Source Books in the History of the Sciences

Edward H. Madden, *General Editor*

From Frege to Gödel: A Source Book in Mathematical Logic, 1879–1931
Jean van Heijenoort

A Source Book in Animal Biology
Thomas S. Hall

Source Book in Astronomy, 1900–1950
Harlow Shapley

A Source Book in Chemistry, 1400–1900
Henry M. Leicester and Herbert S. Klickstein

Source Book in Chemistry, 1900–1950
Henry M. Leicester

A Source Book in Classical Analysis
Garrett Birkhoff

A Source Book in Geology, 1400–1900
Kirtley F. Mather and Shirley L. Mason

Source Book in Geology, 1900–1950
Kirtley F. Mather

A Source Book in Greek Science
Morris R. Cohen and I. E. Drabkin

A Source Book in Mathematics, 1200–1800
Dirk J. Struik

A Source Book in Physics
William Francis Magie

A Source Book in the History of Psychology
Richard J. Herrnstein and Edwin G. Boring

A Source Book in Medieval Science

Edited by Edward Grant

Harvard University Press   Cambridge, Massachusetts   1974

To Marshall Clagett

General Editor's Preface

The *Source Books* in this series are collections of classical papers that have shaped the structure of the various sciences. Some of these classics are not readily available and many of them have never been translated into English, thus being lost to the general reader and in many cases to the scientist himself. The point of this series is to make these texts easily accessible and to provide good translations of the ones that have not been translated at all, or only poorly.

The series was planned to include volumes in all the major sciences from the Renaissance through the nineteenth century. It has been extended to include ancient and medieval science and the development of the sciences in the first half of the present century. Many of these books have been published already and others are in various stages of preparation.

The Carnegie Corporation originally financed the series by a grant to the American Philosophical Association. The History of Science Society and the American Association for the Advancement of Science have approved the project and are represented on the Editiorial Advisory Board. This Board at present consists of the following members:

Alan Anderson, Philosophy, University of Pittsburgh

Marshall Clagett, History of Science, Institute for Advanced Study, Princeton

I. Bernard Cohen, History of Science, Harvard University

Gerald Holton, Physics, Harvard University

Ernst Mayr, Zoology, Harvard University

Ernest Moody, Philosophy, University of California at Los Angeles

Ernest Nagel, Philosophy, Columbia University

Dorothy Needham, Chemistry, Cambridge University

Harry Woolf, History of Science, Johns Hopkins University

The series was begun and sustained by the devoted labors of Gregory D. Walcott and Everett W. Hall, the first two General Editors. I am indebted to them, to the members of the Advisory Board, and to Joseph D. Elder, Science Editor of Harvard University Press, for their indispensable aid in guiding the course of the *Source Books.*

Edward H. Madden

Department of Philosophy,
State University of New York at Buffalo

Preface

That the history of medieval science is eminently worthy of study both for its own sake and for its significant role as transmitter of the achievements of Greek and Arabic science has been established beyond doubt in this century by the heroic research efforts of Pierre Duhem, Anneliese Maier, Marshall Clagett, and others. As scholars have been increasingly attracted to undertake research in the primary sources of medieval science, a growing, though still insufficient, number of carefully edited texts and translations have become available. As yet, however, no source book of reasonably comprehensive scope has appeared to represent the broad spectrum of medieval science to historians of science and to the ever-expanding audience of educated readers keenly interested in the development and fate of the sciences in all periods of history. These volumes attempt to fill such a need and do for the Latin Middle Ages what M. R. Cohen and I. E. Drabkin did for Greek antiquity in their invaluable *Source Book in Greek Science*.

The source readings presented here have been drawn largely from authors in the Latin West, ranging from the Latin encyclopedists of the third to seventh centuries on up to the fifteenth century. Occasionally I have moved outside these temporal and geographic bounds in order to include useful and relevant readings from Greek, Arabic, and seventeenth-century science. Such selections usually reveal either a preliminary development or point of departure of later medieval discussion and controversy, or show the continuity of medieval ideas and concepts and even their termination and repudiation.

The book is divided into parts representing the Early and Later Middle Ages. For the early period, emphasis has been placed upon the general inadequacy of scientific comprehension so typical of the Latin encyclopedists, who nevertheless preserved a modicum of the Greek scientific heritage and thereby exerted a substantial influence on the Later Middle Ages. In this first part only a few representative selections have been included (indeed, two selections from Isidore of Seville were displaced to the section on Medicine in the Later Middle Ages), since the vast bulk of mature scientific discussion falls in the later period, following an interval of translation in the twelfth century. Arabic science, to which a separate volume will be devoted in this series, was omitted entirely except in a few instances involving treatises that were of special significance and influence in the Latin West.

To represent medieval science faithfully it seemed essential that science be construed in a medieval, rather than a modern, sense. It was necessary, therefore, to go beyond the mathematical, physical, and biological sciences and include alchemy, astrology (or, at least, an attack on astrology which also reveals the general attitudes and arguments of astrologers), logic, and theological reactions to science and philosophy. Indeed, even within the more traditional sciences, certain topics have been included which might seem strange to the modern reader but which were of considerable importance and centrality within the broad range of medieval science.

Although approximately 190 distinct selections have been included from about 85 authors, no claim is made for completeness of coverage. Certain topics in the sciences have gone unrepresented, while technology has been deliberately omitted because of the virtual nonexistence of readily available literary sources in a tradition that was largely oral. A determined effort was made to avoid highly abbreviated and distorted selections wrenched out of context. Wherever feasible, reasonably full and substantial discussions of problems and topics have been presented. Indeed, a number of complete, albeit relatively brief, treatises have been included (for instance, the anonymous *Theory of the Planets,* "On Comets" by Albertus Magnus, "On the Formation of Minerals" and "On the Formation of Stones and Mountains" by Avicenna, "On the Causes of the Tides" and "On the Rainbow" by Robert Grosseteste, "Letter on the Magnet" by Peter Peregrinus, and so on) as well as complete, or virtually complete, chapters of parts of larger treatises (for instance, Averroes' Comment on Text 71, Part I of Oresme's *Algorismus proportionum,* Part I of Richard of Wallingford's *De sinibus demonstratis,* and so on).

Of the approximately 190 selections, perhaps as many as 85 appear here for the first time in any vernacular translation. This was made possible by the cooperation of other scholars, whose contributions are acknowledged below. Most of the selections have been annotated, some rather extensively. Where previously published translations were utilized, informative introductions and helpful scholarly notes were frequently included from the original volume, although edited and rearranged to suit the needs of the present work. Many cross-references have been inserted to indicate relevant and interesting connections and to render the book more useful and enjoyable. As a further convenience, biographical data for every author represented by at least one selection, arranged alphabetically, have been added.

In view of the large number of annotated selections, drawn from a wide variety of printed sources, it was deemed futile to attempt a uniform mode of citation. Consequently, translations and footnotes have been left as found. It was, however, often necessary to introduce qualifications, cross-references, or explanations of one kind or another into the translations or accompanying footnotes. All such editorial interpolations are readily detectable, since they alone include the abbreviation *"Ed."* for "editor" at the conclusion of my editorial comment within brackets. Here is an illustration of the form for such editorial comments: [see Albert of Saxony's discussion of this problem on p. 560 —*Ed.*]. Where editorial comments are enclosed within brackets without *"Ed."* they are attributable to the translator or annotator of the particular selection in which they appear. To facilitate location of the large number of references to passages in the works of Aristotle, a standard form of citation, known as Bekker numbers, has been used throughout. In addition to book and chapter numbers, where these are appropriate, all citations consist of page, column, and line numbers from the Berlin Academy edition of the Greek text of Aristotle's works published between 1831 and 1870 by Immanuel Bekker. For example, a reference to *De caelo (On the Heavens)* II. 12.292b.5–10 signifies a passage in Book II, chapter 12, of *De caelo* on page 292, column b, lines 5–10, of the Bekker edition. Since the Oxford and Loeb English translations, as well as virtually all other modern vernacular translations of this century, repeat the page, column, and line numbers from the Bekker edition, the reader can readily find any reference to the works of Aristotle.

The choice and organization of a relatively large number of source readings from virtually the whole range of medieval science is a task that no single scholar could perform with any large measure of confidence. It seemed highly desirable, therefore, to seek the cooperation of as many specialists as possible and assign to them responsibility for certain topics and selections. The response was generous and gratifying. No reader can fail to be impressed with their splendid contributions. It is now a most pleasant duty to acknowledge my debt to Professor John E. Murdoch, Chairman of the Department of the History of Science, Harvard University, who organized the sections on Logic and Atomism, contributed the selection "What is Motion," and collaborated in the translation and annotation of Averroes' Comment on Text 71; to Professor David C. Lindberg of the Department of the History of Science, The University of Wisconsin, who organized the whole of the section on Optics and also translated and annotated the majority of the selections in that section; to William A. Wallace, O.P., who generously translated the most significant parts of Theodoric of Freiberg's *On the Rainbow;* to Professor Michael McVaugh of the Department of History, The University of North Carolina, who assumed full responsibility for the section on Medicine and also translated the selection containing Gerard of Cremona's important list of translations; to Professor Olaf Pedersen of the University of Aarhus, Denmark, for his generous permission to print his previously unpublished translation of the *Theorica planetarum;* and to Professor Victor E. Thoren, my colleague in the Department of History and Philosophy of Science at Indiana University, who not only translated and annotated John of Saxony's difficult introduction to the Alfonsine tables but also provided for the first time an extremely useful sample calculation that reveals the manner in which medieval astronomers used these tables. I am grateful also for his advice and counsel on other astronomical selections.

On the institutional side, I am grateful to the Division of Social Sciences of the National Science Foundation for their support of my research on *The Concept of Void Space in Medieval Physics.* My translations and annotations of the selections on void space constitute a part of that over-all project.

It is also a distinct pleasure to acknowledge the extensive typing assistance contributed by departmental secretaries; their nimble fingers and keen paleographic instincts transformed an illegible scribble into a final typescript. Grateful thanks are owed to Mrs. Jean Coppin, Mrs. Joyce Chubatow, and Mrs. Ina Mitchell, whose Herculean labors produced more than half the total typescript, and to Mrs. Lynore Carnes, who happily brought it to a conclusion.

A source book of this scope and magnitude would not have been possible without the cooperation of many publishers and individuals who generously granted permission to reprint or translate from copyrighted books and articles. Every translation reprinted as a selection has been acknowledged in the selection itself. In addition, I am grateful to the editorial board of the *Journal of the History of Ideas* for permission to incorporate into Selection 44 a translation by Ernest A. Moody from the latter's "Galileo and Avempace: The Dynamics of the Leaning Tower Experiment," Vol. 12 (1951), pp. 184–186, and to Librairie Philosophique J. Vrin (Paris) for permission to use Pierre Thillet, *Alexandre d'Aphrodise De Fato ad imperatores version de Guillaume de Moerbeke édition critique avec introduction et index* (Paris: J. Vrin, 1963), pp. 29–35, as the basis of Selection 8.

The following publishers and institutions kindly permitted translations to be made from Latin texts under their control: Det Kongelige Bibliotek (Copenhagen) (acknowledgment is made not only for permission to translate the *Theorica planetarum* from Ms. Add.447, 2°, fols.49 recto to 56 recto, but also to reproduce the diagrams therein); Dietrich-Coelde-Verlag (Werl/Westfalen); E.J. Brill (Leiden); Edizioni Minerva Medica (Torino); Franciscan Institute, St. Bonaventure University (St. Bonaventure, New York); Franz Steiner Verlag Gmbh (Wiesbaden) (for materials from *Sudhoffs Archiv*); Librairie Delalain Éditeur (Paris); Maisonneuve et Larose, Éditions (Paris); Mediaeval Academy of America (Cambridge, Massachusetts); Pontifical Institute of Mediaeval Studies (Toronto); The University of Chicago Press (Chicago, Illinois); Verlag Aschendorff (Münster/Westfalen).

For permission to reprint passages in introductions, footnotes, and biographical sketches, I am grateful to the following publishers and individuals: Appleton-Century-Crofts (New York); Carnegie Institution of Washington (Washington, D.C.) for material from Publication 376: George Sarton, *Introduction to the History of Science* (3 vols. in 5 parts; published for the Carnegie Institution of Washington by the Williams & Wilkins Company, Baltimore, 1927–1948); Cambridge University Press (Cambridge, England) for extracts from Thomas L. Heath, *The Thirteen Books of Euclid's Elements translated from the text of Heiberg* (3 vols., revised; Cambridge University Press, 1926; reprinted by Dover Publications, Inc., New York, 1956) and J.L.E. Dreyer, *History of the Planetary Systems from Thales to Kepler* (Cambridge University Press, 1906; reprinted by Dover Publications, Inc., New York, 1953, under the title *A History of Astronomy from Thales to Kepler,* Revised with a Foreword by W.H. Stahl); The Clarendon Press (Oxford), especially for passages from *The Works of Aristotle translated into English* under the editorship of J.A. Smith and W.D. Ross (Oxford, 1908–1952); Columbia University Press (New York); Desclée de Brouwer, Éditeurs, S.A. (Brussels); Encyclopaedia Britannica (Chicago, Illinois) for extracts from *Great Books of the Western World* and *Encyclopaedia Britannica;* Harvard University Press (Cambridge, Massachusetts), especially for material from translations of Aristotle's works in The Loeb Classical Library; H.M. Stationary Office and the Director, Royal Botanic Gardens, Kew; Professor Ernest A. Moody and Springer-Verlag (New York); Penguin Books Ltd (Harmondsworth, Middlesex); Petch and Co. (London) and the Executors of the late Mrs. Leyel for quotations from M. Grieve, *A Modern Herbal* (New York: Hafner Publishing Co., 1959); Mrs. Margaret Sherwood Taylor for passages from F. Sherwood Taylor, *The Alchemists Founders of Modern Chemistry* (New York: Henry Schuman, 1949); The University of Chicago Press (Chicago, Illinois); The University of Michigan Press (Ann Arbor, Michigan); and The University of Wisconsin Press (Madison, Wisconsin), especially for material from volumes in the series *Publications in Medieval Science.*

In dedicating this volume to Marshall Clagett, Professor in the School of Historical Studies, The Institute for Advanced Study, Princeton, New Jersey, I have sought to honor my master and friend, to whom I owe so much. Since old and new translations from many fields of medieval science are represented in this source book, which is surely destined to reach a considerably wider public than any monographic study that I might produce, it seemed an appropriate way to honor the name of one of the foremost historians of medieval physical and mathematical thought in this, or any other, century. The numerous selections reprinted here from Professor Clagett's works bear eloquent testimony to his profound and lasting scholarship.

BLOOMINGTON, INDIANA

EDWARD GRANT

Contents

Early Middle Ages

The Latin Encyclopedists

1 ON THE QUADRIVIUM, OR FOUR MATHEMATICAL SCIENCES
Isidore of Seville (d. 636)

Translated by Ernest Brehaut[1]

Revised, expanded, and annotated by Edward Grant

The character of Isidore's *Etymologies* is aptly described by the title. Much of the treatise is given over to fantastic word derivations for all kinds of scientific terms (for instance, see Isidore's chapter 3, where *numerus* is derived). We shall follow William H. Stahl's advice, "The less said about Isidore's word derivations, the better."[2] But Stahl has also provided a concise insight into Isidore's use of authorities in the *Etymologies*:

> Meticulous research has uncovered abundant deceptions in Isidore's references to his authorities. Kettner has pointed out that 36 passages citing Varro were not derived from Varro and that Isidore did not consult Varro's works. Klussman found nearly 70 passages derived from Tertullian's works, and yet Isidore does not cite Tertullian. Mynors' edition of Cassiodorus' *On Sacred and Profane Literature* lists 34 borrowings from that slim work in the first three books of the *Etymologies*, many of them extended passages of close correspondence or verbatim copying; but Isidore withholds acknowledgment. Mommsen, in his edition of Solinus' *Collectanea* lists nearly 600 excerpts drawn by Isidore, but Solinus' name appears nowhere in the *Etymologies*. . . . On the other hand, there is an array of citations of early Latin playrights, satirists, and epic poets, most of whose works had disappeared centuries before Isidore's time: Livius Andronicus, Naevius, Plautus, Ennius, Caecilius, Turpilius, Afranius, Pacuvius, and Lucilius. It has been presumed that Isidore derived most of these excerpts from Servius' Virgil commentaries. . . , but his name is omitted. It is not reassuring to reflect that our knowledge today about the text and attributions of the collected fragments of lost classical works must depend, in many cases, upon the scholarship of such compilers as Isidore. (Stahl, pp. 215–216.)

On the basis of such wholly derivative and unoriginal scholarship, it is depressing to learn that Isidore's *Etymologies* "was one of the most widely read books for the next thousand years; and its serving as a model for encyclopedists as late as Vincent of Beauvais affords a measure of Isidore's importance in the world of scholarship" (Stahl, p. 215). In the selections from Isidore which follow, the reader will see what passed for science, mathematics, and knowledge of the physical world for many centuries. For science, it was truly a dark age. The original treatises of Greek science had, for the most part, been left untranslated. The meager scientific knowledge that came across into Latin was derived from a handbook tradition that went back ultimately to the Hellenistic period. Thus, Isidore was the hapless heir of a drastically diluted scientific heritage. He was rarely better than his sources, and occasionally much worse.

Book III

ON MATHEMATICS

[PREFACE]

Mathematics is called in Latin *doctrinalis scientia* (that is, a theoretical science). It considers abstract quantity. For that is abstract quantity which we treat by reason alone, separating it by the intellect from the material or from other nonessentials, as for example, equal, unequal, or the like. And there are four sorts of mathematics, namely, arithmetic,

---

1. The selection below is drawn from Isidore of Seville's *Etymologies (Etymologiae)*, Book III, as translated in Ernest Brehaut, *An Encyclopedist of the Dark Ages* (New York: Columbia University Press, 1912), pp. 125–152. Brehaut's volume consists of a series of extracts from a few of the twenty books of Isidore of Seville's *Etymologies,* written toward the end of Isidore's life (he died in 636). Since Brehaut's translation is frequently inaccurate and distorted, I have corrected it freely, using the later Latin edition of W. M. Lindsay (see n. 3) and have also translated a number of passages that Brehaut omitted, often without indication.
2. William H. Stahl, *Roman Science* (Madison, Wis.: University of Wisconsin Press, 1962), p. 216.

3

geometry, music, and astronomy.[3] Arithmetic is the science of quantity numerable in itself. Geometry is the science of magnitude and forms. Music is the science that treats of numbers that are found in sounds. Astronomy is the science that contemplates the courses of the heavenly bodies and their figures, and all the phenomena of the stars. These sciences we shall next describe at a little greater length in order that their significance may be fully shown.[4]

CHAPTER 1

ON THE NAME OF THE SCIENCE OF ARITHMETIC

1. Arithmetic is the science of numbers. For the Greeks call number ἀριθμός. The writers of secular literature have decided that it is first among the mathematical sciences since it needs no other science for its own existence.

2. But music and geometry and astronomy, which follow, need its aid in order to be and exist.

CHAPTER 2

ON THE WRITERS

1. They say that Pythagoras was the first among the Greeks to write of the science of number, and that it was later described more fully by Nicomachus, whose work Apuleius first, and then Boethius, translated into Latin.[5]

CHAPTER 3

WHAT NUMBER IS

1. Number is multitude made up of units. For one is the seed of number but not number. *Nummus* (coin) gave its name to *numerus* (number), and from being frequently used, originated the word.

Unus derives its name from the Greek, for the Greeks call *unus* ἕνα, likewise *duo, tria,* which they call δύο and τρία.

2. *Quattuor* took its name from a square figure *(figura quadrata). Quinque,* however, received its name from one who gave the names to numbers not according to nature but according to whim. *Sex* and *septem* come from the Greek.

3. For in many names that are aspirated in Greek we use *s* instead of the aspiration. We have *sex* for ἕξ, *septem* for ἑπτά, and also the word *serpillum* (thyme) for *herpillum. Octo* is borrowed without change; they have ἐννέα, we *novem;* they δέκα, we *decem.*

4. *Decem* is so called from a Greek etymology, because it ties together and unites the numbers below it. For to tie together and unite is called among them δεσμός. . . . [6]

CHAPTER 4

WHAT NUMBERS SIGNIFY

1. The science of number must not be despised. For in many passages of the holy scriptures it is manifest what great mystery they contain. For it is not said in vain in the praises of God (Book of Wisdom 11:21): "but thou hast ordered all things in measure, and number, and weight."[7] For the senarius, which is perfect in respect to its parts,[8] declares the perfection of the universe by a certain meaning of its number. In like manner, too, the forty days which Moses and Elias and the Lord himself fasted are not understood without an understanding of number.

3. So, too, other numbers appear in the holy scriptures whose nature none but experts in this art can wisely declare the meaning of. It is granted to us, too, to depend in some part upon the science of

3. These four mathematical sciences were customarily designated as the *quadrivium* and formed the scientific part of the traditional seven liberal arts. The remaining three, called the *trivium,* consisted of grammar, rhetoric, and dialectic (or logic). The concept of seven liberal arts—that is, *artes liberales,* studies fit for a free man as opposed to a slave—can be traced to the Greeks as far back as the fourth century B.C. It was Martianus Capella (fl. 410–439), however, in his tremendously influential book *The Marriage of Mercury and Philology,* who canonized the seven liberal arts for the Latin medieval tradition. Isidore is but following this tradition, which he also helps to establish.

It should be noted that Brehaut's translation was made from DuBreul's edition of Isidore's works published in Paris in 1601. In the more recent two-volume edition of the *Etymologies* (Oxford: Clarendon Press, 1911) by W. M. Lindsay, music appears before geometry in this sentence and is described immediately after arithmetic a few sentences below. Such trivial differences will be ignored.

4. The substance, and sometimes the very words, of this paragraph were taken from Cassiodorus' section "On Mathematics" in his *An Introduction to Divine and Human Readings,* translated by L. W. Jones (New York: Columbia University Press, 1946), pp. 178–179. Almost all of what Isidore writes on arithmetic was taken from Cassiodorus, who in turn drew largely upon Boethius' *Arithmetic.*

5. This paragraph is drawn directly from Cassiodorus (see Jones, p. 187).

6. In the next few lines Isidore gives etymological derivations for 20, 30, 100, 200, 1,000, and thousands.

7. I have replaced the Latin text of this Biblical quotation with the English translation from the Douay version. This quotation was widely cited as justification for the study of mathematics.

8. Six was considered a perfect number because it equals the sum of all its factors.

numbers, since we learn the hours by means of it, reckon the course of the months, and learn the time of the returning year. Through number, indeed, we are instructed in order not to be confounded. Take number from all things and all things perish. Take calculation from the world and all is enveloped in dark ignorance, nor can he who does not know the way to reckon be distinguished from the rest of the animals.

CHAPTER 5

ON THE FIRST DIVISION INTO EVEN AND ODD

1. Number is divided into even and odd. Even number is divided into the following: evenly even, evenly uneven, unevenly even, and unevenly uneven. Odd number is divided into the following: prime and incomposite, composite, and a third intermediate class *(mediocris)* which in a certain way is prime and incomposite but in another way secondary and composite.

2. An even number is that which can be divided into two equal parts, as II, IV, VIII.[9] An odd number is that which cannot be divided into equal parts, there being one in the middle which is either too little or too much, as III, V, VII, IX, and so on.

3. Evenly even number is that which is divided equally into even number, until it comes to indivisible unity, as for example, LXIV has a half XXXII, this again XVI; XVI, VIII; eight, IV; four, II; two, one, which is single and indivisible.

4. Evenly uneven is that which admits of division into equal parts, but its parts soon remain indivisible, as VI, X, XXXVIII, and L, for presently, when you divide such a number, you run upon a number which you cannot halve.[10]

5. Unevenly even number is that whose halves can be divided again but do not go on to unity, as XXIV. For this number being divided in half makes XII, divided again VI, and again, three; and this part does not admit of further division, but before unity a limit is found which you cannot halve.

6. Unevenly uneven is that which is measured unevenly by an uneven number, as XXV, XLIX; which, being uneven numbers, are divided also by uneven factors, as, seven times seven, XLIX, and five times five, XXV. Of odd numbers some are prime, some composite, some intermediate *(mediocris)*.

7. Simple [or prime] numbers are those which have no other part [or factor] except unity alone, as three has only a third, five only a fifth, seven only a seventh,[11] for these have only one factor.

Composite numbers are those which are not only measured by unity, but are produced[12] by another number, as nine, *(novem)*, XV, XXI, [XXV]. For we say three times three *(ter terni)* [are IX], and seven times three *(septies terni)* [are XXI], and three times five *(ter quini)* [are XV], and five times five *(quinquies quini)* [are XXV].

8. Intermediate *(mediocris)* numbers are those which in a certain fashion seem prime and incomposite and in another fashion secondary and composite. For example, when nine *(novem)* is compared with XXV, it is prime and incomposite because it has no common number except the monad [or unit] only; but if it is compared with fifteen *(quindecim)*, it is secondary and composite since there is in it a common number in addition to unity, that is, the number three (because three times three make nine, and three times five make fifteen).

9. Likewise of even numbers some are excessive, others defective, others perfect. Excessive are those whose factors being added together exceed its total, as for example, twelve. For it has five parts [or factors]: a twelfth, which is one; a sixth, which is two; a fourth, which is three; a third, which is four; a half, which is six. For one and two and three and four and six being added together make XVI, which is far in excess of twelve. And there are many similar kinds, as eighteen, and many such.

10. Defective numbers are those which being reckoned by their factors make a less total, as for example, X, which has three parts: a tenth, which is one; a fifth, which is two; and a half, which is five. . . .

11. The perfect number is that which is equaled

9. In Lindsay's Latin edition of the *Etymologies,* Roman and rhetorical numerals are used indiscriminately. To convey a sense of this inconsistency, I have expressed all numbers as they appear in the text.

10. According to Cassiodorus, an "evenly uneven" (Jones renders it as "even-times-odd") number "is one whose similar division into two equal parts can occur but once; for example, 10, whose half is 5. . ." (Jones, p. 181). Although Isidore's examples agree with this definition, he fails to restrict his version to a single division.

11. Here Isidore has slightly altered and made more cumbersome the remark of Cassiodorus, who says merely that a prime number "is one which can be divided by unity alone; for example, 3, 5, 7, 11, 13, 17, and the like" (Jones, p. 182). Isidore, however, speaks of 1/3 of 3 as the factor of 3; 1/5 of 5, and so on.

12. Cassiodorus properly says that such numbers can be divided by unity and another number. But Isidore says obscurely that they can also be "produced" *(procreantur)* by another number.

5

by its factors, as six, for it has three parts: a sixth, a third, and a half. Now a sixth of it is one, a third is two, a half is three. When these parts are summed —that is, one, two, and three—they perfect and complete the [number] six. The perfect numbers are, under ten, VI; under a hundred, XXVIII; under a thousand, CCCCXCVI.

CHAPTER 6

ON THE SECOND DIVISION OF ALL NUMBER

1. All number is considered either with reference to itself or in relation to something. The former is divided as follows: some are equal; others are unequal. The latter is divided as follows: some are greater, some are less.[13] The greater are divided as follows: into multiples, superparticulars, superpartients, multiple superparticulars, multiple superpartients.[14] The lesser are divided as follows: submultiples, subsuperparticulars, subsuperpartients, submultiple subsuperparticulars, submultiple subsuperpartients.

2. A number is said to be by itself which is said to be without any relation [or ratio], as III, IV, V, VI, and similar others. A number is related to something which is compared relatively to others, as, for example, IV to II, which is called a double [ratio]; also VI to III, VIII to IV, X to V; and III to one is a triple [ratio, as are] VI to II, IX to III, and so on.[15]

3. Those numbers are said to be equal which are equal in quantity, as II to II, III to III, X to X, and C to C. Those numbers are unequal which when mutually compared show an inequality, as III to II, IV to III, V to IV, X to VI; and universally, when a greater is compared to a lesser [number] and a lesser to a greater, they are said to be unequal.

4. A number is greater which contains the smaller number to which it is compared [or related] plus something more; for example, number five is greater than number three because number five contains number three and two other parts of it; and so on for others.

5. A smaller number is one that is contained by the greater to which it is compared [or related] plus some part of it, as three to five, for it is contained by it with its two other parts. . . . .[16]

7. The superparticular number is when a greater number contains in itself a lesser number with which it is compared, and at the same time one part of it.

For example, III when compared with II contains

in itself II and also one, which is the half of two. IV when compared with III contains III and also one, which is the third of three. Likewise V when compared with IV contains the number four and also one, which is the fourth part of the said number four,[17] and so on.

8. The superpartient number is that which contains the whole of a lesser number and in addition II parts of it, or III, or IV, or V, or other parts. For example, when V is compared with III, the number five contains three and in addition to this II parts of it. . . . .[18]

13. Isidore is actually referring to ratios. Thus if $A > B$, where A and B are numbers, then A/B is a ratio of greater inequality and B/A one of lesser inequality.

14. Since only the superparticular and superpartient ratios are discussed below, I shall cite Isidore's examples for the others:

(1) Multiple ratios: $2/1, 3/1, 4/1$; the respective submultiples are $1/2, 1/3, 1/4$. That is, $n/1$ and $1/n$ respectively, where n is any integer.

(2) Multiple superparticular ratios: $5/2$ and $9/4$, where the greater term contains the lesser an integral number of times plus a unit fractional part. The respective submultiples are $2/5$ and $4/9$. In general, $(mn+1)/n$, where m and n are greater than 1 and all terms are integers; the reciprocal will represent the submultiples.

(3) Multiple superpartient ratios: $8/3, 14/6, 16/7, 21/9$, where the greater term contains the lesser an integral number of times plus a proper fraction reduced to its lowest terms. Thus, although Isidore includes $14/6$ and $21/9$, these do not qualify, since $14/6 = 7/3 = 2\frac{1}{3}$ and $21/9 = 7/3 = 2\frac{1}{3}$, which converts them to superparticular ratios. The respective submultiple superpartients would be $3/8, 7/16$. Generally, one may represent multiple superpartient ratios by $P + m/n$, where P is an integer greater than or equal to 1 and m and n are mutually prime integers greater than 1 and $m < n$.

15. I have translated and added this paragraph and paragraphs 3–5.

16. For the remainder of chapter 6, Isidore provides descriptions and examples of the five types of ratios of greater and lesser inequality enumerated in paragraph 1 of this sixth chapter. Only the sections on superparticular and superpartient ratios of greater inequality are translated here (for the rest, see n. 14). All five types of ratios were drawn from Cassiodorus, who derived them from the *Arithmetic* of Boethius.

17. The subsuperparticulars of these respective examples are $2/3, 3/4, 4/5$. Generally $(n+1)/n$, where $n \geq 2$, represents any superparticular; in each case the reciprocal provides the submultiple.

18. Any superpartient ratio can be represented by $1 + (m/n)$, where m and n are mutually prime integers greater than 1 and $m < n$. Once again, the reciprocal of any superpartient ratio represents its subsuperpartient. Although in his brief summary Isidore refers to multiple, superparticular, and superpartient *numbers*, it is ratios that are meant, since in all cases we have a relation between two numbers.

CHAPTER 7

ON THE THIRD DIVISION OF ALL NUMBER

1. Numbers are discrete *(discreti)* or continuous *(continentes)*.[19] The latter are divided as follows: first, lineal; second, superficial; third, solid. Discrete number is that which is made up of discrete units. For example, III, IV, V, VI, and so on.

2. Continuous number is that which is made up of connected units, as, for example, the number three understood in magnitude, that is, it is said to be continuous in a line or space, or solid; similarly for the numbers four and five.[20]

3. A lineal number is one that begins from unity and is written lineally to infinity.[21] For this reason alpha is used to designate lines, since among the Greeks this letter signifies one.

4. A superficial [or plane] number is that which is constituted not only by length but also by breadth, as triangular, square, pentangular, or circular numbers, and the rest that are contained in a plane surface or superficies.[22]

5. A circular number is one that has been multiplied by itself; it begins with itself and returns to itself. For example, five times five is XXV.[23] A solid number is one that is contained by length, width, and depth, as are pyramids, which rise in the manner of a flame.

6. Cube numbers are like dice. Spheres are things that are everywhere equal in rotundity. A spherical number is one that has been multiplied from a circular number; it begins from itself and returns to itself. Five times five is XXV. When this circle has been multiplied by itself, it makes a sphere, that is, five times XXV [makes] CXXV.

CHAPTER 8

ON THE DISTINCTION BETWEEN ARITHMETIC, GEOMETRY, AND MUSIC

1. Between arithmetic, geometry, and music there is a difference in finding the means. In the first place, you do as follows in arithmetic. Join [or add] the extremes and divide and you get the mean. For example, let VI and XII be the extremes and you add them and they make [or equal] X and VIII;[24] you divide the means[25] and get IX, which is the mean of arithmetic *(analogicum arithmeticae)*. Thus the mean exceeds the first term by as many units as it is exceeded by the extreme. For IX exceeds VI by three units and XII exceeds it by three units.

2. According to geometry you find it this way. The extremes multiplied together make as much as the means multiplied *(duplicata)*; for example,

VI and XII multiplied make seventy two *(septuagies dipondius)*; the means VIII and IX multiplied make the same.[26]

3. According to music, you find it this way: the

19. I have altered Brehaut's translation from "abstract or concrete" to "discrete or continuous." These changes have been made systematically in the rest of the chapter; other changes have also been required by Brehaut's misunderstanding of the chapter; I have also added the translation of paragraph 3 and some lines omitted in other paragraphs.

20. Cassiodorus, from whom Isidore drew the last two paragraphs, says (Jones, p. 185): "A *continuous* number is one which consists of connected units; as 3, for example, would be called continuous if it was understood to be the measurement of a magnitude, that is, the measurement of a line or a space or a solid; the same applies to 4 and to 5."

21. In commenting on this passage as it appears in Cassiodorus and from whence Isidore derived it, Jones observes (p. 185, n. 20), "Unity is not included, for a line is the aggregate of two or more points. 2, 3, 4, 5, 6, etc., are linear numbers."

22. As examples of superficial numbers Isidore gives a triangular number, a square number, and a pentagonal number, which were apparently represented by appropriate figures in the manuscripts but were omitted in the Latin edition by Lindsay. Cassiodorus, from whom Isidore derived all this, provided diagrams for these figurate numbers (see Jones, p. 180) but supplied no additional descriptive information. From Nicomachus and Boethius, however, it is apparent that a triangular number is represented by $[n(n+1)] / 2$, so that if $n = 1$ $2, 3, 4, 5, \ldots$, then the successive triangular numbers are $1, 3, 6, 10, 15, \ldots$, where 6, for example, is the sum of $1+2+3$ and could be thought of in the triangular arrangement.

Fig. 1

The successive square numbers may be represented generally as n^2, where $n = 1, 2, 3, 4, 5, \ldots$ and n is the side of each square. Pentagonal numbers consist of the sequence $1, 5, 12, 22, 35, \ldots$ and in general $n^2 + (n/2)(n-1)$, where $n = 1, 2, 3, 4, 5, \ldots$.

23. Thus $5^2 = 25$ is circular, because after the original number is multiplied by itself it terminates with itself; the same applies to $6^2 = 36$.

24. Although it is unclear, Isidore may have intended to use some or all of the terms 12, 10, 8, and 6 to illustrate the three kinds of means. However, the text of the musical mean seems corrupt and the numbers used cannot be exactly determined.

25. Here Isidore divides the sum of 10 and 8, which are means between 12 and 6.

26. Isidore has utterly confused the geometric mean, for although he offers an acceptable definition, his example does not illustrate the definition. A geometric mean is represented as $A/B = B/C$, so that $AC = B^2$; but after inserting means 8 and 9 between extremes 6 and 12, Isidore multiplies $8 \cdot 9 = 6 \cdot 12$ and believes that he has found a geometric mean! In his chapter 13, Isidore repeats this example.

mean exceeds the first term by the same part as the mean is exceeded by the extreme [or last term]. For example, VI, VIII [and XII]; so that the mean [i.e., VIII] exceeds [VI] by two parts, which is one third [of the first term], and the mean is exceeded by the last term [by a third part of it].[27]

CHAPTER 9

THAT INFINITE NUMBERS EXIST

1. It is most certain that there are infinite numbers, since at whatever number you think an end must be made I say not only that it can be increased by the addition of one, but, however great it is, and however large a multitude it contains, by the very method and science of numbers it can not only be doubled but even multiplied.

2. Each number is limited by its own proper qualities, so that no one of them can be equal to any other. Therefore, in relation to one another they are unequal and diverse, and the separate numbers are each finite, and all are infinite.

CHAPTER 10

ON THE INVENTORS OF GEOMETRY AND ITS NAME

1. The science of geometry is said to have been discovered first by the Egyptians, because when the Nile overflowed and all their lands were overspread with mud, its origin in the division of the land by lines and measurements gave the name to the art. And later, being carried further by the keenness of the philosophers, it measured the spaces of the sea, the heavens, and the air.

2. For, having their attention aroused, students began to search into the spaces of the heavens after measuring the earth; how far the moon was from the earth, the sun itself from the moon, and how great a measure extended to the summit of the sky; and thus they laid off in numbers of stades with probable reason the very distances of the sky and the circuit of the earth.

3. But since this science arose from the measuring of the earth, it took its name also from its beginning. For *geometria* is so named from "earth" and "measuring." For the earth is called γῆ in Greek, and measuring, μέτρα. The art of this science embraces lines, intervals, magnitudes, and figures; and in figures, dimensions and numbers.

CHAPTER 11

ON THE FOURFOLD DIVISION OF GEOMETRY

1. The fourfold division of geometry is into plane figures, numerable magnitude, rational magnitude, and solid figures.

2. Plane figures are those which are contained by length and breadth, and which are five in number according to Plato.[28] Numerable magnitude is that which can be divided by the numbers of arithmetic.

3. Rational magnitudes are those whose measures we can know, and irrational, those the amount of whose measurement is not known.

4. Solid figures are those that are contained by length, breadth, and thickness,[29] as [for example], a cube; and there are five species in a plane.[30]

CHAPTER 12

ON THE FIGURES OF GEOMETRY

1. The first of these, the circle, is a plane figure which is called a circumference, in the middle of which is a point upon which everything converges *(cuncta convergunt)*, which geometers call the center and the Latins call the point of the circle.

2. A quadrilateral figure is a square in a plane which consists of four straight lines, thus.[31] A

27. Brehaut's translation and Lindsay's Latin—which differ substantively—do not offer an example of a musical or harmonic mean. Either the text is corrupt or Isidore failed to understand the meaning of musical mean. In the example that I have substituted, the three terms are 6, 8, and 12 such that $12/6 = (12–8)/(8–6)$ and generally $A/C = (A–B)/(B–C)$. Thus 12 exceeds 8 by 4, which is 1/3 of 12; and 6 is less than 8 by 2, which is 1/3 of 6. Hence the difference between each extreme and the mean is 1/3 of that extreme. In Lindsay's Latin text Isidore gives only numbers 6 and 8 and says that the last term exceeds 8 by a ninth ("VII[I] superatur ab ultima nona")! I have added 12 because it seems that Isidore wished to present his examples with two or more of the series of numbers 12, 10, 8, and 6. This gains plausibility from the fact that in his section on music Isidore repeats an example of a harmonic mean using the numbers 6, 8, and 12. See Isidore's chapter 23 and my note 49. For Boethius' discussion of the three types of proportions or means, see Selection 2, chs. 43–47.

28. It is probable that Isidore has corrupted the five regular solids, which Plato discusses in *Timaeus* 53C–55C, into five types of plane figure.

29. Up to this point Isidore has taken the substance of his geometry from Cassiodorus (see Jones, p. 198). Since Cassiodorus has little more than this, Isidore could not have derived the rest of this geometrical section from him.

30. The text seems defective at this point. It is unclear whether Isidore intended to declare here that there are five species of solid figures in a plane(!) or whether, after completing his definition of solid figure and having exemplified it by the cube, he wishes now to take up the five species of plane figure mentioned in paragraph 2.

31. Here again Isidore equates a genus of geometric figure, namely four-sided figure, with a particular type

dianatheton grammon is a plane figure,[32] thus. An orthogonium, that is, a right angle *(rectiangulum)*, is a plane figure, for it is a triangle and has a right angle.[33] The plane figure *isopleuros* is straight and constructed underneath.[34]

3. A sphere is a figure of rounded form equal in all its parts.

4. A cube is a proper solid figure which is contained by length, breadth, and thickness.[35] A cylinder is a square figure with a semicircle above.[36]

5. A cone *(conon)* is a solid figure which narrows from a broad base like a right-angled triangle.[37]

6. A pyramid is a figure which narrows from a broad base to a point like fire.[38] For among the Greeks fire is called πῦρ.

7. Furthermore, just as all number is [contained] below X,[39] so is the outline of all figures contained within a circle.[40] The first figure of this kind is a point, which has no part.[41] The second is a line, which has length besides breadth. A straight line is that which lies evenly in respect to its points. A surface is that which has length and breadth only. Lines are the limits [or boundaries] of surfaces, and the forms [or shapes] in the ten figures mentioned above are not posited because they are found among them.[42]

CHAPTER 13[43]

ON THE NUMBERS OF GEOMETRY

You investigate numbers according to geometry as follows: The extremes being multiplied *(multiplicata)* amount to as much as the means multiplied *(duplicata)*;[44] as for example VI and XII being multiplied make seventy two; the means VIII and IX being multiplied amount to the same.[45]

CHAPTER 15

ON MUSIC AND ITS NAME

1. Music is the practical knowledge of melody, consisting of sound and song; and it is called music by derivation from the Muses. And the Muses were so called ἀπὸ τοῦ μάσαι, that is, from inquiring, because it was by them, as the ancients had it, that

inition of solid figures given in his chapter 11, paragraph 4. In *Elements* XIII, Definition 25, Euclid defines it as follows: "A cube is a solid figure contained by six equal squares." The translation is that of Thomas L. Heath, *The Thirteen Books of Euclid's Elements,* 2d ed. (New York: Dover, 1956), III, 261.

36. How Isidore obtained or arrived at this incredible definition is a mystery. Perhaps it was suggested upon observing the following kind of two-dimensional representation of a cylinder:

Fig. 2

Or perhaps in some manner this is a distortion of Euclid's definition of cylinder (*Elements* XI, Def. 21): "When, one side of those about the right angle in a rectangular parallelogram remaining fixed, the parallelogram is carried round and restored again to the same position from which it began to be moved, the figure so comprehended is a cylinder" (Heath p. 262). Although the "rectangular parallelogram" may have been distorted to a square, it is difficult to see how the semicircle on the square could have been extracted from this.

37. Compare Euclid's definition (*Elements* XIII, Def. 18): "When, one side of those about the right angle in a right-angled triangle remaining fixed, the triangle is carried round and restored again to the same position from which it began to be moved, the figure so comprehended is a cone" (Heath, p. 262).

38. It is again instructive to compare Euclid's definition (*Elements* XIII, Def. 12): "A pyramid is a solid figure, contained by planes, which is constructed from one plane to one point" (Heath, p. 262).

39. Isidore probably means that all numbers can be generated from the numbers 1 through 9.

40. Here Isidore simply means that these figures can be inscribed in a circle. For the solid figures, however, a sphere is required.

41. I fail to see how this sentence relates to what has immediately preceded. Brehaut solved the problem by omitting "The first figure of this kind is. . ." and converting the rest to "A point is that which has no part." Thus, the embarrassment of designating a point as a particular kind of figure is avoided.

42. Isidore's meaning is wholly unclear. Are the shapes of the figures not mentioned because, for Isidore, they are all inscribed in circles? The point is hardly worth further discussion.

43. Brehaut calls this chapter 14; but in this I follow Lindsay's text.

44. Ordinarily *duplicata* means "squared" but this translation would invalidate Isidore's example, which involves the multiplication of two unequal means. Note that Isidore used *multiplicata,* the proper term in this context, at the beginning of the sentence.

45. The same example, with the same terminology, was used earlier in chapter 8, paragraph 2, of the discussion on arithmetic. Chapter 14, the last in the section on geometry, has been omitted. It consists of a paragraph briefly describing additional figures. Thus we conclude this frequently incomprehensible and pitiful remnant of Euclidean geometry.

---

of four-sided figure, a square. Here and elsewhere in this chapter, figures were to be inserted. Lindsay's text does not contain them.

32. The figure intended here is unclear to me

33. Isidore seems to equate right angle with triangle and to commit the same error mentioned above in note 31.

34. Because Isidore's meaning is so obscure, I present the Latin: "Isopleuros figura plana, recta et subter constituta."

35. This definition fails to distinguish a cube from any other kind of solid and is identical with Isidore's def-

the potency of songs and the melody of the voice were inquired into.

2. Since sound is a thing of sense it passes along into past time, and it is impressed on the memory. From this it was pretended by the poets that the Muses were the daughters of Jupiter and Memory. For unless sounds are held in the memory by man, they perish, because they cannot be written.

CHAPTER 16

ON ITS DISCOVERERS

1. Moses says that the discoverer of the art of music was Jubal, who was of the family of Cain and lived before the flood. But the Greeks say that Pythagoras discovered the beginnings of this art from the sound of hammers and the striking of tense cords. Others assert that Linus of Thebes, and Zethus, and Amphion, were the first to win fame in the musical art.

2. After whose time this science in particular was gradually established and enlarged in many ways, and it was as disgraceful to be ignorant of music as of letters. And it had a place not only at sacred rites but at all ceremonies and in all things glad or sorrowful.

CHAPTER 18

ON THE THREE PARTS OF MUSIC[46]

1. There are three parts of music, namely, *harmonica, rhythmica, metrica. Harmonica* is that which distinguishes in sounds the high and the low. *Rhythmica* is that which inquires concerning the succession of words as to whether the sound fits them well or ill.

2. *Metrica* is that which learns by approved method the measure of the different meters, as, for example, the heroic, iambic, elegiac, and so on.

CHAPTER 19

ON THE TRIPLE DIVISION OF MUSIC

1. It is agreed that all sound which is the material of music is of three sorts. First is *harmonica,* which consists of vocal music; second is *organica,* which is formed from the breath; third is *rhythmica,* which receives its numbers from the beat of the fingers.

2. For sound is produced either by the voice, coming through the throat; or by the breath, coming through the trumpet or tibia, for example; or by touch, as in the case of the cithara or anything else that gives a tuneful sound on being struck.

CHAPTER 20

ON THE FIRST DIVISION OF MUSIC, WHICH IS CALLED HARMONICA

1. The first division of music, which is called *harmonica,* that is, modulation of the voice, has to do with comedians, tragedians, and choruses, and all who sing with the proper voice. This [coming] from the spirit and the body makes motion, and out of motion, sound, out of which music is formed, which is called in man the voice. . . .

2. *Harmonica* is the modulation of the voice and the concord or fitting together of very many sounds.

3. *Symphonia* is the managing of modulation so that high and low tones accord, whether in the voice or in wind or stringed instruments. Through this, higher and lower voices harmonize, so that whoever makes a dissonance from it offends the sense of hearing. The opposite of this is *diaphonia,* that is, voices grating on one another or in dissonance.

7. *Tonus* is a high utterance of voice. For it is a difference and measure of harmony which depends on the stress and pitch of the voice. Musicians have divided its kinds into fifteen parts, of which the hyperlydian is the last and highest, the hypodorian the lowest of all.[47]

8. Song is the modulation of the voice, for sound is unmodulated, and sound precedes song. . . .

CHAPTER 21

ON THE SECOND DIVISION, WHICH IS CALLED ORGANICA

1. The second division, *organica,* has to do with those [instruments] that, filled with currents of breath, are animated so as to sound like the voice, as for example, trumpets, reeds, Pan's pipes, organs the pandura, and instruments like these. . . .[48]

CHAPTER 22

ON THE THIRD DIVISION, WHICH IS CALLED RHYTHMICA

1. The third division is *rhythmica,* having to do

46. Isidore derives much of this from Cassiodorus' *Introduction to Divine and Human Readings; Secular Letters,* chapter V, "On Music" (Jones, pp. 189–196). The discussion of music in the *Etymologies* is almost wholly nonmathematical (in contrast to Boethius' *On Music*). Cassiodorus, but not Isidore, enters into a discussion of fifteen Greek tones.
47. For this paragraph, Cassiodorus served as Isidore's source.
48. Brehaut notes that the *pandura* was not a wind but a stringed instrument.

with strings and instruments that are beaten, to which are assigned the different species of cithara, the drum, and the cymbal, the sistrum, acitabula of bronze and silver, and others of metallic stiffness that when struck return a pleasant tinkling sound, and the rest of this sort.

2. . . . . The form of the cithara in the beginning is said to have been like the human breast, because as the voice was uttered from the breast so was music from the cithara, and it was so called for the same reason. For *pectus* is in the Doric language called $\kappa\iota\theta\acute{\alpha}\rho\alpha$. . . .

CHAPTER 23

ON THE NUMBERS OF MUSIC

1. You inquire into numbers according to music as follows: Setting down the extremes, as, for example, VI and twelve, you see by how many units VI is surpassed by XII, and it is by VI units; you square it; six times six make XXXVI. You add those first-mentioned extremes, VI and XII; together they make XVIII; you divide thirty-six by eighteen; two is the result. This you add to the smaller amount, six namely; the result will be VIII and it will be a mean between VI and XII. Because VIII surpasses VI by two units, that is by a third of VI, and VIII is surpassed by XII by four units, a third part [of twelve]. By what part, then, the mean surpasses, by the same is it surpassed.[49]

2. Just as this proportion exists in the universe, being constituted by the revolving circles, so also in the microcosm—not to speak of the voice—it has such great power that man does not exist without harmony. . . .

CHAPTER 24

ON THE NAME OF ASTRONOMY

1. Astronomy is the law of the stars, and it traces with inquiring reason the courses of the heavenly bodies, and their figures, and the regular movements of the stars with reference to one another and to the earth.

CHAPTER 25

ON ITS DISCOVERERS

1. The Egyptians were the first to discover astronomy *(astronomia)*. And the Chaldeans first taught astrology *(astrologia)*[50] and the observance of nativity. Moreover, Josephus asserts[51] that Abraham taught astrology to the Egyptians. The Greeks, however, say that this art was first elabo-

rated by Atlas, and therefore it was said that he held the heavens up.

2. Whoever was the discoverer, it was the movement of the heavens and his rational faculty that stirred him, and in the light of the succession of seasons, the observed and established courses of the stars, and the regularity of the intervals, he considered carefully certain dimensions and numbers, and by limiting and distinguishing them he wove them into order and discovered astrology.

CHAPTER 26

ON ITS TEACHERS

1. In both Greek and Latin there are volumes written on astronomy by different writers. Of these Ptolemy, King of Alexandria,[52] is considered chief among the Greeks. He also formulated rules *(canones)* by which the courses of the stars may be discovered.

CHAPTER 27

THE DIFFERENCE BETWEEN ASTRONOMY AND ASTROLOGY

1. There is some difference between astronomy and astrology. For astronomy embraces the revolution of the heavens, the rising, setting, and motion of the heavenly bodies, and the origin of their names. Astrology, on the other hand, is in part natural, in part superstitious.

2. It is natural astrology when it describes the courses of the sun and the moon or the fixed positions of the stars and the times (seasons?). Superstitious astrology is that which the mathematici

---

49. Here we have an harmonic mean involving the numbers 6, 8, and 12. See Isidore's chapter 8, paragraph 3, and my note 27.

50. Isidore uses the Latin terms *astronomia* and *astrologia* much as we would use their English equivalents (see his chapter 27). However, through much of the later Middle Ages, the two terms were used interchangeably and without meaningful distinction.

51. Cassiodorus, who cites Josephus (*Antiquities,* Bk. I, ch. 9), is Isidore's source for this remark. See *An Introduction to Divine and Human Readings; Secular Letters* (Jones, p. 179).

52. Isidore has confused Claudius Ptolemy, the Greek astronomer of the second century A.D. and author of the *Almagest,* with the Greek dynasty of Ptolemaic kings who ruled in Egypt from around 304 B.C. to 30 B.C. (see a selection by Claudius Ptolemy in the section on Cosmology, and a short biography of him at the end of this source book.

follow who prophesy by the stars and who distribute the twelve signs of the heavens among the individual parts of the soul or body and endeavor to predict the nativities and characters of men from the course of the stars.

CHAPTER 28

ON THE SUBJECT MATTER OF ASTRONOMY

1. The subject matter of astronomy is made up of many kinds. For it defines what the universe is, the heavens, the position and movement of the sphere, the axis of the heavens and the poles, what are the climates of the heavens, what the courses of the sun and moon and stars, and so forth.

CHAPTER 29

ON THE UNIVERSE AND ITS NAME

1. *Mundus* (the universe) is that which is made up of the heavens and earth and the sea and all the heavenly bodies. And it is called *mundus* for the reason that it is always in motion *(motus)*.[53] For no repose is granted to its elements.

CHAPTER 30

ON THE FORM OF THE UNIVERSE[54]

1. The form of the universe is described as follows: As the universe is raised toward the region of the north, so it is inclined toward the south; its head and face are, as it were, the east, and its extreme part the north.

CHAPTER 31

ON THE HEAVENS AND THEIR NAME

1. The philosophers have asserted that the heavens are round, in rapid motion, and made of fire, and that they are called by this name *(coelum)* because they have the forms of the stars fixed on them, like a dish with figures in relief *(coelatum)*.

2. For God decked them with bright lights, and filled them with the glowing orbs of the sun and moon, and adorned them with the glittering images of flashing stars. . . .[55]

CHAPTER 32

ON THE SITUATION OF THE CELESTIAL SPHERE

1. The sphere of the heavens is rounded and its center is the earth, equally shut in on every side. This sphere, they say, has neither beginning nor end, for the reason that being rounded like a circle it is not easily perceived where it begins or where it ends.[56]

2. The philosophers have brought in the theory of seven heavens of the universe, that is, planets moving with the harmony of the spheres, and they assert that all planets are connected to their orbs, and they think that these, being connected and, as it were, fitted to one another, move backward and are borne with definite motions in contrary directions.

CHAPTER 33

ON THE MOTION OF THE SAME SPHERE

1. The sphere revolves on two axes, of which one is the northern, which never sets, and is called Boreas; the other is the southern, which is never seen, and is called Austronotius.

2. On these two poles the sphere of heaven moves, they say, and with its motion the stars fixed in it pass from the east all the way around to the west, with the northern stars near the pole *(iuxta cardinem)* describing smaller circles.

CHAPTER 34

ON THE COURSE OF THE SAME SPHERE

1. The sphere of heaven, [moving] from the east towards the west, turns once in a day and night, in the space of twenty-four hours, within which the sun completes his swift revolving course over and under the earth.

53. As indicated by the title *(Etymologies)*, Isidore is frequently interested in showing the origin of words, motivated by the belief that such knowledge conveys an insight into subject matter. Almost all are false and forced, as is *mundus* from *motus*.

54. Chapter 30 is almost a verbatim repetition of Isidore's words in his earlier work *On the Nature of Things (De natura rerum)*, chapter 9, "On the World." Since the earlier work is exclusively concerned with astronomy, cosmology, and natural phenomena in the upper regions of air and fire (for example, he considers thunder, lightning, rain, clouds), Isidore draws heavily upon it in the *Etymologies*. Indeed, the section on astronomy in the *Etymologies* often seems a highly abbreviated version of the *De natura rerum*. The latter treatise has been edited and translated into French by Jacques Fontaine, *Isidore de Seville Traité de la Nature* (Bordeaux: Féret et Fils, 1960).

55. One sentence involving the derivation of the Greek word for heavens, *ouranos*, is omitted.

56. Here again, Isidore forms this chapter by repeating, almost verbatim, a few lines from chapter 12, paragraph 4, of his earlier *De natura rerum* (see Fontaine, p. 219).

CHAPTER 35

ON THE SWIFTNESS OF THE HEAVENS

1. With such swiftness is the sphere of heaven said to run, that if the planets *(astra)* did not run against its headlong course in order to delay it, it would destroy the universe.[57]

CHAPTER 36

ON THE AXIS OF THE HEAVENS

1. The axis is a straight line north which passes through the center of the globe of the sphere and is called axis because the sphere revolves on it like a wheel, or it may be because the Wain is there.

CHAPTER 37

ON THE POLES OF THE HEAVENS

1. The poles are little circles which run on the axis. Of these one is the northern, which never sets and is called Boreas; the other is the southern, which is never seen and is called Austronotius. . . .

CHAPTER 38

ON THE CARDINES OF THE HEAVENS

1. The *cardines* of the heavens are the ends of the axis and are called *cardines* (hinges) because the heavens turn on them or because they turn like the heart *(cor)*.[58]

CHAPTER 40

ON THE GATES OF THE HEAVENS

1. There are two gates of the heavens, the east and the west. For by one the sun appears, by the other he retires.

CHAPTER 42

ON THE FOUR PARTS OF THE HEAVENS

1. The *climata* of the heavens, that is, the tracts or parts, are four, of which the first part is the eastern, where some stars rise; the second, the western, where some stars set; the third, the northern, where the sun comes in the longer days; the fourth, the southern, where the sun comes in the time of the longer nights.[59]

4. There are also other *climata* of the heavens, seven in number, as if seven lines from east to west, under which the manners of men are dissimilar and animals of different species appear; they are named from certain famous places, of which the first is Meroe; the second, Syene; the third, Catachoras, that is Africa; the fourth, Rhodus; the fifth, Hellespontus; the sixth, Mesopontus; the seventh, Boristhenes.[60]

CHAPTER 43

ON THE HEMISPHERES

1. A hemisphere is half a sphere. The hemisphere above the earth is that part of the heavens the whole of which is seen by us; the hemisphere under the earth is that which cannot be seen as long as it is under the earth.

57. The periodic motion of the planets from west to east in a contrary direction to the daily motion of the heavens from east to west slows the great velocity of the daily motion and prevents the destruction of the heavens.

58. In what sense *cardines* turn like a heart is wholly unclear.

59. Isidore uses the term *climata* in two different senses, both of which derive from Greek antiquity. Here he employs it to represent the four directions of the celestial sphere, a usage which can be found in Cleomedes' treatise *On the Circular Motion of the Heavenly Bodies,* chapter 9, and in Strabo's *Geography* (see D. R. Dicks, *The Geographical Fragments of Hipparchus* [London: University of London, 1960], pp. 155–156). The more usual sense of *climata* is discussed in the next paragraph and in the next note.

60. The seven *climata* mentioned here agree with the seven distinguished by the Greeks except that the latter have Lower Egypt in place of Isidore's Catachoras. The seven *climata* were apparently introduced late (probably in the second century B.C.) and only because "they happened to be the seven parallels which passed through the best-known regions of the inhabited world" (Dicks, p. 157). Originally "the $\kappa\lambda\ell\mu\alpha\tau\alpha$ were narrow belts or strips of land on either side of a parallel of latitude; inhabitants of the same *clima* were assumed to be situated in the same geographical latitude, since, for practical purposes, the celestial phenomena, lengths of the longest and shortest days, and general climatic conditions did not change appreciably within the *clima*. . ." (Dicks, p. 154). Writers like Polybius and Strabo came to use the term in a broader sense to represent a district or region of the inhabited world (Dicks, p. 156). Eventually "the word lost its original scientific meaning and acquired the broader one of 'region' or 'district', but the names of the seven best known parallels were perpetuated" (Dicks, p. 158). In *Almagest,* Book II, chapter 13, Ptolemy drew up tables based upon the seven *climata* using the synonymous terms $\pi\dot\alpha\rho\dot\alpha\lambda\lambda\eta\lambda o\iota$ and $\kappa\lambda\ell\mu\alpha\tau\alpha$ to designate them, a practice which most ancient writers followed. Pliny (*Natural History,* VI, 212–220) also described seven circles or parallels which, however, were based on astrological rather than geographical criteria (Dicks, p. 157).

CHAPTER 44

ON THE FIVE CIRCLES OF THE HEAVENS

1. There are five zones in the heavens, according to the differences of which certain parts of the earth are inhabitable because of their moderate temperature and certain parts are uninhabitable because of extremes of heat and cold. And these are called zones or circles for the reason that they exist on the circumference of the sphere.

2. The first of these circles is called ἀρκτικὸς [the Arctic] because the constellations of the north are seen enclosed within it; the second is called θερινὸς [i.e., summer], which is called τροπικὸς [i.e., summer tropic], because in this circle the sun makes summer in northern regions and does not pass beyond it but immediately returns.

3. The third circle is called ἡμερινὸσ, which is equivalent to *equinoctialis* in Latin, for the reason that when the sun comes to this circle it makes equal day and night. For ἡμερινὸσ means in Latin day equal to the night, and by this circle the sphere is seen to be equally divided. The fourth circle is called ἀνταρκτικὸς [antarctic] for the reason that it is opposite to the circle which we call Arctic.

4. The fifth circle is called the Χειμερινὸς τροπικὸς [i. e., winter tropic, or tropic of Capricorn],[61] which in Latin is *hiemalis* or *brumalis*, because when the sun comes to this circle it makes winter for those who are in the north and summer for those who dwell in the parts of the south.

CHAPTER 47

ON THE SIZE OF THE SUN

1. The size of the sun is greater than that of the earth, and so from the moment when it rises it appears equally to east and west at the same time. And as to its appearing to us about a cubit in width, it is necessary to reflect how far the sun is from the earth, which distance causes it to seem small to us.

CHAPTER 48

ON THE SIZE OF THE MOON

1. The size of the moon also is said to be less than that of the sun. For while the sun is higher than the moon and still appears to us larger than the moon, if it should approach near to us it would be plainly seen to be much larger than the moon. Just as the sun is larger than the earth, so the earth is in some degree larger than the moon.

CHAPTER 49

ON THE NATURE OF THE SUN

1. The sun, being made of fire, heats to a whiter glow because of the excessive speed of its circular motion. And its fire, philosophers declare, is fed with water, and it receives the virtue of light and heat from an element opposed to it. Whence we see that it is often wet and dewy.

CHAPTER 50

ON THE MOTION OF THE SUN

1. They say that the sun has a motion of its own and does not turn with the universe. For if it remained fixed in the heavens all days and nights would be equal, but since we see that it will set tomorrow in a different place from where it set yesterday, it is plain that it has a motion of its own and does not move with the universe. For it accomplishes its yearly orbits by unequal distances on

61. Obviously, the fourth and fifth circles should be reversed. Isidore repeats this very same order of circles in Book XIII (*On he World and Its Parts*), chapter 6. This is surprising, since he seems to have ordered them correctly in his earlier work *On the Nature of Things* (*De natura rerum*), where chapter 10 is titled "On the Five Circles of the World." In the latter treatise, however, Isidore arranges the circles according to the fingers of the hand, seemingly as if these circles were in a single plane. In neither treatise does Isidore identify the zones as lying between two circles, or a pole and a circle; rather he identifies one zone with one circle, rendering his account unclear. Here is my translation of the relevant passage in *On the Nature of Things*, chapter 10 (translated from Fontaine, p. 209):

"In their definition of the world the philosophers have five circles (which the Greeks call parallels, i.e., zones) into which the orb of the earth is divided. In the *Georgics*, Virgil shows these, saying: "The sky has five zones" [*Georgics* I, 233—*Ed.*] But let us fix them in the manner of our right hand so that our thumb is the arctic circle, [which is] uninhabitable because of cold; the second [finger] is the summer (*therinos*) circle, [which is] temperate and habitable; the middle [finger] is the equinoctial (*isomerinos*) circle, [which is] torrid and uninhabitable; the fourth [finger] is the winter (*xeimerinos*) circle, [which is] temperate and habitable; the smallest [finger] is the antarctic circle, [which is] frigid and uninhabitable.

"2. The first of these is the north circle, the second the solstitial circle [i.e., tropic of Cancer], the third the equinoctial circle [i.e., equator], the fourth is the winter circle [i.e., tropic of Capricorn], and the fifth is the south circle [i.e., antarctic circle]."

Despite Isidore's arrangement of the circles in the same plane, it is not likely that he thought the earth flat. He was simply confused.

account of the changes of the seasons *(temporum mutationes)*.[62]

2. For going further to the south, it makes winter, in order that the land may be enriched by winter rains and frosts. Approaching the north, it restores the summer, in order that fruits may mature and what is green in the damp weather may ripen in the heat.

CHAPTER 52

ON THE JOURNEY OF THE SUN

1. The rising sun makes a journey to the meridian; and after it comes to the west and has bathed itself in ocean, it passes by unknown ways beneath the earth and again returns to the east.

CHAPTER 53

ON THE LIGHT OF THE MOON

1. Certain philosophers hold that the moon has a light of its own, that one part of its globe is bright and another dark, and that, turning by degrees, it assumes different shapes. Others, on the contrary, assert that the moon has no light of its own but is illumined by the rays of the sun. And therefore it suffers an eclipse if the shadow of the earth is interposed between itself and the sun. For the sun is farther than it. Hence when the moon is under it [i.e., between earth and sun], the sun lights the farther [or upper] part of the moon, and the nearer [or lower] part, which it holds toward the earth, would be darker.

CHAPTER 56

ON THE MOTION OF THE MOON

1. The moon governs the measures of the months by alternately losing and recovering its light. It advances in its path obliquely, not directly as the sun, lest it should appear in the center of the earth's [shadow] and frequently suffer eclipse.[63] For its orbit is near the earth. The waxing moon has its horns looking east; the waning, west: rightly, because it is going to set and lose its light.

CHAPTER 57

ON THE NEARNESS OF THE MOON TO THE EARTH

1. The moon is nearer the earth than is the sun. Therefore, with a smaller orb, it finishes its course more quickly. For it traverses in thirty days the journey the sun accomplishes in three hundred and sixty-five. Whence the ancients made the months

depend on the moon, the years on the course of the sun.

CHAPTER 58

ON THE ECLIPSE OF THE SUN

1. There is an eclipse of the sun as often as the thirtieth moon reaches the same line where the sun is passing, and, interposing itself, darkens the sun. For it seems that the sun disappears to us when the moon's orb is opposed to it.

CHAPTER 59

ON THE ECLIPSE OF THE MOON

1. There is an eclipse of the moon as often as the moon runs into the shadow of the earth. For it is thought to have no light of its own but to be illumined by the sun, whence it suffers eclipse if the shadow of the earth comes between it and the sun. The fifteenth moon suffers this until it passes out from the center and shadow of the interposing earth and sees the sun and is seen by the sun.

CHAPTER 61

ON THE LIGHT OF THE STARS

1. Stars are said to have no light of their own, but to be lighted by the sun like the moon.

CHAPTER 62

ON THE POSITION OF THE STARS

1. Stars are motionless, and being fixed, are carried along by the heavens in perpetual course, and they do not set by day but are obscured by the brilliance of the sun.

CHAPTER 63

ON THE COURSES OF THE STARS

1. Stars either are borne along or have motion. Those are borne along which are fixed in the heavens and revolve with the heavens. Certain have motion, like the planets, that is, the wandering stars, which go through roaming courses, but with definite limitations.

62. Since the lengths of the seasons are unequal and the sun's motion is always assumed to be uniform, it follows that the distances, which represent each of the four parts of the sun's annual orbit and which correspond to the four seasons, would be unequal.

63. Isidore is here referring to the obliquity of the lunar and solar orbits. If they were in the same plane, eclipses would occur frequently.

CHAPTER 64

ON THE VARYING COURSES OF THE STARS

1. According as stars are carried on different orbs of the heavenly planets, certain ones rise earlier and set later, and certain rising later, come to their setting earlier. Others rise together and do not set at the same time. But all in their own time revolve in a course of their own.

CHAPTER 65

ON THE DISTANCES OF THE STARS

1. Stars are at different distances from the earth and therefore, being of unequal brightness, they are more or less plain to the sight; many are larger than the bright ones which we see, but being further away they appear small to us.[64]

CHAPTER 66

ON THE CIRCULAR NUMBER OF THE STARS

1. There is a circular number of the stars by which it is said to be known in what time each and every star finishes its orbit, whether in longitude or latitude.

2. For the moon is said to complete its orbit every year, Mercury in twenty, Lucifer in nine, the sun in nineteen, Vesper in fifteen, Phaeton in twelve, Saturn in thirty.[65] When these are finished, they return to a repetition of their orbits through the same constellations and regions.[66]

CHAPTER 67

ON THE WANDERING STARS

1. Certain stars are called *planetae,* that is, wandering, because they hasten around through the whole universe with varying motions. Because they wander or produce irregularities, they are called retrograde; that is, when they add and subtract little bits. When they subtract so much of the rest, they are said to be retrograde; when, however, they stand [or rest], they make a station.[67]

CHAPTER 68

ON THE PRECEDING [MOTION] OF STARS

1. *Praecedentia* or *antegradatio* of stars is when a star seems to be making its usual course and [really] is somewhat ahead of it.

CHAPTER 69

1. *Remotio* or *retrogradatio* of stars is when a

star, while moving on its regular orbit, seems at the same time to be moving backward.

CHAPTER 70

1. The *status* of stars means that while a star is continuing its proper motion it nevertheless seems in some places to stand still.

CHAPTER 71

ON THE NAMES OF STARS

3. *Stellae* is derived from *stare,* because the stars always remain *(stant)* fixed in the heavens and do not fall. As to our seeing stars fall, as it were, from heaven, they are not stars but little bits of fire that have fallen from the ether, and this happens when the wind, blowing high, carries along with it fire from the ether, which as it is carried along gives the appearance of falling stars. For stars cannot fall; they are motionless (as has been said above) and are fixed in the heavens and carried around with them.

16. A comet is so called because it spreads light from itself as it if were hair *(comas).* And when this kind of star appears it indicates pestilence, famine, or war.

17. Comets are called in Latin *crinitae* because they have a trail of flames resembling hair *(in modum crinium).* The Stoics say there are over thirty of them, and certain astrologers have written down their names and qualities. . . .[68]

64. Here Isidore appears to assume that the stars are at different distances from the earth, a position that seems incongruous with the notion of a stellar sphere in which the stars are embedded. Only if the sphere of fixed stars was thought to be of enormous thickness could one suppose that stars of varying sizes were scattered about between its widely separated concentric surfaces.

65. In this strange passage Lucifer is Venus, Phaeton is Jupiter, and Vesper, which means evening star, must signify Mars (probably Pyroeis, the Greek name for Mars, was intended), although it was a name used for Venus. Aside from twelve years for Jupiter and thirty for Saturn, which are their respective sidereal periods, it is wholly unclear what orbital phenomena Isidore thought he was representing with his other data.

66. Paragraph 3, which consists of a few lines and concludes chapter 66, has been omitted.

67. This opaque description is improved somewhat in chapter 69. For a discussion of retrogradations and stations see Selection 64.

68. The remainder of chapter 71, the final chapter of Book III, contains approximately two more pages on how various celestial bodies and divisions received their names.

2　ON ARITHMETIC

Boethius (fl. 480–524)

Translation, introduction, and annotation by Edward Grant[1]

The *Arithmetic* of Boethius is actually a Latin paraphrase, bordering on translation, of the *Introduction to Arithmetic* written in Greek by Nicomachus of Gerasa sometime around A.D. 100.[2] The latter is a theoretical and philosophical arithmetic rather than a business arithmetic, or *logistic* as it was called by the Greeks. Among theoretical arithmetics, however, it is necessary to distinguish between those represented by Euclid's *Elements* VII–IX (unfortunately, the only one of its genre extant) and those represented by Nicomachus' *Arithmetic*.[3]

It is a very far cry from Euclid to Nicomachus. In the *Introductio arithmetica* we find the form of exposition entirely changed. Numbers are represented in Euclid by straight lines with letters attached, a system which has the advantage that, as in algebraical notation, we can work with numbers in general without the necessity of giving them specific values; in Nicomachus numbers are no longer denoted by straight lines, so that, when different undetermined numbers have to be distinguished, this has to be done by circumlocution, which makes the propositions cumbrous and hard to follow, and it is necessary after each proposition has been stated, to illustrate it by examples in concrete numbers. Further, there are no longer any proofs in the proper sense of the word; when a general proposition has been enunciated, Nicomachus regards it as sufficient to show that it is true in particular instances; sometimes we are left to infer the general proposition by induction from particular cases which are alone given. . . . Probably Nicomachus, who was not really a mathematician, intended his *Introduction* to be, not a scientific treatise, but a popular treatment of the subject calculated to awaken in the beginner an interest in the theory of numbers by making him acquainted with the most noteworthy results obtained up to date; for proofs of most of his propositions he could refer to Euclid and doubtless to other treatises now lost. The style of the book confirms this hypothesis; it is rhetorical and highly coloured; the properties of numbers are made to appear marvellous and even miraculous; the most obvious relations between them

are stated in turgid language very tiresome to read. It was the mystic rather than the mathematical side of the theory of numbers that interested Nicomachus. If the verbiage is eliminated, the mathematical content can be stated in quite a small compass. Little or nothing in the book is original, and, except for certain definitions and refinements of classification, the essence of it evidently goes back to the early Pythagoreans.[4]

It was this treatise, described so aptly by Heath, that was taken over by Boethius, who, though frequently adding and occasionally rearranging, contributed nothing really new to the original and seems to have been dependent exclusively on Nicomachus. Boethius' Latin version was destined to exert a great influence on subsequent encyclopedic authors of the sixth and seventh centuries and throughout the Middle Ages up to the sixteenth century. From the sixth to the twelfth century, when Greek geometry had almost vanished and science was at its lowest ebb, Boethius' *Arithmetic*, for all its faults, preserved the ideal of a theoretical science. Not until the thirteenth century, when Jordanus de Nemore's *Arithmetic* appeared in ten books,[5] do we have a theoretical arithmetic on the Euclidean model, complete with proofs.

Book I

CHAPTER 2

ON THE SUBSTANCE OF NUMBER

All things constructed from the first nature of things seem to have been formed after the manner

1. My translation is from the Latin edition by G. Friedlein, *Boetii De institutione Arithmetica libri duo; De institutione Musica; accedit Geometria quae fertur Boetii* (Leipzig, 1867).
2. For an English translation of the work, see *Nicomachus of Gerasa: Introduction to Arithmetic*, translated by Martin Luther D'Ooge, with studies in Greek Arithmetic by Frank E. Robbins and Louis C. Karpinski (New York: Macmillan, 1926). This will be referred to hereafter as *Nicomachus: Arithmetic*.
3. Between Euclid and Nicomachus no theoretical arithmetics have survived, but undoubtedly some were written.
4. Thomas L. Heath, *A History of Greek Mathematics* (Oxford: Clarendon Press, 1921), I, 97–99.
5. A few propositions from this treatise are included in Selection 21.

of numbers. Indeed, this was the principal exemplar in the mind of the Creator from which came the multitude of the four changing elements, the alterations in times, the motions of the stars, and the revolution of the sky. And since the status of all things is observed by the union [or association] of numbers, it is also necessary that the number in a particular substance always be preserved equally and that it [the number] not be composed of different things—for what could conjoin [or unite] the substance of a number?—but that it be seen to be composed of itself alone.[6] Moreover, nothing seems to be composed of similar things; nor is anything constituted without any harmonious ratio; and [nothing is constituted of things] that are wholly distinct from each other in substance and nature. It is evident, then, that although number is composed, it is not composed of similar things, nor from things that are mutually unrelated by any kind of ratio [or harmony]. In the first place, therefore, numbers will be united to a substance that always endures and is permanent. For nothing at all can be produced from nonexistent things, but only from dissimilar things [coupled] with the power of composing [these dissimilar things]. Number, then, consists of these things; even and odd, which, though they are dissimilar and contrary, flow together by a certain divine power and are joined in one composition and measure.[7]

CHAPTER 3

THE DEFINITION AND DIVISION OF NUMBER AND THE DEFINITION OF EVEN AND ODD

First, we must define what a number is. Number is a collection of units or a multitude of a quantity brought together from units. The first division of this is into odd and even. Even is that which can be divided into two equals with nothing intervening; odd is that which nothing divides into equals because one [unit] intervenes in the middle. And this is a common definition and is known.

CHAPTER 8

THE DIVISION OF AN EVEN NUMBER

There are three species of even numbers. One is called evenly even *(pariter par)*, another evenly uneven *(pariter impar)*, and a third is unevenly even *(inpariter par)*. Indeed evenly even and evenly uneven seem to be contraries holding the place of extremes. The number which is called unevenly even is a certain middle which shares [something] with each of these [extremes].

CHAPTER 9

ON AN EVENLY EVEN NUMBER AND ITS PROPERTIES

Evenly even is a number that can be divided into two even parts; and one of these parts can be divided into two other even parts; and one of these parts [divided] into another two even parts; and this happens as often as possible until the division of parts arrives naturally to the indivisible unit [i.e., 1]. Thus number[8] 64 has 32 as a half, which in turn has 16 as a half, which has 8 as half; and this is divided into equal parts of 4, which is the double of two. But two is divided in half by the unit, which,

6. As Boethius tells us further on, number is composed of "itself alone" in the sense that it consists of odd and even, which are unified by divine power.

7. Because of the seeming strangeness of Boethius' discussions, it will be helpful to reproduce the corresponding chapter in Nicomachus' Arithmetic (Bk. I, ch. 6; *Nicomachus: Arithmetic,* pp. 189–190):

"All that has by nature with systematic method been arranged in the universe seems both in part and as a whole to have been determined and ordered in accordance with number, by the forethought and the mind of him that created all things; for the pattern was fixed, like a preliminary sketch, by the domination of number preëxistent in the mind of the world-creating God, number conceptual only and immaterial in every way, but at the same time the true and the eternal essence, so that with reference to it, as to an artistic plan, should be created all these things, time, motion, the heavens, the stars, all sorts of revolutions.

"It must be, then, that scientific number, being set over such things as these, should be harmoniously constituted, in accordance with itself; not by any other but by itself. Everything that is harmoniously constituted is knit together out of opposites and, of course, out of real things; for neither can non-existent things be set in harmony, nor can things that exist, but are like one another, nor yet things that are different, but have no relation one to another. It remains, accordingly, that those things out of which a harmony is made are both real, different, and things with some relation to one another.

"Of such things, therefore, scientific number consists; for the most fundamental species in it are two, embracing the essence of quantity, different from one another and not of a wholly different genus, odd and even, and they are reciprocally woven into harmony with each other, inseparably and uniformly, by a wonderful and divine Nature, as straightway we shall see."

Unless it is of unusual interest or significance, no effort will be made to indicate correspondences or differences between Nicomachus and Boethius.

8. Since Arabic (or Hindu) numerals were not yet known in the Latin West, both Isidore of Seville and Boethius used Roman numerals. Although these were retained in the preceding selection from Isidore, they have been altered here to Arabic numerals. This particular example of an evenly even number was used by Isidore of Seville (see previous selection, ch. 5, par. 3).

since it is naturally singular, does not receive any division. . . . The generation of these is as follows: [Starting] from one and taking [successive] double ratios, you will observe that evenly even numbers are generated. Other than this generation, it is impossible that they be generated otherwise. An example of this could be given by a series taking all the doubles starting from one: 1, 2, 4, 8, 16, 32, 64, 128, 256, 512, and so on; and if an infinite progression were made, you would find all such numbers. . . .

CHAPTER 10

ON AN EVENLY UNEVEN NUMBER AND ITS PROPERTIES[9]

CHAPTER 11

ON AN UNEVENLY EVEN NUMBER AND ITS PROPERTIES[10] AND ON THE RELATION OF IT TO EVENLY EVEN AND EVENLY UNEVEN [NUMBERS]

CHAPTER 13

ON ODD NUMBER AND ITS DIVISION

Similarly, odd number, which is distinct from the nature and substance of even number—the one can be divided into two (gemina) equal parts, the other cannot be divided, for it is prevented by the intervention of a unit—has three subdivisions. One of these is a number which is called prime and incomposite;[11] the second (secunda) is called secondary (secundus) and composite;[12] and the third is in between these and is linked to them naturally, drawing something from its relation to each of them, for by itself it is secondary and composite, but compared to others it is found to be prime and incomposite.[13]

CHAPTER 17

ON THE PRODUCTION OF THE PRIME AND INCOMPOSITE [NUMBERS] BY THEMSELVES AND THE SECONDARY AND COMPOSITE BY THEMSELVES [AND] ON THE PRODUCTION OF A SECONDARY AND COMPOSITE [NUMBER] TO ONE THAT IS PRIME AND INCOMPOSITE

The generation and origin of these is derived from a discovery which Eratosthenes called "a sieve" (cribrum).[14] With all the odd numbers

only factor is unity or 1; for when we take 1/3 of 3, 1/5 of 5, 1/7 of 7, and so on, we always obtain 1. In *Elements* VII, Definition 11, Euclid says that a prime number is measured by unity alone. Thus, number 2 satisfies his definition (it is also called prime by Aristotle in *Topics* VIII. 2.157a.39) but not that of Nicomachus, who restricts the notion of prime number to odd numbers alone. In his *Etymologies,* Isidore of Seville includes an abbreviated version of Boethius' fourteenth chapter (see previous selection, ch. 5, par. 7).

12. In chapter 15 (omitted here), where Boethius discusses this class of odd numbers, the numbers 9, 15, 21, 25, 27, 33, 39 are offered as illustrations of odd numbers with factors other than 1. Obviously there are also even numbers with factors other than 1, but these are excluded from the class of composite numbers. For Isidore's resumé of composite numbers, see previous selection, chapter 5, paragraph 7.

13. This is taken up by Boethius in chapter 16 (omitted here), where the numbers 9 and 25 are used as illustrations. The numbers 9 and 25 are considered secondary and composite, since each has factors in addition to 1; but they share no common factors other than 1 and hence are prime to each other. This third category of odd numbers embraces relatively prime numbers. In the previous selection by Isidore, chapter 5, paragraph 8, see his somewhat confused summary presentation of Boethius' sixteenth chapter. In Euclid's *Elements* VII, Definition 13: "Numbers prime to one another are those which are measured by an unit alone as a common measure." The translation is from Thomas L. Heath, *The Thirteen Books of Euclid's Elements,* 2d ed. (New York: Dover, 1956), II, 278. Obviously Euclid's definition includes comparisons between odd and even numbers, which were excluded by Nicomachus and Boethius.

14. Eratosthenes lived in the third century B.C. and was a contemporary of Archimedes. He made contributions to literature and the sciences and also estimated the circumference of the earth by a geometric method. See M. R. Cohen and I. E. Drabkin, *A Source Book in Greek Science* (Cambridge, Mass.: Harvard University Press, 1948), pp. 149–153, and Selection 64.1.

Our information on the "sieve of Eratosthenes" is derived from Nicomachus' Arithmetic, Book I, chapter 13, and was made available to the Middle Ages by Boethius. The rather long-winded account of the "sieve" that follows is nicely summarized by Thomas Heath (*A History of Greek Mathematics,* I, 100):

"The method is this. We set out the series of odd numbers beginning from 3. 3, 5, 7, 9, 11, 13, 15, 17, 19, 21, 23, 25, 27, 29, 31, . . . Now 3 is a prime number, but multiples of 3 are not; these multiples, 9, 15, . . . are got by passing over two numbers at a time beginning from 3; we therefore strike out these numbers as not being prime. Similarly 5 is a prime number, but by passing over four numbers at a time, beginning from 5, we get multiples of 5, namely 15, 25, . . . ; we accordingly strike out all these multiples of 5. In general, if n be a prime number, its multiples appearing in the series are found by passing over $n-1$ terms at a time, beginning from n; and we can strike out all these multiples. When we have gone far enough with this process, the numbers which are still left will be primes. Clearly, however, in order to make sure that the

9. This is described by Isidore of Seville (see previous selection, ch. 5, par. 4).

10. See Isidore of Seville (previous selection, ch. 5, par. 5).

11. In chapter 14 (omitted here), where Boethius discusses this type, he offers as examples the numbers 3, 5, 7, 11, 13, 17, 19, 23, 29, and 31, where in each case the

arranged in order, there can be distinguished by the art [or method] *(artem)* which we are about to describe those numbers which are of the first kind, or the second, or the third.[15] For let all the odd numbers be arranged in sequence from the number three and extended as far as one pleases: 3, 5, 7, 9, 11, 13, 15, 17, 19, 21, 23, 25, 27, 29, 31, 33, 35, 37, 39, 41, 43, 45, 47, and so on. After these have been ordered, we must consider the first number and see which of the numbers in the sequence it can measure. But having passed over two numbers that were placed after it, it then measures the next one; and if the two other numbers which follow the number measured are passed over, the number after them is measured; and in the same way if one should leave two places after it, which would be measured by the first number; and always in the same way, by omitting two numbers, all numbers, proceeding to infinity would be measured by the first number.

But I do not wish to present this confusedly. For example, the first number measures by its own quantity *(per suam quantitatem)* that number which is placed two numbers after it. Thus, after omitting the two numbers 5 and 7, number 3 measures 9. It does this [in a measure equal to its] quantity, that is, three times, for it measures 9, the third number, by three. Then if I should omit two numbers after 9, the number I run into is measured by the first number [i.e., 3] by a quantity [equal to] the second odd number, namely number 5. For if I omit two numbers after 9, namely 11 and 13, the third number is 15, which is measured by the quantity of the second number, that is, it is measured five times, since three measures 15 five times. Again, if, beginning from 15, I omit the two numbers that are placed after it, the first number is the measure of it by [a quantity equal to] the third odd number [i.e., 7]. For if I omit 17 and 19 after 15, [the number] 21 occurs, which is measured by 7 taken three times, for the seventh part of the number 21 is three. And if two numbers are always omitted, by doing this ad infinitum I find that the first number [3] can measure all the numbers following according to the quantity of the odd numbers arranged in sequence.[16]

But if now we take the number 5, the second number in the series, and someone wishes to find the first and next measure of it, it will be the fifth from it and appear after omitting four odd numbers. For let four odd numbers be omitted, namely 7, 9, 11, and 13. The number after these is 15, which the number five measures as many times as the

quantity of the first number, namely 3, for 5 measures 15 by three. And if the next four terms are omitted, 5 will measure the second number after them [i.e., 25] by its own quantity, namely five times. Thus, if after 15, [the numbers] 17, 19, 21, and 23 are omitted, I find 25, which the number five measures several times, for multiplying 5 five times increases to *(succrescunt)* 25. Now if the sequence is preserved and the next four numbers are omitted,[17] the number which follows them [i.e., 35] is measured five times by the sum total of the third number [in the series], that is by seven; and this is an infinite progression.[18]

If, now, the third number [i.e., 7] which can measure is considered, six numbers will be abandoned and the sequence will show that the seventh term would be measured by the quantity *(quantitatem)* of the first number, that is, by 3. And after the next six numbers, a number will occur in the series which is measured five times (that is, by the second number in the series [beginning from 3]) by the third number [in the series, namely 7]. But if another six numbers were eliminated, the number which follows is measured seven times by the same number 7, that is, by the quantity of the third *(per tertii quantitatem)* [number in the series of odd numbers beginning with 3]; and the series progresses in this way to the end.

---

odd number $2n+1$ in the series is prime, we should have to try all the prime divisors between 3 and $\sqrt{(2n+1)}$; it is obvious, therefore, that this primitive empirical method would be hopeless as a practical means of obtaining prime numbers of any considerable size."

15. That is, (1) absolutely prime numbers, (2) composite numbers, or, in this context, multiples of prime numbers, and (3) composite numbers that are relatively prime to each other.

16. That is, all the multiples of the prime number 3 will be measured by it. In the series of odd numbers, these multiples would be located by noting every third number after 3. Moreover, the successive terms 9, 15, 21, 27, 33, . . . are measured by prime number 3 in accordance with the series of odd numbers beginning with 3. That is,

3,	9,	15,	21,	27,	33, . . .
	3(3),	3(5),	3(7),	3(9),	3(11), ...

17. That is, 27, 29, 31, 33.

18. That is,

5,	15,	25,	35,	45, . . .
5(3),	5(5),	5(7),	5(9), ...

Here the second number in the series of odd numbers, namely 5, measures its first multiple, 15, by the first term in the series, namely 3; 5 measures the second multiple term in the series, namely 25, five times, which is equal to the second term in the series of odd numbers; and 5 measures its third multiple, 35, by a number of times equal to the third odd number in the series, namely 7, and so on.

Therefore, the naturally constituted series of odd numbers, just as they are [i.e., in their fixed order], are assumed to measure the changes. Now, if they pass over the intervening odd numbers by a fixed interval [determined] by the even numbers beginning with 2, the measurement would occur in such a way that the first [number of odd numbers omitted] is two; the second is four; the third six; the fourth eight; the fifth ten; or [this could be arrived at] if [the number of] their positions were doubled and the terms [or numbers] to be omitted were made in accordance with this duplication, so that 3, the first number and [therefore represented by] 1—for every first term is 1—would multiply its place twice; and since we get 2, the first number [in this series of odd numbers] would [cause us to] pass over two intervening terms. Again, if the position of the second [number in the series of odd numbers], namely 5, were doubled, it would produce 4, so that 4 places would intervene. Similarly, if 7, which is the third [number in the series], were doubled, it would create six intervening places, for 3 taken twice produces 6.[19] Therefore, 6 [successive places] in the series would be omitted. Also, if the fourth place were doubled, 8 would result and eight [successive places] would be passed over. And this is clear for the other places.

This series will furnish a method of measuring according to the sequence [of the terms] arranged. For the first number that numbers the first number numbers in accordance with the first number, that is according to itself; and the first number that numbers the second number numbers according to the second, and the third number [in the series is first numbered] by the third number, and the fourth by the fourth,[20] and the measurement in the other numbers will be done in a similar way.

Therefore, if you should consider other numbers, either others that measure, or those that are measured by others, you would find that there can be no common measure of all so that any other [number] could not number all at the same time; moreover [you would find] that some of them could be measured by another—indeed they might be numbered by one [number] only; you would find that yet others could be measured by several [numbers], and that for some there would be no common measure other than unity. Those numbers that can be measured by no number other than unity, we judge to be prime and incomposite; those that share some measure other than unity or the designation of a different part,[21] we call secondary and composite. The investigator would

find that [numbers of] the third kind, which taken by themselves are of the secondary and composite type, would be primary and incomposite when compared to each other. For if you should multiply any whatever of these [prime and incomposite] numbers by their own quantity, the numbers that are produced are not connected by any common measure when related to each other. Thus, if you multiply 3 and 5, 3 by 3 makes 9, and 5 times 5 makes 25. Therefore, there is no relation of a common measure. Again, if you relate what 5 and 7 produce, these will also be incommensurable, for, as was said, 5 times 5 makes 25, and 7 times 7 makes 49, and there is no common measure except unity, the generator and mother of all these [numbers].

CHAPTER 18

ON FINDING THOSE NUMBERS THAT ARE SECONDARY AND COMPOSITE WITH RESPECT TO EACH OTHER [AND NUMBERS THAT ARE] PRIME AND INCOMPOSITE RELATIVE TO OTHERS

The method by which we can find such numbers, if someone proposes them to us and declares that it is not known whether they are commensurable in any measure or [whether] the unit alone measures each, is this. Should two unequal numbers be given, it will be necessary to subtract *(auferre)* the smaller from the greater, and if what remains is greater, subtract the smaller from it again; [but] if it is smaller, subtract *(detrahere)* it from the greater [number] that remains, and this should be done until [either] unity finally prevents any further diminution, or, if each of the numbers proposed is odd, some number [is reached that is] necessarily odd; but you will see that the number which is left

19. The Latin text, which I have carefully followed at this point, is very misleading, for it leads one to expect that 7 is to be doubled. But as is obvious, it is only the number 3, representing the location of 7 as the *third* number in the series, that is doubled.

20. That is, 3, the first number in the series of odd numbers, numbers itself; 5, the second number, also numbers itself first; 7, the third number, is the first number that numbers 7 (hence "the third number . . . is first numbered by the third number"); the fourth number 9 is the first number to number or measure 9.

21. With respect to what appears to be the corresponding passage in Nicomachus, Robbins and Karpinski declare that "the numbers referred to are the squares of odd prime numbers. 9, e.g., has ninths, of course, and as it is measured by 3, it will also have thirds, the denomination of which is derived from 3" (*Nicomachus: Arithmetic*, p. 205, n. 2). Thus, *ninths* and *thirds* are the different parts that are shared.

is equal to that [odd] number.[22] And so it is that if this subtraction *(subtractio)* should, in turn, reach one, the numbers are said to be prime to each other necessarily and they are conjoined by no other measure except unity alone. If, however, the end of the subtraction arrives at some [odd] number as was said above, it will be a number that measures each sum, and we call the same number that remains the common measure of each.[23]

Take two proposed numbers with respect to which we do not know whether some common measure measures them; let these be 9 and 29. Now we make an alternate subtraction. Let the smaller be subtracted from the greater, that is, 9 from 29, and 20 is left; let us now again subtract the smaller, that is, 9, from 20, and 11 is left; I again subtract 9 from the remainder [i.e., 11] and 2 remains. If I subtract this from 9, 7 is left, and if I again take 2 from 7, 5 remains; and from this another 2 and 3 remains, which after it is diminished by another 2 leaves only unity. Again, if I subtract one from two, the end of the subtraction is fixed at one, which shows that there is no other common measure of these two numbers, namely 9 and 29. Therefore, we will call these numbers prime to each other.

But should other numbers be proposed in the same situation, that is 21 and 9, they could be investigated since they would be mutually related. Again I subtract the quantity of the smaller number from the greater, that is, 9 from 21, and 12 remains. From 12 I take 9 and 3 remains, which if subtracted from 9 leaves 6; and if 3 were taken from 6, 3 would be left, from which 3 cannot be subtracted for it is equal to it. For 3, which was reached by continually subtracting, cannot be subtracted from 3, since they are equal. Therefore, we shall pronounce them commensurable, and 3, which is the remainder, is their common measure.[24]

Book II

CHAPTER 5

ON LINEAR NUMBER

So it is also in number. Although unity is not a linear number, yet it is the beginning of number extended into length;[25] and linear number, although it is devoid of width, nevertheless shares the beginning of a number extended into another dimension of width. Also a plane number, although it is not a solid body, is the beginning *(caput)* of a solid body added to the width. This will be seen more clearly from examples. Beginning with two,

linear number always increases in one and the same dimension by the addition of 1. And it is this which we consider.

CHAPTER 6

ON PLANE RECTILINEAR FIGURES AND THAT THE TRIANGLE IS THE ORIGIN (PRINCIPIUM) OF THEM

After adding width to the definition a plane number is found in numbers beginning with 3. The angles differ by the multitude of the natural numbers that follow, so that the first is a triangular number;[26] the second a square number; the third, which is contained by five angles, the Greeks call a pentagon; the fourth is a hexagon, that is, it is enclosed by six angles; and, in the same way, the others increase their angles in the plane by one, namely by the definition *(descriptione)* of the figures. Therefore, these begin with number 3, because only 3 is the beginning of width and surface. In geometry the same thing is found more clearly. For two straight lines do not contain a space, and if the lines of every triangular figure, or tetragon, or pentagon, or hexagon, or any such figure whatever that is contained by several angles are drawn from the middle through each angle, they will divide the figure into as many triangles as it has angles. For lines that have been drawn in this way divide

22. That is, the final term reached, or the subtrahend, is equal to the remainder. This is shown in the following example.

23. The procedures here are formally demonstrated by Euclid in *Elements* VII.1 ("Two unequal numbers being set out, and the less being continually subtracted in turn from the greater, if the number which is left never measures the one before it until an unit is left, the original numbers will be prime to one another.") and VII.2 ("Given two numbers not prime to one another, to find their greatest common measure."). The translations are by Thomas Heath, *The Thirteen Books of Euclid's Elements,* II, 296, 298. It should be noted that Euclid's proofs embrace all numbers, odd and even, whereas Boethius and Nicomachus are concerned only with odd numbers.

24. In the remaining chapters of Book I, Boethius considers perfect and imperfect numbers and the five species of ratios of greater and lesser inequality. Since these topics were included in Isidore of Seville's treatment of the quadrivium (see previous selection, ch. 6), they are omitted here.

25. The relation between the unit, linear, plane, and solid numbers in arithmetic is given as analogous to the relation between point, line, surface, and solid in geometry. This was discussed in Boethius's Book II, chapter 4 (not included here).

26. For a discussion of triangular numbers based ultimately upon Boethius but transmitted by Cassiodorus, see note 22 of Isidore of Seville's "On the Quadrivium."

a square into four [triangles], a pentagon into five triangles, a hexagon into six, and divide others into triangles in the mode and measure [or number] of their angles. Thus, this description can be set forth [Fig. 1[27]]: A square is divided into four triangles,

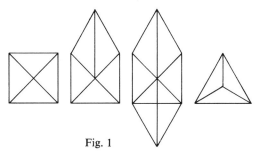

Fig. 1

a pentagon into five triangles, a hexagon into six triangles. But when a triangular figure will have been divided in this way, it will not be resolved into other figures but only into itself. For a triangle is resolved into three. A triangle is divided into three triangles. And thus this is the elemental figure of width so that all surfaces are resolved into it, but it is resolved only into itself, since it is subject to no more basic figures nor does it originate from another extended figure. Now, the same thing can be done in numbers, as will be shown in what follows in this work.

CHAPTER 40

ON PROPORTIONS

And enough has been said about these things.[28] Now some things should be considered about proportions, which could be applied to musical speculations, astronomical subtleties, or to the power of geometrical thought, or even to an understanding of ancient works. Thus it is very fitting that this terminate the arithmetic introduction.

A proportion *(proportionalitas)* is the taking of a whole collection of ratios *(proportionum)*, two or three or however many. It is also commonly defined [in this way]: a proportion *(proportionalitas)* is a similar relation of two or more ratios, even if they are not constituted by the same quantities and differences.[29] A difference *(differentia)* is the quantity between numbers. A ratio *(proportio)* is a certain mutual relation of two terms and to the extent that it is continuous *(continentia)*, the arrangement of the terms which it [i.e., the continuity] produces is proportional and from the ratios that were linked there arises a proportion *(proportionalitas)*. The least proportion is found in three terms. It also occurs in greater terms but in a

longer series. [For example,] 2 to 1 is a double ratio (since there are two terms); but if you compare 4 to 2, there is here also a double ratio. If you consider these three terms continually, there is a proportion of two ratios and the proportion is 1 to 2 and 2 to 4.[30] For, as was said, a proportion *(proportionalitas)* is a collection of ratios in one relation. It also happens in longer series. For if to these four [terms] you wish to join 8; and to these you wish to add 16; and then 32, and then [you add the] successive doubles which follow, a double proportion will result from all these double ratios. Therefore, as often as one and the same term is linked to the two terms around it, so that to one term it is greater *(dux)* and to the other smaller *(comes)*,[31] this is called a continued proportion *(proportionalitas continua)*, as 1, 2, 4. For there is

27. The pentagon and hexagon are reproduced in Fig. 1 as they appear in Friedlein's edition on pp. 91, 92. That they are not regular figures is obvious. One manuscript describes the pentagon as regular, and two call the hexagon regular. These readings, however, were placed among the variants by Friedlein. To allow for the ambiguity, I have translated *medietate,* which occurs two sentences earlier, as "middle" rather than "center."

28. A reference to the preceding chapters involving the formation of various sequences and types of numbers.

29. Here Boethius means that if $A/B = C/D$, the quantities could be of different kinds; for example, A and B may form a ratio of distances and C and D a ratio of times. Moreover, if $A - B \neq C - D$, their differences are unequal.

30. It is curious that Boethius gives the original ratios as double ratios, that is, $2/1$ and $4/2$, but links them in a proportion as subdouble ratios, that is, $1/2$ and $2/4$. Nicomachus (Bk. II, ch. 21; *Nicomachus: Arithmetic,* p. 265) is consistent on this, presenting it all in terms of subdouble ratios.

31. In Book I, chapter 23 (omitted here), Boethius discusses superparticular ratios, which are of the form $(n+1)/n$ where n is any integer (see Isidore of Seville, "On the Quadrivium," ch. 6, par. 7, and my n.17 on that selection). Boethius declares that the greater numbers of such ratios are called *duces* (the plural of *dux*) and the smaller numbers *comites* (the plural of *comes*). Since the greater number in a superparticular ratio is also the first or *antecedent* term (*dux* means leader) and the smaller term is the second or *consequent* term (*comes* means follower), these terms would also be suitable equivalents in the first example, where $4/2 = 2/1$. That the primary signification of *dux* and *comes* would appear to be *greater* and *smaller,* respectively, rather than antecedent (numerator) and consequent (denominator), respectively, is borne out by the converse of the first example of a continued proportion. Thus, when $1/2 = 2/4$, 2 is *dux* in $1/2$ only if it is construed as the greater number but not if sequence, or order, is paramount, for then it is a consequent or denominator; similarly 2 is *comes* in $2/4$ only if *comes* is interpreted as the smaller number and order is ignored, since 2 is also the first or antecedent term in $2/4$.

equality in these ratios: just as 4 is to 2 so is 2 to 1; and conversely, as 1 is to 2 so is 2 to 4. There is equality also in the quantity of the numbers, for as 3 exceeds 2 so 2 exceeds 1; and [conversely,] as 1 is less than 2 so is 2 exceeded by 3. But if another term is related to one term and another to another, it is necessary that this relation be called *disjunct* with respect to the quality of the proportion, as [for example,] 1, 2, 4, 8: for just as 2 is to 1 so is 8 to 4; and conversely, as 1 is to 2 so is 4 to 8; and alternately *(permutatim)*,[32] as 4 is to 1 so is 8 to 2. And with respect to the quantity of number [in a disjunct relation], as 1, 2, 3, 4, [we see that] as 1 is exceeded by 2, by the same amount is 3 exceeded by 4; and as 2 exceeds 1, by so much does 4 exceed 3; and alternately *(permixtum)*[33] by as much as 1 is less than 3 so also is 2 less than 4; or by as much as 3 exceeds 1, by so much does 4 surpass 2.

CHAPTER 43

ON THE ARITHMETIC PROPORTION[34] AND ITS PROPERTIES

We call this an arithmetic proportion: whenever three or more terms are assumed and there is found an equal and same difference between all the terms so arranged. However, the equality of the ratios formed is disregarded as the preservation of the differences is maintained. For example, [take the series of numbers] 1, 2, 3, 4, 5, 6, 7, 8, 9, and 10. If anyone should care to examine the continuous differences between the terms in this natural sequence of numbers, [he would find that] in accordance with an arithmetic proportion the difference between the terms themselves is the same, for the differences are equal, but there is not the same ratio and relation.[35] Therefore, if the discussion concerns three terms, it is called a continued proportion *(continua proportionalitas)*; but if there is another greater term *(dux)* and another smaller term *(comes)* and there are other terms for each [ratio], this will be called a disjunct proportion *(disiuncta medietas)*. Hence, if you consider a continued proportion in only three terms, or a disjunct proportion in four or more terms, you will always observe the same differences between the terms, with only the ratios between the terms altered. . . .

CHAPTER 44

ON GEOMETRIC PROPORTION AND ITS PROPERTIES

Now, the geometric mean, which follows this, is alone, or especially, able to be called a proportion because the greater or smaller ratios between its terms are the same. For an equal ratio is always preserved, although the quantity and multitude of the numbers is ignored, which is opposite to what was described for arithmetic proportion. Thus let there be 1, 2, 4, 8, 16, 32, 64 [in a double ratio], or, in a triple ratio, 1, 3, 9, 27, 81; or numbers arranged in a quadruple, or quintuple, or any whatever multiplicity of numbers extended in this way. You can take any terms of these series and they will form a geometric proportion, for just as one is to the very next term following, so is the one following to the next term; and this would happen if you did this in an alternate way. For if three terms, 2, 4, and 8, were assumed, then just as 8 is to 4 so is 4 to 2; and if you convert them, just as 2 is to 4 so will 4 be to 8. . . .

CHAPTER 47

ON THE HARMONIC PROPORTION AND ITS PROPERTIES

The harmonic proportion consists neither of the same differences nor of equal ratios, but is this: Just as the greatest term is to the smallest, so is the difference between the greatest and middle term to the difference between the middle term and smallest term.[36] As an example, take 3, 4, 6 or 2, 3, 6. . . .

32. See Selection 27, Definition 12, for the Euclidean version of alternated proportion (*Elements* V, Def. 12).

33. Although *permixtum* signifies disordered or mixed, I have rendered it as *alternately* because *permutatim,* the more appropriate term, was so rendered to express the very same relation a few lines before.

34. Although the term *medietas* is used and should be translated as "mean" or "mediety," it is apparent that Boethius is describing and considering the arithmetic proportion, which is formed from a series of successive terms related as arithmetic means. The same is true of the chapters "On the Geometric Proportion" and "On the Harmonic Proportion," which follow. Indeed, the basic discussions of arithmetic, geometric, and harmonic proportions fall under the general rubric of arithmetic, geometric, and harmonic means.

35. That is, although the difference between any two successive terms is the same, the ratios formed from those terms are unequal. Thus, $4-3=3-2=1$ but $4/3 \neq 3/2$.

36. The general form is $A/C=(A-B)/(B-C)$. For Isidore of Seville's much inferior description of arithmetic, geometric, and harmonic (or musical) proportion, see the previous selection, chapter 8.

3 ON THE UNIVERSE AND ITS PARTS

Isidore of Seville (d. 636)

Translated by Ernest Brehaut[1]

Revised and annotated by Edward Grant

Book XIII

[PREFACE]

In this book, as it were in a brief outline, we have commented on certain causes in the heavens, and the sites of the lands, and the spaces of the sea, so that the reader may run them over in a little time and learn their etymologies and causes with compendious brevity.

CHAPTER 1

ON THE UNIVERSE

1. The universe is the heavens, the earth, the sea, and what in them is the work of God, of whom it is said (John 1:10): "And the universe was made by him." The universe *(mundus)* is so named in Latin by the philosophers because it is in perpetual motion *(motu)*,[2] as for example, the heavens, the sun, moon, air, seas. For no rest is permitted to its elements, and therefore it is always in motion.

2. Whence also the elements seem to Varro living creatures since, he says, they move of themselves. The Greeks have borrowed a name for the universe from ornament, on account of the variety of the elements and the beauty of the stars. For it is called among them κόσμος, which means ornament.[3] For with the eyes of the flesh we see nothing fairer than the universe.[4]

CHAPTER 2

ON ATOMS

1. The philosophers call by the name of atoms certain parts of bodies in the universe so very minute that they do not appear to the sight nor admit of τομή, that is, division, whence they are called ἄτομοι (atoms). These are said to flit through the void of the whole universe with restless motions and to move hither and thither like the finest dust that is seen when the rays of the sun pour through the windows.[5] From these certain philosophers of the heathen have thought that trees are produced, and herbs and all fruits, and fire and water, and all things are made out of them.[6]

2. Atoms exist either in a body, or in time, or in number, or in the letters. In a body as a stone. You divide it into parts, and the parts themselves you divide into grains like the sands, and again you divide the very grains of sand into the finest dust, until if you could, you would come to some little particle which is now [such] that it cannot be divided or cut. This is an atom in a body.

3. In time, the atom is thus understood: You divide a year, for example, into months, the months into days, the days into hours, the parts of the hours still admit of division, until you come to such a point of time and small part of a moment that it could not be produced by any pause *(morulam)*, and therefore cannot be divided.[7] This is an atom of time.

---

1. This selection is drawn from various chapters in Book XIII of Isidore of Seville's *Etymologies* and was translated by E. Brehaut in *An Encyclopedist of the Dark Ages* (New York: Columbia University Press, 1912), pp. 234–238. For further details on these works, see Selection 1, note 1.

2. Here Isidore repeats his etymological derivation given in Book III, chapter 29 (see Selection 1).

3. In chapter 9, paragraph 1, of *On the Nature of Things (De natura rerum)*, Isidore had remarked that "the ancients established the communion of man with the structure of the world, for since the world *(mundus)* is called *cosmos* in Greek, man is a microcosmos, i.e., he is called a smaller world" (the Latin text of *De natura rerum* has been edited and translated into French by Jacques Fontaine in *Isidore de Séville, Traité de la Nature* [Bordeaux: Féret et Fils, 1960]). This was a very popular view in antiquity and the Middle Ages.

4. The remaining paragraphs, 3 to 8, of chapter 1 have been omitted because they repeat material included in the earlier selection from Book III ("On the Quadrivium").

5. This analogy is found in Lucretius, *On the Nature of Things* II. 114–141.

6. Atomism was formulated in Greece during the fifth century B.C. by Leucippus and Democritus. Much of what was known about it in the later Middle Ages was furnished by Aristotle by way of his numerous attacks against it.

7. For Isidore, since the atom of time is smaller than any detectable or measurable pause that man can determine or know, it would not be further divisible into any other temporal units; hence it is an atom of time. Because it was misleading and confused, I have altered Brehaut's translation, which says (p. 235) that for Isidore an atom of time is such "that it cannot be lengthened by any little bit and therefore it cannot be divided."

4. In numbers, as for example, eight is divided into fours, again four into twos, then two into ones. One is an atom because it is indivisible. So also in case of the letters. For you divide a speech into words, words into syllables, the syllable into letters. The letter, the smallest part, is the atom and cannot be divided. The atom is therefore what cannot be divided, like the point in geometry. For division is called τόμος in Greek; no division [is called] ἄτομος.[8]

CHAPTER 3

ON THE ELEMENTS

1. The Greeks call ὕλη (hyle) the prime matter of things, which is in no way formed but has a capacity for all bodily forms, and out of it these visible elements are formed. Wherefore they have derived their name from this source. This hyle the Latins called materia, for the reason that everything in the rough from which something is made is always called materia. Then the poets, not inappropriately, called it silva because materials are made of wood.

2. The Greeks moreover call the elements στοιχεῖα, because they are akin to one another in a harmonious association and have a sort of common character, for they are said to be allied with one another in a certain natural way, now tracing their origin from fire all the way to earth, now from earth all the way to fire, so that fire fades into air, air is thickened to water, water coarsened to earth, and again earth is dissolved into water, water refined into air, air rarefied into fire.

3. Wherefore all elements (omnia elementa) are present in all, but each of them has received its name from that which it has in greater degree.[9] And they have been assigned by divine providence to the living creatures that are suited to them, for the Creator himself filled the heaven with angels, the air with birds, the sea with fish, the earth with men and other living creatures.

CHAPTER 5

ON THE PARTS OF THE HEAVENS

1. Ether is the place in which the stars are, and it signifies that fire which is separated on high from the whole universe. Ether is the element itself; and aethra is the glow of the ether and is a Greek word. . . .[10]

CHAPTER 7

ON THE AIR AND THE CLOUDS

1. Air is emptiness, having more rarity mixed with it than the other elements. Of it Virgil says (Aeneid, XII, 354): "Length pursued by emptiness." Air (aer) is so called from ἀέρειν (to raise), because it supports the earth or, it may be, is supported by it. This belongs partly to the substance of heaven, partly to that of the earth. For yonder thin air, where windy and gusty blasts cannot come into existence, belongs to the heavenly part; but this more disordered air, which takes a corporeal character because of dank exhalations, is assigned to earth, and it has many subdivisions. For being set in motion, it makes winds; and being vigorously agitated, lightnings and thunderings; being contracted, clouds; being thickened, rain; when the clouds freeze, snow; when thick clouds freeze in a more disordered way, hail; being spread abroad, it causes fine weather; for it is known that dense air is a cloud and that a cloud rarefied and dissolved is air.[11]

CHAPTER 8

ON THUNDER

1. Thunder (tonitruum) is so called because its sound terrifies (terreat), for tonus is sound. And it

---

8. This is a pitiful description and understanding of the atomic theory as expounded much earlier by Lucretius. Throughout chapter 2 it is not clear whether Isidore himself subscribes to the atomic theory or is merely reporting on it. On the basis of paragraph 1, where he describes what the philosophers say about it, and from the fact that he says "certain philosophers of the heathen" have thought that all things are made out of atoms, one is left with the impression that Isidore probably did not subscribe to the atomic theory. Indeed, in the very next chapter, he goes on to speak of Aristotle's four element theory, which was the great rival theory to the atomist doctrine. That he appears to accept the four element theory may perhaps be argued on the basis of his statement in chapter 3, paragraph 3 of this selection, where he says that the Creator has assigned the four elements to living creatures.

9. Only the visible elements were conceived this way by later authors. See my note 7 to Selection 109.1, Rufinus, on "Simple Medicines."

10. Much of the remainder of chapter 5 repeats material in Book III, chapters 33, 37–39 (see Selection 1). Chapter 6, "On the Circles of the Heaven," is also omitted because it repeats what is substantially given in Book III, chapter 44.

11. The four or five lines of chapter 7, paragraph 2, have been omitted.

sometimes shakes everything so severely that it seems to have split the heavens, since when a great gust of the most furious wind suddenly bursts into the clouds, its circular motion becoming stronger and seeking an outlet, it tears asunder with great force the cloud it has hollowed out, and thus comes to our ears with a horrifying noise.

2. One ought not to wonder at this since a vesicle, however small, emits a great sound when it is exploded. Lightning is caused at the same time with the thunder, but the former is seen more quickly because it is bright and the latter comes to our ears more slowly. Moreover, the light *(lux)* which appears before the thunder is called lightning *(fulgetra)*. And, as we said, it is seen before because it is a clear light *(lumen)*; the thunder reaches the ears more slowly.

4   ON THE ORDER OF THE PLANETS

Macrobius (fl. 400)

Translated by William H. Stahl[1]

Introduction and annotation by Edward Grant

The order of the planets expressed in terms of their distances from the earth, taken as the center of the world, was a much-discussed subject in the handbooks and compendia of antiquity and the Middle Ages. Astronomically, the problem was not solvable with the means available, and other criteria were invoked to resolve the issue. The nature of the problem and the impossibility of a decisive resolution are clearly described by Ptolemy, who also offers his own reasons for selecting one of the two fixed arrangements current in antiquity. In Book IX, chapter 1, of the *Almagest,* we read:

First, then, concerning the order of their spheres, all of which have their positions about the poles of the ecliptic, we see the foremost mathematicians agree that all these spheres are nearer the earth than the sphere of the fixed stars, and farther from the earth than that of the moon; that the three—of which Saturn's is the largest, Jupiter's next earthward, and Mars' below that—are all farther from the earth than the others and that of the sun. On the other hand, the spheres of Venus and Mercury are placed by the earlier mathematicians below the sun's, but by some of the later ones above the sun's because of their never having seen the sun eclipsed by them. But this judgment seems to us unsure since

CHAPTER 10

ON THE RAINBOW AND THE CAUSES OF CLOUDS

1. The rainbow is so called from its resemblance to a bent bow. Its proper name is Iris and it is called Iris, as it were *aeris* (of the air), because it comes down through the air to earth. It comes from the radiance of the sun when hollow clouds receive the sun's rays full in front, and they create the appearance of a bow; and this thing gives various colors because the water is rarefied, the air clear, the clouds darkening. The rays issuing forth create different colors.

2. Rains *(pluviae)* are so called because they flow, as if *fluviae*. They arise by exhalation from earth and sea, and being carried aloft they fall in drops on the lands, being acted upon by the heat of the sun or condensed by strong winds.[12]

these planets could be below the sun and never yet have been in any of the planes through the sun and our eye but in another, and therefore not have appeared in a line with it; just as in the case of the moon's conjunctive passages there are for the most part no eclipses.

Since there is no other way of getting at this because of the absence of any sensible parallax in these stars, from which appearance alone linear distances are gotten, the order of the earlier mathematicians seems the more trustworthy, using the sun as a natural dividing line between those planets which can be any angular distance

12. The remaining chapters (there are twenty-two) of Book XIII are omitted.

1. Reprinted by permission of Columbia University Press from *Macrobius: Commentary on the Dream of Scipio.* Translated with an Introduction and Notes by William Harris Stahl (New York, 1952), pp. 162–164. *The Dream of Scipio* constituted the closing section of Cicero's *On the Republic (De re publica).* Macrobius' two-book commentary on the *Dream* proved instrumental in the preservation of the only known version of it up to the nineteenth century (Stahl, *Macrobius,* p. 10). The present selection is drawn from Book I, chapter 19. My annotations draw heavily upon Stahl's accompanying notes and appendices as well as upon his volume *Roman Science* (Madison, Wis.: University of Wisconsin Press, 1962).

from the sun and those which cannot but which always move near it. Besides, it does not place them far enough at their perigees to produce a sensible parallax.[2]

The issue rested squarely on determinations about the relation between the sun, Mercury, and Venus, which are always seen in close proximity and whose periods appear to be of approximately the same duration (see n. 5). Lacking means for determining sensible parallax, Ptolemy arbitrarily resorts to reasons of convenience and similarity for using the sun as a natural divider between "those planets which have the full range of angular elongation from the sun" (Saturn, Jupiter, and Mars)[3] and "those which do not" (Mercury and Venus). This very order of the planets had earlier been adopted by Cicero, Geminus, Cleomedes, Vitruvius, and Pliny. Macrobius, who was a Platonist, chose to follow Plato (*Timaeus* 38D), as did Aristotle (*Metaphysics* XI.8.1073B) and others: he placed the moon, sun, Venus, Mercury, Mars, Jupiter, and Saturn beyond the earth in that order. (Macrobius later reversed the positions of Mercury and Venus and erroneously ascribed this reversal to Plato; Stahl doubts he ever read Plato [see *Roman Science,* p. 158]).

CHAPTER 19

[1] Next we must say a few things about the order of the spheres, a matter in which it is possible to find Cicero differing with Plato, in that he speaks of the sphere of the sun as the fourth of seven, occupying the middle position, whereas Plato says that it is just above the moon, that is, holding the sixth place from the top among the seven spheres. [2] Cicero is in agreement with Archimedes and the Chaldean System; Plato followed the Egyptians, the authors of all branches of philosophy, who preferred to have the sun located between the moon and Mercury even though they discovered and made known the reason why others believed that the sun was above Mercury and Venus.

Those who hold the latter opinion are not far from a semblance of truth; indeed, their misapprehension arose from the following cause. [3] From the sphere of Saturn, the first of the seven, to the sphere of Jupiter, the second from the top, there is so much space intervening that the upper planet completes its circuit of the zodiac in thirty years, the lower one in twelve. Again, the sphere of Mars is so far removed from Jupiter that it completes its circuit in two years.[4] [4] Venus is so

much lower than Mars that one year is sufficient for it to traverse the zodiac. But the planet Mercury is so near Venus and the sun so near Mercury that these three complete their revolutions in the same space of time, that is, a year more or less.[5] On this account Cicero called Mercury and Venus the *sun's companions,* for they never stray far from each other in their annual periods. [5] The moon, moreover, lies so much farther below these that what they accomplish in a year it does in twenty-eight days. Hence there was no disagreement among the ancients regarding the correct order of the three superior planets, the vast distances between them clearly arranging them, nor about the location of the moon, which is so much lower than the rest. But the proximity of the three neighboring planets, Venus, Mercury, and the sun, was responsible for the confusion in the order assigned to them by astronomers, that is, with the exception of the skillful Egyptians, who understood the reason, here outlined.

[6] The sphere in which the sun journeys is encircled by the sphere of Mercury, which is above it, and by the higher sphere of Venus as well. As a result, when these two planets course through the upper reaches of their spheres, they are perceived to be above the sun, but when they pass into the

---

2. This translation is by R. Catesby Taliaferro in *Great Books of the Western World* (Chicago: Encyclopedia Britannica, Inc., 1952), XVI, 270.

3. This division omits the moon, which is below the sun and has the full range of angular elongation, thus invalidating Ptolemy's reasoning, as Copernicus observed in *De revolutionibus,* Book I, chapter 10 (see "Nicholas Copernicus on the Revolutions of the Heavenly Spheres," trans. Charles Glenn Wallis in *Great Books of the Western World,* XVI, 523).

4. These were the standard approximations. Mars is considerably off, since its sidereal period is 687 days.

5. Stahl observes (*Macrobius,* p. 163, n. 4) that "the orbits of Venus and Mercury are within the earth's orbit and so these planets appear to swing back and forth with respect to the sun. They never go very far from the sun's position in the sky and are seen during the same year as morning and evening stars. The greatest apparent distance of Venus from the sun (maximum elongation) is 48 degrees of celestial arc and of Mercury 28 degrees. At the beginning of a solar year Mercury and Venus might appear ahead of the sun as morning stars and at the end of the year might be behind as evening stars, or vice versa. So Macrobius, who is using a geocentric orientation, says that their revolutions consume 'a year more or less.' The revolutions of Venus and Mercury are actually 225 and 88 days respectively. Macrobius' figures for the duration of the five planets' revolutions are the same as those regularly found in the classical handbooks."

lower tracts of their spheres they are thought to be beneath the sun. [7] Those who assigned to them a position beneath the sun made their observations at a time when the planets' courses seemed to be beneath the sun, which, as we noted sometimes happens; indeed, this position is more noticeable since we get a clearer view at that time. When Mercury and Venus are in their upper regions, they are less apparent because of the sun's rays.[6] As a result, the false opinion has grown stronger, and this order has received almost universal acceptance.[7]

5 ON THE MOTION OF MERCURY AND VENUS AROUND THE SUN

Translations by Sir Thomas Heath

Introductions and annotation by Edward Grant

1. Chalcidius (fl. fourth century)[1]

In the preceding selection it was suggested that Macrobius may have garbled a report about Heraclides of Pontus. In the text that follows, Chalcidius actually mentions Heraclides by name but still misconstrues and misrepresents his views. Instead of describing Mercury and Venus as moving around the sun, the physical center of their orbits, Chalcidius places the sun, Mercury, and Venus on concentric circles that move around a geometric point, their common center, which presumably moves around its center the earth. Thus, Chalcidius has converted Heraclides' limited heliocentric system for Mercury and Venus into a system containing three concentric epicycles. Indeed, he later ascribes to Plato essentially the same view, actually identifying the circles of sun, Mercury, and Venus as epicycles even though Plato lived before the system of epicycles was devised. It should be noted that although Chalcidius mentions only Venus (he calls it Lucifer) and the sun in the passage quoted below, he had implied earlier that Mercury's motion is also included.

Lastly Heraclides Ponticus, when describing the circle of Lucifer as well as that of the sun, and giving the two circles one point *(unum punctum)*[2] and one middle *(unam medietatem),*[3] showed how

discussion about the correct fixed order of the planets, makes it apparent that Macrobius failed to understand what he had read or heard about Heraclides. For the Heraclidean theory is incompatible with a fixed order of the planets. In any event, Macrobius' account is too unclear to have conveyed or transmitted the Heraclidean system, and Stahl is right to reject the views of those historians of astronomy (for instance, Dreyer, Heath, and Duhem) who have interpreted this passage as a proper description of that system (see *Macrobius,* pp. 249–250, and *Roman Science,* p. 159).

1. This selection is reprinted by permission of the Clarendon Press, Oxford, from *Aristarchus of Samos, the Ancient Copernicus* (Oxford: Clarendon Press, 1913), p. 256. Heath's translation was made from Chalcidius' Latin *Commentary on Plato's Timaeus,* chapter 110. This commentary, and Chalcidius' accompanying Latin translation of the *Timaeus,* formed the basis of medieval knowledge of Plato and was a factor of some importance in the history of medieval science.

2. For *unum punctum,* I have replaced Heath's "one center" with "one point." The Latin expressions are my additions, taken from the recent edition of Chalcidius, *Timaeus a Calcidio translatus commentarioque instructus* in societatem operis coniuncto P. J. Jensen edidit J. H. Waszink (London: Warburg Institute; Leiden: E. J. Brill, 1962), p. 157.

3. The "one point" and "one middle" are, of course, the same center for the concentric epicycles. Heath writes *(Aristarchus,* p. 257):

"According to this we are to suppose a point which revolves uniformly about the earth from west to east in a year. This point is the centre of three concentric circles (epicycles) on which move respectively the sun (on the innermost), Mercury (on the middle circle), and Venus (on the outermost); the sun takes, of course, a year to describe its epicycle. That the epicycle for the sun is wrongly imported into Heraclides' true system is confirmed by the next chapter of Chalcidius, with its illustrative figure, where he imports epicycles into Plato's system also . . . . Hence the epicycles must be rejected altogether so far as Plato's system is concerned. Similarly, we must eliminate the sun's epicycle from the account of Heraclides' system, and we must suppose that he regarded Mercury and Venus as simply revolving in concentric circles about the sun."

The system of Heraclides, whose works have not survived, is a reconstruction based on reports of a few authors who lived long after him. Of these authors, Chalcidius and Capella are probably the most important.

---

6. Stahl says *(Macrobius,* p. 164): "The reason given is incorrect. Venus and Mercury are less brilliant when they are nearing superior conjunction because of their greater distance from us. Greatest brilliance of Venus is attained 36 days before or after inferior conjunction, at which time she is $2\frac{1}{2}$ times as brilliant as at superior conjunction."

7. The "false opinion" is the one adopted by Ptolemy. In this paragraph, when Macrobius speaks of Mercury and Venus "coursing through the upper reaches of their spheres" and then passing "into the lower tracts of their spheres" and so lying beneath the sun, he is probably reflecting something he read about the system proposed by Heraclides of Pontus (see the next selection). His vague and obscure account, embedded in the context of a

Lucifer is sometimes above, sometimes below the sun. For he says that the position of the sun, the moon, Lucifer, and all the planets, wherever they are, is defined by one line passing from the centre of the earth to that of the particular heavenly body. There will then be one straight line drawn from the centre of the earth showing the position of the sun, and there will be equally two other straight lines to the right and left of it respectively, and distant 50° from it, and 100° degrees from each other, the line nearest to the east showing the position of Lucifer or the Morning Star when it is furthest from the sun and near the eastern regions, a position in virtue of which it then receives the name of the Evening Star, because it appears in the east at evening after the setting of the sun. . . .

2. Martianus Capella (fl. 410–439)[4]

Although the Heraclidean scheme (see preceding selection by Chalcidius) is inconsistent with a definite fixed order of the planets (Mercury and Venus would at times be on the far side of the sun and at times on the near, or earth, side of the sun), neither Chalcidius nor Capella, the two significant transmitters of this doctrine, seemed to recognize this. They discussed the two traditional fixed orders without in any way taking cognizance of the inconsistency of such arrangements with the Heraclidean view of the motions of Mercury and Venus around the sun.[5] Later authors who found occasion to discuss or repeat the Heraclidean system also did so as part of a general consideration of the fixed order of the planets (for instance, Baudoin de Courtenay in his L'Introductoire d'Astronomie,

written in 1270, and Peter of Abano in his Lucidator astrologiae of 1310; see Stahl, p. 247).

The Heraclidean theory was partly geocentric, partly heliocentric (the sun, with Mercury and Venus circling about it, was described as rotating around the earth, the center of the world). Extension of the Heraclidean scheme to embrace the other planets and the earth would have produced the full heliocentric systems proposed by Aristarchus of Samos in antiquity and Copernicus in the sixteenth century. Indeed, Copernicus himself, in referring to the text reproduced below, emphasized (De Revolutionibus, Bk. I, ch. 10) that his own theory was but an extension of the "ingenious view held by Martianus Capella," which, as we know but apparently Copernicus did not, is nothing other than the system of Heraclides of Pontus, whose name, however, was not mentioned by Capella.

For, although Venus and Mercury are seen to rise and set daily, their orbits do not encircle the earth at all, but circle round the sun in a freer motion. In fact, they make the sun the centre of their circles, so that they are sometimes carried above it, at other times below it and nearer to the earth, and Venus diverges from the sun by the breadth of one sign and a half [45°]. But, when they are above the sun, Mercury is the nearer to the earth, and when they are below the sun, Venus is the nearer, as it circles in a greater and wider-spread orbit. . . .

The wider circles of Mercury and Venus I have above described as epicycles.[6] That is, they do not include the round earth within their own orbit, but revolve laterally to it in a certain way.

6   ON OCEAN AND TIDES

Macrobius (fl. 400)

Translated by William H. Stahl[1]

Annotated by Edward Grant

[1] After giving these matters an examination that is by no means useless, as it seems to me, let us now confirm the statement we made about Ocean, that the whole earth is girt about not by a single but by a twofold body of water whose true and original course man in his ignorance has not yet determined.

4. This selection is reprinted by permission of the Clarendon Press, Oxford, from Aristarchus of Samos the Ancient Copernicus (Oxford: Clarendon Press, 1913), p. 256. The translation was made from Martianus Capella's Marriage of Philology and Mercury (De nuptiis Philologiae et Mercurii), Bk. VIII, 880, 882

5. William H. Stahl, Roman Science (Madison, Wis:. University of Wisconsin Press, 1962), pp. 147, 149, 183–184.

6. Since the circles of Mercury and Venus have an actual body (the sun), rather than a geometric point, as their physical center, they cannot properly be called epicycles.

1. Reprinted by permission of Columbia University Press from Macrobius: Commentary on the Dream of Scipio. Translated with an Introduction and Notes by William Harris Stahl (New York, 1952), p. 214. The present selection is taken from the beginning of Book II, chapter 19.

That Ocean which is generally supposed to be the only one is really a secondary body, a great circle which was obliged to branch off from the original body. [2] The main course actually flows around the earth's torrid zone, girdling our hemisphere and the underside, and follows the circumference of the equator. In the east it divides, one stream flowing off to the northern extremity, the other to the southern; likewise, in the west, streams flow to the north and south, where they meet the streams from the east at the poles.[2] [3] As they rush together with great violence and impetus and buffet each other, the impact produces the remarkable ebb and flow of Ocean,[3] and wherever our sea[4] extends, whether in narrow straits or open coast, it shares in the tidal movement of Ocean's streams. These we now speak of as Ocean proper because of the fact that our sea is filled from Ocean's streams. [4] But the truer bed of Ocean, if I may call it that, keeps to the torrid zone; it follows the circuit of the equator as the streams originating in it follow the circuit of the horizon in their course, thus dividing the whole earth into four parts and making each inhabited quarter, as we previously stated, an island.

2. The theory that two oceans, one running east and west along the equatorial regions and the other north and south at right angles to the equatorial ocean, divide the inhabited portions of the world into four land masses surrounded by ocean and in complete isolation from one another can be traced to Crates of Mallus (second century B.C.). Mankind, as they knew it, was thought to be confined to one of these four island masses. This particular theory was held by Macrobius and Martianus Capella and was influential in the Middle Ages in geographical literature and in map-making. For further details, see John K. Wright, *Geographical Lore of the Time of the Crusades* (New York: American Geographical Society, 1925), pp. 18–19.

3. This theory of tides may have originated with Crates of Mallus (see Stahl, p. 214, n. 2). Apart from fanciful theories, tidal explanations seemed to fall into two major categories: (1) those involving the influence of the moon (Posidonius and Pliny were ancients who invoked the moon in tidal discussions); and (2) those utilizing the physical impact of the two ocean currents colliding near the polar regions and rebounding to cause the ebb and flow of the tides. Obviously, Macrobius belongs in the second category. His opinion was widely known in the Middle Ages. The problem is discussed by John Wright in *Geographical Lore* (pp. 190–196). For a viewpoint upholding lunar influence on the tides, see Selection 82 by Robert Grosseteste.

4. The Mediterranean.

Later Middle Ages

The Translation of Greek and Arabic Science into Latin

7   A LIST OF TRANSLATIONS MADE FROM ARABIC INTO LATIN IN THE TWELFTH CENTURY

Gerard of Cremona (ca. 1114–1187)

Translation, introduction, and annotation by Michael McVaugh

Gerard of Cremona's name will undoubtedly always symbolize that twelfth-century phenomenon, the widespread movement to translate into Latin the philosophical and scientific works possessed by the Arabs.[1] There were many other translators of importance, but none nearly so prolific as he. As Charles Haskins put it, "More of Arabic science in general passed into western Europe at the hands of Gerard of Cremona than in any other way," and the quality of his translations remained remarkably high throughout. We are today able to identify his work with ease, owing to the brief biobibliography presented here, which his students at Toledo attached to his translation of Galen's *Tegni* or *Ars parva*.

As a light shining in darkness must not be set under a bushel, but rather upon a candlestick, so too the splendid deeds of the great must not be held back, buried in timid silence, but must be made known to listeners today, since they open virtue's door to those who follow, and in worthy memorial offer to modern eyes the example of the ancients as a model for life. Thus, lest master Gerard of Cremona be lost in the shadows of silence, lest he lose the credit that he deserved, lest in brazen theft another name be affixed to the books translated by him (especially since he set his name to none of them), all the works he translated—of dialectic as of geometry, of astronomy as of philosophy, of medicine as of the other sciences—have been diligently enumerated by his associates at the end of this *Tegni* just translated by him, in imitation of Galen's enumeration of his own writings at the end of the same book; so that if an admirer of their works should desire one of them, he might find it the quicker by this list, and be surer of it.

Although he scorned fame, although he fled from praise and the vain pomp of this world; although he refused to spread his name in a quest for empty, insubstantial things, the fruit of his works diffused through the world makes plain his worth. For while he enjoyed good fortune, possessions or their lack neither delighted nor depressed him; manfully sustaining whatever chance brought him, he always remained in the same state of constancy. Hostile to fleshly desires, he clove to spiritual ones alone. He worked for the advantage of all, present and future, mindful of Ptolemy's words: when approaching your end, do good increasingly. He was trained from childhood at centers of philosophical study and had come to a knowledge of all of this that was known to the Latins; but for love of the *Almagest,* which he could not find at all among the Latins, he went to Toledo; there, seeing the abundance of books in Arabic on every subject, and regretting the poverty of the Latins in these things, he learned the Arabic language, in order to be able to translate. In this way, combining both language and science (for as Hamet says in his letter *De proportione et proportionalitate,* a translator should have a knowledge of the subject he is dealing with as well as an excellent command of the languages from which and into which he is translating), he passed on the Arabic literature in the manner of the wise man who, wandering through a green field, links up a crown of flowers, made from not just any, but from the prettiest; to the end of his life, he continued to transmit to the Latin world (as if to his own beloved heir) whatever books he thought finest, in many subjects, as accurately and as plainly as he could. He went the way of all flesh in the seventy-third year of his life, in the year of our Lord Jesus Christ 1187.

1. For a general treatment of this activity, in which Gerard of Cremona was the outstanding figure, see C. H. Haskins, *Studies in the History of Medieval Science* (Cambridge, Mass.: Harvard University Press, 2d ed., 1927), ch. 1, from which the following quotation is taken. The translation of Gerard's biography is based upon the texts published by F. Wüstenfeld, *Die Übersetzungen Arabischer Werke in das Lateinische (Abh. der Kön. Gesell. Wiss. zu Göttingen),* XXII (1877), 57–81, and by Karl Sudhoff, "Die kurze Vita und das Verzeichnis der Arbeiten Gerhards von Cremona," *Archiv für Geschichte der Medizin,* XIV (1923), 73–82.

These are the titles of the books translated by master Gerard of Cremona, at Toledo:

On Dialectic

[1] Aristotle, *Posterior Analytics*[2]

[2] Themistius, *Commentary on the Posterior Analytics*[3]

[3] Alfarabi, *On the Syllogism (Liber Alfarabii de syllogismo)*

On Geometry

[4] Euclid, The Fifteen Books [of the *Elements*] *(Liber Euclidis tractatus XV)*[4]

[5] Theodosius, Three Books *On the Sphere*

[6] Archimedes, [*On the Measurement of the Circle*]

[7] [Ahmad ibn Yusuf], *On Similar Arcs (De similibus arcubus)*

[8] Mileus [i.e., Menelaus], Three Books [on Spherical Figures]

[9] Thabit [ibn Qurra], *On the Divided Figure (De figura alchata)*[5]

[10] Banu Musa [i.e., the Three Sons of Moses or the Three Brothers], [*On Geometry*][6]

[11] Ahmad ibn Yusuf, [Letter] *on Ratio and Proportion (Liber Hameti de proportione et proportionalitate)*[7]

[12] [Abu 'Uthman or Muhammad ibn 'Abd al-Bāqī], *The Book of the Jew on the Tenth Book of Euclid*

[13] Al-Khwārizmī, *On Algebra and Almucabala*[8]

[14] *Book of Applied* [or Practical] *Geometry*[9]

[15] Anaritius [i.e., al-Nairizi], [*Commentary*] *on* [*the Elements of*] *Euclid*

[16] Euclid, *Data*

[17] Tideus [i.e., Diocles][10], *On* [*Burning*] *Mirrors*

[18] Alkindi, *On Optics (De aspectibus)*

[19] *Book of Divisions (Liber divisionum)*[11]

[20] [Thabit ibn Qurra], *Book of the Roman Balance (Liber Karastonis)*[12]

On Astronomy

[21] Alfraganus [i.e., al-Farghani], The Book Containing XXX Chapters[13]

[22] [Ptolemy], The Thirteen Books of the *Almagest*

[23] [Geminus of Rhodes], *Introduction to the Spherical Method of Ptolemy (Liber introductorius Ptolomei ad artem spericam)*[14]

[24] Geber [i.e., Jabir ibn Aflah], *Nine Books* [*on the Flowers from the Almagest*][15]

[25] Messehala, *On the Orb (De orbe)*[16]

2. The translation has been edited by L. Minio-Paluello in *Aristoteles Latinus,* Vol. IV, part 3 (Bruges and Paris: Desclee de Brouwer, 1954).

3. Edited by J. Reginald O'Donnell, C.S.B., "Themistius' Paraphrasis of the Posterior Analytics in Gerard of Cremona's Translation," *Mediaeval Studies,* XX (1958), 239–315.

4. Although in the Middle Ages, Euclid's *Elements* were thought to consist of fifteen books, Books XIV and XV were added long after Euclid's time.

5. This is transcribed from the Arabic al-qaṭṭā' meaning "sector," for which reason the title is sometimes given as *De figura sectore.* It seems to derive from a theorem by Menelaus in Book III, Proposition 1, of his *Spherics* or from two lemmas by Ptolemy in Book I, chapter 13, of the *Almagest,* in which a triangle is cut, or divided, by transversals.

6. Also called *Verba Filiorum Moysi Filii Sekir, i.e. Maumeti, Hameti, Hasen (The Discourse of the Sons of Mūsa Ibn Shākir: Muhammad, Ahmad, and Hasan).* This treatise has been edited and translated by M. Clagett, *Archimedes in the Middle Ages* (Madison, Wis.: University of Wisconsin Press, 1964), Vol. I: *The Arabo-Latin Tradition* pp. 238–355. A proposition from this work appears in Selection 32.1.

7. Edited and translated by Sister M. Walter Reginald Schrader, O.P., "The Epistola De proportione et proportionalitate of Ametus Filius Josephi," Ph. D. dissertation, University of Wisconsin, 1961).

8. A selection from Robert of Chester's translation of this treatise will be found in Selection 22.

9. Perhaps identical with Abubacer's (Abu Bakr al-Hasan) *De mensuratione terrarum (On Terrestrial Measurements).*

10. A Greek who probably flourished in the second century B.C.

11. Perhaps Euclid's *Book of Divisions,* mentioned by Proclus, in which a given figure is divided into figures differing from it. The Greek text is lost. See Selection 30.

12. Edited and translated by M. Clagett, in M. Clagett and E. A. Moody, *The Medieval Science of Weights* (Madison, Wis.: University of Wisconsin Press, 1952).

13. A common title used for Alfraganus' treatise was *Book on the Complete Science of the Stars and the Principles of Celestial Motions (Liber de aggregationibus scientie stellarum et principiis celestium motuum).*

14. Since Ptolemy lived about two centuries after Geminus of Rhodes (fl. ca. 70 B.C.), the latter could not have written an introduction to any book by the former. However, the treatise listed above is a medieval compilation from Geminus' extant work, *Elements of Astronomy.* It would seem that the compilation was intended as an aid in the study of Ptolemaic astronomy.

15. Geber, or Jābir ibn Aflah (not to be confused with Jabir the alchemist), wrote "nine books of spherical demonstrations intended primarily as criticisms of many technical doctrines of Ptolemy's *Almagest.*" (F. Carmody, *Arabic Astronomical and Astrological Sciences in Latin Translation, A Critical Bibliography* [Berkeley: University of California Press, 1956], p. 163.) An alternative title is *Elementa astronomica.*

16. Also known as *On the Science of the Motion of the* [*Celestial*] *Orb.*

[26] Theodosius, *On Habitable Places (De locis habitabilibus)*

[27] Hypsicles, [*On the Rising of the Signs (De ascensionibus signorum)*]

[28] Thabit [ibn Qurra], *On the Exposition of Terms in the Almagest (Liber Thebit de expositione nominum Almagesti)*[17]

[29] Thabit [ibn Qurra], *On the Forward and Backward Motion (De motu accessionis et recessionis)*[18]

[30] Autolycus, *On the Moving Sphere (De spera mota)*

[31] Book of the Tables of Jaen with Its Rules *(Liber tabularum iahen cum regulis suis)*[19]

[32] [Abu Abdallāh Muḥammad ibn Mu 'adh], *On the Dawn (De crepusculis)*[20]

On Philosophy

[33] Aristotle, *On the Exposition of Pure Goodness (De expositione bonitatis pure)*[21]

[34] Aristotle, *Physics (De naturali auditu)*[22]

[35] Aristotle, Four Books *On the Heavens and World (Celi et mundi tr. IV)*[23]

[36] Aristotle, *On the Causes of Properties and the Four Elements,* Book I *(Liber Aristotelis de causis proprietatum et elementorum primus)*[24]; he did not translate the second treatise of this work, because he could find only a little bit of its ending in Arabic.

[37] Aristotle, *On Generation and Corruption*

[38] Aristotle, *Meteorology,* Books I-III; the fourth he did not translate, since it was already translated.

[39] Alexander of Aphrodisias, *On Time (De tempore), On the Senses (De sensu),* and another *That Augment and Increase Occur in Form, Not in Matter (Quod augmentum et incrementum fuit in forma et non in yle)*[25]

[40] Alfarabi, *Commentary on Aristotle's Physics (Distinctio Alfarabii super librum Aristotelis de naturali auditu)*

[41] Alkindi, *On the Five Essences (De quinque essentiis)*

[42] Al Farabi, *On the Sciences*[26]

[43] Alkindi, *On Sleep and Vision*

On Medicine (De fisica)

[44] Galen, *On the Elements*

[45] Galen, *Commentary on Hippocrates' Treatment of Acute Diseases*

[46] [pseudo-] Galen, *Secrets [of Medicine] (De secretis)*[27]

17. Also known under the title *On Things Which Require Exposition before Reading the Almagest (De hiis que indigent expositione antequam legatur Almagesti),* which was edited by F. Carmody (Berkeley: University of California Press, 1941), together with the next title on this list.

18. A presentation of Thabit's theory of trepidation, in which he tried to account for an imaginary variability in the precession of the equinoxes. It also carried the title *On the Motion of the Eighth Sphere (De motu octave spere).*

19. Concerning this treatise, George Sarton has the following entry *(Introduction to the History of Science,* Vol. II, Part I [Baltimore: Williams & Wilkins, 1931], p. 342): "Liber tabularum iahen cum regulis suis. Tabulae Jaén (?) (Jaén is an important town or district in Andalusia); Tabulae Gebri (?) al-Jaihānī (?)(which means of Jaén). Supposing this al-Jaihānī to be the author of the tables, it is not easy to identify him. He may be the Abū 'Abdallah Muḥammad ibn Yūsuf ibn Aḥmad ibn Mu'adh al-Juhani (or Jaihānī) of Cordova, born in 989–990, who spent five years in Egypt, 1012–1017; or he may be the wazīr and qādī Abu 'Abdallāh Muḥammad ibn Mu'adh of Seville who wrote treatises on the dawn, and on the total solar eclipse of July 3, 1079. The matter requires investigation." In a recent article, A. I. Sabra conjectures that the second of these authors composed the Table of Jaen (or Jahen) (see p. 85 of the article cited in the next note).

20. This work, concerned with atmospheric refraction, has been traditionally ascribed to the great Muslim scientist Alhazen. Recently, however, A. I. Sabra ("The Authorship of the *Liber de crepusculis,* An Eleventh Century Work on Atmospheric Refraction," *Isis,* LVIII [1967], pp. 77–85) has presented convincing evidence in favor of the authorship of Abū ibn Mu'adh, an Andalusian mathematician who lived in the second half of the eleventh century, and who is the second of the authors mentioned above in note 19.

21. This treatise, which was mistakenly ascribed to Aristotle, was actually an Arabic compilation of extracts from Proclus' *Elements of Theology.* Its more popular title was *The Book of Causes* [*Liber de causis*].

22. *De naturali auditu* means literally "On the Natural Things That Have Been Heard" and was an alternative title for Aristotle's *Physics.*

23. This translation is being edited by Ilona Opelt, who has already summarized certain peculiarities of Gerard's method of translation in "Zur Uebersetzungstechnik des Gerhards von Cremona," *Glotta* XXXVIII (1959), 135–170.

24. Another treatise falsely ascribed to Aristotle but of Arabic origin.

25. All three brief treatises are edited in chapter 3 of G. Théry, O.P., *Autour du décret de 1210: II. Alexandre d'Aphrodise aperçu sur l'influence de sa noétique* (Paris: Le Saulchoir Kain, 1926).

26. Latin and Arabic texts have been published by Angel González Palencia, *al-Fārābī, Catalogo de las ciencias,* 2d ed. (Madrid: Instituto Miguel Asín, 1953).

27. A text describing medical treatments and observations, perhaps collected from Galen's works by the Arabic translator; see Lynn Thorndike, *A History of Magic and Experimental Science* (New York: Macmillan, 1923), II, 758–761.

[47] Galen, *On the Temperaments (De complexionibus)*

[48] Galen, *On the Evils of an Unbalanced Temperament (De malicia complexionis diverse)*

[49] Galen, *On Simple Medicines,* Books I-V *(Liber Gal. de simplici medicina tr. V)* [28]

[50] Galen, *On Critical Days*

[51] Galen, *On Crises*

[52] Galen, *Commentary on Hippocrates Prognostics*

[53] [pseudo-] Hippocrates, *Book of the Truth (Liber veritatis)* [29]

[54] Isaac [Ishāq al-Isrā'ilī], *On the Elements*

[55] Isaac [Ishāq al-Isrā'ilī], *On the Description of Things and Their Definitions (De descriptione rerum et diffinitionibus earum)* [30]

[56] Rhazes [i.e., Abu Bakr Muḥammad ibn Zakariyā al-Rāzī], *The Book of Almansor (Liber Albubatri rasis qui dicitur Almansorius tr. X)* [31]

[57] [Rhazes], *The Book of Divisions,* containing CLIIII chapters

[58] [Rhazes], *Short Introduction to Medicine (Liber. . . introductorius in medicina parvus)*

[59] Abenguefit [i.e., Abū al-Mutarrif 'Abd al-Rahmān ibn al-Wāfid], *Book of Simple Medicines and Foods,* in part *(Pars libri Abenguefiti medicinarum simplicium et ciborum)*

[60] John Serapion [i.e., Yaḥya ibn Sarāfyūn], *Breviary (Breviarium)* [32]

[61] Azaragui [i.e., Abū-al-Qāsim al-Zahrāwī], *Surgery* [33]

[62] Jacob Alkindi, *On Degrees [of Compound Medicines]* [34]

[63] Avicenna, *Canon* [35]

[64] Galen, *Tegni,* with the commentary by Ali ab Rodohan [i.e., 'Alī ibn Ridwān] [36]

On Alchemy

[65] [Jābir ibn Hayyān, attrib.], *Book of Divinity of LXX (Liber divinitatis de LXX)* [37]

[66] [pseudo-Rhazes], *On Alumens and Salts* [38]

[67] [pseudo-Rhazes], *The Light of Lights (Liber luminis luminum)* [39]

On Geomancy

[68] A book on geomancy concerning the divining arts, beginning "Estimaverunt indi" [40]

[69] Alfadhol [de Merengi], *[Book of Judgments and Advice]* (Liber Alfadhol id est arab de bachi) [41]

[70] *Book on Accidents (Liber de accidentibus alfel)* [42]

[71] [Harib ibn Zeid, Calendar] *(Liber anoe)* [43]

28. The remaining six books were eventually translated (from the Greek) in the middle of the fourteenth century by Nicholas of Reggio.

29. Edited by Karl Sudhoff, "Die pseudohippokratische Krankheitsprognostik . . . ," *Archiv für Geschichte der Medizin,* IX (1915), 79–116; it is more often spoken of as the *Capsula Eburnea* (see n. 19 to Selection 101).

30. Edited by J. T. Muckle, "Isaac Israeli Liber de definicionibus," *Archives d'Histoire Doctrinale et Littéraire,* XII–XIII (1937–1938), 299–340.

31. The smaller of Rhazes' two medical compilations, dedicated to the caliph al-Mansur (whence its name); the ninth of its ten books (on general therapeutics) was particularly popular and was often copied and commented separately.

32. More frequently known as the *Practica.* The seventh book, an antidotary, was in wide circulation separately during the Middle Ages.

33. During most of the Middle Ages, the author was known as Abulcasis; this work (illustrated) was in fact only a portion of a larger work (al-Tasrif), the rest of which Gerard ignored. A French translation has been made (1861) by Leclerc.

34. This work develops a mathematical treatment of pharmacy, and was sometimes included by copyists with Gerard's other mathematical translations. It is to be published in M. R. McVaugh, ed., *Thirteenth-Century Mathematical Pharmacy* (University of Wisconsin Press).

35. An extract from the first book of the *Canon* is given in Selection 91.

36. The Galenic work is the *Microtegni* or *Ars parva,* which had earlier been translated by Constantine the African, and which served as the basic text for medical education throughout the Middle Ages. The Galenic text and its commentary are easily confused; the two together may have been important in the later development of a theory of scientific method (see A. C. Crombie, *Robert Grosseteste and the Origins of Experimental Science 1100–1700* [Oxford: Clarendon Press, 1953], pp. 76–78).

37. This is the *Liber de LXX,* a work of seventy sections whose first Latin title is *Liber divinitatis.* Gerard's translation has been edited by M. Berthelot in *Archéologie et Histoire des Sciences. Mémoires de l'Académie des Sciences,* 2d Ser., XLIX (1906), 310–363.

38. Edited by Robert Steele, "Practical Chemistry in the Twelfth Century," *Isis,* XII (1929), 10–46. Julius Ruska attributes the original Arabic text to an (unknown) Spanish alchemist of the eleventh or twelfth century.

39. Perhaps the "Lumen luminum liber Rasis philosophi," partially transcribed by Julius Ruska, "Pseudepigrapher Rasis-Schriften," *Osiris,* VII (1939), 61–65.

40. If the incipit given in these manuscripts is to be trusted, this work (edited in part by Paul Tannery, *Mémoires Scientifiques,* IV [1920], 405–409) cannot have been translated by Hugh of Santalla, as Tannery and others have supposed.

41. On the content and origin of this work, see Lynn Thorndike's articles in *Speculum,* II (1927), 326–331, and *Speculum,* IV (1929), 90.

42. "Alfel" is probably a rendering of the Arabic al-fāl, meaning an omen or augury. The work cannot be certainly identified.

43. Edited by Guillaume Libri, *Histoire des Sciences Mathématiques en Italie,* I (1838), 393–458.

8   A LIST OF TRANSLATIONS MADE FROM GREEK INTO LATIN IN THE THIRTEENTH CENTURY

William of Moerbeke (ca. 1215—ca. 1286)

Arrangement, introduction, and annotation by Edward Grant[1]

The Flemish Dominican William of Moerbeke was not the first to translate from Greek into Latin during the Middle Ages, but he was unquestionably the most prolific of those who did. Although a number of the titles listed were revisions of earlier translations, on the whole it can be said that Moerbeke made available in Latin the scientific works of Aristotle, the mathematical and physical treatises of Archimedes, and other important philosophical and theological treatises. In undertaking the translations of the works of Aristotle, he was probably encouraged by his friend, St. Thomas Aquinas, who was anxious to make available Latin translations that would be closer to the originals, and therefore more reliable, than those current in earlier translations made from the Arabic. Other of his translations may perhaps be accounted for by virtue of his interest in Neo-Platonism, an interest he shared with his famous Polish friend, Witelo, who produced an important commentary on the optics of Alhazen (see biographical sketch of Witelo at the end of this source book and the many extracts from his work in Selection 62).

Despite the extreme literalness of his translations, which drew criticism from Roger Bacon, Moerbeke's impressively large number of serviceable translations (on occasion the literalness has even encouraged modern scholars to attempt reconstruction of original Greek passages) show conclusively that long before the Renaissance a rather impressive number of important Greek authors had been translated into Latin from Greek manuscripts. So well received were Moerbeke's translations that Renaissance authors cited and plagiarized them (Leonardo da Vinci did the former and Niccolo Tartaglia the latter). It was a fitting tribute to Moerbeke that his translations were used in the first printed version of the works of Archimedes made by Guarico in 1503 at Venice.

Alexander of Aphrodisias[2] (fl. A.D. 193–217)

1. Commentary on the *Meteorologica* of Aristotle[3]
2. Commentary on *On Sense and Sensible Objects (De sensu et sensato)* of Aristotle
3. *Short Work on Fate (Opusculum De fato)*
4. *To the Emperors on Fate (De fato ad imperatores)*

Ammonius[4] (fl. A.D. 500)

5. Commentary on *On Interpretation (De interpretatione)* of Aristotle[5]

Archimedes (ca. 287–212 B.C.)

6. *On Spiral Lines*[6]

1. Aside from a few minor rearrangements, the list below follows the compilation of Moerbeke's translations as set forth by Pierre Thillet in his *Alexandre d'Aphrodise De Fato ad imperatores version de Guillaume de Moerbeke édition critique avec introduction et index* (Paris: J. Vrin, 1963), pp. 29–35. Thillet's notes have also been helpful. Although this list cannot be described as definitive or final, a sufficiently large number of the translations cited have been verified beyond doubt and convey a reasonably accurate sense of the magnitude of Moerbeke's great achievement.

2. One of the greatest commentators on the works of Aristotle, who became head of the Lyceum sometime between 198 and 211.

3. This translation was completed at Nicaea, in Greece on April 12, 1260.

4. Ammonius, son of Hermias, was an Aristotelian commentator in Alexandria.

5. Completed in Vieterbo, Italy, on September 12, 1268.

6. Completed in February of 1269. Indeed, all of Moerbeke's Archimedean translations were completed in that year. Prior to Moerbeke's translation of almost all of the works of Archimedes from Greek into Latin during 1269, the Archimedean, or Archimedean-type, treatises that were available were translated from Arabic and limited to a few works (*On the Measurement of the Circle*, some propositions from Book I of *On the Sphere and Cylinder*, and a pseudo-Archimedean work, *De insidentibus in humidum*, based on Archimedes' *On Floating Bodies*). Only a few Greek manuscripts of the works of Archimedes were available to the Latin West:

"The Greek text extant in the Byzantine period consisted of at least three manuscripts, whose fates we can trace. Manuscript A, the principal manuscript, contained all of the works now known except the *On Floating Bodies*, the *On the Method*, the fragmentary *Stomachion*, and the *Bovine Problem*. This was one of the two manuscripts available to Moerbeke. It is the source of all Renaissance copies of Archimedes—but it has been lost. Manuscript B included the mechanical works—*On the Equilibrium of Planes, On Floating Bodies*—and the *On Spiral Lines* and the *Quadrature of the Parabola*. It too was available to Moerbeke. But it drops out of history after a reference to it in the early fourteenth century. Finally, we can mention manuscript C, not available to the Latin West in the Middle Ages, or in modern times

7. *On the Equilibrium of Plane Figures*[7]
8. *Quadrature of the Parabola*[8]
9. *Measurement of the Circle*[9]
10. *On the Sphere and the Cylinder*[10]
11. *On Conoids and Spheroids*[11]
12. *On Floating Bodies*[12]

Aristotle (384–322 B.C.)

13. *Posterior Analytics*[13]
14. *On the Soul (De anima)*[14]
15. *On the Heavens (De caelo)*[15]
16. *Categories*
17. *Nicomachean Ethics*[16]
18. *On the Generation of Animals (De generatione animalium)*[17]
19. *History of Animals (Historia animalium)*[18]
20. *On Interpretation (De interpretatione)*
21. *Metaphysics*[19]
22. *Meteorologica*
23. *On the Motion of Animals (De motu animalium)*
24. *On the Parts of Animals (De partibus animalium)*[20]
25. *On the Short Natural (or Physical) Treatises (Parva Naturalia)*[21]
26. *Physics*[22]
27. *Poetics*[23]
28. *Politics*[24]
29. *On the Progress of Animals (De progressu* [or *incessu*] *animalium)*
30. *Sophistical Refutations (De sophisticis elenchis)*[25]
31. *Rhetoric*[26]
32. *On Colors (De coloribus)*[27]

Eutochius (fl. A.D. 500)

33. Commentary on *The Equilibrium of Plane Figures* of Archimedes[28]
34. Commentary on the *Sphere and the Cylinder* of Archimedes[29]

Galen (A.D. 129—ca. 200)

35. *On Nutriments (De alimentis)*[30]

Heron of Alexandria (fl. A.D. 100)

36. *Catoptrics*[31]
37. *Pneumatics (Liber de aquarum conductibus et ingeniis erigendis)*[32]

University of Wisconsin Press, 1964], Vol. I: The Arabo-Latin Tradition, p. 3.)

7. Dated April 1269.
8. Completed May 8, 1269.
9. For two medieval Latin versions and commentaries on this treatise, see Selection 31.
10. Completed September 25, 1269.
11. Completed November 13, 1269.
12. Completed December 10, 1269.
13. A revision of a previous translation by James of Venice, a twelfth-century Italian translator.
14. Revision of an earlier version by James of Venice; perhaps completed in 1268.
15. 1260 (?) Books I and II constitute a revised translation of an earlier version by Robert Grosseteste who is represented by a few selections in this source book. Books III and IV were translated anew.
16. Perhaps a revision of a version by Grosseteste.
17. 1260(?)
18. Translated in 1260.
19. Books I–X are a revision of an earlier translation from the Greek; from Book XI on, Moerbeke translated anew.
20. Completed at Thebes, in Greece, on January 10, 1260.
21. Moerbeke's translations of the short treatises included under this single title were revisions of earlier translations, the bulk of them by James of Venice. The titles are: *On Sense and Sensible Objects (De sensu et sensato)*; *On Memory and Recollection (De memoria et reminiscentia)*; *On Sleep and Waking (De somno et vigilia)*; *On Dreams (De insomniis)*; *Prophesying by Dreams (De insomniis et de divinatione per somnum)*; *On Length and Shortness of Life (De longitudine et brevitate vitae)*; *On Youth and Old Age, On Life and Death, On Respiration (De iuventute et senectute, De vita et morte, De respiratione)*.
22. A revision of a previous translation from the Greek by James of Venice.
23. 1278.
24. Completed at Orvieto in 1260.
25. A revision of Boethius' earlier translation.
26. 1270(?)
27. Moerbeke's translation of the pseudo-Aristotelian treatise was left incomplete.
28. Completed Novemebr 21, 1269.
29. Completed October 23, 1269.
30. Completed October 22, 1277, This is probably a translation of Galen's περι τροφῶν δυναμεως.
31. Completed December 31, 1269 (see note 137 to Selection 62). This treatise was formerly ascribed to Ptolemy under the title *Claudii Ptolemei de speculis (On Mirrors of Claudius Ptolemy)*.
32. That is, *The Book on the Flow of Water and Construction of Machines*. On the basis of indirect evidence, A. Birkenmajer ascribed such a translation to Moerbeke (*Vermischte Untersuchungen zur Geschichte der mittelalterlichen Philosophie* [Münster, 1922], pp. 19–32). In my judgment, his arguments are unconvincing and the attribution to Moerbeke of a Latin translation of Hero's (or Heron's) *Pneumatics* is unwarranted (see Edward Grant, "Henricus Aristippus, William of Moerbeke and the Two Alleged Medieval Translations of Hero's *Pneumatica*," *Speculum*, XLVI (1971), 662–669.

---

until its identification by Heiberg in 1906 in Constantinople. It contains fragments of *On the Sphere and the Cylinder, On Spiral Lines, Measurement of the Circle, On the Equilibrium of Planes, Stomachion,* most of the Greek text of *On Floating Bodies,* and the brilliant work *On the Method* not present in the other manuscripts." (M. Clagett, *Archimedes in the Middle Ages* [Madison, Wis.:

Hippocrates (b. ca. 460 B.C.)

38. *On Prognostications of Sicknesses according to the Motion of the Moon (De prognosticationibus aegritudinum secundum motum lunae)*[33]

John Philoponus (fl. first half of sixth century A.D.)

39. Commentary on *On the Soul (De anima)* of Aristotle[34]

Proclus A.D. (410–485)

40. *Elements of Theology (Elementatio Theologica)*[35]
41. *Ten Doubts on Providence (De decem circa providentiam dubitationibus)*[36]
42. *The Existence of Evils (De malorum subsistentia)*[37]
43. *On Providence and Fate (De providentia et fato)*[38]
44. Commentary on the *Parmenides* of Plato[39]
45. Commentary on the *Timaeus* of Plato[40]

Ptolemy (fl. first half of second century A.D.)

46. *Book on "Taking up" (Liber de analemmata)*[41]

Simplicius (fl. first half of sixth century A.D.)

47. Commentary on the *On the Heavens (De caelo)* of Aristotle[42]
48. Commentary on the *Categories* of Aristotle[43]

Themistius (fl. A.D. 400)

49. Paraphrase of *On the Soul (De anima)* of Aristotle[44]

33. Probably Hippocrates' *On Prognostics,* concerned with the prediction of the courses of acute diseases.
34. A fragmentary translation of Book I and a separate version of Book III, the latter completed on December 17, 1268.
35. Completed at Viterbo, May 18, 1268.
36. Completed at Corinth, February 4, 1280. The Greek text is only partially preserved.
37. Completed at Corinth, February 21, 1280.
38. Completed at Corinth, February 14, 1280.
39. A translation of the seventh part only.
40. A partial translation.
41. Probably translated at the beginning of 1270. In this treatise Ptolemy was concerned in general with orthogonal projection onto a plane of the various points and arcs of the celestial sphere and in particular with determining the sun's position at any time of the day. Heath (*A History of Greek Mathematics,* II, 287) explains that "the word ἀνάλημμα [that is, *analemma—Ed.*] evidently means 'taking up'. . . in the sense of 'making a graphic representation' of something, in this case the representation on a plane of parts of the heavenly sphere." Although I have followed Heath, this word might also be rendered as "sun-dial," as suggested in the *Greek-English Lexicon* by Liddell and Scott. Heath's statement that Moerbeke's translation was made from an Arabic version of the *Analemma* must be dismissed since there is no evidence that Moerbeke knew Arabic.
42. Completed at Viterbo, June 15, 1271.
43. Completed March 1266.
44. Completed at Viterbo, November 22, 1267.

The Reaction of the Universities and Theological Authorities to Aristotelian Science and Natural Philosophy

9 THE CONDEMNATION OF ARISTOTLE'S BOOKS ON NATURAL PHILOSOPHY IN 1210 AT PARIS

Translated by Lynn Thorndike[1]

Annotated by Edward Grant

Let the body of master Amaury[2] be removed from the cemetery and cast into unconsecrated ground, and the same be excommunicated by all the churches of the entire province. Bernard, William of Arria the goldsmith, Stephen priest of Old Corbeil, Stephen priest of Cella, John priest of Occines, master William of Poitiers, Dudo the priest, Dominicus de Triangulo, Odo and Elinans clerks of St. Cloud—these are to be degraded and left to the secular arm. Urricus priest of Lauriac and Peter of St. Cloud, now a monk of St. Denis, Guarinus priest of Corbeil, and Stephen the clerk are to be degraded and imprisoned for life. The writings of David of Dinant are to be brought to the bishop of Paris before the Nativity and burned.

Neither the books of Aristotle on natural philosophy nor their commentaries are to be read at Paris in public or secret, and this we forbid under penalty of excommunication.[3] He in whose possession the writings of David of Dinant are found after the Nativity shall be considered a heretic.

As for the theological books written in French we order that they be handed over to the diocesan bishops, both *Credo in deum* and *Pater noster* in French except the lives of the saints, and this before the Purification, because [then] he who shall be found with them shall be held a heretic.

1. Reprinted by permission of the Columbia University Press from Lynn Thorndike, *University Records and Life in the Middle Ages* (New York: Columbia University Press, 1944), pp. 26–27. Thorndike's translation was made from the *Chartularium Universitatis Parisiensis*, ed. H. Denifle and E. Chatelain (Paris, 1889–1897), I, 70. The Condemnation of 1210 was issued by the provincial synod of Sens, which included the bishop of Paris as a member.

2. Amaury of Bène (d. 1206 or 1207) was a teacher of logic and theology at Paris, whose excommunication was probably the consequence of his pantheistic views. Of the others mentioned here, only David of Dinant is more

than a mere name since quotations from his treatise *On the Divisions (De tomis, id est de divisionibus)* have been preserved in the works of Albert the Great, Thomas Aquinas, and Nicholas of Cusa. From these quotations, it would seem that he too was charged with pantheism. Of the group in general, Thorndike remarks (p. 27): "Among the heresies of the aforesaid persons—we are told—were that the Father had worked without the Son and Holy Spirit before the incarnation of the Son; that the Father was incarnated in Abraham, the Son in Mary, and the Holy Spirit in us today; that whatever is, is God; that the Son had ruled until now, but that henceforth the Holy Spirit would begin to rule until the end of the world."

3. This prohibition had only local force and was restricted in its application to the arts faculty at the University of Paris, leaving theologians free to read the prohibited books, which, to make matters more perplexing, are not even specified by titles. The effect of all this on the study of Aristotle's physical treatises at Paris is not clear. During the years when the ban was allegedly in force, English scholars, including Robert Grosseteste and Roger Bacon, studied at Paris. Had the prohibition proved effective, they probably would have remained at Oxford, where Aristotle's works could be read and discussed publicly. At the very least, it is a reasonable assumption that the banned works were read in private. By 1240, however, there were already signs that discussion of the disputed books had become public. An interesting consequence of the prohibition at Paris occurred in connection with the University of Toulouse, following its foundation in 1229. A circular, perhaps written by John of Garland or drawn up by representatives of the University, was disseminated which invited masters and students of all lands to come study at Toulouse, where, among other inducements, "Those who wish to scrutinize the bosom of nature to the inmost can hear the books of Aristotle which were forbidden at Paris" (Thorndike, p. 34). Since a quarrel between university and local authorities in Paris culminated in a shutdown of the university between 1229 and 1231, it is likely that the Toulouse circular was intended to capitalize on this situation. The attraction of studying the Aristotelian physical treatises was obviously thought a tempting lure for seducing students to the sunny south of France.

10   THE COMMAND TO EXPURGATE ARISTOTLE'S BOOKS ON NATURAL PHILOSOPHY (1231)

Translated by Lynn Thorndike[1]

Annotated by Edward Grant

To masters W., archdeacon of Beauvais, Symon de Alteis of Amiens, and Stephen of Provins of Reims, canons.

Since other sciences ought to render service to the wisdom of holy writ, they are to be in so far embraced by the faithful as they are known to conform to the good pleasure of the Giver, so that anything virulent or otherwise vicious, by which the purity of the Faith might be derogated from, be quite excluded, because a comely woman found in the number of captives is not permitted to be brought into the house unless shorn of superfluous hair and trimmed of sharp nails, and in order that the Hebrews might be enriched from the despoiled Egyptians they were bade to borrow precious gold and silver vessels, not ones of rusty copper or clay.

But since, as we have learned, the books on nature which were prohibited at Paris in provincial council[2] are said to contain both useful and useless matter, lest the useful be vitiated by the useless, we command your discretion, in which we have full faith in the Lord, firmly bidding by apostolic writings under solemn adjuration of divine judgment, that, examining the same books as is convenient subtly and prudently, you entirely exclude what you shall find there erroneous or likely to give scandal or offense to readers, so that, what are suspect being removed, the rest may be studied without delay and without offense. Given at the Lateran, April 23, in the fifth year of our pontificate.

11   THE NATURAL BOOKS OF ARISTOTLE IN THE ARTS CURRICULUM AT THE UNIVERSITY OF PARIS IN 1255

Translated and annotated by Lynn Thorndike[1]

Additional notes by Edward Grant

In the year of the Lord 1254. Let all know that we, all and each, masters of arts by our common assent, no one contradicting, because of the new and incalculable peril which threatens in our faculty—

1. Reprinted by permission of the Columbia University Press from Lynn Thorndike, *University Records and Life in the Middle Ages* (New York: Columbia University Press, 1944), pp. 39–40. The translation was made from the *Chartularium Universitatis Parisiensis,* ed. H. Denifle and E. Chatelain (Paris, 1889–1897), I, 143–144. On April 13, 1231, Pope Gregory IX issued an important papal bull called *parens scientiarum* ("parent of the sciences," often characterized as the Magna Carta of the University of Paris) in which, without specific mention of Aristotle but with the prohibition of 1210 in mind, he declared that "those books on nature which were prohibited in provincial council for certain cause they shall not use at Paris until these shall have been examined and purged from all suspicion of errors" (Thorndike, p. 38). This selection is dated April 23, 1231, and represents Gregory's effort to implement the declaration of April 13, which had marked a considerable softening in attitude from the harsh prohibition of 1210. The committee chosen for this important task consisted of the distinguished Paris theologian, William of Auxerre (Master W.), Stephen of Provins, presumably the expert on Aris-

totle, and Symon de Alteis. Whatever the reason, the committee never submitted a report (perhaps, as has been suggested, because William of Auxerre died in that very same year of 1231) and the command to expurgate Aristotle's natural philosphy was never carried out. By the 1240's his physical works were read in Paris, even if not publicly discussed, and by 1255 all the questionable physical treatises were prescribed for study and public discussion (see the next selection). Curiously, on January 19, 1263, Pope Urban IV reconfirmed the bull *parens scientiarum* of 1231, including the reference to the natural books of Aristotle condemned in 1210. However, it has been plausibly argued that this may have been an oversight, one that was certainly without impact, since Aristotle's natural books were by that time firmly entrenched in the scientific and philosophical curriculum of the University of Paris.

The argument of despoiling the Egyptians and taking from the enemies of Christianity what was valuable for the study of the Bible was of ancient vintage, for it was employed during the Roman Empire period to justify the appropriation and use of Greek philosophy and science by Christians.

2. A specific reference to the Condemnation of 1210, in which Aristotle was mentioned by name (see the preceding selection).

1. [Reprinted by permission of the Columbia University Press from Lynn Thorndike, *University Records and*

some masters hurrying to finish their lectures sooner than the length and difficulty of the texts permits, for which reason both masters in lecturing and scholars in hearing make less progress—worrying over the ruin of our faculty and wishing to provide for our status, have decreed and ordained for the common utility and the reparation of our university to the honor of God and the church universal that all and single masters of our faculty in the future shall be required to finish the texts which they shall have begun on the feast of St. Remy[2] at the times below noted, not before.

The Old Logic, namely the book of Porphyry, the *Praedicamenta, Periarmeniae, Divisions* and *Topics* of Boethius, except the fourth, on the feast of the Annunciation of the blessed Virgin[3] or the last day for lectures preceding. *Priscian minor* and *major, Topics* and *Elenchi, Prior* and *Posterior Analytics* they must finish in the said or equal time. The *Ethics* through four books in twelve weeks, if they are read with another text; if *per se,* not with another, in half that time. Three short texts, namely *Sex principia, Barbarismus,* Priscian on accent, if read together and nothing else with them, in six weeks. The *Physics* of Aristotle, *Metaphysics,* and

De animalibus on the feast of St. John the Baptist;[4] *De celo et mundo,* first book of *Meteorology* with the fourth, on Ascension day;[5] *De anima,* if read with the books on nature, on the feast of the Ascension, if with the logical texts, on the feast of the Annunciation of the blessed Virgin; *De generatione* on the feast of the Chair of St. Peter;[6] *De causis* in seven weeks; *De sensu et sensato* in six weeks; *De sompno et vigilia* in five weeks; *De plantis* in five weeks; *De memoria et reminiscentia* in two weeks; *De differentia spiritus et animae*[7] in two weeks; *De morte et vita* in one week. Moreover, if masters begin to read the said books at another time than the feast of St. Remy, they shall allow as much time for lecturing on them as is indicated above. Moreover, each of the said texts, if read by itself, not with another text, can be finished in half the time of lecturing assigned above. It will not be permitted anyone to finish the said texts in less time, but anyone may take more time. Moreover, if anyone reads some portion of a text, so that he does not wish, or is unable, to complete the whole of it, he shall read that portion in a corresponding amount of time. . . .

12 STATUTE OF THE FACULTY OF ARTS DRASTICALLY CURTAILING THE DISCUSSION OF THEOLOGICAL QUESTIONS (1272)

Translated and annotated by Lynn Thorndike[1]

To each and all of the sons of holy mother church who now and in the future shall see the present page, the masters of logical science or professors of natural science at Paris, each and all, who hold and observe the statute and ordinance of the venerable father Symon by divine permission

the World); *Meteorology; De anima (On the Soul); De generatione* (that is, *On Generation and Corruption); De causis (The Book on Causes;* spurious); *De sensu et sensato (On Sense); De sompno et vigilia (On Sleep and Waking); De plantis (On Plants;* spurious); *De memoria et reminiscentia (On Memory and Reminiscence); De morte et vita* (perhaps Aristotle's *De Longitudine vitae,* that is, *On Length of Life).*—Ed.]

2. October 1.

3. March 25.

4. June 24.

5. A movable feast, forty days after Easter, usually in May.

6. February 22.

7. The author of this treatise, *On the Difference between Spirit and Soul,* was the Arab Costa ben Luca (tenth century). John of Seville made the translation from Arabic into Latin during the twelfth century.

1. [Reprinted by permission of the Columbia University Press from Lynn Thorndike, *University Records and Life in the Middle Ages* (New York: Columbia University Press, 1944), pp. 85–86. The translation was made from the *Chartularium Universitatis Parisiensis,* ed. H. Denifle and E. Chatelain (Paris, 1889–1897), I, 499–500. —Ed.]

Life in the Middle Ages (New York: Columbia University Press, 1944), pp. 64–65. The translation was made from the *Chartularium Universitatis Parisiensis,* ed. H. Denifle and E. Chatelain (Paris, 1889–1897), I, 277–278. Despite the date of 1254 in the document itself, the editors of the *Chartularium* have dated it March 19, 1255, since the new year began at easter during this period. Since not all of the Aristotelian treatises are readily identified, both the genuine and pseudo-Aristotelian treatises (the latter were thought to be genuine) are now listed: *Praedicamenta (Categories); Periarmeniae (De interpretatione,* or *On Interpretation); Topics; Elenchi (On Sophistical Refutations); Prior and Posterior Analytics; Ethics* (that is, the *Nichomachean Ethics); Physics; Metaphysics; De animalibus (On Animals;* as it stands, at least three of Aristotle's biological works are candidates for this title); *De celo et mundo (On the Heaven and*

cardinal priest of the title of St. Cecilia, legate of the apostolic see, made after separate deliberation of the nations, and who adhere expressly and entirely to the opinion of the seven judges appointed by the same legate in the same statute, greeting in the Saviour of all. All should know that we masters, each and all, from the preceding abundant and considered advice and deliberation of good men concerning this, wishing with all our power to avoid present and future dangers which by occasion of this sort might in the future befall our faculty, by common consent, no one of us contradicting, on the Friday preceding the Sunday on which is sung *Rejoice Jerusalem,* the masters one and all being convoked for this purpose in the church of Ste. Geneviève at Paris, decree and ordain that no master or bachelor of our faculty should presume to determine or even to dispute any purely theological question, as concerning the Trinity and incarnation and similar matters, since this would be transgressing the limits assigned him, for the Philosopher says that it is utterly improper for a non-geometer to dispute with a geometer.[2]

But if anyone shall have so presumed, unless within three days after he has been warned or required by us he shall have been willing to revoke publicly his presumption in the classes or public disputation where he first disputed the said question, henceforth he shall be forever deprived of our society. We decree further and ordain that, if anyone shall have disputed at Paris any question which seems to touch both faith and philosophy, if he shall have determined it contrary to the faith, henceforth he shall forever be deprived of our society as a heretic, unless he shall have been at pains humbly and devoutly to revoke his error and his heresy, within three days after our warning, in full congregation or elsewhere where it shall seem to us expedient. Adding further that, if any master or bachelor of our faculty reads or disputes any difficult passages or any questions which seem to undermine the faith, he shall refute the arguments or text as far as they are against the faith or concede that they are absolutely false and entirely erroneous, and he shall not presume to dispute or lecture further upon this sort of difficulties, either in the text or in authorities, but shall pass over them entirely as erroneous. But if anyone shall be rebellious in this, he shall be punished by a penalty which in the judgment of our faculty suits his fault and is due. Moreover, in order that all these may be inviolably observed, we masters, one and all, have sworn on our personal security in the hand of the rector of our faculty and we all have spontaneously agreed to be so bound. In memory of which we have caused this same statute to be inscribed and so ordered in the register of our faculty in the same words. Moreover, every rector henceforth to be created in the faculty shall swear that he will cause all the bachelors about to incept in our faculty to bind themselves to this same thing, swearing on their personal security in his hand. Given at Paris the year of the Lord 1271, the first day of April.[3]

13 THE CONDEMNATION OF 1277: A SELECTION OF ARTICLES RELEVANT TO THE HISTORY OF MEDIEVAL SCIENCE

Translation, introduction, and annotation by Edward Grant[1]

Ever since the introduction of Aristotle's natural philosophy into the Latin West in the twelfth century, ecclesiastical authorities had feared its impact on theological studies and Christian belief. The Parisian prohibition of Aristotle's natural books in 1210 and the demand for their expurgation in 1231 (see Selections 9 and 10) reflect the apprehension of some Church authorities. But the scientific and philosophic riches in the natural books of Aristotle made inevitable their inclusion into the medieval university curriculum, and by 1255 Aristotle's works formed the core of medieval university education (see Selection 11) which was so heavily oriented toward logic and natural philosophy.

Aristotelian natural philosophy and metaphysics provided students and teachers of the thirteenth century with philosophical tools of analysis that were applied with great fervor to all areas of human

2. [Perhaps a reference to Aristotle's remark in *Physics* I.2.185a.1–3: "For just as the geometer has nothing more to say to one who denies the principles of his science—this being a question for a different science or for one common to all—so a man investigating *principles* cannot argue with one who denies their existence."—*Ed.*]

3. As Easter fell at that time after the first of April, the year is 1272 according to our reckoning.

1. This translation has been made from the *Chartularium Universitatis Parisiensis,* ed. H. Denifle and E.

thought. Aristotle and his Arabian commentators, especially Averroes, who was himself known as "the Commentator," reigned supreme during the middle decades of the thirteenth century and were zealously studied at the University of Paris and elsewhere. It was not long before the views held by some masters of arts on a number of controversial issues became obnoxious to the theologians at Paris. Being thoroughly aroused, they persuaded the bishop of Paris, Étienne Tempier, to condemn in 1270 thirteen articles or propositions, among which were the doctrines of the eternity of the world and the unicity of the intellect (which ascribed to all men a single common intelligence). In 1272 they compelled the masters of arts to swear an oath that they would not treat theological questions (see the preceding selection). The controversy forms the central theme of the treatise of Giles of Rome (Aegidius Romanus), who sometime between 1270 and 1274 published, in defense of his faith, a treatise called *Errors of the Philosophers (Errores philosophorum)*, in which the so-called errors of Aristotle, Averroes, Avicenna, Algazali, Alkindi, and Moses Maimonides were listed.[2] Finally, in 1277, Pope John XXI (formerly Peter of Spain, author of the *Summulae logicales*), concerned over the uproar, instructed the bishop of Paris to investigate the controversies besetting the University of Paris. Not only did Tempier investigate but in only three weeks, on his own authority, he issued a condemnation of 219 propositions drawn from many sources, including, apparently, the works of Thomas Aquinas, some of whose ideas found their way onto the list. The Pope appears to have acquiesced in the actions of the bishop of Paris, and he accepted the penalty of excommunication for all who upheld even a single proposition.

There is little doubt that the condemnation influenced the course of philosophy, for under penalty of excommunication, many deterministic arguments drawn from, or based on, Aristotle's philosophy had of necessity to be modified and qualified. Alternatives, previously thought to be silly or absurd, had now to be entertained as at least possible—even if only by virtue of God's infinite and absolute power. Because of the condemnation, it became a characteristic feature of fourteenth-century scholastic discussion for authors to declare that although something was naturally impossible, it was supernaturally possible. Thus while it was naturally impossible for more than one world to exist, or for a vacuum to exist,

God could achieve both of these effects if He so desired. Theologians expanded the domain where God's absolute and unpredictable power was operative and severely reduced the domain of certain and demonstrated knowledge. The God who lived in obedience to His own laws in the Thomistic interpretation was little in evidence in the fourteenth century.

It was because the condemnation was, in effect, a frontal assault on Aristotelian metaphysics and philosophy that it seemed important and significant to Pierre Duhem, the eminent historian of medieval science, who saw it as an instrument liberating medieval science from bondage to Aristotelian cosmological and metaphysical assumptions and conclusions. It was fortunate, according to Duhem, that propositions denying God's power to move the universe with a rectilinear motion (see Proposition 49) or to create more than one world (see Proposition 34) were condemned, because it was a happy consequence that thereafter discussions on the possibility of a void and the existence of other worlds were forthcoming and served to stimulate scientific imagination and investigation. Indeed, Duhem was moved to declare that "if we must assign a date for the birth of modern science, we would, without doubt, choose the year 1277 when the bishop of Paris solemnly proclaimed that several worlds could exist, and that the whole of the heavens could, without contradiction, be moved with a rectilinear motion."[3] Although this claim is extreme and it is also doubtful whether the new intellectual vistas opened up by the condemnation were helpful to medieval science (Alexandre Koyré, for example, denied it completely), it is a fact that the questions which Duhem thought so monumental were indeed wide-

---

Chatelain (Paris, 1889–1897), I, 543–555. The original order of the propositions as printed in the *Chartularium* has been retained. A complete translation of all 219 condemned propositions following the regrouping and renumbering of P. Mandonnet, *Siger de Brabant et l'averroisme latin au xiiime siècle*, 2d ed. (Louvain: Institut supérieur de philosophie de l'Université, 1908), Part II, has been made by Ernest L. Fortrin and Peter D. O'Neill in *Medieval Political Philosophy: A Sourcebook*, ed. Ralph Lerner and Muhsin Mahdi (New York: The Free Press of Glencoe, 1963), pp. 337–354.

2. This treatise has been edited and translated by Josef Koch and John O. Reidl (Milwaukee, Wis.: Marquette University Press, 1944).

3. *Études sur Léonard de Vinci* (Paris: A. Hermann, 1906–1913), II, 412. The translation is mine from my article "Late Medieval Thought, Copernicus, and the Scientific Revolution," *Journal of the History of Ideas*, XXIII (1962), 200, n. 8.

ly discussed in the fourteenth century. It is also true to assert that fourteenth-century physicists and natural philosophers departed on many specific points from Aristotelian solutions and mechanisms of explanation.

On February 14, 1325, the condemnation was annulled because of a great sentiment in favor of the teachings of Thomas Aquinas, a number of whose ideas and positions had been included in the condemnation. In the document of nullification (printed in *Chartularium*, II, 280–281), the bishop of Paris declares that "on the basis of certain knowledge held at present, we wholly annul the aforementioned condemnation of articles and judgments of excommunication as they touch, or are said to touch, the teaching of blessed Thomas, mentioned above; and because of this we neither approve nor disapprove of these articles, but leave them for free scholastic discussion." From this it would appear that the articles condemned in 1277 no longer carried the penalty of excommunication. It is of interest, however, that, in annulling the condemnation of 1277, the bishop of Paris adopted a position of neutrality toward the articles themselves, which perhaps explains why they continued to be cited in the latter part of the fourteenth century by men of the stature of Jean Buridan, Nicole Oresme, Albert of Saxony, and others.

To all who shall examine these words, Stephen, by divine permission unworthy minister of the Church of Paris, sends greetings in the [name of the] Son of the glorious Virgin. Frequent reports from great and important people aflame with the zeal of faith have made known that some who study at Paris in the arts exceed the proper bounds of the faculty and in the schools presume to write about, and dispute, certain obvious and abominable errors—indeed, rather, "vanities and lying follies"[4] (these are listed on the roll following this letter)—as if they were [merely] doubtful.[5] They seem not to understand what Gregory said, namely that one who endeavors to speak wisely should be very apprehensive lest the unity of his audience be confounded by his discourse, especially since they support the aforesaid errors by the writings of the gentiles, which—oh! for shame—they declare in their ignorance, that these have such force that they do not know how to answer them. However, that they should not seem to assert what they hint at, they declare such feeble responses that while they are eager to avoid Scylla, they fall into Charybdis. For, indeed, they say that things are true according

to philosophy but not according to the Catholic faith; as if there could be two contrary truths,[6] and as if contrary to the truth of Sacred Scripture there could be truth in the statements of the damned gentiles of whom it was written: "I will destroy the wisdom of the wise"[7] because true wisdom destroys false wisdom. Oh that such people might understand the advice of the wise man who says: "If thou have understanding, answer *thy* neighbour: but if not, let thy hand be upon thy mouth, lest thou be surprised in an unskilful word, and be confounded."[8] Therefore, lest dangerous discourse should draw the innocent into error, we strictly forbid, on the advice communicated to us by doctors of sacred theology and other prudent men, that such things and similar things be done, and wholly condemn these things, excommunicating all those who shall have taught some or all of the said errors, or shall have presumed to defend or support them in any way whatever, and also those who listen to these things, unless it be disclosed that within seven days they have come forward to us or to the Chancellor of Paris, notwithstanding other punishments that might be inflicted upon them for the extent of their guilt, as the law shall dictate. By our same judgment, we also condemn the book *De amore*,[9] or *De Deo amoris*,[10] which begins as follows: *Cogit me multum*, etc., and ends with these words: *Cave igitur, Galtere, amoris exercere mandata*, etc.; likewise the book of Geomancy, which begins this way: *Estimaverint Indi*, etc., and ends this way: *Ratiocinare ergo super*

4. Psalms 39:5.

5. Errors that should have been identified unequivocally were instead treated as if they were only doubtful.

6. This is the doctrine of the double truth attributed to the Latin Averroists in the thirteenth century. Not only is it doubtful that Averroes himself ever held the view described here but no blatant adherents of the doctrine can be identified in the Latin West—not even Siger of Brabant, so often alleged to have been its most famous proponent. As presented here, the doctrine implies that two contrary statements could be maintained if it were merely argued that one was true by virtue of natural reason (and could, therefore, be false supernaturally) and the other by virtue of faith and dogma (and perhaps false by natural reason; see Proposition 90). When faced with such contrary statements, an adherent of this doctrine would have argued, presumably, that he accepted the truth of faith as a higher truth, but that this did not conflict with his acceptance of the truth of the contrary statement arrived at through natural reason from the premises of physics and metaphysics.

7. I Corinthians 1:19.

8. *Ecclesiasticus* 5:14.

9. *On Love.*

10. *On the God of Love.*

eum, et invenies, etc.; likewise [we condemn] books, rolls, or pamphlets *(quaternos)* on magic containing experiences of prophesying, invocations of demons, or conjurations on the peril of souls, or [writings] in which such things and similar things are evidently treated in ways that are adverse to the good customs of the orthodox faith. On all those who shall have taught or heard the said rolls, books, [or] pamphlets, we set forth the judgment of excommunication, unless, as expressed above, within seven days they shall have revealed themselves to us or the aforementioned Chancellor of Paris, notwithstanding other punishments that might be exacted according to the seriousness of the crime. Given at the curia of Paris, in the year of the Lord, 1276,[11] on the Sunday in which *Rejoice Jerusalem* is sung.

The letter ends. The errors listed on the roll follow:

. . . . .

6. That when all celestial bodies have returned to the same point—which will happen in 36,000 years—the same effects now in operation will be repeated.[12]

9. That there was no first man, nor will there be a last; on the contrary, there always was and always will be generation of man from man.

21. That nothing happens by chance, but all things occur from necessity and that all future things that will be will be of necessity, and those that will not be it is impossible for them to be; and that upon considering all causes, [it will be seen that] nothing happens contingently. [This is an] error because by definition a concourse of causes occurs by chance, as Boethius says in his book *On Consolation* [*of Philosophy*].[13]

34. That the first cause could not make several worlds.[14]

35. That without a proper agent, as a father and a man, a man could not be made by God [alone].

37. That nothing should be believed unless it is self-evident or could be asserted from things that are self-evident.

38. That God could not have made prime matter without the mediation of a celestial body.

48. That God cannot be the cause of a new act [or thing], nor can he produce something anew.

49. That God could not move the heavens [that is, the world] with rectilinear motion; and the reason is that a vacuum would remain.[15]

52. That that which is self-determined, as God, either always acts or never acts; and that many things are eternal.[16]

66. That there are several first movers.

74. That a motive intelligence of the heavens influences the rational soul just as a celestial body influences the human body.

87. That the world is eternal as to all the species contained in it; and that time is eternal, as are motion, matter, agent, and recipient; and because the world is [derived] from the infinite power of God, it is impossible that there be novelty in an effect without novelty in the cause.

88. That nothing could be new unless the sky were varied with respect to the matter of generable things.

90. That a natural philosopher ought to deny absolutely the newness [that is, the creation] of the world because he depends on natural causes and natural reasons. The faithful, however, can deny the eternity of the world because they depend upon supernatural causes.[17]

91. That the argument of the Philosopher[18] dem-

11. Although the year 1276 is recorded here, the date March 7, 1277, precedes the document published in the *Chartularium*. During this period in France, the new year began at Easter.

12. See Selection 51.2, last paragraph, and note 86, where Nicole Oresme is probably alluding to this very proposition.

13. *De consolatione philosophiae*, Book V, Prose I.

14. This is Aristotle's position in *On the Heavens (De, caelo)*, Book I, chapter 9. A detailed discussion of this widely considered question is given below, in Selection 71, by Nicole Oresme (see also next note). Here we have one of a number of propositions which limited God's power and aroused the ire of the theologians (for similar restrictions on God's power, see propositions 35, 38, 48, 49, and 141).

15. On the basis of this condemned proposition, it was henceforth respectable to argue that a vacuum was possible by divine action, and scholastics would concede this routinely. But it did not produce any proponents of an actually existent void space within the confines of the universe (concerning the existence of void beyond the world, see the dialogue between Hermes Trismegistus and Asclepius, Selection 72. Indeed, acceptance of motion in a hypothetical void antedates the Condemnation, for we find Aquinas asserting it in his *Commentary on the Physics*, quoted in Selection 55.1. Condemnation of this proposition could hardly have had the consequences ascribed to it by Duhem (see the introduction to this selection). However, at least five scholastics (Richard of Middleton, Thomas Bradwardine, Jean Buridan, Nicole Oresme, and Jean Hennon) found occasion to mention it.

16. Cited by Thomas Bradwardine in *De causa Dei contra Pelagium* (ed. Henry Savile [London, 1618], p. 177) and included in Selection 73.1.

17. Here is an illustration of the doctrine of the double truth.

18. That is, Aristotle.

onstrating that the motion of the sky is eternal is not sophistical; and it is amazing that profound men do not see this.

92. That celestial bodies are moved by an internal principle, which is soul; and that they are moved by a soul and by an appetitive power *(per virtutem appetitivam)* [that is, by force of desire] just as an animal; for just as an animal is moved by desire, so also is the sky.

93. That celestial bodies have eternity of substance but not eternity of motion.

94. That there are two eternal principles, namely the body of the sky and its soul.

95. That there are three principles in celestial bodies: (1) a subject of eternal motion, (2) a soul of a celestial body, and (3) the prime mover as that which is desired.—The error concerns only the first two.

98. That the world is eternal because that which has a nature by [means of] which it could exist through the whole future [surely] has a nature by [means of] which it could have existed through the whole past.

99. That the world, though it was made from nothing, was not, however, made anew; and although it came into being from nonbeing, nevertheless nonbeing did not precede being in duration, but only in nature.

100. That theologians who say that the sky [or heavens] sometimes rests argue from a false assumption; and that to say that the sky exists and is not moved is to utter contradictories.[19]

101. That an infinite [number] of celestial revolutions have preceded which it was not impossible for the first cause [that is, God] to comprehend, but [which are impossible of comprehension] by a created intellect.

102. That the soul of the sky is an intelligence and the celestial orbs are not instruments of the intelligences, but organs, just as the ear and the eye are organs of the sensitive power.

106. That the immediate effective cause of all forms is an orb.

107. That the elements are eternal. However, they have been made [or created] anew in the relationship which they now have.

110. That the celestial motions occur because of an intellective soul; but an intellective soul or intellect cannot be produced except by means of a body.

111. That no form coming from outside can become one with matter. For what is separable does not make [or become] one with what is cor-

ruptible.

137. That although the generation of men might become deficient, it does not because of the power of the first orb, which not only moves to generate the elements but also to generate men.

140. That to make an accident exist without a subject is an impossible argument implying a contradiction.

141. That God cannot make an accident exist without a subject nor make more [than three] dimensions exist simultaneously.

143. That from the different [zodiacal] signs of the sky diverse conditions are assigned in men, both with respect to spiritual gifts and temporal things.

145. That no question is disputable by reason which a philosopher ought not to dispute and determine, since arguments *(rationes)* are taken from [or based on] things. Moreover, philosophy has to consider all things according to its diverse parts.[20]

147. That the absolutely impossible cannot be done by God or another agent.—An error, if impossible is understood according to nature.

148. That by nutrition a man can become another numerically and individually.[21]

150. That on any question, a man ought not to be satisfied with certitude based upon authority.[22]

151. That for a man to have certitude of any conclusion, it is necessary that he found it on self-

19. It is tempting to speculate that perhaps this condemned proposition stimulated Buridan and Oresme to argue that the heavens may rest and the earth move (see Selections 67. 4, 5). But neither cites this proposition, and it is more than likely that the fourteenth century discussions derived from Simplicius' *Commentary on Aristotle's De caelo*, which was translated by Moerbeke in 1271 and used by Thomas Aquinas and Nicole Oresme both of whom (Selections 67.3,5) cite Heraclides of Pontus, the ancient Greek proponent of the earth's axial rotation, whose name and views are mentioned by Simplicius.

20. This and other propositions reflect the desire of the philosophers, whose numbers came largely from the teaching masters in the arts faculty at Paris, to subject all knowledge and ideas to philosophical and scientific scrutiny (usually along Aristotelian lines). As far back as 1270 and 1272 (see the preceding selection, on the oath of the arts faculty in 1272) the theologians had sought persistently to control and confine these activities.

21. Among the many problems raised by this preposterous proposition is the obvious one of the fate of souls when one person becomes another in all respects.

22. Although they accepted the articles of their faith, the overwhelming number of scholastics were not content to argue from authority.

evident principles.—An error, because it speaks in a general way about both certitude of understanding and [certitude of] adhesion *(adhesionis)*.[23]

152. That theological discussions are based on fables.

153. That nothing is known better because of knowing theology.

154. That the only wise men of the world are philosophers.[24]

161. That the effects of the stars on free will are hidden.

162. That our will is subject to the power of the celestial bodies.

185. That it is not true that something could be made from nothing, and also not true that it was made in the first creation.

186. That the sky never rests because the generation of the lower things, which is the end purpose of celestial motion, ought not to cease; another reason is because the heaven has its being and power from its mover which things are preserved by its motion. Whence if its motion should cease, its existence would cease.

199. That in efficient causes when the first cause [God (?)] ceases [to act] the second [or secondary] cause does not cease its operation since it could operate in accordance with nature.

201. That He who generates the whole world assumes a vacuum because place necessarily precedes what is generated in that place; therefore, before the generation of the world there was a located place which is a vacuum.[25]

202. That the elements have been made in a previous generation from chaos; but they are eternal.

.   .   .   .   .

Here ends the roll [or list] of errors containing two hundred and nineteen articles condemned at Paris by Stephen, faithworthy minister and bishop of the same place; [issued] in the curia of Paris in the year of the Lord 1276, on the Sunday in which *Rejoice Jerusalem* is sung.

14   AN OBJECTION TO THEOLOGICAL RESTRICTIONS IN THE DISCUSSION OF A SCIENTIFIC QUESTION

John Buridan (ca. 1300—ca. 1358)

Translated and annotated by Edward Grant[1]

In the eighth [question] we inquire whether it is possible that a vacuum exist by means of any power.

It is argued that this cannot be . . . . [After citing a few arguments of others in behalf of this position, Buridan presents his own views.—*Ed.*]

The opposite [position] is argued because God could annihilate everything under the lunar orb with the magnitude and figure of the lunar orb preserved. Then the concave orb of the moon, which is now a plenum in the lower world, would be a vacuum,[2] just as a pitcher would be a vacuum if God annihilated the wine in it while preserving the pitcher and where no other body enters or is made in the pitcher. And thus some of my lords and masters in theology have reproached me on this, [saying] that sometimes in my physical questions

century, however, Thomas Bradwardine himself utilized substantially the same position in his *De causa Dei contra Pelagium* (Savile edition, p. 178) in seeking to show that an imaginary infinite void space exists outside the cosmos. Curiously, in the very section where this condemned proposition is accepted and utilized by Bradwardine, he cites propositions 49 and 52 from the Condemnation of 1277, which he identifies as having been issued by the bishop of Paris. It should be noted, however, that in Bradwardine's version of condemned Proposition 201, God himself, who is ubiquitous, occupies the place in which eventually He created the world. Hence, the place which the world was to occupy was void of matter prior to the creation, but not void of God. In Proposition 201 the impression is given that the pre-existent vacuum was not created by God, but was coexistent and coeternal with Him.

1. This selection constitutes a part of Book IV, Question 8, of John Buridan's *Questions on the Eight Books of the Physics of Aristotle* and has been translated from *Acutissimi philosophi reverendi Magistri Johannis Buridani subtilissime questiones super octo phisicorum libros Aristotelis diligenter recognite et revise a Magistro Johanne Dullaert de Gandavo antea usque impresse* (Paris, 1509; Frankfurt: Minerva, 1964), fols. 73 verso, col. 2, to 74 recto, col. 1.

2. From Buridan's time and thereafter this particular illustration became almost commonplace. For Albert of Saxony's use of it, see Selection 53.1.

23. How *adhesio* should be translated in this context is unclear. But the term seems to be used here by way of contrast with understanding on the basis of self-evident principles, suggesting a reference to articles of faith.

24. The hostility and tension that had developed between theologians and philosophers is well illustrated by propositions 152–154.

25. Who the proponents of this decidedly un-Aristotelian proposition were is a puzzle. In the fourteenth

I intermix some theological matters which do not pertain to the artists [that is, Masters of Arts]. But with [all] humility I respond that I very much wish not to be restricted [with respect] to this, namely that all masters beginning in the arts swear that they will dispute no purely theological question,[3] nor [dispute] on the incarnation; and they swear further that if it should happen that they dispute or determine some question which touches faith and philosophy, they will determine it in favor of the faith and they will destroy the arguments *(rationes)* as it will be seen that they must be destroyed. Now

it is evident that if any question touches faith and theology, this is one of them, namely whether it is possible that a vacuum exist.[4] And so, if I wish to dispute it, it is necessary that I say about it what appears to me must be said according to theology, or to perjure myself and avoid the arguments on the opposite side insofar as this will seem possible for me. But I could not resolve these arguments [on the opposite side] unless I produce them. Therefore, I am compelled to do these things. I say, therefore, that "vacuum" can be imagined in two ways . . . .

15 AN ASSESSMENT OF BURIDAN'S OBJECTIONS

Ludovicus (Luis) Coronel (fl. 1511)

Translation and introduction by Lynn Thorndike[1]

Annotated by Edward Grant

The following passage from the *Perscrutationes physicales* or commentary on the eight books of the *Physics* of Aristotle by Ludovicus Coronel, composed between 1506 and 1511,[2] shows that, while the above statute was still enforced in the days of Buridan in the middle of the fourteenth century, it had become obsolete before the close of the fifteenth century.

After stating that God could supernaturally produce a vacuum, Coronel cites Buridan's commentary on the *Physics,* book IV, question 8.

For he says that many of our masters blamed him for sometimes mingling certain theological matters with physical questions, of which sort is whether this or that which is beyond the power of nature lies within divine power, since to discuss this does not pertain to artists. But he replying humbly said

3. A reference to the Statute of 1272 (Selection 12), which had been in force for many years but which, judging by Ludovicus Coronel's remarks (see next selection), had apparently been abandoned by the end of the fifteenth century.
4. The theological implications of this question on the vacuum were concerned largely with the absolute power of God and the nature of existence prior to the creation of the world (see Selection 73.1 and n. 13). To deny the *possibility* of vacuum was to deny that God could create one supernaturally, a position that no one at Paris would have upheld during the fourteenth century, when the impact of Article 49 of the Condemnation of 1277 was very great indeed. Although no theologians would have quarreled with Buridan's resolution of this question, apparently some were annoyed at his audacity in introducing theological matters into a physical question, since by their interpretation of the Statute of 1272 it was for-

bidden for a Master of Arts to do this. (Actually, in this connection, the statute says only that "if anyone shall have disputed at Paris any question which seems to touch faith and philosophy," it must not be determined contrary to the faith; since Buridan does not seek a defense by appeal to the statute itself, it may possibly have been altered and made more stringent by his time.) Perhaps Buridan was one of the first to challenge the theologians on this matter (see next selection, n.3). Buridan's sensitivity about the reactions of theologians is evident in another part of his *Questions on the Physics* (Bk. VIII, Question 12), where after suggesting the existence of a celestial impetus, he declares, "This I do not say assertively, but [rather tentatively] so that I might seek from the theological masters what they might teach me in these matters as to how these things take place" (see Selection 48 for the context of this quotation). In his *Questions on De caelo,* Book I, Question 20, we observe an even more extreme deference: "Thirdly, I say that there is no body beyond the heaven or world, namely beyond the outermost heaven; and Aristotle assumes this as obvious. But you ought to have recourse to the theologians [in order to learn] what must be said about this according to the truth of faith or constancy" (translated from *Iohannis Buridani Quaestiones super libris quattuor De caelo et mundo,* ed. E. A. Moody [Cambridge, Mass.: Mediaeval Academy of America, 1942], p.93).
1. Reprinted by permission of the Columbia University Press from Lynn Thorndike, *University Records and Life in the Middle ages* (New York: Columbia University Press, 1944), pp. 86–88.
2. Of the edition from which he made his translation, Thorndike says: "In the edition of Lyons, 1530, which I have used, three other forms of the title are given: *Perscrutationes physices, Physica Coronel,* and *Physice perscrutationes egregii interpretis magistri Ludovici Coronel Hispani Segoviensis diligenter castigate.* Prostant Lugduni in edibus Jacobi Giunti in vico Mercuriali, 1530. The passage to be quoted occurs at fols. xcii verso, col. 2—xciii recto, col. 1."

that he would prefer not to be restricted to this. But all the masters, he says, when they incept in arts, swear that they will dispute no purely theological question such as concerning the Trinity and incarnation, and they further swear that, if they chance to dispute or determine a question which touches faith and philosophy, they will determine it in favor of the faith. And they will overthrow the reasons to the contrary as they shall seem to them able to be overthrown. Moreover, it is clear, he says, that if any question touches the faith it is this one whether there can be a vacuum; therefore, if he is going to debate it, he must state what he thinks should be said about it according to theology or perjure himself.

These remarks of Buridan have astonished me, first, that our masters blamed him, for from the declaration of this term, *vacuum,* to conclude that it cannot be produced naturally but can happen supernaturally seems in no way blameworthy.[3] In the second place, Buridan's method of reply, granting that our masters were justified in blaming him, was not satisfactory, for Buridan was not forbidden absolutely to treat of theology, and they did not blame him for this, that where he had treated something concerning the faith he had determined it in favor of the faith, because he was bound to do so apart from his oath. Thirdly, that oath does not seem reasonable when it compels a man to overthrow arguments, for there might be some loyal teacher for whom some argument worked out contrary to the determination of the church, and how would he overthrow it? But to this it seems it should be said that, if the instructor does not know how to overthrow such an argument, he ought not to formulate it in public to his students. But if he should do otherwise, he would remain perjured if he had taken the oath, and will sin although no oath has preceded. Therefore, let Parisian teachers of artists who touch on theological problems look out for themselves. In the fourth

place, I, inadequate and unworthy as I am, do not recall that when I was promoted to the degree in arts I took, or knew of any of my fellows taking, such an oath, but, alas, that laudable custom of the university along with others had become obsolete.[4] In the fifth place, where two catholic masters and teachers disagree about any matter in natural or moral philosophy touching the faith in any way, one of them would not be immune from perjury, which seems to be awkward. For example, if one held that the infinite is possible and the other the opposite; and one that matter can exist without form and the other the opposite; one that, given two things neither of which is God nor either part of the other, God could annihilate one of them and leave the other, and the other, without excepting respective entities, should be unwilling with the Subtle Doctor to concede this. And whether there was an oath or not, I would strive where occasion offered to proceed conformably to the intention of the person who ordered that oath be taken. I therefore assert that God can render any place vacant of content . . . .[5]

3. Indeed, it was common for masters of arts to declare repeatedly that while a natural vacuum was impossible, God could create one supernaturally. In light of the Condemnation of 1277, it would have been foolhardy to deny it at Paris—even, apparently, long after the Condemnation had been set aside. Buridan's complaint may not have been unwarranted, however, for it is quite possible that he was one of the first masters of arts to challenge the Statute of 1272 (Selection 12) forbidding the artists from discussing theological matters. In order to treat many topics properly—especially the possible existence of a vacuum—Buridan fought for the right to introduce theological considerations where necessary. Perhaps it was due to his courageous, and apparently successful, efforts that men like Albert of Saxony, Marsilius of Inghen and other eminent masters who followed aroused no opposition from the theologians when they came to discuss the very same problems.

4. Perhaps because of the efforts of men like Buridan.

5. Interestingly, the locus of Coronel's remarks is his own lengthy discussion on the vacuum.

Classification of The Sciences

INTRODUCTION

Edward Grant

We have already noted (Selection 1, n.3) that the quadrivium was a popular subdivision of the mathematical sciences into arithmetic, geometry, astronomy, and music. But a more significant subdivision that sought to classify all sciences was formulated by Aristotle, who sometimes subdivided the sciences into two categories, theoretical and practical (*Topics* VI. 1 and *Metaphysics* II.1), and sometimes into three categories, adding productive science to the other two (as in *Topics* VI.6. 145a. 15ff. and *Metaphysics* VI.1). In his view theoretical sciences are directed toward knowledge for its own sake; practical sciences are concerned with knowledge that will serve as a guide to conduct and hence embrace ethics and politics; productive sciences seek to organize and discover knowledge pertaining to the production of beautiful or useful things and hence encompass a theory of art. Among the theoretical sciences, the most relevant and significant for the history of science, Aristotle distinguished: (a) *theology or metaphysics,* which investigates being as such—that is, entities that are separable from matter and immovable and unchangeable (this is the highest science); (b) *mathematics,* which is concerned with entities that are associated with bodies and therefore lack separate existence but can be abstracted and treated as unchangeable; and (c) *physics,* which treats of things that have separate existence but are changeable. Logic, which might have been included among the theoretical sciences, was deliberately excluded, since Aristotle conceived it as a general tool or instrument to be mastered before embarking upon a study of the sciences.

Although Aristotle did not elaborate this classification systematically or fully, it nevertheless became a topic of discussion that produced occasional alterations and additions to his general scheme by Augustine, Boethius, Cassiodorus, and others. Boethius took an especially important part in transmitting the distinction between theoretical and practical sciences during the many centuries in which Aristotle's discussions were not directly available (for Boethius' influence on Hugh of St. Victor, see Selection 16). Indeed, both of the selections on classification included here were written in the twelfth century, perhaps before the relevant works of Aristotle were available to either of the authors.

The classification by Hugh of St. Victor in the *Didascalicon,* written in the late 1120's, represents a continuation of the Latin tradition stemming from Boethius and also including as a significant element the direct influence of St. Augustine's *City of God.* This is evidenced by the inclusion of theoretical and practical sciences drawn from Boethius and the addition of logic, which Augustine had included in his threefold division of the sciences into physics, ethics, and logic (*City of God,* Bk. XI, ch. 25), a classification which he ascribed to Plato and which was known as Plato's division until the twelfth century. To theoretical, practical, and logical divisions, Hugh then added mechanical sciences, resulting in a fourfold division that may have been original. In the following selection only the theoretical sciences are included; the three other categories are sketched briefly in footnotes.

The classification scheme presented by Domingo Gundisalvo (Dominicus Gundissalinus, in Latin) in *De divisione philosophiae (On the Division of Philosophy),* written sometime around 1150, drew heavily upon the *De scientiis (On the Sciences)* of the Arabic author al-Fārābī (*d.* 950/51). This Arabic treatise was first translated into Latin sometime around 1140 by John of Seville and Gundisalvo himself (it was later translated brilliantly by Gerard of Cremona; see the list of his translations, Selection 7, no. 42). What Gundisalvo now presented to the Latin West was a far more elaborate and more mature schema than anything found in Hugh's *Didascalicon.* This is explained by the fact that Gundisalvo's primary source, al-Fārābī, had a familiarity with the whole range of Greek science (including Aristotle) and much of Arabic science that had been fashioned by his own day. By virtue of his knowledge of the contents of al-Fārābī's treatise, Gundisalvo's classification reflected the new Greek and Arabic science that was even then in process of becoming available to the Latin West. And although Gundisalvo was himself hardly familiar with the wide range of sciences which he described, al-Fārābī's treatise (as well as works by other Arabic authors such as

Avicenna, al-Ghazzālī, and so on) enabled him to elaborate a classification of the many sciences that would soon be a vital part of the western intellectual heritage. Thus, although no more than approximately twenty years separated the *Didascalicon* and *On the Division of Philosophy,* the former can be said to represent the tradition of the earlier Middle Ages, while the latter, also incorporating aspects of the older tradition, truly heralds the advent into the Latin West of the Greek and Arabic scientific traditions.

16   CLASSIFICATION OF THE SCIENCES

Hugh of St. Victor (d. 1141)

Translated by Jerome Taylor[1]

Annotated by Edward Grant

CHAPTER 1

CONCERNING THE DISTINGUISHING OF THE ARTS

Philosophy is divided into theoretical, practical,[2] mechanical,[3] and logical.[4] These four contain all knowledge. The theoretical may also be called speculative; the practical may be called active, likewise ethical, that is, moral, from the fact that morals consist in good action; the mechanical may be called adulterate because it is concerned with the works of human labor; the logical may be called linguistic from its concern with words. The theoretical is divided into theology, mathematics, and physics—a division which Boethius makes in different terms, distinguishing the theoretical into the intellectible, the intelligible, and the natural, where the intellectible is equivalent to theology, the intelligible to mathematics, and the natural to physics.[5] And the intellectible he defines as follows.

CHAPTER 2

CONCERNING THEOLOGY

"The intellectible is that which, ever enduring of itself, one and the same in its own divinity, is not ever apprehended by any of the senses, but by the mind and the intellect alone. Its study," he says, "the Greeks call theology, and it consists of searching into the contemplation of God and the incorporeality of the soul and the consideration of true philosophy."[6] It was called theology as mean-

ethical conduct, household management, or economic activity, and political management. In Book II, chapter 19 (p. 74 of Taylor's translation), he says: "The practical is divided into solitary, private, and public; or, put differently, into ethical, economic, and political; or still differently, into moral, managerial, and civil. Solitary, ethical, and moral are one; as also are private, economic, and managerial; and public, political, and civil." It is obviously closely related to Aristotle's concept of practical sciences.

3. In Book II, chapter 20, Hugh divides the mechanical sciences into seven particular sciences: (1) fabric making, (2) armament, (3) commerce, (4) agriculture, (5) hunting, (6) medicine, and (7) theatrics. (Each is subsequently allotted a single chapter for further discussion [see Bk. II, chs. 21–27].) After likening these seven mechanical sciences to the seven liberal arts, Hugh says (Taylor, p. 75): "These sciences are called mechanical, that is, adulterate, because their concern is with the artificer's product, which borrows its form from nature. Similarly, the other seven are called liberal either because they require minds which are liberal, that is, liberated and practical (for these sciences pursue subtle inquiries into the causes of things), or because in antiquity only free and noble men were accustomed to study them, while the populace and the sons of men not free sought operative skill in things mechanical."

4. In Book II, chapter 28, Hugh divided logic into grammar and the theory of argument (logic proper), the latter further subdivided (Bk. II, ch. 30), into demonstration, probable argument, and sophistic. "Demonstration consists of necessary arguments and belongs to philosophers; probable argument belongs to dialecticians and rhetoricians; sophistic to sophists and quibblers. Probable argument is divided into dialectic and rhetoric, both of which contain invention and judgment as integral parts. . . . Invention teaches the discovery of arguments and the drawing up of lines of argumentation. The science of judgment teaches the judging of such arguments and lines of argumentation" (Taylor, p. 81). Gundisalvo also mentions "invention" and "judgment" (see Selection 17).

5. The division of theoretical science into theology, mathematics, and physics is, of course, Aristotle's. Boethius equates them with the intellectible, the intelligible, and the natural, respectively, in his *Commentary on the Isagoge of Porphyry,* Book I, chapter 3.

6. *Ibid.*

1. Reprinted with the kind permission of Columbia University Press from *The Didascalicon of Hugh of St. Victor: A Medieval Guide to the Arts,* translated from the Latin with an Introduction and Notes by Jerome Taylor (New York: Columbia University Press, 1961), pp. 62–64, 67–73. There are six books in the *Didascalicon,* but the present selection is drawn wholly from Book II. Many of the references to quotations, as well as other substantive contributions, were supplied by Taylor in the notes to his translation.

2. For Hugh the "practical sciences" serve as guides to

ing discourse concerning the divine, for *theos* means God, and *logos* discourse or knowledge. It is, theology, therefore, "when we discuss with deepest penetration some aspect either of the inexpressible nature of God or of spiritual creatures."[7]

CHAPTER 3

CONCERNING MATHEMATICS

The "instructional" science is called mathematics: *mathesis,* when the "t" is pronounced without the "h," means vanity, and it refers to the superstition of those who place the fates of men in the stars and who are therefore called "mathematicians"; but when the "t" is pronounced with the "h," the word refers to the "instructional" science.[8]

This, moreover, is the branch of theoretical knowledge "which considers abstract quantity. Now quantity is called abstract when, intellectually separating it from matter or from other accidents, we treat of it as equal, unequal, and the like, in our reasoning alone"[9]—a separation which it receives only in the domain of mathematics and not in nature. Boethius calls this branch of knowledge the *intelligible* and finds that "it itself includes the first or *intellectible* part in virtue of its own thought and understanding, directed as these are to the celestial works of supernal divinity and to whatever sublunary beings enjoy more blessed mind and purer substance, and, finally, to human souls. All of these things, though they once consisted of that primary intellectible substance, have since, by contact with bodies, degenerated from the level of intellectibles to that of intelligibles; as a result, they are less objects of understanding than active agents of it, and they find greater happiness by the purity of their understanding whenever they apply themselves to the study of things intellectible."[10]

For the nature of spirits and souls, because it is incorporeal and simple, participates in intellectible substance; but because through the sense organs spirit or soul descends in different ways to the apprehension of physical objects and draws into itself a likeness of them through its imagination, it deserts its simplicity somehow by admitting a type of composition. For nothing that resembles a composite can, strictly speaking, be called simple.

In different respects, therefore, the same thing is at the same time intellectible and intelligible—intellectible in being by nature incorporeal and imperceptible to any of the senses; intelligible in

being a likeness of sensible things, but not itself a sensible thing. For the intellectible is neither a sensible thing nor a likeness of sensible things. The intelligible, however, is itself perceived by intellect alone, yet does not itself perceive only by means of intellect. It has imagination and the senses, and by these lays hold upon all things subject to sense. Through contact with physical objects it degenerates, because, while through sense impressions it rushes out toward the visible forms of bodies and, having made contact with them, draws them into itself through imagination, it is cut away from its simplicity each time it is penetrated by any qualities entering through hostile sense experience. But when, mounting from such distraction toward pure understanding, it gathers itself into one, it becomes more blessed through participating in intellectible substance.

CHAPTER 6

CONCERNING THE QUADRIVIUM[11]

Since, as we have said, the proper concern of mathematics is abstract quantity, it is necessary to seek the species of mathematics in the parts into which such quantity falls. Now abstract quantity is nothing other than form, visible in its linear dimension, impressed upon the mind, and rooted in the mind's imaginative part. It is of two kinds: the first is continuous quantity, like that of a tree or a stone, and is called *magnitude;* the second is discrete quantity, like that of a flock or of a people, and is called *multitude.* Further, in the latter some quantities stand wholly in themselves, for example, "three," "four," or any other whole number;

7. Quoted by Hugh either from Isidore's *Etymologies,* Book II, chapter 24, paragraph 13, or from Cassiodorus' *Introduction to Divine and Human Readings,* Book II, chapter 3, paragraph 6 (on p. 160 of the edition by L. W. Jones [New York: Columbia University Press, 1946]).

8. Taylor notes (p. 197, n. 18) that "at the basis of the distinction are the Greek μάθησις, knowledge, and ματαιότης, vanity."

9. Hugh is here quoting from Cassiodorus' *Introduction to Divine and Human Readings,* Book II, preface 4, and chapter 3, paragraph 6.

10. *Commentary on the Isagoge of Porphyry,* Book I, chapter 3.

11. This chapter was probably drawn from Boethius, *De arithmetica,* Book I, chapter 1 (for excerpts from this treatise, see Selection 2). It differs considerably from Isidore of Seville's brief description in the preface to Book III of the *Etymologies* (see Selection 1). Hugh's discussion of *magnitude and multitude,* and their subdivisions, is also found in Gundisalvo's consideration of the division of mathematics (see Selection 17, n.32).

others stand in relation to another quantity, as "double," "half," "once and a half," "once and a third," and the like. One type of magnitude, moreover, is *mobile,* like the heavenly spheres, another, immobile, like the earth. Now, multitude which stands in itself is the concern of arithmetic, while that which stands in relation to another multitude is the concern of music. Geometry holds forth knowledge of immobile magnitude, while astronomy claims knowledge of the mobile. Mathematics, therefore, is divided into arithmetic, music, geometry, and astronomy.

CHAPTER 7

CONCERNING THE TERM "ARITHMETIC"

The Greek word *ares* means *virtus,* or power, in Latin; and *rithmus* means *numerus,* or number, so that "arithmetic" means "the power of number."[12] And the power of number is this—that all things have been formed in its likeness.[13]

CHAPTER 8

CONCERNING THE TERM "MUSIC"

"Music" takes its name from the word "water," or *aqua* because no euphony, that is, pleasant sound, is possible without moisture.

CHAPTER 9

CONCERNING THE TERM "GEOMETRY"

"Geometry" means "earth-measure," for this discipline was first discovered by the Egyptians, who, since the Nile in its inundation covered their territories with mud and obscured all boundaries, took to measuring the land with rods and lines. Subsequently, learned men reapplied and extended it also to the measurement of surfaces of the sea, the heaven, the atmosphere, and all bodies whatever.[14]

CHAPTER 10

CONCERNING THE TERM "ASTRONOMY"

"Astronomy" and "astrology" differ in the former's taking its name from the phrase "law of the stars," while the latter takes its from the phrase "discourse concerning the stars"—for *nomia* means law, and *logos,* discourse. It is astronomy, then, which treats the law of the stars and the revolution of the heaven, and which investigates the regions, orbits, courses, risings, and settings of stars, and why each bears the name assigned it; it is astrology, however, which considers the stars in their bearing upon birth, death, and all other events, and is only partly natural, and for the rest, superstitious; natural as it concerns the temper or "complexion" of physical things, like health, illness, storm, calm, productivity, and unproductivity, which vary with the mutual alignments of the astral bodies; but superstitious as it concerns chance happenings or things subject to free choice. And it is the "mathematicians" who traffic in the superstitious part.[15]

CHAPTER 11

CONCERNING ARITHMETIC

Arithmetic has for its subject equal, or even, number and unequal, or odd, number. Equal number is of three kinds: equally equal, equally unequal, and unequally equal. Unequal number, too, has three varieties: the first consists of numbers which are prime and incomposite; the second consists of numbers which are secondary and composite; the third consists of numbers which, when considered in themselves, are secondary and composite, but which, when one compares them with other numbers [to find a common factor or denominator], are prime and incomposite.[16]

CHAPTER 12

CONCERNING MUSIC[17]

The varieties of music are three: that belonging to the universe, that belonging to man, and that which is instrumental.

Of the music of the universe, some is characteristic of the elements, some of the planets, some of the seasons: of the elements in their mass, number,

12. The same derivation was given in the ninth century by John the Scot (Scotus Eriugena) in his Annotations on Martianus Capella (*Johannis Scotti Annotationes in Marcianum,* ed. Cora Lutz [Cambridge, Mass.: Mediaeval Academy of America, 1942], pp. 89, 152). For Gundisalvo's mention of the term "ares," see Selection 17.

13. Compare with Boethius' opening remarks of Book I, chapter 2, of *On Arithmetic* (Selection 2).

14. Compare with Isidore of Seville, *Etymologies,* Book III, chapter 10, paragraphs 1–3 (Selection 1).

15. Although Hugh provides more details, his source for this chapter may have been Isidore of Seville, *Etymologies,* Book III, chapter 27 (see Selection 1).

16. Compare with Isidore of Seville, *Etymologies,* Book III, chapter 5, paragraphs 1–2, 8 (Selection 1).

17. Summarized from Boethius, *On Music (De musica),* Book I, chapter 2.

and volume; of the planets in their situation, motion, and nature; of the seasons in days (in the alternation of day and night), in months (in the waxing and waning of the moon), and in years (in the succession of spring, summer, autumn, and winter).

Of the music of man, some is characteristic of the body, some of the soul, and some of the bond between the two. It is characteristic of the body partly in the vegetative power by which it grows—a power belonging to all beings born to bodily life; partly in those fluids or humors through the mixture or complexion of which the human body subsists—a type of mixture belonging to all sensate beings; and partly in those activities (the foremost among them are the mechanical) which belong above all to rational beings and which are good if they do not become inordinate, so that avarice or appetite are not fostered by the very things intended to relieve our weakness. As Lucan says in Cato's praise:

He feasted in conquering hunger;
Any roof from storms served his hall;
His dearest garb, the toga coarse,
Civilian dress of the Roman.[18]

Music is characteristic of the soul partly in its virtues, like justice, piety, and temperance; and partly in its powers, like reason, wrath, and concupiscence. The music between the body and the soul is that natural friendship by which the soul is leagued to the body, not in physical bonds, but in certain sympathetic relationships for the purpose of imparting motion and sensation to the body. Because of this friendship, it is written, "No man hates his own flesh."[19] This music consists in loving one's flesh, but one's spirit more; in cherishing one's body, but not in destroying one's virtue.

Instrumental music consists partly of striking, as upon tympans and strings; partly in blowing, as upon pipes or organs; and partly in giving voice, as in recitals and songs. "There are also three kinds of musicians: one that composes songs, another that plays instruments, and a third that judges instrumental performance and song."[20]

CHAPTER 13

CONCERNING GEOMETRY[21]

Geometry has three parts: planimetry, altimetry, and cosmimetry. Planimetry measures the plane, that is, the long and the broad, and, by widening its object, it measures what is before and behind and to left and right. Altimetry measures the high, and, by widening its object, it measures what

reaches above and stretches below: for height is predicated both of the sea in the sense of depth, and of a tree in the sense of tallness. *Cosmos* is the word for the universe, and from it comes the term "cosmimetry," or "universe-measurement." Cosmimetry measures things spherical, that is, globose and rotund, like a ball or an egg,[22] and it is therefore called "cosmimetry" from the sphere of the universe, on account of the preeminence of this sphere—not that cosmimetry is concerned with the measurement of the universe alone, but that the universe-sphere excels all other spherical things.

CHAPTER 14

CONCERNING ASTRONOMY

What we have just said does not contradict our previous statement that geometry is occupied with immobile magnitude, astronomy with mobile: for what we have just said takes into account the original discovery of geometry, which led to its being called "earth measurement." We can also say that what geometry considers in the sphere of the universe—namely, the measure of the celestial regions and spheres—is immobile in that aspect which belongs to geometrical studies. For geometry is not concerned with movement but with space. What astronomy considers, however, is the *mobile*—the courses of the stars and the intervals of time and seasons. Thus, we shall say that without exception immobile magnitude is the subject of geometry, mobile of astronomy, because, although both busy themselves with the same thing, the one contemplates the static aspect of that thing, the other its moving aspect.

CHAPTER 15

DEFINITION OF THE QUADRIVIUM

Arithmetic is therefore the science of numbers.

18. Lucan, *Pharsalia* II. 384–387.
19. Ephesians 5:29.
20. Boethius, *On Music,* Book I, chapter 34.
21. Based on the introduction *(prenotanda)* to Hugh's own *Practica Geometriae (Applied Geometry),* the text of which has been edited by Roger Baron, *Hugonis de Sancto Victore opera propaedeutica* (Notre Dame, Ind.: University of Notre Dame Press, 1966). See Baron, p. 17. For a selection from a medieval applied geometry, see Selection 33.
22. Taylor observes (p. 203, n. 53) that the measurement of things like a ball and egg in cosmimetry can be found in earlier authors such as Honorius of Autun, Hildegarde of Bingen, and William of Conches, but derives ultimately from the *Saturnalia* of Macrobius. The same threefold division of applied geometry is also found in Gundisalvo's classification (Selection 17).

Music is the distinguishing of sounds and the variance of voices. Or again, music or harmony is the concord of a number of dissimilar things blended into one. Geometry is the discipline treating immobile magnitude, and it is the contemplative delineation of forms, by which the limits of every object are shown. Putting it differently, geometry is "a fount of perceptions and a source of utterances."[23] Astronomy is the discipline which examines the spaces, movements, and circuits of the heavenly bodies at determined intervals.

CHAPTER 16

CONCERNING PHYSICS

Physics searches out and considers the causes of things as found in their effects, and the effects as derived from certain causes:
> Whence the tremblings of earth do rise, or
>   from what cause the deep seas swell;
> Whence grasses grow or beasts are moved with
>   wayward wrath and will;
> Whence every sort of verdant shrub, or rock,
>   or creeping thing.[24]

The word *physis* means nature, and therefore Boethius places natural physics in the higher division of theoretical knowledge.[25] This science is also called physiology, that is, discourse on the natures of things, a term which refers to the same matter as physics. Physics is sometimes taken broadly to mean the same as theoretical science, and, taking the work in this sense, some persons[26] divide philosophy into three parts—into physics, ethics, and logic. In this division the mechanical sciences find no place, philosophy being restricted to physics, ethics, and logic alone.

CHAPTER 17

WHAT THE PROPER BUSINESS OF EACH ART IS

But although all the arts tend toward the single end of philosophy, they do not all take the same road, but have each of them their own proper businesses by which they are distinguished from one another.

The business of logic is with things, and it attends to our concepts of things, either through the understanding, so that our concepts may not be either things or even likenesses of them, or through the reason, so that our concepts may still not be things but may, however, be likenesses of them. Logic, therefore, is concerned with the species and genera of things.

Mathematics, on the other hand, has as its business the consideration of things which, though actually fused, are rationally separated by it. For example, in actuality no line is found without surface and solidity. For no physical entity possesses pure length in such a way that it lacks breadth or height, but in every physical thing these three are found together. And yet the reason, abstracting from surface and from thickness, considers pure line, in itself, taking line as a mathematical object not because it exists or could exist as such in reality, but because the reason often considers actual aspects of things not as they are but as they can exist[27]—exist, that is, not in themselves, but with respect to reason itself, or, as reason might allow them to be. From this consideration derives the axiom that continuous quantity is divisible into an infinite number of parts, and discrete quantity multipliable into a product of infinite size. For such is the vigor of the reason that it divides every length into lengths and every breadth into breadths, and the like—and that, to this same reason, a continuity lacking interruption continues forever.

The business of physics, however, is to analyze the compounded actualities of things into their elements. For the actualities of the world's physical objects are not pure but are compounded of pure actualities which, although they nowhere exist as such, physics nonetheless considers as pure and as such. Thus physics considers the pure actuality of fire, or earth, or air, or water, and, from a consideration of the nature of each in itself, determines the constitution and operation of something compounded of them.[28]

23. After noting that this is a standard definition of Topics *(topica)*, which can be found in Cassiodorus' *Introduction to Divine and Human Readings,* Book II, chapter 3, paragraph 14, and Isidore's *Etymologies,* Book II, chapter 29, paragraph 16, Taylor declares (p. 203, n. 55) that "application of this phrase to geometry may be related to the view of mathematics, geometry in particular, as concerned with visible form, and to the view that form in turn, is the source of a thing's species or genus and name."
24. Virgil, *Georgics* II. 479ff.
25. *Commentary on the Isagoge of Porphyry,* Book II, chapter 3.
26. This is the view ascribed to Plato by St. Augustine in the *City of God,* Book XI, chapter 25.
27. Ultimately derived from Aristotle's conception of mathematics as described in *Metaphysics* XIII.2.1076a. 12–3.1078b.5.
28. Hugh's position is probably derived ultimately from Aristotle, *On Generation and Corruption* II.3.330b. 22–30, where we read (Oxford translation by Harold Joachim) that "fire and air, and each of the bodies we

Nor ought we to overlook the fact that physics alone is concerned properly with things, while all the other disciplines are concerned with concepts of things. Logic treats of concepts themselves in their predicamental framework, while mathematics treats of them in their numerical composition. Logic, therefore, employs pure understanding on occasion; whereas mathematics never operates without the imagination, and therefore never possesses its object in a simple or non-composite manner. Because logic and mathematics are prior to physics in the order of learning and serve physics, so to say, as tools—so that every person ought to be acquainted with them before he turns his attention to physics—it was necessary that these two sciences base their considerations not upon the physical actualities of things, of which we have deceptive experience, but upon reason alone, in which unshakeable truth stands fast, and that then, with reason itself to lead them they descend into the physical order.[29]

Having therefore already shown how Boethius's division of theoretical science fits in with what I gave just before, I shall now briefly repeat both divisions so that we may place their terminologies side by side and compare them.

17 CLASSIFICATION OF THE SCIENCES

Domingo Gundisalvo (fl. 1140)

Translated by Marshall Clagett and Edward Grant[1]

Annotated by Edward Grant

PROLOGUE

. . .Since there is no science which is not some

CHAPTER 18

COMPARISON OF THE FOREGOING

The theoretical is divided into theology, mathematics, and physics; or, put differently, into intellectible, intelligible, and natural knowledge; or still differently, into divine, "instructional," and philological.[30] Thus, theology is the same as the intellectible and the divine; mathematics as the intelligible and the "instructional"; and physics as the philological and the natural.

There are those who suppose that these three parts of the theoretical are mystically represented in one of the names of Pallas, fictional goddess of wisdom. For she is called "Tritona" for *tritoona,* that is, threefold apprehension of God, called intellectible; of souls, called intelligible; and of bodies, called natural.[31] And the name of wisdom by right belongs to these three alone: for although we can without impropriety refer to the remaining branches (ethics, mechanics, and logic) as wisdom, still these are more precisely spoken of as prudence or knowledge—logic because of its concern for eloquence of word, and mechanics and ethics because of their concern for works and morals. But the theoretical alone, because it studies the truth of things, do we call wisdom.[32]

---

have mentioned, are not simple, but blended. The 'simple' bodies are indeed similar in nature to them, but not identical with them. Thus the 'simple' body corresponding to fire is 'such-as-fire,': not fire: that which corresponds to air is 'such-as-air': and so on with the rest of them. . . ." William of Conches (*On the Philosophy of the World* [*De philosophia mundi*], Bk. I, ch. 21), for example, explicitly denied that the fire, earth, air, and water which we see are actually elements; they are already compounded. See Selection 17, n. 9, and Selection 109.2, n. 10.

29. A Platonic conception in sharp opposition to Aristotle's attitude toward experience.

30. Taylor comments (p. 204, n. 62) that "instead of 'philological' one would expect 'physiological' in the sense defined in II.xvi." But he observes that the term *phylologia* was at times used for physics.

31. Taylor cites (p. 204, n. 63) virtually the same three-fold distinction by Remigius of Auxerre (d. 908) in his commentary on Martianus Capella's *Marriage of Mercury and Philology*.

32. Taylor says (p. 205, n. 64): "The present exclusion of logic, ethics, and mechanics from wisdom is at variance with *Didascalicon* II. 17, where Hugh associates logic with mathematics and physics as concerned, though on different levels, with *things,* and with *Didascalicon* I. 8, where Hugh, arguing from premises carefully laid down in earlier chapters of Book I, demonstrates that both practical and mechanical arts must be taken as parts of wisdom."

1. Translated from Dominicus Gundissalinus, *De divisione philosophiae,* edited by Ludwig Baur in *Beiträge zur Geschichte der Philosophie des Mittelalters, Texte und Untersuchungen,* Band IV, Heft 2–3 (Münster Westfalen: Aschendorffsche Verlagsbuchhandlung, 1903), pp. 5–35, 69–72, 74–77, 82–88, 90–94, 103–124. Many of the sources used by Gundissalinus (hereafter cited as Gundisalvo, the Spanish form of his name) were identified by Baur.

part of philosophy, we should first see, therefore, what philosophy is and why it is so called; then what is its purpose and end; then what are its parts and the parts of its parts; finally, what should be considered about each of them.

Philosophers have described philosophy in two ways. One of these is by its properties, the other by its effects. That which has been expressed by its properties is this: "Philosophy is the assimilation of man to the works of the Creator according to the power *(virtutem)* of man." Assimilation to the works of the Creator is indeed the perception of the truth of things, namely the true understanding of them and of their operation, which conforms to the truth. The perception of the truth of things, furthermore, is the perception of these things from the four natural causes: the material cause, the formal cause the efficient cause, and the final cause. . . . [2]

Philosophy is likewise described thus: "Philosophy is the cognition of human and divine things joined with a zeal for living well." Also: "Philosophy is the art of arts and the discipline of disciplines."[3] A description of philosophy taken from its effect is this: "Philosophy is man's total cognition of himself."[4] This is because when a man recognizes himself completely, indeed he recognizes everything that exists *(est)*. For in man there are substance and accident, but substance is twofold: namely spiritual, like soul and intelligence, and corporeal, like a long and wide and thick body.

Similarly accident is twofold: spiritual and corporeal. Spiritual accident is like knowledge and virtue and whatever exists in the soul. Corporeal accident in truth is like "whiteness" and whatever, exists in the body. When a man knows himself perfectly, then he knows whatever exists, because he knows spiritual and corporeal substance and the first substance created out of the power of the Creator without anything intervening, a substance which is of itself subject to diversity. He also recognizes the first general accident, divided into quantity, quality, and relation, and the other six composite accidents, born from the conjunction of substance with the three simple accidents.[5] When, moreover, he comprehends all these things, then he has certainly comprehended every science which is, and he thus deserves to be called a philosopher.

Having recognized what philosophy is, then, it should be seen why it is so called: "Philosophy is the love of wisdom." For *philos* in Greek is translated *amor* (love) in Latin and *sophia* is translated *sapiencia* (wisdom). Whence philosophy is the "love of wisdom" and a philosopher is called a

"lover of wisdom."[6]

. . . . . .

Since we have seen why philosophy is so named, now let us examine its intention. "The intention of philosophy is to comprehend the truth of all things that are, insofar as it is possible for man to do so."[7] But of all things that are, some are from our own work and will—our human works, such as laws, constitutions, religious exercises, wars, and other things of this kind. Others are not from our work or will—such as God, angels, heaven, earth, vegetables, animals, metals, spirits, and all natural things.[8] The totality of all things can be comprehended thus: Every thing which is either comes into being or does not come into being. Everything which has not come into being [is] like God, the Creator of all things: the Father, and the Son, and the Holy Ghost. This is truly eternal, being without beginning and end. Moreover, everything which has come into being is like all creatures. Everything which has come into being either has come "to be" before time, such as *yle* (the principle of matter or first matter) and angelic creatures; or it has come "to be" with time, such as celestial bodies, [invisible and pure] elements, and the [visible] elements made by the first composition [of the invisible elements],[9]

2. Although these four causes were formulated by Aristotle, the material in this paragraph was derived from Isaac Israeli's *Diffinitiones,* a work that had been translated from Arabic to Latin.

3. These two definitions are direct quotes from Isidore of Seville's *Etymologies,* Book II, chapter 24 ("On the Definition of Philosophy"), paragraphs 1 and 9.

4. Drawn from Isaac Israeli's *Diffinitiones.* Indeed, this and the next paragraph are from that source.

5. Including substance, these are the ten categories or types of predicates distinguished by Aristotle in his treatise *Categories.* Nine are predicated of substance, which is primary. Quantity, quality, and relation are taken to be more fundamental than the remaining six (place, date, posture, possession, action, and passivity). According to W. D. Ross (*Aristotle,* 5th ed., rev. [London: Methuen, 1949], p. 23), "The categories are a list of the widest predicates which are predicable essentially of the various nameable entities, i.e., which tell us what kinds of entity at bottom they are."

6. This paragraph is also derived from Israeli's *Diffinitiones.*

7. Quoted from the opening line of Avicenna's *Logica,* chapter 1.

8. The last two sentences were taken almost verbatim from the first part of the *Metaphysics* of al-Ghazali (1058–1111), the Muslim theologian who exerted a great influence on the West. See J. T. Muckle's edition of *Algazel's Metaphysics* (Toronto: Institute of Mediaeval Studies, 1933), p. 1, ll. 22–26.

9. ". . .et elementa et elementata ab eis prima composicione." The distinction between the terms elementa

and these are everlasting without end; or it has come "to be" after [the beginning of] time, such as all other things. Some of these latter are without end, such as the rational soul; others have an end, such as temporal things *(temporalia)*, which begin in time and cease in time. Some of these temporal things are "natural things"; others are "artificial things." Natural things are those which with the motion of nature visibly operating go forth from potency (potentiality) to act (actuality), such as all living things which are born of the earth, as well as species of beasts of burden. They include also inanimate things which arise from the complexion, composition, or conversion of the elements, or which result from phenomena of the air, like snow, hail, rain, and other things of this kind. Artificial things in truth are those things which are made from the art and will of man, such as *subtellares*.[10] Moreover, there are certain things which are at once natural things and artificial things, such as wine, a statue, a sword, a spike, and similar things. But these are natural things with respect to their matter, artificial with respect to the form which makes them what they are. Spiders' webs, birds' nests, and bees' cells are reckoned among natural things.

Out of these [remarks], then, it has become evident why everything that exists is either from our [human] work and our will or is not from our work but from that of God or nature. Since there is no science which does not have a subject of which it treats, there can be nothing which does not arise from one of these two kinds (that is, of man and nature or God). Hence philosophy in the first place is divided into two parts. One part is that by which we know the dispositions of our works; the other is that by which we know all other things which exist. For there is one part of philosophy which makes us know what ought to be done and this is called "practical" *(practica)*; and there is another which makes us know what ought to be understood and this is "theoretical" *(theorica)*. Therefore, one is in intellect, the other is in effect. One consists in the cognition of the mind alone, the other in the execution of work. For since philosophy has been established in order to perfect the mind *(anima)*, there are two ways by which it achieves this, namely by science [or knowledge *(scientia)*] and operation. Therefore, philosophy, which is the order of the mind *(anime)*, is necessarily divided into *science* and *operation*, just as the mind *(anima)* is divided into sense and reason. *Operation* is relevant to the sensible part and

speculation [or *science*] to the rational part. But since the rational part of the mind [or soul] is divided into the cognition of divine things, namely those that are not from our [human] work [or effort], and the cognition of human things, namely those that are from our work [or effort]. Hence the end [or goal] of philosophy is the perfection of the soul, not so that man can know only what he should understand but also so that he may know what he ought to do and [indeed what he] does. For the goal of speculation is the formulation *(conceptio)* of an opinion [or idea] for the understanding [or intellect]; the goal of practice is the formulation *(conceptio)* of an opinion [or idea] for acting.

The parts, therefore, into which philosophy is first divided are theoretical and practical. After this, it remains for us to see what and how many are the parts of these first [two] parts of philosophy. We have said above that theoretical philosophy is the cognition of those things which do not arise out of our work. . . .

. . . . . .

The parts of theoretical philosophy are three: evidently, either speculation [or theory] concerns those things that are not separated from their matters in being *(esse)* or in intellect; or speculation concerns those things that are separated from the matter in intellect but not in being *(esse)*; or speculation concerns those things that are separated from matter both in *esse* and intellect. The first part of this division is called "physical science" or "natural science," which is the first and the lowest; the second is called "mathematical science" or "disciplinal," and is the middle; the third is called "theology" or "the first science" or "first philosophy" or "metaphysics." And because of this Boethius[11] says that physics is unabstracted and with motion, mathematics abstracted and with motion, and theology abstracted and without motion. And these three sciences only are parts of theoretical philosophy in that there can be no more kinds of things other than these three concerning which speculation can be made. Whence Aristotle says:[12] Therefore there are three species of sciences, since one investigates what is moved and corrupted, "natural science"; the second what is moved and not corrupted, the "disciplinal science"; and the

---

and elementata is discussed in Selection 16, n. 28, and Selection 109.1, n. 10.

10. I have failed to locate this term in the standard sources.

11. *On the Trinity (De trinitate)*, chapter 2.

12. See Selection 16, n. 5.

third considers what is neither moved nor corrupted, "divine science."

Moreover, of this tripartite division, the common utility of theoretical science "is to know the dispositions of all existent things so that in our minds the form of the whole could be described according to its order [or arrangement], just as a visible form is described in a mirror. For such descriptions in the soul [or mind] are the perfection of the soul, since the aptitude of the soul for receiving them is a property of it. Thus the fact that it can be described in the soul is the highest nobility and the cause of future happiness."[13]

But since for the pursuit of future happiness it does not suffice that what is be understood by a single science unless a science of doing what is good follows; therefore practical [science], which is also divided into three parts, follows after theoretical [science].

"One of these is the science of building up one's intercourse with all men."[14] Necessary for this are grammar, poetics, rhetoric, and the science of secular laws, among which is the science of ruling cities and the science of understanding the rights of citizens. This latter science is called "political science" and by Tullius (Cicero) is called "civil reason."

"Second is the science of arranging the home and one's own family. One learns from it how he should live with his wife, children, slaves, and with all his domestics."[15] And this science is called "familiar ordination" *(ordinacio familiaris)* [that is, economics in the Greek sense of the word, or household management—*Ed.*].

Third is the science by which a man perceives how to order his very own way of life according to the honesty of his soul, so that he may be uncorrupted and excellent in his manners. This science is called "ethical or moral science."[16]

. . . . .

And in these six sciences is contained whatever can be known and whatever ought to be done, and for this reason it was said that insofar as it is possible, the intention of philosophy is to comprehend whatever is.

. . . . .

Some truth is known, some unknown. An example of known truth is "Two is more than one" and "Every whole is greater than its parts" and similar statements. An example of the unknown truth is "The world began" and "An angel consists of matter and form" and similar statements which lack proof. Moreover, any thing at all which is

unknown does not become known except through something which is known. Therefore, logic is the only science which teaches how by means of the known to arrive at cognition of the unknown. This will be proved later. Wherefore logic naturally precedes all parts of theoretical philosophy and it is necessary to them for the acquiring of truth. But logic indicates the truth by no way except a proposition, and every proposition consists of terms. And furthermore, it is the science of grammar which prepares one for the forming and composing of terms. Therefore, grammar precedes in time logic and all the other sciences. Like a nursemaid it first renders man skillful in speaking correctly.

And thus every science is either a part of philosophy or an instrument, or at the same time a part and an instrument. An example of sciences which are part of philosophy is natural science, or mathematical science, or divine science, as we have said before, while an example of a science as an instrument only is grammar.

But logic is a part and an instrument at the same time. Grammar in truth is an instrument of philosophy as to teaching but not as to learning—for philosophy can be known without words but cannot be taught without them. Logic, however, is in one respect an instrument useful for finding out the truth in itself and in the other sciences, and in another respect is a part of philosophy according as philosophy investigates dispositions (just as it investigates other things) of its subject.

With these three principal parts of theoretical philosophy known—natural, disciplinal, and divine—let us now see what should be investigated concerning each of them. These things ought to be investigated concerning each of the parts: what it is, what is its genus, its matter, its species, its parts, its function *(officium)*, its end, its instrument, who is its artificer, why is it so called, in what order it should be read.

NATURAL SCIENCE

Now, since among all the parts of theoretical philosophy the natural is prior with respect to us,

13. Quoted by Gundisalvo from al-Ghazali's *Metaphysics* (Muckle, p. 2, ll. 4–11).

14. *Ibid.,* ll. 14–15.

15. *Ibid.,* ll. 21–23.

16. Gundisalvo's three practical sciences are in essential agreement with those presented by Boethius in his *Commentary on the Isagoge of Porphyry,* Book I, chapter 3.

let us inquire about it first.[17] Moreover, what natural science is, is defined thus: "Natural Science is the science considering only things unabstracted and with motion." Every form of matter either can be abstracted from matter or not abstracted. Natural science considers matter simultaneously with form which cannot be abstracted, as was said above; therefore it is spoken of as "unabstracted." But since form, while it is in matter, is always varying, therefore it is said to be "in motion."

Its genus, moreover, is that it is the first part of philosophy, since indeed it is first with respect to us. For we apprehend matter simultaneously with form by the senses earlier than we apprehend form without matter by the intellect.

In truth, "the matter of natural science is body, but not according to what is being *(ens)*, nor according to what is substance, nor what is composed out of the two principles which are matter and form, but rather according to what is subjected to motion and rest and change";[18] for the consideration of body can be made in all these ways which are appropriate to the other sciences that treat of bodies. But natural science does not consider bodies except with respect to what is changed and altered.

But since some of the sciences are universal and others are particular, and, moreover, those are spoken of as universal which contain many other sciences under them, then natural science is universal because eight sciences are contained under it. These are the science of medicine, the science of judgments, the science of nigromance according to physics, the science of images, the science of agriculture, the science of navigation, the science of mirrors, the science of alchemy, which is the science of the conversion of things into other species; and these eight are the species of natural science.

Now there are eight parts of this natural science.[19] The first is a consideration about what all corporeal bodies, whether simple or composite, share, namely with respect to principle and the accidents following upon these principles. And this is taught in the book which is called *On Natural Things That Are Heard (De naturali auditu)*.[20]

The second part is a consideration concerning simple bodies: whether they exist, and if they exist what these bodies are and what is the number of them. And this consideration is about the world *(de mundo)*: what it is and how many parts of it there are and that at most there are three or five [parts]. And this consideration is [also] about the heavens *(de celo)* and its division into the remain-

ing parts of the world and that its matter is one. All this is taught in the first part of the first book of what is called *The Books of the Heaven and the World (Libri celi et mundi)*.[21] Next, a consideration about the elements of composite bodies follows, that is, whether in these composite bodies there are simple bodies whose existence was demonstrated— or bodies other than simple bodies. Now, if simple bodies are in composite [or compound] bodies, it is not then possible that they be outside them. Then [one must consider] whether the whole of these [simple bodies] or [only] parts of them [are in the composite bodies]. If only parts of them, then what would there be of them. This investigation [continues] to the end of the first part of the first book of what is called *On the Heavens and World*. Next there follows a consideration of what all the simple bodies share, [for] some [simple bodies] are elements and principles of compound bodies and some

17. The idea expressed here that the subject matter of natural science is first or prior with respect to us is perhaps traceable to Aristotle's description of the investigation of the principles of physics, or natural science, in *Physics* II.1.184a.15–21:

"Plainly therefore in the science of Nature, as in other branches of study, our first task will be to try to determine what relates to its principles.

The natural way of doing this is to start from things which are more knowable and obvious to us and proceed towards those things which are clearer and more knowable by nature; for the same things are not "knowable relatively to us" and "knowable" without qualification. So in the present inquiry we must follow this method and advance from what is more obscure by nature, but clearer to us, towards what is more clear and more knowable by nature." (Oxford translation by R. P. Hardie and R. K. Gaye.)

18. Despite some minor variations, this passage may be legitimately construed as a quotation from Avicenna's *Metaphysics* (or *De philosophia prima*), the beginning of Tract I, chapter 2.

19. Although Gundisalvo makes no mention of it, the description of all eight parts given below is virtually a direct quotation from al-Fārābī's *On the Sciences (De scientiis)*, which has been edited by Angel Palencia (*Catálogo de las ciencias*, 2d ed. [Madrid: Instituto Miguel Asín, 1953]) from the Latin translation by Gerard of Cremona (on Gerard's translation, see Selection 7, no. 42; the eight parts are described on pp. 161–163 of Palencia's edition). It is obvious that al-Fārābī based the eight parts of natural science solely on what were judged to be genuine works of Aristotle. In virtue of this, one can see how Gundisalvo's account of the sciences represented the influx of the new learning and how great was the gulf that separated his account from that of Hugh of St. Victor as represented in the preceding selection.

20. This is an alternative title for Aristotle's *Physics*.

21. Aristotle's *On the Heavens*.

are not elements. This investigation is about the heaven and its parts and is taught in the beginning of the second part of the book called *On the Heavens and the World* and continues on for about two-thirds of it. Then there follows a consideration of the properties of the elements and the nonelements both with respect to the principles and the accidents associated with them. This is taught at the end of the second [and all of] the third and fourth books of the *Book on the Heavens and World*.

The third part is an inquiry about the mixture *(permixtione)*[22] and corruption *(corruptione)* of natural bodies generally and of the things of which they are composed and of the quality of the generation and corruption of the elements; and [it is also an inquiry into] how some [things] are generated from others and how composite [or compound] bodies are generated from them [that is, the elements]; and finally *(in summo)* it teaches the principle of all these things. All this is taught in the book called *On Generation and Corruption (De generatione et corruptione)*.

The fourth part is an inquiry about the principles of actions and passions that are proper to the elements only without *(sine)*[23] the bodies composed of them. This is contained in the first three parts of the book called *On Phenomena of the Upper Regions (De impressionibus superioribus)*.[24]

The fifth part is a consideration of bodies compounded of the elements and of those things [constituted] of similar or dissimilar parts. Of those bodies that are constituted of similar parts there are those which can become parts of different things, as flesh and bone;[25] and there are those which cannot be part of the different parts of a natural body, as salt,[26] gold, and silver.[27] Then consideration follows about what it is that all compound bodies share; next, consideration of what all compound bodies of similar parts share or whether or not they are parts of a compound body of diverse parts. And [all] this is contained in the fourth book of *On Phenomena of the Upper Regions*.

The sixth part is a consideration of what is shared by all compound bodies of similar parts, which are not parts of a compound body of different parts.[28] These are mineral bodies and their species. Then what the properties of these species are is considered next. And all this is taught in the book titled *On Minerals*.[29]

The seventh [part] is a consideration of what species of plants *(species vegetabilium)* share and what the properties of each species are. This is one

of two speculative [or theoretical] parts dealing with compound bodies of different parts, and this is taught in the book *On Plants (De vegetabilibus)*.[30]

The eighth [part] is a consideration of what species of animals share and what the properties of each species are. And this is the second part of speculation on compound bodies of different parts. It is taught in the book entitled *On Animals (De animalibus)*, in the book *On the Soul (De anima)*, and in those books which continue to the end of the natural books.

Therefore, the natural science of every species of all bodies gives the four principles and the accidents of these bodies concomitant with those principles. This, then, is the whole of what is in natural science and its parts and the whole of what is in any part of it.

The function of this art is in the contemplation of natural bodies and accidents—for the latter do not have *esse* [or existence] except through these bodies—and it teaches the things from which, through which, and for which these bodies exist.

. . . . .

And indeed the "matters" and "forms" of bodies and their "agents" and the "ends" through which

22. The corresponding expression used by al-Fārābī is *corruptione et commutatione* (corruption and change).

23. At this point, Baur's edition (p. 22, l. 11) has *et* (and), instead of *sine* (without) as al-Fārābī has it (Palencia, p. 162). It is certain that "without" is intended since the fifth and next part will consider composite bodies while the fourth was to consider only elements.

24. Another title for Aristotle's *Meteorologica*.

25. Flesh and bone are each constituted of similar parts—that is, they are homogeneous—but each can become part of a compounded body as, for example, the arm, which consists of flesh and bone.

26. I have altered Baur's text from *sol* (sun) to *sal* (salt), thereby bringing it into agreement with al-Fārābī's text. It is possible, however, that *sol* was actually intended by al-Fārābī, for it also means "gold" (the alchemist used the term in this way; see Selection 76, n. 13), but then we must explain why he would choose to write *sol et aurum*, using two different expressions for gold.

27. Apparently we are to understand that homogeneous substances in this category do not become components of more complex organic or inorganic bodies.

28. Here only homogeneous bodies such as gold and silver would be considered, because these supposedly are not constituents of more complex compounds involving more than one homogeneous compound.

29. Probably the *Liber de mineralibus* falsely ascribed to Aristotle and discussed in the Introduction to Selection 78.

30. The treatise intended is the pseudo-Aristotelian *De Plantis (On Plants)*. The titles *De plantis* and *De vegetabilibus* are synonymous. In al-Fārābī's Latin version it is titled *Liber plantarum*.

they exist are called the "principles" *(principia)* of bodies. If they relate to the accidents of bodies, they are called "the principles of the accidents which are in bodies."

The "end" of natural science is the cognition of natural bodies. . . . Therefore, this science provides the principles of natural bodies and their accidents. The "instrument" of this science is the dialectical syllogism, which consists of truths and probables. Whence Boethius says: "It is necessary to be versed rationally in natural things."[31] The "artificer" is the natural philosopher who, proceeding rationally from the causes to effect and from effect to causes, seeks out principles. This science, moreover, is called "physical," that is, "natural," because it intends to treat only of natural things which are subject to the motion of nature. Moreover, it is to be read and learned after logic. . . .

ON MATHEMATICS

These same things are also to be sought concerning mathematics. It is defined thus: Mathematics is an abstractive science considering things existing in matter, but without the matter; for example, a line, a surface, a circle, a triangle, and similar things which do not exist except in matter. . . . Whence it is defined by others thus: Mathematics is an abstractive science considering abstract quantity. We call abstract quantity that which we separate from matter in the intellect, or from other accidents—for example: an even number, an odd number, and other things of this kind which we treat by the powers of reason alone.

Now let us see what abstraction is and in how many ways it is done. Abstraction is the apprehension of the form of any kind of a thing whatsoever. "Sense" apprehends form in one way, "imagination" in another, "estimation" (judgment) in a third way, and "intellect" in still another way; for some of them abstract the form of a thing perfectly and others imperfectly. Indeed, those abstract the form of a thing perfectly which apprehend it purely and simply without matter and without all other accidents which are joined in matter. Those abstract the form of a thing imperfectly which apprehend it along with some or several accidents of matter. Thus sight abstracts the form of a thing imperfectly, since it apprehends it only in the presence of matter and with many accidents. Imagination, however, abstracts a little more form from the matter; since form may exist in the imagination itself, it does not require the presence of matter.

But, nevertheless, it [the imagination] does not distinguish [or separate] form from all the accidents of matter. Indeed forms are in the imagination only because they are "sensibles," that is, according to quantity, and some quality and position. For imagination cannot imagine form in such wise that all individuals of that species can come together in it (that is, the form). For when a man is imagined, he can be imagined as one man and it is possible that other men are different than the one imagined.

Estimation *(estimatio)* in truth transcends this order of abstraction, since it apprehends nonmaterial intentions which are not in their matters, although they might exist in matter. Figure, color, position, and things similar to these are things which cannot exist except in corporeal matter; while good and evil, licit and illicit, honest and dishonest, and things similar to these are indeed in themselves nonmaetrial things, although it may happen that they are in matter. For if they were themselves material, never would they be understood except in a body, matter being accepted here as corporeal substance. Therefore, when estimation apprehends material things, it abstracts them from matter and it apprehends nonmaterial intentions, even though they might be in matter. And this abstraction is purer and closer to the absolute [form] than the two earlier ones [that is, sense and imagination]. But it does not wholly separate form from the accidents of matter. . . .

But intellect apprehends forms with perfect abstraction, and abstraction everywhere stripped of matter. As for that which is per se stripped of matter, there is no need that it be abstracted from matter for its apprehension. But as for that which does not have any existence *(esse)* except in matter, either its existence *(esse)* is material or this happens to it: by an absolute [or simple] apprehension, intellect abstracts and apprehends it from matter and from all the additions *(appendices)* to matter, just as, for example, happens with "man," which is predicated of many [men]. For thus the intellect apprehends one nature many times and separates it from every quantity, quality, position, and place. Indeed, unless this nature could be abstracted from these things, it would not be appropriate to be predicated of many things. The human form is a nature in which all the particulars of the species come together equally and there is one definition of them. . . . And so [in a similar manner], mathematics is called abstract, since it completely sepa-

31. *On the Trinity,* chapter 2.

rates the things it treats from matter and its accidents. For since it concerns number or figure, it pays no attention to the matter, color, or position of it but considers it absolutely [or simply,] so that [for example] *all* numbers, *all* triangles, and so on, can be embraced equally in a definition.

The genus of mathematics is the second part of theoretical philosophy abstracted from matter and with motion. For however much things are abstracted from matter by the intellect, nevertheless such entities, which cannot exist without matter—and to the extent that they are not in the intellect—are surely not without motion in matter.

The matter of mathematics is universally quantity but considered separately as *magnitude* and *multitude*. It treats most fully those things happening to magnitude and multitude.

Mathematics is also a universal science, since it contains seven arts under it: arithmetic, geometry, music, astrology, the science of aspects, the science of weights, and the science of devices *(ingenia)*.[32] Moreover, the parts of mathematics are four. Magnitude and multitude [constitute the principal divisions], but (1) some magnitude is mobile; (2) some is immobile. Also (3) some magnitude is per se and (4) some is related. Arithmetic investigates multitude per se; while music avows the science of related multitude. Geometry makes clear properties of immobile magnitude; but astrology reveals knowledge of mobile magnitude. . . .

The instrument of mathematics throughout is demonstration. Demonstration, moreover, is a syllogism arising out of primary and true propositions. Some of the primary propositions are things of the senses, such as the following: "Every fire heats," "All snow is white," and similar statements. Some, on the other hand, are things of the intelligence, as "Every whole is greater than its part" and similar statements. Moreover, among the propositions of intelligence of this kind some are primary, some secondary. Primary propositions are those which when first heard are immediately conceded. Moreover, they cannot be the conclusions of syllogisms, for no propositions are better known than they. And therefore they are spoken of as propositions "known per se" because they cannot be made known through other propositions. Whence they are called "common conceptions of the mind." *(communes animi concepciones)*[33]—ones which anybody approves on hearing them. The secondary propositions of the intelligence are those which are concluded from demonstrations. Of this kind are the theorems of Euclid which,

after they have been proved through primary propositions, are then assumed in a demonstration. Therefore, they are not known per se [or as self-evident] because they are not known in themselves but through other propositions.

Moreover, there are two species of the demonstrative art, namely geometry and logic. The primary propositions of geometrical demonstration are taken from another art which is prior to it, just as that which Euclid said: "A point is that which has no part," and "A line is length without width," "A surface is that which has length and width." Similarly, the primary propositions of logical demonstration are taken from another art which is prior to it, for example when it is said: "Everything which is, with the exception of God, is substance or accident" and "Substance is that which existing in itself is susceptible to contraries." . . . Furthermore, whoever wishes to be skilled in demonstrations of logic ought first to be trained in geometrical demonstrations because they are closer to the understanding and easier to investigate. This is because the examples are things sensible to the sight, although their intentions are things of the intelligence. . . . The artificer is the demonstrator. The function of this art is to prove in the truest fashion everything that is proposed. Its end is the obtaining of certitude from the ambiguity of a proposed question.

Moreover, this science is called "mathematical," that is, "abstractive", for *mathesis* is interpreted as abstraction. Whence it is said there is no second *mathesis,* that is, second abstraction, since from the things that we once abstract from matter by the intellect we cannot abstract other things once again by the intellect; indeed a form is not abstracted from a form, but from matter. But since this science is about things that are understood abstractly, mathematics is therefore called abstractive.

Mathematics is also called doctrinal or disciplinal *(disciplinalis)*[34]—that is, from a discipline—

32. Gundisalvo includes the traditional quadrivium (arithmetic, geometry, music, and astrology [instead of astronomy]) as part of a now expanded conception of mathematics in which the three additions would be classified as applied mathematics. His division into *magnitude* and *multitude,* with further subdivisions into mobile and immobile, is virtually the same as Hugh of St. Victor's in *Didascalicon* II. 8 (see Selection 16, n. 11).

33. This was the expression commonly used to designate the axioms in Euclid's *Elements*. They were assumed to be self-evident.

34. As used here, a *discipline* is something productive of certain knowledge.

because those who are disturbed by different opinions in the preceding sciences find that mathematics. . .holds one opinion only, which is proved by demonstration. . . .

Moreover, mathematics should be read after natural science because whoever considers natural science has considered form together with matter, and after completing what pertains to the accomplishments of that science, it is worthwhile that he should know how to consider form without matter until he is [well] accustomed to consider forms without any matter.

.   .   .   .   .

ON LOGIC[35]

.   .   .   .   .

What [Logic] is, is defined thus: "Logic is the doctrine of diligent discussion, that is, the complete science of disputation. . . ." Moreover, the genus of logic is that it is a part and instrument of philosophy. This is shown as follows: Things exist in two ways—sensibly, that is, according as they are subject to the senses, and intelligibly, that is, according as they are conceived by the mind. The following apply to "things according as they are understood": universality, generality, specialty, accidentality, and similar things. For this reason, then, "things according as they are understood" are genera, species, accidents, and similar things.[36] There is nothing of them in sensible things *(sensibilis)*.

Moreover, when we wish to prove or disprove something with regard to either "sensible things" or "intelligible things," it is necessary that we have recourse to those things which occur to intellects, namely genera, species, and so on, so that through them we might prove those things we intend to prove. Every proof or disproof in the whole philosophic art takes place by means of those things which belong to logic alone. Therefore, according to those people who hold that philosophy treats only of two kinds of things, sensibles and intelligibles, logic will be an instrument. But according to those who hold that philosophy also treats of this third mode which occurs to intellects, logic will be a part of philosophy as well as an instrument in the other arts. . . .[37]

.   .   .   .   .

According to al-Fārābī there are eight parts of logic:[38] "categories, interpretation, prior analytics, posterior analytics, topics, sophistries, rhetoric, poetics. [As it can be seen, then], the names of

[Aristotle's] works are posited for the names of the sciences which are contained in them. . . . Since the fourth part of logic (posterior analytics) is of greater power, it excels all of the other parts in sublimity and dignity. . . .The remaining parts have been discovered only through the fourth. . . ."

.   .   .   .   .

"The sure cognition of truth is not obtained except through demonstration. Therefore, it was necessary that a book be composed in which would be taught how and out of what things demonstration is made. Consequently the book which is called *Posterior Analytics* or the *Book of Demonstration (Liber demonstrationis)* was composed. But since demonstration is made only by means of syllogism and syllogism in truth consists of propositions, it was necessary to have a book in which would be taught the number and kind of propositions, and how according to mode and figure syllogisms should be constructed. For this reason the *Prior Analytics* was written. But propositions cannot compose a syllogism unless they first have been composed out of their own terms. Therefore, it was necessary to have a book which would teach the number of terms and which terms go to make up a proposition. This is fully taught in the work which is called *On Interpretation (Perihermenias)*. Further, a proposition is never well composed from terms unless the signification of each term is first recognized. Therefore, the *Book of Categories (Liber cathegoriarum)* was written to teach how many kinds of terms there are and what is the signification of each of them. . . ."

The species of this art in truth are three kinds of questions: moral, natural, and rational. They are called its species because all of logic is employed in any one of them. The function of this art, according as the art is theoretical, consists in "invention" and "judgment.". . . According as the art is practical, its functions are "division," "definition," and

35. Between mathematics and logic the following sections are omitted here: On Divine Science, On Grammar, On Poetics, and On Rhetoric.

36. These are all concepts in the understanding—that is, "things according as they are understood"—and have no existence in sensible things.

37. In this interpretation, genera, species, accidents, and so on are a third class of things belonging to logic alone and are set by the side of "sensible things" and "intelligible things." Thus, if "things according as they are understood" are construed as part of philosophy, logic will be part of philosophy since this category belongs to logic; if not, logic will be an instrument only.

38. What follows is virtually a quotation from al-Fārābī's *De scientiis* (Palencia, p. 95 ff).

"ratiocination." . . . Ratiocination is the proof or disproof of some doubt. . . . Moreover, ratiocination has three species: (1) dialectical, which is the science of inferring *(colligendi)* through probable things; (2) demonstrative, the science of inferring through things known per se; and (3) sophistic, the science of inferring through things which seem to be true, but are not.

. . . . .

The principal instrument of this art consists in two things: syllogism and induction. There are two secondary ones which come from them by the subtraction of one or more parts: *enthymeme* and example. . . . The artificer of this art is the disputer, who either topically, demonstratively, or sophistically exercises this art on general questions that are moral, natural, or rational. The "logician," moreover, is he who teaches the logical art, [while] the "disputer" is he who exercises that art. . . .

. . . . .

Logic therefore ought to be read after rhetoric but in this order: Since the function *(officium)* of logic is to divide, define, and prove, [and since] a proof is not made except by a syllogism, and a syllogism consists of propositions, and propositions of terms, and terms are words signifying some simple understanding, it is [therefore] necessary to understand the simple terms before the things that are composed of them. Hence in the order of logic, the science of terms, which is taught in the *Categories,* is naturally first. Therefore, that book which should be read first is proposed. In it is treated fully the doctrine of simple words, which are the terms of simple enunciations. But since the *Liber introductorius* of Porphyry is for the understanding of the *Categories (Predicamenta),* then it ought to be read before the *Categories.* But after the cognition of terms it remains [for us] to know how the composition of propositions is made from them. Therefore, after the *Categories (Predicamenta)* comes the *On Interpretation,* which ought to be read immediately. In it is taught how a simple enunciation is constituted out of terms. After we learn how to compose propositions out of terms it is necessary that we compose syllogisms from the propositions and do so according to mode and figure. Consequently, we ought to read the *Prior Analytics* after the *On Interpretation.* In it we are taught how a syllogism is composed from propositions. But syllogism has three species: dialectical, demonstrative, and sophistic. Therefore, the *Topics* is read for dialectical syllogism; the *Posterior Analytics* for demonstrative syllogism; and

Elenchi (On Sophistical Refutations) for sophistical syllogism. It is in this order that Aristotle, the author of Logic, has taught that logic ought to be taught and learned.

ON MEDICINE

But since natural science follows after logic and the first species of natural science is medicine, we must inquire about it all the things stated above, namely:[39] what it is, what is its genus, its matter, its parts, its species, its instrument, who is its artificer, what is its function *(officium)*, its end, why it is so called, what is its use, [and] in what order should it be read.

What medicine is, is shown in this definition: "Medicinal physics is the science of the healthy, the sick, and the inbetween." There are also other definitions according to its function, thus: "Medicine is the science of controlling the natural complexion and of changing that which is foreign (to the body) into the normally existing prior (condition)." Also: "Medicine is that which cares for and restores the health of the human body."

Its genus: it is one of the species of theoretical natural science, for it is unabstracted and with motion.

The matter (subject matter) of medicine is the human body as it becomes diseased and healed. It is through the body that medicine itself has been discovered and on it that medicine is exercised. This science treats of three things: body, symptom *(signum)*, and cause. . . .

Moreover, the efficient cause is primitive, or antecedent, or conjoined *(conjuncta)*. An example of the "primitive" cause is air and foods; of the "antecedent" cause is the humors; and of the "conjoined" is the putrefaction of the humors (that is, body fluids), which is continually followed by fever. Among the natural sciences, therefore, medicine excels in the nobility of its matter, the human body, because the latter excels all natural bodies.

This science is sometimes had by the cognition of the mind alone and sometimes by the exercising of the body. Therefore, according as it is a science of recognizing the principles and precepts of an art, it is theoretical, just as when we say that there are three kinds of fever and nine species of complexions. But insofar as it is a science of operation, it is practical, just as when we say that at the

39. An elaborate program of inquiry of this kind precedes each of the sciences in this treatise; (for example, see the last paragraph of the Prologue.)

beginning there ought to be applied to hot abscesses things that would repel, cool, and thicken; and afterwards we will temporize the repellents with resolvements. . . .[40]

The parts of theoretical medicine are two: the science of conserving health and the science of curing infirmity. These two sciences coming together produce the perfect physician. . . . The parts in truth of practical medicine are three: pharmacy, surgery, and diet.

Pharmatica in Greek is translated in Latin as *medicamentum* (a medicant); *cirurgia* is translated *manus operacio* (manual operation); *dieta* is translated *regula* (rule or reigmen). Pharmacy is therefore a cure by medicants; surgery is incision with instruments; diet is the observation of a law and a way of life. Every cure of disease takes place in these three ways.

The species of this art are the three dispositions of the body: health, sickness, or neutrality. . . . The function of the art is twofold: to conserve by regulation healthy bodies in their state of health and to recall sick or neutral bodies to health. . . . The end [or goal] is twofold: the conservation of health by regimen or the recovery of health through a cure, that is, the expulsion of sickness or neutrality. . . .

Since the instrument is that through which the artificer fulfills his function, therefore diet is the instrument for regulating health. Species of herbs, potions, unguents, and similar things are instruments for the curing of sickness and neutrality. For this reason there are two kinds of those instruments: some are natural and some are artificial. The natural instruments are simple things like herbs, species of fruit, stones, and metals. The artificial instruments are composite, as camphors, potions, unguents, plasters, clysters, pisaria, and surgical instruments.

. . . . .

Why is medicine so called? "Medicine is derived from *modus,* that is, from "moderation," that is, from a due proportion which advises that things not be done immediately *(statim),* but "little by little" *(paulatim).* For nature is pained by surfeit but rejoices in moderation. Whence also those who take drugs and antidotes constantly, or to the point of saturation, are sorely vexed, for every immoderation brings not health but danger."[41] Therefore, medicine is derived from *modus* (mean) in that the *modus* and *mensura* (measure) in all bodies subjected to medicine are conserved by it; or it is called medicine *(medicina)* from "mean"

(mediocritas). For medicine ought to do this: either to conserve the mean state *(mediocritas),* that is, the health, or to repair a fault in some body. Those bodies in which we see elements joined together with equal measure we judge to be temperate. We conserve these bodies in their state by using "similar things" having the same measure of elements. But in distemperate bodies we see the mode and measure of a distemperate body [that is, the measure is off the mean], so that we should cure them by using "contraries" which are distemperate in the same degree.

. . . . .

Since it is in need of all the other arts, it certainly ought to be read after all of them,[42] so that the cognition of those sciences without which it itself is not efficacious would come first.[43]

. . . . .

ON ARITHMETIC

. . . . .

. . . .Arithmetic is defined thus: "Arithmetic is the science of quantity numerable according to itself."[44] Or, "Arithmetic is the science of number." Number, moreover, is considered in two ways: (1) as it is in sensibles and non-sensibles, and (2) as it is considered more abstractly through the intellect apart from everything in which it exists.[45] In sensibles it is like "three" in "three men"; in non-sensibles it is like ["three"] in "three angels." It is considered abstracted from everything through the intellect when we say, "Three is the first odd number [that is, the first number] which cannot be divided into two equal parts." Therefore, in both sensibles and insensibles it is considered when it is numbering [something] or is numbered. But it is

40. Drawn from Avicenna's *Canon of Medicine,* I, fen 1, doct. 1.

41. This is a direct quotation from Isidore of Seville, *Etymologies,* Book IV, chapter 2. The translation of this passage given in Selection 89 has been altered slightly to accommodate the minor differences in Gundisalvo's text.

42. In a preceding paragraph taken almost verbatim from Isidore's *Etymologies,* Book IV, chapter 13, but omitted here, Gundisalvo shows how a physician must have knowledge of grammar, rhetoric, dialectic, arithmetic, geometry, music, and astronomy.

43. Gundisalvo concludes his section on medicine with another unacknowledged quote, taking all of Isidore's *Etymologies,* Book IV, chapters 3 and 4 (these have been translated in Selection 79).

44. Quoted from Isidore's *Etymologies,* Book II, chapter 24, paragraph 15.

45. This twofold division appears based on al-Fārābī's discussion in the third chapter of *De scientiis* (Palencia, p. 145).

abstract when it is neither numbering nor numbered. Number therefore considers those things which are considered in both ways, in itself, or in matter with motion or without motion. . . .Whence according to this (reasoning) some things happen to number from itself; others from its being mixed with matter. The things that happen to it from itself are (for example) that it is odd or even, in excess or diminished, and other things of this kind —things which are pointed out in the Arithmetic of Nichomachus.[46] That which happens to it out of matter is, for example, its being added, subtracted, multiplied, divided, and things of this kind—things which are taught in the *Book of Algorism*.[47] When number is directed to itself, it is called "theoretical" or "speculative." But when it is considered in matter, it is called "practical" or "active." And through this (definition) theoretical arithmetic asserts that no number has been composed out of numbers but only out of units. Nor does it concede that any number is the part or parts of another, because it regards any number at all as a species in itself in that the specific difference, such as "threeness," embraced by the genus which is "number," constitutes this species which is "three." It can be reasoned in the same way concerning other numbers. And therefore [number] divides no number, unless into units only—but not into numbers. Practical arithmetic only considers number in matter which is divisible in many ways. Hence it divides one number into parts and the parts into parts into infinity. It calls its parts fractions, and the parts of parts, fractions of fractions. . . .

The genus of arithmetic is the first of all mathematical disciplines, which considers what is unabstracted and without motion.

The matter of arithmetic is number, since it treats of the accidents of numbers. Although arithmetic is called the science of number, it does not, however, treat of the essence of number, for no science establishes its own subject matter, as Aristotle says.[48] But it does assign properties to number, and it treats those things which happen to number either of itself or from its mixture with matter. It is for another science to treat of the essence of number, namely the divine science (metaphysics) on which it is incumbent to prove the (first) principles of all the sciences.

Moreover, there are some theoretical, some practical, parts of arithmetic. The theoretical parts are three: The first is a consideration of those things which happen to number by virtue of its own essence, such as the fact that some number is even, some odd, and similar things. The second part is consideration of those things which happen to it out of the proportioning of one number to another, such as the fact that some number is a multiple [of another], and another is superparticular to another,[49] and so on. The third is a consideration of what happens to it from a comparison of it to continuous quantities—for example, a certain number is lineal, another two-dimensional (superficial), a third is a cube, still another a solid (that is, any other three-dimensional figure which would be represented by a number), and similar things.

The practical parts are in the main two: the science of conjoining numbers and the science of disjoining numbers. The science of conjoining numbers is the science of adding, the science of duplication, the science of multiplying, [and so on]. The science of disjoining consists of the sciences of subtracting and dividing. However, the science of finding roots of numbers is contained under each of these two, since the root of a number is found in both ways, that is, by multiplying and dividing.

The species of arithmetic also consists of theoretical and practical parts. The species of theory are three: The first is called the arithmetical mean, the second the geometrical mean, and the third the harmonic mean. All of these things are treated in the Arithmetic of Nichomachus.[50] The species of practical arithmetic are in truth the various kinds of businesses in each of which the whole art is employed. . . .All these things (concerning practical species) are treated in the book which is called in Arabic *"Mahamelech."*

The function *(officium)* of theoretical arithmetic is to investigate diligently the natures of numbers. . . .Its end [or goal] is the cognition of all things by the example of numbers. The function of practical arithmetic is fully to conjoin and disjoin numbers of every species [or kind]—integers and fractions. Its end [or goal] is to prevent errors of numbering in every kind of business. The instru-

46. Compare Boethius, *On Arithmetic,* Book I, chapter 2 (Selection 2).

47. Books on algorism considered arithmetic operations; see Selection 20, from Sacrobosco's *Algorism.*

48. Compare *Posterior Analytics* I.1.71a.1–10.

49. For the meaning of superparticular see Selection 1, "On the Quadrivium," chapter 6, paragraph 7, and note 17.

50. For Nicomachus' account of these three "species" as it appears in the paraphrase of Boethius, see Selection 2, chapters 43–45.

ment of theoretical arithmetic in truth is demonstration. . . . While the instrument of practical arithmetic is the table of the abacus *(tabula abachi)*, or the *rithmamachia*, which is the game of contesting numbers according to their proportions *(ludus de pugna numerorum secundum proportiones eorum)*. The artificer is the arithmetician. . . . Arithmetic, moreover, is spoken of as the science of the "virtue" (power) of number, for *"ares"* is translated *virtus*. . . .[51] The "order" of arithmetic is this, that it is prior to all the disciplinal sciences because it is lacking in none of them. . . .[52]

.   .   .   .   .

ON GEOMETRY

.   .   .   .   .

. . . . Certain people have defined geometry thus: "Geometry is the discipline of magnitude and forms which are considered according to magnitude." Others, however, define it in this way: "Geometry is the science of rational magnitudes investigated by reason of probable dimension." It speaks of "rationals" in contradistinction to "irrationals," that is, surds, whose dimensions are not probable; an example of the latter is the quadrature of a circle.

Its genus is that it is one of the four mathematical disciplines, inquiring into things abstracted and with motion, for every discipline of theoretical philosophy either is abstracted or unabstracted, either with motion or without motion. . . .

Its matter *(materia)* is immobile magnitude: line, surface, and solid. It is therefore called "immobile magnitude," since they (that is, line, surface, and so on) are understood as abstracted from matter without motion. Indeed, they are moved in matter, which is always in motion because, as Boethius says, they are changed by contact with a variable thing. Therefore, these things are considered in two ways, as they are outside of matter and inside of matter. Insofar as they are in matter, they are sensible but are not sensed [or perceived] except as they are mixed with other things without which they cannot exist in matter. For they are sensed [or perceived] with visible colors, or with sounds that are heard, or with the smell of an odor or foul smell, or with sweet or bitter taste, or with touch [they are perceived as] hot or cold. But when they are outside matter, they are understood as separated from the things with which they are conjoined in matter.

But since geometry need consider these things in both ways, therefore some of geometry is practical

and some theoretical. Practical geometry accepts these things as they are mixed in matter with other things [as color, sound, and so on,] although it has no regard for these other things with which they are mixed. . . . Theoretical geometry, however, abstracts things from matter by the intellect, and it not only distinguishes them from the things with which they are mixed, but even indeed from themselves, that is, one from the other. It considers these things individually abstracted per se and demonstrates any one of them by itself by defining what it is.

.   .   .   .   .

Now theoretical [geometry] has parts, and practical geometry has parts.[53] There are three theoretical parts: one considers lines, another surfaces, and another bodies. The part concerned with bodies is divided into the number of [types of] bodies, namely into cubes, pyramids, spheres, cylinders, . . . Speculation considers all these in two ways: in one way when it considers any one of them by itself; and in the other way when it compares any one of them in terms of the things which happen to them. Moreover, when it compares them to one another, it considers either their equality or inequality, or some other of their accidents; or it considers how one can be arranged on another—namely how a line can be [placed] on a surface and a surface on a body, or a surface on a surface and a body on a body.

The distinction between natural body and mathematical body ought to be noted. "Natural body" is substance in which can be placed three lines cutting each other at right angles; and thus it is in the category of substance. "Disciplinal (mathematical) body" is extension in three directions: length, breadth, and thickness. . . . and thus it is in the category of quantity.

.   .   .   .   .

The species of geometry also has a theoretical part and a practical part. The species of theoretical geometry are three: operation [or action], knowledge, and discovery. Anything at all which is proposed in this art is in truth proposed for acting, knowing, or finding out. For "acting" there are proposed, for example, the first and second the-

51. Compare Hugh of St. Victor above, Selection 16, n. 12.

52. An elaboration of this point concludes the section on arithmetic, which is followed by a lengthy section on music that is omitted here.

53. The section containing the practical parts is missing from the edition.

orems of Euclid and many others, which are proposed here in order that we might "do" them. For example, it is proposed that we construct an equilateral triangle upon a given straight line and from a given point we draw another line equal to a proposed straight line.[54] At the end of these and similar [propositions for action] it always ought to be said, "and this is what we wished to do." For "knowing," however, propositions like the fifth theorem of Euclid and similar ones are proposed. This fifth theorem is this: If the two angles above the base of any triangle are equal, then the angles which are below the base will be equal. At the end of this and similar propositions it ought to be said, "and this is what we wished to know." For "finding out" are proposed such things as the first theorem of the third book and many others; for example: To find the center with the circle given, for although we know that every circle has a center, we do not know where it is. At the end of this and similar propositions we ought to say, "and this is what we wished to find. . . ."

The end [or goal] of theoretical geometry is to teach something; while the end of practical geometry is to do something. . . .

The species [or types] of practical geometry are also three: altimetry, planimetry, and cosmimetry, [that is, height measurement, surface measurement, and the measurement of solids]. . . .[55]

. . . . The artificer of theoretical geometry in truth is the geometer who has become acquainted thoroughly with all parts of geometry and teaches it. Its instrument is demonstration. The parts of this demonstration are assigned in different ways. According to Boethius they are six: the proposition, the description, the disposition, the distribution, the demonstration, and the conclusion. According to the Arabs, however, there are seven: the proposition, the example, the contrary, the disposition, the difference, the reason, and conclusion.[56] The descriptions of all these [parts of a proposition] are assigned to those who begin to read the book [of *Elements*] of Euclid.

The artificer of practical geometry is he who employs it in working. There are, however, two classes who employ it in working, the measurers and the artisans. The "measurers" are those who measure the height, the depth, or the level surface of the Earth. The "artisans" are those who exert themselves in manufacturing or in working in the mechanical arts, as a carpenter works on wood, an iron worker on iron, a stone mason on cement and stones, and similarly every artificer of the

mechanical arts works according to practical [or applied] geometry. For he forms lines, sufaces, squares, roundnesses, and so on, in the body of the matter which is subject to his art. Now many species, [or types] of these skills are considered by the difference of materials on which and from which they work. Any one of them has its appropriate matter and instruments. For measuring there are the foot *(pes)*, palm *(palmus)*, cubit *(cubitus)*, stade *(stadium)*, pole *(pertica)*, and many others. Of instruments, the carpenters use the hatchet *(securis)*, and ax *(ascia)*, pickax and line, and many others; the iron worker uses the anvil *(incus)*, shears *(forfices)*, hammer, and many others; the stone mason uses line, trowel, and perpendicular, and many others.

. . . . .

Why is geometry so named? This discipline receives its name from measure of the earth, for *ge* is translated "earth" and *metron,* "measure" . . . .[57]

. . . . .

Geometry ought to be read after music. This is reasoned as follows: There are two species of quantity, obviously multitude and magnitude. Moreover, there are two species of multitude: number per se [or absolute number], of which arithmetic treats, and related number, of which music treats. After multitude, magnitude necessarily follows, of which one part of it is immobile, and geometry treats of this, and the other part mobile, astrology treating of it. Moreover, the immobile is before the mobile, since every motion begins from rest. This shows that in the order of the four mathematical disciplines [or sciences], the third, after arithmetic and music, naturally treats geometry. [Hence] by this argument geometry ought to be read after music and before astrology.

. . . . .

54. These are the enunciations of the first two construction theorems of the first book of Euclid's *Elements*.

55. The same threefold division of applied geometry is found in Book II, chapter 13, of Hugh of St. Victor's *Didascalicon* (see Selection 16).

56. Proclus, in his commentary on Book I of Euclid's *Elements,* divided a proposition into six formal parts (see Thomas Heath, *The Thirteen Books of Euclid's Elements,* 2d. ed [New York: Dover, 1956], I, 129–131).

57. The remainder of this paragraph concerning the invention and application of geometry in Egypt is based upon Isidore of Seville's *Etymologies,* Book III, chapter 10, quoted in Selection 1.

ON ASPECTS [OPTICS]⁵⁸

The science of aspects [or optics] inquires about the same things as does geometry, namely figures, magnitudes, positions, order, equality, inequality, and other things, except that [in optics] these entities are in lines, surfaces, and bodies absolutely, so that the science *(speculatio)* of geometry is more common [or universal] than this science [of optics]. But although what it treats is contained in that of which it treats, this is not superfluous but necessary, since what Euclid has proved to be a square necessarily appears to be round when seen from a certain distance; and many things which are equidistant seem to run together, and equals seem to be unequals and conversely. Of things situated in one plane, some appear lower and others higher; and of those things that are before [or in front of] others, some appear behind; and conversely. For this reason, this science was necessary so that it could distinguish between what appears to the sight otherwise than it is and what appears just as it is. Indeed, this science assigns the causes of these things and does this by necessary demonstrations. It also teaches how vision can err so that it might not err but [rather] discover just how everything is that it sees. It also teaches how to detect [or determine] the heights of trees, towers, walls; the widths and depths of rivers and the heights of mountains after the sight falls upon their limits [or boundaries] *(fines)*; then the elongations of celestial bodies and their quantities and everything which can be reached from reflection by the one who sees it; and some of these require an instrument, others do not.

And it teaches that anything seen is seen only by means of a ray penetrating the air and falling continually on the thing at which we are looking. And the rays penetrate transparent [or pervious] bodies until they reach what is seen and then are direct [that is, straight], reflected, reversed, or refracted. Rays are straight when after emerging from the eye *(viso)*,⁵⁹ they are extended in a straight line until they traverse their distance and terminate. Rays are reflected when they begin to emerge from the eye and a mirror obstructs them before they traverse [some distance]; they are reflected from the mirror obliquely toward parts on the sides of the mirror and then they are extended to the sides and go on to the body to which they are reflected. The one who is viewing holds the mirror between his hands in this way:

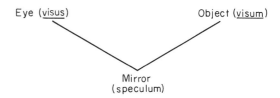

Fig. a

Rays that are reversed *(conversi)* are those which follow the same path returning from the mirror as they followed in previously advancing [toward the mirror]. They return until they fall on the body of the one who is seeing [or looking]. It is for this reason that a man looking at a mirror sees himself with the very same ray. Refracted rays, however, are those which, in returning from the mirror to the side where the viewer is, are extended crookedly from the mirror to one of the viewer's sides and fall on something behind the viewer or to his right or left or above him. For this reason, a man sees what is behind him or what is on one of his sides, as this figure shows:

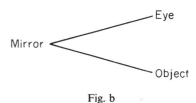

Fig. b

The medium which lies between the eye and what is seen without a mirror consists entirely of transparent bodies as water or air, or celestial bodies or crystalline bodies. Therefore, the science of aspects [or optics] investigates everything that is seen and everything that is viewed with these four rays in any mirror; it investigates everything that happens to the object to which the rays are directed. Whence it is divided into two parts, of which the first is an inquiry about what is seen with straight rays;⁶⁰ the

58. With slight variations and alterations, Gundisalvo adopted the whole of al-Fārābī's section *On Aspects*, probably because this subject was wholly new to him since it had not been considered as part of the earlier Latin tradition on the classification of the sciences (it is absent from Hugh's *Didascalicon*).

59. Gundisalvo has adopted the traditional view that visual rays move from the eye to the object perceived. See Selections 62. 14–17.

60. Baur's text is faulty here. Instead of *cum rectis radiis* (with straight rays), as al-Fārābī has it, Baur has *cum istis radiis*.

second investigates what is seen with non-straight rays. And this is properly called the science of mirrors *(scientia de speculis)*.

ON ASTROLOGY

. . . . .

. . . Astrology is defined as follows: "Astrology is the science of mobile magnitude which seeks out with searching reason the courses of the stars, their figures, and the relations *(habitudines)* of the stars both with respect to themselves and with respect to the Earth."

Its genus, that is, its quality, is that it is mathematical, a doctrinal science. Its matter (material object) is mobile magnitude. Moreover, we call magnitudes mobile which are moved, for example the sun, the moon, and the other planets, the movement of which astrology treats. . . . The matter of astrology is spoken of as "mobile magnitude" not because it considers the magnitude of moving bodies but because it would investigate the motion of the great moving bodies, as for example the seven planets and the ten heavens.

There are three parts to this art.[61] The first is concerned with the numbers and figures of celestial bodies, their order in the world, their mutual quantities, locations, and proportions, and with the quantities of their mutual elongations. And it holds that the whole earth is not in local motion—neither in a place nor from a place. The second part treats of the motions of the celestial bodies—how many there are, that all their motions are spherical [that is, circular], and which of these motions could be communicated to all the celestial bodies, that is, the stars and non-stars—and what is common to all the stars. Next comes [an inquiry into] the motions proper to each star, how many kinds of motions there are for any one of them, and the parts [or directions] to which the stars are moved and in what direction does this motion come to any one of them. It also teaches the way to understand the position of each star in terms of the signs [of the zodiac] in any hour and with all its types of motion. It inquires about all things that happen to celestial bodies and the motions of each of them in the signs [of the zodiac] and what happens to these when they are mutually compared with respect to conjunction, separation, and difference of position. And finally, [astrology inquires about] everything that happens to these bodies from their [own] motions without relating them to the earth, as an eclipse of the sun; but again [it also

inquires about] everything that happens to them because of the position of the earth with respect to them, as an eclipse of the moon. And it states their accidents—how many there are, in what disposition and hour they occur, and in how much time, as [for example] the risings and settings, and other things. The third part inquires about the earth—what is inhabited and what uninhabited; and it shows how much is inhabited and how many great parts [or divisions] of it there are, and what the climes are. It embraces the habitations that exist in any of these climes in that time and where the location is of each habitation and their arrangement in the world. It inquires about what follows necessarily from the revolution of the world that would happen to any of these inhabited climes: there is the revolution of day and night because of the position of the earth in its place [and] on which there are risings and settings, length of day and night, and brevity [of day and night], and other things similar to these. These, therefore, are the parts of astrology.

Its species are two: active and contemplative. Contemplative astrology is the science of all the aforementioned parts. The practical is the science of comprehending through competent instruments the magnitudes, elongations, and mutual comparisons of the celestial bodies.

The function *(officium)* of this art is to assign, by use of the most certain reasons, the magnitudes, differences and proportions of the celestial bodies and the Earth. Its end [or goal] is the sure cognition of those things which are of the visible, incorruptible universe—of those bodies which do have existence neither through generation nor corruption—bodies which are neither [pure] elements *(elementa)* nor [visible] elements *(elementata)* nor natural things nor artificial. For there are three worlds or parts of the world: (1) The first of these is the visible and corruptible one which extends from the center of the Earth up to the circle of the moon. Natural science treats of this world. (2) The second is the visible and incorruptible world, which extends from the moon up to the last heaven. Astrology treats of this one. (3) The third is the

61. These three parts are very nearly a direct quotation from al-Fārābī's discussion of the "Science of the Stars" in chapter 3 of *De scientiis*. It will be noted that al-Fārābī and Gundisalvo reverse our usual understanding of the difference between astronomy and astrology (see Gundisalvo's next section on astronomy), whereas Hugh of St. Victor (Selection 16, chapter 10) retains it. Perhaps this results from the fact that the terms were often used interchangeably.

invisible and incorruptible world, which is neither outside of the heavens nor within them. Theology treats of this last world. Hence, the goal of astrology relates to the cognition of the world of the fifth essence. For the first world is called the "world of nature," the second is called the "world of the fifth essence," and the third is called the "world of intelligence."

The instruments of astrology are the many which Ptolemy teaches how to make in the *Almagest*. . .

Why is astrology so called? Astrology is called the science of the reason of the stars, for *logos* is translated "reason" *(ratio)*. Hence this art is called astrology—the science of the reason of the stars—because whatever it says of the stars, it proves by rational procedures. It ought to be read after geometry but not continually. Nay, the books of certain authors ought to be interposed—introductory books which when known first will lead one easily to the cognition of astrology. The first is Mileus, the second Theodosius, and the third Ascalonita. There then remains to be read after these authors the *Book of the Almagest* in which all of astrology is most fully taught.[62]

.   .   .   .   .

ON ASTRONOMY

There is another science of the stars which is called "astronomy. . . ." What it is, is defined in this way: "Astronomy is the science which describes, according to the belief of men, the courses and position of the stars, for obtaining a knowledge of the times."

Its genus is that it is a science for judging a question deemed necessary according to the position of the planets and signs. For there are many sciences of judging a proposed question, for example, geomancy, which is divination in earth; hydromancy, divination in water; aeromancy, in the air; pyromancy, in fire; chyromancy in the hand, and many others. . . . Thus al-Fārābī says that astronomy is a science of the signification of the stars, namely what the stars signify about the future and many things in the present and the past. Nor are these [sciences of judgment] among the disciplinal [that is, mathematical] sciences but among the forces and powers *(virtutes et potentias)* by which a man can judge of future events just as the power of interpreting visions and the power of auguring in birds and sneezes and other such ways.

Its matter (material object) is mobile magnitude.

Its "parts" are four, the first of which treats of the position and form of the Universe and of the celestial circles. The second treats of the course of the planets and their circles. The third treats of the rising and setting of the signs. The fourth treats of the eclipse of the sun and the moon, and all of astronomy is absorbed into these four parts. The species of astronomy are two: computation and judgment. Computation is constituted in tables, judgment in the distinction of the seasons, the signs, and the planets.

The function of this art is to contemplate the mutual courses, conjunctions, retrogradations and retreats of the planets. The end [or goal] of it is a science for judging about the past, present, and future.

.   .   .   .   .

. . .Astronomy differs from astrology in that the latter relates to the "truth" of the matter: while the former follows the opinion of men. According to Isidore, they differ in this, that astrology embraces the revolutions of the heavens, the rising, setting, and motion of the stars, or from what source (cause) they would be so moved. Astronomy in truth is partly natural, partly superstitious. It is natural while it plots the courses of the sun and the moon or the fixed positions or the times of the stars. The superstitious part is that which is followed by the "mathematicians," who make conjuries by means of the stars and who also arrange the twelve signs by singularities of the soul or members of the body, and who attempt to prejudge the manners and nativities of men by means of the course of the stars.

ON THE SCIENCE OF WEIGHTS (SCIENCIA DE PONDERIBUS)[63]

The science of weights considers weights in two ways: either (1) according to the weights themselves that are being measured or according to what is measured with them and by them; and this is an inquiry about the principles of the doctrine on weights. Or (2) it considers them in so far as they are moved or according to the things with which they are moved; and this is an inquiry about the principles of instruments by means of which heavy

62. Gundisalvo concludes his discourse on astrology by quoting almost the whole of Isidore's *Etymologies,* Book III, chapter 25 (Selection 1).

63. This is virtually a quotation of the whole of al-Fārābī's brief section on the science of weights (Palencia, p. 154).

bodies are lifted and on which they are changed [or carried] from place to place.

ON DEVICES
(DE INGENIIS)[64]

The science of devices is the science of devising ways to make all the things happen whose modes were stated and demonstrated in the sciences *(doctrinis)*. . . . For all other sciences do not consider lines, surfaces, bodies, numbers, and the rest except as they are understood as separated from natural bodies. . . . Therefore the sciences of devices teach the ways of devising and discovering how natural bodies could be joined according to number by any artifice [or device] in this sense: that the use that we seek should arise from things such as the science of numbers called *algebra* and *mucabala*,[65] and other similar things.

This science is common to number and geometry and embraces skillful ways for finding numbers about which Euclid gave the principles of rationals and surds in his tenth book of the *Elements*. For since a proportion of rationals and surds is mutually as a ratio of a number to a number, and every number is relatable *(compar)* to some rational magnitude or surd, those magnitudes are found in a certain way. And for this reason certain rational numbers are assumed so that they could be related to rational magnitudes; and certain numbers are surds so that they could be related to surd [or irrational] magnitudes.[66]

There are many geometric devices, and among them is the art of stone masons concerning a geometrical device for measuring bodies. There are also devices of instruments for lifting, musical instruments, and devices of several active arts as [the making of] bows and weapons.[67]

And there is also among these an aspectual [that is, optical] device in accordance with the art that directs our vision to grasping the truth of things far away when we look at them; and there is the art of mirrors and the science of mirrors concerning places which return the rays—reflecting, converting, or refracting them. And by this device places are also known [and determined] which return [or direct] the solar rays to other bodies, and from this arises the art of burning mirrors. And among devices there is [also] the art of rational weights and instruments according to many arts.

Here, then, are the causes and the sciences of devices *(sciencie ingeniorum)* which are principles of the civil practical arts employed in bodies, figures, arrangements, positions, and mensuration, as in the art of manufacturing [things] among stone masons, and carpenters, and many others.

And these are the sciences [or doctrines] *(doctrine)* and their species [or kinds].

And what has been said about the speculative parts of philosophy should suffice.

64. Taken entirely from al-Fārābī's *De scientiis* (Palencia, pp. 154–156).

65. These two terms are described in Selection 23, n. 3.

66. If I have understood this rightly, Gundisalvo—and of course, al-Fārābī—has declared that certain numbers are irrational. This would represent a novel and important concept in ancient and medieval mathematics, where numbers were almost always assumed to be rational, discrete entities only and never irrational magnitudes, the latter being expressly excluded from the class of numbers (see Selection 27 for Campanus of Novara's comments on Bk. V, Def. 3).

67. The Latin text for this paragraph is difficult and varies somewhat from al-Fārābī's, which is itself somewhat obscure.

Logic

INTRODUCTION

John E. Murdoch

There is perhaps no area of medieval science that enjoyed a more extended history throughout the whole Middle Ages than did logic. For the number of logical texts available—though, to be sure, far from thoroughly appreciated or even read—within the rather intellectually arid earlier medieval period was far more substantial than that in, for example, mathematics, astronomy, biology, or natural philosophy. The early codifications of logical material characteristic of this period—codifications which ran well on into the twelfth century[1]—were constituted basically of Boethius' translations (at least those that were not almost immediately lost) of several of Aristotle's logical works, his commentaries (often in several versions) on these works, and a number of equally Boethian logical opuscula on such topics as categorical and hypothetical syllogisms. As a whole, this collection of logical source material was to become known as the *logica vetus*.[2] However, the rash of translating activity in the twelfth century was soon to swell and complete this basic group of texts through the preparation of Latin versions of Aristotle's *Prior* and *Posterior Analytics, Topics* and *De sophisticis Elenchis.* This, the *logica nova*, when added to the earlier, Boethian-based, *logica vetus*, furnished the Middle Ages with a corpus of texts, a base as it were, that was fundamentally Aristotelian. Yet it was only a base. For when one observes the fully developed medieval logic that was the hallmark of the fourteenth century, the constructions erected upon that base seem larger and more significant than much of the Aristotelian material upon which they rest. Additions to, and transformations of, the legacy of Aristotelian logic had by that time become the order of the day.

We have now to do with the so-called *logica moderna.* The number of issues and problems one finds treated in this logic that are lacking, or at best often only implicitly hinted at, in Aristotle is far from inconsiderable. Still, among the non-Aristotelian advances made by the more mature medieval logican, two are more outstanding than all others in their importance and impact. They are the doctrine of *consequentiae* and the theory of supposition. The following selections are expressly designed to illustrate these two major additions of medieval logic to the inherited Aristotelian base.

The medieval theory of *consequentiae* set forth a systematic consideration of inference in which the basic variables of the system were unanalyzed propositions. In other words, inference or implication proceeded not because of logical connections between universal *terms* (as was the case with Aristotelian syllogistic), but rather because of logical connections between propositions. Such a move, we now know, had already been made within Greek Stoic logic, but the unavailability of Stoic sources in the Latin Middle Ages makes it apparent that the medieval "propositional logic" to be found in the doctrine of *consequentiae* was in all probability an independent development.

Each *consequentia* consisted, as the following selections make amply clear, of an antecedent and a consequent, and the connection between these two elements in a *bona consequentia* was almost always interpreted in the strong sense of implication involved in the *impossibility* of the consequent being false if the antecedent were true. A variety of kinds of *consequentiae,* or consequences, were distinguished by medieval logicians, but what is perhaps most impressive in this particular segment of late medieval logic is the stipulation of all manner of laws of valid inference expressible in these consequences. Here one finds considerable resemblance with much to be found among the theorems of the propositional calculus of modern logic, a factor which has undoubtedly done much to direct the attention of historians to this particular aspect of the medieval doctrine of consequences. And it is

1. A brief account of these codifications can be found in A. Van de Vyver, "Les étapes du développement philosophique du haut moyen-age," *Revue belge de philologie et d'histoire,* VIII (1929), 425–452.
2. The *logica vetus* included Aristotle's *Predicamenta* (that is, the *Categories*) and *De interpretatione,* Porphyry's *Isagoge,* and Boethius' commentaries (often in two versions) on these works, plus Boethius' logical "monographs": *De syllogismis categoricis; Introductio ad syllogismos categoricos,* or *Antepraedicamenta; De hypotheticis syllogismis; De divisione; De differentiis topicis.* Beginning in the later twelfth century, the *De sex principiis* ascribed to Gilbert de la Porrée was also included.

just this aspect which the following selection on consequences from Ockham's *Summa logicae* is meant to exemplify.

The second major logical move beyond Aristotle was the theory of supposition, or better, the supposition of terms *(suppositio terminorum)*. Its concern was the investigation and spelling out of the significative relations of a term, as this term appeared, or was used, *in a proposition*. It is one thing, one might maintain, to ask what a given term may stand for in itself apart from any proposition, but quite another to inquire what it stands for when the answer is determined or limited by some proposition in which the term occurs. The latter is the bailiwick of the doctrine of supposition. Ernest Moody has succinctly expressed the fundamentals of its concerns as follows:

> The doctrine of the *suppositio terminorum* (or, more generally, of the "properties of terms") was an attempt to formulate, on a metalinguistic level, an analysis of the referential function exercised by terms occurring as subjects or predicates of categorical propositions. It presupposed the Aristotelian conception of a simple proposition, whether singular or general, as an expression composed of subject and predicate linked by a tensed verb. The concept of "supposition", indeed, seems to have been derived originally from consideration of the way in which a general term, in subject position, takes the place of, or is used *for,* singular subject-terms denoting the individuals constituting the range of denotation, or of application, of the general term. The standard definition of *suppositio* was that it is the use, or "acceptance", of a term, occurring *in* a proposition, *for* some thing or things—and, in normal cases, for the thing or things signified by, or understood by, the concept or formal content constituting the meaning of the term. [William of] Shyreswood [a logician who flourished around the middle of the thirteenth century] restricted the property of supposition to subject-terms, and said that the terms in predicate position have the property of "copulation" (i.e., adjunction or application) rather than that of supposition. But nearly all the later logicians ascribed supposition to both predicate and subject-terms, arguing that since the rules of conversion allow transposition of subject and predicate, each must be supposed to have the property of standing for individuals. One result of this ascription of supposition to predicate terms as well as to subject terms was

an interpretation of the affirmative copula as a sign of identity, so that both terms were interpreted in extension, and the predicate term made subject to quantification.

The concept of supposition was used for two rather different purposes. One purpose was that of distinguishing between the use and the mention of language expressions, or between discourse about terms and their meanings, on the one hand, and discourse about the things signified or denoted by them, on the other. We may, for example, use the word "man" to denote this very word, as when we say "man is a three letter word"; or we may use it to denote the concept or meaning associated with the word, as when we say "Man is a species of animal". Instead of forming a new term by enclosing the word in quotation marks, or by naming the sense associated with the word by an abstract form such as "humanity", the medieval logicians distinguished three types, or levels, of supposition. When a word-symbol is used to denote itself as a word, it is said to have *material* supposition; when used to denote the concept or meaning which determines its normal use, it is said to have *simple* (or absolute) supposition; but when used for its normal referents, or for the things it was instituted to designate, it is said to have *personal* supposition. The distinctions thus made between word, meaning, and referent, are important, but this method of removing such ambiguities from the use of language was clumsy.

The second main use of the concept of supposition, more pertinent to formal logic, was in the syntactical analysis of propositions whose terms are used in personal supposition for their normal referents. It is in this application that the doctrine constitutes a medieval formulation of quantification theory. An initial distinction was made between *discrete* and *common* supposition, the former being the use of a term for one thing only, which occurs if the term in question is a singular term, or if it is a general term restricted to singular denotation by apposition of the demonstrative pronoun "this". General terms, when not so restricted, were said to have common supposition in one of three ways. The subject term of a particular (or existential) proposition, quantified by the prefix "some," has *determinate* supposition *(suppositio determinata),* and stands for its individual referents in such a manner that the general proposition in which it occurs implies a disjunctive set of singular propositions whose subject terms have discrete supposition for each individual referent of the general term, and whose predicate is that of the general proposition. The subject-term of a universally quantified proposition has *confused* and *distributive* supposition, whereby it stands for its individual referents in such a manner that the "descent to singulars" is effected by a conjunction of propositions with singular subject terms denoting the individuals for which the general term stands. A third mode of common supposition, ascribed to the predicate term of a universal affirmative, was called *merely confused* supposition; the term here stands for its individual referents in such a manner that the reduction to

singulars is effected not by a conjunction or disjunction of singular propositions, but by a proposition of disjunct predicate—e.g., if every man is an animal, then every man is either this animal or that animal or that other animal, etc., but it is not the case that either every man is this animal, or every man is that animal, etc.[3]

In what follows an attempt has been made to document the two important phases of medieval logic one finds in the doctrines of *consequentiae* and *suppositio* through the selection of relevant passages in William of Ockham's *Summa totius logicae.* Several chapters from this work that expound various basic notions concerning terms in general have also been reproduced here.

18 ON TERMS, "SUPPOSITIO," AND CONSEQUENCES

William of Ockham (ca. 1280—ca. 1349)

Translated and annotated by Philotheus Boehner, O.F.M.[1]

[I. GENERAL NOTIONS ON TERMS]

1. On Terms [in General]

All those who deal with logic try to establish that arguments are composed of propositions, and propositions of terms. Hence a term is simply one of the parts into which a proposition is directly divided. Aristotle defines 'terms' in the first book of the *Prior Analytics* by saying: 'I call a term that into which a proposition is resolved (viz. the predicate, or that of which something is predicated) when it is affirmed or denied that something *is* or *is not* something'.

Although every term is or can be a part of a proposition, yet not all terms are of the same kind. Hence to obtain a perfect knowledge of them, we must first get acquainted with some distinctions between terms.

According to Boethius in the first book of the *De interpretatione,* language is threefold: written,

trans. Norman Kretzmann (Minneapolis: University of Minnesota Press, 1968); Peter of Spain, *Summulae logicales,* trans. Joseph P. Mullally (Notre Dame, Ind.: University of Notre Dame Press, 1945); Peter of Spain, *Tractatus syncategorematum and Selected Anonymous Treatises,* trans. Joseph P. Mullally (Milwaukee: Marquette University Press, 1964); Walter Burley, "On Conditional Hypothetical Syllogisms," trans. Ivan Boh in *Franciscan Studies,* XXIII (1963), 4–67; Jean Buridan, *Sophisms on Meaning and Truth,* trans. T. K. Scott (New York: Appleton-Century-Crofts, 1966); Paul of Pergula, "On Supposition and Consequences," trans. Ivan Boh in *Franciscan Studies,* XXV (1965), 30–89. In addition to the works of Moody and Bochenski cited above, the most fundamental recent secondary works on medieval logic are William and Martha Kneale, *The Development of Logic* (Oxford: Clarendon Press, 1962); Philotheus Boehner, *Medieval Logic: An Outline of Its Development from 1250 to c. 1400* (Chicago: University of Chicago Press, 1952); Ernest A. Moody, *Truth and Consequence in Mediaeval Logic* (Amsterdam: North-Holland, 1953). A critical examination of the relation of supposition theory to modern quantification theory can be found in Gareth B. Matthews, "Ockham's Supposition Theory and Modern Logic," *Philosophical Review,* LXXIII (1964), 91–99 (cf. the note by O. P. Henry in *Notre Dame Journal of Formal Logic,* V [1964], 290–293).

3. E. A. Moody, "The Medieval Contribution to Logic," *Studium Generale,* XIX (1966), 447–448. A brief introductory bibliography of medieval logic may be found at the end of Moody's article (p. 452), and a more complete one in I. M. Bochenski, *A History of Formal Logic* (Notre Dame, Ind.: University of Notre Dame Press, 1961). References to the basic Latin texts can be found in these bibliographies. English translations of texts other than those of Ockham which are reproduced in this book are William of Sherwood, *Introduction to Logic,* trans. Norman Kretzmann (Minneapolis: University of Minnesota Press, 1966); William of Sherwood, *Treatise on Syncategorematic Words,*

1. [Reprinted by permission of Thomas Nelson and Sons, Ltd. from *Philosophical Writings, A Selection: William of Ockham,* translated with an introduction by Philotheus Boehner, O. F. M. (Edinburgh and London: Nelson, 1957), pp. 47–58, 64–74, and 84–88; also in paper (Indianapolis: Bobbs-Merrill, 1964), pp. 51–62, 69–79, and 93–97. The three parts following were translated by Boehner from the *Summa totius logicae,* Part I, chapters 1–2, 4, 10–11; Part I, chapters 62–64, 68; and Part III, part 3, chapter 36, respectively.—*Ed.*]

spoken and conceptual. The last named exists only in the intellect. Correspondingly the term is three-fold, viz. the written, the spoken and the conceptual term. A written term is part of a proposition written on some material, and is or can be seen with the bodily eye. A spoken term is part of a proposition uttered with the mouth and able to be heard with the bodily ear. A conceptual term is a mental content or impression which naturally possesses signification or consignification, and which is suited to be part of a mental proposition and to stand for that which it signifies.

These conceptual terms and the propositions formed by them are those mental words which St Augustine says in the fifteenth book of *De Trinitate* do not belong to any language; they remain only in the mind and cannot be uttered exteriorly. Nevertheless vocal words which are signs subordinated to these can be exteriorly uttered.

I say vocal words are signs subordinated to mental concepts or contents. By this I do not mean that if the word 'sign' is taken in its proper meaning, spoken words are properly and primarily signs of mental concepts; I rather mean that words are applied in order to signify the very same things which are signified by mental concepts. Hence the concept signifies something primarily and naturally, whilst the word signifies the same thing secondarily. This holds to such an extent that a word conventionally signifying an object signified by a mental concept would immediately, and without any new convention, come to signify another object, simply because the concept came to signify another object. This is what is meant by the Philosopher when he says 'Words are signs of the impressions in the soul'. Boethius also has the same in mind when he says that words signify concepts. Generally speaking, all authors who maintain that all words signify, or are signs of, impressions in the mind, only mean that words are signs which signify secondarily what the impressions of the mind import primarily. Nevertheless, some words may also primarily signify impressions of the mind, or concepts; these may in turn signify secondarily other intentions of the mind, as will be shown later.

What has been said about words in regard to impressions or contents or concepts holds likewise analogously for written words in reference to spoken words.

Certain differences are to be found among these [three] sorts of terms. One is the following: A concept or mental impression signifies naturally whatever it does signify; a spoken or written term, on

the other hand, does not signify anything except by free convention.

From this follows another difference. We can change the designation of the spoken or written term at will, but the designation of the conceptual term is not to be changed at anybody's will.

For the sake of quibblers, however, it should be noted that 'sign' can assume two meanings. In one sense it means anything which, when apprehended, makes us know something else; but it does not make us know something for the first time, as has been shown elsewhere; it only makes us know something actually which we already know habitually. In this manner, a word is a natural sign, and indeed any effect is a sign at least of its cause. And in this way also a barrel-hoop signifies the wine in the inn. Here, however, I am not speaking of 'sign' in such a general meaning. In another sense, 'sign' means that which makes us know something else, and either is able itself to stand for it, or can be added in a proposition to what is able to stand for something—such are the syncategorematic words and the verbs and other parts of a proposition which have no definite signification—or is such as to be composed of things of this sort, e.g. a sentence. If 'sign' is taken in this sense, then a word is not a natural sign of anything.

2. The Various Meanings of 'Term'

The noun 'term' has three meanings. In one sense, 'term' is the name of everything that can be the copula or one of the extremes in a categorical proposition, namely the subject or the predicate, or any qualification of the subject or predicate or of the verb. In this sense, even a proposition can be a term, as it can be part of a proposition. For it is true to say: ' "Man is an animal" is a true proposition'. In this case, the entire proposition 'Man is an animal' is the subject and 'true proposition' is the predicate.

In another sense, the noun 'term' is contrasted with a sentence. Then, every non-complex expression is called a 'term'. (It was in this sense that I used 'term' in the preceding chapter.)

Thirdly, 'term' in its precise meaning designates everything that in its significative function can be either subject or predicate of a proposition. In this sense a verb or a conjunction or an adverb or a proposition or an interjection is not a term. Even many nouns will not be terms, viz. the syncategorematic nouns[2]; although they may be extremes of a

----

2. Ockham refers to such words as 'all' *(omnis)* which

proposition when taken in material or simple *suppositio,* nevertheless, when taken in their significative function, they cannot be extremes of a proposition. For instance, the sentence ' "Reads" is a verb' makes sense and is true if the verb 'reads' is taken in material supposition.[3] If, however, it were taken in its significative function, the sentence would be unintelligible. The same holds for the following propositions: ' "All" is a noun', ' "once" is an adverb', ' "If" is a conjunction', ' "From" is a preposition'. It is in this [last] sense that the Philosopher understands 'term' when he defines it in the first book of the *Prior Analytics.*

Not only can one simple expression be a term in this sense, but also a composite of two such simple expressions, viz. of an adjective and a substantive, and of a participle and an adverb, or of a preposition with its grammatical case, since such a compound, too, can be subject or predicate of a proposition. For in the proposition 'A white man is a man', neither 'man' nor 'white' is the subject, but only the whole expression 'white man'. Likewise in the proposition '[The one] running swiftly is a man', neither '[the one] running' nor 'swiftly' is the subject but the entire expression '[the one] running swiftly'.

Yet not only the noun in the nominative case may be a term, but also a noun in another case, because such a noun can be subject and also predicate of a proposition. However, a noun which is not in the nominative case cannot be subject in reference to *any* verb. For it is not correct to say 'The man's sees a donkey', although it is correct to say, 'The man's is a donkey'. How a noun in an oblique case may be subject and which verbs can or cannot have such a subject are questions for the grammarian to decide, for his task it is to study the constructions of words.

3. Division into Categorematic and Syncategorematic Terms

There is still another distinction holding both between vocal, and between mental, terms. Some are categorematic, others syncategorematic, terms. Categorematic terms have a definite and fixed signification, as for instance the word 'man' (since it signifies all men) and the word 'animal' (since it signifies all animals), and the word 'whiteness' (since it signifies all occurrences of whiteness). Syncategorematic terms, on the other hand, as 'every', 'none', 'some', 'whole', 'besides', 'only', 'in so far as', and the like, do not have a fixed and definite meaning, nor do they signify things distinct from the things signified by categorematic terms. Rather, just as, in the system of numbers, zero standing alone does not signify anything, but when added to another number gives it a new signification; so likewise a syncategorematic term does not signify anything, properly speaking; but when added to another term, it makes it signify something or makes it stand for some thing or things in a definite manner, or has some other function with regard to a categorematic term. Thus the syncategorematic word 'every' does not signify any fixed thing, but when added to 'man' it makes the term 'man' stand for all men actually, or with confused distributive *suppositio.* When added, however, to 'stone', it makes the term 'stone' stand for all stones; and when added to 'whiteness', it makes it stand for all occurrences of whiteness. As with this syncategorematic word 'every', so with others, although the different syncategorematic words have different tasks, as will be shown further below.

Should some quibbler say that the word 'every' is significant and consequently it signifies something, we answer that it is called significant, not because it signifies something determinately but only because it makes something else signify or represent or stand for something, as we explained before. And just as we say that the noun[4] 'every' does not signify anything in a determinate and limited way, to use Boethius's way of speaking, we must maintain the same of all syncategorematic words and all conjunctions and prepositions.

It is, however, different with some adverbs, because certain of them determinately signify the same things which categorematic words signify, though they do so in a different mode of signification.

4. On the Difference between Connotative and Absolute Terms

Having discussed concrete and abstract terms, we must now speak of another division of names frequently used by the teachers of philosophy.

Certain names are purely absolute, others are connotative. Purely absolute names are those which do not signify one thing principally, and another or even the same thing secondarily; but everything

---

would be classed under the heading of 'nouns' in medieval grammer (as *nomina adiectiva*).

3. Lest it seem that Ockham labours the point of this paragraph unduly, the reader should recollect that in those days there were no quotation marks.

4. See note 2 above.

alike that is signified by the same absolute name, is signified primarily. For instance, the name 'animal' just signifies oxen, donkeys and also all other animals; it does not signify one thing primarily and another secondarily, in such a way that something has to be expressed in the nominative case and something else in an oblique case; nor is there any need to have nouns in different cases, or participles, in the definitions which express the meaning of 'animal'. On the contrary, properly speaking, such names have no definitions expressing the meaning of the term. For, strictly speaking, a name that has a definition expressing the meaning of the name, has only one such definition, and consequently no two sentences which express the meaning of such terms are so different in their parts that some part in the first sentence signifies something that is not signified by any corresponding part in the second. The meaning of absolute names, however, may be explained in some manner by several sentences, whose respective parts do not signify the same things. Therefore, properly speaking, none of these is a definition explaining the meaning of the name. For instance, 'angel' is a purely absolute name, at least if it means the substance and not the office of an angel. This name has not some one definition expressing the meaning of the term. For someone may explain the signification of the name by saying: 'I understand by "angel" a substance which exists without matter'; another thus: 'An angel is an intellectual and incorruptible substance'; again another thus: 'An angel is a simple substance which does not enter into any composition with anything else.' And what is signified by this name is explained just as well by the one as by the other definition. Nevertheless, not every term in each of these sentences signifies something that is signified in the same manner by a similar term in each of the other sentences. For this reason, none is, strictly speaking, a definition expressing the meaning of the name. And so it is with many names that are purely absolute. Strictly speaking, none of them has a definition expressing the meaning of the names. Names like the following are of this kind: 'man', 'animal', 'goat', 'stone', 'tree', 'fire', 'earth', 'water', 'sky', 'whiteness', 'blackness', 'heat', 'sweetness', 'odour', 'taste', and so on.

A connotative name, however, is that which signifies something primarily and something else secondarily. Such a name has, properly speaking, a definition expressing the meaning of the name. In such a definition it is often necessary to put one of its terms in the nominative case and something else in an oblique case. This holds, for instance, for the name 'white'. For it has a definition expressing the meaning of the name in which one expression is put in the nominative case, and another in an oblique case. When you ask, therefore, 'What does the name "white" signify?' you will answer: 'It signifies the same as the entire phrase "Something that is qualified by whiteness", or "Something that has whiteness".' It is manifest that one part of this phrase is put in the nominative case and another in the oblique case. Sometimes it may happen that a verb appears in the definition expressing the meaning of the name. If, for instance, it is asked 'What does the name "cause" signify?' it can be answered that it means the same as the phrase 'Something whose existence is followed by the existence of something else', or 'Something that can produce something else', or the like.

Such connotative names include all the concrete names of the first kind, because such concrete names signify one thing in the nominative case and something else in the oblique case; that is to say, in the definition expressing the meaning of the name, one term signifying one thing must be put in the nominative case, and another term signifying another thing must be put in the oblique case. That becomes evident as regards all such names as 'just', 'white', 'animated', 'human' and the like.

Also to this type belong all relative names. For their definition has to contain distinct parts which either signify the same thing in different ways, or signify distinct things; this is evident as regards the name 'similar'. For if 'similar' is defined, we have to say 'The similar is something that has such a quality as another thing has', or some such definition. However, it does not matter which examples we take.

From this it becomes clear that the common name 'connotative' is a higher genus than the name 'relative', at least if we take the common noun 'connotative' in its broadest sense. For such names include all names pertaining to the genus of quantity, according to those who maintain that quantity is not a different thing from substance and quality. Thus, for them, 'body' has to be considered a connotative name. Hence, according to them, it must be said that a body is nothing else but a thing which has part distant from part in length, breadth and height, and a continuous and permanent quantity is a thing which has part distant from part. This, then, would be a definition expressing the meaning of the name. In consequence, these people have also to maintain that 'figure' or 'shape', 'curvature', 'straightness', 'length', 'height', and the like are connotative names. Further, those who maintain that everything is either a substance or a quality, have to suppose also that all terms contained in the categories other than substance and quality are connotative names, and also some of the genus quality are connotative names, as will be shown later on.

To this group of names belong also such terms as 'true', 'good', 'one', 'potency', 'act', 'intellect', 'intelligible', 'will', 'willable' and the like. The word 'intellect', for instance, has the meaning: 'Intellect is soul able to understand'. Thus the soul is signified by the nominative case and the act of understanding by the rest of the phrase. The name 'intelligible' is also a connotative term. It signifies the intellect, both in the nominative and in the oblique case, since its definition is this: 'The intelligible is something that can be apprehended by the intellect'. In this definition the intellect is signified by the name 'something', and also by the oblique case 'by the intellect'. The same must be said of 'true' and 'good'; for 'true', which is convertible or co-extensive with 'being', signifies the same as 'intelligible'. Likewise 'good', which is co-extensive with 'being', signifies the same as the phrase 'Something which can be willed and loved according to right reason'.

5. On Names of First and Second Imposition

We have thus given the divisions that apply both to terms which signify naturally and to those which are made by convention; we have now to say something about certain divisions which concern only terms made by convention.

A first division is this: Some of the conventional names are names of first imposition and some are names of second imposition. Names of second imposition are names which are applied to signify conventional signs, and also what goes with such signs, but only as long as they are signs.

But the general term 'name of second imposition' can be taken in two senses; one broad, the other strict. In a broad sense a name of second imposition is one that signifies utterances conventionally used, but only as long as they have this conventional use, whether or not such a name be also shared by mental contents, which are natural signs. Such names are 'noun', 'pronoun', 'verb', 'conjunction', 'case', 'number', 'tense' and the like, when used as the grammarian understands them. These names are called names of names, because they are applied

only to signify parts of speech, and only as long as these are significative. For names which are predicated of words both when they are significant and when they are not are not called names of second imposition. Hence such names as 'quality', 'spoken', 'utterance' and the like, are not names of second imposition, though they signify conventional utterances, since they would signify them even if they were not significant as they now are. But 'noun' is a name of second imposition, since neither the word 'man' nor any other word was a noun before it was employed to signify. Likewise 'man's' was of no case, before it was used to signify what it does. The same holds good for the other words of this kind.

In the strict sense, however, 'name of the second imposition' is that which signifies only a conventional sign, and therefore does not refer to mental contents, which are natural signs. Such names are 'figure', 'conjugation' and the like. All other names that are not names of second imposition in one or the other way, are called names of first imposition.

'Name of first imposition', however, can be taken in two senses. In a broad sense all names not of second imposition are names of first imposition. Thus all such syncategorematic signs as 'every', 'none', 'some', 'any' and the like, are names of first imposition. In a strict sense, however, only categorematic names not of second imposition are called names of first imposition, and not syncategorematic names.[5]

Names of first imposition, in the strict sense, are of two classes. Some are names of first intention, others of second intention. Names of second intention are those nouns which are used precisely to signify mental concepts, which are natural signs, and also other conventional signs, or what goes with such signs. All the following are of this kind: 'genus', 'species', 'universal', 'predicable' and the like. For such names signify only mental contents, which are natural signs, or conventional signs.

Hence it can be said that this common term, 'name of second intention', can be taken strictly or broadly. Broadly speaking, that is said to be a name of second intention which signifies mental contents that are natural signs, whether or not it also signifies conventional signs for just such time as they function as signs. In this sense some names of first imposition and second intention are also names of second imposition. Strictly speaking, however, that only is called a name of second

---

5. Or nouns. See note 2 above.

intention which precisely signifies mental contents that are natural signs. In this sense no name of second intention is a name of first imposition.

Names of first intention, on the other hand, are all names that differ from the former; that is, they signify some things which neither are signs nor go with such signs, as for instance, 'man', 'animal', 'Socrates', 'Plato', 'whiteness', 'white', 'true', 'good' and the like. Some signify precisely things that are not signs able to stand for other things; some signify such signs and other things as well.

From all this it may be gathered that certain names precisely signify conventional signs, but only as long as they are signs; some signify both natural and conventional signs; some, however, signify only those things which are not such signs, which are parts of propositions; some indifferently signify both things which are not parts of propositions or speech, and also such signs; of this kind are the following names: 'thing', 'being', 'something', 'one' and the like.

[II. THE THEORY OF 'SUPPOSITIO']

1. 'Suppositio' of Terms [in General]

Up to now we have been speaking about the signification of terms. It remains now to discuss *suppositio,* which is a property belonging to a term, but only when used in a proposition.

We have to know first that *suppositio* is taken in two meanings. In a broad sense, it is not contrasted with appellation; appellation is rather a subclass of *suppositio.* In a strict sense, *suppositio* is contrasted with appellation. However, I do not intend to speak about *suppositio* in this sense, but only in the former. Thus, both subject and predicate have *suppositio.* Generally speaking, whatever can be subject or predicate in a proposition has *suppositio.*

'Suppositio' means taking the position, as it were, of something else. Thus, if a term stands in a proposition instead of something, in such a way (*a*) that we use the term for the thing, and (*b*) that the term (or its nominative case, if it occurs in an oblique case) is true of the thing (or of a demonstrative pronoun which points to the thing), then we say that the term has *suppositio* for the thing. This is true, at least, when the term with *suppositio* is taken in its significative function. Hence, in general, when a term with *suppositio* is the subject of a proposition, then the thing for which the term has *suppositio,* or a demonstrative pronoun pointing to this, is that of which the proposition denotes that the predicate is predicated. But where the term

with *suppositio* is predicate, the thing or pronoun is the one of which the proposition, if formulated, denotes that the subject is a subject. For instance, the proposition 'Man is an animal' denotes that Socrates is truly an animal, so that the proposition 'This is an animal' would be true, if it were formulated while pointing at Socrates. The proposition ' "Man" is a noun', however, denotes that the vocal sound 'man', is a noun. Therefore in this proposition 'man' stands for this vocal sound. Likewise, the proposition 'A white thing is an animal' denotes that this thing which is white is an animal, so that this proposition would be true: 'This'—pointing at that thing which is white—'is an animal'. Hence the subject stands for that thing. Much the same must be said as regards the predicate. For the proposition 'Socrates is white' denotes that Socrates is *that* thing, which has whiteness; and for that reason the predicate stands for *that* thing, which has whiteness. And if no other thing had whiteness but Socrates, then the predicate would stand for Socrates alone.

There is a general rule, namely that in any proposition a term never stands for something of which it is not truly predicated, at least if the term is taken in its significative function. From this it follows that it is false to say, as some ignorant people do, that the concrete term as a predicate stands for a form—for instance to say that in the proposition 'Socrates is white' the term 'white' stands for whiteness. For whichever *suppositio* the terms may have, the proposition 'Whiteness is white' is simply false. Therefore, according to the teaching of Aristotle, a concrete term of this kind never stands for such a form as is signified by the corresponding abstract term. But in regard to other concrete terms of which we have spoken before, this is quite possible. In the same way in this proposition 'A man is God', 'man' truly stands for the Son of God, because He is truly man.

2. The Division of 'Suppositio'

We have to know that *'suppositio'* is primarily divided into personal, simple and material *suppositio.*

Generally speaking, we have personal *suppositio* when a term stands for the objects it signifies, whether the latter be things outside the mind, or vocal sounds, or mental concepts, or writing, or anything else imaginable. Whenever the subject or the predicate of a proposition stands for the object signified, so that it is taken in its significative function, we always have personal *suppositio.*

An example of the first would be 'Every man is an animal', where 'man' stands for the objects it signifies, since 'man' is a conventional sign meant to signify *these* men and nothing else; for properly speaking it does not signify something common to them, but, as St John Damascene says, these very men themselves. An example of the second would be to say 'Every vocal noun is a part of speech'. In this case 'noun' stands only for vocal signs; therefore its *suppositio* is personal. An example of the third would be to say 'Every species is universal', or 'Every mental content is in the mind'. In either case the subject has personal *suppositio*, since it stands for what it conventionally signifies. An example of the fourth would be to say 'Every written expression is an expression'. Here the subject stands only for the thing it signifies, namely for written signs. Hence it has personal *suppositio*.

From this it is clear that personal *suppositio* is not adequately described by those who say that personal *suppositio* occurs when a term stands for a thing. But the definition is this: 'Personal *suppositio* obtains when a term stands for what it signifies and is used in its significative function'.

Simple *suppositio* is that in which the term stands for a mental content, but is not used in its significative function. For instance 'Man is a species'. The term 'man' stands for a mental content, because this content is the species; nevertheless, properly speaking, the term 'man' does not signify that mental content. Instead, this vocal sign and this mental content are only signs, one subordinate to the other, which signify the same thing, in the manner explained elsewhere. This shows the falsity of the opinion held by those who say (as is commonly accepted) that simple *suppositio* occurs when a term stands for the object it signifies. For simple *suppositio* obtains when a term stands for a mental content, which is not properly speaking the object signified by the term, because the term signifies real things and not mental contents.

Material *suppositio* occurs when a term does not stand for what it signifies, but stands for a vocal or written sign, as ' "Man" is a noun'. Here, 'man' stands for itself; and yet it does not signify itself. Likewise in the proposition ' "Man" is written', we can have material *suppositio*, since the term stands for that which is written.

As the three sorts of *suppositio* apply to a spoken sign, so also they can be applied to a written sign. Hence, if the following four propositions are written, 'Man is an animal', 'Man is a species', ' "Man" is a monosyllabic word', ' "Man" is a

written word',[6] each one of these could be true, but each one for a different object. For that which is an animal is by no means a species, nor a monosyllabic word, nor a written word. Likewise, that which is a species is not an animal nor a monosyllabic sign. And so with the others. Yet in the two latter propositions the term has material *suppositio*.

Material *suppositio* could be subdivided according as the subject stands for a spoken sign or for a written sign. If we had terms for them, we could distinguish *suppositiones* for a spoken sign and for a written sign, just as we distinguish *suppositio* for the object signified and *suppositio* for a mental content, calling one 'personal' and the other 'simple' *suppositio*. However, we have no such names.

As such a difference of *suppositio* can apply to a vocal and written term, so also it can apply to a mental term. For a mental content may stand for that which it signifies, or for itself, or for a spoken or written sign.

It should be noted, however, that personal *suppositio* is not called 'personal' because a term stands for a person, nor is simple *suppositio* so called because a term stands for a simple thing, nor is material *suppositio* so called because a term stands for matter; but only for the reasons mentioned. Therefore the terms 'material', 'personal' and 'simple' are being used in an equivocal meaning in logic and in other sciences. In logic, however, they are not used frequently except in conjunction with the term '*suppositio*'.

3. In Whatever Proposition a Term Is Placed, It May Have Personal '*Suppositio*'

In whatever proposition a term is placed, it may have personal *suppositio*, if it is not arbitrarily limited to another *suppositio* by those who use it. In the same manner an equivocal term may stand in any proposition for any one of the objects it signifies, if it is not arbitrarily limited to only one such object by those who use it. However, a term cannot have simple and material *suppositio* in every proposition. It can have these *suppositiones* only when the other term of the proposition, to which it is being compared, relates to a mental content or a spoken or written sign. For instance, in the proposition 'Man is running', 'man' cannot have simple or material *suppositio*, since 'to run' does not refer to a mental content, nor to a spoken or written sign. But it is different in the proposi-

6. Note, again, that in Ockham's day there were no quotation marks; in fact, they partly dispel the ambiguity which he is discussing.

tion 'Man is a species', for 'species' signifies a mental content. For that reason the term can have simple *suppositio*. Hence the proposition must have its various senses distinguished according to the third mode of equivocation, because the subject may have either simple or personal *suppositio*. In the first sense the proposition is true, for then it denotes that a mental content or concept is a species, and that is true. In the second sense, the proposition is simply false, since then it denotes that something signified by 'man' is a species, and· that is manifestly false.

In like manner the following propositions have to have their possible senses distinguished: ' "Man" is predicated of many'; ' "Risible" is a property of man'; ' "Risible" is predicated primarily of man'. These propositions must have various senses distinguished on the side both of the subject and of the predicate. Likewise in the proposition ' "Rational animal" is the definition of "man",' a distinction must be made, for if the subject has simple *suppositio* it is true, if it has personal *suppositio* it is false. Thus it is with many other propositions, such as 'Wisdom is an attribute of God', ' "Creative" is a property of God', 'Goodness and wisdom are divine attributes', 'Goodness is predicated of God', 'Innascibility is a property of the Father', etc.

In a similar manner, if a term is compared with the other extreme [viz. the predicate], which refers to a vocal or written sign, the proposition must be distinguished. For the term may have personal *suppositio* or material *suppositio*. Thus the following propositions have to have their senses distinguished: ' "Socrates" is a name'; ' "Man" is a monosyllable'; ' "Paternity" signifies a property of the Father'. For if 'paternity' has material *suppositio,* this proposition is true; ' "Paternity" signifies a property of the Father'. If, however, its *suppositio* is personal, then the proposition would be false, because paternity is a property of the Father or is the Father Himself [i.e. when we use 'paternity' with personal *suppositio,* we must say it *is* a property of the Father]. In the same manner the following propositions must have their senses distinguished: ' "Rational animal" signifies the quiddity of "man" '; ' "White man" signifies an accidental aggregate'; ' "White man" is a composite term', and so on as regards many other propositions.

Therefore the following rule can be given: When a term capable of this threefold *suppositio* is compared with an extreme [of a proposition] which is a common term for complex and non-complex, spoken and written, terms, the first term can always have material *suppositio,* and therefore with such a proposition a distinction must be made. When, on the other hand, a term is compared with an extreme that signifies a content of the mind, the proposition has to have its possible meanings distinguished, since the term may have simple *or* personal *suppositio*. When, however, the term is compared with an extreme which is a common term for all the aforesaid classes, then its meanings must be distinguished, since it may have simple, material *and* personal *suppositio*. A proposition which must have its meanings distinguished in this manner is ' "Man" is predicated of many'; because if man has personal *suppositio,* then it is false, since then it denotes that something signified by the term 'man' is predicated of many. If it has simple *suppositio,* or material *suppositio* either for the vocal or the written sign, then it is true, because both the common concept and the spoken and written sign *are* predicated of many.

4. Personal 'Suppositio' in Particular

Personal *suppositio* can be divided in the first place into discrete and common *suppositio*. Discrete *suppositio* is that belonging to a proper name of something or to the demonstrative pronoun taken as significative. Such a *suppositio* makes a proposition singular, like these: 'Socrates is a man'; 'This man is a man'; and so with others. It may be objected that the proposition 'This plant grows in my garden' is true, yet the subject has no discrete *suppositio*. To this we have to say that the proposition taken as it stands is false. However, what is understood by it is the proposition 'A plant of this kind grows in my garden', and in this case the subject has determinate *suppositio*. From this it should be noted that when a proposition which as it stands is false, has nevertheless a true sense, then, taking it in this sense, the subject and predicate must have the same *suppositio* as in the proposition which *is* true as it stands.

Common personal *suppositio* is that of a common term, as in 'A man is running', 'Every man is an animal'.

Common personal *suppositio* is divided into confused *suppositio* and determinate *suppositio*. There is determinate *suppositio* when it is possible to make the logical descent to singulars by a disjunctive proposition, as in the correct inference 'A man is running, therefore this man is running, or that man (and so on for every individual)'. It is called 'determinate' *suppositio* because such *suppositio* denotes that a proposition of this kind is true in the case of a determinate singular proposition, which determinate singular proposition by itself, without the truth of any other singular proposition, is sufficient to make such a proposition true. Thus for the truth of the proposition 'A man is running', it is required only that some determinate singular proposition be true; and any one would suffice, even if every other such singular proposition were false. Frequently, however, others also, or even all, of the singular propositions are true. Therefore it is a sure rule that when (1) the logical descent from a common term to its singular inferior terms can legitimately be made by a disjunctive proposition, and (2) from any singular proposition such a proposition is inferable, then this term has determinate personal *suppositio*. Therefore, in the proposition 'A man is an animal', both extremes have determinate *suppositio*; for this follows: 'A man is an animal, therefore, this man is an animal, or this (and so on in regard to each man)'. Likewise this follows: 'This man is an animal'—if we point at any man—'therefore a man is an animal'. Likewise, it follows: 'A man is an animal, therefore a man is this animal, or, a man is that animal, or that animal (and so on with each animal)'. And this correctly follows: 'A man is this animal'—pointing at any animal—'therefore a man is an animal'. Therefore, both 'man' and 'animal' have determinate *suppositio*.

Confused personal *suppositio* is any personal *suppositio* of a common term which is not determinate *suppositio*. This again is subdivided into merely confused *suppositio*, and confused distributive *suppositio*. Merely confused *suppositio* occurs when (1) a common term has personal *suppositio*; (2) we are unable to make the logical descent to the singulars by means of a disjunctive proposition without any change of the other extreme; (3) we can, however, make the logical descent by way of a proposition with a disjunctive predicate; and (4) the original proposition can be inferred from any singular. For instance, in the proposition 'Every man is an animal', 'animal' has merely confused *suppositio*. For (2) it is not open to us to make the logical descent from 'animal' to its inferior terms; for this does not follow: 'Every man is an animal, therefore every man is this animal, or every man is that animal (and so on for every animal)'. However, (3) the logical descent can validly be made to a proposition whose predicate is a disjunction of singular terms. For

this does validly follow: 'Every man is an animal, therefore every man is either this animal or that or that (and so on for every animal)', for the consequent is a single categorical proposition composed of the subject 'man' and the predicate 'this animal or that or that (and so on for every animal)'. And it is manifest that this predicate is truly predicated of every man, and for that reason this universal proposition is simply true. And likewise (4) this proposition is inferred from every thing contained under 'animal'. For this follows correctly: 'Every man is this animal'—pointing at any animal—'therefore every man is an animal'.

Confused distributive *suppositio* occurs when it is licit to make a logical descent in some way to a copulative proposition if the term has many inferiors, but a formal inference cannot be made to the original proposition from one of the instances. So it is with this proposition, 'Every man is an animal'. The subject of this proposition has confused distributive *suppositio*. For this follows: 'Every man is an animal, therefore this man is an animal and that one is (and so on for every individual)'. But this does not formally follow: 'This man is an animal'—pointing at any man—'therefore every man is an animal'.

When I said 'It is licit to make a logical descent in some way', I said so because this logical descent cannot always be made in the same manner. For sometimes it is possible to make this descent without making any change on the part of the propositions, except that in the first proposition a common term is subject or predicate, whereas in the consequent the corresponding singulars are taken, as is clear in the aforesaid example. Sometimes, however, it is allowed to make the logical descent with some variation, even omitting something in one proposition which is put in the other and is not the common term nor contained under the common term. For instance, when we say 'Every man except Socrates is running', it is legitimate to make the logical descent to some individuals in a copulative proposition. For this follows: 'Every man except Socrates is running, therefore Plato is running (and so on for every man who is different from Socrates)'. However, in these singular propositions something is omitted, which was put in in the universal proposition; and that which was omitted is not the common term nor the sign distributing the common term [viz. the quantifier], for what was omitted was the word expressing exception [i.e. 'except'], together with the term to be excepted. Hence the logical descent is not made in the same manner in the two propositions 'Every man except Socrates is running', and 'Every man is running', nor can the logical descent be made altogether to the same things. The first kind of confused distributive *suppositio* is called 'mobile confused distributive *suppositio'*, the second is called 'immobile confused distributive *suppositio'*.

[III. ON CONSEQUENCES]

4. General Rules of Inference

There are many general rules of inference:

(1) From truth falsity never follows.

Therefore when the antecedent is true and the consequent is false, the inference is not valid. This reason is sufficient to prove that an inference is not valid. However, it is to be understood that 'antecedent' means everything that precedes the consequent. Therefore sometimes the antecedent is only one proposition, and sometimes it contains several propositions, as is obvious in a syllogism. In this case, though one of the propositions of the antecedent may be true, the conclusion may be false; but if every one of the propositions is true, the conclusion cannot be false, if it follows from them.

(2) From false propositions a true proposition may follow.

Hence this inference does not hold: 'The antecedent is false, therefore the consequent is false'. But the following inference holds: 'The consequent is false, therefore so is the antecedent'. Therefore, if the consequent is false, it is necessary that the antecedent as a whole is false or that some proposition which is part of the antecedent is false. But it is not required that every proposition which is part of the antecedent should be false. For sometimes from one true and one false proposition a false conclusion follows, as is obvious in the following instance: 'Every man is an animal, a stone is a man, therefore a stone is an animal'.

(3) If an inference is valid, then from the opposite of the consequent the opposite of the antecedent follows.

It has to be noted that when the antecedent is one proposition and the inference is valid, then from the opposite of the consequent the opposite of the entire antecedent always follows. However, when the antecedent contains several propositions, then it is not required that from the opposite of the consequent the opposite of every proposition which is part of the antecedent should follow; but it is

required that from the opposite of the consequent combined with one of the propositions of the antecedent the opposite of the other proposition should follow. Thus this inference follows: 'Every man is white, Socrates is a man, therefore Socrates is white'; but this one does not follow: 'Socrates is not white, therefore Socrates is not a man'. On the other hand, this is correct: 'Every man is white, Socrates is not white, therefore Socrates is not a man'. Thus from the opposite of the conclusion and the major premise the opposite of the minor premise follows, but it does not follow from the opposite of the conclusion alone. Likewise this is correct: 'Every man is an animal, a donkey is a man, therefore a donkey is an animal'; nevertheless from the opposite of the conclusion alone the opposite of the major premise does not follow. For this does not follow: 'No donkey is an animal, therefore not every man is an animal'. But from the opposite of the conclusion and the minor premise, the opposite of the major premise follows in this way: 'No donkey is an animal, a donkey is a man, therefore not every man is an animal'.

(4) Whatever follows from the consequent follows from the antecedent.

For instance, if this follows, 'Every animal is running therefore every man is running', then, whatever follows from the proposition 'Every man is running' follows also from 'Every animal is running'.

From this rule follows another:

(5) Whatever is antecedent to the antecedent is antecedent also to the consequent.

For, if it were not so, then something would follow from the consequent which does not follow from the antecedent.

The following rules, however, are false: *Whatever follows from the antecedent, follows from the consequent.* For this follows: 'Every animal is running, therefore every man is running'; and nevertheless this does not follow: 'Every donkey runs, therefore every man is running'. Likewise this rule is false, for the same reason: *Whatever is antecedent to the consequent, is antecedent to the antecedent.*

From these other rules follow:

(6) Whatever is consistent with the antecedent is consistent with the consequent.

For instance, whatever is consistent with the proposition 'Every animal is running' is consistent with 'Every man is running'. But not everything that is consistent with the consequent is consistent with the antecedent. For with the consequent 'Every man is running', the following is consistent,

'Some donkey is not running'; but this is not consistent with the antecedent 'Every animal is running'. This is the case, when the antecedent does not follow from the consequent either in simple or in factual inference.

(7) Whatever is inconsistent with the consequent is inconsistent with the antecedent.

For instance, whatever is inconsistent with the proposition 'Every man is running' is inconsistent with 'Every animal is running', but not vice versa. For something is inconsistent with an antecedent which is not inconsistent with its consequent. The following propositions are inconsistent: 'Only a man is running' and 'Something different from a man is running'. Yet the following propositions are not inconsistent: 'A man is running' and 'Something different from a man is running'. Therefore, such inferences as these are valid: 'The opposite of the consequent is consistent with the antecedent, therefore the inference is not valid'; and 'The opposite of the consequent is not consistent with the antecedent, therefore the inference is valid'.

However, it is to be understood that such an inference may be valid only at a certain time *(ut nunc)*, because at a certain time the opposite of the consequent may be consistent with the antecedent. But if the opposite of the consequent is or could be consistent with the antecedent, it cannot be a simple inference [which holds at any time].

(8) From something necessary something contingent does not follow.

(9) Something impossible does not follow from something possible.

These two rules have to be understood of simple inference. For in simple inference, from something necessary something contingent does not follow. Nevertheless, it may indeed follow in a factual inference. For instance, this follows: 'Every being exists, therefore every man exists'; but nevertheless the antecedent is necessary and the consequent contingent. Likewise it follows: 'Every coloured thing is a man, therefore every donkey is a man', but nevertheless the antecedent is possible and the consequent impossible. The consequence is valid only as a factual one.

It is to be noted that though, in a simple consequence, from something possible something impossible does not follow, nevertheless it sometimes happens that if something possible is affirmed, something impossible has to be conceded and something necessary has to be denied; but this can be done only in the art of *Obligatio* [that is, in the art of purely logical disputation] and only for the course of a given disputation.

Other rules are given:

(10) From an impossibility anything follows.

(11) What is necessary follows from everything.

Therefore this follows: 'You are a donkey, therefore you are God'. This also follows: 'You are white, therefore God is triune'. But these consequences are not formal ones and they should not be used much, nor, indeed, are they used much.

Mathematics

19   ON THE IMPORTANCE OF STUDYING MATHEMATICS

Roger Bacon (ca. 1219—1292)

Translated by Robert B. Burke[1]

Annotated by Edward Grant

CHAPTER 2

IN WHICH IT IS PROVED BY AUTHORITY THAT EVERY
SCIENCE REQUIRES MATHEMATICS

As regards authority I so proceed. Boetius says
in the second prologue to his Arithmetic, "If an
inquirer lacks the four parts of mathematics, he has
very little ability to discover truth." And again,
"Without this theory no one can have a correct
insight into truth." And he says also, "I warn the
man who spurns these paths of knowledge that he
cannot philosophize correctly." And again, "It is
clear that whosoever passes these by, has lost the
knowledge of all learning." He confirms this by the
opinion of all men of weight saying, "Among all
the men of influence in the past, who have flourished under the leadership of Pythagoras with a finer
mental grasp, it is an evident fact that no one
reaches the summit of perfection in philosophical
studies, unless he examines the noble quality of
such wisdom with the help of the so-called quadrivium." And in particular Ptolemy and Boetius
himself are illustrations of this fact. For since there
are three essential parts of philosophy, as Aristotle
says in the sixth book of the Metaphysics, mathematical, natural, and divine, the mathematical is
of no small importance in grasping the knowledge
of the other two parts, as Ptolemy teaches in the
first chapter of the Almagest, which statement he
also explains further in that place. And since the
divine part is twofold, as is clear from the first book
of the Metaphysics, namely, the first philosophy,
which shows that God exists, whose exalted properties it investigates, and civil science, which determines divine worship, and explains many matters
concerning God as far as man can receive them.
Ptolemy likewise asserts and declares that mathematics is potent in regard to both of these branches.
Hence Boetius asserts at the end of his Arithmetic
that the mathematical means are discovered in civil
polity. For he says that an arithmetic mean is comparable to a state that is ruled by a few, for this
reason, that in its lesser terms is the greater pro-

portion; but he states that there is a harmonic mean
in an aristocratic state for the reason that in the
greater terms the greater proportionality is found.
The geometrical mean is comparable to a democratic state equalized in some manner; for whether in
their lesser or greater terms they are composed of
an equal proportion of all. For there is among all a
certain parity of mean preserving a law of equality
in their relations. Aristotle and his expositors teach
in the morals in many places that a state cannot be
ruled without these means. Concerning these means
an exposition will be given with an application to
divine truths. Since all the essential parts of philosophy, which are more than forty sciences distinct
in their turn, may be reduced to these three, it
suffices now that the value of mathematics has been
established by the authorities mentioned.

Now the accidental parts of philosophy are grammar and logic. Alpharabius makes it clear in his
book on the sciences that grammar and logic cannot be known without mathematics. . . . But not
only does a knowledge of logic depend on mathematics because of its end, but because of its middle
and heart, which is the book of Posterior Analytics,
for that book teaches the art of demonstration. But
neither can the fundamental principles of demonstration, nor conclusions, nor the subject as a whole
be learned or made clear except in the realm of
mathematics, because there alone is there true and
forceful demonstration, as all know and as we shall
explain later. Therefore of necessity logic depends
on mathematics.

What has been said is applicable likewise because

1. Reprinted from *The Opus Majus of Roger Bacon,*
translated by Robert Belle Burke (Philadelphia: University of Pennsylvania Press, 1928; reprinted New
York: Russell and Russell, 1962), pp. 117–127. The
translation was based on the Latin edition of Bacon's
Opus maius by John Henry Bridges (London, 1900).
Precise citations for the numerous references to Aristotle, Averroes, Boethius, and others have not been
furnished. In many instances, this would be superfluous,
since Bacon himself frequently cites his sources accurately.

of its beginning and not only because of its middle and end. For the book of Categories is the first book of logic according to Aristotle. But it is clear that the category of quantity cannot be known without mathematics. For the knowledge of quantity belongs to mathematics alone. Connected with quantity are the categories of when and where. For when has to do with time, and where arises from place. The category of habit cannot be known without the category of place, as Averroes teaches in the fifth book of the Metaphysics. But the greater part of the cagetory of quality contains the attributes and properties of quantities, because all things that are in the fourth class of quality are called qualities in quantities. And all the attributes of these which are absolutely essential to them are qualities, with which a large part of geometry and arithmetic is concerned, such as straight and curved and other essential qualities of the line, and triangularity and other figures belonging to surface or to a solid body; and the prime and non-factorable in numbers, as Aristotle teaches in the fifth book of the Metaphysics, as well as other essential attributes of numbers. Moreover, whatever is worthy of consideration in the category of relation is the property of quantity, such as proportions and proportionalities, and geometrical, arithmetical, and harmonic means, and the kinds of greater and lesser inequality. Moreover, spiritual substances are known by philosophy only through the medium of the corporeal, and especially the heavenly bodies, as Aristotle teaches in the eleventh book of the Metaphysics. Nor are inferior things known except through superior ones, because the heavenly bodies are the causes of things that are lower. But the heavenly bodies are known only through quantity, as is clear from astronomy. Therefore all the categories depend on a knowledge of quantity of which mathematics treats, and therefore the whole excellence of logic depends on mathematics.

CHAPTER 3

IN WHICH IT IS PROVED BY REASON THAT EVERY SCIENCE REQUIRES MATHEMATICS

What has been shown as regards mathematics as a whole through authority, can now be shown likewise by reason. And I make this statement in the first place, because other sciences use mathematical examples, but examples are given to make clear the subjects treated by the sciences; wherefore ignorance of the examples involves an ignorance of the subjects for the understanding of which

the examples are adduced. For since change in natural objects is not found without some augmentation and diminution nor do these latter take place without change; Aristotle was not able to make clear without complications the difference between augmentation and change by any natural example, because augmentation and diminution go together always with change in some way; wherefore he gave the mathematical example of the rectangle which augmented by a gnomon increases in magnitude and is not altered in shape. This example cannot be understood before the twenty-second proposition of the sixth book of the Elements. For in that proposition of the sixth book it is proved that a smaller rectangle is similar in every particular to a larger one and therefore a smaller one is not altered in shape, although it becomes larger by the addition of the gnomon.

Secondly, because comprehension of mathematical truths is innate, as it were, in us. For a small boy, as Tullius states in the first book of the Tusculan Disputations, when questioned by Socrates on geometrical truths, replied as though he had learned geometry.[2] And this experiment has been tried in many cases, and does not hold in other sciences, as will appear more clearly from what follows. Wherefore since this knowledge is almost innate, and as it were precedes discovery and learning, or at least is less in need of them than other sciences, it will be first among sciences and will precede others disposing us toward them; since what is innate or almost so disposes toward what is acquired.

Thirdly, because this science of all the parts of philosophy was the earliest discovered. For this was first discovered at the beginning of the human race. Since it was discovered before the flood and then later by the sons of Adam, and by Noah and his sons, as is clear from the prologue to the Construction of the Astrolabe according to Ptolemy, and from Albumazar in the larger introduction to astronomy, and from the first book of the Antiquities,[3] and this is true as regards all its parts, geometry, arithmetic, music, astronomy. But this would not have been the case except for the fact that this science is earlier than the others and naturally precedes them. Hence it is clear that it should be studied first, that through it we may advance to all the later sciences.

Fourthly, because the natural road for us is from

2. An illustration derived ultimately from Plato's *Meno*.
3. Flavius Josephus, the Jewish historian.

what is easy to that which is more difficult. But this science is the easiest. This is clearly proved by the fact that mathematics is not beyond the intellectual grasp of any one. For the people at large and those wholly illiterate know how to draw figures and compute and sing, all of which are mathematical operations. But we must begin first with what is common to the laity and to the educated; and it is not only hurtful to the clergy, but disgraceful and abominable that they are ignorant of what the laity knows well and profitably. Fifthly, we see that the clergy, even the most ignorant, are able to grasp mathematical truths, although they are unable to attain to the other sciences. Besides a man by listening once or twice can learn more about this science with certainty and reality without error, than he can by listening ten times about the other parts of philosophy, as is clear to one making the experiment. Sixthly, since the natural road for us is to begin with things which befit the state and nature of childhood, because children begin with facts that are better known by us and that must be acquired first. But of this nature is mathematics, since children are first taught to sing, and in the same way they can learn the method of making figures and of counting, and it would be far easier and more necessary for them to know about numbers before singing, because in the relations of numbers in music the whole theory of numbers is set forth by example, just as the authors on music teach, both in ecclesiastical music and in philosophy. But the theory of numbers depends on figures, since numbers relating to lines, surfaces, solids, squares, cubes, pentagons, hexagons, and other figures, are known from lines, figures, and angles. For it has been found that children learn mathematical truths better and more quickly, as is clear in singing, and we also know by experience that children learn and acquire mathematical truths better than the other parts of philosophy. For Aristotle says in the sixth book of the Ethics that youths are able to grasp mathematical truths quickly, not so matters pertaining to nature, metaphysics, and morals. Wherefore the mind must be trained first through the former rather than through these latter sciences. Seventhly, where the same things are not known to us and to nature, there the natural road for us is from the things better known to us to those better known to nature, or known more simply; and more easily do we grasp what is better known to ourselves, and with great difficulty we arrive at a knowledge of those things which are better known to nature. And the

things known to nature are erroneously and imperfectly known by us, because our intellect bears the same relation to what is so clear to nature, as the eye of the bat to the light of the sun, as Aristotle maintains in the second book of the Metaphysics; such, for example, are especially God and the angels, and future life and heavenly things, and creatures nobler than others, because the nobler they are the less known are they to us. And these are called things known to nature and known simply. Therefore, on the contrary, where the same things are known both to us and to nature, we make much progress in regard to what is known to nature and in regard to all that is there included, and we are able to attain a perfect knowledge of them. But in mathematics only, as Averroes says in the first book of the Physics and in the seventh of the Metaphysics and in his commentary on the third book of the Heavens and the World, are the same things known to us and to nature or simply. Therefore as in mathematics we touch upon what is known fully to us, so also do we touch upon what is known to nature and known simply. Therefore we are able to reach directly an intimate knowledge of that science. Since, therefore, we have not this ability in other sciences, clearly mathematics is better known. Therefore the acquisition of this subject is the beginning of our knowledge.

Likewise, eighthly, because every doubt gives place to certainty and every error is cleared away by unshaken truth. But in mathematics we are able to arrive at the full truth without error, and at a certainty of all points involved without doubt; since in this subject demonstration by means of a proper and necessary cause can be given. Demonstration causes the truth to be known. And likewise in this subject it is possible to have for all things an example that may be perceived by the senses, and a test perceptible to the senses in drawing figures and in counting, so that all may be clear to the sense. For this reason there can be no doubt in this science. But in other sciences, the assistance of mathematics being excluded, there are so many doubts, so many opinions, so many errors on the part of man, that these sciences cannot be unfolded, as is clear since demonstration by means of a proper and necessary cause does not exist in them from their own nature because in natural phenomena, owing to the genesis and destruction of their proper causes as well as of the effects, there is no such thing as necessity. In metaphysics there can be no demonstration except through effect, since spiritual facts are discovered through corporeal effects and the creator through the creature, as is clear in that science. In morals there cannot be demonstrations from proper causes, as Aristotle teaches. And likewise neither in matters pertaining to logic nor in grammar, as is clear, can there be very convincing demonstrations because of the weak nature of the material concerning which those sciences treat. And therefore in mathematics alone are there demonstrations of the most convincing kind through a necessary cause. And therefore here alone can a man arrive at the truth from the nature of this science. Likewise in the other sciences there are doubts and opinions and contradictions on our part, so that we scarcely agree on the most trifling question or in a single sophism; for in these sciences there are from their nature no processes of drawing figures and of reckonings, by which all things must be proved true. And therefore in mathematics alone is there certainty without doubt.

Wherefore it is evident that if in other sciences we should arrive at certainty without doubt and truth without error, it behooves us to place the foundations of knowledge in mathematics, in so far as disposed through it we are able to reach certainty in other sciences and truth by the exclusion of error. This reasoning can be made clearer by comparison, and the principle is stated in the ninth book of Euclid. The same holds true here as in the relation of the knowledge of the conclusion to the knowledge of the premises, so that if there is error and doubt in these, the truth cannot be arrived at through these premises in regard to the conclusion, nor can there be certainty, because doubt is not verified by doubt, nor is truth proved by falsehood, although it is possible for us to reason from false premises, our reasoning in that case drawing an inference and not furnishing a proof; the same is true with respect to sciences as a whole; those in which there are strong and numerous doubts and opinions and errors, I say at least on our part, should have doubts of this kind and false statements cleared away by some science definitely known to us, and in which we have neither doubts nor errors. For since the conclusions and principles belonging to them are parts of the sciences as a whole, just as part is related to part, as conclusion to premises, so is science related to science, so that a science which is full of doubts and besprinkled with opinions and obscurities, cannot be rendered certain, nor made clear, nor verified except by some other science known and verified, certain and plain to us, as in the case of a conclusion reached through premises. But mathematics alone, as was shown above, remains fixed and verified for us with the utmost certainty and verification. Therefore by means of this science all other sciences must be known and verified.

Since we have now shown by the peculiar property of that science that mathematics is prior to other sciences, and is useful and necessary to them, we now proceed to show this by considerations taken from its subject matter. And in the first place we so conclude, because the natural road for us is from sense perception to the intellect, since if sense perception is lacking, the knowledge related to that sense perception is lacking also, according to the statement in the first book of the Posterior Analytics, since as sense perception proceeds so does the human intellect. But quantity is especially a matter of sense perception, because it pertains to the common sense and is perceived by the other senses, and nothing can be perceived without quantity, wherefore the intellect is especially able to make progress as respects quantity. In the second place, because the very act of intelligence in itself is not completed without continuous quantity, since Aristotle states in his book on Memory and Recollection that our whole intellect is associated with continuity and time. Hence we grasp quantities and bodies by a direct perception of the intellect, because their forms are present in the intellect. But the forms of incorporeal things are not so perceived by our intellect; or if such forms are produced in it, according to Avicenna's statement in the third book of the Metaphysics, we, however, do not perceive this fact owing to the more vigorous occupation of our intellect in respect to bodies and quantities. And therefore by means of argumentation and attention to corporeal things and quantities we investigate the idea of incorporeal things, as Aristotle does in the eleventh book of the Metaphysics. Wherefore the intellect will make progress especially as regards quantity itself for this reason, that quantities and bodies as far as they are such belong peculiarly to the human intellect as respects the common condition of understanding. Each and every thing exists as an antecedent for some result, and this is true in higher degree of that which has just been stated.[4]

Moreover, for full confirmation the last reason can be drawn from the experience of wise men;[5]

4. At this point, Burke remarks: "The Latin of this sentence is obscure."
5. "Wise men" replaces Burke's "men of science," which is too narrow a rendition of *sapientes*.

for all the wise men in ancient times labored in mathematics, in order that they might know all things, just as we have seen in the case of men of our own times, and have heard in the case of others who by means of mathematics, of which they had an excellent knowledge, have learned all science. For very illustrious men have been found, like Bishop Robert of Lincoln[6] and Friar Adam de Marisco, and many others, who by the power of mathematics have learned to explain the causes of all things, and expound adequately things human and divine. Moreover, the sure proof of this matter is found in the writings of those men, as, for example, on impressions such as the rainbow, comets, generation of heat, investigation of localities on the earth and other matters, of which both theology and philosophy make use. Wherefore it is clear that mathematics is absolutely necessary and useful to other sciences.

These reasons are general ones, but in partic-

ular this point can be shown by a survey of all the parts of philosophy disclosing how all things are known by the application of mathematics. This amounts to showing that other sciences are not to be known by means of dialectical and sophistical argument as commonly introduced, but by means of mathematical demonstrations entering into the truths and activities of other sciences and regulating them, without which they cannot be understood, nor made clear, nor taught, nor learned. If any one in particular should proceed by applying the power of mathematics to the separate sciences, he would see that nothing of supreme moment can be known in them without mathematics. But this simply amounts to establishing definite methods of dealing with all sciences, and by means of mathematics verifying all things necessary to the other sciences. But this matter does not come within the limits of the present survey.

Arithmetic

20 ARABIC NUMERALS AND ARITHMETIC OPERATIONS IN THE MOST POPULAR ALGORISM OF THE MIDDLE AGES

John of Sacrobosco (or Holywood) (d. ca. 1244—1256)

Translated and annotated by Edward Grant[1]

Everything that has proceeded from the earliest beginning of things has been formed in a pattern *(ratione)* of numbers, and the manner in which these exist must be understood as follows: that in the whole comprehension of things, the art of numbering *(ars numerandi)* is concerned with combination [or union] *(cooperativa)*. A certain philosopher named Algus wrote this brief science of numbering, for which reason it is called *Algorismus,*[2] which is understood to be the art of numbering *(ars numerandi)* or the introductory art into number *(ars introductoria in numerum)*.

6. That is, Robert Grosseteste.

7. It is evident throughout this selection that Bacon's interest in mathematics stems from his conviction that it is essential for a proper understanding of natural philosophy and all the sciences. As A. G. Molland has observed, the study and pursuit of geometry for its own sake was little in evidence during the Middle Ages. Few theorems were added to the sum total of pure geometry. The latter was valued primarily for its use in comprehending the world. See A. G. Molland, "The Geometri-

cal Background to the 'Merton School': An Exploration into the Application of Mathematics to Natural Philosophy in the Fourteenth Century," *The British Journal for the History of Science,* Vol. IV, Part II, No. 14 (December 1968), pp. 109–110.

1. Translated from Maximilian Curtze's edition of Sacrobosco's *Algorismus vulgaris (Common Algorism)* in *Petri Philomeni de Dacia Algorismum vulgarem Johannis de Sacrobosco Commentarius una cum Algorismo ipso edidit* (Copenhagen, 1897). During the Middle Ages and into the sixteenth century, this was the most popular treatise describing operations with Arabic numerals. Sometimes known as the *De arte numerandi*, it was translated into Middle English, and somewhat expanded, under the title *The Art of Nombryng* (ed. Robert S. Steele, *The Earliest Arithmetics in English* [London: Oxford University Press, 1922]). Sacrobosco's treatise drew upon al-Khwārizmī's *Algorism* (ninth century) from the Latin translation titled *De numero indorum (On Hindu Numerals)*.

2. "Algus" is probably a corruption of "al-Khwārizmī," the name of the author of a book on Hindu numerals (our Arabic numerals). It is from a corruption of his name that we get the term "algorism." For an extract from al-Khwārizmī's algebra, see Selection 22.

A number is made known *(notificatur)* in two ways: materially and formally. Materially *(materialiter)* a number is a collection of units; formally *(formaliter)* it is the multitude of units extended [or spread out].[3] A unit *(unitas)* is that by which anything is said to be one. Included among numbers are the digit *(digitus)*, article *(articulus)*, and composite number *(numerus compositus)*. A digit is every number smaller than 10; an article is every number divisible into ten equal parts with no remainder;[4] a composite number is mixed, consisting of digit and article.[5] And it must be understood that every number between [any] two proximate [or successive] articles is composite.[6] Moreover, there are nine species of this art [of Algorism]: (1) numeration, (2) addition, (3) subtraction, (4) halving *(mediatio)*, (5) doubling *(duplatio)*, (6) multiplication, (7) division, (8) progression, and (9) extraction of roots, which is twofold, since [it applies] to square numbers and cube numbers. Among these, numeration will be considered first and then the others will follow in order.

ON NUMERATION

The numeration of any number by any suitable figures is an artificial representation. However, figure *(figura)*, difference *(differentia)*, place *(locus)*, and limit *(limes)* suppose the same thing but they are imposed for different reasons. "Figure" is so called because of the drawing of a line: "difference" is shown in this way: how the figure following differs from the figure preceding; "place" is so called by reason of the space in which a number is written; "limit" is the path [or system] that has been organized for the representation of any number. Therefore, it should be understood that alongside the nine "limits" are found nine significative "figures" representing nine digits, which are these:

9. 8. 7. 6. 5. 4. 3. 2. 1.

The tenth [figure], 0, is called *teca*,[7] circle *(circulus)*, cipher *(cyfra)*,[8] or "the figure of nothing" *(figura nichilis)*, since it signifies nothing. It holds a place and signifies for others, for without a cipher or ciphers, a pure article can not be formulated. And so it happens that any number can be represented by these nine significative figures, with a cipher or ciphers added whenever needed, rendering it unnecessary to find more significative figures [than these]. Therefore, it must be noted that every digit must be written with only one figure that is appropriate to it; every article, however, must be represented by a cipher placed in the first position and a digit by which this article is denominated, since every article is denominated by some digit—as ten *(denarius)* by unity, twenty *(vigenarius)* by two, and so on. Indeed, every number that is a digit has to be placed in a first [category of] difference, every article in a second [category of] difference.[9] For every number from ten to one hundred, excluding one hundred, must be written by two figures; and if it is an article *(articulus)*, [it is written] by a cipher, placed in the first position, and a figure written to the left, which signifies the digit by which the article is denominated. If it is a composite number, the digit which is part of it is written first *(praescribatur)*, and as before, the article is placed to the left.[10] Again, every number from one hundred

3. The *material* aspect of number appears to be understood as a collection of units assembled and united to form each number, thereby, in a sense, constituting the "matter" of each number. The formal aspect of number consists of the units themselves conceived as a multiplicity. It is probable that Sacrobosco adopted this twofold definition from Boethius' *Arithmetic,* Book I, chapter 3, where we read: "Number is a collection of units or a multitude of a quantity brought together from units" (see Boethius, "On Arithmetic," Bk. I, ch. 3, in Selection 2. Following Aristotelian philosophical terminology, Sacrobosco then labeled the first part "material," the second part "formal."

4. Thus an "article" would be 10 and any multiple or power of 10.

5. Any whole number can be expressed as a multiple or power of 10 plus digits.

6. For example, the numbers between 100 and 110, and so on.

7. The meaning of this term is unknown to me, but Peter of Dacia, in his commentary on Sacrobosco's *Common Algorism,* says, "*Teca* is iron of round figure," and that the roundness of the cipher was called *teca* by virtue of its similarity to the roundness of the iron (see Curtze, p. 26).

8. A transliteration of the Arabic term *sifr,* meaning "empty."

9. By definition, a "difference" signifies "how the figure following differs from the figure preceding" and appears to be Sacrobosco's way of describing place value in the decimal system.

10. If the passage means what it seems to say, one must conclude that Sacrobosco would write a number like 532 by first writing 2, then to the left of it 3, and finally, 5 to the left of 3. Many Arabs write their numbers in this manner. It must be emphasized, however, that although Arabic text is written and read from right to left, numerals in Arabic are read from left to right. In the early years of the introduction of Arabic numerals, Sacrobosco urged his readers to write the numbers from right to left, believing that he was following the Arabic custom.

The peculiarity in Arabic of reading everything except numerals from right to left may derive from the fact

(centum) to one thousand *(mille)*, excluding one thousand, is written by three ciphers or figures; likewise every number from one thousand to ten thousand is written by four numbers, and so on. It must also be noted that any figure placed in the first position signifies its digit; in the second position ten times its digit; in the third [position] one hundred times its digit; in the fourth [position] one thousand times its digit; in the fifth [position] ten thousand times its digit; in the sixth [position] one hundred thousand times its digit; in the seventh [position] one million times [*millesies millesies*] its digit, and so into infinity by multiplying these three [things]: ten, one hundred, and one thousand. All these things are contained in this maxim: Any figure posited in the place [or position] following signifies ten times as much as in the preceding position. And it must be understood that above any figure in the thousands place, one can conveniently put points for denoting that the last figure ought to represent as many thousands as there are points. Moreover, in this art we write toward the left in the Arabic or Jewish custom, for they are the inventors of this science; or [we do it] for this reason, an even better one, [namely] that in following the usual order in reading, we place a greater number before a smaller number.[11]

ON ADDITION

Addition *(additio)* is the adding *(aggregatio)* of a number, or numbers, to a number. There are two rows *(ordines)* of figures in addition—that is, at least two numbers are necessary, namely the number to which the addition is to be made and the number to be added. The number to which the addition ought to be made is the number that receives the addition and ought to be written above; the number to be added to the other ought to be written below. It is more fitting that the smaller number be written below and added to the greater than contrarily. But whether or not this is done, the same [result] is always produced.

Therefore, if you wish to add a number to a number, write the number to which the addition should be made in the upper row by its differences so that the first number of the lower row is under the first number of the upper row; the second under the second, and so on. After this has been done, the first figure of the lower row is added to the first figure of the upper row. From this addition, then, either a digit, an article, or a composite number results. If a digit, let the resulting digit be written in the place of the upper figure, which is crossed out

(deletae); if an article, let a cipher be written in place of the upper deleted number and let the article be carried to the left and added to the next figure following—if there is a following figure. If, however, there is no figure following, then let it [that is, the article that is carried] be placed in an empty place.[12] However, if it happens that the figure following, to which the addition of the article is to be made, is a cipher [that is, zero], the digit of the article should be written in place of it after it [that is, the cipher] has been deleted. But if there is a figure of nine and a unit is to be added to it, a cipher should be written in place of this nine and the article written to the left of it, as before.[13] After this has been done, the second [number] is added to the second number placed above and the same procedure is followed as before. It must also be noted that in addition and in all the following species [of algorism], when one figure is placed directly above [or over] another, it must be used in this figure and [this is different than] if it were placed by itself.

ON SUBTRACTION

When two numbers have been proposed, subtraction is the finding of the excess of the greater to the lesser term; or subtraction is the taking away of a number from a number so that the sum remaining may be seen. Furthermore, a smaller from a greater or an even from an even [number]

---

that Arabic numerals are of Hindu origin and were arranged to be read from left to right because Sanskrit is written that way. The Arabs, who write in the opposite direction, probably adopted the Hindu numerals basically as they found them; hence the anomalous reading of figurate numbers in a direction opposite to that in which the written language is read.

11. In effect, Arabic numerals, when used by Latins, will be read in the same order as the Latin language. It is odd, however, that Sacrobosco chose to retain the Arabic custom of writing numerals from right to left. Of course, this was eventually reversed. Perhaps he was influenced by the fact that the sum of two numbers, for example, is almost always worked out step by step from right to left (this also happens in the operations of multiplication and subtraction).

12. In the Middle English translation cited in note 1, this example is added (Steele, p. 36), showing that the results were placed directly above the numbers to be added:

The resultant	10
To whom it shalle be addede	7
The nombre to be addede	3
	10

13. That is, $9 + 1 = 10$ would be written $\overset{1}{9}$, with

the result always placed above.

can be subtracted; but a greater from a smaller never. That number is called greater which has more figures so long as the last *(ultima)* is significative. If, however, there are as many [figures] in one number as in the other, [the greater] must be judged by the last, or next to last, [figures], and so on.[14]

In subtraction two significative numbers are necessary, namely the number from which the subtraction is to be made and the number to be subtracted *(numerus subtrahendus)*. The number from which the subtraction is to be made is written in an upper row by its differences *(differentias)*;[15] the number to be subtracted is placed in the lower row by its differences, so that the first [figure] is under the first [figure], the second under the second, and so on. . . .[16]

It must be understood, however, that in both addition and subtraction we can very well begin from the left and go toward the right; but, as was said before, it is more convenient to do it in the manner that has been described. Moreover, if you wish to prove whether you have done it properly or not, add the figures which you have subtracted to the [figures in the] upper row, and if you have done this correctly, the same figures should result which you had before.[17] Similarly in addition, [for] when you will have added all the figures, subtract what you added before, and if you have done this correctly, the same figures which you had before will be produced. For subtraction is the proof of addition, and conversely.

ON HALVING

Halving *(mediatio)* is the finding of half of any number proposed in order to see what and how much is this half. In halving there is only a single row *(ordo)* of figures and only a single number is necessary, namely the number to be halved. Therefore, if you wish to halve any number, let this number be written by its differences and begin from the right, namely from the first figure toward the right, and take it toward the left side. . . .

ON DOUBLING

Doubling *(duplatio)* is the adding *(aggregatio)* of a proposed number to itself so that the resultant sum can be seen. In doubling, only one row of figures is necessary and it must be begun from the left side, namely from the greatest figure—that is, from the figure representing the greatest number.[18]

In the three preceding species, we began from the right side, namely from the smallest figure—that is, [the figure] representing the smallest number. [But] in this species and in all those that follow, we begin from the left side and with the greatest figure. Whence comes the verse:

> You subtract or add or halve from the right
> side,
> On the left side you divide and multiply,
> [And] extract a double root.

The reason for this is that if you begin to double from the first figure it would sometimes happen that the same number would be doubled twice; and although we could begin from the right side, the theory *(doctrina)*[19] and operation *(operatio)* will be more difficult.

If you wish to double any number, let it first be written by its differences and the last figure be doubled.[20] From this doubling there results either a digit, an article, or composite number. If a digit results, it is written in place of the first [figure], which is deleted; if an article results, a cipher [that is, zero] is written in place of the first [figure], which is deleted and the article is carried toward the left;[21] if a composite number results, the digit is written in place of the upper or first deleted figure (which is part of this composite number) and the article is carried toward the left. After this has been done, the next to last number must be doubled

14. Recalling that Sacrobosco would write his numbers from right to left, the last or ultimate numbers would be the first numbers for us. The larger of the "ultimate" numbers would indicate a larger total number; if they are the same, we move to the penultimate number (our second), and so on.

15. That is, in the order of the place value of the numbers.

16. The actual description of the subtractive operation is omitted.

17. If, for example, 16 were subtracted from 21, the
$$5$$
results would be presented as 21. One then adds 16 to the
$$16$$
upper row of figures and obtains the number from which the subtraction was made.

18. We would, of course, do it from the right.

19. Perhaps "teaching" would be a reason.

20. In a number such as 241, the last *(ultima)* figure would be 2, since Sacrobosco orders his terms from right to left (but see the next note).

21. Despite Sacrobosco's insistence on proceeding from left to right, it seems that he is in fact proceeding from right to left, since the transfer of the article—which is really a carrying operation—is to the left. For example, in doubling 25 we first obtain an "article," namely 10; we must then replace 5 and carry the 1 to the left.

and the same thing must be done as before. If, however, a cipher, [or zero,] occurs, it must be left untouched; but if a number is to be added to a cipher, [or zero,] the number to be added must be written in place of what was deleted. The same procedure must be followed for all other numbers until the whole number is doubled. . . . The proof of doubling is halving, and conversely.

ON MULTIPLICATION

Multiplication is the finding of a third number from two proposed numbers which contains one of them as many times as there are units in the remaining number. In multiplication two numbers are principally necessary—the multiplier, [or multiplying number,] *(numerus multiplicans)* and the number to be multiplied *(numerus multiplicandus)*. The multiplier is designated adverbially; the number to be multiplied receives a nominal appellation. A third number to be assigned, which is called the product *(productus)*, arises from leading *(ex ductione)* one number into the other. It should be noted that a multiplying number can always be converted into a number to be multiplied, and conversely, with the product *(summa)*[22] always remaining the same. And this is what is commonly said or alleged by the arithmeticians, namely that every number is converted when it is multiplied by itself.[23]

There are six rules of multiplication.

The first is [for] when a digit multiplies a digit. [After] the smaller digit [has been multiplied] by the difference between the greater [given] digit and ten, it [that is, the product of this multiplication] must be subtracted from the article of its denomination,[24] which has been computed at the same time with ten. For example, if you wish to know how much is eight [multiplied by] four, see how many units there are between eight and ten, while at the same time computing with ten. It is obvious that there are two [units]. Therefore let eight *(quaternarius bis)*[25] be subtracted from forty, and thirty-two will remain, the product *(summa)* of the whole multiplication.

The second rule is this: When a digit multiplies an article, the digit must be multiplied *(ducendus est)* by the digit by which the article is denominated —and this is done according to the first rule. Then any unit [or digit] *(unitas)* will represent ten, and any [multiple of] ten *(denarius)* [will represent] one hundred.[26] And this is true whether the article or composite number increases.

The third rule is this: When a digit multiplies a composite number, the digit, namely the multiplier *(multiplicans)*, must be multiplied into every part of the composite number—the digit by the digit in accordance with the first rule; the digit by the article in accordance with the second rule. The products *(producta)* are added and the sum *(summa)* of the whole multiplication will be obvious.

The fourth rule is this: When an article multiplies an article, the digit which denominates one of them must be multiplied by the digit which denominates the other; and any unit *(unitas)* represents one hundred, any [multiple of] ten *(denarius)*, a thousand.[27]

The fifth rule is this: When an article multiplies a composite number, the digit of the article must be multiplied with each part of the composite number and [then] the products are added and the sum *(summa)* [of the products] will be obvious.

The sixth rule is this: When a composite number multiplies a composite number, each part of the multiplier must be multiplied by each part of the multiplying number, and the products are added; the sum *(summa)* [of the products] will then be obvious. . . .[28]

ON DIVISION

When two numbers have been proposed, division is the distribution of the greater by the smaller into as many parts as there are units in the smaller

22. The term *summa*, usually meaning sum, was occasionally used synonymously with the more appropriate *productus*.
23. Does Sacrobosco mean to confirm the preceding statement by noting that when a number is multiplied by itself it matters not at all which of the terms is designated multiplicand or multiplier?
24. Obtained by multiplying the smaller digit by ten.
25. In this paragraph Sacrobosco uses *octo* and *quaternarius bis* (literally "four twice") for eight.
26. By *unitas* and *denarius* we must understand here, more generally, "digit" and "multiple of ten." For example, in $40 \cdot 2 = 80$, $4 \cdot 2 = 8$ states that 8 is the unit or digit representing the tens column in place value notation; in $40 \cdot 3 = 120$, the multiplication of $4 \cdot 3 = 12$ states that we have a total number in the one hundreds ending in zero—that is, a multiple of ten.
27. For example, in multiplying two articles, $40 \cdot 10 = 400$, multiplication of the digits 4 and 1 yields a product which represents hundreds—that is, 400. But in multiplying 40 by 50, the product of the digits 4 and 5 is 20, a multiple of ten *(denarius)*.
28. A detailed, though general, description of the actual process of multiplication follows. No examples are provided.

[number]. In division, therefore, it must be noted that three numbers are necessary—the number to be divided [that is, the dividend] *(numerus dividendus)*, the dividing number, or divisor *(numerus dividens sive divisor)*, and the number denoting the quotient *(quotiens)*, or number that comes forth. If the division is to be made with whole numbers, the dividend ought always to be greater or at least equal to the divisor number.

Therefore if you wish to divide any number by itself or by another number, write the dividend number by [the order of] its differences[29] in the upper row; and write the divisor by [the order of] its differences in the lower row so that the last [figure] of the divisor would be under the last [figure] of the dividend; the next-to-last [figure] under the next-to-last, and so on, if it can be properly done.[30] For there are two reasons *(causae)* why a last [number] cannot be placed under a last [number]: (1) either because the last number of the lower row cannot be subtracted *(subtrahi)*[31] from the last number of the upper row because [the latter] is smaller than the [number in the] lower row;[32] or because (2) although the last [number] of the lower row could be subtracted from its [corresponding number in the] upper row, the remainder cannot be [subtracted] a certain number of times from the number placed above it; as if the last [number] was equal to the figure above it and the penultimate or antepenultimate [number] is greater. In such a case, the last figure of the divisor must be placed under the penultimate [or next-to-last number] of the dividend.[33]

After arranging the figures, the operation should begin from the last figure of the divisor number. One must see how many times this figure can be subtracted *(subtrahi)* from the figure placed above it; and if something should be left, [see how many times it can be subtracted] from its remainder. But it must be noted that one cannot subtract *(subtrahere)* more than nine times nor less than once. When it has been seen how many times the figures of the lower row could be subtracted from the numbers in the upper row, the number denoting this—that is, the quotient—must be written directly over that figure under which lies the first figure of the divisor, and that figure [indicates how often] the lower figures must be subtracted from their upper figures.

Now when this has been done, the figures of the divisor must be moved forward *(anteriorandae sunt)* toward the right by a single place or difference and the same procedure followed as before.[34]

However, should it happen that after the moving forward *(anteriorationem)* of the divisor, its last figure cannot be subtracted from the figure above it, then directly above the figure under which the first figure of the divisor lies, a cipher, [or zero,] must be written in the row of numbers denoting the quotient and the figures [of the divisor] must be moved forward as before. The same thing must be done wherever it happens that the divisor cannot be subtracted from the number to be divided, [that is,] a cipher must be put in the row of numbers denoting the quotient and the figures [of the divisor] must be moved forward *(anteriorandae sunt)* as before. . . .

29. That is, in the order of their place values.

30. If, for example, we wished to divide 15 into 2352, and recalling that as we face the paper the last figure lies to our extreme left and the first figure to our extreme right, Sacrobosco's arrangement for a division would be:

2352
15

31. The term "subtract" is consistently used for the operation of division. For example, to divide 20 into 63 was apparently conceived as an operation in which 20 is subtracted from 63 as many times as possible. Obviously it can be subtracted three successive times, with a remainder of 3.

32. For example, in dividing 25 into 1352 we cannot follow the arrangement given in note 30, that is,

1352
25

because 2 in the divisor exceeds 1 of the dividend. In this instance we would arrange the numbers in the form

1352
25

33. Illustrating this case would be the division of 1352 by 15. The last numbers are equal but the antepenultimate term of the divisor exceeds its corresponding term in the dividend. Therefore instead of 1352 we would have

15
1352
15

34. After explaining that many medieval Latin writers — including Sacrobosco—advanced their divisors one place to the right each time and called this *anterioratio,* David Eugene Smith *(History of Mathematics* [New York: Dover, 1958], II, 139*)* illustrates how 12 would be divided into 2852:

12	
493	
237	237
2852	8
12	12
12	
12	

The integral quotient is first placed above the dividend, and the complete quotient with fractions is placed to the extreme right (that is, 237 $\frac{8}{12}$). Above all, we observe the successive advancements of the divisor, 12, to the right. For other methods of division used in the Middle Ages, see Smith, pp. 128–144.

ON PROGRESSION

A progression is an aggregation of numbers taken according to equal excesses from unity or from two so that the sum of the whole or of the different numbers is had compendiously. Furthermore, some progressions are natural or continuous, others interrupted or discontinous. A progression is natural or continuous when it is taken from unity and no number is omitted in its ascent, as 1, 2, 3, 4, 5, 6, · · ·, and the number following always exceeds the preceding number by unity only. An interrupted progression occurs when some number is uniformly omitted, as 1, 3, 5, 7, 9, · · · · Similarly, it can begin from 2, as 2, 4, 6, 8, · · · · And the number following always exceeds the preceding by two units. It should be noted, moreover, that two rules are given concerning a natural progression.

The first is this: When a natural progression is terminated in an even number, multiply the next higher number by half [of the even number terminating the series] and you will obtain the total [of the series]. For example, 1, 2, 3, 4. Multiply five by two and you get ten, the sum of the whole progression. Whence the verse comes:

Even, you will multiply a greater by half of an
even number.[35]

The second rule is this: When a natural progression is terminated in an odd number, multiply it by the greater part of it and obtain the total sum. For example, 1, 2, 3, 4, 5. Let five be multiplied by three, and so, three times five results in fifteen, the sum of the whole progression. Whence the verse:

Let odd be multiplied by its greater part.[36]

As for an interrupted progression, two rules are likewise given:

The first rule is this: When an interrupted progression is terminated in an even number, you multiply its half by the next number higher than its half, as [for example,] 2, 4, 6. Let four be multiplied by three, and three times four will result in twelve, the sum of the whole progression. Whence the verse:

If even, multiply the [number] following by its
half.[37]

The second rule is this: When an interrupted progression is terminated in an odd number, multiply its greater part by itself. For example, 1, 3, 5. Let three be multiplied by itself; thus three taken three times produces nine, the sum of the whole progression. Whence the verse:

In odd numbers, the greater mean part is multiplied by itself.[38]

ON THE EXTRACTION OF ROOTS; FIRST IN SQUARE NUMBERS

What follows concerns the extraction of roots, first in square numbers. Therefore, we must see what a square number is; what is the root of a square number; and what it is to extract a root. However, this division must be noted beforehand: some numbers are linear, some superficial [or plane], and some solid.[39] A *linear number* is considered only according to process and not with regard to the multiplication of a number by a number; and it is called "linear" *(linearis)* because it has only a single number, just as a line has only a single dimension, namely length. A *superficial [or plane]* number arises from the multiplication of one number by another number; and it is called "superficial" *(superficialis)* because it has two numbers measuring it, just as a surface has two dimensions, namely length and width. But it must be understood that a number can be multiplied by another number in two ways: either once or twice. If a number is multiplied by another number once, it is either [multiplied by or] into itself or into another [number]; if into itself, it is a square number and is called "square" because written separately by units it will have four equal sides in the manner of a square. If it is multiplied into [or by] another [number], it would be a superficial and non-square number, as two multiplied by three constitutes six, a superficial but non-square number. Thus it is obvious that every square number is superficial, but not conversely.

Moreover, a root *(radix)* of a square number is that number which when multiplied by itself produces the square number, as two taken twice is four. Therefore, four is the first square number and its root is two. However, if a number is multiplied *(ducatur)* twice by a number, it constitutes a solid number; and it is called solid, since it has three dimensions just as a solid body, namely length, width, and depth. Thus this number has three

35. The Latin "verse" reads: "Par, paris media maiorem multiplicabis."
36. In Latin, the "verse" reads: "Impar parte sin maiori multiplicatur."
37. "Si par, per medium se multiplicato sequentem."
38. "Imparibus media pars maior multiplicat se."
39. A threefold subdivision derived from Boethius' *Arithmetic* and ultimately from the Greeks. It was used throughout the Middle Ages.

numbers producing it. But a number can be multiplied twice by a number in two ways: either by itself or by another [number]. If the number were multiplied by itself twice, or [multiplied] once by its square, which is the same thing, it becomes a cube number and is called cube from the name *cubus,* which is a solid. A cube is a certain body having six surfaces, as a die, eight [solid] angles, and twelve edges *(latera).* If, however, any number is multiplied twice by another, a solid and noncube number results, so that twice three twice constitutes twelve. And so it is obvious that every cube number is a solid, but not conversely.

From the aforesaid, it is also obvious that the same number is a root of a square and a cube [number], but the square and cube are not, however, the same. It is also obvious that every number can be the root of a square and cube, but not every number can be a square or cube [number]. Therefore, since the multiplication of a unit by itself once or twice produces nothing but a unit, Boethius says, in his *Arithmetic,* that the unit is every number potentially, but not in act.

It must be noted also that between any two successive square numbers there is a single mean proportional which is produced by the multiplication of one square root into [or by] the root of another; and between any two successive cube numbers there are two mean proportionals—a lesser and a greater. The lesser mean arises from the multiplication of the root of the greater cube [number] by the square of the [root] of the lesser [number]; the greater mean is produced if the root of the lesser cube [number] is multiplied into [or by] the square of the [root of the] lesser [number]. Since, in the present art, no process occurs beyond the sum of solid numbers, only the nine limits of numbers are distinguished. For there is a limit to the continuous sequence of numbers of the same nature contained by extreme terms; hence the first limit is the continuous progression of the nine digits; the second [limit] is that of the nine principal articles;[40] the third is [the limit] of hundreds; the fourth of thousands. Three limits also result in compound [numbers] by the application of digits to any of the three [limits mentioned before, namely articles, hundreds, and thousands]—[that is,] if one is placed before another.[41] But by the multiplication of a final term once beyond itself in the manner of making squares, or twice in the manner of making solids, the penultimate and final limits are produced.[42]

THE EXTRACTION OF THE SQUARE ROOT OF A NUMBER

To find the root of a square number is to find, with respect to some proposed number, the square root of it, if a square number was proposed. If it is not a square, to find the root of the greatest square contained under the proposed number. If, then, you wish to extract the root of some square number, write that number by its differences and count whether the number of figures [in that square number] is even or odd. If even, the operation is to begin under the penultimate [or next-to-last number]; if there is an odd number [of figures], begin from the last figure. To put it briefly, with an odd number [of figures] always begin from the last [of the figures]. Under the last figure[43] placed in the odd position, there must be found a certain digit which when multiplied by itself would offset [that is, equal] *(deleat)* the whole number located above it or come as close to it as possible. When such a number has been found and multiplied by itself, it must be subtracted from the number above it. This digit must then be doubled and placed under the next figure before it to the right and its half must be placed under it. After this has been done, a certain digit must be found under the next figure preceding the doubled number, which multiplied by the doubled number may offset [that is, equal] the whole number located above the doubled number; then this number multiplied by itself offsets, [or equals]—or comes as close as possible to—the whole number located above it . . . .[44]

40. That is, 10, 20, 30 ,40, · · · 90.

41. Thus a fifth limit would be 11, 12, 13, · · · 19; 21, 22, 23, · · · 29; 31, 32, 33, · · · 39; · · 91, 92, 93, · · · 99. Similar operations could be performed on hundreds and thousands.

42. For example, if the 2 in 2000 were repeatedly squared or cubed, the thousands would be increased to 4000; 16,000; and so on; or to 8000; 512,000; and so on. In this manner the numbers generated would carry beyond thousands ad infinitum.

43. That is, the figure on the extreme left.

44. That is, the quotient multiplied by the divisor is then subtracted from the dividend. This complicated verbal description is virtually unintelligible without a concrete example. Here is one furnished by Peter of Dacia in his commentary on Sacrobosco's *Algorism* (Curtze, pp. 81–83):

Find the square root of 9548198. Since 3 is the digit whose square is 9, we can eliminate 9 and proceed as follows:

$$548198$$
$$6$$
$$3$$

Here 6, the double of 3, the first number of the square

21 PROPOSITIONS FROM A THEORETICAL ARITHMETIC

Jordanus of Nemore (fl. 1230—1260)

Translated and annotated by Edward Grant[1]

[DEFINITIONS]

1. A unit is the separate existence of a thing by itself.

2. A number is a collective quantity of separate things.[2]

3. A series [or sequence] of numbers is called natural when its computation is based on the [successive] addition of a unit.[3]

4. That number by which a greater exceeds a lesser number is called the difference between the numbers.

5. Numbers are said to be equidistant (*equedistare*) from other numbers when their differences are equal.

6. A number is [said to be] multiplied by another when it is added to itself as many times as there is a unit in the multiplier; and what arises from the multiplication is called the product (*productus*).

7. A number is said to number [or measure] another when it produces it after being multiplied by a certain number.

8. A smaller number is part of a greater number when the smaller numbers [or measures] the greater; what is numbered is called a multiple of the numbering [or measuring] number.

9. A number denominates when a part of it is taken into its whole.

10. Parts that are denominated by the same number are said to be similar.[4]

11. The number by which another is divided is called a divisor.

12. The things into which a dividend is divided are called parts.

---

root we seek, is placed under 5, and 3 itself is placed under 6. We now seek a number to place under 4, but since 6 cannot be subtracted from 5, we put a zero under 8 and shift the 6 to the right, thus:

$$548198$$
$$60$$
$$30$$

Another appropriate number must now be found under 1. This number is 9 so that 609 multiplied by 9 produces 5481. At this point, Peter of Dacia presents the following:

$$5481$$
$$548198$$
$$609$$
$$30$$

After subtraction, it is put in this form:

$$98$$
$$609$$
$$30$$

Since 9 was the last figure found, it is now doubled and positioned first in this way:

$$98$$
$$6018$$
$$30 \ 9$$

and then after the carrying operation is performed, the numbers are presented this way:

$$98$$
$$618$$
$$309$$

Finally, since the trial divisor 618 is greater than 98, we must add 0, to obtain

$$98$$
$$6180$$
$$3090$$

Hence the square root of 9548198 is 3090, which when squared equals 9548100, leaving a remainder of 98. Thus, 3090 is the nearest integral square root of 9548198.

Since the remainder of this section describes steps already illustrated by this example, it has been omitted; also omitted is the final section on the extraction of cube roots.

1. This selection is translated from my own (unpublished) edition of part of the Latin text of the *Arithmetic* of Jordanus, based primarily upon two thirteenth-century manuscripts (Bibliothèque Nationale, fonds latin, 16644, and Vatican Latin, Ottobonian MSS, 2069). There does not seem to exist any printed edition of this important treatise, although the enunciations of the propositions were published in 1496 by Jacques Le Fèvre, who replaced Jordanus' proofs with his own (see Selection 28, n. 8). The reader should compare the highly formal and Euclidean character of this arithmetic treatise with Selection 2 from Boethius' *Arithmetic* and Selection 20 from Sacrobosco's *Algorism*.

2. We see that, in the Greek tradition, a "unit" or "unity" is not a number, but the latter is a collection of units.

3. This definition is identical with Book VII, Definition 3, of Campanus of Novara's later edition of Euclid's *Elements*. It should be noted that often enough Campanus' definitions, which purport to be those of Euclid, but which frequently differ from other Latin versions of the *Elements* and from the Greek text of Euclid, are either identical or virtually identical with the definitions in Jordanus' *Arithmetic* translated here. It is plausible to suppose that Campanus derived them from Jordanus. These numerous identities or similarities will not be indicated here.

4. If $m,q < n$, then m/n and q/n are called similar parts because each ratio is denominated by n.

13. The first and simple part of a number is a unit *(unitas)*.

14. When two numbers have a common part, [then] as many times as the same part is in the smaller number, just so many times is the smaller number said to be parts of the greater number.[5]

[POSTULATES]

There are three postulates *(petitiones)*:

1. That equals can be taken from any number.

2. That there can be some number greater than any number whatever by however much you please.

3. That the series of numbers can be extended to infinity.

[AXIOMS]

There are seven axioms *(communes animi conceptiones)*, which are these:

1. Every part is smaller which has a greater denomination.[6]

2. Any whatever equimultiples of the same [number] or of equal [numbers] will also be equal.

3. Those [numbers] would also be equal to which the same number is equimultiple or whose multiples are equal.

4. The unit is part of every number that is denominated by itself.

5. Any whole number contains the unit as many times as the unit is part of it.

6. If the unit were multiplied into any number, or the same number [were multiplied] into the unit, it would produce itself.

7. A difference of extremes is composed of differences of the same [extremes] to the mean [term].[7]

[PROPOSITIONS]

BOOK I

1. Every smaller number is either a part or parts of a greater number.

For a smaller number either numbers [or measures] a greater, or not. If it numbers it, it is part of it; [but] if it does not number it and there is some [number] which numbers them in common, that same number will be part of each of them and as often as it numbers the smaller number just so many times will the smaller number be parts of the greater number. However, this same thing could be shown by units since the unit is a part of every number.[8]

2. Every number is [equal to] half [the sum of] the numbers placed around it [on each side] and equidistant from it. And if it is half [the sum of] the numbers [on each side of it], it happens that those numbers are equidistant from it.

Let *A* be the number and let *B*, the greater number, and *C*, the smaller number, be placed around it and *D* be the common difference.[9] Furthermore, let *E*, whose difference with respect to *A* is *T*, be composed of *B* and *C*. Then, since the difference between *E* and *B* is *C*, [it follows] by the last axiom that *T* will consist of *D* and *C*. But since *A* also consists of the same [terms], *T* will equal *A*, so that *E* is double *A*. And if it is assumed [that *E* is double *A*], this will be the reason why *T* equals *A*. But since *T* consists of *C* and *D*, then *A* will consist of the same terms; therefore the difference between *A* and *C* will be *D*. Thus *B* and *C* will be equidistant [*equedistantes*] from *A*, and this is what is proclaimed *(et hoc est quod enunciatur)*.[10]

3. If two numbers have two numbers placed around them equidistantly [on each side], these [pairs of] numbers will be equal when added together; and if they are equal [when added together, the numbers placed around them] would be equidistant from the two given numbers.

The argument in this proposition is the same as

---

5. Essentially, if m and q are two given numbers, with $m < q$, and if n is their common denominator, then $(m:n)/(q:n) = m/q$. For $1/n$ is in m/n m times [that is, $(1/n) \cdot m = m/n$] and m is also m parts of q, that is, m/q.

6. Given $1/n$ and $1/m$, with $n > m$, it follows that $1/n < 1/m$.

7. If m, p, and q are three terms such that $m < p < q$, then if $q - m = h$, it is necessary that $h = (q - p) + (p - m)$.

8. If B and C are any two numbers with $B > C$, then either $B/C = P/1$, where P is an integer; or $B/C \neq P/1$, but some number $Q (\geq 1) < B$ and C is such that $B = QA$ and $C = QD$. Now, since $QD/QA = C/B$, we see that as $C/Q = D$ so does $D/A = C/B$.

9. That is, $D = B - A = A - C$.

10. By concluding the proof with "this is what is proclaimed," Jordanus supplies a Euclidean touch. In summary, the proof is as follows:

If
$$B > A > C \text{ and } D = B - A = A - C,$$

then
$$A = \frac{C + B}{2}; \text{ and conversely.}$$

Let $\quad B + C = E \quad$ and $\quad E - A = T.$

Since $\quad E - B = C,$

∴ $\quad T = D + C$ (by Axiom 7).

But $\quad A = D + C;$

∴ $\quad T = A \quad$ and $\quad E = 2A.$

[and if $E = 2A$, then $T = A$].

But since $\quad T = C + D$ and $A = C + D,$

∴ $\quad A - C = D \quad$ and $\quad B - A = A - C.$

in the one preceding. Thus, let *A* and *B* lie between *C* and *D*, and *A* is greater [than *B*] and *C* is the greatest number, and let *E* be composed of *C* and *D*, the common difference be *T*, and let the difference between *E* and *A* be *Z*. Then since *D* is the difference between *E* and *C*, *Z* will consist of *D* and *T*. But since *B* consists of the same terms, it will equal *Z*, and *E* will consist of *A* and *B*.[11] If the consequent were assumed conversely, [then] since *Z* equals *B*, *B* exceeds *D* by as much as *E* exceeds *A*, that is by *T*.[12] And so the proposition is proved.

4. If a first term is as much a part of a second term, as a third is of a fourth, the first and third terms will be the [same] whole part of the second and fourth terms as the first term is of the second.[13]

Let the first term be *A*, the second *B*, the third *C*, the fourth *D*. And since *A* is the whole part of *B* as *D* is the part of *C*, *B* and *D* are understood to have been divided into similar [that is, an equal number of] parts. But the first of one [pair of terms added] to the first of the other [pair of terms] is as *A* and *C*; similarly the second with the second [is as *B* and *D*]. And since this conjunction [of terms] can be made as often as the first term is contained in the second, it follows that the number equal to [the sum of] *A* and *C* could be taken into [that is, divided into the sum of] *B* and *D* as many times as *A* into *B*. And thus the proposition is proved.

5. If a first term be as many parts of a second term, as a third is of a fourth, the first and third terms will be as many whole parts of the second and fourth as the first is of the second.[14].

As before, let *A* be the first term, *B* the second, *C* the third, and *D* the fourth. And let *E* parts of *A* be taken which are denominated in *B* by *F*; and let one of these parts be *G*;[15] and also let *H* be one of the parts of *C*; and so what is composed of *G* and *H* will, by what has been said before, be part of [the sum of] *B* and *D* denominated by *F* [that is, taken *F* times] and [the sum of *G* and *H*] will be part of [the sum of] *A* and *C* taken *E* times. From all this, it is obvious that [the sum of] *A* and *C* will be as many parts of [the sum of] *B* and *D* as *A* would be of *B*; and this is what we proposed.

6. If there are numbers, however many, that are equimultiple to just as many other [numbers], [then] the number composed of these [equimultiples] will also be equimultiple to the number composed of [all] the other numbers.

For what has been composed of two [numbers] will be equimultiple to what has been composed of two numbers, just as, by the proposition set forth before [that is, Proposition 4], the first [number] is

equimultiple to the first [number in each series]. The same argument will hold for the third number [in the sequence] after it has been added to the term composed of the first [sequence of] numbers. The demonstration is obvious, for, in a similar way, this same argument always applies to the next term when it has been added to the sum of the preceding terms until the last term has been computed.[16]

7. If a unit measures[17] *[or is a part of] a first term as many times as a second term measures [or is a part of] a third term, then as many times as the unit will measure [or be a part of] the second term just so*

11.

Thus

$$C > A > B > D,$$
$$E = C + D,$$
$$C - A = A - B = B - D = T.$$

Assume $Z = E - A.$
Then since $D = E - C,$
∴ $Z = D + T$ (by substituting for *E* and then for *C − A*).

But since $B = D + T$ (from *B − D = T*),
∴ $B = Z.$
And since $Z = E - A,$
∴ $B = E - A.$
And ∴. $E = A + B$ or $C + D$
                   $= A + B.$ Q.E.D.

12. The converse of the consequent was stated as the second part of the enunciation. That is, if $C + D = A + B$, then $C - A = B - D$. Thus, since $Z = B$, where $Z = E - A$ and $B = D + T$, it follows that $T = B - D$, so that $B - D = E - A$, or $T = Z$.

13. That is, if $A = (1/n)B$ and $C = (1/n)D$ then $A + C = (1/n)(B + D)$. Though the proofs differ, this enunciation is basically the same as Euclid's *Elements* VII.5. Heath observes that "this proposition is of course true for any quantity of pairs of numbers similarly related . . ." (*The Thirteen Books of Euclid's Elements*, 2d. ed. [New York: Dover, 1956], II, 304).

14. Let *m* and *n* be integers so that $n > m > 1$. Then if $A = (m/n)B$ and $C = (m/n)D$, it can be shown that $A + C = (m + n)(B + D)$. This proposition is the same as the preceding one and covers the cases involving *m/n* parts (instead of a single part $1/n$). It is much the same as Euclid's *Elements* VII.6.

15. This complicated verbalization reduces to the following: $A = (E/F)B$, where $G = 1/F$.

16. Let us assume that *D, E, F . . .* are equimultiples to *A, B, C, . . .*, in that order. Then if *A* is the first term, *D* the second, *B* the third, *E* the fourth, and so on, then by Proposition 4,

$$A + B = \frac{1}{n}(D + E) \qquad \text{since } A = \frac{1}{n}D.$$

And if $A + B = Q$ and $D + E = R$,

Then $Q + C = \frac{1}{n}(R + F)$... (again by Proposition 4).

17. Although the Latin equivalent of the term "measure" does not appear in the enunciation, this term has been used here to translate "in." Thus instead of "if a unit is in," I have chosen the alternative given in the text.

many times will the first term measure [or be a part of] the third term.[18]

Let the first term be divided into units and the third term divided into just as many [parts], each of which is equal to the second term.[19] But the first unit of the first [term] is as much a whole part of the first of these [terms] as the second term is of the second term.[20] Moreover, by what has been said before *(per premissam)*, the first [term] will be a whole part of the third [term] as many times as the unit [will be part] of one of those [parts of the third term] which is equal to the second [term].[21] Therefore, as many times as the unit measures [or is a part of] the second [term, just so many times will] the first whole [term] be in the third [term].[22]

8. If an alternate multiplication of two numbers is performed, the same number would arise in each case.[23]

By definition *(ex descriptione)*, as many times as the unit goes into one number just so many times does the remaining number go into the product. Therefore, by what has been said before [in Proposition 7], as often as the unit goes into the remaining number, just so many times will the first [of the two numbers] go into the product. You then argue from the conversion of the definition.[24]

9. The [total sum or] result of the multiplication of any number by however many numbers you please is equal to (est quantum) the result of the multiplication of the same number by the number composed of all the others.

Let A be the number multiplied by B and C to produce D and E [respectively]. I say that the composite [or sum] of D and E is produced by multiplying A into the composite of B and C. For it is obvious by Definition [7] that B measures *(numerat)* D A times and that C measures E by the same number, namely A times. By the sixth proposition of this book, you will easily be able to argue this.[25]

10. The [total sum or] result of the multiplication of as many numbers as you please by any [single] number is equal to the result of the multiplication of the number composed of all those numbers by that same [single] number.[26]

By what has already been set forth [in Proposition 8], the result of the multiplication of [each of] those numbers by that [single number] is equal to the result of the multiplication of that single number into each of those numbers. And thus the result of the multiplication of the number that is composed [or is the sum] of those numbers by that [single number] equals the result [derived] from

[the multiplication of] that single number by the composed number. The argument is obvious by what has already been set forth.[27]

11. The results of the multiplication of any numbers whatever into any numbers whatever is equal to the results of the multiplication of the sum (ex composito) [of the one group] by the sum of the other [group].

By what has been said before, if the result of the multiplication of all the numbers [of one group] by the first number of the other [group] is equal to the result of the multiplication of the sum of the same numbers by that same first number, then similarly this will follow for the second [number], and the third, and so on to the last number. Then by what was said before the result of the multiplication of the same sum [of the numbers of the one group] by each [number of the other group] is equal to the

18. This proposition is equivalent to *Elements* VII.15 (VII. 16 in Campanus of Novara's medieval edition of Euclid) and like the latter is a special case of *Elements* VII. 9. It asserts that if four terms 1, A, B, and C are assumed, then, if $1/A = B/C$, it can be shown that $1/B = A/C$.

19. That is, assume that $A/1 = n$ and $C/B = n$.

20. That is, 1, or A/n_1, is the first of n equal parts of the first term A, and C/n_1 is the first of n equal parts of C. Therefore, $(A:n_1)/(C:n_1) = (A:n_2)/(C:n_2)$.

21. This compressed and obscure verbalization declares that $A/C = 1/(C:n)$, where $C/n = B$.

22. That is, $A/C = 1/B$, where B is substituted for C/n in the preceding note.

23. That is, if $A \cdot B = C$, then $B \cdot A = C$.

24. Apparently appealing to Definition 6, Jordanus asserts that if $A \cdot B = C$, then $B/1 = C/A$. Therefore, by Proposition 7, $A/1 = C/B$. Hence if $B/1 = C/A$ produces $A \cdot B = C$, then $A/1 = C/B$ will produce $B \cdot A = C$. Jordanus justifies the last step of this commutation law of multiplication by "conversion of the definition"— namely Definition 6.

25. Jordanus proves that if $A \cdot B = D$, and $A \cdot C = E$, then $D + E = A(B + C)$. By Definition 7, $D/B = A$ and $E/C = A$. And since D and E are equimultiples of B and C, respectively, then by Proposition 6 it follows that $B + C = \left(\frac{1}{n}\right)(D + E)$, and assuming $n = A$, we obtain $A(B + C) = D + E$. I have found no counterpart to this proposition in the Greek text of Euclid or in the medieval edition by Campanus of Novara.

26. That is, $ZA = AZ$, where A represents a single number and Z the composite number such that $Z = B + C + D + E + \cdots + N$.

27. Using the data of note 26, we conclude that by Proposition 8,
$$BA + CA + DA + EA + \cdots + NA = AB + AC + AD + AE + \cdots + AN.$$
Thus
$$(B + C + D + E + \cdots + N)A = A(B + C + D + E + \cdots + N).$$
Therefore, $ZA = AZ$ (see note 26).

result of the multiplication of the same sum by the sum of the other group of numbers.[28] And so the proposition is obvious.

12. Any number that measures (numerat) a whole and what is subtracted from that whole [also] measures the remainder.

Thus if *C* should measure *AB* [that is, *A* + *B*, the whole] and *A*, it will also number *B* [the remainder]. Now let *C* number *AB* by *DE* [that, is *D* + *E* times] and *A* by *D* [that is, *D* times].[29] Since by the ninth proposition, *DE* multiplied by *C* equals *D* multiplied by *C*, and *E* multiplied by *C*,[30] it follows that *A* added to the result [or product] of *E* multiplied by *C* equals *AB* [that is *A* + *B*].[31] Moreover, by an axiom *(per conceptionem)*, *B* will equal the result [or product] of *E* multiplied into *C*.[32] Therefore, since *C* measures one [of the two terms, namely *A*], it will also measure the one remaining, [*B*].[33]

13. The result of multiplying a number by itself is equal to the result of multiplying the same [number] by all its parts.

Let the same number be posited twice, once divided and once undivided. By the ninth [proposition], the multiplication of one by the other is equal to the number that has been divided [multiplied] by all its parts. And the proposition is argued this way.[34]

14. When a number has been divided in two, the result [or product] of the multiplication of the whole number into one of the two numbers is equal to the result of the multiplication of the same [one of those two numbers] by itself and by the remainder.[35]

For by the eighth proposition, the result of the multiplication of the whole number by one [of the two numbers] is the same as the result of the multiplication of that same number by the whole number. Therefore, when the number has been divided twice, you could derive the argument by the ninth [proposition].[36]

Algebra

22 SIX TYPES OF RHETORICAL ALGEBRAIC EQUATIONS

Al-Khwārizmī[1] (fl. 813—833)

Translated and annotated by Louis C. Karpinski[2]

The Book of Algebra and Almucabola[3]
Containing Demonstrations of the Rules of the Equations of Algebra

28. Let us assume one set of numbers $A, B, C, D, \cdots,$ M, such that $A + B + C + D + \cdots + M = N$; and assume another set of numbers $P, Q, R, S, \cdots, T$, such that $P + Q + R + S + \cdots + T = Z$. Now, Jordanus says that if

$$PA + PB + PC + PD + \cdots + PM$$
$$= P(A + B + C + D + \cdots + M),$$

then similarly

$$QA + QB + QC + QD + \cdots + QM$$
$$= Q(A + B + C + D + \cdots + M);$$
$$RA + RB + RC + RD + \cdots + RM$$
$$= R(A + B + C + D + \cdots = M);$$
$$SA + SB + SC + SD + \cdots + SM$$
$$= S(A + B + C + D + \cdots + M);$$
$$TA + TB + TC + TD + \cdots + TM$$
$$= T(A + B + C + D + \cdots + M).$$

Therefore,

$$(P + Q + R + S + \cdots + T)$$
$$(A + B + C + D + \cdots + M) = NZ.$$

29. That is, $(A + B)/C = D + E$, and $A/C = D$.

30. That is, $C(D + E) = CD + CE$.

31. Since $C(D + E) = DC + EC$, and $(A + B)/C = D + E$, we see that $A + B = DC + EC$; and by assumption $A/C = D$ and therefore, $A = CD$. Therefore, $A + B = A + EC$.

32. Whether Jordanus had Axiom 3 in mind is not clear, but it is obvious that since $A + EC = A + B$, then $B = EC$.

33. *C* measures *A* by assumption, since $A/C = D$. Now it is shown that since $B = EC$ (see n. 32), *C* measures *B* *E* times, that is, $B/C = E$.

34. Let *A* and *A* be the two numbers and let $A = C + D + E$ (here *A* is divided) and *A* be left undivided. By Proposition 9, $AC + AD + AE = A(C + D + E)$. But $C + D + E = A$, and $AC + AD + AE = A \cdot A$. This appears to be a special case of Euclid's *Elements* II.1, where the two magnitudes multiplied may be equal or unequal.

35. That is, if $(A + B)$ is the whole number that is divided into *A* and *B*, Jordanus proves that $(A + B)A = A \cdot A + AB$; or $(A + B)B = AB + B \cdot B$.

36. By Proposition 8, $(A + B)A = A(A + B)$; and by Proposition 9, $(A + B)A = A \cdot A + AB$.

1. [In the text and notes Karpinski spells the name as Al-Khowarizmi. I have used the more usual form—*Ed.*]

2. [Reprinted by permission of the University of Michigan from Louis Charles Karpinski and John Garrett Winter, *Contributions to the History of Science, Part I: Robert of Chester's Latin Translation of the Algebra of al-Khowarizmi,* in *University of Michigan Studies,* Humanistic Series (Ann Arbor, 1930), XI, 67–83, 107. The copyright date of Part I is 1916. This treatise and that by Abu-Kamil were the two basic algebraic treatises available in Latin from Arabic sources.

Written some time ago in Arabic by an unknown author and afterwards, according to tradition in 1183,[4] put into Latin by Robert of Chester in the city of Segovia.

The Book of Algebra and Almucabola, concerning arithmetical and geometrical problems.

In the name of God, tender and compassionate, begins the book of Restoration and Opposition of number put forth by Mohammed al-Khwārizmī, the son of Moses.[5] Mohammed said, Praise God the creator who has bestowed upon man the power to discover the significance of numbers. Indeed, reflecting that all things which men need require computation, I discovered that all things involve number and I discovered that number is nothing other than that which is composed of units. Unity therefore is implied in every number. Moreover I discovered all numbers to be so arranged that they proceed from unity up to ten. The number ten is treated in the same manner as the unit, and for this reason doubled and tripled just as in the case of unity. Out of its duplication arises 20, and from its triplication 30. And so multiplying the number ten you arrive at one-hundred. Again the number one-hundred is doubled and tripled like the number ten. So by doubling and tripling etc. the number one-hundred grows to one-thousand. In this way multiplying the number one-thousand according to the various denominations of numbers you come even to the investigation of number to infinity.

Furthermore I discovered that the numbers of restoration and opposition are composed of these three kinds: namely, roots, squares[6] and numbers. However number alone is connected neither with roots nor with squares by any ratio. Of these then the root is anything composed of units which can be multiplied by itself, or any number greater than unity multiplied by itself: or that which is found to be diminished below unity when multiplied by itself. The square is that which results from the multiplication of a root by itself.

Of these three forms, then, two may be equal to each other, as for example:

> Squares equal to roots,
> Squares equal to numbers, and
> Roots equal to numbers.[7]

CHAPTER 1
CONCERNING SQUARES EQUAL TO ROOTS[8]

The following is an example of squares equal to roots: a square is equal to 5 roots. The root of the square then is 5, and 25 forms its square which, of course, equals five of its roots.[9]

Another example: the third part of a square equals four roots. Then the root of the square is 12 and 144 designates its square.[10] And similarly, five squares equal 10 roots. Therefore one square equals two roots and the root of the square is 2. Four represents the square.[11]

In the same manner then that which involves more than one square, or is less than one, is reduced to one square. Likewise you perform the same operation upon the roots which accompany the squares.

CHAPTER 2
CONCERNING SQUARES EQUAL TO NUMBERS

Squares equal to numbers are illustrated in the following manner: a square is equal to nine. Then nine measures the square of which three represents one root.[12]

(For an English translation of Abu-Kamil's *Algebra* made from a Hebrew version, see *The Algebra of Abu-Kamil,* Hebrew Text, Translation, and Commentary. . . by Martin Levey [Madison, Wis.: University of Wisconsin Press, 1966]). What has been reproduced here from al-Khwārizmī's *Algebra* should be compared with Jordanus de Nemore's algebraic treatise (see the next selection).

It should be noted that for convenience in his translation Karpinski has used x and x^2 for the unknown and the square of the unknown. Such letter symbols were not employed by al-Khwārizmī, whose treatise is wholly rhetorical.—*Ed.*]

3. Algebra and almucabola are transliterations of Arabic words meaning "the restoration," or "making whole," and "the opposition," or "balancing." The first refers to the transference of negative terms and the second to the combination of like terms which occur in both members or to the combination of like terms in the same member.

4. The date is given in the Spanish Era, A.D. 1145, according to our reckoning.

5. Mohammed ibn Musa, al-Khwārizmī. The algorism is derived from his patronymic; the spelling and use in the Latin indicate the process of evolution, although the term came into use through al-Khwārizmī's arithmetic and not his algebra.

6. Literally "substances," being a translation of the Arabic word *mal,* used for the second power of the unknown. Gerard of Cremona used *census,* which has a similar meaning.

7. These are the three types designated as "simple" by Omar Al-Khayyami, Al-Karkhi, and Leonard of Pisa. They correspond in modern algebraic notation to the following: $ax^2 = bx$; $ax^2 = n$; and $bx = n$.

8. The chapter headings were not supplied by al-Khwārizmī.

9. $x^2 = 5x$; $x = 5$; $x^2 = 25$.

10. $\frac{1}{3}x^2 = 4x$; $x = 12$; $x^2 = 144$.

11. $5x = 10$; $x = 2$; $x^2 = 4$.

12. $x^2 = 9$; $x = 3$.

Whether there are many or few squares they will have to be reduced in the same manner to the form of one square. That is to say, if there are two or three or four squares, or even more, the equation formed by them with their roots is to be reduced to the form of one square with its root. Further if there be less than one square, that is if a third or a fourth or a fifth part of a square or root is proposed, this is treated in the same manner.[13]

For example, five squares equal 80. Therefore one square equals the fifth part of the number 80 which, of course, is 16.[14] Or, to take another example, half of a square equals 18. This square therefore equals 36.[15] In like manner all squares, however many, are reduced to one square, or what is less than one is reduced to one square. The same operation must be performed upon the numbers which accompany the squares.

CHAPTER 3
CONCERNING ROOTS EQUAL TO NUMBERS

The following is an example of roots equal to numbers: a root is equal to 3. Therefore nine is the square of this root.[16]

Another example: four roots equal 20. Therefore one root of this square is 5.[17] Still another example: half a root is equal to ten. The whole root therefore equals 20, of which, of course, 400 represents the square.[18]

Therefore roots and squares and pure numbers are, as we have shown, distinguished from one another. Whence also from these three kinds which we have just explained, three distinct types of equations are formed involving three elements, as

A square and roots equal to numbers,
A square and numbers equal to roots, and
Roots and numbers equal to a square.[19]

CHAPTER 4
CONCERNING SQUARES AND ROOTS EQUAL TO NUMBERS

The following is an example of squares and roots equal to numbers: a square and 10 roots are equal to 39 units. The question therefore in this type of equation is about as follows: what is the square which combined with ten of its roots will give a sum total of 39? The manner of solving this type of equation is to take one-half of the roots just mentioned. Now the roots in the problem before us are 10. Therefore take 5, which multiplied by itself gives 25, an amount which you add to 39, giving 64. Having taken then the square root of

this which is 8, subtract from it the half of the roots, 5 leaving 3. The number three therefore represents one root of this square, which itself, of course, is 9. Nine therefore gives that square[20]

Similarly however many squares are proposed all are to be reduced to one square. Similarly also you may reduce whatever numbers or roots accompany them in the same way in which you have reduced the squares.

The following is an example of this reduction: two squares and ten roots equal 48 units.[21] The question therefore in this type of equation is something like this: what are the two squares which when combined are such that if ten roots of them are added, the sum total equals 48? First of all it is necessary that the two squares be reduced to one. But since one square is the half of two, it is at once evident that you should divide by two all the given terms in this problem. This gives a square and 5 roots equal to 24 units. The meaning of this is about as follows: what is the square which amounts to 24 when you add to it 5 of its roots? At the outset it is necessary, recalling the rule above given, that you take one-half of the roots. This gives two and one-half, which multiplied by itself gives $6\frac{1}{4}$. Add this to 24, giving $30\frac{1}{4}$. Take then of this total the square root, which is, of course, $5\frac{1}{2}$. From this subtract half of the roots, $2\frac{1}{2}$, leaving 3, which expresses one root of the square, which itself is 9.

Another possible example: half a square and five roots are equal to 28 units.[22] The import of this problem is something like this: what is the square

13. Our modern expression "to complete the square," used in algebra, originally meant to make the coefficient of x^2 equal to unity, i.e., make one whole square.
14. $5x^2 = 80$; $x^2 = 16$.
15. $\frac{1}{2}x^2 = 18$; $x^2 = 36$.
16. $x = 3$; $x^2 = 9$.
17. $4x = 20$; $x = 5$; $x^2 = 25$.
18. $\frac{1}{2}x = 10$; $x = 20$; $x^2 = 400$.
19. Abu Kamil, Omar Al-Khayyami, Al-Karkhi, and Leonard designate these as 'composite' types. In modern notation: $ax^2 + bx = n$; $ax^2 + n = bx$; $ax^2 = bx + n$.
20. $x^2 + 10x = 39$; $\frac{1}{2}$ of 10 is 5; 5^2 is 25; $25 + 39 = 64$. $\sqrt{64} = 8$; $8-5 = 3$. $x = 3$; $x^2 = 9$. For the general type $x^2 + bx = n$, the solution is $x = \sqrt{(b/2)^2 + n} - (b/2)$; the negative value of the square root is neglected, as that would give a negative root of the equation.
21. $2x^2 + 10x = 48$, reducing to $x^2 + 5x = 24$; $\frac{1}{2}$ of 5 is $2\frac{1}{2}$; $(2\frac{1}{2})^2 = 6\frac{1}{4}$; $24 + 6\frac{1}{4} = 30\frac{1}{4}$; $\sqrt{30\frac{1}{4}} - 2\frac{1}{2} = 3$. The general type $ax^2 + bx = n$ is reduced to the preceding by division, giving $x^2 + (b/a)x = n/a$, and the solution is, as before, $x = \sqrt{(b/2a)^2 + n/a} - (b/2a)$.
22. $\frac{1}{2}x^2 + 5x = 28$, reducing to $x^2 + 10x = 56$. $x = \sqrt{5^2 + 56} - 5$, or $x = 4$. Note that the value of x^2 is not given here as it usually is.

which is such that when to its half you add five of its roots the sum total amounts to 28? Now however it is necessary that the square, which here is given as less than a whole square, should be completed.[23] Therefore the half of this square together with the roots which accompany it must be doubled. We have then, a square and 10 roots equal to 56 units. Therefore take one-half of the roots, giving 5, which multiplied by itself produces 25. Add this to 56, making 81. Extract the square root of this total, which gives 9, and from this subtract half of the roots, 5, leaving 4 as the root of the square.

In this manner you should perform the same operation upon all squares, however many of them there are, and also upon the roots and the units.

CHAPTER 5
CONCERNING SQUARES AND NUMBERS EQUAL TO ROOTS

The following is an illustration of this type: a square and 21 units equal 10 roots.[24] The rule for the investigation of this type of equation is as follows: what is the square which is such that when you add 21 units the sum total equals 10 roots of that square? The solution of this type of problem is obtained in the following manner. You take first one half of the roots, giving in this instance 5, which multiplied by itself gives 25. From 25 subtract the 21 units to which we have just referred in connection with the squares. This gives 4, of which you extract the square root, which is 2. From the half of the roots, or 5, you take 2 away, and 3 remains, constituting one root of this square which itself is, of course, 9.[25]

If you wish you may add to the half of the roots, namely 5, the same 2 wihch you have just subtracted from the half of the roots. This gives 7, which stands for one root of the square, and 49 completes the square.[26] Therefore when any problem of this type is proposed to you, try the solution of it by addition as we have said. If you do not solve it by addition, without doubt you will find it by subtraction. And indeed this type alone requires both addition and subtraction, and this you do not find at all in the preceding types.

You ought to understand also that when you take the half of the roots in this form of equation and then multiply the half by itself, if that which proceeds or results from the multiplication is less than the units above-mentioned as accompanying the square, you have no equation.[27] If equal

to the units, it follows that a root of the square will be the same as the half of the roots which accompany the square, without either addition or diminution.[28] Whenever a problem is proposed that involves two squares, or more or less than a single square, reduce to one square just as we have indicated in the first chapter.

CHAPTER 6
CONCERNING ROOTS AND NUMBERS EQUAL TO A SQUARE

An example of this type is proposed as follows: three roots and the number four are equal to a square.[29] The rule for the investigation of this kind of problem is, you see, that you take half of the roots, giving one and one-half; this you multiply by itself, producing $2\frac{1}{4}$. To $2\frac{1}{4}$ add 4, giving $6\frac{1}{4}$, of which you then take the square root, that is, $2\frac{1}{2}$. To $2\frac{1}{2}$ you now add the half of the roots, or $1\frac{1}{2}$, giving 4, which indicates one root of the square. Then 16 completes the square.[30] Now also whatever is proposed to you either more or less than a square, reduce to one square.

Now of the types of equations which we mentioned in the beginning of this book, the first three are such that the roots are not halved, while in the following or remaining three, the roots are halved, as appears above.

23. Attention is called to the force of the expression, "completing the square," as here used with the meaning to make the coefficient of the second power of the unknown quantity equal to unity or making one whole square.

24. $x^2 + 21 = 10x$. For this type of equation both solutions are presented, since both roots are positive. A negative number would not be accepted as a solution by the Arabs of this time, nor indeed was it fully accepted until the time of Descartes.

25. For the general type, $x^2 + n = bx$, the solution is $x = (b/2) \pm \sqrt{(b/2)^2 - n}$, and both positive and negative values of the radical give positive solutions of the equation proposed. In this problem we have $x^2 + 21 = 10x$; $\frac{1}{2}$ of 10 is 5; 5^2 is 25; $25 - 21 = 4$; $\sqrt{4} = 2$; $5 - 2 = 3$, one root; $5 + 2 = 7$, the other root.

26. Another use of the expression, "completing the square." [This expression translates *quam substantiam* 49 *adimplent.—Ed.*]

27. This corresponds to the condition, $b^2 - 4ac < 0$, in the equation $ax^2 + bx + c = 0$; in this event the roots are imaginary.

28. Condition for equal roots, $b^2 - 4ac = 0$.

29. $3x + 4 = x^2$; $\frac{1}{2}$ of 3 is $1\frac{1}{2}$; $(1\frac{1}{2})^2 = 2\frac{1}{4}$; $2\frac{1}{4} + 4 = 6\frac{1}{4}$; $\sqrt{6\frac{1}{4}} = 2\frac{1}{2}$; $2\frac{1}{2} + 1\frac{1}{2} = 4$, the root.

30. The solution of the general type $bx + n = ax^2$, reduced by division to $(b/a)x + n/a = x^2$, is $x = \sqrt{(b/2a)^2 + n/a} + b/2a$, and only the positive value of the radical is taken, since the negative value would give a negative root of the proposed equation.

Geometrical Demonstrations

We have said enough, says al-Khwārizmī, so far as numbers are concerned, about the six types of equations. Now, however, it is necessary that we should demonstrate geometrically the truth of the same problems which we have explained in numbers. Therefore our first proposition is this, that a square and 10 roots equal 39 units.

The proof is that we construct a square of unknown sides, and let this square figure [Fig. 1]

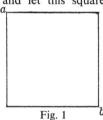

Fig. 1

represent the square (second power of the unknown) which together with its root you wish to find. Let the square, then, be *a b*, of which any side represents one root. When we multiply any side of this by a number (of numbers)[31] it is evident that that which results from the multiplication will be a number of roots equal to the root of the same number (of the square). Since then ten roots were proposed with the square, we take a fourth part of the number ten and apply to each side of the square an area of equidistant sides, of which the length should be the same as the length of the square first described and the breadth $2\frac{1}{2}$, which is a fourth part of 10. Therefore four areas of equidistant sides are applied to the first square, *a b* [Fig. 2]. Of each of

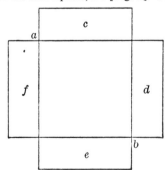

Fig. 2

these the length is the length of one root of the square *a b* and also the breadth of each is $2\frac{1}{2}$, as we have just said. These now are the areas, *c, d, e, f*. Therefore it follows from what we have said that there will be four areas having sides of unequal length, which also are regarded as unknown. The size of the areas in each of the four corners, which is found by multiplying $2\frac{1}{2}$ by $2\frac{1}{2}$, completes that which is lacking in the larger or whole area. Whence

it is that we complete the drawing of the larger area by the addition of the four products, each $2\frac{1}{2}$ by $2\frac{1}{2}$; the whole of this multiplication gives 25.

And now it is evident that the first square figure, which represents the square of the unknown (x^2), and the four surrounding areas (10 x) make 39. When we add 25 to this, that is, the four smaller squares which indeed are placed at the four angles of the square *a b*, the drawing of the larger square, called *G H*, is completed [Fig. 3]. Whence also the

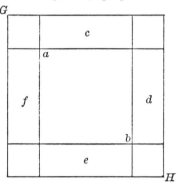

Fig. 3

sum total of this is 64, of which 8 is the root, and by this is designated one side of the completed figure. Therefore when we subtract from eight twice the fourth part of 10, which is placed at the extremities of the larger square *G H*, there will remain but 3. Five being subtracted from 8, 3 necessarily remains, which is equal to one side of the first square *a b*.[32]

This three then expresses one root of the square figure, that is, one root of the proposed square of the unknown, and 9 the square itself. Hence we take half of ten and multiply this by itself. We then add the whole product of the multiplication to 39, that the drawing of the larger square *G H* may be completed;[33] for the lack of the four corners rendered incomplete the drawing of the whole of this square. Now it is evident that the fourth part of any number multiplied by itself and then multiplied by four gives the same number as half of the number multiplied by itself.[34] Therefore if half of the roots is multiplied by itself, the sum total of this multiplication will wipe out, equal or cancel the multi-

31. Evidently meaning a pure number.
32. The proportions of the figures are not correct to scale.
33. This corresponds to our algebraic process of completing the square. The correspondence of the geometrical procedure to the terminology and the methods employed in algebra make it highly desirable to present the geometrical and algebraical discussions together to students of elementary mathematics.
34. $4(a/4)^2 = (a/2)^2$.

plication of the fourth part by itself and then by four.

Another method[35] also of demonstrating the same is given in this manner: to the square $a\,b$ representing the square of the unknown we add ten roots and then take half of these roots, giving 5. From this we construct two areas added to two sides of the square figure $a\,b$. These again are called $a\,g$ and $b\,d$ [Fig. 4]. The breadth of each is equal to

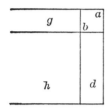

Fig. 4

the breadth of one side of the square $a\,b$ and each length is equal to 5. We have now to complete the square by the product of 5 and 5, which, representing the half of the roots, we add to the two sides of the first square figure, which represents the second power of the unknown. Whence it now appears that the two areas which we joined to the two sides, representing ten roots, together with the first square, representing x^2,[5] equals 39. Furthermore it is evident that the area of the larger or whole square is formed by the addition of the product of 5 by 5. This square is completed and for its completion 25

is added to 39. The sum total is 64. Now we take the square root of this, representing one side of the larger square and then we subtract from it the equal of that which we added, namely 5. Three remains, which proves to be one side of the square $a\,b$, that is, one root of the proposed x^2. Therefore three is the root of this x^2, and x^2 is 9.

. . . . .

Fourth Problem, Illustrating the Fourth Chapter

Multiply $\frac{1}{3}x$ and one unit by $\frac{1}{4}x$ and one unit so as to give as the product 20.[36]

Explanation. You multiply $\frac{1}{3}x$ by $\frac{1}{4}x$, giving $\frac{1}{2}$ of $\frac{1}{6}x^2$, and a unit multiplied by $\frac{1}{4}x$ gives $\frac{1}{4}x$ to be added. Similarly $\frac{1}{3}x$ multiplied by a unit gives $\frac{1}{3}x$ and then a unit by a unit gives a unit. Then this multiplication amounts to $\frac{1}{2}$ of $\frac{1}{6}x^2$, and $\frac{1}{3}x$ and $\frac{1}{4}x$ and one a unit, equal to 20 units. You subtract one unit from 20 units, giving 19 units equal to $\frac{1}{2}$ of $\frac{1}{6}x^2$ together with $\frac{1}{3}x$ and $\frac{1}{4}x$. Now then you complete the square,[37] *i.e.* you multiply throughout by 12. This gives x^2 and $7x$ equal to 228. Then halve the roots, *i.e.* divide them equally, and multiply one-half by itself, giving $12\frac{1}{4}$. You add this to 228, and you will have $240\frac{1}{4}$. From the root of this, $15\frac{1}{2}$, subtract $3\frac{1}{2}$, leaving 12 as the root of the square. Now then this problem has led you to the fourth of the six chapters in which we treated the type, a square and roots equal to numbers.

23   ALGEBRAIC PROPOSITIONS FROM THE TREATISE *ON GIVEN NUMBERS*

Jordanus of Nemore (fl. 1230—1260)

Translated and annotated by Edward Grant[1]

[DEFINITIONS]

1. A given number (*numerus datus*) is one whose quantity is known.

35. A method slightly different from either of these is given by Abu Kamil and also in the Boncompagni version of al-Khwārizmi's algebra, ascribed to Gerard of Cremona. This consists in applying to one side of the square a rectangle with its length equal to 10, while the other dimension is the same as that of the square. The two together represent $x^2 + 10x$, or 39. Bisect the side whose length is 10. Now by Euclid II.6, the square on half the side 10 plus the side of the original square ($x + 5$)^2, equals the whole rectangle (39) plus the square 25 of half the side 10. The rest of the demonstration is similar to that here given.

36. $(\frac{1}{3}x + 1)(\frac{1}{4}x + 1) = 20; \frac{1}{12}x^2 + \frac{1}{3}x + \frac{1}{4}x + 1 = 20; x^2 + 7x + 12 = 240; x^2 + 7x = 228; \frac{1}{2}$ of 7 is $3\frac{1}{2}; (3\frac{1}{2})^2 = 12\frac{1}{4}; 228 + 12\frac{1}{4} = 240\frac{1}{4}; \sqrt{240\frac{1}{4}} = 15\frac{1}{2}; 15\frac{1}{2} - 3\frac{1}{2} = 12$, which is the value of x.

37. The present usage of the expression "to complete the square" is quite different from that of our text. Here it means, of course, to make the coefficient of x^2 unity and this also corresponds to the operation termed by the Arabs *algebra*, as opposed to the operation of *al-muqabala*; see the article al-Djabr wā-'l-Mukābala by Professor H. Suter in *the Encyclopedia of Islam*, Vol. I (Leyden, 1913), pp. 989–990.

1. I have translated from the Latin text of Jordanus' *De numeris datis (On Given Numbers)*, edited by Maximilian Curtze, "Commentar zu dem '*Tractatus De numeris datis*' des Jordanus Nemorarius, "*Historisch-literarische Abteilung der Zeitschrift für Mathematik und Physik*, XXXVI (1891), 1–23, 41–63, 81–95, 121–138. The analytic summaries in the notes are drawn almost entirely from Curtze. Complete proofs of Proposition IV. 16–IV. 35, which contained only enunciations and numerical examples in Curtze's edition, as well as expanded versions of other propositions, were found in MS Codex 4770 of the Austrian National Library (Österreichische Nationalbibliothek) and published by R.

2. [The relation of] one number to another is given when its ratio *(proportio)* to the other is given.

3. A ratio is given when its denomination is known.

Book I

[PROPOSITIONS]

1. *If a given number is divided in two and their difference is given, each of them will be given.*

Since the smaller part *(minor portio)* [or number] plus *(et)* the difference constitute [or equal] *(faciunt)* the greater [number], then the smaller part with [another part] equal to itself and the difference equals *(facit)* the whole [number]. Therefore, when the difference has been subtracted *(sublata)* from the whole, twice the smaller [number] will be given; [and] when this has been divided, the smaller part [or number] will be given, and so also is the greater [number given].[2]

For example,[3] let 10 be divided into two [parts] whose difference is 2 and which, if subtracted from 10, leaves 8 whose half is 4. This is the smaller part; the other is 6.

3. *When a given number has been divided into two [numbers], if what is produced by the multiplication of one by the other is given, it is necessary that each [of the numbers] be given.*

—————————————

Daublebsky von Sterneck, "Zur Vervollständigung der Ausgaben der Schrift des Jordanus Nemorarius: 'Tractatus de numeris datis'," *Monatshefte für Mathematik und Physik,* VIII (Vienna, 1896), 165–179. "In his excellent new edition, translation, and analysis of the 113 propositions of Jordanus' *De numeris datis,* Barnabas Bernard Hughes, O.F.M. (*The "De numeris datis" of Jordanus de Nemore A Critical Edition, Analysis, Evaluation and Translation,* Ph.D. dissertation, Stanford University, 1970; this work reached me too late for use in my translation and analysis, but what follows in the remainder of this note has been drawn from it [pp.iv and 47–52]) observes that by contrast with the few extant medieval algebraic treatises, which taught practical matters useful to lawyers (the division of inheritances was an important part of their duties) and businessmen, the *De numeris datis* is essentially a treatise in algebraic analysis in the tradition established by Greek mathematicians such as Euclid, Diophantus, and Pappus. Although François Viète is credited with the reintroduction of analysis based on a concept of generalized number (see his *Introduction to the Analytical Art* published in 1591 and translated by Winfree Smith in Jacob Klein, *Greek Mathematical Thought and the Origin of Algebra,* translated by Eva Brann, with an appendix containing Vieta's *Introduction to the Analytical Art* trans-

lated by J. Winfree Smith [Cambridge, Mass.: M.I.T. Press, 1968]), Jordanus actually preceded Viète in the application of analysis to algebraic problems. The method employed by Jordanus conforms to the three steps deemed essential by Klein (*Greek Mathematical Thought,* pp. 165–166):

"(1) the construction of the equation, (2) the transformations to which it is subjected until it has acquired a canonical form which immediately supplies the 'indeterminate' solution, and (3) the numerical exploitation of the last, i.e., the computation of unequivocally determinate numbers which fulfill the conditions set for the problem."

That Jordanus regularly proceeded in accordance with these three steps is demonstrated as follows by Hughes (pp.51–52) from Bk.IV, Proposition 6 of the *De numeris datis:*

"IF THE RATIO OF TWO NUMBERS TOGETHER WITH THE SUM OF THEIR SQUARES IS KNOWN, THEN EACH CAN BE FOUND.

Let the ratio of x and y be given. Let b be the square of x and c the square of y; and let $b+c$ be known. Now the ratio of b to c is the square of the ratio of x and y. Hence, the former is known. Consequently b and c can be found.

For example, let the ratio of two numbers be 2 and the sum of their squares 500. Now since the square of one number is four times the square of the other, it follows that 500 is five times the square of the other, which makes it 100. The root of this is 10 for the smaller number, and the larger is 20.

$$x : y = a, x^2 + y^2 = d \tag{1}$$
$$x : y = a, x^2 = b, y^2 = c \tag{2}$$
$$b + c = d$$
$$b : c = x^2 : y^2 = a \tag{3}$$
$$(b/c + 1)y^2 = d \tag{4}$$
$$y = [d/(a^2 + 1)]^{1/2} \tag{5}$$
$$y = [500/(4 + 1)]^{1/2} = 10 \tag{6}$$
$$x = 20$$

(1) is the construction of the equation, the formation of the problem in terms of what is known, a and d, and what is to be found, x and y. (2) – (4) are the transformations to which (1) is subjected until a canonical form (5) is reached. (6) is the numerical exploitation of (5), that is, the computation of unequivocally determinate numbers which fulfill the conditions set for the problem. The answer to our question is affirmative. This is a proposition dealing in problematic analysis."

2. $x + y = s$; $x - y = d$; $y + d = x$; $\therefore 2y + d = x + y = s$, so that $2y = s - d$; $y = (s - d)/2$; $x = s - y$. In the example below, $s = 10, d = 2$; $2y = 8$, $y = 4, x = 6$.

3. All numerals are written verbally or as Roman numerals; for convenience, they are expressed here as Arabic numerals.

Let the given number be *abc,* divided into numbers *ab* and *c,* which when multiplied produce *(fiat) d,* a given number; moreover,[4] *abc* when multiplied by itself produces *e.* Also let quadruple *d* be taken and let this be *f,* which subtracted from *e* leaves *g.* This [*g*] will be the square of the difference between *ab* and *c.* Then the root is extracted from *g,* and let this be *h.* And *h* will be the difference between *ab* and *c;* thus whenever *h* is given, *c* and *ab* will be given.[5]

The procedure will easily be shown in this way. For example, let 10 be divided into two numbers and from the multiplication of one of them with another let 21 be produced, which when quadrupled is 84 and leaves 16 after subtraction from the square of 10—that is, from 100. The root extracted from 16 is 4 and this is the difference, and when this is subtracted from 10, 6 remains and half of it will be 3, which is the smaller part, and the greater part is 7.

16. *That if the difference [between two numbers] has been added to the given result of the multiplication of one by the other [of these two numbers], each of the two numbers will be given.*

Let *ab* be a number that has been divided, and what results from the multiplication of *a* by *b* (*quod fit ex a in b*) when added to the difference is *c,* which after being doubled is *d;* also let *e* be the square of the whole [number *ab*] and *f* what remains after *d* has been subtracted from *e.* If *f* is smaller than *d,* see how much smaller; for if it is smaller by four, the difference [between *a* and *b*] will be 2. If smaller by three, the difference [between *a* and *b*] will be 3 or 1, but this cannot be determined; but if *d* and *f* are equal, the difference will be 4. If, however, *f* exceeds *d,* see by how much it exceeds and let this be *g*; and *g* will be what results from the multiplication of the difference [of *a* and *b*] by itself and its excess over [the difference between *a* and *b* multiplied by] the double of two.[6] Therefore it [that is, *g*] will be obtained and so will the total difference between *a* and *b.*[7]

The procedure is as follows. For example, let 9 be divided in two [parts], and after adding their difference to their product, let this be 21, whose double, 42, is subtracted from 81, leaving 39, which diminished 42 by 3. There can be a difference, therefore, between 1 and 3, and each number can serve. It will be 1 if 9 were divided into 5 and 4 so that 5 by 4 with 1 added equals (*faciunt*) 21; [but] it will be 3 if 9 has been divided into 6 and 3, for similarly 3 multiplied by 6 with 3 added equals 21. Therefore an error might occur.

In a similar manner 9 can be divided [into two numbers whose difference added to their product] produces 19, whose double is 38. If this is subtracted from 81, 43 would remain, which exceeds 38 by 5. The double of 5 could be doubled and produce 20, to which is added the square of 4, which is the double of 2, and 36 is obtained whose root is 6. Subtracting 4 [from 6] and halving the remainder leaves 1, which added to 4 makes 5. And this [namely 5] is the difference of the parts [of number 9], which are 7 and 2.

29. *If a given number is divided in two and the product of the whole into one [of the parts] is equal to the square of the other [number], each [of the numbers] will be given approximately.*

Let the product of *ab* into *b* be as much as [that is, equal to the product of] *a* into itself [that is, *a* squared]; and since *ab* into itself equals [the sum of

4. It is of interest to observe here that Jordanus uses a block of three letters to represent a single number and signifies parts of it by dissociating the letters and using one or two letters to represent parts of the whole number. Indeed, it is significant that letters are used to represent numbers generally.

5. $x + y = s$; $xy = d$; $(x + y)^2 = s^2 = e$; $4xy = 4d = f$; $e - f = g = (x - y)^2$; $x - y = \sqrt{g} = h$. Therefore, by Proposition 1, $2y = s - h$. In the example below, $s = 10$ and $d = 21$ are given; we then get $g = 100 - 84 = 16$, and $h = 4$; therefore $y = 3, x = 7$. In Vienna Codex 4770 (von Sterneck, p. 176), we find "thus whenever *abc* is given," where *abc* replaces *h* in Curtze's edition.

6. This complicated statement is a free translation of "quod fit ex ductu illius, quo differentia excedit duplum binarii, in se et in illud duplum. . . ."

7. $x + y = s$ and $xy + x - y = c$ are the equations, with *s* and *c* given. Now, $2c = d$ and $(x + y)^2 = s^2 = e$; $e - d = f$. Hence $f = x^2 + y^2 - 2(x - y)$ and therefore $d - f = 4(x - y) - (x - y)^2$. Here we must distinguish between the possibilities $d - f \gtreqless 0$.
(1) If $d - f = 0$, then either $x - y = 0$ or $x - y = 4$. Jordanus does not mention the first possibility nor discuss further the second.
(2) If $d - f > 0$, then $d - f$ must equal 3 or 4; if $d - f = 3$, then $x - y = 3$ or 1; if $d - f = 4, x - y = 2$. But if $d - f = 1$ or 2 then $x - y$ equals an irrational; and if $d - f > 4$, imaginary numbers would result.
(3) If $d - f < 0$, the equation takes this form: $f - d = g = (x - y)^2 - 4(x - y)$, from which $x - y$ can be determined with the aid of a corollary of Proposition 7 (omitted in this selection).
In the examples below, for $d - f > 0$, we have $s = 9$, $c = 21$, so that $d - f = 3$ and $x - y = 1$ or 3. If $x - y = 1$, then $x = 5, y = 4$, and $5 \cdot 4 + 1 = 21$; for the second case, $x = 6, y = 3$, and $6 \cdot 3 + 3 = 21$. In the case where $d - f < 0$, we find $s = 9, c = 19$, so that $f - d = 5$ and consequently $(x - y)^2 - 4(x - y) = 5$. By Proposition 7, it follows that $x - y = 5$ and $x = 7, y = 2$; hence $7 \cdot 2 + 5 = 19$.

113

the products of] ab into a and [ab] into b, it will also equal [the sum of the products of] a into itself [that is, a squared] and into ab.[8]

For example, let 10 be divided in two parts so that 10 multiplied into one of the two numbers equals the product of the remaining number multiplied by itself. Now 10 multiplied by itself is 100, and let the double of its double be taken and this will be 400. To this let the square of 10 be added and we will have 500, whose approximate root is now extracted and will be $22\frac{1}{3}$, from which 10 is subtracted and half of the remainder will be $6\frac{1}{6}$, the greater of the parts which must be multiplied by itself.

Book IV

7. *With the mutual relation [or ratio] of two numbers given [and] if the product of their sum (ex composito) and their difference is given, each of them will be given.*

What results from the multiplication of their sum and difference is given, as is the excess of the square of the greater over the square of the lesser— since the ratio of square to square is given; then [the ratio] of this [that is, the product of sum and difference] to that [the difference of the squares]

will be given; therefore, the square is given and then its side, and similarly, the remainder:[9]

For example, let one to the other be a triple [ratio] and their sum and difference equal 32. Therefore, since [the ratio of] square to square is nine to one, and the one will be 8 more than the other, the square of the lesser will be 4, the other will be 2, and the remainder 6.

8. *If a square with the addition of its root multiplied by a given number will make a given number, it [that is, the square] will also be given.*

Let a be the square and b its root multiplied by cd, so that c and d are each half of it [that is, half of cd]; and b multiplied by cd equals e and ae is given. Since bcd multiplied by b equals ae, when the square of d has been added to ae it equals aef. And aef will result from [the multiplication of] bc by itself. Thus when aef is given, bc will be given; and after c has been subtracted, b will remain and be given, and thus a will be given [or determined].[10]

For example, let there be a square whose root if multiplied by 5 and the product be added to it [that is, to the square] makes 36. To this, let the square of $2\frac{1}{2}$ be added, namely $6\frac{1}{4}$, and there results $42\frac{1}{4}$, the root of which is $6\frac{1}{2}$. From this subtract $2\frac{1}{2}$ leaving 4, which is the root whose square is 16.

Number Theory, Probability, and Infinite Series

24 NUMBER THEORY AND INDETERMINATE ANALYSIS

Leonardo of Pisa (Fibonacci; b. ca. 1179; d. after 1240)

Translated by Edward Grant

Annotated by Paul Ver Eecke;[1] additional notes by Edward Grant

Here Begins the Book of Square Numbers Composed by Leonardo Pisano in 1225

Oh most glorious Prince Frederick,[2] when Master Dominic led me to Pisa in the footsteps of Your

8. $x + y = s$, $y(x + y) = x^2$, where s is given. Then $(x + y)^2 = x(x + y) + y(x + y)$, so that $s^2 = sx + x^2$. In the example $s = 10$ and we see that $x = (-10 + \sqrt{500})/2$. Moreover, $\sqrt{500}$ is approximated as $22\frac{1}{3}$, which determines x as approximately $6\frac{1}{6}$ and consequently $y = 3\frac{5}{6}$. Now, from above, $10y = x^2$, but $10 \cdot 3\frac{5}{6} = 38\frac{1}{3}$, while $(6\frac{1}{6})^2 = 38\frac{1}{36}$, which is approximately $\frac{11}{36}$ less than $38\frac{1}{3}$. Curtze observes that this problem actually represents the division of a line according to the golden section (that is, the division of a line into extreme and mean ratio; see Euclid's *Elements* II.11 and VI. 30).

9. If $x/y = m$ and $(x + y)(x - y) = d$, where m and d are given, then we have directly $x^2/y^2 = m^2$ and $x^2 - y^2 = d$. Since $x^2 = m^2y^2$, we obtain $m^2y^2 - y^2 = (m^2 - 1)y^2 = d$. Therefore, $y = \sqrt{d/(m^2 - 1)}$.

In the example $m = 3$, $d = 32$; $m^2 - 1 = 8$, and therefore, $y^2 = 4$, $y = 2$, and $x = 6$.

10. We have here a case of completing the square (compare al-Khwārizmī's first example in chapter 4 of his algebra, in the preceding selection; see also notes 13, 23, 26, and 37 of that selection). The given equation is $x^2 + px = q$, where $p = c + d$ and $c = d = p/2$, so that $x^2 + 2cx = q$. Jordanus then adds $c^2 = d^2$, whereupon $x^2 + 2cx + c^2 = q + d^2$, or $x^2 + px + (p/2)^2 = q + (p/2)^2$. Thus, $x + c = \sqrt{q + d^2}$, or $x + (p/2) = \sqrt{q + (p/2)^2}$ and $x = \sqrt{q + (p^2/4)} - p/2$.

In the example presented in the next paragraph $p = 5$, $q = 36$, $c = d = 2\frac{1}{2}$, $d^2 = 6\frac{1}{4}$, $q + d^2 = 42\frac{1}{4}$, $\sqrt{q + d^2} = 6\frac{1}{2}$; therefore, $x = 4$, $x^2 = 16$.

1. [The propositions included here are drawn from Leonardo Pisano's *Liber quadratorum (Book of Square Numbers)*, written in 1225 and containing twenty propositions which are neither numbered nor distinguished

Highness, John of Palermo, who met me there, proposed to me the question written below, which pertains no less to geometry than to number: to find a square number which when increased or diminished by 5 always produces a square number.[3] After finding the solution of this question and reflecting on it, I saw that the solution took its origin from the many things that happen with square numbers. Moreover, from reports in Pisa and from others returning from the imperial court, I learned recently that Your Sublime Majesty deigned to read the book which I composed on number[4] and that, at times, it pleases you to hear subtleties touching on geometry and number. Recalling that the question written above was proposed to me at your court by your philosopher, I took the subject matter from it and began to construct, in your honor, the work that I wish to call *The Book of Square Numbers*. If its contents are [only] more or less true or necessary, I beg your indulgence, since to have recollection of all things and to err in none is rather a characteristic of divinity than humanity; and no one is devoid of weakness and wholly circumspect.

PROPOSITION 1

I thought about the origin of all square numbers and discovered that they arose from the regular ascent [or increase] of odd numbers. For unity is a square and from it is produced the first square, namely 1; adding 3 to this makes the second square, namely 4, whose root is 2; if to this addition [that is, sum] is added a third odd number, namely 5, the third square will be produced, namely 9, whose root is 3; and so [it is that] the sequence and series of square numbers always takes rise through the regular addition[5] of odd numbers.[6]

Thus when we wish to find two square numbers whose addition produces a square number, I take any odd square number as one of the two square numbers and I find the other square number by the addition *(collectione)* of all the odd numbers from unity up to [but excluding] the odd square number [already selected]. For example, I take 9 as one of the two squares mentioned; the remaining square will be obtained by the addition of all the odd numbers below 9, namely 1, 3, 5, and 7, whose sum is 16, a square number, which when added to 9 gives 25, a square number.[7] And if we wish to employ a geometric demonstration, let any odd numbers be arranged in ascending order from unity, with the last number [of this sequence of terms] a square. Let these numbers be [represented by]

AB, BC, CD, DE, and *EF* [Fig. 1]; and let *EF*

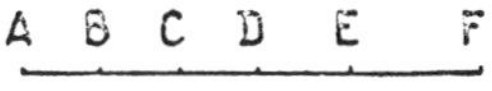

Fig.1

be a square number. Now, since *EF* is a square number and [since] *AE* is a square number—the latter is produced by sequential addition from the odd numbers *AB, BC, CD,* and *DE*—the whole number *AF* is also a square number. And thus from two square numbers *AE* and *EF*, the square number *AF* is formed.[8]

Again, I take any even square number whose half is even, as 36, whose half is 18. Now, if I sub-

in the Latin text as published by Baldassare Boncompagni in his edition of the works of Leonardo titled *Scritti di Leonardo Pisano* (Rome, 1862), II, 253–283. In isolating and numbering the twenty propositions, I have followed Paul Ver Eecke, who performed this task in his French translation of the *Liber quadratorum* (*Léonard de Pise: de Livre des nombres carrés* [Bruges: Desclée de Brouwer, 1952]). His translation has been extremely helpful in the interpretation of difficult passages, and for its useful notes, which have been, for the most part, translated into English and included here. Where it has been necessary to quote Euclidean propositions, I have used Sir Thomas Heath's translation, *The Thirteen Books of Euclid's Elements*, 2d. ed. (New York: Dover, 1956).

The Book of Square Numbers was Leonardo's fifth and final extant mathematical treatise. It was apparently not intended as a systematic treatment of square numbers but rather as a collection of propositions relevant and subservient to two basic problems posed to Leonardo as challenges to his mathematical skill by learned men at the imperial court of Frederick II and which constitute propositions 14 and 20 of the treatise. Although almost wholly ignored and virtually unknown during the Middle Ages, its achievements are considerable, evoking from Ver Eecke the judgment that in the part of it relevant to the theory of numbers it surpasses the level achieved by Diophantus in antiquity and is scarcely surpassed by the work of Fermat and Gauss in modern times (Ver Eecke, p. xxv).—*Ed.*]

2. [Frederick II of Hohenstaufen (1194–1250), Holy Roman Emperor.—*Ed.*]

3. [This problem is considered in Proposition 14—*Ed.*]

4. [It would seem that Leonardo is here referring either to his *Liber Abaci (Book of the Abacus)* or to *Flos super solutionibus quarumdam questionum ad numerum et ad geometriam pertinentium (Flower of solutions of certain questions relevant to number and geometry).—Ed.*]

5. *Collectio,* collection or assemblage, that is, addition.

6. $1 + 3 + 5 + 7 + \cdots + (2n - 1) = n^2$.

7. $9 + (1 + 3 + 5 + 7) = 9 + 16 = 25 = 5^2$.

8. [Although no numbers are expressly assigned to these lettered line segments, it is obvious that Leonardo is simply repeating the preceding example (see n. 7), since he assumes *AB* as unity—*Ed.*]

tract 1 from it and add 1 to it, 17 and 19 are produced, which are odd numbers and successive, *(continui)* since no odd number falls between them. Furthermore, 36, a square number, is produced from their addition [that is, 17 + 19]; and from the remaining odd numbers, which extend from 1 to 15, 64 is produced. From [the sum of] these two square numbers [that is, 36 and 64] comes 100, a square number; but 100 is also produced from the addition of the odd numbers extending from 1 to 19.

Or, I take an odd square number whose third part is a whole number, as 81, whose third is 27. Now I take this 27 along with the two odd numbers which it serves as middle, namely 25 and 29, and these three numbers [25, 27, and 29] make 81, a square number. And from the [addition of the] other [odd] numbers from 1 to 23, there comes 144, whose root is 12. Adding 144 and 81 gives the sum of the addition of the [series of] odd numbers from 1 to 29, namely 225, a square number whose root is 15.

In a similar manner, four and even more successive odd numbers can be found from whose addition a square number can be produced; and from the remaining [odd] numbers that are less than these and extended [in descending order] to unity, another square number will be generated; and [the sum of] these two squares makes another square number.

I also find that any square number exceeds the square immediately preceding it by a quantity equal to the sum of their roots. For example, 121, whose root is 11, exceeds 100, whose root is 10, by a quantity equal to the sum of 10 and 11, namely their roots.[9] Therefore, any square will exceed the second square preceding it by a quantity equal to four times the square root of the intermediate square; as [for example], 121 exceeds 81 by four times 10.[10] In this way, one can find the differences between square numbers by the interval between their roots. And when two successive roots are added and form a square number, the square of the greater root will be equal to two squares.[11] Similarly, when the quadruple of any root is a square, the square of the following root will be equal to two squares, one of which will be that which was created by the previously mentioned quadruple, and the other whose root is 1 less than the quadrupled root. Thus, if 9 were quadrupled, 36 would be produced; then 100, whose root is 10, is equated with [the sum of] 64, whose root is 8, and 36, the quadruple of 9.[12] Now from the quadrupling of any number, it is known that a square number is not

produced unless it was itself a square, because, as Euclid shows, when a ratio of a number to a number is as a ratio of squares, their square is made from their multiplication;[13] and since 4 is a square, it is necessary that that number which it multiplies be a square so that a square number result from [both of] them. There are many ways, then, to find three square numbers of which one will equal the sum of the others.

But how it happens that every square exceeds its [immediately] preceding square by a quantity equal to the sum of their roots will be obvious if we posit the roots in lines *AB* and *BG* [Fig. 2].

$$\text{A} \qquad \text{B} \qquad \text{D} \qquad \text{G}$$

Fig. 2

Now, since *AB* and *BG* are successive numbers, one of them will exceed the other by 1. Let *BG* exceed *AB* by 1 and let *DG*, unity, be subtracted from *BG*, so that *BD*, equal to *BA*, will remain. Since number *BG* has been divided in two, namely into *BD* and *DG*, the multiplication of *BD* by itself with *DG* by itself and with twice *DG* by *BD* will equal the multiplication of *BG* by itself.[14] But the

9. $(121 = 11^2) = (100 = 10^2) + (10 + 11 = 21)$.
10. $(121 = 11^2) = (81 = 9^2) + 4 (\sqrt{100} = 10)$.
11. $4 + 5 = 9$, and $5^2 = 25 = 9 + 16$.
12. $(100 = 10^2) = (64 = 8^2) + (4 \cdot 9 = 36)$.
13. [Although the text reads "a ratio of a number to a number is as a ratio of squares" *(proportio numeri ad numerum est sicut proportio quadratorum)*, Ver Eecke has altered "squares" *(quadratorum)* to "equimultiples" and cited Book V, Proposition 15, as the reference to Euclid's *Elements,* where it is asserted (Heath, II, 163), that "parts have the same ratio as the same multiple of them taken in corresponding order," that is, $A/B = mA/mB$. But "equimultiples" seems unrelated to Leonardo's discussion. Perhaps Book VIII, Proposition 24, is intended, where it is shown that if $A/B = C^2/D^2$, and A is a square number, then B is also square. (In Heath's translation [II, 380]: "If two numbers have to one another the ratio which a square number has to a square number, and the first be square, the second will also be square.") Or, perhaps, *Elements,* Book VIII, Proposition 26, and Book IX, Proposition 1, were intended. In the former Euclid demonstrates (Heath, II, 381) that "similar plane numbers have to one another the ratio which a square number has to a square number." Now, if the "similar plane numbers" are square numbers, then the latter proposition demonstrates (Heath, II, 384) that "if two similar plane numbers by multiplying one another make some number, the product will be square." In this way, we can derive Leonardo's claim that $A^2 \cdot B^2 = C^2.$—*Ed.*]

14. *Elements,* Book II, Proposition 4 (Heath, 379): "If a straight line be cut at random, the square on the whole is equal to the squares on the segments and twice the rectangle contained by the segments." This gives the

multiplication of *BD* by itself is equivalent to the multiplication of *AB* by itself. Therefore, the square made by number *BG* exceeds that made by number *AB* by a quantity equal to the multiplication of *GD* by itself and twice *GD* by *BD*.[15] But the multiplication of *DG* by itself is 1, which is equal to, or the same as, the unit *DG;* and the multiplication of twice *DG* by *BD* makes twice *BD*, since *DG* is 1. Therefore, twice *BD* is *AD* and the square made by number *BG* exceeds the square made by number *AB* by a quantity equal to the sum of their roots, namely *AB* and *BG*, which it was necessary to show.

Otherwise, since number *BD* equals *BA*, the whole will be *AD*, which has been divided in two equal parts at point *B*, to which the unit *DG* is added. Therefore, the multiplication of *DG* by *AG* with the square root of *AB* will equal the square of root *BG*.[16] Hence the square formed by number *BG* exceeds the square formed by number *AB* by the product of *DG* by *AG*. But *DG* multiplied by *AG* produces number *AG*, since *DG* is 1. Therefore, square *BG* exceeds square *AB* by the sum of their roots, namely *AG*. Similarly, it can be shown that every square exceeds every smaller square by [an amount equal to] the multiplication of the excess of those roots by the sum of each root. . . . [17]

PROPOSITION 2

I want to demonstrate why a regular series of squares arises from a regular summation of odd numbers beginning from 1 and going to infinity.[18]

Let as many numbers as *BG*, *GD*, *DE*, and *ZI* be taken in succession from *A*, unity [Fig. 3]; and

A   B   G T   D K   E L   Z M   I N

Fig. 3

let *BG* be composed[19] with unit *A* to produce number *T*; also let each number be composed with its antecedent and consequent. And assume that the sum *(compositus)* of numbers *BG* and *GD* is *K*, [the sum] of numbers *GD* and *DE* is *L*, [the sum] of numbers *DE* and *EZ* is *M*, of numbers *EZ* and *ZI* is *N*.

I say first that, beginning with unity, numbers *T*, *K*, *L*, *M*, and *N* are odd and successive. For number *ZI* is either even or odd. If *ZI* is an even number, *EZ* is an odd number; but if *ZI* is odd, then *EZ* is even, for they are successive numbers. Consequently, the number composed of *EZ* and *ZI*, namely *N*, is odd. Similarly, we can show that the number composed of numbers *DE* and *EZ*, namely *M*, is odd. In the same manner, numbers

L, *K*, and *T* will be shown to be odd. I say, therefore, that the successive numbers *T*, *K*, *L*, *M*, and *N* are odd. Indeed, number *N* was made by the addition *(conjuncto)* of *EZ* and *ZI*; and number *M* was made by the addition *(conjuncto)* of *DE* and *EZ*. Hence by as much as number *ZI* exceeds number *DE*, by just so much does number *N* exceed number *M*. For *ZI* exceeds *EZ* by 1, and this is also the excess of number *EZ* over *DE*. Therefore, number *ZI* exceeds *DE* by 2, so that number *N* exceeds number *M* by 2. In the same manner, it is found that number *M* exceeds number *L*; and number *L*, number *K*; and number *K*, number *T*; and number *T* exceeds *A*, unity. Hence, as was said, numbers *T*, *K*, *L*, *M*, and *N* are successive odd numbers from unity.

And as shown above, the square of number *ZI* exceeds the square of number *EZ* by a number equal to the sum of *EZ* and *ZI*, that is [equal] to number *N*.[20] Similarly, it was shown that the square of number *EZ* exceeds the square of number *DE* by the sum of numbers *DE* and *EZ*, that is, by number *M*;[21] and the square of *DE* exceeds the square of *GD* by number *L*;[22] and the square of number *GD* exceeds the square of number *BG* by *K*;[23] and the square of number *BG* exceeds the

identity $(a + b)^2 = a^2 + b^2 + 2ab$; that is, in our example, $(BG)^2 = (BD)^2 + (DG)^2 + 2 (DG \cdot BD)$.

15. [Since $(BD)^2 = (AB)^2$, Leonardo now substitutes $(AB)^2$ for $(BD)^2$ in the identity described above in note 14.—*Ed.*]

16. *Elements*, Book II, Proposition 6 (Heath, 385): "If a straight line be bisected and a straight line be added to it in a straight line, the rectangle contained by the whole with the added straight line and the added straight line together with the square on the half is equal to the square on the straight line made up of the half and the added straight line." This enunciation, which is translated algebraically by the identity $(2a + b)b + a^2 = (a + b)^2$, gives as the relation of the text: $DG \cdot AG + (AB)^2 = (BG)^2$.

17. That is, we have the identity $x^2 - y^2 = (x - y)(x + y)$, or numerically: $5^2 - 2^2 = (5 - 2)(5 + 2)$. [The remainder of the proposition, some seventeen lines, has been omitted.—*Ed.*]

18. A linear demonstration of the formula $1 + 3 + 5 + \cdots + (2n - 1) = n^2$.

19. *Componere*, to compose, that is, to add.

20. [This was shown above in Proposition 1. Thus $(ZI)^2 = (EZ)^2 + EZ + ZI = (EZ)^2 + N$. Since the numbers are arranged in sequence, $EZ = 5$ and $ZI = 6$, so that $6^2 = 5^2 + 5 + 6 = 36$.—*Ed.*]

21. [That is, $(EZ)^2 = (DE)^2 + DE + EZ = (DE)^2 + M$; or $5^2 = 4^2 + 4 + 5$.—*Ed.*]

22. [That is, $(DE)^2 = (GD)^2 + L$; or $4^2 = 3^2 + 3 + 4$.—*Ed.*]

23. [That is, $(GD)^2 = (BG)^2 + K$; or $3^2 = 2^2 + 2 + 3$.—*Ed.*]

square of unity by number T.[24] For T is 3, and BG is 2; therefore, if number T, by which the square of number BG exceeds the square of unity, is added to the square of unity, that is, to 1, the square of BG will be produced; if one adds number K to this square, the square of number GD will be produced; if one adds number L to this square, the square of number DE will be produced; if one adds number M to this square, the square of number EZ will be produced; and if to this square one adds number N, in which the square of number ZI exceeds the square of number EZ, obviously the square of ZI will be produced. For the numbers A, BG, GD, DE, EZ, and ZI are successive and their squares arise from the successive addition (collectione) of the odd numbers A, T, K, L, M, and N, as it was necessary to show.[25]

PROPOSITION 3

To find two numbers whose squares when added make a square which is itself made by the addition (conjunctione) of two other given square numbers.[26]

Let A and B be two numbers whose squares when joined [or added] form a square number G. It is necessary to find two other numbers whose squares when found equal square number G. Let two other numbers be found whose squares when joined [or added] make a square number and from whose measure straight lines DE and EZ are made and formed into a right angle, namely the angle formed by DEZ [Fig. 4]. The square of side DZ is the

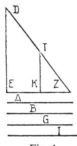

Fig. 4

square of sides DE and EZ, and the square DZ is either equal to number G or not. First, let it be equal. Then two other numbers have been found whose squares when joined [or added] form a square number equal to number G. One of these is equal to straight line DE, the other is equal to straight line EZ. Now, if the square formed by DZ, that is, by number DZ, is not equal to number G, it will be greater or smaller than it.

First, let it be greater. But since the square

formed by DZ is greater than number G, number DZ will be greater than the root of G. Therefore, let the root of number G be taken and let this be I; and take from DZ a number equal to I and let this be TZ. From point T let perpendicular (cathetus) TK be drawn from point T on to line EZ. Therefore, TK is equidistant from straight line DE. Consequently, triangle TKZ is similar to triangle DEZ, so that as ZD is to ZT so is DE to TK. But ratio ZD to ZT is known, for both have been [calculated or] reasoned;[27] therefore, DE to TK will also be known. And since DE is reasoned (ratiocinata) so also will TK be numbered[28] [or calculated]. It will be shown in a similar way that straight line ZK is reasoned (ratiocinatam), since the ratio of it to ZE is as ZT to ZD.[29] Therefore, TK and KZ, the sum of whose squares makes a square formed by TZ, are calculated (numerate). But the square number formed by number TZ is equal to the square formed by number I, since number I is the root of number G. Hence the square formed from TZ is equal to number G, since two numbers, TK and KZ, have been found whose squares when conjoined [or summed] equal square number G.

Now let DZ be smaller than I and extend straight line ZD to L so that ZL is equal to number I [Fig.

24. [That is, $(BG)^2 = (1)^2 + T$; or $2^2 = 1^2 + 1 + 2 — Ed.$]

25. [By conveniently organizing Leonardo's verbal description in this last section, we can readily see how the squares of the successive natural numbers arise from the successive addition of the regular sequence of odd numbers:

1^2 or $A^2 + T = (BG)^2$; or $1 + 3 = 4$
$(BG)^2 + K = (GD)^2$; or $4 + 5 = 9$
$(GD)^2 + L = (DE)^2$; or $9 + 7 = 16$
$(DE)^2 + M = (EZ)^2$; or $16 + 9 = 25$
$(EZ)^2 + N = (ZI)^2$; or $25 + 11 = 36$, etc.—Ed.]

26. A problem of the form $x^2 + y^2 = z^2 = a^2 + b^2$.

27. Raciocinatus, a medieval alteration of ratiocinativus, signifies what is derived from reasoning or calculation, namely that ZD and ZT are known quantities derived by reasoning or calculation from the known quantities DE and EZ given at the beginning. [Since DE and EZ are known, ZD can be calculated. Therefore, ratio DE/ZD is known; but TK/ZT = DE/ZD, so that ratio TK/ZT is also known. Now by Elements, Book V, Proposition 16, DE/TK = ZD/ZT ("If four magnitudes be proportional, they will also be proportional alternately."); and since ZD has already been calculated, ZT can easily be determined, for the ratio ZD/ZT is known. Similarly, the ratio DE/TK is known, and since DE is known, TK can be calculated.—Ed.]

28. Numeratus, an expression analogous to the preceding expression [see n. 27—Ed.], signifying that TK is known by calculation or number.

29. [That is, ZK/ZE = ZT/ZD.—Ed.]

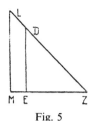

Fig. 5

5]. Similarly, let ZE be extended to M, L be joined to M, and LM be equidistant from [that is, parallel to] line DE. Since triangle DEZ is similar to triangle LMZ and ratio ZD to ZL is known, it follows that numbers ZM and ML will be known. Hence two numbers, LM and MZ, have been found whose squares when conjoined [or summed] form a square equal to number G, since LZ is equal to the root of this number; and this was to be done.

So that all this may be had in numbers, let A be 5 and B, 12, so that G, which is conjoined [or summed] from the squares of numbers A and B, is 169 and its root, namely I, is 13. Now, let the two lines DE and EZ touch and form a right angle DEZ; and let DE be 15 and EZ 8, so that DZ will be 17.[30] But, having taken line ZT equal to I on line DZ, it follows that ZT equals 13. And [now] line TK is produced equidistant from [that is, parallel to] straight line DE;[31] hence as ZD is to ZT so is DE to TK. Therefore, multiply ZT by DE, that is, 13 by 15, and divide the product *(summam)*[32] by DZ, namely by 17, and TK will result, that is, $11\frac{8}{17}$.[33] Also, if you multiply ZT by ZE and divide by ZD, KZ results, namely $6\frac{2}{17}$.[34] Thus it is that two numbers have been found, namely TK and KZ, whose squares conjoined [or summed] form number G, namely the square formed by ZT.[35]

In a similar way one can show this if number DZ is less than I, as in the other figure,[36] where we assume that DE is 4 and EZ is 3, so that DZ is 5; and after extending ZD to L, ZL is equal to I, namely equal to 13. Now, ZD is to ZL as DE is to LM, therefore multiply ZL by DE and divide by ZD to obtain LM, which represents $10\frac{2}{5}$. Also, when the [product of] the multiplication of ZL by ZE has been divided by ZD, namely 39 by 5, you obtain MZ, which will represent $7\frac{4}{5}$. Thus two other numbers have been found, namely $10\frac{2}{5}$ and $7\frac{4}{5}$, whose squares when combined make 169.[37] And so we have shown that this can be done in an infinite number of ways.

PROPOSITION 4

If four nonproportional numbers are proposed

and the first is less than the second and the third less than the fourth, and if the sum of the squares of the first and second is multiplied by the sum of the squares of the third and fourth, and neither of the sums was a square, the number produced will be equal to the square numbers in two ways; and if only one of the sums is a square number, then the number produced will be equal to the two squares in three ways; and if both sums are squares, the number produced will be equal to the two squares in four ways; and this is understood without fractions. . . . [38]

PROPOSITION 5

To find, in another way, a square number equal to two square numbers . . . . [39]

PROPOSITION 6

To find two numbers whose squares when added make a non-square number [which is also] formed from the sum of two squares of given numbers.[40]

30. $(DE = 15)^2 + (EZ = 8)^2 = (DZ = 17)^2$.

31. [See the triangle in Fig. 4, which exactly reproduces the data of this example.—*Ed.*]

32. *Summa* is an expression which signifies a total, that is, a product.

33. We have $DE = 15$, $ZT = 13$, and $ZD = 17$. But $DE/TK = ZD/ZT$, from which we get $TK = (DE \cdot ZT)/ZD = (15 \cdot 13)/17 = 11\frac{8}{17}$.

34. We have $ZT = 13$, $ZD = 17$, and $ZE = 8$. But $ZT/ZD = KZ/ZE$, from which we obtain $KZ = (ZT \cdot ZE)/ZD = (13 \cdot 8)/17 = 6\frac{2}{17}$.

35. We have $(ZT)^2 = (TK)^2 + (KZ)^2$, or $(13)^2 = (11\frac{8}{17})^2 + (6\frac{2}{17})^2$, or $169 = 38{,}025/289 + 10{,}816/289$.

36. [See Fig. 5—*Ed.*]

37. In the second case, we have $DE = 4$, $EZ = 3$, from which it follows that $DZ = \sqrt{16 + 9} = 5$ and $ZL = 13$. But $DE/LM = DZ/ZL$, from which it follows that $LM = (DE \cdot ZL)/DZ = (4 \cdot 13)/5 = 10\frac{2}{5}$. On the other hand, $EZ/ZM = DZ/ZL$, from which it follows that $ZM = (EZ \cdot ZL)/DZ = (3 \cdot 13)/5 = 7\frac{4}{5}$. Then $(ZL)^2 = (LM)^2 + (ZM)^2$, or $13^2 = (10\frac{2}{5})^2 + (7\frac{4}{5})^2$, or $169 = 2704/25 + 1521/25$.

38. This proposition raises an historical question concerning priority, because it enunciates, in disguised terms, identities attributed to Lagrange. In effect, it demonstrates that if four numbers A, B, G, and D are such that $A < B$ and $G < D$, one has the identities ascribed to Lagrange: $(A^2 + B^2) \cdot (G^2 + D^2) = (AD \pm AG) + (BG \pm AD)$; and if one has $A^2 + B^2 = M^2$, we have another: $(A^2 + B^2) \cdot (G^2 + D^2) = (MG)^2 + (MD)^2$; and if $G^2 + D^2 = N^2$, we have $(A^2 + B^2) \cdot (G^2 + D^2) = (AN)^2 + (BN)^2$. [The lengthy proof of this proposition is omitted here.—*Ed.*]

39. A problem of the form $x^2 + y^2 = z^2$.

40. Linear solution of a problem of the form $x^2 + y^2 = a^2 + b^2$, with the condition that $a^2 + b^2$ is not a square number.

Let the two given numbers be G and D, the sum of whose squares make number Z, a non-square number [Fig. 6]. I wish to find two other numbers

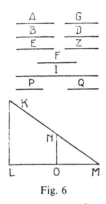

Fig. 6

the sum of whose squares make number Z. Assume two numbers, A and B, the sum of whose squares make number E, a square number; assume also that G is not related to D as A is to B; and multiply E by Z to obtain I. Now find two numbers the sum of whose squares makes [or equals] number I. Let those two numbers be P and Q, whose measure is represented by straight lines KL and LM forming a right angle, namely the angle KLM; and let K and M be joined. Therefore, KM will be the root of number I. Now, take MN, which is equal to the root of number Z, from straight line KM and form a right angle by drawing lines NO and OM, which are rational straight lines the sum of whose squares equals number Z. Since the square of line KM is equal to number I, and number I was formed by the multiplication of E by Z, then, if we multiply the root of number E by the root of number G, the root of number I will be obtained, namely MK. And because the root of number E is rational, as many times as the unit is in this root, so many times is the root of number G, that is MN, in straight line MK. Therefore, F is the root of number E. Hence as unity is to number F so is MN to MK; and just as MN is to MK so is NO to KL and OM to LM. By reason of equality, then, as unity is to number F so is NO to KL and MO to ML. If we divided KL by number F, rational number NO is produced; similarly, if we divide ML by F, OM is produced. Therefore, two numbers, NO and OM, have been found whose squares [when] summed make MN, a non-square number, namely Z, which is composed of the squares of numbers G and D; and this was to be shown.[41]

In order to demonstrate this in numbers, let G be 4 and number D be 5, so that the number composed of their squares, namely Z, is 41. Of the other numbers, let A be 3 and B, 4, the sum of whose squares makes 25, namely number E. The multiplication of E by Z, namely of 25 by 41, produces 1025, and it is possible again to find two other numbers the sum of whose squares makes 1025; one of these is 32, the other 1; or, one number is 31 and the other 8. Therefore, let KL be 32 or 31 and LM, 1 or 8; and let the root of 25 be taken, namely number F, and divide numbers KL and LM by it and we shall have NO and OM; that is, if KL is 32 and LM is 1, then NO will be $6\frac{2}{5}$ and OM will be only $\frac{1}{5}$ of 1. Hence two numbers have been found, namely $6\frac{2}{5}$ and $\frac{1}{5}$, the sum of whose squares equals 41, namely equals number Z. But if KL were 31 and LM, 8, then NO will be $6\frac{1}{5}$ and OM will be $1\frac{3}{5}$. And thus two other numbers are found whose squares make [a sum equal to] 41; and this was to be done.[42]

PROPOSITION 7

If, beginning with unity, any successive numbers whatever, namely odd and even, are arranged regularly, the solid number[43] formed from the

41. We are given the rational numbers G and D such that $G^2 + D^2 = Z$, a non-square number. Let A and B be any rational numbers such that $A^2 + B^2 = F^2$ and assume that $A/B \gtreqless G/D$. We assume $I = E \cdot Z = (A^2 + B^2)(G^2 + D^2) = (A \cdot D \pm B \cdot G)^2 + (A \cdot G \mp B \cdot D)^2 = P^2 + Q^2$, taking $P = A \cdot D + B \cdot G$, and $Q = A \cdot G - B \cdot D$; or better, $P = A \cdot D - B \cdot G$, and $Q = A \cdot G + B \cdot D$. But in the similar rightangled triangles KLM and NOM we have: $KL = P$, $LM = G$, $KM = \sqrt{I}$, $NM = \sqrt{Z}$. But $\sqrt{I} = \sqrt{E \cdot Z} = \sqrt{E} \cdot \sqrt{Z} = F \cdot \sqrt{Z}$; then $\sqrt{Z}/\sqrt{I} = I/F$. But $MN/MK = 1/F$, and $MN/MK = NO/KL = OM/ML = 1/F$; therefore, $NO = KL/F$ and $OM = ML/F$. One then has two numbers such that $(NO)^2 + (MO)^2 = ([KL]^2 + [ML]^2)/F^2 = (KM)^2/E = 1/E = Z = G^2 + D^2$.

42. In the numerical example chosen, it is assumed that $G = 4$, $D = 5$, so that $G^2 + D^2 = 4^2 + 5^2 = 41$, a non-square number. Then $A = 3$ and $B = 4$, so that $A^2 + B^2 = 25 = 5^2$ and $F = 5$. One has, then, $I = (3^2 + 4^2)(4^2 + 5^2) = (4 \cdot 5 + 3 \cdot 4)^2 + (4 \cdot 4 \mp 3 \cdot 5)^2 = 32^2 - 1^2$ for the signs $+$ or $-$; or $= 8^2 + 31^2$ for the signs $-$ and $+$. As the first relation, we shall have $NO = 32/5 = 6\frac{2}{5}$ and $OM = 1/5$; then $(6\frac{2}{5})^2 + (1/5)^2 = 1025/25 = 41$. And as a second solution, we have $NO = 31/5 = 6\frac{1}{5}$ and $OM = 8/5 = 1\frac{3}{5}$, so that $(6\frac{1}{5})^2 + (1\frac{3}{5})^2 = 1025/25 = 41$. Moreover, it is possible to find two other numbers three times whose squares make 1025/25: one [pair] is 32 and 1; another 31 and 8, as above; or even one [pair consisting] of 20 and 25. But these last values are not suitable because 20 is to 25 as 4 is to 5, which is contrary to the condition imposed.

43. *Numerus solidus*, the solid or volumetric number formed by three factors.

[product of the] last number by the number following it [multiplied] by the sum of these two numbers is equal to six times the sum of all the squares formed by all the numbers, namely by unity and all the numbers arranged in this way. . . . [44]

PROPOSITION 8

If, beginning with unity, any odd numbers whatever are arranged regularly, the solid number formed by [the product of] the greatest of them with the odd number following it [multiplied] by the sum [of these two numbers] is equal to twice six times all the squares formed by all the numbers arranged in this way from unity. . . . [45]

PROPOSITION 9

If two mutually prime numbers are composed [or added] and their sum makes an even number, and if the solid number which is formed by [the multiplication of the product of] these numbers by their sum is multiplied by the number by which the greater [of the prime numbers] exceeds the lesser, the number produced will be one whose twenty-fourth part will be a whole number. [46]

Let AB and BG be two mutually prime numbers whose sum is AG, an even number; that is, they are the least numbers in the ratio of AB to BG, with BG greater [Fig. 7]. From number BG, let

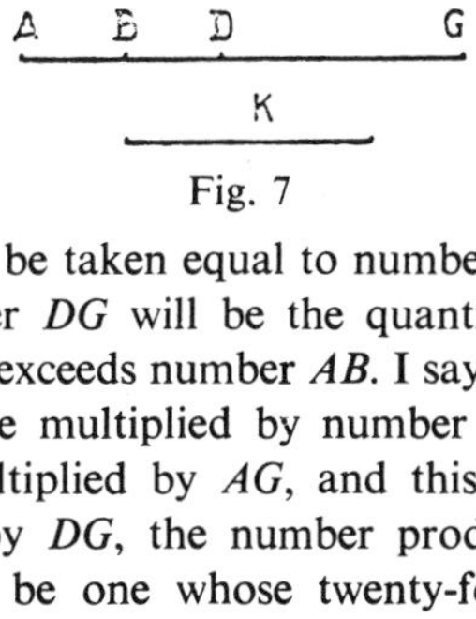

Fig. 7

number BD be taken equal to number AB. Therefore, number DG will be the quantity by which number BG exceeds number AB. I say that if number AB were multiplied by number BG and the product multiplied by AG, and this whole were multiplied by DG, the number produced [by all this] would be one whose twenty-fourth part—that is, a third of an eighth [part] or a fourth of a sixth [part]—will be a whole number.

Now, numbers AB and BG are both odd, for if they were not odd, their sum (compositus) could not be even; nor are both even, for then they could not be mutually prime. Therefore, numbers AB and BG are odd. And since number BD is equal to number AB, number AD is double to AB. Hence number AD is even; therefore the remainder, DG, is even, because if an even number is subtracted from an even number, an even number remains;

and since number DG is even, its half will be either even or odd.

Let it first be odd. Now, half of number AD, namely AB, is odd, so that half of number AD added to half of number DG—that is, with two odd numbers added—makes an even number. Therefore, half of number AG is even and the whole of AG is evenly even (pariter par), [47] so that one-fourth part of it is a whole [number]. By multiplying AG by GD there is produced a number whose eighth part is a whole [number]. But if half of number GD is even, then one-fourth of it will be a whole [number]. Therefore, by multiplying DG by AG a number will be produced whose eighth part is likewise a whole [number]. Hence if the product of the multiplication of AG by BG is multiplied by AB, a number will be produced whose eighth part will be a whole [number]. And since numbers AB and BG are odd, either the third part of one is a whole [number] or not. First, let it be a whole number. Therefore, from the multiplication of the solid number, which is formed by numbers AB, BG, and AG, by number DG, there is produced number K, whose third part is a whole [number] and also whose eighth part will be a whole [number]. As we said, then, one twenty-fourth part of this will be whole.

But if the third part of number AB or BG is not a whole number, then what will remain of each of them after they are divided by 3 will either be equal or unequal. Assume first that what remains is equal. Therefore, number GD is wholly divided by 3, so that from the multiplication of the solid number mentioned above, by number DG, a number is produced whose third part is a whole number. But when numbers AB and BG are divided by 3, their remainders are not equal; indeed, from the

44. A theorem of the form $n(n + 1) [n + (n + 1)] = 6 [1^2 + 2^2 + 3^2 + \cdots + (n - 1)^2 + n^2]$. Numerically the series 1, 2, 3, 4, and 5 gives $5 \cdot 6(5 + 6) = 6(1^2 + 2^2 + 3^2 + 4^2 + 5^2) = 6 \cdot 55$. [The proof is omitted here.—Ed.]

45. A theorem of the form $(2n - 1)(2n + 1)[(2n - 1) + (2n + 1)] = 12[1^2 + 3^2 + 5^2 + \cdots + (2n - 3)^2 + (2n - 1)^2]$. [The demonstration is omitted.—Ed.]

46. It is necessary to demonstrate that if a and b are mutually prime and $a + b$ is an even number, we have $ab(a + b)(a - b) =$ a multiple of 24, a number which is the smallest congruous number resulting from the smallest mutually prime numbers, namely 1 and 3, whose sum is an even number.

47. [According to Boethius' definition of a number pariter par, or "evenly even" (see selection 2), Leonardo could say of AG, which is "evenly even," that "one-fourth part of it is a whole [number]."—Ed.]

one number 1 will remain, from the other 2 will be left. Therefore, from their sum [that is, the sum of AB and BG], namely from number AG, the third part will be whole, so that the third part of the solid number formed by numbers AB, BG, and AG will be a whole number. The multiplication of this solid number by number DG will therefore produce a number whose third part is a whole number; and since the eighth part of this number has also been found to be a whole number, the twenty-fourth part of it will be a whole number, as it was necessary to show. The same thing would occur if numbers AB and BG were not mutually prime.

And if one of the numbers AB and BG were even, their sum (*conjunctus*) will be odd. It could then be shown that if the solid number which is formed by [the multiplication of] the double of each [multiplied] by their sum were multiplied by number DG, a number would also result whose twenty-fourth part will be a whole number, whether or not the numbers are prime to each other. The number obtained in this way, namely the number whose twenty-fourth part is a whole number, I have called a "congruous number."

PROPOSITION 10

If around some number there are arranged as many greater numbers as smaller numbers, and should the multitude [or number] of smaller numbers equal the number of greater numbers, and should it happen that by as much as any one of the greater numbers exeeds this number and that by just so much does this number exceed [one of the corresponding] smaller numbers, then the sum of the numbers arranged in this way will equal [the product of] the multiplication of this number by the number of these numbers.[48]

Let A, B, G, E, Z, and I be numbers arranged around number D; and let the smallest of them be number A, the greatest number I; and by as much as number I exceeds number D, by just so much does D exceed number A; similarly, by as much as number Z exceeds D, by just so much does number D exceed number B; again, by as much as number E exceeds number D, by just so much does number D exceed number G. I say that if D were multiplied by the number numbering all the numbers (*numerum multitudinis numerorum*) A, B, G, E, Z, and I, there would be produced the sum of all these numbers,[49] which is proved as follows:[50] diminish number I by its excess beyond D and add this to

number A, and each of the numbers A and I will equal number D. This can also be done for numbers Z and B, and E and G, so that each of them will equal D. Therefore, there are as many numbers A, B, G, E, Z, and I as there are numbers equal to D in the sum of the series of numbers A, B, G, E, Z, I; and it was necessary to show this.

PROPOSITION 11

To find a number which when added to a square number and subtracted from it always forms a square number.[51]

It is necessary, then, to find three squares and one number so that this number when added to the smallest square forms the second square. If we add this same number to this [second] square, we form the third square, namely the greatest square. And thus having added this number to the second square and diminishing that [second] square by this same number, we always form a square number. . . .[52]

PROPOSITION 12

If a congruous number[53] and its squares are multiplied by some square, the product of the multiplication of the congruous number by this square will be congruous and the remaining squares will be congruent to this congruous number.[54]

48. Symbolically, if n numbers are such that a, b, c $\cdots < N < A, B, C \cdots$ with $2N = A + a = B + b = C + c$, we have $a + b + c + \cdots + A + B + C + \cdots = nN$.

49. [If A, B, G, D, E, Z, and I are the given numbers such that $I > D = D > A$; $Z > D = D > B$; and $E > D = D > G$, then if n represents the number of these numbers, it follows that $Dn = A + B + G + E + Z + I$.—*Ed.*]

50. Numerically, the series 1, 2, 3, 4, 5, 6, and 7 gives $4 \cdot 6 = 1 + 2 + 3 + 5 + 6 + 7 = 24$, and for the same reason, the series 1, 3, 5, 7, 9, 11, and 13 gives $7 \cdot 6 = 1 + 3 + 5 + 9 + 11 + 13 = 42$.

51. A problem of the form $y^2 + N = z^2$ and $y^2 - N = x^2$, or $y^2 = x^2 + N$.

52. [Here we must find a number N and three squares such that $x^2 < y^2 < z^2$; $x^2 + N = y^2$; and $y^2 + N = z^2$. Then $y^2 + N = z^2$; $y^2 - N = x^2$; so that adding N to y^2 and subtracting N from y^2 always results in a square number, namely z^2 and x^2, respectively. The very lengthy proof that follows is omitted, even though it is a problem of indeterminate analysis of the second degree, and as Ver Eecke describes it (p. xvi), "a very remarkable example of geometric algebra." Instead, Proposition 20, also concerned with indeterminate analysis of the second degree, is included.—*Ed.*]

53. [For a definition of "congruous number," see the end of Proposition 9 and the next note.—*Ed.*]

54. Symbolically, if N is a congruous number, that is, a multiple of 24, we have $NK^2 = $ a congruous number.

Let AB be a square number and BG a congruous number, and let GD be equal to number BG [Fig. 8]. Therefore, AG and AD will be square numbers;

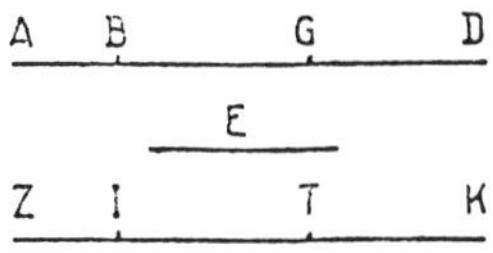

Fig. 8

and let us assume a certain square number E. I say that the product of the multiplication of E by congruous number BG will be congruous; and the numbers produced by the multiplication of E by the squares AB, AG, and AD will be squares that are congruent to the congruent number obtained by the multiplication of E by BG.

Now, let the multiplication of E by AB produce ZI; and the multiplication of E by AG produce ZT; and the multiplication of E by AD produce ZK. Since square E multiplied by the square of AG produces ZT, therefore ZT is also a square number; and ZT equals the [sum of the] two numbers produced by the multiplication of square number E by the square of AB and [E by the] congruous number BG.[55] But the product of square E and square AB is number ZI. The remainder, IT, is therefore produced by the multiplication of E by congruous number BG.[56] And since E and AB are square numbers, their product is a square number; therefore number ZI is a square. Again, since the multiplication of square E by square AD produces number ZK, ZK is equal to the two numbers produced by the multiplication of E by numbers AG and GD.[57] But E multiplied by AG makes square number ZT. Therefore, the remaining number ZK is the product of E by GD. And since number BG is equal to number GD, the product of E by BG will be equal to the product of E by GD. But the product of E by BG is IT. Therefore, IT is equal to number TK.[58] However, if we add number TK to the square of ZT, we get ZK squared; and if from square ZT we take TK, that is, TI, square ZI will remain.[59] Therefore, IT is congruous and the three squares, namely ZI, ZT, and ZK, are congruous with it, which it was necessary to show. In a similar way, the same thing would be obtained if any congrous number and its squares are divided by some square number.

PROPOSITION 13

I wish to find a congruous number whose fifth part is a whole number.[60]

Let one of the numbers set forth be 5 and the other a square that makes their sum a square; and having subtracted the smaller from the greater, there remains a square number. Now, this square number will be 4, which when added to 5 makes 9, a square number. Subtracting 4 from 5 leaves 1, also a square number.[61] I say that the congruous number whose fifth part will be a square number is produced from the numbers arranged in this way, since a congruous number arises from these numbers by multiplying 1 by the double of 5, which whole number is then multiplied by the double of 4 and this product multiplied by 9. And this is, in effect, to multiply the surface formed by 1 and the double of 5 by the surface formed by the double of 4 multiplied by 9, that is, 10 by 72. But the multiplication of 4 by 9 produces a square number, since both are squares. Therefore, the multiplication of the double of 4 by 9 makes the double of this square; and from the multiplication of double this square by double five comes quadruple the square multiplied by five. But quadruple this square makes a square number, therefore quadruple of the square multiplied by five makes the quintuple of this square; and the multiplication of the quintuple of this square by 1, which is a square number, again makes the quintuple of the square, and therefore, the congruous number formed by these numbers. Hence the fifth part [of this congruous number] will be a square number.[62]

PROPOSITION 14

Here now is the question mentioned above in the prologue of this work.

55. [That is, $ZT = E([AB]^2 + BG)$.—Ed.]

56. Let $ZI = E \cdot AB$; $ZT = E \cdot AG$; and $ZK = E \cdot AD$, so that we have $ZT = E(AB + BG) = E \cdot AB + E \cdot BG$, from which it follows that $ZT = ZI + E \cdot BG$, and $ZT - ZI = IT = E \cdot BG$.

57. [That is, $ZK = E(AG + GD)$.—Ed.]

58. We assume that $ZK = E \cdot AD$, so that $ZK = E(AG + GD) = E \cdot AG + E \cdot GD$. But $ZT = E \cdot AG$, and therefore, $ZK - ZT = TK = E \cdot GD$. But $BG = GD$, so that $E \cdot BG = E \cdot DG$. But [see n. 56—Ed.] $IT = E \cdot BG$; therefore, $IT = TK$.

59. $ZT + TK = ZK$, and $ZT - TK = ZT - TI = ZI$.

60. That is, to find a congruous number $24x$ whose fifth part is a whole number. This is a small lemma which will be utilized for the solution of the following proposition.

61. Two successive numbers are assumed, 5 and 4 $= 2^2$, whose sum and difference are, respectively, 3^2 and 1^2.

62. [That is, $4 \cdot 9 = 36 = 6^2$; therefore, $2 \cdot 4 \cdot 9 = 72$, and $2(6^2) \cdot 2(5) = 4(6^2) \cdot 5 = 720$; but $4 \cdot 6^2 = 144 = 12^2$; therefore, $4(12^2) \cdot 5 = 720 = 144 \cdot 5$; and $5 \cdot 144 \cdot 1 = 5 \cdot 144 = 5(12^2)$, or 720, a congruous number. — Ed.]

I wish to find a square number which forms a square number after it has been increased or diminished by 5.[63]

Let us assume a congruous number whose fifth part is a square. This number will be 720, whose fifth part is 144.[64] On this basis, sort out squares congruent to 720, the first of which is 961, the second 1681, the third 2401; and 31 is the root of the first square, 41 of the second, and 49 of the third.[65] From the first square we obtain $6\frac{97}{144}$, whose root is $2\frac{7}{12}$, which arises from the division of 31 by the root of 144, namely by 12; for the second square, that is, for the square sought, we obtain $11\frac{97}{144}$, whose root is $3\frac{5}{12}$, which arises from the division of 41 by 12; and for the last square we obtain $16\frac{97}{144}$, whose root is $4\frac{1}{12}$.[66]

PROPOSITION 15

If any two numbers are added (componantur) and form an even number, the ratio of their sum (compositi) to their difference (residuum)—where the greater [of these two terms] exceeds the smaller—will not be the same as the [ratio which the] greater number has to the smaller. . . .[67]

PROPOSITION 16

I wish to find a square number that forms a square number when its root has been added to itself; and similarly, if this root is subtracted from it, a square number remains. . . .[68]

PROPOSITION 17

Of any three square numbers that are successively odd, the greater square exceeds the middle square by eight more than the middle exceeds the smallest. . . .[69]

PROPOSITION 18

Between three squares, I wish to find two differences in a given ratio.[70]

63. A problem of the form $x^2 + 5 = y^2$; $x^2 - 5 = z^2$, which was put to Leonardo by John of Palermo, a mathematician attached to the court of Frederick II.

64. The congruous number 720 was derived in the preceding proposition.

65. That is, the numbers $961 = 31^2$, $1681 = 41^2$, and $2401 = 49^2$ are separated by the constant difference 720. See the next note for a further discussion of these three square numbers.

66. The proof is given without the details of the calculations. But $961 + 5(720/5) = 961 + 5(144)$; $(961/144) + 5 = 1681/144$; or, $6\frac{97}{144} + 5 = 11\frac{97}{144} = (3\frac{5}{12})^2$. Then $1681 + 5(720/5) = 1681 + 5(144)$; $(1681/144) + 5 =$

2401/144, or $11\frac{97}{144} + 5 = 16\frac{97}{144} = (4\frac{1}{12})^2$. Then the first square is $6\frac{97}{144}$, the second $11\frac{97}{144}$, and the third $16\frac{97}{144}$. As the text says, one could also consider the roots and have $(31^2 + 5[12]^2)/12^2 = (31^2/12^2) + 5 = (41^2/12^2)$, or $(2\frac{7}{12})^2 + 5 = (3\frac{5}{12})^2$; then $(41^2/12^2) + 5 = (4\frac{1}{12})^2$. The author does not explain his choice of the three numbers 31, 41, 49, whose squares are in arithmetic progression in the ratio 720. This, however, is not a puzzle, because we have found that he borrows these numbers from Proposition 7 of Book II of the *Arithmetic* of Diophantus. In Leonardo's time, the work of Diophantus was known neither in the original text nor in Latin translation. It must have been borrowed from an Arabic version or commentary. This proposition of Diophantus seeks to find three numbers with equal differences and such that taken two at a time they form a square, that is, they should be of the form $x - y = y - z$; $x + y = a^2$; $y + z = b^2$; $z + x = c^2$. But the solution demands that the three square numbers with equal differences be found previously and that their half-sum be greater than each of them. This auxiliary problem is then of the form $c^2 - b^2 = b^2 - a^2$, with $(a^2 + b^2 + c^2)/2 > c^2$; $> b^2$; $> a^2$. The solution of Diophantus is summarized as follows: He assumes $a^2 = x^2$ and $b^2 = x^2 + 2x + 1$; then $b^2 - a^2 = 2x + 1$; therefore, $c^2 = (x^2 + 2x + 1) + (2x + 1) = x^2 + 4x + 2$. Adopting the determination $c^2 = (x - 8)^2$, we have $x^2 + 64 - 16x = x^2 + 4x + 2$, so that $x = 62/20 = 31/10$. Consequently, $a^2 = (31/10)^2 = 961/100$; $b^2 = (41/10)^2 = 1681/100$; and $c^2 = (49/10)^2 = 2401/100$. But 100 times these numbers, namely 961, 1681, and 2401, also constitutes a solution, because these numbers have 720 as their common difference; and their half-sum $2521\frac{1}{2}$, exceeds each of them. This problem of Diophantus is of Babylonian origin, as Solomon Gandz, the orientalist, has shown in *Osiris*, VIII (1948), 27.

In proposing to add and subtract the number 5—and not numbers 1, 2, 3, and 4, which do not give solutions—in the problem put forth as a challenge to Fibonacci, John of Palermo must have taken into consideration the nongenerality of the problem. Number 6 could be proposed, the solution for which, $(2\frac{1}{2})^2 + 6 = (3\frac{1}{2})^2$ and $(2\frac{1}{2})^2 - 6 = (\frac{1}{2})^2$, is found in the oldest mathematical work in the New World, published in Mexico in 1556 by Juan Diez, companion of Cortez in the conquest of New Spain. The title of the work is *Sumario Compendioso*. (See the reprinting of this very rare Spanish book in facsimile with an English translation by David Eugene Smith [Boston: Ginn, 1921], p. 43.) Number 7 gives the following solution: $(2\frac{97}{120})^2 + 7 = (3\frac{103}{120})^2$; $(2\frac{97}{120})^2 - 7 = (\frac{113}{120})^2$.

67. In other words, if the sum of two numbers is even, the ratio of their sum to their difference does not equal the ratio of the greater to the smaller number, that is, we never have $a/b = (a + b)/(a - b)$. [The proof is omitted.—*Ed*.]

68. An indeterminate problem of the form $x^2 + x = y^2$; $x^2 - x = z^2$. [The proof is omitted.—*Ed*.]

69. In other words, the greatest of three successively odd square numbers exceeds the middle square by eight units more than the middle square exceeds the smallest. Symbolically, this is expressed by the formula $(2n + 5)^2 - (2n + 3)^2 = 8 + (2n + 3)^2 - (2n + 1)^2$. [The proof is omitted.—*Ed*.]

70. A problem of the form $(x^2 - y^2)/(y^2 - z^2) = m/n$.

Let there be given a ratio of number A to number B, and let numbers A and B be mutually prime, and successive or not. First let them be successive, with B greater than A. Take C as unity and from C arrange as many odd numbers in order as there are units in number B, the greater term. Let these numbers be D, E, F, G, and let squares H, I, and K be taken of E, F, and G. I say that the ratio of the difference between square H and square I to the difference between squares I and K is the same as the ratio of number A to number B, which is proved in the following way . . . .[71]

PROPOSITION 19

I wish to find three square numbers such that the addition of the first and second, and [also the addition] of the three [square] numbers, makes a square number.[72]

First I find two square numbers whose addition produces a square number and which are formed from mutually prime numbers. And let these be 9 and 16, whose addition produces 25, a square number. Then I take the square number that is formed by the sum *(aggregatione)* of all the odd numbers below 25, and this is 144, whose root is half [the sum] of the extreme odd numbers [in this series], namely 1 and 23.[73] From the addition of 144 and 25 comes 169, a square number. And thus three square numbers have been found, two of which make a square number, as do all [three square numbers] added together.[74] If we add to this [latter] square number [that is, 169] the square number obtained from the sum of all the odd numbers arranged from 1 to 167, whose root is 84, that is, half of 168, we obtain 7225, a square number whose root is 85. And so we have found four square numbers of which two or three, and also the sum of all three together, form a square number.[75] We can also add three different square numbers to 7225 and with each of them it would form a square number. The first of these is the square produced by [the sum of] all the odd numbers below 7225 and whose root is 3612;[76] the second is the square number whose square root is 720, which is produced by the sum of all the odd numbers below the fifth part of 7225 diminished by two odd numbers collateral *(collateralibus)* to this same fifth part;[77] the third square, whose root is 132, is produced from [the sum of] all the odd numbers below $\frac{1}{25}$ of 7225 made less by twelve odd numbers collateral ·*(collateralibus)* to this same $\frac{1}{25}$ part.[78] And so an infinity of square numbers can be found

which, when separated or united in this order, form a square number.

PROPOSITION 20

A problem proposed to me by Master Theodore, philosopher of the lord Emperor.

I wish to find three numbers which, when added to the square of the first number, form a square number; if to this square number is added the square of the second number, a square number is produced; similarly if to this square number is added the square of the third number, a square number is produced.[79]

Three square numbers must be found of which the sum of two forms a square, the sum of the three produces a square number, and the smallest of them is greater than the roots of the other two square numbers. Let these numbers be 36, 64, and 576; the root of the second will be 8, of the third 24, and these roots represent the second and third numbers of the three that are sought. For the first number I assume a root, and summing these three numbers I get 32 plus a root,[80] to which I add the

71. [The proof is very lengthy and is omitted here.— *Ed.*]

72. A problem of the form $x^2 + y^2 = z^2$; $z^2 + t^2 = u^2$.

73. The number of odd numbers below 24 is $24/2 = 12$ and their sum is $12^2 = 144$. The root 12 = $(1 + 23)/2$.

74. The three squares that have been found are 9, 16, and 144. We have, then, $9 + 16 = 25 = 5^2$, and $9 + 16 + 144 = 169 = 13^2$.

75. The number of odd numbers from 1 to 167 is $168/2 = 84$ and their sum is $84^2 = 7056$. Therefore, $169 + 7056 = 7225 = 85^2$. We then have four square numbers, 9, 16, 144, and 7225.

76. The number of odd numbers below 7225 is $7225/2 = 3612$ and their sum is $3612^2 = 13,046,544$. This square number when added to square 7225 gives the square number $7225 + 13,046,544 = 3615^2$.

77. The number of odd numbers below $7225/5 = 1445$ diminished by its two collateral, or adjacent, odd numbers, 1443 and 1441, is the number of odd numbers below 1441, namely $1440/2 = 720$; and their sum is $720^2 = 518,400$. The square number added to square 7225 gives the square number $7225 + 518,400 = 525,625 = 725^2$.

78. The number of odd numbers below $7225/25 = 289$ before diminution by twelve odd numbers collateral to 289, that is, [before diminution by] the odd numbers from 265 to 287. The number of odd numbers below 265 is $264/2 = 132$, whose sum is $132^2 = 17,424$. This square number added to square 7225 gives, then, the square number $7225 + 17,424 = 24,649 = 157^2$.

79. An indeterminate problem of the form $x + y + z + x^2 = l^2$; $l^2 + y^2 = m^2$; $m^2 + z^2 = n^2$.

80. That is, we assume as the first number a root as yet unknown, and the sum of the three numbers will be $8 + 24 + \text{root} = 32 + \text{root}$.

square of the root and obtain 32 plus the square of the root plus the root. And I want the whole of this to be equal to the first square assumed, namely 36. I subtract 32 from each side, leaving the square and root equal to four units.[81]

In order to see how to proceed in similar situations,[82] I assume square AC as the square each of whose sides is equal to the root that has been assumed [Fig. 9]; and to it [that is, square AC], I add

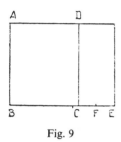

Fig. 9

a rectangular surface DE, which is a root of square AC. Therefore, CE will be 1 and DC a root, since it is one of the sides of square AC.[83] I now divide CE at F, and each of the segments CF and FE will be half of 1. Now, since we found a square and root equal to four units, it will be obvious that rectangular surface AE, which is produced by [the multiplication of] AB by BE, that is, BC by BE, will be 4; and since CE has been divided into two equal parts at F, adding to it in one direction gives a certain straight line CB; now the surface [formed by] BC on BE augmented by the square on line CF will be equal to the square on line BF.[84] But BC [multiplied] by BE produces 4, and if the square of number CF is added to it, which is $\frac{1}{4}$, $4\frac{1}{4}$ will be obtained as the square of number BF. Since this number does not have a root, we say that number BF is the root of $4\frac{1}{4}$. If CF—that is, $\frac{1}{2}$ of 1—is subtracted from this number [that is, BF, the root of $4\frac{1}{4}$], the root of $4\frac{1}{4}$ minus $\frac{1}{2}$ of 1 will be left as the root of BC, which number, although irrational (inratiocinatus), will serve as the first number sought,[85] while the second will be 8 and the third 24. For example, as the sum of these three numbers, we have $31\frac{1}{2}$ plus the root of $4\frac{1}{4}$; if to this sum we add the square of the first number, which is $4\frac{1}{2}$ minus the root of $4\frac{1}{4}$, we obtain 36, a square number.[86] And if to this is added 64, namely the square of the second number, we obtain 100, a square number, whose root is 10; if 576, the square of the third number, is added to this, we obtain 676, a square number, whose root is 26; and this is what we wanted.[87]

In order that the solution of the question raised

above be had in rational numbers, it must be shown first that when a fourth of one whole is added to any number formed by [the product of] two rational numbers, with one exceeding the other by 1, a square number is produced.[88] This is shown in surface AG [Fig. 10], which is formed by two

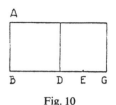

Fig. 10

rational numbers AB and BG where one exceeds the other by 1. Let the greater of the numbers be BG, and subtract GD, unity, from the greater number BG, leaving number BD equal to number AB. GD, unity, should now be divided into two equal parts at E, so that DE will be half of 1. Hence half of DE will be rational (ratiocinata); but number BD is also rational, so that the whole BE is a rational number and its square is also a rational number; and the surface which is equal to it [that is, to BE squared] is formed from [the multiplication of] BD by BG, that is, of AB by BG, plus the square of DE. But the square of DE is $\frac{1}{4}$ of 1; and the number produced from [the multiplication of] AB by BG is formed by two rational numbers, with one exceeding the other by 1. Consequently, when $\frac{1}{4}$ is added to a number formed from [the product of] two numbers, where one exceeds the other by 1,

81. We assume that 32 plus root plus the square of this root equals 36; hence, root plus square of the root equals $36 - 32 = 4$.

82. That is, we move from an abstract numerical procedure to a concrete geometrical procedure.

83. The area $DC \cdot CE$ is taken as the root of square AC, that is, $DC \cdot CE = \sqrt{AC}$; but $AC = (DC)^2$; therefore, $CE = 1$.

84. Elements, Book II, Proposition 5 (Heath, I, 382): "If a straight line be cut into equal and unequal segments, the rectangle contained by the unequal segments of the whole together with the square on the straight line between the points of section is equal to the square on the half." $BE \cdot BC + (CF)^2 = (BF)^2$.

85. $BF = \sqrt{4\frac{1}{4}}$ and $CF = \frac{1}{2}$; therefore, $BF - CF = BC = \sqrt{4\frac{1}{4}} - \frac{1}{2}$.

86. The three numbers sought are $\sqrt{4\frac{1}{4}} - \frac{1}{2}$, 8, and 24, whose sum is $\sqrt{4\frac{1}{4}} - \frac{1}{2} + 8 + 24 = 31\frac{1}{2} + \sqrt{4\frac{1}{4}}$. This sum, increased by the square of the first number, which is $(\sqrt{4\frac{1}{4}} - \frac{1}{2})^2 = 4\frac{1}{2} - \sqrt{4\frac{1}{4}}$, becomes $31\frac{1}{2} + \sqrt{4\frac{1}{4}} + 4\frac{1}{2} - \sqrt{4\frac{1}{4}} = 36 = 6^2$.

87. The first injunction of the text is satisfied by 36, the second by $36 + (8^2 = 64) = 100 = 10^2$, and the third by $100 + (24^2 = 576) = 676 = 26^2$.

88. That is, we have $n(n + 1) + 1/4 = m^2$.

the result is a square number;[89] and I wished to demonstrate this.

And it must be noted that all whole numbers formed by [the multiplication of] two collateral (*collateralibus*), namely successive (*continuis*), numbers [also] arise from the sequence of the regular addition of even numbers. Thus 2, which comes from unity multiplied (*ducta*) by 2, represents the first even number; and 6, which is formed by the multiplication (*ex ductis*) of 2 and 3, is [also] obtained by the sum of the first two even numbers; and 12, which is formed by the multiplication of 3 by 4, is [also] obtained from the sum of three even numbers, namely 2, 4, and 6; and by this same arrangement, the number formed by the multiplication of 10 by 11 is [also] produced by [the sum of] ten even numbers.[90] The same thing must be understood in all other numbers formed by [the multiplication of] two successive whole numbers. And it must be understood that every odd number is the sum (*aggregatio*) of two successive numbers, so that any odd number can be divided into two successive numbers, as 7, which is divided into 3 and 4.

Now, I want to show that when several of the roots of any square number are taken, and the number of these roots is divided into two parts, one of which exceeds the other by 1, and if one of these parts is multiplied (*multiplicetur*) by the other and the product (*quod provenerit*) is added to what remains of the square after its roots have been taken away, there will be produced a number formed by two unequal numbers of which the greater exceeds the smaller by 1.

In order to demonstrate this, let tetragon[91] *AG* be assumed, and take from it some of its roots, which constitute surface *EG* [Fig. 11]. Therefore,

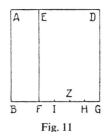

Fig. 11

number *FG* contains as many units as there are roots of square *AG* in surface *EG*. Now let number *FG* be divided into two parts of which the greater exceeds the smaller by 1; and let these parts be *FI* and *IG*, with *IG* the greater. I say that when surface *EG* is taken away from square *AG*, the remainder, namely surface *AF*, plus the surface formed by

[the multiplication of] *IF* and *IG* produces a number constituted of two unequal numbers, the greater of which exceeds the smaller by 1. And this is the same thing as taking from square *AG* the surface *EG* minus the surface formed by [the multiplication of] *IF* by *IG*. If, then, we assume that number *GH* equals number *IF*, leaving *IH* equal to 1, which is then divided into two halves, *ZI* and *ZH*, and the whole number *FG* is divided into two equal numbers at *Z* and into two unequal numbers at *I*, then the multiplication of *FI* by *IG* plus the square of *IZ* equals the square of number *FZ*.[92] Again, since number *FG* has been divided into two equal numbers at *Z* and number *BF* is added to it, the multiplication of *BG* by *BF*, that is, the multiplication of *AB* by *BF*, plus the square of number *FZ* equals the square of number *BZ*. But the square of number *FZ* is equal to the multiplication of surface *FI* by *IG* plus the square of half of *IZ*. Therefore, the multiplication of *AB* by *BF*, that is, surface *AF*, plus the multiplication of *FI* by *IG* plus the square of number *IZ* equals the square of number *BZ*.[93] Again, since the unit *IH* has been divided into two equal numbers at point *Z* and number *BI* is added to it, the multiplication of *BI* by *BH* plus the square of *IZ* will be equal to it. But the surface[94] *AF* and the surface of *FI* by *IG* plus the square of *IZ* are equal to *BZ* squared. Therefore, surface *BI* by *BH* plus the square of *IZ* equals surfaces *AF* and *FI* by *IG* plus the square of *IZ*. Ordinarily, with the square of *IZ* removed, there will remain surface *AF* plus surface *FI* by *IG*, which is equal to surface *BI* by *BH*. But surface *BI* by *BH* consists of two numbers, *BI* and *BH*, one of which exceeds the other by 1, since *IH* equals 1.[95]

89. *Elements*, Book II, Proposition 6 [see n. 16—Ed.]: $BD \cdot BG + (DE)^2 = (BE)^2$; or $AB \cdot BG + (DE)^2 = (BE)^2$. But $DE = \frac{1}{2}$, so that $(DE)^2 = \frac{1}{4}$ and $BG = AB + 1$; therefore, $AB(AB + 1) + \frac{1}{4} = (BE)^2$.

90. $10 \cdot 11 = 110 = 2 + 4 + 6 + 8 + 10 + 12 + 14 + 16 + 18 + 20$.

91. Here "tetragon" signifies a square.

92. *Elements*, Book II, Proposition 5 [see n. 84—Ed.]: $FI \cdot IG + (IZ)^2 = (FZ)^2$.

93. *Elements*, Book II, Proposition 6 [see n. 16—Ed.]: $BG \cdot BF + (FZ)^2 = (BZ)^2$, or: $AB \cdot BF + (FZ)^2 = (BZ)^2$, so that by substitution for the value of $(FZ)^2$ [from n. 92—Ed.], we get $AB \cdot BF + FI \cdot IG + (IZ)^2 = (BZ)^2$.

94. That is, the surface $AB \cdot BF = AF$.

95. For the straight line *IH* we also have according to *Elements*, Book II [see n. 16—Ed.]: $BI \cdot BH + (IZ)^2 = (BZ)^2$, so that we obtain [from relations in n. 93—Ed.] $AB \cdot BF + FI \cdot IG + (IZ)^2 = BI \cdot BH + (IZ)^2$, from which it follows that $AB \cdot BF + FI \cdot IG = BI \cdot BH$. But $BH = BI + IH$ and $IH = 1$; therefore, $BH = BI + 1$.

This can also be shown with numbers. Let AG be 100 and any of its sides 10; let there be removed from square AG 7 of its roots minus [the product of] the multiplication of FI by IG. Now, these roots are [represented by] surface EG, so that 30 will remain as surface AF. If [to this] we add [the product of] the multiplication of FI by IG, that is, of 3 by 4, we obtain 42, the number resulting from [the product of] BI by BH, that is, of 6 by 7. For the whole BG is 10, and if number FG, namely 7, is removed from this, there remains 3 for number BF; and if FI, which is 3, is added to this, the whole number BI will be 6; and if to this is added IH, unity, we will have 7 for number BH.[96]

Now that all these things have been demonstrated, we must return to the question of the philosopher[97] and proceed in the manner stated above[98] until we have a square *(census)*[99] and its root *(radix)*[100] plus 32 equal to the square of 36.[101] Then we see how many roots of 36 are contained in 32, that is, we divide 32 by the root of 36 and obtain $5\frac{1}{3}$ roots. But in order to discover a solution of the aforementioned question in the relation posited for the three squares stated above, namely 36, 64, and 576, it is necessary that we find some square which, after $5\frac{1}{3}$ roots have been taken away from it, leaves a number produced by the multiplication of the unequal numbers mentioned, where the greater exceeds the smaller by 1. If we can, we should find a certain number of roots exceeding $5\frac{1}{3}$, the roots previously mentioned. This can be done in an infinite number of ways. Let us assume 7 roots at will. Divide 7 into two parts, one of which exceeds the other by 1; therefore, they will be 3 and 4. Multiply 3 by 4, producing 12. By what has already been said, we know that when from some square 7 roots of it minus 12 are taken away, there will remain of this square a number produced *(procreatus)* from two unequal numbers, one of which exceeds the other by 1.[102] But we wish to find a square which, after $5\frac{1}{3}$ roots have been removed from it, leaves a number that can be produced from [the multiplication of] two numbers, one of which exceeds the other by 1. Therefore, the $5\frac{1}{3}$ roots of the square that we seek are equal to the 7 roots of the same square minus 12. Hence, if we add 12 to each part, there will result $5\frac{1}{3}$ roots and 12 drachmas[103] equal to 7 roots. Subtract $5\frac{1}{3}$ roots from each side and there will be left $1\frac{2}{3}$ of the root, which is equal to 12 units. Next we triple the whole and 5 roots equal 36, so that if we divide 36 by 5 we will have $7\frac{1}{5}$ as the root of the square we seek, namely of the first [square].[104] The root of

the first square was, indeed, 6; consequently, as 6 is to $7\frac{1}{5}$ so will 8 and 24 be proportionally related to the roots of the second and third squares. But $7\frac{1}{5}$ exceeds 6 by a fifth part of 6, so that if we add to 8 and 24 their fifth parts, we will have $9\frac{3}{5}$ as the root of the second square and $28\frac{4}{5}$ as the root of the third square.[105] And $9\frac{3}{5}$ will be the second of the three numbers sought, and $28\frac{4}{5}$ the third. But unknown as yet is the first number which when added to the second and third of the aforementioned numbers and added to the square of this first number forms the square of $7\frac{1}{5}$, which square is $51\frac{4}{5}+1/5^2$ [that is, $51\frac{21}{25}$].[106] Therefore, we posit a root as the first number and add it to $9\frac{3}{5}$ and $28\frac{4}{5}$ and obtain $38\frac{2}{5}$.[107] And to this we add the square

96. [Let surface $AG = 100$, so that $AB = BG = GD = DA = 10$. Now, $AG - EG = 100 - 7(10) = AF = 30$, and $AG - (EG - [FI \cdot IG]) = 100 - 7(10) + (3 \cdot 4) = 42$. Since $BG = 10$, it follows that $FG = 7$ (this is so because $EG = FG \cdot EF = FG \cdot BG = 7 \cdot 10$). Now, by assumption, $GH = FI$, and $IH = 1$; but $FI + IH + HG = 7$; therefore, $FI = 3$, $IH = 1$, $HG = 3$, so that $IG = 4$ (since $IG = IH + HG$). Furthermore, $BI \cdot BH = 6 \cdot 7 = 42$. That $BI = 6$ follows from the fact that $BG - FG = 10 - 7 = BF = 3$; but by assumption, $FI = BF$, so that $FI = 3$; therefore, $BI = 6$, since $BI = BF + FI$. Similarly, $BH = 7$, since $BH = BI + IH$, where $BI = 6$ and $IH = 1$.—Ed.]

97. That is, the question posed to Leonardo as a challenge by Theodore, the third scholar and probably the astrologer attached to the court of Frederick II.

98. See above, at the beginning of the solution.

99. [*Census* came to signify the square of the unknown, namely x^2. See Selection 22, n. 6.—Ed.]

100. Here 32 is the sum $8 + 24$.

101. We assume, then, the equation $x^2 + x + 32 = 36$, which is that already established at the beginning of the solution [see n. 81—Ed.].

102. The identity $x^2 - (a + b)x + ab = (x - a)(x - b)$ applied to the expressions $3 + 4 = 7$ and $3 \cdot 4 = 12$ gives $x^2 - (3 + 4)x + (3 \cdot 4) = (x - 3)(x - 4)$, or $x^2 - 7x + 12 = (x - 3)(x - 4)$. Therefore, the greater factor, $(x - 3)$, exceeds by 1 the smaller factor, $(x - 4)$.

103. The drachma, a monetary unit, signifies unity here.

104. The text of this passage can be rendered as follows: $5\frac{1}{3}x + 12 = 7x$, so that $12 = 7x - 5\frac{1}{3}x$, or $12 = 1\frac{2}{3}x$. Therefore, $3 \cdot 12 = 3 \cdot 1\frac{2}{3}x$, or $36 = 5x$, from which it follows that $x = 7\frac{1}{5}$ and $x^2 = (7\frac{1}{5})^2$.

105. The three auxiliary squares at the beginning of the proposition are $36 = 6^2$, $64 = 8^2$, and $576 = 24^2$. But the first definitive square found is $(7\frac{1}{5})^2$. But $6 + \frac{6}{5} = 7\frac{1}{5}$; therefore, proportionally $8 + \frac{8}{5} = 9\frac{3}{5}$ will be the root of the second square, and $24 + \frac{24}{5} = 28\frac{4}{5}$ will be the root of the third square.

106. [In the text this is written as $\frac{1}{5}$ $\frac{4}{5}$ 51. Here, however, it has been put into proper order, with a plus sign and exponent added, neither of which were known to Leonardo.—Ed.]

107. $x + 9\frac{3}{5} + 28\frac{4}{5} = x + 38\frac{2}{5}$.

of this root and obtain a square, a root, and $38\frac{2}{5}$, which [together] are equal to $51\frac{4}{5}+1/5^2$ drachmas. Then subtract $38\frac{2}{5}$ from each side and the square *(census)* and root *(radix)* will remain which are equal to $13\frac{2}{5}+1/5^2$.[108] As we did above, let us add $\frac{1}{4}$ to this, namely the square of half the root,[109] and we will have $13\frac{6}{10}+9/10^2$ [that is, $13\frac{69}{100}$], which is the one-hundredth part of 1369.[110] Now divide the root of 1369, namely 37, by the root of 100 and obtain $3\frac{7}{10}$, from which we subtract $\frac{1}{2}$, as half of the root, and $3\frac{1}{5}$ will remain as the first number.[111] And so this question is solved in rational numbers, and in this way it can be solved in an infinite number of ways.

I have also solved this question in whole numbers, of which the first is 35, the second 144, and the third 360, whose sum comes to 539. If to 539 we add the square of the first number, namely 1225, we get 1764, a square number whose root is 42. If to this square we add the square of the second number, which is 20,736, we obtain 22,500, a square number whose root is 150; and if to this square we add the square of the third number, namely 129,600 we obtain 152,100, a square number whose root is 390. I found these numbers by assuming three squares, namely 49, 576, and 3600, of which the sum of two, and [the sum of] all three, make a square number; and I added the roots of the second and third, namely 24 and 60, and obtained 84, which I divided by the root of the first square, namely by 7, to get 12. For this reason, it is necessary that I find a square number from which after taking away 12 roots of it, there would remain a number formed of two unequal numbers, one of which exceeds the other by 1. Therefore, I took 13 and divided it into successive parts, namely into 6 and 7, which I multiplied to obtain 42. It is necessary, then, that I find a square whose 13 roots minus 42 drachmas[112] would be equal to 12 roots of the same square. And having proceeded, then, in the order mentioned before, I have obtained the numbers written above.[113] From these squares I also found these other three numbers, namely $10\frac{2}{5}$, 64, and 160. But not only can three numbers be found in different ways by this method, but also four could be found with four square numbers of which two can be added, or three, and even all together, to form a square number. Therefore, with these four square numbers, namely with . . . and . . . and . . . and . . . ,[114] I found these four numbers, of which the first is 1295, the second $4566\frac{6}{7}$, the third $11,417\frac{1}{7}$, the fourth 79,920, and their sum is 97,199. If to this number we add the square of the first number,

namely 1,677,025, we obtain 1,774,224, a square number whose root is 1332; if we also add to this square. . . .[115]

108. We assume $x^2 + x + 38\frac{2}{5} = 51\frac{4}{5} + 1/5^2$, from which we obtain $x^2 + x = (51\frac{4}{5} + 1/5^2) - 38\frac{2}{5} = 13\frac{2}{5} + 1/5^2$ [As in n. 106, $13\frac{2}{5} + 1/5^2$ is written in the Latin text as $\frac{1}{5}\frac{2}{5}13$ and is equivalent to $13\frac{11}{25}$.—Ed.]

109. That is, $(\frac{1}{2})^2 = \frac{1}{4}$.

110. $x^2 + x + \frac{1}{4} = (13\frac{2}{5} + 1/5^2) + \frac{1}{4} = 13 + \frac{11}{25} + \frac{1}{4} = 13 + \frac{69}{100} = 13\frac{6}{10} + 9/10^2 = \frac{1369}{100}$.

111. $x^2 + x + \frac{1}{4} = \frac{1369}{100} = 37^2/10^2$, or $(x + \frac{1}{2})^2 = (\frac{37}{10})^2$, from which we obtain $x + \frac{1}{2} = \frac{37}{10}$, so that $x = \frac{37}{10} - \frac{1}{2} = \frac{37}{10} - \frac{5}{10} = \frac{32}{10} = 3\frac{1}{5}$.

112. That is, 42 units.

113. The problem is expressed by the equations $x^2 + x + y + z = l^2; l^2 + y^2 = m^2$; and $m^2 + z^2 = n^2$. The author leaves it to the understanding of the reader to apply the method followed in the first part of the proposition to determine the values $x = 35, y = 144$, and $z = 360$, which, substituted in the equations that precede, give, in effect, its solution: $35^2 + 35 + 144 + 360 = 1764 = 42^2$; and $42^2 + 144^2 = 22,500 = 150^2$; and finally $150^2 + 360^2 + 152,100 = 390^2$. In order to arrive at this result, the author could have first let $l = 7$. But $7^2 + 24^2 = 25^2$, from which it follows that he has taken $y = 24$ and $m = 25$. But $25^2 + 60^2 = 65^2$, so that he has taken $z = 60$ and $n = 65$. Adding the values chosen for y and z, we get: $24 + 60 = 84$. The first equation then becomes $x^2 + x + 24 + 60 = 49$, which is unacceptable. Then, in dividing $24 + 60 = 84$ by the value $l = 7$, we get $84/7 = 12$. According to his method, then, the author chooses two neighboring or successive numbers 6 and 7, whose sum is 13 and product 42. Consequently, it is necessary to find a square whose 13 roots minus 42 roots equals 12 roots of this square, so that one root of this square equals 42. Now, if one takes $l = 42$ instead of $l = 7$, the second equation $l^2 + y^2 = m^2$ becomes $42^2 + y^2 = m^2$, or $42^2 + 144^2 = 150^2$. Then $150^2 + z^2 = n^2$, or: $150^2 + 360^2 = 390^2$. In order to find x we have $x^2 + x + 144 + 360 = 42^2$, or $x^2 + x = 1260$, so that $x^2 + x + \frac{1}{4} = 1260\frac{1}{4} = (35\frac{1}{2})^2$, from which it follows that $(x + \frac{1}{2})^2 = (35\frac{1}{2})^2$, so that $x + \frac{1}{2} = 35\frac{1}{2}$, and therefore $x = 35$.

114. These four small lacunae are perhaps not accidents in copying, but may have been left deliberately when one considers that the question was posed as a challenge and that the author left the path to be followed to the wisdom and understanding of his adversary. Thus he inaugurates a custom which one finds among mathematicians of the seventeenth century.

115. Perhaps the solution was left deliberately incomplete for the reason assumed in the preceding note.

The problem, which seeks to generalize the preceding problem, is of the form $x^2 + x + y + z + t = l^2; l^2 + y^2 = m^2; m^2 + z^2 = n^2; n^2 + t^2 = p^2$. The four square numbers given in the text allow the solution to be completed. After a sufficiently lengthy calculation, the four equations become $1295^2 + 1295 + 4566\frac{6}{7} + 11,417\frac{1}{7} + 79,920 = 1332^2$; therefore, $1332^2 + (4566\frac{6}{7})^2 = (4757\frac{1}{7})^2$; so that $(4757\frac{1}{7})^2 + (11,417\frac{1}{7})^2 = (12,386\frac{1}{7})^2$; and finally, $(12,368\frac{4}{7})^2 + 79,920^2 = (80,875\frac{3}{7})^2$. It is not possible to rediscover the value which was given to l, and subsequently to y, z, and t, in order to find $x =$

129

25 A PROPOSITION ON MATHEMATICAL PROBABILITY

Nicole Oresme (ca. 1325—1382)

Translated and annotated by Edward Grant[1]

Proposition X. It is probable that two proposed unknown ratios are incommensurable because if many unknown ratios are proposed it is most probable that any [one] would be incommensurable to any [other].

As found in the first chapter, there are three types of ratios.[2] Some, indeed, are rational ratios, others irrational with denominations, i.e., commensurable to rationals, and perhaps there is a third type, namely irrational ratios which have no denomination because they are not commensurable to any rationals.

Let there be two unknown ratios. Now each might be rational, namely of the first type, and the argument would proceed as follows: with any whatever number of rational ratios forming one or more series of denominations, those which are mutually commensurable are much fewer than those which are incommensurable; and therefore it is likely that any two proposed unknown ratios are incommensurable.

The antecedent is [now] demonstrated.[3] Let 100 ratios of the multiple genus be taken according to the sequence of their denominations, as 2/1, 3/1, 4/1, 5/1, etc., up to 101/1, and let them be as 100 terms mutually compared. Then by comparing any one of these terms to any other of them there would be 4,950[4] ratios which are ratios of ratios and, of these, 25—and no more—are rational and all the others are irrational, as I shall show afterwards.[5] Now if more rational ratios were taken as terms, [for example] 200 or 300, and if ratios of these were taken, the ratio of irrational to rational ratios would be much greater. And if the ratios were taken in a genus other than multiple there will be even fewer mutually commensurable ratios. Indeed, all superparticular ratios are mutually incommensurable, as was seen above.[6] Furthermore, if some ratios are taken from one genus and others

pauca respicientes, edited with Introductions, English translations, and Critical Notes by Edward Grant (Madison, Wis.: University of Wisconsin Press, 1966), pp. 247–255. The notes presented here are based upon my analysis and summary in the introduction to the edition.

In this selection, we may have one of the earliest, if not the earliest, reasonably formal propositions in probability theory.

2. These are described in Selection 29, notes 2, 3, and 4.

3. The antecedent demonstrated in this paragraph declares that when a finite sequence of rational ratios is taken, the number of irrational ratios of ratios will exceed the number of rational ratios of ratios.

4. By taking 100 consecutive rational ratios from 2/1 to 101/1 and relating them two at a time, there would result $100 \cdot 99 = 9900$ possible *ratios of ratios* (for a definition of this expression, see Selection 51. 2, n. 51). Since Oresme is interested here only in relations involving ratios of greater inequality— that is, ratios of the form $n/1$, where n is an integer between 2 and 101, 9900 must be divided by 2, resulting in 4950 ratios of ratios of greater inequality. These may be represented generally by $n/1 = (m/1)^{p/q}$, where n and m are integers and p/q may be rational or irrational.

5. In a practical conclusion, which is really chapter 3, Proposition 11, Oresme provides the details of this particular example (see Grant, *Nicole Oresme: De proportionibus*, pp. 257, 259) and shows that of the 4950 ratios of ratios already mentioned, only 25 are "rational ratios of ratios." These 25 ratios are derived from all the geometric series contained within the sequence of given terms. Thus the series $(2/1)^n$, where $n = 1, 2, 3, 4, 5, 6$, will produce 15 rational ratios of ratios when any two are taken at a time (that is, $(6 \cdot 5)/2 = 15$). For example, one could take the ratios $(2/1)^2$ and $(2/1)^5$, where $(2/1)^5 = [(2/1)^2]^{5/2}$. Thus the rational exponent, 5/2, relates the two rational ratios 32/1 and 4/1. Similarly the series $(3/1)^n$, where $n = 1, 2, 3, 4$, yields 6 rational ratios of ratios; and one each will result from $(5/1)^n$, $(6/1)^n$, $(7/1)^n$, and $(10/1)^n$, where in each case $n = 1, 2$. The total number is therefore 25. The remaining 4925 ratios of ratios will be irrational—that is, the exponent relating the two rational ratios will be irrational. For example, taking 3/1 and 2/1, we see that $3/1 \neq (2/1)^{p/q}$ if p/q is a rational exponent. At the end of his practical conclusion, Oresme declares that the number of irrational to rational ratios of ratios is 197 to 1 (that is, 4925/25). For this reason, he insists, in the next two paragraphs of this selection, that if one were asked to guess whether or not an unknown ratio of ratios is irrational or rational, the mathematical probabilities would clearly dictate the choice of an irrational ratio.

6. That is, $(n + 1)/n \neq [(m + 1)/1]^{p/q}$, where n and m are integers greater than 2 and p/q is a rational exponent.

___

1295, since any arbitrary value could be taken for *l*. However, the presence of a denominator of 7 in the values for y, z, m, n, and p allows one to suppose that the initial value of *l* was 7, or a multiple of 7. However, the four small lacunae cannot be determined with certainty.

1. Reprinted with permission of the copyright owners, The Regents of the University of Wisconsin, from *Nicole Oresme: De proportionibus proportionum and Ad*

from another, this will result in the fewest ratios being mutually commensurable because, as is evident from the third proposition of this [chapter], all ratios in the multiple genus are incommensurable to [ratios of other genera], just as 10/1 is not commensurable to any rational before 100/1, and 7/1 is [commensurable] to [only] one [ratio before 100/1], and 12/1, 13/1, etc., are not commensurable to any below 100/1. For this reason if all the multiple ratios below 100/1 were taken, only 16 would be commensurable to any rational below 100/1, as will be seen afterward.[7] Thus the antecedent has now been stated and all the propositions beyond the ninth of this chapter are designed to make it evident.

I now state the principal consequence[8] With regard to numbers, we see that however many numbers are taken in series, the number of perfect[9] or cube numbers is much less than other numbers and as more numbers are taken in the series the greater is the ratio of noncube to cube numbers or nonperfect to perfect numbers. Thus if there were some number and such information as what it is or how great it is, and whether it is large or small, were wholly unknown—as is the case for the number of all the hours which will pass before antiChrist—it will be likely that such an unknown number would not be a cube number. A similar situation is found in games where, if one should inquire whether a hidden number is a cube num-

ber, it is safer to reply in the negative since this seems more probable and likely.

Now what has been said here about numbers may be applied to ratios of rational ratios, as was shown before, since there are many more irrationals than others, understood in the previous sense [of ratios of ratios]. What is more, [it is even more applicable to ratios, for] if one reflected carefully he would find that between ratios of rational ratios those which are rational are fewer than cube numbers in an aggregate of numbers. Therefore, if any unknown ratio of ratios were sought, it is probable that it would be irrational and its ratios incommensurable. And all this applies if the unknown ratios which are sought should be rational. . . .

And so it is clear that with two proposed unknown ratios—whether they are rational or not— it is probable that they are incommensurable, which was proposed in the first place.[10] Therefore, if many [unknown ratios] are proposed, it is [even] more probable that any one of them would be incommensurable to any other, which was proposed in the second instance. Now the more there are, the more one must believe that any one is incommensurable to any other, for if it is probable that one proposed ratio of ratios is irrational, it is more probable when many are proposed that any one would be irrational, just as could be shown in the example involving cube numbers.

26 INFINITE SERIES

Nicole Oresme (ca. 1325—1382)

Translated and annotated by Edward Grant[1]

QUESTION 1

Concerning the book of Euclid [that is, the

7. The 16 ratios less than 100/1 are all those cited above in note 5 except 10/1 and 100/1.
8. The principal consequence is the enunciation of Proposition X. With the antecedent demonstrated (see n.3), Oresme arouses that any unknown ratio of ratios would probably be irrational.
This is now shown by analogy with perfect and nonperfect numbers and cube and noncube numbers.
9. A perfect number is one whose sum of factors is equal to it (for example, $6 = 1 + 2 + 3$).
10. Oresme claims that his "probability theorem" applies to "ratios of ratios" of all three types that could be formed from the three types of single ratios described at the beginning of this selection—that is, (1) "ratios of ratios" where the ratios that are exponentially related are both rational, as for example, 2/1 and 3/1; or 2/1 and 16/1, and so on; (2) "ratios of ratios" where the two

exponentially related ratios are both irrational but have rational exponents, as for example, $(4/1)^{1/3}$ and $(2/1)^{1/2}$, which are related by a rational exponent (that is, $(4/1)^{1/3} = [(2/1)^{1/2}]^{4/3}$), or $(3/1)^{1/4}$ and $(2/1)^{1/2}$, which are related by an irrational exponent; and (3) "ratios of ratios" where the two exponentially related ratios are irrational with irrational exponents, as for example, $(8/1)^{\sqrt{2}}$ and $(2/1)^{\sqrt{2}}$, which are themselves related by a rational exponent (that is, $(8/1)^{\sqrt{2}} = [(2/1)^{\sqrt{2}}]^{3/1}$), or $(8/1)^{\sqrt{2}}$ and $(5/1)^{\sqrt{2}}$, which are not only irrational ratios with irrational exponents but are also relatable by an irrational exponent.
I have reproduced here only Oresme's discussion of the first type of "ratios of ratios" and omitted his brief discussions of types (2) and (3). By analogy with, and inference from, type (1), however, Oresme declares that for any finite number of ratios related two at a time in categories (2) and (3) there would be more irrational than rational "ratios of ratios." Without additional information about specific sequences of ratios, his claim cannot be properly evaluated.
1. This selection is from Oresme's *Questions on the*

Elements], we inquire first about a certain statement by Campanus asserting that a magnitude decreases into infinity.[2] First we inquire *whether a magnitude decreases into infinity according to proportional parts.*

In the first place, it is argued that it does not [increase into infinity according to proportional parts]. In a continuum there are not infinite parts of the same [finite] quantity, therefore there are not infinite parts of the same proportion. The antecedent is obvious, for then it would be infinite;[3] the consequent is clear because any proportional part is part of the same quantity with any other [proportional part]. Therefore, the parts of the same proportion and of the same quantity are the same.[4]

In the comment by Campanus the opposite [of this] is obvious.

For this question, these things must be noted: firstly, what has [already] been said, namely that proportional parts are[5] parts of the same proportion; secondly, in how many ways can such parts be imagined; thirdly, how something can be divided into such parts; fourthly, assumptions and propositions.

As to the first, it must be noted that proportional parts are said to be in continued proportion and such a proportion is a similitude of ratios, as it is called in the comment [by Campanus] on the ninth definition of the fifth [book of Euclid],[6] where it is said that such a [relation or similitude] is had between at least two ratios; and for this reason Euclid says that the least number of terms in which it [that is, similitude of ratios] is found is three, but he does not give a maximum number because it is a process [that continues] into infinity. It follows from this that it is not proper to speak of [a single] proportional part, nor of two proportional parts, but it is necessary that there be at least three[7] and there can be an infinite number. Proportional parts are said to be continually proportional because the first [proportional part] is related to the second as the second to the third, and so on if more are taken.

As to the second point, I reply that a division can be made into such proportional parts in as many ways as there are continuous proportionals; and there are just as many continuous proportionals as there are ratios, namely infinitely many. For example, it can happen that the first [proportional part] is double the second, and the second double the third, and so on,[8] just as is commonly said about the division of a continuum; and it could happen that the first [proportional part] is triple

the second and the second triple the third, and so on.[9]

As to the third point, it is held that a line and any continuum can be divided into such [proportional] parts. A line [can be so divided] in two ways because there are two extremities of it and such parts

Geometry of Euclid, written perhaps around 1350. The translation, containing all but a few lines of the first two questions, is based upon the Latin text edited by H. L. L. Busard, *Nicole Oresme: Quaestiones super geometriam Euclidis* (Leiden: Brill, 1961), pp. 1–6, and the many emendations and corrections to that text made by John E. Murdoch in his long review of Busard's text in *Scripta Mathematica,* XXVII (1964), 67–91.

The subject of infinite series, with its associated paradoxes, exerted a strong fascination for medieval natural philosophers and mathematicians. Many series were formulated and utilized by Oxford (Merton College) scholastics such as Swineshead, Dumbleton, Heytesbury, and others in the first half of the fourteenth century, and then by Parisian scholastics such as Nicole Oresme and, much later, Alvarus Thomas, to name only two; for some of the infinite series proposed by Alvarus Thomas, see H. Wieleitner, "Zur Geschichte der unendlichen Reihen im christlichen Mittelalter," *Bibliotheca Mathematica,* XIV (1914), 150–168.

2. Immediately following the axioms *(communes animi conceptiones)* and just before Book I of Euclid's *Elements,* Campanus of Novara remarks that "magnitude decreases into infinity, but in numbers this is not so."

3. Presumably, a defender of this position would argue that if a continuous quantity consisted of an infinite number of parts, it would be infinite.

4. Since there cannot be an infinite number of parts in a finite quantity, there cannot be an infinite number of proportional parts, because "the parts of the same proportion and of the same quantity are the same." In the concluding paragraph of this question, Oresme interprets this argument to mean that all proportional parts are equal, an absurd position which he promptly denies without elaboration, for it is contrary to all that has been shown in the question itself.

5. Although the text has *vel* (or), the sense seems to require "are."

6. For a translation of this definition, see Selection 27, Definition 9.

7. If three terms, a, b, and c, are in continued proportion, then $a/b = b/c$. In the same way, a proportion of proportional parts must consist of at least three proportional parts in continued proportion, as, for example, $(1:a)/(1:b) = (1:b)/(1:c)$. Hence at least three such ratios are required before we can speak of proportional parts. Murdoch remarks (p. 68), "The key notion in his analysis is correctly seen by Oresme to be that of 'proportional parts,' parts determined, he adds, by a given continued proportion. Thus, his fundamentals for an investigation of infinite geometric series are in fine order. Moreover, when he probes further his success is, by fourteenth-century standards, considerable."

8. That is, $(1/2)^n$, where $n = 1, 2, 3, 4. \ldots$

9. That is, $(1/3)^n$, where $n = 1, 2, 3, 4. \ldots$

can begin from either one. A surface [can be divided into proportional parts] in infinite ways, and similarly for a body.

As to the fourth point, a first supposition is assumed such that if any ratio were increased to infinity with the greater term unchanged, the smaller term would be diminished into infinity. This is obvious, since a ratio between two [terms] can be increased into infinity in two ways: either by the augmentation to infinity of the greater term or by the diminution to infinity of the lesser term.[10]

A second supposition is that if to any ratio there is added so much, and again so much, and so on into infinity, this ratio will be increased into infinity; and this is common to all quantities.

A third supposition is this: that to any quantity an addition can be made by proportional parts; and by the same [supposition] a diminution can be made by proportional parts.

The first proposition is that if an aliquot part should be taken from some quantity, and from the first remainder such a part is taken, and from the second remainder such a part is taken, and so on into infinity, such a quantity would be consumed exactly—no more, no less—by such a mode of subtraction.[11] This is proved because the whole that was originally assumed, and the first remainder, and the second remainder, and the third, and so on, are continually proportional, as could be proved when arguing with an altered ratio.[12] Therefore, there is a certain ratio, and then so much [of it], and so on endlessly; consequently, such a ratio of the whole to the remainder increases to infinity, because, by the second supposition, it is composed of these [ratios].[13] And the other term, say the whole, is imagined as unchanged, so that, by the first supposition, the remainder is diminished into infinity, and consequently, the whole quantity is consumed exactly.

This corollary follows: If from any foot [length] there should be taken away a half foot, and then half of the remainder of this quantity, and then half of the next remainder, and so on into infinity, the foot length will be removed exactly by this [procedure].[14]

A second corollary is that if one-thousandth part of a foot were taken away [or removed], then [if] one-thousandth part of the remainder of this foot [were removed], and so on into infinity, exactly one foot would be subtracted from this [original foot].[15]

But this is doubted. Since exactly half of one foot and then half of the remainder of that foot, and so

on into infinity, make one foot, let this whole [foot] be a; similarly, by the second corollary, one-thousandth part of this one foot, and one-thousandth of another [that is, the next] remainder, and so on into infinity, make one foot, let this be b. It is then obvious that a and b are equal. But it can be proved that they are not equal because the first part of a is greater than the first part of b; and the second part of a [is greater] than the second [part] of b, and so on to infinity, Therefore, the whole a is greater than the whole b. And this is confirmed, [for] if Socrates were moved over a for one hour and Plato over b [for one hour], and [if] they divide the hour by proportional parts and traverse a and b, respectively, then Socrates is moved quicker than Plato in the first proportional part; and similarly in the second [proportional part], and so beyond. Socrates will therefore traverse a greater distance than Plato, so that a is a greater distance than b.[16] In response to this, I deny the antecedent,

10. Oresme does not say but this supposition can apply only to ratios of greater inequality. He makes this quite explicit in his *On Ratios of Ratios (De proportionibus proportionum,)* ch. 1, *ll.* 106–111 (Grant, *Nicole Oresme: De proportionibus* [Madison, Wis.: University of Wisconsin Press, 1966], pp. 144–147.

11. Murdoch (p. 68) represents this series as:

$$\frac{a}{n} + \frac{a}{n}\left(1 - \frac{1}{n}\right) + \frac{a}{n}\left(1 - \frac{1}{n}\right)^2 + \cdots + \frac{a}{n}\left(1 - \frac{1}{n}\right)^m + \cdots = a.$$

12. By subtracting from a each successive proportional part in the series described in note 11, we obtain $a, a - a/n, a - 2a/n + a/n^2, a - 3a/n + 3a/n^2 - a/n^3, \ldots$, which are the successive remainders in continued proportion. These successive remainders, as Oresme tells us a few lines below, are diminished to infinity and "the whole quantity is consumed exactly."

13. The ratio of a, the whole, to the successive remainders forms a sequence of ratios increasing to infinity. Thus the successive ratios are $a/(a - a/n) < a/(a - 2a/n + a/n^2) < a/(a - 3a/n + 3a/n^2 - a/n^3) < \cdots$.

14. Oresme expresses here the convergent series $1/2 + 1/4 + 1/8 + \cdots + 1/2^n + \cdots = 1$.

15. Here we have the convergent series

$$\frac{1}{1000} + \frac{1}{1000}\left(1 - \frac{1}{1000}\right) + \frac{1}{1000}\left(1 - \frac{1}{1000}\right)^2 + \cdots + \frac{1}{1000}\left(1 - \frac{1}{1000}\right)^n + \cdots = 1.$$

16. This paragraph brings to the fore a doubt that arises from a comparison of corresponding terms in the series described in notes 14 and 15. Since the successive terms of the first series, called a, are greater than the corresponding terms in the second series, called b—and we see this by inspection where $1/2 > 1/1000$; $1/4 > (1/1000) [1 - (1/1000)]$, and so on—it is concluded that the whole magnitude in series a is greater than that of series b, making it impossible for them both to sum at 1. Oresme denies this (see remainder of paragraph) by insisting rightly that eventually the terms of series b will become greater than their corresponding counterparts in series a.

namely that the first part of *a* is greater [than the first part of *b*, and the second part of *a* is greater than the second part of *b*], and so on. The reason [for denying the antecedent] is that although the first part of *a* is greater than the first part of *b*, and the second part of *a* is greater than the second part of *b*, this is not so into infinity, since one part will eventually be reached that will not be greater than the part of *b* to be compared, but smaller. In this way the response to the question is clear, because in the imagination any continuum can have infinite proportional parts; similarly, the first [part] can really be separated from the others in thought[17] and then the second, and so on into infinity.

As for the argument to the contrary, I deny the consequent, and for proof I say that although any proportional part belongs to the same quantity as any other [proportional part], it is nevertheless not of the same quantity with that [part] with which it is of the same proportion. They are not equals, since it follows [that if] they are equals, therefore they are mutually equal to themselves.[18]

QUESTION 2

Next we inquire *whether an addition to any magnitude could be made into infinity by proportional parts*. . . .[19]

For this question, it must be noted in the first place that there is a ratio of equality and it is between equals; another [kind of ratio] is a ratio of greater inequality, which is of greater to smaller, as 4 to 2; and another [kind] is [a ratio] of lesser inequality, which is of smaller to greater, as 2 to 4. And these names differ with respect to relative superposition and subposition,[20] as is obvious in what has been said before; and by this there are three ways in which an addition can be made to any quantity.

Secondly, it must be noted that if an addition were made to infinity by proportional parts in a ratio of equality or of greater inequality, the whole would become infinite;[21] if, however, this addition should be made [by proportional parts] in a ratio of lesser inequality, the whole would never become infinite, even if the addition continued into infinity.[22] As will be declared afterward, the reason is because the whole will bear a certain finite ratio to the first [magnitude] assumed [or taken] to which the addition is made.

Finally, it must be noted that every term smaller than another which bears to it [that is, to the greater] a fixed ratio is called a fraction or frac-

tions, or part or parts; and this is obvious in the principles of the seventh book of Euclid, and it is denominated by two numbers, one of which is the numerator and the other the denominator, as is clear in the same place [that is, in the principles of Euclid's seventh book].[23] For example, one is smaller than two and is called one-half of two and one-third of three, and so on; and two is called two-thirds of three and two-fifths of five, and they ought to be written in this way;[24] and the two is called the numerator, the five the denominator.

The first proposition is that if a one-foot quantity should be assumed and an addition were made to it into infinity according to a subdouble [that is, one-half] proportion so that one-half of one foot

17. Although the text reads *per corrupcionem*, it seems to require emendation to *per concepcionem*, which I have translated as "in thought."
18. See note 4.
19. Eighteen lines, offering a few objections to the affirmative of this question, have been omitted from the translation.
20. I have emended Busard's text from *relativi* <sc. *termini*> *posicionis* to *relative* <*super*> *posicionis* and altered my translation accordingly. The terms 2 and 4 can be related so that one is relatively "superposed" or "subposed" to the other. Thus 4 "superposed" with respect to 2 produces a ratio of greater inequality; if "subposed"—that is, placed under 2—it produces a ratio of lesser inequality. Almost identical terminology was used by Oresme in his *On Ratios of Ratios*, ch. 1, *ll.* 201–204 (Grant, *Nicole Oresme: De proportionibus*, pp. 154, 155).
21. Should the proportional parts be represented as $(a/b)^n$, where $a > b$ and $n = 1, 2, 3, 4, \cdots$, an addition to infinity would produce an infinite whole (that is, a divergent series).
22. Here $a < b$.
23. In Book VII of Campanus' edition of Euclid's *Elements*, fractions are not mentioned. While Campanus speaks of *denomination* of ratios, he does not assert that two numbers are required to denominate a ratio nor does he identify these numbers as numerator and denominator. By "fraction or fractions, or part or parts," Oresme probably meant to distinguish between unit and non-unit fractions. That is, if *a* and *b* are numbers in their lowest terms and $a < b$, then if $a = 1$ and $b > 1$, we have a (unit) fraction or part; but if $b > a > 1$, we have fractions or parts.
24. Apparently Oresme intended to convey the precise manner in which fractions were to be written, but the Latin text, and presumably the manuscripts used to establish that text, fail to reveal this. It seemed best to verbalize them in full. The form desired by Oresme for expressing simple fractions is preserved in Part 1 of his *Algorismus proportionum*, as the reader can see in Selection 28. But just as the manuscripts of the *Questions on the Geometry of Euclid* display no unanimity, so also do the many manuscripts of the *Algorismus* present a bewildering variety of forms for representing irrational ratios.

is added to it, then one-fourth [of one foot], then one-eighth, and so on into infinity by squaring the halves,[25] the whole will be exactly double the first [magnitude] assumed.[26] This is obvious, because if these [very] parts were taken from any quantity, exactly double the first quantity would be taken, as is clear from the first question preceding. By a parity of reasoning, then, if the parts were added [the whole would be exactly double the first].[27]

The second proposition is this: If any quantity were assumed, say one foot, then let one-third as much be added [to it], and then [let] one-third of what was added [be added to the sum], and so on into infinity, the whole will be exactly $1\frac{1}{2}$ feet, namely in a sesquialterate ratio to the first quantity assumed.[28] Furthermore, this rule should be known: We must see how much the second part falls short of the first part, and [how much] the third [falls short] of the second, and so on with the others, and to denominate this [difference] by its denomination, and then the ratio of the whole aggregate to the quantity [first] assumed will be just as a denominator to a numerator. For example, in what was [just] proposed, the second part, which is one-third of the first, falls short by two-thirds of the [quantity of the] first, so that the ratio of the whole to the first part, or what was assumed, is as 3 to 2 and this is a sesquialterate [ratio].[29]

The third proposition is this: It is possible that an addition could be made, though not proportionally, to any quantity by ratios of lesser inequality, and yet the whole would become infinite; but if it were done proportionally, it would be finite, as was said. For example, let a one-foot quantity be assumed to which one-half of a foot is added during the first proportional part of an hour, then one-third of a foot in another [or next proportional part of an hour], then one-fourth [of a foot], then one-fifth, and so on into infinity following the series of [natural] numbers, I say that the whole would become infinite,[30] which is proved as follows: There exist infinite parts of which any one will be greater than one-half foot and [therefore] the whole will be infinite. The antecedent is obvious, since 1/4 and 1/3 are greater than 1/2; similarly [the sum

of the parts] from 1/5 to 1/8 [is greater than 1/2] and [also the sum of the parts] from 1/9 to 1/16, and so on into infinity. . . .[31]

25. "Squaring the halves" translates *duplando subduplos*.

26. That is, $1 + 1/2 + 1/4 + 1/8 + \cdots + 1/2^n + 1/2^{n+1} + \cdots = 2$.

27. If from any magnitude parts were taken in the order $1 + 1/2 + 1/4 + 1/8 + \cdots + 1/2^n + \cdots$, the total magnitude taken would be double the first given quantity (one foot in this case). Reversing the process and reconstituting the whole by adding the successive parts, a quantity double (that is, two feet) the first given quantity (that is, one foot) would be produced.

28. We have here the convergent series $1 + 1/3 + 1/9 + 1/27 + \cdots + 1/3^n + 1/3^{n+1} + \cdots = 3/2$.

29. The difference between the first and second terms is $1 - 1/3 = 2/3$. "To denominate this [difference] by its denomination" is, in medieval mathematics, to represent the quantity two-thirds by a ratio of numbers, namely 2/3. Then, when 2/3 is related as denominator to numerator, we get 3/2, the sum of the series. From the fact that Oresme offered only this example, it would appear that he intended that it apply to any two successive terms in the series. But if the second and third parts—or any other two successive parts—are taken, the concept of "denomination" cannot have its usual meaning for Oresme. For $1/3 - 1/9 = 2/9$, where the difference, two-ninths, is a ratio whose numerical denomination is 2/9, and the ratio of its denominator to its numerator, representing the final step, is 9/2, a result Oresme could not have intended. Obviously, $(2/9):(1/3) = 2/3$ is required where the reciprocal yields 3/2. But "to denominate a difference by its denomination" cannot mean representing 2/9, the difference, by the ratio (2/9):(1/3) unless, perhaps, Oresme extended the meaning of denomination to include the case where the greater of two ratios (in this instance 1/3) can serve as the denominator and denomination of the difference (that is, 2/9). Nevertheless, one is left with the strong impression that Oresme used the expression "to denominate this [difference] by its denomination" correctly in the only instance he gives and improperly assumed that it applied equally to all instances.

30. The series $1 + 1/2 + 1/3 + 1/4 + \cdots + 1/n + \cdots$ is divergent and is offered as illustration of the case in which ratios of lesser inequality are not added proportionally and can constitute an infinitely large whole.

31. Omitted here are the remaining twenty lines of Question 2, in which Oresme replies to the objections mentioned in note 19.

Proportions

27 THE DEFINITIONS OF BOOK V OF EUCLID'S *ELEMENTS* IN A THIRTEENTH-CENTURY VERSION, AND COMMENTARY

Campanus of Novara (d. ca. 1296—1298)

Translated and annotated by Edward Grant[1]

1. A quantity is part of a quantity, the lesser of the greater, when the smaller numbers [or measures] the greater.

Campanus. Sometimes part is taken properly, and this is something which, having been taken a certain number of times, constitutes its whole exactly without diminution or increase and is said to number [or measure] its whole by that [very] number according to which it is taken for the constitution of its whole.[2] Now, here he [that is, Euclid] defines [just] such a part, which we call "multiplicative." Sometimes part is taken commonly and this is any smaller quantity which however many times it has been taken constitutes its whole [only] more or less. We call this part "aggregative" because it would constitute its whole [only] with another different quantity; [but] by itself it could not produce the whole however many times it was taken.[3]

2. The greater is multiple of the lesser when the lesser measures it.

Campanus. A part is said to be relative to a whole and their mutual relation exists in these two extremes. And so having defined the lesser extreme [that is, that part], he [Euclid] defines here the greater extreme. Moreover, he calls it "multiple" because the lesser [term], having been taken a certain number of times, constitutes it. Therefore, they will be called "part" and "multiple" relatively. For every part is a submultiple, as is obvious by its definition.

3. A ratio is a relation, a determinate relation *(certa. . . habitudo)* of one to another of two quantities of the same kind of whatever size.

Campanus. A ratio is a mutual relation of two things of the same kind in the sense that one of them is greater or smaller [than] the remaining one, or equal to it.[4] For indeed ratio is found not only in quantities but in weights, powers, and sounds.

103–112. Of the various differing versions and editions of Euclid's *Elements* which circulated during the Middle Ages, there is little doubt that Campanus' was the most popular, as evidenced both by the large number of extant manuscripts and the frequency with which it was cited by those who had occasion to refer to Euclid's *Elements.* As M. Clagett has shown ("The Medieval Latin Translations from the Arabic of the *Elements* of Euclid," *Isis,* XLIV [1953], 16–42), Campanus drew his enunications of propositions from the twelfth-century translation by Adelard of Bath (from Version II of the three versions distinguished by Clagett) but apparently paraphrased and reworked the proofs from earlier complete versions, also adding material from other sources. This hybrid version of the *Elements* had conferred on it a measure of undying fame when in 1482 it was printed in Venice by Erhardt Ratdolt, thus becoming the first printed edition of Euclid's *Elements* in any language.

All of Campanus' definitions and commentary have been included here. While the reader may at times find it tedious and repetitious, it should be kept in mind that this was one of the most fundamental sources for medieval understanding of Euclidean proportionality theory. Campanus was instrumental in shaping medieval mathematical thought, not only because he popularized certain basic misinterpretations of crucial Euclidean definitions but also because of significant omissions that made more difficult a proper understanding of this vital fifth book. To illustrate and emphasize this, some definitions have been cited from the modern Greek edition in the translation of Sir Thomas L. Heath, *The Thirteen Books of Euclid's Elements,* 2nd. ed., translated from the text of Heiberg with introduction and commentary (New York: Dover, 1956).

2. If *A* and *B* are two quantities and $A > B$, *B* is said to be part of *A* if $A = nB$, where *n* is any integer.

3. Employing the data of note 2, $A \neq nB$, so that $A = nB \pm x$.

4. Since a quantity or magnitude could be made equal to, less than, or greater than another only if both were of the same kind, this conception of ratio went unchallenged in Greek antiquity and was, for the most part, accepted in the Middle Ages. An important exception, if it be that, may have emerged from discussions of motion, where ratios of motive force to resistive power were considered. In the narrow sense, such ratios seem to involve unlike quantities (see Selections 51.1,2, where the ratio F/R is repeated many times) even if only in the weak sense of construing the motive force as an active force and the resistance as a passive force. The use of metric formulations, however, involving ratios with unlike quantities did not become usual until the seventeenth century.

---

1. From *Euclidis Megarensis mathematici clarissimi Elementorum geometricorum libri XV cum expositione Theonis in priores XIII a Bartholomaeo Veneto Latinitate donata, Campani in omnes et Hypsicles Alexandrini in duos postremos* (Basel: Johannes Hervagius, 1546), pp.

In the *Timaeus,* where Plato shows the number of elements, he is of the opinion that there is a ratio in weights and powers.[5] Moreover, [that] there is a ratio in sounds is evident from music, for, as Boethius holds in the fourth [book of *On Music*], if any string [of a musical instrument] were divided into two unequal parts, the proportion between these parts and their sounds would be converse.[6] But all things in which a ratio is found share the nature and property of quantity; for ratio is not found in any two things unless one of them is greater or less than the remaining one, or equal to it. With respect to quantity, moreover, it is proper that it be called equal or unequal, as Aristotle put it in the *Categories*;[7] from this it follows that a ratio is found first in quantity and through this in all other things; nor is there a ratio between any things for which there is not a similar ratio in any [other] quantities. Hence Euclid put it very well when he said that ratio is in quantity absolutely, since he defined it by the mutual relation of two quantities of the same kind. However, this definition, that a ratio is a mutual relation of two quantities, must be understood in the sense that one of them is greater or smaller than the other, or equal to it. From this it is obvious that they must be of the same kind, as two numbers, or two lines, or two surfaces, or two bodies, or two places, or two times. For indeed a line[8] cannot be called greater or smaller than a surface or a body, nor can a time [with respect] to a place; but a line [is called greater or smaller with respect] to a line, and a surface [with respect] to a surface. They are comparable only univocally.

Because he [Euclid] says "determinate relation" *(certa habitudo)*[9] you should not understand this as if it were [necessarily] known or understood *(nota vel scita)* but [only] as if it were determinate [or fixed], so that the sense is, A ratio is a determinate relation of two quantities. And so I say "determinate," because it is this and not something else. Surely it is not necessary that every relation of two quantities be known by us, nor even [known] by nature, for some ratios are between discrete things, as numbers, others between continuous things. In numbers, moreover, the lesser is part or parts of the greater as is shown in the seventh [book], and for this reason in all these [ratios between numbers] there is a determinate and known relation. But in continuous things ratio is more extensive. For in continuous things when a smaller quantity is a part or parts of a greater, the ratio of all such things is known by means of numbers,

and such a ratio is called rational, and all such quantities communicant [that is, commensurable], because one and the same thing necessarily measures them, from which [it follows that] all numbers are communicant, since the unit measures them all. There is also [the case] where the smaller [quantity] is not a part or parts of the greater, and in such [instances] the ratio is known neither to us nor to nature. And this ratio is called irrational[10] and these quantities incommunicant [that is, incommensurable]. Hence any ratio found in numbers would be found in every kind of continuous quantity, as in lines, surfaces, bodies, and times; but not the converse, for there are an infinite number of ratios found in continuous quantities which are not rooted in the nature of numbers. But any ratio found in one kind of continuous quantity is found in all others; for howsoever any line is related to any other, so some surface is related to another [surface], and some body to another [body], and similarly [with] time, but not so for any number to another. It follows that ratio in continuous quantities is more extensive than in discrete quantities. From all this it is obvious that geometric proportion is of greater abstraction than arithmetic proportion, since every ratio with which arithmetic is concerned is rational, but geometry considers rational and irrational ratios equally.

4. Proportion *(proportionalitas)* is a similitude of ratios *(proportionum)*.[11]

5. *Timaeus* 52D–57D.

6. *De institutione musica,* Book IV, chapter 5, in the edition of G. Friedlein (Leipzig, 1867), pp. 314–315. The converse relation would appear to be based on Boethius' assertion that the greater the length of the string, the deeper the sound.

7. *Categories,* chapter 6, where Aristotle says: "Thus it is the distinctive mark of quantity that it can be called equal and unequal" (6a. 34–35).

8. The text has *in ea* where *linea* is clearly intended.

9. This expression appears in the definition itself, where I have translated it as "determinate relation."

10. In asserting that irrational ratios are known "neither to us nor to nature," Campanus stresses that this is not simply a matter of irrational ratios being concealed from human inquiry through some inability or defect on our part, but rather that the very nature of irrational ratios is to be indeterminate and unknowable. Their occurrence in nature itself is indeterminate.

11. Heiberg's edition of the Greek text omits this definition. Here we find Campanus using the Latin term *proportionalitas,* which has been translated as "proportion" but in other places as "proportional," the latter term also serving to render the Latin *proportionalis.* The word "ratio," which appears here with great frequency, translates the Latin *proportio,* since the latter term corresponds precisely to our use of the former term.

Campanus. As if we should say that as ratio *A* is to *B* so is *C* to *D*. The ratio between *A* and *B* is similar to that between *C* and *D*. Moreover, the similitude which results from these ratios is called a "proportion."

5. Quantities which are said to be in continued proportion *(proportionalitatem)* are those whose equimultiples are either equals, or, when taken in sequence *(sine interruptione),* equally exceed or fall short of one another.[12]

Campanus. Having assumed a division of proportionality into continuous and discontinuous, he [Euclid] defines these parts, taking continued proportion first. Indeed, speaking more truly, after having assumed a division of proportionals into continuously and discontinuously proportional, he defines not continued and discontinued[13] proportion but continuously and discontinuously proportional; furthermore, the definition of continued and discontinued proportion is sufficiently clear by a definition of continuously and discontinuously proportional.[14] Now, there is a continued proportion when, for any quantities of the same kind between which there is a ratio, the first [term] antecedes the second in the same [way that] any other antecedes the immediate consequent, as when we say just as *A* is to *B*[15] so is *B* to *C* and *C* to *D*. Any one of these [terms] will be an antecedent and a consequent except the first, which is only an antecedent, and the last, which is only a consequent. In this proportion, it is necessary that all quantities be of the same kind because of the continuation of the ratios since there is no ratio between quantities of different kinds. Continued proportion will be constituted from at least three terms.

There is a discontinued proportion when four, or all, of the quantities are of the same kind; or the first two are of one kind, and the last two of another. In this proportion, the first antecedes the second in the same way as the third antecedes the fourth—as when we say, as *A* is to *B* so is *C* to *D*; and each of these will be either an antecedent only or a consequent only. Nor is it necessary that all four be of the same kind—as was [the case] in continued proportion—because the consequent of the first ratio is not continued to the antecedent of the second, but it is possible for them to be the same or different in kind. For just as it happens that a line is found double or triple to [another] line, so [also can this be found for] a surface to a surface, and a body to a body, and a time to a time, and a number to a number.[16]

Having seen what a continued proportion is,

and what a discontinued proportion is, let us explain the definition of continuous proportionals that was stated before. He [Euclid] says that quantities are continuously proportional whose equimultiples are either equal to one another or, when taken in sequence, equally exceed or fall short of one another.[17] For example, let there be three

12. This definition of continued proportion has no counterpart in the Greek text of the *Elements*. Indeed, it seems to have replaced Definition 4, traditionally ascribed to Eudoxus (ca. 408 B. C.–ca. 355 B. C.), which reads (Heath, II, 114): "Magnitudes are said to have a ratio to one another which are capable, when multiplied, of exceeding one another." Thus, if two magnitudes *A* and *B* $(A < B)$ form a ratio, it must be the case that $nA > B$, where *n* is a positive integer. The designation of ratio would be denied where any of the magnitudes was either infinitesimally small or infinitely large. While other medieval translations and compilations of the *Elements* included Definition 4 in proper form, the most popular medieval version of the *Elements* substituted a quite different definition in which, as we shall see, the concept of "equimultiple" was incorporated, a concept drawn from Definition 5 of the Greek text (see the discussion of this in Def. 6 of Campanus' edition). By omitting Definition 4, Campanus' version lacked important criteria for determining whether or not a ratio could obtain between two magnitudes, for, as already seen, it was the express purpose of Definition 4 to eliminate the possibility of ratios involving the infinitesimally small or the infinitely large. In interpreting Campanus' fifth and sixth definitions, an article by John E. Murdoch ("The Medieval Language of Proportions: Elements of the Interaction with Greek Foundations and the Development of New Mathematical Techniques" in *Scientific Change,* ed. A. C. Crombie, [New York: Basic Books, 1963], pp. 237–271) has been very helpful.

13. The text mistakenly substitutes *continuam* for *incontinuam.*

14. Insofar as there is a genuine distinction between "continued and discontinued proportion" and "continuously and discontinuously proportional," it seems to lie in the application of the concept of "equimultiple" to the one (that is, continuously and discontinuously proportional) and not the other. That is, when equimultiples are taken of quantities in "continued proportion," the relation between the original quantities and the equimultiples would then be called "continuously proportional."

15. The text has *"D."* In Book II, chapter 40, of his *Arithmetica,* Boethius also calls this "continued proportion" (see Selection 2).

16. That is, we can have $L_2/L_1 = S_2/S_1 = B_2/B_1 = T_2/T_1 = N_2/N_1 = 3/1$, where *L* represents line, *S* surface, *B* body, *T* time, and *N* number. Here would be a case of a discontinued proportion involving more than four quantities, where no more than two are of the same kind. In Book II, chapter 44, of his *Arithmetica,* Boethius calls this a "disjunct" proportion in four terms (see Selection 2).

17. That is $A/B = B/C$, if, and only if, $nA/nB = nB/nC$. Although the text at this point does not state ex-

quantities of the same kind, *A, B,* and *C,* to which *D, E,* and *F* are taken as equimultiples, so that as *D* is multiple to *A* so *E* is multiple to *B,* and *F* to *G*; and all will be of the same kind, since multiples and submultiples are of the same kind. And let it happen that *D, E,* and *F* are either mutually equal or related similarly in excess or defect, so that just as *D* exceeds or falls short of *E* so *E* exceeds or falls short of *F*. When these multiples are related in this way, I say that the three quantities *A, B,* and *C* will be continuously proportional.[18] Now, the multiples that are related similarly should not be understood as exceeding or falling short with respect to quantity of excess, but [rather] with respect to proportion, for otherwise the definition would be false, since for any quantities of the same kind that exceed each other by equal differences, the equimultiples also exceed each other by equal differences. In excess or defect, therefore, they are related similarly as to excess of quantity. But the first [or prior] quantities [taken in the sequence] are not continuously proportional; indeed, the greater ratio is always generated from the smaller quantities. This happens because their multiples do not exceed themselves as to proportion but only as to quantity of excess, for the greater ratio is [formed] by the lesser multiples. For example, let three numbers be taken, as 2, 3, and 4, which exceed each other by equal differences, namely in arithmetic proportion *(medietate)*. All equimultiples of the three numbers exceed each other equally, so that double [these numbers produces equimultiples differing] by two; triple [these numbers produces equimultiples differing] by three, and so on.[19] However, 2, 3, 4 are not continuously proportional; on the contrary, a greater ratio is [formed] from the lesser terms, since the ratio of these is a sesquialterate [that is, 3/2]; the ratio of the greater terms is a sesquitertian [that is, 4/3]. Since there is no similitude of ratios between these terms, there will be no continued or discontinued proportion between them. It is obvious, then, that this similitude of excess or defect must not be understood as to quantity of excess but [only] as to proportion, and this will be the sense of the definition given above.

Continuous proportionals are all those whose equimultiples are continuously proportional. But he [Euclid] did not wish to propose this definition under this form because then he would be defining the same thing by the same thing.[20] However, [even] with its definition convertible, it is clear. Moreover, it is necessary that quantities *A, B,* and

C be of the same kind so that their multiples are mutually equal or related similarly in excess or defect. For if *A* and *B* are different kinds [of quantity], their multiples *D* and *E* would be the same kind of different quantities [as *A* and *B*], since

plicitly what has been represented here (Campanus will express it quite clearly later on in his commentary on this definition), it is proper to set it out in this manner because the concept of the equality, excess, and defect of equimultiples described here furnished Campanus with the criteria for determining the continued proportion of equimultiples. We must now inquire what is meant by the strange assertion about equimultiples that are "equal to one another" or "equally exceed or fall short of one another". Thus if $nA/nB = nB/nC$, and $nA = nB$, $nB = nC$, we have equimultiples "equal to one another"; however, should $nA > nB$ and $nB > nC$, these equimultiples are said to equally exceed "as to proportion" *(quantum ad proportionem;* Campanus explains this expression later in this definition); finally, if $nA < nB$ and $nB < nC$, Campanus would say that they equally fall short of one another "as to proportion." In considering equimultiples that "equally exceed or fall short of one another," Campanus explains later in this definition that "equally" must not be construed "as to quantity" *(quantum ad quantitatem)* by which he means, for example, that if $nA > nB$ and $nB > nC$, then $nA - nB = z$ and also $nB - nC = z$. For under these circumstances, it would be impossible to have continuous proportionality of the equimultiples, since $nA/nB \neq nB/nC$.

Until evidence is forthcoming, it may be conjectured that this strangely formulated definition was inherited by Campanus and that he strove valiantly to make sense of it, even to the extent of foisting upon it an interpretation seemingly at variance with the words themselves. For of the crucial but nearly unintelligible words with which Campanus had to contend (that is, "equally exceed or fall short of one another"), Murdoch remarks (p. 256): "Nothing could be farther from Eudoxus' true intentions; only *that* there be a corresponding excess or defect is required; that the excesses or defects be equivalent in amount is not only not specified, but makes no sense."

18. If $D = nA$ and $E = nB$, and $F = nC$ and $D/E = E/F$, then $A/B = B/C$ and *A, B,* and *C* are continuously proportional.

19. Given 2, 3, and 4 and taking $n = 2$, the equimultiples will be 4, 6, and 8 with a constant difference of 2. But since $8/6 \neq 6/4$, therefore $4/3 \neq 3/2$. The same lack of proportion results from any other value of *n*. When the equimultiples differ "as to quantity of excess" there can be neither continued nor discontinued proportion, for this occurs only when the equimultiples are continuously proportional. In Definitions 9 and 10 of the Greek text (corresponding to Defs. 10 and 11 in Campanus' text) of Book V, continuously proportional magnitudes are defined in terms of compounded, or composite, ratios, not equimultiples.

20. That is, he would then have defined the continued proportion of any continuously proportional terms by the continued proportion of their continuously proportional equimultiples.

139

multiples and submultiples are of the same kind, so that D could not be equal to E, nor greater or smaller, for multiples of diverse kinds are not mutually comparable.[21]

6. Quantities said to be in one [and the same] ratio *(proportio)*, the first to the second and the third to the fourth, are those whose equimultiples of the first and third are similar—whether in excess, defect, or equality—to the equimultiples of the second and fourth [quantities] taken in [corresponding] order.[22]

Campanus. Having posited a definition of continuously proportional quantities above [in Def. 5], he [Euclid] now posits a definition of discontinuously proportional quantities. And it is this: After taking equimultiples of the first and third and equimultiples of the second and fourth of any four quantities, and with respect to excess, defect, or equality, the multiple of the first is related to the multiple of the second as the multiple of the third to the multiple of the fourth, the ratio of the first of these quantities will be related to the second as the third to the fourth. For example, let there be four quantities, A, B, C, and D; and to the first and third, which are A and C, let equimultiples be taken that are double and are E and F; and again, let equimultiples be taken that are triple the second and fourth and are G and H. Now let these four multiples that have been taken be mutually compared according to the order of the first four quantities, so that E would be compared to G and F to H, [but] not, however, E to F as G to H. These [multiples] would be similar in excess, defect, and equality; that is, if E exceeds G and F exceeds H, or if E falls short of G and F similarly falls short of H, or if E is equal to G and similarly F is equal to H, then ratio A to B is as C to D.[23] Furthermore, just as in the definition of continuous proportionals, the similitude in exceeding or falling short should be understood here not as to quantity of excess but as to proportion. That he says "taken in [corresponding] order" should be understood as it was interpreted, namely that the multiples are not mutually related according to the order of the quantities in which the equimultiples were assumed, so that a multiple of the first [quantity] would not be related to a multiple of the third, nor a multiple of the second to a multiple of the fourth; but [rather] the multiples are related according to the first sequence of these four quantities, namely a multiple of the first to a multiple of the second, and a multiple of the third to a multiple of the fourth.

And so, consequently, the sense of this definition will be [this]: Four quantities are discontinuously proportional [when] the ratio of the first to the second is as the third to the fourth, and with equimultiples taken to the first and third and likewise to the second and fourth, the ratio of the first multiple to the second multiple is as the third multiple to the fourth multiple. But he [Euclid] did not define it under this form because of the reason stated above,[24] although it would be substantially the same. Now, it is not necessary that the four quantities A, B, C, and D be of the same kind, since B is not in continued proportion with C, but the first two can be of one kind and the two following of another kind. From this it is obvious that it is

21. As, for example, a time and a surface are not comparable.

22. This definition must be taken as equivalent to Book V, Definition 5, of the Greek text (Heath, II, 114): "Magnitudes are said to be in the same ratio, the first to the second and the third to the fourth, when, if any equimultiples whatever be taken of the first and third, and any equimultiples whatever of the second and fourth, the former equimultiples alike exceed, are alike equal to, or alike fall short of, the latter equimultiples respectively taken in corresponding order." Thus $A/B = C/D$ if, for all positive integers m and n, it follows that when $nA \gtreqless mB$, correspondingly $nC \gtreqless mD$. From the very opening words of his commentary on this definition, it will be seen that Campanus interpreted this definition as one of *discontinued proportion*, thus complementing his interpretation of the fifth and immediately preceding definition, which was construed as a definition of *continued proportion*. And so it was that the significance of Euclid's famous fifth definition was interpreted as being merely one of discontinued proportion rather than one in which criteria were laid down for determining equality of ratios for both commensurable and incommensurable magnitudes. Campanus' misinterpretation may have been partially responsible for the fact that Euclid's fifth definition (that is, the present sixth definition) played almost no role in medieval proportionality theory. Instead, another criterion was invoked for determining equality of ratios: ratios were said to be equal if their *denominations* are equal (that is, for rational ratios the denomination is a whole number or fraction representing the ratio in its lowest terms; for the denomination of irrational ratios see Selection 29, notes 3 and 4).

23. The criteria for equality of ratios expounded here by Campanus are in agreement with the customary interpretations of this definition as described near the beginning of note 22.

24. According to Campanus, Euclid did not wish to define the discontinued proportion of any four proportional quantities in terms of the discontinued proportion of their discontinuously proportional equimultiples. As with continuous proportionals (see n. 20), Euclid sought to avoid circularity of argument in his definition of discontinuous proportionals.

necessary that the multiple of the first be related to the multiple of the second and the multiple of the third to the multiple of the fourth; however, not the multiple of the first to the multiple of the third, nor the multiple of the second to the multiple of the fourth, since the multiples of the first and third are not always of the same kind, nor the multiples of the second and fourth. Furthermore, it is necessary to take equimultiples to the first and third, and likewise, equimultiples to the second and fourth, but not equimultiples to the first and second and to the third and fourth, for, unless by the taking of multiples the terms of the first ratio are continued with the terms of the second, it will not be by this means that ratio A to B is as C to D.[25]

7. Quantities whose ratio is one [and the same] are called proportional.[26]

Campanus. After he [Euclid] has defined continuously and discontinuously proportional quantities, he defines proportional quantities absolutely; and the definition is obvious.[27]

8. When there are equimultiples of the first and third and likewise of the second and fourth, and the multiple of the first exceeds the multiple of the second, but the multiple of the third does not exceed the multiple of the fourth, the first is said to bear a greater ratio to the second than the third to the fourth.[28]

Campanus. After defining proportional quantities, he [Euclid] now defines disproportional quantities. Disproportional quantities have no similitude of ratios, which happens in two ways: (1) either because the ratio of the first to the second is greater than that of the third to the fourth; or (2) because it is less. And thus there are two species of disproportionality: first, when the ratio of the first to the second is greater than the third to the fourth, and this is called *greater disproportionality;* second, when the ratio of the first to the second is less than that of the third to the fourth, and this is called *lesser disproportionality.* Therefore, he defines [only] those [quantities] between which the ratio of the first to the second is greater than the third to the fourth, and this is the greater disproportionality. But, because it is obvious from the other [definition], he does not posit a definition of those between which the ratio of the first to the second is smaller than the third to the fourth. Therefore, when there are four quantities and equimultiples are taken of the first and third and equimultiples of the second and fourth, and if the multiples of the first and second will not be similarly related with respect to excess, defect, and equali-

ty to the multiples of the third and fourth, those four quantities will be disproportional.[29] Because if it were true that the multiple of the first was equal to the multiple of the second, indeed, the multiple of the third would be less than the multiple of the fourth;[30] or [if it happened] that the multiple of the first was greater than the multiple of the second, then similarly the multiple of the third [would equal] the multiple of the fourth;[31] however, should the multiple of the first exceed the multiple of the second as to proportion (not as to quantity of excess) more than the multiple of the third [exceeds] the multiple of the fourth,[32] and should the multiple of the first fall short of the multiple of the second as to proportion (not as to quantity of excess) less than the multiple of the

25. In this concluding sentence Campanus explains that in discontinued proportion one cannot take equimultiples of the first and second and equimultiples of the third and fourth terms, because $nA/nB \neq mC/mD$ if $A/B = C/D$. Only in continued proportion, he explains, can equimultiples be taken of the first and second and third and fourth terms, for then $A/B = B/C$ when $nA/nB = nB/nC$ (see n. 17).

26. This corresponds almost identically with Definition 6 of the Greek text.

27. We see here how Campanus interpreted the relations among Definitions 5 to 7: The fifth defines continuously proportional terms; the sixth, discontinuously proportional terms; and the seventh includes the previous two definitions in the broadest sense of the term "proportional."

28. That is, if $nA > mB$ but $nC < mD$, then $A/B > C/D$. This is Euclid's seventh definition in the Greek text. As Heath explains (II, 130), Euclidean commentators distinguished the case where if $nA = mB$ and $nC < mD$, then also $A/B > C/D$. Euclid also failed to mention the case for a smaller ratio. Heath conjectures (II, 130) that "Euclid presumably left out the second possible criterion for a greater ratio, and the definition of a less ratio, because he was anxious to reduce the definitions to the minimum necessary for his purpose, and to leave the rest to be inferred as soon as the development of the propositions of Bk. V enabled this to be done without difficulty." As will be seen, Campanus discusses other possibilities as well as the notion of a lesser ratio.

29. If the criteria for equality of ratios of Definition 6 (this is Def. 5 of the modern edition) do not obtain, we have a disproportionality of the quantities and an inequality of ratios. Now, Campanus goes on to distinguish four ways in which this might occur to produce $nA/mB > nC/mD$.

30. Assuming that $nA/mB > nC/mD$, then we have as Case 1: $nA = mB$ and $nC < mD$.

31. Case 2: $nA > mB$ and $nC = mD$. (The possibility that $nC < mD$ forms part of Euclid's definition [see n. 28], but is ignored by Campanus.)

32. Case 3: $nA > mB$ and $nC > mD$, but $nA/mB > nC/mD$ (rather than, as is possible, $nA/mB < nC/mD$).

third falls short of the multiple of the fourth,[33] [then] in any of these four ways the ratio of the first to the second will be greater than the third to the fourth. Moreover, in the four ways opposite to these, the ratio of the first to the second will be less than the third to the fourth.[34] Examples of all these could obviously be taken from numbers.[35]

Therefore, this excess of the first multiple over the second but not of the third over the fourth, of which the author speaks in the definition, extends to and embraces the four ways mentioned before.[36] Hence the sense of this definition is [this]: When the multiples taken of the first to the second are as [those taken of] the third to the fourth, the ratio of the first to the second will be greater than that of the third to the fourth. However, Euclid did not define [it] under this form because of the common reason stated before.[37] Or we can say that the excess of the first multiple over the second but not of the third over the fourth—and this is spoken of in the definition of greater disproportionality that was set forth—is to be interpreted narrowly, just as the words of the definition signify, and is not extended [or applied] except to the second of the four ways previously mentioned.[38] But, in fact, the ratio of the first to the second is in all four of these ways greater than the third to the fourth. Hence the sense of this [eighth] definition follows: When multiples have been taken as Euclid proposes, if the multiple of the first is greater than the multiple of the second, it is necessary that the multiple of the third not be greater than the multiple of the fourth, then the ratio of the first to the second will be greater than the third to the fourth.[39] For this reason, he could not posit, in the aforementioned definition, the remaining three ways of exceeding, since this [second way] is more intelligible than all those and sufficient for the said definition. For, indeed, the ratio of the first of four quantities to the second is never greater than the third to the fourth, unless it happens that equimultiples of the first and third are found which, when related to equimultiples of the second and fourth, [are such that] the multiple of the first is found to exceed the multiple of the second, but the multiple of the third does not exceed the multiple of the fourth. But this is never found to occur unless the ratio of the first to the second is greater than the third to the fourth, as we shall demonstrate below [in commenting] on the tenth [proposition (?)] of this [book].

Just as quantities that are discontinuously proportional, these disproportional quantities can be of different kinds if there is a discontinued dispro-

portion *(incontinua improportionalitas)* between them, as if a ratio *A* to *B* were said to be greater than *C* to *D*. However, if there is a continued disproportion *(continua improportionalitas)*, all quantities would be necessarily of the same kind —just as they are in continued proportion—as if it were said that ratio *A* to *B* is greater than *B* to *C*.

9. A proportion *(proportionalitas)* is constituted between at least three terms.[40]

Campanus. After the author [Euclid] has defined ratio *(proportio)*, proportion *(proportionalitas)*, and proportional and disproportional quantitites, he shows what is the least number of terms constituting a proportion. He does not posit a maximum number of terms, since a maximum number cannot be taken, for any proportion, whether rational or irrational, can be continued to an infinite number of terms. At least two similar ratios are required for a proportion, since a proportion is a similitude of ratios. Moreover, every ratio has an antecedent and a consequent; therefore, every proportion has at least two antecedents and two consequents—that is, it is impossible to be made in fewer than three terms where the middle term is an

33. Case 4: $nA < mB$ and $nC < mD$, but $nA/mB > nC/mD$ (rather than $nA/mB < nC/mD$).

34. Assuming $A > B > C > D$, it will happen that $nA/mB < nC/mD$ when (1) $nA < mB$ and $nC = mD$; (2) $nA = mB$ and $nC > mD$; (3) $nA > mB$ and $nC > mD$, but now $nA/mB < nC/mD$ (here, the alternative presented in n. 32 is taken); and (4) $nA < mB$ and $nC < mD$, but now $nA/mB < nC/mD$ (here, the alternative presented in n. 33 is taken).

35. Numbers illustrating each case are given in the printed edition but omitted here.

36. Since all four ways produce the same result as mentioned here, namely that $nA/mB > nC/mD$.

37. The significance and interpretation to be placed upon "the sense of this definition" is not readily apparent. But it surely incorporates too much, for in the absence of other data it can be interpreted to mean that if $nA > mB$, so is $nC > mD$, so that $nA/mB \lessgtr nC/mD$ (these are the alternatives given in n. 32); and so on. Euclid's definition avoids this possibility and obtains only the desired result, and no more. Therefore, Campanus' reference to "the common reason stated before" is, perhaps, to the concluding sentence of his commentary on Definition 6, where he remarks that Euclid defined equality of ratios as he did rather than taking "equimultiples to the first and second and to the third and fourth" because the latter approach would have yielded the results sought only for very special conditions, thereby rendering the definition useless (see n. 25 for a brief discussion of this).

38. See Case 2 in note 31.

39. Of the cases outlined above, only Case 2 in note 31 meets these conditions.

40. Its counterpart in the Greek edition is Definition 8.

antecedent and a consequent. Thus the proportion will be continued and a continued proportion is constituted of at least three terms. A discontinued proportion, however, will not consist of fewer than four terms, because any one of them is an antecedent only or a consequent only. The same should be understood about the least number of terms constituting a disproportion, for if it should be continued, it will have at least three terms;[41] if discontinued, at least four terms.[42]

10. If three quantities are continuously proportional, the ratio of the first to the third is said to be the ratio of the first to the second squared (*duplicata*).[43]

Campanus. He [Euclid] defines a ratio constituted from the extremes of a continued proportion in three terms and says that if the ratio of the first to the second is as the second to the third, the ratio of the first to the third will be as the ratio of the first to the second squared (*duplicata*), that is, composed of two such ratios; or, and this is the same thing, the ratio of the first to the third will be as the first to the second squared, that is, multiplied by itself. In numbers, for example, let there be three continuously proportional numbers successively double, as 2, 4, 8. Then the ratio of the first to the third will be as the ratio of the first to the second multiplied by itself. Now the ratio of the first to the second is a double ratio [that is, 2/1, or 8/4], and a double ratio multiplied by itself produces a quadruple ratio [that is, 4/1]. Hence, a ratio of the extremes is a quadruple, namely the double of a double; or in accordance with the previous explanation, the ratio of extreme terms is as the ratio of the first to the second squared, since a quadruple ratio consists of two doubles.

11. When four quantities are continuously proportional, the ratio of the first to the fourth is said to be the ratio of the first to the second cubed (*triplicata*).[44]

Campanus. He [Euclid] defines a ratio constituted from the extremes of a continued proportion in four terms and says that if there were four quantities continuously proportional, the ratio of the first to the fourth would be as the first to the second cubed—that is, composed of three such ratios, since three such ratios are found in it; or, and this is the same thing, the ratio of the first to the fourth will be as the first to the second cubed—that is, multiplied by itself into the product. In numbers, for example, let there be four continuously proportional numbers successively triple, as 1, 3, 9, 27. Then the ratio of the first to the fourth will be as

the ratio of the first to the second multiplied by itself into the product. Now, the ratio of the first to the second is a triple [that is, 3/1] and a triple ratio multipled by itself produces the ratio of 9/1, and a triple [multiplied] by 9/1 produces a ratio of 27/1. Therefore, the ratio of extremes is a ratio of 27/1, which is the triple of a triple ratio; or in accordance with the first explanation, the ratio of extremes is as a ratio of the first to the second cubed, since 27/1 consists of three triple ratios.

But he [Euclid] does not define a ratio of extremes of a continued proportion constituted between more than four terms, because dimensions found in natural things do not exceed three.[45] The denomination of a ratio of two quantities with no mean [term] interposed has the nature of a line; the denomination [of the ratio] of two quantities between which one mean is interposed in continued proportion has the nature of a surface, because it is constituted from the multiplication of the [ratio of the] first two [terms] by itself. For everything produced from the multiplication of one line by another has the nature of a surface; indeed, if the line is multiplied by itself [it has the nature of] a square; if it is multiplied by another [line], one side will be longer than the other. But the denomination of a ratio of those quantities between which two means are interposed in continued proportion has the nature of a solid, since it arises from the multiplication of the denomination [of the ratio] of the first two terms by itself from which multiplication a surface is produced; it [that is, the denomination of the ratio of the first two terms] is then multiplied by this product, and a solid or body arises from this multiplication, since everything produced by the multiplication of a line by a surface grows into a solid. And so it is as if he should say that the ratio of two quantities is a simple interval and has the nature of a simple dimension, as a line; a proportion of three terms is a double interval and has the

41. $A/B \neq B/C$
42. $A/B \neq C/D$
43. If A, B, and C are the terms in question, then $A/C = A/B \cdot B/C$ and $A/C = (A/B)^2$. In the Greek text, this is Definition 9.
44. If A, B, C, and D are the four terms and $A/D = A/B \cdot B/C \cdot C/D$, then $A/D = (A/B)^3$. Omitted here, but added in Definition 10, the corresponding definition of the Greek text, are the words "and so on continually whatever be the proportion" (Heath, II, 114). In the commentary, however, Campanus extends it to any terms whatever.
45. Although Campanus' version of this definition stops at four terms, Euclid extended the definition to embrace any terms whatever, as can be seen in note 44.

nature of a double dimension, as a surface; a proportion of four terms is a triple interval and has the nature of a triple dimension, as a solid. And since dimensions are not produced further, he [Euclid] did not define a ratio obtaining between the extremes of a proportion constituted between five or more terms; or he did not define such a ratio because it could be had from the previously mentioned definitions. For if with three terms, the ratio of extremes consists of the ratio of the first [two terms] squared *(duplicata)*; and if with four terms [the ratio] consists of the same [first two terms] cubed *(triplicata)*; and if with five terms it consists of the same [first two terms raised] to the fourth power *(quadruplicata)*; and if with six terms it consists of the same [first two terms raised] to the fifth power *(quintuplicata)*, then with three terms continuously proportional, the ratio of the extremes contains the ratio of the first [two terms] twice; and with four terms [continuously proportional, the ratio of the extremes contains the ratio of the first two terms] three times; and with five terms [it will contain the first ratio] four times; and with six terms [it will contain the first ratio] five times, and so on, so that the ratio of extremes in continuously proportional terms contains the ratio of the first [two terms] as many times as [the number of] all the terms minus one.[46]

Similarly, if [there is] a ratio of extremes of a continued proportion constituted in three terms, it is produced from the ratio of the first [two terms] multiplied by itself; and [if] in four terms [the ratio of extremes is produced from the ratio of the first two terms] multiplied by itself twice; in five terms the ratio [of extremes] is produced from the ratio of the first [two terms] multiplied three times by itself; and in six terms [multiplied] four times by itself; and so there would always be two more terms than multiplications, or [to put it another way,] the [number of] multiplications would equal the mean terms interposed [between the extremes].[47] And it is known that in a continued proportion the ratio of the extremes is produced from all the intermediate ratios, as is obvious from what has been said before, and that a ratio of extremes of a continued proportion constituted in three terms is denominated by a square [number]; that which is constituted in four terms is denominated by a cube [number]; of these, the side of the square and cube is the denomination of the ratio of the first to the second [term].[48] For example, in numbers, let four numbers be continuously proportional which are successively triple, as 3, 9, 27, 81. The ratio of first to

second is denominated by three, for it is a triple [that is, 81/27];[49] [the ratio] of first to third is denominated by nine, which is the square of three, for it is 9/1; and [finally,] the ratio of first to fourth [that is, 81/3] is denominated by 27, which is the cube of the denomination of the ratio of the first to the second, namely 3, for it is 27/1.

But the ratio of the extremes of a continued disproportion constituted in three terms is denominated by a non-square surface whose sides are the denominations of these ratios; [when] it is constituted in four terms it is denominated by a non-cubic solid whose three sides are the denominations of the three ratios; and this is also obvious in numbers. Let there be four numbers continuously disproportional which are 2, 4, 12, 48. In these numbers, the ratio of first to second is double [that is, 4/2];[50] of second to third triple [that is, 12/4], and the ratio of first to third is 6/1 [that is, 12/2]; of third to fourth is 4/1 [that is, 48/12]. Therefore, 6, which is the denomination of the ratio of first to third, is a surface whose sides are 2 and 3, which are the denominations of the first two ratios;[51] 24, which is the denomination of the ratio of first to fourth, is a solid whose sides are 2, 3, and 4, which are the denominations of the three ratios between those four existing terms.[52]

12. Quantities in one proportion, antecedent to consequent and antecedent to consequent, are said to be converse [when related] as consequent to antecedent and consequent to antecedent; and again, alternately *(permutatim)* as antecedent is to antecedent so is the consequent to consequent.[53]

---

46. If n terms $A, B, C, D, E, \ldots, M, N$ are continuously proportional, then $A/N = (A/B)^{(n-1)}$.

47. Generally, if p is the number of terms in continued proportion and m the number of mean terms, then $p - 2 = m$, and m is equal to the exponent when a ratio of extremes is constituted.

48. In the example immediately following, ratio $81/27 = 3/1$ is the side of the square and cube. For if we take three terms constituting the ratios $81/27 \cdot 27/9 = 3^2$, 3 represents the side of a square; but if we take $81/27 \cdot 27/9 \cdot 9/3 = 3^3$, then 3 is the side of a cube.

49. Despite the order in which Campanus presents the terms, it is obvious that their places are numbered from right to left.

50. The ratios intended are obvious, but the order which Campanus ascribes to the numbers must be disregarded.

51. That is, $4/2 \cdot 12/4 = 2 \cdot 3 = 6$.

52. That is, $4/2 \cdot 12/4 \cdot 48/12 = 2 \cdot 3 \cdot 4 = 24$.

53. If $A/B = C/D$, then $B/A = D/C$ (which represents Def. 13 of the Greek text) and $A/C = B/D$ (which represents Def. 12). Omitted entirely is Definition 11, which reads (Heath, II, 114): "The term *corresponding*

Campanus. He [Euclid] defines six species of proportion, namely converse *(conversa)*, alternated *(permutata)*, separated *(disiuncta)*, conjoined *(coniuncta)*, everted *(eversa)*, and equal *(equa)*.[54] These species are like certain modes of arguing. First, he defines converse and alternated proportions, in which the antecedents and consequents remain the same according to substance (which is not so in separated, conjoined, or everted proportions) and nothing is taken beyond [the originally given] unequals.[55] He calls an "antecedent" the first extreme of a ratio, the "consequent" the second extreme. He means by this definition that [it is true only] if ratio *A* to *B* were as *C* to *D* and that I may conclude from this that *B* is to *A* as *D* to *C*, namely I could make consequents of antecedents and antecedents from consequents, and that this mode of arguing should be called contrary *(econtrario)* or converse proportion. But if I should argue that *A* is to *B* as *C* to *D*, so that *A* would be to *C* as *B* to *D*, that is, both extremes of the first ratio are made antecedents and both extremes of the second [are made] consequents, he would want this mode of arguing to be called an alternated *(permutata)* proportion. In this mode of arguing the antecedent of the second ratio becomes the consequent of the first; and the consequent of the first becomes the antecedent of the second.

13. Conjoined *(coniuncta)* proportion means that as often as an antecedent [is joined] with its consequent [and related] to [its] consequent, so also is the other antecedent [when joined with [its] consequent [related] to [its] consequent.

Campanus. He [Euclid] defines conjoined, separated, and everted [proportions] in which nothing is taken beyond [the original terms], but the terms do not remain the same according to substance. And he means that if it were so that *A* is to *B* as *C* to *D* and then from this I conclude that the whole *A* [plus] *B* is to *B* as the whole *C* [plus] *D* is to *D*, this mode of arguing should be called a "conjoined proportion."[56]

14. Separated *(disiuncta)* proportion means the equal comparison of the antecedents increased beyond the consequents.

Campanus. He means that if the ratio of the whole *A* [minus] *B* to *B* is as the whole *C* [minus] *D* to *D*, and then I conclude from this that *A* is to *B* as *C* is to *D*, this mode of arguing should be called a "separated proportion."[57]

15. Everted *(eversa)* proportion means a similitude of ratios for any antecedents increased beyond their consequents.

Campanus. He [Euclid] means that if *A* [plus] *B* is to *B* as *C* [plus] *D* to *D*, and then I conclude

---

magnitudes is used of antecedents in relation to antecedents, and of consequents in relation to consequents."

54. Of most of the terms mentioned here; which are included in Definitions 12–16 of the Greek text, Heath says (II, 134): "We now come to a number of expressions for the transformations of ratios or proportions. The first is ἐναλλάξ *alternately*, which would be better described with reference to a proportion of four terms than with reference to a ratio. But probably Euclid defined all the terms in Defs. 12–16 with reference to *ratios* because to define them with reference to proportions would look like assuming what ought to be proved, namely the legitimacy of the various transformations of proportions. . . ." What Euclid thoughtfully avoided was adopted in full by Campanus, who defines these six species as *proportions,* not ratios. But not only did Campanus define these as *proportions,* but he later discussed Euclid's demonstrations of four of them ending with formal proofs of what had already been defined (for example, in Bk. V, Prop. 16, he includes Euclid's proof of *alternated proportion:* "If there are four proportional quantities, they will also be alternately proportional"). Although Campanus was fully aware that he had before him definitions that were later given formal proofs (see later, near the beginning of his comment on Def. 16, for his remarks concerning Euclid's proofs of four of these six species of proportion), he seems not to have been disturbed by this.

55. The contrast between converse and alternated proportions on the one hand, and separated, conjoined, and everted proportions on the other, seems to lie in the fact that in the latter group either the consequents or the antecedents are altered by addition or subtraction, whereas in the former, while antecedents and consequents change relative positions, the quantities themselves remain constant, as Campanus tells us when he says "nothing is taken beyond [the originally given] unequals."

56. That is, if $A/B = C/D$, then $(A + B)/B = (C + D)/D$. In Heath's translation of Definition 14 of the Greek text, $(A + B)/B$, for example, is called "composition of a ratio" (see n. 54).

57. If this is a sound translation, then we may represent *separated proportion* as $(A - B)/B = (C - D)/D$. It is indeed what Euclid intended in his definition of "separation of a ratio," as Heath translates it in Book V, Definition 15, of the Greek text. This explains the necessary addition of "[minus]" to Campanus' commentary. But the vagueness and ambiguity of Campanus' version of this definition would allow one plausibly to replace "[minus]" with "[plus]," resulting in an interpretation identical with that of "conjoined proportion," namely $(A + B)/B = (C + D)/D$. On the latter interpretation, it must be argued that Campanus intended to distinguish *conjoined proportion* from *separated proportion* by supposing that in the former if $A/B = C/D$ then $(A + B)/B = (C + D)/D$, whereas in the latter the converse is true, namely if $(A + B)/B = (C + D)/D$ then $A/B = C/D$. In short, the two types of proportion would represent converse arguments. Indeed, both interpretations are consonant with the

from this that A [plus] B is to A as C [plus] D is to C, this mode of arguing should be called an "everted proportion."[58]

16. Equal proportion means that when several quantities have been proposed and an equal number of other quantities have been set forth in one proportion, there will be a similitude of ratios of the extreme terms of each [set of quantities] after an equal number of mean terms have been removed [from each set].[59]

Campanus. He [Euclid] defines equal proportion—it is [also] taken as a proposition to be proved separately[60]—and means that if any quantities, as A, B, and C, are taken, and also several others, as D, E, and F, the latter being either of the same kind as the first [set] or of another kind, and should the second [set of quantities] be in the same proportion as the first, [then] whether they are taken in the same order—as if it were said that A is to B as D to E and B to C as E to F—or in a converse order—as if it were said that A is to B as E to F and B to C as D to E[61]—it could be concluded from this that A is to C as D to F [and] that this mode of arguing should be called an "equal proportion."

Of these six modes of arguing which are said to be species of proportion, the author proves four by means of letters in this fifth book. He proves "alternated proportion" in the sixteenth proposition of this book, "separated" in the seventeenth, "conjoined" in the eighteenth, and "equal proportion" in the twenty-second and twenty-third (in the twenty-second the quantities of the two series are proportional [taken] in the same order; in the twenty-third they are proportional [taken] in the converse order). He does not demonstrate "converse" or "everted" proportions because converse is obvious from the definition of discontinuous proportional quantities, and everted is clear from alternated proportion with the aid of the nineteenth [proposition], as we shall assert [in our commentary] on the same nineteenth proposition.[62] But let us now demonstrate how a converse proportion is manifested from the division of discontinuously proportional quantities. Let ratio A to B be as C to D, so that I want to demonstrate that B will be to A as D to C. Should equimultiples E to A and F to C be taken, and similarly equimultiples G to B and H to D, it will happen, by conversion of the definition of discontinuously proportional quantities, that E and G and also F and H would be related similarly with respect to excess, defect, and equality. I understand, then, that [by conversion]

B is the first, A the second, D the third, and C the fourth. Now, having taken equimultiples G and H to the first and third [quantities] and equimultiples E and F to the second and fourth, and since the multiples of the first and second [quantities], namely G and E, are related similarly with respect to excess, defect, and equality to the multiples of the third and fourth [quantities], namely H and F, the ratio of the first [term], B, to the second, A, will, by the definition stated, be as the third [term], D, to the fourth, C.[63] And this was proposed. In this

enigmatic phrase "antecedents increased beyond the consequents"; for in the first way we must construe it as $(A - B)$, in the second as $(A + B)$.

Despite the alternative possibilities, the interpretation outlined above was chosen because in Book V, Propositions 17 and 18, *separated proportion* is correctly represented. Nevertheless, the ambiguous state of the definition, which is left unclarified in the commentary, leaves legitimate doubt as to which of the interpretations Campanus had in mind.

58. That is, if $(A + B)/B = (C + D)/D$, then $(A + B)/A = (C + D)/C$, and we see that Campanus derived everted proportion from conjoined proportion (see Def. 13). In an additional comment to Book V, Proposition 19, Campanus offers a proof of everted proportion in which he relies on Proposition 19 and alternated proportion. It should be noted that this definition has no counterpart in the Greek text of Euclid.

59. We have here a shortened version of Book V, Definition 17, of the Greek text, which reads (Heath, II, 115): "A ratio *ex aequali* arises when, there being several magnitudes and another set equal to them in multitude which taken two and two are in the same proportion, as the first is to the last among the first magnitudes, so is the first to the last among the second magnitudes; or, in other words, it means taking the extreme terms by virtue of the removal of the intermediate terms." In Heath's estimation (II, 136), "The meaning is clear enough. If $a, b, c, d. . .$ be one set of magnitudes, and $A, B, C, D. . .$ another set of magnitudes, such that

a is to b as A is to B,
b is to c as B is to C,
and so on, the last proportion being, e.g.,
k is to l as K is to L,
then the inference *ex aequali* is that
a is to l as A is to L.
The *fact* that this is so, or the *truth* of the inference from the hypothesis, is not proved until V.22. The definition is therefore merely verbal; it gives a convenient *name* to a certain inference which is of constant application in mathematics. But *ex aequali* could not be intelligibly defined except with reference to two sets of ratios respectively equal." Compare note 54.

60. As mentioned earlier, Campanus seems undisturbed by the fact that Euclid demonstrates this definition. See note 54.

61. The converse relation is demonstrated in Book V, Proposition 23.

62. See note 58.

63. Assuming $A/B = C/D$, Campanus wants to dem-

manner, the mode of arguing called "converse proportion" is established.

The principles of this fifth book seem to be very
difficult for most [people], and, [judging] from
certain propositions which he [Euclid] demonstrates from them, they appear rather remote from
the understanding. Now, nothing seems to stick in
the mind [and understanding] more immediately
than [the proposition] that of any two equal quantities only one [equal] ratio can be formed with any
third quantity; yet the seventh proposition of this
fifth book demonstrates this from the definition of
continued[64] proportion which seems, at first
[glance], very remote from the understanding. For
who would not concede that it is easier [to understand the assertion] that there is the same ratio
between any two equal quantities to any third
quantity than [the assertion] that if equimultiples
of the first and third of four quantities be taken and
related similarly in excess, defect, and equality to
equimultiples taken of the second and fourth,
[then] the ratio of the first to the second would be as
the ratio of the third to fourth?[65] If we regard [or
contemplate] the truth subtly, it will be clearly
established that [the proposition] "the ratio of two
equal quantities to a third forms one [and the same]
ratio" cannot become a part of the understanding
unless [one understands] "what it is to be the [same]
ratio." For if one were ignorant of what it is to have
the same ratio with another, how could he know
that two equal quantities form the same ratio to a
third quantity? Undoubtedly, then, the understanding requires [something] before it comprehends what appears to be an intelligible proposition; and it understands [this proposition] by knowledge of this definition [that is, the definition of
"what it is to be the same ratio"] which is subsequently applied to see whether it supports [the
claim about] two equal quantities compared to a
third. For if the definition were found to apply to
these quantities, one could infer what was proposed
[namely that "the ratio of two equal quantities to a
third forms one and the same ratio."] But if the
opposite [were true], then the proposition, which
had been judged immediate by a superficial comprehension, is not [at all] immediate.

In a similar way also, an initial understanding
might judge that [the assertion]

> of any two unequal quantities the ratio of the
> greater to another [third quantity] is greater than
> [the ratio of] the lesser to that same [third] quan-
> tity

(Euclid demonstrates this in the eighth proposition

of this book) sticks in the mind more immediately
[and directly] than [the claim] that

> of four quantities there is a greater ratio of the
> first to the second than of the third to the fourth
> when, after equimultiples have been taken of the
> first and third and similarly equimultiples of the
> second and fourth, the multiple of the first ex-
> ceeds the multiple of the third but the multiple of
> the third does not exceed the multiple of the
> fourth,

from which the proposition declared before is
demonstrated. But, similarly [as before], the first
assertion cannot be comprehended except by an
understanding of "what it is to be a greater ratio."[66]

Because of this, it was necessary that Euclid define which quantities are to be called proportional
and which disproportional. Those quantities are
proportional whose ratio is one [and equal], and
disproportional whose ratios are different. Moreover, he defined as continuously proportional
quantities those whose ratio is one [and the same]
and in which the extreme terms are not joined by
disconnected means. And he said that this proportion must exist in at least three terms because one
term must be taken twice as a mean. Those [quanti-
ties] in which the means are interrupted are
discontinuously proportional and this proportion
exists in at least four terms because another mean
must be taken. He also defined disproportional
quantities in which one ratio is greater than another.

Now, if every ratio were known or rational, the
mind could easily know which ratios are one [and
equal] and which different. Those having one de-

---

onstrate that $B/A = D/C$. If $E = nA, F = nC$,
$G = mB$, and $H = mD$, then $E/G = F/H$. Should the
proportion given initially be converted, we obtain B/A
and D/C, so that B is the first term, A the second, D the
third, and C the fourth. In the converted order G and H,
which are now equimultiples of B and D respectively, are
therefore equimultiples of the first and third quantities;
similarly, E and F are equimultiples of A and C, the
second and fourth terms. Hence "by conversion of the
definition of continuously proportional," $G/E \gtreqless H/F$,
so that $B/A = D/C$. In Euclid's Greek text, but not in
Campanus, converse or inverse proportion is given as a
porism to Book V, Proposition 7.

64. "Continuae" has been substituted for "incon-
tinuae" (discontinued), since it is obvious from Campanus' exposition of Book V, Proposition 7, that such a
change is required.

65. This is Definition 6 in Campanus' version.

66. The second assertion is the definition of greater
ratio, namely Definition 8, which, while admittedly more
difficult to comprehend, is utilized in Book V, Proposition 8, and, since it is prior to the latter, is deemed
essential to a proper understanding of the latter.

nomination would be one [and equal]; those with different [denominations] would be different [and unequal]. Now, this is easily obvious in arithmetic, since the ratio of all numbers is known and rational. Thus in the second book of his *Arithmetic,* Jordanus,[67] defining which ratios are the same and which different, says that those are the same which receive the same denomination; the greater [ratio that which receives] the greater [denomination]; and the lesser [ratio that which receives] the lesser [denomination]. But there are infinite irrational ratios whose denominations are not knowable, so that when, in this book, Euclid considered proportionals generally [and] not by restricting them to rationals or irrationals—indeed, he understood that ratio is found in continuous quantities, which are common to both rationals and irrationals—he could not, as in arithmetic, define identity of ratios by identity of denominations, because, as was said, the denominations of many ratios are absolutely unknown.[68] But it is necessary that a definition be formulated in terms of things that are known, so that the intractability *(malitia)* of irrational ratios compelled Euclid to posit such definitions. And since, as is obvious from what has already been said, he could not define proportion or identity of ratios by identity of relations or denominations of these terms—[and this was] because of the irrationality of the relations and the unsuitability of the terms—he was compelled to take refuge in the multiples of terms, so that from their relations, considered with respect to excess and equality and taken in equal [numbers or] multitudes, comes the proposed definition by which they are reduced to the nature of irrationality *(ad naturam irrationalitatis).* For in any kind of inequality, nothing is more like the terms than the multiples of the terms—[and this follows] not from the relations between the [original] terms but from the relations between the multiples.[69] Now, because a ratio is a certain relation between two quantities of the same kind in the sense that they are equal, or one is greater than the other, an identity of ratios existing between four quantities, the first to the second and the third to the fourth, is a similar equality of the first to the second and the third to the fourth; or a similar excess *(maioritas)* or defect *(minoritas).* This similar equality, excess, or defect obtains between any four quantities when there are equimultiples between all of them.

That he says in the fifth definition that "quantities which are said to be in continued proportion are those whose equimultiples are either equals,

or, when taken in sequence, equally exceed or fall short of one another," is as if he were to say that I call all these quantities continuously proportional (that is, that they are similarly continuously equal, continuously exceed, and similarly continuously fall short) all of whose equimultiples are either similarly continuously equal mutually to one another, or similarly continuously exceed, or continuously fall short of one another, which is to say that these multiples are continuously proportional because if this [relation] differs anywhere in the multiples, I say that they are not continuously proportional.

Furthermore, that he says in the sixth definition that "quantities said to be in one [and the same] ratio, the first to the second and the third to the fourth," and so on, is as if he were to say that I call all four quantities discontinuously proportional and that the first is related to the second as the third to the fourth (this means that the first is related similarly to the second as the third to the fourth in equaling, exceeding, or falling short) and that all of the equimultiples of the first and third are related similarly either in equaling, or exceeding, or falling short of all equimultiples of the second and fourth. That is, multiples of the first are related in the same ratio to multiples of the second as multiples of the third are related to multiples of the fourth, for if this relation differs anywhere among the multiples, I say that the ratio of first to second would not be as the third to the fourth.

And what he says in the eighth definition is as if he were to say that of four quantities I call the ratio of first to second greater than the third to

67. This constitutes the last of a series of definitions prior to the first proposition of Book II of Jordanus de Nemore's *Arithmetica.*

68. For the denomination of ratios see note 22. In this section Campanus explains that because denominations of irrational ratios are not possible numerically, Euclid was compelled to ignore equality of denomination as a criterion for equality of ratio and instead resort to "equimultiples" for a definition of equal ratio that would embrace both rationals and irrationals. In the fourteenth century, irrational ratios with rational exponents were assigned "mediate" numerical denominations, leaving only those with irrational exponents as wholly intractable to the method of denomination (see Selection 29, notes 3 and 4).

69. That multiples are related exactly as their respective submultiples is not because of the original quantities but only because of the multiples themselves, which happen to bear the same relation to each other. There is no necessary connection between them and no special priority or privileged status need be accorded the originally given quantities.

fourth (that is, the first exceeds the second more than the third exceeds the fourth) when any multiples of the first exceed any multiples of the second, and any multiples of the third [correspondingly equal] in number to the multiples of the first do not exceed any multiples of the fourth [correspondingly equal] in number to the multiples of the second. This is what it means to have a greater ratio of the multiple of the first [term] to the multiple of the second term than of the multiple of the third to the multiple of the fourth.

There are some, however, who demonstrate these definitions and among these is Ametus, son of Joseph, who maintained that he demonstrated them in his *Letter,* which he composed *On Ratio and Proportion.*[70] By way of assumption, he takes three [definitions] as principles which he says are self-evident and do not require proof. The first of these is that if there were four quantities of which the ratio of the first to the second is as the third to the fourth, [then,] conversely, the ratio of the second to the first will be as the fourth to the third, and this is the mode of arguing which Euclid called above "converse proportion."[71] But he [Ametus] has erred, since he said that the proposition is self-evident even though its antecedent and consequent are unknown. For it is not known what it is to be a ratio of the first quantity to the second as the third is to the fourth; and because this has simply been assumed, it is impossible to know what should follow from it.[72] In a similar way, because the consequent is unknown, it is impossible to understand what should antecede it.[73] His second principle is that if there are four quantities of which the ratio of the first to the second is as the third to the fourth, [then] if the first is greater than the second, the third will be greater than the fourth; but if less [the third will be] less [than the fourth]; and if equal, [the third will be] equal [to the fourth].[74] The third [principle assumed by Ametus] is that if there are four quantities of which the ratio of first to second is as the third to the fourth, the first will be related to any multiple of the second as the third to an equal multiple of multiples of the fourth.[75] But in these [final] two principles the same error occurs as occurred in the first [principle], for in all he takes the unknown as well as the known, and therefore has not demonstrated [these definitions if they are based upon these three principles.][76] He erred also in his second demonstration and in the fifth, in each of which he argues from the eighth or tenth [propositions] of this [fifth book], which are proved by the definition of discontinued proportion. For he argues as follows: if ratio *AB* to *E* is greater than

G to *D*, and *NB*, part of *AB*, is to *E* as *G* to *D*, it appears that he assumes from this that of two unequal quantities, *AB* and *NB*, related to *E*, the greater [quantity] bears to it the greater ratio and the smaller the smaller ratio; or that the quantity which will bear to *E* a ratio less than *AB* does will be smaller than *AB*. The first of these [two conclusions] is demonstrated by the eighth [proposition] of this [fifth book], and the second by the tenth [proposition].[77] For since you wish to take a quantity that is related to *E* in ratio *G* to *D*, I will give

70. Ametus, son of Joseph, is the anglicized version of the Latinized form of the Arabic name of Ahmad ibn Yūsuf ibn Ibrahīm ibn al-Dāya al Misri, an Egyptian Arab who probably died sometime around 912/913. At least three of his Arabic works were translated into Latin, among them his *Letter on Ratio and Proportion* (*Epistola de proportione et proportionalitate*), translated by Gerard of Cremona in the second half of the twelfth century (see Selection 7, no. 11). It has recently been edited and translated by Sister M. Walter Reginald Schrader, O.P., under the title "The *Epistola de proportione et proportionalitate* of Ametus Filius Iosephi," Ph.D. dissertation, University of Wisconsin, 1961. The definitions supposedly demonstrated by Ametus are those discussed by Campanus in the section immediately prior to this paragraph, namely Definitions 5, 6, and 8.

71. See Schrader (p. 90) for the translation of this principle and the two following. As Campanus says, Ametus insists that these do not require proof.

72. The first proposition for which Ametus denies the need for proof is this: If one assumes that $A/B = C/D$, then it follows that $B/A = D/C$. Here the antecedent is the initial assumption that $A/B = C/D$. But Campanus insists that we cannot really understand this assumption, since Ametus has not defined or in any way explained what is meant by equality of ratio.

73. For the same reason given in note 72 the consequent, namely $B/A = D/C$, is unknown, because it consists of equal ratios and we do not yet know what equality of ratio is.

74. If $A/B = C/D$ and $A > B$, then necessarily $C > D$.

75. If $A/B = C/D$, then $A/Bm = C/Dm$.

76. In these two assumptions, as in the first (see n. 72), equality of ratios is assumed without our knowing the meaning of equality of ratios. This seems to be the basis of Campanus' criticism.

77. Campanus' representation of Ametus' second (or fifth ?) proposition does not accord with any propositions in that treatise. Assuming that $AB/E > G/D$ and that *NB*, which is part of *AB*, is such that $NB/E = G/D$, Campanus tells us that Ametus concluded from this that if $AB > NB$, then $AB/E > NB/E$ (this conclusion is demonstrated in *Elements,* Bk. V, Prop. 8); and that if $x/E < AB/E$, then $x < AB$ (demonstrated in *Elements,* Bk. V, Prop. 10). In his brief reaction against this procedure, the basis for Campanus' dissatisfaction is not clear but may rest on the fact that if some quantity x is to be related to *E* as G/D, one may take *AB* such that $AB \gtreqqless x$. But if this is what Ametus has demonstrated, it is too imprecise as an aid in finding a quantity x such that $x/E = G/D$.

you *AB* greater, or smaller, or equal, indifferently, as I wish. Therefore, either he does not demonstrate it, or it is circular and the principles are less known than the conclusions. With Euclid, however, the

principles must be assumed as known and they must not be demonstrated from propositions, but [rather] propositions must be demonstrated from them.

28 AN ALGORISM OF RATIOS: MANIPULATION OF RATIONAL EXPONENTS

Nicole Oresme (ca. 1325—1382)

Translated and annotated by Edward Grant[1]

THE PROLOGUE OF MASTER NICOLE ORESME

ON THE TREATISE

ALGORISM OF RATIOS

To the Most Excellent Reverend Phillip of Meaux,[2] whom I would call Pythagoras if it were possible to believe in the opinion about the return of souls, I present this *Algorism of Ratios,* so that if it is agreeable to your Excellency, you may correct that which I put before you. For should it be approved by the authority of so great a man and corrected after his examination [of it], everything that has been revised by your correction would be an improvement. Then, if a disparager should open his mouth and set his teeth to rend [my work] into pieces, he would not find [what he seeks].

PART 1

One half is written as 1/2, one third as 1/3, and two thirds as 2/3, and so on. The number above the crossbar is called the numerator, that below the crossbar the denominator.

A double ratio [2/1] is written as 2^p, a triple ratio [3/1] as 3^p, and so forth. A sesquialterate ratio [3/2] is written as 1^p 1/2, and a sesquitertian ratio [4/3] as 1^p 1/3. A superpartient two-thirds ratio [5/3] is written as 1^p 2/3, a double superpartient three-fourths [11/4] as 2^p 3/4, and so on. Half of a double ratio [$(2/1^{1/2})$] is written as 1/2 2^p and a fourth part of a double sesquialterate [$(5/2)^{1/4}$] as 1/4 2^p 1/2, and so on.[3] But sometimes a rational ratio is written in its least terms or numbers just as a ratio of 13 to 9, which is called a superpartient four-ninths [1 4/9]. Similarly, an irrational ratio such

English translations of these quotations have been included, and unless otherwise specified, are my own).

The present translation is made from my edition of Part I, based on thirteen manuscripts, which appears in "The Mathematical Theory of Proportionality of Nicole Oresme," Ph. D. Dissertation, University of Wisconsin, 1957, pp. 331–339. *The Algorismus proportionum (Algorism of Ratios)* of Nicole Oresme consists of a prologue and three parts; only the prologue and first part— the most important—are translated here. In Part I, Oresme formulates a series of rules for operations with exponents, which he subsequently applies to various physical problems in Parts II and III. All three parts— but not the prologue—were edited by Maximilian Curtze in his *Der Algorismus Proportionum des Nicolaus Oresme; zum ersten Male nach der Lesart der Handschrift R. 4° 2 der königlichen Gymnasialbibliothek zu Thorn* (Berlin, 1868).

The importance of the first part of Nicole Oresme's *Algorismus proportionum* has long been recognized and appreciated. As far as we know now, it appears to be the first extant systematic attempt to describe operational rules for multiplication and division (called by Oresme addition and subtraction) of ratios involving integral and fractional exponents. Despite the apparent popularity of the *Algorismus* (there exist many manuscript copies of it), its subsequent impact, if any, on the history of mathematics is largely unknown.

2. This is the eminent Phillip de Vitri (1291–1361), renowned musical scholar and friend of Petrarch, who was Bishop of Meaux from January 3, 1351, to his death in Paris on June 9, 1361. Since Phillip did not have an ecclesiastical post at Meaux prior to 1351, Oresme's reference to him as "Most Excellent Reverend Phillip of Meaux" (*Reverende Presul Meldensis Phillipe, ...*) indicates that Oresme composed the *Algorismus* sometime between 1351 and 1361, the period during which Phillip was Bishop of Meaux.

3. In writing irrational ratios, Oresme places the exponent first, followed by the ratio that expresses the rational base. Since the manuscripts reveal that Oresme followed the pattern consistently, we have here an early attempt at a mathematical notation. However, the rational bases were written with considerable variation. Thus 1/4 2^p 1/2 might also appear as 1/4 2 2 1/2; 1/4 $p2/12$; 1/4 2^{p1a} 1/2; and so on. Often enough, Oresme followed the usual medieval practice and verbalized the entire expression. With the exception of exponents, the first part of the *Algorismus* is concerned exclusively with ratios of greater inequality—that is, with ratios of the form *A/B*, where $A > B$.

1. This selection and the bulk of the notes were first published under the title "Part I of Nicole Oresme's *Algorismus proportionum*," translated and annotated by Edward Grant in *Isis,* LVI (1965), 327–341. It is reprinted here by permission of the editors of *Isis.* Although the translation has been left intact, the notes have been altered slightly and reduced in extent (for instance, information about manuscripts has been omitted, as have all Latin quotations; however, the

as half of a superpartient two-thirds $[(5/3)^{1/2}]$ is written as half of a ratio of 5 to 3.

Every irrational ratio—and these shall now be considered—is denominated by a rational ratio in such a manner that it is said to be a part or parts of the rational ratio, as [for example] half of a double $[(2/1)^{1/2}]$, a third part of a quadruple $[(4/1)^{1/3}]$, or two thirds of a quadruple $[(4/1)^{2/3}]$.[4] It is clear that there are three things [or elements] in the denomination of such an irrational ratio: [1] a numerator, [2] a denominator, and [3] a rational ratio by which the irrational ratio is denominated, that is, a rational ratio of which that irrational ratio is said to be a part or parts, as, [for example,] in half of a double ratio $[(2/1)^{1/2}]$ the unit is the numerator, or represents the numerator, two is the denominator,[5] and the double ratio is that by which the irrational ratio is denominated,[6] And this can easily be shown for other ratios.

Rule One. [*How*] *to add a rational ratio to a rational ratio.*

Assuming that each ratio is in its lowest terms, multiply the smaller term, or number, of one ratio by the smaller number of the other ratio; and [then] multiply the greater [number of one ratio] by the greater [number of the other], thereby producing the numbers or terms of the ratio composed of the two given ratios. In this way, three or [indeed] any number [of ratios] can be added by adding two of them at a time and then adding a third to the whole composed of those two; then, if you wish add a fourth ratio, and so on.

For example, I wish to add sesquitertian [4/3] and quintuple [5/1] ratios. The prime numbers of a sesquitertian are 4 and 3, of the other 5 and 1. And so, as already stated, I shall multiply 3 by 1 and 4 by 5 obtaining 20 and 3, which is a sextuple superpartient two-thirds ratio $[6\frac{2}{3}]$. In this way a ratio can be doubled, tripled, and quadrupled, as many times as you please.[7] And this can be demonstrated and is [indeed] adequately shown in the sixth proposition of the fifth book of the *Arithmetic* of Jordanus [de Nemore].[8]

Rule Two [*How*] *to subtract a rational ratio from* [*another*] *rational ratio.*

As before, assume any [two] rational ratios in their lowest numbers. The smaller number of one ratio is then cross-multiplied *(ducantur contradictorie)* by the greater number of the other ratio, and the same is done with the remaining numbers, thus producing the terms of a ratio in which a greater term exceeds a smaller term. The greater of the given ratios was that whose greater term when multiplied by the smaller term of the other given ratio produces the greater number [of the resultant ratio].

For example, let a sesquiteritan [4/3] be subtracted from a sesquialterate ratio [3/2]. The prime numbers, or terms, of a sesquitertian are 4 and 3, of a sesquialterate 3 and 2. I shall multiply 4 by 2 and obtain 8, and then multiply 3 by 3 to obtain 9. A sesquialterate ratio is, therefore, greater than a sesquitertian by a ratio of 9 to 8, that is, greater by

4. For Oresme the irrational ratio $(4/1)^{1/3}$ is a *part* of the rational ratio 4/1 because $(4/1)^{1/3} < 4/1$ *and* the exponent, 1/3, is a unit fraction. The irrational ratio $(4/1)^{2/3}$ is *parts* of 4/1 because $(4/1)^{2/3} < 4/1$ *and* the exponent is a proper fraction in its lowest terms, where both integers are greater than 1. In a later work, the *De proportionibus proportionum* (*On Ratios of Ratios;* see Selection 29, n. 3), Oresme related the concept of exponential parts to that of commensurability.

5. Oresme is referring here to the numerator and denominator of the exponent.

6. When an irrational ratio has a rational base, the latter is said to denominate the former. Thus if $(A/B)^{p/q}$ is irrational, A/B rational, and p, q are integers with $p < q$, then A/B will denominate $(A/B)^{p/q}$. In his *De proportionibus proportionum,* Oresme says that such an irrational ratio is *immediately denominated* by the rational ratio A/B and *mediately denominated* by a number—that is, by the exponent p/q when it is a ratio of numbers—Irrational ratios in the form cited above were always expressed by Oresme as exact parts of some rational ratio (see Selection 29, n. 3). The bare distinction, without any elaboration, between immediate and mediate denomination appears as early as 1328 in Thomas Bradwardine's *Treatise On Proportions (Tractatus proportionum seu de proportionibus velocitatum in motibus)*; the published edition is cited in Selection 51.1, n. 1.

7. If we have a rational ratio A^n, where n is an integer, Oresme would expand this ratio by stages. Thus $A \cdot A = A^2; A^2 \cdot A = A^3; A^3 \cdot A = A^4$, and so on until $A^{n-1} \cdot A = A^n$. Note that Oresme says a ratio is "to be doubled" *(duplari)* and "tripled" *(triplari)* when he obviously means "squared" and "cubed." This ambiguous terminology was quite common but usually not troublesome, since context would almost always reveal an author's meaning.

8. Book V, Proposition 6, of the *Arithmetica* of Jordanus de Nemore (propositions from this work are given in Selection 21) is not relevant to the "addition" of ratios in this rule, for in that proposition Jordanus considers how to reduce proportions to their lowest terms (Cambridge, Peterhouse 277 [the codex is in Magdalene College], Bibliotheca Pepysiana 2329, fol. 13 verso, cols. 1–2). But Jacques Le Fèvre d'Estaples (Jacobus Faber Stapulensis [1455–1536]) explains the addition of two ratios in his edition of Jordanus' *Arithmetica* (Paris, 1496; this includes Jordanus' enunciations and Le Fèvre's demonstrations and comments; for a description of this edition see David Eugene Smith, *Rara Arithmetica* [Boston: Ginn, 1908], pp. 62–63]. Before actually commencing the demonstration of Book V, Proposition 3, Le Fèvre says (sig. b6, col. 1; the folios are unnumbered):

a sesquioctavan ratio.[9] This can be shown by the twenty-seventh proposition of the second book of the *Arithmetic* of Jordanus.[10]

Rule Three. If an irrational ratio is parts of any rational ratio, it is possible to designate it as a part

"Before we demonstrate what has been proposed, I wish to show how one ratio is added to another. Now, I say that the ratio of the products of the first term of one ratio by the first term of the other, and of the second [term of one] by the second [term of the other] is composed of these two ratios. For let A to B and C to D be the two ratios that I wish to add and to show what is composed from them. I multiply C by A and get E; and I multiply D by B and get G. I say that ratio E to G is composed of ratios A to B and C to D. Now, again, I multiply C by B and get F, and by the seventh [proposition] of the second [book] ratio A to B is as E to F; and by the eighth [proposition] of the same [book] ratio C to D is as F to G and ratio E to G is composed of ratios E to F and F to G, and, therefore, ratio E to G is composed of ratios A to B and C to D, which was asserted."

We see that "addition" of ratios is simply multiplication. Why, or when, the multiplication of such ratios as $5/1$ and $4/3$ came to be designated "addition" is unknown to me (for instance, it does not appear in Boethius' *Arithmetica,* a very widely used treatise). Where Oresme applies this terminology to exponents a plausible explanation can be offered (see n. 38). In the *De proportionibus proportionum,* Oresme speaks briefly of adding (that is, multiplying) ratios of greater inequality (ch. 1, *ll.* 75–83; Grant, *Nicole Oresme: De proportionibus* · · · [Madison, Wis.: University of Wisconsin Press, 1966], pp. 142, 143). The procedure is couched exclusively in terms of continuous proportionality where extreme terms are assigned and the ratios composed. In the *Algorismus,* however, the addition of ratios is effected by multiplication performed directly on the prime numbers, or numerical denominations, of the ratios; extreme terms are not assigned.

9. $(3/2):(4/3) = 9/8$.

10. The enunciation of this proposition in Jordanus' *Arithmetica* reads (Bibliotheca Pepysiana 2329, fol. 6 recto, col. 1): "If the ratio of the first to the second term is greater than the ratio of the third term to the fourth, then the product of the first and fourth terms is greater than the product of the second and third terms. And if the product is greater the ratio of the first term to the second will be greater."

In his proof Jordanus assumes that $A/B > C/D$, so that $AD = E$, $CB = F$, and $BD = G$. Since $E/G = A/B$ and $F/G = C/D$, it follows that $E/G > F/G$ and, consequently, that $E > F$ (that is, $AD > CB$).

Although this proposition follows the steps outlined by Oresme, nowhere does Jordanus speak of subtracting one ratio from another. But in Book V, Proposition 1, he approximates rather closely the ideas expressed by Oresme, again without using any form of the term *subtrahere.* The enunciation of this proposition reads (Bibliotheca Pepysiana 2329, fol. 13 recto, col. 1): "That ratio by which the ratio of the first to the second term exceeds the ratio of the third to the fourth term is the ratio formed by the product of the second and third terms."

Jordanus assumes that A, B, C, and D are four successive terms, where $A/B > C/D$. Thus if $DA = E$ and $BC = F$, then $(A/B):(C/D) = E/F$, and Jordanus would say that A/B exceeds C/D by ratio E/F just as Oresme says that $3/2$ exceeds $4/3$ by $9/8$, a sesquioctavan ratio. Although Jordanus did not refer to this as a subtraction of ratios, Jacques Le Fèvre, in commenting on this proposition, says (sig. b6, col. 1): "This [proposition] shows how to subtract a ratio from a ratio. And after the lesser ratio has been subtracted from the greater ratio, the ratio that is left, howsoever much it be, is what we call here the difference between one ratio and another."

For Oresme, as for Jacques Le Fèvre and others, "subtraction" of ratios of greater inequality is actually division by cross-multiplication. But why was the quotient of $(3/2):(4/3)$ called a "difference" and the whole process of division designated "subtraction"? In our concluding note we shall see that where exponents are concerned a certain rationale may be supplied for this terminology. But, as with "addition" of ratios, its origin is unknown to me and can only be conjectured (see n. 38). In the *De proportionibus* (ch. 1, *ll.* 72–74) Oresme mentions briefly the subtraction (that is, division) of one ratio of greater inequality from another. In that treatise, subtraction is performed by assigning a mean term between the terms of the greater ratio and subsequently composing the ratios. Thus, if ratio E is to be subtracted from ratio B/C, it follows that $B/C > E$. Assign a mean term, D, between B and C so that either $D > C$ and $D/C = E$ in which event $(B/C):(D/C) = B/D$; or $D < B$ and $B/D = E$, so that $(B/C):(B/D) = B/C \cdot D/B = D/B \cdot B/C = D/C$. Here the *modus operandi* is to produce continuous proportionality by assigning mean terms and composing the two ratios. This is a wholly different emphasis than in the *Algorismus,* where the operation is effected directly by cross-multiplying the prime numbers of the given rational ratios; hence no means need be assigned.

Perhaps with his own *Algorismus* in mind, Oresme informs the readers of the *De proportionibus* that ratios may be added and subtracted in a manner different than that which is appropriate to the latter treatise *(De proportionibus proportionum,* ch. 1, *ll.* 84–89): "If, however, you wish to add a ratio of greater inequality to another by means of algorism *(per artem),* it is necessary to multiply the denomination of one ratio by the denomination of the other. And if you wish to subtract one ratio from another, you do this by dividing the denomination of one ratio by the denomination of the other. The [method] of finding denominations will be taught afterward. Multiplication and division of denominations are done by algorism."

Although we are told here that the addition of ratios can be performed by multiplication of denominations (that is, multiplication of the prime numbers of the two ratios) and that subtraction of ratios can be performed by division of denominations (that is, division of the prime numbers of one ratio into the prime numbers of the greater ratio), Oresme in the ninth rule of his *Algorismus* emphasizes that the multiplication of the denominations of ratios must, strictly speaking, be called addition, not multiplication. Similarly, the division of the numerical denominations of two ratios must be called subtraction, not division (see n. 38 for a possible explanation of this.)

of yet another rational ratio, so that it might more appropriately be called a part rather than parts.

Let *B*, an irrational ratio, be parts of *A*, a rational ratio. Without changing the denominator [of the exponent], I say that *B* will be a part of some ratio [obtained by expanding *A* by the numerator of the exponent] and this [expanded] ratio will be multiple to *A*. Any irrational—and it is of these we speak— will be a part of some rational ratio, since a multiple ratio can be found for every rational ratio.[11]

As an example, let us take a ratio which is two-thirds of a quadruple $[(4/1)^{2/3}]$. Since 2 is the numerator [of the exponent], we shall have one-third of a quadruple ratio squared $[((4/1)^2)^{1/3}]$, namely [one-third of] a sedecuple ratio $[(16/1)^{1/3}]$. The same applies to other ratios of this kind. The justification for this lies in the general truth that one-third of a whole $[(A)^{1/3}]$ equals two-thirds of its half or subdouble[12] $[(A^{1/2})^{2/3}]$, and conversely, two-thirds of a subdouble $[(A^{1/2})^{2/3}]$ equals one-third of a double $[(A)^{1/3}]$. The same reasoning is applicable to any other parts.

Rule Four. [*How*] *to assign the most appropriate denomination of an irrational ratio.*

Here it must be understood that a rational ratio is called *primary* when it cannot be divided into equal rational ratios, and no mean proportional number or numbers can be assigned between its least numbers, as is the case with double [2/1], triple [3/1], or sesquialterate [3/2] ratios. But a rational ratio is called *secondary* when it can be so divided and a mean proportional number or numbers can be assigned between its [least] numbers.[13] As examples, take a quadruple ratio [4/1], which is divisible into two doubles [4/2 · 2/1]; an octuple ratio [8/1], divisible into three doubles [8/4 · 4/2·2/1]; a nonacuple ratio [9/1], into two triples [9/3 · 3/1]; and so on.

From all this we see that if any proposed irrational ratio is denominated by parts, that ratio, by the preceding rule, could be transformed and called a part. However, it must [first] be seen if the rational ratio denominating the irrational is a *primary* ratio. If it is, let it stand, for then the irrational ratio, which is our topic of discussion, is most appropriately denominated, as in one-third of a sextuple $[(6/1)^{1/3}]$ or double ratio $[(2/1)^{1/3}]$, and so forth.

But if the rational ratio that denominates the irrational ratio is *secondary*, one must determine how many primary rational ratios are contained by it, where each primary is an equal part of it. Should the number representing these parts [or

primary ratios] be incommunicant or prime to the denominator of the proposed irrational ratio, the [initial] denomination must be left as it is. The denomination of half of an octuple $[(8/1)^{1/2}]$, for example, is proper because an octuple ratio [8/1] has three equal rational parts, namely three doubles [8/4 · 4/2 · 2/1 or $(2/1)^3$], and 2 is the denominator of the proposed irrational ratio. But 3 and 2^{14} are incommunicant [or prime to one another], so that half of an octuple is not a part of any rational ratio smaller than an octuple, although it could certainly be parts since half of an octuple is three-fourths of a quadruple $[(4/1)^{3/4}]$, but this would not constitute an appropriate denomination.

However, if the number representing the smallest, or primary, parts of such a secondary rational ratio by which an irrational ratio is denominated *and* [the number representing] the denominator [of the exponent] of this irrational ratio were communicant [or mutually nonprime] numbers, the greatest

---

11. If the irrational ratio is $B = A^{m/n}$, where *A* is a rational ratio and *m/n* is a ratio of integers in its lowest terms with $n > m > 1$, then $A^{m/n}$ can be expanded to $D^{1/n}$ so that $D = A^m$. Initially, then, $A^{m/n}$, or *B*, is an irrational ratio that is exponentially *parts* of rational ratio *A*, since the exponent *m/n* is a proper fraction whose numerator is an integer greater than 1. By expanding A^m we obtain a rational ratio *D* which is multiple to *A* in an exponential sense, since $D = A^m$. (Thus when Oresme says that "a multiple ratio can be found for every rational ratio" he means that any rational ratio *A* can be expanded by any integral exponent *m*, and the rational ratio which is the product of that expansion, namely *D*, is called the multiple of *A*.) Hence $A^{m/n} = D^{1/n}$ and has been transformed from an irrational ratio that was *parts* of rational ratio *A*, to one that is a *part*—that is, where the exponent is a unit fraction—of another rational ratio *D*. The most appropriate form of irrational ratio $A^{m/n}$ is, therefore, $D^{1/n}$. This is illustrated by the example in the next paragraph of the text: $(4/1)^{2/3} = [(4/1)^2]^{1/3} = (16/1)^{1/3}$.

12. *Medietas* and *subduplus* are the Latin terms rendered here as "half" and "subdouble" and clearly mean square root. More commonly, however, they signified "half" in the arithmetic sense. The context usually reveals the meaning without ambiguity.

13. In the *De proportionibus proportionum*, Oresme does not use the terms "primary" (*primaria*) and "secondary" (*secundaria*), but, at the conclusion of the first chapter, these concepts constitute the first two of seven conceivable ways in which rational ratios are divisible (see ch. 1, *ll.* 385–389). In the first way rational ratios are divisible into smaller equal rational ratios, which corresponds to secondary ratios in the *Algorismus*; all rational ratios not divisible in the first way constitute the second group, which makes this category equivalent to the primary ratios of the *Algorismus*.

14. The numbers 3 and 2 refer to the exponent in $(2/1)^{3/2}$.

number [or common factor] in which they are communicant must be taken and each of them divided by it. By dividing the number representing the parts of the secondary ratio we arrive at the number of parts or [primary] ratios composing the rational ratio that will most suitably denominate the [proposed] irrational ratio; and by dividing the denominator of the [initially] proposed [irrational ratio] by the same greatest number [or common factor], the most appropriate denominator of the irrational ratio is found.

For example, let a ratio of three-fourths of a quadruple $[(4/1)^{3/4}]$ be proposed. By utilizing the third rule it is clear that it equals one-fourth of $64/1$ $[(64/1)^{1/4}]$. But $64/1$ is composed of six double ratios $[64/1 = (2/1)^6]$, where 6, which signifies the number of primary parts in $64/1$, and 4, which represents the denominator of the proposed ratio,[15] are communicant[16] [or have a common measure] in 2, so that dividing 2 into 6 gives 3, signifying that the proposed ratio is a part of three doubles, namely part of $8/1$.[17] Dividing 2 into 4 yields 2, so that the proposed ratio is one-half—that is, the rule shows that the proposed ratio is one-half of $8/1$, which is written as $1/2$ 8^p $[(8/1)^{1/2}]$, and this is its most proper denomination. In the same way, one-twelfth of four triple ratios $[((3/1)^4)^{1/12}]$, namely $81/1$ $[(81/1)^{1/12}]$, is one-third of $3/1$ $[(3/1)^{1/3}]$; and similarly one-fourth of six triple ratios $[((3/1)^6)^{1/4}]$ is one-half of three triple ratios $[((3/1)^3)^{1/2}]$, namely [one-third of] $27/1$ $[(27/1)^{1/3}]$, and so on.[18]

Rule Five. [How] to add[19] *an irrational ratio to a rational ratio.*

Let B, an irrational ratio, be added to A, a rational ratio, and assume that B is a part[20] of rational ratio D, and in accordance with previous rules, this constitutes the most proper denomination of B.[21] Then, by the first rule, add A to ratio D a number of times equal to the value of the denominator of B, and let C be the total result.[22] I say, therefore, that the ratio composed of [ratios] B and A will be part of C and [it] will be denominated by the same denominator which denominated [or denoted] the part which B was of its denominating ratio, namely D.[23]

For example, let one-third of a double ratio $[(2/1)^{1/3}]$ be added to a sesquialterate ratio $[3/2]$. Now join three sesquialterate ratios with a double ratio to yield a sextuple superpartient three-fourths ratio $[6\frac{3}{4}]$, which is a ratio of 27 to 4. The ratio produced from [the addition of] one-third of a double ratio and a sesquialterate ratio is one-third of the ratio of 27 to 4 $[(27/4)^{1/3}]$, which is written as $1/3$ 6^p $3/4$.[24]

Rule Six. [How] to subtract an irrational from a rational ratio, or the converse.

Let A be a rational ratio and B an irrational ratio

15. Oresme refers here to the denominator of the exponent in the ratio $(2/1)^{6/4}$.

16. In Campanus of Novara's thirteenth-century edition of Euclid's *Elements* (for full title of the edition, see Selection 27, n. 1), Book VII, Definition 8, "communicant" numbers are defined as follows (p. 168): "Numbers are said to be mutually composite or communicant which are measured by a number other than unity, and none of them is prime to any other of them."

17. That is, $[(2/1)^3]^{1/2} = (8/1)^{1/2}$.

18. In the fourth rule Oresme presents criteria for determining the final form of an irrational ratio. If possible, all irrational ratios should have a primary rational ratio as base and a unit fraction for an exponent, as in ratios $(6/1)^{1/3}$ and $(3/1)^{1/4}$. If, however, the base is a secondary rational ratio, Oresme outlines the procedure for determining its most proper denomination or representation, that is, the most proper rational ratio that will serve as base. If $B^{1/q}$ is a given irrational ratio, where B is a secondary rational ratio, one must decompose B into its constituent primary ratios. Let A represent the primary ratios so that $B = A^m$, where m is an integer. Therefore, $B^{1/q} = (A^m)^{1/q}$. Should m and q be unequal mutually prime numbers greater than 1, the original form of the ratio, namely, $B^{1/q}$, will be the most appropriate denomination, since it expresses the irrational ratio as a part of rational ratio B. For example, $(8/1)^{1/2}$ is most properly denominated because the alternatives, $(2/1)^{3/2}$ and $(4/1)^{3/4}$, fail to represent the relation in terms of a single part, or unit fraction.

But if m and q are not mutually prime, they must then be reduced to their lowest terms. Then either $m/q = 1/n$ and the final form is $A^{1/n}$; or $m/q = p/n$, where p and n are greater than 1; and the final form is $D^{1/n}$ (rather than $A^{p/n}$), where $D = A^p$. An example of the first case is $(81/1)^{1/12} = (3/1)^{4/12} = (3/1)^{1/3}$, the last expression being the most suitable final form; an example of the second case is $(64/1)^{1/4}$ (given initially as $(4/1)^{3/4}$ but expanded to $(64/1)^{1/4}$ by the third rule), which equals $(2/1)^{6/4} = (2/1)^{3/2} = (8/1)^{1/2}$, the last expression being the most appropriate. The two examples given by Oresme in which $(8/1)^{1/2}$ is the final form are both represented here.

19. We have already seen that "to add" *(addere)* signifies multiplication in the context of this treatise.

20. Here $B = D^{1/q}$, where q is an integer.

21. That is, $D^{1/q}$ is the most proper denomination of B.

22. The denominator of B is q, so that $D \cdot A^q = C$. In accordance with the first rule, A is expanded in stages. Thus if $q = 3$, then $A \cdot A = A^2$ and $A^2 \cdot A = A^3$.

23. Since $D \cdot A^q = C$, $B < D$, and $A < A^q$, it follows that $(B \cdot A) < C$. Also, $(B \cdot A)$ is an exponential part of C, since $(B \cdot A) = (D \cdot A^q)^{1/q} = C^{1/q}$. But $B = D^{1/q}$, so that $D^{1/q}$ and $C^{1/q}$ have the same denominator in their exponents, namely q. It is in this sense that ratios B and $(B \cdot A)$ are denominated, or represented, by the same denominator.

24. The two given ratios are first cubed, then fully expanded, after which the cube root is taken. Thus $[(2/1)^{1/3} \cdot 3/2]^3 = 2/1 \cdot (3/2)^3 = 27/4$, whose cube root is $(27/4)^{1/3}$ but which Oresme writes as $(6\frac{3}{4})^{1/3}$.

denominated by rational ratio D. Then, whether A is greater or smaller than B, let it be multiplied by the denominator of B. This is the same thing as taking a ratio composed of a number of A's equivalent to [the number representing] B's denominator. The method for accomplishing this is shown in the first rule. Now, let this composed, or produced, ratio be called C. By the second rule one can determine whether C is greater or smaller than D. If it is greater, then A was greater than B; but if D is greater than C, B was greater than A. Whatever the given ratios, namely C and D, the lesser must be subtracted from the greater. This is done by the second rule. Let the remainder be F. And so I say that if A is subtracted from B, or conversely, the remainder will be the same total part of F as B was of D.[25]

For example, subtract a sesquitertian ratio [4/3] from half of a double [$(2/1)^{1/2}$]. I join [or combine] two sesquitertian ratios and obtain a ratio of 16 to 9, which is smaller than a double and must therefore be subtracted from [that is, divided into] a double ratio, leaving a sesquioctavan [9/8] ratio. Hence half of a sesquioctavan ratio [$(9/8)^{1/2}$] will remain after the subtraction of a sesquitertian from half of a double ratio.

Similarly, should you wish to subtract a third part of a double ratio [$(2/1)^{1/3}$] from a sesquialterate [3/2], join [or combine] three sesquialterate ratios [$(3/2)^3$] to obtain a ratio of 27 to 8, from which the double ratio is [then] subtracted, leaving a ratio of 27 to 16. Therefore, if a third part of a double is subtracted from a sesquialterate, there will remain a third [part] of ratio 27 to 16 [$(27/16)^{1/3}$], which is a third part of what would remain if the whole double ratio were subtracted from three sesquialterate ratios.[26] This rule can be demonstrated easily, since, generally, if a first ratio is subtracted from [that is, divided into] a second, a third ratio is left, so that if a third part of the [same] first ratio is subtracted from [that is, divided into] a third part of the [same] second ratio, there is left a third part of the [same] third ratio. This applies to half of any ratio, or a fourth part, or a fifth, and so forth.[27]

General Rule: In the addition of an irrational to an irrational ratio, and in the subtraction of an irrational from an irrational ratio, there are general rules applicable to any quantities whatever.

Let a known part of a known quantity be added to a known part of a known or measurable quantity. For example, let C be part of quantity A, and D part of quantity B; and suppose that C is denominated by number E, and D by number F. I then multiply A by F—that is, I take A continuously F number of times—and produce G; in the same way I multiply B by E and obtain H. Hence C will be a part of G, which is represented by [one part of] the number produced from the multiplication of E by F; and the same number will represent the part that D is of H. Therefore, as C is to G, so is D to H.[28]

25. We are given rational ratio A and irrational ratio $B = D^{1/n}$, where n is an integer (in the enunciation of Rule Six, Oresme does not specify that the numerator of the exponent must be 1; but the rule itself and the examples indicate this unmistakably). Now, whether $A \gtreqless B = D^{1/n}$, we must expand A exponentially by n, the denominator of the exponent of B in $D^{1/n}$. The expansion of A is carried out by the first rule, that is, $A \cdot A = A^2$; $A^2 \cdot A = A^3$; $A^3 \cdot A = A^4$; ... $A^{n-1} \cdot A = A^n$; and let $A^n = C$. We can now determine whether $C \gtreqless D$. Let rational ratio $C = p/q$ (where $p > q$ and both are integers) and let rational ratio $D = s/t$ (where $s > t$ and both are integers). By the second rule, $(p/q):(s/t) = pt/qs$; and if $pt > qs$, then $p/q > s/t$ and $C > D$; but if $qs > pt$, then $s/t > p/q$ and $D > C$. Furthermore, if $C > D$, then $A > B$; and if $D > C$, then $B > A$. This being determined, the lesser ratio must then be subtracted from (that is, divided into) the greater, and this is also carried out by the second rule. If $C > D$, then let us assume that $C:D = F$ (or $A^n : B^n = F$); and since $A = C^{1/n}$ and $B = D^{1/n}$, it follows that $A:B = F^{1/n}$; but if $D > C$, then assume that $D:C = F$ and $B:A = F^{1/n}$. Hence $A:B$ (or $B:A$) $= F^{1/n}$ and $B = D^{1/n}$, which is what Oresme wished to show. Thus $A:B$, or $B:A$, is the same exponential part of F, namely $1/n$, as B is of D.

In general, the original ratios are expanded by a power equal to the denominator n of the exponent of the irrational ratio. The desired ratio is obtained by taking the nth root of the quotient resulting from the division of the expanded ratios. This is illustrated by the first example of the next paragraph, where Oresme subtracts (that is, divides) $A = 4/3$ from (that is, into) $B = (2/1)^{1/2}$. Here $D = 2/1$. Expanding both ratios by the denominator of the exponent, we obtain $[(2/1)^{1/2} : 4/3]^2 = 2/1 : 16/9$, where $16/9 = (4/3)^2 = C$; subtracting (that is, dividing) the lesser from (that is, into) the greater ratio by cross-multiplying (as we are told in Rule Two), we get $2/1 : 16/9 = 18/16 = 9/8$, which shows that $2/1 > 16/9$ (or $D > C$), since $2 \cdot 9 > 16 \cdot 1$, and indicates that the lesser ratio has been properly divided into the greater. Reverting to the original ratios, we see that $(2/1)^{1/2} : 4/3 = (9/8)^{1/2}$.

26. In this example we have $3/2 : (2/1)^{1/3}$. Now, $[3/2 : (2/1)^{1/3}]^3 = 27/8 : 2/1 = 27/16$. Therefore, $3/2 : (2/1)^{1/3} = (27/16)^{1/3}$.

27. Oresme concludes the sixth rule with a general statement that if A, B, and C are ratios and $A/B = C$, then $A^{1/n}/B^{1/n} = C^{1/n}$.

28. The steps outlined in Oresme's general rule are as follows: Let $C = A^{1/E}$, where A is a known quantity or ratio and C is a known part of quantity or ratio A; and let $D = B^{1/F}$, where, similarly, B is a known quantity or ratio and D is a known part of quantity or ratio B. Next take the integral exponent F and expand A so that A^F

Rule Seven. By addition [that is, multiplication], it follows that as C is to G and D to H, so also is the total [or product] *(aggregatum)* of C and D to the total [or product] *(aggregatum)* of G and H. Consequently, the total, [or product,] CD is a part of the total [or product] GH and is represented by [one part of] the number produced by [the multiplication of] E by F.[29]

Rule Eight. By subtraction, however, it follows that if G is subtracted from H, or conversely, and C is subtracted from D, or conversely, the remainder will be the same part of the remainder as part C was of G, or D was of H, and this part is represented by [one part of] the number resulting from the multiplication of E by F.[30]

As an example in the addition of irrational ratios, let half of a double $[(2/1)^{1/2}]$ be added to a third part of a triple ratio $[(3/1)^{1/3}]$. On the one hand, I join or multiply three double ratios just as shown by the first rule. I do this because the other ratio is denominated by a three and is called a third part. On the other hand, and for the same reason and in the same way, I join [or multiply] two triples and then multiply the denominators of the parts, namely 2 by 3, and produce 6. Half of a double is, therefore, a sixth part of three doubles; similarly, a third part of a triple is a sixth part of two triples. Thus the total [or product] of half of a double and a third part of a triple is a sixth part of the total [or product] of three doubles and two triples. By the first rule it is obvious that such a product is a 72^{p1a} ratio, namely 72 to 1. It follows, then, that from the addition [multiplication] of half of a double and a third part of a triple we obtain a sixth part of a 72^{p1e} ratio $[(72/1)^{1/6}]$.

As an example in subtraction [that is, division] of irrational ratios, let half of a double ratio $[(2/1)^{1/2}]$ be subtracted from a third part of a triple $[(3/1)^{1/3}]$. In the first place, and in accordance with the second rule, the product of three doubles $[(2/1)^3]$ should be subtracted from [that is, divided into] the product of two triple ratios $[(3/1)^2]$, leaving a sesquioctavan ratio [9/8]. Then, by subtracting a sixth part from a sixth part—that is, half of a double from a third part of a triple ratio—there will be left a sixth part of a sesquioctavan ratio $[(9/8)^{1/6}]$, since half of a double is a sixth part of three doubles $[((2/1)^3)^{1/6}]$ and a third part of a triple ratio is a sixth part of two triples $[((3/1)^2)^{1/6}]$. Thus subtracting a sixth [part of a ratio] from a sixth [part of another ratio] leaves a sixth [part] of what remains after the subtraction of [one] whole [ratio] from [the other] whole [ratio], and this can be demonstrated easily.

Rule Nine. Now, if the parts should have the same denomination, then it is easier to posit a special rule apart from the general rule already stated, so that if a third part of A $[(A)^{1/3}]$ is added to a third part of B $[(B)^{1/3}]$ we obtain a third part of the result produced by the addition [multiplication] of A and B.[31] In a similar manner, if a third part of A is subtracted from [divided into] a third part of B, there will remain a third part of the remainder that was left after the subtraction of A from B.[32] Thus, [for example,] if a double ratio is added to a triple ratio, a sextuple ratio is produced, so that if half of a double $[(2/1)^{1/2}]$ is added to [multiplied by] half of a triple $[(3/1)^{1/2}]$, the result will be half of a sextuple $[(6/1)^{1/2}]$.[33] Similarly, if a double ratio is subtracted from [divided into] a triple, a sesquialterate ratio is left, so that if a third part of a double ratio $[(2/1)^{1/3}]$ is subtracted from [divided into] a third part of a triple $[(3/1)^{1/3}]$,

---

$= G$; in the same manner take integral exponent E and expand B so that $B^E = H$. Since $A = G^{1/F}$, it follows that $C = (G^{1/F})^{1/E}$, or $C = G^{1/EF}$, where C is an exponential part of ratio G. Similarly, since $B = H^{1/E}$, it follows that $D = (H^{1/E})^{1/F}$, or $D = H^{1/EF}$, where D is an exponential part of H. Therefore, ratio C is the same exponential part of ratio G that ratio D is of H. In the seventh rule, this general rule is applied to addition (that is, multiplication) of two irrational ratios.

29. Since the general rule has shown that C is to G as D is to H, by which Oresme means $C = G^{1/EF}$ and $D = H^{1/EF}$, in the seventh rule he multiplies C and D, obtaining the product $C \cdot D = (G \cdot H)^{1/EF}$. In an example directly following Rule Eight, Oresme adds (that is, multiplies) two irrational ratios. Let $C = A^{1/E} = (2/1)^{1/2}$ and $D = B^{1/F} = (3/1)^{1/3}$. Now, $A^F = G = (2/1)^3$, so that $A = G^{1/F} = [(2/1)^3]^{1/3}$ and $C = A^{1/E} = (G^{1/F})^{1/E} = ([(2/1)^3]^{1/3})^{1/2} = [(2/1)^3]^{1/6}$. Similarly, $B^E = H = (3/1)^2$, so that $B = H^{1/E} = [(3/1)^2]^{1/2}$ and $D = B^{1/F} = (H^{1/E})^{1/F} = (3/1)^{1/3} = ([(3/1)^2]^{1/2})^{1/3} = [(3/1)^2]^{1/6}$. The product of $(2/1)^{1/2} \cdot (3/1)^{1/3} = [(2/1)^3]^{1/6} \cdot [(3/1)^2]^{1/6} = [(2/1)^3 \cdot (3/1)^2]^{1/6}$. Expanding each ratio by the first rule, we obtain $8/1 \cdot 9/1 = 72/1$, which yields as a final product $(72/1)^{1/6}$. In brief, Oresme shows that $A^{1/E} \cdot B^{1/F} = [(A^F)^{1/EF} \cdot (B^E)^{1/EF}] = (A^F \cdot B^E)^{1/EF}$ or $(G \cdot H)^{1/EF}$.

30. Here we have subtraction (that is, division) of the ratios $D/C = (H/G)^{1/EF}$. In an example given immediately before the ninth rule, Oresme subtracts the same irrational ratios that were added in the previous example. Thus $(3/1)^{1/3} : (2/1)^{1/2} = [(3/1)^2]^{1/6} : [(2/1)^3]^{1/6} = (9/1)^{1/6} : (8/1)^{1/6} = (9/8)^{1/6}$. Generally, the eighth rule asserts that $B^{1/F} : A^{1/E} = (B^E)^{1/EF} : (A^F)^{1/EF} = (B^E : A^F)^{1/EF}$ or $(H : G)^{1/EF}$.

31. $A^{1/3} \cdot B^{1/3} = (A \cdot B)^{1/3}$; or generally, $A^{1/n} \cdot B^{1/n} = (A \cdot B)^{1/n}$.

32. $B^{1/3} : A^{1/3} = (B/A)^{1/3}$; or generally, $B^{1/n} : A^{1/n} = (B : A)^{1/n}$.

33. Since $3/1 \cdot 2/1 = 6/1$, it follows that $(3/1)^{1/2} \cdot (2/1)^{1/2} = (6/1)^{1/2}$.

there remains a third part of a sesquialterate ratio $[(3/2)^{1/3}]$;[34] and the same holds in other cases.

Moreover, just as in other things, addition always proves subtraction, and conversely.[35] A ratio can be doubled, tripled, and multiplied—even sesquialterated— or increased proportionally as much as you wish by the addition [multiplication] of a ratio to a ratio. Thus, if someone should want a ratio which is the sesquitertian of a double ratio, it would be necessary, by the fifth rule, to add a third part of a double ratio $[(2/1)^{1/3}]$ to the double ratio so that a third part of a sedecuple $[(16/1)^{1/3}]$ is obtained, which is a ratio that is the sesquitertian of a double ratio.[36] In the same way, subtraction allows a ratio to be subdoubled, subtripled, subsesquialterated,[37] and so on.

Indeed, one ratio cannot be multiplied *(multiplicatur)* or divided *(dividitur)* by another except improperly, as when two doubles are multiplied by two doubles to obtain four doubles. But this is nothing other than a multiplication of numbers, since the multiplication of two doubles $[(2/1)^2]$ by two triples $[(3/1)^2]$ comes to nothing, just as does the multiplication of a man by an ass. The same reasoning applies to division. Thus only addition and subtraction are appropriate types of algorism [that is, operational procedures] for dealing with ratios[38] and we have discoursed sufficiently about them. The first tractate ends here.

34. Since $(3/1) : (2/1) = 3/2$, it follows that $(3/1)^{1/3} : (2/1)^{1/3} = (3/2)^{1/3}$.

35. As shown in the next two notes, this converse relation probably applies to addition and subtraction of exponents having the same base. In *Algorismus de integris,* ascribed to Master Gernardus, Rule 12 reads: "Addition and subtraction are proved mutually." Gernardus tells us that if numbers $a + b = c$, and then if $c - b \neq a$, the addition was necessarily faulty; similarly if $c - a = b$ and $a + b \neq c$, the subtraction was improperly performed. See G. Eneström, "Der 'Algorismus de integris' des Meisters Gernardus," *Bibliotheca Mathematica* XIII (1912/1913), 301–302. Oresme's statement may be an application of this arithmetic relation to addition and subtraction of exponents. Unfortunately, he offers no specific example to substantiate this interpretation.

36. Here Oresme wishes to find ratio $(2/1)^{4/3}$ and invokes the fifth rule, in which an irrational ratio is added to a rational ratio. Thus $(2/1) \cdot (2/1)^{1/3} = (2/1)^{4/3} = (16/1)^{1/3}$; or generally, $A^{n/n} \cdot A^{1/n} = A^{(n+1)/n}$. If we now wish to prove the addition, we would "subtract" $(2/1)^{1/3}$ from $(2/1)^{4/3}$ and obtain $(2)^{4/3} : (2)^{1/3} = (2/1)$, the double ratio to which we initially added $(2/1)^{1/3}$.

37. For subtraction (that is, division) $A^{n/n} : A^{m/n} = A^{(n-m)/n}$, where $m < n$ and both are integers. Addition proves this subtraction of ratios when we take $A^{m/n} \cdot A^{(n-m)/n} = A^{n/n}$. The terms "subdouble," "subtriple," and "subsesquialterate" are, in order, the fractional exponents 1/2, 1/3, and 2/3.

38. In this final paragraph Oresme explains why he believes it proper to speak only of addition and subtraction, rather than multiplication and division, of ratios—and here he actually uses verb forms of *multiplicatio* and *divisio.* The distinction seems connected with the fact that exponents cannot be multiplied—except improperly, as in $(2/1)^2 \cdot (2/1)^2 = (2/1)^4$, where, coincidentally, the correct result is produced. But we cannot multiply the ratios $(2/1)^2$ and $(3/1)^2$ directly, for by analogy with $(2/1)^2 \cdot (2/1)^2$ we ought to obtain four of something. However, such a multiplication would yield something as unintelligible as the union of a man and an ass.

In adopting such terminology, perhaps Oresme was guided by the fact that operations performed with exponents having the same base—as for example, $A^m \cdot A^n = A^{m+n}$ and $A^m : A^n = A^{m-n}$—required the addition and subtraction of exponents. It was then natural to think in terms of adding and subtracting ratios, which are the very entities represented by the exponents. The altered terminology was then applied to cases involving exponents with different bases.

Multiplication, addition, subtraction, and division could be performed on numbers, whether or not those numbers denominated, or represented, ratios. But operations on ratios were restricted to multiplication and division, which are called addition and subtraction respectively. Although evidence is lacking, it would come as no great surprise to discover that Oresme himself originated this terminology and created a new genre of mathematical treatise, an *algorism of ratios,* patterned after, but very different from, an *algorism of numbers.* Treatises of the latter kind, explaining and elaborating the four arithmetic operations as they are performed on numbers in the usual manner, were popular long before Oresme's time (see Selection 20). But I am not aware of the existence of any algorism of ratios in the Oresmian sense prior to Oresme's *Algorismus proportionum.* The absence of such terminology in Jordanus' *Arithmetica* and its use by Jacques Le Fèvre in a much later commentary on that treatise point to a post-Jordanian origin with Oresme himself as the possible originator. He may have derived his new, and peculiar, terminology from addition and subtraction of exponents, and then extended it to the simplest nonexponential cases (represented by the first two rules) in order to make the operational terminology for handling ratios as consistent as possible. If, on the other hand, Oresme inherited this terminology for the simplest cases, he may have extended it to operations involving integral and fractional exponents. The novel manner in which Oresme extended traditional Euclidean terminology to embrace exponential relations in the *De proportionibus* renders such a conjecture perfectly credible (see Selection 29, n. 3).

29 RATIONAL AND IRRATIONAL EXPONENTS DISTINGUISHED

Nicole Oresme (ca. 1325—1382)

Translated and annotated by Edward Grant[1]

Every rational ratio is immediately denominated by some number, either with a fraction, or fractions, or without a fraction.[2] The determination of these denominations will be shown below.

An irrational ratio is said to be mediately denominated by some number when it is an aliquot part or parts of some rational ratio, or when it is commensurable to some rational ratio, which is the same thing, as [for example], the ratio of a diagonal to its side is half of a double ratio.[3]

I say, therefore, that it does not seem true that every irrational ratio is commensurable to some rational ratio.[4] And the reason is that every ratio is just like a continuous quantity with respect to division, which is obvious by the last supposition. Therefore, by the ninth proposition of the tenth [book of Euclid, every ratio] can be divided into two [ratios] any of which is incommensurable to the whole. Thus there will be some ratio which will be part of a double ratio and yet will not be half of a double, nor a third part, or fourth part, or two-thirds part, etc., but it will be incommensurable to a double[5] and, consequently [incommensurable] to any [ratio] commensurable to this double ratio (by the comment on the eighth proposition of the tenth book of Euclid). And further, by the same reasoning there could be some ratio incommensurable to a double and also to a triple ratio and [consequently incommensurable] to any ratios commensurable to these, as [for example], half of a sesquitertian ratio.[6] And the same may be said of other ratios.

And there might be some irrational ratio which is incommensurable to any rational ratio. Now the reason for this seems to be that if some ratio is incommensurable to two [rational ratios], and some

ratio is incommensurable to three rational ratios,

2. That is, $A/B = n$, where n is an integer or ratio of integers.

3. In this treatise Oresme employs the terms *part* (*pars*), *multiple* (*multiplex*), and *commensurable* (*commensurabilis*) in a special and restricted exponential sense. Here is a brief summary of this special usage (Grant, *Nicole Oresme: De proportionibus*, p. 26): "If A and B are two ratios and m and n are integers, then if $A > B$, (1) ratio B is a *part* of ratio A when $B = A^{1/n}$ and (2) A is commensurable and multiple to B, since $A = B^n$; (3) B is *parts* of A when $B = A^{m/n}$, where $n > m > 1$ and m and n are in their lowest terms; so that (4) A is commensurable to B, since they have a common measure in the unit ratio $A^{1/n}$. However, A is not multiple to B, since $A \neq B^n$."

As an example representing this type of proportion, we are given "the ratio of a diagonal to its side" which is "half of a double ratio, that is, $(2/1)^{1/2}$. For Oresme this is a *part* of 2/1 because the latter is its multiple in the special sense that $2/1 = [(2/1)^{1/2}]^2$. Therefore, Oresme would hold that 2/1 is the *immediate denomination* of $(2/1)^{1/2}$, since it is its rational base (indeed, 2/1 would also be the immediate denomination of $(2/1)^{3/2}$, even though the latter is greater than 2/1). But what does Oresme mean when he insists that an irrational ratio is *mediately denominated* by some number? Almost certainly he meant the exponent itself, by means of which we can distinguish between all the different irrational ratios having the same base. Thus the exponent is the mediate numerical denomination of (2/1).

4. In the remainder of this selection Oresme considers a third category of ratio, namely an irrational ratio with an irrational exponent, thereby distinguishing two types of irrational ratio: one with a rational exponent, the other with an irrational exponent.

5. Oresme's "pattern of thought seems to be as follows: Any ratio is like a continuous quantity and therefore divisible in any conceivable mathematically logical manner; now, one conceivable way is to divide a rational ratio into smaller ratios one of which, at least, is irrational and no part, or parts, of the given rational, and hence incommensurable to it; therefore, such a conceivable ratio, logically possible, must correspond to some real category of ratios even though we can not express a single instance of it." (Grant, *Nicole Oresme: De proportionibus*, p. 36.) Because he could not symbolize or otherwise explicitly express an irrational exponent, Oresme could only convey negatively his intuitive sense about such ratios. Thus it could not be $(2/1)^{1/2}$ (that is, "half of a double" ratio), nor $(2/1)^{1/3}$ ("a third part" of a double), nor $(2/1)^{1/4}$ ("a fourth part" of a double), nor $(2/1)^{2/3}$ ("a two-thirds part" of a double), and so on. Indeed, it would be incommensurable to a double ratio because it would be inexpressible in the form $(2/1)^{m/n}$, where m/n is rational and $m < n$. It seems correct to express what Oresme had in mind by $(2/1)^{1/\sqrt{q}}$, since raising this to the qth power will never raise this ratio to equality with 2/1.

6. That is, $(4/3)^{1/2}$

1. Reprinted with permission of the copyright owners, the Regents of the University of Wisconsin, from *Nicole Oresme: De proportionibus proportionum and Ad pauca respicientes,* edited with Introductions, English translations, and Critical Notes by Edward Grant (Madison, Wis.: University of Wisconsin Press, 1966), pp. 161, 163. In this selection from Oresme's treatise *On Ratios of Ratios (De proportionibus proportionum)* we may have the first explicit, though intuitive, sense of an irrational exponent emerging from a persistent analysis of the fourteenth-century concept of a "ratio of ratios" (for the meaning of this expression, see Selection 51.2, n. 51). His treatment of rational exponents can be seen in the preceding selection.

and so on, then there might be some ratio incommensurable to any rational ratio whatever,[7] though this does not follow from the form of the argument as [it does follow when we say that] some continuous quantity is incommensurable to all quantities of one [geometric] series. However, I do not know how to demonstrate this; but if the opposite should be true, it is indemonstrable and unknown. . . .

Geometry

30   ON THE DIVISION OF FIGURES

Leonardo of Pisa (Fibonacci) (b. ca. 1179; d. after 1240)

Translation, introduction, and annotation by Edward Grant[1]

In his commentary on the first book of Euclid's *Elements*, Proclus (410–485) mentions a *Book on Divisions (of Figures)*, which he ascribed to Euclid. In this treatise Euclid apparently divided given figures sometimes into "like" figures (that is, triangles into triangles, and so on) and sometimes into unlike figures (for example, a triangle into a triangle and a quadrilateral). Although the treatise has been lost in Greek, an Arabic text of it containing enumerations of thirty-six propositions but only four proofs was found by F. Woepcke in 1851. Later it was found that in Leonardo Pisano's *Practica geometriae (The Application of Geometry)*, composed in 1220, "no less than twenty-two of Woepcke's propositions are practically identical in statement with propositions in Leonardo; the solutions of eight more of Woepcke are either given or clearly indicated by Leonardo's methods, and all six of the remaining Woepcke propositions (which are auxiliary) are assumed as known in the proofs which Leonardo gives of propositions in Woepcke."[2] The propositions included below appear in Distinction 4 of Leonardo's *Practica geometriae* and are some of those used by Archibald in his reconstruction of Euclid's *Book on Divisions*. It is noteworthy, however, that if Leonardo derived the basis of this section, and a number of its propositions, from Euclid's lost work, he probably supplied additional proofs for some propositions, perhaps rearranged others, and added numerical examples, since he was composing an "applied geometry." Thus, while "his *Practica Geometriae* contains many references to Euclid's *Elements* and many uncredited extracts from this work," and "similar treatment is accorded works of other writers," it seems true to say that "in the great elegance, finish and rigour of the whole, originality of treatment is not infrequently evident."[3]

The popularity of dividing geometrical figures into other figures stems perhaps from the importance of dividing differently shaped plots of land into a variety of figures for purposes of inheritance and crops. It would be surprising, however, if Euclid himself had such concerns when he wrote his lost treatise.

[1.] And if you wish to divide any triangle by a line parallel *(equidistantem)* to one of the sides of the triangle, the manner of doing this will be demonstrated in the following triangle *ABG* [Fig. 1],

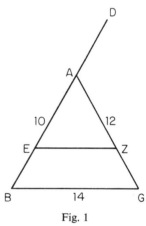

Fig. 1

7. If there is an irrational ratio incommensurable to $(2/1)^n$, there might also be an irrational ratio incommensurable to $(2/1)^n$ and $(3/1)^n$, and so on inductively. By this approach, Oresme was convinced of the existence of one or more irrational ratios that are incommensurable to all irrational ratios, although he readily admitted that he was unable to demonstrate this.

1. The propositions are translated from Leonardo Pisano's *Practica geometriae (The Application of Geometry)* in the edition of Baldassare Boncompagni, *Scritti di Leonardo Pisano* (Rome, 1862), Vol. II, Distinction 4, "On the Division of Fields between Partners," pp. 119–125, 128–129, 138. Translations from Euclid's *Elements* in the following notes are from Thomas Heath, *The Thirteen Books of Euclid's Elements*, 2d. ed. (New York: Dover, 1956).

2. R. C. Archibald, *Euclid's Book on Divisions of Figures, with a Restoration Based on Woepcke's Text and on the Practica Geometriae of Leonardo Pisano* (Cambridge, England: Cambridge University Press, 1915), p. 11.

3. Quotations from Archibald, pp. 11–12. Archibald conjectures (p. 12) that Leonardo may have had a Latin translation of Euclid's treatise made by Gerard of Cremona.

which can be divided into two equal parts by a line parallel to base *BG*.

Let side *BA* be projected *(emictatur)* beyond the triangle to point *D*; and let *AD* equal half of *BA*; and let *AE*, a mean proportional between lines *BA* and *AD*, be taken, that is, as *BA* is to *EA* so is *EA* to *AD*. And let straight line *EZ* be drawn through point *E* and parallel to base *BG*. I say that triangle *ABG* is divided into two equal parts by line *EZ*, one of which parts is triangle *AEZ*, the other quadrilateral *EBGZ*.

This is proved as follows: Triangles *AEZ* and *ABG* are similar, since the two angles *AEZ* and *AZE* are equal to [angles] *ABG* and *AGB*, respectively, namely the exterior angles [are equal] to the interior [angles]; and angle *BAG* is common to each. Now, the similar triangles are related as the double ratio of the corresponding *(homologorum)*, that is similar *(similium)*, sides. Therefore, as *BA* is to *AD* so is triangle *ABG* to triangle *AEZ*. For *BA* is double *AD*, so that triangle *ABG* is double triangle *AEZ*. Therefore, half of triangle *ABG* is triangle *AEZ*.[4]

[2.] Or, [to do it] another way: Since three straight lines are proportional in continued proportion, as the first [line] is to the third so is the figure *(spatium)* on the first [line] to the similarly described figure on the second [line].[5] Therefore, as already stated, as the first line *BA* is to the third line *AD* so is that which is on *AB* double to that which is on *AE*, and therefore triangle *ABG* is double triangle *AEZ*,[6] as I said before.

[3.] This can be demonstrated in another [and third] way: Since *BA* is to *AE* as *AE* to *AD*, it follows that *BA · AD* will be equal to the square *(tetragonum)* on *AE*. Therefore, the square on straight line *AB* is double the square on straight line *AE*. But since *BG* is parallel to *EZ*, [we see that] *BA* is to *AE* as *GA* to *AZ* and therefore as the square of *BA* is to the square of *AE* so is the square of *GA* to the square of *AZ*. [However,] the square on *BA* is double the square on *AE*, and it follows that the square on *GA* will be double the square on *AZ*. Hence *BA · AG* is double *EA · AZ* and therefore triangle *ABG* is double triangle *AEZ* . . . .[7]

HERE BEGINS [A PROPOSITION] ON THE DIVISION OF TRIANGLES INTO THREE PARTS[8]

[1.] If you wish to divide a triangle *ABG* into three equal parts on one of its sides, say on side *BG*, as above [in the preceding proposition], divide this *BG* into three equal parts, which are *BD*, *DE*,

and *EG;* and *A* and *D*, and *A* and *E*, are joined by straight lines [Fig. 2]. I say that triangle *ABG* has

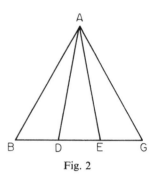

Fig. 2

4. We are to bisect the given triangle by a line parallel to base *BG*. The data are *BA* = 2*AD*; *BA*/*AE* = *AE*/*AD*. Therefore, line *EZ* drawn parallel to *BG* must divide △*ABG* into the two equal parts △*AEZ* and *EBGZ*. Proof:

(1) △ *AEZ* and △ *ABG* are similar (since ∠*AEZ* = ∠ *ABG*, ∠ *AZE* = ∠ *AGB*, and ∠ *BAG* = ∠ *BAG*);

(2) ∴ △ *ABG*/△ *AEZ* = 2/1 = *BA*/*AD* (by *Elements* VI. 19: "Similar triangles are to one another in the duplicate ratio of the corresponding sides" [Heath, II, 232]).

(3) ∴ △*ABG* = 2(△ *AEZ*), and △ABG/2 = △AEZ;

(4) ∴ △*AEZ* = *EBGZ*.

5. Enunciated as a porism to *Elements*, Book VI, Proposition 19. Leonardo, following the usual Greek convention, did not cite the Euclidean theorems supporting the successive steps of a proof.

6. Recalling that *BA* = 2*AD* and that *AB*, *AE*, and *AD* are the three lines in continued proportion (that is, *AB*/*AE* = *AE*/*AD*), the first and third lines are related as the similar figures erected on the first and second lines, that is, on *AB* and *AE*. Therefore, since *AB*/*AD* = 2/1 so does △ *ABG*/△ *AEZ* = 2/1.

7. A summary of the proof follows:

(1) $\dfrac{BA}{AE} = \dfrac{AE}{AD}$;

(2) ∴ *BA · AD* = (*AE*)²; and since *AD* = ½ *BA*,

(3) *BA* · ½ *BA* = (*AE*)², and

(4) (*BA*)² = 2 (*AE*)².

(5) *BG* ∥ *EZ*;

(6) ∴ $\dfrac{BA}{AE} = \dfrac{GA}{AZ}$, and

(7) ∴ $\dfrac{(BA)^2}{(AE)^2} = \dfrac{(GA)^2}{(AZ)^2}$

(by *Elements* VI.22:"If four straight lines be proportional, the rectilineal figures similar and similarly described upon them will also be proportional; · · ·" [Heath, II, 240]).

(8) But (*BA*)² = 2 (*AE*)²;

(9) ∴ (*GA*)² = 2 (*AZ*)²;

(10) ∴ *BA · AG* = 2 (*EA · AZ*), and

(11) $\dfrac{\triangle\,ABG}{\triangle\,AEZ} = \dfrac{2}{1}$.

As in so many propositions, Leonardo concludes with a numerical example.

8. This proposition was not derived from Euclid's *Book on Divisions of Figures.*

been divided into three equal parts, which are triangles *ABD, ADE,* and *AEG*—for they are on equal bases and have the same altitude,[9] making it necessary that they be equal to each other. Now, if I want a third part of a triangle *BGD* on point *Z* [Fig. 3], I will join *B* and *Z;* and if *GZ* is a third

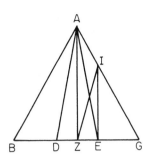

I

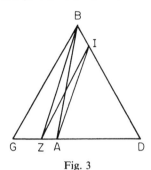

Fig. 3

part of *GD,* surely triangle *BGZ* will be a third part of triangle *BGD.*[10]

[2.] If, however, *GZ* is not a third part of *GD,* then it will be more or less. First, let it be less [than a third part]. Therefore, I take a third part of *GD* and this is *GA.* I join *AI* by a straight line parallel to *ZB* and I join *I* and *Z* by a straight line, and quadrilateral *IBGZ,* which I shall prove to be a third part of the whole triangle *BGD,* is removed from triangle *BGD.* For example, since *GA* is a third part of the whole *GD,* triangle *BGA* will be a third part of triangle *BGD.* And since triangles *BZA* and *BZI* are between parallels *(equidistantes) BZ* and *AI* and on the same base *BZ,* they are equal.[11] Let triangle *BGZ* be added in common [to each triangle, namely *BZA* and *BZI*] and quadrilateral *IBGZ* will be equal to triangle *BGA.* Therefore, quadrilateral *IBGZ* is a third part of triangle *BGD* and the remainder, namely triangle *IZD,*[12] could be divided into two equal parts in such a manner as you would want from the ways described above.[13] Repeating *(rursus),* triangle *ABG* was divided into three equal parts, which are triangles *ABD* and *ADE* and *AEG* [Fig. 2]; therefore, [line] segments *BD, DE,* and *EG* are mutually equal. In the figure above [Fig. 3], however, we placed point *Z* between *B* and *D* [Leonardo assumes the superimposition of Fig. 3 on Fig. 2], which would be the same as if we had placed it between *EG,* since the same procedure would be followed.

[3.] Now, however, let us place point *Z* between *D* and *E* on side *BG* [Fig. 4]; and from point *Z* we want to produce a line cutting a third part of triangle *ABG.* First, let *A* and *Z* be joined and through point *D* or point *E,* let a line be drawn

Fig. 4

that is parallel *(equidistans)* to line *AZ.* Let this line [be drawn through point *E* and] be line *EI*; also let *Z* and *I* be joined. I say that straight line *ZI* cuts off a third part of triangle *ABG,* namely triangle *IZG.*

This is proved in the following manner: Since triangles *AZI* and *AZE* lie between two parallel lines and are on the same base, they are equals.[14] If triangle *ADZ*[15] is added, trapezium *ADZI* will be equal to triangle *ADE;* but triangle *ADE* is a third part of triangle *ABG,* so that quadrilateral *ADZI* is similarly a third part of triangle *ABG.* And triangle *ABD* is another third part of triangle *ABG.* As the remaining third part, there is left triangle

---

9. *Elements* I.38: "Triangles which are on equal bases and in the same parallels are equal to one another" (Heath, I, 333).

10. *Elements* VI.1: "Triangles and parallelograms which are under the same height are to one another as their bases" (Heath, II, 191).

11. *Elements* I. 37: "Triangles which are on the same base and in the same parallels are equal to one another" (Heath, I, 332).

12. My correction of the Latin text from *IZB.*

13. Using Fig. 3, the steps of the proof are as follows: In $\triangle BGD$ assume that $GZ < \frac{1}{3} GD$, $GA = \frac{1}{3} GD$, and $AI \parallel ZB$. Prove that quadrilateral $IBGZ = \frac{1}{3} \triangle BGD$.
(1) $GA = \frac{1}{3} GD \therefore \triangle BGA = \frac{1}{3} \triangle BGD$ (*Elements* VI.1).
(2) Also, $\triangle BZA = \triangle BZI$ (*Elements* I. 38);
(3) $\therefore \triangle BGZ + \triangle BZA = \triangle BGZ + \triangle BZI$.
(4) But $\triangle BGZ + \triangle BZA = BGA$, and
$\triangle BGZ + \triangle BZI = IBGZ$;
(5) $\therefore IBGZ = \triangle BGA$.
(6) $\triangle BGA = \frac{1}{3} \triangle BGD$ [by (1) above];
(7) $\therefore IBGZ = \frac{1}{3} \triangle BGD$.
Now, since $\triangle BGD - IBGZ = \triangle IZD$, it is possible to divide triangle *IZD* into two equal parts (Leonardo demonstrated this at the very beginning of Distinction 4 [Boncompagni, p. 110]), each of which will be $\frac{1}{3} \triangle BGD$. Therefore, when $GZ < \frac{1}{3} GD$, $\triangle BGD$ could be divided into three equal parts—that is, into a quadrilateral and two triangles.

14. By *Elements* I. 37.

15. My correction of the Latin text from *ADE.*

IZG, which was cut from triangle *ABG* by line *IZ*, which it was necessary to do.[16]

[4.] Again, let there be triangle *ABG* which must be divided between three partners, each of whom wants one of the sides of triangle *ABG* in his portion [Fig. 5]. I divide side *BG* into two equal

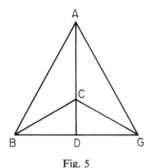

Fig. 5

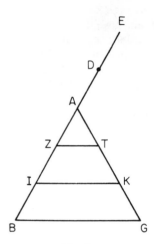

Fig. 6

parts at point *D* and join points *A* and *D* by a straight line. And I mark off *DC*, a third part of line *AD*; and I join points *B* and *C*, and *C* and *G*, by straight lines. I say that triangle *ABG* has been divided into three equal parts, each of which is on one of the sides of triangle *ABG*, for these parts are triangles *ABC*, *ACG*, and *BCG*. I prove this as follows.

Since line *DC* is a third part of line *DA*, line *AC* will be double *CD*; it follows that triangle *ABC* is double triangle *BCD*. For the same reason, triangle *ACG* is double triangle *GCD*. Now, since *BD* is equal to *DG*, triangles *CBD* and *CDG* are equal. Therefore, the whole triangle *CBG* is double each of the triangles *CBD* and *CDG*. But things that are the double of the same thing are equal; therefore, triangles *ACB*, *ACG*, and *BCG* are equal to one another. Therefore, triangle *ABG* has been divided [into three parts, each of which lies on one of the sides of triangle *ABG*].[17]

A demonstration of how a given part is cut from a triangle by a line drawn through a given point within the triangle.[18]

A demonstration of how a given part is cut from a given triangle by a line drawn from a given point outside the triangle.[19]

. . . And if we wish to divide a triangle *ABG* [into three equal parts] by two lines parallel to base *BG* [Fig. 6], I produce side *BA* by a third [of its length] to point *D* and from it extend straight line *DE* equal to line *DA*. The whole [line] *AE* will be two-thirds of line *BA*. Now I posit line *AZ* as a mean [proportional] between *BA* and *AD*; and line *AI* as a mean [proportional] between *BA* and *AE*. Then through points *Z* and *I*, I produce straight

lines *ZT* and *IK* parallel to base *BG*. I say [now] that triangle *ABG* has been divided into three equal parts, one of which is triangle *AZT*, another quadrilateral *ZIKT*, and the third will be quadrilateral *IBGK*. This is proved as follows: Since *BA* is to *AZ* as *AZ* to *AD*, therefore *BA* will be to *AD* as triangle *ABG* to triangle *AZT*, since these triangles are similar. Now *BA* is triple *AD*, so that triangle *ABG* is triple triangle *AZT*. Therefore, triangle

16. Assuming the data of the preceding paragraph and referring to Fig. 4, Leonardo wishes to show that line *ZI* divides △*ABG* in such a way that △*IZG* = ⅓ △*ABG*.
(1) △*AZI* = △*AZE* (by *Elements* I.37);
(2) ∴ △*ADZ* + △*AZI* = *ADZI* = △*ADZ* + △*AZE* = △*ADE*; ∴ *ADZI* = △*ADE*.
(3) But △*ADE* = ⅓ △*ABG* (derived in the first paragraph of this proposition; see n. 9);
(4) ∴ *ADZI* = ⅓ △*ABG*, and
(5) △*ABD* = ⅓ △*ABG* (see first paragraph of proof);
(6) ∴ △*IZG* = ⅓ △*ABG*, since it is the only remaining part.
17. Assuming the data of the preceding paragraph and referring to Fig. 5, this proposition may be represented as follows:
(1) *DC* = ⅓ *DA* (by assumption);
(2) ∴ *AC* = 2*CD*, and
(3) △ *ABC* = 2 △*BCD*, and similarly,
(4) △ *ACG* = 2 △*GCD*.
(5) Also, △*CBD* = △*CDG* (since *BD* = *DG*);
(6) ∴ △*CBG* = 2 △*CBD* = 2 △ *CDG*.
(7) But △*ABC* = 2 △*CBD* = 2 △ *CDG*, and △*ACG* = 2 △*CBD* = 2 △ *CDG*;
(8) ∴ △*CBG* = △*ABC* = △*ACG* (things double the same thing are equal).
18. The discussion on page 121 of the Boncompagni edition is omitted.
19. The proof cited below is included by R. C. Archibald as a proposition that was derived from Euclid's *Book on Divisions of Figures.*

AZT is a third part of triangle *ABG*. Again, because *BA* is to *IA* as *AI* to *AE*, therefore *BA* will be to *AE* as the triangle on *EA* will be to the similarly situated triangle on *AI*. But the triangles *AIK* and *ABG* are similarly described on sides *AI* and *AB*. Therefore, just as *EA* is to *AB*, that is, as 2 is to 3, so is triangle *AIK* to triangle *ABG*—that is, triangle *AIK* is two-thirds of triangle *ABG*. If triangle *AZT*, which is a third part of *ABG*, is subtracted from *AIK*, quadrilateral *ZIKT* will remain, which is a third part of triangle *ABG*. Similarly, the remaining quadrilateral *IBGK* will be the other third part. Therefore, triangle *ABG* has been divided into three equal parts, which was to be done *(quod oportebat facere)*.[20] And so by the modes demonstrated [here], all kinds of triangles can be divided into four or more parts. We come now to the division of quadrilaterals.

HERE BEGINS THE DIVISION OF QUADRILATERALS

There are three kinds of quadrilaterals: (1) the parallelogram *(paralilogramina)*; (2) the trapezium *(caput abscisa)*; and (3) unequal- (or different-) sided *(diversilatera)* figures.[21]

Parallelograms are figures whose sides and the angles opposite them are mutually equal, and there are four species of them: (1) the *square,* which has all sides equal and all angles are right angles; (2) the *rectangle,*[22] which has only the two opposite sides equal and all right angles: (3) the *rhombus,*[23] which consists of four equal sides but has unequal angles; (4) the *rhomboid,* where the two opposite sides are equal and the opposite angles are equal.[24]

Now, since there is only one way of dividing these four species of parallelograms, I have arranged to determine all their figures under the same representations so that what we say in one of these [four] species would seem to apply in the remaining species. Let *ABCD* represent a square, rectangle, rhombus, and rhomboid any one of which could be divided into two equal parts [Fig. 7]. Since parallelograms are quadrilaterals, their diagonals *(dyametri)* would divide them into equal parts . . . .[25]

20. Assuming the data and referring to Fig. 6, this proposition may be described as follows:

(1) $\frac{BA}{AZ} = \frac{AZ}{AD}$ (by assumption);

(2) $\therefore \frac{BA}{AD} = \frac{\triangle ABG}{\triangle AZT}$ (by *Elements* VI. 19, porism; see above and n. 5).

(3) $\frac{BA}{AD} = \frac{3}{1}$ (derived from the initial conditions);

(4) $\therefore \frac{ABG}{AZT} = \frac{3}{1}$, and

(5) $\frac{\triangle AZT}{\triangle ABG} = \frac{1}{3}$.

(6) $\frac{BA}{IA} = \frac{IA}{AE}$ (by assumption);

(7) $\therefore \frac{BA}{AE} = \frac{\triangle \text{ on } EA}{\triangle \text{ on } AI}$ (by *Elements* VI. 19, porism).

But the relation in (7) also obtains between *AI* and *AB*. Since $AB/AI = AI/AE$, it follows by alternated ratio that $AI/AB = AE/AI$. Therefore, the triangles erected on *AI* and *AB* are similarly described as those erected on *AE* and *AI*.

(8) $\therefore \frac{\triangle \text{ on } EA}{\triangle \text{ on } AI} = \frac{\triangle \text{ on } AB}{\triangle \text{ on } AI} = \frac{\triangle ABG}{\triangle AIK}$.

(9) But $\frac{EA}{AB} = \frac{2}{3}$ [by inverting the ratio in (7)];

$\therefore \frac{\triangle AIK}{\triangle ABG} = \frac{2}{3}$;

(10) $\frac{\triangle AZT}{\triangle ABG} = \frac{1}{3}$; $\therefore \triangle AZT = \frac{1}{3} \triangle ABG$;

$\therefore \triangle AIK - \triangle AZT = ZIKT = \frac{1}{3} \triangle ABG$.

(11) $\triangle ABG - \triangle AIK = IBGK = \frac{1}{3} \triangle ABG$.

21. Essentially, the three types contain figures of the following kinds: (1) all sides are parallel; (2) only two sides are parallel; and (3) no sides are parallel.

22. Literally a figure "with one side longer than another" *(parte altera longiora)*. In Campanus of Novara's popular edition of Euclid's *Elements* (for full title, see selection 27, n. 1), the expression for rectangle is *tetragonus longus*.

23. Leonardo uses *rumbus,* a term derived from Greek, whereas in Campanus of Novara's edition of the *Elements* we find *helmuayn,* a direct transliteration of an Arabic term.

24. In his definition of quadrilaterals Euclid (see Bk. I, Def. 22, of Heath's translation) also distinguishes *square, oblong* (that is, rectangle), *rhombus,* and *rhomboid.* All other quadrilaterals are included in the class of *trapezia.* Leonardo, however, restricts trapezium to all quadrilaterals having only two sides parallel. All remaining quadrilaterals would have unequal sides with none parallel.

25. There follows a detailed elaboration of this statement, as well as the application of other divisions. R. C. Archibald has summarized the following two proofs of this proposition, which he includes as part of the reconstruction of Euclid's *Book on Divisions of Figures* (p. 37; a few slight alterations have been made):

"To divide a parallelogram into two equal parts by a straight line drawn from a given point situated on one of the sides of the parallelogram.

Let *abcd* be the parallelogram and *i* any point in the side *ad*. Bisect *ad* in *f* and *bc* in *e*.

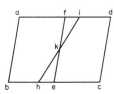

And if you desire to cut any parallelogram into two equal parts by a line drawn from a given point within it—as parallelogram *AG*, within which there is the given point *C* and through which we wish the line to pass that will divide parallelogram *AG* into two equal parts on half of the diagonal *BD*—assume point *F* and join *C* and *F* and produce it to each side in points *E* and *Z*. I say that parallelogram *AG* has been divided into two equal segments, one of which is quadrilateral *EABZ* and the other, quadrilateral *EDGZ*. This is proved as follows: Since the two straight lines *EZ* and *DB* fall on the parallel lines *AD* and *BG*, angle *FDE* will be equal to angle *FBZ*, and angle *FED* to angle *FZB*; and similarly, the angles around *F* are mutually equal [that is, the vertical angles *ZFB* and *EFD*]. Therefore, triangles *FED* and *FBZ* are equiangular; and since line *FB* equals line *FD*, line *DE* will be equal to line *BZ*, and triangle *DFE* will be equal to triangle *BFZ*. Should quadrilateral *DFZG* be taken as common, quadrilateral *EZGD* will be equal to triangle *BGD*. But triangle *BGD* is half of parallelogram *AG*; therefore, quadrilateral *EZGD* is half of parallelogram *AG*, as I said before.[26]

All parallelograms are divided in this way when a line is drawn from a given point lying beyond one of its sides; as if the given point were *E*,[27] it would divide line *DB* into two equal parts at point *F* [Fig. 7], and line *EZ* would be drawn through this

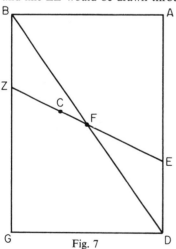

Fig. 7

point and it would divide parallelogram *AG* into two equal parts, as has been proved. And if you wish to divide any parallelogram into two equal parts by a line drawn from a given point outside the parallelogram, as [for example,] *AC*, beyond which lies point *E*, I again draw diagonal *(dy-*

ametrum) *BD* and divide it at point *G* [Fig. 8]

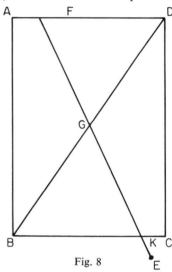

Fig. 8

and I join *E* and *G* and extend it to *F*. The parallelogram *AC* has been divided into two equal parts by the line drawn from given point *E*, and this has been proved by the figure above. [Indeed,] as is obvious in the figure, one half of it is quadrilateral *FABK* and the other *FKCD*.

HERE BEGINS THE DIVISION OF PARALLELOGRAMS INTO MORE [THAN TWO] PARTS.

. . . And if from a given point on one of the

Join *fe*. Then the parallelogram *ac* is divided into equal parallelograms *ae*, *fc* on equal bases.

Cut off *eh* = *fi*. Join *hi*. Then this is the line required.

[Proof] I. Let *hi* meet *fe* in *k*. Then [△s *fki*, *hke* are equal; add to each the pentagon *kfabh*, and so on].

[Proof] II. Since *ae*, *fc* are parallelograms, *af* = *be* and *fd* = *ec*. But

$$fd = \tfrac{1}{2}\,ad$$
$$\therefore fd = af = ec.$$

And since *fi* = *he*, *ai* = *ch*.

So also *di* = *bh*, and *hi* is common.

∴ quadrilateral *iabh* = quadrilateral *ihcd*."

Archibald declares that only the first of these two proofs is Euclidean (p. 37, n. 95).

26. Using Fig. 7 and the given data, this proof proceeds as follows:

To prove that *EABZ* = *EDGZ* and that *EABZ* + *EDGZ* = *ABGD*

(1) ∠*FDE* = ∠*FBZ* (alternate angles are equal by Elements I.29).

(2) ∠*FED* = ∠*FZB* [by the same reasoning as (1)].

(3) ∠*ZFB* = ∠*EFD* (equality of vertical angles by Elements I.15).

(4) ∴ △s *FED*, *FBZ* are equiangular, and since

(5) *FB* = *FD*;

∴ *DE* = *BZ*, and

△ *DFE* = △ *BFZ*.

27. Here point *E* is simply assumed to lie outside the

sides, you wish to draw a straight line which will cut off a third part from a given parallelogram, let us posit the given point to be on side *AD* [Fig. 9].

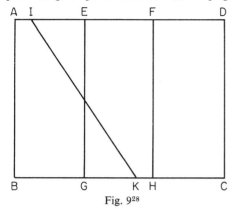

Fig. 9[28]

After dividing the first side, *AD*, at points *E* and *F* in the order stated above,[29] namely that each of the segments *AG*, *EH*, and *FC* is a third part of *AD*, the whole, and if the given point is *E*, I draw line *EG*, and there will be a parallelogram *AG* cut from parallelogram *AC* and drawn from point *E*, which will be a third part of it [that is, of *AC*]. Similarly, if the given point were *F*, parallelogram *FC* will be a third part of parallelogram *AC*.

But if the given point is not at *E* or *F*, it will be between *A* and *E*, between *E* and *F*, or between *F* and *D*. Let the given point be *I* and lie first between *A* and *E*; and because I wish to remove from parallelogram *AC* a third part of it by means of a line drawn from point *I*, I assign the first point, *F*, to the third part of side *AD* and from point *F* draw line *FH* parallel to sides *AB* and *DC*. Therefore, parallelogram *FC* will be a third part of parallelogram *AC*. Hence parallelogram *AH* is double parallelogram *FC*, so that it is necessary to divide it into two equal parts by a line drawn from point *I*. Therefore, by the same distance as *I* is from *A* so far distant do I make *K* from *H*, and I join *I* and *K* to produce quadrilateral *IABK*, which is half of parallelogram *AH*. And so it is that what has been cut by line *IK* is a third part of parallelogram *AC*, which has been divided into [three] equal parts, namely into the quadrilaterals *IABK* and *IKHF*, and into parallelogram *FC*, as is clear in the figure [Fig. 9]. . . .[30] And in the same way, every parallelogram can be divided into four or more equal parts. . . .

HERE BEGINS THE DIVISION OF THE FOUR FIGURES OF TRAPEZIA, WHICH HAVE ONLY TWO PARALLEL SIDES

There are four species of trapezia. . . .and you should know that the things found about some of them could be said about any of the remaining ones. Therefore, let the four related species of trapezia [be represented by] *ABGD* with sides *AD* and *BG* parallel [Fig. 10]. We wish to divide any of

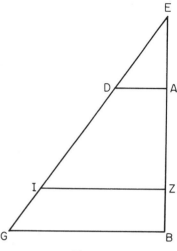

Fig. 10

these [trapezia] by a straight line parallel to their base, which is *BG*; and of the parallel lines, *AD* and *BG*, *AD* is smaller than *BG*. If we produce straight lines *BA* and *GD*, which [now] meet at the sides of *AD*, they would meet at point *E*. Let the square of line *ZE* be half of the squares of straight lines *EB* and *AE*; and through point *Z*, I shall draw straight line *ZI* parallel to base *BG*. I say that trapezium *(trapeziem)*[31] *ABGD* has been divided

figure; but the proof would be identical with that of the preceding paragraph.

28. Since this proposition adds an element to the figure used in the brief proposition immediately preceding but omitted here, I have followed Archibald (p. 38), who considers this a proposition from Euclid's *Book on Divisions,* and combines the two figures, adding line *IK* to the first figure.

29. That is, in the brief proposition omitted here; see note 28.

30. Briefly, the proof may be described as follows:
$$FD = \tfrac{1}{3} AD;$$
$$FH \parallel AB \text{ and } DC;$$
$$\therefore FC = \tfrac{1}{3} AC; \text{ and}$$
$$AH = 2FC.$$
Now, since *EG* divides *AH* into two equal parts (by previous proposition), make *IA = KH*.
$$\therefore IABK = \tfrac{1}{2} AH;$$
$$IABK = IKHF = FC.$$
The same procedure would be used if point *I* were between *E* and *F* or *F* and *D*.

31. Although *caput abscisa* is the usual term used by Leonardo for trapezium, here he uses the term *trapeziem*.

into two equal parts by line *ZI* parallel to line *BG*.

This is proved as follows: Since the [sum of the] square of lines *EB* and *AE* are double the square of line *EZ*, the [sum of the] triangles *EBG* and *EAD* will be double triangle *EZI*, since these triangles are similar. But if we take away triangle *EZI* from triangle *EBG*, quadrilateral *ZBGI* will remain and triangle *EAD* [added to it] will make them equal to triangle *EZI*. And if triangle *EAD* is taken away commonly [from triangle *EZI* and the sum of quadrilateral *ZBGI* and triangle *EDA*], it follows that quadrilateral *ZG* equals quadrilateral *AI*. Therefore, quadrilateral *ABGD* has been divided into two equal parts by line *ZI;* which it was necessary to do. . . .[32]

I shall demonstrate again, in another way, how a quadrilateral of two parallel sides should be divided [by lines drawn] from its angles. Let quadrilateral *ABGD* be a trapezium whose sides *AD* and *BG* are parallel [Fig. 11], with *AD* smaller [than *BG*].

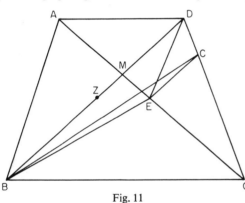

Fig. 11

I draw the diagonals *AG* and *BD*, which intersect at point *M*. Now, since sides *AD* and *BG* are parallel, triangles *AMD* and *BMG* will be similar. Therefore, as *BG* is to *AD* so is *BM* to *MD*, and *GM* to *MA*. But *BG* is greater than *AD*, and therefore *BM* is greater than *MD*, and *GM* than *MA*. Then the diagonals *AG* and *DB* are divided into two equal parts at points *E* and *Z*; and through point *E*, line *EC* is drawn parallel to diagonal *BD*, and *B* and *C* are joined. I say that quadrilateral *ABGD* has been divided from angle *B* by line *BC*.

This is proved as follows: I join *B* and *E*, and *E* and *D*, by straight lines, and since point *E* is in the middle of diagonal *AG*, triangles *ADE* and *DEG* will be equal to each other; and also triangle *ABE* is equal to triangle *BGE*. Therefore, quadrilateral *EDAB* contains half of the whole quadrilateral *ABGD*. And since triangles *BDC* and *BDE*

are between parallels *BD* and *EC* and on base *BD*, they will be mutually equal. After adding both of these triangles to triangle *ABD*, quadrilateral *ABCD* will be equal to quadrilateral *ABED*. But quadrilateral *ABED* is half of quadrilateral *ABGD*, and therefore quadrilateral *ABCD*[33] is half of quadrilateral *ABGD*. . . .[34]

Therefore, it has now been demonstrated how quadrilaterals with two parallel sides are divided from a given point on any side of it; now we discuss the manner in which such quadrilaterals can be divided from a given point outside the figure . . . .

HERE BEGINS THE DIVISION OF UNEQUAL-SIDED QUADRILATERALS

First, I wish to demonstrate how an unequal-sided quadrilateral *(quadrilaterum diversilaterum)* is divided into two equal parts from a given angle. Let quadrilateral *ABCD* [Fig. 12] be the one I wish

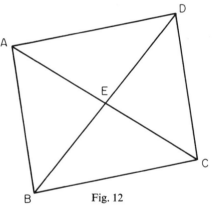

Fig. 12

to divide into two equal parts from a given angle *A*. I draw first the diagonal *BD*, subtending angle *BAD*, and I intersect it by diagonal *AC* at point *E*.

32. Using the given data and Fig. 10, here is a summary of the proof:
(1) $(EB)^2 + (AE)^2 = 2 (EZ)^2$, and $\triangle$ s *EBG, EAD,* and *EZI* are similar;
(2) ∴ $\triangle EBG + \triangle EAD = 2\triangle EZI$ (by *Elements* VI.19);
(3) $\triangle EBG - \triangle EZI = ZBGI$;
(4) ∴ $ZBGI + \triangle EDA = \triangle EZI$.
But $\triangle EZI = \triangle EDA + AZID$. Removing $\triangle EDA$ from both sides, we get:
(5) $ZBGI = AZID$;
(6) ∴ $ABGD = ZBGI + AZID$.
This proof is included in Archibald's reconstruction of Euclid's *Book on Divisions of Figures*.
33. Corrected from *ABGD*.
34. With reference to Fig. 11, demonstrate that *BC* drawn from angle *B* divides *ABGD* into two equal parts:
(1) $\triangle ADE = \triangle DEG$ (since *DE* is a median line in

Straight lines *BE* and *ED* are either equal or not equal. First, let them be equal. Now, since *BE* is equal to *ED*, triangle *ABE* will be equal to triangle *ADE*, as will triangle *EBC* to triangle *ECD*. Therefore, the whole triangle *ABC* will be equal to triangle *ACD*. Hence, quadrilateral *ABCD* has been divided into two equal parts by diagonal *AC* drawn from a given angle; and it was necessary to show this.

But should line *BE* not be equal to *ED*, but *BZ* is equal to line *ZD*, I draw line *ZI* parallel to diagonal *AC*, as is shown in the second figure [of this proposition; see Fig. 13], and I join *A* and *I* by a straight line. I say that quadrilateral *ABCD* has again been divided into two equal parts by line *AI* drawn from given angle *A*, and these two parts are triangle *ABI* and quadrilateral *AICD*.

This is proved as follows: I join *A* and *Z*, and *C* and *Z*, by straight lines and triangles *AZD* and *DZC* will be equal to triangles *ABZ* and *BZC*. Therefore, quadrilateral *AZCD* is half of quadrilateral *ABCD*. And since triangles *ACI* and *ACZ* are on base *AC* and between parallels *AC* and *ZI*, they will be equal to each other. And if triangle

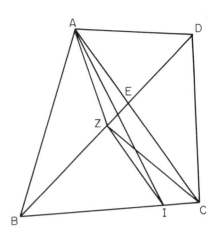

Fig. 13

ACD is added commonly to each, quadrilateral *AICD* will be equal to quadrilateral *AZCD*. But quadrilateral *AZCD* is half of quadrilateral *ABCD*, and therefore quadrilateral *AICD* is half of quadrilateral *ABCD*, which it was necessary to show. . . . .[35]

31 TWO MEDIEVAL VERSIONS OF ARCHIMEDES' QUADRATURE OF THE CIRCLE

Translations, introduction, and annotation by Marshall Clagett[1]

[Sometime before 1187, Gerard of Cremona translated from Arabic into Latin the treatise *On the Measurement of the Circle* by Archimedes (see Selection 7, no. 6).—*Ed.*] Of the Gerard version there are at least twelve manuscripts, and no doubt others which I have overlooked.[2] Furthermore, it was the point of departure for some seven different paraphrases or reworkings of Proposition I, and influenced at least three others. Proposition I, you

35. In Fig. 13, $BE \neq ED$, $BZ = ZD$, and $ZI \parallel AC$. It is necessary to prove that *AI* drawn from angle *A* divides the quadrilateral into two equal parts—that is, $AICD = ABCD/2$.

(1) $\triangle AZD + \triangle DZC = \triangle ABZ + \triangle BZC$ (since $\triangle AZD = \triangle ABZ$, and $\triangle BZC = \triangle DZC$).

(2) $\triangle AZD + \triangle DZC = AZCD$;

$$\therefore AZCD = \frac{ABCD}{2}.$$

(3) $\triangle ACI = \triangle ACZ$ ($ZI \parallel AC$, and *AC* is their common base);

(4) $\therefore \triangle ACD + \triangle ACI = \triangle ACD + \triangle ACZ$.

(5) But $\triangle ACD + \triangle ACZ = AZCD$, and

(6) $\triangle ACD + \triangle ACI = \triangle AICD$;

(7) $\therefore AICD = AZCD$.

(8) But $AZCD = \frac{ABCD}{2}$ [by (2) above];

$$\therefore AICD = \frac{ABCD}{2}.$$

1. [Reprinted with permission of the copyright owners, the Regents of the University of Wisconsin, from M. Clagett, *Archimedes in the Middle Ages* (Madison, Wis.: University of Wisconsin Press, 1964), Vol. I. *The Arabo-Latin Tradition*, pp. 391–397, 407–425. I have arranged the introduction and notes from separate discussions by Clagett in his Archimedes volume.—*Ed.*]

$\triangle ADG$);

(2) $\triangle ABE = \triangle BGE$ (since *BE* is a median line in $\triangle ABG$);

(3) $\therefore EDAB = \frac{ABGD}{2}$ [from (1) and (2), since *EDAB*

$$= \triangle ADE + \triangle ABE].$$

(4) Also $\triangle BDC = \triangle BDE$ (since $BD \parallel EC$, and *BD* is their common base);

(5) $\therefore \triangle ABD + \triangle BDC = \triangle ABD + \triangle BDE$.

(6) $\therefore ABCD = ABED$ (since $\triangle ABD + \triangle BDC = ABCD$, and $+ \triangle ABD + \triangle BDE = ABED$).

(7) But $ABED = \frac{ABGD}{2}$ [from (3)];

$$\therefore ABCD = \frac{ABGD}{2}.$$

Leonardo next demonstrates the same thing for line *GF* drawn from angle *G*.

2. [All of this paragraph is quoted from Clagett, p. 5.—*Ed.*]

will recall, relates the area of the circle to a right triangle whose two sides including the right angle are equal respectively to the circumference and to the radius. These reworkings centered on elaborating by the use of Euclid's *Elements* the geometric steps only implied in the text of Archimedes. For example, by reference to the tenth book of the *Elements* (Proposition I) they demonstrate in a manner similar to that found in Proposition XII.2 of the *Elements* that as we continually double the number of sides of a regular polygon inscribed in a circle, more than half of the unexhausted area of the circle is thereby exhausted at each step—or, to put it in modern terms, they prove that the area of the inscribed polygon converges as close as we like toward that of the circle as we continually double the number of sides. These various reworkings of the Gerard version of Proposition I ought to be classified as a part of our answer to the third principal question on the use of Archimedes.[3] Obviously, at least the first proposition of Archimedes' *Measurement of the Circle* was standard mathematical fare for the geometers of the thirteenth, fourteenth, and fifteenth centuries, since it appears in one form or another in so many of the mathematical codices that date from these centuries. One ought to remark that on occasion the reworkings of the Gerard version appear in more than one manuscript, as the paraphrases themselves became standard for a given school or area.

[Although the two versions reproduced here and a third treatise, the *Pseudo-Bradwardine Version,* have certain characteristics in common, it is true to say that the *Abbreviated Version of Pseudo-Bradwardine* differs from Albert of Saxony's Quadrature of the Circle in (1) using *Elements* X.1 and (2) by its relatively minimal scholastic content as contrasted with the thoroughly scholastic approach employed by Albert of Saxony. Here, now, is Clagett's general discussion and comparison of the three versions in chapter 5 (pp. 368–370) of his Archimedes volume.—*Ed.*]

I have put them together in this chapter because two of them, the Pseudo-Bradwardine Version and the Version of Albert of Saxony, do not directly use Proposition X.1 of Euclid's *Elements* as fundamental to their proofs, as do the emended versions of Chapter Three, while the third version, the so-called *Versio abbreviata*—even though it employs Proposition X.1[4]—is fundamentally a shortened version of the Pseudo-Bradwardine text. Instead of using Proposition X.1, both the Pseudo-Bradwardine and Albert of Saxony versions assert the pos-

sibility (in the first part of the proof) of finding an inscribed regular polygon greater than a given surface (the right triangle with circumference and radius as the sides including the right angle) because of the continuous divisibility of the assumed excess of the circle over the triangle. That is, they do not start with a given regular polygon whose area differs from that of the circle by an amount greater than that by which the circle is said to exceed the

3. [The question is: "What use was made by medieval authors of the works and ideas [of Archimedes] available in Latin translation?" Clagett, p. 2.—*Ed.*]

4. [The role of *Elements* X.1 in the Abbreviated Version reproduced in this selection is much the same as in other versions of the *Quadrature of the Circle* and is summarized by Clagett, pp. 60–62:

"The major point of methodological similarity found in most of the versions is the specific application of Proposition X.1 of the *Elements* of Euclid, which latter proposition holds that 'With two unequal magnitudes proposed, if more than half is subtracted from the greater, and then from the remainder again more than half is subtracted, and this subtraction takes place continuously, there will finally remain a magnitude less than the lesser of the proposed magnitudes.' Archimedes' aim in the first proposition of the *De mensura circuli* (an aim sketched very briefly in the extant Greek text) was to show that if the circle is said to exceed the right triangle, there will result some inscribed polygon which is at the same time greater than and less than the given triangle, an obvious contradiction. A similar contradiction concerning some circumscribed polygon ensues if we assume the circle to be less than the given triangle. . . ." Starting with an inscribed square (see Proposition II of the Abbreviated Version), and continually doubling the sides, they go on to prove that "we are each time extracting more than half of the space remaining between the perimeter of the polygon and the circumference of the circle. Hence, if we continue this process indefinitely, we shall by Proposition X.1 of the *Elements* finally arrive at some remaining quantity that is less than the literally designated quantity by which the circle exceeds the triangle. For the sake of brevity it is said that we have found that remaining quantity when we have inscribed an octagon in the circle. At this point the logical inference is drawn that the regular polygon is greater than the triangle. But, the emended versions go on to say, it can be shown that the polygon is actually less than the triangle, for the polygon is inscribed in a circle and hence the two quantities which double its area—namely, the perimeter of the polygon and the perpendicular drawn from the center of the circle to one of the sides of the polygon—are respectively less than the two quantities which measure double the area of the triangle—that is, the circumference and radius of the circle. Hence the assumption from which the contrary inference was drawn must be false, and so the circle cannot be said to be greater than the triangle. The argument refuting the assumption that the circle is less than the given triangle is also spelled out in some detail in these emended versions."—*Ed.*]

triangle and then proceed on the authority of Proposition X.1 to reach a polygon whose area differs from that of the circle by an amount less than the assumed excess of the circle over the triangle, as was done in the emended versions of Chapter Three. . . . In effect, all of the versions which do not directly use Proposition X.1 rest on the assumption made specific by Albert of Saxony that, given two unequal continuous (and comparable) magnitudes, it is always possible to take from the greater a third magnitude greater than the lesser and less than the greater of the two given magnitudes . . . . The third magnitude is either a regular polygon greater than a triangle (as in the Pseudo-Bradwardine and Albert of Saxony versions) or a straight line (equal to the perimeter of such a regular polygon) greater than a given straight line and less than the circumference of the circle (as in the proof of the *Verba filiorum*). . . .[5]

Other similarities are evident among the three versions of this chapter. They all depend on a basic proposition which asserts that a regular polygon is equal to a right triangle one of whose sides including the right angles is equal to the perimeter

while the other is equal to the line drawn from the center of the polygon to the middle of one of the sides (Pseudo-Bradwardine Version, Proposition I; *Versio abbreviata,* Proposition I; Albert of Saxony Version, Conclusion V). Many of the versions in Chapter Three have a similar proposition, and sometimes they have such a proposition stated as a separate proposition (e.g., see the discussion of the equivalent proposition in the introduction to the Corpus Christi Version). But the versions of this chapter are the only ones to give the area of the polygon as equal to a right triangle; the other versions say the area is equal to the product of one half the perimeter and the line drawn from the center of the polygon to the middle of one of the sides, or something very similar to this. In short, only these versions of this chapter use the expression right triangle to stand for the area in question. I suppose that this was used by the author of the first of these versions as a kind of formal analogy with the statement of the principal quadrature proposition given in the *De mensura circuli* where the circle itself is said to be equal to a right triangle.

1. [The Abbreviated Version of Pseudo-Bradwardine]

[I.] EVERY REGULAR POLYGON IS EQUAL TO A RIGHT TRIANGLE ONE OF WHOSE SIDES INCLUDING THE RIGHT ANGLE IS THE PERPENDICULAR DRAWN FROM THE CENTER OF THE POLYGON TO THE MIDDLE POINT OF ONE SIDE OF THE POLYGON AND THE OTHER IS EQUAL TO THE PERIMETER OF THE POLYGON.

This is proved, assuming as the given polygon a pentagon *ABDEC* [Fig. 1], from whose center *Z*

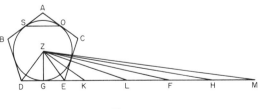

Fig. 1

[line] *ZG* is drawn perpendicular to side *DE*, i.e., at its middle point *G*. Hence let $\triangle DZE$ be formed, which is a fifth part of the pentagon, as is easily deduced: Since all of the sides are equal and since

the lines drawn from the center to the five angles of the pentagon are equal, if follows that these triangles are equiangular and equilateral, and hence equal. Then let line *GE* be protracted rectilinearly as far as is desired. From this protraction in the first place will be cut *EK* equal to *GE*, and let line *ZK* be drawn. It is evident, therefore, that $\triangle ZEK = \triangle ZDG$, since they are on equal bases and are between parallel lines, i.e., are of equal altitude. Therefore, if follows that the whole $\triangle ZGK$ is equal to the whole $\triangle ZDE$, for if equals are subtracted from or are added to equals, the results are the same and equal. Then I shall produce line *ZL*, having cut off from the protracted base a segment (*KL*) equal to base *GK*. And similarly I shall draw lines *ZF*, *ZH*, and *ZM*. From this it is evident that I form five triangles, any one of which is equal to another and consequently to $\triangle ZGK$, which is a fifth part of the pentagon. And consequently all of them taken together equal the given pentagon. But line *GM* is equal to the perimeter of the pentagon. Therefore, that which is proposed is evident.

5. [This refers to a proof in chapter 4 of Clagett's volume from the treatise "The Words of the Sons of Moses . . . (*Verba filiorum Moysi filii.* . .).—*Ed.*]

[II.] ALSO IT IS SUPPOSED IN THE SECOND PLACE THAT WITH SOME SURFACE GIVEN AS SMALLER THAN A GIVEN CIRCLE, THERE CAN BE GIVEN SOME [REGULAR] POLYGON INSCRIBED IN THE SAME CIRCLE WHICH IS GREATER THAN THE GIVEN SURFACE AND LESS THAN THE PROPOSED CIRCLE. AND SIMILARLY WITH SOME SURFACE GIVEN AS GREATER THAN A GIVEN CIRCLE, THERE CAN BE CIRCUMSCRIBED ABOUT THE SAME CIRCLE SOME [REGULAR] POLYGON WHICH IS LESS THAN THE GIVEN SURFACE AND GREATER THAN THE AFORESAID CIRCLE .

Proof: Let there be a circle *ABGD* [Fig. 2], and

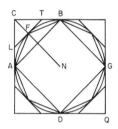

Fig. 2

in the first place let the circle be greater than the given figure. Moreover, I shall construct square *AG* in the circle; and I shall bisect arc *AB* at point *F*, joining *AF* and *BF*. And I shall proceed in the same way in the similar arcs throughout the whole circle. Therefore, there now has been taken away from the segments left over [between the perimeter of the square and the circumference of the circle] more than half of these segments. That [portion thus taken away from the segments] consists in △*AFB* and those [triangles] similar to it. When we have proceeded successively in this manner, there will [sometime] remain segments which will be less than the quantity by which the circle exceeds the given surface. And so the polygon which the circle contains will then be greater than the given surface, which is that which has been proposed.

If, moreover, the circle is less [than the given surface], let there be circumscribed about it square *CQ*, which is tangent to the circle in points *A*, *B*, *G*, and *D*. And I shall bisect arc *AB* as before, in point *F*. And in the same way let all the similar arcs of the whole circle be bisected. And let line *LT* be drawn at the division point *F*—and do the same thing at the [other] similar divisions—so that such lines are tangent to the circle at the points of division. Then from angle *C* of the circumscribed square I shall draw line *CN* to the center, and it will intersect point *F* as it bisects arc *AB*. And con-

sequently it is perpendicular to line *LT* at point *F*. Therefore, there has now been taken away from square *CQ* more than its half by the aforesaid circle. Also, since line *LT* is already bisected at point *F* with *CF* perpendicular to *LT*—and all of the other similar [corners of the square] are treated in the same way—and since *LC* and *CT* are greater than *LT*, therefore their half is greater than its half. Therefore, line *CT* > line *FT*, and consequently line *CT* > line *TB* since *FT* = *TB*. Therefore, △*FCT* is greater than 1/2△*FCB*. Hence a fortiori it is much greater than half the figure contained by [straight lines] *FC* and *CB* and arc *FB*. And I shall say in the same way regarding △ *CFL*, and all the similar figures, that it is much greater than half the figure *CFA* contained by the two [straight] lines *CF* and *CA* and arc *FA*. Therefore, the whole △*LTC* is much greater than half the whole figure *AFBC*; and in the same way the triangles similar [to △*LTC*] are greater than half of the other figures similar [to figure *AFBC*]. Therefore, when we have proceeded in this way successively, there will [sometime] remain figures beyond the circle which when added together will be less than the quantity by which the given surface exceeds the circle; which was the second thing proposed.

With these [two conclusions] supposed, there follows this conclusion:

[III.] EVERY CIRCLE IS EQUAL TO A RIGHT TRIANGLE ONE OF WHOSE TWO SIDES INCLUDING THE RIGHT ANGLE IS EQUAL TO THE RADIUS OF THE CIRCLE AND THE OTHER OF THESE TO THE CIRCUMFERENCE OF THE CIRCLE.

Let there be a circle *Z*, whose center is *Z* [Fig. 3].

Fig. 3

From *Z* let the radius *ZD* be drawn. And from that same center I shall draw as a perpendicular to line *ZD* a line *ZG* equal to the circumference of the circle. And so let *DG* be drawn. I say, therefore, that △*DZG* is equal to the given circle and consequently that the circle can be squared. The consequence is known; and the antecedent is proved. For if not, let it in the first place be greater. Therefore, one can circumscribe about the said circle a polygon less than the given triangle and greater than that circle, by the second supposition. Therefore, let this polygon be a pentagon, for example, one of whose sides BDF has a midpoint D, evidently the point of tangency of the polygon and the circle; and let the radius ZD be drawn. And so then the perimeter of the pentagon is greater than the circumference of the circle; hence it (the perimeter) is greater than line ZG. Therefore, let it be equal to line ZH. Therefore, right $\triangle DZH$ is equal to the pentagon. But that pentagon is less than $\triangle DZG$. Therefore, $\triangle DZH < \triangle DZG$, the whole is less than its part, which is impossible.

If, moreover, it is said that this $\triangle DZG$ is less than the given circle, then by the second supposition there can be inscribed in the same circle a [regular] polygon which is greater than the said triangle and less than that circle, which polygon again is a pentagon with one side NO and radius [i.e., line from the center to the midpoint of one of its sides] ZR. Then in this case the perimeter of the pentagon is less than the circumference of the circle and so less than line ZG. Hence let it be equal to line ZK. And let RK be drawn. Therefore, right $\triangle RZK$ is equal to the given pentagon, by the first supposition. Therefore, this $\triangle RZK$ is greater than right $\triangle DZG$, since the pentagon is greater than it. Therefore, the part is greater than its whole, which as before is impossible, and so on.

2. The Question of Albert of Saxony on the Quadrature of the Circle[6]

It is sought whether it is possible to square a circle. It is argued affirmatively:[7] in the first place by the authority of Antiphon and Bryson, who, as the Philosopher says in the first [book] of the *Elenchi*[8] and in the first [book] of the *Physics*,[9] attempted to square the circle. It is also proved by argument. For if there can be given a square greater than something and a square less than it, there can also be given a square equal to it. But there can be given a square greater than the circle—that is, a square circumscribed about the circle—and also a lesser square—that is, a square inscribed in the circle. Therefore, there also can be given a square equal to the circle.

Second affirmative argument: If there could not be given a square equal to a circle, it would follow that there would take place passage from "greater" to "lesser," or from extreme to extreme, through all the means without ever arriving at "equal" or "middle." But this is false. Therefore, I prove the consequence. For let there be one square inscribed in a circle and let this square begin to be continually and uniformly increased until it becomes larger than the circle. If, therefore, it was at some time equal to the circle, we have the proposition; if not, then passage has been made from "lesser" to "greater" with respect to that circle without ever arriving at "equal."

Third affirmative argument: As a sphere is related to cubing so a circle is related to squaring. But a sphere can be cubed, as is obvious if we pour the water filling up a spherical vase into a squared or cubic vase.

On the opposite side it is argued as follows: If a circle can be squared, a square could be "circled."

The consequence holds, for there appears to be no greater reason for the one than for the other. The falsity of the consequent seems to follow because there does not seem to have been transmitted any art of "circling" the square.

Second negative argument: To square a circle is to find a square equal to a circle. But to find this is impossible, for if the sides of a square are extended equally from the center, the circular surface will be more capacious than this square was before. This

6. [At certain relevant points in Albert's treatise, I have excerpted appropriate discussions from Clagett's summary and interpretation on pp. 398–404, 428, 431, and quoted them as footnotes.—*Ed.*]

7. As a scholastic *questio*, Albert's small tract has all the common features of the fourteenth-century disputative form: affirmative arguments *(rationes quod sic)*, negative arguments *(rationes quod non)*, distinctions *(distinctiones)* and conclusions *(conclusiones)* which constitute a determination of the question, and finally comments on the initial arguments *(rationes principales)*. The scholastic technique is also illustrated by the close attention of the author to the logical form of argument, where the common logical terms are employed: *consequentia* (implication), *antecedens* (antecedent part of the *consequentia*), *consequent* (the conclusion of the *consequentia*), major and minor. . . .

Incidentally, it is something of a breach of scholastic form for Albert to present the affirmative reasons prior to the negative in view of the fact that he will ultimately decide the question in the affirmative. It is far more common for an author to present negative arguments first if he is to decide the question in the affirmative, and the affirmative arguments first if he is to decide it in the negative.

8. *Sophistici Elenchi* 2. 171b. 34—172a. 7.

9. *Physics* I.2.185a.14–17. [For the arguments of Antiphon and Bryson, see Clagett, pp. 426–429; concerning Bryson, see n. 12.—*Ed.*]

is evident from the *Booklet on Isoperimetric Bodies* where it is said that the sphere is the maximum of all isoperimetric bodies. It also follows from another [argument]: Since a square cannot be found whose half is equal to half of a circle, therefore a square cannot be found which equals a circle. The consequence holds, since those whose halves are unequal are themselves unequal. The antecedent is proved, for every rectilinear figure is contained by angles to which the angles of a semicircle cannot possibly be equal; therefore, the half of no circle can be equal to the half of a square. The consequence holds, because Euclid and Campanus prove the equality of figures by the equality of angles, as in proposition I.4 of Euclid and its commentary, which begins: "Of any two, etc." The first part of the antecedent is evident, namely, that every rectilinear figure is contained by angles; for if it were not contained by angles but rather by an angle, it would follow that two straight lines could enclose a surface, which is against the last axiom of [Book] I of Euclid. But the second part is evident, namely, that rectilinear angles and the angles of a semicircle cannot be equal; for this has been demonstrated in III.15 of this book, i.e., [the *Elements*] of Euclid.

In this question we must first make some distinctions regarding quadrature of a circle. Then, secondly, conclusions must be drawn and response made to that which was sought.

As for the first, it should be known that we can understand "quadrature of a circle" in five ways. In one way, to square a circle is to divide it into four equal parts by two diameters which intersect orthogonally in the center. Campanus speaks in this way in his *Theory [of the Planets]*, and [so also do] many other scholars [writing] on quadrature of a circle when they direct one to square (that is, *quarter*) a circle.[10] In a second way we can understand by quadrature of a circle the reduction of a circle to a square or to some figure in some fashion similar to a square by cutting off parts and by transposing them. It is in this way that the ignorant speak of quadrature of a circle. In a third way we can understand by quadrature of a circle the finding of some square which is not equal to the circle but whose sides joined together are equal to the circumference extended in a straight line. And in this way Campanus squared the circle. In a fourth way we can understand by squaring a circle the finding of a square equal to a circle and whose sides joined together are also equal to the circumference of the circle extended in a straight line. In a fifth way we can understand by quadrature of a circle

the finding of a square equal to the circle.

The second distinction is this: Quadrature spoken of in the third and fifth ways is sometimes [considered] with respect to sense and sometimes with respect to intellect, and sometimes with respect to both. To square a circle in the third way with respect to intellect is to prove that there is some square whose sides joined together are equal to the circumference of the circle extended in a straight line. And in this way there is no doubt that a circle can be squared because there is no doubt that there is some straight line equal to the circumference extended in a straight line. If this line is divided into four equal parts and the parts are mutually joined at right angles, they form a square. Now to square a circle in the third way with respect to sense is to find a square whose sides joined together form a straight line so that if one visually compares it with the circumference of the circle extended in a straight line he cannot see any difference between them but would judge one to be as long as the other. In the same way quadrature of a circle is spoken of in the fifth way sometimes with respect to intellect, as in finding a square and demonstratively proving that it is equal to the circle, and sometimes with respect to sense, as in constructing and finding a square such that the sense reveals no difference between it and some circle, and one would not consider one of these to be more capacious than the other; rather he would judge one to be equal to the other. To square a circle in the third way with respect to both [i.e., to sense and intellect] is to find some square and to prove demonstratively that its sides when joined together are equal to the circumference of some circle extended in a straight line, and furthermore the senses could not determine any difference between the sides of the square joined together and the circumference of the circle extended into a straight line. We can speak in a similar way concerning quadrature of a circle in the fifth way with

10. As Heath reports (*History of Greek Mathematics*, Vol. I [Oxford, 1921], pp. 220–221), the recognition of the similarity of the words for "quadrature" and "quartering" is an ancient one and is at the base of the remark made by Aristophanes in the *Birds* (1005) which has Meton the astronomer bringing a rule and compass to make a construction "in order that your circle may become a square." Heath says, "This is a play on words, because what Meton really does is to divide a circle into four quadrants by two diameters at right angles to one another; . . . the word τετράγωνος really means 'with four right angles (at the centre),' and not 'square,' but the word conveys a laughing allusion to the problem of squaring all the same."

respect to both, i.e., to sense as well as to intellect.

As for the second principal [part of our determination], let the first conclusion be this: Quadrature of a circle spoken of in the first way is possible. This is evident to anyone who is attentive; therefore, it is not necessary to demonstrate this now.

The second conclusion is that in speaking of quadrature of a circle understood in the second way, namely, that the parts circumjacent to the circumference form a square equal to the circle, I assert that this is impossible; it is neither knowable nor demonstrable. This is proved: For after the parts are thus made circumjacent to the circumference, a squared figure does not result since not all of its angles are right angles, but in fact no angle of it is a right angle. And also because this figure is not equal to the circle since it is inscribed in the circle. If, however, it is understood [by this] that the circumference can be so divided into four equal parts which when they are transposed form a figure in some way similar to a square, there is no doubt that this is possible, just as it is possible to square a circle in the first way.

Third conclusion: Quadrature of a circle spoken of in the third way is possible with respect both to sense and to intellect. It is proved in the first place with respect to sense by Campanus, who squared a circle in this way, asserting as did many of the philosophers that the circumference of the circle contains the diameter three times and its seventh part. And since the diameter is a straight line, it is evident that if we join together three diameters with a seventh part of the diameter, they will form a straight line equal to the circumference extended in a straight line. If such a straight line is visually compared with the circumference of the circle extended in a straight line, one cannot see any difference between them but would judge them to be completely equal; even if one were superposed on the other, this would be so. And if this line were divided into four equal parts and if a square were formed from these parts, this would be squaring a circle in the third way. And this is possible; therefore, et cetera. That, moreover, quadrature of a circle spoken of in the third way is possible with respect to intellect is evident by the authority of Aristotle in the *Categories* where he says, "quadrature of a circle, if it is knowable, is not yet known." But he has not understood quadrature of a circle in the first way because that is known, nor in the second way because that is impossible, nor in the fourth way because this too is impossible, as I shall immediately prove, nor in the fifth way because

this is known, as will be evident in one of the conclusions to be posited. Therefore, [by exclusion,] he understood quadrature spoken of in the third way. For this way, although perhaps knowable, "is not yet known" since perhaps it has·(had?) not yet been found by art nor demonstrated to the intellect that the circumference of a circle is related in a $3\frac{1}{7}$ ratio to the diameter, nor that any straight line is equal to a circumference. Nevertheless, it is demonstrable to the intellect, although it is difficult. And, therefore, Campanus' quadrature of this circle is with respect to sense not to intellect.

The fourth conclusion: Quadrature of a circle spoken of in the fourth way is impossible, i.e., that some square is equal to a circle [and that at the same time] its sides joined together are equal to the circumference of a circle extended in a straight line. This [conclusion] is evident from the fact that a circular figure is the most capacious of all [isoperimetric figures].

In brief, I do not principally intend quadrature of a circle spoken of in the first, second, third, or fourth way, but rather in the fifth way: that is, to prove demonstratively that some square is equal to a circle.

The fifth conclusion—for proving quadrature of a circle as here intended—is this: Every regular polygon is equal to a right triangle, one of whose two sides containing the right angle is equal to a straight line compounded of all the sides of the polygon joined together, and the other side of those comprising the right angle is equal to a line drawn perpendicularly from the center of the same figure to one of its sides.

For example, let the regular polygon be *ABCD* with center *E* [Fig. 4]. And let a line be drawn

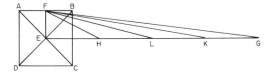

Fig. 4

perpendicularly from the center *E* to one side of the figure, namely, to side *AB*, touching side *AB* in point *F*. And let the right triangle be *EFG* with one of the two sides comprising the right angle, i.e., *EG*, equal to all of the sides of the figure joined together and with the other of the sides comprising the right angle, that is, *EF*, equal to the line, or being the very line, perpendicularly drawn from the center *E* to side *AB*, touching it in point *F*.

Then I say that figure *ABCD* is equal to △*EFG*. This is proved as follows: Those wholes are equal where the parts of similar denomination of the one are equal to the parts of the same denomination of the other. But now it is so that the parts of similar denomination of the regular polygon *ABCD* are equal to the parts of the same denomination of △*EFG*; therefore, et cetera. The major is known by the [first] proposition of [Book] V of [the *Elements* of] Euclid: "If there is any number of quantities which are equal multiples of just as many other [quantities], i.e., the quantities are equal in multitude, it is necessary that just as one of them is related to its comparable term so the whole aggregate of these [quantities of one set] is related to the aggregate of the other quantities [of the second set]." But the minor is proved. For I shall draw a straight line from center *E* to each angle of the aforesaid figure *ABCD* and four equal triangles will be formed, namely, *EBA*, *EBC*, *ECD*, and *EDA*. These triangles are equal because the sides of any one are equal to the sides of any other one, and each of these triangles is one fourth of the regular polygon *ABCD*. Further, I shall divide side *EG* of △*EFG* into four equal parts. Each of these parts will be equal to a side of figure *ABCD*. Let one part of side *EG* be *EH*, the second *HL*, the third *LK*, and the fourth *KG*. Afterwards I shall draw line *FH* from point *F* to point *H* in line *EG*, and line *FL* to point *L*, and line *FK* to point *K*. And [so] △*EFG* will be resolved into four equal triangles because all four triangles fall on equal bases and line *EF* is the altitude of all of them. Hence by VI.1 of [the *Elements* of] Euclid they will be equal. And similarly any of the triangles making up △*EFG* is equal to any of the triangles making up figure *ABCD*, because the base of any triangle of △*EFG* is equal to the base of any triangle of figure *ABCD*, and there is a single altitude for all the triangles, namely, line *EF*. Hence by VI.1 of Euclid it follows that they are equal. For VI.1 of Euclid says: "If the altitude of two parallelograms or triangles is one, then they will be to one another as their bases." Therefore, the four parts, or four fourths, of △*EFG* will be equal to the four fourths of the regular polygon *ABCD*. Therefore, it follows, since the parts of one denomination of the regular polygon *ABCD* are equal to just as many parts of the same denomination of △*EFG*, that the whole surface of figure *ABCD* is equal to the whole surface of △*EFG*; which was to be proved. And it would be the same if the regular polygon were a pentagon, or a hexagon, or a regular polygon of any number

of sides and angles.

The sixth conclusion: Every circle is equal to a right triangle one of whose sides containing the right angle is equal to the circumference extended in a straight line and the other side of those sides containing the right angle is equal to the radius of the same circle.[11]

For the proof of this conclusion I posit in the first place that every figure inscribed in another figure is less than the figure in which it is inscribed, and that its sides when joined together are less than the sides of the figure in which it is inscribed. In the second place, I posit that of any [pair of] right triangles the one is greater whose two sides including the right angle are greater than those same sides of the other triangle, or one of whose [two sides containing the right angle] is equal to one side of the other triangle and the second side is greater than the corresponding side of the other triangle. This is illustrated by triangles *EBC*, *EDF*, and *EDA* [in Figure 5]. Third, I suppose that with two

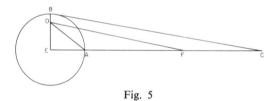

Fig. 5

continuous quantities proposed, a magnitude greater than the "lesser" can be cut from the "greater." This is evident because any excess by which any quantity exceeds another is divisible. With these things posited, I prove the conclusion as follows:

Let there be a circle *AB*, whose center is *E*; and let there be a right △*EBC*, whose side *EC* is equal to the circumference of circle *AB* extended in a straight line and whose other side containing the right angle, namely, *EB*, is the radius, or is equal to the radius, of circle *AB* [Fig. 5]. Then I say that △*EBC* is equal to circle *AB*. This is proved as follows: Since △*EBC* is neither greater than nor less than circle *AB*, therefore it is equal to it. The consequence holds, for any [two] quantities, one of which is neither greater than nor less than the other are equal. But I prove the antecedent, and in the first place, that △*EBC* is not less than circle *AB*. For if so, then by the third supposition a figure greater than △*EBC* could be cut from circle *AB*

11. [This is Proposition I of Archimedes' *On the Measurement of the Circle.—Ed.*]

But this is false. For let there be inscribed in the circle regular polygon *FGHIKLMN*, from whose center *A* is drawn line *AO* perpendicular to side *FG* [Fig. 6]. Then by the fifth conclusion, demon-

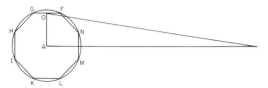

Fig. 6

strated earlier, that regular polygon—an octagon—will be equal to a right triangle one of whose sides containing the right angle is equal to all of the sides of the polygon taken together and the other side is equal to the perpendicular *AO*. But the sides of the aforesaid polygon joined together are less than the circumference of the circle extended in a straight line, as is obvious by the second part of the first supposition. Further, line *AO* is less than the radius. Therefore, the right triangle whose two sides containing the right angle are these same lines is less than right △*EBC* [Fig. 5] by the second supposition—△*EBC* being the triangle you wish to be equal to circle *AB*. Therefore, it also follows that the polygonal figure cut from the circle is not greater than △*EBC*. Therefore, it follows that circle *AB* is not greater than △*EBC*, which was to be proved.

It cannot be said that △*EBC* is greater than circle *AB*, because then there could be cut from △*EBC* a quantity greater than circle *AB*, by the third supposition. But this is false. For if so, then let such a quantity—or a polygon—be cut from it; let this polygon be *APQRSTVX* [Fig. 7]. Since it is

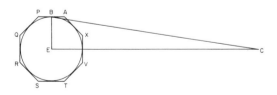

Fig. 7. Rather than triangle *EPC* taken before in Fig. 5 (as formed from the circumference and radius of the circle), one would expect another triangle equal to the circumscribed octagon in the same manner as in Fig. 6.

greater than circle *AB*, then it can be circumscribed about circle *AB*, or about something equal to circle *AB*. But if it is circumscribed about circle *AB*, I prove that it is impossible for it to be less than △*EBC*. For if it were less than △*EBC* and it were

cut from *EBC*, it would follow that the right triangle one of whose sides including the right angle is equal to the perimeter of the circumscribed figure and the other of the sides equal to the line drawn perpendicularly from the center of the figure to one of its sides is less than △*EBC*, which is false. The consequence holds by the fifth conclusion demonstrated earlier. For this triangle is equal to that polygon. The falsity of the consequent is obvious. For the perimeter of the circumscribed figure is greater than the circumference of the inscribed circle extended in a straight line, as is shown by the second part of the first supposition. Now the perimeter of the circumscribed figure constitutes one of the sides including the right angle of the triangle equal to the polygon, and the other side of those sides containing the right angle is equal to the radius. Therefore, by the second part of the second supposition, such a right triangle, which is equal to the polygon circumscribed about the circle, is greater than △*EBC*. Hence, the polygon so circumscribed about the circle is also greater than △*EBC*. Therefore, it follows that it is not less than it, nor is it less than any part cut from △*EBC*. And in the same way I would argue concerning any other polygon greater than that circle. Hence it has thus been proved that △*EBC* is neither less than nor greater than circle *AB*. Therefore, it follows that it is equal to it, which was to be proved.

Seventh conclusion: Quadrature of a circle spoken of in the fifth way is knowable and known, demonstrable and demonstrated.

Proof: [When] it has been demonstrated that a triangle equal to a circle can be designated and it has been demonstrated that a square equal to a triangle can be found, therefore it has been demonstrated and made known that a square equal to a circle can be found. The consequence holds, for wherever there are quantities mutually equal, whatever is equal to the one is also equal to the other. But the first part of the antecedent is evident from what has been said [in the sixth conclusion] and its second part is evident from the last [proposition] of [Book] II of Euclid, which says: "To find a square equal to a given triangle."

Now for the [initial] arguments—first those in the affirmative: For they prove what is intended. Still, those positing these arguments were intending quadrature of a circle in another meaning—in the meaning understood in Aristotle's refutation of Bryson and Antiphon. Bryson perhaps intended that a square equal to a circle can be given where the perimeter of the square is equal to the circumference extended in a straight line;[12] and this is impossible, as was said in one conclusion, although the argument of Bryson might well prove the quadrature of the circle but not in terms of the meaning of Bryson, since it holds and is true that the major of that argument is not universally true, for there can be truly given a rectilinear angle greater than an angle of a segment[13]—as in the case of a right angle, and there can [also] be given a rectilinear angle less than such an angle of a segment; yet there cannot be given a rectilinear angle equal to an angle of a segment. And, therefore, although the first argument follows, yet the major in that argument is not universally true. I answer in the same way regarding the second argument. Although the conclusion is true, yet that which is reputed to be false in it is quite possible. For there can be passage from "lesser" to "greater" by going through every point between "lesser" and "greater" without ever arriving at "equal." For example, some rectilinear angle is less than an angle of a segment and some is greater than that angle of a segment, as for example a right angle, and it is possible that an acute rectilinear angle be continually increased until it becomes equal to a right angle and yet it never becomes equal to the angle of a segment. This is evident in the case where a straight line comprises an acute angle with the diameter of some circle and this line is continually rotated toward a line tangent to the circle until it coincides with that line.[14] And this Campanus clearly posits and deduces [in the commentary] on III.15 of

Euclid.[15] The third argument proves to be true according to the intention of the other arguments.

Now for the arguments that might seem to be against the conclusion as principally intended. To the first argument when it is said: "If a circle could be squared, a square could be 'circled'," I concede that in this meaning [of quadrature] it is possible to find a circle equal to a square just as it is possible to find a square equal to a circle—and this with respect to the meaning posited in the question. And when it is said: "If the sides of some square are extended equally from the center, there will not be a surface equal to the surface before its extension," this is certainly true. But this is not necessary for quadrature of a circle. Nay, it has been said that it is impossible that some square is equal to a circle where all the sides of the square taken together are equal to the circumference extended in a straight line. To the argument where it is said: "It is impossible to find a square where half of it is equal to half a circle," I deny this. And when it was said: "With half of some square given, the angles of this figure are not equal to the angles of the semicircle," I admit this. And when it is said: "therefore half of the square is not equal to half the circle," I deny the consequence. And when it is said: "Euclid and Campanus prove the equality of figures by the equality of angles," I agree completely. From this, however, it does not follow that from the inequality of the angles follows the inequality of the figures, and so on.

32 THE TRISECTION OF AN ANGLE

Translations, introduction, and annotation by Marshall Clagett

1. Banū Mūsā (The Sons of Moses, or Three Brothers) (ninth century)[1]

[XVIII.] AND INDEED IT IS POSSIBLE, WHEN THIS [KIND OF] DEVICE[2] HAS BEEN FOUND, FOR US TO DIVIDE ANY ANGLE WE WISH INTO THREE EQUAL DIVISIONS.

12. [This does not seem to have been Bryson's intent. —Ed.] Heath . . . suggests that Bryson conceived of compressing circumscribed and inscribed figures into one so that they coincide both with one another and with the curvilinear figure in question. However, I wonder if Albert's suggestion to the effect that if there is a square larger and one lesser there must be one equal to the circle [see second affirmative argument—Ed.] is perhaps not close to the intent of Bryson's argument.

13. A mixed angle formed by an arc of a circle and a chord (in this case a diameter of the circle).

14. [The angle formed by the circle and the tangent to

the circle was called a "horn angle."—Ed.]

15. [III. 16 of the Greek text in Heath's translation.—Ed.]

1. [This selection is reprinted with permission of the copyright owners, the Regents of the University of Wisconsin, from M. Clagett, *Archimedean in the Middle Ages* (Madison, Wis.: University of Wisconsin Press, 1964), Vol. I: *The Arabo-Latin Tradition,* pp. 345, 347, 349.

The geometry of the Banū Mūsā (or *Verba filiorum*), which was translated by Gerard of Cremona (see Selection 7, no. 10), was instrumental in conveying to the Latin West a knowledge of Archimedean geometry before the translations of the works of Archimedes by William of Moerbeke (see Selection 8) in the thirteenth century. It influenced Leonardo Fibonacci, Jordanus of Nemore (see the next selection), and others (see Clagett, pp. 224–225). Among nine important contributions to

And so let ∠*ABG* at first be less than a right angle. And I shall take from the two lines *BA* and *BG* two equal quantities *BD* and *BE* [Fig. 1]. And

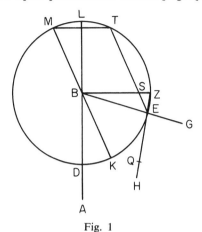

Fig. 1

I shall describe circle *DEL* on center *B* with a radius *BD*. And I shall extend line *DB* up to *L* and erect line *BZ* perpendicularly on line *LD*. Further, I shall draw line *EZ*, extending it to *H*, but without assuming *ZH* to have any fixed length. And I shall cut from line *ZH* a line equal to the radius of the circle, namely, line *ZQ*. Therefore, when we imagine that line *ZEH* is moved in the direction of point *L* and that point *Z* [continually] adheres to the circumference in the course of its motion, and that line *ZH* continues to pass through point *E* of circle *DEL*, and we imagine that point *Z* continues to be moved until point *Q* falls on line *BZ*, then it is necessary for the arc between the point at which *Z* arrives and point *L* to be one third of arc *DE*.

Proof: For I posit point *T* as the place at which point *Z* arrives as point *Q* meets line *BZ*. And I shall draw line *TE* cutting line *BZ* at point *S*. Therefore, line *TS* is equal to the radius of the circle since it is equal to line *ZQ*. And I shall draw through *B* a line parallel to line *TS*, namely, line *MBK*. And I shall draw a line from *T* to *M*. Therefore, lines *MT* and *TS* are [respectively] parallel and equal to the two lines *BS* and *MB*. Therefore, line *MT* is parallel and equal to line *BS*. But line *BS* is perpendicular to the diameter *LD*. Therefore, the chord of arc *TM* forms two right angles with diameter *LD*. Therefore, diameter *LD* bisects chord *MT* and [therefore] it also bisects arc *MT* at point *L*. But arc *ML* = arc *DK*. Therefore, arc *DK* = 1/2 arc *MT*. But arc *MT* = arc *EK*, since line *TE* is parallel to line *MK*. Therefore, arc *DK* = 1/3 arc *DE.* Therefore, ∠*DBK* = 1/3 ∠*ABG*.

And since by means of the device which we have

described in connection with the propositions previously proved and by means of things which are similar to it it is possible for us to move line *ZH* so that point *Z* moves inseparably upon the circumference while line *ZH* in its motion continues to pass through point *E* until point *Q* arrives at line *BZ*, therefore in the same way every angle less than a right angle can be trisected. And by this [technique] we can easily do what we have recounted.

And it is known that if the angle we wish to trisect is greater than a right angle, we bisect it. Then let us trisect each of the two halves in the manner we have described. It is evident, therefore, that we now know the third of an angle greater than a right angle. And this is what we wished to demonstrate. And this is the form of the figure [Fig. 1][3]

mathematics in the Latin West distinguished by Clagett (pp. 223–224) in this treatise of the Banū Mūsā is included "the first solution in Latin of the famous Greek problem of trisecting an angle, a solution to some extent reminiscent of the one found in the so-called *Lemmata* (or *Liber assumptorum*) attributed to Archimedes." Notes 2 and 3 are taken from pages 366–367 of Clagett's volume.—*Ed.*]

2. The reference is to a device or procedure similar to the one used in the second solution of the mean proportional problem [Proposition XVII of the geometry of the Banū Mūsā—*Ed.*] ,i.e., the procedure that allows us to move a line so that its extremity follows a curve while it passes through a certain point, the line being posited as undetermined in length. [See the next note for further elaboration of this device, which reduces the trisection of an angle to a "method equivalent to tracing a conchoid referred to a circular base" (Clagett, p. 666); see also notes 5 and 6 in the second section of this selection. —*Ed.*]

3. This solution of the problem is a mechanical one that serves to solve the problem as reduced to a *neusis* (i.e., to a "verging" problem). [For further discussion, see the introduction and notes to Jordanus' trisection of the angle in the next section—*Ed.*] That is, it gives a mechanical procedure to find the crucial point *S* through which line *ZH* must pass in verging toward point *E*; for with the *neusis* solved, the trisection is solved. This mechanical procedure of moving *ZH* so that *Z* adheres continually to the circumference while *ZH* continually passes through point *E* permits *Q* (chosen on *ZH* so that *ZQ* is equal to the radius) to trace a conchoid which intersects *BZ* at *S*. This is a conchoid referred to a circular base, namely, the circumference of the circle. While this exact solution is not known in antiquity, a similar reduction of the trisection problem to a *neusis* is found in lemma eight of the *Liber assumptorum* (or *Lemmata*) found only in Arabic but attributed to Archimedes. For a discussion of both of these propositions and the subsequent history of the solution of the problem by the Banū Mūsā, see [the introduction and notes of the next section.—*Ed.*]

2. Jordanus of Nemore (fl. 1230–1260)[4]

In the *Verba filiorum* of the Banū Mūsā . . . , we noticed that Proposition XVIII outlined a mechanical method of solving the *neusis* problem to which the trisection of an angle had been reduced, a method equivalent to tracing a conchoid referred to a circular base.[5] While we have no sure evidence of this particular method having been employed in antiquity, it is well known that Nicomedes and others treated conchoids of a line (rather than of a circle) and used them specifically to solve problems that reduce to *neuseis*.[6] Of particular interest to us is the reduction of trisection to a *neusis* found in the eighth lemma of the *Liber assumptorum* (or *Lemmata*), existing only in Arabic but attributed to Archimedes. While there is little doubt that the extant *Liber assumptorum* was not composed by Archimedes, Heiberg and Heath think it reasonable that this particular lemma is derived ultimately from Archimedes, since the *neusis* to which the trisection is reduced in this lemma is exactly similar to the *neuseis* assumed as possible in Propositions VI and VII of the *On Spiral Lines* of Archimedes.[7] Because of its similarities with the solution found in the *Verba filiorum*, let us examine this lemma in more detail. We can translate it as follows:[8] "If we let line *AB* be led everywhere in the circle and extended rectilinearly [Fig. 2], and if *BC* is posited

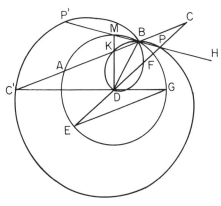

Fig. 2

as equal to the radius of the circle, and *C* is connected to the center of the circle *D*, and the line (*CD*) is produced to *E*, arc *AE* will be triple arc *BF*. Therefore, let us draw *EG* parallel to *AB* and join *DB* and *DG*. And because the two angles *DEG*, *DGE* are equal, $\angle GDC = 2\angle DEG$. And because $\angle BDC = \angle BCD$ and $\angle CEG = \angle ACE$, $\angle GDC = \angle 2CDB$ and $\angle BDG = 3\angle BDC$, and arc *BG* =

arc *AE*, and arc *AE* = 3 arc *BF*; and this is what we wished."

This proposition shows, then, that if one finds the position of line *ABC* such that it is drawn through *A*, meets the circle again in *B*, and its extension *BC* equals the radius, this will give the trisection of the angle. It thus demonstrates the equivalence of a *neusis* and the trisection problem—but without solving the *neusis*. This proposition can be related to the Banū Mūsā solution if we draw the line *DM* perpendicular to *DG* [Fig. 2]. Then the

4. [This selection is reprinted with permission of the copyright owners, the Regents of the University of Wisconsin, from M. Clagett, *Archimedes in the Middle Ages* (Madison, Wis.: University of Wisconsin Press, 1964), Vol. I: *The Arabo-Latin Tradition*, pp. 673, 675, 677. The introduction is drawn from pp. 666–669. The present selection is from Jordanus' *On Triangles (Liber de triangulis)*, Proposition IV. 20.—*Ed.*]

5. The conchoid of a circle (called by Roberval the limaçon of Pascal) has the general equation in rectangular coordinates of $(x^2 + y^2 - 2ax)^2 = b^2 (x^2 + y^2)$ [or in polar coordinates, $r = b + 2a \cos \theta$]. This curve is described as follows: when a rod of length $2b$ moves in such a way that its midpoint describes a circle of diameter $2a$ while the rod is always directed to a fixed point on the circle, the ends of the rod will describe the conchoid of the circle. The particular conchoid traced by the movement of the rule in the Banū Mūsā proposition is such that $a = b$ (see Fig. 2 for such a conchoid). Roberval used this conchoid for the solution of the trisection problem (see "Curves, Special," *Encyclopaedia Britannica*, 14th ed., vol. 6 [Chicago, 1960], p. 889).

6. For a brief history of *neusis* problems and their connections with the trisection of an angle in antiquity, see T. L. Heath, *A History of Greek Mathematics*, Vol. I (Oxford, 1921), pp. 235–41. See his more extended treatment in *The Works of Archimedes* (Cambridge, 1897), pp. c–cxxii. In connection with the *neusis* involved in the Banū Mūsā proposition, we should note the statement defining and illustrating *neusis* given by Pappus (Heath, *ibid.*, p. c): "Two lines being given in position, to place between them a straight line given in length and verging towards a given point. If there be given in position (1) a semicircle and a straight line at right angles to the base, or (2) two semicircles with their bases in a straight line, to place between the two lines a straight line given in length and verging towards a corner *(gōnian)* of a semicircle." It is the first case, with the perpendicular a radius and the given line equal to a radius, that is involved in the Banū Mūsā proposition. Note that most of the pertinent passages from Proclus, Pappus, and Eutocius on *neusis* problems have been collected together and discussed by M. Curtze, *Reliquae Copernicanae* (Leipzig, 1875), pp. 7–21.

7. Heath, *The Works of Archimedes*, pp. ci–cii; J. L. Heiberg, *Archimedis opera omnia*, vol. 2 (Leipzig, 1913), p. 518.

8. Heiberg, *ibid.*

point *K* is equivalent to point *S* in the Banū Mūsā proof and *AK* is equal to the radius (or *TS* in the Banū Mūsā proof). Point *A* can be found by finding *C'*, and *C'* can be found by the intersection of an outer part of the conchoid with the extension of *DG*. If then line *C'AB* were extended so that *BC* also equals the radius, the *neusis* in the form here presented is solved. That is to say, if we had an undetermined line *MH* (equivalent to *ZH* in the Banū Mūsā proof) on which was laid in one direction segment *MP* equal to the radius and in the other direction segment *MP'* also equal to the radius, and *M* continually moved on the circumference of the circle while line *P'MPH* always passed through point *B*, then the motion of *P* would trace the inner loop of the conchoid whose intersection with *DM* determines point *K* and the motion of *P'* would trace the outer part of the conchoid whose intersection with the extension of *DG* determines point *C'*.

Although it is evident, as we have seen, that no solution of the *neusis* was given in the *Liber assumptorum,* a procedure similar to that employed by the Banū Mūsā was attributed by a later Arabic author to an Ancient (i.e., a Greek). Whatever its origin, the technique of the Banū Mūsā was not without influence among the Arabic mathematicians. But our particular concern in this section is its use by Jordanus in the Latin West. If the reader examines the proposition here presented from the *Liber de triangulis,* he will notice that actually Jordanus presents three "solutions" of the trisections (all reducing to the same *neusis*): (1) the solution taken substantially from the *Verba filiorum* (see [the first paragraph of the text following—Ed.]); (2) a slight modification of that solution. . . where, instead of the mobile line being *ZQH* and point *Z* moving to its final position at *T* [Fig. 3], an equivalent line *LON* moves from the contrary direction so that *L* comes to the same final position at *T*: in either case the line *TSE* (with *TS* equal to the radius) is the objective; and (3) a solution where the required line *TSE* is constructed according to the method given in Proposition V.19 of a work entitled *Perspective* (see [final paragraph of text following—Ed.]). In Alhazen's *Optics,* which had already been translated into Latin, Proposition V.34 gives a solution of a *neusis* of the kind represented by the construction of *TSE*, and I suspect it is to this proposition that Jordanus is referring in spite of the difference in proposition numbers. Incidentally, Alhazen's solution is one based on conic sections and would clearly appeal to a ge-

ometer more than the mechanical solution of the Banū Mūsā. Hence we are not surprised at Jordanus' preference for the third solution.

To divide any rectilinear angle into three equal parts.

Let the acute ∠*ABG* be the one to be trisected [see Fig. 3]. With *B* assumed as the center let circle

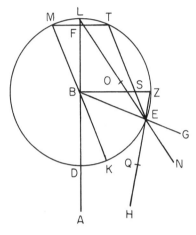

Fig. 3

DZM be described.[9] Let *DB* be extended to *L*, and let *BZ* be erected as a perpendicular to *DL*. Then let line *ZE* be extended to *H*. And I do not pose any determined limit to line *ZH*. And I shall cut *ZQ* equal to radius *DB* from *ZH*. Therefore, let us imagine that line *ZEH* is moved toward *L* in such a way that *Z* during its motion is not separated from the circumference and line *ZH* continues to pass through and adhere to *E*, and *Z* continues to be moved until *Q* falls on *BZ*. And let the terminus of the motion [of *Z*] be *T*. Therefore, a part of line *ZH*—or in other words *ZH*—will coincide with *TE*, and *TS = ZQ* = radius *BD*. I say, moreover, that arc *TL* = 1/3 arc *DE*. From point *B* let us draw *BM* parallel to line *TE*, extending *BM* to *K*, and let points *T* and *M* be connected. Proceed: *TS = MB* and the two lines are parallel. Therefore, *MT* and *BS* are equal and parallel. And *BZ* is perpendicular to *DL*. Therefore, *MT* will cut *DL* at right angles. Therefore, *DL* will bisect chord *MT*. Therefore, arcs *ML* and *LT* are equal. Also, arcs *ML* and *DK* are equal because *MK* and *DL*, intersecting each

9. Points *Z* and *M* are in fact to be determined later. It is not good geometrical form to use these points to designate the circle. Cf. the better procedure of the Banū Mūsā in Proposition XVIII of their *Verba filiorum* [see the preceding section—Ed.]

other at the center B, form mutually equal [vertical] angles. Therefore, by equality twice, arc $KE = 2$ arc DK. Therefore, $\angle KBE = 2\angle KBD$. Therefore, when $\angle KBE$ has been bisected, the proposed $\angle ABG$ will have been trisected.

Now if an obtuse angle is to be trisected, let it first be bisected so that each of its halves will be an acute angle. Then let each of these halves be trisected by the said method. Therefore, that which was proposed is clear.

The same thing will also be proved a little more clearly with only one change made, namely, that instead of HZ, let line LEN be drawn. And since LBZ is a right angle, let $OL =$ line BL. Therefore, let us imagine that NL is so moved toward Z that it always passes through E and that it continues to move until O falls on BZ, and so on as before.

The said demonstration concerning the trisection of an angle does not at all suffice for me, for I can find no certainty in it. To make it suffice for me, I demonstrate the same thing as follows. Let the given acute angle be ABG. Hence, with the foot of a compass placed at B, let a circle be described, and let AB be extended to L in the circumference. And from the center let BZ be erected as a perpendicular to DL. Then by Proposition V.19 of the *Perspective* let a line be drawn from point E through radius BZ so that $TS =$ radius BL. Hence let BM be drawn parallel to line TSE, and [then] extended to K. Since BM and TS are equal and parallel, BS and MT will be equal and parallel. Therefore, since $\angle LBZ$ is a right angle, $\angle BFT$ will be a right angle. Therefore, MT will be bisected by line BF. Therefore, arc $TM = 2$ arc ML. But arc $MT =$ arc KE, because of the parallels. Therefore, arc $KE = 2$ arc ML. Therefore, arc $KE = 2$ arc DK. Therefore, if KE is bisected, that which was proposed is had. If the angle is obtuse, let it be bisected into two acute angles, and let a third part of each be taken, and then that which was proposed is had.

33 CONSTRUCTIONS FROM AN APPLIED GEOMETRY

Dominicus de Clavasio (fl. 1346)

Translation, introduction, and annotation by Edward Grant[1]

Treatises titled *Practica geometriae* (Applied or Practical Geometry) were written by Hugh of St. Victor, Leonardo Fibonacci, and Dominicus de Clavasio, while others, under different titles or anonymously, wrote similar treatises with substantially the same content. As Hugh of St. Victor,[2] Domingo Gundisalvo,[3] and Dominicus tell us, the purpose of applied geometry was to deal with altimetry (*altimetria;* height measurement), planimetry (*planimetria;* surface measurement), and cosmimetry (*cosmimetria;* the measurement of solids; Dominicus uses the term *stereometria*). In each of these parts geometry was applied to determine various measurements in astronomy and optics, as well as to measure heights of mountains, depths of valleys, and in general, lengths, areas, and volumes. The origin of these treatises is perhaps to be found in practical surveying manuals compiled by Roman surveyors *(agrimensores)* during the Roman Empire. With the translation from Arabic into Latin of many Greek and Arabic scientific and mathematical treatises in the twelfth century, this threefold distinction was applied with greater sophistication to problems in optics and astronomy and the general mathematical level was considerably elevated. Problems involving the use of as-

tronomical instruments such as the gnomon, quadrant, and astrolabe were commonly included. Since only a few constructions will be translated here, I shall quote Busard's summary of the contents of Dominicus' treatise (pp. 522–523) in order to convey a proper sense of the nature of a medieval *Practica.*

The *Practica Geometriae* is divided into an introduction and three books. The introduction contains . . . four . . . arithmetical rules and a description of the *instrumentum gnomonicum,* i.e. of the *quadratum geometricum* of Gerbert (Dominicus supposes the astrolabe and the quadrant to be sufficiently known). The three books deal with the measurements of lengths, areas, and volumes, respectively. Particularly to book I Dominicus owes his fame as a good

1. Translated from the Latin text edited by H. L. L. Busard, "The *Practica Geometriae* of Dominicus de Clavasio," *Archive for History of Exact Sciences,* II (1965), 520–575.
2. In the introduction to his *Practica geometriae* and in the *Didascalicon* (although applied geometry is not mentioned in the latter, its threefold division is given; see Selection 16).
3. See Selection 17.

mathematician in his days for in it he gives strict proofs for the rules he set up for mensuration.

. . .

Book I opens with the following construction: *Plani, cuius terminus videtur, cum duabus virgis rectis longitudinem dupliciter mensurare.* In the constructions 2–6 Dominicus applies these methods to measuring the distance between the summits of two mountains (constr. 2), the length of a valley (constr. 3), the distance from the summit to the foot of a mountain (constr. 4), the distance from the summit to a point in the valley, the eye being on the summit (constr. 5), or in the valley (constr. 6). The constructions 7, 8, and 9 show how to measure the distance between the eye and a point in a valley with the *instrumentum gnomonicum* (constr. 7), the astrolabe (constr. 8), and the quadrant (constr. 9). Construction 10 gives a method of measuring the distance between two visible points. This method is applied to find the distance from the summit to the foot, the eye being either on another summit (constr. 11) or in the valley (constr. 12); and to find the distance between two feet (constr. 13). The altitude of a thing is measured in constructions 14–18: with the *radius visualis,* the *radius luminosus,* a mirror, the quadrant, and the astrolabe. It is in construction 15, that Dominicus talks about *sinus rectus arcus, sinus versus arcus, umbra recta,* and *umbra versa* and that he says he intends to write a *Tractatus de umbris et radiis.* In the following constructions are measured the distance from a visible point to an invisible one perpendicularly under it, from the invisible point to the surveyor, and from the surveyor to the visible point, the eye being either in the valley (constr. 19–21), or on the summit (constr. 23–27). Construction 22 shows how to measure the altitude of a tower upon a mountain from the valley; and construction 28 how to measure the altitude of a tower in the valley from a mountain. The constructions 29 and 30 make further inquiries about construction 28. The last constructions of book I (constr. 31–34) deal with measuring the depth of a well in different ways.

In the introduction to book II are discussed the *superficies famosa,* the area, the perpendicular, and the ratio of the circumference of a circle to the diameter ($3\frac{1}{7}$ *vel ea circa,* says Dominicus). Book II opens as follows: if the sides of a triangle are known, determine whether it is right-angled, obtuse-angled or acute-angled (constr. 1), construct a triangle similar to a given

one (constr. 2), and find the perpendicular (constr. 3). It is only in the MS Cod. lat. Monac. 410, that we find another construction of the perpendicular between the constructions 2 and 3, and in this construction there is a reference to the *Almagest* of Ptolemy and to Campanus. The constructions 4–16 show how to find the area of right-angled, obtuse-angled, and an acute-angled triangle, of a triangle with the help of a similar triangle, of a square, an oblong, a rhombus, a rhomboid, a quadrangle, and a regular or irregular polygon. Book II ends with the area of a circle (constr. 17, 19, and 21), a semicircle (constr. 18, 20, 21), a segment of a circle (constr. 22), a sphere (constr. 24), a hemisphere (constr. 25), a cylinder (constr. 26), and a cone (constr. 27). Mathematically speaking, construction 22 is of no value, whereas construction 27 is wrong. It appears from this last construction that Dominicus was not acquainted with Archimedes' *On the Sphere and Cylinder* I, proposition 14.

Finally the constructions of book III give the volume of a cube (constr. 1), a 4-sided prism (constr. 2), an *n*-sided prism (constr. 3), a pyramid (constr. 4), a prismoid (constr. 5), a sphere (constr. 6), a hemisphere (constr. 7), a cylinder (constr. 8, 9, and 12), a cone (constr. 10, 11, and 12), a well (constr. 13 and 14), a wine cask (constr. 15), half a wine cask (constr. 16), and a part of a wine cask (constr. 17). In the constructions 7 and 9, in which the volumes of a hemisphere and a cylinder are compared with the volume of a cube, it is again said explicitly this can be done only approximately (*non relinquendo sensibilem defectum*).

[BOOK I]

(1) To measure any quantity is to find how many times a well-known [or common] quantity is found or contained in it, or what part or how many parts of the well-known quantity there are. (2) A well-known [or common or unit] quantity (*famosa quantitas*) is what is common [or usual] to all or many. (3) A certain quantity is said to be known when it is known how many times a well-known [or common or unit] quantity is found or contained in it, or what part or how many parts there are of some common quantity. These three definitions are quite obvious. But it is known that those are called well-known [or common or unit] quantities which have been used by many, namely the digit, palm, cubit, pace (*passus*), foot(*pes*),the *pertica*, which is a length of 10 feet, the ell *(ulna),* stade(*stadium*),

mile *(milliare)*, league *(leuca)*,and ones that are similar to these. Of other surface and solid quantities no mention is made here, but below, at the beginning of the second part there is a discussion about planimetry, namely about the measurement of quantities by length and width, and in the third part a discussion about stereometry, namely about the measurement of quantities by length, width, and depth. And in the first part only altimetry will be discussed, which deals with the measurement of a quantity according to one dimension only, namely length. . . .

Eighth Construction. To find the length of a plane whose terminus is seen with an astrolabe.

Let *ab* be the plane whose terminus, namely *b*, is seen [Fig. 1]. Suspend the astrolabe in your left

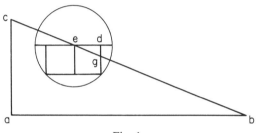

Fig. 1

hand and observe terminus *b* through both openings of the small tables. Then see how many points of the turned shadow *(umbra versa)*[4] the rule cuts, because if the rule should fall on a point of the straight shadow *(umbra recta)*, then the altitude is greater than the length of the plane. After doing this, see what is the ratio of the points of the turned shadow, which the rule cuts, to 12 and this is the ratio of your height to the length of the plane. That is, the distance between your eye and the sole of your foot to the length [or distance] between your foot and terminus *b*. Thus, if the rule cuts two points, which is a sixth part of 12, your height is a sixth part of the plane, and the plane would be six times longer than your height from your eye to the sole of your foot.

Proof: Let *ab* be the plane, *c* the eye, and let the lines of the astrolabe be arranged as they ought to be, let line *cb* be the visual ray which passes through both holes of the small tables, and *gd* the points of the shadow on the astrolabe cut by the rule. Then I imagine two triangles, namely triangle *cab* and triangle *edg*, which I say are equiangular since angle *a* is a right angle, as is angle *d*, making them equal, and angle *deg* is equal to angle *b*, since they are alternate angles between parallel lines *ed* and

ab; by the same argument, angle *dge* is equal to angle *c*, since lines *dg* and *ca* are parallel and angle *dge* and angle *c* are alternates and therefore equal. But if the triangles are equiangular, the sides containing the equal angles are proportional by the fourth [proposition] of the sixth [book of Euclid]. Therefore, as is the ratio of *dg*, the points cut by the rule, to *de,* assumed to be 12, so is the ratio of *ca*, your height, to *ab*, the length of the plane. But ratio

4. The reference here is to the back of an astrolabe. Thus Chaucer, in his *Treatise on the Astrolabe*, Part I, par. 12, describes the *umbra versa* and *umbra recta* as follows (R. T. Gunther, ed., *Early Science in Oxford*, vol. 5, *Chaucer and Messahalla On the Astrolabe* [Oxford: Oxford University Press, 1929], p. 13):
"12. Next to the aforesaid circle of the *abc* [that is, a circle on the astrolabe bearing the names of holidays along with the letters of the alphabet representing the days on which they fall—*Ed.*] and under the crossline is marked a *scale*, like 2 measuring-rules or else like ladders, that serveth by its 12 points and its divisions for full many a subtle conclusion. Of this aforesaid scale, the part from the cross-line to the right angle is called *umbra versa,* and nether part is called *umbra recta,* or else *umbra extensa*. And for more explanation, lo here the figure":

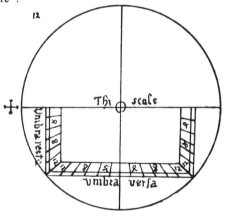

Apparently, it was the Arabs who first made use of the function of "shadows." "It was Ahmed ibn 'Abdallâh, commonly known as Habash al-Hâsib, 'the computer' (c. 860), who constructed the first table of tangents and cotangents, but it exists only in manuscript. The Arab writers distinguished the straight shadow, translated by the later medieval Latin writers as *umbra, umbra recta,* or *umbra extensa,* and the turned shadow, the *umbra versa* or *umbra stans,* the terms varying according as the gnomon was perpendicular to a horizontal plane, as in ordinary dials, or to a vertical wall, as in sundials on a building. They were occasionally called the horizontal and vertical shadows. The shadow names were also used by most of the later Latin authors and by writers in general until relatively modern times, being frequently found as late as the 18th century." (D. E. Smith, *History of Mathematics* [New York: Dover, 1958], II, 620.)

gd to *ed* is known because each line is known;[5] therefore ratio *ca* to *ab* will be known, because their ratio is the same. But *ac* is known, so that by the second supposition[6] *ab*, which is the length of the plane that is sought, will [also] be known.

Sixteenth Construction. To find by means of a mirror the height of an object elevated perpendicularly above a plane whose lower terminus and distance to the one measuring are known.

Let *ab* be the height above plane *bc* [Fig. 2] and

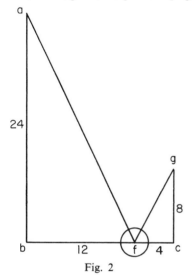

Fig. 2

take a mirror and place it between you and the foot of the height and for a time advance toward or recede from the mirror until you see the summit of the height in the mirror and know the point in the mirror where the summit of the height appears; let this point be *f*. Then see what the ratio of the distance is between your foot and point *f* to your height, namely to the distance that lies between your eye and the bottom of your foot; this will be the ratio of the distance between point *f* in the mirror and *b* to the height *ab*. But the ratio between your foot and *f* to your height is known, since each line is known. Therefore, the ratio of the distance between *f* and *b* to *ab* will be known.

This example *(practica)* is demonstrated by assuming a conclusion from perspective [that is, optics], namely that the angle of incidence and angle of reflection are equal. This conclusion is demonstrated in the fifth book of Witelo's *Perspective,* where in the twentieth conclusion it is said: "In every reflection made by any mirrors whatever, the angle of incidence equals the angle of reflection, from which it follows that the inequality of the lines does not alter the nature of reflection."[7]

Assuming this conclusion, I imagine two triangles, one of which is triangle *abf* formed by (1) the height and the distance between the foot of the height, (2) the mirror, and (3) the incident ray [that is, *af*]; the other triangle is *gfc* formed by (1) the height of the man, which is *gc,* the distance from the eye to the bottom of his foot, (2) the distance between the bottom of his foot and point *f* in the mirror, and (3) the ray of reflection. I prove that these triangles are equiangular, since *b* is a right angle and angle *c* is a right angle and they are equal; moreover, angle *afb* is equal to angle *gfc* because angle *afb* is the angle of incidence and *gfc* the angle of reflection. Therefore, by the conclusion of [Witelo's] *Perspective* declared above, they are equal angles. Hence by Book I, Proposition 32, of Euclid, angle *g* is equal to angle *a*.[8] But if the triangles are equiangular, their sides will be proportional; therefore, as ratio *cf* is to *cg* so is *fb* to *ba*, and so on.

Twenty-eighth Construction. To find the height of an object in a valley when the eye is on a mountain and can see the summit and the lowest terminus of the height.

Let *ab* be the height elevated in a valley and let the eye be on the mountain *cd* [Fig. 3]; you know

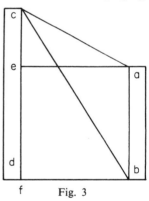

Fig. 3

5. That is each of these two lines is subdivided into a certain number of equal parts which can be read directly from the astrolabe.

6. The second supposition asserts that "if the ratio of two quantities related as a rational ratio is known, and one of these quantities is known, the remaining one will also be known."

7. Dominicus' quotation agrees completely with the enunciation of Book V, Proposition 20, of Witelo's *Optics,* as found on page 200 of Friedrich Risner's edition (Basle, 1572).

8. This is justified by the proposition that if two angles of one triangle are equal respectively to two angles of a second triangle, the third angles are equal. Proposition I.32 of Euclid's *Elements* is concerned with exterior angles and has no application here.

that the height from the mountain where the eye is to the summit of the height which stands in the valley is height *ec;* again, by things that have been said before, you know that the height of the same mountain to the lowest terminus in the valley is height *cf.* Next subtract height *ce* from height *fc,* and line *ef,* which is equal to the height sought, will remain, since *ae,* which has been imagined as drawn parallel to the horizon from the summit of height *ab,* and line *bf* are equal and parallel and lines *ab* and *ef* are drawn between these lines. Thus by Book I, Proposition 33, [of Euclid][9] lines *ab* and *ef* are proved equal. This can be shown from the tenth [construction] of this [part].

BOOK II

Fourth Construction. To find the area of a right-angled triangle.

Let *abc* be a right-angled triangle whose angle *a* is a right angle [Fig. 4]. Multiply one of the sides

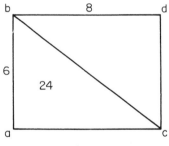

Fig. 4

containing the right angle by the other side containing the same angle, and half of the product is the area of the triangle. This is obvious, because if *ab* is multiplied by *ac,* a surface of parallel sides results which is *abcd,* whose triangle *abc* is half by a corollary of Book I, Proposition 34, of Euclid. It is also obvious from another proposition that a parallelogram is double the triangle, by Book I, Proposition 41, of Euclid, since parallelogram *abcd* and triangle *abc* are between parallel lines *ac* and *bd* and on the same base, so that if side *ab* were 6 feet and *ac* 8 feet, you multiply 6 by 8 and get 48, half of which is 24. I say that triangle *abc* is a surface of 24 feet—that is, it contains 24 surface feet, so that if *f* were a square surface any side of which is 1 foot, 24 such would be contained in triangle *abc*; and if it had multiple fractions, you also know [how] to multiply them.

Fourteenth Construction. To find the area of any equilateral and equiangular polygon.

In what has preceded, we have already discussed triangles and squares. Now we must discuss other figures, namely the pentagon, hexagon, heptagon, octagon, nonagon, decagon, and so on, where it must be understood that a pentagon is said to be a plane figure contained by five sides, and the hexagon by six sides, the heptagon by seven, the octagon by eight, and so on. Let *abcdef* be an equilateral and equiangular hexagon [Fig. 5] and divide two

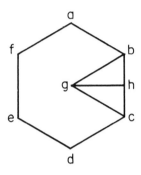

Fig. 5

immediately neighboring angles through the middle, as Book I, Proposition 9, of Euclid[10] teaches. And let there be angles *b* and *c* and draw lines until they meet in point *g,* and then from angle *bgc* drop perpendicular *gh* onto side *bc.* Multiply line *gh* by any side of the hexagon and half of the product is the area. This is proved, because hexagon *abcdef* is divisible into six similar and equal triangles. Now, if line *gh* is multiplied by side *bc,* half of the product is the area of the triangle *gbc.* Therefore, if it [*gh*] were multiplied by all six sides and half of the product taken, the area of the whole hexagon will be had. And if the figure were a pentagon, it is divided into five similar and equal triangles; and if a heptagon, it is divided into seven; and if an octagon, into eight; and if a nonagon, into nine; and so on.[11]

Twenty-sixth Construction. To find the area of a round [column that is, a cylinder].

As Euclid says in the eleventh [book], a round column *(columpna rotunda)* [or cylinder] is generat-

9. "The straight lines joining equal and parallel straight lines (at the extremities which are) in the same directions (respectively) are themselves also equal and parallel." (T. L. Heath, *The Thirteen Books of Euclid's Elements* [New York: Dover, 1956], I, 322).
10. "To bisect a given rectilineal angle" is the enunciation of *Elements* I.9 in Heath's translation.
11. In the sixteenth construction Dominicus shows how to find the area of any polygon, whether regular or irregular.

ed "when with one of the sides containing the right angle in a rectangular parallelogram fixed, this [rectangular] surface is carried round until it returns to its place."[12] Should you know the circumference of the base of this cylinder, you multiply it by its height, and the product is the area of your curved surface; [13] if you add the areas of the two circles you will have the whole area of the cylinder.[14] Thus, if the circumference of the circle *g* of cylinder *abcd* [Fig. 6] were 6 feet and the height

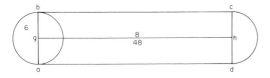

Fig. 6

of the cylinder, which is line *gh,* were 8, you multiply 6 by 8 and obtain 48, and this is the area. And you will imagine that the planes of circles *g* and *h* remain planes and that their circumferences are parallel and equal.

Twenty-seventh Construction. To find the area of a cone.

"A cone *(piramis rotunda)*[15] is [generated by] the transit of a right triangle when one of its sides around the right angle is fixed and the triangle is turned round until it returns to the place from whence it began to be moved."[16] You seek the height of the given cone *(piramidis),* that is, the straight line which falls from the apex of the cone *(cono piramidis)* on to the center of the circular base [Fig. 7]; and you find the height in this way:

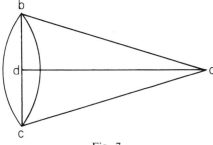

Fig. 7

You know half the diameter of the circular base, which you square, and you also know the straight line from the circumference of the circle to the summit ·of the apex of the cone, which line you similarly square, and from its square you subtract

the square of half the diameter, and the remainder is the square of the height whose square root you seek, because it is the height,[17] since the triangle formed by half of the diameter, the height of the cone, and the straight line extended from the circumference of the circular base to the apex of the cone is a right-angled triangle, and therefore the square of the height with the square of the semidiameter equals the square of the other line. With the height known, multiply it by the circumference of the circular base, and half of the product is the area of your curved surface.[18] If now you will add to it the area of the circular base, the total area of the cone will be had.[19] It is not necessary to mention other angular cones, because the areas of all the surfaces of these cones can be known by the preceding conclusions. If all these surfaces are added, the area of a whole given cone or cylinder will be obtained—if the area of the surface enclosing any cylinder is sought. The nineteenth to the present conclusion have not been demonstrated. But if you do as they teach, the sensible error will not be much. And this is the end of the second book and the third book follows.

BOOK III

A cubic body is enclosed by six equal and square surfaces and has eight angles and twelve sides. A well-known [or common] body is a cubic body enclosed by known surfaces. A cubic body is made just like a die and has six surfaces, eight angles, and twelve sides, so that any die, if it is well made, is called a cubic body or a cube. A body is said to be well-known [or common] when the surfaces con-

12. Quoted directly from Book XI, Definition 11, of Campanus of Novara's edition of Euclid's *Elements* (Bk. XI, Def. 21, in the Greek text.)
    13. That is, $A = ch = 2\pi rh$, where *c* is the circumference and *h* the height of the cylinder.
    14. That is, Total area $= 2\pi r\,(h + r)$, where *r* is the radius. In the next construction Dominicus explains that constructions 19 to 27 are presented without demonstrations.
    15. Literally, "round pyramid."
    16. Quoted directly from Book XI, Definition 9, of Campanus of Novara's edition of Euclid's *Elements* (Bk. XI, Def. 18, in the Greek text.)
    17. If, in the accompanying figure, $bd = dc = r$; $da = h$ (the slant height); and $ba = ca = q$, then $q^2 = r^2 + h^2$, and $h^2 = q^2 - r^2$, and $h = \sqrt{q^2 - r^2}$.
    18. That is, if *A* is the lateral area, $A = \frac{1}{2}(2\pi rh) = \pi rh$, which is correct.
    19. That is, Total area $= \pi rh + \pi r^2 = \pi r\,(h + r)$. I fail to discover the reason which prompted Busard to characterize the twenty-seventh proposition as "wrong" (see the introduction to this selection).

taining this body are known, so that if any surface were 1 foot, it would be called a 1-foot body; and if 2 feet, it would be called a 2-foot body, and so on. Moreover, certain bodies can be measured by a body other than a cube, as a sphere can be measured by a sphere, a cylinder by a cylinder, and a cone by a cone. You should also know that to find the volume *(capacitas)* of some body is to find how many known *(nota)* and common *(famosa)* [or unit] bodies are contained in it; or how many small spheres in a given sphere, or cones in a cone, or cylinders in a cylinder. But if a sphere could be measured by a sphere, and a cone by a cone, nevertheless they are commonly measured by a cubic measure or sometimes by another measure known to us, as [for example,] a pint, or a *sopia,* and other diverse measures among the usual different types.

First Construction. To obtain the volume of a cubic body.

Let *a* be the cubic body [Fig. 8] to be measured

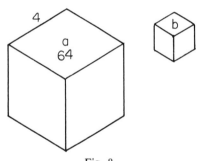

Fig. 8

and *b,* which is the measure, be another cubic body of 1 foot. See how many feet there are in one side of it and cube this number. That is, multiply the number of these feet twice into itself and the product is the number of times that *b* is contained in *a.* For example, if, as was said, *b* is 1 foot, and one side, *a,* is 4 feet, cube 4 by saying 4 times 4 times 4; and the result is 64. I say that *a* contains *b* sixty-four times. If it should please you, this could be proved by a definition of cube number.

Second Construction. To find the volume of a square prism.

A square prism *(columpna quadrilatera)* is a figure that has two equal and parallel square surfaces as its bases and four equal surfaces that are parallel and set up on the sides of those squares [Fig. 9]. If you wish to determine the volume of such a prism, you seek the area of one base by

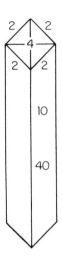

Fig. 9

means of the tenth [construction] of the second [book of this treatise][20] and multiply it by the height of the prism and the product is the volume. Thus in prism *a* if one square, which is a base, is 4 feet and the height of the prism is 10 feet, you multiply 4 by 10 and obtain 40. I say that there are forty 1-foot bodies in *a,* for if the area of the base is 4 feet, four 1-foot bodies cover it by the definition of a 1-foot cubic body; and if the height is 10 feet, it is necessary that ten 4's be assumed. You work in the same way for any other numbers for the base and height.

Tenth Construction. To find the volume of a cone according to [the measure of] a given similar cone.

Cones are said to be similar when the angles on their bases are equal. Let *a* be the cone to be measured [Fig. 10]. Find the area of the circle and see how many times the area of the base circle of

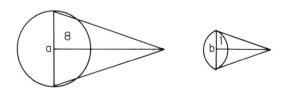

Fig. 10

20. The tenth construction reads: "To find the area of a square."

the small cone is contained in it. With that number multiply the number representing the number of times the height of the small cone is contained in the height of the large cone. The product is the volume of the said pyramid. Or do this: See what is the ratio of the diameter of the base of the one [cone] to the diameter of the base of the other and then triple that ratio and you will have the ratio of cone to cone by Euclid XII. 10.[21] Thus, if the diameter of the base a was double the diameter of base b, a would be eight times [the volume of] b, and b would be contained in a eight times.

Thirteenth Construction. To obtain the volume of a cistern (puteus) according to a given bucket (situla).[22]

Since a bucket ordinarily has one end *(fundum)* greater than another, make the greater end equal to the smaller—thus if the diameter of one end were 6 and the other 4 [Fig. 11],[23] add 6 and 4 to get 10,

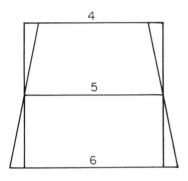

Fig. 11

of which you take half and then you have a common [or average] end. After determining this, see how many times [a bucket] is contained in the area of the mouth of the cistern by the seventeenth [construction] of the second book [of this treatise].[24] Then you can know the depth of the cistern by the third construction of the first book [of this treatise][25] and you can know how many times the height of the bucket is contained in the [line representing the] depth of the cistern. Now you multiply the number representing the number of times the average area of the end of the bucket is contained in the area of the mouth of the cistern by the number representing the number of times the height of the bucket is contained in the depth of the cistern, and the product is the number denoting how many times the cistern contains the bucket. Thus, if the average area of the end of the bucket is contained in the area of the mouth of the cistern 20 times and the height of the bucket is contained

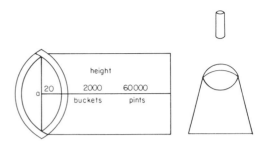

Fig. 12

in the depth of the cistern 100 times, you multiply 20 by 100 and get 2000 [Fig. 12]. I say that the cistern holds 2000 buckets of water. And if you wish to know how many pints [there are in the cistern], you operate in the same way. See how many pints the bucket holds and multiply that number by the number denoting how many buckets the cistern holds. The product is the number which denotes how many pints a cistern holds. Thus, if the bucket which the cistern contains 2000 times contains 30 pints, multiply 30 by 2000 to get 60,000, the number of pints in the cistern. You proceed in the same way if you wish to know how many glasses there are in the cistern; or how many round measures there are, however small they may be.

21. XII. 12 in the Greek text, which Heath (III, 410) translates as: "Similar cones and cylinders are to one another in the triplicate ratio of the diameters in their bases."

22. At the very end of the preceding twelfth construction, Dominicus explains what is to follow: "From what has been said, the manner of determining the volume of cisterns and vessels can be had quite satisfactorily. In order to show this in particular some special conclusions are posited." The remaining conclusions (or constructions) of the treatise (13–17) are devoted exclusively to this purpose. Only the thirteenth is included here.

23. No units are given.

24. I have altered the Latin text from "eighteenth" to "seventeenth" construction. This change seems essential, since Construction 18 seeks to determine how "to find a circular area of a semicircle according to a given circular measure." Thus it employs a circle as the unit area measure of a semicircle. But in our present construction a circular bucket end will be used to measure a circular cistern. Hence 17 is required, since it seeks "to find the circular area of a proposed circle according to a given circular measure." Basic to the seventeenth construction is *Elements* XII.2: "Circles are to one another as the squares on their diameters" (Heath, III, 371).

25. The enunciation of this construction (Busard, p. 533) reads: "To measure the length of a valley when the eye is on the mountain and able to see the foot of the mountain."

Trigonometry

34 TRIGONOMETRY OF THE SINE

Richard of Wallingford (ca. 1292–1336)

Translated by John David Bond[1]
Annotated by Edward Grant

HERE BEGINS THE QUADRIPARTITUM DE SINIBUS
DEMONSTRATIS

Because the canons[2] do not perfectly explain
the use of the sine I intend to do so in this work of
four parts. In the first part I shall clearly set forth
the relation of the circle to its diameter and of any
arc to its chord.[3] In the second I shall show by
means of given numbers the ratio of one sine to
another because the ratio of geometric magnitudes
can not be known except by the things that are
borrowed from arithmetic in EUCLID V. In the
third I shall show that what was true in commensurable numbers is true also in lines; so that to
any chapter of the second part will correspond a
chapter of the third part, because the one is the
origin of the demonstrations of the other. In the
fourth part I shall apply these things to proportion
and show that the ratio of arcs agrees with that
of their chords, so that when the ratio of the chords
is known the ratio of the arcs must also be known.
This completes the outline of the work.

Now we have the *sinus rectus,* the *sinus versus,*
and the *sinus duplatus.* The *sinus duplatus* is the
straight line whose two ends coincide with those of
the arc of the circle: it is the chord of the arc. The
sinus rectus is half the *corda duplata* with respect to
half its arc.[4] The *sinus versus* is always that part
of the diameter lying between the *sinus duplatus*
and the mid-point of its arc, and is the same as the
sagitta (arrow). For example, in the circle *abd*
(Fig. 1) with centre *e* let *bd* be drawn through *e*

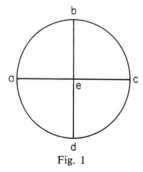

Fig. 1

perpendicular to the diameter *ac.* Then I say that

line *bed* is the *sinus duplatus* of arc *bad, be* the

1. Reprinted by permission of the editors of *Isis* from
"Richard Wallingford's Quadripartitum (English Translation)" by John David Bond, *Isis,* V (1923), 339–358.
Bond also published the Latin text of Part I of Wallingford's *De sinibus demonstratis* (*Isis,* V, 99–115) and
earlier (*Isis,* IV [1921–1922], 295–323) set the stage with
a general article "The Development of Trigonometric
Methods down to the Close of the Fifteenth Century."
Minor alterations have been made wherever necessary,
and the letter designations of the figures have been
changed to numbers and integrated within the propositions.

2. These "canons" *(canones)* were rules for constructing and using astronomical tables (see Selection 65 for
excerpts from the *Canones* of John of Saxony on the Alfonsine tables). Wallingford was familiar with the
Canons to the Toledo tables of Az-Zarqali (or Arzachel,
as he was called by the Latins), who worked in Toledo,
Spain between 1061 and 1080, where he supervised the
construction of the Toledo tables which were widely
known in the Middle Ages.

Although the Arabs and Hindus appear to have identified the six trigonometric functions in one form or
another and to have constructed tables of sines, tangents,
and cotangents, the sine was given the most extensive
development because of its special usefulness in astronomy. What is presented by Wallingford was probably
common knowledge. More noteworthy is the fact that
Wallingford was but one of a rather significant group
of Oxford University scholars in the first half of the
fourteenth century who diligently pursued various
aspects of mathematics (for instance, Bradwardine,
Heytesbury, Swineshead, Maudith, and others).

3. Bond's translation includes only this first part.

4. For the Greeks, trigonometric relations were
based on chords, or the *sinus duplatus* as it is called here.
Hindus and Arabs, however, developed a trigonometry
of the sine, or half chord *(sinus rectus).* But whether the
chord or sine was used, Greeks, Hindus, and Arabs did
not employ the notion of ratio in their trigonometry,
but rather considered chords or sines as lengths expressed as fractions of a total number of parts assigned
to the radius and diameter of a circle.

The Latin term *sinus,* which means bosom, bay, or
curve, was used to translate the Arabic term *jaib* meaning much the same thing. *Jaib* itself was perhaps derived
from the meaningless Arabic word jiba, originally a
phonetic representation of the Hindu word jyā. But the
consonants jb in jiba also allowed for a reading of *jaib,*
which was subsequently accepted by Arabic writers (see
D. E. Smith, *History of Mathematics* [New York: Dover,
1958], II, 616).

sinus rectus of arc *ba,* and *ea* the *sinus versus* both of arc *ba* and of arc *ad,* as is evident in the first circle.

First Proposition: *The sinus rectus of any arc being known, you wish to find its sinus versus.*

In the circle *cgpe* (Fig. 2) let the sinus rectus *fe*

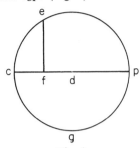

Fig. 2

of arc *ce* be known. I say that from this I can find the sinus versus *cf* of *ce.* Now I prove by[5] EUCLID III 35 that the square of *ef* equals *cf* times *fp.* Hence by II 1 it equals *cf* times *fd* plus *cf* times *cd* since *cd* and *dp* are equal.[6] But by II 1 also *cf* times *cd* equals the square of *cf* plus *cf* times *fd.* Therefore the square of *ef* equals the square of *cf* plus twice *cf* times *fd.* But by II 4 the square of *cd* equals the square of *cf* plus the square of *fd* plus twice *cf* times *fd.*[7] Hence the square of *fe* lacks the square of *fd* of equaling the square of *cd.* Therefore since *cd* is known, for being the radius, it is the sinus totus or 150 minutes,[8] I shall subtract from the square of *cd* the square of *fe,* which is known. The remainder will be the square of *fd.* Then I shall find the root of the remainder, which will be the value of line *fd.* Then I shall subtract *fd* from *cd* and the remainder *fc* will be known. Thus is proved our proposition, to find from the given sinus rectus *fe* the sinus versus *cf* of the same arc.

Second Proposition: *The sinus versus of any arc being known, you ought to know the sinus rectus.*

This is the converse of the preceding proposition. Let the construction of the figure remain as before. I say then that if sinus (versus) *fc* be known sinus *fe* is necessarily known. This is evident as follows. I shall subtract *cf* from *cd* and *fd* will remain. Then since it was shown that the square of *fe* lacks the square of *fd* of equaling the square of *cd,* I shall subtract the square of *fd* from the square of *cd* and find the root of the remainder, which must be *ef,* the sinus rectus of arc *ce.*[9] And this is what I wished to prove.

Third Proposition: *Given the sinus rectus of any arc, it will be easy to find the sinus rectus of half the arc.*

In the given circle *kcde* (Fig. 3) let the sinus

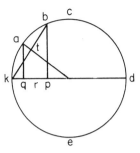

Fig. 3

rectus *pb* of arc *kb* be known. Then I shall find the sinus rectus *qa* of arc *ka,* the half of arc *kb.* For by

5. In Heath, *The Thirteen Books of Euclid's Elements* (New York: Dover, 1956), the second part of *Elements* III.35 says that if two lines intersect in a circle "the rectangle contained by the segments of the one is equal to the rectangle contained by the segments of the other" (Heath, II, 71). Therefore, in Fig. 2, $ef^2 = cf \cdot fp$.

6. *Elements* II.1 enunciates the geometric equivalent of the following algebraic formula: $a(b + c + d + \cdots) = ab + ac + ad + \cdots$. Relating this to Wallingford's proposition, $cf = a$, and since $cd = dp, fp = fd + dp = b + c$. Therefore, $cf \cdot fp = cf(fd + dp) = cf \cdot fd + cf \cdot dp = e^2$.

7. *Elements* II.4 presents the geometric equivalent of the following algebraic formulation: $(a + b)^2 = a^2 + b^2 + 2ab$. As applied here, $cd^2 = (cf + fd)^2 = cf^2 + fd^2 + 2(cf \cdot fd)$. In the remainder of the proof, we see that $ef^2 = cd^2 - fd^2$, and since *cd* and *ef* are known, we find $fd: fd^2 = cd^2 - ef^2$, so that $fd = \sqrt{cd^2 - ef^2}$. Knowing *cd* and *fd,* we can now determine $cf = cd - fd$.

Wherever feasible, I shall quote the analytic summaries which Bond provided for some of the propositions at the conclusion of his translation. "Three forms of each formula will be given. The first is the relation of geometric lines and the second is the functional relation as expressed by the author. The third is the modern equivalent, obtained by setting $r = 1$ and instead of 'sinus *fe* of arc *ce*' (Fig. 2), for instance, reading 'sin $a = fe$,' where $a =$ angle *cde*" (pp. 361–362). The three formulas are as follows:

   I. $cf = cd - \sqrt{cd^2 - fe^2}$;
   II. sinus versus $cf = r - \sqrt{r^2 - (\text{sinus } fe)^2}$;
   III. versin $a = 1 - \sqrt{1 - \sin^2 a}$.

8. Assigning 150 minutes to the *sinus totus,* or radius, and 300 minutes to the diameter can be traced at least to the *Canons* of Az-Zarqali. The significant concept of setting the radius equal to unity, and thereby uniting the trigonometric functions, was probably first proposed by Abu'l Wefa Albuzdjani of Baghdad (940–998). This idea was not exploited until the sixteneth century, by Joost Bürgi.

9. Bond's formulas in the order described above in note 7, and referring again to Fig. 2, are
   I. $fe = \sqrt{cd^2 - (cd - cf)^2}$;
   II. sinus $fe = \sqrt{r^2 - (r - \text{sinus versus } cf)^2}$;
   III. $\sin a = \sqrt{1 - (1 - \text{versin } a)^2}$.

EUCLID III 35 the square of the sinus pb of arc kb equals kp times pd and by[10] I 47 the square of kb equals the sum of the squares of kp and pb. Hence the square of kb equals the square of kp plus the product of kp and pd. Conversely by II 3 the square of kb equals kp times kd. But axiomatically if one whole equals another then a fourth part of the one equals a fourth part of the other. But now the fourth part of kp times kd equals kp times kr since kr is one fourth of kd, as is evident from II 1. And the fourth part of the square of kb equals the square of its half, kt. Therefore since I can find kp from the known pb by Q 1,[11] I shall multiply kp by kr, the fourth part of the diameter, and the square root of this product will necessarily be kt, the half of kb. But by the definition of the sinus, kt equals qa since each is the sinus rectus of arc ka, the half of arc kb. So the proposition is proved.[12] But that lines kt and qa are sinus recti of arc ka will appear if a line be drawn through the center of the circle and the mid-point t to the point a.

As a notable corollary of this proposition it follows that the square of any sinus rectus equals the product of the fourth part of the diameter and the sinus versus of the double arc.[13]

Also observe that it is not my purpose to speak of the sinus of an arc exceeding a quadrant until I have found the sinus of every arc less than a quadrant. And hereafter when I mention the sinus, the sinus rectus is to be understood unless I add something, as *versus* or *duplicis porcionis*.

Fourth Proposition: *The sinus rectus of any arc being known, you can find the sinus of the double arc.*

In the given circle *adef* (Fig. 4) let the sinus hb

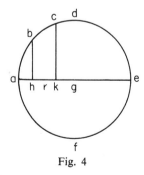

Fig. 4

of arc *ab* be known. I say that I can find line kc, the sinus of arc *ac* which is the double of arc *ab*. This is evident as follows. I shall divide the square of hb by ar, half the radius. By the corollary of Q 3 the quotient will be the sinus versus ak of twice arc *ab*. Knowing the sinus versus ak, by Q 2 I can

find the sinus rectus kc of arc *ac*. This is the sinus rectus of twice arc *ab* and is what I promised.[14]

Fifth Proposition: *The sinus rectus of any arc being known, the sinus rectus of the complementary arc will be known.*

In the circle *cegh* (Fig. 5) let arc *cb*, the half of *ab*,

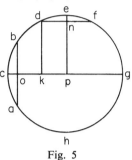

Fig. 5

equal the arc *de*, the half of *df*. I say that if I know the sinus of arc *cb* I can know the sinus of arc *cd*, the complement of *cb* since *cb* is assumed equal to *de*. This proposition is proved as follows. I shall square the sinus rectus *ob* of arc *cb* and subtract

10. The Pythagorean theorem.
11. That is, by Proposition 1 of the *Quadripartitum*.
12. In Fig. 3, we are given pb and arc $ka = \frac{1}{2}$ arc kb, from which we must find sinus rectus qa of arc ka.
Proof:
(1) $pb^2 = kp \cdot pd$ (*Elements* III.35);
(2) $kb^2 = kp^2 + pb^2$ (*Elements* I.47);
(3) $\therefore kb^2 = kp^2 + (kp \cdot pd)$ [by substituting (1) in (2)]; and
(4) $kb^2 = kp \cdot kd$ [from (3), since $kd = kp + pd$].
(5) $\dfrac{(kb)^2}{4} = \dfrac{kp \cdot kd}{4} = kp \cdot kr$ (where it is assumed that $kr = \dfrac{kd}{4}$); and
(6) $\dfrac{kb^2}{4} = kt^2$ ($kt = \dfrac{kb}{2}$, since arc $ka = \frac{1}{2}$ arc kb);
(7) $\therefore kp \cdot kr = kt^2$;
(8) $\sqrt{kp \cdot kr} = kt$; and
(9) $kt = qa$. Q.E.D.
Bond gives the following formulations:
    I. $aq = \sqrt{\dfrac{r}{2} kp}$;

    II. sinus $aq = \sqrt{\dfrac{r}{2}}$ sinus versus kp;

    III. $\sin \dfrac{a}{2} = \sqrt{\dfrac{1}{2}}$ versin $a = \sqrt{\frac{1}{2} (1 - \sqrt{1 - \sin^2 a})}$.

13. $qa^2 = \dfrac{kd}{4} \cdot kp$.
14. With reference to Fig. 4, Bond gives these representations.
    I. $ck = \sqrt{ag^2 - (ag - ak)^2}$
    $= \sqrt{ag^2 - \left(ag - \dfrac{2hb^2}{ag}\right)^2}$;
    II. sinus $ck = \sqrt{r^2 - \left[r - \dfrac{2(\text{sinus } hb)^2}{r}\right]^2}$;
    III. $\sin 2a = \sqrt{1 - (1 - 2\sin^2 a)^2}$.

the result from the square of the radius *cp*. By Q 1 the remainder will be the square of *op*. Therefore the root of the remainder will be *op*. But *op* equals *kd*, the sinus of arc *cd*, which we were trying to find. So the proposition has been proved.[15]

Incidental. But I shall prove that *op* equals *kd*. For line *ab* equals line *df* by[16] III 28 since arc *ab* was assumed equal to arc *df*. Hence by[17] III 3 side *ob* of triangle *obc* equals side *dn* of triangle *den*, by III 28 side *de* equals side *cb*, and by[18] III 26 angle *obc* equals angle *edn*. Therefore by[19] EUCLID I 4 *en* equals *oc*. Consequently *op* equals *pn* by the principle that if equals be taken from equals (the remainders are equal). By[20] I 34 *np* equals *kd*. Hence *kd* equals *op*, which was proposed.

As a corollary of this proposition it is evident that since equal arcs were taken at the end and at the beginning of the quadrant there are equal intersections on the diameters because the *corde verse* and the *sinus recti* of equal arcs are equal.

Sixth Proposition: *It is easy to investigate the sinus rectus of all the kardagas singly and collectively.*

Incidental. Before I proceed to the proposition I shall mention a notable fact, namely, that every circular arc, however small, can have a sinus, and conversely. Then because we can know any sinus only in relation to the diameter, which we can divide into 300 parts, only that sinus can be known exactly whose measure in minutes and seconds of the diameter we can find. We can not do this in the case of a sinus which is incommensurable with the diameter, for example, the sinus of 45 of the 360 degrees or equal parts into which we divide the circle. For of this arc the sinus is the same as the side of a square whose diameter is the radius of the circle. This becomes evident if a square be inscribed in the circle according to[21] IV 6 and a diameter drawn from angle to angle. This diameter is common to the square and the circle. Moreover by[22] X 2 the impossibility (of an exact ratio) is evident. But by the addition of ciphers and the extraction of roots in surds you can make the calculation so accurately by the canon for this that there is no sensible error. For by the addition of ciphers you can approximate the root within one 1111000th (?) part of a physical (sexagesimal) minute.

To the proof of the proposition I now return. But first you shall learn that a *kardaga*[23] is a circular arc which contains 15 degrees. Now let the quadrant *ag* of circle *aghi* (Fig. 6) be divided into six equal parts at the six known points *a, b, c, d, e, f*.

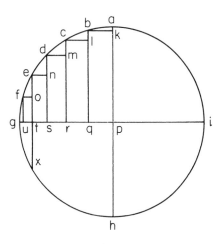

Fig. 6

From point *b* I shall draw line *bq* parallel to the diameter *aph*. Likewise I shall draw lines *cr, ds, et, fu* perpendicular to *pg*. Therefore *gf* is the first kardaga, *fe* the second, *ed* the third, *dc* the fourth, *cb* the fifth, *ba* the sixth. Now when you wish to

15. Bond's three representations (see n. 7) are as follows:

    I. $op = \sqrt{cp^2 - ob^2}$;
    II. sinus $op = \sqrt{r^2 - \text{sinus } ob^2}$;
    III. $\sin(90° - a) = \sqrt{1 - \sin^2 a}$.

In what is surely a misprint, Bond has op^2 for cp^2 in I.

16. *Elements* III.28 states: "In equal circles equal straight lines cut off equal circumferences, the greater equal to the greater and the less to the less" (Heath, II, 59).

17. "If in a circle a straight line through the centre bisect a straight line not through the centre, it also cuts it at right angles; and if it cut it at right angles, it also bisects it" (Heath, II, 10).

18. This is actually III.27 in Heath's translation and is quoted in note 34.

19. The congruence theorem: "If two triangles have the two sides equal to two sides respectively, and have the angles contained by the equal straight lines equal, they will also have the base equal to the base, the triangle will be equal to the triangle, and the remaining angles will be equal to the remaining angles respectively, namely those which the equal sides subtend" (Heath, I, 247).

20. "In parallelogrammic areas the opposite sides and angles are equal to one another, and the diameter bisects the areas" (Heath, I, 323).

21. "In a given circle to inscribe a square" (Heath, II, 91).

22. "If, when the less of two unequal magnitudes is continually subtracted in turn from the greater, that which is left never measures the one before it, the magnitudes will be incommensurable" (Heath, III, 17).

23. Of Hindu origin, *Kardaga* was taken over into Arabic (Az-Zarqali used it) and subsequently into Latin. Although it once signified any value of the *sinus rectus*, Az-Zarqali associated it with an arc of 15°.

find the sinus of the first kardaga *gf*, the twenty-fourth part of the circle, that is 15 degrees, it is very convenient to take a double arc *ex* equal to four times *gf*, that is to the sixth part of the circle. Therefore by the corollary to EUCLID IV 15 line *ex* is half the diameter.[24] Hence by the definition of the sinus the sinus of the arc *eg* of two kardagas is known to us and is *et*, half the radius, that is 75 minutes. Therefore by Q 3 I shall obtain the sinus of the half arc, which is the sinus of arc *fg*, the first kardaga. Having the sinus of arc *fg*, by Q 5 I shall obtain the sinus *bq* of arc *gb*, the complement of *ab*. Then by Q 4 I shall find *et*,[25] the sinus of twice *gf*, that is of *ge* which is two kardagas. And by the same Q 4 and Q 5 I shall find *rc*, the sinus of arc *gc* which is 4 kardagas. Next I shall obtain the sinus rectus maximus *pa*,[26] which is the sinus of 6 kardagas. And by Q 3 I shall find, by taking half this arc, the sinus *ds* of arc *gd*.

And if you wish, take the sinus of any kardaga you please as the second part of this proposition proposes. Subtract the sinus *fu* of the first kardaga from the sinus *et* of two kardagas and the remainder is *oe*,[27] which is called the sinus of the second kardaga or the excess of the sinus of two kardagas over the sinus of one kardaga, which is the same thing hereafter. Subtract sinus *te* from sinus *sd* of three kardagas and there remains sinus *nd* of the third kardaga. Proceed in this way to the sinus *ka* of the sixth kardaga. Then what we set out to prove is clearly established.

Seventh Proposition: *The chords of different arcs being known, it will not be difficult to find the chord subtended by the sum of all the arcs.*

In the circle *adeg* (Fig. 7) let chords *ef, fg, gb* be

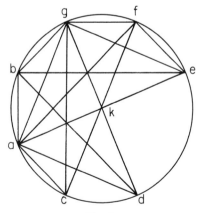

Fig. 7

known. Then I say that I know the chord *be* subtended by the three arcs *ef, fg, gb*. This is evident as

follows. I shall draw the diameters *fkc* and *gkd* of the circle and the chords *ga, ac, fa, gc*. Moreover since by hypothesis line *ef* is known, by I 47 line *fa* must be known, for by[28] III 30 *efa* is a right angle. Also since line *fg* was assumed known, by the same reasoning I shall know line *gc*. Then I consider the inscribed quadrilateral *acfg* and draw its two diameters *gc, fa*. And as I shall at the proper time prove of any inscribed quadrilateral, the product of the diameters equals the sum of the products of its opposite sides.[29] Hence *fa* times *gc* equals *ga* times *fc* plus *gf* times *ac*. Then from the product of the known diameters *fa* and *gc* let the product of the known sides *gf* and *ac* be subtracted. The remainder will be the product of the sides *fc* and *ga*. But since *fc* is the diameter of the circle, that is 300 minutes, I shall divide the product of *fc* and *ga* by *fc* and the quotient will be the value of *ga*. Then I shall draw *ge*, which by I 47 will be known because *ega* is a right angle. In like manner will *ad* be known. So also will *bd* be known from *bg*. Therefore since in the quadrilateral *abgd* I know three sides *gd, gb, ad* and the two diameters *ga, bd* the fourth side *ba* must be known. Then since *ba* is known and *abe* is a right angle, *be* is known. Thus the proposition is proved.

It is to be observed that in the figure accompanying this proposition I found the chord following (that is the chord of the arc supplementary to *efgb*).

24. *Elements* IV.15 shows how, "In a given circle to inscribe an equilateral and equiangular hexagon" (Heath, II, 107), and in a porism to IV.15 we are told, "From this it is manifest that the side of the hexagon is equal to the radius of the circle" (Heath, II, 109). Since *ex* = 4*gf* and subtends an arc of 60°, it forms the side of an inscribed hexagon and equals half the diameter, or radius, of the circle.

25. This was already determined a few lines above as "half the radius, that is 75 minutes."

26. Corrected from *pg*.

27. Corrected from *ce*.

28. *Elements* III.30 is correct for Campanus of Novara's edition of the *Elements* (see Selection 27), which corresponds to III.31 in Heath's translation (II, 61), where we read: "In a circle the angle in the semicircle is right, that in a greater segment less than a right angle, and that in a less segment greater than a right angle; and further the angle of the greater segment is greater than a right angle, and the angle of the less segment less than a right angle." In order that *efa* be a right angle, *ea* must be a diameter, which is nowhere declared. But since Fig. 7 shows that it intersects the two diameters *fkc* and *gkd* at *k*, perhaps this was deemed sufficient to imply that *ea* is a diameter.

29. Ptolemy's theorem, which is proved below in the *Incidental* part of this proposition.

By this proposition, if in a circle two chords having a common extremity are known, I shall know the third chord wich subtends the arc intercepted between the other ends of the given chords. Thus if you know chords *ab, ac* (Fig. 8)

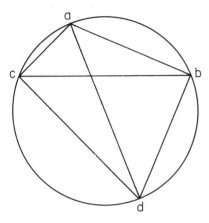

Fig. 8

you will necessarily know the chord of arc *bc*. By reference to the preceding this is evident if you draw from *a* the diameter *ad* of the circle and from *c* and *b* the lines *cd* and *bd*. Since *ad* and *ab* are known *bd* is known.[30] Again from *ac* and *ad*, *cd* is likewise known. Therefore if you know *ab, dc, ac, bd* and the diameter *ad* you will surely know the line *bc*.[31] Hence the proposition is proved.

Therefore if two sides of any inscribed triangle be known the third side must be known. From the figure in the margin you will observe that this is really what I have just proved.

Incidental. But now let us prove the proposition assumed above on which depends the great

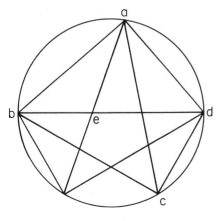

Fig. 9

power of this demonstration, namely, that in an inscribed quadrilateral the product of the two diameters equals the sum of the products of the opposite sides, from which it follows that *ac* (Fig. 9) times *bd* equals *ad* times *bc* plus *ab* times *cd,* and conversely that *ad* times *bc* etc. This theorem is proved as follows.[32] By[33] I 23 I make angle *eab*

30. By *Elements* III.31 (see n. 28) angle *abd* is a right angle and *bd* is determined by the Pythagorean theorem.

31. By Ptolemy's theorem $ad \cdot cb = ab \cdot cd + ac \cdot bd$ (Fig. 8). Since only *cb* is unknown, it can easily be determined. It is presumably in this manner that Wallingford would find the third and only unknown side of any inscribed triangle.

32. Ptolemy, in Book I, chapter 10 ("On the Size of Chords in a Circle"), of his *Almagest,* supplied much that was basic to the later development of trigonometry. A significant part of this contribution was "Ptolemy's Theorem," which was very useful in determining chord lengths. I now quote from the translation of R. C. Taliaferro (pp. 16–17; see Selection 67.1 for an excerpt from the *Almagest*) so that the reader may compare Ptolemy's proof with Wallingford's (they are in essentials much the same). It will be noted that although the latter identified what he believed were the relevant Euclidean propositions justifying different steps in the proof, Ptolemy, in true Greek fashion, omits all this (the

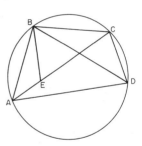

Euclidean propositions given in square brackets were furnished by Taliaferro):

"For let there be a circle with any sort of inscribed quadrilateral *ABCD*, and let *AC* and *BD* be joined.

"It is proved that rect. *AC, BD* = rect. *AB, DC* + rect. *AD, BC.* For let it be laid out such that angle *ABE* =angle *DBC.* If then we add the common angle *EBD*, angle *ABD* = angle *EBC.* But also angle *BDA* = angle *BCE* [Eucl. III, 21], for they subtend the same arc. Thus triangle *ABD* is equiangular with triangle *BCE.* Hence
    BC: *CE*:: *BD*: *AD* [Eucl. VI, 4].
Therefore
    rect. *BC, AD* = rect. *BD, CE.* [Eucl. VI, 16].
Again since
    angle *ABE* = angle *CBD*
and also
    angle *BAE* = angle *BDC*,
therefore triangle *ABE* is equiangular with triangle *BCD.* Hence
    AB : AE :: BD : CD.
Therefore
    rect. *AB, CD* = rect. *BD, AE.*

equal angle *cad* and then consider the two triangles *abe* and *adc.* For angle *a* of the one equals angle *a* of the other by hypothesis. But by[34] III 26 angle *eba* equals angle *dca* because they are constructed over the same arc *da* of the circle. Therefore by[35] I 32 the third angle of the triangle *aeb* equals the third angle of the triangle *adc,* that is, the whole angle *d* equals the partial angle *e.* Hence by[36] VI 4 the ratio of *ab* to *ac* equals the ratio of *be* to *cd.* Then by[37] VI 16 *ab* times *dc* equals *ac* times *be.* The lines are evidently proportional because they face corresponding angles. Then I consider two other triangles *abc* and *aed.* They are equiangular, for by III 26 angles *acb* and *ade,* being over the same arc *ab* of the circle, are equal. And since by hypothesis angle *a* of the one equals angle *a* of the other I shall add to each the angle *cae.* Therefore the whole angle *cab* equals the whole angle *ead.* Then since two angles of one triangle equal two angles of the other, by[38] I 32 the third angle of the one equals the third angle of the other. Hence by[39] VI 4 the ratio of *da* to *ac* equals the ratio of *de* to *cb.* Therefore by[40] VI 16 as before the product of *da* and *bc* equals that of *ca* and *de.* But by[41] II 1 *ac* times *de* plus *ca* times *eb* equals *ac* times *db.* And so the proposition is evident, that is, that the product of the diameters of the quadrilateral equals the sum of the products of the opposite sides. From this it is clearly seen that when three sides and the two diameters of an inscribed quadrilateral are known its fourth side will necessarily be known.

Eighth Proposition: *If you know the sinus rectus of each of two arcs you can find the sinus of the difference of the arcs.*[42]

In the circle *abc* (Fig. 10) let sinus *go* of arc *bo*

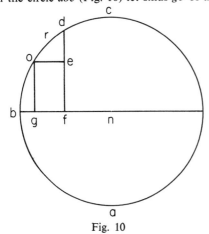

Fig. 10

and sinus *fd* of arc *bd* be known. I say that I shall come to know the sinus of arc *od,* the arc by which

But it was also proved
rect. *BC, AD* = rect. *BD, CE.*
Therefore also
rect. *AC, BD* = rect. *AB, CD* + rect. *BC, AD.*
[Eucl. II,1].
Which was to be proved."
For a detailed summary of Ptolemy's development of a table of chords and its relevance to trigonometry, see M. Clagett, *Greek Science in Antiquity* (New York: Abelard-Schuman, 1955), pp. 200–205.

33. "On a given straight line and at a point on it to construct a rectilineal angle equal to a given rectilineal angle" (Heath, I, 294).

34. "In equal circles angles standing on equal circumferences are equal to one another, whether they stand at the centres or at the circumferences" (Heath, II, 58). (In Heath's translation this is III.27; in the medieval edition by Campanus of Novara, which Wallingford probably used, it is III.26). The reader will note that in his translation of Ptolemy's theorem (n. 32), Taliaferro gives Euclid III.21 as the relevant proposition ("In a circle the angles in the same segment are equal to one another."). Either proposition is applicable.

35. I.32 seems irrelevant, for it proves that the exterior angle of a triangle equals the two interior and opposite angles and that three interior angles of the triangle are equal to two right angles. I.26 would seem to be appropriate.

36. "In equiangular triangles the sides about the equal angles are proportional, and those are corresponding sides which subtend the equal angles" (Heath, II, 200).

37. "If four straight lines be proportional, the rectangle contained by the extremes is equal to the rectangle contained by the means; and if the rectangle contained by the extremes be equal to the rectangle contained by the means, the four straight lines will be proportional" (Heath, II, 221).

38. See note 35.

39. See note 36.

40. See note 37.

41. See note 6.

42. This proposition expresses in terms of the *sinus* of the difference of arcs what Ptolemy, in Book I, chapter 10, of the *Almagest* (immediately after his proof of "Ptolemy's Theorem"), presents as the chord subtending the difference between two arcs. I now quote M. Clagett's summary of Ptolemy's proof (p. 202) and what it expresses in trigonometric terms. The figure is Ptolemy's, with Clagett's addition of angles *a* and *b*:

"From "Ptolemy's Theorem" concerning the diagonals of inscribed quadrilaterals, we get immediately the relationship:

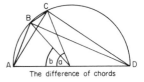

The difference of chords

$$AC \cdot BD = (BC \cdot AD) + (AB \cdot CD)$$
and *thus* $BC = \dfrac{(AC \cdot BD) - (AB \cdot CD)}{AD}$

or $\operatorname{crd}(a-b) = \dfrac{\operatorname{crd} a \cdot \operatorname{crd}(180° - b) - \operatorname{crd} b \cdot \operatorname{crd}(180 - a)}{120}$

bd exceeds *bo*. For if *bo* be subtracted from *bd* there remains only the arc *od*, whose sinus is found as follows. From sinus recti *go* and *fd* I shall know the sinus versi *bg* and *bf*.[43] Therefore I shall subtract sinus versus *bg* from sinus versus *bf* and the remainder *gf* will equal *oe* by reason of the parallelogram according to I 34.[44] Then I shall subtract the sinus rectus *go* of the smaller arc *bo* from sinus rectus *fd* of the larger arc *bd* and the known remainder will be *ed*, for by the same I 34 *fe* equals *go*. Hence by *dulcarnon*, which is I 47,[45] *od* is known because if I square *ed* and *eo* and add the two numbers the root of the sum is the value of line *od*. Therefore I have the sinus duplus of arc *od*. Then by the definition of sinus, half the line *od* is the sinus of half of arc *od*, that is of arc *or*. Therefore by Q 4, I shall obtain the sinus of twice the arc and so I shall have the sinus of the whole arc *od*. And this is what I promised.[46]

Eleventh Proposition:

From the demonstrations in the preceding 10 propositions[47] of the relation of the sinus rectus to its arc and to the whole diameter you will be able to form tables of *corda recta* and of *corda versa* both according to the method of PTOLEMY who used the sinus duplatus of the arc and according to that of ARZACHEL who used the sinus of half that arc, so that what PTOLEMY put in the *Almagest* in his tables opposite any arc ARZACHEL put its half opposite half that arc.[48]

Therefore arrange lines with numbers from one degree to 90 degrees or 180 degrees which is half the circle. Then I shall teach you how to construct tables by ARZACHEL's method. From these you will easily get PTOLEMY's as I shall most certainly show you.

Therefore let this the eleventh in order be the proposition to be demonstrated: *To find the sinus rectus (of every arc) from one degree to the quadrant.*

Supposing then in the beginning that you know incidentally that not every arc has a sinus known by demonstrative knowledge because some sinus are incommensurable with the diameter, I shall proceed in such manner in the demonstration of the proposition that it is not necessary to err from the true value by one 9000th part of one third (sexagesimal). And I shall give you three rules by which you can proceed in this investigation, the second of which is more exact than the first and the third than the second. Here is an example of how to find the sinus of one degree from the degrees of the circle having 360 equal parts and its diameter 300

minutes or 120 parts. Then supposing the diameter to be a quantity of 300 minutes, take half the radius, 75 minutes, as known and square it. The resulting number is 5625. Multiply this by 5 and the product is 28125. By Q 10 the root of this product will contain exactly half the radius and the side of the decagon, that is the corda duplata of 36 degrees of the circle. Then since this root is known, from it I

This is equivalent to the familiar formula
$$\sin(c - d) = (\sin c \cdot \cos d) - (\sin d \cdot \cos c)$$
where $\frac{a}{2} = c$ and $\frac{b}{2} = d$.

With this relationship we can find, e.g.,

crd $(72° - 60°) = $ crd $12° = 12p\ 32'\ 36''$."

The elements of this figure are identified by Clagett (p. 202) as follows:

$AB = $ crd b	$BD = $ crd $(180° - b)$
$AC = $ crd a	$CD = $ crd $(180° - a)$
$BC = $ crd $(a - b)$	$AD = 120$.

It should be noted that where Wallingford follows Azzarqali and divides his diameter into 300 minutes, Ptolemy divides his diameter into 120 equal parts and expresses all chords as fractional parts of 120.

43. By Proposition I of the *Quadripartitum*.

44. See note 20.

45. *Dulcarnon* was one of the names by which the Pythagorean Theorem (I.47 in Heath; I.46 in Campanus' edition) was known. In a discussion of this term, Heath notes (I, 418) that Chaucer used the term in *Troilus and Criseyde*, III, lines 930–933, "where Criseyde says:

'I am, til God me bettre minde send,
At dulcarnon, right at my wittes end.' . . .

Dulcarnon. . . seems to represent the Persian and Arabic *du' lkarnayn*, lit. *two-horned*, from Pers. *du*, two, and *karn*, horn. The name was applied to I.47 because the two smaller squares stick up like two horns and, as the proposition is difficult, the word here takes the sense of 'puzzle;' hence Criseyde was "at dulcarnon" because she was perplexed and at her wit's end." In a note to the Latin text (*Isis*, V, 106) Bond remarks that Alexander Neckham (1157–1217) used an isosceles right triangle to illustrate *dulcarnon*, thereby causing later writers to call *Elements* I.5 *dulcarnon* (which is concerned with isosceles triangles) in place of its usual designation *pons asinorum* ("the bridge of asses").

46. As a modern representation, Bond gives:
$$\sin(a - b) = \sqrt{1 - [1 - 2\sin^2 \tfrac{1}{2}(a - b)]^2} \text{ where}$$
$$\sin \tfrac{1}{2}(a - b) =$$
$\tfrac{1}{2}\sqrt{(\sin a - \sin b)^2 + (\text{versin } a - \text{versin } b)^2}$ where $(\sin a - \sin b)^2 = (\text{sinus } fd - \text{sinus } go)^2$ and $(\text{versin } a - \text{versin } b)^2 = (\text{sinus versus } bf - \text{sinus versus } bg)^2$.

47. Propositions 9 and 10 are omitted. In the ninth, Wallingford determines a line equal to a chord subtended by an arc of 36° (the side of a decagon)—or a tenth part of the circle; and in the tenth proposition he determines a line "composed of half the radius and the side of the decagon of the same circle"—that is, the side of a pentagon subtending an arc of 72°. Ptolemy determines both of these chords in fractional parts of 120 (*Almagest*, Bk. I, ch. 10).

48. See note 4.

shall subtract half the radius, 75 minutes, and the remainder is known: it is the corda duplata of 36 degrees, which is the chord placed in Ptolemy's tables opposite that number of degrees.

Now observe that whenever in this process you find the corda recta of any arc you write it at once opposite that arc in the line of its number in the table you have prepared for the work of Arzachel. And when you have the corda duplata of any arc write it at once opposite the arc in the table you have prepared for the sinus duplati according to Ptolemy's method of procedure. Furthermore observe that you will bisect the already known line which is the sinus duplus of 36 degrees and by the definition of sinus rectus the half is Arzachel's sinus rectus of 18 degrees. Then mark this distinction: the sinus rectus is called the sinus mediatus, and the sinus duplus is the sinus of Ptolemy. Furthermore by Q 6 you will take the sinus rectus of one kardaga and at once by Q 3 I can know the sinus rectus of the half arc, that is of 7½ degrees. Write this down. Moreover by Q 6 you will find the sinus rectus of 30 degrees or two kardagas. From this arc you will subtract 7½ degrees and by Q 8 you will find the sinus rectus of the remaining 22½ degrees. Write it down. Then by Q 3 you can find the sinus rectus of the half, that is of 11 degrees and 15 minutes. Write it. Subtract this arc of 11 degrees and 15 minutes from the arc of 18 degrees and by Q 8 you will find the sinus rectus of the remaining 6 degrees and 45 minutes which is 3/4 of one degree. Furthermore take the arc of 7½ degrees and by Q 3 you will find the sinus of its half, that is, of 3 3/4 degrees, or 3 degrees and 45 minutes. Then subtract this arc of 3 degrees and 45 minutes from 6 degrees and 45 minutes and by Q 8 you will find the sinus rectus of the remainder, exactly 3 degrees. But this can be easily obtained as follows. Bisect the arc of 18 degrees and find the sinus rectus of the half, 9 degrees, either by Q 8 or by Q 3. In the same way by Q 3 you can find the sinus of the half of this, that is of 4 degrees. And if you subtract this from 7 degrees the remainder is 3 degrees, whose sinus rectus you will find by Q 8. From this you can find by Q 3 the sinus of its half, 1½ degrees. With this result, evident is the way to find without sensible error the sinus of one degree, as I said before. And I shall show you, as I promised, first by the method which the commentator used in this case in the last proposition of book I, chapter 9, of the *Almagest*. And it is for you an example of consistent procedure for all incommensurable chords.

Take the corda recta of 1½ degree which, as was seen above, can be accurately determined. For example, let this arc be *ag* (Fig. 11) and let its corda

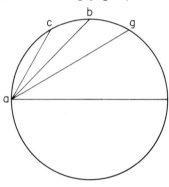

Fig. 11

dupla or simpla, I care not which, be known. According to Ptolemy's demonstration it is one degree, 34 minutes, 15 seconds. Let *ab* be the arc of one degree whose sinus is unknown. Then since as is proved in the *Almagest* I 66 (?) the ratio of arc to arc is greater than the ratio of chord to chord, you will not doubt that arcs and chords are unequal. But by our hypothesis the ratio of arc *ag* to arc *ab* is 3/2. Therefore the ratio of chord *ag* to chord *ab* is necessarily less than 3/2. Hence since it has been demonstrated that chord *ag* is one degree, 34 minutes, 15 seconds from the fact that the diameter is 120 degrees; and since one degree, 34 minutes, 15 seconds is 3/2 times one degree, 2 minutes, 50 seconds, chord *ab* is necessarily greater than one degree, 2 minutes, 50 seconds. Again let us take an arc *ac* of 45 minutes whose sinus can be found by Q 3 since arc *ac* is half *ag*. But arc *ab* is one degree. Therefore the ratio of *ab* to *ac* is 4/3. But it can be easily shown that the chord of arc *ac* is 47 minutes, 8 seconds, to which one degree, 2 minutes, 50 seconds, 40 thirds has the ratio 4/3. So arc *ab* is less than one degree, 2 minutes, 50 seconds, 40 thirds and greater than one degree, 2 minutes, 50 seconds. Therefore the error in placing the chord subtended by an arc of one degree equal to one degree, 2 minutes, 50 seconds, 20 thirds is less than 2/3 of one second and therefore much less than one second. But in the investigation of chords a quantity less than one second is thrown away and hence Ptolemy puts the chord of half a degree equal to 31 minutes and 25 seconds.

Otherwise you can find more exactly the chord of one degree as follows. Having, from the preceding, corda recta of 1½ degree, at once by Q 3 you will find the sinus rectus of half of it, that is of

45 minutes. Moreover by the same Q 3 you will immediately find the sinus of $22\frac{1}{2}$ minutes, 3/8 of one degree, and in the same manner the sinus of the half of this, that is of 11 minutes, 15 seconds. The third part of this is 3 minutes and 43 seconds, that is the sixteenth part of one degree. Then add the arc of 3/16 (degree) to the arc of 3/4 and by Q 7 you will find the sinus of their sum, which is the sinus of one degree less one-sixteenth or 3 minutes and 45 seconds. As is sufficiently evident this is 56 minutes and 45 seconds. Moreover to this arc I shall add the arc of 3/16 and I shall have the sinus of one degree and 2/16 by Q 7. Therefore I shall extract the third part of the difference between the sinus of 15/16 degree and the sinus of one degree and 2/16 and I shall have (by adding to the former) the sinus of one degree.

But it is not difficult to find in another way the sinus of one degree by using the same proposition, Q 7, that you used to find the sinus of 15/16 degree and the sinus of 3/16. In this third case, the most accurate of all and without sensible error, you will proceed as follows. Having the sinus of 3/16, by Q 3 you find the sinus rectus of its half, that is of 3/32. From this by the same Q 3 find the sinus of 3/64. Let this sinus be added to the sinus of 15/16 and by Q 7 you have the sinus of 63/64, as I well know. Further you will take half the arc of 3/64, that is 3/128, and find by Q 3 its sinus. Moreover by the same Q 3 find the sinus of 3/256. Therefore add this arc to the arc of 63/64 and by Q 7 find the sinus rectus of the sum which is one degree less 1/256 part of one degree. Again resume the work, taking the sinus of 3/256. By Q 3 find the sinus of 3/512. Then find the sinus of half this arc, that is of 3/1024. Add this arc to that of 255/256 and by Q 7 find the sinus of the sum and you will have the sinus of one degree less 1/1024 part of one degree. Proceed in this way to 1/9000 part of one degree or as far as you wish in approximating the true value. This is the most accurate procedure but the first method is to be preferred.

Having now the sinus of one degree you will readily make your tables. For by Q 5 you can get the sinus of 89 degrees. At once I shall double the arc of one degree and by Q 4 find the sinus of 2 degrees and by Q 5 the sinus of 88 degrees. Again I shall take the arc of one degree and from its sinus find by Q 3 the sinus of half a degree. I shall subtract this arc from the arc of 2 degrees and by Q 8 find the sinus of $1\frac{1}{2}$ degree. Then by Q 4 I shall find the sinus of its double, that is of 3 degrees, and by Q 5 shall at once know the sinus of 87

degrees. From the sinus of 2 degrees I shall know by Q 4 the sinus of 4 degrees and by Q 5 the sinus of 86 degrees. Then from the arc of 3 degrees I shall subtract the arc of 1/2 degree and by Q 8 I shall know the sinus of $2\frac{1}{2}$ degrees and by Q 4 the sinus of 5 degrees. Then by Q 5 I shall know the sinus of the arc of 85 degrees. And I shall continue this procedure until I complete ARZACHEL's table of the corda recta.

Then according to the corollary to Q 13 I shall take the corda recta of one degree from the radius and obtain the corda versa of 89 degrees, or if I add this corda recta to the radius the sum is the corda versa of 91 degrees. And so I shall do for 2 degrees and 3 degrees and until I have the corda versa of 180 degrees.

This done, I shall construct PTOLEMY's table as follows. I shall take the corda recta and the corda versa of one degree, square them, add the squares, and find the root of the sum. This is the corda duplata of one degree which PTOLEMY places opposite one degree in his table. And finally I shall do this for 2 degrees, 3 degrees, up to the 180 degrees of the semi-circle. Therefore if I have deserved thanks render them to a mind naturally studious.

Twelfth Proposition : *With the sinus of one quadrant completely known, it is not difficult to know the sinus of the other quadrants.*

Since the sinus rectus is the half-chord of the circle and since the chord can not be greater than the diameter nor its arc greater than the semi-circle, the greatest sinus rectus is that of 90 degrees or one quadrant, since to it corresponds the radius. Therefore since in this way one can know the sinus rectus of one quadrant knowledge of the sinus rectus of the other quadrants will be evident as follows.

In the circle *acof* (Fig. 12) let the sinus of the

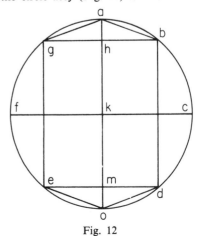

Fig. 12

whole quadrant *ac* be known and let it be required to find the sinus of any arc greater than the quadrant. Then if this arc is less than a semi-circle I shall have to subtract it from the semi-circle, 180 degrees, and find the sinus rectus of the residuum by a preceding (theorem). This is the desired sinus because the general law is that the sinus of any arc less than a semi-circle is the same as the sinus of its supplement, and conversely. Whence if I wish to find the sinus of arc *ad* I shall subtract arc *ad* from arc *ao* and find the sinus *md* of the remainder *do*. But *md* can easily be known since it is equal to *hb*. For if I subtract *ac* from *ad*, *cd* will remain and if I subtract *cd* from *ac*, the remainder will be *ab*. But *ob* equals *ad*. Therefore if their common part *bd* be taken away there will remain equal arcs *ab* and *do*. Hence their sinus recti, and their sinus versi, are equal. This is evident also by III 35. Again if I wish to find the sinus of arc *ace*, from *ace* I subtract *ao* and there remains *oe* whose sinus *em* is equal to *md* and hence to *hb* by a preceding (theorem). Therefore the sinus of *ade* is known. Again if I wish to know the sinus of *adg* I shall subtract it from the whole circle and the remainder is *ag* whose sinus, *gh*, is equal to *hb*. So the proposition is proved. All this about equality can also be proved. If I draw lines *ab, ag, od, oe* the demonstration is evident from right triangles by Q 5.[49]

Thirteenth Proposition : *If equal arcs be taken at the beginning and at the end of the quadrant the sinus rectus and the sinus versus of the one are equal respectively to the sinus versus and the sinus rectus of the other.*

This is evident. For if *ar* (Fig. 13) equals *cd* in the circle *adtg* the corda recta *sr* of arc *ar* equals the line *qc* which is the corda versa of arc *dc*, as is evident enough. From this observe that since the sinus versus of the first kardaga equals the sinus rectus of the last kardaga then the sinus versus of the second kardaga will equal the sinus rectus of the fifth kardaga, and so on for the rest. Observe also that if I wish to find the sinus versus of an arc greater than a quadrant I must take for the quadrant the sinus totus, that is the radius, and for the excess over a quadrant the sinus of the excess measured from the beginning of the quadrant and combine them. To one with insight this operation is sufficiently evident.

From this follows the corollary: The sinus rectus of an arc less than a quadrant and the sinus versus of its complement together equal the radius, and the sinus rectus of the quadrant together with the sinus rectus of the part of the semi-circle in excess of the quadrant equals the sinus versus of the sum of the arcs. In the circle *adtg* (Fig. 13) it is seen that sinus rectus *oc* of arc *ac* and sinus versus *qd* of arc *cd* together equal the radius *md*. Conversely sinus versus *ka* of arc *ab* and sinus rectus[50] *bp* of the complementary arc *bd* together equal the radius *am*. And *md* together with corda (recta) *qt* of arc *dt* equals the corda versa *af* of the whole arc *at*.[51]

49. Bond's representation:
    "sin $(360° - a)$ = sin $(180° + a)$
               = sin $(180° - a)$ = sin a,
only absolute values being considered."

50. The Latin text and Bond's translation have *br*, which is erroneous. I have, therefore, added the letter *p* to Fig. 13 and called *bp* the sinus rectus of arc *bd*.

51. Bond represents the two major points of this proposition as follows (since *r* occurs in Fig. 13, only where *r* is used alone does it represent the radius of the circle):
    I. $sr = r - ao$; $as = r - oc$;
    II. sinus $sr = r -$ sinus versus ao,
            sinus versus $as = r -$ sinus oc;
    III. sin $(90° - a) = 1 -$ versin a,
            versin $(90° - a) = 1 -$ sin a.

    I. $md + qt = af$;
    II. $r +$ sinus $qt =$ sinus versus af;
    III. $1 +$ sin $a =$ versin $(90° + a)$.

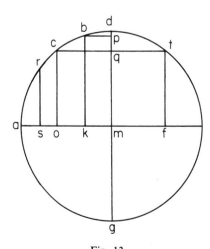

Fig. 13

Typical Scientific Questions Based on Aristotle's Major Physical Treatises

INTRODUCTION

Edward Grant

Questiones, or problems, based on Aristotelian physical treatises constituted a highly significant part of the medieval genre of question literature, which embraced theological, medical, and scientific problems, indeed almost any disputed question. Since the whole or a part of a number of such *questiones* have been translated in this source book, it will be useful to describe the rigidly consistent patterns to which nearly every question was made to conform. Following the enunciation of the question, the author presented an affirmative or negative solution. If affirmative, it was very likely that the author himself adhered to the negative side of the issue. If he presented the negative argument first, he would ultimately underwrite the affirmative position. The central arguments, those the author later rejected, were called the "principal arguments." Following this, the author would often describe his mode of procedure and sometimes further clarify and qualify the question, or define and explain particular terms that occurred in its formulation. He then usually presented a number of conclusions reflecting his opinions and viewpoints. Then he might raise doubts about his own position and subsequently resolve them. Finally, he would respond to the "principal arguments," and, hopefully, by answering them, tie up the remaining loose ends. Although scientific problems had been presented in question form since antiquity (for example, the *Problems* ascribed to Aristotle, and Seneca's *Natural Questions* and in the twelfth century, Adelard of Bath), the form described above was a peculiar product of the late middle ages and was the most significant and widespread form of scientific literature in the thirteenth and fourteenth centuries. While the questions sometimes varied, many were in vogue for centuries, and the long list presented below is wholly typical for each of the four physical treatises represented.

35 QUESTIONS ON THE EIGHT BOOKS OF ARISTOTLE'S *PHYSICS*

Albert of Saxony (ca. 1316–1390)

Translated and annotated by Edward Grant[1]

BOOK I

[1.] In the first [question] we inquire whether *(queritur primo utrum)*[2] natural science considers mobile being as its proper and adequate subject.

[2.] In the second [question] we inquire whether scientific understanding *(cognitio scientifica)* of what has been caused depends on the knowledge of its causes.

[3.] We inquire whether for perfectly understanding something it is necessary to know all its causes.

[4.] We inquire whether the same things are better known to us and nature.[3]

[5.] Next we inquire whether universals are prior to singulars or less universal in the understanding.

[6.] In the sixth [question] we inquire whether every extended thing is a quantity.

[7.] Whether a whole is [composed] of its parts.

1. This complete list of questions is translated from Albert of Saxony's *Questions on the Eight Books of the Physics,* published in *Questiones et decisiones physicales insignium virorum: Alberti de Saxonia in octo libros Physicorum; tres libros De celo et mundo; duos libros De generatione et corruptione; Thimonis in quatuor libros Meteororum. . . . Buridani in librum De sensu et sensato . . . Aristotelis. . . recognitae rursus et emendatae summa accuratione. . . Magistri Georgii Lokert. . .* (Paris, 1518).

2. Here we see the common formula for medieval questions. Often the ordinal number of the question is omitted, leaving only *queritur utrum,* literally "it is sought whether," which I have rendered as "we inquire whether." The cardinal numbers have been added for convenience.

3. Aristotle denied this at the beginning of Book I of the *Physics:* "For the same things are not 'knowable relatively to us' and 'knowable' without qualification. So in the present inquiry we must follow this method, and advance from what is more obscure by nature, but clearer to us, towards what is more clear and more knowable by nature" (184a.18–21).

[8.] In the eighth [question] we inquire whether from the addition of some whole to some whole another whole is made; similarly, [whether] by the removal of some whole from some whole, another whole is made.

[9.] Whether a minimum [piece of] matter can be assigned from whose potentiality a natural form could be brought forth.

[10.] In the tenth [question] we inquire whether in any species a minimum of matter can be assigned under which some individual of this species can exist.

[11.] We inquire whether there are contrary principles of any natural transmutation whatever.

[12.] In the twelfth [question] we inquire whether there are only two or three principles of natural things.

[13.] Whether it would be possible that something be made anew *(de novo)* with no subject presupposed.

[14.] We inquire whether matter could be a principle of any transmutation whatever.

[15.] In the fifteenth[4] [question] we inquire whether prime matter could be knowable per se.

[16.] In the sixteenth, we inquire whether privation is a thing distinct from matter.

[17.] We inquire whether form is the internal principle of any transmutation.

[18.] We inquire whether something of a form that will be generated preexists in matter before its [that is, the form's] generation.

[19.] Finally, we inquire whether every being would like to be permanent.

BOOK II

[1.] We inquire whether every natural being has within itself active and passive principles of natural motion and rest.

[2.] In the second [question] we inquire whether any mixed [that is, compound] *(mixtum)* or simple body could have within itself an internal principle of its alteration.

[3.] In the third [question] we inquire whether a figure is a thing distinct from the figured thing; for example, whether a face and its figure [or shape] are distinct things.

[4.] We inquire whether artificial things could be distinguished from natural things.

[5.] We inquire whether the definition of nature assumed by Aristotle in this second [book] is sound *(bona)*—namely that nature is a source *(principium)* and cause of moving and resting of that in which it is primarily per se and not by accident.[5]

[6.] In the sixth [question] we now inquire whether nature is properly divided into matter and form.

[7.] Next we inquire whether the difference which Aristotle assigns between natural science and mathematics is well conceived—namely that natural science defines by means of motion and mathematics does not.[6]

[8.] In the eighth [question] we inquire whether the middle sciences[7] between pure mathematics and the pure natural sciences are more natural [sciences] than mathematics.

[9.] Next, in the ninth [question] we inquire whether there are four causes—no more nor less—of every natural thing.

[10.] Whether every effecting thing is the cause of that which it is effecting.

[11.] In the eleventh [question] we inquire whether a proportion of effects is mutually similar to the proportion of acting causes [which] mutually [produced those effects].

[12.] Whether an end is a cause.[8]

[13.] We inquire whether Aristotle's definition of *chance (fortuna)* is sound—namely that *chance* is an accidental cause *(causa per accidens)* always external*(extra semper)* and frequently with respect to those things that come to pass for the sake of something.[9]

[14.] We inquire whether anything can happen by chance *(fortuna)* and accident *(casu)*.

[15.] We inquire whether nature endeavors to produce a monster.

[16.] Finally, we inquire whether necessity in natural dispositions and operations comes from the end[10] or from matter.

BOOK III

[1.] Next, with respect to the third book of the *Physics,* we inquire whether it is necessary that ignorance of a motion [entails] ignorance of the nature [of the thing in motion].

[2.] In the second [question] we inquire whether [Aristotle's] definition of motion is sound, namely

---

4. The text has, mistakenly, "sixteenth" *(sextodecimo)*.
5. See Aristotle, *Physics* II.1.192b.21–23.
6. Aristotle, *Physics* II.2.193b.31–35.
7. For example, optics, harmonics, astronomy (see *Physics* II.2).
8. That is, whether the goal or end toward which a thing strives is to be construed as a (final) cause.
9. See Aristotle, *Physics* II.5.197a.32–35 and II.6.197b.18–22.
10. That is, from the final cause, namely "that for the sake of which" the natural dispositions and operations are performed.

that motion is the actualization of what exists in potentiality insofar as it exists in potentiality.[11]

[3.] In the third [question] I inquire *(quero)* whether of all the things that exist, some are permanent and others successive.[12]

[4.] Now, in the fourth [question] we inquire whether a motion of alteration is a thing distinct from the quality that is acquired and from the quality that is lost and [distinct from] the alterable thing which acquires or loses such a quality.

[5.] Next I inquire whether according to Aristotle and his Commentator something that is moved locally requires something that is a certain flux *(fluxus)* [or flow] distinct from the mobile and place.[13]

[6.] Whether upon admitting [certain] divine events, it would be necessary to concede that local motion is something other than the mobile and [its] place.[14]

[7.] In the seventh [question] we inquire whether motion is subjectively in the mobile or in the mover.

[8.] In the eighth [question] we inquire whether every action is a passion.

[9.] In the ninth, we inquire whether determinations about the infinite belong to the natural philosopher [physicist].

[10.] Whether any actual infinite body would be sensible.

[11.] Whether there could be an infinite dimension.

[12.] Whether it is possible for an actual infinite magnitude to exist.

[13.] In the thirteenth [question] we inquire whether there is an infinite [number of] parts in a continuum.

[14.] In the final question we inquire whether for every given magnitude, a greater magnitude could be assigned; and for every given number, a greater number could be assigned.

BOOK IV

Now that the first three books of the *Physics* have been treated, we must next consider the fourth book, and since this book deals principally with *place,* let the first question be this:

[1.] Whether place *(locus)* is a surface.

[2.] In the second [question] we inquire whether place is equal to the thing [or body] that is placed [or located].

[3.] In the third [question] we inquire whether place is immobile.

[4.] In the fourth [question] we inquire whether the definition of place is sound, namely that place

is the terminus of the first immobile containing body.

[5.] In the fifth [question] we inquire whether the natural and proper place of earth is in water or inside the concave surface of water.

[6.] In the sixth [question] we inquire whether the concave [surface] of the moon is the natural place of fire.

[7.] Next, in the seventh [question], we inquire whether every being is in a place.

[8.] In the eighth [question] we inquire whether the existence of a vacuum is possible.[15]

[9.] Next we inquire whether in its downward motion a heavy simple body has an internal resistance; and similarly, [whether] in its upward motion a light [simple] body [has an internal resistance].[16]

[10.] Whether a resisting medium is required in every motion of heavy and light bodies.

[11.] In the eleventh [question] we now inquire whether, if a vacuum did exist, a heavy body could move in it.[17]

[12.] Whether, if a vacuum existed, something could be moved in it with a finite velocity or local motion, or with a motion of alteration.[18]

[13.] Whether condensation and rarefaction are possible.

[14.] Concerning the chapter on time, we inquire whether time is the motion of the sky [or heavens].

[15.] We inquire whether the definition of time is sound, [namely where] time is said to be the number *(numerus)* [or measure] of motion ac-

11. The definition appears in *Physics* III.1.201a.10–11.

12. As Albert expresses it *(Questiones et decisiones Physicales. . .*, fol. 33 verso, col. 1), Aristotle "insists that some things are permanent, as stone and wood, and some successive, as the day and year."

13. For a discussion of this problem by William of Ockham, see Selection 40.

14. The "[certain] divine events" *(casus divinos)* involve the assumption that God annihilates everything except one mobile—say the sphere of the moon—which rotates from east to west. Under these conditions the moon's motion cannot be related to anything external, from which Albert concludes that it must be related in a continually varying manner to something internal or intrinsic to the mobile itself. This he identifies as a "flux," or the motion itself, and argues that motion is therefore distinct from the mobile and the places it occupies (see the enunciation of the immediately preceding, or fifth, question).

15. See Selection 53.1 for a translation of this question.

16. A substantial part of this question is translated in Selection 47.

17. For a translation, see Selection 55.2.

18. *Ibid.*

cording to before and after.[19]

[16.] We inquire whether every motion is measured or measurable by time.

[17.] In the last question we inquire whether rest is measured by time.

BOOK V

[1] We inquire next about the text of the fifth book of the *Physics:* whether every transmutation to a substance is successive.[20]

[2.] In the second [question] we inquire whether generation is motion.

[3.] In the third, we inquire whether every motion is from contrary to contrary.

[4.] In the fourth, we inquire whether there is motion per se according to the three categories of quantity, quality, and place *(ubi)*.

[5.] In the fifth, we inquire whether any motion is one [whole unity].

[6.] Next, in the sixth [question], we inquire whether some motions differ in species.

[7.] In the seventh, we inquire whether a motion could be contrary to a motion.

[8.] In the eighth, we inquire whether rest is contrary to motion.

[9.] In the ninth, we inquire whether contrary qualities could be suffered [or tolerated] in the same subject; for example, whether some degree of hotness could exist simultaneously in the same subject with some degrees of coldness.

[10.] Now, in the tenth [question] we inquire whether in the intension of a quality, the quality that is acquired is acquired all at once [and] simultaneously; or [whether] it is acquired part after part.

[11.] In the final [question], concerning the intension of any quality, we inquire whether what is acquired first remains [or is preserved] with [parts of] the [same] quality acquired afterwards.

BOOK VI

[This is] the sixth book, treating of indivisibles and their motions, in that order, and this difficulty [or problem] is posed:

[1.] Whether a continuum is composed of indivisibles;[21] for example, as a line might be composed of points, time of instants, and motion of mutations [that is, instantaneous changes].

[2.] In the second [question] we inquire whether a continuum is divisible into things that are always [further] divisible.

[3.] Next, in the third [question], we inquire whether an indivisible entity could be moved.

[4.] In the fourth, we inquire whether, in local motion, velocity is measured according to distance traversed.

[5.] In the fifth [question] we inquire in what way velocity of circular motion should be measured.

[6.] In the sixth, we inquire in what sense velocity is measured as if it were an effect in motion of augmentation.[22]

[7.] In the seventh, we inquire whether motion could be accelerated *(velocitari)* to infinity.

[8.] In the eighth, we inquire whether every thing which is moved [or in motion] was moving before.[23]

[9.] In the ninth, we inquire whether an infinite magnitude could be traversed in a finite time; and whether a finite magnitude could be traversed in an infinite time.

[10.] In the final [question] we inquire whether during the time a mobile is moved continuously, it is in a place greater than itself.

BOOK VII

As to the text of the seventh book of the *Physics,* we inquire about this question:

[1.] Whether the same thing could act on itself.

[2.] In the second [question] we inquire whether the mover *(movent)* and the thing moved should be [conjoined] together.

[3.] In the third [question] we inquire whether alteration per se occurs only in qualities of the third species.[24]

[4.] In the fourth [question] we inquire whether it is the case that for motions to be comparable they must be of the same particular species.

19. Aristotle's definition in *Physics* IV.11.220a.24–25.

20. That is, whether every change of substance occurs in a finite time.

21. See Selections 52.1–4 for a discussion of this problem.

22. By "motion of augmentation" is meant the process of increasing a quantity.

23. See Aristotle, *Physics* VI.6.236b.34–35.

24. Albert distinguishes four species of alteration, or change of quality, of which the third is alteration of sensible quality, such as colors, tastes, and so on. His affirmative response to the question agrees with Aristotle's conclusion "that things that are undergoing alteration are altered in virtue of their being affected in respect of their so-called affective qualities, since that which is of a certain quality is altered insofar as it is sensible, and the characteristics in which bodies differ from one another are sensible characteristics: for every body differs from another in possessing a greater or lesser number of characteristics or in possessing the same sensible characteristics in a greater or lesser degree" (*Physics* VII.2. 244b.5–7; in VII.3 Aristotle considers "alterations" that are not really alterations; the four species of alteration discussed by Albert were enumerated by Aristotle, *Categories,* chapter 8).

[5.] In the fifth [question] we inquire whether any motion is comparable to any [other] motion.

[6.] Now, in the sixth [question] we inquire whether a ratio *(proportio)*[25] of velocities in motion varies as a ratio of ratios of the motive powers to their resistances.[26]

[7.] In the final [question] we inquire whether those rules are true which Aristotle assumes in the seventh book on the comparison of mobiles and movers.[27]

BOOK VIII

As to the text of the eighth book of the *Physics,* we inquire about this:

[1.] Whether there always was and always will be motion.

[2.] Next, in the second [question], whether some motion could be eternal.

[3.] In the third, we inquire whether time is eternal.

[4.] Next, in the fourth [question], we inquire whether every body could be moved.

[5.] In the fifth, we inquire whether an animal could be self-moved *(ex se)*.

[6.] We inquire whether inanimate heavy and light bodies could be self-moved.

[7.] In the seventh, we inquire whether a heavy body is self-moved after removal of whatever prevents its motion.

[8.] In the eighth [question] we inquire whether the first mover is immobile.

[9.] In the ninth, we inquire whether local motion is the first of motions.[28]

[10.] Whether it is possible that a mover of finite power move [or act] through an infinite time

[11.] Now we inquire whether an infinite power [or force] could exist in a finite magnitude.

[12.] Whether it is necessary that a moment of rest should intervene between any reflected [that is, converse] motions *(motus reflexos)*.[29]

[13.] In the final [question] we inquire what it is that moves a projected body upwards after separation from what has projected it.[30]

36 QUESTIONS ON THE FOUR BOOKS OF ARISTOTLE'S *ON THE HEAVENS (DE CAELO)*

John Buridan (ca. 1300—ca. 1358)

Translated and annotated by Edward Grant[1]

BOOK I

On the book *On the Heavens and the World (De caelo et mundo)* we inquire:

[1.] Whether the science *(scientia)* of *On the [Heavens and the] World (De mundo)* ought to be distinct from the science of the book of the *Physics.*

[2.] Next we inquire whether in the same body the dimensions length, width, and depth are mutually distinct.

[3.] Next we inquire whether there are three species of magnitude and no more.

[4.] Next we inquire whether every natural body is naturally mobile with respect to place.

[5.] Next we inquire whether there are three simple motions and no more, namely upward motion, downward motion, and circular motion.

[6.] Next we inquire whether by nature there is to one simple body one simple motion only, and also whether by nature there is one simple motion only to one simple body. The sense of this is that by nature, there could not be several simple motions for one simple body; nor by nature could there be one simple motion of several simple bodies.

[7.] Next we inquire whether a mixed *(mixtum)* [or compound] body is moved by the nature of the dominant element.[2]

[8.] Next we inquire whether circular motion has

25. For a brief discussion of the term *proportio,* see Selection 27, n. 11.

26. This is actually a description of "Bradwardine's function" (see Bradwardine's enunciation of it in Selection 51.1 and also n. 29) or "ratio of ratios" (for an explanation of this expression, see Selection 51.1, n. 51), as it was usually designated in the second half of the fourteenth century.

27. In the Middle Ages two rules were usually formulated to represent Aristotle's position. Both are refuted by Oresme in Proposition I, Selection 51.2.

28. That is, whether local motion (or motion from one place to another) is prior to, and more fundamental than, motion of alteration, augmentation, and diminution. See Aristotle, *Physics* VIII.9.265b.16–226a.5.

29. Marsilius of Inghen's response to this question has been translated in Selection 50.1.

30. Among other things, Albert discusses impetus theory here. For medieval impetus theory, see Selection 48.

1. Translated from *Iohannis Buridani Quaestiones super libris quattuor De caelo et mundo,* ed. E. A. Moody (Cambridge, Mass.: Mediaeval Academy of America, 1942).

2. See Albert of Saxony's fifth doubt in Book I, Question 19, of his *Questions on Generation and Corruption* in Selection 77.2 (also n. 41 thereto) and the response to that doubt in the same selection.

a contrary or whether to circular motion there is another contrary motion.

[9.] Next we inquire whether the sky is heavy or light.

[10.] Next we inquire whether the sky [or heaven] is generable and corruptible, augmentable and diminishable, and alterable.

[11.] Next we inquire whether the sky [or heaven] has matter.

[12.] Next we inquire whether the world is perfect.

[13.] If a body that is moved circularly were infinite, we inquire next whether the distance between lines extended from its center would be infinite.

[14.] Next we inquire whether an infinite body, if such existed, would have some active power or even a passive power.

[15.] Next we inquire whether it is possible that a body moved circularly could be infinite.

[16.] Next we inquire whether it is possible that a body moved rectilinearly could be infinite.

[17.] Furthermore, we inquire whether an infinite body is possible.

[18.] This question is next: If there were several worlds, whether the earth of one would be moved naturally to the middle of another.

[19.] Next we inquire whether it is possible that there are several worlds.[3]

[20.] Next we inquire whether there is something beyond the sky [or heaven].[4]

[21.] Next we inquire whether a power [or force] ought to be defined by its maximum capability.[5]

[22.] Next we inquire whether a power's maximum capability can be assigned.

[23.] Next we inquire whether the power to be and not to be of everything which can sometimes be and not be is for a definite, and not infinite, time.

[24.] Next we inquire whether every corruptible thing is corrupted of necessity.

[25.] Next we inquire whether every generable thing will be generated.

[26.] Next we inquire whether *generable* and *corruptible,* as well as *ungenerable* and *incorruptible,* are mutually convertible, so that every generable thing would be corruptible, and conversely; and every ungenerable thing would be incorruptible, and conversely.

BOOK II

Concerning the second book of *On the Heavens and World,* we inquire first:

[1.] Whether or not the sky [or heaven] is moved with any fatigue.

[2.] In the second [question] we inquire whether in the sky there are up and down, front and behind, right and left.

[3.] Next, in the third [question], we inquire whether in the sky up and down should be taken as the extension which stretches from pole to pole, so that one pole ought to be called up and the other down.

[4.] Next, in the fourth, we inquire whether the north *(arcticus)* pole, namely the one appearing to us, is down, and the south *(antarcticus)* pole is up.

[5.] Next, in the fifth, we inquire whether in the sky [or heavens], right and left and front and behind are [part] of the [very] nature of the sky, or are only relative to us.

[6.] In the sixth [question] we inquire whether beyond the heavens that are moved, there should be assumed a heaven that is resting or unmoved.

[7.] Next, in the seventh, we inquire whether the whole earth is habitable.

[8.] Next, in the eighth [question], we inquire whether everything doing work exists for the sake of that work; that is, whether for any working thing, the work is its end [or goal].

[9.] In the ninth, we inquire whether it is appropriate to demonstrate that the plurality of motions and spheres in the heavens is from God.

[10.] Next, in the tenth [question], we inquire whether this consequence is sound: If it is necessary that generations and corruptions be here below, it is [therefore] necessary that there be several celestial motions.

[11.] In the eleventh, we inquire whether the sky [*caelum*] is always moved regularly.

[12.] In the twelfth, we inquire whether natural motion ought to be quicker at the end than at the beginning.

[13.] In the thirteenth, we inquire whether projected things are moved quicker in the middle [of their motions] than in the beginning or end.[6]

[14.] In the fourteenth, we inquire about the discussion of the stars: whether all celestial spheres and all stars are mutually of the same ultimate species *(speciei specialissimae).*[7]

---

3. See Selection 71.

4. For a discussion on infinite space beyond the finite cosmos, see Selections 72 and 73.

5. This and the next question are translated in Selection 57.

6. In terms of the Aristotelian distinction between natural and violent motion, this question is concerned with the latter.

7. "Ultimate species" signifies an individual species not further divisible into other species.

[15.] In the fifteenth [question] we inquire whether by their light *(lumen)* the celestial bodies are generative of heat.

[16.] Next we inquire whether local motion can produce heat.

[17.] Next we inquire whether by its motion the sphere of the sun makes lower things hotter than do other spheres.[8]

[18.] Next we inquire about the chapter [in *On the Heavens and the World*] concerning the motion of stars: whether the stars are self-moved or moved by the motion of their spheres.[9]

[19.] Next we inquire whether spots appearing in the moon arise from differences in parts of the moon or from something external.

[20.] We inquire [now] about the chapter on the order [of the planets];[10] whether the lower spheres of the planets should be moved quicker in their proper motions than the superior spheres.

[21.] Next we inquire whether the sun and moon ought to be moved with fewer motions than other planets.

[22.] Next we inquire whether the earth always rests in the middle [or center] of the world.[11]

[23.] In the next and last question on the second book [of *On the Heavens*] we inquire whether the earth is spherical.

BOOK III

As to the third book of *On the Heavens and the World*, we inquire:

[1.] Whether in terms of heaviness and lightness it could be demonstrated that bodies are not composed of indivisibles.

[2.] In the next and last [question] we inquire

whether after departure from a projector, a stone that is projected, or an arrow that is shot from a bow, and so on in similar cases, [these] are moved by an internal principle or an external principle.[12]

BOOK IV

As to the fourth book of *On the Heavens and the World*, we inquire first:

[1.] Whether there is something absolutely heavy and something absolutely light; and also [whether there is] something [both] heavy and light which is neither absolutely heavy or light.

[2.] Next we inquire whether the natural places of heavy and light bodies are the causes of their motions.

[3.] Next we inquire whether a heavy or light body is moved naturally downward or upward by its generator *(generans)*,[13] or by the thing which removes the impediment [to its motion].[14]

[4.] Next we inquire whether heavy and light bodies are actively moved by heaviness and lightness, [respectively].[15]

[5.] Next we inquire whether heaviness and lightness are substantial forms of heavy and light bodies.

[6.] Next we inquire whether it might suffice to assume two motive natures, and mixtures of them, in order to save [or explain] the appearances of natural motions of heavy and light bodies.

[7.] Next we inquire whether, in its proper region [or place], air is heavy or light; or neither heavy nor light.

[8.] In the last [question] we inquire whether the number of four elements could be derived in terms of [or by arguing from] heaviness and lightness.

37 QUESTIONS ON THE TWO BOOKS OF ARISTOTLE'S *ON GENERATION AND CORRUPTION (DE GENERATIONE ET CORRUPTIONE)*

Albert of Saxony (ca. 1316–1390)

Translated and annotated by Edward Grant[1]

BOOK I

[1.] First we inquire whether it is possible that

8. The question is concerned primarily with the action of the sun on all things between the earth and moon.

9. In Book II, chapter 8, of *De caelo,* Aristotle considers whether the stars are free, self-moving bodies or whether they are attached to spheres which rotate and carry them. It is the latter position that he adopts.

10. See Aristotle, *De caelo* II.10.

11. Buridan's treatment of this very question is given in Clagett's translation in Selection 67.4. Because it was desirable to translate the enunciation of this question in

conformity with the pattern used throughout, my translation differs slightly from Clagett's.

12. This is the locus of one of Buridan's famous discussions of impetus theory. For his views, see Selections 48 and 49.

13. That is, the thing which caused or produced the heavy or light body.

14. For a discussion of this question by St. Thomas Aquinas, see Selection 46.

15. Questions 2, 3, and 4 present three quite different causal explanations of natural motion.

1. This treatise occupies folios 128 verso to 155 recto of the same edition cited in Selection 35, n. 1, which also contains Albert of Saxony's *Questions on the Physics.*

something be generated absolutely or corrupted absolutely.

[2.] In the second [question] we inquire whether generation is alteration.

[3.] We inquire whether, if generation were impossible, alteration would [also] be impossible.

[4.] We inquire whether the generation of one [thing] is the corruption of another.

[5.] We inquire whether in generation there occurs an absolute resolution [or decomposition] into prime matter.

[6.] We inquire whether any quality that was previously in a corrupted [substance] could persist in the [substance] generated [from what was corrupted].

[7.] We inquire whether any generable thing that may be given could be generated from any of several agents. For example, if *A* were a given generable thing which could be produced by agent *B*, [the question is] whether *A* could be generated or produced by an agent other than *B*.

[8.] We inquire whether any generable thing that may be given could be generated indifferently in any one of several instants.

[9.] We inquire whether augmentation is generation.

[10.] We inquire whether that which is increased in augmentation remains the same before and after.

[11.] We inquire whether any part of something that is increased could [itself] be increased.

[12.] We inquire whether augmentation occurs by means of formal, and not material, parts.

[13.] We inquire whether augmentation is a continuous motion.

[14.] We inquire whether every action and passion occurs by contact.

[15.] We inquire whether a similar thing can act on a thing similar to itself.

[16.] We inquire whether in acting [on something] every agent is [also] acted upon *(repatiatur)*.

[17.] We inquire whether an indivisible [thing] could be altered.

[18.] We inquire whether substantial forms of elements can be intended and remitted.

[19.] We inquire whether elements remain [or persist] formally in a compound [or mixed] body.

[20.] We inquire whether a compound is possible.[2]

[21.] We inquire whether a compound *(mixtio)* is natural. By "compound" we understand what Aristotle understands in the first book of this *[On Generation and Corruption]*, namely a "compound" properly so called is that whose every part

is [also] said to be a compound [or mixed] body.

BOOK II

As for the second book, we inquire firstly:

[1.] Whether there are four primary qualities, no more nor less.

[2.] We inquire whether of the four primary qualities (hot, cold, [dry, and moist]) two are active, namely hot and cold, and two passive, namely moist and dry.

[3.] We inquire whether there are four elements, no more nor less.

[4.] We inquire whether fire is primarily hot, air primarily moist, water primarily cold, and earth primarily dry.

[5.] We inquire whether the hotness of fire and the hotness of air are of the same species, differing only by more and less; similarly we ask [the same question] about the dryness of fire and earth, the moistness of air and water, and the coldness of earth and water.

[6.] We inquire whether any element is pure.

[7.] We inquire whether one element could be generated directly from another, so that water could be generated directly from air without something else being generated from it previously; and [the same question can be asked] of the other elements.

[8.] We inquire whether [elements] having a common quality *(symbolum)* are more easily [and quickly] transmuted into one another.[3]

[9.] We inquire whether a third element can be generated from two elements lacking a common quality *(symbolum)*.

[10.] We inquire whether every compound body located around the middle [or center of the world] *(circa locum medium)* is composed of all the simple [or elemental] bodies.[4]

[11.] We inquire whether a compound body can be [equally or] duly proportioned from the [four] elements or qualities of the elements.

[12.] We inquire whether generation is perpetual.

[13.] We inquire whether every corruptible thing has a definite period of duration.

[14.] We inquire whether the process in the generation of things continues into infinity.[5]

2. For translations of Questions 19 and 20, see Selection 77.2.

3. See Aristotle, *On Generation and Corruption* II. 4.

4. See *On Generation and Corruption* II. 8.

5. Aristotle distinguishes between an eternal process conceived linearly and circularly. Rejecting the former, he argues for a cyclical infinite process.

38 QUESTIONS ON THE FOUR BOOKS OF ARISTOTLE'S *METEOROLOGICA*

Themon, Son of the Jew (fl. 1349–1361)

Translated and annotated by Edward Grant[1]

BOOK I

[1.] In the first [question] we inquire whether this whole sensible world is subject to, and continuous[2] with, the superior motions *(superioribus lationibus)*, so that the power of this whole [sensible] world is regulated [and governed] from there.[3]

[2.] We inquire whether the mass *(moles)* of the whole earth—that is, its quantity *(quantitas)* or magnitude *(magnitudo)*—is much less than certain stars.[4]

[3.] We inquire whether the sky has the nature of fire.

[4.] We inquire whether the same opinions are repeated by men an infinite number of times.[5]

[5.] We inquire whether the four elements are continuously proportional or have an equality of a common ratio *(communis analogie)*, that is, whether they form a continuous proportionality.[6]

[6.] In the sixth [question] we inquire whether celestial motions cause hotness here below on lower [or inferior] things.

[7.] In the seventh [question] we inquire whether the middle region of air is always cold.

[8.] In the eighth, we inquire whether every luminous body causes heat by its light.

[9.] In the ninth, we inquire whether the matter of all meteorological phenomena *(impressiones)* is a vapor or exhalation.

[10.] In the tenth [question] we inquire whether burning phenomena *(impressiones ignite)* generated in the upper air occur naturally.

[11.] In the eleventh, we inquire whether clefts *(hyatus)* and chasms *(voragines)* can appear [in the sky] on a clear night.[7]

[12.] In the twelfth, we inquire whether a comet is of a celestial nature or [whether it is] of an elementary nature, say of a fiery exhalation.[8]

[13.] In the thirteenth [question] we inquire whether a comet or stellar comet signifies the death of princes, drynesses, winds, and other evils.[9]

[14.] In the fourteenth, we inquire whether the milky way *(galaxia)* is of a celestial or elementary [that is, sublunary] nature.[10]

[15.] In the fifteenth, we inquire whether the mid-

dle region of air is the place where rain is generated.

[16.] In the sixteenth, we inquire whether moisture *(pluvia ros)* and hoar frost [or dew] *(pruina)* differ as ultimate species *(specie specialissima)*.[11]

[17.] In the seventeenth, we inquire whether hail occurs more in spring and autumn.[12]

[18.] In the eighteenth, we inquire whether a redness in the morning signifies future rains.

[19.] In the nineteenth [question] we inquire

1. My translation was made from the Latin text *(Questiones super quatuor libros Meteororum compilate per doctissimum philosophii professorem Thimonem)* found in folios 155 verso to 214 verso of the same edition cited in Selection 35, n. 1.

2. The text has *contiguus* instead of *continuus*.

3. Aristotle, *Meteorologica* I.2.339a.21–23.

4. *Meteorologica* I.3.339b.7–9.

5. A question based upon Aristotle's claim that "the same opinions appear in cycles among men not once or twice, but infinitely often" *(Meteorologica* I.3.339b.28–29).

6. If we represent the four elements by A, B, C, and D, Themon asks whether the four elements are related in continued proportion so that $(A/B)\cdot(B/C)\cdot(C/D) = (A/D)^3$ and $A/B = B/C = C/D$. In his lengthy discussion of this question Themon cites Euclid's definition of continued proportion (*Elements*, Bk. V, Def. 5), which is translated in Selection 27.

7. See *Meteorologica* I.5. Themon conveys a vague sense of these terms when he observes (fol. 164 recto, col. 1): "The first experience is that sometimes the sky appears perforated or opened so that in these openings a depth appears which is so small and obscure that it is called a cleft *(hyatus)* and if large is called a chasm *(vorago)*."

8. In *Meteorologica* I.7, Aristotle denies that comets are celestial, or supralunar, phenomena, and in I.8 argues that they are fiery elemental exhalations (a second class of comets, however, is associated with the stars). Themon accepts Aristotle's opinions, which are quoted in the introduction to Selection 70.

9. Aristotle says nothing of omens such as the death of princes. His discussion is strictly scientific.

10. Following Aristotle, Themon argues that the milky way is in the celestial regions.

11. That is, whether moisture, on the one hand, and hoar frost or dew on the other, are separate, and not further divisible, species. Themon, once again in accord with Aristotle, holds that all such watery phenomena are of the same species, differing only in possessing more or less water.

12. Discussed by Aristotle in *Meteorologica* I. 12.

whether the waters of springs and rivers are generated in the concavities of the earth.[13]

BOOK II

Next we inquire about the second book of Aristotle's *Meteors*.

[1.] In the first [question] we inquire whether the sea, which is in the natural place of water, is generable and corruptible or perpetual.

[2.] In the second [question] we inquire whether the sea flows at one time and ebbs at another.[14]

[3.] In the third, we inquire whether the whole sea must be salty and bitter.[15]

[4.] We inquire whether river and spring waters could be salty.

[5.] We inquire whether wind is a hot and dry exhalation.[16]

[6.] In the sixth [question] we inquire whether the sun makes the winds cease and moves them.

[7.] In the seventh, we inquire whether a motion of the earth is possible.[17]

[8.] In the eighth, we inquire whether the tranquility of the air is a sign of the earth's motion to come.

[9.] In the ninth, we inquire whether thunder is caused by fire extinguished in a cloud.[18]

[10.] In the tenth, we inquire whether a typhoon and a hurricane are made from a hot and dry exhalation.

[11.] In the eleventh [question] we inquire whether lightning is fire descending from a cloud.

BOOK III

[1.] In the first [question] we inquire whether every visual ray is refracted *(refrangatur)*[19] in meeting a denser or rarer medium.

[2.] We inquire whether a visual ray proceeding from a luminous body of a particular density is refracted *(frangatur)* toward or away from the perpendicular when it meets a medium of a different density.[20]

[3.] In the third [question] we inquire whether every visual ray is reflected *(reflectatur)* when it meets a denser medium.

[4.] In the fourth [question] we inquire whether any visual ray could fall upon mirrors or bodies and be reflected so that it represents color only but not the shape [or figure of the body].[21]

[5.] In the fifth, we inquire whether a halo appears because of the refraction of rays in the vapor interposed between the eye *(visum)* and a luminous body[22] around which it appears.

[6.] In the sixth, we inquire whether a halo could be produced by refraction of visual rays.

[7.] In the seventh, we inquire whether at the time of the appearance of a halo, there is a vapor lying between the eye *(visum)* and a star *(astrum)*.

[8.] In the eighth, we inquire whether the vapor in which a halo appears is of a spherical figure.

[9.] In the ninth [question] we inquire whether a halo must appear as the periphery or circumference of a circle.[23]

[10.] In the tenth, we inquire whether, without the aid of mirrors, a visual ray could be reflected by homogeneous, uncondensed vapor-free air to the same eye [from which the visual ray issued],[24] and which [eye] is [located] in the same air.

[11.] In the eleventh, we inquire whether the colors appearing in the rainbow are where they seem to be and are true colors.

[12.] In the twelfth, we inquire whether a rainbow is a real form impressed on a cloud; or whether it is only an imaginary form.

13. In agreement with Aristotle, Themon adopts the affirmative side of this question (see *Meteorologica* I.13).

14. Based upon *Meteorologica* II.1.354a.6–12. In his Loeb Classical Library translation of the *Meteorologica* (Cambridge, Mass: Harvard University Press, 1962), H. D. P. Lee remarks (pp. 128–129) that "it is not clear exactly what Aristotle means by this ebb and flow (lit. swinging to and fro) of the sea, for he had no real knowledge of the tides." By the Middle Ages, thanks to observations by Bede (ca. 673–735), it was known that tides could be established as roughly constant for a given port (see ch. 29 of his *On the Reckoning of Times [De temporum ratione]*).

15. A topic discussed by Aristotle in *Meteorologica* II.3.

16. *Meteorologica* II.4.

17. Although Themon lists a number of ways in which the earth can be conceived to move (including rotation on its axis, which he rejects), this question is concerned primarily with earthquakes and tremors, the subject matter of II.7 and II.8 of the *Meteorologica*. For a discussion of the earth's possible axial rotation, see Selections 67.3–6.

18. Aristotle, as also Themon, denies this in II.9.

19. In the *Meteorologica*, Aristotle discusses only reflection of visual rays.

20. Since Aristotle did not consider refraction, he could not have raised this question.

21. An affirmative response is given by Aristotle in *Meteorologica* III.2.372a.30—372b.8 and III.4.373b.17–33.

22. Such as sun, moon, or stars.

23. Aristotle answers in the affirmative (*Meteorologica* III.3.372b.34—373a.18).

24. This bracketed addition is warranted from the nature of the question. Although, in the *Meteorologica*, Aristotle speaks as if visual rays issue from the eye, it was not his considered opinion and is denied elsewhere (for example, *Topics* I.14.105b.6). For medieval discussions on visual rays, see Selections 62.14–17.

[13.] In the thirteenth, we inquire whether a rainbow is a translucent *(diaphanalis)* or mirror-like form; or whether it occurs by reflection or refraction, which is the same thing.[25]

[14.] On the supposition that a rainbow can occur by reflection of rays, we inquire whether such reflection occurs in a cloud or whether it occurs in tiny dewdrops or raindrops.[26]

[15.] We inquire whether every rainbow must be three-colored.[27]

[16.] In the sixteenth [question] we inquire whether a rainbow can appear double only, but no more [than double].

[17.] We inquire whether on the appearance of two rainbows, the upper one must always have two colors located in opposite positions [to the colors of the lower bow].[28]

[18.] We inquire whether the colors of the upper or secondary rainbow should necessarily appear weaker *(remissior)* than [the colors] of the principal bow.

[19.] In the nineteenth, we inquire whether a rainbow, if [its formation is] unhindered, ought to appear around the periphery [or circumference] of a circle.

[20.] In the twentieth, we inquire whether, at the time of a rainbow's appearance, it is necessary that the center of the sun, the center of the horizon, the center of the rainbow, and the poles or pole of the rainbow, be in a straight line.

[21.] In the twenty-first [question] we inquire whether at the time of appearance of a rainbow and a halo, the diameter of the rainbow is exactly double the diameter of the halo.[29]

[22.] In the twenty-second, we inquire whether to an observer *(oculo)* placed on the surface of the horizon, some portion of a rainbow would appear during every hour of an artificial [or arbitrary] day, wherever the man or eye of the viewer might be located [on the surface of the horizon].[30]

[23.] In the twenty-third, we inquire whether a rainbow of the moon could appear more than twice in fifty years with colors and properties like the solar rainbow.[31]

[24.] In the twenty-fourth, we inquire whether a rainbow that is seen when the sun or a star is elevated considerably above the horizon appears as a smaller part *(portio)* of a great circle than it would appear if the sun were along the horizon.

[25.] Next we inquire whether rainbows appearing around candles really do exist around those candles.

[26.] In the twenty-sixth [question] we inquire

whether rods *(virge)* and mock suns *(paralleli)* are produced by refraction of rays, just as with a rainbow or a halo.[32]

[27.] We inquire [now] about things made under the earth; whether metals could be produced with the aid of art just as a rainbow and a halo are sometimes made artificially.

BOOK IV

[1.] In the first [question] on the fourth book of the *Meteors* we inquire whether putrefaction is possible, or whether it can be done or exist.

[2.] In the second [question] on this fourth book we inquire whether digestion is a perfection caused by proper and natural heat [acting] on passive opposites.[33]

[3.] In the third, we inquire whether there are only three species of digestion [or concoction].[34]

[4.] In the fourth, we inquire whether digestion

25. For Themon, reflection and refraction differ, but both are required in the formation of a rainbow (see Selection 63.2 for Theodoric of Freiberg's qualitatively correct explanation of the formation of the rainbow), which is simultaneously mirror-like (because reflection occurs) and diaphanous or translucent (because of refraction).

26. Before Themon, Robert Grosseteste (Selection 61.2) had claimed that reflection (and refraction) occurred in a cloud, whereas Theodoric of Freiberg had argued correctly that it (and refraction) occurred in each raindrop (Selection 63.2).

27. For Aristotle, red, green, and blue were the three basic colors of the rainbow (*Meteorologica* III.2.371b. 33—372a.11 and III. 4.374b.7—375b.8). Themon will argue that rainbows can be three-, four-, and even six-colored.

28. Aristotle answered in the affirmative, since he held that the upper or secondary bow consisted of three arcs arranged from outer to inner arc in the order blue, green, and red. The lower, or primary, bow had its colors arranged from outer to inner arc in the order red, green, blue. Hence in the two bows, red and blue are reversed, with green retaining a middle position in each bow.

29. This question is not discussed by Aristotle.

30. Aristotle (and Themon) deny the occurrence of rainbows at any time of day during the year. Indeed, Themon, perhaps relying ultimately on Roger Bacon, denies the occurrence of rainbows whenever the sun rises more than 42° above the horizon.

31. Although Aristotle mentions moon rainbows (*Meteorologica* III.4.375a.18), he does not raise this question.

32. In *Meteorologica* III. 6, Aristotle discusses the formation of rods and mock suns.

33. See *Meteorologica* IV.1.378b.26—379a.2.

34. "The effect of heat is concoction and there are three species of concoction, ripening, boiling, and roasting" (*Meteorologica* IV.2.379b.13–14; trans. H. D. P. Lee).

[or concoction] of blood and organic bodies *(nascentium)* by natural heat is [a process of] ripening *(pepansis)*.

[5.] In the fifth, we inquire whether a simple element could be digested or undergo digestion.

[6.] In the sixth, we inquire whether things that have been boiled are drier than things that have been roasted.

[7.] In the seventh, we inquire whether things that have been fried *(frixata)* are boiled or roasted.

[8.] In the eighth [question] we inquire whether the definition of simple generation is sound, namely where it is said that simple and natural generation is a change *(permutatio)* [effected] by those powers which bear a ratio *(rationem)* [or relation] to the subject matter of any nature.[35]

35. This question is based on Aristotle's discussion of hot and cold as active factors in change; and moist and dry as passive factors. The definition is found in *Meteorologica* IV.1.378b.32—379a.1. If the edition of Themon's text used here is complete, then he chose to terminate his treatise after drawing questions from chapters 1 to 3 only, ignoring chapters 4 to 12.

Physics

Statics, or "The Science of Weights"

INTRODUCTION[1]

Edward Grant

On the basis of extant manuscripts, three types of treatises must be distinguished in the history of medieval statics.

1. Those translated into Latin from Greek or Arabic and originating largely in late antiquity in Alexandria. Included are the *De ponderoso et levi* ascribed to Euclid *(The Book of Euclid Concerning the Heavy and Light)*, the *De insidentibus in humidum* ascribed to Archimedes *(The Book of Archimedes on Floating Bodies)*, the anonymous *Liber de canonio (The Book Concerning the Balance)*, and the *Liber Karastonis (The Book of the Roman Balance)* edited by Thabit ibn Qurra in the ninth century.

2. A group of treatises ascribed to, and some of which were written by, Jordanus de Nemore in the thirteenth century. Jordanus undoubtedly wrote the *Elementa Jordani super demonstrationem ponderum (The Elements of Jordanus on the Demonstration of Weights)* and perhaps the most significant of the treatises ascribed to him, *Liber Jordani de Nemore de ratione ponderis (The Book of Jordanus de Nemore on the Theory of Weight).*[2] A third treatise in this group, *Liber Jordani de ponderibus (The Book of Jordanus on Weights)*, is not by Jordanus.

3. A third group of treatises, written, for the most part, by fourteenth-century authors, were composed as commentaries on, or revisions of, the treatises in the second group.[3]

In the treatise represented by our selection, we find some of the most significant contributions to medieval and early modern statics. Combining the purely Archimedean geometrical tradition in statics with the Aristotelian dynamical tradition derived ultimately from the Pseudo-Aristotelian *Mechanical Problems,* Jordanus and his followers developed not only the form of the principle of virtual displacements but made other conceptual advances as well. Here is Clagett's summation[4] of these medieval contributions:

Heir of Hellenistic mechanics, the medieval mechanician sharpened the concept of virtual displacements, which had had its origin in the Pseudo-Aristotelian treatise *Mechanica*. The form of the principle of virtual displacements used by the thirteenth-century mathematician Jordanus and expressed by one of his thirteenth- or fourteenth-century commentators (in the so-called *Aliud commentum* published in 1533 by Peter Apianus [ed. and trans. J. E. Brown; see n. 2—*Ed.*]) was essentially of this form:

5. *What suffices to lift a weight W through a vertical distance H will lift a weight kW through a vertical distance H/k or a weight W/k through a vertical distance kH.*

This principle was employed by Jordanus to prove the law of the lever as applied to both a straight and bent lever. It was also used in the elegant proof of the proposition concerning the equilibrium of inter-connected weights on oppositely inclined planes whose lengths vary directly as the weights. Furthermore, in the proposition concerning the bent lever, what is essentially the concept of static moment was conceived:

6. *The effective force of a weight on any lever arm, straight or bent, depends on both its weight and its horizontal distance from the vertical line passing through the fulcrum.*

1. The following introduction is drawn from the general introduction by E. A. Moody and M. Clagett to their volume *The Medieval Science of Weights (Scientia de ponderibus), Treatises Ascribed to Euclid, Archimedes, Thabit ibn Qurra, Jordanus de Nemore, and Blasius of Parma* with Introductions, English Translations, and Notes (Madison, Wis.: University of Wisconsin Press, 1952) and from M. Clagett, *The Science of Mechanics in the Middle Ages* (Madison, Wis.: University of Wisconsin Press, 1959).

2. Joseph E. Brown inclines to the view that a pupil of Jordanus was the author of this lengthy treatise, which comprised forty-five propositions. See his *"The Scientia de ponderibus in the Later Middle Ages,"* Ph. D. dissertation, University of Wisconsin, 1967, pp. 64–87. For convenience, we shall ascribe the treatise to Jordanus.

3. A few of the important ones have been edited and translated by Joseph Brown in the work cited in note 2.

4. *The Science of Mechanics in the Middle Ages*, pp. 675–676.

Still another fertile statical concept was used by the author of the *De ratione ponderis*. This was the concept of positional gravity used to determine the component of natural gravity acting along an inclined plane. This concept can be expressed as follows:

7. *The ratio of the effective weight (positional gravity) of a body in the direction of the inclined plane on which it rests to the free natural gravity is equal to the ratio of the vertical component of any given potential trajectory along that plane to that trajectory.*

This concept is equivalent to the modern formulation $F = W \sin a$, where F is the force along the plane, W is the free weight and a is the angle of inclination of the plane. It was employed by the medieval author to establish the law of equilibrium of interconnected weights on oppositely inclined planes. This concept and its use marked an important step in the rise of a vectorial analysis of forces.

Indeed, the medieval achievement in statics is best measured in terms of the four principles of statics distinguished by Ernst Mach in his *Science of Mechanics* and deemed essential to that discipline:

(1) The general lever principle; (2) the inclined plane principle; (3) the principle of composition of forces; and (4) the principle of virtual displacements. The statics of Archimedes, which steered clear of all entanglements with dynamics, attained only the first of these principles; the other three, involving dynamical considerations, characterize modern statics. In the treatises ascribed to Jordanus de Nemore, however, these last three principles appear, and provide a dynamical foundation for the proof of the lever principle.[5]

39 ON THE THEORY OF WEIGHT

Jordanus of Nemore (fl. 1230–1260)

Translated and annotated by Ernest A. Moody[1]

PART 1

Postulates[2]

R1.001. The movement of every weight[3] is toward the center (of the world), and its force is a power of tending downward and of resisting movement in the contrary direction.

5. Moody and Clagett, *The Medieval Science of Weights*, p. 15.

1. [This is a translation of the *De ratione ponderis* ascribed to Jordanus of Nemore (see the introduction, n.2), and edited with introduction, translation, and notes by Ernest A. Moody with the assistance of Raymond Clements, Arthur Ditzel, and Jason Lewis Saunders. It is reprinted with permission of the Regents of the University of Wisconsin from *The Medieval Science of Weights (Scientia de ponderibus), Treatises Ascribed to Euclid, Archimedes, Thabit ibn Qurra, Jordanus de Nemore, and Blasius of Parma*, edited with Introductions, English Translations, and notes by E. A. Moody and Marshall Clagett (Madison, Wis.: University of Wisconsin Press, 1952). The notes, which were cued to line numbers in the Latin text and placed at the end of the book, have been rearranged and assigned as regular footnotes. In what follows, Moody uses the letter *E* to represent *The Elements of Jordanus On the Demonstration of Weights (Elementa Jordani super demonstrationem ponderum)*, a treatise containing nine propositions and undoubtedly written by Jordanus. The present treatise,

De ratione ponderis, referred to by the letter *R*, may have been written by Jordanus himself or a pupil (see the introduction, note 2). It is divided into four parts containing a total of forty-five propositions. Although based on *E*, the *De ratione ponderis* corrects mistakes in *E* and adds much that is original. It is perhaps the finest of the medieval statical treatises and was published in 1565 at Venice from a manuscript once owned by Niccolo Tartaglia.—*Ed.*]

2. The seven postulates are the same as those of the *Elementa Jordani,* and our comments on the latter hold equally for these. [These comments are referenced to the postulates to which they refer.—*Ed.*]

3. The terminology introduced in the Postulates requires a few comments. *Pondus,* which we have translated by "weight," is to be understood in the substantival sense, as that which *has* a natural tendency to move toward the center of the world. This tendency itself is called the *virtus, gravitas,* or *ponderositas* of the heavy body *(grave)* or of the weight *(pondus)*. Thus *pondus* is treated as a scalar quantity, analogous to our notion of mass, while *virtus* or *gravitas* is a vector quantity, analogous to our notion of force. The simple gravitational *virtus* is assumed to be proportional to the scalar weight or *pondus,* just as in Newtonian physics the force of gravity acting on a body in a given region is proportional to its mass. The additional clause, added in the first Postulate by our manuscript *C,* and by most of the manuscripts of the *De ratione ponderis* of Jordanus, makes this plain: *virtutemque ipsius potentiam ad inferiora tendendi et motui contrario resistendi.* This also

R1.002. That which is heavier descends more quickly.

R1.003. It is heavier in descending, to the degree that its movement toward the center (of the world) is more direct.[4]

R1.004. It is heavier in position when in that position its path of descent is less oblique.[5]

R1.005. A more oblique descent is one which, in the same distance, partakes less of the vertical.[6]

R1.006. One weight is less heavy in position, than another, if it is caused to ascend by the descent of the other.[7]

R1.007. The position of equality is that of equality of angles to the vertical, or such that these are right angles, or such that the beam is parallel to the plane of the horizon.[8]

Theorems

R1.01. Between any heavy bodies, the forces are directly proportional to the weights.

―――――――――

indicates that the same *virtus* which measures a body's downward directed force, in falling, measures its resistance to any force acting to lift it vertically upward. This is a basic principle in the consideration of weights in equilibrium, and it is invoked in Postulate *E.06* and in the second part of Theorem *E.1.*

The "heaviness" of a weight, or its *gravitas* or *ponderositas,* is equal to its *virtus* only insofar as it is directed vertically downward. If a weight descends obliquely, as along an inclined plane, its "heaviness" 'in descending along the oblique path (or the force which it exerts along the oblique path) will be less than its natural *virtus* as directed toward the center of the world. This variable "heaviness," or force exerted by a body in descending along a path other than the vertical, is called *gravitas in descendendo.*

4. The second and third Postulates state that the "heaviness" of a body in its descent, along whatever path it descends, is proportional to the speed with which it descends along that path, and both its heaviness and its speed depend on the degree to which its path of descent approaches the vertical. Postulate *E.02* might be given a more general interpretation, as stating that a body of greater natural weight *(pondus)* will fall in free vertical descent faster than a body of smaller natural weight. But in the Aristotelian context, even this thesis would be understood in terms of a descent against a resistance—either a resistant medium, or a resistance such as that of a counterweight on the other arm of a balance, which will be raised more rapidly by a larger weight on the descending arm, than by a smaller weight. Postulates *E.02* and *E.03* seem, however, to be concerned with the restricted case of variability of the force and speed with which a given weight descends along paths which deviate from the vertical—in our language, the concern is with the component of gravity effective in producing motion along a path compatible with a constraint system such as a balance.

5. Postulate *E.04* introduces the expression *gravitas secundum situm,* or "positional gravity." This plays a

major part in the statical theorems which follow. Whereas "gravity in descending" is measured by the obliquity of the path *actually* traversed in the movement of a heavy body along a path compatible with the constraints, "positional gravity" is defined by a merely *virtual* movement along a possible path of descent, in a constraint system which is in equilibrium. In modern language, "positional gravity" is a component in a system of forces whose vector sum is zero. It is measured by the work which would be done if the system permitted the element in question to undergo motion independently of the rest of the system; but as taken in the constraint system, in equilibrium, it actually does no work. The measure of "positional gravity" of a body, relative to its natural gravity or *virtus,* is determined by the ratio of the obliquity of the path it would traverse in a "virtual displacement" of the system, to the vertical path along which its natural gravity is directed.

6. Postulate *E.05* defines "obliquity" as the deviation of a path of descent from the vertical, or as its approach to the horizontal. The contrary of "more oblique" is "more direct," so that the positional gravity varies inversely with the degree of obliquity of the path of virtual movement. Jordanus' definition is precise: the descent is more oblique if the path of descent has a smaller component of the vertical. In Proposition *R1.10* of the *De ratione ponderis* (see below) this notion is explicitly explained, for the case of a rectilinear descent along an inclined plane. Thus, taking an inclined plane as path of possible descent, we project a given length of the inclined plane onto a vertical line, forming a right-angled triangle. Then the ratio of the altitude of this triangle to its hypothenuse varies inversely with the obliquity, or directly with the "positional gravity." The positional gravity, or component of force directed along the inclined plane, is therefore measured by $\sin a \cdot pondus,$ when a is the angle of hypothenuse to base.

7. Postulate *E.06* states that one body has less positional gravity than another, if the descent of the other suffices to cause it to ascend. Obviously it is assumed that the two bodies are connected by a balance beam, or by a pulley arrangement. It is not wholly clear whether the author is also assuming equality of the two weights compared, or whether he is taking into consideration the case where the weights are unequal in their natural gravity or mass. On the first assumption, the postulate would state that if two equal weights are attached so that the descent of one involves the ascent of the other, then if one of them is able to descend and to raise the other, it possesses this advantage by reason of position—i.e., because its path of possible descent, within the constraint system, is less oblique. But if the postulate is understood to hold for unequal weights also, then it asserts that if the descent of one weight causes the ascent of the other, the product of the *pondus* and its "positional" coefficient, in the case of the weight which descends, is greater than the product of the *pondus* and positional coefficient of the other weight.

8. Postulate *E.07* defines the "position of equality," obviously referring to a balance beam capable of rotation on an axis of support. The *situs equalitatis* [that is, position of equality—*Ed.*] is the horizontal position of the beam, such that its arms form equal (and right) angles with the perpendicular to the plane of the horizon passing through the axis of support.

Let there be two weights, *ab* and *c,* of which *c* is the lighter [Fig. 1]. And let *ab* descend to D, and let

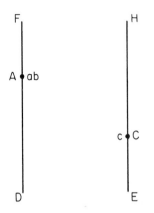

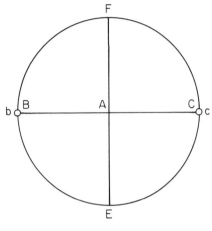

Fig. 1

Fig. 2

c descend to E. Again, let *ab* be raised to F, and *c* raised to H. I then say that the distance AD is to the distance CE, as the weight *ab* is to the weight *c;* for the velocity of descending is as great as the force of the weight. But the force of the combined weight consists of the forces of its components. Let the weight *a* then be equal to the weight *c,* so that *a*'s force is the same as that of *c*. If then the ratio of the weight *ab* to the weight *c* is less than the ratio of the force of *ab* to the force of *c,* the ratio of the weight *ab* to (its component weight) *a* will likewise be less than the ratio of the force of *ab* to the force of *a*. And therefore the ratio of the force of *ab* to that of *b* will likewise be less than the ratio of the weight *ab* to the weight *b*. Consequently the ratio of the same weights will be both greater and less than the ratio of their forces. Since this is absurd, the proportion must be the same in both cases. Hence the weight *ab* is to the weight *c,* as the distance AD is to the distance CE, and conversely as the distance CH is to the distance AF.[9]

R1.02. When the beam of a balance of equal arms is in the horizontal position, then, if equal weights are suspended from its extremities, it will not leave the horizontal position; and if it is moved from the horizontal position, it will revert to it. But if unequal weights are suspended, the balance will fall on the side of the heavier weight until it reaches the vertical position.

A balance is said to be of equal arms, when the arms of the beam, measured from the axis of rotation, are equal. Let the axis, then, be A, and the beam BAC [Fig. 2] and let the suspended weights be *b* and *c,* and let FA be the line of the vertical. If then we describe a circle through B and C, the mid

9. [Since this theorem and its diagram are very similar to Theorem *E.1* and its diagram, Moody's remarks on *E.1* (*The Medieval Science of Weights,* p. 376) in the first paragraph are relevant; what follows is quoted from his discussion of *R1.01* on pages 390–391.—*Ed.*]:

"At first sight the theorem appears to be an assertion that unequal weights, in free fall, descend with velocities proportional to their weight. The diagram normally supplied on the margins of the manuscripts, and which we have reproduced, seems to support such an interpretation. Yet the use of the term 'velocity of descent' [in *R1.01*, *virtus* is substituted for *velocitas* and the equivalent expression is "forces of descent"—*Ed.*], which in postulates *E.02* and *E.03* [these are equivalent to *R1.002* and *R1.003* of the selection reprinted here—*Ed.*] is associated with direction of descent and with "heaviness" directed along an oblique path, suggests that the author is thinking of movements along oblique paths, as between bodies in a connected system, and their "gravities" in the relative sense determined by the paths of movement compatible with the system."

. . . The entire theorem bears on the relative capacities of unequal weights to lift a given counterweight on the other arm of the balance, or to resist being lifted by the given counterweight. Only as so interpreted is the theorem of value in establishing the general lever principle for which it is invoked here (in *R1.06*) just as it was invoked in E.8 of the *Elementa Jordani*. On this interpretation, the theorem is an enunciation of the principle of work, in this form: If a weight w, descending through a vertical distance d, can lift a counterweight x through a certain vertical height h, then another weight $k \cdot w$, by descending the vertical distance d/k, can raise the same counterweight x the same height h; and also, another weight w/k can raise x through the height h by descending the vertical distance $k \cdot d$. The second part of the theorem will then assert that if w, descending the vertical distance d, can raise x through the height h, then $k \cdot w$, descending the same distance d, can raise $k \cdot x$ the same height h, or it can raise x the height $k \cdot h$. The three crucial theorems of Book I of the *De ratione ponderis* (*R1.06* on the general lever principle, *R1.08* on the bent lever, and *R1.10* on the inclined plane) rest on this principle of work; and at least in the case of *R1.06* the

214

point of its lower half being E, it is evident that the descent of both b and c will be along the circumference of the circle, toward E. And since the descents along these paths are of equal obliquity, and since b and c are equal weights, therefore neither of them will move.

Let it now be supposed that the balance is pushed down on the side of b, and elevated correspondingly on the side of c [Fig. 3]. I say that it will revert to

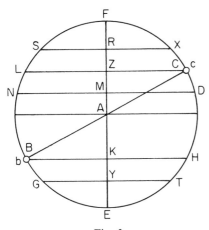

Fig. 3

the horizontal position. For the descent from C toward the horizontal position is less oblique than the descent from B toward E. For let there be taken equal arcs, as small as you please, which we will call DC and BG; and let the lines CZL and DMN, and also BKH and GYT, be drawn parallel to the horizontal. And let fall, vertically, the diameter FRZMAKYE. Then ZM will be greater than KY, because if an arc CX, equal to CD, is taken in the direction of F, and if the line XRS is drawn horizontally, then RZ will be smaller than ZM; and since RZ is equal to KY, ZM will be greater than KY. Since therefore any arc you please, which is beneath C, has a greater component of the vertical than an arc equal to it which is taken beneath B, the descent from C is more direct than the descent from B; and hence c will be heavier in its more elevated position, than b. Therefore it will revert to the horizontal position.[10]

_____

proof is explicitly referred to the principle that the heights through which weights can be lifted by the same counterweight are in inverse ratio to their weights, as a principle which had been established *(ut ostensum est)*. Now it is only this first theorem that could possibly be construed as having established this principle; consequently it seems proper to interpret the first theorem in this sense.

The method of proof here employed is similar to that used in *E.1*, depending on Euclid V, Prop. 30; i.e., if $(a + b)/a < (AE + ED)/AE$, then $(a + b)/b > (AE + ED)/ED$. Here $a + b$ is the greater of the two weights, and its part a is assumed to be the same as the lesser weight c, while the distance of $a + b$'s descent is assumed to be divisible proportionately to its components a and b in such manner that the distance AE corresponding to a's part of the descent is equal to the whole descent of the lesser weight c, and the distance ED represents b's part in the whole descent of $a + b$.

10. Theorem *R1.02*: The first two parts of this theorem correspond exactly to *E.2*, the second part committing the same error in assuming stable equilibrium to obtain. [Moody's remarks on the two parts of *E.2* (*The Medieval Science of Weights*, pp. 378–379) follow.—*Ed.*] "This theorem, like the first, gave rise to repeated discussion and criticism in the later Middle Ages and Renaissance. It is discussed by Roger Bacon in his *Opus Maius* (Part 4, Dist. 3, Ch. 16), a work written between 1266 and 1268. The theorem has two parts, the first of which seeks to demonstrate, by means of the concept of positional gravity, what Archimedes had laid down as his initial postulate—that equal weights on equal lever arms are in equilibrium. Since it was shown in *E.1* that movement (or 'velocity of descent'), as between weights on a balance, is determined either by inequality of *pondera* or by inequality of positional gravities, or by both together, and since the present theorem assumes equality of the *pondera*, the condition determining equilibrium reduces to that of equality of positional gravities—i.e., of the 'obliquities' of virtual descents of the two weights. The first part of the theorem shows that when the balance beam is in the horizontal position, equal arcs cut off just below the horizontal diameter of the circumference of rotation, will have equal components of the vertical; hence the obliquities of virtual descent, for the two weights, will be the same.

"The second part of the theorem, which is erroneous, is what gave rise to the criticisms discussed by Bacon and later writers. Jordanus here attempts to show that the equilibrium of equal weights on equal lever arms is a case of stable equilibrium. The attempted proof involves an erroneous application of the concept of positional gravity. Jordanus first shows that a small finite arc cut off just below the position of the weight on the lever arm that is elevated above the horizontal has a greater component of the vertical than an equal arc cut off just below the position of the weight on the depressed arm. From this he concludes that the elevated weight has greater positional gravity than the depressed weight, so that the balance beam will return to the horizontal position. His error is to compare the positional gravities of the two weights b and c directly, according to the ratios of vertical component to arc of *descent* in both cases—as if the descent of b could be treated independently of the ascent of c. He should have compared the ratio of the obliquity of a small descent of b to the obliquity of an equal *ascent* of c, with the ratio of the obliquity of an equal small descent of c to the obliquity of a corresponding ascent of b; he would then have seen that *these* ratios would remain equal under any displacement one way or the other, so that no advantage would be gained by either side through any displacement. While Jordanus could

Now let *b* be heavier than *c,* and let the balance be in the horizontal position. Then, since the descent on each side is of equal obliquity, it is evident that *b* will descend. For let *b* be placed in any position you please, below, with *c* above [Fig. 4]. I say that in this position *b* will still be heavier.

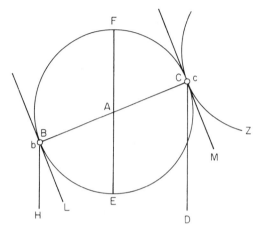

Fig. 4

For let the vertical lines CD and BH be drawn; and let the lines BL and CM be tangents to the circle; and let the arc CZ be drawn, similar and equal to the arc BE, and similarly placed, so that the line CM is tangent to it as well as to the circle. Then the obliquity of the arcs BE and CZ is measured by the angle DCZ, and the obliquity of the arc CE by the angle DCM; and the proportion of the angle DCZ to the angle DCM is smaller than any ratio that can be assigned between a greater and a lesser quantity. Therefore it will also be less than the ratio of the weight *b* to the weight *c*. Since then *b* exceeds *c* to a greater extent than the obliquity exceeds the obliquity, *b* will be heavier in this position than *c*. For this reason *b* will not cease to descend, and *c* to ascend, until the beam is in the position FE.[11]

R1.06. If the arms of a balance are proportional to the weights suspended, in such manner that the heavier weight is suspended from the shorter arm, the weights will have equal positional gravity.

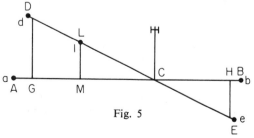

Fig. 5

Let the balance beam be ACB, as before, and the suspended weights *a* and *b* [Fig. 5] and let the ratio of *b* to *a* be as the ratio of AC to BC. I say that the balance will not move in either direction. For let it be supposed that it descends on the side of B; and let the line DCE be drawn obliquely to the position of ACB. If then the weight *d*, equal to *a*, and the weight *e* equal to *b*, are suspended, and if the line DG is drawn vertically downward and the line EH vertically upward, it is evident that the triangles DCG and ECH are similar, so that the proportion of DC to CE is the same as that of DG to EH. But DC is to CE as *b* is to *a;* therefore DG is to EH as *b* is to *a*. Then suppose CL to be equal to CB and to CE, and let *l* be equal in weight to *b;* and draw the perpendicular LM. Since then LM and EH are shown to be equal, DG will be to LM as *b* is to *a*, and as *l* is to *a*. But, as has been shown, *a* and *l* are inversely proportional to their contrary (upward) motions. Therefore, what suffices to lift *a* to D, will suffice to lift *l* through the distance LM. Since therefore *l* and *b* are equal, and LC is equal to CB, *l* is not lifted by *b;* and consequently *a* will not be lifted by *b*, which is what is to be proved.[12]

have avoided his error within the principles and methods of his own theory, it is probable that the complexity introduced into the problem, by the constantly changing obliquities involved in displacements along the circular path, led him to misapply his own principles in this case.''

11. The third part of *R1.02*, not found in *E.2*, shows that if one of the weights is the least bit heavier than the other, its finite excess of natural weight will offset the infinitesimal advantage due to difference in the obliquities of virtual descent, so that the arm on which the heavier weight hangs will continue to fall until it is vertically beneath the axis of support. The argument is of interest because of its glimpse of the infinitesimal aspect of the concept of a virtual displacement. But Duhem's claim (*Les Origines de la Statique*, I, p. 140) that this constitutes ''une refutation concluante'' of the error committed in the second part of the proof, is untenable; for part of the argument in this third section is the assumption that there is an advantage in positional gravity on the side of the weight which is above the horizontal position, which *needs* to be offset by some finite addition to the weight on the other arm in order to keep the balance from reverting to the horizontal.

12. Theorem *R1.06*: This states the general lever principle, and is in all respects identical with *E. 8*. The principle of work, according to Duhem, underlies the proof; and if this is so, we must assume that the first theorem (*R1.01*), here as in the *Elementa*, is Jordanus' statement of the principle of work. [What follows is Moody's comment on *E.8* (*Medieval Science of Weights*, pp. 382–383).—*Ed.*]

''Theorem *E.8:* This is the crucial theorem—the

R1.08. If the arms of a balance are unequal, and form an angle at the axis of support, then, if their ends are equidistant from the vertical line passing through the axis of support, equal weights suspended from them will, as so placed, be of equal heaviness.

Let the axis be C, the longer arm AC, and the shorter arm BC [Fig. 6]. And draw the vertical line

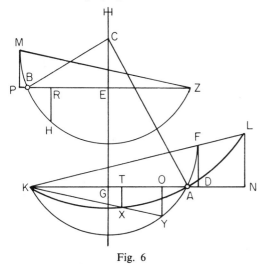

Fig. 6

CEG; and let the lines AG and BE, perpendicular to this vertical, be equal. When, therefore, equal weights are suspended at A and B, they will not change from this position. For let AG and BE be extended by a distance equal to their own length, to K and to Z; and on them let the arcs of circles, MBHZ and KXAL, be drawn; and let the arc KYAF be made similar and equal to arc MBHZ;[13] and let the arcs AX and AL be equal to each other, and similar to the arcs MB and BH. And let the arcs AY and AF also be equal and similar. If then *a* is heavier in this position than *b*, let it be supposed that *a* descends to X and that *b* is raised to M. Then draw the lines ZM, KXY, KFL; and let MP be erected perpendicularly on ZBP, and XT and FD on KAD. And because MP is equal to FD, while FD is greater than XT—on account of similar triangles—, MP will also be greater than XT. Hence *b* will be lifted vertically more than *a* will descend vertically, which is impossible since they are of equal weight. Again, let it be supposed that *b* descends to H and lifts *a* to L; and let HR fall perpendicularly on BZ, and LN and YO on KAN. Then LN will be greater than YO, and consequently greater than HR; so that in the same way the impossible will result.

general principle of the lever. The method of proof used

by Jordanus is wholly different from that of Archimedes, though it is perhaps vaguely implicit in the *Mechanical Problems* ascribed to Aristotle (850a 30—b 10), and also in the Third Proposition of Thabit ibn Qurra's *Liber Karastonis*. But whereas Aristotle, and Thabit as well, had argued that the lesser weight on the longer arm balances the greater weight on the shorter arm, because it would traverse a proportionately greater distance *along the arc* of its movement, Jordanus states that the comparison is determined according to the lengths of *vertical descent or ascent* accomplished in the movements along the curvilinear path. As Duhem pointed out, Jordanus' proof rests on the principle of work, in this form: What can lift a weight W through a height H, can lift a weight W/k through a height $k \cdot$H, or a weight $k \cdot$W through a height H/k. Duhem held, that this principle was 'implicitly' invoked by Jordanus in his proof; and B. Ginzberg, arguing against Duhem, made much of the fact that a man who fails to make his principle explicit is not likely to be very conscious of the fact that his proof depends on that principle. But neither Duhem nor Ginzberg seems to have noticed that Jordanus *does* invoke, as principle of his demonstration, an explicit theorem—namely, the second part of his first theorem *E.1*. This stated that the 'velocities of descent and of ascent' are inversely proportional to the weights. Hence the whole problem of the interpretation of Jordanus' proof of the general lever principle, given here in *E.8*, depends on the meaning to be given to *E.1*. This is clearly seen by the 14th-century commentator whose text was edited by Petrus Apianus, and whose discussion of *E.1* we have printed and translated in our Appendix III. He points out that Jordanus invoked the second part of *E.1* in his proof of *E.8*, and states that *E.1* would only be of value for the proof of *E.8* on condition of being interpreted as a statement that two unequal weights, so placed on a balance that their vertical descents for equal degrees of rotation of the balance on its axis are unequal in inverse proportion, have equal power to sustain or to lift a weight on the other arm of the balance. This interpretation would make *E.1* an explicit statement of the principle of work on which *E.8* depends.

"The whole problem of 'explicit interpretation,' however, is undecidable in an historical sense; for the differentiations which we make among alternative possibilities, through the distinctions developed in Newtonian mechanics, were certainly not made in the thirteenth century. Hence, though we must concede that many of the Newtonian principles were included 'virtually' in the general conception of *potentia motiva* as a function of time, distance, and resistance, the different relationships among these factors such as are expressed by the set of sharply differentiated formulas of Newtonian mechanics, were not distinguished and separated in the thirteenth century, but were included in the general conception in undifferentiated manner. Nevertheless, Jordanus' use of his general dynamic principles, for demonstrating the special theorem of unequal weights on unequal lever arms, contributed to the clarification of the dynamical analysis by determining one explicit sense for the concept of work, as involved in the general lever principle."

13. [I have added the material between the semicolons, which was left untranslated.—*Ed.*]

To make this more evident, let us draw a different figure, as follows [Fig. 7]. Let there be a vertical

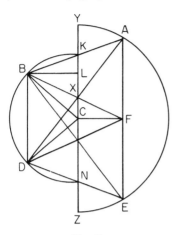

Fig. 7

line YKCNZ, and around the center C let there be drawn two semicircles, YAEZ and KBDN; and let the lines AFE and BD be drawn at equal distances from the diameter, and from these let there be drawn the equal perpendiculars BL and CF. Then draw the lines CB, CA, CE, and CD. If we then suppose that equal weights are suspended at A, B, D, E, and F, they will be of equal positional gravity. For if the lines BA, BXF, BE, DA, DF, and DE are drawn, all of them will be bisected by the diameter—as for instance BXF. For since BL and CF are equal, and the triangles BLX and CFX are similar, BX will be equal to XF. And in the same manner the others will be divided at their mid points. Therefore, since the mid points of all the lines are placed at a common center, so likewise the weights are placed; therefore they will be of equal heaviness.

A more subtle variant may, however, be added, if we suppose that *a* is heavier than *b, b* heavier than *f, f* heavier than *d*, and *d* heavier than *e*. Yet *d* is not able to lift *e;* for the segment of the line DE on the side of E would immediately become greater. But if *a* is given an impulse downward, it is able to raise *b;* and similarly *b* can raise *a;* and *d* can raise *a* and *a* can raise *d;* and *b* can raise *f* and *f* can raise *b;* until they make a complete revolution, and hang in such manner that the angle with the axis is beneath them. For when *b* is moved downward, the segment of the line BA, on the side of B, will become steadily longer, and *b* will become heavier. [14]

R1.09. Equality of the declination conserves the identity of the weight.

Only on a rectilinear path is equality of declination conserved. Let this path be on the line AB, and let the line AC descend vertically [Fig. 8].

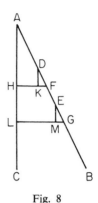

Fig. 8

And let two points, D and E, be assigned on AB. Any weight you please, then, whether it descends from D, or from E, will retain the same heaviness. For equal segments of AB, taken beneath D and E, will have equal components of the vertical. This is shown, if we draw perpendiculars to the line AC, from these points, namely FH and GL, and if we let the lines DK and EM fall perpendicularly on them. Thus, whether a weight is extended along AB, or placed there all at once, it will be of the same heaviness. [15]

14. Theorem *R1.08:* There is no antecedent in the *Elementa Jordani* for this theorem on the bent lever, except that the two erroneous theorems *E.6* and *E.7* involved the problem of equilibrium in a bent lever. The proof given here is rather intricate in its geometrical development, as can be seen by a glance at the diagram. But the proof is valid, and the principle which it invokes is clear enough—namely, that the reason why a balance of bent arms will be in stable equilibrium where the weights are equidistant from the vertical passing through the axis of support, is because in any displacement from this position a weight would be raised some vertical distance, by an equal weight descending less than that distance. Thus the principle of work underlies this demonstration, just as it does in the proofs of the general lever principle *(R1.06)* and of the inclined plane theorem *(R1.10)*.

15. Theorems *R1.09* and *R1.10:* These two theorems, not found in the *Elementa Jordani* or in any of the antecedent literature, Greek or mediaeval, present the first correct statement and proof of the condition of equilibrium of unequal weights on diversely inclined planes, in the history of mechanics. The demonstration is clear and elegant, and requires no comment in explication of the reasoning. The basis of the proof is indicated, in the last two lines, as the principle of work—namely, a force which would suffice to lift the weight *g* through the vertical height ZN, could lift the weight *h* through the height XM, where *g : h = XM : ZN*, or, equivalently,

R1.10. If two weights descend along diversely inclined planes, then, if the inclinations are directly proportional to the weights, they will be of equal force in descending.

Let there be a line ABC parallel to the horizon, and let BD be erected vertically on it [Fig. 9] and

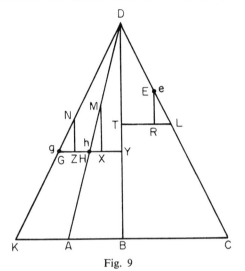

Fig. 9

from D draw the lines DA and DC, with DC of greater obliquity. I then mean by proportion of inclinations, not the ratio of the angles, but of the lines taken to where a horizontal line cuts off an equal segment of the vertical. Let the weight *e*, then, be on DC, and the weight *h* on DA; and let *e* be to *h* as DC is to DA. I say that those weights are of the same force in this position. For let DK be a line of the same obliquity as DC, and let there be on it a weight *g*, equal to *e*. If then it is possible, suppose that *e* descends to L, and draws *h* up to M. And let GN be equal to HM, which in turn is equal to EL. Then let a perpendicular on DB be drawn from G to H, which will be GHY; and another from L, which will be TL. Then, on GHY, erect the perpendiculars NZ and MX; and on LT, erect the perpendicular ER. Since then the proportion of NZ to NG is as that of DY to DG, and hence as that of DB to DK, and since likewise MX is to MH as DB is to DA, MX will be to NZ as DK is to DA—that is, as the weight *g* is to the weight *h*. But because *e* does not suffice to lift *g* to N, it will not suffice to lift *h* to M. Therefore they will remain as they are.[15]

Here ends the first part.

PART 2

R2.01. When there is a balance beam of uniform weight and thickness throughout, and its weight is known, if it is divided into unequal segments and if a body of known weight, suspended from the shorter arm, holds the beam in equilibrium, then the lengths of the arms on each side of the axis of rotation will also be determined.

Let the beam be ABC, of a given weight and of uniform thickness [Fig. 10]. Let a body, *d*, of

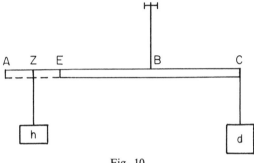

Fig. 10

known weight, hang from the end C, and let BE be equal to BC. From the mid point of AE, designated as Z, let there be suspended a body, *h*, equal in weight to the segment of the beam AE;

where the products of the weights by the heights through which they are raised, are equal. In this way the theorem shows that the force required to sustain a body on an inclined plane (i.e., the force directed upward along the plane) is measured by the weight of the body sustained, multiplied by the ratio of the altitude to the length of incline corresponding to it. And this is of course our modern formula.

What is essential, and apparently original, in Jordanus' whole treatment of the statical problems, is his recognition that it is the *vertical* descent or ascent of a weight, in relation to the length of oblique trajectory, which is to be taken into account. It is this which differentiates Jordanus' definition of "positional gravity" from that offered by the author of the so-called "Peripatetic commentary" (our version "P"), who seems to associate positional gravity with degree of curvature of the trajectory, or with the ratio of vertical to *horizontal* displacement. In the case of the curved trajectory of a weight on a balance arm, Jordanus sees that it is necessary to determine the positional gravity at each point of the curve by the ratio of the tangent to its projection on the vertical; since this is, in effect, a reduction of the more difficult problem of the curved trajectory, to the simpler case of the inclined plane, it seems likely that the inclined plane formulation was present to Jordanus' thought from the start even though it was only in his later work that he introduced a separate theorem in this formulation.

For the history of the attempts to solve the inclined plane problem, from Pappus to Descartes, see Pierre Duhem, *Les Origines de la Statique* [2 vols.; Paris: Hermann, 1905–1906], I, 27–29, 49–51, 182–193, 257–258, 272–279, 303–311, 315–316, and 329–332.

and in this position it will also be of equal heaviness. Since therefore h and d are equally heavy in this position, the proportion of d to h will be that of ZB to BC. And by alternation, the proportion of d to ZB will be that of AE (i.e., of h) to BC. And by composition, the proportion of d plus twice ZB (i.e., AC) to ZB, will be that of AE plus twice BC (i.e., EC) to BC. If therefore the whole weight ABC is multiplied by its half, and the product is divided by the sum of the weights of d and of AC—all these being given—, the weight of the segment BC is thereby determined.[16]

PART 3

R3.01. If the axis is on a perpendicular above the beam of the balance, then, however great a weight be suspended from either of the arms, it will not be possible for it to descend to a position directly below the axis.

Let the balance beam, for example, be ABC [Fig. 11], the perpendicular BDE, with the axis at

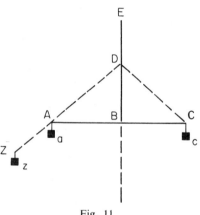

Fig. 11

D, and a greater weight a (at A) than the weight c (at C). Then draw the lines DA and DC, and let DA be extended to Z in such manner that DAZ is to DA as the weight a is to the weight c. Let there be a weight z (at Z) equal to c. When, therefore, the three weights a, c, and z are in this manner attached to the beam ABC, their revolution is around the axis D, just as if they were hanging on the lines DAZ and DC. But as so placed, z will tend to be at the same distance from the vertical passing through D, as c will; and a likewise will be at a proportional distance from the vertical. Therefore a will not be able to descend as far as the vertical.[17]

R3.02. When the ratio is given, of the distance from the axis to the mid point of the beam, to the

length of the beam, and if the ratio of the suspended weights to the weight of the beam is given, then the angle of declination of the perpendicular to the beam, from the vertical, will be determined.

16. Theorem *R2.01*: As the diagram indicates, the argument supposes that the segment AE of the material balance beam is replaced by a weightless line from whose midpoint at Z there is hung a weight h equal to the weight of the material segment AE. In this way the problem of the Roman balance is reduced to that of the ideal balance to which the lever principle, as developed in *R1.06* [above], applies directly. The problem of this theorem is to determine the ratio of the shorter arm BC to the whole beam AC, when the weight of AC and the weight of d are given. Obviously, in dealing with the symmetrical balance beam of uniform material, divisions in its length can be treated as equivalent to divisons of its weight. We suppose that B is the point of division at which the axis of rotation is placed, such that the beam will be in equilibrium when d is suspended from the shorter arm BC. Then we mark off a point E, such that EB = BC. Since EB and BC, considered alone, are in equilibrium (since each part of BC has a corresponding equal part of EB at an equal distance from the axis), it follows that the weight of AE must be what balances the weight d. We then suppose that the weight of AE is replaced by a weight, h, equal to the weight of AE, but concentrated at the mid point of the ideal line AE. The proof then proceeds as follows:

$d : h =$ ZB : BC (by theorem *R1.06*).

Therefore, $d :$ ZB $= h :$ BC (Euclid V, 12: "alternation").

But $h =$ AE (in weight), by hypothesis; therefore $d :$ ZB $=$ AE : BC.

Dividing both sides by 2, $d : 2 \cdot$ ZB $=$ AE $: 2 \cdot$ BC; but $2 \cdot$ ZB$=$AC, and $2 \cdot$ BC $=$ EC; therefore $d :$ AC $=$ AE : EC.

But $d +$ AC : AC $=$ AE $+$ EC: EC (Euclid V, 13; "composition").

Therefore $d +$ AC : (AC/2) $=$ AE $+$ EC : (EC/2); but EC/2 $=$ BC; and AE $+$ EC $=$ AC.

Therefore d $+$ AC : (AC/2) $=$ AC : BC; therefore BC $=$ (AC $\cdot \frac{1}{2}$ AC)/(d $+$ AC).

Since AC and d are known, the value of BC is determined by this equation.

17. Theorem *R3.01*: This demonstration clearly presupposes the principle established in *R1.08* [above] of the first book—namely, that equal weights, suspended from unequal lever arms which form an angle at the axis of support, will be in equilibrium when the extremities of the two arms are equidistant from the vertical passing through the axis of support. Even more than *R1.08*, this theorem represents a correction of the error involved in *E.6* and *E.7* of the *Elementa Jordani*.

The balance beam is initially conceived as a straight beam which is attached rigidly to a rigid perpendicular rod extending upwards, this perpendicular being in turn supported at its upper end by a cord which permits it to swing freely from an axis at D. By drawing the lines DAZ and DC, Jordanus reduces the problem to that of the bent lever of unequal arms supporting equal weights, and invokes *R1.08* to show that a weight at Z, equal to the weight at C, would determine equilibrium

Let the rod which stands vertically be HDLZ, and the rest as before [Fig. 12]. And let the beam be

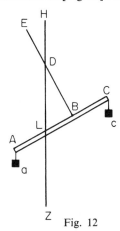

Fig. 12

depressed on the side of A until the line HDLZ cuts it at L. With the beam in this position, it is as if its axis were at L. Since then the suspended weights, and the weight of the balance beam, are given, the segments of the beam, AL and LC, are given. Hence the ratio of the length of each segment, to BD, will be given; and also the ratio of LB to BD. Consequently the angle LDB is determined, and it is the same as the angle EDH. But this is the angle of declination of the perpendicular to the beam, from the vertical.[18]

R3.03. If however the axis is beneath the beam, the weights can scarcely be stabilized in this position.

Let the beam, as before, be ABC, and the perpendicular to it DBE [Fig. 13]. And let E be the

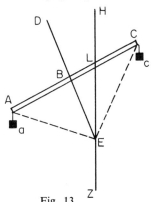

Fig. 13

axis beneath the balance, and let the weights be *a* and *c*. If we draw the lines EA and EC, the weights are so placed as if they were suspended on these lines. If they are in equilibrium in this position, then if any little tilt occurs in either direction,

as for instance on the side of A, the segment of the beam from A to the vertical HLZ will increase, so that the beam will intersect the vertical at L. And this segment will thus become continuously heavier, until the beam rotates to a position beneath the axis E.[19]

in such manner that the horizontal distance of Z from the vertical passing through D, would be the same as the horizontal distance of C from that vertical. Since, by construction, the ratio of DZ to DA is the same as the ratio of the weight *a* to the weight *c*, it follows that the distance from the vertical at which *a* will reach equilibrium will bear the same ratio to the distance from the vertical at which *z* will reach equilibrium, as the ratio of the weight *c* to the weight *a*. Hence, as long as there is some finite weight, however small, on the opposite side, the heavier weight, no matter how great, can never descend to a position vertically below the axis of support. And this is what was to be proved.

18. Theorem *R3.02:* This theorem, and the next, strongly reflect the passage in Aristotle's *Mechanical Problems* (850 a 2–29) dealing with stable and unstable equilibrium. When the axis of rotation is above the center of the straight balance beam, the equilibrium will be stable; if below, it will be unstable. The Aristotelian text involves some obscurities, leading Duhem to state that Jordanus corrected some implicit errors involved in the discussion of the *Mechanical Problems*. In any case, these two theorems constitute a correction by Jordanus of the obvious fallacy involved in the second part of *R1.02* and *E.2,* which had misapplied the theory of positional gravity in an attempt to show that equal weights on an equal armed balance beam (of no thickness) would be in stable equilibrium.

The argument of *R3.02* is as follows: Given the ratio DB : AC, and the ratio of the weights *a* and *c* to the total weight of the beam AC, it follows that since the balance is in equilibrium, *a* : *c* = CL : AL (by *R1.06,* the general lever principle). But since the ratio of DB to AC (which is the same as AL + CL) is given, the ratios of DB to AL, and of DB to CL, are thereby given. But LB = (AC/2) − AL; and consequently the ratio of LB to DB is known. But this ratio LB : DB measures the angle LDB; and this angle is equal to the angle EDH, since they are vertical angles. Therefore the ratio LB : DB measures the declination of the perpendicular ED from the vertical line HZ.

19. Theorem *R3.03:* The proof here used is very similar to that given in the *Mechanical Problems* ascribed to Aristotle (850 a 20–29); i.e., when the material beam is supported from below, any movement of the beam involves an increase in the quantity of beam on one side of the vertical, and hence increases the weight on that side and decreases the weight on the other. Jordanus however uses the method of reduction to the case of weights supported by bent lever arms, considering the weight of the straight balance beam as if it were attached to the ends of the bent arms in such manner that the part of the beam on the left side of the vertical through the axis of support is supported by the left-hand bent lever arm, and the remainder by the right-hand arm.

R3.04. If the beam is in horizontal equilibrium, it is possible to suspend, from either of the arms, as great a weight as you please, in such a way that the balance beam will not depart from its horizontal position.

Let ABC be the beam, B its axis, and DBE a line designating the vertical [Fig. 14]. And suppose

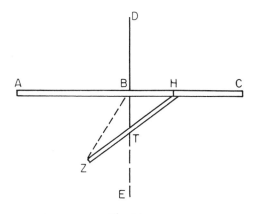

Fig. 14

the beam to be in horizontal equilibrium, by its own weight. Then take some other bar of equal thickness and weight, which we will call HTZ, T being its mid point. And take a segment of the beam, BH, on whichever arm is desired, such that it is of less length than HT. And attach this bar rigidly at H, in such manner that T is vertically beneath B, cutting the line of the vertical at T. I say then that this bar, so suspended, will not cause the beam to change position. For it is placed as if the line BZ were drawn, and as if it were hung on the line BH, in such manner that all its parts equidistant from T would be of equal heaviness. For they are at equal distance from the line of the vertical; hence ZT balances TH, and thus AB plus ZT is as heavy as BC plus TH. Therefore no movement will occur. But beyond this, if any weight desired be suspended from T, it will not cause any inclination.[20]

R3.05. If a heavy bar is raised from a horizontal to a vertical position, it is possible, given the height of the man raising it, to determine how heavily it will weigh on him in each position.

Let the heavy bar be AB [Fig. 15] and let its weight be uniformly distributed throughout its length. Then, the end B remaining in a fixed place, let it be raised at the end A until it reaches the vertical position CB, by its movement describing a

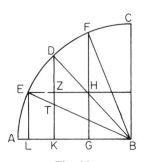

Fig. 15

quarter circle from A to C. And let the horizontal position be the first position, and the vertical position the last. And when the arc AC is divided into equal segments, let this position be called BD, or the middle position. And when the bar is raised to the height of the man lifting it, let this be BE, with the perpendicular EL representing the height of the man lifting it; and let this be the second position. Let the third position be at BF, with the arc FD equal to the arc DE.

I then say that the bar will become continuously lighter from A to E, and then continuously heavier as far as D; then however it will become lighter as far as F, and at F it will be of the same heaviness as at E; and then it will again become continuously lighter as far as C. Yet it is possible for it to be lighter at A than at D, or heavier, or equally heavy, depending on the quantity EL. For let GH, equal to EL, be erected vertically so that it touches DB

20. Theorem *R3.04:* Pierre Duhem, in his *Études sur Léonard de Vinci* (3 vols.; Paris: Hermann, 1906–1913), I, 306–307, discusses this theorem and compares it with a similar one stated by Leonardo da Vinci. It is an application of the basic principle of statical moment as measured by equidistance of weights from the vertical passing through the axis of support, to the somewhat tricky case in which the weight of the bar HTZ, though actually divided evenly on each side of the vertical through the axis, seems to be applied only to one arm because of its rigid attachment to the arm BC at the point H. Children, today, enjoy performing this stunt; e.g., they fix a short stick and a much longer one onto a cork, to form an acute angle, and then show that if the free end of the short stick is set on the very edge of a table, with the other end projecting horizontally out into the air, it will not fall off. Jordanus here seems to take a certain childish delight in the apparently paradoxical character of the theorem; as in his final statement, "But beyond this, if any weight you please is suspended from T, it will not cause any inclination one way or the other." This "experiment" is described in a mediaeval work, the *Secretum philosophorum*, which perhaps comes from an Arabic source; cf. L. Thorndike, *History of Magic and Experimental Science*, Vol. II, N. Y. 1929, p. 789.

at H; and let DK fall perpendicularly on AB. If then G should be at the mid point of AB, in which case GH would be equal to half of AB, then, since the weight DB, at D, is to the weight AB, as the line BK is to the line BA, and since its weight at D is to its weight at H, as BG is to BK, and since BG is to BK as BK is to BA, the weight DB, at H, will be as the weight of AB. But if G is nearer to B, the weight at H will be greater than at A; but if nearer to A, less.

Again, let MN be a perpendicular equal to EL, touching BF at N [Fig. 16]. And draw the line

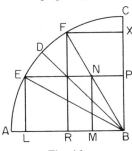

Fig. 16

ENP, and drop the perpendiculars FR and FX to AB and CB respectively. Since therefore the ratio of the weight EB to the weight FB, is as the ratio of LB to RB, or of XB to PB, and since the weight of FB at N is to its weight at F, as FB is to NB, or as RB is to MB, and since XB is to PB as RB is to MB, the weight EB is to the weight FB as the weight of FB at N is to its weight at F. The weight of EB at E is therefore as great as that of FB at N. But that it is lighter at E than at H, is proved, because DH is longer than ET. For DZ is greater than EZ, and the angle BEZ is less than the angle DHZ.[21]

R3.06. A weight not suspended at the middle, makes the shorter part heavier, according to the ratio of the longer part to the shorter part.

Let ABC be that on which the weight *e* is hung [Fig. 17]. Then let the weight *e* be divided into two

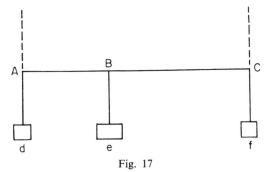

Fig. 17

weights, *d* and *f*, such that *d* is to *f* as the length AB is to the length BC. If then the weight *d* is hung at C, and *f* at A, each of them will be of as much heaviness as the weight *e*, if we imagine in each case that the opposite end is the fulcrum. Hence for those who, at A and at C, are carrying the weight *e* suspended at B, the heaviness at A will be to the heaviness at C, as the length CB is to the length BA.[22]

21. Theorem *R3.05*: [Moody notes that the last two sentences involve reference to Fig. 15—*Ed.*] The kind of experience that this theorem seeks to analyze and explain is illustrated when a man attempts to raise a long ladder from the ground. He first lifts it up at one end, to the extent of his own height, and then he walks toward the end which remains on the ground until he has the ladder standing vertically on that end. A marginal note on our manuscript H gives just such an illustration [the Latin text is omitted here—*Ed.*]:

(Translation): In this fifth theorem the author supposes that the stature of the man who does the lifting, is of the quantity EL; so that when this man commences to raise the weight AB at its end A, he will be bent down; and when he becomes erect he is as the line EL and then, holding up the said weight, he will walk toward B; and when he stands at B, then the said weight will be vertically on end, as the line CB.

There is some obscurity in the text of this theorem, as given in the manuscripts; this led Duhem to conjecture that Jordanus' original theorem had been mutilated by some careless copyist—though he cites this theorem (*Les Origines de la Statique* I, p. 143) as evidence of the author's clear grasp of the principle of statical moment. Presumably the author is concerned to exhibit the way in which the positional gravity of the weight of AB varies in function of two different conditions: (1) with the movement of the supporting bar AB from the horizontal to the vertical position, which steadily decreases its positional gravity or statical moment relative to the axis at B; and (2) with the shift of the point of application of the support EL as it moves from A toward B. What Jordanus wishes to determine is the relative "heaviness" bearing on the support EL as it shifts its point of application to the bar AB. As he formulates the problem, in abstraction from accidental considerations (such as torque) which would introduce considerable complexity into the actual "experiment," the analysis appears to be substantially sound, and the proof offered on lines 116–131 [that is, the second part of the proof associated with Fig. 16—*Ed.*] exhibits an ingenious application of the principle of statical moment to the problem.

22. Theorem *R3.06*: The source of this theorem may have been a passage in the *Mechanical Problems* attributed to Aristotle, where we find (857 b 9) the following:

"Why is it that when two men carry a weight between them on a plank or something of the kind, they do not feel the pressure equally, unless the weight is midway between them, but the nearer carrier feels it more? Surely it is because in these circumstances the plank becomes a lever, the weight the fulcrum, and the nearer of the two carrying the weight is the object moved, and

PART 4

R4.01. Every medium resists what is moved in it.
Let *ab* be that which is moved, and let the medium obstructing it be called *c* [Fig. 18]. And

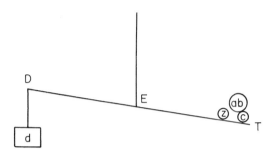

Fig. 18

suppose *c* to be as if on a balance, which we will
call TED. Now if *c* is assumed to be of no gravity,
then, if it does not resist the movement of *ab*
descending, when it is pushed by *ab* it will be forced
to descend; and thus it will be just like something
having weight. Therefore it will be able, as it
descends on the side of T, to raise some weight on
the side D; it will thus turn out that *ab*, in its descent, will be able to raise *d* just as easily as it is
able to push *c* down. Hence what lifts *d* will not be
resisted by *d*'s velocity, which is impossible.

But if *c* should be heavy, then, if it is not in
motion, whatever is impeding it will have to impede
ab to some extent; but if *c* is in motion, then, since
ab follows after it, *ab* will be heavier to the extent
that it is moving faster. Now let *z* be equal in
weight to *ab*. It is possible, then, for *z*, placed on the
side T of the balance, to descend with the movement of *c*, and to raise some weight on the side D.
And then *z* will become just like *c* in weight. If
therefore *ab* encounters no resistance in pushing
c down, it will encounter no resistance in pushing
z down at the same time. Therefore, when *ab* and
z are moved with their natural motion, they will
encounter no resistance in raising *d*, all of which is
impossible.[23]

---

the other carrier is the mover of the weight." (Loeb
Library ed., Aristotle, *Minor Works* I, trans. by W.S.
Hett; London 1936, p. 403.)

This same problem was discussed by Vitruvius, *De
architectura* X, Ch. 3; and by Hero of Alexandria in his
Mechanics. These earlier treatments all contain an erroneous application of the lever principle, just as Aristotle's explanation does. Jordanus, however, corrects the
error involved in treating the weight as the fulcrum, by
treating the opposite end as a fulcrum relative to the man

whose share of the burden is being considered. On this
basis he argues that the positional gravity of *d* at C, is to
that of *e* at B, as that of *f* at A is to that of *e* at B. Duhem
(*Études sur Léonard de Vinci*, I, 299) remarks: "Ce
court passage mérite grandement, par lui-même,
d'attirer l'attention de l'historien des sciences; pour la
première fois, en effet, depuis que les hommes s'occupen
de Mécanique une force de liaison s'y trouve déterminée
par une méthode exacte."

23. Theorem *R4.01*: The most distinctive principle of
Aristotelian dynamics, that motion and velocity essentially involve a finite proportion between the power
by which the body moves through the medium and the
power by which the medium resists this movement, is
involved in this theorem. The proof has special interest
because of its attempt to reduce the case of free fall in a
resistant medium to the case of one weight raising another on the opposite arm of a balance.

The argument, which is somewhat obscure, may be
construed as follows: We suppose the opposite of what
is asserted by the theorem—i.e., we suppose that the
medium offers no resistance to the weight falling through
it. The medium itself, then, will either be of some weight,
or it will be weightless; and if it is of some weight, it
will either be at rest or in motion. We first suppose it to
be weightless, though nevertheless capable of being set
in motion by the heavy body *ab* as it descends in the
medium. The author then argues: If this weightless body
is set in motion, it will be "as if having weight," since a
body in motion has a capacity to do work by reason of
its state of movement. This is supported by supposing
that the weightless body *c* is on one arm of a balance, on
whose other arm is a weight *d*; clearly, if *c* descends, *d*
will be raised, and since *d* has weight, work will be done
in raising it. But if *c*'s descent involves no expenditure of
force on the part of *ab*—as would be the case if the
medium *c* offered no resistance to the fall of *ab* in it—,
it would then occur that the weight *d* would be lifted
without any force being expended in doing this work.
But this is impossible, and in violation of the principle
of work, which is fundamental to Jordanus' whole
treatment of weights.

The second case is where the medium is assumed to
have some weight of its own. Then, if it is assumed to be
at rest when *ab* descends on it, this could only be so if
there is something supporting *c* and preventing it from
falling by its own weight. But in that case, whatever is
resisting *c*'s descent, will resist *ab*'s descent also.

If however the medium is assumed to be in motion (by
reason of *c*'s weight and the absence of any impediment
to its fall), then *ab*, if not held back by any resistance
from *c*, will be accelerated and overtake *c*; and its effective force will increase with its velocity *(erit AB gravius
quo velocius)*. This will then reduce to the same situation
as before—namely, *c* would be able to lift a counterweight *d* in virtue of the velocity given to it by *ab* pushing
against it, so that if *c* offered no resistance to this acceleration by *ab*, work would be done on *d* without the
expenditure of any force.

The last part of the argument, which is rather obscure,
is apparently designed to reinforce the above conclusion.
The author supposes a weight *z*, whose scalar value (at
rest) is the same as that of *ab*; and he then supposes that
z descends with the speed that *c* was assumed to have

R4.02. The heavier the medium through which a body passes, the more difficult is its descent in passing through it.

The kinds of medium through which such passage occurs, are air and water and other fluids. Let the heavier medium, then, be ABC, and the lighter one DEF [Fig. 19], and let the body traversing it

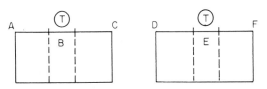

Fig. 19

be T, which, passing through it, comes up against B and E. B, however, is heavier than E; and since these media are also hindered from descending, because when they have to descend they are at rest, that which hinders B is of greater weight than that which hinders E. And because T has to encounter the same impediment, it is hindered more when it is in B. Similarly, if from beneath B and E it is propelled upwards in equal manner, its movement in B will be slower.[24]

R4.06. The longer a heavy body falls, the faster it becomes in descending.

In air, indeed, more, and in water less. For air is related to all movements. When therefore a heavy body is falling, in its first movement it will draw along those parts of the medium which are behind it, and it will move the parts just beneath it; and these, set in motion, move the parts next to them, so that these in turn, being set in motion, offer less resistance to the gravity of the falling body. Hence it becomes heavier, and pushes the receding parts of the medium still more, so that they presently cease to be pushed, and even pull. And so it comes about that the gravity of the falling body is aided by the traction of those parts of the medium, and their movement in turn is aided by the body's gravity. Hence its velocity also is observed to be continuously multiplied.[25]

R4.14. The greater the impulsion given to a body, the more its parts cohere.

This impulsion is produced from behind; and the posterior parts, having been pushed, have to push the front parts ahead; and these, which resist to some extent because of their weight, have to compress the middle parts. Hence these are sometimes driven out at the side. And in this way it also happens that when the lower parts of an object are

attached to the upper ones by being stuck into them, if a downward impulsion is given to the

---

when *ab* was falling against *c* and *c* was merely moving along with it. It is then argued that *z*, falling at *c*'s speed on the side T of the balance, could raise some weight *d* on the other side. But, says Jordanus, if *ab* is not impeded in moving *c* at that speed, it will not be impeded in moving *z* along with it; hence we would have to suppose that when *ab* and *z* are moved by their natural weight, they lift *d* without being resisted, or without expending any force. So again, from the assumption opposed to the theorem, we derive the impossible consequence of work being done without force being expended.

24. Theorem *R4.02*: This theorem expresses the Aristotelian principle that a given heavy body falls more slowly in proportion to the relative "thickness" or density of the medium. But it is interesting to note that Jordanus here assumes that if a body of some mass, such as air or water, is at rest in any place outside the center of the world, this is because some other body is exerting force in holding it up—as if "natural rest" were similar to the "violent rest" of a weight on a balance. He then argues that since whatever is keeping the heavier medium from descending must be of more gravity *(pluris gravitatis)* than what is holding up the lighter medium, the body falling in the heavier medium is as if raising a greater counterweight on a balance. So again we have the effort to reduce the analysis of free fall through a resistant medium, to the case of weights on opposite arms of a balance.

The author seems to make the assumption that all bodies are *per se* heavy, and that "lightness" is to be reduced to an upward pressure caused mechanically by the downward force of other heavy bodies, or of a medium heavier than the body in question.

25. Theorem *R4.06*: This rather fantastic explanation of the natural acceleration of falling bodies had considerable vogue during the last decades of the thirteenth century and at the beginning of the fourteenth century. It represents an intermediate stage between the teleological explanation in terms of increasing "eagerness" due to proximity of the body to its natural place, and the later fourteenth century explanation in terms of cumulative retention of *impetus* or of acquired velocity. A mechanical explanation was being sought, but at first it was sought in a perturbation of the corporeal medium through which the body falls, on the basic assumption that if the medium were at rest it would offer a uniform resistance throughout the fall, and thereby would determine a uniform speed of fall, since the natural gravity of the falling body is a constant.

For the history of attempts to explain the acceleration of falling bodies, from antiquity up to the fourteenth century, cf. P. Duhem, *Études sur Léonard de Vinci*, III, pp. 57–96. As Duhem pointed out in *Les Origines de la Statique*, I, pp. 138–9, the type of explanation offered by Jordanus of Nemore had many defenders in the 16th century. Leonardo da Vinci adopted it, as did Cardanus. It was defended by Card. Contarini in his *De elementis* (published 1548), by Benito Pereira in 1576, and as late as 1640 by Gassendi.

upper parts, the lower ones are driven into them more deeply.[26]

R4.15. If a body, having parts which cohere, is directly obstructed in its motion, it will recoil directly.

This occurs both because of the medium in which the body is carried, whether it be air or water, and also by reason of the rarity of its own parts. Let the medium in which it is moved be B, and the moved body A; and let that against which it strikes be C [Fig. 20]. Since then A moves B,

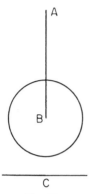

Fig. 20

when A departs from its place and expels B from its place, B has to reverse its motion in order to fill up the places at the rear. Hence it is both pushed ahead, and turned back to the rear, by the same impulsion. But it is carried back with all the more weight, when A strikes C, since it is then unable to proceed further because it is stopped by the mass which is in front of it. And since the impulsive force of A has been broken up on C, and A is now inclined only by its own weight, it must be carried back by B's movement unless its own weight prevails against this; and it will recoil in a direct path, since B recedes equally in every direction.

The rarity of its own parts, however, brings about the same result. For the front parts of A, when they first encounter the obstruction C, are pressed upon by the mass and impulsion of the rear parts, and are compressed on themselves; then, when the impulsion is expended, they revert to their places and in so doing push back the other parts. If the parts are separable, then when they are compressed into a smaller place, they will rebound.[27]

26. Theorem *R4.14:* Here we find the conception of "impulsion" or *impetus* applied to the internal elastic structure of bodies. The argument invites us to contemplate the fall of a piece of wood, into whose under

side nails have been very lightly driven. When the wood falls with great speed, or is given a sudden downward push, the inertial resistance offered by the nails will be sufficient to cause them to be driven more deeply into the wood. We have a vague suggestion here of the conception of the innate resistance of a body to motion, as being a resistance to acceleration; but it is far from explicit. Jordanus is primarily interested in the problem of communication of an impulsive force through the parts of elastic bodies; the example of the nails driving deeper into the wood when the latter moves suddenly, is designed to represent the manner in which the elastic structure of a body permits a compression of its parts generating a recoil—as in a bouncing ball.

27. Theorem *R4.15:* This theorem develops the idea of internal compression, introduced in *R4.14,* in explanation of the rebound of an elastic body. Again the notion of "impulsio" (here also called *impetus*) plays an important part, applied in a quasi-atomistic analysis of elasticity. In the works of Robert Grosseteste we find a somewhat similar use of the concept, in explanation of the vibratory motion of a stretched string in its production of sound. In his treatise *On the Liberal Arts* (L. Baur, *Die philosophische Werke des Robert Grosseteste,* Münster, 1912, pp. 2–3), Grosseteste speaks as follows:

"For when a body is struck violently, the parts which have been struck and compressed recede from their natural positions. But the natural force, tending toward the natural position, makes these parts pass beyond their proper bounds. . . .And thus they again leave their natural position, and by the same natural inclination they return and pass across that position. And in this way, in the smallest parts of the struck body there is generated a vibration, until finally the natural inclination no longer impels them beyond their proper position."

It is of interest to compare this passage from Grosseteste, as well as the present theorem of Jordanus, with another part of the same Question 12, on the eighth book of Aristotle's *Physics,* written by Jean Buridan:

"There are many things such as can be curved back on themselves or compressed by violence, which naturally revert with great velocity to their straightness or natural condition; and, in thus reverting, are able by their impetus to push or pull something else that is attached to them—as is seen in the case of the bow and arrow. Thus also a resilient ball bounced on hard ground is compressed on itself by reason of the impetus of its motion, and immediately after it strikes it reverts with great velocity to its spherical shape, pushing itself upward from the ground; and in so doing it acquires an impetus which moves it for a long distance. In this way also the chord of the zither, tightly strung, and struck hard, remains for a long time in a certain vibration, so that for a perceptible period of time its sound continues. And this happens because, by reason of the initial blow, the string is stretched violently in one direction, and it reverts so rapidly to its straightness that it passes beyond the straight position to the opposite side, and then it returns again, and thus many times."

As these quotations reveal, the fourth book of Jordanus' *De ratione ponderis* takes its place in an historically continuous line of development, leading from early thirteenth century efforts to apply mathematics to phys-

R4.16. A liquid which is continuously poured, forms a narrower stream at its lower end, to the degree that it falls further.

Let the hole through which the liquid escapes be AB, and let the first portion of the liquid be C [Fig. 21] and when C shall have descended as far

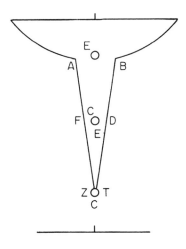

Fig. 21

as DF, let the portion E be at the opening. And then, when E has reached DF, let C be at ZT. Since therefore a body becomes heavier the more it falls, C will be heavier when at DF than when at AB; therefore it will be heavier at DF than E is at AB. But because, when E arrives at DF, C reaches ZT, the distance FZ will be longer than the distance AF; hence the stream will be more slender. And thus it becomes ever thinner, because the front parts are faster, and so, finally, it breaks up into drops.[28]

R4.17. If a body of non-uniform weight is given an impulsion in any given direction, the heavier part will come to be in the front position.

Let AB be the body which is projected, and its heavier part A [Fig. 22]. If then the impulsion is given on the side of A, B will also be impelled;

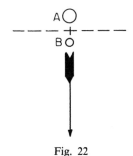

Fig. 22

and since it is lighter, it will more easily yield to the impulsion; and since A does not follow it with such ease, it will be held back in its own motion and will aid the gravity of A. And thus the whole impulsive force will pass back into A; therefore A will move out in front, and with its impetus pull B. If however the part B receives the impulsion from the rear, and A is in front, then B, having received the impulsion, will impel A; and the lightness of B will be retarded in the moving of A, and therefore A will receive more impulsion because it offers more resistance to motion. And being then impelled with the whole of the original impulsive force, it will have to pull B along.[29]

Here ends the fourth part, and with it there ends the book of Jordanus on the theory of weight.

---

ical phenomena, found in such men as Robert Grosseteste, to the developed "impetus mechanics" associated with Buridan, Albert of Saxony, and Nicole Oresme [see Selection 49—*Ed.*]. This line of development is found in the liberal arts tradition, in the disciplines of the *quadrivium;* and it was within this tradition that Jordanus de Nemore lived and worked.

28. Theorem *R4.16:* This little demonstration goes back to a passage in Simplicius' *Commentary on the Physics,* where it is said that Strato of Lampsacus, pupil of Theophrastus, had offered this "experiment" of pouring water from a spout as evidence of the fact that falling bodies are accelerated. Jordanus seeks to explain the fact through the principle that acquired velocity increases the "ponderosity" of a body falling, so that it is able to move continuously faster as it falls. Here again we have a vague anticipation of the "impetus" explanation of acceleration given later on by Buridan; for Buridan states that the effect of the natural gravity of a body, in free fall, is to continuously add new degrees of impetus to the body, thereby continuously increasing the power of the body to overcome the resistance of the medium, and consequently giving it continuously greater speed.

Since Simplicius' commentary had not been translated into Latin in the period when Jordanus wrote the *De ratione ponderis,* we must suppose that this experiment of the poured water, as associated with the problem of acceleration in free fall, had been described by some intermediary source accessible to Jordanus, which had in turn derived it from Simplicius.

29. Theorem *R4.17:* This theorem, though expressed in a most obscure fashion, applies the concept of "impetus" to the case where a body is of unequal composition—i.e., with respect to density, and is projected or thrown from the hand or from a machine. The denser portion of the missile will acquire a greater impulsion or impetus from the initial velocity given to the projectile, than will the rarer portion, as was previously indicated in *R4.10.* Consequently the denser portion will more easily overcome the resistance of the medium, than the rarer portion and it will therefore move out to the front, even if at first it was in the rear. Again we have indication of the connection between Jordanus' dy-

Motion

40 WHAT IS MOTION?

William of Ockham (ca. 1280—ca. 1349)

Translation, introduction, and annotation by John E. Murdoch[1]

The medieval natural philosopher, Aristotelian that he was, expended considerable effort and care in analyzing and expounding fundamentals. The precise nature of all the entities, of all the variables, as it were, that one is to encounter in natural philosophy must, the scholastic felt, be determined with all possible accuracy. And since no one of these variables was more central to the preoccupations of the medieval natural philosopher than motion, it was essential that an adequate understanding of its nature be a primary concern. Indeed, few issues within medieval science received more attention than the problem of the nature or, in other terms, of the definition of motion. And few problems appear to have received a greater variety of answers.

The seeds of the problem are in Aristotle himself, but were more explicitly brought to the fore in Averroes' comment on a particular passage of Book III of the *Physics* (200b.32—201a.3):

> Again, there is no such thing as motion over and above the things.[2] It is always with respect to substance or to quantity or to quality or to place that what changes changes. But it is impossible, as we assert, to find anything common to these which is neither "this" nor *quantum* nor *quale* nor any of the other predicates. Hence neither will motion and change have reference to something over and above the things mentioned, for there is nothing over and above them.

Averroes' presumed resolution of the puzzle[3] he saw in this text did more to excite and extend the discussion of the problem than it did to solve it. For between the absorption of the *Physics* (together with Averroes' commentary) and the fourteenth century one can uncover a considerable spectrum of opinion concerning the nature of motion.[4] Almost all such opinions, however, held that, whatever else might be true, at least motion was certainly something other than the body undergoing motion. It is relative to this point that Ockham's analysis of the problem struck home. For motion was not, he urged, something distinct from the mobile doing the moving. With a single application of his well-known ontological "Razor" he cut away all super-

namics and the "impetus physics" of Buridan; where he says that the heavier part A "will receive more impulsion because it offers more resistance to motion," he is giving crude expression to the truth that a body of greater mass, which is therefore a body that offers greater resistance to an accelerating force, will when set in motion at a given velocity have a greater momentum, which in turn will make it more resistant to the decelerating action of the medium. In Buridan, this insight achieves more explicit formulation though it was not reduced to exact quantitative laws until the 17th and 18th centuries. But the historical continuity of the development of modern mechanics, from the small beginnings found in Jordanus de Nemore through the fourteenth century "impetus physics" and down to Galileo and his contemporaries, is discernible at each stage. In attempting to apply the mathematical method, which he had used so successfully in the field of statics, to the difficult realm of general dynamics, Jordanus took a first step in a direction which led slowly and painfully toward the modern science of mechanics.

1. The following selections were translated from *The Tractatus de successivis attributed to William Ockham*, edited by Philotheus Boehner (St. Bonaventure, N. Y.: Franciscan Institute, 1944), pp. 32–39, 42, 43, 45–49. Although this work was admittedly not composed by Ockham himself, it may nevertheless be considered a true expression of his contentions as it is a compilation drawn directly from his indisputably genuine *Expositio in libros physicorum*.

2. Although Aristotle's τὰ πράγματα, here rendered as "things," does not mean the mobile or the subject undergoing change but rather, as Ross and others have noted, the "respects in which things may change" (which Aristotle specifies in the next sentence), such a refinement was not present in the medieval translations. There τὰ πράγματα was rendered by *res* (for example, the present sentence was rendered as: *Non est autem motus preter res*). Thus the above English translation, at least at this point, adequately reflects what the scholastic had before him. [For the Latin text of this paragraph in Aristotle's *Physics*, see *Aristotelis opera cum Averrois commentariis* (Venice, 1562–1574; Frankfurt: Minerva, 1962), Vol. IV, Book III, Text 4, fol. 86 verso, col. 2.—*Ed.*]

3. This is given in note 11.

4. For a history of some of these opinions see Anneliese Maier, *Die Vorläufer Galileis im 14. Jahrhundert*, 2d ed. (Rome: Edizioni di storia e letteratura, 1966), pp. 9–25, and *Zwischen Philosophie und Mechanik* (Rome: Edizioni di storia e letteratura, 1958), pp. 59–186.

luous entities and reduced motion to, as he puts it, permanent things. In the case of local motion, he tells us, we need only posit a moving body and its continuous, successive occupation of diverse places. Although Ockham's theory is undeniably an improvement upon that of his less perspicacious forebears, we must note, however, that he falls short of being successful in that he fails to account for the relation of the moving body being in, or occupying, the various places. But then his theory of relations itself was unsatisfactory.

Ockham's criticism of the then current views of motion is a particularly good example of a consequence of his belief that the proper subject matter of science consisted in propositions and the terms out of which they were constructed.[5] Motion as a "scientific concept" was a connotative term. To reveal the nature of such a "concept" was to reduce it—or more correctly any proposition in which it occurs—to a proposition containing only absolute terms, terms which required no other thing save what they primarily signify in order to be significant. It is, in effect, this logical reduction that Ockham occupies himself with in establishing his doctrine of the nature, or definition, of motion.

Since it is the common opinion that motion, time, and place are certain things other than the movable body and the placed thing, we must inquire of the opinion of the Philosopher and Commentator, firstly concerning motion, secondly place, and thirdly time.[6]

[THE TREATISE ON MOTION]

With respect to the first of these matters, we must note that according to the Philosopher in Book III of the Physics together with the Commentator in his fourth comment [to that Book]: "Motion does not have a definition stated univocally because it is not named univocally.[7] Rather, it has a definition according to prior and posterior. None the less, such definitions do enter into the natural sciences."[8] In the passage cited, the Commentator says that the Philosopher intends to show that motion belongs to more than one category.[9] The Commentator proves this as follows: Transmutation is only found in that which is transmuted, but a moved thing can be found under more than one category; therefore so can motion.[10] And in the same comment the Commentator brings up the puzzle as to how it can elsewhere be maintained that motion belongs to only one category (namely that of passivity), or even that [motion of itself]

forms a single category. However, it is also said in this very same comment that there are four genera of motion. The Commentator answers [this puzzle] in the passage which begins with the words: "To this however we respond. . . ." See for yourself the solution he gives in this place.[11]

5. See Ockham's discussion of the nature of scientia demonstrativa in his prologue to his Expositio super VIII libros Physicorum as translated by Philotheus Boehner in Ockham, Philosophical Writings (Edinburgh and London: Nelson, 1957), ch. 1.

6. Ockham's discussions of place and time constitute the second and third parts of the treatise at hand. Selections from these subsequent parts have not been included.

7. Averroes is referring to Aristotle's Categories, 1.1a.6–12: ". . . Things are said to be named "univocally" which have both the name and the definition answering to the name in common. A man and an ox are both "animal" and these are univocally so named, inasmuch as not only the name, but also the definition, is the same in both cases."

8. Ockham here quotes Averroes verbatim. For the passage directly preceding this quote, see note 10. [For Book III, Comment 4, see Vol. IV, fol. 87 recto of the edition cited in n. 2 above—Ed.]

9. Specifically, Averroes interprets this intent to reveal that motion belongs to more than one category as the fourth basic point Aristotle sets forth in his attempt to define motion, a task which occupies all the earlier chapters of Book III of the Physics.

10. The text from which Ockham draws this argument is the following: "Motion is not found outside the ten categories. For a thing which has undergone change is always changed either relative to the category of substance, of quantity, of quality, of place, or of some other category; yet change (transmutatio) is only found in a thing which has undergone change, all of which implies that motion exists in more than one category. The argument is constituted in the following manner: Motion exists in the moved thing; but the moved thing exists in more than one category; therefore motion exists in more than one category. 'Moreover,' [Aristotle goes on], 'nothing is found, etc.,' that is to say, it is impossible that some one thing be found which is common to those categories in which motion is found, for example something common to those four categories [mentioned above] which is no one of these four nor of the other categories" (Averroes, Physics III, comment 4 [see Vol. IV, fol. 87 recto of the edition cited in n.2.—Ed]).

11. "To this however we respond that, insofar as it does not differ, save by more or less, from the perfection toward which it is directed, motion necessarily belongs to the genus of the perfection in question. For motion is nothing other than a generation, part by part, of the perfection toward which it is directed, [a generation which continues] until the perfection is reached and actually exists. Thus it is necessary that a motion which takes place in substance is to be found in the genus of substance, and that a motion which occurs in quantity, in the genus of quantity, and, similarly with that which occurs with respect to place and quality. However,

Several things must be examined and proved concerning the grounds for his solution. In the first place one must see and know that motion is sometimes taken in a strict sense insofar as it entails succession, while at other times motion is taken in a general sense to cover all change, be it immediate *(mutatio subita)* or successive. Firstly, therefore, one must prove that immediate change is not another thing, taken by itself as a whole, distinct from the mobile or the moving body and from the terminus acquired or lost [by the change] and from other permanent things. Secondly, one must prove that no motion is some one thing in itself as a whole distinct from all permanent things, and that motion is not some one thing outside the essence of permanent things, and that no permanent thing derives from the essence of motion (just as whiteness is totally outside the essence of substance and substance is totally outside the essence of whiteness).

[THAT IMMEDIATE CHANGE IS NOT SOMETHING DISTINCT FROM PERMANENT THINGS]

Point One: That immediate change is not some thing distinct from permanent things is shown, to begin with, as follows:

When a medium is first illuminated by the sun, I inquire whether or not there might be in this medium several things, distinct from the light itself and from other permanent things, which come into existence in the medium due to the occurrence of the illumination. If not, one has attained that proposed, namely that immediate change is not something other than, and so on. If, on the other hand, there are [such distinct things in the medium], I argue against this as follows: Every single thing is either substance or quality—as has been shown elsewhere and is now assumed.[12] But it is certain that such a thing is not substance, for it is neither matter nor form nor anything composed out of them. Nor is it a quality, as can be inductively proved by running through all species of quality. Nor is it quantity, since it is certain that the quantity of the air [which is the medium in question] is in no way changed. Therefore, and so on, [there is no thing distinct from permanent things in this case]. . . .

Furthermore, it is futile to accomplish with a greater number of things what can be accomplished with fewer *(frustra fit per plura, quod potest fieri per pauciora)*.[13] But, if everything other than matter, form, an agent, and other permanent things be set aside, and if it be assumed that some matter

first does not have a [given] form and then later does have this form (but does not acquire it part by part) then, indeed, that matter is changed. Therefore, in order to account for change, it is vain to assume the existence of some thing other than matter, form, an agent, and other permanent things.

This is confirmed because in order to account for change, it is sufficient that matter have some form it did not previously have. But given the fact that matter has some form, nothing other than the form and the matter is assumed. Nor does the fact that the matter does not previously possess this form necessitate that we assume something in addition to the matter, form, an agent, and other permanent things. Consequently, above and beyond permanent things, there is no other thing which is change. . . .

[OCKHAM'S OWN VIEW]

Therefore, it must be said that immediate change is not some thing distinct from permanent things that is destroyed following the first instant in which the subject suffers immediate change. On the contrary, when matter suffers change toward the accession of a substantial form and the substantial form is introduced into the matter, no thing is in this case afterwards lost, but everything which exists in the first instant exists afterward, or at least can so exist. But for a subject to undergo immediate change involves nothing further than that the same subject have a form which it did not previously have, or that it lack a form which it previously did have. However, [this acquisition or loss of a form] does not occur part by part (that is, in such a way that the subject has one part of the form before another, or loses one part before

insofar as motion is a process *(via)* toward a perfection and is different from that perfection, it is necessary that motion belong to a genus per se. For a process toward a thing is not the same as that thing, and in view of this [motion as a process] has been set down as a category per se. This manner [of treating our problem] is better known, while that referred to above is truer. Therefore, Aristotle has introduced the better known view in the *Categories,* and the true one in this book" (Averroes, *Physics* III, Comment 4 [see Vol. IV, fol. 87 recto of the edition cited in n.2.—*Ed.*]).

12. In Ockham's view quantitative terms signified nothing distinct from substances and instances of qualities (see his *Commentaria sententiarum,* Bk. I, Distinction 2, Questions 1–8, and his *Summa logicae,* Part I, chs. 14–18).

13. This is, perhaps, the most frequent form one finds in Ockham for the expression of his so-called "Razor."

another), but the whole form is simultaneously acquired or simultaneously lost.

And yet [in objection to this view] it may be asked whether or not change is some one thing. If it is not a thing, it is nothing, and therefore nothing is changed by real change. If, however, it is a thing, and is not something other than a permanent thing—as has been proved—then change is a permanent thing, and consequently it is either form or substratum *(subjectum)*. And be it the one or the other, it follows that as long as the substratum or the form would exist, the change would exist, and so something would be changed by such a change.

One must reply [to this objection] that nouns which are derived from verbs, and even nouns derived from adverbs, conjunctions, prepositions, and also from syncategorematic terms (be they syncategorematic nouns or verbs or any other sort or of another part of speech) are only introduced for the sake of conciseness of speech or embellishment of locution. And many of these [nouns], when they are not standing for those words from which they derive, are, in their significative use, equivalent to propositions. And consequently they signify nothing else except those words which they are derived from and the things signified by those words. However, all of the following nouns are of this sort: "negation," "privation," "condition," "perseity," "contingency," "universality," "quantity" (that which is called the "quantity" of a proposition), "action," "passion," "calefaction," "frigefaction," "change," "motion," and in general all verbal nouns derived from verbs which fall under the categories of action and passion, and many other nouns which will not presently be dealt with.

For the noun "change" does not signify some single absolute thing (as is the case with the noun "man" or "ass" or the noun "whiteness"), but it is introduced (1) sometimes for the sake of embellishing locution, as it is sometimes used for the purpose of embellishment quite suitably in a place where the verb from which it is derived would not be used in an embellishing way, (2) sometimes for the sake of brevity, just as this brief sentence, "Every change results from an agent," is equivalent to this whole sentence, "Everything which is changed, is changed by some agent." And so it is with many other sentences, just as this, "Every transmutation is from something and in something and into something," is equivalent to this whole sentence, "Whenever something is transmuted, in such a case there is something which is acquired or lost by

that which is transmuted and there is some other thing which causes it to acquire or lose that something."

And since such nouns are invented for the aforesaid reasons, they are often interpreted equivocally by authors and in different places in different ways. For (1) sometimes such nouns in the writings of various authors *supposit* or stand for the very words from which they are derived, or for the concepts in one's mind corresponding [to such words]. Yet (2) sometimes they *supposit* or stand for that which is signified by one part of that complex expression to which they are equivalent; (3) at other times they stand for that signified by another part; and (4) at still other times for an aggregate of such signified things. Lastly, (5) sometimes they stand for that very thing for which the participle of such a verb as that from which it [that is, such a noun] is derived stands (for in such a case it is used in place of such a participle, together with other elements expressing the *quid nominis*[14] of the verb in question).

Example of the first case: If the following be said: "Change is truly predicated of Socrates," one sense is that the verb "is changed" is truly predicated of Socrates, as in saying "Socrates is changed." Similarly, if one says, "Perseity is suitable to this proposition: 'Man is an animal,' " one sense is " 'Man is an animal' is a proposition per se." And so on with respect to other examples.

Example of the second case: If I say, "Change is capable of change," in this case "change" stands for that signified by the term "something." Whence the definition which expresses the *quid nominis* of the term "to be changed" is the following: "Something, which now has something for the first time which it did not have previously, lacks something which it had previously." And therefore sometimes "change" stands for that which has something which it did not previously have, and so on.

Example of the third case: "Change is really the same as the terminus [involved in the change]," since this is equivalent to this: "That which is acquired or lost by something is really the same as

14. Nominal definitions *(diffinitiones quid nominis)* only tell us what is meant by a term or name. They can be given to all terms whether there is some existing thing for which the terms directly stand or not. On the other hand, a real definition *(diffinitio quid rei)* of a term can be given only if there is some existing thing for which the term primarily stands. Such definitions thus express the essential nature of the thing for which the term in question stands as well as define the term itself.

that which is possessed or is not possessed by that thing."

An example of the fourth case is evident if I say, "Change is both the changeable thing *(mobile)* and the terminus or form existing in this changeable thing," or something of this sort.

An example of the fifth case occurs when I say, "Change is that which proceeds from prior to posterior," which is equivalent to this: "When something is changed, it proceeds from prior to posterior."

Nevertheless, it must be known that the noun "change" is frequently used in place of a complex expression, as are other similar nouns, and that which follows [such a noun] is used in place of a verb. Thus, for example, in saying, "Change is the acquisition or loss of something," the noun "change" is used in place of this complex expression: "When something is changed"; and the nouns "acquisition" and "loss" are used in place of the verbs "it acquires," "it loses"; and the noun (which is in an oblique case) "of something" is used in place of the case which should follow those verbs. For example, the sentence, "Change is the loss or acquisition of something," is used in place of this temporal [sentence]: "When something is changed, it acquires or loses something," or in place of this conditional [sentence]: "If something is changed, and so on". . . .

Therefore, different propositions [involving the term "change"] are to be verified and expounded in diverse ways. Yet it is difficult to discover some one general manner of expounding such propositions, and particularly so because philosophers use the noun "change" equivocally and in different ways. Nevertheless, it must be maintained that, insofar as things are concerned, change is not some thing other than a permanent thing and that the noun "change" does not signify some thing which is not permanent or does not last through time. On the contrary, it signifies that some thing acquires or loses some whole thing simultaneously and not part before part, nor one after another. . . .[15]

[THAT MOTION IS NOT SOMETHING OTHER THAN PERMANENT THINGS]

Now that we have revealed the foregoing with respect to change *(mutatio)*, it remains to prove that no motion *(motus)* is some thing, in itself as a whole, distinct from a permanent thing. . . .

Consequently, it must be said that motion is not some thing in itself wholly distinct from permanent things. For it is vain to accomplish with many

things that which can just as well be accomplished by fewer. And we can save both motion and all things said about motion without recourse to any such [separately existing] entity. Therefore, it is superfluous to posit such a thing. The fact that we can save motion and all things said about it without any such added entity is clear if one enumerates one by one, the various parts of motion.[16]

This is evident with respect to local motion. For if it be assumed that a body is first in one place and then in another place, and that it proceeds in this manner without rest and without there being any thing other than the body and the agent which moves it, it is clear that we have local motion. Thus, it is vain to assume the existence of any additional things.

It may be objected that body and place are not sufficient to account for local motion because, if this were the case, then whenever there were body and place there would be motion and thus a body would always be in motion. To this we must reply that body and place are not sufficient to account for the existence of motion in the sense that the following consequence is not a formal one: "Body and place exist, therefore motion exists." Nonetheless, no other thing is required in addition to body and place. It is only required that a body is first in one place, later in another place, and so on continuously, in such a way that the body never rests at any place throughout the whole time [of its motion]. And it is clear that in addition to all these things, nothing is posited other than permanent things. Thus, by the fact that a body is first in *A*, nothing else except *A* [and the body] is posited; similarly, by the fact that it was not first in *B*, nothing is posited aside from *B* and the body; and similarly by the fact that the body is secondly in *B* nothing is posited other than *B* and the body. And if we proceed with respect to the other [places occupied by the body] in a similar manner, it is clearly evident that above and beyond body and the parts of space and other permanent things it is not necessary to assume the existence of any other thing. On the contrary, it is only necessary to assume that a body at one time is in this place, or in any part of space, and that another time it is not. This is what

15. The qualification that the acquisition or loss in question must be of some whole and must occur simultaneously is required, one will recall, by the fact that Ockham is presently analyzing immediate change *(mutatio subita)*.

16. By "parts of motion" Ockham means the various kinds of motion: local, alteration, and augmentation and diminution.

being moved locally amounts to: for a body first to occupy one place, and with no other thing assumed, at a later time to occupy another place without any intervening rest and without any entity other than place, the body, and other permanent things, and to proceed in this manner continuously. Consequently, beyond these permanent things we need consider nothing else, but alone must add that a body is not simultaneously in [all] these places and does not rest in any of them. By these negative statements no entity aside from permanent things is assumed to exist. Consequently, without any other thing whatsoever, the whole nature of motion can be accounted for by the fact that a body is successively in distinct places and does not rest in any.

The same thing is evident with respect to motion of alteration (that is, that one need not assume anything but permanent things). For, the sole fact that the subject acquires the parts of a form one after another and not simultaneously accounts for motion of alteration. Thus, it is not necessary to posit anything other than the subject and the parts of a form, but it is sufficient to posit only the subject and the parts of a form in such a way that these parts are not simultaneously acquired. However, the fact that these parts are not acquired simultaneously does not mean that some thing other than these parts is posited, but rather merely that some parts of a form exist at different times and not at the same time. Therefore, accordingly, the existence of no other thing is assumed, but rather something is denied, namely that not any one, but that many, parts of a form exist simultaneously.

And if, when we say that "the parts do not exist simultaneously" it be objected that this non-simultaneity of parts is something, we must answer that such an abstract noun fiction constructed out of adverbs, conjunctions, prepositions, verbs, and syncategorematic terms[17] causes many difficulties and leads many into error. For many thinkers imagine that just as there are distinct nouns there are distinct things corresponding to them, so that insofar as there is a distinction between things signified, to such an extent there is a distinction between the signifying nouns. This, however, is not true. For sometimes the same things are signified when there is a difference in the logical or grammatical manner of signifying. Therefore, non-simultaneity is not some thing in addition to those things which can exist simultaneously, but it merely signifies that [such] things do not exist simultaneously. Thus, in modern times, because of the errors arising from

the use of such abstract nouns, it would be better in philosophy, for the sake of simplicity, not to use them but only the verbs, adverbs, conjunctions, prepositions, and syncategorematic terms as they were instituted in the first place. This is preferable to the invention of such abstract nouns and their use. Indeed, were it not for the use of such abstract terms as "motion," "mutation," "mutability," "simultaneity," "succession," "rest," and the like, there would be little difficulty with respect to motion, mutation, time, instants, and so forth.

Moreover, it is clear that no other thing, above and beyond permanent things, is required in the case of motion of augmentation and diminution. For in such a case one can account for motion by the sole fact that there is, successively, more and more, or less and less, quantity. However, when one says that there is more and more quantity it is not necessary to posit anything aside from permanent things, unless someone would wish to form an abstract noun out of the conjunction "and" and thus say that and-ness or and-eity is something other than the parts of quantity. But this is absurd. Therefore, by the fact that in the case of motion of augmentation or diminution there is more and more, or less and less, quantity, one is not required to assume the existence of something in addition to permanent things. Permanent things alone suffice, in such a way that the same [permanent] thing is first of such-and-such a quantity and later is not of this quantity, and thus proceeding without end. Moreover, I have elsewhere said[18] that in this way

17. "Categorematic terms have a definite and fixed signification, as for instance the word "man" (since it signifies all men) and the word "animal" (since it signifies all animals) and the word "whiteness" (since it signifies all occurrences of whiteness). Syncategorematic terms, on the other hand, as "every," "none," "some," "whole," "besides," "only," "insofar as," and the like do not have a fixed and definite meaning, nor do they signify things distinct from the things signified by categorematic terms. Rather, just as, in the system of numbers, zero standing alone does not signify anything, but when added to another number gives it a new signification; so likewise a syncategorematic term does not signify anything, properly speaking; but when added to another term, it makes it signify something or makes it stand for some thing or things in a definite manner, or has some other function with regard to a categorematic term. Thus the syncategorematic word "every" does not signify any fixed thing, but when added to "man" it makes the term "man" stand for all men actually, or with confused distributive *suppositio*." (Ockham, *Summa logicae*, Part I, ch. 4).

18. Ockham appears to be referring to his *Commentaria sententiarum*, Book II, Question 9, which is devoted to an exhaustive analysis of the problem of "whether

motion is composed of affirmations and negations. That is, that in order that there be motion it is sufficient that (1) there be parts or permanent things which do not exist simultaneously in the sense that for the truth of [the proposition] "motion exists" certain affirmative propositions and certain negative propositions [referring to these parts or permanent things] are sufficient, and (2) that no thing other than permanent things is posited or denoted by these affirmative or negative propositions.

In the case of motion of alteration it is sufficient that there is first one thing and not another, and that later there is the other thing ánd not the first, proceeding thus without anything other than permanent things. . . .

Turning to how one speaks [about motion, we should say that] if the noun "motion" is posited in a sentence, we must assert just those things we have asserted above concerning the noun "change." That is, sometimes "motion" stands for the verb "to be moved" and for its moods and tenses; sometimes it stands for the mobile itself; at other times it stands for the terminus or thing acquired when

something is moved; at still other times it is used in place of such expressions as "that it is moved," or "when it is moved," or something of the sort. And similarly, those things which are added to the noun "motion" in a sentence, either in that part of the sentence in which this noun occurs or in the other part of the sentence, are to be expounded in different ways. Hence "motion exists in time" should be expounded as "when something is moved, it does not acquire or lose all that it acquires or loses at one and the same time, but rather part by part." And in such a way it is clear that the nouns "motion" and "time," just as other such abstract nouns, are invented for the purpose of brevity, that is, so that that which is signified by the long expression "that which is moved does not acquire or lose all that it acquires or loses simultaneously, but rather part by part" is also signified by the brief expression "motion exists in time." Similarly, the proposition "motion exists in the moved thing" should be expounded as "that which is moved acquires or loses something." The case is similar with other examples. . . .

Kinematics

41 THE REDUCTION OF CURVILINEAR VELOCITIES TO UNIFORM RECTILINEAR VELOCITIES

Gerard of Brussels (fl. thirteenth century)

Translation, introduction, and annotation by Marshall Clagett[1]

The *Liber de motu* is divided into three short "books." The first attempts to find uniform punctual rectilinear velocities which will make uniform the varying curvilinear velocities possessed by different lines (and perimeters of regular polygons) as they rotate or revolve. The second discovers similar uniform movements for surfaces in rotation, and the last seeks uniform movements for solids in rotation. Gerard's work, then, is probably the first Latin treatise to take the fundamental approach to kinematics that was to characterize modern kinematics, namely, to see in kinematics the basic objective of reducing variations in velocity to uniform velocity.

In proposition I.1 Gerard says that a segment of a rotating radius or the radius is moved "equally" as its middle point. By this he means that if we take the segment or the radius itself and allow it to

move, not in a movement of rotation, but rather in one of translation always parallel to itself so that all of its points are moving rectilinearly with the velocity which the middle point had when it was

motion is a genuine thing outside the soul really different from the mobile and from the terminus [of the motion]."

1. [Reprinted with permission of the Regents of the University of Wisconsin from M. Clagett, *The Science of Mechanics in the Middle Ages* (Madison, Wis.: University of Wisconsin Press, 1959), pp. 185–186, 187–189, 194–195. The introduction has been constructed from relevant passages on pages 185–186; the commentary has been located immediately after the translation, just as in Clagett's volume. Clagett believes that Gerard's *Book on Motion (Liber de motu)* was composed sometime between 1187 and 1260 and represents, among other things, a significant revival of the mathematical works of Euclid and Archimedes. Virtually nothing is known about the life of Gerard.—*Ed.*]

rotating, then the line segment or radius so moved traverses the same area in the same time as it did in revolution. And when Gerard says that the rotating lines move "equally" as its middle point, he means thereby that it is the velocity of the middle point which can convert or transform the varying motion of rotation into uniform motion of translation. This theorem may have stimulated by analogy the discovery at Merton College of the fundamental theorem of uniform acceleration—a theorem which held that with respect to the traversal of space in a given time a uniform acceleration is equal to a uniform motion at a velocity equal to the velocity at the middle instant of the time of acceleration [see Selection 42, Part VI *(continued)*, par. 1, and commentary—*Ed.*]. At any rate, Thomas Bradwardine, the founder of kinematic studies at Merton College, in his *Tractatus de proportionibus velocitatum* of 1328, mentions Gerard's proposition I.1 and uses the treatise as a point of departure for his chapter on kinematics, even while he disagrees with Gerard's conclusions.[2]

Finally, we should notice that, unlike the Greek authors, Gerard throughout his treatise seems to be assigning some given magnitude to velocity, a magnitude distinct from, although proportional to, the distance traversed in a given time. He thereby opens the way for a similar but more varied treatment of the proportionality of velocities in the fourteenth century.

BOOK I

[SUPPOSITIONS:]

[1.] Those which are farther from the center or immobile axis are moved more [quickly]. Those which are less far are moved less [quickly].

[2.] When a line is moved "equally," "uniformly," and "equidistantly" (i.e., uniformly parallel to itself), it is moved equally in all of its parts and points.

[3.] When the halves [of a line] are moved equally and uniformly to each other, the whole is moved equally [fast] as its half.

[4.] Of equal straight lines moved in equal periods of time, that which traverses greater space and to more [distant] termini is moved more [quickly].

[5.] [That which traverses] less [space] and to less [distant] termini is moved less [quickly].

[6.] That which does not traverse more space nor to more distant termini is not moved more [quickly].

[7.] That which does not traverse less space nor to less distant termini is not moved less [quickly].

[8.] The proportion of the movements (i.e., speeds) of points is that of the lines described in the same time.

[Proposition] I.1: Any part as large as you wish of a radius describing a circle, which part is not terminated at the center, is moved equally as its middle point. Hence the radius is moved equally as its middle point. From this it is clear that the radii and the speeds are in the same proportion.

Proceed therefore: I say that *CF* (see Fig. 1) is

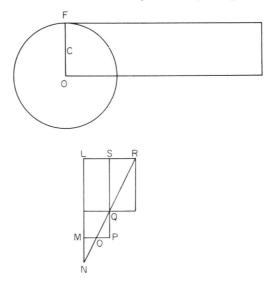

Fig. 1

moved equally as its middle point, it having been proved that the [annular] difference between [concentric] circles is equal to the product of the difference of the radii into half the sum of the circumferences.

For let lines *OF* and *RL* be equal, and let line *LN* be equal to the circumference of the circle with *OF* radius. It is evident by the first proposition of the *De quadratura circuli* [of Archimedes][3] that the circle *OF* and the triangle *RLN* are equal. Also, let lines *SL* and *CF* be equal, as well as lines *ON* and *OQ*. And let lines *SL* and *MP* be parallel. It is necessary, therefore, that triangle *RSQ* be equal to circle *OC* and line *SQ* be equal to the circumference of that circle. For, since triangles *RLN* and *RSQ* are similar, the proportion of *LR* to *SR* is as

2. [This part of Bradwardine's treatise is not included here, but a selection from the earlier chapters of that work appears in Selection 51.1.—*Ed.*]

3. [For two medieval versions of this proposition, see Selection 31.]

the proportion of *LN* to *SQ*. But the proportion of *LR* to *SR* is as that of *OF* to *OC*, and the proportion of *OF* to *OC* is as that of the circumference of *OF* circle to the circumference of *OC* circle, because the diameters and circumferences [of circles] are in the same proportion. Since, therefore, the circumference of *OF* circle is equal to the line *LN*, *SQ* will be equal to the circumference of circle *OC*. And the surface *SLNQ*, which is the difference of the triangles, will be equal to the difference of the circles *OF, OC*. Furthermore, the surface *SLNQ* is equal to the quadrangular surface *SLMP*. This is proved as follows: Triangles *OMN* and *OPQ* are equal and similar because *M* and *P* are right angles, *MP* being parallel to line *SL*, and the opposite angles at *Q* are equal. Hence angle *N* is equal to angle *Q*. Therefore, the sides are proportional. But *ON* is equal to *OQ*, so that *NQ* has been divided into two equal parts at point *O*. Hence *OM* and *OP* are equal, as are *MN* and *PQ*. Therefore, the triangles are equal. Consequently the surfaces *SLNQ* and *SLMP* are equal. But line *LN* plus *SQ* is equal to *LM* plus *SP* because *MN* and *PQ* are equal. But lines *LN* and *SQ* are equal to the circumferences of circles *OF* and *OC*. Hence the lines *LM* and *SP* together are equal to these circumferences together. Therefore, the surface *SLMP* is equal to the product of the difference of the radii by half of the sum of the circumferences, and this is equal to the difference of the circles. This same thing can be proved in another way. But the above proof suffices for the present.

Let *SL*, then, be moved through the surface *SLMP* and the line *CF* through the [annular] difference of the circles *OF* and *OC*. I say, therefore, that lines *SL* and *CF* are equally moved, for they traverse equal spaces and to equal termini, as is now clear from the [following] statements: *SL* is either equally moved as *CF* or it is moved more or less [rapidly]. It is not moved more [rapidly] because it does not describe more space to more [distant] termini [by supposition 5]. Nor is it moved less rapidly, for it does not describe less space to less [distant] termini [by supposition 6]. . . . Since, therefore, *SL* is not moved more [rapidly], and is not moved less [rapidly], it must be moved equally as *CF*.

But *SL* is moved equally as any of its points by the second supposition because it is moved equally and uniformly in all of its parts and points. Hence it is moved equally as its middle point. But the middle point of *SL* is moved equally as the middle point of *CF* because these points describe

equal lines in equal times [by supposition 7]. That moreover, these lines are equal you will prove in the same way that it was proved that line *SQ* is equal to the circumference of circle *OC*. And so therefore, *SL* is moved equally as the middle poin of *CF*. But *SL* is moved equally as *CF*, as was proved. Hence *CF* is moved equally as its middle point. The demonstration is the same for any par you wish of the radius *OF* not terminated at *O* . . . [and also for the radius itself].

COMMENTARY

I.I. The basic intent of this proposition we have already indicated.[4] But we must say something more about the lack of precision in the terminology used. I have purposely left some of the ambiguities since it would be a mistranslation to give precision where precision is lacking. In the first place, the word "motion" *(motus)* is continually used as we would use the more precise term "speed." Thus the verb "to move" is used with the meaning of "to have a speed of motion." And "to move more" means "to have a greater speed of motion."

Similarly there is a lack of precision in the use of the word "equal" as applied to movement. No only is it used to equate one speed to another, i.e. to identify the space traversed in the same time by one line or point with that traversed by another line or point. But it is also used as we use the term "uniform," namely, to describe the character of a single movement. Here it is employed to identify the space traversed by a line or point in any one part of the duration of a movement with the space traversed by a line or point in any other equal par of the duration of the movement. There is an element of both usages in the statement of the first proposition. Thus to say a segment of a radius is moved "equally" as its middle point means tha were the segment to move "uniformly" with a speed "equal" to that of its middle point when rotating so that all of its points were moving uniformly with the speed of the middle point, then it would traverse the same total area in its uniform movement as when rotating.

The proof is simple enough. He finds a rectangular area *SLMP* equal to the area which the rotating segment *CF* describes. Then he lets *SL* equal to *CF* move uniformly over *SLMP* in the same time that

4. [See the introductory remarks at the beginning of this selection. The primary objective of this proposition, and most of the others in Gerard's treatise, is to compare rectilinear and curvilinear velocities.—*Ed.*]

CF sweeps out its annular area. The areas being equal, the "movements" or speeds of the two lines, *SL* and *CF*, are by definition equal. But what is the punctual speed of all the points of *SL*? Since *SL* is moving uniformly, the speeds of all of its points are the same. Hence the speed of all its points is the speed of any one of them, and thus of the speed of its middle point. But the speed of the middle point of *SL* is the same as the speed of the middle of *CF*, since these points describe equal lines in the same time. Therefore, *SL* and *CF* being called equal in movement or speed because they describe equal spaces in equal times, *CF* is said to move equally as its middle point.

2 UNIFORM AND NONUNIFORM MOTION AND THE MERTON COLLEGE MEAN SPEED THEOREM

William Heytesbury (fl. ca. 1350)

Translated by Ernest A. Moody

Introduction and annotation by Marshall Clagett[1]

[From the works of Thomas Bradwardine, William Heytesbury, Richard Swineshead, and John Dumbleton, all of whom were associated with Merton College, Oxford, in the first half of the fourteenth century, there] emerged some very important contributions to the growth of mechanics: (1) A clear-cut distinction between *dynamics* and *kinematics,* expressed as a distinction between the *causes* of movement and the spatial-temporal *effects* of movement.[2] (2) A new approach to speed or velocity, where the idea of an instantaneous velocity came under consideration, perhaps for the first time, and with it a more precise idea of "functionality." (3) The definition of a uniformly accelerated movement as one in which equal increments of velocity are acquired in any equal periods of time. (4) The statement and proof of the fundamental kinematic theorem which equates with respect to space traversed in a given time a uniformly accelerated movement and a uniform movement where the velocity is equal to the velocity at the middle instant of the time of acceleration. It was this last theorem in a somewhat different form that Galileo states and which lies at the heart of his description of the free fall of bodies.

[Of the contributions mentioned by Clagett, the final three appear in the selection from Heytesbury reproduced here.—*Ed.*]

[Part VI. Local Motion]

[PROLOGUE]

There are three categories or generic ways in which motion, in the strict sense, can occur. For whatever is moved, is changed either in its place, or in its quantity, or in its quality. And since, in general, any successive motion whatever is fast or slow, and since no single method of determining velocity is applicable in the same sense to all three kinds of motion, it will be suitable to show how any change of this sort may be distinguished from another change of its own kind, with respect to

1. [Reprinted with permission of the Regents of the University of Wisconsin from M. Clagett, *The Science of Mechanics in the Middle Ages* (Madison, Wis.: University of Wisconsin Press, 1959), pp. 235–237, 270–277. The translation was supplied by Ernest A. Moody to M. Clagett, who made slight alterations so that Moody's translation might better conform to other translations in the same volume. Although the selection reproduced here constitutes a single consecutive portion of Heytesbury's *Rules for Solving Sophisms (Regule solvendi sophismata)*, composed in 1335 or earlier, Clagett published it as two separate selections in successive chapters of his book. Since each selection was followed immediately by Clagett's commentary, I have retained his order and arrangement, omitting only some of the references to other documents in the volume and occasionally adding a note from material in other parts of the volume. All omissions, as indicated by points of ellipsis, were made by Clagett.
The brief introduction is drawn from Clagett's remarks on page 205.—*Ed.*]
2 [This distinction is clearly made by Thomas Bradwardine in his *Treatise on the Proportions of Velocities in Movements* of 1328 (for an extract from this work, see Selection 51.1). "In the third chapter of the work, Bradwardine treats of the 'proportion of velocities in movements in relationship to the *forces (potentias)* of the movers and the things moved,' in short, to the dynamic considerations of velocity. On the other hand, the fourth chapter treats of velocities in 'respect to the magnitudes of the thing moved and of the *space traversed (spatii pertransiti)*,' i.e., to the kinematic measure of movement. . ." (Clagett, pp. 207–208). Authors like Swineshead, and others, also described the distinction as one between the treatment of motion *causally* by forces and resistances (that is, dynamics) and by *effect,* that is, kinematically in terms of distance traversed in a given time (see Clagett, pp. 208–209).—*Ed.*]

speed or slowness. And because local motion is prior in nature to the other kinds, as the primary kind, we will carry out our intention in this section, with respect to local motion, before treating of the other kinds.

[1. MEASURE OF UNIFORM VELOCITY]

Although change of place is of diverse kinds, and is varied according to several essential as well as accidental differences, yet it will suffice for our purposes to distinguish uniform motion from nonuniform motion. Of local motions, then, that motion is called uniform in which an equal distance is continuously traversed with equal velocity in an equal part of time. Nonuniform motion can, on the other hand, be varied in an infinite number of ways, both with respect to the magnitude, and with respect to the time.

In uniform motion, then, the velocity of a magnitude as a whole is in all cases measured *(metietur)* by the linear path traversed by the point which is in most rapid motion, if there is such a point. And according as the position of this point is changed uniformly or nonuniformly, the complete motion of the whole body is said to be uniform or difform (nonuniform). Thus, given a magnitude whose most rapidly moving point is moved uniformly, then, however much the remaining points may be moving nonuniformly, that magnitude as a whole is said to be in uniform movement. . . .

[2. MEASURE OF NONUNIFORM VELOCITY]

In nonuniform motion, however, the velocity at any given instant will be measured *(attendetur)* by the path which *would* be described by the most rapidly moving point if, in a period of time, it were moved uniformly at the same degree of velocity *(uniformiter illo gradu velocitatis)* with which it is moved in that given instant, whatever [instant] be assigned. For suppose that the point *A* will be continuously accelerated throughout an hour. It is not then necessary that, in any instant of that hour as a whole, its velocity be measured by the line which that point describes in that hour. For it is not required, in order that any two points or any other two moving things be moved at equal velocity, that they should traverse equal spaces in an equal time; but it is possible that they traverse unequal spaces, in whatever proportion you may please. For suppose that point *A* is moved continuously and uniformly at *C* degrees of velocity, for an hour, and that it traverses a distance of a foot. And suppose that point *B* commences to move, from

rest, and in the first half of that hour accelerates its velocity to *C* degrees, while in the second half hour it decelerates from this velocity to rest. It is then found that at the middle instant of the whole hour point *B* will be moving at *C* degrees of velocity, and will fully equal the velocity of the point *A*. And yet, at the middle instant of that hour, *B* will not have traversed as long a line as *A*, other things being equal. In similar manner, the point *B*, traversing a finite line as small as you please, can be accelerated in its motion beyond any limit; for, in the first proportional part of that time, it may have a certain velocity, and in the second proportional part, twice that velocity, and in the third proportional part, four times that velocity, and so on without limit.

From this it clearly follows, that such a non-uniform or instantaneous velocity *(velocitas instantanea)* is not measured by the distance traversed but by the distance which *would* be traversed by such a point, *if* it were moved uniformly over such or such a period of time at that degree of velocity with which it is moved in that assigned instant.

[3. MEASURE OF UNIFORM ACCELERATION]

With regard to the acceleration *(intensio)* and deceleration *(remissio)* of local motion, however, it is to be noted that there are two ways in which a motion may be accelerated or decelerated: namely, uniformly, or nonuniformly. For any motion whatever is *uniformly accelerated (uniformiter intenditur)* if, in each of any equal parts of the time whatsoever, it acquires an equal increment *(latitudo)* of velocity. And such a motion is uniformly decelerated if, in each of any equal parts of the time, it loses an equal increment of velocity. But a motion is *nonuniformly accelerated or decelerated,* when it acquires or loses a greater increment of velocity in one part of the time than in another equal part.

In view of this, it is sufficiently apparent that when the latitude of motion or velocity is infinite, it is impossible for any body to acquire that latitude uniformly, in any finite time. And since any degree of velocity whatsoever differs by a finite amount from zero velocity, or from the privative limit of the intensive scale, which is rest—therefore any mobile body may be uniformly accelerated from rest to any assigned degree of velocity; and likewise, it may be decelerated uniformly from any assigned velocity, to rest. And, in general, both kinds of change may take place uniformly, from any degree of velocity to any other degree.

COMMENTARY

Notice in this passage the definitions of uniform velocity, uniform acceleration, and instantaneous velocity. In the definition of uniform velocity, Heytesbury speaks of the traversal of an equal space in an equal part of the time. Thus he failed to say in *any* equal parts of the time. That this would appear to be understood by him is clear, however, from his definition of uniform acceleration in terms of the acquisition of an equal increment of speed in *any* equal parts of the time. Heytesbury's contemporary, Richard Swineshead, was careful to specify that uniform velocity is to be defined by the traversal of an equal distance in *every (omni)* equal period of time. Hence Swineshead, at least, was anticipating Galileo's admonition to include the word "any" in a proper definition of uniform motion.[3]

One final important point should be noticed about this selection from Heytesbury's *De motu*. For him instantaneous velocity is to be measured or determined by the path which *would* be described by a point if that point were to move during some time interval with a uniform motion of the velocity possessed at the instant.[4]

I. Part VI. Local Motion *(continued)*

In this connection, it should be noted that just as there is no degree of velocity by which, with continuously uniform motion, a greater distance is traversed in one part of the time than in another equal part of the time, so there is no latitude (i.e., increment, *latitudo*) of velocity between zero degree [of velocity] and some finite degree, through which a greater distance is traversed by uniformly accelerated motion in some given time, than would be traversed in an equal time by a uniformly decelerated motion of that latitude. For whether it commences from zero degree or from some [finite] degree, every latitude, as long as it is terminated at some finite degree, and as long as it is acquired or lost uniformly, will correspond to its mean degree [of velocity]. Thus the moving body, acquiring or losing this latitude uniformly during some assigned period of time, will traverse a distance exactly equal to what it would traverse in an equal period of time if it were moved uniformly at its mean degree [of velocity].

2. For of every such latitude commencing from rest and terminating at some [finite] degree [of velocity], the mean degree is one-half the terminal degree [of velocity] of that same latitude.

3. From this it follows that the mean degree of any latitude bounded by two degrees (taken either inclusively or exclusively) is more than half the more intense degree bounding that latitude.

4. From the foregoing it follows that when any mobile body is uniformly accelerated from rest to some given degree [of velocity], it will in that time traverse one-half the distance that it would traverse if, in that same time, it were moved uniformly at the degree [of velocity] terminating that latitude. For that motion, as a whole, will correspond to the mean degree of that latitude, which is precisely one-half that degree which is its terminal velocity.

3. [In the Third Day of the *Dialogues Concerning Two New Sciences (Discorsi)*, Galileo defines uniform motion as follows:

"On equable (i.e., uniform) motion (*De motu aequabili*). In regard to equable or uniform *(uniformis)* motion, we have need of a single definition, which I give as follows. I understand by equal *(equalis)* or uniform movement one whose parts *(partes)* gone through *(peracte)* during any *(quibuscumque)* equal times are themselves equal. We must add to the old definition—which defined equable motion simply as one in which equal distances *(spatia)* are traversed in equal times—the word "any" *(quibuscumque)*, i.e., in "all" *(omnibus)* equal periods of time. . . ." Clagett's translation (p. 251), while based in part upon the Crew and De Salvio translation, is made more literal to show the ultimate dependence of Galileo's terminology on the Merton College formulations. Indeed, the Latin terms included by Clagett are frequently the very same terms employed by Galileo's medieval predecessors.—*Ed.*]

4. [As Clagett observes, Galileo's definition of instantaneous velocity is much the same as Heytesbury's. Here, in Clagett's translation (p. 251), is how Galileo expresses it (the omissions are Clagett's):

"Thus we may conceive that a motion is uniformly and continually accelerated when in any equal time periods equal increments of swiftness are added. . . . To put the matter more clearly, if a moving body were to continue its motion with the same degree or moment of velocity *(gradus seu momentum velocitatis)* it acquired in the first time-interval, and continue to move uniformly with that degree of velocity, then its motion would be twice as slow as that which it would have if its velocity *(gradus celeritatis)* had been acquired in two-time intervals."

Of Galileo's definitions of motion, cited in this and the preceding note, Clagett remarks (p. 252): "Note once more that Galileo in this passage compared the instantaneous velocities at the end of the first time-period and at the end of the second time-period (in a uniformly accelerated movement) by imagining that the bodies were moving uniformly over some time-period with these respective instantaneous velocities. This, as we have seen, was precisely what Heytesbury and Swineshead recommended in their treatment of instantaneous velocity. Needless to add also, the definitions of uniform velocity and uniform acceleration given by Galileo have their almost exact Merton counterparts."—*Ed.*]

5. It also follows in the same way that when any moving body is uniformly accelerated from some degree [of velocity] (taken exclusively) to another degree inclusively or exclusively, it will traverse more than one-half the distance which it would traverse with a uniform motion, in an equal time, at the degree [of velocity] at which it arrives in the accelerated motion. For that whole motion will correspond to its mean degree [of velocity], which is greater than one-half of the degree [of velocity] terminating the latitude to be acquired; for although a nonuniform motion of this kind will likewise correspond to its mean degree [of velocity], nevertheless the motion as a whole will be as fast, categorematically, as some uniform motion according to some degree [of velocity] contained in this latitude being acquired, and, likewise, it will be as slow.

6. To prove, however, that in the case of acceleration from rest to a finite degree [of velocity], the mean degree [of velocity] is exactly one-half the terminal degree [of velocity], it should be known that if any three terms are in continuous proportion, the ratio of the first to the second, or of the second to the third, will be the same as the ratio of the difference between the first and the middle, to the difference between the middle and the third; as when the terms are 4, 2, 1; 9, 3, 1; 9, 6, 4. For as 4 is to 2, or as 2 is to 1, so is the proportion of the difference between 4 and 2 to the difference between 2 and 1, because the difference between 4 and 2 is 2, while that between 2 and 1 is 1; and so with the other cases.

Let there be assigned, then, some term under which there is an infinite series of other terms which are in continuous proportion according to the ratio 2 to 1. Let each term be considered in relation to the one immediately following it. Then, whatever is the difference between the first term assigned and the second, such precisely will be the sum of all the differences between the succeeding terms. For whatever is the amount of the first proportional part of any continuum or of any finite quantity, such precisely is the amount of the sum of all the remaining proportional parts of it.

Since, therefore, every latitude is a certain quantity, and since, in general, in every quantity the mean is equidistant from the extremes, so the mean degree of any finite latitude whatsoever is equidistant from the two extremes, whether these two extremes be both of them positive degrees, or one of them be a certain degree and the other a privation of it or zero degree.

But, as has already been shown, given some degree under which there is an infinite series of other degrees in continuous proportion, and letting each term be considered in relation to the one next to it, then the difference or latitude between the first and the second degree—the one, namely, that is half the first—will be equal to the latitude composed of all the differences or latitudes between all the remaining degrees—namely those which come after the first two. Hence, exactly equally and by an equal latitude that second degree, which is related to the first as a half to its double, will differ from that double as that same degree differs from zero degree or from the opposite extreme of the given magnitude.

And so it is proved universally for every latitude commencing from zero degree and terminating at some finite degree, and containing some degree and half that degree and one-quarter of that degree, and so on to infinity, that its mean degree is exactly one-half its terminal degree. Hence this is not only true of the latitude of velocity of motion commencing from zero degree [of velocity], but it could be proved and argued in just the same way in the case of latitudes of heat, cold, light, and other such qualities.

7. With respect, however, to the distance traversed in a uniformly accelerated motion commencing from zero degree [of velocity] and terminating at some finite degree [of velocity], it has already been said that the motion as a whole, or its whole acquisition, will correspond to its mean degree [of velocity]. The same thing holds true if the latitude of motion is uniformly acquired from some degree [of velocity] in an exclusive sense, and is terminated at some finite degree [of velocity].

From the foregoing it can be sufficiently determined for this kind of uniform acceleration or deceleration how great a distance will be traversed, other things being equal, in the first half of the time and how much in the second half. For when the acceleration of a motion takes place uniformly from zero degree [of velocity] to some degree [of velocity], the distance it will traverse in the first half of the time will be exactly one-third of that which it will traverse in the second half of the time.

And if, contrariwise, from that same degree [of velocity] or from any other degree whatsoever, there is uniform deceleration to zero degree [of velocity], exactly three times the distance will be traversed in the first half of the time, as will be traversed in the second half. For every motion as a whole, completed in a whole period of time, corresponds to its mean degree [of velocity]—namely, to the degree it will have at the middle instant of the time. And the second half of the motion in question will correspond to the mean degree of the second half of that same motion, which is one-fourth of the degree [of velocity] terminating that latitude. Consequently, since this second half will last only through half the time, exactly one-fourth of the distance will be traversed in that second half as will be traversed in the whole motion. Therefore, of the whole distance being traversed by the whole motion, three-quarters will be traversed in the first half of the whole motion, and the last quarter will be traversed in its second half. It follows, consequently, that in this type of uniform intension and remission of a motion from some degree [of velocity] to zero degree, or from zero degree to some degree, exactly three times as much distance is traversed in the more intense half of the latitude as in the less intense half.

8. But any motion can be uniformly accelerated or decelerated from some degree [of velocity] to another degree in an endless number of ways, because it may be from some degree to a degree half of that, or to a degree one-fourth of it, or one-fifth, or to a degree two-thirds of that degree, or three-quarters of it, and so on. Consequently there can be no universal numerical value by which one will be able to determine, for all cases, how much more distance would be traversed in the first half of this sort of acceleration or deceleration than in the second half, because, according to the diversity of the extreme degrees [of velocity], there will be diverse proportions of distance traversed in the first half of the time to distance traversed in the second half.

But if the extreme degrees [of velocity] are determined, so that it is known, for instance, that so much distance would be traversed in such or such a time by a uniform motion at the more intense limiting degree [of velocity], and if this is likewise known with respect to the less intense limiting degree [of velocity], then it will be known by calculation how much would be traversed in the first half and also how much in the second. For, if the extreme degrees [of velocity] are known in this way, the mean degree [of velocity] of these can be obtained, and also the mean degree between that mean degree and the more intense degree terminating the latitude. But a calculation of this kind offers more difficulty than advantage.

And it is sufficient, therefore, for every case of this kind, to state as a general law, that more distance will be traversed by the more intense half of such a latitude than by the less intense half—as much more, namely, as would be [the excess of distance] traversed by the mean degree [of velocity] of this more intense half, if it moved in a time equal to that in which this half is acquired or lost uniformly, over that [distance which] would be traversed by the mean degree [of velocity] of the less intense half, in the same time.

9. But as concerns nonuniform acceleration or deceleration, whether from some degree [of velocity] to zero degree or *vice versa,* or from one degree to some other degree, there can be no rule determining the distance traversed in such or such time, or determining the intrinsic degree to which such a latitude of motion, acquired or lost nonuniformly, will correspond. For just as such a nonuniform acceleration or deceleration could vary in an infinite number of ways, so also that motion as a whole could correspond to an infinite number of intrinsic degrees [of velocity] of its latitude—indeed, to any intrinsic degree whatsoever, of the latitude thus acquired or lost.

In general, therefore, the degree [of velocity] terminating such a latitude at its more intense limit is the most remiss degree [of velocity], beyond the other limit (i.e., the most remiss extreme) of the latitude, to which such a nonuniformly nonuniform motion as a whole *cannot* correspond; and the degree [of velocity] terminating that latitude at its more remiss limit is the most intense degree [of velocity] beneath the upper limit of the same latitude, to which such a nonuniformly nonuniform motion *cannot* correspond. Consequently, it is not possible for such a motion as a whole to correspond to such a remiss degree (as that of the lower limit); nor to such an intense degree (as the upper limit).

COMMENTARY

1. An able commentary on this whole section, comparing this passage to similar passages in the *Probationes conclusionum* attributed to Heytesbury, to the section of Swineshead's *Liber calculationum* entitled *Regule de motu locali,* and to the *De motu* of John of Holland, is given by Curtis Wilson in his *William Heytesbury* (Madison, 1956), Chapter 4.

The substance of this first passage is that it makes no difference whether a body is uniformly accelerating from V_o to V_f or uniformly decelerating from V_f to V_o in the same time, the distance traversed, $\varDelta S$, will be the same. The reason given (without proof) is that regardless of whether ΔV is positive or negative, the equivalent uniform speed with which a body would traverse the same distance in the same time is the mean velocity, i.e., the velocity at the middle instant of the period of acceleration. This would appear to be the earliest definitely established statement of the mean speed theorem. . . . No formal proof of this theorem is given.

2. Furthermore, it can be shown that for uniform acceleration, if $V_o = 0$, then $V_m = V_f/2$, where V_f is the final velocity of the acceleration, V_m is its mean velocity and V_o is the initial velocity.

3. Hence if $V_o > 0$, then $V_m > V_f/2$, whether we include or exclude the terminal velocities in ΔV.

4. $S_a = S_m = S_f/2$, S_a being the distance traversed in the course of the uniform acceleration from rest, S_m the distance traversed by a body moving uniformly during the same time with the mean velocity, and S_f being the distance traversed by a body moving uniformly in the same time with the final velocity.

5. When $V_o > 0$, $S_a > S_f/2$.

6. For the next conclusion, Heytesbury feels it necessary to give a fuller treatment. It is clear that Heytesbury here characterizes a uniform acceleration from rest not only by equal increments of velocity in equal time periods [see p.239—Ed.] but also by the fact that if any degree of speed within such an acceleration is taken, we shall always find, on dividing the time period (to that degree of velocity) into continually proportional parts, that the degrees of velocities at the ends of those periods are continually proportional in the same way. . . . Such a definition of uniform acceleration was also assumed by Swineshead in one of his proofs of the mean speed theorem. The meaning of this alternate definition of uniform acceleration can be made clearer if we pretend for a moment that the geometric system of representing movements developed later at Paris by Nicole Oresme was in existence [see, section III.vii of Selection 43—Ed.]. In the accompanying figure (see Fig. 1), AC represents the time of acceleration and the perpendicular lines on AC making up *in toto* the triangle represent velocities at successive instants along AC. Heytesbury's alternate definition is illustrated by taking any velocity QD. Then at time $AD/2$ there is a velocity $QD/2$, at $AD/4$ a velocity $QD/4$, etc. The same would hold true if we took the final velocity FC instead of QD, namely, there would be a velocity $FC/2 (= EG)$ at time $AC/2 (= AG)$.

In Heytesbury's elaboration he indicates that

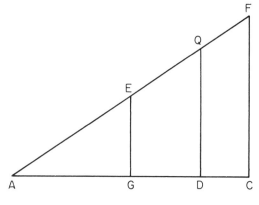

Fig. 1

EG is the mean or middle velocity, and thus there is equal distance in velocity increment (latitude) from the two extremes, namely FC and zero. Hence $FC = 2EG$. While this would appear to be obvious, Heytesbury would like to show this is so without directly reckoning with zero velocity. Assume
$$\Delta V_1 = \Delta V_2, \text{ where } \Delta V_1 = FC - EG, \text{ and}$$
$$\Delta V_2 = \left(EG - \frac{EG}{2}\right) + \left(\frac{EG}{2} - \frac{EG}{4}\right) + \cdots + \left(\frac{EG}{2^{n-1}} - \frac{EG}{2^n}\right)$$
$+\cdots$ and rewriting $\Delta V_2 = \dfrac{EG}{2} + \dfrac{EG}{4} + \cdots + \dfrac{EG}{2^n} + \cdots$

It is clear then that, as n goes to infinity, ΔV_2 sums to EG, and thus $FC = 2EG$.

7. From the special case of uniform acceleration from rest, the mean speed theorem can be proved for uniform acceleration commencing from a finite velocity. Furthermore, from the special case of the theorem for acceleration from rest, it can be proved that the distance traversed in the first half of the uniform acceleration from rest is 1/3 of that traversed during the second half of the period of acceleration. Conversely, the distance traversed during the first half of the time of uniform deceleration to rest is three times that traversed in the second half. The proof is indicated for the case of deceleration. The deceleration of each half corresponds to its mean velocity (for the equivalent traversal of space). Thus in the second half, the mean velocity is 1/4 the initial velocity and in the first half it is 3/4 the initial velocity. Using modern formulas, $S_1 = (3\ V_f/4)\ (t/2)$ and $S_2 = (V_f/4)\ (t/2)$ and thus $S_1/S_2 = 3/1$.

8. No single universal relationship between the distances traversed in the first and second half of the time of uniform acceleration can be established if the acceleration commences with a finite velocity rather than rest. But, if we know the terminal

velocities, we can, by using the mean speed theorem, determine what the relationships are, for the proportion will be as that of the mean velocities of the respective halves of the acceleration.

9. In the case of nonuniform accelerations, no general rule is applicable for determining the equivalent uniform speed allowing for an equal traversal of space in the same time. The only evident conclusion is that the equivalent speed cannot be either of the terminal velocities. (See examples for a uniformly increasing or decreasing acceleration drawn from Swineshead's treatment [on pp. 292–294, 297 of *Science of Mechanics*— *Ed.*].)

43 THE CONFIGURATION OF QUALITIES AND MOTIONS, INCLUDING A GEOMETRIC PROOF OF THE MEAN SPEED THEOREM

Nicole Oresme (ca. 1325–1382)

Translation, introduction, and annotation by Marshall Clagett[1]

[After judging that Oresme probably composed his *Tractatus de configurationibus qualitatum et motuum (A Treatise on the Configuration of Qualities and Motions)* during the 1350's while at the College of Navarre, Clagett launches into a general description of the treatise.—*Ed.*][2]

In its entirety, the tract includes a proemium, a table of chapter titles, and ninety-three chapters divided into three parts. The first and second parts each have forty chapters and the third has thirteen. As I shall show later, there is a strong possibility that Oresme undertook the preparation of this work as an elaboration of the long section on the configuration doctrine appearing in his *Questiones super geometriam Euclidis*. Speaking generally, one can say that the first part of the *De configurationibus* establishes the tenets of the geometry of the figuration doctrine, applies the doctrine to qualities, i.e., to entities which are essentially permanent or enduring in time, and relates it to the intricacies of the internal configurations of qualities. In the course of elaborating the doctrine of internal configurations, he suggests how the theory might explain numerous physical and psychological phenomena. The second part describes how the configuration doctrine can be fruitfully applied to motion, i.e., to entities that are successive. Here again after describing the external, geometrical aspects of the doctrine, he goes on to a detailed analysis of how the actual natures of motions, possessing some kind of essential configuration, may well account for certain sonic and musical effects. And he concludes the second part with a discussion in many chapters of how these essential configurations of motion go far in explaining magical and psychological effects. Thus we see his broadly conceived doctrine of configurations giving him a physical basis to attack magic, just as his elaborate

discussion of the ratios of ratios gave him a mathematical base to attack astrology.[3] Finally, in the third part, Oresme returns again to the external geometrical figures used to represent qualities and motions. He shows how in the comparison of the areas of these figures we have a basis for the comparisons of different qualities and motions.

Initially, I should point out that there are two keys to a proper understanding of the *De configurationibus*. The first is that Oresme uses the term *configuratio* with two distinguishable but related meanings, i.e., a primitive meaning and a derived meaning. In its initial, primitive meaning it refers to the fictional and imaginative use of geometrical figures to represent or graph intensities

1. [Reprinted with permission of the Regents of the University of Wisconsin, from *Nicole Oresme and the Medieval Geometry of Qualities and Motions, A Treatise on the Uniformity and Difformity of Intensities Known as Tractatus de configurationibus qualitatum et motuum,* edited with an Introduction, English Translation, and Commentary by M. Clagett (Madison, Wis.: University of Wisconsin Press, 1968), pp. 165–169, 173–175, 177–183, 191–195, 199–203, 271, 277, 283–285, 393–395, 409–411. The introduction and notes have been brought together from other parts of Clagett's volume.

For the origins and subsequent influence of Oresme's configuration doctrine, see Clagett's excellent chapter "The Configuration Doctrine in Historical Perspective," pp. 50–121. There Clagett describes the important influence of Aristotle and the fourteenth-century Merton College authors, one of whom, William Heytesbury, is represented in the preceding selection. The influence of Oresme's coordinate or graphing system, which he used to represent variations in qualities and motions, is traced through such later authors as Galileo, Thomas Hariot, John Wallis, and Christiaan Huygens.—*Ed.*]

2. [The introduction to this selection is drawn from *Nicole Oresme. . .*, pp. 14–16.—*Ed.*]

3. [For Oresme's discussion of "ratios of ratios" and his use of it to attack astrology, see Selection 51.2, notes 51 and 84, and Selection 69.—*Ed.*]

in qualities and velocities in motions. Thus the base line of such figures is the subject when we are talking about linear qualities or the time when we are talking about velocities, and the perpendiculars raised on the base line represent the intensities of the quality from point to point in the subject or represent the velocity from instant to instant in the motion. The whole figure, consisting of all the perpendiculars, represents the whole distribution of intensities in the quality, i.e., the quantity of the quality, or in case of motion the so-called total velocity, dimensionally equivalent to the total space traversed in the given time. A quality of uniform intensity is thus represented by a rectangle, which is its *configuration;* a quality of uniformly non-uniform intensity starting from zero intensity is represented as to its configuration by a right triangle. Similarly, motions of uniform velocity and uniform acceleration are represented respectively by a rectangle and a right triangle. There is a considerable discussion of other possible configurations.

Now we see that for Oresme differences in configuration taken in its primitive meaning reflect in a useful and suitable fashion internal differences in the subject. Thus we can say that the external configuration represents some kind of internal arrangement of intensities which we can call its essential internal *configuration*. In this way, we arrive at the second usage of the term configuration. Configuration in this second sense abandons the purely spatial or geometrical meaning, since one of the variables involved, namely, intensity, is not essentially spatial, although, Oresme tells us, variations in intensity can be represented by variations in the length of straight lines. . . .

Oresme suggests at great length how differences in internal configuration may explain many physical and even psychological phenomena, otherwise not simply explicable on the basis of the primary elements that make up a body. Thus two bodies might have the same amounts of primary elements in them and even in the same intensity but the configuration of their intensities may well differ, thus producing different effects in natural actions.

The second key to the understanding of the configuration doctrine of Oresme is that which we may call the suitability doctrine, and it pertains to the nature of the configurations in their primitive meaning of external figures. Briefly, it holds that any figure or configuration is suitable or fitting to describe a quality when its altitudes (ordinates, we would say in modern parlance) on any two points

of its base or subject line are in the same ratio as the intensities of the quality at those points in the subject.

CHAPTERS OF PART I

I.i On the Continuity of Intensity

Every measurable thing except numbers is imagined in the manner of continuous quantity. Therefore, for the mensuration of such a thing, it is necessary that points, lines, and surfaces, or their properties, be imagined. For in them (i.e. the geometrical entities), as the Philosopher has it,[4] measure or ratio is initially found, while in other things it is recognized by similarity as they are being referred by the intellect to them (i.e., to geometrical entities). Although indivisible points, or lines, are nonexistent, still it is necessary to feign them mathematically for the measures of things and for the understanding of their ratios.[5] Therefore, every intensity which can be acquired successively ought to be imagined by a straight line perpendicularly erected on some point of the space or subject of the intensible thing, e.g., a quality. For whatever ratio is found to exist between intensity and intensity, in relating intensities of the same kind, a similar ratio is found to exist between line and line, and vice versa. For just as one line is commensurable to another line and incommensurable to still another, so similarly in regard to intensities certain ones are mutually commensurable and others incommensurable in any way because of their [property of] continuity. Therefore, the measure of intensities can be fittingly imagined as the measure of lines, since an intensity could be imagined as being infinitely decreased or infinitely increased in the same way as a line.

Again, intensity is that according to which something is said to be "more such and such," as "more white" or "more swift." Since intensity, or rather the intensity of a point, is infinitely divisible in the manner of a continuum in only one way, therefore there is no more fitting way for it to be imagined

4. The primary reference seems to be to Aristotle's *Metaphysics*, Bk. X, Chap. 1, 1052b. . . . [*Nicole Oresme. . .*, p. 438; on page 54, Clagett remarks that for Oresme it was probably this passage of the *Metaphysics* "that clinched the transference of the concept of ratios between lines to that of ratios between intensities." —*Ed.*]

5. Oresme suggests here an opinion that geometrical entities are really "nichil," i.e., they are mere mathematical fictions useful for the "measures of things" and "for the understanding of their ratios" [*Nicole Oresme . . .*, p. 438—*Ed.*].

than by that species of a continuum which is initially divisible and only in one way, namely by a line. And since the quantity or ratio of lines is better known and is more readily conceived by us—nay the line is in the first species of continua, therefore such intensity ought to be imagined by lines and most fittingly by those lines which are erected perpendicularly to the subject. The consideration of these lines naturally helps and leads to the knowledge of any intensity, as will be more fully apparent in chapter four below. Therefore, equal intensities are designated by equal lines, a double intensity by a double line, and always in the same way if one proceeds proportionally. And this is to be understood universally in regard to every intensity that is divisible in the imagination, whether it be an active or non-active quality, a sensible or non-sensible subject, object, or medium. For example, it is to be understood in regard to the light of the body of the sun, to the illumination of a medium, or to a species in the medium, to a diffused influence or power, and similarly to others, with the possible exception of curvature, concerning which we shall speak in a limited way in chapters twenty and twenty-one of this part [of our work].

Of course, the line of intensity of which we have just spoken is not actually extended outside of the point or subject but is only so extended in the imagination, and it could be extended in any direction whatever except that it is more fitting to imagine it standing up perpendicularly on the subject informed with the quality.[6]

I.iii On the Longitude of Qualities

. . . Let the extension of a quality be called its longitude and intensity its latitude or altitude.[7] But however this might be, it is obvious from the things said that certain moderns do not speak in the best way when they call the whole of the quality its latitude, just as it would be an abuse [of terminology] to understand by the breadth of a surface the whole surface or figure.[8] For just as the breadths of some unequal surfaces or figures are equal, so, as will be seen later, many latitudes of unequal qualities are equal, or vice versa.

I.iv On the Quantity of Qualities

The quantity of any linear quality is to be imagined by a surface whose length or base is a line protracted in a subject of this kind, as the preceding chapter says, and whose breadth or altitude is designated by a line erected perpendicularly on the

aforesaid base in the way that the second chapter proposes. And I understand by "linear quality" the quality of some line in the subject informed with a quality. . . .

Now, reverting to the subject at hand, just as the quality of a point is imagined as a line, and the quality of a line by a surface, so the quality of a surface is imagined as a body whose base is the surface informed with the quality.[9] This will be more fully clarified as we go along. Moreover, since in any kind of a body there is an infinite number of equivalent[10] surfaces and the quality of any one of them is imagined as a body, it is not unfitting but necessary that one body be imagined to be at the same time in the place where another body—or even any other body whatever—is imagined to be. [We can think of this taking place] by penetration or by mathematical superposition or the simultaneous placing of the bodies so imagined. However, this penetration is not real. And although a surface quality is imagined by means of a body and it does not happen that a fourth dimension exists

6. The contrast between *secundum rem* [this expression is translated by "actually"—*Ed.*] and *secundum ymaginationem* [translated as "imagination" in this paragraph—*Ed.*] made here by Oresme is clear evidence that he considers the specific, external figure representing the quality as a fictional entity not existing as such in the matter (i.e., as a specific geometrical figure) [*Nicole Oresme. . . ,* p. 439—*Ed.*].

7. [In I.ii Oresme had argued that—*Ed.*] lines of intensity. . . ought to be called the longitude of the quality, but generally and conventionally they are called the latitude of the quality. Now, the extent of the quality in the subject, if the quality is imagined as existing in a line, is to be imagined as a line on which the lines representing the intensity from point to point are to be erected perpendicularly. This line representing the extension of the quality in the subject, which ought to be called its latitude, is conventionally called its longitude [*Nicole Oresme. . . ,* p. 16—*Ed.*]

8. ["In short, some call the whole quantity of quality 'latitude,' while it is more proper to apply it to one dimension alone (conventionally, intension)." Clagett, *The Science of Mechanics in the Middle Ages* (Madison, Wis.: University of Wisconsin Press, 1959), p. 362. That is, some applied the term "latitude" to the whole figure representing the sum of all the intensions, a practice of which Oresme disapproved.—*Ed.*]

9. That is, Oresme is telling us that the surface quality can be represented by a three-dimensional figure that consists of the totality of perpendiculars rising on the surface informed with the quality [*Nicole Oresme. . . ,* p. 17—*Ed.*].

10. One would suppose that Oresme would have conceived of them as being of infinitely small thickness, syncategorematically speaking, i.e., that they are thinner than any assignable quantity [*Nicole Oresme. . . ,* p. 177, n. 4—*Ed.*].

or is imagined, still a corporeal quality is imagined to have a double corporeity: a true one with respect to the extension of the subject in every dimension and another one that is only imagined from the intensity of this quality taken an infinite number of times and dependent upon the multitude of surfaces of the subject.[11] The suitableness of this imagined concept has been touched upon before and will be more fully apparent in what follows.

I.v On the Figuration of Qualities

Every linear quality is "figured" (i.e., represented in figures) by means of a surface perpendicularly erected upon a subject line. For let AB be a line informed with a quality [see Fig. 1]. And since by

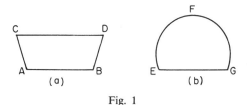

Fig. 1

the preceding chapter this quality is designated by a surface, it is necessary that it be imagined as "figured" by the surface by which it is designated or imagined. The latitude of this surface designates the intensity of this quality. It is necessary also that any point of this surface or figure outside of subject line AB stands perpendicularly above this same line AB, as is obvious in the first chapter, for otherwise the intensity and quality would be [laterally] outside the subject.[12] [This is true] because anything which according to our imagination is above this line is actually in the subject, and vice versa. Accordingly, if anything were imagined as being above the subject but not above it perpendicularly, then it would actually be outside of the subject. Thus it is obvious that no quality is to be imagined by a surface or figure having an angle at the base greater than a right angle, e.g., quadrangle $ABCD$; or by a segment of a circle that is greater than a semicircle, e.g., segment EFG. But some linear quality can be imagined by any other plane figure.

I.vi On the Clarification of the Figures

Although some linear quality can be correctly imagined by any plane figure other than those mentioned before, still not any quality can be imagined by any figure. Indeed no linear quality is imagined or designated by any figure except the ones in which the ratio of the intensities at any points of that quality is as the ratio of the lines

erected perpendicularly in those same points and terminating in the summit of the imagined figure.[13]

For example [see Fig. 2], let line AB be divided

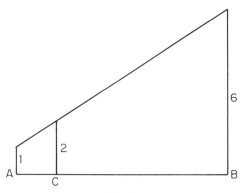

Fig. 2

in point C in any way such that the intensity in point C is double that in point A; and in point B let it be triple that in point C. Therefore, by the first chapter the line imagined as rising perpendicularly above point C and denoting the intensity at that point is double the line imagined as rising above point A, and the line imagined as rising above point B is three times the line imagined as rising above C. Therefore, this quality can be imagined only by the figure which at point C is twice as high as at point A or whose summit at C is double that at point A, and whose summit at point B is triple that at point C—with the further stipulation however that the figure of this sort could be varied in altitude according to the ratio of intensities in the other points of line AB. But

11. Hence, in effect, Oresme is telling us that we can represent a corporeal quality by an infinite number of interlacing bodies, each of which represents the quality of some surface or plane of the body (*Nicole Oresme . . .*, p. 17—*Ed.*].

12. The obvious meaning of this is that if the lines were not perpendicular, not only would they be outside of the subject vertically (and this he was prepared to grant in order to have some geometric representation of intensity which is not itself geometrical) but they would also have to be outside of the subject laterally, in the sense that the surface representing the whole quantity of the quality would not then be directly above the subject line. He had already admitted that it was quite possible to have the intensity lines intersecting the subject line at an angle other than a right angle. But it would be more fitting from the standpoint of what one might call physical intuition to have the quality kept within the lateral bounds of the subject [*Nicole Oresme. . .*, p. 441—*Ed.*].

13. This is the first statement of the suitability doctrine that lies at the heart of Oresme's system. It will be made more precise in the succeeding chapter [*Nicole Oresme . . .*, p. 441—*Ed.*].

from this it is apparent that a quality of this sort cannot be designated by a rectangle or by a semicircle; and similarly concerning an infinite number of other figures.

I.vii On the Suitability of the Figures

Any linear quality can be designated by every plane figure which is imagined as standing perpendicularly on the linear [extension of the] quality and which is proportional in altitude to the quality in intensity. Moreover, a figure erected on a line informed with a quality is said to be "proportional in altitude to the quality in intensity" when any two lines perpendicularly erected on the quality line as a base and rising to the summit of the surface or figure have the same ratio in altitude to each other as do the intensities at the points on which they stand.

For example [see Fig. 3], let there be line *AB* on

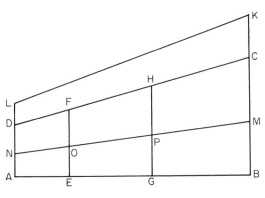

Fig. 3

which surface *ABCD* stands and let the two lines *EF* and *GH* be erected on the base. If, therefore, the ratio of *EF* to *GH* is as the ratio of the intensity in point *E* to the intensity in point *G*, and similarly for the other points and their corresponding lines, then I say that this surface or figure is "proportional in altitude to this quality in intensity," so that the altitude of the surface is similar to the intensity of this quality. Therefore, this quality is most fittingly designated by such a figure or surface.[14] Moreover, since on the same line *AB* a great number of surfaces can be erected which are proportional or similar in altitude—some of which are larger and some smaller than *ABCD*, as for example surface *ABKL* which is larger and surface *ABMN* which is smaller, and any number of others which would be of similar although unequal altitude—it follows that the quality of line *AB* can be designated by any one of them indifferently. There is however this

provision: if the quality is imagined by some one of these designated figures, then with this figuration retained a quality which is double the original one in intensity and similar to it will be designated by a figure of similar altitude but twice as high.[15] The same thing holds proportionally for any greater or lesser quality, notwithstanding the fact that the first quality could have been imagined in the beginning by a greater or lesser surface or figure. Moreover these greater or lesser surfaces are unequal in area, dissimilar in figure and also unequal in altitude, and yet they are similar or proportional in altitude. Hence, if two points *O* and *P* are marked in the intersections as in the accompanying figure, then if *GH/EF = GP/EO* and similarly in regard to any two lines erected in like fashion on the base *AB*, I say that surface *ABCD* and surface *ABMN* are of similar or proportional altitude.

I.xi On Uniform and Difform Quality

And so every uniform quality is imagined by a rectangle[16] and every quality uniformly difform terminated at no degree is imaginable by a right triangle.[17] Further, every quality uniformly difform

14. [Clagett explains (p. 441) that it is from this sentence and the title of chapter vii that he coined the expression "suitability doctrine"—*Ed.*].

15. This constitutes an important proviso to the general statement that any figure whose altitudes (ordinates) were in the same ratio as the intensities would be suitable. The proviso merely holds that when we are *comparing* qualities, we must take as the basis of our comparison some specific figure of the infinite possible figures of the same kind [*Nicole Oresme . . .*, p. 442—*Ed.*].

16. The quality that is uniform in intensity is obviously represented by a rectangle, since the altitude of the rectangle is uniform throughout; i.e., the perpendicular height above any point of the base is the same [*Nicole Oresme. . .*, p. 18—*Ed.*].

17. [Uniformly difform qualities are] qualities in which the intensity of the subject quality varies in a uniform way as it is distributed through the subject (see I. viii–I.ix). Oresme discusses two categories of uniformly difform qualities. The first one is a uniformly difform quality that begins from zero intensity and ends with a given intensity, or which begins at a given intensity and ends with zero intensity. Such a quality is to be represented by a right triangle. For this reason it is called in the title of the chapter (I.viii) a 'right-triangular quality.' Oresme is once more naming the quality from the figure by which it can be represented. Again Oresme applies the suitability doctrine and concludes that this quality could be equivalently represented without any difference by every triangle having a right angle on the base. This is proved by showing that the ratio of any two ordinates on the base of a right triangle on a given base is the same regardless of the height of the triangle. Therefore, some quality is assimilated to any one of these triangles, and, further, the *same* quality can be assimilated to any

terminated in both extremes at some degree is to be imagined by a quadrangle having right angles on its base and the other two angles unequal. Now every other linear quality is said to be "difformly difform" and is imaginable by means of figures otherwise disposed according to manifold variation. Some modes of the "difformly difform" will be examined later. The aforesaid differences of intensities cannot be known any better, more clearly, or more easily than by such mental images and relations to figures, although certain other descriptions or points of knowledge could be given which also become known by imagining figures of this sort: as if it were said that a uniform quality is one which is equally intense in all parts of the subject, while a quality uniformly difform is one in which if any three points [of the subject line] are taken, the ratio of the distance between the first and the second to the distance between the second and the third is as the ratio of the excess in intensity of the first point over that of the second point to the excess of that of the second point over that of the third point, calling the first of those three points the one of greatest intensity.[18]

Let us clarify this first with respect to a quality uniformly difform which is terminated at no degree and which is designated or imagined by △*ABC* [see Fig. 4 (a)]. With the three perpendicular

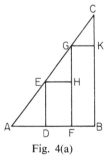

Fig. 4(a)

lines *BC, FG,* and *DE* erected, then let *HE* be drawn parallel to line *DF* and similarly *GK* parallel to line *FB*. Therefore, the two small triangles *CKG* and *GHE* are formed and they are equiangular. Hence, by [proposition] VI.4 of [the *Elements* of] Euclid, *GK/EH* = *CK/GH, CK* and *GH* being excesses. And since *GK* = *FB* and similarly *EH* = *DF*, so *FB/DF* = *CK/GH, FB* and *DF* being the distances on the base of the three points and *CK* and *GH* being the excesses of altitude proportional to the intensity of these same points. Since, therefore, the quality of line *AB* is such that the ratio of the intensities of the points of the line is as the ratio of the altitudes of the lines perpendicularly

erected on those same points, that which has been proposed is evidently clear, namely that the ratio of the excess in intensity of the first point over the second to the excess of the second over the third is the same as the ratio of the distance between the first and second points to the distance between the second and the third, and similarly for any other three points. Hence what we have premised in regard to a quality difform in this way is quite fitting, and so it (this quality) was well designated by such a triangle.

By the same method the aforesaid description or property can be demonstrated for a quality uniformly difform terminated in both extremes at [some] degree, and thus for one which we let be imagined by quadrangle *ABCD* in which line *DE* is drawn parallel to base *AB* forming △*DEC* [see Fig. 4 (b)]. Then let lines of altitude be drawn in the

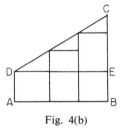

Fig. 4(b)

quadrangle and also transversals parallel to the base in this triangle, thus forming small triangles. And then one can easily argue concerning the excesses and the distances in this triangle just as was argued in the other one. This will be easily apparent to one who is observant.

Further, every quality which is disposed in [any] other way than those described earlier is said to be "difformly difform." It can be described negatively as a quality which is not equally intense in all parts of the subject nor in which, when any three points of it are taken, the ratio of the excess of the first over the second to the excess of the second over the third is equal to the ratio of their distances.

I.xiv On Simple Difform Difformity

We now treat of difform difformity; there are

other one of them and be imagined by it. There is this proviso, however: if some quality is designated by one triangle, another quality of similar but double intensity must be designated by a triangle that is twice as high, and similarly for the proportionally greater [intensities]. [*Nicole Oresme*. . ., p. 19—*Ed.*].

18. This is similar in modern parlance to saying that we have a figure of a uniformly difform quality when the slope of the summit line is a constant [*Nicole Oresme* . . ., p. 20—*Ed.*].

two modes of such difformity: simple and composite. We must first talk of the simple mode. Simple difform difformity is that which can be designated by a figure whose line of summit or line of intensity is a single line, i.e. not composed of several lines. It is necessary, therefore, that the line be a curve; because if it were straight, then it would be simply a uniformity or uniform difformity, as is clear from the preceding chapter. Furthermore, it is necessary that the curvature of the summit line does not attain that of a circular segment greater than a semicircle so that the angle[19] on the base is greater than a right angle, as was clear in chapter five. However, it can happen that the angle on the base is less than a right angle by any amount you please.

Therefore, for example, let there be line *AB*, whose quality can be designated by semicircle *ACB* [see Fig. 5]. This is possible, as is evident

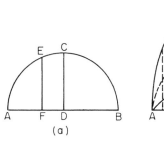

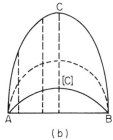

(a)

(b)

Fig. 5

from chapter seven. And so I now say that the same quality of line *AB* is imaginable or can be designated by a figure having an altitude greater or less than that of the semicircle by any amount you please. For let line *CD* be drawn as a perpendicular to center *D* and again let another line *EF* be drawn as a perpendicular to line *AB*. Therefore, since it is possible to construct on the same points two other perpendiculars less than *CD* and *EF* but having the same ratio between them as do *CD* and *EF* and in the same way to construct on all the points of line *AB* perpendiculars which are greater or less than the corresponding perpendiculars in semicircle *ACB* constructed on those points of *AB* and having between any two of them the same ratio as the corresponding perpendiculars on *AB* in semicircle *ACB*, it follows that there can be erected on base *AB* a figure of less height but which will be proportional in altitude to this semicircle and with equal reason a figure of greater height by any amount you wish. Therefore, by chapter seven the quality of line *AB* can be correctly imagined by any of

these figures without it making any difference [which figure is used].[20]

For if it were not so that the quality of line *AB* imaginable by the semicircle could be imagined by a figure greater or less than the semicircle which is proportional [in altitude to the semicircle] it would follow (1) that the intensity of point *D* could not be correctly designated by a greater or lesser line than *DC*, and similarly for all the points, unless the intensity were varied, and thus (2) that any intensity would in itself determine the definite length of the line by which it would be imaginable, and then (3) an intensity would be equivalent and comparable to a line or to quantitative extension, and as a consequence (4) local motion would be comparable in velocity to [qualitative] alteration, all of which seems excessively absurd.

However, any figure by which this quality of line

19. In this case the angle would be a mixed angle composed of the curve and the straight base line [*Nicole Oresme. . .*, p. 199, n. 1—*Ed.*].

20. [In coping with the problem of the quality represented by a semicircle, Oresme altered an earlier view and—*Ed.*] now realized in the *De configurationibus* that such a quality could be represented by higher or lower figures whose altitudes (i.e., ordinates) are in the same ratio as those of the semicircle. He was puzzled as to what kinds of figures these would be. He definitely rejects the possibility [see the final paragraph of I.xiv—*Ed.*] that the figures of higher altitude could be segments of a circle, while he says that he will not consider the case of the figures of lower altitudes. Incidentally, the compositor of MS J of the *De configurationibus* independently proves the case of the figures lower than the semicircle and further suggests that a similar proof can be constructed for the case of the figures higher than the semicircle. [Clagett gives the Latin text and translation in Appendix III of his edition.—*Ed.*]. Unfortunately, Oresme had little or no knowledge of conic sections. For, in fact, the conditions he specifies for these curves comprise one of the basic ways of defining ellipses: if the ordinates of a circle $x^2 + y^2 = a^2$ are all shrunk (or stretched) in the same ratio a/b, the resulting curve is an ellipse whose equation is $x^2/a^2 + y^2/b^2 = 1$. Hence, without realizing it, Oresme has given conditions which show that the circle is merely one form of a class of curves we call elliptical. It should be remarked that even if Oresme had had some knowledge of Greek conic sections, he might well have had difficulty in applying the Greek formulation of the ellipse to his figures, which are like curves in Cartesian geometry. [*Nicole Oresme . . .*, pp. 442–443; in the next chapter Oresme elaborates four kinds of simple difform difformity: (1) rational convex, (2) rational concave, (3) irrational convex, and (4) irrational concave. In I.xv, these four are added to simple uniformity and uniform difformity to constitute six simple figurations of qualitative intensity from which sixty-three species of composite difformity can be formed—*Ed.*].

AB is imaginable is curved. But whether the figure less than a semicircle by which this quality can be imagined is a segment of a circle, I leave aside as a matter to be discussed. But I do say that it cannot be designated by a greater figure which is at the same time a segment of a circle. For this quality can be designated by no figure of which *AB* is not the base or chord. But *AB* cannot be the chord in a circle smaller than circle *ACB* if that circle were completed, for *AB* is the diameter of that circle. Therefore, this quality cannot be imagined by a greater figure which is a segment of a smaller circle than circle *ACB;* nor also of a greater figure which is a segment of a larger circle. [This last is evident,] for then that segment would either (1) be greater than half of its own circle, and therefore no quality could be designated by it, as is clear from chapter five, or (2) it would be less than half its own circle. [But in the case of the second possibility,] since the segment which is less than half of a larger circle would have the same chord as semicircle *ACB,* the segment would be less [in area] and would be a part of semicircle *ACB,* as is easily evident and can be proved by the last [proposition] of the sixth [book] of [the *Elements* of] Euclid. Therefore, this quality cannot be designated by a figure which is a segment of a circle and is [at the same time] greater [in altitude and area] than semicircle *ACB,* and yet it can be designated by [some] greater curved figure, as was proved before. Therefore, the curvature of the greater [curved] figure will not be circular but will bound a figure which in altitude is proportional to that which the circular curvature bounds; and so there will be two figures proportional in altitude, the curvature of one being circular and that of the other being non-circular.

HERE BEGINS THE SECOND PART OF THIS TRACT AND IT TREATS OF THE DIFFORMITY OF SUCCESSIVE THINGS

II.i On the Double Difformity of Motion

Every successive motion of a divisible subject has parts and is divisible in one way according to the division and extension or continuity of the mobile, in another way according to the divisibility and duration or continuity of time, and in a third way—at least in imagination—according to the degree and intensity of velocity. From its first continuity motion is said to be "great" or "small"; from its second, "short" or "long," and from its third, "swift" or "slow.". . .

II.iii On the Quantity of the Intensity of Velocity

Since each uniformity of motion posited in the first chapter consists in equality of intensity and each difformity arises from inequality [of intensity] we ought to set out first [the measure of gradual intensity, i.e. we ought to specify] with what the gradual intensity of the velocity is measured. However, in the matter of velocity three closely related ideas can be considered. One is the total quantity of the velocity taking into account both intensity and extension. I shall speak of this in the third part of this tract, which will be concerned with the measures of qualities and velocities. Another thing to be considered in connection with velocity is the denomination in terms of which a subject is said to become such a kind more quickly or more slowly. I shall also speak of this in the following chapter. Third, there is the gradual intensity [of velocity]. This is the subject which must now be considered. Therefore, I say universally that that degree of velocity is absolutely more intense or greater by means of which in an equal time more is acquired or lost of that perfection according to which the motion takes place. For example, in local motion that degree of velocity is greater and more intense by means of which more space or distance would be traversed. In alteration, similarly, that degree of velocity is greater by means of which more intensity of quality would be acquired or lost; and so in augmentation, by means of which more quantity is acquired, and in diminution, by means of which more quantity or extension is lost. And so generally [our definition would hold] wherever motion would be found.

II.v On Certain Other Successions in Motion

. . . .

There can be imagined one further succession,[21] for every velocity is capable of being increased in intensity and decreased in intensity. Now its continuous increase in intensity is called acceleration, and indeed this acceleration or augmentation of velocity can take place more quickly or more slowly. Whence it sometimes happens that velocity is increasing and acceleration is decreasing, while sometimes both are simultaneously increasing. Similarly acceleration of this sort sometimes takes place uniformly and sometimes non-uniformly and in diverse ways. But since every divisibility or succession which is found in acceleration of this sort is according to parts of the subject, or accord-

21. [The first part of II.v considers "succession according to inception," where successive parts of a body begin to be moved while yet others remain at rest, and succession according to quantitative parts—*Ed.*].

ing to parts of the time, or according to gradual intensity, and from such threefold divisibility arises twofold uniformity or difformity, as was demonstrated in the first chapter, therefore, as before, every uniformity and difformity which can so arise is reduced to the two above-mentioned kinds, that is, to uniformity and difformity according to parts of the subject or uniformity and difformity according to parts of the time. And so let us first speak of that which is according to parts of the subject.

HERE BEGINS THE THIRD PART [OF THIS TREATISE]: ON THE ACQUISITION AND MEASURE OF QUALITIES AND VELOCITIES

III.i How the Acquisition of Quality Is To Be Imagined

Succession in the acquisition of quality can take place in two ways: (1) according to extension, (2) according to intensity, as was stated in the fourth chapter of the second part. And so extensive acquisition of a linear quality ought to be imagined by the motion of a point flowing over the subject line in such a way that the part traversed has received the quality and the part not yet traversed has not received the quality. An example of this occurs if point *c* were moved over line *AB* so that any part traversed by it would be white and any part not yet traversed would not yet be white [see Fig. 6(a)]. Further the extensive acquisition of

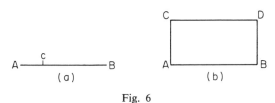

Fig. 6

a surface quality ought to be imagined by the motion of a line dividing that part of the surface that has been altered from the part not yet altered. And the extensive acquisition of a corporeal quality in a similar way is to be imagined by the motion of the surface dividing the part altered from the part not yet altered.[22]

The intensive acquisition of punctual quality is to be imagined by the motion of a point continually ascending over a subject point and by its motion describing a perpendicular line imagined [as erected] on that same subject point. But the intensive acquisition of a linear quality is to be imagined by the motion of a line perpendicularly ascending over the subject line and in its flux or ascent leaving behind a surface by which the acquired

quality is designated. For example [see Fig. 6(b)], let *AB* be the subject line. I say, therefore, that the intension of point *A* is imagined by the motion, or by the perpendicular ascent, of point *C*, and the intension of line *AB*, or the acquisition of the intensity, is imagined by the ascent of line *CD*. Further, the intensive acquisition of a surface quality is in a similar way to be imagined by the ascent of a surface, which (by its motion) leaves behind a body by means of which that quality is designated. And similarly the intensive acquisition of a corporeal quality is imagined by the motion of a surface because a surface by its imagined flux leaves behind a body, and one does not have to pose a fourth dimension, as has been said in the fourth chapter of the first part.

One should speak and conceive of the loss of quality in the same way that we have now spoken of its acquisition, whether that loss is of extension or intensity. For such loss is imagined by movements which are the opposite of the movements described before. Furthermore, one ought to speak of the acquisition or loss of velocity, both in extension and intensity, in the same way we have just spoken of the acquisition or loss of quality.

III.vii On the Measure of Difform Qualities and Velocities

Every quality, if it is uniformly difform, is of the same quantity as would be the quality of the same or equal subject that is uniform according to the degree of the middle point of the same subject.[23] I understand this to hold if the quality is linear. If it is a surface quality, [then its quantity is equal to that of a quality of the same subject which is uniform] according to the degree of the middle line; if corporeal, according to the degree of the middle surface, always understanding [these concepts] in a conformable way. This will be demonstrated first for a linear quality. Hence let there be a quality imaginable by △*ABC*, the quality being uniformly

22. For reasons of symmetry, Oresme also extends his concept to the production of surface and corporeal qualities by the movement of lines and surfaces. But neither here nor in the next chapter on the summits of the configurations representing surface or corporeal qualities does Oresme take up the difficulties of actually using his imagery. His knowledge of geometry was simply inadequate for any detailed discussions of the kind he presents for linear qualities [*Nicole Oresme. . .*, p. 491—*Ed.*].

23. This is, of course, the famous Merton College Rule of uniformly difform [motion—*Ed.*]. And this chapter with its geometric proof constitutes the most significant chapter of the work, historically speaking [*Nicole Oresme. . .*, p. 494—*Ed.*].

difform and terminated at no degree in point *B* [see Fig. 7(a)]. And let *D* be the middle point of the

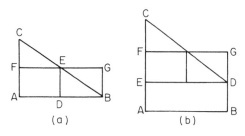

Fig. 7

subject line. The degree of this point, or its intensity, is imagined by line *DE*. Therefore, the quality which would be uniform throughout the whole subject at degree *DE* is imaginable by rectangle *AFGB*, as is evident by the tenth chapter of the first part. Therefore, it is evident by the 26th [proposition] of [Book] I [of the *Elements*] of Euclid that the two small triangles *EFC* and *EGB* are equal. Therefore, the larger △*BAC*, which designates the uniformly difform quality, and the rectangle *AFGB*, which designates the quality uniform in the degree of the middle point, are equal. Therefore the qualities imaginable by a triangle and a rectangle of this kind are equal. And this is what has been proposed.[24]

In the same way it can be argued for a quality uniformly difform terminated in both extremes at a certain degree,[25] as would be the quality imaginable by quadrangle *ABCD* [see Fig. 7(b)]. For let line *DE* be drawn parallel to the subject base and △*CED* would be formed. Then let line *FG* be drawn through the degree of the middle point which is equal and parallel to the subject base. Also, let line *GD* be drawn. Then, as before, it will be proved that △*CED* = □*EFGD*. Therefore, with the common rectangle *AEDB* added to both of them, the two total areas are equal, namely quadrangle *ACDB*, which designates the uniformly difform quality, and the rectangle *AFGB*, which would designate the quality uniform at the degree of the middle point of the subject *AB*. Therefore, by chapter ten of the first part, the qualities designatable by quadrangles of this kind are equal.

It can be argued in the same way regarding a surface quality and also regarding a corporeal quality. Now one should speak of velocity in completely the same fashion as linear quality, so long as the middle instant of the time measuring a velocity of this kind is taken in place of the middle

point [of the subject].[26] And so it is clear to which

24. It is clear, then, that what Oresme has done is to give a rudimentary geometric proof of the so-called Merton Rule of measuring uniform acceleration. [*Nicole Oresme. . . .*, p. 46; see Heytesbury's statement, without formal proof, of the Merton Rule, or mean speed theorem, and Clagett's commentary, in the preceding selection—*Ed.*].

The question has often been raised in connection with this proposition of III.vii and its geometric proof whether Oresme realized that the areas of his figures—that is, the areas of the right triangle and rectangle in this proof—represented, or were in the same ratio as, the distances when the theorem was applied to velocities. I think that there can be little doubt that he did realize this, although he certainly does not make a central point of it. One has to remember that in the Merton Rule of uniformly difform motion, it is the distances of the uniform motion and uniform acceleration that are the same, and indeed this is so specified by Oresme in his *Questions on the Geometry of Euclid* (Appendix I, Question 13, lines 72–77). Hence when it is shown in this chapter of the *De configurationibus* that the areas are equal, it is evident that they must represent distances or, to put it more precisely, whatever ratio one area has to another one distance will have to another. This is even more apparent in III.viii of the *De configurationibus*, where after determining the ratios of the various areas over the proportional parts of the base, Oresme specifically states that the distances traversed are in the same ratios as those already specified for the areas. Thus it would appear that for Oresme, just as the simple denomination of surfaces was to be determined by attention to the dimensions of length and breadth, so the denomination of motions was to be determined by the distances which the areas represented or were proportional to, and thus by attention to the "latitudes" and "longitudes" of motion. Admittedly this is somewhat obscured by the method of calculation by means of the compound ratios of intensities and extensions outlined in III.vi; but this obscurity rises from the medieval and antique practice of using Euclidian proportionality statements in which the ratios of like quantities must be used, so that the ratios of the motions are determined by the form: Total velocity a/Total velocity $b = (v_1/v_2) \cdot (t_1/t_2)$, where v_1 and v_2 are the average velocities, while the modern expression would simply be: Distance a/Distance $b = v_1 t_1/v_2 t_2$. That the techniques of integral calculus are needed for more rigorous proofs concerning "areas under curves," as Oresme's figures would be called in analytic geometry, should not blind us to the apparent fact that for Oresme these figures were representative of, or proportional to, the distances traversed, or that in his terms the "total velocity" imagined by such a figure is strictly proportional to the distance traversed in the time imagined by the base line of the figure [*Nicole Oresme. . . .*, p. 47—*Ed.*].

25. [This case is merely mentioned without discussion by Heytesbury; see above, Part VI *(continued)*, par. 7.—*Ed.*]

26. This statement shows that Oresme realized that the same geometric proof holds for uniformly difform motion, i.e., uniformly accelerated motion. Hence it is

uniform quality or velocity a quality or velocity uniformly difform is equated. Moreover, the ratio of uniformly difform qualities and velocities is as the ratio of the simply uniform qualities or velocities to which they are equated. And we have spoken of the measure and ratio of these uniform [qualities and velocities] in the preceding chapter.

Further, if a quality or velocity is difformly difform, and if it is composed of uniform or uni-

formly difform parts, it can be measured by its parts, whose measure has been discussed before. Now, if the quality is difform in some other way, e.g. with the difformity designated by a curve, then it is necessary to have recourse to the mutual mensuration of the curved figures, or to [the mensuration of] these [curved figures] with rectilinear figures; and this is another kind of speculation. Therefore what has been stated is sufficient.

Dynamics

44 DOES FINITE AND TEMPORAL MOTION REQUIRE A RESISTANT MEDIUM? THE RESPONSES OF AVERROES AND AVEMPACE IN COMMENT 71

Averroes (1126–1198)

Translations, introduction, and annotation by John E. Murdoch and Edward Grant[1]

The comment by Averroes on the text of Aristotle from Book IV of the *Physics* presented here proved to be one of the most important in medieval physics. Here we find Averroes agreeing with Aristotle that finite motion could occur only in a resistant medium and arguing against Avempace (Ibn Bajja, a Spanish Muslim of the twelfth century), who insisted that temporal motion could occur in a resistance-less medium and therefore, presumably, in a void, though strictly speaking he does not mention void in the passages quoted by Averroes. So important was this particular text of Averroes (it was known in the Middle Ages as Comment

71 of Bk. IV) that it gave rise not only to discussions about the possibility of motion in a hypothetical void but also raised the question as to whether inorganic elemental bodies could be self-moved. (Although Averroes denied the self-motion of inorganic bodies, his explanation was ambiguous and seemed to allow it; in any event, Thomas Aquinas criticised him for invoking the form of a body as the mover in natural motion. See below, n. 35). A number of selections in this source book are concerned, wholly or in part, with the issues just described.

1. [The Text of Aristotle][2]

And that which we have asserted is also made apparent in the following way: We see that one and the same body is moved faster because of one or the other of two reasons: either (1) because the medium in which it is moved is different (for example, its motion may occur in water, or in earth, or in air, or in fire), or (2) because the moved thing [itself] is different.

If, therefore, other things, which are due to the difference in heaviness and lightness [of the moved body], are equal, the medium in which the motion will occur will be the cause.[3] For it resists the motion, either to a great extent, when the motion of that [medium] is contrary to the motion of that which is moved in it, or much less so, when it

the Commentaries of Averroes] (Venice, 1562–1574; Frankfurt: Minerva, 1962), Vol. IV, fols. 158 recto, col. 1, to 162 recto, col. 1. Aristotle's discussion, which prompted Averroes' lengthy commentary and which immediately precedes it in the volume cited, has also been translated here from the version rendered from Arabic into Latin in the twelfth century.

2. The text is part of Book IV, chapter 8, of Aristotle's *Physics* (215a.24—215b.20).

3. It is interesting to compare this with the Oxford translation made directly from the modern Greek text of Aristotle. "We see the same weight or body moving faster than another for two reasons, either because there is a difference in what it moves through, as between water, air, and earth, or because, other things being equal, the moving body differs from the other owing to excess of weight or of lightness." For Aristotle, the phrase "other things being equal" (or "are equal") is a qualification establishing the possibility that a difference in weight may be the cause of a greater, or lesser, speed. That Averroes understands this passage to be a qualification which *excludes* the possible differences in weight, and hence refers to the alternative, that a variation in media is the cause, is clear from his gloss below on "other things are equal."

---

clear that this proof is essentially the same as Galileo's first theorem in the "Third Day" of the *Discorsi*. . . [*Nicole Oresme*. . . , p. 495—*Ed.*].

1. Translated from Averroes' Comment 71 on Book IV of Aristotle's *Physics*, as found in *Aristotelis opera cum Averrois commentariis* [*The Works of Aristotle with*

[that is, the medium] is at rest. And the [medium] offers greater [resistance] when it is not easily divided, that is, when it is such as that in which there is a density *(spissitudo)*.

Therefore, let body *A* be moved *(expellatur)* in medium *B* for time *C*, and for time *H* in medium *D*, which is more rare *(subtilius)* and, insofar as it is an impeding [or resisting] body, equal in length to *B*. The purpose of this is the following: Since *B* could be water and *D*, air, then to the extent that air is rarer and of lighter corporeity, to that extent will *A* be moved faster in medium *D* than in medium *B*. And let the ratio of speed to speed be as the ratio of the excess [in rarity] of air over water in such a way that if [the air] is twice as rare, then [*A*] will traverse space *B* in twice the time in which it traverses space *D*. And then the time *C* will be double time *H*. And always by so much as the medium in which there will be motion is of lighter

corporeity and of less resistance and more easy to divide, so much faster will be the movement *(expulsio)*.

Indeed, a void has no ratio *(comparationem)* to a body by which that body exceeds it, just as that which is called zero[4] has absolutely no ratio to that which is called a number. For if 4 exceeds 3 by 1, and exceeds 2 by more, and further exceeds 1 by even more, still it does not have a ratio to zero *(non unum)* by which it exceeds zero. For that which exceeds is necessarily divisible into that which is exceeded and the excess *(augmentum)*. Therefore, 4 is divisible into these two things: into the excess [by which it exceeds zero] and into zero.[5] Thus it is that a line does not exceed a point, since it is not composed of points. Similarly, a void does not have a ratio to a plenum; therefore, there will be no motion in it.

2. [Averroes' *Expositio* on the Text]

This is another demonstration in which he [Aristotle] declares that it is impossible for motion to occur in a medium which is a void. And he has said: *"And that which we maintain is also apparent"*,[6] namely that it is impossible that motion occur in a void. And this is apparent above because we see that two heavy bodies differ in motion according to speed and slowness *(velocitatem et tarditatem)* for two reasons. And these are (1) If the body be the same or if there be two bodies equal in weight, then the difference will be due to that in which they are moved. For example, the same body is moved faster in air than in water (the case is similar with two bodies equal in weight, shape, and size). However, (2) If the bodies are different, and they are moved in one medium, then the difference in motion will be due to the difference in heaviness and lightness *(gravitas et levitas)*, that is, that the heavier of them will be faster. And, after he has asserted this proposition—namely that the cause of the speed and the slowness in movable bodies is either the difference in the density and rarity *(spissitudo et tenuitas)* of the medium, or the difference in heaviness and lightness of the bodies— he says, *"Other things. . . are equal."* That is, if the other things that exist in bodies [and which are] capable of causing speed or slowness (as the difference in weight, shape, and size of bodies) are the same in two different bodies, then, as we see that that which is moved in air is moved faster than that which is moved in water, it is quite clear that the

cause of the diversity in motion is the medium in which the motion occurs. And the same thing is true in the case of one body, that is, we see that its motion in a rarer body [or medium] is faster than in a denser body [or medium].

Then he specifies the reason why this occurs in a medium and says, *"As it resists* [the motion] *to a great extent,* and so on." That is, a retardation is caused because that in which the motion will occur [that is, the medium] will resist the thing moved, and this resistance varies by degrees. For it will resist a great deal when the medium is moved with a motion contrary to that which is moved in it. And it will resist less when it is at rest.

Then [Aristotle] says, *"And* [the medium] *offers*

---

4. What has been translated here as "zero" is literally "that which is not called 1 *(quod non dicitur unum)*.

5. In his exposition of this sentence Averroes makes it clear that this division of 4 into an excess and zero is impossible. Therefore, 4 does not really exceed zero; nor, *mutatis mutandis,* does a plenum exceed a void in density, or a void a plenum in rarity.

6. Here, and in all subsequent instances in this translation, the cue words drawn from the Aristotelian text just translated are italicized in our translation of Averroes' exposition. Frequently, for greater ease of reference and intelligibility, more of the English translation is repeated than is warranted by the few cue words cited by Averroes. It should also be noted that the Latin cue words are not always those actually found in the medieval text of Aristotle and therefore, in those instances, will not be the very same words that appear in our translation.

greater [resistance], and so on." That is, a medium which, by its motion, resists a contrary motion, resists more when it is dense and difficult to divide. Therefore, the cause of the diversity of motions of the same moved thing—or of mobiles like one another—depends on the difference of the media with respect to two factors, namely in the difficulty and facility of the resistance offered by these media, be they in motion or at rest.

Then he tells us what [kind of medium] is difficult to divide and says, *"It is that in which there is great density."* And, when he has asserted this proposition, he begins to explain his contentions by using letters. He says, *"Therefore, let* [body *A*] *be moved and so on."* That is, therefore, let *A* be the body which is moved in a medium, and let *B* be the medium in which it is moved, and let *C* be the time during which it is moved. Moreover, let it be assumed that *A* is moved in another medium more rare than the first, and let this [second medium] be *D*, and the time during which *A* is moved in this medium *H*. And let *A* be moved through equal spaces in the two media, that is, in the more dense *B* and the more rare *D*. Therefore, it follows that the time in which [*A*] traverses these two spaces will be different. For we know that the swifter traverses the same space in less time. Moreover, since the cause of speed and slowness in these two motions [that is, the motion of *A* in *B* and in *D*] is the difference of the media in rarity and density, it follows that the ratio of time to time is as the ratio of the density which exists in one medium to that which exists in the other. Similarly, so stands the ratio of motion to motion.[7]

And since he [Aristotle] gives an example in letters, he [also] gives an example in terms of [elemental] matter so that the demonstration will be clearer and that what seems less [clear] in one example will appear in the other. And he says, *"The purpose of this is* [the following], *and so on"*; for example, water the denser medium, is assumed in place *B*, and air, the rarer medium, in place *D*.[8] Therefore, as air is rarer than water and of less resistance against something that is moved [in it], so will the velocity of motion *A* in it be greater than in water. Hence the ratio of speed to speed will be as the ratio of the excess of the rarity of air over the rarity of water. And indeed he says, *"Therefore the ratio of speed to speed will be, and so on,"* that is, it follows from these propositions that a ratio of the speed of the motion of *A* in *B* to its motion in *D* is as the ratio of the excess of the rarity of air over the rarity of water. Then he says

that *"If* [the air] *is twice as rare, and so on,"* that is, it follows that if we assume that air is twice as rare as water, a thing moved in water traverses in twice the time a distance equal to the distance it traverses in air. Therefore, the time *C*, in which the moved thing traverses a distance in the denser body [or medium], is twice time *H*, in which it is moved an equal distance in the rarer medium, *D*. For it always traverses the same distance slower, [that is,] in a longer time. And although he does not express this proportion, it must, nevertheless, be understood from his discussion.

Then he says, *"And always by so much as the medium, and so on,"* and this is also self-evident. And when he posited these propositions, he also joined to them another proposition, and he says, *"Indeed, a void* [has no ratio to a body by which that body exceeds it], *and so on."* That is, the rarity of a void has no ratio to the rarity of a body that is the medium where a motion occurs. And when he says, *"It has no ratio to a body,"* he means that the void has no ratio—for example, in rarity— by which it exceeds the body that is the medium. For in a void there is no rarity by which it could be

7. Here and elsewhere Averroes' text reads so that one cannot be absolutely certain whether the proportionality between motions and densities or resistances is direct or inverse. That is, if *V* is velocity, or speed, and *R* resistance, does he mean $V/V' = R/R'$ or $V/V' = R'/R$? Surely we would say the latter, inverse proportionality, is intended. However, there was good reason for confusion concerning this matter in the Middle Ages. One could use two kinds of terms to refer to the resistance of a medium: (1) One could speak of its density (*spissitudo* or *crassitudo*) and of one medium being more dense (*spissior* or *crassior*) than another. Thus, interpreting the resistance of a medium as its density, *inverse* proportionality would obtain between resistances and motions: $V/V' = R'/R$. For (other things being equal) an increase in speed follows from a decrease in resistance as density, and vice versa. But (2) one could also refer to the resistance of a medium as its rarity (*subtilitas* or *tenuitas*) and of one medium being more rare (*subtilior*) than another. Hence, following the Aristotelian–Scholastic notion of density and rarity as contraries, one could maintain that *direct* proportionality obtained between motions and resistances when the latter were interpreted as the respective rarities of media: $V/V' = R/R'$. For (other things being equal) an increase in speed follows from an increase in resistance as rarity and a decrease from a decrease. (Averroes says just this in the first paragraph of the exposition of his own opinion below.) This ambiguous use of "proportional" also occurs in the comparison of other quantities, for instance, resistances to retardations. It will be left to the reader to fill in "direct" or "inverse" when such equivocation occurs below.

8. The text mistakenly says *"C."*

said to be rarer than fire, for example, or air. Then he says, "*Just as* [that which is called zero] *has* [absolutely] *no ratio* [to that which is called a number], and so on." That is, there is no ratio between void and the body that is the medium (since there is no rarity in a void; but it is [by means of a ratio] that media are mutually compared with respect to rarity and density), just as there is no comparison or ratio between zero[9] and a number. For numbers are mutually compared because 1 is their common measure *(communicant in uno)*. And, universally, any two things which lack a common measure with respect to the same nature do not have a mutual ratio. And Aristotle intends this when he says, "*For if 4 exceeds 3* [by 1], and so on." That is, numbers exceed each other by the units in which they have their common measure. But zero cannot be compared to a number. Similarly, what lacks rarity cannot be compared to bodies possessing rarity. Indeed, it is in virtue of "more" or "less" that different things necessarily have a common measure with respect to the same nature.

Then he says, "*Still* [it does not have a ratio to] *zero, and so on.*" That is, what is composed of zeroes [that is, things in which 1 is not found] does not bear a ratio to anything that contains 1, however much the excess between them diminishes. Then he says, "*For that which exceeds* [is necessarily divisible], and so on." That is, what is said to exceed something is necessarily divided into that which is exceeded and the excess. For example, since 4 exceeds 2, it is necessary that it be divided into the 2 which is exceeded and the 2 which exceeds. Therefore, when there is something which is not divided into something and there is another thing which is divided into something, it cannot be said that what is divided into something exceeds what is not divided into anything. For example, 4 cannot be said to exceed zero[10] because if it should exceed it, then 4 would be divisible into an excess and into zero, which is impossible. And this is what Aristotle intended when he said, "*Therefore, 4 is divisible* [into these two things], and so on." That is, since it is necessary that 4 be divided into that by which it exceeds something which contains 1 and into zero [that is, that which lacks 1 *(in quo non est unum)*], and because the ratio that he intended to declare is common to numbers and other things, and because he gave an example in numbers, he wished also to give an example in magnitudes, and so he says, "*Thus it is that a line does not exceed a point.*" That is, it is because of this that there is no ratio between a line and a point from

which an excess could arise, because if an excess did arise then that which exceeds and that which is exceeded would be parts of a line so that a line would be composed of points which will afterwards be declared to be impossible.

And when he has asserted this proposition, saying (1) that everything which exceeds something should be divided into an excess and the thing exceeded, and (2) that a void is not divided into that by which it exceeds a body [that is, exceeds it in rarity]—since there is no rarity in a void—he sets down the conclusion which follows from this. That is, that a void has no ratio to a plenum by which it may be said to exceed the plenum in rarity. And he said, "*Similarly, a void* [does not have a ratio to a plenum], *and so on.*" This syllogism is also in the second figure and what he has added[11] is the minor proposition [or premise]. When he has asserted this conclusion he reveals the impossibility that follows from it if something were to be moved in a void, and says, "*Therefore, there will be no motion in it.*" That is, since a void has no ratio to a plenum in rarity, and something is moved in it, it follows that motion in a void has no ratio to motion in a plenum, and thus [the former] will be an indivisible motion occurring in an indivisible time, namely in an instant. But this is impossible, as will be shown later. For the present, he takes for granted what is declared in book VI, to wit, that every motion takes time and that motion is divisible.

[AVEMPACE'S CRITICISM]

Avempace, however, here raises a good question. For he says that it does not follow that the ratio of the motion of one and the same stone in water to its motion in air is as the ratio of the density of water to the density of air, unless we assume that the motion of the stone takes time only because it is moved in a medium.[12] And if we make this assumption, it would imply that motion only takes time because of something resisting it—for the medium seems to impede the thing moved. And, if this were the case, then the heavenly bodies would be moved instantaneously as they have no medium resisting

9. Here again, and throughout the Latin text, what has been translated as "zero" is given literally as "that which is not 1" or "that which lacks 1" [that is, lacks unity] *(quod non est unum)*.
10. Literally, "something in which there is no 1" *(aliquid in quo non est unum)*.
11. "Subticuit" (!) has been altered to *subiacuit*.
12. Thus Avempace rejects the opinion of Aristotle and Averroes.

them. And he says that the ratio of the rarity of water to the rarity of air is as the ratio of the retardation (*tarditatis*) suffered by the moving body in water to the retardation suffered by it in air.

And these are his own words, in the Seventh Book of his work, where he says: "This resistance which occurs between the plenum and the body which is moved in it is that between which and the potency of the void Aristotle made a ratio in his Fourth Book.[13] And what is believed because of his opinion is not so. For the ratio of water to air in density is not as the ratio of the motion of the stone in water to its motion in air. On the contrary, the ratio of the cohesive power (*potentiae continuitatis*) of water to that of air is as the ratio of the retardation suffered by the moved thing due to the medium in which it is moved, namely water, to the retardation suffered by it when it is moved in air.[14]

"For, if what certain thinkers have believed were true, then natural motion would be violent. Thus, if there were no resistance present, how could there be motion? For it would necessarily be instantaneous. Moreover, what then shall be said concerning circular motion [that is, the circular motion of the heavenly spheres]? No resistance is there, since there is absolutely no division there, for the place of a circle [that is, of a celestial sphere] is always the same, so that it does not abandon one place and enter another. Therefore, it is necessary that circular motion occur instantaneously. Yet we observe in it the greatest slowness, as in the case of the motion of the fixed stars, and also the greatest speed, as in the case of the diurnal rotation.[15] And this is only due to a difference in perfection *(nobilitate)* between the mover and the thing moved. Therefore, when the mover is of greater perfection, that which is moved by it will be swifter; and when the mover is of less perfection, it will be nearer [in perfection] to the thing moved, and the motion will be slower."[16]

[AVERROES' CRITICISM OF AVEMPACE]

And these are his [Avempace's] words. And if that which he [Avempace] has said be conceded, then Aristotle's demonstration will be false. For, if the ratio of the rarity of one medium to the rarity of another medium is as the ratio of the retardation suffered by a moved thing in one of them to the retardation suffered by it in the other, and is not as the ratio of the motion itself, it will not follow that what is moved in a void is moved instantaneously. For if this [that is, Avempace's contention]

were the case, then there would be subtracted from the mobile's motion only the retardation which affects it by reason of the medium, and its natural motion would remain. And as every motion involves time, that which is moved in a void is also necessarily moved in time and with a divisible motion.[17] If this is so, nothing impossible follows. This, therefore, is Avempace's question.[18]

13. This is a reference to the next passage (215b.23—216a.7) following the text of Aristotle translated above (see n.2).

14. It is clear that Avempace's formulation gives us only the ratio of retardations for bodies falling in resisting media. It says nothing, however, about the manner in which bodies would fall in resistanceless media or in a hypothetical void. Thus we are unable to determine from this passage whether for Avempace such fall was to be understood as proportional to weight, or to dimension, or perhaps to some indwelling motive power. Avempace's vagueness stands in contrast with Philoponus' discussion of the same problem (it is possible that Avempace derived his view point from Philoponus, a sixth-century Greek commentator on Aristotle), for the latter insists that fall in the void can be determined by the inverse proportional relation between the weights of the falling bodies and their respective times of fall over equal distances (that is, $W_2/W_1 = T_1/T_2$, where W is the weight of a body and T its time of fall) from which it would follow that $W_2/W_1 = V_2/V_1$ when $S_2 = S_1$ (V is velocity and S distance). We see, then, that Philoponus has actually adopted for fall in the void the same relation that Aristotle had employed to represent exclusively the fall of bodies in media. See M.R. Cohen and I. E. Drabkin, *A Source Book in Greek Science* (Cambridge, Mass.: Harvard University Press, 1948), p. 218.

15. Latin scholastics in agreement with Avempace frequently cited celestial motion as the most basic illustration of temporal motion in a resistanceless medium (for example, see Selections 55.1,2.)

16. What Avempace understood by greater or lesser perfection of movers is unclear; nor is it readily apparent from the context of the discussion that he intended this last sentenec as a general assertion applicable to both celestial and terrestrial motions (see also n. 24 and n. 25.)

17. The quotation from Avempace leads Averroes to infer that Avempace's general rule for motion is $V = V' - r$, where V is the velocity of a body in a resistant medium, V' its natural velocity in a void or resistanceless medium, and r the retardation of the natural motion caused by the resistance of the medium. This rule has sometimes been improperly formulated as $V = F - R$, where V is velocity, F the motive power or weight of a falling body, and R the resistance of the medium. Thus when $R = 0$, a body would fall in the void with a natural speed proportional to its motive power or weight (see n. 14 and n. 25). Indeed the confused manner of Averroes' presentation led to this very interpretation of it in the Middle Ages, an interpretation more akin to that of John Philoponus (for Thomas Bradwardine's repudiation of this form of the rule see the discussion of his first erroneous theory at the beginning of Selection 51.1; for

However, we maintain that it is self-evident that if (1) we are given things equal in weight, shape, and size, and (2) these things be moved in one and the same medium, or, indeed, in two media of equal rarity and density, and (3) the cause of the equality of their motions is held to be the equality of the medium, *then* (4) it is also clear that the diversity of the medium, with respect to more, or less, rarity, is the cause of the diversity of motion with respect to fastness or slowness. Moreover, this diversity— whose cause is the diversity of the medium— essentially follows the rarity [of the medium], namely that an increase of this diversity follows from an increase in rarity, and a decrease follows a decrease. Therefore, it is obvious that the ratio of the motions varies as the ratio of the media in rarity and density.

These propositions are self-evidently clear: namely that when two bodies, equal in heaviness and lightness, in shape and size, are moved in one and the same medium, their motion will be equal in speed, and both will traverse an equal space in an equal time. And [secondly], when the two media [in which these bodies move] are of different rarity, their motion will be diversified with a proportional difference.

It is also universally clear that the cause of the diversity, and the equality, of motions, is the diversity and equality of the ratio of the mover to the thing moved.[19] Therefore, whenever there are two

Galileo's acceptance of it and the special Archimedian interpretation which he placed upon it, see Selections 55.3–5 and n. 58. While many discussed it, few in the Latin West adopted and applied this rule to bodies falling in an actual corporeal resistant medium. However, largely as a consequence of this passage, it became commonplace during the Middle Ages to argue that a body would move through a hypothetical void with a finite and temporal motion and not instantaneously, as Aristotle and Averroes had insisted (for example, see Selection 55.1 and the quotation from Albert of Saxony in Selection 47, n. 6).

18. In this translation of Comment 71, the parts under "Avempace's Criticism" and "Averroes' Criticism of Avempace" were previously translated by E. A. Moody in his important article "Galileo and Avempace," *Journal of the History of Ideas*, XII (1951), 184–186. For the most part, our translation follows his.

19. A word should be said about Averroes' various statements about what the "diversity" or "ratio" of motions does follow. He tells us variously that the ratio of motions follows: (a) the ratio of the densities or rarities of the media involved, or (b) the proportion (ratio) of the resistances involved, or (c) the ratio of the ratios of the movers to the things moved. As a reading of

the sequel makes clear, these are not mutually exclusive. For a resistance necessarily arises between a mover and a moved thing (see Averroes' "Resolution of the Objection," following). Thus to assert (c) implies that the ratio of motions follows a ratio between some resistances or other. However, it might be asked whether or not these latter resistances are due to the respective densities or rarities of media; in other words, can we directly relate (c) with (a) and (b)? To answer this question we must determine more exactly the function of media in the motion of bodies for Averroes. To do this we must anticipate something of what he has to say later on in the present selection. The following considerations are relevant: (1) In the case of the violent motion of bodies, the resistance arising from the relation of mover to moved is due either to the moved body—that is, to its tendency to seek its natural place—or to the resistance to penetration offered by the medium through which the body moves (see the beginning of "Exposition of Averroes' Opinion Resumed," following). (2) In the case of the motion of heavenly bodies and animals, the resistance which arises is due solely to the moved body itself. That is, it is due to the action of the form ("Intelligences" and "Souls," respectively) of such bodies acting on the matter of the bodies (see the final paragraph of the translation). In the case of the natural motion of inorganic bodies (both "simple" and compound) the resistance is due to the medium alone (see the final paragraph of the translation). But one remark must be made apropos this third, and most important, distinction. We must note that merely because this resistance is due to the medium itself *(ex ipso medio)* does not mean that there is not also a resistance obtaining between the mover and the thing moved. This puzzling contention may be cleared up if we consider (just after the beginning of "Resolution of the Objection," following) that in no sense are we to take Averroes to mean that the medium in the *natural* motion of inorganic bodies is merely an accidental impediment to their motion. The medium does not *directly resist* the natural motion of simple bodies, yet there is a resistance in such motions *due to* the medium. How is this so? This resistance is due to the medium, it seems, because the medium is a necessary condition in order that resistance obtain between the mover and the moved body. This is why Averroes repeatedly says that there can be no natural motion of simple bodies without a medium. (Further details on this point will be found in n. 30.) Furthermore, Averroes felt that the medium could not offer direct resistance to the natural motion of a simple body because then such bodies would be *observed* to move with something less than their natural motion. For they would then, as Avempace maintained, move with $V' - r$ (since $V' - V = r$, where V' stands for their natural motion, V for their motion in the resistant medium, and r represents the difference in velocities or accidental retardation due solely to the direct resistance of the medium; (see n. 28). To return now to the original problem, we may say that for all bodies the ratio of motions follows both the ratio of resistances involved and the ratio of the ratios between movers and things moved. For it is the relation between the mover and the moved which gives rise to the resistances concerned. Thus, even in the case of the motion of the heavenly

movers, and two things moved, and the ratio of the one mover to one of the things moved is as the ratio of the other mover to the other of the things moved, then the two motions will be equal in speed. And when the ratio is different, the motions are diversified according to that ratio.[20] And this is true in both natural and violent motion. But in the case of violent motion, if the two movers are equal in potency, the diversity [of the resulting motions] will be that of the resistance. In natural motion, however, the diversity, or equality, follows the diversity, or equality, of the ratio of the mover to the thing moved. But from this it does not follow that equality in natural motion follows from the equality of the movers in potency when the ratio [between the mover and moved] is changed—just as this does not occur in the case of violent motion when the resistance is changed. For [in both cases] the ratio will be changed by a change in resistance. Therefore, these propositions are self-evidently clear.

[CRITICISM OF AVEMPACE RESUMED]

If that which Avempace says were true—namely that in such motions the ratio of one motion to another does not follow the ratio of the density [of one medium] to another, but rather the ratio of the retardation [of one motion] to another—then this would, in the case of any violent motion, make what is said in the last chapter of Book VII [of the *Physics*[21]] false, namely that if a given body moves another body through a certain space in a certain time, then it will move half that body through the same space in half the time. For, according to Avempace, this would occur so that the ratio of one time to the other would only be that of one retardation to the other.[22] The cause of this error lies in the judgment that slowness and speed are motions added to, and subtracted from, a motion in the same manner as a line is added to, or subtracted from, a line. Therefore, when it is supposed that speed is an addition to, and slowness a subtraction from, natural motion, the cause of this addition and subtraction would be resistance. It follows from this that the ratio of resistance to resistance is as the ratio of retardation to retardation and not as the ratio of motion to motion.[23] And this is so because when the same quantity is added to, or subtracted from, proportional quantities, they no longer remain proportional.

Avempace is of the opinion that observable motion *(motus sensibilis)* is what remains of natural motion. He has judged that natural motion is, as it were, a single measure from which two measures are subtracted according to the ratio of two other quantities.[24] And he has seen that when this occurs

bodies, where there is no material resistance of a medium, a resistance does come into the picture, a resistance arising solely from the mover-moved relation and having nothing at all to do with a medium. One final point: In of the case the motion of heavenly bodies—and in almost all cases of the motion of animals—it seems that we would have to exclude the possibility that the ratio of such motion could follow a ratio of the densities or rarities of media. For, as we have just seen, no external media are here involved. But, in the natural motion of simple bodies, and at times in their violent motion, the resistance between mover and moved is determined by the resistance of the medium (see Selection 45, n. 12). Thus "resistance" is used ambiguously by Averroes, at times in a general sense to stand for a relation between mover and moved, and at other times to stand for the impedance of a medium which in turn itself determines a resistance in the first sense of the word.

20. That is, the motions have to one another the same diversity of ratio as do the two differing ratios between the respective movers and mobiles. Loosely speaking, if we let A equal the mover, B equal the moved thing, and C equal the resulting motion, then this passage might be formulated as follows: (1) When $A/B = A'/B'$ then $C = C'$; and (2) when $A/B > A'/B'$ or $A/B < A'/B'$ then $C/C' = (A:B)/(A':B')$.

21. *Physics* VII.5.

22. Here Averroes is confused. To begin with, Avempace does not say—to judge from the quotation given above—that the ratio of motions follows the ratio of retardations. On the contrary, he says that the ratio of retardations follows the ratio of resistances, and that the latter does *not* follow the ratio of motions, i.e., $V/V' \neq R'/R = r'/r$ (where V equals motion, R equals resistance, and r equals retardation). The only possible explanation is that here, and again later (see n. 26), Averroes uses retardation to mean the "amount by which the natural motion (V') in a void is slowed down by the accidental impedance (r) of a medium," that is, retardation $= V' - r$. But elsewhere he, and what is more important, Avempace, uses retardation $= r$ alone.

23. Here, the ambiguity referred to in note 7 is particularly striking, since the ratio of resistances is *directly* proportional to the ratio of retardations, but *inversely* proportional to the ratio of motions. But Averroes refers to both in a single sentence without differentiation.

24. Averroes' resumé of Avempace's contentions might be formulated as follows: (1) Natural motion is as a single measure; that is, holding the potency of the mover constant, there is a motion V' which every body moving in a void possesses. It appears, as a matter of fact, that V' is equal to the potency of the mover, even though neither Avempace nor Averroes expressly declares this to be the case. (2) ". . .two measures are subtracted [from this V' as a single measure] according to the ratio of two other quantities." That is, if A's speed (for a given mover) in a void is V', then if it is subsequently moved in medium B of a resistance R and in medium D of a resistance R', the following results: (a) In medium B its speed will be $V' - r$, where r stands

in quantities, it does not follow that the ratio of one of the subtracted quantities to the other is as the ratio of one remaining quantity to the other remaining quantity.[25] Moreover, he is of the opinion that a certain subtraction occurs in natural motion because of resistance; and this subtraction occurs according to the ratio of resistance to resistance. Further, he contends, as we have said, that observable motion is what remains of natural motion after this subtraction, in the same way as something remains of a [mathematical] magnitude after a subtraction. Consequently, he has concluded that it does not follow that the ratio of one observable motion to another is as the ratio of one resistance to another, but rather as the ratio of one retardation to another.[26] Analogously, it does not follow that if quantities are diminished by some ratio the remainders will be in that ratio. On the contrary, they differ, because when some actually existing quantity is subtracted from a mathematical quantity, an actually existing quantity remains. When this subtraction occurs in natural motion, then the natural motion has only potential being.[27] Consequently, observable motion can only be the non-natural motion which is what remains of natural motion [after the retardation is subtracted] and not the natural motion from which this subtraction occurs.[28] And similarly, when some moved thing is moved faster than its natural motion, in that case there will not be natural motion and an addition, but only an addition.[29] Nevertheless, we also say that the ratio of retardation to retardation is as the ratio of resistance to resistance, as Avempace maintains. But in this case there is nothing else but retardation.

The problem at hand is really sophistical, though difficult, for it is all based upon a confusion [of categories]. And this is so because, as he assimilates motion to a line, he contends that what occurs apropos a line, occurs apropos motion.

[EXPOSITION OF AVERROES' OPINION RESUMED]

Therefore, in this passage we have asserted that the ratio of resistance to resistance is as the ratio of motion to motion. This applies to the case of violent motion when the mover is held constant and the resistance is due either to the medium or the moved body itself. But in natural motion, since there is no impediment [or resistance] this varies according to the diversity of the ratios which obtain between movers and things moved; indeed [it varies] according to the diversity of the ratio of the excess of the first motive power over the thing moved to

the excess of the second power over the thing moved.

[OBJECTION]

Yet, someone may object that when we assert that motion will occur in a medium resisted by that medium, it does not follow, if we maintain that the ratio that motions have to one another in that medium is as the ratio of resistance to resistance,

---

for a retardation due to the resistance R of B, and (b) in medium D its speed will be $V' - r'$, r' being the retardation due to the resistance R' of D. The retardations r and r' are the two quantities subtracted from the single measure V' of "natural motion." But they are subtracted "according to the ratio of two other quantities," that is, according to the ratio of the respective resistances R/R'. (3) The motions $V' - r$ and $V' - r'$ are the observable motions of A in two different media. They are what remain of its natural motion V' in those media.

25. Continuing our analysis of the foregoing note, we may say that Avempace realizes the ratio of the subtracted retardations r/r' is not equal to the ratio of the remaining observable motions $(V' - r)/(V' - r')$. This is as good a place as any to make two remarks about our analysis of Avempace's opinions. To begin with, he is not in any sense of the word attempting to discover the appropriate form of proportionality between motions and their respective forces and resistances. That is, he is not bent upon replacing the so-called Aristotelian $V = F/R$ by $V = F - R$. (Moody's article cited in n. 18 strongly suggests this interpretation of his work). On the contrary, it appears that Avempace is merely pointing out that the *natural* motion of terrestrial bodies occurs on the same basis as that of celestial bodies. This brings us to the second point. Avempace seems to be speaking only about the natural motion of bodies. Thus our analyses of Avempace as reported by Averroes should only be taken as aids in clarifying particular passages and not as implying any attempt to formulate a completely general law of motion.

26. In this last clause Averroes seems to misinterpret Avempace. In fact, what Averroes says here is inconsistent with his report in the preceding paragraph. For there he correctly maintains that Avempace holds that $r/r' = R/R'$. But if this is the case, and further, if $R/R' \neq (V' - r)/(V' - r')$, then we cannot conclude, as he does in the present passage, that $r/r' = (V' - r)/(V' - r')$. The first two formulas find support in Avempace's own words; the last does not (see n. 22).

27. That is, observable motion, which actually exists, is $V' - r$; the natural motion V' however, only potentially exists, since it never actually occurs without some r or other being subtracted from it.

28. That, consequently, observable motion would not be natural is unthinkable in view of Averroes' pronounced Aristotelianism, which requires that the natural be identified with the observable (see Moody, pp. 189–190).

29. That is, we observe in this case not a natural motion plus some additional motion but only their sum in which, apparently, the addends are indiscriminable.

that when the same thing is moved in a void it is moved instantaneously and with an indivisible motion. On the contrary, [they would say,] it necessarily follows that [in a void the same thing] is moved in time. And the ratio of this time [a] to the time [b] in which it is moved in a medium is as the ratio of the excess [a'] of the power of the mover over the thing moved insofar as the thing is moved without any extrinsic resistance [that is, in a void] to the excess [b'] of the power of the mover to the thing moved insofar as the thing is resisted [that is, is moved in a medium]. Hence it is even believed that although these propositions which are used may be conceded, one is, nevertheless, not led to the contradiction which one sought to produce.

[RESOLUTION OF THE OBJECTION]

To this, one must reply that those motions of simple bodies in the media of water and air are not motions resisted by media, as appears at first glance. For, if this were the case, natural motions would exist which have not yet been observed, and which never will be observed—unless it were possible that these bodies could move without a medium. But this would only be possible if a void existed. The reason for this is that such moved things require a medium; for a moved thing and its mover necessarily should give rise to some resistance by which the potency of the mover exceeds the potency of the moved thing. This is so in the case of natural motion. If it were not, any mover could move any body.

Furthermore, the motions of the heavenly bodies are absolutely differentiated according to the diversity of the ratio which arises out of this resistance between the mover and the thing moved. Thus certain [heavenly bodies] are swifter than others. And by this resistance the mover only moves the thing moved when its potency exceeds the potency of the thing moved.

[RELATION OF MOVER AND MOVED]

Yet a moved thing is universally similar to the mover in a certain way, and in another way contrary to it. This we must take into account with respect to the perfection (nobilitas) which Avempace speaks of in celestial bodies. For the mover whose potency exceeds [the potency of the thing moved] is more perfect than the mover whose potency is less than [the potency of the thing moved]. It is clear that a resistance is observed between the mover and the thing moved when the thing moved is distinct in itself [from the efficient

mover]. This is the case in the celestial bodies.[30] Yet, in the elements [that is, in the motions of inorganic bodies within the spheres of fire, air, water, and earth], the moved thing potentially exists and the mover actually exists, since [simple, inorganic bodies] are composed out of prime matter and simple forms; the mover is a form, and the thing moved matter. And, since these bodies [that is, simple bodies] are not [actually] distinguished into a thing moved and an actually existing mover, it is impossible that they be moved without a medium being involved. Moreover, Aristotle has declared (in Book VIII of this work) that, accordingly, it is necessary that these bodies are not moved by themselves.[31] Thus, if these simple bodies were moved without there being a medium, there would be no resistance between the mover and the thing moved. On the contrary, the moved thing would in no way essentially exist.[32] Moreover,

30. That is, in celestial and organic bodies (animals) the mover and the moved are actually *distinct within* the body itself. It is the immaterial intelligence actually existent in the celestial sphere that is its mover. As such it is actually distinct from the celestial sphere *qua* moved thing. The case is similar in animals whose bodies are moved by a soul actually distinct from the body as moved. (We must note here that "mover" is taken as an efficient cause; we are not speaking of it as a final cause.) However—and this will become clearer in what follows —this is not the case in the motion of inorganic, or as Averroes calls them, "simple," bodies. Here, mover and moved are not actually distinct entities within the body. Averroes tells us, in the sentence following the present passage, that in such simple bodies the mover, as its form, actually exists, but that the body *qua* moved— that is, as matter—only potentially exists. Thus, within these bodies the form, or "essence," of the body as mover is not actually distinct from the body as a moved thing. But, according to Averroes, in order that there be any motion, there must be an *actual* distinction between the mover and the moved. Therefore, as the form of a simple body in natural motion is not actually distinct from the body *qua* moved, this form cannot be efficient cause of the motion. A simple body cannot, in short, move itself; its form cannot act upon its matter. The form only moves the body *per accidens* by overcoming the resistance of the medium in which the body moves. Thus a medium is required for any such motion. It is the form of the body plus the medium which affords the efficient cause of the motion. This is the substance of the present paragraph (see n. 19).

31. Book VIII, chapter 4.

32. That is, the body *qua* moved thing would not *actually* exist. For as we have seen, the body *qua* moved thing only potentially exists. It is only when a medium is involved and there arises a resistance between the mover and the moved that motion actually occurs. Without such a medium and the consequent resistance, the body would not actually be moved.

if this were the case, [simple bodies] would be moved with any motion whatsoever in no time at all. Because of this Aristotle has said that if these bodies were moved in a void, it would follow that they would be moved with an indivisible motion in an indivisible time (provided, of course, that it is natural for these bodies to be moved in a medium). The reason for all this is that [simple bodies] may be moved and yet not moved by themselves with an unnatural motion as Avempace believes.

[SUMMARY]

Therefore, Avempace raises doubts in this passage in two places. One occurs when he says that the ratio of motion to motion is not as the ratio of the density of one medium to the density of another. The second, however, is that these motions are resisted by the medium and are not natural. Yet nobody before him had arrived at these questions, and thus he was more profound than any others.[33] And in the copy [of Avempace's work] from which we have written, we found a certain page all by itself, and this was written on it.

However, we maintain that it is necessary that there be a resistance between the mover and the thing moved. For the mover moves the thing moved insofar as it is contrary [to it], and the thing moved is moved by it insofar as it is similar. And every motion will be according to the excess of the potency of the mover over the thing moved, and the diversity of motions in speed and slowness follows the ratio which exists between the two potencies.[34]

And this resistance [between the mover and the thing moved is due to one of three things]: (1) It will be due to the thing moved itself when that which is moved by its own will is divided into an actually existing mover and an actually existing thing moved (as is the case in animals and heavenly bodies); or (2) it will be due to the medium itself in which it is moved, and this will occur when the moved thing is not divided into a mover and an actually existing moved thing (as is the case in simple bodies); or (3) the resistance will be due to both, that is, to the thing moved and the medium (as is the case in animals which are moved in water). Moreover, those things in which a self-moved thing is divided into an actually existing mover and an

actually existing moved thing do not necessarily require a medium, but if one is involved, it will be so accidentally. Yet, those things which move themselves[35] and are not divided into a mover and an actually existing moved thing, necessarily do require a medium. Of this sort are heavy and light bodies. And if there were no medium [for such bodies], they would be moved in no time at all. For there would be nothing present to resist the motive power, and it would be impossible that they have a natural motion and be always resisted. And thus it comes about that when we assume these bodies to be moved in a void, they are moved instantaneously, and that bodies of different heaviness are moved with equal speed, which is impossible. Therefore, what Avempace has maintained is that these simple bodies have a natural motion without a medium.

33. However, as Moody himself notes, criticisms of Aristotle's dynamics had preceded Avempace. He refers to the late antique commentator on Aristotle, John Philoponus. For more complete information on these earlier criticisms, see M. Clagett, *Greek Science in Antiquity* (New York: Abelard-Schuman, 1955), pp. 169–177; and Cohen and Drabkin, pp. 217–223.

34. That is, in each case, between the potency of the mover and the potency of the thing moved.

35. It is puzzling what Averroes means by this declaration. He has just cited Aristotle as believing that "it is necessary that these bodies are not moved by themselves." Perhaps Averroes meant to distinguish the natural motion of heavy and light bodies from the violent motion by referring to the former as things "which move themselves." He may have felt this justified because in such cases the indwelling form or essence of the body is *per accidens* the cause of its motion. Yet in order that this cause may actualize itself an *external* medium is needed. Thus, strictly speaking, the bodies do not move themselves since they possess no internal efficient cause of movement.

It is important to realize, however, that although Averroes may not have intended to assert that heavy and light bodies move themselves, this is what he seems to say here, thereby providing sufficient grounds for later Latin authors to believe that he disagreed with Aristotle on this point. The same ambiguity and lack of clarity appear in the next selection. It is probable that in Selection 46 (at the end of paragraph 1035) Thomas Aquinas has Averroes in mind when he says that "it is clearly contrary to the Philosopher's intention to say that there is an active principle in matter, which some maintain is necessary for natural motion."

45 THE MOVER OR CAUSE IN NATURAL MOTION
Averroes(1126–1198)

Translated and annotated by Edward Grant[1]

Then he [Aristotle] says "and so the mover does not accompany the body that is moved with an accidental [or unnatural] motion."[2] That is, it is not necessary that the mover accompany the body in motion until the motion terminates. But although the mover rests, the body in motion will be moved nonetheless, because the air will then move it, since the motion it receives at the beginning from the external mover will remain after it has been separated from its mover. This self-motion [of the air] is in accordance with a natural principle in it, namely a heaviness or lightness. For if this were not so, a stone would fall when separated from its mover since a stone does not have a principle of lightness within itself. . . .[3] And it can be understood that if there were no air, there would be no violent motion. . . .[4]

Moreover,. as Aristotle says, air is also an aid to natural motion in the same manner, necessarily, as in violent motion.[5] But it also seems necessary that in natural motion there should be a proper mover; but there is difficulty in this. For indeed it was asserted about the elements that among them none are self-moved because they are moved to a place only by the agent which brought them into existence. Thus when something is brought [from potentiality] into actuality by an agent [or *generans*], all things, as well as all other proper accidents, which constitute it in its proper place exist in it,[6] unless something should impede it from moving to its proper place. Therefore, just as other accidents exist in a generated thing, they do exist in it only through the form of the generated thing which is produced by the agent that generated it. And this is true of motion. For example, when the element is fire, all the accidents existing in fire and which make it fire exist immediately in the first part of the fire that is produced by the generating fire. One of these accidents is upward motion, because insofar as lightness exists in the generated part [of the fire] it will have the place of a light body; and [the same may be said] about other proper accidents. But, nevertheless, since these accidents from the generating agent exist only in the generated body and motion occurs primarily and essentially by means of a mover, and in no other way, this motion is therefore not produced primarily or essentially from an external mover but [rather] from the form

of the body that is moved.[7] And so this kind of motion [that is, natural motion] has the characteristic that by means of the form a thing is moved by itself when something does not hinder it from motion by violence, namely when it will rest after being hindered. Now, it is in this way that this kind of motion is likened [or compared] to the motion of things that are self-moved. And since a mover is distinguished essentially in being from that which is moved, as in animals, where the mover in them is distinct from the thing moved, since the mover is the soul and the thing moved is the body, so in simple bodies the mover and the thing moved are the same in species but differ in their modes. For a stone moves itself insofar as it is actually heavy, and it is moved insofar as it is a lower potency. The reason that this occurs in one way in [terms of] actuality and in another in [terms of] potentiality is that a stone is composed of matter and form. Therefore, its form moves [that is, is a motive

1. Translated from Averroes' Comment 28 on Book III of Aristotle's *On the Heavens (De caelo)*, as found in *Aristotelis opera cum Averrois commentariis [The Works of Aristotle with the Commentaries of Averroes]* (Venice, 1562–1574; Frankfurt: Minerva, 1962), Vol. V, fols. 198 recto, col. 1, to 199 recto, col. 1.

2. *De caelo* III.2.301b.27–28. In the passage from which this quote was taken, Aristotle explains that the motive force which initially sets a body in motion does not itself accompany that body but rather communicates a motion to the air, which then possesses the capacity to serve as motive power and continue pushing the body.

3. That is, unless air moved a stone after the latter lost contact with its initial mover, the stone would fall immediately to the ground in a vertical downward natural motion.

4. *De caelo* III.2.301b.28–29.

5. *De caelo* III.2.301b.29–30.

6. Thomas Aquinas would have found this statement wholly acceptable. See the introduction to Selection 46.

7. This would appear to be the specific passage to which Thomas Aquinas strenuously objected (see Selection 46, n. 31). For it seems that Averroes posits that the forms of inorganic bodies are movers or efficient causes in natural motion. However, Averroes seems to qualify this elsewhere by insisting that in such natural motions the form alone of a simple body cannot move the body without the concomitant and necessary action of an external medium (see below, near the end of this selection). Nevertheless, his vagueness and lack of clarity on this important issue provoked much later controversy (see Selection 44, notes 30 and 35).

power] insofar as it is form; and it is moved insofar as it is in matter. And thus a stone is not essentially self-moved, but is [only] similar *(assimilatur)* to what is self-moved and therefore requires an external mover in its motion.[8] Nor is it even counted amongst those things that are moved by an external mover, since it is not immediately [or directly] moved with this motion by an external mover but [rather is moved] by means of its form. And this is one of those cases where Plato thought it necessary that the primary mover should be moved essentially—that is, in the sense that the mover in the stone is moved. But although the mover in the stone is identical with the thing moved, they do not, as Plato thought, exist in the same mode but [rather] differently, as it appears in the first demonstration at the beginning of the seventh book of the *Physics,* where it is said that every motion has a mover.[9] Now, as we said, since the stone is not divided into mover and thing moved where each exists in actuality, the stone is moved, then, by a principle that is in it accidentally, since it does not have its being except through that principle.[10] For there is no motion of a stone except as it is in an inferior potentiality and the mover is heavy. Therefore, if there is self-motion essentially insofar as it is heavy, and since we already assume that it [that is, the whole stone] does not move essentially except insofar as it is heavy, then the mover and the moved exist in essentially the same way, which is impossible.[11] And so it follows that the motion of the stone is here not an essential motion, and it does not move itself essentially. And since this is so, it is necessary that everything which is moved accidentally be moved because another thing moves the moved thing essentially by itself. For example, a man does not move himself accidentally in a ship unless he moves the ship essentially. And since this is so, a stone does not move essentially unless it moves the air in which it is, and it moves itself because what moves itself follows [or depends on] the motion of the air, just as the man with the ship.[12] Therefore, since this is so, air and water are necessary in the motion of a stone and in our exposition of the *Physics* this is what we promised to declare here [in our exposition of *De caelo*]; and this is the most appropriate place. . . .

8. It requires an external mover because it is not actually self-moved but is like something that is self-moved. But in all self-moved things (for instance animals) the mover and the thing moved can be distinguished in one and the same thing. Hence, by its similarity to self-moved things, a mover and moved thing must be distinguished in the stone. But Averroes now goes on to make further qualifications and seems to shift back to the form as mover. It is hardly surprising that his position gave rise to controversy.

9. Aristotle begins the seventh book of his *Physics* with this remark (VII.1.241b.24–25).

10. Here Averroes emphasizes that in a stone, mover and thing moved cannot be distinguished in actuality (see Selection 44, n. 30).

11. Here Averroes argues that the heaviness of the stone cannot be the motive cause of its motion, since it is also the case that heaviness is an essential property of the stone itself. In brief, *heaviness* would be common to mover and moved; but he argued above that mover and moved must exist in different modes.

12. The manner in which the rower analogy should be applied to the motion of the stone is left somewhat ambiguous. At least two interpretations seem plausible. First, the rower moves the ship essentially by directly overcoming the resistance of the water with his oars. Once the resistance is overcome, the ship moves and carries the rower with it *per accidens.* In the same way, the form of the stone directly overcomes the resistance of the air thus enabling the stone to move naturally through the air. As the stone moves, it carries its form with it *per accidens.* In this first interpretation, the medium serves only to resist the motion of the inorganic body. But unless this resistance were overcome, the stone could not move at all because *mover* and *moved* are not actually distinct within it. Hence the efficient cause of motion requires both the form of the inorganic body and a resistant medium (see Selection 44, notes 19, 30, 32, and 35). In this first interpretation the medium is construed only as a resistance, not as a mover. But a second interpretation would have the medium serve as the direct contactual mover of the inorganic body. In this view, the rower is said to move the ship essentially by agitating the water, which then moves the ship directly. Hence the rower is again carried by the ship *per accidens.* In the same manner, the form of the stone acts upon the air, which in turn pushes the stone directly with the latter carrying its form *per accidens.* On this second interpretation the medium would function as direct mover (and the form as indirect mover) and the inorganic body would serve as resistance, just as in violent motion. In one place, at least, Thomas Aquinas seems to understand Averroes in this way (see Selection 46, n. 31; see also Selection 46, introduction, where he denies that the natural form is a mover). In the first interpretation, the form of the stone is the (indirect) mover by virtue of overcoming the resistance of the medium. Once this is achieved the body simply moves, since the impediment to its motion has been overcome. Thus the mover is the form, the inorganic body is the thing moved, but the resistance is the medium, not the inorganic body. Because the first interpretation emphasizes the resistive function of the medium, it seems preferable.

46 THE MEDIEVAL ARISTOTELIAN PRINCIPLE OF MOTION: "WHATEVER IS MOVED IS MOVED BY ANOTHER"

St. Thomas Aquinas (ca. 1225–1274)

Translated by Richard J. Blackwell, Richard J. Spath, and W. Edmund Thirlkel[1]
Introduction and annotation by Edward Grant

The Aristotelian principle "whatever is moved is moved by another" (its usual medieval Latin form was *omne quod movetur ab alio movetur*) was applied to all animate and inanimate objects capable of motion. As applied to inanimate objects, it was at first widely held that in cases of violent motion, or motion away from a body's natural place, the mover could be easily identified since it would always be external to the body it moved. For example, when a stone is thrown, the initial cause of its motion would be the thrower and the continuing cause the ambient air, which is continually in contact with the stone. Physical contact between mover and moved is a prime characteristic of violent motion. When, however, many came to reject air as a direct mover in violent motions and invoked instead incorporeal impressed force or impetus, serious problems arose, since now it was no longer meaningful to speak of "physical" contact. Indeed, the agent in violent motion was no longer "obvious." Was it the initial projector, an external object, with the impetus merely an instrumental agent? Or, was it the impetus itself, an internal incorporeal entity?

In natural motion, Aristotle himself emphasized the difficulty in determining the mover when he wrote: "It is in these cases that difficulty would be experienced in deciding whence the motion is derived, e.g. in the case of light and heavy things. When these things are in motion to positions the reverse of those they would properly occupy, their motion is violent: when they are in motion to their proper positions— the light thing up and the heavy thing down—their motion is natural; but in this latter case it is no longer evident, as it is when the motion is unnatural, whence their motion is derived" (*Physics* VIII.4.255a.1–5). But subsequently, at the very end of the chapter, Aristotle suggests that in natural motion light and heavy things ". . . are moved either by that which brought the thing into existence as such and made it light and heavy [the medieval term for this causal agent was *generans*—Ed.], or by that which released what was hindering and preventing it [the scholastics called this *removens prohibens*—Ed.]. . . ." (*Physics*

VIII.4.256a.1–2). Although, in other places, Aristotle would suggest yet other causes of natural motion and free fall, such as an active principle and the gravity or heaviness of a body, in the passage just quoted he did not distinguish a separate motive force or internal principle to explain natural motion but instead accounted for it by a natural spontaneous action of the body, an action that it would always undergo when not otherwise impeded. The capacity for such action was assumed to be a natural property communicated by the agent (that is, the *generans*) that caused or produced it. It is this very position which Aquinas adopts in this selection and which earlier in this very treatise he expressed as a formal principle of motion: "However, in heavy and light bodies there is a formal principle of motion. . . . For just as other accidents are consequent upon substantial form, so also is place, and thus also 'to be moved to place.' However the natural form is not the mover. Rather the mover is that which generates and gives such and such a form upon which such a motion follows" (*Physics* II, Lecture 1, par. 144, on p. 71 of the translation by Blackwell, Spath, and Thirlkel). But even before

1. Included here are Lectures 7 and 8 of Book VIII of the *Commentary on Aristotle's Physics* by St. Thomas Aquinas, translated by Richard J. Blackwell, Richard J. Spath, and W. Edmund Thirlkel, with an Introduction by Vernon J. Bourke (New Haven: Yale University Press, 1963), pp. 503–512; reprinted by permission of the Yale University Press. These two lectures *(lectiones)* incorporate Aquinas' exposition of the whole of Book VIII chapter 4, of Aristotle's *Physics*.

The lucidity and intelligibility of Aquinas' exposition has required almost no additional commentary. Although slightly expanded and occasionally altered, the numerous cross-references and citations to other books supplied by the translators have been included here. The phrases between single quotes represent cue words to the text of Aristotle which Aquinas furnished in lieu of needlessly repeating the full text of Aristotle. They have been taken by the translators from the Oxford English translation of the *Physics* (by R. P. Hardie and R. K. Gaye) and followed in each instance by Bekker numbers, thus conveniently enabling the reader to locate the full Aristotelian discussion in any modern edition or translation. Paragraph numbers have been retained and will also appear in notes supplying cross-references.

Thomas wrote, Averroes, while seemingly arguing the same position which Thomas later adopted, namely that the *generans* is the sole cause of natural motion, appears to have actually distinguished the form of a natural body as an internal mover and the body as the thing moved. And although he still insisted that inanimate bodies are not self-moved in natural motion, his distinction (for Averroes' discussion, see his Commentary 28 on Bk. III of *De caelo,* translated in the preceding selection) may have formed the basis of an even more radical departure initiated by Duns Scotus near the beginning of the fourteenth century. In his *Sentence Commentary,* Book II, Distinction 2, Question 10, Scotus rejected the *generans* as a direct efficient cause of free fall (the latter was now considered as only a remote cause) and argued instead that light and heavy inorganic bodies are self-moved by an internal principle. Though variations on this theme were formulated, it became one of the basic positions in the later Middle Ages. Some who saw the heaviness *(gravitas)* of a body as the internal principle causing it to fall, came to argue that this heaviness produced successive and cumulative increments of impetus which, in turn, caused the falling body to accelerate (this is Buridan's opinion; see Selection 49, section 5). It seems evident that while Scotus' position bears a certain affinity with that of Averroes, it is wholly at variance with Aquinas' viewpoint.

LECTURE 7 (254 B 7–255 A 18)
WHATEVER IS MOVED IS MOVED BY ANOTHER

1021. After the Philosopher has explained[2] his intention, he here begins to develop it; namely, not everything is sometimes moved and sometimes at rest. Rather there is something which is absolutely immobile, and something which is always moved.

This discussion is divided into two parts. First he shows that the first mover is immobile. Secondly, where he says, "And further, if there is. . . ." (259 b 32),[3] he shows that the first mobile object is always moved.

The first part is divided into two parts. First he shows from the order of movers and mobile objects that the first mover is immobile. Secondly, where he says, ". . . but also by considering. . ." (259 a 21),[4] he proves the same thing from the eternity of motion.

The first part is divided into two parts. First he shows that the first mover is immobile. Secondly, where he says, "Since there must always. . ." (258 b 10),[5] he shows that the first mover is eternal.

Concerning the first part he makes two points. First he explains something that is necessary for the proof of what follows; namely, whatever is moved is moved by another. Secondly, where he says, "Now this may come about. . ." (256 a 3),[6] he proves his position.

He has shown above at the beginning of Book VII[7] that whatever is moved is moved by another with a common argument taken from motion itself. But since he has begun to apply motion to mobile things, he shows that that which was proven universally above is universally verified in all movers and mobile objects.

Hence the first part is divided into two parts. First he gives a division of movers and mobile objects. Secondly, where he says, "The fact that a thing. . ." (254 b 24),[8] he explains the proposition in individual cases.

Concerning the first part he makes two points. First he divides movers and mobile objects. Secondly, where he says, "Thus in things that. . ." (254 b 15),[9] he explains this division.

1022. First, therefore, he gives three divisions of movers and mobile objects.

The first of these is that some movers and mobile objects move or are moved *per accidens,* and some *per se.* Here he uses *"per accidens"* in the broad sense, according to which it includes even that which is "according to a part." Hence, to explain the meaning of *'per accidens',* he adds that a thing can be said to move or be moved *per accidens* in two ways. First, those things are said to move *per accidens* which are said to move because they are in certain movers. For example, it is said that music cures, because he who cures is musical. Similarly, those things are said to be moved *per accidens* which are in things which are moved, either as located in a place, as when we say that a man is moved because he is in a ship which is moved, or else as an accident in a subject, as when we say that white is moved because a body is moved. Things are said to move or be moved *per accidens* in another way insofar as they move or are moved according to a part. For example, a man is said to strike or be struck because his hand strikes or is struck. Those things are said to be moved or to

2. Lecture 6, 1016.
3. Lecture 13, 1083.
4. Lecture 13.
5. Lecture 12.
6. Lecture 9.
7. Book VII, Lecture 1.
8. 1024.
9. 1023.

move *per se* which are free of the two preceding modes. For they are not said to move or be moved because they are in other things which move or are moved, nor because some part of them moves or is moved.

Omitting things which move or are moved *per accidens,* he subdivides those which are moved *per se.* First, of those things which are moved *per se* some are moved by themselves, such as animals, and others are moved by others, such as inanimate things.

He gives a third division; namely, some things are moved according to nature, and others are moved outside of nature.

1023. Next where he says, "Thus in things that . . . " (254 b 15), he explains how motion according to nature and motion outside of nature is found in things which are moved by themselves and in things which are moved by another.

He says first that things which are moved by themselves (as are animals, which move themselves) are moved according to nature. He proves this from the fact that they are moved by an intrinsic principle. We say that those things whose principle of motion is in themselves are moved by nature. Hence it is clear that the motion of an animal by which it moves itself, if compared to the whole animal, is natural, because it is from the soul, which is the nature and form of the animal. But if it is compared to the body, it happens that motion of this kind is both natural and outside of nature. For this must be considered according to the diversity of motions and of elements from which the animal is constituted. If an animal is composed of predominantly heavy elements, as is the human body, and if it is moved upward, then this motion will be violent in respect to the body. But if it is moved downward, this motion will be natural to the body. If, however, there are some animal bodies which are composed of air, as some Platonists held, then the contrary must be said of them.

Secondly, he explains how violent and natural motion are found in things which are moved by another.

He says that some of these are moved according to nature; for example, fire is moved upward and earth downward. But some are moved outside of nature; for example, earth upward and fire downward. This latter is violent motion.

Thirdly, he gives another mode of unnatural motion in animals according to which the parts of animals are often moved outside of nature if one considers the natures *[ratio]* and modes of natural

motion in the parts of animals. For example, a man bends his arms forward and his legs backward; but dogs and horses and such animals bend their front feet to the rear and their rear feet to the front. But if a contrary motion occurs in animals, the motion will be violent and outside of nature.

1024. Next where he says, "The fact that a thing . . ." (254 b 24), he proves that whatever is moved is moved by another.

He shows this first in cases in which it is obvious. Secondly, where he says, "The greatest difficulty . . ." (254 b 33),[10] he shows this in cases in which it is open to question.

He omits, however, things which are moved *per accidens,* because such things are not themselves moved, but are said to be moved because some other things are moved. Among things which are moved *per se,* it is especially clear that things which are moved violently and outside of nature are moved by another.

For it is clear from the very definition of violence that things which are moved by violence are moved by another. For as is said in *Ethics,* III,[11] the violent is that whose principle is outside, the patient contributing none of the force.

After he has shown that things which are moved by violence are moved by another, he shows that things which are naturally moved by themselves, as animals are said to move themselves, are also moved by another. For in these cases it is clear that something is moved by another. But there can be a difficulty of how one should designate the mover and the moved in these cases. At first sight it seems (and many think this) that that which moves and that which is moved are distinct in animals, just as they are in ships and in other artificial things which do not exist according to nature. For it seems that the soul which moves is related to the body which is moved as a sailor is related to a ship, as is said in *De Anima,* II.[12] Thus it seems that a whole animal moves itself insofar as one part of it moves another. Whether the soul is related to the body as a sailor to a ship he leaves for investigation in *De Anima.* He will explain later[13] how a thing is said to move itself insofar as one part of it moves and another part is moved.

1025. Next where he says, "The greatest difficulty

10. 1025.
11. *Ethics* III.1.1110a.1–3; St. Thomas, *On the Ethics,* Book III, Lecture 1, 386.
12. *De anima (On the Soul)* II.1.(413a.7–9; St Thomas *In de Anima,* Book II, Lecture 2, 243.
13. Lecture 10.

. . ." (254 b 33), he explains his position in cases in which there is greater difficulty.

Concerning this he makes three points. First in regard to cases in which there is greater difficulty he states that whatever is moved is moved by another. He is referring to heavy and light things when they are moved according to nature. Secondly, where he says, "It is impossible. . ." (255 a 5),[14] he shows that such things do not move themselves. Thirdly, where he says, "It is the fact that. . ." (255 a 19),[15] he explains how they are moved.

He says, therefore, first that it is rather clear that the proposition "whatever is moved is moved by another" applies to things which are moved by violence and to things which move themselves. The main difficulty arises in the remaining alternative of the last division, namely, in things which do not move themselves, but are nevertheless moved naturally.

He says the "last" division; namely, of things which are not moved by themselves but by another, some are moved outside of nature and some on the contrary are moved according to nature. In the latter cases the difficulty is: by what are they moved? For example, heavy and light things are moved to their contrary places by violence and to their proper places by nature, that is, the light is moved upward and the heavy downward. But what moves them is not clear when they are moved by nature as it is when they are moved outside of nature.

1026. Next where he says, "It is impossible. . ." (255 a 5), he proves with four arguments that such things do not move themselves.

The first of these is that to move oneself pertains to the nature [ratio] of life and is proper to living things. For by motion and sensation we distinguish the animate from the inanimate, as is said in De Anima, I.[16] Now it is clear that these things are not living or animated. Therefore they do not move themselves.

1027. He gives the second argument where he says, "Further, if it were. . ." (255 a 7). The argument is as follows.

Things which move themselves can also be the cause of their own rest. For example, we see that by their appetites animals are moved and stopped. Therefore, if heavy and light things move themselves by natural motion, they would be able to stop themselves, just as if a man is the cause of his own walking, he is also the cause of his non-walking. But we see that this is false, because such things are not at rest outside of their proper places

unless some extrinsic cause prevents their motion. Therefore they do not move themselves.

But someone might say that even though such things are not the cause of their own rest outside of their proper places, nevertheless, they are the cause of their own rest in their proper places. Hence he adds a third argument where he says, ". . . and so, since on this supposition. . ." (255 a 9). The argument is as follows.

It is irrational to say that things which move themselves are moved by themselves with respect to only one motion and not a plurality of motions. For that which moves itself does not have its motion determined by another, but determines its own motion. Sometimes it determines for itself this motion, and sometimes another. Hence, that which moves itself has the power to determine for itself either this or that motion. Therefore, if heavy and light things move themselves, it follows that if fire has the power to be moved upward, it also has the power to be moved downward. But we never see this happen except through an extrinsic cause. Therefore they do not move themselves.

It should be noted, however, that these two arguments are probable with respect to those things which are apparent in things which move themselves among us. Such things are found to be moved sometimes by this motion, sometimes by another motion, and are sometimes at rest. And so he did not say "impossible" but "irrational," which is his custom when speaking of probables. For he will show below[17] that if there is some self-motion in which the mover is altogether immobile, then it is moved always and by one motion. But this cannot be said of heavy and light things in which there is not something which is not moved per se or per accidens, since they are generated and corrupted.

1028. He gives the fourth argument where he says, "Again, how can anything. . ." (255 a 12). The argument is as follows.

No continuous thing moves itself. But heavy and light things are continuous. Therefore, none of these moves itself.

He proves as follows that no continuous thing moves itself. The mover is related to the moved as agent to patient. When, however, the agent is contrary to the patient, a distinction is necessary between that which naturally acts and that which is

14. 1026.
15. Lecture 8.
16. *De anima* I.2.(403b.26); St. Thomas, *In de Anima,* Book I, Lecture 3, 32.
17. Lecture 13.

naturally acted upon. Therefore, insofar as some things are not in contact with each other, but are altogether one and continuous both in quantity and form, they cannot be acted upon by each other. Therefore , it follows that no continuous thing moves itself. Rather the mover is separated from that which is moved, as is clear when inanimate thing are moved by animate things, as a stone by hand. Hence in animals which move themselves there is a certain collection of parts rather than a perfect continuity. For one part can be moved by another, but this is not found in heavy and light things.

LECTURE 8 (255 a 19–256 a 2)
HE EXPLAINS HOW HEAVY AND LIGHT THINGS ARE MOVED

1029. After he has shown that heavy and light things do not move themselves, he here explains how they are moved. First he explains how they are moved. Secondly, where he says, "If then the motion. . ."(255 b 31),[18] he concludes to his main point.

Concerning the first part he makes two points. First he shows that they are moved naturally by something. Secondly, where he says, "But the fact that. . ." (255 a 30),[19] he inquires into that by which they are moved.

He says, therefore, first that, although heavy and light things do not move themselves, nevertheless they are moved by something. This may be made clear by distinguishing between moving causes. For just as things which are moved are moved either according to nature or outside of nature, so also in things which move some move outside of nature, for example, a staff does not naturally move a heavy body like a stone, and some move according to nature, for example, that which is actually hot moves that which according to its nature is potentially hot. And the same is true of other such things. And just as that which is in act naturally moves, so that which is in potency is naturally moved, either with respect to quality, or quantity, or place.

He has said in Book II[20] that those things are moved naturally whose principle of motion is in them per se and not per accidens. From this it may seem that when that which is only potentially hot becomes hot, it is not moved naturally, since it is moved by an active principle existing exterior to it. As if to refute this objection, he adds, ". . .when it contains the corresponding principle in itself and not accidentally. . ." (255 a 25). This is as if he

were to say that for motion to be natural, it is sufficient if "the corresponding principle," that is, potency, which he mentioned, be in that which is moved, per se and not per accidens. For example, a bench is potentially combustible, not insofar as it is a bench, but insofar as it is wood.

Explaining what he means when he says, ". . . and not accidentally. . ." (255 a 25), he adds that the same subject may be both quantified and qualified, but one of these is related to the other per accidens and not per se. Therefore, that which is potentially a quality is also potentially a quantity, but per accidens.

Since, therefore, that which is in potency is moved naturally by another which is in act, nothing can be in potency and act with respect to the same thing. It follows that neither fire nor earth nor anything else is moved by itself, but by another. Fire and earth are indeed moved by another, but through violence, when their motion is outside of their natural potency. But they are moved naturally when they are moved to their proper acts to which they are in potency according to their nature.

1030. Next where he says, "But the fact that . . ." (255 a 30), he explains that by which they are moved. And since that which is in potency is moved by that which is in act, he first distinguishes potency. Secondly, where he says, "As we have said. . ." (255 b 17),[21] he explains that by which such things are moved.

Concerning the first part he makes three points. First he shows that it is necessary to know in how many ways a thing is said to be in potency. Secondly, where he says, "One who is learning. . ." (255 a 32),[22] he explains these distinctions. Thirdly, where he says, "But, be it noted. . ." (255 b 14),[23] he answers a certain question.

He says, therefore, first that since being in potency is predicated in many ways, it is not clear what moves heavy and light things in natural motions, for example, fire upward and earth downward.

1031. Next where he says, "One who is learning . . ." (255 a 32), he distinguishes being in potency, first in the intellect, secondly in quality, where he says, "In regard to natural bodies. . ." (255 b 5),[24]

18. 1036.
19. 1030.
20. Book II, Lecture 1, 145.
21. 1035.
22. 1031.
23. 1034.
24. 1032.

and thirdly in local motion, where he says, "So, too, with heavy. . ." (255 b 8).[25]

He says, therefore, first that the potency for science in one who is learning and does not yet have the habit of science differs from the potency for science in one who already has the habit of science but is not using it.

A thing is reduced from first potency to second when something active is joined to its passivity. And then this passivity, through the presence of that which is active, comes to be in this kind of act, which up to this point was in potency. For example, the learner, through the act of the teacher, is reduced from potency to act, to which act is joined another potency. Thus, a thing existing in first potency comes to be in another potency. For one who has science but who is not contemplating it is in a certain way in potency to the act of science, but not in the same way as he was before he learned. Therefore he was reduced from first potency to act, to which is joined a second potency, through some agent, namely, a teacher.

But when one possesses the habit of science, it is not necessary for him to be reduced to second act by some agent. Rather he does this immediately by his own contemplation unless something else prevents him, for example, business, or sickness, or his will. But if he is not impeded so that he cannot contemplate, then he does not have the habit of science but its contrary, namely, ignorance.

1032. Next where he says, "In regard to natural bodies. . ." (255 b 5), he explains the same thing with respect to qualities.

He says that what was said above[26] about the potency for science in the soul also applies to natural bodies. For when a body is actually cold, it is potentially hot, just as one who is in ignorance is in potency for knowing. But when it has been changed so that it has the form of fire, then it is fire in act, having the power of acting. And it operates immediately by burning, unless something prevents it by a contrary action or unless it is in some other way impeded, for example, by a withdrawal of the combustible object, just as it was said that after someone has become a knower by learning, he immediately contemplates, unless something prevents him.

1033. Next where he says, "So, too, with heavy . . ." (255 b 8), he explains the same thing with reference to the local motion of heavy and light things.

He says that a heavy thing becomes light in a way similar to that in which a cold thing becomes hot.

For example, air, which is light, comes from water, which is heavy. The water, therefore, is first in potency to become light, and afterwards it becomes light in act, and then it immediately possesses its operation, unless something prevents it. But the now existing light thing is compared to place as potency to act (for the act of a light thing, as such, is to be in some determined place, namely, up). But it is prevented from being up because it is in the contrary place, namely, down. For it cannot be in two places at once. Hence that which holds the light thing down prevents it from being up. And what is said about local motion must also be said of motion in respect to quantity or quality.

1034. Next where he says, "But be it noted. . ." (255 b 14), he answers a certain question about the foregoing.[27]

Granted that the act of a light thing is to be up, nevertheless some ask why heavy and light things are moved in their proper places. The reason for this is that they have a natural aptitude for such places. For to be light is to have an aptitude for that which is up. And the nature [ratio] of the heavy is to have an aptitude for that which is down. Hence, to ask why a heavy thing is moved downward is nothing other than to ask why it is heavy. The same thing which makes it heavy also makes it to be moved downward.

1035. Next where he says, "As we have said. . ." (255 b 17), he shows from the foregoing what moves heavy and light things.

He says that since that which is in potency is moved by that which is in act, as was asserted,[28] it must be realized that a thing is said to be potentially light or heavy in many ways.

In one way, while it is still water, it is in potency to become light. In another way, when air has already been made from water, it is still in potency to the act of the light, which is to be up, just as one who has the habit of science but is not contemplating it is still said to be in potency. For it happens that what is light may be prevented from being up.

But if that impediment is removed, it rises immediately so that it may be up. As was said[29] with respect to quality, when a quality is in act, it strives immediately toward its own operation, just as he who is a knower immediately contemplates, unless something prevents him. And the same applies to

25. 1033.
26. 1031.
27. 1033.
28. 1029, 1030.
29. 1032.

the motion of quantity. For when an addition of quantity has been made to quantity, extension follows immediately in the body which can be increased, unless something prevents it.

Therefore, it is clear that that which removes that which prevents and restrains in a certain sense moves and in another sense does not. For example, if a column supports something heavy, and so prevents it from falling, he who destroys the column in a certain sense is said to move the weight supported by the column. Similarly, one who removes a stone which stops water from flowing out of a vessel in a certain sense is said to move the water. He is said to move *per accidens,* and not *per se.* For example, if a sphere, that is, a ball, rebounds from a wall, it is moved by the wall *per accidens* and not *per se.* It is moved *per se* by the initial thrower. For the wall did not give it any impetus[30] toward motion, but the thrower did. It is *per accidens* because, when the ball was impeded by the wall, it did not receive a second impetus. Rather because of the same remaining impetus it rebounded with an opposite motion. And similarly, one who destroys a column does not give to the supported weight an impetus or inclination downward. For it has this from its first generator which gave to it the form which such an inclination follows. Therefore, the generator is the *per se* mover of heavy and light things. But that which removes an obstacle is a *per accidens* mover.

He concludes, therefore, that it is clear from what has been said that no heavy or light thing moves itself. Nevertheless the motion of these things is natural because they have a principle of motion within themselves, not, indeed, a motive or active principle, but a passive principle, which is the potency for such act.

From this it is clearly contrary to the Philosopher's intention to say that there is an active principle in matter, which some maintain is necessary for natural motion.[31] This passive principle, which is the natural potency for act, is sufficient.

1036. Next where he says, "If, then, the motion . . . ." (255 b 31), he arrives at the conclusion principally intended in this whole chapter.

He says that if it is true that everything that is moved is moved according to nature or outside of nature and by violence, then it is clear that all things which are moved by violence are moved not only by some mover but by some other extrinsic mover. And further of things which are moved according to nature, some are moved by themselves, in which it is clear that they are moved by

something, not indeed, extrinsic, but intrinsic, and some are moved according to nature but not by themselves, as heavy and light things. These latter are also moved by something, as was shown (because either they are moved *per se* by the generator which makes them heavy and light, or they are moved *per accidens* by that which removes what impedes or prevents their natural motion). Therefore, it is clear that whatever is moved is moved by some mover, either intrinsic or extrinsic, which he calls 'being moved by another'.

30. Although the English term "impetus," which occurs four times in the next few lines, represents a translation of the Latin *impetus,* this is in no sense an early version of the medieval impetus theory. For it is obvious that Thomas is here employing the term in its nontechnical sense of "tendency" or "inclination" to motion, rather than as an incorporeal impressed force, or as some type of separate and distinguishable entity producing or preserving motion. Elsewhere, however, in his *Commentary on the De caelo,* Book III.2, Lecture 7, Thomas did consider and reject impressed forces as motive powers (see M. Clagett, *The Science of Mechanics in the Middle Ages* [Madison, Wis.: University of Wisconsin Press, 1959], pp. 516–517).

31. A probable allusion to the opinion of Averroes in Commentary 28 of Book III of the *De caelo,* which is translated in Selection 45. In his commentary on Book III of *De caelo,* Lecture 7, No. 9, Thomas emphatically repudiates Averroes' alleged position that the natural form is a mover:

"Both arguments stem from the same error. [Averroes] thought that the form of heavy and light bodies is an active principle of motion after the manner of a mover needing some resistance contrary to the tendency of form, and that motion is not immediately due to the agent who conferred the form. But this is absolutely false. The form of heavy and light bodies is not a principle of motion as a generator of motion, but as a means by which the mover moves, just as color, a principle of sight, is a means by which something is seen. . . . Thus movement of heavy and light bodies does not come from the generator by the intervention of another moving power *(mediante alio principio movente).* Nor even is there any need to look for resistance here other than that which exists between generator and generated. Thus it follows that air is not required for natural motion of necessity *(ex necessitate),* as in the case of violent motion, since that which moves naturally has a force *(virtutem)* imparted to it which is a source of motion. Consequently there is no need for a body to be moved by any other force impelling it, as though it were a case of violent motion, having no implanted force from which motion springs."

The translation is by James A. Weisheipl, O. P., in his article "The Principle *Omne quod movetur ab alio movetur* in Medieval Physics," *Isis,* LVI (1965), 41. An examination of Averroes' discussion will reveal that Aquinas ignored the fact that Averroes would allow the form of an inorganic body to be a mover only indirectly, or *per accidens,* denying that it can function as a direct and essential mover.

47 EXTERNAL AND INTERNAL RESISTANCES TO MOTION

Albert of Saxony (ca. 1316–1390)

Translated and annotated by Edward Grant[1]

Next it is sought whether, in its downward motion, a heavy simple body has an internal resistance; and similarly, [whether] in its upward motion a light [simple] body [has an internal resistance]. [This statement of the question is followed by six arguments in favor of internal resistance in simple bodies (all six are subsequently rejected), after which Albert observes that Aristotle denied the idea of internal resistance in simple heavy bodies. It is at this point that Albert enters into an enumeration and description of the different types of resistance.—*Ed.*]

In this question we must first posit the different ways of imagining resistance in the motion of heavy and light bodies; and secondly, one way of imagining the internal resistance of simple bodies must be disproved. Then, in the following question, I shall disprove certain types of internal resistance of simple bodies with respect to their local motion and then we shall respond to what has been sought.

As to the first part [of this question], one must know that resistance is called an inclination [or tendency] not to be moved, [an inclination] to opposite motion, or [an inclination] to rest. For example, if there were a certain body that had four degrees of heaviness and two degrees of lightness, then, when that body should descend, those two degrees of lightness would resist those four degrees of heaviness because they would tend to an opposite motion, namely upward.[2]

In the second place, it must be understood that there is a difference between a simple heavy body and an absolutely heavy body. For having assumed that some [piece of earth] has suffered from fire so that some degrees of lightness have been induced into the earth by this fire, then that [piece of] earth is substantially simple and heavy and is called a simple heavy body. But it is not called absolutely heavy, since it has some degree of lightness [mixed] with degrees of heaviness. Now, this question [under discussion here] should be understood [as being concerned] with a simple heavy body which is absolutely simple.

In the third place, it is known that resistance in the motion of heavy and light bodies can be imagined as external or internal. External [resistance] can be imagined in many ways. In the first way as in balances; thus, having assumed that *a,* pure earth, was moved to one arm of a balance and *b,* [also] pure earth, to the other [arm], then none of them would descend because one would resist the other[3]—and this would be [resisting] externally. External resistance can be imagined in a second way, just as if a light body were connected to a heavy body; then the light body would resist the descent of this heavy body externally. In a third way, that which is connected sometimes resists by virtue of its shape, as if [for example] an extended [of flattened] sheet of paper were connected to some heavy body which would then descend slower than if it were not so connected. In a fourth way, sometimes pulling [or attracting] toward the opposite [direction serves to] resist externally, just as if, for example, a magnet were held up and under it there were posited [a piece of] iron so great that this magnet would not suffice to hold back the iron. Then that [piece of] iron would descend slower than if there were no magnet posited above, because the magnet sometimes attracts iron and by such an attraction the velocity of the descending iron is impeded. In such a case, therefore, the magnet can be called an external resistance. In a fifth way, a medium through which there is motion can be called an external resistance, for because of its thickness it resists a mobile.[4] It is for this reason that any mobile is moved slower through water than air. In a sixth way, a moving power can take

1. From the Latin text in Albert of Saxony's *Questions on the Physics of Aristotle,* Book IV, Question 9, published in *Questiones et decisiones physicales insignium virorum: Alberti de Saxonia in octo libros Physicorum; tres libros De celo et mundo. . . Thimonis in quatuor libros Meteororum. . . Buridani in librum De sensu et sensato. . . Aristotelis. . . recognitae rursus et emendatae summa accuratione. . . Magistri Georgii Lokert. . .* (Paris, 1518). Our selection occupies fols. 49 recto, col. 1, to 49 verso, col. 1.

2. This example illustrates the internal resistance of contrarily inclined elements in a heavy mixed body. Although Albert accepts this as a genuine kind of resistance, it is not discussed in this selection (but see Selection 55.2) because he is concerned here only with the motion of simple elemental—not mixed—bodies. The one type of "internal resistance" considered below is radically different.

3. Presumably because *a* equals *b* and the balance is equal-armed.

4. This was the fundamental and usual sense of resistance in natural motion.

the place of an external resistance, as if, [for example,] a moving power does not wish to move quickly, nor as quickly as it could.[5] A seventh way is the *incompossibility [or incompatibility] of the termini,* namely the termini from which, and the termini to which, can be called an external resistance.[6]

[Now], as to the second [part of this question] concerning internal resistance, some say that any natural agent is of finite power, that is, it is limited to some certain degree of velocity which it does not exceed however much the external resistance should be removed, so that they assume such a limitation as an internal resistance. Whence it is that such people say that if pure earth were assumed in a vacuum, it would descend with a finite velocity, say with that degree of velocity to which it would be limited.[7] Furthermore, they say that heavier bodies would be limited to a more intense degree of velocity than less heavy bodies; moreover, they say that a simple heavy body posited in a vacuum would not be moved in it instantaneously but [rather] successively because of the limitation and finitude of the motive power.

But this imagination [or conception] is contrary to Aristotle, for which reason I posit one conclusion contrary to it, which is that the limitation and finiteness of the power of the agent is not the cause of the succession of motion of a simple heavy body; nor is it a substitute for resistance. This is proved because if any natural agent should act successively, this would be because it is of finite power. It ought to follow, [therefore,] that a luminous body should produce its light in a medium successively. But this is false because illumination of a medium is instantaneous,[8] as has been seen elsewhere. The consequence holds because it is of finite power.[9]

Secondly, let it be proved [in this way]: By the same [reasoning as before], a visible object ought to send its species into the eye successively, and a nearer visible object ought to send its species into the eye more quickly than a visible object more remote from the eye; and thus it ought also to be seen more quickly. But this is false, because if Socrates is 4 feet from me and Plato 8 feet away, I see Plato just as quickly as I see Socrates when opening my eyes. But this would not be so if a visible object could send its species into the eye successively.

Thirdly, if the aforesaid imagination [or con-

5. A motive power which, for whatever reason, fails to

exert its maximum force cannot produce its maximum effect and from this standpoint retards the maximum motion that it is capable of producing. It is not obvious under what part of his threefold definition of resistance Albert would have classified and explained this type of resistance.

6. In the very next question (Question X on fol. 49 verso, col. 2), Albert defines this type of resistance as follows: "There is an imagination [or conception] that if there were no resistance of a medium but a vacuum could be imagined, a simple heavy body would nevertheless descend successively and with a finite velocity because it is impossible that it be simultaneously in the terminus from which, and the terminus to which, [it moves]. And because this is impossible, it is necessary that it be moved successively from the terminus from which [it moves] to the terminus to which [it moves], so that the incompossiblity [or incompatibility] of being simultaneously in different places is the cause of the finite velocity of motion of simple heavy and light bodies when surrounded by the resistance of a medium. And those who are of this opinion say that it is the cause of the finite velocity of celestial motion which is without any of the aforementioned resistances." Albert then proceeds to reject this opinion (as did also John Buridan), which, as we have already seen, was put forth, perhaps for the first time, by St. Thomas Aquinas (see Selection 55.1 and n. 4) as justification for the view that motion in a hypothetical void must take a finite time. Since Thomas did not himself designate this argument as a form of resistance (nor indeed did he name it "incompossibility of the termini"), it must have been so designated sometime after.

7. This form of "internal resistance" may also be traceable to that very passage of Thomas Aquinas cited in the preceding note and containing the substance of the "incompossibility of the termini" argument which is ultimately traceable to Averroes' quotation of Avempace's opinion (see Selection 44.2) on motion in a resistanceless medium. As a consequence of the "incompossibility" argument Thomas insisted that finite motion in the void is possible and that "if motion in a void were assumed, it follows that no retardation would occur beyond the natural velocity;. . . ." However, Thomas did not attribute this natural finite velocity to the finite power of the natural agent and did not conceive of the natural agent as a resistance. These were all later refinements.

8. Here Albert is probably following Aristotle who insisted that light moves instantaneously (*De anima* II. 7.418b.20–27, and *De sensu,* 6.446a.26–446b.3; the latter passage is cited in Selection 55.1, n. 4). Many, in antiquity and the Middle Ages, disagreed with Aristotle (see Roger Bacon's discussion in Selection 62.6).

9. The consequence in this argument declares that a natural agent acts successively because it is of finite power; therefore a luminous body should produce light successively in a medium. The consequence is upheld because every luminous body is of finite power and should diffuse light successively. Albert rejects the consequence, however, since he believed that light is diffused instantaneously, from which it follows that not every finite agent can produce a successive motion. Consequently, it is false to assume that finitude of power is the cause of successive motion.

ception] were true, then a greater force should be less limited and a smaller force more limited. But this is false, for then the stars would not produce their light through the whole medium as quickly as the moon, which, however, is false because any of these produces its light through the whole medium instantaneously. The consequence holds for this reason: Since the moon is of greater power, it would be less limited, and so should produce its effect more quickly; and since the stars are of lesser power, they should be more limited and produce their effect slower.

Fourthly, it follows universally that no mutation is naturally possible in an instant (the opposite of this will be seen afterward). The consequence holds, because any natural being is limited to a certain degree of velocity in producing its effect.[10]

Fifthly, it follows that after every external impediment has been removed,[11] a simple heavy body would not be moved to its natural place until its middle coincided with the middle [or center] of the world.[12] The consequent is false[13] and this is [now] proved: Let *a* be a uniformly [or homogeneously] simple heavy body whose total heaviness [or gravity *(gravitas)*] is as six and its limitation as three.[14] Now *a* descends until a third part of it is under the other side of the center [of the world]; and let this [part of *a*] be *c,* and let the remainder, which is still on the other side of the center, be *d* so that the center [of the world] is between *c* [and] *d*. It is clear, then, that not even the middle of *a* is the middle [or center] of the world. But I prove that it does not descend further, for *c,* which is on the other side of the center, resists *d* with its total limitation. Now because the total limitation is as three and *c* is as two, and since these are together five and its motive power is only as four, it follows that there is more resistance than motive power, and, as a consequence, *a* rests and is not moved further, even though its center of gravity is not the center of the world.[15] Therefore, et cetera, and consequently the conclusion is true.[16].

10. From the assumption that any natural being is limited, and of finite power, it follows universally that no mutation or change is naturally possible in an instant. As a counterinstance, however, Albert has already insisted that light illuminates a medium instantaneously, and says he will show later that there are instantaneous changes.

11. This is the antecedent of this conditional sentence.

12. This is the consequent.

13. The consequent is false because, as Albert shows

below, by assuming the limitation theory as cause of finite successive motion and assuming the conditions just asserted, it can be shown that a body will come to its natural place when its center of gravity does not coincide with the geometric center of the world. For by the "natural place" of a simple heavy body in a universe where all impediments to its motion have been removed, Albert means primarily the place where it will come to rest. But, as his example shows, the place of rest can be such that the center of gravity of the body and the geometric center of the universe fail to coincide.

14. The ratio of motive force to resistance is therefore 6 to 3.

15. A figure may be useful in explaining Albert's example; *F* is motive force, or heaviness; *R* is total limita-

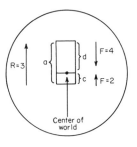

Center of world

tion, or resistance. We see that the total limitation, or resistance, of *a*, which equals 3, acts in opposition to the heaviness of *d*; and since part *c* of *a* has now moved to the other side of the geometric center of the world, and since, furthermore, all heavy things move naturally toward the center of the world, the heaviness, or motive force, of *c*, namely 2, tends to move toward the center of the world in opposition to *d*. Hence *c*'s motive force of 2 must be added to 3, the total limitation, or resistance, of *a*, resulting in a total resistance of 5 to the motive force, or heaviness, of *d*, namely 4. Since a force of 4 cannot move a resistance of 5, it follows that *a* will not descend further and its center of gravity will not coincide with the center of the world.

16. As already enunciated, the true conclusion is that "the limitation and finiteness of the power of the agent is not the cause of the succession of motion of a simple heavy body; nor is it a substitute for resistance." In the next question (Question 10 [fol. 50 verso, col. 1]: "Whether a resisting medium is required in every motion of heavy and light bodies"), where Albert continues to discuss absolutely heavy and light bodies, we discover his own position in the enunciation of these two conclusions: "The first conclusion: simple heavy and light bodies do not have internal resistance in their motion . . . .The second conclusion: an external resistance, as the resistance of a medium or some such thing, is required for the motion of heavy and light bodies. This is proved, because every motion follows a ratio of moving power to resistance, and, as a consequence, a resistance is required for every motion. Therefore, since [simple] heavy and light bodies do not have internal resistance, it follows that an external resistance is required for their motion. . . ."

48 THE IMPETUS THEORY OF PROJECTILE MOTION

John Buridan (ca. 1300—ca. 1358)

Translation, introduction, and annotation by Marshall Clagett[1]

In the same loose sense that Bradwardine was the founder of a school of mechanicians at Merton College, so Buridan was the "founder" of such a school at Paris. But Buridan's principal interest so far as mechanics was concerned was in dynamics. And his exposition of the impressed force (called by him *impetus*)[2] was the starting point of similar discussions of his principal successors at Paris in the second half of the century, namely Nicole Oresme, Albert of Saxony, and Marsilius of Inghen.

BOOK VIII, QUESTION 12.

1. It is sought whether a projectile after leaving the hand of the projector is moved by the air, or by what it is moved.

It is argued that it is not moved by the air, because the air seems rather to resist, since it is necessary that it be divided. Furthermore, if you say that the projector in the beginning moved the projectile and the ambient air along with it, and then that air, having been moved, moves the projectile further to such and such a distance, the doubt will return as to by what the air is moved after the projector ceases to move. For there is just as much difficulty regarding this (the air) as there is regarding the stone which is thrown.

Aristotle takes the opposite position in the eighth[3] [book] of this work (the *Physics*) thus: "Projectiles are moved further after the projectors are no longer in contact with them, either by antiperistasis, as some say, or by the fact that the air having been pushed, pushes with a movement swifter than the movement of impulsion by which it (the body) is carried towards its own [natural] place." He determines the same thing in the seventh and eighth [books] of this work (the *Physics*) and in the third [book] of the *De caelo*.

2. This question I judge to be very difficult because Aristotle, as it seems to me, has not solved it well. For he touches on two opinions. The first one, which he calls "antiperistasis," holds that the projectile swiftly leaves the place in which it was, and nature, not permitting a vacuum, rapidly sends air in behind to fill up the vacuum. The air moved swiftly in this way and impinging upon the projectile impels it along further. This is repeated

continually up to a certain distance. . . . But such a solution notwithstanding, it seems to me that this method of proceeding was without value because of many experiences *(experientie)*.

The first experience concerns the top *(trocus)* and the smith's mill (i.e. wheel—*mola fabri*) which are moved for a long time and yet do not leave their places. Hence, it is not necessary for the air to follow along to fill up the place of departure of a top of this kind and a smith's mill. So it cannot be said [that the top and the smith's mill are moved by the air] in this manner.

1. [Reprinted with permission of the Regents of the University of Wisconsin from M. Clagett, *The Science of Mechanics in the Middle Ages* (Madison, Wis.: University of Wisconsin Press, 1959), pp. 532–538. The introductory paragraph appears on page 522. For convenience, and following his order, Clagett's commentary (pp. 538–540), which is cued to the paragraph numbers of the selection, is reproduced after the translation. The editorial omissions are Clagett's, whose translation was made from the 1509 Paris edition of Buridan's *Questions on the Eight Books of the Physics* (for full title, see Selection 14, n. 1) with the modifications of A. Maier, *Zwei Grundprobleme der scholastischen Naturphilosophie*, 2d ed. (Rome: Edizioni di Storia e Letteratura, 1951), pp. 201–214.—*Ed.*]

2. [An impressed force theory explaining the motion of bodies projected through the air was enunciated at least as early as the sixth century A.D. by the Greek Aristotelian commentator, John Philoponus. In his *Commentary on the Physics of Aristotle*, Philoponus shows the absurdities of two related explanations (one is known as *antiperistasis*) proposed by Aristotle in which the air is invoked as the motive force continuing the motion of the projectile after it has lost contact with its initial mover (this selection is translated in M.R. Cohen and I. E. Drabkin, *A Source Book in Greek Science* [Cambridge, Mass.: Harvard University Press, 1948], pp. 221–223). Instead he argues that it is necessary "*to assume that some incorporeal motive force is imparted by the projector to the projectile,* and that the air set in motion contributes either nothing at all, or else very little to this motion of the projectile" (Cohen and Drabkin, p. 223). Undoubtedly familiar with the views of Philoponus, a number of Arabs discussed, and some even adopted, the concept of an impressed force (for instance, Avicenna and Abū l'Barakāt). Although it is believed that Arab discussions on impressed force were somehow transmitted to the Latin West, the path of transmission has not been discovered. See Clagett, pages 505–525, for a thorough discussion of the history of impetus theory—*Ed.*]

3. This is a statement from the fourth, rather than the eighth, book.

The second experience is this:[4] A lance having a conical posterior as sharp as its anterior would be moved after projection just as swiftly as it would be without a sharp conical posterior. But surely the air following could not push a sharp end in this way, because the air would be easily divided by the sharpness.

The third experience is this: a ship drawn swiftly in the river even against the flow of the river, after the drawing has ceased, cannot be stopped quickly, but continues to move for a long time. And yet a sailor on deck does not feel any air from behind pushing him. He feels only the air from the front resisting [him]. Again, suppose that the said ship were loaded with grain or wood and a man were situated to the rear of the cargo. Then if the air were of such an impetus that it could push the ship along so strongly, the man would be pressed very violently between that cargo and the air following it. Experience shows this to be false. Or, at least, if the ship were loaded with grain or straw, the air following and pushing would fold over *(plico)* the stalks which were in the rear. This is all false.

3. Another opinion, which Aristotle seems to approve, is that the projector moves the air adjacent to the projectile [simultaneously] with the projectile and that air moved swiftly has the power of moving the projectile. He does not mean by this that the same air is moved from the place of projection to the place where the projectile stops, but rather that the air joined to the projector is moved by the projector and that air having been moved moves another part of the air next to it, and that [part] moves another (i.e., the next) up to a certain distance. Hence the first air moves the projectile into the second air, and the second [air moves it] into the third air, and so on. Aristotle says, therefore, that there is not one mover but many in turn. Hence he also concludes that the movement is not continuous but consists of succeeding or contiguous entities.

But this opinion and method certainly seems to me equally as impossible as the opinion and method of the preceding view. For this method cannot solve the problem of how the top or smith's mill is turned after the hand [which sets them into motion] has been removed. Because, if you cut off the air on all sides near the smith's mill by a cloth *(linteamine)*, the mill does not on this account stop but continues to move for a long time. Therefore it is not moved by the air.

Also a ship drawn swiftly is moved a long time after the haulers have stopped pulling it. The surrounding air does not move it, because if it were covered by a cloth and the cloth with the ambient air were withdrawn, the ship would not stop its motion on this account. And even if the ship were loaded with grain or straw and were moved by the ambient air, then that air ought to blow exterior stalks toward the front. But the contrary is evident, for the stalks are blown rather to the rear because of the resisting ambient air.

Again, the air, regardless of how fast it moves, is easily divisible. Hence it is not evident as to how it would sustain a stone of weight of one thousand pounds projected in a sling or in a machine.

Furthermore, you could, by pushing your hand, move the adjacent air, if there is nothing in your hand, just as fast or faster than if you were holding in your hand a stone which you wish to project. If, therefore, that air by reason of the velocity of its motion is of a great enough impetus to move the stone swiftly, it seems that if I were to impel air toward you equally as fast, the air ought to push you impetuously and with sensible strength. [Yet] we would not perceive this.

Also, it follows that you would throw a feather farther than a stone and something less heavy farther than something heavier, assuming equal magnitudes and shapes. Experience shows this to be false. The consequence is manifest, for the air having been moved ought to sustain or carry or move a feather more easily than something heavier. . . .

4. Thus we can and ought to say that in the stone or other projectile there is impressed something which is the motive force *(virtus motiva)* of that projectile. And this is evidently better than falling back on the statement that the air continues to move that projectile. For the air appears rather to resist. Therefore, it seems to me that it ought to be said that the motor in moving a moving body impresses *(imprimit)* in it a certain impetus *(impetus)* or a certain motive force *(vis motiva)* of the moving body, [which impetus acts] in the direction toward which the mover was moving the moving body, either up or down, or laterally, or circularly. *And by the amount the motor moves that moving body more swiftly, by the same amount it will impress in it a stronger impetus.* It is by that impetus that the stone is moved after the projector ceases to move. But that impetus is continually decreased *(remittitur)* by the resisting air and by the gravity

4. [A similar argument was given by Philoponus. See Cohen and Drabkin, pp. 221–222.—*Ed.*]

of the stone, which inclines it in a direction contrary to that in which the impetus was naturally predisposed to move it. Thus the movement of the stone continually becomes slower, and finally that impetus is so diminished or corrupted that the gravity of the stone wins out over it and moves the stone down to its natural place.

This method, it appears to me, ought to be supported because the other methods do not appear to be true and also because all the appearances (apparentia) are in harmony with this method.

5. For if anyone seeks why I project a stone farther than a feather, and iron or lead fitted to my hand farther than just as much wood, I answer that the cause of this is that the reception of all forms and natural dispositions is in matter and by reason of matter. *Hence by the amount more there is of matter, by that amount can the body receive more of that impetus and more intensely* (intensius). *Now in a dense and heavy body, other things being equal, there is more of prime matter than in a rare and light one. Hence a dense and heavy body receives more of that impetus and more intensely, just as iron can receive more calidity than wood or water of the same quantity.* Moreover, a feather receives such an impetus so weakly *(remisse)* that such an impetus is immediately destroyed by the resisting air. *And so also if light wood and heavy iron of the same volume and of the same shape are moved equally fast by a projector, the iron will be moved farther because there is impressed in it a more intense impetus, which is not so quickly corrupted as the lesser impetus would be corrupted. This also is the reason why it is more difficult to bring to rest a large smith's mill which is moving swiftly than a small one, evidently because in the large one, other things being equal, there is more impetus.*[5] And for this reason you could throw a stone of one-half or one pound weight farther than you could a thousandth part of it. For the impetus in that thousandth part is so small that it is overcome immediately by the resisting air.

6. From this theory also appears the cause of why the natural motion of a heavy body downward is continually accelerated *(continue velocitatur)*. For from the beginning only the gravity was moving it. Therefore, it moved more slowly, but in moving it impressed in the heavy body an impetus. This impetus now [acting] together with its gravity moves it. Therefore, the motion becomes faster; and by the amount it is faster, so the impetus becomes more intense. Therefore, the movement evidently becomes continually faster.

[The impetus then also explains why] one who

wishes to jump a long distance drops back a way in order to run faster, so that by running he might acquire an impetus which would carry him a longer distance in the jump. Whence the person so running and jumping does not feel the air moving him, but [rather] feels the air in front strongly resisting him.

Also, since the Bible does not state that appropriate intelligences move the celestial bodies, it could be said that it does not appear necessary to posit intelligences of this kind, because it would be answered that God, when He created the world, moved each of the celestial orbs as He pleased, and in moving them He impressed in them impetuses which moved them without his having to move them any more except by the method of general influence whereby he concurs as a co-agent in all things which take place; "for thus on the seventh day He rested from all work which He had executed by committing to others the actions and the passions in turn." And these impetuses which He impressed in the celestial bodies were not decreased nor corrupted afterwards, because there was no inclination of the celestial bodies for other movements. Nor was there resistance which would be corruptive or repressive of that impetus.[6] But this I do not say assertively, but [rather tentatively] so that I might seek from the theological masters

5. [Of the lines which he italicized, Clagett observes (p. 523) that "Buridan gives a quasi-quantitative definition of *impetus* at the time of its imposition. We must say 'quasi' because there is no formal discussion of its mathematical description. The exact nature of the impetus as conceived by Buridan is difficult to pin down. He spoke of it as motive force and as the reason for the continued movement, . . . It appears that while the impetus is unquestionably a force, it nevertheless seems close to being the effectiveness which the original force has on a particular body, an effectiveness measurable in terms of the velocity immediately supplied to the body and the quantity of matter in the body."—Ed.]

6. ["The characteristic of permanence which Buridan assigned to his impetus made it plausible for him to explain the everlasting movement of the heavens by the imposition of impetus by God at the time of the world's creation. . . . The use of impetus to explain the continuing movement of the heavens is the closest that Buridan comes to the inertial idea of Newton's mechanics. It can scarcely be doubted that impetus is analogous to the later inertia, regardless of ontological differences (Clagett, pp. 524–525)." Although Buridan used impetus to explain terrestrial and celestial motion, Clagett denies that he had sought to formulate a single mechanics for the whole universe. From other passages, it is clear that Buridan accepted the Aristotelian dichotomy between the behavior of celestial and terrestrial bodies. See the next selection, paragraph 7, where Buridan also mentions celestial impetus.—Ed.]

what they might teach me in these matters as to how these things take place. . . .[7]

7. The first [conclusion] is that that impetus is not the very local motion in which the projectile is moved, because that impetus moves the projectile and the mover produces motion. Therefore, the impetus produces that motion, and the same thing cannot produce itself. Therefore, etc.

Also since every motion arises from a motor being present and existing simultaneously with that which is moved, if the impetus were the motion, it would be necessary to assign some other motor from which that motion would arise. And the principal difficulty would return. Hence there would be no gain in positing such an impetus. But others cavil when they say that the prior part of the motion which produces the projection produces another part of the motion which is related successively and that produces another part and so on up to the cessation of the whole movement. But this is not probable, because the "producing something" ought to exist when the something is made, but the prior part of the motion does not exist when the posterior part exists, as was elsewhere stated. Hence, neither does the prior exist when the posterior is made. This consequence is obvious from this reasoning. For it was said elsewhere that motion is nothing else than "the very being produced" *(ipsum fieri)* and the "very being corrupted" *(ipsum corumpi)*. Hence motion does not result when it *has been* produced *(factus est)* but when it *is being* produced *(fit)*.

8. The second conclusion is that that impetus is not a purely successive thing *(res)*, because motion is just such a thing and the definition of motion [as a successive thing] is fitting to it, as was stated elsewhere. And now it has just been affirmed that that impetus is not the local motion.

Also, since a purely successive thing is continually corrupted and produced, it continually demands a producer. But there cannot be assigned a producer of that impetus which would continue to be simultaneous with it.

9. The third conclusion is that that impetus is a thing of permanent nature *(res nature permanentis)*, distinct from the local motion in which the projectile is moved. This is evident from the two aforesaid conclusions and from the preceding [statements]. And it is probable *(verisimile)* that that impetus is a quality naturally present and predisposed for moving a body in which it is impressed, just as it is said that a quality impressed in iron by a magnet moves the iron to the magnet. And it

also is probable that just as that quality (the impetus) is impressed in the moving body along with the motion by the motor; so with the motion it is remitted, corrupted, or impeded by resistance or a contrary inclination.

10. And in the same way that a luminant generating light generates light reflexively because of an obstacle, so that impetus because of an obstacle acts reflexively. It is true, however, that other causes aptly concur with that impetus for greater or longer reflection. For example, the ball which we bounce with the palm in falling to earth is reflected higher than a stone, although the stone falls more swiftly and more impetuously *(impetuosius)* to the earth. This is because many things are curvable or intracompressible by violence which are innately disposed to return swiftly and by themselves to their correct position or to the disposition natural to them. In thus returning, they can impetuously push or draw something conjunct to them, as is evident in the case of the bow *(arcus)*. Hence in this way the ball thrown to the hard ground is compressed into itself by the impetus of its motion; and immediately after striking, it returns swiftly to its sphericity by elevating itself upwards. From this elevation it acquires to itself an impetus which moves it upward a long distance.

Also, it is this way with a cither cord which, put under strong tension and percussion, remains a long time in a certain vibration *(tremulatio)* from which its sound continues a notable time. And this takes place as follows: As a result of striking [the chord] swiftly, it is bent violently in one direction, and so it returns swiftly toward its normal straight position. But on account of the impetus, it crosses beyond the normal straight position in the contrary direction and then again returns. It does this many times. For a similar reason a bell *(campana)*, after the ringer ceases to draw [the chord], is moved a long time, first in one direction, now in another. And it cannot be easily and quickly brought to rest.

This, then, is the exposition of the question. I would be delighted if someone would discover a more probable way of answering it. And this is the end.

COMMENTARY

The reader's attention is first called to the refutation in passages 2 and 3 of the two theories presented by Aristotle. The first point worth noting is that

7. [Elsewhere in the same *Physics* treatise Buridan was more assertive and less deferential to the theologians (see Selection 14).—*Ed.*]

it is largely on the basis of experience that these two theories are shown to be inadequate. The *experientie* adduced against *antiperistasis,* i.e., the mechanical action of the air, are the following: (1) The spinning of a top or smith's wheel takes place without leaving its place of motion and the air can hardly be said to come behind the moving body to continue its motion. (2) The sharpening to a point of the posterior end of a lance does not thereby reduce its speed, as one would expect if this theory were correct. (3) In the course of the continuation of the movement of a ship in a river after the haulers have stopped pulling it, a sailor on deck does not feel the air pushing him from behind but rather feels it resisting him. Nor if behind some cargo would he be pushed against it; and similarly straws in the rear are not bent over. Similar "experiences" are brought against the second theory which held for a successive communication of motive power to the parts of the air.

In passage 4 Buridan states his acceptance of the theory which posited that the projectile motion is continued because the motor impresses in the projectile an *impetus* or motive force. He relates the intensity of impressed impetus to the velocity imparted by the original force to the projectile, i.e., the greater the velocity, the greater the impetus. Quite evidently this velocity is the speed of the projectile immediately after the original force of action has ceased. At the same time, he says that the impetus is made to decrease (i.e., is remitted and corrupted) in the same way that the motion is made to decrease, by the resistance and contrary inclination of the moving body. Perhaps Buridan might hold that the factors of impetus, resistance, and the continuing speed of the projectile are related as Bradwardine held generally for all cases of motion considered dynamically. However, he makes no such statement; and one might suppose from his arguments that, were there no resistance, not only would the *impetus* last indefinitely (as he states) but also that the movement maintained by the impetus would be both finite and uniform. However, such a conclusion would be difficult to fit in with the Aristotelian framework, which Buridan generally accepted. And hence our puzzlement as to just what kind of force this *impetus* really is.

Then Buridan in passage 5 relates the intensity of impetus imparted to the projectile with the quantity of prime matter in the projectile: This "quantity of prime matter" is a kind of analogue of the "mass" of early modern physics.

It is interesting to speculate on how it happened

that Buridan seized upon velocity and quantity of matter as the two factors in determining the intensity of the *impetus*. It seems to be another case of the simultaneous consideration of extensive and intensive factors. The schoolmen had considered the effectiveness of a heat agent (i.e., its *potentia*) to be dependent on both the intensity of heat and the extent of the subject through which the heat is distributed, in short on a quantity of heat. Now *impetus* seems semantically to have the meaning of something like force of impact; its effectiveness, according to Buridan, would depend on how fast it is going (an intensive factor) and how much matter there was in motion (an extensive factor). The analogy is apparent.

Having outlined the measure of impetus in terms of the velocity imparted to the projectile and the quantity of matter in the projectile, Buridan then in passage 6 notes that the continuous impress of impetus in a falling body by gravity can account for its acceleration. This idea is taken up in greater length by Buridan in a question on the *De caelo,* . . .[8] The impetus theory also explains the fact that a broad jumper takes a long initial run before making his jump. Buridan incidentally adds that such a jumper never feels the air moving him along from behind but rather feels it resisting him. In this passage Buridan also presents his tentative hypothesis that the impetus theory could be used to account for the continuing motion of the heavenly bodies, thus eliminating the necessity of positing Intelligences as the movers. . . .

While not discussing here the ontology of movement, Buridan in his first conclusion of this question opposes the identification of impetus and the motion of the projectile, and initially on the ground that, since impetus produces the continuing motion something cannot produce itself. Buridan thus appears to accept the continuance of motion as a new effect for which a cause must be sought. And if the continuing motion is such an effect, it cannot be identical with its cause, namely, impetus. And so, while fundamentally in opposition to William Ockham in conceiving of the continuing motion as a new effect demanding a causal explanation, Buridan does not in this passage appear to be opposing specifically Ockham's view that the motion of a projectile is merely *secundum se,* for Buridan would appear to be arguing against those who would identify impetus with motion as an entity separate from the projectile.

8. [See the next selection.—*Ed.*]

In addition to rejecting the identification of impetus and motion, Buridan rejects the complementary view that impetus is a successive entity of the same kind as motion (see passage 8). Rather impetus must be something permanent, distinct from the local motion of the projectile. However, it is corrupted and diminished in the same way as the motion, i.e., by resistance and contrary inclination.

Finally, according to Buridan (passage 10), the impetus theory gives a satisfactory explanation of the phenomena of rebound, the vibration of a bow string or the string of a musical instrument, and the pendular swing of a bell.

49 ON THE CAUSE OF ACCELERATION OF FREE-FALLING BODIES

John Buridan (ca. 1300—ca. 1358)

Translated and annotated by Marshall Clagett[1]

BOOK II, QUESTION 12.

1. Whether natural motion ought to be swifter in the end than the beginning. . . . With respect to this question it ought to be said that it is a conclusion not to be doubted factually *(quia est)*, for, as it has been said, all people perceive that the motion of a heavy body downward is continually accelerated *(magis ac magis velocitatur)*, it having been posited that it falls through a uniform medium. For everybody perceives that by the amount that a stone descends over a greater distance and falls on a man, by that amount does it more seriously injure him.

2. But the great difficulty *(dubitatio)* in this question is why this [acceleration] is so. Concerning this matter there have been many different opinions. The Commentator (Averroës) in the second book [of his commentary on the *De caelo*] ventures some obscure statements on it, declaring that a heavy body approaching the end is moved more swiftly because of a great desire for the end and because of the heating action *(calefactionem)* of its motion. From these statements two opinions have sprouted.

3. The first opinion was that motion produces heat, as it is said in the second book of this [work, the *De caelo*], and, therefore, a heavy body descending swiftly through the air makes that air hot, and consequently it (the air) becomes rarefied. The air, thus rarefied, is more easily divisible and less resistant. Now, if the resistance is diminished, it is reasonable that the movement becomes swifter.

But this argument is insufficient. In the first place, because the air in the summer is noticeably hotter than in the winter, and yet the same stone falling an equal distance in the summer and in the winter is not moved with appreciably greater speed in the summer than in the winter; nor does it strike harder. Furthermore, the air does not become hot through movement unless it is previously moved and divided. Therefore, since the air resists before there has been movement or division, the resistance is not diminished by its heating. Furthermore, a man moves his hand just as swiftly as a stone falls toward the beginning of its movement. This is apparent, because striking another person hurts him more than the falling stone, even if the stone is harder. And yet a man so moving his hand does not heat the air sensibly, since he would perceive that heating. Therefore, in the same way the stone, at least from the beginning of the case, does not thus sensibly heat the air to the extent that it ought

1. [Reprinted with permission of the Regents of the University of Wisconsin from M. Clagett, *The Science of Mechanics in the Middle Ages* (Madison, Wis.: University of Wisconsin Press, 1959), pp. 557–562. Clagett's commentary, which follows this selection, is reproduced here from pages 562–564. Clagett's translation was made from E. A. Moody's Latin edition of Buridan's *Questions on the Four Books On the Heavens and the World of Aristotle* (Cambridge, Mass:. Mediaeval Academy of America, 1942). The editorial omissions indicated in the selection were made by Clagett.

"Aristotle was by no means clear in his views on either the cause or the measure of the acceleration of falling bodies. He appears to have believed that acceleration depends on the increasing proximity of the body to its natural place, i.e., the center of the world. This increasing proximity produces additional weight, which in accordance to his dynamical rules would bring about the observed quickening. Yet, at the same time in refuting infinite locomotion [*De caelo* I.8.277a.27–277b.8—*Ed.*] Aristotle appears to imply two mutually incompatible ideas, namely that the speed of fall varies with the proximity to the center and that it also varies as the distance of fall" (Clagett, p. 542). Elaborations of this explanation, as well as additional explanations, were devised through the centuries. Some of these are repeated by Buridan. Along with the problem of explaining the continuation of projectile motion (see the beginning of the preceding selection), the problem of accounting for the acceleration of free-falling bodies was the most important problem in the history of physics.—*Ed.*]

to produce so manifest an acceleration *(velocitatio)* as is apparent at the end of the movement.

4. The other opinion which originated from the statements of the Commentator is this: Place is related to the thing placed as a final cause, as Aristotle implies and the Commentator explains in the fourth book of the *Physics*. And some say, in addition to this, that place is the cause moving the heavy body by a method of attraction, just as a magnet attracts iron. By whichever of these methods it takes place, it seems reasonable that the heavy body is moved more swiftly by the same amount that it is nearer to its natural place. This is because, if place is the moving cause, then it can move that body more strongly when the body is nearer to it, for an agent acts more strongly on something near to it than on something far away from it. And if place were nothing but the final cause which the heavy body seeks naturally and for the attainment of which the body is moved, then it seems reasonable that that natural appetite *(appetitus)* for that end is increased more from it as that end is nearer. And so it seems in every way reasonable that a heavy body is moved more swiftly by the amount that it is nearer to [its] downward place. But in descending continually it ought to be moved more and more swiftly.

But this opinion cannot stand up. In the first place, it is against Aristotle and against the Commentator in the first book of the *De caelo,* where they assert that, if there were several worlds, the earth of the other world would be moved to the middle of this world. . . .

Furthermore, this opinion is against manifest experience, for you can lift the same stone near the earth just as easily as you can in a high place if that stone were there, for example, at the top of a tower. This would not be so if it had a stronger inclination toward the downward place when it was low than when it was high. It is responded that actually there is a greater inclination when the stone is low than when it is high, but it is not great enough for the senses to perceive. This response is not valid, because if that stone falls continually from the top of the tower to the earth, a double or triple velocity and a double or triple injury would be sensed near the earth than would be sensed higher up near the beginning of the movement. Hence, there is a double or triple cause of the velocity. And so it follows that that inclination which you posit not to be sensible or notable is not the cause of such an increase of velocity.

Again, let a stone begin to fall from a high place

to the earth and another similar stone begin to fall from a low place to the earth. Then these stones, when they should be at a distance of one foot from the earth, ought to be moved equally fast and one ought not be swifter than the other if the greater velocity should arise only from nearness to [their] natural place, because they should be equally near to [their] natural place. Yet it is manifest to the senses that the body which should fall from the high point would be moved much more quickly than that which should fall from the low point, and it would kill a man while the other stone [falling from the low point] would not hurt him.

Again, if a stone falls from an exceedingly high place through a space of ten feet and then encountering there an obstacle comes to rest, and if a similar stone descends from a low point to the earth, also through a distance of ten feet, neither of these movements will appear to be any swifter than the other, even though one is nearer to the natural place of earth than the other.

I conclude, therefore, that the accelerated natural movements of heavy and light bodies do not arise from greater proximity to [their] natural place, but from something else that is either near or far, but which is varied by reason of the length of the motion *(ratione longitudinis motus)*. Nor is the case of the magnet and the iron similar, because if the iron is nearer to the magnet, it immediately will begin to be moved more swiftly than if it were farther away. But such is not the case with a heavy body in relation to its natural place.

5. The third opinion was that the more the heavy body descends, by so much less is there air beneath it, and the less air then can resist less. And if the resistance is decreased and the moving gravity remains the same, it follows that the heavy body ought to be moved more swiftly.

But this opinion falls into the same inconsistency as the preceding one, because, as was said before, if two bodies similar throughout begin to fall, one from an exceedingly high place and the other from a low place such as a distance of ten feet from the earth, those bodies in the beginning of their motion are moved equally fast, notwithstanding the fact that one of them has a great deal of air beneath it and the other has only a little. Hence, throughout, the greater velocity does not arise from greater proximity to the earth or because the body has less air beneath it, but from the fact that that moving body is moved from a longer distance and through a longer space.

Again, it is not true that the less air in the aforementioned case resists less. This is because, when a stone is near the earth, there is still just as much air laterally as if it were farther from the earth. Hence, it is just as difficult for the divided air to give way and flee laterally [near the earth] as it was when the stone was farther from the earth. And, in addition, it is equally difficult or more difficult, when the stone is nearer the earth, for the air underneath to give way in a straight line, because the earth, which is more resistant than the air, is in the way. Hence, the imagined solution *(imaginatio)* is not valid.

6. With the [foregoing] methods of solving this question set aside, there remains, it seems to me, one necessary solution *(imaginatio)*. It is my supposition that the natural gravity of this stone remains always the same and similar before the movement, after the movement, and during the movement. Hence the stone is found to be equally heavy after the movement as it was before it. I suppose also that the resistance which arises from the medium remains the same or is similar, since, as I have said, it does not appear to me that the air lower and near to the earth should be less resistant than the superior air. Rather the superior air perhaps ought to be less resistant because it is more subtle. Third, I suppose that if the moving body is the same, the total mover is the same, and the resistance also is the same or similar, the movement will remain equally swift, since the proportion of mover to moving body and to the resistance will remain [the same]. Then I add that in the movement downward of the heavy body the movement does not remain equally fast but continually becomes swifter.

From these [suppositions] it is concluded that another moving force *(movens)* concurs in that movement beyond the natural gravity which was moving [the body] from the beginning and which remains always the same. Then finally I say that this other mover is not the place which attracts the heavy body as the magnet does the iron; nor is it some force *(virtus)* existing in the place and arising either from the heavens or from something else, because it would immediately follow that the same heavy body would begin to be moved more swiftly from a low place than from a high one, and we experience the contrary of this conclusion. . . .

From these [reasons] it follows that one must imagine that a heavy body not only acquires motion unto itself from its principal mover, i.e., its gravity, but that it also acquires unto itself a certain impetus with that motion. This impetus has the power of moving the heavy body in conjunction with the permanent natural gravity. And because that impetus is acquired in common with motion, hence the swifter the motion is, the greater and stronger the impetus is. So, therefore, from the beginning the heavy body is moved by its natural gravity only; hence it is moved slowly. Afterwards it is moved by that same gravity and by the impetus acquired at the same time; consequently, it is moved more swiftly. And because the movement becomes swifter, therefore the impetus also becomes greater and stronger, and thus the heavy body is moved by its natural gravity and by that greater impetus simultaneously, and so it will again be moved faster; and thus it will always and continually be accelerated to the end. And just as the impetus is acquired in common with motion, so it is decreased or becomes deficient in common with the decrease and deficiency of the motion.[2]

And you have an experiment [to support this position]: If you cause a large and very heavy smith's mill [i.e., a wheel] to rotate and you then cease to move it, it will still move a while longer by this impetus it has acquired. Nay, you cannot immediately bring it to rest, but on account of the resistance from the gravity of the mill, the impetus would be continually diminished until the mill would cease to move. And if the mill would last forever without some diminution or alteration of it, and there were no resistance corrupting the impetus, perhaps the mill would be moved perpetually by that impetus.

7. And thus one could imagine that it is unnecessary to posit intelligences as the movers of celestial bodies since the Holy Scriptures do not inform us that intelligences must be posited. For it could be said that when God created the celestial spheres, He began to move each of them as He wished, and they are still moved by the impetus which He gave to them because, there being no resistance, the impetus is neither corrupted nor diminished.[3]

You should note that some people have called that impetus "accidental gravity" and they do so aptly, because names are for felicity of expression. Whence this [name] appears to be harmonious with Aristotle and the Commentator in the first [book] of this [work, the *De caelo*], where they say that gravity would be infinite if a heavy body were moved infinitely, because by the amount that it is

2. [For Galileo's explanation, see note 41 of the next selection.—*Ed.*]

3. [See paragraph 6 of the preceding selection, where Buridan also discusses this; also note 6.—*Ed.*]

moved more, by that same amount is it moved more swiftly; and by the amount that it is moved more swiftly, by that amount is the gravity greater. If this is true, therefore, it is necessary that a heavy body in moving acquires continually more gravity, and that gravity is not of the same constitution *(ratio)* or nature as the first natural gravity, because the first gravity remains always, even with the movement stopped, while the acquired gravity does not remain. All of these statements will appear more to be true and necessary when the violent movements of projectiles and other things are investigated. . . .

COMMENTARY

This selection (passages 3–5) tells us of three common explanations of the cause of the acceleration of falling bodies in addition to the fourth, the impetus explanation which Buridan is supporting: (1) a heating of the medium which decreases its resistance and thus increases the velocity; (2) proximity to natural place which acts by some virtue or other (like that of the magnet) as a moving cause, this virtue being increased as the body comes closer. . . ; (3) as the body falls there is continually less air beneath it acting as resistance; hence the velocity increases. . . ;[4] (4) the impetus explanation, i.e., gravity continually introduces an impetus which acting as a supplementary increasing cause of movement, and acting with the gravity, produces the acceleration. Among the other explanations not mentioned by Buridan were two which centered around the supplementary action of the medium: (5) one which held that air stirred up by the movement is able to get behind the falling body and give it supplementary pushes (a theory taken over from Aristotle's explanation of the continuance of projectile motion. . .); and (6) the falling body not only draws the air behind, but in pushing the air beneath it, it sets it in motion, and this air sets other air in motion, and the drawing action of the air makes it less resistant and helps the gravity of the body (a theory found in the *Liber de ratione ponderis* attributed to Jordanus[5]. . .). One other opinion drawn from Simplicius (and supported later by Galileo in his early work *De motu*) held (7) that when a body starts to fall, it still has unnatural lightness which is a holdover from its previous violent movement, and this lightness slows up the body until it is dissipated by the action of the gravity.

Like Albert of Saxony later. . ., Buridan seems to believe that the increasing velocity is a simple function of the increasing distance of fall (see the second paragraph of passage 5). But he does say in one place, you will notice, that the increasing velocity is dependent on the "length of movement" (passage 4, last paragraph), which is somewhat ambiguous but sounds as if he is talking of time. Furthermore, the adverb he uses in his exposition of the impetus theory (passage 6) and its application to falling bodies is temporal, e.g., he speaks of the velocity *continually (continue)* increasing. I believe actually he made no clear distinction between the mathematical difference involved in saying that the velocity increases directly as the distance of fall and saying that it increases directly as the time of fall. The reader should compare this discussion with the discussions of Nicole Oresme. . . , Albert of Saxony. . . and Galileo. . . .[6]

The source of the continually increasing impetus acquired by the falling body clearly appears to be the continually acting gravity, and, furthermore, the increasing velocity caused by the gravity and continually impressed impetus is a measure of the increased impetus. Hence this whole passage seems to confirm his previous description of impetus as varying directly with the velocity it maintains.

The strongly empirical and observational character of Buridan's refutation of the "erroneous" opinions as well as of the exposition of his own theory should be noticed. For example (see passage 4), if the theory that held that the cause of acceleration lay in the proximity of the body to its natural place were tested by dropping a body from a high place and another body from a low place, according to the theory when these stones "should be at a distance of one foot from the earth, [they] ought to be moved equally fast and one ought not to be swifter than the other if the greater velocity should arise only from nearness to [their] natural place because they should be equally near to [their] natural place. Yet it is manifest to the senses that the body which should fall from the high point would be moved much more quickly than that which should fall from the low point, and it would kill a man while the other stone [falling from the

4. [See M. R. Cohen and I. E. Drabkin, *A Source Book in Greek Science* (Cambridge, Mass.: Harvard University Press, 1948), p. 210, where this is discussed in Simplicius' *Commentary on Aristotle's De caelo.—Ed.*]

5. [See above, Theorem *R4.06*, Selection 39. Later in the same treatise, Jordanus employed a crude version of impetus; see Theorems *R4.14–17* and notes 26–29.—Ed.]

6. [The omissions are references to these discussions given by Clagett on p. 563.—Ed.]

low point] would not hurt him." Notice then that, like Strato,[7] Buridan takes as his practical measure of velocity of fall the force of impact, as also did Leonardo and Galileo later.

50 IN OPPOSITION TO ARISTOTLE: CONTRARY MOTIONS CAN BE CONTINUOUS WITHOUT AN INTERVENING MOMENT OF REST

Introduction by Edward Grant

In *Physics,* Book VIII, Chapter 8, Aristotle argues that the only possible single and continuous infinite motion must be rotatory not rectilinear. After a lengthy discussion involving, among other things, one of Zeno's paradoxes, Aristotle presents the following passage which aroused considerable controversy and disagreement in the Middle Ages.

Everything whose motion is continuous must, on arriving at any point in the course of its locomotion, have been previously also in process of locomotion to that point, if it is not forced out of its path by anything: e.g. on arriving at *B* a thing must also have been in process of lo-comotion to *B*, and that not merely when it was near to *B*, but from the moment of its starting on its course, since there can be no reason for its being so at any particular stage rather than at an earlier one. So, too, in the case of the other kinds of motion. Now we are to suppose that a thing proceeds in locomotion from *A* to *C* and that at the moment of its arrival at *C* the con-tinuity of its motion is unbroken and will remain so until it has arrived back at *A*. Then when it is undergoing locomotion from *A* to *C* it is at the same time undergoing also its locomotion to *A* from *C*: consequently it is simultaneously under-going two contrary motions, since the two mo-tions that follow the same straight line are con-trary to each other. With this consequence there also follows another: we have a thing that is in process of change from a position in which it has not yet been: so, inasmuch as this is impossible, the thing must come to a stand at *C*. Therefore the motion is not a single motion, since motion that is interrupted by stationariness is not single.

Further, the following argument will serve better to make this point clear universally in respect of every kind of motion. If the motion undergone by that which is in motion is always one of those already enumerated, and the state of rest that it undergoes is one of those that are the opposites of the motions (for we found that there could be no other besides these), and moreover that which is undergoing but does not always undergo a particular motion (by this I mean one of the various specifically distinct motions, not some particular part of the whole motion) must have been previously undergoing the state of rest that is the opposite of the motion, the state of rest being privation of motion; then, inasmuch as the two motions that follow the same straight line are contrary motions, and it is impossible for a thing to undergo simultaneously two contrary motions, that which is undergoing locomotion from *A* to *C* cannot also simultaneously be undergoing locomotion from *C* to *A*; and since the latter locomotion is not simultaneous with the former but is still to be undergone, before it is undergone there must occur a state of rest at *C*; for this, as we found, is the state of rest that is the opposite of the motion from *C*. The foregoing argument, then, makes it plain that the motion in question is not continuous. (264a.7–35)

As will be seen in the notes, it would seem that some of the basic anti-Aristotelian arguments were passed on from Arabic sources to the Latin West, where they were utilized and added to by Marsilius and others in the fourteenth century, eventually passing in the sixteenth century to Benedetti and Galileo, who apparently found the question signi-ficant because it provided them with one more

7. [Clagett's reference is to Simplicius' *Commentary on Aristotle's De caelo,* where Simplicius attributes the following argument to Strato (fl. ca. 287 B.C.): "If one drops a stone or any other weight from a height of about an inch, the impact made on the ground will not be per-ceptible, but if one drops the object from a height of a hundred feet or more, the impact on the ground will be a powerful one. Now there is no other cause for this powerful impact. For the weight of the object is not greater, the object itself has not become greater, it does not strike a greater space of ground, nor is it impelled by a greater [external force]. It is merely a case of accelera-tion. And it is because of this acceleration that this phe-nomenon and many others take place." The translation appears in Cohen and Drabkin, pp. 211–212, and is quoted by Clagett, p. 546. Since William Moerbeke translated Simplicius' *Commentary on De caelo* from Greek to Latin in 1271 (Selection 8, no, 47), Buridan probably derived his argument in Passage 4 from Sim-plicius.—*Ed.*]

point of attack against Aristotle. The question is of interest because, incorporated into the responses,

1. Marsilius of Inghen (ca. 1340–1396)

Translated and annotated by Edward Grant[1]

BOOK VIII, QUESTION 7

Whether an intervening [moment of] rest (quies media) is necessary between any contrary [or reverse] (reflexos) motions.

Now, it is argued affirmatively *(quod sic)* that between things resting in opposite places there is an intervening motion; therefore, between any contrary motions there is an interval of rest *(quies media)*. The consequence holds by similarity [or analogy];[2] the antecedent is obvious by experience.[3]

Secondly, [on the assumption of an intervening moment of rest,] Aristotle, in this eighth book, argues and proves that a rectilinear motion is not continuous and perpetual. Since a continuous and perpetual rectilinear motion could not occur except by breaking the motion,[4] it is necessary that a moment of rest intervene between any motions and the breaks *(reflexos)* in those motions. But motions which have an intervening [moment of] rest are not continuous.

Thirdly, unless this is so [namely that a moment of rest intervenes], it would follow that contrary motions are one continuous motion, which is impossible. The consequence[5] is proved, because there is no moment of rest between the contrary motions.

Fourthly, it would follow that something could be moved with contrary motions. This consequent is impossible, as is evident in the fourth book of [Aristotle's] *Metaphysics*.[6] [But] the consequence can be proved,[7] for if something should ascend and afterwards descend without an intervening moment of rest, then in the middle instant [of the motion], it ought to ascend with as much reason as descend. Therefore, either the body is moved [simultaneously] with each motion [that is, up and down] and the consequent is proven;[8] or it moves with neither motion, so that the proposition is upheld and the body rests there.[9]

Fifthly, it would follow that something *could* both *be* and *not be* in some place. This consequent is implied and the consequence is proved: After assuming that *A* is moved from *B* to *C* and then returned *(reflectátur)* over the same line, I inquire whether or not *A* recedes from *C* at the middle instant[10]

we find dynamic considerations involving impressed forces *(impetus)*.

[of the motion]. If not, then it rests in the same place and what has been proposed is had; if it does [recede from *C*] then it is not in *C*, because everything which recedes from any terminus is no longer in that terminus, as is evident in this eighth book; and yet it was previously posited in *C*.

Sixthly, with the exception of circular motion, every motion tends to come to rest, as if that were

1. From Book VIII, Question 7, of *Questiones subtilissime Johannis Marcilii Inguen (sic) super octo libros Physicorum secundum nominalium viam*. . . (Lyon, 1518); reprinted as *Johannes Marsilius von Inghen Kommentar zur Aristotelischen Physik* (Frankfurt: Minerva, 1964), fols. 83 verso, col. 2, to 84 recto, col. 1. The lengthy quotation from Aristotle's *Physics* in the Introduction was drawn from the Oxford translation of that work by R. P. Hardie and R. K. Gaye.
2. See Marsilius' response to the first principal argument below.
3. The antecedent declares that "between things resting in opposite places there is an intervening motion." For Marsilius it is only by experience that we know that a body can rest in two different places at two different times and only by virtue of some kind of intervening motion.
4. In a finite world, a body could move with continuous and perpetual rectilinear motion only if it oscillated between two fixed points separated by a finite distance. The text of this second argument appears corrupt in at least one place. In the first sentence, I have added "non" before "esse" in the statement "probat motum rectum esse continuum et perpetuum."
5. Namely, that without a moment of rest contrary motions are continuous.
6. Probably a reference to Aristotle's remark that it is impossible for contrary attributes—in this case up and down motion—to belong to the same subject at the same time (*Metaphysics* IV.3.1005b.25–26; also IV.4.1006a. 3–5).
7. As so frequently happens in scholastic argumentation, Marsilius rejects the consequent—namely that something could move with contrary motions—as impossible, but then shows further that if the antecedent is true—namely if a moment of rest does not intervene—the rejected consequent must follow logically. For this reason the antecedent must be rejected and there must be a moment of rest.
8. Namely that a body can move with contrary motions, which is impossible,
9. The proposition upheld is the claim that a moment of rest intervenes.
10. That is, the point at which the motion becomes oppositely directed.

its end, [or goal, or terminus] *(terminus)*. But if over the very same path *(per eandem lineam)* there occurred an upward motion and then a downward motion of the same mobile, it follows that rest will occur at the upward [terminus of the motion], and what has been proposed is shown.[11]

Seventhly, let it be assumed that *A* is moved from *B* to point *C* and then is returned *(reflectatur)* over the same line, I inquire whether it was in *C* at some time or never. If never, then it was not moved through the whole distance, which is contrary to what was assumed. Furthermore, *A* recedes from some terminus in which it was not. [But] if *A* was once in *C* through some [period of] time, *A* consequently rested through that time.[12]

Eighthly, let it be assumed that a certain magnitude *D* is conjoined with point *C*. I ask, then, whether or not *A* touches *D* at the point lying between the contrary motions, [namely point *C*]. If it does, and since every motion is temporal, it follows that *A* touches [*D*] and consequently rests in point *C* through [some interval of] time. [But] if *A* never touches *D* [in point *C*], it follows that it was not moved through the whole distance, which is contrary to the assumption. . . .[13]

[I argue for] the opposite position[14] because if a heavy body were projected upward, it would rest prior to descending, and then it would follow that a heavy body would rest [while high] up without being detained violently.[15] Secondly, [I argue for the opposite position] because there appears to be no reason why it should rest more [that is, through a greater length of time (?)] at one time than another, and thus it would rest either through no time at all or through any time whatever.[16]

It must be noted that it is possible that an upward motion occurs, and afterward a downward motion, without [an exact] return [over the same path], as [for example] if a motion occurs over a curved line, as is obvious in the motion of an arrow through the air. In another way there can first be ascent and then a descent by returning through an angle. And this can happen in two ways: in one way by returning *(reflectendo)* through an acute angle in a rectilinear or curvilinear path, or [a path] composed of a straight line and a curved line, or even [returning] through an obtuse or right angle by coming over one line [or path] and returning over another; in the other [or second] way, there is a return at a straight angle in such a way that the mobile comes over the same line [or path] by which it returns, and this motion is properly composed of two contrary motions and this question must be

understood to be concerned [only] with such motions that turn back *(reflexos)*.[17]

Now these conclusions are posited. The first is that it is not necessary that there be a temporal moment of rest between any motions that turn back *(reflexos)* [over the same path]. The proof is that if a bean *(fabba)* were projected upward against a millstone *(molarem)* which is descending, it does not appear probable that the bean could rest before descending, for if it did rest through some time it would stop the millstone from descending, which

11. Here the upward motion and downward motion are assumed to be independent motions, each having rest as its goal. Obviously, then, on this argument, since a heavy body thrown into the air will have reached its goal or terminus at the uppermost point of its motion, it will come to rest there "naturally." This would turn Aristotelian physics upside down, with heavy bodies "naturally" suspended at rest, unsupported above the surface of the earth.

12. The point of this argument is that the first alternative—that *A* was never in *C*—is absurd and impossible. The second—that *A* was in *C* for some time—must be true, and we conclude that *A* rested momentarily in *C*.

13. Point *C* is the midpoint of the whole motion, being the last point of the upward leg and the first point of the downward movement. If a magnitude *D* is assumed to be in *C*, then when *A* is thrown upward either it touches *D* in *C* or not. If it does, then their contact in *C* constitutes a temporal interval of rest; if *A* does not touch *D*, it never reached *C* and its motion is incomplete, contrary to the assumption that it completed its up-down motion. This seems but a variation on the preceding (seventh) argument.

14. That is, that no temporal moment of rest intervenes.

15. This would be contrary to a fundamental tenet of Aristotelian physics, for it would make "up" the natural resting place of a heavy body. See also note 11.

16. In the absence of further details, two interpretations seem plausible here, both employing the principle of sufficient reason. If the body rests at the turning point of the motion, it ought to be able to rest at any other point of its up-down path. Furthermore, if it rests only through a durationless instant, there is no moment of rest; but if it rests through some temporal period, however small, then it ought to do so at any point of the path (since none are privileged), in which event it could possibly rest "through any time whatever." A second, and perhaps more likely, intepretation, takes the "interval of rest" as applicable only to the turning point of the motion where, if rest occurs at all, one must inquire about the length of the interval, which, in the absence of any reasons or explanations, may range from a durationless interval to one of any temporal duration, however great. Hence the body "would rest either through no time at all or through any time whatever." My own uncertainty is revealed by the query in square brackets.

17. The problem of an interval of rest is thus restricted to cases of up and down motion where the path is represented by a single straight line.

seems impossible.[18] Secondly, let it be assumed that Socrates *(Sortes)* is moved toward the west in a ship that is at rest. Then it is possible that Socrates might cease moving in any instant. Now let it be assumed that in the [very] same instant in which Socrates should cease to be moved [toward the west], the ship, with all its contents, begins to be moved toward the east. Hence, immediately before, Socrates was moved to the west, and immediately after, will be moved toward the east. Therefore, previously he was moved with one motion and afterward with another, and contrary, motion without a moment of rest.[19]

But we must see now the manner in which response is made to the first two arguments. To the first, it is said that a millstone forcefully pushes the air before it and that this air by its pushing stops the bean before the millstone reaches it.[20] But [I argue] against this because either the impetus of the air, which you say stops the bean, is equally as forceful *(fortis)* as the impetus of the bean for moving upward against the resistance of the medium, or it is stronger or weaker. If stronger, then it would make the bean descend and consequently the bean would not rest there; if weaker, the bean would ascend upward even further, and as a consequence, the air will not stop the bean. And if [the air] is equally as forceful, this will endure exactly for an instant[21] because as the millstone approaches closer [to the bean] so much more does it push the air. There is, consequently, no temporal rest in the middle [or midpoint of the bean's motion].

To the second argument, it is customary, in the first place, to respond that Socrates is self-moved and is moved [only] accidentally with the motion of the ship. Since one of them is self-moved and the other is moved accidentally, we do not have contrary motions and hence no rest at the middle of the motion. But concerning two self-moved [contrary] motions, it is impossible [that there should be no temporal rest]. Another response is that although it is possible that Socrates might cease to be moved in the very instant in which the ship begins to be moved, nevertheless it is impossible that these two [events] occur at the same instant. I argue against the first solution because our conception of motion requires that the same judgment be made about self-motion and accidental motion, even though there is a difference with respect to the movers [in each of these motions]. If, therefore, one of the contrary motions is self-moved and the other is accidental and there is no interval of rest,

it follows that there is also no interval of rest between contrary motions when each motion is self-moved *(per se)*. Secondly, since any argument *(ratio)* drawn from Aristotle, and any other argument which might prove something about self-motion, can be utilized to derive the same thing about accidental motion, [it follows] therefore that what is conceded about the one [kind of motion] must be conceded about the other.

I argue against the second solution, because it does not seem [to offer] a [good] reason why the ship could not begin to be moved in the same instant in which Socrates ceases to be moved, so that each can occur in the same instant. In the second place, this can also be proved by varying the case. Let it be assumed that the ship is moved uniformly toward the east and that Socrates is moved difformly (that is, non-uniformly) on the ship with a proper motion toward the west, but quicker than the ship in the first half of the hour and slower in

18. In this example, we have the most fundamental experiential appeal of those who rejected Aristotle's interval of rest. The example itself, and its numerous trivial variations, can be traced back at least as far as Abū l'Barakāt al-Baghdâdî (ca. 1080—ca. 1165) who, in order to reject Avicenna's arguments in defense of Aristotle's position, formulated it in terms of a falling millstone striking an ascending date pit. (See S. Pines, "Études sur Ahwad al Zamân Abû l'Barakāt al-Baghdâdî," *Révue des études juives,* New Ser., IV [1938], 5.) Indeed, since Avicenna sought to repudiate it with an argument which Marsilius repeats (see n. 20), it probably antedates Avicenna. Marsilius and many others would utilize it, including Johannes Versor, Pico della Mirandola, Hasdai Crescas, who made a mountain out of a millstone, Giovanni Benedetti, and, finally, Galileo, who by that time could cite the argument "as the well-known one about a large stone falling from a tower" and striking an ascending pebble (see Selection 50.2).

19. Abū l'Barakāt offers analogous arguments where a motion becomes oppositely directed without a moment of rest. Since his works (in Arabic) were not translated into Latin, the manner in which his arguments may have reached the Latin West—if indeed they were not independently formulated—is unknown to me. Continuing with the first conclusion, Marsilius presents five more brief arguments (fol. 84 recto, col 1) which are omitted here. They are analogous to the two already presented but involve the motion of spheres, light, heat, augmentation and diminution. Immediately after, he returns, for further consideration, to the first two arguments translated above.

20. This very argument was formulated long before by Avicenna. The air pushed downward by the descending millstone would stop the ascent of the bean, which would remain motionless through an interval of time and then descend (see Pines, p. 5, n. 265).

21. As we see from the very next sentence, the instant is of no temporal duration.

the next half hour. In the middle instant of the time,[22] it is true to say that immediately before, Socrates was moving toward the west, and immediately after, will be moved toward the east. Therefore, previously he was moved with one motion and afterward with another and contrary motion without an interval of temporal rest.[23]

A second conclusion is this: It is not necessary that between any contrary motions there be rest during an indivisible or intervening instant. This is demonstrated because there is no rest in an instant, as was proved in the sixth [book] of this [treatise]. Secondly, since nothing can be moved locally in an instant,[24] therefore there can be no rest [in an instant] because there is no other measure [of time] in which rest can be produced but not motion.[25] Thirdly, because in no instant intervening between contrary motions is it true to say that a mobile is related otherwise [or differently] than it was immediately before and than it will be immediately after.[26] Therefore the mobile does not rest [in the intervening instant between contrary motions]. Fourthly, an argument that could prove that in this instant rest occurs, could also prove that in any instant of motion the mobile is at rest.[27] Obviously, then, it is not necessary that rest intervene either instantaneously or temporally between any contrary motions.[28]

A third conclusion is that between any *contrary natural motions,* it is necessary that there be an interval of rest. The proof of this is that since every natural motion, except celestial motion, is directed toward coming to rest as if that were its goal or end, therefore, if it does not rest when it reaches its terminus [or goal], it will have been frustrated [or prevented] from attaining its end [or goal]. Secondly, what is moved naturally has not only a natural inclination [or tendency] to motion, but also to rest in the terminus [or goal] of the motion. But this natural inclination [for rest] in the terminus of the motion cannot be lost instantaneously *(subito)*. It is necessary, therefore, that there be a time [lapse] before this inclination to contrary motion be acquired [again]. Consequently, it will rest through that interval of time.[29]

A fourth conclusion is that when something is moved upward violently and then descends downward over the same path, there is, necessarily, an interval of rest between those motions, unless it should be hindered by the resistance of something external. Let this be proved [as follows]: For just as long as anything is moved upward after it has been projected, just so long does its motive power

exceed the natural heaviness of the projected body plus the resistance of the medium. But when the impetus moving the projected body is equal to the natural heaviness of the stone, the stone ceases to be moved upward. Nevertheless, it does not yet begin to be moved downward. This is proved because the natural heaviness scarcely exceeds the

22. That is, at the end of the first half hour.
23. During the first half hour the resultant or net motion of Socrates is westward, in the direction of his own proper motion. But at the very instant when Socrates' westward motion becomes slower than the eastward motion of the ship, his resultant motion will become eastward, although he continues to walk westward on the ship. Hence while his direction of motion changed instantaneously from west to east, he did not rest during the change. Marsilius has shown that, by varying the conditions of the first argument against an interval of rest, it is not essential to rely on the mere coincidence that Socrates ceases to be moved exactly at the instant when the ship begins to be moved in the opposite direction. Now Socrates continues walking westward while his direction of motion is changed to eastward. Much the same type of argument was used by Galileo (see the next selection).
24. In Book VI, Question 10, Marsilius asks "whether motion could occur in an instant"; it cannot, he argues.
25. Neither rest nor motion can be measured by "instants"; that is, if motion is unmeasurable by an "instant," so also is rest.
26. The intention here is to show that if the mobile's behavior is identical during the three successive instants of time embracing (1) the instant immediately prior to the intervening instant, (2) the instant intervening between the contrary motions, and (3) the instant immediately following the intervening instant, it follows that the mobile did not rest in the instant represented by (2).
27. If rest can be proven for one instant, it can be proven for any instant, and consequently, for every instant. Therefore, the mobile will be at rest in every instant.
28. Throughout this question, Marsilius interweaves these two notions of an "instant"—one durationless, the other of temporal length.
29. In the example above of the millstone and bean, the sequence of motions was violent to natural (the bean ascends by a violent motion, and after having been struck by the millstone, descends by natural motion) with no intervening moment of rest. But here the sequence is reversed, where a natural motion is followed by a violent motion and, it is now argued, an interval of rest must occur. This can be illustrated by a rubber ball (a heavy earthy body) falling naturally toward the earth. Since it has a natural tendency to rest when it reaches its natural place, and since "this natural inclination [for rest] in the terminus of motion cannot be lost instantaneously," we must suppose that before it bounces upward in a violent motion, it will rest for a temporal moment. Thus the curious expression "contrary natural motions" must not be taken literally (for then it would be unintelligible in Aristotelian terms; one and the same body cannot have "contrary natural motions") but rather must signify a natural motion followed by a violent one.

medium yet and it is necessary, therefore, that time should pass before it does.[30] Consequently, the body will rest in this time.

And so it is clear in what manner it is not necessary that an interval of rest occur between any contrary motions. However, between *contrary natural motions* of one and the same mobile, it is necessary that the mobile rest for an interval of time and [also] when one of the motions is natural and the other violent, except when there is an obstacle by the interference of another [external body].[31]

Now [let us turn] to the [principal] arguments. To the first, I deny the consequence, which is not similar [or analogous], since things resting in *opposite places* can do so only in different places, whereas *opposite motions* can occur over the same

space.[32] Furthermore, by denying the antecedent, one can declare otherwise, for if the earth did not rest naturally and if by the power of the sun it became hot and light on one side, then there is now another center of gravity than before. But previously the earth rested naturally, but now it rests violently, and yet there is no intermediate motion [between the natural and violent rest].[33] That the earth now rests is obvious, because the lightness produced [on one side of the earth] is not sufficient to move the earth locally because of the resistance of the surrounding medium. To the second argument, I say that Aristotle understands that there is an interval of rest between any contrary natural motions and this is sufficient for his purpose, because if any motion were continuous and eternal it would necessarily be natural. . . .[34]

2. Galileo Galilei (1564–1642)

Translated by I. E. Drabkin[35]
Annotated by Edward Grant

CHAPTER [20]

In which, in opposition to Aristotle and the general view, it is shown that at the turning point [an interval of] rest does not occur.

Aristotle and his followers believed that two

30. Since the natural heaviness is balanced by the impetus pushing the body upward, the body will not actually move down until the balancing impetus, or incorporeal impressed force, will have become further corrupted, after which it is exceeded by the natural gravity or heaviness of the body. This fourth conclusion presents again the case of a violent motion followed by a natural motion, just as with the bean and millstone. But now Marsilius assumes a temporal interval of rest because there is no external body, such as the millstone, to interfere with the smooth flow of the up-down movement. For this same case, Avicenna argued for an interval of rest by insisting that the upward impressed force had to be wholly dissipated before the natural force could start the body downward. The interval of rest occurs between the disappearance of the one and the advent of the other. Abū l'Barakāt, on the other hand, denied Avicenna's arguments and the interval of rest for this case.

31. The cases distinguished by Marsilius may be summarized as follows:

I. *No interval of rest*
   1. Vertical Contrary Motions: Violent motion followed by natural motion plus an external interference (millstone-bean)
   2. Horizontal Contrary Motions (east-west) (Socrates on the ship)

II. *Interval of rest*

   1. Vertical Contrary Motions: Natural motion followed by violent motion
   2. Vertical Contrary Motions: Violent followed by natural motion with no external interference.

32. Because the proposition "between things resting in opposite places there is an intervening motion" is true, it does not follow, argues Marsilius, that the proposition "between any contrary motions there is an interval of rest" is true by analogy or similarity. Such an argument merely trades on the superficial similarity between the forms of the statements.

33. The antecedent of the first principal argument (on p. 285) is that "between things resting in opposite places there is an intervening motion." Marsilius denies this by positing a contrary case, namely by supposing that the earth does not rest naturally and that its center of gravity continually shifts because of the sun's heating action. But the earth does rest, therefore it rests violently as its center of gravity shifts. Hence the earth rests violently in different places but does so without intervening motion, since it was at violent rest—but rest nonetheless—throughout.

34. The remaining counterarguments, brief and unilluminating, have been omitted.

35. Reprinted with permission of the Regents of the University of Wisconsin from *Galileo Galilei "On Motion" and "On Mechanics" Comprising "De Motu"* (ca. 1590), translated with introduction and notes by I. E. Drabkin, and *"Le Meccaniche"* (ca. 1600), translated with introduction and notes by Stillman Drake (Madison Wis.: University of Wisconsin Press, 1960), pp. 94–100. The basic similarity between some of Galileo's arguments and those of Marsilius of Inghen in the preceding selection will be obvious. On this question, Galileo merely continues and elaborates a well-established medieval tradition.

contrary motions—he defines contrary motions as those which tend toward opposite goals—could in no way be continuous with each other. And therefore they believed that, when a stone is projected upward and then falls back over the same path, it must necessarily remain at rest at the turning point. The chief argument with which Aristotle tries to prove this is as follows. "Whatever moves by approaching some point and then moving away from that same point, using it as an end and as a beginning, will not move away from it, unless it has first stopped at it. But that which moves to the farthest point of a line and then moves back from that point, uses that point both as an end and as a beginning. It must therefore remain stationary between the motion toward and the motion away from the point."[36] Aristotle proves his major premise thus: "Whoever treats something both as beginning and as end, makes what is one in number two in logic, just as the person who in thought takes a point, which is one and the same numerically, and makes it two in logic, namely the end of one thing and the beginning of a second thing. But if something uses one thing as two it must necessarily remain stationary there; for there is [an interval of] time between the two."

Such is Aristotle's argument. But how weak it is will soon be clear. For, as he himself holds, the moving body makes use of a point on the line of its motion, i.e., one point, numerically, for what are two things in logic, for a beginning and an end. Yet there is no line between these two things, since they are only one in number. And why, similarly, will the same body not use the same instant (one, in number) as two in logic, namely, for the end of the time of moving toward [the turning point] and for the beginning of the time of moving away [from it], so that between these instants that are two in logic, no time intervenes, since they are only one in number?

There is no compelling reason why this should not be the case, especially since Aristotle himself holds that what is true of a line is true also of time and motion. If, then, on the same line the same point, numerically, is both the end of one motion and the beginning of a second, and if, nevertheless, it is not necessary that a line form a connection between this beginning and that end, then, in the same way, the same instant numerically, will, in logic, be the end of one time and the beginning of a second time, and it will not be necessary for time to intervene between the two. It is clear, then, that the refutation of Aristotle's argument can be neatly

derived from the propositions of that same argument. Hence, since the argument no longer has compelling force for us, let us see whether we can construct arguments for the opposite view that are more sharply convincing.

So much in opposition to Aristotle. But in order for us to show by other arguments that [an interval of] rest does not occur at the turning point, and that there need not be such rest between contrary motions, consider these additional arguments.

Secondly, suppose that some continuum, such as the whole of line ab, moves in the direction of b in a motion like a forced motion which becomes continuously slower. And while the line so moves, suppose that a body, say c, moves on the same line in the opposite direction, from b to a. But let this motion be like a natural motion, that is, one that is accelerated. And let the motion of the line at the beginning be faster than the motion of c at the beginning. Now it is clear that at the beginning c will move in the same direction as that in which the line moves, because the motion by which it is carried in the opposite direction is slower than the motion of the line. And yet, since the motion of the line becomes slower and the [leftward component of the] motion of c becomes faster, at some moment c will actually move toward the left, and will thus make the change from rightward to leftward motion over the same line. And yet it will not be at rest for any [interval of] time at the point where the change occurs. And the reason for this is that it cannot be at rest unless the line moves to the right at the same speed as body c moves to the left. But it will never happen that this equality will continue over any interval of time, since the speed of one motion is continuously diminished, and that of the other continuously increased. Hence it follows that c will change from one motion to its contrary with no intervening state of rest.[37]

My third argument can be drawn from a certain rectilinear motion which Nicholas Copernicus in his *De Revolutionibus* compounds from two circular motions.[38] There are two circles [the center of] each of which is carried on the circumference of the other. When one circle moves more swiftly than the other, a point on the circumference of the first circle moves in a straight line continuously back and forth over the same path. And yet it can-

36. *Physics* VIII.8.262b.2–8 and also the quotation at the beginning of this selection.
37. This argument is fundamentally the same as Marsilius of Inghen's, in the preceding section.
38. Book III, chapter 4.

not be said that the point is at rest at the extremities, since it is carried continuously by the circumference of the circle.[39]

My fourth argument is the well-known one about a large stone falling from a tower.[40] A little pebble is forcibly thrown up from below against it, but the stone will not be sufficiently blocked by the pebble so as to allow the pebble to be at rest for any interval of time. Hence the pebble will surely not remain at rest at the farthest point of its upward motion, and, despite what Aristotle said, it will use that farthest point as two termini, namely, of upward motion and of downward motion. And the last instant is taken twice, viz., as the end of one interval of time and as the beginning of the other.

But in order to escape from this argument my adversaries declare that the large stone is at rest, and so they believe that they have answered the argument.[41] But, so that they may not believe this in the future (unless they are thoroughly obstinate), I shall add the following to my argument. Suppose that those stones which move with contrary motions move not up and down but on a plane surface parallel to the horizon, one with great impetus, and the other more slowly. And suppose that they move in opposite directions from opposite parts and meet in the middle in an interacting motion. In that case, there is no doubt that the weaker will be thrust back by the stronger and forced to move back. But how can they say that at that point of impact an interval of rest occurs? For if once they remained at rest, they would thereafter always be at rest, since they would not have reason for moving. In the case of the large stone falling from a high point, even if it were stopped by the pebble, yet after the interval of rest, both would fall together, moving down by reason of their own weight. But when they are in a plane parallel to the horizon there exists no cause of motion after the [supposed] interval of rest.

Before expounding my last argument, I make these two assumptions. My first assumption is that only then is it possible for a body to be at rest outside its proper place, when the force that opposes its fall is equal to its weight, which exerts pressure downward. Surely this is clear: for if the impressed force was greater than the resistant weight, the body would continue to move upward; and if it were smaller, the body would fall. Secondly, I assume that the same body can be sustained in the same place over equal intervals of time by equal forces.

I then urge the following. If a state of rest lasting for some interval of time occurs at the turning point, e.g., when a stone changes from forced upward motion to [natural] downward motion, then over the same interval of time there will exist equality between the projecting force and the resisting weight. But this is impossible, since it was proved in the previous chapter that the projecting force is continuously diminished.[42] For the motion in which the stone changes from accidental lightness to heaviness is one and continuous, as when iron moves [i.e., changes] from heat to coldness. Therefore the stone will not be able to remain at rest.

Furthermore, suppose that the stone moves forcibly from *a* to *b*, and naturally from *b* to *a*. If, then, the stone is at rest at *b* for some interval of time, suppose that this time has as its end moments *c* and *d*. If, then, the body is at rest for time *cd*, the external projecting force will, through time *cd*, be equal to the weight of the body. But the natural

39. Drabkin notes (p. 97, n. 5) that "there is an analogous and simpler example in Benedetti, *Diversarum speculationum. . .liber*, p. 183 (cited by A. Koyré, *Études Galiléennes*, I, 51), in which one end of a rigid bar is attached to a point (other than the center) of a continuously rotating circle, and the other end moves back and forth along the same straight line." Abū 1' Barakāt had formulated much the same argument long before (see my n. 19).

40. See Inghen section and note 18.

41. Probably a reference to the type of argument cited by Marsilius of Inghen but formulated long before by Avicenna (see Inghen section and n. 20). However, Galileo's version is distorted. It was not claimed that the large stone would rest, but rather that the pebble would have had its upward direction reversed (by the force of the air pushed downward by the large falling stone) before the large stone reached it. Perhaps Galileo's version represents a Renaissance variation on the traditional arguments.

42. Actually demonstrated in Chapter 18 ("In which it is shown that the motive force is gradually weakened in the moving body"). In chapter 19 upward deceleration and downward acceleration of bodies is explained in terms of a self-expending impressed force. (Also see Drabkin, p. 89). "In Galileo's exposition, the notion of a residual force plays the leading role. Initially, the mover imparts an impressed force to a stone that is hurled aloft. As the force diminishes, the body gradually decreases its upward speed until the impressed force is counterbalanced by the weight of the stone at which moment the stone commences to fall, slowly at first and then more quickly as the impressed force diminishes and gradually dissipates itself. The acceleration arises from the continual increase of the difference between the weight of the stone and the diminishing impressed force." (Edward Grant, "Bradwardine and Galileo: Equality of Velocities in the Void," *Archive for History of Exact Sciences*, II [1965], 360.)

weight is always the same. Therefore the [projecting] force at moment c is equal to that force at moment d. Now it is the same stone and the same place: hence the stone will be held there over equal intervals of time by equal forces. But the force at moment c sustains the body throughout time cd. Hence the force at moment d will sustain the same stone throughout an interval of time equal to interval cd. The body will therefore be at rest throughout twice time cd. But this is inconsistent: for it was assumed to be at rest only through time cd. Indeed, by continuing the same form of argument, we could prove that the stone would always be at rest at b.

But do not be confused by the argument that, if the weight and projecting force are equal at some time, then the body must be at rest for some time. For it is one thing to say that the weight of the body at some time comes to be equal to the projecting force; but it is another thing to say that it remains in this state of equality over an interval of time. This becomes clear from the following consideration. While the body is in motion, since (as has been shown) the projecting force is always being diminished, but the intrinsic weight always remains the same, it must follow that, before they arrive at a relation of equality, countless other ratios occur. Yet it is impossible for the force and the weight to remain in any of these ratios over any interval of time. For it has been proved that the projecting force never remains at the same level over an interval of time, since it is always diminishing.

And so, it is true that the [projecting] force and the weight pass through ratios of, let us say, 2 to 1, 3 to 2, 4 to 3, and countless other ratios; but it is false and impossible that they should remain for any interval of time in any one of these ratios. So, too, they arrive at equality at some moment, but they do not remain at equality. This being so, since local motion upward and downward is a consequence of that alterative motion of change from light *per accidens* to heavy *per se,* in such a way that upward motion flows from an excess of impressed force, downward motion from a deficiency thereof, and rest from equality, and since this equality does not persist over an interval of time, it follows that neither does the state of rest persist.

51 MATHEMATICAL REPRESENTATIONS OF MOTION

1. Thomas Bradwardine (ca. 1290–1349): "Bradwardine's Function" and the Repudiation of Four Opposition Theories on Proportions of Motion

Translated by H. Lamar Crosby, Jr.[1]
Annotated by Edward Grant

CHAPTER 2

Having looked into these introductory matters,[2] let us now proceed with the undertaking which was proposed at the outset. And first, after the manner of Aristotle, let us criticize erroneous theories, so that the truth may be the more apparent.

There are four false theories to be proposed as relevant to our investigation, the first of which holds that: *the proportion between the speeds with which motions take place varies as the difference whereby the power of the mover exceeds the resistance offered by the thing moved.*[3]

This theory claims in its favor that passage from Book I of the *De caelo et mundo* (in the chapter on the "infinite") in the text which reads:

1. Reprinted with permission of the copyright owners, the Regents of the University of Wisconsin, from *Thomas of Bradwardine: His Tractatus de Proportionibus; Its Significance for the Development of Mathematical Physics,* edited and translated by H. Lamar

Crosby, Jr. (Madison, Wis.: University of Wisconsin Press, 1955). The following selection is from chapters 2 and 3, pp. 87–117. Bradwardine wrote the *Tractatus de proportionibus (Treatise on Proportions)* in 1328.

2. The introduction and chapter 1 were devoted to a description of the various types of ratios, proportionalities, and the enunciation of a number of axioms and theorems necessary for subsequent developments. Many of the preliminaries were drawn from the *Arithmetica* of Boethius and the *Elements* of Euclid (especially Bk. V).

3. This first erroneous theory is representable by $V_2/V_1 = (F_2 - R_2)/(F_1 - R_1)$, where V is velocity, F motive force, and R the resistance of the moving body (or, should the body fall with a natural downward motion, R would represent the resistance of the medium). It is likely that Bradwardine's version of this theory was derived ultimately from Avempace, a conjecture that gains support from the fact that a few lines further on, we find mentioned Averroes' Comment 71 on Book IV of Aristotle's *Physics,* the very place where Avempace's views were reported. If so, this could only have occurred by reading into Avempace's words a great deal more than was warranted, for it was noted (Selection 44, n. 17) that Avempace did not represent the natural motion of bodies by the formulation $V = F - R$.

"It is necessary that proportionally as the mover is in excess, etc.," together with Averroes' Comment[4] 71 on Book IV of the *Physics,* in which he says: "Every motion takes place in accordance with the excess of the power of the mover over that of the thing moved." In Comment 35, on Book VII of the *Physics,* he further states that: "The speed proper to any given motion varies with the excess of the power of the mover over that of the thing moved," and in the final Comment, Comment 39, he says that: "The speed of alteration and the quantity of time will vary in accordance with the amount whereby the power of that which is causing the alteration exceeds the resistance offered by what undergoes the alteration." Many other passages afford similar remarks.

The present theory may, however, be torn down in several ways:

First, according to this theory, it would follow that, if a given mover moved a given *mobile* through a given distance in a given time, half of that mover would not move half of the *mobile* through the same distance in an equal time, but only through half the distance. The consequence is clear, because, if the whole mover exceeds the whole *mobile* by the whole excess, then half the mover exceeds half the *mobile* by only half the former amount; for, just as 4 exceeds 2 by 2, half of 4 (namely, 2) exceeds half of itself (that is, 1) by 1, which is only half of the former excess.

That such a consequence is false is apparent from the fact that Aristotle proves, at the close of Book VII of the *Physics,* that: "If a given power moves a given *mobile* through a given distance in a given time, half that power will move half the *mobile* through an equal distance in an equal time" [250a.4–6—*Ed.*]. Aristotle's reasoning is quite sound, for, since the half is related to the half by the same proportion as the whole is to the whole, the two motions will, therefore, be of equal speed.

Secondly, it follows from this theory that, given two movers moving two *mobilia* through equal distances in equal times, the two movers, conjoined, would not move the two *mobilia,* conjoined, through an exactly equal distance in an equal time, but, instead, through double that distance. This consequence follows necessarily because the excess of the two movers, taken together, over the two *mobilia,* taken together, is twice the excess of each of them over its own *mobile;* for, just as anything having a value of 2 exceeds unity by 1, so two such "2's" (which make 4) exceed two "1's" (which make 2) by 2, which is twice the excess of

2 over 1. The foregoing holds in all cases in which two subtrahends are equally exceeded by two minuends.

That the above consequence is false is evident from the foregoing argument of Aristotle, in Book VII of his *Physics,* where he demonstrates the following conclusion: "If two powers move two *mobilia,* separately, through equal distances in equal times, those powers conjoined, will move the two *mobilia* conjoined, through an equal distance in a time equal to the former one" [250a.25–28—*Ed.*]. This argument of Aristotle is sufficient proof that the relation between a single mover and its *mobile,* and a compound mover and its *mobile,* is a proportional one.

In the third place, it would follow that a geometric proportion (that is, a similarity of proportions) of movers to their *mobilia* would not produce equal speeds, since it does not represent an equality of excesses; for, although the proportions of 2 to 1 and 6 to 3 are the same, the excess of the one term over the other is 1 in the first case and 3 in the second case.

The consequence to which we are thus led is, however, false and opposed to Aristotle's opinion, as expressed at the close of Book VII of the *Physics* and in many other places, where, from an equality of proportions of movers to their *mobilia,* he always argues equal speeds. Averroes supports the same view in his remarks on the passages just mentioned and also in his Comment 71 on *Physics* IV, Comment 63 on *De caelo et mundo* I, and in many other places.

Nor can it be legitimately maintained that, in the passages cited, Aristotle and Averroes understand, by the words "proportion" and "analogy," arithmetic proportionality (that is, equality of differences), as some have claimed. Indeed, in Book VII of the *Physics,* Aristotle proves this conclusion: "If a given power moves a given *mobile* through a given distance in a given time, half that power will move half the *mobile* through an equal distance in an equal time, because, 'analogically,' the relation of half the mover to half the *mobile* is similar to that of the whole mover to the whole *mobile.*"[5]

---

4. References to the comments of Averroes on the *Physics* and *De caelo (On the Heavens)* of Aristotle may be found in the *Aristotelis opera cum Averrois commentariis* (Venice, 1562–1574; Frankfurt: Minerva, 1962), Vols. IV and V. Comment 71 is translated above, in Selection 44.

5. 250a.4–9. In this quotation, Bradwardine has omitted an inessential line.

Such a statement, interpreted as referring to arithmetic proportionality, is discernibly false (as has already been made sufficiently clear in the first argument raised against the present theory). Moreover, regarding this same passage, Averroes says that the proportion will be the same "in the sense that geometricians universally employ in demonstrations."

The above thesis of Aristotle may be demonstrated geometrically as follows: As is the whole mover to half the mover, so is the whole *mobile* to half the *mobile*. Therefore, permutatively (by Axiom 7 of Chapter I):[6] As is the whole mover to the whole *mobile*, so is half the mover to half the *mobile*. And this is what was to have been proved.

The reading of Aristotle proposed by the present theory does not stand up, moreover, because, in Book VII of the *Physics,* Aristotle also proves this conclusion: "If two movers, separately, move two *mobilia* through equal distances in equal times, those two movers, conjointly, will move those two *mobilia,* conjoined, through a distance and in a time equal to the former,"[7] if the medium through which the motions take place remains the same, "for the motions are 'analogous'." By "analogous" he means a proportional, but not in the sense of arithmetic proportionality, for the simple mover does not exceed the simple *mobile* by the same amount that the compound mover exceeds the compound *mobile,* (as was made clear in the second argument against this theory).

Averroes, commenting on this passage, proves that, "although the proportion will be the same, the excess will not." He does this by employing Proposition [Crosby has "Theorem"—*Ed.*] I of Book V of Euclid's *Elements,* which states that: "If the members of a given set of quantities are either equal multiples of, equally greater or less than,[8] or exact equals of the members of a corresponding set, it follows that the relations between corresponding individual terms of the two sets will be the same as the aggregate relation of the two sets." Therefore, the above-mentioned reading of Aristotle cannot be taken as valid.

The fourth criticism is as follows: It would follow, on the basis of the present theory, that a mixed body, possessing internal resistance, could move faster through a medium than through a vacuum. Let *A,* for example, be a heavy mixed body (possessing within itself both motive and resistive power)[9] and imagine it to descend of itself through some medium, *B.* Let *C* represent a quantity of pure earth, possessing less power than the excess

of motive power over resistance in *A. A* will, of itself, move at a determinate speed in a vacuum. Now let the medium, *B,* be rarified to the point at which *C* moves in it with a speed equal to that of *A* in the vacuum. If *A* is now placed in the same medium with *C,* it should move faster than *C* (for it possesses a greater excess of motive power over resistance). *C* will move in that medium with a speed equal to that of *A,* moving in the vacuum. Therefore *A* will move faster through the medium than through the vacuum.[10]

6. The seventh axiom or assumption declares that if four quantities are related as $A/B = C/D$, then permutatively $A/C = B/D$.

7. 250a.25–28. This quotation is meant to be identical with one given above even though the Latin texts, and hence the translations, differ somewhat, most noticeably in the replacement of "powers" *(potentiae)* by "movers" *(motores).*

8. The phrase "equally greater or less than" does not appear in the modern Greek text nor in the popular medieval Latin version by Campanus of Novara.

9. A heavy mixed body was one compounded of both heavy and light elemental bodies, where the light elements, with their natural upward inclination, were assumed to function as internal resistance to the prevailing heavy elements, whose natural downward inclination functioned as the motive force actually moving the body downward. Every such body was held to be capable of moving through a hypothetical vacuum with finite speed—rather than instantaneously, as Aristotle held—since its internal resistance guaranteed that the motion would take time and consequently be theoretically measurable. The innovator of the concept of internal resistance, a radical departure from Aristotle, is unknown to me. Mixed bodies and internal resistance play a fundamental role in Selections 55.2,4,5.

10. Let $(f - r)$ represent the difference between motive force and internal resistance in *A,* a heavy mixed body that can fall in a resistant medium *B* and also fall in the void with a determinate speed. Furthermore, let *C* be a quantity of pure elemental earth whose motive power is represented by F_C (because *C* is a pure elemental body, it has no internal resistance). Finally, it is assumed that $(f - r) > F_C$ (although the motive power of a body *C* is always relative to the resistive power of a corporeal medium, in this example Bradwardine seems to compare $(f - r)$ to the absolute power of *C* without reference to a medium; indeed, it is as if *C* were falling in a void).

Since *A* is taken as falling with a finite velocity in the void, it is possible to produce an equal velocity for *C* falling in medium *B* by properly adjusting the density of *B.* In terms of the first erroneous theory, we may represent this as: $f - r = F_C - B$, so that V_A in void $= V_C$ in *B.* But now if *A* is let fall in the same medium *B,* it will fall with a speed greater than *C,* since $(f - r) - B > F_C - B$ (for by assumption $f - r > F_C$) so that V_A in $B > V_C$ in *B.* Drawing together these two consequences, we see that V_A in void $= V_C$ in B, and V_A in $B > V_C$ in *B,* from which it follows that V_A in $B > V_A$ in void.

That it is, in fact, possible to rarify B to a point at which C would move in it with the speed just specified is evidently true, for, by rarefaction of the medium, local motion can be accelerated to any desired degree. This is shown to be the case in Book IV of the *Physics* (in the chapter on the "void" [215b.1–11—*Ed*.]), where it is stated that, with the moving power remaining constant, it is possible to arrive at any given speed of local motion by rarefying the medium.

Thus (positing that a local motion could take place in a void), we find that the same body could move at the same velocity in both a medium and a vacuum.[11]

Our fifth criticism is that it would also follow that, if a given mover exceeded its resistance by a lesser amount than another mover exceeded its resistance, the former motion would be the slower one. Let a large quantity of earth, possessing a given resistance which its downward force greatly exceeds, be supposed to fall. Let also another quantity of earth, possessing a lesser such excess of power, be supposed to fall. Letting the larger quantity of earth and the resistance associated with it remain constant, let the medium in which the smaller quantity of earth moves now be rarefied to the point at which its speed becomes equal to that of the larger quantity. The smaller quantity now moves its resistance with the same speed that the larger moves its own, and yet exceeds it by a smaller amount.[12]

Sixthly, it would also follow that, if a bit of pure earth were moving in some medium whose resistance its power exceeded by a ratio of two to one, or more, it could not move at double that speed in any other medium. It could not exceed any medium by double the first excess, for, in that case, the entire moving power would be excess, and, with the moving power remaining constant, it would consequently not be possible to increase the speed of the motion indefinitely by rarefaction of the medium. Such a consequence has already been established as false.[13]

In the seventh place, another consequence would be that, if a given mover were to exceed its resistance by a greater amount than another mover exceeded its own resistance, the former motion would be the faster. Then, since a strong man exceeds anything he moves by a greater excess of power than a weaker mover (such as a boy, or a fly, or something of that sort) exceeds what it moves, he should move it more rapidly.

Experience, however, teaches us the contrary,

for we see that a fly carrying some small particle flies very rapidly, and that a boy also moves a small object rather rapidly. A strong man, on the other hand, moving some large object which he can scarcely budge, moves it very slowly, and even if there were added to what he moves a quantity larger than either the fly or the boy can move, the man will then move the whole not much more slowly than he did before.

From all these considerations, therefore, the

By the assumption of a subtractive relation between the motive force and internal resistance of A (that is, $f - r$), one derives the absurd consequence that one and the same heavy mixed body will fall more quickly in a plenum than in a vacuum.

11. A lapse in Bradwardine's thought seems to have occurred here. For the proof just presented (and summarized in the preceding note) demonstrates that the same body would move faster through a medium than through a void, and not that it would move with equal speeds in medium and vacuum (in *Physics* IV.8.215b.20—216a.7, Aristotle showed that one and the same body would, in the same time, fall equal distances in void and plenum).

12. It must be understood that Bradwardine is here speaking of heavy mixed bodies of earth possessing internal resistances as well as being opposed by an external resistance in the form of an ambient medium. Let A_2 and A_1 be heavy mixed bodies, with $A_2 < A_1$. If $f_2 - r_2 < f_1 - r_1$, where f is motive force and r internal resistance, then, as a consequence of the erroneous theory, $V_2 < V_1$ (where V_2 is the velocity of A_2 and V_1 of A_1). Taking R_1 as the initial resistance of the external medium and R_2 as the resistance of the medium after it has been rarefied to the point where $f_2 - r_2 - R_2 = f_1 - r_1 - R_1$, it follows that the velocities of A_2 and A_1 are equal—that is, the two forces f_2 and f_1 now move A_2 and A_1 with the same speeds even though $f_2 - r^2 < f_1 - r_1$. That Bradwardine should have taken seriously this criticism is surprising. An adherent of the theory in question would undoubtedly have replied that it is the total resistance—not just internal resistance—which must be subtracted from the motive force, and if the remainders are equal, so also are the resultant velocities.

13. On the assumption that a medium is infinitely rarefiable (this was assumed above), Bradwardine now argues that if $F - R = 2 - 1$, where F is the motive force of a piece of pure earth, say A_i and R the resistance of the medium, then A could not double in speed in any whatever corporeal medium, but could do so only in void space, where $F - R = 2 - 0$. But if, as all believed, a medium is infinitely rarefiable, it ought to be possible to double any speed by holding the force constant and suitably rarefying the medium. Hence a universally accepted Aristotelian position is violated and the theory is held to collapse. Bradwardine's argument is wide of the mark, since a partisan of this theory was committed to a rejection of the Aristotelian position that velocity is inversely proportional to the density of the medium.

following negative conclusion is sufficiently well established:

The proportion of speeds in motions does not vary with the amount whereby the power of the mover exceeds that of the thing moved.

Objections to this conclusion are not difficult to dissolve, for Aristotle and Averroes, when they say that the speed of a motion varies in accordance with the amount by which the power of the mover excels or exceeds that of the thing moved, understand by "excellence," or "excess," a proportion of greater inequality whereby the power of the mover excels, or exceeds, the power of the thing moved.[14]

CHAPTER 2, PART 2
[THEORY II]

Let us now turn to the second erroneous theory, which supposes *the proportion of the speeds of motions to vary in accordance with the proportion of the excesses whereby the moving powers exceed the resisting powers.*[15] This idea is evidently based on Averroes' Comment 36 on Book VII of the *Physics,* for he there states that the speed of a motion is determined by the proportion whereby the power of the mover exceeds that of the thing moved.

This theory should, however, be refuted as false. For just imagine the case in which the excess of the power of the mover over that of the thing moved is equal to the power of the thing moved. No mover will be able to move any *mobile* either faster or slower than the speed produced by this proportion, because no other proportion can be either greater or smaller than that whereby the excess of power of this mover over its *mobile* is related to the power of the *mobile* as a whole (as is demonstrable by Theorem VII, Chapter I).[16]

In the second place, a moving power moves a whole *mobile* primarily by means of its total strength, and not by means of a residuum of its strength. A motion and its speed vary primarily and essentially, therefore, with the relation, or proportion, between the entire power of the mover and that of the thing moved, and not (except accidentally and secondarily) according to a proportion of excess.

This negative conclusion is therefore evident: *The proportion of the speeds of motions does not vary in accordance with the proportion[17] of the excess of the motive power over the power of the thing moved.* The above-mentioned statement by Averroes may, if anyone were so to desire, be interpreted

in the same way as were the other authorities cited in support of Theory I.

CHAPTER 2, PART 3
[THEORY III]

There follows the third erroneous theory, which claims that: *(with the moving power remaining constant) the proportion of the speeds of motions varies in accordance with the proportion of resistances,[18] and (with the resistance remaining constant) that it varies in accordance with the proportion of moving powers.[19]*

With respect to its first part, this theory is seen to be founded on many passages of Aristotle's

14. That is, velocity varies as F/R, where $F > R$ (rather than as $F - R$, where $F > R$).

15. We may represent this as $V_2/V_1 = (F_2 - R_2) : R_2/(F_1 - R_1) : R_1$. I know of no supporters of this theory prior to Bradwardine, and can cite only Giovanni Marliani (d. 1483) as one who adopted it after Bradwardine. In Latin, the language describing this theory is almost indistinguishable from that of the previous theory. It is only by means of specific examples that one can tell them apart with any assurance.

16. If $V \propto (F_1 - R_1)/R_1$ and $F_1 - R_1 = R_1$, then $(F_1 - R_1)/R_1 = 1$ and we have a ratio of equality, which, by Theorem VII of chapter 1, can bear no exponential relation to ratios of greater or lesser inequality. In short, a ratio of equality, $(1/1)^n$, cannot be made equal to p/q, where $p \neq q$ and p and q are integers. Thus $(F_1 - R_1)/R_1$, where $F_1 - R_1 = R_1$, cannot be made equal to, or greater than, or be compared in any way with, $(F_2 - R_2)/R_2$, where $F_2 - R_2 > R_2$. As a consequence of this theory, then, it follows that whatever the speed with which F_2 moves R_2, it is not comparable to the speed with which F_1 moves R_1. While this consequence is an important feature of Bradwardine's "true" theory (see later), it seems strange to us that he should invoke Theorem VII of chapter 1 against this "erroneous" theory, where such a theorem would have been wholly irrelevant. However, it was fairly typical of medieval scholastic argumentation to repudiate opposing theories by deriving consequences from the basic presuppositions of one's own theory, which were then used against opposition theories subscribing to a quite different set of fundamental assumptions. (Nicole Oresme does much the same thing; see n. 59 below.)

17. I have substituted the remainder of this sentence for Crosby's version which reads: "whereby the excess of the moving power over its mobile is related to the power of that mobile." While this presents an accurate verbal description of the theory Bradwardine repudiates, it adds elements that do not appear in the Latin text, which differs very little from the first rejected theory.

18. $V_2/V_1 = R_1/R_2$ when $F_1 = F_2$.

19. $V_2/V_1 = F_2/F_1$ when $R_2 = R_1$. The two parts of this third theory are representable as $V \propto F/R$, which actually represents Aristotle's verbal formulation despite Bradwardine's attempt to associate Aristotle with his own theory.

writings. In Book IV of the *Physics* (in the chapter on the "void") he speaks as follows: "Let *B* represent a given quantity of water and *D* a given quantity of air. Now, by however much air is thinner and more incorporeal than water, by so much will *A* (that is, the moving body) move faster through *D* than through *B*. Let the one speed bear the same ratio, or proportion, to the other as that whereby air differs from water, and then, if air is twice as thin, the body will traverse *B* in twice the time required to traverse *D*" [215b.4–8—*Ed*.]. Furthermore, the text immediately following manifestly makes the supposition that, with the moving power remaining constant and the medium being varied, the proportion of the speeds of motions varies in accordance with the proportion of media, and that, conversely, the proportion of the times measuring those motions varies also in accordance with the proportion of media (namely, that the longer time corresponds to the motion through the denser medium and the shorter time to the motion through the rarer medium).

Further, in Book I of the *De caelo* (in the chapter on the "infinite") he speaks as follows: "It is held to undergo a greater and less effect by the action of the same agent, in a longer and shorter time, any such effects are divided proportionally to the time" [275a.32—275b.2—*Ed*.].

And, at the end of Book VII of the *Physics,* Aristotle wishes it to be understood that, if a given power moves a given *mobile* through a given distance in a given time, the same power will move half the same *mobile* through twice the distance in an equal time, and through the same distance in half the time [249b.30—250a.2—*Ed*.].

Thus much in favor of the first half of the present theory.

In support of the second part of this position, Aristotle holds, in Book IV of the *Physics* (in the chapter concerning the "void") that, other conditions remaining constant, heavy and light bodies differing in quantity will move through a given distance in the same medium more swiftly and more slowly in accordance with the proportion of the heavy and light bodies to each other [216a.14–16—*Ed*.].

According to Averroes' exposition, at the close of Book VII of the *Physics,* Aristotle intends that, if a given power moves a given *mobile* through a given distance in a given time, double the power will move that *mobile* through double the distance in an equal time.

At the close of Book VIII of the *Physics,* Aris-

totle maintains that a motive power which is double another such power will move the same *mobile* in half the time required by the lesser power [266b.10–11—*Ed*.], and that, universally, a motive power greater than another will move the same *mobile* which the smaller power moves in a time that is less by converse proportion (that is, that less time is required by the larger power and more time by the lesser power) [266b.17–19—*Ed*.].

Moreover, Aristotle intends the same thing in Book I of the *De caleo* (in the chapter on the "infinite"), where, speaking of heavy bodies that fall equal distances in the same medium, he writes as follows: "The 'analogy' (that is, the proportion) between the weights will be the contrary of that between the times. For example, if the whole weight in a given time, then double the weight in half that time" [273b.31—274a.2—*Ed*.].

Further, in Book III of the *De caelo,* where Aristotle proves every body to possess a rectilinear gravity or levity, he states that heavy bodies unequal in power will traverse, in the same medium and the same time, distances proportional to those powers [perhaps 301b.11–15—*Ed*.].

The same is evident from Theorem I of the *De ponderibus,* which states the following: "The proportion between the speeds of descent of any given heavy bodies is the same as that between their respective weights."[20]

The theory may also be set forth by the following reasoning: If one mover has exactly twice the power possessed by another, it can move the same *mobile* exactly twice as much, or move twice the *mobile* the same amount, for if it is exactly twice the power, it can accomplish exactly twice as much. If it could accomplish more than twice as much, it would be of more than twice the power; and if it were not capable of twice, but only of less, it would be of less than twice the power.

So much for the second part of this position, and thus we have seen what are the foundations of both parts of the theory.

The theory is, however, refutable on two grounds: first, on that of insufficiency, second, because it yields false consequences.

It is insufficient, because it does not determine the proportion of the speeds of motions except in cases where either the mover or the *mobile* are constant. Concerning motions in which the moving

20. This is from the *Liber Jordani de ponderibus (The Book of Jordanus on Weights),* the third treatise mentioned in section 2 of the introduction to Selection 39. Despite the title, it is not by Jordanus of Nemore.

forces, as well as the *mobilia,* are varied, it tells us almost nothing.

The theory is, on the other hand, to be refuted on the ground of falsity, for the reason that a given motive power can move a given *mobile* with a given degree of slowness and can also cause a motion of twice that slowness. According to this theory, therefore, it can move double the *mobile.* And, since it can move with four times the slowness, it can move four times the *mobile,* and so on *ad infinitum.* Therefore, any motive power would be of infinite capacity.[21]

A similar argument may be made from the standpoint of the *mobile.* For any *mobile* may be moved with a given degree of slowness, with twice that degree, four times, and so on without end; and, therefore, by the given mover, and by half of it, one fourth of it, and so on, without end. Any *mobile* could, therefore, be moved by any mover.[22]

Nor is it legitimate to object that slowness of motion cannot be doubled indefinitely, for, supposing this to be true, let *A* represent some slowness of a *mobile* that cannot be doubled. Now imagine a sphere or cylinder, revolving about a fixed axis; then, at some point near the pole of the sphere, or the axis of the cylinder, there is a degree of slowness double that of A, as is quite clear and easy to demonstrate. Now, at this point let there be attached a strong, long cord, at whose end is affixed a given weight, *B.* Then the slowness of the motion of *B* is twice that of *A,* and this is what we wished to demonstrate.

Nor can some quibbler properly claim that the motion of *B* is accidental motion, merely motion *in potentia,* and that it really has no relevance to the question; for this motion possesses a mover, a thing moved, initial and final limits, a time and a space traversed. . . all *in actu.* Therefore, the motion is a real one. Nor can it be maintained that the mover is not *in actu,* but only *in potentia,* inasmuch as it is part of the sphere, or cylinder, and because it is the whole that moves primarily and the part by consequence. In that case, if a man were to pull that weight by hand, by means of the cord, it would move accidentally, because by virtue of a part of the man; it would then follow that no motion, extrinsically caused, could be a "real" motion, (one which is *in actu),* since no mover can apply itself wholly to the thing moved, but can only do so by means of a part.

Thirdly, the present theory is to be refuted on the ground of falsity, because sense experience teaches us the opposite. We see, indeed, that if,

to a single man who is moving some weight which he can scarcely manage with a very slow motion, a second man joins himself, the two together can move it much more than twice as fast.[23] The same principle is quite manifest in the case of a weight suspended from a revolving axle, which it moves insensibly during the course of its own insensible downward movement (as is the case with clocks). If an equal clock weight is added to the first, the whole descends and the axle, or wheel, turns much more than twice as rapidly (as is sufficiently evident to sight).

Since the situation regarding retardation is closely similar, whether the thing moved be constant and the mover be diminished or the reverse we arrive at this negative conclusion: *With the mover remaining constant, the proportion of the speeds of motions does not vary in accordance with the proportion of resistances, nor, with the resistance remaining constant, does it vary in accordance with the proportion of movers.*

As for the reasons which seemed to support this theory, it should be pointed out that all authorities claiming that, with the mover remaining constant, the proportion of the speeds of motions varies in accordance with the proportion of resistances, really mean that the proportion of speeds

21. If $F/R \propto V$, where $F > R$ and produces motion, then $F/nR \propto V/n$, where $n = 2, 4, 8, 16, 32, . . .$ Thus, when $R > F$, and however great R becomes, F will nevertheless be capable of producing a positive velocity in accordance with the ratio it bears to nR. Hence F will be of "infinite capacity," able to move any resisting weight, however large (Oresme draws the same consequence and then demonstrates that, although it should follow from the false theory, it actually does not; see Selection 51.2). The theory is rejected by Bradwardine as physically impossible, since it violated the generally accepted axiom that in order for motion to occur a motive power must exceed the resistance opposing it, whether this resistance be another body or a corporeal medium. Bradwardine's new theory remedied this defect (see n. 30). Aristotle, who first formulated and accepted this third theory, actually anticipated Bradwardine's criticism and would have countered by insisting that if after successive halvings of a velocity, R became equal to, or greater than, F, motion would cease and mathematical rules of proportionality would no longer apply (see his shiphauler argument in *Physics* VII.5.250a.10–20).

22. Here $F:nR \propto V/n$, where $n = 2, 4, 8, 16, 32, . . .$ As the motive force, F, is successively halved, it will happen eventually that $R > F:n$. But however small and feeble F becomes it can yet produce a proportionate velocity in R. See note 30.

23. As can be seen further on (Theorem VI, chapter 3), Bradwardine's theory will mathematically yield such results.

varies with the proportion of the things affected to the things affecting them.[24]

As a matter of fact, in the case of the first authority cited (that of Book IV of the *Physics*) it should be realized that what Aristotle means is that, to whatever extent the proportion of air to a given body which moves through it is smaller than that of water to that same body (due to the greater thinness and incorporeality of air), to that extent the body will move faster through air than through water; for to whatever extent the proportion of air to a given body is smaller than the proportion of water to that body, to the same extent is the proportion of the body to air larger than its proportion to water, and, as will later be shown, the proportion of the speeds of motions varies in accordance with the proportion of movers to things moved.

The second authority cited should also be construed in the same sense, together with that passage, from Book I of the *De caelo et mundo,* which reads: "It is held to undergo a greater and less effect, by the action of the same agent, in a longer and shorter time, and any such effects are divided proportionally to the time." In other words, the proportions of any given effects are divided proportionally to the time.

In the case, moreover, of the theorem drawn from Book VII of the *Physics* (which states that, if a given power moves a given *mobile* through a given distance in a given time, the same power will move half of it through double the distance in an equal time), Aristotle understands, by "half the *mobile*," a part of the *mobile* possessing a proportional relation to the given moving power which is half that of the whole *mobile* to that power.

This quite clearly appears to be Averroes' interpretation, for, regarding the passage in question, he proves the above-mentioned theorem as follows: "When we divide the motion (or the thing moved), it follows necessarily that the proportion of the power of the mover to the motion (or thing moved) is twice the former proportion. This would, nevertheless, be untrue, unless understood in the sense previously indicated, for, although a given whole may bear a given proportion to some other whole, it does not follow that it bears to half of the latter whole a proportion double the first proportion [after "bears" Crosby has "a proportion to half of the latter whole that is half the original proportion"—*Ed.*] (as will be shown in what follows)." It is in this sense that the following authorities should be interpreted.

The theorem drawn from the *De ponderibus* must be read in the same way. "Between whatever heavy bodies, etc. . . . .," that is, between whatever heavy bodies the proportion of speed of descent and that of the proportion of weight to resistance are taken in the same order. (And with the proviso that the resistance remains equal.)

The author of this work, however, proposes no principle in proof of this theorem, and it may well be objected against the above interpretation, that neither does any commentator prove the theorem in this sense, but rather in another, which the present theory supports (namely, that there is neither any proportion nor excess of motive power over resistive power) and which is, in fact, what the words of this theorem really mean.

The fact is that no commentator whom we have seen either proves this theorem according to our interpretation, or according to any other. One commentator, for example, takes two unequal weights and two unequal lines representing their descents, and then, first taking it as granted by his adversary that the proportion of the larger weight to the smaller is greater than the proportion of the longer line to the shorter, he argues from this that the proportion of the smaller weight to the larger is less than the proportion of the shorter line to the longer. From this he concludes that the proportion of weights is less than that of descents (the opposite of which had been stipulated).

This however, presents no obstacle, for it was admitted that the proportion between the weights was greater than that between the descents, in the first place, and it not only does not invalidate the theorem, but it follows (conversely) that the proportion of the smaller weight to the larger will be

---

24. Bradwardine argues here that partisans of the third erroneous theory have misinterpreted the authors cited in their favor. The references and quotations from Aristotle, Averroes, and the *Liber de ponderibus* do not, Bradwardine insists, uphold the position that when $F_2 = F_1$ the ratio of velocities will vary inversely as the ratio of resistances ($V_2/V_1 = R_1/R_2$) but rather that the ratio of velocities will vary as the relation between the whole ratio F_2/R_2 to the whole ratio F_1/R_1. These authors do not take V to vary with R alone when F is held constant; nor indeed do they take it to vary with F alone when R is held constant; but it will vary with alterations in the whole ratio F/R. Bradwardine sees these authors as proponents of a viewpoint that is akin to his own, although, as he declares, none of them have demonstrated it. In truth, none of these authors subscribed to the position attributed to them by Bradwardine.

smaller than that of the shorter descent, or line, to the longer.[25]

Another commentator, also taking two unequal weights and their unequal descents, adds to the smaller weight a second, such that the two together are equal to the larger. He now posits that the descent of the added weight, by itself, through a time equal to that of the previous descents, when added to the descent of the lesser weight, is equal to the descent of the larger weight.

This had neither been previously proved, nor is independently known, nor follows logically, nor is universally true. In many cases it is, in fact, false (as will appear from what follows).

From this supposition, at any rate, he concludes that the proportion of the larger weight to the weight that was added is less than that of the descent of the larger weight to the descent of the additional weight (the opposite of what he states to have been given). Yet this is not the case, for it was not previously laid down as universally true that, "of any given unequal weights, the proportion of the larger to the smaller is in the proportion of their descents, taken in the same order," but only specifically that, "of these two weights, which have been chosen, and of their descents." This is, therefore, not incompatible with the thesis that, in the case of certain other weights, the proportion of the larger to the smaller should be less than that of their respective descents; for, in the case of some weights, the proportion of the larger to the smaller is equal to that of their respective descents, in other cases it is greater, and in yet other cases less (as will be clearly demonstrated, later on).[26]

As for the argument which is most convincing to writers on this subject, it should be replied that, although the causal principle adduced in its proof is true, the first consequence drawn from it does not hold. Fundamentally, a power which is double another power can move a mobile which is twice that moved by the lesser power through an equal distance and time. Instead of it following from this that the greater power can move the lesser *mobile* twice as fast, it rather follows that the greater power can move the lesser *mobile* at a speed which is as much greater as that expressed by the proportion of double the resistance to the former speed, and that it requires twice the power to do it. That speed will, in some cases, be exactly double the former speed, in some cases more than double, and in some cases less than double (as will be clear from later portions of this work).[27]

CHAPTER 2, PART 4
[THEORY IV]

A fourth theory declares there to be *neither any proportion nor any relation of excess between motive and resistive powers*. It holds that, instead of there being some proportion or excess of motive power over thing moved, motions take place in accordance with some sort of "natural dominance" of mover over moved.

This contention may be seen as founded on the authority of Averroes, who (Comment 79, *Physics* VIII) in solution of the same problem, says that an incorporeal power is neither to be called finite nor infinite, because only bodies may be referred to in this way. Furthermore, one incorporeal power may not be referred to as greater than another, since the terms, greater and less, may be applied only to quantities. Moreover, powers separate from body can neither be proportionals nor possess proportional relations, since a proportion can only be between one magnitude and another.

From the above it is seen to follow that no motive power can be either finite or infinite, greater or lesser, or in any way proportional to the power of

25. If $W_2 > W_1$ (W is the weight of a body) and $A > B$, where A and B are lines representing the speeds of descent of the weights, and if it is further assumed that $W_2/W_1 > A/B$, it follows that $W_1/W_2 < B/A$. But Bradwardine insists that this does not controvert the initial assumption that $W_2/W_1 > A/B$, since the anonymous commentator has merely converted a relation between ratios of greater inequality to one between ratios of lesser inequality.

26. If W is the weight of a body and V its velocity, the commentator assumes initially that $W_2 > W_1$ and $V_2 > V_1$. He then says that $W_1 + w = W_2$ (w is the added weight) and assumes, therefore, that $V_1 + v = V_2$ (where v is the velocity of w), from which he apparently concluded that $W_2/w < V_2/v$. Although this conclusion does not seem to follow from the data, Bradwardine accepts it as proper but argues that he sees in it nothing that is incompatible with his own view, since this consequence does not invalidate any general rule to the effect that $W_2/W_1 = V_2/V_1$. In fact, there is no such universally proven rule, so that it cannot be said that this commentator has disproved it. Indeed his consequence is compatible with consequences that are derivable from Bradwardine's own theory, where generally $F_2/R_2 = (F_1/R_1)^{V_2/V_1}$. For example, in Theorem VI of chapter 3 (see later), $F_2/F_1 = 2/1$ (that is, the ratio of motive forces or weights are related as $2/1$) and $V_2/V_1 > 2/1$, so that $F_2/F_1 < V_2/V_1$. Therefore, just as in the consequence drawn by the commentator, the ratio of forces or weights are not related as the ratio of velocities taken in the same order.

27. See in Theorems I–VII of chapter 3.

the thing it moves, because no motive power is a body, but is rather a form (either extensive within the body or separate from it).

This theory, together with Averroes' opinion, may be further confirmed by the definition of proportion, for a proportion is a comparison of two things of the same kind (as is evident on the basis of the definition of proportion set forth in Chapter I). It is obvious, however, that active and passive powers are not of the same kind.

Furthermore, if active and passive powers were to bear a proportion to each other, they would then be comparables. They would, therefore, be of exactly the same species and would consequently, have a subject or substance of exactly the same species. This consequence is false, because powers are divided into active and passive, as is a genus, by incompatible differences. That it nevertheless follows is shown by Aristotle, at the end of Book VII of the *Physics,* where he expresses the opinion that all things which are to be compared must be of the same individual species and entirely without difference with respect to the subject or substance of comparison, as well as to that regard or those regards in which the comparison is made.

Further, if there were some proportion of a motive power to the thing it moves, it would have to be one of greater inequality, since it would have to exceed the power of the thing moved. And since everything which exceeds something else is divisible into what exceeds and what is exceeded (as appears from Book IV of the *Physics,* in the chapter on the "void" [215b.12–18—*Ed.*]), it follows that any motive power may be so divided. This is false, for all incorporeal motive powers are fundamentally indivisible, and an embodied motive power is smaller in extension than the power of the thing moved.

Nor can it be claimed that Aristotle is here speaking of excesses only in the strictest sense (the sense in which they are found among quantities), for he is actually referring to the excess or rarity of one medium over another. To the same effect, in Book I, Chapter 7 of Aristotle's *Rhetoric,* where relative good and relative utility are under discussion, there appears the following: "So let the thing that surpasses be as much as and more than the exceeded thing contained within it." Aristotle's dictum is, therefore, not true merely of the strict usage of "excellence."

This theory is, however, capable of disproof, for if there were no proportion between powers, for the reason that they are not quantities of the same kind, neither could there be such a proportion

between musical pitches, and the entire science of harmonics would collapse, accordingly. For the *epogdoös* or "tone" is constituted in the proportion of nine to eight, the *diatessaron* in the proportion of four to three, the *diapente* in the proportion of three to two, the *diapason* (composed of *diatessaron* and *diapente*) in the proportion of two to one, the *diapason* and *diapente* combined in the proportion of three to one, and the double *diapason* in the proportion of four to one. This is sufficiently evident from various passages of the *Music* [probably the *De Musica* of Boethius—*Ed.*].

Furthermore, Averroes, in Comments 36 and 38 on Book VII of the *Physics,* proves certain theorems concerning the proportion of the speeds of motions by means of geometric theorems, as has already appeared in the third argument against Theory I. In Comment 65 on Book I of the *De caelo,* he also proves this theorem: It being granted that the infinite can move the finite in a finite time and that a finite agent can, in the same time, move part of this finite resistance, no infinite can move a finite. What he does is take a second finite mover which bears the same relation to the former as the whole finite resistance bears to this part. He then argues permutatively, from Definition xii of Book V of Euclid's *Elements,* that the proportion of the larger finite mover to the whole resistance is equal to that of the lesser mover to the part. And from this he concludes that the larger finite mover moves the whole resistance in a time equal to that in which the lesser finite mover moves the part and also equal to that in which an infinite mover moves the whole resistance.[28]

Further, according to the present theory and

28. Averroes does little more than repeat the argument given by Aristotle in Book I of *De caelo* (275a.14–22), where the latter assumes that A is an infinite power, BF (mistakenly called B at the outset) a finite resistive power, and C the time of A's acting on BF. Now, if A should produce a finite motion in BF during time C, then during the same time a finite power, D, should be able to cause the same motion in F, a finite resistive power or mobile less than BF. Now, because $D/F = A/BF$, where $D < A$ and $F < BF$, it follows that $V_F = V_B$ in time C (V is velocity). Next, take another power E such that $E/D = BF/F$, so that by permutation of ratios $E/BF = D/F$, from which it follows that $E/BF = A/BF$ and, as Aristotle concludes, "the finite and the infinite effect the same alteration in equal times. But this is impossible; for the assumption is that the greater effects it in a shorter time." The relevance of this argument for Bradwardine is that Averroes (and, of course, Aristotle) has accepted as proper the relation between the finite movers and their respective resistances ($E/D = BF/F$) and would obviously oppose this fourth theory.

(as a matter of fact) in reality, the power of the mover "dominates" that of the thing moved. In many passages Averroes says that the power of the mover exceeds that of the thing moved, and that the mover is of greater power than the thing moved. If, therefore, it "dominates" and exceeds and is of greater power, this must necessarily take place according to some proportion, whether strictly or generally understood, and both Aristotle and Averroes, in many passages, suppose there to be some proportion of the power of the mover to that of the thing moved.

We thus arrive at this affirmative conclusion: *A proportion is found to exist between any motive power and the resistive power of the thing it moves.*

The first reasons which were brought forward in favor of the present theory are easily countered by means of the definition of proportion already given in Chapter I, for the proportion found to exist between motive and resistive power is not a strict, but only a general one.

The next argument concerning comparison is overthrown by a similar distinction in the meaning of the word, "comparison." The authority which was cited is to be understood as speaking of strict rather than general comparison. It is to be noted, indeed, that comparison is made: (1) within a genus (as, for example, indicated by the terms: "more virtuous," "wiser," etc.), (2) within the most general genus (for example, "form is more [Crosby omits "more" *(magis)—Ed.*] substance [after "substance" Crosby has "rather," which is omitted here—*Ed.*] than matter or a compound of the two"), and (3) also in transcendence of every genus (for example, "substance is more [Crosby omits "more" *(magis)* and has "rather" after "being"—*Ed.*] being than accident").

As for the final argument, it is to be replied that it is true that there is, in the general sense, a proportion of excess between motive and resistive power. To the authority cited as saying that everything that exceeds is divided into what exceeds and what is exceeded, it must be replied that, just as "what exceeds" may be taken in two senses, so also "to be divided into what exceeds and what is exceeded" may be taken in two senses (i.e., generally and strictly). Everything that exceeds is, therefore, strictly divisible in this manner.

In the general sense, on the other hand, what exceeds may be divided as follows: In a general sense everything that exceeds may be reduced or lessened until equal to what was exceeded; and thus may be understood the entire latitude whereby

it exceeds and likewise the similarity or equality which is contained virtually and potentially in the thing that exceeds.

Or it may be carried out thus: Everything exceeding something else is divided, in the general sense, into excess and what is exceeded, not, of course, in itself, but in comparison to something else outside itself (for example, action, and passion or resistance). In this sense, the powers of mover, moved and resistance can be compared to each other in every way in terms of excess and what exceeds. And if one were to take the example of a motive power equal to a resistive power, that motive power is twice half of the resistive power, not because it can produce twice the motion, but because twice the halved resistance has precisely the power of resistance that the motive power has of moving. Concerning every other proportion of mover to moved this is proportionally true.

This theory, therefore, together with the former ones, is pronounced false.

CHAPTER 3

Now that these fogs of ignorance, these winds of demonstration, have been put to flight, it remains for the light of knowledge and of truth to shine forth. For true knowledge proposes a fifth theory which states that the proportion of the speeds of motions varies in accordance with the proportion of the power of the mover to the power of the thing moved.[29]

29. This fifth and "true" theory is Bradwardine's function, which we have already represented as $F_2/R_2 = (F_1/R_1)^{V_2/V_1}$. In general, to obtain n times any velocity arising from a ratio F/R, the latter must be raised to the nth power. So vague is the language used here and in Theorem I to describe this theory that without reliance on the subsequent theorems it would be almost impossible to discern Bradwardine's meaning. In what immediately follows, Bradwardine associates Averroes, and to a lesser extent Aristotle, with his own theory, but there is little doubt that their views are quite properly represented by Bradwardine's third erroneous theory.

The exponential function described here, and commonly called "Bradwardine's Function," probably had its origin in a pharmaceutical rule which sought to calculate the effects of hot and cold parts in compound medicines. On the formulation of this rule by al-Kindī and Arnald of Villanova, see Michael R. McVaugh, "Arnald of Villanova and Bradwardine's Law," *Isis* LVIII (1967), 56–64. For the suggestion that Bradwardine might have derived his function directly from Walter of Odington's *Icocedron*, in which the rule is applied to the quantitative composition of intensive qualities, see Donald Skabelund and Phillip Thomas, "Walter of Odington's Mathematical Treatment of the Primary Qualities," *Isis* LX (1969), 331–350.

This is what Averroes intends when he says, in Comment 71 on Book IV of the *Physics*: "It is manifest that, universally, the cause of the diversity and equality of motion is the equality and diversity of the proportion of mover to thing moved. If, therefore, there are two movers and two things moved, and the proportions between these movers and the things which they respectively move are equal, then the two motions are of equal speed. If the proportion is varied, the motion is also varied in that proportion."

Further on in the same comment, he also says: "The difference between motions with respect to slowness and fastness varies in accordance with the proportion between the two powers (namely, motive and resistive)."

In Comment 36, on Book II of the *De caelo*, he says: "Fastness and slowness do not occur otherwise than in accordance with the proportion of the power of the mover to that of the thing moved. By however much, therefore, the proportion is greater, by so much will the motion be faster; and by however much the proportion is less, by so much will the motion be slower."

In Comment 35, on Book VII of the *Physics*, from a doubling of the proportion of mover to moved he argues a doubling of the speed of the motion, as follows: "If we divide the *mobile* in two, it necessarily comes about that the proportion of the power of the mover to the thing moved becomes double the former proportion, and thus the speed will be twice what it was before."

Further on, in the final comment, he remarks that: "These two (that is, the speed of alteration and the quantity of time) vary in accordance with the proportion between that which causes the alteration and that which undergoes it. If, therefore, the proportion is great, the speed will be great and the time short, and conversely."

Concerning this same problem, both Aristotle and Averroes (as is evident in the third argument against Theory I) express, in many passages, the opinion that, from an equality of proportion between mover and moved, there follows equality of speed. Equality of the proportion of movers to *mobilia* is, therefore, the causal condition which, when fulfilled, posits an equal speed of motions and which, when not fulfilled, makes impossible an equality of speeds. Equality of the proportion of movers to *mobilia* is, thus, the primary and precise cause of equality of the speeds of motions, and to the variation of this cause there directly cor-

responds the variation of proportion between different motions.

Furthermore, there does not seem to be any theory whereby the proportion of the speeds of motions may be rationally defended, unless it is one of those already mentioned. Since, however, the first four have been discredited, therefore the fifth must be the true one.

We, therefore, arrive at the following theorem:

Theorem I. *The proportion of the speeds of motions varies in accordance with the proportion of motive to resistive forces, and conversely.* Or, to put it in another way, which means the same thing: *The proportion* [the words "of the proportion," which immediately follow, are omitted here, since they are not represented in the Latin text— Ed.] *of motive to resistive powers is equal to the proportion of their respective speeds of motion, and conversely.* This is to be understood in the sense of geometric proportionality.[30]

Theorem II. *If the proportion of the power of the mover to that of its mobile is that of two to one, double the motive power will move the same mobile exactly twice as fast.* This may be demonstrated by means of an example. Let A be a motive power that is twice B (its resistance), and let C be a motive power that is twice A. Then, (by Theorem I, Chapter I) the proportion of C to B is exactly double that of A to B. Therefore (by the immediately preceding theorem), C will move B exactly twice as fast as A does. This is what was to be proved.[31]

Theorem III. *If the proportion of the power of the*

30. This is Bradwardine's function, already cited in the preceding note. No formal proof of it is offered, but its truth is assumed on grounds that only five possibilities exist, and since four have already been "proven" false, Bradwardine's theory stands confirmed by a process of elimination. We can now see how Bradwardine avoided the major criticism which he directed against the third erroneous theory (see n. 21 and n. 22). If initially motion occurs, then necessarily $F > R$ and it is thereafter impossible for R to become equal to, or greater than, F. This is obvious, since in this function, successively halving a velocity arising from a given ratio F/R is achieved by taking $(F/R)^{1/n}$, where $n = 2, 4, 8, 16, 32, \ldots$. In this way Bradwardine remained faithful to contemporary Aristotelian physics by insisting that velocity is determined by a ratio of force to resistance, but avoided the mathematical difficulties.

31. Once again using F's, R's, and V's instead of A, B, and C, we see that if $F_1/R_1 = 2/1$, $F_2 = 2F_1$, and $R_2 = R_1$, then $F_2/R_2 = (F_1/R_1)^{2/1}$, so that $V_2 = 2V_1$, since 2/1, the exponent, represents the ratio of velocities V_2/V_1. Therefore, F_2 moves the same resistance ($R_2 = R_1$) with twice the velocity with which it is moved by F_1.

mover to that of its mobile is two to one, the same power will move half the mobile with exactly twice the speed. This you may demonstrate by an argument like that used for Theorem II.[32]

Theorem IV. *If the proportion of the power of the mover to that of its mobile is greater than two to one, when the motive power is doubled the motion will never attain twice the speed.* This may be demonstrated by means of Theorem IV, Chapter I; and Theorem I, Chapter III.[33]

Theorem V. *If the proportion of the power of the mover to that of its mobile is greater* [Crosby has "less" although the Latin reads *maior*—Ed.] *than two to one, when the resistance of the mobile is halved the motion will never attain twice the speed.*[34] This may be demonstrated by means of Theorem III, Chapter I and Theorem I, Chapter III.

Theorem VI. *If the proportion of the power of the mover to that of its mobile is less than two to one, when the power moving this mobile is doubled it will increase the speed to more than twice what it was.*[35] This is likewise easily demonstrable, from Theorem VI, Chapter I and Theorem I, Chapter III.

Theorem VII. *If the proportion of the power of the mover to that of its mobile is less than two to one, when the same mover moves half that mobile the speed of the motion will be more than doubled.*[36] This may be demonstrated clearly, from Theorem V, Chapter I and Theorem I, Chapter III.

Theorem VIII. *No motion follows from either a proportion of equality or one of lesser inequality, between mover and moved.*[37] With the addition of the following axiom, independently known:

Axiom 1. *All motions of the same species may be compared to each other with regard to slowness and fastness;* this theorem may be proved by means of Theorems VII and VIII of Chapter I and Theorem I, Chapter III.

Theorem IX. *Every motion is produced by a proportion of a greater inequality, and from every proportion of greater inequality a motion may arise.* The first part of this may be proved by Theorems I and VIII of Chapter III and the axiom just given. The second part is demonstrable from the fact that every excess of mover over moved suffices to produce motion, as will be shown elsewhere.

Theorem X. *Given any motion, one twice as fast and one twice as slow can be determined.* This may be proved by Theorem I and Theorem IX (Part 2) of Chapter III, with the help of the following axiom, independently known:

Axiom 2. *A proportion of greater inequality of mover to moved may be halved or doubled indefinitely.*[38]

Theorem XI. *An object may fall in the same medium both faster, slower, and equally with some other object that is lighter than itself.*

Let, for example, *A* represent a heavy mixed body composed of heavy and light and having a certain weight, and let *B* represent some pure heavy body, as small as you please. Now let a given medium be rarefied to the point at which *B* bears to it a proportion equal to, or greater than, that of the heaviness to the lightness in *A*. Then let both bodies be placed in the same medium. The heaviness of *A* will now be in a lesser proportion to its total intrinsic and extrinsic resistance than *B* is to its resistance. Therefore, by Theorem I, Chapter III, *A* moves more slowly than *B*.[39]

32. If $F_1/R_1 = 2/1$, $F_1 = F_2$, and $R_2 = R_1/2$, then $F_2/R_2 = (F_1/R_1)^{2/1}$, where $2/1 = V_2/V_1$, and it is shown that the same power moves half the original resistance with twice the speed of the whole resistance.

33. If $F_1/R_1 > 2/1$, $F_2 = 2F_1$, and $R_2 = R_1$, it follows that $F_2/R_2 < (F_1/R_1)^{2/1}$, so that $V_2 < 2V_1$.

34. If $F_1/R_1 > 2/1$, $F_2 = F_1$, and $R_2 = R_1/2$, it follows that $F_2/R_2 < (F_1/R_1)^{2/1}$, so that $V_2 < 2V_1$.

35. If $F_1/R_1 < 2/1$, $F_2 = 2F_1$, and $R_2 = R_1$, it follows that $F_2/R_2 > (F_1/R_1)^{2/1}$, so that $V_2 > 2V_1$.

36. If $F_1/R_1 < 2/1$, $F_2 = F_1$, and $R_2 = R_1/2$, it follows that $F_2/R_2 > (F_1/R_1)^{2/1}$, so that $V_2 > 2V_1$.

37. When $F \leq R$ no motion can occur. In medieval physics this was accepted as almost axiomatic, since it was held that motion could occur only when the motive force exceeded the resisting body or medium (that is, when $F > R$; this is made explicit in the next theorem). The proof of Theorem VIII is based on Theorems VII and VIII of Chapter 1, where it is shown that a ratio of equality, A/A, is not relatable by any exponent whatever to a ratio of greater inequality A/B $(A > B)$ in the sense that $(A/A)^n$, where n may have any value however great, always remains equal to A/A and consequently cannot be made equal to, or greater than, A/B; similarly, a ratio of lesser inequality D/C $(D < C)$ is not relatable by means of any exponent to a ratio of greater inequality A/B, since whatever the value of n in the expression $(D/C)^n$ it will always remain less than A/B.

Although Bradwardine might have appealed to the accepted axiom that no motion could be produced when $F \leq R$, he chose instead to "demonstrate" this mathematically by showing that in the context of his function even if velocities could arise when $F \leq R$, such motions were not comparable to motions arising when $F > R$.

38. When $F > R$, the motion produced can always be doubled or halved by squaring or extracting the square root of F/R—that is, determining $(F/R)^2$ or $(F/R)^{1/2}$.

39. Let f/r represent the ratio of heavy to light (or motive force to internal resistance) in body A, and let B represent a pure heavy body as small as you please; and assume also that R is an external medium which is rarefied so that the ratio of $B/R \geq f/r$. Upon placing both bodies in R, we see that $B/R > f/(r + R)$ and therefore $V_A < V_B$ in R.

Conversely, let the medium be condensed to the point at which the proportion of B to it is less than the proportion of the heaviness of A to its entire intrinsic and extrinsic resistance. Then, by Theorem I of this chapter, A moves faster than B.[40]

Thirdly, let the medium be so determined that the proportion of B to it is equal to the proportion of the heaviness of A to its entire intrinsic and extrinsic resistance. Then, by Theorem I of the present chapter, A and B will move at equal speeds.[41]

Alternatively, let A be supposed to have a determinate speed, C, in a vacuum, and let some medium be rarefied until B falls in it with speed C or faster; then A, placed in the same medium, will fall more slowly than B.[42] Conversely, let the medium now be condensed as required, and the remaining two consequences will follow.[43]

Corollary 1. It is manifest, from the foregoing, that *the fastness and the slowness of any pure body and the slowness of any mixed body may be doubled indefinitely, but that the fastness of a mixed body may not be so doubled by rarefaction of the medium.*[44] This corollary is sufficiently well established on the basis of what has been said above.

Theorem XII. *All mixed bodies of similar composition will move at equal speeds in a vacuum.* In all such cases the moving powers bear the same proportion to their resistances. Therefore, by Theorem I of this chapter, all such bodies move at the same speed.[45] From this you must also understand that:

Corollary 2: *If two heavy mixed bodies of unequal weight, but similar composition, were balanced on a scale within a vacuum, the heavier would descend.*

Let A and B represent two such heavy bodies, A greater and B less; let C and D represent the heaviness and lightness of A, respectively, and let E and F represent the heaviness and lightness of B, respectively. Then C, D, E, and F are four proportionals, C being the greatest and F the smallest.[46] Therefore (by Axiom [Crosby has "Theorem" —*Ed.*] VIII, Chapter I) C and F, combined, exceed D and E, combined.[47] Since C and F tend to raise B and only D and E resist, therefore (by Part ii of Theorem IX of the present chapter) B ascends and A descends.[48]

40. Here $B/R < f/(r + R)$, so that $V_A > V_B$.
41. Here the density of the medium is such that $B/R = f/(r + R)$ and $V_A = V_B$.
42. In heavy mixed body A let f/r produce speed C in a vacuum and assume the density of medium R to be such that pure elemental body B is related to it as $B/R \geq f/r$, so that $V_B \geq V_A = C$. Subsequently dropping A

in R makes V_A (or C) $< V_B$, since $B/R > f/(r + R)$.
43. By properly adjusting the density of R one can arrange to have A and B fall in R so that V_A (or C) $>$ V_B, and $V_A = V_B$.
44. The speeds of a pure elemental body and a heavy mixed body can be halved indefinitely because theoretically a corporeal medium can be made infinitely dense. But only a pure elemental body can have its velocity doubled indefinitely, since it was almost universally assumed that a corporeal medium is rarefiable ad infinitum. But rarefaction of a medium ad infinitum will not permit the infinite doubling in speed of a mixed body, for when R becomes zero or void in the formulation $f/(r + R)$, a maximum finite speed would be determined by the ratio of motive force to internal resistance, namely f/r.
45. Since the two unequal mixed bodies are of homogeneous composition, the ratio of motive force to internal resistance (f/r) per unit of matter is equal in each body and consequently their velocities of fall in the void will be equal. Bradwardine relies here on a specific factor (namely equality of ratio per unit of matter) rather than utilizing an extensive factor, or gross weight, as did Aristotle. Indeed the distinction is analogous to that between specific and gross weight. This important conclusion, which was rejected by Aristotle as an absurdity (*Physics* IV. 8.216a. 14–21), is akin to that enunciated in the sixteenth century by Galileo in his *De motu (On Motion)* (and even earlier by Giovanni Benedetti), where it is declared that homogeneous bodies of different weight or size would fall in the void with equal finite velocities. Galileo, however, abandoned the medieval distinction between pure elemental bodies (for Bradwardine such bodies, devoid of internal resistance, could not fall in the void with finite velocities) and mixed bodies, and the dichotomy between absolute heaviness and lightness, so that his arguments in favor of essentially the same conclusion were quite different, despite a few striking similarities (see Selection 55.5 and n. 79 thereto).
46. C, D, E, and F are four proportionals, because A and B are homogeneous bodies.
47. Axiom VIII, Chapter 1, states that if "four quantities are proportionals and the first the largest and the last the smallest, the sum of the first and last will be necessarily found greater than the sum of the other two" (trans. Crosby).
48. Upon suspending heavy mixed bodies A and B (with $A > B$) from a balance placed in a vacuum, we see

from the figure (added here for convenience; it is not in the text) that C and F (that is, the downward acting motive power of A and the upward acting resistive power or lightness of B) act conjointly to move B upward, while D and E (the upward acting resistive power or lightness of A and the downward acting motive power of B) act conjointly to move A downward. Since we have a constrained system and $C + F > D + E$ (by Axiom VIII, Chapter 1, quoted in n. 47), we can form a ratio of greater inequality $(C + F)/(D + E)$, where $C + F$ acts as total motive power and $D + E$ as total resistance.

2. Nicole Oresme (ca. 1325–1382): Extended Application of "Bradwardine's Function"

Translated and annotated by Edward Grant[49]

Proposition 1. That the following rules are false: If a power moves a mobile with a certain velocity, double the power will move the same mobile twice as quickly. And this [rule]: If a power moves a mobile, the same power can move half the mobile twice as quickly.[50]

The falsity of the first [rule] is obvious: Now, let *B* be a power that moves mobile *C* with a certain velocity, and let *A* be a double power. Then if ratio *B* to *C* were a double ratio, it follows, by the third noteworthy point of the first chapter, that ratio *A* to *C* will be [equal to] ratio *B* to *C* squared. It certainly follows, by the first supposition, that the velocity with which *A* moves *C* is double the velocity with which *B* moves *C,* since a ratio of velocities is just like a ratio of ratios.[51]

But now, on the contrary, if ratio *B* to *C* were less than a double when ratio *A* to *B* is a double by assumption, it follows, by the second supposition,[52] that ratio *A* to *C* would be greater than ratio *B* to *C* squared because it is composed of ratio *B* to *C,* the lesser, and ratio *A* to *B,* the greater, namely a double. Consequently, by the first supposition, *A* will move *C* more than twice as quickly as *B* will move *C.*[53]

Again, if ratio *B* to *C* is greater than double and *A* to *B* is assumed to be a double ratio, it follows by the second supposition, that ratio *A* to *C* will be less than ratio *B* to *C* squared. Hence, by the first supposition, *A* will move *C* with a velocity that is less than twice the velocity with which *B* moves *C.*[54]

For example, let *A* be 8 and *B,* 4. Therefore, if ratio *B* to *C* is related as ratio *A* to *B,* so that *C* would be 2, then ratio *A* to *C* will be ratio *B* to *C* squared, and thus the velocity is doubled. Furthermore, if ratio *B* to *C* were less than double, so that *C* might be 3, then ratio *A* to *C* will be greater than ratio *B* to *C* squared, and the velocity is more than doubled. If, however, ratio *B* to *C* were greater than a double ratio, so that *C* might be a unit, then ratio *A* to *C* [will be less than the square of *B* to *C*], and similarly, the velocity will be less than doubled.

Thus it is clear that from a doubling of the power, a doubling of the velocity does not follow except in one case, namely when the power is first taken to be double the mobile. Hence the rule is false, because it is conditional and ought to be necessary. And the antecedent ought to be in-

capable of being true without the consequent [being true], but the truth of the consequent does not agree with the truth of the antecedent except in one case.

The falsity of the second rule can be shown by the same principles, because if a power moves a mobile with a certain velocity, the same power will not move half [the mobile] twice as quickly unless the first velocity should arise from a double ratio. But

---

Now, by the second part of Theorem IX, Chapter 3 "from every proportion of greater inequality a motion may arise," so that motive power *C* + *F* will cause *A* to descend, while *D* + *E,* the total resistance, will cause *B* to ascend.

49. The translation is reprinted with permission of the copyright owners, the Regents of the University of Wisconsin from *Nicole Oresme: De proportionibus proportionum* and *Ad pauca respicientes,* edited with Introductions, English Translations, and Critical Notes by Edward Grant (Madison, Wis.: University of Wisconsin Press, 1966). The propositions included here are taken from chapter 4 of his *De proportionibus proportionum (On Ratios of Ratios)* on pages 269–275, 279–283, 287–289, and 299–309. The annotations have been specially prepared for this volume but are based almost wholly on material in the edition and translation.

50. "The first false rule asserts that if $F/R \propto V$, then $(2F/R) \propto 2V$; in the second false rule, if $F/R \propto V$, then $F/(R:2) \propto 2V$" (p. 269). As in this instance, citations drawn directly from my edition will be placed within quotation marks, followed by the page number in parentheses. The two false rules under attack here are substantially the same as those represented by the third erroneous theory in Bradwardine's *Treatise on Proportions* (see chapter 2, Part 3, in the preceding selection).

51. "Using *F*'s, *R*'s, and *V*'s for *A*'s, *B*'s, and *C*'s, we see that if $F_1/R = 2/1$ and $F_2/F_1 = 4/2$, then $F_2/R = (F_1/R)^{2/1}$, where the exponent $2/1 = V_2/V_1$" (p. 269). By the expression "ratio of ratios" (Oresme calls his treatise "On Ratios of Ratios" from the technical expression *proportio proportionum*), Oresme means an exponent (expressed always as a ratio) relating two ratios. If the exponent is rational, we have a "rational ratio of ratios"; if irrational, the exponent is called an "irrational ratio of ratios." Since any ratio of velocities, V_2/V_1, serves as an exponent relating two ratios of force and resistance, it "is just like a ratio of ratios."

52. A reference to the second supposition of Chapter 4, where it is stated that if $A/C = A/B \cdot B/C$, and $A/B > B/C$, then $A/C < (A/B)^2$, and $A/C > (B/C)^2$.

53. Again using *F*'s, *R*'s, and *V*'s, we may summarize this as: "If $F_1/R < 2/1$, and $F_2/F_1 = 4/2$, then $F_2/R < (F_1/R)^{2/1}$ (where $2/1 = V_2/V_1$), since $F_2/R = F_2/F_1 \cdot F_1/R$" (p. 269).

54. "If $F_1/R > 2/1$, and $F_2/F_1 = 4/2$, then $F_2/R < (F_1/R)^{2/1}$, where $2/1 = V_2/V_1$" (p. 271).

sometimes what this [power] moves twice as quickly would be exactly half the first [mobile], sometimes greater than half, [and] sometimes less [than half].

In the second place, I argue against the second rule thus: If it were true, it follows that any power, however feeble, can move any mobile, whatever its resistance.[55]

Now, let a power A be taken, which can move C. And let D be a mobile double to C, and E another mobile double to D, and F double to E, and G to F, and so on. Then this consequence is proved: A can move C; therefore it can move D. And similarly this will be proved: A can move D; therefore it can move E. And likewise it can move E, and therefore F, and so on. Moreover, since it is possible, let B be a power which could move D exactly twice as slowly as A can move C.

Thus if the rule were true, B can move C twice as quickly as this very same B can move D, since C is subdouble to D and half of it. Therefore, B can move C exactly as quickly as A can move C, since A moves C twice as quickly as B moves D, and B moves C twice as quickly as B can move D. Hence A and B can move C equally quickly, by the ninth [proposition] of the fifth [book of Euclid, which says]: If two [quantities] bear the same ratio to a third, they are equal. Consequently, by the third supposition,[56] A and B are equal powers. But by assumption, B can move D: therefore, by the fourth supposition, A can move D, which was to be proved.

And this consequence will be proved in the same way: A can move D; therefore it can move E by taking a power that can move E twice as slowly as A can move D. It can then be proved as before that A and this given power that can move E are equal, from which the proposed consequence follows just as it was deduced before.[57]

However, if A should move C in a quadruple ratio, then in virtue of this it surely follows that A moves C. Then that power, namely power B, which moves D twice as slowly, that is, in a double ratio, can move C, the half of D, exactly as quickly as A moves C. Thus they are equal powers and A can move D because, similarly, A, as well as B, is related to D in a double ratio. And this is so because it follows that B moves D in a double ratio; therefore, B moves C twice as quickly, as conceded before, because, like A, it moves it in a quadruple ratio.[58]

For this reason it was stated in the corollary to the fourth proposition of the third chapter that a ratio of a quadruple to a double ratio is just like a ratio of [their] denominations, but this is not found in other ratios. And because of this it does not follow further in the aforementioned case that [if] A moves D in a double ratio, therefore that power that can move E twice as slowly can move

55. As will be seen, Oresme uses a radically different approach than Bradwardine in repudiating this false rule. Where Bradwardine was content to reject the false rule because it yielded this consequence, which he deemed physically absurd, Oresme shows that although the physical consequence enunciated here is entailed by the false rules, it does not in fact follow mathematically. The major criticism of Oresme's procedure will be described below.

56. Supposition III asserts that "all powers are equal which can move the same mobile, or equal mobiles, with equal velocity" (p. 263).

57. Up to this point Oresme has sought to show that any motive power, say A, can move any resisting body whatever. This is accomplished by showing that if A moves a body D, it can also move $2D$, $4D$, $8D$, and so on ad infinitum. The basic move involves the postulation of another motive power that can move $E = 2D$ with half the speed with which A moves D. Should this other motive power be P, it is taken as obvious by Oresme that if P moves $2D$, it can also move D with twice the velocity with which it moves $2D$. But A also moves D with twice the velocity with which P moves $2D$, from which it follows that $A = P$ (by the third supposition quoted above in n. 56). Obviously, A will also move $E = 2D$. By use of this procedure, it can be shown that A can move $4D$, $8D$, and so on. The power of A is therefore of potentially infinite capacity (compare Bradwardine's discussion in the preceding selection. This completes the first part of the demonstration. Now Oresme seeks to demonstrate that despite the entailment of A's infinite capacity as a motive power, this consequence does not in fact follow mathematically.

58. If $A/C = 4/1$, $D = 2C$, and if $B/D = 4/2$, it follows that B can move C, so that $A/C = B/C$ and we see that $A = B$. Therefore, $A/D = B/D = 2/1$, and $A/D = (1/2)$ (A/C) so that $V_D = V_C/2$ when A is the motive power. Thus if A can move C, it can move $2C = D$. But, as Oresme will emphasize at the beginning of the next paragraph, A can move D with half the velocity with which it moves C only because the initial conditions of this particular case are $A/C = 4/1$ and $B/D = 4/2$, so that $A/C = 2(B/D)$, which signifies that A moves C with twice the velocity with which B moves D. This result, arising from the false rule, agrees with the result that would be produced from what Oresme accepts as the true theory (Bradwardine's function) where for the same data $A/C = (B/D)^{2/1}$, which equals $A/C = 2(B/D)$ in the false rule. In this case Oresme would say that the "ratio of denominations" (that is, 2/1, where 4 is the denomination of A/C and 2 of B/D) equals the "ratio of ratios" (that is 2/1, where 2/1, the "ratio of ratios," is the exponent in $4/1 = [2/1]^{2/1}$). But in all other cases there would be no agreement, as, for example, if we were to relate 9/1 and 3/1, where the ratio of denominations is $9/3 = 3/1$, while their ratio of ratios is 2/1, since $9/1 = (3/1)^{2/1}$.

D exactly as quickly as A moves D. On the contrary, it follows [that it will move D] more quickly, because that power would be B, and then ratio B to E will be half of a double ratio; and since ratio E to D is a double ratio it follows that ratio B to D will be composed of a double and half of a double ratio. It will, consequently, be greater than double, so that, by the first supposition, B will move D quicker than A moves D, since A was double to D. Thus it no more follows that [because] A can move D, therefore it can move E. But, as was shown, this would nevertheless follow if the rule were true.[59]

What, then, should we say to Aristotle, who seems to enunciate the repudiated rules in the seventh [book] of the *Physics*? It must be said that they are false unless [the following] is added to the first rule: If a power moves a mobile *in a double ratio*, a double power will move [the same mobile twice as quickly]. And likewise [this must be added] to the second rule: If a power [moves a mobile] *in a double ratio*, the same power will move [half of the mobile twice as quickly]. And so we can gloss [Aristotle] and say that these rules ought to be understood [in this way]. Perhaps Aristotle said this but has been poorly translated. But if he did not say it, perhaps he failed to understand [the rules] properly.

Proposition 3. The ratio of a power to each of [two] mobiles can be made known when [both] the ratio of the two mobiles is known and the ratio by which the same power moves the lesser mobile more quickly than the greater mobile.

Let A be the power, B the greater and C the lesser mobile, and since I assume that every motion arises from a ratio of greater inequality, A will be greater than B. And let E be ratio A to B, F ratio B to C, and D ratio A to C.

Then ratio D is composed of intermediate ratios, which are E and F.[60] Furthermore, let G be ratio D to E, so that G will be a ratio of ratios from which the velocities arise.[61] Now, since the ratio of velocities is known by hypothesis, the ratio of ratios, namely G, will be known, by the first supposition. [The first supposition is actually a statement of Bradwardine's function; see chapter 3, Theorem I, in the preceding selection.—*Ed.*] We will then have four ratios, D, E, F, and G, of which two are known, namely G and F; and two are unknown, namely D and E, which we wish to make known.[62]

Let it be argued as follows: The ratio of D, the whole, to E, its part, is known, and therefore the ratio of the same part, namely E, to the remainder,

F, will be known, by the first part of the eighth supposition. And similarly, the ratio of the whole, namely D, to the remainder F, will be known, by the second part of the same eighth supposition.[63] Then, continuing, ratio E to F is known, and F is a known ratio or quantity, so that E will be known, by the ninth supposition.[64] Likewise, ratio D to F is known, and F is a known ratio or quantity, and consequently D will be known. Thus the two ratios, namely E and D, are now known, and this is what I desired.[65]

59. I quote here my own summary of this argument from the edition cited in note 49 above: "Thus according to the second false rule, if A moves D in a double ratio, i.e., $A/D = 2/1$, and $E = 2D$, then should another force B move E with half the velocity with which A moves D, it follows that $B/D = A/D$. But by Supposition I, which is Bradwardine's function, this is false, for according to that function to say that B moves E with half the velocity with which A moves D is to say that $B/E = (A/D)^{1/2}$ rather than $B/E = (1/2)(A/D)$, as required by the false rule. Now, $E/D = 2/1$ and $A/D = 2/1$. Consequently, $B/E = (A/D)^{1/2} = (2/1)^{1/2}$, and $B/D = B/E \cdot E/D = (2/1)^{1/2} \cdot (2/1)$, so that $B/D = (2/1)^{3/2}$. It is now evident that $B/D > A/D$, since $(2/1)^{3/2} > (2/1)$. Therefore, $B > A$ and $B/E > A/E$ (where $E = 2D$). Thus from the fact that $A < B$, it does not follow that because A can move D it can move $E = 2D$, even though B can move E and *a fortiori* can move D, which is half of E. Indeed, A may be incapable of producing any motion whatever in E. But the second false rule requires that $B/D = A/D$, making $B = A$, in which event A must necessarily be capable of moving D and E.

The success of Oresme's demonstration depends entirely on setting $B/E = (A/D)^{1/2}$ rather than $B/E = (1/2)(A/D)$. This move enables him to show that $B/D > A/D$ instead of $B/D = A/D$. The entire demonstration is improper, since Oresme has substituted his own function for the false rule and merely demonstrated that one obtains quite different results from the rival theories. He has produced no good reasons for repudiating the false rule, but has only shown that in terms of Bradwardine's function the consequence of the false rule—namely that any power, however weak, can move any resistance, however great—is contradicted."

60. $D = E \cdot F$, or $A/C = A/B \cdot B/C$.

61. That is, $D = E^G$.

62. That is, $D = E^G \cdot F$, where G and F are known.

63. In $D = E^G$, G is known by assumption, thus allowing us to determine that $E = (D)^{1/G}$, and $F = (D)^{(G-1)/G}$, from which it follows that $E = (F)^{1/(G-1)}$, and $D = (F)^{G/(G-1)}$. This is all derived from the eighth supposition of the fourth chapter.

64. Since F is known by assumption and exponent $1/(G - 1)$ is known (since G is also known by assumption), $F^{1/(G-1)}$ is completely known, and consequently, so is E.

65. Here also F and exponent $G/(G - 1)$ are both known, so that $F^{G/(G-1)}$ is completely known, and consequently, D is also known (see n. 63.) In the relation $D = E^G \cdot F$, all four quantities are now known.

For example, let B be double C and let the velocity with which C is moved be triple the velocity with which B is moved. By the first supposition, therefore, the ratio of ratios, namely G, will be a triple so that D will be triple to E. Hence E will be one third of D, and F, the remainder, two thirds of D. The ratio of D to F will then be as a denominating number to its numerator, namely 3 to 2, a sesquialterate. Now, by hypothesis, F is a double ratio, and therefore D would be composed of a double and half of a double ratio and will be three fourths of a quadruple ratio. And since D is composed of E and F, and F is a double ratio, E will be half of a double; or E will be one third [of D] and F two thirds [of D]. Thus their ratio is as a ratio of [their] numerators, so that E is related to F as 1 to 2.[66]

It must, however, be noted that if B were double C, it does not follow for this reason that A could move C twice as quickly as it moves B, because this is disproved by the first proposition of this chapter.[67]

It must also be understood that just as the ratio of a power to each of [two] mobiles can be ascertained [if] the ratio of those two mobiles and [the ratio of] the respective velocities produced by that same power are known,[68] so, contrarily, with the ratio of two powers known, as well as the [ratio of] velocities they produce with respect to the same mobile, the ratio of each power to that mobile can be made known. And thus the ratios from which velocities arise will be known, so that if it were assumed that C is the greater power, B the lesser power, and E the mobile which is moved by each [power] successively, and if the other things remain as before, one can carry on the argument exactly as above.[69]

Proposition 5. When any velocity is given and the ratio from which it arises is known, then the ratio that gives rise to any [other] velocity can be known, provided that the ratio of the velocities is known.

Let A be a known ratio from which a given velocity arises, and let there be another velocity, commensurable to this given velocity, which arises from ratio B. I say that B can be found, and argue as follows: A ratio of ratios, namely A to B, is like a ratio of velocities, by the first supposition; and the ratio of velocities is known, as is assumed; therefore, ratio A to B is known. But A is known by hypothesis, and consequently B can be made known, by the ninth supposition [of this chapter]; and this is what has been proposed.[70]

In this way we can find the ratio of a power to a

resistance, namely the ratio from which a velocity arises, provided the power is not separated from its mobile or resistance, or [simultaneously] applied to diverse mobiles.

For example, let A be a double ratio from which a given velocity arises, and let some other velocity which arises from ratio B be quadruple to it. Therefore, ratio B is quadruple to A, namely to a double ratio. Then, as was taught in the first chapter, I shall increase the double ratio to quadruple of itself and shall get a sedecuple ratio. Therefore, B was a sedecuple ratio.[71]

Proposition 6. If a ratio from which a velocity arises is known and rational, give its prime numbers; but if irrational, find two lines the greater of which is as the power of the mover, the lesser as the resistance of the thing moved. . . .

66. To summarize this example, let $D = A/C$, $E = A/B$, and $F = B/C$, so that $D = E \cdot F$ (where A is motive force and B, C are resistances). If $B/C = 2/1$, and $A/C = (A/B)^{3/1}$, then $D = (E)^{3/1}$. Therefore, $E = (D)^{1/3}$, and $F = (D)^{2/3}$, so that $D = (F)^{3/2}$, and $E = (F)^{1/2}$. By hypothesis $F = B/C = 2/1$, which determines that $E = (2/1)^{1/2}$, and $D = E \cdot F = (2/1)^{1/2} \cdot (2/1) = (4/1)^{3/4}$. Ratios D and E are now known and are $(4/1)^{3/4}$ and $(2/1)^{1/2}$, respectively; furthermore, $E = (F)^{1/2}$. When applied to the ratios of force and resistance, we see that $A/C = (4/1)^{3/4}$ and $A/B = (2/1)^{1/2}$, from which it follows that the ratio of velocities, or "ratio of ratios," is $3/1$, since $(4/1)^{3/4} = (2/1)^{3/2}$, and $(2/1)^{3/2} = [(2/1)^{1/2}]^{3/1}$, or $A/C = (A/B)^{3/1}$, where $3/1 = V_C/V_B$ (V_C representing the velocity of C and V_B the velocity of B). Using F's, R's, and V's for force, resistance, and velocity, this proposition utilizes the basic relation $F_2/R_2 = F_1/R_1 \cdot R_1/R_2$, where $F_2 = F_1$, $R_1 > R_2$, and $F_2/R_2 = (F_1/R_1)^{V_2/V_1}$. Since ratios R_1/R_2 and V_2/V_1 are assumed known, it is easy to determine ratios F_2/R_2 and F_1/R_1.

67. Oresme insists here that simply because the ratio of mobiles or resistances is $B/C = 2/1$, it does not follow that A will move C with twice the velocity with which it moves B. This will happen only if $A/B = 4/2$ when $B/C = 2/1$, in which event $A/C = A/B \cdot B/C = 4/2 \cdot 2/1 = 4/1$ (see n. 58).

68. An example representing this part of the proposition is summarized in note 66.

69. Replacing C, B, and E with F's, R's, and V's (see n. 66), we are told here that if there is only one mobile or resistance and two motive powers or forces, the two unknown ratios of force to resistance can be determined. For if $F_2/R_2 = F_2/F_1 \cdot F_1/R_1$, with $R_2 = R_1$, $F_2 > F_1$, and $F_2/R_2 = (F_1/R_1)^{V_2/V_1}$, then should V_2/V_1 and F_2/F_1 be known, the two ratios F_2/R_2 and F_1/R_1 can be determined.

70. This can be formulated as $A = (B)^{m/n}$, where m/n and A are known, so that B is easily found. Since A and B represent ratios of force to resistance, the exponent, m/n, which is a "ratio of ratios," is also the ratio of velocities.

71. If $A = 2/1$, and $B = (A)^{4/1}$, then $B = (2/1)^4 = 16/1$.

Proposition 7. If any velocity should arise from a rational ratio that has no mean proportional number or numbers between its prime numbers, then every lesser velocity that is commensurable to it arises from an irrational ratio; and similarly every greater [velocity that is commensurable but] not multiple to it [arises from an irrational ratio].

Since, by the first supposition, a ratio of velocities is like a ratio of ratios, and, by the fifth proposition of the second chapter, any ratio [that has no mean proportional number or numbers] between its prime numbers is incommensurable to any lesser [rational] ratio and to any greater that is not multiple to it, then any velocity which arises from such a ratio will be incommensurable to any lesser velocity which arises from a rational ratio.[72] Therefore, if any smaller velocity is commensurable to such a velocity, it will have been produced by an irrational ratio;[73] and this also applies to any greater velocity which is not multiple to it.

For example, let *B* be a velocity produced by a double ratio. I say that every velocity commensurable to it arises from an irrational ratio; and every lesser velocity arising from a rational ratio is incommensurable to it. However, not every lesser [velocity] which is incommensurable to it arises from a rational ratio, since some irrational ratio less than a double ratio may be commensurable to it and another incommensurable.[74]

Furthermore, every velocity greater than *B*, but not multiple to it, that arises from a rational ratio is incommensurable to *B*; and every [greater velocity] that is commensurable, but not multiple, to *B* arises from an irrational ratio. For example, if a league were traversed in one day by velocity *B*, and a league-and-a-half by velocity *C*, I say that *C* arises from an irrational ratio.[75] For indeed these [velocities] are related as ratios of ratios, which is clear from the fifth proposition of the second chapter and other places in the third chapter. Thus, by means of the first supposition of this chapter and the propositions given in the third chapter, one who understands can demonstrate many propositions about velocities.

For example, every velocity that arises from a multiple ratio is incommensurable to any other velocity not arising from a multiple ratio.[76] But even if each velocity were produced by a multiple ratio, it does not follow that they are commensurable.[77] This is obvious from the first supposition and third proposition of the third chapter. Moreover, by the fifth proposition of the same third chapter, it is clear that every velocity that arises

from a superparticular ratio is incommensurable to any other arising from a superparticular ratio.[78]

And so, in a like manner, one or more [additional] propositions about velocities can be demonstrated from any proposition of the third chapter,[79] but in order to be briefer so that I may move on, I omit all the propositions up to the tenth of the third chapter, from which latter proposition, however, such a proposition is [now] elicited.

72. In Proposition 5 of chapter 2, Oresme had shown that if A is a rational ratio that has no mean proportional number between its prime numbers and B is a smaller rational ratio, then $B \neq (A)^{m/n}$, where m/n is a rational exponent. As applied to ratios of force and resistance, let $A = F_2/R_2$, and $B = F_1/R_1$, so that $F_2/R_2 \neq (F_1/R_1)^{V_2/V_1}$, where $V_2/V_1 = m/n$. In the same fifth proposition, Oresme had also shown that if $C > A$, and $C \neq (A)^n$ (n is any integer), ratios C and A must be incommensurable, since they would be related by an irrational ratio. For Oresme, two ratios (both may be rational or irrational; or one rational, the other irrational) are commensurable when they can be related by a rational exponent; they are incommensurable when they cannot be related by a rational exponent (see also n. 51).

73. If $F_2/R_2 = (F_1/R_1)^{V_2/V_1}$, where V_2/V_1 is a rational exponent and F_2/R_2 is a ratio lacking a mean proportional number between its terms, then F_1/R_1 is irrational. For example, if $F_2/R_2 = 2/1$, then should $F_1/R_1 = (2/1)^{1/2}$, it follows that $V_2/V_1 = 2/1$, since $2/1 = [(2/1)^{1/2}]^{2/1}$.

74. In note 73 we saw that $F_1/R_1 = (2/1)^{1/2}$ was commensurable to $F_2/R_2 = 2/1$. But if $F_1/R_1 = (2/1)^{m/n}$, with m/n irrational, then F_1/R_1 is incommensurable to F_2/R_2.

75. Since $S_C/S_B = V_C/V_B = 3/2$ when $T_C = T_B$ (where S is distance, V, velocity, and T, time), it follows that $V_C = (3/2)(V_B)$. Now, in the preceding paragraph it was assumed that V_B arises from a ratio of force to resistance of $2/1$, so that V_C must be generated by the irrational ratio $(8/1)^{1/2}$, since $(8/1)^{1/2} = (2/1)^{3/2}$.

76. A multiple ratio is of the form $n/1$, where n may be any integer. Here Oresme is saying that if p/q is a ratio of numbers in its lowest terms and $p > q > 1$, then $n/1 \neq (p/q)^{m/r}$, where m/r is a ratio of integers, that is, a rational exponent. Thus any two ratios of force to resistance related in this manner will produce a ratio of velocities that is irrational.

77. For example, 3/1 and 2/1 are multiple ratios but incommensurable, since no rational exponent can relate them. Any ratios of force to resistance related as these two ratios will produce incommensurable velocities.

78. A superparticular ratio is of the form $(n + 1)/n$, where n is any integer (see p. 6, n. 17). If in one superparticular ratio $n = 2$ and in another $n = 3$, those two ratios are not relatable by a rational exponent, and therefore any ratios of force to resistance related in the same way will give rise to incommensurable velocities represented by an irrational ratio.

79. There are ten strictly mathematical propositions in chapter 3 plus an eleventh practical proposition illustrating, by way of example, the tenth proposition.

When two velocities have been proposed whose ratio is unknown, it is probable that their ratio is irrational and that these velocities are incommensurable. And when more velocities are proposed, it is exceedingly probable that any [one of them] would be incommensurable to any [other of them]. And as more are proposed it must be considered even more probable, since it has frequently been said, with reference to the first supposition, that a ratio of velocities is as a ratio of ratios. But when one unknown ratio of ratios has been proposed, it is probable that it is incommensurable and that those ratios would be incommensurable, because if more ratios of ratios are proposed, it is most probable that any one of them would be irrational, since there are fewer rationals among ratios of ratios, just as there are fewer cube numbers among numbers, which was stated in that [very] tenth proposition of the third chapter. For this reason, the same thing must be said about ratios of velocities, namely that when two velocities have been proposed [whose ratio is unknown, it is probable that their ratio is irrational and that these velocities are incommensurable], which has been proposed.[80]

Now, any ratio of magnitudes [or distances] would be just like the ratio of velocities with which those magnitudes [or distances] were traversed in the same time or in equal times.[81] And a ratio of times is just like a ratio of velocities when it happens that equal distances are traversed in those times, and conversely, which is clear from the sixth [book] of the *Physics* [of Aristotle]. [From what has been said] this proposition follows: When there have been proposed any two things whatever acquirable [or traversable] by a continuous motion and whose ratio is unknown, it is probable that they are incommensurable. And if more are proposed, it is more probable that any [one of them] is incommensurable to any [other]. The same thing can be said of two times and of any continuous quantities whatever.

For example, let there be two unequal motions which last through an equal time and whose ratio is unknown. I say that it is probable that the magnitudes [or distances] traversed would be incommensurable,[82] as would any other magnitudes [or distances] that are traversed or traversable by these motions.

Now, if the ratio of times were unknown between two motions traversing equal magnitudes [or distances] in unequal times, it is probable that the times of this ratio would be incommensurable;

and if there were more times [one would argue] as before. Therefore, it must be said that it is probable that a day and the solar year are incommensurable times.[83] And if this be so, it is impossible to discover the true length of the year, for it is just as if the year should last through a certain number of days and one part of a day which is incommensurable to a day. And one can make similar remarks about other things.

From all the things which have been said, this proposition also follows: When two motions of celestial bodies have been proposed, it is probable that they would be incommensurable, and most probable that any celestial motion would be incommensurable to the motion of any other [celestial] sphere,[84] but if the opposite of this were true, it could not be known. And this seems especially true since, as I shall declare afterward, harmony comes from incommensurable motions.

Now that I have declared that any celestial motion might be incommensurable to any other celestial motion, many very beautiful propositions

80. Since this paragraph represents a straightforward application of Proposition 10, chapter 3, to ratios of velocities, the reader is urged to examine that proposition in Selection 25. Oresme is saying here that in the proportional relation $F_2/R_2 = (F_1/R_1)^{V_2/V_1}$, if V_2/V_1, the ratio of velocities, is initially unknown, it is probable that it would be irrational and the velocities incommensurable, so that the ratios F_2/R_2 and F_1/R_1 would also be incommensurable. The mathematical basis for assuming the probability of their incommensurability was presented by Oresme in Proposition 10, chapter 3, of this same treatise.

81. In the basic relation $F_2/R_2 = (F_1/R_1)^{V_2/V_1}$ it is obvious that $V_2/V_1 = S_2/S_1$ when $T_2 = T_1$ (where S is distance and T, time), so that what was said about the probable incommensurability of the ratio of velocities (see n. 80) applies also to the ratio of distances traversed in equal times. At the end of this paragraph, Oresme extends his argument to any continuous magnitudes whatever.

82. Since $V_2/V_1 = S_2/S_1$ when $T_2 = T_1$, it follows that if the ratio of velocities is probably irrational so also is the ratio of distances traversed.

83. Oresme stated this in a number of other treatises.

84. Here Oresme extends to celestial motions what was said in preceding paragraphs about terrestrial motions. His mathematical arguments in support of the probable incommensurability of celestial motions were utilized to discredit astrological prediction, since he could argue that if two celestial motions are probably incommensurable, it would be impossible for astrologers to make exact predictions of celestial positions, and all consequences based upon imprecise knowledge of these positions would be necessarily uncertain and questionable. Not even the most accurate astronomical data could remedy this inherent mathematical indeterminacy.

that I arranged at another time follow, and I intend to demonstrate them more perfectly later, in the last chapter,[85] among which will be these.

One [of them] is: If a perfect [or total] eclipse of the moon should occur only once—and this could happen—it is impossible that a like conjunction should have happened at another time and impossible that it happen again during an eternal future time to come. And I always understand this "naturally speaking," and have even assumed an eternity of motion and the principles put forth by Aristotle in the second [book] of his *On the Heavens* and in other places.

Another proposition is: If two planets, with respect to longitude and latitude, should be conjuncted once in a point, they will never again be conjuncted.

Another proposition is: If three planets were conjuncted with respect to longitude so that they were on the same meridian, it is impossible for them to conjunct again even if they were moved eternally. Thus they were in conjunction in only one way through perpetual times.

Another proposition—I shall not relate any more for now—will be this: In any instant it is necessary that celestial bodies be so related that in any moment there will be a configuration such that there never was a similar one before, nor will

there be one after in all eternity, just as it has been written in the twelfth [chapter] of Daniel: "And a time shall come such as never was from the time that nations began even until that time." And the Lord disposeth through these celestial bodies, [for] as the poet says: "The First [Cause] rules all things through the heavens as He wishes; [and the heavens] are the organs of the Prime [Being], the instruments of the Supreme [Being]."

Indeed, along with these I shall demonstrate many other no less beautiful propositions based on the same principle, [but] with a few other more probable principles assumed from those which have [already] been demonstrated. Many errors about philosophy and faith could be attacked by the use of these [propositions], as [for example], that [error] about the Great Year, which some assert to be 36,000 years, saying that celestial bodies were in an original state and then return [to it in 36,000 years] and that past aspects are arranged again as of old;[86] and other errors of this kind which people have been accustomed to reject not by demonstrations but rather by strife and verbosity. But it is better to attack philosophers with philosophy and mathematicians with mathematics, just as Goliath was struck dead by a suitable weapon, and so also truth is made manifest and falsity destroyed. This, then, ends the fourth chapter.

52 THE DEVELOPMENT AND CRITICISM OF ATOMISM IN THE LATER MIDDLE AGES

Translations, introductions, and annotation by John E. Murdoch

To Aristotle's critical eye, the atomism of Leucippus and Democritus was, although admittedly among the most encouraging of the pre-Socratic systems of natural philosophy, inadequate to the task of yielding a satisfactory explanation of the physical world. Yet his opposition to these two of his predecessors was not Aristotle's only criticism of atomistic conceptions. For in Book VI of the *Physics* he clearly demands that no continuum, be it space, time, or motion, be composed of indivisibles; the atomistic structure, not of sensible bodies but of magnitude in general, was, in Aristotle's view, not permissible. Thus, if in rejecting the hypothesis of Leucippus and Democritus, Aristotle

Celestial Motions), which are two treatises devoted wholly to the subject of celestial commensurability and incommensurability (indeed there are also brief discussions of this subject in other of his works). For some propositions drawn from his *De commensurabilitate* see Selection 69. Oresme's objective in enunciating these four propositions in the following paragraphs was to emphasize that if the celestial motions are incommensurable, no celestial configuration or event can ever repeat exactly in all respects.

86. The concept of a Great Year as an unending sequence of equal time intervals in each of which the same exact celestial phenomena—and for some of its proponents even the very same terrestrial events—are repeated was rather popular (especially among some Stoics) during the Hellenistic and Roman Empire periods. Oresme calls it an error in philosophy and faith because in 1277 the bishop of Paris condemned the following proposition (one of 219 condemned; see Selection 13, no. 6). "That when all the celestial bodies have returned to the same point—which will happen in 36,000 years—the same effects now in operation will be repeated." To Christians, the unending cyclical repetition of events was repugnant, since it was contrary to their belief in a unique creation that must run its course but once.

85. These were never demonstrated as part of the *De proportionibus proportionum*, which terminates with the fourth chapter. However, the four specific propositions mentioned in the following paragraphs all appear in Oresme's *Ad pauca respicientes* and all but one in his *De commensurabilitate vel incommensurabilitate motuum celi* (*On the Commensurability or Incommensurability of the*

had disavowed a physical atomism by arguing in Book VI of the *Physics* for the absolute continuity of all extensive magnitudes, he had also, in effect, rigorously disassociated himself from the possibility of a mathematical atomism, from, to cite the most crucial examples, the possible composition of lines out of points or of time intervals out of instants. It was against this, against, as it were, the very substance of Book VI of the *Physics,* and not against Aristotle's dissatisfaction with Democritus, that the atomism of the later Middle Ages reacted. Moreover, given such a context for the development of atomism in the earlier fourteenth century, it was quite natural that this atomism be, in essence, a mathematical one. Mathematical not in the sense that atomistic conceptions were employed in anything like a mathematical system of analysis, but mathematical merely in the philosophical sense of the atoms at hand always being points or extensionless indivisibles. Thus it is not surprising that this late medieval atomism was not intended, as was that of Democritus or Epicurus, as any kind of general system which might cover or explain the natural world [Nicholas of Autrecourt was among the few scholastics, if not the only one, to propose physical atomism as a more plausible alternative to Aristotle's explanation of the world; see Selection 56.3, n.12—*Ed.*]. It was intended rather as a single facet of natural philosophy, designed simply to explain the structure of magnitudes, and specifically of space, time, and motion as magnitudes. In medieval terms, this atomism proposed an anti-Aristotelian answer to the question of the composition of continua. Indeed, Aristotle's opposition to the views the medieval atomist wished to espouse made the latter's task a predominantly defensive one.

The first object of his dissent was, of course, to render lame the highly effective (so they were usually regarded) arguments of Aristotle against the indivisibilist composition of continuous magnitudes. For in the opening chapter of Book VI of the *Physics,* Aristotle had made it clear that the constitution of continua out of indivisibles must necessarily falter, not merely because indivisibles cannot be continuous but chiefly because there is no manner in which they can touch or be in contact. In other terms, if one is to maintain the indivisibilist structure of a continuum, then one must be able to explain exactly how the component indivisibles are "connected to" one another. But, Aristotle cautioned, no "connecting relation" is possible which will, at the same time, also account for the extended continuum such indivisibles are supposed to com-

pose.[1] Viewed in such a light, the initial move of the fourteenth-century atomist was to circumvent Aristotle's argument by discovering and establishing a "relation of connection" for his indivisibles which did not fall prey to the by now traditional gambit of the Aristotelian attack.[2]

Should he be so fortunate as to have survived this stage of the battle (at least in his own eyes, if not those of his critics), the medieval atomist had no cause to rest or to expect eventual victory. Other arguments that were yet more formidable remained to be repulsed, and these were, at bottom, mathematical. For the scholastic, their origins could be found in the *Metaphysica* of Algazel. But they gained immeasurably in force and prestige when John Duns Scotus employed such mathematical arguments for the primary bastion of his criticism of indivisibilism in the second book of his *Opus Oxoniense*. From Scotus on, mathematical arguments against the atomist increased, in complexity as well as in number, at an almost exponential rate. In fact, the core of the substantial body of fourteenth-century literature comprising the treatises of these newly born atomists and their critics exemplifies the history of the debate, over such mathematical arguments and over the strictures pronounced by Aristotle in the sixth book of his *Physics* against mathematical atomism. The following selections are meant to illustrate this debate.[3]

1. This is, in effect, the substance of Aristotle's argument in *Physics* VI. 1.231a. 26–231b. 6. A medieval reformulation of just this argument of Aristotle is given below in the selection from William of Alnwick as the first of five arguments against the atomistic position of Henry of Harclay. The "confirmation" constructed by Averroes of this argument is also given in the passage from Alnwick, an addition which became as standard as the original argument itself in all medieval treatises dealing with the composition of continua.

2. Such a move is precisely what Harclay was concerned with when he replied to Aristotle's argument (see section 4 below).

3. Further details concerning other participants, both atomist and anti-atomist, in the fourteenth-century continuum debate, as well as with respect to the whole history of late medieval atomism, may be found in the following publications by John E. Murdoch: *Rationes Mathematice: Un aspect du rapport des mathématiques et de la philosophie au Moyen Age* (Paris: Université de Paris, 1962); "Superposition, Congruence and Continuity in the Middle Ages" in *Mélanges Alexandre Koyré* (2 vols.; Paris, Hermann, 1964), Vol. I: *L'aventure de la science,* pp. 416–441; "Two Questions on the Continuum: Walter Chatton (?), O.F.M., and Adam Wodeham, O.F.M.," *Franciscan Studies,* XXVI (1966), 212–288 [with Edward A. Synan]. Critical editions of the complete texts of the works from which the following selec-

1. Thomas Bradwardine (ca. 1290–1349)

Easily the most impressive and the most exacting refutation of the incipient atomism of the fourteenth century was Bradwardine's *Tractatus De continuo* (written, it would appear, sometime between 1328 and 1335). In form the *Tractatus* is geometrical, modeled—as was Bradwardine's later major theological work *De causa Dei*—on Euclid's *Elements*. It opens with 24 definitions and 10 suppositions or postulates and then moves on to demolish atomism of all sorts in the formulation and demonstration of 151 conclusions. Unfortunately, the work is far too systematic and interconnected to allow selection of brief passages illustrative of the substance of its criticism of indivisibilism. Yet the very brief selection given below is instructive in revealing Bradwardine's awareness of the then rising current of atomism and his view concerning the various forms which such an atomism could, and did, assume. The present translation has been made from my own as yet unpublished edition of the *Tractatus De continuo*.

TRACTATUS DE CONTINUO

If one continuum is composed in a certain manner out of indivisibles, any [continuum] is so composed, and if one is not composed of atoms, none is.

2. Algazel (1058–1111)

Although the repetition and refinement of Aristotle's arguments against atomism naturally formed the basis of any criticism of the rising indivisibilism of the later Middle Ages, a second group of arguments that became standard fare in any such criticism were six drawn from the *Metaphysics* of Algazel (as he was known to the Latins). The *Metaphysics* was a Latin translation of part of al-Ghazzālī's *Maqāsid*, a work which did not mean to give his own views but rather a summary of those of Avicenna. Thus, the arguments given below derive initially from Avicenna, although the Latin West knew them only under the name of *rationes Algazelis*. My translation has been made from J. T. Muckle's edition of *Algazel's Metaphysics* (Toronto: Institute of Medieval Studies, St. Michael's College, 1933), pp. 10–13, as well as from additional manuscripts.

Metaphysics
First Treatise

CHAPTER 3. ON THE DIVERSE VIEWS OF THE COMPOSITION OF A BODY

In order to understand this conclusion we must realize that there are five famous opinions concerning the composition of continua among ancient and modern philosophers. For certain [philosophers], like Aristotle, Averroes and most of the moderns, hold that a continuum is not composed of atoms, but rather of parts divisible without end. Others, however, hold with its composition out of indivisibles. But there are two variants [of this position]. For Democritus maintains that a continuum is composed of indivisible bodies. Others claim [it is composed] of points, and these [thinkers] fall into two groups. For Pythagoras—the father of this sect—Plato, and Walter the modern[4] contend that a continuum is composed of a finite number of indivisibles. Others, however, [believe in its composition] out of an infinite number [of indivisibles], and these [indivisibilists] are [again] twofold. For certain of them, like Henry the modern,[5] say a continuum is composed of an infinite number of indivisibles immediately joined [to one another]; others still, like [the Bishop of] Lincoln[6] [side with] an infinity of [indivisibles] which are mediate to one another.

Men have been of diverse opinions concerning the composition of bodies, and we can only learn the truth about this problem after we have revealed which of these opinions—and there are three of them—is the truest. Some have said that a body is composed of parts that are indivisible, both conceptually and physically, and are called atoms. Indeed, these thinkers have called these [indivisible

---

tions are drawn, as well as of other works directly relevant to this particular segment of fourteenth-century science, have been prepared by Murdoch and will shortly appear in a volume setting forth the history of the debate about continuity and atomism in the fourteenth century.

4. That is, the Franciscan Walter Chatton, who, presumably in 1323, included an exhaustive examination of the problem of atomism and continuity in the second book of his *Commentary on the Sentences*.

5. That is, Henry of Harclay; see section 4 below.

6. That is, Robert Grosseteste, who did not, strictly speaking, maintain the atomistic composition of continua, but who, on the grounds of several important implications within his work, was clearly held to have been an atomist by such thinkers of indivisiblist sympathy as Henry of Harclay.

parts] units and odd substances[7] and have claimed that a body is composed of them. However, others have said that a body is in no manner a composite thing but that it is a being that is one both in essence and by definition and that contains no multiplicity. And still others have held that a body is composed of matter and form.

[*Refutation of the First View*]

The first opinion of those who have said that a body is composed of odd substances is destroyed in six ways.

[1] The first[8] is that if one of these substances is placed between two others, then each of the two extreme ones [will either touch the same things in the middle one as the other [extreme one] does or something different [from that which the other extreme touches]. If it touches something different, then [the middle substance] is divisible, since that which one of the extremes occupies [in the middle substance] is not that which the other extreme occupies. On the other hand, it is, without doubt, impossible for [the two extremes] to touch the same thing [in the middle substance]. For if they did, it would follow that each of the extremes would penetrate the mean substance totally insofar as each has touched the whole [mean] because the latter has no totality [of parts] due to the fact that it is one. Yet one of the extremes has touched something belonging to the mean; therefore, it has touched the whole of it. Moreover, the other extreme has similarly touched the whole of it. Consequently, it follows that there is one place for three things and the mean substance. Alternatively, the mean substance lies separated between the two extremes and touches one of the extremes differently than it does the other, since it can only touch one in the same thing as it does the other if one of the [extremes] has penetrated the other. The same thing will occur if a third, fourth, or more [substances] be added. It necessarily follows, then, that a distance of a thousand parts is no greater than a distance of one part. But it is clear that this is inconsistent; for, on these grounds, nothing could be composed out of these [indivisible substances].

[2] The second [way of refuting this indivisibilist position] is to arrange five of the aforesaid substances in a line and to place two [other substances] over the two extremities of the line in this fashion:

—————   —————

————— ————— ————— ————— —————

There is no doubt that the intellect can conceive that these two superposed parts simultaneously

begin to move toward one another with equal speeds [and continue to do so] until they meet. If this be assumed, then each of them will cut a part from the middle substance [in the line]. Consequently, the middle substance is divided. And all this is true because if it is not, then it follows that the substances moving with equal speeds cannot meet, but rather, after they have begun to move and one of them has reached the second part of the line, this [moving substance] will remain at rest in that place until the other one reaches the third part of the line. But from whence comes the variety in this unequal motion, and why does this occur on the right rather than on the left when either one [of the moving substances] is equally capable of accomplishing this and both are equally capable of receiving motion?

[3] The third [manner of refutation] is to take two lines, both composed of six parts, one of them *AB*, the other *CD*, and let them be parallel and opposite one another in this fashion:

A		E	Z		B
C		H	T		D

Next, assume that two parts [of these lines] begin to move: one from *A* to *B*, the other from *D* to *C* until they are opposite one another, one over the other. Surely, [our two mobile parts] will first be opposite one another and one will be over the other; then, one will pass the other. Moreover, it is possible for them to move with equal speed. Then, if [each] substance [composing our two lines] is odd,[9] they cannot possibly meet. For their meeting can occur in only one of three ways: [a] They will

7. The Latin text reads *substantias impares* which is a mistranslation of what in Arabic was the standard term for atom or indivisible: *al-jawhar al-fard*. This should have been rendered *substantia simplex* (and hence whenever "odd substance" appears below, one should read "simple substance," or merely "atom"), but I have let the mistranslation stand, since it appears in an overwhelming majority of extant manuscripts, and since, consequently, it was just such an incorrect text that the Latins used and which, on occasion, in its erroneous form gave rise to criticism that would otherwise not have been possible. Thus, the Franciscan Gerard of Odo (apparently writing about 1326) had, as an atomist, grounds to object to the fourth argument (see below) of Algazel by claiming that: "He seems to contradict his own assumptions. For he explicitly says that an odd number of substances, that is, indivisibles, may be taken, and then goes ahead to take sixteen, an even, not odd, number of them."

8. This argument derives ultimately from Book VI of Aristotle's *Physics* (see n. 1).

9. See note 7.

meet in the two points *E* and *H*, and hence one traverses three parts of the line, the other two; or [b] they meet in the two points *T* and *Z*, and hence [again] one traverses three parts, the other two (and [in either of these two cases their] motion will not be equal); or [c] one will be in point *T* and the other in point *E* and both will traverse two parts; but *E* and *T* are not opposite. Therefore, their direct opposition and traversal by equal motion is destroyed, which is completely impossible. However, this impossibility arises only because we have assumed the substances to be indivisible and [our mobiles] come into opposition over a half of each of two [such indivisible substances]. For every length is susceptible of division into two equal halves. Therefore, the bisection of our line occurs in the middle of the substances and it is [only] in this way that [the mobiles] come to be directly opposed, one over the other.

[4] The fourth [manner of refuting this indivisibilist position] is to arrange sixteen continuously joined odd substances into a square figure, four in [each of] four lines, in this way [Fig. 1]:

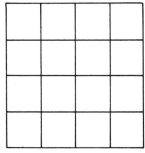

Fig. 1

Furthermore, though we illustrate these [substances] as being separate from one another, we nevertheless conceive them to be joined to each other in such a fashion that no space falls between them. Moreover, there is then no doubt that the four sides are equal; for each side is composed of four [equal] parts. Yet the diagonal is also composed of four. Hence, it follows that the diagonal is equal to each of the sides, which is impossible. For every diagonal cutting a square into two equal triangles is always greater than any of the sides. This is sensibly evident in all squares and is proved to be the case in geometry. Yet this becomes impossible if we assume the existence of odd substances.

[5] The fifth manner [of refutation] is that if we have placed a stick perpendicularly [in the earth] facing the rays of the sun, it will undoubtedly cause a shadow; then, by means of a ray, a straight line is projected from the end of the shadow cast by the top of the stick to the sun. Therefore, it is then necessary either for the shadow to move when the sun moves or it is not. A ray, moreover, is extended only in a straight line. If, then, the shadow does not move when the sun moves, a straight line will have two ends in a single direction, one in the place previously occupied by the sun and the other in the place to which the sun has subsequently been moved. But this is impossible. On the other hand, if the shadow does move, then, when the sun moves the distance of a single atom, the shadow will move less than an atom, which means an atom is divided. If, finally, the shadow moves the same extent as the sun, this is certainly an inconsistency. For the sun traverses a million miles while the shadow moves only so much as a hair's breadth.

[6] The sixth manner [of refutation] is the following: When a wooden or stone wheel revolves, the parts at its center clearly move less than the exterior parts because of the fact that a circle at the center [of a wheel] is smaller than a circle at its perimeter. However, when the exterior circle moves one atom, the center circle will either move less than an atom and hence an atom will be divided, or it will not move at all and it will follow that all parts of the wheel are separate so that some move, while others do not, but remain at rest. But our senses tell us this is false, for the parts of the wheel are not separated in any way at all. . . .

3. John Duns Scotus (ca. 1266–ca. 1308)

The following two geometrical arguments against the composition of a continuum out of indivisibles are drawn from Scotus' rather lengthy examination of the problem of "whether an angel can move from place to place with a continuous motion." Once proposed by Scotus, these two arguments never failed to grasp the attention of all concerned, both atomist and anti-atomist, with the structure of continua. The translation has been made from my own edition, based largely on MS Assisi, Biblioteca communale, 137 and occasionally corrected by MS Padua Anton. 173 and the Luke Wadding edition of Scotus published at Lyons in 1639.

ON BOOK II OF THE SENTENCES: THE OPUS OXONIENSE
DISTINCTION II, QUESTION 9

Nevertheless, this antecedent[10] is even more effectively proved by a pair of geometrical arguments or propositions. The first of them is the following: [1] In accordance with Postulate 2 of Book I of Euclid "one can construct a circle of any size one wishes about any center." Let there be constructed, then, two [unequal] circles about some given center A [Fig. 2]; call the smaller circle

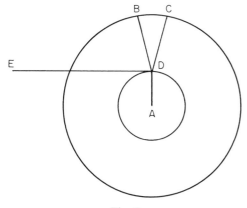

Fig. 2

D, the larger one B. If the larger circumference is composed of points, take two of them immediately next to each other and call them B and C. Next, draw straight lines from A to B and C in accordance with Postulate 1 of Book I of Euclid (which reads "To draw a straight line from point to point, etc."). The straight lines so drawn will pass through the circumference of the smaller circle at right angles. I ask, then, whether they will cut it in one and the same point or in different ones. If in different ones, there will then be as many points in the smaller circle as in the larger one. Yet it is impossible for two unequal quantities to be composed of parts equal in both magnitude and number. [And this applies to our case,] since one point does not exceed another in magnitude and there are the same number of points in the smaller circumference as there are in the circumference of the larger circle. Therefore, the smaller circumference is equal to the larger one, and consequently, the part to the whole. If, on the other hand, the two straight lines AB and AC cut the smaller circumference in the same point, call it D; then erect a straight line DE over AB, cutting it at point D, which, by III.15 of Euclid,[11] also touches the smaller circle. Now, by I.13 of Euclid,[12] DE forms two right angles, or [two angles] equal to two right angles, with AB; furthermore, by the same I.13, DE will also form two right angles, with line AC (which is assumed to

be a straight line). Consequently, $\angle ADE$ together with $\angle BDE$ is equivalent to two right angles. By a parity of reasoning, $\angle ADE$ together with $\angle CDE$ is equivalent to two right angles. Yet any pairs of right angles are equal, on grounds of Postulate 3 of Book I of Euclid; therefore, if we take away [the angle these two pairs] have in common (that is, ADE) the remainders will be equal; consequently $\angle BDE$ will be equal to $\angle CDE$ and thus the part will be equal to the whole.

[*Objection*]

An opponent might reply to this that DB and DC do not enclose an angle, since if they did, a base could be drawn between B and C under the angle, which is contrary to that assumed, since B and C were held to be immediate points. Therefore, the claim that $\angle CDE$ functions as a whole relative to $\angle BDE$ as part is to be denied, since nothing from $\angle CDE$ is added to $\angle BDE$ because of the fact that there is no angle between B and C about their juncture at point D.

[*Reply to the Objection*]

This response seems absurd on two counts. First, because in denying that there is an angle at the point where two [straight] lines—which are extended over a surface and not directly applied to each other—meet, it contradicts the definition of an angle given in Book I of Euclid. Secondly, because in denying that a line can be drawn from one point to another, it simultaneously denies Postulate 1 of Book I of Euclid. Nonetheless, even though these absurdities ensue, they are not held to be inconsistencies, because they follow from that proposed [by our opponent]. In view of this, I argue against the response in another way: $\angle CDE$ includes the whole of $\angle BDE$ and, though you impudently claim that it adds no angle, adds at least a point [to BDE]; yet a point is, on your grounds, a part; therefore, $\angle CDE$ adds some part to $\angle BDE$ and functions, consequently, as a whole relative to the latter. Our assumption is evident no matter which way one interprets an angle. For either: (1) an angle

10. Namely, that a continuum cannot be composed of indivisibles.

11. That is III.16 of the Greek Euclid, the relevant portion of which reads: "The straight line drawn at right angles to the diameter of a circle from its extremity will fall outside the circle. . . ."

12. Which reads: "If a straight line set up on a straight line make angles, it will make either two right angles or angles equal to two right angles."

is held to be the space intercepted between the lines [enclosing it] without counting these lines themselves, and then the first point of line *DB* beyond the smaller circumference will not belong to ∠*BDE* but will belong to ∠*CDE*; or (2) an angle includes, in addition to this intercepted space, the enclosing lines as well, and then the first point of line *DC* beyond the smaller circumference will not belong to ∠*BDE* but will belong to ∠*CDE*. Consequently, from either point of view, ∠*CDE* adds a point to ∠*BDE*.

Nor can the principal demonstration be opposed by pretending that the two [straight] lines do not begin to diverge from one another in the smaller circumference, but rather at some other point closer to, or further from, the center. For I shall draw the smaller circumference wherever you assume this occurs.

The second part [of our argument]—to wit, that the smaller circumference is not cut in just one point if it is cut by both of the lines—would need no proof were it not for the arrogant obstinacy of our opponent. For it is evident enough that if one continuously and directly extends one and the same line, it will never end in two points in a single direction. And if this is conceded to be manifestly true, that which is proposed is immediately evident from the chain of reasoning in the first part [of our argument].

[2] The second [geometrical] proof is based on Propositions 5 and 7 of Book X of Euclid. For the fifth proposition says that the ratio of all commensurable quantities to one another is as the ratio of some number to some number, and consequently, as the seventh proposition asserts, if several lines are commensurable, their squares will stand to one another as one square number to some other square number. However, the square of a diagonal [of a [square] does not stand to the square of the side as some square number to another square number; therefore, the line which was the diagonal of the square will not be commensurable to the side of the square. The minor premise is evident from the penultimate proposition of Book I [of Euclid's *Elements*], because the square of the diagonal is twice the square of the side insofar as it is equal to the squares of two sides; yet no square number is twice another square number, as is evident if one runs through all squares arising from any roots whatsoever.

From the foregoing, the following conclusion is evident: The diagonal is asymmetrical, that is, incommensurable, with the side. However, if these

[two] lines were composed of points, they would not be incommensurable, for the points of one would not stand to the points of the other in any ratio of numbers. And it would not only follow that the lines would not be incommensurable, but that they would even be equal, which plainly runs contrary to our senses.

Proof of the consequence: Take [Fig. 3] two

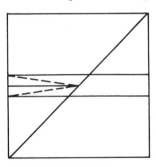

Fig. 3

immediate points in one side and two other immediate points directly opposite them in the facing side; draw two straight lines between these two pairs of points parallel to the base [of the square]; these lines will cut the diagonal. I ask, then, whether [they cut it] in immediate points or mediate ones. If in immediate points, then there are no more points in the diagonal than in the side and, hence, the diagonal is no greater than the side. If [they cut the diagonal] in mediate points, then I take a mean point falling between these two mediate points of the diagonal; it will belong to neither of the two originally given [parallel] lines. Then, by I.31 [of Euclid],[13] I draw a line from that mean point parallel to both of the given lines. This latter parallel is drawn continuously and directly in accordance with the second part of Postulate 1 of Book I [of the *Elements*]. It will cut the side, yet will do so in neither of its given points but rather between them both. If this were not the case, it would meet a line with which it is, by hypothesis, parallel, which is contrary to the definition of parallel (the last definition set down in Book I [of Euclid]). Therefore, there is a mean point between those two points in the side that were assumed to be immediate. This follows from the fact that it was claimed that there was a mean point between the points in the diagonal [arising from its intersection with out first two parallels]. Thus, the denial of the consequent implies the denial of the antecedent; therefore,

13. Which reads: "Through a given point to draw a straight line parallel to a given straight line."

and so on. Indeed, one should even say in general that the whole of Book X of Euclid overthrows the composition of lines out of points. For, given such, there would be no irrational or surd lines at all, which nevertheless, are Euclid's principal occupation in Book X, as is evident from the many species of irrational lines he points out in that place. . . .

4. Henry of Harclay (ca.1275–1317) and William of Alnwick (ca. 1270–1333)

Henry of Harclay, Chancellor of Oxford University in 1312, was, it appears, the first thoroughgoing atomist and, one can add, adherent of the existence of an actual infinite, in the later Middle Ages. Although his long-lost treatise on the infinite and the continuum has recently been discovered, his views concerning the latter are given below not from that work but rather from the second socalled *Determinatio* of one William of Alnwick, Franciscan, editor of the works of his teacher Duns Scotus and, it seems, first critic of Harclay's atomist utterances. Inasmuch as Alnwick reports Harclay's arguments almost verbatim (he even takes some of his objections to Harclay word for word from possible objections Harclay himself formulated), the substance of the latter's contentions are most accurately reported in the work of his Franciscan critic.

Both Harclay's treatise and Alnwick's *Determinatio II* are rather lengthy *quaestiones,* so only a brief portion of the latter has been presented in what follows. However, the selection has been made with a view to giving: (1) Harclay's two major positive arguments for his atomistic stance; (2) three of the most often employed arguments against any atomist doctrine; (3) Harclay's replies to these three frequently occurring arguments; and (4) Alnwick's criticism of the initial two positive arguments of his opponent. Harclay's replies to these criticisms of Alnwick, as well as Alnwick's counter-replies have been omitted; so also Alnwick's criticism of Harclay's response to the three arguments noted in (2) above. The translation has been made from my own as yet unpublished edition, based directly upon manuscript versions.

Determinatio II

ARTICLE TWO OF THE QUESTION

In the second article of the question[14] it remains to see whether continuous quantity is composed of indivisibles. And I shall proceed as follows in this article: First, I shall set forth the arguments of the aforesaid doctor [that is, Henry of Harclay] by means of which he holds a continuum to be composed of indivisibles, for example, a line of points.

Secondly, I shall offer arguments which prove with greater evidence and certainty that a continuum is not composed of indivisibles. Thirdly, I shall give the replies on which he bases his refutation of these arguments and show that these refutations are unsatisfactory. Fourthly, I shall refute the arguments which he constructs to support his point of view.

The Opinion of Master Henry, Chancellor of Oxford: That a Continuum Is Composed of Indivisibles

(1) Relative to the first point [of our investigation], this doctor has used two arguments to show that a continuum is composed of indivisibles. These arguments have [even] moved him to asserting this fact. The first of them has the same force as the second argument given above among those set forth above in the [introductory] dispute. It is this: God actually perceives or knows the first inchoative point of a line and any other point capable of designation in the same line. Therefore, God either perceives that a line can fall between this inchoative point of the line and any other point in the same line, or He does not. If not, then He perceives point immediate to point, which is that proposed. If He does, then, since points can be assigned in the mean line [which falls between the first inchoative point and the other, random, point], these mean points would not be perceived by God, which is false. The consequence is evident: For, by hypothesis, a line falls between the first point and any other point of the same line perceived by God, and consequently there exists some mean point between this [first] point and any other point perceived by God; therefore, the mean point is not perceived by God.

(2) His second argument is the following: God can annihilate the end point of a line without corrupting the line (as will be proved). This done, the line actually possesses a terminus between which and the corrupted point there was no mean line.

14. The question itself reads: "Whether there are potentially more parts in a greater continuous quantity than in a smaller one," while the first article dealt with the problem of whether one infinite could be greater than another.

For otherwise some line would have been corrupted, the opposite of which will be shown. In
the same way, He could corrupt another point
without corrupting any line or part thereof, and it
will necessarily follow that point is immediate to
point. Proof of the major premise: Let some continuous line be broken at the middle, [the two
halves] then create two actual points; let [these two
halves] be joined to one another, then [these two
points] will exist potentially. This accomplished, it
is possible that numerically the same point which
actually existed before and was the beginning of
the right part of the line is now the end of the left
of the line. Proof: A line can be broken at any one
of its points in such a way that any point can indifferently be the beginning of one or the other [of the
resultant parts], though not of [both] at the same
time. Therefore, the point which was the beginning
of some other line can now be the end of this line,
and conversely. Hence, as one line can be corrupted
while the other remains an integral whole, God can,
if He corrupts one line, corrupt the point which
was the beginning of the other line without any
part of this [other] line being corrupted.

[*Five Arguments against Henry's Opinion*]

Secondly, we must set forth contrary arguments
which prove that a continuum is not composed of
indivisibles. There are five arguments.[15]

(1) The first of these is that of Aristotle in [Book]
VI of the *Physics;* it is the following: If a continuum were composed of indivisibles, it would be
necessary that the addition of indivisible to indivisible would cause an increase in size. The consequent is impossible, thereore [so also is] the antecedent. The consequence is self-evident, since, if
the addition of indivisible to indivisible would not
cause an increase in size, there would [upon such
addition] result but a single indivisible and such a
procedure would never yield a continuum composed of the [indivisibles in question]. Therefore,
opposite follows from opposite. Aristotle proves
the consequent to be false as follows: If indivisible
added to indivisible were to create an increase in
size, it would be necessary that one indivisible
touch another; therefore, [they will touch] either
according to their wholes or according to a part.
Not according to a part, since [an indivisible] does
not have a part (for if it did, it would be divisible
and not indivisible). If [they touch] according to
their wholes, then no increase in size will occur,
since whole touching whole causes no increase in
extension. And we must argue in the same fashion

with respect to an infinity of additional indivisibles.

This reasoning is corroborated by Averroes'
argument in the same place in [Book] VI of the
Physics, Comment 2, where he argues as follows:
"When one thing wholly touches another, superposition occurs, and no magnitude that did not
exist previously results from superposition. Hence
no magnitude in breadth is caused by the superposition of line to line, as they are only superposed
breadthwise, in which [dimension] a line is indivisible; similarly, the superposition of surface to
surface does not create depth. And since a point is
indivisible in every dimension, [a point wholly
touching, or superposed to, a point] does not
create something of greater size that possesses
parts." These are the words of Averroes.

(2) The second argument [against Harclay's
view] is this: If a line were to be composed of
points, the diagonal of a square would be equal to
the side, which is false and contrary to our senses.
Proof of the consequence: The sides of a square
are equal and, you maintain, composed of points,
hence there are an equal number of points in each.
Therefore, let lines be drawn from each and every
point (infinite in number) of one side to each and
every facing point of the opposite side. These lines
will be mutually parallel, and neither point nor
line can fall between them since they are drawn
parallel from points [in the side] that are immediately next to one another. Then [one argues] as follows: These lines pass through the diagonal and
cut it. Hence, they either cut it in every one of its
points, or they do not. If in every one, then there
are an equal number of points in the diagonal and
the side, and consequently the side and diagonal
are equal. It may be claimed that the lines which
pass through the diagonal do not touch every one
of its points, and that thus some point of the diagonal falls between these lines. Yet, since these
lines are parallel, it follows that howsoever much
falls between them in one place just as much falls
between them in another place, and hence a point
can fall between these lines where they are joined to
the side: but this is contrary to that assumed, since
these lines exhaust all the points of [each] side.
Therefore, no point can fall between two lines
drawn [parallel] from two immediately situated
points in the side.

(3) The third argument has to do with the indivisible instants in a continuous, successive time

15. The fourth and fifth arguments, here omitted, are
those of Algazel dealing with the rotating wheel and the
shadow-casting stick (see section 2 above, p. 316).

which is not composed of instants; for if it were, one 1-foot quantity would be greater than another 2-foot quantity, and a part would be greater than, or equal to, the whole. Proof of the consequence: Take a slow mobile which is moved for two days through a space of 1 foot, and let a quick mobile be moved for one day through a space of 2 feet. Then take the assertion of Aristotle in [Book] VI of the *Physics:* When a mobile is moved, it is in a space equal to itself in any designated or designatable instant, and from one instant to another it is in one and another space, for otherwise it would be at rest and not be continuously moved. Therefore, as many instants as one can assign in the time during which a mobile is moved through some space, so many spaces equal to the mobile can be assigned in the same magnitude, and so many points terminating these spaces. Hence, if time be composed of instants, then, as you claim there are more instants in two days than in one, there are more points in a 1-foot space traversed in two days than in a 2-foot space traversed in one day, which is impossible. . . .

Replies of the Aforesaid Doctor to these Arguments

The third task left to fulfill in this article is to note this doctor's replies to the aforesaid arguments and to show that these replies are unsatisfactory. . . .

[*To the first.*] To the first argument drawn from [Book] VI of the *Physics*—that the addition of indivisible to indivisible causes no increase in size, since whole touches whole—he replies that whole touching whole can occur in two ways: either in the same local position, or in distinct positions immediately next to one another. In the first way superposition occurs, and no increase in size results, as Averroes [himself] argues. This is also evident [from the following consideration]: An infinite number of points terminating an infinity of lines running from the circumference [of a circle] to its center do not, since they are superposed at the center in the same local position, cause an increase in size. In the same way, the points of two intersecting lines occupy the same position at the point of intersection and do not cause an increase in size. Hence, this [lack of increase in size] is not due to the fact that an indivisible is added to an indivisible in just any manner, but rather because they are added according to the same position. However, when they are added according to distinct positions, then [the addition] can create an increase in extension. Whence the points of two lines which touch in a common boundary of contact have distinct positions and therefore cause an increase in size.

He first explains this reply by means of an example dealing with numbers: A unit, even though it be indivisible, nonetheless causes an increase in numerosity when added to another unit, not insofar as the second unit is added to the first relative to the same numbered thing (for in such a case an infinite number of units do not make more than one), but when [one unit] is added to another relative to another numbered thing. But points are related to an increase in the genus of continuous quantity just as units are to an increase in the genus of discrete quantity. And thus point added to point in the same position causes no increase in size, but [if added] in a distinct position it does.

He also explains this by an appeal to the fact that if a surface or a body be added to another [surface or body] according to the same position, no increase in size results, but only [when the addition takes place] according to diverse positions. For a single body which possesses every dimension can be added or conjoined to another body without there occurring an increase in size. For he explains at length that "quantitative dimension no more prevents one body from simultaneously existing [in a single position] with another body than a point with a point." From these considerations he infers that not merely a point but also a line or a body creates an increase in extension only when it is applied [to another thing] according to diverse positions. Thus he claims that if two indivisibles—points, for example—be applied to one another according to diverse positions, they will cause an increase in size. "Yet the points or lines of two touching lines or surfaces, or the surfaces of two touching bodies, have corresponding distinct positions, just as do the bodies of which they are the surfaces [or the surfaces and lines of which they are, respectively, lines and points]. For just as the places and positions of bodies are distinct and separate, so are the termini of these bodies." Therefore, he says that "point added to point according to a distinct position does cause an increase in size, but that such addition of point to point will not attain a quantity perceptible or comprehensible by us unless there occurs an addition of an infinite number of points (just as not every part in a continuum creating an increase in size is perceptible to us)." With this, the reply to the argument of Aristotle and Averroes in [Book] VI of the *Physics* is evident. . . .

To the second argument. To the second argument about the side and diagonal of a square, this doctor replies that a point falls according to oblique position, and not according to right-angled position, between the parallel lines cutting the diagonal, and that therefore a point [in the side] does not fall between the lines drawn between the sides, because these points [in the side] have right-angled positions [relative to the parallel lines]. And this is the way in which there will be more points in the diagonal than in a side, since a point in the diagonal, but not in the side, falls between the [parallel] lines. . . .

To the third argument. He replies to the third argument dealing with time and the. instants of time by making a distinction: that time can be considered as it is discrete, or as it is continuous. If [it be considered] as continuous, he claims that instants are multiplied in time according to the *mutata esse*[16] of a mobile [moving] over a magnitude. And, since there are more points in a greater magnitude than in a smaller one, it follows that if a magnitude of 2 feet is traversed in one hour, and a magnitude of 1 foot in a hundred years, there will be more instants in one hour than in a hundred years. But the converse holds in time [considered] insofar as it is discrete. . . .

[Replies] to the Arguments Supporting [Henry's] View

Fourthly, it remains in this article to reply to the arguments supporting his view.

(1) To the first—where, in arguing, it is asked whether or not there is a mean line between the first point of a line and every other point known by God. If not [it is replied] then God perceives this [first] point to be immediate to the other [points]. If there is, then, since points can be assigned in the [mean] line, these mean points will not be perceived by God, which is false. To this argument, I have replied at length in the question: "Whether God knows an infinity of things," while excluding the unsatisfactory replies of others who are ignorant of logic. However, I reply in brief that this is true: "Between the first point of the line and every other point of the same line known by God there is a mean line." For any singular [of this universal] is true, and moreover, its contradictory is false. And this is so because the term "mean line" in the predicate mediately following the universal sign has merely confused supposition.[17] On the other hand, this is false: "There is [some one] mean line between the first point and every other point of the same line perceived by God," since there is no [one] mean

line between the first point and every other point

16. A *mutatum esse,* or alternatively *motum esse* (literally a "having been changed" or a "having been moved"), was what an indivisible of motion was called, just as, the Scholastics often explain, a point is an indivisible of space and an instant is an indivisible of time.

17. The doctrine of supposition was one of the most important, and certainly one of the most used, additions made by medieval logicians to the logic of Aristotle as they knew it. It is impossible to explain in brief compass the precise significance of this doctrine [the theory of supposition is discussed in Selection 18—*Ed.*], but for present purposes it is perhaps sufficient to note that, very roughly, the doctrine has to do with the logic of the discrimination of what a term, *within a proposition,* "stands for" *(supponit pro)*. Thus, William of Ockham defines supposition (*Summa logicae,* I, ch. 62) as follows: "It remains to speak of supposition, which is a property of the term, but only when it is in a proposition. . . . It is called supposition, in the sense that it is a positing for other things, such that when a term in a proposition stands for something, we use the term for that of which (or of a demonstrative pronoun indicating it) that term is verified." Ockham's meaning becomes clearer if we follow his discussion and division of supposition further. Thus, one of the major types (for our purpose the most significant one) of supposition is that called *personal,* when the term in question "stands for the objects it signifies." Personal supposition can, however, itself be divided into a variety of distinct types, each type being characterized, among other things, by the kind of "logical descent" one can, or cannot, make from the initial proposition containing the term whose supposition is in question to another proposition involving singular terms. Thus, Ockham tells us (ch. 68) that "*confused distributive* supposition occurs when it is licit to make a logical descent in some way to a *copulative* proposition if the term has many inferiors. . . . So it is with this proposition, 'every man is an animal.' The subject of this proposition has *confused distributive* supposition. For this follows: 'Every man is an animal, therefore this man is an animal and that one is (and so on for every individual)'." On the other hand, in the proposition "some man is an animal," the term "man" has *determinate* supposition, which means that a licit logical descent can be made to a *disjunctive* proposition, that is, to the proposition: "This man is an animal, or that man is an animal (and so on for every individual)." Similarly, if a term has, as in the passage in Alnwick now in question, *merely confused* supposition, then we know that, among other things, a logical descent to a disjunctive proposition *cannot* be made. Therefore, since the term "mean line" in the proposition "Between the first point of the line and every other point of the same line. . . there is a mean line" does have *merely confused* supposition, we cannot make a disjunctive descent to the proposition "Between the first point of the line and every other point of the same line. . .there is this mean line or that mean line (and so on for every individual)." This is, then, the point which Alnwick wishes to make, which is but another way of saying that although there is a mean line between the first point of the line and every other point, there is no *one* mean line between the first point and every other point.

perceived by God. For there cannot be any such mean line, for if there were, it would fall between the first point and itself; nor would that line be perceived by God. And therefore, when it is inferred: "If there is [such a mean line], then, as points can be assigned in the line, etc.," the term "line" there has particular supposition. And hence an inference is made affirmatively from a superior to an inferior, and thus the fallacy [of affirming] the consequent is committed.[18] Similarly, an inference is made from a term having merely confused supposition to the same term having determinate or particular supposition,[19] and *quale quid* is changed into *hoc aliquid,* and a fallacy of a figure of speech occurs.[20]

To the proof [of the consequence in Harclay's argument]: When he takes as that assumed "a line falls between this first point and any other point of the same line perceived by God," we must reply that this is not what is assumed or conceded. But this has been conceded if "mean line" is placed in the predicate, namely [if we assert]: "between the first point and any other point perceived by God there is a mean line." The other [proposition above] does not follow from this, because of the reason stipulated, as is evident. And thus it is clear that this doctor, howsoever subtle, has cozened by [committing] fallacies of the consequent and of a figure of speech.

(2) To the second argument—when it is argued that God can annihilate the point at the end of a line without corrupting the line or a part of it—I reply that this is not so. For that which only exists positively through the existence of another thing is only annihilated by means of annihilation of the existence of that [other] thing through which it exists. Hence, as a point possesses existence only through the existence of a line, it is only annihilated by the line's annihilation. And when you prove [your point] by breaking a line and in turn joining [the parts so obtained, and argue]: "then [the previous end points of the parts] exist potentially; this accomplished, it is possible that numerically the same point which actually existed before and was the beginning of the right part of the line is now the end of the left part of the line," I reject [the reasoning]. Not only because an actually existent point reduced to that which exists potentially does not return as numerically the same thing but [also because] if it be given that by divine power it were to return as numerically the same thing, it would return as the beginning or end of the same line to which it belonged before and cannot wander from

one line to another, which this argument would prove [to be possible] even by means of the ability of a created being to divide [lines].

Furthermore, to the proof: When it is argued "a line can be broken at any one of its points in such a way that any point can indifferently be the beginning of one line or another, even though not of [both] at the same time," I reply: When it is claimed that a line can be broken at any one of its points, one may concede this with respect to any one of its points continuing the line, but one cannot take some [point] immediate to that [point] which existed before. And therefore that [point] which was previously at the end of one line, even though it were to be agreed that it can exist in another line, will nonetheless not be immediate [to the previous end point]; nay, rather there will be a mean line between it and the point which actually existed. Secondly, I say that when it is assumed that "a line can be broken at any one of its points," this is true of any one of its points which remains in the line according to the actuality

18. That is, X is inferior relative to Y as superior when X is related to Y as, for example, species to genus. Now, there is a valid consequence when one argues from an inferior to a superior (for instance, A man is bald, therefore an animal is bald), but not from a superior to an inferior (for instance, An animal is bob-tailed, therefore a man is bob-tailed). But Alnwick here claims that Harclay has the term "mean line" in his initial proposition as a superior, and "mean line" (or "line") in that which he *infers* from this proposition as an inferior, which means that he infers an inferior *from* a superior. Yet if inference *from* an inferior *to* a superior is that which is valid (call this, "if I, then S"), then to argue as Harclay does is to commit the fallacy of affirming the consequent (that is, given "if I, then S" to claim "S, therefore I").

19. As we have seen in note 17, if a term has *determinate* supposition, then we can make a logical descent from the proposition in which such a term occurs to a corresponding *disjunctive* proposition. This means that Alnwick is charging Harclay with fallaciously moving from a true proposition from which no disjunctive descent can be made to a false proposition where the corresponding disjunctive descent can be made. For, since in the proposition "There is a mean line between the first point and every other point of the same line" the term "mean line" has *determinate* supposition, one can licitly descend to the disjunctive proposition "there is this mean line or that mean line (and so on for every individual) between the first point and every other point of the same line." But this latter is clearly false.

20. That is, Harclay argues, claims Alnwick, from the same term ("line") functioning as a species (the *quale quid,* or a "what kind of a what") to this same term functioning as singular, to "that line" (the *hoc aliquid,* a "this something"), which is one kind of a "fallacy of a figure of speech," that is, in general, a fallacy involving the use of the same term to cover really different things.

of its presence, just as a part exists remaining in a whole. But a point which actually was before the beginning or end of a line does not remain in a line once again joined and united to another [line] unless it does so in the potency of matter. Therefore, it cannot be broken at that point.

As to the fact that it is also assumed that "any point can indifferently be the beginning of one line or another, even though not of [both] at the same time," we must reply that this is not true. For a point in an undivided line does not remain after the division of the line occurs at that point, since the point which exists in a continuous line would, if the line were divided at that point, have no greater reason to remain in one [resultant] line after division rather than in the other. And therefore it would either remain in both [resultant lines], which is impossible, or in neither, which is true. And therefore that point ceases to be a continuing [point] and withdraws into the potency of matter, and two points actually result each of which existed only in the potency of matter before division. . . .

On Vacuum

53 NATURE ABHORS A VACUUM

1. Albert of Saxony (ca.1316–1390): A Natural Vacuum Denied

Translated and annotated by Edward Grant[1]

QUESTIONS ON THE PHYSICS OF ARISTOTLE, BOOK IV, QUESTION 8

In the eighth [question] we inquire whether the existence of a vacuum is possible. . . .[2]

In this question, some things must be premised and, secondly, conclusions posited.

As to the first [part, namely the assertion of premises], it must be understood that this name "vacuum" is a privative term and valid in signifying only this sentence: "a place not filled by body." Therefore, when "vacuum" is mentioned it is "a place not filled by body." The said definition is posited as it signifies. But it does not follow that because "a vacuum is a place not filled by body," therefore "a vacuum exists. . . ."[3] Secondly, it must be understood that just as "place" is to be imagined [or conceived] in two ways, so also "vacuum" is to be imagined in two ways. Thus some have imagined that "place" is equal to all the dimensions of what is placed [or located] in it; others, however, as Aristotle and his followers, have imagined that place is not such a space but [rather] a body external to the body placed [or located and] equal to it in two dimensions, namely width and length. And "vacuum" must also be imagined in this twofold way: in one way as a separate space in which there is no other body conjointly; in the second way as a body between whose mutually closest sides there is nothing. As for the first way of speaking of a vacuum, it must be conceded that a vacuum would exist within the concave orb of the moon if

God were to annihilate all local bodies within the concave orb of the moon;[4] however, it must not be conceded that the heaven or concave orb of the moon is a vacuum, although this would be conceded in the second way declared above. . . .[5]

1. Translated from Albert of Saxony's *Questions on the Physics of Aristotle,* Book IV, Question 8, published in *Questiones et decisiones physicales insignium virorum; . . . tres libros De celo et mundo. . . Thimonis in quatuor libros Meteororum. . . Buridani in librum De sensu et sensato. . . Aristotelis. . . recognitae rursus et emendatae summa accuratione. . . Magistri Georgii Lokert. . .* (Paris, 1518), fols. 48 recto, col. 2, to 48 verso, col. 2.

2. Since he will argue for a highly qualified affirmative position, Albert first presents unqualified affirmative arguments, which he will later repudiate. These have been omitted, and our translation commences with Albert's personal response to the question.

3. Merely because "vacuum" can be defined does not imply that it exists.

4. This constitutes perhaps the most popular illustration in discussions on the existence of vacuum (see Selection 14).

5. At this point, there is a hole or tear in the page of my copy of this book, rendering the next four lines unintelligible.

For the proper comprehension of the remainder of this question, the two ways in which "vacuum" is distinguished must be kept in mind. In the first, it is the emptiness or space *between* certain containers or boundaries. In the second way, however, it is the container itself which is taken to be the vacuum, provided that it is a container between whose sides there is only emptiness. To use Albert's example of the concave orb of the moon after God has annihilated everything within it, it would be the intervening space itself that is taken as the vacuum

As to the second [part of this question], let the first conclusion be that it is impossible for a vacuum to exist, when speaking of a vacuum in the first way.[6] This can be proved, because in the first way a vacuum is a separate dimension and this cannot be posited because a separate accident must not be assumed to exist without a subject.[7] Secondly, if such a separate space were posited, it follows that because it is a body (for it would have length, width, and depth)[8] an interpenetration of bodies would occur when it received something placed in it, which is impossible.[9] Thirdly, if such a vacuum were possible, it would follow that if it existed, the vacuum which it implies would not be a vacuum. The consequence is proved, because when all the locatable things within the [concave] surfaces of the sky have been annihilated, a separate void space would exist there, according to those who hold that place is such a space; hence, according to them, a vacuum exists there. But I prove that it does not exist there because *something* would be there, namely such a separate space.[10]

A second conclusion is that a vacuum is not possible by [the action of] any natural power, when taking "vacuum" in the second way. This is proved by [appeal to] certain experiences. First, if all the openings in a bellows were stopped up, no power could raise one handle *(asserem)* from the other[11] unless a break occurred somewhere through which air could enter. After this had been done, one of the handles could easily be raised [or parted] from the other, for then there would be something that could be received between the sides of the bellows. This [experience] seems to be a sign that nature abhors a vacuum. This same thing could be shown for a clepsydra.

A third conclusion, however, is that, in the second way, a vacuum is possible by means of supernatural power. This is obvious, because God could annihilate everything that is within the sides [or concave surface parts] of the sky, after which the sky would be a vacuum;[12] and in this case, the vacuum would be a good [and proper] circularly mobile thing. But, on the contrary, this is doubted, for having assumed the case just mentioned, either the sides of the sky would be distant [that is, separated] or not. If they are not separated by a distance, they would be conjoined as two leaves of a book or an empty purse; but then it ought not to be conceded that the sky is a vacuum, because in the aforesaid case the sides of the sky would be immediate [that is, in contact]. If, however, it should be stated in this case that the sides of the sky are yet distant

[or separated], there would then be some dimension between them by which they would be distant [or separated]; hence there would be no vacuum. To this, one might respond that the sides [or concave surface parts] of the sky are not distant [or separated] by a straight line, although they may well be separated by a curved line. But if it is said that then the sides of the sky would be conjoined, I deny this, for the sky would remain spherical, just as now; and its sides would not be in direct contact, even if not separated by a [rectilinear] distance.[13]

---

in the first way, whereas in the second way the concave orb itself is the vacuum, for it is the container of that empty space. In the remainder of the question, Albert denies the first way and allows the second.

6. That is, as a separate and continuous space.

7. Although a separate vacuum is treated as a dimension, it is here also dissociated from a body or subject—that is, the dimension of the vacuum is a property or attribute of vacuum, but since vacuum is "nothing," the dimension cannot be said to inhere in anything. Although it was generally assumed that no accident or quality could exist naturally without inhering in a body or subject, it was also conceded that by his absolute power God could create an accident without a subject. To deny this would have meant running afoul of Articles 140 and 141 of the Condemnation of 1277 (see Selection 13).

8. In the preceding note we saw that Albert treated dimension as a quality or attribute without a subject or body in which it inheres. Now, however, for the sake of a new argument, he seems to shift ground and consider the dimension of a vacuum as a body, which would qualify it as a subject. Thus in one argument it serves as a free-floating quality and in the next as a body or subject.

9. This argument is drawn from Aristotle, *Physics* IV.8.216a.27–216b.11.

10. Here again, Albert trades on the "dimension" of a vacuum. For if it has dimension, it is a "something." Hence "vacuum" is a "something" and does not exist in the sense of "nothing."

11. That is, separate the sides of the bellows.

12. Since it would fuaction as the corporeal container of the emptiness that remains after the annihilation.

13. Although this conclusion and its arguments are hardly obvious, Albert does allow that if God annihilated all things within the concave surface of the outermost heaven, a vacuum would exist. Drawing upon the first two conclusions, Albert first formulates a counterargument in which the vacuum is denied under the conditions described and then proposes a solution to the counterargument. In essence, Albert argues that a vacuum of the second kind would exist, since the spherical shape of the celestial container would remain after the annihilation. He agrees that rectilinear distances cannot be measured across, or through, this void, for if they could be measured, this empty space would be a dimension and hence a "something" and not a vacuum; the first kind of vacuum is therefore impossible (see the third part of the first conclusion above). But it is nonetheless

A fourth conclusion is that by assuming the previously mentioned case [where God annihilates everything within the concavity of the sky], one pole of the world would not be near the other, and yet would not be distant [or separated] from it by a rectilinear distance. This is obvious, because distance is nothing but intervening (intermediam) dimension. But since, in the case assumed before, there could be no intervening rectilinear dimension, it follows because of this that the poles are not separated by a rectilinear distance; therefore, [one pole is not near the other, although they are not separated by a rectilinear distance].[14]

A fifth conclusion is that the poles may well be distant [or separated] by a curved distance. This is obvious, because a curved dimension would come between them, namely half the circumference of a great circle imagined in the sky. Therefore, and so on [that is, they may well be separated by a curved distance]. . . .

2. John Buridan (ca. 1300–ca. 1358): Experiments Demonstrating that Nature Abhors a Vacuum

Translated and annotated by Edward Grant[15]

Again, every universal proposition in natural science ought to be conceded as a principle which can be proved by experimental induction, just as in many particular [occurrences] something is clearly found to be so and in no instances did it fail to appear. For Aristotle puts it very well [when he says] that many principles must be accepted and known by sense, memory, and experience.[16] Indeed, at some time or other, we could not know that every fire is hot [except in this way].

[Now,] by such experimental induction it appears to us that no place is a vacuum, because everywhere we find some natural body, namely air, or water, or some other [body]. But now let us show by experience that we cannot separate one body from another unless another body intervenes. Thus if all the holes of a bellows were perfectly stopped up so that no air could enter, we could never separate their surfaces. Not even twenty horses could do it if ten were to pull on one side and ten on the other; they would never separate the surfaces of the bellows unless something were forced or pierced through and another body could come between the surfaces.[17]

[Another experiment] is with a [hollow] reed, one end of which you place in wine, and the other end in your mouth. By drawing up the air standing in the reed, you [also] draw up the wine by moving it above [even] though it is heavy. This happens because it is necessary that some body always follow immediately after the air which you draw upward in order to prevent the formation of a vacuum. And thus there are many other experiences of a mathematical kind.[18] We ought, therefore, to concede that a vacuum is not naturally possible, as is [indeed] known by that method [deemed] adequate for assuming and conceding principles in natural science. And by such induction it can be shown that there is no vacuum in any of the two ways mentioned be-

possible for the second kind of vacuum to exist, because it is defined as the physical container of an empty space, in this instance the outermost sphere itself. Distances between any two points on this spherical physical surface are indeed potentially measurable, albeit only by curvilinear distances traced along the material surface, since, as we have already seen, it is not possible to measure the distances between these points rectilinearly through the void space.

14. This and the fifth conclusion are merely extensions of his argument in the third conclusion (see n. 13).

15. Translated from John Buridan's *Questions on the Eight Books of the Physics of Aristotle*, Book IV, Question 8, fol. 73 verso, col. 1. The full Latin title is cited in Selection 14, n. 1.

16. *Posterior Analytics* II.19.100a.4–9.

17. In 1654 Otto von Guericke performed experiments demonstrating his air pump before the Imperial Diet at Ratisbon. These were subsequently described in Book III of von Guericke's *New (as they are called) Magdeburg Experiments on Void Space (Experimenta Nova (ut vocantur) Magdeburgica de Vacuo Spatio)* (Amsterdam, 1672); a selection from this treatise is given in Selection 73.4. "The most impressive of these was that of the celebrated 'Magdeburg Hemispheres' [Bk. III, ch. 23—*Ed.*]. Two hollow bronze hemispheres were fitted carefully edge to edge, and the interior was evacuated through a stop-cock in one of them which was then closed. A team of eight horses was harnessed to each hemisphere and the two teams were driven in opposite directions, but they were unable to pull the hemisphere asunder so long as the stop-cock was left closed." A. Wolf, *A History of Science, Technology and Philosophy*, 2d ed. (London: George Allen and Unwin, 1950), p. 101. Although no air pump is involved, Buridan's account bears an obvious physical resemblance to von Guericke's experiment. But where one team of horses labored mightily to demonstrate nature's abhorrence of a vacuum, the other labored with equal energy to demonstrate the creation of an artificial vacuum and the powerful pressure of the atmosphere.

18. Does Buridan mean to classify the experiments cited above as "mathematical"?

fore,[19] since we always see that natural bodies follow one another by touching [or contact] without there being any space between them devoid of a natural body such as air, or water, or some such body.

3. Marsilius of Inghen (ca. 1340–1396): Experiments Demonstrating that Nature Abhors a Vacuum

Translated and annotated by Edward Grant[20]

. . . In the second place, one may declare against the vacuum by experiments given in the treatise *On Emptiness and Void (De inani et vacuo)*,[21] since natural bodies are moved against their natural inclinations lest a vacuum occur.[22] The first experiment *(experientia)* is that of water rising to extinguish a candle covered by a vessel.[23]

A second experiment is this: If a vessel is made with two arms, one longer than the other, and the shorter is assumed to be in water, then if air is withdrawn through the longer arm the water rises through the shorter arm which it would not do unless it were preventing a vacuum.[24]

A third experiment is that a heavy body existing upwards without any obstacle [to its descent] does not descend because a vacuum might occur. Therefore, a vacuum is not possible in lower things by means of a natural agent. The antecedent is proved, because if a vessel having many small openings below and one large opening above was filled with water and the opening above was obstructed, the water would not descend through the lower openings because there would not be any other means of preventing a vacuum.[25]

Fourthly, we see that light elements, which descend into the depths of cisterns, descend downward. Here is an argument against this conclusion: Let there be a body that is everywhere concave within. . . and let this body be filled with air. This

19. A reference to an earlier section of the same question (fol. 73 recto, col. 2). The first way assumes that a vacuum is an entity or corporeal dimension having length, width, and depth. When this dimension is filled with a natural body it becomes a full space; when not occupied by any natural body it is a void place. A second kind of void is not itself an entity or dimension but can be imagined when a body occupying a place is annihilated and the place retains its shape or figure. There would be "nothing" in that place, where place is understood as the innermost surface of the containing body. Buridan's two-fold conception of vacuum is much like Albert of Saxony's in the preceding section.

20. I have translated from Book IV, Question 13 ("Whether it is possible that a vacuum exist"), of Marsilius' *Quesions on the Eight Books of the Physics* as published in *Questiones subtilissime Johannis Marcilii Inguen (sic) super octo libros Physicorum secundum nominalium viam*. . . (Lyon, 1518); reprinted as *Johan-*

nes Marsilius von Inghen Kommentar zur Aristotelischen *Physik* (Frankfort: Minerva, 1964), fol. 55 verso, col. 1.

21. The identity of this treatise is uncertain. Perhaps it is a version of the *De ingeniis spiritualibus (Pneumatics)* of Philo of Byzantium (ca. third century B.C.), which exists in a medieval Latin translation. Another candidate is the *Pneumatics* of Hero of Alexandria (first century A.D.), which is somewhat similar to Philo's treatise and of which there appears to have been a fragmentary Latin version available in the Middle Ages. Or perhaps the *Tractatus de inani et vacuo* was a compilation from Arabic sources, as Pierre Duhem believed.

22. In this statement and the four "experiments" that follow, one sees the extremes to which medieval authors would go in order to deny a vacuum. They would readily sacrifice the fundamental Aristotelian principle that a heavy body (for instance, water) must descend in a lighter medium (for instance, air).

23. In the Latin text published by Wilhelm Schmidt (*Heronis Alexandrini Opera*. . . , Vol. I: *Pneumatica et Automata*. . . *accedunt*. . . *Philonis De ingeniis spiritualibus* [Leipzig, 1899]) this experiment is described in section VIII on page 476. It is translated in M. R. Cohen and I. E. Drabkin, *A Source Book in Greek Science* (Cambridge, Mass.: Harvard University Press, 1948), p. 256.

24. A similar experiment appears in section IX of Philo's *Pneumatics* (p. 478 of Schmidt's edition).

25. This effect is produced in a clepsydra and is described in the *Problems* XVI.8 attributed to Aristotle but probably compiled by his followers. It was apparently

a device for transferring liquids and "consists of a globular vessel [see fig.—*Ed.*] with small holes at the bottom and a pipe handle open at the top. When the clepsydra is placed in a vessel containing liquid, this liquid enters the clepsydra through the holes in the bottom. Then, if the hole at the top is covered with the thumb, the clepsydra may be removed without the loss of the liquid in it. When the thumb is taken from the opening, the liquid flows out of the holes at the bottom" (Cohen and Drabkin, pp. 245–246, n. 3). The same illustration appears in section XI of Philo's *Pneumatics* (pp. 480, 482 of Schmidt's edition), in Hero's *Pneumatica* I.7 (see Cohen and Drabkin, pp. 326– 327; the figure has been reproduced from p. 327), and is also described by Nicholas of Autrecourt (see Selection 56.2).

body is then filled with air and placed in water or in something that is made intensely cold so that the air contained in the vessel will be condensed, since condensation follows upon coldness. Hence the air will occupy a smaller place than before, and as a consequence, there will be a vacuum. [But to this]

it must be replied that it is impossible that the air in this body be condensed by coldness, however cold it may be, unless the vessel is broken or there is an opening in it through which another body enters it.

4. Galileo Galilei (1564–1642): Experiments Demonstrating that Nature Abhors a Vacuum

Translated by Henry Crew and Alfonso de Salvio[26]
Annotated by Edward Grant

[59]

Salv. A truly ingenious device![27] I feel, however, that for a complete explanation other considerations might well enter; yet I must not now digress upon this particular topic since you are waiting to hear what I think about the breaking strength of other materials which, unlike ropes and most woods, do not show a filamentous structure. The coherence of these bodies is, in my estimation, produced by other causes which may be grouped under two heads. One is that much-talked-of repugnance which nature exhibits towards a vacuum; but this horror of a vacuum not being sufficient, it is necessary to introduce another cause in the form of a gluey or viscous substance which binds firmly together the component parts of the body.[28]

First I shall speak of the vacuum, demonstrating by definite experiment the quality and quantity of its force [*virtù*]. If you take two highly polished and smooth plates of marble, metal, or glass and place them face to face, one will slide over the other with the greatest ease, showing conclusively that there is nothing of a viscous nature between them. But when you attempt to separate them and keep them at a constant distance apart, you find the plates exhibit such a repugnance to separation that the upper one will carry the lower one with it and keep it lifted indefinitely, even when the latter is big and heavy.[29]

This experiment shows the aversion of nature for empty space, even during the brief moment required for the outside air to rush in and fill up the

(Selection 56.3) Galileo did accept the existence of interstitial vacua but nevertheless maintained the traditional viewpoint that nature abhorred a separate vacuum and sought as quickly as possible to occupy it with matter. Not until Pascal explained the role of atmospheric pressure was this long-held opinion emphatically denied (for Pascal's exposition, see next selection).

27. A reference to a device for descending from a window at a controlled rate, described by Sagredo, one of the interlocutors in the dialogue.

28. Later, Galileo identifies this other cause as interstitial vacua, which he suggests function as a binding or attractive force *(forza del vacuo)* holding together the particles of a solid body (see Selection 56.3).

29. The opinions of Salviati and Sagredo represent Galileo's viewpoint on this subject. Admitting a momentary vacuum between the plates (see the last two paragraphs) represents a radically different response than was given in a typically medieval exposition by Roger Bacon to explain much the same phenomena.

After demonstrating the impossibility of void by rational argument, Bacon offers a series of supporting illustrations and experiments, the first of which is basically that presented by Galileo. "Let two plane tables be taken and joined together so that nothing lies between them. If one were raised above the other, the air would enter between the tables, first reaching the extreme parts, and then the central point. But in this raised position, and before the air reaches the central point, a void will exist there." In order to deny the formation of a void, Bacon says that one might suppose that when the tables are parted, air reaches all parts of the tables instantaneously. Finding this repugnant, he proposes another solution. "I reply to this by saying that the two round tables could not be raised if one were above the other unless there should be an inclination of some part. Hence it would be necessary that some part be inclined before it could be raised, for otherwise a void would be produced. This applies to all of nature. . . . It is evident, then, that anyone who wishes to raise [anything] must incline one part beforehand. . . . For this reason a vacuum should not be assumed." (My translation is from Bacon's *Questions on the Eight Books of the Physics of Aristotle*, Book IV ("Whether a vacuum can be assumed below the heavens"), in *Opera hactenus inedita Rogeri Baconi*, Fasc. XIII: *Questiones supra libros octo Physicorum Aristotelis*, edited by F. M. Delorme, O.F.M., with the aid of Robert Steele [Oxford: Clarendon Press, 1935], pp. 225–226.)

26. Reprinted with the kind permission of the Macmillan Company from the First Day of the *Dialogues Concerning Two New Sciences by Galileo Galilei*, translated from the Italian and Latin into English by Henry Crew and Alfonso de Salvio with an Introduction by Antonio Favaro (New York: Macmillan, 1914), pp. 11–12. As we shall see here and in another brief selection

region between the two plates. It is also observed that if two plates are not thoroughly polished, their contact is imperfect so that when you attempt to separate them slowly the only resistance offered is that of weight; if however, the pull be sudden, then the lower plate rises, but quickly falls back, having followed the upper plate only for that very short interval of time required for the expansion of the small amount of air remaining between the plates, in consequence of their not fitting, and for the entrance of the surrounding air. This resistance which is exhibited between the two plates is doubtless likewise present between the parts of a solid, and enters, at least in part, as a concomitant cause of their coherence.

[60]

Sagr. Allow me to interrupt you for a moment, please; for I want to speak of something which just occurs to me, namely, when I see how the lower plate follows the upper one and how rapidly

it is lifted, I feel sure that, contrary to the opinion of many philosophers, including perhaps even Aristotle himself, motion in a vacuum is not instantaneous. If this were so the two plates mentioned above would separate without any resistance whatever, seeing that the same instant of time would suffice for their separation and for the surrounding medium to rush in and fill the vacuum between them. The fact that the lower plate follows the upper one allows us to infer, not only that motion in a vacuum is not instantaneous,[30] but also that, between the two plates, a vacuum really exists, at least for a very short time, sufficient to allow the surrounding medium to rush in and fill the vacuum; for if there were no vacuum there would be no need of any motion in the medium. One must admit then that a vacuum is sometimes produced by violent motion [*violenza*] or contrary to the laws of nature, (although in my opinion nothing occurs contrary to nature except the impossible, and that never occurs).

54 NATURE DOES NOT ABHOR A VACUUM

Blaise Pascal (1623–1662)

Translated by I. H. B. Spiers and A. G. H. Spiers[1]
Annotated by Edward Grant

CONCLUSION OF THE TWO PRECEDING TREATISES

I have recorded in the preceding Treatise all the general effects which have been heretofore ascribed to nature's effort to avoid a vacuum, and have shown that it is utterly wrong to attribute them to that imaginary cause. I have demonstrated, on the contrary, by absolutely convincing arguments and experiments that the weight of the mass of the air is their real and only cause. Consequently, it is now certain that nature nowhere produces any effects in order to avoid a vacuum.

It is not difficult to demonstrate, furthermore, that nature does not abhor a vacuum at all. This manner of speaking is improper, since created nature, which is the nature under consideration, is

A. G. H. Spiers, with Introduction and Notes by Frederick Barry (New York: Columbia University Press, 1937), pp. 67–75. The present selection, translated from the original French, constitutes the complete conclusion in a volume by Pascal edited and published in 1663, one year after his death, by his brother-in-law F. Perier. The two treatises in the volume, referred to in the heading of the conclusion, are titled "On the Equilibrium of Liquids" and "On the Weight of the Mass of Air." As Frederick Barry remarks in a foreword (pp. v–vi), "to him we owe the first conclusive proof of the pressure of the atmosphere, and the final banishment from the minds of natural philosophers of that ancient and persistent conception of *horror vacui* which had long inhibited investigation in this field; a complete mechanical correlation of all the diverse phenomena of fluid equilibrium, without recourse to that or any other imaginary conception; and finally, the establishment of detailed analogies between the effects of pressure in liquids and in air which affected the unification of hydrostatics and aerostatics as one deductively organized and logically coherent discipline." In the preceding chapter of this source book the selections from John Buridan and Marsilius of Inghen typically represent the medieval conception of "abhorrence of a vacuum." Even Nicholas of Autrecourt, who argued for minute, interstitial vacua, reaffirmed nature's abhorrence of a separate or extended vacuum. Pascal's exposition is so simple and clear that it requires little comment.

30. Some scholastics argued that motion in a hypothetical vacuum would not be instantaneous (Selection 47, n. 6, and Selection 55.1). Of great interest here is the fact that Galileo holds that an actual vacuum would exist momentarily before nature rushed to fill it.

1. Reprinted with the kind permission of the Columbia University Press from the *Physical Treatises of Pascal: The Equilibrium of Liquids and the Weight of the Mass of the Air,* translated by I. H. B. Spiers and

not animated, and can have no passions. Such language is in fact metaphorical, and means nothing more than that nature makes the same efforts to avoid a vacuum as if she abhorred it. Those who use this phrase mean that it is the same thing to say that nature abhors a vacuum as to say that nature makes great efforts to prevent a vacuum. Now, since I have shown that nature does nothing at all to avoid a vacuum, the conclusion is that nature does not abhor it. To carry out the metaphor: just as we say of a man that a thing is indifferent to him when his actions never betray any movement of desire for, or of aversion to, that thing, so should we say of nature that it is supremely indifferent to a vacuum, since it never does anything either to seek or to avoid it. (I am here still using the word "vacuum" to mean a space empty of all bodies which our senses can apprehend.)[2]

It is perfectly true (and this is what misled the ancients) that water rises in a pump when the air has no access to it, that a vacuum would result if the water did not follow the piston, and that the water ceases to rise as soon as any cracks develop by which the air can get in to fill the pump. Thus it looks as though the water rose merely for the purpose of preventing a vacuum, since it rises only when otherwise there would be one.

Similarly it is a fact that a bellows is hard to open when its apertures are so carefully sealed that no air can enter it; and it is true that its opening would produce a vacuum. This resistance ceases when air can enter to fill the bellows, and since it is met with only when a vacuum would otherwise result, it seems to be due to nothing else than the fear of a vacuum.[3]

Finally, it is a fact that all bodies in general make great efforts to follow one another and to keep together whenever their separation, and nothing else, would produce a vacuum between them. This is why it has been inferred that this close adhesion is due to the fear of a vacuum.

To reveal the weakness of this reasoning the following example will serve. When a bellows is placed in water in the manner we have often described, with its nozzle at the end of a tube assumed to be twenty feet long which projects out of the water into the air, and with all its side apertures sealed so as to exclude the air, everyone knows that it is hard to open, and the more so the greater the amount of water above it; whereas if the vents in one of the wings are unsealed so that they admit the water freely, the resistance disappears.

If one wished to reason out this effect like the others, he might say: When the side vents are closed and when, therefore, if the bellows is to be opened, the air must enter through the tube, there is difficulty in opening it; but when water instead of air can enter to fill it, the resistance ceases. Therefore, since there is resistance only to the entrance of air, the resistance arises from an abhorrence of the air.

There is no one who will not laugh at this inference, seeing there may well be another cause of the resistance. It is evident indeed that the bellows cannot be opened without raising the water, because the water that would be pushed aside in the act of opening cannot enter the body of the bellows, and, compelled to find room for itself elsewhere, raises the whole body of the water and causes the resistance. This does not occur when the bellows has vents through which the water can enter; for whether it is opened or shut the water neither rises nor falls in consequence, since water enters the bellows just as fast as it is pushed aside, and thus offers no resistance to its opening. All this is clear, and consequently we must believe that the bellows cannot be opened without two results: first, the air does really enter, or second, the level of the water is raised. It is the latter action that causes the resistance and with this the former has nothing to do, although it occurs simultaneously.

Let us give the same explanation for the difficulty experienced in opening in the air a bellows sealed on all sides. If it were forced open two things would occur: first, a vacuum would really be formed; second, the whole mass of air would be raised and upheld. It is the latter action that causes the resistance felt; the former has nothing to do with it. This resistance also increases or diminishes in proportion to the weight of the air, as I have shown.

The same facts explain the resistance to separation offered by all bodies between which there is a vacuum: air cannot filter in, otherwise there would be no vacuum, and this being so, they cannot be separated except by raising and upholding the

2. The usual assumption made by Aristotle and his followers was that a vacuum, by definition, was absolutely devoid of matter. However, in discussing imaginary void space beyond the cosmos it was assumed that, although devoid of corporeal entities, it was wholly occupied by God or spirit (see Selection 73, introduction). Pascal, here, seems to allow for the possibility that nonperceptible, or non-sensible, matter might, in some way, occupy, or partially occupy, a vacuum.

3. Compare to Buridan's discussion of the bellows in the preceding selection.

whole mass of the air. It is this which occasions the resistance.

Such then is the real cause of the adhesion of bodies between which there exists a possible vacuum. It was for a long time not understood because erroneous opinions were entertained which were discarded only by degrees. There have been three different periods during which different opinions of this character were held; and these involved three generally prevailing errors which made it absolutely impossible to understand the cause of this adhesion of bodies.

The first is that, in nearly all times, it was believed that the air has no weight; for the ancients said so,[4] and their professed disciples followed them blindly. They would have remained forever wedded to that theory had not keener thinkers rescued them by the force of experimental evidence;[5] for it was impossible to believe that the weight of the air causes such adherence, so long as it was held that air had no weight.

The second error lay in the belief that the elements have no weight in themselves,[6] for the sole reason that the weight of water is not felt by those who are in it, that a bucket full of water immersed in the water is easy to lift so long as it stays there, and that its weight begins to be felt only when it is lifted out. As if these effects could not be due to another cause—or rather, as if this one was not wholly beyond all probability! For there is no sense in believing that water in a bucket has weight when out of the water, but has no weight left after it is poured back into the well; that it loses its weight when mixing with the rest, and recovers its weight when lifted above the surface. Strange are the means men employ in order to cloak their ignorance! Because they could not understand why the weight of water is not felt, and were loath to confess their ignorance, they declared it had no weight, for the satisfaction of their vanity and to the ruin of truth. Their views prevailed; and, of course, the weight of the air could not be accepted as the cause of these effects so long as this vain imagining had currency. Even had it been known that air has weight, the claim would still have been made that it has no weight when contained within itself, and consequently the belief would have persisted that it can effect nothing by its weight. That is why I have shown in the Equilibrium of Liquids that water weighs the same within itself as outside, and I have explained there why, in spite of that weight, a bucket is not hard to raise while it is in the water and its weight is scarcely felt.

And in the Treatise on the Weight of the Mass of

4. On the basis of an experiment in which an inflated and uninflated bladder were found to be of equal weight, the Greek commentator, Simplicius, in his *Commentary on Aristotle's De caelo*, concluded that air has no weight in air but apparently believed that it would have weight in the natural place of fire. Aristotle (*De caelo*, 311b.6–10), however, on the basis of the same experiment insisted that an inflated bladder weighs more than an empty one and concluded that air would have weight in its own natural place as well as in the natural place of fire. The measurements were too delicate for the instruments available and the issue could not be resolved in this manner. See M. R. Cohen and I. E. Drabkin, *A Source Book in Greek Science* (Cambridge, Mass.: Harvard University Press, 1948), pp. 247–248.

5. Frederick Barry observes (pp. 27–28, n.3): "The famous Milanese mathematician and philosopher of nature, Girolamo Cardano (1501–1576) appears to have been the first to demonstrate that air has weight and to attempt a determination of its specific gravity. After him, and before it was known how to produce a vacuum, many experiments were made to establish his principal conclusion as indubitable fact. The best of these were carried out by Galileo and described in his *Discorsi e dimonstrazioni mathematiche* (1638), in the dialogue of the First Day. With a syringe he forced into a properly valved bottle, previously weighed, an amount of air which at ordinary pressure would have been of two or three times its volume, and determined with precision an actual increase of weight; then he allowed the excess air to escape without loss through a tube which led into another bottle completely filled with water and so constructed that the water displaced by the entering air could be caught in a suitable weighed vessel, the increase in the weight of which, finally, he determined. Galileo remarked that if air were specifically light, the first flask when filled with condensed air should be lighter than before, not heavier; that air consequently had weight (as, indeed, previous experience had already proved); and that since the volume of the water displaced was that of the excess air introduced, in its normal condition, the ratio of the determined weight of this air to that of the displaced water was the specific gravity sought. In a second, more elegant experiment, he forced water, instead of air, into the valved flask first used and thus compressed the air within it; weighed it in this condition; then opened the valve, thus allowing to escape a volume of air which, in its normal condition, had previously occupied the space now filled by the water; and finally weighed again. In this case, the increase in weight after the water had been introduced was the weight of this water, the subsequent loss in weight being that of the same volume of air in its normal condition. See the English translation of the *Discorsi* by Henry Crew and Alfonso de Salvio, *Dialogues Concerning Two New Sciences by Galileo Galilei* (New York: Macmillan, 1914)."

6. Pascal is in error here, for Aristotle insisted that earth, water, and air have weight in their own natural places (*De caelo*, 311b.6–10; fire is an exception, for it was deemed to be weightless or absolutely light). See note 4.

the Air I have given the same demonstration in the case of the air, to clear up all doubts.

The third error is different in kind. It does not appear in connection with [the weight of] the air, but in connection with the effects which were ascribed to the abhorrence of a vacuum. Concerning these the most erroneous theories were entertained. For it had been imagined that a pump raises water not only to ten or twenty feet, which is true enough, but still farther—to a height of fifty, one hundred, or one thousand feet, or as high as you please, without limit. Likewise the belief was held that it is not only difficult, but actually impossible, to separate two polished bodies in close contact;[7] that not even an angel, or any created force, could do so, with hundreds of exaggerations which I scorn to repeat. And so with the rest.

This is an error of observation so ancient that it cannot be traced back to its source. Heron himself, who is one of the oldest and best of the authors who have written on the raising of water, states as a positive and uncontrovertible fact that the water of a river may be made to pass over a mountain ridge and to flow into the valley beyond, provided this valley be somewhat lower down, by means of a siphon placed on the summit with its legs stretching along the slopes, one into the river and the other on the farther side; and he asserts that the water will rise from the river over the mountain and drop down again into the other valley, however high the ridge between may be.[8]

All writers on the subject have said the same thing; and even at the present time our fountain builders guarantee that they can make suction pumps which will raise water as much as sixty feet if it be desired.

Neither Heron, nor those other writers, nor the artisans, and still less the natural philosophers, can have carried their tests very far; for had they tried to draw water to the height of forty feet, they would have failed. They had only seen suction pumps and siphons six, ten, or twelve feet high, which worked beautifully; and in all the experiments they had occasion to make, had observed no case in which water failed to rise. They never imagined, consequently, that there was a limit beyond which water behaved otherwise. They conceived that the facts they had noticed were the results of an invariable natural necessity; and since they believed that water rose by an invincible abhorrence of a vacuum, they concluded that as it rose at first, so it would continue to rise without limit, applying their interpretation of what they did

observe to what they did not observe and declaring both statements to be equally true.

So positively was this believed that philosophers have made it one of the most general principles of their science and the foundation of the treatises on the vacuum. It is and has been didactically asserted every day in all the schoolrooms in the world, ever since books were written. Everyone has firmly believed it, and it has remained uncontradicted down to our own time.

This fact perhaps may open the eyes of those who dare not doubt an opinion which has always been universally entertained; for simple workmen have been able to prove in this instance that all the great men we call philosophers were wrong. Galileo declares in his dialogues[9] that Italian plumbers taught him that water rises in pumps only to a certain height: whereupon he himself confirmed the statement as others did also, afterward, first in Italy and later in France, by using quicksilver, which is easier to handle but provides merely several other ways of making the same demonstration.

Before men gained that knowledge, there was no incentive to prove that the weight of the air was the cause of water rising in pumps; since, the weight of the air being limited, it could not produce an unlimited effect.

But all these experiments were insufficient to show that the air does produce those effects: they had rid us of one error but left us in another. They taught us, to be sure, that water rises only to a certain height, but they did not teach us that it rises higher in low-lying places. On the contrary, the belief was held that it always rises to the same height, in every place on the earth. And since the weight of the air never entered anybody's head, it was vaguely thought that the nature of the pump

7. The problem of two plane surfaces in contact was discussed by Galileo (see Selection 53.4 and n. 29 thereto).

8. Barry remarks (pp. 71–72, n. 3) that the modern Greek edition of Hero's *Pneumatica* and the editions available to Pascal contain no such explicit statement and that Pascal either misread Hero or relied on hearsay. "On the other hand, Heron, in discussing the siphon . . . makes no mention of any limitation to its efficacy; neither does Cardan (*De subtilitate*, 1560: I.3.364a); and Galileo's discussion of pumps in the *Mathematical Discourses* of 1638. . . implies that it was then still taken for granted by the philosophers, if not by all workmen, that a siphon of any height would operate."

9. See the First Day of Galileo's *Dialogues Concerning Two New Sciences* in the translation of Crew and de Salvio, pp. 16–17.

was such that it lifted water to a limited height and no further. Indeed Galileo took that to be the natural height of a pump, and called it *la altessa limitatissima*. How indeed could it have been imagined that that height was different in different places? Certainly, it would seèm improbable. Yet that last error again put out of the question the proof that the weight of the air causes these effects; since, because this weight would be greater at the foot than at the top of a mountain, its effects, obviously, would be proportionately greater there.[10]

That is why I decided that the proof could be obtained only by experimenting in two places, one some four or five hundred fathoms above the other. I chose for my purpose the Puy de Dôme mountain in Auvergne, for the reasons that I have set forth in a little paper which I printed as early as the year 1648, immediately after the experiment had proved successful.[11]

This experiment revealed that fact that water rises in pumps to very different heights, according to the variation of altitudes and weathers, but is always in proportion to the weight of the air. It perfected our knowledge of these effects, and put an end to all doubting; it showed their real cause, which was not abhorrence of a vacuum, and shed on the subject all the light that could be wished for.

Try now to explain otherwise than by the weight of the air why suction pumps do not raise water so high by one-quarter on the top of Puy de Dôme in Auvergne as at Dieppe; why the same siphon lifts water and draws it over at Dieppe and not at Paris; why two polished bodies in close contact are easier to separate on a steeple than on the street level; why a completely sealed bellows is easier to open on a house-top than in the yard below; why, when the air is more heavily charged with vapors, the piston of a syringe is harder to withdraw;[12] and lastly why all these effects are invariably proportional to the weight of the air, as effects are to their cause.

Does nature abhor a vacuum more in the highlands than in the lowlands? In damp weather more than in fine? Is not its abhorrence the same on a steeple, in an attic, and in the yard? Let all the disciples of Aristotle collect the profoundest writings of their master and of his commentators in order to account for these things by abhorrence of a vacuum if they can. If they cannot, let them learn that experiment is the true master that one must follow in Physics; that the experiment made on mountains has overthrown the universal belief

in nature's abhorrence of a vacuum, and given the world the knowledge, never more to be lost, that nature has no abhorrence of a vacuum, nor does anything to avoid it; and that the weight of the mass of the air is the true cause of all the effects hitherto ascribed to that imaginary cause.

10. Barry supplies the following note (pp. 73–74. n, 5): "Pascal here refers, in the first place, to ideas current up to the time of Galileo, which were consistent with the theory of a limitless *horror vacui;* then to Galileo's observations on the pump which, though they permitted the retention of the general notion of a *horror vacui,* nevertheless proved that it was a finite and measurable force. But Galileo did not dogmatically assert, as Pascal implies, that it was an invariable force: it was contrary to his habit of thought, in the absence of experimental evidence, to do anything of the sort; indeed, for lack of sufficient knowledge, he left the phenomenon unexplained. Pascal's omission of all reference to the several physicists who, after Galileo, quite definitely conceived that this force was external and due in fact to atmospheric pressure—among whom were Mersenne and Descartes, whose ideas were almost certainly known to him—and particulary his failure to mention Torricelli, who verified this theory, leaves a false impression too favorable to himself. It was not his work that invalidated the conception of *horror vacui:* and his contradictory hypotheses were not only suggested but were almost completely verified by his predecessors. It is sufficient honor to him that he so brilliantly completed their work.

11. Perier included this paper in a later part of the same volume which contained the conclusion reproduced here. It describes Pascal's most famous experiment.

12. In an earlier note (pp. 51–52, n. 1) Barry explains, "although the density of moist air is less than that of dry air under the same conditions, barometric pressures at the same altitude and like temperatures are usually greater when the air is moist; primarily on account of its greater compressibility at ordinary temperatures in this condition, since this may occasion an increase of density that overcompensates, at the moderate altitudes where measurements are usually made, the decrease due to the mere presence of water vapor—especially when, as is usual, the upper air is dry. Perier observed this effect in a long series of observations made with the mercury barometer at one station in 1649–51 . . . ; and Pascal, in his experiment with the bellows. . .not only noted it, but made it the basis of several acute meteorological inferences. . . . In 1753 Bouguer remarked especially the strikingly greater difference of pressure between two fixed altitudes when the air is moist—an informing fact which had previously been missed by Perier for lack of a sufficient number of observations; but it was not until 1784 that Deluc, as the result of a long series of measurements made under widely variable conditions and carefully compared, finally explained the phenomenon satisfactorily. See Deluc: *Recherches sur les modifications de l'atmosphère,* nouvelle ed. 1784, III, 715, 716.

55 MOTION IN A HYPOTHETICAL VOID

1. St. Thomas Aquinas (ca.1225–1274): A Kinematic Argument for Finite Motion in a Hypothetical Void

Translated and annotated by Edward Grant[1]

But several difficulties arise against the opinion of Aristotle. The first of these is that it does not seem to follow that if a motion occurs in a void it would bear no ratio in speed to a motion made in a plenum.[2] Indeed, any motion has a definite velocity [arising] from a ratio of motive power to mobile—even if there should be no resistance. This is obvious by example and reason. An example is that of the celestial bodies, whose motions are not impeded by anything, and yet they have a definite speed in a definite time.[3] An appeal to reason is this: Just as there is a prior and posterior part in a magnitude traversed by a motion so also we understand that in the motion [itself] there is prior and posterior. From this it follows that motion takes place in a definite time.[4] But it is true that in virtue of some impediment [or resisting medium] something could be subtracted from this speed. It is not necessary, therefore, that a ratio of speeds be related as a ratio of resistance to resistance, for then, if there were no resistance, motion would occur instantaneously. But it is necessary that the ratio of retardation to retardation be as the ratio of resistant medium to resistant medium.[5] Thus if motion in a void were assumed, it follows that no retardation would occur beyond the natural velocity,[6] and it does not follow that motion in a void would bear no ratio to motion in a plenum.

1. My translation was made from St. Thomsa' *Commentary on the Eight Books of the Physics of Aristotle (Commentaria in octo libros Physicorum Aristotelis)*, Book IV, Lectio 12, in Vol. II of *Sancti Thomae Aquinatis . . . opera omnia* iussu impensaque Leonis XIII, P. M. edita (Rome, 1884), p. 186, col. 1. It first appeared in my article "Motion in the Void and the Principle of Inertia in the Middle Ages," *Isis*, Vol. LV, Part 3, No. 181 (1964), p. 269. The notes have been written especially for this source book. This brief selection is Thomas' comment on Text 71 of Book IV of Aristotle's *Physics*, and it is obvious that he was thoroughly familiar with the disagreement between Avempace and Averroes as set forth in Averroes' comment on Text 71 (see Selection 44). Without mentioning either Avempace or Averroes, Thomas emphatically sides with the former against Averroes and Aristotle.

2. A reference to Aristotle's argument in *Physics* IV.8.215b.12–216a.7.

3. This is one of Avempace's principal arguments (see Selection 44.2).

4. Later authors named this argument "the incompos-

sibility [or incompatibility] of the termini" *(incompossibilitas terminorum* or *distantia terminorum)*, and some even designated as a resistance the distance traversed in a finite time (see Albert of Saxony's discussion of this in Selection 47 and n. 6 thereto). Perhaps St. Thomas was influenced in the formulation of this particular kinematic proposition by Aristotle's report in *De sensu* (446a.26–446b.3) of Empedocles' argument for the finite velocity of light. For just as Thomas denies instantaneous velocity in the void because intervening parts of a distance must be traversed before the *terminus ad quem* is reached, so Empedocles says that "light from the sun arrives first in the intervening space before it comes to the eye, or reaches the earth." In response, Aristotle concedes, "This might plausibly seem to be the case. For whatever is moved [in space] is moved from one place to another; hence there must be a corresponding interval of time also in which it is moved from the one place to the other." But Aristotle immediately rejects Empedocles' argument because "any given time is divisible into parts; so that we should assume a time when the sun's ray was not as yet seen, but was still traveling in the middle space." Since we cannot physically distinguish the sun's rays in the prior parts of its path from the posterior parts, Aristotle concludes that the transmission of light is instantaneous. However, Thomas, who commented on the *De sensu* and obviously knew Empedocles' argument, may have been sufficiently impressed to extend its application against another claim for instantaneous motion—this time in void space. A similar argument in favor of the successive and finite motion of light (and other bodies) in a void was given by Roger Bacon (see Selection 62.32; for yet another instance, see Selection 56.2 and n. 29 thereto).

5. Thus Aquinas rejects Averroes' opinion that $V/V' = R/R'$, where V is velocity and R is the *density* of the resistant media; or $V/V' = R/R'$, where R is the *rarity* of the resistant media (see Selection 44.2, n. 7). In place of it, he adopts Avempace's formulation that $r/r' = R/R'$ (here r is the retardation of motion resulting from the resistance of the density of medium R, and $r = V' - V$, where V' is the natural speed of a body in a resistance-less medium and V is its speed in a resistant medium (see above, Selection 44.2, n. 26).

6. For a brief discussion of this statement, see Selection 47 and n. 7 thereto. It may have prompted others to explain the finite velocity of a body in void space as the direct consequence of a natural limitation—a limitation that would sometimes be called a resistance—of the falling body.

Elsewhere Aquinas argued that while external forces were necessary in violent motion, they were superfluous for natural motion. In rejecting Averroes' claim that a separate motive force is necessary in natural motion, he insists that "movement of heavy and light bodies does not come from the generator by the intervention of another moving power. Nor even is there any need to

2. Albert of Saxony (ca. 1316–1390): Dynamic Arguments Justifying Motion in a Hypothetical Void

Translated and annotated by Edward Grant[7]

BOOK IV, QUESTION 11

In the eleventh [question] we now inquire whether if a vacuum did exist, a heavy body could move in it.

And it is argued yes, because if a vacuum did exist and a heavy mixed [or compound] body *(mixtum)* were assumed in it, it would descend successively because of the internal resistance[8] it has. Secondly, every body that is outside its natural place is moved, or begins to be moved, toward it. Hence if any heavy body were assumed to be in a vacuum and outside its natural place, it would be moved toward it. Consequently, if a heavy body were moved in a vacuum—if there were a vacuum —it would move to its natural place if unimpeded.

Thirdly, everything that seeks some end [or goal] follows it unless impeded. But when a heavy body is outside its natural place, it seeks to be in it as an end. Therefore, it is moved to it if unimpeded, since it cannot pursue its end except by motion.

Fourthly, a heavy body is moved through air more quickly than through water because air is subtle and resists less than water, and since in a vacuum there would be absolutely no resistance, it seems that a heavy body ought to be moved most easily *(optime)* in a vacuum. Fifthly, if a heavy body were not moved in a vacuum—assuming a vacuum existed—it follows that this would be solely because of a lack of resistance. But this cannot be, since any luminous body could easily illuminate any space successively. This is obvious with respect to a luminous body moved locally, for nothing resists it. It also seems, therefore, that a simple heavy or light body could be moved in a vacuum without any resistance to it; similarly, the sky is moved successively and yet the sky offers no resistance to the mover, namely the Intelligence.[9] Therefore, and so on.

Aristotle seems to take the opposite position, arguing that if a vacuum existed, a heavy simple body could not be moved in it because there would be no reason why it should be moved to one different [place] than to another.[10] Furthermore, even if a vacuum existed, a heavy body assumed in it would be moved in an instant.[11] But that motion be made in an instant implies a contradiction. To this question we must first respond as it is expressed *(sicut verba sonant)*, and [then], secondly, by reforming the question we must respond to the conception [and purpose] of those who ask it.

As for the first [part, concerning the question as expressed], let this be the first conclusion: This [proposition] is false: if a vacuum existed, a heavy body could be moved in it. [The falsity of] it is proved because it is a conditional proposition whose antecedent can be true when the consequent is false; consequently, it is false. The antecedent is obvious, [for] if by the power of God the whole which lies between the sides of the sky were annihilated and no heavy body existed, then this [that is, the antecedent] is true: a vacuum exists; but yet it is false [to assert] that a heavy body is moved in it.

A second conclusion [is this]: Similarly this [proposition] is false: if a vacuum existed, a heavy body could not be moved in it. For it is also obvious that this is a conditional [statement] whose antecedent can be true when the consequent is false. Nor do such conditional propositions contradict [each other].[12]

---

look for resistance here other than that which exists between generator and generated. Thus it follows that air is not required for natural motion of necessity, as in the case of violent motion, since that which moves naturally has a force *(virtutem)* imparted to it which is a source of motion. Consequently there is no need for a body to be moved by any other force impelling it, as though it were a case of violent motion, having no implanted force from which motion springs." (This paragraph is repeated from note 31 to Selection 46).

This passage would seem to qualify the present selection. Kinematically, both violent and natural motion would require finite times for completion in a hypothetical void. Dynamically, however, violent motion is an impossibility without air or an external medium to maintain and continue it. Hence not even in a hypothetical void could violent motion be sustained. In contrast, a body in natural motion requires nothing for its realization beyond itself and could readily move through an actual void, if such existed.

7. This is virtually the whole of Book IV, Questions 11 and 12, of Albert of Saxony's *Questions on the Eight Books of the Physics* appearing on fols. 50 recto to 51 verso of the edition cited in note 1 to Selection 35.

8. For a discussion of *mixtum* and internal resistance, see Selection 51.1, n. 9, and Selection 77.2, n. 22.

9. Probably based on Avempace's argument that despite a lack of resistance in the celestial regions, the fixed stars and planets move successively and with finite speeds. See Selection 44.2 and note 15 thereto.

10. *Physics* IV.8.214b.28–33.

11. Probably a reference to *Physics* IV. 8.215b.12–24.

12. Toward the beginning of his *Questions on the Phys-*

As to the second [part of this], the question can be formed [in this way]: "If a vacuum exists, whether a heavy body could be moved in it."[13] But then concerning the vacuum, we must make the distinction about it which we declared above in the first article of the eighth question.[14]

Let this be the first conclusion: By taking "vacuum" in the second way,[15] a heavy simple body could be moved in it. This is obvious, for with earth and water having been destroyed, and if some heavy body were posited at the sphere of fire, it would descend successively through the air [perhaps "fire" is intended here—Ed.] to the concavity of the air, because in the air it would have a sufficient resistance [which it would not have in the] vacuum.

A second conclusion [is this]: By taking "vacuum" in the first way,[17] as it is commonly taken in this question, a heavy mixed body is easily moved in it successively. This is clear, for let there be a heavy mixed body whose heaviness *(gravitas)* is as 2 and lightness *(levitas)* as 1. And let it reach the concave [surface] of air and descend successively until its center of gravity *(medium gravitatis)* is the middle [or center] of the world *(medium mundi)*. [This will happen] because it has an internal resistance, for it has one degree of lightness inclining [or tending] upward and two degrees of heaviness inclining downwards.

A third conclusion [is this]: In some [particular] case a heavy mixed [or compound] body *(mixtum)* could be moved quicker in a plenum than in a vacuum. This can be demonstrated, for let a mixed body be composed of earth, which is as 3, and air, which is as 2. Now, should everything below the sphere of fire be annihilated, let us then assume that heavy body is at the sphere of fire which resists it with [a resistance of] 1.[18] Then, because both the earth and air [of the mixed body] are outside their natural places, and because both seek to descend through the fire, the motive power is as 5, since the earth is as 3 and the air as 2; and the resistance of the medium is as 1, as has been assumed in this case. Hence, this body is moved through fire, or through a plenum, in a quintuple ratio.[19] Now, when it reaches to where the natural place of air would be if there were no vacuum, the air [in the

of a vacuum, which is possible, perhaps there is no heavy body in it; or, although there could be a heavy body in it, perhaps it rests either because of the divine power or for another reason. A third conclusion is that this is also not a good consequence: 'A vacuum exists, therefore a heavy body is not moved in it,' because it is possible that it is moved at least by the divine power, as was said. And so these two conditional statements must be denied: (1) 'If a vacuum exists, a heavy body could be moved in it:' [and] (2) 'If a vacuum exists, a heavy body could not be moved in it.'"

13. In the first part Albert showed that both conditional propositions are false because logically possible situations are imaginable which falsify them. Now he explores whether a body could move in a vacuum on the assumption that vacuum exists *and* that a heavy body is actually located within it.

14. Albert refers here to two types of vacuum that are distinguished in this eighth question, which is reproduced in Selection 53.1.

15. In this sense, a vacuum is conceived as a body between whose sides there is nothing (see Selection 53.1).

16. Assuming a correct text, Albert's intent is wholly unclear. Under the conditions described here, it seems that the body does not even enter the vacuum formed where earth and water formerly existed but remains in the sphere of air. In what sense, then, are we to conceive of it as having moved in a vacuum? Are we to suppose that the entire sublunar region, including the concavities of fire and air as well as the concavities where earth and water formerly existed, is conceived as a vacuum? This would indeed be a strange interpretation of the second type of vacuum. That Albert believes a heavy simple body would move instantaneously if it fell freely in a vacuum as usually understood emerges in what follows (see n. 17 and the fifth conclusion, below).

17. That is, as a separate space devoid of body (see Selection 53.1). As Albert tells us, this is the usual sense of vacuum. Simple (or pure elemental) heavy bodies would, it seems, move with infinite velocity in this kind of separate vacuum, since they do not have internal resistance and the void itself lacks external resistance. According to almost all adherents of Aristotelian physics in the Middle Ages, if such a body moved freely it would have an infinite velocity, since no ratio of motive force to resistance would be possible. This seems to be Albert's position in the fifth conclusion of this question. But as will be seen in the fourth conclusion of this question and especially in the fourth conclusion of the next (twelfth) question, Albert believes that force and resistance relations are distinguishable in suitably devised thought experiments, so that motion of simple heavy bodies in a vacuum would be possible. Heavy mixed bodies, however, which are compounds of varying proportions of the four elements, have internal resistances (understood in the sense described in Selection 51.1, n. 9) and could move in a hypothetical vacuum with finite velocities, as Albert shows in this conclusion.

18. In this imaginary situation, the concentric spheres of earth, water, and air are assumed to be annihilated, with only the outermost sphere of fire remaining as an external resistance.

19. That is, $F/R = 5/1$, where F represents the motive power and R the resistance of the fire.

---

ics, Bk. IV, Question 10, fol. 77 recto, col. 1 (for full title, see Selection 14, n. 1), John Buridan offers virtually the same two conclusions. "And so for the second conclusion, I say that this is not a good consequence: 'A vacuum exists, therefore a heavy body is moved in it.' The reason for this is that having assumed the existence

mixed body], which is as 2, begins to resist the earth, which is as 3, because the air strives to remain there. But since this mixed body is dominated by earth, it descends further so that only the earth in it moves while the air resists, whereas previously in the full medium ·[of fire] both were moving.[20] Thus the mixed body descends further into the vacuum by a ratio of 3 to 2, which is smaller than a quintuple ratio, as is the motion it produces. And so, the same mixed body has a much slower motion in a vacuum than in a plenum, namely fire. It follows from this that the natural motion of some heavy body can be quicker in the beginning than in the end. For example, if a mixed [or compound] body of four elements should have one degree of fire, one of air, one of water, and four of earth and if everything were annihilated within the sides of the sky except this mixed body, and if the mixed body were placed where the fire was, then this mixed body would descend more quickly through the vacuum of fire *(vacuum ignis)* than through the vacuum of air *(vacuum aeris),*[21] and so on, as can easily be deduced from this case. But you [now] say, What should be said, therefore, about the common assertion that natural motion is quicker in the end than at the beginning? One can say that is universally true of the motion of heavy and light bodies but not of the motion of heavy and light mixed [or compound] bodies.

A fourth conclusion [is this]: If a vacuum existed, a heavy simple body could be moved in it successively. For example, assuming that everything within the concave [surface] of the lunar orb were annihilated except for a small uniformly heavy body 1 foot [in length], of which one fourth is on one side of the middle [or center] of the world *(medii mundi)* and the remainder is on the other side [of the center]. Then this heavy body would be moved until its middle [or center] becomes [or coincides with] the middle [or center] of the world and through the whole motion its smaller part, which is on one side of the center of the world, would resist the remainder.[22] All these things are understood on the assumption that in heavy bodies there is encompassed a tendency [or striving or appetite] *(appetitus)* by means of which they would seek to fill a vacuum if such existed and that they possess *(obtineant)* a natural tendency for this, namely a tendency for being in their proper places.

And let this be a fifth conclusion:

Assuming what has already been said and that everything within the sides of the sky is annihilated except one heavy simple body which touches the concave surface of the sky, [then] *in infinitum* quickly it [that is, the heavy simple body] would descend *(infinite velociter descenderet)* [because of what follows in this paragraph, I have translated *infinite* by *in infinitum*—Ed.]. This is proven, because [it will fall] more than twice as fast as any body descending with a definite [or finite] velocity; and more than four times as fast; and more than eight times as fast; and so on into infinity, because it would descend without any internal or external resistance. Secondly, if it should descend with any definite [or finite] velocity, it would fall with a definite ratio of motive power to resistance. But this is false. The falsity of it is obvious, because nothing would resist it; but the consequence holds, because every definite [or finite] velocity varies as

20. The concept of the potency of "natural place," which was accepted by virtually all who discussed physics in the Middle Ages, is well illustrated here. Although all is void space below the sphere of fire, that portion of it formerly containing air causes the body to behave differently than those portions which formerly contained water or earth. Thus, although in one sense it would be absurd to say that one portion of void space "differs" from any other, it is yet true to say that for Albert the different intrinsic cosmic properties of each natural portion of sublunar space continue to exert their customary influence on the behavior of elements. The potency and behavior of each natural place is thus unalterable, and the air in the mixed body behaves as would any quantity of air in that portion of space— that is, it resists the motion of heavy bodies and acts as if it were an external medium.

21. That is, the respective vacua where formerly there were fire and air.

22. Compare Albert of Saxony's discussion at the end of Selection 47 and note 15 thereto. In this example we have a homogeneous body situated as in Fig. 1. Let *c*

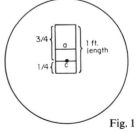

Fig. 1

represent the center of the world and *a* the center of the body. The body will move in a vacuum, because the fourth part of it on one side of the center (or diameter) of the world will resist the motion of the other three fourths, since both parts of this single body will seek the exact center of the world, thereby giving rise to contrary motions. But that part which is three fourths of it has a greater motive power than the resisting one-fourth part and will consequently cause the entire body to move until *a* coincides with *c*, at which time motion ceases.

some ratio of mover to resistance, at least in things moving involuntarily. But you say that if a heavy simple body would descend *in infinitum* quickly *(descenderet. . .infinite velociter)* in a vacuum, then a heavy simple body could be moved in a vacuum. But if it were moved, then it would be moved in time [that is, in a finite time] in a vacuum, since every motion occurs in time. Consequently, it would not be moved in an instant, which is the opposite of what Aristotle says. In replying to this, I concede the consequent, but deny the antecedent, for I did not say in the question that "it would descend *in infinitum* quickly," but I said that "*in infinitum* quickly it would descend."[23] The first of these statements is false and the second true; nor does the first imply the second, because in the first the words "it would descend" are taken determinately, but in the second they have merely confused supposition, because in this second statement the syncategorematic term "infinite" precedes [the words "it would descend"], while in the first statement it follows them.[24]

Yet, you might say, if a heavy simple body were moved infinitely quickly *(in infinitum velociter moveretur)* in a vacuum, it would follow that any resistance, however small, would be of infinite power. However, this is false. But the consequence is proved, because a very small resistance could reduce *(demere)* an infinite velocity with which a body is moved in a vacuum by taking from it a [single] simple degree [of velocity]. But to do this is to be of infinite power, because to diminish any degree of velocity by some [amount] is to be of some [resistive] power; and to diminish it by twice that amount is to be of [even] greater [resistive] power, and so on. Therefore, to take from it an infinite velocity is to be of infinite power. As the solution of this argument, I say that if something were moved infinitely quickly with a resistance, then I readily concede that that which could make it be moved with a finite speed would be of infinite power. If, however, something could be moved infinitely quickly without a resistance, then I say that that which makes it move with a finite speed need not necessarily be of infinite power, and from this to the argument. . . .[25]

BOOK IV, QUESTION 12

Whether, if a vacuum existed, something could be moved in it with a finite velocity or local motion, or with a motion of alteration.[26]

In the first place, it is argued that local motion [does] not [occur in a vacuum], for if it did this

would happen especially in the local motion of mixed [or compound bodies]. But this would not

23. The antecedent which is denied is the assertion that a heavy simple body "would descend *in infinitum* quickly" in a vacuum; the conceded consequent declares that "a heavy simple body could be moved in a vacuum."

The intelligibility of this difficult paragraph depends on the distinction between the expressions (1) *descenderet in infinitum velociter,* or *descenderet infinite velociter,* and (2) *in infinitium velociter descenderet,* or *infinite velociter descenderet.* By the rules of Latin grammer and syntax, these phrases are equivalent and would ordinarily be translatable by the same expression. Not, however, in this paragraph, where word order becomes crucial. That *descenderet* ("it would descend") precedes *in infinitum* (or *infinite,* which is equivalent to *in infinitum*) in expression (1) and follows in expression (2) requires two quite different interpretations, which will be described in the next note. To emphasize the fundamental importance of word order, the equivalent Latin terms "in infinitum" and "infinite" have been rendered as "in infinitum" in the translation (ordinarily, they would be translated as "infinitely")." The words that precede and follow will thus be readily detectable.

24. In the first expression, *descenderet in infinitum velociter* ("it would descend *in infinitum* quickly"), the term "infinitum" follows the term it modifies, namely *descenderet* ("it would descend"), and thereby gives determinate supposition to *descenderet.* Under these circumstances, the term *infinitum* was often considered, as it is here, a categorematic infinite, which signifies that the heavy simple body descending from the concave surface of the sky would fall with a single, actual infinite speed that is greater than any other assignable speed. Albert rejects this in favor of the second mode of expression, *in infinitum velociter descenderet.* Here the term "infinitum," or "infinite," precedes the term it modifies, namely *descenderet,* and thus gives merely confused supposition to *descenderet.* In this instance, the term *infinitum* is considered a syncategorematic infinite, which signifies that the heavy simple body will fall infinitely quickly in the sense that however large the assigned speed, the body can always descend with yet a greater speed. On this interpretation, Albert holds that a heavy simple body could be conceived to move temporally in a vacuum only with a syncategorematic infinite speed. For the distinction between *determinate* and *merely confused suppositio,* see Selection 18, [II] 4, and Selection 52.4, reply to the first argument supporting Henry of Harclay's views and notes 17 and 19 thereto. I am indebted to John E. Murdoch for the interpretation in this and the preceding note, and for that part of the translation which they elucidate.

25. A paragraph containing Albert's reply to the five principal arguments enunciated at the beginning of this question, is omitted. Although the infinite speed discussed in this final paragraph of Question 11 seems to be syncategorematic, I have not indicated this by word order.

26. "Motion of alteration" signifies change of quality, as, for example, from hot to cold, white to black, and so on.

occur, because mixed bodies are moved only with the motion of the predominating element. But no element is moved with a finite velocity in a vacuum, as declared in another question,[27] because of a lack of internal and external resistance. Therefore, it would seem that no mixed body [could be moved in a vacuum]. This can be confirmed [as follows]: Take a portion of pure earth and take a portion of lead heavier than this earth. Now, this [piece of pure earth] would not descend in the vacuum with a finite velocity but [rather] with an infinite velocity. The piece of lead, therefore, could not descend with a finite velocity.

Secondly, if [there could be motion with a finite velocity in a vacuum], it would seem especially true with respect to motion of alteration [that is, change of quality], since no external resistance is required for this. But this is not so, because local motion, which is the primary motion, cannot occur in a vacuum.[28] Therefore, it seems that motion of alteration would not occur. Therefore, and so on [that is, no motion of finite velocity can occur in a vacuum].

I argue against this. For if a vacuum existed, men could, nonetheless, walk around on the earth—although birds could not fly. It is true that men could not walk around for long, since they could not live long without breathing. Similarly, if some heated water were assumed to exist in a vacuum, it would change itself to its prior state of coldness. It seems, therefore, that local motion as well as motion of alteration could be made in a vacuum, for as stated in the other question,[29] mixed [or compound] bodies could be moved in a vacuum because they have internal resistance.

In brief, I say that we have never experienced the existence of a vacuum, and so we do not readily know what would happen if a vacuum did exist. Nevertheless, we must inquire what might happen if it existed, for we see that natural beings undergo extraordinarily violent actions to prevent a vacuum. Indeed we see that if some tube (*fistula*) is put in water, and air is drawn from the tube, the water follows by ascending, striving to remain contiguous with the air lest a vacuum be formed.[30]

And so, let the first conclusion be that if God should annihilate everything between two walls, those two walls would become rarefied infinitely quickly [*in infinitum velociter*] in order to avoid a vacuum. This can be proven, because it has already been declared that natural things seek to avoid a vacuum and are inclined toward this action by

their very natures. Now, they strive for, or are inclined to, this finitely or infinitely. If the latter, the sides [or walls] would meet infinitely quickly and the intent of this [conclusion] would be had. However, if the striving is only finite, it follows that a finite power (*virtus*) that is less than the power and inclination of these walls would be sufficient to hold them and prevent their meeting. Hence a finite power could conserve a vacuum and even increase it if it were a greater power [than the one that only conserves it] and could condense those walls. But this is false, because only God could conserve, increase, and make a vacuum.[31]

But you say [now] that there is no infinite striving (*appetitus infinitus*) in these lower beings, so that it does not follow that if a vacuum did exist these natural beings would seek to avoid it infinitely. I reply by conceding that in fact there is no infinite striving [or inclination] in these lower beings. Nevertheless, if God should make [or form] a vacuum, natural beings would then strive infinitely to avoid it if their natures could be directed; and this is what is commonly said: "Nature abhors a vacuum infinitely." Indeed, before the sky would allow a vacuum to remain, it would descend and fill it.

A second conclusion [is this]: If there were a vacuum between sky and earth, and if there should be a plane surface equidistant from the center on which there are two heavy spheres A and B, with A heavier than B, then any power (*virtus*) [or force], however small, could, with infinite ease, move any

27. Probably a reference to Aristotle's argument cited near the beginning of Question 11. As will be evident in some of the conclusions to follow in this selection, Albert does not accept this unqualifiedly, since he believes that under properly imagined circumstances pure elemental bodies could move with finite velocities in a vacuum (see also n. 16 and n. 17). Here, however, he is formulating an opposition argument which he will subsequently reject.
28. Proponents of this argument, which Albert will reject, would insist that a change of quality depends on spatial motion of the material parts of the body undergoing the change of quality.
29. See the second and third conclusions of Question 11, Selection 55.2.
30. Substituting wine for water, the same illustration is cited by Buridan (Selection 53.2); compare also Marsilius of Inghen's second experiment (Selection 53.3).
31. That only God could make a vacuum was a standard feature of medieval discussions of the void. Except for Nicholas of Autrecourt, and perhaps a few others, none believed that God had made, or would make, a vacuum within the physical cosmos.

of these spheres on this surface.[32] This can be proved, because any of these spheres would touch this surface at a point and thus any half would counterbalance equally the other half just as two weights in a balance.[33] But since any excess would suffice to produce motion, it follows that any [motive] power (*virtutem*) [however small] could move any one of these spheres with infinite ease.[34] However, it is true that this could not occur in a plenum, because then not only would one [half] of the sphere resist the pushing force [*pellenti*] but also air. And so in air not every force that is as small as you please could push and move these spheres.[35]

[Here is] a third conclusion: Although a small force (*virtus*), however small, could move these spheres in a vacuum, no force could move the large and the small heavy body with equal velocity. Let this be set forth in numbers. For it happens that in the greater sphere, namely in *A*, one half of it pulls against the other half, as [for example] 2 against 2; and in the small sphere, say *B*, one half pulls against the other, [say] as 1 against 1. And should there be a power as 1, it would move the small body in a greater ratio—and therefore more quickly—than the larger body. Thus it moves the small body with a double ratio [that is, $F/R = 2/1$] and the large body with a sesquialterate ratio [that is, $F/R = 3/2$]. For when this power begins to move the small body, one of those halves in moving the other [half] moves itself with it, and when any one of those halves is as 1, the whole power, the total motive power of the small body will be as 2 and the resistance as 1. But 2 to 1 is a double ratio. And in the motion of the great sphere, the total motive power is as 3, the resistance as 2, and the ratio of 3 to 2 is a sesquialterate ratio.

32. Since no figure accompanies this conclusion, the manner of conceiving this demonstration is not obvious. For, what does Albert mean by "a plane surface equi-

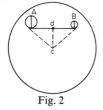   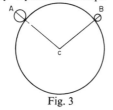

Fig. 2   Fig. 3

distant from the center on which there are two spheres *A* and *B*" ("aliqua superficies plana eque distans a centro super quam essent due sphere graves, una *A* et alia *B*"; fol. 51 recto, col. 1)? Obviously all the points on a plane cannot be equidistant from the center of a sphere. Perhaps something like Fig. 2 is meant, where the two points of contact of sphere and plane are equidistant from *c*, and *d* is the midpoint of the surface. If, however, *plana*

signifies merely an even or uniform, rather than a flat, surface, perhaps a curved surface is intended, as in Fig. 3, where the circumference of the circle would make all the points of the surface equidistant from the center *c*. Indeed any representation in which the two heavy spheres are virtually equidistant from the center of the world would be satisfactory, since "heaviness" of a body in Aristotelian physics varies with distance from the center of the world.

33. That is, one half of sphere *A* will counterbalance the other half of *A*, just as one half of *B* counterbalances the other half of *B*. This interpretation gains support from the illustration in the next (third) conclusion. Thus Albert imagines the two halves of each sphere as constituting a force-resistance relation. Since in each sphere $F = R$, where *F* is motive force and R resistance, neither will move, for it was an accepted principle in medieval physics that motion occurs only when $F > R$ but not when $F \leq R$.

34. In this interesting conclusion Albert remains faithful to Aristotelian physics in one sense and drastically departs from it in another. Should any force, however small, be applied to spheres *A* or *B*, a motion would be produced, since by assumption $F = R$ in each sphere, so that $F + f > R$, where *f* is any applied force. For, as stated in note 33, motion is possible only when the force is greater than the resistance, as required by the commonly accepted interpretation of Aristotelian physics. But if we ignore the arbitrary division of each sphere into halves representing motive force and resistance, and focus on the fact that a force as small as you please is deemed able to move either sphere, we realize that from this standpoint Albert daringly allows that motion in a hypothetical vacuum can arise when $F < R$. Generally, as long as the force is applied, a force as small as you please can move a body as large as you please. This second interpretation may be compared to Galileo's early discussion in *De motu (On Motion)*. There, in chapter 14, Galileo truly abandons the Aristotelian doctrine that motion can arise only when $F > R$. If a perfectly smooth, hard spherical body rests on a perfectly smooth and hard plane, and "if everything is arranged in this way, then any body on a plane parallel to the horizon will be moved by the very smallest force, indeed, by a force less than any given force"; *Galileo Galilei, On Motion*, trans. I. E. Drabkin (Madison, Wis.: University of Wisconsin Press, 1960), pp. 65–66. In addition to the fact that Galileo's discussion is not hypothesized in a void, the "plane parallel to the horizon" follows the earth's surface and circumference, so that the motion would be circular if continued indefinitely. Although it does not appear to be inertial in the *De motu*, it is indeed inertial (though still circular) in *Dialogues Concerning Two New Sciences (Discorsi)*, written many years later. Albert does not describe the path which *A* or *B* will follow. Also, their motions could not be inertial, since removal of *f* would make $F = R$ and the motion would cease immediately. Marsilius of Inghen, in his *Questions on the Eight Books of the Physics*, Book IV, Question 12, has virtually the same example (fol. 54 verso, col. 2, of the edition cited in Selection 53.3, n. 20).

35. That is, if $F = R$, and *f* is the applied force, however small, and *r* the resistance of the air, then motion arises in air only when $F + f > R + r$, and $f > r$.

A fourth conclusion [is this]: Similarly if any balances were suspended in a vacuum, the greater weight—even if it were a heavy simple body—would descend successively. This is obvious, for although it would not have internal resistance nor external resistance in the form of a medium, yet it would have an external resistance [in the form] of a counterweight.

A fifth conclusion [is this]: After suspending balances in a vacuum and hanging two equal weights, any heavy body, however small, could be added to [either] one [of the suspended weights] and cause it to descend. The reason for this is that any excess suffices to produce motion. But this would not be true in a plenum, as was said.

A sixth conclusion [is this]: If there were a table placed transversely, [or crosswise] in a vacuum and on it is posited a heavy simple spherical body, the body would descend on the table with a successive finite velocity. This is obvious, because it would not descend by a straight line [but] would descend by an uncertain path (natando); and one part [of the table(?)] would raise the other, and then that part which is raised violently would take the place of a resistance.[36]

[Here is] a seventh conclusion: Although a lead mass should weigh more on a balance than simple water, nevertheless if both were posited in a vacuum, water would move or descend infinitely quickly; lead, however, would descend successively. This is obvious, because lead, although it has no external resistance, yet because it is a mixed [or compound] body, has internal resistance;[37] water, however, has neither internal nor external resistance.[38]

An eighth conclusion [is this]: Mixed [or compound] bodies of homogeneous composition [consimilis compositionis] are moved with equal velocity in a vacuum but not in a plenum. The first part [concerned with fall in a vacuum] is obvious, because they are of homogeneous mixture. The ratio of motive power to total resistance in one body is the same as in another homogeneous [consimilis] body, because they both have only internal resistance.[39] The second part [that is, dealing with the plenum] is [also] obvious, because if there is a mixed [or compound] body, say A, whose heaviness (gravitas) is as 8 and lightness (levitas) as 4, and there is another mixed body, B, whose heaviness is as 4 and lightness as 2, then A and B are homogeneous mixed bodies.[40] [Finally] these bodies are posited in a resistant medium taken as 1. I [now] prove A descends quicker than B. This is obvious from [A's]

greater ratio of moving power to resistance. For by computing together the internal and external resistances, the total resistance of A is as 5 and the motive power as 8; and in B the total resistance is as 3 and the motive power as 4. Now, a ratio of 8 to 5 is greater than a ratio of 4 to 3.[41]

A ninth conclusion [is this]: In a vacuum—if it existed—an alteration would be possible. This is obvious, for if a vacuum existed and heated water were assumed in it, it would be altered from outside to its prior coldness.

36. In this ambiguous and fuzzily presented illustration of finite and temporal motion in a vacuum, it seems that the sphere is rolling down a table lying crosswise with respect to an observer at the center of the world (see Fig. 4). The force-resistance dichotomy is either

Fig. 4

conceived in the spherical body itself, as in the second and third conclusions, or the part of the table which rises as the sphere rolls down serves as the resistance to the sphere. In the first possibility, if half the sphere opposes the other half and $F = R$, how will the motion arise? Perhaps, then, the second alternative is more plausible. Realizing that a sphere would roll down an inclined table, Albert feels compelled to distinguish a force and resistance, so that he could meet the Aristotelian requirements for a finite motion.

37. In an obvious mistake, the text has "external" (extrinseca).

38. See note 17.

39. Earlier, Bradwardine had formulated the same conclusion and undoubtedly for the very same reason (see Theorem XII in Selection 51.1 and note 45 thereto), whereas Galileo, who also reached the same conclusion about two centuries later, argued on a quite different basis (see the introduction to Selections 55.4, 5 and Selection 55.6, by Galileo.

40. Since their ratios of heaviness to lightness—or motive force to resistance—are equal (that is, $8/4 = 4/2$), Albert concludes that they are necessarily bodies of the same homogeneous composition.

41. While two unequal homogeneous mixed bodies would fall with equal speed in the void, their behavior in a plenum is made to depend on which particular equal ratio is assigned to it. In the example cited here A descends with greater speed than B because the ratio $8/4$ was assigned to it and the equal ratio $4/2$ was assigned to B. Thus A might be smaller than, equal to, or greater than B, and it would, under the conditions set forth, descend more quickly then B. It is painfully evident how bewildering were the consequences of the doctrine of internal resistance when interpreted in this arbitrary manner

A tenth [and final] conclusion [is this]: If a vacuum existed, an alteration would not be possible in it from the outside unless the agent and patient are immediate [that is, in direct contact], for otherwise one of them could not act on the other. And this could happen because something intermediate that could serve to receive and transmit (*deferentis*) the action is lacking. Thus, if a vacuum existed between us and the sky, we could not see the stars nor the sky. And this could be the result of a lack of a medium by means of which the visible species could be multiplied.[42] Similarly, if there were a vacuum between you and me, then however much I shouted you would neither hear nor see me; and all this would be the result of a lack of a medium.

[I now reply] to the principal arguments.[43] To the first, I say that although a mixed [or compound]

body is moved with the motion of the predominant element in it, nevertheless it does not move with so great a velocity as that element [itself] if it were separated from the compound. The confirmation of this is obvious from the question. To the second [argument] I say that indeed any local motion could occur in a vacuum—if it existed or could be made. And so, what was assumed in this argument must be denied.[44] But also [it must be denied] that every local motion is the cause and first motion of all other [types of] motions. But [this can be said of] the circular motion of the sky only.[45] It is [certainly] not necessary that if a motion of alteration [or change of quality] could occur in a vacuum, that [therefore] any other local motions could occur. It follows, and so on [that the second principal argument is false].

3. Galileo Galilei (1564–1642): His Earliest Law of Motion and His Arguments for Finite Velocity in a Void

Translated and annotated by I. E. Drabkin;[46] additional notes by Edward Grant[47]

CHAPTER [10]

IN WHICH, IN OPPOSITION TO ARISTOTLE, IT IS PROVED THAT, IF THERE WERE A VOID, MOTION IN IT WOULD NOT TAKE PLACE INSTANTANEOUSLY, BUT IN TIME.

Aristotle, in Book 4 of the *Physics,* in his attempt to deny the existence of a void adduces many arguments. Those that are found beginning with section 64[48] are drawn from a consideration of motion. For since he assumes that motion cannot take place instantaneously, he tries to show that, if a void existed, motion in it would take place instantaneoulsy; and, since that is impossible, he concludes necessarily that a void is also impossible. But, since we are dealing with motion, we have decided to inquire whether it is true that, if a void existed, motion in it would take place instantaneously. And since our conclusion will be that motion in a void takes place in time,[49] we shall first examine the contrary view and Aristotle's arguments.

In the first place, of the arguments adduced by

45. Following Aristotle, it was generally accepted that the circular motions of the heavens were essential for all other terrestrial motions, since the annual motion of the sun determined the changes of season, the growth of crops, and life itself.

46. [Reprinted with permission of the Regents of the University of Wisconsin from *Galileo Galilei "On Motion" and "On Mechanics" Comprising "De Motu" (ca. 1590),* translated with Introduction and Notes by I. E. Drabkin, and *"Le Meccaniche" (ca. 1600),* translated with Introduction and Notes by Stillman Drake (Madison, Wis.: University of Wisconsin Press, 1960), pp. 41–50. Although possibly the earliest extant scientific essay by Galileo, it was left unpublished. It reveals an interesting mixture of traditional (that is, medieval) and new concepts.—*Ed.*]

47. [Since additional notes are interspersed with Drabkin's, the numbers of the notes as they appear in his translation have been altered. Four simple marginal figures, which contribute in no way to the understanding of the arguments, have been omitted.—*Ed.*]

48. [214b. 10ff. I have replaced Drabkin's citation of 214a.27ff., since it does not correspond to Galileo's reference to Text 64 of Book IV of Aristotle's *Physics.* Here, and in other places below, Galileo is using the traditional medieval system of reference to the texts of Aristotle. In all such cases Drabkin has converted the medieval citation to the modern system of reference based upon Bekker numbers.—*Ed.*]

49. [As Galileo seems to have acknowledged (see n. 69), this viewpoint was widely adopted in the Middle Ages and, as already stated, is traceable to Avempace and ultimately to Philoponus. See Selection 44.2 and n. 17 thereto.—*Ed.*]

42. That is, transmitted. For a discussion of multiplication of species, see Selection 62.1–4.

43. These are the two responses to the two opening arguments of Question 12.

44. In the second principal argument it was assumed that no local motion could occur in a vacuum, and consequently no motion of alteration, or change of quality, was possible.

Aristotle there is none that involves a necessary conclusion, but there is one which, at first sight, seems to lead to such a conclusion. This is the argument set forth in sections 71 and 72,[50] in which Aristotle deduces the following inconsistency—that, on the assumption that motion can take place in time in a void, then the same body will move in the same time in a plenum and in a void. In order to be better able to refute this argument, we have decided to state it at this point.

Thus, Aristotle's first assumption, when he saw that the same body moved more swiftly through the rarer than through the denser medium, was this: that the ratio of the speed of motion in one medium to the speed in the second medium is equal to the ratio of the rareness of the first medium to the rareness of the second. He then reasoned as follows. Suppose body a traverses medium b in time c, and that it traverses a medium rarer than b, namely d, in time e. Clearly, the ratio of time c to time e is equal to the ratio of the density of b to the density of d. Suppose, then, that there is a void f and that body a traverses f, if it is possible, not in an instant, but in time g. And suppose that the ratio of the density of medium d to the density of some new medium is equal to the ratio of time e to time g. Then, from what has been established, body a will move through the new medium in time g, since [the density of] medium d has to that of the new medium the same ratio as time e to time g. But in the same time g body a also moves through the void f. Therefore a will in the same time move over two equal paths, one a plenum, the other a void. But this is impossible. Therefore the body will not move through the void in time; and therefore the motion will be instantaneous.

Such is Aristotle's proof. And, indeed, his conclusions would have been sound and necessary, if he had proved his assumptions, or at least if these assumptions, even though unproved, had been true. But he was deceived in this, that he assumed as well-recognized axioms propositions which not only are not obvious to the senses, but have never been proved, and cannot be proved because they are completely false. For he assumed that the ratio of the speeds of the same body moving in different media is equal to the ratio of the rarenesses of the media. But that this is false has been fully proved above.[51] In support of that proof, I shall add only this. Suppose it is true that the ratio of the rareness of air to the rareness of water is equal to the ratio

of the speed of a body moving in air to the speed of the same body in water. Then, when a drop or some other quantity of water falls swiftly in air, but does not fall at all in water, since the speed in air has no ratio to the speed in water, it follows, according to Aristotle himself, that there will be no ratio between the rareness of air and the rareness of water. But that is ridiculous.[52]

Therefore, it is clear that, when Aristotle argues in this way, we must answer him as follows. In the first place, as has been shown above, it is not true that differences in the slowness and speed of a given body arise from the greater or lesser density and rareness of the medium. But even if that were conceded, it is still not true that the ratio of the speeds of the motion of the body is equal to the ratio of the rarenesses of the media.

And as for Aristotle's statement in the same section that it is impossible for one number to have the same relation[53] to another number as a number has to zero, this is, of course, true of geometric ratios [viz., a/b], and not merely in numbers but in every kind of quantity. Since, in the case of geometric ratios, it is necessarily true that the smaller magnitude can be added to itself a sufficient number of times so that it will ultimately exceed any magnitude whatever, it follows that this smaller magnitude is something, and not zero. For zero, no matter how often it is added to itself, will exceed no quantity. But Aristotle's conclusion does not apply to *arithmetic* relations [viz., the difference $a - b$]. That is, in these cases, one number can have the same relation to another number as still another number has to zero. For, since [two pairs of] numbers are in the same arithmetic relation when the difference of the [two] larger is equal to

50. [215a.24–216a.3. Drabkin gives 215b–216a, which does not precisely embrace Texts 71 and 72.—*Ed.*]

51. [In chapter 8. In the next few lines, Galileo describes Aristotle's law in terms of the rareness, or subtlety, of media, that is, $V_2/V_1 = R_2/R_1$, where R is the rareness of a medium. See Selection 44.2, n. 7.—*Ed.*]

52. There is the following marginal addition at this point: "If this were true, the ratio of the speed of motion in air to the speed of motion in water would be the same for all bodies. Hence the ratio of the speed of lead in air to its speed in water would be the same as the ratio of the speed of wood in air to its speed in water. But who cannot see that this is false? For lead sinks in water, wood does not."

53. Galileo's point depends on two senses of this word *(proportio)*. Aristotle's conclusion is sound for what Galileo here calls a geometric *proportio* (i.e., the ratio a/b), but not for an arithmetic *proportio* (i.e., the difference $a—b$).

the difference of the [two] smaller,[54] it will, of course, be possible for one number to have the same [arithmetic] relation to another number, as still another number has to zero. Thus, we say that the [arithmetic] relation of 20 to 12 is the same as that of 8 to 0: for the excess of 20 over 12, i.e., 8, is equal to the excess of 8 over 0.

Therefore, if, as Aristotle held, the ratio of the speeds were equal to the ratio, in the geometric sense, of the rarenesses of the media, Aristotle's conclusion would have been valid, that motion in a void could not take place in time. For the ratio of the time in the plenum to the time in the void cannot be equal to the ratio of the rareness of the plenum to the rareness of the void,[55] since the rareness of the void does not exist.[56] But if the ratio of the speeds were made to depend on the aforesaid ratio, not in the geometric, but in the arithmetic sense [i.e., as a ratio of differences], no absurd conclusion would follow. And, in fact, the ratio of the speeds does depend, in an arithmetic sense, on the relation of the lightness of the first medium to that of the second. For the ratio of the speeds is equal, not to the ratio of the lightness of the first medium to that of the second, but, as has been proved, to the ratio of the excess of the weight of the body over the weight of the first medium to the excess of the weight of the body over the weight of the second medium.

So that this may be clearer, here is an example. Suppose there is a body a whose weight is 20, and two media unequal in weight, bc and de. Let the volume of b be equal to that of a, and the volume of d also equal to that of a. Since we are now discussing downward motion that takes place in a void, let the media be lighter than the body a, and let the weight of b be 12, and of d 6. It is clear, then, from what was proved above, that the ratio of the speed of body a in medium bc to the speed of the same body in medium de will be equal to the ratio of the excess of the weight of a over the weight of b to the excess of the weight of a over the weight of d, that is, as 8 is to 14. Thus if the speed of a in medium bc is 8, its speed in medium de would be 14. Now it is clear that the ratio of the speeds, 14 to 8, is not the same as the ratio (in the geometric sense) of the lightness of one medium to the lightness of the other. For the lightness of medium de is double that of medium bc (for since the weight of b is 12, and of d 6, i.e., since the weight of b is double the weight of d, the lightness of d will be double the lightness of b); but a speed of 14 is less than twice

a speed of 8. Yet the speed 14 has to the speed 8 the same relation, in the arithmetic sense, as the lightness of d to the lightness of b, since the difference between 14 and 8 is 6, and 6 is also the difference between the lightness of d (12) and the lightness of b (6).[57]

Furthermore, if medium de should be lighter, so that the weight of d is 5, the speed f will be 15 (for 15 will be the difference between the weight of body a and the weight of the medium d). And again the relation [i.e., arithmetic difference] of speed 15 and speed 8 will be the same as between the weight of medium b (12) and the weight of medium d (5), that is, the same as the relation of the lightness of d and the lightness of b. For the difference in each case will be 7. Furthermore, if the weight of d is only 4, the speed f will be 16: and the relation of speed 16 and speed 8 (with a difference of 8) is the same arithmetic relation as between the weight of b (12) and the weight of d (4), i.e., between the lightness of d and the lightness of b, the difference being also 8. If, again, medium de becomes lighter, and the weight of d is only 3, the speed f will now be 17. And between the speed f (17) and the speed 8 (a difference of 9), the difference is the same as between the weight of b (12) and the weight of d (3), i.e., as between the lightness of d and the lightness of b. If, again, medium de becomes lighter, and the weight of d is only 2, the speed f will now be 18. And the arithmetic difference between that speed and the speed 8 will be the same as the difference between the weight of b (12) and the weight of d (2), i.e., between the lightness of d and the lightness of b. In each case the difference will be 10. If, again, medium de becomes lighter, and the weight of d is only 1, the speed f will now be 19. And there will be the same arithmetic difference between this speed and the speed 8 as between the weight of

54. Or, what is equivalent, the difference between the two members of each pair is equal.

55. There seems to be a slip here. Galileo may have meant to say "density" instead of "rareness" (see the next note). The same effect would have been achieved by saying "ratio of the rareness of the *void* to the rareness of the *plenum*" or by speaking of the ratio of *speeds*, not of times.

56. In the sense of being greater than any finite number. But Galileo may have meant to speak of density, in which case *nulla* would be "zero."

57. If the weights of d and b are 6 and 12, respectively, Galileo here speaks of their lightnesses as 12 and 6, respectively, with difference 6. And similarly in what follows.

b (12) and the weight of d (1), i.e., between the lightness of d and the lightness of b. In each case the difference will be 11. Now if, finally, the weight of d is 0, so that the difference between the weight of body a and of the medium d is 20, the speed f will be 20; and the arithmetic difference between the speed f (20) and the speed 8 will be the same as that between the weight of b (12) and the weight of d (0), the difference in each case being 12.

It is clear, therefore, that the relation of speed to speed is the same as the relation of the lightness of one medium to the lightness of the other, not geometrically [i.e., as a quotient] but arithmetically [i.e., as a difference]. And since it is not absurd for this arithmetic relation [i.e., difference] to be the same between one quantity and a second quantity as between a third quantity and zero, it will similarly not be absurd for the relation of speed to speed to be the same, in this arithmetic sense, as the relation of a given lightness [of medium] to zero.

Therefore, the body will move in a void in the same way as in a plenum. For in a plenum the speed of motion of a body depends on the difference between its weight and the weight of the medium through which it moves. And likewise in a void [the speed of] its motion will depend on the difference between its own weight and that of the medium.[58] But since the latter is zero, the difference between the weight of the body and the weight of the void will be the whole weight of the body. And therefore the speed of its motion [in the void] will depend on its own total weight.[59] But in no plenum will it be able to move so quickly, since the excess of the weight of the body over the weight of the medium is less than the whole weight of the body. Therefore its speed will be less than if it moved according to its own total weight.

From this it can clearly be understood that in a plenum, such as that which surrounds us, things do not weigh their proper and natural weight, but they will always be lighter to the extent that they are in a heavier medium. Indeed, a body will be lighter by an amount equal to the weight, in a void, of a volume of the medium equal to the volume of the body. Thus, a lead sphere will be lighter in water than in a void by an amount equal to the weight, in a void, of an aqueous sphere of the same size as the lead sphere. And the lead sphere is lighter in air than in a void by an amount equal to the weight, in a void, of a sphere of air having the same size as the lead sphere. And so also in fire, and in other media. And since the speed of a body's motion depends on the weight the body has in the medium in which it moves, its motion will be swifter, the heavier the body is in relation to the various media.

But the following argument is invalid: "A void is a medium infinitely lighter than every plenum; therefore motion in it will be infinitely swifter than in a plenum; therefore such motion will be instantaneous." For it is true that a void is infinitely lighter than any [nonvacuous] medium; but we must not say that such a [nonvacuous] medium is of infinite weight. We must instead understand [the applicability of the term "infinite"] in this way, that between the lightness of air, for example, and a void there may exist an unlimited number of media lighter than air and heavier than a void. And if we understand the matter in this way, there may also exist, between the speed in air and the speed in a void, an unlimited number of speeds, greater than the speed in air and less than the speed in a void. And so also between the weight of a body in air and its weight in a void, an unlimited number of intermediate weights may exist, greater than the

58. [Galileo's law of motion is $V = W_{body} - W_{medium}$, where V is velocity and W is weight per unit volume, or specific weight. Hence a difference in specific weights determines velocity, so that it does not matter whether a body falls in a plenum or a void. The subtractive relation is such that when $W_{medium} = 0$, velocity is proportional to the specific weight of the body. In this Galileo is apparently following the tradition stemming from Avempace (and beyond him probably to Philoponus; for references, see n. 49), where velocity was thought to be determined by the difference—not the ratio as Aristotle would have it—between the weight of a body and the resistance of the medium through which it fell. Resistance of a medium was something to be subtracted, since it served only to retard motion. In the absence of a resistant medium bodies were held to fall freely in the void with a natural finite velocity. Although numbers were arbitrarily assigned and subtractions made (see Bradwardine's criticism of the first erroneous theory at the beginning of Selection 51.1), to my knowledge no one in the Middle Ages, Avempace included, specified how this subtraction could be interpreted physically. This is not surprising when one realizes that it was usual to conceive of bodies in terms of gross, or total, weight and the resistance of a medium by its density (often vaguely understood). Under these circumstances, no proper subtraction was possible. By utilizing specific weight as the criterion for measuring the difference between body and medium, Galileo, and Giovanni Benedetti before him, made Avempace's law theoretically precise and intelligible even though it was as erroneous as the dynamic theories with which it contended.—Ed.]

59. I.e., its weight in the void, undiminished by any weight of medium.

weight of the body in air, but less than its weight in a void.

And the same is true of every continuum. Thus between lines a and b, of which a is greater, an unlimited number of intermediate lines, smaller than a, but greater than b may exist (for since the amount by which a exceeds b is also a line, it will be infinitely divisible). But we must not say that line a is infinitely greater than line b, in the sense that even if b were to be added to itself without limit, it would not produce a line greater than a. And by similar reasoning, if we suppose a to be the speed in a void, and b the speed in air, an unlimited number of speeds, greater than b and smaller than a, will be able to exist between a and b. Yet we must not conclude that a is infinitely greater than b, in the sense that the time in which [the motion with] speed a is accomplished, when added to itself any number of times without limit, can still never exceed the time corresponding to speed b, and that, therefore, the speed corresponding to time a is instantaneous.

It is therefore clear how the argument is to be understood. "The lightness of a void infinitely exceeds the lightness of a [nonvacuous] medium; therefore the speed in the void will infinitely exceed the speed in a plenum." All that is conceded. What is denied is the conclusion: "Therefore the speed [i.e., the motion] in the void will be instantaneous." For such motion can take place in time, but in a shorter time than the time corresponding to the speed in a plenum; so that between the time in the plenum and the time in the void an unlimited number of times, greater than the latter and smaller than the former, may exist. Hence it follows, not that motion in a void is instantaneous, but that it takes place in less time than the time of motion in any plenum.

Therefore, to put it briefly, my whole point is this. Suppose there is a heavy body a, whose proper and natural weight[60] is 1000. Its weight in any plenum whatever will be less than 1000, and therefore the speed of its motion in any plenum will be less than 1000. Thus if we assume a medium such that the weight of a volume of it equal to the volume of a is only 1, then the weight of a in this medium will be 999. Therefore its speed too will be 999. And the speed of a will be 1000 only in a medium in which its weight is 1000, and that will be nowhere except in a void.

This is the refutation of Aristotle's argument. And from this refutation it can readily be seen that motion in a void does not have to be instantaneous.

The other arguments of Aristotle are without force or cogency. To say, for example, that in a void the body will not move in one direction rather than in another, or up rather than down, because the void does not give way upward or downward but equally in all directions,[61] is childish. For I could say the same thing about air. That is, when a stone is in air, how does the air give way downward rather than upward, or to the left rather than to the right, if the rareness of the air is everywhere the same? At this point someone, quoting Aristotle,[62] might say that air has weight in its own place and therefore helps downward motion more. We shall examine these fantasies in the next chapter, where we shall investigate whether elements have weight in their own proper places. And similarly, when they say that in a void there is neither up nor down, who dreamt this up? If the air were a void, would not the void near the earth be nearer the center than the void which is near [the region of] fire?

Similarly lacking in force is the argument which Aristotle makes about projectiles when he says:[63] "Projectiles cannot move in a void, for projectiles, when they have left the hand of the thrower, are moved by the air or by some other corporeal medium that surrounds them and is set in motion. But this is not present in a void." For Aristotle assumes that projectiles are carried along by the medium; and in the proper place we shall show that this is false. And what he adds to his argument, about different bodies moving in the same medium, is also false. For he assumes that in a plenum heavier bodies move more swiftly because they cleave the medium more forcibly, and that this is the only reason for their speed; but since that resistance is not present in a void, he supposes that all motions in a void will take place in the same time and with the same speed—and this, he asserts, is impossible.[64]

Now, in the first place, Aristotle errs in that he does not prove that it is absurd for different bodies to move in a void with the same speed. But he makes an even greater error when he assumes that the speeds of different bodies depend on an ability of heavier bodies to divide the medium better. For, as we showed above, the speed of moving bodies does not depend on this, but on the size

60. I.e., its weight in a void.
61. Cf. *Physics* 215a.22–24.
62. Cf. *De Caelo* 311b.9.
63. Cf. *Physics* 215a.14–19.
64. *Physics* 216a. 12–21.

of the difference between the weight of the bodies and the weight of the medium. For the speeds are in the ratio of these differences. But the difference between the weights of different bodies and the weight of the same medium is not the same (for otherwise the bodies would be equally heavy). Therefore the speeds will not be equal. For example, in the case of a body whose weight is 8, the excess over the weight of the void (which is 0) is 8; hence its speed will be 8. But if the weight of a body is 4, the excess over the [weight of the] void will, in the same way, be 4; and hence its speed will be 4. Finally, using the same method of proof in the case of the void as we used in the case of the plenum,[65] we can show that bodies of the same material but of different size move with the same speed in a void. So much for that.

Such[66] is the force of truth that learned men, even Peripatetics, have recognized that Aristotle's view on this subject was mistaken, though none of them could properly refute his arguments. And, as for what is contained in *Physics* 4.71–72,[67] certainly no one was ever able to refute that argument, for up to now the fallacy in it has never been noticed. And though Scotus, Saint Thomas, Philoponus,[68] and some others hold a view opposed to Aristotle's, they arrive at the truth by belief rather than real proof or by refuting Aristotle.[69] And, indeed, if one were to accept Aristotle's assumption about the ratio of the speeds of the same body moving in different media, one could scarcely hope to be able to refute Aristotle and upset his proof. For Aristotle assumes that the speed in one medium is to the speed in the other, as the rareness of the first medium is to the rareness of the second. And no one up to now has ventured to deny this relation.[70]

Nor is there any validity in the assumption made by the aforesaid writers, namely, a twofold resistance to the motion of the body—one external, resulting from the density of the medium, the other internal, by reason of the determinate weight of the body.[71] For there is something artificial about this; since those two resistances do not, if we look at the matter carefully, differ from each other. For, as has been made clear above, the density or (to use a better term) the weight of the medium makes for lightness of the moving body, and the lightness of the medium is responsible for the heaviness of the body; and the same body is now heavier, now lighter, according as it is in a lighter or heavier medium.[72] And so, these writers add nothing new when they assume this twofold resistance, since it is

65. [For Galileo's proof, see section 6 below.—*Ed.*]

66. This section to the end of the chapter is on a separate sheet. Galileo merely indicated that it belongs in this chapter; and Favaro [the editor of the National Edition of Galileo's works—*Ed.*] placed it at the end of the chapter.

67. [215a.24—216a.3. Drabkin repeats the numbers cited in note 50.—*Ed.*]

68. Ioannes Duns Scotus, *Opera Omnia* III (Paris, 1891), 102–3 (on Arist. *Phys.* 4. Quaest 12.18,19); D. Thomas Aquinas, *Opera* III (Rome, 1884), 186. *Comm. on Arist. Phys.* IV. ch.8. lectio 12, no. 8; Ioannes Philoponus, *Comm. in Phys. Arist.*, pp. 678–84 (Vitelli).

69. [Duns Scotus, Aquinas, Philoponus, and Avempace (the last-named is mentioned in note 72) are cited because they held, in opposition to Aristotle, that bodies would move in a hypothetical void with finite, not instantaneous, velocities. None of these authors, however, held or asserted Galileo's other conclusion, namely that homogeneous bodies would fall with equal speed in the void. But a view similar to Galileo's was held by Thomas Bradwardine (see Theorem XII, Selection 51.1) and Albert of Saxony (see section 2 above, Question 12), and very likely by others.—*Ed.*]

70. But Philoponus, cited above by Galileo, *does* deny the relation as it is expressed in the form "time required in one medium is to the time required in the other medium as density of first medium is to density of the second." Cf. *Commentary on Aristotle's Physics*, p. 682.30–32 (Vitelli). [This section from Philoponus is translated in M. R. Cohen and I. E. Drabkin, *A Source Book in Greek Science* (Cambridge, Mass.: Harvard University Press, 1948), p. 219; Avempace, whom Galileo does not mention in this connection, agreed with Philoponus on this point.—*Ed.*] G. B. Benedetti also denies the relation (*Diversarum speculationum mathematicarum et physicarum liber* [Turin, 1585], p. 172; see A. Koyré, *Études Galiléennes*, I, 50). From what Galileo says one may question how much direct and detailed knowledge he had, at this time, of either Philoponus' or Benedetti's work. [The remainder of this note is omitted.—*Ed.*]

71. [The internal resistance described here by Galileo is not that frequently accepted in the Middle Ages, which was held to arise either from light elements acting in an opposite direction to the predominant heavy elements of a body, or vice versa (see Selection 51.1 n. 9). Nor is it properly attributable to Duns Scotus or to Thomas Aquinas, as Galileo declares. Indeed, since no explicit statement is made by either Philoponus or Avempace (see n. 72), it is also highly questionable whether it is proper to interpret them as construing the weight (or force) of a body as a kind of resistance which is instrumental in producing finite motions in the void. There is, however, little doubt that such a position was distinguished in the Middle Ages (see Albert of Saxony's discussion in Selection 47). It is worth noting that those who accepted the type of internal resistance arising from contrary tendencies of light and heavy elements would have rejected the assumption that a pure elemental body, say earth, could move with a finite speed in a hypothetical void. However, this assumption would probably have been adopted by those who accepted the internal resistance described by Galileo.—*Ed.*]

72. Galileo's meaning may be clarified by the memo-

merely increased or decreased according to the decrease or increase[73] in the heaviness or density of the medium. But if, on the other hand, they admit that the resistance increases or decreases in the *ratio* in which the weights of the medium vary, their attempts to upset Aristotle's argument will be in vain.

4. Thomas Bradwardine (ca. 1290–1349) and
5. Albert of Saxony (ca. 1316–1390): Unequal Homogeneous Bodies Fall with Equal Velocity in a Vacuum

Introduction by Edward Grant[74]

To arrive at the conclusion that two bodies of similar or homogeneous composition but unequal volume and weight will fall with equal speed in a vacuum, Thomas Bradwardine, Albert of Saxony, and Galileo, in his early work *De motu* (*On Motion*), used intensive rather than extensive factors of a-nalysis. Since the bodies were assumed to be homo-geneous, it followed that the parts of each body, however small or large, were identical. Consequent-ly, their behavior was identical. The whole body, which was merely an aggregate of identical units, would behave exactly as every one of its constit-uent parts behaved. The smallest unit would fall with the same speed as an aggregate of such units formed into a large body. Subtraction or addition of identical units could neither diminish nor in-crease the speed of the body or its units. By con-trast, Aristotle was concerned exclusively with the body as a gross or extensive object. Homogeneous bodies of different size, and therefore of unequal weight, would fall with different speeds because their total weights differed.

In coming to the same conclusion, however, the two medieval authors, Bradwardine and Albert, employed a ratio of force to resistance per unit of matter, whereas Galileo used weight per unit vol-ume. For Galileo, the concept of effective weight rather than gross weight was the determinant of velocity. But the effective weight, or heaviness, of a body in a medium depended on the difference in the specific weights of the body and the medium through which it fell. Hence, it is actually a differ-ence in specific weights which determines velocities. The free fall of a body may be represented as $V \propto$ *specific weight of body* minus *specific weight of medium,* where V is speed;[75] the velocity of a rising body would be $V \propto$ *specific weight of medium* minus *specific weight of body.* Clearly, then, if the specific weights of two unequal bodies are the same, they must fall with equal speeds in the same medium or in a vacuum. This reasoning underlies Galileo's argument at the conclusion of section 6 below,

namely that if bodies *a* and *b* were equal in size and of the same composition they would fall with equal speed, but if joined together would not double that speed, as Aristotle held, but would fall with the same speed as when they fell individually, since bodies of the same material should fall with the same speed.

Despite a significant similarity in reliance on intensive rather than on gross or extensive factors, our two medieval authors formulated their interest-ing conclusions within the framework of Aristote-lian physics, whereas, already in *De motu*, Galileo has moved outside of that framework and into early modern physics.

Where Bradwardine and Albert of Saxony accept without question the notion that heaviness and lightness are absolute properties of bodies and believed they operated as contrary forces—one as motive force, the other as internal resistance—in mixed or compound bodies, Galileo has completely abandoned absolute heaviness and lightness. He

randum (410.21–26): "Philoponus, Avempace, Avicen-na, Saint Thomas, Scotus and others who try to main-tain that motion takes place in time [i.e., not instantane-ously] in the void, are mistaken when they assert a two-fold resistance in the moving body, viz., one accidental and due to the medium, the other intrinsic and due to the body's own weight. These two resistances are clearly one, for the medium, insofar as it is heavier, both offers more resistance and [by that very fact] renders the body lighter."

That is, though one may, by abstracting, analyze the weight of a body into two factors (weight of the body in a void, and weight of the medium in a void), these factors do not act independently and are not separable in the actual case of a body in a medium.

73. "Decrease or increase" seems to be a slip for "in-crease or decrease."

74. This introduction is partly a rearrangement and partly a verbatim reproduction from pages 356–359 of my article "Bradwardine and Galileo: Equality of Velocities in the Void." *Archive for History of Exact Sciences,* II (1965), 344–364.

75. For the relation of Galileo's law to the tradition stemming from Avempace, see note 58 and Selection 44, n. 17.

insists that everything has weight and that things are only less heavy, equal to, or heavier than other things; void alone is weightless, and it is not a substance. Adopting the notion of relative density, Galileo utilized it in its more precise form of specific weight. While the true weights of bodies could only be determined in the void, their effective weights in a given medium could be determined by comparing the difference in specific weights between each body and the medium.

By abandoning absolute heaviness and lightness and adopting the concepts of relative density and specific weight, Galileo could avoid many of the strange consequences, distinctions, and explanations that were unavoidably associated with the use of the absolute heaviness-lightness dichotomy. Where Bradwardine, Albert, and medieval Aristotelians generally, explained direction of motion in terms of absolute lightness and heaviness functioning as motive qualities, Galileo relied on the relation between the specific weight of body and medium. No longer was it necessary to distinguish the behavior of simple or pure elemental bodies from that of mixed bodies. The distinction forced upon most medieval physical speculators that only certain bodies (mixed) could fall with finite speed in a hypothetical void space and that all others (pure elemental bodies) could not became utterly meaningless in the physics of Galileo's *De motu*, where, by virtue of the concept of specific weight, *all* bodies were to be treated alike, regardless of composition, and *all* bodies would fall with finite speed in void and plenum. With Galileo, homogeneous Archimedean magnitudes replaced the simple and heavy mixed bodies of the later Middle Ages. The

emptiness of this distinction between bodies in the analysis of motion and the rejection of absolute heaviness and lightness destroyed the need for the medieval concept of internal resistance. Internal resistance, which, in the Middle Ages, had been invoked to permit an explanation in dynamic terms for finite motion in the void, depended upon the contrary tendencies of elements distinguishable as light and heavy in a mixed body. In downward motion heavy and light functioned as motive force and internal resistance respectively; in upward motion their roles were reversed. Galileo, however, required neither internal nor external resistance for the production of finite speed in a void, where the speed of a falling body would be directly proportional to its specific weight.

Thus where specific weight provided Galileo with a consistent standard of comparison for the motions of bodies in void and plenum, those who relied on internal resistance in mixed bodies were led to peculiar and often bizarre conclusions. Thus Albert of Saxony would argue that a heavy mixed body might fall more slowly at the end of its natural motion than at the beginning;[76] that one and the same heavy mixed body could be moved quicker in a plenum than in a vacuum;[77] and that even if one compound heavy body were smaller than, or equal to, another of the same homogeneous composition, it would descend more quickly in a plenum.[78] Bradwardine deduced similar paradoxical consequences.

For the remarks of Bradwardine and Albert of Saxony on this subject, see Theorem XII in Selection 51.1, and the eighth conclusion of the second section above.

6. Galileo Galilei: Unequal Homogeneous Bodies Fall with Equal Velocity in a Vacuum

Translated by I. E. Drabkin[79]
Annotated by Edward Grant

Similarly, if two bodies move downward [in natural motion], one more slowly than the other, for example, if one is wood and the other an [inflated] bladder, both falling in air, the wood more swiftly than the bladder, our assumption is as follows: if they are combined, the combination will fall more slowly than the wood alone, but more swiftly than the bladder alone. For it is clear that the speed of the wood will be retarded by the slowness of the bladder, and the slowness of the bladder will be accelerated by the speed of the wood; and, as before, some motion will result

intermediate between the slowness of the bladder and the speed of the wood.

76. *Questions on the Physics of Aristotle,* fol. 50 verso, col. 1., of the edition cited in Selection 35, n. 1.

77. See the third conclusion from Book IV, Question 11, Selection 55.2.

78. See the second part of the eighth conclusion of Book IV, Question 12, Selection 55.2.

79. Reprinted with the kind permission of the Regents of the University of Wisconsin from *Galileo Galilei "On Motion" and "On Mechanics" Comprising "De Motu" (ca. 1590),* translated with Introduction and Notes by I. E. Drabkin, and *"Le Meccaniche" (ca. 1600),* trans-

On the basis of this assumption, I argue as follows in proving that bodies of the same material but of unequal volume move [in natural motion] with the same speed. Suppose there are two bodies of the same material, the larger *a*, and the smaller *b*, and suppose, if it is possible, as asserted by our opponent, that *a* moves [in natural motion] more swiftly than *b*. We have, then, two bodies of which one moves more swiftly. Therefore, according to our assumption, the combination of the two bodies will move more slowly than that part which by itself moved more swiftly than the other. If, then, *a* and *b* are combined, the combination will move more slowly than *a* alone. But the combination of *a* and *b* is larger than *a* is alone. Therefore, contrary to the assertion of our opponents, the larger body will move more slowly than the smaller. But this would be self-contradictory.

What clearer proof do we need of the error of Aristotle's opinion? And who, I ask, will not recognize the truth at once, if he looks at the matter simply and naturally? For if we suppose that bodies *a* and *b* are equal and are very close to each other, all will agree that they will move with equal speed. And if we imagine that they are joined together while moving, why, I ask, will they double the speed of their motion, as Aristotle held, or increase their speed at all? Let us then consider it sufficiently corroborated that there is no reason *per se* why bodies of the same material should move [in natural motion] with unequal velocities, but every reason why they should move with equal velocity. Of course, if there were some accidental reason, e.g., the shape of the bodies, this will not be considered among causes *per se*. Moreover, as we shall show in the proper place, the shape of the body helps or hinders its motion only to a small extent.

56 ON INTERSTITIAL VACUA

1. Marsilius of Inghen (ca. 1340–1396): Explanation of Condensation and Rarefaction, and Denial of Interstitial Vacua

Translated and annotated by Edward Grant[1]

In the fourteenth [question], we inquire whether it is possible that anything be rarefied and condensed.

And it is argued first that it is not possible. . . .

Aristotle argues the opposite of this at the end of the fourth book [of the *Physics*] in the treatise on the vacuum. There are many ways of responding to this question. One opinion assumes that condensation occurs by the entrance of the parts of a condensible body into certain separate vacuities within the parts of the condensible body; and rarefaction occurs by the generation of such vacuities. Against this opinion can be adduced all the arguments by which it was proved that it is impossible for a vacuum to exist.[2]

A second way assumes that condensation occurs by the departure of a more subtle body as appears with respect to a sponge, which is condensed by the

a plenum the speed of motion of a body depends on the difference between its weight and the weight of the medium through which it moves. And likewise in a void [the speed of] its motion will depend on the difference between its own weight and that of the medium. But since the latter is zero, the difference between the weight of the body and the weight of the void will be the whole weight of the body. And therefore the speed of its motion [in the void] will depend on its own total weight." Since, throughout this passage, "weight" signifies "specific weight," and since bodies move in the void in the same way as they do in a plenum, Galileo by the same method described here could easily have demonstrated equality of speed for homogeneous bodies falling in a void. Deeming it superfluous, he remained content with the declaration that "using the same method of proof in the case of the void as we used in the case of the plenum, we can show that bodies of the same material but of different size move with the same speed in a void" (Drabkin, pp. 48–49). In the First Day of the *Dialogues Concerning Two New Sciences (Discorsi)*, published in 1638, Galileo modified and corrected this opinion when he argued that *all* bodies of whatever composition and specific weight fall with equal uniformly accelerated speeds in the void, but not in a plenum. See the translation by Henry Crew and Alfonso de Salvio (New York: Macmillan, 1914), pp. 72, 74.

lated with Introduction and Notes by Stillman Drake (Madison, Wis.: University of Wisconsin Press, 1960), ch. 8, pp. 29–30. See also the end of section 3 above.

Although, in the argument reproduced here, the homogeneous but unequal bodies are falling in a uniform medium, there is no doubt of Galileo's conviction that the same argument could demonstrate that homogeneous bodies would also fall with equal speeds in a void. For he says later (Drabkin, p. 45) that a "body will move in a void in the same way as in a plenum. For in

1. Translated from Book IV, Question 14, on folios 55 verso to 56 recto of Marsilius of Inghen's treatise cited in Selection 50.1, n. 1.

2. Marsilius argued this in the preceding, or thirteenth, question, from which an excerpt has been included in Selection 53.3.

explusion of air and rarefied by the entrance of air or water. I argue against this opinion, for if every condensation and rarefaction occurs in this way, then it follows that no body would be uniform [or homogeneous] except the most simple or subtle body. The consequent is false, because many stones or metals are uniform bodies. The consequence is proved,[3] because then such a [subtle and simple] body could not be condensed by the departure of a more subtle body, since there is no subtler [or rarer] body in it, for it is uniform throughout.

A third opinion is that condensation occurs by the penetration of parts of a body by the intrusion of the bodies [or parts] at its circumference [or surface] into the parts of the body situated around its center. And yet the body cannot be condensed to any extent whatever, because in every species there is a certain degree beyond which no individual of the species can be condensed. Therefore, it happens that bodies cannot mutually interpenetrate to any extent whatever but only to the extent that condensation can take place. And [in this opinion] rarefaction occurs by departure of one condensed body [or part] from another part. But one can argue against this opinion, because then it would follow that air, no part of which penetrates another part, could not be rarefied. The consequent is false, because fire can be made from air;[4] and the consequence is proved, because rarefaction occurs only by the departure of one part of a rarefiable body from another part, where these parts [had been assumed] previously to be mutually interpenetrable. A second argument is that if it happens that by condensation two bodies can mutually interpenetrate, it follows, by the same argument, that three, or four, or indeed an infinite number of bodies can mutually interpenetrate, as the Philosopher argues in this fourth book.[5]

After dismissing these opinions it must be noted that "condensation" is said to occur when any body is made smaller without the withdrawal of any part of it; and "rarefaction" is said to occur when a body is made larger without any other body entering from outside. In this respect, condensation and rarefaction differ from increase and decrease, for increase (augmentatio) occurs when some body is added, and decrease (diminutio) occurs when body is removed, as is clear in the first book of On Generation. Secondly, it must be noted that condensation is found in three ways. Sometimes it is violent, and this occurs only by compression of the condensed body—as when wax is compressed. In another way condensation occurs by

means of a previous alteration. And in a third way it occurs from a previous generation.[6] Now these conclusions are posited.

The first is this: it is possible that something be condensed by compression alone without any other change having been made prior to the condensation. This is proved, for unless it were true, it would follow that no rectilinear local motion could be made. The consequent is contrary to experience,[7] but the consequence is proved, because the medium in which such a motion is made yields to the mobile, for otherwise there would be a penetration of bodies. Therefore, either the condensation occurs by the compression of the mobile—and the proposition is had—or another body yields to it, and a third body yields to that body, and so on up to the sky, which is impossible, because either the sky would yield to that body [which pressed against it] or that body would penetrate the sky. Each of these alternatives is absurd.

A second conclusion is this: It is possible that a body could be condensed by alteration without generation. For if a glass vessel were taken and heated in fire and the mouth of the vessel were then immediately placed in water with its bottom

---

3. The consequence declares that if condensation occurs by the departure of a more subtle body, then only the most subtle or simple bodies could be homogeneous. Since it follows that homogeneous bodies could not be condensed, the theory is absurd. The role of a "subtle matter" in increase and decrease is discussed by Nicholas of Autrecourt in section 2 (see also n. 40).

4. The consequent referred to is the statement in the preceding sentence that air could not be rarefied. This is simply false, Marsilius argues, presumably by appeal to the universally accepted opinions that fire can be made from air and that fire is rarer than air. Thus when air is converted to fire it is rarefied. In the very next sentence Marsilius goes on to show that although the consequent is false, it follows from the rejected opinion. For if rarefaction occurs only by the exit of one part from another part of previously interpenetrated parts, and the parts of air do not mutually interpenetrate, it follows that air will not rarefy.

5. In Physics IV.8.216b. 3–11, Aristotle argues that the volume of a cube placed in a void will occupy an equal volume of void, which will penetrate it. Thus the dimensions of cube and void will occupy the same place. "And if there can be two such things," declares Aristotle, "why cannot there by any number coinciding? This, then, is one absurd and impossible implication of the theory."

6. Marsilius illustrates each type further on.

7. That is, the consequent that "no rectilinear motion could be made" is contrary to our experience and hence is false. Therefore, it must be true that condensation could be made by compression alone, and Marsilius goes on to show this.

upward, it would appear obviously that the air is condensed from the cold by alteration;[8] and as the air is condensed so does the water rise, lest a vacuum be produced.

A third conclusion is that it is possible for a condensation to occur by means of a previous alteration *and* generation. This is proved, because it would otherwise follow that a denser element could not be made from a rarer element, which is false,[9] as is obvious in the second book of *On Generation*.

But there might be a doubt, first about the mode of condensation, i.e., how it happens, whether by the penetration of bodies or in some other way. Let the reply be that condensation occurs in this way: The parts [of the body] lie closer together than before. This is not true of the immediate parts,[10] but any of the mediate parts lie closer than before. Thus any two points that may be

assigned in a condensed body are mutually closer than before condensation.

A second doubt might arise about whether or not something is lost by condensation; the same doubt may arise as to whether something is acquired in rarefaction. The reply to this could be made in different ways depending on the different opinions about quantity. If quantity is assumed to be a thing distinct from substance and quality, then quantity is a dimension by means of which something is extended, as whiteness *(albedo)* is a certain quality by means of which something is white *(album)*. In this event, it must be said that a new quantity is acquired in rarefaction and lost in condensation. But if every quantity were assumed to be a substance or quality, then it is not necessary that something be lost in condensation and acquired in rarefaction. . . .[11]

2. Nicholas of Autrecourt (b. ca. 1300; d. after 1350): The Existence of Interstitial Vacua Affirmed

Translated and annotated by Edward Grant[12]

ON VACUUM

We have probably destroyed many discussions. Now we want to take into consideration certain matters about which we have to say some unusual things, much of it in the previous manner. And [so] let us begin a discussion on the vacuum: whether this description that has been assumed reveals the "quid" of the word ["vacuum," namely] that by the word "vacuum" we understand that in which there is no body [but in which], nevertheless, a body can be [or exist]. And it seems that it does, because if not it would follow (1) that rectilinear

not possess more or less of whatever substance it is, a substance that will retain its identity; nor would it lose or gain with respect to quality, since its qualities would remain unaltered if its identity were preserved, regardless of the size it would attain as a consequence of rarefaction or condensation.

A minor objection has been omitted as well as very brief responses to the six principal reasons that were enunciated at the very beginning of the question and which were also omitted.

12. Translated from Autrecourt's *Exigit ordo executionis*, edited from a single known manuscript by J. Reginald O'Donnell, C.S.B., in "Nicholas of Autrecourt," *Mediaeval Studies*, I (1939), 179–280. The section on vacuum extends over pages 217–222.

In advocating the probable existence of indivisible atoms and interparticulate vacua Nicholas of Autrecourt either stands alone in the Middle Ages or belongs to a very small group whose members are not yet known. Believing that things and qualities do not undergo change (that is, are neither generated nor corrupted), Nicholas sought to account for apparent change by explaining generation as the coming together of indivisible, eternal, and unalterable atoms and corruption of bodies by the dissociation of these atoms. In the present selection he seeks not only to justify the existence of interstitial vacua as essential for a proper explanation of motion but attacks Aristotle's physics of motion based upon the existence of a plenum and the total denial of vacua of any kind. Unlike the ancient atomists Nicholas does not argue for an infinite universe and a plurality of worlds.

A thorough study and analysis of Nicholas' point of view appears in Julius R. Weinberg, *Nicolaus of Autrecourt: A Study in 14th Century Thought* (Princeton, N. J.: Princeton University Press, 1948).

8. The alteration thus involves a change of quality from hot to cold.

9. The conversion of air to water, for example, would be conceived as a substantial change (or generation), since one element has been generated from another. It is also a qualitative change (or alteration), since air possesses the qualities hot and moist and water moist and cold. The change from air to water requires that hotness replace coldness.

10. Immediate parts are already in contact and cannot lie closer together.

11. Thus Marsilius allows that condensation and rarefaction can be properly interpreted on the basis of two different conceptions of quantity. If quantity is merely dimension or extension, then condensation, which is obviously a loss of extension, is necessarily a loss of quantity; and rarefaction will be a gain of extension and consequently of quantity. But if quantity is thought of as a substance or a quality, then nothing is lost or gained, presumably because the condensed or rarefied body will

local motion could not be; or (2) that two bodies could exist simultaneously [in the same place]; or (3) that with one thing moved, all things would necessarily be moved and change place. Thus in nothing can we persist so probably as in a vacuum; therefore vacuum must be assumed.

That all these positions are more absurd than the assumption *(positio)* of a vacuum seems to be sufficiently known and it should be conceded to us that they would follow. This is shown, because when this body changes place either (1) the place in which it was received was a vacuum and thus what has been proposed [is had], or (2) it is filled with a body and then either (a) the body remains and there will be two bodies [in the same place] simultaneously, which is another absurdity, or (b) it withdraws [from that place] and crosses to another place; and then I inquire in the same way as before.[13] By proceeding in this way, either you concede that the process always goes to infinity, or that two bodies will exist [in the same place] simultaneously; or you must concede a vacuum.

But here it might be replied that finally it [that is, the process of motion from one place to another] will terminate because a condensation will occur.[14] But we cannot relate *(dicere)* this discussion with the negation [or denial] of the vacuum, although, perhaps, Aristotle could according to his principles. For we do not assert that "dense" occurs by means of the generation of some new quality which did not exist before,[15] but [rather we declare] that "dense" occurs only by the retreat [or compacting] of bodies, as in wool; or because the parts come together, that is, because more [parts] are related more closely than before. And "rare" will not come to be unless the parts of this body are more separated than before. Thus "dense" or "rare" do not come to be except by the local motion of the parts. And so it is that another interpretation of rare and dense is of no help to us.[16]

There is another response which I heard from a certain worthy master which seems more probable with respect to Aristotle and our conclusion.[17] Let it be imagined that in a rectilinear local motion the situation is as follows: The motion continues [or endures] because it is conceived as if it were on a circle, so that a body located in *A* will be moved to place *B* and the body located in *B* will be moved to place *A* and so it will continue [in this way].[18] But this does not seem to suffice, [for] let us [now] imagine that around this space there are differences of position [or direction, namely] in front *(ante)*, behind *(retro)*, and so on. Now let this mobile be moved in a straight line. It seems that what is [directly] behind it, say air, could reach the place previously occupied by the mobile before what was in front of the mobile could reach that [very same] place; and what was behind it would enter that place; and so [it would happen that] the state

13. That is, if a body departs to make room for another, we must pose the same alternative about the new place which it will occupy.

14. If matter becomes more dense somewhere in this process, space will become available to accommodate all bodies, and the process will not be infinite.

15. That is, Aristotle could deny a vacuum and accept condensation or compression of matter because his explanation of the way bodies become dense and rare involves the generation of a new quality and does not, as Nicholas describes in the following sentences, involve compression and extension of particles in vacua (see Weinberg, pp. 160–161). For Aristotle, dense and rare are contrary qualities possessed at different times by one and the same matter. When a rare matter like air is generated from a dense matter like water, this is a qualitative change which one and the same quantity of matter undergoes (see *Physics* IV.9.217a.20—217b.26).

16. Perhaps Nicholas is making the point that Aristotle's interpretation of rare and dense is of no use to him because "rare" and "dense" involve local motion of parts or atoms. Hence even if a condensation occurred to terminate the process of motion from one place to another, the same problems described in the preceding paragraph would arise with respect to the parts of the body: Either the process goes to infinity or two bodies would exist in the same place simultaneously. If these alternatives are unpleasant, one ought to concede the existence of vacua.

17. What follows "will be more probable with respect to Aristotle and our conclusion" because it will show the absurdity of Aristotle's explanation and be favorable to Nicholas's position.

18. This explanation of rectilinear motion was frequently called "mutual replacement" or *antiperistasis*, the Greek term used by Aristotle in his brief mention of it in *Physics* IV.8.215a.15 and VIII.10.267a.15–21. It is clear that Aristotle did not utilize it to explain continuous violent rectilinear motion produced by a motive power in media such as air or water, because in "mutual replacement" "all the members of the series are moved and impart motion simultaneously, so that their motions also cease simultaneously: but our present problem concerns the appearance of continuous motion in a single thing, and therefore, since it cannot be moved throughout its motion by the same movent, the question is, What moves it?" (*Physics* VIII.10.267a.18–21). Thus while Aristotle may have accepted "mutual replacement" for nonrectilinear motion or, perhaps, for natural up or down rectilinear motion, he does not regard it as an appropriate explanation for violent rectilinear motion (his explanation of the latter appears in the section immediately preceding our quotation [VIII.10.266b.28–267a.15]). It seems reasonable to conclude that if Nicholas were criticizing Aristotle—and we note that his name is not explicitly associated with this argument—his

described above would never exist[19] unless a vacuum were conceded. Now, a body which is in front of a mobile does not appear to be moved unless it is pushed and yields. Now I ask whether in the same instant in which what is in A is [also] in B, that which is in B is [also] in A, and then it does not appear that this [that is, what is in front] yields more to this mobile than conversely [that is, than the mobile yields to what is directly in front of it]. Then how does it yield as if pushed?[20] Or, [if what is in B will be in A, or vice versa] in another [and different] instant and then, in this first instant, either two bodies were simultaneously [in the same place] or there was a vacuum. . . .[21] What I have said about the immediate instant you should understand in terms of the conclusion posited above, where it was said that a continuum is composed of indivisibles and time of instants.[22]

Let another argument be adduced with respect to new wine in a jar. When it becomes rarefied, there appears to be a greater quantity, not because of the generation of a new quality as above, nor because of the advent of new bodies, but only, it seems, because the parts, having [previously] come together as if there were a certain trampling down, now separate and are more distant from each other. What seems more intelligible in the description of rare and dense supports this argument. The usual description of dense is that "dense" is something whose parts lie closer together, so that water does not seem denser than air unless it is because its parts are mutually closer.[23] And so it is that the parts in a rare body are in some way further separated so that a vacuum intervenes—that is, the parts could be mutually closer [yet]. Hence the vacuum is something where there is no body [but] where, nevertheless, there could be a body. And this is how I expound this matter, for those who urge the opposite position seem to frighten people merely by the term ["vacuum"].

It is true that others use a description of "dense" in another signification, declaring that that is dense which yields less to the touch and "rare" is that which yields [readily to the touch]. But nothing of what the understanding seeks in this— namely why this body is denser than another—is determined by this approach. Moreover, they fall into a description of other things, namely of "the solid" (solidi) and "the light" (levis). Whence it is that the concept of dense is grasped by the understanding by the sight of the body alone immediately upon seeing the body, because when we see it we say that it is dense. But "solid" is grasped

by the understanding by means of touch, so that it could be said that a solid is that which yields to the touch with difficulty, and "light" is the opposite. And if we could say this, the investigation would be over, based on a solution of contraries. But I do not wish to determine this fully now.

And first, we should see how the assumption of a vacuum destroys the objection posited above, from which it was concluded that there is either no [rectilinear] local motion, or with one thing moved, all things are moved, and so on. It must be said that [on the assumption of a vacuum] this will be understood in local motion. For when a body is moved locally in air, the parts of the air are in some way compressed and come together, and thus it is not necessary that such a process continue to infinity or that two bodies exist simultaneously [in the same place]. And you might imagine that what was [just] stated could be explained in a manner similar to this: Assume that the subtle bodies

---

criticism could only be relevant to natural rectilinear motion. From the general tenor of the discussion it seems likely that Nicholas was rejecting the "mutual replacement" interpretation for both natural and violent rectilinear motion. At best, then, only that part which is relevant to natural motion might appropriately be used against Aristotle.

19. The situation just described requires that when body a in place A moves to place B, body b in place B must move to A. But, according to Nicholas, this will not occur. For if b is "in front" of a and a moves into B, a unit of air, say c, directly behind A will move into A before b itself can come around and occupy A. Thus a and b will be incapable of exchanging places. Only the assumption of a vacuum could circumvent this dilemma.

20. Another difficulty with "mutual replacement" arises in distinguishing between mover and moved. For if at the same instant, body a moves to B and body b moves to A, how can pusher and pushed be differentiated?

21. Now the bodies are assumed to exchange places not simultaneously but in different instants. Thus, if body b moved into place A in a first instant and body a moved into place B during the next, or second, instant, bodies a and b would be in place A simultaneously during the first instant. (A brief but difficult sentence, which plays no role in the argument, has been omitted.)

22. Discussed in a preceding section, "On Indivisibles."

23. Are we to assume that the only difference between air and water lies in the degree of separation of their constituent atoms? Or, do atoms of water differ from those of air? Considerations of this kind were not raised by Nicholas, perhaps because the main thrust of his discussion was to demonstrate the general plausibility and probability of the existence of atoms and void and to discredit traditional Aristotelian explanations. A detailed and comprehensive application of atomism to the physical world was apparently not attempted.

which are said to exist in wool are not interstitial[24] and that someone is moved in a room filled with wool; the parts [of the wool] would come together more and a greater proximity [of parts] would occur. If there are any good reasons against this conclusion, they would seem to be those of Aristotle in the fourth book of the *Physics,* for although they are few, they are by no means trivial. Against those who assume an internal [or interstitial] vacuum Aristotle adduces certain arguments in the fourth book of the *Physics.* But these arguments were directed more against the causes [or reasons] which led them to assume a vacuum than against the conclusion itself. But we do not use these causes [or reasons]. And the argument which follows seems very difficult and is thought to be the most subtle of his arguments, at least against one way of assuming a vacuum. [The argument is:] If a vacuum existed, it would follow that local motion would occur in an instant. The consequence is proved, because as is a ratio of medium to medium so is a ratio of motion to motion in speed and slowness, since it seems that a slower motion results from a greater resistance of a medium. But there is no ratio of a full medium to a void medium; therefore [there is] no [ratio] of motion to motion with respect to speed and slowness. Now if motion through a vacuum occurred in [some] time, there would be a certain ratio between the times; [but this does not happen]. Therefore, there will be motion in an instant.[25]

We ought to see first that this argument does not tell against the kind of vacuum which I have assumed, nor did Aristotle wish to adduce [it against this kind of vacuum][26] for we do not posit a pre-existing separate vacuum *(vacuum separatum praeexistens)* through which a motion could occur, but [rather we assume a vacuum only] between the parts of a body, and when the body must be moved locally those parts come together and are nearer each other. Now, if you wish to speak about the motion of this body whose parts come together [or compress] and which is first assumed to be moved through a medium–although it could [also] be received into a vacuum; nevertheless it was first assumed to be in a plenum—then, other things being equal, to the extent that these parts are compressed [or come together] faster so is the body moved faster, just as if we were to assume that it is moved through a pre-existing vacuum.[27] But although this [that is, Aristotle's] argument does not oppose us on this, it does oppose us in another way concerning the parts of the body which are

compressed; for according to this argument, the parts come together within [the body] where there was a pre-existing vacuum [that is, pre-existing interstitial vacua], and thus it seems that the motion of these parts occurs in an instant.[28]

It can be shown from other ways that this argument does not conclude. As to whether the consequent is absurd, see above in the treatise [or section] on motion, for a mobile can be for an instant in any place in which it was not previously, and if by motion you understand something composed of several instantaneous changes *(ex pluribus mutatis),* then motion would not occur in a single instant but in several. And whatever there might be about this consequent that we did not say is absolutely absurd above in the treatise [or section] on motion, I say that the consequent does not hold if one understands properly the manner in which the speed and slowness of one mobile is measured with respect to another as stated above in the treatise [or section] on the continuum when I replied to Aristotle's arguments.

Thus, leaving aside the agent, it [that is, a mobile] is not said to be slower unless in some way it rests.

24. The compression of wool would have been described by Aristotelians as the squeezing out of "subtle bodies" (presumably air) between the particles of wool. Nicholas suggests that if the existence of such "subtle bodies" between the particles of wool be denied, no good reasons could be suggested for supposing that the particles will not simply squeeze close together if someone walks into a room filled with wool. A similar explanation would be applicable to account for the compression of air.

25. A considerably altered representation of Aristotle's discussion in *Physics* IV.8.215b.20–216a.11.

26. The argument just described was intended by Aristotle for use against the existence of a separate vacuum and not against interparticulate vacua.

27. For Nicholas, the speed of a body is a direct function of the speed of its constituent atoms as they come together and make the body denser. How this is to be understood is not made clear. Is the body's speed greater because it becomes denser as a whole, or is the speed solely a function of the speeds of the individual particles? Since the speed of a body is made to depend primarily on the motion of its constituent particles, it follows, as Nicholas says, that a body could move in a hypothetically pre-existing separate vacuum (although he did earlier deny its existence) as well as in a material medium. Presumably, the external resistance of a material medium would act to reduce its speed.

28. Nicholas recognizes that although Aristotle's claim of instantaneous motion in a separate vacuum is inapplicable to his position, since he rejects the existence of such vacua, it might apply to atoms moving in the interstitial vacua. Thus it might be relevant on the microlevel of atomic motion.

One must therefore reflect that there is a certain essential [characteristic] *(essentialitas)* in motion, [namely] that just as one part of space is before another, so it is traversed before;[29] and this essential [characteristic] cannot be removed [or ignored]. And [now] consider slowness, which occurs because a mobile rests, so that to say that one mobile is slower than another is to say nothing other than that one mobile rests more than another.[30] Thus either rest belongs to [or inheres in] a mobile because of an intrinsic property *(ex determinatione)* of the mobile's nature, or [it occurs] from the resistance of a medium so that the parts of a medium do not come together easily. Now, therefore, should there be motion in a vacuum, why does it not follow that it should have a ratio to motion made in a vacuum in the sense that it had the essential [characteristic] of motion, [namely] that it traversed one part of the space before another, Hence you should not conclude that the motion made over the whole distance was made in an instant. Nor, again, [should you conclude that] there is no rest [for motions in the void]. For it is true to say [that there would be rest] if we do not assert that the total cause of rest in such a motion is the resistance of a medium but [rather assert that] the cause of rest is an intrinsic property *(determinatio)* of the moving thing.[31] And the understanding of succession which I assumed above, [namely] that it should be considered with respect to the order of parts in space, seems [also] to have been the understanding of Avempace.[32]

Other arguments are produced for denying a vacuum—as, [for example,] certain natural experiences *(experientiae)* which do not seem to have any rationale *(habere locum)* unless it be to show the impossibility of a vacuum. The first of these is the clepsydra, which is a vessel having an aperture below and above and filled with water. If a finger is held over the upper aperture, the water does not go forth even though by its [very] nature it should descend; should the finger be removed, the water descends. But that it does not descend in the first case does not seem explicable except by reason of an "abhorrence of a vacuum" *(fuga vacui)*, since a vacuum would exist if the water went forth, because another body could not enter through the top, since it is stopped up [by the finger]; nor could it enter through the bottom, because then two bodies would exist in the same place simultaneously, which is impossible.[33] Let it be said that this argument is not made against the

conclusion which we hold, because we hardly suppose that such a separate vacuum *(vacuum separatum)*, which would be left in the vessel after the water had been removed, exists.

Now I have heard a subtle response [to this]: That the water does not flow forth is not because of any "abhorrence of a vacuum" *(fuga vacui)*, but it is imagined that the whole universe has its mode of fullness *(modum plenitudinis)*, so that to the extent that something exists it is able to enter into the universe of existent beings, which is considered to be good. Now, if this body should flow forth, another could not enter, and the whole universe would not be a total plenum and [yet] two bodies would [not] be [in the same place] simultaneously.[34] Now I choose the first explanation.[35]

---

29. This kinematic argument, used to justify motion in hypothetical separate vacua, was formulated by Thomas Aquinas (see Selection 55.1) and frequently repeated. As we see, Nicholas invokes it here in order to justify temporal motion of atoms through interstitial vacua and subsequently credits Avempace with having understood this argument. But in the form in which it is given here, it was probably derived ultimately from Thomas Aquinas (see Selection 55.1, n. 4).

30. An explanation which Nicholas probably adopted from the description of Arab atomism by Moses Maimonides (see *Guide of the Perplexed*, Bk. I, ch. 73). Earlier in the treatise (O'Donnell, p. 207), after assuming that space and time are composed of indivisible units, Nicholas argued that if a body moves through three units of space, say *a, b, c,* in three instants of time, and another body traverses them in six instants, the first is said to be moved with twice the speed of the second because it moves through each unit of space for only one instant and rests in none, whereas the second also rests for one instant in each unit. Obviously, all this is meant to apply to the motion of individual atoms. Thus not only will atoms move temporally through interstitial vacua but a ratio of the motions of any two atoms can be formed.

31. By assuming that rest is an intrinsic property of atoms and not dependent on the action of an external resistant medium, Nicholas can argue that motions in vacua are comparable. Why one atom should rest for one instant, another for two or three instants, and so on, is not explained.

32. See note 29.

33. The clepsydra illustration is also cited by Marsilius of Inghen (see Selection 53.1 and n. 25 thereto).

34. This strange argument assumes that matter enclosed in a vessel will seek to depart in order to join the totality of existent matter outside the vessel. Some form of the concept of a "community of matter" seems to be operative here.

35. Nicholas opts for nature's abhorrence of a separate vacuum as a better explanation than the one offered in this paragraph.

There is, [however,] doubt about the way in which we assume that a vacuum occurs. Suppose that this body is moved locally; you say that the parts of the body, say air, which were in the place to which the body moved, come together and occupy a smaller place than before.[36] But with regard to this, I inquire about the place from which the air withdraws, and [I ask] whether something could enter there. For if we deny it, then we shall hold a conception of [the term] vacuum that is held by others and we would be assuming a separate vacuum (vacuum separatum); but if we do not deny that something could enter, and something should enter, say air, then, as before, we ask about the place from which this air withdraws. And so the reason which caused us to posit the abhorrence of a vacuum, does not allow us to avoid the other absurdity [of an infinite process].[37] Now, perhaps an appropriate response with respect to this should be to say that the parts of air which were moving behind the body are sometimes extended [or rarefied] so that fewer [parts] coming together fill the place up to a certain magnitude but not into infinity. Now, although we do not have to assume a vacuum on the side behind the body, nevertheless we must assume it for another [reason] lest we fall into an absurdity about the bodies which are in front of the body, [namely] that from one motion all things are moved. Thus, as we stated briefly, we do not approve of this, because [it would happen] that the parts of space would be pushed together in the first instant.[38] But perhaps we can consider this [problem] at another time.

Against what we say about vacuum there is the experience about the inflated bladder which cannot be compressed. Now this would not happen [those who oppose us say] if an interstitial void (vacuum interceptum) [actually existed,] enabling the parts to come together. With respect to this [inflated bladder] one of two explanations must be declared:[39] (1) Either [this is explained by the fact that] the parts of the body contained in the bladder cannot easily be compressed further, so that although the parts of air come together and motion occurs this way, they are not, however, so compressed but that between them any [separate] vacuum would remain. Therefore, the parts could be brought together [or compressed] up to a certain point (ad aliquem modum), which does not [however, leave an empty] place in what is in the bladder. [Or] it might be explained in another way by saying (2) that this [lack of further compressibility of the bladder] happens because otherwise there would

remain a wholly separate vacuum (vacuum separatum) in a part of the bladder, since another body could not enter [to occupy it].

The ancients produced one argument for the assumption of a vacuum, arguing that if there were no vacuum, local increase [or augmentation] could not occur. They used to prove this [by arguing that] since augmentation would not occur unless something were added to a body, either what is added is received in a plenum or in a vacuum. If in a plenum, two bodies would exist simultaneously [in the same place]; if in a vacuum, what has been proposed [is had]. [For] it is not valid to say that [the increase is received] into something that was previously filled by a certain subtle matter; in the first place because it seems to be nothing but [a type of explanation similar to] the abhorrence

36. Under these circumstances Nicholas seems to concede that a vacuum would not occur. For if there is a contraction of the air—whatever the reason—room will be made for the newly arrived body. But now he shifts ground and, in the next few lines, asks about the space which the moving body has vacated. Thus even if one mass of air, *a*, contracts in its place, allowing another mass of air, say *b*, to enter we must ask about the place which *b* has vacated. Either it is a separate vacuum, which Nicholas deems absurd, or since nature abhors a separate vacuum, as Nicholas believes, a body *c* will enter and occupy this place. But then the same question must be asked about the place vacated by *c*. We are therefore involved in an infinite regress.

37. That is, despite the fact that nature abhors a separate vacuum we cannot avoid the absurdity of an infinite process for every motion in a plenum. Only by assuming interstitial vacua can we avoid the dilemma, as Nicholas suggests in the next few lines.

38. If the world is a plenum and body *a* moved into place *B*, body *b* must vacate *B* and move to place *C*, from which body *c* must vacate, and so on. This infinite chain reaction must occur in an instant or no motion would ever cease. To avoid this absurdity, one should assume interstitial vacua.

39. These two explanations are offered by Nicholas as possible responses to his opponents. In each alternative he seeks to explain the absence of compressibility by retaining interstitial vacua and also denying the formation of a separate or continuous vacuum. The essence of such "solutions" lies in the assumption that particles of air will compress up to a certain point or not at all, stopping short at some hypothetical critical point at which a separate vacuum might form. In these ad hoc explanations, nature, acting through its atomic particles, seems to "know" the critical point. Unlike ancient atomism Nicholas' system falls short of a thoroughgoing mechanical determinism and relies on nature to "know" the critical point at which the normal behavior of atoms is suspended and no further compression would occur despite the fact that the interstitial vacua would allow further contraction.

[of a vacuum];[40] secondly because it [that is, the subtle matter] could not occupy a place greater than the body itself. With these [subtle] bodies interspersed, a whole arm was at one time but a magnitude of 1 foot, but it is now a magnitude of 2 feet.[41] Aristotle does not respond to this argument but says that these arguments refute themselves, for let there be a vacuum and then it follows that there will be no augmentation [or increase] because either it follows that any part of the increase will not be an increase or that augmentation does not occur by addition to some body, or that two bodies will exist simultaneously [in the same place], or that the whole body will be increased.[42] The proof of the consequent is this:[43] because if there are vacuities in an increased body, either these vacuities are everywhere—and then the whole will be a vacuum and nothing of it will be body, which is one of the absurdities—or a certain part of the increased body is void and a certain part full, and then either the full part lacking in void could be increased or not. If it be declared that it could be increased, then I ask whether this will occur by adding body or not. If not, the other absurdity holds;[44] if it is increased by adding body, then, with this part having been assumed absolutely full, two bodies, according to them, will then be in the same place [simultaneously].

But saving the reverence of Aristotle, this argument does not tell against them. First let us see what will be understood in augmentation. Assume a certain body that is composed of four indivisibles A, B, C, D. I say that there are vacuities between these indivisibles. Let us assume that one indivisible is placed between A and B. I say that with the advent of a suitable thing, they spread or extend themselves, as it were, so that they are separated; [for example,] say that A and B are separated by an extent of vacuum equal to two indivisibles and [then] they [namely the vacuities] receive within themselves what is appropriate to their nature.[45] And this is how an augmentation occurs. And the understanding of such a proposition as "Every part of an increased [or enlarged] body is increased" does not at all include the prime parts of the increased body, so that it be understood that the indivisible [part] is increased.[46] But the understanding [of all this] is that [only] after the augmentation of any composite part is obvious to the sense and has been compared to that which preceded the increase is it true to say that it is greater, as a hand is larger (major) after an increase than before; and so it is in other things.

Moreover, the ancients argued about a vessel filled with ashes which received as much water as if there were no ashes [whatever in the vessel].[47] Firstly, the Commentator, Averroes, who wished to solve this, says that he has not experienced it;[48] secondly, he says that this is because some parts of the water are corrupted and some parts of the ashes are corrupted. Now, he asserts this because if the ashes are afterward dried they do not appear in so great a quantity [as before]. With respect to the first [statement by Averroes], it seems that he could not be much of a lover of truth to neglect experience so readily;[49] [and concerning the second assertion] it seems dreadful (fuga) that he did not oppose what he said [with what the ancients proposed,] because the ancients said that if a smaller quantity appears, this is because the parts are nearer than before.

The ancients argued in [yet] another way[50]

40. By analogy with the explanation that "nature abhors a vacuum," where bodies rush in to fill spaces that might otherwise be left void, this explanation would have grosser matter rush in to replace "subtle matter," where the latter constituted no part of the bulk of the body and the former will now increase the quantity of matter. "Subtle matter" is also mentioned by Marsilius of Inghen (see the beginning of selection 1 above).

41. If someone argues that augmentation is merely the replacement of subtle matter by gross matter, then how could a 1-foot magnitude interspersed with subtle matter become a 2-foot magnitude? That is, if a mere replacement is involved, the size of the body should remain constant. It would be as if the grosser matter were merely filling a vacuum. It is because of this that Nicholas likens this explanation to that of "nature abhors a vacuum."

42. See Physics IV.7.214b.3–10.

43. Here Nicholas shows how Aristotle's consequences follow from his assumptions. But in the next paragraph, he will reject them.

44. Namely that augmentation does not occur by the addition of matter to a body.

45. It would appear generally that the atomic interpretation of increase advocated here is such that whenever appropriate atoms enter the vacuities between the atoms of a body, immediately neighboring atoms will disperse sufficiently to allow for the subsequent intrusion of additional atoms, at which time another expansion will occur. In this way, a body is always potentially able to increase in size and quantity.

46. Individual particles or atoms do not increase; rather vacuities between atoms are occupied, producing an increase of body.

47. An illustration described by Aristotle in Physics IV.6.213b.20–22.

48. See Averroes, Physics, Book IV, Text 56 (folio 150 recto of the edition cited in Selection 44.1, n. 1).

49. The charge would seem to be that Averroes failed to test what could have been readily tested.

50. What follows is probably based on Aristotle, Physics IV.9.217a.10–19.

against those who denied a vacuum and assumed generation in things, for suppose that from the body of a lesser quantity a body of greater quantity is made, then either it is necessary that there be a vacuum or that a necessary condensation of bodies occur,[51] or that necessarily in another place there be generated [a body] of lesser quantity from a greater [body] in the same proportion as here [in this place] a body of greater quantity is generated from a smaller [body]. Now in order to avoid this,

they concede [the existence of a vacuum]. The consequent follows, because the body that has been generated occupies a greater place than it did before.[52]

Again, when the argument concerns a body which is moved locally, it is either received in a plenum or in a vacuum. If in a plenum, two bodies will exist [in the same place] simultaneously; if in a vacuum, what has been proposed is had, [namely that a vacuum exists]. . . .[53]

3. Galileo Galilei (1564–1642): The Existence of Interstitial Vacua Affirmed

Translated by Henry Crew and Alfonso de Salvio[54]

Annotated by Edward Grant

SAGR. It still remains for you to tell us upon what depends the resistance to breaking, other than that of the vacuum; what is the gluey or viscous substance which cements together the parts of the solid?[55] For I cannot imagine a glue that will not burn up in a highly heated furnace in two or three months, or certainly within ten or a hundred. For if gold, silver and glass are kept for a long while in the molten state and are removed from the furnace, their parts, on cooling, immediately reunite and bind themselves together as before. Not only so, but whatever difficulty arises with respect to the cementation of the parts of the glass arises also with regard to the parts of the glue; in other words, what is that which holds these parts together so firmly?

[66]

SALV. A little while ago, I expressed the hope that your good angel might assist you. I now find myself in the same straits. Experiment leaves no doubt that the reason why two plates cannot be separated, except with violent effort, is that they are held together by the resistance of the vacuum; and the same can be said of two large pieces of a marble or bronze column. This being so, I do not see why this same cause may not explain the coherence of smaller parts and indeed of the very smallest particles of these materials. Now, since each effect must have one true and sufficient cause and since I find no other cement, am I not justified in trying to discover whether the vacuum is not a sufficient cause?

SIMP. But seeing that you have already proved that the resistance which the large vacuum offers to the separation of two large parts of a solid is really very small in comparison with that cohesive force

which binds together the most minute parts, why do you hesitate to regard this latter as something very different from the former?

SALV. Sagredo has already answered this question when he remarked that each individual soldier was being paid from coin collected by a general tax of pennies and farthings, while even a million of gold would not suffice to pay the entire army.[56] And who knows but that there may be other extremely minute vacua which affect the smallest

51. That is, either space is made for the increased body by means of vacua or neighboring bodies must be compressed.

52. The implications and alternatives just described follow from the fact that the new body would occupy a greater place than before.

53. Approximately the last twelve lines of the section "On Vacuum" are omitted. They embody Nicholas' response to an objection raised by someone who might have been a contemporary (he is referred to as "the latest of the expounders of Aristotle") and are concerned with the relative or absolute nature of vacua.

54. Reprinted with the kind permission of the Macmillan Company from the First Day of *Dialogues Concerning Two New Sciences by Galileo Galilei*, translated from the Italian and Latin into English by Henry Crew and Alfonso de Salvio. . . (New York: Macmillan, 1914), pp. 18–19. Galileo's straightforward physical explanation should be compared to the preceding selection, where Nicholas of Autrecourt gives a complicated scholastic defense of interstitial vacua.

55. A reference to an earlier remark, quoted in Selection 53.4.

56. In connection with increasing the coherence of materials Sagredo had earlier declared (Crew and De Salvio, pp. 13–14): "I was wondering whether, if a million of gold each year from Spain were not sufficient to pay the army, it might not be necessary to make provision other than small coin for the pay of the soldiers." In terms of the present discussion, if a large single force were inadequate, perhaps a great number of minute forces acting together would be more effective.

particles so that that which binds together the contiguous parts is throughout of the same mintage? Let me tell you something which has just occurred to me and which I do not offer as an absolute fact, but rather as a passing thought, still immature and calling for more careful consideration. You may take of it what you like; and judge the rest as you see fit. Sometimes when I have observed how fire winds its way in between the most minute particles of this or that metal and, even though these are solidly cemented together, tears them apart and separates them, and when I have observed that, on removing the fire, these particles reunite with the same tenacity as at first, without any loss of quantity in the case of gold and with little loss in the case of other metals, even though these parts have been separated for a long while, I have thought that the explanation might lie in the fact that the extremely fine particles of fire, penetrating the slender pores of the metal (too

small to admit even the finest particles of air or of many other fluids), would fill the small intervening vacua and would set free these small particles from the attraction which these same vacua exert upon them and which prevents their separation. Thus the particles are able to move freely so that the

[67]

mass *[massa]* becomes fluid and remains so as long as the particles of fire remain inside; but if they depart and leave the former vacua then the original attraction *[attrazzione]* returns and the parts are again cemented together.

In reply to the question raised by Simplicio, one may say that although each particular vacuum is exceedingly minute and therefore easily overcome, yet their number is so extraordinarily great that their combined resistance is, so to speak, multipled almost without limit. . . .

Measurement of Forces

57 ON MAXIMUM AND MINIMUM POWERS

John Buridan (ca. 1300–ca. 1358)

Translated and annotated by Edward Grant[1]

Questions on the *De caelo [On the Heavens]* of Aristotle

BOOK I, QUESTION 21

Next, we inquire whether a power [or force] ought to be defined by its maximum capability.[2]

This question is very difficult. And with respect to the intention of Aristotle and the Commentator,

1. From *Iohannis Buridani Quaestiones super libris quattuor De caelo et mundo,* edited by Ernest A. Moody (Cambridge, Mass.: Mediaeval Academy of America, 1942), pp. 95, 96–100, 102–103, 109–112.
2. The expression "maximum capability" has been used to render the Latin phrase *maximum in quod ipsa potest.* Questions 21 and 22, large parts of which are translated here, are based upon a passage in Aristotle's *De caelo* (I.11.281a.7–27), which I now translate from a Latin translation possibly based on William of Moerbeke's medieval Latin translation from the Greek, made in 1271 and published in *Aristotelis opera cum Averrois commentariis* (Venice, 1562–1574; Frankfurt: Minerva, 1962), Vol. V, fol. 78 verso, col. 1, to 79 recto, col. 2. What follows is embedded in a lengthy discussion by Aristotle as to whether or not the world is eternal.
"Therefore if something can move through one hundred stades or lift a weight, we always speak of the most [or the maximum] that it can do, as to lift one

hundred talents or to walk one hundred stades and it can accomplish any parts below these. And if there is a maximum *(excessum)* it would be necessary that the power be defined to the limit of the maximum [*excessus;* throughout this passage *excessus* has been used for 'maximum']. It is necessary, therefore, that what someone can do according to the maximum he can accomplish everything below that maximum, just as if someone who can raise one hundred talents can also lift two; and if someone walks through one hundred stades, he can walk through two stades. Moreover, a power is of the maximum, and what cannot do as much as the maximum cannot do more, just as what cannot walk a thousand stades obviously cannot walk through a thousand and one stades. Furthermore, nothing should trouble us [about this], for that which is properly possible is determined by the limit of the maximum. Now perhaps someone will insist that what has been said is not necessary, for someone seeing a stade will not necessarily see a smaller magnitude. On the contrary, he who can see a dot or hear a slight sound will [also] sense greater things. But this does not bear on the argument. For the maximum is determined either in the power or in the thing. What has been said is obvious, for vision which surpasses is of the smaller object, [whereas] a velocity which surpasses [another velocity] is of the greater body." The very last point should perhaps be understood in accordance with Aristotle's belief that a heavier body falls more quickly than a lighter body in natural motion. The passage just quoted occasioned a considerable

it must be noted that their words in opposition [to the arguments set forth above] should not be understood as pertaining to definition or description in a proper sense, because neither an active nor passive power is defined by a maximum or minimum but [rather] by a principle for transmuting another or [being transmuted] by another, just as the reasons given at the beginning of this question argue.[3] But coming now to Aristotle's intention, it must be understood that we do not know how strong a power is other than by its effect, and therefore we judge a motive power to be greater because it can move a greater mobile, [or] other things being equal, [it can move a mobile] quicker or through a greater distance or through a longer time; and so concerning other powers.

It must be said, therefore, that according to Aristotle an active power must be determined by its maximum capability, in the sense that we know how much strength an active power has by knowing its maximum capability or at least by knowing the maximum below which it is able to do everything. And I assume this distinction because of a great difficulty which will be investigated afterward. For some say that it is not a matter of giving the maximum of which a power is capable but of giving the minimum of which it is incapable; and that minimum could be the maximum, not of what it is able to do but below which it is able to do everything. This conclusion is asserted: When[4] we take the lifting power of Socrates or Plato, or of another, we see that that power is greater or stronger which can lift 100 pounds than that which can lift only 50. And we assume this. Then we also assume that any power that can lift 100 pounds can also lift 50 or 40, but not contrarily.

And then we argue as follows: We know by this [that is, the assumptions of the preceding paragraph] how strong a power is, and we can distinguish it from a stronger and weaker power [or force]. But this is by means of its maximum capability, or at least by its capacity to overcome everything below that point, but not otherwise. Therefore, and so on. The major [premise] is obvious. The minor [premise] is asserted, because to know a maximum capability includes two things, namely to know that it is able to do so much and to know that it is not able to do more. That we know it can do so much allows us to distinguish it from a lesser power [or force,] because that [lesser power] cannot do as much. And that we know it can do no more allows us to distinguish it from a greater power, since a greater power

can do more. And thus we can know definitely how much it [that is, the maximum capability] is.

Nevertheless, I say, moreover, that not every active power can be determined in this way by its maximum capability or [the point] below which it can do everything, because the power of God, which according to the truth of faith is of infinite strength, cannot be determined in this way. For, indeed, no maximum can be given which God can move or do—neither a maximum velocity with which God can move the sky, nor an effect so perfect that God could not make it [even] more perfect. And this is because the infinite is not properly more in some degree, nor is it measurable or proportionable. Nevertheless, it must be noted that if we should want, in some way, to say how much is the motive or creative power of God, we should say that it is so much that in every mobile or

---

literature on maximum and minimum powers—indeed, the discussions were usually called *On Maximum and Minimum (De maximo et minimo)*—which had, as we shall see, both physical and mathematical aspects. In the present selection, which is intimately associated with the Aristotelian passage, it is the physical discussion that predominates, but, as Curtis Wilson explains in his treatment of this subject, for others, as, for example, Heytesbury, "The problem of the purely physical and qualitative character of different potencies recedes into the background and the discussion assumes. . .an extremely hypothetical, logicomathematical character. A set of rules is proposed as covering all imaginable cases of action-passion; the rules are couched in language borrowed from the *logica moderna,* and the center of interest appears to have shifted from that of answering strictly physical questions to the logicomathematical problem of setting extremes to classes or aggregates." (Curtis Wilson, *William Heytesbury: Medieval Logic and the Rise of Mathematical Physics* [Madison, Wis.: University of Wisconsin Press, 1956], ch. 3, "De maximo et minimo," p. 69).

At the outset, Buridan presents four arguments (omitted here) in defense of the opinion that a power ought not to be defined by its maximum capability. Then, after remarking that Aristotle and the Commentator, namely Averroes, say that "every power ought to be defined by its maximum capability," Buridan continues his discussion as translated here.

3. In the first of the four reasons, we read (*Iohannis Buridani . . .*, p. 95): "And it is argued in the negative [namely that a power ought not to be defined by its maximum capability] because in the fifth [V.12.1019a. 15–1019b.14] and ninth [IX.1.1064a.4–15] books of the *Metaphysics* powers are defined as active as well as passive, but not by a certain maximum. Indeed, it is said that an active power is a principle of transmuting another and a passive power is a principle of being transmuted by another."

4. Although the text has *quia,* I have used "when" rather than "since" or "because."

creatable thing it can move [that thing] greater [that is, with a greater velocity] or more perfectly or even to create [it more perfectly]. Indeed, it is by this that we believe Him to be infinite.

Next, concerning a passive power, it should also be understood that one must speak in a contrary way to [that in which one would speak about] an active power. For just as any active power that can act on a greater [thing] can also act on a lesser [thing], so also a passive power that can endure [or be acted upon by] a smaller active [power] can also be acted upon by a greater active power, but not the contrary—just as if Socrates can lift a great stone, so also can he lift a smaller stone; and if this stone can be lifted [or raised] by a strong force, it can also be lifted [or raised] by an even stronger force. And so, just as an active power should be determined by a maximum capability, so a passive power should be determined by the minimum which it can endure [or by which it can be stimulated], for by so much as the passive is said to be more passible [or receptive] by so much more does it suffer [or undergo change] from a lesser active power. Thus, if a passion should be [considered] good, a passive power is said to be better, stronger, and greater—not in acting or resisting, but in suffering and receiving—which receives more from a smaller object.

And by this [approach], the argument made about vision is solved.[5] For a visual power is passive and receptive of species from an object. Thus a visual power is called stronger, better, and more perfect, that receives from a smaller visible object a species sufficient to cause vision.

And so it appears that the [four] reasons given at the beginning of this question follow in their manner in accordance with the things said above, although a great doubt remains as to which part of the disjunctive [proposition] posited before is to be held;[6] but this will be discussed in another question. And it is obvious that it must be expressed in this manner.

BOOK I, QUESTION 22

Next, we inquire whether a power's maximum capability can be assigned.[7]

And it is argued "yes" on the authority of Aristotle and the Commentator, who say that a power or virtue must be determined by its maximum capability, which would not be true if that maximum could not be assigned.

The opposite [of this] is argued by the authority of many others, who say commonly that it is not

[a question] of giving a maximum capability but the minimum of which it is incapable.

This question is very difficult because of the diversity of powers, both active and passive, and because of the diverse ways of acting or suffering and of moving or transmuting.

In the first place, therefore, I shall say something about the first power, namely the divine power. And, in accordance with faith, I posit this conclusion: No maximum mobile can be assigned which God cannot move, because, as stated before, I assume that an infinite mobile cannot be assigned; and yet, nevertheless, for every finite mobile that is assigned or assignable, God, by means of His infinite power, could make a greater finite mobile and move it. And so it may also be said that no minimum mobile can be assigned which God could not move, since God can move every mobile, great or small. Thus the power of God must not be determined by its maximum capability nor by the minimum of which it is incapable. I also say that there is no maximum velocity with which God is able to move [a mobile], since, indeed, whatever velocity may be assigned or assignable, He can move it [even] more quickly.

But now we must report what Aristotle would

5. Here Buridan refers to the fourth and final principal reason proclaimed at the beginning of the question (p. 96): "Again, Aristotle assigns the cause whereby a power must be determined by a maximum, namely because what is able to do a greater thing can do a lesser, but not conversely. Thus, if this cause should not be universally true, it must not be universally conceded that a power should be determined by its maximum capability. But this cause is not universally true, since it does not follow that if you can see a millet seed you can [also] see the thousandth part of it; for, indeed, many can distinguish and see greater things who cannot, however, distinguish smaller things." In the lines that follow, Buridan responds to this objection with the observation that vision is a passive—not active—power, so that any visual power capable of stimulation by a smaller object is a greater passive power than one whose threshold of visual stimulation can only be activated by a larger object.

6. A reference to the original question. At this juncture Buridan admits that he cannot decide the question "whether a power [or force] ought to be defined by its maximum capability." We are told that this will be considered in another question but apparently not in this treatise.

7. While admitting the difficulty in deciding whether a "maximum capability" can define a power, Buridan seeks to determine in this question whether a maximum power of any kind can be assigned. As will be seen, no maximum can be assigned to God's capacity for producing velocity, nor indeed to the action of an agent or power on a patient or resistance.

say about this. And it seems to me that Aristotle would say that a maximum velocity, or even a maximum slowness, can be assigned with which God can move the last [or outermost] sphere; and this maximum velocity, or even this maximum slowness, is that with which God does in fact move the last sphere. For Aristotle would say that God, because of His immutability and will and because of the unchangeability of the last sphere, cannot move this sphere quicker or slower than He now moves it.

However, all this notwithstanding, it must be said against Aristotle[8] that no maximum speed or maximum slowness can be assigned with which God can move, or even with which God does in fact move, [the last sphere]. On the contrary, whatever the velocity with which He moves something, He can move [it] with yet a greater velocity— I do not say greater extensively, but greater intensively. That is, in the last sphere what is moved cannot be assigned as quickest, for this whole sphere is not moved [with a] quickest [speed], since some parts of it, say near a pole, are moved more slowly than other parts.

Now if something were difformly [that is, non-uniformly] white or black so that from one end [apex? cone?; how the Latin *conus* ought to be rendered in this context is problematic] to the other end [apex? cone?] there would be a gradual diminution in whiteness part after part, this whole body would not, however, be denominated as a highest [or greatest] whiteness, even though the remission would begin from the end with the highest [or greatest degree of] whiteness. Indeed, by balancing one part against another, it ought more to be denominated by the mean degree between the greatest white part and the least white part.[9] Thus it could be said that if a stick were disposed in this way [that is, denominated by its mean degree], some part of the stick would be whiter than the [whole] stick and another would be less white than the [whole] stick. It is the same with respect to velocity in the issue proposed here. For the more remote parts of a sphere are moved more quickly and the parts nearer the poles are moved more slowly. It follows from this that the whole sphere is not moved most quickly or most slowly; on the contrary, a part remote from the pole is moved more quickly than the whole sphere and a part near the pole is moved more slowly than the whole sphere; and as a part is more remote from a pole— indeed from both poles—so much more quickly is it moved.

Now since this is so, it [that is, the whole sphere] ought not to be said to have a quickest motion. Indeed, by balancing one part against another it ought to be denominated more by a velocity mediate between the quickest and slowest moved parts.[10] And from this being so, it follows that no most intense velocity can be assigned in the sky unless there can be assigned a part most remotely distant from the poles. But such a velocity cannot be assigned unless you assume something without width, for [only] then would it be true that the circumferential line of the equinoctial circle would be most remote from the poles and would be moved most quickly. But we assume that such lines cannot be assigned.[11] Moreover, if between the poles you assign length, width, and depth to any part, it cannot be most distant from the poles. On the contrary, if it were divided into three parts according to latitude, the middle part of it would be more distant from the poles; therefore, no quickest motion can be assigned.

And the same appears [to be true] about slowness, because as a part around a pole is smaller so

8. As required by the sense of the argument, the Latin *secundum Aristotelem* has been rendered as "against Aristotle" rather than "according to Aristotle," its usual meaning.

9. The terminology and concepts embodied in this example are drawn from the subject matter of intension and remission of forms, which is described in Selection 43.

10. By analogy with the balancing of parts of whiteness in arriving at a mean degree of whiteness in the preceding paragraph, in this example concerning velocity Buridan implicitly assumes that the velocities of the points along the radius increase uniformly as they recede from the center. It is assumed intuitively that for every velocity less than the mean speed by whatever amount, there is a corresponding velocity greater than the mean speed by the same amount and that the points involved are equidistant from the point moving with the mean speed. All this is drawn from, and associated with, the mean speed theorem originating at Oxford in the 1330's (see Selections 42 and 43). It should be noted that in his *Treatise on Proportions (Tractatus de proportionibus)*, chapter 4, Part II, written in 1328, Thomas Bradwardine insisted, contrary to the position adopted by Buridan and most fourteenth-century scholastics, that the speed of a rotating radius be measured by the speed of its fastest moving point (see M. Clagett, *The Science of Mechanics in the Middle Ages* [Madison, Wis.: University of Wisconsin Press, 1959], p. 221).

11. For terminists or nominalists like Buridan, only physical—not mathematical—points, lines, and planes were conceded reality. Thus, only if the equinoctial circle were taken as mathematical—that is, lacking in width and depth—could a quickest point be admitted. Buridan, however, rejects such a move.

much the slower would it be moved; and thus a slowest motion cannot be assigned unless a minimum [velocity] can be assigned, which is impossible. . . .[12]

Having spoken of the powers of God and the intelligences, we must now consider natural and corporeal powers. It must be noted that some [natural powers] act without the resistance of a patient, as [for example,] light illuminating a medium, or a visible object multiplying its species in a diaphanous [or transparent] medium,[13] or into the eye; and so with many other [natural powers]. Other powers, however, act with the resistance of a patient, as [for example,] heating, or lifting a stone, or the downward motion of a heavy body, and so with many other [natural powers]. For it is a medium which resists. But first we shall speak about those things that act without resistance, [for it is here] that the major difficulty lies. In the first place we shall consider vision, but since illumination is required for vision, we shall speak first about illumination. . . .[14]

12. Although I have omitted the next section of the discussion, where Buridan considers the power of incorporeal intelligences (*Iohannis Buridani*. . . , pp. 100–102), I include Curtis Wilson's summary of it (pp. 64–65; Wilson has provided a brief description of the entire question):

"SECOND: The powers of the celestial intelligences.—Here again it is probable that Aristotle would assign a maximum body which each intelligence can move, namely that sphere which it does in fact move, and also a maximum velocity with which each intelligence can move this body, namely that velocity with which it does in fact move its sphere. For the intelligences are finite and immutable in power, and the bodies which they move unalterable; hence the relation between mover and moved will remain constant, and each intelligence in moving its sphere will be acting according to its maximum power. But a doubt arises, supposing that God can augment the sphere of the moon, and further that the power of the mover must exceed the resistance of what is moved. For the power of the lunar motor will exceed the resistance of the lunar sphere by a divisible excess; let God then increase the resistance of the sphere by half this excess, and the lunar intelligence will still be able to move it. Hence the present lunar sphere is not the maximum which the lunar intelligence can move. Buridan answers that the reasoning is not demonstrative, because the celestial spheres offer no resistance to their movers, but have a pure inclination to their proper movement. In such a case, although the active power will exceed the passive in nobility and perfection, it is not necessary that its activity be greater than the passivity of the passive. The distinction between pure passivity and resistance is drawn more precisely in the next section."

13. For the concept of "multiplication of species," see Selections 62.1–4.

14. A lengthy discussion of illumination and vision (*Iohannis Buridani*. . . , pp. 103–109) has been omitted. Wilson's summary of these sections is offered instead (pp. 65–67):

"THIRD: Light which illuminates a transparent medium.—To begin with, we must decide whether the medium offers resistance to the light which illuminates it. It would seem that it does insofar as it is opaque, for the less transparent and more opaque medium is illuminated less intensely and less far by the same light than the more transparent and less opaque. In opposition, however, it is objected that if opacity constituted a resistance, the illumination of the medium would become more intense and extend to a greater distance over a period of time, as we see in the case of heat-action. But this is false; the medium is illuminated instantaneously to its maximum potentiality. Buridan decides, therefore, that the medium is without resistance; the cause of its being more or less transparent is a greater or less passivity to light. Resistance is through a contrary or an inclination to a contrary; but light has no contrary.

"It may now be asked whether there is a maximum distance to which a given light can propagate its species in a given medium. If the light be of finite potency, and if neither the light nor the medium alter, then the light will illuminate part of the medium and part it will not; and between the two parts there will be only an indivisible or mathematical point. Hence the distance from the lightsource to this point will be the maximum distance to which the light propagates its species. It will also be the maximum distance to which the light *can* propagate its species; for the illumination does not improve with time, and the light cannot illuminate more than it in fact illuminates. A further consequence is that there is no minimum distance to which the light cannot propagate its species; for such a *minimum quod non* could not differ by a divisible magnitude from the *maximum quod sic* already assigned, the entire part of the medium beyond the dividing point being unilluminated.

"Again, as a corollary it follows that there is a maximum space through which a given visible body multiplies the species of color; since the action of light and that of color are entirely similar with respect to the multiplication of species. Thus it is possible for a given colored body to multiply its species to the surface of the eye and no further, or the point which is geometrically at the center of the eye and no further.

"FOURTH: A given medium to be illuminated by a light.—Buridan now turns to the case of the corresponding passive potency. Given a determinate medium, spherical in form, is there a minimum or maximum light which can illuminate the whole of it? Buridan answers that there is a *minimum quod sic;* and it will be precisely that light the maximum range of which extends throughout the given medium.

"FIFTH: Vision.—Here Buridan does not consider, as had Averroes, the case of different objects seen at a fixed distance, but rather keeps the object fixed and allows the distance to vary. Given a determinate power of vision *a,* a visible body *b,* and a uniform medium, is there a maximum distance through which *a* can see *b*? The common and probable answer, Buridan states, is in the negative. For suppose there were such a distance; through this distance *a* would see *b* with a certain intensity of

Now, finally, we must speak briefly about agents [acting] with [or against] the resistance of a patient. Concerning these, it is assumed as a principle that if a resistance is stronger than the active virtue or power, there is no action. But if the active power is stronger than the resistive power, there is an action and the action will be so much the greater or quicker as the excess of the active power to the resistance will be greater. These things are assumed. Then again we assume that the strength of the agent is not changed.

This first proposition is now posited: There cannot be assigned a maximum resistance by means of which an agent can act or with which it can act. For if A ought to move B, and you wish to assign the maximum resistance with which A moves B, let this maximum resistance be CD. Then I take a resistance equal to the active power and it [that is, the power] will be unable to move with this resistance. But it was able to move with resistance CD, therefore that equal resistance is greater than resistance CD; and let this be CE. But then the difference between D and E is divisible,[15] and let I represent the division. Then resistance CI is smaller than resistance CE, and consequently the active power exceeds it and so can move it.[16] But resistance CI is greater than resistance CD, so that resistance CD was not the maximum [resistance] with which A could move B, even though our adversary assumed it was a maximum.

From this, a second proposition that has been customarily assumed follows immediately, namely that a minimum resistance can be assigned with which A cannot move B, and this is the resistance that is equal to the active power. For it has been said that there is no action with such a resistance, but that however much smaller the resistance may be, the active power would exceed it and move it. Therefore, there is no smaller resistance that it could not move; but as there is nothing smaller [that it cannot move,] this is the minimum. Therefore, this is the minimum with which it was unable to move.[17]

vision (*visio aliquantae intensionis*). Now a form or quality which is naturally capable of remission or diminution in intensity is not corrupted all at once from a certain intensity to zero intensity, but remits in a continuous manner to its total corruption; hence if the distance between a and b is increased, a will continue to see b, albeit less and less intensely, until vision is finally reduced to zero intensity. Thus the original distance assigned was not a *maximum quod sic*, and no such can be assigned.

"And yet, although this reasoning appears demon-strative, a grand difficulty arises, which shows how the problem of assigning limits to potencies could become hopelessly involved with the problem of the physical nature of particular potencies. It was previously admit-ted (third section) that there is a maximum distance through which the given visible object b can multiply its visible species. Suppose this distance extends from b precisely to that point of the eye, say the center of the eye, at which vision becomes possible; will not this be the maximum distance through which the eye can see b? Thus we have two arguments: one based on the nature of vision as a remissible form, and leading to the assign-ment of a *minimum quod non*; the other based on the mode of propagation of the visible species of an object, as extending a definite maximum distance, and leading to the assignment of a *maximum quod sic* for vision as well. The two arguments cannot both stand.

"Buridan attempts to extricate himself from the diffi-culty by means of a distinction. There is a difference between 'the space in which b can be seen' and 'the space *through which* b can be seen'; b is in the first, but not in the second. Buridan then assigns a maximum space *in which* the body b can be seen; and a minimum space *through which* b cannot be seen, equal to the maxi-mum *in which* it can be seen. Apparently—although Buridan does not say as much—this space is also equal to the maximum distance through which b can multiply its visible species. Thus if b is just outside the given space, it cannot be seen; if it approaches the eye by as little a distance as you please, it comes into the range of vision; and when it is altogether in the given space, it can be seen as a whole. However ingenious this distinction, it may be questioned whether it meets the central difficulty, which arises from the problem of the nature of light and of vision."

15. That is, between CD and CE. The Latin reads in-correctly *inter D et C* ("between D and C") rather than *inter D et E*.

16. That is, $CE > CI > CD$.

17. The principles embedded in the two propositions of the last two paragraphs were used earlier in this very question, and although omitted from the translation were summarized in Wilson's account quoted above in note 12. Indeed, much the same proof appears also in an anonymous work entitled *Probationes conclusionum;* it is summarized by Wilson on page 58 of his volume on Heytesbury.

That no maximum resistance is possible Buridan dem-onstrated in the first proposition by utilizing one of the most fundamental axioms in medieval physics, namely that motion is impossible when the resistive force is equal to, or greater than, the motive force. Thus, if F is a force and R a resistive body, then if $F > R$, motion can occur. But it is obvious that R cannot represent F's maximum capacity, since another resistance, say R_1, can be found such that $F > R_1 > R$; nor does R_1 repre-sent F's maximum capability, since R_2 can be found such that $F > R_2 > R_1$, and so on. Hence in the second proposition Buridan concludes that *the maximum capacity of any motive force must be measured by the least resistance which it is unable to move,* and this will equal the motive force (that is, $F = R$).

Of this proof as it appears in the *Probationes con-clusionum,* Wilson remarks (pp. 58–59): "The proof is

And now more particularly, a third proposition is concluded, [which is] that one cannot assign [or specify] a maximum heavy body that Socrates can lift, but one can assign [or specify] a minimum heavy body that he cannot lift; and this is what is equal to his power.[18]

Similarly,[19] it must also be said if we posit a medium resisting the downward motion of a heavy body by its thickness [or density], we do not assign the thickest [or most dense] medium through which heavy body A can be moved downward but the least thickness [or denseness] through which it cannot be moved downward. Nor should a maximum weight be assigned which weight A can incline or raise in an [equal-arm] balance, but [rather] a minimum weight which cannot be inclined [or raised], and that is a weight equal to A. Therefore,

_____

rigorous, if two assumptions are granted: (1) that the rule according to which no action or motion accrues from a ratio of equality between power and resistance is correct; and (2) that the sequence of all possible weights [where the anonymous *Probationes* speaks of weights, Buridan speaks of resistance in general—*Ed.*] forms a continuum. From this first illustration, indeed, it is possible to generalize, and to say that the medieval discussion of maxima and minima presents two major aspects of interest: (1) a physical or dynamical aspect, involving the relations between powers and resistances; and (2) a purely mathematical aspect, involving the concepts of continuity and limit.

"As physicists— or we might say 'speculative physicists,' since the theory here was seldom such as to yield conclusions admitting of empirical verification— the Schoolmen introduced numerous distinctions between types of powers and resistance, distinctions which necessitated a variation in the manner of assigning limits to potencies. For example, how is the power to be bounded if it is active or passive, debilitable (subject to weakening) or indebilitable, finite or infinite? If a power divides a resisting medium, how will the boundary be assigned when the medium is uniform; when it is difform (nonuniform)? What alterations in the form of the analysis must be introduced when the velocity and duration of the movement effected by a power are taken into account?

"The mathematical interest in the discussion lies, first of all, in the use of two kinds of boundaries for continuous sequences: one, in which the element serving as a boundary is itself a member of the sequence of elements which it bounds (*maximum quod sic* or *minimum quod sic*); and the other, in which the element serving as a boundary stands outside the range of elements which it bounds (*maximum quod non* or *minimum quod non*). In the preceding chapter we have noted (1) that this distinction is not capable of empirical verification; (2) but that, in recent times, it has proved of importance in the theory of functions and in the definition of irrational numbers; (3) that the medieval Schoolmen, on the contrary, were concerned with it only in connection with magnitudes which are, at least hypothetically, physical.

Once more, however, we find that the medieval discussion, particularly where it enters the context of logic, develops in extremely imaginative directions; becomes, in effect, a discussion of the ways in which aggregates may be bounded, and of the ways in which quantities may vary. The application of the two types of boundaries to finite and infinite aggregates and the incisive use of the notion of 'any' in connection with variable quantities permit us to speak of a logicomathematical interest in the Scholastic discussion."

In this physical discussion, and in the two examples following, Buridan has divided the continuum of resistances into two classes, one containing all resistances that can be moved, which includes all resistances smaller than the particular resistance that is equal to the motive force; the other class of resistances containing all those which the motive force cannot overcome and move, and embracing all resistances greater than the resistance which equals the motive force. Thus the R that equals F divides the continuum of resistances into two distinct classes, a division which also meets the requirements of Dedekind's postulate or cut enunciated in 1872 and which I quote in the translation of Wooster W. Beman:

"III. If a is any definite number, then all numbers of the system R fall into two classes, A_1 and A_2, each of which contains infinitely many individuals; the first class A_1 comprises all numbers a_1 that are $< a$, the second class A_2 comprises all numbers a_2 that are $> a$; the number a itself may be assigned at pleasure to the first or second class, being respectively the greatest number of the first class or the least of the second. In every case the separation of the system R into the two classes A_1, A_2 is such that every number of the first class A_1 is less than every number of the second class A_2." (Richard Dedekind, *Essays on the Theory of Numbers. I: Continuity and Irrational Numbers. II: The Nature and Meaning of Numbers* [Lasalle, Ill.: Open Court Publishing Co., 1948], p. 6.)

Where for Dedekind the assignment of the dividing point or number, a, was wholly arbitrary—that is, it could be assigned as the maximum term of the first class or the minimum term of the second class—it was hardly an arbitrary matter for Buridan and the scholastics who were concerned with real physical continua. Thus, for Buridan, the resistance that equals the motive force must necessarily be the minimum force which F cannot move, and hence the first and least term of the second class. As for the first class, it has no maximum resistance, since its successive resistances only approach the minimum resistance of the second class as a limit. Thus each physical case or example would have to be analyzed in order to determine to which of the two classes the dividing quantity is to be assigned. Despite this rather important difference between the Dedekind cut and its medieval counterpart, their similarity is nonetheless striking.

18. The results of the first two propositions are combined to form this third proposition.

19. Since the two Latin manuscripts used in establishing this edition diverge radically in their accounts of this paragraph, Moody included verbatim texts of both versions in parallel columns. Only the longer and seemingly more complete version in MS Bruges 477 has been translated here.

in any things whatever, [although] the greater mobile would resist more, no maximum can be assigned which *A* can move, but a minimum can be assigned which it cannot move.

But since some also speak about a maximum velocity, therefore I should speak of natural and not voluntary agents. Assuming that an active power is neither increased nor diminished, and that the resistance is not [increased or diminished], and [assuming further] that the resistance is such that *A* could move *B*, I could assign a maximum velocity and even a minimum velocity according to which *A* can move *B*. And it is that by which *A* does in fact move *B*, because in the case posited, *A* moves *B* with a certain velocity and cannot move *B* with a greater or smaller velocity. Indeed, from the aforesaid observations, it will always move it uniformly, since the same ratio of active force to resistance will always remain.

From[20] this one must conclude that when the medium remains uniform in the downward motion of a heavy body, the moving force is continually increased. This is due either to the attraction of the place or the impetus acquired by the motion,[21] or any other reason whatever about which I shall speak later. And this is obvious, because the natural motion of this heavy body downward is continually

quicker and quicker, while the resistance remains the same, since we assumed a uniform medium. What I have said about local motions ["motions" renders *motibus,* which I have substituted for the textual reading *mobilibus*] must also be understood of alterations.

However, someone might object, because a drop of water thrown on a great fire acts [on it] and in a very short time extinguishes a glowing coal on which it falls. However, this great fire is of greater resistance than is the active power of the drop of water. Therefore, an agent acts even though the resistance is greater than its active power. Only it must be asserted that not the whole fire suffers from this drop but [only] a certain part of it whose resistance is exceeded at the beginning by the active power of the drop.

You ought to speak in this way about all other natural agents by considering that if an active power is increased in the downward motion of a heavy body, or if it is diminished, as in violent motions and in fatigable agents, then the velocity is increased or diminished. And so for all these cases it is necessary to reflect and respond according to the requirements [of each case].[22] And so it has been said.

Magnetism

58 AN ENCYCLOPEDIST'S DESCRIPTION OF THE MAGNET

Bartholomew the Englishman (fl. 1220–1250)

Translated and annotated by Edward Grant[1]

ON THE MAGNET

The magnet is an Indian stone of an iron-like color. It is found in India among the Troglodytes and attracts iron, for, as Isidore says, it makes a chain of rings so that the vulgar call it "living iron" *(ferrum vivum).* It is believed that it also attracts clear glass, just as it does iron. As Augustine says, the force *(vis)* of it is so great that if anyone held the same magnet under a gold or bronze vessel and placed iron above it, the iron above will be moved by the motion of the [magnetic] stone underneath and follows [the magnet] in the same place. Whence, in a certain temple, a statue was made of

20. In the next two paragraphs I again translate from the longer version in MS Bruges 477, for the same reason given in the preceding note.

21. For the attraction of place and the role of impetus on bodies falling with accelerated motion, see Selection 49.

22. "The conclusion of Buridan's discussion is thus that a single rule cannot be given for assigning maxima and minima, but it is necessary to have regard for the particular exigencies of each case" (Wilson, pp. 68–69). For the most part, Buridan rejects Aristotle's position that a power is measured by its maximum capability (this is true of all the sections translated here). Only in vision does he partially accept Aristotle's viewpoint, but even here his attempt to reconcile it with the concept of a minimum which cannot be affected or altered failed (see n. 14).

1. Translated from *Bartholomaei Anglici De genuinis rerum coelestium terrestrium et inferarum proprietatibus libri XVIII.* (Frankfurt, 1601; Frankfurt, Minerva, 1964) pp. 746–747. This inferior, meager, and fantastic account should be compared to Peter Peregrinus' splendid exposition in the next selection.

iron which, by the power of the magnet, was seen to hang in air. Indeed, there is another species of magnet in Ethiopia which repels *(respuit)* iron and flees from itself. Furthermore, the magnet sometimes attracts iron from one angle and repels it from itself in another [angle]. Moreover, the darker [or bluer] a magnet is, the better it is. Isidore, a second *Dionysius,* and the *Lapidarius* add to this that this kind of stone restores husbands to wives and increases elegance and charm in speech. Moreover, along with honey, it cures dropsy *(hydropsim),* spleen, fox mange *(alopecia),*[2] and burns *(arsura).* From the dizziness and confusion of the brain [caused by] the powder [of this magnet] when it is sprinkled over burning coals lying through the four angles of a house, it will seem that the house tumbles down immediately. And the

magnet, just like the loadstone, when placed on the head of a chaste woman causes its poison *(virum suum)* to surround her immediately;[3] [but] if she is an adulteress she will instantly remove herself from bed for fear of an apparition. According to Plato, the Magi especially used this stone. A magnetic stone is hot and dry in the third degree, and has power to attract iron. There are mountains made of such stones and they attract and dissolve ships made of iron. Its dust, perfected with *apostolico,* an application for wounds, is especially valuable for wounds. Indeed the powder of a magnet in a quantity of two drams attracts iron [and] with the juice of fennel is valuable against dropsy, spleen, and fox mange, as Avicenna says, for it draws phlegm and black bile *(melancholia).*

59 THE FIRST SYSTEMATIC DESCRIPTION IN EUROPE OF THE PROPERTIES OF THE LODESTONE

Peter Peregrinus (fl. 1269)

Translated by Brother Arnold (Joseph Charles Mertens)[1]

Annotated by Edward Grant

The Letter of Peregrinus

PART I

Chapter 1
Purpose of this Work

Dearest of Friends:[2]

At your earnest request, I will now make known to you, in an unpolished narrative, the undoubted though hidden virtue of the lodestone, concerning which philosophers up to the present time give us no information, because it is characteristic of good things to be hidden in darkness until they are brought to light by application to public utility. Out of affection for you, I will write in a simple style about things entirely unknown to the ordinary

one by H. D. Harradon ("Some Early Contributions to the History of Geomagnetism -I," *Journal of Terrestrial Magnetism and Atmospheric Electricity* [subsequently retitled *Journal of Geophysical Research*], 48 [1943], 6–17).

For descriptions of knowledge about magnetism in Greek and Roman antiquity, see M.R. Cohen and I.E. Drabkin, *A Source Book in Greek Science* (Cambridge, Mass.: Harvard University Press, 1948), pp. 310–314, and Duane H. D. Roller, *The "De Magnete" of William Gilbert* (Amsterdam: Menno Hertzberger, 1959), pp. 11–25 (for the Middle Ages, see pp. 25–42, which include a section on Peregrinus).

Although certain magnetic effects were known, for instance, chain-like attraction (magnetic induction), repulsion, and the action of magnets through certain media, no systematic description of the magnet's properties was given nor was magnetic polarity recorded. Indeed, no artificial magnets were made, and the application of magnetism to compasses was unknown. In this selection Peter Peregrinus reveals knowledge on all these matters. It would also seem that his letter was of some influence in the history of magnetism, for it is cited in William Gilbert's *De magnete* (1600) and in Athanasius Kircher's *De arte magnetica* (1641) and it was plagiarized by Jean Taisnier in the latter's *De natura magnetis* (1562). References to Gilbert's *De magnete* will be to the translation of Silvanus P. Thompson, made in 1900 and re-issued in 1958 in *The Collector's Series of Science,* published by Basic Books.

2. The letter was written to one Siger of Faucaucourt (or some variant of this place name), who was also from Picardy.

---

2. For Isidore of Seville's description of fox mange, see Selection 89.

3. Presumably to ward off potential attackers.

1. *The Letter of Petrus Peregrinus On the Magnet, A.D. 1269,* translated by Brother Arnold (Joseph Charles Mertens), with Introductory Notice by Brother Potamian (M. F. O'Reilly) (New York: McGraw, 1904), pp. 3–34. The translation was made from a Latin edition published by Achilles Gasser in Augsburg, 1558. An earlier English translation was made by Silvanus P. Thompson. (*Epistle of Petrus Peregrinus of Maricourt, to Sygerus of Foucaucourt, soldier, concerning the magnet* [London: privately printed at the Chiswick Press, 1902] and a later

individual. Nevertheless I will speak only of the manifest properties of the lodestone, because this tract will form part of a work on the construction of philosophical instruments. The disclosing of the hidden properties of this stone is like the art of the sculptor by which he brings figures and seals into existence. Although I may call the matters about which you inquire[3] evident and of inestimable value, they are considered by common folk to be illusions and mere creations of the imagination. But the things that are hidden from the multitude will become clear to astrologers and students of nature, and will constitute their delight, as they will also be of great help to those that are old and more learned.

Chapter 2
Qualifications of the Experimenter[4]

You must know, my dear friend, that whoever wishes to experiment, should be acquainted with the nature of things, and should not be ignorant of the motion of the celestial bodies. He must also be skilful in manipulation in order that, by means of this stone, he may produce these marvelous effects. Through his own industry he can, to some extent, indeed, correct the errors that a mathematician would inevitably make if he were lacking in dexterity. Besides, in such occult experimentation, great skill is required, for very frequently without it the desired result cannot be obtained, because there are many things in the domain of reason which demand this manual dexterity.

Chapter 3
Characteristics of a Good Lodestone

The lodestone selected must be distinguished by four marks—its color, homogeneity, weight and strength. Its color should be iron-like, pale, slightly bluish or indigo, just as polished iron becomes when exposed to the corroding atmosphere. I have never yet seen a stone of such description which did not produce wonderful effects. Such stones are found most frequently in northern countries, as is attested by sailors who frequent places on the northern seas, notably in Normandy, Flanders and Picardy.[5] This stone should also be of homogeneous material; one having reddish spots and small holes in it should not be chosen; yet a lodestone is hardly ever found entirely free from such blemishes. On account of uniformity in its composition and the compactness of its innermost parts, such a stone is heavy and therefore more valuable. Its strength is known by its vigorous attraction for a large mass of iron; further on I will

explain the nature of this attraction. If you chance to see a stone with all these characteristics, secure it if you can.

Chapter 4
How to Distinguish the Poles of a Lodestone

I wish to inform you that this stone bears in itself the likeness of the heavens, as I will now clearly demonstrate. There are in the heavens two points more important than all others, because on them, as on pivots, the celestial sphere revolves: these points are called, one the arctic or north pole, the other the antarctic or south pole. Similarly you must fully realize that in this stone there are two points styled respectively the north pole and the south pole. If you are very careful, you can discover these two points in a general way. One method for doing so is the following: With an instrument with which crystals and other stones are rounded let a lodestone be made into a globe and then polished.[6] A needle or an elongated piece of iron is then placed on top of the lodestone and a line is drawn in the direction of the needle or iron, thus dividing the stone into two equal parts. The needle is next placed on another part of the stone and a second median line drawn. If desired, this operation may be performed on many different parts, and undoubtedly all these lines will meet in two points just as all meridian or azimuth circles meet in the two opposite poles of the globe. One of these is the north pole, the other the south pole.[7]

3. Apparently Siger's inquiries on magnetism furnished the incentive for this important letter.

4. For Roger Bacon's praise of Peregrinus' talents as an experimenter, see the biographical sketch of Peregrinus, at the end of this source book.

5. S. P. Thompson remarks that Peregrinus' "reference to the navigation of the northern seas, . . . is of significance in the disputed question whether the use of the compass began with the Arab traders in the Levant or with the Northmen in the Baltic" ("Petrus Peregrinus De Maricourt and His Epistola de Magnete," *Proceedings of the British Academy*, II [1905–1906], 385). Actually, it may have been employed first by the Chinese (see n. 9).

6. Peregrinus may have been the first to actually shape a magnet, or, at least, the first to speak of an artificially shaped magnet. It appears, then, that the first shaped magnets were spherical. In his *De magnete,* William Gilbert called the spherical magnet a *terrella,* a term which gained general acceptance. It was so called because the spherical *terrella* was conceived as a miniature spherical earth, for Gilbert considered the earth to be a great magnet.

7. This method of locating the poles is also described by William Gilbert in Book I, chapter, 3 of his *De magnete* (see p. 13 of Silvanus P. Thompson's translation of the *De magnete* cited in n. 1).

Proof of this will be found in a subsequent chapter of this tract.

A second method for determining these important points is this: Note the place on the above-mentioned spherical lodestone where the point of the needle clings most frequently and most strongly; for this will be one of the poles as discovered by the previous method. In order to determine this point exactly, break off a small piece of the needle or iron so as to obtain a fragment about the length of two fingernails; then put it on the spot which was found to be the pole by the former operation. If the fragment stands perpendicular to the stone, then that is, unquestionably, the pole sought;[8] if not, then move the iron fragment about until it becomes so; mark this point carefully; on the opposite end another point may be found in a similar manner. If all this has been done rightly, and if the stone is homogeneous throughout and a choice specimen, these two points will be diametrically opposite, like the poles of a sphere.

Chapter 5
How to Discover the Poles of a Lodestone and How to Tell Which is North and Which South

The poles of a lodestone having been located in a general way, you will determine which is north and which south in the following manner: Take a wooden vessel rounded like a platter or dish, and in it place the stone in such a way that the two poles will be equidistant from the edge of the vessel; then place the dish in another and larger vessel full of water, so that the stone in the first-mentioned dish may be like a sailor in a boat. The second vessel should be of considerable size so that the first may resemble a ship floating in a river or on the sea. I insist upon the larger size of the second vessel in order that the natural tendency of the lodestone may not be impeded by contact of one vessel against the sides of the other. When the stone has been thus placed, it will turn the dish round until the north pole lies in the direction of the north pole of the heavens, and the south pole of the stone points to the south pole of the heavens.[9] Even if the stone be moved a thousand times away from its position, it will return thereto a thousand times, as by natural instinct. Since the north and south parts of the heavens are known, these same points will then be easily recognized in the stone because each part of the lodestone will turn to the corresponding one of the heavens.

Chapter 6
How One Lodestone Attracts Another

When you have discovered the north and the south pole in your lodestone, mark them both carefully, so that by means of these indentations they may be distinguished whenever necessary. Should you wish to see how one lodestone attracts another, then, with two lodestones selected and prepared as mentioned in the preceding chapter, proceed as follows: Place one in its dish that it may float about as a sailor in a skiff, and let its poles which have already been determined be equidistant from the horizon, i.e., from the edge of the vessel. Taking the other stone in your hand, approach its north pole to the south pole of the lodestone floating in the vessel; the latter will follow the stone in your hand as if longing to cling to it. If, conversely, you bring the south end of the lodestone in your hand toward the north end of the floating lodestone, the same phenomenon will occur; namely, the floating lodestone will follow the one in your hand. Know then that this is the law: the north pole of one lodestone attracts the south pole of another, while the south pole attracts the north. Should you proceed otherwise and bring the north pole of one near the north pole of another, the one you hold in your hand will seem to put the floating one to flight. If the south pole of one is brought near the south pole of another, the same will happen.[10] This is because the north pole of one seeks the south pole of the other, and therefore repels the north pole. A proof of this is that finally the north pole becomes united with the south pole. Likewise if the south pole is stretched out towards the south pole of the floating lodestone, you will observe the latter to be repelled, which does not occur, as said before, when the

8. This method is also described by Gilbert in Book I, chapter 3, of the *De magnete* (Thompson, p. 14). Gilbert probably derived these two methods of locating poles from Peregrinus. He added a third using a *versorium*, "a piece of iron touched with a loadstone, and placed upon a needle or point firmly fixed on a foot so as to turn freely about. . ." (Thompson, *De magnete*, p. 13).

9. In describing the magnetic compass—it had been in use long before Peregrinus wrote his letter—Alexander Neckham listed this property. Knowledge of the magnetic compass was had by the Chinese no later than 1088 (Joseph Needham, *Science and Civilisation in China* [Cambridge, Eng.: Cambridge University Press, 1954–1965], Vol. IV, Part I, p. 249), which is at least a century earlier than its mention by Neckham.

10. Repulsion between the like poles of a magnet is described by Roger Bacon in his *Opus Majus* (see Thompson, "Petrus Peregrinus. . . ," p. 378). The phenomenon of magnetic repulsion was known in Greek and Roman antiquity (for instance, Lucretius, *On the Nature of Things* VI) but was not associated with polarity.

north pole is extended towards the south. Hence the silliness of certain persons is manifest, who claim that just as scammony attracts jaundice on account of a similarity between them, so one lodestone attracts another even more strongly than it does iron, a fact which they suppose to be false although really true as shown by experiment.

Chapter 7
How Iron Touched by a Lodestone Turns Towards the Poles of the World

It is well known to all who have made the experiment, that when an elongated piece of iron has touched a lodestone and is then fastened to a light block of wood or to a straw and made float on water, one end will turn to the star which has been called the Sailor's star because it is near the pole; the truth is, however, that it does not point to the star but to the pole itself. A proof of this will be furnished in a following chapter. The other end of the iron will point in an opposite direction. But as to which end of the iron will turn towards the north and which to the south, you will observe that that part of the iron which has touched the south pole of the lodestone will point to the north and conversely, that part which had been in contact with the north pole will turn to the south. Though this appears marvelous to the uninitiated, yet it is known with certainty to those who have tried the experiment.

Chapter 8
How a Lodestone Attracts Iron

If you wish the stone, according to its natural desire, to attract iron, proceed as follows: Mark the north end of the iron and towards this end approach the south pole of the stone, when it will be found to follow the latter. Or, on the contrary, to the south part of the iron present the north pole of the stone and the latter will attract it without any difficulty. Should you, however, do the opposite, namely, if you bring the north end of the stone towards the north pole of the iron, you will notice the iron turn round until its south pole unites with the north end of the lodestone. The same thing will occur when the south end of the lodestone is brought near the south pole of the iron. Should force be exerted at either pole, so that when the south pole of the iron is made touch the south end of the stone, then the virtue in the iron will be easily altered in such a manner that what was before the south end will now become the north

and conversely. The cause is that the last impression acts, confounds, or counteracts and alters the force of the original movement.

Chapter 9
Why the North Pole of One Lodestone Attracts the South Pole of Another and Vice Versa

As already stated, the north pole of one lodestone attracts the south pole of another and conversely; in this case the virtue of the stronger becomes active, whilst that of the weaker becomes obedient or passive. I consider the following to be the cause of this phenomenon: the active agent requires a passive subject, not merely to be joined to it, but also to be united with it, so that the two make but one by nature. In the case of this wonderful lodestone this may be shown in the following manner: Take a lodestone which you may call *AD*, in which *A* is the north pole and *D* the south; cut this stone into two parts, so that you may have two distinct stones; place the stone having the pole *A* so that it may float on water and you will observe that *A* turns towards the north as before; the breaking did not destroy the properties of the parts of the stone, since it is homogeneous; hence it follows that the part of the stone at the point of fracture, which may be marked *B*, must be a south pole; this broken part of which we are now speaking may be called *AB*. The other, which contains *D*, should then be placed so as to float on water, when you will see *D* point towards the south because it is a south pole; but the other end at the point of fracture, lettered *C*, will be a north pole; this stone may now be named *CD*. If we consider the first stone as the active agent, then the second, or *CD*, will be the passive subject. You will also notice that the ends of the two stones which before their separation were together, after breaking will become one a north pole and the other a south pole.[11] If now these same broken portions are brought near each other, one will attract the other, so that they will again be joined at the points *B* and *C*, where the fracture occurred. Thus, by natural instinct, one single stone will be formed as before. This may be demonstrated fully by cementing the parts together, when the same effects will be produced as before the stone was broken. As you will

---

11. Peregrinus observes here the fact that a magnet can be broken into pieces, each of which is a complete magnet with north and south poles. Its omission here is of little consequence, since the discussion is easily followed.

perceive from this experiment, the active agent desires to become one with the passive subject because of the similarity that exists between them. Hence *C,* being a north pole, must be brought close to *B,* so that the agent and its subject may form one and the same straight line in the order *AB, CD* and *B* and *C* being at the same point. In this union the identity of the extreme parts is retained and preserved just as they were at first; for *A* is the north pole in the entire line as it was in the divided one; so also *D* is the south pole as it was in the divided passive subject, but *B* and *C* have been made effectually into one.[12] In the same way it happens that if *A* be joined to *D* so as to make the two lines one, in virtue of this union due to attraction in the order *CDAB,* then *A* and *D* will constitute but one point, the identity of the extreme parts will remain unchanged just as they were before being brought together, for *C* is a north pole and *B* a south, as during their separation. If you proceed in a different fashion, this identity or similarity of parts will not be preserved; for you will perceive that if *C,* a north pole, be joined to *A,* a north pole, contrary to the demonstrated truth, and from these two lines a single one, *BACD,* is formed, as *D* was a south pole before the parts were united, it is then necessary that the other extremity should be a north pole, and as *B* is a south pole, the identity of the parts of the former similarity is destroyed. If you make *B* the south pole as it was before they united, then *D* must become north, though it was south in the original stone; in this way neither the identity nor similarity of parts is preserved. It is becoming that when the two are united into one, they should bear the same likeness as the agent, otherwise nature would be called upon to do what is impossible. The same incongruity would occur if you were to join *B* with *D* so as to make the line *ABDC,* as is plain to any person who reflects a moment. Nature, therefore, aims at being and also at acting in the best manner possible; it selects the former motion and order rather than the second because the identity is better preserved. From all this it is evident why the north pole attracts the south and conversely, and also why the south pole does not attract the south pole and the north pole does not attract the north.

Chapter 10
An Inquiry into the Cause of the Natural Virtue of the Lodestone

Certain persons who were but poor investigators of nature held the opinion that the force with which a lodestone draws iron, is found in the mineral veins themselves from which the stone is obtained; whence they claim that the iron turns towards the poles of the earth, only because of the numerous iron mines found there. But such persons are ignorant of the fact that in many different parts of the globe the lodestone is found; from which it would follow that the iron needle should turn in different directions according to the locality; but this is contrary to experience.[13] Secondly, these individuals do not seem to know that the places under the poles are uninhabitable because there one-half the year is day and the other half night. Hence it is most silly to imagine that the lodestone should come to us from such places. Since the lodestone points to the south as well as to the north, it is evident from the foregoing chapters that we must conclude that not only from the north pole but also from the south pole rather than from the veins of the mines virtue flows into the poles of the lodestone. This follows from the consideration that wherever a man may be, he finds the stone pointing to the heavens in accordance with the position of the meridian; but all meridians meet in the poles of the world; hence it is manifest that from the poles of the world, the poles of the lodestone receive their virtue. Another necessary consequence of this is that the needle does not point to the pole star, since the meridians do not intersect in that star but in the poles of the world. In every region, the pole star is always found outside the meridian except twice in each complete revolution of the heavens.[13a] From all these considerations, it

12. Peregrinus explains the attraction of unlike poles in terms of an agent, or active principle, and a patient, that which is acted upon (the agent–patient relationship was a popular type of medieval explanation invoked frequently in physical treatises). Thus a magnetized pole, whether north or south, which could cause another pole to move toward it would be labeled as active, its opposite as passive. Such a distinction is wholly arbitrary and of no value. See Cohen and Drabkin, p. 313, for Lucretius' atomistic explanation of magnetic attraction, which is devoid of any concept of polarity.

13. In dismissing this false theory that magnetic deposits in the earth attract the iron, Peregrinus was to turn away completely from the fruitful concept of associating magnetic properties with the earth. It remained for later authors, especially William Gilbert, to treat the earth as a large spherical magnet.

13a. Thus Peregrinus was aware that the pole star, Polaris, is not at true north but rotates around that point. Samuel Eliot Morison (*Admiral of the Ocean Sea, A Life of Christopher Columbus* [2 vols.; Boston: Little, Brown and Co., 1942], Vol. I. p. 271) remarks that Columbus discovered for himself the rotation of Polaris,

is clear that the poles of the lodestone derive their virtue from the poles of the heavens.[14] As regards the other parts of the stone, the right conclusion is, that they obtain their virtue from the other parts of the heavens, so that we may infer that not only the poles of the stone receive their virtue and influence from the poles of the world, but likewise also the other parts, of the entire stone from the entire heavens. You may test this in the following manner: A round lodestone on which the poles are marked is placed on two sharp styles as pivots having one pivot under each pole so that the lodestone may easily revolve on these pivots. Having done this, make sure that it is equally balanced and that it turns smoothly on the pivots. Repeat this several times at different hours of the day and always with the utmost care. Then place the stone with its axis in the meridian, the poles resting on the pivots. Let it be moved after the manner of bracelets so that the elevation and depression of the poles may equal the elevation and depressions of the poles of the heavens of the place in which you are experimenting. If now the stone be moved according to the motion of the heavens, you will be delighted in having discovered such a wonderful secret;[15] but if not, ascribe the failure to your own lack of skill rather than to a defect in nature.[16] Moreover, in this position I consider the strength of the lodestone to be best preserved. When it is placed differently, i.e., not in the meridian, I think its virtue is weakened or obscured rather than maintained. With such an instrument you will need no timepiece, for by it you can know the ascendant at any hour you please, as well as all other dispositions of the heavens which are sought for by astrologers.

PART II

Chapter 1
The Construction of an Instrument for Measuring the Azimuth of the Sun, the Moon, or any Star on the Horizon

Having fully examined all the properties of the lodestone and the phenomena connected therewith, let us now come to those instruments which depend for their operation on the knowledge of those facts. Take a rounded lodestone, and after determining its poles in the manner already mentioned, file its two sides so that it becomes elongated at its poles and occupies less space. The lodestone prepared in this wise is then enclosed within two capsules after the fashion of a mirror. Let these capsules be so joined together that they cannot be

"a fact which many late medieval and renaissance astronomers had denied. Practical seamen had assumed for centuries that the North Star marked true north.

14. Gilbert (*De magnete*, Bk. III, ch. 1; Thompson, p. 116), scorns this and similar explanations of the causes of magnetism, declaring that "the crowd of philosophizers, in order to discover the reasons of the magnetical motions, called up causes lying remote and far away." Among those whose explanations he rejected are Peter Peregrinus (who thought that "the direction arises from the poles of the sky"), Cardano (who thought that "the turning of iron was caused by a star in the tail of the Great Bear"), and Marsilio Ficino (who argued that "the loadstone follows its own Arctick pole"). (See also Thompson, *De magnete*, p. 153).

Had Peregrinus been aware of magnetic declination and dip, it is not likely that he would have adopted the position that the poles of a magnet receive their virtue from the celestial poles. The phenomenon of magnetic declination is mentioned in a Chinese work by Shen Kua written around 1088. He observes that "magicians rub the point of a needle with the lodestone; then it is able to point to the south. But it always inclines slightly to the east, and does not point directly at the south" (Needham p. 249). Gilbert attributes the discovery of magnetic declination to Robert Norman, who described it in his book *Newe Attractive* in 1576.

15. Such a pivoted spherical magnet would be a perpetual motion machine exactly simulating the motion of the celestial sphere (see further on for Peregrinus' effort to construct a perpetual motion wheel) and would serve as a clock, as we are told a few lines further on. William Gilbert, taking cognizance of this claim, remarks (*De magnete*, Bk. VI, ch. 4; Thompson, pp. 223–224) that "Peter Peregrinus constantly affirms that a terrella suspended above its poles on a meridian moves circularly, making an entire revolution in 24 hours: which, however, it has not happened to ourselves as yet to see; and we even doubt this motion on account of the weight of the stone itself, as well as because the whole Earth, as she is moved of herself, so also is she propelled by other stars: and this does not happen in proportion (as it does in the terrella) in every part." In the Third Day of his *Dialogue Concerning the Two Chief World Systems*, Galileo criticized Gilbert for not vigorously rejecting the rotation of a spherical lodestone. For if the earth has a diurnal axial rotation, spherical lodestones already possess this motion "and to assign them a motion around their own centers would be to attribute to them a second movement quite different from the first. Thus they would have two motions; that is, a rotation in twenty-four hours about the center of the whole, and a revolution about their own centers. Now this second motion is arbitrary, and there is no reason whatever for introducing it" (trans. Stillman Drake [Berkeley, Calif.: University of California Press, 1962], pp. 413–414).

16. Convinced of the truth of his theory, Peregrinus anticipates the response of all who will try and fail by attributing their failure to lack of skill. There could be no experimental or experiential grounds for rejecting such a theory. In this it resembles the conviction of the alchemists that base metals are transmutable to gold. Every failure was attributable to faulty procedure, lack of knowledge, or misunderstanding.

separated and that water cannot enter; they should be made of light wood and fastened with cement suited to the purpose. Having done this, place them in a large vessel of water on the edges of which the two parts of the world, i.e., the north and south points, have been found and marked. These points may be united by a thread stretched across from north to south. Then float the capsules and place a smooth strip of wood over them in the manner of a diameter. Move the strip until it is equally distant from the meridian-line, previously determined and marked by a thread, or else until it coincides therewith. Then mark a line on the capsules according to the position of the strip, and this will indicate forever the meridian of that place. Let this line be divided at its middle by another cutting it at right angles, which will give the east and west line; thus the four cardinal points will be determined and indicated on the edge of the capsules. Each quarter is to be subdivided into 90 parts, making 360 in the circumference of the capsules. Engrave these divisions on them as usually done on the back of an astrolabe. On the top or edge of the capsules thus marked place a thin ruler like the pointer on the back of the astrolabe; instead of the sights attach two perpendicular pins, one at each end.[17] If, therefore, you desire to take the azimuth of the sun, place the capsules in water and let them move freely until they come to rest in their natural position. Hold them firmly in one hand, while with the other you move the ruler until the shadow of the pins falls along the length of the ruler; then the end of the ruler which is towards the sun will indicate the azimuth of the sun. Should it be windy, let the capsules be covered with a suitable vessel until they have taken their position north and south. The same method, namely, by sighting, may be followed at night for determining the azimuth of the moon and stars; move the ruler until the ends of the pins are in the same line with the moon or star; the end of the ruler will then indicate the azimuth just as in the case of the sun. By means of the azimuth may then be determined the hour of the day, the ascendant, and all those other things usually determined by the astrolabe.[18] A form of the instrument is shown in the following figure [Fig. 1].

Chapter 2
The Construction of a Better Instrument for the Same Purpose

In this chapter I will describe the construction of a better and more efficient instrument.[19] Select a

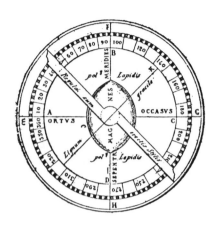

Fig. 1. Azimuth compass

vessel of wood, brass or any solid material you like, circular in shape, moderate in size, shallow but of sufficient width, with a cover of some transparent substance, such as glass or crystal; it would be even better to have both the vessel and the cover transparent. At the centre of this vessel fasten a thin axis of brass or silver, having its extremities in the cover above and the vessel below. At the middle of this axis let there be two apertures at right angles to each other; through one of them pass an iron stylus or needle, through the other a silver or brass needle crossing the iron one at right

17. S. P. Thompson calls this floating compass "the earliest compass with proper divisions" ("Petrus Peregrinus. . . ," p. 388).
18. Peregrinus conceives of this instrument as an astronomical one capable of performing the tasks of an astrolabe. That it was also intended to function as a mariner's compass is clear from the description of the second instrument, described in his next chapter, which was devised as an improvement over this one.
19. This instrument eliminates the need for water. It is a dry, pivoted compass placed in a circular box with a glass lid and provided with sights. Shortly after 1300— probably in southern Italy, perhaps in Amalfi—a light card with the compass points marked on it was affixed to the pivoting needle (see S. P. Thompson, "The Rose of the Winds; the Origin and Development of the Compass-Card," *Proceedings of the British Academy*, VI [1913], 181). This was the origin of the compass card, which was used effectively in the dry compass. The helmsman could now keep on course much more easily. The Chinese did not become aware of the dry, pivoted compass or the compass card until introduced to it by Dutch or Portuguese ships in the latter part of the sixteenth century (see Needham, p. 290). In truth, Peregrinus' pivoted compass of the cross-needle type with a graduated circle on which were marked the cardinal points was a major advance which he seems to have been the first to announce. Although the compass card was important, it is a mere refinement.

Fig. 2. Double-pivoted needle

angles.[20] Divide the cover first into four parts and subdivide these into 90 parts, as was mentioned in describing the former instrument. Mark the parts north, south, east and west. Add thereto a ruler of transparent material with pins at each end. After this bring either the north or the south pole of a lodestone near the cover so that the needle may be attracted and receive its virtue from the lodestone.

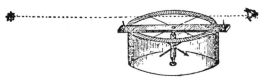

Fig. 3. Pivoted compass

Then turn the vessel until the needle stands in the north and south line already marked on the instrument; after which turn the ruler towards the sun if day-time, and towards the moon and stars at night, as described in the preceding chapter. By means of this instrument you can direct your course towards cities and islands and any other place wherever you may wish to go by land or sea, provided the latitude and longitude of the places are known to you. How iron remains suspended in air by virtue of the lodestone, I will explain in my book on the action of mirrors.[21] Such, then, is the description of the instrument.[22]

Chapter 3
The Art of Making a Wheel of Perpetual Motion

In this chapter I will make known to you the construction of a wheel which in a remarkable manner moves continuously. I have seen many persons vainly busy themselves and even becoming exhausted with much labor in their endeavors to invent such a wheel. But these invariably failed to notice that by means of the virtue or power of the lodestone all difficulty can be overcome.[23] For the construction of such a wheel, take a silver capsule like that of a concave mirror, and worked on the outside with fine carving and perforations, not only

20. When the magnetic needle lies north and south, the silver or brass needle will point west and east.

21. I am unaware of the existence of such a book on mirrors. Did he ever write one? Roger Bacon's statement about Peregrinus indicates that the latter was interested in optics (see the biographical sketch of Peregrinus at the end of this source book).

22. I have slightly truncated the last sentence, which Brother Arnold gives (p. 31) as "Such, then, is the description of the instrument illustrated below. (See Figs. 2 and 3.)" But an emended text by Timoteo Bertelli ("Sulla Epistola di Pietro Peregrino di Maricourt e sopra alcuni trovati e teorie magnetiche del secolo XIII. Memoria seconda" in *Bullettino di bibliografia e di storia delle scienze matematiche e fisiche,* I [Rome, 1868], 87) has only "Et hec est iam dicti Instrumenti descriptio." The two figures on p. 30 of Brother Arnold's translation appear neither in Gasser's edition nor in the manuscripts but were copied from diagrams supplied by Bertelli in a fold-out plate at the end of Volume I of the *Bullettino*. Since the text is sufficiently clear, Bertelli's figures have been omitted.

23. This account may represent the first genuine scientific attempt to construct a perpetual motion machine. Somewhat earlier in the thirteenth century Villard de Honnecourt (fl. 1225–1250) sketched a perpetual motion machine and supplied a brief legend which says: "Often have experts striven to make a wheel turn of its own accord. Here is a way to do it with an uneven number of mallets and with quicksilver" (quoted from Theodore Bowie, *The Sketchbook of Villard de Honnecourt* [Bloomington, Ind.: Indiana University Press, 1959], p. 134). Since no further elucidation is provided, Villard's concept cannot be readily compared with that of Peregrinus, who relied on the repulsive and attractive forces of magnetic poles and applied what he had learned about the behavior of magnets to construct a perpetual motion machine. In the fourteenth century scholastics generally rejected perpetual motion on theoretical grounds, arguing that inexhaustible forces *(vires infatigabiles)* did not exist in nature. Many would simply have supported Aristotle's contention in *De caelo* I.11.281a.26–27 that whatever had a beginning must eventually cease to be. Nicole Oresme also agreed that such perpetual motions did not occur in nature, but he added that there is no contradiction in conceiving a motion that has a beginning but no end (*Le Livre du ciel et du monde,* Bk. I, ch. 29; pp. 201–203 of A. D. Menut's translation [Madison, Wis.: University of Wisconsin Press, 1968]). After devising an illustration of a perpetual motion involving four wheels and incommensurable speeds, Oresme declares that "although such a series of events cannot occur in nature, nor be shown by material art or in destructible matter, nor endure so long [as forever], nevertheless, it contains or implies no contradiction whatever, nor is it within its own frame of reference incongruous to reason, ..." Despite the healthy skepticism about perpetual motion prevailing in the fourteenth century, many believers are recorded through the centuries (see Henry Dircks, *Perpetuum mobile* [London, 1870]). Even so great a figure as Gottfried Leibniz sought to devise a perpetual motion machine (see A. Maier, *Metaphysische Hintergründe der spätscholastischen Naturphilosophie* [Rome: Edizioni di Storia e Letteratura, 1955], p. 269, n. 50).

for the sake of beauty, but also for the purpose of diminishing its weight. You should manage also that the eye of the unskilled may not perceive what is cunningly placed inside. Within let there be iron nails or teeth of equal weight fastened to the periphery of the wheel in a slanting direction, close to one another so that their distance apart may not be more than the thickness of a bean or a pea; the wheel itself must be of uniform weight throughout. Fasten the middle of the axis about which the wheel revolves so that the said axis may always remain immovable. Add thereto a silver bar, and at its extremity affix a lodestone placed between two capsules and prepared in the following way: When it has been rounded and its poles marked as said before, let it be shaped like an egg; leaving the poles untouched, file down the intervening parts so that thus flattened and occupying less space, it may not touch the sides of the capsules when the wheel revolves. Thus prepared, let it be attached to the silver rod just as a precious stone is placed in a ring; let the north pole be then turned towards the teeth or cogs of the wheel somewhat slantingly so that the virtue of the stone may not flow diametrically into the iron teeth, but at a certain angle; consquently when one of the teeth comes near the north pole and owing to the impetus of the wheel passes it, it then approaches the south pole from which it is rather driven away than attracted, as is evident from the law given in a preceding chapter. Therefore such a tooth would be constantly attracted and constantly repelled. In order that the wheel may do its work more speedily, place within

the box a small rounded weight made of brass or silver of such a size that it may be caught between each pair of teeth; consequently as the movement of the wheel is continuous in one direction, so the fall of the weight will be continuous in the other. Being caught between the teeth of a wheel which is continuously revolving, it seeks the centre of the earth in virtue of its own weight, thereby aiding the motion of the teeth and preventing them from coming to rest in a direct line with the lodestone. Let the places between the teeth be suitably hollowed out so that they may easily catch the body in its fall, as shown in the diagram above. [Fig. 4.]

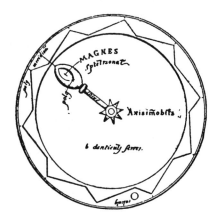

Fig. 4. Perpetual-motion wheel

Farewell: finished in camp at the siege of Lucera on the eighth day of August, Anno Domini MCCLXIX.

Optics

60 THE ENCYCLOPEDIC TRADITION IN OPTICS

1. Adelard of Bath (fl. 1116–1142): Natural Questions

Translated by Hermann Gollancz[1]

Annotated by David C. Lindberg

CHAPTER 23. WHAT THEORY IS TO BE HELD CONCERNING SIGHT?

NEPHEW: . . . We have heard what you have said about hearing; now let us see what you have to say about seeing.

ADELARD: Many different views have been expressed about the nature of sight, and it will be perhaps convenient first to set them forth, and then

1. The following selection is reprinted by kind permission of the heirs, Mr. Rodney Gollance and Mrs. Phyllis Simon, from *Dodi Venechdi (Uncle and Nephew)*, *the Work of Berachya Hanakdan, now edited from MSS. at Munich and Oxford, with an English Translation, Introduction etc., to which is added the first English Translation from the Latin of Adelard of Bath's Quaestiones Naturales* by Hermann Gollancz (London: Oxford University Press, 1920), pp. 114–121. Adelard was both a translator of scientific works from Arabic to Latin and a writer of scientific treatises in his own right; the *Natural Questions* were written, by his own testimony, after he had received some exposure to Arabic science.

enquire which of them is the most reasonable.

NEPHEW: Then, if you approve, it shall be my task to state these various theories, and yours to state any objections there are to them when I have done so.

ADELARD: Your suggestion is a reasonable one.

NEPHEW: The theories I have been able to collect in various quarters about sight fall into four different groups. Some say that the mind, sitting in the brain as its chief seat, and looking forth upon outer things through open windows, viz., the eyes, gets knowledge of the shapes of things, and when it has got knowledge of them, judges them; it being always understood that nothing from the mind passes to the outside, and nothing from the shapes outside makes its way to the mind. Others, again, maintain that sight takes place through the approach of shapes, saying that the shapes of things give shape to the air that intervenes between themselves and the eyes, and that in this way the materials for judgment pass to the mind. Very many also assert that something is sent forth by the mind, i.e., visible breath, and that the shapes of the things that are to be seen meet it in mid-air: having taken shape from these, the breath returns to its seat, and presents the shape to the mind for it to exercise judgment upon. A fourth party maintains that no shapes of objects approach the eye, but that something which they call "fiery force," and which is produced in the brain by means of concave sinews, passes first through the eyes, and then to the objects to be seen, and by returning to its point of origin brings back to the mind, with the same quickness as it went, the shape impressed upon it as though by a potter.

Let us deal first with the first theory. The use of all the senses belongs, as Boethius declares in his "*De musica*," to all living creatures; but what is their strength, and what are the limits to their use, is not perfectly clear except to the intellect of the philosopher; for both the effects of things follow upon the antecedent causes according to a most subtle nexus, and causes along with their effects differ from one another by the most subtle differences, so that even philosophers themselves often fail in their understanding of nature. Thus in the present case, if they mean that the mind placed in the brain as a seat regards external objects through the windows of the eyes, then when they assert this they will either be attributing a sort of power to the windows of the eyes, or else assigning no means of contemplating external things to these windows. But by calling the eyes windows, they are attribut-

ing to them a certain faculty for seeing external objects: do they then mean that the mind is a corporeal thing? In that case it would require a free and unimpeded means of egress without risk of injuring itself by coming into contact with other things; or do they think that in that conception of shapes which we call sight, there is some corporeal power needed for the mind to prevent it itself from being checked by contact with any obstacle, and that hence windows are required? But neither the mind, nor its consideration of external things, are corporeal things, and therefore they need have no fear of contact with corporeal things, and consequently, so far as the things perceived by them are concerned, do not require any corporeal opening. The formula then of these people includes no force, nor do they assume any path for the mind through the windows of the eyes, since they do not require this. This being so, why do they say that it sees external things through the eyes, especially as the eyes too are not perforated, but are more solid than the other instruments of sense, so that if the mind required any corporeal power, the ears or the nostrils (which have passages through them) would supply it. Further if, as is asserted, the mind should see the shapes of things by looking at external things, how could the human mind, by regarding a mirror, see the shape of its own face? The mirror, being opposite to its windows, it would be able to see; but its own face, though opposite to it, could not be seen according to this idea. This theory therefore lacks consistency, and we must now discuss the second view previously stated.

ADELARD: By all means, nor do I see how the upholders of this first view are to get the better of your objections. It was on this account that a theory was propounded by the Stoics, that external forms make their way to the mind itself, and that it is imprinted by them as wax might be. This it was that made Boethius in his "*De Consolatione*" say, "in bye-gone days the Stoic school introduced us to a set of out-of-date old men who hold that perceptions and images are imprinted on our minds by external bodies, just as at times it is people's habit to make and leave marks on the smooth paper with a swift pen." Although Boethius in the same passage attacks this idea as being pointless, I should like to tell you my objections to it as follows:—The shapes of things to be seen approach the mind of the seer either in their own subject and along with it, or in another subject and along with that. Now it is impossible that it should do so in its own subject, and not along with it; nor does the body that is seen approach me: therefore it must be in another subject, and along with that.

NEPHEW: In air and with air. For the body, as they say, gives shape to air, and that air to other air, till it comes to that in the brain which acts upon the mind itself.

ADELARD: If approach to the mind therefore were to be by such a transmission of shapes how could this be accomplished should a glassy body be interposed? Would not the advance be prevented through such an obstacle? Would not the air in the neighbourhood of the glass be acted upon by it? Will it receive shape from that of the more distant and less splendid body? This is of course absurd, and absurd too is the theory we are discussing. However, let us, as Boethius says in the "*Topica,*" admit it for the sake of argument, that we may see what follows. Let us grant that the progression of shapes as far as the eye, is made by the eye even to the brain, or the air of the brain; but let us note that passage which they call a sort of potter's impression. If the air of the brain imprints its shape on the mind, it gives either that shape actually and essentially, or another like it. But that it should be the same is impossible, for no individual shape can pass from one subject to another, as in the passing the shape would be without a subject, which is impossible. On the other hand, if it imprints another shape, it will at once require local parts in the mind just as it had in the air, and not finding them in it, it will not be able by moulding to give shape to it. Since then this theory is obviously untenable, let us pass to the next. This one—the third—is held by those who grant that a visible truth issues from the brain, but most certainly does not go so far as the body, which is relatively very far from it, for the reason that it finds the shapes it seeks in the intervening air.

Hence it has no need to do more than this, and consequently having been impressed returns, bringing back the shape to the mind by which it was sent forth. Those who say this do not realise what is the necessary consequence of their words; for if, as they assert, between the seer and the things to be seen, the air, by reason of the progression of similar shapes, offers to the visible breath that of which it was in search, then let us assume, first, a man looking from the east westwards, with the object of discerning the shape of a white body (and this, they say, he will find in the intervening air); next, at the same moment, assume another

looking from the west eastwards, in order to behold a black body; and let him consequently by the same progression find the shape of the black body in the same intervening air: from this it follows, that the same or indivisible air is the subject alike of blackness and whiteness. Thus two contrary things, acting in opposition, are found in the same subject. But this is impossible, and we are therefore bound to reject the views of those who assert it. Finally, and beyond doubt, there is another consideration which, even if we put other objections on one side, is sufficient to upset all the views we have mentioned. It is this: we are familiar enough with the sight of our own shapes in a mirror; but this, though its reality is established by everyday life, does not agree with the theories we have recounted. It will be better, therefore, to deal with the view of which philosophy approves, and dismissing other theories as lacking strength, put our faith in this academic truth. This theory is as follows: In the brain there is generated a certain air of the most subtle nature, and made of fire, and consequently exceedingly light: this makes its way from the mind along the nerves, whenever it so pleases, and necessity arises, to see things outside. Hence it is called by physicists "visible spirit": being a body, it naturally requires a local exit, which exit it finds through the different concave[2] nerves, which the Greeks call "optic," extending from the brain to the eyes: then travelling to the body to be seen, it makes its way with wondrous speed, and being impressed with the shape of the body, it both receives and retains the impression, and then returning to its original position, it communicates the shape it has received. Now this spirit is called by philosophers "fiery force," and this force, when it finds a mirror opposite to it, or any other light-giving body, being reflected by it, returns as a result of the reflection to its own face, and still retaining the shape, when it enters, reveals it to the mind. You are not, however, to suppose that this fiery force found the shape of the face in the mirror; but, being reflected from the surface which is too smooth for it to abide there, received the shape while returning, and having received it, brought it back. This then is the divine theory which Plato has adopted, among other things, in his "*Timaeus.*" "There are, in my opinion," says Plato, "two virtues in fire, one consuming and destructive, the other soothing and endowed with harmless light. With this one, therefore, in virtue

2. That is, hollow.

of which light bringing in the day unfolds itself, the divine powers are in harmony, for it has been their pleasure that the intimate fire of our bodies, own brother of the fire which is a passing bright, clear, and purged fluid, should flow through the eyes, and issue from them in order that through the eyes, slight, cramped, and affrighted, as it were, by the stouter substance, but yet offering a narrow medium, the more subtle clear fire might flow down through the same medium. Hence, when the light of day lends itself to the diffusion of sight, then no doubt the two like lights meeting in turn cohere into the appearance of a single body, in which the flashing brightness of the eyes meet, while the intimate brightness of the diffusion as it spreads is reflected by meeting with the image at close quarters. All this then goes through one and the same experience, and the result of that same experience, when either it touches something else, or is touched by it; moved by this contact, it spreads itself through the whole body, and making its way through that body to the mind and produces the sense which is called 'sight.' " . . .[3]

CHAPTER 24. WHETHER VISIBLE BREATH IS A SUBSTANCE OR AN ACCIDENT?

NEPHEW: I understand, and nothing could give me greater pleasure. For alike on philosophical grounds, from the physical effects of causes, and from ethical considerations, I am inclined to give my whole-hearted assent to Plato rather than to you.

I have now a task to impose both on Plato and on you, and I think it will annoy you both: for if, as has been previously explained, anything issues from the brain, whether with the physicists we call it visible breath, or—as Plato insists—the fire of our own bodies, it must necessarily be either substance or accident.

ADELARD: It is corporeal substance; for, as the philosopher says, fire is a most subtle substance, composed of the four elements.

CHAPTER 25. HOW CAN THIS SAME BREATH IN SO SHORT A TIME TRAVEL TO THE STARS AND RETURN FROM THEM?

NEPHEW: Certainly; for if it were an accident, it would not be able to pass to outer things without a subject. Since then you understand that it is a body, it is easy to understand what inconvenience would follow thence. The firmament, which by some is called *aplanos*,[4] and by others the further sphere, contains all body and place: and when

I speak of stars, I mean those upper bodies which not seeming to move are fixed in the plane. Well then, this fiery breath with a single glance of the eye, so to speak, sees the stars; and we have to admit, that a body in so short a time traverses and returns through as great a density of air as there is between us and the moon and the infinite space that lies between us and the moon, and between the moon and the sun up to the very aplanos itself: and this is mere madness and quite impossible, since the actual breadth[5] of the whole earth bears no sort of proportion to the infinite diameter of the sky.

ADELARD: That is the sort of difficulty that a man gets into who knows nothing of nature: do you then pay careful attention and store up in your mind what I am going to say. Just as of spaces some are wide, some wider, and some widest of all, so of bodies some I hold are swift, others swifter, and others swiftest of all. This, however, not everyone can understand. Just as the extent of the sky and its shape as laid down by geometers is barely or not at all clear to the vulgar mind, so to it such a great speed in so short a time cannot be clear, though the eye in such people is swifter than the mind; for they measure, or rather mis-measure, everything according to the fallacious evidence of their senses in terrestrial matters. They think that the size of the sky exactly coincides with the earth, and that the bulk of the moon and sun and other things, which true reason would show them to exceed in size the earth, are not one whit greater than they seem to their bleary eyes. Those, however, who in matters of this sort are more apt to use reason, the incorporeal eye of the mind as their guide, just as they see clearly the magnitude of the external continent, or boundary, and the almost infinite extent of its diameter, so also see clearly the revolution of the heavens and the unutterably swift movement of the visible breath— and this in both cases, thanks to the use of reason. Just as the mind of the external sphere excels all created beings in the execution of virtues, so also this visible breath is more subtly perfected by the wonderful energy of creative virtue than all things compounded of elements. Just as the travelling forth of mind is the cause of that swift revolution, so the travelling forth of body is the reason of this swift going and returning, and it is not strange that

3. *Timaeus* 45B–45D.
4. That is, fixed.
5. Gollancz's edition has "breath," obviously a typographical error.

the man who is ignorant of this should also be ignorant of and wonder at its effects. One thing more I must tell you, and that is, that not all the physicists agree with you in saying that the upper stars are fixed in the aplanos, for they assert that they move inside the sky at a far lower altitude—this point, however, I will deal with later on, if I live long enough. For the present, now that impossibility and space are dwindling, let doubt dwindle also.

CHAPTER 26. HOW IS IT THAT WHILE THE EYE IS SHUT, THE VISIBLE BREATH IS NOT LEFT OUTSIDE?

NEPHEW: Nay, doubt grows from more to more, and I fancy that I myself shall dwindle away to utter nothingness before I give up doubting. Just imagine that while the visible breath is touching a star, the eye is shut. This is bound to happen, frequently, and the breath will then be left outside.

ADELARD: If only you would remember what I said previously, you would not raise this objection. The breath is very swift, and, as said above, it is established that it is sent by the mind, and through the mind. It is clear that the eye can be closed only by the mind, by which all voluntary motion is communicated to the body; let the consequence then of this also be clear, that that which sends also receives; and when it wills, shuts the door in such a way as to involve no injury to itself; and so that nothing of its own may remain outside, especially too when the thing itself is of such speed that it immediately receives it back again. Idle then would it be to grieve over a disaster of this sort when the thing sent obeys nature which sends it, and a door is no resistance to one who obeys.

CHAPTER 27. WHY THE BREATH DOES NOT HINDER ITSELF IN GOING AND RETURNING.

NEPHEW: I only hope that you, who while following Plato are tumbling into a pit, may have no causes for grief. I grant that in our discussion we should admit anything that has a possibility of truth, yet I think my next question will land you in difficulties. Assume that this breath, or if you will, fire, makes its way as far as to what is to be seen by it: is sight effected before it returns or afterwards?

ADELARD: It brings no message until its return.

NEPHEW: If therefore sight takes place immediately after its return, when we look at anything for a long while with fixed gaze and see it, it follows that it has both gone and returned at the same time, if we see the thing continuously and without any interval: hence both its going and returning will be simultaneous and without division, and therefore it hinders itself while going and returning.

ADELARD: Nay, it is you who hinder yourself, for as a result of your not understanding you make foolish objections; it does not follow that because we await anything with a continuous gaze that the breath when going is also returning, or hindering itself. When we say that it goes and returns so quickly, that when one return is accomplished, another starting takes place without a word, because there is no delay perceptible to any sense, since there is nothing that is manifest to sense, i.e., to the intellect, there is therefore no great need that it should go with the same speed after its return as it returned with after its going.[6] As, however, this interval is not discernible to sense, the journey wrongly seems continuous.

2. Alexander Neckam (1157–1217): Concerning the Natures of Things

Translated and annotated by David C. Lindberg and Greta J. Lindberg[7]

CHAPTER 153. CONCERNING SIGHT.

It is generally conceded that the more remote a thing, the smaller it appears. However, vapor can and commonly does prevent this general occurrence, for the body of the sun appears larger toward dawn on account of the remains of the nocturnal vapors than when it shines at midday. Moreover, a fish or anything placed in water seems larger in the water than out of it. Thus a dog swimming in water holding a piece of meat in its mouth is deceived by seeing a shadow and lets go of the meat that it was holding in its mouth, hoping to secure

a larger piece for itself, but in vain. Let the waters represent tribulations; martyrs placed in tribulations were greater than in time of peace. The sun stands for power, which seems greater the more

6. That is, the interval between return of the breath and its re-emission may be longer than the interval between the original emission of the breath and its return.

7. Translated from *Alexandri Neckam De naturis rerum libri duo,* edited by Thomas Wright (London, 1863), pp. 234–240. The two chapters translated here constitute all the optical material contained in Neckam's encyclopedia. They are valuable chiefly as illustrations of the low level of optical knowledge in the twelfth century.

remote it is. Something worthy of admiration is found also in geometrical investigations: there is something that appears larger the more remote it is; for the closer the angle of tangency,[8] the smaller it appears to be. The cause of this is as follows: The larger the circle, the smaller is the angle of

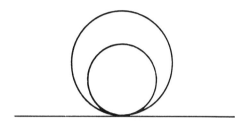

Fig. 1

tangency, and the smaller the circle, the larger is the angle of tangency, as can be seen in the adjacent figure [Fig. 1]. But the further away the circle, the smaller it appears to be; for the more remote it is, the smaller is the angle under which it is seen. However, the smaller the circle appears, the greater appears the angle of tangency; therefore, the more remote the angle of tangency, the larger it appears to be. Similarly, the further the acquaintance of a powerful man is from being achieved, the more worthy of praise is he considered to be. [However,] having become the friend of the powerful man, so much less desirable will his friendship appear to you.

Nevertheless, to some people it seems that this argument is invalid. The smaller the circumference, the larger the angle of tangency; therefore, the smaller the circumference appears to be, the larger the angle of tangency appears to be, since as the angle under which the circle appears decreases because of its elongation from the eye, so the angle under which the angle of tangency appears decreases as it is removed [from the eye]. But how? Under what angle is the angle of tangency viewed? Likewise, a straight rod appears bent in water, which is customarily attributed to reflection[9] of the rays from the surface of the water. [Now,] waters represent tribulations and the straight rod good works. Thus the works of the just, who are vexed by tribulations, are often regarded as bent, although they are [actually] straight. Furthermore, the man who is in a dark place sees a man standing in the light, but not vice versa; in the same way, unimportant people, whose fortune is dark, perceive the deeds of important people, but not vice versa.

A marvelous thing can be observed [even] by somebody poorly instructed in such things, since no parallel lines appear to be parallel.

A denarius placed in the bottom of a basin filled with water will be seen by one standing far away; when the water is drawn off, it will not be seen. Likewise, as long as there is tribulation in the soul, many things are observed and perceived which are not perceived when the tribulation has been removed. And as too much prosperity blinds the mind, so tribulation restores keener sight to the mind. Some one wishing to maintain that the denarius is not seen will contend that the water itself represents such a form to the eyes of the observer. To another, it will seem that a shadow appears because of the reflection made in the lucid body. But I think Aristotle is more to be believed than the masses.

To a person looking into water a tower not far removed from the water will appear to be inverted, although it is actually straight and erect; rejoice in the vanities of things, and they who do well will seem to you not to be standing upright, but inverted. An image is not formed in turbulent water but in quiet water; so also a disturbed mind is not attentive; anger hinders the soul, so that it cannot discern truth.

Although the body of the sun is round, it does not appear to have a convex surface, which occurs because of the impotence of sight and the excessive splendor of the sun. Similarly, the status of important men does not appear as it really is because of the splendor of the vainglory of conspicuous men; it is subject to the vicissitudes of life, although to the less prudent it appears otherwise. Furthermore, many are lifted in spirit who appear perfectly flat. Is it more necessary to ignore their thoughts than [those of] this powerful man? It is necessary that he fear many, who is feared by many.

Notice how certain animals see at night but not during the day on account of the whiteness of their eyes and the subtlety of their visual spirit.[10] For during the day the visual spirit (inasmuch as it is subtle and bright) is easily dispersed by the brightness of the air striking against the eyes, the whiteness of the eyes cooperating in this separation; and thus the power of sight of the spirits is

8. On angles of tangency or contingent angles, see note 10 on Grosseteste's *Concerning Lines, Angles, and Figures*, Selection 61.1.

9. Although the cause of the rod's bent appearance is refraction, Neckam uses the term *reverberatio*.

10. On the visual spirit, see Adelard of Bath in the preceding section.

darkened by perdition. But at night, aided and strengthened by the darkness of the air, the visual spirit illuminates (by its own brightness) the air between the seeing thing and the thing to be seen, and thus it prepares the way for discerning the thing opposite; and this is evident in owls. In other animals, illuminated by the brightness of the day, the visual spirit produces sight during the day, because the eyes [of those animals] are black and the visual spirit is gross and dark; but at night, darkened by much greater obscurity, it does not perfect sight, but confuses it; and this is seen clearly in men. However, in bats, because their instruments [of perception] are varied and their visual spirit is midway between darkness and brightness, sight is achieved only at dawn or at dusk, because of the mediocrity [at those times] of the air; for during the day, because of the excessive brightness dividing the spirit, and at night, because of the excessive darkness obscuring the spirit, vision is as far removed from the bat as activity. So bright and subtle is the visual spirit in a lynx that by its irradiation, illuminating the dark air contained in the pores of a wall, the spirit is moved to see something placed on the other side of the wall. The eagle, however, although it has an excellent organ of vision (both because its visual spirit is strong and because its pupil is small, on which account very little of the spirit is scattered outside), exposes its eyes to the rays of the sun and [yet] directs its eyes without injury to seeing. But in us, because there is not so much visual spirit and the pupil of the eye is large, we are unable to look at the sun for very long on account of the [resulting] excessive dissolution of the spirits.

The basilisk[11] is said to kill man by sight alone, for it infects the air with the badness of its ray. The visual ray of a wolf is productive of stupor, and a man appearing unexpectedly and seen by a wolf is struck dumb for a time. Wherefore they say that bewitching is produced by the badness of the ray of the seeing thing; and this is why nurses lick the face of a bewitched little boy.

But if a ray issues forth, it is marvelous how it perceives the surface of the solar body, for the sun is 166 times and a fraction larger than the whole earth. This is why it appears to very inquisitive men that the sun is not [really] seen but that air adjacent to the earth and subject to the sun is seen. For how is the ray issuing forth turned back? How also can the ray be emitted [both] yesterday and today, or is a new ray emitted by the eye at every single instant? Moreover, when a certain man is sitting in a different place from others, how do those men see him if the rays emitted from their eyes [that is, his as well as theirs] encounter one another? Also, will one of the rays withdraw from the other, or will there be two rays [in the same place] at the same time?[12] Also, since the stars are much more distant from us than is the sun, how does a small ray extend itself to such a remote place without becoming thinner as it approaches the more remote obstacle? However, to him who is well instructed in physics, it is not doubtful that the visual ray is sent from the lining of the brain through the optic nerve to the window of the eye. And note how the two rays sent to the eyes intersect each other transversely in the form of a cross;[13] just as without faith in the cross the inner man does not see well, so neither is exterior vision achieved without the form of the cross. It should be noted also that a gross vapor sometimes unites itself to a visual ray, whence certain small black bodies sometimes appear to flit before human eyes. . . .[14]

CHAPTER 154. CONCERNING THE MIRROR.

When a mirror is whole, there being but one observer, only one image is produced; but should the glass be broken into several pieces, there appear as many images as there are breaks. And thus in Holy Scripture, as many understandings shine forth as there are expositions. But, marvelous thing!, remove the lead placed behind the glass; immediately no image is visible to the observer. [Likewise,] remove the foundation of the faith, and immediately you will no longer see yourself clearly in Holy Scripture. The lead can be understood to represent sin; thus in the mirror of Holy Scripture you will discern yourself less clearly unless you confess yourself to be a sinner. For if we say that we have not sinned, we deceive ourselves, and the truth is not in us.[15]

11. A fabulous lizard.
12. One of the rays is from the solitary man, the other from the group of men or from one member of the group. "Ray," as Neckam uses the term, denotes not a geometrical line or a narrow beam, but the entire radiation issuing from the eye or eyes of an observer or observers.
13. The optic nerves intersect in the lower forehead, most of the fibers proceeding to the opposite eye. In Neckam's view, rays pass from the brain to the eyes through the optic nerves, and thus they intersect in the form of a cross at this optic chiasma. On the chiasma, see Pecham's *Perspectiva communis,* I, Proposition 32 (Selection 62.7).
14. In the brief remaining portion of this chapter Neckam discusses the effects of a lunar ray on a horse's sore.
15. I John 1:8.

In a concave mirror, the image of the observer appears inverted; in a plane and convex mirror, erect. Who will provide a sufficient explanation of this?[16]

Also the pupil, which is a very small substance, is a mirror, in which the image of the observer shines back. However, for three days before the death of a man, the brightness of the pupil is already darkened, so that for [those] three days the image of an observer does not appear in it.

The image appearing in a mirror appears to conform itself to that of which it is the image. It laughs at a laughing person, and when the observer cries, the image seems to cry [also]. Thus the soul is the mirror of its Maker and ought to sympathize with the suffering Christ and rejoice with the risen and rejoicing Christ. Therefore, at different times different countenances are to be assumed, so long as we always have the countenance of one proceeding toward Jerusalem. . . .[17]

3. Bartholomew the Englishman (fl. 1220–1250): Concerning the Properties of Things

Translated and annotated by Bruce S. Eastwood[18]

BOOK VIII, CHAPTER 40. CONCERNING *lux* AND ITS PROPERTIES.

According to Basil, light *(lux)* is completely homogeneous. Authorities speak diversely on the question whether light is substance or accident. For Aristotle, light *(lumen)* is not corporeal, nor is it an emission by a body. Damascene likewise says that light *(lumen)* has no substance of its own. In *Super Genesim ad litteram,* Augustine maintains that light *(lux)* is a corporeal being; that of corporeal genera it is the simplest; that in efficacy it is the most diverse; that it is most mobile and penetrating while being least resistant; that it most highly unifies diverse and contrary things; that it is most convertible and is ultimately the origin and principle of natural motion. Light *(lux)*, he continues, is most perfectible, most pleasing, and most communicable. Whence there can be discovered in bodies nothing more useful, convenient, beautiful, swift, subtle, impassible, or virtuous than light.

Between *lux* and *lumen* there is a difference. *Lumen* is a certain emission, or irradiation, by the substance of *lux. Lux* is the original substance from which *lumen* arises. Now, if *lux* be considered in itself as an accident, as such it must be caused by the form of its subject. Thus, if *lux* in air be an accident, true *lux* is caused by the form of air—which cannot be. *Lux* changes its subject. This is manifest, because *lux* is naturally to the east at first, but subsequently it is in the west. The eastern light engenders more light immediately beside it, which new generation continues the process until light appears in the west. But accidents do not change their subject, nor do they act outside it, so that *lux* seems not to be an accident. Also, if it be an accident of the air, the air must move instantaneously *(subito)* from east to west in the same

manner as light rays. However, such motion is incompatible with the air, as with any other element. Again, nothing is nobler than *lux,* but accident is less noble than substance, from which it seems that light is no accident, for air is far more ignoble than light.

If light be corporeal, it becomes difficult to understand how it may be in the air or any other transparent body, for instance, in crystals, since two bodies cannot simultaneously occupy the same place. Yet it is neither impossible nor difficult to posit light simultaneously as corporeal substance and as accompanying some body. We see water and ashes similarly mixed together, preserving the corporality and position of each component; the surface of each remains distinct from the surface of the other, and the parts of each are contained within their proper surface. . . . Light can likewise be in air or another body while retaining

16. Note the primitive understanding of mirrors and their properties: Neckam knew about erect and inverted images but regarded them as incomprehensible.

17. This chapter contains nothing further of optical interest.

18. This translation is based on a comparison of the Latin editions of Strassburg, 1485; Nürnberg, 1519; Frankfurt, 1601 [for the title of this edition, see Selection 58, n. 1—*Ed.*]; and the English version of London, 1582. The sections on vision and geometrical optics in this encyclopedia are extremely general and elementary—much less informative than Grosseteste's *De lineis.* However, Bartholomew's section on *lux,* excerpted here, nicely summarizes the main points of light metaphysics as found in the first half of Grosseteste's *De luce.* Bartholomew's encyclopedia represents a less technical but widely read version of this doctrine, which authors like Bacon and Witelo considered relevant to the science of optics. The authorities cited here by Bartholomew are Neoplatonic, Basil being the most important. While Grosseteste is not mentioned in this chapter, "Rupertus [Robertus] Linconiensis" is listed prior to Book I as an authority.

its own corporality and the continuity of its own substantial parts, . . . although the simplicity of light may prevent its being directly perceived in the medium. How much more wonderful is it that many lights may be brought together and united as one, still preserving unmixed the substantial form of each, by which each differs from the others. Now does any *lux* exist materially or formally by virtue of another. Dionysius teaches this openly in *De divinis nominibus,* as follows. When lamp lights *(lumina)* are in a house and gather together in common, each light remains discrete by one distinction and by distinct unity.[19] We observe many lights as one illumination and one clarity. Nor can anyone, I believe, separate one of these lights from the others, or from the air containing them all at once; nor can anyone discern one of these lights from another, as they exist confusedly among themselves. If one of the lamp lights is removed, it carries with it nothing from the others and relinquishes nothing to them. Indeed, there is a perfect unity, naturally commingled, and no part is confused, and the unity is such in corporeal air and in material light. Thus Dionysius teaches that while luminaries may be united, the substantial and accidental properties of each are preserved, either in increase or in decrease of illumination. . . .

Lux is diffused from the highest heaven to the center of the earth; this light is unitary and simple in its root and substance, but varied according to the diversity of its recipients, whether the sun or the other orbs, superior or inferior, having one substance, virtue, and operation of light, though without luminosity. This is the light created on the first day of Creation, according to Basil, and it preceded the sun and the stars made on the fourth day. The sun and stars are the vehicles of the primeval *lumen,* and because of the support of incorporeal *lux,* they are sufficient to shine perpetually without loss of their own substance. This

lux reaches everywhere and perfects and influences all bodies, more or less. *Lux* therefore is the root and basis of every illumination; it is one in substance and is not contained accidentally by any lower body but rather contains all bodies. In comparison with them *lux* is quite formal, although it is in itself material and has position. So far as *lux* is material, it has part separate from part potentially. Because *lux* is in truth formal, it has part separate from part actually, for what matter has in potency, form has in actuality. Thus, when *lux* in a kind of bodies has a minimum of matter and a maximum of form, this light is closer to form than to matter. Therefore as matter it is at a minimum, that is, at a point; as form it is everywhere. . . . The form of *lux,* being the noblest of corporeal forms, extends its matter maximally. One point of *lumen* or of *lux* suffices to illuminate the whole world, because of the nobility of its matter and the highest actuality of its form, according to Algazel. Therefore *lux* is unitary and simple as it is uniform in essence, though various luminaries are distinct among themselves. . . .

Lux, as Basil says, is extremely mobile, incessantly moving and generating itself rectilinearly and circumferentially, diffusing itself in every direction, and moving instantaneously *(subito).* *Lux* is most active and without resistance to internal penetration of anything. *Lux* engenders diversities and brings into unity and accord the contradictions in elements and mixed bodies. By the projection of its rays—direct, intersecting, refracting, and reflecting rays—*lux* brings things into being and conserves them, or it dissolves their existence, as Chalcidius says in commenting on the *Timaeus. Lux* rules and directs the life and permanence of everything. It naturally multiplies itself, generating *lux,* which likewise generates successive *lux.* This process occurs in an instant from a single point, filling the whole world.

61 ROBERT GROSSETESTE AND THE REVIVAL OF OPTICS IN THE WEST
Translations, introduction, and annotation by David C. Lindberg

Prior to the thirteenth century, Western optical knowledge was exceedingly limited. The best Greek and Islamic sources, including Ptolemy and Alhazen, were as yet unknown. Encyclopedists like Adelard of Bath and Alexander Neckam were thus unable to present more than a smattering of general knowledge gathered from traditional Western sources, such as Plato, Augustine, Boethius, Sol-

inus, Chalcidius, and Isidore, and a few treatises of Greek and Islamic authorship that were just beginning to appear.

But Western optics was transformed in the first

19 In other words, as explained further on, the lamp lights remain quantitatively distinct as individuals ("by one distinction") and as integral entities ("by distinct unity").

half of the thirteenth century through the efforts of Robert Grosseteste (d. 1253). Grosseteste had at his disposal several newly translated treatises on optics, including Euclid's *Optica* and *Catoptrica,*[1] Aristotle's *Meteorologica,* and Alkindi's *De aspectibus.* (It is possible, but unlikely, that he also knew Ptolemy's *Optica.*)[2] The profound influence of these treatises on the level of optical discussions is revealed by the two works of Grosseteste that follow. In another treatise, *De luce* (not included here, but available in English translation),[3] Grosseteste presents a metaphysics of light, according to which the world was created by the self-diffusion of a point of light into a spherical form. Optics thus became for him the fundamental science of

nature, revealing not only the manner of creation of the material world but also the mode of all natural actions.

Grosseteste's influence was exercised not only through his written works but also through his association with Oxford University, first as lecturer in theology to the Franciscans and later as Bishop of Lincoln. His metaphysical conceptions influenced the next generation of writers on optics, and his stress on the value of optics and other scientific subjects served to justify a great deal of later scientific endeavor. Indeed, through his influence a scientific tradition was initiated among the Oxford Franciscans, which flowered later in the century in the works of Roger Bacon and John Pecham.[4]

1. Robert Grosseteste (ca. 1168–1253): Concerning Lines, Angles, and Figures[5]

The usefulness of considering lines, angles, and figures is very great, since it is impossible to understand natural philosophy without them. They are useful in relation to the universe as a whole and its individual parts. They are useful also in connection with related properties, such as rectilinear and circular motion. Indeed, they are useful in relation to activity and receptivity, whether of matter or sense; and if the latter, whether of the sense of vision, where activity and receptivity are apparent, or of the other senses, in the operation of which something must be added to those things that produce vision.

Since we have spoken elsewhere of those things that pertain to the whole universe and its individual parts and of those things that relate to rectilinear and circular motion,[6] we must now consider universal action insofar as it partakes of the nature of sublunary things *(inferiorum)*; this universal action is a subject receptive to diverse activities, insofar as it descends to operation in the matter of the world; and some things can be brought in as intermediaries, which are able to bring to perfection that which is advancing toward greater things *(maiora).*

Now, all causes of natural effects must be expressed by means of lines, angles, and figures, for otherwise it is impossible to grasp their explanation. This is evident as follows. A natural agent multiplies its power from itself to the recipient, whether it acts on sense or on matter. This power is sometimes called species, sometimes a likeness, and it is the same thing whatever it may be called; and the agent sends the same power into sense and into matter, or into its own contrary, as

heat sends the same thing into the sense of touch and into a cold body. For it does not act by deliberation and choice, and therefore it acts in a single

1. The latter is no longer attributed to Euclid. Both works have been translated by Paul Ver Eecke in *Euclide, L'Optique et la catoptrique* (Paris: Blanchard, 1959).

2. This is Albert Lejeune's position in his *L'Optique de Claude Ptolémée dans la version latine d'après l'arabe de l'émir Eugène de Sicile* (Louvain: Bibliothèque de l'Université, Bureaux du recueil, 1956), p. 31*. The alternative but, I believe, less easily defended position is adopted by A. C. Crombie, *Robert Grosseteste and the Origins of Experimental Science, 1100–1700* (Oxford: Clarendon Press, 1953), pp. 116, 120.

3. Robert Grosseteste, *On Light,* translated by Clare C. Riedl (Milwaukee: Marquette University Press, 1942).

4. On the Franciscan school, see A. G. Little, "The Franciscan School at Oxford in the Thirteenth Century," *Archivum Franciscanum Historicum,* XIX(1926), 803–874. However, note the cautions set forth by Stewart Easton in an appendix to his *Roger Bacon and His Search for a Universal Science* (Oxford: Blackwell, 1952), pp. 206–209.

5. I have used the text from *Die philosophischen Werke des Robert Grosseteste, Bischofs von Lincoln,* edited by Ludwig Baur in *Beiträge zur Geschichte der Philosophie des Mittelalters,* Vol. IX (Münster/Westfalen: Aschendorffsche Verlagsbuchhandlung, 1912), pp. 59–65. Frequent reference has been made to the prior, unpublished translation of this treatise by Bruce S. Eastwood in "The Geometrical Optics of Robert Grosseteste" (Unpublished Ph.D. dissertation, University of Wisconsin, 1964). A glance at Baur's variant readings reveals that a more justifiable title would be "Concerning the Refraction and Reflection of Rays." Baur gives both titles, but the former has become so firmly established in the literature that a change seems inadvisable.

6. Baur, *Philosophischen Werke,* p. 79*, identifies the works of Grosseteste treating these other matters as *De sphera, De motu corporali et luce,* and *De motu supercelestium.*

manner whatever it encounters, whether sense or something insensitive, whether something animate or inanimate. But the effects are diversified by the diversity of the recipient, for when this power is received by the senses, it produces an effect that is somehow spiritual and noble; on the other hand, when it is received by matter, it produces a material effect. Thus the sun produces different effects in different recipients by the same power, for it cakes mud and melts ice.

The power from a natural agent reaches the recipient either along a shorter line—and then the power is more active, since the recipient is less distant from the agent—or along a longer line— and then it is less active, since the recipient is more distant. In any case, the power comes from the surface of the agent with or without mediation. If without mediation, the power comes either by a straight line or by a bent line. But if by a straight line, the action is stronger and better, as Aristotle proposes in Book V of the *Physics*,[7] since nature acts in the briefest possible manner; and a straight line is the shortest of all lines, as he says in the same place. Moreover, a straight line possesses equality because it has no angle; but the equal is superior to the unequal, as Boethius says in his *Arithmetic*.[8] However, nature acts in the briefest possible manner; consequently it acts best along a straight line. Similarly, every united power acts more strongly. But there is greater oneness and unity in a straight line than in a bent line, as is asserted in Book V of the *Metaphysics*;[9] therefore action in a straight line is stronger.

Now, a straight line falls either at equal angles (that is, perpendicularly) or at unequal angles. If at equal angles, the action [along it] is stronger for the three aforementioned reasons, since such a line is shorter and possesses equality, and the power comes uniformly along it to the parts of the recipient. But a line falling at equal angles on some body falls at right angles when it is incident on a plane, at acute angles on a concave body, and at obtuse angles on a sphere. This is illustrated as follows: If an incident line is drawn passing through the center of a sphere, it is perpendicular to the line of tangency, and between the line of tangency and the sphere, contingent angles are formed on both sides.[10] Therefore, such a line incident on a sphere forms two angles with the surface of the sphere, each of which is larger than a right angle, since it equals a right angle plus a contingent angle. Therefore, when power is incident at angles that are not merely equal but right

angles, its action is seen to be the strongest, since complete equality and uniformity exist.

If there should be a line that is not straight (or one that is bent), then since it is not circular because a natural agent does not emanate its power along a circular path but along the diameter of a circle on account of the brevity of the latter, it is evident that such a line has angles. And this would not occur in the presence of a single medium or when a single body is encountered; but [in order for the line to be bent] there must be two media, since the power is multiplied along certain straight lines in the first medium and along others in the second. However, this can occur only in two ways, namely when the recipient body is dense so as to impede the passage of the power, chiefly in the case of our senses—and then it is called a reflected line, because the power returns—or when the opposing body is rare, thereby permitting passage of the power. If in the first way, then the power incident on a dense body falls either at equal angles (that is, perpendicularly) or at unequal angles. If the first of these, the power returns into itself along the same path by which it approached [the dense body]. The explanation of this is that lines incident on and reflected from a body form similar angles of the same size. Therefore, the power must be reflected at an angle equal to the angle of incidence and [thus] return in the same way [as it approached]. For if it should return at another angle or by another path, deviating to the right or to the left, it could not return at a [reflected] angle equal to the angle of incidence; rather, the angle would be larger or smaller. But if the power is incident nonperpendicularly, it returns along a path by which it is able to form an angle with the surface of the resisting body equal to the angle of incidence (that is, the angle formed by that body and the incident line) for the aforementioned reason. Thus the angles of incidence and reflection are always equal, which may now be assumed.

Since reflection occurs in these two modes, it is to be understood that a power reflected into itself is stronger than a power reflected along some other path, on account of the doubling of the power

7. Baur (p. 61) suggests that the statement may actually have been drawn from Aristotle's *Metaphysics* V.6.

8. I.32 and II.1, according to Baur, p. 61.

9. V.6.1016a ff., according to Baur, p. 61.

10. Medieval scientists regarded "contingent angles" (or angles of tangency) as having a finite magnitude. On the history of this view, see T. L. Heath, *The Thirteen Books of Euclid's Elements,* 2d ed. (New York: Dover, 1956), II, 39–42.

in the same place. Nevertheless, however much this may [seem to] follow from the theory of reflection, the action [of a power] is weaker when reflection is along the same path [as that of the approach,] because, since every reflection weakens a power, that reflection is weakened more which makes a power deviate completely [that is, 180°] from the direct progress that is proper to it if it is to pass through the body; and this is reflection along the path of incidence. And such a path is altogether contrary and opposite to the direct progress proper to the power.

When reflection occurs from a polished body possessing the nature of a mirror, then the reflection is best and the action strongest; but when reflection occurs from rough bodies, the species is dissipated and the action is weak. The Commentator assigns the cause of this in his treatise on sound,[11] saying that in the reflection of a species all parts of a polished body and a smooth surface concur in one action, on account of their evenness and uniformity; therefore the whole species is reflected from a polished body integrally, just as it approaches the body. But the parts of a rough body are uneven and the highest parts reflect the species first, and thus the parts do not unite in one action; for this reason the species is broken into parts, and therefore its action is not strong.

When reflection takes place from concave bodies, its action is stronger than when reflection takes place from plane or convex bodies, because rays reflected from a concave surface converge together; but this is not true of the other surfaces.

Now, if the body encountered by a power should not [entirely] prevent passage of the power, then a ray incident at equal angles (that is, perpendicularly) maintains rectilinear progress, and this is the strongest ray. But the ray incident at unequal angles deviates from the rectilinear path that it had in the first substance, which would be maintained if the medium [in which the power is propagated] were uniform. And this deviation is called refraction of the ray. This refraction is twofold: If the second substance is denser than the first, the ray is refracted to the right[12] and passes between the path of rectilinear progress[13] and the perpendicular drawn to the second body at the point of refraction; if, on the other hand, the second substance is subtler than the first, the ray is refracted toward the left, receding from the perpendicular on the other side of the path of rectilinear progress. And since these things are so, it should be understood that a power incident along

a refracted line is stronger than a power incident along a reflected line, since a refracted line deviates slightly from a rectilinear course, which is strongest, while a reflected line deviates a great deal, [passing] in the opposite direction;[14] therefore reflection weakens a power more than refraction. Regarding the two modes of refraction, it can be asserted that a power refracted toward the right is stronger than that refracted toward the left, since the power refracted toward the right more nearly approaches the perpendicular course, whether we refer to the perpendicular drawn from the point of refraction or that drawn from the agent, from the same point of which issue both a perpendicular and a refracted line.[15]

Besides these three essential lines, there is a fourth accidental line, along which an accidental and weak power proceeds. However, this power does not come directly from the agent but from a power multiplied along any of the three aforementioned lines; for example, accidental light comes to all corners of a house from a ray falling through a window. This power is the weakest of all, since it does not issue directly from the agent but from a power of the agent directed along a straight, reflected, or refracted line. These things are asserted regarding lines and angles.

Regarding figures, there are two kinds that must be considered here. One of these, namely the sphere, is required by the multiplication of a power; for every agent multiplies its power spherically, since it does so in every direction and along all diameters—upward, downward, before, behind, to the right, and to the left. This is evident, since, for the same reason that a line extends in one direction from a centrally located agent, lines will extend likewise in every direction according to

11. Two manuscripts give the reading "chapter on sound in the second book of *De anima*." Crombie gives the reference as II.4.vi, fol. 143 recto of the Venice, 1551, edition of Averroes' commentary on Aristotle's *De anima*.

12. Grosseteste obviously had a figure in mind, although none appears in any of the manuscripts. Nevertheless, his intent is clear: in passing from a less dense to a more dense medium, the ray is refracted toward the perpendicular.

13. That is, the rectilinear extension of the incident ray.

14. It is, of course, not true that a reflected ray must deviate from rectilinearity more than a refracted ray.

15. This conclusion appears to contradict the earlier statement that bending weakens, for here Grosseteste seems to imply that a ray gains strength by deviating from its rectilinear course and approaching the perpendicular.

all differences of position—and thus necessarily in spherical fashion. This is asserted by the Commentator in the second book of *De anima*.[16] Moreover, wherever an organ of sense is placed, it is able to perceive such an agent at a suitable distance;[17] but this is possible only by means of a species or power coming from the agent. Therefore, that power is multiplied in all directions.

The other figure required for natural actions is the pyramid.[18] For if power should issue from one part of an agent and terminate at one part of the recipient, and if this should be true of all powers so that power always comes from one part of the agent to a single part of the recipient, no action will ever be strong or good. But action is complete when the power of the agent comes to every point of the recipient from all points of the agent or from its entire surface. But this is possible only by means of a pyramidal figure, since powers issuing from the single parts of the agent [which constitutes the base of the pyramid] converge and unite at the apex of the pyramid; and therefore all are able to act strongly on the part of the recipient encountered. Thus infinitely many pyramids are able to issue from one surface of an agent; as base of all these pyramids there is one thing, namely the surface of the agent, and there are as many apices as there are pyramids. The apices are located everywhere, at all the different points of the medium or recipient; and infinitely many pyramids issue in any one direction, some shorter and some longer. But there is no diversity [of power] in those pyramids of equal length or shortness, since, for their part, they all act identically, although diversity can originate with the recipient matter.

When one pyramid is shorter than another and both issue from the same agent, there is great difficulty [in determining] whether the apex of the shorter pyramid acts more strongly on the recipient. It must be asserted that a shorter pyramid

acts more strongly, since its apex is less distant from its source, and consequently more power reaches the apex than in a longer pyramid; thus the recipient is more closely joined to the agent by the shorter pyramid, and therefore it is more strongly altered by the power [of the agent]. Moreover, if rays from the right side of the lateral surface of the shorter pyramid are extended beyond the apex in a continuous straight line, they form smaller angles[19] with rays from the left side of the lateral surface of the pyramid than do similar rays in a longer pyramid, as is evident from Proposition 21 of the first book of Euclid's *Geometry* and also to sense. In the same way, rays coming from the left side of the pyramid extended beyond the apex in a continuous straight line are more completely united with rays from the right side of the lateral surface of the pyramid than similar rays of a longer pyramid. Consequently, since every congregation and union is more active, the apex of a shorter pyramid acts more strongly and also alters the recipient more than the apex of a longer pyramid. Nevertheless, if it should be objected rationally that power comes to the apex of a longer pyramid from the whole surface of the agent,[20] that the power congregates there more completely because that apex is more acute than the apex of a shorter pyramid, that every united power acts more strongly, and moreover that the rays of a longer pyramid are closer to the perpendicular rays drawn from the extremities of the diameter of the agent, and therefore that longer pyramids are stronger because a perpendicular path is the strongest, it can be asserted that those reasons are quite conclusive insofar as they suffice; therefore, they prevail unless the stronger reasons mentioned above are to the contrary.

Here ends the treatise by [the Bishop] of Lincoln on the refraction and reflection of rays.

2. Robert Grosseteste: On the Rainbow[21]

Investigation of the rainbow is the concern of both the student of perspective and the physicist. It is for the physicist to know the fact and for the student of perspective to know the explanation. For this reason Aristotle, in his book on *Me-*

16. Averroes, as cited in note 11.

17. That is, neither too close nor too far.

18. Grosseteste uses the term "pyramid" to denote all figures having a plane or curved base and straight lines extending from every point on the base to a common apex; the base need not be a regular polygon.

19. That is, the angles supplementary to the angle of the apex.

20. Grosseteste evidently has in mind a convex surface, such as the surface of a sphere, so that a longer pyramid includes rays from a larger portion of the surface than do shorter pyramids.

21. I have translated this treatise from the Latin text edited by Ludwig Baur, pp. 72–78. Portions of this treatise have also been published in translation by Crombie, chapter 5. The majority of the manuscripts collated by Baur contain the title *De iride*; however, the British Museum MS contains the title *De iride et speculo*, and Baur gives both titles.

teorology, has not revealed the explanation, which concerns the student of perspective; but he has condensed the facts of the rainbow, which are the concern of the physicist, into a short discourse. Therefore, in the present treatise we have undertaken to provide the explanation, which concerns the student of perspective, in proportion to our limited capability and the available time.

In the first place, then, we state that perspective is the science based on visual figures and that this is subordinate to the science based on figures containing radiant lines and surfaces, whether that radiation is emitted by the sun, the stars, or some other radiant body. Nor is it to be thought that the emission of visual rays [from the eye] is only imagined and without reality, as those think who consider the part and not the whole. But it should be understood that the visual species [issuing from the eye] is a substance, shining and radiating like the sun, the radiation of which, when coupled with radiation from the exterior shining body, entirely completes vision.

Wherefore natural philosophers, treating that which is natural to vision (and passive), assert that vision is produced by intromission. However, mathematicians and physicists, whose concern is with those things that are above nature, treating that which is above the nature of vision (and active), maintain that vision is produced by extramission. Aristotle clearly expresses this part of vision that occurs by extramission in the last book of *De animalibus,* saying: "A deep-set eye sees from a distance; for its motion is neither divided nor destroyed, but a visual power leaves it and goes directly to the objects seen."[22] Again in the same book [Aristotle writes]: "The three senses referred to, namely vision, hearing, and smell, issue from the organs [of perception] as water issues from pipes, and for this reason long noses have a strong power of smell."[23] Therefore, true perspective is concerned with rays emitted [by the eye].

Perspective has three principal subdivisions, according to the triple mode in which rays are transmitted to the visible object. For either transmission of the ray to the visible object is in a straight line through a transparent medium of a single kind interposed between the observer and the object; or in a straight line to a body having a spiritual nature which is such as to make it a mirror, from which the ray is reflected to the object; or through several transparent media of different kinds, at the junctions of which the visual rays are refracted to form angles, so that the ray

reaches the object not by direct approach but by a path of several straight lines joined at angles.

The first subdivision comprises the science called *"De visu,"* the second subdivision the science called *"De speculis."* The third part has remained untouched and unknown among us until the present time. Nevertheless, we know that Aristotle completed this third part, which surpasses the other parts in its exceeding subtlety and the greatly admired profundity of its natures.[24] This part of perspective, if perfectly understood, shows us how to make very distant objects appear close, how to make nearby objects appear very small, and how to make a small object placed at a distance appear as large as we wish, so that it would be possible to read minute letters from incredible distances or count sand, seeds, blades of grass, or any minute objects.[25] The way in which these astonishing things occur is made manifest as follows. The visual ray penetrating through several transparent substances of diverse natures is refracted at their junctions, and its parts, in the different transparent media existing at those junctions, are joined at an angle. This is revealed by that experiment which is considered fundamental in the book *De speculis:*[26] If an object is placed in a vessel and the observer stations himself at a position from which the object cannot be seen, the object will become visible when water is poured in. The same thing is also revealed by the fact that the subject of a continuous ray is a substance possessing a single [homogeneous] nature. Therefore, at the interface between two transparent media of different kinds, a visual ray must be interrupted. However, since the entire ray is generated by a single principle, and its continuity cannot be entirely destroyed unless its generation ceases, the discontinuity of the ray at the interface between the two transparent media must be incomplete. But the only mean between complete continuity and complete discontinuity is a point touching the two parts of one ray, [joined] not in a straight line but at an angle.

22. *De generatione animalium,* V.1.781a.1–2.
23. *Ibid.,* V.1.781b.2–13.
24. Grosseteste apparently has in mind an ancient tradition to the effect that Aristotle wrote a book on optics; see Hero of Alexandria, *De speculis,* in *Heronis Alexandrini Opera quae supersunt omnia,* ed. L. Nix and W. Schmidt, Vol. 2, fasc.1 (Leipzig, 1900), p.318; compare Carl B. Boyer, "Aristotelian References to the Law of Reflection," *Isis,* XXXVI (1945–1946), 93, n. 14.
25. These views, long marveled at in Roger Bacon, were obviously borrowed by the latter from Grosseteste.
26. Pseudo-Euclid, *Catoptrica,* Definition 6.

The amount of divergence from rectilinearity of rays joined at an angle can be represented as follows. Imagine a ray from the eye incident through air on a second transparent medium, extended continuously and rectilinearly [into the second medium], and a line perpendicular to the interface drawn into the depth of the medium from the point at which the ray is incident on the [second] transparent medium. I say, then, that the path of the ray in the second transparent medium is along the line bisecting the angle enclosed by the ray which we have imagined to be extended continuously and rectilinearly and the perpendicular line drawn into the depth of the second transparent medium from the point of incidence of the ray on its surface.[27]

That the size of the angle of refraction of a ray may be thus determined is evident from an experiment similar to those by which we have learned that reflection of a ray by a mirror occurs at an angle equal to the angle of incidence. The same thing is revealed to us by the following principle of natural philosophy, namely that every operation of nature takes place in a manner as limited, well-ordered, brief, and good as possible.[28]

An object seen through several transparent media does not appear as it really is but appears to be situated at the intersection of the ray emitted by the eye, extended in a continuous straight line, and the line drawn perpendicularly from the visible object to the surface of the second transparent medium (that is, the one nearer the eye). This is revealed to us by similar reasoning and the same experiment as that by which we know that objects seen in mirrors appear at the intersection of the visual ray, extended in a straight line, and the perpendicular drawn [from the visible object] to the surface of the mirror.

Thus far we have considered the size of the angle of refraction at the interface between two transparent media and the place of appearance of an object seen through several transparent media. To these we add the following principles, which the optical theorist appropriates from the natural philosopher, namely that the size, position, and order of the visible object are determined from the size of the angle and the position and order of the rays under which the object is seen and that it is not great distance that makes an object invisible, except accidentally, but rather the smallness of the angle under which it is observed. From these principles, it is perfectly evident by geometrical reasoning how an object of known distance, size, and position will appear with respect to location, size, and position. From the same principles it is evident how one must shape transparent substances so that they will receive the rays emitted by the eye, according to the desired angle formed in the eye, and will draw the rays back [toward convergence] in any desired manner onto visible objects, whether these objects are large or small, far or near. Thus all visible objects are made to appear to the observer in the desired position and of the desired size; and very large objects, if so desired, are caused to appear very small; and conversely, very small objects placed far away are caused to appear large and very easily perceptible by sight.

The science of the rainbow is subordinate to this third part of perspective [that is, to the science of refraction]. Now, a rainbow cannot be formed by means of solar rays falling in a straight line from the sun into the concavity of a cloud,[29] for they would produce a continuous illumination of the cloud in the shape of the opening toward the sun through which the rays enter the concavity of the cloud rather than in the shape of a bow. Nor can a rainbow be formed by the reflection of solar rays from the convexity of the mist descending from a cloud as from a convex mirror, in such a way that the concavity of the cloud would receive the reflected rays and the rainbow would thus appear; for if that were so, the rainbow would not always be in the form of an arc, and as the sun rose the rainbow would become proportionately larger and higher, and as the sun set the rainbow would become smaller; the contrary, however, is evident to sense. Therefore, the rainbow must be formed

27. That is, the angle of refraction equals half the angle of incidence. Diagrams illustrating this conception appear in Crombie, page 121. This half-angle law of refraction is remarkably primitive by contrast with Ptolemy's elaborate attempts to find, by empirical means, the true mathematical law of refraction; see Albert Lejeune, *Recherches sur la catoptrique grecque* (Bruxelles: Palais des académies, 1957), pp. 152–166. It seems doubtful that Grosseteste would have formulated his half-angle law had he known Ptolemy's *Optica*.

28. This paragraph reveals that Grosseteste's half-angle law of refraction was determined not through measurement but rationally, on grounds of symmetry and brevity of action. See Bruce S. Eastwood, "Grosseteste's 'Quantitative' Law of Refraction: A Chapter in the History of Non-Experimental Science," *Journal of the History of Ideas*, XXVIII (1967), 403–414.

29. This is roughly the Aristotelian view, expressed in the *Meteorologica* III.2–5. On the history of the theory of the rainbow, see Carl B. Boyer, *The Rainbow: From Myth to Mathematics* (New York: Yoseloff, 1959), and Selection 63.2 from the *De iride* of Theodoric of Freiberg.

by the reflection of solar rays in the mist of a convex cloud. I maintain that the outside of a cloud is convex and the inside concave, as is evident from the nature of light and heavy. And that which we see of a cloud must be less than a hemisphere, although it may look like a hemisphere; and since the mist descends from the concavity of the cloud, it must be pyramidally convex at the top, descending to the earth, and therefore more condensed near the earth than in the higher part.[30]

Therefore, there are, in all, four transparent media through which a ray of the sun penetrates: [first,] pure air containing the cloud; second, the cloud itself; third, the higher and rarer mist coming from the cloud; and fourth, the lower and denser part of the same mist. Therefore, in accordance with what was said before about the refraction of rays and the size of the angle of refraction at the interface between two transparent media, solar rays must be refracted first at the interface between the air and the cloud and then at the interface between the cloud and the mist. By these refractions the rays converge in the density of the mist and, being refracted there once more as from the vertex of a pyramid, spread out not into a round pyramid but into a figure like the curved surface of a round pyramid[31] expanded opposite the sun. Therefore it assumes the shape of an arc, and in these [northern] regions the rainbow does not[32] appear in the south. And since the vertex of the aforementioned figure is near the earth and it is expanded opposite the sun, half the figure or more must fall on the surface of the earth and the remaining half or less onto a cloud opposite the sun. Therefore, near sunrise or sunset the rainbow appears semicircular and is larger, and when the sun is in other positions the rainbow appears as part of a semicircle; and the higher the sun, the smaller the [visible] part of the rainbow. For this reason, in regions where the sun closely approaches the zenith, the rainbow never appears at noon. Aristotle's claim that the variegated rainbow is of small measure at sunrise and sunset is not to be interpreted as smallness of size, but as smallness of luminosity, which occurs because the rays pass through a greater multitude of vapors at this time of day than at other times. Aristotle himself indicates this later, when he says that this [smallness] occurs because of the diminution of that which shines from the solar ray in the clouds.

However, since color is light mixed with a transparent medium—the transparent medium being diversified according to purity and impurity, while light is divided in a fourfold manner (according to brightness and darkness and according to multitude and paucity), and all colors being generated and diversified according to the combinations of these six differences—the variety of colors in different parts of one and the same rainbow occurs chiefly because of the multitude and paucity of solar rays. For where there is a greater multiplication of rays, the color appears clearer and more luminous, and where there is a smaller multiplication of rays, the color appears more bluish and obscure. And since the multiplication of light and the diminution determined by this multiplication result only from the splendor of the illumination on the mirror or from the transparent medium, which collects light in a certain place because of its shape [that is, the shape of the medium] and, after it has come together, diminishes it by separation, and since this disposition for the reception of light is not fixed, it is manifest that painters do not have the ability to represent the rainbow; nevertheless, its representation according to a fixed[33] disposition is possible.

Actually, the difference in color between one rainbow and another arises both from the purity and impurity of the recipient transparent medium and from the brightness and darkness of the light impressed on the medium. For if the transparent medium is pure and the light is bright, the color of the rainbow will be whitish and similar to light. But if the recipient transparent medium should contain a mixture of smoky vapors and the light is not very bright, as occurs near the rising or setting [of the sun], the color of the rainbow will be less brilliant and darker. Similarly, [the production of] all variations in color of the variegated bow is sufficiently apparent from other combinations of brightness and darkness of light and purity and impurity of the transparent medium.

Here ends the treatise on the rainbow by [the Bishop] of Lincoln.

30. For debate on the interpretation of the scheme described in this paragraph, see D. C. Lindberg, "Roger Bacon's Theory of the Rainbow: Progress or Regress?," *Isis*, LVII (1966), 240; Bruce S. Eastwood, "Robert Grosseteste's Theory of the Rainbow," *Archives internationales d'histoire des sciences*, XIX (1966), 323–324. Grosseteste's theory has also been analyzed by Crombie, chapter 5.

31. That is, a hollowed-out cone.

32. Baur's text does not include the negative, but Crombie (p. 126) argues for its inclusion and adduces manuscript support for his position.

33. According to Baur, the manuscripts collated for his text are unanimous in reading *non fixam*. However, omission of the negative seems to be required if sense is to be made of the sentence.

Translations, introduction, and annotation by David C. Lindberg[1]

Although Robert Grosseteste greatly stimulated European interest in optics, his own investigations were hampered by the lack of many important sources. Ptolemy's *Optica* was just becoming known in the West, and it is doubtful that Grosseteste was familiar with it; Alhazen's *Perspectiva,* although translated late in the twelfth or early in the thirteenth century, had not come to Grosseteste's attention by the time he composed his optical works. Consequently, Grosseteste's knowledge of optics, though more complete than that possessed by any Western predecessor or contemporary, was primitive by comparison with the optical achievements of Islam; thus his works lack the scope and depth of such a work as Alhazen's great optical treatise, as even the most casual inspection will reveal.

However, by the second half of the century, the most advanced treatises of Greek antiquity and medieval Islam had been rendered into Latin. The most important of these for the development of optics were the *Perspectiva* of Alhazen, the *Optica* of Ptolemy, a number of works of Avicenna, and (toward the end of the period) the *Catoptrica* of Hero of Alexandria. These new riches, especially Alhazen's *Perspectiva,* occasioned a dramatic surge in the study of optics, which found its best expression in the works of Roger Bacon, Witelo, and John Pecham. Thus, whether in Alhazen's *Perspectiva* (which not only served as the principal source for Western writers but also circulated widely itself) or in the works of Bacon, Witelo, and Pecham which it inspired, the West possessed for the first time treatises in which almost the entire gamut of problems now classified as optical was systematically treated; moreover, in these treatises we find an enormous increase in the sophistication of mathematical techniques employed. Indeed, we find what must be considered a grand synthesis of optical knowledge, comprehensive in scope and incorporating the very best learning of Greece, Islam, and the Latin West. It speaks for the vigor and excellence of this synthesis, that the progress of optics throughout the remainder of the Middle Ages came about as scientists extended the application of principles set forth late in the thirteenth century[2] or as questions raised in the thirteenth century, but incompletely or unsatisfactorily answered, became the objects of inquiry and debate.

The selections appearing in this chapter are an attempt to convey the principal doctrines and techniques of this late thirteenth-century synthesis. They are drawn from the works of Roger Bacon, Witelo, and John Pecham, all of whom wrote on optics during the 1260's and 1270's. In addition, it has been deemed essential to include selections from the Latin text of Alhazen's *Perspectiva* to illustrate further the optical knowledge available in Europe after about 1250 and the origin of many of the ideas expressed by Bacon, Witelo, and Pecham.[3] Since Alhazen and his three thirteenth-century followers were in such complete agreement on most fundamental issues, it has been unnecessary to present the view of each individual on every question; rather, those passages have been selected

1. Except for section 18, which was translated by Robert B. Burke.
2. Set forth, that is, by Alhazen's *Perspectiva* as well as by those thirteenth-century treatises based upon it. One could legitimately argue that this synthesis was actually Islamic and occurred in the eleventh century in the works of Alhazen and Avicenna. Nevertheless, their works did not become available to the West until the thirteenth century, whereupon they inspired further efforts by Bacon, Witelo, and Pecham; thus, so far as the optical knowledge of the West is concerned, the synthesis occurred in the late thirteenth century.
3. Unless otherwise noted, the selections have been translated or reprinted from the following editions: Alhazen, *Perspectiva,* from *Opticae thesaurus Alhazeni Arabis libri septem,* edited by Friedrich Risner (Basel, 1572); Witelo, *Perspectiva,* bound with Alhazen's *Perspectiva* in the *Opticae thesaurus,* but separately paginated; Roger Bacon, *The Opus Majus of Roger Bacon,* edited by J. H. Bridges (London, 1900); Roger Bacon, *De multiplicatione specierum,* included in Volume II of Bridges' edition of the *Opus maius.* The propositions from John Pecham's *Perspectiva communis* (both revised and unrevised versions) are reprinted by permission of the copyright owners, the Regents of the University of Wisconsin, from *John Pecham and the Science of Optics: Perspectiva communis* edited with an introdution, English translation, and critical notes by David C. Lindberg (Madison, Wis.: University of Wisconsin Press, 1970). The order in which the propositions appear in this source book and their page numbers in my Pecham volume are as follows: Part I, Proposition 27, revised version (p. 109); Propositions 29–34 (pp. 111–119); Propositions 43, 28, 37, 33, 38 (pp.127, 109–110, 121, 119, 121–123); Propositions 44–46 (pp. 127–131); Part II, Propositions 6, 20, 30 (pp.161–163, 171–173, 183–185); Part III, Proposition 4 (p.215); Proposition 16 (pp.229–231). Henceforth abbreviated citations will be used.

that most clearly and concisely present the shared theory. In cases of major disagreement, as on the question of visual rays, alternative views are presented.

Finally, it should be noted that the magnitude of the thirteenth-century optical synthesis has made it both possible and necessary to omit all but two selections illustrating optical progress during the later Middle Ages: possible, because the thirteenth-century synthesis provided the framework within which later optics was pursued, so that the views presented here largely represent later medieval views as well; necessary, because late medieval discussions were of a highly particular sort impossible to treat adequately in a source book.[4]

1. Roger Bacon (ca. 1219–1292): The Nature and Multiplication of Light or Species

OPUS MAIUS,[5] IV, DISTINCTION 2, CHAPTERS 1–2

Every efficient cause acts through its own power, which it exercises on the adjacent matter, as the light *(lux)* of the sun exercises its power on the air (which power is light [*lumen*] diffused through the whole world from the solar light [*lux*]). And this power is called "likeness," "image," and "species" and is designated by many other names, and it is produced both by substance and by accident, spiritual and corporeal, but more by substance than by accident and spiritual than corporeal. This species[6] produces every action in the world, for it acts on sense, on the intellect, and on all matter of the world for the generation of things, since one and the same thing is produced by a natural agent in whatever it acts upon because it does not have the power to deliberate; therefore it produces the same action in whatever it encounters. But if it acts on sense and the intellect, it produces species, as everybody knows; and if it acts on the contrary or on matter, it [also] produces species. And in things that have reason and intellect, although they do many things with deliberation and choice, nevertheless this operation (that is, the generation of species) is natural for them as for other things. Wherefore the substance of the mind multiplies its power inside the body and outside the body, and every body produces its power outside itself; angels move the world through powers of this kind. . . . Therefore, such powers of agents produce every action in this world. Now, there are two things to be considered with respect to these powers: one is the multiplication itself of the species and power from the place of its generation; the other is the varied action [of this species or power] in the world because of the generation and corruption of things. Since the second cannot be known without the first, the multiplication itself must first be described.

Every multiplication is according to lines, angles, or figures. As long as a species advances through a medium possessing a single rarity, as entirely in the heavens, entirely in fire, entirely in air, or entirely in water, it always maintains a rectilinear path, since Aristotle says in the fifth book of the *Metaphysics* that nature always acts in the shortest possible way, and a straight line is the shortest of all lines. . . .

But when there is a second body of different rarity and density, so that it is not altogether dense but alters the passage of the species in some way—like water, which is partly rare and partly dense, and similarly, crystal and glass and such substances through which we can see—then the species is incident on the second body either perpendicularly, and continues along the same straight line as before, or if it is not incident perpendicularly its rectilinear path is necessarily altered, and it forms an angle upon entering the second body. And this deviation from rectilinear progress is called refraction of the ray and species. The explanation of this is that the perpendicular is stronger and shorter and therefore nature acts in a better way along it, as geometrical demonstrations teach, of which further mention will be made below in the proper place. But this refraction is double, since if the second body is denser (as is the case in descending from the heaven into the lower regions), all stellar powers which are not perpendicularly incident on the sphere of the elements are refracted

4. On the history of late medieval optics, see Graziella Federici Vescovini, *Studi sulla prospettiva medievale* (Turin: Giappichelli, 1965).

5. Bridges, I, 111–117. Similar views on species appear in Bacon's *De multiplicatione specierum,* II, chapters 1–2 (Bridges, II, 457–465); however, as is frequently the case, Bacon expresses himself more concisely in his later and more refined *Opus maius.* Note the similarities between this passage and portions of Grosseteste's *Concerning Lines, Angles, and Figures,* included in the preceding selection.

6. Neither Witelo nor Alhazen employs the term "species," as do Bacon and Pecham. Nevertheless, as I have argued at length in my "Alhazen's Theory of Vision and Its Reception in the West," *Isis,* LVIII (1967), 321–341, Alhazen's and Witelo's "forms" are identical, in most respects, to Bacons and Pecham's "species."

between the direct path and the perpendicular erected at the place of refraction. And if the second body is subtler (as is the case in ascending from water to higher regions), then the direct path falls between the refracted ray and the perpendicular erected at the place of refraction.

When the second body is so dense that it in no way permits passage of a species—I speak of sensible passage, as concerns the judgment of human sight—then we say that the species is reflected. However, according to Aristotle and Boethius, the sight of a lynx penetrates walls. Therefore, in fact the species passes through, and this is true; however, human sight is unable to judge regarding this, but [only] regarding reflection, which necessarily takes place.[7] For because of the difficulty of passage through the dense body, since it has easy passage in the air through which it arrives, it multiplies itself more copiously in the direction from which it came [than straight ahead through the dense body]. . . .

But the fourth [kind of multiplication] is more necessary to the world, although it is called accidental multiplication. For light *(lumen)* is called accidental with respect to the principal light *(lux)*[8] coming from the thing, since the former does not come [directly] from the agent but from the principal multiplication, as the principal multiplication falls from the sun into a house through a window, but to a corner of the house comes accidental light from a ray passing through the window. The bodies

of mortals could not always be exposed to the principal species without undergoing corruption, and therefore God has moderated all things through such accidental species.

The fifth [kind of multiplication] is different from the others, for it violates the common laws of nature and claims a special privilege for itself. And this multiplication occurs only in an animated medium as in the nerves of the senses; for a species follows the tortuous path of the nerve and does not concern itself with rectilinear passage.[9] And this takes place by the power of the mind directing the passage of the species, as the operations of an animated being require. . . .

OPUS MAIUS,[10] V.1, DISTINCTION 9, CHAPTER 4

But a species is not body, nor is it moved as a whole from one place to another; but that which is produced [by an object] in the first part of the air is not separated from that part, since form cannot be separated from the matter in which it is unless it should be mind; rather, it produces a likeness to itself in the second part of the air, and so on.[11] Therefore there is no change of place, but a generation multiplied through the different parts of the medium; nor is it body which is generated there, but a corporeal form that does not have dimensions of itself but is produced according to the dimensions of the air; and it is not produced by a flow from the luminous body but by a drawing forth out of the potentiality of the matter of the air. . . .

2. John Pecham (ca. 1230–1292): The Nature and Multiplication of Light or Species

PERSPECTIVA COMMUNIS, REVISED VERSION, I, PROPOSITION 27

Every natural body, visible or invisible, diffuses its power radiantly into other bodies. The proof of this is by a natural cause, for a natural body acts

outside itself through the multiplication of its form. Therefore, the nobler it is, the more strongly it acts. And since action in a straight line is easier and stronger for nature, every natural body, whether visible or not, must multiply its species in a continuous straight line; and this is to radiate . . .

3. Roger Bacon: The Nature and Multiplication of Light or Species

DE MULTIPLICATIONE SPECIERUM,[12] II, CHAPTER 1

It is to be understood that lines along which multiplication [of species] occurs do not have length alone, extended between two points, but all of them [also] have width and depth, as the authors

plicatione specierum, I, ch. 1; Bridges, II, 409) that the two terms are usually employed interchangeably. [Compare Bartholomew the Englishman's discussion in Selection 60.3.—*Ed.*]

9. As will be seen further on, Bacon and his contemporaries regarded the optic nerve as hollow and believed that the species of objects were propagated to the brain through the spirits filling the nerve.

10. Bridges, II, 71–72.

11. For a full discussion of the nature and propagation of species, see my "Alhazen's Theory of Vision and Its Reception in the West," *Isis,* LVIII (1968), 321–341.

12. Bridges, II, 459–460.

7. That is, species pass through, but not in sufficient quantity to produce sight; however, they are reflected sufficiently to be observed.

8. Grosseteste was careful to distinguish between *lux* and *lumen* in his *De luce.* Here Bacon observes Grosseteste's distinction, but remarks elsewhere (*De multi-*

of books on sight determine. Alhazen demonstrates in his fourth book that every ray coming from a part of a body necessarily has width and depth as well as length.[13] Similarly, Jacob Alkindi says[14] that an impression is similar to that which produces it; now, the impressing body has three dimensions, the corporeal property of which [dimensions] is possessed by the ray. And he adds that a ray does not consist of straight lines between which are intervals, but multiplication is continuous, and therefore it will not lack width. And thirdly, he says that whatever lacks width, depth, and length is not perceived by sight; therefore a ray [if it lacks width and depth] is not seen, which is false. And we know that a ray must pass through some part of the medium, but every part of the medium has three dimensions.

4. Witelo (fl. 1250–1275): The Nature and Multiplication of Light or Species

PERSPECTIVA,[15] II, THEOREM 3

Every line along which light passes from a luminous body to a facing body is a natural sensible line, having a certain width, within which may be supposed an imaginary mathematical line.

Light proceeds only from body, since it exists only in body. Hence it is evident that in the least light that can be supposed there is width, since we call "the least light" that which, if divided, no longer has the reality *(actus)* of light, since it would not be visible; but both parts [in case of division] would be destroyed, since neither of them would be light or would appear to sense. Therefore, in a radial line along which light is diffused, there is some width, and in the middle of that line is an imaginary mathematical line, parallel to which are all the other mathematical lines in that natural line. And since the least light proceeds toward the least part of a body that light can occupy, it must proceed along the mathematical line which is in the middle of the sensible line and along the extreme lines parallel to the middle line. And the least light is not incident on a mathematical point of the facing body but on a sensible point corresponding to all the indivisible mathematical points on which the mathematical lines of the sensible line can terminate. Therefore, we employ the fantasy of advancing mathematical lines to demonstrate the properties of light.

5. Witelo: The Speed of Propagation of Light

PERSPECTIVA,[16] II, THEOREM 2

Unimpeded light is necessarily moved in an instant through the whole of a medium proportioned to it.[17]

Let there be a line proportioned to the diffusion of a strong light, as the diameter of the world in the case of solar light. Let this line be *ABCD*, and let there be a strongly luminous body at point *A* [Fig. 1]. If it should be said that the light is moved through line *ABCD* in time and not in an instant, then in part of that time it is moved through line *AB*, and in the least sensible time through the least sensible line *AB*; for if it were moved through an

13. Alhazen, *Perspectiva,* IV, sec. 16; Risner, p. 112.

14. Alkindi, *De aspectibus,* edited by A. A. Björnbo and Sebastian Vogl in *Abhandlungen zur Geschichte der mathematischen Wissenschaften,* Vol. XXVI, Part III (Leipzig: Teubner, 1912), ch. 11, p. 13.

15. Risner, pp. 63–64. Witelo wrote the *Perspectiva* no earlier than 1270.

16. Risner, p. 63.

17. Witelo's view that light is propagated instantaneously is opposed to the opinions of Alhazen and Roger Bacon. See Alhazen, *Perspectiva,* II, sec. 21; Risner, pp. 37–38. Also see Bacon, section 6 immediately following.

Fig. 1

insensible space in a sensible time, it would follow that a sensible space is composed of insensible [parts], just as the time measured after that space [*AB*] is composed of partial times that are sensible.[18] Therefore, light is moved in the least sensible time through the least sensible space. But in the same time, the form of a luminous body weaker than that more strongly luminous body is moved through the same space, since there is no sensible space smaller than the least sensible space and no sensible time less than the least sensible time. Therefore, stronger and weaker lights are of equal strength,[19] which is impossible, since they involve contradictories. Therefore, it is impossible for light to be diffused in time through a medium proportioned to it, and thus it is necessary for this diffusion to take place in an instant, which is the proposition. . . .

6. Roger Bacon: The Speed of Propagation of Light or Species

OPUS MAIUS,[20] V.1, DISTINCTION 9, CHAPTER 3

But here arises a great doubt concerning the species of the eye and of the visible object, whether they are produced suddenly and in an instant or in time; and if in time, whether in a sensible and perceptible time or not. Alkindi endeavors to demonstrate in his book, *De aspectibus*,[21] that a ray traverses [some space] in an altogether indivisible instant, and he adduces quite a curious and probable argument when he says that if a certain species, such as the light of the sun when it rises, is produced in time in the first part of the air, then if that time is doubled in the second equal part and tripled in the third part, when it reaches the west it would be a time composed of many parts, great in proportion to the first time; and although the first time would be insensible, nevertheless the whole time will be sensible because of its magnitude, which is nearly incomparable with respect to the first time. And Aristotle says in the second book of *De anima* that although the multiplication of light over a small distance can be hidden from our sense, it cannot be hidden in the case of a distance as great as that between east and west.[22] Therefore, if a species were produced in a certain time, this would be perceptible to sense. But we do not perceive it; therefore it is not produced in time but in an instant. And Aristotle says in his book *De sensu et sensato* that the reasoning regarding light is different from that regarding the other sensibles; and concerning these others he teaches that their multiplications are in time.[23] Therefore, the multiplication of light takes place in an instant.

And all authors express the same view except Alhazen, who tries to refute this view in his second book, arguing as follows.[24] Consider the last instant in which light is at the *terminus a quo* and the first in which it is at the *terminus ad quem*. Since the instants are different, as he endeavors to prove by experiences, there will be a middle time between them. And he says that every alteration takes place in time, and the medium and the eye are altered by species. But these arguments of Alhazen have no efficacy. The first is refuted elsewhere; for it is not always necessary to specify the last instant of a thing's existence at the *terminus a quo*, as universally occurs in the generation of permanent things, but it is necessary to specify the first instant of the *terminus ad quem*, as Aristotle teaches in the eighth book of the *Physics*.[25] Thus, when Socrates, from nonwhite, becomes white, it cannot be said that now finally he is not white, taking now as an instant, but [it can be said that] now he is first white; for he is not white during the whole time that measures the change, and he becomes white at the end of that time, namely at the instant which is its *terminus*, as Aristotle teaches and as is certain, although it is too difficult to understand unless carefully explained; but this is inquired into elsewhere. Alhazen's second argument is of no value [either], for all those who suppose the opposite deny that the multiplication of light is successive and temporal change.

Nevertheless, an irrefutable argument for the opinion of Alhazen can be drawn from his statements in the seventh book. For there he teaches that from the same origin a perpendicular ray arrives sooner at the end of the space than a non-

18. That is, as the sum of the sensible partial times is equal to the whole time required to traverse space *AD*, so the sum of the insensible spaces corresponding to those partial times must equal the sensible space *AD*; but, as Witelo implies, it is absurd to hold that a sensible space is composed of insensible parts.

19. That is, if any light (or form) is moved through the least sensible space in the least sensible time (the view Witelo is attempting to refute), then strong and weak lights move with equal rapidity; but this, Witelo apparently thinks, violates the assumption of unequal strengths.

20. Bridges, II,68–71.

21. Björnbo and Vogl, chapter 15, pp. 25–27.

22. *De anima* II.7.418b.20.

23. *De sensu et sensato*, 6. 446b. 25ff.

24. Alhazen, *Perspectiva*, II, sec. 21; Risner, pp. 37–38.

25. *Physics* VIII. 8.

perpendicular ray. But sooner and later do not exist without time, as Aristotle maintains in the fourth and sixth books of the *Physics*.[26] And this is demonstrated without possibility of contradiction, for no finite power acts in an instant, as Aristotle says in the sixth book of the *Physics;* and he proves this, since if it did, a greater power would act in less than an instant, which is impossible. But the power of the eye and of its species and of all created things is finite; therefore it cannot act in an instant. And at the end of the eighth book of the *Physics,* he holds that a finite and an infinite power cannot act in the same or equal durations, since then they could have equal effects, and thus they would be equal.[27] But it is the property of an infinite power to act in an instant. Therefore, a finite power cannot do anything in an instant but only in time.

Moreover, as an instant is to time, so is a point to a line. Therefore, by permutation, as an instant is to a point, so is time to a line. But the traversal of a point occurs in an instant; therefore the traversal of every line occurs in time. Therefore, species traversing a linear space, however small, traverse it in time. Furthermore, prior and posterior in space are the cause of prior and posterior in traversal of the space and in duration, as Aristotle says in the fourth book of the *Physics*.[28] Therefore, since the space through which a species is moved has a prior and posterior [part], its traversal must have a prior and posterior part in itself and in its duration; but prior and posterior in duration do not exist without time, since they cannot exist in an instant. And if it should be declared that this is true of those things that are measured by parts of space but that this excludes species (as is supposed), still it is nothing, since this second statement is applicable only to spiritual existence. Since the species of a corporeal thing has a real corporeal existence in the medium and is a true corporeal thing, as we have shown above, it must necessarily be dimensional and therefore adapted to the dimensions of the medium.

On the other hand, if in a single instant the species were produced throughout the whole medium, then it would simultaneously be at the *terminus a quo,* in the intermediate space, and at the *terminus ad quem.* But this is impossible for many reasons. First, it follows from this that a created thing would be simultaneously in several places, and by the same reasoning in an infinite number of places, as has already been considered in the chapter on matter. Therefore, it would have infinite power and would be God or equal to God. Secondly, it is argued from this that when a thing is at its *terminus a quo,* it is altogether at rest and is not moved in any way; and when it is at its *terminus ad quem,* the motion has already occurred, and the motion occurs between these *termini.* Therefore, the species is simultaneously at rest before translation, has completed its motion, and is actually being moved through the whole space. Therefore, it is simultaneously moved and not moved, which is contradictory, as Aristotle argues (reducing the matter to impossibility) in another case in the sixth book of the *Physics*.[29]

The final explanation of this is that inasmuch as the multiplication of light does not depend on any other motion, we may assume that the heaven is at rest and not moved, for the multiplication of light can occur readily with the heaven at rest, and this will take place at the end of the world if the heaven stands still, as is believed. Therefore, if the multiplication of light is instantaneous and requires no time, there will be an instant without time, since there is no time without motion. But it is impossible for an instant to exist without time, as it is for a point to exist without a line. Therefore, the remaining possibility is that light is multiplied in time, and likewise all species of a visible thing and of the eye. Yet this time is not sensible, perceptible by sight, but insensible, since everybody knows by experience that he does not perceive the time required for light to travel from east to west.

7. John Pecham: The Structure of the Eye

PERSPECTIVA COMMUNIS,[30] I, PROPOSITIONS 29–34

Proposition 29. The eye would be unsuited for the perception of size if it were not round.[31]

26. *Physics* IV. 11.
27. *Physics* VIII. 10.
28. *Physics* IV. 11.
29. *Physics* VI. 6.
30. The anatomical scheme described by Pecham in

these propositions is very close to that of Alhazen, Bacon, and Witelo. Compare Alhazen, *Perspectiva*, I, secs. 4–13 (Risner, pp. 3–7); Bacon, *Opus maius*, V, Distinctions 2–4 (Bridges, II, 12–30); Witelo, *Perspectiva*, III, Theorems 3–11 (Risner, pp. 85–90). For a more complete discussion of Pecham's views, see pp. 37 and 247–250 of my edition of the *Perspectiva communis,* cited in note 3. [See Mundinus' description of the eye, Selection 96.—*Ed.*]

31. The revised version of this proposition differs

Because of the many things that must be quickly perceived, the eye must be round so that it can easily move and turn. Furthermore, if the part through which the eye is acted upon were not spherical, the eye could see at a single glance only things equal [in size] to itself. This is evident, because vision occurs through straight lines incident perpendicularly on the eye[32] and intersecting at its center, as will be taught below. If the eye had a flat surface, only perpendiculars from a surface of equal size could reach it. For example, although impossible, let the eye have a plane surface *AB,* and let *CD* be the visible object [Fig. 2]. Then drop

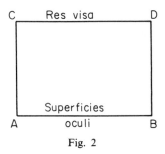

Fig. 2

a perpendicular from *B* to *D*. Likewise, draw another perpendicular from point *A* to point *C*. Since *AB* and *CD* are parallel (this may be assumed, since no absurdity follows from it), line *AC,* drawn perpendicularly by hypothesis, will be equal to line *BD;* consequently, line *BA* will equal line *DC.* That is, the thing seen cannot exceed the breadth of the eye. However, since this is false, it follows that the eye is not flat but more nearly spherical, and into its center rays can fall perpendicularly from a magnitude far greater [than the eye]. Moreover, roundness accords with interior capaciousness, because [the sphere] is the most capacious of all isoperimetric (that is, commensurable) figures.

Proposition 30. Some of the bodies constituting the eye must depart from complete sphericity.

For instance, if the consolidativa (that is, the white fat) that surrounds the eye were to surround it entirely, the eye would see nothing because the consolidativa lacks transparency. The uvea, too, has an opening in front, and the glacial humor[33] also departs from sphericity.

Proposition 31. Bodies of various dispositions are required for the constitution of the eye.

This is evident, for the part in which the power of sight resides[34] is very delicate and sensitive because it is transparent and watery and of the most delicate composition. If this were not so, it would be unsuited to the subtlety of the visual

spirits coming from the brain; yet neither would the species in any way admit of being subject to the immaterial and purified. Furthermore, the influence of species can be perceived only by a most subtle body. This humor, however, would be easily corrupted if it were not surrounded by other stronger ones. Hence the eye is so disposed that there is an outermost tunic called the consolidativa,[35] strong and thick so as to maintain the entire eye in its proper arrangement. Within the consolidativa is a tunic called the cornea, because it is like horn. The cornea is hard, [as it must be] because of its exposure to the air, and transparent in order to be pervious to species.

Within the cornea is the tunic called the uvea. The uvea is black like a grape to darken the humor in which [the power of] sight resides, for unless that humor is darkened, the visible species will not appear in it. This is a strong tunic to prevent the enclosed humor from exuding, and so that species may pass through, it has a circular aperture in front,[36] the diameter of which is about the size of the side of the square that can be inscribed within the sphere of the uvea.

Within the uvea is the albugineous humor,[37] which is like the white of an egg. It is transparent, so that species may be conducted through it freely, and damp, to moisten the glacial humor so that the web surrounding it will not be corrupted by dryness.

The glacial humor,[38] which is like glass, is the innermost humor. It is moist in order to be sensitive to light—by virtue not only of transparency but also of the possibility of sense—and it is subtle so that it might be easily stimulated [by species]. It is also somewhat dense to permit the retention of species within it, for they would otherwise die away. This humor is divided into two parts: It has an anterior part[39] that is a portion of a larger

---

somewhat from the unrevised text presented here; however, since Pecham's revised version experienced so little circulation, it has seemed appropriate to print the text of the more popular unrevised version.

32. See I, Proposition 28, in section 13 of this selection.
33. That is, the crystalline lens.
34. The glacial humor, or crystalline lens; see note 38.
35. That is, the sclera. For a medieval drawing of the eye, see p. 115 of my edition of Pecham's *Perspectiva communis,* cited in note 3.
36. That is, the pupil.
37. That is, the aqueous humor.
38. The term "glacial humor" was employed before to denote the crystalline lens. Here it denotes the combined crystalline lens and vitreous humor.
39. The crystalline lens.

sphere, and this part is concentric with the whole eye and parallel to the front of the eye; and it has a posterior part called the vitreous, more subtle than the anterior part. These two parts are surrounded by a certain fine web like a spider web, called the aranea;[40] the function of the aranea is to contain this moist humor. And so, according to the Physicist,[41] the eye has three humors and four tunics.

Others who investigate anatomy more attentively, however, consider (as it is observed in the book *De elementis*) that the uvea takes its origin from the pia mater and the cornea from the dura mater—the pia mater and dura mater being two webs that surround the cerebrum. This book adds that the eye is composed of three humors and seven tunics, the first of which is the conjunctiva or consolidativa; and it divides the cornea into two parts, calling the anterior part the cornea and the interior part the schrosim. It divides the uvea similarly, calling the anterior part the uvea and the posterior the secundina. Likewise the aranea is divided, the anterior part being termed the aranea and the interior part the retina. Thus to divide the eye, however, is not the concern of physics, which considers only what pertains to eccentricity or concentricity, refraction, and direction [of the rays].

Proposition 32. The duality of the eyes must be reduced to unity.

The benevolence of the Creator has provided that there should be two eyes so that if an injury befalls one, the other remains. However, their source is as follows. Two hollow nerves originate in the anterior part of the cerebrum and are directed toward the face. At first these are joined and form a single nerve, but then they branch into two [nerves] at the two hollow openings beneath the forehead; here they are spread apart, and on the ends of these nerves the eyes are formed. Accordingly, the species of visible things are received by both eyes, so that, if the species were not united, one thing would appear as two. This is evident also when a finger is placed under the eye and one eye is elevated from its [customary] position: one thing then looks like two because the species received through the two eyes are not joined in the common nerve. Therefore, [if vision is to be single,] species must be brought together in the common nerve and united there.

Proposition 33. It is necessary for some of the spheres constituting the eye to be mutually eccentric.

This is evident as follows. Since the species of the visible object is incident on the eye in the form of a pyramid, the vertex of which is to be imagined at the center of the eye, the rays converging at that center would intersect and proceed beyond if there were no change in transparency; and right would appear left, and left right. Hence nature has arranged for the interior glacial humor to have the same center as the cornea and albugineous humor, so that species traversing them should not be refracted before they reach the sensitive power residing in the glacial humor. . . .

Proposition 34. The centers of all the tunics and humors are contained on one line.

This is proved by an effect, for otherwise light could not enter all the tunics and humors in a regular manner, nor could any ray remain unrefracted. Consequently, there could be no certification[42] by the conveyance of the eye over the visible object from one end to the other, and that is false.

8. Alhazen (ca. 965–ca. 1039): The Lens as the Sensitive Organ of the Eye

PERSPECTIVA,[43] I, SECTION 16

And we would say in the first place that sight occurs only by means of the glacial humor, whether sight takes place through forms coming from the visible object to the eye or in some other way. Sight does not occur through one of the other tunics in front of it, since those tunics are merely instruments of the glacial humor; for if injury should befall the glacial humor, the other tunics remaining sound, sight is destroyed; if the other tunics should be corrupted, their transparency and the health of the glacial humor being retained, sight is not destroyed. And also, if there should be an obstruction in the opening of the uvea,[44] and the transparency of its humor were destroyed, sight would be destroyed along with the health of the cornea; and if the obstruction were removed, sight would return. Similarly, if a dense opaque fragment were placed within the albugineous humor in front of the glacial humor, between it and the opening of the uvea,

40. That is, the retina.
41. Alhazen.
42. On pecham's doctrine of certification, see I, Proposition 38, in section 13 below.
43. Risner, p. 8. Note that the headings of the sections (or enunciations of the propositions, if you will) in the Risner edition of Alhazen's *Perspectiva* did not originate with Alhazen but were added by Risner in the sixteenth century; therefore, they have not been translated.
44. That is, the pupil.

sight would be destroyed; and when that fragment is removed or is displaced in some direction from the direct line between the glacial humor and the opening of the uvea, sight will be restored. And medicine attests all these things. Therefore, the sense of sight is destroyed when the glacial humor is corrupted, the tunics in front of it remaining healthy. And this is the argument [to the effect that] the sense of sight is associated only with that [glacial] humor and not with the other tunics in front of it.[45]

9. Roger Bacon: The Lens as the Sensitive Organ of the Eye

OPUS MAIUS,[46] V.1, DISTINCTION 4, CHAPTER 2

The anterior glacial humor has many properties. The first and principal of these is that the visual power resides only in it, according to Alhazen and others. For all other humors anterior to it are its instruments and exist for its sake. For if the glacial humor should be injured, the others being preserved, sight is destroyed; and if the glacial humor is preserved, injury befalling the others (provided their transparency remains), sight is not destroyed but is still of service; for as long as the transparency within the glacial humor remains continuous with the transparency of the air, sight is not destroyed, provided the anterior glacial humor itself is preserved. And the anterior glacial humor is moist, so that it is more readily affected by the species of light and color, for very dry substances do not easily receive impressions. It is also subtle, since subtlety of body serves subtlety of feeling. And it is somewhat transparent, so that it can receive the forms of light and color, and they can pass through it to the common nerve. And it is [also] somewhat dense, so that it can retain species within itself for a long time, and they can appear to the visual power, and it can form a judgment. For if it were too transparent, species would pass through it [too readily] and would not remain so that a judgment could be formed. Therefore it must be somewhat dense, so that it may undergo a kind of pain on account of the species; for we see that strong lights and colors narrow the [pupil of the] eye and injure it and induce pain. But every action of light is of one nature (and the same is true of color), except that one [action] is stronger and another weaker; therefore, the eye always undergoes a kind of pain, although this is not always perceptible, as for example when the species are moderate. But painful effects cannot occur in a body unless it is quite dense, since if it is too rare, the species does not remain long enough to produce the effect of pain. . . .

10. Witelo: The Lens as the Sensitive Organ of the Eye

PERSPECTIVA,[47] III, THEOREM 4

And so it is evident from these things that the glacial humor is properly the organ of the visual power, for the transparency of only this humor is receptive of the forms of visible objects, and it is located in the middle of all the humors and tunics. And if any tunic or humor whatever, saving the glacial humor, should be injured, the eye always receives care by the aid of medicine and is healed, and sight is restored. However, when the glacial humor itself has been damaged, all of sight is destroyed without hope of restoration through the aid of medical treatment. And so principally the crystalline or glacial humor is the organ of the visual power, on which account it is carefully preserved.

11. Alhazen: The Act of Sight

PERSPECTIVA,[48] I, SECTIONS 25–26

Also it has been declared that this [glacial] humor

45. Knowledge of the ancient technique of couching for the removal of cataracts should easily have refuted Alhazen's claim that the visual power resides in the lens. However, in the Middle Ages cataracts were not regarded as corruptions of the lens but as obstructions intervening between the lens and the pupil; indeed, Alhazen's remark about displacement of an opaque fragment from behind the uvea may be a reference to couching. Alhazen's argument, though based on a false premise, does appear to demonstrate that the visual power cannot reside in any humor anterior to the lens; but it does not prove (as Alhazen seems to have realized) that a humor or tunic behind the lens cannot possess the visual power; nevertheless, Alhazen does not claim simply that the visual power resides in the glacial humor or some humor posterior to it, but that it resides in the glacial humor. Probably because of this stress on the glacial humor as the sensitive organ, it was never (during the Middle Ages) regarded as a device for focusing rays on the retina.
46. Bridges, II, 27–28.
47. Risner, p. 87.
48. Risner, pp. 15–16.

is somewhat transparent and somewhat dense; and in this respect it is like glass. Now, since it possesses some transparency, it receives forms, and these pass through it because of its transparency. And since it [also] possesses some density, it prevents forms from passing through it because of its density; and [thus] forms are fixed in its surface and body, although weakly. Similarly, when light is incident on any transparent body in which there is some density, the light passes into it because of its transparency and is fixed in its surface because of its density.[49] And also, the glacial humor is suited for the reception of those forms and for perceiving them. Therefore, forms pass into the glacial humor on account of its sensible power of reception; and when the form reaches the surface of the glacial humor, it acts on it, and the glacial humor suffers because of the form, since it is a property of light to act on the eye and a property of the eye to suffer because of the light. And this effect, which light produces in the glacial humor, passes through the glacial body. . . . And when light passes through the glacial body, color passes through with it, for color is mixed with the light; and the glacial humor receives that action [of light and color], and the color passes through. And the glacial humor perceives on account of this action and suffering, because of the forms of visible things that are on its surface and pass through its whole body; and it perceives through the ordering of the parts of the form on its surface and throughout its body. . . .[50]

And this action produced in the glacial humor by light is a kind of pain; for certain pains are capable of being received, and [yet] the member is not injured because of them. Such pains are not manifest to sense, nor does the latter judge them a painful affliction. And the meaning of this is that light induces pain, since strong lights offend the eye and manifestly injure it (as the light of the sun when the observer looks at its body, or the light of the sun reflected to the eye from smooth bodies); for those lights induce manifest pain in the eye. And the action of all light on the eye is of the same kind and is diversified only according to more and less. Since all are of one kind, and the operation of strong lights is a kind of pain, all actions of light are a kind of pain. . . .

12 Witelo: The Act of Sight

PERSPECTIVA,[51] III, THEOREM 6

Sight occurs through the action of visible forms on the eye and the suffering of the eye on account of this form.

That visible forms act on the eye is evident from the second and third suppositions [of this book]; for the eye is injured by strong light, as in observation of the solar body or some other strong light (such as light reflected to the eye from a polished body or from some other very white body). In these cases, the eye is disabled in such a way that it fails in its operation until it is restored by its natural intrinsic power. But sight also suffers from sensible forms, since it sometimes retains in itself their strong impressions. For after the eye gazes for a long time at bright light or color, if it should then look at a dark or weakly illuminated place, it would [still] perceive very well the visible object that it observed directly before, along with its light, color, and shape.[52] And sometimes a strong color impressed on sight is mixed with the colors of things seen in [relative] darkness, and those things appear colored, mixed with the other color, as a bright green object that is observed makes white objects seen afterwards in a darker place appear mixedly green; and if the eye is closed, the form seen previously nevertheless appears to sight. Therefore, visible forms act on sight, and sight suffers on account of them. And since the visibles *per se* are light and color, and light is the hypostasis of colors—and light is always diffused spherically to every different position—it is plain that colors are diffused similarly. Thus, when the eye is opposite an illuminated or colored object, light is multiplied either by itself or with the color of the object,

49. A substance without any density would offer no resistance to the passage of light; however, all real transparent substances possess some density.

50. This point is crucial to Alhazen's theory of vision. Clear vision results when the form of the visible object is ordered on the surface and throughout the interior of the glacial humor exactly as on the object itself; or to express the idea more precisely, an object becomes visible when the species representing its individual points (or small parts) fall on the glacial humor in the same order as the points they represent and from which they issued. A one-to-one correspondence is thus established between points on the object and points on the surface of the glacial humor. Compare Pecham, *Perspectiva communis*, I, Proposition 37, in section 13 below.

51. Risner, pp. 87–88.

52. Witelo is referring to the after-image. Alhazen also stressed the after-image at the beginning of his *Perspectiva* and elsewhere.

and arriving at the surface of sight,[53] it acts on sight, and sight suffers from it. When light and color come simultaneously to the surface of sight and act on it, and sight suffers because of them, and the power of the soul achieves understanding

because of the union of the visible forms and the soul's organ, then sight takes place on account of the presence of the visible forms acting on [the organ of] sight. . . .

13. John Pecham: The Act of Sight

PERSPECTIVA COMMUNIS, I, PROPOSITIONS 43, 28, 37, 33, 38

Proposition 43. The action of the visible object on the eye is painful.[54]

This is proved as follows. Because the action of the visible object on the eye is of a single kind and the action of bright lights on the eye is sensibly painful and injurious, it follows that all actions of light are painful and injurious, even if they do not seem to be. This is the argument of the Physicist[55] on the properties of sight, and it is seen to follow necessarily, since there is no visible object so delightful to the eye that when continuously observed it does not cause fatigue, the cause of which is clearly the preceding observation [of the object]. Now, the Physicist construes it this way, but philosophers who treat natural things say otherwise, [insisting that] since the sensible object is the perfection of sense, nothing induces pain in the act of perception unless it is excessive [in its action]. But the argument that if a superior sensible body induces pain, an ordinary one must do so also, is not compelling, for a violent action distresses while a moderate action delights and pleases. Therefore, what has been said here is limited to all prolonged observation and does not apply to every brief glance.[56]

Proposition 28. Sight occurs through lines of radiation perpendicularly incident on the eye.

This is obvious, for unless the species of the visible object were to make a distinct impression on the eye, the eye could not apprehend the parts of the object distinctly; and no distinction could be made among the partial species representing the parts of the object except by perpendicular lines, for otherwise the species would mingle together and would exhibit the object to sight confusedly. Besides, when perpendicular lines between the visible object and the eye have been intercepted, vision ceases. Accordingly, the opposite effect requires the opposite cause.[57]

Proposition 37. Vision takes place by the arrangement of the species on [the surface of] the glacial humor exactly as [the parts] of the object [are arranged] outside.

The possibility of this is obvious, notwithstanding the smallness of the glacial humor, because there are as many parts in the smallest of magnitudes as in the greatest. Species, however, are received without matter. Therefore, regardless of the size of the visible object, its species can be received distinctly and in [proper] order on [the surface of] the glacial humor. Unless this were so, the eye would not see the object distinctly; for if the species of two parts of the visible object were received in the same part of the glacial humor, the parts of the object would not be perceived distinctly by reason of the confusion of forms acting on the same part of the eye.[58]

53. *visus superficiem.* This could be translated "at the surface of the eye," but it appears to me from the context that Witelo has in mind the surface of the glacial humor, which contains the power of sight; hence, "the surface of sight."

54. This proposition is from the unrevised version of the *Perspectiva communis;* the revised version varies slightly.

55. That is, Alhazen. Pecham's reference is to the passage translated in section 11 above.

56. Here Pecham disagrees with Alhazen's conclusion that *every* action of light is painful.

57. In this proposition Pecham expresses one of the most original aspects of Alhazen's theory of sight (cf. Alhazen, *Perspectiva,* I, sec. 18; Risner, p. 9). A problem was created by the requirement that clear vision occurs only if species from the object fall on the glacial humor in the same order as the points from which they issued (see n. 50); for it was recognized that species issue in all directions from every point on the surface of the visible object, so that species or rays from every point on the object are incident on every point of the glacial humor. No one-to-one correspondence can thus be established. However, the problem is solved if all rays are ignored except those perpendicular to the surface of the eye and glacial humor: there is one such ray from each point on the visible object, and the complete set of such rays has the same arrangement on the surface of the glacial humor as on the visible object; moreover, these rays are rectilinear, since they are perpendicularly incident on the tunics of the eye and thus penetrate without refraction. Though not stated here, the principle which allows all nonperpendicular rays to be ignored is simply that refraction weakens, so that only perpendicular rays are powerful enough to influence the power of sight.

58. In this proposition Pecham deals with a further problem raised by Alhazen's intromission theory of

Proposition 33. . . . . Upon meeting the interior glacial humor, which is eccentric to [the cornea and albugineous humor], or the vitreous humor, which is more subtle than the anterior glacial humor, the rays are separated and refracted away from the perpendicular. From there the species is carried to the place of interior judgment by the way of spirits.[59]

Proposition 38. Visible objects are perceived by means of the pyramid of radiation; perception is certified by the axis [of the pyramid] being conveyed over the visible object.

To be sure, the pyramid of radiation, impressed on the eye by the visible object, manifests the object to the eye; but the visible object is certified by a turning of the eye [all about] over the object, the latter being the base of the pyramid. For although the whole pyramid is perpendicular to the center of the eye, that is, the anterior glacial humor, it is not perpendicular to the whole eye. Therefore, only that perpendicular called the axis, which is not refracted,[60] manifests the object efficaciously, and other rays are correspondingly stronger and better able to manifest [the object] as they are closer to the axis. Therefore, the eye is turned about so that the object, which is perceived under the pyramid all at once, is discerned efficaciously by appearing along this perpendicular successively. . . .

14. Alhazen: The Debate about Visual Rays

PERSPECTIVA,[61] I, SECTION 23

It is permissible for us to assert that a transparent body receives something from sight and transmits it to the visible object and that perception occurs through the unbroken succession of that thing to the visible object, between the eye and the visible object. And this is the view of those who suppose that rays issue from the eye. Therefore, [for the sake of argument,] let it be supposed that this is true, that rays issue from the eye and pass through the transparent medium to the object of sight and that perception occurs by means of those rays. And since, [according to the view now under consideration,] perception occurs in this way, I inquire whether those rays return something to the eye. If perception occurs [only] through [such] rays, and they do not return anything to the eye, then the eye does not perceive. But the eye does perceive the object of sight, and [we have supposed that] it perceives only by the mediation of rays. Therefore, those rays that perceive the visible object [must] transmit something to the eye, by means of which the eye perceives the object. And since the rays transmit something to the eye, by means of which the eye perceives the object, the eye does not perceive the light and color in the visible object unless something comes to the eye from the light and color in the object; and this is delivered by the rays.

Therefore, according to all possibilities, sight does not occur unless something of the visible object comes from the object, whether or not rays issue from the eye. It has already been shown that sight is achieved only if the body intermediate between the eye and the visible object is transparent, and it is not achieved if the medium be-

sight: If vision takes place through the action of perpendicular species from every point on the surface of the object converging towards an apex in the eye, how is it possible for these species to squeeze together as they approach the apex? Or, in mathematical rather than material terms, can there possibly be as many points (or small parts) on the tiny surface of the glacial humor as on the surface of a large visible object? The answers to these questions are, first, that species "are received without matter," so that squeezing is no problem, and second, that there are indeed as many points (or parts) in a small as in a large surface.

59. It is most important, in the theory of Alhazen and his thirteenth century followers, that the perpendicular species responsible for vision do not actually reach an apex and then reverse their order as they continue beyond the apex; for if such were the case, the visual impression would be inverted and the world would be seen upside down. (It should be noted that although the chief organ of vision is the glacial humor, vision is "completed" in the brain; thus it is necessary for the rays or species to maintain the proper order even after incidence on the surface of the glacial humor.) To explain why rays do not form an apex in the eye, Alhazen and his followers argued that the surface separating the glacial from the vitreous humor is situated anterior to the center of the eye, so that the species are refracted away from the center, thereby failing to reach an apex, and thus maintain their proper order as they progress through the vitreous humor and optic nerve to the "place of interior judgment" in the brain.

60. All rays perpendicular to the surface of the eye remain unrefracted until after they have passed through the surface of the glacial humor. However, only the central perpendicular ray (that is, the axis of the visual pyramid) is also perpendicular to the surface separating the glacial and vitreous humors, and this ray alone passes through the entire eye without refraction.

61. Risner, pp. 14–15. On the question of visual rays, see also Grosseteste's *On the Rainbow*, Selection 61.2, and Adelard of Bath's *Natural Questions*, Selection 60.1.

tween them is opaque. And it is evident that a transparent and an opaque body are distinguished only in the aforesaid way.[62] Since, as we have said and as has been demonstrated, the forms of the light and color in the visible object reach the eye (if they were [originally] opposite the eye), that which comes from the visible object to the eye (through which the eye perceives the light and color in the visible object no matter what the situation [with respect to visual rays]) is merely that form, whether or not rays issue [from the eye]. Furthermore, it has been shown that the forms of light and color are always generated in air and in all transparent bodies and are always extended to the opposite regions, whether or not the eye is present. Therefore, the egress of rays [from the eye] is superfluous and useless. Consequently, the eye does not perceive the light and color of the visible object unless form comes from the light and color [to the eye]. . . . .

It remains only to consider the view of those who maintain that rays issue from the eye and to indicate what in this view is false and what true. Therefore, let us assert that if sight is due to something issuing from the eye to the visible object, that thing is either corporeal or incorporeal. If it is corporeal, it follows that when we look at the sky and see the stars in it, corporeal substance issues from our eye in that time and fills the whole space between heaven and earth, and [yet] the eye is in no way destroyed; and this is [obviously] false. Therefore, sight does not occur through the passage of corporeal substance from the eye to the visible object. But if that which issues from the eye is incorporeal, it will not perceive the object, since there is no perception except in corporeal things. Therefore, nothing issues from the eye to the visible object to perceive that object.[63] Now, it is evident that sight occurs through the eye; and since this is so, and the eye perceives the visible object only when something[64] issues from the eye to the visible object, and since that which issues forth does not perceive the object, therefore that which issues from the eye to the visible object does not return anything to the eye, by which the eye perceives the object. And [therefore,] that which issues from the eye is not sensible but conjectural, and nothing ought to be believed except through reason or by sight.[65]

However, those who assume that rays issue from the eye suppose this because they ascertain that the eye perceives the visible object and that between the eye and the object is space; and it is known[66] by

mankind that there is no perception except by contact. Consequently, they have conjectured that sight does not occur unless something issues from the eye to the visible object, so that the thing issuing forth perceives the object in place of the eye, or rather receives something from the visible object and returns it to the eye; then the eye perceives that thing. And since corporeal substance cannot issue from the eye to perceive the object and nothing perceives a visible object unless it is corporeal, nothing remains except to conjecture that the thing issuing from the eye to the object of vision receives something from the object and returns it to the eye. And since it has been demonstrated that air and transparent bodies receive the form of the visible object and transmit it to the eye and to every facing body, that which they conjecture to return something from the visible object to the eye is nothing but air and the transparent bodies between the eye and the object of vision. Since air and transparent bodies transmit something from the visible object to the eye, regardless of the time and according to all arrangements (when the eye is opposite the visible object), without requiring that something issue from the eye, the reasoning that led from positing rays to maintaining their existence is unnecessary; for that which led them to say that rays exist is their supposition that sight cannot be completed except through something extended from the eye to the visible object, which returns something from the object to the eye. And since air and transparent bodies do this without requiring that something issue from the eye, and

62. That is, one allows the transmission of rays and thus the achievement of sight, while the other does not.

63. Note Alhazen's care in asserting that thus far he has proved only that visual rays (if they exist) are not responsible for visual perception, not that they are nonexistent. Alhazen, of course, was convinced that visual rays did not exist, as he makes clear later on, but the caution he displayed in this argument was to provide Western writers with room to maneuver. Thus Bacon, in an argument appearing in Selection 62.15, affirms the existence of visual rays and even assigns them a minor role in vision, while agreeing with Alhazen that the principal agents of vision are rays issuing from the observed object. This question is discussed in more detail in my "Alhazen's Theory of Vision," cited in note 6.

64. The Latin term *aliquid,* here translated, does not appear in the Risner text but is found in British Museum Royal MS. 12.G. VII, fol. 7 verso.

65. The last three words in this sentence *(vel a visu)* are included in the British Museum manuscript cited in note 64 but not in the Risner text.

66. The Risner text reads *magnum* but the British Museum MS reads *cognitum.*

moreover, since the air and transparent bodies are extended between the eye and the visible object without defect, it is useless to suppose that something else returns something from the object of vision to the eye. Therefore, it is useless to say that [visual] rays exist.

Moreover, all mathematicians who say there are

15. Roger Bacon: The Debate about Visual Rays

OPUS MAIUS,[68] v. 1, DISTINCTION 7, CHAPTERS 3–4

And if Alhazen, Avicenna (in the third book of *De anima*), and Averroes (in his little book *De sensu et sensato*) are cited as opposed to this view [that species issue from the observer's eye], I reply that they are not opposed to the generation of the species of the eye or to its role in sight but [only] to those who have supposed that some body, as a visible species or something similar, extends from the eye to the visible object, by means of which the eye perceives the object, and that it seizes the species of the visible object and returns it to the eye . . . . Therefore, it must be maintained that the aforesaid philosophers, namely Alhazen, Avicenna, and Averroes, attack only this view, as is evident from their texts. . . .[69]

The explanation of this view [that species issue from the eye] is that every natural thing completes its action solely through its own power and species, as the sun and other celestial bodies produce the generation and corruption of things through their powers dispatched into the things of the world; inferior things do likewise, as fire dries and consumes and does many things by its power. And therefore the eye must perform the act of sight through its own power. But the act of sight is the perception of a visible object at a distance, and therefore the eye perceives the visible object through its own power multiplied to the object. Furthermore, the species of the things of the world are not immediately suited of themselves to bring to completion an action on the eye because of the nobility of the latter. Therefore, they must be assisted and excited by the species of the eye, which proceeds through the space occupied by the visual pyramid, altering and ennobling the medium and rendering it commensurate with sight. Thus the species of the eye prepares for the approach of the species of the visible object and, moreover, ennobles the species of the object so that it is wholly conformable to and commensurate with the nobility of the animate body (that is, the eye). But since this view is doubtful to many, I shall adduce,

rays use nothing in their demonstrations except imaginary lines, and they call them radial lines; . . . and the belief of those who consider radial lines to be imaginary is true, and the belief of those who suppose that something [really] issues from the eye is false. . . .[67]

besides the verifications already presented, various true and certain experiences, according as they are relevant to other conclusions (which necessarily attend this view) in various places below.

Now, concerning the multiplication of this species, it must be recognized that it occurs in the same place, between the eye and the thing seen, as the species of the visible object; and it takes place along a pyramid, the apex of which is in the eye and the base of which is on the visible object. And just as the species of the object proceeds rectilinearly in a single [homogeneous] medium and is refracted in different ways when it encounters a medium of different transparency and is reflected when it meets the obstacle of a dense body, so the species of the eye proceeds along the same path as the species of the visible object. And although the species of the eye has the form of a pyramid, the apex of which is in the eye and the base of which is on all parts of the visible object, from the surface of the glacial humor [as base] proceed an infinite number of pyramids, all having the same base, and

67. There has been some confusion among historians about whether or not Alhazen really denied the existence of visual rays. The heading inserted by Risner (Alhazen's sixteenth-century editor) before the next section of Alhazen's text has caused most of the confusion: "Vision seems to occur through συναύγειαν, that is, rays simultaneously received and emitted." Directly beneath this heading, Alhazen's text reads: "It has been asserted on account of this that both schools of thought speak the truth and that both beliefs are correct and consistent; but one does not suffice without the other, and there can be no sight except through that which is maintained by both schools of thought" (I, sec. 24; Risner p. 15). However, what Alhazen means by this remark is simply that mathematicians, who are concerned to account for the phenomena mathematically, can legitimately use visual rays to represent the geometrical properties of sight. He does not compromise his denial of belief in the real existence of visual rays.

68. Bridges, II, 50–53.

69. With great insight Bacon observes that the attack on visual rays by the three Muslim philosophers mentioned actually overthrows only ancient visual ray theories like those of Plato and Ptolemy and that it does not apply to the theory he is about to propose.

their apices fall on individual points of the visible object,[70] so that all parts of the visible object appear as strongly as possible. Nevertheless, one pyramid is principal, namely the one whose axis is the line passing through the center of all parts of the eye (that is, the axis of the whole eye); for [the ray proceeding along] that [line] certifies everything, as has been said above and as will be explained more fully [below].

And although the species of visible objects, as the species of light and color, are mixed in a medium—that is, several lights are united and several colors are mixed, as has been said—and this species of the object and the species of the eye occupy the same common place, nevertheless there is no confusion of those species, nor mixing, nor do they unite; for they are not of the same kind nor of the same genus, since the pupil has no color, and color and light do not possess the power of soul. But the species of the eye is the species of an animate body, in which the power of soul rules, and therefore there is no comparison between it and the species of an inanimate thing so that they should be united. . . .

16. John Pecham: The Debate about Visual Rays

PERSPECTIVA COMMUNIS, I, PROPOSITIONS 44–46

Proposition 44. By assuming that sight occurs through rays issuing from the eye, mathematicians exert themselves unnecessarily.

For the manner in which vision occurs is adequately described above, by which [description] all the phenomena of vision can be saved. Therefore, it is superfluous to posit such rays. I say this following in the footsteps of the author of the *Perspectiva*.[71] Nevertheless, Alkindi teaches differently [in the treatise] *De aspectibus,* the Platonists have judged otherwise, and philosophers are seen in many places to understand differently. Augustine, who declares that the power of the soul has an effect on the light of the eye in a manner different from any that has hitherto been investigated, also teaches otherwise.

Proposition 45. Rays issuing from the eye and falling on a visible object cannot suffice for vision.

If it should be supposed that rays issue from the eye and fall on the visible object as if to seize it, either they return to the eye or they do not. If they do not return, vision is not achieved through them, since soul does not issue from body. If they do return, how do they do so? Are they animated? Are all visible objects mirrors by [virtue of the property of] reflecting rays? Furthermore, if the rays return to the eye with the form of the visible object, they go out in vain, since light itself (or the form of the visible object through the power of light) diffuses itself throughout the whole medium. Therefore, the visible object need not be sought out by rays as by messengers. Moreover, how would any power of the eye be extended all the way to the stars, even if the whole body were transformed into spirit?

Proposition 46. The natural light of the eye contributes to vision by its radiance.

For, as Aristotle says, the eye is not merely the recipient of action but acts itself, just as shining bodies do.[72] Therefore, the eye must have a natural light to alter visible species and make them commensurate with the visual power, for the species are emitted by the light of the sun but moderated with respect to the eye by mixing with the natural light of the eye. This is why Aristotle said that sight occurs when the outward action [of the eye] is strong, [for] when the action entering the eye is strong, as with a solar ray, it overpowers sight; nor, of itself, can this inward action be made commensurate with sight. It is evident, therefore, that there is some kind of emission of rays, but not of the Platonic type such that rays emitted by the eye are, as it were, immersed in the visible form and then returned to the eye as messengers. But rays [emitted by the eye] do have some effect on sight in the manner described above. This is evident as follows. Since vision is of the same kind in all animals, and certain animals are able to bestow a multiplicative power on colors by the light of their eyes so as to see them at night, it follows that the light of the eye has some effect on [external] light. Whether it goes beyond this, I do not determine, save only by following in the footsteps of the Author,[73] as I have said before.

70. These pyramids are reversed with respect to the principal visual pyramid, which has its apex in the observer's eye and its base on the visible object. That is, every point on the object serves as the apex of a pyramid of rays, the base of which is the surface of the glacial humor.

71. That is, Alhazen, in the passage translated in section 14 above.

72. *De generatione animalium* V.1.780a.5–15, 781a.1–10.

73. Alhazen. But Pecham was also influenced by Bacon, as his reconciliation of the existence of visual rays with Alhazen's attack on visual rays reveals.

17. Witelo: The Debate about Visual Rays

PERSPECTIVA,[74] III, THEOREM 5

It is impossible for sight to be applied to the visible object by rays issuing from the eyes.[75]

If from the eyes should issue certain rays, by which the visual power is united with external objects, those rays are either corporeal or incorporeal. If corporeal, then when the eye sees stars and the sky, something corporeal issuing from the eye necessarily fills the entire space of the universe between the eye and the visible part of the sky, without diminution of the eye itself. But it is impossible that this should occur and also that it should occur so swiftly, the substance and size of the eye being preserved. If it should be said [instead] that the rays are incorporeal, then those rays do not perceive the visible object, since perception exists only in corporeal things. Therefore, the corporeal eye cannot perceive by the mediation of this insentient incorporeal ray. Nor do such incorporeal things return something to the eye, by which sight could perceive the visible object, since

sight occurs only through contact between the eye and the visible form, because there is no action without contact. Therefore, if rays issuing from the eye return nothing to the eye, those rays do not produce sight. But if they return something to the eye, these are lights or colors, which appear by themselves [without the mediation of rays issuing from the eye] and which are multiplied to the eye among the [visual] rays. Therefore, [visual] rays do not cause sight to be applied to the things seen; but something else, which is multiplied to the eye, is by itself the cause of vision. It is therefore impossible for rays of themselves to cause vision, unless perhaps the lines drawn through the points of the forms multiplied from the surfaces of the visible object to the eye are called rays; for, as is evident by Theorem 2 of this book, in order that the object may actually appear, it must be possible to draw straight lines between any point on the surface of the visible object and a given point on the surface of sight. But such rays do not issue from the eyes.[76] Therefore, the proposition is evident.

18. Roger Bacon: Psychology of Visual Perception

Translated by Robert B. Burke[77]

OPUS MAIUS, V.1, DISTINCTION 1, CHAPTERS 2–4

Chapter 2. Concerning the internal faculties of the sensitive soul, which are imagination and the common sense (sensus communis).[78]

Since the optic, that is, the concave, nerves causing vision have their origin in the brain, and writers on optics ascribe, to a discriminative faculty through the medium of vision, the formation of judgments concerning twenty kinds[79] of visible things, which will be considered later, and since it is not known whether that discriminative faculty is among the powers of the soul, the organs of which are distinct in the brain, and since many other things to be treated later suppose a definition of the faculties of the sensitive soul, therefore we must begin with the parts of the brain and the faculties of the soul, in order that we may discover

74. Risner, p. 87.

75. In the thirteenth century only Witelo followed Alhazen in denying the existence of visual rays.

76. Witelo here distinguishes between rays actually issuing from the observer's eye and imaginary lines that can be drawn between points in the eye and points on the visible object. This echoes Alhazen's similar distinction; see note 67.

77. *The Opus Majus of Roger Bacon* (Philadelphia: University of Pennsylvania Press, 1928; reprinted, New York: Russell and Russell, 1962), II, 421–427. For this one long selection, it seemed preferable to reproduce Burke's translation of the *Opus maius* rather than to retranslate the Bridges text. However, a number of changes have been made in the interests of accuracy and clarity. First, the paragraphing and punctuation have been slightly modified. Second, a number of expressions have been given different translations; the most common changes are as follows: *virtus distinctiva* has been translated "discriminative faculty" instead of "distinct function"; *virtus* has been translated "power" or "faculty" instead of "function" or "impression"; *species* has been rendered as "species" instead of "form"; *sensibile,* which Burke renders as "sensation," "impression," "attribute," "property," or "quality," has in every case been rendered as "sensible"; and *collocantur sub* has been translated as "are subordinate to" instead of "are placed under" or "are under." Finally, Burke's translations of Bacon's citations of other works have been rephrased. Other modifications are noted as they appear.

78. On the internal faculties of the soul, see Harry A. Wolfson, "The Internal Senses in Latin, Arabic, and Hebrew Philosophic Texts," *Harvard Theological Review* XXVIII (1935), 69–133.

79. Instead of "kinds," Burke has "species," which is ambiguous because of its double meaning in optical contexts.

those things that are necessary for vision. Writers on optics give us a means to this end by showing how the visual nerves descend from the membranes of the brain and from the lining of the cranium, but no one explains all things necessary in this matter.

I say, then, as all writers on the subjects of nature, all physicians and authorities on optics agree, that the brain is enfolded in a double membrane, one of which is called the pia mater, which enfolds the brain by direct contact; and the other, the dura mater, which adheres to the concave side of the bone of the head called the cranium. For this latter membrane is harder, so that it may resist the bone, and the other is softer and tenderer owing to the softness of the brain, the substance of which is like marrow and ointment, with phlegm as the chief constituent, and with three distinctions, which are called chambers, cells, parts, and divisions. In the first cell there are two faculties: The one in the anterior part is the common sense, as Avicenna states in the first book of *De anima,* which is like a source[80] with respect to the particular senses, and like the center with respect to the lines extending from that same point to the circumference, according to Aristotle in the second book of *De anima.*[81] This common sense judges concerning each particular sensible. For the judgment is not completed in regard to what is seen before the species comes to the common sense, and the same is true in regard to what is heard and to the other senses, as is clear from the end of the work *De sensu et sensato*[82] and in the second book of *De anima.* This common sense forms a judgment concerning diversity in the sensibles[83] as, for example, that in milk whiteness is different from sweetness, a distinction which sight cannot make, nor taste, because they do not distinguish things in other categories, as Aristotle maintains in the second book of *De anima.* It judges concerning the operations of the particular senses, for vision does not perceive that it sees, nor hearing that it hears, but another faculty does, namely, the common sense, as Aristotle maintains in the second book of *De somno et vigilia.*[84] But the final action of this faculty is to receive the species coming from the particular senses and to complete a judgment concerning them. But it does not retain these impressions, owing to the excessive slipperiness of its own organ, according to the statement of Avicenna in the first book of *De anima.*

Therefore, there must be another faculty of the soul in the back part of the first cell, the function of which is to retain the species coming from the

particular senses, owing to its tempered moistness and dryness, which is called imagination and is the coffer and repository of the common sense. Avicenna cites as an example a seal, the image of which water readily receives but does not retain, owing to its superabundant moistness; wax, however, retains the image very well, owing to its tempered moistness with dryness. Wherefore he says that it is one thing to receive and another to retain, as is clear from these examples. Such is the case in the organ of the common sense and of imagination.

The whole faculty, however, composed of these two, namely that which occupies the whole first cell, is called phantasia or the *virtus phantastica.* For according to the second book of *De anima* and *De somno et vigilia* and the book *De sensu et sensato,* it is evident that phantasia and the common sense are the same according to subject but differ according to being, as Aristotle says, and that phantasia and imagination are the same according to subject but differ according to being. Wherefore phantasia includes both faculties and does not differ from them except as the whole from the part. Therefore, since the common sense receives the species, and imagination retains it, a complete judgment follows regarding the thing, a judgment formed by phantasia.

Chapter 3. Concerning the sensibles that are apprehended by special senses, by the common sense, and by imagination.

We must note that imagination and the common sense and any particular sense form judgments by[85] themselves only concerning twenty-nine sensibles; as sight concerning light and color; touch concerning heat and cold, moistness and dryness; hearing concerning sound;[86] smell concerning odors; taste concerning savor. These are the nine proper sensibles[87] that belong to their own senses, as I have named them, of which no other particular sense can form a judgment. There are, moreover, twenty

80. The Latin term is *fons,* which Burke translates "fountain."

81. The proper reference, as indicated by Bridges, is *De anima* III.1–2.

82. See *De sensu et sensato,* 7.449a.

83. ". . . de diversitate sensibilium," which Burke translates "concerning differences of impressions on the senses."

84. *De somno et vigilia,* 2.455a.15–25.

85. Burke has "of," which is misleading.

86. Burke has failed to translate the words *auditus de sono.*

87. . . . *propria sensibilia,* translated "special properties" by Burke.

other sensibles, namely, distance, position, corporeity,[88] figure, magnitude, continuity, discreetness or separation, number, motion, rest, roughness, smoothness, transparency, thickness, shadow, darkness, beauty, ugliness, also similarity and difference in all these things and in all things composed of them. Besides these qualities there are some that are subordinate to one or more of these qualities, as order to position, and writing and painting to figure and order. Further examples are straightness and crookedness, and concavity and convexity, which are subordinate to figure; multitude and fewness, which are subordinate to number; equality, augmentation, and diminution, which are subordinate to similarity and difference; eagerness, laughter, and sadness, which are apprehended from the form of the face together with the shedding of tears; and moistness and dryness, which are subordinate to motion and rest. . . . The same is true concerning many other sensibles, which are reduced to the species and principal modes enumerated above belonging to visible things. All these matters are explained in the first book of Ptolemy's *Optica* and in the second book of Alhazen's *De aspectibus*,[89] and in other authors on optics.

There are, moreover, common sensibles, some of which Aristotle defines in the second book of *De anima* and in the beginning of his work *De sensu et sensato*, as, for example, magnitude, figure, motion, rest, and number.[90] These are not the only common sensibles but also all those mentioned before, although most writers on the subjects of nature do not consider this fact, because they are not expert in the science of optics. For common sensibles are not so called because they are perceived by the common sense but because they are determined by all the special senses or by several of them, and particularly by sight and touch, since Ptolemy states in the second book of his *Optica* that touch and sight participate in all these twenty. These twenty-nine, with those that are reduced to them, are apprehended by the special senses, and by the common sense, and by imagination, and these faculties of the soul cannot judge of themselves concerning other sensibles except by accident.

Chapter 4. Concerning the investigation of the estimative, memorative, and cogitative faculties.

But there are other sensibles *per se*, for animals use sense alone, since they do not possess intellect. The sheep, even if it has never seen a wolf, flees from it at once; and every animal experiences fear

at the roaring of a lion, although it has never heard a lion before nor seen one. The same is true in regard to many things that are hurtful and contrary to the constitution or complexion of animals. The same principle holds good as regards what is useful and in conformity with their natures. For although a lamb may never have seen another lamb before, it runs to one and willingly remains with it, and the same is true concerning other animals. Brutes, therefore, have some perception in things advantageous and in things harmful. There is, then, something sensibly in them besides the twenty-nine sensibles mentioned above and besides those that are reduced to them. For there must be something more active and productive of change in the sentient body than light and color, because it not only causes comprehension[91] but also a state of fear or love or flight or delay. This is the property of the complexion belonging to each object by which it is assimilated to others in a nature special or general. Through this quality things agree, are strengthened and invigorated, or differ and oppose one another and are mutually harmful. Wherefore not only do light and color produce their species and powers, but to a far greater degree do the complexional qualities,[92] nay, the very natures of things as regards their substance, agreeing or disagreeing with one another, produce strong species, which change greatly the sensitive soul, so that it is moved to states of fear, horror, flight, or the opposites.

These species or powers coming from things, although they change and alter special senses and the common sense and imagination, just as they do the air through which they pass, yet no one of these faculties of the soul judges concerning these impressions, but of necessity a far nobler and more powerful faculty of the sensitive soul does, which is called estimation or the estimative faculty, as Avicenna states in the first book of *De anima*, a faculty which he says perceives the insensible forms connected with sensible matter. Sensible matter is spoken of here as that which is apprehended by the special senses and by the common sense, as are

88. Burke fails to translate *corporeitas*.
89. Ptolemy discusses these matters at the beginning of Book II of the *Optica*. See Alhazen, *Perspectiva*, II, sec. 15; Risner, p. 34.
90. *De anima* II.6.418a.15–20; *De sensu et sensato*, 1.437a.5–10.
91. Burke has "apprehension," which is ambiguous in the present context.
92. . . . *qualitates complexionales*, which Burke has rendered as "properties of complexions."

the aforesaid twenty-nine. We call insensible form that which is not taken cognizance of by those senses of themselves, since they are commonly called senses, although other faculties of the sensitive soul may equally well be called senses, should we wish so to name them, because they are parts of the sensitive soul. For every part of the sensitive soul can be called a sense, because it is in truth a sense, and a sensitive faculty. The statement, therefore, that qualities belonging to complexions are not apprehended by a sense must be understood as applying to a special sense and the common sense and imagination; but they can readily be apprehended by estimation, which although not called a sense is, however, a part of the sensitive soul.

But estimation does not retain a species, although it receives it like the common sense, and it therefore requires another faculty in the remotest part of the posterior cell to retain species coming to the estimative faculty and to be its storehouse and repository, just as imagination is the storehouse of the common sense. This is the memorative faculty, as Avicenna states in the first book of *De anima.*

Cogitation or the cogitative faculty is in the middle cell and is the mistress of the sensitive faculties. It takes the place of reason in brutes, and is therefore called the logical, that is, the rational faculty, not because it employs reason but because it is the ultimate perfection of brutes, just as reason

is in man, and because the rational soul in man is united directly with it. By this faculty the spider weaves its geometrical web, and the bee makes its hexagonal house, choosing one of the figures that fill out space, and the swallow its nest. The same is true of all the works of brutes that are similar to human art. Man by means of this faculty sees wonderful things in dreams, and all the faculties both posterior and anterior of the sensitive soul serve and obey it, because they all exist on account of it. For the species[93] that are in the imagination multiply themselves into the cogitative faculty, although they exist in the imagination according to their nature primarily because of phantasia, which uses those species; but the cogitative faculty holds those species in a nobler way, and the species of the estimative and memorative faculties exist in the cogitative faculty in accordance with a nature nobler than that existing in those faculties, and therefore the cogitative faculty uses all the other faculties as its instruments.

In man there is in addition from without and from creation the rational soul, which is united with the cogitative faculty primarily and immediately, and uses this faculty chiefly as its own special instrument. Species are formed in the rational soul by this faculty. Wherefore, when this faculty is impaired, the judgment of reason is especially perverted, and when it is in a healthy condition the intellect functions in a sound and rational way.

19. John Pecham: The Geometry of Reflection

PERSPECTIVA COMMUNIS, II, PROPOSITIONS 6, 20, 30

Proposition 6. The angles of incidence and reflection are equal, and the incident and reflected rays are in the same plane as the line erected [perpendicularly] at the point of reflection.

The angle formed by the incident ray and either the surface of the mirror (on one side) or the imaginary line erected [perpendicularly] at the point of reflection (on the other side) is called the angle of incidence; the angle formed by these [that is, the surface of the mirror and the imaginary perpendicular] and the reflected ray is called the angle of reflection. Equality of the angles is gathered from experience and proved by reason in any of several ways; for if an incident ray could advance into the depth of a mirror, it would form (with the perpendicular extended into the depth [of the mirror] at the point of reflection) an angle equal to the angle of incidence because, according

to Euclid, vertical angles are equal.[94] Therefore, the ray recoils in the same mode as that in which it would be transmitted [if it were not deflected by a reflecting surface], and consequently it must rebound at an equal angle [to the angle of incidence].[95] Accordingly, if it is incident on the mirror perpendicularly, it is reflected back on itself; if it is incident obliquely, it is reflected obliquely toward the other side. The same thing is evident in the motion of a body, since a heavy body descending vertically onto a solid body or

93. Burke has "forms or species," but the Latin text simply reads *species.*
94. *Elements* I.15.
95. Note that although Pecham indicates that equality of angles can be gathered by experience, he argues from the principle of symmetry. Doubtless the experiences or experiments to which he referred were those described by Alhazen in the *Perspectiva,* IV, secs. 7–12; Risner pp. 104–109.

projected perpendicularly along a line is driven back along the same line; if projected obliquely, it rebounds along a similar line on the other side. . . . [96]

It is evident also that the three lines [that is, the incident ray, the reflected ray, and the perpendicular] are in the same plane, because a ray conforms as closely as possible to a rectilinear path, since straightness is natural to light. However, if a ray were to forsake that plane, it would be departing doubly from straightness, both by rebound and by deviation.

Proposition 20. In plane mirrors, and for the most part in others, the images appear at the intersection of the ray and the cathetus.

The cathetus is the line dropped perpendicularly from the visible object to the surface of the mirror, whether plane or spherical. Indeed, that which is seen in a mirror appears to be located at the imaginary intersection of the ray under which the object is seen and the perpendicular dropped from the object to the surface of the mirror. This can be explained by reference to Proposition 67 of Part I, since [by that proposition] the lengths of rays are perceived by the eye. But because the reflected part of the ray acts directly on sight and sight occupies itself with that part, through its mediation the part of the ray incident on the mirror is so perceived that the whole ray is apprehended by the eye as though proceeding without interruption in a straight line; for reflection cannot be discerned by the eye, which perceives nothing except the part of the ray that [actually] conveys the qualities [of the visible object] to sight. Therefore, an object seen in a mirror, if above the mirror, must appear beneath it at the imaginary intersection of the ray and the cathetus. [97]

For example, let *ABG* be the mirror, *CK* the visible object, and *D* the eye of the observer [Fig. 3]. Rays *KA* and *CB* fall from the visible object and are reflected to the eye along rays *AD* and *BD*. Consequently *KA* appears to be the [direct] continuation of *AD*, and *CB* of *BD*; it follows that *KA* and *CB* seem to be directed behind the mirror under the same angles at which they are reflected, because vertical angles are equal, and *DA* falls to *E* and *DB* to *F*. Furthermore, the object appears on the aforementioned perpendicular, that is, the cathetus, exactly as in its true location. . . .

Proposition 30. In convex spherical mirrors the image appears at the intersection of the ray and the cathetus, that is, the line dropped [from the object] to the center of the sphere.

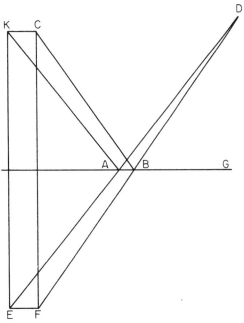

Fig. 3

This can be proved by experience and [can be shown to result] from natural causes, as it was for plane mirrors above. Nevertheless plane mirrors and convex spherical mirrors differ in that an object always appears as far below a plane mirror as it actually is above it, whereas in a convex spherical mirror the image appears sometimes on the surface of the mirror, sometimes inside the mirror, and sometimes outside it. [98] For example, let *E* be the visible point, *G* the eye, *N* the point of reflection, and *D* the center of the sphere [Fig. 4]. It is clear that the image is located at *K*. However, if the visible object should be placed at *B*, the image would appear at *O*; if the visible object should be placed still closer to the sphere, the image would

96. It must be asserted emphatically that Pecham is not implying that light is corpuscular; the mechanical example employed here is intended to elucidate only the equality of angles, not the nature of light.

97. The essence of Pecham's argument is that the eye knows the length of the ray by which the object is seen but is not aware of the bend; consequently, the object appears to be along the backward rectilinear extension of the reflected ray at its intersection with the cathetus. This makes ray *ED* in Fig. 3 equal to the sum of rays *KA* and *AD*, which is in accord with the eye's ability to determine the length of the entering ray. It should be noted, however, that this equality holds only for plane mirrors.

98. Although it is impossible for the image to be located anywhere except *behind* a convex mirror, Pecham

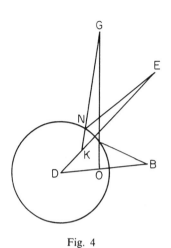

Fig. 4

appear outside the sphere, as will be evident by careful investigation.

The point of reflection is easy to find in these [mirrors] when the eye and visible object are at exactly equal distances from the sphere. Otherwise the length [of time required] to find the point is greater than the difficulty or utility, as can be seen by examining the section [in Alhazen's *Perspectiva*] on images.[99] From that section also it is apparent that the image in convex spherical mirrors is nearer the mirror than is the visible object, [a relationship] which is in contrast to [that for] plane mirrors as shown above.

20. Witelo: A Problem of Image Formation by Reflection

PERSPECTIVA,[100] VIII, THEOREM 53

When the lines of incidence [drawn from the endpoints of some object] intersect each other in concave spherical mirrors, oblique lengths situated inside the point of intersection[101] appear as they [truly] are; but those that are outside the intersection of the lines appear as reversed images.

Let there be a concave spherical mirror *AG*, with center *M;* let *B* be the center of the eye, and let *DE* be a line situated obliquely above the surface of the mirror [Fig. 5]. Let the form of point *D* of this line be reflected to the eye *B* by point *A* of the mirror, and the form of point *E* by point *G*. Let the lines of incidence (that is, *DA* and *EG*) intersect one another at point *I;* and inside point *I,* situated obliquely with respect to the surface of the mirror, let there be line *KC,* point *K* of which is reflected from point *G* of the mirror and point *C* from point *A* of the mirror.[102] Thus, from point *D* to the center of the mirror, draw line *DM*, which, because of the obliquity of line *BA* with respect to the surface of the mirror (since line *DM* is perpendicular to the

is correct in asserting that the image may be inside, outside, or on the surface of the mirror. Consider the convex spherical mirror in the figure, with the visible object situated at *O* and the observer's eye at *E*. The image is located at the intersection of the cathetus and the rectilinear extension of the reflected ray, *I*, which is outside, though *behind,* the mirror.

99. V, sec. 39; Risner, pp. 150–151. This problem of determining the point of reflection, given the locations of the visible point and the observer's eye, is known as "Alhazen's problem." For a good discussion of "Alhazen's problem," see Paul Bode, "Die Alhazensche Spiegel-Aufgabe in ihrer historischen Entwicklung nebst einer analytischen Lösung des verallgemeinerten Problems," *Jahresbericht des physikalischen Vereins zu Frankfurt* (1891–92), pp. 63–107; cf. Sabetai Unguru, "Witelo as a Mathematician: A Study in XIIIth Century Mathematics, including a Critical Edition and English Translation of the Mathematical Book of Witelo's *Perspectiva*" (unpublished doctoral dissertation, University of Wisconsin, 1970), pp. 368–375. Much of the rest that has been written on the problem is misleading.

100. Risner, p. 355.

101. That is, between the point of intersection (*I* in Fig. 5) and the mirror. The collection of points such as *I* for all positions of the object and the eye constitutes the focal plane of the mirror, and although Witelo does not employ modern terminology in describing this plane, he is correct in asserting that objects situated outside it appear inverted or reversed, while those situated inside the focal plane appear erect or unreversed. This proposition well illustrates the advanced level to which geometrical optics was brought in the thirteenth century.

102. For ease of demonstration, Witelo has arranged lines *DE* and *KC* so that both are observed by means of the same incident and reflected rays (Fig. 5). But he need not have done so; any line situated above *I* (that is, outside the focal plane) will have an inverted image, and any line situated below *I* will have an erect image.

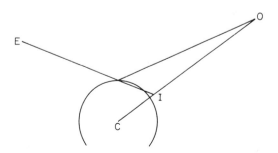

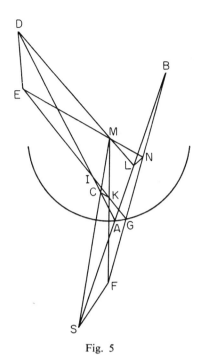

Fig. 5

cident obliquely on the surface of the mirror, as Book I, Theorem 14, of this treatise is able to reveal; and let the point of intersection be *L*. Similarly, line *EM* intersects line *BG*; let the point of intersection be *N*. Thus it is evident, by Book V, Theorem 37, of this treatise, that the image of the form of point *D* is located at point *L*, and the image of the form of point *E* at point *N*.[103] Draw line *NL*, which will be the image of the whole line *DE*. Now image *NL* has reversed itself with respect to line *DE*, since point *N*, the image of the lower point *E*, is higher than point *L*, the image of the higher point *D*.

Also, let line *MK* be extended until it intersects the extension of line *BG*—they intersect because of the obliquity of line *BG* with respect to the surface of the mirror and the perpendicularity of line *MK*—and let the point of intersection be *F*; extend line *MC* until it intersects the extension of line *BA*, and let the point of intersection be *S*. Let line *FS* connect [these points of intersection]; then line *FS* will be the image of line *KC*, and as point *K* is higher than point *C*, so point *F* is higher than point *S*. Thus image *FS* has a position conforming to that of the visible object *KC*, situated in front of the mirror inside the point of intersection, *I*, of the incident lines. Therefore, the proposition is evident.

same surface of the mirror by Book I, Theorem 72, of this treatise, because it passes through the center *M* of the mirror), intersects line *BA*, which is in-

21. Alhazen: Paraboloidal Burning Mirrors

ON BURNING MIRRORS[104]

One of the most exquisite things invented by geometers, about which the ancients were concerned, in which the most excellent properties of geometrical figures are apparent, and which follows (according to them) from natural things, is the construction of mirrors that produce combustion through the reflection of a solar ray. Now, in inventing these mirrors, they employed various means. That is, they discovered that a ray is reflected from the surfaces of plane mirrors and also from the surfaces of spherical mirrors, and that the places to which the ray is reflected vary with the size of the various dimensions of the mirror. However, it became clear to them that a ray [proceeding from a given point on an object and] reflected from a plane mirror is reflected to one point from only one point [of the mirror] and that a ray reflected from a spherical mirror is reflected to one point from the circumference of only one of the circles lying in that sphere. And demonstrations of this are presented in their books. Thus some of them attempted to employ many plane mirrors

joined together, from all of which rays would be reflected to one point, so that combustion would be strong. And those who invented these mirrors

103. Note that *DM* is the cathetus, since it is dropped from the visible point *D* perpendicular to the surface of the mirror, and *AB* is the reflected ray; the image of *D* is thus situated at intersection *L* of these two lines. The same principle is used to determine the location of the images of points *E*, *C*, and *K*.

104. Translated from the text edited by J. L. Heiberg and Eilhard Wiedemann, "Ibn al Haitams Schrift über parabolische Hohlspiegel," *Bibliotheca Mathematica*, Ser. 3, X (1910), 201–237. In the same article Heiberg and Wiedemann present a German translation based on both the Arabic and Latin texts, which texts, as their translation reveals, do not diverge significantly. In translating the excerpts that follow, I have consulted both this German translation and the English translation of an Arabic manuscript by H. J. J. Winter and W. 'Arafat, "Ibn al-Haitham on the Paraboloidal Focussing Mirror," *Journal of the Royal Asiatic Society of Bengal*, Ser. 3; Vol. XV, Nos. 1–2 (1949), pp. 25–40. This shorter work by Alhazen was translated into Latin probably by Gerard of Cremona or one of his school; in any case, it was known by the second half of the thirteenth century and became very popular during the later Middle Ages.

were famous men, such as Archimedes, Anthemius,[105] and others.

Then it occurred to them to consider the properties of the figures from which the ray is reflected. Thus they investigated the properties of conic sections[106] and discovered that the rays incident on the entire concave surface of a paraboloid are reflected to one and the same point. It thus became evident that the combustion produced by a mirror of this shape is stronger than the combustion produced by a mirror of any other shape. Nevertheless, they did not present an adequate demonstration of these facts, nor of the method by which they discovered them. Therefore, since in this matter there is great benefit and general utility, I have taken care to explain it and to reveal it, so that he who desires to know truth will have knowledge of it, and he who is concerned with the swiftness *(velocitatibus)*[107] of things will understand it. . . .

Premises agreed upon: A solar ray issues from the body of the sun to the surfaces of all kinds of mirrors along straight lines. All rays incident on plane mirrors are reflected from the surfaces[108] at equal angles. All rays incident on concave or convex mirrors are reflected from the surfaces at equal angles, formed with the plane surfaces tangent to those surfaces at the points of incidence. By the ray reflected at equal angles,[109] I mean that the reflected ray forms two equal angles with the straight line that is the common difference[110] between the surface of the two straight lines (that is, the incident and reflected rays) and the plane surface (that is, the surface of a [plane] mirror or the [plane] surface tangent to a concave or convex mirror). The rays that proceed along the straight lines reaching the surfaces of all mirrors and that are reflected at equal angles (whether from the surfaces of plane mirrors or from the [plane] surfaces tangent to the surfaces of concave and convex mirrors), that is, the lines that are reflected according to the manner of a reflected ray, are reflected in turn along those lines.[111] By the plane surface tangent to the concave surface, I mean that [surface] which has only one point [in common] with the concave surface. By the surface of the reflected line or ray, I mean the surface in which those two lines lie, that is, the line itself and the reflected line with which it forms an angle.[112]

[Consider] a section of a parabola, the axis of which is drawn, and [on which] a distance equal to one-fourth of the latus rectum is marked off from the end of the axis. In every [such] section, all lines

drawn parallel to the axis, arriving at the section and variously reflected to the point that marks off the fourth part, form two equal angles with the line tangent to the section at the point [of reflection]. For example, let *ABG* be a section of a parabola, and let its axis be *AD* and its latus rectum *L* [Fig. 6]. From *AD*, I mark off line *AE* equal to one fourth of line *L*, and I draw line *TB* parallel to line *AD*. I connect *B* and *E* and draw tangent *KBH*. I say, then, that angle *TBK* equals angle *EBH*.

First, let angle *BEH* be acute. Then by the method of resolution, I suppose that angle *TBK* equals angle *EBH*.[113] And since line *TB* is parallel to line *DA*, angle *TBK* will equal angle *BHE*. But angle *TBK* equals angle *EBH* by assumption; therefore, angle *EBH* equals angle *BHE*. Therefore, line *BE* is equal to line *EH;* consequently, the square of *BE* equals the square of *EH*. Now I draw *BZ* perpendicular to the axis. Then the two squared quantities *EZ* and *BZ* [summed] are equal to the square of *EH*. But the square of *BZ* is equal

105. On Anthemius, see *Les opuscules mathématiques de Didyme, Diophane et Anthemius suivis du fragment mathématique de Bobbio,* translated by Paul Ver Eecke (Paris/Bruges: Desclée de Brouwer, 1940); Sir Thomas L. Heath, "The Fragment of Anthemius on Burning Mirrors and the 'Fragmentum mathematicum Bobiense,' *Bibliotheca Mathematica,* Ser. 3, VII (1907), 225–233. On Archimedes, see A. Rome, "Note sur les passages des catoptriques d'Archimède conservés par Théon d'Alexandrie," *Annales de la Société scientifique de Bruxelles,* Ser. A, LII (1932), 30–41.

106. The Latin text reads *sectionum pyramidum,* but cone is meant.

107. This makes absolutely no sense; the Arabic text reads "application."

108. . . . *ex superficiebus angulorum.* I have not translated *angulorum,* which makes no sense and is lacking in the Arabic text.

109. The text reads "right angles," but emendation is obviously required.

110. That is, the straight line formed by the intersection of the two planes.

111. Alhazen appears to mean that rays and the geometrical lines employed to represent them—rays have width and are thereby unlike geometrical lines—follow identical paths; that is, the rays move along the geometrical lines.

112. That is, the reflected line to which it is joined angularly.

113. Alhazen assumes that which is to be demonstrated; then, following the method of resolution (or analysis), he deduces from it some known mathematical relation. In the reverse process of composition (or synthesis) which follows, he begins with the known mathematical relation and deduces that which is to be demonstrated, namely the equality of angles.

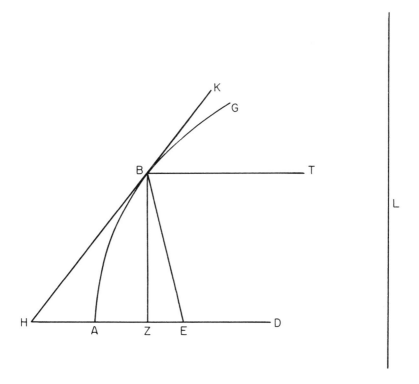

Fig. 6

to the product of *AZ* and *L*,[114] which is the latus rectum,[115] as the great Apollonius demonstrates in his book on conics.[116] Therefore, the square of *EZ* added to the product of *AZ* and *L* is equal to the square of *EH*. But *AE* is one fourth of *L;* thus four times the product of *AZ* and *AE* is equal to the product of *AZ* and *L*. Consequently, four times the product of *AZ* and *AE* added to the square of *EZ* is equal to the square of *EH*. Therefore, *AH* is equal to *AZ*.[117] But this is so because *BH* is tangent[118] and *BZ* is as arranged [that is, perpendicular to the axis].

And by the method of composition, I suppose everything as before, and I say that angle *TBK* equals angle *EBH*. The proof of this is as follows. I draw *BZ,* as before [perpendicular to the axis]. Since *BH* is tangent to the section [of the parabola] and *BZ* is as arranged, *AZ* will be equal to line *AH;* therefore four times the product of *AE* and *AZ* added to the square of *EZ* is equal to the square of *EH*. But *AE* is one fourth of *L;* consequently, four times the product of *AE* and *AZ* is equal to the product of *L* and *AZ*. Therefore, the product of *L* and *AZ* added to the square of *EZ* is equal to the square of *EH*. But the product

of *L* and *AZ* is equal to the square of *BZ,* since *BZ* is as arranged [that is, perpendicular to the axis]; therefore, the square of *BZ* added to the square of *EZ* is equal to the square of *EH*. But the two squared quantities *BZ* and *EZ* [summed] are equal to the square of *BE;* thus the square of *BE* is equal to the square of *EH,* and *BE* is equal to *EH*. Therefore, angle *EBH* equals angle *BHE*. And again, *TB* is parallel to *DA;* therefore, angle *TBK* equals angle *BHE*. Therefore, angle *EBH* equals angle *TBK*. And similarly, all lines drawn parallel to the axis[119] are reflected variously to

114. By the defining equation of a parabola, expressed in Cartesian notation as $y^2 = Lx$, where L is the length of the latus rectum.

115. The Latin text reads "straight line," but latus rectum is obviously intended.

116. *De pyramidibus.*

117. A number of steps have been omitted, but it does indeed follow from $4AZ \cdot AE + EZ^2 = EH^2$ that $AH = AZ$.

118. Apollonius had demonstrated that for any parabola $AH = AZ$ if BH is tangent; see Heiberg and Wiedemann, p. 209, n. 1.

119. The text reads "diameter," but Alhazen obviously means the axis.

point E and form acute angles with AE. And that is what we wished to demonstrate. . . .[120]

When any concave reflecting surface having the same concavity as a paraboloid is placed opposite the body of the sun so that its axis is in a direct line with the body of the sun, rays issue from the body of the sun to its whole surface; and all of these rays are reflected to the point on the axis located a distance from the pole of the surface

body of the sun. Consequently, this line will always be incident near point T; but point T is within the body of the sun. Therefore, it is incident on the sun; thus let it be incident at point K.

Therefore, the ray that proceeds from point K to point G proceeds along line KG.[121] Similarly, for every line extended from a point on the reflecting surface parallel to the axis and reaching the body of the sun, the ray that issues from that

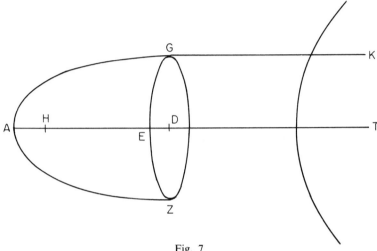

Fig. 7

equal to one fourth of the latus rectum of the section that generated the surface. For example, let there be a concave reflecting surface having the same concavity as a paraboloid, the pole of which is point A, its base the circle GEZ, its axis AD, and the distance of point H from point A equal to one fourth of the latus rectum of the section that generated the surface [Fig. 7]; and let the body of the sun (circle T) be placed opposite it, so that axis AD, when extended in a straight line, reaches point T within the body of the sun. I say, then, that rays issue from the whole body of the sun to this entire surface, and all are reflected to H.

This is demonstrated as follows. Since a ray proceeding from the body of the sun to the reflecting surface proceeds along straight lines, the ray proceeding from point T to point A proceeds along line AT. However, I choose a point at random on the circumference of the base of the reflecting surface, namely point G; and I imagine a line proceeding from point G parallel to line AT, such as line GK. Therefore, line GK, when extended rectilinearly, is incident on the body of the sun, since the separation between it and line AT is small, and insignificant by comparison with the

point [on the body of the sun] to the point on the reflecting surface proceeds along that line. But it has already been declared that rays proceed from the body of the sun to the entire surface of the reflecting body along lines parallel to the axis. Therefore, I maintain that all of them are reflected to one point. And since surface AGEZ is a concave surface having the same concavity as a paraboloid, when lines parallel to the axis encounter the surface and are reflected to point H, they form equal angles with the straight lines drawn in the same planes as the incident and reflected rays and tangent to the concave surface, as is evident in the figure above [Fig. 7]; and the rays that proceed along the straight lines drawn to the reflecting surfaces are reflected at equal angles with the lines [lying] in the same planes as the reflected lines and tangent to the reflecting surfaces. Therefore, the rays that proceed to the entire concave

120. The same demonstration is then repeated for the cases when angle BEH is right and obtuse.

121. Here Alhazen shifts from a discussion of geometrical lines to the rays propagated along them, in accordance with the distinction made in the fifth premise, in the third paragraph of this section.

surface along lines parallel to the axis are reflected along those lines that reach the point. And it has already been demonstrated that rays proceed from the body of the sun to the entire reflecting surface along lines parallel to the axis; therefore [these] rays, which are parallel to the axis, drawn to the

entire concave reflecting surface having the same concavity as a paraboloid, are all reflected to point *H*. And this is the point located a distance from the pole of the surface equal to one fourth of the latus rectum. And this is what we wished to demonstrate. . . .[122]

22. Witelo: Paraboloidal Burning Mirrors

PERSPECTIVA,[123] IX, THEOREM 43

When a concave mirror, of the same concavity as a portion of a parabola, is placed opposite the sun so that its axis points directly toward the body of the sun, all rays parallel to the axis and incident on the mirror are reflected to one point of the axis, removed from the surface of the mirror a distance equal to one fourth of the latus rectum of the portion of the parabola generating the surface of the mirror. It is evident from this that it is possible to kindle fire by means of the surface of such mirrors.

Let there be a concave mirror having the same concavity as a portion of a parabola [Fig. 8]. Let

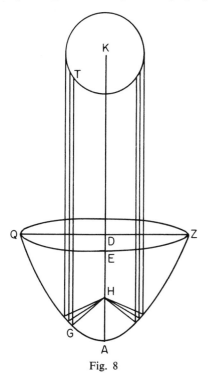

Fig. 8

its vertex be point *A*, its base circle *QEZ*, its axis *AD*; and let the distance of point *H* on the axis from the vertex of the mirror (point *A*) be equal to one fourth of line *QZ*, the latus rectum of the por-

tion of the parabola *AGQ*, which generates the surface of the concave mirror by its motion about axis *AD*. Let the mirror be placed opposite the sun according to its axis *AD*; that is, let *K* be the center of the solar body, and let the mirror be situated so that its axis *AD*, if extended, reaches point *K* in the center of the sun. I say that all solar rays parallel to ray *KA* incident on the surface of the proposed mirror are reflected to point *H* of line *AD*, the axis of the mirror.

Since all rays proceeding from any point whatever of the solar body to any point of the surface of the mirror proceed along straight lines, as is evident by Book I, Theorem 2, of this treatise, it is clear that line *KA* is a straight line. Thus on the periphery of any portion of the parabola of the mirror—let this be *GAZQ*—mark point *G* at random; and from point *G* of the mirror to some point of the solar body—let this be *T*—draw line *GT* parallel to ray *AK*, which is incident on the surface of the mirror along axis *AD*, by Book I, Proposition 31 [of Euclid's *Elements*]. Now, it is necessary for the line from any point of the mirror extended parallel to ray *AK* to encounter the surface of the body of the sun, since the ratio between the surface of the mirror and the surface of the solar body is either zero or small; thus let there be point *T*, which is the end of line *GT*, on the surface of the solar body. Thus all lines that can be drawn from the surface of the mirror parallel to its axis *AD* encounter the solar body; and, like the ray incident along the axis, the incidence of all rays parallel to the axis takes place along these lines. This is true of all rays incident on any point whatsoever of the whole surface of the mirror, since (by Book I, Proposition 31 [of Euclid's *Elements*,]) from any given point, near or far, we know how to draw a line parallel to any given line, such as axis *AD* in the proposition.

Thus I say that all those rays are reflected from

122. The remainder of the treatise deals with the fabrication of paraboloidal mirrors.

123. Risner, pp. 401–402. This proposition abundantly reveals Witelo's close reliance on Alhazen.

the whole surface of the mirror to one point on the axis of the mirror, point H. Since all those rays are straight lines, it is evident from what has been asserted above that they form equal angles with the lines drawn from all their points of incidence to point H; therefore, by Book V, Theorem 20, of this treatise, all those rays are reflected along the lines passing through point H. From this it is clear that all rays incident on the periphery of a portion [of the paraboloid] and parallel to the ray incident along the axis[124] of that portion are reflected to the point on the axis that cuts a line segment equal to one fourth of the latus rectum of portion $GAZQ$ [of the paraboloid] from the end of the axis at the periphery of that portion,[125] since every reflection from all regular polished bodies takes place according to equality of the angles formed between the incident and reflected lines and the line tangent to the surface of the mirror at the point where reflection occurs. And since all those lines intersect at point H, it is clear that the convergence of all those rays is in point H; therefore, every power of all the rays incident on the entire surface of the mirror is assembled at that point. And since every small beam brings with it some of the active power of the solar body, it is evident that the whole power is assembled at that point, namely the power of all rays parallel to axis AD incident on the surface of the mirror. Thus it is obvious that if combustible material should be placed at point H, fire can be kindled. And this is the best and strongest of all figures for assembling solar rays at one point, since solar rays are assembled at one point by its whole surface and its every point. Therefore, the proposition is evident.

23. Alhazen: Causal Analysis of Reflection

PERSPECTIVA,[126] IV, SECTION 18

The reason why light is reflected along a line having the same slope as the line by which the light approaches the mirror is that light is moved very swiftly, and when it falls on a mirror it is not allowed to penetrate but is denied entrance into that body. And since the original force and nature of motion still remain in it, the light is reflected in the direction from which it came, along a line having the same slope as the original ray. We can see the same thing in natural and accidental motion, for if we allow a heavy spherical body to descend perpendicularly onto a smooth body from a certain height, we will see it reflected along the same perpendicular by which it descended. In accidental motion,[127] if a mirror is raised to the height of a man and firmly fastened to a wall and a sphere is fixed to the tip of an arrow, and if the arrow is [then] projected by a bow at the mirror in such a way that the arrow and mirror have equal elevations and the arrow is horizontal, it is evident that the arrow approaches the mirror perpendicularly; and it will be seen to recede from the mirror along the same perpendicular. But if the motion of the arrow should be oblique relative to the mirror, the arrow will be reflected not along the line of incidence, but along a line horizontal (like the line of incidence) and of the same obliquity with respect to the mirror and the perpendicular. And it is clear that a motion of reflection is experienced by light because of the opposition of a polished body, since the stronger the repulsion or opposition, the stronger is the reflection of light.[128]

But the reason why the motions of approach and reflection are the same is as follows. When a heavy body descends along the perpendicular, the opposition of the polished body and the motion of descent of the heavy body are directly opposite: there is no motion except perpendicular motion, and opposition takes place along the perpendicular. Therefore, the body is opposed along the perpendicular, and thus it returns perpendicularly. But when the body descends along a sloping line, the line of descent falls between the perpendicular to the surface of the polished body (passing through that body) and the line in the surface orthogonal to this perpendicular. And if the motion were to penetrate beyond the point of its incidence, finding free passage, this line would fall between the per-

124. The text reads "diameter," but Witelo obviously means the axis; see note 119.

125. The complicated phraseology at this point tends to obscure Witelo's meaning, which is simply that the point is situated on the axis at a distance from vertex A equal to one fourth of the latus rectum.

126. Risner, pp. 112–113.

127. That is, violent or forced motion.

128. Note the mechanical terms in which Alhazen discusses reflection. Nevertheless, it should not be thought that he regarded light as corpuscular; like his Western followers, he thought of light as a power or form propagated through a material medium. For a fuller analysis of Alhazen's explanation of reflection, see my introduction to the forthcoming facsimile reprint of the Risner edition.

pendicular passing through the body and the line in the surface orthogonal to the perpendicular, and it would maintain the same measure of position with respect to the perpendicular passing [through the body] and the other line orthogonal to that perpendicular. For the measure of position of this motion is composed of a position along the perpendicular and a position along the orthogonal.[129] But since repulsion occurs along the perpendicular, because the body cannot repel motion according to the measure which it has along the perpendicular passing through the polished body and because a moderate thing would not enter, the body repels according to the measure of position along the perpendicular which it has along the orthogonal.[130] And when the regression of the motion has the

same measure of position along the orthogonal which there was formerly along the orthogonal on the other side [of the point of reflection], it will also have the same measure of position along the perpendicular passing [through the body] as it had before.

But in the rebound of a heavy body, when the motion of repulsion ceases, the body descends because of its nature and tends toward the center [of the world]. However, light, which has the same nature of reflecting [as a heavy body], does not by nature ascend or descend; therefore, in reflection it is moved along its initial line until it meets an obstacle which terminates its motion.[131] And this is the cause of reflection.

24. Roger Bacon: Causal Analysis of Reflection

DE MULTIPLICATIONE SPECIERUM,[132] III, CHAPTER 1

If one speaks of the rebound of species from a body, it is evident from what has been said that this does not occur by violence; but when it is for-

bidden to pass through [the body] by the density opposing it, the species generates itself in a direction open to it. For if it were to be driven back like a ball from a wall, it would be necessary for it to be body.[133]

25. Witelo: Causal Analysis of Reflection

PERSPECTIVA,[134] V, THEOREM 18

Everything seen by means of any mirror whatever is perceived by the eye under the shortest [possible] lines.

Let there be a mirror, in the surface of which is a straight or curved line *ACB*,[135] and let *D* be a point of the visible object and *F* the center of the eye; point *D* is seen to be reflected from point *C* of the mirror [Fig. 9]. I say that lines *FC* and *DC*

point on the surface of the mirror; let this be *E*. These lines are neither shorter than, nor equal to,

129. Alhazen resolves the motion of an oblique ray into components perpendicular and parallel to the surface of the mirror.

130. This and the next sentence are most obscure; consequently, I have provided a very literal translation. It does seem clear, however, that Alhazen regards both the parallel and perpendicular components of motion as being preserved in reflection. This conclusion is consistent with the claim, earlier in the argument, that after the reflection of light "the original force and nature of motion still remain in it"; combined with the knowledge that the angles of incidence and reflection are equal, this implies that the perpendicular and parallel components of the reflected ray possess "the original force and nature of motion" of those components of the incident ray. Thus Alhazen has analyzed the reflection of light in a fully Cartesian manner.

131. Although the behavior of heavy bodies can elucidate the reflection of light, Alhazen here takes care to point out that light is not itself a heavy body.

132. Bridges, II, 505.

133. In this brief passage Bacon reasserts the point he has made elsewhere that light is not corpuscular and that reflection is not an instance of mechanical rebound; reflection is rather to be explained by the self-diffusive properties of species.

134. Risner, p. 198.

135. The meaning seems to be that line *ACB* is straight in a plane mirror, curved in a curved mirror.

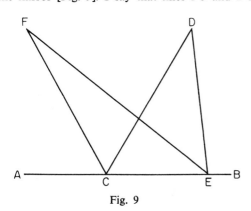

Fig. 9

[summed] are shorter than [the sum of] lines drawn from points *D* and *F* to any other points of the mirror. Let lines *ED* and *EF* be drawn to another

lines *CD* and *CF*, but longer. Now, nature always acts along the shortest lines, as is evident from Theorem 5 of this book, and the multiplication of forms to the surfaces of mirrors is natural, since it, like every other diffusion of forms, takes place by the action of nature, as we have shown in [our] *Philosophia naturalis*,[136] in the chapter on natural action. Similarly, the reflection of forms from the surfaces of mirrors to the eye is purely natural, since it, like every other visual perception, takes place by the action of nature and is completed by the action of the soul, as is evident from the whole fourth book of this our book of science; and the soul is, as it were, the nature of animals. Therefore, it is evident that this diffusion, reflection, and comprehension of a form, which takes place of itself, is truly natural. Consequently it takes place along shorter lines, which was the proposition; for in vain would it take place along longer lines, since it could take place better and more surely along shorter lines.[137]

26. Alhazen: An Instrument for Investigating Refraction

PERSPECTIVA,[138] VII, SECTION 2

The construction of an instrument of refraction.[139]

That light indeed passes through air and is extended along straight lines was declared in the first book of this work; however, air is but one of the transparent substances. Light [also] passes through water, glass, and transparent stones, and [in them] it proceeds along straight lines. But this is learned by experience. If one would like to experience it, let him take a round plate of copper, not less than one cubit in diameter, thick enough to be rather strong. This plate should have a circular rim perpendicular to its surface, not less than two finger-breadths in height. In the middle of the back of the plate, perpendicular to its surface, there should be a small cylindrical body not less than three finger-breadths long. Now place this instrument in a lathe with which the turner fashions instruments of copper; place one point of the lathe in the middle of the plate and the other in the middle of the end of the body mounted on the back of the plate. Then scrape this instrument by revolution, with suitable abrasion, until the roundness of its rim, inside and out, is ascertained, and the interior and exterior surfaces are made even, and the two surfaces are made parallel. We also scrape the body on the back [of the plate] until it becomes round.

When this instrument has been completed by abrasion, we mark two diameters on its interior surface, crossing through its center and intersecting each other perpendicularly [Fig. 10]. Then we mark a point at the bottom of the rim of the instrument, one finger-breadth away from the end of one of the two intersecting diameters. From this point we draw a third diameter passing through the center of the plate and extended to [the limits of] its whole surface. From the two ends of this

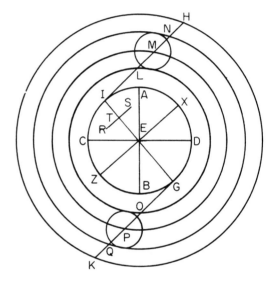

Fig. 10

136. This work has been lost.

137. Witelo's use of the principle of least action or minimum distance to explain the law of reflection is doubtless drawn from chapter 4 of Hero of Alexandria's *De speculis*, translated into Latin by William of Moerbeke at Witelo's request and completed on December 31, 1269. The date is given at the end of the holograph of this treatise, in Vat. Ottob. Lat. MS 1850, fol. 61v. Witelo's demonstration that lines forming equal angles with the "reflecting" surface satisfy the principle of least distance appears in Bk. I, Theorem 17 and Bk. V, Theorem 19.

138. Risner, pp. 231–233. Figure 11 has been corrected. Eilhard Wiedemann has published a German translation of portions of this passage in his "Ueber den Apparat zur Untersuchung und Brechung des Lichtes von Ibn al Haitam," *Annalen der Physik und Chemie*, n. s., XXI (1884), 541–544.

139. The same instrument is described by Witelo; however, there is no evidence that Witelo actually used it.

[third] diameter[140] we draw two lines[141] on the surface of the rim of the instrument, perpendicular to the surface of the plate. Then out of one of those lines we divide three small equal parts,[142] the first of which is next to the surface of the plate; and the length of each of these line segments is equal to the size of half a grain of barley. Thus, along this perpendicular line there are three points, which are the ends of those line segments. Then we return this instrument to the lathe and we mark on it three parallel circles,[143] passing through the three points on the perpendicular line at the end of the diameter. Thus the other perpendicular, which is perpendicular to the other end of this diameter, is also cut by those three circles, and three points are marked on it; [thus] there are, on each of the three circles, two opposite points, which are the ends of certain of their diameters. We then divide the middle of those three circles into 360 degrees and, if possible, into minutes. We form a round opening in the rim of the instrument, the center of which is the middle of the three points that are on one of the two lines perpendicular to the ends of the diameter of the plate, and its radius equal in size to the distance between circles. Thus the circumference of the opening extends between the two outermost parallel circles.[144]

After this, we take a thin rectangular plate of moderate thickness, the length of which is equal to the height of the rim of the instrument and the width of which is nearly that [Fig. 11]. Its

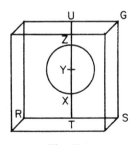

Fig. 11

[largest] surface is made as even as possible, and also [the surface of] its thickness, which falls along one of its ends, until the common difference between the surface of its face and the surface of its thickness is a straight line.[145] We divide this line into two equal parts, and from the middle we draw a straight line on the surface of its face, perpendicular to the straight line [RS] which is the common difference. Then we divide out of this perpendicular line, from the end that is above the

common difference, three segments equal to each other and to one of the small lines marked along the perpendicular line on the rim of the plate; thus there are three points[146] on the perpendicular line on the face of the small plate. Then we pierce this small plate with a round hole, the center of which is the middle point of those that mark off the lines in it, and the radius of which is equal to any one of those small lines; consequently, this opening is equal to the opening in the rim of the instrument.

On the diameter of the [circular] plate, at the ends of which are the two perpendicular lines, we mark a point[147] in the middle of the line that connects the center of the plate and the end of the diameter that extends toward the opening; through this point we draw a line perpendicular to the diameter. Then we place the base of the small plate on this line so that the common difference of the small plate is superimposed on this line perpendicular to the diameter; and the point that divides the common difference of the small plate into two equal parts is placed over the point [T] marked on the diameter of the plate. When this has been done, the small plate is attached to the large plate by means of a complete joining and consolidation; the opening in the small plate will then be opposite the opening in the rim of the instrument.[148] And there is an imaginary line, which joins the centers of the two openings, in the plane of the intermediate of the three circles that are on the inside of the rim of the instrument; and it will be parallel to the diameter of the plate. The

140. Points *I* and *G* in Fig. 10.

141. *IH* and *GK*. Perspective has been distorted in Fig. 10 in order to make all parts of the apparatus visible. The concentric circles passing through points *L*, *M*, and *N* represent circular grooves or scratches made in the rim mounted perpendicular to the disk, which is represented by the innermost circle. Thus lines *IH* and *GK* are actually perpendicular to the disk and parallel to one another.

142. I have translated the passage rather literally. Alhazen means that the line is divided into four equal segments (*GO*, *OP*, *PQ*, and *QK* in the figure). Apparently segment *QK* does not count—hence three parts are "divided out."

143. *LO*, *MP*, and *NQ*.

144. There are two such circular openings, represented in Fig. 10 by the small circles centered on points *M* and *P*.

145. Line *RS* in Fig. 11.

146. *X*, *Y*, and *Z*. When Alhazen talks about "dividing out" three equal segments of the perpendicular line, he means dividing the line into four equal parts; see n. 142.

147. That is, point *T* in Fig. 10.

148. These two apertures, of the same size and carefully aligned, serve as a sighting device.

small plate, which is applied to the point, is like the rim of an astrolabe. When this has been completed, from the four distinct quarters between the first two diameters that intersect perpendicularly, we cut away that quarter adjacent to the quarter containing the opening;[149] and the place of this portion [of the rim] is evened out until it becomes one with the surface of the plate.

We then take a copper rod, not less, but more, than one cubit long and of square section, which is enclosed by four equal surfaces two fingerbreadths wide; and its surfaces are made as even as possible, until they become equal and have right angles [Fig. 12]. Then we pierce one of its surfaces

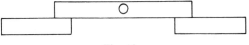

Fig. 12

with a round hole of such a size that it can receive the body on the back of the instrument and allow it to revolve, not easily, but with difficulty; and this hole should be perpendicular to the surface of the rod and should pass through to the other side. We place the instrument on the rod, and we put the body on the back of the instrument into the hole in the middle of the rod until the surface of the instrument rests on the surface of the rod. When this has been done, we cut from the ends of the rod those parts that extend beyond the diameter of the plate, for the rod is longer than the diameter of the plate, since we ordained it so. When we have cut the two excess lengths from the two ends of the rod, we replace them on the two ends of the rod in such a way that the two ends of the excess lengths are placed on the two ends of the remaining part of the rod.[150] And we place the surface of the excess lengths on the surface of the back of the instrument; and the amount of the two excess lengths that is placed upon the remaining part of the rod will equal one finger-breadth. This position having been noted, the two excess lengths will project over the two ends of the rod. Also, it will be better if that part of the body on the back of the instrument that sticks out [through the hole in the rod] is pierced and an iron peg inserted to keep the body from coming out. This having been done, the instrument is complete.

Then let the experimenter take a copper rule of small width (double the diameter of the opening in the rim of the instrument), of thickness equal to the diameter of this opening, and of length not less than half a cubit; this rule is perfected until it

becomes very straight and true, and its surfaces become equal and parallel. We cut one of the sides of its width obliquely until the edge of its length and the edge of its width enclose an acute angle, so that one can thus easily turn and move it anywhere he wishes; and he makes its width at the other end perpendicular to the edge of its length. Then we divide this [latter] width into two equal parts, and from the place of division we draw a line on the surface of the face of the rule; this line is extended along the length of the rule and is perpendicular to its width. When this rule is placed on the surface of the plate, its upper surface will be in the plane of the intermediate of the three circles marked on the inside of the rim of the instrument; for the thickness of this rule is equal to the diameter of the opening, and the diameter of the opening is equal to the perpendicular extending from the center of the opening in the rim of the instrument to the surface of the plate, since the diameter of the opening is equal to two of the three small line segments marked off on the perpendicular line on the inside of the rim of the instrument. Thus, when this rule is mounted on the rim, and the surface of its width is on the surface of the plate, the line marked on its middle will be in the plane of the aforesaid middle circle, since the perpendicular that extends from any point on this line to the edge of the length of the rule is equal to the perpendicular extending from the center of the opening to the surface of the plate; for both of those perpendiculars are equal to the diameter of the opening.[151]

149. The purpose in removing this section of the rim was to provide the observer with unobstructed vision of the inside of the remainder of the rim: ". . . *inspiciat ad fundum aquae, ex quarta, cuius orae sunt abscisse . . .*" (*Perspectiva*, VII, sec. 3, p. 233).

150. See Fig. 12. The purpose of this rod is to support the disk in a vertical plane, thus allowing it to be inserted into a tank of water. The two "excess lengths," which extend beyond the circular disk, rest on the sides of the tank; since the hole in the rod is below these supports, it is possible to suspend the disk in such a way that its center is exactly at the surface of the water.

151. This copper rule is used to mark various positions of the light passing through the apertures in the course of actual experiments. For example, it can be used to determine the point on the surface of the water at which light passing through the upper pair of apertures is refracted or to position the disk with its center precisely at the surface of the water. Unfortunately, Alhazen's account of the use of the disk and associated equipment is far too verbose to be included here, but from his description of the apparatus, here presented, it is quite clear how one would proceed in general to measure angles of

27. Roger Bacon: The Geometry of Refraction

DE MULTIPLICATIONE SPECIERUM,[152] II, CHAPTER 3

It must be recognized, secondly, that multiplication [of species] is diversified along refracted lines; and refraction occurs in two ways. When the second medium is denser [than the first], refraction of the species at the surface of the second substance takes place between the direct path and the perpendicular extended into the second body from the point of refraction;[153] thus the species deviates from the direct path into the depth of the second body and divides the angle between the direct path and the perpendicular extended into the second body from the point of refraction. Nevertheless, it does not always divide that angle into two equal parts, though some think it does,[154] since greater deviation and less refraction from the direct path occur as a result of different densities of the second medium, according to the various differences in the angles of refraction determined by Ptolemy in Book V of his *De aspectibus* and Alhazen in his Book VII.[155] For the denser the second body, the more refraction deviates from the direct path, on account of the resistance of the denser medium, for density resists the ray, as Alhazen says.[156] But when the second substance is subtler [than the first], refraction of the species at the surface of the second body takes place beyond the direct path, by recession from the perpendicular drawn at the point of refraction; that is, the direct path falls between the refracted ray and the perpendicular. And the recession of the refracted ray and the size of the angle vary as the subtler [second] substance is more or less subtle.

Therefore, I will illustrate with diagrams the manner in which rays are understood to be refracted or unrefracted at plane and spherical surfaces. First, in plane surfaces, it is declared concerning a second substance of plane surface that the whole power [of refraction] is determined by its shape. Therefore, if the second substance should have a plane surface, a ray incident perpendicularly is not refracted, as has been said; but all nonperpendicular rays are refracted. And if the second body (which is plane) is denser, then, as has been said,

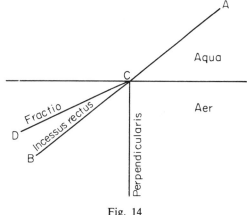

Fig. 14

incidence and refraction: If sunlight passes through the upper pair of apertures and is refracted at the surface of the water, a spot of light will appear on the submerged portion of the rim; the angles can then be determined from the markings on the middle of the three circles engraved on the inside of the rim. For additional detail, see A.C. Crombie's description of the use of Witelo's apparatus, which is identical to that of Alhazen, in *Robert Grosseteste and the Origins of Experimental Science, 1100–1700* (Oxford: Clarendon Press, 1953), pp. 220–223.

Alhazen adds a number of impressive refinements, not mentioned in the above account or by Crombie, which leave no doubt about his competence as an experimenter. Thus, for example, he suggests that a needle be placed over the middle of the aperture through which sunlight enters the apparatus; the resulting shadow in the middle of the spot of reflected light on the inside of the rim permits greater precision of measurement than the relatively large spot of light.

152. Bridges, II, 466–468.
153. That is, refraction is toward the perpendicular.
154. Chiefly Robert Grosseteste; cf. his treatise *On the Rainbow*, Selection 61.2.
155. Ptolemy's work on refraction has been analyzed by Albert Lejeune in *Recherches sur la catoptrique grecque* (Bruxelles: Palais des académies, 1957), pp. 153–166. On Alhazen, see *Perspectiva*, VII, secs. 10–12; Risner, pp. 243–247.
156. Alhazen, *Perspectiva*, VII, sec. 8; Risner, p. 240.

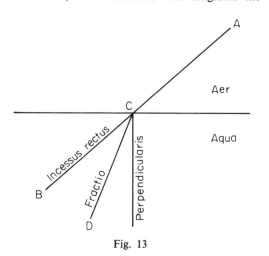

Fig. 13

the ray is refracted between the direct path and the perpendicular extended into that second body from the point of refraction, in this manner [Fig. 13]. But if the second substance is subtler, direct ingress falls between the refracted ray and the perpendicular drawn from the point of refraction; for example: [see Fig. 14]. If, however, the second body is spherical and denser than the first, refraction occurs at its surface between the direct path and the perpendicular leading from its center to the point of refraction, in this manner [Fig. 15]. But if the second body should be subtler, the direct

path falls between the perpendicular drawn from the point of refraction and the refracted ray itself, as here [Fig. 16]. And Ptolemy, in Book V of *De aspectibus,* and Alhazen, in his Book VII, agree regarding this double refraction; nor can it be otherwise. And they prove this by means of instruments, which they there show to be designed for this purpose, so that one can see with the senses the manner in which nonperpendicular rays are refracted. Likewise, they examine these refractions through causes and experiments.[157]

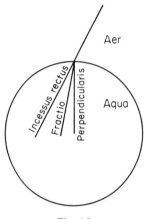

Fig. 15

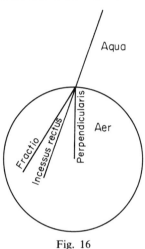

Fig. 16

28. Witelo: The Geometry of Refraction

PERSPECTIVA,[158] X, THEOREM 8

The angles of all refractions are revealed by means of tables.

Having observed by means of an instrument,[159] as closely as we were able, the angles of all refractions for all known transparent substances, by turns (as from air to water and glass, and from water to glass; and conversely, from water and glass to air, and from glass to water), we have discovered that the angles of refraction[160] are always the same [in passing] from any rare transparent substance to a denser transparent substance and [in passing] from that same dense substance [back] into the same rare substance. Accordingly, we have constructed these tables, of which this is the form [Table 1]. Out in front, in the first column, we have placed the angles of incidence; then we have added the other angles according to the modes of their circles, which we have set forth at the heads of their columns. And so by means of these tables, experimentally formulated by means of the aforementioned instrument, the

diligent investigator can learn all the angles of refraction for any and all media of diverse transparency.[161]

157. Note that Bacon makes no claim of having performed experiments himself, in contrast to Witelo in the following section.

158. Risner, pp. 412–413.

159. This is the instrument described by Alhazen, in section 26 above, and also by Witelo, II, Theorem 1 (Risner, pp. 61–63); X, Theorems 4–7 (Risner, pp. 407–412).

160. Witelo's "angle of refraction" is what is now usually referred to as the "angle of deviation," that is, the angle formed between the incident and refracted rays; Alhazen employs similar terminology. Hence this is a statement of the reciprocal law—the angle of deviation of a ray passing from substance A to substance B is equal to the angle of deviation of another ray traversing the same path from substance B to substance A.

161. Here and in the opening sentence of the theorem Witelo claims to have formulated these tables of refraction personally by experimental investigation. But there is ample evidence that this is untrue: In the first place, the values are not those given by experiment, but sets of numbers conforming to a regular progression—the

[Table 1.]

Table of the quantity of the angles of incidence common to all the following	Refracted angles[a] from air to water		Angles of refraction of the same		Refracted angles from air to glass		Angles of refraction of the same		Refracted angles from water to glass		Angles of refraction of the same	
	deg.	min.	deg.	min.	deg.	min.	deg.	min.	deg.	min.	deg.	min.
10	7	55[b]	2	5	7	0	3	0	9	30	0	30
20	15	30	4	30	13	30	6	30	18	30	1	30
30	22	30	7	30	19	30	10	30	27	0	3	0
40	29	0	11	0	25	0	15	0	35	0	5	0
50	35	0	15	0	30	0	20	0	42	30	7	30
60	40	30	19	30	34	30	25	30	49	30	10	30
70	45	30	24	30	38	30	31	30	56	0	14	0
80	50	0	30	0	42	0	38	0	62	0	18	0

Table of the quantity of the angles of incidence common to all the following	Refracted angles from water to air		Angles of refraction of the same		Refracted angles from glass to air		Angles of refraction of the same		Refracted angles from glass to water		Angles of refraction of the same	
	deg.	min.	deg.	min.	deg.	min.	deg.	min.	deg.	min.	deg.	min.
10	12	5	2	5	13	0	3	0	10	30	0	30
20	24	30	4	30	26	30	6	30	21	30	1	30
30	37	30	7	30	40	30	10	30	33	0	3	0
40	51	0	11	0	55	0	15	0	45	0	5	0
50	65	0	15	0	70	0	20	0	57	30	7	30
60	79	30	19	30	85	30	25	30	70	30	10	30
70	94	30	24	30	101	30	31	30	84	0	14	0
80	110	0	30	0	118	0	38	0	98	0	18	0

[a]The expression "refracted angle" refers to the angle between the refracted ray and the perpendicular, that is, to what is now generally termed the "angle of refraction." For Witelo's "angle of refraction," see note 160.
[b]Printed editions of Witelo's *Perspectiva* give this angle as 7°45'. However, three early manuscripts that I have checked all give the value appearing here.

It is evident from these tables that the angles of incidence of the form of the same point [which successively assumes different positions] nearer the ray [that issues] from a point of the visible object incident perpendicularly on the surface of the transparent body, which [body] gives rise to refraction, are smaller; and those angles more remote from that ray are larger. For when a larger angle has been subtracted from the right angle of that [perpendicular] ray, the angle that remains is smaller than the remaining angle when a smaller angle is subtracted from the right angle. And in a transparent body denser than the first, the angle of refraction corresponding to the greater angle of incidence will be larger than the angle of refraction corresponding to the smaller angle of incidence.[162] Also, the excess of the greater angle of refraction over the smaller angle of refraction will be smaller than the excess of the greater angle of incidence over the smaller angle of incidence; and the ratio between the angle of refraction corresponding to the greater angle of incidence and that greater

angle [of incidence] will be larger than the ratio between the angle of refraction corresponding to the smaller angle of incidence and that smaller angle [of incidence].

And the refracted angle by which the larger angle of incidence exceeds its angle of refraction is larger

---

differences between successive angles of refraction form an arithmetic progression with a common difference of one-half degree. Secondly, the upper half of the table, with the exception of the "angles of refraction" (which Witelo computed simply by subtracting the "refracted angles" from the angles of incidence) and the refracted angle of 7° 55', was taken from Ptolemy's *Optica*. Finally the lower half of the table was computed from the values in the upper half by erroneous application of the reciprocal law; consequently, it includes preposterous results, such as "refracted angles" (measured from the perpendicular) larger than 90° and no recognition at all of total internal reflection.

162. The reader is reminded that "angle of refraction" denotes the angle between the incident and refracted rays. The expression "angle of incidence," however, denotes the angle between the incident ray and the perpendicular.

425

than the refracted angle by which the smaller angle of incidence exceeds its angle of refraction. Thus the refracted angle, in a second transparent substance denser than the first, will always be smaller than the angle of incidence; and the ratio of those refracted angles to equal angles of incidence varies with the different densities of the [second] medium. For when refraction takes place in water and in glass, [the forms having come] through the same air and at equal angles of incidence, the refracted angles are more acute in the glass than in the water; thus the angles vary with the different transparencies. But if the second transparent medium is rarer [than the first], then the refracted angle is always larger than the angle of incidence; and the ratio of these angles to the other angles, since they are formed in the opposite manner

to the aforesaid angles, will be as though the tables set forth above were arranged in the reverse manner. And the ratio of these refracted angles and angles of refraction to the same angle of incidence are varied according to greater and less rarity of the transparency of the second medium. For when refraction takes place from glass to water or to air, the angles formed in the air are larger than the angles formed in the water; accordingly, the ratio of the angles of refraction to the angles of incidence is varied.

These are the things that occur to lights and colors and universally to all forms in their diffusion through transparent bodies and in the refraction that occurs in all of them, both according to themselves and with respect to sight. And so what has been inquired into is evident.

29. John Pecham: Image Formation by Refraction

PERSPECTIVA COMMUNIS, III, PROPOSITION 4

Proposition 4. The image is located at the intersection of the pyramid under which the object is seen and the perpendiculars that one can imagine dropped from the visible object to the surface of the adjacent transparent medium.[163]

As was shown above,[164] everything that is viewed appears [as though] in a straight line; and by the apprehension of the ray through which the eye perceives the object, the object is judged to be at the end of the straight-line extension of the ray. Therefore, just as the appearance of the object at the intersection of the perpendicular and [the rectilinear extension of] the ray is considered basic to mirrors, so in the case at hand it [that is, the image] is at the intersection of the ray and the perpendicular dropped from the visible object. For example, let A be the eye, B the visible object, BC the bent ray (refracted at C, from which [point] proceeds [ray] CA) that presents the object to the eye, and BLD the perpendicular [Fig. 17]. I say that point B appears at L.

30. Roger Bacon: Image Formation by Refraction

OPUS MAIUS,[165] V.3, DISTINCTION 2 CHAPTERS 2–3

Chapter 2. Concerning different places of appearance of the image by refraction at plane surfaces.

It should be understood that vision by refraction is at the intersection of the visual ray and the

163. As in the case of reflection, the image of any given point is located at the intersection of the *ray* under which

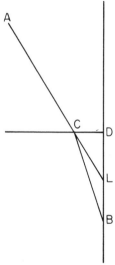

Fig. 17

the point is seen and the *perpendicular* dropped from the point to the refracting surface. However, an object is composed of many points, and the collection of the rays under which those points are seen constitutes a pyramid. Consequently, in the enunciation of this proposition, Pecham speaks of the intersection of the *pyramid* and the *perpendiculars*.

164. See *Perspectiva communis*, II, proposition 20, in section 19 above.

165. Bridges, II, 148–153.

cathetus, as has been stated for reflection. But this can occur in various and marvelous ways. However, in order that we may understand every diversity of such appearances, it is necessary to consider how such diversity occurs in plane, concave, and convex substances, when the eye is in the subtler or denser medium and the visible object conversely. Now, if the eye is in the subtler transparent medium, and between the eye and the visible object there is a denser medium, such as water or crystal or glass of plane surface or some other such transparent medium, then the object appears much larger than it is; for it is seen under a larger angle and is somewhat nearer than if there were no change in medium. The demonstration of this is evident in this figure [Fig. 18]. For the visible

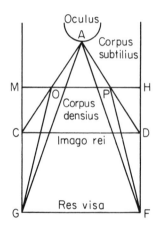

Fig. 18

point F will appear at D, where visual ray AD intersects the cathetus FH; similarly G will appear at C, where visual ray AC intersects cathetus GM. Therefore, the whole object GF will appear in position CD nearer to the eye and will be seen under a larger angle than if the substance [of the medium] were one; for it is seen under angle OAP through these two substances, but through one medium without refraction it would be seen under angle GAF.

But if the eye is in the denser medium and the visible object in the subtler medium, the contrary occurs; for the object will appear smaller both because it is seen under a smaller angle and because it appears more remote [Fig. 19]. Thus O is seen at H, and F at K, beyond the visible object, so that OF appears at KH, since visual ray AB intersects cathetus HC at H, and visual ray AD intersects cathetus PFK at K. And it is seen under a smaller angle than if it were seen through a

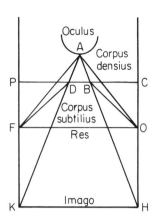

Fig. 19

single medium, since now the whole object is seen under angle DAB because of the refraction, whereas without refraction it is seen under the larger angle FAO.

Chapter 3. Concerning the different locations of the image in spherical media.

If the substances through which the eye sees are not plane, but spherical, there is great diversity. Now, either the concavity or the convexity of the substance is toward the eye. If the concavity, then there are four cases, two if the eye is in the subtler medium and two if the eye is in the denser medium. If the eye is in the subtler medium and the concavity of the medium is toward the eye, the eye can be between the center [of the concave surface] of the medium and the visible object, or the center can be between the eye and the visible object. And here there should be no notion of the center of the denser medium or of the center of the subtler medium, since the same thing is the center of both, and the concavity of each is toward the eye; for the same thing is the center of the containing sphere and the contained sphere. First, then, I shall state all the cases and then illustrate them with figures; it is necessary to proceed thus because of the brevity of the rules covering individual cases and the great number of figures. And all these things are revealed in the following figures, which are presented in the order of the eight aforementioned cases.[166]

166. I have had to correct all of Bridges' figures except the eighth; in each case, my correction has embodied principles clearly recognized by Bacon, and for most of the corrections it is possible to adduce either textual or manuscript support. For example, Bridges mistakenly interchanged point C and the eye in the first figure; my rearrangement is confirmed by the reference to angle

If the eye is in the subtler medium, and the concavity is toward the eye, and the eye is between the center and the visible object [Fig. 20], the

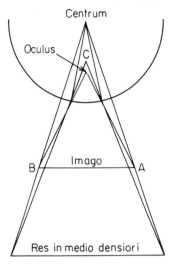

Fig. 20

object appears nearer than it [actually] is and [is seen] under a larger angle, for thus[167] the visual angle is larger than if straight lines are drawn from the eye to the ends of the object without refraction; and yet the image is smaller [in absolute terms] than the object itself. If the eye is in the subtler medium, and the concavity is toward the eye, and the center [of the concavity] of the denser substance is between the eye and the object [Fig. 21], the object still appears nearer; but the angle is smaller,

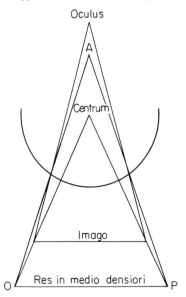

Fig. 21

and the image is smaller. But if the eye is in the denser medium, and the concavity is toward the eye, and the eye is between the center of the concave substance and the object [Fig. 22], then the object

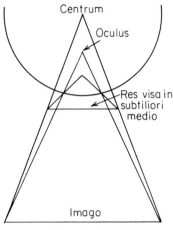

Fig. 22

appears beyond its [true] place, more remote [from the eye], and under a smaller angle; and the image will be larger [in absolute terms]. If the center of the concave substance is between the eye and the visible object, other conditions remaining the same [Fig. 23], the visible object still appears more

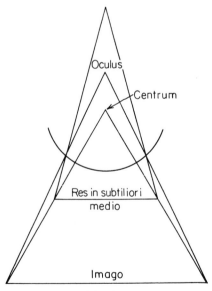

Fig. 23

BCA of this figure near the end of the chapter. Compare the figures in A. C. Crombie, *Augustine to Galileo*, 2nd ed., rev. (Cambridge, Mass.: Harvard University Press, 1961; reprinted, 1963, 1967, as *Medieval and Early Modern Science*), I, 106–109.

167. That is, when the object appears nearer.

remote and under a larger angle; and the image is larger.

However, if the convexity of the substance is toward the eye, there are likewise four cases, two if the eye is in the subtler medium and two if the eye is in the denser medium. Now, if the eye is in the subtler medium, and the convexity of the medium in which the object is situated is toward the eye, then the visible object can be between the center and the eye, or the center can be between the eye and the visible object. If the object is between the eye and the center [Fig. 24], the image

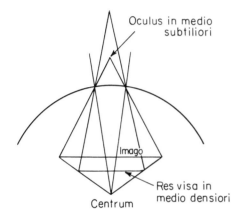

Fig. 24

is nearer and larger, and the angle is larger. If the center is between the eye and the visible object [Fig. 25], the image is still larger and the angle is

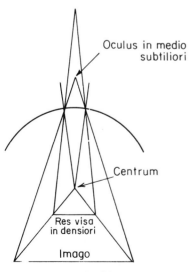

Fig. 25

larger; but the location of the image is more re-

mote. However, if the eye is in the denser medium, and the visible object is between the eye and the center [Fig. 26], the image is more remote and

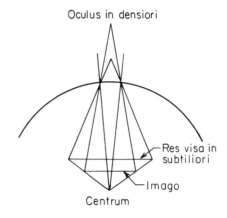

Fig. 26

smaller, and it is seen under a smaller angle. If the eye is in the denser medium, and the center is between the eye and the object [Fig. 27], the image

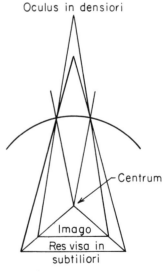

Fig. 27

is nearer and smaller and is seen under a smaller angle; and the size of the angle under which the object is seen is recognized to be smaller than it would be if the medium were one [substance]. And this is also the case whenever it [that is, the angle under which the object is seen] encloses the angle formed by the lines of direct progress, and they are terminated inside[168] it at another point; but [the

168. The Bridges text reads "outside," but the passage makes no sense unless emended.

angle under which the object is seen] is called greater, as is evident in the first figure [Fig. 20], when it converges at another point below [the angle formed by the lines of direct progress]; but then the angle formed by the lines of direct progress is smaller than the angle under which the object is seen, and therefore the angle under which the object is seen is larger than if there were one medium.[169] For then the object is seen under angle

31. John Pecham: The Burning Glass

PERSPECTIVA COMMUNIS, III, PROPOSITION 16

Proposition 16. Fire can be kindled by the convergence of refracted rays.

It is evident from Proposition 17 of Part II, and the next to last proposition of the same part, that [fire can be kindled] by reflected rays in mirrors. The same thing happens when round transparent bodies are exposed to solar rays; but mirrors and transparent bodies are different in that fire is kindled between a mirror and the sun, whereas with transparent bodies, on the contrary, it is the transparent body that is between. For example, let there be a round crystal of diameter *YAZ*, and let rays *XC, XS, XR, XQ*, and *XP* fall on it from the sun [Fig. 28]. It is certain that only *XR* falls to the center *A* and proceeds unrefracted to *H*. Therefore, the others are refracted toward the perpendicular and fall from *C* to *B*, from *S* to *G*, from *Q*

32. Roger Bacon: Causal Analysis of Refraction

DE MULTIPLICATIONE SPECIERUM,[171] II, CHAPTER 3

However, they [that is, Ptolemy and Alhazen] assign the cause of refraction as follows. Since the descent of a perpendicular species is strong (as is evident in a falling stone, provided its descent is not diverted from the vertical), if something should impede perpendicular descent and make the stone deviate from a perpendicular course, it is manifest to sense that its ability to penetrate is weakened. Wherefore a man falling perpendicularly from a high place is killed by the fall. But if something should divert him from direct approach as he descends, he is spared insofar as he diverges from perpendicular approach. Similarly, if a sword or axe or some other instrument designed to cut is applied to a rod perpendicularly by the hand of the one wielding the instrument, it penetrates and divides the rod; [however,] if the instrument is applied obliquely, either it does not cut at all or it does not cut as much. But if the substance that

BCA enclosed under the direct lines. And in the following figure [Fig. 21] the object would be seen under angle *OAP*, enclosed by the direct lines, if the medium were one [substance], and thus under a larger angle than the angle formed by the refracted lines under which the object is seen through two media. And it is to be understood in this way in all the other figures that follow.

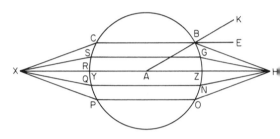

Fig. 28

to *N*, and from *P* to *O*. Accordingly, ray *CB* arriving at the concave surface of the air does not proceed directly to *E* but is refracted away from perpendicular *BK* to *H*, and similarly for the other rays. When these rays have been assembled and the air rarefied, fire is kindled [just] beyond the terminus of their species.[170]

receives an object falling perpendicularly should

169. Bacon is referring to the angles situated above the interface in each of the figures. In Fig. 20, for example, the lines of direct progress terminate at *C*, which is outside the angle under which the object is seen (terminating at the eye); consequently, the angle under which the subject is seen is larger than the angle formed by the lines of direct progress.

170. The closest approach by a medieval author to a geometrical analysis of lenses was discussion of the burning sphere (or cylinder). It is true that Alhazen and Witelo described a plano-convex (that is, semicylindrical) piece of glass or crystal, to be used in connection with their refraction apparatus, but this was so arranged that refraction occurred only at its plane surface. It was not employed as a lens, and the paths of rays through it were not traced.

171. Bridges, II, 468–470. For an analysis of medieval discussions of the cause of refraction, including Alhazen's, see my article, "The Cause of Refraction in Medieval Optics," *The British Journal for the History of Science,* Vol. IV, Part I, No. 13 (June 1968), 23–38, from which this selection (on pp. 36–38) is reprinted with the kind permission of the editors.

resist it altogether, the falling object will return along the same path by which it approached because of the strength of its approach, as a ball thrown against a wall or something that altogether resists [penetration]. And a ball falling obliquely deviates toward the other side of the perpendicular in accordance with [the direction of] its oblique fall, as is evident to sense; and it does not return along the path by which it approached [the wall] because of the weakness of its fall. Experiences of such things are without number.

But a transparent substance does not altogether resist [penetration by] species, causing them to return, for only a dense substance, which produces reflection, does that. Wherefore species incident on any transparent substance penetrate it, whether they are incident obliquely or perpendicularly. But since a perpendicular species is stronger and an oblique species weaker,[172] both cannot proceed in the same manner in a transparent substance. But there is no path except a rectilinear and a bent course. Now, since the rectilinear course is the stronger, it is suited to perpendicular species, allowing them to proceed along the same straight line in the second substance as in the first; and a bent path is suited to oblique species. Therefore, a perpendicular species does not bend and is not refracted, since bending is refraction; but an oblique species is refracted at the surface of the second substance, since it is more strongly resisted by the transparent substance than is a perpendicular species because of the weakness of the oblique species and the strength of the perpendicular species; for every transparent substance possesses some coarseness, by which it is able to resist somewhat and to impede the passage of species. As Alhazen says in the seventh [book of his *Perspectiva*],[173] transparency in natural things is limited so that it does not extend infinitely, and there is some coarseness even in celestial substance. Consequently, both perpendicular and oblique species are somewhat impeded, but oblique species more so. For, on account of its strength, the perpendicular is not diverted from a rectilinear path, though any coarseness whatever of the medium introduces a greater succession into its [that is, the perpendicular species'] traversal [of the medium]. Indeed, if there were a void medium without any natural coarseness, and if light or any body whatsoever could pass through it, there would be some succession caused by the medium simply as a result of the prior and posterior [character] of space itself,[174] quite apart from any resisting coarseness of the medium. But

in a natural plenum, which has some natural coarseness, it is necessary that the coarseness (which neither altogether yields nor altogether impedes, but partly one and partly the other) should produce a new succession. Therefore, by its coarseness a transparent substance impedes both oblique and perpendicular species and introduces succession into both; and oblique and perpendicular species share this in common. But beyond this, a transparent substance impedes an oblique species more [than a perpendicular species] on account of the weakness [of the former] and does not allow it to continue in a straight line along the same path as it had formerly in the higher substance.[175]

Now, the reason why a species passing from a subtler into a coarser substance is refracted toward the perpendicular leading from the place of refraction (that is, between the rectilinear path and the perpendicular) is that just as the perpendicular course is strongest, so every course close to the perpendicular is stronger than any course more remote from the perpendicular, as reason declares and authors assert. Therefore, since the species [in the case under consideration] was moved much more swiftly in the more subtle substance than it could be in the second denser substance because of the great resistance offered by the coarseness of such a substance, the natural power generating the species desires the easier passage and chooses it;[176]

172. Supposedly, this has been demonstrated by the mechanical examples with which the discussion began: a sword cuts more deeply if applied to a rod perpendicularly, and a man's chances of survival increase as he is diverted from perpendicular descent. However, Bacon's claim is that perpendicular *traversal of a medium* is stronger than oblique traversal of the medium, whereas the examples demonstrate merely that perpendicular *incidence on a resisting surface* is stronger than oblique incidence on that surface.

173. VII, sec. 16; Risner, p. 252.

174. [The same justification for successive motion in a void was given by Thomas Aquinas; see Selection 55.1— Ed.]

175. Here Bacon argues that species incident obliquely on a transparent interface are refracted, while species incident perpendicularly are not, because oblique species are weaker and more easily deviated. The implication is that the medium resists equally in all directions and that variation is introduced only by the varying strength of the species. However, in the following paragraph, Bacon clearly asserts that the denser medium resists less in a direction perpendicular to its surface than in directions oblique with respect to its surface; a given species chooses a path closer to the perpendicular, because it finds easier passage in that direction.

176. Note Bacon's animistic terminology. Pecham,

and this is toward the perpendicular. Wherefore it must be refracted between the rectilinear path and the perpendicular leading from the place of refraction.

And since contraries are causes of contraries and the effects of contrary causes are contrary, it is necessary that when a species passes from a denser substance into a subtler substance, it bends away from the perpendicular, so that the rectilinear path falls between the line of refraction and the perpendicular drawn from the place of refraction. For since there is great resistance in the former dense substance (because its coarseness surpasses the coarseness of the second substance, which is very subtle so that species entering it act with ease), it is unnecessary for the species to seek an easier course (that is, toward the perpendicular), since every path is easy for it by comparison with the difficulty it encountered in the first substance. Therefore, the species is able to assume a direction away from the perpendicular, and it must necessarily choose this direction, since insofar as possible it acts uniformly and in one mode.[177] And therefore it passes reluc-

tantly toward that which is violently contrary to its disposition. Accordingly, since in the first substance the species had great difficulty or [only] moderate ease of passage, on account of its uniformity (which it always maintains insofar as possible), it ought not choose the superior contrariety, according to which it would deviate toward the direction of easiest passage (that is, toward the perpendicular); but to a certain degree, as much as it can and ought, it will maintain [its] difficulty or moderate ease of passage; and this occurs as the species deviates from the rectilinear path in the direction away from the perpendicular, since such a course would suffice for it and be able to fulfill it. For example, when a species passes from a subtler into a coarser substance, it maintains its ease of traversal in the second substance, so that its passage in each substance is, insofar as possible, proportional and uniform; however, superfluous coarseness in the second substance [as when a species passes from a rare to a dense medium] excites the generative power of a species, so that it bends toward the direction of easier traversal.

33. Witelo: Causal Analysis of Refraction

PERSPECTIVA,[178] II, THEOREM 47

Although a perpendicular ray penetrates any transparent substance [without refraction], a ray obliquely incident on a second transparent medium of greater density [than the first medium] is refracted toward the perpendicular to the surface of the second transparent medium (drawn from the point of incidence); and [when incident] on a second transparent medium of greater rarity [than the first medium, the obliquely incident ray] is refracted away from the same perpendicular.

We [now] intend to support by a natural demonstration this [proposition], which has been proved previously by particular experiments employing an instrument. For all natural motions occurring along perpendicular lines are stronger [than others], since they are aided by the universal celestial power, which influences every body situated below according to the shortest straight line.[179] Moreover, the impulses of objects projected perpendicularly are stronger than of those that are projected obliquely. Similarly, perpendicular blows are stronger than all oblique blows, and among oblique blows, those which more closely approach the perpendicular are stronger.

And so, since the density of every body impedes the passage of light, it must be conceived that light

is thrust back from passage by the resistance of a dense body, and even more by the resistance of a denser body. And we understand, by this resistance of a passive quality (that is, density) to an active quality (that is, light), a certain mode of the motion of light through the midst of resisting bodies, which are more or less susceptible to the impression of light; this is not to maintain that there would be some motion in the local transmutation of light itself (as is evident from Theorem 2 of this book), but that in the same instant light restrains or diffuses itself more, according to the diversity of

who in other respects presented a similar explanation, was to rebuke him for this; see *Perspectiva communis,* I, Proposition 15.

177. Here is the fundamental principle employed by Bacon in his explanation of refraction. Species choose a direction in the second medium that will most nearly maintain uniformity of strength or action. Consequently, species passing from a less dense to a more dense medium deviate toward the perpendicular, where the greater ease of traversal (by comparison with traversal of more oblique paths) compensates for the greater resistance of the denser medium; species entering a less dense medium choose a more resistant path for the same reason.

178. Risner, pp. 81–83.

179. Witelo introduces a novel element into discussions of refraction: motion along the vertical (which Witelo takes also to be the perpendicular to the refracting surface) is stronger because a celestial power aids it.

the media.[180] And here we call this the motion of light itself. Thus every light passing through a transparent body traverses it with an exceedingly swift and insensible motion. And yet the motion occurs more swiftly through more transparent bodies than through less transparent bodies;[181] for every transparent body resists penetration by light more or less according as it participates more or less in transparency, since the coarseness of bodies is always resistant to penetration by light.

Therefore, when light passes obliquely through a transparent body and encounters a denser transparent body, the latter impedes the light more vigorously than the former rarer body impeded it; consequently, the resistance of the denser body must alter the motion of the light. And if the resistance should be strong, that motion would be refracted in the opposite direction;[182] and, in fact, since light does not strongly resist [deviation], it would not [be permitted to] proceed in its original direction. But if the resistance should be weak, in accordance with the greater rarity of a more transparent body,[183] the incident light would not be refracted in the opposite direction, nor would it be able to proceed in its original direction; rather, it would be altered in direction. (Nevertheless, it is not altered when incident perpendicularly on transparent bodies possessing any variety of transparencies, but penetrates all the bodies in a straight line, because the perpendicular [ray] is stronger than all others, and oblique [rays] nearer the perpendicular are stronger than those more remote from the perpendicular.) Therefore, when light is incident obliquely on a denser transparent body, it proceeds [after refraction] along a straight line approaching the perpendicular issuing from the point at which the light encounters the surface of the dense transparent body and extended above the surface of the denser body, since motion along the perpendicular line is easiest. Thus, if a ray of light is incident along a perpendicular line, it will continue in a straight line because of the strength of motion along the perpendicular. And if the ray is incident obliquely, it will not be able to continue directly because of the weakness of motion along oblique lines; consequently, it deviates toward some direction in which passage is easier than in the original direction.[184] But the easier motion and the one more aided by celestial influence is along the perpendicular, and that which is closer to the perpendicular has easier passage than that which is more remote from the perpendicular.

Accordingly, let many rays from point A of a luminous body be incident through medium AB, on the surface of another transparent body containing line $BCDE$; and let BF be the depth of the other transparent body and line AB the perpendicular to its surface [Fig. 29]. It is evident from the foregoing

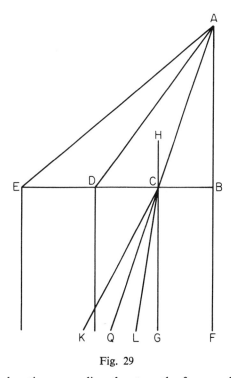

Fig. 29

explanations regarding the strength of perpendicular [rays] and from the experiments with instru-

180. In this theorem, translated in section 5, Witelo argues that light is propagated instantaneously; consequently, one cannot speak of its local motion or transmutation.

181. This appears to contradict previous arguments on the instantaneous propagation of light. While it might be maintained that Witelo's earlier arguments *succeed* in demonstrating merely that the time required for the propagation of light is insensibly small, it appears that his *intention* was to demonstrate that the propagation of light over a finite space does not require any time at all.

182. That is, the motion would be turned back on itself, as in reflection. However, here Witelo prefers to continue to speak of the process as refraction. This is legitimate, since (as the present argument suggests) reflection is simply a more intense degree of refraction.

183. Rarity and transparency were commonly equated in medieval optics.

184. It is apparent that Witelo's explanation of the direction of refraction is essentially the same as Bacon's.

ments described in Theorems 42 and 44 of this book that a perpendicular ray incident along line *AB* penetrates the entire body *BEF*. However, if the ray incident along line *AC* were to traverse body *BEF* without deviation, [this would imply that] there is no difference in transparency between bodies *ABE* and *BEF*, contrary to hypothesis; accordingly, because of the difference in resistance [of the two substances], line *AC* will not be extended directly. But if a ray has been moved freely along line *AC* through a body offering little resistance, it cannot be moved along the same line in a body resisting more or less. Therefore, if body *BEF* is denser than body *ABE,* it is evident from the foregoing that passage through the former is more difficult. If, therefore, line *AC* is refracted away from the perpendicular drawn to surface *BCDE* of the body from point *C*—let this perpendicular be *CG*—it will be weakened and will fail to achieve its full effect; therefore, it has been thwarted. But it was accepted at the outset that nature is not thwarted;[185] therefore, as was demonstrated experimentally in Theorem 43 of this book, line *AC* must be refracted toward perpendicular *CG*, so that its action may be strengthened. The same is true of rays incident along lines *AD* and *AE*.

But if the body containing line *BCDE* in its surface should be of rarer transparency (like body *ABE* [in the foregoing case]), perpendicular ray *AB* would still penetrate without refraction, because of the strength of action [along the perpendicular]; but the ray passing through the denser body along line *AC* and incident on the surface of the rarer body at point *C* would not encounter resistance as before. And since forms have the property of always diffusing themselves according to the amplitude of every material suited to [receiving] them, it is evident that ray *AC* does not continue along line *AC*, for this would be the case only if the transparent bodies were equivalent in their resistance to the reception of light, contrary to hypothesis. Therefore, ray *AC* is refracted, but not toward perpendicular *CG*, since refraction in such a direction would not result from the resistance of the material, but from victory of the acting form over matter more suited [to receiving the form] than the first [denser matter]. Therefore, by its own power, the form diffuses itself away from its initial advance along line *AC* and in the direction away from perpendicular *CG* and its parallel *BF*. The same is true of all other oblique rays, such as *AD* and *AE*.

And so the motion of rays incident obliquely on a body of denser transparency, such as *BEF*, is composed of a motion in the direction of perpendicular *AB* (crossing through body *BEF* in which the motion takes place) and of a motion along line *CB*, which is perpendicular to line *CG*. Now, since perpendicular passage is the strongest and easiest of motions and the density of the body opposes [the reaching of] the goal toward which the motion was directed, line *AC* is necessarily moved toward perpendicular *CG* issuing from point *C*, where ray *AC* encounters the surface of the denser body. And since it is prohibited from that [initial] motion because of the grossness of the medium and the nature of the other motion (that is, along line *CB*), which is not altogether destroyed but only impeded by the resistance of the medium,[186] the light will deviate toward point *B,* always approaching perpendicular *ABF*. Thus, in a second medium of grosser transparency than the first, refraction of ray *AC* occurs along line *CL*, which is closer to perpendicular *CG* issuing from point *C* (where ray *AC* encounters the denser body) than is line *AC* (along which the ray was incident on the surface of the denser body) extended beyond point *C* to point *Q*, to the same perpendicular extended beyond point *C* to point *H*. Consequently, angle *ACH* is greater than angle *LCG*. Nevertheless, [refracted ray *CL*] does not intersect perpendicular *BF* in the direction of point *F*, but in the direction of point *A*, by Book I, Theorem 2, of this treatise, since *CL* intersects line *CG* (which is parallel to *BF*) at point *C*.

But if ray *AC* issues from the grosser body to the more subtle, then since the latter has less resistance, motion of the ray will be swifter and more diffusive in it than in the former. And since the resistance of a denser medium always occasions an oblique light, so that it unites with the perpendicular line drawn to the surface of that body at the point of incidence (that is, line *CG*), and in a medium of rarer transparency the resistance is less than [the resistance of] the first [denser medium], it is evident that the light is moved toward the region from which the greater motion was prohibited by the resistance.

185. This is much like Bacon's principle of uniform action.

186. Here Witelo attempts an explanation of refraction in terms of the perpendicular components of the incident and refracted rays. Apparently, the ray assumes a direction closer to the perpendicular in the denser medium because (or partly because) its component parallel to the interface is weakened (or slowed?). It is not clear, however, what is happening to the perpendicular component at the same time. Nevertheless, the resemblance to Descartes' later analysis is noteworthy and suggests some reservations regarding Descartes' originality.

Thus, in a rarer transparent body, light is moved away from the perpendicular, so that angle *GCK* is larger than angle *ACH*. However, when light *AC* is refracted by a second body of rarer transparency than that of the first body, motion of the light always takes place between lines *CG* and *CE*, since when angle *GCE* is right, angle *GCK* cannot be right. Therefore, the proposition is evident.

63 LATE MEDIEVAL OPTICS

Introduction by David C. Lindberg

Fourteenth- and fifteenth-century optics developed within the framework (broadly defined) established in the thirteenth century. Indeed, the most notable optical treatises of the fourteenth century were several long technical commentaries on the *Perspectiva communis* of John Pecham. The extent to which this late medieval tradition continued to debate questions raised in the thirteenth century is illustrated by the first section, which reproduces the enunciations of the fifteen questions discussed by Henry of Hesse in his *Questions* on Pecham's

Perspectiva communis.[1]
Few problems were debated more urgently in the thirteenth century than the cause of the rainbow, treated unsuccessfully by every major thirteenth-century author. Perhaps the most dramatic development of fourteenth-and fifteenth-century optics was the first correct geometrical analysis of this problem, by Theodoric of Freiberg early in the fourteenth century. The second section, from Theodoric's *De iride*, embodies all the essential geometrical elements of his theory.

1. Henry of Hesse (1325?–1397): Questions Concerning Perspective

Translated and Annotated by David C. Lindberg[2]

1. Whether light is multiplied by means of rays.
2. Whether a luminous body terminates a pyramid of its light in every point of the medium.
3. Whether light incident through a triangular aperture is reduced to roundness by nature.
4. Whether a luminous body appears larger from afar than from nearby.
5. Whether every ray incident or issuing forth under an obtuse angle is oblique.[3]
6. Whether a perpendicular ray is strongest.
7. Whether an opaque body smaller than the luminous body always casts a pyramidal shadow.
8. Whether the moon increases in light [in passing] from conjunction to opposition.
9. Whether everything seen is viewed by means of straight lines.
10. Whether every visual perception takes place under an angle.
11. Whether all visible intentions are apprehended or revealed through the species of colors.
12. Whether reflection takes place from the surface of every denser body.
13. Whether everything seen by reflection appears at the intersection of the reflected ray and the cathetus.
14. Whether rays are refracted in a difform medium.
15. Whether the rainbow appears by means of circular reflection from a cloud.

2. Theodoric of Freiberg (b. ca. 1250; d. shortly after 1310): On the Rainbow

Translated and annotated by William A. Wallace, O.P.[4]

PART II.

Chapter 25

In general, concerning the matter or subject [body]

1. All but one or two of these questions had been explicitly raised by Pecham or his contemporaries, Bacon and Witelo. It has been impossible, in this *Source Book,* to include more than the enunciations of Henry's *Questions.*
2. Translated from an unpublished edition by H. L. L. Busard.

3. The meaning of this question is clarified by the contrasting case of a perpendicular ray, which is incident or issues forth under two right angles.
4. From *Theodoricus Teutonicus de Vriberg: De Iride et Radialibus Impressionibus* (Dietrich von Freiberg: Über den Regenbogen und die durch strahlen erzeugten Eindrücke), edited by Joseph Würschmidt, *Beiträge zur Geschichte der Philosophie des Mittelalters,* Vol. XII, pts. 5–6 (Münster/Westfalen: Aschendorffsche Velagsbuchhandlung, 1914), pp. 113–114, 119–120, 133–138, 151, 153–154, 156–159. I have corrected Würschmidt's reading, as noted below, by reference to the Leipzig manuscript, Cod. Lips. 512, fols. 47 recto to 72 verso.

in which the rainbow is generated, its efficient cause, and its mode of radiation.[5]

The [radiant] impression that we call the rainbow extends above the horizon to the upper part of the airy [that is, atmospheric] region in the form of an arc, with the arch on top and with both extremities touching the plane of the horizon when it is fully formed.[6]

The matter or subject body in which this impression comes to exist is twofold. One is a dewy cloud, that is, a cloud resolved into drops or watery droplets of dew shaped into spherical form, located in or around the place of their generation, and detained and suspended there partly by the power of the generator and alterator, which has not yet been separated from it, and partly from the nature of a certain levity, from the fact that its substance shares markedly in the concrete nature of an airy property, as has been noted in Part I.[7] The other subject body or matter of the aforementioned impression or rainbow is a collection of falling drops released from a cloud as rain. Although these drops, which are like watery spherulets, continually move downward in their fall, the fact that some drops succeed others in the same places causes this impression to appear to sight as being in the same place.[8]

These [notes], therefore, have been made with respect to the matter or subject body in which the aforementioned radial impression that we call the rainbow comes to exist. Its efficient cause is radiation from a star that is shining brightly, such as the sun and the area immediately surrounding it—for it is not the radiation of a single form that causes the described four colors, as has been noted above.[9] This radiation is effected after the third mode of radiating among those that have been enumerated in the preceding Part I,[10] a radiation, I say, incident on some one of the described bodies at a place on the spherical particles fixed by the configuration between this place, the star that is radiating, and the sight to which such radiation is reflected. This incidence of radiation and its reflection to sight occur only in an area that is fixed geometrically by nature.[11]

Chapter 29

Concerning the incidence and reflection of solar radiation whereby the colors of the [lower] rainbow are generated.

Another mode of solar radiation takes place in some of the described spherical droplets, which the foregoing altitude circle *ADB* intersects at right

angles [Fig. 1].[12] This mode, I say, is different from

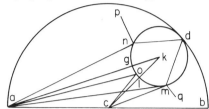

Fig. 1. Diagram from Theodoric of Freiberg's *On the Rainbow,* showing the path of solar radiation through an individual drop when generating the colors of the lower rainbow. The sun is at *a*, the observer at *c*, and the raindrop at *d*.

5. The title of this introductory question reveals Theodoric's basic Aristotelian methodology, wherein he is seeking the various causes (material, efficient, and so on) of the rainbow. For a full discussion of this method, see W. A. Wallace, O. P., *The Scientific Methodology of Theodoric of Freiberg.* Studia Friburgensia, New Ser., No. 26 (Fribourg: The University Press, 1959).

6. In Part I, chapter 6, of this treatise Theodoric classified the rainbow as one of fifteen "radiant impressions" produced in the upper atmosphere by light from the sun or other heavenly body. Part II of the work is concerned with the lower rainbow; Part III, with the upper rainbow; and Part IV, with the remaining radiant impressions.

7. Theodoric's views of gravity and levity, while Aristotelian, differed in several respects from those of his contemporaries. For details, see W. A. Wallace, O.P., "Gravitational Motion according to Theodoric of Freiberg," *The Thomist* XXIV (1961), 327–352; reprinted in *The Dignity of Science,* edited by J. A. Weisheipl, O.P., (Washington, D.C.: The Thomist Press, 1961), pp. 191–216.

8. Albert the Great (d. 1280), like Theodoric a German Dominican and a Prior Provincial of Teutonia, is usually credited with being the first to have recognized the role of the individual drops in the production of the rainbow. See Carl B. Boyer, *The Rainbow: From Myth to Mathematics* (New York: Thomas Yoseloff, 1959), pp. 94–99, especially p. 96.

9. Theodoric had worked out his own peculiar theory of color formation, whereby two forms (more and less bright) are required to generate the four colors. For details, see *The Scientific Methodology of Theodoric of Freiberg,* pp. 188–205, 152–173.

10. Five different modes of radiation are described by Theodoric in chapters 7–11 of Part I; these are summarized in *The Scientific Methodology of Theodoric of Freiberg,* pp. 181–182.

11. This conclusion is the result of extensive experimentation by Theodoric that has been well described by A. C. Crombie, *Robert Grosseteste and the Origins of Experimental Science, 1100–1700* (Oxford: Clarendon Press, 1953), pp. 233–259; also by Boyer, pp. 110–128. See note 24.

12. The diagram of Fig. 1 is based on the faulty Aristotelian geometry of a "meteorological sphere" and its "altitude circle." See the detailed discussion in Boyer, pp. 119–122.

the foregoing [mode] both with respect to the place and with respect to the mode of incidence and reflection. For in the first mode the incidence of the radiation and its reflection take place on the surface of the spherical droplet that is on the convex side in the direction of the sun and the [person] viewing. In the mode of which we now speak, however, the incidence and reflection to sight take place from the interior of the droplet, or from the concave spherical surface in the direction away from the sun and the [person] viewing, and this according to the third mode of radiation explained above in Part I. For the solar ray is incident on the surface of the sphere facing the sun; it is incident, I say, at a place higher than the point where the perpendicular ag falls [on the surface]; and let this place of incidence be n and the incident ray an. For in this place the described ray, entering the body of the drop, is refracted toward the perpendicular of its incidence, because such a drop is a translucent body of greater termination or density and, as a consequence, of lesser transparency than the air through which the described ray passes on its way to the drops.[13]

According to this mode of refraction and of penetration, therefore, the ray passing through the body of the drop arrives at its concave [surface] at point d over line nd, and the described ray is again reflected from this place of incidence, or d— as by a concave mirror, since it has the nature of a mirror—and is reflected toward the surface of the sphere on which it was first incident. Passing through the body of the sphere it arrives at point m, at a place below the perpendicular ag coming from the sun; moreover, the same place of incidence will be below the perpendicular cl, coming from the [person] viewing to the indicated drop. The ray so reflected from the concavity of the foregoing surface at the place already described is dm, and the radiation leaving this place on the body of the sphere is refracted, on leaving, away from the perpendicular of its incidence, because the air which it enters on leaving the body of this sphere is of more subtle substance and, as a consequence, of greater transparency and translucency. Thus the radiation passing from the point of such refraction arrives at the [person] viewing over line mc, and in a place and at a configuration fixed by the nature of the star, of the drop, and the [position of the] viewer. Again, the colors of the rainbow appear at the places of incidence, refraction, and reflection, as will now be explained; this is apparent in the description of the foregoing figure.

Chapter 38

Concerning the mode whereby the colors of the lower rainbow are transmitted to the viewer.

Now we must consider the mode whereby the colors that appear in the rainbow are transmitted to the viewer, and this for the case of the lower rainbow. To do so we must review some knowledge from physics that is required for this purpose, and first this [fact], which has been demonstrated above, that radiation of a luminous object, precisely as luminous, when obliquely incident and traveling through a watery or crystalline drop—especially if the drop is of very small size—such radiation, I say, projects the color red from its part that is incident around the sides, angles, and angular sides of such a drop, then [the color] yellow, then green, and finally blue in the portion that on entrance is farther from the angles and the angular sides of the sphere and stays more toward the middle of the drop.[14] Also supposed is this [fact], already indicated, that the place of elevation of the drops on the altitude circle from which the rainbow appears is not a point, but occupies a distance on the altitude circle that corresponds to the width of the entire rainbow and also of each of its component colors.

Another [factor] to be similarly noted is that the place of incidence of the radiation on any drop from which the colors of the rainbow shine on the surface of the drop is not a point but has a width, the different parts of which reflect different colors of the rainbow in such a way that each width of these parts contains a proportionate amount of each of the said colors.

Likewise the place on the opposite surface in the interior of the drop on which such [radiation] is incident has a width that is proportioned to the aforementioned [partial] widths, as has been noted. As a consequence, moreover, the place on the drop from which the foregoing colors depart and are transmitted to the viewer has a width that is proportioned to the width of incidence already mentioned.

13. Theodoric obtained his knowledge of the law of refraction from Alhazen (ibn al-Haitham), whom he cites fifteen times in *De iride* as the "auctor *Perspectivae.*"

14. According to Theodoric's theory of color formation, the color red would be generated in the part of the drop where there is greater clarity and boundedness, that is, by rays passing closest to the drop's surface, whereas the other colors would be generated by rays passing more toward the middle of the drop; see *The Scientific Methodology of Theodoric of Freiberg*, pp. 188–196.

Again, this too is to be remembered, that the more obliquely radiation is incident on one side at a position more remote from the perpendicular, the more obliquely it departs toward the viewer on the other side and at a position more remote from the same perpendicular.

Moreover, with the foregoing, [the point] that has been mentioned above is not to be omitted, namely that with one drop receiving such radiation in this manner and transmitting it to the viewer, all of the mentioned colors seen at the drop do not

appear be *defgh;* moreover, let it suffice that each aggregate or collection of drops so located as to [produce] each of the mentioned colors be represented by one drop. Then, let there be marked off on each of the mentioned drops a place of incidence having a width, as has been said, and let this be the arc *ln,* and similarly a place at which the said radiation is incident in the interior of the drop, since it has a width, *os,* in the foregoing manner; [then,] the place of exit toward the viewer *tz,*[16] the partial positions within the same width *tvxyz,* the

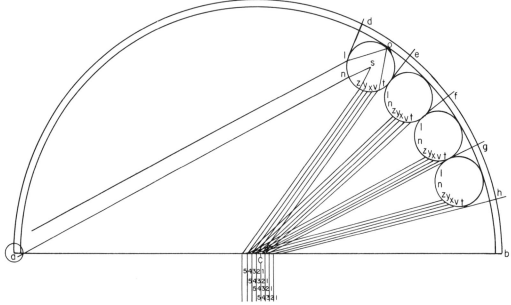

Fig. 2. The formation of the lower rainbow, showing the four drops (or collections of drops) that produce the four colors Theodoric held were present in the bow. The sun is at *a,* the observer at *c,* and the four drops at *de, ef, fg,* and *gh.*

come to the viewer in one and the same spot, but, depending on the various positions of the viewer with respect to this drop, different colors come to view. Hence, if the viewer sees all the colors simultaneously, as happens in the rainbow, this must come from different drops that are differently situated with respect to the viewer, and the viewer with respect to them.

I will therefore draw a horizon represented by the line *ab,* which also will represent the diameter of a sphere whose center is *c* [Fig. 2].[15] Let the altitude circle cutting the horizon at right angles be *adb.* Let the position on the altitude circle to which the spherical drops are elevated, and in which the rainbow appears, be the arc *dh.* Since this is not a point, but has a width, as has been said, let the parts of the position in which the individual colors

place where the total radiation carrying the colors of the rainbow falls on the diameter of the sphere on the horizon, *51,* [and] the partial places at which the individual colors fall, *54321.* And let the sun be on the surface of the horizon so as to provide an example that is easily seen.

With the foregoing [points] that are here called to mind, this too should not be forgotten, as has been shown above, that the arc described on the drop that is included between the place of inci-

15. This diagram is based on the same faulty geometry as Fig. 1; see note 12.

16. Würschmidt's text gives the letters *SZ* here; I have emended his reading on the basis of Cod. Lips. 512, fol. 65 recto. Crombie (p. 253, n. 1) notes a similar change on the basis of the Basel manuscript, Cod. Basil. F. IV. 30, fol. 33 verso.

dence, which is *ln,* and the place of departure toward the viewer, *zyxvt,* is larger or smaller according as the spherulet is more or less elevated above the horizon on the altitude circle. Therefore, [considering] the spherical droplet at the highest place on the altitude circle from which radiation incident from the sun can be transmitted to the viewer, namely, at position *de,* the [incident] radiation falls on the surface of the drop at arc *ln* and, entering the body of the drop there and being refracted toward the perpendicular, falls on the opposite internal surface of the drop at arc *os;* then it returns, as if reflected by a mirror, to the first surface and falls on the arc *tz.* Then, in leaving, it is refracted away from the perpendicular and, arriving at the viewer, represents the color red in the portion that when passing through the body of the drop is incident around the angles and sides and confinement of the boundaries of the drop.[17] Thus the portion *vt* of the aforementioned radiation comes to the viewer in the center of the [meteorological] sphere with the color red at the farther limit *21* on the plane of the horizon. The other three colors, yellow, green, and blue, fall on the diameter of the sphere[18] *acb* between the viewer and the sun, namely yellow at position *32,* green at position *43,* and blue at position *54* at the other limit of these markings. These do not come to the viewer, but the red color alone comes to the viewer from this drop and, in fact, from the entire collection of spherical droplets that are located on the altitude circle at position *de.*

The radiation that is incident on drops that are lower on the altitude circle, namely at position *ef,* so arrives at the viewer that the color yellow from the portion *xv* of the arc *zyxvt* comes to the viewer at the second position on the horizon toward the lower limit, *32,* from each of the drops located at the position *ef* on the altitude circle. The color red, on the other hand, falls on the diameter of the sphere[19] [*acb*] in front of the viewer, namely between the viewer and the place where the rainbow appears. The other two [colors], namely green and blue, fall on the diameter of the sphere [*acb*] behind the viewer, namely between the viewer and the sun, and none of these come to the viewer except the yellow.

Considering the drops located at the third position lower on the altitude circle already described, namely *fg,* from each of these the color green comes to the viewer in the portion *yx* of the arc *zyxvt.* The red and the yellow, on the other hand, fall on the diameter of the sphere[20] [*acb*] in front of the viewer, namely between the viewer and the place where the rainbow appears. The fourth [color], namely blue, falls on the diameter of the sphere behind the viewer, [that is,] between the viewer and the sun, at the positions noted in the preceding [account].

From the drops at the fourth and lowest of the described positions on the altitude circle, namely at *hg,* the fourth color, blue, comes to the viewer at the portion of *zy* of the arc *zyx;* the other three colors, red, yellow, and green, fall on the diameter of the sphere in front of the viewer or between the viewer and the rainbow,[21] and none of these colors come to the viewer except the color blue.

Thus all the colors of the rainbow [are seen] at the same time, and the entire rainbow appears on the altitude circle in different spherical droplets according as they are more or less elevated along the arc *hd* at various positions, from each of which the individual colors come to the viewer in the manner described above.

On the other hand, no incident radiation comes to the viewer from the drops that are located on the altitude circle above the mentioned arc *hd.* Those [drops] that are located below the mentioned arc, however, transmit some incident radiation to the viewer, but not with the colors of the rainbow; rather [they transmit] a kind of white light, unmixed with colors, as has been treated above and will be explained more fully below.[22]

Therefore, according to what has already been said, since the color red is radiated from the highest position on the altitude circle, yellow from the [position] next to this, then green from the third [position], and blue from the lowest [position],

17. This passage again presupposes Theodoric's theory of color formation; see note 14.

18. Würschmidt's text gives the reading *sphaerulae,* meaning spherulet or drop; I have emended his reading to *sphaerae* on the basis of Cod. Lips. 512, fol. 65 recto. The same error occurs with some frequency, as noted below. Crombie (p. 254, nn. 3–5, and so on) has made similar corrections on the basis of the Basel manuscript, Cod. Basil. F. IV. 30, fol. 33 verso ff.

19. Cod. Lips. 512, fol. 65 verso.

20. *Ibid.*

21. I have corrected Würschmidt's text to read *ante visum sive inter visum et iridem* in place of *ante visum et iridem* on the basis of Cod. Lips. 512, fol. 65 verso; see Crombie, p. 254, n. 6, for the same emended reading based on the Basel Manuscript.

22. This paragraph provides Theodoric's explanation of "Alexander's band" and the "Aphrodisian paradox" unaccounted for by previous Aristotelian explanations of the rainbow; for a detailed explanation, see Boyer, pp. 64, 119–123, 215.

as a result the highest and outermost periphery [of the rainbow] is red, the next under [it is] yellow, then follows green, and the lowest and innermost [periphery is] blue.

PART III.

Chapter 3

In general, concerning the difference between the [modes of] radiation whereby the two rainbows are variously generated.

The mode of radiation by which the upper rainbow is generated, to speak only in general, is this [Fig. 3]. Just as in the lower rainbow solar

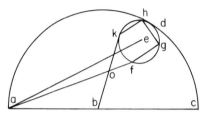

Fig. 3. The path of solar radiation through an individual drop when generating the colors of the upper rainbow. The sun is at *a*, the observer at *b*, the raindrop at *d*.

radiation falls on any drop on the part that is higher or more remote from the diameter of the sphere[23] of the world, and this either higher or more to the right or to the left, as has been explained above, and is reflected to the viewer from the lower part of the drop, or that closer to the diameter of the sphere of the world, so correspondingly in the present case, or in the generation of the upper rainbow, solar radiation falls on the drop at the part that is lower or closer [for instance, point *f*] to the diameter of the sphere of the world [that is, *ac*], whether the drop is elevated above the horizon on the altitude circle at the summit of the rainbow or is located to the right or to the left of the viewer; and thus this radiation, entering the body of the drop and penetrating to the opposite concave surface of the same drop [for instance, at points *g* and *h*], is reflected from it to the viewer [at point *b*] from the part of the drop that is higher or more remote [for instance, point *k*] from the diameter of the sphere of the world [that is, *ac*].

The mode of this type of radiation can be verified by experiment, if one uses a translucent crystalline stone, which is called a beryl, or any clear spherical drop that is so situated with respect to the sun and the viewer, namely with the viewer located between the sun and such a drop located off to one side of the straight line that goes from the sun to the viewer.[24] But this suffices in general.

Chapter 4

In particular, concerning the place or quantity of elevation on the altitude circle. . . of the upper rainbow.

Now to be considered more in particular is [the fact] that the place of generation of this rainbow determined on the altitude circle is eleven [degrees] or approximately this degree of elevation higher on the altitude circle than the place of elevation of the lower [rainbow]. . . .[25]

Chapter 6

Concerning the four conditions or modes of this upper rainbow, which are deduced from the foregoing, with certain introductory notes that are inserted for considering them in particular.

From the foregoing mode of radiation whereby this rainbow is generated, therefore, we may deduce its four modes or conditions: the first, namely that the situal arrangement of these four colors in the upper rainbow is just the opposite of what it is in the lower rainbow.[26] The second also we deduce from physics, namely its shape; the third also, namely why this rainbow is more obscure than the lower [rainbow]; [and] the fourth, why it appears with less frequency than the lower [rainbow]. . . .[27]

Chapter 7

In detail, concerning the cause of the arrangement of the colors of the upper rainbow as these appear to sight.

23. See note 18; Cod. Lips. 512, fol. 68 recto; Crombie p. 256, n. 2.

24. This passage only hints at the experimental methodology used by Theodoric to verify his theory. His ingenious coupling of theory and experiment, in Boyer's expression, "represents one of the greatest scientific triumphs of the Middle Ages. It is difficult to overestimate the significance of his step in immobilizing the raindrops. For the first time the rainbow in the laboratory could be studied more thoroughly than the astronomer in the observatory scans the heavens" (Boyer, p. 124).

25. By his own account Theodoric used the astrolabe to measure such angles; see Part II, chapter 34 (Würschmidt, p. 127, ll. 5–6). The angle he reports in this passage is correct, whereas the angle he measured for the radius of the lower bow is not; for details, see Boyer, pp. 120–121.

26. Theodoric was the first, it may be noted, to provide an essentially correct explanation for the phenomenon of color reversal in the secondary rainbow. See the translation from Part III, chapter 7, following.

27. The explanations for these remaining properties of the upper rainbow are given in *The Scientific Methodology of Theodoric of Freiberg*, pp. 218–223; compare Boyer, pp. 117–118.

Therefore, considering the spherical drop or, in fact, the entire collection of drops elevated on the

viewer and the center of the rainbow k . . . .

Therefore, [after having discussed the drops at

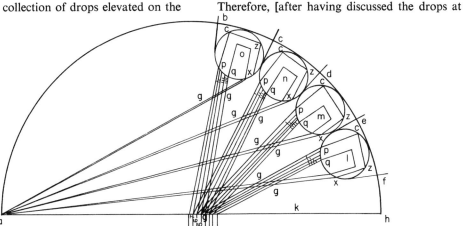

Fig. 4. The formation of the upper rainbow, showing the four drops (or collections of drops) that produce the four colors that Theodoric held were present in the bow. The sun is at a, the observer at g, and the four drops at bc, cd, de, and ef.

altitude circle at arc fb, [let us examine] the lowest portion of the arc, fe [Fig. 4]. So as to have a very clear example, let the sun be at the horizon [for instance, at a]; then solar radiation falls on each one of the mentioned drops at its lower side, namely at arc x, and there entering the body of the drop rotates, through its incidences and reflections, around the opposite concave surface of the drop until it arrives at its place of exit, namely at arc pq. On emerging from this position it is refracted away from the perpendicular because it meets a body of greater transparency, namely air; thus, by considerable circumgyration it arrives at the line of original incidence, ax, and, intersecting this at gg,[28] comes to the viewer [at] g. Therefore, since for the particular portion or part of the radiation that emerges near p the path of rotation comes closest and nearest the angles, arcs, and angular sides of the drop, it is certain,[29] from the foregoing, that this same part of the radiation, which is designated in the diagram by one mark [|], will present the color red to the viewer. The other three colors, namely yellow, green, and blue, which are designated by the remaining marks [||, |||, ||||], fall on the diameter of the sphere[30] between the

ed, dc, and cb], since in the generation of this [upper] rainbow the color red comes to the viewer from the lowest position within the arc fb, which is determined by nature to produce this type of rainbow in view of the elevation of the spherical drops within it, and since yellow is reflected to the viewer next to the red from the next higher portion of the mentioned arc, while the color green radiates from the third portion of the said arc, two [places] above [the red], and finally blue comes from the topmost [part], it is clear that the colors of this upper rainbow are arranged in an order that is the reverse of that seen in the lower rainbow. . . .

28. Cod. Lips. 512, fol. 69 verso, gives a single g here, although there are two g's on the diagram in the manuscript corresponding to Fig. 4. I have inserted gg so as to avoid the ambiguity otherwise introduced by Theodoric's using g also to designate the viewer at the center of the meteorological sphere. Würschmidt gives the reading f, following the Basel manuscript; compare Crombie, p. 258, n. 1.

29. I have replaced Würschmidt's *tertium*, meaning third, by *certum*, following Cod. Lips. 512, fol. 69 verso, where the "c" is written like a "t," as in many fourteenth-century manuscripts. The conclusion again presupposes Theodoric's theory of color formation; see note 14.

30. See note 18. Cod. Lips. 512, fol. 69 verso.

Astronomy, Astrology, and Cosmology

Astronomy

64 THE TWO MOST POPULAR MEDIEVAL HANDBOOKS OF THE ELEMENTS OF ASTRONOMY

1. John of Sacrobosco (d. ca. 1244–1256): On The Sphere

Translated by Lynn Thorndike[1]

Annotated by Edward Grant

PROEMIUM

Contents of the Four Chapters[2]

The treatise on the sphere we divide into four chapters, telling, first, what a sphere is, what its center is, what the axis of a sphere is, what the pole of the world is, how many spheres there are, and what the shape of the world is. In the second we give information concerning the circles of which this material sphere is composed and that supercelestial one, of which this is the image, is understood to be composed. In the third we talk about the rising and setting of the signs, and the diversity of days and nights which happens to those inhabiting diverse localities, and the division into climes. In the fourth the matter concerns the circles and motions of the planets, and the causes of eclipses.

CHAPTER 1

Sphere Defined.

A sphere is thus described by Euclid:[3] A sphere is the transit of the circumference of a half-circle upon a fixed diameter until it revolves back to its original position. That is, a sphere is such a round and solid body as is described by the revolution of a semicircular arc.

By Theodosius[4] a sphere is described thus: A sphere is a solid body contained within a single surface, in the middle of which there is a point from which all straight lines drawn to the circumference are equal, and that point is called the "center of the sphere." Moreover, a straight line passing through the center of the sphere, with its ends touching the circumference in opposite directions, is called the "axis of the sphere." And the two ends of the axis are called the "poles of the world."

Sphere Divided.

The sphere is divided in two ways, by substance and by accident. By substance it is divided into the ninth sphere, which is called the "first moved" or the *primum mobile;* and the sphere of the fixed stars, which is named the "firmament"; and the seven spheres of the seven planets, of which some are larger, some smaller, according as they the more approach, or recede from, the firmament. Wherefore, among them the sphere of Saturn is the largest, the sphere of the moon the smallest, as is shown in the accompanying figure.[5]

1. This selection, which includes all of chapters 1, 2, and 4 of John of Sacrobosco's *On the Sphere (De sphaera)* as translated by Lynn Thorndike, is reprinted here by permission of the University of Chicago Press from Lynn Thorndike, *The Sphere of Sacrobosco and its Commentators* (Chicago: University of Chicago Press, 1949), pp. 118–129, 140–142.

This nontechnical account, written primarily as a textbook for use in the universities, was probably the most popular astronomical and cosmological treatise of the Middle Ages and served as the basis of many later commentaries. For instances of its considerable influence on Pierre d'Ailly's *Ymago Mundi,* see Selection 81.

2. This caption, and the many which follow, have no basis in the Latin text and were supplied by Thorndike, presumably for the convenience of the reader.

3. In Book XI, Definition 14, of the modern Greek edition (see Thomas Heath: *The Thirteen Books of Euclids Elements,* 2d. ed. [New York: Dover, 1956], III, 261). Sacrobosco probably derived it from one of the versions of Adelard of Bath's translation of Euclid's *Elements,* perhaps from Campanus of Novara's popular edition (see Selection 27, n. 1.).

4. From Theodosius of Bithynia's (b. ca. 180 B.C.) *On Spheres (De spheris),* a work originally written in Greek but translated into Latin in the twelfth century by Gerard of Cremona (see Selection 7, no. 5).

5. Although Sacrobosco mentions an accompanying figure, none appears here or elsewhere in Thorndike's edition and translation. Perhaps Sacrobosco never supplied diagrams, or, if he did, they may have been omitted in the subsequent manuscript copies. Whatever the explanation, the figure intended here was probably one that contained all the celestial spheres and their associated planets arranged concentrically.

By accident the sphere is divided into the sphere right and the sphere oblique. For those are said to have the sphere right who dwell at the equator, if anyone can live there. And it is called "right" because neither pole is elevated more for them than the other, or because their horizon intersects the equinoctial circle and is intersected by it at spherical right angles. Those are said to have the sphere oblique who live this side of the equator or beyond it. For to them one pole is always raised above the horizon, and the other is always depressed below it. Or it is because their artificial horizon intersects the equinoctial at oblique and unequal angles.

The Four Elements.

The machine of the universe is divided into two, the ethereal and the elementary region. The elementary region, existing subject to continual alteration, is divided into four. For there is earth, placed, as it were, as the center in the middle of all, about which is water, about water air, about air fire, which is pure and not turbid there and reaches to the sphere of the moon, as Aristotle says in his book of *Meteorology.* For so God, the glorious and sublime, disposed. And these are called the "four elements" which are in turn by themselves altered, corrupted and regenerated. The elements are also simple bodies which cannot be subdivided into parts of diverse forms and from whose commixture are produced various species of generated things. Three of them, in turn, surround the earth on all sides spherically, except in so far as the dry land stays the sea's tide to protect the life of animate beings. All, too, are mobile except earth, which, as the center of the world, by its weight in every direction equally avoiding the great motion of the extremes, as a round body occupies the middle of the sphere.

The Heavens.

Around the elementary region revolves with continuous circular motion the ethereal, which is lucid and immune from all variation in its immutable essence. And it is called "Fifth Essence" by the philosophers. Of which there are nine spheres, as we have just said: namely, of the moon, Mercury, Venus, the sun, Mars, Jupiter, Saturn, the fixed stars, and the last heaven. Each of these spheres incloses its inferior spherically.

Their Movements.

And of these there are two movements. One is of the last heaven on the two extremities of its axis,

the Arctic and Antarctic poles, from east through west to east again, which the equinoctial circle divides through the middle. Then there is another movement, oblique to this and in the opposite direction, of the inferior spheres on their axes, distant from the former by 23 degrees. But the first movement carries all the others with it in its rush about the earth once within a day and night although they strive against it, as in the case of the eighth sphere one degree in a hundred years.[6] This second movement is divided through the middle by the zodiac, under which each of the seven planets has its own sphere, in which it is borne by its own motion, contrary to the movement of the sky, and completes it in varying spaces of time—in the case of Saturn in thirty years, Jupiter in twelve years, Mars in two, the sun in three hundred and sixty-five days and six hours, Venus and Mercury about the same, the moon in twenty-seven days and eight hours.

Revolution of the Heavens from East to West.

That the sky revolves from east to west is signified by the fact that the stars, which rise in the east, mount gradually and successively until they reach mid-sky and are always at the same distance apart, and, thus maintaining their relative positions, they move toward their setting continuously and uniformly. Another indication is that the stars near the North Pole, which never set for us, move continuously and uniformly, describing their circles about the pole, and are always equally near or far from one another. Wherefore, from those two continuous movements of the stars, both those that set and those which do not, it is clear that the firmament is moved from east to west.

The Heavens Spherical.

There are three reasons why the sky is round: likeness, convenience, and necessity. Likeness, because the sensible world is made in the likeness of the archetype, in which there is neither end nor beginning; wherefore, in likeness to it the sensible world has a round shape, in which beginning or end cannot be distinguished. Convenience, because of all isoperimetric bodies the sphere is the largest and of all shapes the round is most capacious. Since largest and round, therefore the most capacious. Wherefore, since the world is all-containing, this

6. The value of the precession of the equinoxes as given in Ptolemy's *Almagest* and subsequently cited in other astronomical treatises, for example, in Alfraganus' popular astronomical work (see n. 8).

shape was useful and convenient for it. Necessity, because if the world were of other form than round—say, trilateral, quadrilateral, or many-sided—it would follow that some space would be vacant and some body without a place, both of which are false, as is clear in the case of angles projecting and revolved.[7]

A Further Proof.

Also, as Alfraganus says,[8] if the sky were flat, one part of it would be nearer to us than another, namely, that which is directly overhead. So when a star was there, it would be closer to us than when rising or setting. But those things which are closer to us seem larger. So the sun when in mid-sky should look larger than when rising or setting, whereas the opposite is the case; for the sun or another star looks bigger in the east or west than in mid-sky. But, since this is not really so, the reason for its seeming so[9] is that in winter and the rainy season vapors rise between us and the sun or other star. And, since those vapors are diaphanous, they scatter our visual rays so that they do not apprehend the object in its true size, just as is the case with a penny dropped into a depth of limpid water, which appears larger than it actually is because of a like diffusion of rays.

The Earth a Sphere.

That the earth, too, is round is shown thus.[10] The signs and stars do not rise and set the same for all men everywhere but rise and set sooner for those in the east than for those in the west; and of this there is no other cause than the bulge of the earth. Moreover, celestial phenomena evidence that they rise sooner for orientals than for westerners. For one and the same eclipse of the moon which appears to us in the first hour of the night appears to orientals about the third hour of the night, which proves that they had night and sunset before we did, of which setting the bulge of the earth is the cause.

Further Proofs of This.

That the earth also has a bulge from north to south and vice versa is shown thus: To those living toward the north, certain stars are always visible, namely, those near the North Pole, while others which are near the South Pole are always conceal-ed from them. If, then, anyone should proceed from the north southward, he might go so far that the stars which formerly were always visible to him now would tend toward their setting. And the

farther south he went, the more they would be moved toward their setting. Again, that same man now could see stars which formerly had always been hidden from him. And the reverse would hap-pen to anyone going from the south northward. The cause of this is simply the bulge of the earth. Again, if the earth were flat from east to west, the stars would rise as soon for westerners as for orien-tals, which is false. Also, if the earth were flat from north to south and vice versa, the stars which were always visible to anyone would continue to be so wherever he went, which is false. But it seems flat to human sight because it is so extensive.

Surface of the Sea Spherical.

That the water has a bulge and is approximately round is shown thus: Let a signal be set up on the seacoast and a ship leave port and sail away so far that the eye of a person standing at the foot of the mast can no longer discern the signal. Yet if the ship is stopped, the eye of the same person, if he has climbed to the top of the mast, will see the signal clearly. Yet the eye of a person at the bot-tom of the mast ought to see the signal better than he who is at the top, as is shown by drawing straight lines from both to the signal. And there is no other explanation of this thing than the bulge of the water. For all other impediments are excluded, such as clouds and rising vapors.[11]

7. Aristotle's *De caelo* (II.4.278a.12–23) is the source of the argument for the *necessary* sphericity of the uni-verse.

8. A reference to the *Differentie scientie astrorum* (or *Liber de aggregationibus,* as it was also known) of Al-Farghani (known as Alfraganus in Latin) (fl. 813–833), an Arab astronomer. Translated twice into Latin (by John of Seville and Gerard of Cremona), it was influen-tial in the history of medieval astronomy. For a modern Latin edition, see Francis J. Carmody, *Al Farghani, Differentie scientie astrorum* (Berkeley, Calif., 1943).

9. The explanation which follows is ultimately—if not directly—derived from Ptolemy's *Almagest,* Book I, chapter 3. There, Ptolemy explains that "what makes the apparent size of a heavenly body greater when it is near the horizon is not its smaller distance but the vapor-ous moisture surrounding the earth between our eye and the heavenly body. It is the same as when objects im-mersed in water appear larger, and in fact, the more deeply immersed, the larger" (trans. Cohen and Drab-kin, *A Source Book in Greek Science* [Cambridge, Mass.: Harvard University Press, 1948], p. 124, and repeated on p. 283, where, in a note, they explain that "the apparent increase of size of sun and moon at the horizon is largely due to an optical illusion, not to refraction").

10. In the next two paragraphs the substance of the arguments for the earth's sphericity are derived ultimate-ly from Ptolemy's *Almagest,* Book I, chapter 4.

11. In a similar manner, but using a different specific

Also, since water is a homogeneous body, the whole will act the same as its parts. But parts of water, as happens in the case of little drops and dew on herbs, naturally seek a round shape. Therefore, the whole, of which they are parts, will do so.

The Earth Central.

That the earth is in the middle of the firmament is shown thus. To persons on the earth's surface the stars appear of the same size whether they are in mid-sky or just rising or about to set,[12] and this is because the earth is equally distant from them. For if the earth were nearer to the firmament in one direction than in another, a person at that point of the earth's surface which was nearer to the firmament would not see half of the heavens. But this is contrary to Ptolemy[13] and all the philosophers, who say that, wherever man lives, six signs rise and six signs set, and half of the heavens is always visible and half hid from him.

And a Mere Point in the Universe.

That same consideration is a sign that the earth is as a center and point with respect to the firmament, since, if the earth were of any size compared with the firmament, it would not be possible to see half the heavens. Also, suppose a plane passed through the center of the earth, dividing it and the firmament into equal halves. An eye at the earth's center would see half the sky, and one on the earth's surface would see the same half. From which it is inferred that the magnitude of the earth from surface to center is inappreciable and, consequently, that the magnitude of the entire earth is inappreciable compared to the firmament.[14] Also Alfraganus says that the least of the fixed stars which we can see is larger than the whole earth. But that star, compared with the firmament, is a mere point. Much more so is the earth, which is smaller than it.

The Earth Immobile.

That the earth is held immobile in the midst of all, although[15] it is the heaviest, seems explicable thus. Every heavy thing tends toward the center. Now the center is a point in the middle of the firmament. Therefore, the earth, since it is heaviest, naturally tends toward that point.[16] Also, whatever is moved from the middle toward the circumference ascends. Therefore, if the earth were moved from the middle toward the circumference, it would be ascending, which is impossible.

Measuring the Earth's Circumference.

The total girth of the earth by the authority of the philosophers Ambrose, Theodosius, and Eratosthenes is defined as comprising 252,000 stades, which is allowing 700 stades for each of the 360 parts of the zodiac (*sic*).[17] For let one take an astrolabe on a clear starry night and, sighting the pole through both apertures in the indicator, note the number of degrees where it is. Then let our measurer of the cosmos proceed directly north until on another clear night, observing the pole as before, the indicator stands a degree higher. After this let the extent of his travel be measured, and it will be found to be 700 stades. Then, allowing this many stades for each of 360 degrees, the girth of the earth is found.

And Diameter.

From these data the diameter of the earth can be found thus by the rule for the circle and diameter. Subtract the twenty-second part from the circuit of the whole earth, and a third of the remainder— that is, 80,181 stades and a half and third part of one stade—will be the diameter or thickness of the terrestrial ball.[18]

CHAPTER 2

OF THE CIRCLES AND THEIR NAMES

Celestial Circles.

Of these circles some are larger, some smaller,

illustration, Ptolemy, in Book I, chapter 4, of the *Almagest,* also asserts the sphericity of the surface of the earth's waters when he declares that "whenever we sail towards mountains or any high places from whatever angle and in whatever direction, we see their bulk little by little increasing as if they were arising from the sea, whereas before they seemed submerged because of the curvature of the water's surface" (trans. R. C. Taliaferro in *Great Books of the Western World* [Chicago: Encyclopaedia Britannica, 1952], XVI, 9).

12. Earlier Sacrobosco had observed that the celestial bodies *appear* larger (though they are not actually so) when rising or setting and attributed this to refraction (see the previous paragraph "A Further Proof" and n. 9). Here, in apparent contradiction, he declares flatly that the stars *appear (stelle apparent)* exactly the same size in every part of the sky.

13. See *Almagest,* Book I, chapter 5 ("That the Earth is in the Middle of the Heavens"; Taliaferro, p. 10).

14. Ptolemy is again the ultimate source of this argument in *Almagest,* Book I, chapter 6 ("That the Earth Has the Ratio of a Point to the Heavens").

15. In his lengthy review of Thorndike's volume on Sacrobosco, Edward Rosen (*Isis,* XL [1949], 260, col. 1) rightly insists that in this context *cum* be translated as "because" rather than "although."

16. This Aristotelian argument could have been derived from Ptolemy's *Almagest,* Book I, chapter 7 (see Selection 67.1).

17. See Selection 81, chapter 5 and n. 19.

18. See Selection 81, chapter 5 and n. 22.

as sense shows. For a great circle in the sphere is one which, described on the surface of the sphere about its center, divides the sphere into two equal parts, while a small circle is one which, described on the surface of the sphere, divides it not into two equal but into two unequal portions.

The Equinoctial.

Of the great circles we must first mention the equinoctial. The equinoctial[19] is a circle dividing the sphere into two equal parts and equidistant at its every point from either pole. And it is called "equinoctial" because, when the sun crosses it, which happens twice a year, namely, in the beginning of Aries and in the beginning of Libra, there is equinox the world over. Wherefore it is termed the "equator of day and night," because it makes the artificial day[20] equal to the night. And 'tis called the "belt of the first movement."

The Two Movements Again.

Be it understood that the "first movement" means the movement of the *primum mobile,* that is, of the ninth sphere or last heaven, which movement is from east through west back to east again, which also is called "rational motion" from resemblance to the rational motion in the microcosm, that is, in man, when thought goes from the Creator through creatures to the Creator and there rests.

The second movement is of the firmament and planets contrary to this, from west through east back to west again, which movement is called "irrational" or "sensual" from resemblance to the movement of the microcosm from things corruptible to the Creator and back again to things corruptible.

The North and South Poles.

'Tis called the "belt of the first movement" because it divides the *primum mobile* or ninth sphere into two equal parts and is itself equally distant from the poles of the world. It is to be noted that the pole which always is visible to us is called "septentrional," "arctic," or "boreal." Septentrional" is from *septentrio,* that is, from Ursa Minor, which is derived from *septem* and *trion,* meaning "ox," because the seven stars in Ursa move slowly, since they are near the pole. Or those seven stars are called *septentriones* as if *septem teriones,* because they tread the parts about the pole. "Arctic" is derived from *arthos,* which is Ursa Maior, for 'tis near Ursa Maior. It is called

"boreal" because it is where the wind Boreas comes from. The opposite pole is called "Antarctic" as opposed to "Arctic." It also is called "meridional" because it is to the south, and it is called "austral" because it is where the wind Auster comes from. The two fixed points in the firmament are called the "poles of the world" because they terminate the axis of the sphere and the world revolves on them. One of these poles is always visible to us, the other always hidden. Whence Virgil:

This vertex is ever above us, but that
Dark Styx and deep Manes hold beneath our
           feet,[21]

The Zodiac.

There is another circle in the sphere which intersects the equinoctial and is intersected by it into two equal parts. One half of it tips toward the north, the other toward the south. That circle is called "zodiac" from *zoe,* meaning "life," because all life in inferior things depends on the movement of the planets beneath it. Or it is derived from *zodias,* which means "animal," because, since it is divided into twelve equal parts, each part is called a sign and has its particular name from the name of some animal, because of some property characteristic of it and of the animal, or because of the arrangement of the fixed stars there in the outline of that kind of animal. That circle in Latin is called *signifer* because it bears the "signs" or because it is divided into them. By Aristotle in *On Generation and Corruption* it is called the "oblique circle," where he says that, according to the access and recess of the sun in the oblique circle, are produced generations and corruptions in things below.[22]

The Twelve Signs.

The names, order, and number of the signs are
set forth in these lines:

There are Aries, Taurus, Gemini, Cancer, Leo,
           Virgo,

19. That is, the celestial equator.
20. In his review (p. 259, cols. 1–2) Rosen observes that "the 'natural day' had traditionally been defined as the period from sunrise to sunset, in contradistinction to the full period of twenty-four hours from sunrise to sunrise. Sacrobosco reversed the terminology, applying the term 'natural day' to the twenty-four hour period, and 'artificial day' to the interval from sunup to sundown."
21. *Georgics* I. 242–243.
22. *On Generation and Corruption* I. 10.336a.32— 336b.24.

Libra and Scorpio, Architenens, Caper, Amphora, Pisces.

Morevoer, each sign is divided into 30 degrees, whence it is clear that in the entire zodiac there are 360 degrees. Also, according to astronomers, each degree is divided into minutes, each minute into 60 seconds, each second into 60 thirds, and so on. And as the zodiac is divided by astronomers, so each circle in the sphere, whether great or small, is divided into similar parts.

While every circle in the sphere except the zodiac is understood to be a line or circumference, the zodiac alone is understood to be a surface, 12 degrees wide of degrees such as we have just mentioned. Wherefore, it is clear that certain persons in astrology lie who say that the signs are squares, unless they misuse this term and consider square and quadrangle the same. For each sign is 30 degrees in longitude, 12 in latitude.

The Ecliptic.

The line dividing the zodiac in its circuit, so that on one side it leaves 6 degrees and on the other side another 6, is called the "ecliptic," since when sun and moon are on that line there occurs an eclipse of sun or moon. The sun always moves beneath[23] the ecliptic, but all the other planets decline toward north or south; sometimes, however, they are beneath the ecliptic.[24] The part of the zodiac which slants away from the equinoctial to the north is called "northern" or "boreal" or "Arctic," and those six signs which extend from the beginning of Aries to the end of Virgo are called "northern." The other part of the zodiac which tips from the equinoctial toward the south is called "meridional" or "austral," and the six signs from the beginning of Libra to the end of Pisces are called "meridional" or "austral."

Extended Uses of "Sign."

When it is said that the sun is in Aries or in another sign, it should be understood that *in* is taken for *beneath* according as we now accept sign. In another meaning a sign is called a "pyramid," whose quadrilateral base is that surface which we call a "sign," while its apex is at the center of the earth. And in this sense we may properly say that the planets are *in* signs. "Sign" may be used in a third way as produced by six circles passing through the poles of the zodiac and through the beginnings of the twelve signs. Those six circles divide the entire surface of the sphere into twelve parts, wide in the middle but narrower toward the

poles, and each such part is called a "sign" and has a particular name from the name of that sign which is intercepted between its two lines. And according to this usage stars which are near the poles are said to be "in signs." Also think of a body whose base is a sign in this last sense which we have accepted but whose edge is on the axis of the zodiac. Such a body is called a "sign" in a fourth sense, according to which usage the whole world is divided into twelve equal parts, which are called "signs," and so whatever is in the world is in some sign.

Colures.

There are two other great circles in the sphere which are called "colures," whose function is to distinguish solstices and equinoxes.[25] "Colure" is derived from *colon*, which is a member, and *uros*, which is a wild ox, because, just as the lifted tail of the wild ox, which is its member, describes a semicircle and not a complete circle, so a colure always appears to us imperfect because only one half of it is seen.

The colure distinguishing the solstices passes through the poles of the universe and through the poles of the zodiac and through the greatest declinations of the sun, that is, through the first degrees of Cancer and Capricorn. Wherefore, the first point of Cancer, where that colure intersects the zodiac, is called the "point of the summer solstice," because, when the sun is in it, the summer solstice occurs and the sun cannot approach further toward our zenith. The zenith is a point in the firmament directly above our heads. The arc of the colure which is intercepted between the point of the summer solstice and the equinoctial point[26] is called the "sun's greatest declination" and is according to Ptolemy, 23 degrees and 51 minutes, according to Almeon,[27] 23 degrees and 33 minutes. Similarly, the first point of Capricorn is called the "point of the winter solstice," and the arc of the colure intercepted between that point and the

23. Although the expression is *sub ecliptica,* Sacrobosco means that the sun moves along the ecliptic line as its immediate background. It is not inclined (or declined) to the ecliptic as are the other planets.

24. Presumably, when they cross the ecliptic.

25. See Selection 81, chapter 6 and n. 32.

26. Rather than "equinoctial point," *equinoctialem* should be translated here as "celestial equator" (see Rosen, p. 259, col. 1).

27. Thorndike notes that Almeon probably represents the Abbasid caliph al-Ma'mun, who ruled from 813 to 833. It was during his reign that a value of 23 degrees 33 minutes was determined for the inclination of the ecliptic.

equinoctial is called the "sun's greatest declination" and is equal to the former.

The other colure passes through the poles of the universe and through the points of Aries and Libra where are the two equinoxes, whence it is called the "colure distinguishing the equinoxes." Those two colures intersect at the poles of the world at spherical right angles. The signs of the solstices and equinoxes are stated in these verses:

These two solstices make, Cancer and Capricorn
But Aries and Libra equal the nights to days.[28]

The Meridian.

There are yet two other great circles in the sphere, namely, the meridian and the horizon. The meridian is a circle passing through the poles of the world and through our zenith, and it is called "meridian" because, wherever a man may be and at whatever time of year, when the sun with the movement of the firmament reaches his meridian, it is noon for him. For like reason it is called the "circle of midday." And it is to be noted that cities of which one is farther east than the other have different meridians. The arc of the equinoctial intercepted between two meridians is called the "longitude" of the city.[29]. If two cities have the same meridian, then they are equally distant from east and from west.

The Horizon.

The horizon is a circle dividing the lower hemisphere from the upper, whence it is called "horizon," that is, "limiter of vision." It is also called the "circle of the hemisphere." Moreover, the horizon is twofold—that is, right, and oblique or slanting. Those have a right horizon and right sphere whose zenith is on the equinoctial, since their horizon is a circle passing through the poles of the world cutting the equinoctial at right angles, wherefore it is called "right horizon" and "right sphere." But those to whom the pole of the world is raised above the horizon have an oblique or slanting horizon, since their horizon intersects the equinoctial at unequal and oblique angles and is called "oblique horizon" and the sphere "oblique" or "slanting." Moreover, the zenith over our heads is always the pole of the horizon.

Elevation of the Pole.

From these things it is evident that the elevation of the pole of the world above the horizon is as great as the distance of the zenith from the equator, which is shown in this way. Since in every natural day either colure twice joins or becomes identical with the meridian, whatever is true of one holds for the other. Take, then, a fourth part of the colure distinguishing the solstices, which is from the equinoctial to the pole. Take another fourth part of the same colure, which is from zenith to horizon. Since the zenith is the pole of the horizon, those two quarters, since they are quarters of one and the same circle, are equal. But if equals are subtracted from equals, or the same thing common to both is subtracted, the remainders will be equal. Therefore, if we subtract the common arc, namely, that between the zenith and the pole, the remainders will be equal, namely, the elevation of the pole above the horizon and the distance of the zenith from the equinoctial.

Tropics of Cancer and Capricorn.

Having told of the six great circles, we must speak of the four smaller circles. Be it noted, then, that the sun, when in the first point of Cancer of the summer solstice, as it is carried by the firmament describes a circle, which is the one last described by the sun in the direction of the Arctic pole. Wherefore it is called the "circle of the summer solstice" for the reason aforesaid, or the "summer tropic" from tropos, which is "turning," because then the sun begins to turn toward the lower hemisphere and to recede from us. The sun again, when in the first point of Capricorn or winter solstice, as it is carried by the firmament describes another circle which is the one last described by the sun in the direction of the Antarctic pole, whence 'tis called the "circle of the winter solstice" or the "winter tropic," because then the sun turns toward us.

Arctic and Antarctic Circles.

Since the zodiac slants from the equinoctial, the pole of the zodiac will decline from the pole of the world. Therefore, since the eighth sphere and the zodiac, which is a part of it, are moved about the axis of the world, the pole of the zodiac, too, will move about the pole of the world. And that circle which the pole of the zodiac describes about the Arctic pole of the world is called the "Arctic circle." And that circle which the other pole of the zodiac describes about the Antarctic pole is called the "Antarctic circle."

28. The author of these lines is unidentified.
29. Since the Latin term is *civitatum*, "of the city" should be translated "of the cities," that is, the intercepted arc provides a measure of arcal distance between the cities.

As great as is the maximum declination of the sun, so great is the distance of the pole of the world from the pole of the zodiac, which is shown in this way. Take the colure distinguishing the solstices which passes through the poles of the world and the poles of the zodiac. Since all quarters of one and the same circle are equal, the quarter of this colure between equator and pole is equal to the quarter of the same colure from the first point of Cancer to the pole of the zodiac. Then, if we subtract from those equals the common arc from the first point of Cancer to the pole of the world, the remainders will be equal, namely, the maximum declination of the sun and the distance from the pole of the world to the pole of the zodiac. Moreover, since the Arctic circle at every point is equidistant from the pole of the world, it is evident that that part of the colure which lies between the first point of Cancer and the Arctic circle is almost double the maximum declination of the sun or the arc of the same colure intercepted between the Arctic circle and the Arctic pole, which is equal to the maximum declination of the sun. Since that colure, like other circles in the sphere, has 360 degrees, a quarter of it will be 90 degrees. Then, since the maximum declination of the sun according to Ptolemy is 23 degrees and 51 minutes and of as many degrees is the arc which is between the Arctic circle and the Arctic pole, if those two combined, which make about 48 degrees, are subtracted from 90, the remainder will be 42 degrees, as is the arc of the colure which lies between the first point of Cancer and the Arctic circle. So it is clear that that arc is almost double the maximum declination of the sun.[30]

It is also to be noted that the equinoctial with the four small circles are called "parallels," as it were equidistant, not that the first is as far from the second as the second is from the third, because this is false, as has already been shown, but because any two taken together are equidistant at every point. They are called the "equinoctial parallel," the "parallel of the summer solstice," the "parallel of the winter solstice," the "Arctic parallel", and the "Antarctic parallel." It is further to be noted that the four minor parallels, namely, the two tropics and the Arctic parallel and Antarctic parallel, distinguish five zones or five regions in the heaven. Wherefore, Virgil:

Five zones possess the sky, of which one is ever
Red from blazing sun and ever burnt by fire.[31]

Also a like number of zones is distinguished on earth directly beneath the said zones. Wherefore, Ovid:

. . .and just as many zones are marked on earth.[32]

The Five Zones.

That zone which lies between the tropics is said to be uninhabitable[33] because of the heat of the sun, which ever courses between the tropics. Similarly, the zone of earth directly beneath it is said to be uninhabitable because of the fervor of the sun, which ever courses above it. But those two zones which are described by the Arctic circle and the Antarctic circle about the poles of the world are uninhabitable because of too great cold, since the sun is far removed from them. The same is to be understood of the zones of earth directly beneath them. But those two zones of which one is between the summer tropic and the Arctic circle and the other between the winter tropic and the Antarctic circle are habitable and tempered from the heat of the torrid zone between the tropics and from the cold of the extreme zones which lie about the poles. The same is to be understood of the stretches of earth directly beneath them.

CHAPTER 4[34]

Movement of the Sun.

It should be noted that the sun has a single circle in which it is moved in the plane of the ecliptic, and it is eccentric. Any circle is called "eccentric"[35] which, like that of the sun, dividing the earth into equal parts, does not have the same center as the earth but one outside it. Moreover, the point in the eccentric which approaches closest to the firmament is called *aux* or *augis,* meaning

30. Compare to Pierre d'Ailly's discussion in Selection 81, chapter 6. Oresme, on whose *Traitié de l'espere* d'Ailly depended directly, probably drew upon Sacrobosco's account of the four smaller circles and the sun's maximum declination.

31. *Georgics* I. 233–34.

32. *Metamorphoses* I. 48.

33. Compare d'Ailly's account in Selection 81, chapter 6, and see also notes 35 and 38. In the latter note it will be seen that Columbus disagreed with Sacrobosco and d'Ailly by insisting upon the habitability of the torrid zone.

34. In chapter 3, which has been omitted here, Sacrobosco discusses risings and settings, ascensions, inequalities of days, the number of times the sun passes directly overhead during a year for those whose zeniths are at the four smaller circles and the zones between them, and the seven climes.

35. Sacrobosco's definitions of "eccentric," "aux" and other technical astronomical terms in this fourth chapter should be compared to those given in the *Theorica planetarum* in the next section.

"elevation." The opposite point, which is farthest removed from the firmament, is called the "opposition" of the *aux*.

Moreover, there are two movements of the sun from west to east, one of which is its own in its eccentric, by which it is moved every day and night about 60 minutes. The other is the slower movement of the sphere itself on the poles of the axis of the circle of the signs, and it is equal to the movement of the sphere of the fixed stars, namely, 1 degree in a hundred years. From these two movements, then, is reckoned the sun's course in the circle of the signs from west to east, by which it cleaves the circle of the signs in 365 days and a fourth of one day, except for a small fraction which is imperceptible.

Of the Other Planets: Equant, Deferent, and Epicycle.

Every planet except the sun has three circles, namely, equant, deferent,[36] and epicycle. The equant[37] of the moon is a circle concentric with the earth and in the plane of the ecliptic. Its deferent is an eccentric circle not in the plane of the ecliptic —nay, one half of it slants toward the north and the other toward the south—and the deferent intersects the equant in two places, and the figure of that intersection is called the "dragon" because it is wide in the middle and narrow toward the ends. That intersection, then, through which the moon is moved from south to north is called the "head of the dragon," while the other intersection through which it is moved from north to south is called the "tail of the dragon."[38] Deferent and equant of each planet are equal, and know that both deferent and equant of Saturn, Jupiter, Mars, Venus, and Mercury are eccentric and outside the plane of the ecliptic, and yet those two are in the same plane. Also every planet except the sun has an epicycle. An epicycle is a small circle on whose circumference is carried the body of the planet, and the center of the epicycle is always carried along the circumference of the deferent.

Stationary, Direct, and Retrograde.

If, then, two lines are drawn from the center of the earth to include an epicycle, one on the east and the other on the west, the point of contact on the east is called the "first station," while the point of contact to the west is called the "second station." And when a planet is in either of those stations it is called "stationary." The upper arc of the epicycle intercepted between those two

stations is called "direction," and when the planet is there it is called "direct." But the lower arc of the epicycle between the two stations is called "retrogradation," and a planet existing there is called "retrograde." But the moon is not stationary, direct, or retrograde because of the swiftness of its motion in its epicycle.[39]

Cause of Lunar Eclipse.

Since the sun is larger than the earth, it is necessary that half the sphere of earth be always illuminated by the sun and that the shadow of the earth, extended into the air like a cone, diminish in circumference[40] until it ends in the plane of the circle of the signs inseparable from the nadir of the sun. The nadir is a point in the firmament directly opposite to the sun. Hence, when the moon at full is in the head or tail of the dragon beneath the nadir of the sun, then the earth is interposed between sun and moon, and the cone of the earth's shadow falls on the body of the moon. Wherefore, since the moon has no light except from the sun, it actually is deprived of light and there is a general[41] eclipse, if it is in the head or tail of the dragon directly but partial if it is almost within the bounds determined for eclipse. And it always happens at full moon or thereabouts. But, since in every opposition—that is, at full moon—the moon is not in the head or tail of the dragon or beneath the nadir of the sun, it is not necessary that the moon suffer eclipse at every full moon.

---

36. In the *Theorica planetarum* these circles are called "eccentric equant" and "eccentric deferent" (see next section, par. 32).

37. Sacrobosco's use of the term "equant" here is at the least unfortunate, and probably confused. For the relation of the lunar circles, see paragraphs 10–16 and 29–30 of the *Theorica planetarum*, in the next section.

38. It is not clear whether the two intersecting circles described here, namely the "equant" and deferent circles of the moon, are identical with the two intersecting circles described in the *Theorica planetarum*, paragraph 29, namely the eccentric circle of the moon and the eccentric circle of the sun, or ecliptic. Perhaps Sacrobosco's "equant" corresponds to the eccentric circle carrying the nodes 3 minutes daily from east to west, described in paragraph 30 of the *Theorica*.

39. A more faithful translation of this last sentence would be: "However, the moon is not assigned a station, direction, or retrogradation and therefore is not said to be stationary, direct, or retrograde because of the swiftness of its motion in its epicycle."

40. "Roundness" would be a better rendering of *rotunditate* than "circumference."

41. In this context *generalis* should be translated as "total" rather than "general."

Cause of Solar Eclipse.

When the moon is in the head or tail of the dragon or nearly within the limits and in conjunction with the sun, then the body of the moon is interposed between our sight and the body of the sun. Hence it will obscure the brightness of the sun for us, and so the sun will suffer eclipse—not that it ceases to shine but that it fails us because of the interposition of the moon between our sight and the sun. From these it is clear that a solar eclipse should always occur at the time of conjunction or new moon. And it is to be noted that when there is an eclipse of the moon, it is visible everywhere on earth. But when there is an eclipse of the sun, that is by no means so. Nay, it may be visible in one clime and not in another, which happens because of the different point of view in different climes. Whence Virgil most aptly and concisely expresses the nature of either eclipse:

> Varied defects of the moon, and of the sun
> travails.[42]

Eclipse during the Passion Miraculous.

From the aforesaid it is also evident that, when the sun was eclipsed during the Passion and the same Passion occurred at full moon, that eclipse was not natural—nay, it was miraculous and contrary to nature, since a solar eclipse ought to occur at new moon or thereabouts. On which account Dionysius the Areopagite is reported to have said during the same Passion, "Either the God of nature suffers, or the mechanism of the universe is dissolved."

2. Anonymous: The Theory of the Planets

Translation and introduction by Olaf Pedersen

Annotated by Edward Grant

From the middle of the thirteenth century the primary source of theoretical astronomy was Ptolemy's *Almagest,* which was accessible in various Latin translations. This text was, however, much too advanced and difficult to be of any use to students at an elementary level, and comparatively few manuscripts of these translations are known. For teaching purposes in the Liberal Arts course in the universities it was therefore usually replaced by one of a number of elementary manuals, the most popular of which was the *Theorica Planetarum* which is still extant in hundreds of manuscripts dating from the thirteenth to the sixteenth centuries. In astronomical codices from the Middle Ages it was often placed immediately after the equally popular works of John of Sacrobosco, the *Tractatus de Sphaera,* the *Compotus,* and the *Algorithmus.* It was usually followed by a set of astronomical tables and one or two treatises on instruments such as the quadrant or the astrolabe, the whole forming a kind of astronomical compendium which, with some additions, remained essentially unchanged throughout the centuries.

Unfortunately neither the author nor the date of the *Theorica Planetarum* can be determined with certainty. The great majority of the manuscripts are anonymous, from which it follows that the author was unknown to most medieval scholars. A few among them ascribed the text to one of a dozen authors, among whom was the great translator Gerard of Cremona (d. 1187), and in the many early printed editions it was commonly listed among his works. In the nineteenth century Boncompagni [*Atti dell'Accad. Pont. de'Nuovi Lincei* IV (1852), 449 ff] rejected this ascription and tried to establish a case for the Italian astrologer Gerard of Sabbioneta (about 1250) as the author, whereas Duhem [*Le Système du Monde* (Paris: Hermann, 1913–1959), III, 219 ff.] sought to give it back to Gerard of Cremona. However, in the light of modern scholarship none of these assumptions can be maintained and the question remains unsolved. The most probable hypothesis is that the *Theorica Planetarum* was composed by an unknown teacher of astronomy as a manual destined to supplement the meager treatment of planetary theory in the fourth book of the *Sphaera* of Sacrobosco. There are some reasons to suppose that this happened in Paris sometime during the period 1260 to 1280.

The translation given below is based upon one of the oldest and best preserved manuscripts in the Royal Library in Copenhagen as MS. Latin Add. 447, 2°, 49r–56r, dating from perhaps a little before 1300. The division into eight chapters is found in most manuscripts, and also the chapter headings given here. For the sake of easy reference the translator has found it convenient to divide the

42. *Georgics* II.478.

text into short paragraphs numbered consecutively from 1 to 121. As for the translation, some of the technical terms, for instance, *aux* or *motus,* have to be used in their Latin form because in their Ptolemaic sense they have no precise equivalents in modern English. The reader should consult Professor D. J. de Solla Price's *The Equatorie of the Planetis* (Cambridge, Eng.: The University Press, 1955) for a more detailed discussion of such problems.

Here begins the theory of the seven planets.

I. THEORY OF THE SUN'S MOTION

1. A circle whose center is not at the center of the world is called an eccentric circle, or a circle with a displaced cusp, or an outgoing center.

The part of the eccentric which has the greatest distance from the center of the world is called aux, or the greatest distance. The part nearest to the center is called the opposite aux, or the least distance. Therefore, the two parts of the circle situated in the middle between the aux and the opposite aux are called the mean distances.

2. The Sun has its own motion in its eccentric circle, where it every day moves 59 minutes 8 seconds with uniform speed, so that it must perform a nonuniform motion through the zodiac from west to east; but the universe rotates from east to west with a uniform motion.

3. The mean motus of the Sun is an arc of the zodiac cut off between two lines, one from the center of the Earth to the place of Aries, and another from the same center to the firmament, parallel to a line from the center of the eccentric through the center of the Sun.

4. The true motus of the Sun is defined as an arc of the zodiac lying between the Head of Aries and a line from the center of the Earth through the center of the Sun.

5. The equation of the Sun is defined as an arc of the zodiac lying between the mean motus and the true motus of the Sun. The equation of the Sun is zero when the Sun is at the aux or the opposite aux, but when it is at the mean distances the equation is maximum.

6. The argument of the Sun is defined as an arc of the zodiac lying between the aux and the line defining the mean motus.

7. The aux of the Sun in the second sense is defined as an arc of the zodiac lying between the Head of Aries and the line through the aux of the eccentric circle.

8. Throughout one half of the heavens the mean motus is greater than the true motus so that the equation must here be subtracted; throughout the other half it is smaller, so that the equation must be added.

9. To find the mean motus of the Sun means to find a certain arc of the zodiac which has the same ratio to the whole zodiac as the arc of the eccentric circle described by the Sun has to the whole eccentric circle. This is found by the parallel line as seen in the following figure [Fig. 1].

II. THEORY OF THE VARIOUS MOTIONS OF THE MOON

10. The Moon's epicycle, or orbit of revolution, or small orbit, is defined as the small circle whose center moves from west to east upon the circumference of the eccentric.

11. The eccentric circle of the Sun is immovable apart from the motion of the eighth sphere, but the eccentric circle of the Moon moves every day about 11 degrees from east to west. The center of the eccentric describes a certain small circle around the center of the world. The center of the epicycle of the Moon moves about 13 degrees per day from west to east, and the aux of the eccentric moves 11 degrees per day from east to west. The center of the Sun moves almost 1 degree per day in the opposite direction to the firmament.

12. From this it appears that if the center of the Sun, and the aux of the eccentric of the Moon, and the center of the epicycle of the Moon were at a certain place of the heavens at a certain hour, then the following day the Sun would have moved 1 degree towards the east from the said place. The aux of the eccentric would have moved 11 degrees towards the west and the center of the lunar epicycle 13 degrees towards the east, wherefore the Sun will then be in the middle between the aux and the epicycle center. If, therefore, the distance between the Sun and the center of the lunar epicycle is doubled, we get the distance between the aux and the epicycle. This distance is called the Moon's centrum, or the double interval.

13. This makes it clear that these three points are either at the same place or that the Sun is in the middle between the two others, or at their opposite point.

14. It is also seen that the center of the lunar epicycle goes round the eccentric circle twice in a month, and that it is at the aux when in conjunction with, or in opposition to, the Sun. In quadrature it is at the least distance.

15. When the Moon is in the upper part of the

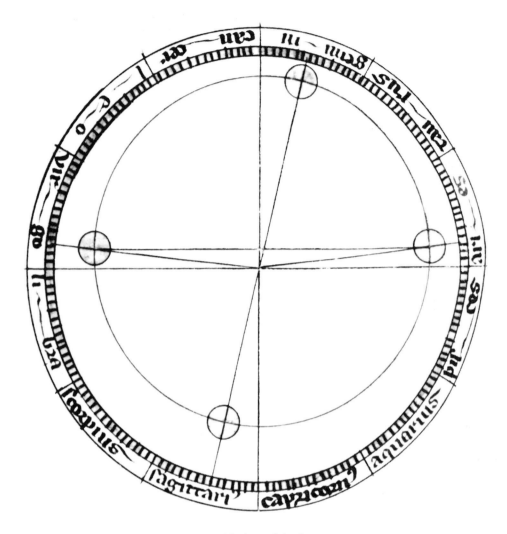

Fig. 1. Motions of the Sun.

circumference of the epicycle it moves from east to west and therefore has a slow motion. In the lower part the motion is in the opposite direction and therefore fast. The other planets behave in the opposite manner.

16. Just as the Sun moves uniformly around the center of its eccentric, the center of the lunar epicycle moves uniformly around the center of the world.

17. The mean motus of the Moon is an arc of the zodiac beginning at Aries and continuing according to the order of the signs, that is, Aries, Taurus, and so on, and ending at a line from the center of the Earth through the center of the epicycle.

18. The true place of the Moon is cut off by a line from the center of the world through the center of the Moon's body.

19. The mean aux of the epicycle is defined as the point of intersection between the upper part of the epicycle and a line passing through the center of the epicycle from a certain point opposite to the center of the Moon's eccentric, and with the same distance from the center of the Earth as this center of the eccentric.

20. The true aux is defined as the point of the epicycle where it is cut by a line from the center of the world through the center of the epicycle to the upper part of the latter.

21. The equation of center is a small arc of the epicycle lying between the mean and the true auges.

22. The mean argument of the Moon is defined as an arc lying between the mean aux and the center of the Moon's body and reckoned in accordance with the motion of the Moon upon the epicycle.

23. The true argument is defined as an arc between the true aux and the center of the Moon's body.

24. The equation of center is zero, that is, the true aux and the mean aux are the same, when the center of the epicycle is at the aux of the eccentric (that is, in mean conjunction and mean opposi-

tion) and also when it is at the opposite aux (that is, in quadratures between the Moon and the Sun).

25. The lunar equation of argument is defined as an arc of the zodiac lying between the mean motus and the true motus. This equation is zero when the epicycle center is at the aux or the opposite aux of the eccentric, the Moon being at the same time at the greatest or least distance of the epicycle. The equation of argument is maximum when the epicycle center is at the mean distances of the eccentric and the Moon at the mean distances of its epicycle.

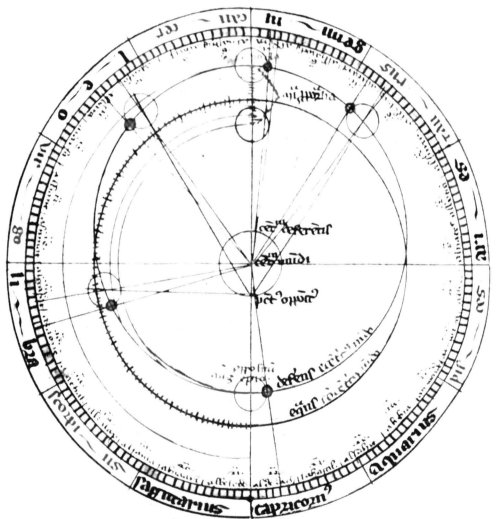

Fig. 2. Motions of the Moon. [(1) is the center of the world (*centrum mundi*), namely the Earth (concentric with it is the celestial sphere divided into its 12 familiar zodiacal signs; also shown are the 28 numbered houses of the Moon, whose names were transliterated from Arabic); (2) is the center of the deferent (*centrum deferentis*), that is, the center of the circle "bearing" the epicycle; (3) is the equant point (*punctus oppositus*) or center of mean motion; and (4) represents proportional minutes (*minuta proportionalia*). The abbreviated forms of the Latin terms used in the figure have been expanded in this caption, which was supplied by V. E. Thoren.—*Ed.*]

When the Moon is on the western half of the epicycle, the mean motus is greater than the true motus; here, therefore, the equation of argument must be subtracted. Upon the eastern half the opposite is true.

26. You must know that when the epicycle center is at the opposite aux of the eccentric, the equations of argument are greater than when it is at the aux. The difference between these equations of argument, with the epicycle center at the aux and the opposite aux, is called the equation of the diameter variation of the small circle [that is, of the epicycle].

27. The equations of argument which are recorded in the tables presuppose that the epicycle center is always at the aux of the eccentric; but when the epicycle center is found at other places, the equations of argument increase as the epicycle approaches the opposite aux, since the epicycle center comes nearer to the center of the Earth.

28. The equations of argument are found by means of proportional parts. These proportional parts are defined as sixty small parts of a line twice as long as the line between the center of the Earth and the center of the eccentric, and divided into sixty parts. The line from the center of the Earth to the opposite aux does not contain any of these parts, but the line to the aux contains them all. Other lines to other points contain a number of those parts, corresponding to their nearness or remoteness from the aux and the opposite aux, as seen from the figure [Fig. 2].

III. THEORY OF THE MOTION OF THE DRAGON'S HEAD

29. In the following we consider the Head and Tail of the Dragon. The eccentric circle of the Moon deviates from the path of the Sun in two directions, namely to the North and to the South, intersecting the eccentric circle of the Sun at two places which are always opposite. These intersections are called the Head and Tail of the Dragon. The intersection where the Moon begins to turn to the north is called the Head, and the other is the Tail.

30. These points of intersection move daily about 3 minutes from east to west. They are carried by a certain eccentric circle existing in the heaven of the Moon and equal to the eccentric in magnitude. It lies in the plane of the zodiac, that is, in the path of the Sun, and this motion is unlike the motion of the planets, which is from west to east.

31. In order to obtain a similarity in motions,

the Head of Genzahar is said to have a mean motion contrary to the firmament and equal to that which it actually has along with the firmament. It follows that the mean motus of the Head subtracted from twelve signs is equal to the true place of the Dragon's Head, reckoned according to the order of the signs, which appears from the following figure [Fig. 3].

IV. THEORY OF THE MOTIONS OF THE THREE SUPERIOR PLANETS

32. We shall now proceed with the three superior planets. You must therefore note that any of the three superior planets has two eccentric circles lying in the same plane and immovable, apart from the motion of the eighth sphere and the diurnal motion around the Earth from east to west. One of them is called the eccentric deferent, along the circumference of which the epicycle center is carried from west to east. The other is called the eccentric equant, about the center of which the center of the epicycle moves uniformly, describing equal angles in equal times.

33. These two eccentrics are equally elevated towards the same place in the heavens, and the one whose center is nearest to the Earth is the deferent. The other one is the equant. The deferent center has the same distance from the center of the Earth as from the center of the equant because it is in the middle. And these three centers are upon one line, and both eccentric circles have the same magnitude.

34. In the lunar theory the diameter of the epicycle points towards the center of the Earth when the epicycle center is at the aux or the opposite aux, and later, at other places of the eccentric, it inclines towards a certain point opposite to the center of the eccentric and at the same distance from the center of the world as the center of the eccentric. In the three superior planets the diameter of the epicycle likewise points to the center of the world when the epicycle center is at the aux or the opposite aux of the eccentric, but when the epicycle center is at other points of the eccentric this diameter points towards the center of the equant. We call this "reflection".[43]

35. On the superior part of the epicycle the Moon moves from east to west, but the opposite way on the inferior part. However, these three planets

43. The term "reflection" (reflexio) has a quite different application here than it does in the section translated in note 47, where it signifies terrestrial parallax.

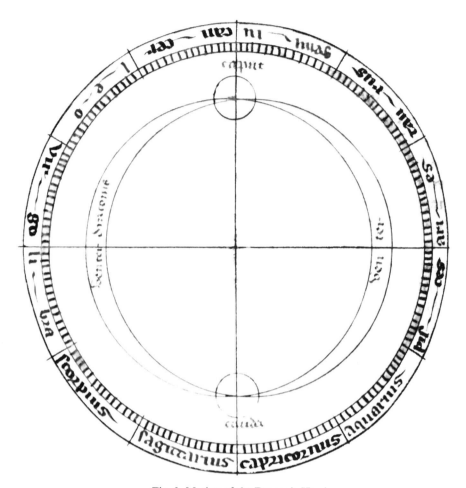

Fig. 3. Motion of the Dragon's Head.

move from west to east on the superior and the opposite way on the inferior part.

36. When any of these planets is in conjunction with the Sun with respect to the mean motus, it will always be upon the superior part of its epicycle, that is, at the mean aux. But in mean opposition to the Sun it will be at the lowest point of its epicycle, and in quadrature with the Sun it is at the mean distances of the epicycle. Therefore, it will go around the epicycle in the same time as the Sun needs for returning to a conjunction.

37. The mean aux of the epicycle is defined as a point upon its superior part where a line from the center of the equant through the center of the epicycle intersects the latter. This aux does not change.

38. The true aux is a point defined by a line from the center of the Earth through the center of the

epicycle, and this aux changes with the increase or decrease of the equation of center of the epicycle.

39. The equation of center of the epicycle is an arc of the latter lying between the mean and true auges.

40. The equation of center in the zodiac is an arc of the latter lying between the mean motus and true motus of the epicycle, and the ratio between the equation and its corresponding circle is the same in the two cases. This can be proved by a line lying between the two parallels. Consequently, if one equation is found in a table, we also have the other.

41. The mean motus of any of these planets, that is, of their epicycles, is an arc of the zodiac lying between Aries and a line from the center of the Earth parallel to the line passing from the center of the equant through the center of the epicycle.

42. The true place of the epicycle is an arc of the

zodiac between Aries and a line from the center of the Earth through the center of the epicycle.

43. The true place of the planet is determined by a line from the center of the Earth through the center of the planet.

44. The aux of the planets in the second sense is defined, as in the solar theory, as an arc of the zodiac beginning at Aries and ending at a line drawn from a certain point of the Earth just beneath the auges of the eccentric circle.

45. The mean centrum of the planet is defined as an arc of the zodiac between the aux of the eccentric and the mean motus of the epicycle. It is this arc which is called "argument" in the solar theory and centrum, double interval, or double distance in the Moon.

46. The true centrum is defined as an arc of the zodiac lying between the aux of the eccentric and the true place of the epicycle.

47. The mean argument is defined as an arc of the epicycle lying between the mean aux and the center of the planet.

48. The true argument is defined as an arc of the epicycle lying between the true aux and the center of the planet.

49. In one half of the heavens the equation of center in the zodiac must be subtracted from the mean centrum, whereas the equation of center in the epicycle must be added to the mean argument in order to obtain the true centrum and the true argument. In the other half the opposite is true, which is easily seen from a figure [Fig. 4]. When

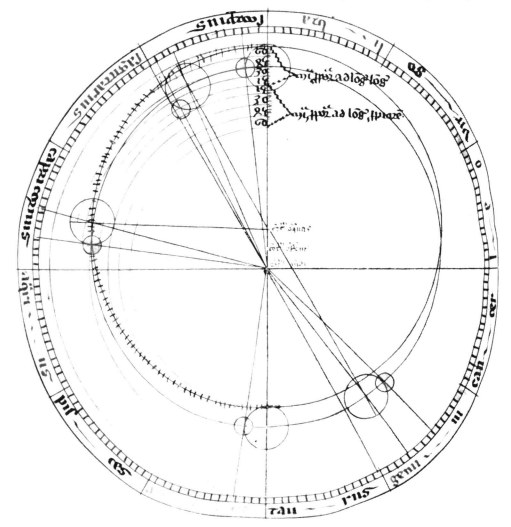

Fig. 4. Motions of the three superior planets.

the epicycle center is at the aux or the opposite aux, the said equations are zero.

50. The planet's equation of argument is defined as an arc of the zodiac between the true place of the planet and the true place of the epciycle.

51. If the true argument of the planet has the same value, it is clear that the more the epicycle center approaches the center of the Earth, the more the equation of argument will increase. The reason is that the equations of argument are greater when the epicycle center is at the least distance of the eccentric circle than when it is at the mean distances, and that the equations are greater at these mean distances than at the aux.

52. In tables the equations of argument are written as if the epicycle center were always at the mean distances of the equant, not because it moves upon the equant but because it has a uniform motion about the center of the equant. The epicycle center is said to be at the mean distance of the equant when the diameter of the epicycle is perpendicular to the diameter of the world passing through the center of the eccentric circles.

53. The differences between the equations at the mean distance and at the aux are called the diameter variations of the small circle [that is, the epicycle] corresponding to the greatest distance. The differences between the equations at the mean distance and at the opposite aux are called the diameter variations of the small circle corresponding to the least distance, and these variations are recorded in the tables.

54. The excess of a line drawn from the center of the Earth to the aux of the equant over another line from the same center to its mean distance is divided into sixty parts, called proportional parts corresponding to the greatest distance. The excess of a line drawn to the mean distance over a line drawn to the least distance is divided into sixty parts, called proportional parts corresponding to the least distance. Therefore, the diameter variations corresponding to the greatest distance must be subtracted from, or those corresponding to the least distance added to, the equation of argument, which is clearly seen from a figure [Fig. 4] by the attentive reader.

55. As mentioned above, the more the epicycle center approaches the center of the Earth, the more the equation of argument will increase, as seen from the following figure [Fig. 4].

V. THEORY OF THE MOTIONS AND CIRCLES OF VENUS AND MERCURY

56. We continue with Venus and Mercury. Mercury has two eccentric circles of the same magnitude and lying in the same plane, of which the equant is nearest to the center of the Earth. For the deferent center is twice as far away from the equant center as the latter is from the center of the Earth, because a certain small circle must pass through these two centers of the eccentrics.

57. The deferent center of Mercury moves upon the circumference of this small circle from east to west with a daily motion equal to that with which the Sun moves contrary to the firmament, carrying the aux with it. For in the same time in which the Sun goes round the firmament, the aux of the eccentric deferent and also any other point of it will go round the equant, and the center of the eccentric deferent will go round its small circle. Sometimes it therefore happens that the equant and deferent centers are at the same place, so that the two circles here are one and the same. Except for this moment, the deferent will be nearer to the firmament than the equant. The deferent moves so that it describes equal angles in equal times around the center of the equant.

58. The equant is immovable apart from the motion of the eighth sphere.

59. The epicycle of Mercury moves upon the circumference of its eccentric from west to east with the same speed with which the Sun moves contrary to the firmament. It therefore appears that the epicycle center of Mercury goes twice around its eccentric in one year, once because of its own motion and once because of the motion of the aux, just as the epicycle of the Moon goes twice around the circumference of its eccentric in one month.

60. Similar to the epicycles of the other planets, this epicycle also has two motions, that is a mean and a true one. The mean motus is determined by a line from the center of the Earth parallel to the line from the center of the equant through the center of the epicycle, but this mean motus is the same as the mean motus of the Sun.

61. Sometimes these three lines are parallel, that is, the line from the center of the Sun's eccentric through the center of the Sun, and the other line from the center of the eccentric equant of Mercury through the center of its epicycle, and the line from the center of the Earth parallel to the two others. And since these three lines move equally, they will always be parallel, or they will, from time to time, be two lines only, or even one and the same line. From this it does not follow that

the epicycle center of Mercury and the center of the Sun are at the same place, or that the line from the center of the Sun's eccentric through the center of the Sun is the same as the line from the center of the equant through the center of the epicycle.

62. The true place or motus of the epicycle is determined by a line from the center of the Earth through the center of the epicycle.

63. As already mentioned, the center of the epicycle and the aux of the eccentric move in opposite directions at the same rate, equal to that of the mean motus of the Sun. In equal times the center of the epicycle and the aux and any other point of the eccentric deferent always describe equal angles about the center of the equant. The center of the deferent also moves at the same rate upon the small circumference in such a way that it describes equal arcs upon the equant in equal times, thus describing unequal arcs upon the small circumference. Thus, when it has moved through a quarter of its small circle, it will not have moved through a quarter of the equant.

64. From what is said above, it can also be shown that the aux of the deferent cannot be at every point of the equant, since the lines passing from the center of the Earth through the center of the deferent and pointing to the aux of the latter will always lie inside an arc of the equant, confined by two lines tangential to the small circle upon which the deferent center moves and passing through the center of the Earth.

65. From this it follows that the aux of the deferent is always confined to this arc, that is, that it now approaches and now recedes from the aux of the equant and this to both sides of the immovable aux of the equant; and although the aux of the deferent cannot be outside the said arc, nevertheless the point of the deferent, which in a way is the aux point, can be at any point of the equant, that is, in line with it. The reason is that the aux is continually changing.

66. When the aux of the deferent is moving away from the aux of the equant towards the west, or inward, then the intersection between the deferent and the equant following after the aux of the latter according to the order of the signs will approach the aux of the equant, and the other intersection will recede from it. The opposite is true when the aux of the deferent moves away from the aux of the equant towards the east.

67. Each time the epicycle center of Mercury coincides with the aux of the deferent, that is, falls upon it, then the aux of the deferent will be at the

aux of the equant. It then follows that the centers of the Earth, the equant, and the deferent, and the auges and the opposite auges, and the center of the epicycle will lie upon the same diameter of the world. And when the epicycle center is at the opposite aux of the deferent, then the points mentioned above are upon the same diameter and the two eccentric circles become one.

68. It also follows from what is said above that although the epicycle center goes round its deferent twice in a year, it will nevertheless be at the aux only once, for when the epicycle center is upon one half of its equant the center of the eccentric will move upon the opposing half of its small circle. And when the epicycle center is at the aux it is at the point of the deferent which can be farthest away from the Earth. Thus it can be at the point (of the deferent) farthest away from the Earth, but it cannot be at the point nearest to the Earth, for when it is at the opposite aux, then the two circles are one and the same and the epicycle center is at the opposite aux of the equant.

69. But there are two places in which the epicycle center comes as near to the center of the Earth as possible, whereas it cannot come so near at other places. These two places are points opposite to the aux of the equant and endpoints of the lines tangential to the small circle carrying the center of the deferent and passing through the center of the Earth.

70. Therefore, when the epicycle center is at the aux, it immediately begins to move towards the east upon its deferent, and likewise the center of the deferent begins to move towards the west upon its small circle. And when the epicycle center is at the head of the tangents in the neighborhood of the opposite aux of the equant, then the deferent center will be at the point of contact between this line and the small circle. The aux of the deferent is then as far as possible from the aux of the equant, and the epicycle center is at the opposite aux of the deferent.

71. At this moment, therefore, it (that is, the epicycle center) is as near as possible to the Earth, because the deferent center now descends more on its small circle and the opposite aux is more removed from the center of the Earth, which is easily seen by anyone who regards the figure [Fig. 5] and understands the motion. And as long as the epicycle center is upon the arc of the deferent lying between the points or heads of the tangents, the heads being near to the opposite aux of the equant, it will always be at the opposite aux of the

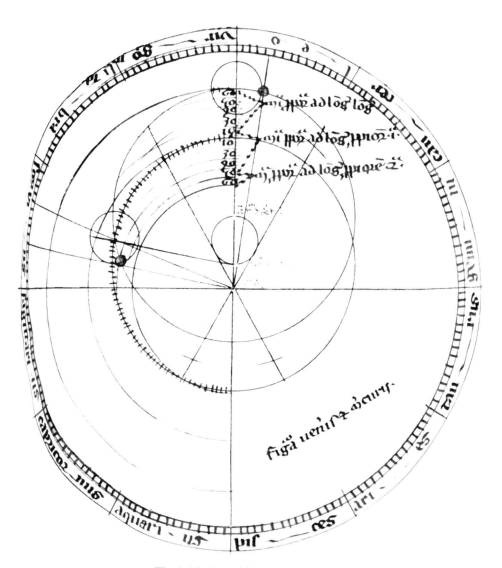

Fig. 5. Motions of Venus and Mercury.

deferent. This follows from the fact that as the epicycle moves in one direction, the deferent center will move as much in the other. Therefore, they will always be upon the same line, passing through the center of the Earth and the center of the deferent. And yet the epicycle center never comes so near to the center of the Earth as it does at the heads of the tangents.

72. The mean motus of Venus and Mercury, the mean and true centrums, the mean and true arguments, the equations of center in the zodiac and in the epicycle, the mean and true auges, and the equation of argument—all these things are defined in Venus and Mercury as in the three superior planets.

73. As in the three superior planets, the equations of argument of Mercury increase as the epicycle center approaches the Earth. The equations which are written in the tables presuppose that the epicycle center is at the point of intersection between the equant and the deferent.

74. For when the epicycle center is at the aux, it immediately begins to move towards the east, and the point of intersection which is before it moves towards it, so that they will be conjoined. And the equations of argument corresponding to this

460

position are written one after the other in the tables. The same thing takes place at the other point of intersection. When the epicycle center is at other places of the deferent, the equations of argument are found by proportional parts.

75. There must therefore be three sets of proportional parts. The first are the proportional parts corresponding to the greatest distance and defined as sixtieth parts of the difference between a line from the center of the Earth to the center of the epicycle, the latter being at the aux of the deferent, and another line to the intersection of the circles. The proportional parts corresponding to the least distance are sixtieth parts of the difference between a line from the Earth to the intersection, and the tangents [from the Earth] to the points where the epicycle center comes nearest to the Earth. Or again, from the same place to the opposite aux of the equant. As this is a moving line, the proportional parts will vary.

76. The diameter variation of the small circle can be described in two ways, as in the three superior planets.

77. Venus has a deferent and an equant situated as in the three superior planets, and with their highest points at the same place as that of the eccentric circle of the Sun. Its epicycle center moves with the same speed as the Sun, so that the mean motus of the Sun is also the mean motus of Venus, because there are two lines parallel to the line from the center of the Earth, one of which goes from the center of the Sun's eccentric circle through the center of the Sun, and the other from the center of the equant through the center of the epicycle.

78. The deferent and the equant are immovable, apart from the motion of the eighth sphere, and apart from the fact that the deferent moves in latitude to the south and to the north, so that when the deferent and equant sometimes lie in the plane of the ecliptic, it will soon move away from this line towards each pole, and the deferent will incline relative to the ecliptic. We shall treat of this motion in the chapter on latitudes.

79. All other things concerning Venus are similar to the three superior planets. And this will be sufficient for Venus and Mercury.

VI. ON THE RETROGRADE MOVEMENTS OF THE PLANETS

80. In the following we deal with the retrograde movements of the planets. A planet is called direct when its motion is supported by the motion of the epicycle against the firmament. It is called retrograde when its motion is not supported by the motion of the epicycle.

81. The first station is defined as the point of the epicycle where the planet begins its retrograde motion. The second station is the point where it begins to be direct.

82. The Moon cannot be said to exhibit any of these three phenomena, although it has an epicycle, because this epicycle always moves faster than the Moon itself. Yet in the superior part of its epicycle the Moon is called slow and in the inferior part fast in its motion.

83. The first station in the second sense is defined as the arc of the epicycle between the true aux of the latter and the first station in the first sense.

84. The second station in the second sense is defined as the arc of the epicycle lying between the true aux and the second station in the first sense, and stretching through the first to the second station.

85. The arc of retrogradation is an arc of the epicycle lying between the first and the second station. This arc decreases as the epicycle center approaches the center of the Earth, so that the stationary points are changing.

86. If the arc of the first station is subtracted from the second station, the arc of retrogradation will remain, and if the first station is subtracted from the whole circle, the second station remains, because the are *ABC* is equal to the arc *ACB*.

87. The mean motus of any planet corresponding to a given time not appearing in the tables is found in this manner: Take the radix[44] corresponding to a collected year. Then take the mean motus from the expanded years corresponding to the intermediate years between the first of the collected years and the year you seek. Subtract this motus from the radix if it be possible. If not, add one full revolution of twelve signs to the radix. The remainder is the desired mean motus of the planet.

88. A mean conjunction or a mean opposition is defined with reference to the mean motus, a true conjunction with reference to the true motus, and a visible one with reference to the visible [planets] and this with attention to sign, degree, and minute.

89. An eclipse digit is defined as the twelfth part of the diameter of the lunar or solar body.

90. The minutes of half duration are the minutes (of arc) through which the Moon moves from the beginning of an eclipse until the middle of it, if it is

---

44. For the concept of *radix,* see Selection 65.

not totally obscured, or until the beginning of the total eclipse, if it is totally obscured. In the Sun, too, the minutes of half duration are the minutes through which the Moon moves from the beginning to the middle of a solar eclipse.

91. The minutes of half the time of total obscuration are the minutes through which the Moon proceeds from the beginning of a [total] eclipse until the middle. If, therefore, these minutes are divided by the average motion of the Moon in one hour the result is the time used for the motion between these points.

92. The residual minutes or points are defined as the part of the diameter of the epicycle remaining between the aux and the direction of the Moon.

93. The numbers written in the columns of the eclipse tables are arguments of equivalent latitudes, or equivalent latitudes.

94. The great years are computed to the number of terms and the least years to the smallest revolutions. Some are intermediate between the great and least years of the Sun and the Moon. The great years belong to their great revolutions, the least years to their least revolutions, and the mean years to their mean revolutions.

95. The planets are said to be slow and of least speed when they are retrograde, but fast, that is, with increased speed, when they are direct. The planets are said to be numerically increased when the equation of argument is added to the mean motus, and to be numerically decreased when the opposite is true. They are called increased as to their light when they recede from the Sun or the Sun from them, but decreased as to their light when they come nearer to the Sun or the Sun to them.

96. The natures, properties, and operations of the planets and the signs are determined in such a way that one always begins with the Sun, because it is the most noble of the planets, and with Aries as the noblest of the signs, and from noon as the nobler part of the day, and from the site of the equator, which is in the middle of the world.

VII. ON THE LATITUDES OF THE PLANETS

97. The latitude of a planet is defined as its distance from the path of the Sun. The declination of a planet is its distance from the equinoctial line. It follows that the Sun has no latitude but only a declination. This declination of the Sun is found by means of the distance of the Sun's center from the first intersection between the circle of the Sun and the equinoctial line, that is, from the beginning of Aries.

98. The latitude of the Moon is found by means of the distance of the center of the Moon's body from the first intersection between the circles of the Moon and the Sun, that is, from the Dragon's Head, and thus we also find the declination of that degree of the zodiac where the Moon is, and for the same time we find the latitude of the Moon from the zodiac, that is, from the path of the Sun. And if both of them, namely the latitude and the declination, are to the north or to the south, we combine them, and the result is the Moon's declination from the equinoctial; but if they are different, we must subtract the smaller from the greater. In this way the declination is found in the other planets too.

99. It must be noted that the eccentric circle of the Moon always deviates from the path of the Sun in the same way and that its epicycle always lies in the plane of the eccentric, so that the Moon has only one latitude. The other five planets have a double latitude, one by which the epicycle deviates from the eccentric and another due to the eccentric itself, which deviates from the path of the Sun.

100. The latitude corresponding to the epicycle is found by means of a binary table, and the latitude corresponding to the eccentric by means of a quaternary table. A table is called binary[45] because it has two entrances, and quaternary because it has four. A binary table is computed for six signs, that is, half a circle, and a quaternary for a quarter circle. This means that a binary table is computed for six signs, namely a particular table for each; but a quaternary table for three signs, that is, a particular table for one sign, but serving for four signs by equivalence.

101. Because a binary table corresponds to the epicycle, it is entered by the argument, and because a quaternary table corresponds to the eccentric, it is entered by the distance from the ascending node (the Dragon's Head). The latitude recorded in a binary table is the distance of the various parts of the circumference of the epicycle from the circumference of the eccentric, this distance being computed towards the path of the Sun, which is called ecliptica because eclipses of the Sun and the Moon occur upon this line or in its neighborhood.

102. The epicycle is inclined to the eccentric in such a way that the planet is always between the ecliptic and the center of the epicycle, except when the latter is at the Head or the Tail of the Dragon. For then the epicycle lies in the plane of the eccen-

45. Binary tables appear in Selection 65.

tric. But when the planet is at the aux of the epicycle it has its maximum distance from the path of the Sun, so that we here find the maximum latitude in the table. At places in between, a smaller latitude is found, because the planet has a smaller distance.

103. The planetary latitude recorded in the quaternary table is the distance of the circumference of the eccentric from the path of the Sun. This distance is very small in the neighborhood of the nodes, but greatest at places three signs from the latter. In a quaternary table we thus find a small latitude at the beginning and the maximum latitude at the end, with intermediate latitudes in between.

104. Thus it is clear that when these two latitudes have been found, one of them must always be subtracted from the other.

105. In order to prove his mastery, the computer of the tables has refrained from recording the true values of the said latitudes, one of which is always subtracted from the other. Instead he gives equivalent values by means of which you get the same result by dividing one by another, as that resulting from the subtraction of one of the true values from another. For to every subtraction corresponds a certain division, and conversely. For the same thing results if you divide 60 by 30 as if you subtract 2 from 4, and consequently one latitude is divided by another.[46]

106. When the epicycle center is at the nodes the latitudes are zero. Then the epicycle lies in the plane of the eccentric and its center on the path of the Sun. That the values quoted in the tables are not the true ones is seen from the fact that the table of Mercury contains a latitude greater than 6 degrees, so that it would sometimes be outside the zodiac. But its true latitude appears by dividing one of these numbers by the other.

107. What I have said is corroborated by the tables of the real latitudes, if such tables are at hand, in which subtraction or addition takes the place of division. And it is stated that if you divide the second latitude by the first, you are reckoning the latitude from the path of the Sun, but if you divide the first by the second, you are reckoning from the limb of the zodiac, for a latitude of 2 degrees according to one way of reckoning is the same as the latitude of 4 degrees according to the other.

108. The Head and the Tail of the three superior planets are immovable, but the Head and Tail of Venus and Mercury are moving, so that the argument of latitude of the latter is found in a different way from that of the former.

109. The Head of Venus (and Mercury) moves at such a rate that its true place always has the same distance from it [that is, the planet] as the true place of the Head quoted in the canon to the table deviates from the place determined by the mean motus of the Sun and their equated argument. Consequently, we must add their equated arguments to the mean motus of the Sun.

110. The true places of the Heads are by definition computed from Aries according to the order of the signs, that is, Aries, Taurus, Gemini, and so on.

111. The mean places of the Heads are by definition computed in the other direction, that is, Aries, Pisces, Aquarius, and so on, so that the mean motus of the Head, together with its true motus, makes twelve signs. If the mean motus of the Head is subtracted from twelve signs, the true motus remains.

112. The computers of the tables based upon Arim[47] are said to have been Nembroth, Hermes, Ycominus, Ptholomeus, Albategni, Albumazar, Algorismus. Arim has the same distance from the gates of Alexander and Hercules projected upon the equator. It has, namely, a distance of 90 degrees from the gates of Alexander in the east and also of 90 degrees from the gates of Hercules in the west and a distance of 90 degrees from each pole.

113. If you wish to adjust the tables to other localities, you will have to subtract the mean motus of the stars in as many hours as there are between these localities and Arim, that is, subtract or add it in the collected years only.

114. When the planets are equated before or after noon, you can find the ascendant and the hour for the said time. If you wish to do so, you must place the degree of the Sun upon the meridional line [of an astrolabe] and note the place of the almuri upon the degrees of the limb. Let it go 15 degrees forwards or backwards for each of the hours you had before or after noon, and you will find the ascendant for the present hour.

115. If the Moon is in the middle of the heavens and you equate it by means of a table for a certain region, you will know the longitude between the two regions by the difference between the places of the Moon without having to wait for an eclipse.

116. If you take the altitude of the lower limb of

46. The description in this paragraph is of an Indian method foreign to Ptolemaic astronomy. See A. S. Kennedy and W. Ukashah, "Al-Khwārizmi's Planetary Latitude Tables," *Centaurus*, Vol. 14 (1969), 86–96.

47. Arim was the Indian city, Ujjain, which served as the standard meridian for earlier astronomical tables.

the Sun and Moon with the dorsum of an astrolabe, and then the altitude of the upper limb, and notice the displacement of the alidade, you will know the magnitudes of the diameters of the Sun and the Moon, but you must look at the Sun through a cloth.

117. Note that Thabit says that the auges are said to move 7 degrees towards the east during nine hundred years, and the same number of degrees to the west during another nine hundred years.[48] Moreover, they are, according to Albategni, said to move 1 degree in sixty years and four months, and always towards the east. But Alfraganus states that they move 1 degree in one hundred years, and also always towards the east.

118. Note also that as long as the Sun is on that half of its eccentric which is farthest away from the Earth, that is, at the greatest distance, the alidade on the dorsum of the astrolabe is higher than the degree of the Sun upon the rete placed upon the almucantarat at noon. The opposite is true upon the other half of the eccentric. And upon that day when the distance between these two altitudes is greatest, the Sun will be at the aux upon that half of its eccentric first mentioned above. The value of this distance is the same as the eccentricity of the Sun, that is, about 2.5 degrees.

119. If the degree of the Sun is placed among the almucantarat upon the number corresponding to that by which the nadir of the Sun falls below the place beyond which it rests, the hour determined here will be wrong by so much. If, in this way, we have considered the day when the Sun at noon is at the aux in one year and have measured its altitude by the alidade, and if we do the same the next year at the same hour and measure how much the altitude is increased, we will know how much the aux has moved in one year.

120. In this way Albategni is said to have found how much the planetary auges move per year, per month, and per day, and he computed tables of this motion, and he had a great astrolabe of 3 cubits or more.[49]

VIII. THE DETERMINATION OF THE PLANETARY ASPECTS

121. The aspect of the planets can be found in this way. Enter the table of right ascensions of the signs which begins from Aries at the equated length of any planet. The number of degrees found under the sign containing this planet must be retained. Thereupon enter at the equated length of any other planet, and the number of degrees found

under its sign must be taken and retained. Of these two numbers taken in this way, the smaller deviates from the greater. If it amounts to the

48. This is probably a reflection of Thabit ibn Qurra's theory of trepidation, as given in his work *On the Motion of the Eighth Sphere*. See J. L. E. Dreyer, *A History of Astronomy from Thales to Kepler,* 2d. ed. (New York: Dover, 1953), pp. 276–277.

49. Some manuscripts of the *Theorica planetarum* contain an additional passage that is lacking in the Copenhagen manuscript from which the present translation was made. Because Professor Pedersen considers this passage to be a later interpolation, I have relegated to this note my translation of it (made from MS. Florence, Ashburnham 208, p. 51, col. 2, to p. 52, col. 2, and an edition of the *Theorica* published in 1531):

"A line drawn from the center of the earth through the center of the planet's body [and on] to the sky reveals the true place of the planet, as is obvious from the radiation of the sun.

"A line drawn from our position through the center of the planet's body [and on] to the sky shows the apparent place of the planet; but it is there only when the planet is in our zenith. For then [that is, when the planet is in our zenith] the line drawn from the center of the earth to the center of the planet's body and the line drawn from our position are the same, and the place in which this line is terminated is the same; at other places, [however,] it is not the same.

"Therefore, the arc cut off between the true place and the apparent place of the planet is called the reflection *(reflexio)* [that is, terrestrial parallax] or difference of aspect *(diversitas aspectus)*. The said reflection occurs sometimes in latitude, sometimes in longitude, sometimes diverse [that is, in both latitude and longitude; and] sometimes it is greater, sometimes smaller. It must be understood that in the planets above the sun, namely Saturn, Jupiter, and Mars, there is no reflection perceptible to the sense. It occurs only in the sun, Mercury, Venus, and the moon. In the sun, a reflection of about 3 minutes is found; and when the sun is near the horizon it [reflection] is a maximum; further this [maximum is also] perceived in the moon [that is, when the moon is at the horizon]. Furthermore, the reflection is wholly in longitude when a planet is on the meridian of the observer; and it is wholly in latitude when a planet is on the circle of altitude; and it is diverse, that is, partly in longitude, partly in latitude, when a planet is neither on the meridian nor on the circle of altitude. And it is sometimes greatest because of one, or both, of two causes. For there are two causes which produce a greater reflection as well as a smaller reflection. One of these is when a planet is more distant from our zenith; therefore, reflection would be greatest near the horizon. The other cause of reflection is when a planet is nearer the earth. Hence, when the moon is at the least distance [from the earth] and near the horizon, the total reflection is 1 degree 44 minutes, which is the maximum of all reflections. When, however, it is at the greatest distance [from the earth], the total reflection is 54 minutes. Moreover, during hours of eclipses, the greatest reflection is 1 degree 4 minutes. . . ."

Approximately three lines of garbled text are omitted.

sixth part of a circle, those two planets have the sixth aspect, because there is a sixth part of a circle between them. If there remains the fourth part of a circle, it will be the fourth aspect; [if there remains half a circle, it will be the second aspect.] And if there remain so many degrees as to make up the third part of a circle, it will be the trine aspect. But if there remains half of the circle, it will be the

aspect of opposition. If there remain more or less degrees, the planets have no aspect. If nothing remains, then these planets will be in corporal conjunction. If there remain so many degrees as to make up another splendor, then these planets will be conjoined as to their light and not as to their body.

Explicit.

65 EXTRACTS FROM THE ALFONSINE TABLES AND RULES FOR THEIR USE

John of Saxony (fl. 1327–1355)

Translated by Victor E. Thoren and Edward Grant

Introduction and annotation by Victor E. Thoren

The *Alfonsine Tables* were prepared during the third quarter of the thirteenth century as part of an extensive scientific and literary program sponsored by Alfonso X of Castile. From Spain they passed to Paris, where, in 1327, they were drawn up in an essentially new form and provided with a set of Latin precepts by John of Saxony. In that form they soon spread throughout Europe, to enjoy—first in manuscript, then in numerous printed editions—a reign of three hundred years as the Latin West's authoritative astronomical work.

From a purely technical standpoint the *Alfonsine Tables* are of little interest. They are essentially the tables of the *Almagest,* shorn of their explanatory and justificatory apparatus and recomputed in accordance with minor adjustments in a few of the constants. There is one respect, however, in which the *Tables* are quite distinctive; namely their peculiarly complete utilization of the sexagesimal system. Instead of the usual grouping of days into 365's, they employ powers of 60; and instead of the traditional grouping of degrees into twelve signs of 30 degrees each, they adopt six "physical" signs of 60 degrees each. How truly remarkable these innovations are has emerged only comparatively recently, through the discovery that they are not really Alfonsine at all—that they constitute a departure from the form of the original tables themselves! By whom the modifications were introduced has yet to be ascertained. All that is known is that the Spanish introduction to the original *Tables* (which have not survived in Spain) describes something quite different from what left Paris fifty years later to circulate Europe under Alfonso's name; and that the author of that version, John of Saxony, credits the difference neither to himself nor to his teacher John of Linières. As for the scheme itself, in addition to providing a very tidy symmetry, it

allows a most ingenious abbreviation of the task of tabulating the various mean motions—a crucial consideration in an era when all tables had to be hand-copied.

The immediate source of the material presented here is the first printed edition of the *Tables,* published at Venice in 1483.[1] The portion of the work selected is the minimum required for the prediction of a lunar eclipse: about 30 percent of the introductory *canones* and about 20 percent of the complete set of tables. To facilitate comprehension of the steps involved in an eclipse prediction, a sample prediction of eclipses has been included between the translated precepts and the appended tables. Boldface numbers in parentheses inserted within the introductory *canones* refer to illustrative operations in the sample calculation. Obvious misprints have been corrected without recourse to the usual scholarly apparatus.

VIII. *To set out what is to be understood by the term radix,*[2] *when the radix of this or that motion is mentioned.*

It is to be known that the radix of any *motus*[3] is

1. *Alfontij regis castelle illustrissimi celestium motuum tabule necnon stellarum fixarum longitudines ac latitudines Alfontij tempore ad motus veritatem mira diligentia reducte. Ac primo Joannis Saxoniensis in tabulas Alfontij canones ordinati incipiunt faustissime* (Venice: Erhard Ratdolt, 1483).

2. This is the first of several technical terms which seem to be better left untranslated. As with most of them, it is defined in the text, according to which it would have to be rendered as "epoch value" if it were to be translated. It has been pluralized in the Latin form, *radices.*

3. The Latin term *motus* is used to signify both "movement" and the position resulting from a particular movement. Its most common appearance in the latter sense is as a technical term indicating angular distance from the first point of Aries (see pars. 3, 17, and 41 of

nothing other than the place in the circle of the signs in which that motus stood at the beginning of the era of which it is the radix. For example, in the table of radices[4] of the sun, the radix of the Incarnation of Christ is 4 signs 38 degrees 21 minutes. This means that where this number falls in the zodiac (beginning the reckoning from Aries), there the line of mean motus of the sun was situated at noon of the last day of December[5] (or the beginning of January) at the time of Christ. This should also be understood for all [the rest of] the mean motuses. [On the other hand,] the radix of any argument (when I speak of "argument," I always refer to motion in an epicycle) is the distance of the body of the planet from the mean aux[6] of the epicycle at the beginning of the era of which it is the radix. For example, in the table of radices for the argument of the moon, the radix at the Incarnation of Christ is 3 signs 19 degrees, and so on. This means that in the time of Christ at the beginning of January, as I said, the distance of the body of the moon from the mean aux was this much; for just as the mean motuses were reckoned from Aries in the circle of the signs, so the mean arguments are reckoned from the mean aux in the epicycle. . . .

XI. *To find by means of tables the mean motuses of the sun, moon, and the rest of the planets; the mean arguments of the moon, Venus, and Mercury; and the motion of access and recess of the eighth sphere; the motus of the auges and the fixed stars; the argument of latitude of the moon; and the motus of the head of the dragon.*

From the beginning of whatever era you use as reference, up to the hour you wish to consider, reduce the total time to fourths, thirds, seconds, and units of days (**1**); likewise, reduce the hours and minutes of hours, if you have them, to minutes and seconds of days. This done, write first the radix (**2**) of that motion which you wish to know for the time of the era in terms of which you are working. For example, if you are seeking the mean motus of the sun, write the radix of the sun; and if you are working in years of Christ, write the radix of the Incarnation. If you want to work to some other years, write the radix corresponding to them. With the radix written aside, enter with the number of fourths into the table of that motion you seek, and what you find in the fourth line after the line of the number (namely in the line entitled "4ths") are signs, then degrees, next minutes, and so forth, according to the order of the fractions (**2**). Write them down under the chosen radix, each under its

own kind: signs under signs, degrees under degrees, and so on. Then enter the same table with the number of thirds, and what you find in the line entitled "3ds" are the signs, next degrees, and so on in order: Write them down under their own kinds in the usual order. Then enter with the number of seconds, and what you find in the line entitled "2ds" are the signs, then degrees, and so on. Then enter the same table with the number of units, and what you find in the line entitled "units" are signs, then degrees, and so on: Write them under the others already written. This done, if you have minutes of the day, enter into the same table with them, and what you find in the first line are degrees, then minutes, and so forth: Write them down under the others, each under its kind. Then enter the same table with the seconds of days, if you have them, and what you find in the first line are minutes of a degree, and so forth: Write them under the others already written. Continue in the same fashion if you have more fractions of time. Now, that these fractions of degrees correspond to any fraction of time is obvious from [the introductory paragraph?][7] of this table.

This done, add everything together column by column, beginning from the smallest fraction. As often as the sum surpasses 60, one unit is to be substitued for them in the place of the next larger fraction. This procedure is to be followed up to

---

the *Theorica planetarum* in Selection 64.2). Where it is being used in this way, it has been left untranslated and given an anglicized plural, "motuses." Where it is intended in its more general sense, it has been rendered as "motion" or "movement."

4. In the table of radices cited (as well as for each of the other sixteen motions he tabulates), Alfonso provides radices for ten different eras; for instance, those of the Flood, Alexander, Julius Caesar, and the Hegira. For the purpose of the present translation, the radix of the Incarnation of Christ has been extracted from each of the tables relevant to the sun or moon, and all have been combined in the table of radices (Table [A]).

5. Note that Alfonso reckons his year from half a day *before* the beginning of the civil year rather than half a day *after*, as was the usual astronomical custom.

6. In general terms, the *aux* is the point farthest from the earth on any circle—apogee, in modern parlance (see pars. 1, 19, and 20 of the *Theorica planetarum* in Selection 64.2). Adapted from the Arabic, it has been pluralized in the Latin form, *auges*.

7. The text reads *per titulum inferiorem ipsius tabulae*. The treatise opens with "Time is the measure of the motion of the prime mover, as Aristotle puts it in the fourth [book] of the *Physics*. Therefore, when we wish to know a motion, it is necessary that we know the times beforehand; so that with the quantity of time known, we can know the motion corresponding to it."

signs, since 60 degrees equal 1 sign. 6 signs make one revolution or one circle, so if by adding the signs together, more than 6 signs emerge, subtract 6 as often as you can. Set the remainder aside,[8] for it is what you have been seeking.

It should be understood that the radices taken in these tables are for the meridian of Toledo (53), so that by this procedure you have motions only for the meridian of Toledo. Now, if you wish to find the radices of motion for your meridian or any others, you need to know whether your city is farther east or west. And for however much time, seek the motion you want,[9] and subtract it from the motion that has been found if the city is farther east, but add it if the city is farther west. What emerges after the addition or subtraction is the motion you seek for the meridian of your city. Note that you could abbreviate the work considerably by reducing all the radices to your meridian at one time, and it would serve you as long as you live. Note also that you can seek all the mean motuses and mean arguments in the beginning of any year and place them on one list and take that year as your radix. This will greatly reduce your work, because for the whole of this year it will be necessary to enter only with seconds and units (and perhaps, minutes and seconds of days, if you should have minutes and seconds of days).

XII. *To find the time of mean conjunction and opposition of the sun and moon by the tables made for this purpose.*

For the day during which you estimate that conjunction or opposition will occur, reduce the appropriate time interval to fourths, thirds, seconds, and units according to the method already given. This done, write first the radix of the mean elongation (2) of the moon, working in the same way as was taught concerning the mean motions. When you have made all the entries and added the numbers, if exactly 6 signs result for that time, then there is a mean conjunction; if 3, there is a mean opposition. For mean conjunction is defined as the situation of the lines of mean motus in the same place of the zodiac, while mean opposition is when they are in opposite places.

If neither precisely 6 nor precisely 3 signs result, then, if you wish to find conjunction, subtract your result from 6 signs and set the remainder aside; if you wish to find opposition, subtract your result from 3 signs and set the remainder aside. This done, seek the remainder in the table of mean elongations. If there are any signs, seek them in the first line: they are days (3). If, however, you do not find

a whole number of signs or degrees exactly, take the nearest smaller number, find the number standing opposite in the line of numbers, and write it aside; for these are days. Then subtract what you took in the table (namely the signs, degrees, and so on) from the number which you had (namely from the signs, degrees, and so on), and with the remainder enter again into the same table. If there are degrees in the remainder, seek them in the first line after the line of numbers; and if you find them there, what was written in the line of numbers are minutes of days: Write them down after the days you have already noted. If, however, there are so many degrees that you cannot find them in the first line after the line of the numbers, seek them (or the nearer lesser number) in the second line after the line of the numbers, in that part where there is zero or a cipher in the first line. Then what is written in the line of the numbers are days: Write them aside under the others already written. This done, if after subtracting that with which you entered the table from your number, there should be a remainder, seek it again in the same table. If there are minutes in the remainder and you can find them in the first line after the line of the numbers, then what is written in the line of the numbers are seconds of days. But if there are so many minutes as cannot be found in the first line after the line of the numbers, then seek them in the second line (namely, in that part where the zero is); and what is written in the line of numbers are minutes of days. The small table written below shows these denominations. Enter in the same manner until the number is used up: It is very much the same operation as where you sought the years in fourths, thirds, seconds, and units from the table.

This done, add the days, minutes of days, and the rest of the fractions that you found by the operations to the time with which you [first] sought the mean elongation. What results from the addition will be the time of mean conjunction or opposition.

If, now, having conjunction, you wish to get opposition in an easier way, or, having opposition, you want to get the following conjunction, then, since both intervals are equal, add 14 days, 45 minutes of a day, 55 seconds, 3 thirds, and 48 fourths of a day to the time of conjunction if you want the

8. The imperative *serva* occurs frequently. It is rendered throughout as "set aside."

9. That is, for the time equivalent to the difference in longitude, find the distance traveled by the planet involved.

following opposition, or to the time of opposition if you want the following conjunction. From this it is clear that, having the first conjunction of the year, all the conjunctions and oppositions of that year can be found by the addition of numbers alone. If you wish to work with this method, seek the first conjunction of the year in this way: Reduce to fourths, thirds, seconds, and units the total time from the beginning of your era up to the beginning of the year whose first conjunction you seek. With them, enter the table in the way already described in this chapter. Subtract what results from 6 signs, and with what remains, enter in the way already given, taking days, minutes of days, and so forth: Set these aside, for these are the days and minutes of days, and so on, from the beginning of the year to the conjunction immediately following. Now add 29 days, 31 minutes of a day, 50 seconds, 7 thirds, and 36 fourths of a day (for that is the time from one mean conjunction to another) to the [time of] conjunction just set aside (5). And if you add this interval to the second conjunction, there results a third; and in the same way, others. If you wish to know oppositions, add to the time of any conjunction the time previously given, namely 14 days, 45 minutes of a day, 55 seconds, 3 thirds, and 48 fourths of a day, since this is the time from conjunction to opposition and from opposition to mean conjunction. If you also want to find quadratures, that is, the distance of a quarter circle between the mean motus of the moon and the mean motus of the sun, add 7 days, 22 minutes of a day, 57 seconds, 31 thirds, and 54 fourths of a day to the time of [mean] conjunction and you will have first quadrature; add this same amount to the time of opposition and you will have second quadrature. You can write these numbers in the margins of the tables so you can find them quickly.

XIII. *To determine whether an eclipse is possible during any conjunction or opposition.*

Seek the argument of latitude of the moon for the time of mean conjunction, according to the rules of Alfonso and John of Linières (as appears in the thirty-third proposition of his work), if you are inquiring as to the possibility of a solar eclipse, or for the time of opposition, if you are checking for a lunar eclipse (4). If the argument of latitude of the moon turns out to be no signs and less than 12 degrees, or 5 signs and more than 48 degrees, or 2 signs and more than 48 degrees, or 3 signs and less than 12 degrees, say that an eclipse would be possible (6). But if you get a result outside these limits, an eclipse is impossible.

XIV. *To master the means of correcting the tables of uniform motion and even constructing them anew when the motion for any day whatever is known.*

You know the motion of one day of the table which you wish to correct, for it is the motion of any given day which is written in the first line of the table. So when you doubt whether some line in the table is correctly copied, look at the line immediately preceding the one whose truth you doubt. Write it down, then write the motion of one day (namely the first line of the table) and add the two together: What results will be the value of the line you doubt. But suppose it is the rectitude of the first line itself that is in doubt. Then take any line whatever in the table, write it down, and take the immediately preceding line and write it under the one previously written. Subtract it from the upper, and what remains is the motion of one day, namely the first line of the table. If they agree, fine; if they do not agree, however, take two successive lines in another section, subtract one from the other, and check agreement until you have certified the motion of one day, since it is not likely that the table will be erroneous in many places. And when you have certified it, you will be able to correct the whole table by the said motion of one day; if it were all false, you could even make a whole table from the beginning, in the following manner. Write the motion of a day first, then double it: the doubled value is the motion of two days. To this, add the motion of one day, and there will result the motion of three days; add again the motion of one day, and there will result the motion of four days. Do this until you complete sixty lines and you will have finished the whole table.

XV. *To find the true place of the aux of any planet by the tables.*

Seek first the motus of the auges and fixed stars according to the method given for the mean motuses, but do not take any radix: Set it aside (8). Then seek the motion of access and recess of the eighth sphere and add it to the radix of the eighth sphere. This done, if there are any signs here, resolve them into degrees and add the degrees to the others, if there are degrees in addition to the signs. Then enter the table of equations of access and recess, and find the given number of degrees in the line of the numbers (9), if there are less than 90 degrees. If there are more than 90, however, reckon in backward or contrary order; that is, if you have 91, take 89, and if 92, take 88, if 93, 87, and so forth, back to the beginning of the table, where the equation is null for 180 degrees. If there are more than 180 degrees, begin at the beginning of the table; thus 181 will be 1, 182 will be 2, and 183, 3, and so forth up to 90 again, which will be 270 degrees. But if there are more than 270 degrees, work backward again from the end of the table towards the beginning; that is, for 271 take 89, and for 272 take 88, and so forth right to the beginning of the table, which will be 360, where the equation is again null. But you will not need this retrograde computation in our lifetime, because it will be many years before 90 degrees accumulate.

Seek, then, the number of degrees in the lines of the numbers, and see how many degrees, minutes, and seconds stand opposite it. Write them down in two places. This done, if you have any minutes beyond the degrees, enter in the same way again with one more than the number of degrees with which you entered. That is to say, if you entered first with 67 degrees, and if there are some minutes beyond 67 degrees, enter with 68 degrees. See what equation stands opposite, and write it down. This done, see whether the first equation (namely, that which you took first) is greater than the second, or the contrary. Subtract the smaller from the greater, and what remains is called the difference between the two equations. Of this difference, take the proportional parts according to the proportion of the minutes beyond degrees to 60.

The proportional parts are found in this fashion. If there are minutes and seconds in the difference between the two equations, reduce them to the same denomination (namely to seconds): This is the first number. Then, in the same way, reduce those minutes and seconds beyond the degrees, with which you did not enter the tables, to seconds: This is the second number. Then multiply the second number by the first, divide what results by 60, and the quotient will be the proportional parts. Add these proportional parts to the first equation taken if the second was greater; but subtract them from the first if the second was smaller. What emerges after the addition or subtraction is the equation of the motion of the eighth sphere. This done, add this equation of the eighth sphere to the motion of the auges and fixed stars (which I directed to be set aside) if the degrees with which you sought the equation were fewer than 180, but subtract if they were more. Add what results after the addition or subtraction to whatever radix of the aux you choose, and you will have its true place (10). Note that this operation is necessary for determining the motion of the planets, because for no planet except the moon can the true place be found without the motion of the aux, as will appear from what follows.

XVI. *To find the true place of the sun by the tables.*

Seek first the mean motus of the sun and set it aside (11). You know the aux of the sun from the preceding chapter. Subtract it from the mean motus of the sun if you can (that is, if the mean motus is greater than the aux). But if you cannot subtract it (that is, if the aux is greater than the mean motus), add one revolution, or 6 signs, and then subtract the aux. What emerges after the subtraction is the argument of the sun (12). Note here that in this chapter and in all succeeding ones, whenever the rules dictate the subtraction of such a number from such a number, it will always be necessary to subtract in that way, whether the number from which the subtraction is to be made is greater or smaller than that which is to be subtracted. If that from which the subtraction is supposed to be made is smaller, then one revolution must be added and the subtraction made. This should be noted very carefully, since the situation will often arise in your work.

Having the argument of the sun, enter with it into the table of equations of the sun, and seek the same number in the lines of numbers (13). If the whole number of the argument can be found exactly (that is, if there are only degrees and signs, without minutes and so forth), you will find the equation of the sun opposite. Subtract the equation from the mean motus if the argument of the sun (by which you found the equation) is less than 3 signs, but add it to the mean motus if the argument is more than 3 signs. What emerges after the addition or subtraction is the true place of the sun in the ninth sphere. If great precision is not necessary, you can proceed in this fashion. See whether there are any minutes and seconds beyond the signs and degrees in the argument of the sun. If there are more than 30, add 1 degree in place of them to the degrees of the argument: if there are less than 30, do not worry about them. Thus, with one entry, you have the equation, without the necessity of computing proportional parts. The method is sufficiently precise for knowing the degree of the sun's position and can be used in the same way for the rest of the planets. But since this does not suffice if you wish to work precisely, it is necessary to give the complete method. . . . [Details of interpolation procedure follow.]

XVII. *To find the true place of the moon by the tables.*

Seek first the mean motus of the moon (26) and set it aside. Seek also the mean motus of the sun and subtract it from the mean motus of the moon. Double what remains, and what results will be the mean center of the moon (27); for the *Theorica* says that if the distance between the sun and moon is doubled, the distance between the aux of the eccentric of the moon and the mean motus of the moon is obtained, and this distance is called the mean center of the moon. Set it aside with the mean motus. Then seek the mean argument of the moon (30) and set it aside also. These things found and set aside, enter into the table of equations of the moon with the mean center, and seek the equation of center (28), with a double entry [by interpolation] if there are any minutes in the [mean] center beyond the signs and degrees, as set out in the directions regarding the sun. In the same place (namely where you entered with the mean center), look up the proportional minutes (29) which stand opposite, and set them aside. Then see whether the mean center is less than 3 signs, or more; if less, add the equation of the center to the mean argument, but if more than 3 signs, subtract the same equation from the mean argument. What results from the addition or subtraction will be the corrected[10] or true argument of the moon (31), which is the distance of the body of the moon from the true aux of the epicycle.

This had, enter with it into the same tables, seeking the corresponding number in the lines of numbers, and you will find the equation of the argument standing opposite. Set it aside: it is called the first equation. Then enter with one degree more, and again take the equation standing opposite, which is called the second equation. Subtract the lesser from the greater, and the difference of the two equations will remain, of which you take the proportional parts (namely, the proportion of the minutes beyond the degrees to 60). Add the result to the first equation if the second is greater, but subtract from the first if the second is smaller. What remains after the addition or subtraction is the first examined[11] equation of the argument (33).

In the same place extract also the variation of the diameter of the small circle (32), making a double entry if it is necessary (that is, if there are any minutes, and so forth, beyond the degrees in the corrected argument). Then take the proportional part of the variation of the diameter according to the proportion of the proportional minutes

(which I earlier told you to set aside) to 60 (34). You know how to do this from the rules for the sun. Add the proportional parts to the first examined equation of the argument, to get the second examined equation of the argument, which you add to the mean motus of the moon if the corrected argument is more than 3 signs, and subtract if less. For the *Theorica* says that if the moon is in the half of the epicycle which one regards as west on the right-hand side, the mean motus is greater than the true, so that the equation of the argument is to be subtracted; in the other half the contrary obtains, so that it is to be added. Add or subtract as necessary, then, and what results will be the true place of the moon (35) in the ninth sphere.

XVIII. *To find the true place of the head of the dragon by the tables.*

It should be known that the motion of the intersections of the lunar eccentric with the ecliptic (which is called head or tail of the dragon) is not like the motion of the planets but is in the opposite direction, as the *Theorica* says. The head of Genzahar is said to move as much in mean motus against the firmament as it (in reality) goes with the firmament,[12] so that by subtracting the mean position of the head of the dragon from 6 signs, there remains its true place computed according to the succession of the signs. Therefore, when you want the true place of the head of the dragon, find its mean motus in its table, subtract it from 6 signs, and what results from the subtraction is the true place of the head of the dragon reckoned according to the succession of the signs from Aries. Having, then, the true place of the moon by a preceding rule, and the true place of the head of the dragon by this rule, you can determine the true

10. The phrase is *argumentum equatum sive verum.* "Corrected" seems to be the sense of *equatum,* which appears frequently in astronomical literature up through the seventeenth century.

11. The phrase is *equatio argumenti primo examinata.* "Corrected" seems to be the sense of *examinata,* but if that were meant, *equata* might have been used. Perhaps "approximated" is what is intended.

12. Thus, the ascending node (Ω) actually slides *backward* (in the direction of the nightly movement of the stars) toward the vernal equinox (Υ). This means that the longitude of the node ($\Omega - \Upsilon$) is a *decreasing* quantity. The straightforward way of handling it would be to establish an epoch value for ($\Omega - \Upsilon$), sum up the regression during the time interval involved, and *subtract* the (backward) movement of the node ($\triangle \Omega$) from the epoch value to get the current value ($\Omega - \Upsilon - \triangle\Omega$). John does it by using $360° - (\Omega - \Upsilon)$ as the epoch value, *adding* $\triangle\Omega$, then subtracting the whole [$360° - (\Omega - \Upsilon) + \triangle \Omega$] from $360°$, to get ($\Omega - \Upsilon - \triangle \Omega$).

place of the moon's latitude in this way: Subtract the true place of the head of the dragon from the true place of the moon, and the true argument of latitude of the moon will remain.[13] The true argument of latitude of the moon is defined as the distance of the body of the moon from the head of the dragon. It is called the moon's argument of latitude because by means of it the latitude of the moon from the ecliptic is found. . . .

XXVI. *To find the extent and duration of a solar eclipse from the tables.*

According to the teaching of the rules established for the purpose, seek a mean conjunction of the sun and moon for which the possibility of an eclipse exists (6). This found, seek the mean motus of the sun (11), the mean motus of the moon, and the mean argument of the moon (14) for the time of mean conjunction. You do not need to worry about the argument of latitude yet, whatever the other rules say. Seek also the mean motion of the eighth sphere, so that you can get the aux of the sun (10) according to the rules established for this purpose. Upon finding these things, subtract the aux of the sun from the mean motus, so that the argument of the sun remains (12). With it, enter the table of solar equations, seeking the number matching the argument in the lines of numbers, and finding the equation opposite it (13). Subtract [it] from the mean motus of the sun if the argument is less than 3 signs, but add if it is more. What emerges from the addition or subtraction will be the true place of the sun for the time of mean conjunction: Set it aside. Then enter the table of lunar equations with the mean argument of the moon, seeking its like in the lines of numbers (15). After finding the equation of the argument standing opposite, subtract [it] from the mean motus if the argument is less than 3 signs, but add if it is more than 3 signs; and what results is the true place of the moon.[14]

Examine the true places of the sun and moon. If they agree in signs, degrees, and minutes, true and mean conjunction occur at the same time. But if they do not agree, subtract the lesser from the greater and set [what is termed] the distance [*longitudo*] aside (16). See of what it is the distance: It is the distance of that [body] which precedes the other[15] in the order of the signs (16). To this distance add a twelfth part of itself, and divide the whole in half. Add the half to the mean argument of the moon if the distance is "of the sun", but subtract from it if the distance is "of the moon." What results after the addition or subtraction is

the argument of the moon,[16] corrected for the purpose of finding the motion of the moon in 1 minute of the day (17): Set it aside. Then enter with the argument of the sun into the table of the motion of the sun in 1 minute[17] of a day, and what you find standing opposite is the motion of the sun in 1 minute of a day (19). Likewise, enter with the argument of the moon (corrected, as I said) into the table of the motion of the moon in 1 minute of a day, and what you find standing opposite is the amount the moon moves in 1 minute of a day (18). Then subtract the motion of the sun in 1 minute of a day from the motion of the moon in 1 minute of a day, and there will remain the fraction or "excess" per minute of a day (20). For every degree of distance between the sun and moon, add to this excess 1 minute of motion in 1 degree, and for

13. That is, subtract $\Omega - \Upsilon$ from $\mathbb{)} - \Upsilon$ to get $\mathbb{)} - \Omega$.

14. The involved computation of the moon's true place (related in sec. XVII and shown in the sample calculation) is here legitimately abbreviated by the fact that the computation is for syzygy. Since twice the elongation is 0 degrees, steps (27) to (29) and (31) are eliminated.

15. At another place John elaborates: "Thus the distance is *of the sun* if the moon has not yet reached the sun in conjunction or [if it has not yet reached] the nadir of the sun in opposition. It is *of the moon* if the moon precedes the sun in conjunction or [precedes] the nadir of the sun in opposition."

16. The ultimate object of this paragraph is to estimate the time from mean to true conjunction (or opposition). During that time t, the moon, traveling at a velocity V, will move not only through the "distance" D, but also through an extra angle d resulting from the velocity of the sun v during t.

Thus, $Vt = D + d$, where $d = vt$.

Then, $D = Vt - vt = t(V - v)$; so $t = \dfrac{D}{V - v}$.

With D known, all that is necessary to get t is a knowledge of the velocities of the sun and moon at the time in question. Both vary with position in the epicycle; and the moon moves sufficiently fast in the epicycle to necessitate an attempt at determining its *average* position (and, hence, its average velocity) through the interval involved. The *longitude* through which the moon moves is $D + d$; in the short term it can be used as an approximation of the *argument* through which the moon moves. From the equations above, $d = vt = v[D/(V - v)]$. With V and v equaling roughly 13 degrees and 1 degree, it is clear that the approximation $d = D/12$ is a very reasonable one. Then half of $d = D/12$ added to (or subtracted from) the mean argument of the moon will approximate the position of the moon in the epicycle halfway through the overtaking period.

17. Omitted for the sake of abbreviating the tables. The *hourly* velocities (that is, motions in one twenty-fourth rather than one sixtieth of a day) given in Table [P] are used in the sample calculation.

every second, add a third[18] if the moon is in the lower part of its epicycle; but subtract if it is in the upper part of its epicycle. What remains will be the corrected excess, which is to be divided into the distance between the sun and moon (21): The quotient will be the number of minutes of the day between mean and true conjunction. (If the distance is smaller than the excess, multiply the distance by 60, then divide by the excess, and what results will be seconds of the day. If anything is left over, multiply by 60 and divide by the excess, and what results will be [units of] the next smaller fraction. You can do this as many times as you wish.) The minutes and seconds are to be added to the time of mean conjunction if the distance [you found] was of the sun, but subtracted if it was of the moon. What results will be the time of true conjunction, very closely approximated indeed, if you have done the work well.

For this time, seek the mean motus of the sun (24), and the mean motus (26), mean center (27), and mean argument (30) of the moon. Then correct the sun and moon with all the precision you can; and if they agree in signs, degrees, and minutes, that will suffice. If they do not agree, however, subtract the lesser from the greater, leaving the distance (35), which you set aside. Then, to the mean motus of the sun that you just now found, add the movement of the sun in 1 second of a day;[19] do the same to the argument and correct the sun as before. This done, subtract the place [motus] of the sun found previously from the place found just now, and what results will be the motion of the sun in 1 second of a day. Do the same thing with the moon: That is, correct it by 1 second of a day after the time for which you first figured it, then subtract the first value from the second to get its motion in 1 second of a day. This done, subtract the motion of the sun in 1 second of a day from the motion of the moon in 1 second of a day to get the excess (37), by which you divide the distance to get seconds of a day (38). If anything remains, multiply by 60 and divide again by the same number as before to get thirds of a day, and so on as far as you wish. When this is finished, add the time of this division to the time of true conjunction already found if the last distance was of the sun; but subtract it if it was of the moon. What results from the addition or subtraction will be the time of true conjunction uncorrected for the equation of days.

For this [time], then, seek the places of the sun and moon precisely corrected (39); and seek also

the mean motus of the head of the dragon (42) for this time, which you add[20] to the true place of the moon (41). What results will be the second corrected argument of latitude of the moon (42): Set it aside. Then reduce the fractions of a day that you have for the time of truest conjunction to hours and fractions of an hour (7), and add them (46) to the equation of days, which is found in this manner. Enter the table of the equation of days with the degrees of the sun, using a double entry if it is necessary (that is, if there are any minutes beyond complete degrees in the place of the sun). And for every degree found here, take 4 minutes of an hour (40) and for every minute, 4 seconds; and for every second, 4 thirds. Add[21] them to the hours, minutes, and so on, of truest conjunction, and you will have the time of truest conjunction corrected for the equation of days. . . . [The rest of the details of solar eclipse prediction follow. Only what has already been presented is necessary for understanding the procedure for computing lunar eclipses.]

XXVIII. *To find the extent and duration of lunar eclipses.*

Seek first mean opposition (namely the one for which you found the possibility of a lunar eclipse) and then true opposition, working in exactly the same way as was said for conjunction, except that

18. The rule of thumb given here is not only garbled in the text, but is also somewhat different from that given in an earlier (omitted) discussion. What is intended is a correction of 1 part in 3600, less 1; for example, 4 seconds for a distance of 5°. It is copied from the Toledan Tables, in which the original rule of al-Battani (simply 1 part in 3600) has been slightly corrupted. Obviously, the correction is so small as to be scarcely worth considering, but it is theoretically valid. Because of the eccentricity of Ptolemy's lunar orbit, the moon is very slightly closer to the earth (and hence moves somewhat faster) when just outside of syzygy than in syzygy itself.

19. This operation is a purely academic exercise. It is unlikely to produce a prediction differing by as much as 1 second from one derived from use of the table of the sun's motion per minute of the day. The value used in arriving at (37) is one sixtieth of the hourly velocity tabulated in [P].

20. In section XVIII (and see n. 12) the mean motus of the head of the dragon is defined as the *retrograde* arc from the first point of Aries to the ascending node: $\Upsilon - \Omega$ instead of the usual $\Omega - \Upsilon$. This means that it must be added to, rather than subtracted from, the place of the moon ($\mathbb{C} - \Upsilon$) to get the argument of latitude ($\mathbb{C} - \Omega$).

21. In order to render all the corrections additive, the table has been augmented by a small portion of what would rightfully be the radix of the sun.

in finding conjunction it is the distance between the sun and moon that is sought, while for opposition it is the distance between the nadir of the sun and the body of the moon that is sought. The work differs in no other way. If I described here the method of working, I would do nothing but repeat what was said.

Having the time of truest opposition, then, seek for that time the mean motus of the head of the dragon, and add it to the true place of the moon found for the time of truest opposition. What results is the second corrected argument of latitude. Then seek the equation of days by entering with the position of the sun into the table made for this purpose. Reduce the equation to fractions of hours, as was said concerning conjunction, and add these fractions of hours to the time of true opposition. What results is the time of true opposition corrected by the equation of days. This you take as the time of mid-eclipse (46).

With the second corrected argument of latitude (42), enter the table for lunar eclipses at greatest distance (43) and extract the points of eclipse which stand opposite: Use a double entry if it is necessary, following the same method given for solar eclipses. At the same time, take the minutes of *casus* and the minutes of *mora*,[22] if there is any mora; or only the minutes of casus if there are no minutes of mora (for if 00 appears in the line entitled "mora," it has no mora). Now, it is to be understood that mora is defined as the time from the beginning of total obscuration (namely when the whole moon has entered the shadow) up to the beginning of the return of light. And what is called the minutes of mora is the space of the heavens traversed by the moon's own motion from the beginning of mora up to mid-eclipse. The minutes of casus are defined as the space which the moon traverses from the beginning of eclipse up to the beginning of mora. Having the points of eclipse, the minutes of casus, and the minutes of mora for the greatest distance, then set them aside as what was to have been found if the moon was in the true aux of its epicycle. But if the moon was not in the aux of its epicycle, enter with the same argument of latitude into the table for lunar eclipses at the least distance (44), and take here the points of eclipse, minutes of casus, and minutes of mora, by working in the same manner as you did with the greatest distance. Set them aside as the things to have been found if the moon was in opposition to the true aux of its epicycle. In short, if the moon was in the true aux of its epicycle, enter only at the greatest

distance; while if it was in opposition to the aux of the epicycle, enter only at the least distance. But if it was neither in the aux nor in opposition to the aux, enter both distances in the way already said. Then subtract the points of eclipse found for the greatest distance from the points found for the least distance, and the minutes of casus from minutes of casus, and minutes of mora from minutes of mora, and set the differences aside, each by itself. Then, with the corrected argument of the moon for the hour of true opposition, enter the table of proportions tabulated in increments of 2 degrees, and (using a double entry if necessary) take the proportional minutes standing opposite (45). Multiply these proportional minutes by each difference previously set aside, and divide by 60 to get the proportional part of each of the quantities. Add these to the points of eclipse, minutes of casus, and minutes of mora found for the greatest distance, each to its own kind (namely to the proportional parts of the difference of points, add the points; and to the proportional parts of the minutes of casus, add the minutes of casus; and to the proportional parts of the difference of the minutes of mora, add the minutes of mora). You will then have all these quantities corrected according to the distance of the moon from the earth.

Now, divide the minutes of casus (as just corrected) by the excess of the moon in 1 hour, and the result will be hours. And if there is any remainder, multiply by 60 and divide by the same thing as before to get minutes of an hour: and if something still remains, multiply again by 60 and divide as before to get seconds of an hour. This done, you will have the hours, minutes, and seconds from the beginning of eclipse up to the beginning of mora (47), if there is going to be any mora; or from the beginning of eclipse up to the middle, if there will be no mora. In the same way, divide the minutes of mora by the excess of the moon in 1 hour, and there will result the hours, minutes, and seconds (or minutes and seconds, alone, if there are no hours) from the beginning of mora up to mid-eclipse (48). This done, if you do not have any mora, subtract the hours from the beginning of eclipse up to the middle from the time of truest opposition corrected by the equation of days. What remains will be the time of the beginning of the eclipse, and if you add these same hours to the time of truest opposition you will get the time of the end

22. As will appear from the definitions accompanying them, *casus* and *mora* refer to the periods of partiality and totality of the eclipse.

of the eclipse. However, if the eclipse will have mora, add the hours, minutes, and seconds from the beginning of eclipse to the beginning of mora to the hours, minutes, and seconds from the beginning of mora to the middle of the eclipse. Subtract the whole from the time of truest opposition to get the time of the beginning of the eclipse (**49**); and if you add the same thing, there will result the time of the end of the eclipse (**50**). Then subtract the time from the beginning of mora to mid-eclipse from the time of truest opposition, and there will remain the beginning of mora; and if you add the same, that will be the end of mora.

If you double the time from the beginning of eclipse to mid-eclipse, the result will be the total duration of the eclipse from beginning to end. If you double the time from the beginning of mora to mid-eclipse, the result will be the total mora (that is, the length of time the moon stays completely in the shadow). Then, if you wish to make a diagram, multiply the hours from the beginning of eclipse up to mid-eclipse by the motion of the moon in 1 hour (**51**), and subtract what results from the true place of the moon found for the time of truest opposition, and also from the argument of latitude (**52**), so that the true place of the moon for the time of the beginning of the eclipse will emerge. From the argument of latitude, subtract the motus of the head of the dragon[23] to get the argument of latitude corrected for the time of the beginning of the eclipse; and if you add the same thing that you just now subtracted, you will get these things for the time of the end of the eclipse. Moreover, you find the latitude of the moon for these three times (**52**) by entering the table of latitudes of the moon with these three arguments of latitude (namely for the beginning, middle, and end of the eclipse). Then you can make a diagram according to the teaching of Master John of Linières. . . .[24]

Prior to the seventeenth century, *canones* such as the foregoing constituted the ultimate form of help for the would-be calculator. The greatest concession found in such theoretical works as the

Almagest or *De Revolutionibus* was an occasional set of precepts for carrying out an isolated operation. The following sample exercise is thus an anachronism. As seventeenth-century astronomers were to admit, however, even the experienced calculator can benefit from an explicit display of the various conventional usages adopted by a particular tabulator[25] and for the uninitiated such a picture is, indeed, worth a thousand words. In order to maximize its value while minimizing the commentary involved, the sample calculation (see Table 2) has been harnessed with a rather complicated reference system. The first element is a bold face marginal number labeling the various operations and results, by means of which the reader can move from the textual description of a given procedure to the illustration of it. The second element enables the reader to move in the opposite direction: If the short comments in the lefthand column are not sufficient to convey the drift of the computation, the last previous roman numeral in the same column will direct him back to the appropriate section in the text. The third element indicates the source of the numbers actually involved in the calculation: A letter in brackets identifies the table from which the corresponding number in the fourth (computation) column derives, while a parenthetical boldface number signifies that the number opposite it in the fourth column has simply been carried forward from some previous operation. In the computation column the prevalent operation is addition. Where a subtraction is involved, a minus sign has been entered for the appropriate number. Other operations (except rounding and interpolating) have been indicated in the explanatory columns.

23. While strictly accurate, this correction is so small that it has been omitted from the sample calculation.

24. The printed edition omits the concluding tribute "*a quo habeo scientiam meam*" found in most manuscripts.

25. See, for example, *Histoire de l'Astronomie du Moyen Age* (Paris, 1819), pp. 251–252, where Delambre, intending to provide illustrative computations for September 20, 1476, actually obtains positions for the following year.

Table 2

Sample Investigation of Eclipses for the Year 1511

	XI		[B]	1000^y		1	41	27	30
				500			50	43	45
				10			1	0	52
(1)	days: epoch to beginning 1511					2	33	12	7^d

			[A]	Radix		3	24°	25	50
			[J]	2 4ths		53	23	15	44
				33 3rds		17	40	53	49
				12 2nds		26	17	20	20
				7 units		1	25	20	7
(2)	elongation of moon at new year					0	11°	15	50

	XII		3^ss – (2)		2	48°	44	10
			[J]		2	38	28	47
						10	15	23
						10	9	32
(3)	time to first mean opposition				5	51	13^d 50 29
						5	53

	XIII		Rx		3	34°	28	43
			[K] 2		31	18	44	52
			33		34	6	39	20
			12		38	45	7	52
			20^d		4	24	35	13
			50			11	1	28
			29				6	24
(4)	argument of latitude at opposition				4	20°	43	52

			[K] 29^d		6	23°	39	4
			31			6	50	7
			50				11	1
(5)	monthly increase arg. lat.				0	30°	40	12

In three months the argument of latitude will fall within the eclipse limits. As it happens, however, the resulting eclipse occurs during daylight hours in Europe. Six months later, the moon approaches the other node: that is,

(6)   (4) + 9 × (5) = 2^ss57°.

		(3)		13^d	50	29
	(29^d 31 50) times 9 mos.		4	25	46	30
(7)	since 4 33^d = end of Sept.		4	39^d	37	or
	[C₂]		14^h 48^m	of	7	Oct.
	time of mean opposition	2 33	16	46^d	37	·

	XV		[F] 2		0	8°	41	23
			33			2	23	23
			16				1	10
			46					3
(8)	auges and fixed stars					11°	5	59

			Rx		5	59°	12	34
			[E] 2		1	0	49	38
			33			16	43	39
			16				8	7
			46					23
	eighth sphere				1	16°	54	21

			[D]			+ 8°	45	49
(9)	equation	(8)				11	5	59
			Rx		1	11	25	23
(10)	aux ☉				1	31	17	11

Table 2 (continued)

XVI		Rx	4	38°	21	1
		[G] 2	16	39	14	38
		33	31	34	47	32
		16	15	46	13	14
		46ᵈ		45	20	23
		37			36	28
(11)	motus ☉		3	24°	33	16
		(10)	− 1	31	17	11
(12)	arg. ☉		1	53°	16	5
(13)	eq. ☉	[M]		− 2°	1	9
XXVI		Rx	3	19°	0	15
		[I] 2	47	55	0	42
		33	8	40	37	42
		16	29	2	23	20
		46ᵈ	10	0	59	22
		37		8	3	24
(14)	arg. ☽		3	6°	4	45
(15)	eq. ☽	[N]		+ 0°	34	15
		(13)		− 2°	1	9
(16)	"distance," "of the moon"			2°	35	24
		(16) ÷ 12		+ 12		57
				2°	48	21
				− 1	24	10
		(14)	3	6°	4	45
(17)	corrected arg. ☽		3	4°	40	35
(18)	vel. ☽	[P]			36	20
(19)	vel. ☉	[P]			2	29
(20)	excess				33	51
(21)	correction (16) ÷ (20)			− 4ʰ	35½ᵐ	
		[C₁]	0ᵈ	11	29	
		[G] 11′ 29″		0°	10	51
						29
(22)				− 0°	11	20
		(12)	1	53	16	5
	corrected arg. ☉		1	53°	4	45
(23)	eq. ☉	[M]		− 2°	1	13
		(11)	3	24°	33	16
		(22)		− 0°	11	20
(24)	mean ☉		3	24°	21	56
		(23)		− 2°	1	13
(25)	true ☉		3	22°	20	43
XVII		Rx	2	2°	46	50
		[H] 2	10	2	30	22
		33	49	15	41	21
		16	30	49	20	20
		46ᵈ	10	6	6	51
		37		8	7	32
			0	24°	33	16
		[H] 11′ 29		− 2	24	56
				−	6	22
(26)	mean ☽		0	22°	1	58
	mean ☉	(24)	− 3	24°	21	56
			2	57°	40	02
(27)	2 (☽ − ☉)		5	55°	20	04
(28)	eq. center	[N]			− 0°	42

476

Table 2 (continued)

(29)	prop. parts	[N]					00
		[I] 11′	0	2°	23	43
		29			6	19
				− 2°	30	2
		(14)	3	6	4	45
(30)	mean arg. ☽		3	3°	34	43
		(28)		− 0°	42	
(31)	true arg. ☽		3	2°	53	
(32)	div. diam.	[N]		0°	11	
(33)	eq. arg.	[N]		+ 16′	17″
(34)		**(29) × (32)**		00	00
		(26)	0	22°	1	58
	true ☽		0	22	18	15
	XXVI	**(25)**	− 3	22	20	43
(35)	"distance"			− 0°	2	28
(36)	vel. ☽	[P]			36″22 per^m
	vel. ☉	[P]			2″29 per^m
(37)	excess				33″53 per^m
(38)	correction	**(35) ÷ (37)**	+ 4^m	or	0^d 0 10

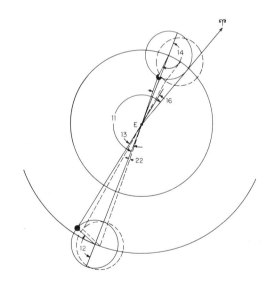

477

Table 2 (continued)

				+ 0°	0	10
	movement ☉	(25)		3	22	20	43
(39)	longitude mid-eclipse			3	22°	20	53

(40) eq. of days $\quad$ [Q] $\qquad$ 7° $\quad$ 36′ $\quad$ or $\quad$ 30½ᵐ

(41)	longitude ☽ mid-eclipse	(39)		0	22°	21
		Rx	1	31	56
	XXVIII	[L] 2	21	16	15
	mean motus	33	44	50	58
	of the head	16		50	50
	of the dragon	46ᵈ		2	26
		26			1
(42)	arg. lat.		2	54°	47

	distance in epicycle	points		casus		mora	
(43)	greatest [S]	10	50	44	19	0	0
(44 a, b, c)	least [T]	13	04	45	20	11	57

(45) From (**31**) the table of proportions (R) indicates (59ᵖ 57′) that no further interpolation is necessary. The ☽ is essentially at its least distance, so the values of (**44**) obtain.

	mean opp.	(**7**)	7 Oct.	14ʰ	48ᵐ
		(**21**)		− 4ʰ	35½ᵐ
		(**38**)		+	4ᵐ
		(**40**)		+	30½ᵐ
(46)	MID-ECLIPSE		7 Oct.	10ʰ	47ᵐ

(47)	half-duration of partiality	(**44b**) ÷ (**37**)	1ʰ	20ᵐ
(48)	half-duration of totality	(**44c**) ÷ (**37**)	0ʰ	21ᵐ

(49)	BEGIN ECLIPSE		7 Oct.	9ʰ	6ᵐ
		(**47**)		1	20
	BEGIN TOTALITY			10ʰ	26ᵐ
		2 × (**48**)		0	42
	END TOTALITY			11ʰ	8ᵐ
		(**47**)		1	20
(50)	END ECLIPSE			12ʰ	28ᵐ
(51)	half-distance of eclipse	(**47 + 48**) × (**36**)	1°	1′

		BEGINNING	MIDDLE	END
	longitudes of eclipse	0 21° 20	0 22° 21	0 23° 22
(52)	args. of latitude	2 53° 46	2 54° 47	2 55° 48
	latitudes [0]	32′ 32″	27′ 15″	21′ 56″

Copernicus (*De revolutionibus orbium coelestium* IV.5) "painstakingly observed" this eclipse and found that it began 1⅛ hours before midnight and ended 2⅓ hours after midnight.

(53) When corrected by 1ʰ36ᵐ, to account for the 24° difference in the longitudes of Frauenburg and Toledo, these times become 9ʰ16ᵐ and 12ʰ44ᵐ.

Table [A] of radices for the incarnation of Christ.

Radix	S	G	m	s
vement of the eighth sphere	5	59	12	34
vement of the aux of the sun	1	11	25	23
tus of the sun	4	38	21	1
tus of the moon	2	2	46	50
gument of the moon	3	19	0	15
ument of latitude of the moon	3	34	28	43
ngation of the moon	3	24	25	50
vement of the head of the dragon	1	31	55	53

Table [B] for converting years and months into days.

Years	Days 4ths	3ds	2ds	Units	Years	Days 3ds	2ds	Units	*		Normal Years		Leap Years	
40	0	4	3	30	1	0	6	5	15	Januarius	0	31	0	31
60	0	6	5	15	2	0	12	10	30	Februarius	0	59	1	0
80	0	8	7	0	3	0	18	15	45	Martius	1	30	1	31
100	0	10	8	45	4b	0	24	21	0	Aprilis	2	0	2	1
200	0	20	17	30	5	0	30	26	15	Maius	2	31	2	32
300	0	30	26	15	6	0	36	31	30	Junius	3	1	3	2
400	0	40	35	0	7	0	42	36	45	Julius	3	32	3	33
500	0	50	43	45	8b	0	48	42	0	Augustus	4	3	4	4
600	1	0	52	30	9	0	54	47	15	September	4	33	4	34
700	1	11	1	15	10	1	0	52	30	October	5	4	5	5
800	1	21	10	0	11	1	6	57	45	November	5	34	5	35
900	1	31	18	45	12b	1	13	3	0	December	6	5	6	6
1000	1	41	27	30	13	1	19	8	15					
2000	3	22	55	0	14	1	25	13	30					
3000	5	4	22	30	15	1	31	18	45					
4000	6	45	50	0	16b	1	37	24	0					
5000	8	27	17	30	17	1	43	29	15					
6000	10	8	45	0	18	1	49	34	30					
7000	11	50	12	30	19	1	55	39	45					
8000	13	31	40	0	20b	2	1	45	0					

*This column for tabular symmetry only; not to be used in computation

Table [C₁] for finding minutes of a day and their fractions from minutes of an hour and their fractions.

S		m	M S
2;30		31	1 17 30
5;0		32	1 20 0
7;30		33	1 22 30
10;0		34	1 25 0
12;30		35	1 27 30
15;0		36	1 30 0
17;30		37	1 32 30
20;0		38	1 35 0
22;30		39	1 37 30
25;0		40	1 40 0
27;30		41	1 42 30
30;0		42	1 45 0
32;30		43	1 47 30
35;0		44	1 50 0
37;30		45	1 52 30
40;0		46	1 55 0
42;30		47	1 57 30
45;0		48	2 0 0
47;30		49	2 2 30
50;0		50	2 5 0
52;30		51	2 7 30
55;0		52	2 10 0
57;30		53	2 12 30
1 0;0		54	2 15 0
1 2;30		55	2 17 30
1 5;0		56	2 20 0
1 7;30		57	2 22 30
1 10;0		58	2 25 0
1 12;30		59	2 27 30
1 15;0		60	2 30 0

Table [C₂] for finding hours and fractions of hours from minutes of a day and their fractions.

M	H m	M	H m
1	0 24	31	12 24
2	0 48	32	12 48
3	1 12	33	13 12
4	1 36	34	13 36
5	2 0	35	14 0
6	2 24	36	14 24
7	2 48	37	14 48
8	3 12	38	15 12
9	3 36	39	15 36
10	4 0	40	16 0
11	4 24	41	16 24
12	4 48	42	16 48
13	5 12	43	17 12
14	5 36	44	17 36
15	6 0	45	18 0
16	6 24	46	18 24
17	6 48	47	18 48
18	7 12	48	19 12
19	7 36	49	19 36
20	8 0	50	20 0
21	8 24	51	20 24
22	8 48	52	20 48
23	9 12	53	21 12
24	9 36	54	21 36
25	10 0	55	22 0
26	10 24	56	22 24
27	10 48	57	22 48
28	11 12	58	23 12
29	11 36	59	23 36
30	12 0	60	24 0

Table [D] of equations for the access and recess of the eighth sphere.

	G m s		G m s		G m s
1	0 9 25	31	4 37 17	61	7 51 49
2	0 18 49	32	4 45 18	62	7 56 19
3	0 28 11	33	4 53 14	63	8 0 41
4	0 37 32	34	5 1 5	64	8 4 56
5	0 46 52	35	5 8 51	65	8 9 2
6	0 56 12	36	5 16 30	66	8 12 58
7	1 5 31	37	5 24 4	67	8 16 45
8	1 14 48	38	5 31 33	68	8 20 23
9	1 24 4	39	5 38 57	69	8 23 52
10	1 33 20	40	5 46 16	70	8 27 11
11	1 42 34	41	5 53 26	71	8 30 23
12	1 51 46	42	6 0 29	72	8 33 24
13	2 0 57	43	6 7 26	73	8 36 15
14	2 10 6	44	6 14 17	74	8 38 56
15	2 19 13	45	6 21 2	75	8 41 28
16	2 26 17	46	6 27 40	76	8 43 50
17	2 37 16	47	6 34 10	77	8 46 2
18	2 46 11	48	6 40 33	78	8 48 5
19	2 55 2	49	6 46 49	79	8 49 59
20	3 3 49	50	6 52 58	80	8 51 44
21	3 12 47	51	6 59 0	81	8 53 19
22	3 21 36	52	7 4 53	82	8 54 44
23	3 30 20	53	7 10 38	83	8 55 55
24	3 36 57	54	7 16 15	84	8 57 0
25	3 47 27	55	7 21 44	85	8 57 55
26	3 55 54	56	7 27 7	86	8 58 40
27	4 4 17	57	7 32 21	87	8 59 15
28	4 12 38	58	7 37 27	88	8 59 40
29	4 20 55	59	7 42 21	89	8 59 55
30	4 29 10	60	7 47 10	90	9 0 0

Table [E] of the mean motion of access and recess of the eighth sphere.

#	Units	2ds	3ds	4ths			
1	0	0	0	0	30	24	49
2	0	0	0	1	0	49	38
3	0	0	0	1	31	14	27
4	0	0	0	2	1	39	16
5	0	0	0	2	32	4	5
6	0	0	0	3	2	28	54
7	0	0	0	3	32	53	43
8	0	0	0	4	3	18	32
9	0	0	0	4	33	43	21
10	0	0	0	5	4	8	10
11	0	0	0	5	34	32	59
12	0	0	0	6	4	57	48
13	0	0	0	6	35	22	37
14	0	0	0	7	5	47	26
15	0	0	0	7	36	12	15
16	0	0	0	8	6	37	4
17	0	0	0	8	37	1	53
18	0	0	0	9	7	26	42
19	0	0	0	9	37	51	31
20	0	0	0	10	8	16	20
21	0	0	0	10	38	41	9
22	0	0	0	11	9	5	58
23	0	0	0	11	39	30	47
24	0	0	0	12	9	55	36
25	0	0	0	12	40	20	25
26	0	0	0	13	10	45	14
27	0	0	0	13	41	10	3
28	0	0	0	14	11	34	52
29	0	0	0	14	41	59	41
30	0	0	0	15	12	24	30
31	0	0	0	15	42	49	19
32	0	0	0	16	13	14	8
33	0	0	0	16	43	38	57
34	0	0	0	17	14	3	46
35	0	0	0	17	44	28	35
36	0	0	0	18	14	53	24
37	0	0	0	18	45	18	13
38	0	0	0	19	15	43	2
39	0	0	0	19	46	7	51
40	0	0	0	20	16	32	40
41	0	0	0	20	46	57	29
42	0	0	0	21	17	22	18
43	0	0	0	21	47	47	7
44	0	0	0	22	18	11	56
45	0	0	0	22	48	36	45
46	0	0	0	23	19	1	34
47	0	0	0	23	49	26	23
48	0	0	0	24	19	51	12
49	0	0	0	24	50	16	1
50	0	0	0	25	20	40	50
51	0	0	0	25	51	5	39
52	0	0	0	26	21	30	28
53	0	0	0	26	51	55	17
54	0	0	0	27	22	20	6
55	0	0	0	27	52	44	55
56	0	0	0	28	23	9	44
57	0	0	0	28	53	34	33
58	0	0	0	29	23	59	22
59	0	0	0	29	54	24	11
60	0	0	0	30	24	49	0

Table [F] of the mean movement of the auges and the fi[xed] stars.

#	Units	2ds	3ds	4ths					
1	0	0	8	0	4	20	41	17	12
2	0	0	0	0	8	41	22	34	24
3	0	0	0	0	13	2	3	51	36
4	0	0	0	0	17	22	45	8	48
5	0	0	0	0	21	43	26	26	0
6	0	0	0	0	26	4	7	43	12
7	0	0	0	0	30	24	49	0	24
8	0	0	0	0	34	45	30	17	36
9	0	0	0	0	39	6	11	34	48
10	0	0	0	0	43	26	52	52	0
11	0	0	0	0	47	47	34	9	12
12	0	0	0	0	52	8	15	26	24
13	0	0	0	0	56	28	56	43	36
14	0	0	0	1	0	49	38	0	48
15	0	0	0	1	5	10	19	18	0
16	0	0	0	1	9	31	0	35	12
17	0	0	0	1	13	51	41	52	24
18	0	0	0	1	18	12	23	9	36
19	0	0	0	1	22	33	4	26	48
20	0	0	0	1	26	53	45	44	0
21	0	0	0	1	31	14	27	1	12
22	0	0	0	1	35	35	8	18	24
23	0	0	0	1	39	55	49	35	36
24	0	0	0	1	44	16	30	52	48
25	0	0	0	1	48	37	12	10	0
26	0	0	0	1	52	57	53	27	12
27	0	0	0	1	57	18	34	44	24
28	0	0	0	2	1	39	16	1	36
29	0	0	0	2	5	59	57	18	48
30	0	0	0	2	10	20	38	36	0
31	0	0	0	2	14	41	19	5	
32	0	0	0	2	19	2	1	1	
33	0	0	0	2	23	22	42	2	
34	0	0	0	2	27	43	23	4	
35	0	0	0	2	32	4	5		
36	0	0	0	2	36	24	46	4	
37	0	0	0	2	40	45	27	36	
38	0	0	0	2	45	6	8	53	
39	0	0	0	2	49	26	50	10	
40	0	0	0	2	53	47	31	21	
41	0	0	0	3	58	8	12	4	
42	0	0	0	3	2	28	54		
43	0	0	0	3	6	49	35	14	
44	0	0	0	3	11	10	16	36	
45	0	0	0	3	15	30	57	54	
46	0	0	0	3	19	51	39	11	
47	0	0	0	3	24	12	20	23	
48	0	0	0	3	28	33	1	4	
49	0	0	0	3	32	53	43		
50	0	0	0	3	37	14	24	26	
51	0	0	0	3	41	35	5	32	
52	0	0	0	3	45	55	46	54	
53	0	0	0	3	50	16	28	11	
54	0	0	0	3	54	37	0	21	
55	0	0	0	3	58	57	50	46	
56	0	0	0	4	3	18	32	1	
57	0	0	0	4	7	39	13	52	
58	0	0	0	4	11	59	54	32	
59	0	0	0	4	16	20	35	54	
60	0	0	0	4	20	41	17	12	

Table [G] of the mean motus of the sun.

#	Units	2ds	3ds	4ths					
1	0	0	59	8	19	37	19	13	56
2	0	1	58	16	39	14	38	27	52
3	0	2	57	24	58	51	57	41	48
4	0	3	56	33	18	29	16	55	44
5	0	4	55	41	38	6	36	9	40
6	0	5	54	49	57	43	55	23	36
7	0	6	53	58	17	21	14	37	32
8	0	7	53	6	36	58	33	51	28
9	0	8	52	14	56	35	53	5	24
10	0	9	51	23	16	13	12	19	20
11	0	10	50	31	35	50	31	33	16
12	0	11	49	39	55	27	50	47	12
13	0	12	48	48	35	5	10	1	8
14	0	13	47	56	34	42	29	15	4
15	0	14	47	4	54	19	48	29	0
16	0	15	46	13	13	57	7	42	56
17	0	16	45	21	33	34	26	56	52
18	0	17	44	29	53	11	46	10	48
19	0	18	43	38	12	49	5	24	44
20	0	19	42	46	32	26	24	38	40
21	0	20	41	54	52	3	43	52	36
22	0	21	41	3	11	41	3	6	32
23	0	22	40	11	31	18	22	20	28
24	0	23	39	19	50	55	41	34	24
25	0	24	38	28	10	33	0	48	20
26	0	25	37	36	30	10	20	2	16
27	0	26	36	44	49	47	39	16	12
28	0	27	35	53	9	24	58	30	8
29	0	28	35	1	29	2	17	44	4
30	0	29	34	9	48	39	36	58	0
31	0	30	33	18	8	16	56	11	56
32	0	31	32	26	27	54	15	25	52
33	0	32	31	34	47	31	34	39	48
34	0	33	30	43	7	8	53	53	44
35	0	34	29	51	26	46	13	7	40
36	0	35	28	59	46	23	32	21	36
37	0	36	28	8	6	0	51	35	32
38	0	37	27	16	25	38	10	49	28
39	0	38	26	24	45	15	30	3	24
40	0	39	25	33	4	52	49	17	20
41	0	40	24	41	24	30	8	31	16
42	0	41	23	49	44	7	27	45	12
43	0	42	22	58	3	44	46	59	8
44	0	43	22	6	23	22	6	13	4
45	0	44	21	14	42	59	25	27	0
46	0	45	20	23	2	36	44	40	56
47	0	46	19	31	22	14	3	54	52
48	0	47	18	39	41	51	23	8	48
49	0	48	17	48	1	28	42	22	44
50	0	49	16	56	21	6	1	36	40
51	0	50	16	4	40	43	20	50	36
52	0	51	15	13	0	20	40	4	32
53	0	52	14	21	19	57	59	18	28
54	0	53	13	29	39	35	18	32	24
55	0	54	12	37	59	12	37	46	20
56	0	55	11	46	18	49	57	0	16
57	0	56	10	54	38	27	16	14	12
58	0	57	10	2	58	4	35	28	8
59	0	58	9	11	17	41	54	42	4
60	0	59	8	19	37	19	13	56	0

Table [H] of the mean motus of the moon.

#	Units	2ds	3ds	4ths					
1	0	13	10	35	1	15	11	4	35
2	0	26	21	10	2	30	22	9	10
3	0	39	31	45	3	45	33	13	45
4	0	52	42	20	5	0	44	18	20
5	1	5	52	55	6	15	55	22	55
6	1	19	3	30	7	31	6	27	30
7	1	32	14	5	8	46	17	32	5
8	1	45	24	40	10	1	28	36	40
9	1	58	35	15	11	16	39	41	15
10	2	11	45	50	12	31	50	45	50
11	2	24	56	25	13	47	1	50	25
12	2	38	7	0	15	2	12	55	0
13	2	51	17	35	16	17	23	59	35
14	3	4	28	10	17	32	35	4	10
15	3	17	38	45	18	47	46	8	45
16	3	30	49	20	20	2	57	13	20
17	3	43	59	55	21	18	8	17	55
18	3	57	10	30	22	33	19	22	30
19	4	10	21	5	23	48	30	27	5
20	4	23	31	40	25	3	41	31	40
21	4	36	42	15	26	18	52	36	15
22	4	49	52	50	27	34	3	40	50
23	5	3	3	25	28	49	14	45	25
24	5	16	14	0	30	4	25	50	0
25	5	29	24	35	31	19	36	54	35
26	5	42	35	10	32	34	47	59	10
27	5	55	45	45	33	49	59	3	45
28	6	8	56	20	35	5	10	8	0
29	6	22	6	55	36	20	21	12	55
30	6	35	17	30	37	35	32	17	30
31	6	48	28	5	38	50	43	22	
32	7	1	38	40	40	5	54		
33	7	14	49	15	41	21	5		
34	7	27	59	50	42	36	16		
35	7	41	10	25	43	51	27		
36	7	54	21	0	45	6	38		
37	8	7	31	35	46	21	49		
38	8	20	42	10	47	37	0		
39	8	33	52	45	48	52	11		
40	8	47	3	20	50	7	23		
41	9	0	13	55	51	22	34		
42	9	13	24	30	52	37	45		
43	9	26	35	5	53	52	56		
44	9	39	45	40	55	8	7		
45	9	52	56	15	56	23	18		
46	10	6	6	50	57	38	29		
47	10	19	17	25	58	53	40		
48	10	32	28	1	0	8	51		
49	10	45	38	36	1	24	2		
50	10	58	49	11	2	39	13		
51	11	11	59	46	3	54	24		
52	11	25	10	21	5	9	35		
53	11	38	20	56	6	24	47		
54	11	51	31	31	7	39	58		
55	12	4	42	6	8	55	9		
56	12	17	52	41	10	10	20		
57	12	31	3	16	11	25	31	2	
58	12	44	13	51	12	40	42		
59	12	57	24	26	13	55	53		
60	13	10	35	1	15	11	4		

Table [I] of the mean argument of the moon.

2ds	3ds	4ths							Units	2ds	3ds	4ths					
13	3	53	57	30	21	4	13	31	6	45	0	52	42	40	53	10	43
26	7	47	55	0	42	8	26	32	6	58	4	46	40	11	14	14	56
39	11	41	52	31	3	12	39	33	7	11	8	40	37	41	35	19	9
52	15	35	50	1	24	16	52	34	7	24	12	34	35	11	56	23	22
5	19	29	47	31	45	21	5	35	7	37	16	28	32	42	17	27	35
18	23	23	45	2	6	25	13	36	7	50	20	22	30	12	38	31	48
31	27	17	42	32	27	29	31	37	8	3	24	16	27	42	59	36	1
44	31	11	40	2	48	33	44	38	8	16	28	10	25	13	20	40	14
57	35	5	37	33	9	37	57	39	8	29	32	4	22	43	41	44	27
10	38	59	35	3	30	42	10	40	8	42	35	58	20	14	2	48	40
23	42	53	32	33	51	46	23	41	8	55	39	52	17	44	23	52	53
36	46	47	30	4	12	50	36	42	9	8	43	46	15	14	44	57	6
49	50	41	27	34	33	54	49	43	9	21	47	40	12	45	6	1	19
2	54	35	25	4	54	59	2	44	9	34	51	34	10	15	27	5	32
15	58	29	22	35	16	3	15	45	9	47	55	28	7	45	48	9	45
29	2	23	20	5	37	7	28	46	10	0	59	22	5	16	9	13	58
42	6	17	17	35	58	11	41	47	10	14	3	16	2	46	30	18	11
55	10	11	15	6	19	15	54	48	10	27	7	10	0	16	51	22	24
8	14	5	12	36	40	20	7	49	10	40	11	3	57	47	12	26	37
21	17	59	10	7	1	24	20	50	10	53	14	57	55	17	33	30	50
34	21	53	7	37	22	28	33	51	11	6	18	51	52	47	54	35	3
47	25	47	5	7	43	32	46	52	11	19	22	45	50	18	15	39	16
0	29	41	2	38	4	36	59	53	11	32	26	39	47	48	36	43	29
13	33	35	0	8	25	41	12	54	11	45	30	33	45	18	57	47	42
26	37	28	57	38	46	45	25	55	11	58	34	27	42	49	18	51	55
39	41	22	55	9	7	49	38	56	12	11	38	21	40	19	39	56	8
52	45	16	52	39	28	53	51	57	12	24	42	15	37	50	1	0	21
5	49	10	50	9	49	58	4	58	12	37	46	9	35	20	22	4	34
18	43	4	47	40	11	2	17	59	12	50	50	3	32	50	43	8	47
31	56	58	45	10	32	6	30	60	13	3	53	57	30	21	4	13	0

Table [J] of the mean elongation of the moon.

	Units	2ds	3ds	4ths							Units	2ds	3ds	4ths					
1	0	12	11	26	41	37	51	50	39	31	6	17	54	47	40	33	47	10	9
2	0	24	22	53	23	15	43	41	18	32	6	30	6	14	12	11	39	0	48
3	0	36	34	20	4	53	35	31	57	33	6	42	17	40	53	49	30	51	27
4	0	48	45	46	46	31	27	22	36	34	6	54	29	7	35	27	22	42	6
5	1	0	57	13	28	9	19	13	15	35	7	6	40	34	17	5	14	32	45
6	1	13	8	40	9	47	11	3	54	36	7	18	52	0	58	43	6	23	24
7	1	25	20	6	51	25	2	54	33	37	7	31	3	27	40	20	58	14	3
8	1	37	31	33	33	2	54	45	12	38	7	43	14	54	21	58	50	4	42
9	1	49	43	0	14	40	46	35	51	39	7	55	26	21	3	36	41	55	21
10	2	1	54	26	56	18	38	26	30	40	8	7	37	47	45	14	33	46	0
11	2	14	5	53	37	56	30	17	9	41	8	19	49	14	26	52	25	36	39
12	2	26	17	20	19	34	22	7	48	42	8	32	0	41	8	30	17	27	18
13	2	38	28	47	1	12	13	58	27	43	8	44	12	7	50	8	9	17	57
14	2	50	40	13	42	50	5	49	6	44	8	56	23	34	31	46	1	8	36
15	3	2	51	40	24	27	57	39	45	45	9	8	35	1	13	23	52	59	15
16	3	15	3	7	6	5	49	30	24	46	9	20	46	27	55	1	44	49	54
17	3	27	14	33	47	43	41	21	3	47	9	32	57	54	36	39	36	40	33
18	3	39	26	0	29	21	33	11	42	48	9	45	9	21	18	17	28	31	12
19	3	51	37	27	10	59	25	2	21	49	9	57	20	47	59	55	20	21	51
20	4	3	48	53	52	37	16	53	0	50	10	9	32	14	41	33	12	12	30
21	4	16	0	20	34	15	8	43	39	51	10	21	43	41	23	11	4	3	9
22	4	28	11	47	15	53	0	34	18	52	10	33	55	8	4	48	55	53	48
23	4	40	23	13	57	30	52	24	57	53	10	46	6	34	46	26	47	44	27
24	4	52	34	40	39	8	44	15	36	54	10	58	18	1	28	4	39	35	6
25	5	4	46	7	20	46	36	6	15	55	11	10	29	28	9	42	31	25	45
26	5	16	57	34	2	24	27	56	54	56	11	22	40	54	51	29	23	16	24
27	5	29	9	0	44	2	19	47	33	57	11	34	52	21	32	58	15	7	3
28	5	41	20	27	25	40	11	38	12	58	11	47	3	48	14	36	6	57	42
29	5	53	31	54	7	18	3	28	51	59	11	59	15	14	56	13	58	48	21
30	6	5	43	20	48	55	55	19	30	60	12	11	26	41	37	51	50	39	0

Table [K] of the argument of latitude of the moon.

Units	2ds	3ds	4ths							Units	2ds	3ds	4ths					
0	13	13	45	39	22	25	53	45	31	6	50	6	35	20	35	22	46	15
0	26	27	31	18	44	51	47	30	32	7	3	20	20	59	57	48	40	0
0	39	41	16	58	7	17	41	15	33	7	16	34	6	39	20	14	33	45
0	52	55	2	37	29	43	35	0	34	7	29	47	52	18	42	40	27	30
1	6	8	48	16	52	9	28	45	35	7	43	1	37	58	5	6	21	15
1	19	22	33	56	14	35	22	30	36	7	56	15	23	37	27	32	15	0
1	32	36	19	35	37	1	16	15	37	8	9	29	9	16	49	58	8	45
1	45	50	5	14	59	27	10	0	38	8	22	42	54	56	12	24	2	30
1	59	3	50	54	21	53	3	45	39	8	35	56	40	35	34	49	56	15
2	12	17	36	33	44	18	57	30	40	8	49	10	26	14	57	15	50	0
2	25	31	22	13	6	44	51	15	41	9	2	24	11	54	19	41	43	45
2	38	45	7	52	29	10	45	0	42	9	15	37	57	33	42	7	37	30
2	51	58	53	31	51	36	38	45	43	9	28	51	43	13	4	33	31	15
3	5	12	39	11	14	2	32	30	44	9	42	5	28	52	26	59	25	0
3	18	26	24	50	36	28	26	15	45	9	55	19	14	31	49	25	18	45
3	31	40	10	29	58	54	20	0	46	10	8	33	0	11	11	51	12	30
3	44	53	56	9	21	20	13	45	47	10	21	46	45	50	34	17	6	15
3	58	7	41	48	43	46	7	30	48	10	35	0	31	29	56	43	0	0
4	11	21	27	28	6	12	1	15	49	10	48	14	17	9	19	8	53	45
4	24	35	13	7	28	37	55	0	50	11	1	28	2	48	41	34	47	30
4	37	48	58	46	51	3	48	45	51	11	14	41	48	28	4	0	41	15
4	51	2	44	26	13	29	42	30	52	11	27	55	34	7	26	26	35	0
5	4	16	30	5	35	55	36	15	53	11	41	9	19	46	48	52	28	45
5	17	30	15	44	58	21	30	0	54	11	54	23	5	26	11	18	22	30
5	30	44	1	24	20	47	23	45	55	12	7	36	51	5	33	44	16	15
5	43	57	47	3	43	13	17	30	56	12	20	50	36	44	56	10	10	0
5	57	11	32	43	5	39	11	15	57	12	34	4	22	24	18	36	3	45
6	10	25	18	22	28	5	5	0	58	12	47	18	8	3	41	1	57	30
6	23	39	4	1	50	30	58	45	59	13	0	31	53	43	3	27	51	15
6	36	52	49	41	12	56	52	30	60	13	13	45	39	22	25	53	45	0

Table [L] of the mean motion of the head of the dragon.

	Units	2ds	3ds	4ths							Units	2ds	3ds	4ths					
1	0	0	3	10	38	7	14	49	10	31	0	1	38	29	41	44	39	24	10
2	0	0	6	21	16	14	29	48	20	32	0	1	41	4	19	51	54	13	20
3	0	0	9	31	54	21	44	27	30	33	0	1	44	50	57	59	9	2	30
4	0	0	12	42	32	28	59	16	40	34	0	1	48	1	36	6	23	51	40
5	0	0	15	53	10	36	14	5	50	35	0	1	51	12	14	13	38	40	50
6	0	0	19	3	48	43	28	55	0	36	0	1	54	22	52	20	53	30	0
7	0	0	22	14	26	50	43	44	10	37	0	1	57	33	30	28	8	19	10
8	0	0	25	25	4	57	58	33	20	38	0	2	0	44	8	35	23	8	20
9	0	0	28	35	43	5	13	22	30	39	0	2	3	54	46	42	37	57	30
10	0	0	31	46	21	12	28	11	40	40	0	2	7	5	24	49	52	46	40
11	0	0	34	56	59	19	43	0	50	41	0	2	10	16	2	57	7	35	50
12	0	0	38	7	37	26	57	50	0	42	0	2	13	26	41	4	22	25	0
13	0	0	41	18	15	34	12	39	10	43	0	2	16	37	19	11	37	14	10
14	0	0	44	28	53	41	27	28	20	44	0	2	19	47	57	18	52	3	20
15	0	0	47	39	31	48	42	17	30	45	0	2	22	58	35	26	6	52	30
16	0	0	50	50	9	55	57	6	40	46	0	2	26	9	13	33	21	41	40
17	0	0	54	0	48	3	11	55	50	47	0	2	29	19	51	40	36	30	50
18	0	0	57	11	26	10	26	45	0	48	0	2	32	30	29	47	51	20	0
19	0	1	0	22	4	17	41	34	10	49	0	2	35	41	7	55	6	9	10
20	0	1	3	32	42	24	56	23	20	50	0	2	38	51	46	2	20	58	20
21	0	1	6	43	20	32	11	12	30	51	0	2	42	2	24	9	35	47	30
22	0	1	9	53	58	39	26	1	40	52	0	2	45	13	2	16	50	30	40
23	0	1	13	4	36	46	40	50	50	53	0	2	48	23	40	24	5	25	50
24	0	1	16	15	14	53	55	40	0	54	0	2	51	34	18	31	20	15	0
25	0	1	19	25	53	1	10	29	10	55	0	2	54	44	56	38	35	4	10
26	0	1	22	36	31	8	25	18	20	56	0	2	57	55	34	45	49	53	20
27	0	1	25	47	9	15	40	7	30	57	0	3	1	6	12	53	4	42	30
28	0	1	28	57	47	22	54	56	40	58	0	3	4	16	51	0	19	32	40
29	0	1	32	8	25	30	9	45	50	59	0	3	7	27	29	7	34	20	50
30	0	1	35	19	3	37	24	35	0	60	0	3	10	38	7	14	49	10	0

Table [M] of the equation of the sun.

Common numbers S	G	S	G	Equations G	m	s
0	1	5	59	0	2	10
0	2	5	58	0	4	19
0	3	5	57	0	6	27
0	4	5	56	0	8	36
0	5	5	55	0	10	44
0	6	5	54	0	12	53
0	7	5	53	0	15	2
0	8	5	52	0	17	10
0	9	5	51	0	19	19
0	10	5	50	0	21	28
0	11	5	49	0	23	36
0	12	5	48	0	25	45
0	13	5	47	0	27	53
0	14	5	46	0	30	1
0	15	5	45	0	32	8
0	16	5	44	0	34	16
0	17	5	43	0	36	23
0	18	5	42	0	38	30
0	19	5	41	0	40	37
0	20	5	40	0	42	43
0	21	5	39	0	44	49
0	22	5	38	0	46	55
0	23	5	37	0	48	59
0	24	5	36	0	51	4
0	25	5	35	0	53	4
0	26	5	34	0	55	2
0	27	5	33	0	57	1
0	28	5	32	0	58	59
0	29	5	31	1	0	57
0	30	5	30	1	2	54

Common numbers S	G	S	G	Equations G	m	s
0	31	5	29	1	4	46
0	32	5	28	1	6	37
0	33	5	27	1	8	28
0	34	5	26	1	10	19
0	35	5	25	1	12	9
0	36	5	24	1	13	58
0	37	5	23	1	15	41
0	38	5	22	1	17	24
0	39	5	21	1	19	6
0	40	5	20	1	20	48
0	41	5	19	1	22	29
0	42	5	18	1	24	10
0	43	5	17	1	25	50
0	44	5	16	1	27	29
0	45	5	15	1	29	8
0	46	5	14	1	30	46
0	47	5	13	1	32	23
0	48	5	12	1	33	59
0	49	5	11	1	35	30
0	50	5	10	1	37	0
0	51	5	9	1	38	30
0	52	5	8	1	39	58
0	53	5	7	1	41	27
0	54	5	6	1	42	54
0	55	5	5	1	44	14
0	56	5	4	1	45	34
0	57	5	3	1	46	53
0	58	5	2	1	48	10
0	59	5	1	1	49	28
1	0	5	0	1	50	44

Common numbers S	G	S	G	Equations G	m	s
1	1	4	59	1	51	51
1	2	4	58	1	52	56
1	3	4	57	1	54	0
1	4	4	56	1	55	6
1	5	4	55	1	56	9
1	6	4	54	1	57	11
1	7	4	53	1	58	11
1	8	4	52	1	58	52
1	9	4	51	1	59	41
1	10	4	50	2	0	26
1	11	4	49	2	1	16
1	12	4	48	2	2	2
1	13	4	47	2	2	41
1	14	4	46	2	3	21
1	15	4	45	2	3	59
1	16	4	44	2	4	36
1	17	4	43	2	5	16
1	18	4	42	2	5	48
1	19	4	41	2	6	17
1	20	4	40	2	6	45
1	21	4	39	2	7	12
1	22	4	38	2	7	37
1	23	4	37	3	8	2
1	24	4	36	2	8	27
1	25	4	35	2	8	45
1	26	4	34	2	9	1
1	27	4	33	2	9	17
1	28	4	32	2	9	32
1	29	4	31	2	9	45
1	30	4	30	2	9	57

Common numbers S	G	S	G	Equations G	m	s
1	31	4	29	2	9	59
1	32	4	28	2	10	0
1	33	4	27	2	10	0
1	34	4	26	2	10	0
1	35	4	25	2	9	57
1	36	4	24	2	9	51
1	37	4	23	2	9	36
1	38	4	22	2	9	20
1	39	4	21	2	9	2
1	40	4	20	2	8	45
1	41	4	19	2	8	25
1	42	4	18	2	8	6
1	43	4	17	2	7	41
1	44	4	16	2	7	14
1	45	4	15	2	6	46
1	46	4	14	2	6	18
1	47	4	13	2	5	48
1	48	4	12	2	5	13
1	49	4	11	2	4	42
1	50	4	10	2	4	5
1	51	4	9	2	3	27
1	52	4	8	2	2	37
1	53	4	7	2	1	45
1	54	4	6	2	0	51
1	55	4	5	1	59	53
1	56	4	4	1	58	55
1	57	4	3	1	57	57
1	58	4	2	1	56	57
1	59	4	1	1	55	57
2	0	4	0	1	54	57

Common numbers S	G	S	G	Equations G	m	s
2	1	3	59	1	53	46
2	2	3	58	1	52	35
2	3	3	57	1	51	24
2	4	3	56	1	50	12
2	5	3	55	1	48	59
2	6	3	54	1	47	40
2	7	3	53	1	46	20
2	8	3	52	1	44	53
2	9	3	51	1	43	26
2	10	3	50	1	41	57
2	11	3	49	1	40	27
2	12	3	48	1	38	57
2	13	3	47	1	37	25
2	14	3	46	1	35	53
2	15	3	45	1	34	20
2	16	3	44	1	32	46
2	17	3	43	1	31	12
2	18	3	42	1	29	37
2	19	3	41	1	27	50
2	20	3	40	1	26	3
2	21	3	39	1	24	16
2	22	3	38	1	22	28
2	23	3	37	1	20	40
2	24	3	36	1	18	51
2	25	3	35	1	17	0
2	26	3	34	1	15	8
2	27	3	33	1	13	13
2	28	3	32	1	11	16
2	29	3	31	1	9	10
2	30	3	30	1	7	7

Common numbers S	G	S	G	Equations G	m	s
2	31	3	29	1	5	1
2	32	3	28	1	2	54
2	33	3	27	1	0	47
2	34	3	26	0	58	40
2	35	3	25	0	56	33
2	36	3	24	0	54	25
2	37	3	23	0	52	17
2	38	3	22	0	50	9
2	39	3	21	0	48	1
2	40	3	20	0	45	53
2	41	3	19	0	43	44
2	42	3	18	0	41	35
2	43	3	17	0	39	26
2	44	3	16	0	37	16
2	45	3	15	0	35	6
2	46	3	14	0	32	51
2	47	3	13	0	30	35
2	48	3	12	0	28	19
2	49	3	11	0	26	1
2	50	3	10	0	23	42
2	51	3	9	0	21	22
2	52	3	8	0	19	1
2	53	3	7	0	16	40
2	54	3	6	0	14	19
2	55	3	5	0	11	58
2	56	3	4	0	9	36
2	57	3	3	0	7	12
2	58	3	2	0	4	48
2	59	3	1	0	2	24
3	0	3	0	0	0	0

Table [N] of the equations of the moon.

Common numbers (S G S G)	Equation of center (G m)	Prop. parts (m)	Variation of the diameter (G m)	Equation of the argument (G m s)
1	5 59	0 9	0	0 3	0 4 46
2	5 58	0 18	0	0 5	0 9 31
3	5 57	0 27	0	0 7	0 14 15
4	5 56	0 36	0	0 10	0 19 0
5	5 55	0 45	0	0 12	0 23 44
6	5 54	0 53	0	0 14	0 28 28
7	5 53	1 2	0	0 17	0 33 11
8	5 52	1 11	0	0 19	0 37 54
9	5 51	1 20	0	0 21	0 42 35
10	5 50	1 29	0	0 24	0 47 19
11	5 49	1 38	0	0 26	0 52 0
12	5 48	1 46	1	0 28	0 56 41
13	5 47	1 55	1	0 31	1 1 20
14	5 46	2 4	1	0 33	1 5 59
15	5 45	2 13	1	0 35	1 10 38
16	5 44	2 22	1	0 38	1 15 15
17	5 43	2 31	1	0 40	1 19 51
18	5 42	2 39	1	0 42	1 24 27
19	5 41	2 48	1	0 45	1 29 0
20	5 40	2 57	2	0 47	1 33 32
21	5 39	3 5	2	0 49	1 38 3
22	5 38	3 14	2	0 52	1 42 33
23	5 37	3 23	2	0 54	1 47 1
24	5 36	3 31	2	0 57	1 51 27
25	5 35	3 40	2	0 59	1 55 52
26	5 34	3 49	2	1 1	2 0 15
27	5 33	3 57	3	1 3	2 4 37
28	5 32	4 6	3	1 6	2 8 57
29	5 31	4 15	3	1 8	2 13 14
30	5 30	4 23	3	1 10	2 17 29
0 31	5 29	4 32	3	1 12	2 21 43
0 32	5 28	4 41	4	1 14	2 25 55
0 33	5 27	4 49	4	1 16	2 30 5
0 34	5 26	4 58	4	1 19	2 34 12
0 35	5 25	5 7	4	1 21	2 38 17
0 36	5 24	5 15	4	1 23	2 42 21
0 37	5 23	5 24	5	1 26	2 46 22
0 38	5 22	5 33	5	1 27	2 50 19
0 39	5 21	5 41	5	1 29	2 54 14
0 40	5 20	5 50	5	1 31	2 58 7
0 41	5 19	5 59	6	1 33	3 1 58
0 42	5 18	6 7	6	1 35	3 5 46
0 43	5 17	6 16	6	1 37	3 9 31
0 44	5 16	6 25	7	1 39	3 13 13
0 45	5 15	6 33	7	1 40	3 16 51
0 46	5 14	6 42	7	1 42	3 20 26
0 47	5 13	6 50	8	1 44	3 23 59
0 48	5 12	6 58	8	1 45	3 27 30
0 49	5 11	7 7	8	1 47	3 30 57
0 50	5 10	7 15	9	1 48	3 34 20
0 51	5 9	7 23	9	1 49	3 37 40
0 52	5 8	7 32	9	1 51	3 40 57
0 53	5 7	7 40	10	1 53	3 44 10
0 54	5 6	7 48	10	1 54	3 47 20
0 55	5 5	7 56	10	1 56	3 50 26
0 56	5 4	8 4	11	1 58	3 53 29
0 57	5 3	8 12	11	1 59	3 56 30
0 58	5 2	8 20	11	2 1	3 59 26
0 59	5 1	8 28	12	2 2	4 2 17
1 0	5 0	8 36	12	2 3	4 5 4

Common numbers (S G S G)	Equation of center (G m)	Prop. parts (m)	Variation of the diameter (G m)	Equation of the argument (G m s)
1 1	4 59	8 44	13	2 5	4 7 47
1 2	4 58	8 52	13	2 6	4 10 27
1 3	4 57	8 59	14	2 7	4 13 3
1 4	4 56	9 7	14	2 9	4 15 35
1 5	4 55	9 15	15	2 10	4 18 3
1 6	4 54	9 22	15	2 11	4 20 27
1 7	4 53	9 30	15	2 13	4 22 47
1 8	4 52	9 37	16	2 14	4 25 2
1 9	4 51	9 44	16	2 15	4 27 12
1 10	4 50	9 52	17	2 16	4 29 18
1 11	4 49	9 59	17	2 17	4 31 20
1 12	4 48	10 6	18	2 18	4 33 18
1 13	4 47	10 13	18	2 19	4 35 11
1 14	4 46	10 20	19	2 20	4 36 59
1 15	4 45	10 27	19	2 21	4 38 43
1 16	4 44	10 34	20	2 22	4 40 23
1 17	4 43	10 41	20	2 23	4 41 58
1 18	4 42	10 48	21	2 24	4 43 28
1 19	4 41	10 55	21	2 25	4 44 53
1 20	4 40	11 2	22	2 26	4 46 13
1 21	4 39	11 8	22	2 27	4 47 26
1 22	4 38	11 15	22	2 28	4 48 35
1 23	4 37	11 21	23	2 29	4 49 38
1 24	4 36	11 27	23	2 30	4 50 41
1 25	4 35	11 33	24	2 31	4 51 38
1 26	4 34	11 39	24	2 32	4 52 28
1 27	4 33	11 44	25	2 33	4 53 11
1 28	4 32	11 50	25	2 34	4 53 50
1 29	4 31	11 55	26	2 35	4 54 25
1 30	4 30	12 0	26	2 36	4 54 54
1 31	4 29	12 5	27	2 37	4 55 18
1 32	4 28	12 10	27	2 37	4 55 37
1 33	4 27	12 15	28	2 38	4 55 49
1 34	4 26	12 20	28	2 38	4 55 55
1 35	4 25	12 24	29	2 38	4 56 0
1 36	4 24	12 28	30	2 38	4 55 56
1 37	4 23	12 32	30	2 38	4 55 43
1 38	4 22	12 36	31	2 39	4 55 25
1 39	4 21	12 39	31	2 39	5 55 4
1 40	4 20	12 43	32	2 39	4 54 44
1 41	4 19	12 45	32	2 39	4 54 12
1 42	4 18	12 48	33	2 39	4 53 38
1 43	4 17	12 51	33	2 40	4 52 59
1 44	4 16	12 54	34	2 40	4 52 14
1 45	4 15	12 56	35	2 40	4 51 22
1 46	4 14	12 58	35	2 40	4 50 22
1 47	4 13	13 0	36	2 40	4 49 17
1 48	4 12	13 2	36	2 40	4 48 10
1 49	4 11	13 4	37	2 40	4 46 54
1 50	4 10	13 5	37	2 39	4 45 33
1 51	4 9	13 6	38	2 39	4 44 7
1 52	4 8	13 7	38	2 38	4 42 34
1 53	4 7	13 8	39	2 38	4 40 56
1 54	4 6	13 9	39	2 37	4 39 15
1 55	4 5	13 9	40	2 36	4 37 29
1 56	4 4	13 8	40	2 36	4 35 37
1 57	4 3	13 7	41	2 35	4 33 41
1 58	4 2	13 6	41	2 34	4 31 34
1 59	4 1	13 5	42	2 33	4 29 20
2 0	4 0	13 4	43	2 31	4 27 0

Table [N] of the equations of the moon.

Common numbers S G S G	Equation of center G m	Prop. parts m	Variation of the diameter G m	Equation of argument G m s	Common numbers S G S G	Equation of center G m	Prop. parts m	Variation of the diameter G m	Equation of argument G m s
2 1	3 59	43	2 30	4 24 38	2 31	3 29	9 8	56 1 32	2 34 52
2 2	3 58	43	2 29	4 22 16	2 32	3 28	8 53	56 1 29	2 30 6
2 3	3 57	12 59	3 27	4 19 38	2 33	3 27	8 38	56 1 26	2 25 16
2 4	3 56	12 56	2 26	4 16 58	2 34	3 26	8 22	56 1 24	2 20 23
2 5	3 55	12 53	2 25	4 14 13	2 35	3 25	8 5	57 1 21	2 15 26
2 6	3 54	12 50	2 23	4 11 23	2 36	3 24	7 48	57 1 18	2 10 26
2 7	3 53	12 46	2 22	4 8 28	2 37	3 23	7 31	57 1 16	2 5 22
2 8	3 52	12 41	2 21	4 5 31	2 38	3 22	7 14	57 1 13	2 0 17
2 9	3 51	12 36	2 19	4 2 30	2 39	3 21	6 56	57 1 10	1 55 9
2 10	3 50	12 30	2 18	3 59 20	2 40	3 20	6 39	58 1 8	1 49 58
2 11	3 49	12 23	2 17	3 56 5	2 41	3 19	6 21	58 1 5	1 44 44
2 12	3 48	12 16	2 15	3 52 47	2 42	3 18	6 3	58 1 2	1 39 27
2 13	3 47	12 9	2 14	3 49 23	2 43	3 17	5 45	58 0 59	1 34 9
2 14	3 46	12 2	2 12	3 45 52	2 44	3 16	5 27	58 0 56	1 28 49
2 15	3 45	11 54	2 10	3 42 17	2 45	3 15	5 8	59 0 52	1 23 26
2 16	3 44	11 46	2 9	3 38 37	2 46	3 14	4 49	59 0 49	1 18 1
2 17	3 43	11 38	2 7	3 34 53	2 47	3 13	4 30	59 0 46	1 12 34
2 18	3 42	11 29	2 5	3 31 3	2 48	3 12	4 11	59 0 42	1 7 6
2 19	3 41	11 20	2 3	3 27 10	2 49	3 11	3 52	59 0 39	1 1 36
2 20	3 40	11 11	2 1	3 23 12	2 50	3 10	3 32	59 0 36	0 56 5
2 21	3 39	11 2	1 58	3 19 9	2 51	3 9	3 12	59 0 32	0 50 32
2 22	3 38	10 53	1 56	3 15 2	2 52	3 8	2 52	60 0 29	0 44 58
2 23	3 37	10 43	1 54	3 10 50	2 53	3 7	2 32	60 0 25	0 39 23
2 24	3 36	10 33	1 51	3 6 35	2 54	3 6	2 11	60 0 21	0 33 47
2 25	3 35	10 22	1 49	3 2 15	2 55	3 5	1 50	60 0 18	0 28 10
2 26	3 34	10 11	1 46	2 57 53	2 56	3 4	1 29	60 0 15	0 22 33
2 27	3 33	10 0	1 43	2 53 23	7 57	3 3	1 7	60 0 11	0 16 56
2 28	3 32	9 48	1 41	2 48 51	2 58	3 2	0 45	60 0 8	0 11 18
2 29	3 31	9 35	1 38	2 44 15	2 59	3 1	0 23	60 0 4	0 5 40
2 30	3 30	9 22	1 35	2 39 35	3 0	3 0	0 0	60 0 0	0 0 0

Table [O] of the latitude of the moon.

Common numbers	Ascending latitudes 0	1	2	Descending latitudes 3	4	5
1 29	0 5 13	2 34 24	4 22 21	4 59 58	4 17 7	2 25 17
2 28	0 10 27	2 38 52	4 24 51	4 59 50	4 14 22	2 20 40
3 27	0 15 40	2 43 57	4 27 14	4 59 35	4 11 34	2 16 2
4 26	0 20 53	2 47 39	4 29 34	4 59 15	4 8 37	2 11 21
5 25	0 26 7	2 51 57	4 31 49	4 58 51	4 5 38	2 6 40
6 24	0 31 19	2 56 10	4 33 59	4 58 21	4 2 37	2 1 56
7 23	0 36 31	3 0 21	4 36 4	4 57 45	3 59 28	1 57 8
8 22	0 41 42	3 4 29	4 38 4	4 57 4	3 56 16	1 52 17
9 21	0 46 52	3 8 35	4 40 0	4 56 17	3 53 0	1 47 23
10 20	0 52 1	3 12 39	4 41 52	4 55 25	3 4 40	1 42 27
11 19	0 57 9	3 16 39	4 43 38	4 54 28	3 46 17	1 37 29
12 18	1 2 16	3 20 35	4 45 18	4 53 25	3 42 49	1 32 31
13 17	1 7 23	3 24 26	4 46 52	4 52 17	3 39 17	1 27 33
14 16	1 12 30	3 28 15	4 48 20	4 51 3	3 35 41	1 22 35
15 15	1 17 36	3 32 0	4 49 44	4 49 44	3 32 0	1 17 36
16 14	1 22 35	3 35 41	4 51 3	4 48 20	3 28 15	1 12 30
17 13	1 27 33	3 39 17	4 52 17	4 46 52	3 24 26	1 7 23
18 12	1 32 31	3 42 49	4 53 25	4 45 18	3 20 35	1 2 16
19 11	1 37 29	3 46 17	4 54 28	4 43 38	3 16 39	0 57 9
20 10	1 42 27	3 49 40	4 55 25	4 41 52	3 12 39	0 52 1
21 9	1 47 23	3 53 0	4 56 17	4 40 0	3 8 35	0 46 52
22 8	1 52 17	3 56 16	4 57 4	4 38 4	3 4 29	0 41 42
23 7	1 57 8	3 59 28	4 57 45	4 36 4	3 0 21	0 36 31
24 6	2 1 56	4 2 37	4 58 21	4 33 59	2 56 10	0 31 19
25 5	2 6 40	4 5 38	4 58 51	4 31 49	2 51 57	0 26 7
26 4	2 11 21	4 8 37	4 59 15	4 29 34	2 47 39	0 20 53
27 3	2 16 2	4 11 34	4 59 35	4 27 14	2 43 57	0 15 40
28 2	2 20 40	4 14 22	4 59 50	4 24 51	2 38 52	0 10 27
29 1	2 25 17	4 17 7	4 59 58	4 22 21	2 34 24	0 5 13
30 0	2 29 52	4 19 47	5 0 0	4 19 47	2 29 52	0 0 0
	11	10	9	8	7	6 b

Table [P] of hourly movements of the sun and moon.

Common numbers S	G	S	G	Movements Moon m	s	Sun m	s
0	1	5	59	30	21	2	23
0	2	5	58	30	21	2	23
0	3	5	57	30	21	2	23
0	4	5	56	30	21	2	23
0	5	5	55	30	22	2	23
0	6	5	54	30	22	2	23
0	7	5	53	30	22	2	23
0	8	5	52	30	22	2	23
0	9	5	51	30	23	2	23
0	10	5	50	30	23	2	23
0	11	5	49	30	24	2	23
0	12	5	49	30	24	2	23
0	13	5	47	30	25	2	23
0	14	5	46	30	25	2	23
0	15	5	45	30	26	2	23
0	16	5	44	30	26	2	23
0	17	5	43	30	27	1	23
0	18	5	42	30	27	2	23
0	19	5	41	30	28	2	23
0	20	5	40	30	28	2	23
0	21	5	39	30	29	2	23
0	22	5	38	30	29	2	23
0	23	5	37	30	30	2	24
0	24	5	36	30	31	2	24
0	25	5	35	30	32	2	24
0	26	5	34	30	33	2	24
0	27	5	33	30	34	2	24
0	28	5	32	30	35	2	24
0	29	5	31	30	36	2	24
0	30	5	30	30	37	2	24
0	31	5	29	30	38	2	24
0	32	5	28	30	39	2	24
0	33	5	27	30	40	2	24
0	34	5	26	30	41	2	24
0	35	5	25	30	42	2	24
0	36	5	24	30	43	2	24
0	37	5	23	30	44	2	24
0	38	5	22	30	45	2	24
0	39	5	21	30	46	2	24
0	40	5	20	30	48	2	24
0	41	5	19	30	49	2	25
0	42	5	18	30	50	2	25
0	43	5	17	30	52	2	25
0	44	5	16	30	54	2	25
0	45	5	15	30	56	2	25
0	46	5	14	30	58	2	25
0	47	5	13	31	0	2	25
0	48	5	12	31	2	2	25
0	49	5	11	31	4	2	25
0	50	5	10	31	6	2	25
0	51	5	9	31	8	2	25
0	52	5	8	31	10	2	25
0	53	5	7	31	12	2	25
0	54	5	6	31	14	2	25
0	55	5	5	31	16	2	25
0	56	5	4	31	18	2	25
0	57	5	3	31	20	2	25
0	58	5	2	31	22	2	25
0	59	5	1	31	24	2	25
1	0	5	0	31	26	2	25
1	1	4	59	31	28	2	26
1	2	4	58	31	30	2	26
1	3	4	57	31	32	2	26
1	4	4	56	31	34	2	26
1	5	4	55	31	36	2	26
1	6	4	54	31	39	2	26
1	7	4	53	31	42	2	26
1	8	4	52	31	44	2	26
1	9	4	51	31	46	2	26
1	10	4	50	31	48	2	26
1	11	4	49	31	50	2	26
1	12	4	48	31	53	2	26
1	13	4	47	31	56	2	26
1	14	4	46	31	59	2	26
1	15	4	45	32	1	2	26
1	16	4	44	32	3	2	26
1	17	4	43	32	5	2	26
1	18	4	42	32	8	2	26
1	19	4	41	32	11	2	26
1	20	4	40	32	14	2	27
1	21	4	39	32	17	2	27
1	22	4	38	32	20	2	27
1	23	4	37	32	23	2	27
1	24	4	36	32	26	2	27
1	25	4	35	32	29	2	27
1	26	4	34	32	32	2	27
1	27	4	33	32	35	2	27
1	28	4	32	32	38	2	27
1	29	4	31	32	41	2	27
1	30	4	30	32	44	2	27
1	31	4	29	32	47	2	27
1	32	4	28	32	50	2	27
1	33	4	27	32	53	2	27
1	34	4	26	32	56	2	27
1	35	4	25	32	59	2	27
1	36	4	24	33	2	2	27
1	37	4	23	33	5	2	27
1	38	4	22	33	8	2	28
1	39	4	21	33	11	2	28
1	40	4	20	33	14	2	28
1	41	4	19	33	17	2	28
1	42	4	18	33	20	2	28
1	43	4	17	33	23	2	28
1	44	4	16	33	26	2	28
1	45	4	15	33	29	2	28
1	46	4	14	33	32	2	28
1	47	4	13	33	35	2	28
1	48	4	12	33	38	2	28
1	49	4	11	33	41	2	28
1	50	4	10	33	44	2	28
1	51	4	9	33	47	2	28
1	52	4	8	33	50	2	29
1	53	4	7	33	53	2	29
1	54	4	6	33	56	2	29
1	55	4	5	33	59	2	29
1	56	4	4	34	2	2	29
1	57	4	3	34	5	2	29
1	58	4	2	34	8	2	29
1	59	4	1	34	11	2	29
2	0	4	0	34	14	2	29

Common numbers S	G	S	G	Movements Moon m	s	Sun m	s
2	1	3	59	34	17	2	29
2	2	3	58	34	20	2	29
2	3	3	57	34	23	2	29
2	4	3	56	34	26	1	30
2	5	3	55	34	29	2	30
2	6	3	54	34	32	2	30
2	7	3	53	34	35	2	30
2	8	3	52	34	38	2	30
2	9	3	51	34	40	2	30
2	10	3	50	34	42	2	30
2	11	3	49	34	44	2	30
2	12	3	48	34	46	2	30
2	13	3	47	34	48	2	30
2	14	3	46	34	50	2	30
2	15	3	45	34	52	2	30
2	16	3	44	34	55	2	30
2	17	3	43	34	58	2	30
2	18	3	42	35	0	2	30
2	19	3	41	35	3	2	30
2	20	3	40	35	5	2	30
2	21	3	39	35	7	2	30
2	22	3	38	35	9	2	30
2	23	3	37	35	11	2	31
2	24	3	36	35	13	2	31
2	25	3	35	35	15	2	31
2	26	3	34	35	18	2	31
2	27	3	33	35	20	2	31
2	28	3	32	35	22	2	31
2	29	3	31	35	24	2	31
2	30	3	30	35	26	2	31
2	31	3	29	35	29	2	31
2	32	3	28	35	32	2	31
2	33	3	27	35	34	2	31
2	34	3	26	35	36	2	32
2	35	3	25	35	38	2	32
2	36	3	24	35	40	2	32
2	37	3	23	35	42	2	32
2	38	3	22	35	44	2	32
2	39	3	21	35	46	2	32
2	40	3	20	35	48	2	32
2	41	3	19	35	50	2	32
2	42	3	18	35	52	2	32
2	43	3	17	35	54	2	32
2	44	3	16	35	57	2	32
2	45	3	15	35	59	2	33
2	46	3	14	36	1	2	33
2	47	3	13	36	3	2	33
2	48	3	12	36	5	2	33
2	49	3	11	36	7	2	33
2	50	3	10	36	9	2	33
2	51	3	9	36	11	2	33
2	52	3	8	36	13	2	33
2	53	3	7	36	15	2	33
2	54	3	6	36	17	2	33
2	55	3	5	36	19	2	34
2	56	3	4	36	21	2	34
2	57	3	3	36	22	2	34
2	58	3	2	36	23	2	34
2	59	3	1	36	24	2	34
3	0	3	0	36	25	2	34

485

Table [Q] of the equation of days

G	Capricorn m	s	Aquarius m	s	Pisces m	s	Aries m	s	Taurus m	s	Gemini m	s
1	3	41	0	37	0	12	2	15	4	41	5	33
2	3	33	0	33	0	15	2	20	4	44	5	33
3	3	25	0	30	0	17	2	25	4	48	5	33
4	3	18	0	26	0	20	2	30	4	52	5	33
5	3	11	0	23	0	22	2	35	4	55	5	33
6	3	4	0	20	0	25	2	41	4	59	5	32
7	2	57	0	18	0	28	2	46	5	2	5	29
8	2	50	0	15	0	31	2	52	5	5	5	26
9	2	43	0	13	0	35	2	57	5	8	5	23
10	2	37	0	10	0	38	3	3	5	10	5	20
11	2	31	0	8	0	42	3	8	5	13	5	17
12	2	24	0	6	0	45	3	14	5	16	5	15
13	2	17	0	5	0	49	3	19	5	18	5	12
14	2	10	0	4	0	54	3	24	5	20	5	10
15	2	3	0	3	0	58	3	29	5	22	5	8
16	1	56	0	2	1	3	3	34	5	23	5	5
17	1	49	0	1	1	7	3	39	5	25	5	3
18	1	43	0	0	1	12	3	44	5	27	5	0
19	1	37	0	0	1	16	3	49	5	28	4	56
20	1	31	0	1	1	21	3	54	5	29	4	53
21	1	25	0	1	1	25	4	0	5	30	4	49
22	1	19	0	1	1	30	4	5	5	31	4	46
23	1	14	0	2	1	34	4	11	5	31	4	43
24	1	9	0	2	1	39	4	16	5	32	4	40
25	1	4	0	3	1	44	4	20	5	32	4	36
26	0	59	0	4	1	49	4	23	5	32	4	33
27	0	54	0	6	1	54	4	27	5	32	4	30
28	0	49	0	7	2	0	4	31	5	33	4	27
29	0	44	0	8	2	5	4	34	5	33	4	24
30	0	41	0	10	2	10	4	39	5	33	4	20

Table (Q) of the equation of days

G	Cancer m	s	Leo m	s	Virgo m	s	Libra m	s	Scorpio m	s	Sagittarius m	s
1	4	16	3	4	3	48	6	9	7	49	6	59
2	4	14	3	4	3	51	6	14	7	50	6	55
3	4	10	3	4	3	54	6	18	7	51	6	50
4	4	7	3	4	3	57	6	22	7	52	6	44
5	4	3	3	4	4	1	6	27	7	52	6	38
6	4	0	3	4	4	6	6	32	7	53	6	32
7	3	57	3	4	4	12	6	36	7	53	6	25
8	3	54	3	4	4	17	6	41	7	54	6	19
9	3	51	3	4	4	21	6	45	7	54	6	12
10	3	48	3	5	4	26	6	49	7	53	6	6
11	3	45	3	6	4	31	6	53	7	52	5	59
12	3	42	3	7	4	36	6	57	7	51	5	53
13	3	39	3	8	4	41	7	1	7	50	5	48
14	3	36	3	9	4	46	7	6	7	49	5	44
15	3	33	3	10	4	51	7	10	7	47	5	39
16	3	30	3	11	4	56	7	15	7	45	5	32
17	3	27	3	13	5	1	7	19	7	43	5	25
18	3	25	3	14	5	6	7	23	7	41	5	18
19	3	23	3	16	5	11	7	27	7	39	5	11
20	3	21	3	17	5	16	7	31	7	37	5	4
21	3	19	3	19	5	21	7	34	7	34	4	57
22	3	17	3	21	5	26	7	36	7	31	4	49
23	3	15	3	24	5	31	7	38	7	27	4	42
24	3	13	3	27	5	36	7	39	7	23	4	34
25	3	11	3	30	5	41	7	41	7	20	4	27
26	3	9	3	33	5	45	7	42	7	16	4	19
27	3	8	3	36	5	50	7	44	7	12	4	12
28	3	7	3	39	5	55	7	45	7	9	4	4
29	3	6	3	42	5	59	7	47	7	5	3	56
30	3	5	3	45	6	4	7	48	7	3	3	49

Table [R] of proportions.

Deg.	Prop. min.	Deg.	Prop. min.	Deg.	Prop. min.
0 2	0 2	1 2	14 52	2 2	45 0
0 4	0 6	1 4	15 45	2 4	46 0
0 6	0 12	1 6	16 41	2 6	47 7
0 8	0 20	1 8	17 38	2 8	47 46
0 10	0 30	1 10	18 36	2 10	48 57
0 12	0 42	1 12	19 36	2 12	49 30
0 14	0 57	1 14	20 36	2 14	50 19
0 16	1 15	1 16	21 36	2 16	51 6
0 18	1 34	1 18	22 36	2 18	51 50
0 20	1 55	1 20	23 36	2 20	52 32
0 22	2 18	1 22	24 36	2 22	53 11
0 24	2 42	1 24	25 36	2 24	53 48
0 26	3 5	1 26	26 38	2 26	54 24
0 28	3 25	1 28	27 40	2 28	54 59
0 30	3 54	1 30	28 32	2 30	55 34
0 32	4 21	1 32	29 44	2 32	56 8
0 34	4 50	1 34	30 46	2 34	56 42
0 36	5 21	1 36	31 48	2 36	57 15
0 38	5 57	1 38	32 50	2 38	57 43
0 40	6 34	1 40	33 52	2 40	58 8
0 42	7 13	1 42	34 54	2 42	58 31
0 44	7 52	1 44	35 56	2 44	58 50
0 46	8 32	1 46	36 58	2 46	59 7
0 48	9 15	1 48	38 0	2 48	59 21
0 50	10 0	1 50	39 0	2 50	59 33
0 52	10 46	1 52	40 0	2 52	59 43
0 54	11 33	1 54	41 0	2 54	59 51
0 56	12 21	1 56	42 0	2 56	59 56
0 58	13 10	1 58	43 0	2 58	58 58
1 0	14 0	2 0	44 0	3 0	60 0 m

Table [S] for lunar eclipses at the greatest distance in the epicycle.

Arguments of latitude				Digits	Minutes	
Northerly		Southerly			Casus	Mora
0 11 0	1 49 0	3 11 0	5 49 0	0 0	0 0	0 0
0 10 30	2 49 30	3 10 30	5 49 30	0 40	12 10	0 0
0 10 0	2 50 0	3 10 0	5 50 0	1 40	19 30	0 0
0 9 30	2 50 30	3 9 30	5 50 30	2 40	24 32	0 0
0 9 0	2 51 0	3 9 0	5 51 0	3 35	28 7	0 0
0 8 30	2 51 30	3 8 30	5 51 30	4 32	31 13	0 0
0 8 0	2 52 0	3 8 0	5 52 0	5 30	34 10	0 0
0 7 30	2 52 30	3 7 30	5 52 30	6 25	36 27	0 0
0 7 0	2 53 0	3 7 0	5 53 0	7 23	38 42	0 0
0 6 30	2 53 30	3 6 30	5 53 30	8 21	40 28	0 0
0 6 0	2 54 0	3 6 0	5 54 0	9 20	42 11	0 0
0 5 30	2 54 30	3 5 30	5 54 30	10 17	43 36	0 0
0 5 0	2 55 0	3 5 0	5 55 0	11 14	44 52	0 0
0 4 30	2 55 30	3 4 30	5 55 30	12 11	41 4	0 0
0 4 0	2 56 0	3 4 0	5 56 0	13 9	36 42	10 21
0 3 30	2 56 30	3 3 30	5 56 30	14 7	34 1	13 47
0 3 0	2 57 0	3 3 0	5 57 0	15 4	32 44	15 48
0 2 30	2 57 30	3 2 30	5 57 30	16 2	31 38	17 38
0 2 0	2 58 0	3 2 0	5 58 0	17 0	30 31	19 14
0 1 30	2 58 30	3 1 30	5 58 30	17 57	30 3	20 12
0 1 0	2 59 0	3 1 0	5 59 0	18 53	29 52	20 52
0 0 30	2 59 30	3 0 30	5 59 30	19 50	29 19	21 16
0 0 0	3 0 0	3 0 0	6 0 0	20 46	29 16	21 32

Table [T] for lunar eclipses at least distance in the epicycle.

Arguments of latitude				Digits	Minutes	
Northerly		Southerly			Casus	Mora
0 13 0	2 47 0	3 13 0	5 47 0	0 26	12 25	0 0
0 12 30	2 47 30	3 12 30	5 47 30	1 13	20 52	0 0
0 12 0	2 48 0	3 12 0	5 48 0	2 2	26 7	0 0
0 11 30	2 48 30	3 11 30	5 48 30	2 50	30 23	0 0
0 11 0	2 49 0	3 11 0	5 49 0	3 36	34 27	0 0
0 10 30	2 49 30	3 10 30	5 49 30	4 34	37 0	0 0
0 10 0	2 50 0	3 10 0	5 50 0	5 29	41 27	0 0
0 9 30	2 50 30	3 9 30	5 50 30	6 10	43 26	0 0
0 9 0	2 51 0	3 9 0	5 51 0	6 54	45 21	0 0
0 8 30	2 51 30	3 8 30	5 51 30	7 41	47 25	4 0
0 8 0	2 52 0	3 8 0	5 52 0	8 31	49 28	0 0
0 7 30	2 52 30	3 7 30	5 52 30	9 26	51 6	0 0
0 7 0	2 53 0	3 7 0	5 53 0	10 11	52 44	0 0
0 6 30	2 53 30	3 6 30	5 53 30	10 54	54 9	0 0
0 6 0	2 54 0	3 6 0	5 54 0	11 43	55 20	0 0
0 5 30	2 54 30	3 5 30	5 54 30	12 35	47 14	9 7
0 5 0	2 55 0	3 5 0	5 55 0	13 27	43 53	14 4
0 4 30	2 55 30	3 4 30	5 55 30	14 25	49 54	17 35
0 4 0	2 56 0	3 4 0	5 56 0	15 0	59 9	19 57
0 3 30	2 56 30	3 3 30	5 56 30	15 50	37 50	21 57
0 3 0	2 57 0	3 3 0	5 57 0	16 38	36 51	23 32
0 2 30	2 57 30	3 2 30	5 57 30	17 24	36 0	24 49
0 2 0	2 58 0	3 2 0	5 58 0	18 15	35 31	25 47
0 1 30	2 58 30	3 1 30	5 58 30	19 5	35 5	26 32
0 1 0	2 59 0	3 1 0	5 59 0	19 54	34 49	27 1
0 0 30	2 59 30	3 0 30	5 59 30	20 43	34 40	27 16
0 0 0	3 0 0	3 0 0	6 0 0	21 31	24 35	27 27

Astrology

Nicole Oresme (ca. 1325–1382)

Translated and annotated by G. W. Coopland[1]

Here begins Master Nicole Oresme's Book against Divination.

It is my aim, with God's help, to show in this little book, from experience, from human reason, and from authority, that it is foolish, wicked, and dangerous even in this life, to set one's mind to know or search out hidden matters or the hazards and fortunes of the future, whether by astrology, geomancy, nigromancy, or any other such arts, if they can correctly be called arts; and, further, that such things are most dangerous to those of high estate, such as princes and lords to whom appertains the government of the commonwealth.

Hence I have written this little book in French so that laymen may understand it for I have heard that many of them are overmuch given to such stupidities. At a former time I wrote in Latin on this matter and if any man wishes to attack what I shall say let him do it openly and with set argument, not in mere slander, and let him write against it and I shall reply so far as I am able, for in such wise we may arrive at truth. In any case what I say is submitted to the correction of those whom it concerns. And I beg that I may be excused for my rough manner of expression because I have never learned or been used to set forth or write anything in French.

Here follow the chapters of the book.

The first; of the arts by which men enquire into hidden and secret things.

The second; as to how much truth there is in the various branches of astrology.

The third; as to what truth there is in the other arts aforesaid.

The fourth; a reply to an objection.

The fifth; on the arguments in favour of princes applying themselves to these sciences.

The sixth; the arguments in favour of the possibility of knowing the future by these methods.

The seventh; the arguments to show that divination is a possible and profitable thing.

The eighth; true proof of the contrary from experience.

The ninth; on my proposition from authority.

The tenth; proof of my proposition by reason.

The eleventh; that there is no certainty in such arts.

The twelfth; on how men are deceived by such arts.

The thirteenth; on the fitting attitude of princes to such sciences.

The fourteenth; replies to the arguments of the fourth chapter.

The fifteenth; replies to the arguments of the fifth chapter.

The sixteenth; replies to the arguments of the sixth chapter.

The seventeenth; recapitulation and conclusion of the whole work.

Here ends the Prologue.

CHAPTER 1

There are many arts or sciences by means of which men are accustomed to enquire into the future or into things occult, secret, and hidden, or which can be applied to such uses. One of these is astrology which appears to me to have six principal parts. The first of these has to do chiefly with the movements, the signs, and the measurements of the heavenly bodies, so that by means of tables, constellations, eclipses and suchlike things in the future can be known.[2] The second is concerned with the qualities, the influences, and physical powers of the stars, with the signs of the zodiac, with degrees, with the heavenly signs, and so on; as, for instance, that a star in one quarter of the sky signifies or has power to cause heat or cold, dryness or moisture, and similarly with other physical effects. This part of astrology is introductory to the making of predictions. The third part deals with the revolu-

1. [Reprinted with the permission of the Harvard University Press from G. W. Coopland, *Nicole Oresme and the Astrologers: A Study of His Livre de Divinacions* (Cambridge, Mass.: Harvard University Press, 1952), pp. 51, 53, 55, 57, 61, 71, 75, 87, 89, 91, 93, 109, 111, 113. The present selection was translated from Oresme's *Livre de Divinacions*.

As will be seen, although Oresme was tolerant of certain features of astrology, he was utterly opposed to its determinism and wholly sceptical of its ability to predict future events. Indeed, in his *On Ratios of Ratios* and *Treatise on the Commensurability or Incommensurability of Celestial Motions*, he offered mathematical reasons for denying astrology's predictive claims (see Selection 69, Part III).—*Ed.*]

2. [Here we have astronomy proper.—*Ed.*]

tions of the stars and with the conjunctions of the planets, and is applied to three kinds of predictions; first, that we may know from the major conjunctions the great events of the world, as plagues, mortalities, famine, floods, great wars, the rise and fall of kingdoms, the appearance of prophets, new religions, and similar changes; next, that we may know the state of the atmosphere, the changes in the weather, from hot to cold, from dry to wet, winds, storms, and such movements in nature; third, that we may judge as to the humours of the body and as to taking medicine and so on.

The fourth part has to do with nativities, and especially with decisions as to a man's fortune, from the constellation which is in the heavens at the moment of his birth.

The fifth deals with interrogations, that is, decides and answers a question according to the constellation which is in the heavens at the time when the question is asked.

The sixth is of elections by which the time to start a journey or to undertake a task is ascertained and in this part is included the branch which teaches how to make images, carettes, rings, and such things.

The other sciences are geomancy, hydromancy, and similar devices, palmistry, experiments, and such auspices as sneezing, encounters, arguments drawn from the song or flight of birds, from the members of dead animals, magic art, nigromancy, interpretation of dreams, and many other vanities which are not sciences properly speaking.

CHAPTER 2

The first part of astrology is speculative and mathematical, a very noble and excellent science and set forth in the books very subtly, and this part can be adequately known but it cannot be known precisely and with punctual exactness, as I have shown in my treatise on the Measurement of the Movements of the Heavens and have proved by reason founded on mathematical demonstration.[3]

The second part is a part of natural science and is a great science and it too can be known so far as its nature is concerned but we know too little about it and in particular the rules in the books are false, as Averroes says, and have either slight proof or none. And some of them which were fulfilled in the place or at the time when they were laid down, are false in other places or at the present time: for the fixed stars which according to the ancients have great influence are not now in the position they

were in then and these same positions are used in making predictions.

And of the three subdivisions of the third part of astrology the first, which is concerned with the great events of the world, can be and is sufficiently well known but only in general terms. Especially we cannot know in what country, in what month, through what persons, or under what conditions, such things will happen, or the other particular circumstances. Secondly, as regards change in the weather, this part by its nature permits of knowledge being acquired therein but it is very difficult and is not now, nor has it ever been to any one who has studied it, more than worthless, for the rules of the second part are mostly false as I have said, and are assumed in this branch. And, similarly, the detailed rules bearing on this part are false, so that we see every day that sailors and husbandmen can prophesy changes in the weather better than the astronomers. In the third place, so far as medicine is concerned, we can know a certain amount as regards the effects which ensue from the course of the sun and moon but beyond this little or nothing. All this third part of astrology has to do chiefly with physical effects; the parts which follow, with the effects of fortune.

The fourth part, of nativities, is not in itself beyond knowledge, so far as the complexion and inclination of the person born at a given time are in question, but cannot be known when it comes to fortune and things which can be hindered by the human will, and this section has to do with those things rather than with physical effects. And one often sees in practice that two people are born at intervals of time so minute as cannot be recorded and yet their fortunes are quite different, so I say that this part of astrology cannot be known and the rules written down on it are not true.

The fifth part, of interrogations, and the sixth, of elections, have no reasonable foundation and there is no truth in them and Averroes says in his commentary on the XIIth book of Metaphysics that the part about images sprang from the corruption of philosophy and the fables of the pagans; and any one who has read the works of Hyginus[4] and Aratus,[5] who deal with this matter, can un-

3. [At least three candidates for this reference come to mind: his *Treatise on the Commensurability or Incommensurability of the Celestial Motions*, Part III (Selection 69); *On Ratios of Ratios*, where the mathematical theory is formulated in favor of the probable incommensurability of celestial motions (see Selection 51.2 and notes 80 and 84 thereto); and the *Ad pauca respicientes* (see Selection 69).—*Ed.*]

derstand that such images can have no effect unless it be by magic art or by nigromancy.

CHAPTER 4

Some may say that the aforesaid sciences, and especially astrology, are said to be of very ancient authority and are set forth by prophets, by reasonable authors, and in authentic books, and yet I attack some portions and some rules of these sciences and the arts that have been named without giving reasons, like a man who hates them and yet knows nothing about them. In reply I say, first, that it would be too long a business to demonstrate in detail the falsity and weakness of the principles on which the sciences that I call worthless are based. Again, it is not easy to deal with the matter in French or in such a way as laymen could easily understand. Further, I am prepared, if necessary, to defend what I shall say by set argument in Latin against those who wish to maintain the contrary, in a fitting way. And it is a matter of every-day experience that such divinations fail to speak truth. And, indeed, the real reply is that my main aim in this work is not to prove the falsity of such pursuits but, leaving on one side that question, to demonstrate that it is foolish and perilous to make use of such arts so far as enquiring into particular fortunes is concerned. And to prove this does not require that one should know the said arts perfectly, just as a man who condemns tables or dice need not be a master of those games.

CHAPTER 8

Notwithstanding all the above arguments, I wish to demonstrate with certainty my main proposition; and first by experience and by induction from past history. And I say that princes and others who have set themselves to such matters have come to a bad end and have had evil fortune in this world. And consider Zoroaster, King of the Bactrians, who was such a great astronomer, and who was the first to invent the arts of magic; he was slain by Ninus who deprived him of his life and of his kingdom. . . . [6] Hence it is clear that the conclusion may be drawn from this great induction, from so many experiences, observations, instances, that divination was never at any time propitious.

CHAPTER 9

Now I will prove my proposition by argument from authority. First, Moses in the 18th chapter of Deuteronomy teaches and commands the people of Israel, when they come into the promised land, to beware lest they follow the abominations of the poeple of that land and forbids them to have among them any manner of diviners, or to put questions to them, for our Lord abominates such things, and for such manner of iniquity would take away their land and would give it to the people of Israel. For he had said previously, in the 9th chapter, that they should not possess that land because of their righteousness, but that the inhabitants would be destroyed for their iniquities. Surely, any man who considers this text of the Scriptures, so ancient, so authentic, so approved, received, and commended, by the whole world, whether Christian, Jew, or Saracen, must be in great fear and horror of listening to such divinations; for the Scripture gives no reason but this for the destruction of those peoples and for the ruin of those kingdoms. . . . [7]

CHAPTER 11

Although sometimes a person may, by nature, be able to see what is absent or to come, in dreams, or sickness, or when he has been put out of his senses by magic art, as I have shown elsewhere, yet such visions are often false or occult, without certainty and dangerous of acceptance, and there is still less certainty in the other arts aforesaid. For, in the case of astronomy, which has the most appearance of truth, a philosopher named Baren says that a diligent interpreter or expositor of dreams will investigate and will find the truth more easily than an astronomer.[8] "For," says he, "the institutes of astronomy are too defective," and he goes on to say that he saw in his own day many who knew astrology more than excellently yet in their predictions they were variable and discordant. Also Avicenna, in his Metaphysics, expresses the opinion that the disposition of the heavens and of the stars cannot be completely known, and even supposing we did know it we could know nothing

4. See *C. Julii Augusti Liberti Fabularum ejusdem Astronomicon libri quattuor,* etc., Basle, 1535, for illustrations of this.

5. Aratus (c. 270 B.C.) the author of the *Phoenomena et Prognostica.* Oresme is probably indebted to Haly's *Commentary* [that is, Haly Eben Rodan's Commentary on Ptolemy's *Tetrabiblos* or *Quadripartitum,* as it was known in Latin—*Ed.*] for the reference.

6. [Many similar "historical" instances are cited.— *Ed.*]

7. [References to other "authorities" fill out the remainder of the chapter.—*Ed.*]

8. Long search and enquiry in many quarters have revealed no clue to this reference.

of absent things.[9] And he says further that many of the rules of astrology are based on poetry and on rhetoric, that is to say on fables and imaginings that cannot be accepted in natural science. Hence he says that we must not believe the predictions of the astrologers although of our good pleasure we may allow their writings to be true, out of respect for their science. Again, according to Ptolemy in the Quadripartitum, those who live towards the south are more apt for the knowledge of astrology than those who live towards the north,[10] and to this Haly adds that those who live towards the east are similarly more fitted for it than those who live further west. And this is true, and more especially so in the case of judicial astrology. And, according to the opinion expressed by Ptolemy in his Centiloquium[11] and also according to human reason, those men can profit little in astrology who are not inclined thereto by nature; hence it follows that Frenchmen, and still more Englishmen and those further away yet, who live near the confines of the habitable north or of the west, can commonly have little profit from judicial astrology. And, in particular, as I indicated briefly in the second chapter, the part of astrology which deals with the shapes of the constellations is nearly all founded on the stories of the poets who pretended that good princes and certain others were changed into stars. And just as old women to-day speak of St. Martin's Staff,[12] or the True Cross, or St. James' Road,[13] so the poets gave to the stars the names of persons or things that in ancient times were held in common memory or reverence. Thus Andromeda, Perseus and Hercules were named and the astronomers accepted those names because it was convenient to have separate names to distinguish the stars, and every star or constellation did not possess a shape or outline which suggested a name like the Crown, the Wain and others. This one may know from the old histories and books which deal with the origin of these names, as, for instance, Aratus, Ovid,[14] Fulgeratus,[15] and many others. And certain of the rules of judicial astrology are based on the names of the heavenly shapes, and from this it is clear that there is no certainty in these matters. And St. Ambrose stresses this and relates how they say that if a man is born under the sign of the Ram he will be generous, because the sheep parts willingly with its fleece, and if he is born under the sign of the Bull he will be a tiller of the ground, and such-like nonsense.[16] Also we have books of "experiments of images", such as that of Olbriza-balis,[17] which, similarly, are based on the manners

or deeds or doings of the personages after whom these shapes are named, just as if one were to construct an "image" to cure fevers when St. Martin's Staff was in the ascendant. And some wish to connect the images in geomancy with these shapes, but without justification, for they have no resemblance to them, either in number of stars, or outline or shapes of the figures. And, even if they were alike, it would not be possible to draw any conclusion as to their predictions so we can see plainly that the whole thing is worthless. But it would be waste of time to delay to attack it, for the man who is willing to listen to such things is lacking in

9. See *Liber Avicennae de philosophia prima sive scientia divina*, ed. 1492 (Bodleian Lib.), especially the first chapter of the 10th tractatus, f. 108 recto, col. 2 [the same folio reference applies to a 1508 edition of Avicenna's works reprinted in Frankfurt am Main by Minerva G. m. b. H. in 1961—*Ed.*]. For Avicenna's views on astrology see also A. F. Mehren, *Vues d'Avicenne sur l'Astrologie et sur le rapport de la responsabilité humaine avec le destin*, in *Homenaje a D. Francisco Codara*, Zaragoza, 1904, pp. 235–250. The article is based on examination of Avicenna's *Réfutations des Astrologues* and *Traité sur le destin* as existing in Arabic MSS. at Leyden. Avicenna compared judicial astrology with the quack side of medicine, and criticised the fact that the astrologers confined influence to the planets and neglected the millions of stars of the Milky Way; he dealt, too, with many of the traditional dilemmas and difficulties.

10. Ptolemy, *Quadripartitum*, II, cap. 2.

11. ["The astrological *Centiloquium* or *Karpos*, and other treatises on divination and astrological images ascribed to Ptolemy in medieval Latin manuscripts are probably spurious, but there is no doubt of his belief in astrology" (L. Thorndike, *A History of Magic and Experimental Science*, [New York: Macmillan, 1929], I, 111). The *Tetrabiblos*, or *Quadripartitum*, is believed to be Ptolemy's.—*Ed.*]

12. This is no longer well known in France, though it may still be used in certain districts.

13. The Milky Way is still so called in France. In England it was once called Watling Street and Walsingham Way; see on this Skeat's edition of Chaucer (1894), the notes to the *Hous of Fame*, II, 1.939.

14. e.g. *Fasti*, II and V.

15. [Coopland mentions a *Fulgentius*, cited by Pierre d'Ailly (in his *Principium in Cursum Bibliae*) as the author of *Mythologies*; and also cites a work *Fabii Planciadis Fulgentii Mythologicon*, Amsterdam, 1681.—*Ed.*]

16. [St. Ambrose, *Hexameron*, Bk. IV.—*Ed.*] The work of St. Ambrose here used by Oresme is one of the fullest and most interesting of orthodox condemnations of divination.

17. [Identified by Coopland as Albohaly (Yahyā ibn Ghālib, Abu Ali Al-Khaiyat). Thorndike (*History of Magic*, II, 75) mentions that in 1153, John of Spain, or John of Toledo, translated the *Nativities* of Albohali.—*Ed.*]

reason, so that it would be trouble wasted, and, if he listens to reason, he need only consider the matter for himself and he will soon find that so far as predictions are concerned, they are mere foolishness without base or reason. Further, it appears to me to be ridiculous to believe that an astrologer or geomancer or other of the sort can make sure predictions as to things in the future, which are fortuitous and under control of the changeable human will, when he can't say what sort of a day it will be to-morrow or prophesy a change in the weather or in the wind. Yet these latter things happen independently of fortune and result from the influence of the heavens and cannot be impeded except by divine miracle. Also, supposing a fortuitious result were known to the man to whom it is to happen, the which event depends on free will, I pose this question. Either it can be prevented or it cannot. If it cannot then it follows that all things happen of necessity and it is not worth while to ascertain them beforehand. And if it can be prevented then it cannot be said to be known. So that it seems to be impossible to know the future, precisely and certainly, by any human art. And that is what the Sage says in Ecclesiastes, namely, that man has many afflictions and cannot know by any message what is to come.[18] And, on this, it is written in a letter of King Alexander that when he asked his fate of the sacred trees the reply was that he would die poisoned. And when he asked at whose hand, he was told that he could not know this, for if he did he might impede the decrees of the fates.[19] And to any man who studies attentively the books of predictions it is plain, whatever they may say to the contrary, that the writers believe everything happens of necessity. And Julius Fermacus in his book, which includes almost the whole of judicial astrology, states this explicitly,[20] but it is false and contrary to the Faith and is condemned and attacked by every law. And Tully wrote a book on Divinations in which he gives plenty of stories and examples to illustrate the fact that there is no certainty in these matters;[21] and following his example I have been pleased to name this treatise, "On Divination". Further, besides the arts I named in the first chapter, there are others which are more or less promising and yet entirely uncertain, such as Saturn's head,[22] the notary art, the contrenotary art, pyromancy, spatulomancy, the art of sparks, divination by means of metals, wax, bread, sneezes, thunder, and similar sorceries. But I won't stop to reprove them for it is not my main purpose. At the

same time, I do not say that some things that are absent, or still to come, cannot be known by prophecy, or by divine revelation, or by the use of human reason, and without the aid of the sciences in question; but such visions come to men of sober and peaceful life, whose souls are like clear and shining mirrors, clean from all worldly thoughts. So that Rabbi Moses the Egyptian says that concupiscence and melancholy prevent prophecy[23] and Isaiah speaks to the same effect.[24] And sometimes, wars, plagues, and death can be conjectured from the stars, from comets, from monstrosities, from extraordinary appearances in the atmosphere, as we see in the second book of Machabees,[25] and in the history of the downfall of Jerusalem, and in many other places, especially in the history of Rome, but these conjectures can be only general and without certainty. With regard to that, Seneca says that a comet appeared in the reign of Octavian which was followed by good and not evil and he says this removed the ill repute of comets, for previously all the authorities attributed an evil significance to them.[26]

CHAPTER 15.

The aim of the sixth chapter was to show that it is possible to know the future by means of art. My reply is that it is possible only by the means set

18. *Ecclesiastes* VIII, 6, 7.
19. [Coopland's reference is to a work by Valerius on the deeds of Alexander the Great of Macedon (*Valerii Alexandri Polemi Res Gestae Alexandri Macedonis*, p. 190; no date or place of publication is supplied).—*Ed.*]
20. [That is, Julius Firmicus.—*Ed.*] See *Julii Firmici Materni Matheseos Libri VIII*, ed. Kroll and Skutch, 1897, 1913. . . p. 17.
21. *De Divinatione*, II, generally.
22. I have found no mention of this in any authority; the only famous head in this connection seems to be that credited to Roger Bacon.
23. Moses Maimonides, called by the Latins Rabbi Moyses of Cordova. [A selection from Maimonides' *Guide of the Perplexed* is printed in Selection 68.1.—*Ed.*] I have read only his *Director Dubitantium*, and that only as translated by Friendlander, New York, 1904. . .
In the *Director Dubitantium* there are many passages justifying Oresme's reference, e. g. p. 228 (*ed. cit.*), "You will find therefore that prophets are deprived of the faculty of prophesying when they mourn or are angry, or are similarly affected. Inspiration does not come upon a man when he is sad or languid and prevalence of sadness and dullness was undoubtedly the direct cause of the interruption of prophecy during the exile."
24. Possibly, though the passage is not conspicuous.
25. II Mach. III, 24 *et seq.*; X, 29, 30, *etc.*
26. *Nat. Quaes*, VII, Sect. 17. [Coopland observes that Oresme expands Seneca's passage.—*Ed.*]

forth in the eleventh chapter. Now, as regards the argument drawn from the dumb animals who have this foreknowledge, I say that Pliny, in the seventh book of his Natural History,[27] describes the miseries of human nature in comparison with, and in regard to, the other animals, and says that man is born dumb, helpless, impotent, able to do nothing without teaching except weep, and that this is not true of the other animals; does it not ensue, then, that if nature has given a gift to another animal that she has given it to man? But God and nature have given to man something better, namely, reason and memory, by use of which he can acquire human prudence by consideration of things past, present, and possible in the future, according to reasonable conjecture, by good counsel, and by legitimate industry, better than by listening to such divinations; and this is Seneca's opinion, as given in chapter ten. The second argument on human prudence may be met similarly by what I say here. And in this matter of provision for what may happen in the future, great assistance may be obtained from consideration of the history of times past. This is the view of Sallust,[28] and lords who have the task of government should sometimes read such histories, or hear them read, instead of the impossible fables to be found in many romances. And because the most reliable histories that we possess are in Latin, it is very desirable that princes should know Latin. As to a further argument, namely, that Ptolemy laid down that some men have a natural tendency towards power to foretell the future and can assist that power by art, my reply is that, if the tendency is towards prophesying changes in the atmosphere or weather, it may be assisted by astronomy or still better by observation of changes in the atmosphere, as Ptolemy says, and he calls such changes the secondary stars.[29] But, if the tendency is towards prophesying by the help of visions, one can help this by leading a sober and peaceful life, as I said in the eleventh chapter, and in no other way unless it be by magic art or divine miracle, and, even here, not in regard to fortunes but in the respects treated of in the eleventh chapter. As for the remaining arguments, there are two points and not more that call for an answer. The first is Bernard's when he says that the sky is a book wherein are written the fortunes of kings. I reply that what is to happen in the future is not written in the sky, except in so far as from congruent movements we may know future constellations which are, or will be, cause or sign of various inclinations and diverse fortunes, saving

always the freedom of the human will. The second point is that astrology would be worthless and God would have displayed this book, the sky, to us, without significance, if we cannot derive such knowledge from it. My reply is that this is not so, and that astrology has three very noble ends. The first is to have knowledge of such great matters, for to this, according to the philosophers, is human nature naturally inclined. And, as Aristotle says in a work that we no longer possess,[30] if it is a delectable and gracious thing to hear or know the ordinance and description of a noble royal palace, or a beautiful mountain, how much more excellent is it and more desirable to have knowledge of the form of the heavens, which exceeds and surpasses in beauty all things visible; and Tully says this in his book on the Nature of the Gods.[31] And Aristotle on this point says in the book of the Heavens and the Earth that it is better to know imperfectly a great matter than much more completely a matter of less moment, for, according to the philosophers, the natural perfection of the human understanding is attained in the study of such things. The second end, and the chief, is that it gives great aid in the knowledge of God the Creator. And Isidore says that it was ordained last of secular sciences so that by means of it the thought of man should be withdrawn from earthly, though praiseworthy, things to contemplation of the divine,[32] on which Tully says, in the work named above, that the argument that had most weight in bringing ancient philosophers to a knowledge of God was the marvellous quantity and most ordered movement of the heavens and the stars.[33] And the Scriptures speak similarly in the Book of Wisdom[34] and the Apostle in the Epistle to the Romans;[35] and Saint Denis, in an Epistle that he wrote, confesses that what he knew of astrology was the occasion of his being

27. The famous passage which introduces the book.
28. *De Bello Jugurthino* IV; the sentiment is there expressed only by implication, as also in the opening of the *De conjuratione Catilinae*.
29. [Apparently drawn from Haly's commentary on No. XIII of the *Centiloquium* attributed to Ptolemy. See n. 11.—*Ed.*]
30. [Perhaps a reference to some pseudo-Aristotelian treatise which was lost, or unavailable, in Oresme's day.—*Ed.*]
31. [*De natura deorum*, Bk. II, 40, 104.—*Ed.*]
32. Isidore, *Etymologies*, III, 71.
33. The thought is not expressed in this form in Cicero, but *De natura deorum*, II, 61, has the substance of it. . . .
34. XIII, 2, 4, 5.
35. I. 20.

converted to the faith of Jesus Christ.[36] The third end of astrology and the least important is to ascertain certain dispositions of this lower and corruptible nature, whether present or to come, and nothing beyond that, as I have laid down in several passages and especially in the second chapter. I say then that astrology is not worthless and it is not without purpose that God and Nature have displayed the heavens to us.

Cosmology

67 ON THE POSSIBLE DIURNAL ROTATION OF THE EARTH

Introduction by Edward Grant

Prior to Copernicus very few had accepted the daily rotation of the earth as a part of their astronomy and cosmology. The most notable proponents of its rotation were the Greeks Heraclides of Pontus and Aristarchus of Samos, whose influence in subsequent centuries was extremely minimal, as the physical and cosmological opinions of Aristotle and Ptolemy dominated Greek, Arabic, and medieval thought. But reports by Ptolemy, Simplicius, and even Aristotle (whose report on Plato seemed to classify Plato as an advocate of the earth's motion) preserved knowledge of these earlier views about the earth's rotation and gave rise in the Middle Ages to important discussions by John Buridan and Nicole Oresme, who, after formulating arguments in behalf of the earth's diurnal rotation, accepted the traditional position for quite different reasons. Although it has never been determined whether either one influenced Copernicus, it will be evident that a number of their arguments in favor of the earth's rotation appear in strikingly similar fashion in Book I of the *De revolutionibus*.

1. Ptolemy (fl. 127–151): The Immobility of the Earth in the Center of the World

Translated by R. Catesby Taliaferro[1]

Annotated by Edward Grant

7. THAT THE EARTH DOES NOT IN ANY WAY MOVE LOCALLY

By the same arguments as the preceding it can be shown that the earth can neither move in any one of the aforesaid oblique directions, nor ever change at all from its place at the centre. For the same things would result as if it had another position than at the centre. And so it also seems to me superfluous to look for the causes of the motion to the centre when it is once for all clear from the very appearances that the earth is in the middle of the world and all weights move towards it. And the easiest and only way to understand this is to see that, once the earth has been proved spherical considered as a whole and in the middle of the universe as we have said, then the tendencies and movements of heavy bodies (I mean their proper movements) are everywhere and always at right angles to the tangent plane drawn through the falling body's point of contact with the earth's surface. For because of this it is clear that, if they were not stopped by the earth's surface, they too would go all the way to the centre itself, since the straight line drawn to the centre of a sphere is always perpendicular to the plane tangent to the sphere's surface at the intersection of that line.

All those who think it paradoxical that so great a weight as the earth should not waver or move anywhere seem to me to go astray by making their judgment with an eye to their own affects and not to the property of the whole. For it would not still appear so extraordinary to them, I believe, if they stopped to think that the earth's magnitude compared to the whole body surrounding it is in the ratio of a point to it. For thus it seems possible for that which is relatively least to be supported and pressed against from all sides equally and at the same angle by that which is absolutely greatest and homogeneous. For there is no "above" and

36. See *Acta Sanctorum* [*Acts of the Saints—Ed.*], IV, Oct. IX, cols. 774–5, for the wonders of the heavens which influenced St. Denis.

1. Reprinted with the permission of the *Encyclopaedia Britannica* from Ptolemy's *Almagest*, Book I, chapter 7, as translated by R. Catesby Taliaferro in *Great Books of the Western World* (Chicago: Encyclopaedia Britannica, 1952), XVI, 10–12.

"below" in the universe with respect to the earth, just as none could be conceived of in a sphere. And of the compound bodies in the universe, to the extent of their proper and natural motion, the light and subtle ones are scattered in flames to the outside and to the circumference, and they seem to rush in the upward direction relative to each one because we too call "up" from above our heads to the enveloping surface of the universe; but the heavy and coarse bodies move to the middle and centre and they seem to fall downwards because again we all call "down" the direction from our feet to the earth's centre.[2] And they properly subside about the middle under the everywhere-equal and like resistance and impact against each other. Therefore the solid body of the earth is reasonably considered as being the largest relative to those moving against it and as remaining unmoved in any direction by the force of the very small weights, and as it were absorbing their fall. And if it had some one common movement, the same as that of the other weights, it would clearly leave them all behind because of its much greater magnitude. And the animals and other weights would be left hanging in the air, and the earth would very quickly fall out of the heavens. Merely to conceive such things makes them appear ridiculous.

Now some people, although they have nothing to oppose to these arguments, agree on something, as they think, more plausible. And it seems to them there is nothing against their supposing, for instance, the heavens immobile and the earth as turning on the same axis from west to east very nearly one revolution a day,[3] or that they both should move to some extent, but only on the same axis as we said, and conformably to the overtaking of the one by the other.

But it has escaped their notice that, indeed, as far as the appearances of the stars are concerned, nothing would perhaps keep things from being in accordance with this simpler conjecture, but that in the light of what happens around us in the air such a notion would seem altogether absurd.[4] For in order for us to grant them what is unnatural in itself, that the lightest and subtlest bodies either do not move at all or no differently from those of contrary nature, while those less light and less subtle bodies in the air are clearly more rapid than all the more terrestrial ones; and to grant that the heaviest and most compact bodies have their proper swift and regular motion, while again these terrestrial bodies are certainly at times not

easily moved by anything else—for us to grant these things, they would have to admit that the earth's turning is the swiftest of absolutely all the movements about it because of its making so great a revolution in a short time, so that all those things that were not at rest on the earth would seem to have a movement contrary to it, and never would a cloud be seen to move toward the east nor anything else that flew or was thrown into the air. For the earth would always outstrip them in its eastward motion, so that all other bodies would seem to be left behind and to move towards the west.

For if they should say that the air is also carried around with the earth in the same direction and at the same speed, none the less the bodies contained in it would always seem to be outstripped by the movement of both.[5] Or if they should be carried around as if one with the air, neither the one nor the other would appear as outstripping, or being outstripped by, the other. But these bodies would always remain in the same relative position and there would be no movement or change either in

2. Unlike Aristotle, Ptolemy seems here to deny an absolute "up" and "down" with respect to the earth. These directions are only relative.

3. This interpretation was held by Heraclides of Pontus and Aristarchus of Samos (see sections 3 and 5 of this selection and n. 22). By reporting this important minority opinion, Ptolemy joins Simplicius *(Commentary on Aristotle's De caelo)* as one of three primary sources introducing this conception to medieval authors (the third is Aristotle's *De caelo,* where Aristotle seems to attribute to Plato a belief in the axial rotation of the earth at II.13.293b.30–32 and to mention it again, rather vaguely, in II.14.296.b.1–4). Aquinas probably derived it from Simplicius' commentary in the translation made available by William Moerbeke (see number 47 of Moerbeke's list in Selection 8); Oresme probably knew all three sources (see section 5 of this selection).

4. It is of interest to note that Ptolemy concedes that the earth's diurnal rotation is compatible with the purely astronomical phenomena, while holding it absurd in terms of physics.

5. Buridan was unquestionably familiar with this argument, for he reports that adherents of diurnal rotation say that "the earth, water, and the air in the lower region are moved simultaneously with diurnal motion (see pars. 7 and 8 of section 4 of this selection). While Buridan appears to accept this, he rejects diurnal rotation by arguing that bodies moved violently ought to resist the motion of the air as it accompanies a moving earth. Experience shows that they do not (see par. 9). Copernicus explicitly reports and rejects Ptolemy's opinion on the disruptive effects of a terrestrial motion (see Bk. I, ch. 7 and ch. 8, of Copernicus' *De revolutionibus,* in section 6 of this selection).

the case of flying bodies or projectiles. And yet we shall clearly see all such things taking place as

if their slowness or swiftness did not follow at all from the earth's movement.[6]

2. Thomas Aquinas (ca. 1225–1274): The Immobility of the Earth in the Center of the World

Translated and annotated by Edward Grant[7]

1. Inasmuch as the Philosopher [that is, Aristotle] has described the opinions of others concerning the earth, he here considers it according to the truth [of the matter]. First he solves the problem of the earth's place and rest; and secondly, where he says "Its shape must necessarily be spherical"[8] (297a.9), he considers its shape. He makes two points concerning the first part: First he determines the truth by means of natural reasons; and secondly, where he says "This view is further supported. . . " (297a.3), he resorts to astronomical appearances. With regard to the first of these [two parts], he makes two points: First he shows that it is impossible for the earth to be moved; secondly, where he says ["From what we have said the explanation of the earth's immobility] is also apparent" (296b.28), he gives the true explanation for the earth's immobility from premises.

Concerning the first part,[9] he makes three points. First he directs attention to what must be discussed first, namely whether the earth moves or rests. If it moves, then we must move on to other things that have to be considered about the earth. And thus he posits this first[10] in order to assume it as a beginning for what is to follow.

Secondly, where he says "For, as we said, [there are some who make it one of the stars]. . . " (296a.25), he shows the necessity for the aforesaid inquiry.[11] For, as stated above,[12] some, namely the Pythagoreans, assumed that the earth was moved around the center of the world—as if it were one of the planets;[13] others, indeed, assuming that the earth is in the center, as it is written in the *Timaeus,* say that it is rotated around the middle of the pole—that is, around the axis dividing the sky down the middle.[14]

2. Thirdly, where he says "That both views are

them this way or that—for a wind in the air is as a current in the sea" (Bk. I, ch. 8, in section 6).

7. Translated from Aquinas' *Commentaria in libros Aristotelis De caelo et mundo,* Lib. II, cap. XIV, Lectio 26, in *Sancti Thomae Aquinatis Opera Omnia iussu impensaque Leonis XIII* (Rome, 1886), Vol. III, p. 218, col. 1, to p. 221, col. 2. Wherever Thomas quotes the words of Aristotle, I have consulted the Oxford English translation, *The Works of Aristotle Translated into English under the Editorship of W. D. Ross,* Vol, II (1930; reprinted 1953).

8. Thomas comments upon this in Lectio 27.

9. That is, Aristotle's intention to determine the truth by natural reasons.

10. The motion of the earth.

11. The inquiry into the motion of the earth is necessary, because those who assume that the earth moves disagree as to the kind of motion it has.

12. Lectio 20, note 3.

13. Since that aspect of Aristarchus' view which assumed the earth's revolution around the sun was apparently unknown in the Middle Ages, Aristotle's account (293a.20—293b.6) of the Pythagorean view (to which Thomas alludes here)—that the earth is like a planet moving round a central fire—was the only other opinion available in which the earth was given an orbital motion in contradistinction to an axial rotation. This Pythagorean view was mentioned by Copernicus (see ch. 5 of section 6 of this selection).

14. It has been argued whether Aristotle meant to attribute to Plato (in the *Timaeus*) the view that the earth had an axial rotation. This was discussed earlier in Lectio 21, par. 5 (p. 204, col. 2, to p. 205, col. 2), where Thomas comments upon Book II, chapter 13 (293b.30–32). Thomas is quite explicit that Plato believed in the absolute immobility of the earth at the center of the universe and that Aristotle did not mean to attribute such a view to Plato. All Aristotle meant to say, according to Thomas, is that in the *Timaeus* it is assumed that the earth is packed, so to speak, around the axis of the world. Numerous scholars, ignorant of Thomas' opinions, adopted the same interpretation. It was J. L. E. Dreyer who first claimed that "this simple explanation [that is, that the expression "as it is written in the *Timaeus*" in 293b.32 refers only to the earth lying in the center about the axis of the universe and not to any motion of the earth—*Ed.*] was brought forward already six hundred years ago by good old Thomas Aquinas, a fact which neither Zeller nor Martin nor any other modern writer seems to have noticed" (*A History of Astronomy from Thales to Kepler,* 2d. ed. [New York: Dover, 1953] p. 78). Sir Thomas Heath, it should be noted, rejects Thomas' interpretation (*Aristarchus of Samos* [Oxford: Clarendon Press, 1913], p. 177).

Rather than Plato, Thomas believed that Aristotle had in mind Heraclides of Pontus ("Moreover we can say briefly that a certain Heraclitus Ponticus—whose

6. This last argument, that there would be no detectable motion if the air and everything in it rotated with the earth, seems to have been answered indirectly by Oresme and Copernicus. Thus Oresme says (in section 5) that "according to this opinion, not only the earth moves, but also with it the water, and the air, as we stated above, although the water and air here below may be moved in addition by the winds or other forces." Perhaps influenced by Oresme, Copernicus conjectures that despite the rotation of the earth, "the air nearest Earth, with the objects suspended in it, will be stationary unless disturbed by wind or other impulse which moves

untenable. . . ." (296a. 27), he shows by means of four reasons that it is impossible for the earth to be moved in this way. In the first of these [reasons] he takes it as a starting point that if the earth were moved circularly—whether in the center of the world or outside of it—such a motion would be necessarily violent. Now, it is obvious that circular motion is not a proper and natural motion for the earth; for if it were, every particle of earth should also have this motion, since, as stated above, a natural motion is the same for whole and part. But we see that this is false, because all parts of the earth are moved with a rectilinear motion toward the center of the world. If, however, the motion of the earth were circular, it would be violent and contrary to nature, and [therefore] could not be eternal, since, as was said earlier, nothing violent is eternal. But if the earth were moved circularly it would be necessary that such a motion be eternal—assuming, [of course,] that the world is eternal. In his [that is, Aristotle's] opinion this is so, because it is necessary that the universal order be eternal, and the motion or rest of the principal parts of the world [that is, heavens and earth] are part of this order. It follows, therefore, that the earth is not moved with a circular motion.

3. Where he says "Again, everything [that moves with the circular movement] . . ." (296a.34), he posits the second reason, which is this. All bodies that are moved circularly seem to drop back, that is, do not maintain a constant position, because every one of them—except the first sphere, which according to Aristotle is the sphere of the fixed stars—is moved with several motions and not by one only. Therefore, if the earth—whether it is in the center or outside the center—has a circular motion, it would necessarily be moved with several motions, namely with the motion of the first sphere around the equinoctial poles [that is, the poles of the equator], and with some other proper motion around the poles of the zodiac. But this cannot be, for if it were so, the fixed stars would exhibit changes and turnings with respect to the earth, which, in virtue of its proper motion, would lag behind and not return to the same point together with a fixed star—just as happens with the planets. This applies either to the whole earth or any part that may be designated. And so it would follow that the fixed stars would not always be seen to rise and set at the same part of the earth. But this does not happen, because the fixed stars rise and set at the same designated places. Therefore, the earth is not moved circularly.[15]

4. Where he says "Further, the [natural] movement [of the earth] . . ." (296b. 7), he asserts the opinion Aristotle presents here—assumed that the earth was moved in the center [of the world] and that the sky rested." Lectio 21, par. 5; p. 205, col. 2.). Heath (*Aristarchus of Samos*, p. 240) believes it implausible that Aristotle was representing the view of Heraclides, who assumed a daily axial rotation:

"It seems likely, as Dreyer suggests, that in speaking of a motion of the earth 'at the centre itself,' Aristotle is not thinking of a rotation of the earth in *twenty-four hours*, i.e. a rotation replacing the apparent revolution of the fixed stars, as Heraclides assumed that it did; for he does not mention the latter feature or give any arguments against it; on the contrary, he only deals with the general notion of a rotation of the earth, and moreover mixes up his arguments against this with his arguments against a translation of the earth in space. He uses against both hypotheses his fixed principle that parts of the earth, and therefore the earth itself, move naturally towards the centre. Whether, he says, the earth moves away from the centre or *at* the centre, such movement could only be given to it by *force;* it could not be a natural movement on the part of the earth because, if it were, the same movement would also be natural to all its parts, whereas we see them all tend to move in straight lines towards the centre; the assumed movements, therefore, being due to force and against nature, could not be everlasting, as the structure of the universe requires."

15. This unclear and quite confused passage must be examined against the background of the specific Aristotelian passage (296a.34—296b.6) which it purports to explain. Thomas Heath translates it as follows (pp. 240–241):

"Further, all things which move in a circle, except the first (outermost) sphere, appear to be left behind and to have more than one movement; hence the earth, too, whether it moves about the centre or in its position at the centre, must have two movements. Now, if this occurred, it would follow that the fixed stars would exhibit passings and turnings. This, however, does not appear to be the case, but the same stars always rise and set at the same places on the earth."

Heath supplies an important commentary to this passage (p. 241), which is worthy of full citation:

"The bodies which appear to be 'left behind and to have more movements than one' are of course the planets. The argument that, if the earth has one movement, it must have two, is based upon nothing more than analogy with the planets. Aristotle clearly inferred as a corollary that, if the earth has two motions, one must be oblique to the other, for it would be the obliquity to the equator in at least one of the motions which would produce what he regards as the necessary consequence of his assumption, namely that the fixed stars would not always rise and set at the same places. As already stated, Aristotle can hardly have had clearly in his mind the possibility of one single rotation about the axis in *twenty-four hours* replacing exactly the apparent daily rotation; for he would have seen that this would satisfy his necessary condition that the fixed stars shall always rise and set at the same places, and therefore that

third reason, which depends upon the motion of parts of the earth and of the whole. He makes three points concerning this. In the first place, he proposes what the natural motion of the earth and its parts must be; secondly, where he says "but it might be [questioned] . . ." (296b. 9), he raises a certain doubt concerning this; and thirdly, he proves what he proposed.

In the first place, then, he says that motion of parts of the earth—in accordance with the nature of earth—is toward the center of the whole world; and, similarly, if the whole earth were outside the center of the world, it would, by its very nature, be moved to the center of the world, because the natural motion of the whole is the same as that of a part.

5. Then, when he says "but it might be [questioned]. . . " (296b.9), he raises a certain doubt. First, he proposes the doubt, saying that if the earth were assumed in the middle or center of the world—he means by this that the center of the whole world and of this earth would be the same—one can ask to which of these are heavy bodies—especially parts of the earth—moved naturally. That is, are they moved to the center because it is in the center of the world, or because it is the center of the earth?

Secondly, where he says "The goal, surely. . . " (296b.13), he resolves the doubt, saying that heavy bodies are moved to the center because it is the center of the whole world. For the motion of heavy bodies is contrary to the motion of light bodies; but light bodies, especially fire, are moved to the extremity of the [innermost] celestial body.[16] Therefore, heavy bodies, especially earth, are moved to the center of the earth, not, however, per se but accidentally, because the center of the earth and the center of the world coincide. . . .

6. Thirdly, where he says "That the center of the earth is the goal of their movement. . ." (296b.18), he proves what he assumed, namely that heavy

lowing Aristotle, tells us that if the earth moved at all it would necessarily possess these two motions—whether or not it is at the center of the world. If it were at the center of the world, it would have an axial rotation—but no translatory motion—and simultaneously rotate around the poles of the zodiac—that is, ecliptic—which is inclined to the earth's axis by approximately $23\frac{1}{2}$ degrees. However, despite the obliquity of the second motion to the first, and assuming uniformity of motion, the risings and settings would, it seems, follow regular patterns. And if we assume that the earth is not in the center of the universe, the two motions attributed to the earth would be those applicable to any planet—that is, a daily motion around the equinoctial poles as it is carried along with the other planets by the daily rotation of the sphere of the fixed stars (hence no axial rotation peculiar to the earth is involved), and a proper motion, that is, an annual periodic motion stemming from the motion of the sphere in which the earth—here treated as any other planet—would be embedded. On this interpretation, then, the earth would be assumed to have two translatory motions but no daily axial rotation—and this would be consistent with Thomas' comparison of the earth's motion to that of a planet. If Thomas had this scheme in mind, and the two motions are assumed uniform and commensurable, his consequence that "the fixed stars would not always be seen to rise and set at the same part of the earth" would again not follow. The phenomena of risings and settings would proceed in regular patterns.

Finally, it might be asked whether we do not have here, perhaps, a reflection of the heliocentric system enunciated by Aristarchus, where one motion of the earth would be a daily axial rotation and the other a revolutionary translatory motion around the sun in the course of a year. Such an interpretation seems unwarranted, since Thomas makes no mention of the sun as center. From his all too brief description, one is only justified in assuming that the two motions are of the same kind for each of the two possibilities. Thus, as we have seen, if the earth is at rest in the center of the universe, the two motions are axial; if it is not in the center, both are translatory. There is no justification for taking one motion as axial and the other as rotational. But just this combination of motions would be required if we are to see in this discussion the principal assumptions enunciated by Aristarchus. However, on the remote possibility that this is what Thomas intended, his consequence that such motions of the earth could not account for stellar risings and settings would be false.

Perhaps the underlying difficulty in all this lies in an implicit, but rash, inference. For Aristotle and Aquinas seem to have inferred that because a planet appears to move irregularly against the background of fixed stars, an observer on that planet would necessarily see only irregular risings and settings of these stars. Then, by analogy, they concluded that if the earth had the two motions of a planet, observers on earth would also see variations in the risings and settings of stars.

16. Fire would rise no higher than the concave surface of the moon, beyond which lay the celestial region that was filled with a fifth element, aether, extending all the way to the outermost sphere (that is, the sphere of the fixed stars).

---

he would have to get some further support from elsewhere to his assumption that the earth must have *two* motions. Still less could he have dreamt of the possibility of Aristarchus's later hypothesis that the earth has an annual revolution as well as a daily rotation about its axis, which hypothesis satisfies, as a matter of fact, both the condition as to two motions and the condition as regards the fixed stars."

It is immediately evident that Thomas has made explicit the two motions left unspecified by Aristotle. We have both a motion around the poles of the earth (equinoctial poles) and the poles of the zodiac. Thomas, fol-

bodies and parts of the earth are moved to the center. . . .[17]

7. Where he says "It is clear, then. . ." (296b. 21), he concludes what he has proposed. But he infers two propositions. The first of these is that the earth is in the center of the world, which is deduced from premises in the following way: [1] All heavy bodies are self-moved toward the center of the world; [2] furthermore, as was [just] demonstrated, they are also moved toward the center of the earth; therefore, the center of the earth is the center of the world. And so the earth is in the center of the world.

The second proposition is that the earth is immobile. This is deduced from premises in the following way: [1] Nothing is moved [when it is] in a place to which it is moved naturally, since it is natural for it to rest there; [2] but earth is moved naturally toward the center of the world; therefore, it is not moved in the center. Moreover, as has been shown, the earth cannot but be in the center of the world; therefore, the earth is in no way moved.

8. He posits the fourth reason where he says "because heavy bodies forcibly thrown. . . " (296b.23). We see that if a stone were suspended above some table, it would fall down in a straight line; and as long as its motion is downward it will fall along the same straight line. Now, if the table were not moved, the stone would fall to the [very] same place where it fell before; but if the table were moved, the stone would fall to another place

that would be as far removed [from the previous place] in proportion as the height from which the stone is dropped increases, since a greater time will elapse between the beginning and the end of the fall. However, we see that heavy bodies fall downward in a regular manner—that is, in a straight line—and that they fall to the same place on earth when dropped repeatedly from a particular place. Now, someone might say that because of the slowness of the earth's motion, there is an imperceptible distance between each of these places. But this overlooks the fact that if someone were to cause the stone to fall downward time after time for an infinite number of times, the same thing would occur, so that the extent of the time [involved] should make the distance between the places perceptible. And so it is obvious that the earth is not moved.

Then, in conclusion, he deduces from premises that the earth can neither be moved nor located outside the center of the world. . .

11. That the earth is not movable from place to place with a translatory motion results from the fact that it is always in the center. But should it be moved with any motion whatever, it would follow, again, that in virtue of its velocity all other motions—whether of clouds or animals—would be concealed from us. For that does not appear to be moved which is moved more slowly near a body that is moved more quickly.[18]

Therefore, in closing, the Philosopher concludes that enough has been said on the relation between the position, motion, and rest of the earth.

3. Thomas Aquinas: Heraclides of Pontus, and Aristarchus Mentioned as Proponents of the Earth's Diurnal Rotation

Translated and annotated by Edward Grant[19]

1. After the Philosopher [that is, Aristotle] shows what sort of nature the stars have, he takes up their motion here. First he shows how the stars are moved; secondly, where he says "From all this it is clear. . . " (290b. 12),[20] he shows whether sounds could be produced by their motion.

On the first point he offers three reasons to show that the stars are not self-moved, but [rather] that

the earth moved (translatory or rotatory), its speed would be greater than that of any other body on, or near, its surface. In this event, bodies on the earth would appear to be unmoved on the strange principle that a body will appear stationary when it is "moved more slowly near a body that is moved more quickly." It is possible that Thomas has garbled one of Ptolemy's arguments against the earth's axial rotation which we have quoted elsewhere (see section 1 of this selection).

19. Translated from Aquinas' *Commentaria in libros Aristotelis De caelo et mundo*, Lib. II, cap. VIII, Lectio 11, in *Sancti Thomae Aquinatis Opera Omnia*, Vol. III, p. 162, col. 1. Thomas is here commenting upon the first part of II.8 (289b.1—290a.7) of Aristotle's *On the Heavens*. For the translation of Thomas' quotations of Aristotle, I have consulted the Oxford English translation.

20. This is a reference to the opening words of II.9 of Aristotle's *On the Heavens*.

17. The "proof" according to Aristotle is "indicated by the fact that heavy bodies moving toward the earth do not move parallel but so as to make equal angles, and thus to a single centre, that of the earth" (296b.19–21). That is, wherever heavy bodies fall toward a point on the surface of the earth, they fall at right angles to a tangent drawn at that point.

18. Brevity and lack of clarity make this passage difficult to comprehend. Thomas seems to argue that if

they are carried by the motion of the orbs [that is, spheres]. The first of these reasons is put forth by way of a comparison of stars to orbs. In this reason, indeed, he presupposes one thing that is evident to the senses: We see that the stars and the entire sky are moved. Now, of necessity, this could happen in [only] three ways:[21] in one way, each— namely both star and sphere—is at rest; in another way, each of them is moved; in the third way, one of them rests and the other is moved. This division being assumed, he takes up the three parts just mentioned.

2. He first takes up the first way, where he says "That both should be at rest. . . " (289b.4). On this point he says that it is impossible to say that each—namely star and sphere—rests if it is also

assumed that the earth rests. For the apparent motion of the stars could not be saved if the stars, which appear to be moved, and the men who view the stars should both be at rest. Indeed, the appearance of motion is caused either by the motion of the thing seen or by the motion of the one who sees it. For this reason some people, assuming that the stars and the whole sky rest, have posited that the earth on which we dwell is moved once daily from west to east around the equinoctial poles. Thus by our motion, it seems to us that the stars are moved in a contrary direction, which is what Heraclides of Pontus and Aristarchus are said to have posited.[22] However, in the present discussion, Aristotle assumes that the earth rests; but later he will prove it.

4. John Buridan (ca. 1300—1358): The Compatibility of the Earth's Diurnal Rotation with Astronomical Phenomena

Translation and commentary by Marshall Clagett[23]

Annotated by Edward Grant

BOOK II, QUESTION 22

1. It is sought consequently whether the earth always is at rest in the center of the universe. . . .

This question is difficult. For in the first place there is a significant doubt as to whether the earth is directly in the middle of the universe so that its center coincides with the center of the universe. Furthermore, there is a strong doubt as to whether it is not sometimes moved rectilinearly as a whole, since we do not doubt that often many of its parts are moved, for this is appearent to us through our senses. There is also another difficult doubt as to whether the following conclusion of Aristotle is sound, namely, if the heaven is by necessity to be moved circularly forever, then it is necessary that

translate the relevant passage in Simplicius' commentary as follows: "There have been some, among them Heraclides of Pontus and Aristarchus, who thought that the phenomena could be accounted for by supposing the heaven and stars to be at rest, and the earth to be in motion about the poles of the equator from west [to east] making approximately one complete rotation each" (*A Source Book in Greek Science* [Cambridge, Mass.: Harvard University Press, 1948], pp. 106–107). Heraclides (Thomas spells it Heraclitus) lived ca. 388 B.C. to ca. 310 B.C.; Aristarchus of Samos ca. 310 B.C. to ca. 230 B.C. This translation also serves as a very accurate rendition of Moerbeke's Latin translation on folio 71 verso, col. 1, of the Venice edition of 1540. Also in this same translation of Simplicius' commentary are two other references to Heraclides of Pontus as a proponent of the earth's diurnal rotation (see fols. 83 recto, col. 1, and 87 verso, col. 1). Medieval discussions of the earth's daily axial rotation were, for the most part, probably derived ultimately from Simplicius via Moerbeke's translation rather than from Ptolemy's *Almagest,* where it is also mentioned (see section 1 of this selection). This is likely not only because few would have read so formidable a work as the *Almagest* but also because Ptolemy makes no mention of Heraclides or Aristarchus, thus indicating that where one or both of these names are associated with a discussion of the earth's diurnal rotation, Simplicius is the probable source (see n. 3).

21. After some discussion Aristotle concludes that "since . . . we cannot reasonably suppose either that both are in motion or that the star alone moves, the remaining alternative is that the circles should move, while the stars are at rest and move with the circles to which they are attached" (289b.31–34). By circles Aristotle means orbs or spheres. Almost without exception in the Middle Ages it was held that the stars and planets are not self-moved but are carried around by the spheres in which they were thought to be embedded.

22. Thomas almost certainly gets this from Simplicius' commentary on Aristotle's *On the Heavens,* which was translated from Greek into Latin by William of Moerbeke in 1271. Indeed, Simplicius mentions Heraclides and Aristarchus as proponents of the earth's diurnal rotation in commenting on the very same passage as St. Thomas (*De caelo* II.8.289b.5–7). Cohen and Drabkin

23. Reprinted with permission of the Regents of the University of Wisconsin from Marshall Clagett, *The Science of Mechanics in the Middle Ages* (Madison, Wis.: University of Wisconsin Press, 1959), pp. 594–599. Clagett's translation was made from E. A. Moody's Latin edition of Buridan's *Quaestiones super libris quattuor De caelo et mundo* (Cambridge, Mass.: Mediaeval Academy of America, 1942), pp. 226–232. The omissions, as indicated by ellipses, are Clagett's.

the earth be at rest forever in the middle. There is also a fourth doubt whether, in positing that the earth is moved circularly around its own center and about its own poles, all the phenomena that are apparent to us can be saved (*possent salvari omnia nobis apparentia*). Concerning this last doubt let us now speak.

2. It should be known that many people have held as probable that it is not contradictory to appearances for the earth to be moved circularly in the aforesaid manner, and that on any given natural day it makes a complete rotation from west to east by returning again to the west—that is, if some part of the earth were designated [as the part to observe]. Then it is necessary to posit that the stellar sphere would be at rest, and then night and day would take place through such a motion of the earth, so that that motion of the earth would be a diurnal motion (*motus diurnus*). The following is an example of this [kind of thing]: If anyone is moved in a ship and he imagines that he is at rest, then, should he see another ship which is truly at rest, it will appear to him that the other ship is moved. This is so because his eye would be completely in the same relationship to the other ship regardless of whether his own ship is at rest and the other moved, or the contrary situation prevailed.[24] And so we also posit that the sphere of the sun is everywhere at rest and the earth in carrying us would be rotated. Since, however, we imagine that we are at rest, just as the man located on the ship which is moving swiftly does not perceive his own motion nor the motion of the ship, then it is certain that the sun would appear to us to rise and then to set, just as it does when it is moved and we are at rest.

3. It is true, however, that if the stellar sphere is at rest, it is necessary to concede generally that the spheres of the planets are moving, since otherwise the planets would not change their positions relative to each other and to the fixed stars. And, therefore, this opinion imagines that any of the spheres of the planets moved evidently like the earth from west to east, but since the earth has a lesser circle, hence it makes its rotation (*circulatio*) in less time. Consequently, the moon makes its rotation in less time than the sun. And this is universally true, so that the earth completes its rotation in a natural day, the moon in a month, and the sun in a year, etc.

4. It is undoubtedly true that, if the situation were just as this position posits, all the celestial phenomena would appear to us just as they now appear. We should know likewise that those persons wishing to sustain this opinion, perhaps for reason of disputation, posit for it certain persuasions. . . . The third persuasion is this: To celestial bodies ought to be attributed the nobler conditions, and to the highest sphere, the noblest. But it is nobler and more perfect to be at rest than to be moved. Therefore, the highest sphere ought to be at rest. . . .

The last persuasion is this: Just as it is better to save the appearances through fewer causes then through many, if this is possible, so it is better to save [them] by an easier way than by one more difficult. Now it is easier to move a small thing than a large one. Hence it is better to say that the earth, which is very small, is moved most swiftly and the highest sphere is at rest than to say the opposite.[25]

5. But still this opinion is not to be followed. In the first place because it is against the authority of Aristotle and of all the astronomers (*astrologi*). But these people respond that *authority does not demonstrate,* and that it suffices astronomers that they posit a method by which appearances are saved, whether or not it is so in actuality. Appearances can be saved in either way; hence they posit the method which is more pleasing to them.

6. Others argue [against the theory of the earth's diurnal rotation] by many appearances (*apparentiis*). One of these is that the stars sensibly appear to us to be moved from the east to the west. But they solve this [by saying] that it would appear the same if the stars were at rest and the earth were moved from west to east.

7. Another appearance is this: If aynone were moving very swiftly on horseback, he would feel the air resisting him. Therefore, similarly, with the very swift motion of the earth in motion, we ought to feel the air noticeably resisting us. But these [supporters of the opinion] respond that the earth, the water, and the air in the lower region are moved simultaneously with diurnal motion. Consequently there is no air resisting us.[26]

24. See the similar discussions of relative motion in a ship by Oresme (section 5) and Copernicus (section 6).

25. Oresme presents the same argument (section 5) as did Copernicus later in rebutting Ptolemy near the beginning of Book I, chapter 8, of *De revolutionibus* (see section 6). Galileo, in turn, would repeat Copernicus' arguments in the *Dialogue Concerning the Two Chief World Systems* (see p. 115 of Stillman Drake's translation [Berkeley, Calif.: University of California Press, 1962]).

26. Ptolemy had already formulated this counterargument against his own position. However, Buridan makes no mention of Ptolemy's response to it (see section 1 and n. 6).

8. Another appearance is this: Since local motion heats, and therefore since we and the earth are moved so swiftly, we should be made hot. But these [supporters] respond that motion does not produce heat except by the friction (confricatio), rubbing, or separation of bodies. These [causes] would not be applicable there, since the air, water, and earth would be moved together.

9. But the last appearance which Aristotle notes is more demonstrative in the question at hand. This is that an arrow projected form a bow directly upward falls again in the same spot of the earth from which it was projected. This would not be so if the earth were moved with such velocity. Rather before the arrow falls, the part of the earth from which the arrow was projected would be a league's distance away. But still the supporters would respond that it happens so because the air, moved with the earth, carries the arrow, although the arrow appears to us to be moved simply in a straight line motion because it is being carried along with us. Therefore, we do not perceive that motion by which it is carried with the air. But this evasion is not sufficient because the violent impetus of the arrow in ascending would resist the lateral motion of the air so that it would not be moved as much as the air. This is similar to the occasion when the air is moved by a high wind. For then an arrow projected upward is not moved as much laterally as the wind is moved, although it would be moved somewhat. . . .[27]

10. Then I come to the other doubts. One would be whether the earth is situated directly in the middle of the universe. It should be answered in the affirmative. For we suppose that the place [designated] absolutely (simpliciter) as "upward," insofar as one looks at this lower world, is the concave [surface] of the orb of the moon. This is so because something absolutely light, i.e., fire, is moved toward it. For since fire appears to ascend in the air, it follows that fire naturally seeks a place above the air, and this place above the air is at the concave [surface] of the orb of the moon; because no other element appears to be so swiftly moved upward as fire. Now the place downward ought to be the maximum distance from the place upward, since they are contrary places. Now that which is the maximum distance from the heaven is the middle of the universe. Therefore the middle of the universe is absolutely downward. But that which is absolutely heavy—and earth is of this sort—ought to be situated absolutely downward. Therefore, the earth naturally ought to be *in* the

middle of the universe or *be* the middle of the universe.

11. But it is a significant difficulty as to whether the center of magnitude in the earth is the same as the center of gravity (medium gravitatis). It seems according to some statements that it is not. This is because if a large region of the earth is not covered with waters due to the habitation of animals and plants, and the opposite part is covered with waters, it is clear that the air which is naturally hot, and the sun, make hot the noncovered part, and thus they make it to some degree more subtle, rarer, and lighter. The covered part remains more compact and heavier. Now if one body in one part is lighter and in an opposite part heavier, the center of gravity will not be the center of magnitude. Rather with the center of gravity given, the greater magnitude will be in the lighter part, just as in the case of balances if on one side a stone is placed and on the other side wool [and they balance], the wool will be of a much greater magnitude.

12. With this understood, it ought to be seen which of those centers is the center of the universe. It should be answered immediately that the center of the universe is the center of gravity of the earth. This is because, as Aristotle says, all parts tend toward the center of the universe through their gravity,[28] and a part which is heavier would displace another, and thus finally it is necessary that the center of the universe coincide with the center of gravity. From these arguments it follows that the earth is nearer to the heaven in the part not covered with waters than in the covered part, and thus at the covered part there is greater declivity, and so the waters flow to that part. So, therefore, the earth, with respect to its magnitude, is not directly in the center of the universe. We commonly say, however, that it is in the center of the universe, because its center of gravity is the center of the universe.

13. By this another doubt is solved, evidently, whether the earth is sometimes moved according to its whole in a straight line. We can answer in the affirmative because from this higher [part of] the earth many parts of the earth (i.e., debris) continually flow along with the rivers to the bottom of the sea, and thus the earth is augmented in the covered part and is diminished in the uncovered part. Consequently, the center of gravity does not remain the same as it was before. Now, therefore,

27. See Clagett's Commentary below.
28. *On the Heavens* II.14. 296b.7–25.

with the center of gravity changed, that which has newly become the center of gravity is moved so that it will coincide with the center of the universe, and that point which was the center of gravity before ascends and recedes, and thus the whole earth is elevated toward the uncovered part so that the center of gravity might always become the center of the universe. And just as I have said elsewhere, it is not apparent how it could be saved unless the mountains were consumed and destroyed sometimes, nay infinite times, if time were eternal. Nor is any other way apparent by which such mountains could be generated. This was spoken of elsewhere, so I shall now desist. . . .

COMMENTARY

The first passage outlines the various "doubts." In passage 2 the relative nature of the perception of motion is introduced: "If anyone is moved in a ship and he imagines that he is at rest, then, should he see another ship which truly is at rest, it will appear to him that the other ship is moved." In the same way, an observer on a moving earth would think he was at rest and the heaven, truly at rest, he would believe to be in motion. On the other hand (passage 3) the planets change their positions relative to each other and the fixed stars. Hence they are in movement. But all celestial phenomena (passage 4) would appear to us as they do regardless of whether the heavens or the earth rotate.

Furthermore, according to Buridan, certain persuasions can be introduced to support the diurnal rotation of the earth. After presenting some that involve the natural superiority of the heavens over the earth, Buridan in the last persuasion gives the principle of economy or simplicity; saying that it is better to save appearances by the simpler and easier movement, i.e., the rotation of the earth.

In passage 5 Buridan notes that the authority of

Aristotle is against the diurnal rotation of the earth. But the holders of the theory of the earth's rotation would however answer rightfully, "Authority does not demonstrate." Furthermore, astronomers are only interested in finding the simplest way to save appearances. They are not concerned with whether the theory is so in actuality. Here Buridan is assigning the same sort of role to astronomers as did the Greek astronomer Geminus, whom Simplicius quotes as saying that the astronomer merely seeks hypotheses by the assumption of which the phenomena will be saved; it is not his business, but rather that of the physicist, to determine which hypothesis is actually more suitable to nature.[29]

In passage 7 Burdian presents the argument that, as the result of the diurnal rotation of the earth, one would expect a continually resisting wind. But the answer given is that the earth, water, and air are moved together. This would also explain why there is no heat generated by the friction between the air and the moving earth (passage 8). Finally, Buridan presents as an answer to the apparent rectilinear ascent and descent of arrows or projectiles by saying that the arrows are carried along with the air in the same diurnal motion as the earth. Buridan says this is not true, for the vertical motion of the arrow would resist the air. Perhaps Oresme, later, realized that it is not the air that causes the arrow to move laterally but rather the fact it is a part of the mechanical system of the rotating earth, just as a man moving his hand vertically up and down a mast is part of the mechanical system of the boat.[30]

In passage 11–13, Buridan argues that geological processes are always causing a redistribution of the matter of the earth; hence they are continually changing its center of gravity. But the center of gravity always strives to be at the center of the universe. Hence the earth is constantly shifting about near the center of the universe.

5. Nicole Oresme (ca. 1325–1382): The Compatibility of the Earth's Diurnal Rotation With Astronomical Phenomena and Terrestrial Physics

Translated by Albert D. Menut[31]

Annotated by Edward Grant

T. There are others who hold that the earth is at the center of the world and that it revolves and

was expressed by Simplicius in Book III, chapter 1, of his commentary on Aristotle's *De caelo* (the passage is translated in E. Grant, "Late Medieval Thought, Copernicus, and the Scientific Revolution," *Journal of the History of Ideas,* XXIII [1962], p. 198.)

30. See Oresme's discussion in the next section.

31. Reprinted with permission of the copyright owners, the Regents of the University of Wisconsin, from

29. This explicit distinction reported by Simplicius (the passage is translated in Cohen and Drabkin, pp. 90–91) appears in his commentary on Aristotle's *Physics,* which was not available in Latin translation during the Middle Ages. However, a similar sentiment

moves in a circuit around the pole established for this purpose, as is written in Plato's book called *Timaeus*.[32]

G. This was the opinion of a philosopher named Heraclides Ponticus,[33] who maintained that the earth moves circularly and that the heavens remain at rest. Here Aristotle does not refute these theories, possibly because they seemed to him of slight probability and were, moreover, sufficiently criticized in philosophical and astrological writings.

However, subject, of course, to correction, it seems to me that it is possible to embrace the argument and consider with favor the conclusions set forth in the above opinion that the earth rather than the heavens has a diurnal or daily rotation. At the outset, I wish to state that it is impossible to demonstrate from any experience at all that the contrary is true; second, that no argument is conclusive; and third, I shall demonstrate why this is so. As to the first point, let us examine one experience: we can see with our eyes the rising and setting of the sun, the moon, and several stars, while other stars turn around the arctic pole. Such a thing is due only to the motion of the heavens, as was shown in Chapter Sixteen, and, therefore, the heavens move with daily motion. Another experience is this one: if the earth is so moved, it makes its complete course in a natural day with the result that we and the trees and the houses are moved very fast toward the east; thus, it should seem to us that the air and wind are always coming very strong from the east and that it should make a noise such as it makes against the arrow shot from a crossbow or an even louder one, but the contrary is evident from experience.[34] The third argument is Ptolemy's—namely, that, if someone were in a boat moving rapidly toward the east and shot an arrow straight upward, it would not fall in the boat but far behind it toward the west. Likewise, if the earth moves so very fast turning from west to east and if someone threw a stone straight upward, it would not fall back to the place from which it was thrown, but far to the west; and the contrary appears to be the case.[35] It seems to me that what I shall say below about these experiences could apply to all other theories which might be brought forward in this connection. Therefore, I state, in the first place, that the whole corporeal machine or the entire mass of all the bodies in the universe is divided into two parts: one is the heavens with the sphere of fire and the higher region of the air; all this part, according to Aristotle in Book I of *Meteors*,[36] moves in a circle or revolves each day.

The other part of the universe is all the rest—that is, the middle and lower regions of the air, the water, the earth, and the mixed bodies—and, according to Aristotle, all this part is immobile and has no daily motion. Now, I take as a fact that local motion can be perceived only if we can see that one body assumes a different position relative to another body. For example, if a man is in a boat *a*, which is moving very smoothly either at rapid or slow speed, and if this man sees nothing except another boat *b*, which moves precisely like boat *a*, the one in which he is standing, I maintain that to this man it will appear that neither boat is moving.[37] If *a* rests while *b* moves, he will be aware that *b* is moving; if *a* moves and *b* rests, it will seem to the man in *a* that *a* is resting and *b* is moving, just as before.[38] Thus, if *a* rested an hour and *b* moved, and during the next hour it happened conversely that *a* moved and *b* rested, this man would not be able to sense this change or variation; it would seem to him that all this time *b* was moving. This fact is evident from experience, and the reason is that the two bodies *a* and *b* have

---

Nicole Oresme, Le Livre du ciel et du monde, edited by Albert D. Menut and Alexander J. Denomy, C.S.B. †, translated with an Introduction by Albert D. Menut (Madison, Wis.: University of Wisconsin Press, 1968), pp. 519–539.

The present selection constitutes virtually the whole of Book II, chapter 25, of Oresme's translation from Latin into French of Aristotle's treatise *On the Heavens (De caelo)*, along with Oreme's accompanying commentary (see Selection 71, n. 1). The letters *T* and *G* represent Aristotle's text and Oresme's commentary (or gloss), respectively.

32. These words of Aristotle's are found in *On the Heavens* II.13.293b.30–32. The chapter which follows consists of Oresme's lengthy commentary on these few words relating to the possible diurnal axial rotation of the earth.

33. See note 3.

34. The same point was mentioned and refuted by Buridan (see par. 7 of section 4).

35. Although Ptolemy mentions neither boat nor stone, these arguments reproduce faithfully the substance of his counterargument in the penultimate paragraph of Book I, chapter 7, of the *Almagest* (see section 1).

36. Probably *Meteorologica* I.3.340b.10–12, where, however, there is some dispute as to whether Aristotle intended to subdivide it quite as Oresme presents it.

37. This particular illustration of the relativity of motion was not given by Buridan in paragraph 2 of section 4, but was included by Copernicus near the beginning of Book I, chapter 5 of *De revolutionibus* (see section 6).

38. This is virtually the same point made by Buridan in paragraph 2 of section 4.

a continual relationship to each other so that, when *a* moves, *b* rests and, conversely, when *b* moves, *a* rests. It is stated in Book Four of *The Perspective* by Witelo that we do not perceive motion unless we notice that one body is in the process of assuming a different position relative to another.[39] I say, therefore, that, if the higher of the two parts of the world mentioned above were moved today in daily motion—as it is—and the lower part remained motionless and if tomorrow the contrary were to happen so that the lower part moved in daily motion and the higher—that is, the heavens, etc.—remained at rest, we should not be able to sense or perceive this change, and everything would appear exactly the same both today and tomorrow with respect to this mutation. We should keep right on assuming that the part where we are was at rest while the other part was moving continually, exactly as it seems to a man in a moving boat that the trees on shore move. In the same way, if a man in the heavens, moved and carried along by their daily motion, could see the earth distinctly and its mountains, valleys, rivers, cities, and castles, it would appear to him that the earth was moving in daily motion, just as to us on earth it seems as though the heavens are moving. Likewise, if the earth moved with daily motion and the heavens were motionless, it would seem to us that the earth was immobile and that the heavens appeared to move; and this can be easily imagined by anyone with clear understanding. This obviously answers the first experience, for we could say that the sun and stars appear to rise and set as they do and that the heavens seem to revolve on account of the motion of the earth in which we live together with the elements. To the second experience, the reply seems to be that, according to this opinion, not only the earth moves, but also with it the water and the air, as we stated above, although the water and air here below may be moved in addition by the winds or other forces.[40] In a similar manner, if the air were closed in on a moving boat, it would seem to a person in that air that it was not moving. Concerning the third experience, which seems more complicated and which deals with the case of an arrow or stone thrown up into the air, etc., one might say that the arrow shot upward is moved toward the east very rapidly with the air through which it passes, along with all the lower portion of the world which we have already defined and which moves with daily motion; for this reason the arrow falls back to the place from which it was shot into the air. Such a thing could be possible in this way,

for, if a man were in a ship moving rapidly eastward without his being aware of the movement and if he drew his hand in a straight line down along the ship's mast, it would seem to him that his hand were moving with a rectilinear motion; so, according to this theory it seems to us that the same thing happens with the arrow which is shot straight down or straight up.[41] Inside the boat moved rapidly eastward, there can be all kinds of movements—horizontal, criss-cross, upward, downward, in all directions—and they seem to be exactly the same as those when the ship is at rest.[42] Thus, if a man in this boat walked toward the west less rapidly than the boat was moving toward the east, it would seem to the man that he was approaching the west when actually he was going east; and similarly as in the preceding case, all the motions here below would seem to be the same as though the earth rested. Now, in order to explain the reply to the third experience in which this artificial illustration was used, I should like to present an example taken from nature, which, according to Aristotle, is true. He supposes that there is a portion of pure fire called *a* in the higher region of the air; this fire, being very light, rises as high as

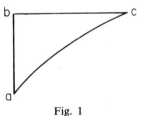

Fig. 1

39. *Vitellonis Thuringopoloni Opticae libri decem*, Bk. IV, par. 110, p. 167, of the edition published by Friedrich Risner (Basel, 1572). This declaration of the relativity of motion was taken by Witelo from Book II, paragraph 49, of the *Optics* of Alhazen, which, although separately paginated, was published by Risner in the same volume that included Witelo's *Optics;* Alhazen's statement appears on p. 60.

40. See note 6.

41. Oresme is here proposing a very different explanation than Buridan for the phenomenon of the arrow shot vertically into the air and falling vertically down to approximately the place from which it was shot. Where Buridan invoked the impetus theory (see par. 9 of section 4) to explain the arrow's return and inferred from it that the earth and its ambient air did not rotate, Oresme argues that the arrow might share the rotational motion of earth and air and hence would naturally fall to the same place from which it was shot.

42. In *De revolutionibus*, Book I, chapter 8, Copernicus says the very same thing (see section 6), as does Galileo in the *Dialogue Concerning the Two Chief World Systems* (Drake, p. 142).

possible to a place called *b* near the concave surface of the heavens [see Fig. 1]. I maintain that, just as with the arrow above, the motion of *a* in this case also must be compounded of rectilinear and, in part, of circular motion, because the region of the air and the sphere of fire through which *a* passed have, in Aristotle's opinion, circular motion. If they were not thus moved, *a* would go straight upward along the line *ab;* but because *b* is meanwhile drawn toward *c* by circular and daily motion, it appears that *a* describes the line *ac* as it ascends and that, therefore, the movement of *a* is compounded of rectilinear and of circular motion, and the movement of the arrow would be of this kind of mixed or compound motion[43] that we spoke of in Chapter Three of Book I. I conclude, then, that it is impossible to demonstrate by any experience that the heavens have daily motion and that the earth does not have the same.

With regard to the second point, if it could be demonstrated by rational arguments, in my opinion they would be the following, to which I shall reply in a manner that could be employed to refute all other pertinent argument. First, every simple body has a single simple motion, and the earth is a simple element which has, according to its various parts, natural rectilinear movement downward. So, it can have no other motion, a fact fully explained in Chapter Four of Book I. Circular motion is not natural to the earth for it has another motion, as already noted; if circular motion is violent to it, the earth could not be perpetual, as appears in several passages of Book I. All local motion is relative to some body at rest, as Averroes states in Chapter Eight,[44] from which he concludes that the earth must be at rest in the center of the heavens. Now, all motion is produced by some motive power or force, as shown in Books Seven and Eight of the *Physics*,[45] and the earth cannot move circularly because of its weight; if it is so moved by an external force, this movement would be violent and not perpetual. If, in reality, the heavens did not have diurnal motion, all astronomy would be false as well as a large part of natural philosophy throughout which such motion is taken for granted. It would, moreover, contradict Holy Scripture which states: The sun riseth and goeth down and returneth to his place; and there rising again, maketh his round by the south and turneth again to the north; the spirit goeth forward surveying all places round about and returneth to his circuits.[46] And it is also written of the earth that God made it motionless: Etenim firmavit orbem

terre, qui non commovebitur.[47] The Scripture states that the sun stopped its course in Joshua's time and returned in King Hezekiah's;[48] if, as is posited in this theory, it is the earth that moves and the heavens that remain motionless, then this stopping would have been a turning backward, which would have been more than a stoppage. And this is contrary to the statement in the Scriptures. As for the first argument where it is stated that every simple body has a single simple motion, I say that the earth, which as a whole is a simple body, has no movement, according to Aristotle in Chapter Twenty-two. Against the interpretation of anyone who maintained that Aristotle means that this body has a single simple motion not proper to itself as a whole, but applying only to its parts when they are out of their proper place, we can cite the case of air which moves downward when it is in the region of fire and upward when it is in the region of water, both being simple movements. Therefore, we can say with an ever greater show of reason that each simple body or element of the universe, with the possible exception of the sovereign heaven, moves in its proper place with circular motion. If any part of such a body is out of its place or outside the main body, it returns to it as directly as it can, once the hindrance is removed; this would surely happen if some part of the heavens were to get outside. It is not necessary that a simple body have its own simple motion in its proper place and another motion in its parts when they return to their proper place, and, according to Aristotle, we shall have to grant this assumption, as I shall do a little later. To the second

43. Like Oresme, Copernicus would also argue (Bk. I, ch. 8, of *De revolutionibus;* see section 6) that on the assumption of the earth's rotation, rising and falling objects would have a motion that is the resultant of two motions, rectilinear and circular.

44. Averroes, Commentary on *De caelo,* Book II, Comment 18, on folio 107 verso, paragraph 1, in Volume V of the edition cited in Selection 44, n. 1.

45. That is, Aristotle's *Physics.*

46. Ecclesiastes 1:5–6.

47. Psalms 92:1.

48. Joshua 10:12–14. This Scriptural account was also used against the Copernican theory, as Galileo reports in the Third Day of the *Dialogue Concerning the Two Chief World Systems* (Drake, p. 357), and even long before, in his 1615 *Letter to Madame Christina of Lorraine Grand Duchess of Tuscany Concerning the Use of Biblical Quotations in Matters of Science* (translated into English by Stillman Drake in *Discoveries and Opinions of Galileo* [New York: Doubleday, 1957], pp. 211–215), where Galileo attempts to reconcile the Joshua passage with the Copernican theory (see n. 51).

argument, I say the motion is natural to the earth as a whole and in its place; however, its parts have a different natural motion, rectilinear upward and downward, when they are out of their natural place. According to Aristotle, we must admit the same with respect to fire, parts of which move naturally upward when out of their proper place, and besides, also according to him, the entire element of fire in its sphere and in its place moves perpetually with diurnal motion, which could not be a true statement if its movement were violent. Now, in the theory we are discussing, it is not the element of fire, but the earth that moves in this manner. I say no to the third argument which states that all motion requires some body to be at rest, unless the motion must be perceptible to the senses; to make such motion apparent, it would suffice that the first body be moved in a different manner. But it is not required that there be a second body in order that this motion should exist, as was explained in Chapter Eight. Assuming that the heavens have diurnal movement and that the earth is moving in the opposite direction or imagining even that the earth were annihilated, we would note that the heavens had not stopped moving on this account; nor would they move faster or more slowly because neither the intelligence which moves the heavens nor the moving body of the heavens as a whole would be disposed to do otherwise. Besides, if it is assumed that circular motion did require another body at rest, such a body would not be situated in the middle of the one moving; in the middle of a millstone of a flour mill or of any similar moving body, nothing is at rest save a single mathematical point which is not a body, and the same is true at the center of the movement of the polar star. Thus, it could be said that the sovereign heaven rests or moves differently from the motion of the other bodies because its movement requires the existence of the other motions or requires that they be perceptible to the senses. To the fourth argument, we can say that the force causing this lower region of the world to move in a circle is its nature or form; and this same force—similar in nature to that which draws iron to the magnet—moves the earth to its proper place when it gets outside. Besides, I ask Aristotle what force it is that moves fire in the diurnal movement of its sphere, for we cannot say that the heavens pull it thus or seize it violently not only because this motion is perpetual, but also because the concave surface of the heavens is so highly polished, as noted in Chapter Eleven, that it passes over the

sphere of fire without rubbing, pulling, or pushing, as stated in Chapter Eighteen. So, we must say that fire is moved circularly by its own nature and form or by some intelligence or celestial influence. Exactly the same could be said by one who maintains that the earth has diurnal rotation and that the sphere of fire remains at rest. I say to the fifth argument, where it is held that, if the heavens did not make a rotation from day to day, the whole of astronomy would be false, that such a statement is not true, because all heavenly aspects, conjunctions, oppositions, constellations, figures, and influences would be exactly as they are in every respect, as is apparent from what was stated in reply to the first experience; and the astronomical tables of the heavenly motions and all other books would remain as true as they are at present, save that, with respect to diurnal motion, one would say that it is *apparently* in the heavens, but *actually* in the earth; no other effect would follow or result from one theory more than from the other. Aristotle's statement in Chapter Sixteen is pertinent in this connection, namely, that the sun seems to us to turn and twist and the stars to flicker and twinkle and that whether the thing we see moves or whether our vision moves makes no difference; and in the present case one could say that our vision is affected by diurnal movement. One could answer the sixth argument, which concerns the reference in Holy Scripture about the sun's turning, etc., by saying that this passage conforms to the customary usage of popular speech just as it does in many other places, for instance, in those where it is written that God repented, and He became angry and became pacified, and other such expressions which are not to be taken literally.[49] And more pertinent to our present subject, we read that God covers the heavens with clouds: Qui operit celum nubibus,[50] while the fact is that the heavens cover the clouds. Thus, we could say that the heavens, rather than the earth, appear to move with diurnal motion, while the truth is the exact opposite. And we could say that, in reality, the earth does not move from its place, nor apparently within its place, but it does actually move within its place. To the seventh argument, we could reply in much the same manner that in the time of Joshua the sun stopped and that in the time of

49. To illustrate Oresme's point that Scripture is not to be taken literally when it speaks of God repenting, and so on, Menut cites Genesis 6:6, Isaiah 47:6, Psalms 59:3, I Par. 13:10, Psalms 105:40.

50. Psalms 147:8.

Hezekiah it returned, but only apparently so; for, in fact, it was the earth which stopped moving in Joshua's time and which later in Hezekiah's time advanced or speeded up its movement; whichever occurrence we prefer to believe, the effect would be the same.[51] The latter opinion seems more reasonable than the former, as we shall make clear later.

Regarding the third point of this discussion, I want to present several opinions or reasons favorable to the theory that the earth moves as we have stated. In the first place, everything that requires another thing for its natural existence must aim at receiving the good it derives from the other through the motion or action natural to it. In this way we can see that each element moves to its natural place where it is conserved and that it goes to its place rather than its place coming to it. Thus the earth and the elements here below which require the heat and influence of the heavens round about them must needs be disposed by their movements to receive these benefits in due degree, just as, to speak familiarly, the meat being roasted before the fire receives around it the heat of the fire by being turned and not by the turning of the fire around the meat. If neither experience nor reason indicates the contrary, it is much more reasonable, as stated above, that all the principal movements of the simple bodies in the world should go or proceed in one direction or manner. Now, according to the philosophers and astronomers, it cannot be that all bodies move from east to west; but, if the earth moves as we have indicated, then all proceed alike from west to east—that is, the earth by rotating once around the poles from west to east in one natural day and the heavenly bodies around the zodiacal poles: the moon in one month, the sun in one year, Mars in approximately two years, and so on with the other bodies.[52] It is unnecessary to posit in the heavens other primary poles or two kinds of motion, one from the east to the west and the other on different poles in the opposite direction, but such an assumption is definitely necessary if the heavens move with diurnal motion. . . . Although Averroes says in Chapter Twenty-two that motion is nobler than rest,[53] the contrary seems true, because, again on Aristotle's authority in Chapter Twenty-two, the noblest thing possible achieves its perfection without movement, and this is God Himself. Rest is the end purpose of motion, and so Aristotle holds that the bodies here below move to their natural places in order to rest there.[54] A further sign that rest is best is that we pray for the dead that God may

give them rest: Requiem eternam, etc. Therefore, to rest or to be moved less is a better and nobler condition than to be moved or to be moved farther and farther from rest.[55] From this, it seems that the position we have taken above is very reasonable, for it could be said that the earth, the vilest element, along with the other elements here below make their rotation very fast, that the sovereign air and fire move less fast—as can be observed in the case of the comets—and that the moon and lunar sphere move still more slowly, for it moves in a month only the distance the earth travels in a natural day. Proceeding in this manner, the higher heavens make their revolution more slowly yet, although there is some variation, and this process continues up to the heaven of the fixed stars, which is motionless or makes its revolution very slowly, according to some in thirty-six thousand years or one degree in one hundred years.[56] In this way and no other can we solve the question proposed by Aristotle in Chapter Twenty-one, with only slight additions. It is not necessary to assume so many degrees of things nor such obscure difficulties as Aristotle introduces in his reply in Chapter Twenty-two. It is indeed very reasonable that the bodies that are larger or farther from the center should make their circuit or revolution in longer

51. Oresme's solution to the "Joshua problem" is more daring than Galileo's, since it was more at variance with the literal sense of the text. Galileo argued that the sun controls all celestial motions, which would cease immediately if the sun's motion ceased. "The sun, then," says Galileo, "being the font of light and the source of motion, when God willed that at Joshua's command the whole system of the world should rest and should remain for many hours in the same state, it sufficed to make the sun stand still. Upon its stopping, all the other revolutions ceased; the earth, the moon, and the sun remained in the same arrangement as before, as did all the planets; nor in all that time did day decline towards night, for day was miraculously prolonged. And in this manner, by the stopping of the sun, without altering or in the least disturbing the other aspects and mutual positions of the stars, the day could be lengthened on earth—which agrees exquisitely with the literal sense of the sacred text"(from Galileo's *Letter to Madame Christina*, pp. 213–214 of Drake's translation).

52. Buridan says virtually the same thing in paragraph 3 of section 4.

53. Averroes, *Commentary on De caelo*, Book II, Comment 61, on folio 140 recto, paragraph A, in Vol. V of the edition cited in Selection 44, n.1.

54. *Physics* V.6.230b.25–27.

55. Buridan offers much the same argument in favor of the immobility of the heavens and the motion of the earth (see the third persuasion in par. 4 of section 4).

56. Ptolemy's value for the precession of the equinoxes.

time than those nearer the center, because, if they made their circuit in the same or equal time, their movements would have to be excessively fast. So we could say that nature compensates by ordaining that the rotations of the bodies farther from the center shall be accomplished in much longer time. Accordingly, because of its great size, the sovereign or primary heaven takes a very long time to make its circuit or rotation although it moves very fast. But the earth, which has a very small circuit, can cover the distance in one diurnal movement, while the other bodies intermediate between the highest and lowest heavens accomplish their circuits in time periods midway between the extremes, although these periods are not proportionate. In this way, a constellation near the north, i.e., the Great Bear which we call the Chariot, does not move backward, the chariot in front of the oxen, as it would if moved with diurnal motion; but it actually goes forward in the right direction. All philosophers say that an action accomplished by several or by large-scale operations which can be accomplished by fewer or smaller operations is done for naught.[57] And Aristotle says in Chapter Eight of Book I that God and nature do nothing without some purpose.[58] Now, if it is true that the heavens move with diurnal motion, it becomes necessary to posit in the major bodies of the universe and in the heavens two contrary kinds of movement: one from east to west and the other from the opposite direction, as we have often stated. And with regard to diurnal motion, we must assume an excessively great speed; for, if we consider thoughtfully the height or distance of the heavens, their magnitude, and the immensity of their circuit, mindful that this circuit is traveled in but one day's time, no man could imagine or conceive how marvelously swift and excessively great, how far beyond belief and estimation their speed must be.[59] Since, then, all the effects we see could be produced and all appearances saved by substituting for the diurnal movement of the heavens a smaller operation, namely, the diurnal motion of the earth, a very small body as compared with the heavens, and by so doing avoid the multiplication of operations so diverse and so outrageously great, then it follows that God and nature must have created and arranged them for naught; and this is an inadmissible conclusion, as we have often said. Assuming the entire heavens to move with daily motion and, in addition, assuming the eighth sphere to have a different motion, as the astronomers believe, then we must necessarily posit a ninth sphere moving

only with diurnal motion. However, if we assume that the earth moves as stated above, then the eighth heaven moves with a single slow motion and it is consequently unnecessary to imagine a ninth natural sphere invisible and starless; for God and nature would have made this ninth sphere for naught since by another method, i.e., assuming the earth to move, everything can remain exactly as it is. Also, when God performs a miracle, we must assume and maintain that He does so without altering the common course of nature, in so far as possible. Therefore, if we can save appearances by taking for granted that God lengthened the day in Joshua's time by stopping the movement of the earth or merely that of the region here below—which is so very small and like a mere dot compared to the heavens—and by maintaining that nothing in the whole universe—and especially the huge heavenly bodies—except this little point was put off its ordinary course and regular schedule, then this would be a much more reasonable assumption. And appearances can be saved in this way, as is evident from the reply to the seventh argument, presented against this opinion. As much could be said with regard to the return of the sun in Hezekiah's time. Thus, it is apparent that one cannot demonstrate by any experience whatever that the heavens move with diurnal motion; whatever the fact may be, assuming that the heavens move and the earth does not or that the earth moves and the heavens do not, to an eye in the heavens which could see the earth clearly, it would appear to move; if the eye were on the earth, the heavens would appear to move. Nor would the vision of this eye be deceived, for it can sense or see nothing but the process of the movement itself. But if the motion is relative to some particular body or object, this judgment is made by the senses from within that particular body, as Witelo explains in *The Perspective*;[60] and the senses are often deceived in such cases, as was related above in the example of the man on the moving ship. Afterward, it was demonstrated how it cannot be proved conclusively by argu-

57. This principle was expressed by Aristotle as nature doing nothing in vain (see n. 58, and compare to Ockham's "razor," Selection 40, n. 13).

58. Aristotle, *De caelo* I.4.271a.33; see also II.8.290a.31.

59. The same substantive argument was made by Buridan (par. 4 of section 4) and Copernicus (ch. 7 of section 6). See note 25.

60. See note 39.

ment that the heavens move. In the third place, we offered arguments opposing their diurnal motion. However, everyone maintains, and I think myself, that the heavens do move and not the earth: For God hath established the world which shall not be moved, in spite of contrary reasons because they are clearly not conclusive persuasions. However, after considering all that has been said, one could then believe that the earth moves and not the heavens, for the opposite is not clearly evident. Nevertheless, at first sight, this seems as much against natural reason as, or more against natural reason than, all or many of the articles of our faith.[61] What I have said by way of diversion or intellectual exercise can in this manner serve as a valuable means of refuting and checking those who would like to impugn our faith by argument.

6. Nicolaus Copernicus (1473–1543): The Compatibility of the Earth's Diurnal Rotation with Astronomical Phenomena and Terrestrial Physics

Translated by Charles Glenn Wallis[62]

Annotated by Edward Grant

PREFACE AND DEDICATION TO POPE PAUL III

. . . .

Accordingly, when I had meditated upon this lack of certitude in the traditional mathematics concerning the composition of movements of the spheres of the world, I began to be annoyed that the philosophers, who in other respects had made a very careful scrutiny of the least details of the world, had discovered no sure scheme for the movements of the machinery of the world, which has been built for us by the Best and Most Orderly Workman of all. Wherefore I took the trouble to reread all the books by philosophers which I could get hold of, to see if any of them even supposed that the movements of the spheres of the world were different from those laid down by those who taught mathematics in the schools. And as a matter of fact, I found first in Cicero that Nicetas[63] thought that the Earth moved. And afterwards I found in Plutarch that there were some others of the same opinion: I shall copy out his words here, so that they may be known to all:[64]

Some think that the Earth is at rest; but Philolaus the Pythagorean says that it moves around the fire with an obliquely circular motion, like the sun and moon. Herakleides of Pontus and Ekphantus the Pythagorean do not give the Earth any movement of locomotion, but rather a limited movement of rising and setting around its centre, like a wheel.

Therefore I also, having found occasion, began to meditate upon the mobility of the Earth. And although the opinion seemed absurd, nevertheless because I knew that others before me had been granted the liberty of constructing whatever circles they pleased in order to demonstrate astral phenomena, I thought that I too would be readily permitted to test whether or not, by the laying down that the Earth had some movement, demonstrations

61. In the final analysis, then, Oresme denies the diurnal rotation of the earth. He argues that although one might plausibly believe in the earth's rotation—for there are no persuasive arguments to deny it conclusively—it seems contrary to natural reason, even more so than do some articles of the faith. Thus Oresme acquiesces in tradition, custom, and "natural reason," to conclude in favor of the earth's immobility. He sought to humble reason and show that physical arguments could not establish a relatively simple physical problem. If reason is impotent in the solution of a physical problem, how much more impotent would it be in coping with articles of the faith, which some men had tried to demonstrate by reason. Thus Oresme's elaborate and brilliant arguments were designed to protect the faith from demonstrations based on applications of reason and science, both of which seem incapable of deciding straightforward physical problems. By showing that it was impossible to know which alternative is *really* true, Oresme, the theologian, succeeded in using reason to confound reason.

62. Reprinted by permission of Encyclopaedia Britannica from *Nicolaus Copernicus on the Revolutions of the Heavenly Spheres,* translated by Charles Glenn Wallis in Great Books of the Western World (Chicago: Encyclopaedia Britannica, 1952), XVI, pp. 508, 514–521. This selection contains an excerpt from Copernicus' Preface and chapters 5 to 9 of Book I of the *De revolutionibus orbium coelestium.* Another, but inferior, translation of Preface and Book I has been made by John F. Dobson, assisted by Selig Brodetsky, in *Occasional Notes of the Royal Astronomical Society,* Vol. II, No. 10 (May 1947).

63. Cicero's *Academica,* Book II, chapter 29 (the quotation is given in n. 67). Copernicus uses the name Nicetus (not Nicetas, as given here, or Hicetas, as used by Wallis, in Bk. I, ch. 5) for Hicetas, the form employed by Cicero.

64. *On the Opinions of the Philosophers (De placitis Philosophorum),* Book III, chapter 13. Copernicus actually quoted the Greek text.

less shaky than those of my predecessors could be found for the revolutions of the celestial spheres.

. . . . . .

BOOK I

5. Does the Earth Have a Circular Movement? And of Its Place

Now that it has been shown that the Earth too has the form of a globe, I think we must see whether or not a movement follows upon its form and what the place of the Earth is in the universe. For without doing that it will not be possible to find a sure reason for the movements appearing in the heavens. Although there are so many authorities for saying that the Earth rests in the centre of the world that people think the contrary supposition inopinable and even ridiculous; if however we consider the thing attentively, we will see that the question has not yet been decided and accordingly is by no means to be scorned. For every apparent change in place occurs on account of the movement either of the thing seen or of the spectator, or on account of the necessarily unequal movement of both. For no movement is perceptible relatively to things moved equally in the same directions—I mean relatively to the thing seen and the spectator.[65] Now it is from the Earth that the celestial circuit is beheld and presented to our sight. Therefore, if some movement should belong to the Earth it will appear, in the parts of the universe which are outside, as the same movement but in the opposite direction, as though the things outside were passing over. And the daily revolution in especial is such a movement. For the daily revolution appears to carry the whole universe along, with the exception of the Earth and the things around it. And if you admit that the heavens possess none of this movement but that the Earth turns from west to east, you will find—If you make a serious examination— that as regards the apparent rising and setting of the sun, moon, and stars the case is so. And since it is the heavens which contain and embrace all things as the place common to the universe, it will not be clear at once why movement should not be assigned to the contained rather than to the container, to the thing placed rather than to the thing providing the place.

As a matter of fact, the Pythagoreans Herakleides and Ekphantus were of this opinion[66] and so was Hicetas[67] the Syracusan in Cicero; they made the Earth to revolve at the centre of the world. For they believed that the stars set by reason of the interposition of the Earth and that with cessation

of that they rose again. Now upon this assumption there follow other things, and a no smaller problem concerning the place of the Earth, though it is taken for granted and believed by nearly all that the Earth is the centre of the world. For if anyone denies that the Earth occupies the midpoint or centre of the world yet does not admit that the distance [between the two] is great enough to be compared with [the distance to] the sphere of the fixed stars but is considerable and quite apparent in relation to the orbital circles of the sun and the planets; and if for that reason he thought that their movements appeared irregular because they are organized around a different centre from the centre of the Earth, he might perhaps be able to bring forward a perfectly sound reason for movement which appears irregular. For the fact that the wandering stars are seen to be sometimes nearer the Earth and at other times farther away necessarily argues that the centre of the Earth is not the centre of their circles. It is not yet clear whether the Earth draws near to them and moves away or they draw near to the Earth and move away.

And so it would not be very surprising if someone attributed some other movement to the earth in addition to the daily revolution. As a matter of fact, Philolaus the Pythagorean—no ordinary mathematician, whom Plato's biographers say Plato went to Italy for the sake of seeing[68]—is supposed to have held that the Earth moved in a circle and wandered in some other movements and was one of the planets.

Many however have believed that they could show by geometrical reasoning that the Earth is in the middle of the world; that it has the proportionality of a point in relation to the immensity of the heavens, occupies the central position, and for this reason is immovable, because, when the universe

65. The same statement was made earlier by Oresme (see section 5 and note 37).
66. See the quotation from Plutarch in the Preface.
67. See the Preface and note 2. Here is the quotation from Cicero's *Academica*, Book II, chapter 29 (trans. H. Rackham in the Loeb Classical Library, p. 627):
"The Syracusan Hicetas, as Theophrastus asserts, holds the view that the heaven, sun, moon, stars, and in short all of the things on high are stationary, and that nothing in the world is in motion except the earth, which by revolving and twisting round its axis with extreme velocity produces all the same results as would be produced if the earth were stationary and the heaven in motion; and this is also in some people's opinion the doctrine stated by Plato in *Timaeus*, but a little more obscurely."
68. See Plutarch, *De placitis philosophorum*, III.13.

moves, the centre remains unmoved and the things which are closest to the centre are moved the most slowly.

6. On the Immensity of the Heavens in Relation to the Magnitude of the Earth[69]

It can be understood that this great mass which is the Earth is not comparable with the magnitude of the heavens, from the fact that the boundary circles—for that is the translation of the Greek ὁρίζοντες—cut the whole celestial sphere into two halves; for that could not take place if the magnitude of the Earth in comparison with the heavens, or its distance from the centre of the world, were considerable. For the circle bisecting a sphere goes through the centre of the sphere, and is the greatest circle which it is possible to circumscribe.

Now let the horizon be the circle *ABCD* [Fig.2],

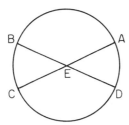

Fig. 2

and let the Earth, where our point of view is, be *E*, the centre of the horizon by which the visible stars are separated from those which are not visible. Now with a dioptra or horoscope or level placed at *E*, the beginning of Cancer is seen to rise at point *C*; and at the same moment the beginning of Capricorn appears to set at *A*. Therefore, since *AEC* is in a straight line with the dioptra, it is clear that this line is a diameter of the ecliptic, because the six signs bound a semicircle, whose centre *E* is the same as that of the horizon. But when a revolution has taken place and the beginning of Capricorn arises at *B*, then the setting of Cancer will be visible at *D*, and *BED* will be a straight line and a diameter of the ecliptic. But it has already been seen that the line AEC is a diameter of the same circle; therefore, at their common section, point *E* will be their center. So in this way the horizon always bisects the ecliptic, which is a great circle of the sphere. But on a sphere, if a circle bisects one of the great circles, then the circle bisecting is a great circle. Therefore the horizon is a great circle; and its centre is the same as that of the ecliptic, as far as appearance goes; although nevertheless

the line passing through the centre of the Earth and the line touching to the surface are necessarily different; but on account of their immensity in comparison with the Earth they are like parallel lines, which on account of the great distance between the termini appear to be one line, when the space contained between them is in no perceptible ratio to their length, as has been shown in optics.

From this argument it is certainly clear enough that the heavens are immense in comparison with the Earth and present the aspect of an infinite magnitude, and that in the judgment of sense-perception the Earth is to the heavens as a point to a body and as a finite to an infinite magnitude. But we see that nothing more than that has been shown, and it does not follow that the Earth must rest at the center of the world. And we should be even more surprised if such a vast world should wheel completely around during the space of twenty-four hours rather than that its least part, the Earth, should. For saying that the centre is immovable and that those things which are closest to the centre are moved least does not argue that the Earth rests at the centre of the world. That is no different from saying that the heavens revolve but the poles are at rest and those things which are closest to the poles are moved least. In this way Cynosura [the pole star] is seen to move much more slowly than Aquila or Canicula because, being very near to the pole, it describes a smaller circle, since they are all on a single sphere, the movement of which stops at its axis and which does not allow any of its parts to have movements which are equal to one another. And nevertheless the revolution of the whole brings them round in equal times but not over equal spaces.

The argument which maintains that the Earth, as a part of the celestial sphere and as sharing in the same form and movement, moves very little because very near to its centre advances to the following position: therefore the Earth will move, as being a body and not a centre, and will describe in the same time arcs similar to, but smaller than, the arcs of the celestial circle. It is clearer than daylight how false that is; for there would necessarily always be noon at one place and midnight at another, and so the daily risings and settings could not take place, since the movement of the whole and the part would be one and inseparable.

69. Virtually the same chapter title was used by Ptolemy in Book I, chapter 6, of the *Almagest*.

But the ratio between things separated by diversity of nature is so entirely different that those which describe a smaller circle turn more quickly than those which describe a greater circle. In this way Saturn, the highest of the wandering stars, completes its revolution in thirty years, and the moon which is without doubt the closest to the Earth completes its circuit in a month, and finally the Earth itself will be considered to complete a circular movement in the space of a day and a night. So this same problem concerning the daily revolution comes up again. And also the question about the place of the Earth becomes even less certain on account of what was just said. For that demonstration proves nothing except that the heavens are of an indefinite magnitude with respect to the Earth. But it is not at all clear how far this immensity stretches out. On the contrary, since the minimal and indivisible corpuscles, which are called atoms, are not perceptible to sense, they do not, when taken in twos or in some small number, constitute a visible body; but they can be taken in such a large quantity that there will at last be enough to form a visible magnitude. So it is as regards the place of the earth; for although it is not at the center of the world, nevertheless the distance is as nothing, particularly in comparison with the sphere of the fixed stars.

7. Why the Ancients Thought the Earth Was at Rest at the Middle of the World as Its Centre[70]

Wherefore for other reasons the ancient philosophers have tried to affirm that the Earth is at rest at the middle of the world, and as principal cause they put forward heaviness and lightness. For Earth is the heaviest element; and all things of any weight are borne towards it and strive to move towards the very centre of it.

For since the Earth is a globe towards which from every direction heavy things by their own nature are borne at right angles to its surface, the heavy things would fall on one another at the centre if they were not held back at the surface; since a straight line making right angles with a plane surface where it touches a sphere leads to the centre. And those things which are borne toward the centre seem to follow along in order to be at rest at the centre. All the more then will the Earth be at rest at the centre; and, as being the receptacle for falling bodies, it will remain immovable because of its weight.[71]

They strive similarly to prove this by reason of movement and its nature. For Aristotle says that the movement of a body which is one and simple is simple, and the simple movements are the rectilinear and the circular. And of rectilinear movements, one is upward, and the other is downward. As a consequence, every simple movement is either toward the centre, i.e., downward, or away from the centre, i.e., upward, or around the centre, i.e., circular. Now it belongs to earth and water, which are considered heavy, to be borne downward, i.e., to seek the centre: for air and fire, which are endowed with lightness, move upward, i.e., away from the centre. It seems fitting to grant rectilinear movement to these four elements and to give the heavenly bodies a circular movement around the centre. So Aristotle.[72] Therefore, said Ptolemy of Alexandria, if the Earth moved, even if only by its daily rotation, the contrary of what was said above would necessarily take place. For this movement which would traverse the total circuit of the Earth in twenty-four hours would necessarily be very headlong and of an unsurpassable velocity. Now things which are suddenly and violently whirled around are seen to be utterly unfitted for reuniting, and the more unified are seen to become dispersed, unless some constant force constrains them to stick together. And a long time ago, he says, the scattered Earth would have passed beyond the heavens, as is certainly ridiculous; and a fortiori so would all the living creatures and all the other separate masses which could by no means remain unshaken. Moreover, freely falling bodies would not arrive at the places appointed them, and certainly not along the perpendicular line which they assume so quickly. And we would see clouds and other things floating in the air always borne toward the west.[73]

8. Answer to the Aforesaid Reasons and Their Inadequacy

For these and similar reasons they say that the Earth remains at rest at the middle of the world and that there is no doubt about this. But if someone opines that the Earth revolves, he will also say that the movement is natural and not violent. Now things which are according to nature produce effects contrary to those which are violent. For things to which force or violence is applied get

70. The immobility of the earth was also the subject of Book I, chapter 7, of the *Almagest* (see section 1).
71. Based on Aristotle, *De caelo* II.14.296b.
72. Aristotle, *De caelo* I.2–3; III.3–5.
73. Most of this paragraph is drawn from the *Almagest*, Book I, chapter 7 (see section 1).

broken up and are unable to subsist for a long time. But things which are caused by nature are in a right condition and are kept in their best organization. Therefore Ptolemy had no reason to fear that the Earth and all things on the Earth would be scattered in a revolution caused by the efficacy of nature, which is greatly different from that of art or from that which can result from the genius of man. But why didn't he feel anxiety about the world instead, whose movement must necessarily be of greater velocity, the greater the heavens are than the Earth? Or have the heavens become so immense, because an unspeakably vehement motion has pulled them away from the centre, and because the heavens would fall if they came to rest anywhere else?

Surely if this reasoning were tenable, the magnitude of the heavens would extend infinitely. For the farther the movement is borne upward by the vehement force, the faster will the movement be, on account of the ever-increasing circumference which must be traversed every twenty-four hours: and conversely, the immensity of the sky would increase with the increase in movement. In this way, the velocity would make the magnitude increase infinitely, and the magnitude the velocity. And in accordance with the axiom of physics that *that which is infinite cannot be traversed or moved in any way*, then the heavens will necessarily come to rest.[74]

But they say that beyond the heavens there isn't any body or place or void or anything at all; and accordingly it is not possible for the heavens to move outward:[75] in that case it is rather surprising that something can be held together by nothing. But if the heavens were infinite and were finite only with respect to a hollow space inside, then it will be said with more truth that there is nothing outside the heavens, since anything which occupied any space would be in them; but the heavens will remain immobile. For movement is the most powerful reason wherewith they try to conclude that the universe is finite.

But let us leave to the philosophers of nature[76] the dispute as to whether the world is finite or infinite, and let us hold as certain that the Earth is held together between its two poles and terminates in a spherical surface. Why therefore should we hesitate any longer to grant to it the movement which accords naturally with its form, rather than put the whole world in a commotion—the world whose limits we do not and cannot know? And why not admit that the appearance of daily revolution belongs to the heavens but the reality

belongs to the Earth? And things are as when Aeneas said in Virgil: "We sail out of the harbor, and the land and the cities move away."[77] As a matter of fact, when a ship floats on over a tranquil sea, all the things outside seem to the voyagers to be moving in a movement which is the image of their own, and they think on the contrary that they themselves and all the things with them are at rest.[78] So it can easily happen in the case of the movement of the Earth that the whole world should be believed to be moving in a circle. Then what would we say about the clouds and the other things floating in the air or falling or rising up, except that not only the Earth and the watery element with which it is conjoined are moved in this way but also no small part of the air and whatever other things have a similar kinship with the Earth? whether because the neighbouring air, which is mixed with earthly and watery matter, obeys the same nature as the Earth or because the movement of the air is an acquired one, in which it participates without resistance on account of the contiguity and perpetual rotation of the Earth. Conversely, it is no less astonishing for them to say that the highest region of the air follows the celestial movement, as is shown by those stars which appear suddenly— I mean those called "comets" or "bearded stars" by the Greeks. For that place is assigned for their generation; and like all the other stars they rise and set. We can say that that part of the air is deprived of terrestrial motion on account of its great distance from the Earth. Hence the air which is nearest to the Earth and the things floating in it will appear tranquil, unless they are driven to and fro by the wind or some other force, as happens. For how is the wind in the air different from a current in the sea?[79]

But we must confess that in comparison with the world the movement of falling and of rising bodies is twofold and is in general compounded of the

---

74. Although both Buridan and Oresme used the argument that it is more appropriate for the relatively small earth to rotate daily than for the vast heavens to do so (see section 4, par. 4; n. 25; and section 5), neither includes mention of a vehement motion *(motus vehementia)* or force expanding the heavens to infinity.

75. Aristotle, *De caelo* I.9. For those who maintained the existence of an infinite void beyond the cosmos, see Selection 73.

76. "Philosophers of nature" translates *physiologi*.

77. *Aeneid* III.72.

78. Oresme had already made this point (see section 5 and n. 42).

79. See note 6.

rectilinear and the circular.[80] As regards things which move downward on account of their weight because they have very much earth in them, doubtless their parts possess the same nature as the whole, and it is for the same reason that fiery bodies are drawn upward with force. For even this earthly fire feeds principally on earthly matter; and they define flame as glowing smoke. Now it is a property of fire to make that which it invades to expand; and it does this with such force that it can be stopped by no means or contrivance from breaking prison and completing its job. Now expanding movement moves away from the centre to the circumference; and so if some part of the Earth caught on fire, it would be borne away from the centre and upward. Accordingly, as they say, a simple body possesses a simple movement—this is first verified in the case of circular movement—as long as the simple body remains in its unity in its natural place. In this place, in fact, its movement is none other than the circular, which remains entirely in itself, as though at rest. Rectilinear movement, however, is added to those bodies which journey away from their natural place or are shoved out of it or are outside it somehow. But nothing is more repugnant to the order of the whole and to the form of the world than for anything to be outside of its place. Therefore rectilinear movement belongs only to bodies which are not in the right condition and are not perfectly conformed to their nature—when they are separated from their whole and abandon its unity. Furthermore, bodies which are moved upward or downward do not possess a simple, uniform, and regular movement—even without taking into account circular movement. For they cannot be in equilibrium with their lightness or their force of weight. And those which fall downward possess a slow movement at the beginning but increase their velocity as they fall. And conversely we note that this earthly fire—and we have experience of no other—when carried high up immediately dies down, as if through the acknowledged agency of the violence of earthly matter.

Now circular movement always goes on regularly, for it has an unfailing cause; but [in rectilinear movement] the acceleration stops, because, when the bodies have reached their own place, they are no longer heavy or light, and so the movement ends. Therefore, since circular movement belongs to wholes and rectilinear to parts, we can say that the circular movement stands with the rectilinear, as does animal with sick. And the fact that Aristotle

divided simple movement into three genera: away from the centre, toward the centre, and around the centre, will be considered merely as an act of reason, just as we distinguish between line, point, and surface, though none of them can subsist without the others or without body.

In addition, there is the fact that the state of immobility is regarded as more noble and godlike than that of change and instability, which for that reason should belong to the Earth rather than to the world[81] I add that it seems rather absurd to ascribe movement to the container or to that which provides the place and not rather to that which is contained and has a place, *i.e.,* the Earth. And lastly, since it is clear that the wandering stars are sometimes nearer and sometimes farther away from the Earth, then the movement of one and the same body around the centre—and they mean the centre of the Earth—will be both away from the centre and toward the centre. Therefore it is necessary that movement around the centre should be taken more generally; and it should be enough if each movement is in accord with its own centre. You see therefore that for all these reasons it is more probable that the Earth moves than that it is at rest—especially in the case of the daily revolution, as it is the Earth's very own. And I think that is enough as regards the first part of the question.

9. Whether Many Movements Can be Attributed to the Earth, and Concerning the Centre of the World

Therefore, since nothing hinders the mobility of the Earth, I think we should now see whether more than one movement belongs to it, so that it can be regarded as one of the wandering stars. For the apparent irregular movement of the planets and their variable distances from the Earth—which cannot be understood as occurring in circles homocentric with the Earth—make it clear that the Earth is not the centre of their circular movements. Therefore, since there are many centres, it is not foolhardy to doubt whether the centre of gravity of the Earth rather than some other is the centre of the world. I myself think that gravity or heaviness is nothing except a certain natural appetency implanted in the parts by the divine providence of the universal Artisan, in order that they should unite with one another in their oneness and wholeness and come together in the form of a globe. It is believable that this affect is present in the sun,

80. See note 43.
81. See section 4, paragraph 4.

moon, and the other bright planets and that through its efficacy they remain in the spherical figure in which they are visible, though they nevertheless accomplish their circular movements in many different ways. Therefore if the Earth too possesses movements different from the one around its centre, then they will necessarily be movements which similarly appear on the outside in the many bodies; and we find the yearly revolution among these movements. For if the annual revolution were changed from being solar to being terrestrial, and immobility were granted to the sun, the risings and settings of the signs and of the fixed stars—whereby they become morning or evening stars—will appear in the same way; and it will be seen that the stoppings, retrogressions, and progressions of the wandering stars are not their own, but are a movement of the Earth and that they borrow the appearances of this movement. Lastly, the sun will be regarded as occupying the centre of the world. And the ratio of order in which these bodies succeed one another and the harmony of the whole world teaches us their truth, if only—as they say—we would look at the thing with both eyes.

68 ON SAVING THE PHENOMENA AND THE REALITY OR UNREALITY OF EPICYCLES AND ECCENTRICS

Introduction by Edward Grant

The following selections from Moses Maimonides, Bernard of Verdun, and John Buridan reflect a controversy about the role of epicycles and eccentrics in astronomy and cosmology that is traceable to the Arabs and beyond them to the Greeks, especially Posidonius, Geminus, and Ptolemy. Here is Duhem's characterization of the problem:[1]

One could conceive the hypotheses of astronomy as simple mathematical fictions which the geometer combines in order to render the celestial movements accessible to his calculations. Or one could view them as the description of concrete bodies, of movements actually realized. In the first case, a single condition is imposed on these hypotheses, that of *saving the appearances*; in the second case, the freedom of the one who imagines these hypotheses is very narrowly limited. If he is one who is versed in a philosophy that pretends to know something of the celestial essence, he will necessarily express his hypotheses in accordance with the precepts of that philosophy.

On the subject of astronomical hypotheses, the author of the *Almagest*, and the Hellenic thinkers who came after him, adopted the first of these two opinions. They could, therefore, construct their geometric theories without concerning themselves about the different physics about which they disputed; they could choose their suppositions without troubling themselves if agreement was lacking between the results of their calculations and the data of observation.

On the contrary, however, after the author of the *Hypotheses [of the Planets]*,[2] with Thabit ibn Qurrah and Ibn al Haitham [Alhazen], Arab astronomers desired that the hypotheses imagined by them should correspond to the true movements of really existing solid or fluid bodies. Hence they rendered these hypotheses amenable to the laws assumed by physics.

But the physics professed by most of the philosophers of Islam was peripatetic [that is, Aristotelian] physics, the physics that Xenarchus and Sosigenes had long ago opposed to the astronomy of eccentrics and epicycles showing that the reality of the latter could not be reconciled with the truth of the former. The realism of Arab astronomers would necessarily provoke the Peripatetics of Islam into an ardent and merciless quarrel against the doctrines of the *Almagest*.

In the twelfth century, we see this quarrel—led vigorously by the most distinguished Arab thinkers, Ibn Bâdja (Avempace), Ibn Tofail (Aboubekr), Ibn Roshd (Averroes), Moses ben Maimon (Maimonides)—produce an astronomical system, the system of al-Bitruji (Alpetragius), which, until the sixteenth century, will attempt unceasingly to be substituted for the system of Ptolemy.

1. My translation from *Le Système du monde* (Paris: Hermann, 1913–1959), II, 130–131.
2. Ptolemy himself was author of the *Almagest* and *Hypotheses of the Planets*. Thus in the former treatise he was primarily saving the astronomical phenomena by use of eccentrics and epicycles in a purely geometric manner without concern for the mechanical relations between these geometric devices. In the *Hypotheses*, however, Ptolemy is concerned exclusively with actual physical epicyclic and eccentric spheres and the manner in which they are physically located one within the other.

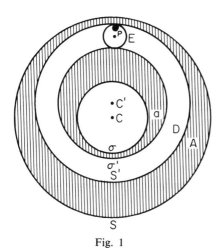

Fig. 1

Figure 1[3] conveys something of Ptolemy's effort to conceive physical eccentrics and epicycles. Almost all subsequent efforts of a similar kind were based on Ptolemy's account. The two spherical surfaces S and σ have as their center C, the center of the world. These constitute the *sphere of the planet*. The two spherical surfaces S' and σ' have as their common center C', the center of the eccentric. What lies between S' and σ', namely D, comprises the *deferent orb*. The epicyclic sphere, E, is located in D between S' and σ'. Thus the epicyclic sphere has a diameter equal to the thickness of the deferent orb. Set in the sphere of the epicycle is the planet P. The solid masses A and a, which are enclosed by the surface S, S' and σ', σ, respectively, are each moved with two rotations around axes passing through C, the center of the world. One of these rotations, the daily motion, is effected from east to west around the poles of the center of the world;

the other rotation is very slowly from west to east around the poles of the ecliptic. Thus double rotation is communicated by A and a to D, the deferent orb, which is itself carried in a third motion around C', the center of the eccentric planet. All these motions are conceived as communicated to E, the epicyclic sphere, which, in turn, rotates around an axis passing through its center. The great circles, epicycles, and eccentrics that are treated in Ptolemy's *Almagest* are derived from this three dimensional cosmological description as follows: (1) By the rotation of the epicyclic sphere, the planet would describe a great circle on it which is the epicycle circle; (2) an eccentric circle, or deferent, is described as the center of the epicyclic sphere carried round by the deferent orb; and (3) the epicyclic sphere is subject to an oscillatory movement by means of which one can explain the inclination of the plane of the epicycle on the plane of the eccentric circle.

In the following selections it will be seen that Maimonides presents a thoroughly Aristotelian viewpoint, Bernard of Verdun argues for the existence of epicycles and eccentrics but partially justifies them by use of Aristotelian principles, whereas Buridan, in a more qualified position, emphatically denies the existence of epicycles but seems to argue for the physical existence of eccentrics. It will be evident that in their reports of the arguments of others and in their attacks against epicycles and eccentrics, Maimonides and Averroes strongly influenced the history of this problem and the responses to it. Helping to prolong the dispute was the fact that epicycles and eccentrics had not yet been devised in Aristotle's lifetime, thereby allowing the disputants to justify their positions by direct appeal to Aristotelian principles.

1. Moses Maimonides (1135–1204): The Reality of Epicycles and Eccentrics Denied

Translated by Shlomo Pines[4]

Annotated by Edward Grant

You know of astronomical matters what you have read under my guidance and understood from the contents of the "Almagest." But there was not enough time to begin another speculative study with you. What you know already is that as far as the action of ordering the motions and making the

tion and Notes by Shlomo Pines with an introductory essay by Leo Strauss (Chicago: University of Chicago Press, 1963), pp. 322–327. *The Guide of the Perplexed*, of which the present selection constitutes the whole of Book II, chapter 24, was written originally in Arabic between 1187 and 1190 and translated into Hebrew by Samuel ibn Tibbon at Arles (France) in 1204. From the latter translation, a Latin translation was made during the first half of the thirteenth century. The *Guide* was well known and influential in the Latin West, being cited by Thomas Aquinas and very likely conveying some of the arguments and opinions to which Bernard of Verdun and John Buridan reacted.

3. Reproduced from Duhem, *Le Système du monde*, II, 91. The description follows Duhem.

4. Reprinted with the kind permission of the University of Chicago Press from *The Guide of the Perplexed, Moses Maimonides*, translated with an Introduc-

course of the stars conform to what is seen is concerned, everything depends on two principles: either that of the epicycles or that of the eccentric spheres or on both of them. Now I shall draw your attention to the fact that both those principles are entirely outside the bounds of reasoning and opposed to all that has been made clear in natural science. In the first place, if one affirms as true the existence of an epicycle revolving round a certain sphere, positing at the same time that that revolution is not around the center of the sphere carrying the epicycles—and this has been supposed with regard to the moon and to the five planets—it follows necessarily that there is rolling, that is, that the epicycle rolls and changes its place completely. Now this is the impossibility that was to be avoided, namely, the assumption that there should be something in the heavens that changes its place. For this reason Abū Bakr Ibn al-Sā'igh[5] states in his extant discourse on astronomy that the existence of epicycles is impossible. He points out the necessary inference already mentioned. In addition to this impossibility necessarily following from the assumption of the existence of epicycles, he sets forth there other impossibilities that also follow from that assumption. I shall explain them to you now.

The revolution of the epicycles is not around the center of the world. Now it is a fundamental principle of this world that there are three motions: a motion from the midmost point of the world, a motion toward that point, and a motion around that point. But if an epicycle existed, its motion would be neither from that point nor toward it nor around it.

Furthermore, it is one of the preliminary assumptions of Aristotle in natural science that there must necessarily be some immobile thing around which circular motion takes place. Hence it is necessary that the earth should be immobile. Now if epicycles exist, theirs would be a circular motion that would not revolve round an immobile thing. I have heard that Abū Bakr has stated that he had invented an astronomical system in which no epicycles figured, but only eccentric circles.[6] However, I have not heard this from his pupils. And even if this were truly accomplished by him, he would not gain much thereby. For eccentricity also necessitates going outside the limits posed by the principles established by Aristotle, those principles to which nothing can be added. It was by me that attention was drawn to this point. In the case of eccentricity, we likewise find that the circular motion of the spheres does not take place around the midmost

point of the world, but around an imaginary point that is other than the center of the world. Accordingly, that motion is likewise not a motion taking place around an immobile thing. If, however, someone having no knowledge of astronomy thinks that eccentricity with respect to these imaginary points may be considered—when these points are situated inside[7] the sphere of the moon, as they appear to be at the outset—as equivalent to motion round the midmost point of the world, we would agree to concede this to him if that motion took place round a point in the zone of fire or of air, though in that case that motion would not be around an immobile thing. We will, however, make it clear to him that the measures of eccentricity have been demonstrated in the "Almagest" according to what is assumed there. And the latter-day scientists have given a correct demonstration, regarding which there is no doubt, of how great the measure of these eccentricities is compared with half the diameter of the earth, just as they have set forth all the other distances and dimensions. It has consequently become clear that the eccentric point around which the sun revolves must of necessity be outside the concavity of the sphere of the moon and beneath the convexity of the sphere of Mercury. Similarly the point around which Mars revolves, I mean to say the center of its eccentric sphere, is outside the concavity of the sphere of Mercury and beneath the convexity of the sphere of Venus. Again the center of the eccentric sphere of Jupiter is at the same distance—I mean between the sphere of Mercury and Venus. As for Saturn, the center of its eccentric sphere is between the spheres of Mars and Jupiter. See now how all these things are remote from natural speculation! All this will become clear to you if you consider the distances and dimensions, known to you, of every sphere and star, as well as the evaluation of all of them by means of half the diameter of the earth so that everything is calculated according to one and the same proportion and the eccentricity of every sphere is not evaluated in relation to the sphere itself.

Even more incongruous and dubious is the fact that in all cases in which one of two spheres is inside the other and adheres to it on every side, while the centers of the two are different, the smaller sphere can move inside the bigger one without

5. That is, Ibn Bajja, or Avempace (see Selection 44).
6. This view is discussed by Buridan (see section 3 below).
7. That is, beneath.

the latter being in motion, whereas the bigger sphere cannot move upon any axis whatever without the smaller one being in motion. For whenever the bigger sphere moves, it necessarily, by means of its movement, sets the smaller one in motion, except in the case in which its motion is on an axis passing through the two centers. From this demonstrative premise and from the demonstrated fact that vacuum does not exist and from the assumptions regarding eccentricity, it follows necessarily that when the higher sphere is in motion it must move the sphere beneath it with the same motion and around its own center. Now we do not find that this is so. We find rather that neither of the two spheres, the containing and the contained, is set in motion by the movement of the other nor does it move around the other's center or poles, but that each of them has its own particular motion. Hence necessity obliges the belief that between every two spheres there are bodies other than those of the spheres. Now if this be so, how many obscure points remain? Where will you suppose the centers of those bodies existing between every two spheres to be? And those bodies should likewise have their own particular motion. Thābit[8] has explained this in a treatise of his and has demonstrated what we have said, namely, that there must be the body of a sphere between every two spheres. All this I did not explain to you when you read under my guidance, for fear of confusing you with regard to that which it was my purpose to make you understand.

As for the inclination and deviation that are spoken of regarding the latitude of Venus and Mercury, I have explained to you by word of mouth and I have shown you that it is impossible to conceive their existence in those bodies. For the rest Ptolemy has said explicitly, as you have seen, that one was unable to do this, stating literally: No one should think that these principles and those similar to them may only be put into effect with difficulty, if his reason for doing this be that he regards that which we have set forth as he would regard things obtained by artifice and the subtlety of art and which may only be realized with difficulty. For human matters should not be compared to those that are divine.[9] This is, as you know, the text of his statement. I have indicated to you the passages from which the true reality of everything I have mentioned to you becomes manifest, except for what I have told you regarding the examination of where the points lie that are the centers of the eccentric circles. For I have never come across

anybody who has paid attention to this. However this shall become clear to you through the knowledge of the measure of the diameter of every sphere and what the distance is between the two centers as compared with half the diameter of the earth, according to what has been demonstrated by al-Qabīsī[10] in the "Epistle Concerning the Distances." If you examine those distances, the truth of the point to which I have drawn your attention will become clear to you.

Consider now how great these difficulties are. If what Aristotle has stated with regard to natural science is true, there are no epicycles or eccentric circles and everything revolves round the center of the earth. But in that case how can the various motions of the stars come about? Is it in any way possible that motion should be on the one hand circular, uniform, and perfect, and that on the other hand the things that are observable should be observed in consequence of it, unless this be accounted for by making use of one of the two principles[11] or of both of them? This consideration is all the stronger because of the fact that if one accepts everything stated by Ptolemy concerning the epicycle of the moon and its deviation toward a point outside the center of the world and also outside the center of the eccentric circle, it will be found that what is calculated on the hypothesis of the two principles is not at fault by even a minute. The truth of this is attested by the correctness of the calculations—always made on the basis of these principles—concerning the eclipses and the exact determination of their times as well as of the moment when it begins to be dark and of the length of time of the darkness. Furthermore, how can one conceive the retrogradation of a star, together with its other motions, without assuming the existence of an epicycle? On the other hand, how can one imagine a rolling motion in the heavens or a motion around a center that is not immobile? This is the true perplexity.

However, I have already explained to you by word of mouth that all this does not affect the astronomer. For his purpose is not to tell us in which way the spheres truly are, but to posit an astronomical system in which it would be possible for the motions to be circular and uniform and to

8. Thabit ibn Qurra, a pagan Arab astronomer and mathematician who died in 900.
9. *Almagest* XIII.2.
10. A tenth-century astronomer.
11. Epicycles and eccentric circles are the two principles.

correspond to what is apprehended through sight, regardless of whether or not things are thus in fact.[12] You know already that in speaking of natural science, Abū Bakr Ibn al-Sā'igh expresses a doubt whether Aristotle knew about the eccentricity of the sun and passed over it in silence— treating of what necessarily follows from the sun's inclination, inasmuch as the effect of eccentricity is not distinguishable from that of inclination—or whether he was not aware of eccentricity. Now the truth is that he was not aware of it and had never heard about it, for in his time mathematics had not been brought to perfection. If, however, he had heard about it, he would have violently rejected it; and if it were to his mind established as true, he would have become most perplexed about all his assumptions on the subject. I shall repeat here what I have said before.[13] All that Aristotle states about that which is beneath the sphere of the moon is in accordance with reasoning; these are things that have a known cause, that follow one upon the other, and concerning which it is clear and manifest at what points wisdom and natural providence are effective. However, regarding all that is in the heavens, man grasps nothing but a small measure of what is mathematical; and you know what is in it. I shall accordingly say in the manner of poetical preciousness: *The heavens are the heavens of the Lord, but the earth hath He given to the sons of man.*[14] I mean thereby that the deity alone fully knows the true reality, the nature, the substance, the form, the motions, and the causes of the heavens. But He has enabled man to have knowledge of what is beneath the heavens, for that is his world and his dwelling-place in which he has been placed and of which he himself is a part. This is the truth. For it is impossible for us to accede to the points starting from which conclusions may be drawn about the heavens; for the latter are too far away from us and too high in place and in rank. And even the general conclusion that may be drawn from them, namely, that they prove the existence of their Mover, is a matter the knowledge of which cannot be reached by human intellects.[15] And to fatigue the minds with notions that cannot be grasped by them and for the grasp of which they have no instrument, is a defect in one's inborn disposition or some sort of temptation. Let us then stop at a point that is within our capacity, and let us give over the things that cannot be grasped by reasoning to him who was reached by the mighty divine overflow so that it could be fittingly said of him: *With him do I speak mouth to mouth.*[16] That is the end of what I have to say about this question. It is possible that someone else may find a demonstration by means of which the true reality of what is obscure for me will become clear to him. The extreme predilection that I have for investigating the truth is evidenced by the fact that I have explicitly stated and reported my perplexity regarding these matters as well as by the fact that I have not heard nor do I know a demonstration as to anything concerning them.

2. Bernard of Verdun (fl. late thirteenth century): The Reality of Epicycles and Eccentrics Affirmed

Translated and annotated by Edward Grant[17]

Distinction III
On Explaining the Causes of the Apparent Irregularities in the Sun's Motion

CHAPTER 2

The second chapter will set forth some things that are necessary for showing that in many planets *(stellis)* and the sun it is necessary to assume an eccentric orb or a concentric [orb] with an epicycle whose radius is equal to the aforementioned eccentricity [Fig. 2]. According to *On the Heavens,* Book II,[18] the irregularity of motion is not in the celestial bodies [themselves], for if an irregularity

12. An attitude that was widely prevalent in the Middle Ages, perhaps wholly, or in part, the result of this statement, but which goes back ultimately to Simplicius' *Commentary on Aristotle's Physics,* and even beyond, to Geminus and Posidonius (see Cohen and Drabkin, *A Source Book in Greek Science* [Cambridge, Mass.: Harvard University Press, 1948], pp. 90–91). Buridan repeats it (see near the end of Question 10 in section 3.)

13. Compare to Book II, chapter 22.

14. Psalms 115:16.

15. In a note Pines explains (p. 327, n. 12), "In Ibn Tibbon's translation the passage has a different meaning: 'The general proof from them is that they indicate the existence of their Mover, but the knowledge of other matters concerning them cannot be reached by human intellects.' "

16. Numbers 12:8.

17. Translated from Bernard of Verdun's *Tractatus super totam astrologiam* edited by P. Polykarp Hartmann, O.F.M. (Werl, Westphalia: Dietrich-Coelde Verlag. 1961), pp. 67–72.

18. Aristotle, *De caelo* II.6 (288a. 13—289a.10). In this chapter Aristotle shows that there are no irregularities in celestial motions.

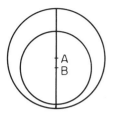
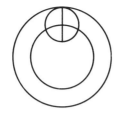

Fig. 2

were in them, there would be an intension and remission [of motion][19] in the beginning and end; indeed a natural motion is intensified at the end. But this is repugnant [when applied] to celestial circulation. Or such irregularity will result from a variation in the motive power, which is not true, because such movers are separate and do not undergo any change of power, nor does the substance of a [celestial] mobile change, since it is wholly immune from change. Furthermore, an intension and remission [of motion] occurs because of an impotence. Now, an impotence is contrary to nature, for nature intends what is best. Nothing is contrary to nature there [in the celestial regions], since celestial bodies are simple and in their proper region and nothing is contrary to them.

A second thing that is set forth is that in the planets, and especially in the moon, there seems to occur an advance *(accessus)* toward the center of the earth and a withdrawal *(recessus)* from the center by the same [celestial body] according to a greater or lesser amount. . . .

A third thing that is set forth is that in the motion of the sun and all the planets there appears an irregularity in speed and slowness, and all admit this, so that it is unnecessary to delay on this point.

CHAPTER[20] 4

In this fourth chapter a twofold way of saving the previously mentioned appearances is posited; the first is excluded and the second is inferred to be necessary.[21] All who speak reasonably on this subject say there is only a twofold way of saving the aforesaid phenomena. One way is touched upon by Averroes in Book XI of the *Metaphysics*,[22] where he says that true astronomy is based upon natural principles that destroy eccentrics and epicycles, which can happen if one orb is moved on many poles so that these motions are composed of many motions. Yet he confesses that he cannot explain this astronomy. Thus it seems to some that Alpetragius (al-Bitruji), following him and perhaps aroused by these things, developed the roots of

Averroes into beautiful branches, flowers, and fruits.[23] For he assumes that the planets are moved with particular motions on poles other than the poles of the first mobile, namely on the poles of the zodiac. He assumes that they are moved on inclined circles whose poles are moved on the circumference of a small circle around or near the poles of the zodiac. Thus the planets are moved against the order of the signs through one half [of the circle] and with the order of the signs in the other half. And when the arc that runs counter to the order of the signs has been traversed and it corresponds to more of the zodiac than the orb of the planet will have traversed in revolving around the poles of the zodiac, a planet will appear retrograde; when these two are equal, the planet will appear stationary; and when the arc that has been traversed in the contrary order of the signs is less than the part traversed by the planet along the zodiac, the planet will have a direct movement but at a slow speed. However, when it moves in the order of the signs it appears to move quickly. And for every planet each of these events occurs in the very same point of the small circle.[24]

19. That is, the velocity of celestial bodies would increase and decrease. This is denied by Aristotle.

20. This is actually chapter 3, but an error lists it as chapter 4. The next will be listed as 5, but is actually 4.

21. Rashly assuming that his two alternatives are exhaustive, Bernard believes that he has arrived at a necessary truth by elimination. This mode of reasoning is frequently encountered in medieval treatises (see Selection 51, n. 30; after eliminating four theories, Bradwardine considers a fifth as necessarily true).

22. Actually Book XII, Comment 45, of the *Metaphysics*. These arguments are mentioned in the next section (see n. 49).

23. Duhem noted (*Le Système du monde*, III, 447) that all of chapter 4 to this point was taken substantially, though not verbatim, from Roger Bacon, *Epistola ad Clementem IV in Opus maius*, Part II, ch. 2.

24. The *De motibus celorum (On the Celestial Motions)* of al-Bitruji, in which this interpretation is found, was translated from Arabic into Latin by Michael Scot in 1217. Apparently following the path of Averroes, al-Bitruji opposed Ptolemy and denied epicycles and eccentrics. It will be useful to quote part of J. L. E. Dreyer's description of this system (*A History of Astronomy from Thales to Kepler*, 2d ed. [New York: Dover, 1953], pp. 265–266):

"The object of this system was to explain the constitution of the universe as it really is, and not merely to represent the motions of the planets geometrically, so as to be able to foretell their places in the heavens at any time; and the author (be he Ibn Tofeil or Al Betrugi *alias* Alpetragius) specially disclaims any intention of testing the theory by comparing it with observations or of accounting for minor details of the motions. The

This position is seen to be impossible in many ways. For when any point of the sphere has been moved, the whole sphere and any point of it have been moved except the two points which are the poles of motion. Therefore, when the poles will have traversed a fourth part of the small circle on which they are revolved, according to Alpetragius, the whole sphere will have traversed a fourth of the distance of its revolution; similarly when the whole [small] circle is traversed by the poles, the whole sphere will traverse its whole distance. For when the line coming from the center of the said small circle through the pole of the sphere, with whose motion the planet is moved, has terminated at the planet, not only when it has traversed the small circle does it also traverse the arc of the zodiac intercepted between the lines touching the small circle and coming from the pole of the world and terminating in the zodiac, but it traverses the whole zodiac. From all this, the great laughability of this position is obvious.[25]

Again, in this theory, direct and retrograde arcs would always be equal, and stationary points always the same. All these things are wholly false.

Again, direct and retrograde motions would be quickest when a planet would be at its maximum latitude; the opposite of this is certain and is shown below. Again, the sun would have a latitude, which is wholly false.

Again, if this way were true, the things we premised in the preceding chapter, which are very obvious to any one capable of reasoning, would be destroyed. Although many arguments could be brought against this position, these should suffice.

From all this, the first way is impossible and insufficient for saving the things shown before, which anyone capable of reasoning necessarily has to assume. The second way is left and therefore is necessary, namely the way which assumes eccentric and epicycle at the same time and a plurality of orbs around different poles of revolution in the planets other than the sun. In the sun [there is needed] only an eccentric, which is more likely, or a concentric with an epicycle. By these means, all the aforementioned absurdities are avoided and all the things mentioned in the preceding chapter are saved. Maintaining this from the beginning [of astronomical investigation], all the things which we happen to know about celestial motions and the distances and magnitudes of celestial bodies were investigated truly up to this day. If the basis of all this was false, this could not have happened. For, in any thing whatever, a small error at the beginning is great at the end.

Moreover, all the celestial phenomena are consonant with this way and in contradiction with the

leading idea is that of the homocentric spheres, each star being attached to a sphere, and the motive power is the ninth sphere, the sphere outside that of the fixed stars. The Spanish philosopher ought therefore to have been content with the system of Eudoxus or its modification by Aristotle (whom he never mentions by name, but only as "the sage"), but unfortunately he became possessed with the notion that the prime mover must everywhere produce only a motion from east to west, and he had therefore to reject the independent motion of the planets from west to east, and revert to the old Ionian idea that the seven planets merely perform the daily revolution with a speed slightly slower than that of the fixed stars. The true speed of the *primum mobile* is a little faster than this; the eighth sphere performs a revolution in a slightly longer period (24 hours), and the effect of the prime mover is gradually weakened more and more, with increasing distance, until we find the sphere of the moon, being furthest from the prime mover, taking nearly twenty-five hours to complete a revolution. This was the old primitive Ionian idea, but Al Betrugi (or his teacher) saw that this was not sufficient, as not only is the pole of the ecliptic different from that of the equator, which prevents the planets from moving in closed orbits, but the planets do not even keep at the same distance from the pole of the ecliptic but have each their motion in latitude, as well as a variable velocity in longitude; and all this had yet to be accounted for. The ninth sphere has but one motion, but the eighth has two, that in longitude (precession) and another which is caused by the pole of the ecliptic describing a small circle round a mean position thereby producing the supposed oscillation or trepidation of the equinoxes. Similarly, the pole of each planet describes a small circle round a mean position (i.e. the pole of the ecliptic), thereby producing inequalities in longitude and motion in latitude. Whenever the actual orbit-pole of a planet is on the parallel of the mean pole, it is obvious that the planet will perform its daily revolution with its mean velocity, while the velocity is increased or lessened when the actual pole is respectively at its minimum or maximum distance from the pole of the heavens (the motion of the pole of the orbit being added to or subtracted from the motion of the planet), so that the epicycle is hereby rendered superfluous."

Dreyer calls this a retrograde step in the history of astronomy. Duhem believed that Bernard of Verdun did not fully understand al-Bitruji's system and made unjust criticisms (see *Le Système du monde,* III, 449,450). No evaluation of al-Bitruji's system or Bernard's criticism will be attempted here. The Latin text has been edited by F. J. Carmody, *Al-Bitruji De motibus celorum* (Berkeley: University of California Press, 1952).

It is of considerable interest to realize that the only other contending theory is not the vague, qualitative Aristotelian system of homocentric spheres described in the *Metaphysics,* but al-Bitruji's mathematically formulated alternative.

25. The substance and meaning of this criticism elude me.

first way. Therefore, it is necessary that the premises stand and be conceded with the same necessity as the celestial motions and the whole of nature, and so on. To deny the things that are very certain in every argument *(omni ratione)* because of certain sophistic arguments is as foolish as are the follies of the ancients who falsely denied motion, all transmutations, and a plurality of beings, the contrary of which is apparent to sense alone. For these things cannot be demonstrated, just as one cannot demonstrate that fire is hot or that substance and accident are in existent things; but we accept this by the senses. Thus the Philosopher [Aristotle] says that we do not have to dignify every argument, and he says it is absurd to seek a reason for these things. All our reasoning always presupposes the senses.[26]

CHAPTER 5

In this fifth chapter are presented the arguments of those who attack the way stated above and the response to them. For they argue that if there is an eccentric or epicyclic orb then either the body of the sky will be divisible or rarefiable or there will be a vacuum, or two bodies in the same place. Indeed the part of the eccentric farthest from the earth will take the place of the part of the eccentric nearest the earth, and conversely. Similarly an epicycle that is revolved through an orb will cut it.

But it should be replied that the first difficulty is avoided because an eccentric is not assumed to be revolved around the center of the world, except with an accidental motion, but around its own center, so that its different parts succeed themselves continually in the points or places of the farther and nearer longitude that are imagined in the convexity of the orb surrounding the eccentric.

Similarly, the second difficulty is avoided because we do not assume that the center of the epicycle is moved on the circumference of the deferent in reality *(secundum veritatem),* but we assume that the epicycle is a small sphere located between the thickness of the [eccentric] orb and is moved with the motion of the orb in which the planet is immediately located.[27] And, as will be seen, this sphere embraces several spheres required for saving the different appearances of such a motion. Nor does it seem absurd if the motion of superior [planets] is distributed to the lower spheres. I say this about the motion of the center of the epicycle inclined in the orb of the eccentric, [that is,] the motion of such lower eccentrics is the

aggregate of their own proper motions and the motions of all the higher [or upper] eccentrics; and the inclination *(declivitas)* of their motions [is the aggregate of] their own proper inclinations and that of the higher [spheres]. For example, if in the sun, in which no inclination or latitude appears, a latitude or inclination were to appear, an epicycle carrying the sun should be assumed revolving in latitude in order to compensate for the latitude that is to be offset. Or a multitude of great spheres could be assumed, making the sun accord with its appearances; or if this does not seem suitable, immobile spheres could be posited [between the different orbs] that are not of the same thickness in their parts, so that it would not be necessary to assume that the motion of the lower spheres is composed of the motions of all the higher spheres[28] and it would not be necessary to assume a division of the spheres and a penetration or vacuity. Nor are these things assumed superfluously, since the motions of the other [spheres] makes them necessary. Thus we can understand that in all natural things, some things are in them for their own sake, and some because of other things only. And it is the same way in artificial things; for all the parts of a cither do not produce sound, and yet if the parts that produce no sound were not in the cither, the others would not produce sound. Also in animals there are bones that are immobile per se, in which other bones are revolved.

Furthermore, even the beauty and perfection of creatures consists more in variety, numerousness, and proportionality than in complete uniformity; and just as the beauty *(formositas)* of the starry orb consists in the variety and numerousness of the stars, so [does the beauty] of the lower [celestial bodies or planets consist] in the variety and numerosity of their motions and orbs, as is declared in Book II of *On the Heavens.*[29] In the second place, one may argue this way: If the moon and any planet are rotated by means of the motion of an epicycle, then the same part of the body of that planet does not always face the earth, the contrary of which is shown by Aristotle regarding the moon, whose spot always appears to us in the same shape [or form].[30]

26. The dependence of induction on the senses is emphasized in *Posterior Analytics* I.18.81a.38—81b.9.
27. See the figure in the Introduction to this selection.
28. The immobile spheres between the orbs would not transmit motions from the outer orbs to the inner orbs.
29. 292b.25—293a.14.
30. Aristotle argues (*On the Heavens* II.290a.25–27) that since the moon always shows its same face to us, it

I reply that if this were so, it [that is, the moon or any planet] is countered [or brought back *(reducitur)*] by the motion of some of the orbs containing the epicycle, in whatever way this would be required.[31] Moreover, granting that the same part of the body of a planet does not always face the earth, yet it could, nevertheless, always appear to us in the same shape [or form] because we assume that we see every part of the moon's body, since it is like a solid spherical body of great transparency which extends to every part of its depth and in the interior of it there is a spot of spherical shape. Therefore, however this body may be revolved, the spot always appears to us in the same shape [or form].[32]

Furthermore, every natural motion is away from a center, toward a center, or around a center. It must be said that since the motion of eccentrics and epicycles are circular, they move around a center, since they move around their own centers. Thus the Philosopher, at the beginning of *On the Heavens*[33] divides local motion in two ways, namely into rectilinear and circular motion, since every motion is straight or circular or a mixture of these. For these are the only two simple motions, since these magnitudes[34] are the only simple magnitudes. And this includes the circle, which is around a center; and thus he calls circular motion and motions around a center the same. Hence the center *(medium)* around which the prime motion [of the outermost sphere] occurs, and takes with it all the orbs, is the absolute center [of the world]. And the centers *(media)* around which the lower orbs are moved are centers in a relative sense *(secundum quid)*. At present, what we have said about these matters is sufficient.

3. John Buridan (ca. 1300—ca. 1358): An Intermediate Position: Epicycles Denied, Eccentrics Affirmed

Translated and annotated by Edward Grant[35]

Book XII

QUESTION 10

In the tenth question, we inquire whether epicycles are to be assumed in celestial bodies.

First it is argued that they are not to be assumed,[36] by the authority of the Commentator [Averroes], who expressly tries to disprove them. Next by the authority of Aristotle, who, in enumerating the celestial motions and spheres in this twelfth book, did not enumerate epicycles. Indeed even in the second book of *On the Heavens and World*, in saving the irregularities *(difformitates)* of the planetary motions that are apparent to us, he says that these irregularities arise because several motions are brought together in the same mobile. Thus he does not assume that they arise by eccentrics or epicycles.

Likewise, the Commentator argues that every celestial motion ought to be a simple motion and every simple motion is (1) away from the center of the world, or (2) toward the center of the world, or (3) around the center of the world, as is taught in the first book of [Aristotle's] *On the Heavens*.[37] Therefore, every celestial motion must be one of these [three possibilities]. But if the motion of an epicycle were assumed, it would be none of these. For it would not be away from the center, because cannot be said to roll or revolve. This argument is ex-

tended to all celestial bodies. This second objection against epicycles may have been drawn from Roger Bacon's *Opus Tertium* or the *Communia naturalium*, Book II.

31. Bernard argues that the planets could be carried on epicycles and yet always present the same face to us if only we understand that in the total machinery of spheres and orbs for each planet there are spheres that would counter the epicycle's motion and allow us to see only the same face. Bernard offers no technical details for achieving this.

32. In this interesting response, Bernard assumes that the moon is a transparent sphere with a spherical opaque spot lying at its center. Thus however the moon may turn or rotate, we will see the spot embedded in the transparent moon and, to a terrestrial observer, always possessed of the same shape.

33. I.2.268b.17f.

34. That is, straight line and circle.

35. From Buridan's *Questions on the Metaphysics of Aristotle* as published in *In Metaphysicen Aristotelis Questiones argutissimae Magistri Ioannis Buridani in ultima praelectione ab ipso recognitae et emissae* . . . (Paris, 1518; Frankfurt: Minerva G.M.B.H., 1964), fols. 73 recto to 74 recto.

36. By first stating the arguments for the negative position and subsequently adopting it, Buridan diverges from the usual pattern of a *questio* (for a similar instance, see Selection 31, n. 7). Despite a rejection of epicycles in this question, Buridan allows for a qualified acceptance of them at the conclusion of the next question (see n.54).

37. This threefold Aristotelian analysis of motion is also repeated by Maimonides (see the beginning of section 1) and Bernard of Verdun (near the end of section 2).

that motion would be the motion of light bodies, namely an upward motion; nor would they move toward the center, because that motion is the motion of heavy bodies. Nor would it move around the center of the world, because this epicycle is not around the earth [that is, does not have the earth as its center]. Indeed it is above [the earth] in an orb and in this orb it has both its own center and circumference. Therefore, such an epicycle must not be assumed. Moreover, every celestial motion ought to be circular and, consequently, must be around something [that is, around a physical center]. But the motion of this epicycle would not be around anything, since it does not have a center—except an indivisible [center or point] around which it is moved. Such an indivisible [entity] is not a natural center. Indeed it is nothing. Hence this motion is not around anything.

Nevertheless, Ptolemy and all modern astronomers (astrologi) assume the opposite. For otherwise the appearances of the planets could not be saved, especially their approaches (approximationes) and recedings (elongationes) from the earth which we obviously and notably experience. Thus it is that sometimes the moon is so high that it cannot be totally eclipsed, but the cone of the earth's shadow falls onto the middle of it, so that while it is seen to have been eclipsed in the middle, the circumference remains in light. And sometimes the moon is so near to the earth that it falls deeply into the earth's shadow and remains totally eclipsed.

It must be understood that in the world the natural center is this earth. Therefore, an indivisible center is not assumed except in the imagination. But nevertheless a point can be imagined in the center of the earth as if it were the center of the world, and then all spheres having their centers in the center of the earth are called concentric. But in all the spheres of the planets, the moderns assume a sphere which they call eccentric and which is wholly around the earth; it does not have its center in the center of the world, but outside it. And for this reason, it is called eccentric. Furthermore, they assume little spheres (spherulas), which they call epicycles, that are fixed in the width of the eccentric. The planet is fastened on these epicycles. But then it is necessary that the whole sphere of the planet be concentric, for otherwise there would necessarily be a vacuum between the total spheres of two planets. Indeed it is necessary to imagine two spheres of very nonuniform magnitude, [one] above the eccentric, the other below the eccentric,

and each of them is very wide on one side and on the other side of virtually no magnitude.[38] Then it is necessary to imagine that this whole sphere composed of all the things just mentioned is moved simultaneously with a daily motion from east to west, carrying the eccentric with it; but the eccentric is also moved on the poles of the zodiac between the two parts [or sides of the whole sphere] in a motion contrary to the daily motion, namely from west to east. Then, thirdly, the epicycle, carrying the planet, is moved on its [own] proper center in this eccentric.

After saying all this, we must let it be known how this appears to us. It seems to me very probable that epicycles should not be assumed, because if an epicycle is not posited in the orb of the moon it ought not to be posited in the orb of the other planets, since all the reasons which apply to the other planets should also apply to the moon; and should anyone assume epicycles on the spheres of the other planets, he should similarly assume an epicycle on the sphere of the moon. Therefore, if it were shown that an epicycle should not be assumed in the [orb of the] moon, it could be concluded that no such epicycles should be assumed.

Then I argue that it ought not to be assumed for the moon, because then it would follow that in that spot (macula) of the moon which appears as if it were an image of a man whose feet always appear to be below [or toward the bottom], the feet would sometimes appear above [in the upper part of the moon]. But we experience that this consequent is false, since this image always appears situated in the same way with respect to us, on the assumption that it appears to us in the same part of the sky, for example, on the meridian. But I prove the principal consequent,[39] because if the feet appear to us below when the moon is in the aux [that is, apogee] of the epicycle, it should follow that when it reaches the opposite of of the aux [that is, perigee] the feet should appear above, since by the motion of the epicycle the moon has traversed [a distance such that] the part of the moon which previously was above is now below.

It must be understood that there is no escape in

38. The figure in the introduction to this selection illustrates Buridan's description.

39. Buridan wishes to prove that if a lunar epicycle were assumed, it must follow that the feet of the man in the moon (or the spot on the moon) will sometimes appear in the upper part and sometimes in the lower part of the moon. For Bernard of Verdun's denial of this argument, see near the end of section 2, and note 32.

resolving this argument except in one way only: namely by saying that just as this epicycle is moved around its proper center, so also is the body of the moon moved around its proper center, in a motion contrary to that of the epicycle and with an equal speed. Thus the epicycle completes its circulation in the same time as the moon completes its circulation. It is certain that if this conception *(imaginationem)* were true, the argument would be solved, since the part of the moon which was previously above will always remain above. But there is an objection against this conception, because if the body of the moon had a special motion, by a parity of reasoning the same would seem to apply to the other planets and stars, since any star *(stella)* [or planet][40] is a spherical body like the moon. No reason can be offered for the moon's motion which ought not to apply to any other star [or planet]. But then, according to Aristotle, it would be necessary to assume as many intelligences as there are stars in the sky because each star would require a special *(proprium)* mover for its special motion. Aristotle did not assign [or concede] such a multitude [of motions and intelligences].

Moreover, according to Aristotle, these different motions of the stars and planets are required for explaining the difference of species and transmutations in lower [or inferior] things. Therefore, motions must not be posited in the sky that are not associated [directly] with a star. Otherwise there would not follow any diverse action in lower things, but by such a motion of the sun or Jupiter no diversification would appear to us, because it is probable that the sun is uniform and that the upper part of it is just like the lower part. And thus there is no force *(vis)* which a part of the sun would reveal to us. And so it is that if the motion of the sun were around a special center [other than the earth], it would be in vain, conferring nothing on this lower world. But if the sun does not have such a motion, it does not seem reasonable that the moon should have it, since the sun is much nobler than the moon.

To the authority of the astronomers, the Commentator replies that this way of assuming or imagining epicycles and eccentrics is quite valid for computation and for knowing the positions and dispositions of the planets with respect to each other and with respect to us; and astronomers seek nothing more than this. Thus, although such conceptions are useful for them, they do not exist in the thing [that is, in the celestial spheres].[41] And when it is also said that the appearances [of the

planets] cannot be saved, I readily concede that they cannot be saved without the assumption of epicycles or eccentrics. But all the appearances can be saved by eccentrics [alone] without epicycles. And this will be seen in another question.

QUESTION 11

In the eleventh question, we inquire whether eccentric orbs are to be assumed in the sky.

By the authority of the Commentator, who attempts to reject them, it is argued that [eccentric orbs are] not [to be assumed in the sky]. Furthermore, it follows that the motions of these eccentrics would not be simple; and yet all celestial motions ought to be simple, since they arise from simple movers. But the first consequence is proved,[42] because that eccentric motion which participates in circular motion would also participate in somewhat of an upward motion and somewhat of a downward motion, so that sometimes a planet would appear nearer the earth—and therefore [participate in] more of a downward motion—and sometimes farther from the earth—and thus [participate in] more of an upward motion.[43] Now, such motions, which participate in circular, upward, and downward movements, are not simple, and so on.

Furthermore, in the second book of *On the Heavens,* Aristotle does not save the motions and appearances by such eccentrics but [rather] by combining several motions in the same orb. It was not Aristotle's intention that there be such eccentrics.

Again, the world ought to be one and as uniform as possible, since it depends on a single and simple principle. But it would detract from its

40. In this context it seems best to understand *stella* as "planet" rather than "star."

41. An opinion also held by Maimonides. However, in at least one instance, Averroes seems to cast slight doubt on the appropriateness of epicycles and eccentrics for computation. In *Metaphysics* XII.8, Comment 45, he says with respect to epicycles and eccentrics that "present-day astronomy does not deal with realities, and is suitable merely for computing unrealities" (translated by Francis Carmody, "The Planetary Theory of Ibn Rushd," *Osiris,* X [1952], 572).

42. That is, Buridan now demonstrates that the motion of eccentrics is not simple, but composite. He argues (near the end of section 3), however, that it is sufficient "that an eccentric motion is simple, uniform, and regular with respect to its center, although not with respect to the earth."

43. Since the center of the eccentric circle differs from the earth's center, the eccentric will allow for variation in the distance of a planet from the earth.

unity if it did not have the same circumference and center for all the spheres composing it: for example, that the natural circumference would be the highest sphere and the natural center would be the earth. Now, assuming eccentrics would destroy the unity of the world, since the earth would not be the center of all the celestial spheres.

Again, it would follow that only in the sky would there be bodies whose function it is to fill a vacuum, which seems absurd and against Aristotle, who says that every celestial sphere is disposed for carrying a [single] planet. . . .[44]

Again, not only is an intelligence the cause of celestial motion but it is also the cause of the orb, since it should not be said that the sky does not have a cause. Now, as Aristotle says, it is imagined that from a simple intelligence there ought to arise only one simple and regular motion. Thus it is probable that an orb caused by a simple intelligence ought to be one and therefore regular. But those who posit eccentrics assume very irregular orbs, as it seems, for those planets that are said to have an eccentric. Therefore, eccentrics should not be assumed.

These are the arguments of the Commentator against eccentrics in the twelfth book of the *Metaphysics* and in the second book of *On the Heavens*.[45]

Ptolemy and all modern astronomers argue for the opposite position, and the argument is based on calculations *(rationes)*, because no way has yet been found, nor does it appear that it could be found, for saving the appearances—especially concerning the receding and approach of the planets [away from and] toward the earth—except by assuming eccentrics or epicycles.

It must be noted that we observe many apparent irregularities in the planets and stars. First there are the different retardations of the planets from the daily motion; and in this they are assumed to be moved with a proper motion contrary to the daily motion, so that with its proper motion the sun completes one circulation in one year and similarly Venus and Mercury; and the moon in one month, Mars in two years, Jupiter in twelve years, and Saturn in thirty-two years.[46] The retardation is such that if the sun and a fixed star rise simultaneously today, they will not rise simultaneously tomorrow, but the fixed star would rise before. Thus in one year a fixed star rises more often than the sun, namely more often in one period.

A second irregularity is the common withdrawal from our tropic to the other tropic and back

again. . . ,[47] and because of this nonuniformity an oblique circle is assumed, namely the zodiac, under which the planets move about on poles other than the poles of the daily motion.

A third irregularity is the inclination *(declinatio)* of a planet from the middle line of the zodiac, which is usually called the ecliptic. It is an inclination toward the south or north, depending on the width of the zodiac.

A fourth irregularity is speed and slowness. For it is noticeably appearent that a planet moves sometimes quicker and sometimes slower. Thus it is that sometimes the moon passes through a sign [that is, 30 degrees] in less time and sometimes in a considerably greater time.

A fifth irregularity is retrogradation. For some planets in moving from west to east contrary to the daily motion not only appear to be moved sometimes quicker and sometimes slower but even appear to regress, so that they return from west to east more quickly than the fixed stars. And you ought to know that by the association of several different motions in the same sphere—as Aristotle says in the second book of *On the Heavens*—all these irregularities are easily saved without imagining epicycles or eccentrics.[48]

---

44. A few lines are omitted here.

45. The arguments of the Commentator, Averroes, in both the *Metaphysics* and *On the Heavens* have been translated by F. Carmody ("The Planetary Theory of Ibn Rushd," *Osiris*, X [1952], 556–586). Some of the relevant sections will now be quoted:

"We claim that eccentrics and epicycles are unnatural, and that epicycles are impossible. A body moved in a circle can only be moved about the center of the universe; if there were circular motion not associated with this center, we would have to suppose another and different center, hence another and different earth, which is impossible, as stated in the Physics.

"The same perhaps holds for Ptolemy's eccentrics: if there were many centers, there would be many solid bodies different from the earth, and the center would not be unique, for it would have breadth and be divided; all this is impossible. Furthermore, accepting eccentrics, we would have to find among heavenly bodies some superfluous ones, useful merely to fill spaces, as in the bodies of animals. Nothing in planetary motion leads us to suppose epicycles or eccentrics" (p. 568; from Averroes, *Metaphysics* XII.8, Comment 45).

"A body moving in a circle can only move about a center; hence eccentrics are impossible unless between celestial bodies there is either a vacuum or filling-in bodies which are neither naturally round nor naturally moved" (p. 570; *On the Heavens* II.6, Comment 35).

46. Thirty years would be more nearly correct.

47. Two lines are omitted here.

48. According to Buridan, the first five irregularities

527

A sixth irregularity is the nearness of a planet to the earth and at times its remoteness or distance from it. This irregularity is well experienced and demonstrated. It seems to me that this irregularity cannot be saved without eccentrics or epicycles, since any concentric sphere in the world would be everywhere equidistant from the earth. By virtue of the motion of such a sphere a planet could never be nearer or farther. The Commentator tried briefly to save this appearance in another way, but he could not, as he confessed.[49] In the second place, it seems to me that these appearances could be saved by eccentrics without epicycles, or even by epicycles without eccentrics. An eccentric has one of its sides nearer the earth and the other farther away, so that when the planet is on the side of the eccentric more distant from the earth it would be farther away from us; but on the other side it would be nearer to us, although no epicycle was assumed. And similarly, even if there were no eccentric, yet an epicycle could have one side nearer to us and the other more distant from us.[50]

But some argue against the previously stated conclusion, because when the planet approaches or recedes there appears a very great irregularity, since sometimes it approaches very close before it begins to withdraw and sometimes it approaches less than half as much before it begins to withdraw.[51] But this could not be saved by assuming an eccentric alone or an epicycle alone. To this argument it could be replied that if anyone wished to assume epicycles without eccentrics it would be necessary, in order to save this appearance [or phenomenon], to posit an epicycle on an epicycle, and then every irregularity which is saved by an eccentric and an epicycle could be saved by these two epicycles.[52] Also, whoever wished to assume eccentrics without epicycles would have to assume two eccentrics, one on the other, just as we assumed an eccentric on a concentric. Then everything that was saved by an eccentric and an epicycle could be saved by these two eccentrics. But, to put it briefly, it does not seem to me that any of these ways is really demonstrable or disprovable.

To the arguments put forth by the Commentator one can say this: that an eccentric motion is simple, uniform, and regular with respect to its center, although not with respect to the earth. And this suffices. It is not necessary to assume the world as one being but as a union of many things having a mutual order and one principle; nor is it necessary to assume that these [celestial] bodies would serve only to fill a vacuum. Rather it should be assumed that the whole [world] is made up of these things and is moved on an eccentric simultaneously with the daily motion and that the different parts of the eccentric are related to the earth in different ways. For God and the intelligences intend such irregularities for the purpose of better explaining all the different material species in this world, and it is not necessary that the sphere moved by one intelligence be itself one [that is, undivided], because there is no discontinuity there, since the whole world contains many spheres and it [the world] is moved all at once by the simplest prime mover. Nor [indeed] is there any irregularity of magnitude with respect to the moving intelligence, since the whole sphere [of the world] is concentric and of uniform magnitude and it is moved all at once by its proper intelligence. But, again, an eccentric has a special intelligence by which it is moved with a certain other motion; and this eccentric is also of a uniform magnitude and

can be explained without epicycles or eccentrics, but not the next, or sixth, irregularity.

49. Probably based on Averroes' remarks in *Metaphysics* XII.8, Comment 45 (Carmody, p. 572): "But Ptolemy did not understand that the ancients postulated spiral motion because epicycles and eccentrics are impossible; wherefore we must return to a study of the true theories based on rational precepts. I have set up a system of motion on a single sphere but about two or more different poles, as the appearances required; by this motion one can, for the asters, explain speed, slowness, advance and recession, and other motions for which Ptolemy was able to set up no theory; and from this, also, lesser and greater distance, as for the moon. In my youth I hoped to complete this study, but in my old age I despair of this; yet perhaps these statements will lead someone to further study." Perhaps al-Bitruji was motivated by these early efforts of Averroes (see n. 24).

50. Underlying Buridan's statement is the demonstrated equivalence in representing the apparent irregular motion of a planet from west to east (especially the sun's annual motion) either by (1) an eccentric circle whose center is not the earth, or (2) a concentric circle with the earth at its center but on which an epicycle is assumed whose radius equals the eccentricity of the eccentric circle described in (1). A demonstration of this equivalence appears in Ptolemy's *Almagest*, Book III, chapter 3 (see pp. 86–91 of R. Catesby Taliaferro's translation of the *Almagest* in Vol. XVI of the *Great Books of the Western World* [Chicago: Encyclopaedia Britannica, 1952]).

51. Buridan observes that the perigee of a planet varies, a phenomenon that was not representable by an epicycle or a concentric or an eccentric circle alone.

52. That is, a concentric circle with an epicycle on an epicycle can be made equivalent to an eccentric with an epicycle.

moved uniformly around its center.[53] And so, by assuming eccentrics, these arguments are avoided. As for those who wish to assume epicycles *(epi-* *ciclos)*, there is no strong argument against them except for the spot *(macula)* on the moon.[54] And so Question 11 is terminated.

69 ON THE COMMENSURABILITY OR INCOMMENSURABILITY OF CELESTIAL MOTIONS

Nicole Oresme (ca. 1325–1382)

Translation, introduction, and annotation by Edward Grant[1]

Nicole Oresme's fascination with the subject of commensurability and incommensurability in mathematics, physics, and cosmology is evidenced by a number of treatises in which he saw fit to discuss, or at least mention, it. Sometime after 1351, and perhaps between 1351 and 1360, he composed *On Ratios of Ratios (De proportionibus proportionum)*, in which the propositions of the first three chapters were concerned with commensurable and incommensurable relations between various types of ratios, and the fourth chapter developed the application of these propositions to terrestrial and celestial motions.[2] Even prior to this, however, Oresme seems already to have written *Ad pauca respicientes* (these are the opening words of the treatise),[3] in which he concerned himself almost exclusively with the kinematics of circular motion.

[His] purpose was to determine the consequences deriving from the motions of points or bodies moving with different uniform circular velocities. These bodies, usually two or three, are, in some propositions, assumed to move with mutually commensurable velocities and in others with incommensurable velocities.

In contrast to astronomers, and especially in Part I, Oresme is generally interested in precise punctual relations between the moving bodies. That is, he wishes to specify the exact point or points at which the bodies will meet or oppose one another, and if the bodies move commensurably, to determine the precise time intervals between such occurrences. On the assumption of incommensurable speeds, such precision is impossible.

In Part 2 Oresme shifts his point of emphasis and concerns himself with general dispositions and relative positions of two or more mobiles before and after they enter into actual conjunction or opposition. The particular points in which these events occur are now of little significance.

Of great interest is Supposition II of Part 1 (p. 385), where Oresme assumes that any two quantities are probably incommensurable. Although this supposition is not utilized until

the very end of the treatise, its application is the highlight of the *Ad pauca respicientes*. Having formulated numerous abstract propositions involving points or bodies moving with commensurable or incommensurable speeds, Oresme, in Proposition XVII of Part 2 (p. 423), invokes Supposition II and applies it to celestial motions in order to demonstrate that no configuration or relationship of celestial bodies can ever repeat. This constitutes the basis of his repudiation of astrological prediction in Proposition XIX (pp. 425, 427). If every celestial configuration is unique—as it must be on the assumption of incommensurable speeds, which by Supposition II are probable—astrological prediction is hopeless for it must rely on cumulative observations of precisely recurring events.[4]

53. It seems obvious from the preceding that Buridan argues here for a physically real eccentric sphere which is moved around *a center other than the earth* by its own special intelligence. This represents a total break with a central theme in Aristotelian cosmology.

54. Probably a reference to Question 10, where Buridan argued that if epicycles existed, the spot or man in the moon should appear sometimes in the upper part, sometimes in the lower part, of the moon. Since such a phenomenon was not experienced, this was taken as a weakness of the epicycle theory. Now Buridan says that if one were ready to accept this imperfection as an inherent defect of epicyclic theory, the latter would be serviceable in all other respects. See also note 39.

1. Reprinted with permission of the copyright owners, the Regents of the University of Wisconsin, from *Nicole Oresme and the Kinematics of Circular Motion: "Tractatus de commensurabilitate vel incommensurabilitate motuum celi,"* edited with an introduction, English translation, and commentary by Edward Grant (Madison, Milwaukee, and London: University of Wisconsin Press, 1971), pp. 179, 193, 201, 215–217, 237–243, 249–253, 261–265, 273–277, 279, 319–321. Where it seemed relevant, I have incorporated into the notes material from the introduction and commentary. Occasionally these have been altered.

2. Excerpts from this treatise appear in Selections 25 and 51.2.

3. See pages 76–78 of my edition of Oresme's *De proportionibus proportionum* (for full title, see Selection 25, n. 1).

4. *Ibid.,* pp. 121–122; the page numbers given in parentheses are references to my edition.

There is reason to believe that Oresme intended to add a revised version of the *Ad pauca respicientes* to the *De proportionibus* as an illustration of one important application of incommensurability. In the process of revision, however, Oresme probably became dissatisfied with its organization and content and subsequently reworked and enlarged the propositions and format so thoroughly that a wholly new treatise emerged, which he called *Treatise on the Commensurability or Incommensurability of the Celestial Motions (De commensurabilitate vel incommensurabilitate motuum celi)*.[5] Now unsuitable as an addition to the *De proportionibus proportionum,* it was issued as a separate and rather lengthy treatise.

The *De commensurabilitate vel incommensurabilitate motuum celi* consists of a prologue and three parts. After the presentation of a few preliminary definitions and assumptions, the remainder of Part I is devoted to 25 propositions in which bodies, or mobiles, are assumed to move with commensurable speeds on concentric circles (in Proposition 20 the circles are eccentric). Part II contains 12 propositions in each of which at least two of the motions are incommensurable. In these two parts, Oresme's objective is to derive various consequences from the motions of two or more bodies whose speeds are first assumed to be commensurable and then incommensurable. In most propositions, these kinematic consequences are applied arbitrarily to astronomical aspects—conjunction is used as the paradigm case—to determine the times in which, and places where, these can occur in terms of the specific data and assumptions. Oresme concludes the treatise by presenting Part III in the form of a debate presided over by Apollo and involving, as antagonists, personifications of Arithmetic and Geometry, who argue whether the celestial motions are commensurable (Arithmetic) or incommensurable (Geometry).[6] The setting of the debate is a dream in which Apollo and his Muses appear before Oresme. Following the debate, and just as Apollo is on the verge of rendering a judgment as to whether the celestial motions are commensurable or not, Oresme awakens and the conclusion is left in doubt. The attentive reader, however, would have had good reason to suppose that Oresme really considered that the celestial motions were probably incommensurable, for it was only this position that enabled him to discredit exact astrological predic-

tion with mathematical certainty (see Part III of this selection and notes 80 and 84).

Although Oresme had predecessors who discussed the incommensurability of celestial motions and some of its consequences, no one is yet known who gave it such full and subtle mathematical development, nor who exerted as great an influence on subsequent discussants (for instance, Henry of Hesse, Marsilius of Inghen, Pierre d'Ailly, and probably Jerome Cardano).

PART I

. . . .

I take the commensurability and incommensurability of circular motions in terms of the magnitude of the angles described around the center or centers, or in terms of the circulations, which is the same thing. Thus, things are moved commensurably when, in equal times, they describe commensurable angles around the center, or when they complete their circulations in commensurable times. Circulations are incommensurable when they are completed in incommensurable times, and when, in equal times, incommensurable angles are described around the center. Accordingly, conjunctions, oppositions, aspects, and all the motions ascribed to the heavens by astronomers are to be measured in this way, since a ratio of velocities varies as a ratio of the circular lines described by the mobiles. Whether or not it ought to be taken in this way is not relevant to what is proposed here. Moreover, incommensurability can be found in every kind of continuous thing, and in all instances in which continuity is imaginable, either extensively or intensively. For a magnitude can be incommensurable to a magnitude, an angle to an angle, a motion to a motion, a speed to a speed, a time to a time, a ratio to a ratio, a degree to a degree, and a voice to a voice, and so on for any similar things.

The return of *one mobile* along a circular path from any point to that same point, I call a *circulation*. The return of *several mobiles* from any state to a wholly similar state, or aspect, I call a *revolution*.[7]

5. *Ibid.,* pp. 79–80.
6. *Nicole Oresme and the Kinematics of Circular Motion*, p. 6 ("Summary and Analysis of the *De Commensurabilitate*") of my edition of Oresme's *De commensurabilitate*.
7. A *circulation* of a single body is analogous to a sidereal period in astronomy, but a *revolution* has no direct counterpart in astronomy, since it requires that the bodies in motion, or mobiles, return to the very same point or configuration taken as absolutely fixed in

In this little book, my purpose is to consider *exact* and *punctual aspects* of mobiles that are moved circularly. I do not, however, propose to deal with aspects near a point which is usually the intention of astronomers, who care only that there be no sensible discrepancy—even though a minute, undetectable error would produce a perceptible discrepancy when multiplied over a long [period of] time.

. . . .

Proposition 4. If two mobiles are now in conjunction, it is necessary that they conjunct in the same point at other times.

In this part of the book the discussion is always about things moved commensurably. And so let *A* and *B* be in conjunction in point *g*, since, as far as what is considered here, a conjunction in a point is the same as on a line or surface. Since the motions are commensurable, it follows by the fifth [proposition] of the tenth [book] of Euclid[8] that a ratio of these two motions is as a ratio of two numbers. Therefore, let *A*'s motion be as number *C*, and *B*'s motion as number *D*. Then, in the same time in which *A* makes *C* circulations, *B* makes *D* circulations, so that at the end of this time *A* and *B* will have completed an exact number of circulations and each of them will be where it is now. Consequently, they will be in conjunction in the point in which they are now.

For example, let *A*'s velocity be as 5 and *B*'s as 3. Then, when *A* will have made 5 circulations, *B* will have made 3 and they will be where they are now. A similar argument could be made for past time, namely, that they had conjuncted at other times.

Proposition 8. [How] to determine the place of the first conjunction following the present conjunction of these two mobiles.

Since, by the sixth proposition,[9] it has already been shown [how to find] the time required for such a conjunction, there only remains to be seen how great a distance either of these [two] mobiles traverses in that [very same] time, and then to subtract the whole circle from that distance as many times as possible. This is the same as dividing the whole circle by that distance so that what has been proposed will be found.

For example, from the velocity given in the case presented before,[10] it is obvious that in one day *B* traverses $\frac{1}{5}$ of its circle. But since by the sixth proposition, $7\frac{1}{2}$ days represents the time elapsed before the first conjunction following [the present conjunction], it is evident that during that time *B*

traverses $\frac{7}{5}$ of a circle plus $\frac{1}{2}$ of $\frac{1}{5}$ of a circle[11]—that is, the whole circle plus half of it. Consequently, the [very next] conjunction following will occur in a point opposite the point in which *A* and *B* are now in conjunction. The same result will be obtained by operating with the motion of *A*.

Proposition 12. If there were more than two mobiles, it is possible that no more than two will ever be in conjunction at the same time.

Let *A*, *B*, and *C* be three mobiles. By the preceding propositions,[12] *A* and *B* can conjunct in only a fixed number of points, or places, and nowhere else; let any such point be called *d*. Similarly, *B* and *C* conjunct in a certain number of fixed points, and nowhere else; let any such point be called *e*. Then, if no *d* is an *e*, which is possible, these three mobiles will never be in conjunction at the same time.

---

space. Thus a revolution is more like a Great or Perfect Year in ancient and medieval astrology, a concept in which all planetary bodies return simultaneously to the same celestial configuration from which they started, an event which was thought to occur at regular intervals. In the *De commensurabilitate*, Oresme even includes as instances of a Great Year the return of any two planets to the same point, and even includes the case of a single celestial body with two or more simultaneous motions, if, in equal time intervals, these motions always brought it back to the very same point from which it started (see Part I, Proposition 22). The Great Year was mentioned by Plato (*Timaeus* 39D) and other ancient authors. See also selection 51.2, n. 86.

8. "Any two commensurable magnitudes have a ratio of a number to a number" (*Omnium duarum quantitatum communicantium est proportio tanquam numeri ad numerum*). See p. 247 of the edition by Campanus of Novara cited in Selection 27, n. 1.

9. In Part I, Proposition 6, Oresme shows "[How] to find the time of the first conjunction that will follow a [present] conjunction of two mobiles whose velocities have been given." In this proposition, two bodies are assumed to be in conjunction at the outset and the time that will elapse before their very next point of conjunction is then determined.

10. The data are drawn from Part I, Proposition 6, where body *B* traverses 1/5 of its circle per day and body *A* traverses 1/3 of its circle per day. To determine the place of their very next conjunction, one can work with the daily distance traversed by either *B* or *A*; Oresme chooses the former.

11. That is, in 7 days *B* will traverse 7/5 of its circle and in the remaining 1/2 day it will traverse $(1/2) \cdot (1/5) = 1/10$ of its circle. Adding, $(7/5) + (1/10) = 3/2$. Thus in $7\frac{1}{2}$ days, *B* will have traversed its circle $1\frac{1}{2}$ times and meet *A* in conjunction in a point diametrically opposite their first conjunction. That *A* will also be in the same point in $7\frac{1}{2}$ days is shown by multiplying $7\frac{1}{2} \cdot \frac{1}{3} = 2\frac{1}{2}$. Thus, while *B* has traversed its circle $1\frac{1}{2}$ times, *A* has traversed its circle $2\frac{1}{2}$ times, and they will conjunct.

12. Propositions 7, 8, 10, and 11 are relevant here.

In a like manner, if there were four or five, or any number of mobiles, it can happen that only two will ever conjunct at the same time since it is possible that when the mobiles are taken two at a time, they do not share any common places of conjunction; but it is also possible that they could share some common points, and that [for example], of six mobiles there might be simultaneous conjunctions of three, or four and no more, or five, or all; and the same reasoning applies to other sets of mobiles.

For example, assume that *A* and *B* are in conjunction in point *d* [Fig. 1], and mobile *C* precedes

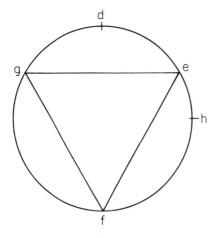

Fig. 1

them by $\frac{1}{8}$ part of the circle—i.e., by $1\frac{1}{2}$ [zodiacal] signs. And let the velocity of *A* be as 4, that of *B* as 2, and *C* as 1; let point *e* be separated from point *d* by $\frac{1}{6}$ part of the circle—i.e., by two signs—and let *f* and *g* be two other points such that *e*, *f*, and *g* divide the circle into three equal parts. [Finally] assume that *h* is another point separated from point *d* by $\frac{1}{4}$ part of the circle—i.e., by three signs.

Having now set out the data, we shall demonstrate, by [means of] the ninth proposition,[13] that *A* and *C* will conjunct first in point *e*;[14] and by the same [ninth] proposition it will be obvious that *B* and *C* will conjunct first in point *h*;[15] and by the tenth, or even eleventh, proposition it will be shown that *A* and *B* will conjunct only in point *d*,[16] and, by the same proposition, that *B* and *C* will conjunct only in point *h*. It follows immediately from this that *A*, *B*, and *C* will never conjunct simultaneously. Indeed, by the same [tenth] proposition it will be seen that *A* and *C* conjunct only in the other three points *e*, *f*, and *g*, from which it could again be argued that *A*, *B*, and *C* will never conjunct

simultaneously. When more mobiles are involved it can happen that 3, or 4, or any number of them can be in conjunction simultaneously. How this happens will be considered later.

Proposition 22. [How] propositions [that are] similar [to those demonstrated previously for two or more distinct mobiles] can be applied to one and the same mobile moving with several [simultaneous] motions.

For example, let it be assumed that the sun has only two motions, namely, proper and diurnal; and on the ninth sphere, let *A* be the first point of Cancer, which describes the same circle every day, namely, the summer tropic; and let *B* be the center of the sun describing the ecliptic in one year with a proper motion [Fig. 2]. [Finally] assume that

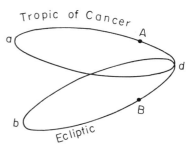

Fig. 2

A and *B* are now in point *d*, which is imagined as motionless. Now on the assumption that these motions are commensurable, the ratio of the motions will be rational. If this ratio were multiple, it follows that *A* and *B* will never be together simultaneously except in point *d*. This can be made clear at once by an example, for if *A* should make 100 circulations in the same time that it takes *B* to make one circulation, then at the termination of

13. Part I, Proposition 9, reads: "[How] to determine the place and time of the first conjunction following [a given conjunction] when a distance between the two mobiles has been assigned."

14. From the data, $V_a/V_c = 4/1$, where *V* is a scalar velocity. Initially *A* and *B* are in conjunction in *d*, and *C* precedes them by 1/8 part of a circle or $1\frac{1}{2}$ zodiacal signs. Now, since $V_a/V_c = 4/1$, as *A* moves 2 signs from *d* to *e*, *C*, which precedes *A* by $1\frac{1}{2}$ signs, will move $\frac{1}{2}$ sign and conjunct with *A* in *e*. Indeed, it is evident that *A* and *C* can conjunct only in points *e*, *f*, and *g*.

15. Since *C* precedes *B* by $1\frac{1}{2}$ signs and $V_B/V_C = 2/1$, it follows that *C* and *B* will conjunct in *h*, which is separated from *d*, and hence from *B*, by 3 signs and from *C* by $1\frac{1}{2}$ signs. Thereafter, *B* and *C* will conjunct only in *h*.

16. Since $V_A/V_B = 2/1$ and *A* and *B* are initially in conjunction in *d*, it follows that *A* and *B* will conjunct only in *d*.

these 100 circulations—and not before—*A* and *B* will be in point *d;* and this will always occur every 100 days.

But if the ratio of the motions is not multiple, a fraction must be assumed in its denomination. Furthermore, the denominator of the fraction indicates the number of points in which *A* and *B* can meet. Thus, if *A* makes $100\frac{1}{2}$ circulations in the same time as *B* makes one circulation, it follows that they would be found in a point opposite *d,* the first point; and after the next sequence [of circulations], they would be found in *d* [itself]. [In this example] there are only two fixed places where *A* and *B* could meet. But *B* could meet *A* in exactly three fixed points, if *B* makes one circulation when *A* makes $100\frac{1}{3}$; and if [*A* makes] $100\frac{1}{4}$ [circulations], there would be four [points of contact], and so on. Also, if *A* should complete $100\frac{2}{5}$ circulations while *B* makes one circulation, there would be five places where they could meet, since in that time *A* would traverse its circle a certain number of times plus $\frac{2}{5}$ of it, and during another [equal] time [period] would again make just as many circulations plus $\frac{2}{5}$ [of one]; and this applies to all [such cases].[17]

The number and arrangement of the celestial points on which the sun could enter the first point of Cancer—as well as other degrees of the zodiac— become apparent from all this. Thus if the sun were moved with only two motions and the solar year were measured by an exact number of integral days, it is obvious that the sun could enter [the first point of] Cancer only when it coincided with one [particular] meridian; and after another [equal interval of] time it would again [enter Cancer] on the same meridian, and never on another. But if the year contained a certain number of days exactly plus some $\frac{1}{4}$ part, or parts, of a day—as, for example, $365\frac{1}{4}$ days—there would be only four equidistant points on this tropical circle on which the sun could enter the first point of Cancer; and the same applies to other degrees [of the zodiac].

If, however, three motions were involved, then, by the method already outlined, they will be taken two at a time and their combinations mutually compared in quite the same way as in our previous discussion of a conjunction of three mobiles;[18] the same procedure would be followed for four or more motions.

From all this, it follows that whatever the number of motions involved—provided they are all mutually commensurable—there is always, for any

given disposition, a certain number of fixed places—and no more—in which that mobile can be related in the same way. And when it has proceeded through all these places maintaining this same relationship, it will begin, once again, to run through them exactly as before.

17. To explain all this refer to Fig. 2, where circle *a* is the tropic of Cancer (summer tropic) and *A* is the first point of Cancer. Let A_n, where *n* can be any integer, represent any point on circle *a*. Circle *b* is the ecliptic on which the center of the sun, *B*, moves with a uniform motion and completes one circulation in a year. Finally, *d* is the only point of contact between circles *a* and *b*. Therefore, any point, A_n, will become the first point of Cancer when it is in *d* simultaneously with *B*, the center of the sun.

Initially, assume that *B* and some point on the tropic of Cancer, say A_1, are in *d* simultaneously. Let V_a and V_b represent the commensurable velocities of circles *a* and *b*, respectively. Now if $V_a/V_b = n/1$, where *n* is an integer, then, since A_1 is carried around by circle *a*, and *B* by circle *b*, it follows that when *a* makes *n* circulations and *b* one circulation, A_1 and *B* will meet again in *d*. Thus, in Oresme's first example, whenever circle *a* completes 100 circulations while *b* completes one circulation, A_1 and *B* will meet in *d*.

But if the ratio of velocities of the circles is not exactly a multiple ratio, but includes some fractional part of a circulation, the denominator of the fraction will indicate the exact number of points of circle *a* that *B* could meet in *d*. Thus, if $V_a/V_b = P\frac{m}{n}/1$, *n* will represent the total number of points on circle *a* that will meet *B* in *d*. In the mixed fraction, *P* is the integral number of circulations that circle *a* will complete in some given time and m/n represents an additional fractional part of a circulation. Since the *n* points on circle *a* will divide it into equal parts, it is only necessary to determine the distance that separates two successive meetings in *d* in order to arrange the sequence of the points on circle *a* through which the sun can enter the first point of Cancer. For example, should circle *a* complete $100\frac{1}{4}$ circulations to one for *b*, there would be four equidistant points on *a* that would meet *B* in *d* at equal intervals. If A_1 and *B* were initially in *d*, then, after $100\frac{1}{4}$ circulations, A_2 will meet *B* in *d;* after the next $100\frac{1}{4}$ circulations of *a*, A_3 will meet *B* in *d*; and finally, A_4 will arrive in *d* simultaneously with *B* after the next $100\frac{1}{4}$ circulations of circle *a*. This will be followed, once again, by a meeting of A_1 and *B* in *d*, after which the cycle will repeat ad infinitum.

If, however, circle *a* completes $100\frac{2}{5}$ circulations to every one for circle *b*, there will be five equidistant points on *a* that will meet *B* in *d*, but the order of the points meeting *B* will not be A_1, A_2, A_3, A_4, and A_5, where each point is separated from its immediate neighbor by 1/5 of the circle. For if at the beginning, A_1 and *B* are together in *d*, after the next $100\frac{2}{5}$ circulations of circle *a*, point A_3—not A_2—will meet *B* in *d*; and after the next series of circulations A_5 will meet *B*; then A_2, and finally, A_4. The order of points which meet *B* in *d* is, therefore, A_1, A_3, A_5, A_2, and A_4, after which the cycle is repeated ad infinitum.

18. As, for example, in Part I, Proposition 12.

For example, let *d* be one of only four fixed points on which the sun can enter [the first point of] Cancer. I say that if the sun begins from point *d* and moves continuously for four years describing daily a new spiral until it returns to point *d* at the end of the fourth year, it would then resume its course from the beginning; and so on, endlessly. In the same manner, this would also happen if there were more points of this kind, but then such a revolution would be completed in a greater time. The same thing could be said about the moon and any of the other planets.

Therefore, if all the motions by which the sun is moved were commensurable, the sun's center, moving within a space imagined as immobile, would describe a finite path, or line, that is like a circular line without terminating points. And yet it does not describe a circular path. Indeed, its path is composed of many circular-like or spiral lines that are [interlaced or] interwoven in a certain way. Some of these lines begin toward the tropic of Cancer and proceed in the direction of the tropic of Capricorn, which is the path the sun follows if beginning its motion there; and at other times the sun returns in the opposite direction describing lines that intersect those made previously—as if weaving a pattern.[19] Also the number of circular-like lines, or spirals, between the two tropics equals the number of days that it takes the sun to depart from and return to the point which is imagined as fixed in the heavens. The [end] product of such an interweaving between the two tropics is similar to a quadrangle, oblong, or rhombus.[20] But throughout all eternity, the center of the sun never was, nor will ever be, inside the area or inner space of these [figures]. Furthermore, once this path has been traversed by the sun in its period, it begins again to move over the same path as before. The same thing could be said of the moon and the other planets.

Thus, any mobile taken independently, but having several motions, has a fixed period upon the completion of which it begins again; and this goes on an infinite number of times. This period can be called the Great Year of that mobile,[21] just as, in a similar way, any two celestial mobiles that might be taken complete their courses simultaneously in a fixed period of time, after which they begin over again as before; and this can be said about three mobiles, or [indeed] any number of them. These mobiles can also be said to have a Great Year, just as some say that 36,000 solar years constitute a Great Year of the sun and the eighth sphere. But

a Great Year of all the planets and the eighth sphere is very much longer.[22] In brief, if all the celestial motions are mutually commensurable,

19. As the sun moves between the two tropics or solstices with two simultaneous motions, diurnal and annual, it traces out a spiral line which results from the fact that its two motions are oppositely directed. For each day of the year there is a corresponding spiral line, and at the end of every year a pattern is traced which is similar to a fabric woven with criss-crossing lines. Oresme may have read about this pattern of spiral lines in Chalcidius' Latin translation of Plato's *Timaeus*, where in 39A, B, Plato discusses the simultaneous motions of the circle of the Same (that is, the celestial equator which rotates daily from east to west carrying all the planets with it) and the circle of the Different or Other (that is, the ecliptic, or more properly, the zodiacal band, on which each planet moves in its period of revolution from west to east). The generation of Plato's spiral is summarized by Sir Thomas Heath as follows (*Aristarchus of Samos* [Oxford, Clarendon Press, 1913], p. 169; the figure appears on p.160.):

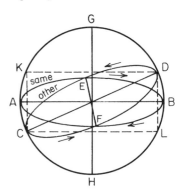

Suppose a planet to be at a certain moment at the point F. It is carried by the motion of the Same about the axis *GH*, round the circle *FAEB*. At the same time it has its own motion along the circle *FDEC* [i.e. the ecliptic]. After 24 hours accordingly it is not at the point *F* on the latter circle, but at a point some way from *F* on the arc *FD*. Similarly after the next 24 hours, it is at a point on *FD* further from *F*; and so on. Hence its complete motion is not in a circle on the sphere about *GH* as diameter but in a spiral described on it. After the planet has reached the point on the zodiac (as *D*) furthest from the equator it begins to approach the equator again, then crosses it, and then gets further away from it on the other side, until it reaches the point on the zodiac furthest from the equator on that side (as *C*). Consequently the spiral is included between the two small circles of the sphere which have *KD*, *CL* as diameters."

20. In the figure in note19, we can produce a right angled quadrilateral figure if we join lines *KD* (the diameter of the tropic of Cancer) and *CL* (the diameter of the tropic of Capricorn) by joining the points *K* and *C* and *D* and *L* by straight lines.

21. See note 7.

22. See note 7.

it is necessary that there be one maximum [or perfect] period for all of them at the same time, [a period] which is repeated after its termination. But if the world were eternal, the perfect period [or Great Year], although it would always be similar, would not be repeated in the same place, but, rather, in infinite [different] places.[23]

PART II

In the second part of this work, it is assumed that some of the celestial motions are incommensurable. We must now show what follows from that assumption.

Proposition 1. If two mobiles have moved with incommensurable velocities, and are now in conjunction, they will never conjunct in that same point at other times.

Assume that A and B are in point d. If, at some later time, they should meet in point d, it follows that during that time each will have made a certain number of exact circulations. Let E be the number of circulations that will have been made by A, and G the number that will have been made by B. Since a ratio of velocities varies as the ratio of the distances traversed [by these velocities] during the same time, it follows directly that the speeds of these mobiles are related as the two numbers E and G. Therefore, the velocities are commensurable, which is contrary to what was assumed.

For example, if A and B are now in point d, and at some later time they are again in conjunction in the same point, assume that A will have made 5 circulations and B will have made 3. It is obvious, then, that their velocities are related as the numbers 5 and 3, and that A is moved more quickly than B by a superpartient two-thirds ratio.

And one can show in the same manner that they were never before in conjunction in point d. All this is applicable, in just the same way, to any other aspect. Thus if A is in point d, and B is in the point opposite [to d], they will never be in opposition at other times in these very same points. This could be demonstrated in the same manner as before. Hence, the [period of] revolution of these mobiles is infinite; indeed, it might be truer to say that there is no period of revolution at all.

Proposition 2. If two mobiles are now in conjunction, they will never conjunct at other times in any point separated from their present point of conjunction by a part of the circle commensurable to the whole circle.

From the eighth and ninth [propositions] of the tenth [book] of Euclid, it follows at once that if something is added to each of two commensurable quantities, and is commensurable to each of them, the wholes will be commensurable. Therefore, if to each of two sets of circulations—there may be any number of circulations in each set—there is added a part of a circulation commensurable to a whole circulation, the wholes will be commensurable. Now if A and B are in conjunction in some point separated from point d, then, in the time that has intervened [since their conjunction in d], each will have completed an exact number of circulations plus some part of a circulation commensurable to a whole circulation.[24] Their motions must, therefore, be commensurable, since the distances traversed in the same period of time would be commensurable.

For example, if A and B are in conjunction in some point that is separated from point d by $\frac{1}{4}$ part of the circle, and if A had by then completed $5\frac{1}{4}$ circulations and B, $3\frac{1}{4}$ circulations, the ratio of the motions of A and B would be as $5\frac{1}{4}$ to $3\frac{1}{4}$—i.e., as 21 to 13, which is a ratio of $1\frac{8}{13}$. The same may be said for any other aspect. Thus if A is presently in point d and B is in the point opposite d, then, by taking a part of the circle commensurable to the whole circle, one can demonstrate by the same method that they will never again be in opposition in points that are separated from their present points of opposition by distances that are commensurable to the whole circle. Any other disposition can be treated in this manner. And the same argument applies to the past as well as the future. One may, therefore, conclude that the distance, or sector of a circle, that separates two successive conjunctions is incommensurable to the whole circle; and this may be said about any other aspect. And similarly can the time between

23. If the world were eternal without beginning, we would be ignorant of the unique point or configuration from which all the motions commenced and to which a Great Year could be uniquely referred. Without such knowledge, an infinite number of points or configurations would be equally privileged to serve as reference points.

24. That is, each of the bodies, A and B, will have completed an integral number of circulations with respect to d plus the same angular part of the circle. If that angular part of the circle is commensurable to the whole circle, it follows that the distances traversed by A and B, as expressed in whole numbers of circulations plus the commensurable fractional part, will be commensurable. Since the speeds are proportional to the distances traversed, the velocities will be commensurable, which is contrary to the assumption that they are incommensurable.

two such conjunctions be related to the time it takes each mobile to complete one circulation; and these times of circulation are likewise mutually incommensurable.

Proposition 7. It is possible that three or more mobiles whose motions are mutually incommensurable are now in conjunction but can never conjunct in another place. Therefore, in one case of incommensurability it follows that the mobiles can be in conjunction an infinite number of times, but in another case this does not follow.[25]

As before, let *A, B,* and *C* be in point *d*. Let us now reconsider something that was demonstrated in the second proposition of the first part—namely, that if a continuum were divided into infinity by one rational [geometric] proportionality, it could also be imagined as divisible by [yet] another rational [geometric] proportionality; and these proportionalities are incommunicant [or prime to each other].[26] If this continuum were a circle, then, except for one point only, no point in one division is a point in the other division. However, if the proportionalities were communicant, there would be points in common between the two divisions.

The same distinctions apply to divisions made by proportionalities consisting of irrational ratios. Thus if one proportionality is based on half of a double ratio [i.e., $(2/1)^{1/2}$], and the other is based on half of an octuple ratio [i.e., $(8/1)^{1/2}$], these are communicant; but if the proportionalities are based on half of a double ratio and half of a triple ratio [i.e., $(3/1)^{1/2}$], they are incommunicant [or prime to each other].[27] Hence it is possible for the continuum to be divided by points that are separated by incommensurable distances, so that no point in one division is also a point in the other division—except, perhaps, one point. If the continuum is circular, two such divisions are possible in a circle. Now since these two series of infinite numbers will begin in sequence from point *d* and they are separated from each other by incommensurable distances, the two series will not meet in [or share] any point. It is possible, then, that [beginning] from point *d*, all the other infinite number of points in which *A* and *B* will conjunct are incommunicant with those in which *B* and *C* will conjunct, so that *B* and *C* will never conjunct in the same places where *A* and *B* conjunct. Therefore, *A, B,* and *C* will never conjunct at other times.

For example, assume that *A* and *B*, which are now in *d*, are afterward in conjunction in *e*, then in *f*, and so on [Fig. 3]. Then, because of the

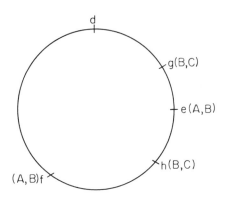

Fig. 3

25. In Part II, Proposition 6, Oresme had shown that three mobiles whose motions were respectively uniform but mutually incommensurable could conjunct again in other points. Indeed they could conjunct an infinite number of times but never twice in the same place. In Proposition 7, however, Oresme establishes conditions which would deny the results of Proposition 6. For now, three mobiles in conjunction will never again conjunct elsewhere.

26. In Part I, Proposition 2, it was demonstrated that if, for example, a continuum were divided into 2^n successive parts (where *n* is 1, 2, 3, 4, . . .) and then into 3^n successive parts, these two geometric proportionalities would not divide the continuum into parts which had common points of division (except for 1, which would represent the whole continuum before any divisions were made). This follows from the fact that the numbers following unity in each proportionality are mutually prime to each other (that is, 2 and 3). Where this is not the case, there will be points in common. Oresme asserts that statements true of rectilinear continua apply also to circular continua, the only difference being that where a single point can divide a rectilinear continuum, two points are required for division of a circle. In Part II, Proposition 1, Oresme considered only rational proportionalities, but he now concentrates on irrational proportionalities.

27. They are "communicant" or commensurable because $(2/1)^{1/2}$ and $(8/1)^{1/2}$ can be related by a rational exponent—that is, $(8/1)^{1/2}=[(2/1)^{1/2}]^{3/1}$. Hence these two irrational ratios constitute a "rational ratio of ratios." But the irrational ratios $(2/1)^{1/2}$ and $(3/1)^{1/2}$ form an "irrational ratio of ratios" since no rational exponent can relate them—that is, $(3/1)^{1/2}\neq[(2/1)^{1/2}]^{p/q}$, where p/q is a rational ratio. These distinctions were elaborated by Oresme in his earlier work, *On Ratios of Ratios (De proportionibus proportionum)*, especially Part I, Proposition 9 (see pp. 200–204 of my edition; also see Selection 51.2, n. 51).

If a circle were divided from the same point by the "commensurable" proportionalities $(2/1)^{1/2}$ and $(8/1)^{1/2}$, it is clear that all points represented by $[(2/1)^{1/2}]^{3n}$ and $[(8/1)^{1/2}]^n$, where $n = 1, 2, 3, 4, . . . $, are common to both proportionalities, since for any value of *n* both ratios would be equal and represent the very same point. Thus when $n = 2$, both ratios equal 8/1.

regularity of the motions, points *d, e, f,* and so on, are separated by equal distances. Let the whole circle be related to arc *de* as half of a sesquialterate ratio.[28] Also, assume that *B* and *C,* which are in point *d,* are afterward in conjunction in *g,* then in *h,* and so on. [Finally] let the whole circle be related to arc *dg* as half of a sesquitertian ratio [namely, as $(4/3)^{1/2}$].[29]

Now since half of a sesquialterate and half of a sesquitertian are incommensurable ratios, as is shown in the *De proportionibus proportionum,* it follows that the aforementioned divisions are incommunicant. Thus, *A, B,* and *C* will never conjunct at other times; this applies also to the past and to more [than three] mobiles.[30]

It should be noted, however, that the circle ought not to be divided by such points in accordance with a ratio of velocities, or any ratio commensurable to it. This is immediately obvious in rational ratios, for if *A* were moved four times quicker than *B,* the points of conjunction would divide the circle into three, not four, parts; and the whole circle would be triple any one of these parts. This is shown by the eleventh proposition of the first part.

Proposition 11. . . . On the assumption of the incommensurability and eternity of motions, it is truly beautiful to contemplate how such a configuration as an exact conjunction occurs only once through all of infinite time, and how it was necessary through an eternal future that it occur in this [very] instant with no conjunction like it preceding or following. One cannot even find a reason as to why it happens at that time [rather] than at another time,[31] unless it be because the velocities of motions and the unalterable inclinations of moving bodies are [simply that way].

And if [celestial] configurations are the continuous causes of effects in the lower regions, then, whenever extraordinary aspects concern a whole species, there would occur such a disposition that never again will there be one like it in this world. Speaking naturally, it does not seem inconceivable that a great conjunction of the planets, different from anything that happened before, could produce some individual unlike any other which would begin as a new and previously unseen species in either substance or accident, just as Pliny, in the twenty-sixth book [of his *Natural History*], says that with regard to sickness "the face of man has been afflicted with new and unknown diseases in every age."[32] And, if the world were eternal, perhaps it is even possible that once such a species

has come into being, it would never cease to exist; or it might at some time cease to exist because of the power of another configuration.[33] Similar things may be said about similar corollaries that are deducible from what has been said.

Proposition 12. Propositions enunciated previously [for two or more distinct mobiles] will now be applied to [one and] the same mobile moving with several [simultaneous] motions.

Now just as in the twenty-second proposition of the first part of this work, the motions discussed were commensurable, so in this proposition they are to be taken as incommensurable. For example, assuming that the sun is moved with only two incommensurable and perpetual motions—namely, proper and diurnal—let *A* be the first point of Cancer which daily describes one and the same circle, namely the summer tropic; and let *B* be the center of the sun's body which describes the ecliptic in one year.[34] And [finally] let *A* and *B* be in point *d* imagined motionless in space—i.e., *A* and *B* are in conjunction there.

I say, then, that *A* and *B* will never be simultaneously in point *d* at other times, and this can be demonstrated exactly as was the first proposition [of this part]. Thus, if the sun should enter the first point of Cancer on some [particular] meridian, it will never enter, nor has it ever entered, this sign

28. That is, circumference/arc $de = (3/2)^{1/2}/1$.

29. That is, circumference/arc $dg = (4/3)^{1/2}/1$.

30. That *A, B,* and *C* will never conjunct anywhere else except *d* follows from the fact that irrational ratios $(3/2)^{1/2}$ and $(4/3)^{1/2}$ would not share any common points on the circle except *d,* which they share as a common starting point. The motions of *A* and *B* are such that after leaving *d* their points of conjunction will always differ from those of *B* and *C.* For the circumference of the circle is related to the arcs separating the successive conjunctions of *A* and *B* in the ratio of $(3/2)^{1/2}/1$; and the circumference of the circle is related to the arcs separating the successive conjunctions of *B* and *C* as $(4/3)^{1/2}/1$. In effect, arcs *de* and *dg* are incommensurable, so that when *B* conjuncts with *A* it cannot simultaneously conjunct with *C.*

31. Oresme raised this same question in his *Ad pauca respicientes,* Part II, Proposition 18 (see p. 425 of my edition cited in Selection 25, n. 1); *Questiones super De celo,* Book I, Question 24 (see pp. 419–420 of Claudia Kren's Ph.D. dissertation "The 'Questiones super De celo' of Nicole Oresme," University of Wisconsin, 1965); and his *Le Livre du ciel et du monde,* Book I, chapter 34, 57a–57c (on p. 243 of Albert Menut's translation [Madison, Wis.: University of Wisconsin Press, 1968]).

32. Pliny, *Natural History* XXVI, 1, 1 in the translation of W. H. S. Jones (Loeb Classical Library).

33. This is also discussed in the three other Oresme treatises cited in note 31.

34. Figure 2 of Part I, Proposition 22, illustrates this.

at other times when on this same meridian; nor, indeed, as shown in the second proposition of this part, will it enter when it is on any meridian separated from this one by a distance that is commensurable [to the whole circle of the summer tropic]. There is also an infinite set of points scattered everywhere on circle *a* in each of which *B* has entered Cancer; and there is [also] another infinite set of points in each of which *B* will enter this same sign in the future. Then, as was the case for conjunctions in the fourth proposition of this part, no part [or arc] of this circle is so small that it does not contain some meridian on which *B* appeared [in the past] when it was in the first point of Cancer; nor can it be so small that it does not contain some meridian on which *B* will appear in the first point of Cancer sometime [in the future].[35] One should understand that what has been said about the first point of Cancer applies also to any point of the zodiac.

It follows from all this that *B* describes daily a new spiral—imagined as motionless in space—which it has never before described; and it traverses a path which it has never before traversed. And so, by its track, or imagined flow, *B* seems to extend a spiral line that has already become infinite [in length] from the infinite spirals that were described in the past. This infinite spiral line is constituted from the spirals that were described between the two tropics, and from the intersections between the earlier and later spirals. Consequently, through an eternal time, *B* appears twice in any point of intersection [between these spiral lines][36]—and only once, or never, in any of the other points.[37]

In accordance with what has been imagined here, the whole celestial space between the two tropics is traced by *B*, leaving behind a web-or net-like figure expanded through the whole of this space. The structure of this figure was already infinitely dense through the course of an infinite past time, and yet, nonetheless, it will be made continually more dense, since it produces a new spiral every day. Throughout the extent of such a space, *B* has appeared between any two assignable points and yet, despite this, in the very same space [between any two such points] there are an infinite number of points in which *B* never was, nor ever will be. There is also another infinite [set of points] in each point of which *B* enters only once through all eternity; and there is yet another infinite set of points through each of which *B* passes twice during the course of an infinite time; but there is no point through which it could pass more than twice.

Furthermore, from the incommensurability mentioned above, it might be [the case] that the mean solar year contained a certain number of days and a part of a day incommensurable to a whole day. If this were assumed [to be true], the precise length of the year could not possibly be expressed in numbers, nor could a perpetual almanac be established, or a true calendar found.[38]

35. The enunciation of Part II, Proposition 4, reads: "No sector of a circle is so small that two such mobiles could not conjunct in it at some future time, and could not have conjuncted in it sometime [in the past]." This proposition, which was concerned with two bodies moving with incommensurable speeds, is now applied to the motion of a single body, the sun, moving with two simultaneous, but incommensurable, motions.

36. In this paragraph Oresme turns, once again, to the "Platonic spiral." In Part I, Proposition 22, the simultaneous motions of the sun were assumed commensurable and the spiral line traced by its path was of finite length (though not terminating in any one point), which the sun perpetually retraced as it moved between the tropics (see n. 19). But with its simultaneous motions assumed incommensurable, the sun's path would be an infinite spiral having neither beginning nor end, for it already has described a new spiral every day through an infinite past and will describe a new daily spiral through an infinite future. Although Oresme does not elaborate, it is clear that the spiral must be infinite, since the sun has always entered the tropic of Cancer at a different point and then spiraled down toward the tropic of Capricorn (the formation of the spiral is explained in n. 19), which it always entered, and always will enter, at a different point. It then spirals upward once again toward the tropic of Cancer entering at a new point. Under these circumstances and assuming an infinite past, as Oresme does (of course, as a Christian, he believed that the world had a temporal beginning), the spiral will be of infinite length without beginning or end. Furthermore, since the sun never retraces a previously described spiral, its path can intersect any other spiral line at only one point, and hence through all eternity the sun will have been in every one of these points of intersection twice, but no more.

37. The class of points through which *B*, the sun, passes only once includes all points on the infinite spiral line which it has traced, exclusive of points of intersection. Points through which it never passes are presumably points on either tropical circle that cannot serve as the first point of Cancer or Capricorn. For example, by Part II, Proposition 2, the sun will never enter Cancer on any point that is separated from one which it previously entered by an angular distance that is commensurable to the whole circle.

38. The same opinions were expressed by Oresme in his *De proportionibus proportionum* (see Selection 51.2), *Ad pauca respicientes*, Part II, Proposition 16 (see pp. 420–422, ll. 192–197, of my edition cited in full in Selection 25, n. 1), and his *Questions on the Geometry of Euclid (Quaestiones super geometriam Euclidis)* (edited by H. L. L. Busard [Leiden: Brill, 1961], p. 25, ll. 5–10).

PART III

. . . . . .

[*The Oration of Geometry*][39]

. . . . . .

Furthermore, those who object that man would be ignorant [of celestial events if these motions were incommensurable] are unconvincing. For it is enough to know beforehand that a future conjunction, or eclipse, of this mobile falls below a certain degree, minute, second, or third; nor is it necessary to predict the exact point or instant of time [in which these will occur], since, as Pliny says, "the measure of the heavens is not reducible to inches."[40] And according to Ptolemy, we are unable to determine the exact truth in such matters.[41] In these matters, then, anyone who announces results that are free of noticeable error would seem to have determined things adequately and beautifully. But if men knew all motions exactly, it would not be necessary to make further observations, or to record the celestial revolutions with attentive care. As far as such excellent things are concerned, it would be better that something should always be known about them, while, at the same time, something should always remain unknown, so that it may be investigated further. For such an inquiry, acquired with a sweet taste, would divert noble minds from terrestrial things and, with their desire continually aroused, totally engage and engross those [already] occupied in so respectable an exercise of high-minded endeavor. If, however, these motions were known punctually, and this Great [or Perfect] year were possible, then, surely, all things to come, and the whole order of future events, could be foreseen by men, who could then construct a perpetual almanac based on all the effects of the world. They would become like the immortal gods. But it is the gods, not men, who know future times and moments, which are subject to the divine power alone. Indeed, it would be very repugnant that men should come to know about future events beforehand. It seems arrogant of them to believe that they can acquire a foreknowledge of future contingents, only some of which are subject to celestial powers.

It seems better, therefore, to assume the incommensurability of the celestial motions, since these difficulties do not follow from that [supposition]. Indeed, incommensurability is shown in yet another way, for, as demonstrated elsewhere,[42] when any two unknown magnitudes have been designated, it is more probable that they are incommensurable than commensurable, just as it is more probable that any unknown [number] proposed from a multitude of numbers would be nonperfect rather than perfect. Consequently, with regard to any two motions whose ratio is unknown to us, it is more probable that that ratio is irrational than rational—provided that no other consideration intervenes that was not taken into account in what has already been discussed.

70 ON COMETS

Albertus Magnus (ca.1193–1280)

Translated by Lynn Thorndike[1]

Introduction and annotation by Edward Grant

In this selection Albertus Magnus not only reports a number of explanations of comets which Aristotle had refuted but adds a few others proposed subsequently.[2] A basic issue was whether comets are celestial or sublunar phenomena. For

39. Part III consists of an introduction in which Oresme dreams he is conversing with Apollo, of whom he inquires whether the celestial motions are commensurable or incommensurable. Arithmetic orates first on behalf of commensurability, followed by Geometry, who argues that the celestial motions must be incommensurable. The excerpt which follows is but a part of Geometry's lengthy discourse, which represents Oresme's true opinion.
40. Pliny, *Natural History* II, 21, 87.
41. This purported remark of Ptolemy was drawn by

Oresme from Al-Battani, *De scientia stellarum (On the Science of the Stars)*. See chapter 52, folio 81 recto, of the edition published in 1537.
42. Undoubtedly a reference to Oresme's *De proportionibus proportionum*, Chapter III, Proposition 10, which is included in Selection 25.
1. Reprinted with permission of the University of Chicago Press from *Latin Treatises on Comets Between 1238 and 1368* A.D., edited by Lynn Thorndike (Chicago: University of Chicago Press, 1950), pp. 62–76. Thorndike's translation was made from Book I, Tractate III, of Albert's commentary on Aristotle's *Meteorologica* as printed in the Borgnet edition, IV, 499–508.
2. A good summary account of cometary theory to the end of the fourteenth century is provided by C. Doris Hellman, *The Comet of 1577: Its Place in the History of Astronomy* (New York: Columbia University Press, 1944), ch. 1, pp. 13–65.

Albert, following Aristotle, they are sublunar, occurring in the uppermost regions of the terrestrial world. Since Aristotle's explanation in Book I of the *Meteorologica* became the standard account of comets in the Middle Ages, it will be useful to quote the most relevant and substantive portions of it.

> When the sun warms the earth the evaporation which takes place is necessarily of two kinds, not of one only as some think. One kind is rather of the nature of vapour, the other of the nature of a windy exhalation. That which rises from the moisture contained in the earth and on its surface is vapour, while that rising from the earth itself, which is dry, is like smoke. Of these the windy exhalation, being warm, rises above the moister vapour, which is heavy and sinks below the other. Hence the world surrounding the earth is ordered as follows. First below the circular motion comes the warm and dry element, which we call fire, for there is no word fully adequate to every state of the fumid evaporation: but we must use this terminology since this element is the most inflammable of all bodies. Below this comes air. We must think of what we just called fire as being spread round the terrestrial sphere on the outside like a kind of fuel, so that a little motion often makes it burst into flame just as smoke does: for flame is the ebullition of a dry exhalation. So whenever the circular motion stirs this stuff up in any way, it catches fire at the point at which it is most inflammable. The result differs according to the disposition and quantity of the combustible material. If this is broad and long, we often see a flame burning as in a field of stubble: if it burns lengthwise only, we see what are called "torches" and "goats" and shooting-stars. Now when the inflammable material is longer than it is broad sometimes it seems to throw off sparks as it burns. (This happens because matter catches fire at the sides in small portions but continuously with the main body.) Then it is called a "goat." When this does not happen it is a "torch." But if the whole length of the exhalation is scattered in small parts and in many directions and in breadth and depth alike, we get what are called shooting-stars. (*Meteorologica* I.4.341b.7–35 in the Oxford translation of E. W. Webster)

In chapter 7 (344a.9—344b.17) Aristotle pursues the matter further with reference to comets:

> We know that the dry and warm exhalation is the outermost part of the terrestrial world which falls below the circular motion. It, and a great part of the air that is continuous with it below, is carried round the earth by the motion of the circular revolution. In the course of this motion it often ignites wherever it may happen to be of the right consistency, and this we maintain to be the cause of the "shooting" of scattered "stars." We may say, then, that a comet is formed when the upper motion introduces into a gathering of this kind a fiery principle not of such excessive strength as to burn up much of the material quickly, nor so weak as soon to be extinguished, but stronger and capable of burning up much material, and when exhalation of the right consistency rises from below and meets it. The kind of comet varies according to the shape which the exhalation happens to take. If it is diffused equally on every side the star is said to be fringed, if it stretches out in one direction it is called bearded. We have seen that when a fiery principle of this kind moves we seem to have a shooting-star: similarly when it stands still we seem to have a star standing still. We may compare these phenomena to a heap or mass of chaff into which a torch is thrust, or a spark thrown. That is what a shooting-star is like. The fuel is so inflammable that the fire runs through it quickly in a line. Now if this fire were to persist instead of running through the fuel and perishing away, its course through the fuel would stop at the point where the latter was densest, and then the whole might begin to move. Such is a comet—like a shooting-star that contains its beginning and end in itself.
>
> When the matter begins to gather in the lower region independently the comet appears by itself. But when the exhalation is constituted by one of the fixed stars or the planets, owing to their motion, one of them becomes a comet. The fringe is not close to the stars themselves. Just as haloes appear to follow the sun and the moon as they move, and encircle them, when the air is dense enough for them to form along under the sun's course, so too the fringe. It stands in the relation of a halo to the stars, except that the colour of the halo is due to reflection, whereas in the case of comets the colour is something that appears actually on them.
>
> Now when this matter gathers in relation to a star the comet necessarily appears to follow the same course as the star. But when the comet is formed independently it falls behind the motion of the universe, like the rest of the terrestrial

world. It is this fact, that a comet often forms independently, indeed oftener than round one of the regular stars, that makes it impossible to maintain that a comet is a sort of reflection, not indeed, as Hippocrates and his school say, to the sun, but to the very star it is alleged to accompany—in fact, a kind of halo in the pure fuel of fire.

The Aristotelian interpretation of comets as sublunar phenomena was exceptionally long-lived. Not until observation of the comet of 1577 by Tycho Brahe was their supralunar and celestial locale determined by parallax. Indeed, "no claim can be made that immediately after the appearance of that comet all who wrote on astronomy completely dropped Aristotelian and similar theories and immediately recognized comets as celestial bodies moving according to the general laws governing the motion of such bodies. These laws, indeed, as applied to the solar system, were first stated in their present generalized form by Kepler in the beginning of the seventeenth century" (C. D. Hellman, *The Comet of 1577*, p. 307). But it was not until the 1680's that Dörfel determined a parabolic path for some comets and Halley an elliptical path for others.

CHAPTER 1
OF THE ERROR OF THOSE WHO SAID THAT A
COMET IS A CONJUNCTION OF SEVERAL STARS

Now let's speak of comets or stars having tails, and of what sort they are with respect to efficient form, and how they vary in kind and figure. Some physicists who deviated from nature have said —and, as Seneca says in the book of *Natural Questions,* Apollonius, too, was of that opinion— that stars having tails are many stars gathered together in the lower spheres, which are mobile like the planets and not fixed like the stars of the eighth sphere.[3] Every planet may be in conjunction with every other planet in two ways, truly and not truly. Truly indeed when one eclipses another, and then the lower one is seen and not the upper. Not truly, moreover, again in two ways, namely, that they are in the same part of the meridian, although they do not touch or eclipse; and this is called being on the same meridian but not on the same circle of altitude. Moreover, the meridian is that arc which starts from the North Pole, goes straight overhead through the zenith, crosses the equator, and passes through the South Pole. The circle of altitude is that which comes from the east through the star under observation and goes on to the west.

So it often happens that two planets lie on the same meridian but not on the same arc of the circle of altitude, because one planet may be farther north and the other farther south, and then they are said to be in conjunction so far as the meridian is concerned, but not truly, because they're not in the line of the same circle of altitude. Sometimes, moreover, by their motion they are so close on the meridian that their bodies touch or so nearly touch that the eye cannot discern any of the arc of the meridian between the body of one and the body of the other, and then it may be that one is below the other, yet the eye sees them as a single star made from two or three. And this they said was a comet, and the reason for the deception was that it undoubtedly is like a comet, for, when two or more planets are thus gathered, a person looking at them sees a long continuous light like a tail. This, then, is the opinion of the said physicists as to what a comet is.

CHAPTER 2
OF THE OPINION OF THOSE WHO SAID THAT A
COMET IS VAPOR CLINGING TO A PLANET, JUST AS
THE MORNING SUN IS SEEN WITH THE COLOR OF
VAPOR

Certain Pythagoreans among the Italian philosophers said that a tailed star is only one star, and not many gathered together, although by one star they did not mean just one in number, which is always tailed, because many in turn might be tailed according to them; for this tailed star is one of those that move, that is, one of the planets which sometimes seems to have a tail.[4] For when it rises, it looks small, since the eye, surveying the horizon of our habitable hemisphere, sees only a little of it because of the thick vapor close to the ground between our vision and the rising star. But when it has risen some space above the horizon, the vapor is thinner, and then the star is seen in the vapor, and sometimes the vapor is spread out like long hairs and the planet shines in it, and that diffusion of vapor seems to be its tail, and the star changes color as the vapor changes. For in pure vapor the tail will seem white, and in moist it will appear red; and if what is between us and the star is very watery, the tail will seem golden-red or green; and if

3. Probably based on Aristotle's remark that "Anaxagoras and Democritus declare that comets are a conjunction of the planets approaching one another and so appearing to touch one another" (*Meteorologica* I.6. 342b.25–29).
4. Compare *Meteorologica* I.6.342b.30–343a.20.

there are dark clouds, the tail will appear black. This, then, is the reason, as the Pythagoreans say, why sometimes a tailed star appears.

CHAPTER 3
OF THE OPINION OF THOSE WHO SAID THAT A
TAILED STAR IS THE IMPRESSION OF A STAR ON
HUMID AIR AS IN A MIRROR

Certain others talked similarly concerning a comet so far as matter is concerned, since they said it was an impression on moist vapor, although they differ from these as to the location of that vapor. And they are imitators of Hippocrates of Cos and were his disciples, Nichius and Paul. For they said that, although it was one of the planets which seemed tailed, yet the tail was not an essential part of it.[5] But rather, when one of the five planets moved in its varied course, now rising from the east, now descending to a western sign, now in latitude north of the equator, now carried southward in the opposite direction beyond the equator, then our vision would sometimes see it as if it had a shining tail. This came about because this star impressed its light as in a mirror on the humid and pure air in the upper region of the air, and the air which first received the light reflected it in the humid air adjoining, just as a mirror reflects the form which it receives; and so the air kept reflecting it until a long light appeared about the star, and this is called the tail of the star. Moreover, they call humid air a moist vapor of subtle substance and clear, which they say is drawn to the upper air by the force of the sun, which, according to the Egyptian physicists, draws that humor to nourish itself and the other planets even beyond the sphere of fire. Moreover, such a star, according to them, is [not] seen in the north, where every star has more of its orbit above than below the earth because of the obliquity of the horizon; and so the sun, when in the northern signs, makes the days longer than the nights; and so they say that that star stays above earth longer than another.

Some give a different explanation,[6] that that moist vapor which is attracted for nourishment, although diminished, remains standing about the star until consumed and so is seen to rotate with it for some time. Those supporting this view give the reason that in the west, where is completed the revolution of the star above the hemisphere, there is abundant humor from cold and vapors and that there it frequently appears tailed. They add that in the north, where humor abounds because of the cold, and in the south, where humor abounds

which the planets attract as they cross those regions, a tailed star is more often seen; and when a star is in the air where there is no humor, north and south, it is not seen to have a tail. But, although this sometimes happens in the circle farthest south, as in the small circle which the movement of the pole of the zodiac describes to the south, yet it does not appear to us there because of the too great remoteness from our vision. But when it is in the north over our habitable world, then its tail is visible to us, since then our sight can comprehend it because it is near and overhead. These, then, are the views which Aristotle found had been expressed as to comets.

CHAPTER 4
A DIGRESSION TO GIVE THE OPINIONS OF
SENECA AND JOHN OF DAMASCUS AND OF
CERTAIN MODERNS AS TO COMETS

There are still some false views which Aristotle did not bother to note which we give here in order to overthrow them at the same time. For John of Damascus says (II, 7):

Often comets are formed, signs signifying the deaths of kings, which are not of those stars which were generated from the beginning, but by divine command are formed to suit the time being and again are dissolved.

Seneca, moreover, in the book of *Natural Questions* thus expresses his opinion concerning comets, saying:

A comet is a star created together with the works of nature, whose nature it may be is ignored, yet that it is not transient fire is proved thus. Whatever things air produces are short-lived. How then can a comet last long in air, when air itself is not long permanent? For they are born in what is flighty and mobile, and it cannot be that fixed fire should have its seat and so pertinaciously adhere in a wandering body.[7]

And later he adds in the same place:

A comet has its seat and so is not quickly expelled but traces its measured course, nor is extinguished but passes on.[8]

There are also certain learned moderns who say that a comet is the impression of one of the five

5. Based on *Meteorologica* I.6.343a.23–30.
6. The following explanation is drawn from *Meteorologica* I.6.343a.1–20.
7. Seneca, *Natural Questions*, VII.22. Thorndike notes that Albert's version of this passage differs considerably from the modern edition.
8. *Natural Questions*, VII.23.

planets, not merely in fire, nor merely in air, but rather at the meeting line of fire and air, where the convex surface of the air mingles with the concave of fire, for there there are certain obscure lines of air and certain luminous, as they say, lines of fire which are further illuminated by the light descending from one of the five planets, and so the tail seems to them to be made up of obscure lines and luminous lines intermixed. These have four arguments for themselves, which I will briefly touch upon mediately rather than argumentatively. For they say that it seems unfitting, if the fixed stars are the causes of some impression and the planets of none, since it is evident that the planets are much stronger in moving inferiors than the fixed stars are. If, therefore, they are causes of some impression, it is proved inductively that no impression suits them except tailed stars. The second line of argument is that comets are generated either in the way aforesaid or from dry terrestrial vapor. But if from dry terrestrial vapor, either that vapor is uniform or not uniform. If it is uniform, either it is raised all at once or successively. Not all at once, because, if all at once, then it would also be inflamed and illuminated all at once, and so would appear in a round form like *assub,* as will be shown below and so the comet would not be seen in elongated form. But if not raised all together, then it will be long and inflamed, standing erect in air; therefore, it would look like a lance or perpendicular fire. But if it is not uniform, either it is raised all at once, or not all at once. If it is raised simultaneously, then the first subtle part will be raised and set on fire, and the lower because of its coarseness will be spread out, and thus it will seem perpendicular and broad below; and so again it will not be in the figure of a comet. But if it is not raised simultaneously, then the subtle part will rise without the coarse; therefore, it will also be set afire without it; therefore, the coarse will never follow the subtle; therefore, the figure of that fire will never be elongated by the coarse part set afire and joined to the subtle; and so again it will not have the figure of a comet. The third argument is that if it is of terrestrial vapor set afire, it seems that the same should happen in space as happens in the case of *assub,* namely, that it descend that it may straightway ascend, and this we do not see, since often it long moves in a circle. Therefore, it is not of inflamed terrestrial vapor. The fourth argument is that no inflamed terrestrial vapor, which at the same time rises to the hot upper region of the air and secondarily descends to earth,

appears to rise in the air, since either it is consumed in that upper region in which fire triumphs, as said above, or it is weighed down with cold and depressed to earth. But a comet often appears to rise in the air; therefore, it is not from terrestrial vapor inflamed in or below the upper region of air. These, then, are the views about comets which Aristotle did not touch upon.

CHAPTER 5
A DIGRESSION GIVING THE CORRECT VIEWS OF PHYSICISTS, AVICENNA, ALGAZEL, PTOLEMY, AND MANY OTHERS, AS TO COMETS, WHO ALL AGREE ON THE SAME POINT.

Having excluded all these erroneous views, as our rebuttal of them below will show, we will give the correct view as to comets, and we will confirm it by the authority of many physicists and even by the arguments which those philosophers advance for themselves. I say, then, that a comet is nothing else than coarse terrestrial vapor, whose parts lie very close together, gradually rising from the lower part of the upper region of the air to its upper part, where it touches the concave surface of the sphere of fire and there is diffused and inflamed and so often seems long and diffuse. I say "terrestrial vapor" to denote the material of the vapor. And I say "coarse" because, if it were subtle, it would quickly evaporate and be dissipated. And I say "whose parts are close together," since it is well mixed and viscous in so far as such vapor which is not actually humid and viscous can be. And it is said "ascend gradually," since in rain-producing vapors are intermingled some ignited terrestrial parts which do not all descend with the rain, and those which do descend for the most part reascend with evaporation. And they escape beyond the middle cold region of the air because of their state and stay there and multiply. And from that multiplying, as from some reservoir, they gradually ascend because of the heat of the upper region of the air, because they have much constancy in many respects. At first, the mass is diffused by the heat of the fire and afterward inflamed; yet in the middle it always remains dense, since it is supplied by its reservoir below it, and so its flame is very white and thick. But that which is diffused from it at a distance to the sides is tenuous and has a thin flame like a white cloud, and this is called the "tail." Moreover, it lasts through all the time until its supply is exhausted. This is the view of the philosopher Constantinus, who is said to have commented in Greek on the *Liber meteororum* of Aristotle, where he speaks of comets.[9]

Moreover, that it is so, testify the illustrious philosophers Avicenna and Algazel. For Avicenna speaks thus: "A star which is called tailed is made of thick ignited fumes, which, since it is quickly converted, sometimes revolves circularly with the sphere of fire." Algazel in his *Physics* says:

> If fire prevails over elevated fumes, it will purge them of darkness and then all the fumes will be converted into fire. And only one of two things can take place there: either it will be ignited, or it will become pure fire, because there is no frigidity there. But if it is thick, it will indeed ignite; but because it is not such as is readily altered, it will so remain for some time and will seem a tailed star. Which revolves with the sky, because the parts of fire are continuous with the concave surface of the heavens; so it revolves because of its relationship.

From these words is shown concerning comets what we have said before, and so Algazel says that the matter of the coarse vapor revolved with the sky seems threefold. For sometimes it is flame, as was just said. But sometimes coarse and black, and then it is ignited like coal, and that comet looks red. Sometimes, moreover, the fire with it is extinguished, because the matter is too coarse, and it remains smoky, and then it looks like coal that is black and extinct; and so it is that dark apertures are seen in the sky, which phenomenon is called by the people "perforation of the sky."

Alfraganus, moreover, in his *Astronomy* proves that fire is moved in a circle from the fact that the vapor set on fire adhering to it rotates in a circle beneath it, saying:

> Circular motion is derived from the sky apart from natural motions; and this is so, since we see in the element fire a circular motion like the motion of the sky, as in the case of scintillations of stars apart from natural motions, of flaming vapors in the upper air, in evening twilight, so that the spectator thinks them stars and sees them moved with the motion of the stars and following these until they are hidden.

From this it is inferred that they are like stars and not stars and are vapors ignited in the air. Ptolemy implies the same in his *Centiloquium,* third proposition, where he says that third judgment is from second stars, on which three commentators, namely, Haly and Abraham and Bugaforus, remark that second stars are the effects of the stars upon vapor elevated and ignited, as in comets and *assub* and the like. Albumasar also implies this in the book on *Conjunctions of the Planets,* where he discusses the quality of the science of the generation of superior individuals apart from inferior accidents and says that one part of that signification is from superior meteors, such as fire and *assub* and comets. And he adds that they are generated from one matter by the power of Mars alone, without any other star. So it is clear that this view, which we have stated, is that of all the better philosophers.

Reason, too, rallies to the support of this opinion, since it is evident that flame is nothing but kindled fumes; now a comet is a sort of flame, as is apparent to the sight; therefore, it is kindled fumes. Further, every other impression which is luminous, having a distinct light by itself, is from kindled vapor; therefore, this too. Besides, let us ask our modern doctors, since a comet is light kindled by one of the five planets at the boundary of air and fire and since the planets always are moved above this kind of boundary where air is mixed with fire, why don't comets always appear? Also why don't several comets appear at the same time, since several planets may rise simultaneously above this mixture of air and fire? Also why is diverse form seen in comets? Since, as Seneca says, sometimes it revolves, pointing its tail upward, sometimes casting it downward, and sometimes sideways. It seems that they would be unable to make any reasonable reply. Furthermore, if a comet is always produced by one of the five planets, then it should never be seen outside the path of the planets. But the path of the planets is in the zodiac or very little outside; therefore, a comet should always be seen in the zodiac or near it; and yet this is false, since Aristotle says that we see comets to the north and south and in every part of the sky. Moreover, I with many others in Saxony in the year 1240 from the Incarnation of the Lord saw a comet close to the North Pole, and it projected its rays between east and south, but more toward the east, and it is evident that there was not the path of any planet. But, since also the other three opinions which Aristotle mentioned say that a comet is generated by the five planets, therefore this opinion together with them in this respect will receive disproof.

9. Thorndike observes that "the allusion seems to be to Konstantinos Psellos, usually known as Michael."

CHAPTER 6
A DIGRESSION DISPROVING THE OPINION
OF SENECA AND OF JOHN OF DAMASCUS

That the statement of Seneca is an error is patent, since what vanishes without setting but within the hemisphere, that is corrupted not completing its course through the hemisphere. Moreover, a comet thus vanishes before the eyes of men; therefore, it is extinguished and disappears. Moreover, that it vanishes before the eyes of men is proved by two experiences which Seneca himself records in the book of *Questions*,[10] saying:

After the death of Demetrius, king of Syria, shortly before the Achaean War, a comet shone forth not smaller than the sun, at first fiery and rubicund and emitting a clear light; then gradually its size was restricted and its clarity vanished; finally it totally died out.

From this, then, is evident what has been said. He recorded the second in the same place, saying:

While Attalus was ruling in Greece a moderate-sized comet appeared; then it spread out and reached the equator so that it equaled in extent that part of the sky called the Milky Way, and there within an instant it ceased to be.[11]

The opinion of John of Damascus, that God makes a new star, is not probable, since there is no reason why He should make it in different figures and colors. But a comet is seen in different figures and colors; therefore, it is not a work of God alone. That it has different figures is evident from the words of Seneca, who says:

The form of a comet is not single, for there were many comets from which flame hung like a beard, and those which spread their hairs in every direction about them, and those in which fire was diffuse but tending to a point: yet all are of the same nature and are rightly called comets. Whose forms, since they appear at long intervals, are difficult to compare. Indeed, at the very time they appear, observers do not agree as to their appearance, but as each is sharp or dull sighted, so he says they are more brilliant or rubicund or obscure.[12]

Since, therefore, this variety of form is because of some pre-existing matter from which nature works, it is evident that a comet is not then produced by divine work only. So then it is evident, with regard to the three opinions of Seneca, John of Damascus, and certain doctors of our time, that they are false, and that that which is true is to be held.

CHAPTER 7
A DIGRESSION TO DISPROVE THE FIRST OPINION WHICH
ARISTOTLE DISPROVES

What Anaxagoras and Democritus said,[13] and Apollonius after them, that a comet is many planets congregated, Seneca disproves by four reasons, of which the first two are taken from experiences. For if a comet appeared before the Achaean War as large as the sun, whereas no other stars, albeit gathered together, ever equal the sun, it is evident that a comet is not generally a congregation of stars. The second case is that the comet which appeared while Attalus was ruling in Greece so extended itself that it equaled in immense extent that region of the sky known as the Milky Way. For the planets cannot come together so that they occupy continuously so long a tract of the sky. Two other arguments are based on reason, one from the position of the planets, which he states in these words: "Apollonius says that not one comet is made up of planets but many planets are comets." But if this were so, comets would move within the limits of the zodiac. The second argument is from the nature and property of the stars, which he states thus: "A star never shines through a star, since otherwise one would not eclipse another; through a comet sometimes, as through a cloud, objects beyond are seen, whence it appears that a comet is not a star but a light and tumultuous fire." On the same point Seneca further remarks: "It is certain that comets are not seen in one part of the sky only but both in the east and west, even more frequently to the north."

CHAPTER 8
DISPROOF OF THE SECOND AND THIRD OPINION BY
REASONING OF ARISTOTLE

This being so, let us overthrow the other two opinions by the arguments of Aristotle. Since we have already repeated their utterances concerning comets, we ought now to ask whether these are true or false. I say, then, that often we see comets outside the eighth sphere of the fixed stars, and so it cannot be that a comet is one or more of the fixed stars, since then it would be in the first belt, which is the eighth sphere which encloses all the starry sphere, which is false because we see comets

10. *Natural Questions,* VII.15.
11. *Ibid.*
12. *Natural Questions,* VII.11.
13. See note 3.

outside that sphere far below it. But if it were one of the planets, then it would never be seen except in the spheres of the planets, which is false because it is seen outside these, beneath them, and outside the circle of the signs. But if comets were stars seen through humid air, as those of the second opinion said, then when the star comes to mid-sky, where our vision observes it without the interposition of humid air, it ought to appear without a tail. And this again is false, since a comet sometimes appears anywhere in the revolving sky and sometimes for many days. Furthermore, if the tail were caused by humid air, then every star seen through humid air would be a comet; so comets would be as numerous as the stars and would not appear only in one of them but in many at once. And this is false, since we do not find stars having tails except five, and and these not at once but successively, as we shall state below. And their assertion that a comet is not found except in the north is false, since one is often seen elsewhere than in the north, for a comet is sometimes seen with the sun preceding it as it rises in summer. Moreover, the summer rising of the sun is the sign in which the sun rises in summer, such as the northern signs in which it moves from the beginning of Aries to the beginning of Libra, because this summer rising is interpreted broadly for any rising in which the day exceeds the night, since then the sun regards us more directly and warms our habitation. And likewise tailed stars are in the south as time varies. And this happens especially in the city of Arim, which is on the equator at the center of the world, ninety degrees from the Arctic pole and ninety from the Antarctic, and ninety to the east and ninety to the west of the habitable world. For the region at the equator is more temperate than any other spot in this world, as Ptolemy says and Avicenna and many other philosophers. Comets are also seen in the north in the morning, rising before the sun in its winter rising, and this cannot be otherwise than when the sun rises in winter in southern signs, since then the comet was rising in the north in northern signs. All this that has been said shows us that tailed stars are other than the fixed stars and planets which they have said, and are seen in summer and in winter in various parts of the sky and world. So then is shown the error of their statements.

CHAPTER 9
ARISTOTLE'S OPINION CONCERNING COMETS

I say, then, that stars having tails effectively and by the place of their generation are air inflamed by the vicinity of fire and which contains the matter from afar and the comet itself. I say air close to fire which contains in itself the heat of fire, since, when dry coarse vapor is inflamed in it, then a continuation is made with the light of the flame of the comet from the lower air where stands vapor of the same nature, and so it is elongated from this extension continuous with it; and such is a comet, as is clear from the aforesaid. Some expound otherwise this opinion of Aristotle, but I do no violence to their exposition, since almost all philosophers who have followed Aristotle agree with our interpretation.

CHAPTER 10
A DIGRESSION EXPLAINING WHY THERE ARE SAID TO BE FIVE COMETS AND NO MORE

Someone may ask according to that interpretation of Aristotle why there are said to be only five comets. For if the number is not taken from the planets, it will seem to depend merely on difference in material, but differences in material are infinite and so should not be stated as five. Moreover, those stars are called "secondary" by the philosophers; therefore, there should be some primary stars which are their causes; it is evident that these are not fixed stars, it seems therefore that they are planets; they are not the two luminaries, so they should be the five planets. But, according to this, they should be only in or near the zodiac, whereas it was held above that they are seen in every part of the sky.

To this without prejudice I think it should be answered that in truth, according to astronomers who base judgments upon them, five comets are not reckoned as five planets, as we will state below, but either according to five differences of the vapor composing them or according to five forms of stars having tails. The differences of vapors are that all cometary vapor, though of a coarse and sticky variety, yet may be relatively coarse or subtle or halfway between. Moreover, where there is a midway mean, there is also a mean near either extreme, and so there will be three means, namely, halfway, and between it and subtle, and between it and coarse. Nor can there be greater diversity of vapor in variety, and so comets number five. Also they are five by figure, for either the vapor surrounds it in a circuit, or from lightness is above the comet, or hangs downward, or to one side, in which case it is evidently either from both sides or from one side: and so they are again five. It may be, however, that the primary difference is one of matter, since it causes difference in form, since thin vapor takes on one shape, and thick another. That they are called "secondary stars" is because they signify the force of the planets in diverse places, as we shall soon state, yet not on this account are they numbered according to the planets. It may be, too, that difference in vapor as coarse or subtle is a difference in matter; yet it is not purely material but rather ordained for form, since by intrinsic and extrinsic heat such matter takes on such formality, and this question of diversity is one which properly belongs to the physicist.

CHAPTER 11

A DIGRESSION WHY COMETS SIGNIFY THE DEATH OF POTENTATES AND WARS

Now it must be asked if we can comprehend why comets signify the death of magnates and coming wars, for writers of philosophy say so. The reason is not apparent, since vapor no more rises in a land where a pauper lives than where a rich man resides, whether he be king or someone else. Furthermore, it is evident that a comet has a natural cause not dependent on anything else; so it seems that it has no relation to someone's death or to war. For if it be said that it does relate to war or someone's death, either it does so as cause or effect or sign. Evidently it does not have sufficient relation to anything as cause, since it is neither efficient cause, nor formal, nor final, nor material. Similarly it is proved not to be an effect, since neither necessarily follows the other, but an effect necessarily follows its cause. Nor does it relate even as a sign, since it has not conformity, but every sign has conformity with what it signifies. Furthermore, as some say, the distinction of five comets depends on the five planets, and, since not all the planets have such bad effects, since Jupiter signifies fortune and similarly Venus and Mercury do not signify misfortune, it seems that comets should not as a rule be signs of evil.

Consider, then, things like them, for as Albumasar says in the seventh tractate *On the Conjunctions of the Planets,* the event of fires and *assub* and

comets is not inferred from any planet except in the air, especially when its rays were in terrestrial or aerial signs, the moon not impeding, since it stirs up humid, aqueous vapor which might prevent such fiery phenomena. Hence it is clear that it is not stated that comets are interpreted according to the five planets, since such phenomena are only of the character of Mars and so are from it as prime mover, except perhaps occasionally, as in the conjunction of Jupiter and Mars, since from that conjunction scintillations and coruscations and running fires are also moved through the air.

In answer to the objection as to land of poor and rich, it should be said that it indicates the destruction of both according to the diversity of period of which we have spoken at the end of the second book *On Generation and Corruption.* But the death of kings is noticed more because of their fame, since their periods have more planetary dignity, and so greater significations are referred to them. In answer to the assertion that it has no natural cause related to anyone's death, it should be said that it is not the immediate cause of the death of kings; yet its first moving cause denotes the force of Mars, which is a sign of death especially by violence, and then Mars dominates the commotion of the elements; also it is lord in the period and shows the cause of death, so far as the prime mover may be called the cause, which is not necessary but rather inclinatory. As to the question whether it be sign or cause, it should be said that it is properly sign, since, as Aristotle says in the book *On Sleep and Waking,* such are like a counselor whose advice can be changed by finding better. So the rule of Mars signifies wars and death of inhabitants and disturbances of rulers, as inclining to this by wrath and heat and drought, from which comes animosity or exciting of peoples against one another. And so the significance of comets is primarily connected with Mars, which is the cause of war and destruction of peoples; and so a comet is said to signify these, although its immediate significance is not connected with them.

71 THE POSSIBILITY OF A PLURALITY OF WORLDS

Nicole Oresme (ca. 1325–1382)

Translated by Albert D. Menut[1]

Introduction and annotation by Edward Grant

Oresme's analysis of the possible existence of other worlds was perhaps the most cogent and

1. *Nicole Oresme: Le Livre du ciel et du monde,* edited by Albert D. Menut and Alexander J. Denomy, C.S.B.†;

penetrating discussion of this question during the Middle Ages. Not only did he incorporate many of the standard arguments, but he went considerably beyond, adding new insights and subtleties. In his response, he followed a somewhat usual pattern by distinguishing different possible interpretations of a "plurality of worlds," taking up each in turn. Of special interest is the third interpretation involving the simultaneous existence of many worlds. Here Oresme's objective is to demonstrate the inconclusiveness of Aristotle's arguments that only one world is possible. Although writing one hundred years later, Oresme's attitude reflects the influence of Article 34 of the Condemnation of 1277, which denounced those who would deny that God could create as many worlds as He pleased (see Selection 13). Not only does Oresme advocate significant cosmological ideas, but he attempts to cope with some of the physical problems that would have immediately arisen for Aristotelians. Indeed, he seeks to apply and extend Aristotelian physical principles to these other possible worlds, whose existence Aristotle had denied categorically. In the end, however, Oresme upholds the traditional Aristotelian and Christian position that there is only one world. It was never his intention to overturn tradition, but merely to demonstrate the possibility—and no more—that other worlds could exist quite readily if God but chose to create them.

Of the three interpretations discussed here, the first and third were of ancient vintage—namely the concept of a succession of single worlds and a plurality of simultaneous, separate, existing worlds scattered about in space. The second interpretation—where the plurality is simultaneous but the worlds are contained within each other—appears to have been a medieval invention. Albert of Saxony—and Oresme himself in his earlier *Questiones super de celo (Questions on De caelo)*, Bk. I, Question 18—even conceived of these worlds as arranged concentrically or eccentrically. As Albert explains it (*Questions on De caelo*, Bk. I, Question 11, fol. 95r, col. 1, of the edition cited in n. 1 to Selection 35):

Several worlds can be imagined simultaneously which are either concentric or eccentric. Indeed, they would be concentric if beyond the last heaven of our world we should imagine an earthy orb, and beyond that an aqueous orb, and beyond that an airy orb, and beyond that a fiery orb, and beyond that celestial orbs like the celestial orbs of our world; or, if inside our earth, it could be imagined that the aggregate of celestial

orbs was contiguous with the concavity of the orb of our earth and within this aggregate of supercelestial orbs, four elements could be imagined, just as in our world. And so, by ascending and descending, an infinite number of concentric worlds can be imagined.

Similarly, we can imagine several eccentric worlds: either (1) one lies wholly outside the other, and this could be [imagined] in the way several globes are placed in a sack, or [it might be imagined that they are inside but] do not touch; or (2) that one does not lie wholly outside the other, but that there is another world in some part of our world, as if, [for example,] another world were imagined in the moon, or in the sun, and in the other planets.

Now we have finished the chapters in which Aristotle undertook to prove that a plurality of worlds is impossible, and it is good to consider the truth of this matter without considering the authority of any human but only that of pure reason. I say that, for the present, it seems to me that one can imagine the existence of several worlds in three ways. One way is that one world would follow another in succession of time, as certain ancient thinkers held that this world had a beginning because previous to this all was a confused mass without order, form, or shape. Thereafter, by love or concord, this mass was disentangled, formed, and ordered, and thus was the world created. And finally after a long time this world will be destroyed by discord and will return to the same confused mass, and again, through concord, another world will then be made. Such a process will take place in the future an infinite number of times, and it has been thus in the past. But this opinion is not touched upon here and was reproved by Aristotle in several places in his philosophical

translated by Albert D. Menut (Madison, Wis.: University of Wisconsin Press, 1968), pp. 167–179. Reprinted with permission of the copyright owners, The Regents of the University of Wisconsin. Of Oresme's *Le Livre du ciel et du monde (Book on the Heavens and the World)*, which is a French commentary on Oresme's own French translation of Aristotle's *De caelo (On the Heavens)*, the present selection includes most of Book I, chapter 24, and is an actual commentary on I.9.279a.12–279b.4 of Aristotle's *De caelo*. Completed in 1377, this represents Oresme's last extant work before his death in 1382. It was one of his best. Apart from a few minor changes and the omission of a few cross-references and folio numbers to one of the French manuscripts, the translation is faithfully reproduced.

works.[2] It cannot happen in this way naturally, although God could do it and could have done it in the past by His own omnipotence, or He could annihilate this world and create another thereafter. And, according to St. Jerome,[3] Origen used to say that God will do this innumerable times.

Another speculation can be offered which I should like to toy with as a mental exercise. This is the assumption that at one and the same time one world is inside another so that inside and beneath the circumference of this world there was another world similar but smaller. Although this is not in fact the case, nor is it at all likely, nevertheless, it seems to me that it would not be possible to establish the contrary by logical argument; for the strongest arguments against it would, it seems to me, be the following or similar ones. First, if there were another world inside our world, it would follow that our earth is where it is by constraint, because inside this earth and beneath its circumference toward its center would be another heaven and other elements,[4] etc. Also, the earth of the second world would be absolutely massive and at the center of both worlds; and the earth of our world would be empty and concave and neither the whole earth nor any part of it would be at the center. Thus, since their natural places are different, it follows from what is said in Chapter Seventeen that these two worlds are of different form so that the world beneath us and this our world would be dissimilar, etc. Also, all natural bodies are limited in bigness and smallness, for the size of a man could diminish or grow so much that he would no longer be a man, and the same with all bodies. So, the world we have imagined inside our own world and beneath its circumference would be so small that it would not be a world at all, for our sun would be more than 2,000 times the size of the other and each of our stars would be larger than this imaginary world. To pursue our thought, one could dig in the ground deep enough to reach the earth of the other world beneath ours. This is an untenable absurdity. Also, we should have to posit two Gods, one for each world, etc. Likewise, we might assume another world like our own to exist in the moon or some other star, etc. Or we could imagine another world above and another beneath the one which is our world, etc. To show that these and similar speculations do not preclude the possibility of such a thing, I will posit, first of all, that every body is divisible into parts themselves endlessly divisible, as appears in Chapter One; and I point out that *large* and *small* are relative, and not

absolute terms, used in comparisons. For each body, however small, is large with respect to the thousandth part of itself, and any body whatsoever, however large, would be small with respect to a larger body. Nor does the larger body have more parts than the smaller, for the parts of each are infinite in number.[5] Also from this it follows that, were the world to be made between now and tomorrow 100 or 1,000 times larger or smaller than it is at present, all its parts being enlarged or diminished proportionally, everything would appear tomorrow exactly as now, just as though nothing had been changed. And, if a stone in a quarry had a small opening in it or a concavity full of air, it is not necessary to say that this stone is outside its natural place. Likewise, if there were a concavity the size of an apple full of air at the earth's center, it would not follow that the earth was out of its natural place nor that it was there by violence. Also, if such concavity were to become a bit larger and then still larger until it became very large, we could not place a limit upon this growth at which point one could say the earth would be out of its natural place, precisely because large and small are relative terms, as we have already said. Therefore, for the earth to be in its natural place, it is enough that the center of its weight should be the center of the world, regardless of the concavity inside the earth, provided that it be held firmly together. And this is the answer to the first argument; for, if a world were enclosed within a concavity inside our earth, nevertheless our earth would be in its natural place since the center of the world would be the middle or center of its weight.[6] A propos, I say further that, according to

2. This is the view of Empedocles (fifth century B.C.). Among other places, see Aristotle's *Physics* VIII.1.252a. 7–10; VIII.9.265b.19–21; *On the Heavens* I.10.279b. 15–17; II.13.295a.30–-295b.1; III.2.301a.16–20; *Metaphysics* I.4.985a.22–30.

3. In a letter to Avitus (see letter 124 in Vol. 3 of I. Hilberg's edition of St. Jerome's letters, *Sancti Eusebii Hieronymi epistulae* [Vienna: F. Tempsky, 1918]).

4. By extension of Aristotelian principles to this hypothetical situation, it could only be concluded that the earth above the heaven would be out of its natural place and could be maintained there only by force.

5. We have here the assertion of a one-to-one correspondence between an infinite set and its subset. This important concept was expressed at least as early as the Greek Stoics and repeated in the Middle Ages by many scholastics, including Roger Bacon. Galileo later applied this notion to numbers, showing a one-to-one correspondence between the infinite set of natural numbers and its subset of square numbers.

6. That is, the center of the world would correspond to

Scripture, water is above the heavens or the firmament; whence the psalm says:[7] Who stretchest out the heavens, etc., Who coverest these heavens with water. And, elsewhere: Bless the Lord, ye waters that are above the heavens.[8] And if this water were not heavy in substance if not in fact, then it would not be water. For this reason it is said to be solid and as though frozen or solidified and is called the glacial or the crystalline heavens. Accordingly, this heaven or this water is in its natural place, in spite of the fact that all the other heavenly spheres and elements are enclosed within the concavity of this sphere, for it is solid and the center of its weight is the center of the world. To the second argument I reply that, even if this earth were hollow and concave, nevertheless it would be in the center of the world or worlds, just as though this were its proper place, taking *place* in the sense of the second member of the distinction made in reply to an argument in Chapter Seventeen.[9] From this it appears that our earth and the earth of the other world within it would be in the same place. To the third argument, which stated that all natural bodies are limited in quantity, I say that in this world they are limited to one quantity or size and that in another world they would be fixed at other limits, for large and small, as we have said, are relative terms which do not mean variation or difference in form. Accordingly, we see men—all of the same form—larger in one region and smaller in another. To the fourth argument, where it was stated that one could dig deep enough into the earth, etc., I answer that nature would not permit this, any more than one could naturally approach the sky close enough to touch it. To the fifth argument, regarding the possibility of two Gods, it does not follow; for one sovereign God would govern all such worlds, but it is possible that additional intelligences would move the heavenly bodies of one world and other intelligences the heavens of the other world. To the sixth argument, where it was said that by analogy one could say there is another world inside the moon, and to the seventh, where it was posited that there are several worlds within our own and several outside or beyond which contain it, etc., I say that the contrary cannot be proved by reason nor by evidence from experience, but also I submit that there is no proof from reason or experience or otherwise that such worlds do exist. Therefore, we should not guess nor make a statement that something is thus and so for no reason or cause whatsoever against all appearances; nor should we sup-

port an opinion whose contrary is probable; however, it is good to have considered whether such opinion is impossible.[10]

The third manner of speculating about the possibility of several worlds is that one world could be entirely outside the other in an imagined space,[11] as Anaxagoras held. This solitary type of other world is refuted here by Aristotle as impossible. But it seems to me that his arguments are not clearly conclusive, for his first and principal argument states that, if several worlds existed, it would follow that the earth in the other world would tend to be moved to the center of our world and conversely, etc., as he has loosely explained in Chapters Sixteen and Seventeen.[12] To show that this consequence is not necessary, I say in the first place that, although *up* and *down* are said with several meanings, as will be stated in Book II [see ch. 4], with respect to the present subject, however, they are used with regard to us, as when we say that one-half or part of the heavens is up above us and the other half is down beneath us. But up and down are used otherwise with respect to heavy and light objects, as when we say the heavy bodies tend downward and the light tend upward. Therefore, I say that up and down in this second usage indicate nothing more than the natural law concerning heavy and light bodies, which is that all the heavy bodies so far as possible are located in the middle of the light bodies without setting up for them any

the center of gravity of the earth, but not to the actual center of the earth.

7. Psalms 103:2–3.

8. See Daniel 3:60.

9. In Book I, chapter 17 (Menut translation, p. 139), Oresme distinguishes between (1) place as the immediate container of a thing—as, for example, the concave surface of a barrel of wine is the container of the wine—and (2) place "as that which defines the proper state of a body seated in its unique natural location; in this sense the center of the world is the place natural and proper to the earth and to all the multitude of objects possessing weight, . . ." Thus, even if the earth were hollow and concave, its center would still coincide with the geometric center of the world. With this approach, Oresme is satisfied that the earths of every world are in their natural places, for every one of their centers would coincide with the geometric center of the universe.

10. And so Oresme shows that a series of worlds arranged one inside the other is not a contradictory concept and is therefore logically possible. For this particular interpretation of a plurality of worlds, Albert of Saxony adopted much the same attitude as Oresme.

11. In this sentence I have altered slightly Menut's translation.

12. Aristotle makes this point in *On the Heavens* I.8.276b.10–18.

other motionless or natural place. This can be understood from a later statement and from an explanation in the fourth chapter, where it was shown how a portion of air could rise up naturally from the center of the earth to the heavens and could descend naturally from the heavens to the center of the earth.[13] Therefore, I say that a heavy body to which no light body is attached would not move of itself;[14] for in such a place as that in which this heavy body is resting, there would be neither up nor down because, in this case, the natural law stated above would not operate and, consequently, there would not be any up or down in that place. This can be clarified by what Aristotle says in Book Four of the *Physics*, namely, that in a void there is no difference of place with respect to up or down.[15] Therefore, Aristotle says that a body in a vacuum would not move of itself.[16] In the eleventh chapter of this first book it appears, according to Aristotle, that, since nothing is lower than the center of the earth, nothing is or can be higher than the circumference or the concavity of the lunar sphere,[17] the place proper to fire, as we have often said. Thus, taking up in the second sense above, beyond or outside of this circumference or heaven there is no up nor down. From this it follows clearly that, if God in His infinite power created a portion of earth and set it in the heavens where the stars are or beyond the heavens, this earth would have no tendency whatsoever to be moved toward the center of our world. So it appears that the consequence stated above by Aristotle is not necessary. I say, rather, that, if God created another world like our own, the earth and the other elements of this other world would be present there just as they are in our own world. But Aristotle confirms his conclusion by another argument in Chapter Seventeen and it is briefly this: all parts of the earth tend toward a single natural place, one in number; therefore, the earth of the other world would tend toward the center of this world. I answer that his argument has little appearance of truth, considering what is now said and what was said in Chapter Seventeen. For the truth is that in this world a part of the earth does not tend toward one center and another part toward another center, but all heavy bodies in this world tend to be united in one mass such that the center of the weight of this mass is at the center of this world, and all the parts constitute one body, numerically speaking. Therefore, they have one single place. And if some part of the earth in the other world were in this world, it would tend toward the center of this

world and become united with the mass, and conversely. But it does not have to follow that the portions of earth or of the heavy bodies of the other world, if it existed, would tend to the center of this world because in their world they would form a single mass possessed of a single place and would be arranged in up and down order, as we have indicated, just like the mass of heavy bodies in this world. And these two bodies or masses would be of one kind, their natural places would be formally identical, and likewise the two worlds.[18]

13. In Book I, chapter 4, Oresme set up certain hypothetical conditions to show how one might disagree with Aristotle's contention that a simple or elemental body has only one simple motion—that air, for example, rises in the region of water and always falls when in the region of fire, in both instances seeking its natural region of air. Oresme counters this with an illustration: "I imagine the case of a tile or copper pipe or other material so long that it reaches from the center of the earth to the upper limit of the region of the elements, that is, up to the very heavens. I say that, if this tile were filled with fire except for a small amount of air at the very top, this air would drop down to the center of the earth for the reason that the less light body always descends beneath the lighter body. And if this tile were full of water save for a small quantity of air near the center of the earth, this air would mount up to the heavens, because by nature air always moves upward in water. From these it appears that air can, by reason of its nature, descend and move upward to the distance of the semidiameter of the sphere of the elements. Now, these two motions are both simple and contrary, and thus a simple body is by its nature capable of moving in two simply contrary motions." Oresme immediately qualifies this example, but it is the object of the reference above.

14. From the discussion that follows, Oresme illustrates this situation by assuming a heavy body in void space where presumably "no light body is attached" or in contact with the heavy body. Under these circumstances, the heavy body would remain at rest since there is no up or down in void and no lighter medium to fall through, as happened in the example to which he referred in note 13.

15. *Physics* IV.8.215a.6–13. Aristotle's fundamental point is that in void one part of emptiness is like every other, so that no absolute directions such as up or down can be identified or isolated.

16. Lacking directional places like "up" or "down", Aristotle denied that natural motion could occur in a vacuum, since, for him, all natural motion was necessarily directed toward a place.

17. Aristotle does not actually mention the center of the earth and the concavity of the lunar sphere in the chapter referred to by Oresme. But he does discuss the upper and lower contrary directions associated with the contrary motions of up and down. See *De caelo* I.6.273a. 7–18.

18. Oresme considers each world a closed system where every bit of matter acts in accordance with Aristotle's principles of natural motion and remains in its own

In Chapter Twenty Aristotle mentions another argument from what was said in the *Metaphysics*—namely, that there cannot be more than one God and, therefore, it seems there can be only one world. I reply that God is infinite in His immensity, and, if several worlds existed, no one of them would be outside Him nor outside His power; but surely other intelligences would exist in one world and others in the other world, as already stated. And my reply to this argument is given more fully in Chapter Twenty. He argues again in Chapters Twenty-two and Twenty-three of which the purport is briefly this: this world is composed of all the matter avilable for the constitution of a world, and outside this world there can be no body or matter whatsoever. So it is impossible that another world exists.[19] In reply, I say in the first place, that, assuming that all the matter now existing or that has ever existed is comprised in our world, nevertheless, in truth, God could create *ex nihilo* new matter and make another world. But Aristotle would not admit this. Thus, I say, secondly, that, assuming that nothing could be made save from matter already existing and considering the replies we have given to Aristotle's first arguments regarding this problem—arguments whose substance he repeats and employs here in the present case—nonetheless he does not prove that another or more than one world besides our own could not now exist or may not always have existed, just as he states this world of ours to exist without beginning or end. He argues again in Chapter Twenty-four that outside this world there is no place or plenum, no void, and no time;[20] but he proves this statement by saying that outside this world there can be no body, as he has shown by the reasoning above to which I have replied; so it is unnecessary to answer this argument again. But my position could be strengthened or restated otherwise; for, if two worlds existed, one outside the other, there would have to be a vacuum between them for they would be spherical in shape; and it is impossible that anything be void, as Aristotle proves in the fourth book of the *Physics*.[21] It seems to me and I reply that, in the first place, the human mind consents naturally, as it were, to the idea that beyond the heavens and outside the world, which is not infinite, there exists some space whatever it may be, and we cannot easily conceive the contrary. It seems that this is a reasonable opinion, first of all, because, if the farthest heaven on the outer limits of our world were other than spherical in shape and possessed some high elevation on its outer surface

similar to an angle or a hump and if it were moved circularly, as it is, this hump would have to pass through space which would be empty—a void—when the hump moved out of it.[22] Now, if we assumed that the outermost heaven was not thus shaped or that nature could not make it thus, nevertheless, it is certainly possible to imagine this and certain that God could bring it about.[23] From the assumption that the sphere of the elements or of all bodies subject to change contained within the arch of the heavens or within the sphere of the moon were destroyed while the heavens remained as they are, it would necessarily follow that in this concavity there would be a great expanse and empty space. Such a situation can surely be imagined and is definitely possible although it could not arise from purely natural causes,[24] as Aristotle

world. Should it be removed to another world, it would become a part of that world and move with respect to the absolute places of that world. John Buridan devoted a special question to this problem and came to the same conclusion as did Oresme—*Questions on De caelo*, Book I, Question 18: "Whether, if there were several worlds, the earth of one would be moved naturally to the middle of another world" (edition of E. A. Moody [Cambridge, Mass.: Mediaeval Academy of America, 1942], p. 83). Others, including Albert of Saxony, also adopted this position. Although none believed in the actual existence of more than one world, they were compelled to acknowledge that God could create more if He wished. On the assumption that he did create more and they existed simultaneously, many scholastic natural philosophers seemed interested in pursuing the seeming cosmological and physical consequences that would follow from a hypothetical plurality of worlds. Most concluded that no physical absurdities would follow and seemed satisfied that they could save the hypothetical phenomena which they had imagined.

19. *On the Heavens* I. 9.279a.7–11.

20. *On the Heavens*, I.9.279a.12–19.

21. *Physics* IV.6–9. For much of the remainder of this section Oresme considers the existence of an infinite void space lying beyond our finite cosmos. This topic is discussed in the next selection.

22. With only slight alterations, Oresme is here utilizing an illustration which Aristotle employs to demonstrate the sphericity of the universe in *On the Heavens* II.4.278a.12–23.

23. Appeals to God's absolute power to do a multiplicity of things deemed absurd by Aristotle were commonplace as a consequence of the Condemnation of 1277 (Selection 13).

24. Imagining that God annihilates all or part of the elemental spheres, thereby leaving behind a void space, was commonplace in the fourteenth century. The various consequences relevant to problems of motion that would have followed upon such an action were frequently described and discussed (for an illustration, see Selection 53.1).

shows in his arguments in the fourth book of the *Physics*, which do not settle the matter conclusively, as we can easily see by what is said here. Thus, outside the heavens, then, is an empty incorporeal space quite different from any other plenum or corporeal space,[25] just as the extent of this time called eternity is of a different sort than temporal duration, even if the latter were perpetual, as has been stated earlier in this chapter. Now this space of which we are talking is infinite and indivisible, and is the immensity of God and God Himself,[26] just as the duration of God called eternity is infinite, indivisible, and God Himself, as already stated above. Also, we have already declared in this chapter that, since our thinking cannot exist without the concept of transmutation, we cannot properly comprehend what eternity implies; but, nevertheless, natural reason teaches us that it does exist. In this way the Scriptural passage, Job 26: [7], which speaks about God can be understood: Who stretchest out the north over the empty place. Likewise, since apperception of our understanding depends upon our corporeal senses, we cannot comprehend nor conceive this incorporeal space which exists beyond the heavens. Reason and truth, however, inform us that it exists. Therefore, I conclude that God can and could in His omnipotence make another world besides this one or several like or

---

25. On the basis of other statements made in *Le Livre du ciel et du monde,* Oresme conceived of this void space as infinite and extensionless, because God, who is identified with it (see the next few lines of this selection) is infinite and extensionless (though, paradoxically, omnipresent). Not only did Oresme view this infinite, incorporeal void space as the container of many possible worlds but in Book II, chapter 8, also envisioned it as the spatial backdrop against which one could conceive of an absolute motion arising if God chose to move our whole spherical finite cosmos in a straight line. In the quotation which follows, we find an appeal to God's absolute power as well as an actual reference to Article 49 of those condemned in 1277:

"But perhaps someone will say that to move with respect to place is to change one's position in relation to some other body which may, or may not, be in motion itself. Yet I say that this is not valid primarily because there is an imagined infinite and immobile space outside the world, as was stated at the end of the twenty-fourth chapter of the first book, and it is possible without contradiction that the whole world could be moved in that space with rectilinear motion. To say the contrary is an article condemned at Paris. Now assuming such a motion, there would be no other body to which the world could be related with respect to place, and the description given above would be invalid." (The translation is my own from Menut's French text on pp. 368, 370.)

In the famous Clarke-Leibniz correspondence of 1715–1716, Samuel Clarke found this unusual illustration useful in defending Newton's absolute space against Leibniz's relational concept of space. Clarke argued that "if space was nothing but the order of things coexisting [as Leibniz maintained], it would follow that if God should remove the whole material world entire, with any swiftness whatever, yet it would still continue in the same place" (Par. 4 of Clarke's Third Reply to Leibniz in 1716, quoted, with altered punctuation, from H. G. Alexander, ed., *The Leibniz-Clarke Correspondence* [New York: Philosophical Library, 1956], p. 32). The point of Clarke's argument is to show that if space is but a relation between coexistent things—as Leibniz held— and God moved the world taken as a single thing, the world could not be said to have undergone any motion, since there would be no other existent thing to which it could be related. For Clarke, as for Oresme, the hypothetical conditions under which God moves the world would indeed produce a motion, one that is absolute.

26. Oresme here identifies the infinite extramundane void space with God Himself (in this he differs from Bradwardine; see Selection 73.1, n. 25). Despite a lack of explicitness, and even some ambiguity, it seems that Oresme would have denied physical extension to the extramundane void. He does explain that "God in His infinite grandeur without any quantity and absolutely indivisible, which we call immensity, is necessarily all in every extension or space or place which exists or can be imagined. This explains why we say God is always and everywhere . . ." (Bk. II, ch. 2; Menut translation, p. 279). Thus God, though dimensionless, would be in every physically extended space. Is it not plausible, then, to assume that the infinite void beyond the cosmos is also physically extended and occupied wholly by God, just as He would occupy any finite physical space? This seems unlikely, however, for as the passage above indicates, Oresme identified the extramundane void with God Himself, who is dimensionless; hence this particular space would also be dimensionless. But nowhere does Oresme identify "God Himself" with actual physical or dimensional spaces; he rests content to say only that God is in them because He is everywhere.

It would seem, then, that Oresme's identification of God with an infinite void beyond the cosmos conferred upon the latter a transcendent and nondimensional character (this is not to say, however, that Oresme did not conceive of bodies moving through this "transcendent" infinite space, or that it might not serve as the space of other worlds; see n. 25). Many medieval authors would have construed it in much the same manner (not, however, Johannes de Ripa; see Selection 73, introduction). Some authors in the seventeenth and eighteenth centuries—for example, Henry More, Joseph Raphson, and Isaac Newton—would come to associate God with a physically extended infinite void and conceive of God as actually extended in physical space. Indeed, Raphson, Newton, and Samuel Clarke conceived of physical space as an attribute of God (see my article "Medieval and Seventeenth-Century Conceptions of an Infinite Void Space Beyond the Cosmos," *Isis,* LX [1969], 57–59). Thus, as Leibniz was to note in his controversy with Samuel Clarke, God was in danger of becoming a physically extended being.

unlike it. Nor will Aristotle or anyone else be able to prove completely the contrary. But, of course, there has never been nor will there be more than one corporeal world, as was stated above.

72 ON THE EXISTENCE OF AN IMAGINARY INFINITE VOID SPACE BEYOND THE FINITE COSMOS

Hermes Trismegistus to Asclepius: Void beyond the Cosmos Lacks Matter but not Spirit

Translated by Walter Scott[1]

Introduction and annotation by Edward Grant

The Latin dialogue between Hermes Trismegistus and Asclepius was translated from a Greek original now lost. The text, which for a variety of reasons is faulty, has required considerable emendation based on careful study. Scott explains in his translation, "I have aimed at expressing what I suppose to have been the meaning of the original Greek, rather than the meaning—or, too frequently, the absence of meaning—of the Latin" (*Hermetica*, I, 51). *Asclepius* was subdivided into three parts by Scott because it seemed to consist of that many distinct and unrelated parts. The following selection is from *Asclepius* III, which Scott believes to have been composed in the original Greek in Egypt by an Egyptian sometime between 260 and 310, indeed perhaps between 268 and 273. Its translator into Latin is unknown, but Scott suggests C. Marius Victorinus, who died soon after 362. The treatise was reasonably well known in the Latin Middle Ages, and a number of manuscripts of it exist.

This selection on void may have influenced Bradwardine (see Selection 73.1 and n. 7 thereto) and other Christians to suppose that beyond the cosmos there existed a space void of matter but not God. Christians could easily have placed such an interpretation on Hermes Trismegistus' declaration that "I hold that not even the region outside the Kosmos is void, seeing that it is filled with things apprehensible by thought alone, that is, with things of like nature with its own divine being." Since all space is occupied by matter or spirit, Hermes Trismegistus denies the existence of void. But void in the usual sense of a place devoid of matter may indeed exist.

But as to Void, which most people think to be a thing of great importance, I hold that no such thing as void exists, or can have existed in the past, or ever will exist. For all the several parts of the Kosmos are wholly filled with bodies of various qualities and forms, each having its own shape and magnitude; and thus the Kosmos as a whole is full and complete. Of these bodies, some are larger, some are smaller; and they differ in the greater or lesser firmness of their substance. Those of them which are of firmer substance are more easily seen, as are also those which are larger; whereas smaller bodies, and those which are of less firm substance, are almost or quite invisible, and it is only by the sense of touch that we are made aware of their existence. Hence many people have come to think that these bodies do not exist, and that there are void spaces; but that is impossible. And the like holds good of what is called the "extramundane," if indeed any such thing exists; for I hold that not even the region outside the Kosmos is void,[2] seeing that it is filled with things apprehensible by thought alone, that is, with things of like nature with its own divine being. And so our Kosmos also—the sensible universe, as it is called—is wholly filled with bodies, and living bodies, suited to its character. The shapes presented by these bodies to our sight differ in magnitude; some of these shapes are very large; others are very small, when the distance of the objects makes them appear small to us; and some things, on account of their extreme minuteness of tenuity, are wholly invisible to us, and are consequently supposed by many people to be nonexistent.

And so, Asclepius, you must not call anything void, without saying what the thing in question is void of, as when you say that a thing is void of fire or water or the like. For it is possible for a thing to be void of such things as these, and it may consequently come to *seem* void; but the thing that

1. In *Hermetica, the Ancient Greek and Latin Writings Which Contain Religious or Philosophic Teachings Ascribed to Hermes Trismegistus,* edited with English Translation and Notes by Walter Scott (Oxford: Clarendon Press, 1924), I, 317–321.
2. Certain Stoics (notably Posidonius) believed in the existence of an infinite void beyond the visible and finite cosmos (but denied the existence of void within the cosmos). The author of *Asclepius* III, however, speaks of an extramundane region in hypothetical terms only.

seems void, however small it be, cannot possibly be empty of spirit and of air.

And the like must be said of Space.[3] The word "space" is unmeaning when it stands alone; for it is only by regarding something which is in space, that we come to see what space is; and apart from the thing to which it belongs, the meaning of the term "space" is incomplete. Thus we may rightly speak of the space occupied by water, and fire, and so on (but not of space alone). For as there cannot be a void, so it is impossible to determine what space is, if you regard it by itself. For if you assume a space apart from something which is in it, it will follow that there is a void space; and I hold that there is no such thing as that in the universe. If void has no existence, then it is impossible to find any real thing answering to the word "space" taken by itself.

73 ON A GOD-FILLED EXTRAMUNDANE INFINITE VOID SPACE

Introduction, translations, and annotation by Edward Grant

In these selections on the existence of an infinite extramundane void space, we shall see an essentially Christian and theological reaction to Aristotle's claim that there can be "neither place, nor void, nor time outside the heaven,"[1] a conclusion reached in the process of rejecting the existence of other worlds. Despite Aristotle's position, subsequent authors could not refrain from inquiring about the hypothetical disposition of objects, say a lance or an arm, pushed through the outermost celestial sphere. Where Aristotle would have deemed such questions absurd, arguing nothing whatever existed beyond the world, Stoics and some others who subscribed to a finite spherical universe came to hold that an infinite void lay beyond (see the previous selection, n. 2). Thus a lance or an arm pushed beyond the universe would emerge into an infinite void. This type of example and Stoic advocacy of an extramundane infinite void came to be widely known in the Middle Ages through William of Moerbeke's Latin translation of Simplicius' *Commentary on Aristotle's De caelo* (see Selection 8, no. 47), where a brief summary of Stoic arguments was presented.[2] This, along with the general availability of *Asclepius III,* where, as we have seen, it was emphasized that if extramundane void existed it would be filled with spirit,[3] came to constitute one of the two major sources for medieval discussions of extramundane void. When we add to all this the theological concept of God's absolute power which emerged from the Condemnation of 1277 (see Selection 13) we have the major historical influences that shaped Bradwardine's interesting arguments. And the other authors from whom excerpts have been included in this selection[4] would argue, as did Bradwardine, for the existence of an infinite extramundane void filled with an omnipresent God, a void that was very different indeed from the spiritless and truly empty void proposed by the Stoics. This medieval conception seems ultimately to have influenced the Cambridge Platonists, (for instance, Henry More), Otto von Guericke, Isaac Barrow, and Sir Isaac Newton, whose absolute space is ontologically strikingly similar to the medieval God-filled extramundane void space.

In the selections from Bradwardine, Oresme, and the Coimbra Jesuits, we appear to have a medieval

3. Throughout this paragraph Scott used the term "space" (instead of "place") to translate *locus,* not *spatium.* However, "place"—in the restricted sense of the container of a body—would have been preferable. Although *locus* and *spatium* appear sometimes to be used interchangeably in the sense of a container, the term *spatium* is more usually introduced when distance or extent is intended.

1. *De caelo* I.9.279a.12–13, 17–18.

2. Simplicius tells us (fol. 44 verso, col. 2, of the edition published in Venice, 1540): "The Stoics, however, thinking that there is a vacuum beyond the sky, prove it by this kind of assumption: Let it be assumed that someone standing motionless at the extremity [of the world] extends his hand upward. Now, if his hand does extend, they take it that there is something beyond the sky to which the hand extends. But if the arm could not be extended, then something will exist outside that prevents the extension of the hand; but if he then stands at the extremity of this [obstacle that prevents the extension of his hand] and extends his hand, the same question as before [is asked], since something could be shown to exist beyond that being." I have quoted this translation from my article "Medieval and Seventeenth-Century Conceptions of an Infinite Void Space beyond the Cosmos," *Isis,* LX (1969), 41. Thomas Aquinas discussed these arguments in his *Expositio in De caelo.* They seem also to have been known to John Buridan and Nicole Oresme.

3. See Selection 72, introduction. In Bk. XI, ch. 5 of the *City of God,* St. Augustine says much the same thing, and was probably familiar with the Latin version of *Asclepius* III.

4. Summaries and analyses of their discussions are given in my article in *Isis* cited in note 2. Some of the notes in this selection are based upon my article.

tradition in which an extramundane infinite void was conceived as either dimensionless (Bradwardine and Oresme are hardly clear on this point; for an explicit statement from the Coimbra Jesuits see Article 4, section 3 below), or dimensional in a special transcendent sense associated with the "dimensions" of a Divine Being who is Himself extensionless. Bradwardine and the Coimbra Jesuits, and probably Oresme, are quite emphatic in characterizing this infinite extramundane void as an existent entity—for it is the immensity of God Himself, who cannot be confined to a finite world— that has no real and positive being of its own. These authors seem to have conceived of infinite extramundane void space—it is also characterized as imaginary—as coextensive (in a transcendental sense) with God's total immensity (or "extent") and therefore eternal, because it is God Himself.

But another important view (not represented in this selection) was proposed around 1350 by Johannes de Ripa. He conceived an imaginary infinite extramundane void space as being less "extensive" than the immensity of God Himself. Indeed, God's infinite immensity was said to circumscribe and infinitely exceed the immensity of the imaginary infinite void beyond the world (see "Jean de Ripa, I Sent. Dist. XXXVII: De Modo Inexistendi Divine Essentie in Omnibus Creaturis. Édition critique par André Combes et Francis Ruello, présentation de Paul Vignaux. Immensité Divine et Infinité Spatiale," *Traditio,* XXIII (1967), 235, 238 (text), 199 (Vignaux's discussion). Perhaps because he radically distinguished between the dimension of the infinite void space and God's infinitely more extensive immensity, Johannes conceived of the infinite void as a possible three dimensional enity. Of great interest, however, is the fact that he too denied that this imaginary infinite extramundane space was a positive and real thing (Vignaux, pp. 232, 233, 234).

1. Thomas Bradwardine (ca. 1290–1349)[5]

[COROLLARIES]

1. First, that essentially and in presence, God is necessarily everywhere in the world and all its parts;

Whether Otto von Guericke conceived of imaginary extramundane infinite void as coextensive with God Himself or inclined rather to a position more in accord with the view of Johannes de Ripa is difficult to determine. In one instance he declares that God can be contained by no place, since "He Himself is the place for Himself, or the thing that is empty *(vacuum)* of every creature, or the space, or universal container of all things." Elsewhere, however, he asserts that "the space and vacuum of the whole creation is in Him, by Him, and for Him" (see section 4). On a conjectural basis, it would seem appropriate to regard von Guericke as closer to Bradwardine and the Coimbra Jesuits than to de Ripa in believing that total infinite void space is coextensive with God Himself.

On one significant point von Guericke breaks completely with his medieval predecessors just mentioned: He insists that the imaginary space beyond the world is indeed a *real and positive* thing (see near end of Bk. II, ch. 6, in section 4). Perhaps this judgment was generated in part by seventeenth-century experiments and discoveries concerning atmospheric pressure, achievements in which von Guericke had himself played a leading role. Not only had it been demonstrated that nature did not abhor a vacuum (see Selection 54) but production of three-dimensional artificial vacua might have encouraged acceptance of the vacuum as in some sense a real three-dimensional entity. Moreover, ancient atomism, with its three-dimensional void extending to infinity, and Stoic cosmology, with its sealed finite world surrounded by a real three-dimensional infinite void, were well known, and each had its partisans. For some, or all, of these reasons, it is perhaps not surprising that Otto von Guericke and others in the seventeenth century came to regard three-dimensional void space as positive and real.

5. This selection is from Thomas Bradwardine, *De causa Dei contra Pelagium et De virtute causarum. . .* (London: Henry Savile, 1618), Bk. I, ch. 5, pp. 177–180. This theological treatise was apparently written in 1344. In a manner similar to Spinoza in the *Ethics,*

Bradwardine employed a quasi-mathematical form, announcing propositions and deriving corollaries from them. A French translation of many of the passages relevant to infinite void space was made by Alexandre Koyré: "Le vide et l'espace infini au XIVe siècle," *Archives d'Histoire Doctrinale et Littéraire du Moyen Age,* 24th Year (1949), pp. 45–91. My translations and annotations in this and the following selections on extramundane void space were an outgrowth of a research project on the concept and role of void space in the history of medieval physics supported by the Division of Social Sciences of the National Science Foundation.

2. And also beyond the real world in a place, or in an imaginary infinite void.[6]

3. And so truly can He be called immense and unlimited.

4. And so a reply seems to emerge to the old questions of the gentiles and heretics—"Where is your God?" And, "Where was God before the [creation of the] world?"

5. And it also seems obvious that a void can exist without body, but in no manner can it exist without God.[7]

. . . .

That God is necessarily everywhere in the world follows obviously.[8] For if He were not in any [particular] place in the world, but could be in it—since, by the seventh part of the [corollary] of the first [chapter], He is omnipotent—He cannot do this by His motion, nor through the motion of any creature, since there is a creature there now, and there was always one there previously. And the same reasoning applies to any other creature. This suffices to show the existence of God there, but not by his motion, as the chapter has shown. The same thing is plainly shown by a corollary of the second [chapter] of this book and by its demonstration.

In order to demonstrate the second part,[9] I assume that *A* is the fixed, imaginary place of this world, and *B* is simultaneously an imaginary place beyond the world that is separated from *A*; and I assume that God moves the world from *A* to *B*, placing it in *B*. Then, by the first part [or corollary] of this [fifth] chapter and by a corollary of the second [chapter] and its demonstration, God is now in *B*.[10] Therefore, either He was there [in *B*] before, or not. If He was there, then, by the same reasoning, He was there before and can now be imagined as everywhere outside the world.[11] If He was not there before, then, by the same reasoning, He is not now in *A*, but [must be] in *B*. Therefore, God was in *A* before, but not in *B;* and He is not now in *A* but in *B*, which is completely separate. Consequently, He withdrew from *A* and came to *B*, so that He was moved locally or with respect to position, just as our soul is moved with the motion of our body.

Some of those who follow the Philosopher [Aristotle] in Book I of *On the Heavens* reply to this. They assume that every local motion is necessarily upward, downward, or circular—that is, away from the center [of the world], toward the center, or around the center.[12] But they say that if this motion [from *A* to *B*] were assumed, it could not be any one of the ways just mentioned. For

this reason, they say that it is impossible for the world to be moved. But these [followers of Aristotle] seriously diminish and mutilate the divine—indeed, omnipotent—power. For, in the beginning, God could have created this world in *B*. Why, then, is He unable to put it in *B* now? Furthermore, He can now create another world in *B*. Why, then, is He unable to put this world in *B*? This reply [that the world cannot be moved] is condemned by Stephen, Bishop of Paris, in these words: "That God could not move the heavens [that is, the world] with rectilinear motion; and the reason is that a vacuum would remain."[13] But this response does not avoid the difficulty. For it could be assumed that without [resort to] local motion, God could create another world in *B* and annihilate the world in *A*. [On this assumption,] the difficulty returns.[14]

6. Bradwardine nowhere explains what he means by "imaginary infinite void." Perhaps it was "imaginary" because of its incomprehensibility, which arose from its necessary lack of dimensionality stemming from the fact that God, who filled it, is dimensionless. Or, perhaps he rather inclined toward Oresme's indirect suggestion (see Selection 71, at the end) that infinite incorporeal space is "imaginary" because it is imperceptible to the senses and comprehensible—if at all—by the reason alone. A few authors discuss this term explicitly (see Question 2, Article IV, of section 3 and Bk. I, ch. 35 of section 4).

7. The influence of *Asclepius* III is detectable here, for, as we saw in the preceding selection, its author insisted that if void existed, it would of necessity be filled with spirit. This treatise was known in the Middle Ages as *De aeterno verbo (On the Eternal Word)*. Although Bradwardine does not mention it in the section translated here, he cites it elsewhere in the *De causa Dei*. Indeed, he calls Hermes Trismegistus "a brilliant prophet and glorious philosopher" ("Hermes Mercurius Trismegistus, clarus propheta et philosophus gloriosus"; *De causa Dei*, p. 98). A quote from *Asclepius* appears in Question 2, Article II, of section 3.

8. The first corollary is now demonstrated. Bradwardine argues that without undergoing movement, God is omnipresent in the world.

9. That is, the second corollary, namely that God exists beyond the world in an imaginary infinite void.

10. After moving the world from *A* to *B*, God must be in *B*, the place of the world, since by the first corollary God is everywhere inside the cosmos.

11. Obviously, place *B* can be rightly conceived as representing each and every place outside, or beyond, the world. Hence God is everywhere.

12. Aristotle, *De caelo* I.2.268b.16–24.

13. This is Article 49 of the Condemnation of 1277 (see Selection 13 and n. 15 thereto). Reliance on God's absolute power to perform acts contrary to Aristotelian physics and metaphysics is a vital component of Bradwardine's over-all approach.

14. Thus, whether the world is transferred from place

Other followers of the Philosopher would, it seems, respond to this by saying that there is no place or void outside of [or beyond] the world;[15] therefore God is not there and the world cannot move there. But as a consequence these [followers of the Philosopher] must then say that God, for all His omnipotence, could not have made the world greater or smaller in any thing, which would very much restrict and confine His omnipotence. For they must say that God necessarily makes the world in place A, and that place A, and no other, existed before there was a world.[16] But why this place, and no other? Why was the world just this size and no greater or smaller? For, indeed, either this [that is, the size of the world] was fixed by God, or by itself and not by God. If by God, then, in virtue of his infinite power, He could make a place greater and greater; and [He could] make another place, and yet another place, without end. If, however, the size is determined by the world itself —and not by God—what power determines this? What is the reason for it? What nature has fixed the world at this [particular size], and inviolately determined the limit beyond which it cannot pass?[17]

Moreover, as no one can ignore, this imaginary place (situs imaginarius) could have no positive nature, for otherwise there would be a certain positive nature which is not God, nor from God, the opposite of which was shown in the second and third [chapters] of this [book]; such a nature would be coeternal with God, something no Christian can accept. It was for this reason that Bishop Stephen of Paris condemned an article asserting that "many things are eternal,"[18] and that even if He wished, God could not destroy them; therefore, He is deprived of omnipotence. The argument of these [Aristotelians] tells against them. For if, according to the assumption of the Philosopher and his followers, there could be no void, nor any imaginary place not filled with body, [then] the world is eternal, which is heretical, and which these people [themselves] deny;[19] or, prior to the creation of the world, there was an imaginary void place, unoccupied by any body. The previously cited article [of Bishop Stephen of Paris] condemned this reply and its irrational argument.[20] Since these arguments do not serve to hinder us in any way, it appears that the reasoned premise stands unscathed.[21]

On the contrary, creating the world and every part of it in the beginning, God was simultaneously with the world and every part of it. For . . . in every motion and action, the mover and agent exists simultaneously with what is moved and suffers the action. Thus Augustine in his second Homily on John, 1, [where John says] "He was in the world and the world was made by Him,"[22] asks, "How was He in the world?" And he replies, "As the artist, ruling what he makes." For He does not make things as does a carpenter who is outside the box that he makes, a box which, during its construction, is also located in another place [than the carpenter]. And though he is very near to it, yet he who makes it is located in another place, and is external to what he makes. God, however, is infused into the world He makes, which is placed wherever He makes it. And He does not withdraw from anything and is not external . . . [to what He

---

A to B by local motion, a possibility denied by Aristotelians, or by the divine power annihilating the world in A and recreating it in B, the same problem arises: Was God in place B before the creation of the world?

15. Aristotle, De caelo I.9.279a.12–13, 17–18.

16. No proper Aristotelian could have subscribed to the consequence drawn by Bradwardine, since it was a fundamental tenet of Aristotelian cosmology and physics that the universe was not itself in a place (Physics IV.5.212b.7–11). Perhaps Bradwardine thought that by restricting God's omnipotence with the claim that He "could not have made the world greater or smaller," the Aristotelians were committed to the consequence that "place A, and no other, existed before there was a world." That is, A was a pre-existent place into which God had to fit the world exactly; therefore, He could not tinker with its size.

17. Apparently deeming the alternative opinion absurd, Bradwardine adopts, without further comment, the position that God determined the size of the world and the size of its place and could have made it any size whatever.

18. Probably a reference to Article 52 of the Condemnation of 1277.

19. The sense of this argument would appear to be that if no void space or place ever existed, then it follows that a plenum has always existed. Hence matter—and consequently, the world—is eternal. As Christians, the Aristotelians, whom Bradwardine here opposes, would have been committed to the repudiation of such a position.

20. The alternative argument, that "prior to the creation of the world, there was an imaginary void place, unoccupied by any body," is also rejected by appeal to Article 52 of those condemned in 1277. Such a pre-creation void place conceived as mere emptiness devoid of matter and God would be a thing coeternal with God. As Bradwardine will argue in what follows, God himself must necessarily exist in such a void place— indeed in all such void places, for He is omnipresent.

21. The "reasoned premise" is that "this imaginary place could have no positive nature"; therefore, no positive thing is coeternal with God.

22. John 1:10.

makes]. The presence of [His] majesty makes what He makes; [and] His presence governs what He has made. Therefore, He was in the world even as the world was made by Him. . . .

God is a foundation without a foundation—that is, He is based in no prior thing but is rather the first and original foundation of all other things, on which all other things rest, since they are unstable themselves, while the foundation by which they are always maintained and continuously supported is itself fixed, stable, and immobile, as was shown in many ways in the second [chapter]. However, the contrary of this is not true. God can be in any place He wishes without the need of a creature; and He is not there anew, since He does not get there by His motion, as the fifth [chapter] has shown;[23] therefore He is there in eternal rest. Hence God persists essentially by himself in every place, eternally and immoveably everywhere. Furthermore, it is more perfect to be everywhere in some place, and simultaneously in many places, than in a unique place only; and a spirit that can do this is more perfect than a body unable to do this. But God is an infinitely perfect spirit, as the first assumption and its demonstration[24] reveal. Hence this is infinitely suitable for Him, and [it is so] without the need of any creature, as shown above, and without any mutability whatever. God is, therefore, necessarily, eternally, infinitely, everywhere in an imaginary infinite place, and so truly omnipresent, just as He can be said to be omnipotent. In a certain sense, and for a similar reason, He can also be called infinite, [as for example], infinitely great, or of infinite magnitude, even though extensive [is taken in] a metaphysical and improper sense. For He is infinitely extended without extension and dimension.[25] For truly, the whole of an infinite magnitude and imaginary extension, and any part of it, coexist simultaneously, for which reason He can be called immense, since He is unmeasured, nor is He measurable by any measure; and He is unlimited, because nothing surrounds Him fully as a limit; nor, indeed, can He be limited by anything, but [rather] He limits, contains, and surrounds all things. Thus Sixtus[26] the Pythagorean says in his aphorisms (*in sententiolis suis*) that "you could not discover the magnitude of God even if you could fly with wings." This opinion is attested to by the second of the twenty-four definitions of God posited by the twenty-four Philosophers,[27] which says, "God is an infinite sphere whose center is everywhere and circumference nowhere";[28] and the eighteenth definition,

which says, "God is a sphere that has as many circumferences as points"; and the tenth definition, which says, "God is that whose power is not numbered, whose being is not enclosed, [and] whose goodness is not limited. . . ."

. . . .

. . . Indeed, if against the possibility of the rectilinear motion of the whole world,[29] which was touched on above, anyone should oppose the opinion of the Philosopher mentioned before, namely that every motion is from the center, or around the center of this world, it must be said that he speaks of species of natural motions of natural bodies and parts of this world. But if anyone objects because of the vacuum [that would be left behind], it must be said that perhaps he under-

23. See the discussion immediately after the Corollaries at the beginning of this section, and note 8.

24. A reference to the first supposition of the first chapter, which declares that "God is the highest perfection and good because there can be nothing more perfect or better." *De causa Dei*, p. 1; this is demonstrated on pages 1–2.

25. Since the infinite void prior to the creation of the world was occupied by an omnipresent God who is "infinitely extended without extension and dimension," it seems plausible to conclude that the infinite extra cosmic void was thought to be dimensionless—especially since Bradwardine had earlier argued that the void possessed no independent existence, for then it would be coeternal with God. Although Bradwardine did not identify infinite void space with God or call it an attribute of God, Oresme did make such an identification but, like Bradwardine, seems to have denied dimensionality to this void. Others would not only confer dimensionality upon it (see Introduction; near end of Book II, ch. 6, of section 4; and n. 71), but would also conceive of God as in some sense physically extended within this space, which was even described as His attribute (see Selection 71 and note 26 thereto).

26. Is this Sextus, or Sextius, the Pythagorean, who lived in the reign of Augustus or Tiberius, sometime between 29 B.C. and A.D. 37? I have not located the quotation.

27. The three definitions following are drawn from the *Book of the XXIV Philosophers*, a popular Pseudo-Hermetic treatise. The Latin text was edited by Clemens Baeumker, "Das pseudo-hermetische 'Buch der vierundzwanzig Meister' (Liber XXIV philosophorum)" in *Beiträge zur Geschichte der Philosophie des Mittelalters*, Band XXV, Heft 1–2 (Münster, 1927), pp. 194–214. The three definitions appear on pages 208, 210, 212.

28. This definition became widely known in the Renaissance and influenced, among others, Nicholas of Cusa, Marsilio Ficino, Giordano Bruno, and Robert Fludd, as well as the anonymous Jesuit commentator at Coimbra, from whose treatise we have included a selection on infinite void (see Question 2, Article II, of section 3, and n. 43).

29. See note 13.

stands that vacuum must not be posited necessarily and naturally in the world nor even outside the world, because of the effects which the Pythagoreans, Democritus and Leucippus, and many ancient philosophers, assumed from it, namely because of local motion, motion of augmentation and diminution, condensation and rarefaction, motion of alteration, and for distinguishing beings lest they be one, or because of respiration of the heavens from an infinite void beyond it, as is obvious in [Book] 4 of the *Physics,* [Books] 3 and 4 of *On the Heavens and the World,* and [Book] 2 of *On the Soul,* and also in other places.[30]

And it seems that vacuum can be assumed in two ways: in one way only privatively, namely as pure privation itself or the pure lack of a plenum; in another way positively, and this can be assumed in two ways. In one way, it would only be a corporeal dimension separated from natural forms, as if it were a mathematical body, or a quantity by itself separated from other natural things; it is sometimes taken in this way in the fourth [book] of the *Physics.* In another way it would also have an active power and be moved according to place.[31] Vacuum is also

defined by him [Aristotle] as a place that is not filled, but is capable of being filled. Moreover, according to him [Aristotle], place is an ultimate [surface], namely the last surface of the containing body.[32] Furthermore, such a void is not outside the heavens; nor is there a place [beyond the heavens] that could be filled by natures formed [or constructed] from natural things. Nevertheless, these arguments do not at all convince with an irrefutable necessity. Indeed, by means of His absolute power, God could make a void anywhere that he wishes, inside or outside of the world. Truly, even now, there is in fact an imaginary void place outside of the world, which I say is void of any body and of everything other than God, as shown by what has preceded. . . .

However, one might argue against [my position by saying] that if such a void place existed outside this world, another world would be possible there. This is contrary to the same Philosopher in his *On the Heavens,* [Text] 76,[33] and afterward where he rejects this in many places. But who would deny or doubt this conclusion, except, perhaps, one who denies the omnipotence of God[34]. . . .

2. Nicole Oresme (ca. 1325–1382)

See Selection 71.[35]

3. Jesuit Commentators at the College of Coimbra, Portugal (second half of sixteenth century)[36]

Question 2

Whether or Not God Exists Beyond the Sky [or Celestial Heavens]

ARTICLE I

The Argument On the Negative Side of the Controversy

Those who have followed the negative part of this question are Scotus[37] in a unique question of

30. Here Bradwardine has brought together from a number of Aristotelian works phenomena which the atomists sought to explain by the motion of atoms in void space. He argues that one ought not to posit the existence of vacuum for the reasons which they offered, namely that it could explain the above-mentioned phenomena.

31. Aristotle, *De caelo* IV.2.309b.15–19.

32. Both definitions appear in Book IV of the *Physics.*

33. Aristotle's preliminary statement against a plurality of worlds appears in Book I, Text 76, of the medieval text of Aristotle's *De caelo;* the argument does not conclude until Book I, Text 85. In the modern editions and translations of *De caelo,* see I.8.276a.18—277a.12.

34. Here, again, Bradwardine rejects Aristotle's restriction on God's absolute power to create more worlds if He pleased. In so doing, Bradwardine abides by the spirit which motivated the Condemnation of 1277, probably with Article 34 in mind.

35. This selection properly begins on page 552 with the words: "He argues again in Chapter Twenty-four that outside this world there is no place or plenum, no void, and no time; . . ."

36. This selection is translated from the Commentary on Aristotle's *Physics* published in Cologne, 1602, under the title *Commentariorum Collegii Conimbricensis Societatis Iesu: In octo libros Physicorum Aristotelis Stagiritae . . . .* It is drawn from Book VIII, chapter 10, Question 2, Articles I, II (part), and IV, on cols. 514, 518–519.

37. In this question John Duns Scotus (ca. 1265–1308) denied the necessity for God's omnipresence in an infinite extramundane void on grounds that God's presence in a place was not a necessary precondition for His acting in that place. God's will, not His omnipresence, was taken as the basis of His actions. Hence, He could act in a place remote from His presence. With "action at a distance" as God's *modus operandi,* Scotus denied the necessity of God's presence in the empty place where He came to create the world and, a fortiori, rejected the necessity for God's omnipresence in an infinite void

[Book] I, Distinction 37 [of his *Sentence Commentary*]; St. Bonaventure in Article II, Question 1, Number 22 [of his *Sentence Commentary,* and] Capreolus,[38] in a unique question, Article III, and others.

In behalf of this position, some have put forth this most powerful argument: (1) There is nothing beyond the sky; but God cannot be in nothing; therefore He is not beyond the sky.

(2) Again, if God were in a space, which we can imagine beyond the encompassing sky, [and] since this space is, in truth, nothing positive or real, then by a parity of reasoning *(pari ratione)* it would be possible to say that God is in darkness, and in other privations, which is absurd.

(3) Next, just as we conceive such a space only in the mind, so it is a consequence that God is not in the thing itself but is said to exist in it only in our conception. Therefore, God does not really exist beyond the sky [or heaven].

(4) It happens that substances devoid of matter are not in place except by operation. Indeed, God operates nothing beyond [or outside] the world, just as He did not operate anything before the world was founded, which St. Augustine says in Confessions, Book XI, chapter 12: "[How, then, shall I respond] to him who asks, 'What was God doing before he made heaven and earth?' I do not answer, as a certain one is reported to have done facetiously (shrugging off the force of the question), 'He was preparing hell.'" And a bit later on [Augustine says,] "Before God made heaven and earth, he did not make anything at all. For if he did, what did he make unless it were a creature?"[39] Finally, St. Augustine holds the same opinion on this in his book *Against Maximinus,* chapter 21, where to those wishing to know where God was before He made the world, he [Augustine] replied that He was in Himself.[40] This is repeated in Psalm 120[41] in these words: "I have lifted up my eyes to the mountains." There is no mention here of an imaginary space in which God existed.

ARTICLE II

AN AFFIRMATIVE PART CONSISTS OF THE TESTIMONY OF PHILOSOPHERS

The opposite part of the question, however, is true, especially what Trismegistus teaches in *Asclepius,* where he says: "God who dwells above the summit of the highest heaven, is present everywhere, and from all around He watches all things. For there is beyond the sky a space *(spatium)* without stars removed from all things

corporeal."[42] To these statements, he adds this definition of God, which is said to have been furnished by the same person: "God is an intelligible[43] sphere whose center is everywhere and circumference nowhere." In this description, the greatest perfection of the divine nature is signified by the sphere, the most absolute *(absolutissima)* [or perfect] of all figures. . . .

Moreover, that the same opinion concerning the existence of God beyond the world was followed by certain other ancient philosophers, as well as by Plato and Aristotle,[44] is testified to by Eugubinus in Book IV of *De perenni philosophia (On Perennial Philosophy)*, chapters 1 and 2. . . .

ARTICLE IV

WHAT IS IMAGINARY SPACE? WHEN THIS HAS BEEN DETERMINED [WE SHALL INQUIRE WHETHER] GOD EXISTS IN IT.

The solution to the arguments of the first article. For the most thorough explication both of our opinion and the arguments, certain things must be considered about space, and we shall do this with respect to space that is within and beyond the sky.

space. See Scotus' *Questio unica* in Book I, Distinction 37, of his *Questions on the Sentences* in *Joannis Duns Scoti . . . opera omnia editio nova juxta editionem Waddingi XII tomos continentem . . .* (Paris: Vivès edition, 1893), Vol. X, p. 597, col. 2.

38. That is, Johannes Capreolus (d. 1444), sometimes called "The Prince of Thomists" *(Thomistarum princeps)*.

39. These two passages from Augustine's *Confessions* are quoted here from the translation of Albert C. Outler, *Augustine: Confessions and Enchiridion* (Philadelphia: Westminster Press, 1955), p. 253.

40. I was unable to locate this statement in the work and chapter cited.

41. The text has 122.

42. Since the text of *Asclepius* III possessed by our author at Coimbra differs somewhat from that edited by Walter Scott in *Hermetica . . .* (Oxford: Clarendon Press, 1924), I, 324, I have altered the latter's translation accordingly. As the most significant variation, it should be noted that where Scott's text has *"huic est enim ultra caelum locus"* (translated as "his abode is beyond heaven"), the Coimbra text has *"Est enim ultra caelum spatium,"* which I have rendered as "For there is beyond the sky a space." See also note 7.

43. Although the text has *intelligibilis,* it should read *infinita*—that is, "God is an infinite sphere . . . ." This definition, which is not from *Asclepius* but is found in the pseudo-Hermetic *Book of the XXIV Philosophers,* was also quoted by Bradwardine and others (see Selection 73.1 and n. 28 thereto).

44. This is not true for Aristotle, unless one is prepared to argue that the prime mover was located beyond the world.

The first is that this [imaginary] space is not a true quantity possessed of three dimensions;[45] moreover, bodies cannot be received in it, since several such dimensions cannot be in the same position simultaneously by [any actions of] the forces of nature *(naturae viribus)*. Furthermore, it cannot be another real and positive being, since, beside God, no such thing could exist from eternity.[46] But this space always existed and always ought to be.

A second thing about it is that this space is not a being [or entity] of the reason,[47] since by means of this thing itself bodies are received within the world without the action of the intellect; and they can [also] be received outside [or beyond] the world if they were created there by God. Therefore, the dimensions of this space are not usually said to be imaginary because they are fictions *(fictitiae)* or depend solely on a mental conception or that they are thought to be beyond the understanding *(nec extra intellectum dentur)*, but because we imagine them in space, in a certain relation corresponding to the real and positive dimensions of bodies.[48]

We assert, therefore, that God is actually in this imaginary space, not as in some real being but through his immensity, which, because the whole universality of the world cannot [accommodate it], must of necessity also exist in infinite spaces beyond the sky.[49] But God Himself exists in one way in Himself, in another way in created things, and still another way outside the world. For, indeed, He is in Himself because He needs the support of no one; He is in created things by essence, presence, and power; He is beyond the sky because He cannot be excluded from any place, whether true or imaginary. And yet, He is not present in the imagination by coexisting with any creature, but [rather He is there] according to the real and infinite existence of His being.

Now we must explain the arguments which we set forth at the beginning. To the first [see beginning of Article I], we deny that God cannot be in nothing—that is, in a space which is not a real and positive being; otherwise no stone could exist beyond the world by [an act of] the divine power.[50]

To the second [argument], we must deny that the same reasoning applies to space as to darkness and other privations. For just as space is appropriately disposed to receive bodies, this is not true of other negations or privations. And so it is also fitting that God should be said to be in this space, but that He should not be said to be in other negations or privations. To the third [argument we say] that

that space beyond the world should be conceived by us as if it were truly there, but not as a real positive being. To the fourth [argument we say] that God is not properly said to exist in a place, since He is not enclosed by any spaces. But created substances devoid of matter do not exist in a place by operation but by their substances considered in a certain way;[51] however, this cannot be explained more concisely in this place. To the last [and fifth argument we say that] although St. Augustine makes no mention of imaginary space in his book, he, nevertheless, made [allusion to it] with clear and lucid words, in a place cited by us in *The City of God*, Book II.[52] Since these things are so, it is clear that Aristotle,[53] in Book I of *Metaphysics*, Explanation [that is, Text] 14, was not right to label as an error and find fault with this opinion, which

45. Insofar as this discussion applies to space beyond the world, our Jesuit author agrees with Bradwardine and Oresme that imaginary space lacks dimensions (see n. 25 above and Selection 71, n. 26).

46. Both Bradwardine and Johannes de Ripa held this opinion (see Selection 72, n. 20, and the introduction to this selection).

47. That is, it is not created by the reason.

48. Otto von Guericke argued (near end of Bk. II, ch. 6, in section 4) that imaginary space is not just *like* a positive and real thing but actually *is* a positive and real thing.

49. After our author ascribes an essentially unreal but distinct status to imaginary space—that is, its dimensions merely *correspond* to the positive and real dimensions of bodies, for in itself it is nothing—he insists that God is actually in this imaginary space, a space which always existed and in which the world was created. Here we ought, presumably, to understand that since this infinite space is eternal (see preceding part of Question 2, Article IV in section 3) it is assumed to be God himself, for otherwise it would be an eternal thing coexistent with, but separate from, God. In this, the Coimbra Jesuits agree with Bradwardine and Oresme that infinite space is God Himself and, therefore, coextensive and eternal.

50. Since no one would deny God's power to create a stone in the unreal space beyond the cosmos, there is no good reason for supposing that God Himself could not be in this space.

51. A spiritual substance is not properly locatable. Thus God exists in this unreal space only in a transcendent sense.

52. Earlier, our author cited *The City of God*, Bk. II, ch. 5, presumably the reference intended here. In this chapter Augustine actually expresses disbelief in the existence of infinite places or spaces beyond the world. He is convinced that there is no place beyond this one world. But he insists that if infinite spaces existed beyond the world, God would be omnipresent in them, since no reason could be found for confining God to our relatively small finite world.

53. The text reads *Aquarium* instead of *Aristotelem*.

we follow, concerning the existence of God beyond

4. Otto von Guericke (1602–1686)[54]

BOOK I

Chapter 35

Because according to Aristotle, as the prince of philosophers, and as the more common opinion of both philosophers and theologians, there are no more worlds beyond this world, this provides a good reason that the world must ultimately have a limit, since it is also against God and nature that something (with the exception of God) be infinitely extended. And so finally, the human mind asks, What is outside [or beyond] the world (or its limits: either the so-called prime mobile, or the empyrean heaven, or those waters believed to be beyond the heavens, or something else)? Or what contains or circumscribes the world? Aristotle, especially in Book III, chapters 8 and 9, of *De caelo*, expressly declares that beyond the last [or outermost] sky [or heaven,] (which, for him, is the starry sky or natural body or substance of the last revolving thing *[conversionis]*) there is neither place *(locum)*, nor time *(tempus)*, nor vacuum *(vacuum)*, but absolutely nothing *(nihil)*;[55] indeed, by his arguments, he zealously seeks to confirm that no natural body could be there, and so on.

Now, some state that there is a certain imaginary space which they call "possible locations" *(ubicationes possibiles)*. They offer an occasion for imagining or conceiving the different ways [of thinking about] these various interpretations.

The Coimbra group *[Conimbricenses]*, in commenting on Aristotle's *Physics*, Book VIII, chapter 10, Question 2, Article IV, declares "concerning the space which is found within and beyond the sky. . . ."[56] And in their explication of Book I, chapter 9, of *De caelo*, they confirm this when they write: "On the other hand, although there is no body beyond the sky, there is, nevertheless, a space *(spatium)* or receptacle *(capedinem)* for receiving bodies."[57]

Others say that imaginary space is nothing other than nothing.

Others hold that it is empty of all reality.

Others proclaim that it is the negation of all being.

Others respond [to these last three descriptions as follows]: (1) As for its being "nothing," if it were nothing, space could not be conceived, for "nothing" could have neither extent, nor width,

the sky.

nor length, nor depth. With respect to [description] (2), they infer that if imaginary space is space that is empty of all reality and being, and since the same thing is said of vacuum, therefore vacuum or imaginary space are one and the same thing. As for [description] (3), they reply that if imaginary space is the negation of every being, then it cannot also be called "space."[58]

Furthermore, some desire that imaginary space be merely something fictitious. From this they conclude that before the creation of the world there was no thinking [or conceiving] intellect [and] therefore no imaginary space. And since God cannot be in that which does not exist (as in chimeras), but only in that which does exist,

54. This selection is from *Ottonis de Guericke Experimenta Nova (ut vocantur) Magdeburgica De Vacuo Spatio primum a R. P. Gaspare Schotto, e Societate Jesu, et Herbipolitanae Academiae Matheseos Professore. Nunc vero ab ipso Auctore Perfectius edita, variisque aliis Experimentis aucta . . .*(Amsterdam, 1672; Aalen: O. Zeller, 1962), pp. 51, col. 1, to 52, col. 2; pp. 61, col. 1, to 62, col. 2; pp. 63, col. 1, to 65, col. 2.

Probably through the work of Gaspar Schott, Otto von Guericke was thoroughly familiar with the different viewpoints on imaginary void space, and was especially influenced by the opinions of the Coimbra Jesuits (see section 3), who transmitted a variety of medieval opinions. In this way von Guericke probably played a significant role in disseminating the medieval position represented by Bradwardine and Oresme (indeed, he may have even read Bradwardine's *De causa Dei* in the 1618 edition).

In the history of science von Guericke's treatise, *New Magdeburg Experiments on Void Space,* is most famous for its description of pneumatic experiments, especially the celebrated "Magdeburg Hemispheres," in which two hollow bronze hemispheres were fitted together and subsequently evacuated. Despite the efforts of two teams of eight horses harnessed to each hemisphere and pulling in opposite directions, the evacuated hemispheres could not be pulled apart. See Selection 53.2 for Buridan's similar experiment.

55. See introduction to this selection. Like his medieval predecessors, von Guericke's point of departure is Aristotle's denial of void or any kind of existence outside the finite world.

56. Since von Guericke now quotes much of the substance of the first four paragraphs of Article IV, translated in section 3 above, I have omitted that section here. The edition from which he quoted differs slightly from the one I used.

57. The Coimbra Jesuits also commented upon Aristotle's *De caelo* (and virtually all the physical treatises).

58. Presumably, for these critics, space is a positive thing.

imaginary space does not exist, according to their view. Therefore, God is not in imaginary spaces.

Others understand imaginary space as some merely possible, immense, corporeal mass that is diffused everywhere into infinity; or [imaginary space is] a possible location of such a corporeal mass [or quantity].

Some say (as does Lessius in Book 2 of *De perfectione divino*): "Imaginary space is God Himself, who, in accordance with His immensity is necessarily everywhere, or infinitely, diffused."[59] For if someone could transcend all the [celestial] heavens and seek what is in the space which he forms in his imagination, he would surely find God.

Some wholly reject imaginary space for this reason: because what is conceived *only* by our imagination or mind is not something about a real thing. For example, someone could conceive or imagine that he has 1000 gold pieces, when, indeed, he does not have a single one. Therefore, there is no imaginary space.

A certain more recent philosopher, René Descartes, in his *Principles of Philosophy,* Part II, Number 21, observes that this imaginary space is something truly real when he says:[60] "We likewise recognise that this world, or the totality of corporeal substance, is extended without limit, because wherever we imagine a limit we are not only still able to imagine beyond that limit spaces indefinitely extended but we perceive these to be in reality such as we imagine them, that is to say, that they contain in them corporeal substance indefinitely extended." "For," as has already been shown in his [Descartes's] opinion "the idea of extension that we perceive in any space whatever is quite evidently the same as the idea of corporeal substance.[61] It is thus not difficult to infer from all this that the earth and heavens are formed of the same matter, and that even were there an infinitude of worlds, they would all be formed of this nature *(natura)*,[62] from which it follows that there cannot be a plurality of worlds, because we clearly perceive that the matter whose nature consists in its being an extended substance only, now occupies all the imaginable spaces where these other worlds could alone be, and we cannot find in ourselves the idea of any other matter.[63] There is, therefore, but one matter in the whole universe, and we know this by the simple fact of its being extended," and so on. And this is the understanding or observations of the said René [Descartes].[64]

Finally, if imaginary space is conceded, it follows that it is infinite and immense or infinitely ex-panded in all its parts, that is, in length, width, and depth;[65] it also follows that it is incorruptible, sempiternal, immobile, fixed, and permanent, so that by no force or reason could it be destroyed, so that it could be the receptacle [or container] of any body whatever, whether great or small. On this matter, see several places in Book II, following.

. . . .

BOOK II

Chapter 6
Whether Space, or the Universal Container of All Things, Is Finite or Infinite

[After reviewing, in the first half of this chapter, the different opinions on whether an imaginary space exists beyond the world, von Guericke rejects Descartes's opinion (quoted in the preceding book), declaring that he will demonstrate its falsity throughout the treatise, for it is "against God to assert something infinite and immense; for only God could know no limits to his extension." At this point, von Guericke sets forth his own opinion.—*Ed.*].

Now let us proceed to what has been proposed,

---

59. A conception of imaginary space that is almost identical with Oresme's (see Selection 71). The Lessius referred to here may be Leonard Lessius (1554–1623), a Jesuit professor in theology at the University of Louvain.

60. Von Guericke's quotations from Descartes are given here in the translation by Elizabeth S. Haldane and G. R. T. Ross, *The Philosophical Works of Descartes* (Cambridge, England: Cambridge University Press, 1911; reprinted, with corrections, 1931), I, 264–265.

61. Part II, Principle 21, terminates here, but von Guericke continues and also quotes Principles 22 and 23.

62. Descartes has *materia,* not *natura.* Except for this, and *constitit* for *consistit* (a few words after), von Guericke's quotation agrees with the corresponding passage in the edition by C. Adam and P. Tannery, *Oeuvres de Descartes* (Paris: L. Cerf, 1905), VIII, 52.

63. Principle 22 ends here, followed by the first few lines of Principle 23.

64. Descartes' "imaginary space" is an indefinitely extended plenum, not a void.

65. In light of later remarks, this appears to represent von Guericke's own interpretation of extramundane infinite void space. For him, then, it is three-dimensional, despite the fact that earlier, in Book II, chapter 4 (p. 57 of the 1672 ed.), he declared that he is not considering space three-dimensionally, as is commonly done, but rather as a universal container of all things "which is not to be conceived according to quantity, or length, width, and depth." For evidence that he conceived extramundane space as three-dimensional, see page 565 and notes 69 and 71; and page 567 and note 84.

namely whether there is a space beyond the world and, [if so, whether] it is finite or infinite.

Since it is certain that the world has a limit,[66] the human mind inquires especially about the last terminus of the world. What indeed is it (whether it ends, or the first mover is a fiction, or [whether] one ought to believe there are waters beyond the heavens, or some other thing)? What contains or circumscribes all these things?

Assuming, according to the more common opinion, that there is "nothing" *(nihil)* beyond the world (taking "nothing" in the Aristotelian sense)[67] then if someone should reach the last confines of the world, he either could, or could not, extend his arm beyond the last surface of the last heaven or hurl a spear and a stone into this nothingness.[68]

In the first case, if one could [do this], there would be nothing to obstruct it and so it would be true that "nothing" is there; but nevertheless *space* is there (for if no space were there, a stone could not be projected beyond, nor an arm be extended because it cannot be in something into which it cannot be sent; nor could anything be in it, because what is not has neither extent nor width but is absolutely incapable of receiving anything), and consequently, there would be something [out there].

In the second case, if one could not [do this], something must necessarily impede [or obstruct] it, [since] one [body] could not occupy the place of another. Now, what obstructs it will be, necessarily, something hard or corporeal. Therefore, something will exist outside the last boundary [or terminus] of the world, since [mere] "nothing" cannot obstruct. Nor can what obstructs be called "nonbeing." [But] you say: *It will not be a body or some other [thing] that resists [or opposes] but this nothing (nihil), or lack of space and place, resists and thus opposes nature as if the spear or stone were repelled by a hard body.* But who will be able to understand this? That what is not could obstruct and have a certain nature of resisting. For appearances teach us that lack of place or space is nothing other than one body impeding another, not that one body occupies the space of another [simultaneously]; however, with the one body removed, another can occupy that space. Indeed, since it cannot be denied that the Divine Essence, infinite [and] immense, is beyond the world, then all the more reason it is appropriate that it [that is, the space beyond the world] is immense in extent, width, depth, that is, according to space or expanse;[69]

but it is not to be considered an absolutely infinite nothing (indeed, according to what will be said in the following chapters, we know that it is not [an infinite nothing]).

It follows, moreover, than unless the spear or stone were impeded by another body, it could be projected beyond and such a projection *(jactus)* could be continued into infinity.

Furthermore, since the projection of a spear or stone could be continued into infinity beyond the world and there is nothing to stop it, while, nevertheless, such a projection could not be accomplished except in space, it follows:

1. That the *nothing (nihil)* beyond the world and *space (spatium)* are one and the same; and so-called *imaginary space* is true space, for imaginary space (in the common opinion of philosophers) is nothing and nothing is space, and the space which they call imaginary is true space *(spatium verum.)* Nor can it be objected that whatever is imaginary is not a positive and real thing[70] even if it is not always efficacious. For it is necessary that at least we imagine everything we never see and which is beyond our grasp, as, for example, it is necessary that one who never sees Rome, or a spirit, or an exotic animal, or some other thing, must imagine them; and because no one can comprehend the infinite, it is grasped at least in some way, by the imagination. Meanwhile, it does not follow that Rome, or spirit, or the infinite, are not real *(verum)* or positive things.

2. If the projection of a spear or stone could be continued into infinity and no terminus for resting is revealed, it follows that this space beyond the world is extended or expanded without end toward all parts and thus is infinite and immense.[71] Moreover, if it is "infinite" and "immense" what do we mean by these? Before we can respond to this question, it will be necessary to inquire about *those things that exist (ea quae sunt)*.

66. Von Guericke accepted and defended the truth of the Copernican system.

67. That is, as absolute privation.

68. This statement, and the two cases that follow, derive ultimately from Stoic arguments, probably from Simplicius' version, which is quoted in note 2.

69. In this statement von Guericke indicates that extramundane space is three-dimensional.

70. Here von Guericke disagrees with the Jesuit authors at Coimbra, who, as we have seen (end of section 3), argued that imaginary space is not a real and positive being. See also the introduction to this selection.

71. Three-dimensionality seems to be implied here. This is a description of an inertial motion which had already been given earlier on this page.

Chapter 7
On that which is and that which is said not to be.

. . . .

Thus, if it were asked, "What is to be understood [as existing] before the world was established," and someone responds, "What is Uncreated," and another responds, "Nothing," each would be a proper response. For the one who says "Uncreated" answers as rightly as the one who says "Nothing." Indeed, he [that is, the one who answers "Uncreated"] is thinking of what was created, because it [the "Uncreated"] was surely the "Nothing" of it [that is, of the Created]. Nevertheless, it was and yet is Uncreated.

In the same way, we say that the heaven and earth were made from Nothing, that is, from the Nothing [or negation] of what was created, that is, from the Uncreated.[72] And just as all things have been created from Nothing, so have all things been received and fixed in Nothing—that is, in the Uncreated. For where created things subsist and are now, Nothing was there before and they were received in Nothing, that is, in the Uncreated, since they could not be received in any created thing, both because it was not, and because if it was it was not a Nothing but a created something. Moreover, if it was [or did exist], something could not be received in something. Therefore, it was Nothing in which all things were received, that is, in the Uncreated.

Therefore, everything is in Nothing *(nihilo)* and if God should reduce the fabric of the world *(machinam mundi)*, which he created, into Nothing *(nihilum)*, nothing would remain of its place other than Nothing *(nihil)*, that is, the Uncreated *(increatum)*. For the "Uncreated"[73] is that whose beginning does not pre-exist; and Nothing, we say, is that whose beginning does not pre-exist. Nothing contains all things. It is more precious than gold, free of origin and distinction, more joyous than the appearance of beautiful light, more noble than the blood of kings, comparable to the heavens, higher than the stars, more powerful than a stroke of lightning, perfect and blessed in every part. Nothing always inspires. Where Nothing is, there ceases the jurisdiction of all kings. Nothing is without any mischief. According to Job[74] *[Hiob]*, the earth is suspended over Nothing. Nothing is outside the world. Nothing is everywhere. They say the vacuum is Nothing; and they say that imaginary space and space itself is Nothing.

Chapter 8
Whether Space, or the Universal Container of All Things, Is a Created or an Uncreated Something?

Since vacuum, and imaginary space, or space itself, is Nothing, [and] Nothing is the negation of one thing and the affirmation of another (as was demonstrated in the preceding chapter),[75] it follows that vacuum, or imaginary space, or space itself is one of these two things that are—it is either created or uncreated, for no third thing is given. But that vacuum, or imaginary space, or space itself, is the Nothing [or negation] of the created is acknowledged by Jacques du Bois, an ecclesiastic of Leyden, in his *Dialogus Theologico-Astronomicus*[76] (written against Galileo to the Galileans Philip of Lansberg[77] and others, who declare that the sun is in the middle of the world and that the earth revolves) when he writes on page 39: "Who denies the vacuum beyond the extreme heavens contradicts *(contradicit)* the infinity of the divine essence, for they do not take the God of the Heaven of Heavens as Solomon witnesses in I Kings, 8, verse 27.[78] Therefore, beyond these heavens there is some place *(ubi)* in which there also is this divine essence, which cannot be contained by any limits. We are accustomed to call this place *(ubi)* "vacuum" because it contains no body. Accordingly, before the creation of the universe, God was somewhere; indeed He was everywhere infinitely in this infinite abyss which could not be filled by a finite body, however large, which is what the su-

---

72. That is, the Nothing from which the world was created is the Uncreated.

73. What follows can only be described as an *Ode to Nothing*. The contrast with the Stoic conception of infinite void is striking. No longer is the infinitely extended void an inefficacious, impotent, and completely empty three-dimensional space. It has become a positive, real, all-powerful divine force—God Himself.

74. Job 26:7.

75. Omitted here.

76. A reference to Du Bois' *Dialogus Theologicus–Astronomicus in quo ventilatur quaestio An Terra in centro universi quiescat, . . . an vero, sole quiescente, terra circa eam feratur (Theological-Astronomical Dialogue in which is aired the question whether the earth rests in the center of the universe, . . . or whether, with the sun at rest, the earth is carried around it)*, published in Leyden, 1653.

77. 1561–1632.

78. "But will God indeed dwell on the earth? behold, the heaven and heaven of heavens cannot contain thee; how much less this house that I have builded?" (King James Version).

preme [or outermost] heaven is (note: here the author [Jacques du Bois] declares with Aristotle that the heaven [or sky] is a body).[79] For to declare with René Descartes that the whole of corporeal substance has no limits to its extension is, in reality, to declare an infinite creation, which God and nature abhor, and so on"[80]

It is clearly obvious from these words that Jacques du Bois states that God is infinitely everywhere in some void place *(in aliquo ubi, vacuo)*. But this statement was not well put: "Therefore, there is some place in which there also is this divine essence." and so on. But [instead] he ought to say: "Therefore, there is a place or space, not *in which* the divine essence is, but which is itself the divine essence." For God can be contained by no place *(ubi)* or vacuum or space, because He Himself is the place *(ubi)* for Himself, or the thing that is empty *(vacuum)* of every creature, or the space, or universal container of all things.

On the other hand, Father Athanasius Kircher[81] of the Society of Jesus, a most famous mathematician, although he attacks the world system of Copernicus and concedes no void space in nature, nevertheless, where he takes up imaginary space in his *Itinerarium Ecstaticum*,[82] Dialogue 2, chapter 9, Number 4, he uses the following words:

"Since God is infinite in actutality, he also necessarily fills every void and empty space (which everyone conceives as immense and infinite) with his substance and presence, eliminating every nothing, every vacuum, emptiness, and nonbeing. And so, outside of God, it is necessary that neither Nothing, nor emptiness, nor vacuum be left. For if there is a nothing outside of God, you must acknowledge that God is not infinite in the least, inasmuch as the divine substance would be confined and contained by this nothing, which is impossible. Hence, it follows again that the world does not subsist in Nothing, but is received in God. Therefore, when you imagine this imaginary space beyond the world, do not imagine it as nothing, but conceive it as a fullness of the Divine Substance extended into infinity. For He who has perfected and created all things by Himself and His fullness receives, confers, carries, encompasses, contains, preserves, and so on."

From these words, nothing other follows than that *space (spatium)*, vacuum *(vacuum)*, and emptiness *(inane)*, which we conceive as immense outside the world, is not Nothing or not-being as understood in the opinion of Aristotle, but it is

the Nothing [or negation] of the created, that is, the uncreated. But Father Kircher speaks improperly when he says "that God fills all imaginary space, vacuum, or emptiness, by His substance and presence," and so on, because he himself [Kircher] denies their existence in the nature of things. How can God fill what is not? He ought to have said with Lessius (who is quoted above) that "imaginary space, vacuum, or the Nothing beyond the world, is God Himself," as he finally does when he announces that the space beyond the world is not Nothing, but is the fullness of the Divine Substance.

Now, we ought to consider that the infinite essence of God is not contained in space, or vacuum, since God, who is present everywhere, is not contained in space or vacuum, but the space and vacuum of the whole creation is in Him, by Him, and for Him. For He is an essence other than all created things [and] contained by the periphery of none of them nor (as I would say) comprehended by any enclosure, but infinite and therefore as much outside all things as within all things. Accordingly, he is also present in all our actions and brings us to judgment for them. "Am I a God at hand, saith the Lord and not a God far off? Can any hide himself in secret places that I shall not see him? saith the Lord. Do not I fill heaven and earth? saith the Lord" (Jeremiah, chapter 23, verse 23).[83]

Thus in the same manner as the said authors consider the space beyond the world as the Divine Essence Himself, so also is it necessary that they consider the world within in the same way or quality. For while they assert that the space beyond is infinite, immense, and full of the divine essence, it follows that within the world it must be the same way, because this world is contained in this infinite space (indeed, in relation to the immensity of this space, it does not have the likeness of a point or atom)[84] and cannot be excluded from it so

79. This parenthetical note within the quotation from Du Bois was supplied by von Guericke.

80. The quotation from Du Bois terminates here.

81. 1601–1690. After teaching mathematics and philosophy at the Jesuit college in Würzburg, Kircher settled in Rome, where he taught mathematics, physics, and oriental languages.

82. Published in 1656.

83. This quotation embraces verses 23 and 24 of the King James Version, the translation used here.

84. A comparison between two physical magnitudes seems implied here, once again suggesting that von Guericke thought of extramundane infinite void space as three-dimensional.

that space withdraws [or diminishes] to the extent that this world system has been received in it; but as the space also contains, it penetrates and thus receives, confers, carries, encompasses, contains, preserves, and so on, all the bodies existing in the world system. Nor does it matter whether the world is located here or elsewhere by an infinite number of parasangs,[85] as we shall see in a follow-ing chapter. The Coimbra [*Conimbricenses*] discussions confirm all these things, as can be seen [above] in [Book I], chapter 35.[86]

85. A parasang is a Persian league containing 30 stades. Its exact measure is irrelevant, since von Guericke speaks of an infinite number of them.
86. For the references, see page 563.

Alchemy and Chemistry

74 ON THE FORMATION OF MINERALS AND METALS AND THE IMPOSSIBILITY OF ALCHEMY

Avicenna (980–1037)

Translated and annotated by E. J. Holmyard
and D. C. Mandeville[1]

The time has now arrived for us to give an account of the properties of mineral substances. We say, therefore, that mineral bodies may be roughly divided into four groups, *viz.* stones, fusible substances, sulphurs and salts.[2] This is for the following reason: some of the mineral bodies are weak in substance and feeble in composition and union, while others are strong in substance. Of the latter, some are malleable and some are not malleable. Of [the former, *i. e.*] those which are feeble in substance, some have the nature of salt and are easily dissolved by moisture, such as alum, vitriol, sal-ammoniac and *qalqand*,[3] while others are oily in nature and are not easily dissolved by moisture alone, such as sulphur and arsenic [sulphides].[4]

Mercury is included in the second group, inasmuch as it is the essential constituent element of malleable bodies or at least is similar to it.[5]

All malleable bodies are fusible, though sometimes only indirectly, whereas most non-malleable substances cannot be fused in the orthodox way or even softened except with difficulty.

The material of malleable bodies is an aqueous substance united so firmly with an earthy substance that the two cannot be separated from one another. This aqueous substance has been congealed by cold after heat has acted upon it and matured it. Included in the group [of malleable bodies], however, are some which are still quick[6] and have not congealed on account of their oily nature; for this reason, too, they are malleable.[7]

1. [This is the third part ("On the Four Species of Mineral Bodies") of Avicenna's *"De congelatione et conglutinatione lapidum"* being sections of the Kitâb al-Shifâ', edited . . . by E. J. Holmyard and D. C. Mandeville. For a full description and discussion of this treatise see Selection 78, introduction and note 1.

2. [Holmyard and Mandeville cite a classification in the Kitâb al-Asrâr of al-Râzî (d. 924) in which terrestrial elements are divided into the following six classes: (1) spirits (four kinds); (2) bodies (seven kinds); (3) stones (thirteen kinds); (4) vitriols (five); (5) niters (six); (6) salts (eleven). Other schemes of classification were also proposed. On al-Râzî, see Selection 76, n. 129—*Ed.*]

3. qalqand, χάλκανθος, green vitriol, FeSO₄. See later

in this selection.

4. zarnîkh, realgar, orpiment, As₂S₂ and As₂S₃.

5. This is not merely an expression of Ibn Sînâ's natural caution but a reference to general alchemical opinion, *viz. Our mercury is not the mercury of the vulgar.*

6. In the sense of *alive,* or as in *quicksilver.* This passage proved very troublesome to the Latin translator, or at least to the copyists,the general rendering being: *Et erit exemplum a vino quod nondum gelavit propter suam unctuositatem: et ideo non est ductile.* ["And an example is wine, which scarcely congeals because of its oiliness; and so it is not malleable."—*Ed.*] The Trinity MS. 1122 [see Selection 78, n. 3.—*Ed.*], however, seems to read *a vivo* for *a vino*, so that it is possible that Alfred the Englishman was not far out. [That is *a vivo* would give "And an example is the living, . . . " which would bring it quite near the sense given the passage by Holmyard and Mandeville.—*Ed.*]

7. On the theory that metals are composed of an aqueous substance and an earthy substance, *cf.* Aristotle, following Heraclitus (W. D. Ross, *Aristotle,* London, 1923, p. 109):—"There are two 'exhalations' produced by the sun's rays acting on the surface of the earth. When the sun's rays fall on dry land, they draw up from it an exhalation which is hot and dry, and which Aristotle likens for the most part to smoke but also to fire and wind. When they fall upon water, they draw up an exhalation which like water is moist and cold, and is called the vaporous in opposition to the smoky exhalation. The dry exhalation consists of minute particles of earth on the way to being fire, and exhibiting already, though in a weaker degree, the properties of fire—heat and dryness. The moist exhalation consists of minute particles of water on the way to becoming air, but exhibiting in the main the qualities of water—coldness and moisture. . . . Aristotle turns next to the effects produced by the exhalations when 'imprisoned' in the earth, *i.e.* the minerals. These are divided into the metals, which are formed by the moist exhalation, and the 'fossils', formed of the dry; most of the latter are said to be either 'coloured powders' or stones formed out of such."

According to Jâbir ibn Hayyân there are two constituents of metals, an earthy smoke and a watery steam. Condensation of these exhalations in the bowels of the earth gives rise to sulphur and mercury; combination of sulphur and mercury results in the formation of metals. [Little is known of Jâbir ibn Hayyân, an Islamic alchemist of the eighth century who acquired so great a reputation that long after his death many alchemical treatises written in the tenth century were ascribed to him by their authors. These treatises, many of which were written by members of a sect known as the Brethren

As regards the stony kinds of naturally-occurring mineral substances, the material of which they are made is also aqueous, but they have not been congealed by cold alone. Their congelation has, on the contrary, been brought about by dryness which has converted the aquosity into terrestreity. They do not contain a quick, oily humidity and so are non-malleable; and because their solidification has been caused mainly by dryness, the majority of them are infusible unless they are subjected to some physical process which facilitates fusion.

Alum and sal-ammoniac belong to the family of salts,[8] though sal-ammoniac possesses a fieriness in excess of its earthiness, and may therefore be completely sublimed.[9] It consists of water combined with a hot smoke, very tenuous and excessively fiery, and has been coagulated by dryness.

In the case of the sulphurs, their aquosity has suffered a vigorous leavening with earthiness and aeriness under the leavening action of heat, so far as to become oily in nature; subsequently it has been solidified by cold.

The vitriols[10] are composed of a salty principle, a sulphureous principle and stone, and contain the virtue of some of the fusible bodies [metals].

---

of Purity, came to constitute what is called the "Jâbir corpus" (in Latin "Geber") of alchemical literature. An integral part of this literature is the Sulphur-Mercury theory, which had a long subsequent history. Of Jâbir's theory, Holmyard writes (*Alchemy* [Middlesex: Penguin Books, 1957], pp. 72–73):

"He believed that, under the influence of the planets, metals were formed in the earth by the union of sulphur (which would provide the hot and dry 'natures') and mercury (providing the cold and moist). This theory, which appears to have been unknown to the ancients, represents one of Jâbir's principal contributions to alchemical thought; it may have been wholly original, though perhaps Jâbir found the germs of it in Apollonius of Tyana. It was generally accepted by later generations of alchemists and chemists, and survived until the rise of the phlogiston theory of combustion in the concluding years of the seventeenth century.

"A proviso should be made concerning the character of the sulphur and the mercury of which Jâbir supposed metals to be formed. He knew quite well that when ordinary sulphur and mercury are heated together the product obtained is a non-metallic stony substance; in fact he describes this very experiment and says that the resulting solid is cinnabar. The sulphur and mercury composing metals were, then, not the substances commonly known by those names, but hypothetical substances to which ordinary sulphur and mercury formed the closest available approximations.

"The reasons for the existence of different kinds of metal are that the sulphur and mercury are not always pure, and that they do not always unite in the same proportion. If they are perfectly pure, and if also they

combine in the most complete natural equilibrium, then the product is the most perfect of metals, namely gold. Defects in purity and, particularly, proportion result in the formation of silver, lead, tin, iron, or copper; but since these inferior metals are essentially composed of the same constituents as gold, the accidents of combination may be rectified by suitable treatment. Such treatment, according to Jâbir, is to be carried out by means of elixirs." From all this, it is apparent that in the present selection, Avicenna was strongly influenced by Jâbir.—*Ed.*]

Abu'l-Qâsim al-Irâqî (see his *Kitâb al-'Ilm al-Muktasab*, edited and translated by E. J. Holmyard, Paris, 1923, p. 13) says:—"The moistness and dryness of which minerals are composed are nothing but watery steam and earthy smoke; if compounded together in right proportion, they give rise to the six metals, while if the dryness, that is, the smoke, is in too great proportion, then are formed brittle stones such as the marcasites, magnesias, tutias, and the stones related to the mineral substances from the *kuhl* and *zarnîkh*, etc. If the moistness, that is, the steam, is in too great proportion, mercury, and nothing else, will result."

8. Al-Râzî classifies alum as a stone and sal-ammoniac as a spirit. Sal-ammoniac is practically universally classed as a spirit by the Arabs, Ibn Sînâ being the only exception so far as we are aware.

9. ["Sublimation, which consists in heating a substance until it vaporizes and then condensing the vapour directly back to the solid state by rapid cooling, is not of general application, since most vapours pass through a liquid phase before solidifying. However, many of the substances used in alchemy happen to produce sublimates very easily: they include sulphur and many sulphides, amber and other resins, and camphor" (E. J. Holmyard, *Alchemy* pp. 44–45). For Albertus Magnus' description, see Selection 76.—*Ed.*]

10. In his book *Kitâb al-Burhân fi Asrâr 'Ilm al-Mîzân*, Al-Jildaki [*fl.* 1339–42] has a section on vitriols, of which the following is a summary:—"There are seven kinds of vitriol, *viz.* the yellow, the green, the red, *qalqatâr, qalqand, qalqadîs* and *shahîra;* they are all naturally-occurring minerals, and from any one of them all the others may be prepared. *Qalqadis* is white vitriol, *qalqand* yellow, shading off into green and black, *qalqatâr* is yellow vitriol in which are shining golden eyes. *Al-sûrî* is red vitriol; *al-shahîra* is yellow, shading off into green and pure blue. Al-Râzî says that there are seven sorts of vitriol [but compare note 2—*Ed.*] but mentions only six, and one of these is alum, which is not a vitriol; yet in another place he groups the alums and the vitriols separately in spite of what he said previously. In still another place he says that *qalqâdis* is a white vitriol, which is correct, and that *qalqand* is green vitriol, but this is only partially true for it implies that every green vitriol is *qalqand,* and this is not so. He says too that *qalqatâr* is yellow vitriol and that *al-sûrî* is red vitriol, but does not say how they may be distinguished. Finally, he says that the vitriol mines are in Cyprus and that vitriols are formed from [crude] vitriols and alums, which the waters dissolve and carry down with them into the hollows of the earth, after which the heat of the sun coagulates them together. This, however, is only partially true, for the alums are different from the

Those of them which resemble *qalqand and qalqaṭâr* are formed from crude vitriols by partial solution, the salty constituent alone dissolving, together with whatever sulphureity there may be. Coagulation follows, after a virtue has been acquired from a metallic ore. Those that acquire the virtue of iron become red or yellow, *e. g. qalqaṭâr,* while those which acquire the virtue of copper become green. It is for this reason that they are so easily prepared by means of this art.

Mercury seems to be water with which a very tenuous and sulphureous earth has become so intimately mixed that no surface can be separated from it without something of that dryness covering it. Consequently it does not cling to the hand or confine itself closely to the shape of the vessel which contains it, but remains in no particular shape unless it is subdued.[11] Its whiteness is derived from the purity of that aquosity, from the whiteness of the subtle earthiness which it contains, and from the admixture of aeriness with it.[12]

A property of mercury is that it is solidified by the vapours of sulphureous substances[13] it is therefore quickly solidified by lead[14] or by sulphur vapour. It seems, moreover, that mercury, or something resembling it, is the essential constituent element of all the fusible bodies, for all of them are converted into mercury on fusion.[15] Most of them, however, fuse only at a very high temperature, so that their mercury appears red. In the case of lead, an onlooker does not doubt that this is mercury, since it melts at a lower temperature, but if during the fusion it is heated to the high temperature [mentioned above], its colour becomes the same as that of the other fusible bodies, *i. e.* fiery-red.

It is for this reason, *viz.* that it is of their substance, that mercury so easily clings to all these bodies.[16] But these bodies differ in their composition from it by reason of variation in the mercury itself—or whatever it is that plays the same part—and also through variation in what is mixed with it and causes its solidification.[17]

If the mercury be pure, and if it be commingled with and solidified by the virtue of a white sulphur which neither induces combusion nor is impure, but on the contrary is more excellent than that prepared by the adepts,[18] then the product is silver. If the sulphur besides being pure is even better than that just described, and whiter, and if in addition it possesses a tinctorial, fiery, subtle and non-combustive virtue—in short, if it is superior to that which the adepts can prepare—it will solidify the mercury into gold.

Then again, if the mercury is of good substance, but the sulphur which solidifies it is impure, possessing on the contrary a property of combustibility, the product will be copper. If the mercury is corrupt, unclean, lacking in cohesion and earthy, and the sulphur is also impure, the product will be iron. As for tin, it is probable that its mercury is good, but that its sulphur is corrupt; and that the commingling [of the two] is not firm, but has taken place, so to speak, layer by layer, for which reason the metal shrieks.[19] Lead, it seems likely, is formed from an impure, heavy, clayey mercury and an impure, fetid and feeble sulphur, for which reason its solidification has not been thorough.[20]

---

vitriols in nature, reactions and properties. According to Ibn Sînâ, the vitriols are partially soluble bodies, which have been dissolved and then solidified [*cf.* text]; *qalqatâr* is yellow vitriol, *qalqadîs* is white, *qalqant* is green and *al-sûrî* red. Dioscorides and Galen do not mention *qalqant* among the species of vitriol, but mention *qalqadîs* alone, saying that its name in Greek is *chalkanthos.* It is clear to anyone who examines their writings that *qalqant* with them is *qalqadîs* itself, and that what they call *vitriol* is green vitriol, which Ibn Sînâ calls *qalqant.*"

11. Probably the meaning is: unless it is amalgamated or sublimed or fixed, *i.e.* converted into a compound.

12. *Cf.* Geber, *Sum of Perfection,* Third Part, Chap. VI (London, 1678):—"Argentvive, which is also called Mercury by the Ancients, is a viscous water in the bowels of the earth, by most temperate Heat united, in a total Union through its least parts, with the substance of white subtile earth, until the humid be contempered by the dry, and the dry by the humid, equally. Therefore it runs easily upon a plain superficies, by reason of its Watery Humidity; but it adheres not, although it hath a viscous Humidity, by reason of the Dryness of that which contemperates it, and permits it not to adhere. It is also, as some say, the Matter of metals with Sulphur." [This quotation is from a medieval Latin alchemical treatise translated into English in 1678.—*Ed.*]

13. *I.e.* converted into sulphides.

14. It is rather difficult to understand why lead should be given as an example of a sulphureous substance which coagulates mercury. Ibn Sînâ may be confusing elementary lead with *galena,* PbS, or perhaps lead is an error for *zarnîkh,* arsenic sulphide.

15. This naive interpretation of the nature of a fused metal is common in alchemy.

16. *I.e.* amalgamates with the metals.

17. *I.e.* sulphur or something resembling it.

18. *I.e.* the alchemists.

19. A reference to the well-known *cry of tin. Cf.* Geber, *ed. cit.,* p. 163:—"That there is a twofold substance of argentvive in tin, whereof one is not fixed, and the other is fixed, is proved, because it makes a crashing noise before its calcination, but after it hath been thrice calcined, that crashing is not; the reason of this is, because the fugitive substance of its argentvive, making that crashing, is flown away." [Modern chemistry ascribes the cry to friction of the crystalline particles; Albertus Magnus mentions a "crackle" of tin; see Selection 76.—*Ed.*]

20. According to Geber, *op. cit.,* p. 166, "lead differs

There is little doubt that, by alchemy, the adepts can contrive solidifications in which the qualities of the solidifications of mercury by the sulphurs are perceptible to the senses, though the alchemical qualities are not identical in principle or in perfection with the natural ones, but merely bear a resemblance and relationship to them.[21] Hence the belief arises that their natural formation takes place in this way or in some similar way, though alchemy falls short of nature in this respect and, in spite of great effort, cannot overtake her.[22]

As to the claims of the alchemists,[23] it must be clearly understood that it is not in their power to bring about any true change of species. They can, however, produce excellent imitations, dyeing the red [metal] white so that it closely resembles silver, or dyeing it yellow so that it closely resembles gold. They can, too, dye the white [metal] with any colour they desire, until it bears a close resemblance to gold or copper; and they can free the leads[24] from most of their defects and impurities. Yet in these [dyed metals] the essential nature remains unchanged; they are merely so dominated by induced qualities that errors may be made concerning them, just as it happens that men are deceived by salt, qalqand, sal-ammoniac, etc.[25]

I do not deny that such a degree of accuracy may be reached as to deceive even the shrewdest, but the possibility of eliminating or imparting the specific difference has never been clear to me. On the contrary, I regard it as impossible, since there is no way of splitting up one combination into another. Those properties which are perceived by the senses are probably not the differences which separate the metals into species, but rather accidents or consequences, the specific differences being unknown. And if a thing is unknown, how is it possible for anyone to endeavour to produce it or to destroy it?

As for the removal or imparting of the dyes [above-mentioned], or such accidental properties as odours and densities, these are things which one ought not to persist in denying merely because of lack of knowledge concerning them, for there is no proof whatever of their impossibility.

It is likely that the proportion of the elements which enter into the composition of the essential substance of each of the metals enumerated is different from that of any other. If this is so, one metal cannot be converted into another unless the compound is broken up and converted into the composition of that into which its transformation is desired.[26] This, however, cannot be effected by

fusion, which maintains the union and merely

----

not from tin . . . except that it hath a more unclean substance, commixed of the two more gross substances, viz. of sulphur and argentvive, and that the sulphur in it is burning, and more adhesive to the substance of its own argentvive; and that it hath more of the substance of fixed sulphur to its composition than Jupiter hath." [For another brief discussion on the formation of metals, perhaps derived from Avicenna, see Selection 76.—Ed.]

21. The passage beginning There is little doubt is not easy to render literally, though the meaning is clear, viz. that the alchemists can artifically prepare substances which to all appearances are metals, though the apparent qualities are not absolutely identical with those of real metals. The Latin translation evades the difficulty of exactness with Et artifices gelationem fere similem artificialiter facinut, quamvis artificialia non eodem modo sunt quo naturalia. ["And the adepts make a similar solidification artificially, although artificial things are not just like natural things."—Ed.] It will be seen from a later passage that Ibn Sînâ was of opinion that the sensible qualities—in the Aristotelian sense of the word —were perhaps not those which in reality distinguish one metallic species from another.

22. But the general view of the alchemists was that expressed in a marginal note to one of the Trinity MSS. —Natura nonnunquam operatur arte iuvante de qua hic non loquitur. ["Nature sometimes operates with the aid of art, but this is not discussed here."—Ed.]

23. Here begins the passage so famous in the Middle Ages. [Al-Jildaki offers a good summary of Avicenna's famous anti-alchemical views (Holmyard and Mandeville, p. 7): "Avicenna considered that each of the six metals was a distinct species of one genus, just as the genus plant includes different species, and the genus animal likewise. And in the same way that it is impossible to convert a horse into a dog or a bird into a horse, or a man into a bird, so it is impossible to convert silver into gold or copper into silver or lead into iron. . . . He believed, however, that it was possible to dye copper white, and thus to give it the appearance and colour of silver, but it would still be copper, only dyed; and to dye silver red and thus to give it the appearance and colour of gold, but it would still be silver, only dyed, and not gold. The white dye, he believed, was extracted from arsenic, mercury and silver, and the red from sulphur, gold and sal-ammoniac, and also from certain plants and animals."—Ed.]

24. I.e. lead and tin.

25. Ibn Sînâ seems to imply that artificial sal-ammoniac, qalqand, etc. are not identical with the natural products—a view which is paralleled to-day among the general public, who usually imagine that synthetic indigo, for example, is not veritable indigo but only a very good imitation.

26. Cf. the Latin version of this passage. [Hec compositio in aliam mutari non poterit compositionem nisi forte in primam reducantur materiam. . . . ("This composition cannot be changed into another composition unless, perhaps, they are reduced into prime matter . . . ."—Ed.] It will be seen that the Arabic says nothing of converting the substance into its prime matter, though perhaps this is implied.

causes the introduction of some foreign substance or virtue.

There is much I could have said upon this sub-ject if I had so desired, but there is little profit in it nor is there any necessity for it here.[27]

75 TWENTY-SIX ARGUMENTS AGAINST ALCHEMY AND THE RESPONSES THERETO

Petrus Bonus (fl. 1330)

Translated by A. E. Waite[1]

Introduction by E. J. Holmyard; annotation by Edward Grant

[Of this treatise, and especially of the extract presented below, E. J. Holmyard says the follow-ing:—Ed.][2]

An alchemical work highly esteemed by the alchemists themselves was 'The New Pearl of Great Price' (*Pretiosa Margarita Novella*), written in 1330 or thereabouts by Petrus Bonus. The author cannot be identified with certainty, but he may have belonged to the Avogadrus family of Ferrara; the book itself was written at Pola, then a city of the Italian province of Istria but now, under the name of Pulj, included in Yugoslavia. It was first edited in 1546 by Janus Lacinius Therapus in an abridged and parapharsed form, and a further abridged edition in English was published by A. E. Waite in 1894.

A remarkable feature of the book is that Bonus, though declaring that the whole secret of transmu-tation can be learned in a single day, or even in a single hour, admits at the end that he himself had never been successful in the Art. 'This unusual candour in an alchemical writer', says Lynn Thorndike pertinently, 'rather disarms our criti-cism, and makes us feel that we have to do with a genuinely first-hand document which reflects the relation of alchemy to the thinking world of a particular past period rather than with the forgery of some quack or romancer who directed his appeal to gullible and unthinking followers of a current fad and delusion.'

Although Bonus asserted that essential alchemi-cal knowledge could be transmitted in a very short time, he goes on to explain that the search for that knowledge is very difficult, partly because the ad-epts use words not only in their ordinary sense but in allegorical, metaphorical, enigmatical, equivocal, and even ironical ways. Moreover, alchemical writers often contradict one another, and working alchemists use different practical methods.

In the manner of the Schoolmen, Bonus first marshals the arguments that can be advanced against the truth of alchemy, and does so very cogently; then, later in the book, he shows how these arguments may be refuted. Some of the rea-sons militating against the reality of the Art are as follows. The metals are composite substances, but the alchemists do not know their exact composi-tion, therefore cannot produce them. Neither are the peculiar manner and mode of metallic composi-tion known. In the production of metals, Nature uses a mixed heat, derived partly from the Sun and partly from the centre of the Earth; this cannot be imitated by alchemists. In Nature, the generation of metals takes thousands of years, and occurs in the bowels of the Earth. This process cannot be hasten-ed appreciably by heat, because excessive heat would hinder development, and neither can glass or earthenware vessels replace the natural womb of metals in the ground. Alchemy cannot produce animal life, yet animals are easily decomposed and putrefied; much less therefore can it produce metals, which are of a much stronger composition. It is true that metals are generically alike, but they are different specifically; now it is impossible to change one species into another. Metals are formed under the movements and influences of the stars; but these movements no human mind can fix or direct to any given spot. It is easier to destroy things than to make them; but we can hardly destroy gold, so how should we make it? The ancient philosophers were in the habit of teaching all the arts and sciences they knew to their disciples,

27. The Latin version omits the last sentence: no doubt the sarcasm of it was unpalatable.

1. This selection is taken from the translation by A. E. Waite of *The New Pearl of Great Price. A Treatise Concerning the Treasure and Most Precious Stone of the Philosophers. Or the Method and Procedure of this Divine Art: With Observations Drawn from the Works of Arnol-dus, Raymondus, Rhasis, Albertus, and Michael Scotus, first published by Janus Lacinius. the Calabrian, with a Copious Index. The Original Aldine Edition translated into English* (London, 1894), pp. 49–77, 152–183.

2. Reprinted with the kind permission of Penguin Books Ltd. from E. J. Holmyard, *Alchemy* (Middlesex: Penguin Books, 1957), pp. 138–140. Although twenty-six refutations of alchemy are given, no response is in-cluded to the twenty-first.

and of writing them in their books; but of alchemy they mention never a word. The Stone is supposed to harden lead and tin into gold, and to soften silver, iron, and copper into gold; but it is impossible that one and the same thing should produce opposite effects. It is not correct to call metals other than gold and silver imperfect, since in their own way they are complete. Critics also say that alchemists can merely alter and not transmute, and that alchemical gold and silver are not the same as those precious metals naturally derived.

This would seem a powerful case, but Bonus thinks it may be easily refuted. He does not consider it necessary to go back to ultimate constituents, for the proximate constituents of all metals are known to be sulphur and mercury; the base metals are imperfect and diseased but were ordained by Nature to become gold eventually and are already well on the way. It therefore only remains for the alchemist to cure the diseased metals by ridding them of superfluity of sulphur. Astrological influences need not be taken into account, since the celestial power is a constant factor in the growth or transmutation of metals; there is no virtue in working under a particular constellation or sign. As to the length of time required, Bonus is of the opinion that, to have reached even the base-metal stage, the process of metallic generation must have been going on so long that the final stage may be accomplished very quickly—in fact instantaneously on the addition of the Stone. Transmutation is merely the work of Nature aided by the Art and directed by the divine will. The inconsistencies that critics find in the Stone cannot be explained on natural grounds but must be accepted by faith, which has no difficulty in accepting the Christian miracles. In alchemy, work without faith is foredoomed to failure; the Art is a divine secret, and this explains also why it was not revealed by the ancient philosophers.

The New Pearl of Great Price
Being a Concordance of the Sages on the Great Treasure, the Stone of the Philosophers, the Arcanum, the Secret of All Secrets, and the Gift of God.

Both among ancients and moderns the question whether Alchemy be a real Art or a mere imposture has exercised many heads and pens; nor is it possible for us entirely to ignore the existence of such a dispute. A multiplicity of arguments has been advanced against the truth of our Art; but men like Geber[3] and Morienus,[4] who were best fitted to come forward in its defence, have disdained to answer the cavilling attack of the vulgar. They have not, as a matter of fact, furnished us with anything beyond the bare assertion that the truth of Alchemy is exalted beyond the reach of doubt. We will not follow their example, but, in order to get at the foundation of the matter, we will pass in review the arguments which have been or may be, set forth on both sides of the question.

In the case of a science which is familiarly known to a great body of learned men, the mere fact that they all believe in it supersedes the necessity of proof. But this rule does not apply to the Art of Alchemy, whose pretensions, therefore, need to be carefully and jealously sifted. The arguments which make against the justice of those claims must be fairly stated, and it will be for the professors of the Art to turn back the edge of all adverse reasoning.

Every ordinary art (as we learn in the second book of the Physics) is either dispositive of substance, or productive of form, or it teaches the use of something. Our Art, however, does not belong to any one of these categories; it may be described indeed as both dispositive and productive, but it does not teach the use of anything. It truly instructs us how to know the one substance exclusively designed by Nature for a certain purpose, and it also acquaints us with the natural method of treating and manipulating this substance, a knowledge which may be either practically or speculatively present in the mind of the master. There are other crafts which are not artificial, but natural, such as the arts of medicine, of horticulture, and glass-blowing. They are arts in so far as they require an operator; but they are natural in so far as they are based upon facts of Nature. Such is the Art of Alchemy. Some arts systematise the creations of the human mind, as, for instance, those of grammar, logic, and rhetoric; but Alchemy does not belong to this class. Yet Alchemy resembles other arts in the following respect, that its practice must be preceded by theory and investigation; for before we can know how to do a thing, we must understand all the conditions and circumstances under which it is produced. If we rightly

3. For remarks on Geber, that is, Jâbir ibn Hayyân, see Selection 74, n. 7.
4. According to E. J. Holmyard (*Alchemy*, pp. 62–63), Morienus, or Marianos, was a Christian alchemist who revealed the secrets of alchemy to the Islamic Ummayyad Prince Khalid ibn Yazid, who lived at Damascus about 660–704.

apprehend the cause or causes of a thing (for there often is a multiplicity or complication of causes), we also know how to produce that thing. But it must further be considered that no one can claim to be heard in regard to the truth or falsity of this Art who does not clearly understand the matter at issue; and we may lay it down as a rule that those who set up as judges of this question without a clear insight into the conditions of the controversy should be regarded as persons who are talking wildly and at random.

REASONS APPARENTLY MILITATING AGAINST THE REALITY OF OUR ART.

It was usual among the ancients to begin with a destructive argument. This custom we will now follow.

Reason First.

Whoever is ignorant of the elements of which any given substance is composed, and of the quantities of each element in such composition, cannot know how to produce that substance. Now, the alchemists are necessarily ignorant of the exact composition of metals: therefore, as the metals are composite substances, it is not possible that the alchemists should know how to produce them.

Reason Second.

Again, if you are unacquainted with the determinate proportion of the elements entering into the composition of any given substance, you cannot possibly produce that substance. I allude to the exact degree of digestion which has taken place in, and the peculiar manner and mode of composition which constitute the specific essence, or form, of any assigned substance, and make it what it is. This specific form of metals can never become known to a human artist. It is one of Nature's own secrets, and the Art of Alchemy must, therefore, be pronounced not only unknowable, but utterly impossible.

Reason Third.

We are also ignorant of the proper or specific instrument, or means, which Nature uses to produce those peculiar substances defined as metals. We are aware that Nature, in the production of every different substance, uses a certain modified form of digestive heat. But in the case of metals, this digestive heat is not derived either from the sun, or, exclusively, from any central fire, for it is inextricably mixed and compounded of the two,

and this in a manner which no man can imitate. Therefore, Alchemy is impossible.

Reason Fourth.

Moreover, we know that the generation of metals occupies thousands of years. This is the case in Nature's workshop in the bowels of the earth: hence we see that even if this Art were possible, man's life would not be long enough for its exercise. Everything requires for its generation a certain predetermined period of time; and we find in the case of animals and vegetables that this period of generation and development cannot be hastened to any considerable extent. It might indeed be said that Art can do in a month what Nature requires a thousand years to accomplish—by intensifying and exalting the temperature of the digestive warmth. But such a course would defeat its own object, since a greater degree of heat than is required for the development of metals (*i.e.,* an unnatural temperature) would hinder rather than accelerate that development.

Reason Fifth.

Again, the generation of metals, as of all things else, can only be accomplished in a certain place specially adapted to the purpose. Definite peculiar local conditions must be fulfilled if a seed is to spring up and grow; an animal can only be generated and developed in its own proper womb. Now, glass, stone, and earthenware jars and vessels can never take the place of the natural womb of metals in the bosom of the earth. Hence, Alchemy is nothing but a fraudulent pretence.

Reason Sixth.

Once more, that which is effected by Nature alone, cannot be produced artificially; and metals belong to this class of substances. Generation and corruption are the effect of an inward principle, and this inward principle is Nature, which creates the substantial forms of things. Art, on the other hand, is an outward principle, which can only bring about superficial changes.

Reason Seventh.

If Art cannot produce that which is of easy separation, and, therefore, of easy composition, it cannot produce that the separation and composition of which are more difficult. Now, a horse or a dog are easily decomposed, while the putrefaction of metals requires a great length of time. But yet

Art cannot produce a horse or a dog; hence it can still less produce metals.

Reason Eighth.

Metals do indeed belong to the same genus or kind; they are all metals, just as a horse and a man are both animals. But as horse and man are specifically different, and as one species cannot be changed into another, so the various metals are specifically different; and as a dog can never become a man, so neither can one metal be changed into another.[5] This reason and its solution are advanced by Geber.

Reason Ninth.

The principles which stir up the vital spark slumbering in metals are necessarily unknown to the student of Nature. For these principles are supplied by the movements and influences of the stars and heavenly bodies, which are overruled by the Supreme Intelligence, and preside over the generation, corruption, and conservation of species, imparting to everything its own peculiar form and perfection. These influences which determine whether a certain metallic substance shall be gold, silver, etc., no human mind can possibly fix or direct to any given spot. Therefore, etc.

Reason Tenth.

Artificial things bear the same relation to natural things which Art bears to Nature. But as Art is not Nature, neither are artificial things the same as natural things: and artificial gold, even if produced, would not be the same thing as natural gold. For the methods of Nature are inward, they are always one and the same, and never vary; but the methods of Art, on the other hand, vary with the idiosyncrasies of the artist.

Reason Eleventh.

It is easier to destroy than to make things: but we can hardly destroy gold: how then can we make it?

Reason Twelfth.

The ancient philosophers were in the habit of teaching all the arts and sciences they knew to their disciples, and of declaring them in their books; but of this Art they never mention a word, which proves that it was unknown to them. Moreover, Aristotle tells us that if a man knows a thing he can teach it: but the books of the so-called Alchemistic Sages are full of obscurities and a wantonly perplexing phraseology. This shews that their boasted knowledge was an impudent pretence.

Reason Thirteenth.

Many ancient Sages, as well as kings and princes, who had hundreds of profound scholars at their beck and call, have sought the knowledge of this Art in vain; now, this would not have been the case if it had any real existence.

Reason Fourteenth.

Alchemists say that their one Stone[6] changes all metals into gold; this would mean that it hardens lead and tin, which are softer than gold, and that it softens silver, iron, and bronze, which are harder than gold. But it is impossible that one and the same thing should produce opposite effects. If, indeed, it could produce two such mutually exclusive effects, it would have to do the one *per se* and the other *per accidens*—and either that which is hardened or that which is softened would not be true gold. We should thus have to assume the existence of two Stones, one which hardens and colours *per se,* and one which softens and colours *per se;* but this would be in flat contradiction to one of the few clear statements of the Alchemists themselves. And even if there were two different Stones, their difference would be reproduced in their effects, and there would thus result two different kinds of gold, which is impossible.

Reason Fifteenth.

If gold and silver could be evolved out of any metallic substance, they could be prepared most easily out of that which is most closely akin to them; but as it is impossible to prepare them out of their very first principles, viz., quicksilver and sulphur,[7] they cannot be evolved out of metals specifically different from them. For it is clear that out of these two matters all metals are derived and generated; orpiment, sal armoniac, and secondary spirits like marcasite, magnesia, and tutia, being all reducible to these two primary forms. There are seven spirits of Alchemy, the four principal ones, quicksilver, sulphur, orpiment, and sal armoniac, and the three secondary and composite spirits, marcasite, magnesia, and tutia; but sulphur and quicksilver include them all. The Stone would

5. Compare note 23 of Selection 74.
6. That is, elixir. For a discussion of the "philosopher's stone," see Selection 76, n. 54.
7. An allusion to the Sulphur-Mercury theory, which is discussed in Selection 76, n. 18, and Selection 74, n. 7.

have to be obtained either from the metals or from these spirits. But the Sages represent the Stone as bearing the same relation to the metals which is borne by form to substance, or, soul to body: hence, it cannot be extracted from such gross things as metals. They do indeed say that by calcining, dissolving, distilling, and coagulating those bodies they purge out all that is gross, and render the metals spiritual and subtle. But they know well enough that any fire violent enough to perform this would kill or destroy the vital germ of the metal.

Nor can so highly spiritual a substance as the Philosopher's Stone is represented be obtained from the metallic spirits (sulphur and quicksilver). For they must either be fixed or volatile. If they are volatile they are useless: they evaporate when exposed to the action of fire, and leave bodies still more impure and defiled than they were before; or they even cause other bodies to evaporate along with them. If, on the other hand, the spirits in a fixed state are to represent the Stone, they will not be able to accomplish any of those things which the Stone is supposed to encompass. For, in that case, they are hard and petrine, like earth or flint, and thus are unable to enter other bodies and pervade them with their own essence. If they are subjected to the violent action of fire they become like glass, *i.e.*, they undergo a process of vitrification, and, with their metallic humour, they lose their malleability and all their other metallic properties. Even lead and tin become glass when their metallic humour is burnt out of them, and it is rank absurdity to say that the vitreous substance is malleable, or ever can become so; for it is the metallic humour which renders metals malleable and fusible. Moreover, glass, or anything vitrified, in melting does not amalgamate with other metals, but floats on the surface like oil. Besides, quicksilver in its natural state adheres to all metals, but it does not adhere either to marcasite (which resembles it too closely for such a purpose), nor to glass: this shews, incontrovertibly, that glass is no metal, whether such glass be natural, or some other substance vitrified. Again, glass, or any vitrified substance, when it has been dipped in cold water, or otherwise refrigerated, can be broken, pounded, and converted into powder; but all metals will bend rather than break, because of their greater malleability and the metallic humour which is in them. You can also either engrave or stamp any image upon cold metals and it will retain that image; but glass (unless in a state of fusion) will do nothing of the kind. Thus, it appears that malleability is a prop-

erty which belongs to metals, and to metals only; and in the various metals this property, with the property of fusibility, exists in different degrees, according to the grade of their digestion and sulphureous admixtion. In glass, too, there are different proportions of fusibility, perspicuity, opacity, and colouring, which depend upon differences of the material used in its manufacture. Only metals in a cold state capable of a certain degree of liquefaction; glass, on account of its great viscosity, may be liquefied when it is melted in a fiercely heated furnace, but not after refrigeration, because then the aforesaid viscosity disappears. When metal is cold or red hot its viscosity is greatest, and in such a state it can be examined; but fusion separates its different parts, and then much of this viscosity is lost. With glass the very opposite is the case. Therefore, if by calcination a metallic spirit becomes vitrified, it is not capable of any further change; and, being fixed, it cannot enter other bodies, or convert them. Therefore, also, if metallic spirits, which are the very vital principles of gold and silver, cannot evolve them out of metals, nothing else can.

Reason Sixteenth.

Again, the Alchemists appear to say that they do not create metals, but only develop those which are imperfect; they call gold and silver perfect metals, and the rest imperfect. We reply that this is an impossibility. The fact is that everything which has its own substantial form, and all its peculiar properties, is specifically perfect. A horse is perfect as a horse, though it has not the rational nature of man; and tin and lead are as perfect in their way as gold and silver. Whatever is perfectly that which it was designed to be, the same also is bound so to remain; thus, lead and tin are fully as permanent and enduring as gold and silver.

Reason Seventeenth.

Again, whatever is multiplied by Nature after its kind, in its own species, may be regarded as permanently belonging to that species. And tin and lead, etc., are of this class. They are not an imperfect form of that which we behold perfected in gold and silver. They are base metals, while gold and silver are precious; and a base thing can never develop into a precious thing, just as a goat can never become a horse or a man.

Reason Eighteenth.

Where there is not the same ultimate disposition

of elements, there cannot be the same substantial and specific form. Now alchemistic gold and silver cannot exhibit the same ultimate arrangement as natural gold and silver; consequently, they are not the same thing. Hence, if there be such a thing as alchemistic gold, it is specifically different from ordinary gold.

Reason Nineteenth.

Again, those things which have not the same generation, must be, so far, different from each other. Now, gold of Art, if any, is generated by a different process from that which Nature employs. It follows that the gold of Alchemy is not true but fantastical gold.

Reason Twentieth.

Anything that is contingent, and liable to chance, cannot be the subject matter of science: for science deals with the necessary, incorruptible, and eternal. The Alchemists themselves say that the secret of their Art seldom becomes known to any one: hence they themselves put their own claim to scientific accuracy out of court.

Reason Twenty-first.

Again, Aristotle (Meteor. iv.)—according to the ancient version—expressly denies the truth of Alchemy, calling it a sophistical and fantastical pretence—though some say those words were interpolated by Avicenna (which, however, we do not believe). We beg leave to transcribe Aristotle's very words:[8] Let me tell the Alchemists that no true change can take place between species; but they can produce things resembling those they desire to imitate; and they can tinge (*i.e.,* colour) with red and orange so as to produce the appearance of gold, and with white so as to produce the appearance of silver (tin or lead). They can also purge away the impurities of lead (so as to make it appear gold or silver); yet it will never be anything but lead; and even though it look like silver, yet its properties will still be those of lead. So these people are mistaken, like those who take armoniac salt for common salt[9]—which seem the same and yet are in reality very much diverse. But I do not believe that the most exquisite ingenuity can possibly devise any means of successfully eliminating specific difference (*i.e.,* the substantial form) of metals. The properties and accidents which constitute the specific difference are not such as to be perceived with the senses; and since the difference is not cognizable (*i.e.,* not sensuously perceptible),

how can we know whether they have had it removed or not? Moreover, the composition of the various metallic substances is different, and, therefore, it is impossible that one should be changed into another, unless they be first reduced to their common prime substance. But this cannot be brought about by mere liquefaction, though it may appear to be done by the addition of extraneous matter.

By these words the philosopher seems to imply that there can be no such thing as a pure Alchemistic Art, that which passes current under the name being mere fanciful and deceptive talk. From his remarks we elicit five reasons which (apparently) militate against the truth of our Art.[10]

Reason Twenty-second.
[The First of Aristotle.]

He who only changes the accidents of things, does not change them specifically, and, as the substantial form remains the same, we cannot say that any real alteration has been effected. Now, the transformation (if any) which takes place in Alchemy is of this kind; therefore, we may confidently assert that it is not real. Alchemists may, as it were, wash out the impurities of lead and tin, and make them look like gold and silver; but in their substantial form they are still neither better nor worse than lead and tin. Certain foreign ingredients (colouring matter, etc.) may make people fancy that they see real gold and silver before them. But those are the same people who could not tell the difference between common salt and salt of ammonia. Nevertheless, these two, though generically the same, exhibit considerable specific differences, and no skilled master of chemistry could possibly confound them.

Reason Twenty-third.
[The Second of Aristotle.]

Any transformation that does not involve the destruction of the substantial and specific pre-existent form, is no real transformation at all, but a mere juggling pretence. Now this exactly describes the performances of Alchemy.

8. The quotation attributed to Aristotle was probably drawn from a spurious work or fabricated by Petrus Bonus (see Selection 76, n. 22). That Avicenna actually subscribed to the opinion expressed in the preceding sentence is clear from his remarks near the end of Selection 74.

9. See Selection 76 and note 69 thereto.

10. Although some theoretical notions of Aristotle are incorporated into the five following reasons, the specific arguments are not found in Aristotle's genuine works.

Reason Twenty-fourth.
[The Third of Aristotle.]

It is impossible for us to know whether a thing which in itself is incapable of being perceived by our senses has been removed or not. Now, the specific differences of metals belong to this category: therefore, Alchemy falsely claims the power of accomplishing a thing which in reality transcends all human possibility and knowledge. The external characteristics with which we are acquainted in metals are not those which constitute their inward and essential nature, but their accidents, and properties, and passivities, which are alone subject to the cognizance of our senses. If this mysterious and deeply hidden something could be touched and handled, we might hope to destory, or abolish, and change it. But, as it is, such an attempt must be considered utterly hopeless.

Reason Twenty-fifth.
[The Fourth of Aristotle.]

Thing which are not mixed in the same elementary proportions, and are not compounded after the same manner, cannot be regarded as belonging to an identical species. Now, this relation does not exist between natural gold and the metals which Alchemy claims to transmute into that metal. Consequently, they cannot become real gold. The fact is that we are ignorant of the true composition of the precious metals —and how can we bring about a result the nature of which is not clear to us?

Reason Twenty-sixth.
[The Fifth of Aristotle.]

One species can only be transmuted into another by returning into the first substance common to both, before each was differentiated in the assumption of its own substantial form. This first substance must then be developed into the other species. But such a complicated operation the Alchemists fail to achieve. They do not reduce the metals to the first substance; hence there is with them no true generation, nor is there any genuine corruption, but only a spurious manipulation of accidents. They melt the metal in their furnace, and then add to it certain prepared chemical substances which change its appearance; but no one can say that there has been a true transformation. So long as they do not reduce the metal to its first substance, and then introduce into it another substantial form, it will still be the same metal, whatever alterations they may seem to effect in its outward appearance. The

original substance and first principle of gold and silver are quicksilver and sulphur. To this substance they cannot reduce any metal by bare liquefaction. Hence their transmutation of metals is never true, but always sophistical. If you wish to generate a man out of meat and vegetables, and other food that is eaten, this food will first have to become blood, and the blood will have to undergo a chemical change into seed, before it can be available for purposes of generation. In the same way, if any metal is to become gold and silver, it must first become quicksilver and sulphur. The Alchemists may indeed say that there is between metals no specific, but only an accidental difference. They suppose that the base metals are in a diseased condition, while gold and silver exhibit the healthy state of the metallic substance: and thus they contend that lead and tin can be converted into gold and silver by a mere alterative motion, just as an alterative motion (produced by some medicine) may convert a diseased into a healthy man. But this is equivalent to the affirmation that, apart from the morbid matter which they contain, all metals are actually gold, and here is an assertion which it is impossible to substantiate. If all metals have the same substantial form, they have the same properties and passivities; for properties and passivities are directly the outcome of the substantial form. Hence all metals would have the same properties and qualities (whether active or passive) as gold. But this is not the case; for they do not abide the test of fire as gold does, nor have they the same comforting medicinal effect, which proves that the difference between them is not merely accidental, but specific. Yet they might again advance that, though all metals have the same substantial form as gold, they have not the same qualities and properties, because these are kept inactive or obscured by the morbid matter; as, for instance, when a man suffers from epilepsy, or apoplexy, or madness, he cannot perform the operations of a complete man; and if a woman suffer from contraction of the womb, or syncope, she may have the substantial form of a woman, and yet she cannot exercise all the functions of a woman. They further say that, as in the human subject this incapacity is removed by the alterative action of some medicine, so in metals, the full effects of the substantial form (which is that of gold) may be brought out by alchemistic action. But the substantial form is not complete until the development is fully accomplished, and if the base metals are not fully developed, they can have no real substantial form, let alone that of gold and silver. And if a thing have not the same substantial form with anyting else, it cannot have, even in a latent condition, the properties and qualities characteristic of that thing. Nothing can have the peculiar qualities of a man that has not the form (*i.e., the essential characteristics*) of a man. The form of gold consists in the brightness which the sulphur receives from the purifying quicksilver in digestion. This brightness belongs only to gold and silver, or even to gold exclusively, as will be shewn. It is a sign that the development of these precious metals is complete, and the fact that the other metals do not possess it also shews that they cannot have the substantial form or essential characteristics of gold. Hence the comparison of the base metals to diseased bodies is false and misleading. We have thus demonstrated that the claims of Alchemy are frivolous, vain, and impossible. We might adduce other reasons, but we believe those already given to be sufficient.

. . . .

REFUTATIONS OF THE OBJECTIONS TO ALCHEMY

The foregoing distinctions and declarations having been set forth, to the great elucidation of the whole subject, we will now proceed to refute the arguments alleged against the truth of our Art, at the same time giving such illustrations and explanations as may suggest themselves.

It is hoped that what has been said has supplied the reader with all desirable information with regard to the scope and bearing of our Art. We now propose to say something in refutation of the arguments intended to discredit Alchemy in the eyes of those who suppose themselves to be learned.

Refutation of the First Five Objections.

The fact is that, in producing gold, the Art of Alchemy does not pretend to imitate the whole work of Nature. It does not create metals, or even develop them out of the metallic first-substance; it only takes up the unfinished handiwork of Nature (*i.e., the imperfect metals*), and completes it (transmutes metals into gold). It is not then necessary that Nature's mode of operation, or the proportion of elements, or their mixture, or the proper time and place, should be so very accurately known to the Artist. For Nature has only left a comparatively small thing for him to do—the completion of that which she has already begun. Moreover, our Artists do not, as a matter of fact, set to work without having first investigated

Nature's method of procedure. Nature herself is set upon changing these metals into gold; the Artist has only to remove the cause which hinders this change (*i.e., the corrupting sulphur*), and then he can depend upon Nature for the rest. This matter will, however, be more clearly explained below in our chapters on the generation of metals. As to the brief space of time required for the conversion in our Art, it must not be thought that we bring this about by exposing metals in the furance to the sudden operation of fierce heat. If we did so, their metallic moisture would, of course, be destroyed and dried up. But we only just melt the imperfect metals over the fire, and then add to them the Philosopher's Stone, which, in a moment of time, imparts to them the form of gold, thus changing and ennobling their nature, and conserving their own proper metallic humour. It would not be possible for us to evolve gold and silver out of the metallic first-substance; but with the help of our Stone, in a fire sufficient for liquefaction, preserving the moisture and removing the superfluity, do we generate that volatile Stone which we seek, to which we unite our fixed Red Stone, and then we can very easily hasten and facilitate an inward action which Nature has already set going, which alone has been brought to a standstill by the presence of impure sulphur.

It is a frequent cause of error to reason about some particular fact or facts in vague and general terms. Where particulars are concerned, you ought to confine your syllogism to the same category, or we may be logically compelled to admit what we know to be nonsense. Now, if you look at the five first arguments directed against our Art, you will find that they are all couched in the most indefinite language; and, therefore, until our opponents descend to matter-of-fact particulars, we cannot consent to regard their arguments as deserving of a refutation.

Refutation of the Sixth Objection.

In our Magistery there are two things to be taken into account—the action of Nature, and the ministration of our Art. In respect of the first consideration—the indwelling natural agent—the whole work from beginning to end is, of course, brought about by it, and by it alone; the digestion, conjunction, generation, and formation of our Blessed Stone are due to it. Nevertheless, there is another point of view, in which our Magistery may be termed an artificial process; without its aid the action of Nature could either not go on at all, or would not be accomplished with so great rapidity. But the moving principle in our Art is undoubtedly natural, and the same must be true of its products. In a word, generation and combination are natural, but the ministration is the work of art, being in Alchemy even as in the cooking of food.

Refutation of the Seventh Objection.

In this argument of our opponents the conclusion is invalid because the form, which is the perfection of a thing, is twofold, one in so far as it is mixed, and one in so far as it has the principle of life and development, or has such a principle introduced from without by means of the quintessence, or in some other way. In the case of animate objects, the nobler part of the composition is often this vital principle; with inanimate objects, indeed, the reverse is naturally the case. For this reason we cannot form a lion, a goat, or a man; for though we might know the exact composition of their bodies, yet it is impossible for us ever to understand the evolution of the soul. In like manner, though we are familiar with the generation of some minerals, vegetables, and animals, yet we are ignorant of their specific forms. But in the generation of gold we know the specific form or composition, separated from the perfectible matter, and the methods of perfection and conjunction, according to Nature. The specific form of the common metals is, as a matter of fact, the same as that of gold and silver. There is no need for us to create metals; we only remove certain impurities which stand in the way of their development, and they then become commuted into gold and silver of their own accord.

Refutation of the Eighth Objection.

This objection is not conclusive because the metals, as has been said, differ not specifically, but only accidentally. But this ojection will be more irrefragably refuted below, when we deal at some length with the argument advanced by Aristotle.

Refutation of the Ninth Objection.

It is true that the generation of some earthly things is dependent on the influences and movements of heavenly bodies, for the introduction of their form, but it is not needful for us to know of them, nor indeed is it possible, except in a confused way, as, for example, in the seasons of the year which are caused by the movements of the sun, and determine the sowing and the growth of plants,

with the sexual commerce of horses, asses, hawks, falcons, etc., which are capable of producing offspring only at certain periods of the solar year. But the rule does not apply to men, pigeons, and fowls. If we wish to generate worms in a putrefying body, we need not attend to the season of the year, but only to certain conditions of warmth, etc., which it is easy for us to bring about by artificial means. In the same way, a certain degree of equable warmth will always hatch the eggs of the domestic hen. The same principle may be observed in the generation of lime, vitriol, salt, and so on. To operations of this kind the heavenly influences appear to be always favourable; and all Sages are unanimous in saying that our Magistery belongs to this class, because it may be performed at any time or period of the year. It is only indispensable, says Rhasis, that all other necessary conditions should be properly fulfilled, and then the stellar influences will not be wanting. And this dictum is substantially confirmed by Lilium and others. So also Plato states that the celestial influences are poured down according to the value of the matter. Wherever, indeed, it is necessary to infuse a new accidental form, the sites, aspects, and conjunctions of the heavenly bodies must be carefully observed. But as the Art of Alchemy makes no demand of this kind, the knowledge required for such an operation is not needed.

Refutation of the Tenth Objection.

Forms are either natural or artifical; and natural forms are either substantial or accidental. The substantial form is that which makes a thing what it is, and differentiates it from all other objects of the same genus. It is also called the specific form. The accidental form embraces all the proper manifestations of the substantial or specific form, such as the active and passive qualities of any given object, and its colour, smell, taste, and shape. Artificial forms are entirely accidental, and are nothing but the shapes and qualities imparted to anything by art through the will of the artist, such as the shape of a house, or ship, or coin. Some of these artificial accidents are permanent, as, for instance, a house or a ship; some pass away with the act in which they consist, as, for instance, dancing and singing, and all successive actions. The generation of the Philosopher's Stone is brought about through the mediation or agency of Nature, using the natural instrument of fire, with the natural colour, smell, and shape thereof, which are its accidental forms, following its determined substantial form, but at the same time by means of the artist's aiding hand. Its form is necessarily natural and substantial, and is known by its natural accidental qualities, like everything else in the world. Some assistance is indeed given to the development of the inherent principle; but the inward agent is natural, and the form which is brought into existence by it is also natural, and not artificial, as is falsely asserted by our opponents. Hence the gold which is obtained by means of our Stone, differs in no respect from natural gold, because its form is natural and not artificial.

In order fully to understand the refutation of this tenth objection, we should further consider that natural forms are evolved in two ways. Either Nature supplies the substance and works it up into a given specific form in the absence of any aid from without, or natural substances are combined and prepared in a certain way by art, and then attain to perfection by means of a natural operation. To this latter class belong most chemical compounds. Though here Nature cannot herself prepare and combine the requisite ingredients, yet the result could never be brought about by a merely artificial operation, and is due to Nature alone. Health is restored to the body by Art, but the real agent is Nature, Art only supplying the necessary conditions under which Nature is to work. There is all the difference in the world between an artificial product of this kind and a real artificial product, such as a house, a ship, and the like. Natural products admit of but little variety, and the gold which is produced by Nature, either in the one or the other of the ways indicated above, will always be the same gold. Hence the gold of Alchemy, which is due to a natural process, rendered possible and assisted by Art, is evidently not wanting either in the specific form or the accidental properties of gold found in mines. The principle of Art is Nature, and, after all, the works of Nature are the operations of Supreme Intelligence, and natural conditions may be established by the intelligent mind of the Artist.

Refutation of the Eleventh Objection.

Our assailants say that it is easier to destroy than to construct. But Geber tells us that what is difficult to construct is also difficult to disintegrate; the stronger the composition of anything, the more difficult is also its decomposition. The making or construction of a thing may be considered in a twofold aspect. There is the initial development of a thing out of its first principles, as, for instance,

the blood in the uterine veins of the hen, out of which the egg is formed; then the development of the chicken out of the egg by subjecting it to the warmth of the hen for a certain period, when all necessary conditions of this development already pre-exist in the egg. We may also distinguish a third operation, viz., the laying of the egg by the hen. The change brought about by Alchemy is of the second description. For in the common metals all the necessary conditions of gold are already found, just as the chicken is already contained in the egg. It is not the business of the Alchemist either to know or to put together the component elements of gold. Rather, we may say that he has them in an unfinished state, and commutes them into gold by a process similar to that which changes an egg into a chicken. The twelfth and thirteenth objections are already met by what has been advanced in our previous arguments.

Refutation of the Fourteenth Objection.

To the fourteenth objection, which asserts that it is impossible that the same thing should operate in two contrary ways, we answer that this is true of the same thing, but not true of different things; and this diversity depends on the thing receiving rather than on the thing received. In the human body, for instance, the same agent changes very different foods into the chyme and blood of exactly the same composition, hard food being softened and soft food hardened. Galen tells us that both cold and warm foods ultimately produce animal heat in the body. Considered as foods, all these substances are different, yet they are all turned to the same use by the one agent which we call the vital power. In the same way, the common metals, which are dug out of mines, differ from each other as to the hardness or softness of their composition, and the degree of their purity, etc.; yet they are all subject to the same natural digestion and the inherent action of the same specific form is developing them all in the same direction. In this case, too, through the operation of one and the same force, the hard substances are softened and the soft substances rendered hard, so that both are reduced to one intermediate degree of consistency. Would it not be absurd to say, as is nevertheless asserted by some who are wise in their own conceits, that it is impossible for our Stone to change both copper or lead and iron into gold, because the one is hard and the other soft. It is the digestive power of metals, and it deals with them as the digestive power of the human stomach is able to deal with food. There is, then, as Geber says, in our Magistery only one thing which changes all metals into the same precious substance, viz., the Red Tincture, and this assertion involves no contradiction in terms, as has been supposed on account of the diversity of the common metals. This one medicine hardens that which is soft, and softens that which is hard, fixes that which is fugitive (or volatile), and glorifies them all with its own magnificent brilliancy and splendour. The true artist knows the causes of the hardness of metals, as well as of their softness, the causes of their fusibility, whether that process be quickly or slowly accomplished, and the causes of their fixation and volatility; he is acquainted also with the causes of the perfection of metals, and of their corruption, of all their defects and superfluities; and, therefore, has all the knowledge which our Magistery requires and presupposes.

Refutation of the Fifteenth Objection.

The refutation of this argument is sufficiently patent from what has already been said.

Refutation of the Sixteenth Objection.

It is advanced that common metals are perfect in their own species, and that it is, therefore, impossible to bring them to any higher degree of perfection, just as a horse can never be perfected into a man. But there is such a thing as specific completeness which, nevertheless, admits of a higher development. An egg, for instance, as far as it goes, is specifically complete in itself; and yet it is not perfect as regards the intention of Nature, until it has been digested by means of natural heat into a bird. It would be absurd to say that an egg must always remain an egg, because as such it has certain well-defined properties and a substantial form of its own. The same holds good with regard to the seeds of plants, which are specifically complete as seeds, yet Nature nevertheless designs them to be perfected into living plants. In the same way, tin, lead, and iron, are perfect in their own species, yet in another sense are not perfect, are at once noble and ignoble, and still have not yet achieved the highest possibilities of their nature. The delay in their development is caused by Nature for the sake of man, because the common metals can be turned into uses for which gold and silver could not be employed.

Refutation of the Seventeenth Objection.

The solution of this difficulty is patent from that which has already been said.

Refutation of the Eighteenth and Nineteenth Objections.

Here, too, we may partly refer the reader back to what has already been proved, and partly we must ask him to wait until we deal with the five arguments of Aristotle.

Refutation of the Twentieth Objection.

We are told that the subject matter of this Art must be contingent, and dependent upon chance rather than upon the strict sequence of natural causes and effects, because the Sages themselves admit that it has never become known to any very considerable number even of its most diligent students. Hence it is asserted that our Art cannot aspire to be a science, and can never, at the very best, be more than a system of haphazard guess work. But it is a mistake to suppose that that which happens only seldom, must therefore necessarily be subject to chance. If our objectors only knew our Art, they would readily admit that it is governed by as rigid a system of unchangeable laws as the most exact science in the world.

We Will Now Proceed to Answer the Five Arguments of Aristotle.

As to the first, it has already been met by our proof that the transmutation of metals in Alchemy is brought about by a natural process.

The same remark holds good of the second objection. A solution of his third difficulty has also been given when we proved that it is not necessary for the artist to be acquainted with the exact composition, or substantial and preexistent form, either of the common metals or of gold and silver, since the necessary process of change is brought about, not by the artist but by the inward natural principle, which strives to fulfil the intention of Nature with regard to it. It is enough to be acquainted with their accidents, properties, and passions, which are the consequence of their form. When any transmuted metal is found to have the properties and passions of mineral gold, without superabundance and without deficiency, we conclude of necessity that it has also the form of gold. It is, of course, impossible, and always will be impossible, for any one to know things by means of their forms, because they do not fall within the cognisance of our senses. That which does the work, and performs the functions, of an eye is an eye, but nothing else really deserves this name; hence a stone or a wax eye is not an eye, but only the similitude of an eye, because it does not perform the functions of an eye. I affirm, notwithstanding, that, among composite things, the form of gold and the Stone of the Philosophers alone can be properly known through the perfect knowledge or cognition of the immediate matter which underlies the visible accidents, which, if the same do not subtend, then is the form unknown and inoperable, as in other composites. There is, however, no need for us to know the forms of common metals; for us it is sufficient to be aware that all metals are in course of development into gold, through the properties and accidents in the immediate first matter, and are capable of being endowed with the form of gold. Whosoever is ignorant of the form in a given matter is ignorant of the possibility of its transmutation, and must judge by his knowledge of accidents and qualities; and, seeing that the gold of the mine and the gold produced by alchemy have precisely the same properties in appearance, and endure the same tests, we conclude that they are both real gold, and are impressed with the same form. The fourth objection of the philosopher has already been met by what has been said above concerning the proportion in which the elements are mixed in any given thing.

We Will Now Attempt to Answer the Fifth Objection of Aristotle.

Aristotle obliges us to confess that metals differ not only in their accidents, but specifically, and therefore his argument requires to be answered at some length. Now, there is this difference between potentiality and actuality, that the one is related to the other as non-existence to existence. The potential becomes the actual, the imperfect the perfect, and substance becomes form; but the process is never reversed. Seed is never potentially blood, nor blood potentially food, nor food potentially the four elements. Not everything that is changed into something else is called the substance of that other thing; a living body is not the substance of a corpse, nor wine of vinegar. In the generation of metals all common metals are potentially what gold is actually, they are imperfectly what gold is perfectly; they are substantially what gold is formally. This is evident from the fact—which shall be proved later on—that Nature changes all metals into gold, while gold is never changed into any of the other metals. Hence, if

our Art is to succeed it must follow the course of Nature, and do as it is taught by Nature.

It must be further distinguished that in this connection potentiality is of two kinds disposition towards the form and the faculty of receiving form. The first may be divided into approximate, remote, and remotest. The second is also duplex. Now, complete goodness or perfection is one, and amongst the metals there is only one which is good and perfect, namely gold, and gold does not need to go through any change to make it good and perfect. To be perfect is for anything to have realised the ultimate intention of Nature concerning it; the common metals have not yet realised this ideal; hence it still remains for them to be changed into gold. And, as that which is nearest to perfection is the best among imperfect things, silver comes next after gold, then bronze, then tin, then copper, then lead, then iron—as appears from what has been said above.

Gold alone among metals has, therefore, reached the highest stage of actual perfection. All other metals are only potentially perfect. Some of them, however, have left behind the more remote grades of potentiality, and the change they require to undergo is inconsiderable, because their distance from the highest stage of metallic actuality is not very great. We do not affirm, with other writers, that the intention of Nature has been frustrated or arrested in the imperfect metals. We affirm that they are produced in accordance with her intention, and that they are in course of development into gold. This operation is performed either by Nature in the bowels of the earth, or, in an infinitely shorter space of time, by our most glorious Magistery.

There are also three kinds of perfection and imperfection:—(1) Among things which have the same substantial form; (2) among things which have different substantial forms; (3) among things which are in course of development into the same form. The first kind of perfection belongs to a man who has the complete use of all his organs, senses, and faculties; a man who suffers from any defect in these particulars is not so perfect a specimen of humanity. The second kind of perfection is comparative, when we place two things, which are complete in their own species, side by side. So, for instance, a man is a more perfect creature than a horse, and a horse is more perfect (or noble) than an ass. The third kind of perfection we find only amongst those things of the same kind which are in different stages of development towards a certain highest point. This is the species of perfection we refer to when we speak of metals. Each metal differs from all the rest, and has a certain perfection and completeness of its own; but none, except gold, has reached that highest degree of perfection of which it is capable. For all common metals there is a transient and a perfect state of inward completeness, and this perfect state they attain either through the slow operation of Nature, or through the sudden transformatory power of our Stone. We must, however, add that the imperfect metals form part of the great plan and design of Nature, though they are in course of transformation into gold. For a large number of very useful and indispensable tools and utensils could not be provided at all if there were no copper, iron, tin, or lead, and if all metals were either silver or gold. For this beneficent reason Nature has furnished us with the metallic substance in all its different stages of development, from iron, or the lowest, to gold, or the highest state of metallic perfection. Nature is ever studying variety, and, for that reason, instead of covering the whole face of the earth with water, has evolved out of that elementary substance a great diversity of forms, embracing the whole animal, vegetable, and mineral world. It is, in like manner, for the use of men that Nature has differentiated the metallic substance into a great variety of species and forms.

Nevertheless, the great process of development into silver and gold is constantly going on. This appears from the fact that miners often find solid pieces of pure silver in tin and lead mines, and also from the experience of others who have met with pure gold in metallic veins of iron—though this latter occurrence is more rarely observed, on account of the great impurity of iron. In some silver mines, again, quantities of solid gold have been discovered, as, for instance, in Servia; at first, the whole appears to be silver, but in the refiner's crucible the gold is subsequently separated from the less precious metal. Thus it is the teaching of experience that Nature is continually at work changing other metals into gold, because, though, in a certain sense, they are complete in themselves, they have not yet reached the highest perfection of which they are capable, and to which Nature has destined them—just as the human embryo and the little children are complete and perfect as far as they go, but have not yet attained to their ultimate goal of manhood. Gold is found in different forms, either mixed with a coarse rocky substance, or in a solid condition, or amongst the sand in the beds of rivers, being washed out of the

mines by water. Golden sand is also found in the deserts of India, where there are no rivers. Silver is never found mixed with the sand of rivers, but mostly in the shape of ore in mines, or like a vein running through a rock. Lead and tin occur mostly in the shape of ore, and sometimes they are mingled with earth. The same facts have been commonly observed with regard to iron and the other metals. When different metals are discovered in the same mine, the less pure of the two will generally be found uppermost, because in the digestion of the metallic substance the impure elements have a tendency to ascend and leave what remains more force to develop in the right direction. The difference between metals, then, may be called specific; but it is not the same difference as that which exists between a horse and a man; it is rather a difference of development, or of the degree of digestion. The common metals have the same metallic form as gold; but the digestion of gold is complete, while that of the others is still more or less imperfect. Thus, there is nothing left for us to do except to continue the digestive process until gold is reached, and so finish it: there is no need for us to reduce the common metals to their first substance, to revert them to the principle of digestion, or to accomplish any other difficult feat of the kind. If, indeed, a horse were to be changed into a man, it would be necessary, by corruption and distinegration, to convert the lower animal form into the first substance, and from this first substance to evolve human seed. Such an operation is, of course, impossible, and to attempt it would be to court failure. Art, therefore, follows Nature in that which it would accomplish after the manner of Nature, and it extols Nature wonderfully, not by violating Nature, but by governing her. But far different is the case of metals, which are all naturally in a state of transition and development into gold. In our Art the metals are not, indeed, changed back into their first substance; but by the juxtaposition and influence of the Blessed Stone, and its subtle mingling with all, even the smallest, parts of the base metal, the Stone, which is the substantial form of gold, impresses this form on every atom of the lead or copper, and thereby transmutes it into gold. This mingling cannot take place, however, without a preliminary melting or liquefaction, which renders the base metal accessible in all its parts to the subtle influence of the Stone, and to the transmuting power of the transmuting medicine. The form which is thus introduced is not accidental, but substantial; and, therefore, the gold which results is not artificial, but natural and real.

Even if it be true, as is generally assumed, that all things are evolved out of the four elements, this theory in no way conflicts with the claims of our Art. For this first substance is not available for any special purpose, unless it has first been changed into a suitable and specifically differentiated form. Thus it is impossible for us to generate a man out of the four elements: for this purpose we must have them in the more specific form of human seed. But where there is human seed, a man may be generated from it without first changing it back into the four elements; rather, the digestion of those four elements, which has already begun, must be continued until the substance assumes the human form. So we cannot produce metals out of the four elements; we must have a viscous, heavy, intermediate water mingled with subtle sulphureous earth, which is the special metallic first substance—that is, quicksilver. This substance, then, through the agency of the sulphur, is developed into gold, or into some common metal, and then into gold. In order to effect this ultimate change, there is no need to reduce the common metals to their first matter, for they already contain that proximate first matter, which may, by comparison, be called the seed of gold, which also has in itself the principle of ultimate development into gold. In the working of Nature there is no regres-

sion; we cannot change the embryo back into the seed, nor the seed into the four elements. The common metals are a substance intermediate between gold and silver, on the one hand, and quicksilver and sulphur on the other. Seeing, then, that the middle must always be nearer to the end than is the beginning, therefore the imperfect metals are nearer to gold than is the first matter; and, consequently, it must be easier to obtain gold from the common metals than from a more remote, or less developed substance, like quicksilver and sulphur.

If we say that the common metals are an intermediate substance, and represent the different stages of transition from quicksilver to gold, this remark must be understood to apply to the natural aspect of the process. As far as our Art is concerned, there is a difference both in the arrangement and in the time. Our Stone perfects the quicksilver of the common metals by purifying and partly eliminating their sulphur; and this process of digestion, which may occupy ages in the bowels of the earth, is accomplished by our Stone in a moment of time, on account of the high degree of digestion possessed by our Stone. This elaborate discussion of the arguments for and against our Art was composed by Master Peter Bonus, of Ferrara, in the year of our Lord 1338. The Master was at that time residing in Pola, a township of Istria.

76 A DESCRIPTION OF ALCHEMICAL OPERATIONS, PROCEDURES, AND MATERIALS

Albertus Magnus (ca. 1193–1280)

Translated and annotated by Sister Virginia Heines, S.C.N.[1]

PREFACE

"All wisdom is from the Lord God and hath been always with Him, and is before all time."[2] Let whoever loves wisdom seek it in Him and ask it of Him, "who gives abundantly to all men, and does not reproach."[3] For He is the height and the depth of all knowledge and the treasure house of all wisdom, "since from Him and through Him and unto Him are all things":[4] without Him nothing can be done; to Him be honor and glory forever. Amen.

Therefore, at the beginning of my discourse I shall invoke the aid of Him Who is the Fount and Source of all good to deign, in His goodness and love, to fill up by grace of His Holy Spirit my small knowledge so that I may be able by my teaching

to show forth the light which lies hidden in the darkness and to lead those who are in error to the pathway of truth. May He Who sitteth on High deign to grant this. Amen.

Though I have laboriously traveled to many

1. [This selection is reprinted by permission of the Regents of the University of California from *Libellus de alchimia ascribed to Albertus Magnus*. Translated from the Borgnet Latin edition; introduction and notes by Sister Virginia Heines, S.C.N., with a foreword by Pearl Kibre (Berkeley: University of California Press, 1958), pp. 1–4, 7–14, 15–16, 18–19, 21–26, 29–30, 35–36, 39, 40–46, 48–51, 58–60, 65–66, 66–69, 71–74. The footnotes have been slightly edited and their numbers changed to fit the needs of the present work.—*Ed.*]
2. Ecclesiasticus, I, 1.
3. James, I, 5.
4. Romans, XI, 36.

regions and numerous provinces, likewise to cities and castles, in the interest of the science called Alchemy, though I have diligently consulted learned men and sages concerning this art in order to investigate it more fully, and though I took down their writings and toiled again and again over their works, I have not found in them what their books assert. Therefore, I examined books pro and con and I found them to be worthless, devoid of all profit and of usefulness. I found, moreover, many learned men of wealth, abbots, bishops, canons, natural philosophers, as well as unlettered men, who expended much money and great effort in the interest of this art, and yet failed because they were not capable of tracking it down.

Yet I have not despaired, but rather I have expended infinite labor and expense, ever going from place to place observing, considering, as Avicenna[5] says, "If this is so, How is it? If it is not, How is it not?" I persevered in studying, reflecting, laboring over works of this same subject until finally I found what I was seeking, not by my own knowledge, but by the grace of the Holy Spirit. Therefore, since I discerned and understood what was beyond nature, I began to watch more diligently in decoctions[6] and sublimations,[7] in solutions and distillations,[8] in cerations[9] and calcinations[10] and coagulations[11] of alchemy and in many other labors until I found possible the transmutation[12] into Gold and Silver,[13] which is

5. Also known as ibn-Sīnā. Abu-'Ali al-Husain ibn-'Abdallah Sīnā (980–1037), the great Arabian philosopher and physician. Among Avicenna's numerous writings is the alchemical tract De anima, frequently quoted by thirteenth-century writers such as Albertus Magnus, Roger Bacon, and Vincent of Beauvais. This work was first printed in Basel in 1527. A great many alchemical treatises have been attributed to Avicenna, but most of them are probably spurious. See Lynn Thorndike, A History of Magic and Experimental Science, II, 471, n. 3; see also ibid., V, App. 3, for a list of the writings attributed to Avicenna. [For extracts from one of Avicenna's genuine works, see Selections 74 and 78.—Ed.]

6. decoctionibus, digestion; decoquere, to boil away. That is: the action of decocting or boiling water or other liquids in order to extract the soluble parts or principles of substances; to boil down or concentrate.

7. See below, chap. 30, for a discussion of the process of sublimation. [Also see Selection 74, n. 9. —Ed.]

8. See below, chaps. 34 and 35.

9. [Chapter 36, omitted here, is titled "What is Ceration and How is it Done?" In a note Sister Virginia describes "Ceration, a word now obsolete," as "the softening of a hard substance or reducing it to a waxlike state."—Ed.]

10. See below, chaps. 31, 53, and 54. [The last chapter, omitted here, is titled "How are Copper Plates Cal-

cined?"—Ed.]. The words calx, calcining, and calcination occur frequently in the Alchimia. Alchemists sought to form a calx by roasting, thereby expelling volatile matter and reducing the substance to a more friable state.

11. See below, chaps. 32 and 50. [Chapter 50 (on p. 65 of Heines's translation), titled "On the Coagulation of all Solutions," is so brief that it can be quoted here in full: "Coagulation of all solutions is accomplished with heat and dryness. Put the liquid which you wish to dry in a glass vessel with a narrow neck, and place it in ashes over a slow fire, and it will coagulate within six hours into a white or red sheet."—Ed.]

12. transmutationem. The dream of all the ancient workers in alchemy was to convert the base metals, such as iron, lead, and copper, into gold and silver. They were thoroughly convinced of the reality of transmutation. Albertus, in the Alchimia, although skeptical, maintains that transmutation may be possible. [Avicenna denied the possibility altogether (see Selection 74). —Ed.]

13. in Solem et Lunam, into Sun and Moon. This use of planetary signs to designate metals dates from the infancy of alchemy. The Liber de mineralibus (III, i, 6; Borgnet ed., V), attributed to Albertus Magnus, gives the reasons for the planetary names of the metals. The Jammy (1651) and the Borgnet (1898) editions of the Alchimia frequently employ: Sol for gold, Luna for silver Venus for copper, Mercury for mercury, Saturn for lead, Jupiter for tin. When the planetary name is used in the Borgnet edition, the present translation capitalizes the name of the metal; for example, Venus is translated as Copper, but cuprum as copper. On the planetary designation of metals, see J. R. Partington, "Report of Discussion upon Chemical and Alchemical Symbolism, Ambix, I (1937), 61; and Pearl Kibre, "The Alkimia minor Ascribed to Albertus Magnus," Isis, XXXII (2) (June, 1949), 270. [Even more interesting than the replacement of the true names of the metals by those of the planets is the fact that the symbols representing the planets came to represent the metals, just as "we write H for an atom of hydrogen, [and] K for an atom of potassium, . . ." (F. Sherwood Taylor, The Alchemists [New York: Schuman, 1949], p. 51). This convenient shorthand goes back to the earliest Greek alchemical manuscripts, dating from around A.D. 250. "We have considerable lists of the signs in the oldest Greek manuscripts. Some of them are derived from the signs of the planet with which the metals were associated, others from the pictorial representations of the things symbolized, others from the initial letters of the name.

"The connection of the planets and metals is certainly ancient, and it persists throughout the whole of alchemy.

"The metals have all received planetary signs. Gold receiving the sign ⊙, representing the sun; silver the sign of the waxing moon ☾; mercury that of the waning moon ☽ (Hermes speaks of "that which drips from the waning moon"); copper has the sign of Venus (Aphrodite–Isis–Hathor) ♀; lead has the sign of Saturn ♄; iron has the sign of Mars ♂. There remain the signs of electrum and tin. Tin has in these old lists the symbol of Hermes ☿, and electrum that of Zeus ♃. In later times (between 500 and 700 A.D.) the symbol of Hermes was given to mercury in place of that of the waning moon. Electrum was no longer considered a separate metal, and its symbol was then given to tin.

better than the natural [metal] in every testing[14] and malleation.[15]

I, therefore, the least of the Philosophers, purpose to write for my associates and friends the true art, clear and free from error; however, in such a way that seeing they may not see, and hearing they may not understand. Therefore, I beg and I adjure you by the Creator of the world to hide this book from all the foolish.[16] For to you I shall reveal the secret, but from the others I shall conceal the secret of secrets because of envy of this noble knowledge. Fools look down upon it because they cannot attain it; for this reason they consider it odious and believe it impossible; they are, therefore, envious of those who work in it and say that they are forgers. Beware, then of revealing to anyone our secrets in this work. A second time, I warn you to be cautious; persevere in your labors and do not become discouraged, knowing that great utility will follow your work.

2

HOW DO METALS ARISE?

Alchemy is an art invented by [the] Alchemist: the name is derived from the Greek *archymo*,[17] which in Latin is *massa*. Through this art, corrupted metals in minerals are restored and the imperfect made perfect.

It should be noted that metals differ from one another only in their accidental form, not in their essential form; therefore the stripping of accidents in metals is possible. Hence, it is also possible, through this art, to bring about a new body, since all species of metals are produced in the earth from a commixture of sulphur and quicksilver[18] or because of foetid earth. Just as a boy in the body of his mother, contracts infirmity from a diseased womb by reason of the accident of location and of infection, though the sperm is healthy, yet, the boy becomes a leper and unclean because of the corruption of the womb. Thus it is in metals which are corrupted, either because of contaminated sulphur or foetid earth; thus there is the following difference among all the metals, by which they differ from one another.

When pure red sulphur comes into contact with quicksilver in the earth, gold is made in a short or long time, either through the persistence [of the contact] or through decoction of the nature subservient to them. When pure and white sulphur comes into contact with quicksilver in pure earth, then silver is made, which differs from gold in this, that sulphur in gold will be red, whereas in silver

it will be white. When, on the other hand, red sulphur, corrupt and burning, comes into contact with quicksilver in the earth, then copper is made, and it does not differ from gold except in this, that in gold it was not corrupt, but here [in copper] it is corrupt. When white sulphur, corrupt and burning, comes into contact with quicksilver in the earth, tin is made, [as is indicated from the fact that] it crackles between the teeth[19] and quickly liquefies, which happens because the quicksilver was not

---

"This system of metals and planets remained unaltered throughout the subsequent development of alchemy and indeed until Dalton suggested his new chemical symbols based on the atomic theory. But the form of some of the signs was altered, and the forms found in printed works are: Gold ☉, silver ☽, copper ♀, lead ♄, iron ♂, tin ♃, mercury ☿.

The system could be extended to express the composition of alloys by putting together the symbols of the metals composing them very much in the modern style." (Taylor, pp. 51–53).—*Ed.*]

14. *examinatione*.

15. *malleatione*.

16. This phraseology was commonly used by the alchemists. Marcellin Berthelot quotes from a Greek manuscript (Paris, Bibl. Nat., Gr. MS. 2419): "Wishing to write for my friends, so that seeing they may not see, and hearing they may not understand, I adjure you, in the name of God, to hold this book hidden from the ignorant." See M. P. E. Berthelot, *Introduction à l'étude de la chimie des anciens et du moyen âge*, pp. 205–207.

17. Gr. μάζα, Lat. *massa*, a lump, mass. Footnote (Borgnet ed., p. 547) reads: *Vel melius ex articulo arabico al et verbo graeco* χυμός. ["Or better, [the name *archymo* is derived] from the Arabic article *al* and the Greek word *chymos*."—*Ed.*] This note does not appear in the Jammy edition.

18. *argento vivo*. The Sulphur-Mercury theory of the origin of metals was important to mediaeval alchemists. "The mercury and sulphur of this theory were not held to be identical with the common substances bearing these names. . . . The names stood rather for combinations of properties or quantities: for example, sulphur was sometimes held to typify visible properties, such as colour, while mercury represented the invisible or occult properties." John Read, *Prelude to Chemistry*, p. 25. "Geber postulated an intermediate formation of sulphur and mercury, from the exhalations, in the interior of the earth. Finally, sulphur and mercury, by combining in different proportions and in different degrees of purity, gave rise to the various metals and minerals." *Ibid.*, p. 18. [See Selection 74, n. 7, for information on Geber (or Jâbir) and a further elaboration of the Sulphur-Mercury theory; and for Avicenna's opinion (which opposes Albert's) that metals are not mutually transmutable because they are of different species and essentially different, see Selection 74 and n. 23 thereto.—*Ed.*]

19. [Probably a reference to "the cry of tin." See Selection 74 and n. 19 thereto.—*Ed.*]

well mixed with the sulphur. When white sulphur, corrupt and burning, comes into contact with quicksilver in foetid earth, iron is made. When sulphur, black and corrupt, comes into contact with quicksilver, lead is made. Aristotle says of this that lead is leprous gold.[20]

Now sufficient has been said about the origin of metals and how they differ from one another in accidental but not in essential form. It remains now to examine the proofs of the philosophers and authorities, to see how they demonstrate that this is the true art, so that we may be able to contend with those who maintain that it is not true.

3

THE PROOF THAT THE ALCHEMICAL ART IS TRUE

Some persons, and they are many, wish to contradict us, especially those who neither know anything about the art nor are acquainted with the nature of metals, and who are ignorant of the intrinsic and extrinsic properties of metals, understanding very little about their dimensions and densities.[21] To these, when they set against us the words of Aristotle, who says, "let the masters of Alchemy know that the species of things cannot be changed,"[22] we must answer that he said this about those who believe in and wish to effect the transmutation of metals that are still corrupt, but this, without doubt cannot be done. Let us, therefore, listen to the words of Aristotle which say the following: "It is true that experiment destroys the form of the species, and especially in metals, and this is the case when some metal is calcined and hence is reduced to ashes and calx, which can be ground, washed, and softened with acid water until made white and natural: and thus these bodies through calcinations and various medicines[23] may lose the brown corrupt vapor,[24] and acquire an airy, vivifying vapor, and the whitened calx will be reduced to a solid mass, which can be colored white or red." For this reason, Hermes[25] says that spirits cannot enter bodies unless they are purified, and then they enter only through the instrumentality of water. Aristotle says: "I do not believe that metals can be transmuted[26] unless they are reduced to prime matter, that is, purified of their own corruption by roasting in the fire."

To those still dissenting and unbelieving, I wish to make myself clearer because we know whereof we speak and have seen what we are asserting: we see different species receive different forms at different times; thus it is evident that by decoction,[27] and persistent contact, what is red in

arsenicum will become black and then will become white by sublimation; this is always the case.

If, by chance, someone should say that such species can easily be transmuted from color to color, but that in metals it is impossible, I will reply by citing the evident cause through various indications and proofs, and will thoroughly destroy their error.

For we see that azure,[28] which is called transmarinum,[29] is produced from silver; since, as is more easily seen, when it is perfected in nature losing all corruption, the accidental is destroyed rather than the essential. We see, furthermore, that copper receives a yellow color from calamine stone,[30] and yet neither the copper nor the calamine stone is perfect, since fire acts on both.

We see that litharge[31] is made from tin, but tin through too much decoction turns a golden color; however, it is possible to convert it to a species of silver, since it is of this nature.

We see iron converted to quicksilver, although this may seem impossible to some; why it is possible I have already stated above; namely, that all metals are made from quicksilver and sulphur;

20. [Compare Avicenna's discussion of the formation of metals in Selection 74.—Ed.]

21. profunditates.

22. [Since alchemy did not exist in Aristotle's time, the sorts of quotations ascribed to him in this chapter were not derived from his genuine works, but probably formed part of an alchemical literature in which quotations were frequently fabricated and attributed to great figures of the past.—Ed.].

23. medicinas.

24. humiditatem corruptam et adustivam.

25. All the ancient Egyptian writings relating to the different sciences such as mathematics, music and medicine were attributed to Hermes. Neoplationsts, mystics, alchemists, and others regarded the Egyptian god, Thoth, as more or less identical with the Greek god, Hermes. "Among the Arabs and in medieval Latin learning the reputation of Hermes continued not only as an alchemist, but as a fountain of wisdom in general." Thorndike, op. cit., II, 219. [For a sample of Hermetic literature, see Selection 72.—Ed.]

26. transubstantiari.

27. perdecoctionem.

28. azurum. This may be the lapis lazuli stone, which is a complex silicate commonly called the sapphire by the ancients. The most important mineral in the stone is lazurite, which is a rich azure blue in color.

29. Perhaps ultramarine.

30. This word was used by the ancients to designate ores of zinc.

31. Lead was known to the Egyptians and was sometimes confused with tin by the early alchemists. Lead monoxide or litharge is generally made by cupellation of lead. See below, chap. 28, n. 92.

wherefore, since quicksilver is the origin of all metals, it is possible also for iron to be reconverted to quicksilver. Do you not perceive, for example, that water solidifies in the winter time through excess cold, and becomes ice, and that ice melts by the heat of the sun and returns to water as before? Thus from quicksilver, wherever it is in the earth, and from sulphur, if this also is present, a union of these two comes about and through a very mild decoction over a long period of time, in which they are combined and hardened to a mineral stone, from which the metal may be extracted.

Likewise, we see that cerussa[32] is made from lead, minium[33] from cerussa, and lead from minium.

Behold, now, it has already been sufficiently proved how species are changed[34] from color to color even to the third or fourth form. From this it must not be doubted at all, that corrupted metals can become pure by their own medicines.[35]

Since the foundation for this art has now been laid, let us see what we shall build upon. For if we build upon hay or wood or straw, fire will consume all. Therefore, let us procure stones, which are neither destroyed by fire nor by decay; then we will be free from all anxiety.

From what we have said concerning the difficulties of the art—its principle, and, finally, concerning its proof—it is evident that we have established that it is the true art.[36] Now it remains to be seen how to proceed, and at what time and in what place.

First, at the outset, certain precepts are to be laid down.

The first precept is that the worker in this art must be silent and secretive and reveal his secret to no one,[37] knowing full well that if many know, the secret in no way will be kept, and that when it is divulged, it will be repeated with error. Thus it will be lost, and the work will remain imperfect.

The second precept is that he should have a place and a special house, hidden from men, in which there are two or three rooms in which are carried on the processes for sublimating and for making solutions and distillations, as I will show later.

The third one is that he should observe the time in which the work must be done and the hours for sublimations and solutions; because sublimations are of little value in the winter; but solutions and calcinations may be made at any time: All these things, however, I will show clearly in [the discussion of] these operations.

The fourth is that the worker in this art should

be careful, and assiduous in his efforts, and not grow weary, but persevere to the end. For, if he begins and does not persevere, he will lose both materials and time.

Fifth, it should be done according to the usage of the art: first in collecting [supplies], second in sublimations, third in fixations,[38] fourth in calcinations, fifth in solutions, sixth in distillations, seventh in coagulations, and so on in order. If he should wish to color besides subliming, and to both coagulate and distill, he will lose his powders, because when they will have been volatilized[39] he will have nothing left of them whatever, but they will be very quickly dispersed.[40] Or, if he wishes to color with fixed powders which are neither dissolved nor distilled, they will neither penetrate nor mix with the bodies [to be colored].

The sixth is that all vessels in which medicines may be put, either waters or oils, whether over the fire or not, should be of glass or glazed. For, if acid waters[41] are placed in a copper vessel, they will turn green; if placed in an iron or lead one, they will be blackened and corrupted;[42] if placed in earthenware, the walls will be penetrated and all will be lost.

The seventh is that one should be on one's guard before all else against [associating oneself] with princes or potentates in any [of these] operations,

32. [In Chapter 27 ("How and from What Is Cerussa Made?"), Albert explains that cerussa is obtained from lead plates suspended "above the vapors of strong vinegar in a stout jar." On page 38, note 148, Sister Virginia quotes from J. M. Stillman's *Story of Early Chemistry*, pp. 19–20: "Cerussa (our white lead), used as a pigment, and externally in medicine, was obtained by submitting lead to the action of the fumes of vinegar in closed vessels for ten days, after which time the 'rust' was scraped off, and the process repeated." Sister Virginia remarks that "this is the centuries-old Dutch process still used for making white lead."—*Ed*.]

33. See below, chap. 28.

34. *permutantur*.

35. [See below, n. 54.—*Ed*.]

36. [The preceding selection in this volume is devoted to an elaborate defense of alchemy and its objectives.—*Ed*.]

37. [The emphasis on secrecy borders on the paranoiac in alchemical treatises. See the Preface to this selection and note 16.—*Ed*.]

38. Probably a stable solid material was formed.

39. *projecti*.

40. *sed revolabit citissime*.

41. Arabian chemists did not attempt to distinguish by their choice of words the specific character of vinegar and acid water. J. M. Stillman, *Story of Chemistry*, p. 282.

42. *inficiuntur*.

because of two dangers: If you have committed yourself, they will ask you from time to time, "Master, how are you succeeding? When will we see some good results?" and, not being able to wait for the end of the work, they will say that, it is nothing, it is trifling, and the like, and then you will experience the greatest dissatisfaction. And if you are not successful, you will suffer continued humiliation because of it. If, however, you do succeed, they will try to detain you permanently, and will not permit you to go away, and thus you will be ensnared by your own words and caught by your own discourses.

The eighth precept is that no one should begin operations without plenty of funds, so that he can obtain everything necessary and useful for this art: for if he should undertake them and lack funds for expenses then he will lose the material and everything.

· · · · · ⬦ ·

5

ON THE QUALITY AND QUANTITY OF FURNACES

Take common clay[43] and to four parts add a fifth part of potter's clay and grind well, and add a little sand, grind again (some prudently add manure or salt water in which manure will have been dissolved); after doing this make a wall, as mentioned before,[44] above the pit, two feet high or a little less, one span thick, and permit to dry. Then have a disc made of potter's clay, which can sustain strong fire, everywhere perforated with fifty or sixty holes, according to the size of the disc [with the perforations] made like a finger, the upper part narrow and the lower wider so that ashes can easily descend. Below, in the earth, make a canal[45] through earth and wall before the disc has been put in place; this should be narrow at the pit end, while outside, at the wall, it should be wider, about one span in width, so that the wind may enter. This canal should be lined with clay; then the disc should be placed on top, in such a way that the wider openings of the perforations are on the underside. Next a wall is built upon the first wall and the disc, to the thickness of one span, but the wall should be above the disc to about the distance of one arm. The furnace should have a hole in the middle above the disc where the coals[46] will be laid. At the top there should be a hole through which calcining vessels may be placed: this hole is to be covered over afterwards with a tight cover. The furnace may also have beneath four or five small holes about three digits[47] wide.

This is the general plan of the furnace.

Note also that a clay tripod should be placed above the disc, upon which are to be placed the calcining vessels, and under which the coals.

· · · · ·

10

THE FOUR SPIRITS OF METALS WHICH COLOR

Note that the four spirits of metals[48] are mercury sulphur, auripigmentum or arsenicum, and sal ammoniac. These four spirits[49] color metals white and red, that is, in Gold and Silver: yet not of themselves, unless they are first prepared by different medicines for this, and are not volatile,[50] and when placed in the fire burn brilliantly. These spirits fashion Silver from Iron and Tin, or Gold from Copper and Lead.

Thus, as I shall say briefly, all metals may be transmuted into Gold and Silver, which are like all the natural metals, except that the iron of the Alchemist is not attracted by adamantine stone[51] and the gold of the Alchemist does not stimulate the heart of man, nor cure leprosy, while a wound made from it may swell, which does not happen with natural gold.[52] But it is evident that in all other operations, as malleation, testing, and color, it will last forever. From these four spirits the

---

43. *argilla simplici*. Good fire clays are made from a mixture of sand and clay with traces of iron, lime, and magnesia, and can stand a very high temperature. Potter's clay *(argilla figulorum)* was no doubt added for its great plasticity.

44. [In chapter 4, Albert says that the furnace should be built near a wall.—*Ed.*]

45. *meatus*.

46. *carbones*.

47. Measures and weights have been Anglicized in this translation. The H. C. and L. H. Hoover translation of Georgius Agricola's *De re metallica* gives (App. C) the following values of the Roman long measure: 1 digitus = 0.726 inches; 4 digiti = 1 palmus = 2.90 inches.

48. The four spirits of metals were frequently mentioned by the early alchemists. "I say therefore first, that spirits (spiritus) are four. . . . The four spirits are quicksilver, sulphur, orpiment, and sal armoniacum." This quotation (Stillman, *op. cit.*, p. 245) appears in a Latin manuscript published by Berthelot and attributed to Djaber [that is, Jâbir, or Geber, as he was called in the Latin West.—*Ed.*]

49. Each of these substances is discussed in separate chapters. See below, chaps. 13–17.

50. *et non effugiant ignem*, and they do not flee the fire.

51. That is, magnetite, an important iron ore.

52. Albertus here shows his distrust of the gold produced by the alchemist.

tincture[53] is made, which in Arabic is called *elixir*,[54] and in Latin, *fermentum*.

11

WHAT IS ELIXIR, AND HOW MANY OF THE METALS ARE TRANSMUTED THROUGH THESE FOUR SPIRITS?

Elixir is the Arabic name and *fermentum* is the Latin: because, just as bread is leavened and raised through good yeast, so the matter of metals may be transmuted through these four spirits into white and red, but especially through mercury, because it is the source and origin of all metals.

. . . . . .

13

WHAT IS MERCURY AND WHAT IS ITS ORIGIN?

Mercury[55] is viscous fluid united in the interior of the earth with a white subtile earth, through the most moderate heat until there is equal union of the two. It rolls on a flat plane with ease and, despite its fluid nature, does not stick to it, and it may possess a viscous form because of its dryness, which tempers it, and prevents adherence [to a surface].

It is the matter of metals when combined with sulphur, that is, as a red stone[56] from which quicksilver[57] can be extracted; and it occurs in the mountains, especially in old drains, in great quantities.

By nature mercury is cold and moist and is the source of all metals, as has been said above. It is created with all metals, is mixed with iron, and without it no metal can be gilded.[58]

Addition.

Quicksilver and sulphur, sublimed with sal ammoniac is converted into a brilliant red powder,[59] but when burned in the fire returns to a fluid and humid substance.

14

WHAT IS SULPHUR, ITS PROPERTIES, AND ITS OCCURRENCE?

Sulphur,[60] the fatness of the earth, is condensed in minerals of the earth through temperate decoction, whereby it hardens and becomes thick; and when hardened it is called sulphur.

Sulphur has a very strong action,[61] and is a uniform substance throughout; for this reason its oil cannot be separated from it by distillation, as from other substances having oil, but rather by means of acute waters, by boiling sulphur in them. It occurs in the earth, sometimes in the mountains and sometimes in the marshes.[62] There are many varieties; namely white, red, green, yellow, or black: and

besides it occurs in the dead form.[63] It is living when extracted from fusible earth, and is effective

53. *tinctura.*

54. Ar. *al-iksēr*, Gr. ξηρίον, powder, Lat. *fermentum.* The alchemists understood by the words elixir, tincture, magisterium, medicine, or philosopher's stone, a compound that was supposed to possess the power of transmuting baser metals into gold and silver. "Of this substance they recognized two sorts—a white powder transmuting metals into silver and a red powder transmuting them into gold." J. C. Brown, *History of Chemistry*, p. 185. See also Thorndike, *op. cit.*, III, 74–75, 81–82. [F. Sherwood Taylor's brief description of the elixir is helpful here (*The Alchemists*, p. 82): "The Greeks, and especially Galen, whose works were very well known to the Arabs, attributed many diseases to excess of one of these qualities [that is, hotness, coldness, moistness, and dryness—*Ed.*]. If a patient suffered from an excess of, e.g., the hot element, he was given a medicine made of substances in which the cold elements were supposed to preponderate. Jâbir tried to do the same—to cure the baseness of metals with medicines, which he termed *elixirs*. The Greek alchemists likewise talked of the 'medicine' *(pharmakon),* which was to be added to a mixture in order to cause a transmutation. Jâbir greatly developed the idea of the 'supreme elixir,' the medicine of the metals." See also note 142.—*Ed.*]

55. Mercury as a metal dates from ca. 1500 B.C. E. R. Caley ("Mercury and Its Compounds in Ancient Times," *Journal of Chemical Education,* V [1928], 419), states that Aristotle (*Meteorologica* IV, 8, 11) gives us the first written evidence of the metal. He called it fluid silver.

56. That is, red sulphide of mercury—our cinnabar. The alchemist often confused the red color of cinnabar with red lead oxide.

57. *argentum vivum.*

58. The use of mercury to amalgamate gold and silver is an ancient practice.

59. When mercury is heated in the air for a long time just below its boiling point, in a long-necked flask, it forms a red crystalline powder: mercuric oxide. Mercuric sulphide, the pigment vermilion formed by sublimating mercury with sulphur, may be the product mentioned here.

60. "Sulphur is a fatness of the earth, thickened until hardened and made dry, and when it is hardened it is called sulphur." Geber in the *Summa Perfectionis Magisterii,* as quoted in Stillman, *op. cit.,* p. 280.

61. This may signify the properties of some of its compounds; for example, sulphur dioxide, a suffocating gas, or sulphuric acid, a highly corrosive liquid.

62. H. Casey, O.P. ("The Scientific Works of St. Albert the Great," *Irish Ecclesiastical Record,* XXXIX [1932], 378) points out that St. Albert showed keen powers of observation, for sulphur does not occur in large quantities in the marshes, but is produced by the action of "sulphur bacteria—*beggiatoa alba* which infest water containing protein matter. Probably St. Albert observed this sulphur in some stagnant marsh which had been contaminated by organic refuse."

63. *Living* carries the connotation of natural, native, crude; and *dead,* that of being already worked, that is, burned by exposure to the fire or treated chemically.

against the itch. It is dead when it is poured into cylinders, as it is found among apothecaries.

Addition.

Sulphur has a fiery nature, liquefies as gum and is entirely smoky.

15

WHAT IS AURIPIGMENTUM AND WHAT IS ITS ORIGIN?

Auripigmentum[64] is a mineral stone and is made thus. Earthen dung pits[65] in the bowels of the earth through long processes of decoction transform[66] it into the substance of auripigmentum. Its viscosity is twofold: one is fine and the other coarse: one is freed through washing and decoction in urine; another through sublimation, as stated below.

Addition.

Auripigmentum is active and burning, unless whitened. After sublimation it may whiten copper into a species of silver: this is done by adding two parts of sal ammoniac to four of rock salt, placing the latter on top of the former, and repeating the process three times until you are satisfied [with your work].

16

WHAT IS ARSENICUM?

Arsenicum[67] is a subtle substance of a sulphurous color and occurs as a red stone. By nature it is like auripigmentum: the flowers are white and red. It is easily sublimed and is whitened in two ways:[68] through decoction and sublimation.

17

SAL AMMONIAC IS TWOFOLD

Sal ammoniac[69] is of two kinds, natural and artificial. The natural variety occurs in the earth as both white and red; both may be extracted from a hard and crystalline mineral, with a very salty taste, which is by nature warm and dry. It is used in washing, for cleansing, and for the refining [of metals]. The artificial kind is better than the natural for our work and it is more noble than all other salts. It converts Mercury into a liquid when it is roasted and ground with it. When [it is] set in some damp place to dissolve: it dissolves iron and lead. It is also an oil which the dryness of fire congeals. By its nature it is warm and moist and serves as a subtle spirit for the elixir: for without it, substances cannot be dissolved nor well mixed.

Note that this spirit neither whitens, reddens,

nor transmutes bodies like other spirits; but it aids the penetration of other bodies[70] purging and cleansing the bodies of blackness. Thereupon it leaves behind spirits mixed with these bodies, aiding their reaction and itself passing away.[71]

64. The early alchemists thought this mineral contained gold, hence the name [that is, *aurum* (gold) + *pigmentum* (color). Vitruvius, Celsus, and Pliny used the term.—*Ed.*]. It is our orpiment, As_2S_3, a yellow sulphide of arsenic. The Greek name, αρσενικόν, *arsenikon*, was often used by the ancients.

65. Casey (*op. cit.,* p. 386), gives an explanation of the use of dung pits: "What appears to be absurd crudities in laboratory methods . . . is that of his [Albertus Magnus] custom of putting things into 'sterquilinium' for varying periods. This is strictly scientific. Saint Albertus Magnus would have had the occasion to require a warm temperature which would remain more or less constant over a prolonged period. Possessed of no thermostat with electric heating and fire controls . . . Saint Albert's practical turn of mind cast about for a suitable means of attaining this end. Nature provided him with 'sterquilinium' (dung pit). It contains thermophilic bacteria which are capable of producing a temperature of 50-70° by their biological activity. The temperature is automatically maintained about this figure, for should it rise higher than the optimum temperature of the organisms, their activity is diminished and, consequently, the temperature falls. The result is a natural incubator which will even hatch eggs."

66. *transit.*

67. Agricola, *De re metallica* (Hoover trans.) V, 111, n.: "Metallic arsenic was unknown although it has been maintained that a substance mentioned by Albertus Magnus *(De Rebus Metallicis)* was the metallic form. Agricola, who was familiar with all Albertus' writings, makes no mention of it, and it appears that the statement of Albertus referred only to the oxide from sublimation."

68. The red stone here is our red realgar (As_2S_2). The disulphide becomes yellow when exposed to light, from the formation of the mixture, As_2S_3 (yellow), and As_2O_3 (white). Both sulphides burn to white trioxide and sublime when heated in the air.

69. " 'Sal armoniacum' is with Albertus, as with ancient writers, classed as a variety of common salt, though he refers to a salt of which he has heard, that is prepared from human urine, chiefly of young boys, prepared by the operations of alchemy, by sublimation and distillation. As he characterises this salt no further, it leaves a doubt as to whether he considered this as essentially different from common salt, though Arabian writers had previously made the distinction clear." (Stillman, *op. cit.,* pp. 250–251). "It is often difficult however to tell in the case of medieval Latin writers whether in writing sal ammoniacum, sal hammoniacum, sal armeniacum, etc., they mean with Pliny the superior grade of common salt, or sal ammoniac, as the confusion of spellings and signification is great and often no clue is given as to properties of the salt alluded to." (*Ibid.,* p. 246). Sal ammoniac (NH_4Cl) was often used by the early alchemists as a flux in soldering, just as it is today.

70. *sed ipse dat aliis introitum.*

71. *recedit.*

Addition.

A solution of sal ammoniac, sublimed or not, but filtered or passed several times through a thick, fine-textured cloth, aids the penetration of any calcinated body into another fusible body, provided the calcinated body is given opportunity to be absorbed in it, and afterwards it is itself congealed and finally dissolved.[72] When saffron iron[73] is prepared in this way, it provides a tincture of the best gold for silver.

. . . . .

20

WHAT IS THE USE OF SAL ALKALI, AND HOW IS IT PREPARED?

Sal alkali[74] is important in this art and, when it has been well prepared, frees all the calxes of bodies as a solid mass. By nature it is warm and moist. It is prepared in this manner: take a large quantity of putrid oaken ashes, or better clavellated ashes,[75] which are used for washing garments, grind very finely, add a sixth part of quicklime, mix once and put a closely woven cloth over a *tina*[76] and upon it as much of the ashes mixed with the calx as it will hold, and pour hot water over the whole from above. Then filter into the lye[77] until all the bitterness has been extracted.[78] Remove this solution and replace it by a fresh one, and repeat [the procedure] as before. Put all the filtrates into the same vessel until morning, and then distill through a filter. Heat in a small cauldron[79] until the solution evaporates and does not fume. Allow to cool and a hard stone will remain which is called alkali, that is, dregs of bitterness. Half fill an earthen jar with this salt and set it uncovered in the furnace. Apply a slow fire at first, heating gently so that it does not boil over (or bump); afterwards increase the heat until the alkali reddens and liquefies as wax; then, using tongs, pour at once into another jar, for if you delay it quickly gets too hard to pour. Place this white alkali salt in a glass vessel in a warm, dry place, since it dissolves in a moist one.

Addition.

Or this alkali salt can be made in another way. Take ashes of certain herbs, called Soda,[80] crush well, [and] boil in a jar with water. Pass through a mesh,[81] as with claret[82] once or twice, [and] then distill through a filter. Afterwards place [the salt] in a new earthern pot and congeal with a slow fire at first, [then] increase the heat until the salt solidifies. Place the salt in a clean dry place.

Plant alum[83] is called alkali alum, alkali salt,[84] and clavellated ashes, or [it] is made from them.

Crush and dissolve alum of Yemen in three pounds of distilled urine water, distill through a new filter, then harden white, and, when this is done, crush on a marble slab.[85] Sprinkle another marble slab with distilled vinegar, place the ground alum from the first slab upon it, raising the slab on one side so that the clear liquid may be drained into a glass vessel, while the residue remains on the slab as a white earth. This should be done in a cold and moist place, and [the solution] should be collected in a well-stoppered glass vessel. This liquid can be hardened in a slow, moist fire [water bath]. With this alum a spirit may be fixed, and with this liquid calcinated bodies may be washed.

. . . . .

25

HOW AND FROM WHAT IS CINNABAR MADE?

Cinnabar is a noble substance: it pertains to this

---

72. *et postmodum congeletur, et ultimo pulverizetur et dissolvatur.* The translation is an alternate reading. In the second interpretation the subject would be the calcinated body and not the sal ammoniac.

73. *crocus ferri.* Iron oxide is used here for treating silver to make it resemble gold. This process is given in more detail in chapter 53, below.

74. This salt is "a white solid, closely resembling soda in many of its properties, [it] was prepared by extracting the white ash of burnt wood with water in earthenware pots. It was known as potash (originally 'pot-ash') or 'pearl ashes.' During the Middle Ages, until the time of Boyle [1627–1691], potash was the only alkali used in Europe; afterwards, however, soda was prepared from the ash of sea weeds and was used on a large scale, under the name of 'barilla,' for the manufacture of soap." (T. M. Lowry, *Inorganic Chemistry,* p. 51).

75. Clavellated ashes were burned lees or dregs of wine, from which potash was obtained.

76. A vessel for holding wine.

77. *lixivium.*

78. The method described here is the modern fundamental procedure for preparing caustic soda.

79. *caldarium.* A large tub used as a water bath.

80. Agricola, *op. cit.* (Hoover trans.), XII, 558: "The actual difference between potash and soda—the nitrum of the Ancients, and the *alkali* of Geber . . . was not understood for two hundred years after Agricola. . . ."

81. *caligam.*

82. *claretum.*

83. *Alumen faseolum. Faseolus, phaseolus,* an edible bean; the kidney-bean.

84. *alofer;* probably should be *alafur,* alkali salt. See Carbonelli, *Sulle Fonti Storiche della Chimica e dell' Alchimia in Italia,* p. 196.

85. Marble slabs were used then as today in the laboratory for exposing materials to the air.

art [of Alchemy] and is called *usifur*.[86] It is made from quicksilver and sulphur thus: take two parts of mercury, and a third part of sulphur, and grind as with vermilion.[87] Place [the mixture] in a glass vessel with a narrow neck, smear all over with clay to the depth of one digit and, after drying, place above a tripod; close the mouth of the vessel well and apply fire slowly for half a day. Afterwards increase the fire, roast through the entire day, until you see red fumes appear above, then allow to cool, break the vessel, and take it out; you will find good and fine cinnabar. Therefore, work hard, and you will arrive at the truth.

As you watch the cinnabar, note when moisture begins to escape from the vessel, and when you see yellow fumes come out, open very cautiously. After less than an hour these will become red, that is, [the color of] cinnabar. Then, put a stick into the *ampulla*[88] and draw out some of it to examine. If it is satisfactory, allow to cool.

Addition.

But mercury should first be washed with ashes and salt and passed through a loose-textured cloth.[89] In like manner, sulphur may be boiled[90] in urine and in vinegar to remove supernatant cloudiness. After it is dried, it should be immersed[91] for one day in vinegar and a second day in urine. But I have seen this [recipe] in writings: take two parts of mercury, three parts of sulphur, and four parts of sal ammoniac. This is according to Hermes.

. . . . .

28

HOW IS MINIUM MADE FROM CERUSSA?

Minium from cerussa[92] is made thus: crush cerussa well with water upon the stone, and form into the shape of lozenges[93] and place in an earthen vessel shaped like a basin[94]—neither round nor long. [For support] first place a large jar on a stone or on two clay walls, each one span high; place the jar on them so that its mouth lies on one wall, and the bottom on the other. Then put a *vasculus*[95] containing cerussa inside the jar and close it with another vessel fitted to it.[96] Now apply a slow fire, then increase it through half a day; allow to cool, and you will have cerussa, moderately red.[97] Again repeat, grinding for half a day, then take it out, and you [will] have good minium.

. . . . .

30

WHAT IS SUBLIMATION AND IN HOW MANY WAYS CAN IT BE DONE?

Sublimation is the volatilizing of a dry substance by fire, causing it to cling to the sides of the vessel.[98] Sublimation in fact is diversified according to the diversity of the spirits of those things to be sublimated. One kind [is accomplished] by ignition, as with marchasita,[99] magnesia or tuchia;[100] another with moderate ignition as with mercury

86. *Usifur, uzifur,* from Arabic *zanjifur,* cinnabar.
87. Here vermilion means minium. See below, chap. 28.
88. A glass or earthenware flask of globular shape. Today's Erlenmeyer flasks are similar in shape.
89. Before attempting to make cinnabar the writer here recommends purifying the mercury and the sulphur.
90. *bulliatur.*
91. *decoquatur.*
92. Casey ("The Scientific Works of Albert the Great," *Irish Ecclesiastical Record,* XXXIX [1932], 386) says of this procedure, that by heating the cerusite, litharge, the yellow oxide of lead was obtained. By carefully regulating the heat the litharge was converted to red lead or minium, otherwise, strong heat would have produced litharge again. Lead was, of course, known in antiquity, and cerusite or native white lead was used then as today for a pigment. [For the method of obtaining cerussa, see note 32.—*Ed.*]
93. *pastilla.*
94. *bacinus.* A small basin.
95. A vase-like vessel.
96. The alembic. [See note 120.—*Ed.*]
97. This is very close to our modern method. "When white lead, massicot, or finely ground litharge is heated in air at about 340° it absorbs oxygen and forms a scarlet crystalline powder, red lead or minium . . . . Above 450° this decomposes again into lead monoxide and oxygen." (Partington, *op. cit.,* p. 839.)
98. [For a description of sublimation, see Selection 74, n. 9.—*Ed.*]
99. [On page 20, note 79, Sister Virginia quotes Stillman's (*Story of Early Chemistry,* p. 253) translation from the medieval *Book on Minerals (Liber de mineralibus)* in which it is said that "Marchasita, or Marchasida, as some call it, is a stone in substance and there are many species, wherefore it takes the color of any metals whatsoever and is thus called silver or gold marchasita."—*Ed.*]
100. [On page 20, note 76, Sister Virginia translates a description of tuchia from the *Liber de mineralibus:* "Tuchia, which is frequently used in the transmutation of metals, is an artificial and not a natural mixture, for tuchia is made from the smoke which rises and is coagulated by adhering to hard bodies, when the brass is purified from the stones and tin which are in it; but the best kind is from that which is sublimed. . . . There are many kinds of tuchia: white, and [ranging] from yellow to red."—*Ed.*]

and arsenic; and still another with a low fire as with sulphur. Indeed, in one type of sublimation of mercury the separation of its earth will result and there will be a change in its fluidity. On the other hand, it is natural that superfluous earth very often is mixed with things with which it has no affinity,[101] hence its sublimation has thus to be repeated more often. Examples of these are the calx of eggshells and of white marble, and finely ground glass, and every kind of prepared salt. From these latter, it [the earth] is cleansed, from others it is not, unless the bodies are [in a state of] perfection; however, they are rather more corrupt, because all such things have sulphureity which, ascending with it in sublimation, corrupts the work. Because of this, if you sublimate from tin or lead you will note that after the sublimation it is contaminated with blackness. Therefore, sublimation is better accomplished with those things with which it does not agree [in nature]. However, sublimation, in general, would be more readily accomplished with those things with which it [the substance to be sublimated] agrees [in nature] if it were not for the sulphureity [in any of the components] with which it does not agree [in nature].[102] A method of removing moisture is to mix and grind with calxes—with which the sublimation should be done—until the metal can no longer be detected, and then the moisture is removed by slow heating. As [the moisture] of [the mixture] recedes, the moisture of the mercury will recede with it, as I shall teach you in the following sublimations of spirits.

31

WHAT IS CALCINATION AND IN HOW MANY WAYS CAN IT BE DONE?

Calcination of any kind is the pulverizing of substances by fire to remove the moisture uniting the parts.[103] Bodies diminished of their own perfection are calcined.

There are also different kinds of calcinations. Bodies are calcined so that the sulphureity corrupting and defiling them may be removed. In fact, each sulphureity may be burned from the substance with which it is combined, but which without calcination cannot be removed. Soft bodies are, indeed, particularly hardened by it, but they [also] take an impression more clearly and harden more readily.[104] Spirits are calcined the better to fix them and bring them more quickly into solution. Every kind of calcined body is more fixed, and more easily sublimed than the uncalcined;

hence, soft bodies can be easily calcined through fire; hard bodies need very strong fire [to be calcined], as I shall teach you at the end [of this book].

Addition.

Silver[105] may be calcined thus: take an ounce of purest Silver, or more if you wish, and from this make plates thin as the [finger] nails of the hand. Add a third part of common salt, from the preparation commonly prepared and calcined, and a fourth part of sublimated mercury, making a powder of said mercury and salt by grinding. Afterwards cement the plates together in the sublimatory, by placing first a layer of the powder, then a second layer of the sheets, and follow layer by layer; then sublime with a slow fire until all the moisture of the mixture evaporates. Close well the opening and increase the fire through the natural day; take care not to remove the vessel from the fire immediately, but let it cool [for] three hours. Do not open the vessel until it is cold, because the spirits will evaporate. When the vessel is cold, take out the sublimed mercury, clear as a crystal, and set [it] aside; then take out the silver that remains half-calcined with the common salt. If possible,

101. Albertus Magnus "introduced the word affinity to designate the cause of the sublimation of the metals with sulphur." (F. P. Venable, *History of Chemistry,* New York, 1922, p. 24). This is the only occurrence of the word *affinitas* in the text of the Borgnet edition.

102. *Et propter hoc si sublimas a stanno vel plumbo, post sublimationem ipsum conspicies nigredine infectum: ergo melior est sublimatio per ea cum quibus non convenit: melius autem esset cum eis cum quibus convenit, si sulphureitatem non convenit.* This is a knotty sentence to untangle. The writer seems to be saying that sublimation would be more easily effected with a mixture or compound of substances of a like nature, if it were not for the sulphureity which is of a contrary nature.

103. [A useful general description of calcination is furnished by Holmyard (*Alchemy,* pp. 43–44): "Calcination implied the reduction of a solid to the state of fine powder, principally by means of heat and generally with a change of composition. Thus although gold could be obtained as a powder, and was then known as calx of gold, it was in fact still in the metallic state; but when lead is heated in air it is converted into a yellowish-brown powder of a non-metallic nature—in this case the calx of lead is a lead oxide. For calcining by means of heat many different kinds of furnaces were used, as they were for most of the other operations; hence large furnaces, small furnaces, and furnaces of intermediate size formed the major part of the equipment of an alchemical laboratory."—*Ed.*]

104. This probably refers to some kind of stamping test; the susceptibility for the impression being related to the hardness of the body.

105. That is, Luna. [See note 13.—*Ed.*]

crush the salt and the half-calcined Silver at once above the *porphyry*.[106] If it cannot be ground, put it into a glass *cassola*[107] and separate the whole salt with fervent waters,[108] until you perceive no salty taste; dry the remaining calx in the bottom of a *paropsis*,[109] and calcine once again with new salt and mercury sublimed five or six times. Alternate the calcining and washing of the Silver calx until you detect no salty taste. Your calcined Silver will then be the whitest and cleanest [kind], like the rays of the stars, so that if you melt part of the said calx with borax, or with good sal nitrum or sal alkali, you will find your Silver converted to white gold.

32

WHAT IS COAGULATION AND WHY IS IT USED?

Coagulation is the reduction of liquid substances to a solid mass by deprivation of their vapors.[110] It was devised to harden mercury and purify medicinal solutions of moisture mixed in them. Mercury is coagulated by its frequent precipitation with violence to the dryness of the fire.[111] The dryness of the fire removes the moisture. This is accomplished in a long narrow vessel.

33

WHAT IS FIXATION AND IN HOW MANY WAYS ARE BODIES FIXED?

Fixation[112] is the appropriate tempering of a volatile substance in fire. It was devised so that every coloring, and every alteration is perpetuated in another and is not changed: for bodies, whose perfection has been diminished[113] through calcination, are fixed when they are freed from corrupting and volatile sulphureity. Sulphur and arsenicum are fixed in two ways: one method is the repetition of their sublimation from one state to another,[114] or until they achieve stability. Spirits are also fixed in another way, either with the solutions of metals or with oil of tartar,[115] as I shall say below.

Addition.

Take sublimed mercury, an equal amount of sal ammoniac, and sublime seven times, or until melted, [then] let the stone remain at the bottom; crush it and expose to damp air so it will become a liquid. Soak metallic arsenicum in this water, dissolve in distilled vinegar, and distill seven times, or congeal, and dissolve, and a stone will result.

Metallic arsenicum[116] is made by melting one part of arsenicum with two parts of white soap. Another [procedure] is given in Geber's *[Liber]*

Fornacum:[117] where you may read [it], if you wish.

Either sublime mercury, or sulphur, or prepared arsenicum, or several of these, at the same time, along with sal tartarum or saltpeter, or sal ammoniac. Do this many times until they remain fixed, then extract [them] with warm water.

34

WHAT IS SOLUTION AND IN HOW MANY WAYS IS IT DONE?

Solution is the resolution of any calcined substance into water. It was devised so that the intrinsic qualities of substances might become extrinsic and vice versa, and so that they might be made suitable for distilling; thus they are freed from every contamination. Solution is achieved either by heat and moisture or by cold and moisture, as I shall teach in the following [chapters].

Addition.

Some [substances] dissolve after being calcined

106. A purple stone found in Egypt; often used by alchemists instead of marble.
107. A kind of saucepan with a handle.
108. Perhaps alcohol was used here to precipitate the salt. "Fervent water" might also mean boiling water.
109. A small dish.
110. [In his glossary of alchemical terms Holmyard (*Alchemy*, pp. 271–272) defines "coagulation" as "crystallization, conversion of a liquid to a solid."—*Ed.*]
111. In Aristotle's concept of the four elements both fire and earth were dry (earth: cold and dry; fire: hot and dry) but fire represented dryness pre-eminently.
112. Fixation to the alchemist meant the stabilizing of a substance by removing impure vapors by heat or by converting them into another form.
113. See above, chap. 30.
114. *super illa in aliis,* from the nonsublimed to the sublimed state.
115. Stillman says (*op. cit.,* p. 293), in speaking of the recipes of alchemists, "The preparation of a concentrated and purified syrup (oleum) of potassium carbonate from the ignition of tartarum (argol from wine) is given with elaborate detail."
116. "Although the Greeks and the Romans used a substance which they called 'arsenic' this was not the metal itself. The so-called 'arsenic' of the ancients consisted of the poisonous sulphides, orpiment and sandarac, mined with heavy loss of life by slave labor. No one knows who first isolated the metal, but this honor is sometimes accredited to Albert the Great . . . who obtained it by heating the orpiment with soap." (M. E. Weeks, *Discovery of the Elements,* p. 10.)
117. [*The Book of Furnaces.* This treatise and a number of others were written first in Latin and attributed to Geber (Jâbir) to confer upon them prestige and authority.—*Ed.*]

with an equal weight of sulphur, with water or the juice of limes,[118] in a closed crucible.[119]

35

WHAT IS DISTILLATION AND HOW IS IT DONE?

Distillation is the rising of the vapors of a liquid in its own container. There are different methods: with and without fire, that with fire is of two kinds; one, through rising vapors, as with an alembic;[120] the other through a descensory,[121] as with a pipe, and through fire superimposed on vessels.

The general purpose of distillation is [the] purification of a liquid from its dregs. We can see that the distillate is rendered purer [than the original liquid]. The special purpose of pure water is the imbibition of spirits and clean medicines, so that we can have a pure solution when we need one, for the dregs that can contaminate our medicines and purified spirits will have been removed. Distillation was invented to extract, through a descensory, an oil pure in its nature, whenever we cannot [evidently] have an oil combustible in its nature, as is true of petroleum. However, distillation, through filtration, is devised solely to obtain a clear liquid.

Addition.

Mercury is sublimed thus: take an ounce of dry vitriol and one of previously calcined common salt, pulverize, mix, and [in turn] add to this time and time again, one ounce of mercury. Grind well together by sprinkling with some of the partly distilled vinegar, so that it may be better fortified; [or even better] use a little *aqua fortis*,[122] which will be even stronger, then place it in the sublimatory. If this be made of glass, [place it] in the midst of ashes and let it be luted with Cretan earth pulverized with flour mixed with egg albumin. If the sublimatory be of earth, lute it with potter's clay and quicklime, moistened with horse dung and salt water [according to] the best papyrus [manuscript], and place it above the coals.

· · · · · ·

37

HOW IS MERCURY PREPARED AND WHITENED LIKE SNOW?

Take a pound of mercury, crush on a stone thus: take the calx of eggshells, or of white marble, or of green copper, and put on a stone; pour over enough strong vinegar to make a paste; add a little mercury, rub until it disappears,and once again add a little mercury and rub as before [or rub] all to-

118. *succo limonum.*
119. *crucibulo.*
120. [F. S. Taylor observes (*The Alchemists,* pp. 37–38) that "nothing that can really be called distillation was known before the time of the alchemists. It seems that a sort of sublimation of liquids was occasionally practiced. Thus sea water was heated in covered cauldrons and the drops condensed on the lids were shaken off and used as drinking water; again "oil of pitch" was made by heating pitch and condensing the vapor on fleeces. Mercury was made by heating cinnabar on an iron saucer in a pan covered by a pot called an "ambix," on which the vapor of mercury condensed (Fig. 1), but none of these pieces of apparatus could be

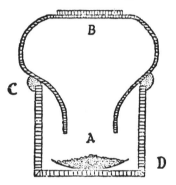

Fig. 1. Probable appearance of the mercury still of Dioscorides. (Courtesy of *Annals of Science*).

called a still. A still or alembic [the word comes from the Arabic, *al-inbiq—Ed.*] consists of three parts: a vessel in which the material to be distilled is heated, a cool part to condense the vapor, and a receiver. The traditional form of still is, indeed, that shown in Figure 2.

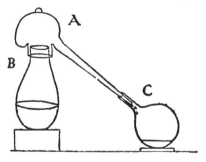

Fig. 2. The still. (Courtesy of *Annals of Science*).

A is usually called the still-head, *B,* the body and *C,* the receiver—though many other terms were used.

"This piece of apparatus was invented by the Greek alchemists, or, at least, is first described in their writings, and it continued to be listed in chemical catalogues as late as 1860! It is first described by Mary the Jewess, though we do not know that she invented it."
Mary may have lived around A.D. 100.—*Ed.*]
121. *per descensum.*
122. Possibly *nitric acid.*

gether until no mercury is seen.[123] Make the paste into small lozenges and place in a vessel until [no mercury] is seen. Make lozenges of this, and put on [a container], made in the form of a *scutella*,[124] and dry in the furnace with moderate heat, lest the mercury volatilize by overheating. Take one pound of mercury and one of the calx, a measure of [sal] ammoniac, so that there is always as much calx as mercury, and grind until the mercury disappears. When well dried, [grind] with vinegar until the mercury cannot be seen. A sign of perfect grinding is to moisten it with a little saliva, smear it on a silver coin,[125] and [observe whether] the mercury will fail to cling to it. If it does, then it is not well rubbed. When it is satisfactory, grind it very finely and evenly, and place in the sublimatory, completely covered.

When you put the medicine [in a vessel], do not fail to brush its surface [level] with a feather. Close the container with potter's clay, smeared well in the joints to prevent leaks, and place on the sublimation furnace and apply a slow fire through half a day until the moisture evaporates. Test with a plate until it is no longer moist; then, when dry, close well with potter's clay and apply a stronger fire, and finally the greatest heat. When night comes allow [the vessel] to cool. Open [it] in the morning and [note that] what has not reacted appears in the upper part of the aludel;[126] the other part [lies], like snow, on the dregs, pulverizable around the lower walls of the vessel, and sometimes on the upper part. Collect and save this. Now, as before, grind the unreacted part with the dregs, and sublime until all is reacted and is pulverizable. Do not add the least liquid to it, because it will regain its vivacity, and your labor will be in vain. Then, take one part of prepared salt, as I have taught, well purified and dry, add to it one-half part of sublimed mercury, mix together with the hand, and put into a sublimation vessel; level off, close and sublime as described above. In the morning take out what is sublimed, collect and test [it]. If anything remains in the dregs, note this: quickly take the salt from the dregs and put it above coals; if it fumes, sublime, as before, [for] a second day; afterwards diligently collect all the sublimate. On the third day sublime with new salt, and continue as before, and you will find it whiter than snow. Then test to see if anything remains in the dregs; [if] so, sublime until all is recovered. Then, once again, take new salt and do as before, four or more times up to fixation, and set aside.

Addition.

Also, it is best calcined from talc.[127] Master Joide Meun[128] in his great work adds green copper; this, however, is contrary to Geber who, in his chapter on sublimation of mercury, says that it should be sublimed with those substances which have no sulphureity, and so on.

A better way to sublime mercury is given by Rhases[129] in his book of *Divination,* the twenty-third [chapter] of the seventieth [book]. Crush well one part of rock salt, [and] one of Egyptian atramentum; then place above them a quantity of quicksilver equal to the others and crush all of it well. Place [the substance] at the bottom of a pergamen aludel,[130] and above this put roasted rock salt, to which a medicine may be added after it is dried, because no fluid remains in it. [Now] light a small fire under it, then increase the fire until the quicksilver is sublimated; then collect, grind well, and sublime it. Let the aludel head be broad in the first part, then successively smaller and narrower in

123. When mercury is triturated with powders, such as chalk or eggshells, it is formed into very minute globules.

124. A salver, tray for vessels.

125. *denarius.*

126. A pear-shaped pot open at both ends so that the neck can be fitted into the bottom of a similarly shaped second pot; used as a condensing tube in a sublimatory.

127. *talco.* Talc is a magnesium silicate.

128. Jean de Meun (1280–1365), French poet, satirist, translator of Boethius and others, and master of the literary and scientific knowledge of his time, has had many apocryphal writings in alchemy attributed to him. (See Sarton, *Introduction to the History of Science,* II, 929). He lived, of course, at a later time than Albertus Magnus, hence this paragraph represents another late addition to the text of the *Alchimia.*

129. Abu-Bakr Muhammad ibn Zakariyā al-Rāzi (commonly, Rasis, Rhazes), a Persian physician who lived in the late ninth and early tenth centuries. He was the author of numerous medical works several of which were translated into Latin in the twelfth century. As with Aristotle, Djaber, and Avicenna, many treatises on alchemical matters written in the twelfth and early thirteenth centuries were ascribed to Rasis. According to Steel, "Practical Chemistry in the Twelfth Century," *Isis,* Vol. 12 (1929), p. 10, the *De aluminibus et salibus* of Rasis was translated into Latin by Gerard of Cremona (1114?–1187) from a text altered and interpolated by Moorish alchemists. See also Thorndike, *History of Magic and Experimental Science,* I, 667–669; II, 751–808 *passim.*

130. Pertaining to Pergamum (Pergamon), chief city of ancient Mysia in Asia Minor. Under the Attalid dynasty in the third century it was the capital city and a notable literary and art center.

the next part down to a palm's width. It [the mercury compound] will then sublime in crystals under the shield head of the aludel. That [substance formed] above the shield will not be made a stone, but will be a white powder. Do this seven times. If it becomes solid above the fire, and makes a burning slab,[131] this will be well, and if not, repeat the sublimation with atramentum and salt, so that it will be confined above the head: for, in fact, it may flow (or cling) and make a white slab and be pure silver.

If you sublime mercury more often, you should, each time, diminish a fifth part of the dregs. If you sublime with vitriol and salt, as above, they should be roasted.

The clay used for closing the cracks may be [made] from ashes, potter's clay, and common salt dissolved in urine. Some persons [use] egg albumin and quicklime.

· · · · ·

44

THE REVELATION AND TEACHING OF THE SECRETS OF THIS ART BEGIN HERE

Now I have already taught you how to collect various flowers full of the fine fragrances, redolent with health and beauty, and the glory of this world: this is the flower of flowers, the rose of roses, and the lily of the valley. Rejoice therefore, O Youth, in thy adolescence and gather the flowers, since I have introduced you into the garden of Paradise; make from these a wreath for your head, that you may rejoice and enjoy the delights of this world.

I have disclosed to you the meaning, now I will help you to understand the secrets of this art, and what was hidden for such a long time, I shall now bring to light.

Previously, I taught you how to sublimate and to collect the flowers of these substances, therefore, now I shall teach you how to plant them so that they may bear much fruit, and their fruit may last forever. I shall teach you how to fix the powders sublimed, that they may remain in the fire, be combined and mixed with bodies, and [I shall show you that this may be done] in two ways.

45

THE FIXATION OF POWDERS, SO THAT THEY CAN MIX WITH BODIES, IS TAUGHT HERE

Take as much [powders] of these as you wish, one pound or two without anything else, and place in the vessel of fixation, and shape off the opening with good clay, not glazed, of the glassmakers, one digit thick, and close the cracks with good clay,

namely, clay of wisdom. When this is done, put [it] upon a sublimation furnace, and apply fire for a whole day. Now if done in summer, the amount of heat is as of sublimed mercury after mid-day; however, if it is done in the morning, turn the upper layer underneath, alternate two times at least, then open and see in this way if the powder is fixed; place a little of it over the coals: if it smokes, it is not yet fixed, but if it does not smoke, then it is fixed, and this is the sign of every spirit. If, however, it is not fixed, return to the furnace, closing the vessel as before, and apply fire for five days or until at length you hear a sound in the vessel like falling stones, as very often happens, when it is dried up too much. (Another direction says that it may be tested over a burning plate to see if it melts or flows, or fails to give off smoke),

A second way [to fix powders] is with the imbibition of oil of tartar. However, you can do it this way: take sublimed arsenicum or sulphur or auripigmentum, and crush over the stone with oil of tartar, until all becomes liquid. Then place in a glass phial in ashes, which have been sifted through a fine sieve, and place the vessel with the ashes over a distillation furnace, and apply the fire at first very slowly as [is done] in masticating, lest the vessel be broken. After heating the glass, increase the fire; then dry the medicine in an open vessel, if you wish, but it is better [to do it] in a closed one. Place above it an alembic which collects the water distilled from it, because [this distillate] is useful for many things. When the medicine is dry, the vessel has to be broken, since you cannot empty it otherwise, and you will find the powders hardened like stone. This has to be well ground as before with the distilled oil [of tartar]. Using the same procedure, again break the glass, remove, grind well, place in another ampulla, and set [it] in a warm dung pit for seven days, and then it will be dissolved into a liquid. Then place the vessel in warm ashes and heat with a slow fire, then you will have the spirits fixed; and the color will remain firm and lasting. And of this powder, add one part to fifty parts of calcined Iron or Copper, and this will be good for every malleation and testing.

· · · · ·

51

HOW CAN GOLD AND SILVER BE CALCINED?

The calcination of all metals must now be noted.

---

131. *tabulam ignitam.* The term is not susceptible of modern interpretation, since the substances used were impure.

First, take the calcination of Gold and Silver. Place the filings of either one you wish in vinegar for nine days. Then remove and, when dried, crush into dry powder; afterwards add water [and] sal ammoniac, crushing and drying six times. Then place over a stone, as I have taught for dissolving, and distill and put aside; and from this liquid take the powder for the solution.

Note this, however, that you should use liquids of Gold for making red solutions and [liquids] of Silver for white [ones].

· · · · ·

53

HOW CAN OTHER METALS BE CALCINED?

Take plates and heat them, afterwards you may rub them with salt water, because the salt eats up all the impure moisture in the bodies, and select the earthen vessel prepared for this. Fill the plates, place on a tripod in the furnace for calcination, put coals under it, and make a strong fire and have the furnace closed. When the plates glow, lower the fire so they will not melt but permit [them] to stay at such heat until [they are] entirely consumed by the fire. Take out in the morning, scrape and preserve what is calcinated; what remains, moisten again with salt water and continue until [all] is calcinated.

Additions.

The calcination of Gold and Silver[132] is done through amalgamation of either, with equal parts of mercury and common salt, by placing over a slow fire and continually stirring until the mercury volatilizes. Wash [the residue] with lukewarm water, and distill through a filter, and you will find your substance reduced to a calx. Another [method of] calcination of Gold and Silver and of other substances may be found in Aristotle's *De Perfecto magisterio.*[133]

Likewise, the double calcination of Gold and Silver is found in Brother Roger Bacon's book, entitled, "A method of compounding medicine through equalization of the elements,"[134] in the chapter on "the calcination of bodies," where he discusses the calcinations of the various bodies.

Calcination of gold. Liquefy it, make of it thin sheets, fill an iron crucible, cover, and let it melt with common salt, or partly prepared [sal] ammoniac and acid of pomegranates,[135] because it will be crushed to powder by the sharpness of the salt and acid. Then place in a baker's oven for a day and

night, and draw off the yellow-red calx to which there is no equal.

Calcination of silver. Melt it and make of it sheets; fill an iron crucible; then put into the crucible common salt, slightly ground, fill, cover, and melt—it soon melts by the beneficence of the salt—then crush. This, too, the salt helps to do. Put [it] into a baker's oven for a day and night, and draw off the Silver calx to which there is no equal.

Calcination of lead. Melt the lead, cover with a solution of common salt and the sourest of white vinegar, so that the salt may be partly prepared and dissolved. Then immerse the melted lead in this water forty times, and place (pulverized) in a new, crude clay-closed jar. Dry [it] in the oven for a day and a night, and in the morning take out the snow-like calx, heavy as salt.

Calcination of tin. Melt and immerse ten times in the liquid of common salt, made by dissolving it in the strongest white vinegar, mixed with two ounces of the honey of bees.[136] Then crush and place in a jar in a baker's oven for a day and night; then extract the tin calx.

Calcination of iron. Crush and imbibe with a solution of common salt, sorb-apple vinegar,[137] and pomegranates, in which the salt is dissolved, and dry ten times in ashes. Afterwards, regenerate by triturating with aqua fortis to the measure of five digits, then place under dung ten days and it will be dissolved. Then congeal the liquid for a day

132. "One of the methods of purifying or perfecting silver from lead was repeated calcination, quenching in sal ammoniac or vinegar, distilling, and calcining again. Now, although it is ridiculous to suppose that calcining removed any of the impure sulphur, or the impure mercury, or the lead, leaving a purer product and more noble metal, and thus effecting its transmutation into silver, the process was probably founded upon chemical fact. Many lead ores are argentiferous, and hence much of the lead in use was also argentiferous." (J. C. Brown, *A History of Chemistry,* p. 178).

133. [A spurious work attributed in the Middle Ages to Aristotle and also to Arnald of Villanova (see the biographical sketch at the end of this book). Indeed it may have been written after the death of Albertus Magnus. References and citations from it would then be later interpolations into Albert's original text.—*Ed.*]

134. *Modo componendi medicinam per aequationem elementorum.* This title does not appear in the lists of Roger Bacon's authenticated writings. For a full bibliography of Bacon's works, see the appendix to A. G. Little's *Roger Bacon Essays on the Occasion of the Commemoration of the Seventh Centenary of His Birth* (Oxford, 1914).

135. *mali granati,* a pomegranate.

136. *mellis apum.*

137. *aceto-sorbarum.*

and it will be colored; then crush and put into a baker's oven for a day and night, and you will have a calx as red as blood. This is the calx of Solis [the Sun or Gold], which is called [Saffron iron].

Calcination of copper. Make copper plates, suspend them in a jar with chains, in which is strong vinegar, and place [the jar] in the dung for ten days. Withdraw the plates and scrape off the deposit from them, and repeat this until the plates are finished. Wash [the corroded plates] with vinegar, rub gently, filter through a cloth, and throw away the dregs, which are its thickest part. Then let [the vinegar water] settle, and you will get the subtlety of green copper.

Calcination of quicksilver. [This method] will be found in the chapter on sublimation.[138]

. . . . .

56

HERE BEGINS THE FIRST OF THE OPERATIONS

Take, in the name of the good God, one part of white mercury, a second part of sulphur, and a third part of arsenicum. Mix these all together [and] add one-half part of liquid silver, [and] put into a glass vessel and heat above strongly with a glowing iron until the glass melts and congeals. Then place in ashes above the fire, and congeal; when it is congealed, a fixed persevering brown color will develop throughout the coagulated mixture. Place therefore one part of this colored tincture over a hundred parts of iron ore or purified copper and it will always, without doubt, be good, and in every malleation and testing will endure forever.

Addition.

Note this, that the colored substance should be dissolved and mixed with the bodies for calcination and converted to liquid. When the liquid is mixed with water, [the bodies] will never be separated from each other, [in the same way] as a little red wine will tint a great amount of water.

I saw other tracts, where this chapter was omitted, and only the [material in the] following [chapter treated].

57

HOW CAN GOLD AND SILVER BE OBTAINED ACCORDING TO ALL THE ABOVE?

Take one part of sublimed and fixed mercury, another of fixed arsenicum, and a third of Silver calx. Triturate well above a stone and saturate with a solution of sal ammoniac. Do this three or four

times, grinding, saturating, and roasting as above; set aside similarly for dissolving, and keep the solution. If it does not dissolve, grind more, and add a little sal ammoniac, and thus the whole will dissolve. When all has been dissolved, place [it] for distilling into warm water, and the entire [solution], as I have taught, will be distilled. Take care not to put into ashes for distilling, for it will harden for the most part, and it will have to be redissolved as at first. When the whole distillation is finished, put into a glass cucurbite, coagulate, and there will be a white substance, hard and clear as a form of crystal, liquefying above the fire as wax, a permeating and stable [substance]. Take of it one part to a hundred parts of any purified and calcined metal, and it will be good for all time. Take care not to touch with any unpurified metal, because the color will vanish after two or three testings.

Additions.

Aristotle's book *De Perfectionis magisterio*,[139] speaks of the sublimated and calcined mercury, which I think is the same as fixed, because it cannot be calcined unless it is first fixed; and unless it is calcined it cannot be dissolved. Near the end [of the experiment] some [writers] say that a certain white oil of philosophers should be added for the softening of this medicine. If fixed spirits cannot be penetrated,[140] add a similar amount that is not fixed, dissolve and coagulate and penetration[141] will take place, etc. Thus, also, if the calcined body cannot be reduced to a solid, add some of its own melt and it will be reduced. Divide the egg of the philosophers into four parts[142] of which each will

138. [See chapter 30.—*Ed.*] The general method of calcining was to react the melted metal with a substance to dissolve it; on further ignition it was converted to a dry powder. In the text "additions" amalgamation took place between mercury and gold, and between mercury and silver.

The "tin" here must have been an alloy. For example, Pliny confused tin with alloys of lead and tin. The calcination of iron presented no difficulty, but the copper was perhaps the alloy of brass.

139. [See note 133. The Latin title of this pseudo-Aristotelian work varies slightly.—*Ed.*]

140. *ingredi.*

141. *intrabit.*

142. The "philosopher's stone" was a mythical stone, substance, or chemical which was supposed to cause the transmutation of base metals to gold or silver. It first appears in alchemical writings in the seventh century. "In the early centuries of alchemy, there was also developed a mass of symbolism which lost nothing of complexity and obscurity with the development of alchemy. Thus the egg, symbol of the round universe, or

have one nature, then bring together equally and proportionately, so that it has no inconsistency, and you will have achieved that which was proposed, the Lord willing.

This is a universal method. However, I will explain it to you in specific operations divided into four. Two of the methods work well and without corruption. When, therefore, you will have water from air and air from fire, then you will have fire from the earth. Dispose, therefore, of the airy substance through separation, and of the earthy substance through heat and moisture, until they come together and unite and do not differ nor divide. Then you may add to it two effective virtues: namely, water and fire, and the work will be completed.[143] For if you mix water alone, Silver will be made; if you add fire, it will make it redden, the Lord willing.

77 HOW ELEMENTS PERSIST IN A COMPOUND

1. Thomas Aquinas (ca. 1225–1274)

Translated by Vincent R. Larkin[1]

Annotated by Edward Grant

ON THE COMBINING OF THE ELEMENTS

1. Many men are in doubt as to the manner of existence of elements in a compound.[2]

2. Some think that the substantial forms[3] of the elements remain, while the active and passive qualities of the elements are somehow placed, by being altered, in an intermediary state;[4] for if they did not remain, there would seem to be a kind of corruption of the elements, and not a combination.

3. Again, if the substantial form of the compound is the act of matter, without presupposing the forms of the simple bodies, then the simple bodies would lose the nature of elements; for an element is that out of which something is in the first instance formed, and which remains in it, and is by its nature indivisible;[5] for if the substantial forms are withdrawn, the compound will not be formed from the elements in such a way that they remain in it. Now this cannot be the case; for it is impossible for the same portion of matter to receive the forms of the different elements. If then the substantial forms of the elements are preserved in the compound, they must be in different parts of matter. But it is impossible for different parts of matter to receive them, unless we assume that quantity is present in matter; for if quantity is not presupposed, the substance would still be indivisible, as is made clear in the first book of the *Physics*.[6] Now a physical body is composed of matter that is subject to dimensions and of a substantial form united to it. Therefore the different parts of the matter that support the forms of the elements

are sometimes said to typify the four metals which form the basis for transmutation, copper, tin, lead and iron." (Stillman, *op. cit.,* p. 170). [See also note 54.—*Ed.*]

143. In this chapter the writer gives the final recipes for preparing gold and silver and treats again of the four elements: earth, water, fire, and air. This Aristotelian doctrine of the composition of matter was generally accepted until the time of Paracelsus, 1493–1541 (Theophrastus Bombastus von Hohenheim) who, in his writings on medicine and chemistry, profoundly influenced the work of his contemporaries and succeeding generations. Paracelsus introduced the doctrine that all matter is composed of three principles: salt, sulphur, and mercury.

The treatise here concludes with the notion that matter could assume different forms, and that these forms could be removed. Perhaps the author wished here to emphasize the idea of the transmutation of metals.

1. This is a complete translation of Thomas Aquinas' *De mixtione elementorum (On the Combining of the Elements)* made from *Opuscula philosophica* of the Marietti edition of Aquinas' works edited by R. M. Spiazzi, O.P., and is reprinted by permission of the editors of *Isis* from *Isis,* Vol. LI, Part I, No. 163 (1960), pp. 68–72. The years 1270/1271 and 1273 have been conjectured as dates of composition. As does Albert of Saxony in the next section, Aquinas is here elaborating a problem arising from a discussion in Aristotle's *On Generation and Corruption (De generatione et corruptione),* Book I, chapter 10. The paragraph numbers are given as they appear in Larkin's translation.

2. The doubt was largely the result of Aristotle's vague discussion in Book I, chapter 10 (see n. 19).

3. On substantial and accidental forms, see note 23.

4. For a further description of this position, which was Avicenna's solution to the problem, see note 25.

5. Drawn from Aristotle, *Metaphysics* V.3.1014a. 26–27 (see note 40).

6. In *Physics* I.2.185b.2–4, Aristotle says: "If, then, Being is both substance and quantity, it is two, not one: if only substance, it is not infinite and has no magnitude; for to have that it will have to be a quantity." Thus, if Being is substance only, it will have no magnitude and will consequently be indivisible.

___

of eternity; the "egg of the philosophers" consisted, like the physical universe, of four components, white and yolk, a skin and shell. These four constituents again

receive the natures of several bodies. Now it is impossible for many bodies to be [in a compound] at the same time. The four elements will not then be in each part of the mixture, and hence there will not be a true mixture,[7] but only an apparent one, as is the case when bodies, which are invisible or imperceptible due to their minuteness, are clustered together.

4. Furthermore, every substantial form demands a special disposition in matter, without which it cannot exist; as a result, alteration precedes generation and corruption. Now it is impossible that the special disposition which is demanded by the form of fire and which is demanded by the form of water should be found in the same [portion of matter], because it is on account of such dispositions that fire and water are contraries. It is indeed impossible for contraries to be entirely present in the same thing at the same time. Therefore it is impossible for the substantial forms of fire and water to be in the same part of a compound. If then a compound is formed while the substantial forms of the elements remain, it follows that it is not a true compound, but only an apparent one, as when parts, indiscernible because of their smallness, are placed next to one another.

5. Some men, wishing to escape both arguments, have fallen into a greater difficulty.

6. For, in order to distinguish the combinations of elements from their corruption, they said that the substantial forms of the elements indeed remain somehow in the compound; but, lest they should be forced to admit that this is an apparent combination, and not a true one, they maintained that the forms of the elements do not remain in the compound in their entirety, but are reduced to some intermediary state; for they say that the forms of the elements admit of more or less, and are related to one another as contraries. But because this plainly contradicts the common opinion of men and the words of the Philosopher who says in the *Praedicamenta*[8] that substance has no contrary and that it does not admit of more or less, they go further, and say that the forms of the elements are the least perfect of all, inasmuch as they are nearer than others to prime matter; hence they stand midway between substantial and accidental forms; and thus, inasmuch as they approximate the nature of accidental forms, they can admit of more or less, even though they are related to one another as contraries.[9]

7. Now this position is improbable for many reasons.

8. First, because it is altogether impossible for something to stand midway between substance and accident, for then there would be a mean between affirmation and negation. For it is the nature of an accident that it be in a subject but it is the nature of a substance that it be not in a subject. Now substantial forms are indeed in matter, but not in a subject; for a subject is something individual. Now a substantial form is that which causes the individual subject. It does not, on the contrary, presuppose it.

9. Likewise, it is ridiculous to say that there is something midway between things which do not belong to the same class; for the mean and the extremes must belong to the same class, as is proved in the tenth book of the *Metaphysics*.[10] Therefore there can be no mean between substance and accident.

10. Furthermore, it is impossible for the substantial forms of the elements to admit of more or less. For every form that admits of more or less is accidentally divisible inasmuch as the subject can participate in it more or less. Now we find continuous motion in that which is divisible essentially or accidentally, as is made evident in the sixth book of the *Physics*.[11] We have as examples change of place, and growth and decay with respect to space and quantity which are essentially divisible, and alteration with respect to qualities, such as hot and white, that admit of more or less. If then the forms of the elements admit of more or less, there will be continuous motion in both the generation and the corruption of the elements. But this is impossible. For continuous motion exists only in three genera, namely quantity, quality, and place, as is proved in the fifth book of the *Physics*.[12]

11. Furthermore, every difference in substantial

---

7. Under the circumstances described here, if four elements constitute a given compound, each element will fully retain its identity and represent a part of the compound in complete isolation from the other three elements. Thus a compound would consist of four elements each occupying a different part of the compound but unmixed in any manner. Hence it is not a true mixture or compound. For the distinction between a *physical mixture* and a *true mixture* see note 19.

8. See Aristotle's *Categories* [or *Praedicamenta*, as it was known in the Middle Ages], 5.3b.24, 33.

9. The argument summarized in this paragraph was enunciated by Averroes in *De caelo*, Book III, Comment 67 (see note 26).

10. *Metaphysics* X.7.1057a.20.

11. *Physics* VI.2.233b.15, 31.

12. *Physics* V.1.225b.9.

form results in a change of species. [13] Now what admits of more or less differs from that which is less, and is in some way contrary to it, as in the case of the more white and the less white. If then the substantial form of fire admits of more or less, it will result in a change of species according as it is more or less realized, and it will not be the same form but another one. Hence the Philosopher says in the eighth book of the *Metaphysics* that just as the species is changed in the case of numbers by addition and subtraction,[14] so also is it changed in the case of substances.

12. Therefore some other explanation must be found by which the truth that a combination is effected and that the elements are not wholly destroyed but remain in some way in the compound may be safeguarded.

13. Let us consider then that the active and passive qualities of the elements are related to one another as contraries, and admit of more or less. From these contrary qualities which admit of more or less there can be formed a mediant quality which partakes of the nature of each extreme as does grey which lies between white and black and as does warm which lies between hot and cold.[15] Thus, when the perfections of the qualities of the elements are modified, there is formed from them

some kind of mediant quality which is the quality characteristic of the compound and which differs in different compounds according to the different proportions of the combinations; and this quality is, in fact, the disposition that belongs to the form of the compound, just as the elementary quality is the disposition that belongs to the form of an element. Therefore, just as extremes are united in the mean which participates in the nature of both, so the qualities of elements are united in the quality that is characteristic of the compound. The quality, however, of an element is indeed distinct from its substantial form; however, it acts by virtue of the substantial form; otherwise, heat would merely warm, and not by its power would a substantial form be brought to actual existence; for a thing's activity cannot transcend its nature.

14. In this way then the powers of the substantial forms of the elements are retained in compounds.[16] Therefore the forms of the elements are present in compounds not actually but virtually; and this is what the Philosopher says in the first book of the *De generatione et corruptione.* "The elements do not remain actually in the compound, as body and white do, and neither one of them nor both of them are destroyed or altered; for their power is preserved."[17]

2. Albert of Saxony (ca. 1316–1390)

Translated and annotated by Edward Grant[18]

BOOK I

Question 19

We inquire whether elements remain [or persist] formally in a compound [or mixed] (mixto)body.[19]

13. This argument probably formed the basis of Albert of Saxony's rebuttal of Averroes (see Section 2 and note 32).

14. In the course of arguing that "definition is a sort of number," Aristotle says: "And (2) as, when one of the parts of which a number consists has been taken from or added to the number, it is no longer the same number, but a different one, even if it is the very smallest part that has been taken away or added, ..." (*Metaphysics* VIII.3.1043b.36—1044a.1).

15. Albert of Saxony's position was derived from this opinion (see section 2 and note 34). Compare Avicenna's discussion in Selection 91.

16. That is, through the elemental qualities which united to form the mediant quality, which becomes the characteristic quality of the compound.

17. *On Generation and Corruption* I.10. 327b.30–32. See note 19, where these few lines are given in the Oxford translation.

18. The selection here is from Albert of Saxony's *Questions on the Two Books of On Generation and Corruption (Questiones in duos libros De generatione et corruptione)*, Book I, Questions 19 (virtually the whole text) and 20 (less than half), published in *Questiones et decisiones physicales insignium virorum: Alberti de Saxonia in octo libros Physicorum; tres libros De celo et mundo; duos libros De generatione et corruptione; Thimonis in quatuor libros Meteororum. . .Buridani in librum De sensu et sensato. . .Aristotelis recognitae rursus et emendatae summa accuratione. . .Magistri Georgii Lokert. . .*(Paris, 1518), fol. 142 verso, col. 2, to fol. 144 verso, col. 2.

19. During the fourteenth century this was one of the most widely discussed questions. In his *On Generation and Corruption (De generatione et corruptione)*, Book I, Chapter 10, Aristotle had distinguished between what the scholastics were to call a *mixtio ad sensum*, which corresponds to a physical mixture (or "composition," as it is described in the Oxford translation) and a *mixtio secundum veritatem*, which corresponds more closely to our notion of a chemical compound and which Aristotle defined as follows: " 'Combination' is unification of the 'combinables,' resulting from their 'alteration.' " (A few lines below, this definition appears as *"mixtio est mixtibilium alteratorum unio"* and is trans-

First it is argued yes on the authority of Aristotle, who says that two things are neither corrupted nor altered, but that their powers *(virtutes)* are preserved *(salvantur)*. In the second place, Aristotle also argues that a compound is a union of mixable things that have been altered *(mixibilium alteratorum unio)*.[20] Now, if in the generation of a body mixed from elements, the elements are corrupted and do not remain, then the compound *(mixtio)* ought to be called a union of mixable things that have been corrupted *(mixibilium corruptorum unio)* and ought not to be called [a union] of [mixable] things [that have been] altered. Now, in the third place, in the syllable *ab,* just as the letter *a* remains formally, as does the letter *b,* although the sound of each is weakened *(remittatur),* so it seems that each of the mixable things in a compound body remains, although weakened *(remissum)* [somewhat].

Aristotle is absolutely opposed to this, for he says that two things that have been altered cannot be compounded because each is preserved and remains.[21] It is as if he wished to say about this that if each thing were altered it would be required that each of them be preserved; but they do not remain and it follows, therefore, that elements do not remain, nor are they preserved in a compound body *(mixto).*[22] In the second place, if they should

lated as "a compound is a union of mixable things that have been altered.") But how elements were contained in a compound was answered by Aristotle only very briefly and sketchily: "Since, however, some things *are-potentially* while others *are-actually,* the constituents combined in a compound can 'be' in a sense and yet 'not-be.' The compound may *be-actually* other than the constituents from which it has resulted; nevertheless each of them may still *be-potentially* what it was before they were combined, and both of them may survive undestroyed. . . .The constituents, therefore, neither (*a*) persist actually, as 'body' and 'white' persist: nor (*b*) are they destroyed (either one of them or both), for their 'power of action' is preserved" (I.10.327b.24–31; see par. 14 of section 1 above). It was left to the scholastics to explain the mechanism by means of which an element could be said to have its power preserved in a compound and yet not actually persist. A number of solutions were proposed, the most important of which are mentioned in this selection.

20. This definition is given at the very end of Book I (that is, at the end of ch. 10) of *On Generation and Corruption* (see n. 19).

21. Perhaps this is an inference from Aristotle's statement that *alteration* "occurs when the substratum, which is perceptible, persists, but there is change in its properties. . ." (*On Generation and Corruption* I.4.319b. 11–12). Since, in alteration, a substance persists with only accidental changes, it follows that if two things

have been merely altered, they cannot have lost their identity and, consequently, cannot have been compounded. But Aristotle also maintained that a compound "is a union of mixable things that have been altered" (see n. 19). In this definition, the term "altered" cannot be understood in the sense of the definition of "alteration" given a few lines above, since the perceptible substrata of combining elements do not persist in a compound. Aristotle is clear on this. It appears that "alteration" as applied here to the combining units must be construed in the broadest sense of change. Aristotle himself tells us how the "mixable things" must be altered before their union can be effected. It is necessary that (*a*) elements suffer reciprocal action and therefore be reciprocally active and passive; hence "only those agents are 'combinable' which involve a contrariety— for these are such as to suffer action reciprocally" (I.10. 328a.33–34); (*b*) combining entities be easily divisible ("liquids are the most 'combinable' of all bodies— because of all divisible materials, the liquid is most readily adaptable in shape, unless it be viscous." [I.10. 328b. 3–4]); and (*c*) the quantities of the combining units or pieces be reasonably equal (for best results they should also be small), for otherwise "the effect produced is not 'combination,' but increase of the dominant: for the other material is transformed into the dominant. (That is why a drop of wine does not 'combine' with ten thousand gallons of water: for its form is dissolved, and it is changed so as to merge in the total volume of water)" (I.10.328a.26–29).

22. The various forms of the participle *mixtum* and the noun *mixtio* have been rendered here as "compound body" (and occasionally also as "mixed body") because their meaning seems to approximate most closely to our notion of a chemical compound, for which reason it was also used in the Oxford translation, where Aristotle says that if " ' combination' has taken place, the compound *must* be uniform in texture throughout—any part of such a compound being the same as the whole, just as any part of water is water: whereas if 'combination' is 'composition of the small particles,' nothing of the kind will happen" (I.10.328a.11–13).

As was the case with other authors, Albert elsewhere, especially in discussions of internal resistance and falling bodies, used the very same terms. These usages, however, have been rather consistently rendered in this source book as "mixed body" rather than "compound," for although chemical homogeneity is presupposed, the bodies are treated as if they actually contained independently active component elements in a physical mixture, contrary to the interpretation adopted by Albert in this selection. For example, in his *Questions on the Physics,* Book IV, Question 11 (fol. 50 verso, col. 1 of the work cited in n.18), Albert considers internal resistance in a heavy mixed body and assigns to it a heaviness of 2 degrees and a lightness of 1 degree; and a few lines below, he assumes a "heavy mixed body" consisting of earth, with 3 degrees of heaviness, and air, with 2 degrees of lightness. In these contexts "mixed bodies" are assumed to behave as if their elemental components persist with a capacity for independent activity, an assumption that is in seemingly sharp contrast to Albert's discussion in this selection, where he argues that the elements are wholly corrupted in a *mixtum,* although

remain, it follows that several forms could be in the same matter. But this is false, as was seen elsewhere. The consequence is proved, because in the same matter the form of fire would remain along with the newly created form of the compound [or mixed body] *(forma mixti)*. In this question, I shall first present some opinions on this matter with their disproofs; secondly, I shall reply to the question; and thirdly I shall remove doubts against this opinion and resolve them.

As the first [of these three tasks], it must be known that some have said that in the generation of a compound body the forms of the elements remain without a new generation of a mixed and super-added form. Hence they imagined that in a mixed [or compound] body there were no forms except the substantial forms[23] of the elements, each of which had a mutual relation with the others. According to this opinion, a mixed body would be of one species and then another.[24] Others, however, assumed that the substantial forms of the elements remain in a compound body and, in the generation of the compound body, add a new form, the form of the compound body. But these were divided into two groups, since some of them assumed that the substantial forms, but not the qualities, of the elements remained unweakened *(non refractas)* in a compound body; Avicenna was one of them.[25] Others, however, assumed that the elements remain formally in a compound body, but weakened both as to their substantial and accidental forms; and among these was the Commentator [Averroes], who posits this in the third [book] of the *De caelo*,[26] reproving Avicenna [and] assuming that the substantial forms of the elements are intensible and remissible.

Against these three opinions, I posit these three conclusions: The first is that in a compound body it is necessary to assume a form distinct from the forms of the things that are mixed; the second [conclusion] is that substantial forms of elements do not remain in a compound body in their intense [and complete] being; the third conclusion is that substantial forms of elements do not remain in a

23. The scholastics distinguished between "substantial" and "accidental" forms. The latter were generally made dependent on the former, for when a substantial form was introduced into matter it was assumed that accidents—and therefore accidental forms—would be associated with it (compare Averroes' remarks on form and accidents in Selection 45). Aristotle, however, did not sharply distinguish particular substantial forms. In his explanation, when the qualities cold and dry combined with prime matter, the element earth was the direct result. For most scholastics this would have produced only an accidental form and the element earth would not have materialized until a substantial form was introduced into the matter.

24. In this first opinion the elements are assumed to persist and a compound is thought to consist of the substantial forms of the component elements. Apparently, each particular association of substantial forms in a compound was held to represent a different species of compound because the proportional relations obtaining between the varying quantities of the elements were unique for each compound.

25. Two major discussions of this problem by Avicenna appear in Book I, chapter 10, of his *Sufficientia*, which is the title of the medieval Latin translation of his commentary on Aristotle's *Physics* (see the beginning of the next selection and n. 2 thereto), and in his commentary on Book I of Aristotle's *On Generation and Corruption*. Avicenna argued that the substantial or essential forms of the combining elements persist unaltered in a compound, with only their qualities altering and weakening. Contrary qualities blend into a *complexio*, or mean quality (see Selection 91). But the new complexions, or mean qualities, do not produce a new substantial form in the newly forming compound. Rather, they prepare the matter of the compound to receive a new substantial form that is infused directly by the "giver of forms" *(dator formarum)*, namely the Agent Intellect. This new substantial form is simply added to the four substantial forms of the elements already present in the compound. The properties or accidents of the compound are then finally determined by the new substantial form. Though Avicenna's theory was almost without influence in the Latin West, it was usually mentioned and included amongst the theories to be rejected (see par. 2 of section 1 above, where Aquinas mentions it).

26. In Comment 67 on *De caelo*, Averroes held that both the qualities (accidents) and substantial forms of the component elements are diminished in intensity in the new compound. Indeed, the weakened forms blend and constitute the new "form of the compound" *(forma mixti)*. In order to conform with Aristotle's position that substantial or essential forms are invariant—that is, a man is no more a man than another man—Averroes maintained that elemental forms are not really substantial forms but lie midway between substance and accident and can therefore experience intension and remission. On this last point, Latin scholastics were in complete disagreement with Averroes, for they denied that elemental forms could be mean forms between substance and accident, since means arise only from contraries—but substance and accident are not contraries.

---

they maintain a "virtual" or "potential" existence. To reconcile these apparently conflicting interpretations, we are forced to assume that the behavior of "mixed bodies" as discussed in the *Physics* is the direct and sole consequence of the "virtual" presence of the former elemental components, despite appearances to the contrary when Albert and others speak as if the elements are actually distinguishable and capable of separate existence and activity in a "mixed body."

compound body in a weakened state. The first conclusion is against the first opinion; the second is against Avicenna; and the third is against the opinion of the Commentator.

The first conclusion is proved as follows: For unless a form other than the forms of the elements were assumed in a compound [or mixed] body to make, govern, and shape it, all heavy bodies in a compound would indeed descend and the light bodies ascend. But this is false, for if it were so, it would follow that in a compound [or mixed] body such as a man or horse, bone would lie [lowest and] below, and everything above [or in the upper part of the body] would be spirit and heat, and everything in the middle ought to be flesh. But this is false, as is self-evident. The consequence *(consequentia)* holds, because bones are generated from heavier things, spirit and heat from lighter things, and flesh and blood are generated in a middling way.[27] Briefly, then, in a compound body, it is necessary to assume one form, other than the form of an element, which governs the aforesaid things. Secondly, if the conclusion were false, it follows that heated *(ignitum)* iron and heated stone would be more similar in substance than heated *(ignitum)* iron and cold *(frigidus)* iron. But this is false and the consequence holds, because qualities of elements have a greater mutual appropriateness [or affinity] in heated iron and heated stone than they would have in heated iron and cold iron.[28] Thirdly, one may argue in this way: It follows that an alteration would be a generation, wihch is contrary to what was said before. This consequence *(consequentia)* is proved, for inasmuch as there would be one mutual proportion of elements for one compound body and another for another, one compound body would be of one species and another of another species, so that a specific name could be changed by the mere mutual alteration of the elements.[29] But, as stated before, when there is a change in the name of a species, there is a generation of substance *(generatio substantialis)*.[30]

The second conclusion is [now] proved. For if the forms of the elements remained in existence in their most intense and perfect state in a compound [or mixed] body, it follows that in a compound body the form of fire would persist without corruption under a greater density and heaviness than would the form of water. But this is false and the consequence obvious, since fire would persist in a marble stone. But the falsity of the consequent is proved: For if it were true, it follows that water

could never be generated from fire—and this is false. . . . .[31]

The third conclusion is proved, for with the opposite of the conclusion asserted, the substantial forms of the elements would be intensible and remissible, which is contrary to what was said before. Secondly, it follows that just as a mean quality results from hotness and coldness, and similarly from wetness and dryness, so also by the mutual remission of the substantial form of fire and substantial form of water there would result a mean substantial form, and similarly from the substantial forms of air and earth there would

27. In this strange example, Albert wishes to show that if compounds, or true mixtures, were mere aggregates of two or more of the four independent elemental forms, then the elements in digested food ought to act independently, the heavy elements tending to the lower extremities, the lighter to the upper extremities, and the middling elements collecting around the center. From this the impossible stratification described above ought to result. On the basis of this absurd consequence, Albert regards his first conclusion as demonstrated and assumes that nonelemental forms inhere in compound bodies.

28. On the assumption that compound bodies have as many substantial forms as constituent elements, Albert believes that a necessary, but absurd, consequence of this would enable one to argue that heated iron and stone are more closely related substances than heated iron and cold iron. Presumably, this could be maintained on grounds that the quality "hotness" would be common to the iron and stone, while heated iron and cold iron possess the contrary qualities "hotness" and "coldness." Hence, despite the fact that hot and cold iron are the very same substance, this position separates them into very different categories and distorts their proper relation.

29. See note 24.

30. Albert's argument against the first of the three positions trades on the distinction between an alteration *(alteratio)*, or change of property without perceptible change of substance, and generation *(generatio)*, the coming-to-be of a new perceptible substance from a previously different perceptible substance. If compounds differ only in the proportion of the four elements (earth, air, water, and fire) which comprise them, and since we apply different specific names to these different compounds, it would follow that a mere *alteration* of the proportions between the elements would generate a new substance requiring a new specific identity. In this theory the usual distinction between "alteration" and "generation" would be obliterated (see also n. 21).

31. That water is generable from fire is stated by Aristotle in *On Generation and Corruption* II.4.331b.6–10. In the few lines that are omitted here, Albert "proves" this. Also omitted are brief second, third, and fourth arguments in favor of the second conclusion. These arguments actually constitute a series of arguments against Avicenna's opinion (see n. 25).

result one other mean form distinct from the first [mean form] in species. But if this is so, then every compound body would consist of different species, which is absurd. The consequence holds, because two substantial forms distinct in species, yet not subordinated one to the other, would be extended through all the matter of the body.[32]

As to the second part,[33] let this be the first conclusion: that the elements do not remain in a compound according to their substantial forms. The second conclusion is that [the elements do] not [remain in a compound] according to their accidental forms or qualities, which are hotness and coldness. The third conclusion is that in the generation of a compound body two qualities are generated, one of which is a mean between the active qualities, namely hotness and coldness, and the other a mean between the passive qualities, namely wetness and dryness. And the mean between the active qualities in a compound body heats a colder body and makes a hotter body cold. And similarly for the mean between passive qualities: [The mean quality] dries a wetter [body] and wets a drier [body]. And in this way such mean qualities differ from the primary qualities of the elements, since the qualities of the elements are [weaker or] more remiss. [For example], if the hotness of [the element] fire were weakened, it could not, because of this, make something cool, as could an active quality of a compound body.[34]

The first conclusion is proved, because elements do not remain in a compound body according to their strongest (intensas) or weakest (remissas) substantial forms, as was proved. Therefore, in no way [whatever do substantial forms remain in compound bodies]. Secondly, if substantial forms did remain, it would follow that in every generation an analysis (resolutio) could be made right down to the prime matter [of the body]. But this is false and the consequent is demonstrated, for so much as the substantial forms of the elements remain with the form of a compound body after generation, to that extent were they present in the matter before generation. The falsity of this consequent is obvious, for then the subject of generation would be a being in act, namely a composite of the matter and form of the element, and consequently generation would be alteration,[35] which is contrary to what has been said.

The second conclusion is [now] demonstrated. For if the qualities of the elements remained, but not their substantial forms, it follows that an accident could pass from subject to subject. But

this is false. The consequence[36] seems to hold, because an accident that was first in an element as in a subject would be in a compound afterward.

The third conclusion is [now] demonstrated. We

32. In demonstrating this third conclusion, Albert is, of course, attacking the position of Averroes described above in note 26. The major argument depends upon consequences that follow from blending of substantial forms. Every compound body would constitute a distinct species, since its component substantial forms would blend in different and unique proportions, a condition contrary to experience. Here, for the sake of argument, Albert implicitly concedes the variation of substantial forms (see n. 26 and par. 11 of section 1 above).

33. At this point, Albert considers the question directly.

34. Taken together, these three conclusions of the second part approximate most closely to the position formulated by Thomas Aquinas (see par. 13 of section 1 above), a position widely accepted in the fourteenth century and usually described as the "opinion of the moderns." Along with the interpretations of Avicenna and Averroes, it formed the basis of fourteenth-century discussions concerning the formal persistence of elements in a compound. For Thomas, mean qualities in a compound are most appropriate and constitute the proper disposition for the new form of the compound, just as the simple elemental qualities constitute the "proper disposition" of the elemental forms. But because Thomas did not explain the disposition of the elemental qualities in the new mean quality—that is, did they actually exist in the mean quality? or did they persist only virtually or potentially? or were they, perhaps, not even represented in the new mean quality?— there was much subsequent discussion on this point. (Although Thomas did not explain the disposition of the elemental qualities, he did hold that the elements themselves were virtually, though not actually, present in compounds [see par. 14 in section 1 above].) The dominant opinion which emerged was that the mean qualities are not caused by the extreme elemental qualities, nor do they actually contain them; rather, they bear a certain similitude to the extreme elemental qualities. This is the position which Albert appears to adopt in this paragraph and in his demonstration of the third conclusion three paragraphs further on. For my information on Avicenna, Averroes, and Aquinas, I am indebted to Miss Anneliese Maier's perceptive analysis in Part I ("Die Struktur der materiellen Substanz") of her volume An der Grenze von Scholastik und Naturwissenschaft, 2d ed. (Rome: Edizioni di storia e letteratura, 1952).

35. If the substantial forms of the elements persist in the formation of a compound, then the generation of a compound, or new substance, could involve no more than a change of accidental qualities, that is, an alteration (alteratio), since the matter and form of the component elements would remain unaltered (see n. 21).

36. That is, if the qualities of elements remained without the substantial forms of the elements, an accident could pass from subject to subject.

say that a compound body is hot and cold, wet and dry. But since this does not happen by virtue of the qualities of the elements—this is obvious from the second conclusion, since those qualities would be contrary, and contraries cannot be in the same subject simultaneously—it seems that this would happen through qualities that are mean between the active and passive [qualities], as was declared in the conclusion.

As for the third [part of this question concerning doubts],[37] let this be the first [doubt]: If elements do not remain in a compound, it follows that a compound would be as simple as an element. But this is false. The consequence holds, [however,] because in a compound there would be nothing more than prime matter and form, just as in a simple element; and this would [indeed] happen if the elements did not remain in a compound. Secondly [there is this doubt]: If the elements did not remain [in a compound], it follows that the forms of compounds would not be received in matter that was mediated by forms of the elements, for otherwise the elements would not seem to be principles of compounds more than conversely [that is, more than compounds would seem to be principles of elements]. A third doubt is this: Aristotle says, here in the first book [of *On Generation and Corruption*], that a drop of water cannot be mixed in a vessel of wine because the water would be corrupted;[38] but he also wishes to say that the mixable things in a compound ought not to be corrupted, but [rather] remain.[39] A fourth doubt: In the fifth book of the *Metaphysics*, it is said that an element is that of which something consists when it contains it.[40] But if the elements do not remain in a compound, this would seem to be false. A fifth [doubt is this]: In the first book of *On the Heavens (De caelo)*, it is stated that a compound is moved with the motion of the predominant element.[41] But this is false unless the elements—or something of them—should remain in a compound. A sixth doubt: If the elements did not remain, but were corrupted, either all would be corrupted—but this cannot be, for then a compound would be compounded of nonbeings; and thus it does not appear that the predominant element is corrupted—or some would be corrupted and some not; but this cannot be because this would not be a compound [or mixed body] but rather a joining and generation *(alligatio et generatio)* of one element from several, or of several [elements] from several [elements]. [Finally,] a seventh doubt is based on experiences, for we see

in green wood that has been placed in a fire that a liquid *(humor)* appears at the extremities lying outside the fire. Now, this could not occur if the liquid were not present in the compound. Similarly, we see that a water, called rose, is extracted from roses, which are a certain compound. Also similarly, fire issues forth when a stone is struck with iron, which ought not to happen, it seems, unless this fire were present in any one, or both, of these bodies.

[I now respond] to these doubts. To the first, I say that an element is something as simple as a compound with respect to simplicity, which is opposed to [the notion of] the composition of matter and substantial form. For as to its substantial composition, a compound body, like an element, has nothing more than its matter and form. But to the understanding, however, a compound is itself more composite in its accidental composition than any element, since it has qualities which can make something hot or cold, wet or dry;[42] but no element has qualities which can do all these things. Therefore, a compound body is called a compound because, firstly, it does not have any primary quality in its pure state *(in summo)*, just as it does not have any element [in its pure state]. Secondly, [it is called a compound] because a compound has mean qualities. For if it were constituted of the purest primary qualities, this would be contrary

37. The doubts raised here by Albert are objections against his own position, which he will resolve later. In this particular question all the doubts are first listed in numerical sequence and then resolved in this same order; but in the very next question each doubt is followed immediately by its resolution. Posing doubts and subsequently resolving them was a fairly characteristic feature of medieval *questiones*, a form of scholastic literature concerned largely with problems related to, and drawn from, the physical treatises of Aristotle.

38. For Aristotle's words, see note 21. Where Aristotle places a drop of wine in water, Albert puts a drop of water in wine.

39. The passage to this effect is cited in note 19.

40. Aristotle, *Metaphysics* V.3.1014a.26. In the Oxford translation we read: " 'Element' means (1) the primary component immanent in a thing, and indivisible in kind into other kinds."

41. "Necessarily, then, movements also will be either simple or in some sort compound—simple in the case of simple bodies, compound, in that of the composite—and in the latter case the motion will be that of the simple body which prevails in the composition." *De caelo* I.2.268b.30–269a.1.

42. Albert is speaking here of "mean qualities" and their capacity for performing the actions of each of the two contraries of which they are in some sense a blend. See the third conclusion of the second part and note 34.

to the nature of a compound; and conversely, if a compound could be reduced, [it would be reduced] to its original mean qualities, so that it is not [made] from elements, for if an element has a mean quality it would seek neither to remain with it nor to be reduced to it, but rather would seek to move from the mean quality to its own quality, as is obvious in the case of warm water.[43] And in this way we can know when something which has been qualified by a mean quality is a substantial compound or a simple compound. Thus, if we should see that it remains close to this mean quality after every external thing or aid has been removed, it is called a substantial compound; but if we do not see it remain close to this mean quality but rather tend to some quality in its purest state—as we see [for example] in hot water [that tends to become cold]—we say that this is not a substantially compounded body, although it might well be accidentally [compounded]. Thirdly, a compound is [also] known to be a substantial compound in this way: because it has the sensible qualities color, odor, taste, and the like, which a pure element does not have.[44] It is for this reason we say that water that is more tasteless is nearer to pure water than water that is more tasteful; and it is for this reason that salted water, as [for example] sea water, is not as close to the pure element water as is water that has less taste. This will be seen in the second book.[45]

To the second doubt, I say since the Commentator assumes that the elements persist in a compound, he says that the forms of the compounds are received in their [respective] matters with the forms of the elements mediating. What the Commentator says about this can be denied. Nevertheless, we can assert that the forms of compound bodies are received in their matters by mediating forms of the elements, that is, by certain mediating dispositions left behind in the prime matter through the mutual action and passion of the elements. But then it was also said that the elements ought no more, and so on [that is, ought no more to be principles of compounds than conversely].[46] This is denied because for the generation of any compound it is necessary, or was necessary, that the elements should come together mediately or immediately. For the generation of an element, however, this is not necessary. Now, I say "mediately or immediately" because sometimes a compound body is generated from a compound body; but sometimes, however, in tracing the formation of a compound body, it will have been generated

directly from the elements.[47] It is for this reason that I said any compound body is generated "mediately or immediately" from the elements. But because it may well be the case that an element that was never generated from a compound may have been generated from another element, we ought to say that elements are principles of compounds rather than conversely.[48]

To the third doubt, where it is said that a drop of water, and so on, I say that by this [statement] Aristotle did not wish to maintain that in a compound it is necessary for the mixable things to persist [or endure]; but rather he wished [to claim] that when one of the mixable things is corrupted and the remaining one persists, then there is no perfect compound,[49] but, instead, there is a generation and joining. When, however, all mixable things are corrupted, then a perfect compound exists, which is not the case in a mixture of a drop of water with a jug of wine.

43. The primary qualities of water are moistness and coldness. Warm or tepid water, however, has a mean quality in the sense that this quality lies somewhere between hot and cold and is presumably a blend of these qualities. But warm water will eventually cool and Albert interprets this as a mean quality seeking to reacquire its own primary quality, in this instance "coldness." This illustration seems to represent a general rule that all elements with mean qualities will seek to reattain their primary quality.

44. From this it would seem to follow that all perceptible physical bodies in the world are substantial compounds, whereas elements are hypothetical abstractions arrived at analytically. But in Book II, Question 6, of the same treatise, Albert denies this for earth and fire (the former, we are told, is pure near the very center of the earth, and the latter in the sphere of fire just below the moon) and suggests that, despite appearances, perhaps all water and air are not impure and that somewhere pure specimens of them exist (fols. 149 verso, col. 2, to 150 recto, col. 1 of the edition cited in n. 18).

45. A reference by Albert to Book II, Question 6, of this very treatise (see n. 44).

46. An explicit statement of this is made in the second doubt above.

47. A compound is said to be generated "mediately" from elements when it is generated from another compound, and generated "immediately" when formed directly from elements.

48. The priority of element over compound is based ultimately on the argument that whereas elements can be generated independently of compounds—that is, an element can be generated from another element—compounds must be generated from elements, whether this be "mediate" or "immediate."

49. Strictly speaking, although Aristotle seems to accept this view, it is one that he attributes to "some thinkers" (*On Generation and Corruption* I.10.327a.35 and 327b.2–5).

To the fourth doubt, I say that in the fifth book of the *Metaphysics,* Aristotle understands prime matter, and so his authority is not against us.[50]

To the fifth doubt, where it is said "that a compound is moved with the motion of the predominant element," [I say] it is understood in this way: A compound [or mixed body] is moved according to a disposition that was left behind *(secundum dispositionem derelictam)* in it by the predominant element in the compound.

To the sixth doubt, I say that all things are corrupted. But when it is said therefore something is compounded of nonbeings, this consequence is denied, because from this compound there is an ampliative term *(terminus ampliativus)*[51] of another term for assuming from this what is or was, just as this term must be generated. Thus, this [proposition] must be conceded: "Something is compounded of nonbeings which are or have been," just as it is conceded that a man is born from a seed, not which is but which was.[52] Later on, [in the sixth doubt,] where it is said that "it does not appear that the predominant element is corrupted," I say that if the predominant element is predominant over the other three [elements] to such an extent that those three cannot corrupt it, [then] it is not a compound, but the predominant element transmutes the other three into its own nature. If, however, this element is not predominant to such an extent, but is [itself] corrupted with the others through a mutual action and reaction of all the other elements, then there is a compound and the generation of a compound, not an element.

To the seventh doubt concerning experiences, I say to the first of these that the liquid *(humor)* that emerges from the wood and is drawn from the rose is not pure elemental water, which is in these things [namely wood and rose] only proportionately, just as with blood in the body of an animal. That such a liquid is not an element is obvious from the fact that it has color, odor, and taste, which do not occur in pure elemental water. As for the other [experience], where iron is considered, I say that the fire was not in the bodies that were mutually struck, but that some small particles are separated from the stone and ignited by motion and mutual friction *(confricationem).*[53]

As for the principal arguments *(rationes principales),* I respond to the first by asserting that Aristotle understands that both of them are not corrupted and that the one element is not changed into the other, but rather, that a compound is

newly generated from these corrupted elements. And Aristotle puts it well when he says that these elements persist in a compound potentially but not

50. There is nothing to suggest that Aristotle is speaking about prime matter in *Metaphysics,* Book V, chapter 3 (see n. 40). Actually, Albert is following the commentary of Averroes on this passage, where the latter says *(Metaphysics,* Bk. V, Comment 31, Vol. VIII, fol. 106 verso, col. 1, of the edition cited in Selection 44, n. 1) that "a true element is common to all composed things; it is that which is first and from which all things are composed, and it exists in everything and all things are resolved into it. And this element ought to be the cause of other elements, and this is what he [that is, Aristotle] calls *prime matter."* The basis for Averroes' interpretation is perhaps Aristotle's discussion in *Physics,* Book I, chapters 6 and 7 (or so it seems from a remark he makes in Comment 31 of Bk. III *De caelo* [Vol. V, fol. 201 recto, col. 1, of the edition cited in Selection 44, n. 1], where he cited Bk. I of the *Physics).* This fourth doubt leads one to suspect that the concept of an "element" was confounded by the fact that the four pure physical elements (earth, water, air, and fire; though these might exist, they were not physically perceptible in a pure state, since they were colorless, odorless, and tasteless; see n. 44) were further analyzable into the fundamental, but abstract, principles of prime matter and form (in the elements, the forms were associated with the primary qualities hot–cold and dry–moist). Albert is arguing here that prime matter persists in a compound, not the four pure corporeal elements, and implies that prime matter should not be interpreted as an "element."

51. According to Philotheus Boehner *(Medieval Logic* [Chicago: University of Chicago Press, 1952], p. 11), " 'Ampliatio' is the property of a common or universal term of which the personal supposition is extended to signify not only significates or objects of the present, but also of the past or future, or of the realm of possibility. In other words, the number of individuals signified by the terms is enlarged or "amplified." The ampliatio has to be expressed by an appropriate term, usually the verb, as, for instance, in the proposition: 'Every man will run,' the term 'man' is extended or, at least, may be extended or amplified so that it not only supposits for the actually existing man, but also supposits for all future men." See note 52.

52. The point of Albert's argument is that his opponent has made it seem as if the "nonbeings" presently in a compound were always "nonbeings" in the past. For this reason, Albert restates the proposition and concedes only that the compound is *now* made up of "nonbeings," but these did have existence before they were corrupted, namely when they were actually existent elements. Thus his opponent has improperly "amplified" the sense of "nonbeings" to apply to the whole past (see n. 51).

53. It is of interest that Albert chose to explain the origin of the fire as a consequence of motion and friction, rather than appeal to a disposition of fire left behind in these two compounds by the element fire before it was corrupted. This latter explanation would have been justified by what he says in response to the first principal argument (see next paragraph).

actually, that is, their matter, which is a certain potentiality, persists, but not their forms, which are actualities, for he said that their power *(virtus)* is preserved *(salvatur)*, that is, because of the mutual action and reaction of the elements certain dispositions of them are left behind in matter. To the second [principal reason], I say that a compound is first a union of thinge that have been altered and afterward a union of things that have been corrupted. In order that the matter of the [respective] elements should be mutually united and continued by the form of the compound *(forma mixti)*, it is first necessary that the elements be altered mutually and then corrupted and the form of the compound generated. To the third principal reason, I say that it is indeed similar in the sense that just as in the syllable *ab* each of the letters, *a* and *b,* loses its perfect sound, so in a compound each of the things that are mixed is weakened in the quality which determines it. But it is dissimilar, because in the syllable *ab* both the letter *a* and the letter *b* persist, so that it is unlike the things mixed in a compound.

Question 20
We inquire whether a compound is possible.[54]

First, it is argued in the negative, for if it were possible, it would either be by juxtaposition of the things compounded [or mixed] or by penetration. But it can be done in none of these ways, therefore, and so on [a compound is not possible]. In the first place, it cannot be by juxtaposition, for if it were, then not every part of the compound would be a compound. Nor is it by penetration, for then several bodies could be in the same place, which is contrary to what Aristotle says in the fourth book of the *Physics.*[55] In the second place, it is argued. . . .[56]

Aristotle takes the opposite position [namely that a compound is possible]. The opposite is also obvious to the senses, for we see that wine and water and other things can be mutually united. In this question I shall first assume certain distinctions; secondly, I shall respond to the question itself; thirdly, I shall propose certain problems [or doubts] concerning this question and resolve them.

As for the first [of these parts, namely the assumption of distinctions], it must be understood that in one way a compound *(mixtio)* [or mixture] is taken as a juxtaposition of several things which retain their form and shape, as when we say that barley and wheat are mixed.[57] In another way, a compound is understood as a juxtaposition of some

very small sensible parts, as in a mixture of wools. In a third way, it is taken as a juxtaposition of insensible parts, as happens in the grinding of several species into mortar.[58] Fourthly, a compound is properly so called when any part of it is [also] a compound, whether that part be sensible or insensible.[59]

A second distinction is that in such a mixture a compound is said to be mixed homogeneously when any part of it is mixed in the same way as any other part. Another compound is said to be heterogeneous when any part of it is not compounded like every other part. And here we must understand whole parts, not essential parts as matter and form.

A third distinction is that something is said to be a compound either from the standpoint of its *terminus ad quem* or its *terminus a quo.* An example of the first is a compound which is generated from elements; an example of the second is the four elements.[60]

As for the second part [of this question], I posit this conclusion: A compound *(mixtio)* in its most proper sense is possible. This is obvious, for unless every part of a compound, whether sensible or insensible, is [also] a compound, a compound could not have one continuous form.

As for the third part [of this question], I pose some problems, the first of which is, "How is a compound made?" I reply that it happens as follows: The things that are compounded, having been divided into small particles, act and react mutually until the substantial elemental forms of all these particles are corrupted. Then a substantial form of the compound is produced *(educitur)* from the whole composite mass of these particles; and by the production *(eductionem)* [of this substantial

54. Aristotle asks this very question in *On Generation and Corruption* I.10.327a.35.
55. *Physics* IV.8.216a.27–216b.11.
56. Principal arguments two to five, along with Albert's response to them at the end of the question, are omitted here.
57. This is Aristotle's example in *On Generation and Corruption* I.10.328a.1–5.
58. By this analogy, we are presumably to understand that the several ingredients of mortar are so finely ground and mixed together that the individual ingredients are no longer discernible.
59. This is the sense of compound which Albert takes as proper and legitimate.
60. As the end product of a generation, a compound can be construed as generated from elements. Looked at from the standpoint of its originative or generating units, a compound can be equated with the four elements that will contribute to its formation and constitution.

form] these matters, which previously were only contiguous, since they were controlled by their elemental forms, are made continuous. It follows from this that the juxtaposition of parts alone is not sufficient for a compound, or the generation of a compound, but continuation [of the parts] is also required. Secondly, such a division of a mixable thing or compound does not continue into infinity but will reach minute parts. Thirdly, it follows that every compound of things is a generation, but not conversely. This is obvious, because a compound connotes that many things are corrupted and made one and whenever this happens there is a generation. But there is not always a corruption of many things when there is a generation, for sometimes there is a generation but the things that were mixed are not corrupted except one of them, as happens when a compound is generated from a compound.[61]

A second doubt is this: Whether a compound could be made without the division of mutually mixable things. I reply that it can; for when there is a division into small parts, these small parts are mixed [or compounded] without a [further] division.[62] Similarly, if *a* is juxtaposed to *b,* it is possible that from their mutual action and reaction both are corrupted and a mean thing prepared *(temperatum)* . . . . A thirteenth and last doubt is this:

What is the reason *(causa)* that any part of a compound is a compound, since fire applied to any compound reduces *(resolvat)* one part to ashes, another to smoke, and another to flame, as when fire is applied to wood. But in another compound, it [that is, fire] reduces every part to flame, as when fire is applied to oil. I respond that this is due to the heterogeneity of the compound of wood, and the homogeneity of the compound of oil.

[I reply now] to the principal arguments. To the first, I say that it is first by juxtaposition, then by their corruption, and then, with the advent of the substantial form of the compound, that any part of the compound is a compound. To the second, [I say]. . . .[63]

---

61. Are we to understand that the compound itself, taken as a single homogeneous entity, is corrupted to form a new compound? Or is only one ingredient of the compound corrupted, thereby producing a new compound?

62. A claim based upon Aristotle's assertion (*On Generation and Corruption* I.10.328a.34–36) that bodies "combine more freely if small pieces of each of them are juxtaposed. For in that condition they change one another more easily and more quickly; whereas this effect takes a long time when agent and patient are present in bulk."

63. The remaining few lines of the question are omitted (see n. 56).

Geology, Geography, and Oceanography

Geology

78 ON THE FORMATION OF STONES AND MOUNTAINS

Avicenna (980–1037)

Translated and annotated by E. J. Holmyard and D. C. Mandeville[1]

Introduction by Edward Grant

This selection and Selection 74 form a treatise that "is partly a direct translation and partly a *résumé* of sections of a genuine work of Avicenna, *viz.*, the *Kitâb al-Shifâ',* the 'Book of the Remedy,' which he composed in response to his friend Al-Juzjânî's request that he should write a general commentary on Aristotle's works. He was too busy to write a formal commentary, but compromised by writing a plain exposition free from any attempt at refutation of adverse views. He had already written the first book of his great *Canon* of Medicine, and thereafter worked at the *Shifâ'* and *Canon* simultaneously" (Holmyard and Mandeville, p. 8). The section on Physics, to which the two selections published in this source book belong, was completed between 1021 and (probably) 1023.

Sometime during the last thirty years of the twelfth century, Alfred of Sareshel, or Alfred the Englishman, translated this treatise from Arabic into Latin. In the form in which it has been preserved, it is divided into three paragraphs: (1) *De congelatione et conglutinatione lapidum (On the Congealing and Hardening of Stones)*, which frequently served as the title of the treatise;(2) *De causa montium (On the Cause of Mountains;* these first two paragraphs or sections constitute this particular selection and are translations of section 1 of Discourse II of Subject *(fann)* V of the Physics portion of the *Shifâ')* ; (3) *De quatuor speciebus corporum mineralium (On the Four Species of Mineral Bodies;* this brief part is included in Selection 74, and is a translation of section 4 of Discourse II, Subject V. While a few authors (for example, Albertus Magnus) were fully aware of Avicenna's authorship, confusion soon developed as the treatises came frequently to be ascribed to Aristotle, often under the title *Liber de mineralibus Aristotelis (The Book on Minerals of Aristotle)*. The basis for the attribution was a promise made

by Aristotle at the conclusion of the third book of the *Meteorologica* (6.378b.5), where, after a brief summary on the formation of minerals and metals, he declares that "this is the general theory of all these bodies, but we must take up each kind of them and discuss it separately." But no such discussions appear in the fourth and final book of the *Meteorologica,* so that it was quite plausible to mistake Avicenna's treatise as the fulfillment of Aristotle's promise, for it deals "with topics which Aristotle no doubt would have treated in the fourth book of his *Meteorologica* if he had ever completed it" (Holmyard and Mandeville, p. 9). And so it was that the *De mineralibus* was often appended as an additional chapter (in three paragraphs) to the fourth book of the *Meteorologica.*

Of the treatise itself, which they translate directly from the better Arabic text, Holmyard and Mandeville observe (p. 11), "From a scientific point of view, Ibn Sînâ's [he is more commonly referred to as "Avicenna," the Latinized form of his name— *Ed.*] opinions upon the formation of stone, rocks and mountains are remarkably interesting, in that they show an astonishingly accurate insight into geological phenomena. They are so clearly and concisely expressed that we have felt it unnecessary to give a synopsis, preferring that Ibn Sînâ should be allowed to speak for himself. Similar remarks apply to his theories upon the nature of minerals,

1. [*In Avicennae "De congelatione et conglutinatione lapidum" being sections of the 'Kitâb al-Shifâ.' "* The Latin and Arabic texts edited, with an English translation of the latter and with critical notes, by E. J. Holmyard and D. C. Mandeville (Paris: Librairie Orientaliste Paul Geuthner, 1927), pp. 17–32. Of the numerous notes supplied by Holmyard and Mandeville, only those (or parts of those) have been included which illuminate the text substantively (thus all strictly philological notes have been eliminated). The notes have been renumbered and, where necessary, slightly altered to meet the needs of the reader.—*Ed.*]

and particularly to his ruthless criticism of the alchemists [see Selection 74—*Ed.*] and their attempts to transmute the base metals into gold."

THE BOOK OF THE REMEDY[2]

In the name of God, the Merciful, the Compassionate!

THE FIFTH SUBJECT OF THE PHYSICS, consisting of two discourses upon meteorological phenomena. This subject comprises the secondary causes of the inanimate creation such as minerals, meteorological phenomena and the like.

The First Discourse, upon those things which occur upon the earth.

Section 1. Upon Mountains.

We shall begin by establishing the condition of the formation of mountains and the opinions that must be known upon this subject. The first [topic] is the condition of the formation of stone, the second is the condition of the formation of stones great in bulk or in number, and the third is the condition of the formation of cliffs and heights.

We say that, for the most part,[3] pure earth does not pertrify, because the predominance of dryness over [i.e. in] the earth endows it not with coherence but rather with crumbliness. In general, stone is formed in two ways only (a) through the hardening[4] of clay, and (b) by the congelation [of waters].[5] Many stones in fact, are formed from a substance in which earthiness predominates, and many of them are derived from a substance in which aquosity predominates. Often a clay dries and is changed at first into something intermediate between stone and clay, *viz.* a soft stone, and afterwards is changed into stone [proper]. The clay which most readily lends itself to this is that which is agglutinative, for if it is not agglutinative it usually crumbles before it petrifies. In my childhood I saw, on the bank of the Oxus, deposits of the clay which people use for washing their heads;[6] subsequently I observed that it had become converted into a soft stone, and that was in the space of approximately 23 years.[7]

Stone has also been formed from flowing water in two ways (a) by the congelation of the water as it falls drop by drop or as a whole during its flow,[8] and (b) by the deposition from it, in its course, of something which adheres to the surface of its bed and [then] petrifies. Running waters have been observed, part of which, dripping upon a certain spot, solidifies into stone or pebbles of various

colours, and dripping water has been seen which, though not congealing normally, yet immediately petrifies when it falls upon stony ground near its channel. We know therefore that in that ground there must be a congealing petrifying virtue which converts the liquid to the solid. Thus the bases of the formation of stone are [either] a soft clayey substance or a substance in which aquosity predominates. Congelation of the latter variety must be caused by a mineralizing, solidifying virtue, or earthiness must have become predominant in it in the same way in which salt is coagulated, *i.e.* earthiness becomes predominant in it by reason of its [peculiar] virtue and not of its amount. If indeed the earthy quality is not like that in salt, but is of a different kind, nevertheless the two must be similar in that they are transformed by heat, and in that the advent of heat coagulates them. Or it may be that the virtue is yet another, unknown to us. Alternatively, the converse may be true—that its earthiness has prevailed merely by a cold dry virtue.

In short, it is in the nature of water, as you know,

2. *Shifâ',* remedy, the becoming free from disease, or the giving of health. In medieval Latin the word was wrongly translated *Sufficientia.*

3. The Latin version begins here. [The Latin text published by Holmyard and Mandeville is drawn from a fifteenth-century manuscript found in Trinity College, Cambridge. For comparison, variant readings are given from another fifteenth-century manuscript in Trinity College, Cambridge, as well as from two printed editions (Bologna, 1501, and Lyons, 1528). Occasionally, the Latin text and its variants will be mentioned.—*Ed.*]

4. [Holmyard and Mandeville remark that "hardening" represents the Latin word *conglutinatio,* which, in turn, renders an Arabic term that "is unknown to the dictionaries. It appears to mean the conversion of clay into the hard form which it assumes when baked."—*Ed.*]

5. Latin *congelatio.* The reference is to the deposition of solids from water, interpreted by Ibn Sînâ (and others) as a solidification of the water itself.

6. [In notes generously made available to the translators, Professor H. E. Stapleton of Calcutta says:—*Ed.*] "This clay, which is known in Baghdad as *Ṭîn Khâwah,* presumably contains sodium carbonate, but we have been unable to obtain a sample in Calcutta. In India the naturally occurring *Sajji Matti* (crude sodium carbonate) is used for the same purpose." It is green in colour and of a creamy consistency. [The Latin text has no equivalent for the words "deposits . . . heads."—*Ed.*]

7. The Latin version says 33 years, but this is clearly an error. [Actually, this is a variant reading; the text printed by Holmyard and Mandeville says 100 years.—*Ed.*]

8. Literally, with its rope, i.e. altogether, as a whole. The Latin translator omitted the latter part of this sentence ["or . . . flow"—*Ed.*], probably because he did not understand it.

to become transformed into earth through a predominating earthy virtue; you know, too, that it is in the nature of earth to become transformed into water through a predominating aqueous virtue. In this connection, there is a substance used by those folk who have lost their way amid their artful contrivances[9] which, when they are so minded, they call *Virgin's Milk;*[10] it is compounded of two waters which coagulate into a hard solid. This is an indication of the truth of [what I have said above]. They have also many things which they use in liquefaction and coagulation which bear witness to the soundness of these judgments.

Stones are formed, then, either by the hardening of agglutinative clay in the sun, or by the coagulation of aquosity by a desiccative earthy quality, or by reason of a desiccation through heat. If what is said concerning the petrifaction of animals and plants is true, the cause of this [phenomenon] is a powerful mineralizing and petrifying virtue which arises in certain stony spots, or emanates suddenly from the earth during earthquakes and subsidences, and petrifies whatever comes into contact with it. As a matter of fact, the petrifaction of the bodies of animals and plants is not more extraordinary than the transformation of waters.

It is not impossible for compounds to be converted into a single element if the virtue of the latter gets the mastery over them, for each of the elements they contain may be converted into that element. For this reason anything which falls into salt-pans is converted into salt, while objects which fall into the fire are converted into fire. As for the swiftness or slowness of the conversion, that is a matter which necessarily varies according to the variation in the strength of the virtues; if they are very violent they perform the conversion in a short time. In Arabia there is a tract of volcanic earth[11] which turns to its own colour everyone who lives there and every object which falls upon it. I myself have seen a loaf of bread in the shape of a *raghîf*[12]— baked, thin in the middle, and showing the marks of a bite—which had petrified but still retained its original colour, and on one of its sides was the impression of the lines in the oven. I found it thrown away on a mountain near Jâjarm,[13] a town of Khurâsân, and I carried it about with me for a time. These things appear strange only on account of their infrequent occurrence; their natural causes, however, are manifest and well-known.

Certain varieties of stone are formed during the extinction of fire, and frequently ferreous and stony bodies originate during thunder-storms, by reason of the accidental qualities of coldness and dryness which fieriness acquires when it is extinguished. In the country of the Turks[14] there fell, amid thunder and lightning, coppery bodies in the

9. A thrust at the alchemists. Stapleton says: "The full rendering of the Arabic of these last two sentences is as follows: 'There is a thing that certain people, who have gone astray, prepare in their experiments, called Virgin's milk, which is composed of two waters. This coagulates into a dry substance, which clearly proves the truth of what we have stated above (i.e. that water can be turned into stone). They have likewise many other things that they prepare during the processes of liquefaction and coagulation that also proves these laws.' The version of the Latin translator betrays the fact that he himself is numbered amongst those who 'have gone astray.' "

10. *Lac virginis* is a name applied to many substances in alchemical literature. . . .
Berthelot (*La Chimie au Moyen Âge,* Vol. 2, p. 29) says [my translation—*Ed.*]: "Virginal milk is a symbolic name frequently used by the alchemists. In particular, it is the name which the alchemists, following Bar Bahloul, gave to mercury. However, one should understand by it a chloride of mercury rather than metallic mercury.
It seems clear that Ibn Sînâ's reference is merely to two liquids which when mixed give a white precipitate, such as the solutions of lead acetate and potassium carbonate. . . . That a white solid is thrown down from the mixture of these solutions was known to Jâbir ibn Hayyân." [*Lac virgineus* and *lac virginis* are the forms found in the Latin text.—*Ed.*]

11. *Harra.* Lane describes a *harra* as a stony tract of which the stones are black and worn and crumbling, as though burned with fire. Stapleton has the following note: "Harrah is an outcrop of volcanic rock often found in Arabia. Burton (*Pilgrimage to Al-Madinah and Meccah,* Memorial ed. Vol. I, p. 421) says: 'Harrah, from Harr (heat) is the generic name of lava, porous basalt, scoriae, greenstone, schiste, and other [stones] supposed to be of igneous origin. It is also used to denote a ridge or hill of such formation.' Several famous ones exist close to Madinah."
Doughty, *Arabia Deserta,* 1923, I, 419, says: "In the train of Harras we see a spectacle of the old vulcanic violence that tormented this border of the Arabian peninsula. I have followed these Harras almost to Mecca; that is through nearly seven degrees of latitude." Doughty has numerous other references to *harras,* for which see the index to his book.

12. Lane: A *raghif* is "a round cake of bread, such as is thick, or not thin, generally about a span, or less, in width and from half an inch to an inch in thickness."

13. Latin *Lurgeam!* Jâjarm lies approximately 56° by 37° N (in modern Persia). [Neither the Latin text nor the variants contain *Lurgeam.* Indeed the sentence "I found it thrown away . . . for a time" is not translated in the Latin version. Perhaps Holmyard and Mandeville were merely indicating that in other treatises it was usual to use *Lurgeam* to transliterate Jâjarm.—*Ed.*]

14. Latin version: *in Persia.*

shape of arrowheads with a projection turned back towards the top. A similar one fell in Jil and Dailam,[15] and when it fell it penetrated into the earth. The substance of all these was coppery and dry. I myself undertook, in Khwarazm,[16] the difficult task of fusing a head of that kind, but it would not melt; a greenish fume continued to come off from it until at length an ashy substance remained.[17]

I regard as true, on unexceptionable evidence, an event which happened in Jûzjânân[18] in our own time: a ferreous body, which perhaps weighed 150 mana,[19] fell from the sky, penetrated the ground, and then rebounded once or twice like a ball which has been thrown against a wall; afterwards it penetrated into the ground again. People heard a tremendous, terrifying, noise and when they investigated the matter they took possession of the object and carried it to the Governor of Jûzjânân. He wrote about it to the Sultan of Khurâsân, contemporary with us, the Amîr Yamîn al-Daula wa 'Amîn al-Milla Abu'l-Qâsim Maḥmûd ibn Sabaktagîn al-Muzaffar al-Mughallib,[20] who ordered him to send him the object or a part of it. Its removal, however, proved too difficult on account of its weight, so they attempted to break off a piece of it, but the tools could work it only with difficulty, every drill and chisel used upon it breaking. At length, however, they cut off a fragment of it and sent it to the Sultan, who ordered a sword to be struck from it, but this proved too difficult. It is said that the substance was entirely composed of small rounded granular particles adhering to one another.[21] All this was seen by my friend the lawyer Abû 'Ubaidallah 'Abdu'l-Wâḥid ibn Muḥammad al-Jûzjânî.[22] I am told that many of the beautiful swords of the Yaman are made from this kind of iron only[23] and that the poets of the Arabs have described the fact in their poems.[24]

This, then, is one kind of way in which stone is formed. A trustworthy man from among the Shaikhs of the kingdom of Iṣfahân, Abû Manṣûr Hormuz Diyâr ibn Mashakzâr, one in close relation with the illustrious Amir Abû Ja' far Muḥammad ibn Dushmanzâr (may God have mercy upon him!)[25] told me that there fell from the sky, in the mountains of Ṭabaristân, an object the fall of which resembled the fall of the above-mentioned mass of iron, except that in this case it was a huge stone. This completes the discourse upon the formation of stones.[26]

As for the formation of large stones, this may occur all at once, by intense heat acting suddenly upon a large mass of clay, or little by little with the passage of time.

15. Two Persian provinces on the south-west shores of the Caspian Sea.

16. Modern Khiva.

17. Mention of the greenish fume confirms Ibn Sînâ's identification of the substance as "coppery," though perhaps it may have been nickel as is the case with many meteorites.

18. Jûzjânân is in Herat (Khurâsân).

19. A mana is about two pounds. Among the Chaldaeans and Egyptians the mana was the sixtieth part of the talent, which itself weighed 93.65 lbs. It is an extremely ancient weight. . . .
One would not have thought that a weight of rather less than 3 cwt. was very difficult to transport (see further on in this paragraph). The Latin version has 100 marks, i.e., approximately 100 lbs.

20. S. Lane-Poole, Mohammadan Dynasties, 1925, pp. 286 and 287: "Yamin al-Daula Maḥmûd ibn Sabaktagîn came to the throne in A.D. 998 (A.H. 388)."

21. Cf. Encycl. Britannica, XVIII, p. 264 (Eleventh Edition): About eleven out of every twelve of the known meteoric stones belong to a division to which Rose gave the name chondritic χόνδρος, a grain); they present a very fine-grained but crystalline matrix or paste, consisting of olivine and enstatite or bronzite, with more or less nickel-iron, troilite, chromite, augite and triclinic feldspar; through this paste are disseminated round chondrules of various sizes and generally with the same composition as the matrix; in some cases the chondrules consist wholly or in part of glass. Some meteorites consist almost solely of chondrules; others contain only a few; in some cases the chondrules are easily separable from the surrounding material.

22. Ibn Sînâ's faithful companion for many years.

23. Sir Robert A. Hadfield, F. R. S., the eminent authority on iron and steel, whom we consulted upon this point, writes: "As regards the sword made from meteoric iron, as the latter is usually a compound of Iron and Nickel, with Cobalt and some other elements, yet at the same time hardly any Carbon, it is quite to be understood that this material, although forgeable, would only give a product which on quenching would be tough but quite soft, and not carry a cutting edge."

24. Having been unable to confirm this statement, we asked the help of Professor R. A. Nicholson, who writes: "I cannot find any authority for the statement of Ibn Sînâ that the Arab poets refer to meteoric iron as the material from which the best swords were made. Possibly Ibn Sînâ may have misunderstood such passages as that in which Muslim b. al-Wahid describes a flashing sword as 'a deadly meteor.'"

25. S. Lane-Poole, Mohammadan Dynasties, 1925, p. 145: "Mohammad b. Dushmanzâr, known as Ibn-Kâkwayh, was first cousin to Majd-al-dawla the Buwayhid, of Hamadhân, whose dominions he annexed by the deposition of Samâ-al-dawla in 1023 A.D. He had previously taken Ispahân in 1007 A.D."

26. In the Encyclopaedia Britannica, Vol. XVIII (Eleventh Edition), p. 261, it is stated that an abundant meteoric display took place on October 15, 1002 A.D., a date which accords well with the observations of Ibn Sînâ.

The formation of heights is brought about by (a) an essential cause and (b) an accidental cause. The essential cause [is concerned] when, as in many violent earthquakes, the wind which produces the earthquake raises a part of the ground and a height is suddenly formed.[27] In the case of the accidental cause, certain parts of the ground become hollowed out while others do not, by the erosive action of winds and floods which carry away one part of the earth but not another. That part which suffers the action of the current becomes hollowed out, while that upon which the current does not flow is left as a height. The current continues to penetrate the first-formed hollow until at length it forms a deep valley, while the area from which it has turned aside is left as an eminence. This may be taken as what is definitely known about mountains and the hollows and passes between them.[28]

Very often both water and wind would be in-effectual except for the fact that the earth is not uniform, some parts of it being soft and others stony. The soft, earthy parts become hollowed out and the stony parts are left behind as elevations. With the passage of time, the channel is excavated and widened more and more, while the raised portion is left, becoming relatively higher and higher as more earth is hollowed out from [beside] it. These, then, are the principal causes of the three changes [mentioned at the beginning of the *faṣl*, viz. the formation of stone, the formation of stones great in bulk or in number, and the formation of cliffs and heights].[29]

Mountains have been formed by one [or other] of the causes of the formation of stone, most prob-ably from agglutinative clay which slowly dried and petrified during ages of which we have no record. It seems likely that this habitable world was in former days uninhabitable and, indeed, sub-merged beneath the ocean.[30] Then, becoming ex-posed little by little, it petrified in the course of ages the limits of which history has not preserved; or it may have petrified beneath the waters by reason of the intense heat confined under the sea. The more probable [of these two possibilities] is that petrifaction occurred after the earth had been exposed, and that the condition of the clay, which would then be agglutinative, assisted the petri-faction.[31]

It is for this reason [*i.e.* that the earth was once covered by the sea] that in many stones, when they are broken, are found parts of aquatic animals, such as shells, etc.[32]

It is not impossible that the mineralizing virtue[33]

was generated there [*i.e.* in the petrifying clay] and aided the process, while the waters also may have petrified. Most probably, mountains were formed by all these causes.

The abundance of stone in them is due to the abundance, in the sea, of clay which was afterwards exposed. Their elevation is due to the excavating action of floods and winds on the matter which lies between them, for if you examine the majority of mountains you will see that the hollows between them have been caused by floods. This action, however, took place and was completed only in the course of many ages, so that the trace of each individual flood has not been left; only that of the most recent of them can be seen.

At the present time, most mountains are in the stage of decay and disintegration, for they grew

27. Cf. the opinion of Aristotle, in *Meteorologica* [Bk. II, ch. 8—*Ed.*], that earthquakes are due to the formation of winds within the earth.

With the views expressed here and later in this section, it is interesting to compare those which Ovid ascribes to Pythagoras (*Metamorphoses,* XV, trans. by F. J. Miller, 1916, pp. 383 etc.): "I myself have seen what was once solid land changed into sea; and again I have seen land made from the sea. Sea-shells have been seen lying far from the ocean, and an ancient anchor has been found on a mountain-top. What once was a level plain, down-flowing waters have made into a valley; and hills by the force of floods have been washed into the sea. What was once marsh is now a parched stretch of dry sand, and what was once dry and thirsty now is a marshy pool . . . . Near Troezen, ruled by Pittheus, there is a hill, high and treeless, which once was a perfectly level plain, but now a hill; for, horrible to relate, the wild forces of the winds, shut up in dark regions underground, seeking an outlet for their flowing and striving vainly to obtain a freer space, since there was no chink in all their prison through which their breath could go, puffed out and stretched the ground, just as when one inflates a bladder with his breath, or the skin of a horned goat. That swelling in the ground remained, has still the appearance of a high hill, and has hardened as the years went by."

28. That is, so far Ibn Sînâ has merely been stating what was generally accepted; his own views upon the details and methods of the processes follow.

29. See p. 616.

30. A suggestion which goes back at least as far as Xenophanes of Colophon (614 B.C.) and was accepted by Herodotus, Strabo, Aristotle and others.

31. See p. 616.

32. Ibn Sînâ was not alone in this anticipation of the conclusions of Nicolas Steno, for Xenophanes and the other philosophers mentioned in note 30 based their views upon the occurrence of shells in regions far from the sea. After the death of Ibn Sînâ, however, orthodox belief repudiated the natural deduction from the ob-served facts, although Leonardo da Vinci was courage-ous enough to reaffirm it.

33. See p. 616.

and were formed only during their gradual exposure by the waters. Now, however, they are in the
grip of disintegration, except those of them which
God wills should increase through the petrifaction
of waters upon them, or through floods which
bring them a large quantity of clay that petrifies on
them. I have, I believe, heard that this has been
observed on certain mountains. As for [the similar
phenomenon] which I witnessed upon the banks
of the Oxus, that place cannot properly by called a
mountain.[34]

Of the land which was exposed by the retreat of
the waters, those parts which were of harder clay
or more strongly petrified or of greater bulk than
the rest remained as elevations and heights when
the other parts had been carried away.

As for the veins of clay that are found in mountains, it is possible that these were formed not from
the main substance which has undergone petrifaction, but from *débris* of the mountains that turned
into dust and filled the valleys and ravines. It then
became moistened by streams which flowed upon
it, and was covered by the layers of stone forming
the mountains, or interlaid with the good clay of
the latter. It is possible also that the ancient clay
of the sea was not uniform in substance, and that
in succession some of it petrified thoroughly, while
some did not petrify at all, and some was converted
only into a soft stone through a certain quality
predominant in it or by reason of some one of
innumerable other causes.

It is also possible that the sea may have happened
to flow little by little over land consisting of both
plain and mountain and then have ebbed away
from it; and so it came to pass that the plain was
turned into clay without the same befalling the
mountain. Once converted into clay, it was in a
fit state to undergo petrifaction when it became
exposed, and its petrifaction would be complete
and strong. When exposure of the matter which
was petrifying took place, it must frequently have
happened that the old petrified portions [*i.e.* the
mountains] were in a state fit for disintegration,
and so would suffer the converse of what was
happening to the earth. That is, they became moist
and soft and turned into dust again, which itself
is in a fit state for petrifaction. For example, when
you soak a brick, some earth and some clay in
water, and then expose each of them to the fire,
the soaking will increase the tendency of the brick
to be disintegrated again by the fire, and will also
increase the tendency of the earth and the clay to
petrify strongly.

It is possible that each time the land was exposed
by the ebbing of the sea a layer was left, since we
see that some mountains appear to have been piled
up layer by layer, and it is therefore likely that the
clay from which they were formed was itself at one
time arranged in layers. One layer was formed
first, then, at a different period, a further layer
was formed and piled [upon the first, and so on].
Over each layer there spread a substance of different material, which formed a partition between
it and the next layer; but when petrifaction took
place something occurred to the partition which
caused it to break up and disintegrate from between
the layers.[35]

As to the bottom of the sea, its clay is either
sedimentary or primaeval, the latter not being
sedimentary. It is probable that the sedimentary
clay was formed by the disintegration of the
strata of mountains.[36]

Such is the formation of mountains.[37]

---

34. See p. 616.

35. The text of the first sentence of this passage is
unsatisfactory in all three MSS we have used. We have
therefore adopted Stapleton's rendering, as he seems to
have had a text more correct than ours in this passage.
He has, however, not translated the passage from *piled
up layer by layer* to *between the layers*, remarking that
the Arabic original of the latter portion of *Fasl* I is full
of mistakes, and that if the mediaeval translator had a
similar text before him, it is easy to understand the
difficulty he evidently experienced in rendering the
Arabic into Latin.

Except for the sentence above-mentioned, the passage
in question does not present any peculiar difficulty if our
text is correct, though it is true that what Ibn Sînâ means
to convey about the partitions between the layers is not
as clear as could be desired. It is evident that he wished
to explain (*a*) why the various layers of clay did not mix
while they were still in a viscous state, which he does by
postulating the formation of partitions; and (*b*) why the
partitions are not to be discovered in actual practice.

36. Cf. *Encyclopaedia Britannica*, XI, p. 656: "The
sedimentary or stratified rocks form by much the larger
part of the dry land of the globe, and they are prolonged
to an unknown distance from the shores under the
bed of the sea. They include those masses of mineral
matter which, unlike the igneous rocks, can be traced
back to a definite origin on the surface of the earth.
Three distinct types may be recognized among them: (*a*)
By far the largest proportion of them consists of different kinds of sediment derived from the disintegration of
pre-existing rocks. In this 'fragmental' group are placed
all the varieties of shingle gravel, sand, clay and mud,
whether these materials remain in a loose incoherent
condition, or have been compacted into solid stone. (*b*)
Another group consists of materials that have been
deposited by chemical precipitation from solution in
water. The white sinter laid down by calcareous springs
is a familiar example on a small scale. Beds of rock-salt

79 ON THE MOTION OF THE EARTH'S CENTER OF GRAVITY AND THE FORMATION OF MOUNTAINS

John Buridan (ca. 1300—ca. 1358)

Translated and annotated by Edward Grant[1]

QUESTION 7

We inquire next whether the whole earth is habitable.

(1) In the first place, it is argued yes because it is commonly said that a fourth part of the earth is habitable, and yet there seems no reason why one quarter ought to be more habitable than another quarter. Therefore, all quarters ought to be conceded to be habitable, and consequently, the whole earth. And this also seems obvious by a concession of Aristotle, who concedes that the earth opposite to us is as habitable as this part; concerning these people, he says that they live upward and to the right and that we are downward and to the left.[2]

(2) Next, it is argued that no part of the earth ought to be habitable because the earth is spherical and in the middle of the world as its center, just as is had in the second [book] of this treatise.[3] Now water is naturally located above the earth and always flows down toward a lower place, causing a sphere to be produced naturally, as is had in this second book. It follows from this that water ought to surround the whole earth and that because of the waters no part of the earth should be habitable. Nor is it valid to say that there are mountains and heights on the earth to which the surrounding water does not reach, making these heights habitable. There is a strong objection against this if the world is eternal, as Aristotle assumes. For at all times, many of the higher parts of the mountains descend to the valleys, and no parts, or few, ascend; thus through an infinite time these mountains ought to be wholly consumed and reduced to a plain.

Similarly, if the earth is more elevated where there is no sea and more depressed or sloped where there is a sea, then it is obvious that through the whole of time many of the more elevated parts of the earth were carried by the rivers into the depths of the sea, for which reason it happens that the sea is made gross and salty. But those parts of the earth are not returned from the depth of the sea to this earth, for what is elevated from the sea through exhalation or evaporation is nothing but a subtle watery matter and not a gross earthy substance. Through an infinite time, then, it would seem that

and dolomite have, in some regions, been accumulated to a thickness of many thousand feet, by successive precipitations of the salt contained in the water of inland seas. (*c*) An abundant and highly important series of sedimentary formations has been formed from the remains of plants and animals. . . . As the sedimentary rocks have for the most part been laid down under water, and more especially on the sea floor, they are often spoken of as "aqueous," in contradistinction to the igneous rocks. . . . It is from the sedimentary rocks that the main portion of geological history is derived. They have been deposited one over another in successive strata from a remote period in the development of the globe down to the present time. From this arrangement they have been termed 'stratified,' in contrast to the unstratified or igneous series. They have preserved memorials of the geographical revolutions which the surface of the earth has undergone; and above all, in the abundant fossils which they have enclosed, they furnish a momentous record of the various tribes of plants and animals which have successively flourished on land and sea."

37. Chapter II [or paragraph II—*Ed.*] ends here.

1. The present selection is drawn from Book II, Question 7 ("Whether the Whole Earth Is Habitable"), of Buridan's *Questions on the Four Books of the De caelo et mundo*, on pp. 154–155, 157–160 of E. A. Moody's edition (Cambridge, Mass.: Mediaeval Academy of America, 1942). Only that part of the question relevant to geology is included here. It may be properly described as a brief and succinct summary of Buridan's more extensive treatment in his *Questions on the First Three Books of the Meteors*—that is, Aristotle's *Meteorologica*. Parts of the latter have been translated into French by Pierre Duhem, *Le Système du Monde*, (Paris: Hermann, 1913–1959), IX, pp. 293–305. Although Duhem (p. 293) ascribed to Buridan the formulation of a new mechanical geology as distinguished from an astrological geology which had been in vogue earlier, it must be remembered that Avicenna had also presented a mechanical geology that was well known in the Latin West (see the preceding selection from Avicenna). Though Buridan's name is not often mentioned, his theories were accepted in the fourteenth and early fifteenth centuries by Albert of Saxony; Themon, son of the Jew; Marsilius of Inghen; Pierre d'Ailly; and others, and were discussed into the seventeenth century, especially by Jesuits. Leibniz himself rejected it, but thought it an ingenious theory (Duhem, pp. 317–323).

2. *De caelo* II.2.285b.23–25.

3. In Book II, Question 22, Buridan discusses "Whether the earth always is at rest in the center of the universe" (see Selection 67.4), and in Book II, Question 23, he considers "Whether the earth is spherical." These topics are treated by Aristotle in *De caelo* II.14.296b.7—298a.20.

the whole depth of the sea ought be to filled with earth, thus consuming the [portion of] earth that was elevated. In this way, water ought to surround the earth naturally, leaving no uncovered elevations. And this is also confirmed in another way, for just as in the proportion of the elements, the sphere of air greatly exceeds the sphere of water in magnitude and depth, so ought the sphere of water notably exceed the sphere of earth in magnitude and depth. But if this is so, the sphere of water ought to be elevated above the earth more than the highest mountains. Therefore, nothing ought to remain habitable.[4]

     .   .   .   .   .   .

Now, it remains to speak of uninhabitation caused by waters. There are three great opinions on this.[5] Some assume that only one quarter is habitable, while others assume that all quarters are habitable. And this [latter] opinion will be discussed first.

They say that both earth and water are concentric to the world, so that the center of the world is the center of both of them. However, they say that in any quarter of the earth there are many regions not covered by waters because of many protrusions of earth and elevations of mountains projecting above the waters. And they [also] say that many other parts of the earth are covered by waters because of their depressions, such as valleys between the aforementioned elevations. They say that is so in any quarter of the earth. The sign of this is that from one very large uncovered region, we cross a great and long sea and come to [yet] another very large uncovered region. It is probable that this would be the case as one went round the whole earth.[6]

But there are great doubts against this opinion.

The first is that all the seas which can be crossed and all the habitable lands which can be found are contained in this quarter of the earth which we inhabit. And some have tried to cross the sea into other quarters but they could never reach any habitable land. It is said, therefore, that Hercules fixed pillars at the limits of this quarter as a sign that beyond them there was no habitable land and no passable sea.

The other doubt, which has been previously stated, is more difficult, because if the world is eternal, this opinion cannot explain how the earth's elevations can be saved through eternity, since from these elevations many parts of earth always flow with the rivers into the depths of the sea. Through an infinite time the depths of these seas

ought to be filled and the elevations of the earth ought to be consumed, which is not a convenient thing to tell those who wish to maintain the eternity of the world in a condition as favorable to animals and plants as at the present time.

Hence, there was another opinion which assumed that for the well-being of animals and plants, God and nature ordained that the water be eccentric so that the center of the earth is the center of the world but the center of the [sphere of] water is outside the center of the world. And so they say that water always flows down toward a lower place, not with respect to the center of the earth or world but with respect to the proper center of the water. Thus it can be the case that one part of the earth, say a fourth, would not be covered with waters, while all the others exist covered with waters. And so they save the phenomena, because approximately one fourth of the earth only would be uncovered and habitable.

But, again, there remain two doubts against this opinion. The first is that although the world is governed by God, yet, speaking naturally [that is, in accordance with the laws of nature alone], it is by means of the heavens that the world is governed. Therefore, it would be necessary to assign the cause of this eccentricity to the heavens. For it could not be assigned to the earth, whose parts are like and homogeneous; nor to water, whose parts are also like and homogeneous. Nor indeed can it be assigned to the motion of the heavens, because the heavens are turned everywhere around earth and water indifferently. Therefore, no reason can be posited to explain why the center of water should lie more to one side than to another outside the center of the earth.

The second doubt concerns the manner in which one could save the mountain through an infinite time, since at all times many parts descend from higher places to lower places and few parts ascend or could be carried from low places to high places, especially if we wish to speak of very high mountains. In an infinite time, then, these mountains ought to be consumed.

For this reason, there is a third opinion, which

4. After discussing the five climatic zones, Buridan takes up the problem of this effect of water on the earth's surface.

5. As will be seen, the three opinions are that (1) the spheres of earth and water are concentric, (2) eccentric, and (3) concentric but with the earth's center of magnitude distinguished from its center of gravity.

6. Therefore, all quarters of the earth should be habitable.

seems probable to me and by means of which all appearances could be perpetually saved. Let both the earth and water be assumed concentric to the world so that the whole earth naturally collects around the center of the world and all water naturally flows down toward a lower place with respect to the center of the world. But there is much water in the bowels of the earth and through evaporation there is also much water mixed together with air. Hence it is not necessary that there be so much water in the sea that it should exceed the elevations of the earth.

But now we must ask how these elevations will be saved eternally. If, according to Aristotle, the world is assumed eternal, it could be replied that from eternity the world has been ordained for the well-being of animals and plants because one part of the earth, nearly a fourth, has not been covered by waters and rises above the waters; and it always remains, and will remain, naturally uncovered despite the concentricity and even though we might confine [or restrict] the mountains.

And there is a conception *(imaginatio)* that in the uncovered part the earth is altered by air and the sun's heat, and much air is mixed with it, so that this earth becomes rarer and lighter and has many pores filled with air or subtle bodies. However, the part of the earth covered with waters is not altered by air and sun and therefore remains denser and heavier. And therefore, if the earth were divided through the middle [or center] of its magnitude, one part would be much heavier than another, but that part which is uncovered would be much lighter. It seems, then, that there is one center of magnitude of the earth and another center of gravity. For the center of gravity is where the heaviness is just as much on one side as on the other, but as was said, this is not in the middle of the magnitude. Furthermore, since by its heaviness the earth tends to the middle of the world, its center of gravity is the middle of the world and not its center of magnitude. It is because of this that the earth is raised above the water on one side and is wholly under water on the other side.

But since the parts of the earth that are elevated flow with the rivers into the depth of the sea to the other side, then when you ask further how this elevation can be saved, it should be replied that if many parts of the elevated earth are moved to the depth of the sea at the other side, then they make heavier that side to which they flow; and the sun and air always act on what is left uncovered and render this the lighter part. Thus the part which

was previously the center of gravity will no longer be the center of gravity, and it will be necessary that the whole earth be moved and elevated toward the uncovered region. From this it follows that through a process of time the parts which are in the center of the earth will come to the surface of the habitable earth because parts which flow to the opposite side are continually removed from the earth. In this way the elevation of the earth is always saved.[7]

The generation of the highest mountains is thus also saved, because there are very dissimilar parts within the earth—as diggings show, for some parts are stony and hard, others are more soft and easily divisible. Therefore, since these interior parts of the earth are raised to the earth's surface in the way mentioned before, those which are soft and divisible are again moved to the depth of the sea by winds, rains, and rivers; the others which are harder and stony cannot be so divided and made to flow and so they remain and through the longest times are continually raised by the whole elevation of the earth and thus it is that the highest mountains can be made.[8] And if there were no mountains now, yet they could be formed in this way in

7. The parts of earth which are continually carried from the mountains to the depths of the sea by the rivers leave the elevated regions of the earth lighter and the regions covered by water heavier. Thus the earth's center of gravity continually shifts within the heavier parts. Since the center of gravity seeks to coincide with the geometric center of the world, the earth shifts toward the lighter, uncovered regions and in so doing pushes previously covered regions out of the water. Thus new lands are elevated. This unending process guarantees that the parts of earth near the center of gravity at the present time will be pushed by degrees toward the surface, eventually rising above the waters to the heights of the mountains that were previously eroded. Thus, as the center of the earth's gravity continually shifts to coincide with the geometric center of the world, the process of dissolution is always counterbalanced by the re-emergence of a small part of the earth.

8. As the soft stones weather away and are carried back to the sea, the hard stones remain and are repeatedly elevated as the center of the earth's gravity continually shifts. Over a long period of time, these aggregations of hard rocks will form high mountains. That part of Buridan's account of mountain formation which relies on weathering of soft stones and the relative permanence of the hard stones bears a resemblance to Avicenna's explanation of the accidental cause of mountain formation, for the latter stressed erosion of soft stones and the subsequent formation of valleys that leave the hard stones as eminences which finally develop into mountains. But Avicenna also emphasized earthquakes in mountain formation, a feature which is wholly lacking in Buridan's theory (see Selection 78). By way of contrast, the earth's

the future. There do not appear to be other ways by which such mountains could be generated and remain as mountains. Indeed, some assume that mountains are generated from motions of the earth by means of exhalations. But if this were true of some small mountains, yet it could not be true for the highest and tallest mountains because where

could so great an exhalation be confined which could elevate so much of the earth? This is hardly apparent. And [even] if so much earth could be elevated, yet after the exhalation had gone forth, it [that is, the elevated part of the earth] would again fall into its hole.

And thus the question is clear.

80 ON THE MATERIAL, HARDNESS, AND FISSILITY OF STONES

Albertus Magnus (ca. 1193–1280)

Translated and annotated by Dorothy Wyckoff[1]

BOOK I, TRACTATE I

Chapter 1: The Plan of the Book, and the Divisions, Method, and Order of Things To Be Discussed

This chapter is a general introduction, stating Albert's intention—to fill a gap in the existing series of books on natural science, coming after Meteorology and before the biological works[2]—and outlining his plan for the whole treatise. The authorities mentioned are cited again in later books as writers of lapidaries or alchemical works.

Mixing and hardening, and likewise solidification and liquefaction, and all the other ways in which things are acted upon, have already been discussed in the book on *Meteorology*.[3] Among natural things the first in which such effects appear are the stones and metals, and intermediates between these, like *marchasita* and alum and other things of that kind. And since these are the first compounds naturally formed from the elements, inasmuch as they come before the combinations[4] that are alive, they are the next subject to be discussed after the

its many errors of fact or interpretation of fact, there is something here that we recognize: the introductory exposition of general principles (the origin, physical and chemical properties of minerals), followed by descriptions of individual minerals (appearance, mode and place of occurrence, uses, etc.). This general pattern is still to be seen in our own textbooks (Wyckoff, p. xxxv).—*Ed.*]

2. [On p. xxvii, Wyckoff reproduces the order of Albert's commentaries on Aristotle's scientific treatises, an order which Albert himself formulated near the beginning of his commentary on the *Physics*. "Those marked with an asterisk (*) are directly based on corresponding works in the Aristotelian *corpus*. But we must remember that Albert never had a 'complete edition' of Aristotle. . . . *The Nature of Places* and *The Book of Minerals* were put together by Albert himself, when he failed to find any Aristotelian treatises on geography and mineralogy" (p. xxviii).

    Physics (Physica)
    The Heavens (De caelo et mundo)
    The Nature of Places *(De natura locorum)*
    Properties of the Elements *(De causis proprietatum elementorum)*
    Generation and Corruption (De generatione et corruptione)
    Meteorology (Meteora)
    The Book of Minerals *(Mineralia)*
    The Soul (De anima)
    Life and Death (De morte et vita)
    Youth and Age (De iuventute et senectute)
    Nourishment *(De nutrimento et nutribili)*
    Sleep and Waking (De somno et vigilia)
    The Senses (De sensu et sensato)
    Memory and Recollection (De memoria et reminiscentia)
    Movement of Animals *(De motibus animalium)*
    Breath and Breathing (De spiritu et respiratione)
    The Intellect *(De intellectu et intelligibili)*
    Plants (De vegetabilibus)
    Animals (De animalibus)

Albert's many commentaries on, and paraphrases of, Aristotle's scientific works were intended to assist his fellow Dominicans to obtain a total grasp of Aristotelian science.—*Ed.*]

3. *Meteor.* IV, which has been called the 'chemical treatise'.

4. *complexionata.* Albert (*Phys.* II, ii, 1) defines the terms used here: minerals are *commixta,* simple 'mix-

---

center of gravity, which plays so prominent a role in Buridan's description of geological processes, finds no counterpart in Avicenna's discussion.

It is obvious that Avicenna's account of mountain formation is far superior to Buridan's.

1. [Reprinted with the permission of the Clarendon Press, Oxford, from *Albertus Magnus Book of Minerals,* translated by Dorothy Wyckoff (Oxford: Clarendon Press, 1967), pp. 9–10, 11–14, 46–49, and 76–78. The translator has introduced each chapter with a brief introductory comment, which is retained along with her notes, renumbered here to maintain a successive order.

The Book of Minerals, written sometime after 1248, is divided into five books, dealing with stones (Bks I–II), metals (Bks III–IV), and "intermediates" (Bk V), which are "neither stones nor metals, but have some characteristics of both" (Wyckoff, p. xxxii), such as salt, alum, soda, and so on. "Taken as a whole, the *Book of Minerals* is an impressive attempt to organize a science of mineralogy. Despite its background of medieval thought,

Meteorology: for they seem to contain little except a simple mixture of elements. We have not seen Aristotle's books about these [minerals], but only some excerpts from them; and what Avicenna says about [minerals] in the third chapter of the first book which he wrote about them is not sufficient.

First, then, we shall investigate stones, and afterwards metals, and finally substances intermediate between these; for in fact the production of stones is simpler and more obvious than that of metals. Many things come to mind which ought to be said about the nature of stones in general, and these we shall put first. And then of course we shall discuss particular stones, such of them as have names. But we shall make our discussion of these brief, since the causes of many of the things that must be mentioned have already been determined in the *Meteorology.*

In treating of stones in general, we shall investigate their material and the immediate cause that makes them, and the place where they are produced; and then, the way in which they are mixed, and the cause of the variety of their colours and of the other accidental properties found in them—such as greater and lesser hardness, fissility and non-fissility, porosity and compactness, heaviness and lightness, and so on; for stones seem to have no small variety, not only in the specific nature and number, but even in the general character, of such properties.

There are indeed some men of the highest authority in philosophy who have treated of some kinds of stones, although not of all. Among these are Hermes, [Evax],[5] King of the Arabs, and Diascorides, Aaron, and Joseph; but they have treated only of precious stones, not stones in general. Even less satisfactory is the account given by Pliny in his *Natural History:* for he does not offer an intelligent explanation of the causes common to all stones. But we do not need to introduce the opinions of all these men, because knowledge of this subject is not so occult that we have to extract it from among the errors of many authorities. The nature and constitution of stones will be sufficiently well understood when we understand the material that is peculiar to them, the immediate cause that produces them, their forms, and their accidental peculiarities, according to the method of inquiry outlined in the fourth book of the *Meteorology.*[6]

· · · · · · ·

Chapter 2: The Material of Stones

Here Albert begins his discussion of the causes of

stones, taking first the material cause, that is, the matter of which they are made. His 'chemistry' is that of Aristotle's *The Heavens, Generation and Corruption,* and *Meteorology,* but its application to stones is based on Avicenna's *De congelatione et conglutinatione lapidum.* The formation of stones from Earth (gravel, sand, or dust) requires a 'gluing together' (*conglutinatio*) of the dry particles by moisture. Thus any stone, however hard and apparently dry, must contain some of this cementing Water: otherwise it would simply fall apart.

To begin, then, with our treatment of the nature of stones: we say in general that the material of all stone is either some form of Earth or some form of Water. For one or the other of these elements predominates in stones; and even in stones in which some form of Water seems to predominate, something of Earth is also important. Evidence of this is that nearly all kinds of stones sink in water: and so they must be rich in the material of Earth, as we have said in the science of *The Heavens.*[7] For if the lighter[8] elements were predominant in them, undoubtedly they would float on water. Now no kind of stone floats, unless it is spongy, or burnt and made spongy by burning, like pumice and the stone spewed out by hot springs and the fire of a volcano; and even of these, if they are reduced to powder, the powder sinks in water. Furthermore, if in transparent stones there were not something earthy mixed with the Water and imposing a boundary[9] on the moisture, they would not sink in water,

---

tures' of elements. Plants are *complexionata,* 'combinations' of humours. Animals are *composita,* organisms 'composed' or built up of tissues and organs. But his usage is not always strictly consistent with the definitions.

5. *Cuates* (or *euates* in texts of 1495, 1518) is an error for *Evax,* later cited several times, once as 'Evax King of the Arabs' (II, ii, 1, *Agathes*).

6. *Meteor.* IV, 8, 385 a 12ff. gives a classification of physical properties which served Albert as a model in discussing the 'accidental' properties of stones (I, ii) and metals (III, ii).

7. *The Heavens,* IV, 4, 311 a 15: heaviness and lightness are explained by the doctrine of 'natural places' and 'natural motions'. The natural place of Earth is at the center of the world and its natural motion is downwards towards the center. Therefore Earth, or anything composed mostly of Earth, will sink through all other elements; or, conversely, anything that does so sink must be composed mostly of Earth.

8. *superiora,* literally 'upper', since Fire and Air have their natural places above Water and Earth.

9. *Gen. and Corr.* II, 2, 329b 31 distinguishes between solid and liquid: a solid is *terminatum,* 'determined by its own boundary', but a liquid (or anything that is very

as rock crystal and beryl do; for ice and the other things that are entirely or chiefly made up of Water do not sink. And likewise, all stones that are produced in the kidneys and bladders of animals are made of a viscous, gross, and earthy moisture; and therefore something of the sort must be the material of stones.

In speaking in particular of those stones which are made of Earth, it is perfectly clear that in these Earth is not the only material, for this would not cohere into solid stone. For we say that the cause of coherence and mixing is moisture, which is so subtle that it makes every part of the Earth flow into every other part; and this is the cause of the thorough mixing of the parts of the material. And in that case, if this moisture were not soaked all through the earthy parts, holding them fast, but evaporated when the stone solidified, then there would be left only loose, earthy dust. Thus there must be something viscous and sticky, so that its parts join with the earthy parts like the links of a chain.[10] Then the earthy dryness holds fast to the moisture, and the watery moisture existing within the dryness gives it coherence.

Avicenna testifies to this when he says that pure Earth does not become stone,[11] since on account of its dryness Earth does not produce coherence, but rather a tendency to break into little pieces; for the dryness predominating in it prevents it from sticking together. The same philosopher explains that sometimes clay is dried out and becomes something intermediate between stone and clay, and then after a while it becomes stone. And again he says that the clay most suitable for transmuting into stone is unctuous, and the reason why that kind does not break into little pieces or crumble into dust is that its moisture is not easily separable from it.

Evidence of this is that in the stones themselves there frequently remain layers of Earth; it is hard dry Earth, and if it is compressed or pounded it becomes dust. And the cause of this simply that its moisture, which was not unctuous or viscous enough, evaporated when the stone solidified; and so the Earth was left hard and easily broken, because of the solidifying power of the surrounding stone. And there is still another evidence of this: for when stones are produced not in one continuous mass, but like timbers,[12] one above another, the earth in the intervening layers is not firmly united, but breaks into pieces, if subjected to pressure or a blow, and yet it is hard. And the cause of this we have stated above.

And that it is the viscous and unctuous moisture which gives chorenece to the material of stone is indicated by the fact that the animals called shell-fish[13] are very commonly produced with their shells in stones. These are extremely common in the stones found [at Paris],[14] in which there are many small holes shaped like the shells which some people call moon-shells.[15] For the cause of this is the moisture which has evaporated there; and being confined by the surrounding material, it rolled itself up, hardening first on the outside and coiling inwards, and received vital spirit, as we have said in the fourth book of the *Meteorology*.[16]

This then is the common material of those stones which are not transparent or nearly so. But there are many different kinds, as will appear in the following chapters.

BOOK I, TRACTATE II

Chapter 4: The Cause of the Differences in Hardness in Stones

Hard and soft are defined by Aristotle (*Gen. and Corr.* II, 2, 330 a 8) in terms of dry and moist, resistant or unresistant to pressure.

The accounts of the various stones are somewhat reminiscent of Vitruvius (*On Architecture*, II, 7, 1–3) on the hardness of building stones, and (II, 8, 2–3) on the failure of walls. But Albert is not really

---

moist) has no such 'boundary' of its own, but takes the shape of its container.

10. *Meteor.* IV, 9, 387a 12 uses the analogy of a chain to explain the consistency of viscous things: they do not fall apart into drops (like a liquid) nor into grains (like a friable solid).

11. *Terra pura lapis non fit* is the beginning of Avicenna's *De congelatione;* the next few sentences are also quoted or paraphrased from this. [For excerpts from Avicenna's treatise and a discussion of it, see Selections 74 and 78.—*Ed.*]

12. *asser*, a pole or beam: the outcrop, on a cliff or quarry face, of alternating strata of sandstone and shale is aptly compared to timber-work with the interstices filled with clay.

13. *testudo*, in classical Latin, is a tortoise; but Albert uses it for any kind of shell; cf. *Animals*, XXIV, 32, where *testudines* are snail shells.

14. *parvis:* evidently an error for *Parisiis*. Cf. Albert's *Properties of the Elements* (II, ii, 5) for another mention of these fossil moulds or impressions of shells *in lapidibus Parisiensibus*. The Paris Basin is made up of Cretaceous and Tertiary sediments, many of them richly fossiliferous.

15. *lunares:* perhaps the species of *Natica* still called 'moonshells'; perhaps less specific, since there was a general belief that some marine animals were directly affected by the moon (cf. Albert, *Animals*, XXIV, 32).

16. *Meteor.* IV, 1, 379b 7: spontaneous generation in putrefying material; but a closer parallel, relating to shellfish, is *Generation of Animals*, III, 11, 762a 19.

quoting from Vitruvius, nor from Pliny or Isidore, who repeat many of the same statements. Probably Albert, since he lived in the great age of cathedral building, acquired some of this information directly, by watching workmen and talking to them, either at Cologne or during his travels. He recorded one such incident at Venice (II, iii, 1).

We shall speak next about the differences in hardness, in which stones are found to differ greatly. All kinds of precious stones are so hard that the file removes nothing from them, and if struck forcibly against each other or against hardened steel, they emit fire. On the other hand, nearly all kinds of tufa are found to have so little hardness that they can be cut with an ordinary tool. And certain white stones which the common people call chalk,[17] and some which are even softer and whiter, are found to have less hardness than any other kind of stone. All kinds of flint[18] are very hard; and after these come the marbles, according to their kinds. Freestone,[19] however, is of intermediate hardness among stones, and yet some variation in hardness is found in this, too. But it commonly happens that the harder stones, if exposed to the air for a long time in cold weather, are later, in the sunshine, broken into many pieces. On the other hand, some which are less hard—unless they are poorly mixed, like quicklime solidified by baking heat *(optesis)*— the longer they are exposed to the air[20] in buildings, the better and harder they become and they are not broken by cold.

It is [the task] of natural science to assign causes for these accidental properties, based on the material and efficient causes, in the manner described elsewhere. Let us say, therefore, that the general cause of hardness is dryness. For since what is hard has a natural tendency to resist anything that touches it, and what is soft does not have this tendency, the cause can only be dryness, which stands firm and does not yield to anything else. Dryness is caused by two things in the nature of stones, as has already appeared: for either heat has dried the moisture out of earthy material, leaving it hard; or else very cold dryness has intensely attacked transparent moisture and in converting it to its own properties [of cold and dryness] has expelled the moisture, and by intensely compressing the material has hardened and highly compacted it. [This is the case] in transparent stones, and so they are very hard, and when struck emit fire; and they resist the file and must be polished, as it were, by grinding and rubbing.

But in the other stones made of the material of Earth, the cause of greater hardness is nothing but greater dryness, which is due to greater or lesser heat, acting as the efficient cause, and moisture, which is separable from the material with greater or lesser ease, acting as the material cause. For if the moisture is very unctuous it coheres easily; but if it is entirely watery, it evaporates easily. And therefore stones like chalk, or those softer than chalk, [which are] very white and leave a white streak on whatever they touch, have surely been mixed with a moisture highly susceptible to evaporation, and have been burnt by a heat exceeding [that which merely causes] solidification, and have already begun to be calcined.

Therefore they are not durable in walls. For because their dryness has been calcined they are always rough on the surface, which tends to separate from the grip of the cement, so that the stone as a whole is not held fast by the cement; and so these stones fall out of walls, and after a while, a wall made of them becomes like an earth wall. But flints are very hard because their moisture is not separable from the material and is thoroughly dried out and hardened by intense earthy dryness. And so they do not hold the cement well, because their pores are contracted and do not absorb it. And this is why stonemasons rarely use them in construction, and say that these stones ruin walls. Marbles are also very well mixed and intensely baked; and therefore they are hard and suitable for walls. But freestones are best of all for buildings; and when they are intensely hard they have a great deal of dryness and [only] a little moisture holding them together. And when [the moisture] is hardened by cold, it leaves the outside and retreats

17. *creta,* includes a number of white, soft, almost earthy materials. Pliny (*Nat. Hist.* XXXV, 57, 195–9) mentions varieties like fuller's earth and polishing powders. Isidore (*Etym.* XVI, 1, 6) abbreviates Pliny, adding (as usual) the origin of the name—from the island of Crete.

18. *silex* is always a hard stone. Vitruvius (*On Architecture,* II, 5, 1) says it can be burnt for lime, so some *silex* must have been hard limestone. But other kinds were probably lavas, since Pliny (*Nat. Hist.* XXXVI, 49, 168) says 'the fire does not harm them', a statement that could not apply to limestone.

19. *quadrum* would include a variety of sandstones and limestones, having a considerable range of hardness and durability. Some building stones of this type weather very badly, splitting along bedding planes if improperly laid.

20. Hardening in the air after quarrying is characteristic of many rather soft limestones, like those of northern France.

to the inside; and such moisture is not well incorporated with the parts of the stone and therefore it is easily transferred inwards or outwards; and so, after it has been forced into motion by cold, it is later dried out by exposure to the heat of the sun, and then the stone breaks into separate pieces. On the other hand, stones which are slightly moist, with the moisture firmly incorporated in their composition, are intensely dried out in the air; and therefore in buildings they become harder and better with the passage of a long time. In tufas the cause of their softness is moisture, which is neither completely removed nor very well mixed with the Earth; and therefore tufa is soft; and if exposed to fire, it is not baked hard like brick but is transformed into earthy ash.

These, then, are the statements made about the hardness of stones; and from these, other differences, too, can easily be understood.

Chapter 5. Fissility and Non-fissility in Stones

Fissility (*dolabilitas*) is the tendency of some rocks to break into thin slabs, either along bedding planes in sediments, or along planes developed by metamorphism in slates and schists. Albert does not seem to include here the cleavage of crystals, although he later mentions the fact that mica and gypsum can be split into thin pieces (II, ii, 17, *Specularis*). Rocks that are not fissile are *comminuibiles*—tending to break into little bits—brittle, friable, or crumbly.

The explanation, based on the role of 'pores', is taken from *Meteor.* IV, 9, 386 a 9–17; 386 b 26–387 a 3.

On this [same basis], the cause of fissility and non-fissility can be explained. For stones that are extremely hard are not fissile but have a tendency to break into little pieces; and since they do not have their pores arranged in rows they do not split straight. For just as in wood the knottiness depends upon variations in the flow of the [sap][21] by which the body of the wood is produced, so [the same thing] in stones depends upon variations in their mixture, and disorder in their materials; and that knottiness causes the stone to break irregularly and not straight. Nevertheless the hardest and driest stones, whether they are knotty or not, have a tendency to break into little pieces rather than to be fissile; for compaction or baking heat *(optesis)* has compacted and distorted the pores in them so that their capacity for splitting and parting

has been removed. But stones that have not been excessively compacted or hardened are fissile and can be cut straight; though, to be sure, they cannot be cut like wood, but rather by chipping away a little at a time, meanwhile leaving the rest of the stone untouched. This, then, is what fissility and non-fissility are.

And the very procedures of the stonemasons' art show this: for stonemasons cleave fissile stones straight, parallel to the whole surface,[22] but non-fissile stones, which tend to break into little pieces, they do not cleave parallel to the whole surface; but it is enough if the projections of the surface, which is not flat but rough, are in line. This is what the stonemasons of Lesbos are said to do; because in the island of Lesbos the only stones are those that tend to break into little pieces.

. . . .

BOOK II, TRACTATE II

Chapter 2. Those Beginning with the Letter called B

BERYLLUS

This is beryl. Pliny (XXXVII, 20, 76–79) correctly described it, noting several pale colours, yellow, green, and 'the colour of sea water' (that is, the variety now called aquamarine). Medieval writers, however, took the last to mean 'clear as water' and thus confused beryl with rock crystal ('water clear' quartz). Both minerals occur in hexagonal prisms, some of which are large enough to be carved into cups, balls, or simple lenses. Spectacles were invented in the thirteenth century and seem to have been named after beryl (German, *Brille*) though they were probably never made of beryl.

Beryllus (beryl) is a stone of a pale, clear, transparent colour; and therefore we have said above that when it is turned about, water can be seen moving inside it.[23] Mostly it is produced in India, as many other gems are. There are many kinds and varieties of this stone; but the better kind is said to be paler and to have more drops of water that

---

21. *sicci*, evidently an error for *succi*: cf. Albert's work on *Plants* (I, ii, 1) for discussion of the role of sap or juice *(succus)* in forming the grain and knots in wood. [This discussion is given in Selection 87.—*Ed.*]

22. That is, along a regular cleavage like that of slates.

23. Liquid inclusions with movable gas bubbles are sometimes found in beryl, but much more commonly in quartz. Albert had probably seen such a specimen, since this point is not in anyone else's description of beryl.

can be seen [moving]²⁴ inside it. It is said to be effective against peril from enemies and against disputes, and to give victory. It is also said to cause mildness of manner and to confer cleverness. Some medical men also say that it is good against sloth, and pains of the liver, and against shortness of breath and belching, and that it is good for watery eyes. For it is known by experience that when it is shaped into a ball²⁵ and is placed in direct sunlight, it burns, and kindles fire. The goldsmiths also say that it makes husband and wife agree in marriage.

Chapter 3: Those Beginning with the Letter C

CARBUNCULUS

Carbuncle includes almost all brilliant red, transparent gems. Latin *carbunculus* (diminutive of *carbo*) is a translation of Greek *anthrax*, 'a red-hot coal.' But these words came to be used in two different senses: (1) literally, a stone that burns (whence our anthracite); (2) figuratively, a fiery-red stone (carbuncle). A fusion of the two ideas may account for the ubiquitous tales of stones that shine in the dark. Of course some minerals do exhibit luminescence or phosphorescence after being rubbed, heated, or exposed to sunlight; and it is possible that Albert had seen 'such a one'; but it is also possible that he was deceived by some trick of coating a stone with phosphorescent material from fish or fungi.

Another interesting point is the attempt to subdivide carbuncle into subspecies. Pliny (XXXVII, 25–30, 92–104) names several varieties of *carbunculus,* one of which, *alabandicus,* had already achieved independent status in medieval lapidaries (see II, i, 1, *Alamandina*). Thomas of Cantimpré, like Albert, recognizes *rubinus* or *rubith* (our ruby), and *balagius* or *balastus* (our balas ruby, red spinel). Albert adds a third, *granatus,* our garnet (see II, ii, 7, *Granatus*); but his remarks about it are partly based on a misunderstanding of his sources (see note 29).

Carbunculus (carbuncle), which is *anthrax* in Greek, and is called *rubinus* (ruby) by some, is a stone that is extremely clear, red, and hard. It is to other stones as gold is to the other metals. It is said to have more powers than all other stones, as we have already said.²⁶ But its special effect is to disperse poison in air or vapour. When it is really good it shines in the dark like a live coal, and I myself have seen such a one. But when it is less good, though genuine, it shines in the dark if clear limpid water is poured over it²⁷ in a clean, polished black vessel. One that does not shine in the dark is not of perfect, noble quality. It is mostly found in Libya; and although there are several varieties, so that Evax says that there are eleven²⁸ kinds, nevertheless Aristotle, according to Constantine,²⁹ says that there are [only] three kinds which we have enumerated above—namely, *balagius, granatus,* and *rubinus.* And—what surprises many people—he says that *granatus* is the most excellent of these; but jewellers consider it less valuable.

24. *manere,* 'staying inside it'? More likely the reading should be *movere* (as above), since it was the *movement* of the bubbles that excited curiosity.

25. Thomas says 'round as an apple', so evidently a sphere was used as a burning glass.

26. In II, i, 2 where the colour of these stones is related to the Sun and Mars.

27. This statement certainly requires elucidation, and there are three possibilities: (1) Albert may simply be recalling an old story coming by way of Pliny (XXXVI, 34, 141–2, *gagates*) from Theophrastus (*On Stones,* 13), of a stone that burns when wet—possibly by spontaneous combustion in lignite or soft coal. (2) But if this is part of what Albert himself had seen, it was probably a trick with a 'doctored' stone. (3) He seems also to be alluding to a well-known optical trick: a small object is put into an empty basin, and the observer is placed so that the rim of the basin barely hides the object from his sight. Then the basin is filled with water, and the object can be seen (this was inexplicable to anyone who knew nothing of the refraction of light). Albert mentions this trick in *The Soul* (II, iii, 12), where he is again discussing this very topic of things that shine in the dark, and seems to think that the transparency of the water somehow 'retains the light' and makes the object visible.

28. But Evax (Marbod) says *twelve* kinds; probably a scribe wrote XI for XII.

29. Constantine (*Opera,* p. 352) was quoting the *Lapidary* of Aristotle, which distinguishes three kinds of *hyacinthus* (corundum gems), red, yellow, and blue, of which the red (*granatus,* 'like pomegranate seeds'), that is, rubies, are said to be the best. But Albert takes *granatus* to be the best of the *red* stones; and since his *granatus* seems to be garnet, of course it is surprising that it should be considered more valuable than ruby. . . .

Geography

81 THE IMAGE OR REPRESENTATION OF THE WORLD *(YMAGO MUNDI)*

Pierre d'Ailly (1350–1420), with the Marginal Comments of Christopher Columbus

Translated and annotated by Edward Grant[1]

The image of the world, or imaginary description of it representing it as in a physical mirror,[2] is not without use for the elucidation of Holy Scriptures, which frequently make mention of its parts and especially about the places of the habitable earth. It is because of this that I was led to write this treatise and thought it worthwhile to have gathered, briefly and truthfully, things that have been written diffusely by scholars on this subject.

[TABLE OF CONTENTS]

The treatise is divided into sixty chapters.

1. Pierre d'Ailly completed the *Ymago Mundi* on August 12, 1410, and it was printed at Louvain between 1480 and 1483 in a volume containing fifteen other treatises by d'Ailly and five by Jean Gerson. A copy of this volume was subsequently annotated in Latin by Columbus before his first voyage of discovery (this volume is today preserved in Seville). Although most of the annotations are mere repetitions of d'Ailly's text, they

will be translated here, so that the reader can see not only what Columbus deemed worthy of emphasis but also something of his views about world geography and thereby acquire a sense of the background to Columbus' historic voyages. It is noteworthy that of the books which Columbus read and annotated between 1485 and 1490, the *Ymago Mundi* was one of the most important. Indeed, it "seems to have been Christopher's bedside book for a number of years . . ." (Samuel Eliot Morison, *Admiral of the Ocean Sea* [Boston: Little, Brown, 1942], I, 120). These annotations will be presented as notes rather than in the margins, where Columbus wrote them.

The translation has been made from the Latin text in Vol. I of the edition by Edmond Buron, *Ymago Mundi de Pierre d'Ailly* (Paris: Maisonneuve Frères, 1930), which is accompanied by a facing French translation. Buron's notes have been extremely helpful in identifying d'Ailly's sources.

2. Many authors before d'Ailly had compared descriptions of the world to an image in a mirror, signifying by this that while the descriptions are not physically real, they can be taken to provide a perfect representation of reality.

The traditional and encyclopedic character of this treatise is nicely characterized by Buron (I, 152, n. 3; the translation is mine): "The beginning of this treatise is well known to all who have frequented the ancients even slightly. We do not venture to name any writers before d'Ailly who describe the world in the same manner as he did; for it would be necessary to enumerate some hundreds of Greek, Arabic, and Latin authors. What Isidore of Seville wrote had been said by twenty authors; what the Venerable Bede, or Honorius of Autun, or Sacrobosco, or Albert the Great, or Michael Scot, or Vincent of Beauvais, or Bacon has compiled has served for d'Ailly and all have reproduced Isidore, or Pliny, or Solinus, or Ambrose, or Jerome, or Macrobius, or Capella, or Aristotle, Ptolemy, Strabo, Hegesippus, etc. How can one recognize the contribution of each in these encyclopedic accumulations? Some idea, or even some textual phrase, of d'Ailly is found simultaneously in Pliny, Solinus, Josephus, Isidore, Bacon, Vincent, etc. Did our author borrow his phrase from Solinus or Isidore? We can only know this if d'Ailly, or some witness, informs us of it. In the present case, this is fortunately what happens. Our author does not confide all his secrets to us, although one must recognize in him much honesty and candor; but sometimes we can disclose his neglect by analysis, by comparison of texts, by fortunate errors committed by medieval translators or editors." It may be added that sometimes d'Ailly was an unscrupulous plagiarist (see n. 10).

3. I have altered the ordinal numbers of the Latin text to cardinal numbers.

CHAPTER 1

ON THE WORLD AND ITS PARTS IN GENERAL

The world has a spherical[4] or round figure with a variety of many parts. First, it is composed of four elements: earth, water, air, and fire. Then [it is composed] of nine heavens [or spheres] *(ex nonem celis)*, namely of the heaven of the Moon *(ex celo Lune)*, Mercury, Venus, the Sun, Mars, Jupiter, Saturn, the firmament, and the first mobile.[5] Beyond this certain philosophers assume a tenth immobile heaven; and beyond this there is

4. In the margin Columbus writes: [5] "A sphere is called a space, thus 'sphere of a sphere' *(spera spere)*, that is, a rotundity or body round on every side. And a sphere is [formed] when a semicircle is moved completely around a diameter that remains [at rest]; and a sphere is called a round space, and therefore the roundness of the world is called the sphere of the world because its spheres have been formed in the round. And everything round and of a round form is called a sphere because of its round form and roundness, just as the balls with which children play. 'Sphere' comes from the Greek, for the Greeks call something that is round a sphere from whence is derived *spericus, sperica, spericum,* that is round and solid like a sphere. And they say that the sphere of the sky is a certain kind formed in roundness whose center is the earth enclosed equally on all sides. This sphere is turned from east and west once in a day and night with such a swiftness that if the stars did not move against [or contrary to] its swift course and retard it, it would destroy the world. They [also] say that the sphere has neither beginning nor end. And thus it is not easy to understand where what is round, as a circle, begins and ends." [The next few words are incomprehensible, followed by a few lines of little interest, which are omitted.]

The number "[5]" preceding Columbus' comment signifies that this is his fifth marginal comment. All the comments, in this and other volumes annotated by Columbus, have been assigned numbers in order of occurrence. These will be repeated here before each comment. It should be noted that comment [5] is meant to apply to much of the material in this paragraph. Most of Columbus' marginal notes merely repeat phrases or sentences from the text of d'Ailly.

5. The sixty chapters of the *Ymago Mundi* are preceded by a lengthy preface containing diagrams and tables which are relevant to various chapters. A figure (not reproduced here) showing nine concentric spheres is presented as relevant to this first chapter. In his description of the planets and spheres in this diagram, d'Ailly

said to exist a crystalline heaven [or sphere] and then comes the empyrean heaven [or sphere], the outermost of all [spheres], where the abode of God is and the dwelling of the saints.[6] But, speaking naturally, [that is, in terms of natural phenomena alone] these last two do not fall within the scope of philosophers and astronomers. Few astronomers concern themselves with the tenth immobile heaven, but are rather concerned in their speculations with the other nine. In the first place, they assume that the ninth sphere, or first mobile [the sphere that is moved first], exists where there are no stars and that its motion is regular and uniform, moving around the whole earth in a natural day and carrying with it all the other spheres from east to west. After this heaven [or sphere] is the eighth sphere, which is called the firmament, that is, the starry sky, whose proper and natural motion is contrary to that of the first mobile and all the other spheres, namely from west to east. But it does not move [from west to east] absolutely, but moves in a contrary way with a certain declination. According to Ptolemy, this motion of the firmament is 1 degree in one hundred years made around the poles of the zodiac; and according to Thabit [ibn Qurra, this degree is acquired] on the first points of Aries and Libra. Thus its complete revolution is made in 36,000 years, which is called a Great Year.[7] Then come the seven spheres of the planets named above, which complete their revolutions in different times with a proper motion from west to east, as Saturn in around thirty years, Jupiter in almost twelve years, Mars in around two years, the sun, Mercury, and Venus in one year, and the moon in 27 days and 8 hours.[8]

Of the aforementioned spheres, astronomers have been chiefly concerned with the eighth sphere because its influence is powerful and strong, and in it, or under it, are found all the stars. Ancient wise men assumed that there were 1029 principal stars, namely the seven planets, which are called wanderers because their motions are diverse and on different circles, and 1022 fixed stars. . . .[9]

· · · · · · · ·

CHAPTER 4

ON THE FOUR ELEMENTS AND THEIR LOCATION[10]

After treating to some extent of the heaven, we must now consider the things that lie under it.

_____

seems to draw upon Nicole Oresme's *Traitié de l'espere* (edited by Laura McCarthy in a master's thesis, University of Toronto, 1942), a treatise written in Old French and on which d'Ailly relied heavily in the *Ymago Mundi*

(see n. 10). Since Oresme incorporated cosmological opinions of Aristotle, these are also reflected in the following description (Latin text in Buron, I, 126, 128):

"This figure is relevant to the first chapter. Only the nine celestial spheres which conform to the opinions of astronomers are described here; for Aristotle assumes only eight [spheres].

"Saturn is naturally cold and dry in its effect [on other things]; it is pale and of an evil disposition. Jupiter is hot and wet, clear and pure, thus tempering the maliciousness of Saturn. Mars is hot and dry, fiery and radiant, thereby harmful and provoking to war. The Sun is hot and luminous, distinguishing the times [and seasons], illuminating the stars, and greater than any [star]. Venus is hot and wet, most splendid amongst the stars, and always companion to the sun, called Lucifer when it precedes the sun [it is then a morning star—*Ed.*] and Vesper when it follows the sun [it would now be an evening star—*Ed.*] Mercury is radiant and keeps pace with the sun, being never more distant than 24 degrees. Thus it is rarely perceptible. The moon is cold and wet and the mother of waters; illuminated by the sun, it illuminates the night.

"The heaven [or sky] is not of the nature of the four elements, nor does it have any of their qualities, since it is not generable or corruptible; nor is it called hot except virtually *(virtualiter)*, since by its power *(virtute)*, it makes [things] hot. Nor is it properly colored, for it is clear [and lucid]; nor is it properly light or heavy; soft or hard; rare or dense. Only improperly is it called hard because it is unbreakable and impenetrable; and only improperly is it called dense or thick because a star is said to be the densest part of its sphere. Nor can it be moved slower or quicker than it is moved. Moreover, the spheres of the sky are unequal in width, not proportionally according to their magnitude, however, but [rather] according as the stars or planets [embedded] in them are greater or smaller."

Some of the special properties that were conferred upon the celestial region from Greek antiquity to the seventeenth century are described here by d'Ailly. It should be noted that in his fixed order of the planets, Venus and Mercury are placed below the sun, an order which agrees with Ptolemy (*Almagest,* Bk. IX, ch. 1) and Cicero (*Republic,* Bk. VI). See Selection 4, Macrobius.

6. Sacrobosco (*On the Sphere,* ch. I) lists the same nine spheres (see Selection 64.1) but ignores the tenth, the crystalline, and the empyrean heavens. By way of contrast to the view that located God at the empyrean heaven see the cosmological discussions in Selections 72 and 73, where it is assumed that He is everywhere inside the cosmos as well as beyond in an infinite void space.

7. That is, $100 \times 360 = 36,000$. On the Great Year, see Selection 51.2, n. 86, and Selection 69, n. 7.

8. The periods of the planets are probably derived from Sacrobosco, *On the Sphere,* chapter 1; see Selection 64.1.

9. The total of 1022 fixed stars derives ultimately from Ptolemy's *Almagest,* Book VIII. Since it was repeated frequently, d'Ailly could have derived it from many possible sources.

10. Although inverting the order and introducing some slight modifications, d'Ailly took this chapter from Nicole Oresme's *Traitié de l'espere (Treatise On the Sphere),* chapter 1. In his *Tractatus de legibus et sectis*

Immediately after [or below] the sphere of the moon, the philosophers place the sphere of fire, which is most pure there and invisible because of its rarity. Just as water is clearer than earth and air than water, so this fire is rarer and clearer than air, and so is the heaven [or sky] rarer or clearer than fire, except for the stars, which are thicker [or denser] parts of the sky so that the stars are lucid and visible. Afterwards is the sphere of air, which encloses water and earth. This is divided into three regions, one of which is the outermost (next to fire) where there is no wind, rain, or thunder, nor any phenomenon of this kind, and where certain mountains, such as Olympus, are said to reach. Aristotle says[11] that starry comets appear and are made there and that the sphere of fire and this supreme region of air with its comets are moved simultaneously with the heaven [or sky] from east to west.[12] The middle region [of air], however, is where the clouds are and where various phenomena occur,[13] since it is always cold. The other [and third] region is the lowest, where the birds and beasts dwell.[14] Then follow water *and* earth, for water does not surround the whole earth, but it leaves a part of it uncovered for the habitation of animals. Since one part of the earth is less heavy and weighty than another, it is, therefore, higher and more elevated from the center of the world. The remainder [of the earth], except for islands, is wholly covered by waters according to the common opinion of philosophers.[15] Therefore, the earth, as the heaviest element, is in the center or middle of the world, so that the earth's center is the center of its gravity; or, according to some, the center of gravity of the earth and also of water is the center of the world.[16] And although there are mountains and valleys on the earth, for which reason it is not perfectly round, it approximates very nearly to roundness. Thus it is that an eclipse of the moon, which is caused by the shadow of the earth,[17] appears round. They say the earth is round, therefore, because it approximates to roundness.

CHAPTER 5

ON THE SIZE OF THE EARTH AND ITS MEASUREMENT

Having assumed the roundness of the earth, one must presuppose for the measurement of its size that in going from one side of the earth to the other, one part of the sky after another will appear and that the horizon varies. If anyone should go from the north directly toward the south, the north pole would appear to him continually less elevated; and if he went the other way, it would appear to him

more elevated. And if he should go directly southward along one meridian such that the north pole were less elevated on the horizon by a twelfth part of the meridian, [then] he would have traversed a twelfth part of the earth's circuit; and if he went so that the pole appeared less elevated to him by 1 degree, he would then have traversed 1 degree of the earth's circuit, and there are 360 such parts in its circuit. It is by this method that the measure of the earth was found, because the elevation of one pole or one star is easily found by means of an astrolabe or quadrant.[18] Now, it is found by experience that if one goes toward the north so that the pole is more elevated by 1 degree, or directly toward the south so that it is less elevated by 1 degree, he will have traversed 700 stades of the earth. And according to the author of the *Sphere,* it is necessary to multiply this number by 360, so that we shall have 252,000 stades for the whole circuit [or circumference] of the earth,[19] which equals 15,750 leagues, since 1 league equals 2 miles,

---

contra supersticios astronomos (Treatise on the Laws and Sects against the Superstitious Astronomers), which is the third treatise in Columbus' edition of the *Ymago Mundi,* and one which he annotated, d'Ailly plagiarized much of Part III of Oresme's *On the Commensurability or Incommensurability of the Celestial Motions,* part of which is included in Selection 69.

11. *Meteorologica* I.7.344a.9–23.

12. Columbus: [8]"Mount Olympus, where comets are made."

13. Columbus: [8]"Where phenomena occur."

14. Columbus: [8]"Where the birds dwell."

15. Columbus: [8]"Water and earth together make a round body."

16. Columbus: [9]"The center of gravity of the earth and water together is the center of the world."

17. Columbus: [9]"An eclipse of the moon is caused by the shadow of the earth."

18. Columbus: [10]"The measure of the earth was found by this method."

19. Sacrobosco, *On the Sphere,* chapter 1 ("Measuring the Earth's Circumference"). See Selection 64.1. This estimate is derived ultimately from Eratosthenes (third century B.C.), who first arrived at a figure of 250,000 stades, which he subsequently altered to 252,000, probably because 1 degree would then be equal to 700 stades, a round number. The section from Cleomedes, *On the Orbits of the Heavenly Bodies,* Book I, chapter 10, which describes the methods of Posidonius and Eratosthenes in determining the measurement of the earth's circumference, appears in M. R. Cohen and I. E. Drabkin, *A Source Book in Greek Science* (Cambridge, Mass.: Harvard University Press, 1948), pp. 149–153. It should be apparent that Columbus did not discover the rotundity of the earth. Aside from a few trivial exceptions, Aristotle, Ptolemy, and almost all educated men in Greek and Roman antiquity and the Middle Ages believed in its sphericity.

and 1 mile equals 8 stades; 1 stade equals 125 paces;[20] 1 pace equals 5 feet; 1 foot, 4 palms; and 1 palm, 4 digits. It is obvious, then, that if one traveled 10 leagues per day, he could go round the earth in 1570 days, which equals 4 years, 16 weeks, and 2 days.[21]

Knowing the size of the earth's circumference, there is a given rule by which the size of its diameter can be known: subtract from the circle its twenty-second part and divide the remainder into three parts and the diameter equals one of these thirds. And so it is that the diameter equals 80,181 stades plus 1/2 and 1/3 of a stade,[22] which equals 2505 leagues plus 1/2 and 3 stades. Knowledge of this, although not completely precise, is, however, wonderful and useful, since it has no notable deficiency[23] and especially since Ptolemy and other wisemen have shown the size of the stellar bodies by means of the size of the earth; and by the size of the earth's radius (*dimidii dyametri*) they have shown the distance of the stars from the pole of the earth (*cuspide terre*).[24] But it is not our present intention to describe these quantities. It is sufficient here to note that the sun is much greater than any other star and that some say that the magnitude [or size] of its body is equal to more than 166 times that of the earth,[25] although some say, as does Isidore, that it is only 8 times greater than the earth.[26] After the sun, there come [in order of size] the greatest of the fixed stars, namely those in the first magnitude of stars, after which Jupiter is the greatest, then Saturn. The rest of the fixed stars follow in order of their degrees [of magnitude] until we reach the stars in the sixth magnitude; then come Mars, Venus, the Moon, and finally, Mercury.[27] Each of these stars is perceptibly larger than the earth, except for the last three, which are smaller than it.[28] This is obvious, because among all the stars perceptible to the sight, these three are smallest, since all the others are said to be greater than the whole earth. Any one of the stars in the first magnitude is approximately equal to 107 times the size of the earth; and any one of the stars in the sixth magnitude is [approximately] equal to 18 times the size of the earth.[29] From all this, it is also obvious that the philosophers say with good reason that the size of the earth is as a point in relation to the size of the heaven [or sky].[30]

CHAPTER 6

ON THE DIVISION OF THE WHOLE EARTH[31]

The earth is divided proportionally, just as the sky is divided into four quarters, by two colures and the equinoctial [circle or celestial equator];[32]

20. The value of the stade, in terms of feet, varied according to the different foot measurements employed (see Cohen and Drabkin, p. 150, n. 2).

21. This should be 4 years, 15 weeks, and 4 days.

22. That is, 80,181$\frac{5}{6}$ stades, the very figure given by Sacrobosco in chapter 1 of *On the Sphere* (see Selection 64.1). Using d'Ailly's figures, where π is equivalent to $3\frac{1}{7}$, the calculations yield 80,181$\frac{9}{11}$ stades.

23. Up to this point chapter 5 is, for the most part, a translation from French into Latin of chapter 26 of Oresme's *Traitié de l'espere*. Oresme had drawn some of his data from Sacrobosco's *On the Sphere.*

24. Columbus:[11] "They have shown the size *(quantitatem)* of the celestial bodies by the size (*quantitatem*) of the earth. By the size of the earth's radius they have shown the distance of the stars."

25. Probably derived from Roger Bacon's *Opus maius,* Part IV, where Bacon records that Thabit ibn Qurra "says that the sun is equal to the earth one hundred and sixty-six times." See Robert B. Burke's translation *The Opus Majus of Roger Bacon* (1928; New York: Russell and Russell, 1962), I, 256.

26. I failed to locate this reference in Isidore's *Etymologies* or *On the Nature of Things,* but the 8 to 1 ratio of sizes between sun and earth was given, perhaps for the first time, by Macrobius in his *Commentary on the Dream of Scipio,* Book I, chapter 20, [32]. In his translation of Macrobius' *Commentary* (New York: Columbia University Press, 1952), William H. Stahl observes that this estimate of the sun's size was widely adopted in the Middle Ages and cites as instances Helpericus of Auxerre (*De computo XX*) and Honorius of Autun (*De philosophia mundi* II. xxxii).

27. This particular order of the sizes of the celestial bodies was probably derived from Roger Bacon's *Opus maius,* Part IV ("The Usefulness of Mathematics in Divine Things"). It is identical with Bacon's arrangement except that between Mars and Earth, Bacon includes, and d'Ailly omits, all fixed stars that are perceptible to the sight but smaller than any in the first six magnitudes. Of these Bacon writes: "Then there are other stars in infinite number, the size of which cannot be ascertained by instruments and yet they are known by sight, and therefore have sensible size with respect to the heavens, like the part with respect to the whole" (Burke, I, 258).

28. That is, Venus, the Moon, and Mercury.

29. These size relations are identical with those given by Roger Bacon, *Opus maius,* Part IV (Burke I, 258).

30. Columbus:[12] "The size of the earth is as a point in relation to the size of the heaven [or sky]." D'Ailly probably took this from Sacrobosco, *On the Sphere,* chapter 1, where we are told that "the earth is as a center and point with respect to the firmament . . ." (see Selection 64.1). Ultimately it is traceable to Ptolemy's *Almagest,* Book I, chapter 6 ("That the Earth has the Ratio of a Point to the Heavens").

31. The bulk of this chapter is drawn from Oresme's *Traitié de l'espere.*

32. Columbus:[13] "The earth is divided in the same manner as the sky." The colures are two great circles intersecting at right angles at the poles; one great circle passes through the equinoctial points (that is, the points of intersection between the ecliptic and celestial equator) and the other through the solstitial points. The effect

and these circles on the earth are similarly divided by degrees and minutes, so that the whole circumference is 360 degrees and a fourth part is 90 degrees. Just as the sky, the earth is divided proportionally by four smaller circles into five unequal parts, namely by two circles, arctic and antarctic, and by two tropics. And the five zones on earth are said to be [related] proportionally just as [the zones] in the sky, because certain districts or regions on earth correspond to those zones. The first zone is located between the arctic pole and arctic circle;[33] the second between the arctic circle and summer tropic; the third between the summer tropic and winter tropic; the fourth between the winter tropic and antarctic circle; the fifth between the antarctic circle and antarctic pole.[34] Furthermore, the distance of these zones is unequal, because the distance of the pole of the zodiac [that is, the pole of the ecliptic] from the pole of the world [that is, the pole of the celestial equator] is 23 degrees, 51 minutes.[35] The maximum declination of the sun is the same as the distance between the pole of the zodiac and the pole of the world. Indeed, the arc cut off between the maximum declination of the sun and pole of the zodiac is 42 degrees, and 18 minutes.[36]

According to some, the first and last of these zones or regions of the earth are uninhabitable because of excessive cold, since they are so far from the sun.[37] The third zone, which is the middle zone, lies under the sun's path and is very near it; and therefore it is called the torrid zone and is said to be uninhabitable because of excessive heat.[38] The other two, namely the second and fourth, are neither excessively close nor far from the sun and so are temperate in heat and cold, and from this standpoint they are habitable unless something else should hinder them. In this division of the earth, I called that which is in the direction of the arctic pole, which always appears to us, the first zone, even though Aristotle, in assigning the differences of position in the sky says that the antarctic pole is the top of the world;[39] from this standpoint we are at the bottom and left side of the world because

of this intersection is to divide the sky into four quarters.

33. Columbus: [14]"There are five zones on the earth, as in the sky."

34. In order, the zones are (1) the North Frigid Zone; (2) North Temperate Zone; (3) Torrid Zone; (4) South Temperate Zone; (5) South Frigid Zone. This discussion of the five climatic zones is a great improvement over that given by Isidore of Seville in his *Etymologies,* chapter 44, and in chapter 10 of his *On the Nature of Things* (see Selection 1 and note 60 thereto). In his preface

d'Ailly includes a diagram to illustrate the five zones.

35. Columbus: [15] "The distance between the pole of the zodiac and the pole of the world is 23 degrees, 51 minutes." This is a value for the obliquity of the ecliptic —that is, the angle formed by the intersection of the poles of the ecliptic and celestial equator—which causes both the change of seasons and inequality of the five climatic zones previously described. Since the value for the obliquity of the ecliptic is approximately 23½ degrees (more precisely 23 degrees, 27 minutes; the value given by d'Ailly is 23 degrees, 51 minutes; see Sacrobosco *On the Sphere,* ch. 2, p. 447), the frigid zones lie within 23½ degrees of their respective poles (that is, between 90 degrees N and S latitudes and 66½ degrees N and S latitudes, respectively); the temperate zones lie within 66½ degrees N and S latitudes and 23½ degrees N and S latitudes, respectively, so that each temperate zone extends over an arc of the earth's circumference subtended by an angle of approximately 43 degrees; finally, the torrid zone, extending from the tropic of Cancer, which is 23½ degrees N latitude, to the tropic of Capricorn, which is 23½ degrees S latitude, embraces 47 degrees of latitude.

36. Columbus: [15]"From the tropic of Cancer to the arctic circle there are 42 degrees, 18 minutes." Using 23 degrees, 51 minutes, d'Ailly's figure for the obliquity of the ecliptic, and observing that 23 degrees, 51 minutes, would be the extent of the North Frigid Zone and 23 degrees, 51 minutes, the angle subtending the arcal distance between equator and tropic of Cancer, we would, upon adding these two figures, obtain 47 degrees, 42 minutes. Subtracting this from 90 degrees, we get 42 degrees, 18 minutes, the angle subtending that portion of the earth's circumference falling in the temperate zone, or the angular distance between the arctic circle and the tropic of Cancer.

37. Columbus: [15] "The Frigid Zones." For emphasis, Columbus enclosed this remark in a rectangular box.

38. Columbus: [16]"The Torrid Zone is not uninhabitable *(non est inhabitabilis),* because today the Portuguese navigate through it; indeed it is very populated. Below the equator there is the fort of Mine belonging to the Most Serene King of Portugal and which we have seen." Perhaps in order to emphasize his disagreement with d'Ailly and traditional accounts of the uninhabitability of the torrid zone, Columbus drew a rectangular box around this marginal comment. Apparently oblivious to his opinion here, d'Ailly remarks later on that experience shows that the region below the first clime is inhabited. See beginning of chapter 1 and note 60, where Columbus notes d'Ailly's opinion without drawing attention to the conflicting judgment given here.

39. Columbus: [16]"Aristotle. The antarctic pole is at the top of the world." Aristotle argued that the celestial sphere had a top and bottom and a right and left. The celestial pole furnished top and bottom, and in a seemingly strange argument, Aristotle insisted that the north (or visible) pole is the bottom and the south (or invisible or antarctic) pole the top. "Clearly therefore the invisible pole is the upper, and those who live in the region of it are in the upper hemisphere and to the right, whereas we are in the lower and to the left" (*On the Heavens [De caelo],* II.2.285b.23–25). For an analysis of this rather unimportant point, see Thomas L. Heath, *Aristarchus of Samos* (Oxford: Clarendon Press, 1913), pp. 231–232.

we dwell toward the west and the arctic [or north] pole. But if we consider the proper motion according to the zodiac, which is carried along in a contrary motion, as has been said, then the arctic [or north] pole is above, because this motion is perhaps more noble, and the spheres which are so moved are more noble, and the greater multitude of noble stars is in this half of the sky towards the arctic pole, at least with regard to the half [of the sky] that appears visible to us. In accordance with this, we should be in the superior and nobler part and to the right. Thus the first zone should be towards the arctic pole.[40]

. . . .

CHAPTER 8

ON THE SIZE OF THE HABITABLE EARTH

In order to investigate the size of the habitable part of the earth, it must be understood that "habitable" is taken in two ways. In one way it is taken with regard to the sky, namely how much can, and how much cannot, be inhabited because of the sun.[41] This was discussed above in general and enough was said. In the other way, "habitable" is considered with respect to water, namely as water limits [or prevents] habitability;[42] and we must now consider this on which wise men offer various opinions.

For Ptolemy, in his book *On the Arrangement of the Sphere (De dispositione sphere)*[43] would have it that about a sixth part of the earth is habitable because [of an absence] of water and all the rest is covered by water. But in his *Almagest,* Book II, he posits the known habitation extends over only one fourth of the earth, namely over the part on which we dwell, which extends in longitude from

the basic objective of his first voyage. As we shall see, chapters 8, 10 and 11 were quite useful in this regard.

Columbus: [24] "Note that in this year of [14] 88, in the month of December, Bartholomew Diaz, captain of three caravels which the Most Serene King of Portugal had sent to Guinea to take land, landed in Lisbon. He reported to the Most Serene King that he had sailed beyond the Yan, having traveled 600 leagues, namely 450 to the south and 250 to the North, and that he had reached a promontory to which he gave the name of "Cape of Good Hope," which we thought to be in Agensinba. By means of an astrolabe he found that in this place he was 45 degrees beyond [or south of] the equinoctial line [or equator] and that this place was 3100 leagues from Lisbon. He pictured and wrote about this voyage league by league on a navigation chart so that he could show it to the eyes of the Most Serene King. I was present at all these things.

"This agrees with the statement of Marinus [of Tyre], whom Ptolemy corrected with respect to his voyage to the Garamantes, where Marinus said he had travelled 27,500 stades beyond the equator. Ptolemy rejects and corrects this. He agrees with Pierre d'Ailly that the waters do not cover three fourths of the earth.

"This agrees that the sea is wholly navigable and that excessive heat is no impediment."

John K. Wright (*The Geographical Lore of the Time of the Crusades* [New York: American Geographical Society, 1925], p. 41) suggests that the country of the Garamantes is the more modern Fezzan, located in West Central Libya and forming part of the Sahara desert. He goes on to say that "Ptolemy mentions two expeditions that had been made at an unknown period to the south from the land of the Garamantes, one under Septimius Flaccus, who arrived at the country of the Ethiopians after three months' journey, and the other under Julius Maternus and the king of the Garamantes, a four months' journey to a country called Agisymba [probably identical with "Agensinba" mentioned above by Columbus—*Ed.*], abounding in rhinoceroses." It seems that Columbus believed that Bartholomew Diaz's voyage had penetrated as far south as the expedition from the Garamantes to Agensinba (thought by Columbus to include the Cape of Good Hope) described by Ptolemy. Marinus of Tyre (ca. 90–130), a Greek geographer from whom Ptolemy drew much detailed information, had made the earth excessively large, so that Ptolemy found it necessary to reduce its extent on the basis of an estimate of 180,000 stades as the earth's circumference. In this marginal comment Columbus cites one instance in which Marinus had greatly exaggerated the estimate of the distance that he had traveled to the Garamantes and where Ptolemy had rightly corrected him. Indeed, the Garamantes, if they really lived in Fezzan, would have been at 27 degrees N latitude, well above the equator.

The broad significance of all this for Columbus lies in the fact that Diaz's voyage not only confirmed the ancient reports of habitation beyond the equator but also reinforced his belief that the earth is a relatively small and navigable sphere.

43. I have not found such a title amongst the known works of Ptolemy. Perhaps it was some part of one of his treatises which circulated independently.

40. Columbus: [17]"The author. The arctic pole is above." D'Ailly's disagreement with Aristotle is based wholly on arguments presented by Oresme in *Traitié de l'espere.*

41. See the preceding chapter. Although slightly rearranged and altered, this eighth chapter is based almost wholly on Roger Bacon's *Opus maius,* Part IV (Burke, I, 310–312).

42. Columbus was greatly interested in arguments demonstrating that the sea between Spain and India was a relatively small one which could be navigated in a short time. The *Ymago Mundi,* which was a world geography used widely prior to the reintroduction of Ptolemy's *Geography* in the fifteenth century, was invaluable to Columbus because it furnished arguments which he could subsequently use to convince others of the feasibility of his plans to sail westward and reach the Orient,

east to west, embracing half of the equinoctial circle [that is, equator], and the whole latitude extends from the equator to the [north] pole, covering one fourth of the colure.[44]

But at the end of his book *On the Heavens and World,* Aristotle says that it covers more than one fourth, and Averroes confirms this.[45] Aristotle says that a small sea lies between the end of the western side of Spain and the beginning of the eastern part of India.[46] But he was not speaking of Hither Spain, which is now commonly called Spain, but [rather] of Farther Spain, which is now called Africa [and] of which certain authors, as Pliny, Orosius, and Isidore have spoken.[47] Moreover, Seneca, in the fifth book of *Natural Things* [that is, *The Natural Questions*] says that if the wind is favorable, this sea is navigable in a few days. And Pliny, in the second book of the *Natural History,* teaches that some have navigated from the Arabian Gulf to the Pillars of Hercules [that is, Straits of Gibralter] in a time that is not very great.[48] For these and other reasons, which I shall consider further when I speak of the ocean, some conclude, apparently, that the sea is not so great that it could cover three fourths of the earth. To this can be added the authority of Esdras, who in his fourth book says that six parts of the earth are inhabited and the seventh is covered by water.[49] The saints hold in reverence the authority of this book and the sacred truths confirmed by it. And so it seems that although the inhabited part known to Ptolemy and his followers was confined below one fourth, more than that is habitable.[50] But

---

44. For Ptolemy's estimate that only one fourth of the world is inhabitable (extending over 180 degrees of longitude and 90 degrees of latitude), see the *Almagest,* Book II, chapters 1 and 6. Columbus sought to discredit Ptolemy's estimate, since it implied that the oceans would be too large and formidable to navigate.

45. Neither Aristotle nor Averroes explicitly declares that the inhabitable part of the world occupies more than one fourth of the earth's surface.

46. Columbus: [24]"Aristotle. Between the end of Spain and the beginning of India there is a small sea navigable in a few days." Actually Aristotle says only that some hold that the Pillars of Hercules (Straits of Gibraltar) and India are connected, since elephants are found at both ends. Nothing is said about the sea being small or navigable. See *On the Heavens,* II.14.298a.9–15.

Columbus eagerly seized upon all mentions of a narrow ocean in the *Ymago Mundi.* These he annotated or copied into the margin. See S. E. Morison, *Admiral of the Ocean Sea,* I,122–123, for a list of such instances. See also notes 70, 71.

47. According to Roger Bacon, "We must know that Spain in this locality is considered not as hither but as farther Spain, of which certain authors speak, as Pliny in the *Questions on Nature* [i.e., *Natural History*], and Merlinus in his prophecy, and Orosius in his book *Ormesta Mundi,* and Isidore in the fourteenth book of the *Etymologies*" (Burke, I, 312). In his comment quoted in the preceding note Columbus (like Aristotle; see n. 46) says nothing about the distinction between Hither and Farther Spain. According to Bacon (Burke, I, 312), "Between the Spain so called at present and the Africa so named at this time water did not flow, but the land was continuous in times past, but later the ocean broke through in the low parts of the earth and joined the Tyrrhenian Sea [i.e., the Mediterranean], which flows by the coast of the province of Arragone and of Italy. Therefore Hither Spain extends from the Pyrenees Mountains to Carthage: but Farther Spain crosses the Strait of Gibralter up to the provinces of Africa. Hence it extends beyond Gibralter and reaches the Atlas range." Bacon goes on to explain that although Aristotle did not distinguish between Hither and Farther Spain and spoke only of Spain in general (actually he spoke only of the Straits of Gibralter or Pillars of Hercules), his allusion to elephants applies only to Farther Spain (see n. 46), or Africa, since "it is a fact that elephants abound near the Atlas range." In this manner, Bacon and his predecessors tried to make sense of "elephants in Spain."

48. Columbus: [24]"Pliny. In a not very great time, the Arabian Gulf to the Pillars of Hercules *(Gades)* has been navigated."

49. Columbus: [24]"Esdras. Six parts are inhabited and the seventh is covered with waters" (III Esdras 6:42). The third and fourth books of Esdras have been considered apocryphal. Since this statement by Esdras was very favorable to Columbus' claims that a voyage westward would be easily feasible, he sought to confer great respectability and reliability on Esdras (see the next note).

50. Columbus: [24]"Note that blessed Ambrose and Aurelius Augustine and several others took Esdras as a prophet and approved his book, as the extracts from his books (below) reveal; for they do not seem to be apocryphal.

"And with respect to the water, the opinion of Peter the Eater *(Petrus Comestor)* agrees; for if we consider what we know at present through navigations, this is found to be true. [The last two paragraphs were enclosed by Columbus in a single rectangular box.—*Ed.*].

"St. Augustine; *The City of God.*

"Franciscus de Mayronnis in his [book on] [*Theological*] *Truths* says in the twelfth truth that after the return of the Jews from Babylon they did not have any prophets after Malachi, Haggai, Zechariah, and Esdras, until the coming of the Savior unless [one counts] Zechariah, father of John, old Simeon, and the most glorious John the Baptist. Finally, in this chapter [Mayronnis says that] a noteworthy thing is understood, [namely] that Esdras was a prophet and his prophecies, although not canonical, seem authentic. And this is confirmed, since blessed Ambrose admits it by accepting the utterance of this prophecy: My son Jesus died and the world is converted.

Aristotle, with the aid of Alexander, was able to know more about this, as was Seneca, with the aid of Nero; for they had been stimulated to investigate doubtful things about this world. Pliny, in his eighth book,[51] and Solinus[52] also testify about Alexander; and Seneca tells about Nero in his book on *Natural Things*.[53] It seems that they ought more to be believed than Ptolemy or even Albategni [al-Battani], who assumes that even less of the earth is habitable, namely only a twelfth part;[54] but this lacks proof, as could be shown, but by reason of brevity I pass over this, because even from what follows this will be obvious.

Thus, from what has already been said and what will be said below, it appears that the habitable earth is not round as a circle, as Aristotle says, but is as the fourth part of the surface of a sphere, of which fourth the two extreme parts are somewhat cut off, namely those which are not habitable because of excessive heat or cold. But this cannot be as suitably figured on a plane [map] as on a spherical [one].[55]

· · · · · · ·

CHAPTER 10

ON THE LONGITUDE AND LATITUDE OF THE CLIMES[56]

Every clime has more longitude than latitude, because longitude proceeds from east to west and latitude from the south toward the north. According to astronomers, the first clime contains in longitude half the circumference of the earth, which some measure by stades, as was said above. But Alfraganus and certain others measure it by miles[57] and say that any degree of the earth's circumference contains $56\frac{2}{3}$ miles,[58] so that the whole circumference contains 20,400 miles and half would be 10,200 miles, which, if 1 league is reckoned as 2 miles, equals 5100 leagues, the length of the first clime. The second clime would be shorter than the first and the third less than the second, and so on, in ascending northward. But authors do not assume certitude for the measure of longitude of these climes. The longitude [or length] of these climes has been found by the aspect at [the time] of eclipse of the moon.

The latitude [or width] of these climes is measured from south to north. The beginning of the first clime is 12 degrees and some minutes this side of the equator, which according to Alfraganus, is equivalent to $712\frac{1}{2}$ miles, which equals $356\frac{1}{4}$ leagues. . . .

CHAPTER 11

ON THE ANTECLIMES AND POSTCLIMES

From what has been said, the climes when taken together are terminated by two lines, one of which begins the first clime and is $12\frac{1}{2}$ degrees from the equator,[59] the other terminates the seventh clime. Some designate as "anteclimes" areas

---

"Peter the Eater, master of ecclesiastical histories says: The third day God brought together the waters into one place under the firmament even though they could acquire more places. However, since all were continuous in the bowels of the earth, it could be said that they were capable of being brought together, so that the waters, which occupied the whole space of air when in a vaporized state occupied a small space in a solid state; or the earth rises little by little until it encloses the waters as in a vessel; and so the earth appears dry." I have omitted the last section, which continues in the same vein. For the various references to Franciscus Mayronnis (a scholastic at the University of Paris who died in 1325), Peter the Eater, and others, see the appropriate footnotes in Buron I, 210–215.

51. *Natural History* VIII, 17.

52. *Polyhistor seu de mirabilibus mundi IV*. The *Polyhistor* of Solinus, who lived in the third century A.D., was a geographical work based largely on Pliny.

53. *Natural Questions* VI, 8.

54. Here again, d'Ailly aids Columbus by arguing that the waters on the earth are less extensive than some prominent authors had allowed. In *Traité de l'espere*, chapter 37, Oresme had also cited Albategni's theory.

55. This last paragraph is taken from Oresme, *Traitié de l'espere*, chapter 31.

56. This chapter is largely a translation of Oresme's *Traitié de l'espere*, chapters 33, 34. In chapter 9, d'Ailly had discussed the division of the inhabited part of the earth into seven climes. This was accepted by almost all authors and is found, for example, in Alfraganus (*Elementa*), Sacrobosco (*On the Sphere*), Roger Bacon (*Opus maius*) and Oresme (*Traitié de l'espere*). Here d'Ailly continues the discussion of the seven climes. See also Selection 1, notes 58 and 59.

57. For instance, Sacrobosco and Oresme.

58. Columbus: [28] "Every degree contains $56\frac{2}{3}$ miles; therefore the whole circuit [or circumference] of the earth is 20,400 miles." S. E. Morison observes that "Columbus notes every mention of the length of a degree, and sometimes writes in the margin 'Not so. A degree is $56\frac{2}{3}$ Roman miles,' or words to that effect." (*Admiral of the Ocean Sea*, I, 122). In arguing for the feasibility of a voyage westward, this value of a degree produced a sufficiently small earth to lend credibility to Columbus' claims. In a part of the present chapter omitted here, Columbus, in marginal comments 30 and 31, emphasizes that $56\frac{2}{3}$ miles per degree is an exact measurement (interestingly, these comments are enclosed in boxes). This particular figure was arrived at by the Arabs in the ninth century.

59. That is, north of the equator.

inhabited beyond the first line and as "postclimes" those areas beyond the second line. Now, it has been proven by experience and the authority of many that not only are the seven climes inhabited, but more than one fourth of the earth contains nations of men, so that authors assume more inhabited areas beyond [or below] the first line.[60] According to Pliny, in the second, fifth, and sixth books,[61] and Ambrose in the fourth book of his *Hexameron*,[62] there are, in parts of the south, some who are without shade for two days of the year because they have the sun over their heads and are illuminated everywhere by its revolution. For this reason they are called "Ascii",[63] that is, without shade, and "Amphiscii,"[64] that is, surrounded by shadows. And these people are below the equator and, according to Pliny,[65] are called "Orestes," "Menodes," and "Symarians." Located among them is a mountain named Malk on which the shadows are changed every six months. And yet beyond the equator, Haly places two cities,[66] one of which is called Deleyt, whose longitude is 122 degrees and whose latitude is 3 degrees; and the other is called Baraya, whose longitude is 125 degrees and whose latitude 3 degrees. Moreover, according to Pliny, we find that there is habitation under the tropic of Capricorn and beyond. Thus there is a region called the island of Pathalis possessing a famous port where shadows fall only to the south. Therefore, those who dwell in this place have the sun always to the north.[67] And in the sixth book [of his *Natural History,* Pliny] says[68] the same thing about the island of Taprobane [Ceylon], whose inhabitants who came to Rome during the principate of Claudius were amazed that their shadows fell to the north and that they rose to the south.[69] According to Albategni, this island is opposite India toward the east. . . .[70]

As for the line which terminates the climes toward the north, there is much inhabited [land] beyond it, as England, Scotland, Dacia, Norway, and several other regions, the last of which, according to some, is the island of Thule.

And according to Aristotle and Averroes at the end of the second book of *De caelo et mundo,* the end of the habitable earth toward the east and the end of the habitable earth toward the west are very close and there is a small sea between;[71] but its latitude over the earth extends over a distance greater than half the circumference of the earth. Thus, if, as some authors say, the climes are extended to the western extremity but do not extend

in length beyond half the circumference of the earth, it follows, according to Averroes, that the climes do not extend to the eastern extremity but that there are large inhabited areas beyond, namely beyond the climes toward the east. It is necessary, therefore, that the climes, or some of them, be longer than the astronomers suppose.[72] And this briefly is what I want to say about those places which are beyond the climes.[73]

60. Columbus: [33]"The author proves that there is habitation under the equator and beyond the tropic of Capricorn." See chapter 6 and n. 38, where d'Ailly denies the habitability of the torrid zone. D'Ailly's source for much of this chapter was Roger Bacon, *Opus maius,* Part IV. Oresme, *Traitié de l'espere,* chapter 38, seems also to have been utilized.

61. *Natural History* II, 8; II, 75, 76; VI, 24, 34, 39.

62. *Hexameron* IV, 5.

63. Ambrose gives the Greek term *askioi,* which means literally "those who are without shade."

64. Here again, Ambrose gives the Greek term *amphiskioi,* which, in the Greek lexicon of Liddell and Scott, is defined as "throwing a shadow both ways, sometimes north, sometimes south, of those who live within the tropics." Thus the inhabitants of this region are sometimes shadowless (see n. 63) and for the rest of the time cast shadows sometimes north and sometimes south; hence are "surrounded by shadows."

65. Pliny, *Natural History* II.112.5; II.75.3.

66. Haly, *De judiciis astrorum (On the Judgments of the Stars),* IX, 37; page 406 of the edition by Peter Liechtenstein, Basle 1551 or 1571.

Columbus: [34]"Two cities lie beyond [or below] the equator, one is called Deleit, the other Baraya."

67. Colubmus: [35]"Beyond [or below] the tropic of Capricorn, the region of Pathalis is found which has a port where shadows fall only to the south and the sun is always to the north."

68. VI, 22 and 24.

69. Columbus: [35]"Men came from Taprobane to Rome."

70. Columbus: [36]"Taprobane is opposite India; [and] since it is toward the east opposite [India], it has nations [or peoples]."

[37]"Note that if Taprobane is situated as described above, it would be distant from true west by 58 degrees to the west; for this reason we approve [the opinion] that the sea which separates Spain and India is small" (see n. 46 and n. 71).

71. Columbus: [43]"The limit of the habitable earth in the east and the limit of the habitable earth in the west are quite close and there is a small sea between" (see n. 46 and n. 70).

72. Columbus: [44]"The climes are longer than the astronomers suppose."

73. Condensed and somewhat rearranged, the last two paragraphs are closely akin to chapter 38 of Oresme's *Traitié de l'espere.*

Oceanography

82 ON THE CAUSES OF THE TIDES

Robert Grosseteste (ca. 1168–1263)

Translated by Richard C. Dales[1]

Annotated by Edward Grant

AN INQUIRY INTO THE CAUSES OF THE TIDES

I

In our investigation of the tides, let us first determine the material cause, which is twofold, namely general and specific.

The general cause is this: The spheres of the four elements are so arranged that earth is in the center and fire on the periphery, while water and air occupy places between these two. These elements are of such a nature that they can be moved by rarefaction and condensation. But water and air share this capacity to a greater extent than fire and earth; therefore they are better suited to be moved. The minor premise is proved thus: Condensation is the motion of the parts of matter toward the center of the universe, and rarefaction is the motion of the parts of matter toward its periphery. Therefore, that which is in the very center, such as earth, is not further condensable; and that which is at the periphery, such as fire, cannot be further rarefied. However, those things which are between the center and the periphery, such as water and air, can be rarefied and condensed.

Now for the specific material cause: Every motion of a generable or corruptible thing takes place in water and air. Let us omit the motions of the air and speak about the motions of water.

This investigation has three parts. The first part concerns the material and efficient cause of such motions. The second part concerns the reasons for the increase and decrease of the tides. The third part concerns the three kinds of seas: those which have tides, those which do not have tides, and those which have very small tides but do not seem to have any.

Turning our attention to the first part, let us first determine the efficient cause. This efficient cause must necessarily be a power of the sky or the power of a star in the sky. For an element is not moved by itself or by another element, but we ought to reduce every motion of the elements to an efficient cause which is an immaterial species.

In this connection, Alpetragius[2] says that all the lower spheres, as far as the sphere of water, are moved from east to west by the power of the outermost sky. But the lower any particular sphere is, the less power it will receive from the outermost sky, because this power is the power of a body and diminishes with distance. The earth, however, both because it is farthest away from the outermost sphere and because of its heaviness, remains completely immobile. Therefore, the water of the sea is moved by the power of the outermost sphere from east to west, and from this, Alpetragius says, a crashing together of the waters occurs and consequently a high tide. But the water returns to its original place because of its heaviness, and when it has all returned it begins to be moved again and to rise as far as its heaviness will permit, and then again it returns. And these two high and low tides take place in a time greater than one day and night.

That this is false is clear thus: In the outermost sky, the parts which are moved most rapidly are the parts existing on the equinoctial circle, directly under which, on the surface of the earth, is the great sea called Ocean, surrounding the land; and this is the source of all seas.[3] Therefore, if in one part of the sky there were a motive force and in

1. Reprinted with the kind permission of the editors of *Isis* from Richard C. Dales, "The Text of Robert Grosseteste's *Questio de fluxu et refluxu maris* with an English Translation," *Isis,* LVII (1966), 468–473.

2. That is, al-Bitruji (fl. 1185), a Spanish Muslim whose Arabic astronomical treatise was translated into Latin by Michael Scot in 1217 with the title *On the Celestial Motions (De motibus celorum)*. Grosseteste is here summarizing chapter 4, paragraph 5, page 81 of the edition published by Francis J. Carmody (Berkeley: University of California Press, 1952). *On the Celestial Motions* was an attempt "to explain the constitution of the universe as it really is, and not merely to represent the motions of the planets geometrically, so as to be able to foretell their places in the heavens at any time; . . . The leading idea is that of the homocentric spheres, each star being attached to a sphere, and the motive power is the ninth sphere, the sphere outside that of the fixed stars" (J. L. E. Dreyer, *A History of Astronomy from Thales to Kepler,* 2d ed. [New York: Dover, 1953], p. 265; for the full context of this quotation, see Selection 68, n. 24).

3. Compare Macrobius' views, Selection 6.

another part there were not, when this part of the sky rose one part of the sea would be moved by it and another would not. If, therefore, one part of the sea should be moved and another not, the one part would resist the other, and one part would have to be elevated above the other and occupy a greater place. But since the sea is circular, and the motive power operates equally in every part of the equinoctial circle, therefore all the parts of the sea under each single part of the equinoctial circle will be moved equally in the same direction. Therefore, one part of the sea will not resist the motion of the other part, and there will be no reason why the water should occupy a greater place. Likewise, if the sea were moved in this manner, only local motion would result from such a cause. But we perceive by experience that the rise and fall of the sea results from its condensation and rarefaction. For ships in the sea draw less water at high tide than at low tide, and this is because the low tide is caused by rarefaction and the high tide by condensation. Also, the water is found to be hotter at high tide than at low, and this can only be because of the greater subtlety of its parts. These arguments and experiments are sufficient to prove Alpetragius' explanation false.

That the moon alone, and not some planet or image of the fixed stars, is the cause of this motion is clear for this reason, that the motion of the sea follows the motion of the moon more closely than that of any other planet;[4] and there is a definite proportion between these two motions, as will be made clear below. The arguments of the astronomers lead us to a similar conclusion. They point out that there are two great luminous bodies, the sun and the moon. Of these two, the sun is primarily responsible for motions which take place in the air, and the moon for those which take place in water, since in these two spheres [i.e., air and water] the generation and corruption of all living things takes place, and these two luminous bodies are the principal causes of every generation and corruption. Also we know by experience that, of all the heavenly substances, the moon exercises the greatest control over moist and cold bodies.[5] Thus certain people are called lunatics because, when the moon wanes, they suffer a diminution of the cerebrum, since the cerebrum is a cold and moist substance. These arguments are enough to show that the moon is the sufficient and efficient cause of the tides.

But how it causes them, we must still discover. When the moon rises on the horizon of any sea, it first casts its luminous rays on the center of that sea and, strongly impressing its power, it moves this sea, and this motion increases until the moon arrives at the meridian. But when it passes over the meridian its effective power is diminished, and the sea recedes toward its original place until the moon has set. When the moon again passes over from the west to the middle of the sky under the earth, the sea is increased; and when the moon passes from the middle of the sky under the earth, the sea is decreased until the moon again begins to rise. And thus in one revolution of the moon from rise to rise, two high tides occur in that place over whose horizon the moon has risen.

However, the explanation of why the high and low tides correspond to these four quarters of the sky is very difficult. It is clear enough from what has been said why the high tide occurs between the moon's rise and its arrival at the meridian and why the low tide occurs when the moon is passing from the meridian to the western horizon. For while it is rising it impresses its power more strongly than when it is setting. But the other high and low tide

4. See Selection 6, n. 3. Abū Ma'shar's *Introduction to Astronomy (Introductorium in Astronomiam)* may have served as Grosseteste's source for the moon's role in causing tides. "Here, in the chapters on the moon, a full description is given of the various characteristics of the tides together with copious speculations regarding their causes. The actual observations of fact were exact and careful. Abū Ma'shar explains with not a little accuracy the relation of the tides to the moon's rising and setting, to her phases, and to the position of the sun; he understood that winds might cause exceptionally high water; he recognized the influences of local topographic features, that some seas display different tidal phenomena from others and that the flood waters may be retained by reefs, or valleys, or deep bays. On the other hand, Abū Ma'shar's treatment of the causes of the tides was less successful. Though he believed firmly that the moon produces the ebb and flood, he failed to account for the presence of the high tide at the time of the moon's opposition. His explanation of the moon's attraction of the waters was in keeping with astrological methods of reasoning. Our satellite was supposed by astrologers to be of peculiarly aqueous nature and for that reason exceptionally capable of governing the movements of the liquid element of the earth (J. K. Wright, *The Geographical Lore of the Time of the Crusades* [New York: American Geographical Society, 1925], pp. 84–85.

Wright's account is drawn from Pierre Duhem's lengthy description of Abū Ma'shar's views in *Le Système du Monde* [Paris: Hermann, 1913–1959], II, 369–386. According to Duhem, Abū Ma'shar's explanation of lunar causation of tides was widely adopted in the Latin west. Two translations of it were made in the twelfth century, first by John of Seville and then by Hermann of Carinthia in 1140.

5. See note 4.

take place when the moon is in the two quarters below the horizon, and its light cannot act on the sea. Since heavenly bodies can only act on lower bodies by their light, it is doubtful how the moon can be the cause of the motion of the sea. The astronomers answer this by saying that opposite quarters of the sky have similar effects, but whether this is true remains to be proved and is in need of further investigation.

Therefore the time taken by two complete high and low tides exceeds the time of one day and night by the same amount as does one revolution of the moon from rise to rise. You must find out, therefore, by how many hours the rise of the moon follows or precedes the rise of the sun. We know by how many hours the beginning of the rise of the sea follows or precedes the beginning of the day and that one lunation contains twenty-nine days and a few minutes. Therefore, in seven and a quarter days and a few minutes the moon will be one-quarter of a circle east of the sun; whence in one-third of this time, when the sun is beginning to rise, the moon will be in the middle of the sky under the earth, and then the tide will begin to recede, since at the beginning of this time the tide began to rise. In twice this time, however, when the sun is in the east and the moon is in the west, the tide will begin to rise.[6] But if the moon should be more than three-quarters away from the sun, when the sun begins to rise, the moon will be in the middle of the sky above the earth, and the tide will begin to recede.[7] However, when the moon is again in conjunction with the sun, the beginning of the day and the beginning of the rise will occur at the same time. And this is what sailors say, that if the beginning of the rise coincides with the beginning of the day, seven days later the beginning of the recession will coincide with the beginning of the day. But this rise and fall in the same times does not appear on the shores of seas which are far away from the middle of that sea where the rise and fall of the tides originate. And you should know that on the shores nearer to the south and east, the rise takes place more quickly. Let this suffice concerning the efficient cause.

Now it remains to discuss the specific material cause, and it is this: Waters tend to come together in a deep and broad place, and in these waters is much matter of vapors and winds. Whence the moon, rising and impressing its power, generates many vapors in these waters and stirs up the winds. But this water, because of the heaviness and viscosity of its parts, will not let them escape, but

rumbles and becomes swollen because of the trapped vapors. But since fresh water is subtle and its parts are penetrable, when any vapor is generated in it, it is expelled at once; hence it does not, strictly speaking, have tides.

Also, if a heavenly body acts on lower bodies only by its light rays, since these light rays are in some manner incorporated with the elements, they intersect themselves at one point when they are reflected, and thus they generate heat by scattering the parts of matter in various directions, as is clear from the last proposition of Euclid's *Catoptrica*. Therefore the subtler the matter, the less heat will be generated by the ray. For this reason, snow remains longer on the tops of mountains than in valleys because the air is subtler there. Therefore if fresh water is much more subtle than sea water, the lunar rays will be much less incorporated in it than in sea water, and thus they will generate less heat and a smaller effect. Let these remarks suffice concerning the first part of our investigation.

II

Let us go on to the second part, in which we are to seek the reason why sometimes the tide increases and becomes strong and at other times it decreases and becomes weak. There are eight reasons for this.

The first is that when the sun and moon are in conjunction, the power of the moon becomes stronger, and the tide increases and becomes strong.[8] But as the moon recedes from the sun, the tide decreases until the moon is one quarter of a circle away from the sun.[9] From this point until the moon is a half-circle away from the sun, its power increases, not because of the moon's nearness to the sun but because of the increase of light in it. And again, from this point until the moon is three quarters away from the sun, the tide decreases; and then it increases again until the sun and moon are in conjunction. And this occurs proportionately according to the four parts of one month corresponding to the four parts of one day.

The second reason is the comparison of the diverse motion of the moon with the mean motion

6. Under these conditions a spring tide would occur where the sun and moon are in opposition (or in conjunction), and solar and lunar tides reinforce each other. It is not clear, however, that Grosseteste is describing this phenomenon.

7. Is Grosseteste here describing a neap tide, which occurs when the moon is at quarter phase and the solar and lunar tides partly neutralize each other?

8. Spring tide (see n. 6).

9. Neap tide (see n. 7).

of the moon. For if the motion of the sea is caused by the motion of the moon, then the mean motion of the sea should correspond to the mean motion of the moon, and the diverse motion of the sea to the diverse motion of the moon. And so, when the moon's true motion varies from the mean motion in one direction, the tide is increased; and when it varies in the other direction, the tide is lessened.

The third reason is the approach of the moon to the *aux* of its circle, that is to its longitude farthest from earth, for then its power on the sea is diminished because of its greater distance from earth. And when it approaches the point opposite the *aux,* that is its longitude nearest the earth, its power increases, and then the rise of the sea is strong.

The fourth reason is the southward declination of the moon in latitude from the zodiacal circle, for then it approaches the middle of the Ocean in which the strong rise begins. However, when the moon declines north of the zodiac, the high tide is lower because of the moon's greatest distance from the middle of the Ocean.

The fifth reason is the existence of the moon in the southern or northern signs of the zodiac, and this is a particular cause of the high tide. For when the moon is in the southern signs, it increases the high tide in the corresponding place on earth. But when it is in the northern signs, it increases the tide in the northern sea.

The sixth reason is the days which are called "Egyptian days" by the ancients, because the Egyptians first discovered them. But let us omit this reason for the present because many effects lie hidden in these days.

The seventh reason is the help the sun gives the moon. From the spring equinox to the summer solstice, the sun increases the tide. But from that point to the autumnal equinox, it diminishes it. From the autumnal equinox, however, until the winter solstice, it increases it again. And from there to the spring equinox, it diminishes it. And there are these four quarters of the year, just as earlier we noted the four quarters of the month corresponding to the four quarters of the day.

The eighth reason is the wind. When it blows in the direction of the tide, it increases it; but when it blows in the opposite direction, it decreases it.[10]

If, therefore, we make a classification of the reasons for the increase of the tide, we shall see that each cause will be either accidental or essential. The essential cause is general, and the accidental cause is specific. This eighth cause is general, but the seven others are specific. For every star has a special operation as a result of its special nature and a general operation in common with other stars. And let these remarks suffice concerning the second part.

III

For a knowledge of the third part of our investigation, we must be aware that certain rivers and springs have tides not because of their own natures but because they are continuous with the waters of the sea; and it does not matter whether this continuity is in underground channels or in channels appearing on the surface of the earth.

There are three kinds of seas. There are some that do not have tides, some that have tides but do not seem to, and some which have tides and seem to. If a sea does not have tides, this will be because of the absence of a material or efficient cause (and remember that the material cause is twofold). One instance of this is that waters tend to come together in a deep and broad place. I say a deep and broad place since, when the waters come together there in the winter, they cause no apparent increase; nor, when they leave it in a summer because of their consumption of the sun's heat, is there an apparent diminution. For this reason, that sea which is called broad, in which there is no sensible increase or decrease, does not have tides. Another example of this kind of sea is that in which the hardness of the earth is not sufficient, and the earth is spongy. So when the moon makes the waters more subtle there, the water enters the pores of the earth and does not rise to the earth's surface. A third reason is the great distance from the motion of the moon of a certain northern sea.

Another kind of sea is that which has tides and does not seem to; and there are similarly three kinds of these. One is a sea which is very wide so that one of its shores is near the moon's motion, and on this shore there are tides. But they are not seen because no one lives on that shore. On the other shore, however, where there are inhabitants, no tide will appear because it is too far away from the motion of the moon. Another kind is when there are inhabitants on both shores, but because of its great width, the waters which are near the bottom of the sea and far distant from its center are somewhat rarefied. So when they are moved by the moon, they let the vapors escape, just as fresh water does, and no tide is apparent in them although there is a very slight one. The third kind

10. See note 4.

is that which has somewhat cavernous earth, in whose pores and caverns are heavy vapors. The water does not enter these pores and caverns because of its heaviness and its retention of vapors. But when the vapors are moved by the moon, the water becomes more subtle; these vapors are expelled and the waters enter the lower parts. And in the upper part it expels them so that the sea is more subtle than other seas, and thus the waters flow together by contrary motions both upwards and downwards. No tide is apparent in seas of this kind, although there is a slight one.

Other seas which have tides, and seem to, behave similarly to a greater or lesser degree.

Biology
Zoology

83 DESCRIPTIONS OF ANIMALS FROM A TWELFTH-CENTURY BESTIARY

Translation, introduction, and annotation by T. H. White[1]

A Bestiary is a serious work of natural history, and is one of the bases upon which our own knowledge of biology is founded, however much we may have advanced since it was written.

There is no particular author of a bestiary. It is a compilation, a kind of naturalist's scrap-book, which has grown with the additions of several hands. Its sources go back to the most distant past, to the Fathers of the Church, to Rome, to Greece, to Egypt, to mythology, ultimately to oral tradition which must have been contemporary with the caves of Cromagnon. Its influence has extended throughout literature, and, as has been seen in the Notes, country people are still repeating some of its saws.

A full history of the subject would have to begin with oral tradition in various parts of the world, and to continue through Herodotus, Aristotle, and Pliny, for many hundred years, before it reached the Physiologus which is the immediate ancestor of the work with which we are dealing. Some idea of this history can be gathered from the Family Tree which is printed on page[2] 645 and which the present translator has adapted from a very interesting diagram by Dr Ansell Robin.[3] Unfortunately, Dr Robin's version does not mention the bestiaries as such.

The immediate ancestor of our manuscript is the Physiologus. His information may have come from any or all of the sources shown before him in the family tree, but the fact remains that an anonymous person who is nicknamed 'the Physiologus' appeared between the second and fifth centuries A.D., probably in Egypt, and wrote a book about beasts, possibly in Greek. The book was a success and was translated north and south into Syriac, Armenian, and Ethiopian. The earliest Latin translation of him which we have is of the eighth century.

It has to be remembered that the circulation of ancient books was by manuscript. So popular was the work of Physiologus that, like a stone thrown into a pool, it proceeded to spread itself over the surface of the literate world in a series of concentric rings, as it was copied and translated from one language into another, century by century. 'Perhaps no book,' says E. P. Evans, 'except the Bible, has ever been so widely diffused among so many people and for so many centuries as the Physiologus. It has been translated into Latin, Ethiopic, Arabic, Armenian, Syriac, Anglo-Saxon, Icelandic, Spanish, Italian, Provençal, and all the principal dialects of the Germanic and Romanic languages.' The last hand-written manuscript known to the present translator, except his own, was copied in Iceland in 1724.[4]

Meanwhile, as the versions of the Greek Physiologus spread across the globe, other naturalists, some of whom may themselves have been influenced by him, were writing other natural histories.

There was no reason why a scribe, setting out to

1. [Reprinted by permission of T. H. White and Jonathan Cape, Ltd., from *The Book of Beasts being a translation from a Latin Bestiary of the Twelfth Century* made and edited by T. H. White (London: Jonathan Cape, 1954). Because of its interest and relevance, I have reproduced a portion of White's Appendix (pp. 231–236) to serve as introduction to the descriptions that follow.—*Ed.*]
2. [I have replaced White's reference to "p. 233" with "page 645."—*Ed.*]
3. [Presumably in Dr Robin's book, *Animal Lore in English Literature* (London: John Murray, 1932).—*Ed.*]
4. T. K. Abbott's *Catalogue of MSS in the Library of Trinity College, Dublin* (1900), p. 175, no. 1017.

THE FAMILY TREE

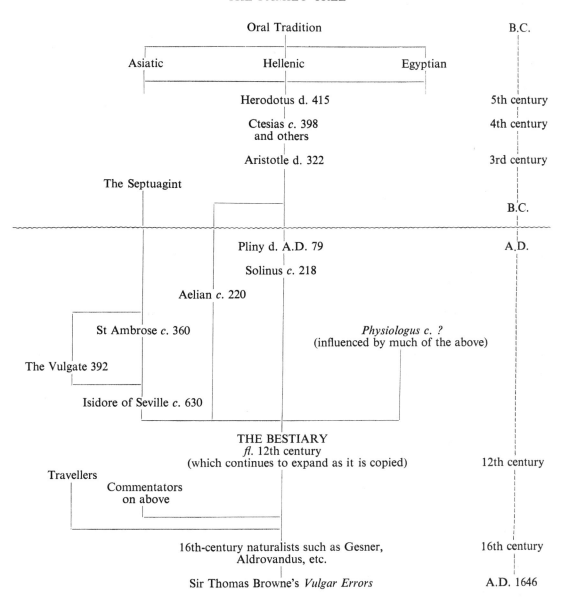

Oral Tradition

Asiatic   Hellenic   Egyptian   B.C.

Herodotus d. 415   5th century

Ctesias *c.* 398
and others   4th century

Aristotle d. 322   3rd century

The Septuagint

B.C.

Pliny d. A.D. 79   A.D.

Solinus *c.* 218

Aelian *c.* 220

St Ambrose *c.* 360   *Physiologus c. ?*
(influenced by much of the above)

The Vulgate 392

Isidore of Seville *c.* 630

THE BESTIARY
fl. 12th century
(which continues to expand as it is copied)   12th century

Travellers

Commentators
on above

16th-century naturalists such as Gesner,
Aldrovandus, etc.   16th century

Sir Thomas Browne's *Vulgar Errors*   A.D. 1646

copy the original Greek but being aware of a later authority, should not expand his copy with interpolations from that authority. This in fact happened, and the Bestiary, as distinct from the Physiologus, began to grow like a living tree. There are forty-nine beasts in the Greek original: in the manuscript here translated there are between 110 and 150. The scribe himself says 144 in the sea.

It was in the twelfth century that the scrap-book as such reached its finest foliage in Latin prose, and that happened particularly in England. 'Indeed,' says Dr M. R. James, 'the Bestiary may be reckoned as one of the leading picture-books of the twelfth and thirteenth centuries in this country. It ranks in this respect with the Psalter and the Apocalypse. Leading, and influential: for researches such as those of Cahier and Martin in the last century and of Mr G. C. Druce in this, have shown how widely images and ideas taken from it have permeated medieval art.'⁵ Dr James examined thirty-four English copies. In this connection, it is perhaps a surprising fact that our own translation is the first full version to be published in English, although translations of much shorter works in Latin verse and Norman French have appeared. The version of Trevisa is not of the true bestiary. Those of Sylvester's *Du Bartas* and Topsell are very far from it.

Dr James, in editing this manuscript for the Roxburghe Club, took the trouble to trace every statement to its proximate source. Except for thirteen quite insignificant additions (from Rhabanus Maurus, the *Aviarum* of Hugo, the Pantheologus and a few untraceable) he found that the immediate ancestors of this version were (*a*) the Physiologus, (*b*) Solinus, (*c*) St. Ambrose, and (*d*) Isidore of Seville. All of them, of course, were liable to quote from the Septuagint, and from anybody who preceded them.

In short, if we are asked for 'the author' of the bestiary at its fullest, as here reproduced, we can only place a finger in the middle of the family tree⁶ where the previous sources converge.

In adapting Dr Robin's Zoological Pedigree to the particular tradition of the bestiary itself, the present translator has omitted Theophrastus, Nicander, Oppian, Capella, Cassiodorus, Neckham, Vincent of Beauvais, English Bartholomew, Albertus Magnus, Marco Polo, Odoric, the supposed Mandeville, Belon, and one or two general categories. These were associated with the remoter sources or with the subsequent influence, but the

tree is complicated enough without them.

This brings us to the branches of the Bestiary in general, rather than to its roots.

Bartholomew Glanville (English Bartholomew) wrote his *De Proprietatibus Rerum* in the thirteenth century, as Neckham had written his *De Naturis Rerum* in the twelfth, and these books were extensions of the subject as we have seen it here.

Bartholomew was translated into English by Trevisa in 1397. A host of other writers had begun to appear, all of whom, with Bartholomew and Neckham, are included in our diagram under the comprehensive heading 'Commentators on the Above.'

In all probability, it would have been from people like Pliny and the commentators that Du Bartas (1578) and Topsell (1607), the translator of Gesner, derived some of their facts about animals, and from these latter that the poets like Spenser and Milton took theirs.

Meanwhile, the travellers and serious observers had begun to add their quota to the subject, people like Marco Polo and still more modern voyagers— there was even a Baron Munchausen of the subject, the pseudo-Mandeville, who resembled his predecessor Ctesias the Cnydian—until, with the sixteenth and seventeenth centuries, the voluminous sages like Gesner and Aldrovandus had vast stores of learning to collect. They were already beginning to have their doubts about the wilder statements.

In 1646, in England at any rate, which was the home of the bestiaries,⁷ Sir Thomas Browne subjected what he called their 'tenents' to rational criticism in his *Pseudodoxia Epidemica,* and began to raise the subject of biology to a scientific level, for the first time since Aristotle.

This meant the end of the Physiologus as a serious authority, and memories of him declined as the factual approach of the eighteenth century gained in popularity.

5. [Although White cites neither book nor page number, this is almost certainly a reference to M. R. James, *The Peterborough Psalter and Bestiary,* edited for the Roxburghe Club, Oxford: 1921).—*Ed.*]

6. [The words "on page 233" have been omitted.—*Ed.*]

7. The present translator has, for the sake of clarity, confined himself to the English side of the story. A fuller survey would deal particularly with Norman-French writers like Guillaume le Clerc, whom Mr Druce translated, with Jacob van Maerlant's *Der Naturem Bloeme* (1280) and with the *Bestiare d'amour* of Richard de Furnival.

PART I BEASTS[8]

TIGRIS the Tiger gets his name from his speedy pace; for the Persians, Greeks and Medes used to call an arrow 'tygris'. The beast can be distinguished by his manifold specklings, by his courage and by his wonderful velocity. And from him the River Tigris is named, because it is the most rapid of all rivers.

Hyrcania is his principal home.

Now the Tigress, when she finds the empty lair of one of her cubs that has been stolen, instantly presses along the tracks of the thief. But this person who has stolen the cub, seeing that even though carried by a swiftly galloping horse he is on the point of being destroyed by the speed of the tigress, and seeing that no safety can be expected from flight, cunningly invents the following ruse. When he perceives that the mother is close, he throws down a glass ball, and she taken in by her own reflection, assumes that the image of herself in the glass is her little one. She pulls up, hoping to collect the infant. But after she has been delayed by the hollow mockery, she again throws herself with all her might into the pursuit of the horseman, and, goaded by rage, quickly threatens to catch up with the fugitive. Again he delays the pursuer by throwing down a second ball, nor does the memory of his former trick prevent the mother's tender care. She curls herself round the vain reflection and lies down as if to suckle the cub. And so, deceived by the zeal of her own dutifulness, she loses both her revenge and her baby.

UNICORNIS the Unicorn, which is also called Rhinoceros by the Greeks, is of the following nature.

He is a very small animal like a kid, excessively swift, with one horn in the middle of his forehead, and no hunter can catch him. But he can be trapped by the following stratagem.

A virgin girl is led to where he lurks, and there she is sent off by herself into the wood. He soon leaps into her lap when he sees her, and embraces her, and hence he gets caught.

Our Lord Jesus Christ is also a Unicorn spiritually, about whom it is said: 'And he was beloved like the Son of the Unicorns'. And in another psalm: 'He hath raised up a horn of salvation for us in the house of his son David'.

The fact that it has just one horn on its head means what he himself said: 'I and the Father are

One'. Also, according to the Apostle: 'The head of Christ is the Lord'.

It says that he is very swift because neither Principalities, nor Powers, nor Thrones, nor Dominations could keep up with him, nor could Hell contain him, nor could the most subtle Devil prevail to catch or comprehend him; but, by the sole will of the Father, he came down into the virgin womb for our salvation.

It is described as a tiny animal on account of the lowliness of his incarnation, as he said himself: 'Learn from me, because I am mild and lowly of heart'.

It is like a kid or scapegoat because the Saviour himself was made in the likeness of sinful flesh, and from sin he condemned sin.

The Unicorn often fights with elephants, and conquers them by wounding them in the belly.

There is an animal called an ELEPHANT, which has no desire to copulate.

People say that it is called an Elephant by the Greeks on account of its size, for it approaches the form of a mountain: you see, a mountain is called 'eliphio' in Greek. In the Indies, however, it is known by the name of 'barrus' because of its voice—whence both the voice is called 'baritone' and the tusks are called 'ivory' (ebur). Its nose is called a proboscis (for the bushes), because it carries its leaf-food to its mouth with it, and it looks like a snake.

Elephants protect themselves with ivory tusks. No larger animals can be found. The Persians and the Indians, collected into wooden towers on them, sometimes fight each other with javelins as if from a castle. They possess vast intelligence and memory. They march about in herds. And they copulate back-to-back.[9]

Elephants remain pregnant for two years, nor do

8. [The bestiary is divided into three parts: I. Beasts, II. Birds, III. Reptiles and Fishes. In order of appearance the selections included here are on pp. 12–13, 20, 24–28, 28–29, 45–47, 49–51, 82–83, 119–120, 123–124, 165–167, 191–194, 209, 210, 210–212. Notes by the translator have been included only where they furnish information about the animal described. The charming illustrations which depict many of the beasts described here could not, alas, be included.—Ed.]

9. The copulation of elephants was a matter for speculation in the dark ages, and still is, as it is rarely witnessed. Solinus quotes Pliny to the effect that their genitals, like those mentioned by Sir Thomas Browne in his note on hares, were put on backward. It was supposed that, being modest, they preferred to look the other way

they have babies more than once, nor do they have several at a time, but only one. They live three hundred years. If one of them wants to have a baby, he goes eastward toward Paradise, and there is a tree there called Mandragora,[10] and he goes with his wife. She first takes of the tree and then gives some to her spouse. When they munch it up, it seduces them, and she immediately conceives in her womb. When the proper time for being delivered arrives, she walks out into a lake, and the water comes up to the mother's udders. Meanwhile the father-elephant guards her while she is in labor, because there is a certain dragon which is inimical to elephants. Moreover, if a serpent happens by, the father kills and tramples on it till dead. He is also formidable to bulls—but he is frightened of mice, for all that.

The Elephant's nature is that if he tumbles down he cannot get up again. Hence it comes that he leans against a tree when he wants to go to sleep, for he has no joints in his knees. This is the reason why a hunter partly saws through a tree, so that the elephant, when he leans against it, may fall down at the same time as the tree.[11] As he falls, he calls out loudly; and immediately a large elephant appears, but it is not able to lift him up. At this they both cry out, and twelve more elephants arrive upon the scene: but even they cannot lift up the one who has fallen down. Then they all shout for help, and at once there comes a very Insignificant Elephant, and he puts his mouth with the proboscis under the big one, and lifts him up. This little elephant has, moreover, the property that nothing evil can come near his hairs and bones when they have been reduced to ashes, not even a Dragon.

Now the Elephant and his wife represent Adam and Eve. For when they were pleasing to God, before their provocation in the flesh, they knew nothing about copulation nor had they knowledge of sin. When, however, the wife ate of the Tree of Knowledge, which is what the Mandragora means and gave one of the fruits to her man, she was immediately made a wanderer and they had to clear out of Paradise on account of it. For, all the time that they were in Paradise, Adam did not know her. But then, the Scriptures say: 'Adam went in to his wife and she conceived and bore Cain, upon the waters of tribulation'. Of which waters the Psalmist cries: 'Save me, O God, for the waters have entered in even unto my soul'. And immediately the dragon subverted them and made

them strangers to God's refuge. That is what comes of not pleasing God.

When the Big Elephant arrives, i.e., the Hebrew Law, and fails to lift up the fallen, it is the same as when the Pharisee failed with the fellow who had fallen among thieves. Nor could the Twelve Elephants, i.e., the Band of the Prophets, lift him up, just as the Levite did not lift the man we mentioned. But it means that our Lord Jesus Christ, although he was the greatest was made the most Insignificant of All the Elephants. He humiliated himself, and was made obedient even unto death, in order that he might raise men up.

The little elephant also symbolizes the Samaritan who put the man on his mare. For he himself, wounded, took over our infirmities and carried them from us. Moreover, this heavenly Samaritan is interpreted as the Defender about whom David writes: 'The Lord defending the lowly ones'. Also, with reference to the little elephant's ashes: 'Where the Lord is present, no devil can come nigh'.

It is a fact that Elephants smash whatever they wind their noses round, like the fall of some prodigious ruin, and whatever they squash with their feet they blot out.

They never quarrel with their wives, for adultery is unknown to them. There is a mild gentleness

---

while they were about it. Albertus Magnus held that they copulated like other quadrupeds, but that, owing to the great weight of the husband, he either had to dig a pit for his wife to stand in or else he had to float himself over her in a lake, where his gravity would naturally be less. In fact, they copulate in the ordinary way and, according to Lieut-Colonel C. H. Williams, more gracefully than most. [Compare Albertus Magnus' description of the elephant in the next chapter.—*Ed.*]

10. Mandrake. There is still a genus of plants called the Mandragora, said to be emetic, narcotic and 'fertilizing'. 'The mandrakes which Reuben found in the field were used by his mother Leah for venereal purposes (Gen. xxx. 14–16), and this precious peculiarity is enlarged upon in rabbinical literature. The Greeks spoke of them as anthropomorphic; and according to popular superstition they spring from human sperm spilled on the ground, and are so full of animal life and consciousness that they shriek when torn out of the earth, so "that living mortals, hearing them, run mad." ' E.P. Evans. [See a description of the Mandrake by Rufinus in Selection 109.1.—*Ed.*]

11. Julius Caesar, in his *Gallic War*, relates the same fable about the elks *(alces)*. The present translator is informed by Captain A.A.F. Minchin of the Indian Forest Service that elephants do have, as it were, scratching posts at their watering places, and that the hunter is able to forecast the size of his quarry by the height of the mud rubbed off against these trees.

about them, for, if they happen to come across a forwandered man in the deserts, they offer to lead him back into familiar paths. If they are gathered together into crowded herds, they make way for themselves with tender and placid trunks, lest any of their tusks should happen to kill some animal on the road. If by chance they do become involved in battles, they take no little care of the casualties, for they collect the wounded and exhausted into the middle of the herd.

This is an animal called CASTOR the Beaver, none more gentle, and his testicles make a capital medicine. For this reason, so Physiologus says, when he notices that he is being pursued by the hunter, he removes his own testicles with a bite, and casts them before the sportsman, and thus escapes by flight. What is more, if he should again happen to be chased by a second hunter, he lifts himself up and shows his members to him. And the latter, when he perceives the testicles to be missing, leaves the beaver alone.[12]

Hence every man who inclines toward the commandment of God and who wants to live chastely, must cut off from himself all vices, all motions of lewdness, and must cast them from him in the Devil's face. Thereupon the Devil, seeing him to have nothing of his own about him, goes away from him confused. That man truly lives in God and is not captured by the Devil who says: 'I shall persevere and attain these things'.

The creature is called a Beaver (Castor) because of the castration.

URSUS the bear, connected with the word *'Orsus'* (a beginning), is said to get her name because she sculptures her brood with her mouth *(ore)*. For they say that these creatures produce a formless foetus, giving birth to something like a bit of pulp, and this the mother-bear arranges into the proper legs and arms by licking it. This is because of the prematurity of the birth. In short, she pups on the thirtieth day, from whence it comes that a hasty, unformed creation is brought forth.

A bear's head is feeble: the greatest strength is in the arms and loins, for which reason they sometimes stand upright.

Nor do they neglect the healer's art. Indeed, if they are afflicted with a serious injury and damaged by wounds, they know how to doctor themselves by stroking their sores with a herb whose name is Flomus,[13] as the Greeks call it, so that they are cured by the mere touch.

A sick bear eats ants.

Numidian bears excel others so far as the thickness of their shaggy hair is concerned, but the creature itself is the same wherever they breed.

They do not make love like other quadrupeds, but, being joined in mutual embraces, they copulate in the human way. The winter season provokes their inclination to lust. The males respect the pregnant females with the decency of a private room, and, though in the same lairs for their lying-in, these are divided by earth-works into separate beds. The period of gestation is short, since the thirtieth day relieves the womb. This is why the precipitate childbirth creates shapeless fruits. They bring forth very tiny pulps of a white colour, with no eyes. They gradually sculpture these by licking;[14] and meanwhile they cherish them to their bosoms so as to draw up the animal spirit, being warmed by this careful incubation. During this time, with absolutely no food for the first fourteen days, the sleepless she-bears get so deeply drowsy that they cannot be woken up, even by wounds, and they lie hid after bearing for three months. Then, after coming out into the free daylight, they suffer so much from being unaccustomed to the light that you would take them to be struck blind.

Bears look out for the hives of bees and long for the honeycombs very much. Nor do they grab anything else more greedily than honey.

When they have eaten the fruits of the mandrake they die—unless they hurry off, for fear that the

12. The medicine was called 'castoreum'. It was situated not in the testicles, but in a different gland. The testicles of a beaver are internal and cannot be bitten off. 'The originall of the conceit,' says Sir Thomas Browne, 'was probably Hieroglyphicall, which after became Mythologicall unto the Greeks and so set down by Aesop, and by process of tradition stole into a totall verity'. [See the next selection for Albertus Magnus' description of the beaver, and note 3 thereto for a discussion of "Castoreum."—*Ed.*]

13. φλόμος, mullein.

14. There was a reasonable substratum of truth for the medieval belief. Bear cubs really are born blind and hairless, and remain so for no less than five weeks. Naked, born in a jealous seclusion which made observation almost impossible, and constantly licked by their dam as bitches clean their puppies, the idea that the cubs were produced as a kind of mola was not absolutely preposterous. 'The cub comes forth involved in the Chorion,' says Sir Thomas Browne, 'a thick and tough membrane obscuring the formation, and which the Dam doth after bite and teare asunder, (so that) the beholder at first sight conceives it a rude and informous lumpe of flesh, and imputes the ensuing shape into the mouthing of the Dam'.

poison should grow strong enough to destroy them, and eat ants to recuperate their health.

If ever they attack bulls, they know by what parts to bring them down most readily, nor will they go for anything but the horns or the nostrils—the nostrils, because the sharper pain comes in the tenderer place.

This is called a COCODRYLLUS[15] from its crocus or saffron colour. It breeds in the River Nile: an animal with four feet, amphibious, generally about thirty feet long, armed with horrible teeth and claws. So great is the hardness of its skin that no blow can hurt a crocodile, not even if hefty stones are bounced on its back. It lies in the water by night, on land by day.

It incubates its eggs in the earth. The male and female take turns. Certain fishes which have a saw-like dorsal fin destroy it, ripping up the tender parts of its belly. Moreover, alone among animals, it moves its upper jaw, keeping the lower one quite motionless.[16] Its dung provides an ointment with which old and wrinkled whores anoint their figures and are made beautiful, until the flowing sweat of their efforts washes it away.[17]

Hypocritical, dissolute and avaricious people have the same nature as this brute—also any people who are puffed up with the vice of pride, dirtied with the corruption of luxury, or haunted with the disease of avarice—even if they do make a show of falling in with the justifications of the Law, pretending in the sight of men to be upright and indeed very saintly.

Crocodiles lie by night in the water, by day on land, because hypocrites, however luxuriously they live by night, delight to be said to live holily and justly by day. Conscious of their wickedness in doing so, they beat their breasts: yes, but with use, habit always brings to light the things which they have done.

The monster moves his upper jaws because these people hold up the higher examples of the Fathers and an abundance of precepts in speech to others, while they show in their lower selves all too little of what they say.

An ointment is made of its evil dung because bad people are often admired and praised by the inexperienced for the evil they have done, and extolled by the plaudits of this world, as if beautified by an ointment. But when the Judgement, sweated out by the evils perpetrated, moves its anger to the striking, then all that elegance of flattery vanishes like a smoke.

The ONAGER is said to be the Wild Ass. The Greeks, to be sure, called the Ass *'on'* and they called wild *'agra'*. Africa breeds these creatures—large and untameable and wandering about in the desert.

One male at a time presides over the herds of females. When little males are born, the fathers get jealous of them and remove their testicles with a bite—for fear of which, the mothers hide them in secret places.

Physiologus says of the Wild Ass that, when twenty-five days of March have passed, it brays twelve times in the night and the same number in the day. From this the season is recognized as the 'Equinox'[18] and people can tell, hour by hour, the time of day or night by counting the brays of the ass.

Now the Devil is symbolized by this animal, for he brays about the place night and day, hour by

15. [This is the Latin word for crocodile which appears in the manuscript instead of *crocodilus,* the usual form. See p. 242 of White's volume, where *cocodryllus* is the first word of a paragraph of Latin text reproduced from the manuscript.—*Ed.*]

16. This was the common belief of the ancients. 'Cuvier thus accounts for the error: "The lower jaw being prolonged behind the skull, the upper seems to be movable, and the ancients have so recorded it; but it moves only with the whole head".' Dr W. Ogle. [The title of Ogle's work in which the quotation appears was omitted by White. Perhaps it is in Ogle's translation of Aristotle's *On the Parts of Animals.—Ed.*] Aristotle himself fell into this error. [The same error was repeated by Albertus Magnus; see his description of the crocodile, in the next selection.—*Ed.*]

17. Francis Bacon, who held that 'some Putrefactions and Excrements do yield excellent Odors,' pointed out that: 'We finde also, that places where men Urine commonly have some smell of Violets. And Urine, if one hath eaten Nutmeg, hath so too'.

The scented dung was thought to come from a special kind of crocodile living in 'the Province of Xanagarra'. It was better than musk. As the sweet-smelling Panther also inhabited those parts, it was conjectured that perhaps the habitat was responsible for the smell. Galen considered that crocodile dung was good for freckles. Aetius recommended that it should be lighted and the smoke puffed into snake-holes. Kiranides held that the teeth were aphrodisiac, but they had to be taken out alive.

18. The earliest text of *Physiologus,* upon which the bestiaries are based, here names an Egyptian month, according to the *Encyclopaedia Britannica.* If the date is there stated with precision, it seems just possible that an astronomer who was acquainted with the history of the calendar and with the alterations of the equinox might be able to give an approximate date at which the *Physiologus* was written—a date which at present can only be debated between the second and fifth centuries A.D.

hour, seeking his prey. He does this when he knows night and day to be equal, i.e. when he knows that the number of those who walk in darkness is equal to the number of the sons of light. For the Wild Ass does not bray unless it wants its dinner. As Job says, 'Doth the Wild Ass bray when he hath grass?' Wherefore the Apostle also: 'Our adversary the Devil, as a roaring lion, walketh about, seeking whom he may devour'.

PART II BIRDS

There is a bird called the YBIS (Ibis) which cleans out its bowels with its own beak. It enjoys eating corpses or snakes' eggs, and from such things it takes food home for its young, which comes most acceptable. It walks about near the seashore by day and night, looking for little dead fish or other bodies which have been thrown up by the waves. It is afraid to enter the water because it cannot swim.

This bird is typical of Carnal Man, who goes in for deadly dealings as if they were good spiritual food—by which his miserable soul gets nourished for punishment.

You, on the other hand, good Christian fellow—who are born again by water and the Holy Spirit to enter into the spiritual oceans of God's mysteries—on you he bestows the very finest good which he mentioned to the apostles, saying: 'The fruit of the Spirit moreover is affection, praise, peace, forbearance, long suffering, etc.'

Now if the sun and moon do not throw out their cruciform rays, they do not shine: if the birds do not spread their wings like a cross, they cannot fly. Just so, Man, if you do not protect yourself by the sign of the cross and spread your yard-arm wings of love, you will not be able to go through the tempests of this world to the quiet haven of the heavenly land.

Finally, when Moses raised his arms like a bird, Israel prevailed: When he lowered his arms, Amalech conquered.

The ALTION (halcyon) is a seabird, which hatches its young on the beaches. It lays its eggs in the sand, in the very middle of winter—for that is the season which destiny has chosen for the business—and it produces them just when the sea normally rises to its roughest on the shores, and when the waves are most boisterously dashed against them.

Now the object of all this is, that the mother-love of the halcyon may be illustrified by an unexpected celebration of kindness.

For when the full-waved sea reaches the position of the eggs it suddenly moderates, and the violent winds fall, and the booming of the breeze softens, and the sea stands calm in the light airs, while the halcyon warms her eggs.

What is more, there follow seven quiet hatching days, until, when these are over, the mother lets her chicks out and stops sitting. Then there are seven more days during which she feeds the young, while they are beginning to grow up. You may perhaps wonder at such a short infancy, when the brooding time was also so short.

People say that during those fourteen days sailors take it for granted that there will be fine weather. They call these days 'the Halcyon Days', in which there is no tide and the sea is not rough.[19]

PART III REPTILES AND FISHES

DRACO the Dragon[20] is the biggest of all serpents, in fact of all living things on earth. The Greeks call it 'draconta' ($\delta\rho\acute{\alpha}\kappa\omega\nu$) and hence it has been turned into Latin under the name 'draco'.

19. It will be noticed that this is not the classical story of Halcyone, whose nest should properly be on the sea itself, thus accounting for the seven halcyon days when 'birds of calm sit brooding on the charmed wave'. Here the nest is on the beach, like that of the ringed plover. Perhaps the medieval version is more interesting than the classical one, for it seems to link the Greek legend with a real phenomenon, the neap tides. The mid-winter neaps are the tides at which the rise and fall between high-tide and low-tide would be at a minimum, thus causing less disturbance to a bird nesting near the tide line, and the numbers seven and fourteen are, moreover, evidently connected with lunar calculations, hence with tides.

The tides of the Mediterranean, from which sea the legend originated, are not pronounced like ours, however, and Oppian could place the incubation of these birds between sea and shore. 'They are so fond of the sea that they put their nest alongside the waves, and though they wet their breasts, they lay their tails on dry land.' (Auc. Bk. 2).

Birds of the tide-line do have a very short infancy, like many others which are hatched at ground level.

20. Aldrovandus [1522–1605; White omits the title of the particular work in which Aldrovandus discusses dragons.—Ed.] gives fifty-nine folio pages to dragons, and turns up much interesting material in the process. He deals with humans of the name of Draco, with sea-serpents, tarantulas, plants, trees, stars, devils, quicksilver, mountains, traps, fistulae, sirens, Hydras, anacondas, whales, leviathan, fossils, hieroglyphs and even with an early form of aircraft called a Dragon, though not manufactured by De Havilland, which flew. He adds that it is possible for unscrupulous people to forge a dragon, by plastic surgery on the cadaver of a Giant Ray. But his main point is that the words 'dragon' and 'serpent' are interchangeable. He points out that the reptile which attacked Laocoön is called by Virgil a serpent in one place and a dragon in another. 'Why',

When this dragon has come out of its cave, it is often carried into the sky, and the air near it becomes ardent. It has a crest, a small mouth and a narrow gullet through which it draws breath or puts out its tongue. Moreover, its strength is not in its teeth but in its tail, and it inflicts injury by blows rather than by stinging. So it is harmless as regards poison. But they point out that poisons are not necessary to it for killing, since if it winds round anyone it kills him like that. Even the Elephant is not protected from it by the size of its body; for the dragon, lying in wait near the paths along which the elephants usually saunter, lassoes their legs in a knot with its tail and destroys them by suffocation.

They are bred in Ethiopia and India, in places where there is perpetual heat.

The Devil, who is the most enormous of all reptiles, is like this dragon. He is often borne into the air from his den, and the air round him blazes, for the Devil in raising himself from the lower regions translates himself into an angel of light and misleads the foolish with false hopes of glory and worldly bliss. He is said to have a crest or crown because he is the King of Pride, and his strength is not in his teeth but in his tail because he beguiles those whom he draws to him by deceit, their strength being destroyed. He lies hidden round the paths on which they saunter, because their way to heaven is encumbered by the knots of their sins, and he strangles them to death. For if anybody is ensnared by the toils of crime he dies, and no doubt he goes to Hell.

VERMIS the Worm is an animal which is mostly germinated, without sexual intercourse, out of meat or wood or any earthly thing. People agree that, like the scorpion, they are never born from eggs. There are earth worms and water worms and air worms and flesh worms and leaf worms, also wood worms and clothes worms.

A Spider is an air worm, as it is provided with nourishment from the air, which a long thread catches down to its small body. Its web is always tight. It never stops working, cutting out all loss of time without interruption in its skill.

The earth Millipede gets its name from the number of its legs, and this is the one which hides in large numbers under flower-pots, rolled up in a ball.

Sanguinea the Leech is a water worm, so called because it sucks blood. It also lies in wait for people who are taking a drink of water and, when it slips down their throats or manages to catch on any-

where, it sups their gore. When it is gorged with blood it spews it up again, so that it can start once more with fresh.

The Scorpion is a land worm which we classify with worms rather than with snakes. It is a stinging creature, and is called the Archer in the Greek Language because it plunges in its tail and injects its poisons with a curving wound *(aculeus: arcuatus)*. The oddest thing about a scorpion is that it will not bite you in the palm of your hand.[21]

Bombocis the Silkworm is a worm of leaves, from whose productions silken garments are made *(bombycina)*. It is also called this *(bombus* = a hollow sound) because it empties itself out while it is spinning the threads, and only air remains inside it.

Eruca the Caterpillar[22] is another worm of leaves. After rolling up young vine shoots, it is believed to fatten itself by eating them away *(ab erodendo)*. Plautus makes mention of it as follows: 'It curls itself round the young shoots like a nasty, malignant beast, nor does it buzz off like the locust, which wanders from place to place when it has half dined, but it sticks to the same growing fruits and gobbles the whole thing up with a tardy slither and slothful nibbles.'

The Greeks call wood worms *'Teredonas'*, because they eat by grinding *(terendo)*. We call them Termites. The Latins call them Vermes. Trees which are felled in the wrong season breed them.

The clothes worm is called Tinea because it gnaws *(terat)*, and whenever this one gets in, it eats the thing away. Hence it is a pertinacious creature, for it sticks to the same thing all the time.

Flesh worms include Emigramus, Lumbricus, Ascaridae, Costae, Pediculi, Pulices, Lendex, Tarmus, Ricinus, Usia and Cimex.

wrote Kingsley in 1849, 'should not these dragons have been simply what the Greek word dragon means—what . . . the superstitions of the peasantry in many parts of England to this day assert them to have been—"mighty worms", huge snakes?' This is the proper way to regard them. 'Dragon' was simply the medieval word for a large reptile, and the more one regards it as not being a joke from the fairy stories, the more interesting the following pages may prove to be. The very first definition of 'dragon' in the N.E.D. is 'A huge serpent or snake; a python. *Obs.*' In modern zoology the dragon is a flying lizard forming the genus *Draco*, belonging to the family *Agamidae*, and there are twenty species.

21. This fancy derives from Pliny. But since the blow is inflicted downwards, the palm of the hand and the sole of the foot do seem less likely to be wounded than their opposite sides.

22. The word 'caterpillar' seems to be derived from *'chatepelose'* (hairy cat) or *'chat-pilour'* (plundering cat).

Emigramus is known as a head worm.

Lumbricus is a stomach worm, *quasi 'lumbicus'*, because it is slipped in; or else because it lives in the loins *(in lumbis)*.

Pediculi the Lice are flesh worms which are named from 'feet' *(a pedibus)*—and lousy people *(pediculosi)* also get their name from this root. They swarm with them.

Fleas are rightly called fleas *(pulices)* because they are mainly nourished by dust *(ex pulvere)*.

Tarmus is a lard worm.

Ricinus the Tick is a worm called after the dog, because it clings to dogs' ears. (The pun seems to be, *Ri-cinus: Auriscanis*.) Also, *Kunos* is the Greek for dog.

Usia (probably *'urica'*, the canker) is known as a pig worm, because its bite burns *(urit)*. The place where it has bitten gets so fiery that you make water on the spot, when it has bitten you. (There is an obscure pun here about 'urine'.)

Cimex the Bug gets its name from its similarity to a certain herb, whose offensive smell it has.[23] Appropriately enough, this worm is born on rotting meat.

You get the Tinea in clothes, the Eruca in vegetables, the Teredo in wood and the Tarmus in earth.

Worms do not crawl by plain wriggles like snakes, nor by pressing on their scales, because they do not have the same rigidity of the spine that snakes have.[24] Instead, by contracting the parts of their small bodies into a straight line and then expanding them, their movement develops by dragging the stretched parts along and by crawling with this concertina action.

ANGUILLAE the Eels are similar to snakes, and thus they derive their name *(anguis* = serpent). They are engendered from mud and for this reason, if you catch hold of an eel, the creature is so slippery that the harder you press it the more quickly it slips away. They say that in the eastern River Ganges eels are born measuring thirty feet. If eels are drowned in wine, those who drink of it get a loathing for liquor.

TORPEDO the Electric Ray or Eel gets its name because it makes one's body grow numb *(torpescere)* if one touches it when alive. Pliny the younger relates: 'If the electric ray out of the indigo sea is touched, even from a distance, even with a spear or a stick, the arm-muscles, howsoever mighty, will be paralysed and the feet, however swift in running, will be struck still'. So great is its power, moreover, that even the radiation of its body affects the limbs.

CANCER the Crab goes in for a cunning stratagem, due to his greed. He is very fond of oysters and likes to get himself a banquet of their flesh. But, although eager for dinner, he understands the danger, since the pursuit is as difficult as it is hazardous. It is difficult because the inner flesh of the oyster is contained within very strong shells, as if Nature its maker had by her imperial command fortified the soft part of the body with walls. She feeds and cherishes this flesh in a kind of arched dome in the middle of the shell: disposes it, as it were, in a sort of hollow. For this reason, the handling of oysters has to be done carefully, because nothing can open the closed oyster by force, and thus it is dangerous for the crab to insert his claw. Betaking himself to artfulness, therefore, the crab lays an ambush with a new plot of his own. Because all species delight in relaxing themselves, the crab investigates to find out whether at any time the oyster opens that double shell of his in places remote from all wind and safe from the rays of the sun, or whether it unlocks the fastenings of its gates, so that it may pleasure its internal organs in the free air. Then the crab, secretly casting in a pebble, prevents the closing of the oyster, and thus, finding the lock forced, inserts his claws safely and feeds on the internal flesh.

Now is not that just like Men—those corrupt creatures who follow the habit of the crab, creep into the practise of unnatural trickery and eke out the weakness of their real powers by a sort of cunning! They join deceit to cruelty and are fed upon the distress of others. Do you, therefore, be content with your own things and do not seek the injury of your neighbours to support you.

The simple fare of a man who does no harm is the right food. Having his own property he knows not how to plot against his fellow man's, nor does he burn with the flames of avarice. Covetousness is to him only a loss of virtue and an incentive to greed. And so, blessed is that poverty which

23. Either 'cummin' or 'cyma', a young cabbage.

24. The ancient category of 'worms' was more extensive than the modern one. They included the serpent which stung Cleopatra, the maggots which bred in corpses, the malignities which destroyed Herod, and many other creepers or crawlers. In Golding's translation of Solinus we find the sunshun: 'It is a verie little Worme and like a Spyder in shape, and it is called a *Shunsunne* because it shunneth the daie light. It lyeth most in Sylver Mynes. . . . It creepeth privily, and casteth the plague uppon such as sitte upon it unawares.'

truthfully sticks to its own goods, and meet it is to be preferred above all riches. 'Better is little with the fear of the Lord than great treasure and trouble therewith: better is a dinner of herbs where love is, than a stalled ox and hatred therewith.'

Let us then devote ourselves to acquiring merit and to maintaining what is wholesome, not to the cheating of another's innocence. Let it be left to us to make use of the marine example in perfecting our own well-being, not in the undoing of our neighbour.

84 AN ATTEMPT AT A SCIENTIFIC DESCRIPTION OF ANIMALS

Albertus Magnus (ca.1193–1280)

Translated and annotated by Edward Grant[1]

BOOK XXII

Beaver (Castor)

The Beaver is an animal having goose-like [that is, web-like] hind feet for swimming and fore-feet like a dog because it frequently walks around on land.[2]

He is called castor from castrating *(castrando)*— not [however,] because he castrates himself, as Isidore [of Seville] says, but because he especially seeks castration. Indeed, as has often been learned in our parts, it is false that the beaver, having been excited by a hunter, castrates himself by his own teeth and projects an odorous secretion [or musk][3] *(castoreum)* and that if he should be aroused, after having been castrated in some way, he shows that he lacks testicles by lifting himself up.[4]

With his teeth, this animal fells trees of considerable size and constructs dwellings on the banks of waters before the burrows *(antra)* in which he frequently lives.[5] He makes these dwellings two-chambered or three-chambered with terraces *(cum solariis)*, so that when the water rises or falls he can ascend or descend. It is said that in gathering the trees *(ligna)* for such structures beavers from many parts come together for [common] service[6] and lie on their backs with the trees properly placed between their feet, and by means of their tails they diligently pull the trees over their bellies to the place of construction. In this way they construct their homes.

The hide of the beaver is ash-colored, inclining toward black. At one time it was valuable but is now of little value. It consists of thick, short hair. Its food is fish and it also eats the barks of trees.

Within the body the beaver has an odorous musk *(castoreum)* that is hot and dry, which strengthens the nerves and works very well against tremors of the parts and paralysis, for it exerts a divisive and drying effect on the phlegmatic humor.[7]

1. My translation has been made from Albertus

Magnus' treatise *Twenty-Six Books on Animals (De animalibus Libri XXVI)*, edited by Hermann Stadler in Vols. XV and XVI of the *Beiträge zur Geschichte der Philosophie des Mittelalters* (Münster Westfalen: Aschendorffsche Verlagsbuchhandlung, 1921), pp. 1370–1371, 1376–1377, 1528, 1548. The first nineteen books paraphrase Michael Scot's translation of Aristotle's zoological works titled *De animalibus* (see Selection 86, n. 1). Books XX to XXVI constitute Albert's own zoological discussion plus a great many animal descriptions. Since the animal descriptions translated here are also included in the immediately preceding selection representing a twelfth-century bestiary, a comparison will quickly reveal that Albert's descriptions are generally superior, being genuine attempts to portray the animals in a faithful and scientific manner without moralizing and sermonizing. True, Albert also repeats many fantastic tales (see especially the section on the elephant), but he brings a critical spirit to his descriptions and occasionally mentions that he personally observed this or that animal.

2. Albert's description is accurate. The beaver's fore-feet are quite short and differ from the webbed hindfeet.

3. In both sexes of the beaver there are "musk glands in the lower abdomen that produce a musk-smelling liquid, castoreum, once valued for its medicinal properties, but now used in perfumes" (article on "Beaver" in *Encyclopaedia Britannica*, III [1968], 348, col. 2). "Castoreum is an oily, viscid glandular secretion contained in two pairs of membraneous sacs between the anus and external genitals of both sexes of the castor beaver, . . . It is yellow or yellow-brown in colour, of a butterlike consistency, and has a highly disagreeable odour due to the presence of an essential oil. After the death of the animal, the sacs are removed and dried in smoke or in the sun; drying darkens and hardens the castoreum and reduces its odour. Castoreum is soluble in alcohol and is prepared for use as a tincture. In this form castoreum has a pleasant scent and is used in the blending of perfumes" (article on "Castoreum," *Encyclopaedia Britannica*, V [1968], 42, col. 2).

4. Albert's rejection of this fabulous tale should be compared to its ready acceptance by the author of the bestiary in the preceding selection (see especially n. 12 thereto).

5. "In rivers and lakes beavers often burrow into banks" (article on "Beaver" in *Encyclopaedia Britannica*).

6. "Of placid disposition, beavers often work cooperatively" *(ibid.)*.

7. See note 3, and the preceding selection, n. 12.

It has a wide, very fatty tail, like a scaly hide; and it seems as if it always wants to let it hang into the water. Moreover, the tail is very suitable for swimming, like a rudder in a ship.[8] But it is not true that it never withdraws it from very cold, icy water. Hence, it is false that this animal compacts mud *(luterem)* around [its tail] so that in the winter it can move its tail around, lest it be frozen. But it is superior to the otter *(luthery)* and expels it or kills it, for it has a very sharp bite. Besides the tail, all its flesh is abhorrent.

Moreover, the odorous fluid [or musk] *(castoreum)* which has been extracted with the humor around it and prepared in a homogeneous quantity *(quantitate aequale)* and then dried ought to be of a subtle substance, inclining not toward the color black but toward a dark blood color.[9] However, if it is mottled *(varium),* inclining toward blackness, it is poisonous and sometimes kills on the same day and sometimes transforms to another terrible sickness. But when it is good, it is hot and dry with a most subtle dryness and is beneficial to the nerves and for paralytics when it is drunk hot with saliva *(quando cum saliva calidum bibitur).* In child birth, after the bleeding of the saphena vein *(post minutionem saphenae),*[10] which is a vein coming from the liver into the lower part of the arm *(brachii),* and after having drunk it with an aromatic herb *(calamento)* and two ounces of honey water, it easily expels the fetus and afterbirth; it also readily produces menstruation, because if the vein is not bled *(si vena non minueretur),* it would have little effect. After it has been drunk in this way, it also causes men's testicles to be aroused.

Elephant

The Elephant is the largest animal among the quadrupeds. We have already spoken sufficiently of his manner in coitus and generation.[11] Having a trunk 10 cubits in length, he uses it in place of a hand[12] in both war and in [obtaining] food, as well as in other things. And sometimes it makes a sound [or cry] through its mouth and that is a terrible sound; but at other times it sounds through its trunk and it is a sweet sound, just as the sound made in the hollow of a great tube.

Moreover, it is said that this animal fears a mouse[13] and hates the grunts of pigs, so that sometimes he flees from them.[14] It is said to be disturbed by bulls in the woods *(thauris silvestribus).* Some say that dragons with great bodies fight with elephants and, subduing them, drink their blood,

which, they say, cools them from heat. But I believe this is incredible *(fabulosum).*

On account of the weight of its body, it has testicles on the inside.[15] However, in order that coitus might be quicker, it sends for [brings down?] the testicles at the time of coitus. It has a black, pumice-like, and rough skin into which flies fall; by contracting its skin, it kills them within the wrinkles. They are also said to be chaste animals and not to know adultery;[16] but they join together at the time of passion. So strong are they that they carry twelve men, and more, in wooden towers placed upon them; indeed some large ones are said to be capable of carrying forty men. There are said

8. The beaver has "a broad, flattened, scaly tail, about one foot long, which is used as a rudder in the water" (article on "Beaver" in *Encyclopaedia Britannica*).

9. See note 3.

10. According to the *Oxford English Dictionary,* "*saphena* is the distinctive name of two veins in the leg: (1) the *long* or *internal saphena,* which extends from near the ankle-joint along the inner surface of the leg, and ends in the femoral vein; (2) the *short, posterior,* or *external saphena,* which extends from the foot along the calf of the leg, and finally joins the popliteal vein." Albert, however, clearly locates the *saphena* in the arm, not the leg.

11. In Books II, VII, VIII, and XII.

12. See Aristotle, *On the Parts of Animals* IV.12.692b. 15.

13. According to the description in the *Encyclopaedia Britannica* (article on "Elephant," VIII [1968], 274, col. 1), "An elephant is not inherently afraid of mice, but is concerned that these and other small animals might enter its trunk. Of course any such foreign object that does enter is forcibly expelled."

14. This claim may have a sound basis in the historical use of elephants in warfare. "Owing to the tendency of the elephant to be frightened by unusual sounds, attempts were made by the enemy to disturb them by the use of firebrands and musical sounds. The squealing of a pig seemed to be particularly disastrous to the peace of mind of the elephant. Frequently the animals were more dangerous to their own armies than to those of the enemy, owing to their disposition to stampede" (Francis G. Benedict, *The Physiology of the Elephant* [Washington, D.C.: Carnegie Institution of Washington, 1936], p. 9). Although the elephant may have reacted to the squeals of pigs, Richard Carrington argues that although "most naturalists from Pliny to Buffon have referred to the elephant's dislike of any member of the pig family, . . . it is extremely doubtful if there is any truth in the belief" (*Elephants* [London: Chatto and Windus, 1958], p. 77).

15. Albertus is here probably following Aristotle, who correctly reported (*Generation of Animals* I.12.719b.15) that the testes of the male elephant are not visible externally, but lie within the animal's body (see Carrington, p. 52).

16. This is untrue, for elephants change partners frequently (Carrington, *Elephants,* p. 57).

to be two kinds of them, but the more noble of them reveal an enormity of body.

Its legs are large and the lower and upper parts are nearly equal to [the height of] a column. Although the foot is divided into many parts, yet nature joined together the toes *(digitos)*, so that the foot might be strengthened.[17] And for this reason also [that is, in order to strengthen the leg], they are said not to have flexion in their legs at [above?] the knees *(post genua)*; perhaps it does not bend in a relaxed manner but rigidly. Thus, without such experience, it is believed that it does not have a regulated [or methodical] step, because if it cannot bend, it cannot have a regulated [or methodical] step.[18]

Its flesh is cold, dry, and abominable. Whoever partakes of it with water and salt and fennel seeds is cured from a chronic cough. If a pregnant woman should partake of this after it has been boiled and liquefied with vinegar and fennel seed, it would bring forth whatever she has in the uterus. Its bile *(fel)* placed in the nostrils near a weight of gold *(ad pondus aurei)* is said to work against the falling sickness [epilepsy]. When the extremity of its liver has been eaten with water and the leaves of a citrus fruit *(foliis citranguli)*, it is helpful for pains of the liver. If the excrement of an elephant should be smeared on skin in which lice appear and left until it dries upon the skin, the lice will not remain on it but will depart immediately. If the fat *(sepum)* of an elephant is smeared with it [elephant excrement], it is said to cure the pain of one who suffers a headache; it is even said that if an ounce of elephant bone is drunk with ten ounces of wild mountain mint from something which a leper first touched, it does the most for a headache. If a house or place in which there are stinging insects is fumigated with the excrement of an elephant, they flee and die.

BOOK XXIV

Crocodile

The crocodile is [included] among aquatic animals and has the shape of a lizard *(lacerta)* in all respects except that its tail is not as rounded[19] and has scales(?) *(pinnas)*.[20] It reaches a length of 20 cubits when it attains its complete size. But I saw two, one of 16 feet, the other of 18 feet. Its skin is wrinkled but so firm that it is like a shield. At night it remains in water, but in the daytime it comes forth on land[21] for food. Because of its short feet, it is slow. Its mouth opening extends to the location of the ears—if it had ears; its teeth are very strong, and it lacks at tongue.[22] And though there are many kinds of crocodiles, all animals move the lower jaw except the crocodile *(praeter tencheam)*

17. Where Albert says that nature has joined the elephant's toes, Aristotle observes that "this animal has somewhat indistinctly articulated toes, and its forelegs are much larger than the hind ones. Still, it has five toes . . ." (*History of Animals* II.1.497b.22–25 in the Loeb Classical Library translation of A. L. Peck; see also *Generation of Animals* IV.4.771b.9 and *On the Parts of Animals* II.16.659a.24–27). This confusion may be resolved in terms of the following explanation by Carrington (p. 42): "More exactly and scientifically expressed, man is plantigrade (or walks on the soles of his feet) and the horse is digitigrade (or walks literally on tiptoe). The elephant represents a mid-way stage, having a digitigrade skeleton, but being more properly considered as semi-plantigrade by reason of its supporting cushions. All elephants have five skeletal toes on each foot, but the number of nails externally visible varies between the different species and subspecies as described in Chapter One."

18. It seems that Albert, who probably never saw an elephant, denies that elephants bend at the knee and in this fails to follow Aristotle's correct account, where we read that "the elephant does not behave as some used to allege, but settles down and bends its legs. . . . Its hind legs it bends just as a human being does" (*History of Animals*, II.1.498a. 9–13). Perhaps Albert was following "the old account" as described by Aristotle (*Progression of Animals* 709a.9–12 [Loeb Classical Library]), where Aristotle says that "it is possible, however, to move even if the leg has no bend in it, as happens when children crawl. (The old account attributed such motion to elephants, but it is untrue.) Movement of this kind takes place through a bending in the shoulders or hips. But no creature could walk erect in this way continuously and safely. . . ."

19. The tail of the crocodile is indeed vertically flattened.

20. *Pinna* signifies a kind of mussel and *penna* (also spelled *pinna*) a feather or wing. None of these meanings seems appropriate.

21. The source of this statement is probably Aristotle, *History of Animals* II.10.503a.9–15.

22. Here Albert is following Aristotle, who argues that while the crocodile has no separately articulated tongue, it does have a "paltry sort of tongue" which can only be seen when its mouth is opened wide. He says that all blooded, oviparous animals "have a tongue inside the mouth—all except the river crocodile, which apparently has none, but only a space for it; and the reason is that in a way he is both a land-animal and a water-animal. In virtue of being a land-animal, he has a space for a tongue; as a water-animal, he is tongueless. This agrees with our previous statement, that some fishes appear to have no tongue unless you pull the mouth very well open, others have one which is not distinctly articulated" (*Parts of Animals* IV.11.690b.20–27 [Loeb Classical Library]; see also II.17.660b.12–34, and *History of Animals* II.10.502b.35—503a.2; see the next note).

and thus it has a very strong bite.[23] And so it lies immobile in the sun and seems to be dead, and then opens its mouth and small birds *(aviculae)* come and clean its teeth for it; sometimes the mouth closes and the birds are swallowed. It ambushes all animals, especially gazelles. The gazelle, on the other hand, tramples the crocodile in the sand. When it can, the crocodile also kills men, but afterward some say it laments this. Prostitutes make an ointment from the dung of crocodiles which they spread over the wrinkles of their faces and after the face has been washed, the original wrinkles are less evident.[24]

These beasts live on the Nile river, the rivers of India, and certain other rivers. Glutted with fishes, they come on the banks like sleepyheads. Now, the bird which cleans its teeth and which it sometimes swallows is called *crochilos* in Greek and *regulus* in Latin. As Pliny says, on an island in the Nile called Tyntirus,[25] certain men with small bodies learned a method by which they could swim out and mount the backs of these beasts and, after bridling them with pieces of wood tied in their mouths, lead them captive to the banks and sometimes compel them to give up prey that had been taken recently.

Torpedo[26]

The torpedo is a fish which, in the earlier books, we called "stupefier" *(stupefactor)*. It hides in the mud and seizes and devours approaching fish. And one who touches it is stupefied, however slight [the touch] and however quickly [the hand is] withdrawn. Thus one of our colleagues touched it only with the extremity of his finger, and after half a year of bathing with warm water and [applying] ointments, sensation was barely restored to his arm. Pliny and Isidore say that the touch of a spear stupefies and that a breeze blowing from the torpedo stupefies those who stand nearby. It has a very delicate liver *(epar tamen habet tenerrimum)*.

85 ON THE STRUCTURE AND HABITS OF BIRDS

Frederick II of Hohenstaufen (1194–1250) (Holy Roman Emperor, 1220–1250)

Translated and annotated by Casey A. Wood and F. Marjorie Fyfe[1]

Introduction by Edward Grant

The selection below is drawn from one of the great scientific treatises of the Middle Ages, whose author would have been deemed extraordinary in any age. Although most of our knowledge of Frederick II comes from his enemies, the inevitable distortions have not dimmed his image. Described by an awe-filled contemporary as *stupor mundi,* the wonder of the world, Frederick has been aptly described by Ernst Kantorowicz as "statesman and philosopher, politician and soldier, general and jurist, poet and diplomat, architect, zoologist, mathematician, the master of six or it might be nine languages, who collected ancient works of art, directed a school of sculpture, made independent researches in natural science, and organized states. . . ."[2] Living most of his life in Sicily and Italy, Frederick was wholly oriented toward the Mediterranean and Latin civilization, rarely visiting Germany, though he was Holy Roman Emperor

the immobility of its lower jaw, to which the tongue is naturally joined. We must remember, however, that the crocodile's jaws are topsy-turvy; the bottom one is on top and the top one below; this is clearly so, because in other animals the top jaw is the immovable one. The tongue is not fixed to the upper jaw (as one might expect it to be) because it would get in the way of the food as it entered the mouth, but to the lower one, which is really the upper one in the wrong place. Furthermore, although the crocodile is a land-animal, his manner of life is that of a fish, and this is another reason why he must have a tongue that is not distinctly articulated" *(Parts of Animals* II.17.660b.26–34).

24. This fantastic account bears a strong resemblance to that given in Selection 83.

25. See Pliny, *Natural History* VIII, 92.

26. Aristotle discusses the torpedo, or electric ray, in a number of places; see also the description in Selection 83.

1. [Reprinted by permission of the Stanford University Press from *The Art of Falconry being the De arte venandi cum avibus of Frederick II of Hohenstaufen,* translated and edited by Casey A. Wood and F. Marjorie Fyfe (Stanford, Calif.: Stanford University Press, 1943). —*Ed.*]

2. [*Frederick the Second 1194–1250,* translated from the original German by E. O. Lorimer (London: Constable, 1931), p. 669 (reprinted, New York: Ungar, 1957).— *Ed.*]

23. Here again Aristotle was probably Albert's source for this mistaken view (see Selection 83 and n. 15 thereto). Aristotle argues that "among the factors which contribute to the deformity of the crocodile's tongue is

and half-German by birth. Based in cosmopolitan Norman Sicily, the meeting ground of Greek, Arabic, and Latin cultures, Frederick had readily available the knowledge of many learned contemporaries (for example, the work of Pisano [Fibonacci] was personally known to him; see Selection 24) as well as numerous scientific treatises of Greek and Arabic origin. The intellectual life of his kingdom centered in his court rather than in the universities under his jurisdiction. Science was one of his serious interests, to which he devoted some of the resources of the state. He was a patron of the sciences as well as a practitioner, bestowing his largesse on Jew, Moslem, and Christian alike. He listened intently to philosophial, mathematical, and scientific discussions and sought out the company of scientists and philosophers wherever he went. In order to resolve troublesome scientific questions it was his practice to draw up questionnaires and dispatch them to the learned men of the Muslim world. Such questions on philosophy, optics, and geometry have been preserved. He was deeply interested in animals and collected a great menagerie, which followed him about through Italy and which included panthers, elephants, camels, gerfalcons, lions, leopards, and so on.

As a consequence of his restless and controversial nature and an intense desire to learn things firsthand and by experience, a host of barbaric and bizarre stories of legendary proportions have clustered around his name through the centuries. His enemies made it seem that a natural cruelty coupled with an insatiable desire to learn by experience led him to drown a man in a wine cask in order to prove that the soul died with the body (this would have made him a heretic as well) and to perpetrate other horrors, such as opening men up to observe the digestive process for the sake of obtaining similar knowledge. In truth, he was a firm believer in testing traditional explanations of problems of direct and deep interest, as was the case, for example, with birds.

The emperor not only tested the artificial incubation of hens' eggs, but, on hearing that ostrich eggs were hatched by the sun in Egypt, he had eggs and experts brought to Apulia that he might test the matter for himself. The fable that barnacle geese were hatched from barnacles he exploded by sending north for such barnacles, concluding that the story arose from ignorance of the actual nesting-places of the geese. Whether vultures find their food by sight or by smell he ascertained by sealing their eyes while their

nostrils remained open. Nests, eggs, and birds were repeatedly brought to him for observation and note, and the minute accuracy of his descriptions attests the fidelity with which his observations were made.[3]

On the Art of Hunting with Birds was probably completed in the form we have it sometime between 1244 and 1250. The edition from which the present translation was made is in six books (there is also a two-book version) and was intended as a purely scientific account of the habits and structures of birds, in which the first part (Bk. I) presented the theory, the second (Bk. II?) practice, and a third part was to embrace theory and practice.[4] It is indeed a masterpiece of objective and concise scientific description. Although Frederick used Aristotle's zoological works for zoological data (and also utilized earlier treatises on falconry as well as citing Pliny, Avicenna, the pseudo-Aristotelian *Mechanica,* and the *Aphorisms* of Hippocrates), he relied mostly on his own personal experience for the details of bird habits and structure. Indeed, in the preface to his treatise, he was quite critical of Aristotle, declaring,

We discovered by hard-won experience that the deductions of Aristotle, whom we followed when they appealed to our reason, were not entirely to be relied upon, more particularly in his descriptions of the characters of certain birds.

There is another reason why we do not follow implicitly the Prince of Philosophers: he was ignorant of the practice of falconry—an art which to us has ever been a pleasing occupation, and with the details of which we are well acquainted. In his work, the *Liber animalium* [*Book of Animals*[5]—*Ed.*], we find many quota-

3. [Charles H. Haskins, *Studies in the History of Mediaeval Science* (Cambridge, Mass.: Harvard University Press, 1924; 2d ed., 1927), pp. 263–264.—*Ed.*]

4. [In a thirteenth-century Vatican manuscript of the two-book version there is a magnificent array of approximately 900 marginal illustrations, almost all in color, depicting birds, falconers, equipment, and various other animals. Although it is recorded that Frederick could draw, and Kantorowicz believes that he made hundreds of bird drawings (*Frederick the Second,* p. 362), the drawings in this manuscript were made after Frederick's death, probably at the command of his son, Manfred. They served as exemplars for later copies.—*Ed.*]

5. [Probably a general title applied to the translations of Michael Scot (1217) containing Aristotle's *History of Animals, Parts of Animals,* and *Generation of Animals.* Of the three works, it would seem that the *History of Animals,* where birds are discussed throughout, served as Frederick's primary source.—*Ed.*]

tions from other authors whose statements he did not verify and who, in their turn, were not speaking from experience. Entire experience of the truth never follows mere hearsay.[6]

BOOK I THE STRUCTURE AND HABITS OF BIRDS

Chapter 1: Falconry Is an Art More Noble than Other Forms of Hunting

Since falconry is undoubtedly a variety of the chase, and as the art of hunting has numerous branches, each with its peculiar practices, we might consider in what both the art of venery, with all its subdivisions, and the actual practice of hunting consist. Setting aside all else, we shall at the present time discourse mainly on falconry.

The art of hunting is the sum total of experience by which men have learned to capture wild creatures of all sorts for their use by means either of force or of skill.

Hunting itself is nothing else but a form of bodily exercise and practices employed to capture animals. There are, in fact, three kinds of venery: that in which inanimate instruments are employed; that in which live animals are trained to catch other live animals; and that in which combinations of the two are used.

The art of hunting with inanimate objects is a greatly diversified one and includes the employment of nets, snares, slings, bows, arrows, and numerous other instruments.

Examples of venery of the second class are seen in the use of such living animals as dogs, leopards, and other four-footed beasts, as well as birds of prey. What birds are to be considered rapacious and what nonrapacious we shall shortly determine.

As we intend to confine the present work to hunting by means of birds, we shall now take up the employment for that purpose of trained raptores and in this chapter give our reasons for believing it to be an occupation more worthy than other forms of hunting and explain why we select it for discussion.

There are many arguments that can be advanced to demonstrate the noble character of falconry, as the discriminating reader of this book will soon discover; and he will in this way learn more about the secrets of nature than if he followed other kinds of venery. It is true that the latter are more popular, because their technique is crude and easier to learn; falconry, on the other hand, is less familiar and does not commend itself to the majority because skill in it is difficult to acquire and because it is more refined.

Moreover, as regards other forms of hunting, which so many follow with enthusiasm, they are less noble because they depend merely upon the use of artificial implements, such as nets, snares, traps, huntings spears, javelins, bows, and slings, or they are carried on by means of four-footed animals, both tame and wild, such as various sorts of leopards, dogs, lynx (male and female), ferrets, and other beasts.

On the other hand, the art of falconry is not dependent upon such auxiliaries as artificial tools or four-footed animals but is almost entirely conducted with the aid of birds of prey that are indeed more noble instruments of the chase than inanimate objects or trained quadrupeds.

It is also true that it is far more difficult and requires more ingenuity to teach raptorial birds the stratagems of hawking than to instruct dogs or wild quadrupeds to hunt, because birds of prey are more afraid of man than are other birds or such four-footed animals as are used in the chase.

Moreover, raptorial species do not eat grain or similar food cultivated by man as do many other birds. As a result they do not associate with men and do not easily become domesticated. It is also well known that raptorials avoid man more than do other avian species and certain quadrupeds. Again, birds of prey frequent localities inhabited by man less than do the last-named animals. It may be added that wild and shy quadrupeds that shun mankind are difficult to tame and train for hunting and these difficulties are still more marked in the case of birds of prey. It is to be remembered, also, that the habitat of quadrupeds is limited to the earth's surface, their movements are not very rapid and they generally run along in an upright position, whereas birds fly quickly through the air. Consequently the former are more easily brought under human subjection than are the latter, and they are readily caught by the use of force or are trapped by other means because they remain on the ground. Fully fledged birds, on the other hand, can be captured and trained only by finesse.

It is evident not only that the art of falconry presents greater difficulties but requires more unusual skill than do other forms of venery.

By means of this noble art most raptorial birds can be taught to hunt and capture even such birds as cranes, bustards, geese, and other large game birds that are bigger and heavier than those they capture alone in their wild state, as well as to take

6. [Wood and Fyfe, pp. 3–4.—*Ed.*]

smaller quarry not only in their natural fashion but more often than is effected by other methods.

Although it is true that birds of prey display an inborn antipathy to the presence and company of mankind, yet by means of this noble art one may learn how to overcome this natural aversion, to win their confidence, and to induce them even to seek those they previously avoided.

By the proper exercise of falconry, raptorial birds are taught to tolerate the society of human individuals and their associates for hunting purposes, to fly after quarry, and to behave (without control) just as they would in their wild state. Any dabbler in venery can readily hold in leash or let loose dogs or other quadrupeds; but in the pursuit of falconry no tyro can so easily join in the chase, either to carry his birds or to throw them off at the quarry. Falcons and other hawks are rendered clumsy or entirely unmanageable if placed under control of an ignorant interloper. By using his hearing and eyesight alone an ignoramus may learn something about other kinds of hunting in a short time; but without an experienced teacher and frequent exercise of the art properly directed no one, noble or ignoble, can hope to gain in a short time an expert or even an ordinary knowledge of falconry.

Here it may again be claimed that, since many nobles and but few of the lower rank learn and carefully pursue this art, one may properly conclude that it is intrinsically an aristocratic sport; and one may once more add that it is nobler, more worthy than, and superior to other kinds of venery.

Let it then be the first one discussed. To other forms of venery, especially those patronized by the nobility, we shall return (our life being spared) when we have completed this present treatise.

Chapter 2: Of the Division of Birds into Waterfowl, Land Birds, and Neutral Birds

In this first section of our work we shall discuss those aspects of bird life it is necessary to understand before investigating the art of falconry. Birds, like all other animals, may be divided into various species in accordance with their sexual relations, their parturition, their methods of procuring food, their variations in residence at different seasons of the year (chiefly as a result of alterations in temperature), and their change in diet.[7]

Consider, then, a general division of birds based on the different activities they employ in securing food, the great varieties of that food, the flights they make to localities both near and far in search

of heat or cold, the variety of and support given them by their limbs, the peculiarities of their feathers, their art of flying, their contests, and their moulting. It is, therefore, quite obvious, in so far as our plan permits, that we, as practitioners of falconry who hunt with birds of prey, should discuss these avian traits so that we can better understand where, when, and how hunting birds can be taught to catch their prey. All that we do not include on the nature of birds can be found in Aristotle's book *On Animals*.[8]

All birds may be divided as follows:

Waterfowl are birds that habitually live in or near aquatic areas and whose organs are so fashioned that they may remain for indefinite periods immersed in water.

Land birds are those that prefer a continuous life on land, an existence for which their bodies are especially constructed.

Neutral birds are those that may change from one habitat to another, from earth to water and vice versa, as shown both by their preferences and by their bodily structure.

Although Aristotle declares that every creature may be classed either as a water resident or as a terrestrial animal, and that only fish are truly aquatic, and includes under the term of land animals all those that progress both by walking and by flying, yet he does not make the mistake of classifying all winged creatures as birds. We however, following the usage of falconry experts and adopting its terms, divide birds (in whose mingled constituents the lightest of elements predominate and who are the lightest and most agile of the winged inhabitants of the air) into water, land, and neutral birds; and of all these we shall give examples. We find that they may also be divided into various genera and these again into a number of species.

Water birds rarely leave and prefer to remain in

7. This is largely Aristotelian, or Stagirite doctrine. [In Book VIII, chapter 12, of his *History of Animals*, Aristotle observes that "the habits of animals are all connected with either breeding and the rearing of young, or with the procuring of a due supply of food; and these habits are modified so as to suit cold and heat and the variations of the seasons. For all animals have an instinctive perception of the changes of temperature, and, just as men seek shelter in houses in winter, or as men of great possessions spend their summer in cool places and their winter in sunny ones, so also all animals that can do so shift their habitat at various seasons" (Oxford translation by D'Arcy Wentworth Thompson). Much of this relatively brief chapter is subsequently devoted to birds.—*Ed.*]

8. [See note 5.—*Ed.*]

the water.[9] They do not leave it in search of food or for other purposes except when they fly from one body of water to another or during their seasonal migrations. These species include mergansers, cormorants, swans, and those birds that Aristotle in his *Natural History* calls pelicans and which the Apulians dub "cofani."[10] The last-named are as large and almost as white as swans; they have a long, broad beak that has a pouch attached that they open and shut like a fishnet.[11] They also have a [swimming] web between their feet that extends to the hind toe; the latter condition is not found in other waterfowl with webbed feet.[12]

These and many others rarely leave aquatic areas. Others occasionally do so for the reasons aforementioned, as well as in search of food, and then return to it according to their natural impulses. This is the rule with some species of geese, ducks, and similar birds. Certain land birds visit bodies of water, but only for drinking and bathing, among them quail, partridge, pheasants, [common] bustards, and lesser bustards[13] (that are like the former only much smaller). To this class belong also peacocks and birds like them.

Certain birds live most of the time on land but resort to water not only for drinking and bathing but also to secure their food, like aquatic fowl, returning to the land for rest. These include the sea eagles, which dive into the sea, rivers, and swamps to catch fish, after which they return to the cliffs and trees where they dwell. It is proper to class them with land birds because they are birds of prey and as such ought not to be called waterfowl.

Of neutral birds one may distinguish three types. Some of them (like the curlew) prefer water to land, in response to the demands of their bodily structure. Then there are others that have a changeable residence but seem to prefer land to water, such as plover, lapwings, and snipe. Of these, plover love the water less than lapwings, and lapwings more than snipe. The latter more nearly approach the character of land birds than lapwings or plover, because, though both these birds often sleep on land, snipe slumber there more frequently. There are still other birds that remain as much in the water as on land, like the cranes, both large and small,[14] also both kinds of storks, the white and the black. The latter are frequently seen wandering about, fishing in water and in swamps and other wet places, returning afterward to dry land.

Into these three classes (land, water, and neutral) are all birds divided.

It may be added that those neutral birds who spend the greater part of the time in the water are customarily called waterfowl by bird hunters and these as well as true aquatics are termed shore birds.[15] Those neutral birds that follow their natural bent and live mostly on land are often called terrestrial; and both these and land birds may be styled field birds, or birds of the plain. Some of those species that live equally on land and water may be termed either land birds or water birds; but those neutrals that come twice a day to feed on dry land but return each time to an aqueous resort for safety and rest, although they frequent the meadows, are nevertheless to be classed as aquatic birds, since they most resemble true waterfowl in their habits and seek the water as a permanent place of refuge. Among such birds are the cranes. Those neutral birds, however, such as snipe, plover, and lapwings, that live in meadows but go to bathe and drink in the water and take refuge there when pursued by birds of prey, may well be called land birds; for, as will be observed in the chapter on bird defense, many land birds take to the water when chased by their raptorial foes, for example, the crows.

9. The mergansers make use of tall trees and terrestrial nests in which to raise their young: at any rate they prefer the land for this purpose. As for cormorants, their nests are found on trees or cliffs. Swans patronize islands and often make floating nests. Pelicans nest in canebrakes or among reeds. All the birds mentioned by Frederick II leave the water for various purposes.

10. One is struck by the evident reference to the pelican's pouch. It. *cofano*, a basket or small trunk; Lat. *cophinus*; Gr. κόφινος. [Perhaps a reference to Aristotle's *History of Animals* VIII.12.597a.9–13.—*Ed.*]

11. The common white pelican, *Pelicanus onocrontalus*, changes from brownish to a light rose color after its third year, and looks white at a distance, especially when old and just before moulting.

12. The same palmated formations seen in pelicans occur also in common cormorants, darters, gannets, tropic birds, and frigate birds, that constitute a separate sub-genus of *palmipeds* or web-footed birds.

13. The little or lesser bustard *(Otis tetrax)* is common enough in the Far East and in Southern and Southeastern Europe but is rare in the North. Frederick II gives these birds a nickname—*Anates quae dicuntur campestres, quae similis sunt bistardis*—bustards being called by him *bistardae*.

14. Probably the "small" cranes were the young of adult birds and wore their nestling garb. They associated with older individuals; or they may have been "demoiselle cranes" *(Anthropoides virgo)* that occasionally visited Italy and are certainly smaller and more delicate than the common crane *(Grus cinerea)*.

15. *aves de rivera.*

Chapter 3

Of the Division of Birds into Raptorial and Non-raptorial Species

Birds may be classified in still another manner—as raptorial and nonraptorial species.[16] We call raptorial all those birds who, employing their powerful flight and the special fitness of their members, prey upon any other bird or beast they are able to hold and whose sole sustenance is the flesh of such animals. These are the eagles, hawks, owls, falcons, and other similar genera. They feed only on their prey—never upon dead flesh or carrion *(carnibus cadaverum neque residuis)*—and are therefore called rapacious birds. Aristotle calls them "greedy-clawed" birds or sometimes "birds of the hooked claws";[17] but this nomenclature seems to us unsuitable, since it is erroneous in so far as birds such as jackdaws, the larger swallows, and vultures have hooked claws and yet may not properly be called raptores, as they do not feed upon their own quarry.

[It was the habit of Aristotle and the philosophers to classify objects into positive and negative groups and to begin their discussions with the positive.[18] Since it is our purpose to give special attention to raptorials, we shall first consider the nonrapacious (or negative) varieties; afterward we shall consider at length raptorial birds.]

Nonraptorial species are those (whether aquatic, land, or neutral birds) that do not live entirely by robbery; in fact they cannot be regarded as true birds of prey if they subsist in part by plunder and partly on grains and fruit, like some ravens, crows, and magpies—less frequently certain species of vultures, and the so-called "bone-breakers" (lammergeiers)—also some ignoble eagles[19] that never plunder other birds or quadrupeds but feed on dead bodies and scraps.

It is,[20] therefore, evident that all birds may be included under the two categories of rapacious and nonrapacious, and that birds of prey differ from harmless species not only in their method of securing food, as is herein described, but also in many other ways, as, for example, in the form of their members, in their behavior, and in the thickness or sparseness of their plumage.

Among the characteristic forms of their organs may be mentioned: the beak, which in birds of prey is generally curved, strong, hard, and sharp; claws that are bent inward and are hard and needle-pointed; retracted eyes;[21] a short neck, short legs, and the posterior toe of each foot very strong. The female is larger than the male. Not all of the foregoing is true of nonraptorial birds.

Functionally also they differ in that raptorials are more keen-sighted and have more acute hearing than other birds. They are strong in flight but walk badly. They dislike water and drink little, fly alone,

16. The remainder of this paragraph (given here as it appears in the Bologna MS.) is missing entirely from the Vatican Codex. In its place there has been substituted the next paragraph (placed in brackets) which does not appear in the Bologna MS. [The Bologna MS. is Bologna, University Library, MS. Lat. 419(717), written in the thirteenth century. The Vatican MS., also of the thirteenth century, is in the Vatican Library, MS. Palatine Lat. 1071. Although there exist other manuscript versions of Frederick's treatise, these two formed the basis of the Wood and Fyfe translation.—*Ed.*]

17. *aves unquis rapaces et aliquando aves uncorum unguium.* Bologna MS., fol. 3. [The reference to Aristotle is perhaps to the *History of Animals* IX.32.619b.7–9.—*Ed.*]

18. [An obvious reference to classification by dichotomy, which was used by Plato and initially by Aristotle. In this system one differentia was intended to define a species, and these differentiae were expressed in terms of pairs of opposites. Thus one might divide all animals into feathered and unfeathered (just as Frederick divides bird into rapacious and nonrapacious). Since unfeathered is not further divisible, one could then further dichotomize "feathered" into barbed and unbarbed, and so on. Aristotle eventually rejected this method after formulating a number of criticisms of it. Thus, in the example above, since unfeathered is not further subdivisible, it is a differentia and should represent some species. But "unfeathered" is a characteristic of many species and hence useless as a classifying characteristic. "The method of dichotomy is either impossible," Aristotle came to argue, "(for it would put a single group under different divisions or contrary groups under the same division), or it only furnishes a single ultimate differentia for each species, which either alone or with its series of antecedents has to constitute the ultimate species." Instead, Aristotle insisted that it was essential "to recognize the natural groups, following the indications afforded by the instincts of mankind, which led them for instance to form the class of Birds and the class of Fishes, each of which group combines a multitude of differentiae, and is not defined by a single one as in dichotomy" (both quotes are from *Parts of Animals* I.3.643b.10–17 in the Oxford translation by William Ogle). In the *History of Animals,* which Frederick seems to have known best, Aristotle employed the method of dichotomy, but eventually utilized criteria such as reproduction, levels of vital heat, and so on.—*Ed.*]

19. *modus aquilarum ignobilium,* probably vultures other than the *Geier.* Schöpffer remarks that all the German eagles are predatory, and, although they do not eat carrion, they rob other birds.

20. The text from here to the sentence beginning "It is also to be noted," on page 663, is entirely lacking in the Vatican Codex.

21. *oculos concavos.*

and live long. They drive their young early from the nest and then abandon them; and this behavior is not that of nonrapacious birds.

As to plumage, it varies among raptores; the first year after hatching (when they are called sorehawks *[saure]*) they moult only once, while other birds *(generally)* shed their feathers twice. The large quill feathers of the wings and tail are limited to a definite number; this is not true of other birds.

In numbers also the two classes differ, for there are fewer rapacious birds than nonrapacious; and there are no raptores among aquatic and neutral birds, but only among land birds, and even here they are few in number; so that all water and neutral birds and the greater part of land birds are nonrapacious.

Rapacious birds (which are universally warmer and drier than aquatics and neutrals) dislike water for two reasons, one active and the other passive. Since they have not members and plumage of a suitable form, they do not live in the water, nor can they do so, because they cannot continue to stand in deep water, lacking long legs like those of herons and cranes, nor can they swim about with ease, as their feet and toes are not webbed like those of geese, ducks, coots, and nearly all aquatic birds. Were a raptorial bird overturned, or submerged, in water her feathers and quills would be more inclined than those of aquatic birds to become soaked, so that she could hardly fly, and her claws would become so softened that she would be unable to wound or hold her prey. For these reasons, birds of prey dread remaining in the water, since they are extremely feeble in that element. There are certain birds, however, similar to eagles but smaller, that perch above bodies of water (or on high banks) and, when they perceive fish in the water, suddenly drop on them, draw them out alive, and feed on them. They are, therefore, called fish eagles. Their members and plumage are better adapted for this purpose than are those of other raptores.

The genera into which raptores are divided, and the species in each genus, will be discussed more fully in another treatise[22] and in other parts of this work.[23] It is also to be noted that the same genera and species are given different names by diverse authors. Sometimes the same bird may have a variety of synonyms; and the same name applied to diverse birds that are so dissimilar that one cannot establish the true identity of a species simply by its name. In consequence of this multiplicity

of terms, a description of the essential characters of individual birds is more difficult to furnish, whether they resemble or are different from another in the shape of the limbs, the movements they make, the way they feed, the care of their young, their mode of flight, and their style of defense.[24] Let it, however, be remembered that, in general, their bodily conditions and their other peculiarities are due to definite causes.

Very different localities may possess the same species and genera not found elsewhere; or a single region may be the habitat of birds of a species found nowhere else; while, on the other hand, in one region may be seen a genus found elsewhere but of a different color, or varying in other respects, but which may be identified by the characters of its members, its feeding habits, and other essentials. Therefore when we give one example of a genus and speak of others as similar, it is not necessary to repeat all the identification marks, lest we be guilty of tiresome prolixity. Indeed there is a multitude of birds, aquatic, terrestrial, and neutral; and so many genera are in each class that it would take too long even to mention them. However, when birds presenting essentially the same (or entirely different) characteristics are encountered, they may be described as belonging to the same or to a different genus, and this will be true even of birds not seen elsewhere and not previously identified.

Chapter 7
Of the Return of Aquatic Birds to Their Water or Swamp Refuges

The home localities to which waterfowl return may be lakes, ponds, swamps, or some of the longer or smaller streams, and other water-covered areas sometimes called flats; but they prefer to swim about and plunge into some rocky and extensive body of water possessing islands and other advantages. Their home-coming takes place during the day, that they may guard against such wild animals as fish otters,[25] foxes, and birds of prey. The last-named do not pursue waterfowl very much in the water, because swimmers can easily

22. Probably in the work on hawks that was often promised but was presumably never written.
23. At this point ends the passage missing in the Vatican Codex.
24. A difficulty not confined to medieval ornithology.
25. Schöpffer properly says that the fish otter (*Entra vulgaris*) may now and then catch birds for food but only when very hungry and unable to secure his proper food, viz., fish.

escape from them by diving. Even weak waterfowl more successfully defend themselves in this way than many other stronger birds that live away from water. Aquatic birds[26] return to the water in the daytime for the purpose also of drinking and resting. This is especially true in summer. At night they remain standing in the water not only for protection against otters, foxes, wolves, and other wild animals that may harm them while sleeping but also that they may pass the night in peace, sleeping and resting.

Chapter 8
Of the Positions Assumed by Birds when Asleep or Awake on or in the Water

During sleep swimming birds pass part of the time floating on the water, part of it near the shore with one or both feet on the bottom. Nonswimmers, aquatic birds or neutrals, keep either both feet on the ground under the water, or rest on one foot with the water up to the knee or above that point, depending upon its depth.

The larger number of water birds, like some land birds, when sleeping, turn round the head and rest it on the back between their shoulders. Indeed both swimmers and nonswimmers usually sleep on one foot, holding their heads on their backs below the shoulders so that the head as well as the cold and horny beak may be kept warm. This position not only provides warmth but prevents wetting the head (keeps it dry) and thus avoids possible freezing of the parts in very cold weather. Disease in the head, called *gipsus,*[27] may easily set in from exposure to wet and cold.

Not only waterfowl but almost all other birds sleep on one foot in order that they may sleep lightly and be easily alarmed and readily awakened to meet approaching danger. Water birds asleep with one foot in the water readily sense any disturbance of the surrounding fluid and are thus warned of the approach of beasts of prey or other enemies. Since[28] water is liquid and its limits are ill-defined, it recedes from the point of disturbance with a circular motion that extends to the leg of the bird, who, feeling the motion, is instantly alarmed and put on guard. Moreover, a body that has several points of support stands more firmly and is more difficult to move than one resting less securely. Hence a bird resting on one foot only is easily roused. This, added to the fact that any person sleeping under threat of danger is easily disturbed, makes it doubly plausible that birds

sleeping normally under these conditions will be quickly awakened.

[Addition by King Manfred:[29]

There are other reasons for the foregoing which Manfred, former King of Sicily, son of the divine, august, and imperial author of this treatise, appended when he ordered it read to him.

When birds are fatigued after their search for food they seek nocturnal rest by sleeping for a time first on one foot and then on the other, just as tired-out four-footed animals do.

There is in this connection another consideration—the feet of birds are almost entirely devoid of flesh, being composed mostly of bones, nerves, ligaments, skin, and horny substance (toenails) that in their nature lack warmth. Birds meet this defect by drawing them up alternately under their feathers to heat them.]

When a storm blows up during sleep, birds turn their breasts toward the wind, so that their feathers may remain smooth and not be ruffled, as they would otherwise be were the wind to come from behind, and so disturb them during their slumbers.

Chapter 17
Of Which Birds Migrate and at What Season

As migrants, first in order come the waterfowl, including those that live on fish alone or on fish and other aquatic food, as well as those that prefer terrestrial products or that feed on both; the last-named are, however, not good fliers. Next in order come the neutrals, especially those that subsist on worms dug out of the earth. Land birds, both the raptorial and harmless species, large or small, are also birds of passage.

Not all water, shore, or land birds migrate; those who are unable to take long flights, inclusive of weak or disabled individuals, cannot make this journey, particularly to distant lands. They make another change in place of it, moving to neighboring localities; in winter, perhaps, they fly from hills to valleys near by and in summer come back again from the valleys to the mountains. This is the habit of such land birds as partridges, red-

26. The text to the end of this chapter, missing in the Vatican Codex, is from Bologna MS. fol. 5, col. 2.
27. Acute catarrh, i.e., inflammation of the cranial passages.
28. The material from this point to the end of the paragraph not found in the Vatican Codex, is from Bologna MS., fol. 5, col. 1.
29. See Vatican Codex, fol. 8ᵛ, col. 1, l. 6, to col. 2., l. 9. [Manfred was Frederick's favorite son and was much like his father (see n. 4).—*Ed.*]

legged partridges, pheasants, peacocks, bustards, and, generally speaking, most of those birds that do not pair off during the breeding season. In winter some aquatic birds seek (neighboring) warm waters, either a large river, the sea near by, deep thermal springs, or lakes the waters of which have a summer temperature. This is the habit of such limited fliers as mergansers and rails. Superior fliers, who have become weak or lame, also make a change of this sort instead of attempting to migrate to far-off regions. These handicapped birds, though true migrants, are dubbed, in the neighborhood where they remain behind, "natives," since they become, as it were, permanent residents.

Of course, birds born and living in India and in regions below the equator or near it (the Tropics) build their nests there and rear their offspring until they are mature. They need not migrate to procure food elsewhere or to avoid excessive cold, since in those regions the sun reaches the zenith twice in the year and they have really two annual springs, two summers, two winters, and two autumns. Consequently all avian foods—fruits, herbs, seeds, locusts, and beetles—are found in abundance. Moreover, no cold of winter does injury to bird life, since the sun is always overhead except for 23° or less. Since tropical birds feed their young in spring and summer, like other avian species, if they wish to migrate and avoid excessive heat above or below the equator they can do this in a direction reverse to that of the solar movements. If they wish to avoid cold, they can easily find a hot climate. However, they are much more likely to be forced to escape excessive heat, particularly at the time they build their nests and rear their young; but even at that time they do not migrate to a great distance. Like other nonmigratory birds they easily make a short change from valleys and plains to mountains or woods near by, from neighboring warm to cool waters, and thus avoid extreme heat, and follow the contrary rule to avoid cold.

Chapter 18
Of the Reasons Why Birds Migrate

The reasons why birds desert their usual resorts are numerous; but, as we have said, it is chiefly to avoid excessively cold or very hot weather. For birds, like other living creatures, depend for their existence and survival upon a certain balance of fundamental conditions, and for this reason (since due proportion is conserved by moderate and destroyed by excessive heat or cold) birds require

for their well-being a moderation of the atmosphere and other environmental conditions; so they take refuge from extremes of either heat or cold. Although they can pass spring and summer comfortably in cold climates, northern resorts become disagreeable in autumn and winter. Birds, therefore, seek warmer haunts even though these are to be found only in remote countries.

Other reasons for migration are less important. One such explanation is their search for food; when water freezes and herbage dries up, and the surface of the soil solidifies so that grass seeds, fruits, and other foods that fall to the ground cling to and become part of the hardened earth, and when even the worms beneath the soil cannot be reached, it is with the greatest difficulty that birds obtain their necessary fare—flesh, fish, plants, or insects—with the result that they fly off to warmer localities where neither earth nor water freezes and there is no scarcity of food.

For rapacious birds it is no less needful to seek change of climate, although they can withstand temperature variations better than most other birds; but as they feed on avian species it is necessary to follow the latter on their excursions to milder climates.

Also, smaller birds—goldfinches, blackbirds, starlings, redwings, and thrushes—migrate for the reasons mentioned above. Cranes, herons, geese, and ducks also change their temporary habitat because of intemperate weather conditions. Storks, curlews, plovers, lapwings, and all those birds with long bills who cannot bore into the frozen earth for food, as well as snipe, which live on beetles and other insects, join the migrating flocks.

Among other reasons for a change of climate are the heavy rains, storms, and snowfalls of cold countries, which make bird flight and life generally very difficult. Still another explanation of avian migration in the case of waterfowl and some neutral species is that the freezing over of their aquatic resorts renders these localities useless as a refuge from predatory beasts and birds of prey. So they seek a climate where watery sanctuaries do not freeze.

Chapter 19
How Birds Prepare for Migration

Birds do not prepare for migration as soon as they leave their nests, for they have neither the strength nor feathers firm enough to transport them for great distances. Moreover, the cold weather does not, generally, set in to compel them to migrate, and there is still plenty of food available. The young birds fly about their birthplace from early spring to late summer, sometimes with their own, sometimes with other species, alone or in numbers, without apparent plan or purpose. As summer wanes and winter approaches, both young and adults leave their breeding grounds and gather with their own kinds, young and old (indiscriminately), waiting, feeding, and preparing, for a day favorable to their long journey.

This takes place in the case of all migratory water, neutral, and land birds with the exception of certain land birds. Raptores, however, never prepare for migration by mingling with other birds of prey, whether of the same or of different species, because it it their habit to rob each other of any quarry that is caught. They therefore travel alone in fear that another rapacious bird may carry off their prey.

As noted, most neutral, land, and water birds assemble, species with species. It must be remembered that, although different species that closely resemble one another may feed in the same localities and may mix freely, rarely or never will they form flocks with each other, no matter how closely related they are. They will gradually join their fellows, as may be seen in a promiscuous gathering of geese, from which red geese, gray geese, barnacle geese, and small geese gradually withdraw and finally make up distinct, specific groups.

As cold weather looms in one locality after another, birds migrate little from one vicinity to another. In moderately cold countries preparations for migration are more prolonged than in climates where severe winters are the rule.

Chapter 20
Of the Seasons and Weather That Favor Migration

With a prophetic instinct for the proper time to migrate, birds as a rule anticipate the storms that usually prevail on their way to and from a warmer climate. They are conscious of the fact that autumn follows summer (when they are strongest and their plumage is at its best) and that after these seasons comes the winter—the time they dread most. They are instinctively aware of the proper date of departure for avoiding the winds to which they may be exposed in their wanderings and for eluding the local rains and hailstorms. They usually are able to choose a period of mild and favoring winds. North winds, either lateral or from the rear, are favorable, and they wait for them with the same sagacity that sailors exhibit when at sea. With

such helpful breezes progress and steering in the air are made easy. With these to help them on their way, they reach, with comparative comfort, the distant lands of heart's desire. When they fly before the wind they can rest on an even keel, still maintaining progress, especially when propelled in a proper direction. When becalmed they do not fly so satisfactorily, for they must exert themselves all the more. With head winds there is a threefold difficulty in attempting to float, to fly forward, and to overcome direct aerial obstacles.

Among flight obstructions there are also to be considered not only contrary winds but local rains, hailstorms, and other forms of bad weather that may affect both air and sea, so that some birds fall into the ocean and others, when possible, fly on board a ship (where they are easily caught), preferring that fate to certain death or to continued exposure to the rigors and dangers of oceanic storms.

We notice also that when a favoring wind springs up, whether by day or night, migrating birds generally hasten to take advantage of it and even neglect food and sleep for this important purpose. We have observed that migrating birds of prey, that have begun to devour food we have thrown to them, will abandon it to fly off if a favorable wind begins to blow. They would rather endure hunger and travel day and night than forego such an advantageous opportunity.

The calls of migrating cranes, herons, geese, and ducks may be recognized flying overhead even during the night, and not, as Aristotle claims, as a part of their efforts in flight; they are the call notes of one or more birds talking to their fellows. For example, they understand wind and weather so thoroughly that they know when meteorologic conditions are favorable and are likely to remain so long enough to enable them to reach their intended haven. Weak flyers postpone their journey until they are sure of a prolonged period of good weather sufficient for their migrating venture, but hardy aviators take advantage of the first propitious period to begin their flight.

Chapter 21
Of Early and Late Departure in Bird Migration, and of Their Flight Formations

The slower migrants begin their departure early. For example, the smaller birds, as well as storks and herons, remain until the end of summer and leave the last of August so that they may not be embarrassed by changeable weather or early (autumn) storms. The more robust species and better fliers remain until the beginning of harvest (about the middle of September). Among the latter are the larger and smaller cranes. At that date strong fliers can readily defy the early winds and rains. There are, moreover, still better and swifter fliers who postpone their departure to the end of the autumnal season, say, until November. These include certain ducks and geese who do not fear high winds and heavy rains because of their skill in flight and because their plumage protection against cold is adequate. This rule applies also to the smaller geese who may remain behind in the sixth and seventh climatic zones[30] the whole winter through, inasmuch as they can find there the herbage on which they feed. The larger geese also possess unusual meteorological instincts and avian alertness. In years when there are short summers, i.e., when the winter threatens to set in early, they migrate much sooner than usual.

Certain birds, cranes for example, who pass the summer in the far north (where winter comes on early) on account of the longer journey before them, migrate sooner than others of their species who, having nested farther south, prolong their northern visit, since their winter comes later and they have a shorter journey to make. When autumnal winds are favorable, these birds resume their southern flight and, traveling without intermission, quickly accomplish the voyage. Inclement weather, however, may delay the flight of species that have hatched their young in more southern localities until the storm has passed. Those nearest the equator begin their migration last.

The order of avian departure may be summed up as follows: not all shore birds depart pell-mell, like the disorderly land birds; the latter do not seem to care what birds lead the van or which form the rear-guard of the migrating flocks. Water birds, on the contrary, preserve the following order: one forms the apex of advance, and all the others in the flock follow successively in a double row, one to the left and one to the right. Sometimes there are more in one series than in the other, but the two rows, meeting at an angle, form a pyramidal figure. Occasionally there is a single line.

This order they maintain not only when migrating to distant points and returning but, as has been explained, in going to and from their local feeding grounds.

One member of the flock continually acts as leader and, especially in the case of cranes, does this not because he alone knows the goal they seek but that he may be ever on the lookout for danger, of which he warns his companions; he also notifies them of any change to be made in the direction of flight. The whole flock is thus entirely under control of their leader or guide. When the latter becomes fatigued from the performance of this important work, his place in front is taken and his duties are assumed by another experienced commander, and the former leader retires to a rear rank. It is not true, as Aristotle asserts, that the same leader heads the migrant column during the whole of their journey.

Chapter 23–D
Of the Mating of Birds

It is not our intention to enter into a full discussion of the subject of avian reproduction, which generally takes place after the return of birds to their native haunts. It may be said, however, that the pairing of male and female is for the purpose of breeding. Pairing is a preliminary to coitus, which, in its turn, is followed by breeding. Mating is preceded by the abandonment of the now full-grown young that resulted from a previous union. This procedure is necessary because last year's fledglings, mindful of the food furnished by the parents, are not inclined to leave them until they themselves reach the age for mating or until they are otherwise prevented from following their parents or are actually driven off by them. Raptorial species not only expel their young in this way as soon as they can fly but drive them from the neighborhood, as will be explained later.

This method is not, as a rule, adopted by harmless species, who do not desert their offspring until they are again affected by the instinct to breed; meantime the young birds endeavor to follow and

30. Abu Abdallah Mohammed Idrisi (ca. 1099–1154), the Arabic geographer, attached for twenty-five years to the court of Roger II (1101–1154) at Palermo, was the author of *Al Rajori*, or *The Going Out of a Curious Man to Explore the Regions of the Globe, Its Provinces, Islands, Cities and Their Dimensions and Situations*. In this work Idrisi not only describes journeys to such places as Scandinavia, the African coast, Egypt, Syria, etc., but he recognizes the rotundity of the earth. He divides the known world into seven climates between the equinoctial line and the point where the earth becomes uninhabitable because of extreme cold. [Information about the seven climes could also have reached Frederick through various other sources (see Selection 1, n. 60.—*Ed.*] Longitudinally he makes eleven divisions between the most westerly point of the African continent and the eastern coast of Asia. Two manuscripts of his work are found in the Bodleian Library and two in the Bibliothèque Nationale.

associate with their parents. Even those species that breed several times a year are accompanied by their fledglings, who are driven off only at the last moment.

In the case of birds that breed once only during the year, particularly cranes, the young birds follow their elders and are not expelled by them for a whole year. But in every instance fledglings are driven away and, if they are unwilling to go, the parents use force and even beat their offspring to prevent them from interfering with the new brood by destroying the nest, breaking the eggs, or killing the occupants.

Having abandoned their young and all others of their kind among whom they have been living, the full-grown birds now give themselves up to the business of mating, each male choosing a single female of his own species. This applies, of course, only to those species in which birds pair off for purposes of reproduction. There are, however, many races of birds where the male does not mate with a single female but is polygamous. The males of these species do not, as a rule, assist the females to build the nest, to incubate the eggs, or to feed the young; for example, those dark-colored birds called wild peacocks, a second black-hued species resembling a pheasant,[31] and many others, including certain varieties of quail, peacocks, pheasants, and bustards (nearly all nonmigratory varieties), make little effort to help the female in the tasks of nesting and of rearing the young. A few of these resident species are, however, monogamous—for example, the partridges.

All migratory birds, when they attain the proper age and have the desire to mate, pair off in the manner described and the male bird invariably assists his mate in some part of the many phases of brooding and bringing up the young birds.

We say "when they attain the proper age," because some birds (like the cranes who breed only once a year) do not undertake the business of procreation until after they are at least a year old, sometimes waiting until the second year of life; others breed at the end of their first year, and these birds usually select their mates before the end of that year. These last are those that, like the pigeons, often breed several times a year.

We have emphasized the avian urge to mate because there are individuals who, though they are ripe as to age for reproductive activities, neither mate nor reproduce. They are prevented by some abnormal impediment from satisfying this impulse to perpetuate the race. Such birds are, as a rule,

solitary; more rarely they are found in colonies.

As soon as birds feel the impulse to mate and reproduce, they select their partner in the following manner: They return to their native country, or to a locality that closely resembles it, if for any reason it is not available. At this time more than at any other, birds give voice to various cries, frequently singing and whistling. By these calls males and females recognize each other, and particular cries are indicative of a desire for coitus. Not only do males distinguish the notes of females, and vice versa, but each individual bird can identify its mate by the sound of his or her voice. When birds have made their choice, they pair off and usually forsake all others.

The behavior of raptores is somewhat different from that of other migrants. One of a pair will be seen waiting for the other in the vicinity of their permanent nest or eyrie. Sometimes it is the male who first makes his appearance near the nest, where he may await the female for many days. In this same fashion the female may be seen watching for her mate; or occasionally they arrive simultaneously at the eyrie.

Chapter 23–E
Of Coitus in Birds

Inasmuch as the pairing of birds at the period of the return ultimately ends in coitus, the laying and brooding of eggs, and the reproduction of the species, we would now discuss the minute details of the whole process and the attendant avian behavior if it were essential or in any way served our purpose; but as it is not relevant to the subject in hand we shall omit further discussion of it. There is, however, one fundamental fact that must not be overlooked. Nature in her endeavor to preserve the race by the continuous multiplication of individuals has decreed that every species of the animal kingdom, whether it progresses by the use of wings or walks on the ground, shall take pleasure in sexual union so that they may seek instinctively to bring about such enjoyment. Birds take such delight in this natural function that even birds of prey, who at no other time seek the companionship of their kind, not only come together at the mating season but even exhibit definite signs of mutual affection.

Chapter 23–G
Of the Laying and Incubation of Eggs

We pass now from the subject of nest building

31. Probably the capercailzie and the blackcock, both species of European grouse.

to the laying and incubation of birds' eggs. To begin with, we cannot give here exact data as to the number of eggs laid by a given species, since the figures would be too numerous and too diverse to record. The following facts must not, however, be overlooked. Cranes and other large birds, such as swans, pelicans, bustards, and vultures, whether land or water species, lay few eggs, since the greater part of their food is consumed in flesh-building, required by the bulk of their bodies. Vultures in particular lay a small quota of eggs. In several instances we have inspected nests in which there was but one solitary egg which was being incubated alone. Aristotle, however, in his book on animals, states that neither the nest nor the young of vultures have ever been seen.

Smaller birds lay a large number of eggs in proportion to their size. Being slight of build, only a small percentage of their food is needed for bodily support, while the remaining and larger portion is changed into sperm and egg-building material. This rule applies also to the breeding of quadrupeds; the larger ones produce small litters, whereas the smaller animals beget a numerous progeny.

It happens occasionally that birds lay eggs without direct contact with the male, but such eggs are sterile and none of them reach fruition. Illustrations of this fact are seen in hens, pigeons, geese, ducks, and peacocks, both domestic and wild. It frequently happens, also, that raptorials lay sterile eggs, especially those falcons that have been confined to the mews and have not been flown during the winter but allowed to remain idle and rest until spring.

The shells of birds' eggs are of infinite variety in color (depending upon their species). Some are all white, or white with spots; and the spots vary in color and shape. Eggshells may be grayish, green, blue, and many other shades. Expert ornithologists can often identify a bird as soon as they see its eggs.

The eggs of those aquatic birds that nest near and feed in the water and along the shores differ from those that live and procure their nourishment in dry meadows in this respect—they contain double the amount of yolk in proportion to the white. This statement does not hold true in the case of land birds.

The shape of birds' eggs also varies with the species; some eggs are more nearly ellipsoidal than others that are rounder. Most eggs correspond to the corporeal lines of the bird that laid them and of the bird that will ultimately be hatched from them

—elongated eggs for the slender birds, round ones for the more stocky species.

As regards the actual incubation of eggs, in certain species the female alone sits on and keeps them warm, and never the male. This is, as a rule, true of species that do not pair prior to reproduction, and when the male does not assist the female in nest building. Among such birds are fowls, bustards, and peafowl. In some species, e.g., pigeons, both male and female take turns in sitting on and keeping the eggs at an even temperature. This is generally true only of species who pair to propagate the race and where the male joins in nest building. Birds of a warm nature, who are aided by fine weather and other favorable circumstances, brood their eggs for a shorter period than do those of a cold temperament who do not enjoy such advantages. Furthermore, males of certain species, although they do not directly help the females with incubation, supply her with food and so render unnecessary her forced absence from the nest.

We have already alluded to the fact that the cuckoo does not incubate its own eggs. The same is true of the ostrich, whose failure to perform this function we ascribe to fear of breaking the eggs by subjecting them to the weight of their large bodies. Moreover, the extreme heat of the sun, warming the sand in which the eggs are deposited, is sufficient to hatch them. A similar phenomenon is to be observed in Egypt, where eggs of the barnyard fowl are kept warm and the young hatched out independent of the mother bird. We ourselves saw this, and we arranged to have it repeated in Apulia by experts whom we summoned from Egypt.

Chicks are not hatched from every incubated egg, for a number of the latter are destroyed by wind, lightning stroke, and other agencies and accidents. Sterile eggs, of course, bring forth no young, since they lack the activating male spermatozoa.

A discussion of the manner in which the embryonic chick develops within the egg, the order of appearance, and the formation of its various organs and members, the most suitable season and the length of time required for incubation, as well as numerous other constant factors, we omit here because they all have been adequately discussed in the *Liber Animalium*;[32] nor are they relevant to our main subject, which deals chiefly with methods

32. [See Aristotle, *History of Animals*, Book VI, chapter 3, and *On the Generation of Animals*, Book III, chapter 2.—*Ed.*]

of teaching full-grown raptorial birds to capture well-developed nonraptorial species.

Chatper 23–I
Of the Functions of Avian Organs

We have discussed, to the extent required of us, the reproduction of birds. We shall now consider various external and internal organs that distinguish birds as members either of a species or of a genus.

Structural differences the young bird inherits chiefly from his ancestors. If all birds were uniformly constructed, their members would exhibit in detail a corresponding uniformity of function, no matter how many species were represented; but avian organs show a great diversity in form and appearance, so much so that individuals may be distinguished one from another. These variations are at times so marked that they at once divide bird life into various categories.

The avian body, like that of any other aquatic and terrestrial animal, may be divided into cellular (homogeneous) and organic (functional) parts.[33] The cellular parts are those that are constructed of similar elements, like bones, cartilages, nerves, the cardiac ligaments, blood vessels, flesh and fat, and the tissues of the skin, feathers, and nails. We shall say little concerning each of these substances, merely mentioning them when we discuss the nature of the organs of which they form a part. The internal organs of birds do not vary greatly from one species to another in their component tissues.

The functioning organs of birds are constructed from various cellular groups. They are, however, distinguishable from one another in build, size, number, and location. Some organs, chiefly external, are of a sensitive (nerve-possessing) nature, while some internal organs are insensitive. It is by means of both these organic structures that birds perform the functions esesntial to their well-being, to the preservation of the species, indeed to the very existence of the individual..

One should not conclude that the functions of the members determine their conformation, since that would be to attribute the cause *a posteriori* rather than *a priori*. Organs come first, according to their nature; then their characteristics, which are manifested through action and function, just as action depends upon the objective. As functions are determined by characteristics, and characteristics are derived from members, obviously functions depend upon organs.

It must be remarked that creative power has allocated to the proper place material naturally adapted to the formation of various organs and has endowed each one with a construction resembling that of the parent bird. Hence every organ is made of material suitable to its function as well as having a functional purpose. Moreover, if productive Nature had formed organs to fulfill the functions for which they are appropriate, it might be predicated that she made one bird that it might destroy another, viz., a predatory bird that would destroy and live on a species that is nonrapacious; in other words it would follow that Nature has created one species for the annihilation of another, and, according to this axiom, Nature is not only benevolent in one species and malevolent in another but, what is more important, exhibits her two opposite aspects at the same time, for each species finds in another what is harmful to it.

It must be held, then, that for each species and each individual of the species, Nature has provided and made, of convenient, suitable material, organs adapted to individual requirements. By means of these organs the individual has perfected the functions needful for himself. It follows, also, that each individual, in accordance with the particular form of his organs and the characteristics inherent in them, seeks to perform by means of each organ whatever task is most suitable to the form of that organ.

The external parts of birds that are of a sensitive nature are chiefly the head, eyes, ears, nasal cavities, mandibles, shoulder blades, joints, sides, belly, rump, hips, shinbones, feet, toes, back, thighs, external breathing apparatus, tail, oil glands and other related parts.

The internal organs are the meninges, the brain, the vertebral canal, the tongue, other parts of the mouth, the bronchial tubes, lungs, heart, cardiac ventricles,[34] diaphragm, esophagus, larynx and vocal cords, intestines, stomach, spleen, liver, kidneys, testes, uterus, and many other organs.[35]

33. The terms *consimilia* and *officilia* conform closely to the modern "tissues" and "organs." Tissues, *consimilia*, are made up of primitive cells disposed in various ways to form bone, cartilage, connective tissue, etc. From these *consimilia* are constructed the organs of the body, the *officilia*. [The use of the term "cellular" is anachronistic and unfortunate.—*Ed*.]

34. *casula cordis*.

35. The clear-sighted, remarkable, and corrrect explanations given by the Emperor of the terms *consimilia* and *officilia* are in marked contrast to some subsequent applications of these two terms when he, as may have been noticed, confuses the one with the other.

We shall now discuss this list of both internal and external avian organs (by which birds consume their food and digest it and by which they avoid dangers, live in their dwelling places, fly about in space, and change their habitats) and shall include matters that it is necessary to study for the purpose of writing about the treatment of their diseases.[36]

Indeed, birds have particular organs for definite functions, examples of which are many. One organ may serve a single or more than one purpose; or several organs may be required to carry on but one function. That we may avoid needless prolixity in our discussion of these topics we shall mention only those organs and functions that are pertinent to our thesis, beginning with the bird's head.

The head contains the brain and the organs of special sense. It is constructed of many bones that form the skull-cap. It includes tissues and nerves that govern the sense organs and motor apparatus. The head contains also the coverings and other parts of the organs of special sense, the eyes for sight, the ears for hearing, the nose for smelling, and the mandibles and tongue for taste. It occupies the chief and most important situation in the body because it is the seat of and controls its most important, finest, and most essential functions.

Various species, often the same genus, present great diversity in size, appearance, number, and position of the parts composing the head. Long- and short-eared owls, plover, lapwings, and some other birds have large heads in proportion to the rest of their bodies. Vultures, bustards, pelicans, swans, and many others have comparatively small heads. Other birds, again, have heads well proportioned to the size of their bodies, for example, hens and pigeons. Certain birds, like geese and swans, have elongated heads, and there are individuals that have longer heads than others of the same species. Other species such as rapacious birds and those related to them have more or less short heads. Some birds have a nearly round head, like curlew, lapwings, and plovers, as well as those that dig worms out of the ground, who, though they possess long bills, are none the less round-headed.

Some species have the head bare, others are well clothed with feathers, while in other instances the head has neither down nor feathers, as is the case with the black vulture *(galeranus niger campester)*, which has a black beak and black legs, and the piebald (black and white) vulture *(galeranus varius ex albo et nigro)*, which is frequently found in Syria, Egypt, and the Far East. This last-named bird has the whole head and throat naked, having neither down nor feathers. In some white species of carrion eaters that have black feathers at the extremities of their wings the saffron yellow of the mandibles extends to the middle of the head.[37]

Cranes have no true head feathers except a few hirsute appendages on the crown. Certain vultures, particularly the white variety, have some down on the head but no developed feathers. The majority of birds with feathered heads have no crests; others are crested. These head ornaments may be thin, fleshy, or tuberculated; as in fowls that have red combs and are provided with wattles at the throat like beards, or such as one sees on the heads of certain water birds called Pharaoh's ducks.[38] These are white and black geese, with red beaks, legs, and feet. Between nose and forehead they have a small, red, and fleshy excrescence, but it is not so protuberant nor so notched as is the cock's comb. The same condition is noticed in the male of a certain species of swan[39] with red mandibles that, between the nasal openings and the forehead, shows a red, fleshy swelling, like a hazel nut,[40] also called a "comb." Other birds sport a crest of feathers, among them the hoopoe, the horned lark, and certain parrots imported from India.[41] One of the latter was sent to us by the Sultan of Babylon; it had white feathers and quills, changing to yellow under the sides.

Some birds, such as herons, have projecting from the center of the head long feathers that reach to and lie like thick hairs along the back. A kind of

36. Frederick never, so far as we know, contributed a chapter on avian ailments and their care. We have several times referred to this unfulfilled promise.

37. This is the Egyptian carrion vulture *(Neophron Percnopterus)*. Other vultures cannot be identified.

38. *anates de faraone*, the *Vulpanser tadorna*, according to Schöpffer. The coloration changes in this species. The male develops during the breeding season a brilliant excrescence on his beak. The expression may easily refer to the Red-crested Pochard *(Netta Rufina)*, a species common in Lower Egypt.

39. Swans are difficult to identify, but perhaps the red-beaked and red-knobbed swan may be the spur-winged goose *(Plectropterus rupelli, Sclat.)*, whose habitat is North Africa. The male has a rather well-marked knob on the beak. The bird does not present the usual anserine form, although he has a rather long throat. Doubtless during his many journeys in the Orient the Emperor had often seen this goose.

40. *avellana*.

41. These cannot be modern Indian birds, since the "parrots" mentioned are cockatoos and they are not found in India. The Sunda Islands is the nearest habitat of these birds. Legendary descriptions of these Indian birds are very doubtful. The title "India" was (in medieval times) often applied to any unknown country.

bustard found in the desert has a high crest on its head extending to the back, like a mane. Other avian heads have tufts of feathers to the right and left like horns, as seen in some pheasants, the great horned owl, and the short-eared owls; and still others sport tufts of this sort and in addition other feathers pendant from either side and from the throat, like a certain species of diver.[42] There are still others that have three feathers that hang down from the middle of the head around the throat like a beard. These feathers are more plainly displayed during the breeding season and when the bird is frightened or angry, for then they bristle. These bridal plumes are more abundant and in marked evidence during the spring.

There are also birds that have a horny protuberance (or helmet) on their heads, like Indian fowl; or they present in the same situation an indurated growth like their own spurs; sometimes (though rarely) there are two of these horns. Other peculiarities about the head call for little or no consideration in this short review.

Chapter 24
Of the Eyes of Birds

The eyes are the organs and instrument of vision. In the *Liber Animalium*[43] it is fully explained why birds have two of them, why they lie in front[44] of the head, why they are placed higher in the body than other sense organs, and why they are composed of three humors and seven coats. By means of eyesight an animal correctly perceives dangers, sees objects he needs, and is able to avoid the one and to search for the others. The eyes vary greatly in size, color, and shape. Some birds have large eyes in comparison with their bodies, some small, some of medium size. Certain birds have quite black eyes, like the falcons. The blackness of the pupil is deeper than that of the margins of the iris; some dark pupils have a blue-gray border, as in the genera of hawks and sparrow hawks.[45] The color of the eyes may vary with the age of the bird.

Among hawks and sparrow hawks there are some individuals that have very dark eyes, like falcons. Often both eyes are black, but occasionally one only. Falcons, however, never have either one or both eyes the same color as hawks' eyes. It appears that the black pupils of the latter bird's eyes are enlarged when they are fixed upon some object and diminished when not staring at something that draws their attention. Some birds, moreover, see better than others.[46]

Most birds shut the eyes by the aid of the lower lids, which are long and adjusted to the purpose; in some others both lids take part in closing the eyes. Avian lids are devoid of true hairs or lashes.[47]

For cleaning the eyeball there is provided a peculiar membrane that is quickly drawn across its anterior surface and rapidly withdrawn.[48] The many peculiarities of the eye cannot (all) be described here. The superciliary shield that projects forward as a protecting wall over the eyes is, in birds of prey, bare and lacking in feathers; it is so stiff and thin as to have almost a cutting edge; in certain other birds it is less prominent but feathered.

Chapter 27
Of the Wings of Birds

Wings take the place in birds of human arms and the front legs of lower animals. They are constructed of several bones joined end to end and linked in various ways. Wings also have many joints (to insure flexibility), in addition to the necessary bones, nerves, veins, arteries, muscles, tendons, and other ligaments.

The wing bones are arranged as follows: to the bone called by the doctors the arm *(armus)*, which articulates with the shoulder, are attached at its lower extremity two other bones that are called *focilia*[49] [bones of the forearm]. They present many differences in size and structure. In their middle portion they are separated and placed one above the other, but their ends are joined together. At their inferior extremities they are connected with

42. The grebe *(Colymbus vel Podiceps cristatus)*.
43. Schneider says that this *Liber Animalium* was probably written by the Emperor himself and then lost. It was not a reference to the work of Aristotle, since the latter has not given us any definite description of the avian eye.
44. The author says *"in prora capitis."* Schneider remarks that most of them are laterally placed; rarely (as in the owls) do they look directly forward.
45. The Emperor is astray in his description of the color of the eyes of rapacious birds. Falcons have usually brown irises, and common hawks and sparrow hawks have yellow, not blue-gray, eyes. The iris is often gray in the brightly colored mouse-bustards and silver-gray, verging on yellow, in the red kite. The author is referring, of course, to the European sparrow hawk, which resembles our Cooper's hawk but is smaller.
46. The whole of the foregoing paragraph (the statements of which are not quite accurate) is found only in the Bologna and other six-book manuscripts.
47. Birds' lashes are not hirsute but are in reality small feathers.
48. This skin is called the nictitating membrane.
49. *focilia,* the *ulna* and the *radius.*

that part of the wing that in all birds [anatomi-cally] corresponds to the hand in mammals. This member is composed of an assemblage of small bones closely articulated and arranged in the following order: first of all are some quite small bones that articulate with the bones of the forearm and fill up the cavities in the joints but are inde-pendent of them. One of these, separated from the others, stands alone and (in birds that habitually strike with their wings) at some distance from the wing proper. It is compact, very hard, and is employed for both offense and defense. It is quite prominent in geese and swans, where it stands out from the wing. In little birds it is small, in larger ones large.

In the remaining parts one finds other isolated bones, especially one that seems to be the analogue of our [human] thumb and that is, on the outer side of its extremity, hard and sharp like a talon. Under this lies another weak and delicate bone on which the "thumb" rests; beside this is a large bone that is fused into a single bone at the ends,[50] but is fenestrated and separated into two bones in its midportion. At the extremity of this are two short bones, one (the external) broader than the other and overlying it. To this wider bone the smal-ler (very small) bone is attached; and these are the last of the true wing bones.[51]

The folding of the wings takes place as follows: when they are drawn in and made to rest against the body, the upper arm descends from the shoul-der and lies close along the side of the body and the ilium. The next segment, which articulates with the extremity of the upper arm at the elbow, rises alongside the upper arm; and the third portion of the wing settles down in place from the point called the *impulsorium*.[52] This last part of the wing is call-ed by the French *le bout,* "the tip."

It is well that there are several joints and cor-responding divisions in the wing and not merely one. In the latter case the single segment would have to be either long or short; if it were short, the wing would not be big enough to sustain the weight of the avian body, or adapted to the requirements of bird flight; and if it were long, this single joint could not readily hug the birds body; it would project behind the tail and prevent the bird from roosting or setting, and when it was stretched for-ward it would project in front of the head and the wing feathers would open and not remain close together. They would not be able to fulfill their intended purpose in that situation, because the bird must move his wings in all directions—

right, left, forward, or backward, up, or down—and a single articulation of the wing would not be sufficient to enable him to accomplish this. On the other hand, were alar appendages provided with more joints, bird flight would be weaker than it is. The present number of articulations fulfills all the requirements of diverse wing motions. Any mem-ber is proportionately weakened by each additional joint. Moreover, such an arrangement would interfere with the bird's ability to close the wing feathers one over the other. Were there even a fourth joint the pinions could not be properly drawn close to the body but would lie exposed and unprotected by the other wing feathers to which they would from an impediment; likewise the quill feathers (remiges) attached thereto would not be properly arranged and every one of the external pinions that ought to lie close to the body would stick out in an absurd fashion.

The largest of the flight feathers are those at the posterior extremity, and they have above and below them smaller cover feathers which will be described in our chapter on plumage.

The wings, whether expanded or contracted, are capable of movements in all directions at the will of the bird. These (complicated) motions are con-trolled by various muscles; indeed their move-ments—forward, backward, up, and down—are due to muscular forces operating in the desired direction. Likewise there are special muscles that provide for stretching and bending the wings. During flight the bird moves his wings both upward and downward, lifting them high over his back so that he can more effectively sweep them downward on the return stroke and thus be lifted up and driven forward.

This forceful beating of the wings is more evi-dent in birds whose breast muscles (which control it) are thick and strong and whose flight is there-fore powerful and fleet. From which one might draw the following syllogism: All birds that have thick and strong muscles from the breast to the wings have strong and rapid movements of the wings. All birds that have large and firm breasts have thick and strong muscles extending from the breast to the wings. It follows, therefore, that

50. The metacarpal bones of the larger and small "wing-fingers."

51. The wing bones of birds have but three (true) "fingers," viz., thumb and larger and smaller finger bones. The toes are as a rule four, but sometimes three. The African ostrich has only two.

52. The wrist joint, the portion of the wing that strikes the air with the greatest force.

birds with large and firm breasts are capable of strong and rapid wing movements and are powerful and swift flyers.

When in motion a wing describes an arc of a circle; all the largest flight feathers follow the same movement and describe shorter or longer arcs depending on whether they are near to or remote from the body of the bird, and these arcs are all parallel with each other. Those feathers that are farthest from the body of the bird and describe in flight the greatest arcs have the greatest power to lift and carry the bird forward. About this Aristotle says in his book on the art of raising weights that the greater the orbit, the greater the power to lift a weight.[53]

Wings are useful in that they enable the bird to move about in search of food, to seek a better climate, and to escape danger; some birds employ them as instruments of offense and defense. Also they protect the sides and part of the back from the evil effects of cold. Many birds are recognized (classified) by the character of the wings. In proportion to their bodies birds' wings may be long, short, or of medium length. In addition to this the wing feathers of long-winged birds may be either long or short, and the pinions of short-winged birds also may vary in length. Some birds have pointed and scythelike wings, like those of the falcon, and others have them broad and rounded. There are still other differences that might be mentioned, but what has been said will suffice for the present.[54]

Chapter 41
Concerning the Stomach

The stomach is the organ to which (by means of the esophagus) food and drink are conveyed. It extends downward to the intestines. In some birds the stomach is very fleshy, thick, and compact, as in ducks, geese, and fowls, in others, birds of prey for example, it is more tendonous (or sinewy) and consists mostly of an inner and an outer coat. Food and drink are conducted from the mouth through the esophagus to the stomach. After digestion, the waste matter is carried from the stomach to the intestines. The latter form numerous convolutions and extend as far as the anus, situated beneath the tail.

Chapter 42
Of the Liver

The liver is a blood-making organ, divided into two segments, a right and a left lobe. Attached to the liver of some birds one finds a gall bladder, which is absent in other birds—pigeons and turtledoves, for example.

Chapter 43
Of the Kidneys[55]

Birds have two kidneys, one on the right, one on the left side. They lie close to the vertebrae under the iliac bones and extend toward the anus. The urine is excreted from the kidneys by way of the anus through the uriniferous tubules, which pass below and in close contact with them. As the urine is passed with the feces, the bird does not require and does not, indeed, possess a urinary bladder.

Chapter 44
Of the Testes and the Ovaries[56]

At the root of the kidneys, in the male bird,

53. [The Aristotelian treatise translated here as "the art of raising weights" is cited in the Vatican MS. Palatine Lat. 1071, fol. 24 recto, as *Liber de ingeniis levandi pondera* and is a reference to the important pseudo-Aristotelian treatise *Mechanica,* written in Aristotle's Lyceum during the fourth century B.C. No manuscripts of a Latin translation of the *Mechanica* have yet been found, nor is mention made of it in the standard medieval catalogues. However, there seems reason to believe that such a medieval translation was made, since some propositions in the *De ratione ponderis* (see Selection 39) seem derived from the *Mechanica.* Moreover, "in the early fifteenth century, a license for the export of books from Bologna, dated August 18, 1413, includes the title *Reportorium super mechanica Aristotelis(!) [Report on the Mechanics of Aristotle—Ed.*]. The fact that a reportorium had been composed certainly suggests that a prior translation had been made. And it is not improbable that the *Liber mechanicorum [Book of Mechanics—Ed.*] listed among a group of books at Padua in 1401 is the translation we seek. In view of these facts, we must keep an open mind on the question of a possible translation of the *Mechanica*" (M. Clagett, *The Science of Mechanics in the Middle Ages* [Madison, Wis.: University of Wisconsin Press, 1959], pp. 71–72).

In citing the *Mechanica,* Frederick probably had these lines in mind: "The further one is from the fulcrum, the more easily will one raise the weight; the reason being that which has already been stated, namely, that a longer radius describes a larger circle" (3.850b.1–4; Oxford translation by E. S. Forster). The lines translated as "that the greater the orbit, the greater the power to lift a weight" might be rendered more precisely as "a greater circle causes a weight to be raised more [easily]" (. . . *quod magis facit levari pondus maior circulus).—Ed.*]

54. For comparison, the student is recommended to read Elliot Coues' discussion of the bird's wing in his *General Ornithology.* The remarkable accuracy of the Emperor's description then becomes apparent.

55. Aristotle denies that birds possess these organs.

56. Chapters XLIII and XLIV are all too short. It must be added to this chapter that it is only the left

there project over the vertebrae two testes, one on the right, one on the left. In the female are the organs in which eggs are created, which in consequence may be called the matrices (ovaries). These sexual organs exhibit but little variation.[57]

Chapter 54
Of Modes of Flight in Birds

After a consideration of certain organs in birds, of their modifications and their functions, not to mention their plumage, there still remains the study of flight itself, which is founded on the subjects already discussed.

Although flight is progress through the air by means of wings, comparable to walking along the ground with the aid of feet, it is not correct to regard every winged creature as a bird; for many animals that do not belong to an avian species have wings. Yet every bird has wings. But to be called a bird it must have also well-marked down and flight (quill) feathers.[58]

Some birds move their wings in a well-defined, monotonous manner during flight and do not alter these movements except from sheer necessity; others use their wings with ease in all directions and change their motion without apparent purpose. The former may use their flying apparatus either much or very little.

In comparison with the size of their bodies some birds have short wings with relatively short primary flight feathers, while others have long wings bearing short primaries. Both types of bird are unable without repeated and hurried wing strokes to fly forward or to remain easily suspended in midair. They require to flap their wings all the more frequently if the flight feathers are thin and delicate or have natural or accidental vacant spaces between them. If the pinions are stiff and coarse and without gaps there is less need for frequent wing movements.

The necessity for rapid wing strokes is seen among water birds—ducks, geese, goosanders, and coots, and such neutrals as plovers—also among francolins, partridges, pheasants, quail, and little (lesser) bustards. Also, in every species one notices variations in the wing beats of individuals.[59]

Certain birds have longer wings and primary flight feathers than their body measurements seem to demand; others, again, have short wings with long quills. These are all able to float in the air and fly quickly forward without oft-repeated wing strokes. When, as stated, such birds have strong and well-fashioned quills with no vacant interspaces, they need for long and effective flight com-

paratively few wing beats. But if, with a similar arrangement of wings and pinions, the latter are soft and slender or with broken spaces, the bird cannot maintain a forward motion or remain long in the air without wing motion.

Heavy birds with short, relatively soft wings must naturally employ rapid wing beats in flying, otherwise they would fall to the ground, as all weights are attracted to the center of the earth. Examples of birds that do not make frequent strokes with their wings are the herons (both white and gray), the *albani,* screech owls, kites, and many other aquatic, land, and neutral birds.

Large birds with a long wing sweep do not need rapid wing strokes, as they describe with their flight feathers a large semicircle covering a wide space; also their wings are slowly drawn against the body to repeat the movement, avoiding the need for frequent motion. On the contrary, birds with short wings and flight pinions, because of their short wing sweep, require an oft-repeated motion. And once more it may be stated that the frequency of wing motion varies with individuals of the same species.

[Addition by King Manfred:

Birds with long wings and perfect flight feathers sail along by backward strokes of their wings as if to set the air in motion. The longer the wings and the more nearly perfect the pinions, the more support is given by the spreading out of the wings and these backstrokes of the latter on the air, thus reducing the necessary number of beats. It has, in this connection, a resemblance to the relative movement of a broad and of a pointed piece of lead; the downward fall of the former being slower

---

ovary of the avian female that functions (important to remember in sterilizing domestic and other birds). The right Fallopian tube is, as a rule, more plainly seen than that on the left. Whether it acts as a substitute bladder or as a canal, its reproductive functions are certainly as negligible as that of the right ovary.

57. During the pairing season the testes are greatly enlarged, although in undeveloped young birds they are barely visible. This is important for collectors to keep in mind while determining the sex of a bird. One must not mistake the suprarenal gland, that lies close to them, for the male testes—a common error.

58. Struthious birds—the ostrich, rhea, emu, apteryx, cassowary, *et al.*—have contour feathers and down indistinguishable from one another; but of course the Emperor knew nothing of South American or Australian species, and may not have closely examined the African ostrich.

59. Schöpffer thinks this abnormality may be due to a temporary paresis, to a permanent paralysis, or to weakness attending an advanced moult.

because of the greater resistance of the air, as noted by the philosopher in his work entitled "On the Heavens and the Earth."][60]

Now when birds have their wings and flight quills in proper proportion to their body (neither too short, nor too long), their wing beats describe, in correspondence, neither a wide nor a narrow arc, nor are they of a rare or great frequency. When they have perfect, stiff quill feathers, they do not require to flap their wings as often as must those birds whose pinions are not so well developed. The wider and stiffer the flight feathers, the less frequent is the wing motion. The reverse is true in birds that possess soft, narrow pinions. Birds displaying moderate wing movements of this sort frequently belong to a genus of falcons. On the whole, it may be said that the longer, more complete, and rigid the bird's flight feathers, the better it flies; and the opposite holds true.

From all this the mode adopted in flying by certain birds is made clear, and it may, in addition, be stated that in case of desperate need they may change their manner of flight; for example, when they are urged by excessive fear, birds beat their wings as fast as possible; and this rule holds true of birds of prey engaged in chasing their quarry. On the other hand, when they have a long journey (as during migration) before them, they husband their strength, both coming and going, and do not tire themselves by wide wing beats; hence they are able to complete such excursions with ease. It is to be noted that raptorial birds, expecially those that have interspaces in their wing feathers, like the hawks and the sparrow hawks (*Accipiter nisus*), after an exhausting hunt for their prey which they have failed to catch, soar about in the air with outstretched but apparently motionless wings. On these occasions they seem to be taking a rest. When the same species wish to rise high in the air by making a spiral ascent (or by "ringing up"),[61] they change their mode of flight. Hawks and sparrow hawks by rapid movement of the wings can gain a great altitude; then, after a period of frequent wing beats, they rest, circling about on outstretched motionless wings. If they wish to rise higher, they repeat the motion, then circle again and rest. They do this because they tire from too prolonged and rapid flight. Thus it seems that quick movements of the wings are induced by fatigue and fatigue itself precludes a protracted period of rapid movements. This forces the bird to change its mode of flight and to take a rest. In going to perch the wings are as a rule held extended and motionless in order

to effect a safe landing.

Birds that change their style of flight for no imperative reason, but because it is a normal process with them, do so in various ways. Some attain speed by rapid wing beats, then, folding their pinions, dive through space, and thus reach their objective. Among such birds are the long-eared owls (*noctua*) and some small birds. There are others that in flying beat their wings sometimes in rapid succession, sometimes not so often, and then with folded wings fly forward, and thus continue their flight in (interrupted) uneven lengths. Examples of this style of flight are the magpies, jays,[62] and hoopoes. These birds cannot, because of their short wings and primaries, and the interspaces (lacunae) in their flight feathers, long sustain rapid progression and uniform wing beats; so they must vary them, as explained.

There is a Syrian pigeon[63] which in the course of its usual flight forward suddenly turns over two or three times as if it were bewildered, and then resumes its usual course. We do not know why it does this—perhaps because of good spirits and cheerfulness. Other examples of unusual forms of flight we do not mention because they are too numerous and a recital of them would be tedious.

To sum up—some birds have a rapid flight, by which we mean they cover a long distance in short time; others, contrariwise, fly slowly. Strong birds with a well-developed muscular (*carnosum et lacertosum*) breast, as well as compact, stiff, and long primary flight feathers, fly rapidly. Those that exhibit infrequent wing beats are the eagles, bustards, pigeons, and the baldheaded vultures, as well as some water birds. These birds have large wings and long flight feathers with which they are able to describe wide circles in the air and to make rapid progress after the manner of galleys furnished with long oars. Provided in this way, certain birds fly fast and with rapid wing beats, as do geese, ducks, plover, the lesser bustards, francolins, pheasants, partridge, and quail.

Weak birds with undeveloped breastbones and soft, flexible flight feathers exhibit a corresponding lack of flying powers, and have a languid, leisurely kind of flight. These include all the herons, the *albani,* and the screech owls, especially those that

60. This passage by King Manfred is taken from the Vatican Codex, fol. 38, col. 1, l. 25.
61. The falconer's term for this rapid, spiral ascent.
62. *jahyus,* a word of which the origin is unknown to us; Pacius suggests "nutcracker."
63. The tumbler pigeon.

sail along by means of occasional wing beats. On the other hand, among those that utilize frequent strokes of the wings are mergansers, coots, land and water rails, the so-called birds of paradise from the Orient, and some other waterfowl and land birds.

Other birds, in their efforts to avoid capture, turn and twist in their flight when followed by predatory enemies. They have a slow wing beat and small bodies, so that they are not prevented by the rate of wing movement or the size of their bodies from turning quickly and dodging. These birds are herons, *albani,* screech owls, crows, hoopoes, kites, lapwings, magpies, and other small and medium-sized birds. Other species continue their original direction when pursued by birds of prey, and these are as a rule such as have command of rapid wing movements and are prevented thereby from effecting rapid changes of direction, e.g., geese, ducks, plover, francolins, partridge, and quail. Birds that are unable because of their large size and heavy weight to fly away in the distance and escape (despite rapid and increased flappings of their wings) are the swans, pelicans, bustards, cranes, and similar large and heavy birds.

The duration of flight varies in bird species. Some are able to continue it for long periods; others are soon exhausted, so much so that they can be captured, if not by the hand alone, by the help of dogs. The latter class includes pheasants, partridges, quail, and francolins. On the whole, it may be said that those species that persist in flight for short periods only move their wings rapidly; the reverse, however, is not true, for certain birds with rapid wing beats fly continuously for long periods; but no birds with a slow wing beat are found whose flying powers are limited to short flights.

The explanation of this abbreviated power of flight is that in all birds exhibiting that variety of weakness the wings are small compared with the size of the body; their primary flight feathers are slender, with many lacunae in them, requiring repeated strokes to get results; so they are easily fatigued and they cannot remain in the air. Owing to the disproportion between wings, plumage, and body, they are obliged to return to the earth.

Birds that are able to sustain long, continuous flights never become so exhausted that they can be caught by men or dogs, examples of which are seen in herons, eagle owls, and kites. In comparison with the size of their bodies they have long wings and well-developed pinions. Among those birds

possessed of a slow, infrequent wing movement those that move their wings most rarely can maintain the longest flights and accomplish the most distant journeys. This applies to individuals as well as to species. By the aid of favoring winds, such birds as quail, whose usually short flights are made by rapid wing strokes, undertake long journeys (as when migrating or returning to their nesting places). On such occasions, however, they take their time and husband their wing power as they fly from island to island.

Birds with long wings and good flight feathers fly better with the wind than those with short wings and pinions, even when they both fly equally fast without the aid of the wind. However, short-winged birds fly better against the wind than those with long wings, even if both species progress equally fast in the absence of a stiff breeze.

Also, birds with rapid wing strokes fly better against the wind than those with a slow wing movement; and there are small birds of every species that make better time in defiance of an adverse wind than do larger birds.

Waterfowl suffer least from heavy showers of rain; in wet weather the bird with its first supply of feathers is more affected than after it has moulted. The second plumage, if well developed and still fresh, is less likely to be affected by dampness than the older covering.

Birds who fly only by day may be placed in two categories. Magpies, pigeons, sparrows, and most small birds have no definite time schedule, whereas others fly only in the early morning and toward sunset. In this second group we find those who feed in the early morning, aquatic birds who feed on land in the evening hours, and many neutrals and land birds who are afraid to fly at midday [because they feel heavy after a full meal and are sluggish and less able to escape from eagles and other birds of prey that are in the habit of soaring aloft, especially on hot summer days when they cool themselves and ventilate their wings.

There is another obvious reason why most birds take a rest in the daytime: After digesting a full meal during the night, in the early morning they are again hungry and fly to their feeding grounds, where they enjoy a comfortable meal that demands undisturbed quiet for its digestion.][64]

Some species find frequent excursions at night (or in the gloaming) more to their taste, as do long-

---

64. This passage (in brackets) is taken from the Vatican Codex, fol. 40^v, col. 2, and is substituted for a shorter one in the Bologna MS., fol. 32, col. 1.

eared and eagle owls; they fly about like bats, not so much because they see best in the dark but because they fear the persecution of birds who dislike and worry them.[65]

Other birds, such as snipe, fly around both day and night. Birds of all sorts take advantage, either by day or night, of favoring winds during their migration, as has already been noted. Domestic fowl, partridge, pheasants, peacocks, quail, and similar birds (that are heavy weights and poor fliers) always keep near the ground so that they may quickly reach a refuge in time of need. Raptorial as well as peaceful birds that are light on the wing and good fliers ascend high into the air on the lookout for food. This habit coincides in birds of prey with their sharp eyes. Birds that soar to great altitudes fly either straight ahead, like the screech owls, or in circles, ringing up, like herons. If birds wish to take advantage of a favoring breeze they often rise high in the sky; but if the wind is adverse they generally remain lower down, because at a great altitude wind has greater force. Some birds choose a solitary flight like the birds of prey. The reason for this is that raptorial birds attempt to seize the captured quarry of another. They therefore fear one another and fly alone in order to have uninterrupted leisure in which to consume their prey. They do not swallow it in large portions, but first deplume and then eat it bit by bit.

As a rule birds fly in flocks, especially those whose food consists of single grains that they pick up one by one and swallow forthwith. These associated birds pursue their journeys either in orderly fashion, like cranes, geese, ducks, and waterfowl, or in disorder, like sparrows, starlings, and pigeons, which to secure safety not only consort with their own species but keep company with others of a similar kind. Certain birds, like turtle— and ringdoves, fly sometimes alone, sometimes in flocks. During the mating season all birds pair off, a male with a female. When they migrate to avoid heat or cold and when they are not nesting they gather in larger flocks.

Chapter 57
Of the Moult of Birds

While it is true that birds shed their feathers every year, we must not leave it at that but must inquire further into the cause of this moulting, why it happens regularly each year, at what season it occurs, and whether it is completed quickly or

gradually, and last, but by no means least, discuss the alterations in plumage before and after the moult.

It is both necessary and useful that the bird should change its down, contour, and flight feathers, because they are reproduced by an overflow of humors and are composed of perishable matter subject to many hazards, and because this material can last only a year at most; if the moult is delayed any longer, degeneration of the plumage would set in and the feathers would then be so worn out and broken down as to be of little or no use to the bird. For this reason Nature has provided a substitute for them and brought about an exchange of the old feathers. The appearance of new feathers depends upon an abundance, even superfluity, of the bird's bodily humors regularly supplied. As evidence of this, should a falcon complete her moult earlier than usual, she may again shed some pinions (which had appeared in the premature moult).

This accident may be a result of the heat of the season, which opens the pores, as well as from too rich food, or from excessive humors.

It may be further remarked that when hairs of the human head and body are broken or cut, they grow again to their former or even greater length, for under the continuous flow of generative matter they develop to an indeterminate length. But in the case of feathers and quills this is not so, for these adjuncts have a definite size and form, and if broken cannot regain in toto the portion that has been lost. In time degeneration would extend to the barrel, and if other feathers were not substituted for the damaged plumage the bird would be without feathers or quills and entirely exposed to the inclemency of the weather and rendered incapable of flight.

Moulting takes place every year; indeed feathers are not of much use longer than that period. In harmless birds the first plumage, after it has fully developed, is exchanged for another supply which develops continuously and uniformly with the growth of the body. This second supply of feathers is completed during the first year in pheasants, partridge, and quail, and moulted at the end of that period.

The question why moulting occurs only once during the first year in raptorial species and twice

65. Owls and bats also choose twilight hours for catching night moths and butterflies, of which they are very fond.

during the same period in harmless birds has already been discussed in the chapter on plumage.[66] An answer to this query is that more dangers threaten harmless birds during the moulting season than birds of prey; for that reason the former need constantly both their down and flight feathers to insure safety and to procure food; and this is particularly the case when the parents do not supply their fledglings with nutriment. Again, these birds are by nature richer in humors than are raptorial birds and their flight feathers develop more rapidly than the remainder of the body and are thus somewhat weak and soft. With continued development of the body these early feathers and pinions are unable to lift the increased weight and are moulted. Nature then provides stronger plumage capable of carrying the bird for the first year. The second supply of feathers then falls out and a third lot begins to grow and this provision lasts throughout the succeeding year.

Birds of prey are not subject to as many dangers, nor are they of such full habit, as are harmless birds; hence their feathers develop more slowly, grow more gradually, and, consequently, thrive more regularly and are in better proportion to the size of the body whose weight they are able to support until the end of the first year. Thus it is that the raptores moult only once during their first year, afterwards changing their plumage annually, like most inoffensive species.

The season for beginning a moult is, with most birds, the springtime, when there is on hand a good supply of food. During the following summer season they do not suffer from the cold and their flight feathers are so full grown that they are prepared for the autumn migration. Spring is, in fact, the only suitable time for moulting, as is explained in the chapter on migration. There are, however, some birds that do not begin to lose their feathers until the summer—following the rule that moulting largely depends upon the date they issue from the shell. When a bird moults it may do so at the end of the year that begins with the date of its birth. The same phenomenon is evident in the appearance of new leaves on trees and shrubs and a new growth of grass from the roots, which first sets in after the lapse of about a full year.

There is a regular as well as an irregular form of avian moult. To the first class belong those birds in whom a number of feathers fall out at one time, followed by their immediate renewal, making them fit for flying before others are lost. The irregular kind includes those cases in which many feathers fall out, then another lot before the first loss is repaired.

Raptores *(Falconiformes),* that are in constant need of their flight feathers to aid in capturing their prey, have a regular form of moult, so that they never entirely lose their flying ability. Harmless birds that are not in such urgent need of wing power to gain a living (i.e., those whose provender does not fly away from them) moult in less orderly fashion; but as they require flying power to secure shelter and to avoid dangers, the moult is not entirely without plan. Waterfowl (i.e., swimmers), on the other hand, make a complete and unusual moult influenced by the fact that they do not escape dangers nor obtain their sustenance by flight. By living in the water they attain both objectives.

A further study of the moulting process we shall defer to a later chapter on the moulting of birds of prey, concerning which we are much better informed than regarding that of other birds.

Here we remark, in general, how wisely it has been arranged that birds begin to moult their flight feathers before their other contour plumage. The former, much larger and longer, demand for their development and growth a greater expenditure of vitality than the latter; if the other feathers (or accompanying plumage) were the first to fall out they would have completed their moult long before the pinions had been replaced, and the renewal of the entire plumage would not have been completed at the same time, so that the flying powers of the bird would be lessened.

Water birds moult earlier than other birds, owing to the richness of their nutritional humors and fat; then come the harmless land birds, which also possess a large amount of adipose tissue; while birds of prey, the driest (i.e., those provided with the smallest amount of humors), are correspondingly the slowest to lose their feathers.

The length of the period from the beginning to the end of the moult, as well as the order observed in changing their flight and tail feathers, will be later discussed more fully and in its proper place.

The plumage of unmoulted birds differs from that of moulted fledglings in that, as a rule, the feathers after the moult are in better condition and are of a different color. They have more down around them and the new pinions do not get wet so readily, nor are they so fragile or so easily frayed.

66. [In chapter 45, which has been omitted.—*Ed.*]

BOOK II

OF FALCONS USED IN HUNTING, THEIR FURNITURE,
CARE, AND MANNING

Author's Preface

That falconry is a gentle art was maintained in
the previous book, and its nature was discussed
there. It was shown to consist chiefly of hunting
nonrapacious birds with birds of prey trained for
the purpose. It was claimed that this sport is a
nobler form of venery than any other and, moreover, that a work of this kind in its very nature
involves a study of waterfowl, neutral birds, and
land birds, raptores as well as harmless species.

In the second part of this treatise—and in
others—we arrive at the essential portion of our
research, true falconry, including everything desirable for a man to know and to learn who wishes
to understand and practice this art.

The requisite traits of the novice, who must
begin his education under an experienced teacher,
will be discussed later. It is self-evident that, having
secured his falcons, he must first of all gain experience in the art. However, it is not everyone who
can qualify for the practice of falconry; only an
indefatigable, enthusiastic lover of it, who is fitted
for the sport at the same time by instinct and by
training, can succeed in it.

Although the birds of prey with which one hunts
are mere instruments in the hands of a master, yet
the skilled falconer should give his entire attention
to them and to their equipment. He must be tireless
in watching and directing his falcons and in studying their deportment. These birds include not only
falcons and hawks but also other raptorials whose
performance we shall thoroughly discuss. For the
art of falconry varies greatly according to whether
one chooses to practice it with falcons or with the
large and small hawks.

The art and science of falconry are further divided into *theory* (the general mental survey and understanding of the principles of our subject without
reference to is practical application) and *practice*
(the knowledge of how to put into operation the
rules that we shall expound).

When a man without either theoretical knowledge (which must ever be a preliminary consideration) or practical experience undertakes to pursue
the art of falconry, although he may secure good
results in his hunting (just as in the gymnasium
the inexperienced boxer may strike a good blow,
or the unskilled archer may even hit the mark),

we may well say of him that he is merely testing his
luck and not his skill.

The falconer's primary aspiration should be to
possess hunting birds that he has trained through
his own ingenuity to capture the quarry he desires
in the manner he prefers. The actual taking of prey
should be a secondary consideration. He will, in
this way, secure that threefold advantage that we
have discussed at the beginning of our first book.
The falconer's postulates that he should ever bear
in mind are set forth in this book of ours as examples and precepts covering the entire range of
falconry. In stating them we shall observe such
brevity as is consonant with our duty to preserve
a constant clarity of expression.

The eye of envy will not influence us to be prolix,
or guilty of repetitions or superfluities, or to talk
on immaterial subjects. We intend to confine ourselves to matters that are relevant to the subject
in hand.

The tasks of the falconer are numerous and of
great variety. To begin with, he must make a study
of birds of prey, then secure and carefully tame
his hunting birds so that they will stay with him.
To this end he must so tame them that they will
lose their innate fear of man and of intercourse
with him. Also, as a part of their education, falcons
should become accustomed to be slipped from the
first and learn to return obediently to the hand
when summoned.

As we have said, the falconer must also teach
his falcons that they are to chase only birds of his
choice and in the manner he desires. This is not
an easy task, as it runs counter to their natural
inclinations. To this end he should utilize items of
information he has already gathered while engaged
in other forms of venery. There are many other
important duties incident to his career; he must,
for example, insure the good health of his falcons,
because their usefulness deepnds upon it; and he
must learn how to treat birds that become ill.

After making a sufficient study of the foregoing
tasks, as set forth in the various sections of this
work, one may with confidence own falcons, since
they will then be in safe keeping and be well tamed.

Among the various implements included in the
expert falconer's outfit are devices for trapping his
falcons, or hawks, such as nets and snares; and in
this connection it must not be forgotten that some
birds of prey may be used to catch others; for
instance, falcons are captured by hawks.

Other accessories are employed for the retention
and accommodation of captured birds, such as the

falcon sock,[67] jesses,[68] leash, and proper perches, as well as stools or blocks.[69]

Other devices are employed in taming wild birds; for example, the "tiring,"[70] the hood,[71] and the bathtub.[72] There are also certain contrivances used in teaching the falcon to fly from and return to the falconer's fist. These include the line, or creance, and the lure[73] (with or without meat attached thereto) that is used to recall the falcon.

There is still another device whereby the falconer summons the falcon that he is teaching to capture certain birds, namely, the train,[74] a form of the lure—a decoy prepared either with a crane, a heron, or some other bird, or even with the straw-filled skin of a hare; for there are many birds that will "wait on"[75] the train.

Among other contrivances used in flying the trained falcon are the small hand drum and the falconer's glove (or gauntlet).[76]

For the conservation of the health of his hunting birds there are several provisions to be made by the falconer, especially while they are moulting their flight and down feathers. They ought to be kept in a small moulting-house or mews that should be well stocked with remedial agents. In addition to remedies for healing the sick there should be provided suitable receptacles for administering them. A further description will, in its proper place,[77] be given of all these implements.

Abandoning our prologue, we must now return to our main theme. As a starting point we must define a bird of prey, for by definition and description all existing things may be fully explained. We shall also explain later why they are called rapacious, or birds of prey.

86 PHILOSOPHICAL AND THEORETICAL ZOOLOGY

Albertus Magnus (ca. 1193–1280)

Translated and annotated by Edward Grant[1]

BOOK I

Question 3
Whether an Organic Member That Has Been
Cut Off Can Be Restored

We inquire whether an organic member that has been cut off can be restored.

(1) It seems that it can, because in animals

67. *maleolus;* a close-fitting linen bag into which the recently captured falcon is thrust, leaving her tail, head, and feet free.

68. Footgear, made of a leather strap attached to the lower portion of the falcon's leg.

69. *sedilia;* truncated cone of wood, or stone blocks, to be placed outdoors when the hawk is to be "weathered" or set out at hack.

70. *tiratoria;* pieces of meat, e.g., the fresh or dried leg of a chicken for a short (emergency) meal.

71. *capellum;* this most important, useful, as well as ornamental part of the falcon's dress is thoroughly described in Book II, chapters lxxvii and lxxviii.

72. *tina balneatoria;* Book II, chapter lxix.

73. *loyrum;* cf. Book III, chapters i ff.

74. *trahina;* cf. Book III, chapters xxii ff.

75. "Wait on," a falconer's term meaning to hover on outstretched wings high above the falconer until quarry is flushed or the lure is thrown out.

76. *chirotheca;* used to protect the falconer's hand from the bird's sharp talons when resting on his fist.

77. From this point to the beginning of chapter xxxi, the translation is entirely from the Vatican Codex, for these chapters were inserted in the treatise by King Man-

fred from notes found by him among Frederick's papers, as is explained by the King in chapter xviii.

1. Translated from Albert's *Questions "On Animals"* (*Quaestiones super De animalibus*), edited by Ephrem Filthaut, O.P., in *Alberti Magni Ordinis Fratrum Praedicatorum Liber de natura et origine animae . . . edidit Bernhardus Geyer; Liber de principiis motus processivi . . . edidit Bernhardus Geyer; Quaestiones super De animalibus* primum edidit Ephrem Filthaut, O.P., pp. 80–82, 92, 122, 170–171, 229-231, 238–239, 241–242, 242–243, 259–260, 266, 269, 285, 294, 295. Monasterium Westfalorum in Aedibus Aschendorff, 1955. These *Questions on "On Animals"* were disputed by Albert in 1258 and edited by one Conrad of Austria sometime after 1260. Consisting of nineteen books, Albert's *Questions* are based on the translation of Aristotle's zoological works which were translated (no later than 1220) from Arabic by Michael Scot under the title *"On Animals"* (*De animalibus*). This lengthy Aristotelian treatise comprised three of Aristotle's zoological works: *History of Animals* (*Historia animalium*), *On the Parts of Animals* (*De partibus animalium*), and *On the Generation of Animals* (*De generatione animalium*). Although later in the century William Moerbeke translated these and the other of Aristotle's zoological works directly from Greek (see Selection 8), Michael Scot's translation was used in the universities and was by far the more popular version. It served as the basis for Albertus Magnus' lengthy *Twenty-Six Books on De animalibus* and the later and briefer *Questions on "On Animals"* in nineteen books, from which this selection has been translated. The sources from which Albert drew most frequently are the whole corpus of Aristotelian treatises, the works of Avicenna (especially the *De animalibus, Canon of Medicine,* and the *De anima*), and scattered works of Galen.

organs are as branches in plants; but a branch that has been cut can be regenerated. Therefore, by the same reason, organic members in animals [can be regenerated].

(2) Again, organic members are composed of similar parts; but similar parts can be regenerated, as is seen in flesh. Therefore, and so on.

(3) Again, nutriment is converted into the substance of the one nourished. Therefore, it is possible that any member that is lost can be restored by nutriment.

The opposite [of this] is obvious to the senses; a hand that is cut is not regenerated, nor an eye that has been torn out.

It must be said that organic or functional parts in animals cannot be restored, because in proportion as a thing is more noble so nature is more concerned about its production. Thus the seed of an animal is nobler and more subtle than the seed of a plant, so that nature has ordained a fixed place for the seed of an animal, that is, [in] the testicles, and not for the seed of a plant. Now, the organic members are generated from spermatic seed by a formative power *(virtus)* assigned to such parts. If, then, such a member should be cut, no matter would be left in the body from which such a member could be produced, since the productive power of a hand is in the hand and of a foot in the foot; and when the hand has been generated, this productive power disappears. Hence, such parts cannot be regenerated, both because of a lack of matter and a lack of an agent.

[Response] to the [principal] reasons. (1) To the first it should be said that branches that have been cut can be regenerated because among animate things a plant is closer to matter and inanimate things. Therefore it is begotten in whole and in part from a single parent. Hence for the production of a branch nothing is required for the matter but a nutritional principle and the generation of branches is the same. But this is not so for organic members in animals.

(2) To the second, it should be said that certain parts, as nerves, bones, and similar [or homogeneous parts] *(consimilia)* cannot be restored any more than organic parts. For these have more form and less matter, so that those which occur mostly from spermatic seed cannot be restored. But those which possess more the characteristics of matter, or are closer to matter, and yet are not so much made from spermatic fluid *(ex humido spermatieo)* but more from a nutritional fluid *(ex humido nutrimentali)* can be restored—as flesh,

hair, and nails. However, there is some flesh, as the flesh of the face, which cannot be restored; and this is so because such flesh is made from spermatic fluid.

Or [this can be put] otherwise. Flesh is twofold: In one sense it is [taken] according to species, and in another sense according to matter. The first is not regenerated, but the second can be. The first [type] is, [for example,] the flesh of the foreskin, lips, jaws, and so on.

(3) To the third, it should be said that although nutriment can be converted into the substance of the one who is nourished, nevertheless, when a hand has been cut off, the power which might convert the nutriment into a likeness of the hand is lacking. And so, as has been declared, it is plain that such members cannot be restored by nutriment.

Question 4
Whether Every Animal Breathes Air

"And the modes of animals," and so on.[2] We inquire here whether every animal breathes air.

(1) It seems that this is not so, because breathing is for the cooling of the heart and lungs; but air is warm and moist, water cold and moist. Therefore, a breathing animal has to draw more water than air.

(2) Again, an animal drawing water lives from water. The sign of this is that if he is outside water, he immediately dies. But an animal cannot live from air; therefore, he is not able to draw air.

The Philosopher [Aristotle] says the opposite.[3]

It must be said that there are certain animals that have a very warm heart, and nature has given them a lung which is like a fan for the heart, for by its dilation air is drawn in and by its compression air is emitted, as can be seen in a fan or bellows. But the heart is an impassible member "for it is not susceptible to infirmity," according to the Philosopher in [Book] III of *On the Parts [of Animals]*.[4] But water and earth concern things materially; and thus neither water nor earth are drawn in for the disposition of the heart, because they are too material; nor is fire drawn in, because it would increase the heat of the heart. But air is

2. The cue words refer to Aristotle, *History of Animals* I.1.487a.11–12. In the Oxford translation by D'Arcy W. Thompson the full sentence reads: "Animals differ from one another in their modes of subsistence, in their actions, in their habits, and in their parts."
   3. *History of Animals* I.1.487a.28–32.
   4. *On Parts of Animals* I.4.667a.33.

taken in because of its subtlety and conformity. Indeed, animals that do not properly have a heart or lung but have some material thing like a heart, as certain fishes, take in water because of their disposition just as those having lungs take in air. And the Philosopher says in his work that "some [animals] take in air and emit air"[5] and "some, as fishes, take in and emit water through their gills."[6]

[Response] to the [principal] reasons.

(1) To the first, it should be said that although air is warm and moist, it is less intensely warm than the heart.[7] And it is obvious that very hot water can be tempered by mixing in or infusing tepid water. Furthermore, the air in which we live is the coldest because of the mixture of vapors and its nearness to earth and water.

(2) To the second, it should be said that an animal which takes in water does not live from that water, because, according to Aristotle in the book *On Generation,* "we are nourished from the same things of which we are constituted."[8] Therefore, since an animal is a mixed body, it is necessary that his food be mixed. Those who assume that the chameleon *(gamaleon)* and mole, small fish *(allec)* and salamander, live on pure elements do not speak the truth, because food must be in the closest relation to the one eating it; but a simple [body or element] is not closest to a mixed body in potentiality but [far] removed. For this reason fishes do not live on sea water or water alone but on some earthy matter mixed together with the water.

Question 17
Whether Touch is in the Flesh or in Something Like It

Now we inquire about the members [or parts] devoted to the senses; and first [we ask] whether touch is in the flesh or in something like it.[9]

. . . .

[In response] to the question it should be said that touch can be in something in four ways. [In the first way] as in an organ and foundational thing, so that "touch is in something that is like the heart," as is said in *On Sense and Sensible Objects (De sensu et sensato).*[10] In another way touch is in something which makes its operation evident; and so it is in the brain. In a third way it is in something which is a carrier of the tactive power; and so it is in the nerves. In a fourth way it is in something which senses as a medium; and so it is in flesh. Therefore, when the Philosopher says "touch is in the flesh,"[11] he does not understand that it is in the

flesh as in some organ, unless the term "flesh" were extended to blood, nerve, cartilage, and such things.

. . . .

BOOK II
Question 31

Whether It Is Necessary That Every Animal Have Special Parts Assigned for Generation

We inquire whether it is necessary that every animal have special parts, namely testicles, assigned for generation.

(1) It seems that not. A power that resides in every part of a body is not required in a definite [and fixed] part. But a generative power *(virtus generativa)* is in every part of the body, otherwise what has been generated could not be assimilated to the generator in the whole and in particular parts. For unless a generative force were in the hand of the one who generates, the thing that is generated would not have a hand; and so for other parts.

. . . .

The opposite is obvious from the determination of the Philosopher.[12]

It must be said that some animals are generated by means of propagation and some by means of putrefaction. Among those generated by putrefaction, no parts are assigned for generation because they are not generated from semen. But in those generated by propagation, this is necessary, because in such animals the male emits semen beyond himself, which would not occur unless semen were gathered and distributed in some definite part. And thus males have testicles, in which semen is organized and fermented, and a rod [or penis] by means of which the semen is emitted at the time of generation. And a female has a matrix, where it receives and informs the seed; and the matrix has an opening *(os)* through which it receives the semen when the latter is discharged *(evomitur)* from the private parts [of the male].

[Response] to the [principal] reasons. (1) To the

5. *History of Animals* I.1.487a.28–29.
6. Probably *History of Animals* I.1.487a.17–18.
7. And, therefore, it can serve to cool the heart.
8. *On Generation and Corruption* II.8.335a.10–11.
9. In this question, and in most of those to follow, the principal reasons, and the responses to them, have been omitted. Only Albert's direct reply to the question will be included.
10. Aristotle, *De sensu et sensato,* 439a.1–2.
11. *History of Animals* I.4.489a.23–25.
12. See *History of Animals* III. 1.

first, it should be said that semen is derived from an excess of nutriment *(alimentum)*, which is very close to conversion into a part [of this body] but which is not converted because there is too much of it. For this reason, semen has the potential disposition of all the parts [or members of the body] which it would have in actuality if it were converted into those parts. Thus, when the semen receives the form of semen *(semen)*, or sperm *(sperma)*, or generative seed *(genitura)*—all these are the same—from the power of that in which it is, it receives a power to produce a thing similar [or like] that of which it is the semen. And although semen is distributed or gathered in a definite part, as in the testicles, it is nevertheless an excess of nutriment which is in potential likeness to the whole body. Therefore, not only is there produced from the semen a similar part but also a likeness to the whole generating body.

.   .   .   .

BOOK VII

Question 2
Whether There Is a Mean between the Living and the Nonliving

Furthermore, we inquire whether there is a mean between the living and the nonliving.

(2) And it seems not. . . .

The Philosopher holds the opposite.[13]

To this it should be said that "mean" *(medium)* is taken in many ways. In one way by the denial of each extreme, just as a mean is assumed by the Philosopher in the *Categories (Predicamenta)*[14] between health and sickness, good and evil. In another way by mixing together the extremes, just as gray *(fuscum)* is a mean between white and black. In a third way mean can be taken as equidistant between extremes, just as a virtue is a mean between two vices, or a point between the two extremes of a line. In a fourth way a mean can exist by the participation of properties, so that if something shares the nature of one thing in a certain way and the nature of another in another way, it can be said to be a mean between them.

And it is in this [last] way that the Philosopher assumes the existence of means mixed among living and nonliving things.[15] Thus plants are like means between living and nonliving things. For they are immobile with respect to place, just as nonliving things and are materially unchanged; nevertheless they are nourished and increased just as living things. Compared to nonliving things, then, the

genus of plants is living, and compared to animals it is nonliving.

But more to the point, there is a certain mean between nonliving things and plants, as mushrooms and fungi[16] which are abundant in the woods of Cologne. Similarly, between plants and animals there is something like a mean, as the sea sponge which is moved just as an animal by way of dilation and constriction, but nevertheless it has leaves like a plant, which we see with our eyes in the sea. In a similar way, a child is like a mean between brute and man, because through the whole day he eats and drinks like a brute. Therefore, children, in living, share the nature of a brute—for this reason, drunken and intemperate men are said to have puerile sins, as is said in Book III of the *Ethics;*[17] nevertheless, they differ from the brute in participating in a rational soul.

.   .   .   .

BOOK XII

Question 7
Whether the First Concoction Occurs in the Mouth

We inquire whether the first concoction *(digestio)* or first concoctive power occurs in the mouth.

(1) And it seems that it does. . . .[18]

The Philosopher says the opposite.[19]

To this it should be said that here there is controversy between the Philosopher and many physicians. For some physicians assume that the first concoction *(prima digestio)* occurs in the mouth and the second in the stomach. But according to the Philosopher, mastication, which occurs in the mouth is preparatory for the operation of the stomach, for it disposes and prepares the food so that it may be more easily concocted in the stomach. Therefore, it must be said that concoction does not properly occur in the mouth because the attractive and appetitive powers precede concoction; but the first appetite occurs in the upper part of the stomach. Therefore, there is no concoction before the nutriment reaches this part.

.   .   .   .

13. *History of Animals* VII.1.588b.4–6.
14. *Categories* 11b.38—12a.25.
15. *History of Animals* VIII.1.588b.4–27.
16. Compare Albert's evaluation of fungi and mushrooms in the next selection, and see note 26 thereto.
17. Aristotle, *Nichomachean Ethics* III.15.1119a. 33–34.
18. Two principal arguments follow, both of which are later rejected.
19. *History of Animals* II.3.650a.8–29.

Question 8
Whether the First Concoction, which Occurs in the
Stomach, Changes the Nutriment into a Different
Species

. . . .

It should be said that the first concoction, which occurs in the stomach, changes the species of the nutriment when it proceeds in a natural way. I say "in a natural way" because if there were a natural power *(naturalis virtus)* beyond the natural disposition, so that if the retentive power were sufficiently weak and the expulsive power sufficiently strong, then it could happen that the food would be discharged [from the stomach] as the same species as the one that entered [the stomach]. But when concoction proceeds in a natural way it is otherwise, because the nutriment is dissimilar at the beginning and similar at the end. But this could not occur unless the nutriment were changed into a contrary disposition. By the power of natural heat, then, the nutriment in the stomach is separated into several parts, the purer of which is called chyle *(chylus)*, the more impure waste *(egestio)*. Moreover, the chyle is changed into blood in a second concoction; and this blood is changed into members [or parts of the body] in a third concoction. Therefore the species [of the nutriment] is changed.

. . . .

Question 9
Whether the Four Humors Are Generated from the
Same Nutriment

Further, we inquire whether the four humors are generated from the same nutriment.

(1) And it seems that they are not. . . . .

And the Philosopher says the opposite.[20]

It should be said that the four humors can be generated from the same nutriment, because any nutriment, however uniform it may be, is a mixture, since simple bodies do not nourish, as the Philosopher says in [Book] I of *On Generation*[21] and in *On Sense and Sensible Objects (De sensu et sensato)*.[22] Hence in any nutriment there is the power of four things mixed together. Thus the four elements can be separated [or sorted out] from anything whatever; and, similarly, the four humors, which are related *(proportionantur)* to the four elements, can be separated from anything. For just as in milk there are four substances and by a process of coagulation three substances can be separated from it, one of which is related to earth, as cheese, another [is related] to air, as butter, and the third

to water, (some say this third substance is related to fire) as the watery part of curdled milk. And so, by a natural power, different things can be separated from nourishment, since the more earthy parts are converted into black bile *(melancholia)*, the watery parts into phlegm, the airy parts into blood, and the fiery parts into yellow bile *(cholera)*.

. . . .

Question 10
Whether Blood Is the Final Food of the Parts [of
the Body]

We inquire, further, about blood; and first whether blood is the final food of the parts [of the body].

(1) And it seems that it is not. . . . .

The Philosopher says the opposite.[23]

It should be said that the final food of the parts is twofold: common and appropriate. If we speak of the final common food, blood is the final food; but if we speak of the final appropriate food, a certain moisture is more ultimate than blood. For blood that has been delivered to any part undergoes a further concoction in that part and is changed into moisture which is in the closest proximity to the part. Nevertheless, the Philosopher understands this in the first way, while the physicians speak according to the second way.[24]

20. *On the Parts of Animals* II.3.649b.9—4.651a.19. On the four humors, see Selections 89 and 90.
21. *On Generation and Corruption* II.8.335a.11–15.
22. *On Sense and Sensible Objects* 4.441b.26–27.
23. *On the Parts of Animals* II.3.650a.33–35.
24. Aristotle's views on nutrition are admirably summarized by William Ogle in the latter's Oxford translation of Aristotle's *On the Parts of Animals*, note 2 to III.5.668a.8 (in lieu of pagination, Bekker numbers are employed): "The food masticated in the mouth, but not otherwise altered (ii.3.650ᵃ 11), reaches the stomach, where it is concocted; the heat for this purpose, which is not common heat but a heat with special powers, being supplied by the liver and spleen, which are hot organs in close contiguity with the stomach (iii.7.670ᵃ 21). The solid and indigestible portion passes off by the lower bowel, but the fluid portion, which alone can be serviceable in nutrition (ii.2.647ᵇ 26), is absorbed by the blood-vessels of the stomach and intestine (iv.4.678ᵃ 10), over the surface of which they are spread like the roots of a plant (ii.3.650ᵃ 25). These blood-vessels open by very minute and invisible pores into the intestine, pores like those in jars of unbaked clay that let water filter through (G.A.ii.6.743ᵃ 9). The matter thus absorbed passes up to the heart in the form of vapour, not as yet being blood, but only (ii.4.651ᵃ 17) an imperfect serum. In the heart and vessels (*De Somno* 3.456ᵇ 4) it undergoes a second concoction, these being the hottest parts of the body, and by this second concoction the serum is converted into blood (H. A. iii.19.521ᵃ 17), the ultimate

BOOK XIII

Question 1
On the Composition (complexio) of the Heart

We have next to consider the teeth, and so on.[25]
In this thirteenth book we shall first inquire about
the heart, because we have previously inquired
about teeth and claws. And first we inquire about
the composition of the heart.

. . . .

. . . the heart can be considered in two ways:
either materially or formally. If considered materially, the composition of the heart is melancholic
(melancholica),[26] because the heart is very hard and
compact and, according to Isaac in [the book] *On
Diets,* very difficult to nourish. If the heart is considered formally, it is of a hot and dry composition
and choleric *(cholericum).* That it is hot is obvious,
because natural heat and spirit thrive [and flourish]
in the heart. That it is also dry is obvious, because
heat cannot be conserved for long in moist matter.
But heat is conserved in the heart during life, and so
it is necessary that it be in dry matter.

Moreover, the heart is the beginning of motion
and is not subject to pain, according to the Philosopher in the first book.[27] It is necessary, therefore, that it be capable of resisting injuries, and
moistness cannot do this as can dryness. Therefore, and so on.

Briefly, then, it could be said that the heart, by
reason of its mixture, is of a melancholic composition, and inasmuch as it is the beginning of the
generation of the blood and spirits, it is of a
choleric composition. Hence, in speaking properly
about the heart, just as we now inquire about the
heart, it is hot and dry, but is made wet by the
blood contained in it, which is sent to it from the
liver.

. . . .

Question 7
Whether Blood Is Generated First in the Heart or in the Liver

We inquire, further, whether blood is generated
first in the heart or in the liver.

And it seems that it is generated in the liver. . . .

The Philosopher says the opposite.[28]

It should be said that blood can be generated in
two ways: either from something moist, as when it
is generated from phlegm; or it is generated from
something that is not moist or wet, as when it is
generated from chyle. And this second generation
can be made from something in two ways: either

virtually or materially. And each of these occurs
in two ways: with respect to the whole or the part.
In the first way the blood can be generated in any
member, as when it is generated from something
wet; in the second way it can be generated both in
the heart and in the liver. But with respect to the
whole [body] and materially, it is generated in the
liver; and with respect to the part and formally, or
virtually, it is generated in the heart, because the
blood generated in the liver is thick *(grossus)* and
not completely concocted, while the blood generated in the heart is rare *(subtilis)* and perfectly
concocted. Hence, according to the physicians
(medici), the first generation of blood occurs in
the liver, but according to the Philosopher the first
generation occurs in the heart, because to the
Philosopher the heart is the source and origin
(principium et origo) of the veins[29] (according to
the physicians, it is the liver). But this controversy
between the Philosopher and the physicians can be
resolved by [resorting to] the distinction [involved]
in "prior," because "prior" is taken in two ways:
with respect to "generation" and with respect to

food of all the organs. The amount of blood thus formed
is extremely small, as compared with the original materials (G.A.i.18.725ª18). The blood when made passes
from the heart by the vessels (arteries and veins alike),
being mingled with air inhaled by the lungs and thence
conveyed to the heart, and is carried to all parts of the
body. Each organ selects from the common stock those
materials which it requires. The nobler parts, such as the
flesh and the organs of sense, take the choicer elements,
while the inferior parts, as bones and sinews, are fed on
the inferior elements or leavings of the former (G.A.ii.6.
744ᵇ 15). This nutrition of the parts goes on most actively at night (*De Somno* I. 454ᵇ 32).

"Thus every part of the blood that can be turned to
account is utilized; but such as from its quality is unfit
for use, for instance any bitter substance, is excreted as
bile, urine, sweat, etc., in company with the matter
which results from the decay of the parts themselves.

"Such surplus of nutritious matter as there may be,
after all parts are satisfied, is either stored up in the body
as fat or the like, or passes out to form hairs, scales,
feathers, and other cutaneous appendages."

25. These are the opening words of Book III of Aristotle's *On the Parts of Animals* (III.1.661a.34—36). Aristotle begins his discussion of the heart in chapter 4 of the
same book.

26. Since it is black bile that makes something melancholic, does Albert mean that the heart is composed of
black bile?

27. Although Aristotle discusses the heart in Book I of
History of Animals (I.17.496a.4—496b.9), this remark
appears in *On the Parts of Animals* III.4.667a.33–34.

28. *On the Parts of Animals* III.5.667b.15–17 and
History of Animals III.19.521a.8–10. See also note 24.

29. *On the Parts of Animals* III.5.667b.15–21.

"perfection," just as a child is prior with respect to the generation of a man but is not prior with respect to perfection. Hence, in relation to what has been proposed here, blood in the liver is prior in generation and time, but in the heart it is prior in perfection.

. . . .

Question 9
Whether Venal Blood Is Thicker than Arterial Blood

It seems that it is not. . . .

The Philosopher says the opposite.[30]

It should be said that blood in the artery is rarer than in the vein. The reason for this is that the veins are properly vessels of the blood, while the arteries are the carriers of spirit and heat. Hence there is very little blood in the arteries, indeed just enough to nourish [or sustain] the spirit and vital heat with respect to the veins— where there is an abundance of nutrimental blood—both because motion and heat have the power to rarefy and make [the blood] more subtle. By virtue of the continuous motion of vital spirit carrying vital power or life to all the parts of the body and continually drawing air and natural heat, the blood in the artery is rarefied and dilated and consequently made more subtle. But the blood in the vein is in much greater abundance and is intended to nourish the parts of the body, so that this blood is thicker, just as parts of the body [are thicker or denser] than spirits. Therefore, the spirits are nourished from the first kind [of blood, namely arterial,] and the solid parts from the second kind [or venal blood]. . . .

. . . .

BOOK XV

Question 1
Whether Sex Is Necessary for the Generation of Animals

We have now determined above, and so on.[31] In this fifteenth book we inquire first about sex, whether sex is necessary for the generation of animals.

And it seems that it is not. . . .

The Philosopher says the opposite.[32]

It should be said that generation is twofold: one way by the transmutation of one thing from another, and in this the generator arranges the matter of another and induces its form; such a generation is found in the elements. The other way is generation by division or separation *(descissio)* of a part from a whole, or of one thing from another, and in this generation not only is the form induced by the generator but the matter is also provided. And since nature always intends what is best, it separates what is best and more noble from the most ignoble in things capable of feeling; and an agent is more noble than a patient, just as form [is more noble than matter]. Thus, in the generation of animals, nature separates the male from the female as nobler, since perfect animals generate by separation [or departure] of semen, in which the male is as the agent and the female as the patient. And so sex is necessary for the generation of such things.

Furthermore, sperm is not always suitable everywhere for generation but only in a proper place. Moreover, the place that is receptive of sperm, rendering it fit for generation, exists only in the female or in a part discharged by the female. And he [Aristotle] says this for fishes, because the female discharges a creature *(genitura)* or egg and the male then distributes his sperm over it. Therefore, sex is necessary for the generation of animals.

. . . .

Question 12
Whether Sperm Comes from the Whole Body

We inquire further about the beginning of generation; and first whether sperm comes from the whole body.

It seems that it does. . .

The Philosopher says the opposite.[33] It is argued by reason, because what proceeds from the whole body does not have a definite receptacle in the body, as is obvious with sweat. But sperm has a definite receptacle; therefore it does not proceed from the whole body.

To this it should be said that sperm can proceed from any part of the body and also simultaneously from a definite part of the body, because sperm is the excess of the final food.[34] But the final food of

---

30. Although it is likely that Aristotle considered venal blood thicker than arterial blood, I was unable to locate an explicit statement to that effect.

31. These are Aristotle's opening words of the treatise *On Generation of Animals (De generatione animalium)* I.1.715a.1.

32. In *On Generation of Animals* I.1.715a.18–24, Aristotle believes that almost all sanguineous animals are produced from a sexual union.

33. *On the Generation of Animals* I.17.721b.7–722a.16.

34. It seems that Albert disagrees with Aristotle. Compare Isidore of Seville's opinion in Selection 93, par. 139.

any part can have an excess, and therefore sperm can be derived from any part.

Moreover, it can also arise from a definite part, because there is nothing to prevent one part from being of perfect power and another to be imperfect. A perfect part [of the body] can convert the final food wholly into its [own] nature, with no excess remaining; but a part of imperfect power can be hindered in this. Therefore, nothing prevents sperm from being derived from one part and not another.

. . . .

Question 16
Whether Sperm Could Be Generated Directly from Blood

We inquire further whether sperm could be generated directly from blood.

It seems that it can. . . .

To this it should be said that sperm cannot be generated directly from blood. Indeed, excess blood flowing through the veins is changed into rarer moisture in the extremities of the veins. And this moisture passes through the sponginess of the parts and becomes [even] rarer with the pure [part] separated from the impure. What is impure is expelled through sweat, or abscesses (apostemata), or through sediment (hypostasim) in the urine. But what is pure becomes like those parts, and what remains after a sufficient transformation of this moisture into parts [of the body] becomes sperm. The universal beginning of the generation of sperm is blood, but the direct beginning of its generation is the moisture which is generated from blood. Now, this moisture is of another nature in different parts of the body, because it is of another nature in bone and flesh; but this depends on its potency.

. . . .

BOOK XVI

Question 15
Whether the Heart Is Generated First

We inquire further whether the heart is generated first.

(1) It seems that it is not . . . .

The Philosopher says the opposite.[35]

To this it should be said that a part is twofold. Sometimes it is necessary to a thing passing into the very constitution of it; and sometimes a part does not pass into the constitution of a thing but is called "part" because it is necessary for generation and is formed first from the matter of genera-

tion. If we speak in the first way, then the heart is what is generated first; if in the second way, then it is the afterbirth (secundina), which is a certain film [or membrane] (pellicula) containing the semen, until it is concocted and perfected and the form of the fetus is induced. This afterbirth occurs from a certain particle of the semen, but finally the afterbirth departs with the fetus and does pass into the constitution of the fetus. Hence the afterbirth is the first thing generated from the semen, but it is not the first part of the animal generated from the semen, unless it be the heart itself. The reason for this is that the first force (virtus) is a vital force; for nothing is nourished, augmented, or capable of sensation, unless it is alive. And so all forces [or powers] are based on the vital [or life] force. Now, the liver does not function except by means of heat; nor can sensation occur without heat or motion, as is stated in [Book] III of On the Soul (De anima) or [Book] II.[36] But the source (radix) of heat is in the heart. Hence the heart is the first part generated in an animal, after which comes the liver. Thus Avicenna said[37] that in the semen there first appear two swellings (ampullae), from one of which the heart is made and from the other the liver. Nevertheless, the soul is first sustained in the heart, for which reason the heart is said to be the seat of the soul. The Philosopher proves this [namely that the heart is generated first] many times in his works.[38]

BOOK XVII

Question 12
Whether All the Elements Come Together for the Generation of an Animal Generated by Putrefaction

We inquire further about the generation of animals by putrefaction. And first whether all the elements come together for the generation of an animal generated by putrefaction.

(1) And it seems that they do not . . . .

It should be said that for the generation of an animal all the elements come together either "in

35. On the Generation of Animals II.6.741b.15–19. Without mentioning Aristotle, Isidore of Seville reports this view as well as another in which the fetus is held to begin with the head (see Selection 93, par. 143).

36. Aristotle, On the Soul III.1.425a.6 and perhaps II.5.416b.32—417a.6.

37. Canon of Medicine (Liber canonis), III, fen 21, tr. 1, ch. 2.

38. In addition to the discussion cited in note 35, see also On the Generation of Animals II.4.740a.3–4 and II.6.743b.25–26; On the Parts of Animals III.3.665a. 10–14.

themselves" or "in their effects." "In themselves" as when the four elements, after having been divided, are in such a close proportion that no one of them is wholly denominated by another; but by a reciprocal action and passion the forms of each are expelled, and in the matter which was previously under four distinct forms one form is induced, having the power of the four in a certain mixture (*confusio*).[39]

"In their effects," the elements arrive at the constitution of an animal just as blood is generated from one nutriment and semen from blood. Therefore, since such an animal would be a mixed body, the four elements come together for its generation either "in themselves" or in another. This can be stated briefly: Generations are made from contraries; and for the generation of all mixed things earth and water are required; therefore their contraries are required; but air is the proper contrary to earth, and fire to water. Therefore, and so on [that is, the four elements come together for the generation of an animal generated by putrefaction].

Question 13

Whether Putrefaction Is a Path [or Means] to Generation

(1) It seems that it is not . . . .

The Philosopher says the opposite.[40]

To this it should be said that an animal can be generated from putrefaction. For just as in concoction the semen in the womb (*matrix*) is concocted and perfected by an internal power, and what is impure is expelled, and the pure part which remains is converted into the matter of the fetus, so [also] is it proper that in the bowels of the earth putrified matter can be concocted by the heat of the celestial body [the sun?] and the terrestrial or containing body; and that what is impure is expelled; and what is pure and remains can be in potency to the form of an animal. For just as internal heat disposes the matter of an animal for the generation of life, so external heat can dispose matter for the generation of an imperfect animal.

.   .   .   .

Botany

87 PHILOSOPHICAL AND THEORETICAL BOTANY

Albertus Magnus (ca. 1193–1280)

Translation, introduction, and annotation by Edward Grant[1]

Botany as a theoretical and classificatory science was represented par excellence in Greek antiquity by the treatises of Theophrastus (ca. 372–ca. 288 B.C.) called *Enquiry into Plants* and *Causes of Plants*. During the first century B.C. the Aristotelian commentator Nicholas of Damascus, utilizing the works of Theophrastus, apparently composed a brief theoretical treatise in Greek called *On Plants (De plantis)*, the text of which was lost, though not before it had been translated into Arabic. The Arabic text, in turn, was apparently also lost, sometime after it had been translated into Latin by the Englishman Alfred of Sareshal in the twelfth century. By virtue of its Aristotelian character, it was considered a genuine work of Aristotle's throughout the Middle Ages and well into the nineteenth century, and for this reason it is usually cited as Pseudo-Aristotle *On Plants* (as it will be called here for convenience). While this brief two-book treatise (it has been translated in the Oxford English translation by E. S. Forster) formed the basis of Books I and IV of Albertus Magnus' lengthy seven-book theoretical botanical

treatise *On Plants (De vegetabilibus)*, Albert not only elaborated upon much that he included from Pseudo-Aristotle but with a fine instinct for morphology[2] included much that was original. The

39. Compare the discussions by Aquinas and Albert of Saxony on the status of elements in a compound in Selections 77.1,2. Albertus Magnus' position appears to differ somewhat from theirs.

40. *Meteorology* IV.1.379a.2–18. In the course of his discussion Aristotle defines putrefaction as "the destruction of the peculiar and natural heat in any moist subject by external heat, that is, by the heat of the environment."

1. From *Alberti Magni ex ordine praedicatorum De vegetabilibus libri VII* . . . a critical edition begun by Ernest Meyer and completed by Carl Jessen (Berlin, 1867), pp. 20–21, 23–27, 56–62, 71–77, 138–142, 148–153, 254–255, 312.

2. Agnes Arber, *Herbals Their Origin and Evolution,* 2d ed. (Cambridge, England: The University Press, 1938), pp. 4–5, remarks on Albert's intuitive sense of morphology "in which he was probably unsurpassed during the next four hundred years. He points out, for instance that, in the vine, a tendril sometimes occurs in place of a branch of grapes, and from this he concludes that the tendril is to be interpreted as a bunch of grapes

first five books are theoretical and Aristotelian, while the sixth is devoted to a series of plant and herb descriptions, and the seventh, to agriculture and horticulture. The emphasis on theory and philosophy of biology is made evident at the start of Book VI, where before presenting a sequence of alphabetized descriptions of a variety of individual plant species Albert apologetically explains that particulars of this kind do not constitute philosophy or theory but are presented for the purpose of satisfying those who are curious about such matters. Ironically, "after the lapse of nearly seven centuries, it is his descriptions of species that are important, both for identifying the plants concerned and for ascertaining the precise connotation of the terms employed by him. His general philosophy of the Plant World, on the other hand, is now of relatively little interest; his conception of the plant being zoomorphic, he constantly endeavoured to explain its organs and mode of life by means of false analogies drawn from the Animal World. In fact, he seems to have regarded the plant as an imperfect animal."[3]

Albertus Magnus is justly celebrated for his attention to detail and his descriptive powers, but his heroic effort in this treatise to present botany in theoretical and philosophical terms must not be scorned despite its imperfections, which after all are but the shortcomings of all of theoretical botany—Greek, Arabic, and Latin—until the Renaissance.[4] His treatise represents virtually the sole theoretical effort in botany during the course of the Middle Ages. Indeed, since the great works of Theophrastus were unknown during the Latin Middle Ages, only the relatively short Pseudo-Aristotelian *On Plants* was available as a purely theoretical work (a nontheoretical botanical description is represented in the next selection; numerous herbals, oriented toward drugs and medicine, were also available; see Selection 109.1). While using herbals as a point of departure, Albert went considerably beyond them, presenting, as we have said, much that was original.

BOOK I, TRACT I

Chapter 1
[Objective of Book]

. . .We must begin with the bodies of plants. Concerning this, we intend to present in this book the usual things appropriate to plants according to the whole and its parts, since particulars are infinite and no science *(disciplina)* can be made of them, as Plato put it very well. Now, since the first

common principle appropriate to all plants and their parts is life [itself], which is found in plants, we shall first inquire about the life of plants. . .[5]

incompletely developed. He distinguishes also between thorns and prickles, and realizes that the former are of the nature of stems, while the latter are merely a surface development."

3. T. A. Sprague, "Plant Morphology in Albertus Magnus," *Bulletin of Miscellaneous Information, Royal Botanic Gardens* (1933), pp. 431–432. This article, and another following immediately after ("Botanical Terms in Albertus Magnus," pp. 440–459), define and illustrate in alphabetical order the many botanical terms that appear in Albert's *De vegetabilibus*. It has been of great value in this translation. In the second article mentioned here, Sprague notes the pitfalls in interpreting medieval usage of botanical terms (pp. 440–441):

"One of the principal difficulties encountered in reading mediaeval botanical works is the strange yet deceptively familiar terminology employed in them. Many botanical terms still current nowadays were then used in very different senses, and certain words possessed a more general signification, covering several morphological categories now recognized as distinct. The word *cotyledo* illustrates both difficulties: now the equivalent of cotyledon, in Albertus it meant a stalk of various kinds—petiole, peduncle, pedicel, filament or funicle. Other modern terms used by Albertus with different or additional meanings were: *arillus, arista, folliculus, legumen, nodus, pomum, racemus, siliqua, spica, stipula, testa, theca, virgula.* The word *siliqua*, now applied to the capsular fruit characteristic of *Cruciferae*, was used by Albertus to denote a spathe, glumes of a grass spikelet, a calyx, a dry seed-vessel or pericarp, and the cartilaginous "endocarp" of a pome. A further source of difficulty is the employment of two or three different words in practically the same sense: thus the three terms *folliculus, siliqua* and *theca* were used more or less indiscriminately with the general meaning of closed covering. Similarly *camera, casa* and *cellula* were equivalents of loculus.

"The gradual change of meaning of certain terms from the first through the seventh to the thirteenth century is illustrated by the word *arbustum*. This was used by Pliny in its classical sense of plantation of trees, on which vines were trained; by the time of Isidorus it had acquired the additional meaning of sapling; and in Albertus Magnus it meant shrub—the equivalent of the French word arbuste. Thick, as applied to leaves, was *spissus* (not *crassus*) in Albertus, and narrow was *strictus* (not *angustus*). *Reticulatio* denoted the reticulation of veins in a leaf—previous writers do not seem to have paid much attention to this character, and Albertus was possibly the first to use the word in this sense.

"Caducous members whose function was not understood by Albertus were naturally considered by him as excreta: thus the male catkins of *Alnus, Corylus* and *Juglans* were called *purgamenta* or *superfluitates*. Similarly pollen *(formale croceum)* was regarded by Albertus as a secretion of bile *(cholera)*."

4. Not until the works of Andrea Cesalpino in the late sixteenth century was Albert's botanical work surpassed (Arber, p. 6).

5. In the following chapters Albert discusses whether or not plants have life and concludes that they do.

Chapter 6
Concerning the Reasons of Aristotle by Which He Proves that Plants Do Not Sense or Sleep

Now, according to our intent, we must investigate a problem about which we made mention in our preceding discussion, namely whether plants have desire, animal motion, and a soul, and concerning what was said about plants being relaxed by sleep and wakefulness, just as sleep in an animal is caused by what is released by evaporation from the place of digestion. That plants do not have such a spirit, which is drawn by inhaling and exhaling, as Anaxagoras said, we can prove in this way: because we find that there are many animals that do not inhale and exhale; however, such a spirit is more appropriate to animals than plants, because animals are, by their natures, hotter and participate in [that is, contain more of] the higher elements than plants, which are earthy and cold. For this reason animals require animal spirits more than plants.

Similarly, we find that plants do not sleep or wake, since in them evaporation does not occur from the place of digestion to any cool place located in their bodies. Therefore, the evaporation which is released in them descends and immobilizes the exterior parts of the plant, which by nature always exist immobile and insensible. This is especially so since, as was made clear in *On Sleep and Waking (De somno et vigilia)*,[6] to be awake is a certain faculty and force and effect of the senses; sleep, on the contrary, is a certain weakness and loss of the senses. But since plants wholly lack sense they cannot participate in accidents of the senses which involve a force or power [of the senses] and a lack of the senses. . . .

Chapter 7
On the Sex of Plants according to the Statements of the Ancients

Since sex is an accident of living things, let us speak of the sex of plants, conveying about this things that were said by the ancients according to reason. For it seems that in this science it must be especially investigated whether a male sex and a female sex are found in plants or whether they are found in common or mixed together, as was said by Protagoras, whom they called Abrutulus. If we consider the definitions of these sexes, then a male is that which generates from its semen an individual in another of its species, playing the role of agent in generation; a female is that which generates in itself after receiving the seed from another individual of its species, playing the role of matter and patient in generation. If we consider these as the definitions, then, without doubt, sex is not found in plants. But perhaps it is possible to discover some properties of participating sexes in plants. For properly taken, the sexes are mutually separated in subject and place, but some common properties of each sex are also found in plants. [Properties of] the male are found in generation in the sense that there is something forming and, as it were, shaping [or figuring the plant]; a female property is found in the sense that something is formed and shaped and has properties opposite to those [of the male], as softness, moistness, and smoothness, since these receive a formation very well. And in plants we also find those that are called masculine, because everything that is generated from them is harder and rougher, and what is generated from those that are called female is softer and smoother. And the female plant is more fruitful and has thicker fruit because of its abundance of moisture. We shall show this below. Nevertheless, by what has been established, it is obvious that there is no separate [or distinct] sex in plants.

Let us now inquire whether sex is mixed [or joined] together in plants, as Protagoras said. By taking the meaning of "mixture" *(commixtio)* truly and properly it ought not to be thought that sex is mixed in plants.[7] For any things that are mixed were previously separated and would be again separable after mixture, as was proved at the end of the first [book] *On Generation (peri geneseos)*.[8] Therefore, if the sexes in plants were said to be mixed, it would be necessary that previously in matter there should be in nature or in a genus of plants a separate male [plant] per se and a separate female [plant] per se which would be mixed together afterward by generation, since nothing that is mixed is produced from mixable things except by generation in which the mixable things are mutually altered. But if this were true, there should be found a plant [wholly] distinct in sex before such a mixture [or mingling] of the sexes, which is utterly absurd. There is also another absurdity which

6. Presumably Book I, Tract I, chapter 9, of Albert's commentary on this Aristotelian treatise.
7. In what follows, Albert elaborates and adds to the discussion in the Pseudo-Aristotelian *De Plantis*, Book I, chapter 2.
8. Aristotle's *On Generation and Corruption*, Book I, chapter 10 (see Selection 77,n.19). Albert gives the Greek, rather than the Latin, form of the title of this Aristotelian treatise.

seems to follow from this. Now, since the male is the agent and the female is the patient [or passive entity], if the sexes were mixed together in a plant, the same thing will appear to be agent and patient, which is impossible, as was shown in the first book *On Generation (peri geneseos)*.

Furthermore, that male cannot be mixed with female in plants can be shown in this way. For if it were true, a plant would not require anything outside itself for generation. Therefore, since it would be perfect [that is, complete], lacking nothing, a plant would be more perfect in generation than an animal, since among the latter the female, which is imperfect [or incomplete] for [the process of] generation, desires the male, just as matter desires form and the base seeks the good. Moreover, the male [also] seeks the female to help in generation, but he is not imperfect except *per accidens*. . . just as a form seeks matter *per accidens* so that it may have its material existence in it. Furthermore, it is wholly improbable, since we see that plants lack many things, as [for example] they require a special time of year and a special place in which to sprout [or grow]. The reason for this is that the beginning of the nutriment of plants is from the earth, from which they generate an excess, and the first digestion of this nutriment occurs in the earth; and the active principle of the generation of the fruits of a plant is from the sun . . . . Moreover, Lycophron, whom the Arabs call Lecineon, said that the earth is the mother of plants, ministering them as a patient, and the sun is the father, generating as an agent [or active cause]. . . . Truly, therefore, there is no sex in plants, neither separate nor combined [or mixed].

But if we wish to assign to plants a combination of the sexes, it must be conceived in another way. For sex as it is found in individual complete substances is not in plants, either separated or combined. But the forces or powers of the sexes are in the seeds, embryos, or impregnations. For I say that embryos or impregnations are seeds that have been conceived. For in them the power of the male *(vis masculi)* is what makes the form and is the agent; the seed of the female mixed with it is what is formed and is the patient [or passive entity]. And the seed of the male is as the artificer, the seed of the female is as the thing of which it is to be made, the thing in which the form is produced . . . . In all embryos the power of each sex is by way of agent and patient. It is in this way that we should imagine that the powers of the sexes are combined in plants, since it cannot be doubted

that there is an agent and form-giving principle in them and [also] something that is matter which receives the action and formation. . . .

Chapter 1
On the Absolute Diversity of the Parts of a Tree That Are Most Prominent in the Genus of Plants

In inquiring about the body of plants it is necessary that we proceed by the path of nature, according to which the principles of compound bodies are the parts of which they are composed. The bodies of plants must be understood through the parts, because, as Aristotle put it so elegantly, we know what is compounded when we know from how many things and of what kinds it is compounded. Hence, just as the bodies of animals are known through anatomy *(anatomia)* which is called division *(divisio)*, so is the nature of the bodies of plants known by the division of the bodies of plants. And so, with respect to parts, one must consider first the whole diversity of the parts. First, however, we shall only cite these differences and afterward assign the causes of all the differences. If we did not follow Aristotle, however, but others, we would surely proceed otherwise.

We therefore say, with Aristotle,[9] that certain plants called trees *(arbores)* have gum *(gummas)*— as the pine tree *(pix de abiete)*, resin and almond gum, myrrh, frankincense *(thus)*, and gum arabic, and others about which we shall present a special tractate below—and certain others do not have gum, or have very little of it, as the box tree *(buxus)* the oak *(quercus)*, and others of this kind having very compact wood. And this is the first difference taken from the nutriment of the tree itself.

Now, the second [difference] is that some trees have nodes *(nodos)*, veins *(venas)* [or vascular bundles], tissues,[10] wood, bark, and inner pith *(medullam)*, all of which parts are certain organs in which nourishment is completed. Some trees lack some of these parts, or in many of them the bark constitutes most of the tree.

I call the connection of the joints nodes, just as there are nodes in vines; and these nodes are called

9. The ultimate source of this chapter is Theophrastus, *Enquiry into Plants* (London, 1916; translated into English in the Loeb Classical Library by Arthur Hort). Throughout this chapter Albertus greatly elaborates brief statements in the Pseudo-Aristotelian *On Plants*.

10. Through a mistranslation from the Arabic, the text has *venter* instead of *caro* (flesh or tissue).

malleoli.[11] What is between the nodes[12] is more subtle [matter] but is made [or becomes] gross *(ingrossatur)* at the nodes. And the node is where the nutriment that has collected receives, by means of digestion, a greater assimilation than before to the wood around it.

Certain plants, however, are of a rare and very similar substance and wholly lack such nodes—as [for example] the scirpus *(cirpus)* plant. This is the reason for the proverb [in which it is said that] a node is sought in the scirpus when someone seeks for something in vain[13]—for the opposite is obvious.

Veins *(vena)*[14] are not truly in any plants as perfectly structured veins. But there are straight paths [or channels] in which the nutriment flows, as is seen in the leaves of the plantain[15] when they are separated slightly. For certain threads *(fila)* appear that are paths [or channels] in which the nutriment flows in the manner of straight and curved lines extended through the plants. And sometimes these channels are separated and sometimes congregated, just like veins. No plant is found that lacks these unless those which consist almost wholly of bark; but even in these something similar to channels is found, but less than in the others. These veins are found to arise in plants from two sources. In those plants that are not full of nodes *(nodosae)* and are of rarer substance these channels are extended up from the roots through the stem, through the branches, and through the leaves that are divided and spread. And sometimes they are derived from the pith *(medulla)* and extended to the surface of the plant. This is known, because when the plants are cut, white lines appear in the form of a star extended from the middle of the pith to the bark of the plant; as in vines and other node-filled plants, especially in the node and near the node.

"Tissue" [or "flesh"][16] of plants is appropriate to plants [only] metaphorically. In one way earth is the general flesh of all plants, as will be shown below. But some plants are soft on the inside and have empty spaces in which the nutriment that has been drawn there is cooked [or concocted] so that afterward it can be distributed through the whole plant, as *cassia fistula* [mezereon?][17] and the plant which bears *zucarum,* and many others.

The bark *(cortex)* is twofold in large plants, namely a harder exterior, and softer and juicier interior; and the exterior is dry, the interior frequently sticky [or viscous].

The pith *(medulla)* is sometimes like a crumbling powder, [varying from] purple to black, succeeding [or transforming] into a slight humor. Sometimes it is more humid and soft, or perhaps sometimes [even] fluid; but sometimes it is of a rare and white substance, or of another color.

This, then, is the second difference among plants and is found especially among trees, since trees alone are the most perfect plants, just as animals with the largest bodies are found to be more perfect than other animals. And thus, also, a difference among the plants that are trees is especially set forth here, because the greater difference is found among them and their parts are better known. And so it is that by analogy with them one must understand the parts of small plants, whose parts are not so well known to us.

With respect to the nodes of which we spoke before, it must be understood that nodes are found in trees and plants in three ways. One is that which was declared above, and only this way is wholly natural and intended by nature. Another way in which nodes are found is caused by the dryness and earthiness of the tree and the nutriment which nourishes it. In this, the pores of the tree are constricted because of the firmness and thickness of its parts, and since the nutriment is earthy and thick, it closes the direct path [or channel] through the veins of the wood and diverts it to another part, where it is pressed together into a node. And all wood is caused to be filled with nodes and resistant to division [or cleavage], as we said in the fourth book of the *Meteors*.[18] A third way of producing nodes is from a certain growth which appears on the outside of trees and is converted into wood that is net-like and seems like an interweaving of wood fibers *(ex nervis ligneis)* because of a manifold turning by means of which the humor [or sap] of the wood is turned into itself. This node is called

---

11. That is, "hammer-headed."
12. The internodes?
13. The proverb was cited by Isidore of Seville, *Etymologies,* Book XVII, chapter 9, paragraph 97.
14. That is, vascular bundles.
15. A plant of the genus *Plantago. Plantago maior* is a low herb with broad flat leaves spread close to the ground.
16. As mentioned in note 10, although the Latin term is *venter,* it should have been *caro,* and it is here translated as "flesh" or "tissue."
17. The Harper Latin dictionary defines *cassia* as "a tree with an aromatic bark" and "a fragrant shrub-like plant, mezereon, ..."
18. Book IV, Tract III, chapters 12 and 13, of Albertus' *Meteorologica.*

murra [that is, a burr on tree trunks] in Latin....[19] Therefore, the first kind of node is called *malleolus* ["hammer-headed"]; the second is properly called node *(nodus)* and the wood in which it exists is called node-filled *(nodosum)*; and the third node is called by the proper name *murra*.

A third difference [among trees] is that certain trees have their fruit between wood and bark—that is, in certain woody barks. These have wood in the center of their fruits *(pomis)*[20] and a woody bark, as the fir *(abies)* and pine *(pinus)* and certain other trees that have woody fruits *(poma lignea)*. Furthermore, some trees bring forth their fruits in other places and in many ways, as will be shown below.

A fourth difference is in the parts of the plants mutually compared. For some of these parts are simple, since they retain the name and structure of the whole after they have been divided into parts by a continuous division—as [for example] humor [or sap], which is found in the vein of a plant, and even nodes, although it is less appropriate in nodes than in other parts. Furthermore, of parts some are composed of these [simple] parts, as [ordinary] branches *(rami)* and slender branches *(virgae)*,[21] and similar parts, anyone of which has nodes, veins, and humor [or sap] within itself. However, [it should be observed that] humor is a potential part, while others are parts in actuality. Composite parts, moreover, are divided by a heterogeneous division into the things of which it is composed— that is, into nodes, veins, and other such parts. All the parts described here are found together in all plants, but only in the most perfect and largest, as we said above.

And a fifth difference [among trees] is that these more perfect plants have other functional *(official-es)* parts than those which were declared and which are roots *(radices)*; slender branches *(virgas)* of roots [twigs?] and leaves in the covering of fruits; branches [pedicels?] that are like the supports of the slender branches; and flowers [catkins?] that are the excretions of menstruations; and sprouts *(pullulationes)* of branchlets in different parts of their bodies, as certain water shoots which do not bear fruit in the first year or even in several years, but when they have grown will produce fruit. And with respect to the type of parts there are, there is a roundness of shape which is found in the large plants [that is, trees]; and similarly, the bark, which covers the body of the plant from external harm and through which must be expelled what should be purged from the body of the tree;

and for the same reason, similarly with respect to the bark which surrounds the fruit.[22]

Chapter 2
How the Parts of Large Plants are Compared to Animals, and How Not

Just as there are in animals certain members [or limbs] of similar parts and of dissimilar parts called organic or dissimilar members, so also [are these to be found] in perfect plants. For the parts of such plants are compared by a certain imperfect imitation to the members of animals. Thus the bark in a plant is compared to the skin of an animal. And roots are comparable to the mouth of an animal; nodes to the nerves of an animal, though not to those that are sensitive and motive nerves but [only] to those that are called ligaments, which are stretched from one end of a bone to the joint of the other end....[23]

Chapter 4
On the Diversity of the Essential and Principal Parts of a Plant

19. Approximately three lines are omitted here.

20. Two terms for "fruit" were used by Albertus and are distinguished by T. A. Sprague (pp. 447, 453) as follows: (1) *fructus*—"fruit in a general sense, including drupes, berries, legumes, capsules, caryopses, nuts;" and (2) *pomum*— "fruit, used to denote both fleshy and woody 'fruits,' e.g. of *Pinus, Musa, Pyrus, Citrus, Cupressus,* ..."

21. *Virgae* is a term applied to root (see below, the fifth difference) or stem.

22. As a fair indication of how extensively Albertus added to, and elaborated upon, the briefer remarks of the Pseudo-Aristotelian *De Plantis*, I quote from Book I, chapter 3, of the latter treatise (in E. S. Forster's translation) the entire passage that served as the basis for Albertus' lengthy discussion of the five differences among trees:

"Some trees contain a gummy substance, such as resin and almond-gum and myrrh, and frankincense, and gum arabic. Some trees have fibres and veins and flesh and wood and bark and marrow within them; some trees consist almost wholly of bark. In some the fruit is underneath the bark, that is, between the bark and the wood. Some parts of the tree are simple, such as the moisture found in it and the fibres and veins; other parts are composite, such as the branches and twigs and the like. These are not all found in all plants; for some have composite and some simple parts, while others do not have them. Some plants possess various other parts as well (roots, twigs), leaves, pedicels, flowers, catkins, tendrils, and bark surrounding the fruit."

23. This is based on Pseudo-Aristotle, Book I, chapter 3, and is cited here only to reveal the unfortunate but understandable analogies that were drawn between plant and animal. Albertus goes into much greater detail than Pseudo-Aristotle.

The most special difference is that some plants produce fruit above the leaves (as pears, apples, bullaces (*cini*) [that is, *Prunus insititia—Ed.*]), plums (*pruni*), and almonds, in which the leaf is found at the root of the peduncle of the fruit [*a radice cotyledonis fructus*]); other plants produce fruit below the leaf, but these are rare in our climes; and some produce fruit opposite to the leaf, as in vines. . . .

There are parts of all plants . . . which grow and increase . . . and these are the essential parts of plants: These are roots, branches of roots (*virgae*), stems (*stipites*), and branches. For these parts in plants are like the parts of animals. . . .

The root of a plant is organically the mediator between the plant and food. . . .

The stem of a plant is what alone is produced by the earth from the root and is erected perpendicularly at right angles [to the root?]. It is like the height of the upright plant.

Suckers (*surculi*) [or grafts] are properly twigs [or shoots] from the root of a tree sprouting around the stem. Taken improperly, suckers are called branches (*palmites*) which sprout from the stem and branches in different parts of the plant.

Branches (*rami*) are those parts which arise immediately from the stem—not in place of suckers—above and beyond the stem. And such branches are not found in all plants but usually only in he large plants. . . .

Barks, wood, and pith, and all such essential parts of this kind arise from the seminal humor [or sap] and from the food of the plant. Some call the pith (*medulla*) the mother of the tree because in it the seed (*semen*) seems to be conceived and formed. . . .

Nodes, veins, and the woody or herbaceous tissues (*caro*) of plants, which fill what is between the veins and nodes, are all constituted of the four elements. Indeed, in these things the powers of the elements are found more than in the parts of animals, because plants are more material and nearer to the elements and in them the elements are altered less than in the bodies of animals. . . .

Chapter 5
In Which the Genus of Plants is Divided into Species and the Reason for This

If we now consider the plant according to the community and range of its predication, parts of it are designated univocally by name and structure. Of these, some are trees; some lie between trees and herbs and their name shows their intermediate

being, for in Greek they are called *ambragyon*[24] and in Latin are called by the common name *arbusta*, bushes. Moreover, some are bushy herbs (*olera virentia*). Almost every plant falls under these names.[25] I say almost because the arborescent shrub (*fructicem magnum*) lies between shrub and tree, inclining, however, more toward the tree. And according to this, the highest of the plants is the tree, the lowest the herb. But since nature does not pass instantaneously from extreme to extreme, but passes through all means that have been fitted between them, thus it [nature] has made many means, of which only one is equidistant and middle; and there are others which are closer to the extremes. This will be obvious in the example and definitions of the names introduced.

A tree is that which has a strong stem from its root, and on this stem arise many branches, and on the branches many slender branches (*virgae*), and on the latter arise what are called ultimate branchlets (*flagra*), as in the olive, cypress, and fig trees.

Plants that are mean between trees and herbs are called *ambragyon* in Greek and *arbusta* (bush) in Latin. From their roots they send forth many branches in the manner of long branches. And there are three kinds. . . .

There is another mean plant with much the property of an herb which is called *olus virens* (bushy herb) by authors. This plant projects many stems from one root and different branches on the stems, but it has little or nothing woody, although, in later age, some of them grow hard like wood. Among this kind are rue (*ruta*), cabbage (*caulis*), and many others.

Herbs seem to retain the lowest place among plants; they have one root, from which nothing is produced but leaves, as a clump of scallions (*rapa porri*) and some others.

In this division of plants there are no fungi (*fungi*), truffles (*tuberes*), or mushrooms (*boleti*) of any kind, because they seem to possess the least of the life and powers of a plant, so that they lack the force to form leaves but are said to be certain

24. In his translation of Pseudo-Aristotle, where the version of this term is *ambrachion* (Bk. I, ch. 4) E. S. Forster remarks that this term is unknown but that its meaning is clear—namely "bush."

25. This fourfold division of plants is found in Pseudo-Aristotle (Bk. I, ch. 4) and its ultimate source is Theophrastus, *Enquiry into Plants*, Book I, chapter 3. In his translation of the latter work Hort names the four as *tree, shrub, under-shrub*, and *herb*. Thus in our translation Theophrastus' shrub is a "bush" and undershrub a "bushy herb."

plant-like things that have been exhaled and evaporated from other plants. For this reason they are rarely found except between other plants and they endure for a short time. . . .[26]

. . . .

BOOK II, TRACT II

IN WHICH THESE THINGS ARE INVESTIGATED: WHAT HAPPENS NATURALLY TO PLANTS WHICH CONTRIBUTES TO THEIR PRODUCTION OF FRUIT OR GENERATION

Chapter 1
In Which We Discuss What Will Be Said in this Tractate and also about the Nature of Leaves.

We shall now examine the common things that happen essentially and naturally to plants with respect to their generation, according as nature intends to save each of their species. We shall not inquire here about the place or mode of generation, but we only desire to know the nature of those parts which remain on them when they are in generation. Now, these parts are the *leaves,* which have the function of covering the fruits, the *flowers,* which are the signs of fruits, and the *fruits* and *seeds.* It is indeed necessary to know the natures which these have in common and their differences, so that after these things have been considered, the causes of the same parts of the body are assigned more carefully and in an orderly manner. For a cause is sought in vain if the reason for which the cause is sought is not known before. Indeed, to philosophize is to investigate the certain, manifest, and true cause of a known effect and to show how it is the cause of it and that it is impossible that it be related otherwise.

Therefore, we say of the leaf in general that the matter of a leaf in all plants is a watery humor mixed somewhat with, and not well separated from, dried earth—[that is,] not purged from the dregs of earthy matter. The sign of this is that the leaves of large plants having broad and thin leaves suck up moisture quickly and thus almost all leaves are produced full of holes.[27] But where the humor is viscous and more watery, being agitated from the heat that has been retained within its viscosity, the leaves are thicker and cling more to those plants. But the reason why leaves are frequently thick in herbs and bushes is that the humor is abundantly distributed and multiplied in them because of their nearness to, and continuity with, their roots.

Now, a final purpose [or goal] of leaves is to cover the fruit; [moreover,] nature requires a purging from an excess of watery humidity.[28] Now, since nature is wise and ingenious, it uses the same purging for the protection of the fruits. Thus in many plants nature produces the leaves under the fruit, so that the spread leaf is extended beyond the fruit. However, the production and location of leaves is more general, because the leaf sprouts at the base of the peduncle[29] of the fruit *(ad basim cotyledonis fructus).* This is more fitting for the matter and purpose of the leaf. There are two kinds of material vapors in the flesh of the plant and in all other vapors, namely a *humid vapor* and a *dry, windy vapor.*

The humid vapor is the matter of the leaf and the dry, windy vapor is the matter of the fruit. For this reason the fruit is judged to be windy by virtue of its nature. The dry vapor is sharper and distributed in the flesh of the tree; the humid vapor is duller [or weaker]. With its sharpness it cuts the body of the plant so that each of these vapors bursts forth. The vapor of the fruits ascends higher and the vapor of the leaves emerges under it. Nevertheless, since these vapors are mixed together in the flesh of the plant, it is necessary that the humid vapor should have a motion by virtue of the windy vapor mixed with it. And so it happens that because the leaf is generated by a humid vapor it comes forth near the fruit, close to the base of the fruit as in many [plants].

The location of a leaf is varied in three ways. Sometimes it is under the fruit at the base of the peduncle of the fruit, as in pear, apple, bullace (*cino*) [that is, *Prunus insititia—Ed.*], and plum trees, and in many others; sometimes it is opposite the fruit, as in vines; and sometimes it is above the fruit, as in the violet *(viola)* and many other herbs. The cause of the leaf being at the base of [or below] the fruit has been assigned above in terms of the fitness [or suitability] of the matter. This location is also appropriate to its goal [or end], because the leaf that comes forth underneath is extended toward the flower and covers it more usefully and protects the flower.

26. Although Heinrich Balss (*Albertus Magnus als Biologe* [Stuttgart: Wissenschaftliche Verlagsgesellschaft, 1947], p. 83) asserted that Albertus had added mushrooms as a fifth class or type of plant, it appears rather that he excluded them from the plant world as but a by-product of plant physiology.
27. For Albert the quick absorbtion of water by leaves is evidence of their "earthiness," since earth absorbs water quickly.
28. As will be seen, the leaf is thought to form from an expulsion of watery vapor.
29. A peduncle is a primary flower stalk supporting a cluster of flowers or only one.

When the leaf is opposite the fruit, it comes forth in plants, drawing much humor, and especially does the fruit of these plants fill very much with a subtle vaporous and windy humor. From this vapor the watery vapor, which is less distributable and less subtle—just as if it were a contrary—is expelled to the opposite side by a formative power. It is for this reason that the leaf comes forth in an opposite place. This is also appropriate for the purpose of the leaf, since the fruit of such a plant requires a great boiling [action] on the part of the sun because of the abundance of humor. But if the leaf was spread over the fruit it would hinder [the necessary boiling action of the sun on the fruit]. . . .

Leaves that emerge above the fruit especially cover the fruit, so that it is almost always in the shade. And the cause of this is matter, because the fruits have much of earthiness and therefore, when the windiness has been closed off [or hindered], the water humor in them rises higher, and the vapor, from which the fruit is generated, comes forth in the lower parts of it as if drawn with violence. The utility of this effort of nature can be judged by its end [or goal], since such fruits are cold and humid, so that frequently the shade of the leaves helps preserve these two things [namely the coldness and humidity of the fruit].

These, then, are the three general locations of leaves, and from their causes the variation of the location of the leaves is easily known. Nevertheless, perhaps sometimes ways other than those declared here may be found in certain plants.

Chapter 4
On the Nature and Generation of Flowers

The flower, which is the sign of fruit, is found in plants. As in several things, the generation of flowers is of the same substance with fruit. For this reason, also, the flower very frequently adheres to the upper part of the fruit, as in trees; or the fruit is formed in the middle of a flower, as in bushes and herbs.[30] And what we have said about trees to whose fruit the flower adheres when first formed is especially apparent in the flowers of the pomegranate *(in balaustiis malorum punicorum)* and in pears and apples. But it is not generally suitable, because in almost all fruits having "stones" *(ossa)* inside, the flower is formed around the fruit and the fruit is formed in the middle of the flower, as in all plums and acacias. And this is proper for all bushes, as the poppy, and in almost all herbs whose seed vessels *(siliquae)*, where seeds are

formed, come forth from the middle of the flower.

From all this one may easily infer [things about flowers], since the flower is produced from the nature of a subtle watery humor well mixed with earth. In substance, then, a flower is of a very solid and smooth *(planae)* substance. Therefore, when it is immersed in water their solidity and compactness prevent the entrance of air and prevent their rise above the water.

That flowers are changed to another color than green happens because of the moist transparency in them and because of the earth that is well distributed and mixed with it. For all the differences of colors are caused by different flotations of earth expanded as vapor in the moistness, or in fiery or clear smoke, or in burned earth, as we said in our discussion of the generation of sensible things.[31]

Generally, the substance of flowers is made from very subtle moisture which appears first from heat and, because of the abundance of water in it, is extended like a leaf. Therefore, since the moisture has a great distribution, the flower is almost universally of good odor. But this would not happen unless the moisture were very well distributed and very subtle [or rare] and the earth in it were very rare and very well mixed with the moisture. For since the creation of fruits is from windy earthy vapor, there is in this vapor something rarer and wetter and of a lesser earthiness which does not easily show, and it is thickened by a distributive heat. And since this is more vaporizable than the rest of what is in the substance of the plant where the bud *(gemma)* is located and where the fruit emerges, it comes forth immediately with the first heat and is formed into a flower.

For this reason, the dew, which produces honey *(mel)* and wax *(ceram)*, clings to the flower. And these are found deep in the interiors of the flowers, for when nature forms a subtle moisture which has been effected by a rare and very well mixed dryness, there flows from it this rare and well-boiled watery moisture—in the manner of a sweet phlegm in the creation of humors in animals—and this [substance], gathered and warmed by the effort of bees, is converted into the nature of honey. . . .[32]

30. "Albertus distinguished 'flowers' *(flores)* of two kinds: those situated on top of the fruit, e.g. *Pyrus, Malus, Cucurbita, Purnica* and *Rosa;* and those beneath or round the fruit, e.g. *Prunus, Amygdalus* and *Papever*" (T. A. Sprague, p. 437).

31. In Albertus' *De sensu et sensato*, Book I, Tract II.

32. Approximately three lines are omitted from the translation.

Wax, which is in the lowest parts, is like a purge of yellow bile *(purgamentum cholerae)* that trickles from the ears of animals in the purging of the brain of animals. While the flower is formed, what it has of earth is rejected with the easily inflammable fat and like a powder is sprinkled over the insides of the flower, because what is inflammable cannot endure, by natural or alchemical means, the action of natural heat determining and forming the being of things. Indeed, before it can be formed it burns and is converted to the yellowness of yellow bile. It appears to be a formal yellow, especially in the flowers of the poppy, the linden tree, and the white birch *(miricae)*. But it is in all flowers more or less, and it clings to the rear legs of bees when they gather honey. Indeed, they build their hives from it for the preservation of the honey. . . .[33]

From all these things it is established that the substance of flowers is of a subtle [or rare] wateriness mixed together with subtle earth, which by reason of its subtlety [or rareness] is rather formable into the figure of a leaf than into the denseness of fruits. And thus with the first heat the flowers come forth, because of the subtlety of their material substance, and it is for this very reason that they are more easily injured by cold than are leaves or fruits. Many odors are emitted by flowers because of the rare moisture which has been suitably extended by dryness and which is released into their substance as if it were spiritual. Furthermore, the moisture of the leaves is of a gross undistributed water and the moisture of the fruits is at the beginning more astringent and earthy, requiring much by way of digestion. For this reason the fruit is completed last, after the leaves and flowers.

. . . .

From all the things that have been said, it is obvious that those trees which do not flower, but produce fruit, as the fig tree and certain apple trees, lack flowers for one of two causes or both together. Either they have a very thick [or viscous] moisture whose parts lie together . . . so that the subtle water cannot be released from the thick moisture and earth that is in it, just as in the fig tree. For this reason its fruits are very sweet, because its moisture has been decocted a long time with sweetness, which remains in them the whole time. Or [and this is the second cause] this happens because of the rarity of the substance of trees, through which each moisture flows simultaneously through the enlarged pores. In this way the fruit is formed without a flower, as happens in certain apple trees. But such trees are rare.

Trees that have a very bitter sap with much earthiness have either a small flower or none at all because of a lack of subtle moisture—as nuts . . . .[34]

These are the statements about the nature of flowers and generation.

BOOK IV, TRACT III
ON THE PRINCIPLES OF GENERATION AND THE FECUNDITY OF PLANTS

Chapter 1
On the Five Principles of Generation and Growth of Plants and the Doubts Emerging about Them

Herbs and whatever is planted and grows from a root fixed in the earth require one or more of five things. These are a seed *(semen)*, putridity [rottenness] *(putredo)*, humor [or sap] *(humor)*, water *(aqua)*, and planting *(plantatio)*. Of these five, the first has the formative power of the plant in itself and matter and the power to effect is in it at the same time, as we said in the second book of the *Physics*.[35] The second thing (putridity) receives a formative power from the power of the stars. The humor, which is mixed from the elements, is the food and matter both of generation and of the plant that has been generated; indeed the plant draws this [material] purified from the earth with its first digestion. As in all nutriments, water plays no role except that it is the vehicle of the food. Nor does food flow to the parts of the plant except by means of the motion of water. . . .[36] Planting [or transplanting] *(plantatio)* another plant is twofold: either (1) it is by grafting, or (2) it is done in the manner described above, namely when a plant having a putrid humor within exhales it by the [action of the] power of the sun, and this putrid humor, impressed into a plant, develops into another plant differing in shape from the plant that exhaled the humor.

Therefore, three of these serve directly for the generation of the plant, while the other two contribute to the nourishment of the plant. The seed, putridity, and planting [or transplanting] contribute toward generation. To contribute toward generation it is necessary that something formative should be in it leading to a species of plant. Now, what is formative is produced either by a lower power, as the seed, or by a universal higher power, as putri-

33. Approximately four lines are omitted here.
34. Approximately eight lines are omitted here.
35. Presumably Book II, Tract II, chapter 3, of Albert's commentary on the *Physics* of Aristotle.
36. Two lines are omitted here.

dity,[37] which [as we have seen above] contributes to the generation of a plant in two absolute ways. The third and remaining one, which is the transplantation *(plantatio)* of a plant into another plant, contributes simultaneously to the generation of it and the transmutation of it in shape.

The things which contribute to the food of a plant have a twofold difference: for the humor is the substance of the food, and the water contributes the proper motion to the food toward the parts that are nourished. . . .

. . . . . .

BOOK V TRACT I

Chapter 7
On Five Ways of Transmuting One Plant into Another[38]

On the transmutation of plants, the works of nature are found to be amazing. Five ways are known in which plants are mutually transmuted. One of these ways is the transmutation of their seeds. A second occurs when a plant is cut off and another species sprouts from its putrefaction. A third way, associated with a place, has not been found among those plants in our clime,[39] which is the seventh, except in one place and that is the sucker *(surculus)* of one plant, after it has been pulled from it and fixed in the earth, is altered into a plant of another species. A fourth way results from the putrefaction of a whole plant with respect to its natural humor, which upon reaching the surface is changed into a plant of another species. A fifth and final way is by grafting, when a shoot of one species of plant is implanted into the stem [or trunk] *(truncus)* of another species and is altered into a plant of a third species, just as we will show by way of example. And [when] a graft *(sceptrum)* of a certain species is implanted into the stem [or trunk] of the same species, it will be changed into another species, as when a pear or apple tree is healthy and a shoot is cut from it above, and after that the stem is cut and the shoot that was cut off before is implanted into it, then it is changed into a pear or apple tree of another species. We shall give examples and teach the modes and physical causes of all these things.

88 AN ILLUSTRATION OF DESCRIPTIVE BOTANY: THE OAK TREE

Albertus Magnus (ca. 1193–1280)

Translated by Charles Singer[1]

Annotated by Edward Grant

OF THE OAK AND ITS QUALITIES

The oak is a very large and tall tree with broad branches; it has many roots which go deep down and when old has a very rough bark, but the young tree is smooth. It has great breadth and size in its branches; when it is thriving its leaves are thickly set, broad, and hard. The leaves are wholly surrounded by triangles, the bases of which are upon the leaf and the angle at the exterior. Many leaves are attached to it, but they fall off. When they are dried up, however, they still sometimes cling in large numbers. Its wood increasing by straight layers is composed of straight pores, can be split to a line and can be hewn and retains well the shapes of large incisions, but in this box-wood surpasses it. The outer zone is of a pale colour but towards the center it shades into a reddish tint. If it is put in water, at first it swims, but at last sinks owing to its earthy nature and then grows black. Its fruit is called acorn *(glans)*; it is not joined by a stalk of its own to the branch on which it grows, but small cups spring out from the branches and in these the acorn is formed. The acorn also has outside a hard shell in which it is enclosed; this resembles well-polished wood, shaped like a column except that its apex is not a plane superficies but a hemisphere and has a point in the middle to represent the pole. Below is the base of the acorn through which it draws nourishment from the little cup; that also is

37. This is formed ultimately by the heat of the sun, a universal higher force.
38. Acceptance of the transmutation of animal and plant species was common in the Middle Ages.
39. On climes, see Selection 1, n. 60.
1. Reprinted by permission of the Clarendon Press, Oxford, from Charles Singer, "Greek Biology and its Relation to the Rise of Modern Biology," in *Studies in the History and Method of Science*, edited by Charles Singer (Oxford: Clarendon Press, 1921), II, 74–75. Singer's translation is from Albert's *De vegetabilibus*, Book VI, Tractate I, chapter 31, on pages 440–443 of *Alberti Magni ex ordine praedicatorum De vegetabilibus libri VII* . . . a critical edition begun by Ernest Meyer and completed by Carl Jessen (Berlin, 1867). In the appropriate places I have indicated omissions in the translation..

not simply a plane superficies but is somewhat flattened at the pole; this depression is formed by the weight of the acorn, for if it were an exact hemisphere there would be no place for the reception of nourishment except the point, and through that it could not receive enough. The acorn within the sheath is surrounded by rind, not hard but soft, which is formed from the excretion of the acorn; the acorn is twisted round itself and divided down the middle as a column might be cut lengthways by a plane surface. At the apex, however, is the life germ and what is beneath is of a floury substance and is to be regarded as matter and food for the germ. The little cup in which the acorn itself sits is concave and evenly formed as if it were smoothed on a lathe inside. The bottom is somewhat levelled since from it the acorn draws its nourishment; on the outside it is rough because of its own earthy nature which is expelled from the material of the acorn. It is not joined by any sort of stem to the branch but sits immediately upon it. This is to prevent the acorn from being too far distant from the branch, because if it had to draw its food a long way, it would become hard and cold and would do no good, especially as the juice of this tree is very earthy.

On the leaves of the oak often grow certain round ball-like objects called galls, which after remaining some time on the tree produce within themselves a small worm bred by the corruption of the leaf. If the worm exactly reaches the midst of the gall apple weather prophets foretell that the coming winter will be harder: but if it is near the edge of the gall they foretell that the winter will be mild. . . . Galls have a juice pure in itself as long as the apple is green and moist; but when it is rubbed against a flat clean piece of iron it immediately is transformed into a kind of very black encaustic. . . .

. . . The leaves of the acorn are extraordinarily astringent but less dry. The acorn resembles the chestnut in that both are astersive and cause flatulence in the lower bowel; both strengthen the limbs and both are good food especially for pigs. Galen says that the acorn as well as the chestnut is good for nourishment and deserves more praise than all the fruits of growing trees: but the chestnut is more nutritious than the acorn on account of its greater sweetness. But the food they afford lacks the praise of men because it is too astringent, but if chestnuts are mixed with sugar they make good food; taken otherwise they will be of slow digestion but the acorn[2] is even slower. The astringency in the inner bark of the acorn is greater than in the acorn itself. The leaves of the oak ground to powder and laid upon wounds make the flesh unite. The acorn is of as much value as the chestnut as a remedy against poison.

. . . The juice of galls darkens the hair: powder of them gets rid of superfluous flesh and warts. Galls are helpful also if placed on decayed spots of the teeth and in many other medical operations which must be determined in (books on) simple medicines.

Medicine

Theory

89 EARLY MEDIEVAL MEDICINE

Isidore of Seville (ca. 570–636)

Translated and annotated by William D. Sharpe, M. D.[1]

Book IV On Medicine

CHAPTER 1

MEDICINE

1. Medicine, *medicina,* is that which either protects or restores bodily health: its subject matter deals with diseases and wounds.

2. I have replaced "oak" with "acorn," since the Latin term is *glans.*

1. [Translated from the *Etymologies,* Book IV (On Medicine), in *Isidore of Seville: The Medical Writings. An English Translation with an Introduction and Commentary* by William D. Sharpe in *Transactions of the American Philosophical Society,* New Ser., Vol. LIV, Part II (1964), pp. 55–61, 64; reprinted with the kind permission of the American Philosophical Society and the translator, William D. Sharpe, M. D. I have eliminated most of the references to Greek and Latin medical texts as well as to philological discussions; for the most part only substantive explanatory information has been retained.—*Ed.*]

2. There pertain to medicine not only those things which display the skill of those to whom the name physician, *medicus,* is properly applied, but also food and drink, shelter and clothing. In short, it includes every defense and fortification by which our body is kept [safe] from external attacks and accidents.

CHAPTER 2

ITS NAME

The name of medicine is thought to have been given it from "moderation," *modus,*[2] that is, from a due proportion, which advises that things be done not to excess, but "little by little," *paulatim.* For nature is pained by surfeit but rejoices in moderation. Whence also those who take drugs and antidotes constantly, or to the point of saturation, are sorely vexed, for every immoderation brings not health but danger.

CHAPTER 3

THE FOUNDERS OF MEDICINE

1. Among the Greeks, the discoverer and founder of the art of medicine is said to have been Apollo. His son, Aesculapius, added to it by his fame and works.

2. However, after Aesculapius died from a lightning bolt, it is said that the healing art was forbidden and that the art died along with its founder, and was hidden for nearly five hundred years, until the time of Artaxerxes, King of Persia. At that time, Hippocrates, born on the Island of Cos and whose father was Aesculapius, restored it to light.

CHAPTER 4

THE THREE SECTS OF PHYSICIANS

1. These three men founded as many sects. The first, or Methodist, *Methodica,* was founded by Apollo,[3] whose remedies are also discussed in poems. The second, or Empiric, *Enpirica,* that is, the most fully tested, was established by Aesculapius and is based upon observed factual experience alone, and not on mere signs and indications. The third, or Logical, *Logica,* that is, rational sect, was founded by Hippocrates.

2. For having discussed the qualities of the ages of life, regions, and illnesses, Hippocrates thoroughly and rationally investigated the management of the art; diseases were searched through to their causes in the light of reason [and their cure was rationally studied]. The Empirics follow only experience; the Logical join reason to experience;

the Methodists study the relationships of neither elements, times, ages, nor causes, but only the properties of the diseases themselves.

CHAPTER 5

THE FOUR HUMORS OF THE BODY

1. Health, *sanitas,* consists in an integrity of the body and a harmonious proportion in its nature as regarding the hot and moist qualities embodied in the blood, wherefore it is called "health" as though the state of the blood, *sanguis.*

2. The word "disease," *morbus,* is a general term which includes all bodily afflictions; our elders called it "disease" to indicate by this name the power of death, *mors,* which could arise therefrom. Healing, *curatio,* consists in a middle course between health and disease for, unless congruent with the disease, it does not conduce to health.

3. All diseases arise from the four humors: that is, from blood and yellow bile, from black bile and phlegm. [Healthy people are maintained by them and the ill suffer from them. When any of the humors increase beyond the limits set by nature, they cause illnesses.] Just as there are four elements, so also there are four humors, and each humor imitates its own proper element: blood the air; yellow bile fire; black bile earth; and phlegm water. Thus, there are four humors as well as four elements which preserve our bodies.

4. Blood, *sanguis,* derives its name from a Greek etymology, because it grows and sustains and lives. The Greeks call yellow bile [such] because it is bounded by the space of one day, whence also it is called *choler,* that is, *fellicula,* for this is an effusion of bile, *fellis effusio.* For the Greeks call bile CHOLÊ.

5. Black bile, *melancholia,* is so named because it is a mixture of the dregs from black blood with an abundance of bile: the Greeks call "black," MELAS and "bile," CHOLÊ.[4]

6. Blood, *sanguis,* is named in Latin because it is pleasant to the taste, *suavis;* whence also men in whom the humor blood predominates are pleasant and agreeable.

2. [As so often, Isidore suggests a bizarre etymological derivation.—*Ed.*]

3. [The founder of the Methodist medical sect is usually said to be Asclepiades of Bithynia, who lived in the first century B.C.—*Ed.*]

4. Galenic physiology generally held that the yellow bile was separated from the blood of the portal vein, passing to the gall bladder and thence to the duodenum. The heavy impurity, the black bile, passed to the spleen. Phlegm was often the residue of digested food.

7. Phlegm, *phlegma,* however, is named because it is cold, for the Greeks call coldness PHLEGMONÊ.[5] By these four humors healthy people are maintained and the ill suffer from them, for when they increase beyond the limits set by nature, they cause illnesses. The acute diseases, which the Greeks call OXEA, arise from blood and yellow bile, but the longstanding disorders, which the Greeks call CHRONIA, come from phlegm and black bile.

CHAPTER 6
ACUTE DISEASES

1. OXEIA is [the name applied to] an acute disease which either passes away quickly or kills speedily, such as pleurisy, phrenitis. The word for "acute" in Greek is OXUS, which means "rapid." CHRONIA is [the name applied to] a protracted disease of the body which tarries for long periods of time, such as gout, phthisis. Among the Greeks, CHRONOS is said for "time." Certain diseases, however, are named after their own proper causes.

8. Pleurisy, *pleurisis,* is a sharp pain of the side accompanied by a fever and a bloody sputum: in Greek, the side is called PLEURA, whence [also] the pleuritic disease derives its name.

9. Pneumonia, *peripleumonia,* is a disease of the lung marked by sharp pain and a sighing respiration. The Greeks call the lung PLEUMÔN, whence also this disease is named.

10. Apoplexy, *apoplexia,*[6] is a sudden effusion of blood by which those who die are suffocated. It is called apoplexy because it is caused by a sudden lethal blow. The Greeks call a blow APOPLÊXIS.

15. HUDROPHOBIA, that is, fear of water, for the Greeks call water HUDÔR, fear PHOBOS, whence also considering this fear of water the Latins called this disease "madness," *lymphaticus.* It arises [either] from the bite of a rabid dog, or from spume falling from the air to the ground. Should man or beast touch this, he straightway becomes demented or is also made rabid.

16. A carbuncle, *carbunculus,* is named because in the beginning it is red like fire and in the end, black like a dead coal, *carbo extinctus.*[7]

17. Plague, *pestilentia,* is a contagion which, when it takes hold of one person, quickly spreads to many. It arises from corrupt air, *corrupto aëre,* and by penetrating into the the viscera settles there. Even though this disease often springs up from air-borne potencies, *per aërias potestates,* nevertheless it can never come about without the will of Almighty God.

CHAPTER 7
CHRONIC DISEASES

1. *Chronia* is a protracted disease which tarries for long periods of time, such as gout, phthisis, for CHRONOS is said in Greek for "time."

5. Epilepsy, *epilemsia,* is named because hanging over the mind, it equally also possesses the body. The Greeks term anything "weighing upon" EPILÊPSIA. It arises whenever the black bile happens to develop in excess and is turned in its course to the brain. This disease is also called the "falling sickness," *caduca,* because the sick man, falling down, *cadens,* suffers convulsions.

6. These the common people call "lunatics" since under the influence of the moon's cycle, *lunae cursus,* the snare of demons snatches them up.[8] This likewise accounts for the term "ghost-ridden," *larvaticus.* It is the same thing as the comitial disease, that is, a greater and divine disease by which those who have fallen to the ground are held—it is powerful enough to make a strong man fall down and froth at the mouth.

7. It is called the comitial disease because if, among the pagans, this seized anyone on a meeting day set for the Comitia, this assembly was dismissed. However, there was a solemn day for the meeting of the Comitia at Rome on the kalends of January.

9. *Melancholia* is named from black bile: the Greeks call black MELAS, and bile CHOLÊ. Epilepsy arises in the phantasy; melancholia in the reason; mania in the memory.[9]

5. *Phlegma* is the word for fire, heat or flame and, by derivation, for inflammation in the medical sense, but may stand for the humor phlegm. *Phlegmonê* does not seem to have been used for *rigor,* "coldness" or "stiffness," but for inflammation.

6. If *apoplexia* here refers to cerebral hemorrhage or thrombosis, the English "stroke," this passage reflects ancient knowledge that paralysis and death from respiratory failure can follow such cerebral accidents. The Greek word refers to this and not to a sudden pulmonary hemorrhage, a much less likely alternative interpretation of this passage. Galen, *Definitiones medicae* 254, clearly refers to coma and paralysis as characteristic of *apoplexia.*

7. Cassius Felix, *De Medicina* 22, ascribes carbuncles to a collection of hot blood and black bile.

8. Ancient medical thought generally assigned the moon a part in controlling epileptic attacks; see Owsei Temkin, *The Falling Sickness* (Baltimore, 1945), who quotes Isidore at p. 94.

9. This classification of psychiatric disturbances agrees with post-classical views on cerebral localization. Francis Adams (*The Seven Books of Paulus Aegineta,* Vol. 1, p. 90) notes: "The later Greek authorities, as, for

10. Intermittent fevers, *typi,* are chilling fevers incorrectly called *tipi* after a certain herb which grows in water. In Latin, "form and state," *forma atque status* are said for the disease is an accession and recession of the fever which recurs over regular intervals of time.[10]

15. Pneumonia, *peripleumonia,* gets its name from the lungs, since it is a swelling, *tumor,* of the lung marked by an effusion of blood-stained sputum.

23. Dropsy, *hydropis,* takes its name from a watery humor of the skin, for the Greeks call water HUDÔR. It is a subcutaneous accumulation of fluid, marked by a turgid swelling and a fetid, labored breath.[11]

32. Calculus, *cauculus,* is a stone which arises in the bladder, whence also it is named: however, [the stone] is composed of phlegmatic matter.

35. Diarrhea, *diarria,* is a continuous flow from the bowel not accompanied by vomiting.

36. Dysentery, *disinteria,* is a breach in the continuity, that is, an ulceration of the intestine, for *dis* is a division, *intera,* the intestine. It arises following a flux, which the Greeks call DIARROIA.[12]

CHAPTER 8

DISEASES WHICH APPEAR ON THE BODY'S SURFACE

1. Mange, *alopicia,* is a loss of the hair in patches

---

example, Theophilus Protospatharius and Nemesius, adopt a division of the brain as regards its connexion with mind, to which Galen and his immediate followers appear to have been strangers. According to it, fantasy is connected with the anterior part of the brain; cogitation, or the *discursus mentis,* with the middle; and memory with the posterior."

Isidore derived the word "bad," *malus,* from (Bk. X., paragraph 176 of the *Etymologies*) the word for black bile, concluding, "for this reason, those men are called melancholy who flee from human society and suspect even their dearest friends." (It is likely that many people now considered "nervous" would, in ancient times, have been classified as "melancholy.") Mental deficiency is briefly mentioned (X. 79) by Isidore: "Demented, *demens,* means the same thing as 'mindless,' *amens,* that is, without a mind, or indicates that such have diminished mental power. Foolish, *desipiens,* in that they begin to understand less than was their wont."

10. [In an appendix on undulant fevers (p. 67) Sharpe says that *typi,* or intermittent fevers, constituted] a group of fevers which probably resembled typhoid, malaria, and various undulant fevers now extinct or unidentifiable. In ancient clinical discussions of fever, four stages are commonly identified:

Beginning	. . . . *Initium*	. . . . ARCHÊ, APHORMÊ
Increasing phase	. . . . *Augmentum*	. . . . AUXESIS, EPIDOSIS
Highest stage	. . . . *Status*	. . . . AKMÊ
Decline	. . . . *Declinatio*	. . . . ANESIS, PARAKMÊ

These stages are simply defined at Pseudo-Soranus, *Quaestiones medicinales* [edited by Valentin Rose in *Anecdota graeca et graecolatina,* Vol. 2, p. 257] and at Galen, *Definitiones medicae* 136 [in Kühn's edition of Galen's works, Vol. 19, p. 338]: the names of these stages of a febrile illness are self-explanatory. Their actual time sequence would vary from disease to disease, and would be helpful both in differential diagnosis and in predicting the outcome of any particular patient's case; contrast, e.g., the differing fever patterns between an acute bacterial penumonia and a chronic tuberculosis abscess. Ancient physicians attached great importance to the exact numerical relationships involved, and this is one of the reasons why Isidore felt (Bk. IV, chap. 13, par. 2 [quoted below—*Ed.*]) that the physician should study mathematics.

The ancient *typhus* (the word TUPHOS is conventionally derived from the smoke of a burning funeral pyre: the *Iliad* begins during such an epidemic) was apparently a collective term which included several kinds of fevers accompanied by stupor and insensibility at some time during their course or termination. The *typi* which Isidore describes seem to be intermittent or remittent fevers, and although it is now impossible to identify them, the regular paroxysms of malaria would certainly suggest some relationship. Aurelius, *De acutis passionibus* 1 [see C. Daremberg, "Aurelius de acutis passionibus," *Janus,* Vol. 2 (1846), p. 489] describes the stages of a fever: the beginning is marked by moderate subjective awareness of fever; the increasing phase leads to the highest stage, which varies in duration and character from disease to disease and which increases until the beginning of the decline, unless the patient dies. There are periodic fevers which return daily or every third or fourth day, and Aurelius' description clearly refers to quartan and tertian malaria.

. . . If Isidore's reference to herbs which grow in water can be stretched to imply that there is a relationship between the *typi,* undulant or remittent fevers, and the marshy waters, the identification of these *typi* as malarial would be stronger, but this is conjectural.

11. In ancient medical writers generally dropsy included ascites and anasarca, and Celsus distinguishes between three varieties of dropsy: ascites, anasarca, and tympanites.

12. In ancient nosology generally, diarrhea is looseness of the bowels with frequent but relatively painless passage of watery stools; dysentery is painful, prostrating, and often marked by the passage of blood from mucosal ulceration. Typhoid fever, various parasitic infestations, and colonic malignancy were doubtless lumped together with the dysenteries. Aurelius describes *diarria* as above, but notes that it may be followed after some days by dysentery.

The ancient *cholera* was an entity now difficult to identify: Aurelius (*De acutis passionibus* 14 [Daremberg's edition of 1846, p. 715]) notes that it is accompanied by

circumscribed by reddish-yellow hairs having the color of copper, called by this name from a resemblance to the animal, the fox, which the Greeks call ALÔPÊX.[13]

5. Herpes, *serpedo,* is a redness of the skin accompanied by prominent pustules, and takes its name from "creeping," *serpere,* because it creeps gradually along the members.[14]

12. The disease *elefantiacus* is named from its similarity to the elephant, the skin of which is naturally hard and rough, and the name has been given to the disease among human beings because the surface of the body becomes similar to the skin of elephants,[15] or because the pain is immense, like the animal itself from which its name has been taken.

13. Jaundice, *hicteris,* the Greeks have named after a certain animal because it has the color of yellow bile. The Latins term this the arcuate disease, from its resemblance to the rainbow, but Varro says that this disease is "gold-bearing," *auriginis,* because of its golden color. Some think that the same disease is termed that of kings in that it is the more readily cured by a regimen of good wines and rich foods.

14. *Cancer* is named from its resemblance to a certain marine animal: physicians say that it is a wound which admits of cure by no medications, and for this reason they usually cut off the member in which cancer has arisen from the body, so that [the patient] may live a little longer. In truth, yet death will come from it, although a little later.[16]

CHAPTER 13

THE STUDY OF MEDICINE

1. Some ask why the art of medicine is not included among the other liberal disciplines. It is because whereas they embrace individual subjects, medicine embraces them all. The physician ought to know literature, *grammatica,* to be able to understand or to explain what he reads.

2. Likewise also rhetoric, that he may delineate in true arguments the things which he discusses; dialectic also so that he may study the causes and cures of infirmities in light of reason. Similarly also arithmetic, in view of the temporal relationships involved in the paroxysms of diseases and in diurnal cycles.

3. It is no different with respect to geometry be-

cause of the properties of regions and the locations of places. He should teach what must be observed in them by everyone. Moreover, music ought not be unknown by him, for many things are said to have been accomplished for ill men through the use of this art, as is said of David who cleansed Saul of an unclean spirit through the art of melody. The physician Asclepiades also restored a certain insane man to his pristine health through music.[17]

4. Finally also, he ought to know astronomy, by which he should study the motions of the stars and the changes of the seasons, for as a certain physician said, our bodies are also changed with their courses.[18]

5. Hence it is that medicine is called a second philosophy, for each discipline claims the whole of man for itself. Just as by philosophy the soul, so also by medicine the body is cured.[19]

---

diarrhea, nausea, vomiting, and prostration. Colic was distinguished by swelling and pain of the viscus.

13. Ancient folklore held that the vegetation upon which a fox urinated was forthwith killed leaving burnt patches, and this is often suggested as the origin of the word *alopecia,* "fox mange."

14. This seems to be herpes zoster.

15. This *elefanticus morbus* refers to a cutaneous infection of some kind, and not to any of the modern diseases called elephantiasis. That due to lymphedema was called *elephantiasis Arabum; elephantiasis Graecorum* in *Plinii Secundi Iunioris de medicina* (ed. Valentine Rose [Leipzig, 1875] p. 33) is a cutaneous infection spreading from one nidus.

16. Cancer of the female breast was well known in ancient times.

17. Rudolf Allers, "Microcosmus," *Traditio,* Vol. 2 (1944), pp. 375–376, observes: "The relation between human music, inner harmony, and the harmony of the spheres, therefore the cosmic law, explains the high esteem in which music was held throughout medieval times."

Aristotle, *De caelo* 2.9 [290b.12] explains that man does not hear the music of the spheres since its sound is continuous [Aristotle rejected outright the Pythagorean doctrine that the celestial motions produced music.— *Ed.*], and man perceives sounds only when interrupted by periods of silence. Cassiodorus, *Institutes* 2.5.9, has a discussion of the medical usefulness of music.

18. In Hippocrates, *Airs, Waters, and Places* 2, we read: "Astronomy contributes not a little, but a very great deal indeed to medicine."

19. The notion that philosophy, i.e., philosophical and metaphysical theology, would cure some of the disorders of the soul and the art of medicine some of those of the body developed early in Christian thought. [This notion is traceable to Plato.—*Ed.*].

90 THE GALENIC SYSTEM

Joannitius (Hunain ibn Ishāq) (809–877)

Translated and annotated by H. P. Cholmeley,[1] with additional notes by Michael McVaugh

Isagoge

THE BEGINNING OF THE INTRODUCTION OF
JOANNITIUS TO MEDICINE.

Medicine is divided into two parts, namely theoretic and practical. And of these two the theoretic is further divided into three, that is to say, the consideration of the naturals, the non-naturals, and the contra naturals. From the consideration of these arises the knowledge of sickness, of health, and of the mean state, and their causes and significations; of when the four humours increase in an abnormal manner, or of what may be the causation (*occasio*) or significance of sickness.

OF THE NATURALS.

The naturals are seven in number: elements, qualities *(commixtiones)*, humours *(compositiones)*, members, energies, operations, and spirits. But some add to these four others: namely, age, colour, figure, and the distinction between male and female.

The Elements.

There are four elements: fire, air, water, and earth. Fire is hot and dry; air is hot and moist; water is cold and moist; earth is cold and dry.

The Qualities.

There are nine qualities, eight unequal and one equal. Of the unequal, four are simple: namely, hot, cold, moist, and dry. From these arise four compound: namely, hot and moist, hot and dry, cold and moist, cold and dry.

The equal is when the body is so disposed that it is in good condition and in a mean state, when it has a proper amount of all four.

Of the Humours (De Humoribus).

The humours *(compositiones)* are four in number: namely, blood, phlegm, reddish bile, and black bile. Blood is hot and moist, phlegm is cold and moist, reddish bile is hot and dry, black bile is cold and dry.

OF PHLEGM

There are five varieties of phlegm. There is the salt phlegm, which is hotter and drier than the rest and is tinged with the biliary humour. There is the sweet phlegm belonging to hotness and dampness, which is tinged with the sanguine humour. There is the acrid phlegm belonging to coldness and dryness, which is tinged with the melancholic humour. There is the glassy phlegm, which arises from great coldness and coagulation such as occurs in old people who are destitute of natural warmth. And there is another which is cold and moist; it has no odour, but retains its own coldness and moistness.

OF REDDISH BILE

Reddish bile exists in five different fashions. There is reddish bile which is clear or pure and hot, both by nature and substance, of which the origin is from the liver. There is another which is straw-coloured, from which the origin is from the watery humour of the phlegm, and pure reddish bile, and therefore it is less hot. Another is vitelline. It is similar to the yolk of an egg, and it has its origin from a mixture of coagulated phlegm and clear red bile, and this is less hot. Another one is green bile, like the green of a leek *(prasium)*, and it arises generally from the stomach or the liver; and there is another which is green like verdigris, and which burns after the fashion of a poison, and its origin is from too much *adustio*, and it possesses its own proper colour and its own energies, both good and evil.

OF BLACK BILE

Black bile exists in two different fashions. In one way it may be said to be natural to the dregs of the blood and any disturbance of the same, and it can be known from its black colour whether it flows out of the body from below or above, and its property is cold and dry. The other kind is altogether outside the course of nature, and its origin is from the *adustio* of the choleric quality, and so it is rightly called black, and it is hotter and lighter, and having in itself a most deadly quality and a pernicious character.

1. [H. P. Cholmeley, *John of Gaddesden and The Rosa Medicinae* (Oxford: Clarendon Press, 1912), 136–166. The *Isagoge,* written by Hunain as an introduction to Galenic medicine, was available in the West quite early, towards the end of the eleventh century; it was the only systematic exposition of medical theory known in this period before the translation of the bulk of the Greco-Arabic medical corpus. Its conciseness kept it in use as a standard text for over two hundred years, long after European physicians had attained a vastly more detailed knowledge of Galenic theory.—M. M.]

Of the Various Kinds of Members.

There are four kinds of members. Some of them are principal, and are as it were the substance and fundamentals of the body, as, for instance, the brain, the heart, the liver, and the testicles; and there are others which do service to the aforesaid principals, such as the nerves, which minister to the brain, and the arteries, which minister to the heart, and the veins, which minister to the liver, and the spermatic vessels, which minister to the testicles, and bring the sperm to them. Some of the members, again, have their proper energy whence the members are ruled and in which their particular qualities consist. Such are the bones, and the cartilages, and the membranes which lie between the skin and the flesh and the muscles and the fat and the flesh. Others there are which work by the energy proper to them, but yet they obtain their origin and vigour from the principals and fundamentals. Such are the stomach, the kidneys, the intestines, and the muscles *(lacerti)*. For these by their own proper energy pick up the food and commute it, and they do their actions according to their nature, and they have other energies of their own arising from the principals and fundamentals, in which principals consist sense and life with voluntary motion.

Of the Number and Division of the Energies.

The energies are divided into three. There is the animal energy and the spiritual and the natural.

OF THE NATURAL ENERGY

There is one natural energy which does service, and one to which service is done. But the natural energy to which service is done at one time generates, at another time nourishes, and at another feeds.

But the energy which does service and is not done service to, in the same way desires, retains, and digests, and it expels those matters which are subject to the feeding energy, just as the feeding energy is subject to the nourishing energy. And the natural energy in its generating function is served by two others, one which transmits the food and the other which shapes it; and these two differ, the one from the other, for the first changes the food and ministers to the generating energy, without the shaping, but the second does the same thing with shaping. And the operations of the informative energy are five—Assimilativa, Concava, Perforabilis, Aspera, Lenis.

OF THE SPIRITUAL ENERGY

From the spiritual energy proceed two others, one the operative, and the other the result of the operation. The operative energy is that which at one time dilates and at another contracts the heart and the arteries, and the results of this are as follows—Indignation, Victory, Domination, Astuteness, and Anxiety.

OF THE ANIMAL ENERGY

The zodiacal energy embraces three things. There is one which arranges and puts together and classifies. The second is that which is moved by voluntary motion, and the third is that which is called sensible. From the first proceed Imagination in the front of the head, Cogitation or Reasoning in the brain, and Memory in the occipital region. The second moves the muscles *(lacerti)*, by which the other members are moved, that is to say by voluntary motion. And the sensible energy resides in sight, hearing, taste, touch, and smell.

The Operations.

Operations are of two kinds: there are some each one of which individually performs that which is its own (function). Such, for instance, is Appetite by means of heat and dryness; Digestion by means of heat and moisture; Retention by cold and dryness; Expulsion by cold and moisture. There are also compound operations which are of a double nature: such are desire *(desiderium)* and carrying off *(deportatio)*. Desire is compounded of a double energy: the one longs for *(appetit)*, and the other feels; for the stomach is conscious of its own place ('stomachus enim suam mansionem sentit'). Carrying off is of two or more energies, one casts out, another attracts or feels, and a third longs for.

The Spirits.

The spirits are three in number: the first the natural spirit, having its origin from the liver; the second the vital spirit, having its origin from the heart; the third the animal spirit, having its origin from the brain. Of these three the first is diffused throughout the body by means of the veins which have no pulse; the second is diffused throughout the body by the heart; and the third is diffused throughout the body through the agency of the nerves by the brain. These are the matters which come under the heading of spirit in the seventh division of the seven naturals.

Of the Ages.

There are four ages; namely, adolescence, the prime *(juventus)*, decline *(senectus)*, and decay *(senium)*. The period of adolescence is hot and moist, during which the body increases and grows up to the twenty-fifth or thirtieth year. The prime follows which is hot and dry, during which the body remains in perfection without any diminution of bodily force, and it lasts from the thirty-fifth to the fortieth year. Next comes decline, which is cold and dry, and during this period the body begins to lessen and decrease, although the bodily force is not abated, and it lasts to the fiftieth or sixtieth year. Finally succeeds decay, which is cold and moist, with appearance of the phlegmatic humour, and during this period the bodily forces are abated, and the period ends with the end of life.

Of the Colours of the Skin and Their Divisions.

The colours of the skin are of two kinds; namely, those due to internal causes and those due to external. And the internal causes again are two in number; namely, excess or equality of humours. From equality comes that tint which is composed of white and red; from inequality proceed black, yellow, reddish *(rubeus)*, greyish *(glaucus)*, and white. The reddish, black, and yellow set forth the ruling humour of the body: yellow by itself signifies reddish bile; black by itself, black bile; reddish by itself, abundance of blood. White and greyish signify an excess of coldness ruling the body; greyish arises from black bile *(melancholia)* and white from phlegm.

Certain colours arise from external circumstances such as from cold among the Scots and from heat among the Ethiopians. And there are many others from other causes.

There are also special or spiritual colours, due to fear, anger, grief, or other affections of the mind.

Of the Colours of the Hair.

There are four colours of the hair—black, reddish, greyish, and white. Black is due to an excess of over-heated bile or blood; reddish to a superfluity of a rather lower heat *(caloris non adusti)*— this is always the cause of reddish hair; greyish arises from an excess of black bile, and white from a deficiency of the natural heat and the operation of putrid phlegm, and is therefore chiefly found in the aged.

Of the Coats of the Eye.

The eye has seven coats and three humours.

The first coat is the retina, the second the secundine, the third the sclerotic, the fourth the spider's web *(tela aranea)*, the fifth the uvea, the sixth the cornea, and the seventh the conjunctiva. And of the humours the first is the vitreous, the second the crystalline, and the third the albugineous which is in front of the uvea.

Of the Qualities of the Body.

The qualities of the body are five in number; namely, excess or grossness; thinness or tenuity; 'sinthesis' (wasting), squalidity, and the mean state *(equalitas)*. There are two kinds of grossness, the one consisting in excess of flesh, and the other in fat. Excess of flesh arises from excess of heat and humours; but fatness from cold and intense humidity; loss of fat or thinness arises from heat and intense dryness. Sinthesis arises from cold and intense dryness; squalidity either from cold and intense humidity, or from an intensity of both together. And the mean state arises from a proper proportion of the humours. These are the appearances of the body.

Of the Difference between Male and Female.

The male differs from the female in that he is hotter and more dry; she, on the contrary, is colder and more moist.

THE BEGINNING OF THE TREATISE ON THE NON-NATURALS.

And First of the Changes of the Air.

Changes of the air come about in five different ways; from the seasons, from the rising and setting of the stars, from the winds, and from the different countries and their exhalations *(fumositas)*.

Of the Seasons.

There are four seasons; namely, Spring, which is hot and moist; Summer, which is hot and dry; Autumn, which is cold and dry; Winter, which is cold and moist.

The nature of the air is also changed by the stars, for when the sun approaches a star or a star the sun, the air becomes hotter. But when they separate the coldness of the air is altered, viz., either increased or diminished.

Of the Number and Properties of the Winds.

There are four winds; the East *(subsolanus)*, the West, the North, and the South *(auster)*. And of these the nature of one is cold and dry and of another hot and moist. The two others are of an equal nature, for the East is hot and dry and the West is cold and moist. The South is slightly hotter and moister and the North colder and dryer.

Of Varieties of Places and Their Qualities.

There are four varieties of places; namely, height, depth, nearness to mountains or to the sea, and those particular qualities in which one district differs from another. Height produces cold and depth the contrary.

The relation to mountains is as follows: if the mountains are to the south, the locality will be the cooler, for the mountains keep off the hot winds, and so the north winds seek it out with their cool breath. But if the mountains are to the north of the locality the reverse is the case.

As regards relation to the sea: if the sea is on the south the locality will be hot and dry, if to the north it will be cold and dry.

Soils differ among themselves. Stony land is cold and dry; fat and heavy land is hot and moist; clay lands are cold and moist. Exhalations from marshy land or other places where decay is going on also change the air and give rise to disease and pestilence.

Of Exercise.

Exercise has an effect on the body. To a mean amount it causes a mean amount of heat, i.e. exercise in moderation maintains the normal bodily heat. Violent exercise first of all heats the body, but afterwards cools and dries it.

Rest also affects the body; if excessive it increases cold and moisture, if of a normal amount it maintains the normal amount of coldness and moisture.

Of Baths.

Baths are either of fresh water or not fresh. Fresh-water baths soften the body, and if hot they warm it, but if cold they cool it. But a fresh-water bath does not dry up the body. Baths of salt or bitter or sulphurous waters heat and dry up the body. Aluminous or lime *(gipsea)* baths cool and dry up the body.

Of Foods.

Foods are of two kinds. Good food is that which brings about a good humour, and bad food is that which brings about an evil humour. And that which produces a good humour is that which generates good blood; namely, that which is in the mean state as regards quality *(commixtio)* and working. Such is clean, fresh, fermented bread,

and the flesh of lamb or kid. Bad food brings about the contrary state, and such is old and bran *(opirus)* bread, or the flesh of old beeves or goats. Foods producing good or evil humours may also be heavy or light. Of the first kind are pork and beef, of the second chicken, or fish. And of these the flesh of the middle-sized and more active kinds is better than that of the fatter and scaly varieties.

Certain kinds of vegetables produce evil humours; for instance, nasturtium, mustard, and garlic beget reddish bile. Lentils, cabbage, and the meat of old goats or beeves produce black bile. Pork, lamb, purslain, and attriplex beget phlegm. Heavy foods produce phlegm and black bile, light food produces reddish bile, and either of these is evil.

Of Drinks.

Drinks are of three kinds: firstly, drink which is nothing but drink, as water; secondly, drink which is both drink and food, as wine; and thirdly, drink partaking of the nature of both of these, called *potio,* which is given to counteract the evil of some disease. Such are mellicratum, mulsa, and conditum.

The use of the food is to restore the wholeness of the body, the use of the drink is to distribute the food throughout the body. But that kind of drink which we have above called *potio* converts the nature of the body to itself. ('Sed illius potus quem potioni diximus pertinere, corporis naturam ad se convertit.')

Of Sleep.

Sleep changes the nature of the body in that it cools it exteriorly and warms it interiorly. If it be prolonged it cools and moistens the body.

Waking also changes the body, for it warms it exteriorly, while interiorly it cools and dries it.

Of Coitus.

Coitus is good for the body; it dries the body and diminishes the natural forces, and so cools it, although sometimes the body is warmed by much friction *(concussione)*.

Of Affections (Accidentibus) of the Mind.

Sundry affections of the mind produce an effect within the body, such as those which bring the natural heat from the interior of the body to the outer parts or the surface of the skin. Sometimes this happens suddenly, as with anger; sometimes gently and slowly, as with delight and joy. Some affections, again, withdraw the natural heat and conceal it either suddenly, as with fear and terror, or again gradually, as poverty. And again some affections disturb the natural energy both internal and external, as, for instance, grief.

OF THE CONTRA NATURALS.

There are three contra naturals; namely, disease, the cause of disease, and the concomitants or sequels of disease. Disease is that which primarily injures the body, without the aid of any intermediary, as, for instance, heat in continued (*succedente*) fever.

Of Fevers.

Fever is unnatural heat, i.e. heat which overpasses the normal course of nature. And it proceeds from the heart into the arteries, and is harmful by its own effects.

And of it there are three kinds: the first in the spirit (*anima*), which is called ephemeral; the second arises from the humours which putrefy, and which is therefore called putrid; and the third affects for ill the solid portions of the body, and this is called ethic (i.e. hectic).

Of these three the ephemeral variety arises from non-essential causes (*ab accidenti occasione*). Putrid fever arises from putrid matters, and these are simple and uncombined, and they are four in number.

The first is that which arises from putridity of the blood and burns up both the interior and exterior of the body; such, for instance, is continued fever (*sinochus*).

The second is that which arises from putridity of reddish bile; such, for instance, is tertian fever (*tritheus*).

The third arises from putridity of phlegm; such, for instance, is quotidian fever.

And the fourth arises from putridity of black bile; this attacks the sick man after an interval of two days, and it is called quartan.

In addition there are three kinds of fevers occurring from putridity. First there is the fever which lessens day by day; such, for instance, as that called *peraugmasticus*, i.e. decreasing ($\pi\alpha\rho\alpha\kappa\mu\alpha\sigma\tau\iota\kappa\acute{o}s$).

Secondly, that which increases until it departs; such as that called *augmasticus* ($\grave{\alpha}\kappa\mu\alpha\sigma\tau\iota\kappa\acute{o}s$).

Thirdly, that which neither decreases nor increases until it again (*iterum*) departs; such, for instance, as that called *homothenus* ($\acute{o}\mu\acute{o}\tau\sigma\nu\sigma s$).

Continued fever arising from putridity in the veins begins to decline by departing from out the veins into other parts of the body.

Goose-skin or shivering (*horripilatio*) occurs in

fevers from an infusion of putrid matter into the sensitive members, which gnaws (*mordens*) and makes them cold.

And, therefore, goose-skin occurs in these fevers which are characterized by remissions (*anesim*) or variations (*interpolationem*), for the putrid matters are outside the veins.

Of Swellings.

There are four simple kinds of swellings; those which arise from the blood and are called phlegmons; those which arise from reddish bile and are called *erysipelas;* those which arise from coagulated phlegm and are called *undimiae*[2] or *cimiae*, that is to say tumour; and finally, those which arise from black bile and are called cancerous phlegmons.

The signs of a swelling arising from the blood are these: redness, a hard pulse, pain, heat, swelling.

And the signs of those arising from bile are these: heat, a reddish yellow colour, great pain of a darting character, and rapid increase. And the signs of those arising from phlegm are these: a white colour and softness, so that if the finger be pressed thereon it makes an impression; moreover it is painless.

And the signs of those arising from black bile are these: great hardness, a black colour, and absence of feeling.

Of the Natural Condition (Res Naturalis) in the Human Body.

In the human body, if each and singular natural condition maintains its proper quality, such a condition makes for health. If any one of them fails, either sickness follows or else the neutral state.

Of the Classes of Sickness.

There are three classes of sickness: (1) the similar, (2) the official, and (3) the universal.

An *aegritudo consimilis* is one affecting the similar members (tissues), and they receive names of like nature to the suffering; such, for instance, as an aching (head). ('Est quidem egritudo consimilis similibus membris contingens quae similia sortiuntur vocabula cum eadem passione; ut est caput dolens.')

And an *aegritudo officialis* is one which occurs in special members, such as the feet, the hands, the tongue, or the teeth. This takes its name from the accompanying infirmity (*ex accidenti infirmitate*), such as podagra in the foot or chiragra in the hand.

2. *Undimiae* is a corruption of $o\check{\iota}\delta\eta\mu\alpha$. An English translation of Lanfranc, circa 1400, has 'vdemia'.

Or again, it may take the name from the member in which they occur *(ex membris quibus accidunt)*, as podagra, chiragra.[3]

And finally an *aegritudo universalis* is one which is associated with the two aforesaid, as separation of the limbs and paralysis.

Of Diseases in the Similar Members.

Diseases of the similar parts are eight in number, four simple, and four compound.

The simple arise solely from heat, from moisture, from cold, or from dryness.

And these four may be combined so as to be compound, such as cold and moist, cold and dry, hot and moist, hot and dry.

Each of the eight kinds may be of two varieties, for either it is of a simple quality *(ex qualitate simplici)*, or it is combined with one or other of the humours.

For example, a disease of a simple quality is one affecting the solid members, such as the Greeks call *ethica* (i.e. hectic).

A hot disease arising from combination with some humour is a putrid fever, as has been said above.

Chilling *(algor)* due to very cold air or snow is a simple cold disease without admixture of any humour. But a cold disease, with an admixture of humour, is paralysis, either complete or partial.

The mark of a moist disease is that it has an admixture of humour; for instance a cold or ulcerated *(vacuum)* wound; or again a very foetid wound accompained by wasting of the body, as, for instance, the puffed-up flesh of dropsical persons, which flesh is inactive and in an unprofitable condition.

A moist disease is one which attracts to itself foreign humours, as, for instance, dropsy.

A dry disease with an admixture of humour is, for instance, a hard and dry cancer.

Of Diseases in the Official Members.

Diseases occurring in the official members are four in kind: they concern shape, size, number, and position.

(1) As to shape. Abnormalities of this are unbecomingness of a member, e.g. (*a*) a very long head; (*b*) absence of the normal concavity, as when the hollow of the foot or hand is filled up with flesh; (*c*) variations in the size of canals, as stricture or dilatation; (*d*) roughness, as of the throat or of the trachea and bronchi; (*e*) smoothness, as of the womb or the stomach.

(2) As to size. Abnormalities of size arise from overplus of sperm, owing to which the member grows to a greater size than it should do. And so we sometimes see a very large head or tongue.

Also a member may be unbecomingly small, as we sometimes see in the case of the head, the stomach, or the liver.

(3) As to number. Abnormalities of this kind occur either by augmentation or diminution. And those due to augmentation are either according to the course of nature or outside it. Of the first kind are extra fingers; of the second are round worms *(lumbrici)*, thread worms *(ascarides)*, warts, and acrocordines, that is to say large fleshy growths or large spreading warts or fistulae *(pori)*.

Those due to diminution are either universal or particular. Of the first kind is absence of all the fingers; of the second kind absence of one finger.

(4) As to position. Abnormalities of position are due either to a removal of the member from its proper place or to some defect in its relation to neighbouring parts. Such abnormalities we see in the fingers and the lips. The fingers may be double *(conglutinantur)*, or webbed *(vel adherent)*, or the lips may be separated and not joined (? harelip).

Separation in a member which ought to be whole happens both in similar members and in official members. The similar members are the bones, the nerves, the flesh, the veins, the muscles, and the skin.

When separation occurs in a bone it is termed a fracture; when occuring in the flesh (if recent) it is termed a wound. But if the injury is of old standing it is not called simply a wound, but a putrid wound.

Separation in veins, nerves, or arteries is sometimes called by one name and sometimes by another.

If the injury occurs in the middle of a muscle it is called a contusion or a bruise *(contritio)*.

If it occurs in the skin, it is termed an excoriation; but this, if it be of long standing, may become a putrid wound.

Separation occurring in the official members may be lasting, as, for instance, loss of the hand or foot.

---

3. [Of this passage the translator says: "This passage is very obscure, but it seems to mean that an *aegritudo consimilis* receives its name from the kind of suffering, as, for instance, aching, burning, and the like. The *aegritudo officialis* is so called from the organ or member which it affects. Thus 'toothache' is an *aegritudo officialis*, while aching, burning, or darting pains in the head would be *aegritudines consimiles*."—M. M.]

Of the Qualities of the Body.

The qualities of the body are three in number; namely, health, sickness, and the mean state.

Health is that condition in which the temperament of the body and the seven naturals are working according to the course of nature.

Sickness is defect in temperament outside the course of nature, and injuring nature, whence arises an efficient condition of harm which may be felt. The mean state is that which is neither health nor disease. And there are three kinds of this mean state: (*a*) when health and disease co-exist in the same body; which may happen in different members, as in the blind or the lame; (*b*) in the bodies of the aged, in whom no one member remains that is not in evil case or suffers; (*c*) in those who are well at one season and sick at another. For instance, persons of a cold nature are sick in the winter and well in the summer; and those of a moist nature are sick in childhood, but well in youth and old age. Those of a dry nature are well in childhood, but sick in youth and old age.

Health, sickness, and the mean state are evident in three ways; (1) in the body in which any one of them occurs; (2) in the cause which produces, which governs, and which preserves them; (3) in their indicating signs.

Of the Causes (Occasiones).

Causes are of two kinds, either natural or outside the course of nature.

Natural causes either produce health or preserve it.

And the preservative causes pertain to the maintenance of health, but the productive causes to the expulsion of sickness.

The non-natural causes pertain to sickness or to the mean state. Causes of sickness produce sickness, and they also maintain it. And that which pertains neither to health nor to sickness brings about and maintains the mean state.

Of the Causes of Health and of Sickness.

The causes which have a relation to health and sickness are six in number,[4] and of these the first is the air which surrounds the body. Then follow food and drink, exercise and rest, sleep and waking, fasting and fullness, and finally affections of the mind. All these, if in moderation as to quantity, quality, time, function, and order, tend to preserve health. But if in excess in one of these matters they tend to produce sickness and to maintain it. The causes which bring about sickness are of three kinds; those which are called primitive and affect the body externally, such as cold and heat; those which are accidental and act within the body, such as fullness or fasting; and those which are called conjoint, because when they are present disease is present, and vice versa, such as, for instance, putridity in fevers.

Of the Varieties of Sickness.

Sickness may again be classified under two heads; namely, common and proper.

Those under the first heading occur either accidentally, such as striking, burning, biting, catching in a trap (?) *(deceptio)*, or other harmful effects.

Or they occur of necessity; for example, those just mentioned as having a relation to health and sickness. And these are really proper sicknesses and they occur in the similar members, where they maintain sickness, or in the official members, and in cases of permanent separation, i.e. in loss of the hand or foot.

Of the Diseases Arising from Heat.

Disease may arise from heat in five different ways.

Firstly, from disturbance of the spirits or of the body. An example of the former is anger, of the latter, fatigue, or sexual excitement *(superbia)*.

Secondly, the direct action and obvious effects of heat, such as hypanthasis, i.e. sunburn or sunstroke.

Thirdly, heating of the body by some substance which has an accompanying faculty of heat, such as the use of 'acrumina' [e. g. onions and garlic].

Fourthly, from the shutting up of the pores, as from cold in winter.

Fifthly, from putridity of the humours, as in fevers.

Of the Diseases Arising from Cold.

Disease may arise from cold in eight different ways.

Firstly, from the direct action and visible effect of cold, such as from the coldness of snow.

Secondly, from cold drugs such as opium, which strongly affect the human body.

Thirdly, from excess of food, which fills up the body and extinguishes the natural heat.

Fourthly, from deprivation of food, which also extinguishes the natural heat.

4. I.e. the six non-naturals.

Fifthly, from excess of cold or of cold humours, which block up the pores, so that the natural heat is lessened.

Sixthly, from purging and opening of the body, so that the natural heat is purged away and evacuated.

Seventhly, from violent exercise with profuse sweating, whereby the body is weakened.

Eighthly, from sleep and plenty of leisure (*occasione multa*).

Of the Diseases Arising from Dryness.

Disease may arise from dryness in four different ways.

Firstly, from the direct action and visible effect of dryness, such as the dryness of poison.

Secondly, from the presence in the body of some dry harsh substance, as for instance, vinegar, salt, or mustard.

Thirdly, from deficiency of food or drink.

Fourthly, from over-exercise.

Of the Causes of Diseases Arising from Moisture.

Disease may arise from moisture in four different ways.

Firstly, from the direct action and visible effect of moisture, such as a bath.

Secondly, from the presence in the body of some moist substance, such as fresh fish.

Thirdly, from excess of food or drink.

Fourthly, from sleep and leisure.

Of the Modes of Disease.

There are four ways in which disease may occur from an abnormal motion of some humour to a weakly part.

Firstly, vigour of the impelling, and weakness of the receiving, member.

Secondly, an abundance of humour.

Thirdly, weakness of the nutrient energy.

Fourthly, an abnormal largeness of the pores.

Of the Evil Quality ('Malitia') of Sickness.

The evil quality of sickness attacks and resides in a similar member in five different ways.

Firstly, in the uterus.

Secondly, at the time of birth.

Thirdly, from the infant being too tightly swaddled.

Fourthly, from defective nutrition.

Fifthly, from any sickness which may occur at the aforesaid times or afterwards.

Sickness of the embryo or foetus arises from a defect in the sperm, which may be either too thick and rich or too thin and watery.

If the child be not born rightly it may be affected for ill, as, for instance, if it be delivered looking upwards (? face presentation) or with the knees bent (? breech presentation). If it be too tightly swaddled it may be injured by being doubled up (*duplicatus debilitatur*).

Or it may be defectively nourished by not being able to suck or to take milk.

And at any of these seasons or afterwards sickness may occur in a consimilar member by incision of any tendon or nerve (*nervus*).

Or some accident may occur, or a wound or a swelling (*apostema*).

Of Sicknesses of the Consimilar Members.

Sickness may affect a similar member in seven different ways.

Firstly, from a midwife who holds the child improperly.

Secondly, if the child be allowed to walk too soon.

Thirdly, from an ignorant physician if he puts to right, or bandages, deformed (*contrafacta*) or bruised limbs unskilfully.

Fourthly, from the patient himself, should he move a broken or injured limb, after being put up by the surgeon, before it has properly healed or set.

Fifthly, from fracture, as, for instance, when the hip is twisted above the muscle of the hip-bone which is on the femur.[5]

Sixthly, from a blow as, for instance, if the nose be driven in and a 'chimus' is the result.

Seventhly, from some evil humour, as is the case in lepers, or from some deficiency of the humour as is the case of those who are phthisical.

Of Constriction or Dilatation of the Pores: Of Smoothness and of Roughness.

Constriction of the pores happens in three ways. Firstly, from uncomplicated constriction; secondly, from fleshiness; thirdly, from narrowing (*coartatione*).

Uncomplicated constriction is caused (*a*) by excess of the retentive energy; or (*b*) by deficiency of the expulsive energy; or (*c*) by excessive cold; or (*d*) by tight constriction of any part of a limb, as often happens from a tight bandage; or (*e*) from excess of dryness.

5. This passage would seem to be a reference to some dislocation at the hip-joint.

Secondly, fleshiness contracts the pores, as in the case of swellings *(apostema)* or in the seat of an old wound (? scar tissue).

Thirdly, narrowing contracts the pores, when anything is deposited in them such as a humour, or a stone, or a blood clot, or again anything which lies hid therein, such as proud flesh or 'scabies'.

Dilation of the pores may be due to four causes.

Firstly, excess of the expulsive energy; secondly, to deficiency of the retentive energy; thirdly, to excess of the colours of humours. Fourthly, to aperient medicines.

Smoothness may occur after two fashions; internal and external.

If internal it may be due to liquefied and viscous humours, if external to ointments.

Finally, roughness may occur after two fashions: internal, due to excess of sharp humour, or external from smoke and dust.

Of Excess of the Number of Members.

Excess of the number of members happens in two ways. If natural, it is due to an excess of the natural and normal humour, or from excess of the informative energy.

If outside the course of nature it is due to an unnatural and abnormal humour, or to an excess or deficiency of energy.

Of Diminution in the Number of the Members.

In like manner, diminution of the number of the members happens in two ways, internal and external.

If internal it may be due to diminution of humours. If external, to burning, or to cold, or to putridity, or to cutting.

Putridity is due either to some poisonous draught which causes mortification or brings about putrefaction, or to constriction and retention of the humour which is thereby broken down.

Of the Size of the Members.

Bigness of the members occurs in three ways. Firstly, from an excess of the humours; secondly, from an excess of the (formative) energy; thirdly, from an admixture of the two.

Smallness also occurs in three ways. Firstly, from deficiency of the energy; secondly, from cutting *(incisione)*; thirdly, from burning by fire or from excessive cold.

. . . .[6]

Of Separation of Parts Normally Joined.

Separation of parts normally joined is due to either an intrinsic or to an extrinsic cause.

Intrinsic causes are the invasion of *(a)* an acute humour; and *(b)* a ventosity,[7] which distends and weakens the parts.

Extrinsic causes are cutting, fracture, rupture due to muscular effort *(exercitium nimium)*, sword cuts, anything which stretches *(distendit)*, as a rope, or which bruises, as a stone.

Of the Kinds and Number of Symptoms.

There are three genera of symptoms, which refer respectively to health, to sickness, and to the mean state.

And each genus is divided into two species; namely, those which have to do with the official members, and those which have to do with the similar members.

Again, the symptoms of similar members are of two kinds, i.e. substantial and accidental.

The substantial are heat, cold, dryness, and moisture. The accidental are those which show their significance either by touch, as hardness or softness; or by sight, as colour. Some, again, are obvious by action of the energies, as, for instance, when the functions are well and fully performed.

Of Symptoms in the Official Members.

Symptoms in the official members are likewise divided into substantial and accidental. The first are four in number, namely, number, position, *ars* (?function), *modus* (?arrangement). The accidental are likewise four: namely, good, bad, perfect, imperfect.

Of the Genera of Symptoms.

There are three genera of symptoms: (1) those which show what has happened, and they are called cognitive or agnitive. For instance, when we find the body wet we know that sweating has gone before. (2) Those which show what is present and are called by Galen demonstrative, as, for instance, when we find a large and quick pulse we understand that fever is present. (3) Those which show what will happen and the perception of those precedes the event, as, for instance, if we see the lower lip tremble we judge that vomiting will occur, which after it has happened is called *precessio significativa*.

But between symptoms and signs *(accidentia)*

6. [In a note the translator explains that "I have not translated this passage from sheer inability to do so; . . ." He includes only the Latin text of approximately twelve lines with the heading *Of the Displacement of a Member.*—M. M.]

7. Cf. the old term 'Spina Ventosa'.

there is a difference, and there is a gradually widening division between them. If you carefully examine each one of the differences respectively it will have one particular signification; but some things which are signs to the patient are symptoms to the physician.

Of Signs and Their Number.

Significant signs are of three kinds: a change in the operative energy, as, for instance, indigestion; a change in bodily quality, as, for instance, jaundice; a change in excretion, as, for instance, black urine.

Of Change in the Operative Energy.

Changes in the operative energy are three in kind: total, as indigestion; or partial, as obscurity of the eyes, or slow digestion; or from one quality to another, as when good digestion is changed into a turbid or acid digestion, or when specks like flies or chips *(ligna)* appear before the eyes, or partial obscurity of the sight.

Of Changes in Quality.

Changes in quality are four in kind. Those obvious to sight, as jaundice, or morphea,[8] or black tongue; or known by smell, as foetid breath or sweat or polypus; or again those which are known by taste, as salt, bitter and acid; and finally those which are known by touch, as hard and soft.

Of Excretions.

There is a double significance in the excretions. For some come forth with noise, such as eructations from the mouth, rumbling in the intestines, and wind from the anus.

And those which come forth without sound may be abnormal in three different ways: in quantity, or quality, or in both. An example of the first is lientery, of the second black urine, of the third a flux of blood.

Of Alterations in Members.

Alterations in members are primarily divided into two classes, namely intrinsic and extrinsic.

Intrinsic changes are six in number: (*a*) change of the operative energy of a member; (*b*) changes in the excretions; (*c*) changes from pain in the neighbourhood of a member; (*d*) changes from pain in the member itself; (*e*) changes by abnormal mobility; (*f*) changes gathered from the opinion of the patient. And intrinsic changes are three in number: (*a*) those obvious to the sight, as whiteness or blackness; (*b*) those obvious to the touch, as

hardness or softness, heat or cold; (*c*) those which are obvious to both senses, as greatness or smallness, increase or decrease.

Of the Cause of Sickness.

The causes of sickness are three in number: Firstly, a change of nature; secondly, *habitudo inconveniens* of an official member; thirdly, a separation of continuity.

Of the Operation of Medicine.

The operation of a medicine has the following threefold effect: it preserves health, after its many different kinds *(secundum multitudinem suam)* ; or, out of illness it produces health; or finally it acts in the contrary fashion.

Of the Regimen of Health.

The regimen of health is of three kinds according as it deals with those prone to illness, those just beginning to be ill, and weakly persons.

The first classes are treated by proper regulation of the aforesaid six things, i.e. the non-naturals.

Those in the second class are treated in two ways: first, by removal of the excess of humour *(chimus)* ; secondly, by repairing any defect in nature and by counselling adherence to the proper observance of the non-naturals.

'Weakly persons' are infants, old persons, and convalescents.

Of the Divisions of Medicine in General.

All Medicine comes under one of two heads, general or particular.

General Medicine concerns itself with the right ordering of the non-naturals.

Particular Medicine has three divisions, according to whether it is concerned with the similar members, the official members, or with solution of continuity.

Abnormal changes in the similar members are treated by being brought back to their original condition or state, and retained therein by bandaging.

If hollow organs (?channels) are over-dilated we bring them back to their proper size and keep at rest. If they are too small we do the contrary.

If there be defect of the retentive energy, we mollify the place with fomentations and cataplasms.

If there be defect of the expulsive energy we use diaphoretics and carminatives *(confortativa)* .

8. 'Morphea is a spice of lepre that sitt in the skyn.' Lanfranc, *Science of Cirurgie,* English translation about 1380, *Early English Text Society,* Orig. Series, No. 102.

If the cause be of a styptic nature use a softening remedy; if dry, a moist one; if the disease arises from constriction apply some remedy which will break it down.

In cases where there is a change from the natural order of things, we restore them to their proper working. If, for instance, there is an apostema, we cure it by bringing it to a head. If the trouble arises from an adhesion *(naturali junctura)* we either use an aperient medicament or else we alter it by surgery. If there is some new growth we take it away; if roughness be present we use smoothness, and vice versa.

Of the Removal of Overgrowth.

Overgrowth we remove either in part, as in the case of scrofulous swelling, or totally, as in the case of cancer.

'Separatio' we cure as follows: if from overplus of blood, we take it away (? bleed) at all ages.

If a limb be too small we increase its size by exercise and fomentation, if too big we lessen it by rest and bandaging.

Of Displacement of Members.

Displaced members are restored to their normal condition in two ways: (1) by joining what is separated; (2) by separating what is joined.

In the former case there are four essentials: (*a*) to join the separated parts; (*b*) to keep the joined parts in place; (*c*) to prevent displacement occurring again; and (*d*) to maintain the seat of injury in a wholesome condition ('naturam loci custodire').

Of the Sufficiency and Division of Medicine.

The practice of medicine deals with the right ordering of the non-naturals, with giving of drugs *(potio)*, and with surgery. Drugs are administered internally by the mouth, by the ears, by the nose, by the anus, and by the vulva. Externally by means of poultices, plasters, and stupes.

Medicines administered internally act in three ways; they loosen or they bind, or they bring about an alteration in quality, as does cold water in a fever.

Sometimes they act in four ways: they reduce over-excess; or they supply deficiency, as, for instance, flesh or blood when administered; or they bind what is loose, as does a styptic; or they bring about an alteration in quality, as does water in fever.

Of Surgery.

Surgery deals with two tissues, i.e. with the flesh and with the bones.

When dealing with the former it cuts, sews, and heals *(coquere)*.

When dealing with the latter it consolidates, unites, and scrapes.

Of the Judging of Drugs (Species).

The right judging of drugs takes into consideration five different matters: quality, quantity, season (of gathering), arrangement; and the question of whether they be good or bad.

The end of the Book of the Introduction of Joannitius and Praise be to God.

91 CANON

Avicenna (980–1037)

Translated by O. Cameron Gruner; modified and annotated by Michael McVaugh)

Book I, Fen I

DOCTRINE I

Chapter 1: The Definition of Medicine.

Medicine is the science by which we learn the various states of the human body, when in

1. Reprinted by permission of Luzac and Co., Ltd., from O. Cameron Gruner, *A Treatise on the Canon of Medicine of Avicenna* (London: Luzac and Co., 1930), pp. 25–26, 29–31, 57–59, 62–63, 77–80, 83–85, 88–92, 107, 110–111. This translation has been considerably modified in certain places in order to present a more accurate rendition of the Latin version of Venice, 1569.

The *Canon*, like the *Isagoge*, is a summary of the Galenic medical system, but on a very different scale; Avicenna's book is an exhaustive compilation and sys-

tematization of all the scattered knowledge contained in Galen's writings, and it is so successful a clarification that even the modern reader has comparatively little difficulty in following the orderly arrangement and substance of its doctrines. Its first book surveys medical theory generally; its second examines the properties and virtues of all simple (that is, uncompounded) medicines; the third and fourth treat of diseases, of individual members and of the whole body, respectively; and the fifth is an antidotary. Such treatises helped set the tone of Western medicine in the thirteenth and fourteenth centuries; however much they might emphasize the importance of practical knowledge, or the nature of medicine as "art," the effect of their very structure was to favor the logical element over the clinical. Only in surgery—which in any case was tending to separate from medicine during this period—did the empirical approach retain some strength (though see below, pp. 802–806).

health and when not in health, whereby health is conserved and whereby it is restored, after being lost. Although some divide medicine into a theoretical and a practical part, you have assumed that it is wholly theoretical "because," you say, "it is pure science." But in fact there are some arts that have both a theoretical and a practical aspect; philosophy has a theoretical and a practical side, and so has medicine. In each of these areas we mean one thing by theory and another by practice. The difference between the two need be explained only in the case of medicine. Thus, when, in regard to medicine, we say that practice proceeds from theory, we do not mean that there is one division of medicine by which we know, and another, distinct therefrom, by which we act—as many, examining this problem, suppose. We mean instead that these two aspects are both sciences—but one deals with the basic principles of knowledge, the other with the mode of operation of these principles. The former is theory; the latter is practice. Theory is that which, when mastered, gives us a certain knowledge, apart from any question of treatment. Thus we say that there are three forms of fever and nine complexions. The practice of medicine is not the work which the physician carries out, but is that branch of medical knowledge which, when acquired, enables one to form an opinion upon which to base the proper plan of treatment. Thus it is said that in case of hot apostemes, the first agents to employ are infrigidants, inspissants, and repellants; then we temper these with mollificants; and, finally, when the process is subsiding, resolvent mollificants will accomplish the rest. But if the diseased apostemes contain matter which depends for its expulsion on the integrity of the principal members, such treatment is not applicable. Here the theory guides to an opinion, and the opinion is the basis of treatment. Once the purpose of each aspect of medicine is understood, you can become skilled in both theoretical and applied knowledge, even though there should never come a call for you to exercise your knowledge. . . .

Chapter 2: The Subject Matter of Medicine.

Since medicine considers the human body from the standpoint of how it is made healthy and how it sickens, and since we can have knowledge of neither unless it is known through its causes, we must in medicine know the causes of health and of sickness. Now as health and sickness and their causes are sometimes evident to the senses and sometimes only perceived by means of the evidence afforded by the various symptoms *(accidentia),* we must in medicine gain a knowledge of the symptoms of health and sickness. It is a dictum of the exact sciences that knowledge of a thing is attained only through a knowledge of the causes and the origins of the causes—assuming there to be causes and origins. Consequently our knowledge (of health and sickness) cannot be complete without an understanding both of symptoms and of the principles of being.

There are four kinds of cause: material, efficient, formal, and final. The material cause is the subject in a state of health or disease—the immediate subject is the members and the spirit *(spiritus);* the more remote subject is the humors; the most remote is the elements. The latter two are in composition when they are subjects, and are liable to vary. But a certain unity is achieved in the composition and alteration of any such thing composed, and this unity to which the many things are reduced is called either the complexion or the form: complexion in regard to alteration, form in regard to composition. The efficient causes are such as change or maintain the states of the human body. Namely: the air and affiliated agents; comestibles, potables, and the like; evacuation and retention; locale, cities and habitable places; motion and rest, sleeping and waking; the changes at the different periods of life, and in occupations, in habits and customs; and in those things affecting the human body by contact, whether contrary to nature or not. The formal causes are the complexions, the faculties *(virtues)* proceeding from the complexions, and the compositions. The final causes are the actions *(operationes).* A knowledge of these presupposes a knowledge of the faculties and the spirits, which are the subjects of the faculties, as we shall show.

These, then, are the subjects which pertain to medicine. Familiarity with them gives one insight into how the body is maintained in a state of health, and how it becomes ill. A full understanding of how health is conserved, or ill-health removed, depends on understanding the underlying causes of each of these states and of their "instruments": for example, the regimen in regard to food, drink, choice of climate, regulations regarding labor and repose, the use of medicines, operative interference. Physicians treat of all these points under three headings, as will be referred to later—health, sickness, and a state intermediate between the two. But we say that the state which they call intermediate is not really a mean between the other two.

Now that we have enumerated these groups of

causes we may proceed to discuss whatever medicine has to say concerning the elements; the complexions; the humors, or fluids of the body; the bodily members, simple and composite; the spirits and their natural, animal and vital faculties; the functions; the states of the body—health, sickness, intermediate conditions; and their causes—food, drink, climate, water, localities of residence, exercise, repose, age, sex, occupation, customs, race, evacuation, retention; the external accidents to which the body is exposed from without; the regimen, in regard to its food, and drink; and medicines and manual operations, for conserving health and curing every illness. . . .

DOCTRINE III

Chapter 1: The Complexions.

The complexion is that quality which results from the mutual interaction and interpassion of the four contrary primary qualities residing within the elements. These elements are so minutely intermingled as each to lie in very intimate relationship to one another. Their opposite powers alternately conquer and become conquered until a quality is reached which is uniform throughout the whole: this is the complexion. Inasmuch as the primary powers in the aforesaid elements are four in number (namely, heat, cold, moisture, dryness), it is evident that the complexions in bodies undergoing generation and destruction accord with these powers. A simple rational classification is into two modes. One is the equable or balanced, in which the contrary qualities are present in the complexion in equal quantities—neither of them being in excess or deficiency. This complexion is the quality which is exactly the mean between the two extremes. The other mode is when the complexion is not the absolute mean between the contraries, but tends a little more to one—whether between hot and cold, between moist and dry, or both.

A complexion, as understood by medicine, is never strictly equable or strictly inequable. The physician should abide by the philosopher who is aware that the really equable state we have defined cannot be found, especially in a human complexion or member. The term "equable," used by doctors in their treatises, does not refer to weight but to an equity of distribution *(iustitia in divisione).* It is this distribution which is the primary consideration in the complexion—whether of the body as a whole, or of some individual member—so that the measure of the elements in it, as to quantity and quality, is that which human nature ought to have—both

in best proportion and in equity *(equitatem)* of distribution. As a matter of fact, this distribution of qualities, such as is characteristic of man, actually is very close to true equality . . . .

In saying a medicine is temperate (that is, of equable complexion or temperament), we do not use this expression in the absolute sense, because that would be an impossibility. Nor do we mean that it is attempered correspondingly to the complexion of the human temperament, for in order to be that the medicine would have to be actually composed of human substance. We mean this—that when the medicine is exposed to the action of the innate heat within the human body, its quality will not overreach either of the limits of the range of equable temperament proper to the human being. Consequently it will not produce an effect beyond those limits. Therefore, in regard to its actions within the human body it is of equable ⟨complexion⟩. Similarly, when we say a drug is hot or cold, we do not mean an absolute heat or coldness of substance, or that it is hotter or colder in substance than is the human body; for if it were so, a drug whose complexion was like the human complexion would be temperate. What we mean by the statement is that through the drug hotness or coldness comes to the body, over and above that which it has itself. Consequently a medicament may be at the same time cold—that is, compared with the human body—and hot—that is, compared with the body of a scorpion; it may be at the same time hot—that is, compared with the human body—and cold—that is, compared with the body of a serpent. More than that, a medicament may be hotter towards the body of Peter than it is to the body of Paul. It is therefore essential for those who wish to alter a complexion to abandon any medicine which cannot have the effect desired.

Now that we have explained the subject of the temperate complexion sufficiently we pass on to consider the distemperate complexions. . . .

DOCTRINE IV

Chapter 1: What a Humor Is, and What Kinds There Are.

A humor is a fluid moist body into which our aliment is transformed. A good humor is such as has the capacity for becoming transformed into actual body-substance, either by itself or in combination with something else. In short, it is that which replaces the loss which the body substance undergoes. The residue from such, the bad humor, does not do this, and is only exceptionally convertible into good humor. It is proper that it should be expelled from the body instead. . . . The humors are of four kinds: blood (the best of them all), phlegm, red bile, and black bile.

In nature the blood is hot and moist, and is either normal or abnormal. Normal blood is red in color, has no unpleasant odor, and has a very sweet taste. When blood is abnormal, it is either because its good complexion has become intrinsically altered or vitiated—i.e. has become colder or hotter—but not from admixture with any foreign matter; or because bad humor is mixed with it. This may happen by a humor coming to it from without, penetrating and corrupting it, or by the production of another humor within it—for example, a part of it might decay, the rarefied product becoming red bile and the denser product becoming black bile, and either one, or both together, might remain in the blood. . . .

Phlegm can also be either normal or abnormal. The normal is such as is capable of transformation into blood at any time, seeing that it is in fact an imperfectly matured blood. It is a sort of sweet phlegm which is not very cold; that is, it is cold compared with the blood and red bile, but hardly at all cold compared with the body as a whole. There is also an abnormal sweet phlegm, a tasteless phlegm, which we shall describe presently; which when it occurs is produced by admixture of normal blood—as occurs often in rheums and in saliva. Galen has said that nature prepared no special member as a receptacle for the sweet normal phlegm, as she did for the two biles; for phlegm resembles blood closely in that it is equally necessary for all the members, and they therefore receive it along with the blood. It is necessary for two reasons, one essential and the other accessory. The essential function is twofold: one is that it must be near the members in case they should be deprived of their habitual nutriment (which is healthy blood) by reason of retention of the material, created in the stomach or liver, from some cause. Phlegm is normally acted upon by the natural faculties, which change and digest it and are themselves maintained thereby. The transformation of phlegm into blood is achieved by the innate heat; external heat only putrefies the material and decomposes it. The two biles do not have this function, since unlike phlegm they are not converted into blood by the innate heat; but they resemble phlegm in undergoing putrefaction and decomposition under the influence of external heat. It is secondly essential that phlegm mix with blood so

as to prepare it to nourish members of phlegmatic complexion. The phlegm must be present in a definite proportion in the blood which nourishes such members, e.g. the brain. The same is true of the two biles. The accessory function is that of moistening the joints and members which are frequently in motion, lest the heat and friction of the movement produce dryness. . . .

Red bile too can be either normal or abnormal. The normal form is the foam of blood, red and clear in color; it is light and pungent. The redder its color, the hotter it is. It is formed in the liver, and then pursues one of two routes—it passes either into the blood or into the gallbladder. That which moves with the blood serves both a necessary and an accessory function: its necessary function is to mix with the blood for the proper nourishment of those members in whose complexions red bile is present in a dispersed form, e.g. the lung. Its accessory function is to attenuate the blood so as to enable blood to traverse the very minutest channels of the body. The part which passes to the gallbladder is also both necessary and useful: necessary because it cleanses the entire body of superfluity and nourishes the gallbladder, useful because it cleanses the intestines of fecal matter and viscous phlegm as well as stimulates the muscles of the intestine and anus, so that they may perceive what they should, and emit wastes. Thus colic very often occurs due to a blockage of the duct from the gallbladder to the intestine. . . .

Black bile can also be either normal or abnormal. The normal form is the lees of good blood, its sediment or residue; in taste it is between sweetness and bitterness. It is formed in the liver and then divides into two portions, one of which enters the blood, and the other goes to the spleen. The portion which moves with the blood serves both a necessary and useful purpose: necessary because it mixes with the blood as required to nourish those members in whose complexions black bile is present, e.g. the bones; useful because it draws the blood together, thickens and strengthens it. The portion which passes to the spleen is that which the blood does not require, and there serves both a necessary and a useful purpose: necessary because it cleanses the entire body of superfluity and nourishes the spleen, useful in that by travelling to the mouth of the stomach by a sort of milking movement it tightens, strengthens, and thickens it, and by its bitterness causes a disturbance there, exciting it to hunger and creating an appetite. You must remember that the red bile which passes to the gallbladder is something not needed by the blood, and that the part which emerges from the gallbladder is something not needed by that either. It is much the same with the black bile. That part which goes to the spleen is such as is not needed by the blood, and that part which emerges from the spleen is such as is no longer needed by the spleen. And just as the red bile then arouses expulsive forces below, so the black bile then arouses appetitive forces above. . . .

Chapter 2: The Production of the Humors.

Aliment undergoes a certain digestion in mastication. The lining of the mouth is continuous with that of the stomach, there being as it were one surface, and therefore contains a digestive force *(virtus)*. When it comes in contact with something masticated it produces a certain change in it, aided by the saliva acquired in digestion, containing innate heat. That is how it is that when wheat is masticated it brings about the maturation of furuncles and abscesses, but has no such effect when simply rubbed with water, or even if boiled with water. Some say that a sign showing us that food is already somewhat digested in mastication is that previously there is neither odor nor taste in it. Once the aliment has entered the stomach, true digestion goes on—not so much by reason of the heat of the stomach as by reason of the heat of the enveloping members: on the right, the liver; on the left, the spleen (the spleen warms not in virtue of its own substance, but in virtue of the many arteries and veins within it); in front, the abdomen, whose fat easily takes up heat and reflects it back to the stomach; above, the heart, which warms the stomach by way of the diaphragm.

The first stage of digestion yields the essence of the aliment, which, in many animals, becomes "chyle" by the help of admixture with the fluid which one has consumed; this is a liquid substance, of the consistency of a ptisan. The portion of this chyle which is thus diluted is drawn from the stomach into the intestines, and then is caused to enter into the mesenteric veins: fine, firm vessels which are found all along the intestinal tract. Through these it comes to the vein called the gateway to the liver [portal vein], enters the liver, and travels along finer and ever finer hairlike channels until it comes to the roots of the vein [vena cava] emerging from the convexity of the liver. The passage of the nutriment through these very narrow channels could not take place were it not admixed with water consumed in excess of the strict require-

ments of the body. When it is distributed through these channels, the liver is almost completely in contact with the whole of the chyle, and for this reason its action is then more violent, stronger, and quicker, and the chyle is digested. In every digestion of this sort there is to be found a foam and a sediment, and perhaps as well something burnt (if the digestion was extreme) or something uncooked (if the digestion did not proceed far enough). The foam is red bile; and the sediment is melancholy [black bile], both normal. The attenuated portion of the overcooked product is bad red bile, and the denser portion bad black bile, both abnormal; and the uncooked material is phlegm. But if the digestion is a proper one, what is formed is blood. As long as it stays in the liver, the blood is more attenuated than it should be, because of the excessive wateriness which is necessary for the reason we have given. But when the blood leaves the liver it is freed from this excessive wateriness, which was needed only for a reason no longer valid. The wateriness is taken from the blood into the veins which go to the kidneys, carrying with it the blood quantitatively and qualitatively proper for their nutrition. The fatness of these fluids nourishes the kidneys; then what is left passes to the bladder and finally to the penis. The good blood passes by the vein leading to the convexity of the liver [vena cava] into the veins branching off from it . . . and eventually on to the hairlike vessels; then the blood sweats out through their orifices and bathes the members, by God's doing. . . . You must remember that hot and cold, among other things, are causes of the production of the humors. When the heat is equable, blood forms; in excess, it forms red bile; in very great excess it forms black bile, due to the burnt residue. Cold produces phlegm and in great excess produces black bile, because of the excessive condensation. . . .

You must know that the blood and that which flows with it undergoes a third digestion in the veins; and when it passes into the members, so that each has received its nutriment, undergoes a fourth digestion. The residue from the first digestion, in the stomach, passes out by way of the intestines; that from the second digestion, in the liver, passes out chiefly by the urine, while what is left goes to the spleen and gallbladder. The residues from the other two digestions are discharged partly as insensible transpiration *(resolutionem)* and perspiration, and partly in material evacuations: through visible orifices, the nostrils and ears; through the invisible orifices, or pores;

through unnatural orifices, such as open sores *(apostemata)*; or in corporeal growths like the hair and nails. . . .

DOCTRINE VI

Chapter 1: On the Faculties.

Faculties *(virtutes)* and functions *(operationes)* are to be distinguished from each other. Every faculty is the source of function, and every function originates in a faculty. For this reason we can treat both in one chapter. Physicians recognize three kinds of faculties and of functions deriving from them: the vital, the natural, and the animal. It is held by many philosophers and all physicians, foremost Galen, that each faculty has its own principal member, which is its seat, and from which its functions emerge. They hold that the seat of the animal faculty is the brain, and that its functions originate there; that the natural faculty is two-fold—one aspect concerned with the welfare and preservation of the individual and securing nourishment to it to the end of life (the seat of this aspect and the source of its functions being the liver), and one concerned with the preservation of the race, governing generation and separating out from the bodily humors the spermatic substance (the seat of this aspect and the source of its functions being

the testicles); and that the vital faculty is that which conserves the spirit, which is the vehicle of sensation and movement, and makes it able to receive these impressions when it reaches the brain, and makes it capable of imparting life wherever it spreads (the seat of this faculty and the source of its function being the heart). The greatest of all philosophers, Aristotle, holds that the heart is the source of all these functions, though they are manifested in the aforesaid organs—while for physicians the brain is the chief seat of sentience, and each sense then has its own member in which its function appears. But if they consider and judge as they should, they will find things to be as Aristotle said, and not as they thought; and will find their writings to have been drawn from sufficient propositions rather than necessary ones, in which conclusions follow only from appearances. It is not for the physician, as a physician, to decide which of these two judgments is the truth; the philosopher or the natural scientist *(physicus)* should decide that. It being agreed that these above-mentioned members are the sources of the faculties, the physician is not concerned, while considering medical treatment, to know whether they derive from other sources prior to them or not. However, his ignorance of these matters would not be tolerated in the philosopher.

Scientific Method

92 COMMENTARIES ON GALEN'S *TEGNI*

Jacopo da Forlì (ca. 1350–1414) and Haly Rodohan ('Alī ibn Ridwān) (ca. 998–1061)

Expanded by Michael McVaugh from the translation of John Herman Randall, Jr.[1]

Annotated by Michael McVaugh

I. *In all the ways of teaching which follow a definite order there are three orders of procedure. . . .* [Galen]

II. *One of them is that which follows the way of conversion and resolution.* [Galen]

In this you set up in your mind the thing at which you are aiming, and of which you are seeking a scientific knowledge, as the end to be satisfied. Then you examine what lies nearest to it, and nearest to that without which the thing cannot exist; nor are you finished till you arrive at the principle which satisfies it. . . . [Haly]

III. *The second follows the way of composition, and is the contrary of the first way.* [Galen]

In this you begin with the thing at which you

have arrived by the way of resolution, and then return to the very things resolved, and put them

1. During the fourteenth century physicians concerned with medical theory began to show an increasing interest in the problem of proper scientific method. Such an interest is most apparent in the many commentaries written during the fourteenth and fifteenth centuries on Galen's *Tegni*, or *Ars parva*. This work was translated by Gerard of Cremona, together with a commentary upon it by Haly Rodohan ('Alī ibn Ridwān) [see Selection 7, no. 64—*Ed.*]; the first two or three phrases of the *Tegni* and the remarks of Haly associated with them comprised an obvious starting point for any medieval treatment of scientific methodology. Out of a number of commentaries on this material there developed a very sophisticated theory of scientific method which, it has been argued, culminated in sixteenth-century Padua with "a clear formulation of the structure of a science of hypothesis and demonstration, with the dependence of

together again in their proper order, until you arrive at the last of them. This is the second way of teaching which follows a definite order; and from what I have said about its opposite it is clear what resolution is. Demonstrations are carried out in these two ways; but demonstration *quare* is effected by composition, and demonstration *quia* by resolution. [Haly]

* * *

Secondly, he says that the second way is by composition of things found by resolution. Resolution is twofold, natural or real, and logical. Real resolution, though taken improperly in many senses, is strictly the separation and division of a thing into its component parts. Logical resolution is so called metaphorically. The metaphor is derived in this fashion: Just as when something composite is resolved, the parts are separated from each other so that each is left by itself in its simple being, so also when a logical resolution is made, a thing at first understood confusedly is understood distinctly, so that the parts and causes touching its essence are distinctly grasped. Thus, when you first consider a fever, you understand the concept of fever in general and confusedly; you then resolve the fever into its causes, since any fever comes either from the heating of the humor or of the spirits or of the members; and again, the heating of the humor is either of the blood or the phlegm, etc.; until you arrive at the specific and distinct cause and knowledge of that fever. Here, then, we have spoken of the subject taken logically. Composition can be dealt with analogously in every way.

Note that on the subject there are many contradictory interpretations to be found. The first of these is of the *plusquam commentator* [Torrigiano, 1270–ca. 1350], holding that by the first knowledge, that which comes by dissolution or resolution, is meant knowledge *propter quid,* i.e. from cause to effect, or the way in which an effect is demonstrated from the cause. The opposite is true of the compositive way, *quia,* by which we prove cause from effect, i.e. proceed from effect to causes. According to this interpretation, the text should be interpreted as follows. The first way is from knowledge of the end, i.e. it proceeds from a knowledge of the ultimate cause to the immediate effect, namely [one] between which and the effect there is no intermediate cause. For by "end" he means the immediate or ultimate cause, beginning with the furthest and advancing to the effect. Note that this cause is immediate to the effect by elimina-

tion of intermediate causes by means of dissolution (that is, resolution), for by resolution of intermediate or remote cause we can proceed until we arrive at the cause immediate to the effect. Thus from a blockage we can conclude an inhibition of transpiration, and from that an inflammation of heat and of the humors, and from that the ascent of putrid vapor to the heart, and from that, fever; thus we have resolved the remote cause, the blockage, into the immediate cause, the fever. The second way then is by composition, that is by effect, of what has been discovered, i.e. the things found by resolution (that is, the resolution spoken of above), because in this way we proceed from the effect to the causes which we found in the resolution.

The second earlier interpretation is that of Haly in his commentary, holding that by the way of resolution we are to understand demonstration *quia,* that is, the knowledge of an effect proceeding to a knowledge of its causes; and conversely, by the way of composition we are to understand demonstration *propter quid.* The first is from the "notion," that is, from the knowledge of the end, that is, of the effect. This comes from the "dissolution," that is, from the resolution of the effect into its causes. The second way comes from the composition of what has been discovered, that is, of the causes discovered by resolution. For those things that are discovered by resolution in demonstration *quia* are afterwards put and joined together in a demonstration *propter quid,* until we arrive at the immediate cause, and conclude the

its first principles upon empirical investigation plainly set forth." For discussions of this tradition, see John Herman Randall, Jr., "The Development of Scientific Method in the School of Padua," in *The School of Padua and the Emergence of Modern Science* (Padova: Editrice Antenore, 1961), pp. 15–68; A. C. Crombie, *Robert Grosseteste and the Origins of Experimental Science* (Oxford: Clarendon Press, 1953), pp. 74–78, 296–300; Neal W. Gilbert, *Renaissance Concepts of Method* (New York: Columbia University Press, 1960), pp. 13–24, 98–107.

This selection is from the commentary of Jacopo da Forli, who taught medicine and natural philosophy at Padua; it is apparently in his work that the identification of medical resolution and composition with Aristotelian demonstration *quia* and *propter quid,* respectively, crystallizes. Much of it has been drawn from fragments translated by Randall, pages 31–32, 35–36; I have filled in the rest. I have tried to follow the sense of Crombie's judgments (p. 77) in distinguishing between Galen's text and Haly's commentary; and in one place I have altered a few words of Randall's translation, in accordance with the Latin edition of Venice, 1519, so as to make the meaning clear.

effect. This exposition is approved by the Conciliator, Diff. 8, and by Gentilis, and by other moderns
since them. [Jacopo]

IV. *And the third way follows the way of analysing
the definition.* . . . [Galen]

Anatomy

93 ON ANATOMY

Isidore of Seville (ca. 570–636)

Translated and annotated by William D. Sharpe, M.D.[1]

Book XI
On Man and Monsters

CHAPTER 1

MAN AND HIS PARTS

116. We say *viscera* not only with reference to the
intestines, but to whatever is under the skin, from
the "gum," *viscus,* which is between skin and
flesh. Likewise, the viscera are "vitals," *vitalia,*
that is, placed all around the heart, as though
"vigor of the heart," *viscora,* since life, that is, the
soul, is centered there.

117. Similarly, the heads of the sinews, made
of blood and nerves plaited together, are called
viscera. Again, [muscles] are termed "lizards,"
lacertus, or "mice," *mus,*[2] because these animals
have the equivalent of a heart all throughout their
several members so that the heart is the center of
the whole body, and they are called by the name of
similar animals which hide themselves under the
earth. Hence muscles, *musculus,* are named from a
similarity to sea-mice, *mus,* and are likewise called
torus, because the organs seem twisted, *tortus,*
in these places.

118. Heart, *cor,* is derived from the Greek name,
because they say KARDIA, or from "care," *cura,*
for therein resides all solicitude and concern for
knowledge. It is therefore near the lung so that
when it flames up with wrath, it may be tempered
by the lung's humor. This has two channels,
arteria, of which the left has more blood, the right,
more vital spirit. Whence also we feel the pulse in
the right arm.

119. The *praecordia* are the places near the heart
by which sensation is perceived, and are called the
praecordia because there is located the source, *principium,* of the heart and of cogitation.

120. The pulse, *pulsus,* is named because it pal-
pitates: from its indications we learn concerning
sickness or health. Its motion is duplex, either

simple or composite. The simple, which is marked
by one pace; the composite, which is irregular and
unequal because of many motions. This motion
has definite intervals: striking dactylically so long
as there is no disorder, but whenever it is more
rapid, as DORKADAZONTES, or more feeble, as
MURMIZONTES, it is a mortal sign.[3]

121. Veins, *vena,* are named because they are
channels, *via,* of flowing blood, and are divided
into streams throughout the entire body, by which
all the members are irrigated.

122. Blood, *sanguis,* gets its name from the Greek
etymology, since it is so active, sustaining and
giving life. It is, however, blood while in the body,
but once spilled out it is called "gore," *cruor.* For
it is gore because it gushes down in a stream, or
because it sinks to the ground in a flux. Others
think that gore refers to corrupt blood which is
discharged; still others say that blood is so named
because it is sweet, *suavis.*

123. Except in youth, the blood is not unexhausted, *integer,* for physicians say that the blood supply

1. [Translated from the *Etymologies,* Book XI (Man
and His Parts), in *Isidore of Seville: The Medical Writings. An English Translation with an Introduction and
Commentary* by William D. Sharpe in *Transactions of
the American Philosophical Society,* New Ser., Vol. LIV,
Part II (1964), pp. 46–47, 48–49; reprinted with the kind
permission of the American Philosophical Society and
the translator, William D. Sharpe, M.D. In editing the
footnotes, I have followed the procedure described in
Selection 89, note 1.—*Ed.*]

2. The *mus* from which *musculus,* "muscle," is derived
was an eel of some kind, not the familiar rodent, hence
the use of *lacertus,* "lizard," for muscle.

3. [In an appendix (p. 65) Sharpe remarks that after the
classical period there was extensive study of the pulse,
which was analyzed from ten aspects and its disorders
recognized. "Galen divided the pulse into four parts:
diastole and systole, separated by two periods of rest.
The *pulsus dorcadizon* is discussed at Galen, *De pulsuum
differentiis* I.28, and the *pulsus formicans* at I.26: these
are the disordered pulses mentioned by Isidore at XI, 1,
120.—*Ed.*]

is diminished with age, whence also tremor occurs in the elderly; but properly, blood is controlled by the soul, for which reason women customarily scratch their cheeks in grief, whence also purple clothing and purple flowers are furnished the dead.

124. The lung, *pulmo,* derives a name from the Greek, for the Greeks call the lung PLEUMÔN, since it is a fan for the heart,[4] in which the PNEUMA, that is, the [vital] spirit, is centered, by which they are both set in motion and moved; whence also the lungs are named. For in Greek, the [vital] spirit is called PNEUMA, which by blowing up and stirring, sends out and takes in the air by which the lungs are moved and palpitate, and opening themselves to catch a breath, contracting to expel it. It is an organ of the body.

125. The liver, *iecur,* is so named because the fire, *ignis,* which flies up to the brain has a seat there. It diffuses from there to the eyes, the other senses and members;[5] and by its own heat changes the juice, *sucus,* which it draws to itself from food into blood, which it provides for the nourishment and sustenance of each member. Pleasure and sensual desire have their seats in the liver, according to those who dispute concerning scientific matters.

126. The ends of the liver are pointed, as [also] the ends of endive leaves, or like tongues sticking out. They are called "entrails," *fibra,* because among the pagans, in rites at the altars of Phoebus, responses were given by soothsayers, *ferebantur ab ariolis,* who received their answers when the livers were sacrificed and burned.

127. The spleen, *splen,* is named since it serves as a supplement in the part opposite the liver, lest the space be empty. Some also think that it was made as a source for laughter,[6] for we laugh with the spleen, are angry with the bile, understand with the heart and love with the liver. The whole animal is made up of these four fixed elements.

128. Gall, *fel,* is so named since there is a little bag, *folliculum,* which contains the humor called bile, *bilis.* The mouth, *stomachus,* is called *os* in Greek since it is the door to the belly, and it takes the food and passes it into the intestine.

129. The intestines, *intestinum,* are so named because located in the inner parts of the body, arranged in long circular entwinings, so that the food swallowed can gradually be digested, and not impeded by food swallowed later.

134. Only women have a womb, which resembles the stalk of a plant, *cauliculum,* and within which they conceive. Nevertheless, some authors—not just poets, but others as well—would like a "womb" for both sexes, and many write "belly" for this organ.

135. It is called the *uterus* because it is double and divides itself into two parts which spread in differing directions opposite each other and bend around very much like a ram's horn,[7] or because inside it is filled with the fetus. Hence also [it is called] the "pouch," *uter,* because it has something inside, members and viscera.

136. The "paunch," *aqualiculus,* properly pertains to the pig, hence its transposition to "belly," *venter.* It is called the *matrix* since the fetus arises therein: it carefully cherishes the semen which it has received, and fashions what it has warmed into a body, drawing this embodied mass out into members.

137. The *vulva* is named by analogy to a folding door, *valva,* that is, the door of the belly, because it receives the semen, or because the fetus proceeds from it. The bladder, *vesica,* is named as though a "water vessel," *vas aqua,* thus, it is filled by urine collected from the kidneys, and is distended by fluid. It does not have this function in birds.[8]

138. Urine, *urina,* is named either because it burns, *urere,* or because it is collected from the kidneys. Its indication is a prognosis of both future health and disease.[9] The urine is vulgarly called *lotium* because by its use clothing may be made *lotum,* that is, clean.

139. Seed, *semen,* is that which once sown, is taken up either by the earth or by the womb for the generation of fruit or fetus. It is a liquid made through a decoction of food and of the body and spread through the veins and spinal cord whence,

4. William Harvey, *De motu cordis* 36, has similar views [that is, he accepted the cooling effect of the lungs.—*Ed.*]

5. The liver is the warmest organ of the body at rest, and is exceeded in temperature only by actively contracting skeletal muscle.

6. The function of the spleen was poorly understood in ancient times, but was assigned a role in the production of black bile.

7. This probably refers to a *bicornuate* uterus, not an uncommon developmental anomaly in human beings and the rule in lower animals; a "ram horn" is an appropriate description of a uterus with attached adnexa involved by chronic gonorrheal salpingitis, a disease not rare in women whose bodies come to public dissection.

8. Isidore's comment that the avian air-bladder does not have the same function as the human urinary bladder is correct.

9. Uroscopy was an important diagnostic tool, although in ancient times more used in prognosis than in what would now be considered diagnosis.

sweated out in the manner of bilge-water, it condenses in the kidneys. Ejaculated during coitus and taken up in the woman's womb, it is shaped in the body of a certain visceral heat and the humidity of the menstrual blood.

140. The menstrual flow is a woman's superfluous blood: it is termed "menstrual," *menstrua,* because of the phase of the light of the moon by which this flow comes about. The moon is called MĒNĒ in Greek. These are also called the "womanlies," *muliebria,* for woman is the only menstrual animal.

141. On contact with this gore, crops do not germinate, wine goes sour, grasses die, trees lose their fruit, iron is corrupted by rust, copper is blackened. Should dogs eat any of it, they go mad. Even bituminous glue, which is dissolved neither by iron nor by [strong] waters, polluted by this gore, falls apart by itself.[10]

142. After many menstrual days, however, the semen is no longer germinable because there is no menstrual blood by which the ejaculate can be irrigated. Thin semen does not adhere to the female parts; lacking this power to adhere, it is lost. Likewise, thick semen also lacks the power of growth, being unable to mix with the female blood because of its own excessive thickness. This is why men or women become sterile: from excessive thickness of semen or blood, or from excessive thinness.[11]

143. They say that the human heart is the first part of the body to be formed, because in it is all life and wisdom; then, by the fortieth day, the whole task is made up. This has been learned, they say, from abortions. Others say that the fetus begins to develop at the head, whence also we see in eggs that the eyes are the first parts to be formed in the fetus of birds.[12]

144. The fetus, *foetus,* is named since it is still nourished, *fovere,* in the womb; its afterbirth, *secundae,* is called the "sac," *folliculus,* which is born at the same time as the infant and which encloses him: it is so called because when he comes forth, it also follows.

145. They say that children resemble their fathers if the paternal seed be stronger; the mother if the maternal seed be the stronger. This is the reason faces are formed to resemble others; those with the likeness of both parents were conceived from an equal admixture of paternal and maternal semen. Those resembling their grandparents and great-grandparents do so since, just as there are many seeds hidden in the soil, seeds also lie hidden in us which will give back the figures of our ancestors. Girls are born from the paternal semen and boys from the maternal, because every birth consists of two seeds. When its greater part prevails, it produces a similarity of sex.[13]

146. In our body, certain things are made only for utility, such as the viscera; certain for both utility and beauty, as the senses on the face or the hands and feet on the body, the usefulness of these members is great, and their appearance is most becoming.

147. Certain things are for appearance's sake only, as breasts in men, and the umbilicus in either sex. Some are for the purpose of differentiation, as the genitalia, the long beard, and the wide chest in men; in women, a gentle face and a narrow chest, but for conceiving and carrying young, their loins and sides are wider. What pertains to man and to the parts of his body has, in part, been said, and I shall now discuss the ages of his life.

94 ANATOMICAL DEMONSTRATION AT SALERNO (*THE ANATOMY OF THE PIG*)

Anonymous (first half of twelfth century)

Translated and annotated by George W. Corner,[1] with additional notes by Michael McVaugh

Because the structure of the internal parts of the human body was almost wholly unknown, the ancient physicians, and especially Galen, undertook to display the positions of the internal organs

10. Pliny, *Historia naturalis* 7.15, reports the same views on menstrual blood. "Waters" refers to various acids and solvents.

11. See Lucretius, *De rerum natura* [*On the Nature of Things—Ed.*] 6. 1239–1247, for a review on ancient theories on sterility, and see also Hippocrates, *Aphorisms* 5. 62–63. Owsei Temkin, *Soranus' Gynecology* (Baltimore, 1956) reviews the subject thoroughly.

12. Alcmaeon of Croton [probably flourished in the early fifth century B.C.—*Ed.*] taught that the fetal head was first formed in the uterus so that it might partake of nourishment, but the early formation of the fetal heart was common knowledge from the time of the Corpus Hippocraticum downward. Isidore here contradicts XI.1. 96 [where he said the sacral spine is the first part which develops in an infant—*Ed.*].

13. See Lucretius, *De rerum natura* 4. 1209–1232, which very strongly suggests this passage. Belief in a female semen, necessary for fertilization, persisted until the discovery of the ovum.

1. [Reprinted with the kind permission of the Carnegie Institution of Washington from *Anatomical Texts of the*

by the dissection of brutes. Although some animals, such as monkeys, are found to resemble ourselves in external form, there is none so like us internally as the pig, and for this reason we are about to conduct an anatomy upon this animal.

The term *anatomy* signifies "correct division," which is performed as follows: place the pig on its back and incise its throat in the middle. The first thing which presents itself is the vocal organ,[2] which is bound on the right and the left by certain nerves, called *motivi* [laryngeal nerves]. Also there come to the vocal organ, from below, certain nerves which are called *reversivi* [recurrent laryngeal nerves], because after proceeding from the brain to the lung they return to the vocal organ, by which means it is moved in producing the voice. Nearby there are fleshy masses called *pharynges,* and the same term is applied to swelling of these structures. There are also in this region large glands in which humors collect and cause tumor of the throat.[3] At the base of the tongue arise two passages, namely, the *trachea arteria* [trachea], through which air passes to the lung, and the oesophagus, through which food is transported to the stomach. The *trachea arteria* lies in front of the oesophagus, and upon it there is a certain cartilage known as epiglottis, which at times closes to prevent the entrance of food and drink, opening at other times to allow entrance and exit of air.

Between the *trachea arteria* and the oesophagus is a space known as the isthmus, between two walls, in which humors may collect and cause an abscess called angina; sometimes this is partly internal and partly external and is called quinsy; sometimes it is wholly external and is then called *synanche* [sore throat].

Next separate the *trachea arteria* from the oesophagus in order to expose the lung and heart. The heart is placed somewhat to the left side, and the lung on the right, and each is enclosed in its own capsule. In the capsule of the heart there may be an accumulation of matter, causing syncope; matter which gathers in the capsule of the lung causes peri-pneumonia.

Next you will observe a vein, called *vena concava* [here probably the portal vein], which comes from the liver through the middle of the diaphragm and enters, from below, the inferior auricle of the heart. It then becomes an artery, and from it arise all the other arteries which proceed to the members; in these the pulse occurs. By means of these vessels the heart is connected to the lung and draws air from the cavities of the lung. In these cavities

fluid sometimes gathers and causes *sansugium;* at times it gathers outside and causes *anhelitus,* sometimes both inside and outside and causes *orthomia.* You can show the lung to be hollow by inflating it with a quill. Below these organs, and dividing them from the organs of nutrition, is a membrane called diaphragm, which is attached to the soft parts of the ribs, where the fluid gathers in pleurisy.

Now turn to the organs of nutrition and measure the length of the oesophagus as far as the seventh vertebra. The *os stomachi* begins above the diaphragm, but the stomach itself lies below. After the stomach comes that part of the intestine called *portanarium* [pylorus]; next is the duodenum, after the duodenum the jejunum, after the jejunum the *orbum,* after the orbum the *saccus* [caecum], after the saccus, the *longaon* [rectum], and in the region of the latter there are small intestines called *lateralia,* which are the seat of the iliac passion.

At the right side under the pouch of the stomach the liver is placed. In its substance there is a sac called the gallbladder, and above the liver are two membranes, *zirbus* and *siphac,*[4] which are folded together like a net. The one which appears thick and loaded with fat is called *zirbus,* but the one which is delicate is called *siphac.* These membranes reach as far as the spleen, and are traversed by veins through which black biliary humor *(melancholia)* is transmitted from the liver to the spleen.

The spleen is an oblong organ located in the left side. The regions under the spleen and liver are

Earlier Middle Ages, A Study in the Transmission of Culture with a Revised Latin Text of Anatomia Cophonis and Translations of Four Texts, translated by George W. Corner (Washington, D. C.: Carnegie Institution Publication No. 364, 1927), pp. 51–53.

This text, associated accidentally with the name Copho (*Anatomia Cophonis*), was probably composed between 1100 and 1150; it should be remarked that it is clearly meant to accompany the actual dissection of a pig, and is not a purely analytical textbook description. —M. M.]

2. The term *Lingua* evidently signifies the larynx rather than the tongue.

3. The word *branchos* is defined as tumor of the throat several times in the Salernitan clinical literature. Dr. Singer has called my attention to the following passage in Aristotle's *Historia Animalium* (D'Arcy Thompson's translation): "the pig suffers from three diseases, one of which is called *branchos,* a disease attended with swelling about the wind-pipe and jaws. It may break out in any other part of the body," etc.

4. [These words are Arabic, and refer respectively to the omentum and peritoneum. Corner points out (p. 28) that their presence here indicates that the *Anatomia Cophonis* in this form postdates the activity of Constantine the African at Salerno.—M. M.]

called hypochondria. Next remove all the intestines, whereupon the great artery will be visible in the middle of the spine; under it the great vein is found. The artery is formed from all the arteries of the head, which unite to make up one great artery, and this descends to the loins and sends off various arteries in a downward direction, both on the right and on the left. The great vein is made up of all the veins of the head, and reaches to the kidneys, where it divides; and there also is found the *vena chilis*,[5] into which enter the capillary veins, too small to be seen, through which the urine with the four humors is transmitted to the kidneys. The kidneys are oblong hollow organs situated at the upper part of the loins.

Two passages called *uritides pori* [ureters] are also found in this region, through which the urine oozes into the bladder, passing through a kind of membrane by which all the intestines are enclosed except the *longaon*, and which when broken allows the intestines to fall into the scrotum. Above this is a fatty structure called *omentum*.[6] Above the omentum is the *siphac*.

It is next necessary to discuss the anatomy of the uterus. It must be recognized that nature has contrived this organ in women in order that whatever superfluities are generated during the course of the month may be sent to this organ as if to form the bilge-water of the whole body; this is the nature of the menses which women have. This organ is also nature's field, which is cultivated that it may bear fruit; in which, when seed is sown, it remains as on good ground and through the cooperative action of natural warmth, and the mediation of vital spirits, it becomes implanted like a germinating seed, and sends out twigs through certain roots or mouths by which it is attached to the uterus, and through which nutriment is delivered to it and to the future foetus. Thus, later on, by the action of the bodily forces (as I have often told you, you may recall) the foetus-to-be is generated and augmented. The uterus is located above the intestine; above its neck is the bladder, and under it the *longaon*. Below is the vulva. Next cut the uterus through the middle of its os; you will find two testicles attached above it, by which the female seed is transmitted to the uterus and joins the male seed to form the foetus. The uterus has seven cells, and if the animal is pregnant, you will find the foetuses in these chambers. Over them you will

find a kind of tunic, like a chemise, which is called *secundina* [placenta]. This is broken when the foetus strives for exit. It is attached to the uterus and to the foetus by veins which run in it, and it carries nutriment to the uterus and to the foetus. Those openings by which the foetus is attached are called cotyledons. There is also a large channel, called umbilicus, which is broken (near the uterus when the foetus is delivered; midwives tie it) at a distance of four fingers from the foetus. When it is ligatured this causes phlegmons of the umbilicus.

Now let us examine the structure of the brain. You are to make an incision in the top of the head, and when the skin is removed you find certain layers which are called muscles, under which you discover a bone, called cranium. Under the cranium is a thin layer called dura mater; under this is the pia mater, then the brain. In the eyes there is a tunic called conjunctiva, which appears white. Next to it is the cornea, which is rather grayish; the layer under that is black and slight; it is called uvea. Next cut the eye through the center. The first humor which appears is called *albugineus*, the next is coagulated like a crystal and is called *crystallinus*, and the last is called vitreous.[7] The nerve which comes from the interior to the eye, which is white and slender, is called *opticus*, and that which goes to the ears is called the auditory nerve.

5. [A word not of Arabic origin, as Corner thought, but Greek. "The *vena chilis* . . . lies close to the dorsal spine from above downward. It is called vena chilis from *kilos*, which means "juice." Through this vein juice is borne to the entire body. The vena chilis breaks up into three branches, two small and one large. The two smaller reach and enter the kidneys, to which they carry the urine with the blood. . . . The larger branch of the vena chilis goes to individual members, including the stomach, heart, brain, and the lower members." From the *Anatomia Magistri Nicolai*, a Salernitan text of ca. 1200; translated by G. W. Corner, p. 82. The *vena chilis* apparently may thus refer either to the vena cava or to the renal veins; indeed, because it is understood to originate in the liver, it may sometimes refer to the portal vein. (The physiology assumed here, which treats the veins as afferent vessels, is of course still Galenic.)— M. M.]

6. [Corner noted (p. 28) that "Copho uses both *omentum*, a classical Latin word, and the Arabic *zirbus*, without any clear evidence that the writer was aware of their equivalence."—M. M.]

7. [Compare Selection 96, "The Animal Members: of the Eyes."—*Ed.*]

Master Nicholas (?) (ca. 1200)

Translated and annotated by George W. Corner[1]

As Galen advises, if anyone desires to know the arrangement of the internal and external members of the human body, he must undertake the practice of anatomy. Therefore let us consider what anatomy is and why it is so called, upon what animals it was practiced by the ancients and upon which by the modern, and what are their species and divisions. As we have it from Galen, anatomy is the correct division of the animal members. The word is derived from *ana,* meaning straight, and *thomos,* meaning division; whence *anatomia,* that is to say, correct division of the members. Among the ancients dissection was practiced upon both the living and the dead. The anatomists went to the authorities and claimed prisoners condemned to death; they tied their hands and feet, and made incisions first in the animal[2] or major principal members, in order to understand fully the arrangement of the pia mater and dura mater, and how the nerves arise therefrom. Next they made incisions in the spiritual members, in order to learn how the heart is arranged and how the nerves, veins, and arteries are interwoven. Afterward they examined the nutritive organs and finally the genitalia or subordinate principal members. This was the method practiced upon living bodies. A different method was used upon the dead; they took a crucified corpse and fastened it with rope to a stake in a running stream, with hands and feet tied, so that after a time the skin, flesh, fat, and other superfluous parts were removed by the action of the water and the arrangement of the internal members could clearly be observed. Thus Galen, a remarkable physician of those times, composed two books of anatomy of the dead and one of the living; but such treatment of the human body came to be considered inhuman, especially by Catholics, and the practice of dissection was transferred to animals. Now, some kinds of animals are much like man, especially in outward aspect, for instance, monkeys and bears, while others, such as the pig, are similar to man internally; and therefore the anatomists chose the latter kind, and in particular the female pig, which shows the greatest likeness to the human structure in all internal organs, including the uterus.

Next let us discuss the species and divisions of anatomy; and because its classification depends upon members, let us see first what a member is. A member is defined thus: It is a part of an animal which is firm and solid, composed of similars or dissimilars, and assigned to some special function. "Part of an animal" is specified in order to exclude branches, which are not parts of animals but of trees, and also to exclude parts of wood and

1. [Reprinted with the kind permission of the Carnegie Institution of Washington from *Anatomical Texts of the Earlier Middle Ages* . . . translated by George W. Corner (Washington, D. C.: Carnegie Institution Publication No. 364, 1927), pp. 67–70. Corner's translation of the *Anatomia Magistri Nicolai Physici* was made on the basis of the Latin text edited by Franz Redeker (Leipzig, Diss., 1917).

This work, like the *Anatomia Cophonis,* is of Salernitan origin, but was written perhaps as much as a century later (ca. 1200). As Corner points out (p. 32), it is quite different from the earlier text. "It is in the first place a systematic descriptive anatomy, dealing with all parts of the body in logical sequence rather than in an order dictated by convenience of dissection. It purports to describe the human body, rather than the pig. Moreover, the subject-matter is handled in a highly organized style, much space being devoted to definition, classification, and arrangement of the material. The whole tone of the book suggests the philosophical method of scholasticism." Enough is given here to show the increased emphasis on citation of authority, on elaboration of conceptual language, and on formalized structure.— Michael McVaugh.]

2. A modern translator cannot preserve both the spirit and the letter of his original when dealing with the humoral physiology and the doctrine of vital spirits. A newer analytic physiology has driven the very terms out of the language, or given them altered connotations. Not only, for instance, has the word "animal" acquired a special meaning different from that of its Graeco-Latin significance, but the very idea of the *anima* is foreign to minds accustomed for reasons both religious and scientific to separate soul from mind, and both from life. The *virtutes spirituales* have ceased to be "spiritual" since Harvey and Mayow, having been analyzed into circulatory, respiratory, and thermic functions. However, to avoid circumlocutions, I have retained the terms animal, spiritual, and natural members and forces throughout these translations, leaving it to the reader to make the interpretation. [This passage may reflect the report of Celsus (first century A.D.) in his treatise *On Medicine,* where he relates that the Greek physiologists Herophilus and Erasistratus practiced vivisection on condemned prisoners in Alexandria in the third century B.C. For Celsus' account, see M. R. Cohen and I. E. Drabkin, *A Source Book in Greek Science* (Cambridge, Mass.: Harvard University Press, 1948), pp. 471–472.— *Ed.*]

stone. Note also that the term "part of an animal" may denote either *act* or *aptitude*. If it denotes act, it is accidental, because according to this it is a constituent part of the animal; if it denotes aptitude it is substantial, because according to this it is a constitutive part of the animal; for instance, the hand of the embryo before the infusion of life is a constitutive but not a constituent part of the animal. The term "firm" is specified to exclude spirit, which is not a firm part. "Solid" is specified to exclude humors, which are not solid parts. "Composed of similars or dissimilars" refers to consimilar and official members. The phrase "assigned to a special function" is specified to distinguish the idea "part of an animal" in this sense from mere fractional parts such as a third or a quarter of an animal.

The classification of members is threefold. First, they are either consimilar or official. A consimilar member is one which is of the same nature throughout, that is to say, of the same species or complexion. The latter phrase is specified to allow for the case of the arteries, which consist of two layers, one of which is not of the same species, although of the same complexion, because all arteries are of one complexion, for every part of an artery is arterial. The same applies to other members. The consimilar members are twelve in number, namely, skin, flesh *(caro)*, fat, muscle *(musculus)*, lacerta,[3] bones, nerves, veins, arteries, ligaments, tendons, and medulla.

An official member is one which consists of various consimilars, such as the hand, which consists of skin, flesh, veins, nerves, and the like. Again, some members are principal, some on the other hand are derived from the principal members; some have innate powers, others have both innate and inflowing powers. The principal members are four, brain, heart, liver, and testes. The first three, that is to say, brain, heart, and liver, are called principals *(principalia)*, because they maintain the individual in existence. No animal is to be found which has not these three members, unless it be the acephalic animal which is said to have no head and consequently lacks a brain. The testes are called principal members because they maintain the species in existence by material transmission. If there were not generative power in the testes, the human species would perish, and the same holds good for other species.

The members derived from the principals are the nerves, veins, arteries, and seminal ducts. All the nerves arise directly or indirectly from the brain, all the arteries from the heart, all the veins from the liver. The seminal ducts, both superior and inferior, arise from the testes; the superior are called *didymi,* or dubious, because it is doubtful whether they are derived from the kidneys or the testicles. The inferior are the emunctory vessels, which receive the sperm and carry it through the penis into the mint of conception. The members having innate virtues are consimilar members, which have four natural forces, namely, appetitive, digestive, retentive, and expulsive. The appetitive is that which digests, the expulsive that which expels. The members having both innate and inflowing powers are the official members, which have the innate or natural powers of the consimilars of which they are composed. They are said to possess also inflowing powers, not meaning that anything flows from place to place, but because there are forces in them performing their function, such as the animal force and the spiritual force. For example, the animal spirits are carried by the nerves to the hand to produce sensation and voluntary motion, the vital force is carried by the arteries to confer vitality, and so forth.

The third classification of the members is as follows: Some members are animal, others spiritual, others nutritive, others generative. The animal members are the brain, pia mater, dura mater, and the like. They are situated above the epiglottis. The spiritual members, namely, the heart, the lung, and so forth, are situated between the epiglottis and the diaphragm. The nutritive members, namely, the liver, spleen, stomach, and the like, are between the diaphragm and the kidneys. The genital members, namely, the testes and seminal vessels, are below the kidneys.

The brain, being the most important of the animal members, is surrounded by other structures, either protective, expurgative, or subservient. It has as protective structures the pia mater, dura mater, cranium, and overlying skin. The pia mater immediately surrounds the brain with its arterial net, and shields it from the dura mater, which is cartilaginous and hard as cardboard. This membrane is called pia mater because it surrounds the brain like a devoted mother embracing her child. The dura mater guards the brain from the hardness of the cranium. The name *miringae* is also applied to these two membranes. The skull and overlying skin protect the brain from external injury.

---

3. The author appears to have listed muscular tissue thrice, under *caro, musculus,* and *lacerta.*

As expurgative members the brain has the eyes, ears, nostrils, and palate. Through the eyes it is drained of black-biliary humors as rheumy discharges; through the ears its biliary humors are discharged as yellow fluid; and through the nostrils and palate the brain is purged of phlegmatic humors as *rascationes* (?). The brain has also subservient members, namely, the nerves; for the animal spirits are carried by the nerves to all the members, endowing them with sensation, motion, and what not.

The heart, being the most important of the spiritual members, has certain members which are protective, others expurgative, and others subservient. As defensive members it has a kind of capsule which surrounds it and protects it against the hardness of the ribs; and in turn it has the ribs, which protect it against external injuries. For expurgative members it has the canals of the lung and the trachea. The pulmonary canals drain it of superfluities engendered by frequent ebullition of the blood, carrying away the foam to the trachea, by which in turn it is carried to the mouth and thus ejected with the sputum. As subservient members the heart has the arteries, which take the vital spirits and blood which it generates and carry them to the members to give them heat and life.

The liver, which is the most important of the organs of nutrition, is provided with certain members which are protective, others expurgative, others subservient. As protective members it has the *zirbus* and the *siphac*,[4] which envelop it and protect it from the hardness of the ribs; in turn the ribs protect it against external injuries. It has several organs which drain off the various superfluities formed in it, namely, the gallbladder draining it of excess bile; the spleen draining it of excess black bile; the brain, heart, and stomach draining off excess of phlegm; and the capillary veins and ureters, which carry off urine into the bladder. As subservient members the liver has the veins, which provide it with nutrient blood.

The testes also, as the most important of the generative organs, are provided with certain members which are protective, others expurgative, others subservient. As protective members they have the scrotum, the little pouch in which they hang, which protects them against the pressure of the thighs, and in turn the thighs protect them from external injury. As expurgative members they have the emunctory vessels, which carry the sperm from the testes by the way of the penis into the mint of conception. As subservient members they have the *didymi,* which carry to the testes the raw materials from which they generate the sperm.

The brain, which is the most important among the animal members, is, according to some, of hot complexion; according to others, cold; according to others, moist; in substance, subtle, thin, and soft; in color, white; in constitution, hollow and spongy; in form, oblong with a degree of rotundity. It is covered by the scalp and the skull, and is enclosed in two membranes. It has much of spirit and much of marrow. It is mobile with a twofold mobility and divided into three cells. According to Aristotle, it is hot of complexion, as Isaac testifies in his work, *De dietis universalibus,* but according to Hippocrates it is cold (at least as he seems to hint in the *Prognostics,* where he says that cold harms the brain, spinal medulla, nerves, arteries, and consimilar members in general). All agree that the brain is of moist complexion.

96 ANATOMY BASED ON HUMAN DISSECTION: *THE ANATOMY OF MUNDINUS*

Mondino de' Luzzi (ca. 1265–1326)

Translated and annotated by Charles Singer,[1] with additional notes by Michael McVaugh

I. INTRODUCTION

Of the Whole

As Galen, following the authority of Plato, hath

4. [See Selection 94, n. 4.—*Ed.*]

1. [Reproduced by permission from *The Fasciculo di Medicina. Venice, 1493, with an Introduction, etc., by Charles Singer* (Florence: R. Lier & Co., 1925), I, 59–60, 75–77, 82–84, 94–96. (I wish to express my gratitude to the executors of the late Professor Charles Singer.) Apart from renumbering in sequence, I have retained, for the most part unchanged, Singer's explanatory notes, pp. 100–101, 104–106, and 109–110.

Mundinus (Mondino de' Luzzi), professor of anatomy at Bologna from 1290 until his death in 1326, is universally given credit for the reintroduction of systematic human dissection into anatomy. His is the first anatomical text we have that so clearly and consistently reflects personal investigation. This new concern may have been an outgrowth of the surgical renaissance at Bologna towards the end of the thirteenth century (see Selection 111, n. 1), or perhaps of the postmortem examinations beginning to be performed there (for a later example, see the next selection). But how influential his *Anatomy* was is debatable. Though a considerable number of manuscripts of the book exist, it apparently never became a standard university text. Moreover, while dis-

said in the seventh book of his *Methodus Medendi*[2] "a work in any Science or Art is published for three reasons: first for the satisfying of friends, second for the useful exercise of the faculties, and third as a remedy for the forgetfulness which doth come with lapse of time." Moved by these I have projected a work for my pupils in Medicine. Now in Medicine a knowledge of the parts of the subject— that is of the human body—and the naming and relations of those parts, form a division of the Science—as Averroes hath it in the first book of his *Colliget* in the chapter on *The Definition of Medicine*.[3] I therefore purpose to give, among other topics, some of that knowledge of the human body and of the parts thereof which doth come of anatomy. In doing this I shall not look to style but shall merely seek to convey such knowledge as the chirurgical usage of the subject doth demand.

Having placed the body of oné that hath died from beheading or hanging in the supine position, we must first gain an idea of the whole, and then of the parts. For all our knowledge doth begin from what is known. For though the known is oft vague and though our knowledge of the whole is of a surety vaguer than that of the parts, we yet begin with a general consideration of the whole.

Now considering this whole we ought to know in what man doth differ from the brutes. He differeth then in three ways, to wit, in the form or position of his members, in his manners or arts, and in possession of certain parts.

First in form. Man, we note, is of upright stature and was so made for four reasons:

(1) For that the human body is wrought of matter which is ethereal and airy and is the lightest among all the animals; wherefore it doth ever upward strive.

(2) For that, compared with other animals of the same bigness, man hath a greater degree of heat and the trend of heat is also upward.

(3) For that man hath a most perfect form which he shareth with the Angels and Intelligences that rule the Universe. Thus are all his senses of right in the upper part of his body.

(4) For the end to which he was made. For he is thus upright that he may understand, and for this there serve the senses and, notably, that of sight as is seen in the preface of the *Metaphysics*.[4]

Wherefore in man sight and brain, and by consequence the head, must of need be so placed as to receive the divers impressions of the senses. Moreover the range of vision is wider when raised aloft. Wherefore governors of cities do place their watches on

high, as in towers and the like, that they may see

section did become a more or less regular feature of medical education during the fourteenth century, professors of anatomy did not continue, like Mundinus, to teach from the cadaver, but gave formal lectures instead, with an assistant illustrating on the dissected subject. Mundinus' work—with all its flaws, and its willingness to accept tradition—is still the outstanding medieval achievement in anatomy.—M.M.]

2. The *Methodus medendi*, θεραπευτικὴ μέθοδος, of Galen (130–200 A.D.) is a vast work which occupies the whole of Vol. X of C. G. Kühn's *Claudii Galeni Opera Omnia*, 22 volumes, Leipzig, 1821–1833, the edition to which references to Galen are usually made. Extremely popular throughout the Middle Ages, the *Methodus medendi* formed the basis of a large part of mediaeval therapeutics. It was often spoken of as the *Megatechne* or *Macrotechne* to distinguish from the *Microtechne*, τέχνη ἰατρική, or *Ars medica* a smaller work contained in Vol. I of Kühn's edition.

The *Methodus medendi* was translated several times into Latin. It had been turned into Arabic by Hunain ibn Ishak (809–873 [he died in 877—*Ed.*]). A translation was made from the Arabic by Constantine the African (died 1087). This translation survives in numerous MSS and has been printed. (Opera Isaaci, Lyon 1515, p. 189.) Another translation made in the late XIIIth century was direct from the Greek. It also is quite common in MSS but it has not, I think, been printed. Probably it was this Graeco-Latin translation and not the Arabic-Latin version that was used by Mundinus.

The passage quoted by Mundinus is a greatly abbreviated paraphrase of the opening of the *Methodus medendi*.

3. Averroes is the mediaeval Latin form of the name of the heretical Spanish writer Ibn Roschd of Cordova (died 1198). Averroes takes a very important place in the history of scholasticism by reason of his commentaries on the works of Aristotle. In medical matters he influenced the Latin West chiefly through his work *kètâb al kolligât* i.e. "universal book of medicine" which was known to the Latin West as the *Colliget*. This work was widely read in the later Middle Ages in a translation into Latin made by a Jew named Bonacosa at Padua in 1225 [actually 1285—*Ed.*]. MSS of the version of Bonacosa are well known. Bonacosa's version was printed at Venice in 1482 and appeared in several later editions.

4. The *Metaphysics* of Aristotle to which Mundinus refers he had probably read in the Arabic-Latin version of Michael Scot (1175?–1234?). The *Metaphysics* opens with the sentence "All men by nature desire to know. An indication of this is the delight we take in our senses; for even apart from their usefulness they are loved for themselves; and above all others the sense of sight." Mundinus in these paragraphs proceeds to exhibit the characteristic mediaeval application of the Aristotelian doctrine of the four elements and their arrangement, along with Ether, in a series, Earth, Water, Air, Fire, Ether, of which Earth, the densest, is lowermost and Ether, the most tenuous, is uppermost. This doctrine is set forth especially in the Aristotelian works *De generatione et corruptione,* the *Meteorologica* and the *De caelo et mundo* all of which were commonly available in Latin translations in the time of Mundinus.

afar, as Galen doth say in the ninth part of his *De juvamentis membrorum.*[5] So also Avicenna in the *Canon,* Book III at the beginning, saith that the need that the head of man should be uplifted did not come from the brain, nor from the ears, no, nor from the mouth nor nostrils, but from the eyes for the reason stated.[6] Thus for all four reasons was man of upright stature. Wherefore he is named *Antropos* or *Plantenus,* that is *with turned sole,*[7] and *Microcosm* that is the *smaller world* because, like the world the *Macrocosm,* he hath an upper and a lower.[8] This then is the first difference which marketh him off from brutes.

The second cometh of Manners or of Art; for among animals he hath the gentlest manners, for is he not an animal that is political and civil? Nor hath he any Art implanted in him by Nature, as hath the spider, the bee and their like. Yet this is to the end that he may understand every art, for had he any one Art so implanted in him he would have been deprived of every other, as Galen hath it in the fourth part of his *De juvamentis membrorum.*

Thirdly Man doth differ also from other animals in his parts, for he is without many internal parts which they have,[9] nor hath he parts that are given to them as a defence by Nature as are horns, beaks, claws. He is without these for that he hath the best of all, the hand to wit, whereby he may work for himself every manner of weapon for defence, as Galen also saith in the first part of his *De juvamentis membrorum.* Thus having none of these parts he yet hath all. For which same reason he hath not such parts as fur, feathers or scales. And another cause wherefore he is deprived of these is that he hath not overmuch superfluous earthy matter which is the substance of these parts. Thus also he hath no tail, for being upright in his gait he taketh rest by sitting and a tail preventeth thereunto.

Let this suffice for the anatomy of the whole.

Of the Parts

Although the parts be of two kinds, simple and composite, yet shall I not make a separate anatomical discussion of those that are simple, for their anatomy is not to be seen in a dissected body but rather in one decomposed in streams of water.[10] Notwithstanding, in setting forth the anatomy of the organical members, I shall speak of the simple parts according to that which doth predominate in the organical member under discussion. Thus I shall treat of flesh in the anatomy of the thigh, of bones in that of the back and feet and of nerves in

that of the brain and spinal cord.

As regards the official members of the body[11] know that in most of them as far as anatomy of a

5. The *De juvamentis membrorum* of Galen is an abbreviated Latin translation of an Arabic version of the first nine books of the *De usu partium corporis humani* (K III, I–IV, 366 ["K" is used throughout by Singer to refer to Kühn's edition of Galen's works—M. M.]) which is in seventeen books. The *De usu partium* itself was not available in its entire form in Latin until after the death of Mundinus. He therefore had at his disposal only the Arabic-Latin version of the *De juvamentis* which was prepared by an unknown writer in the XIIIth century and is commonly encountered in MSS of the XIIIth, XIVth and XVth centuries. This abbreviated *De juvamentis* is one of the main sources of Mundinus. The passage referred to here is to be found in VIII § 5 of the *De usu partium* (K III, 703).

6. The so-called *Canon* (in Arabic *Alkanûn fil tebb*) of the Bokhariote Ibn Sina (980–1037) known to the West as Avicenna was by far the most influential medical work among the Latins of the later Middle Ages. It was translated by Gerard of Cremona (died 1185 [actually 1187 —*Ed.*]). This translation was probably among his last works and possibly was not completed by him. It is found in innumerable MSS. The Latin translation of the *Canon* was first printed in 1473 and continued to appear in a whole series of editions down to 1658. Avicenna is one of the main literary sources of Mundinus and most of the Arabic anatomical terms used by Mundinus are taken from this Latin translation of the *Canon.* For those who care to compare the anatomical sections of the *Canon* with Mundinus, the French translation of the anatomical section in P. de Koning's *Trois traités d'anatomie Arabes,* Leyden 1903 will be found convenient.

7. The word *plantenus* is obviously taken from *planta,* the sole. The ridiculous derivation of *anthropos,* represented as *antropos,* and its impossible association with τρόπος, a turning, is a mediaeval commonplace of considerable antiquity.

8. I have discussed the nature of this theory and its origin and implications in my article on "Hildegard" in *Studies in the History and Method of Science,* Vol. I, Oxford, 1917, and in a contribution on "The Dark Ages and the Dawn" in F. S. Marvin's *Science and Civilization,* Oxford, 1923. [See Selection 3, note 3.—*Ed.*]

9. This contrast of the anatomy of animals and of man came quite naturally to a mediaeval writer, more so perhaps than it would to many modern medical men. In the time of Mundinus such dissection as was carried on was largely on the bodies of animals. Mundinus himself refers in several places to animal dissections.

10. By *simple* parts the mediaeval anatomists meant something like what we mean by *tissues* and what Aristotle meant by *homoiomeria.* Among the anatomical processes applied both to human and animal bodies by the mediaeval anatomists was maceration. This was carried on to a degree that left the ligaments and tendons but removed the softer parts. There are several descriptions and many references to this process among mediaeval anatomical writings [compare the beginning of Selection 95—*Ed.*].

11. The *official* members are those which subserve the

dead body is concerned, there are six points worthy to be observed, for so the Alexandrine commentator doth note in his Commentary on the *Book of Sects*,[12] to wit their (1) position, (2) substance, and following therefrom their "complexion," (3) size, (4) number, (5) figure and (6) relations.

As for the anatomy of the living, two points need be considered which are also to some extent evident in dead anatomy.[13] The first is as to purposes and workings. The second is as to distempers that may fall thereto and the showing forth of the proper cure, if such there be.

Now there is division and number in the parts of the body, for there are those which are extreme and are called *extrinsic* or the *extremities,* and those that are deep and are called *intrinsic.* Of these latter, some there are ordained for the preservation of the kind and some for the preservation of the individual. Of the first sort are the reproductive members, of the second are those which be contained in the cavities or *venters.* Of these cavities or *venters* there are again three. The uppermost that holdeth the *animate members* is the head. The lowermost holdeth the *natural members.* The middle holdeth the *respiratory members.*[14] I shall begin with the anatomy of the lowermost venter so that the organs there, being most corruptible, may be the first cast aside.

· · · · · · ·

III. THE GENERATIVE MEMBERS

· · · · · ·

Of the Womb

To continue this discourse. If thou dost dissect a female body, after the *vasa spermatica* thou shouldest see the anatomy of the womb. Examine it as thou hast other members, first as to place and connections, second as to shape, third as to size, fourth as to substance, fifth as to parts, sixth as to purpose and possible injuries.

The position of the womb thou wilt see as being situated in the concavity of the *alchatim* [sacrum], which hollow is surrounded behind by the spondyles of the *allanis* [sacral vertebrae] and of the tail, and in front by the part called *pecten* or *femur* [symphysis pubis]. Thus it is placed immediately between the rectum, which is, as it were, his keel posteriorly, and the bladder anteriorly. It is specially linked as to his neck to the neck of the bladder which is placed upon the neck of the womb even though the cavity of the womb may be higher than the cavity of the bladder. As between right and left, the womb is placed exactly in the middle.

Next thou mayest see the connections of the womb which are very many as being joined, so to say, with all the organs above it. Thus it is linked with heart and liver by veins and arteries, with the brain by many nerves, and consequently with the stomach by both, with the organs which are in the middle of the body as the diaphragm, kidneys and mirach, and specially with the breasts, as already described. It may be also that it hath connection with these organs by means of other veins which have their origin in the *vena chilis* and arise therefrom below the *furcula* [sternum] as will be stated below. It is also linked with the lower organs, as the bladder by means of its neck, and with the colon. It is joined to the *anchae* and to both sciatic joints by two thick and strong ligaments. These ligaments are broad and thick next the womb, and next the anchae are slender and proceed like horns from the head of an animal. They are therefore called the *cornua matricis.*[15]

Second as to shape. The womb is squarish with

needs of the higher *organic* or *principal* members. Thus Andrew Boorde writes in 1547 in his *Breviary of Health* "Princypal members be foure, the herte, the brayne, the lyver, and the stones. . . . All other members be officiall members, and dothe offyce to the pryncypall members."

12. The *Book of Sects* to which Mundinus refers is doubtless the περὶ ἁιρέσεων τοῖς εἰσαγομένοις, *De sectis ad eos qui introducuntur* (K. I, 64). There was an "Alexandrian Commentary" on this which exists in Arabic (represented by the British Museum MS 1356). There are a number of MSS of the XIIIth and XIVth centuries that claim to be translations into Latin of the *De sectis.* Some of these would doubtless be found to be translations of this Arabic version of the Alexandrian commentary.

13. The titles *Anatomia mortuorum* and *Anatomia vivorum* were attached to two pseudo-Galenic Latin tracts on anatomy that were widely diffused in the days of Mundinus. Concerning the source and authorship of these works a considerable literature has arisen. The last contributor to the subject is F. Redeker (*Die Anatomia magistri Nicolai phisici und ihr Verhältnis zur Anatomia Chophonis und Ricardi,* Leipzig, 1917). Those interested in this complicated and difficult question can trace the literature back from the full bibliography in his work.

14. The classification into animate or *animal* members associated with the brain, respiratory or *vital* members associated with the heart, and *natural* members associated with the liver is based on the Galenic physiology with its threefold system of spirits, *animal, vital* and *natural.*

15. [There was a persistent medieval tradition of a horn-shaped structure arising on both sides near the cervix—probably the result of a mistaken interpretation of the cut edge of the vagina. See M. Holl, "Über die sogenannten Hörner des Uterus," *Archiv für Geschichte der Medizin,* XIII (1921), 107–115.—M. M.]

a sort of rotundity and it hath a long neck below. The reason for this shape was the narrowness of space and the use or need for which it was created which will afterward appear. It oweth this shape to being divided into seven cells as will be stated below.[16]

Third as to size. This is moderate and is about that of the bladder: but it varieth for other causes, since it waxeth or waneth by coitus and by pregnancy. A woman who hath borne hath a womb larger than she who is barren, and she that is wed hath larger organs than a virgin or than one that is continent. So also is it with men for use increaseth the size of an organ, according to Galen in the sixth part of the *De interioribus*.[17] Also by reason of age, for a wife hath a larger womb than a maiden or old woman. Also by reason of the complexion and constitution of the body as a whole; this thou canst gather from the *Canon* of Avicenna, Fen II in the third chapter. For these reasons a woman I anatomized last year, that is in the year of Christ 1315, in the month of January, had a womb double as big as her that I anatomized in March of the same year.[18] Yet a further cause for this may have been that which Avicenna doth put forward, to wit that the first had just had her monthly courses and the womb waxeth thicker and stouter at that time. The womb also doth differ in bigness according to the amount of generative power, for the womb of an animal that beareth several at a birth is larger than that of one that beareth but one. It is for that reason that the womb of a sow that I anatomized in the year 1316 was a hundred times greater than I ever saw in the body of a woman. There might be yet another reason in that the sow was actually pregnant and had in her womb thirteen porkers and in it I demonstrated the anatomy of the foetus and of pregnancy which I shall relate to you.

Fourth as to substance. The substance is nervous and membranous that it may expand to contain the foetus; also it is of cold and dry complexion. Also the substance is very thick because it must become thinner during expansion.

Fifthly thou shouldst see the tale of the parts thereof which are both external and internal. The external parts are first the sides on which are fastened the testicles and the *vasa seminaria* already discussed, and second the *cornua* and the neck, the extremity of which is the vulva which clasps the neck. Note that the vulva is the length of a palm's breadth as is the penis. The vulva is broad and capable of expansion being of wrinkled membrane.

These wrinkles are like to leeches; their purpose is that during intercourse it may be sensitive. Towards the upper part anteriorly and about two or three fingers' breadth within the vulva is the orifice of the bladder. At the orifice of the vulva itself are two membranes which may be raised or lowered over the orifice. These hinder entrance of air and external matter into the neck of womb or bladder as the skin of the prepuce guardeth the penis. Therefore in the place already cited Haly Abbas calleth them *praeputia matricis*.[19]

The inner parts thou canst see by cutting the womb in the midst. Then thou wilt see the *os* and cavity. The os is very nervous made like the mouth of a new-born kitten or, to speak more properly, like the mouth of an old tench. In virgins his surface is covered with a thin veil which in the violated is broken and so doth bleed.[20]

His cavity doth number seven cells, three to right and three to left and one in the midst at the top. These cells are nothing but certain cavities in the womb in which the semen may coagulate along with the menses and be there contained and united to the orifices of the veins.[21]

From this we can pass to the purposes of the womb. It is made chiefly for conception and consequently to cleanse or purge the whole body from superfluous, undigested blood. This is the case in human beings only. Other animals do not menstruate and in them such superfluities are consumed by the production of hide, fur, claws, beaks and feathers and the like of which man is deprived.

From this it is evident that the womb must be subject to a multitude of diseases, and many organs suffer in sympathy with it. What are the proper sufferings and accidents thereof and their cause and cures would indeed take long to tell and would be beyond our intention, but seek in the appropriate

16. The extraordinary idea of the division of the uterus into seven cells doubtless came to Mundinus from the very widely read work of Michael Scott *De secretis naturae* § 7. . . . The notion of the uterus as divided in a number of parts goes back to works of the Hippocratic Collection.

17. *De locis affectis* VI § 6 (K VIII, 441).

18. Jan. 1315 old reckoning is Jan. 1316 of our calendar. Thus the date of composition of the work of Mundinus may be 1316 or 1317.

19. Haly Abbas *Liber regalis* Theorice III § 34 in Stephen of Antioch's translation (Ed. Lyons 1523 fo. 41). The *praeputia matricis* are the *labia minora*.

20. In mediaeval anatomies the *cervix* is repeatedly confused with the *vagina* and the *os* with the *vulva*. This was still the case with Vesalius in 1543.

21. See note 16.

places already mentioned as in Avicenna *Canon III*, Fen 21, and in Serapion,[22] Rhazes[23] and our Johannes.[24]

The diseases which arise in sympathy with the womb are indeed as numerous as the organs to which it is connected or locally attached. What these are hath been already said and seen. One, however, thou canst diagnose from anatomy. It is described by Galen in *De interioribus* VI § 4. It is suffocation of the womb. There is suffocation not because the womb moveth in the material sense to the neck, throat or lung, for this cannot be; but because, being unable to expel the vapours downward, it is moved and contracted below so that it driveth them upward. Now if these vapours, by means of one of the connections heretofore described, should reach the stomach then they often cause hickup or vomiting and women say that they have their "womb in their stomach." If, however, the vapours reach the lung and hinder the action of the diaphragm they cause gasping and women say they have their "womb in their throat." But if they reach the heart, which happeth rarely, suffocation doth cause syncope and then they say their "womb hath reached the heart."[25] Yet always suffocation is from sympathy: owing to the connection the womb hath with the part. For the womb doth not reach these organs but only vapour goeth up.

Now how and by what ways it could reach them thou hast seen. What is the cure thereof seek in the authorities, for anatomy dealeth principally with the observation of the facts.

. . . .

IV. THE SPIRITUAL MEMBERS

. . . .

Of the Heart

Having removed the membranes thou wilt see the lung, and in the midst the heart, covered by the lobes of the lung, that the heat and the breath generated therein may be cooled by the air drawn into the lung, and thus be tempered.

First the place and position of the heart is evident, for it is in the midst between front and back, between right and left. The point doth incline to the left but the root to the right, that it may blow in heat and breath to the right side which ought to be warmer than the left. It is also in the midst between upper and lower and removed from the extremities. It was so placed as the source and ultimate root of all the organs.[26]

Second the connections appear as a consequence,

as also the size thereof. It is not too large nor too small, yet is greater in man than in any other animal of the same bigness since he hath more heat.

22. The work referred to under the name Serapion is probably the *Breviarium* of the Syrian Jahiah ibn Serabi (circa 900 A.D.) known as Serapion senior or Serapion Damascenus. This work was translated by Gerard of Cremona (perhaps from Hebrew) before 1185 and printed at Venice in 1497 and often subsequently. Almost equally popular was the work of Serapion junior who lived about a hundred years later. He produced a work *De medicinis simplicibus*. This book was translated into Latin during the youth of Mundinus (about 1290) by Simon of Genoa working in conjunction with the Jewish interpreter Abraham of Tortosa. The book was printed at Milan in 1473 and several times afterwards.

23. Rhazes is the Persian Abu Bekr Muhammad ibn Zakhariah al Razi (died 932) the greatest of all Arabic writing clinicians. The name Rhazes (al Razi) means *man of Ray*, a town near Tehran. The book referred to by our author is probably the vast work known as *Continens* translated by the Sicilian Jew Faradj ben Salem (Farragut) of Salerno during the lifetime of Mundinus (about 1280) and first printed at Brescia in 1486.

24. The writer referred to as *our John* is that puzzle of medical historians known as *Mesue junior*. The identity of this writer is still obscure but it appears not improbable that part of the work which passed under his name was *turned into Arabic from Latin*—perhaps in the XIIth century by a Sicilian Jew—*thence into Hebrew and thence back again into Latin!* The work as it stands is doubtless a compendium gathered from many sources. There are a great number of MSS of it and it is the most frequently printed of all early medical books, the *editio princeps* being that of Venice, 1471. . . . The book was very familiar to the Bologna professors of the XIVth century and was part of the curriculum; hence his description as *our John*.

25. The idea of the womb as mobile was very widespread in the Middle Ages and may still be encountered among the ignorant. To it we owe our word *hysteria* (ὑστέρα = womb). Other names for that disease are *suffocation of the womb*, the *hysteric passion*, the *vapours* and in German *Mutterweh*. The idea was kept before the public and the profession by the very wide circulation of the work of Moschion *De morbis mulierum*. This had been adapted into Latin from Soranus in the Vth or VIth century and a MS of the VIth century has survived. (Leyden Voss 4° 9*). It was translated into most vernaculars—English among them. Curious to relate it was also translated back again into Greek. Its ideas penetrated into many mediaeval works on diseases of women, e.g. the "Cleopatra," pseudo-Trotula and others. The work of Galen ἐι ζῶον τὸ κατὰ γαστρός, *An animal sit quod est in utero* (K XIX, 158) appears to have been unknown in the Middle Ages.

26. The idea of the central position of the heart, a mediaeval commonplace, comes from Aristotle, e.g. *De partibus animalium* III § 4, 665 b 15. The conception was long dominant in physiology and still influenced William Harvey in the preface to the *De motu cordis*, Frankfurt 1628.

Third the shape will be evident to thee, for it is of the form of a pine or pyramid. Everything that is excessively hot must needs be of that form for the proper shape of the prime heat is the pyramid. There is also another reason: the division of the ventricles and of those things that arise therein.[27]

Fourth thou must see the tale of the parts thereof, from which will be evident to thee the substance, complexion and purpose. Some of the parts are outside the substance thereof, but some are within. Outside the substance are the *capsule* and the organs that have their origin therein which will hereinafter be seen. The *pericardium* is nervous or membranous and much dilated. It is made to guard the heart from injuries and to prevent it coming into contact with the other organs during expansion. The reason it is not joined to the heart was to prevent the heart being hindered in movement by weight. There was also another reason, to wit that it might contain a watery substance by which the heart should be moistened and bedewed, lest owing to its great and continuous movement it be dried up: so watery substance is ever found in the capsule. If the capsule were without such aqueous matter it would shrink and become corrupt, and if it be too abundant the heart doth fall into violent beating, trembling and heart disease as Galen doth state in his *De interioribus*.[28]

Of the essential parts of the heart some are outer and some inner. The outer are fat and the *additamenta* or *auricles* of the heart and what ariseth therefrom.

The fat is found in the outer surface of the heart, near the extremity rather than in the active part: for fat is formed by cold or diminished heat, and since the heart is very warm it is generated there in order that the heart may not be dried up owing to his great and constant movement. The *additamenta cordis* are certain membranous parts suited for expanding and contracting. These are formed for the purpose that when excess of blood is formed in the heart or excess of spirit is generated in the left ventricle, it may yet be contained by dilatation.[29]

But thou wilt demand, as did Galen in the place above-quoted, wherefore did not Nature make the heart large enough to contain the full amount of blood and spirit instead of supplying these *additamenta cordis?* I answer that had the heart been very large the force thereof had been weakened by reason of the dispersion of the spirit. Wherefore animals of large heart are timid as the hare and the stag. Moreover since a large quantity of spirit and blood is not at all times to be found in us, had

the heart been over-large the cavity had been oftenmost part empty; but as for these auricles, they can easily contract when not full and therefore there is no empty space therein. Also weight is avoided. The organs that spring from the heart will be seen afterward.

The inner parts of the heart are ventricles, right, left and middle.[30] Cut then the heart first in the

27. In general the description of the heart by Mundinus is borrowed from Avicenna who in turn took it from Aristotle. The absurd idea that "what is hot must be pyramidal" is a mediaeval commonplace of Peripatetic origin. It does not occur in the chapter on the heart in the *Canon* of Avicenna.

28. The pathology of heart disturbance is described by Galen in the *De locis affectis* (K VIII, 302 ff.) which is the reference of our author.

29. Mediaeval anatomists, following Galen, always regard the auricles as of secondary importance, mere *additamenta*.

30. The idea of the *middle ventricle,* about which Mundinus makes such pother, has its origin in a misunderstanding. It results from an attempt to combine the views of Aristotle and Galen. Aristotle, who never dissected a human body, derived his anatomical conceptions largely from cold-blooded animals in some of which the heart is provided with three cavities. He considered that the heart had three chambers, the largest being on the right, the smallest on the left, and one of intermediate size between the two. As far as they can be identified, the largest was the right ventricle plus the right auricle, the smallest or left chamber was the left auricle, while the intermediate cavity appears to have been the left ventricle.

Galen's description differed altogether from that of Aristotle. He tells us, expressly and somewhat contemptuously, that "it is no marvel if Aristotle erred in many anatomical matters, a man who thought, forsooth, that the heart in the larger animals had three chambers." (*De anatomicis administrandis* VII § 157, K II, 62.) Galen always describes the heart as having but two chambers, the right and left ventricles, a wholly subordinate part being assigned to the auricles. These latter were regarded as safety-overflows, expanding to hold superfluous blood when the chambers of the heart to which they correspond became too full.

No third ventricle is described by Rhazes or Haly Abbas and its existence is expressly denied by the latter. Avicenna, in his *Canon*, however, makes an effort to combine the views of Aristotle and Galen. Speaking of the anatomy of the heart (Lib. III, Fen.XI, § 1) he describes the ventricular portion as follows: "In the heart are three cavities, two large, and a third as it were central in position. So that the heart has firstly a receptacle [the right ventricle] for the nutriment with which it nourishes itself—this nutriment is thick and firm like the substance of the heart; secondly a place where the pneuma is formed [the left ventricle], being engendered of the subtil blood; and, thirdly, a canal between the two." A somewhat similar account is given in Constantine's translation of Isaac. The idea soon crept into European medicine, for in a Pisan MS dating from

right side beginning from the point, yet guarding thyself against reaching the opposite wall, but dividing at the side of the middle ventricle. Thou wilt at once come upon the right ventricle and see two orifices therein. One of these orifices openeth toward the liver and the *vena chilis* entereth thereby. This is the largest orifice, for through it the heart draweth blood from the liver and expelleth blood unto all the other organs. Moreover as the heart taketh in through this orifice more than it doth expel, Nature ordained that, at the time of contraction of the heart, it should be shut when it hath to expel and should open again when the heart expandeth. Wherefore it hath three *ostiola* [cardiac valves] which open from without inward. By the same orifice also the expulsion of the perfectly prepared blood doth take place. Yet is not the whole expelled, for a certain portion is sent out to the lung and another portion crosseth to the spirit. Nature therefore ordained that these *ostiola* should not be too greatly depressed lest they should entirely close the orifice.

Observe here two points. First that sense plainly showeth thee that *vena chilis* hath her origin from the heart, since she is united to the substance of the heart and doth not go through it[31] but is greatest next her base and root, like to the stock of a tree. Second that from this vein, before that she entereth the hollow of the heart, there goeth off a vein which compasseth the root of the heart round about, and from which come forth branches dispersed through the substance of the heart. It is from the blood of this vein that the heart is fed.

Next is another orifice toward the lung, that of the *vena arterialis* [pulmonary artery], which carrieth the blood from the heart to the lung, for lung serveth heart in a manner to be stated. In repayment whereof heart doth send blood to lung by this *vena arterialis,* named *vena* for that it carrieth blood, named *arterialis* for that it hath two tunics. These two tunics it hath, first, because it goeth to an organ in continual motion, and second, because it carrieth very subtile and choleric blood. Wherefore, that the blood be not evaporated nor the vein ruptured, it hath these two tunics. In the orifice of this vein are three *ostiola,* which open from within outward and shut from without inward with perfect accuracy. Thus by this orifice the heart doth only expel at the time of its contraction and yet doth not bring in aught at the time of expansion.

When thou hast seen this, cut open the left ventricle, leaving whole the middle wall where is the third ventricle. At once will be evident the cavity

of the left ventricle, the wall of which is thicker and more dense than that of the right. Nature wrought it so for three reasons. First this ventricle hath to contain the spirit, but the right only blood. Now blood is heavier than spirit. Therefore by reason of its contents the right had been heavier than the left and so the heart had not been evenly balanced. But that it might be of equal weight, the left wall was made thicker to compensate thereby the weight of the blood in the right ventricle. Another reason is that the spirit which it hath to contain is easily dissipated and against this the wall was made thick. A third reason is that this ventricle must form spirit from blood. Now spirit is formed from blood by a strong subtilizing and evaporating heat; and heat is the stronger when in a thicker substance.[32]

In the cavity and near the base are two orifices. One is the orifice of the *arteria adorti* [aorta] called *adorti* because it sprang *(orta)* immediately from the heart,[33] or because from it spring, as from the chief source, all the arteries of the body. By it the heart, as it contracteth, sendeth the spirit generated

---

the first half of the thirteenth century (Roncioni MS 99) a crude figure of a three-chambered heart is to be found.

The first translator of the *Canon of Avicenna,* Gerard of Cremona, whose work appeared towards the end of the twelfth century, improved on his original. "In the heart" he said "are three ventricles; two are large and the third as it were between, which Galen called the fovea or non-ventricular meatus, so that there may be a receptaculum for the thick and strong nourishment, like to the substance of the heart, with which it is nourished, and also a storehouse for the pneuma *(spiritus)* generated in it from the subtil blood. And between the two are channels or meatuses." Henri de Mondeville (died about 1320), a contemporary of Mundinus, by going direct to the current Arabico-Latin text of Galen *(De juvamentis,* Book VI) avoided some of the errors of Avicenna, with whom, however, he still describes three ventricles. Mundinus does little but copy Avicenna. There is an attempt to show the central chamber of the heart in a wretched little figure in the edition of Mundinus by J. A. Adelphus, Strassburg 1513. A drawing by Leonardo shows the channels of communication which were supposed to exist between the two ventricles.

31. The idea that certain of the vessels go right through the heart is encountered in Plato's *Timaeus* which was much read in the Middle Ages in the ancient paraphrase of Chalcidius. . . . The view is here opposed by Mundinus. . . .

32. No early figures of the heart are satisfactory or indeed lucid. . . . The fact is that neither Mundinus nor his commentators nor the Arabian writers on which they all drew understood fully the physiological system of Galen.

33. The derivation is a false one. *Adorta* appears to be a mere corruption of *Aorta* which is a word found as early as the Hippocratic writings and is probably related to ἀείρειν, to raise.

therein to all the organs. Therefore Nature ordained that at his root and orifice were to be placed three very dense *ostiola* which shut straitly from without inward and open from within outward.

There is another orifice, that of the *arteria venalis* [pulmonary vein]. It is named *arteria* as conveying vapour and *venalis* as having but one tunic, for Nature was not over solicitous in guarding it. What goeth through it is a smoke-like vapour or air which the heart draweth from the lung. Now the heart doth both attract and expel through this vein; wherefore Nature placed here in the orifice thereof no more than two ostiola which do not shut perfectly. These ostiola are raised high up that they may be of service to the wall of the heart when it expelleth and transmitteth the spirit [through the *arteria adorti*] so that the spirit may not be expelled by the *arteria venalis*. These ostiola are wondrous works of Nature.

No less wondrous is the middle ventricle. For this ventricle is not one cavity but many small cavities, extended rather toward the right than the left, to the end that the blood which crosseth from the right ventricle to the left may be continually subtilized so as to be turned into spirit. This subtilizing prepareth it for being generated into spirit. Now Nature, when transmitting something through the organs or by any path, never doth so idly but ever prepareth it for the form which it shall receive, as she doth when she transmitteth chyle through the meseraics to the liver. Thus doth Galen oft indicate in his book *De juvamentis membrorum* for example in part V *On the meseraic veins*.

So much for the parts of the substance of the heart. There are other parts which spring from it and they are four, to wit the *vena chilis,* the *vena arterialis,* the *arteria adorti,* and the *arteria venalis.* Two of these go to the lung, the anatomy of which thou mayest now see.

. . . . . .

V. THE ANIMAL MEMBERS

. . . . . .

Of the Eyes

Afterward cut both the bones of the eyes and thou wilt see the place of the eye and the connections with the optic nerve and with the motor nerves.[34]

The position of the organ is now clear. It was not placed very deep, since it hath to receive the *species* into the midst,[35] nor was it much raised above the face, that it may not be hurt by exterior things, for the eyes are very soft and easily affected.

For their protection in man Nature ordained eyebrows to guard them from things coming from above, eyelids to guard them from things coming from in front, tuberosities of the jaws to guard them from things coming upward or from the side, while the nose doth guard them at the other side.

34. A great deal of attention was paid by the Arabians to the structure and diseases of the eye, and the essentials of the description by Mundinus are to be found in Rhazes, Haly Abbas and others. The tradition presented by these writers passed early to the West, and is reproduced, for example, in the works of Constantine Africanus and in the anatomy to which the name of Richardus Anglicus is attached. Avicenna's description of the eye is somewhat different, and gave rise to another tradition reproduced in the works of John of Peckham and of Roger Bacon, and influencing the views of Leonardo and Vesalius. The views on the anatomy of the eye expressed by Rhazes and Haly Abbas were, on the whole, more widely accepted than those of Avicenna.

The treatment of the eye was always felt to be hardly within the range of the ordinary practitioner of surgery, and its structure, as we learn from Guy de Chauliac (1300–1370) was not usually included in the general course of anatomy. The custom was rather to refer the student to special works such as that of Alcoatim. The *Ophthalmologia* which Salome ibn Alcoati wrote in 1159 was translated by an unknown worker in the XIIIth century. (The Latin text was printed by J. L. Pagel in his *Neue literarische Beiträge zur mittelalterlichen Medizin,* Berlin, 1896. See also J. Hirschberg in Graefe–Saemisch *Handbuch der gesamten Augenheilkunde* XIII p. 70, Leipzig, 1908).

The description of the eye by Mundinus is that still generally accepted at the end of the XVth and beginning of the XVIth century and closely resembles that of Alcoatim. It is distinctly better than that of his contemporary Henri de Mondeville (1260–1320) or than the pseudo Richardus Anglicus, and superior also to most of the descriptions of the eye dating from the fourteenth and fifteenth centuries brought to light by Sudhoff. [Compare Pecham's description of the structure of the eye, in Selection 62.7.—*Ed.*]

In reading any early description of the eye, it is to be remembered that until the nineteenth century the "emanation theory" prevailed. Light was regarded as of the nature of a stream emitted from the object seen, and the act of vision was considered as a collision of this emanation with an emission of something from the eye itself, called in mediaeval writings the "visual spirit." The emanations took, in the view of some, the form of thin shells, of the things seen. These passing into the eye became the *species.* [See "The Act of Sight" Selections 62.11–13—*Ed.*]

35 Into its *midst.* In mediaeval physiology the actual site of visual sensation was a *central* body, the *humor crystallinus.* This structure was, in fact, the crystalline lens, but even Leonardo and Vesalius placed it in the centre of the globe of the eye. The first to displace it therefrom was Felix Plater (1536–1614) in his work *De partibus corporis humani structura et usu* (Basel, 1583). Plater recognised the real nature of the retina.

Now in the eye are seven tunics and three humours. Thou wilt see the tunics by the straight and careful division of the eye into two, a fore and a hind part. In the fore part are four tunics of which three are joined to the three tunics of the hind part, while for one of them, to wit the *cornea,* there is no continuation in the depth or interior.

(1) The first tunic is the *cornea,* so named as resembling horn in substance and colour. It is transparent and solid. Transparent that, being of no colour, it hinder not the reception of colour, and solid as marking it off from the adjoining external world.

(2) The second tunic is the *conjunctiva,* so named because it doth join, cover and veil the whole of the eye save the *cornea.*

(3) To this is joined behind the *tunica sclerotica.* It surroundeth the whole eye within.

(4) Under the *tunica conjunctiva* in the fore part is the *uvea,* so named as resembling the dark skin of a black grape. In the midst, toward the cornea, is a hole named *pupilla,* through which the *species* of visible things may reach the *crystallinus* and not be hindered by the opaque *uvea.* And the reason why the eye was not entirely covered by this *tunica uvea* is threefold. First that by his colour, green, purple or azure, it doth strengthen the visual spirit since it is the medium between extreme colours. Second that, had it not been there, the visual spirit had been too much dissipated by exterior light. Therefore, that this spirit be held united in one place, to wit in the *crystallinus,* this tunic had an opening, the *pupilla.* For if it happen that this hole is expanded, by Nature or otherwise, and even more if it be contracted, the sight is hindered. Third that every *species* of thing seen reach the eye in a pyramidal figure, the base being the thing seen and the apex of its cone the eye, or its angle in the *crystallinus.* It is therefore necessary that his hole be narrow.

This tunic is called *uvea* because it doth contain the *humor uveus* or *albugineus* formed to moisten the eye, so as to be an internal medium receptive of the *species* of things seen. Now such a medium must be under the sway either of air or water and, since air was impossible, water was put there.

Moreover since the *spiritus visualis* doth run thither from the domination of the air, this humour was put in the *uvea* to separate the *humor crystallinus* from the exterior air, and to separate the *humor crystallinus* from the *cornea.* It also serveth the purpose of keeping the *foramen pupillae* expanded, for it doth come out of the hole in the

uvea to swell the *cornea.* So it is that in those at the point of death, in whom this humour is dried up, the cornea doth sink and become flat. Then it is said by the common people that "a web cometh before the eyes," which is a certain sign of death. On account of such drying up the pupilla also doth become contracted and for many causes there is also dilatation of the pupilla. In fact I have been able to perceive indications of obscure diseases of the eye by means of the *uvea.*

There remaineth only to speak of cataract. Now there is cataract when vapour is generated which falleth from the brain or riseth from the stomach and passeth through the pupil to the humour between the *crystallinus* and the *pupilla.* It doth hinder the reception of *species,* and, because this vapour hath motion, at first the *species* do seem to move. Likewise, since it hath colour, it doth affect the *crystallinus* therewith. And since it is an external object from which *species* are wont to be formed, the power of vision doth judge that it is external things that move. So it doth appear as though flies or bugs or ants were walking across his field of vision.

Later this vapour is changed into water which, condensing and passing into confirmed cataract, covereth the whole of the pupil and doth entirely prevent sight. If it so fall that it doth not wholly obstruct, it may be confined to the side or middle of the pupilla. If at the side he doth see and judge the thing to be smaller in size. If in the middle, he doth judge the thing to have a hole in it, since the part which he cannot see doth seem like to a hole.

From this observe the method of cure of confirmed cataract which is by chirurgical operation.[36] It is not, however, done by taking the cataract

36. We have here an interesting account of couching cataract. The operation had been described by Celsus (1st century B.C.) whose work, however, was unknown to Mundinus being rediscovered only in the XVth century. Galen mentions the operation but does not describe it. Paul of Aegina (625–690) gives a good description of it but the part of his work in which it occurs had not been translated into Latin in the time of Mundinus. There is an excellent and elaborate description of the operation by Avicenna but the procedure is different from that of Mundinus. A similar operation to that of Mundinus is, however, described by Albucasis (died 1106) whose work on surgery was translated by Gerard of Cremona from 1180 and became known in Northern Italy from about 1230 onward. It is highly probable that it would have been studied by Mundinus who might also have read of it in the work of Alcoatim (see note 34). Arabian writers in general paid much attention to the cataract operation, and the impress of their influence on Western ophthalmology is encountered as early as the work of Benvenut-

entirely away for, before it could be drawn off, the whole of the humour *albugineus* would come out. This, however, is the manner. They force a needle into the *cornea* far from the pupil, the *cornea* being pierced deeply in an oblique direction as far as the pupil. Then they press the needle to the place of the cataract or condensed humour and they drive the cataract down as far as possible. Then by pressure and other means they prevent it from rising again. They do not call such an operation a "cure" of cataract but they speak of it as *abatere* that is "to press down" the cataract.

So much for the fourth tunic called the uvea.

(5) Next in the posterior or interior part of the eye is the fifth tunic, called the *secundina* either because it is the "second" from the sclerotic or because it resembles the "afterbirth" *(secundina)*.

(6) After this is the *tunica aranea* surrounding the *crystallinus* in the anterior part since in posterior part the *retina* (7) is joined thereto. Between the two is contained the *humor vitreus* and in the very midst of it the *humor crystallinus*. This is round or spherical in shape somewhat flattened as to his fore part. This humour doth lie more towards the fore part than doth the *humor vitreus* in which it is placed. Now this humor was made to be a place for the *crystallina* and to nourish it.[37]

So is the anatomy of the eye ended.

Of the Ear

Having finished these, turn to the ear.[38] It is placed at the side of the head for sound doth come from right and left, from before and behind, from above and below. So the instrument thereof should be placed right and left but not in the fore part, because there the instruments of the other senses were placed.

Now the ear in man was made of round form and cartilaginous. It was round to be very capacious. It was cartilaginous to be safe from external injury and sonorous. The entrance thereof is a long passage ending at the *os petrosum* in the cavity of which is implanted the *spiritus audibilis* which is the instrument of hearing. The entrance to this cavity is covered by a fine membrane woven of the fibres of the nerves of hearing as hath been said above.

The bones which are below the *os basilare*

cannot be well seen unless they are removed and boiled, but owing to the sin involved in this I am accustomed to pass them by.[39] Thou canst, however, see the beginning and end of the jawbones. These begin from the suture or *adorea* which is between the skull and the *os basilare* which is at the extremity of the eyebrow and forehead. The jaw-bone doth pass hindward by the *os petrosum* and endeth at the ear and at the teeth the anatomy of which I have told above.

us Grapheus who was a student of Salerno at the end of the XIIth century and whose work exists in many MSS. Benvenuto describes the same operation as Mundinus . . . .

37. The technical terms here used for the eye may be conveniently referred to in J. Hirschberg's *Wörterbuch der Augenheilkunde,* Leipzig, 1887.

38. This account of the anatomy of the ear is probably taken from the Pseudo-Galenic *De anatomia vivorum.*

39. [The official attitude of the medieval Church towards human dissection is still a matter of some debate, but there can be no doubt that many anatomists, like Mundinus, were afraid that a sin was involved. The immediate reason is probably the bull issued by Pope Boniface VIII in 1300, concerning the disposition of those "who, when dying in foreign lands, have expressed a desire to be buried in their own country. The custom consists of disemboweling and dismembering the corpse, or chopping it into pieces and then boiling it so as to remove the flesh before sending the bones home to be buried—all from a distorted respect for the dead. Now, this is not only abominable in the sight of God, but extremely revolting under every human aspect. Wishing, therefore, as the duty of our office demands, to provide a remedy for this abuse, by which the custom, which is such an abomination, so inhuman and so impious, may be eradicated and no longer be practiced by anyone, We, by our apostolic authority, decree and ordain that no matter of what position or family or dignity they may be, no matter in what cities or lands or places in which the worship of the Catholic faith flourishes, the practice of this or any similar abuse with regard to the bodies of the dead should cease forever, no longer be observed, and that the hands of the faithful should not be stained by such barbarities." (The full text is given in James J. Walsh, *The Popes and Science* [New York: Fordham University Press, 1911], pp. 32–33.) The immediate addressees of the bull were probably the Crusaders—St. Louis of France, for example, had been treated this way upon his death in Tunis in 1270—but anatomists must have felt that the phrase "any similar abuse" applied to their own activities. Intentionally or not, the bull seems to have considerably restricted the practice of human dissection for at least fifty years.— M. M.]

97 A FIFTEENTH-CENTURY AUTOPSY

Bernard Tornius (1452–1497)

Translated by Lynn Thorndike,[1] annotated by Michael McVaugh

Worshipful Judge, I grieve over thy sad lot, for to lose one's offspring is hard, harder to lose a son, and hardest [to lose him] by a disease not yet fully understood by doctors. But, for the sake of the other children, I think that to have seen his internal organs will be of the greatest utility. Now, therefore, I will not hesitate to state as briefly as I can what we have seen and draw my honest conclusion and adduce the remedies which in my judgment are advantageous.

In the first place, the belly appeared quite swollen, although the abdomen was thin. But after dividing according to rule the abdomen and peritoneum,[2] we saw the intestines and the bladder, which was turgid and full of urine. Removing further the colon and caecum, there appeared in them more gross wind than filth. Then when the ileum and jejunum and duodenum were removed, two worms were found, quite large and white, showing phlegm rather than any other humor. After the intestines had been cut off from the mesentery, since nothing notable was found therein, seeing that the bladder was turgid, I had it cut open and a great quantity of urine appeared, although before he died, as they reported, he discharged a large amount of urine. Afterwards we examined the liver, which was marked with certain spots like ulcers and somewhat swollen about the beginning of the chilic [i.e., portal] vein.[3] But what is more remarkable, there appeared around the source of the emulgent veins in the hollow of the chilic vein an evident obstruction by which the whole cavity was filled with viscous humor for the space of the thickness of a finger, beyond which humor no blood was seen beneath, while the emulgent veins were full of blood, quite watery in character, and the swollen kidneys were also full of this sort of blood, or perhaps of much urinal wateriness admixed with it. Moreover, the ascending chilis [vena cava] had the branch to the heart filled with much blood, and the heart was much swollen, and so the auricles too appeared swollen beyond measure. When these were cut open, a great part of the blood came out, and so almost all the blood was found near the heart. But the vein which carries the nourishing blood to the lungs was also full of similarly viscous humor and seemed wholly free from blood. Having seen this much, I did not examine further concerning anything else, since the cause of his death was apparent in my judgment.

From these facts I infer, first, that this lad had contracted a great oppilation [obstruction] either from birth or in course of time, and it is safe to assume that matter of this sort was accumulated by gradual congestion rather than brought by a deflux from another member.

Second, I infer that those worms were generated after the beginning of his principal illness and were in no way the cause of his death.

Third, I infer that when transmission of blood through the chilic vein and the pulmonary vein was prevented, ebullition and fever resulted. And because in that blood there was much phlegm, that fever was like a phlegmatic [quotidian] one in many of its accidents, though from the manner of its oncoming and development it seemed like a double tertian. For every third day it came on worse in the night, as those present reported and I infer clearly from his restlessness and perceived from his pulse.

I infer, fourth, that those spots of the liver were generated after the oppilation.

Fifth and last, I infer that any son of yours of the same constitution is to be preserved to his twelfth year with the usual medicines which I will mention in closing.

The first corollary quickly follows from what I saw with my own eyes. For Galen remarks in the sixth book of his *Therapeutic*,[4] first, those matters which require consideration should be considered,

1. Reprinted by permission of Columbia University Press from Lynn Thorndike, *Science and Thought in the Fifteenth Century* (New York: Columbia University Press, 1929), pp. 126–131. The Latin text is published for the first time on pages 290–294. The author taught medicine at Pisa from 1478 until his death in 1497; the identity of the boy, even of the father, remains unknown.

Thorndike interprets the case as follows (p. 126): "Expressed in modern medical phraseology, the autopsy seems to have revealed that the boy suffered from multifold metastatic abscesses of the liver, the result of septicemia or pyleophlebitis."

2. The text reads *mirach et sifach;* see Selection 94 and n. 4 thereto.

3. "Vena chilis"; see Selection 94 and n. 5 thereto.

4. C. G. Kühn, ed., *Claudii Galeni Opera Omnia* (Leipzig, 1821–1833), X, 393.

and then certified by experience, so that reasoning may be confirmed by experiment. Moreover, before his death there appeared many signs of obstruction of the liver and veins, since the hue of his face while the fever slackened was discolored, and he had difficulty in breathing, and lassitude of the body, and slowness in his motions, and sometimes I had seen egestion partly chimosic and partly chilosic. So I was assured on seeing the chilic vein that there was an obstruction in it as I had suspected. Moreover, his pulse, since it appeared to vary greatly, indicated obstruction in the veins adjoining the heart, especially those serving the lungs, according to the example in Galen's fourth book on internal members,[5] of the physician whose pulse displayed diversities of all sorts, who afterwards died like those who die of cardiac complaint. But your Worship is witness that that lad had the greatest diversity of pulse, and to such a degree, as I reported, that I was always in much perplexity because I could not administer the requisite medicines.

The first part of the second corollary seems clear, for while long worms, according to the opinion of Avicenna in the chapter "De speciebus vermium," are generated from humidity over which division and separation do not prevail apart from the attraction of the liver and force of putrefaction, and are more harmful than small worms and harder to get rid of, yet it does not seem likely that they would have remained in the duodenum for so many days while medicines were being taken which are good against worms, both to expel and to kill them. And the second part of the corollary is manifest from the opinion of Lord Avicenna, first chapter concerning worms, where he says, "And on this account are produced worms and flies and crawling things from humid putrid substances, since that which is taken away from that matter in order to receive form, when it is rectified becomes a worm-like or fly-like animal, for this is better than that it should remain pure putrefaction."[6] By which words Lord Avicenna seems to suggest that a few worms found in the intestines are not harmful, since, as he says, the worms spring from putrefactions and feed on them because of their homogeneity and take them from the body. While that may be the case, yet I do not assert that, as many believe, worms in our body are a good thing, since, again from the authority of the prince of medicine, it is not their nature to help without harm, since from worms are generated epilepsy and canine hunger and *bolismus*[7] and putre-

faction which is the cause of fever. Nay more, as some have reported, worms sometimes have bored holes in the belly and come out, as Lord Avicenna testifies in the aforesaid chapter.

The third corollary is patent from the proof there given, for the fever was without chill and burning, although at the beginning, because of a great rising of ascending vapors to the head, sleep was produced as if in hectic fever, with some coldness of the extremities while the [diseased] matter was moved about, which perhaps happened, as I have often observed in other cases, from withdrawal of heat to the inner parts when injury is felt about the principal members. Moreover, the fever was continuous, having proportionate exacerbations of cholera,[8] and, since the [diseased] matter was in the parts near the heart, there was a tremulous movement of the heart, where especially was manifested frequent constriction, which, like putrid fever, revealed a great lack of the emission of vapors, nor were the accidents those of *causodes*,[9] since, on account of his humid age and the abundance of phlegm, the impression of cholera was strongly repressed. Nor does it matter that there were two tertians within the veins, with two accumulations in different veins variously moved to putrefaction, as I have shown in other cases, for very likely the matter which was putrefying in the ascending chilic vein [vena cava] made a tertian of greater exacerbation, and that which was in the descending chilic vein made one of lesser exacerbation.

But this is a matter of grave doubt, how the chilic vein, which is so large, especially the descending one, could be so stopped up, for it does not seem that that vein could be clogged unless first the other veins of the liver were clogged with the greatest oppilation, which cannot happen while life lasts. And this is the strongest argument to me for believing that oppilation of this sort was contracted from the disposition of natural principles [i.e., from a fundamental disposition of nature]. Never-

5. Kühn, VIII, 293–296.

6. Thorndike says: "The reference is of course to the then accepted belief that worms, flies, and other insects were spontaneously generated from filth and putrefaction. The chapter on worms occurs in Avicenna, III Canon., Fen 16, Tract. 5, cap. I. . . ." (pp. 839ff. in the Venice, 1564, edition).

7. *Bolismus*=voracious hunger.

8. That is, the humor choler; red or yellow bile.

9. Thorndike says: "A particular variety of fever." The text in fact reads *causonides;* this may refer to a severe fever of bilious origin

theless it is quite true that the matter which was mixed with urinal aquosity in the bladder may have produced oppilation in the veins of the hump of the liver and been expelled by nature and filled those veins with watery fluid simultaneously with death and produced a cathimic crisis, according to that saying of Galen's in the first book on critical days, "Two sons were gladiators, who, fighting together, killed each other":[10] so nature may have expelled the matter and have been overcome by it. But I am more inclined to think that the matter producing the fever, retained, as I have said, in the vein which leads to the heart, was the cause of that suffocation than was anything else, since neither in the meseraic [veins] nor in the liver was any obstruction apparent, but only, as I have said, in the chilic vein. Unless it is held that the matter in the small veins of the liver is so small that it cannot be seen, as can that which was in the chilic vein—gross and viscous in appearance and adhering to its linings—and that nature expelled the subtle [matter] through the urine but could not expel the gross, and so death resulted. And on that point hangs the whole force of this investigation. So, while you are in Florence, you will consult on this point Georgius Ciprius [George of Cyprus?], an exceedingly learned man.

The fourth corollary follows from what has gone before, for from the heat of the heart and the ebullition of that blood was generated salt phlegm causing mordication [necrosis] and ulceration, as happens to many about the gums from salt phlegm descending from the head. As to why his heart was large, I don't think it came from audacity inborn in him, for he seemed timid, rather, when he was in good health, but the heart was filled with a great quantity of blood, which made it turgid and inflated, but it may be he naturally had a large heart, which in man is a sign of audacity, though in hares it signifies timidity, as may be learned from the statements of Lord Avicenna, eleventh Fen of the third part, first chapter.[11]

The fifth corollary is known [or, noted] because after the twelfth year the natural heat becomes acute, exciting all the virtues strongly to expel superfluities, and humidity is lessened, and hence your sons are less liable after that age to incur such a disease. Therefore they are to be preserved by this preservative, that is, each year in April you ought to administer daily the following syrup: Take a dram each of

Water of Hops	[Aque lupulorum]
Fumitory	[Fumiterre]
Maidenhair	[Capilli veneris]
Agrimony	[Eupatrii]
Betony	[Bettonice]

Mix in two doses and with each use one dram of pure vinegar and make a potion and spice it with a little cinnamon and give it hot in the morning. But I think that this prescription should be varied according to their years, as the doctors may judge. After this syrup let them take this medicine: Of choice rhubarb take one dram and let it soak in water of milk taken from goats, and water of endive and absinth. Take in equal parts, with the eight grains of spikenard, through the night. Then let it be strained out and give the juice with cassia or with manna or with dyasena or with a dose of trochees of agaric as the doctor in charge judges best. And as I think that the head, because of the rising of vapors, is one cause of trouble, at least to the extent that when heavy vapors rise in the head beyond measure they descend through the nerves and produce softening, as appeared manifestly, therefore at least once a week when they go to bed I would give them one or two pills of aloes soaked in water of endive, but in winter aloes not soaked might be better.

These are briefly what it seems should be prescribed, but I will visit Your Worship often and shall be able to keep you informed concerning the other children. Farewell, Your Worship, to whom I commend myself with all my heart.

Practice

98 GENERAL INSTRUCTIONS FOR THE PRACTITIONER

Archimatthaeus (first half of twelfth century)

Translated and annotated by Henry E. Sigerist[1]

ADVICE FOR THE PHYSICIAN

Physician! When you shall be called to a sick man, in the name of God seek the assistance of the

10. Thorndike says: "I fail to find such a passage in Kühn's edition of Galen."

11. Page 663 in the Venice, 1564, edition.

1. [Reprinted by permission of the editor of the

Angel who has attended the action of the mind and from inside shall attend departures of the body. You must know from the beginning how long the sick has been laboring, and in what way the illness has befallen him, and by inquiring about the symptoms, if it can be done, ascertain what the disease is. This is necessary because after having seen the faeces and urine and the condition of the pulse you may not be able to diagnose the disease, but if you can announce the symptoms the patient will have confidence in you as in the author of his health and therefore one must devote greatest pains to knowing the symptoms.

Therefore, when you come to a house, inquire before you go to the sick whether he has confessed, and if he has not, he should confess immediately or promise you that he will confess immediately, and this must not be neglected because many illnesses originate on account of sin and are cured by the Supreme Physician after having been purified from squalor by the tears of contrition, according to what is said in the Gospel: "Go, and sin no more, lest something worse happens to you."[2]

Entering the sickroom do not appear very haughty or overzealous, and return, with the simple gesture, the greetings of those who rise to greet you. After they have seated themselves you finally sit down facing the sick; ask him how he feels and reach out for his arm, and all that we shall say is necessary so that through your entire behavior you obtain the favor of the people who are around the sick. And because the trip to the patient has sharpened your sensitivity, and the sick rejoices at your coming or because he has already become stingy and has various thoughts about the fee, therefore by your fault as well as his the pulse is affected, is different and impetuous from the motion of the spirits. When it has quieted down on both parts, you shall examine the pulse in the left arm because, although the right side would be satisfactory, yet it is easier to diagnose the motion of the heart in the left arm on account of its vicinity to the heart. Be careful that the patient does not lie on the right side because the compression would hinder the sense motion, nor should he stretch the fingers or make a fist. While you apply the fingers of your right hand you shall support with the left the patient's arm, because from greater sensibility you will distinguish the different and various motions more easily, and also because the patient's arm being so to say weak requires your support. If the arm is very full and fleshy you must press your fingers hard so as to get into the depth; if it is weak and lean you can feel the pulse sufficiently on the surface. You must examine the pulse to a hundred beats at the very least, so that you may form an opinion on the various kinds of pulses, and the patient's people should receive your words as the result of a long examination of the heart beat.

Finally you request to have the urine brought, and if the change in pulse indicates that the individual is sick, the kind of disease is still better indicated by the urine, but they will believe you to indicate and diagnose the disease not only from the urine but also from the pulse. While you look at the urine for a long time you pay attention to its color, substance, and quantity, and to its contents from the diversity of which you will diagnose the different kinds of diseases, as is taught in the *Treatise on Urines,* whereupon you promise health to the patient who is hanging on your lips. When you have left him say a few words to the members of the household, say that he is very sick, for if he recovers you will be praised more for your art; should he die his friends will testify that you had given him up.

Let me give you one more warning: do not look at a maid or a daughter or a wife with an improper or covetous eye and do not let yourself be entangled in woman affairs—for there are medical operations that excite the helper's mind; otherwise your judgment is affected, you become harmful to the patient, and people will expect less from you. And so, be pleasant in your speech, diligent and careful in your medical dealings, eager to help. And adhere to this without fallacy.

When you have been invited for dinner you should not throw yourself upon the party and at the table should not occupy the place of honor although it is customary to assign the place of honor to the priest and the physician. Then you

Northwestern University Medical School Magazine from Henry E. Sigerist, "Bedside Manners in the Middle Ages: The Treatise *De Cautelis Medicorum* Attributed to Arnold of Villanova," *Quarterly Bulletin,* Northwestern University Medical School, XX (1946), 135–143. It was later reprinted in *Henry E. Sigerist on the History of Medicine,* ed. Félix Martí-Ibañez (New York: MD Publications, 1960), pp. 131–140. This is the third section of the *De cautelis medicorum* ascribed to Arnold of Villanova (see Selection 100.2, n. 8); according to Sigerist, it is actually "an abstract of a large treatise *De Instructione Medici* of the Salernitan physician Archimattheus."—Michael McVaugh.]

2. See Paul Diepgen, *Die Theologie und der ärztliche Stand,* Berlin, 1922, p. 49ff.

should not disdain certain drinks, nor find fault with certain dishes nor be disgusted perhaps because you are hardly accustomed to appease your hunger with millet bread in peasant fashion. If you act thus your mind will feel at ease. And while the attention is concentrated on the variety of dishes, inquire explicitly from some of the attendants about the patient or about his condition. If you do this the sick will have great confidence in you, because he sees that you cannot forget him in the midst of delicacies. When you leave the table and come to the sick, you must tell him that you have been served well, at which the patient greatly rejoices because he was very anxious to have you well served.

If it is the time and place to feed the sick you will feed him. It is necessary, however, that you set the time for the patient's meals, namely in intermittent fevers when the sick have a real remission; in continuous fevers when there happens to be a quiet moment because a decline of their fever does not occur before the crisis. In intermittent fevers they must be fed before the attack and so early that when the attack comes the entire food be digested, because otherwise nature will have to fight a war on two fronts and it will not be strong enough to digest what has been offered at the wrong moment nor will it be able to defeat the enemy disease. When the attack of fever has begun, wait until it has ceased and then wait for two more hours or for one at least, because the organs are exhausted from the preceding battle and the attack of the enemy and do not want that a burden be imposed on them in the form of food, but after having so to say triumphed over the enemy they wish to have a rest.

You shall feed the sick according to the season of the year and according to the change of seasons and of the disease; and quantity and quality of food must be varied according to the diseases, for you shall give the patient ampler food in intermittent than in continuous fever, and colder food in a continuous than in intermittent fever, more food in winter and spring, less in summer and autumn because they stand it very badly. The age must be considered, and you will restore children more often than youths, because their consumption is greater on account of the liquidity of the humors and because they must grow, for it is according to nature to restore where there is a daily loss. Old people you will restore with less food because they have little heat and vigor; and also according to what they are accustomed to eat, because if they are

accustomed to use an ampler and coarser diet you will not give them the same kind of food but rather prescribe a liquid or moderate diet. You must fear constipation of the bowels or flux, and if there is flux you must start out with coarser food such as quinces, sorb-apples, and medlars because they constipate through their thickness. If, however, there is constipation you will start out with lighter liquid foods. Thus you will give prunes and the cooked juice of Damascene prunes because they quickly eject through their heaviness. If the condition of the bowels is between the two, you will begin with a lighter and more liquid diet because this is very useful to the sick and protects against greater harm. If the bowels have moved, give such a diet because it relieves the various organs. Thus you shall give first prunes cooked in water, or pomegranates or almond milk that you shall prepare in the following way: almonds removed from the shells shall be put in hot water, whereupon they shall be ground thoroughly and a little cold water shall be added; the whole shall be stirred, strained through a clear linen cloth and given to drink. If however a little bread, that is the soft part of it, is cooked in the pot the almond milk is better digested than if it is drunk pure. After it has been prepared, a small amount must be poured off, and then one must remove by blowing or with a feather the oily substance that is on the surface because it is a hot matter. After this has been done, give several times chicken broth to drink or water in which the soft part of bread has been dissolved.

You shall also give barley flour and make it in the following way: first wash the barley in cold water, pour it over a stone and rub it so that it loses the skins, whereupon it must be rubbed and ground in a mortar or ground between millstones; then have the finer parts very well cooked and toward the end of the cooking add a little almond milk and present it to the sick. If, however, you wish to have ptisane, cook the coarser parts of the barley in water and give him ptisane in a drink, or water in which bread has been soaked, cooked or not. And remember that while there is food in the stomach you shall not give diuretic water with syrup because such drinks force the food out of the stomach undigested or retard digestion.

Remember, furthermore, that in the beginning of the disease the physician endeavors to oppose it with digestive remedies, for he is the helper of nature and must aid it. Nature namely proceeds to making the crisis, to the triumph over the disease; she wishes to reduce the forces of the disease by changing the condition and quality of the matter and by dispersing it among the organs so that the parts be separated from each other and she may reach her end and more easily than expected with complete results in one weak expulsion. In the same way, the physician in order to drive out the matter that must be driven out, must be prepared to treat the digested matters, according to the aph-

orism in the first book of Hippocrates. Consideration of the cause of the disease determines the choice of remedies that digest the humors; for if the patient suffers from cholera you will give vinegar syrup; if he suffers from a cold humor you will give oxymel, and everything else as I have said in another chapter. Oh physician, thanks be given to God.

Methods of Diagnosis

99 INTERPRETATION OF THE PULSE

Anonymous (thirteenth century)

Translated and annotated by Michael McVaugh[1]

Of all the indications of the internal disposition of the body, two are most reliable, the pulse and the urine. Now, there are two principal members in the structure of the body, which sustain or transform the substance of the whole: these are the heart and the liver.[2] According to Galen, all the powers of the body are founded on these two members; they are alike in composition, and cause the whole body to be uniformly disposed. The heart and the liver fill the same places in the microcosm as the sun and moon do in the greater world.[3] Just as innate heat passes from the sun through the air as a gift to all living things upon the earth, so too heat progresses from the heart, its source, via the vital spirit; it consumes wastes and is propagated in many ways in generously supplying [the body] from the middle to the extreme [parts] and from the near to the remote [parts]. And the liver is analogous to the moon: for just as the moon communicates moisture to the regions next it, and as the full moon is adorned in fullest roundness when the sun casts its own rays upon it, just so the liver generates humors and bestows a suitable humor on every member, while the strengthened substance of the members is sustained with the aid of the heart. Should the sun not regard the moon, the moon would suffer an eclipse, and the substance of the air be wholly darkened. Likewise, when the aid of the heart is lacking, and with it the members' nutriment, the relation of the distant parts to their neighbors is destroyed, weak associations are broken, and the marvelous union of the soul with the body is undone in many ways.

The urine, then, indicates the state of the liver and proclaims any hindrances to the internal members, as well as proclaims their functions.

The pulse, on the other hand, reveals the condition of the heart and the general tenor of life, and by elucidating the passions of the soul, manifests the mind's secrets. Since, therefore, the life of the body and nutrition of the members is most excellent, a knowledge of the pulses far outweighs an understanding of urine. For this reason we have written this our epitome on pulses, fully and clearly compiled from Galen's great work on pulses, his epitome on the same, and the *Summa pulsuum* of Archbishop Alphanus, skilled in both Greek and Latin.[4] Archbishop Alphanus studiously pursued

1. Freely translated from *Summa pulsuum,* the text published by Rudolf Creutz, "Der Frühsalernitaner Alfanus und sein bislang unbekannter 'Liber de pulsibus,' " *Sudhoffs Archiv für Geschichte der Medizin,* XXIX (1936), 57–83. The text itself occupies pages 60–74, and the material translated here pages 60–66, 73–74. A slightly different text was published by Pietro Capparoni, *Il "Tractatus de Pulsibus" di Alfano 1° Arcivescovo di Salerno (Sec. XI)* (Rome: Istituto nazionale medico farmacologico serono, 1936).

This compilation of the thirteenth century is, like most medieval writings on the pulse, faithful to the Galenic view of its cause and correct interpretation. It claims Galen and Alphanus (see n. 4) as authorities; it could perhaps also have claimed Philaretus, to whose *Liber de pulsibus* it seems to owe a certain debt. The essentials of this Galenic doctrine were very little affected by the thirteenth-century influx of Arabic medical writings.

2. There are, of course, four principal organs in the body (see Selection 90), but these two are of particular importance for diagnosis because of their close operational relation to humoral physiology.

3. On the microcosm and macrocosm, see Selection 3, n. 3; and Selection 96 and n. 8 thereto.

4. This is the Archbishop of Salerno who died in 1085, having written medical works on a number of topics. His *Summa pulsuum* has not survived, although it was known

a knowledge of the pulses, led on by envy of the Jew who (as is told in the legend of St. Basil) accurately foretold the saint's moment of death from a knowledge of his pulse.[5]

THE DEFINITION OF THE PULSE

The pulse is the motion of the heart and arteries in diastole and systole so as to cool the natural heat and to expel vaporous waste.[6] It is a motion of the heart, and from the heart to the arteries; but it is not a reversible motion, since it is not a motion of quality or alteration, nor is it an essential or general motion. It is thus a local motion,[7] either direct or circular, and since it is not circular, it is direct. It is not simple but compounded of diastole and systole, of motion upwards and downwards. Its function is to cool the innate heat and to expel the vaporous waste, so that the spirit may be purified and tempered; this allows the establishment of bodily harmony, namely the conjunction and union of several distant parts. For since the soul (anima) is incorporeal and distant from the body, it could in no way inhere in bodily substance did not the spirit (spiritus), like an incorporeal body, serve as a middle term, to bring together distant substances and associate them harmoniously.[8] Although the soul is simple in nature and neither inheres in the greater members to a greater degree nor in the lesser members to a lesser, it resides principally in the brain and heart, which it moves by its essence. But the agents of the soul differ naturally in competence. The heart is hollow, formed like a smith's bellows, containing heat, spirit, and humor. When the heat resolves the blood into spirit and vapors, the spirit is taken up into the concavity of the heart, so that the substance of the heart is enlarged; a contraction follows, due to heaviness [of earthy material?] and the exhalation of spirit. The soul thus moves the pulse, like an artificer; and in the same way, with the artificer absent, his instruments are stilled.

THE DIFFERENT KINDS OF PULSES

The varieties of pulses are differentiated by the physician in a number of ways, in particular according to five considerations: (1) motion of the arteries; (2) condition of the artery; (3) duration of diastole and systole; (4) strengthening or weakening of pulsation; (5) regularity or irregularity of the beat.[9] Ten varieties of pulse derive from these five considerations.

From the first consideration several varieties are derived. They are based on the quantity of arterial

motion, of necessity either great, little, or in-between. These quantities can be classified as long, short, and intermediate; broad, narrow, and intermediate; obvious, hidden, and intermediate. A long pulse is that which can be felt over a space of four fingers or more; a short pulse does not occur over a full four fingers; an intermediate one is neither greater nor less. A broad pulse is felt over a width of four fingers or more; a narrow one occurs as if in a taut cord; an intermediate one does not surpass either limit. An obvious pulse is manifest to the touch; a hidden pulse escapes the sense of touch; and an intermediate pulse is felt to a moderate extent.

A long pulse indicates a plenitude of spirits filling the artery lengthwise: a short pulse, a deficiency of spirits. A long pulse signifies hotness; a short one, coldness; an intermediate one, temperateness. A broad pulse indicates moistness; a narrow one, dryness; an intermediate one, temperancy between the two. An open pulse indicates strong virtue and a healthy organ; a hidden pulse, weak virtue and an

---

to Giles of Corbeil (at the end of the twelfth century) as well as to this author. Creutz believes that the present work is substantively as well as textually quite different from Alphanus' original.

5. The story is that on hearing the Jew's prognosis, St. Basil prayed God for a slight delay, and in fact survived for several hours longer than had been predicted; the Jewish physician was so struck by the power of the Christian faith that he allowed St. Basil to baptize him before he died.

6. This definition follows very closely that of Philaretus.

7. These distinctions between types of motions are surely a reflection of natural-philosophical discussions.

8. The idea suggested here, that "spiritus" is something more than breath but less than soul, is pursued by Boyd H. Hill, Jr., "The Grain and the Spirit in Mediaeval Anatomy," Speculum, XL (1965), 63–73.

9. The first consideration is in structure quite typical of the others, although it is longer than most, and it is the only one translated here. The contents of the others can be summarized quite briefly. The second consideration examines the significance of the condition of the artery—whether hard or soft, full or empty, hot or cold. The third begins with a definition of diastole and systole as the respective dilation and contraction of the vessel (which is of course not what we would emphasize; we recognize the arteries' "dilation" as a consequence of the heart's systole). Only the dilation can be directly perceived, however. Systole must be determined by the lag between two dilations, and this consideration concludes by explaining the implications of systoles of various lengths. The fourth discusses the meaning of pulses which grow either stronger or weaker (the rising or failing pulses referred to in the final portion of the translation), and the last treats certain specific irregular pulses—the goat-like, the hammer-like, and so on.

unhealthy organ; and an intermediate pulse, the mean between the two. A long, broad, open pulse is called great; a short, narrow, hidden pulse is called small; and an in-between pulse is created out of the means between these. A pulse can be long either naturally or unnaturally: naturally long means a lavish nature; unnaturally long means abundance and intensity of heat. A naturally short pulse indicates weakness and inadequacy for great undertakings; an unnaturally short one indicates weakness of heat and of the heart. A broad pulse and abundancy of moistness can occur either naturally or unnaturally: if naturally, it indicates a pliant nature and weakness in action; if unnaturally, it indicates dissolution of the humors and members. A naturally narrow pulse indicates constancy, an unnaturally narrow one aridity of the members and material solidity. An open pulse, natural or unnatural, indicates prodigality, effusiveness, and officiousness; also strength of heat and ease of arterial dilation. An unnaturally open pulse reveals a plenitude of vapors and excess heat, as in the drunk or feverish. A hidden pulse can also be either natural or unnatural: natural, it indicates taciturnity, secretiveness, and scanty heat or vapor; unnatural, it indicates a weakness of natural heat and the natural virtues. A long pulse can be either long and broad or long and narrow; similarly, a short pulse can be short and broad or short and narrow. Long indicates hotness, broad moistness; therefore long and broad shows that it results from a hot and moist humor, namely blood. Long and narrow indicates hotness and dryness and thus derives from a hot and dry humor, namely choler. Short and broad indicates the frigidity of earth and moistness and is thus derived from cold and moist [humor], namely phlegm. Short and narrow indicates coldness and dryness and derives from a cold, dry humor, namely melancholy. An unnaturally long and broad pulse thus indicates the excessive domination, qualitatively or quantitatively, of blood. In quantitative excess, the contraction of the pulse to expel waste vapors from the heart is greater than its dilation to attract cold air into the heart; in qualitative excess, however, the dilation of the pulse to attract the frigid air is greater than its contraction to emit the vapors. Thus, when blood is in excess qualitatively and quantitatively, the pulse is unnaturally long and broad. When it produces fever, the pulse is hot and biting; when it does not, the pulse is neither distempered in hotness nor biting. [A long, narrow pulse indicates] excess of choler, quantitative or qualitative. If the excess is quantitative, the pulse is not biting and contraction is greater than dilation; if it exceeds qualitatively, the pulse is biting and dilation is greater than constriction. A short, broad pulse indicates excess of phlegm, either qualitatively or quantitatively. If qualitatively, the pulse is biting and dilation greater than contraction. If quantitatively, the pulse is not biting and contraction is greater than dilation. A short, narrow pulse, which indicates melancholy, indicates qualitative excess when it is biting and dilation is greater; if it is not biting but slow, and the contraction is greater, it indicates quantitative excess.

A second variety of pulse derived from the first consideration, namely the motion of the artery, is taken from its quickness, slowness, or intermediate speed. A rapid pulse is a dilation that ends more quickly than it begins; a slow one is a dilation that ends more slowly than it begins; and an intermediate one is midway between the two in rapidity. A naturally rapid pulse indicates capacity for resolution and sharpness of dissolving heat, as in a person with quick breathing or quick speech. An unnaturally rapid pulse indicates overabundance, while a naturally rapid pulse shows the man to be quick in all the works he undertakes.

A third variety of pulse is taken from its strength or weakness. A naturally strong, great, rapid pulse indicates a strong man; a weak pulse indicates failure of strength.

. . . . . .

ON FORETELLING LIFE AND DEATH BY THE PULSE

The hour of death can be foretold by the failing pulse on a critical day, and the hour of convalescence determined by a rising one. Suppose, for example, that a patient has a failing pulse; whether he will die or not from this sickness is determined thus. You must reckon from the first failure of his pulse to the second, [finding] the time of day when it happens and the number of beats between the two—say you find thirty beats between the first and second failures at the third hour of the day. But since it is improper for you to stay there in continual calculation, come back the next day at the same hour and count the beats of his pulse again. The first day you counted thirty beats between one failure of the pulse and the next; if now the failure comes on the fifteenth beat, so that you have fifteen strong beats where before you had thirty, it is a sign that when the same number of hours have passed once more, having lost [a further] fifteen beats, the patient will die at that moment.

You should treat a rising pulse in the same way. Suppose someone has a faint pulse, almost failing: you should reckon from the first slightly stronger pulse to the second, determining both the time of day and the number of beats between the two strong pulses. Suppose between the first and second slightly stronger pulses you find thirty beats to occur, so that the thirtieth beat is stronger than the other beats—this at the third hour of the day; and suppose the next day at the same hour you count his pulse again. The first day you found thirty beats between the first and second stronger pulses; now on the fifteenth beat you find a stronger one, so that the patient has picked up fifteen beats in those twenty-four hours. This is a sign that in the next twenty-four hours the patient will not lose but will gain another fifteen beats, and that at that moment the crisis will pass and the patient be relieved and improved.

If you cannot return to the patient at the same hour when you first counted his pulse, and must come back earlier, or later, then, taking his pulse, determine precisely the number of beats, and the number of hours that have passed, and carefully calculate according to the above procedure, taking into account the large or small time difference; you will then be able to make marvelous predictions.

100 INTERPRETATION OF THE URINE

1. Giles of Corbeil (d. between 1220 and 1224)

Translated and annotated by Michael McVaugh

ON URINES

[PART I]

Urine may be so called because it is unified in the kidneys *(fit renibus una)*, or from the Greek *urith*, which means demonstration, or else because it corrodes, dries, and burns *(urit)* whatever it touches. As a clear humor separates from serous milk, so the fluid urine leaves the bulk of the blood. Urine is the serum of blood, the subtle residue of the humors, which are created by the directive force ruling the second digestion, formed by the filtration of aliment, when the pure part is separated from the impure.

The physician who wishes to be considered an expert judge of urine must consider with care the following things: what sort, what, what is in it, how much, how often, where, when; age, nature, sex, exercise, anger, diet, anxiety, hunger, movement, baths, food, ointment, drink. But the first four, which are particularly important, form the best basis of judgment. Health or illness, strength or debility, deficiency, excess, or balance, are determined with certainty by examination in this way.

Urine can be of twenty colors, which you can learn from the descriptions below. . . .

A large quantity of urine, darkened throughout by a black cloudiness, and muddied with sediment, if produced on a critical day [of an illness] and accompanied by poor hearing and insomnia, portends a flux of blood from the nose; depending on whether the other signs are ominous or favorable, the patient will die or will recover. . . .

If the urine is livid, the lividity is partial or total. If total, it means the mortification of a member or of its humors. Livid near the surface, it suggests various things: a mild form of *hemitriteus* fever;[2] falling sickness; ascites; synochal[3] fever; the rupture of a vein; catarrh; strangury; an ailment of the womb; a flux; a defect of the lungs; pain in the joints; consumptive phthisis; the extinction of [natural] heat. These are the causes of lividity—interpret them according to [other] symptoms.

A very limited quantity of urine, passed with difficulty, livid and oily, foreshadows death; livid, passed frequently but in scanty quantity, points to strangury; lividity coupled with minute, distinct particles consistently indicates respiratory trouble; a grainy lividity foretells affliction of the joints and rheumatism. If the womb presses upon the spine or diaphragm, it gives the surface of the urine a livid tinge.

Thin urine, white in color, is a sign of spleen,

1. Translated from the text published in Camille Vieillard, *L'urologie et les médicins urologues* (Paris: Rudeval, 1903), pp. 272–300, where it is accompanied by a French translation. It is a poem of perhaps 350 hexameter lines, something over half of which are translated here. The work was written about 1200 and continued to be widely used during the thirteenth and fourteenth centuries despite its obvious Salernitan (rather than Arabist) derivation.
2. A particularly severe type of tertian fever (see n. 4).
3. *Synochus* (see n. 5).

dropsy, intoxication, nephritis, delirium, diabetes, rheumatism, black bile, epilepsy, dizziness, chill of the liver, or (with a bilious fever) death; in the old it is a sign of debility or childishness; in those suffering in the neck or shoulders, of lipothymia; in women, it is a sign of a number of complaints of the womb; and it also signifies hemorrhoids and condylomata. . . .

Thick urine, whitish, milky, or bluish-white, indicates dropsy, colic, the stone, headache, excess of phlegm, rheum in the members, or a flux; it also portends the disappearance of the factors bringing on apostemes. Milky or bluish-white urine is never found to be thin.

Urine is often composed of separate layers, differing in appearance, so that the middle or upper layer is quite different from the lowest. The upper surface, the middle, and the thick lower level represent the upper, middle, or lower part [of the body], each region standing for its counterpart. This general rule applies to all others.

Pale or weak yellow urine, of fatty substance, portends the periodic chill of a phlegmatic fever. Pale thin urine announces adust choler; pale and less thin, a fever from bitter phlegm. Pale urine moderately thickened is the regular sign of a phlegmatic complexion.

. . . .

Clear, red urine can indicate a double tertian fever, an overheated liver, or a quartan fever.[4] Several things—passage of time, long or short, cloudiness, season, health, age—determine the illness, its nature and place; all these are to be considered in making your judgment, so that it may have a firm foundation.

Turbid, slightly thick urine, red in color, is a certain sign of a continued tertian fever. If green at the surface, or marked with a livid cloudiness, it portends a mild [fever] or severe pleurisy. Thick and red, livid at the surface, means *synochum;* without the paleness, *synacham.*[5] If it is thin rather than thick, it indicates a bilious fever; if the reverse, a synochal fever. Subtle and fiery, it indicates a bilious fever. The rubicund color is the more certain indication.

If the urine is wine-colored, it means danger to health when it accompanies a continued fever; it is less to be feared if there is no fever. [It can be produced when] a caustic humor inflames the kidneys and liver; or when the renal vein(?)[6] ruptures; or when, its vessel broken, the menstrual blood passes from a woman's body. Dancing, overmuch coitus, running, and immoderate exercise produce

the same signs in a healthy body. Blue-black urine follows this pattern too.

Urine tinged green indicates jaundice, spasm, severe fever, and finally death.

[PART II]

The color of urine often misleads the physician in his assessment of it; but there is an exact law, a definite rule for judgment, to be found in its contents. Hippocrates, an author knowledgeable about nature, deferred other considerations and drew the seeds of his true doctrine from these things. Now we will enumerate, in order, the things contained and will explain the significance of each.

[The possible contents are] the circle; bubbles; grit; cloudiness; spume; pus; grease; chyme; blood; sand; hair; bran; lumps; scales; specks; sperm; ash; sediment; and rising vapor.

If there is a thick, watery circle in the urine, the posterior part [of the head] is afflicted with phlegm. Purplish and thick, the front of the head is afflicted with blood; pale and thin, the left side is afflicted with black bile. If the circle is reddish and thin, fiery choler rages on the right side of the head. A leaden color means that the root of the senses is attacked, and that the affliction is passing on to the branches. If it turns from livid to red, the natural state of the brain is improving, and strength restored. If it takes on a greenish color, in fever, the bilious humor is bringing on delirium. If it is tremulous, the spinal members are affected; if black, it means either mortification or adustion, depending on whether it has previously been greenish or livid.

4. "We now explain that the origin of the doctrine [of periodic fevers] was related to the cycles in the life-history of the malarial parasites. Benign tertian fever is set up by the Plasmodium vivax, which has a cycle of 48 hours and which occasions a recurrence of the fever every third day; quartan fever is due to the Plasmodium malariae, with a cycle of 72 hours, the fever recurring every fourth day; . . . besides there are a variety of permutations, whether owing to a repetition of inoculation by the same parasite, or to a mixed infection, etc. Thus two generations of the tertian parasite, maturing on succeeding days, produce a quotidian fever [which here might equally well be called a double tertian fever]; two generations of the quartan parasite cause a fever on two successive days, with one day of freedom from fever [which would have been called a double quartan fever by the medieval physician]." Celsus, *De Medicina,* tr. W. G. Spencer (Cambridge: Harvard University Press, 1940), I, 238–239.

5. *Synochus* and *synocha* are two types of fever: the former is caused by the putrefaction of blood, the latter by an excessive quantity of blood.

6. *Vena chilis;* on this vessel, see Selection 94 and note 5 thereto.

Swollen, aery, persistent bubbles [rising to the top] indicate crudeness of the humors causing the illness; also prolongation of the illness, with nephritis, headache, *rugitum,* vomiting, and diarrhea.

Grit is always the sign of rheumatic ills. If the material remains in the circle, the source of the illness dwells in the brain; if it lies a little below, and does not return after agitating the urine, the humor is affecting a site near the head, with a flux tending toward the weaker members.

. . . .

Pus in the urine indicates ulceration of the bladder, or of the kidneys, if a strong odor is present. A sanious deposit can also derive from the liver, as indicated by pain and impaired functioning of the member.

A greasiness like oil indicates consumption of the entire body if accompanied by fever; if not, of the loins only.

[Crude] humor can be found at the top, middle, or bottom of urine; it indicates ills, respectively, of the upper, intermediate, or lower members. At the bottom, it signifies strangury, nephritis, and tenesmus; in the middle, harm to the nutritive members; and at the top, to the diaphragm.

The urination of blood betrays a failure of the liver, bladder, kidneys, or renal vein(?); you can make a full diagnosis from other signs. If it comes from the bladder, it is clotted, heavy, speckled, inert, and somewhat fetid. Pure [blood] in the urine derives from rupture of the renal vein(?) and is also indicated by severe pain in the seventh vertebra. If it derives from the kidneys, the blood appears pure and unclotted, and an accompanying pain in the kidneys confirms its origin. If the flow is from the liver, there is pain on the right side, and its passage relieves the pain.

Sand testifies to stone in the bladder or kidneys—the kidneys when reddish, the bladder when white, the contents taking the hue of their container.

What looks like hair is either a strip of flesh, or humoral material dried out by heat: from this you will recognize an affliction of the kidneys or of the whole body.

Thin scales and bran in the urine indicate ulceration of the bladder; and if accompanied by fever, indicate a general bodily debilitation.

Lumps are a certain indication that the third form of hectic fever has deeply invaded the body.

Gout is revealed by tiny white flecks. In the case of a woman, you will not be mistaken if you diagnose either pregnancy or an undesirable debility.

Dryness of the womb, dormancy of desire, ces-sation of the menses, a change of appetite, vomiting, pressure on the genital organ, swollen breasts, disturbance of the stomach by honey-water, all these are signs of conception. A shiny face, lightness of the members, congelation of a thick milk, and swelling of the belly on the right side, indicate the conception of a boy; the contrary signs indicate the feminine sex.

Urine clearly containing sperm indicates recent coitus, damage to the genitalia from venereal ferment, spermatorrhea, gonorrhea, or paralysis of the genital organ.

Ashy bodies sinking to the bottom indicate that the lower members are weakened by disease. They often arise from irritated hemorrhoids, from tumor of the spleen, or when the menstrual humor does not appear at its regular periods.

Color, consistency, duration, form, and place are the determining characteristics of urinary sediment. The color should be white, the consistency continuous, the form conical, the place the bottom of the vessel, and the duration marked. If the sediment is all these—white, continuous, lasting, conical—digestion will be good, virtue strong, and natural action will flourish with a triple activity: it whitens, it cleanses, and it condenses, concentrates, and unites. [This shows that] the natural heat is dissipating ventosity and absorbing vapor; that the natural force is sustaining what is beneficial and resisting what is not; that nature's actions are not interrupted or stopped. The sure indication of health derives from these signs. Let the fever have a favorable crisis, and the illness will be of short duration.

. . . .

Now pause, Muse, with my work complete; brake your wheels, check your course; hold back the flow of Musandinus' teachings. Spread no further the treasures of Salernus, the achievements of Urso, the sayings of Maurus.[7] Quietly withdraw to yourself, for fear of the envious—lest an opposing sect be moved to challenge you. Spiteful and mordant, violent, clamorous, and vain, feeding on sterile weeds, the Montpellier physician swells himself with undigested fodder and makes a fool of himself. I fear lest he lash out with insults, lest he cloud and stain the splendor of this doctrine while secretly embracing it. Do not reveal your pure self to that corrupt horde; let it see only the hem of your teachings, a hem which it is not worthy to touch.

---

7. These men had all been Salernitan masters—even Salernus, called "equivocus" by Giles here and elsewhere because he bore the same name as the school.

2. Arnald of Villanova (ca. 1240–1311)

Translated by Henry E. Sigerist[8]

Annotated by Michael McVaugh

ON THE PRECAUTIONS THAT PHYSICIANS MUST OBSERVE

We must consider the precautions with regard to urines, by which we can protect ourselves against people who wish to deceive us. The very first shall consist in finding out whether the urine be of man or of another animal or another fluid; and if it is human urine it is diagnosed in four ways.

The second precaution is with regard to the individual who brings the urine. You must look at him sharply and keep your eyes straight on him or on his face; and if he wishes to deceive you he will start laughing or the color of his face will change, and then you must curse him forever and in all eternity.

The third precaution is also with regard to the individual who brings the urine, whether man or woman, for you must see whether he or she is pale, and after you have ascertained that this is the individual's urine, say to him: "Verily, this urine resembles you," and talk about the pallor, because immediately you will hear all about his illness. It commonly happens with poor people and those of moderate means that they go to the doctor when they are afflicted very seriously.

The fourth precaution is with regard to sex. An old woman wants to have your opinion. You inquire whose urine it is, and the old woman will say to you: "Don't you know it?" Then look at her in a certain way from the corner of your eye, and ask: "What relation is it of yours?" And if she is not too crooked, she will say that the patient is a male or female relation, or something from which you can distinguish the sex. Should she say: "We are not related," then ask what the patient used to do when he was in good health, and from the patient's doing you can recognize or deduce the sex.

The fifth precaution is that you must ask if the patient is old. If the messenger says yes, you must say that he greatly suffers from the stomach, and that he spits a lot, and in the morning more than at any other time, for old people have by nature a cold stomach.

The sixth precaution: whether this illness has lasted for a long time or not. If the messenger says that it has, you must say that the patient is altogether irritable and that one can help him, or some such talk. If he says no, you must say that the patient is altogether oppressed because in the beginnings of diseases there is much matter that oppresses the organ.

There is a seventh precaution, and it is a very general one; you may not find out anything about the case, then say that he has an obstruction in the liver. He may say: "No, sir, on the contrary he has pains in the head, or in the legs or in other organs." You must say that this comes from the liver or from the stomach; and particularly use the word, obstruction, because they do not understand what it means, and it helps greatly that a term is not understood by the people.

The eighth precaution is with regard to conception. An old woman consults you because the patient cannot become pregnant. Perhaps you do not know the cause but say that she cannot hold her husband's sperm which she could have done very well if she had been well disposed.

The ninth precaution is with regard to a woman, whether she is old or young; and this you shall find out from what is told you. Should she be very old, say that she has all the evils that old women have, and also that she has many superfluities in the womb. Should she, however, be young say that she suffers from the stomach, and whenever she has a pain further down, say that it comes from the womb or the kidneys; and whenever she has it in the anterior part of the head, then it comes from the stomach; and whenever on the left side, then it comes from the spleen; whenever to the right, then it comes from the liver; and when it is worse and almost impedes her eyesight, say that she has pains or feels a heaviness in the legs, particularly when she exerts herself.

8. Reprinted by permission of the editor of the *Northwestern University Medical School Magazine*, from Henry E. Sigerist, "Bedside Manners in the Middle Ages: The Treatise *De Cautelis Medicorum* Attributed to Arnald of Villanova," *Quarterly Bulletin, Northwestern University Medical School*, XX (1946), 135–143.

The treatise entitled *De Cautelis Medicorum* included in Arnald of Villanova's *Opera* really consists of four distinguishable sections, only the first of which (the text given here) can have been written as late as his lifetime [see Selection 98, n. 1.—*Ed.*]. Some of Arnald's admirers have been a little shocked at the ethics displayed here and have inclined to doubt his authorship on those grounds, but in fact the work is perfectly in accord with medieval medical ethics (as well as with what we know of Arnald's character). It throws an interesting light on the limits of a physician's faith in urological theory, such as Giles's in the first section of this selection.

The tenth precaution: You must keep yourself very busy spitting or blowing your nose and if the old woman pesters you with the urine say quite casually: "What concern is this of yours?" or "Why do you pester me so much?" If she says: "Yes, it concerns me," then you shall know the sex. If she says, no, ask as has been explained under the fourth precaution.

The eleventh precaution is taken with regard to white or yellow wine. If you have any doubts in this respect, be cautious and put the lid of the urinal down and pour out a little of the content in such a way that the wine in being poured out touches your finger. Then you must give her the urinal and act as if you were going to blow your nose whereby you put the finger that has been dipped in, on or next to your nose; then you will smell the odor of wine, whereupon you must take the urinal again and say to her: "Get away and be ashamed of yourself!"

The twelfth precaution is taken with respect to fluid made from figs and also nettles. Although you could recognize this under the first precaution, yet you will clearly see that the residue extends in the form of a circle touching the urinal and does not make a rotundity or pine cone like a true sediment.

The thirteenth precaution: whenever the old woman asks what disease the patient has, you must say: "You would not understand me if I told you, and it would be better for you to ask what he should do." And then she will see that you have judgment in the matter and will keep quiet. But perhaps she will say: "Sir, he is very hot; therefore he seems to have a fever."—"Thus it seems to you and other lay people who do not know how to distinguish between fever and other diseases."

The fourteenth precaution: When you have been called to a patient, feel the pulse before you examine the urine and make them talk so that the condition of the animal virtue becomes apparent to you. After having recognized these factors you will be able to evaluate the urine better and with more certainty and you may proceed thus.

The fifteenth precaution is: should the patient be in a bad condition so that you think that he may die the following day, do not go to him but send your servant to bring you the urine, or tell them to bring it to-morrow in the early morning because you wish that they prepare for the meal and that after you have seen the urine you will tell what they shall administer. And so from the report of the person who brings the urine you will be able to form an opinion about the patient, whether he is in good or bad condition.

The sixteenth precaution is that when you come to a patient you should always do something new lest they say that you cannot do anything without the books.

The seventeenth precaution is that if by hard luck you come to the home of the patient and find him dead and somebody perhaps says: "Sir, what have you come for?" You shall say that you have not come for that, and say that you well knew that he was going to die that night but that you wanted to know at what hour he had died.

The eighteenth precaution is that if you have a competitor whom you believe to be a shameless crook, be careful when you go to the house of the patient; perhaps he will stir up the urine for you and you will not be able to form a certain judgment from it.

The nineteenth precaution is the following: if two urines of the same patient are presented to you and you wish to know which was the first, ask at what time of the night he got up, for if he did at dawn or after digestion had taken place that urine which is more digested and red will be the first, if it has sediment. If, however, he got up before midnight or around that time, you may judge that the less digested and less red urine is the first.

No other deception can occur outside of these, but these points must be kept in mind, and you must be cautious because the physician is greatly honored if he knows how to be cautious, for he is asked questions many times.

101 INTERPRETATION OF PARTICULAR SYMPTOMS

1. Gilbert the Englishman (fl. 1250): The Symptoms of Leprosy

Translated and annotated by Michael McVaugh[1]

BOOK VII

The General Symptoms of Leprosy.

It is important to recognize the disease of leprosy,

1. Translated from *Compendium medicine Gilberti Anglici* (Lyons, 1510), fols. 339–340.

 Gilbert's discussion of leprosy has been called the first correct Western description of the disease, although individual components can be found in earlier Latin

together with its antecedent and conjoint causes, from its proper symptoms; first in general, then after analysis according to its species, for a clearer understanding.

One symptom is a permanent loss of sensation, coming from within, particularly of the last digits of both the hands and feet, namely the smallest digit and the one next to it; and of the muscle, from the little finger to the elbow or even the shoulder, and from the toe to the knee and sometimes above. A coldness in these places named above is another sign. . . . This coldness sometimes occurs without any clear external cause and is in some cases permanent. Should there be an external cause of cold, these places are easily chilled; from such causes there may be produced a brief numbness and prickling in the forehead, palate, tongue, eyelids, and brows—first as if from ants, then as if from needles, and finally as if from large spines. Prickling is an equivocal symptom, though, and may mean leprosy or paralysis. Leprosy is in the muscles and flesh and the external parts; paralysis is in the nerves and is accompanied by debility of the nerves. Lucidity of the skin is also a symptom, seen because the natural folds of the skin are not present; it is instead stretched into a similitude to very thin polished leather. There is consumption of the muscles, leaving empty space, but this is equivocal and may also indicate a wasting disease. Similarly, distortion of the joints of the foot and hands, and of the mouth and nose [is a symptom]; this is preceded by a tickling sensation, as if some living thing were fluttering about within the body, the thorax, the arms, or the lips. A motion can be felt there which is even apparent to the sight; it sometimes affects the eye and distorts it. This is a very sure symptom. There will be fetidity of the breath, the sweat, and the skin, although this indication can be erroneous. The site [of the disease] will lose its hair, and the hair that grows back will be very fine, so fine that it cannot be seen unless looked at against the sun. Sometimes none grows back at all. When the eyebrows and lashes lose their hair, it is the worst of all signs. Hoarseness and obstruction of the nostrils are also general signs, not deriving from any other cause. When the affected parts are washed, water will run off the skin, just as if it had been oiled; this is a very bad sign. So too if the place is rubbed briskly and the water disappears. The corners of the eyes grow round and brilliant. The skin rises into little pimples, like those that appear on the skin of a plucked goose, even if there has been no touch or

chilliness to cause this. The blood drawn off in phlebotomy is greasy and contains sand. . . . Tumors accompany the loss of hair from the eyebrows. Lepers search for sexual pleasure more than usual and more than they should; they are ardent in the act, yet find themselves weaker than usual. Their skin is tormented by a persistent itching; they suffer inordinately now from heat and now from cold. They do not often come down with fever nor suffer with a quartan fever. When they do, it comes on them only once or twice; or if it comes on more often, it indicates the resolution and cure of the leprous matter. . . . They will suddenly feel a coldness, as if cold water were passing between their skin and their flesh or over their skin; it sometimes seems to them as though drops of rain were striking them on the face or on some other part of the body. They become enraged more easily than usual. Their blood, washed, is lumpy and has a fetid odor. Their eyes become distorted, rimmed with red, striking horror into those who see them. Grains will be found under their tongues, as in leprous swine. Also, if blood is rubbed in the palm of the hand and squeaks or is too greasy, or if the blood passing into clear still water remains on the surface, it is a sign of leprosy. . . . If urine poured over blood mixes easily with it, it is a bad sign. If vinegar poured over blood bubbles and does not mix with it, as it would mix with something dry, it is a sign of corruption; so is fetidity of the blood. The urine of lepers is thin of substance, sometimes containing hairs and sand.

The Symptoms Proper to Each Kind of Leprosy.

Now that we have described the symptoms [of leprosy] in general, we will discuss them in relation to their antecedent causes, namely the four humors. The leprosy caused by adust melancholic blood, or blood infected with adust melancholy, is called *elephantia;* that caused by adust choleric blood, or choler scorching and infecting blood, is called *leonina;* that caused by adust phlegmatic blood, or blood infected with adust phlegm, is called *tyria;* and that caused by blood adust and corrupt in its own right, or corrupted by something external, is called *allopicia.* These varieties are rarely found uncombined; indeed, we frequently find two or

sources. While it is often said that the medieval physician interpreted many different skin disorders as leprosy, the text here clearly identifies a number of the symptoms of true leprosy: muscular wasting, thickening of the mucous membranes of nose and throat, loss of sensation, and so on; of course, it includes wholly imaginary ones as well.

three or four forms occurring together. The combination is to be interpreted according to the symptoms of the individual varieties and a compound medicine administered according to the combination of symptoms.

Let us speak first of *elephantia,* which takes its name from the elephant, not without reason: for just as the elephant surpasses other animals in size and strength and ugliness, so this variety [of leprosy] is greater and stronger than the others, as regards both its cause and its cure. Likewise, as the elephant is a spotted *(maculosum)* animal,

so in *elephantia*—but this is common to all forms of leprosy: spots appear at the beginning and quickly fade away. This [variety] also surpasses the others in harmfulness and ugliness. It is inevitably incurable but under treatment may be diminished or relieved. It is brought on by melancholic foods, such as stale goat's flesh, venison, or beef; beans; strong or spoiled wine; garlic; spoiled food or drink; intercourse after infection *(post leprosum)*; speaking with lepers or seeing them or living with them. . . .[2]

2. Jordan of Turre(?) (fl. 1313–1335): The Symptoms of Lepers

Translated and annotated by Michael McVaugh[3]

Lepers can be recognized by five signs: by the urine, by the pulse, by the blood, by the voice, and by the different members. Whoever wants to determine whether someone has leprosy should first make him sing: If his voice is harsh, it may be a sign of leprosy, but if it is clear, it is a good sign. (Proceed as follows: take a tablet and write the good signs on one side and the bad signs on the other, and you will not become confused.)

It is also possible to judge by the urine, and this in four ways. First, the urine of lepers must be white with a certain transparency. Moreover, it must be clear and thin. As for the contents, they will be clotted and should look like flour or ground bran. Finally, if the urinal is shaken, it should make a sound; the reason is that while with fevered patients there would be no noise, because of the oiliness removed from the body, with lepers there will be, thanks to the earthiness and dryness of the matters contained in it.

It is also possible to recognize lepers by the pulse, which should be weak. Avicenna gives the reason for this, that it has little force because of the resistance of the artery, which is almost entirely dried up.

Similarly, they can be recognized by the blood. You should bleed the patient from a vein in any part of the body so that the blood is collected in a clean vessel; leave it until a residue is formed, which should be put into a linen cloth and shaken in fresh water; squeeze it gently until the water is no longer appreciably tinted. Then take what is left in the cloth after the squeezing, and if you see brilliant white corpuscles there, looking like millet or panicum, it is a sign of leprosy. Or put a large lump of salt in the liquid extracted from the blood;

if it melts or dissolves, this is a good sign, but if not, if it remains whole, it is a sign of leprosy. The reason for this is that in leprous blood there is no good warm moistness able to dissolve it, only a thick earthiness by which it cannot be dissolved. Or pour strong vinegar onto the blood; if the vinegar foams, it is a sign of leprosy, just as when you pour it on the ground it foams, because of its dryness. Finally, you can pour urine onto the blood; if it sinks and mixes, it is a sign of leprosy, and if not, not. The reason for using urine and not some other liquid is that urine is a very subtle, penetrating substance and also has a greater suitability for use with blood, being its filtrate.

It is also possible to recognize lepers by examining the different members. First, because lepers have thin, fine hairs, so that they thicken at the

2. Gilbert goes on to speak of the causes and symptoms of each of the four varieties of leprosy he has just mentioned before turning to the problem of treatment.

3. Translated from *Opera Arnaldi de Villanova* (Venice, 1505), fol. 248 recto–verso. In manuscript this work is widely attributed to Jordanus de Turre, a master at Montpellier in the 1320's, and there are internal reasons for doubting Arnald's authorship. At any rate, it surely represents the teaching of Montpellier in the early fourteenth century.

It is interesting to compare this text with Gilbert's, written a hundred years before. The Arabic authorities now being cited have not changed the description of the disease; all the characteristic features singled out by Jordan can be found in the *Compendium*. But the two treatments clearly reflect different purposes. The later text is intended as a clinical manual rather than as a medical encyclopedia. It does not bother to go through the repetitive description of each of the supposed varieties of leprosy and emphasizes instead the details of the procedure a physician should follow in evaluating a patient's condition.

roots; moreover, their hairs are pale and grey. This is due to the fact that the material from which the hairs are formed is lacking and cannot be exhaled through the pores, and because the pores are [almost] closed. It is advisable to examine the hairs in sunlight, to see if they are thin and straight, like pigs' bristles.

They are also recognized by the skin of the head, which is lumpy, so that one area is higher than the next; and by the lack of hair on the eyebrows, since lepers have no hair there, particularly at the corners; their brows have a curvature, seeming almost round and spherical. They are recognized, too, by the roundness of the eyeballs, which seem to be starting from their sockets, so that a leper's face is horrible to see; its natural expression being distorted, it is a terrible sight. These are the most obvious signs.

But it can also be recognized by a wound in the nostrils, which can be examined in the following manner. Cut a small wooden wand and fork it like tongs and introduce it into the nose, expanding it; then examine the interior with a lighted candle. If you see within an ulceration or excoriation in the deepest part of the nose, it is a sign of leprosy itself; this is a sign which is known not to all but to the wise only. Or it can be recognized by the bridge of the nose, where there should be a depression like a thread stretched lengthwise along it; for the cartilage which joins the two parts is eaten away, leaving a furrow.

They can also be recognized by the veins around the eyes and by the veins of the chest, which are very red, since the veins and arteries are bleached out by desiccation, while the blood is red, so that their red color is seen; for contraries stand out when juxtaposed. Or they can be recognized by the tongue. Draw out the tongue, using a cloth, and if you see white corpuscles at its root, like grains of millet, it is a sure sign of leprosy.

Having done all these things, you must examine the patient completely naked, to see whether his skin is darkened, and to see if its surface feels rough with a certain smoothness at the same time; if so, it is a sign of leprosy. Then sprinkle cold water on the shoulders; if it is not retained, it is a sign of leprosy. The reason is that the oiliness of the skin is weakened by the excess heat beneath the skin [in healthy people?], and the water will not stay on this oiliness; and so it is a sign.

Or make the patient cover his eyes so that he cannot see, and say, "Look out, I'm going to prick you!" and do not prick him. Then say, "I pricked you on the foot"; and if he agrees, it is a sign of leprosy. Likewise, prick him with a needle, from the little finger and the flesh next to it up to the arm. The reason for doing it in these fingers rather than others is that they are the weakest and therefore the ones which are first lost to the natural state.

These are the universal signs of leprosy, which are not dependent on any of its forms. But any of the latter can be known, [for example,] whether it is leonine, by using the signs listed above with those given by Avicenna in Book I, Fen 2, of the *Canon*: "Cuiuslibet humor, scil. sanguinis, flegmatis, etc." If you find signs of blood, you are dealing with the variety called "alopecia," for there their hair falls and they seem flayed like a fox, whence the name. If you find signs of phlegm, it is the variety called "tyria," whose name comes from the serpent "tyro." If you find signs of choler, it is the leonine variety, which gets its name because the face [of the patient] looks like that of a lion. But if symptoms of melancholy are found, it is the variety called "elephantia," in which the skin seems flayed away.

3. John of Mirfeld (d. 1407): Danger Symptoms

Translated and annotated by Percival Horton-Smith Hartley

and Harold Richard Aldridge[4]

PART I, DISTINCTION 6, CHAPTER 19
CONCERNING THE SIGNS OF EVIL PORTENT
APPEARING IN FEVERISH AND OTHER TYPES OF
PATIENTS

When you visit your patient, you must bear in

4. [Reprinted by permission of Cambridge University Press from Percival Horton-Smith Hartley and Harold Richard Aldridge, *Johannes de Mirfeld of St. Barthol-* *omew's, Smithfield* (Cambridge, England: Cambridge University Press, 1936), pp. 55, 57, 59, 61, 63, 65, 67, 69. Mirfeld's work, of the late fourteenth century, suggests much better than the previous specialized treatises the syncretic and rather shallow nature of many of the texts employed by the average practitioner. Mirfeld was himself a secular priest rather than a doctor and compiled the *Breviarium* "so that poor and unlearned men who do not possess a plenty of books . . . [might find] not a few remedies for very many diseases."—Michael Mc-Vaugh]

mind first of all the matters to which you should give heed, such as paying due regard to his countenance, and opening the mouth and looking at his tongue; moreover, his eyes and nails must be inspected, and consideration given to many other matters connected with him which will be set forth below. Wherefore, indeed, it should be known that of those signs of evil portent which manifest themselves in the eyes and features of the sick, some appear within the three first days of the onset of the disease, and others appear later.

Of such evil symptoms, those which sometimes appear within the first three days are these:[5] the eyes are hollowed, the temples sunken, the forehead dry and tense; the nostrils pinched; the ears cold, with the lobes turned outward and contracted; the colour livid, greenish or black, or something similar. If these symptoms appear at the commencement of the disease, and by reason of the virulence of the malady itself, then death is signified, especially if the stools be very loose and the urine oily: but this is not the case if they appear later. If however, the appearance of these signs in the opening stages is due, not to the severity of the disease itself, but rather to its accidents, such as sleeplessness, or frequency of stools, lack of food, and the like, then evil is not thus portended. Moreover, if all these aforesaid symptoms appear after the first three days, and result from the disease itself, or if they develop a good deal later during its course, then they are not of ill omen; because if they do not make an appearance until after a considerable lapse of time, then this is a matter neither of wonderment nor of concern. It must, however, be understood, that if such symptoms do appear after the commencement of the disease and accompanied by pronounced weakness, death may thus be signified, although not so effectually as when the appearance occurs at the beginning: for it is reasonable to assume that the body of the patient has been wasted away owing to the process of time; and it would, indeed, be a bad sign if such were not the case.

Of those signs of evil import, which appear in the features of the patient, there are some which show themselves later than the outset of the disease, such as on the fourth day or afterwards. They are as follows:[6] The patient cannot bear to gaze upon a lighted candle and he sheds involuntary tears, whilst the eyes appear to squint, and one seems smaller than the other: or the whites of the eyes appear bloodshot, and the veins black, swarthy, or sallow, the eyes inflamed, and the eyeballs protruding or sunken, whilst the whole aspect of the face is unsightly and horrible to look upon. These symptoms, I say, all signify death in most cases, and they are of worse import than those mentioned in the first place; they can also appear at the outset; but this, however, rarely happens.

We may therefore, it would seem, conclude that whenever and wherever, at the outset or later, all these last-mentioned symptoms appear on account of the fury of the disease and the weakness of the constitution, then most assuredly death is betokened. And to the aforesaid we may add some other indications of death; namely, if the whites of the patient's eyes appear during sleep,[7] without his being able to see anything; if he has lost all his sense of taste and is totally unable to discern those things, which he was once accustomed to perceive; and if he can smell nothing; and if the teeth also, as well as the tongue, appear dry. These and similar tokens portend not merely an approaching death, but death within the doors. With such cases therefore do not meddle. Let your prognosis be given, and then retire.

Take note also of the following verses[8] indicative of a fatal termination:

When all these various symptoms in the patient
    do appear
Then you may safely prophesy that death at hand
    is near!
Dusky glows the brow, first the soles of the feet
    wax cold,
The dreadful hour of death is by the unwished
    tears foretold![9]
Downwards sags the chin; then pales to its tip
    the nose,
The belly rigid turns; the left pupil smaller grows.
Watch too for signs in sleep; for you can surely
    tell,
That Death doth strike a grey-beard who sleeps
    too sound and well,

---

5. The following passages are largely taken from a medieval Latin translation of the *Prognostics* of Hippocrates, perhaps that by Gerard of Cremona of the Arabic version of Hunain ibn Isḥāk. For a direct translation from the Greek Text see that by W. H. S. Jones (William Heinemann, 1923, Loeb Classics), *Prognostic* II.

6. Hippocrates, *Prognostic* II.

7. Idem, *Prognostic* II.

8. *School of Salerno*, De Renzi, *Coll. Salernit.* Vol. I, p. 491, ll. 1400 *et seq.*

9. See Hippocrates, *Aphorisms*, VII, lxxxiii: "When in illnesses tears flow voluntarily from the eyes it is a good sign, when involuntarily, a bad sign."

Likewise the youth who night and day gains no
    repose at all,[10]
Or if perchance Priapism holds the sufferer in its
    thrall!
If white the urine be, when tertian, synocha, or
    causus[11] rages,
In pleurisy or phrenzy too, this death, alas,
    presages!
And when the sick be stricken with paralysis or
    dropsy,
With consumption, apoplexy, or with dysentery
    as well,
Then if the urine red appear of patients thus
    afflicted,
Alas, their plight is hopeless; you can merely
    death foretell!

Moreover, if the patient lies prostrate on his
back, it is the worst possible sign, and indicates a
vitality at its lowest ebb.[12] Likewise it is a bad sign
to find the patient lying face downwards, for this
indicates either excessive pain or delirium. And if
he should lie with arms and feet rigid, this is an evil
omen because it points to the presence of spasm.
And if, after a sleep, or at other times, he suddenly
turns himself about in a disorderly fashion, changing for example from the head to the foot of the
bed, this shows delirium and bodily weakness; and
if he lies stretched out with his mouth open and his
feet twisted, this is grave, because it indicates
spasm. And if he is suddenly constrained to rise
up or to stand erect, this is grave, because it is due
to suffocation arising from constriction of the
respiratory passages. And if his hands are continually trying to pull straws or hairs, or the like,
out of the bedclothes or from the wall[13]—especially
if his reason be affected, so that he is unaware of
what he is doing—then this is the worst possible
sign, for it signifies the final consuming and mortification. Again, a drawing up of the lips, hands, or
penis is bad, because this indicates spasm.[14] Also,
since the nails are generated from the vapours
coming forth from the heart, the vigour or decline
of the latter organ may be ascertained most readily
by an examination of the former of these members:
for if the heat in the heart be insufficient, then the
nails grow black, pale, or livid; so that by their
change or contraction they present and show forth
the withdrawal of the heat, and the departure of
life. Wherefore it is said that a black or greenish
colour in the nails, accompanied by other grave
symptoms, and occurring before the signs of maturation, is an evil thing and denotes death.

Again, if the respiration be feeble and shallow, or

if it be laboured and hurried, or accompanied by
sounds within the chest, this is a bad symptom and
points to death. When, however, a black vomit
appears at the beginning of a disease, most assuredly death is signified. All vomiting of whatever description, if it occur after the symptoms of the
maturation of the disease, and which brings allevia-
tion to the patient, is good and praiseworthy,
since it comes as a cleansing agent; therefore any
natural vomiting, which thus alleviates, should not
be checked, unless it be excessive. Vomiting of
blood, however, accompanied by continual fever,
is the worst possible symptom, and is also mortal.
Again, if live worms appear in the stools at the
commencement of a disease, this is an evil symptom,
because it signifies that the corruption of the body
is so great that the worms have been unable to bear
with it, and have taken to flight; if the worms be
dead, then this is a good sign, because it would
seem that their appearance is the result of a cleansing. If, however, the worms arrive dead at the close
of the illness, then this is a grave matter, for they
would seem to have been done to death by the
malignity of the disease: if they be alive in such a
case all is well, for they would seem to burst forth
as a result of the cleansing of the patient. Again,
perspiration supervening in a case of fever before
the signs of maturation,[15] is of evil import; likewise death is portended by a cold sweat which
appears in an acute disease: if it occur in a lengthy

10. These lines would appear to be based on the
Hippocratic *Aphorisms* VII, lxxii: "Both sleep and
sleeplessness, when beyond due measure, constitute disease." See also *Aphorisms* II, iii.

11. "Synocha" and "causus" are continued fevers.

12. See *Prognostic* III, upon which the following
paragraph is based.

13. This description of "Carphology" is taken from
Hippocrates, *Prognostic* IV: "In acute fevers . . . if
they (the arms) move before the face, hunt in the empty
air, pluck nap from the bedclothes, pick up bits, and
snatch chaff from the walls—all these signs are bad, in
fact deadly."

14. See Hippocrates, *Prognostic* IX: "Testicle or
member being drawn up is a sign of pain or death."

15. "Digestio" in the Latin. We are obliged to Mr.
Fulton of the British Museum for the information that
the corresponding word in the Arabic text is "Nudj,"
which has the meaning of "gestation," "maturity," and
"ripeness" (as applied to fruit, etc.), thus proving that
the word "digestio" is a translation of the Greek πέψις.
Mirfeld, following the Galenists, held that fever was
produced by the presence in the body of morbid matter
(the "materia peccans" of the Proemium), the expulsion
of which, on maturity, constituted the crisis. "Digestio"
was the stage during which the fever was ripening, and
was terminated by the crisis.

illness, it signifies that death is afar off.

Jaundice occurring before the seventh day is serious; but if it make an appearance after the seventh day,[16] and the bile be digested on the day of crisis, it is always of good import, especially when it brings relief to the patient. Again, when a blackish abscess appears suddenly on the tip of the nose, accompanied by considerable pain, then this is an evil sign, and is mortal. Again, when black blisters resembling the bitter vetch occur on the fingers of either hand accompanied by vehement pain, then the patient will die on the fourth day. Again,[17] foetor of the breath and mouth in acute diseases is a mortal sign. Again, frequent tossing about in bed, and changing from one posture into another, is an evil sign. Similarly, when the sick man can obtain no rest whilst lying upon either side but always desires to lie upon his back, and, moreover, slips down to the foot of the bed, then this is a mortal sign.

Another mortal sign is the appearance upon the patient's knee of a black pustule, that is, one with a blackish edging. Moreover, it is significant of evil when those actions which normally occasion embarrassment, such as the uncovering of the pudenda, or the audible expulsion of flatus from the bowel by the patient, occur (without perturbing him). Quinsy, again, with very acute fever, is most harmful.

Again, another symptom of death is when the eyes are dull and the eyelids not properly closed. Another sign of ill omen is the uttering by the patient of the names of those no longer living; and if hiccoughs follow, the result will be lamentable. Moreover, if the eye suddenly turns yellowish or black; or if the patient is choked so that he cannot swallow his saliva; then these symptoms are fateful. And when grains the size of a chick-pea appear on the tongue in conjunction with very acute fever, the patient will die at dawn on the following day.

Moreover, according to Avicenna,[18] when a patient feels pains in the region about the hip-joint, and then is affected with marked but painless erythema in that region extending over an area the size of three fingers, then if that spot begins to itch violently, so that the patient longs for the application of soothing oils, he will die on the twenty-fifth day. Again, it is not a good sign if the patient frequently looks at his hands, or claws at the bed-linen with his fingers or his hands. Again,[19] it should be known that when a large and virulent abscess appears on either knee,[20] the patient will die within eight days, especially if he should per-

spire excessively at the commencement of his illness. Again, if there be on any one of the fingers a small black and painful blister similar in appearance to the bitter vetch, then the patient will die on the second day. And the symptom of that will be that he will be very much pained in his body during his illness. Again, when a small blister, dark or pallid in hue, appears on the thumb of the left hand without causing pain to the patient, then it should be known that death will ensue on the sixth day of his illness, and an indication of it is that at the commencement of his illness he went to stool profusely and frequently. Again, when the finger-nails are dark or pallid in colour, and a blood-coloured blister appears on the forehead, then know that the patient will die on the fourth day: and an indication is the occurrence of considerable sneezing and yawning. Again, when blood, rather pallid in colour, flows from the nostrils, and a painless white blister appears upon the right hand, then know that the patient will die on the third day of his illness; and the sign is that he has lost all desire for food. Again, if in any disease, the patient perspires from the base of the thorax upwards, flee from him: if he perspires all over his body it is well. Again, whosoever has fever, whatever you will, if he has spasm in that fever, then know that he will die.

16. Hippocrates, *Aphorisms*, IV, lxiv.

17. At this point commences an extract from the "Signs of Evil Import in Fevers" from Rhazes, *Liber Ad Regem Almansorem*, Lib. X, cap. 21.

18. Much of this treatise on prognosis appears to be based upon haphazard excerpts from the *Canon* of Avicenna, Lib. IV, Fen. 2, Tract. 1, C. 26 onwards.

19. At this point commences a series of extracts from the pseudo-Hippocratic *Capsula Eburnea*, one of the spurious productions assigned to Hippocrates in order to give it currency, but perhaps Byzantine or even Arabic in origin. It derived its name from the account, with which it opens, of the supposed discovery of the original MS. within an ivory box placed in the tomb of Hippocrates. It was printed in the 1497 Latin edition of the works of Rhazes, but appeared frequently in manuscripts during the Middle Ages. We are indebted to Mr. Fulton of the British Museum for the information that in the Arabic text (entitled *'Alāmāt al-ḳaḍāyā*, or *al-Rasā'il al-ḳabrīyah*) it is ascribed to Hippocrates, who, feeling death near, ordered the twenty-five dicta to be set down in writing, placed in an ivory box, and buried with him; and that it was found by the Byzantine emperor who opened the tomb. According to the Arabic text, it was translated into Arabic from the Greek by Hunain ibn Ishāk during the reign of the Caliph Ma'mūn (A.D. 813–833). The Arabic text was translated into Latin by Gerard of Cremona (1114–1187).

20. "Both knees" in Arabic, according to Mr. Fulton.

Again, when there is a blister below the nape of the neck, and another white blister in the lower eyelid of the left eye, know that the patient will die on the eleventh day of his illness. And the indication of this is that at the beginning of his illness it happened that he most ardently desired sweet things.[21] Again, if the patient bends himself so that his hands touch his feet, then on no account can he escape.[22] Again, if the patient suffers from insomnia and somewhat frequently turns towards the door, it signifies that he will pass away. Moreover, in acute diseases if the faeces are seen to be black with blood, or even if without fever they are still black, or there be black vomit, then this is mortal. Again, lividity of the lips is a sign of the loss of strength and natural heat, just as, on the contrary, red lips show good health and internal strength.

Again, cold sweats in a man which drench the face or breast signify that a cure will take place. Again, natural philosophers say that those who are about to die, for three days beforehand, lack natural pupils; and, if these be invisible, the case is certainly hopeless. Again, in acute diseases, uncovering and restlessness of the arms is a mortal sign.

Also, Avicenna advises the physician that, when he visits his patient, he should carry in his hand a twig of vervain and enquire of the patient, "How does this feel to you?" If he shall answer, "It is pleasant to me", then he will recover; but if, on the contrary, he shall say, "It is bad", then he will not escape. Also, the Experimenter[23] states that if, unknown to the patient, Artemisia (mugwort)[24] be placed underneath his head, then, if he shall sleep, he will live; if he do otherwise, he will die.

Again, if the hand of the patient be rubbed with yeast, and then this be offered to a dog, if the dog eats it, then the patient will be restored to health; if not, he will die.

Coldness of the extremities, thirst, heat in the
inmost part,
Should'st thou perceive, retire thyself:
to others take thine Art![25]

Again, carefully anoint all over the sole of the patient's right foot with lard, and then throw the lard to a dog. If the dog eats it without vomiting then the patient will live, but if the dog returns it or makes no attempt to eat it, then the patient will die. Again, mix the patient's urine with the milk of a woman who has brought forth a male child: if the milk and urine mix well together, then the patient will live, but if on the contrary they remain separate, then he will die. Also, if the urine of the patient be poured over a green nettle, and if the nettle remain green on the next day, then the patient will live; if not, he will die very shortly.

Again, if the right eye of a sick man sheds tears, then this is the worst possible sign and is mortal: and in the case of a woman this applies to the left eye. Moreover, if a plaster be made from Musa Aenea[26] and this be applied to the places where the pulses are situated, on a day which is not a critical one, then if the patient shall perspire he can be judged as likely to escape. If otherwise, the labour is in vain.

Likewise take gum ladanum[27] and aloes, grind up the aloes and mix with the ladanum, and then make therewith a candle, and place it to the nostrils of the dying patient; who, when he feels the heat penetrate through to his brain, will open his eyes: then ask him whatever you desire, and he will speak to you, whilst he retains consciousness. But of this I have not made proof.

Again, if it is desirable that it should be known whether a person will quickly become feverish, let phlebotomy be performed on him in May or April; allow the blood to coagulate in a dish, and then sprinkle roasted salt upon it: if, after an hour, the salt gives forth an unpleasant odour, the man is a feverish subject; otherwise not. Again, if your patient becomes thin, take a little of his blood and pour it into a cup full of cold water; if it all falls to the bottom in one mass, he will not die during the same year; but if it dissolves and appears, as it were, a drop of oil floating on the surface, then the patient is in danger.

Moreover, if there is any doubt as to whether a person is or is not dead, apply lightly roasted onion to his nostrils, and if he be alive, he will immediately scratch his nose.

Furthermore, if the first vowel of the name of a patient, who has lost an eye or is otherwise mutilated, be A or O, then he will suffer in his right side, and if it be E, I, or U, then in the left. . . .

21. End of *Capsula Eburnea* extracts.
22. The *"emprosthotonos"* of tetanus described by Aretaeus.
23. I.e. Rhazes.
24. Marginal note in Lambeth MS. "Mother of Herbs."
25. *School of Salerno,* De Renzi, ll. 2098, 2099.
26. Musa Aenea is a "compound medicine" and a sedative, containing among other ingredients Opium and Henbane. The prescription is given in full in the *Breviarium,* Part XIII, Dist. 2.
27. Resinous juice obtained from the lada shrub, *Cistus creticus.*

A Method of Medical Practice

Ugo Benzi (Hugh of Siena) (1376–1439)
Translated by Dean P. Lockwood[1]
Annotated by Michael McVaugh

[THE CASE]

The distinguished and noble gentleman, Messer Mariscoto of Nullano (?), about sixty years of age, of a complexion naturally tending to hot and moist, suffers from gout in his whole body and likewise at times from arthritis, whose matter is mixed, although at present, as usual, phlegm predominates. And at intervals he is so free from discomfort that he can easily walk or ride horseback, although in some of his joints a certain degree of stiffness persists. And I will give him here a brief regimen, supported by medical treatment, to the end that his health be preserved as well as possible.

[REGIMEN]

[Habitation or "Air"]

First, then, let him occupy a good chamber between two roof-terraces or balconies *(solaria)*. The chamber should be warmed in cold weather by a fire of dry wood. And in general he should be amply protected against the cold by clothes, shoes, and other appropriate means

[Sleep]

And let him sleep seven or eight hours per night, and when that is not sufficient, let him sleep in the morning before tierce, with his head well elevated and his body well covered.

[Regulation of the Bowels]

And let him be sure that he has a movement every day, and if nature does not respond, let him use a clyster.

[Exercise]

And for his exercise let him walk or ride horseback every day before eating, but not if it gives him much pain; and after eating let him refrain from effort.

[Emotions]

And let him forego sexual intercourse as much as possible.

[Food and Drink]

In regard to his food, he should always incline to moderation, making it a rule to leave the table before his appetite is completely satisfied. Similarly in regard to drink. And let him secure a vessel of seventeen ounces and fill it half full of water of honey, and fill the remaining half with wine, and let him not consume more than the contents of this vessel at lunch or at dinner. His wine should be red, clear, and of moderate strength.

For food let him eat the meat of chickens, partridges, pheasants, larks, and other small birds; also squab, doves, and quail; and similarly kid, veal, and the flesh of young sucklings and of castrated animals; also young rabbits, and young deer *(capriolorum iuuenum)*. He should refrain from waterfowl, domestic pork, lamb, and beef; also from dried meats and all other forms of the flesh of the above-mentioned animals.

1. Reprinted by permission of the University of Chicago Press from Dean Putnam Lockwood, *Ugo Benzi: Medieval Philosopher and Physician 1376–1439* (Chicago: University of Chicago Press, 1951), pp. 54–56.

The *consilium* is exceptional as a form of medical literature that shows something of the medieval physician's actual methods of practice: It was "a physician's written report, in traditional form, prescribing care and treatment for an individual patient on a specific occasion" (Lockwood, p. 47). The more famous academic physicians often had their *consilia* collected and published (in manuscript); we have about a hundred by Ugo (Hugh of Siena), who taught medicine and natural philosophy at half a dozen Italian universities in the early fifteenth century. The form is supposed to have originated with Taddeo Alderotti of Bologna something over a hundred years before.

The "traditional form" of the *consilium* is brought out here by the translator's bracketed headings. After a description of the patient's condition, Ugo prescribes treatment according to the two standard modes of activity available to the physician, regimen and drugs (the surgeon was master of a third—see Selections 111 and 113). It will be remarked that "regimen" is further analysed into six headings, which prove to be the "six non-naturals" (see Selection 90), external factors that help determine the condition of the body. Of the six, diet is almost always the most minutely detailed, as is the case here.

Let him use bread made of good flour, well cooked and well leavened; and let him avoid cheese. Fresh eggs, however, are good, i.e., the yolks lightly cooked, in the shell or poached (*deperditi in aqua*) or otherwise dished up, or in a tender omelet fried in sweet oil, or prepared in other ways, provided they be not overdone nor mixed with cheese. Farinaceous foods, such as pancakes, rolls (*fermentinos*), and so forth he should forego.

As for vegetables, none are good, but he can make a dish [a salad] of spelt, barley, panic, millet, spinach, blite, borage, bugloss, balm, fennel, anise, parsley, marjoram, and savory herbs, such as sage and thyme, along with bread crumbs and eggs—singly or jointly.

And he should eat but little fish; the less harmful sort are those which are small and scaly, living in clear water and of good odor. Crayfish, however, are not good.

And of fruits the following are suitable: raisins, figs either fresh or dried, almonds, pine-nuts, filberts. And for dessert he can have pears stewed with wine and anise and fennel; but from other moist fruits he had better refrain, such as apples and cherries and peaches and so forth, save that a small quantity, e.g., of melons or plums, may be taken before a meal in summer.

And as for condiments, he must forego vinegar, but he may use verjuice and cinnamon and pepper and good spices. And he shall not use raw herbs.

So much for his regimen.

[MEDICINAL TREATMENT]

Coming now to medicinal treatment, at the middle of February or a little earlier he should start taking the following syrup:

℞ of compound syrup without vinegar, ℥ iiii
　　of oxymel of squills, ℥ ii

of sage-water, parsley-water, fennel-water, and marjoram-water, each, ℥ iii
Strain and flavor with ginger and muscat nut. This will be sufficient for four doses. To be taken hot at dawn.

Item, ℞ of fetid pills and of greater hermodactyl (?), each, ℥ ½.
To be divided into seven doses. To be taken two hours before sunrise and after the syrup.

And on this day he should use great precaution against cold and not eat before the eighteenth hour. And on the following morning let him use a cleansing clyster, and take ℥ ii of theriac with ℥ ii of honey-water seven hours before lunch. And a similar purge should be made at the beginning of October. During the intervening periods he should be purged without supervision at least once a month, taking ℥ i of the following pills at midnight or thereabouts:

℞ of pills of the eightfold antidote, ℥ ½
　　of larch-fungus, ℥ ½
　　of sal-gem, ℈ ½
Make into pills with sage-water, and they are for one dose only, but he may have them prepared in advance for many doses.

Item, in winter once a week he should take ℥ ii of mithridate five hours before eating.

Item, in cold weather he should wear on his feet at night little bags of ground salt.

And often in the morning before breakfast he should wash his legs and feet and knees with the following water:

of sage, camomile, and laurel, each, M ii
of water, lb xx
of salt, lb v
To be boiled over a fire, until the salt dissolves, and used frequently as a hot footbath.

Treatment of Particular Ailments

103 GYNECOLOGY

Trotula (eleventh century)

Translated by Elizabeth Mason-Hohl[1]

Annotated by Michael McVaugh

PROLOGUE

SINCE God, the author of the universe, in the first establishment of the world, distinguished the individual natures of things each according to its

1. Reprinted by permission of Elizabeth Mason-Hohl, M. D., from Trotula of Salerno, *The Diseases of Women*, translated by Elizabeth Mason-Hohl, M.D. (Los Angeles: Ward Ritchie Press, 1940), pp. 1–3, 5–8, 16–25.

The "Trotula" to whom this work is ascribed was

own kind, He differentiated the human race above
the other creatures by means of extraordinary
dignity. To it, beyond the condition of other animals, He gave freedom of reason and of intellect.
Moreover, desiring its generation to subsist perpetually, He created it male and female in different
sexes that by means of their fertile propagation
future offspring may never cease to come forth.
Blending their embraces with a pleasing mixture,
He made the nature of the male hot and dry and
that of the female cold and wet so that the excess
of each other's embrace might be restrained by the
mutual opposition of contrary qualities. The man's
constitution being hot and dry might assuage the
woman's coldness and wetness and on the contrary her nature being cold and wet might soothe
his hot and dry embrace. Likewise that the male
having the stronger quality might pour seed into
the woman as into a field and the woman endowed
with a weaker quality, subject as it were to the
function of the man, might naturally take unto her
bosom the poured out seed. Since then women are
by nature weaker than men it is reasonable that
sicknesses more often abound in them especially
around the organs involved in the work of nature.
Since these organs happen to be in a retired location, women on account of modesty and the
fragility and delicacy of the state of these parts
dare not reveal the difficulties of their sicknesses
to a male doctor. Wherefore I, pitying their mis-
fortunes and at the instigation of a certain matron,
began to study carefully the sicknesses which most
frequently trouble the female sex. Since in women
not so much heat abounds that it suffices to use up
the moistures which daily collect in them, their
weaknesses cannot endure so much exertion as to
be able to put forth that moisture to the outside
air as in the case of men. Nature herself, on account of this deficiency of heat, has assigned for
them a certain specific purgation namely the
menses, commonly called flowers. For just as trees
do not produce fruit without flowers so women
without menses are deprived of the function of
conception. This purgation occurs in women just
as 'pollutio' occurs in men. Nature always, bur-
dened by certain moistures, whether in man or in
woman, strives to lay off its yoke and diminish its
exertion. Now a purgation of this sort usually
befalls women about the 13th or 14th year or a
little earlier or later according to whether heat or
cold abounds in them more. It lasts up to about
the 50th year if she is lean; sometimes up to the
60th or 65th if she is moist; in the moderately fat

up to about the 45th. If such purgations have been
of normal time and regularity, Nature sufficiently
unloads women of superfluous moisture. If the
menstruation has taken place too copiously various
sicknesses arise from it. The appetite for food and
drink is diminished, sometimes vomiting occurs,
and often they have an appetite for earth, coals,
chalk, and the like. At times from the same cause
they feel pain around the neck, the back, and
around the head. There may be acute fever, sharp
pains in the heart, dropsy, and dysentery. These
conditions appear either because the menstrual
periods are missing for a long time or because they
do not have them at all. For this latter reason not
only dropsy, dysentery, and heart attacks but also
other more serious illnesses occur. Diarrhoea
occurs too on account of excessive coldness in the
womb, either because the veins are very slender as
in thin women since in this case thick and excessive
fluids do not have free channels through which
they can break forth, or because the liquids are
thick and viscous and because of clotting their egress
is hindered. Sometimes it is because the women eat
too luxuriously or because from some exertion
they sweat profusely. Galen says: "A woman
who does not exercise much must necessarily be
abundant in many menstrual periods in order that
in this respect she may be in good health." Sometimes a woman's periods are lacking because the
blood is clotted in the body or it is emitted through
other parts as through the mouth in the spit,
through the nostrils, or through hemorrhoids.

already spoken of in the twelfth century as a woman
who had practiced medicine at Salerno perhaps a hundred years before. The present text does, in its original
form, go back to eleventh-century Salerno, and we have
undoubtable evidence that there were women practicing
there by the fourteenth century at the latest. But the
historical existence of Trotula herself (traditionally
supposed to be the wife of the Salernitan physician
Johannes Platearius) and her authorship of this work
remains moot: For a balanced judgment, see H. P.
Bayon, "Trotula and the Ladies of Salerno," *Proceedings of the Royal Society of Medicine*, XXXIII (1939–
1940), 471–474. Whoever the author, the treatise "taught
the professional handling of gynaecological and ob-
stetrical conditions in a manner . . . more enlightened
than was usual at that time" (compare the little work of
Constantine the African in the following selection);
nor was it superseded in the later Middle Ages. Yet *The
Diseases of Women* remains quite revealing of the first
stage in the developing medieval medical tradition:
Heavily practical, it preserves disorganized scraps of
classical knowledge (attributed to Galen or Hippocrates)
but is largely ignorant of and unconcerned with theoretical explanations.

Sometimes the periods fail because of excessive grief or anger or excitement or fear. If they have ceased for a long time there is a suspicion of serious future illness. Often the urine is changed into a red color or into a color like the washings from fresh meat; sometimes the woman's appearance is changed into a gray or leaden color, or into the color of grass.

CHAPTER 1
ON THE RETENTION OF THE MENSES

If the menses are lacking and the woman is thin in body, the vein that is under the inner arch of her foot, the internal saphenous, should be lanced. On the first day from one foot, on the following day from the other foot—the blood to be drawn as the case demands. In every sickness general care must be taken and regard must be shown that the patient be not too much weakened. Galen reports the case of a woman whose menses had been lacking for nine months and her body had been tense and thin and her appetite had failed completely; he himself drew out her blood from the aforesaid vein for three consecutive days. He took from one foot on the first day one pint, on the second day one pint from the other foot, on the third day eight ounces from the first foot again and in a short interval her color and lively warmth returned with improved condition of the whole body. Moreover since the bowels are constipated as often happens in suppressed menses, one should make small pills of a certain efficacious medicine and sharpen them to the extent that she will be able to support their sharpness and give them to her. Then let blood be drawn from the internal saphenous vein. Let her be bathed and after the bath let her drink *calamento,* or catnip, or mint cooked with honey in such proportion that there be seven parts of water and nine of honey. Let her bathe often and after the bath let her drink a *diatesseron* with the decoction of honey and water. Diuretics are helpful also—fennel, wheathead, parsley, *ciminum, ameos, carui,* rock parsley, and the like. All these herbs or each separately cooked in wine and drunk are a help. Galen's prescription is "wormwood rubbed in wine or cooked in it, if drunk, helps greatly." It is some help too if in the bath catnip is drunk, or if it be cooked in the bath; or let a trodden green plant be tied over the abdomen either below or above the navel. Or let it be cooked in a pot covered with a perforated gall bladder and let the woman sit over it covered in all directions or let the steam go out through a reed or a tube so that

the steam being received within may penetrate to the womb. Wormwood also is efficacious mixed with *thapsia,* skirwort, salvia, wild marjoram, *ciminum, ameos,* nettle, honey, pennyroyal, dill, *betonica,* anise, savory, and privet. Let all of these or any of them be cooked in water, then take a little wool shredded so fine that it is like powder, dip it in that water, and while hot apply it all to the abdomen. Do this frequently. The best remedy for producing the menses is to take an equal quantity of *flammula,* hemlock, *castorei,* myrrh, century, and salvia; let a powder be made and let one dram be given in water in which nettle and myrrh have been cooked. Let her drink this in her bath, one scruple to be taken at a time. If however the womb has hardened so much that by these means the menses cannot be produced take the gallbladder of a bull or some other gallbladder and powdered soda and let them be mixed with juice of parsley or of hyssop. Let it be soaked into shredded wool and pressed so that it is firm and long and of a size to be inserted into the vulva, and let it be inserted. Or let a hollow pessary be made in the shape of a man's penis and in it let the medicine be placed so that it can be injected.

CHAPTER 2
CONCERNING SCANTINESS OF MENSES

There are women who, when they come to the time of the menses, have none or a very slight amount. These patients we help in this way: take the red roots of willow, the kind of which baskets are woven and crush them after cleaning them well of their outer bark. When crushed blend them by cooking with wine or water and the next day give a warm draught of the decoction for drinking. If she is suffering very severely you will give her food prepared in the following way: grate a rather large carrot and a mallow, mix with barley flour and whites of eggs, and of all this make small curls or noodles. Herbs of this sort are also efficacious when suffumigated for producing menstruation, or add to the same mixture verbena and rue crushing them vigorously, or cook them with bacon fat and give to the patient for food. Also grate the root of a delicate willow and a rather large carrot and give the juice with wine to the patient. If women have scanty menses and emit them with pain, take one dram each of *betonica,* pennyroyal, *centonica,* and wormwood; let them be cooked down to one-half in wine or water. Strain this through a cloth and let it be drunk steaming hot. If the menses have been absent for a long time, make a powder of two drams of rhubarb and one dram each of wormwood and pepper; let her take this morning and evening for three days and let her cover herself so that she sweats. Also take one dram each of mint, pennyroyal, and rue, four drams of grain salt, five drams of red *caulis,* and three heads of leek. Cook all of these together in a clean pot and let her drink it in her bath. Another remedy is this: take a root of iris, catnip, rue, *coloquintidam,* and fennel; clean them well, cook them in wine, and give the wine to the woman to drink. Or cook the following in wine: juniper, parsley, fennel, rock parsley, lovage, and catnip, and let it be drunk. Also take tansy, clover, and wormwood cooked with butter and place over the navel. A certain doctor in the region of Francia did this: he took leaves of laurel and ginger and ground them together in a clean pot; he put this mixture over live coals under a perforated seat and over this he let the woman sit. She took in the smoke from below and thus made the menses begin. It may be necessary to do this thrice or more times. However let the woman who habitually practices fumigations of this sort anoint her vulva inside with cold ointments lest she be irritated. For the aforementioned fumigation the following are also efficacious: *ciminum,* fennel, dill, calmento, mint, and nettle either all mixed together or one alone. For bringing on the menses massage is helpful, and likewise coitus. However, bloodletting is injurious. Let her eat, if she be without fever, leeks, onions, pepper, garlic, *ciminum,* and scaly fish. Let her drink strong wine, if she be without pain in the head and without weakness of muscles and without fever, because in all fever wine is injurious.

. . . .

CHAPTER 11
ON THE HINDRANCES TO CONCEPTION AND OF THE
THINGS WHICH MAKE FOR IMPREGNATION

Certain women are useless for conceiving either because they are too thin and lean or because they are too fat. In these latter the flesh folded around the opening of the womb binds it and does not permit the seed of the man to enter it. Some have a womb so soft and slippery that the seed having been received cannot be retained in it. Sometimes this happens through a defect of the male who has seed so thin that when it is poured into the vagina it slips out because of its own liquidness. Some men also have testicles cold and dry; these men rarely or never beget, because their seed is useless for procreation. It is evident therefore that con-

ception is hindered as often by a defect of the man as of the woman. If it is by a defect of the woman it happens either from excessive warmth or from excessive moistures of the womb. Sometimes on account of its natural softness the womb cannot retain the seed injected into it, and often because of its excessive moisture it suffocates the seed. Often because of its excessive heat the womb burns the seed up and she cannot conceive. If for these reasons she cannot conceive these will be the signs: The lips of the vulva appear ulcerated and as if chapped by the north wind. There will be red spots, continuous thirst, and falling out of the hair. When we have observed these symptoms and the woman has passed thirty years we judge her incurable. If she is very young and the ailment not of long standing you may help her thus: Take mallow and mugwort cooked in water and douche the patient three or four times with this decoction. Between these douchings you will use suppositories and also pessaries for the vulva with oil of weasel and a small quantity of moss that the womb may be relieved. On the seventh day after the purgation or douching take from a great plant which bears thrice a year a fruit in the shape of an acorn and wrap it in silk. Make it into a suppository for the vulva that by its beneficient action and by the many douches she may receive some relief, assuagement, and softness. On the following day you will arrange that she have intercourse with her husband. It will be expedient to employ the same procedure the following week—the douchings and the other pleasant things we have mentioned. Do this until symptoms are relieved and tell her to have intercourse two or three times a week because thus she will be able to become pregnant more quickly. If she cannot conceive because of excessive moisture of the womb these will be the signs: Her eyes will be continuously tearful for since the womb is frequently tied up with the sinews it is necessary that the brain suffer with the womb. If the womb is too moist the brain is filled with water and the moisture running over to the eyes compels them involuntarily to shed tears. Since the brain suffers with the womb, the estrangement of the womb is indicated by the retention of the menses. Therefore purge her first with broken doses of Theodorics, or have Paulinius' made and wrapped in silk so that they may not dissolve. Insert as carefully as possible through the private parts. If the womb has not been well purged you should the next day make a pessary from a thrice bearing plant and add a small quantity of musk. Continue this until you know that the excess moisture has been evacuated. Later take a little musk with oil of weasel or some other strong smelling oil and insert it through the vulva. If she has been well purged she will taste the odor and if one kissed her he would think she were holding musk in her mouth. She will know it too because she will feel thirsty. When she is thus well purged let her have intercourse frequently and she will conceive.

If conception be hindered because of a defect of the male it would be from a lack of force impelling the sperm, a defect of the organ, or a defect of heat. If it be from a defect of heat, the sign is that he is not eager for copulation. Hence he ought to anoint his loins with *arrogon* or take seed of colewort and euphorbia and reduce them to a fine powder. Then mix them with the oils of fleabane and of weasel and with this anoint his loins. If it happens through a defect of the spirit the sign will be that he has desire but the penis is not erected. We aid him with an ointment that generates spirit. If it happens through a defect of the sperm the sign is that when he copulates he emits either none or too little seed. We aid him with things that increase sperm such as orris, domestic parsnips, and the like.

If the woman or the man be sterile you will ascertain it by this method: take two jars and into each put bran. Into one of them put the urine of the man and into the other put the urine of the woman and let the jars be left for nine or ten days. If the barrenness be from a defect of the woman you will find many worms and the bran foul in her jar. On the other hand you will have similar evidence from the other urine if the barrenness be through a defect of the man. But if you have observed such signs in neither urine neither will be the cause of the barrenness and it is possible to help them to conceive by the use of medicines. If they wish to have a male child let the man take the womb and vulva of a hare and have it dried and pulverized; blend it with wine and let him drink it. Let the woman do the same with the testicles of the hare and let her be with her husband at the end of her menstrual period and she will conceive a male. Another remedy is this: Take the livers and testicles of a little pig—one which the sow bore alone—and let them be dried; make a powder of this and give it in a drink to the man and woman who cannot conceive and they will procreate. Also if the woman wishes to become pregnant let her take the dried testicles of a boar pig or wild boar reduced to powder. Mix this with wine and let the woman drink it at the end of her menstrual period

and when she copulates with her husband she will conceive.

"Note," says Galen, "that women who have narrow vulvas and tight wombs ought not to have husbands lest they die if they conceive." But since they cannot all abstain they need our help. If one of them for fear of death dare not conceive let her carry on her naked flesh the womb of a she-goat which has never had offspring. A certain stone is found called *Galgates* which, worn on the neck or even tasted, prevents conception. Also remove the testicles from a weasel and let it be left alive. Let the woman carry these testicles with her on her bosom tied in the skin of a goose or in some other skin and she will not conceive. If she has been injured in childbirth and for fear of death does not wish to conceive again, let her lay on the last afterbirth as many grains of *cataputia* or of barley as the years which she desires to remain sterile. If she wishes to remain barren forever let her lay on a handful.

CHAPTER 12
ON THE FORMATION OF THE SEED WHEN CONCEIVED

In the first month occurs a small clot of blood. In the second occurs the formation of the blood and of the body; in the third the nails and hair are produced. In the fourth motion and therefore women are nauseated. In the fifth the foetus receives the likeness of father or mother. In the sixth, the binding together of the sinews. In the seventh, the bones and sinews are strengthened; in the eighth nature helps and the child puts on flesh. In the ninth, it proceeds from darkness into light.

CHAPTER 13
ON THE POSITION OF THE FOETUS IN THE MOTHER'S WOMB

Galen gives the report that the foetus is fastened in the womb just as the fruit is on the tree, which when it comes forth from the blossom is very tender and falls from any occasion whatsoever. When it has become full grown, riper, and established, it clings to the tree and will not fall on slight occasion. When it has become completely ripe it will fall of itself and not of any other occasion. Thus when a child is first produced from a conceived seed the ligaments by which it is fastened to the womb are tender and unfirm and therefore it is easily let fall by abortion. On account of a cough, diarrhoea, dysentery, excessive activity or anger, or loss of blood, a woman can lose her foetus. But when a soul or life has been infused into the child it clings a little more firmly and does not slip quickly. When the child has ripened it is quickly let out by the office of nature. Hippocrates says that if a woman requires bleeding or purgation, you should not do these things before the fourth month. In the fifth or sixth months she can be bled or purged cautiously, if there be necessity, with a mild cholagogue or decoction according as the strength of the patient shall be able to tolerate. Beyond and before this time an evacuation will be dangerous. When the time for parturition has arrived the child moves more violently and struggles toward the exit. Nature in its own time causes the vulva to be opened, the foetus finds its own exit and thus it is expelled by the force of nature from its own resting place, the afterbirth.

CHAPTER 14
ON SIGNS OF PREGNANCY

For knowing whether a woman is carrying a male or female child take water from a spring and let the woman draw out two or three drops of blood or of milk from the right breast. Let them be poured into the water and if they seek the bottom she is bearing a male; if they float on top she is bearing a female. Hippocrates said that the woman who is bearing a male is well colored and has the right breast larger; if she is pale she is bearing a female and has the left breast larger.

CHAPTER 15
ON REGULATIONS FOR PREGNANT WOMEN

When a woman is first pregnant care must be taken that nothing be named in her presence which cannot be had because if she shall ask for it and it not be given to her she has occasion for miscarrying. But if she should seek to have potter's earth or chalk or coals, let beans cooked with sugar be given to her. When the time for parturition is imminent the woman should be bathed often; anoint her abdomen with olive oil or oil of violets and let her eat light and digestible foods. If her feet have swollen, let them be anointed with oil of roses and with vinegar. Instead of heavy foods let her eat quickly digested things like citrons and pomegranates. If her abdomen is distended with flatulence take three drams each of parsley seed, *ameos,* mint, mastic, *garyophyllous,* cardamon, roots of carrots, coffee, *galangale,* iris, and five drams of sugar; make a very fine powder and cook it all with honey; give three scruples to her with wine for this substance removes flatulence and prevents abortion if properly taken.

CHAPTER 16
ON THE REGULATIONS FOR THE WOMAN ABOUT TO GIVE BIRTH

When the time for giving birth is imminent, let the woman prepare herself as the custom is, and the midwife likewise. Let sneezing be done with great caution, holding tightly the nostrils and the mouth, in order that the greatest part of the strength and spirits may tend toward the womb. Give her a decoction of chick peas, a paste of flax and psyllium as an antidote, or another remedy is a mixture of four ingredients with a decoction of mugwort in wine. Likewise make troches of *galbanum* with asafoetida and myrrh or rue; from this make a fumigation for the nostrils. Above all things let her guard herself from cold. Let an aromatic fumigation be made below the nostrils; it can also safely be applied at the mouth of the womb because then a fragrant womb follows and an ill smelling one is avoided. For this purpose fragrant kinds of substances avail as musk, amber, wood of aloe and the like for rich patients, and fragrant herbs as mint, pennyroyal, *calamentum,* wild *marjoram* and the like for the poor. It is to be noted that there are certain physical remedies whose virtues are obscure to us, but which are advanced as done by midwives. They let the patient hold a magnet in her right hand and find it helpful. Likewise they let her drink a powder of ivory or they find that coral suspended on the neck is helpful. In similar fashion that white substance which is found in the dung of an eagle, when given in drinks is advantageous. Likewise give the dung of babybirds which is found in the swallow's nest. Washings of this are serviceable for this and for many other purposes.

CHAPTER 17
ON DIFFICULTY OF PARTURITION

There are, however, certain women so narrow in the function of childbearing that scarcely ever or never do they succeed. This is wont to happen for various reasons. Sometimes external heat comes up around the internal organs and they are straightened in the act of giving birth. Sometimes the exit from the womb is too small, the woman is too fat, or the foetus is dead, not helping nature by its own movements. This often happens to a woman giving birth in winter. If she has by nature a tight opening of the womb, the coldness of the season constricts the womb still more. Sometimes the heat all goes out of the woman herself and she is left without strength to help herself in childbearing.

In the first place and above all things when there is difficulty in childbirth one must have recourse in God. Descending then to lower means, it is helpful to the woman in difficult labor to be bathed in water in which has been cooked mallow, chick peas, flaxseed, and barley. Let her sides, abdomen, hips, and flanks be rubbed with oil of roses or oil of violets. Let her be rubbed vigorously and let vinegar and sugar be given her as a drink, and powdered mint and a dram of absinth. Let sneezing be provoked by placing dust of incense in the nostrils, or powder of *candisium,* or pepper or euphorbia. Let the woman be led with slow pace through the house. Do not let those who are present look in her face because women are wont to be bashful in childbearing and after the birth. If the child does not come forth in the order in which it should, that is, if the legs or arms should come out first, let the midwife with her small and gentle hand moistened with a decoction of flaxseed and chick peas, put the child back in its place in the proper position. If the child be dead take rue, mugwort, absinth, and black pepper and give this pulverized in wine or in water in which lupins have been cooked. Or let savory be mashed and bound over the abdomen and the foetus, whether dead or alive, will come forth. Verbena drunk with wine, water, or vinegar does the same thing. Or let sea water or rose water be drunk with an equal quantity of ass's milk. Or let butter be mixed with honey or wine and given to her to drink. If the birth is still delayed or if the foetus is dead in her and is not released, take rue, mugwort, *opoponacum,* and absinth with a little oil and let a small quantity of sugar be mixed with it. Place it over the loins and above the navel and it will be very efficacious. Likewise let the woman be girt with the skin of a snake—one which a snake has cast off. The stone *"Aetiteo,"* tied to the thigh, produces the same effect. Or let the root of a gourd be tied to her kidneys and have the foetus be taken away as soon as it comes out lest the womb come out after the emergence of the child. Also those who are in

difficult labor must be aided in the following manner: Let a bath be prepared and the woman put in it; after she has come out let a fumigation be made of wheat and similar aromatics for comforting and relaxing. Let sneezes be produced with white hellebore well reduced to powder. Colphon says to let the limbs be shaken to break the bag of water and in this way the foetus will come forth. Thus also those may be aided who are laboring much to bring forth a dead foetus: Let the patient be placed in a linen cloth stretched by four men at the four corners with the patient's head somewhat elevated. Let the four corners be strongly drawn this way and that by the opposite corners and she will give birth immediately, God favoring her. If the afterbirth has remained within there is need of haste that it shall come out. Let sneezing be provoked with mouth and nostrils shut. Or take lye made from ash tree ashes and mix it with one dram of powdered mallow seed. Give this to the woman to drink and she will immediately vomit. Or give mallow seed powder alone in a drink of hot water and if she vomits it will be a good thing. Also let her be fumigated below with bones of salted fish or with horses hoofs, or with the dung of a cat or lamb. These things bring out the afterbirth. Also let those things be done which have been mentioned before for bringing forth menstruation. If difficulty in childbirth should result from tightness of the mouth of the womb, the cure of this is more difficult than anything else, therefore we subjoin this advice: let the woman take care the last three months in her diet that she so use light and digestible foods that through them the limbs may be opened. Such foods are egg yolks, the meat and juice of chickens and small birds—partridges and pheasants, and scaly fish of good flavor. Let her often take a bath in fresh water to which has been added herbs of softening character such as *matura* and the like. Let her avoid a bath tinctured with copper and calcium. When she comes out of the bath, let her be anointed with hot ointment such as oil of laurel, oil of flaxseed, or the grease of goose, duck, or hen. Let this anointing be done from the navel down.

104 HOW TO COMBAT SPELLS PREVENTING INTERCOURSE

Constantine the African (d. 1087)

Translated and annotated by Henry E. Sigerist[1]

PART II, BOOK VIII, CHAPTER 29

A short treatise about the people who, impeded

1. [Translated in "Impotence as a Result of Witchcraft," reprinted in *Henry E. Sigerist on the History of Medicine,* ed. Félix Martí-Ibañez (New York: MD

by spells, are unable to have intercourse with their wives.

There are people who, impeded by diabolical spells, are unable to have intercourse with their wives. We do not want to deprive our book of their applause, for the remedy, if I am not wrong, is most sacred.

Now, if this should happen to somebody, he must set his hope in the Lord and He will be merciful. Since, however, there are many kinds of spells, it is necessary that we discuss them. Some spells are made of animated substances such as the testicles of a cock. If they are put under the bed with blood of the cock, they bring it about that the people lying on the bed cannot have intercourse. Some are made of letters written with the blood of a bat. Some are made of inanimate substances, for instance if a nut or an acorn is divided in two, and one half is put on one side, the other on the other side, of the road along which the bride and bridegroom must proceed.

There are others also which are made from beans which are not softened with hot water nor cooked on the fire. This spell is very bad if four such beans are placed on the roof or on the road or over or under the door.

There are others also which are of metal, such as those that are of iron or lead, for instance, the iron ones made of the needle with which the dead men or women have been sewn. And because these spells are devilish and are particularly in women, they are sometimes cured by divine, sometimes by human measures.

If therefore bridegroom and bride are disturbed by the above-mentioned spells, it is better to talk about them than to keep silent, for if the victims are not succored they are separated and thus disgraced, and doing this evil they seem to sin not only against their relatives but also against the Holy Ghost.

If we wish to extirpate the spell properly, we must look out: if the above-mentioned spell is under the bed, it must be removed. But if the author of this spell removes it in daytime and puts it back at night, or vice versa, then bridegroom and bride must acquire another house and lie down there.

If the spell is made of letters, which is recognized by the fact that bridegroom and bride do not love each other, one must search above and under the threshold of the door, and if something is found it must be taken to the bishop or priest. If not, one must do what is indicated below.

If a nut or an acorn are the cause of this spell, the woman shall take a nut or an acorn and divide it in two. And with one half the man shall proceed on one side of the road and deposit it there; the woman, however, shall put the other half on the other side of the road. Thereupon bridegroom and bride shall take both parts of the nut without having removed the shell. And then the nut shall thus be made whole again and shall be kept for seven days. Having done this they shall have intercourse.

If, however, it happens on account of beans, it can be cured with divine rather than human means. If it is on account of the needles for the dead, the spells must be sought either in the pillow or in the mattress. If they are not found, the victims shall lie together in another house.

Bile of a male dog purifies the house and brings

Publications, 1960), pp. 146–152, from Henry E. Sigerist, *Essays in Biology* (Berkeley: University of California Press, 1943), pp. 541–546. Reprinted by permission of the Regents of the University of California. Sigerist introduces the text with the remarks below.—Michael McVaugh]

In the majority of all cases impotence is due to nervous causes. It is a phobia. The fear of being impotent prevents individuals from having normal erections. It is not astonishing that in the Middle Ages such a condition was frequently attributed to witchcraft. The question was important because it had legal consequences. Inability to consummate marriage was a reason for declaring it null and void *ab initio*.

The first who brought up the question of witchcraft in this connection was Hincmar, Archbishop of Rheims, who lived in the ninth century. In discussing a definite case he came to the following conclusion: if a marriage has been annulled on account of the impotence of the husband, he cannot marry again if his impotence was due to natural (physical) causes. If his condition, however, was the result of witchcraft and the marriage had been declared void after the customary treatments had failed, he was permitted to marry again.

Gratian, who in the twelfth century codified canon law, accepted Hincmar's view, and so did Peter the Lombard in the same century in his *Liber Sententiarum,* where he has a special chapter: "De His Qui Maleficiis Impediti Coire Non Possunt." Peter's book was commented upon by the leading scholastic theologians.

Since impotence resulting from witchcraft was to be treated, not only was the Church interested in the question but the physicians also. In the eleventh century Constantine of Africa devoted a chapter of his *Pantechne* [from which the present selection is taken—M. M.] to it . . . . Constantine's text with or without additions is frequently found in medical manuscripts as an independent anonymous treatise. It has been published and discussed by Gerda Hoffmann in an excellent dissertation ["Beiträge zur Lehre von der durch Zauber verursachten Krankheit und ihrer Behandlung in der Medizin des Mittelalters," *Janus,* XXXVII (1933), 129–144, 179–192, 211–220.—M. M.].

it about that no evil remedy be brought to the house.

Sprinkle the walls of the house with dog's blood, and it will be liberated from every spell.

If bridegroom and bride carry bile of a fish and particularly of zangarinus (that is lucius) along in a box made of juniper, and if when they go to sleep they pour some on burning coals and are fumigated therefrom, all spells mentioned above vanish.

Similarly, if quicksilver is taken and put into a reed pen and the pen is sealed with wax and lead while the bridegroom and bride know nothing about it, no spell will harm them at the place.

If, however, on account of impending sins, the above-mentioned measures did not help at all, they shall go to a priest or the bishop. And if the bishop has permitted it and no remedy is found,

after having confessed to the bishop or an ordained priest they shall take Holy Communion on the day of the Holy Resurrection or Ascension in Whitsuntide. Having received the Body and Blood of the Lord, bridegroom and bride shall give each other the kiss of peace.

And after they have received the benediction of the bishop or priest, the bishop or priest shall give this verse of the prophet written on paper: The voice of the Lord is upon the waters: the great Lord is upon many waters [Psalms 28:3]. Thereafter they shall go home and shall abstain from intercourse for three days and three nights. Then they shall perform it, and thus all diabolical power is destroyed.

The little treatise on spells has come to an end. Thanks be to God. Amen.

105 DISEASES OF THE SKIN

Theodoric, Bishop of Cervia (1205–1298)

Translated by Eldridge Campbell and James Colton[1]

Annotated by Michael McVaugh

THE DEAD DISEASE [A SPECIES OF LEPRA]

A certain disability occurs on the legs and arms which is called the dead disease, for there are dry and livid ulcers which generate sanies slightly. Sometimes they are made of pure black bile, and sometimes of black bile with an admixture of salty phlegm.

If the disease arises from pure black bile, it is recognized by black pustules without pruritus. But if salty phlegm is mingled with this, the area grows livid, as it were, and is accompanied by pruritus and stinging.

Considering the particular characteristics, therefore, if it arises from pure black bile, digest the matter with oxymel of squills and the syrup of fumitory.

If salty phlegm is part of the mixture, digest the matter with two parts of oxymel, the third part of oxysaccharum, and the fourth part of syrup of fumitory.

The digested matter is recognizable by the coloration and thickening of the urine; and a greater quantity may be purged by the use of two parts of hieralogodion, the third of hierarufinum, and the fourth of an imperial cathartic.

On the following day in each case, make a stupe of calamint, oreganum, rue, mint, marrubium,

laurel leaves, and every type of parsley, cress, scabiosa, fumitory, and spatula fetida. As the patient emerges from soaking in this stupe, offer him greater theriac with hot juice of fumitory, if the cause was pure black bile. But if it was mixed with phlegm, add a third part of madder trochees.

On the following day if the dead disease has spread all over, exercise phlebotomy from the hepatic vein. But if it is on the legs and arms, exercise phlebotomy from the interior and exterior saphenous veins successively; but sometimes place the cups over the kidneys as well.

Likewise make ashes of fennel, spatula fetida, lapathum acutum, fumitory, leaves of elder and dwarf elder; then take some lye water in which the aforesaid herbs have been cooked; and with the aforesaid herbs and the same lye water soak the diseased parts. Then make an inunction upon the

1. Reprinted with the kind permission of Appleton-Century-Crofts from Book III, chapter 49, of *The Surgery of Theodoric ca. A.D. 1267,* tr. Eldridge Campbell and James Colton (New York: Appleton-Century-Crofts, 1955), II, 145–148.

The use of mineral as well as vegetable remedies became frequent only after the introduction of Arabic medical writings. Mercury (used primarily for skin ailments, as here) is the most important of these, and it by no means supplanted the traditional use of vegetable simples.

shoulder blades with this ointment: the skins of tapsus barbatus, ground very fine, and added to these an equal weight of powdered litharge, the whole modified with oil of roses and vinegar, and make an ointment.

Likewise take juice of celadony, lapathum acutum, scabiosa, spatula fetida; and with these mix together in equal quantities, add old pork fat, and add a small amount of powder of hellebore, yellow arsenic, flowers of sulphur, frankincense, mastic, litharge, and incorporate them thoroughly with a pestle. Finally, add some triturated quicksilver and make an ointment.

Likewise another which is effective against scurf, scabies, and the dead disease: Take quicksilver, flowers of sulphur, hellebore, cummin, titymal, and stavesacre; put them together with six ounces of old pork fat, or bear fat, which is better, or fox fat, which is better still.

Likewise take the skins of nettle, tapsus barbatus, celadony, one measure of each; a half measure of sage; grind these up together. And if a salty mixture ought to be made therefrom, then add to a bit of pork fat and incorporate it well with a pestle; and add a bit of triturated quicksilver; and add powdered hellebore, common salt, and ink, and make an ointment.

Macer says that lolium, ground up with radish and a bit of salt, cures livid pustules and lepra.

. . . .

Likewise a Saracenic ointment, which cures scabies, cancer, the dead disease, or salty phlegm by bringing forth the matter through the mouth. And it is said to cure also the leprous in their initial phase, and to be good for arthritis and gout. Take: saltpeter, pyrethrum, burnt lead, euphorbium, two and a half drachms of each; of marine pumice, and chameleon thistle, one and a half drachms of each; of ceruse and quicksilver, a half

drachm of each; put them together with pork fat and oil, first having triturated the quicksilver with old pork fat and with old oil; and when it has been well triturated, add the powder of the types mentioned before and incorporate it well. Next make two fires, and between them place a table on which the patient is located; and anoint him from below the knees as far as the feet and for three fingers above the knees; similarly from the elbows down to the hands, and three fingers above the elbows. This inunction should be made twice a day. And let the patient beware of cold. But when the phlegm begins to dissolve, you may discontinue the inunctions. His diet should be delicate and easily digestible.

. . . .

Others make this ointment thus and it is better: take of euphorbium, litharge, and aurea, a half-pound of each; one pound of quicksilver; one and a half pounds of old pork fat; cleanse the fat very carefully; and then incorporate the quicksilver (which has been triturated with saliva) with this, so that it does not remain in globules; then mix together the other things which have been pulverized and incorporate them well. From this ointment let the patient anoint himself as has been said, and keep his feet and arms well covered as long as the inunction lasts. Let him make this inunction morning and evening between two fires, until his teeth have begun to hurt. Then let him cease straightway from the inunction, and keep himself good and warm, and stay so, until the flow of phlegm from his gums has ceased.

When this has been done let him not wash himself for up to forty days. Let him take gum arabic, bear fat, and rose oil, and a little balsam, and mix them all together, and anoint his hands, face and temples with it until the color which had been darkened, as it were, returns.

106 SMALLPOX

John Gaddesden (ca. 1280–1361)

Translated and annotated by Michael McVaugh

There are a number of things that must be done in curing this disease [smallpox]. First, it is necessary that evacuation take place, with phlebotomy and mild purge, cooling and compressive rather

1. Translated from *Rosa anglica practica medicine* (Venice, 1502), fol. 41 recto to 42 recto. A few emendations of this printed text seemed necessary to me.

Gaddesden has had his critics since the fourteenth

century (see Selection 111), but his *Rosa anglica* is not really so discreditable a work as they have claimed. It is explicitly a compilation of practical medicine which still displays considerable judgment and clear thought—and, unlike Mirfeld's work of fifty years later (see Selection 101.3), it often reflects personal experience, in the light of which Gaddesden has selected or abridged the teachings of others. It is incidentally the first English synthesis of the continental medical traditions.

than attractive or solutive. Second, an internal alteration should be brought about, using acidic substances (if constriction of the chest does not prevent it) or else cold and digestive materials, strengthening stipticity in some way. Third, drugs which lead to the easy eruption of the pox ought to be administered. Fouth, the eyes, chest, nose, mouth, and intestines ought to be protected, because of the nobility of these members, and a diversion of matter from the inside to the outside should be effected. Fifth, there should be a drying out through the use of infusions, which cannot, however, be done before complete maturity [of the pox]. Sixth, when the pox burst they must be cleaned and healed; or if they do not burst by themselves, they must be opened with a gold needle. Seventh, traces of the pox should be healed and eliminated.

In connection with the first, it should be noted that if the body is plethoric or if blood is predominant (always assuming that the patient's forces are strong and that age and other particulars are suitable), phlebotomy should be carried out first from the median vein and then from the bridge of the nose, especially in children; such a blood-letting preserves the higher parts from the malignity of the pox, and is easy for children. If the pox are small and unimportant and the body is replete with blood, phlebotomy can again be performed, until the fourth day; but when they persist or enlarge for two or three days, then it is wrong to administer a bloodletting, because this would draw material inside and prevent the formation of the pox, unless the body experiences a tremendous fullness and a dominance of matter, when it should be withdrawn in the quantity which will reduce that repletion. For it is best that blood-letting be done in this illness, and therefore corruption of the extremities is to be feared if bleeding is not done when it should be. Then purge the stomach, using cassia fistula, prunes, tamarinds, violets, fresh rose-sugar (made with fresh roses or rose juice with sugar and tamarinds); this should not be strongly laxative, since this illness can easily pass into a flux.

In connection with the second, it should be noted that cold, stiptic, and sour things somehow help when they [the pox] first appear; and therefore sour juice, pomegranate wine, verjuice, coriander, the juice of sour prunes, apples, and other fruits, nenufar, camphor, and sandalwood are useful; likewise the confection *triasandali* with water of nenufar, in which a little camphor has been put against fever,

aerial pestilence, and venemosity of matter. But excessively cold things should not be given internally, such as cold water or water chilled with snow; very cold things are of no use after the second day of the appearance of the pox, because they prevent their expulsion as well as the ebullition of the blood that causes their expulsion. Likewise cold things should not be applied externally, lest they divert poisonous matter back inside; nor should the air be cold and dry, lest the pores close. . . . The patient's diet should tend towards the cool foods, such as almond milk, preparations of barley or oats, or lentils (called "tiles" in English) cooked in these two waters. If a high fever accompanies the pox, then a broth of hepatica, lettuce, and scariola should be used; if the stomach is weak, then plantago can be added to these in small quantity. Other broths can be prepared in which sumach, borage, and bugloss are cooked, for the last two herbs excellently purify the blood, while sumach and lentils thicken it. If it is too thin and sharp, make up a decoction of grapes cleaned of their seeds, dry figs(?), and shelled almonds. Patients should not eat flesh nor fish if their strength is weak; until the illness moderates, they may not drink wine nor strong beer, only a ptisan or almond milk. Figs are especially valuable in cases of constriction of the chest or of scantness of breath. . . .

Third, this potion helps bring out the pox easily:
dry figs 1/2 oz.
shelled lentils 3 oz.
tragacanth seed 2 dr.
fennel 2 dr.
saffron 1 dr.

If you are afraid of syncope, grind these up and boil them in two pounds of water until it is half gone; strain, sweeten with sugar, and administer. This expels hotness from the heart, as Avicenna says in [the *Canon,* Book] 4. Fennel and celery taken with sugar, or their juice, and also roots with savory saffron, and a decoction of yellow figs, expels matter very well. Then take a scarlet cloth and wrap whoever is suffering from pox completely in it, or in some other red cloth. I did this when the son of the King of England was suffering with the pox, making sure that everything about his bed was red.[2]

2. This passage has been cited repeatedly as the earliest description of the red-light, or Finsen, treatment for smallpox; in this form, however, the treatment goes back even further, to Gilbert the Englishman at least (who attributes the discovery to "old countrymen"). It is uncertain who Gaddesden's patient was, or indeed whether his father was Edward I or Edward II.

This treatment succeeded for me very well, for he was restored to health without traces of the pox. You must take the greatest care, however, that when the pox appear, their site not be anointed with oil, since this obstructs the pores; nor should the patient be exposed to cold air, unless the air be very hot(?), in which case the air of the dwelling may be tempered with willow leaves and aspersed with rosewater, camphor, and camphorated water. Bind the limbs of the patient, or make him continuously wear gloves on his hands, so that he cannot scratch or touch himself, for this could make an ugly pit in the skin, especially on the face. Then take warm juice of celery and fennel, and moisten a cloth with this; wrap the cloth around the entire body, for this draws the matter out and helps consume it. Or cook celery in water with lentils and dry figs; moisten the cloth in this decoction, wring it out, and wrap the patient in it. Likewise anoint the patient (though not at the outset) not with oil but with the blood of a warm animal, such as a chicken, a dove, or a sheep; and then wrap him in the aforesaid cloth. Do this frquently, but be careful not to remove it incautiously, especially in the case of children.

Fourth, the eyes must be defended and strengthened with a light bandage, because of their sensitivity; use cotton or an old soft cloth dipped in rosewater. Make a collyrium from coriander, rosewater, sumach, and egg-white, shaken together for a long time, and apply it frequently to the eyes. Rosewater by itself and mother's milk are also good; powdered sugar candy is wonderful inasmuch as it prevents filming over, as we have said elsewhere. The juice of verbena with egg-white is best for the film that remains. Should perhaps the physician have been summoned too late and spots remain on the eye after the pox, try this remedy: Take the whitest tragacanth and put it entire in rosewater to dissolve; put some of the solution into the eye once a day, and the spot will disappear, particularly if sugar candy is sometimes alternately placed there in turn, or ordinary white powder. If you want to treat the nostrils, anoint them inside and out with verjuice, vinegar, and rosewater. It helps greatly to sniff pure vinegar. To treat the mouth and throat you should get pomegranates and gargle with their wine; likewise mulberry wine—that is, their juice, which is red like wine—mixed with hot barley-water. To preserve the

lungs poppy seed is good, with almond-milk, rose-sugar, and diapapaver, all of the best. To preserve the intestines, cook sumach and coriander in sourdock juice, or cook sourdock in water, strain, sweeten, add 1/2 oz. of spodium, and administer as a potion. If the stomach is purged, cook quince-flesh with sugar and add 2 dr. of spodium or a troche of spodium, as Isaac (*On Fevers,* Book 5: *De cura synoche*) recommends, for these things are effective.

Fifth, note that before complete maturation you should take rosewater in which saffron has been cooked and sprinkle the pox with the warm water; then cook tamarisks, roses, and sandalwood in water and add a little camphor; sprinkle the patient with this hot water. . . . This way of acting speedily matures and dries the pox and causes them to disappear.

Sixth, if maturation occurs but the pox do not burst, let them be opened with gold or silver needles; dry the moisture that comes forth—first with cotton and then with litharge, burnt lead, and rotten wood (that is, that which is left after worms have eaten it away). Or if you want to dry them before they burst, let the patient sleep on flour made from rice, barley, or beans—or use these things to fill the mattress on which the patient sleeps. If an ulcer should appear, dry it with Rhazes' white ointment described by me above, or bean- and rice-powder, sprinkled on. Rice flour and starch are effective after maturation. To generate flesh, for scabbing and healing, a powder made of *thus,* mastix, and sarcocolla sprinkled on or mixed with honey is good.

Seventh, when the pox have developed and pus has been formed, should they leave scars and other traces, anoint with an ointment of white lead, made in this fashion: Take filtered wax[3] and four times its quantity of oil of roses; mix these, add 1 dr. white lead and stir it around. When this begins to cool, mix egg-white with it, stirring without pause until it thickens; set this aside for use. Once you have applied this ointment, take powdered rice flour and sprinkle the scar with it; do this every night, and first thing in the morning wash the face with bean-flower water. . . .

---

3. A numerical weight of wax to be used seems to have been lost from the text.

107 BUBONIC PLAGUE

Guy de Chauliac (d. ca. 1368)

Translated and annotated by Michael McVaugh[1]

. . .Internal apostemes,[2] especially those near the principal members, are dangerous. We saw this clearly in the great and unprecedented mortality which appeared to us in Avignon in the year of our Lord 1348, in the sixth year of the pontificate of Pope Clement VI, in whose service, by his favor, I unworthily then was. Let it not displease you if I tell of it, since it was so remarkable, and as a guide in case it comes again.

The mortality began with us in the month of January, and lasted for seven months. It was of two kinds. The first lasted for two months; it was characterized by a continuous fever and a spitting of blood, and men died of it within three days. The second lasted for the rest of the time; it too was characterized by continuous fever, and by apostemes and carbuncles and tumors in the external parts, mainly the armpits and groin; and men died of it within five days. It was so contagious (especially that which involved spitting of blood) that one man caught it from another not just when living nearby but simply by looking at him; so much so that people died without servants and were buried without priests. Father would not visit son, nor son, father; charity was dead, and hope prostrate.

I call it great because it engaged all the world, or nearly. For it began in the East, and shooting out into the world, passed through our region toward the west. It was so extensive that it left scarcely the fourth part of the population alive. And it was unprecedented, since, although we read in the *Epidemics* of the mortality in the city of Thrace, and in Palestine, and of another which occurred in Hippocrates' time, and (in the *Euchymia*) of that which afflicted the Romans' subjects in Galen's time, and of that of the city of Rome in the days of Gregory [IX], none of these was as great as this one. For they affected only one area; this one, the whole world; they were curable in one fashion or another; this one, in none. Because of this physicians felt useless and ashamed, inasmuch as they did not dare visit the sick for fear of infection; and when they did visit them they could do very little and accomplished nothing, for all the sick died except for a few towards the end, who escaped with matured buboes.

Many have speculated on the cause of this great mortality. In some places they believed that the

Jews had poisoned the world, and so they killed them. In others, they believed that it was the mutilated poor, and so they drove them away. And in still others, that it was the nobles, so that they feared to travel in the world. Finally it came to the point that a watch was kept in cities and towns, forbidding entry to anyone who was not well known. And if they found anyone with powders or ointments, they made him swallow them, fearing that they might be poisons.

But whatever the people said, the truth is that the cause of this mortality was twofold: one active and universal, one passive and particular. The active, universal cause was the disposition of a certain important conjunction of three heavenly bodies, Saturn, Jupiter, and Mars, which had taken place in 1345, the 24th day of March, in the fourteenth degree of Aquarius.[3] For (as I have said in my book on astrology) the more important conjunctions

1. Translated from Tractate II, Doctrine II, Chapter 5, of Guy's Latin treatise on surgery published as *Ars chirurgicalis Guidonis Cauliaci medici* (Venice, 1546), fols. 21 verso to 22 recto.

2. Here and elsewhere I have used "aposteme" to translate *apostema,* a word which "covers all swellings 'against nature,' which are due to accumulations of humors, either singly or admixed, and in normal or altered form" (Leo M. Zimmerman and Ilza Veith, *Great Ideas in the History of Surgery* [New York: Dover Publications, 1967], p. 155).

3. The ascription of the Black Death to this astrological cause is not unique with Guy but received European acceptance. It was apparently first proposed by the Faculty of Medicine at Paris in October 1348, which had been asked by Philip VI of France to explain what was known of the epidemic. Anna Montgomery Campbell, *The Black Death and Men of Learning* (New York: Columbia University Press, 1931), summarizes the Parisian medico-astrological doctrine (p. 40): "On 20 March, 1345, at one o'clock in the afternoon, occurred an important conjunction of three higher planets in the sign of Aquarius, which, with other conjunctions and eclipses, is the cause of the pernicious corruption of the surrounding air, as well as a sign of mortality, famine, and other catastrophes not connected with the present subject. Aristotle is quoted and other, unnamed, ancient philosophers referred to for the fact that the conjunction of Saturn and Jupiter brings about the death of peoples and the depopulation of kingdoms, great accidents occurring on account of the change of the two stars themselves; and Albert of Cologne is adduced in support of the assertion that the conjunction of Mars and Jupiter causes great pestilence in the air, especially when it takes place in a warm and humid sign, as occurred in

presage marvelous, mighty, and terrible events, such as changes of rulers, the advent of prophets, and great mortalities; and they depend on the [zodiacal] sign and the aspects of the bodies in conjunction. It should not amaze you, therefore, that such an important conjunction signified a marvelous, awful mortality, for it was not just one of the greater ones but one of the greatest. Because the [zodiacal] sign was a human one, it foretold grief for humanity; and because it was a fixed sign, it signified long duration. For [the mortality] began in the East, a little after the conjunction, and was still abroad in the West in [13]50. It so informed the air and the other elements that, as the magnet moves iron, it moved the thick, heated, poisonous humors; and bringing them together within the body, created apostemes there. From this derived the continuous fevers and spitting of blood at the outset, when this corrupt matter was strong and disturbed the natural state. Then, as this lost its strength, the natural state was not so troubled, and expelled what it could, mainly in the armpits and groin, and so caused buboes and other apostemes, so that these external apostemes were the effects of internal ones. The particular, passive case was the disposition of each body, such as cachochymia, debility, or obstruction, whence it was that the working men and those living poorly died.

Men looked for a preventative treatment before the attack and for a curative one thereafter. For prevention nothing was better than to flee the area before becoming infected; and to purge oneself with pills, to diminish the blood by phlebotomy, to purify the air with fire, and to strengthen the heart with tyriac, fruits, and good-smelling things; to fortify the humors with Armenian *bolus,* and to resist decay with sharp[-tasting] things. For a cure men tried bleedings and evacuations, electuaries and cordial syrups. The external apostemes were brought to a head with figs and cooked onions mixed with yeast and butter; then they were opened and treated as ulcers. The tumors were cupped, scarified, and cauterised.

I myself did not dare to leave lest I lose my good name, but in continuous dread kept myself healthy as well as I could, using the remedies above. Nevertheless, towards the end of the mortality I fell into a continuous fever, with an aposteme in the groin; I was ill for six weeks, in such great danger that all my friends believed I would die; but the aposteme ripened, and was treated as I have described, and by God's will I survived.

Afterwards, in [13]60, the eighth year of the pontificate of Pope Innocent VI, the mortality returned to us on its way back from Germany and the northern lands. It began toward the feast of St. Michael [September 29], with swellings, fevers, carbuncles, and tumors, growing little by little, sometimes even dying away, until the middle of [13]61. Then it raged for the next three months, so furiously that in several places it left less than half the population. It differed from the earlier one in that more of the populace died in the first, while in the second more of the rich and noble died, innumerable children, and few women.

During this [second mortality] I made up the following tyriacal electuary, from the teachings of Master Arnald of Villanova and of the masters of Montpellier and Paris:

℞ : iuniperi dr. 2½; gariophylli, macis, nucis muscati, zingiberis, zedoari an. dr. 2; utriusque aristolochie, radicis gentiane, tormentille, radicis herbe tunicis, diptami, radicis enule campane, an. dr. 1½; salvie, rute, balsamite, mente, polemenon. . .an. dr. 1; bacarum lauri, doronici, croci, seminum acetose, citri, ozymi, masticis, olibani, boli armenici, terre sigillate, spodii, ossis de corde cervi, rasure eboris, margarite fragmentorum, sapphyri, smaragdi, coralli rubei, ligni aloes, sandali rubei et muscatellini, an. dr. 1/2; conserve rose, conserve buglosse, conserve nenufaris, theriace probate, an. oz. 1; panis zuc. lib. 3. Fiat electuarium cum aqua scabiose et rose, modicum camphorata.

I took it like tyriac[4] and was preserved by the will of God, whose name be blessed for ever and ever, Amen.

---

this instance. For, the Paris doctors explain, Jupiter, a warm and humid planet, drew up evil vapors from earth and water, and Mars, being excessively hot and dry, set fire to those vapors. Whence there were in the air flashes of lightning, lights, pestilential vapors, and fires, especially since Mars, a malevolent planet generating choler and wars, was from the sixth of October, 1347, to the end of the May of the present year [1348] in the lion together with the head of the dragon. Not only did all of them, as they are warm, attract many vapors, but Mars, being on the wane, was very active in this respect, and also, turning toward Jupiter its evil aspect, engendered a disposition or quality hostile to human life. From this were generated strong winds, which, according to Albert, Jupiter has the property of raising, particularly from the south, giving rise in lower countries to very great heat and dampness; in regions about Paris the dampness was greater than the heat.

4. See Selection 110.2.

Tools Employed in Treatment

108 DIET AND REGIMEN

Regimen Sanitatis Salernitanum (thirteenth century)

Translated by Sir John Harington[1]

Annotated by Michael McVaugh

The *Salerne Schoole* doth by these lines
    impart
All health to *Englands King,* and doth advise
From care his head to keepe, from wrath his
    heart,
Drinke not much wine, sup light, and soone arise,
When meate is gone, long sitting breedeth smart:
And after-noone still waking keepe your eyes.
When mov'd you find your selfe to *Natures Needs,*
Forbeare them not, for that much danger breeds,
Use three Physicions still; first Doctor *Quiet,*
Next Doctor *Merry-man,* and Doctor *Dyet.*

Rise earely in the morne, and straight remember,
With water cold to wash your hands and eyes,
In gentle fashion retching every member,
And to refresh your braine when as you rise,
In heat, in cold, in *Iuly* and *December.*
Both comb your head, and rub your teeth
    likewise:
If bled you have, keep coole, if bath'd keepe
    warme:
If din'd, to stand or walke will do no harme.
Three things preserve the sight, *Grasse, Glasse,*
    & fountains,
At Eve'n springs, at morning visit mountains.

    . . . . . .

To keepe good dyet, you should never feed
Until you finde your stomacke cleane and void
Of former eaten meate, for they do breed
Repletion, and will cause you soone be cloid,
None other rule but appetite should need,
When from your mouth a moysture cleare doth
    void.
All *Peares and Apples, Peaches, Milke and*
    Cheese,
Salt meates, red Deere, Hare, Beefe and Goat:
    all these
Are meates that breed ill bloud, and Melancholy,
If sicke you be, to feede on them were folly.

    . . . . . .

Fine *Manchet* feeds too fat, *Milk* fils the veines,
New cheese doth nourish, so doth flesh of Swine:

The Dowcets of some beasts, the marrow,
    braines,
And all sweet tasting flesh, and pleasant wine,
Soft Egges (a cleanely dish in house of Swaines)
Ripe Figs and Raysins, late come from the Vine:
Chuse wine you meane shall serve you all the
    yeere,
Well-savor'd tasting well, and coloured cleere.
Five qualities there are, wines praise advancing,
Strong, Beautifull, and Fragrant, coole, and
    dancing.

    . . . . . .

If wine have over night a surfet brought,
A thing we wish to you should happen feeld:
Then early in the morning drinke a draught,
And that a kind of remedie shall yeeld,
But gainst all surfets, vertues schoole hath taught
To make the gift of temperance a shield:
The better wines do breed the better humors,
The worse, are causes of unwholesome tumors.
In measure drinke, let wine be ripe, not thicke,
But cleere and well alaid, and fresh and quicke.

The like advice we give you for your Beere,
We will it be not sowre, and yet be stale:
Well boild, of harty graine and old and cleare,
Nor drinke too much nor let it be too stale:
And as there be foure seasons in the yeere,

1. *The Englishmans Doctor. Or, the School of Salerne*
(London, 1608; reprinted New York, 1920).
    The most widely known of all medieval books on
science or medicine may be the *Regimen Sanitatis Salernitanum;* the translation used here testifies to its continuing popularity into early modern times. Like Harington's
version, the original is in verse: A core of perhaps 360
verses seems to have come into existence in the middle
of the thirteenth century (Karl Sudhoff, "Zum Regimen
Sanitatis Salernitanum," *Archiv für die Geschichte
der Medizin* VII–XII [1913–1920]).
Most of the poem is concerned with diet, but the proper
regulation of the other five "non-naturals" is touched on
at the beginning [for the non-naturals, see Selection 90
and Selection 102, n.1.—*Ed.*]. The verses quoted here
reflect pretty accurately the weight given to the several
aspects of regimen in the poem as a whole.

In each a severall order keepe you shall.
In *Spring* your dinner must not much exceed,
In *Summers* heate but little meate shall need:
In *Autumne* ware you eate not too much fruite:
With *Winters* cold full meates do fittest suite.

If in your drinke you mingle *Rew* with *Sage*.
All poyson is expeld by power of those,
And if you would withall Lusts heat asswage,
Add to them two the gentle flowre of Rose:
Would not be sea-sicke when seas do rage,
Sage-water drinke with wine before he goes.
Salt, Garlicke, Parsly, Pepper, Sage, and Wine,
Make sawces for all meates both course and fine.
Of washing of your hands much good doth rise,
Tis wholesome, cleanely, and relieves your eyes.

Eate not your bread too stale, nor eate it hot,
A little Levend, hollow bak't and light:
Not fresh of purest graine that can be got,
The crust breeds choller both of browne and
    white,
Yet let it be well bak't or eate it not,
How e're your taste therein may take delight.
Porke without wine, is not so good to eate
As *Sheepe* with wine, it medicine is and meate
Tho Intrailes of a beast be not the best,
Yet are some intrailes better than the rest.

. . . .

The choyse of meate to health doth much availe,
First Veale is wholesome meat, & breeds good
    bloud
So Capon, Hen, and Chicken, Partridge, Quaile,
The Phesant, Woodcock, Larke, & Thrush be
    good,
The Heath-cocke wholesome is, the dove, the
    raile,
And all that doe not much delight in mud.
Faire swans such love your beauties make me
    beare you,
That in the dish I easily could forbeare you.
Good sport it is to see a *Mallard* kil'd,
But with their flesh, your flesh should not be fil'd.

As choyce make of Fowle, so make of Fish,
If so that kinde be soft, the great be best,
If firme, then small, and many in a dish:
I need not name, all kinds are in request.
Pike, Trowt, and *Pearch,* from water fresh I wish,
From Sea, *Bace, Mullet, Brean,* and *Souls* are
    best:
The *Pyke* a ravening tyrant is in water,
Yet he on land yeelds good fish ne're the later,
If *Eeles* and *Cheese* you eate, they make you
    hoarse,

But drinke apace thereto, and then no force.

Some love at meals to drink smal draughts and
    oft,
But fancie may herein and custome guide,
If *Egges* you eate, they must be new and soft.
In Pease good qualities and bad are tryed,
To take them with the skinne that growes aloft,
They windie be, but good without their hide.
In great consumptions learn'd Physicions thinke,
'Tis good a *Goat* or *Camels* milke to drinke,
Cowes-milke and *Sheepes* doe well, but yet an
    Asses
Is best of all, and all the other passes.

. . . .

Although you may drinke often while you dine,
Yet after dinner touch not once the cup,
I know that some Physicions doe assigne
To take some liquor straight before they sup:
But whether this be meant by broth or wine,
A controversie 'tis not yeat tank up:
To close your stomack well, this order sutes,
Cheese after flesh, *Nuts* after fish or fruits,
Yet some have said, (beleeve them as you will)
One *Nut* doth good, two hurt, the third doth kill.

Some *Nut* 'gainst poyson is preservative:
Peares wanting wine, are poyson from the tree,
But bak't Peares counted are restorative,
Raw Peares a poyson, bak't a medicine be
Bak'd Peares a weake dead stomack doe revive,
Raw Peares are heavie to digest we see,
Drinke after Peares, take after Apples order
To have a place to purge your selfe of ordure.
Ripe *Cherries* breed good bloud, and help the
    stone,
If *Cherry* you doe eate and Cherry-stone.

Coole *Damsens* are, and good for health, by
    reason
They make your intrailes soluble and slacke,
Let *Peaches* steepe in wine of newest season,
Nuts hurt the teeth, that with their teeth they
    crack
With every *Nut* 'tis good to eate a *Raison*.
For though they hurt the spleen, they help the
    back,
A plaister made of Figges, by some mens telling,
Is good against all kernels, boyles and swelling,
With *Poppy* ioyn'd, it drawes out bones are
    broken
By *Figges* are lice ingendred, Lust provoken.

. . . .

If unto *Choller* men be much inclin'd

'Tis thought that *Onyons* are not good for those,
But if a man be flegmatique (by kind)
It does his stomack good, as some suppose:
For oyntment iuyce of *Onyons* is assign'd
To heads whose haire fals faster than it growes:
If *Onyons* cannot helpe in such mishap,
A man must get him a *Gregorian* cap.
And if your hound by hap should bite his master,
With *Hony, Rew,* and *Onyons* make a plaster.

. . . .

Cleane *Hysop* is an hearbe to purge and clense
Raw flegmes, and hurtfull humors from the brest,
The same unto the lungs great comfort lends,
With hony boyl'd: but farre above the rest,
It gives good colour, and complexion mends,
And is therefore with women in request:
With *Hony* mixt, *Cinquefoyle* cures the Canker,
That eates out inward parts with cruell ranker.
But mixt with wine, it helpes a grieved side,
And staies the vomet, and the laske beside.
Ellecompane strengthens each inward part,
A little loosenesse is thereby provoken,
It swageth grief of minde, it cheeres the heart,
Allaeth wrath, and makes a man faire spoken:
And drunke with Rew in wine, it doth impart
Great help to those that have their bellies broken,
Let them that unto choller much incline,
Drinke *Penny-royall* steeped in their wine.
And some affirm that they have found by tryall
The paine of Gowt is cur'd by *Penny-royall.*

. . . .

Our hearing is a choyce and dainty sense,
And hard to men, yet soone it may be mard,
These are the things that breed it most offence,
To sleepe on stomacke full and drinking hard,
Blowes, fals, and noyse, and fasting violence,
Great heate and sodaine cooling afterwards;
All these, as is by sundry proofes appearing,
Breed tingling in our eares, and hurt our
    hearing:
Then thinke it good advice, not idle talke,
That after Supper bids us stand or walke.

You heard before what is for hearing naught,
Now shall you see what hurtfull is for sight:
Wine, women, Bathes, by art to nature wrought,
Leekes, Onyons, Garlicke, Mustard-seed, fire and
    light,
Smoake, bruises, dust, Pepper to powder brought,
Beanes, Lentiles, strains, Wind, Tears, and
    Phoebus bright,
And all sharpe things our eye-sight do molest:
Yet watching hurts them more than all the rest.

Of *Fennells, Vervin, Kellidon, Roses, Rew:*
Is water made, that will the sight renew.

. . . .

Foure humors raigne within our bodies wholly,
And these compared to foure Elements,
The *Sanguine, Choller, Flegme,* and *Melancholy,*
The latter two are heavie, dulle of sence,
Th' other two are more Joviall, quicke and Jolly,
And may be likened thus without offence,
Like ayre both warme and moist, is *Sanguine*
    cleare,
Like fire doth *Choler* hot and drie appeare.
Like water cold and moist is *Flegmatique,*
The *Melancholy* cold, drie earth is like.

Complexions cannot vertue breed or vice,
Yet may they unto both give inclination,
The *Sanguine* game-some is, and nothing nice,
Love Wine, and Women, and all recreation,
Likes pleasant tales, and news, playes, cards & dice,
Fit for all company, and every fashion:
Though bold, not apt to take offence, not irefull,
But bountifull, and kinde, and looking cheerefull;
Inclining to be fat, and prone to laughter,
Loves mirth, & Musick, cares not what comes after.

Sharpe *Choller* is an humour most pernitious,
All violent, and fierce, and full of fire,
Of quicke conceit, and therewithall ambitous,
Their thoughts to greater fortunes still aspire,
Proud, bountifull ynough, yet oft malicious
A right bold speaker, and as bold a lyar,
On little cause to anger great enclin'd,
Much eating still, yet ever looking pin'd:
In yonger years they use to grow apace,
In Elder hairie on their brest and face.

The *Flegmatique* are most of no great growth,
Inclining to be rather fat and square:
Given much unto their ease, to rest and sloth,
Content in knowledge to take little share,
To put themselves to any paine most loth.
So dead their spirits, so dull their sences are:
Still either sitting, like to folke that dreame,
Or else still spitting, to avoid the flegme:
One qualitie doth yet these harmes repaire,
That for the most part *Flegmatique* are faire.

The *Melancholly* from the rest doe varie,
Both sport and ease, and company refusing,
Exceeding studious, ever solitary,
Inclining pensive still to be, and musing,
A secret hate to others apt to carry:
Most constant in his choise, tho long a chusing,
Extreme in love sometime, yet seldom lustfull,

Suspitious in his nature, and mistrustfull,
A wary wit, a hand much given to sparing,
A heavy looke, a spirit little daring.
. . . .
If *Sanguine* humor doe too much abound,
These signes will be thereof appearing cheefe,
The face will swell, the cheekes grow red and
     round,
With staring eyes, the pulse beate soft and breefe,
The veines exceed, the belly will be bound,
The temples and the fore-head full of griefe,
Unquiet sleepes, that so strange dreames will
     make,
To cause one blush to tell when he doth wake:
Besides the moisture of the mouth and spittle,
Will taste too sweet, and seeme the throat to
     tickle.
. . . .
Of Bleeding many profits grow and great,
The spirits and senses are renewed thereby:
Tho these men slowly by the strength of meat,
But these with wine restor'd are by and by.
By bleeding, to the marrow commeth heat,
It maketh cleane your braine, relieves your eye,
It mends your appetite, restoreth sleepe,
Correcting humours that do waking keepe:
All inward parts and senses also clearing,
It mends the voyce, touch, smell & tast, & hearing.

Three speciall Months *(September, April, May)*
There are, in which 'tis good to ope a veine;
In these 3 Months the Moone beares greatest
     sway,
Then old or yong that store of bloud containe,
May bleed now, though some elder wizards say
Some dayes are ill in these, I hold it vaine:
September, April, May, have dayes a peece,
That bleeding to forbid, and eating Geese,

And those are they forsooth of May the first,
Of other two, the last of each are worst.
. . . .
Make your incision large and not too deepe,
That bloud have speedy issue with the fume,
So that from sinewes you all hurt do keepe,
Nor may you (as I toucht before) presume
In six ensuing houres at all to sleepe,
Lest some slight bruise in sleepe cause an
     apostume;
Eat not of milke, nor ought of milk compounded,
Nor let your braine with much drink be
     confounded
Eate no cold meats, for such the strength
     impaires,
And shun all misty and unwholesome aires.

Besides the former rules for such as pleases,
Of letting bloud to take more observation,
Know in beginning of all sharpe diseases,
'Tis counted best to make evacuation:
Too old, too yong, both letting bloud displeases.
By yeares and sicknesse make your computation.
First in the Spring for quantity you shall
Of bloud take twise as much as in the Fall:
In Spring and Summer let the right arme bloud,
The Fall and Winter for the left are good.

The *Heart* and *Liver,* Spring & Summers bleeding,
The Fall and Winter, hand and foot doth mend,
One veine cut in the hand, doth help exceeding
Unto the spleene, voyce, brest, and intrailes lend,
And swages griefes that in the heart are breeding.
But here the *Salerne Schoole* doth make an end:
And here I cease to write, but will not cease
To wish you live in health, and die in peace:
And ye our Physicke rules that friendly read,
God grant that Physicke you may never need.

109 DRUGS: SIMPLE MEDICINES

1. Rufinus (fl. 1287–1300): Simple Medicines

Translation, introduction, and annotation by Edward Grant[1]

The identity of Rufinus is uncertain, and information about him comes largely from the preface to his *Herbal*. There "he tells us that he had pursued the liberal arts in the cities of Naples and Bologna and that, after studying astronomy and astrology, using the *Tables of Toledo,* he turned to the science concerning herbs, which he intended to investigate fully and perfectly" (Thorndike, *The Herbal of Rufinus,* p. xiv). Thorndike conjectures

that the work was written between 1287 and 1300.

The organization of the work is of considerable interest. For each entry Rufinus cites the opinions of one or more authorities like Dioscorides, *Macer*

1. This selection is translated from the Latin text published in *The Herbal of Rufinus edited from the Unique Manuscript* by Lynn Thorndike, assisted by Francis J. Benjamin, Jr. (Chicago: University of Chicago Press, 1946).

Floridus, Circa instans, and so on, but frequently terminates with his own opinion. Thorndike estimates that about one-fifth of the text consists of the personal opinions and descriptions of Rufinus, and he remarks (p. xvii) that the descriptive botany of Rufinus is for particular plants "more specific and discriminatory than that of any previous author, ancient or medieval." William Stearn,[2] however, does not believe Rufinus represents an advance on Theophrastus,[3] and in what seems a general warning against ancient and medieval botanical descriptions, he says that "identification of the plants concerned is almost impossible from the descriptive notes of these authors, unless the plant has very well-marked characters" (as an illustration of an identifiable plant Stearn mentions *centaurea maior,* a description of which is translated in this selection).

Not until the invention of printing was it possible to accompany verbal descriptions with pictorial representations of the plants and be assured that readers in widely scattered areas could study and see brought together the same verbal description and appropriate accompanying figure. As long as it had been necessary to represent plants separately for each manuscript copy, when they were included at all, this was not possible. The problem of identification was very formidable indeed in ancient and medieval botany.

During the Middle Ages herbals were intimately liked with medicine, since herbs were primarily grown and described for their medicinal and healing powers. Except for Albertus Magnus (see Selection 87), and perhaps a few others who were interested in theoretical botany, people associated botanical knowledge primarily with herbals and with medicine rather than considering it an independent science. Although the *Herbal* of Rufinus was hardly known, its inclusion here as a sample of medieval botany is justified because Rufinus quoted from some of the best-known herbals in the Middle Ages and brought together important authorities for a large number of herbs. (For additional translations of herb descriptions, see Thorndike's introduction to *The Herbal of Rufinus,* pp. xvii–xviii.)

Here begins the book on the powers and composition of herbs compiled by the great doctor Master Rufinus from the statements of the greatest philosophers: Dioscorides,[4] *Circa instans,*[5] *Macer,*[6]

2. *Botanical Latin* (New York: Hafner, 1966), p. 24.

3. For selections from Theophrastus see M. R. Cohen and I. E. Drabkin, *A Source Book in Greek Science*

(Cambridge, Mass.: Harvard University Press, 1948).

4. A Greek author of the first century A.D. who compiled one of the most important works in the history of botany. His *De materia medica* in five books describes some six hundred plants and was used for about fifteen hundred years. For an English translation see *The Greek Herbal of Dioscorides . . .* Englished by John Goodyer A.D. 1655; edited and first printed A.D. 1933 by Robert T. Gunther (reprinted, New York: Hafner, 1959); selections are also included in M. R. Cohen and I. E. Drabkin, *A Source Book in Greek Science* (Cambridge, Mass.: Harvard University Press, 1948). Thorndike observes (p. xxviii) that up to the letter O Rufinus quotes most frequently from Dioscorides. The Latin version of Dioscorides used by Rufinus has not been identified, but it bears little resemblance to Goodyer's translation made directly from the Greek (see n. 18)

5. *Circa instans* are the opening words of a treatise titled *On simple medicine (De simplici medicina)* written in the twelfth century by Matthaeus Platearius (d. 1161) at Salerno, a flourishing medical center. *Circa instans,* by which title the treatise was almost always cited, is actually a commentary on the *Antidotarium Nicolai,* a treatise containing about 150 recipes for the preparation of drugs and also written at Salerno, presumably by an otherwise unknown Salernitan physician named Nicolas. The *Antidotarium,* which was enormously popular throughout the Middle Ages, was also probably written in the twelfth century and already reveals Arabic influences. The *Circa instans,* known in longer and shorter versions, was arranged alphabetically (as was the *Herbal* of Rufinus) by the Latin names of the drugs. According to George Sarton, *Introduction to the History of Science* (Baltimore: Williams and Wilkins, 1927–1948), Vol. II, Part I, p. 241: "These names are given also in Greek, Italian, and French. Then for each drug we are offered the following information: description, action, mode of application, signs of purity, falsifications, distinctions of various kinds. The *Circa instans* is especially important from the botanical point of view. For example, of the 273 chapters of the minor text, 229 treat of medicinal plants. It marked a great progress over Dioscorides and other herbals. During the thirteenth and fourteenth centuries it gradually superseded the Herbarius ascribed to Apuleius, and it became one of the prototypes of our Western pharmacopoeias. Illustrated manuscripts also influenced botanical iconography; of 379 illustrations in the Gart der Gesundheit, 94 can be traced back to the Platearius manuscripts."

6. This is the widely known *Macer Floridus,* about which Gösta Frisk provides the following very useful summary:

"*Macer Floridus De Viribus Herbarum* is the title of an old Latin poem written in hexameters. It comprises 2269 lines, in which 77 plants are discussed. This book is supposed to have been written sometime between 849 and 1112. . . .

"The proper and original title of the book is believed to have been *De Viribus Herbarum.* Where and when the secondary title *Macer Floridus* appeared is not quite certain, but *Macer* is mentioned as early as 1120–30 and a little later *Floridus* is added. Then the question arises as to why the work was called *Macer Floridus.* The general theory is that Macer is a pseudonym and that the

779

Alexander,[7] [the Master of] Salerno,[8] Isaac,[9] and many other doctors.

For with unutterable mercy, God arranged the world as a perfect body undiminished in anything, which is constituted of an incorruptible substance, namely the heavens and the things in it; and a corruptible substance, namely the four elements in visible [or composed] bodies *(quatuor elementis in elementatis)*[10] and the four qualities, which results from elements made into bodies by a certain harmony and mixing. And all these will be located inside the corporeal limit, within which all bodies live; beyond the heavens, however, there is an in-corporeal limit, where nothing can exist. And so, I, Rufinus, in the city of Naples and in the city of Bologna, directed my mind in many ways in order to take the treasure of learning [or science] from the seven [liberal] arts. Nevertheless, when, by means of the Toledan Tables[11] I had learned perfectly the revolutions of the planetary orbs *(orbium stellarum)*, [learning] as much about the orbs of the equants as about the [orbs of the] eccentrics and epicycles and the bodies being moved on their circles with a proper motion, I understood the two motions in the outermost [sphere], namely the first motion and the second motion. The first motion moves all things and by virtue of it night and day occur; and it moves from east to west and again to the east. The second motion is opposite to the first motion, that is, from west to east and again to the west and this motion is in the bodies of the planets on their epicycles.[12] These superior [or highest] motions move the lower bodies and rule [them] in accord-ance with nature *(secundum viam naturam)*, not by free will *(secundum arbitrium)*. And so, having learned from the Masters about the science of the stars, I learned to investigate perfectly and fully the science of the lower things, that is, about herbs in which human life is preserved and even death is found; [that is,] life is preserved, as in the spelt grass of dry grain *(herba frumenti sicalis spelta)*[13] and others; and death [is found], as in spurge of

thought to be Odo de Meung (Odo Magdunensis), this name being added to a Macer MS at Dresden. Odo de Meung lived during the first part of the 11th century and was from Meung on the Loire. He was most probably a layman, for no phrases and words typical of monks and clergymen are to be found in the book. This work cannot be said to be unique in any way. It is quite similar to numerous medico-literary works, but among them it has an important place." (*A Middle English Translation of Macer Floriuds de Viribus Herbarum,* ed. Gösta Frisk [Cambridge, Mass.: Harvard University Press, 1949], pp. 13–14.) Many recipes and healing plants are arranged in a rough and confused manner and drawn partly from such authorities as Hippocrates, Dioscorides, Pliny, Galen, and so on.

7. Thorndike speculates that by "Alexander," Rufinus is referring to a commentary on the *Antidotarium Nicolai* (see n. 5). Apparently no work in the genre has yet been identified as authored by an Alexander.

8. *The Masters of Salerno (Magistri Salerni)* is prob-ably a twelfth-century medical work written at Salerno. It has been published by Salvatore de Renzi in *Collectio Salernitana* (Naples, 1852–1859).

9. Isaac Israeli the Elder (or Isaac Judaeus), who died in Tunis about the middle of the tenth century. His *Universal and Particular Diets,* from which Rufinus quotes, was translated from Arabic into Latin in 1087 by Constantine the African.

10. Already in the twelfth century a distinction was made between the terms *elementa,* which signified the real but invisible four elements earth, water, air, and fire, and *elementata,* which represented the actual but seemingly already compounded visible elements earth, water, air, and fire. (For alerting me to this distinction, I am indebted to my student Miss Joan Cadden.) These terms were used by William of Conches in his *Glosses on Plato* (*Glosae super Platonem,* LIX; p. 130 of the edition by Edward Jeauneau [Paris: Vrin, 1965]) and by Domingo Gundisalvo (see Selection 17 and note 9 thereto). Rufinus continues this tradition, which is perhaps traceable to Aristotle (see Selection 16, n. 28).

11. The Toledan Tables were compiled in the eleventh century by al-Zarqālī and translated into Latin by Gerard of Cremona in the twelfth century. It is unlikely that Rufinus was referring to the Alfonsine Tables, also compiled in Toledo, for as Thorndike says (p. xiv, n. 7): "The Alfonsine Tables drawn up about 1270 or 1272, although employing the year 1252 as their point of departure would not yet have been available during Rufinus' student days. Indeed, it is difficult to trace their influence outside the Spanish Peninsula before the four-teenth century."

12. The west to east motion is the sidereal period rep-resented by the motion of the eccentric circle which carries the epicycle on which the planet is located. See Selections 64.1, chap. 4, and 64.2, paragraph 10, for definitions of epicycle. Rufinus' interest in astronomy and cosmology is evident by a lengthy discussion of such matters in the *Herbal* under the entries "Virtus" (see Thorndike, pp. 321–326) and "Zodiacus" (pp. 350–351).

13. My translation is uncertain at this point. According to the *Oxford Dictionary,* "spelt" is a species of grain (*Triticum spelta*) related to wheat and once cultivated widely in southern Europe.

---

author wished to hide behind the name of a Latin author, Aemilius Macer, who died in 15 B.C. and who was a friend of Virgil and Ovid. The reason for this is perhaps that the author aimed at imitating Aemilius Macer's didactic poem of similar content, which poem, however, is lost. On the other hand, this medieval poem may have been written in Aemilius Macer's honor or in an attempt to exploit his name and popularity.

"The real author of the book is nowadays generally

tithymal *(napello titimallis)*[14] and other poisonous herbs.

I shall begin with *A,* by composing *A* with *b;* then *A* with *c;* then with *d;* then with *e* and *f* and *g,* and so on, for all the letters of the alphabet.[15] And by describing the powers of herbs and their operations in inferior bodies from the statements of the ancient wise men, I shall assemble what they had experienced and were able to discover about these things. And first I shall produce the statements of Dioscorides; second, of *Circa instans;* third, of *Macer;* fourth, of Alexander the philosopher; fifth, of the Masters of Salerno; sixth, of Isaac; seventh, of *Synonyms;*[16] and eighth, I shall speak.

(ABSINTHIUM)[17] WORMWOOD

Dioscorides:[18] Hot and dry in the first degree. It strengthens *(confortat)* the stomach and liver *(epar);* having been drunk beforehand, it cures [or prevents?] hysterics, strengthens the appetite, [and] is diuretic.[19] Its juice when squeezed into the ears stops pain and when made into a poultice *(facto cathaplasmate)* makes bearable a pain of the stomach or spleen.[20]

14. In his text Rufinus has no entry under *napello titimallis* but does have an entry *anabula titimalli* (Thorndike, p. 25) which includes quotes from Dioscorides, *Circa instans,* and *the Masters of Salerno.* Perhaps *napello* is a corruption or alteration of *anabula,* since Rufinus' quotation from Dioscorides declares that women often die from it and warns against giving it to a pregnant woman.

Seven kinds of tithymal, often called spurge, are distinguished by Dioscorides, *De materia medica* (pp. 559–566 of the Goodyer translation cited in n. 4). In a fifteenth-century Oxford manuscript, containing a medico-botanical glossary of terms, which seems to be a version of an earlier anonymous treatise called *Alphita* composed sometime in the late fourteenth century, we find the term *Anabulla maior,* which is described as "spurge, mezereon, taking life, making widows, the lion of the earth, the species is tithymal . . . (spurga, mezereon, rapiens vitam, faciens viduas, leo terre, species [est] titimalli . . .)"; in *Alphita Medico-Botanical Glossary from the Bodleian Manuscript, Selden B.35.,* edited by J. L. G. Mowat in *Anecdota Oxoniensia,* Mediaeval and Early Modern Ser., Vol. I, Part II (Oxford, 1887), p. 9, col. 2. The plant was also called "widow wail" because it made widows. Mowat identifies it as Daphne Laureola or Daphne Mezereum, which, according to one authority, "in large doses is an instant poison, causing vomiting and hypercatharsis. The berries have proved fatal to children" (M. Grieve, *A Modern Herbal* [New York: Hafner, 1959], II, 532, col. 2).

15. Thorndike observes (p. xxxvi) that although Rufinus here announces his intention to follow a strictly alphabetical order, he fails to do so; Thorndike remarks

further that he has never found a medieval manuscript which followed a strictly alphabetical order. "In some cases this may have been because the original alphabetical arrangement was based upon the Greek alphabet. More often it is, I suspect, to be attributed to the lack of, or expensiveness of, small movable slips of parchment or paper for purposes of filing and to the use, instead, of a notebook which was not loose-leaf and in which a page or two was reserved for each letter of the alphabet, but with entries on each page made haphazard and according to the exigencies of space left for further entries" (pp. xxxvi–xxxvii).

16. Perhaps the work of Stephen of Antioch (fl. ca. 1127). To his translation of Galen's *Pantegni* (to use its Medieval Latin title) Stephen added a glossary of technical terms from Dioscorides in Greek, Arabic, and Latin. This glossary was later called "Stephen's Synonyms" (see Sarton, Vol. II, Part I, pp. 236–237). In the late thirteenth century a contemporary of Rufinus, Simon of Genoa, compiled a treatise called *Synonyma medicinae,* which contained some 6000 articles on linguistic and botanical terms and was used until the sixteenth century (see Sarton, Vol. II, Part II, p. 1085). It seems unlikely, however, that Rufinus knew Simon's work (Thorndike, pp. xii–xiii).

17. The Latin terms are given first in parentheses in order to emphasize that the entries are alphabetized and also because the reader may find this information of interest.

18. This brief account bears no relation to the lengthy and elaborate discussion of the different varieties of *absinthium* described by Dioscorides. Indeed, the fact that it is characterized as "hot and dry in the first degree" is strong evidence that the treatise ascribed to Dioscorides and used extensively by Rufinus was not actually by the former, but probably represents a drastically altered version. For nowhere in his description of wormwood does Dioscorides assign degrees in terms of hotness and coldness. Indeed, this was not a technique used by Dioscorides but was rather characteristic of Galen and very popular with medical authors in the Middle Ages. For a discussion of the use of degrees in medicine, see Selection 110.3.

19. In describing the powers of *Artemesia pontica* (*absinthium,* wormwood), Dioscorides says: "But it hath a warming, binding digestive facultie, and of taking away ye cholerick matter sticking to ye stomach, and ye belly. It is also ureticall and keeps from surfetting, being drank beforehand" (Bk. III, ch. 26; p. 259 of Goodyer's translation cited in n. 4). This seems to be the ultimate source of Rufinus' version. Wormwoods belong to the genus *Artemesia,* comprising some 180 species. "The whole family is remarkable for the extreme bitterness of all parts of the plant: 'as bitter as Wormwood' is a very ancient proverb" (Grieve, II, 858).

20. These few lines probably represent a jumbled version of this portion of Dioscorides' description: "But for Sugillata with honey and for dullness of sight, and likewise for mattering ears, and the vapour thereby of ye decoction for ye ear's-pain, and for ye toothache. And being sodden with passum it is a cataplasme of very much pained eyes. It is also applied to ye Hypochondria, and ye liver, and a pained stomach. . ." (Bk. III ch. 26; p. 259 of Goodyer's translation cited in n. 4).

Circa instans says that it is hot in the first degree and dry in the second; and that there are two kinds, one a very bitter green wormwood *(ponticum)*, the other white and less bitter. Against intestinal worms *(lumbricos)* the juice of the wormwood is given with pulverized betony *(bethonice)*[21] or centaury *(centauree)*[22] or the [pulverized] nuts of peaches.

Against obstruction of the spleen and liver its juice is given and a syrup is made and given with hot [water]. To bring on menstruation its juice is crystallized(?) *(pessarizetur)* or a suppository is made from wormwood *(ex absinthio et arthemisia)* after it has been boiled [or cooked] in common oil.

Against drunkenness its juice is given with honey and warm [water].

Against hardness of the spleen wormwood *(absinthium)* cooked in oil is made into a poultice.

Against pain and bruises of limbs from having been struck, a plaster *(emplastrum)* is made of the juice of wormwood and pulverized [or ground] Cumin[23] and honey-water.

Macer:[24] . . . Of wormwood it is said: This herb strengthens the stomach in any manner that one takes it; but it is better if it should be cooked [or boiled] in rain water and cooled in the open air before it is drunk. . . . And this also banishes intestinal worms and constitutes a laxative [literally: "softens the bowels" *(mollit et alvum)*], and taking a small quantity of it stops pain. . . . It cures the liver if French nard *(Gallica nardus)*[25] is joined [or pounded together] with it; and with vinegar joined [or mixed] with it, it will help the spleen [or milt]. It also expels the poison of the toadstool. . . . And when it is boiled, the vapor opens stopped ears. . . . Joined [or mixed] with honey-water and soda, it helps with snake bites(?) *(anguine)*. . . . A poultice of wormwood resolves the hardness of the spleen.

Alexander: Nothing *(nichil)*.[26]

The Master of Salerno: Pontic wormwood *(Absinthium ponticum)*[27] binds [or constipates] more; bitter wormwood relaxes *(laxat)*. Therefore, if one should need the stomach to be bound *(constipatum)*, it [the wormwood] would bind; if it should be relaxed [or loosened], it would loosen it. Tempered with honey and placed in the uterus, it brings on menstruation.

Isaac: Wormwood is medially hot and extremely dry, and its sap [or juice] is much hotter than an herb. It strengthens the stomach, draws a reddish bile through the bowels *(per ventrem)* or urine. If it is drunk, it purges the veins of the liver and con-

stricts the liver.

Synonyms: Abscenalium, that is, wormwood *(absinthium)*.

Tables of Salerno:[28] Red. There are two[29] kinds

21. Called *Vettonica* by Pliny *(Natural History,* Bk. XXV, ch. 8), *Betonica officinalis* by Linnaeus, *Stachys Betonica* more recently, and frequently known as bishopswort, this plant, in a number of species, is common in Europe, where it was widely used against a great variety of ailments and evil spirits during the ancient and medieval periods. It was also used for tea and smoking. For a full description, see Grieve, I, 97–99. Dioscorides describes two types of betonica (Bk. IV, ch. 1 and ch. 2).

22. Terms synonymous with "centaury" are gentian, century, red centaury, filwort, centory, Christ's ladder, and feverwort. It is an annual with numerous variations in different regions. It is extremely bitter and "of all the bitter appetizing wild herbs which serve as excellent simple tonics, the Centaury is the most efficacious. . . ." It is "aromatic, bitter, stomachic and tonic. It acts on the liver and kidneys, purifies the blood, and is an excellent tonic. The dried herb is given in infusion or powder, or made into an extract. It is used extensively in dyspepsia, for languid digestion with heartburn after food . . ." (Grieve, I, 183).

In the medico-botanical glossary of the *Alphita,* a *centaurea minor* (smaller centaury) is distinguished from a *centaurea maior* (larger centaury). See page 37 of Mowat's edition cited in note 14, and Rufinus' descriptions of centaury later in this selection.

23. *ciminum.* Perhaps this is cumin (designated *Cuminum cyminum* by Linnaeus), and more particularly *Cuminum* (or *ciminum*) *dulce* or anise, as it is identified in the *Alphita* glossary under *Anisum* (p. 10 of Mowat's edition cited in n. 14). It is a small annual herb cultivated from early times around the Mediterranean. "From Pliny we learn that the ancients took the ground seed medicinally with bread, water or wine and that it was accounted the best of condiments" (Grieve, I, 242).

24. I have translated all of *Macer* that was included by Rufinus. The omissions, represented by ellipses, indicate places where, according to Thorndike, Rufinus omitted lines, something like fifty-one in all for this entry.

25. Nard *(Nardostachys Jatamansi)* is divided into Indian and Syrian by Dioscorides (Bk. I, ch. 6), who also distinguishes Celtic nard *(Valeriana celtica;* Bk. I, ch. 7) and mountain nard *(Valeriana tuberosa;* Bk. I, ch. 8). They were all apparently deemed very useful against stomach pains and ailments.

26. It would seem that "nothing" signifies the complete omission of a particular entry by the author whose name is cited rather than a decision by Rufinus to exclude what that author said on the herb under discussion.

27. *Artemesia pontica* is called "Roman wormwood" by Grieve (II, 860), who describes it as "the most delicate though the least strong of the wormwoods." According to Dioscorides, it was called *Absinthium rusticum* by the Romans (Goodyer, p. 259).

28. Although not mentioned by Rufinus in his introduction, this was another work written at Salerno.

29. Apparently only one is described.

of wormwood: pontic, which has wider leaves [and] in which the force of pontic bitterness and dryness are balanced *(ponderantur)*; therefore it can be given in place of *yerapigra*.[30] And it strengthens the head and clarifies vision.

Rufinus: This herb is spoken of as if clinging and drinking, for unless it clings to the earth and absorbs [or drinks] the moisture from the earth, it would be nothing.

(ACRIMONIA OR AGRIMONIA)[31] AGRIMONY

Dioscorides: It is also called *sarutium.* It is an herb known in many places whose powers counteract poisons. Its root is called *eupatorium* by the Greeks and it is very diuretic. Moreover, agrimony that has been dried, pounded, sifted, and placed near the nostrils *(naribus)* causes one to sneeze a great deal. Again, when you see it pounded, it causes you to sneeze. Again Dioscorides:[32] *sarutium* is called *acrimonia.* By its roughness it provokes urine and menstruation and causes one to sneeze a great deal after it has been pounded and sifted and placed near the kidneys.[33]

Rufinus: Agrimony has hairy leaves that are cut between [or into] quite a bit *(multum intercisa)* and in July it produces pointed yellow flowers.[34] After softening in lye *(in lixivio macerta)* for several days, it produces golden hairs *(capillos aureos)*; or boil it in oil until its substance [or matter] is consumed [and then] clean and oil the hairs.

(CENTAUREA) CENTAURY OR FEVERWORT[35]

Dioscorides: It is similar to *origanum,*[36] one greater, another smaller. It has relaxing powers with some roughness, and it purges the bowels and brings forth black bile and causes menstruation and expels the embryo [that is, causes abortion] *(partum expellit)* and makes for difficult breathing and coughing.

Circa instans: Centaury is hot and dry in the third degree; moreover, it is a very bitter herb, so that it is called "gall of the earth" *(fel terre).* Now, there is a "greater centaury" *(centaurea maior),* which is of a greater efficacy, and a "smaller centaury" of lesser efficacy. Constantine says that the root of the "greater" is hot and dry in the second degree and has a bitterness with some sweetness. It also has astringency *(ponticitatem)* and therefore has the power of joining by virtue of its astringency, and from bitterness it has a diuretic power. The greater centaury has efficacy in its leaves and in its flowers; therefore, when it begins to produce flowers, they ought to be gathered and sus-

pended and dried in a dark place. Through a year it is preserved with its power largely intact. It has a diuretic, attractive, and consumptive power. Note

30. Under a special entry for *yerapigra* (Thorndike, p. 124; see also p. 347) Rufinus says of it: "*yera,* i.e., sacred;*pigra,* i.e., bitter. It produces different effects *(passiones)* of the head, ears, and eyes; it corrects distemperature and causes of the head, and removes or reduces hardness and thickness of the spleen. It also purges the stomach best and alters a disproportionate womb *(matricem distemperatam).* . . ."

31. Perhaps *Agrimonia Eupatoria,* or common agrimony, which "has an old reputation as a popular domestic medicinal herb, being a simple well known to all country-folk. It belongs to the Rose order of plants . . ." (Grieve, I, 12). "The name Agrimony is from *Argemone,* a word given by the Greeks to plants which were healing to the eyes, the name *Eupatoria* refers to Mithridates Eupator, a king who was a renowned concoctor of herbal remedies" (I, 13).

Although in addition to his own description Rufinus here gives only a quotation he ascribes to Dioscorides, I have not located this herb in Goodyer's translation of the *De materia medica.* However, in Book IV, chapter 41, of the latter treatise (pp. 434–435) we find *Eupatorium cannabinum* (common hemp agrimony), one of a number of plants which although "not actually related botanically to the Common Agrimony . . . were given the same name by the older herbalists because of their similar properties" (p. 14). The description of Dioscorides' *Eupatorium cannabinum* bears no actual similarity to the agrimony described by Rufinus' Dioscorides. It should be noted that Pliny (*Natural History,* Bk. XXV, ch. 6, 29) speaks of *eupatoria,* which is also called *agrimonia.*

32. Thorndike comments (p. 8, n. "d") that "this second excerpt from Dioscorides does little but repeat the first. It would seem either that two versions of Dioscorides are being cited or that two chapters from one version of *Acrimonia* and *Sarutium,* respectively, are here combined."

33. The effects described here are consequences of a histamine-like reaction. So powerful is agrimony that, if brought into contact with the mucous membranes, it could produce anaphylactic shock, and perhaps death. Hence the instruction to place it "near the nostrils." The urination and menstruation caused by placing agrimony on the skin near a kidney could also be produced by placing it on any other part of the body. In specifying the kidney, the author is probably reflecting a traditional medical belief that proximity of the medicinal agent to the bodily part or organ affected is partially responsible for producing the observed effect.

34. Grieve observes (I, 12) that "its slender spikes of yellow flowers, which are in bloom from June to early September, and the singularly beautiful form of its much-cut-into leaves, make it one of the most graceful of our smaller herbs."

35. Other names by which it was known are cited in note 22. For a full description, see Grieve, I, 182–184.

36. Wild marjoram (*origanum vulgare*), a perennial herb used by the Greeks for fomentations.

that when centaury is mentioned simply [that is, by itself], the greater should be understood. . . .[37]

Alexander: Centaury is hot and dry. It has the power to dissolve and consume excess humors; it is diuretic. The centaury herb is bitter, in the manner of gall, and is called by another name, "gall of the earth" *(fel terre)*. Moreover, there is a greater and a smaller. When "centaury" is found alone in a recipe, the greater should be assumed, namely the one which produces flowers in the manner of *hypericum*. After removal of the stems, the leaves and flowers should be used in medicine. When this herb produces flowers, it ought to be gathered and dried in a dark place. It retains much of its power throughout the year.

Master of Salerno (Magister Salernus): Greater centaury. The smaller is hot in the third degree. It is not called "smaller" or "greater" because of the smallness or greatness of its effect but [rather because of the size] of its leaves. The smaller purges yellow bile *(coleram)*; the greater, black bile *(melancholiam)*. The smaller can be given for centonic against stomach worms.

Synonyms: Greater centaury joins the leaves near the branch; smaller centaury, *Febrifuga,* or "gall of the earth" *(fel terre)* are the same, and we use the whole herb.

Copho:[38] Centaury is twofold: greater and smaller. Some say that it is hot and dry in the second degree. Galen says that the root of the greater centaury has diverse qualities according to taste and therefore has diverse actions. Indeed, it has a sharp and astringent *(ponticitatem)* taste with a considerable sweetness. And so it is that with its sharpness it provokes menstruation and expels the dead fetus; with its astringency it closes a wound and restricts the flow of blood. . . .

Rufinus: Greater centaury *(centaurea maior)* has a round stem, very bright and green, and leaves like *matersilve,* although small. And the stem passes through the middle of the leaf, and the distance between leaf and leaf is four fingers, and at the top of the stem it has many yellow flowers. The height of the stem is approximately a cubit, and its taste is very bitter, but in its flower there are eight very yellow leaves.

There is also a "middle centaury" *(centaurea media)* which is not mentioned by the philosophers cited above but about which all the herbalists *(herbolarii)* in Bologna and Naples were in agreement. Perhaps the above-mentioned philosophers associated the "middle" with the "smaller" [centaury] as a single type, since these two have a certain similarity in their leaves and flowers. Nevertheless, this "middle centaury" produces many stems on one root and has leaves similar to *maiorane*. Its leaves are separated by three fingers and are sharp and narrow; from two leaves two small branches come forth, and it produces many small purple flowers at the top.

The "smaller centaury" springs forth at the end of May and produces only one stem on one root. Its leaves are similar to middle centaury, and its flower is similar in form to middle centaury; but the smaller centaury has a white and purple flower, for the greater part just like middle centaury; and it is very small and acrid.[39]

(MANDRAGORA) MANDRAKE[40]

Dioscorides: This is an herb whose bark *(cortex)* when mixed with wine is drunk so that those may sleep and feel no pain whose bodies are about to

37. Rufinus' citation from *Macer Floridus* is omitted.

38. Probably not the Copho to whom has been ascribed the *Anatomy of the Pig (Anatomia porci),* written sometime during the first half of the twelfth century (for this work, see Selection 94, n.1). Thorndike, who does not identify Copho, says (p. xxx) that the quotations for which Rufinus cites Copho "almost always correspond closely to passages in the *Liber graduum* [*Book of Degrees—Ed.*], which is found among the works ascribed to Constantinus Africanus in the 1536 edition of his *Opera.* . . ."

39. Perhaps this is red centaury *(Erythraea centaurium),* which is described by Grieve (I, 182) as "an annual with a yellowish, fibrous, woody root, the stem stiff, square and erect, 3 to 12 inches in height, often branching considerably at the summit. The leaves are of a pale green colour, smooth and shiny, their margins undivided. The lowest leaves are broader than the others, oblong or wedge-shaped, narrowed at the base, blunt at the end and form a spreading tuft at the base of the plant, while the stalkless stem-leaves are pointed and lance-shaped, growing in pairs opposite to one another at somewhat distant intervals on the stalk, which is crowned by flat tufts *(corymbs)* of rose-coloured, star like flowers, with five-cleft corollas. . . . A variety is sometimes found with white corollas."

40. *Atropa mandragora.* "It has a large, brown root, somewhat like a parsnip, running 3 or 4 feet deep into the ground, sometimes single, sometimes divided into two or three branches. Immediately from the crown of the root arise several large, dark-green leaves, which at first stand erect, but when grown to full size—a foot or more in length and 4 or 5 inches in width—spread open and lie upon the ground. They are sharp pointed at the apex and of a foetid odour . . ." (Grieve, I, 511).

The leaves had a cooling effect and were used in ointments and poultices; the roots were powerful emetics and purgatives. In large quantities it could produce delirium. When the root was chewed, it could serve as an anesthetic in surgical operations.

be cut [surgically] for the purpose of effecting a cure. There are two species of it: a female [species] with leaves like lettuce *(lactuce)* and cheeks in the likeness of prunes; and a male species with leaves like those of a beet. The mandrake is an herb which some call *appolinaris* [that is, henbane].[41] It has cooling powers and can make one very faint, so that it is believed to be soporific. For it suppresses pain and sleeplessness in men.

Circa instans: The mandrake is of a cold and dry complexion. Its excess is not determined by authors [of this subject]. There are two species of it, male and female. The leaves of the female species are rough [or uneven], and some say that it is more suitable for use as a medicine; [however,] we use them indifferently. Some say that the female has been shaped in the form of a woman and the male in the form of a man, which is false. For nature never assigns the human form to an herb.[42] We have heard from farmers, however, that some [of these herbs] assume such forms.

As for use in medicine, the bark of the root is most suitable, secondly the fruit, and thirdly the leaves. The barks of the root that are gathered are preserved and are very efficacious over [a period of] four years. They have a power to constrict and cool, and a harmless power for causing a small bit of death, that is, for causing sleep. For inducing sleep in fevers a powder of its bark is prepared with the milk of a woman and the white of an egg and is placed on the brow and temples. Against a head-

ache from heat its ground leaves are placed on the temples *(tympora)*. They [the temples] could also be smeared with mandrake-ish *(mandragoraceo)* oil, which is made in this way: After grinding fruits of the mandrake, let them be softened for some time in common oil; afterwards boil it a little and strain [or filter] it. This oil is called *Mandragoraceum* (mandrake-like or mandrake-ish) and would be powerful for producing sleep and for [reducing] a headache from heat if the brow and temples were smeared with oil. Indeed, if the pulse were smeared [with this oil], it would check [and repress] a hot fever. For use against a hot abscess one could spread this oil at the beginning, since it repels matter. . . .[43]

Alexander.[44] Mandrake is cold and dry in the fourth degree. It has a gross substance and can easily be resolved into a great deal of thick smoke. It has a constrictive and enotic *(enoticam)* power, that is, it induces sleep. Its root is medicinal. It should be gathered in summer and dried; it ought to be chosen for heaviness and should be continuous and unperforated. The leaves and fruits of it should be used in unguents, and the root in other medicines. Its powder sucked up with [that is, absorbed by] an egg could cure dysentery from above, and applied in a clyster with barley water, it could cure dysentery from below. Furthermore, a poultice made from its root with the proper juice [or sap], and applied to feverish temples, induces sleep quickly.

2. Taddeo Alderotti (ca. 1223–ca. 1295): The Preparation and Medicinal Use of Alcohol

Translated and annotated by Michael McVaugh[45]

CLXXX. HOW TO MAKE AQUA VITE [ALCOHOL]

Some *aqua vite* is simple, some is compound. The simple is that which is produced from wine alone, without admixture of anything else, and it is called spirit of wine; it is to be taken undiluted by water or wine, just as it is produced from pure wine. Its glory is inestimable; it is the parent and lord of all medicines, and its effects are marvelous against

the old Herbals we find them frequently figured as a male with a long beard, and a female with a very bushy head of hair. Many weird superstitions collected round the Mandrake root. As an amulet, it was once placed on mantelpieces to avert misfortune and to bring prosperity and happiness to the house. Bryony roots were often cut into fancy shapes and passed off as Mandrake. . . . In Henry VIII's time quaint little images made from Bryony roots, cut into the figure of a man, with grains of millet inserted into the face as eyes, fetched high prices. They were known as *puppettes* or *mammettes,* and were credited with magical powers." (Grieve, II, 511–512).

43. Approximately three lines are omitted.

44. This is the second and final entry on the mandrake; Rufinus himself added nothing to his authorities.

45. Translated from Thaddaeus Florentinus, *I Consilia* edited by G. M. Nardi (Torino: Minerva Medica, 1938), pp. 236–238.

The process of distillation of alcohol was known in the twelfth century; Alderotti, however, was apparently the first person to use it extensively for medicinal purposes.

41. Probably *hyoscyamus niger,* which like the mandrake was used to induce sleep and to alleviate pain. Dioscorides (Bk. IV, ch. 69) distinguishes three varieties: *niger* (black), *albus* (white), and *aureus* (golden or yellow) (see Goodyer, p. 464).

42. Although the author of *Circa instans* denied this belief, the Mandrake was the object of much superstition: "The roots of Mandrake were supposed to bear a resemblance to the human form, on account of their habit of forking into two and shooting on each side. In

all cold affections *(passiones)*. It is made in this way: Take the strongest and best red or white wine (red is better, on account of its potency and strength, for it produces this most precious fluid more copiously). Place ten measures of this wine in the vessel mentioned, seal it carefully, and distill it over a slow fire. Collect the fluid which comes out after the initial vapor; it is this which burns. When this is no longer being distilled, discard the rest, for this is of no use in medicine, as I know. Of the ten measures, perhaps a tenth part will be drawn off; if the wine was particularly strong, two [measures], the fifth part, will be enough, if you take so much. If you distill it a second time in this way, you will get half of the two measures which you placed in the still. If this distillate is distilled a third time, you should get four measures of every five; and if you distill a fourth time and perform it correctly, you will find what you placed in the still not at all, or very little, diminished.

The most perfect compounding of this water: If you want to achieve the highest degree of perfection, take a tenth part in the first distillation and distill in this way for ten times, taking just the tenth part in every distillation. After the seventh distillation, the water is called perfect, since it acts so marvelously; after the tenth it is called most perfect, since it perfects and delights with its utter nobility. But because of the difficulty and expenses of such an operation, for most medicines we make the water in the first way mentioned above. If it is distilled three times, or certainly four, it is able to cure virtually all cold afflictions, if you know how to use it.

The effects of *aqua vite:* It picks up the properties of all herbs, save only violets, from their flowers and roots, if they are steeped in it for three hours. The effects of this water are universal within the human body, for it quickly and marvelously cures all the cold afflictions that are curable, particularly ailments of the brain, the nerves, and joints, when taken in drink and applied externally. If the ailment involves a corrupt humor, a purge comes first; otherwise [the patient] should not be purged. In drink, the dose is from one to two drams, more or less according to the strength of the patient. It is good for chronic headaches [that are] cold in origin, if a caputpurge is made with it. It prevents grey hairs and preserves youth. It cures tinea, sores of the head, and every sort of scabies. Washing with it every morning is especially good for the memory. It is good against cerebral epilepsy if it is drunk, if the head is anointed with it, if sniffed, or

if a caputpurge [made with it] is administered; if it is mixed with a little tyriac[46] and taken two or three times on an empty stomach, a few days apart, it cures completely. Administered in the same way, it is good against apoplexy. It is good against palsy and tremors of the limbs. It consumes phlegm when drunk or sniffed or when rubbed on the face and diseased members. In cases of palsy of the tongue, let the patient drink it as discussed above, and let him keep a morsel soaked in this water on his tongue, not just once but repeatedly until he is cured, and he will recover completely. For a facial tic from some disturbance, as occasionally happens to some, the sufferer should quickly rub his face with it, and drink it as described, and he will be cured. Against melancholy and sadness, a half spoonful every morning, taken on an empty stomach, together with a small dipper of fragrant wine, makes a man glad, merry, and happy, and strengthens all the animal virtues against all thick and turbid spirits. It is good by itself against dimming of the eyes and against every failing of sight, especially cataract; rubbed on the temples and eyelids, it dries tears amazingly, dissolves film, and is wonderful for all cold rheums of the eyes. It is good for deafness when drunk or put into the ears on cotton.

It is good for toothache, decayed teeth and rotten gums, when drunk or rubbed on; it clears up bad breath. Drunk with a decoction of figs and licorice, it is good for asthma. It is good for phlegmatic ills of the stomach, for instance, windiness, when drunk or used in a clyster. It relieves tense nerves and cures every cold pain and rheum. It cures cold dropsy when taken repeatedly, to the amount of half an eggshell full three times a day.

It quickly cures every quartan fever, when it has taken up those herbal properties from flowers and roots that the member requires; and administered before the fever's onset, it speedily frees the patient from it. Drunk and administered in a pessary, it is good for sterility of cold origin. It cures dysentery and splenetic ailments. It is good for bronchitis, sciatica, and gout of cold origin, when drunk following the appropriate purge. It is very good for wounds, if they are washed with it. It cures cancer, fistula, *noli me tangere,* and other like afflictions. It expels poison. A man afflicted by Mercury, or *malitia mercurii,*[47] should take *aqua vite* and laurel oil to drink and to anoint the affected place.

---

46. See Selection 110.2.
47. I do not understand this reference.

Its external effect on the body is that anything rubbed with it burns; it restores corrupt, vinegary wine and brings it back to its initial goodness, if a little of it is added to the wine.

Prepared by the second method, in ten distillations, this accomplishes all those things which Aristotle, the supreme master, attributes to *aqua vite* and oil in his book on physics.

110 DRUGS: COMPOUND MEDICINES

Translated and annotated by Michael McVaugh

1. Matthaeus Platearius (d. 1161): The Rationalization of Pharmacy[1]

Some medicines are simple, others compound. A simple medicine is one which is as nature produced it, or which has been artificially prepared without anything else added: examples are pepper, scammony, and many others of this sort. Since medicinal virtue is present in these simples in the highest degree, it may properly be asked, What is the reason for compounding medicines? The reasons are manifold. The first and principal one is for greater efficacy; the second for action against a combination of illnesses, of which one may be quite the opposite of another; the third is for the repression of harmful properties; the fourth is for the preservation of the medicine; the fifth [to disguise] the horrible taste.

Greater efficacy is a reason [for compounding medicines], since some illnesses are compound and cannot be cured with one medicine alone; and because the lesser virtue of one medicine can be increased by the greater virtue of another; it can then do twice as much, which it could not by itself. Combination of illnesses is a reason, since some illnesses are hot, others cold, yet both can exist together in the human body; but one medicine is unable to have different effects on different imbalances *(dyscrasiis)*. A compound medicine can be useful as a laxative when it contains scammony, which primarily purges hot, but secondarily cold, humors. Opposition of illnesses relates to the members. Sometimes the afflicted member is noble, such as the liver, which (being porous) is responsive to dissolution; for if it is chilled, it needs solutive medicines to purge it, so that the pores may be opened and the cold material dissolved. But because violently solutive materials are harmful to the substance of the liver, styptic medicines are to be mixed with the solutive ones, and this will strengthen the weakness of the liver. [Compounding is needed] in order to repress harmful properties, since some medicines, such as the solutive ones, are harmful and sharp, and cannot be taken internally by themselves unless they have

previously been mixed with others to repress their sharpness and harmfulness: for example, mastix, scammony, hellebore, and gariophyllus—for black hellebore and scammony cannot be given by themselves. Compounding is necessary to preserve medicines, since some are naturally humid and quickly decay, so that unless they are mixed with others they cannot be used; examples are ginger, jujubes, and myrobolans. Green jujubes have great efficacy, and lest they decay they are administered with syrups and other things. Compounding is necessary because of a horrible taste, for some medicines, like aloes, are so loathsome that the sick cannot take them, and if they manage to take them, they do not have the required effect; therefore sweet things must be mixed with them to repress their abominable, horrible taste, such as honey and sugar.

Antidotum means "given against," from *anti*, which means "against," and *dosis*, which means "a giving." Those compound medicines are called *antidota* which are compounded from selected medicines, chosen with great care, so that all medicines can really be called *antidota*.

---

1. Translated from the *Expositio super Antidotarium Nicolai* in *Opera Mesuae* (Venice, 1549), (II), fol. 169 recto.

"Mattheus Platearius, who is a tangible historical person of the early twelfth century, . . . appears as the author of a commentary upon the *Antidotarium Nicolai*, which is not only the earliest commentary upon that work but also the earliest extant example of a Salernitan commentary. The appearance of the form of the commentary marks the definite transition from practical to theoretical instruction at Salerno" (Paul Oskar Kristeller, "The School of Salerno: Its Development and its Contribution to the History of Learning," in *Studies in Renaissance Thought and Letters* [Rome: Edizioni di storia e letteratura, 1956], p. 513). Platearius' commentary suggests a transformation of medical education in substance as well as in form, when its analysis of the reasons for a compound medicine's effect is compared with the *Antidotarium's* simple description of the same.

2. *Antidotarium Nicolai* (twelfth century): Traditional Empirical Pharmacy[2]

SOPORIFIC SPONGE[3]

℞ : opii Thebaici oz. 1; succi iusquiami, succi more immature, rubi, seminis lactuce, succi cicute, codii (id est papaveris), succi mandragore, succi hedere arbore an.[4] oz. 1.

Put all these together in a vessel, and in it a marine sponge just as it was taken from the sea, so that fresh water has not touched it. Put this in the sun during the dog days until it is entirely consumed, and when you want to use it, moisten it slightly with very hot water, and put it to the nostrils of the patient, and he will quickly fall asleep. When you want to arouse him, put juice of fennel root to his nostrils, and he will quickly awaken.

· · · · · · ·

THE GREAT THERIAC OF GALEN[5]

Theriac is called the chief of the Galenic medicines, since it was compounded by him. It is good for the most serious afflictions of the entire human body: against epilepsy, catalepsy, apoplexy, headache, stomach ache, and migraine; for hoarseness of voice and constriction of the chest; against bronchitis, asthma, spitting of blood, jaundice, dropsy, pneumonia, colic, intestinal wounds, nephritis, the stone, and choler; it induces menstruation and expels the dead fetus; it cures leprosy, smallpox, intermittent chills, and other chronic ills; it is especially good against all poisons, and the bites of snakes and reptiles, although the dose varies according to the quantity and quality of the different ailments, as is written at the end; it clears up every failing of the senses(?), it strengthens the heart, brain, and liver, and makes and keeps the entire body incorrupt.

℞ : Trochiscorum scilliticorum dr. 2 scr. 2; piperis longi dr. 2 minus gr. 7; trochiscorum tyri, diacoralli an. dr. 2; xylobalsami scr. 2 gr. 7; opii, agariridis, rose, scordei, seminis rape sylvestris, cinnamomi, opobalsami, an. scr. 1 gr. 14; reubarbari, croci, spice, costi, schenanthi, zingiberis, cassie lignee, storacis calamite, myrrhe, terebinthe, thuris masculi, calamenti, dictami, stichados polii, radicis pentaphylon, petroselini, piperis albi, an. scr. 1 gr. 7; folii gummi arabici, acori, chalacanthi usti, sagapeni, terre sigillate seu boli, succi hypocystidos, celtice, chamedreos, gentiane, meu, carpobalsami, amomi, apii, marathri, carui sylvestris, siseleos, nasturcii, seminis anisi, hyperici, an. scr. 1; mumie, castorei, opoponacis, asphalti, galbani, centauree minoris, aristologie longe, dauci sylvatici, an. scr. 1; mellis despumati as needed. Grind up those which require it, and mix the gums, dissolved in wine, with this powder and sufficient honey, or else grind them up with the spices.

It is given in a pill the size of a hazelnut *(in modum avellane)* with warm [water?] for apoplexy, dizziness, headaches, hoarseness of voice, and constriction of the chest; with honey or tragacanth, so that it is held in the mouth, to asthmatics; with a decoction of wild sage *(elelisphaci),* to those spitting blood from the chest and for failings of the lungs with a ptisan; it is given for chronic illnesses with an elixir of hyssop; for jaundice, with a decoction of hazelwort *(asarum);* for dropsy, with oxymel or oxysaccharum; for pneumonia, with henbane *(iure: iusquiamus?)* or a decoction of marrubium; for colic, with an elixir of celery *(apium);* for those suffering from intestinal wounds, with a decoction of sumach; for those with kidney trouble, the stone, and biliousness, it is given with an elixir of gromwell *(granorum solis)* and wild or cultivated celery; for bronchitis, with henbane(?) or a decoction of rue *(pigani);* for use against poison, or for expelling the menses or a fetus, it is given with hot wine, or in honey mixed with water in which mint has been

2. *Opera Mesuae,* II (Supplementum); Venice, 1581, fol. 185.

Like Trotula's work (see Selection 103), the *Antidotarium Nicolai* is a Salernitan work partially traceable to the eleventh century but in its present form dating to the twelfth. It is a collection of some hundred and fifty empirical, time-tested recipes, some perhaps of some value, others not. The two given here suggest both extremes.

3. The effects of opium were known to the classical medical writers and were not forgotten in the Middle Ages. The soporific sponge described here was already in use in the ninth century; from the twelfth century on, it is frequently found included in surgical texts (for instance, the Salernitan "Bamberg surgery," Theodoric) as well as in pharmaceutical manuals—though how powerful an anesthetic it really was is still a matter of question.

4. "an. oz. 1" means 1 ounce of each.

5. The base or focal ingredient of theriac was *tyrus,* viper's flesh, apparently first introduced into antidotes in the first century A.D. to make them homeopathically effective against snakebite. For a discussion of the general principles behind the compounding of theriac in classical medicine, see Gilbert Watson, *Theriac and Mithridatium* (London: Wellcome Historical Medical Library, 1966), p. 71ff.

cooked, or basil *(ozimum);* and for intermittent

chills and all illnesses, with warm [water?].

3. Bernard of Gordon (fl. 1283–1308): Mathematical Pharmacy[6]

CHAPTER 8

CONCERNING URINE OF PALE COLOR, TENDING TO
THICKNESS IN SUBSTANCE, AND CONCERNING THE
CURE OF PHLEGMATIC MATTER

A color tending to paleness, with substance tending to thickness, indicates the dominance of natural phlegm without fever.[7] The paleness of the urine is due to indigestion, the indigestion due to coldness; and the thickness is due to moistness mixed substantially with wateriness. Such a color tends towards milky or greyish or yellowish-green or yellowish, depending on the different degrees of indigestion, the different mixtures, and the various types of phlegm; by itself, however, it indicates nothing but abundance of phlegm. It is possible that there may be fever, but this cannot be discerned from the urine. The symptoms which commonly accompany indigestion and abundance of phlegm are heaviness of the stomach, little thirst, nausea and little appetite, loathing [for food?], heavy sleep with dreams of water, heaviness of the eyes and head, lassitude, indolence, a slow and feeble pulse, and poor memory and understanding (since according to Galen, part three of *De interioribus,* coldness in the first degree and moistness in the second accompany mental confusion, so that such minds are weak and easily changed).

Abstinence is most suitable for the cure of such cases, for they carry food and drink with them: according to Galen, phlegm is changed into blood in time of need. Let them thus avoid all things moist and watery, all fruits, all oils, all fish, broths, egg-yolk, and bones. Wakefulness, exercise, dry and fried things are good for them. The [phlegmatic] matter is digested with those things that heat and dry, and with those things bearing on the stomach, since it is the part of the patient [involved?]; we must therefore bear in mind the dyscrasia[8] and the part of the patient. Let them also avoid completely, from the very beginning, those things which are very hot, lest the matter heat up and so decay.

First of all, therefore, hyssop is prescribed, since it heats and dries in the third degree. The phlegm in the body is colder than it is, so that hyssop is suitable with respect to the primary qualities; since it renders subtle, while phlegm is gross, viscous, and ventose, it is suitable with regard to the secondary qualities, and it is also suitable with regard to the tertiary qualities, since it evacuates the crude humors. Thus, because it has many profitable effects and none harmful, we make hyssop the base; and since according to Mesue we can use a great quantity of such things, of hyssop take one-half pound. Secondly, mint is prescribed, for according to Avicenna it is hot and dry in the second degree, and it has the property of strengthening the stomach and preventing decay; it soothes phlegmatic nausea, strengthens the appetite, and renders subtle. It is thus marvelously suitable. But since (according to Avicenna) it has excess moistness, reduce its quantity, and use four ounces. Thirdly absinth is prescribed, which is hot in the first degree and dry in the second. It has many varieties, but the whitish one is not suitable in this case, since many of its virtues are remitted. Absinth stypticifies, opens, and softens; by virtue of its stypticity it fortifies, but it opens and softens by virtue of its bitterness, and thus that which is not bitter is not suitable here. The stypticity of this variety is helpful, but many of its other properties due to its earthiness are not, especially when it is mixed with earthy substances, so that its quantity should be reduced; and since it is made abominable by its bitterness, we reduce it further and use three ounces. Fourthly calament is prescribed, hot and dry in the third degree; and because it does not fortify and is of scant virtue, we reduce its quantity in comparison to the base, and use two ounces. Then aromatic seeds may be

6. Reprinted by permission of the copyright owner, the Johns Hopkins Press, from Michael R. McVaugh, "Quantified Medical Theory and Practice at Fourteenth-Century Montpellier," *Bulletin of the History of Medicine,* XLIII (1969), 409–413. The translation was made from Bernard of Gordon's *Tractatus de urinis* in *Opus lilium medicinae inscriptum* . . . .(Lyons, 1550), pp. 826–831. A system for mathematically determining the degree of a compound medicine and for preparing one precisely suited for a distemperancy of known degree was described by Arnald of Villanova at Montpellier in the 1290's. His Montpellier colleagues (including Bernard) and his successors there were fascinated by the prospect of an exact pharmacy and produced numerous works explaining Arnald's system in their own words. Bernard's discussion of it in his work *De urinis (On Urines)* is the strongest evidence we have that the theory was ever applied to practice.

7. A perfectly orthodox diagnosis; see Selection 100.1.

8. That is, imbalance of the patient's complexion.

added, hot and dry, such as the seeds of fennel and anise; they are exceedingly aperitive and subtiliative in relation to the material in the stomach, so we reduce their quantity and use one ounce of each. Because all these medicines are hot, we weaken them with roses. Use one ounce of roses to weaken them; and since they (roses) strengthen and are aromatic, add another ounce, which makes two. If it is summer, this medicine is made up with loaf sugar; if winter, with honey. Thus,

R : hyssop—one-half pound
    common mint—four ounces
    green-leaved absinth—three ounces
    calamint—two ounces
    anise ⎫
    fennel⎭—one ounce each
    red roses—two ounces
    loaf sugar—one pound
    and make up a syrup.

This can also be made into an electuary, or a condiment, an oil, a plaster, or a decoction, with a little honey in an elixir if the patient is poor; and if he is naturally weak, remove the absinth and replace it with lemon rind, for lemon rind is hot in the first degree and dry in the third; and the odor of lemon, according to Avicenna, rectifies the corruption and pestilence in the air. Thus in the composition of a medicine we must consider the errant matter, the dyscrasia, and the part of the patient involved.

Because the intensity of the medicine must be increased according to the degree of the dyscrasia, we must therefore know in what degree and fraction of a degree the medicine's intensity lies; if not precisely, at least approximately near the truth. For this we must turn to my book *On Degrees,*[9] where the nature of this gradation is perfectly explained. But in order to make an approximation it is necessary to reduce the medicine to prime doses,[10] and to set the qualities of one kind in one column and the others in another column; then compare them by the twofold method *(secundum viam duplicem)* according to the rules given in the book *On Degrees.*[11]

Let us say, for example, that the prime dose of hyssop in the fifth climate[12] is three ounces; then, since there is one-half pound of hyssop in this recipe, it contains two prime doses. From the first prime dose it has eight hot parts and one cold, and from the second the same; we have therefore sixteen hot parts, predominant, and two halves [=parts] cold. In the second place came mint: let us suppose that four ounces is its prime dose,

so that upon first being administered it can heat to the second degree. It will provide therefore four hot parts and one cold. We join these with the first, and in the column of hot there are twenty parts, while in the column of cold there are only three. In the third place came absinth: let us suppose that its prime dose is three ounces. It will have two parts hot and one cold, and if we join these with the others, there will be 22 parts hot and four cold. In the fourth place came calamint: supposing that two ounces is its prime dose, we have eight parts hot and one cold; adding, we get 30 parts hot and 5 cold. Fifth came the seeds: supposing that the dose of each is an ounce, there will be two prime doses. Thus from fennel, which is hot in the second degree, we get four parts of hot and one of cold; but since anise is in the third degree, supposing that one ounce is its prime dose, this gives us eight parts of hot and one of cold. If we now add, we have 42 parts hot and 7 cold. Sixth were the roses, and supposing that their prime dose is one ounce, there being two ounces used, they will provide four parts of cold and two of hot; add, and there will be 44 parts of hot and 11 of cold. Now suppose that this medicine is compounded with sugar, let us say one pound of it; sugar being in the first degree of hotness, suppose its prime dose to be four ounces, and there will then be three prime doses; we get from the first dose two parts hot and one cold, from the second the same, and from the third likewise, and thus we will have six parts hot and three cold. Add, and there are 50 parts hot and 14 cold. (Although according to Galen, *De simplici medicina,* book I, water is cold, we do not include it;

9. The *Tractatus de gradibus,* completed in 1303.

10. According to Bernard's theory, the "prime dose" is that lowest quantity of a medicine that will produce the medicine's specific effect upon the human body.

11. Briefly, Bernard assumes that the degree of a medicine is related to the ratio of hotness to coldness within the medicine according to the following table:

Medicinal Degree	Hot/Cold Ratio
temperate	1 : 1
hot in the first degree	2 : 1
hot in the second degree	4 : 1
hot in the third degree	8 : 1
hot in the fourth degree	16 : 1

For the corresponding cold medicines, these ratios are of cold to hot.

The Montpellier system bears some formal resemblance to the functional relation assumed in Thomas Bradwardine's dynamic law for local motion (see Selection 51.1, n. 29) and may have suggested it.

12. [For the *climata,* see Selection 1 and note 60 thereto.—*Ed.*]

for according to Galen's *Regimen* it is a constituent of foods and medicine, and so, being a constitutive material, is hot when in hot things and cold in cold things. We therefore disregard it here.)

[6 oz. hyssop	3° H	16	:	2
4 oz. mint	2° H	4	:	1
3 oz. absinth	1° H	2	:	1
2 oz. calamint	3° H	8	:	1
1 oz. fennel	2° H	4	:	1
1 oz. anise	3° H	8	:	1
2 oz. red roses	1° C	2	:	4
1 lb. loaf sugar	1° H	6	:	3
		50	:	14]

That we may see the degree resulting from following the twofold method, let us see what ratio the dominant qualities have to the dominated. It appears that they are more than double, for then there would be only twenty-eight parts hot—if it were so, it would be hot precisely in the first degree. Nor are they quadruple; then there would be fifty-six, and it would be precisely in the second degree. Therefore, since they are more than double and less than quadruple, the medicine is hot between the first degree and the full second degree. . . . This gradation is quite rough, but it suffices for novices, since the doctrine is exactly and precisely communicated in the book *On Degrees*.

Let it be understood that the prime dose should be reduced for the inhabitants of the fourth climate, and increased for inhabitants of the sixth and seventh. I do not claim that these are the precise doses, but rather approximations of a sort; nor do I claim that the syrup is gradated to absolute perfection, only approximately. And it is to be understood that if the syrup is made with honey, the hotness is greater. It should also be understood that as has been said concerning the gradation of the active qualities, so too the gradation of the passive qualities proceeds by the double and by the half. And it should be further understood that in what follows I do not intend to assign a degree to each medicine, since that would be onerous for me and tedious for you. Rather, what I have said is to be taken as an example in accordance with the method of my book, since the theory of gradation is not enough by itself.

Surgery

111 HISTORY OF SURGERY

Guy de Chauliac (d. ca. 1368)

Translated by James Bruce Ross[1]

Annotated by Michael McVaugh

CAPITULUM SINGULARE

The workers in this art, from whom I have had knowledge and theory, and from whom you will find observations and maxims in this work, in order that you may know which has spoken better than the other, should be arranged in a certain order.

The first of all was Hippocrates [5th century B.C.] who (as one reads in the *Introduction to Medicine*) surpassed all the others, and first among

1. From *La grande chirurgie de Guy de Chauliac*, ed. E. Nicaise (Paris, 1890), pp. 12–19, translated by J. B. Ross in *The Portable Medieval Reader*, edited by James Bruce Ross and Mary Martin McLaughlin (New York: Viking Press, 1949), pp. 640–649. Reprinted by permission of The Viking Press, Inc. I have compared the translation with the Latin text in *Ars Chirurgica Guidonis Cauliaci Medici . . .* (Venice, 1546), fols. [viii–ix], and made three short additions.

Guy de Chauliac's historical sketch makes up the larger part of the *capitulum singulare* that begins his *Surgery*, and shows a considerable sophistication in its treatment of the historical development of his profession. Much current history of medicine still follows his interpretations: his feeling that surgery was growing away from medicine per se and that it had lost contact with its theoretical rationale (a common complaint of the surgeons: see Selections 113 and 114); his analysis of the various schools of wound treatment (see Selection 115); or his recognition that the character of European surgery was significantly affected by the introduction of the Arabic authors Abulcasis [whose name is spelled "Albucasis" in this selection—*Ed.*] and Avicenna, about 1200. These authors were first used extensively at Bologna by Bruno Longoburgo and Theodoric, and it is this North Italian tradition which, through Lanfranc and Henry of Mondeville, comes eventually to have a European vogue. With the exception of the selection immediately following this, a severely practical excerpt from the twelfth century, all the material in this section shows the influence of this Arabic surgical literature.

the Greeks led medicine to perfect enlightenment. For according to Macrobius and Isidore, in the fourth book of the *Etymologies* (and as is also related in the prologue to the *Continens*), medicine had been silent for the space of five hundred years before Hippocrates, since the time of Apollo and Aesculapius, who were its first discoverers. He lived ninety-five years, and wrote many books on surgery, as it appears from the fourth of the *Therapeutics* and many other passages of Galen. But I believe that on account of the good arrangements of the books of Galen the books of Hippocrates and of many others have been neglected.

Galen [2nd century A.D.] followed him, and what Hippocrates sowed, as a good labourer he cultivated and increased. He wrote many books, indeed, in which he included much about surgery, and especially the *Book on Tumors Contrary to Nature,* written in summary; and the first six *Books on Therapeutics,* containing wounds and ulcers, and the last two concerning boils and many other maladies which require manual operation. In addition, seven books which he arranged, *Catageni* (that is, about the composition of medicaments according to kinds), of which we have only a summary. Now he was a master in demonstrative science in the time of the Emperor Antoninus [Marcus Aurelius], after Jesus Christ about one hundred and fifty years. He lived eighty years, as is told in *The Book of the Life and Customs of the Philosophers.* Between Hippocrates and Galen there was a very long time, as Avicenna says in the fourth of the *Fractures,* three hundred and twenty-five years, as they gloss it there, but in truth there were five hundred and eighty-six years.

After Galen we find Paul [of Aegina, 7th century] who (as Rhazes attests in the *Continens,* and Haly Abbas in the first of the *Royal Disposition*) did many things in surgery; however, I have found only the sixth book of his *Surgery.*

Going on we find Rhazes [d. c. 923], Albucasis [d. c. 1013], and Alcaran, who (whether they were all one and the same, or several) did very well, especially in the *Books for Almansor* [by Rhazes] and in the *Divisions,* and in the *Surgery* said of Albucasis. In these as Haly Abbas [late 10th century] says, he put all his particulars, and in the *Continens* (which is called *Helham* in Arabic) he repeated the same things, and he collected all the sayings of the ancients, his predecessors; but because he did not select and is long and without conclusion, he has been less prized.

Haly Abbas was a great master, and besides what he sowed in the books on the *Royal Disposition,* he arranged on surgery the ninth part of his *Second Sermon.*

Avicenna [980–1037], illustrious prince, followed him, and in very good order (as in other things) treated surgery in his fourth book.

And we find that up to him all were both physicians and surgeons, but since then, either through refinement or because of too great occupation with cures, surgery was separated and left in the hands of mechanics. Of these the first were Roger [of Salerno, fl. c. 1170], Roland [of Parma, fl. c. 1200], and the Four Masters [anonymous], who wrote separate books on surgery, and put in them much that was empirical. Then we find Jamerius (fl. c. 1230–1252] who did some rude surgery in which he included a lot of nonsense; however, in many things he followed Roger. Later, we find Bruno [of Longoburgo, fl. c. 1252], who, prudently enough, made a summary of the findings of Galen and Avicenna, and of the operations of Albucasis; however, he did not have all the translation of the books of Galen and entirely omitted anatomy. Immediately after him came Theodoric [Borgognoni, 1205–1298], who gathering up all that Bruno said, with some fables of Hugh of Lucca [d. c. 1252–1258], his master, made a book out of them.

William of Saliceto [c. 1210—c. 1280] was a man of worth who composed two compendia, one on medicine and the other on surgery; and in my opinion, what he treated he did very well. Lanfranc [d. before 1306] also wrote a book in which he put scarcely anything but what he took from William; however, he changed the arrangement.

At that time Master Arnald of Villanova [d. 1311] was flourishing in both skills, and wrote many fine works. Henry of Mondeville [d. c. 1325] began in Paris a very notable treatise in which he tried to make a marriage between Theodoric and Lanfranc, but being prevented by death he did not finish the treatise.

In this present time, in Calabria, Master Nicholas of Reggio [d. 1350], very expert in Greek and Latin, has translated at the order of King Robert many books of Galen and has sent them to us at court; they seem to be of finer and more perfect style than those which have been translated from the Arabic. Finally there appeared a faded English rose [*Rosa Anglica* of John of Gaddesden, d. 1361] which was sent to me, and I have seen it. I had thought to find in it sweetness of odour, but I have found only the fables of the Spaniard, of Gilbert, and of Theodoric.

In my time there have been operating surgeons, at Toulouse, Master Nicholas Catalan; at Montpellier, Master Bonet, son of Lanfranc; at Bologna, Masters Peregrin and Mercadant; at Paris, Master Peter of Argentière; at Lyons (where I have practised for a long time), Peter of Bonant; at Avignon, Master Peter of Arles and my companion, Jean of Parma.

And I, Guy of Chauliac, surgeon and master in medicine, from the borders of Auvergne, diocese of Mende, doctor and personal chaplain to our lord the pope, I have seen many operations and many of the writings of the masters mentioned, principally of Galen; for as many books of his as are found in the two translations, I have seen and studied with as much diligence as possible, and for a long time I have operated in many places. And at present I am in Avignon, in the year of our Lord 1363, the first year of the pontificate of Urban V. In which year, from the teachings of the above named, and from my experiences, with the aid of my companions, I have compiled this work, as God has willed.

The sects which were current in my time among the workers in this art, besides the two general ones, which are still in vigour, namely, that of the Logicians or rationalists, and that of the Empiricists (condemned by Galen in the *Book of Sects* and throughout the *Therapeutics*), were five.

The first was of Roger, Roland, and the Four Masters, who, indiscriminately for all wounds and boils, produced healing or suppuration with their poultices and cataplasms; relying for this on the fifth of the *Aphorisms*, "The loose are good, and the hard are bad."[2]

The second was of Bruno and Theodoric who indiscriminately dried up all wounds with wine alone, relying for this on the fourth of the *Therapeutics*, "The dry comes closer to the healthy, and the wet to the unhealthy."[3]

The third sect was of William of Saliceto and of Lanfranc, who wishing to hold the middle ground between the above, cared for or dressed all wounds with unguents and sweet salves, relying for this on the fourth of the *Therapeutics*, that there is only one way to healing, namely, that it be done safely and painlessly.[4]

The fourth sect is composed of all the men at arms or Teutonic Knights and others following war; who treat all wounds with conjurations and liquors, oil, wool, and cabbage leaves, relying on this, that God has put his efficacy in words, in herbs, and in stones.

The fifth sect is composed of women and many ignorant ones who entrust those sick with all maladies only to the saints, relying on this: "The Lord has given it to me as it has pleased Him; the Lord will take it from me when it shall please Him; blessed be the name of the Lord, Amen."

And because such sects will be refuted in the course of this book, let them be put aside for the present. But I am amazed at one thing, that they follow each other like cranes. For one says only what the other has said. I do not know whether it is from fear or love that they scorn to hear anything except what is traditional or proved by authority. They have read Aristotle badly in the second of the *Metaphysics* where he shows that these two things most of all prevent the sight and knowledge of the truth. Let us abandon such friendships and fears, for Socrates or Plato is our friend, but truth is still more a friend. It is a holy and worthy thing to honour truth in the first place. Let them follow the dogmatic doctrine of Galen approved in the *Book of Sects* and throughout the *Therapeutics*, which is entirely composed of experience and reason, in which one seeks things and scorns words. And he himself has taught the means of seeking truth, in the book on *The Constitution of the Dogmatic Art*, chapter seven, which under a certain epilogue he put in the third book of the *Natural Faculties*, chapter ten, in this way: he who would know something better than the others must suddenly be very different from the others (that is, at the beginning and by nature, and by first knowledge). And when he is a boy, or at the age of puberty, he must be seized with a certain amorous rage for the truth; he must not cease to study day and night, to learn all that has been said by the most famous of the ancients. And when he shall reach the flower of his age and learning, then he must judge what he has learned by examining it very well for a long time and perceive all that agrees with the things which are clearly apparent, and all that disagrees with them, and thus choose the one, and reject the other. To such a one, I hope my observations will be very useful; but to the others, these writings will be as superfluous as a fable told to an ass.

I do not say, however, that it is not a very good thing to cite witnesses in one's discourse, for Galen in many places, besides reason and experience, which are for all men the two instruments of judg-

2. *Aphorisms* V. 67.
3. Kühn X. 278.
4. I cannot tell to what part of Book IV Guy (and the others) refer.

ing (as he says in the first of the *Therapeutics,* chapter three), brings in the third, by witnesses. Of which in the first of the *Miamir* or *Compositions according to Plans,*[5] he says that the belief in things which one writes increases with the agreement of those who repeat them; and therefore he says that he will write down all the medicaments which have been made up by expert doctors. And thus I shall do (as I have said) in my procedure, with the aid of the glorious God.

Let us return to our theme, and put down the conditions which are requisite to every surgeon, who wishes by art to exercise on the human body the aforesaid manner and form of operating, which conditions Hippocrates, who guides us well in everything, concludes with a certain subtle implication, in the first of the *Aphorisms:* life is short, and art prolix, time and chance sharp or sudden, experience fallacious and dangerous, judgment difficult. But not only the doctor must busy himself in doing his duty but also the sick person and the attendants, and he must also put in order external things.

There are then four conditions which are valued here, according to Arnald, a very eloquent Latinist. Some are required of the surgeon, others of the sick, others of those present, others in those things which come in from outside.

The conditions required of a surgeon are four: the first is that he be educated; the second, that he be skilled; the third, that he be ingenious; the fourth, that he be well behaved. It is then required in the first place that the surgeon be educated, not only in the principles of surgery, but also of medicine, both in theory and practice.

In theory he must know things natural, non-natural and unnatural. And first, he must understand natural things, principally anatomy, for without it nothing can be done in surgery, as will appear below. He must also understand temperament, for according to the diversity of the nature of bodies it is necessary to diversify the medicament (Galen against Thessales, in all the *Therapeutics*). This is shown by the virtue and strength of the patient. He must also know the things which are not natural, such as air, meat, drink, etc., for these are the causes of all sickness and health. He must also know the things which are contrary to nature, that is sickness, for from this rightly comes the curative purpose. Let him not be ignorant in any way of the cause; for if he cures without the knowledge of that, the cure will not be by his abilities but by chance. Let him not forget or scorn acci-

dents; for sometimes they override their cause, and deceive or divert, and pervert the whole cure, as is said in the first to Glauconius.

In practice, he must know how to put in order the way of living and the medicaments; for without this surgery, which is the third instrument of medicine, is not perfect. Of which Galen speaks in the *Introduction:* as pharmacy has need of regimen and of surgery, so surgery has need of regimen and pharmacy.

Thus it appears that the surgeon working in his art should know the principles of medicine. And with this, it is very fitting that he know something of the other arts. That is what Galen says in the first of his *Therapeutics* against Thessales, that if the doctors have nothing to do with geometry, or astronomy, or dialectics, or any other good discipline, soon the leather workers, carpenters, smiths, and others, leaving their own occupations, will run to medicine and make themselves into doctors.

In the second place, I have said he must be skilled and have seen others operate; I add the maxim of the sage Avenzoar [12th century], that every doctor must have knowledge first of all, and after that he must have practice and exprience. To the same testify Rhazes, in the fourth *Book for Almansor,* and Haly Abbas on the testimony of Hippocrates, in the first of his *Theory.*

Thirdly, he must be ingenious, and of good judgment and good memory. That is what Haly Rodan [11th century] says in the third of his *Techni;* the doctor must have good memory, good judgment, good motives, good presence, and sound understanding, and that he be well formed, for example, that he have slender fingers, hands steady and not trembling, clear eyes, etc.

Fourth, I have said, he should be well mannered. Let him be bold in safe things, fearful in dangers, let him flee false cures or practices. Let him be gracious to the sick, benevolent to his companions, wise in his predictions. Let him be chaste, sober, compassionate, and merciful; not covetous, or extortionate, so that he may reasonably receive a salary in proportion to his work, the ability of his patient to pay, the nature of the outcome, and his own dignity.

The conditions required of the sick man are three: That he be obedient to the doctor, as the

5. This is the *De compositione medicamentorum secundum locos* (Kühn XII.378–XIII.361). Galen's *De compositione medicamentorum per genera* (Kühn XIII. 362–1058) is the *Catageni* to which Guy refers above.

servant is to the master, in the first of the *Therapeutics;* that he have faith in the doctor, in the first of the *Prognostics;* that he be patient, for patience conquers malice, as it is said in other writing.

The conditions for the attendants are four, that they be peaceable, polite or agreeable, faithful, and discreet.

The conditions for things coming from outside are many, all of which ought to be arranged for the advantage of the sick, as Galen says, at the end of the commentary of the *Aphorisms* mentioned above.

112 SALERNITAN SURGERY

Roger of Salerno (fl. 1170)

Translated and annotated by Michael McVaugh[1]

I. ON INJURIES TO THE HEAD

Chapter 1.

The head may suffer several kinds of injury. Sometimes the injury involves fracture of the skull, sometimes not. Sometimes such a fracture associated with the injury is considerable and obvious; sometimes it is small. Both large and small fractures may occur with a large, wide wound; or they may occur with a small and narrow one. But whatever sort of fracture of the skull is involved, we must always be alert for injury to the cerebral membranes; sometimes the pia mater is injured, and sometimes the dura mater. Injury to the dura mater is known by these symptoms: the patient suffers pain in the head, is red of face with inflamed eyes, is deranged, and has a blackened tongue. Injury to the pia mater can be recognized by these symptoms: a failing of strength, loss of voice, appearance of pustules on the face, flow of blood and pus from the nostrils, constipation, and chills three or four times a day, which [last] is a certain sign of death. If all or most of these symptoms are to be seen, death will follow or can be expected during the next hundred days at most. And in particular, if there is injury to the cerebral membranes, it usually happens that the patient dies at the next full moon. Thus, since there is considerable danger from fracture of the skull, we will proceed to explain systematically how we can treat it.

Chapter 2.

When the fracture of the skull is considerable and obvious, with a broad, long wound, as if made with a sword or something similar, and bone has to be withdrawn (unless there would be a great gush of blood, or unless something else interferes), the bone to be withdrawn is removed and a very fine linen cloth is carefully introduced as it were obliquely between the skull and the dura mater, using a feather. At the opening of the fracture, a linen or silk cloth, preferably long enough for both ends to pass under the head, will prevent corrupt matter from flowing from the outside onto the dura mater, which would bring about still greater harm to the brain. A clean, dry marine sponge is also used, for this thirstily soaks up the corruption deriving from the surface. The external wound should be carefully packed with linen soaked in egg-white and slightly pressed out, a little feather placed on top [for drainage], and the whole bound up carefully, following the contour of the head. The dressing should be changed twice [a day] in winter, three times in summer; the patient should be placed to lie with the injured part downward [for drainage]. This treatment is to be maintained until the skull is fully healed.

Chapter 3.

If excess flesh should grow upon the dura mater before the skull has healed, keep a clean, dry marine sponge there until the superfluous flesh has been eaten away. And when, after the skull has healed, superfluous flesh has grown up at the site of healing, we have been accustomed to employ a powder of hermodactyl at the site [to eat it away]. The external wound we cure simply with lint and bandage. After the wound has closed, we use the surgical apostolicon: 1/2 pound of naval pitch; 1 pound of Greek pitch; 1/2 ounce each of galbanum, serapinum, ammoniacum, and opoponax; of wax, 3 ounces; 1/2 pound of vinegar; and make it up as follows. Put the vinegar in a tin vessel together with the gums, which should not be ground up; that is to say, with the galbanum, ammoniacum, serapinum, and opoponax. These things should be placed over a fire to liquefy, and when they have

1. Translated from *Rogerri Medici Celeberrimi Chirurgia,* in *Ars Chirurgica Guidonis Cauliaci Medici . . .* (Venice, 1546), fols. 362 verso to 363. I have not translated the chapter titles, which are no more than restatements of their introductory sentences.

melted, a little of the liquid should be put in cold water. When this thins out and changes color, add the above-mentioned quantity of Greek pitch, finely powdered together with mastix and olibanum, 1/2 ounce of each. Put this with the others in the tin vessel, and, continually stirring with a spatula, mix all well together. When it turns whitish, almost lemon-colored, it is a sign that it is ready; take it from the vessel, and add 1/2 ounce of terebinth; mix well. Strain this mixture through a bag into cold water; anoint your hands with *oleum laurinum* or some other such, and, taking it from the water, knead it so that the water may be forced out of it. Then make it into plasters. This surgical apostolicon is good for spleen, and, after the bone is brought together, helps solidify bone and flesh. It is also good for pain from injuries to the chest, such as arise from contusion, percussion, or falling.

Chapter 4.

If the fracture of the skull is large, but the wound small in area, so that you cannot fully determine the extent of the fracture, introduce your finger into the wound and carefully probe it; for there is no better way to determine the nature of a skull fracture than by your sense of touch. After you have generally determined the extent of the fracture, cut the narrow wound with a razor in the form of a cross, and with a scraper *(cum rugine)*—that is, an iron instrument—separate the flesh from the skull. And unless a large quantity of blood, or something else, should prevent it, bone or anything else that has to be taken out should be removed with forceps. But if there is an efflux of blood, wait until it ceases, although it means postponing your treatment; just remove it as soon as you can. Then carefully introduce a cloth between the dura mater and the skull with a feather, and pursue the treatment of the skull that we have outlined above. The four quarters of the incision then having been drawn tightly together, the whole wound is to be bound up with a linen cloth soaked in white of egg, together with a little feather, and wrapped according to the contour of the head; let it be unbound from morning to evening, or vice versa. If on coming back to the wound you find the four quarters swollen, it is a good sign; while if they are shrunken and show signs of mortification, it is

a bad one. Continue this external treatment until you know that the skull is fully healed, and then you can reduce the bandage and bring the quarters back to their proper place; you should not use lint, or another cloth, to the very end. For in wounds involving fracture of the skull we only use one cloth, unmoistened [by egg-white] after the second or third day. Indeed, we entirely avoid using an ointment or anything greasy; but we do put some of the surgical apostolicon on the skin at the end.

If you want to apply something of your own, make up the following ointment, which you can safely put on the external lips of the wound: take saffron and put it in water, and let it stand long enough for the water to become well colored; then strain it, and add wheat flour to the strainings, mixing it in well. Boil this briefly over a fire, stirring continually. This, kept for use, eases pain and soothes.

Chapter 5.

It may happen that the skull suffers a hairline fracture, so that one side seems neither higher nor lower than the other; and it cannot be told whether such a fracture runs through to the inside or not. In order to tell, let the patient hold his mouth and nostrils shut, and exhale strongly, and if he forces something out through the crack, the skull is split down to the brain itself. We cure this in the following way. If the wound is narrow, enlarge it, and unless there is too much blood, make a space on either side of the crack. Then, with a trepan (an iron instrument), using the greatest care, make as many holes as you think necessary, and with a chisel *(spathumen)* cut the skull itself from one hole to another, so that the incision runs to the ends of the crack; by this means you can carefully withdraw the putridity that has collected above the brain, using cotton or a very fine cloth inserted with a feather between the brain and the skull. To treat the wound from this point on, follow those instructions that we have given above. If the skull is so fractured that one side is slightly depressed, so that you cannot easily remove a fragment from the side where it adheres, you should begin to cut; make as many openings as seem necessary to you; then treat as described above, with the chisel and other instruments.

113 DEFINITION AND OBJECTIVES OF SURGERY

Theodoric, Bishop of Cervia (1205–1298)

Translated by Eldridge Campbell and James Colton[1]

Annotated by Michael McVaugh

INTRODUCTION

To the venerable father and my dearest friend, that excellent gentleman Master Andres Albolat, by God's grace bishop of Valencia, Brother Theodoric, through God's indulgence an unworthy minister of the church of Botonto, offers this work so long in preparation.

Some little time ago, dearest father, while we were both living in Rome, you asked me affectionately (I was at that time your chaplain and a courtier of His Holiness the Pope) to describe clearly and briefly to explain certain matters concerning the art of medicine and surgery which are rather obscure and involved, and incompletely treated by the ancients, making a book about this according to the system of medicine of the excellent Hugo of Lucca,[2] a most expert man in the aforementioned science. Willing to comply in part, if not wholly, with your wishes, I produced a book which you, in your impulsive way, carried off to Spain, though you knew at the time that it was unfinished and uncorrected. Then you asked, and afterward by messenger required in frequent letters that I should send over the same book corrected, with the omissions supplied. For a long time I put it off. But now, with old age approaching, invoking the help of Christ and His guidance, with every evidence of envy left aside, I am laying bare the secrets of the art of surgery.

Take then, dearest father, this little work, remarkable indeed for its small size, yet full in content.

Because I was with the aforementioned Master Hugo for a very short time, and was able neither to see nor to grasp nor to learn to the full his most skillful cures, I shall therefore present what was unfinished, in that portion, from my own experience. And I shall take care to complete the accounts of the ancients, especially of Galen, who we know differs with the aforementioned excellent man in no respect. Moreover, in order that the approach to this book may be easier, I shall briefly point out what surgery is, whence the work comes, what sort of operators it requires, and indeed what should be their objective, and into how many classes it may be divided.

Surgery, then, is a manual operation upon the body of an animal and endeavors to cure. Or thus: Surgery is the ultimate resource of medicine. Now the resources of medicine are three, with which treatments the doctor may attack the causes of diseases: Thus; diet, draught [i.e., medication] and surgery.

Diet is the first method, and is the best, so they say: Galen, in his commentary on the *Regimen in Acute Diseases* [*Hippocrates, De regimine acutorum*], saying: "If we can cure a man with diet, we should not cure him with any draught"; and Damascenus on the *Aphorisms* [Hippocrates],[3] "If you can cure a man by diet, you will find it fortunate: however, this resource or regimen is required to counteract the cause of the disease and promotes digestion."

The second method or resource is the draught, concerning which the same Galen says, "If with a draught we can remove the sickness, one should not resort to surgery. If, however, by the two aforementioned methods or resources it cannot be removed, let it be removed by the aid of the third method, to wit, surgery."[4]

Now "surgery" is said to come from cyros [χείρ] which is *hand,* and agia [ἔργω], which is action: thus surgery [χειρουργία], an operation by

---

1. Reprinted by permission of Appleton-Century-Crofts from *The Surgery of Theodoric ca. A.D. 1267, translated from the Latin by Eldridge Campbell and James Colton* (New York: Appleton-Century-Crofts, 1955), I, 3–7.

2. Theodoric insists throughout his book that the treatments he recommends are those which he has learned directly from Hugo; as his translators put it, "Theodoric was possibly the actual son and certainly the professional and spiritual son of Hugo of Lucca" (p. xv). This background adds to the very considerable interest of Theodoric's technical accomplishments, since his *Surgery* is thus a symbol of the increasing importance of textual instruction—as opposed to Hugo's oral instruction of Theodoric and his other sons—in thirteenth-century surgery.

3. Not the aphorisms of Hippocrates, but of Damascenus himself: the *Aphorismi Johannis Damasceni* (for instance, Bologna, 1489) is a translation of a work of Yūḥarrā ibn Māsawaih (d. 857), called both Mesue and Damascenus in the Latin West.

4. This refers to the *Isagoge.*

the hand, as it were. It behooves practitioners of surgery, as Haly Abbas says in his commentary, to frequent the places where skilled surgeons operate, and to attend their operations diligently, and to commit them to memory. There is no need for them to be rash or daring, but let them be foresighted, gentle, and circumspect, in order that with the greatest deliberateness and gentleness under all circumstances they may operate with what gentleness they can, and especially around cerebral membranes, sensitive parts, and other ticklish places; because all the things which are necessary to the art cannot be included in books, cannot be easily foreseen, and many of these frequently happen to the operator. Damascenus indeed highly commends native talent in a physician. He says, the native talent of a physician aids his skill, and on the other hand, his control of natural forces. They must needs be well-read, and even if they be aided sometimes by experience, yet frequently will they fall into error and into confusion. I scarcely think that anyone can understand surgery without schooling. Now Almansor [Rhazes][5] contends that the practitioners of this art are indeed for the most part uncouth and unfeeling ignoramuses, and by reason of their stupidity most unfortunate things can happen to people. Generally, also, when their operations are not performed with a sure diagnosis, nor done with a plan, people are killed by reason of their lack of skill. They do not really recognize the causes nor yet the names of the diseases which they claim to cure.

The objective of surgeons' operations is concerned with three things, thus: the joining together of broken parts; the separation of parts unnaturally joined; and the removal of whatever is superfluous.

The forms of surgery, as Joannitius bears witness, are two: for he says, "Surgery is of two kinds, either upon flesh, or upon bone, and according to the contrasting nature of the underlying parts. For one type may be done upon soft parts, as upon flesh, muscles, veins, and the like; the other may be performed on hard parts, as on bones and the like. And just as these differ in the contrasting nature of their qualities (that is to say in softness and hardness), so also is there this difference in every surgical operation."[6]

The surgery which may be performed upon soft parts is subdivided into three categories: either it may be done on flesh only, which is termed scarification or incision, or suture; or it may be performed upon veins with no pulse, or upon pulsing ones, and is called phlebotomy or section; or it may be done across the muscles, and then is called incision, or lengthwise and then is called puncture, or splitting. In the case of the diaphragm, however, and of membranes, it is named breaching [rupture]. But the surgery which may be done upon hard parts is of two kinds, for it may be performed upon hard bones, and is called repairing, or the setting of fractures; or it may be done upon bones dislocated from their proper position, and is named the joining of the disjointed.

However, in order that it may be possible for each to discover more easily what he wishes, we shall divide this book into four sections in orderly fashion, separating the individual portions by chapters. Thus, then, in the first part of this book we intend to discuss these things not by appearance, but under a certain general classification.

This book is divided, then, into four parts.

In the first part there is a complete discussion of wounds, ulcers, hemorrhages and the medicines for these; of the signs of death, wounds of the sinews, the cure of hot abscess and of convulsions, as well.

In the second part is a discussion of the same things, according to what is necessary for surgical practice on each part, from the head down to the feet.

In the third will be discussed fistulas, cancer, itch [mange], cancerous growths, scabies, and the other abscesses, and the infections which take hold upon the human body in plain view.

In the fourth and last part we shall end the book by writing some things approved by us for pains in the head and for migraine and by discussing diseases of the eyes, the ears, the nostrils, the mouth, and of the teeth, of arthritis and of gout. And thus, if life last and God grant it, we shall finish this book.

5. *Almansor* or *Liber Almansoris* was the title usually given to the Latin translation of the medical compilation of Rhazes, *Kitāb al-Mansūrī.*
6. The reference is to the *Isagoge*: compare with Selection 90.

114 BLOODLETTING

Lanfranc (d. before 1306)

Translated and annotated by Michael McVaugh[1]

TRACT III, DOCTRINE 3, CHAPTER 16

Phlebotomy is an artificial diminution of the blood contained in the veins. You know that, although because of our pride the office of phlebotomy is today left to barbers, it was once the work of physicians, particularly when surgeons carried it out. Oh Lord, why is there such a distinction made today between the physician and the surgeon? Perhaps because physicians abandoned manual operation to laymen? Or because, as some say, they disdain to work with their hands? Or, as I believe, because they do not understand the method of operation that is necessary? This false distinction is so entrenched, because of surgery's earlier disuse, that some people believe that it is impossible for one man to have a mastery of both. Let everyone know, therefore, that a man who is wholly ignorant of surgery cannot be a good physician. Moreover, a surgeon who is ignorant of medicine ought to be held as nothing—rather, it is necessary for him to know every aspect of medicine well.

We use phlebotomy, generally, for three things: to preserve health, to protect it from possible sickness, and to remove an existing illness in one of the first two ways (which seem the same, but are different). Phlebotomy is called elective, since we can choose time and hour, climate, or disposition of the patient, considering many particulars and waiting for the ideal time and moment. Although we consider many particulars when need be, nevertheless sometimes, omitting other particulars, we insist on nothing except the strength of [the patient's] forces. Actually, necessity may sometimes compel us to bleed him when his forces are weak, but then we do in three or four successive operations what could be done with stronger [patients] in one. We employ phlebotomy with those who eat considerable meat and drink good wine, and whose faculties (virtutes) produce considerable blood, even though it be healthy blood, and who take little exercise; particularly the young or certain old people who are used to it. This method of treatment is allied to the conservation of health. Secondly, we bleed those who suffer with a sanguine pain in the joints, or a constant fever (synocha), or quinsy, or pleurisy, or an intermit-

tent pain. All these things can be avoided by treatment before the normal time of onset of the sickness, and this method is called preventative. Thirdly when a man suffers from a strong pain in the head without fever,[2] from quinsy, pleurisy, pneumonia, hot apostemes of the internal members, or any other illness which derives from overabundance of blood, phlebotomy is performed to remove the incipient or established illness, and this method of treatment is called curative. The necessity for bleeding is common to these three general treatments; still, it is much better for a man to be governed by a regimen including adequate exercise and temperate food and drink, and better to live moderately and abstemiously when he feels himself replete, than to live in such fashion that he comes to require phlebotomy. I would say that this is equally true of laxative medicines.

Let me now propose three universal headings for phlebotomy, under which I will arrange my teachings on the subject. First will be what sort of a man the bloodletter should be, and how he should carry out his office; second, who those people are who require phlebotomy; and third, which those veins are on which bloodletting should be carried out, and how each vein should be opened.

A bloodletter should be a young man, neither a boy nor old; he should have steady hands and be wholly strong of body; he should have good and subtle sight and should be able to recognize the veins and to differentiate them from nerves and arteries; and he should know the different places in which the veins that are to be bled may be opened, and should know how to avoid all danger in those places where the veins come near the nerves and arteries. He should also possess several lancets [phlebotomos] of steel, bright and clean, of different shapes, some of which should be fine and some

1. Translated from *Practica Magistri Lanfranci de Mediolano quae dicitur ars complete totius chirurgiae,* in *Ars Chirurgia Guidonis Cauliaci Medici . . .* (Venice, 1546), fols. 249–250. In preparing the translation I have profited from the Old English version published as *Lanfrank's "Science of Cirurgie,"* Early English Text Society, Original Ser., No. 102 (1894), Part I, pp. 298–303. I have also added or summarized some of the editors' remarks.

2. This follows the reading established by the editors of the EETS version.

somewhat bigger, some short and some longer, so that when it is necessary to open a large vein and make a large incision he may use a larger lancet, and the smaller one in the contrary case. He should hold the lancet with the thumb and index finger of his right hand and boldly palp the vein beforehand to find the best place for bleeding; and then, with these two fingers, insert the lancet in whatever vein he wishes.

As to the second topic, you should know that children should not be bled before puberty, save in urgent necessity—for example, if you saw someone suffocating from too much blood, as evidenced by shortness of breath, by the fullness of the jugular veins in his neck, by the redness of his whole face, or its fullness and humors, and by the bearing of his body. In such a case, before you bleed him, speak to his parents and friends as follows: "Do not say that I advised you that he should be bled, because I am not saying this; but I will say that phlebotomy is necessary in order for him to live. If he is bled and should die, it will not be because of the phlebotomy; but if he is cured, it can only be by bleeding; and if he were my son, I would have him bled, seeing no other way. Still, you choose whatever you or others think best." You should speak thus in all uncertain cases, where something dangerous is to be done. Likewise old men should not be bled, particularly the more decrepit—although this rule is sometimes contravened, since there are certain old people whose forces (virtutes) are stronger than those of some young people. And those recovering from illnesses should not be bled, particularly when they have had a good, full crisis; nor should pregnant women, particularly in the first three and last three months, even though they may be accustomed to being bled regularly, before and after the life of the embryo, without it harming them. Youths whose color is white and pale and whose beard is scanty and thin, and who have tiny veins, difficult to find, are not suited to bleeding; nor are those whose bodies contain crude and melancholic humors, and who have little good blood, for that little blood should be guarded like a treasure. In this respect the French do much harm, since they bleed themselves when they are filled with gross, cold, corrupt humors; they see the putrid, corrupt blood [drawn off] and think they have done very well to remove such blood. The barber will say, "See how much you needed that bloodletting; it will be necessary to draw some more soon"; but a man who goes to the barber all the time will be destroyed by [blood] letting. It is

thus better for him to conserve his blood and to evacuate the corrupt humors by other means. Phlebotomy is also not proper for those who have the first stages of cataract, nor indeed is any diminution of blood, especially with cupping-glasses (ventosis).[3] But phlebotomy is proper in all the cases named above, for if a man eats much meat and drinks much wine, other things being in accord, he is not safe: Sanguine illnesses may be generated in him, or perhaps he may drop dead suddenly. But if he is bled, his health can be preserved longer. Likewise, he who habitually suffers from a sanguine rheumatism will no longer have pain if he is adequately bled before the time of its onset, unless you do something else harmful. This is true of all illnesses which are due to too much blood. In the case of one who suffers from a constant fever, phlebotomy is so necessary that to bleed him till he faints, before the fourth day, will either remove the fever or will so far lessen the matter that the remainder will not putrefy; but if he is not phlebotomized, the blood sometimes boils up into his chest because of the heat and collects there in such great quantity, given the size of the place, that the patient suffocates. Sometimes a vein bursts in the chest or in the lung, and if its flow is not restrained, the patient will die—unless his natural strength moderates the flow, as sometimes happens. I say "moderates," because some-

3. Cupping was another, equally valuable, method for evacuation of humors. Lanfranc treats it in the next chapter of his *Greater Surgery*; Mondeville's description of the technique, however, is more succinct: "The *ventosa* is a glass vessel, round and flat, with a narrow mouth but a large base; it is of course well known. The site where you wish to apply it should previously have been warmed, scrubbed, and slightly oiled, so that the skin will be soft and the application of the *ventosa* less painful. Within the *ventosa* you should place a little torn-up cloth waste, to which you set fire with a straw or candle; then quickly apply it, so that it can quickly catch hold of the skin [as the flame goes out and the air within cools]. Do not remove it until it falls off of its own accord. If it should be useful, you may successively apply several just like the first. Or, if you want to apply a *ventosa* over an incision of the skin, scratch the site several times and then apply the *ventosa* containing its kindled tow. When it has sucked out about a full eggshell of blood, remove it, empty it, and apply it again, as before, until it has sucked out a little more than it did the first time; and you can replace it several times until you have drawn off enough, each time increasing the quantity of blood [removed] over that taken off the previous time. When the operation is complete, wipe the wound dry several times with a hot sponge." (Henry of Mondeville, *Surgery*, Tr. III, Doct. 1, ch. 4, sec. 6; tr. E. Nicaise [Paris, 1893], pp. 559–560.)

times the blood of unbled patients has such force that it bursts through the nose with such force that it brings on the patient's death. There is therefore no avoiding bloodletting in such fevers, if the patient can be bled in any way. If not, you must extinguish the fever with drinks of much cold water, which cannot do anything harmful save prolong the duration of the fever; as Rhazes points out, this is better than to leave the patient to die without it. It is also good for a constant fever to bleed the patient before the fourth day, since the fever and its concomitants are thereby reduced. In this case the blood should be diminished within the body in order to maintain its strength, so that the natural heat may be the food which defends the members and body from non-natural heat; Avicenna discusses this very well. In pleurisy, phlebotomy may be carried out if necessary, since this removes the illness and its concomitants, which as Galen says is the best medicine. It is also useful for all hot apostemes, both external and internal; it is useful, indeed necessary, in cases of sanguine quinsy; it is also useful for spasm caused by repletion and in such cases phlebotomy should be performed without using medicines, especially if it has resulted in a wound or aposteme. Generally, therefore, it is useful in all sanguine illnesses, which can never or rarely be cured without it.

Third, while the veins which branch from the ramose vein which comes from the curve of the liver[4] continue to branch many times over, and although many people have extensively described their possible incisions, in the interests of brevity I will discuss here only the opening of those veins which are in use. There are three particular sites in either arm which are customarily bled. First, the cephalic vein, which is bled in two places, near and slightly above the bend of the elbow. You must bind the arm above the site and compress it slightly until the vein appears, and you should cut it broadly *(large)* since its narrow *(stricta)* incision can create numerous apostemes. You must beware lest you touch the muscle, which lies very near. This vein is also bled between the thumb and index finger, at and a little above their conjunction. It is bled for hot ailments of the head, and of the neck and the members from the breastbone up, generally. The basilic vein is bled in the lower part of the arm, where it too lies near the bend of the elbow; it is right over a large artery, or quite close to it, so that you must pay the greatest attention lest you touch it. It is also bled between the little and ring finger of either hand. Its bleeding is good for all

ailments from the breastbone down, and is also valuable in pursuing a preventative regimen against ailments of the head. The median vein is formed from the basilic and the above-mentioned cephalic veins and likewise opens in the lower arm near the elbow; you must be very careful not to touch one of the two nerves between which it lies. Bleeding it is useful when you mean to evacuate the body, and therefore assists against ailments of the heart and of all the pectoral members, once these ailments are well-established—for before they establish themselves it is more useful to bleed from the basilic vein. [It is also good] against incipient ailments of the head, those which have not yet become established, since after their establishment bleeding is better from the cephalic vein, as has been said. Bleeding from the vein of the thumb is good for head ailments; so is bleeding from the cephalic vein, from which the vein of the thumb arises, but bleeding from the thumb weakens the patient less, and no mistake can arise from its bleeding. The vein which is between the ring and little fingers, proceeding from the basilic vein, is bled for ills of the liver on the right hand, and for ills of the spleen on the left. In the head, the vein of the forehead can be bled for head ailments, particularly of the back of the head, both before and after their establishment, as I myself have verified; this sometimes does away with frenzy, especially when all the material is already present in the head(?). I had as a patient a woman who had an almost unbearable pain in the head, whom I bled from the hand and also purged, to no avail; there was still more material in the front of the head, and the vein [in the forehead] was more prominent than usual, so that I ordered that it be opened, and she was soon well. When you want to bleed from this vein, constrict the neck, and cut the vein lengthwise. Sometimes this vein is also cut at its highest point; this is good for hardened ulcers of the head, for instance *saphati*,[5] and for sanguine *tinea*, particularly if the head is bathed with the blood drawn off.

The veins in both temples may be bled for headache and eye ailments. I once treated a young man who was suffering from a persistent hot headache, repeatedly purged and bled him, and although he

---

4. Portal vein and vena cava.

5. The editions of Lanfranc read *saphati*, Add. MS. 26,106, fol. 89 *saffira. Matth. Silv.:* "Saphiros, sephiros idest apostema durum cui non associatur sensus." *Vigo Interpr.:* "Sephiros is an arabike word, and it is called in Greke scirros, in latyne durities, that is hardenes" (EETS).

found some relief he continued to be troubled with the same thing. Then I cut the vessel *(arteriam)* on the afflicted side, cauterized it (lest it consolidate further), and thereafter his cure was lasting. Likewise, it is good for facial sores and headache to bleed from the veins in back of the ears. Incision of the veins of the tongue is good for quinsy, apostemes of the tongue or throat[6] (when preceded by a bleeding from the cephalic vein), for acute rheum of the eyes, prurigo, nasal sores, and dizziness, whenever these things are of a sanguine origin. Sometimes the jugular veins in the neck are bled when suffocation from excessive blood is feared, and sometimes too in cases of leprosy. The veins in the lower lip are bled for hot sores in the mouth, and for apostemes and hot ills of the gums. In the feet, three veins in each foot are frequently bled: one under the curve of the knee, bled for illnesses of the womb and to bring on menstruation—this vein strongly evacuates the entire body; another is between the heel and the ankle, on the inner side (called the saphenous vein), which is bled for diseases of the womb in women, and for apostemes of the testicles in men—always preceded by bleeding of the basilic vein on the opposite side; and on the outside lies the sciatic vein, which is similarly bled between the heel and the ankle, on the outside, for sciatica—this will cure it, as I have said.

Note that when you mean to bleed a second time at the same place, as sometimes happens (namely when you need to remove a great deal of blood but dare not do it all at once), then make the wound very broad, so that it will not consolidate; [and when you wish to bleed again,] bind the limb and rub the wound to make the place bleed again. Likewise, should you want to use phlebotomy to draw matter to the opposite side, when you have drawn off one third of the blood you wish to remove, put your finger over the wound so that no blood be lost, and make the patient spit; then let the blood flow again. Do this three or four times, for in this way the blood is better diverted to the opposite side and the patient's strength better conserved. Again, note that if the patient usually faints when he is bled, let the bleeding be done with him lying down; it is also good for him to eat beforehand a mouthful of dry bread in pomegranate wine. All veins should be opened along the length of the limb, although some men hold the opposite, except that if the veins of the foot are so fine that blood will not flow through a longitudinal incision, they should be cut laterally. When the veins of the head running from the neck are to be bled, as required above, the neck should be constricted until they stand out, and held tight until the desired quantity of blood is available. When the veins of the arm are to be bled, the arm should be ligatured four fingers above the place to be bled; but it should not be bound so tightly (as is done by some) that all feeling be lost to the arm. If you wish to bleed the veins of the hands or the feet, they should be put in hot water to warm for an hour, and constricted above the wrist or ankle; continue to keep the hand in hot water until you have withdrawn as much blood as you wish.

115 THE TREATMENT OF WOUNDS

Introduction by Michael McVaugh

One of the most momentous issues debated by medieval surgeons concerned the proper treatment of wounds. Orthodox medical theory then held that pus was composed of waste materials, corrupted humors, and that it should therefore be brought to the surface and expelled; to accomplish this, most surgeons advocated probing the wound and applying suppurative medicines to it. Almost alone, Theodoric and his pupil Henry of Mondeville argued for the superiority of treatment that avoided the formation of pus. It is significant that they do not try to defend it primarily on empirical grounds but rather in terms of the standard theory of pus as waste matter, looking (like their opponents) to Galen and Hippocrates for authoritative confirmation. Their doctrine may also have been rejected for reasons of theory. It was looked on critically by Guy de Chauliac, and the principle of "laudable pus" remained extremely strong down to the time of Lister.

6. *Brancus,* med. Lat. Dufr., a throat-disease. *Isid.,* Lib. 4, Cap. 7: "branchos est praefocatio faucium a frigido humore." The editions of Lanfranc read *brachiis* (EETS) [see Selection 95, n. 3].

1. Theodoric, Bishop of Cervia (1205–1298)

Translated by Eldridge Campbell and James Colton[1]

Annotated by Michael McVaugh

BOOK II, CHAPTER 11

Wounds Anywhere in the Body

In whatever part of the body a cut may have occurred, let everything be done in order, according to the rules laid down for wounds in the scalp and face. Indeed, above all else a wound must be made clean. Secondly, having brought the lips of the wound together, they should be replaced accurately in the position which they had in their natural state; if necessary, they should be held there by stitches taken in accordance with the size of the wound. Let the size and depth of the wound determine the closeness and depth of the stitches. For let the physician make no mistake; as has already been said many times, he should be prudent and attentive to every detail. After the suturing has been properly done and the dressings have been carefully arranged, let the wound be bound up skillfully as the position and condition of the part require, that is to say, so that neither the stitches nor the dressings can be disturbed at all. And, just as we have often said before, do not undo the dressing until the third, or fourth, or fifth day if no pain occurs. Afterwards let the dressings be changed every third day unless too much putridity should occur in the wound, in which case it should be changed every day, observing the aforementioned directions. And always, whenever the dressing is changed, by pressing gently upon the wound with a little wine-soaked tow you may express any retained bloody matter. Afterwards let it be bound up according to the aforesaid method and let it be kept thus until the patient has completely

recovered. And if proud flesh should become excessive on a wound, as has been said before, put on the green ointment or something similar, for as long as you see that it is necessary.

Do Not Drain a Scalp Wound

For it is not necessary, as stupid men do, to place a wick at the end of a suture line; nor under such conditions, as was said in the first book, is it necessary to generate bloody matter in a wound, as Roger and Roland and many of their disciples teach, and as almost all modern surgeons continue to do.[2] For there can be no error greater than this, and nothing else which impedes nature so much and prolongs the sickness, prevents uniting and consolidating of a wound, deforms the part and impedes cicatrization. And what is more deleterious, with their treatments the unskilled frequently make wounds become quite eroded and develop sinuses, things which rarely can occur except from lack of care and stem from the great inexperience of the physician. And anyone who reads carefully that book which I entitled "The Daughter of the Prince"[3] will be able, by following the authority of the ancients, and the clearest reasoning, and this present doctrine, to refute a large measure of those things which have been written in the surgical texts of the moderns. But still I fear that we are ploughing in the sands, because, with all this, they will not withdraw from their errors; for it is difficult to relinquish the things to which one is accustomed; and perhaps it is better to let those who are in error continue to err in their own stupidity.

2. Henry of Mondeville (d. ca. 1326)

Translated and annotated by Michael McVaugh[4]

TRACT II, NOTULA 21

It is extremely risky for a little-known surgeon to treat any case other than as his colleagues generally do, for example to treat wounds as

1. Reprinted by permission of Appleton-Century-Crofts from *The Surgery of Theodoric ca. A.D. 1267, translated from the Latin by Eldridge Campbell and James Colton* (New York: Appleton-Century-Crofts, 1955), I 137–139.

2. For an excerpt from Roger's surgery, see Selection 112; see also Selection 111.

3. Campbell and Colton note that this work exists in MS. Bodl. Ashm. 1427, fols. 39–130.

4. The translation has been made from a French version of Mondeville's *Surgery: Chirurgie de Maitre Henri de Mondeville,* tr. E. Nicaise (Paris, 1893), pp. 187–188, 308–311.

Mondeville's *Surgery,* unpublished until the nineteenth century (ed. Pagel, 1890–1891), is the first work by a native Frenchman in the new surgical tradition of Bologna and draws particularly heavily on the work of Theodoric and Lanfranc: Mondeville, as this selection shows, was one of the principal agents in the European dissemination of Theodoric's radical new "dry" treat-

Theodoric instructs in the first part of his *Great Surgery*. . . . Master Jean Pitart[5] and I, who were the first to bring this method into France and first to use it in the treatment of wounds at Paris and on a number of military campaigns, did so against the will and advice of everyone, in particular the physicians. We had to endure scorn and contemptuous words from laymen, and menaces and threats from our colleagues, the surgeons. From some laymen and from physicians, every day and at each new treatment, we suffered such violent attacks that, nearly exhausted by so much opposition, we were about to give the treatment up, and would have given it up entirely, God knows. But the most serene prince Charles, Count of Valois [the brother of Philip IV, and founder of the Valois line], came to our aid, as did several others who had previously seen us in the camps curing wounds by this method. Moreover, we were sustained by the truth, for which man should sooner accept death than yield to error. Is not God the Truth, and was He not willing to suffer death for it? Yet if we had not been strong of faith, royal physicians upheld by the king, of some little literacy, we would surely have had to abandon this treatment. . . .

TRACT II, DOCTRINE I, CHAPTER 3

Throughout this book we have been supposing that every simple wound can be cured without producing a notable quantity of pus, on condition that it be treated according to our own doctrine, that of Theodoric, without deviation. We must now see if this is possible.

It may be said that this is not possible, since in every member which is nourished, whether it be large or small, healthy or unhealthy, the third digestion[6] takes place; and in every digestion wastes are produced, especially within wounded members, and these wastes compose the matter of pus. The weakened [natural] heat is the agent of this transformation; and when an agent and the object upon which it acts are both present, it is impossible that its action should not take place; and it is thus impossible that pus should not be engendered in a wound. The fact is proved on the authority of all the authors in medicine and surgery, and by all the practitioners. The contrary is shown by Theodoric throughout his *Great Surgery,* and we ourselves can also attest it from experience.

It must be concluded that every wound treated thus can be cured without producing a notable quantity of pus. This can be proven in two ways,

by experience and by reasoning: by experience, since we see that it often happens thus; by reasoning, because where the cause is lacking, the effect will be lacking too—and in every simple wound treated by our method we can avoid all the causes of the formation of pus, therefore, and so on. The major [premise] is proved on the authority of the Philosopher; the minor results from the fact that according to experience and to authorities, there are just five causes of the formation of pus in wounds (although Haly gives only three: excess of nourishment, unhealthy quality, and application of unhealthy medicine). The first cause is the alteration caused in the wound by the air; we can avoid this by rapidly closing the wound and maintaining this closure. The second cause is a too violent flux of humors toward the wound. We avoid this by an evacuation,[7] which will divert the humors, by raising the injured member, by an artfully made bandage, by fomentations of hot wine and application of wine to the wound, all things that dissolve a part of those humors that have already reached the member and that repel those that would have come to it; this is because they strengthen the member and expel the humors by constriction, as the press squeezes juice from the grapes. The third cause is the weakness of the injured member, which receives wastes from other parts; we can avoid this by a suitable bandage and by employing wine and other temperate remedies,

---

ment of wounds. The book was written between 1306 and 1320, during much of which time he was engaged in the service of Philip IV of France and of his brother, Charles of Valois; it very clearly reflects his civilian and military experience. But equally clearly it reflects his academic background and commitment. It is indicative of this, though in the long run not very significant, that anatomical illustrations which he had used initially in the lecture room were included (as increasingly corrupt miniatures) in the manuscripts of the *Surgery* (see Loren C. MacKinney, "The Beginnings of Western Scientific Anatomy: New Evidence and a Revision in Interpretation of Mondeville's Role," *Medical History,* VI (1962), 233–239); it is indicative and highly significant that he should be so concerned with theory, authority, and logical form, in what might be expected to be a severely empirical treatise, like Theodoric's. Henry's work reveals quite well the strength and attractiveness of the scholastic approach in even the most empirical branches of medicine, in surgery as well as in anatomy.

5. Surgeon to Philip IV after 1298. For a short summary of what is known of Pitart, see Ernest Wickersheimer, *Dictionnaire Biographique des Médecins en France au Moyen Age* (Paris, 1936), p. 465.

6. For Avicenna, the fourth digestion; see Selection 91.

7. For example, with a purge or by bloodletting.

internally and externally, in moderate quantity—
that is, sufficiently to sustain the organic force;
these medicines strengthen the natural complexion
of the member by their aromaticity. The fourth
cause is an excess of nourishment taken, or its
unhealthy quality, or both; we avoid this by a light
regimen, scanty, easy to digest, and one that
forms good dry blood, not burnt. The fifth cause is
the application of a suppurative medicine; but the
wine and the bandages which we use do not have
this effect, rather they dry and dissolve; thus, and
so on. If any other cause of this sort is given by
authorities under another name, I believe that it
will be reducible to one of these I have just given.

It follows that the minor is proven and that in
every simple wound we can avoid the causes of the
formation of pus; moreover, from this results the
primary conclusion that it is possible to cure every
wound, as a wound, without creating a significant
quantity of pus, when treated in this manner. As
for the reasoning which maintains the contrary, it
should be said that it concludes correctly that pus
is formed in every wound where wastes are produced in sufficient quantity to engender a great deal
of pus, but this reasoning does not prove that pus
is formed in considerable quantity in those wounds
where scant wastes are produced. To the authorities it must be replied that their conclusions are
valid for those wounds for which one prescribes a
cold, humid, suppurative regimen, or the like.

Moreover, once proven and agreed that it is
possible to cure all wounds thus treated without
producing pus there in significant quantity, we can
ask, Which of the two treatments is the more
healthy, that in which the formation of pus is
produced or provoked, or that in which this formation is avoided completely or as much as possible?
It may be argued that the treatment in which the
formation of pus takes place or is provoked is
preferable to that in which it is completely avoided,
since the treatment by which we free nature from
wastes appears preferable to that which does not
have this result—that is the case here, therefore,
and so forth. The major premise is evident; the
minor premise is proven by the fact that nature
discharges herself by suppuration; therefore, and
so on. It is also supported by the authority of
Galen, commenting on the *Aphorisms,* Part V:[8]
In serious wounds the hard is bad, the soft is good;
therefore, and so on.

The contrary is supported by Avicenna's
authority, Book I [of the *Canon*], fen 4, chapter
29[9]—"On treatment by solution of continuity, and

of the kinds of ulcers"—where he says that one
should set himself three goals in the treatment
of the fleshy members, of which the third is to
prevent suppuration as much as possible; therefore, and so forth. This can also be supported by
the authority of Galen, who says the same thing
(in Book IV of the *Megatechni,* chapter 4[10]):
Desiccative medicines are suitable for all wounds,
from first to last, save only those which involve a
contusion, that is to say, an old contusion. And
desiccatives do not engender suppuration; therefore suppuration must not be provoked in wounds,
and so on. Moreover, in the *Techni* (treatise on
causes, chapter 34, beginning "Sufficit autem mani-
feste"[11]), explaining the treatment of apostemes,
Galen says that after having first purged the body,
it is necessary to try to repel [wastes]; then, if that
is impossible, to dissolve them; and finally, if that
does not succeed, he says that it is necessary to
ripen and provoke the suppuration. It is clear that
Galen is trying first to cure by the best method of
treatment; therefore, and so on.

It must be concluded that the treatment in which
no pus is formed, in which one avoids it as much
as possible, is better, surer, and more healthy
than that in which it is produced or is provoked.
The reason for this is as follows: The treatment
which least troubles the patient or the surgeon; in
which there is no loss of substance; in which the
least spirit and vital heat is lost, and by which the
least external cold penetrates (both are in fact
contrary to the principles of life); which involves
neither hot aposteme nor fever; by which the lips of
the wound can be more exactly joined—that treatment is preferable to one which does just the contrary. But this is the case here; therefore, and so
forth. The major is self-evident; the minor can be
proved by going over each of the parts of the argument; therefore, and so on. Besides, it is useless
to do with more, and so on; thus, it is in vain that
we provoke suppuration in wounds, because
according to the opinion of Galen and Avicenna,
the desiccatives are appropriate for all wounds
from first to last, therefore, and so forth.

To the first argument, which says that nature is
relieved by suppuration, we should rather answer
that to provoke suppuration is to injure nature, on

8. Kühn XVII (2). 882; compare Selection 111 and
note 2 thereto.

9. Actually chapter 28 in the edition of Venice, 1564
(pp. 221–222).

10. Kühn X.281ff.

11. Kühn I.401ff.

the authority of Hippocrates (aphorism of Part II on the formation of pus). It is however true that once pus is formed, its expulsion is a relief, and true that it is necessary that it then be expelled; but it would be better if it were never engendered nor expelled, for wounds heal more easily before suppuration than after. Likewise, as there are more people who know how to induce suppuration than how to dry it up, it can happen that suppuration thus provoked cannot be stopped. Then the surgeons announce that St. Eloi's sickness has infected the wound, or something of this sort, so that the people will not criticize them—indeed, they withdraw with honor and no longer concern themselves with the treatment—when neither the patient nor even the saint to whom they ascribe it ever suffered from the disease.

As for Galen's statement, "The hard is bad, etc.," it should be said that Galen meant it not for simple wounds, but for complicated wounds, involving apostemes, contused and of long-standing, whose development is so advanced that they cannot be cleared up by [the normal] evacuation, repercussion, and solution,[12] without provoking suppuration. From the moment a wound develops complications, the sooner it becomes soft and fluid, the better it will be; and the longer it remains hardened, the worse it will be. Some, misunderstanding this opinion of Galen's and applying it to simple wounds, have harmed many people by provoking suppuration in them. Perhaps this statement has been more harmful than useful, since a statement misunderstood leads to error—but all this was contrary to Galen's own intentions.

3. Guy de Chauliac (d. ca. 1368)

Translated and annotated by Michael McVaugh[13]

TRACT III, DOCTRINE II, CHAPTER 1

. . . *Contusion* [of the head] *with extensive fracture*. If the fracture is a considerable one, we must turn to surgery and to dilation of the fracture; this is approved by Galen in Book 6 [of the *Therapeutics*] and by Avicenna in Book 4 [of the *Canon*], for three reasons:

We must open what we cannot bind up. But by using a bandage we cannot preserve the head free of apostemes and exudation of matter (which is our first goal in every contused fracture), since the shape of the head is not suited to binding up, as is brought out there. Therefore, and so on.

Besides, if it is true of what is less [necessary], it is also true of what is more so. And it seems less necessary to open the arms and other bones in order to withdraw fluids than the head; yet it is sometimes necessary [in such cases]. Thus it is still more necessary in the case of the head, because of the brain.

Likewise, if anything could obviate this [necessity], it would be medicines; but they can do nothing without a bandage, as has been well said.

Therefore, in extensive contusions we must expose and dilate a part of the fracture, so that we may dry the cerebral membranes and clear them of their fluids. So do not listen to the claims of the followers of Theodore and Januensis,[14] who promise that they can cure any injury to the head with their spices and potions, without surgery and removal of bones. For while it may be possible in

minor injuries, as I have said, I have never yet seen

12. Repercussion and solution involve, respectively, diverting or dissolving the unwanted waste material.

13. Translated from *Ars Chirurgica Guidonis Cauliaci Medici* . . . (Venice, 1546), fol. 37 verso. In the French translation of E. Nicaise (*La grande chirurgie de Guy de Chauliac* [Paris, 1890]) the passage is to be found on pages 263–264.

There has been some attempt to revise the traditional interpretation of Guy de Chauliac as retrogressive in his treatment of wounds on the grounds that while he did employ suppurative methods, he did not always do so, or did so only when appropriate (for instance, see Leo M. Zimmerman and Ilza Veith, *Great Ideas in the History of Surgery* [New York: Dover Publications, 1967], pp. 152–154). But it is reasonably clear that Guy is basically unsympathetic to the new approach represented by Theodoric and Mondeville. He differs most seriously from them in accepting the classical view that pus is a normal concomitant of certain broad classes of wounds, thereby rejecting their doctrine that the avoidance of suppuration should be a universal goal. The text given here is one of many (see Guy's historical conspectus in Selection 111) to suggest that the disagreement between Chauliac and Mondeville is not really a practical question, one of how to treat particular types of injury—indeed, in this respect there is usually very little to choose between them; rather, it is a question of orientation: Should the physician conceive of suppuration as inevitable and natural, or not? In the fourteenth century Henry's outlook may have had no more practical advantages than Guy's, but it would be difficult to argue that it has not ultimately been a more fruitful one.

14. Simon of Genoa, physician to Pope Nicholas IV (1288–1292), and co-translator of pharmacological works of Abulcasis and Serapion.

it so in serious ones. The *Conciliator's* view[15] that strong medicines can bring up [fluids] from deep within and expel them is worthless; for such medicines are suspect, inasmuch as their excessive strength can induce apostemes in those predisposed, as our bodies usually are—just as Dino [de Garbo] says in his commentary on Book 4 of Avicenna. Nor is Henry's view [valuable] that all wounds can be cured without notable pus—for serious contusions inevitably involve notable pus, according to the fourth book of the *Therapeutics;*[16] nor is what they claim about "the potency of Nature"—for that may be true of a small amount of material, but when it is copious, it is imperative that we open to remove it, as we do beneath the fourth rib for empyema, as will be described below. I am the more amazed when they say that their potion is worthless after the fourth day—I would have thought it to be more effective after that time, when the wound has been saved from flux, pain, and aposteme. Nor do they say what is to be done if their potion is ineffective. I suspect that they would act like the ship captain who (as is told in Book 5 of the *Therapeutics*), losing his ship through his carelessness, then handed a plank to each of the sailors to save themselves if they could.

116 PLASTIC SURGERY

Heinrich von Pfolspeundt (fl. 1460)

Translated by Martha Teach Gnudi[1]

Annotated by Michael McVaugh

TO MAKE A NEW NOSE FOR ONE WHO LACKS IT ENTIRELY, AND THE DOGS HAVE DEVOURED IT

Likewise a master craft is proper. If a person comes to you, whose nose has been devoured, and you wish to make a new nose for him, let no one watch, and make him promise you solemnly to conceal how you wish to heal him. And after the vow, tell him your opinion. If he wishes to risk it with you and endure the pain, approach him with reason and tell him how you must cut and also bind him and how long he must be in bed. And therefore allow no one to watch, so that he may not learn the art from you. Then either you or he must find a trustworthy person who will also promise to conceal the matter, and whom you may allow to observe, provided he be of aid to you and also help the patient to eat and drink and bring him other necessary things. And the room wherein he lies must also be locked. Let no one have access except the person who is to serve you and him.

The Art

Take a piece of parchment or leather and shape and cut it according to the nose wound, as broad and long as the nose was previously. And you must bend it a little at the upper part of the nose so that the nose may not be [too] broad at the top. Then take the same parchment or leather and lay it on the arm a little behind the elbow, where it is thick, and paint around it with ink or other coloring matter, as wide and long as the same pattern was.

Take a good sharp cutting blade or scissors and strike or cut through the skin and take a little of the flesh with it. Cut no further than you have marked with ink or paint, and begin to cut from the back to the front, and when you have matched it with the pattern then cut [still a little farther]. You will do this with one stroke, two fingers' breadth or more. Allow the same flap that you have cut to hang from the arm and do not cut it off. Now raise the arm up to the head and suture the self-same flap directly over the nose, in the same measure that it was before. For this reason you must cut the flap so much longer that you may the better come to the nose. Then you must bind his arm on the head and below the elbow and must keep him

15. The *Conciliator differentiarum philosophorum et precipue medicorum* was the principal work of the Paduan Averroist, Peter of Abano (ca. 1250–1316).

16. Kühn X.280–281; apparently a different use of the text refuted by Henry (see n. 10).

1. Translated from *Buch der Bündth-Ertznei von Heinrich von Pfolsprundt* [sic] *Bruder des deutschen Ordens, 1460,* edited by H. Haeser and A. Middledorpf (Berlin, 1868), pp. 29–31; reprinted by permission of Jerome Pierce Webster from Martha Teach Gnudi and Jerome Pierce Webster, *The Life and Times of Gaspare Tagliacozzi, Surgeon of Bologna* (New York: Herbert Reichner, 1950), pp. 114–115.

Rhinoplasty, together with other forms of plastic surgery, was practiced by Hindu physicians as early as the sixth century B.C. and independently developed by Greek and Roman physicians. In the Middle Ages, Theodoric describes a technique for reuniting a partially severed nose (Bk II, ch. 10).

thus in bandages so that the arm may lie more immovable and may not become so tired. Make that many more bandages from cloths. For he must lie strapped up until the nose has united with the flap. That usually requires eight or ten days' time, or until you see that it is united and healing. Then cut the flap off, but not too short, so that it may afterwards extend somewhat beyond the nose. Now the nose has only one opening. Then cut the flap in such a length and breadth that you can suture it on below, too. You must likewise cut away the skin a little but in addition make the flesh raw in that place [i.e., freshen the margins] and suture the same flap on below where the raw flesh is. Thus the nose becomes two-fold again on the outside, though not within. Then heal it with the wound potion and with the oil and red salve. But before you cut him, lay the arm on his head [several times], higher and lower, so that you may see well where you should cut. And when you have sutured it thus entirely and are ready to heal it, and all the time that you are healing it, arrange the nose and bandage it and take care also with such bindings that it will become narrow, high or low. If the nose is too broad, tie two small sacks, one on either side of the nose. You must insert quills bound with flax in the nose and fill up the space at the front of the nose with it well, thus the nasal openings will not be too narrow and will remain equally wide. If he becomes tired of lying down, you must help him from time to time with pillows and cloths in the bed. These you must tie and arrange so that they help him and through them he can find rest. And from time to time he must rest so that his head is up. At certain times he sits up, at other times he walks about in the room where he lies. And with whatever and in whatever way he can best rest, help him with the same. And it turns out well if only you carry out the cutting correctly and with judgment. By cutting the flap long enough you make it better for him to go about and to rest, and it does no harm. I heartily recommend this for anyone who is without a nose. An Italian taught it to me, one who had helped many people with it and earned much money therefrom.

Should a person come to you and should his nose have been struck off and should it have healed, then cut the skin off, well and widely enough, down to the raw flesh and carry out the above. After that heal it also in the same way. Without a doubt this works and is often worthwhile.

Brief Author Biographies

Included in this part are biographical sketches of all authors from whose works at least one selection has been reproduced in this source book. They are arranged alphabetically by the authors' *first* names or, in the case of Arabic authors, the name by which they are best known. These sketches are solely for the convenience of the reader and are intended to present in brief compass salient features about any author, that is, his time and place of birth, profession, and his more significant scientific works. Although in some instances it has been necessary to compress and condense biographies and bibliographies of well-known figures (for instance, Albertus Magnus, Thomas Aquinas, and so on), it is all too often the case that a paucity of data has allowed presentation of all that is known about some authors; indeed, in all too many instances not even the identity of the author has been properly established. In compiling these biographies, the editor acknowledges his large debt to George Sarton's monumental *Introduction to the History of Science* (Baltimore: Williams and Wilkins, 1927–1948). With the appearance of the first five, of a projected thirteen, volumes of the *Dictionary of Scientific Biography* (New York: Charles Scribner's Sons, 1970–), the reader will soon have available scholarly biographical summaries, with critical bibliographies, for many of the authors included here.

Adelard of Bath (fl. 1116–1142)

Not only was Adelard an early seeker after the new Arabic learning but, upon learning Arabic, distinguished himself as one of the earliest translators. He traveled in France, southern Italy, Sicily, and the Near East. After his early philosophical treatise *On Identity and Difference (De eodem et diverso)* and a work on the abacus he composed the *Natural Questions (Questiones naturales)* in seventy-six chapters, in which he claimed to be propounding the new Arabic learning, although no Arabic author is cited by name. Treatises on the astrolabe, on falconry, and a revised edition of the *Mappae clavicula,* a chemical treatise in which we find one of the earliest references to alcohol, are also ascribed to him. He translated from Arabic into Latin the astronomical tables of al-Khwārizmī and the *Elements* of Euclid. A version of his Euclid translation became the most widely used text of the *Elements* in the Middle Ages.

Albert of Saxony (ca. 1316–1390)

Albert was probably born in Lower Saxony of a burgher family. He received his master's degree in 1351 at the University of Paris, where he also served as rector in 1353 and in various capacities in the English Nation. In 1365 he became the first rector of the University of Vienna, having helped found it approximately one year earlier. In 1366 he was also named bishop of Halberstadt in Saxony, a title he retained until the end of his life. Albert left commentaries or questions on most of the works of Aristotle and also composed a number of logical treatises. His mathematical works include *On the Quadrature of the Circle* and a *Treatise on Proportions (Tractatus proportionum).* Although not the equal of John Buridan or Nicole Oresme, Albert of Saxony was an outstanding and influential scholastic thinker who grappled with most of the major scientific and philosophic issues of his time.

Albertus Magnus (ca. 1193–1280)

A German born in Lauingen, Bavaria, Albertus studied in Padua and became a Dominican in 1223. From 1228 to 1254 he taught in many Dominican schools, primarily in Germany but also in Paris between 1245 and 1248. During the years 1253–1256 he was provincial of the German province of the Dominican Order, and after resigning was made bishop of Regensburg in 1260. He resigned from this ecclesiastical post in 1262, preferring the life of a scholar. He returned to Cologne, where he had resided for short periods before. Although he seems to have ventured from Cologne on occasion (he is said to have gone to Paris in 1277 to defend his own and Aquinas' doctrines against the Condemnation issued in 1277 by Bishop Étienne Tempier of Paris), it was his residence until his death in 1280.

A prolific author, Albert wrote many commentaries on Aristotle for the instruction of his fellow Dominicans. Among these are commentaries on the *Physics, Meteorology, Generation and Corruption,* and *De caelo.* He wrote works in zoology, botany, alchemy, astrology, and theology. The list of his authentic works reveals an impressive range of achievement encompassing a wide variety of scientific topics. Indeed, a glance at the selections included in this source book is sufficient proof of this.

Alexander Neckam (1157–1217)

Alexander Neckam was an Englishman born in St. Albans, Hertfordshire, where he received his early education, which was continued later in Paris. He became an Augustinian canon after returning to England in 1186. In 1213 he became abbot of Cirencester. He died in Kempsey, Worcestershire, in 1217. His relevance to the history of medieval science is based on his five-book treatise *On the Natures of Things (De naturis rerum),* a popular encyclopedia of science in which, among other things, we find the earliest mention in Europe of the magnetic needle for nautical use.

Algazel (1058–1111)

Abū Ḥāmid Muhammad al-Ghazzālī, known to the Latins as Algazel, was born in Tūs, Khurāsān, in 1058, and after some travel to Nishabur, Baghdad, and Alexandria, he died in Tūs in 1111. He was the greatest theologian of Islam. In his autobiography he explained how, beginning at the age of 33, he began a painful transition and by the grace of God's light passed from despair and scepticism to faith and certitude. "The main philosophical works of Algazel belong to the earlier period when he was examining philosophies that failed to satisfy him. One of the most important of these writings is a work that was written in two parts, as though it were two volumes to one set of books. Part I was called *Maqasīd al-Falāsifa (Tendencies, Intentions, or Essential Views of the Philosophers).* It included a prologue and an objective summary of philosophical knowledge of logic, metaphysics, and physics, as held by al-Fārābi and Avicenna. Part II, the *Tahafut al-Falāsifa,* was intended as a refutation or destruction or a means of showing the incoherence in the philosophers' handling of twenty principal theses.

Through a curious historical accident Algazel was to become known to most of the thinkers of the Christian Middle Ages as the disciple of Avicenna, a man with whom he vigorously disagreed. For Part I, the Maqasīd, became separated from the Tahafut which was to complete it, and it also became separated from its own prologue and conclusion," where Algazel declared that his purpose was to present summaries of the sciences of the philosophers (namely logic, physics, and metaphysics) in order subsequently to refute them. Without the prologue, however, "the Maqasīd was so clear and objective a statement of the positions of al-Fārābi and especially, Avicenna, that it is not surprising that St. Albert, St. Thomas, and Giles of Rome, who knew it alone in the Latin translation [in the *Metaphysica,* as it was called; see Selection, 52.2—*Ed.*], should have regarded its author as a follower of Avicenna" (the two quotations are from *Averroes' Destructio Destructionum Philosophiae Algazelis in the Latin Version of Calo Calonymos,* edited with an introduction by Beatrice H. Zedler [Milwaukee: Marquette University Press, 1961], pp. 5–6; the titles transliterated from the Arabic have been slightly altered). Part II, *Tahafut al-Falāsifa (The Incoherence of Philosophy)* was attacked by Averroes in the latter's *Tahafut al-Tahafut* (that is, *The Incoherence of the Incoherence,* which was translated into Latin in the fourteenth century under the title *Destructio Destructionum*), a treatise of some influence in Italy during the Renaissance. Algazel was also the author of mystical treatises and other anti-philosophical tracts.

Alhazen (ibn al-Haitham) (ca. 965–ca. 1039)

Alhazen, whose name was Latinized from ibn-al-Haitham, was born in Basra but spent his productive scientific years in Cairo, where he died. He was undoubtedly the greatest of Muslim scientists, devoting himself to physics, especially optics, astronomy, mathematics, and medicine, as well as commenting upon Galen and Aristotle. His major work was his lengthy *Optics,* in which experiment and a high level of mathematics, including conic

sections, were employed. Translated into Latin, Alhazen's *Optics* exerted a great influence in the history of optics in Europe.

al-Khwārizmī (fl. 813–833)

Abū 'Abdallāh Muhammad ibn Mūsa al-Khwārizmī was born in Khwārizm, south of the Aral Sea. Much of his scientific work was accomplished during the reign of the Caliph, al-Ma'mūn, a generous patron of science. His impact on the Latin West was overwhelming. His arithmetic presented to the Arabs the Hindu numeral system, which we call Arabic numerals. Apparently lost in its original Arabic version, the Latin translation of it, *De numero Indorum,* is extant and was

instrumental in disseminating knowledge of the Arabic number system to medieval western Christendom. His *Book of Algebra* was equally influential after its translation into Latin, for it presented analytic solutions to linear and quadratic equations and formed the basis for algebraic studies in the West. Adelard of Bath's translation of his astronomical tables, as revised by Maslama al-Majrītī (d. ca. 1007), made available sine and tangent functions for the first time in the West.

Archimatthaeus (first half of twelfth century)

Archimatthaeus was a Salernitan physician, perhaps identical with a Matthaeus de Archiepiscopo. He was the author of a few medical treatises,

including the *De instructione medici,* concerned with the conduct of a physician.

Arnald of Villanova (ca. 1240–1311)

Born near Valencia, Spain, Arnald studied medicine in France and then traveled widely, remaining for periods in Paris, Montpellier, Barcelona, and Rome. He died at sea in 1311 while sailing from Naples to Genoa. He was an extraordinary and controversial figure who was not only a physician but also an alchemist, astrologer, translator, diplomat, and reformer. Kings and popes

sought his medical advice. A large number of medical, alchemical, astrological, and theological treatises—as many as 123—have been attributed to him, although many may be spurious. In medicine, he sought to systematize knowledge on Galenic foundations. Learned in Arabic, he translated a few Arabic treatises into Latin, including Avicenna's *De viribus cordis.*

Averroes (1126–1198)

Abū-l-Walīd Muhammad ibn Ahmad ibn Muhammad ibn Rushd, or Averroes as he was known in Latin, was born in Cordova, Spain. After studying law and medicine he served as a judge in Seville and Cordova, and in 1182–1183 the caliph called him to Marrākush, Morocco, to become his personal physician. The next caliph banished him to Lucena near Cordova and ordered all but his strictly scientific works burned. Shortly before his death in 1198 he was pardoned and recalled to Marrākush, where he died. Although one of the greatest—if not the greatest—of Muslim philosophers, Averroes had virtually no influence in the Muslim world. But his impact on Jewish and Christian philosophy and science was very great. To Latin scholastics he was known respectfully as "The Commentator" because of his important

Aristotelian commentaries, many of which had been translated into Latin. He customarily wrote three commentaries on each work of Aristotle: (1) little commentaries, which were largely paraphrases or compendia presenting only Aristotle's conclusions: (2) the lesser, or "middle," commentaries, in which he frequently intermingled Aristotle's doctrine and his own explanations, making it often difficult to distinguish one from the other; (3) the greater commentaries, in which successive sections of Aristotle's text are cited, each followed by Averroes' explanation (these were numbered by the Latin scholastics for convenient references to the text of Aristotle). Latin translations of all three categories exist for his *Posterior Analytics, Physics, De caelo, De anima,* and *Metaphysics.* The logical works exist in Latin only in little and middle commentaries. Other Aristotelian commentaries exist in only one version. His *Tahafut al-Tahafut* (*The Incoherence of the Incoherence;* translated into Latin in 1328 with the title *Destructio Destructionum;* it was translated too late to be of significant influence on scholastic thought) was an attack against al-Ghazzāli's assault on philos-

ophy *(The Incoherence of Philosophy [Tahafut al-Falāsifa]).* In addition to many other works in astronomy, grammar, logic, and other subjects, many of which were not translated into Latin, Averroes wrote a medical encyclopedia, called *Colliget* in its Latin translation (from a corruption of the Arabic term kullīyāt meaning "generalities").

Avicenna (980–1037)

Abū Alī al-Ḥusain ibn 'Abdallāh ibn Sīnā, or Avicenna as he was called in Latin, was born at Afshana, near Bukhara (now in the Soviet Union), and was at least partly Persian. In his autobiography Avicenna relates that by the age of eighteen he had, by his own genius and diligence, mastered logic, metaphysics and all philosophy, medicine, mathematics, astronomy, and, indeed, all the sciences. So thoroughly had he learned these subjects that for the rest of his life he could add nothing but maturity to his understanding of them. Serving numerous rulers whose fortunes frequently declined in the Persia of his time, Avicenna was compelled to wander from the court of one petty ruler to another. He died in Hamadhān.

Although many of his works were unknown in the West, his great fame rests on two monumental works, both of which were translated into Latin. His *Canon of Medicine* was a comprehensive medical treatise approximately one million words in length. After its translation into Latin by Gerard of Cremona, it became the chief textbook for the study of medicine during the next five centuries. The second work is the *Kitāb al-shifā'* (*The Book of the Remedy* or *Book of the Healing*), consisting of four parts (logic, physics, mathematics, and metaphysics). The translation by Domingo Gundisalvo and ibn Daoud (Avendeath) was incomplete, omitting the mathematical part and including only a small part of the logic. Although portions of the

physics were eliminated in the Latin version, the latter was divided into eight parts, given by E. Gilson as follows (*History of Christian Philosophy in the Middle Ages* [London: Sheed and Ward, 1955], p. 641, n.12): "1, Generalities (*Sufficientia,* or *Communia naturalibus,* or *Communia naturalium*); 2, Heaven and the world, probably spurious; 3, On coming to be and passing away *(De generatione et corruptione),* not included in the 1508 ed.; 4, On elements, action and passion *(De actione et passione),* same remark [that is, not in 1508 ed.?—Ed.]; 5, On meteors *(De rebus congelatis)* [our selections in the sections on Geology and Alchemy in this source book are drawn from this fifth section of the physics—Ed.], same remark; 6, On the soul (*De anima* or *Liber sextus naturalium* [that is, the sixth book of natural philosophy in the section on physics—Ed.]; 7, On plants *(De plantis),* not included in the 1508 ed.; 8, On animals (1, *De animalibus;* 2, *De partibus animalium),* translation ascribed to Michael the Scot by the 1508 edition." Sometimes the whole of the physics was called *Liber sufficientiae,* after the title of the first part. Including the metaphysics, the whole of the *Book of the Healing,* with its interwoven Neoplatonic and Aristotelian strands, powerfully shaped and influenced medieval Latin philosophy and science, especially in the thirteenth century, before Aristotle's *Metaphysics* came to be widely and carefully studied.

Banū Mūsā (ninth century)

The Banū Mūsā "were sons of Mūsā ibn Shākir, who is variously described as a reformed bandit and an astronomer or astrologer of the Caliph al-Ma'mūn. He is said to have given his sons to the astronomer Yahya ibn Abī Mansūr for instruction in mathematics. At any rate, they joined the circle of mathematicians that grew up in Baghdad at the time of al-Ma'mūn (fl. 813–33) and his successors.

The names of the three brothers were Abū Ja'far Muḥammad, Abū 'l-Kāsim Aḥmad, and al-Ḥasan. They devoted their energies and resources to the acquisition of Greek scientific manuscripts and the propagation of their contents either by translation or by independent works" (M. Clagett, *Archimedes in the Middle Ages* [Madison, Wis.: University of Wisconsin Press, 1964] I, 226).

Bartholomew the Englishman (fl. 1220–1250)

A Franciscan, Bartholomew lived in Oxford, Paris, and Magdeburg. Although little is known of him, his *On the Properties of Things (De proprietatibus rerum)* in nineteen books was enormously popular for about three centuries, being translated into Italian, French, English, and Spanish during the fourteenth and fifteenth centuries. As was usual for encyclopedic treatises, Bartholomew ranged over a wide variety of subjects, including medicine, natural history, astronomy, cosmology, and geography.

Bernard of Gordon (fl. 1283–1308)

Bernard was a French physician born at one of a number of possible French places called Gordon. He was a well-educated physician with a good knowledge of Aristotle as well as of Greek and Arabic medical science. In 1307 he wrote *On the Preservation of Human Life (De conservatione vitae humanae)* in four parts, each of which sometimes circulated as an independent treatise; their titles are (1) *De phlebotomia,* (2) *De urinis,* from which a selection is included in this book, (3) *De pulsibus,* and (4) *De regimine sanitatis.* His major work, however, was the *Lilium medicinae.* Written at Montpellier in 1303 and ranging over many topics, it was a widely used textbook of practical medicine presented in the scholastic manner.

Bernard of Verdun (fl. late thirteenth century)

Bernard of Verdun was a Franciscan astronomer who wrote an astronomical work entitled *Treatise on the Whole of Astronomy (Tractatus super totam astrologiam).* In it he compared al-Bitruji's theory of homocentric spheres with Ptolemy's theory of eccentrics and epicycles and clearly upheld the latter. Indeed, the treatise is patterned after Ptolemy's *Almagest.*

Bernard Tornius (1452–1497)

A natural philosopher and a physician, Tornius was associated with the University of Pisa for virtually all of the last twenty-two years of his life, first as a teacher of philosophy (1475–1478) and then as a teacher of medicine (1478–1496). In addition to many brief medical treatises, including the *Relatio anatomica (An Anatomical Relation),* reproduced in Selection 97, Bernard wrote a letter on motion suggested by certain problems raised by Giovanni Marliani and Swineshead, and in 1494 published annotations on William Heytesbury's *On Local Motion (De motu locali).*

Blaise Pascal (1623–1662)

The son of a government official of the lesser nobility, Pascal was born in Clermont in Auvergne. He was educated by his father; a child prodigy, he mastered Greek and Latin by the age of twelve and completed an essay on conic sections (*Essai pour les coniques,* published in 1640) by the age of sixteen. Until 1654, and occasionally thereafter, Pascal devoted much time to mathematics, producing contributions in geometry, number theory, and probability. His highly significant contributions demonstrating the pressure of the air and the existence of vacuum were made during the years 1646–1648. On November 23, 1654, Pascal had such an intense religious experience that he devoted the remaining years of his life to religion. He defended Jansenism and attacked the Jesuits in *Lettres provinciales,* a work which Pascal wrote cooperatively with two others and which was placed on the Index in 1657. During the last years of his life Pascal wrote his famous philosophical works, *Pensées,* and *De l'Esprit géométrique,* both left unfinished at his death.

Boethius (480?–524)

Member of an illustrious Roman family, Boethius was an enormously important figure in intellectual history. His knowledge of Greek enabled him to translate into Latin the elementary logical works of Aristotle and Porphyry's *Intro-duction to Aristotle's Categories.* To these he added Latin commentaries and original logical treatises which were instrumental in shaping the philosophical language and terminology of the Middle Ages. Also extant are his manuals *On Arithmetic*

and *On Music,* both based on Greek sources and both used as textbooks throughout the Middle Ages. His most famous work, *On the Consolation of Philosophy,* was written in prison while awaiting execution on a charge of treason against his emperor, Theodoric.

Campanus of Novara (ca. 1205–1296)

From frequent designations as "Magister Campanus," it has been assumed that he was a graduate of a university, with Paris and Bologna as probable candidates. Although he is mentioned here and there, and seems to have been on familiar and friendly terms with Pope Urban IV (1261–1264), who made him rector of a church in Arles in 1263 and a canon in the church of Toledo in 1264, very little about him is actually known. On the basis of the benefices awarded him (he also had one in England), it is probable that he died reasonably affluent. Campanus was primarily a mathematician (in 1267, Roger Bacon cited him as an excellent mathematician), astronomer, and apparently also an astrologer. In addition to his widely used edition of, and comentary on, Euclid's *Elements,* his

important treatises worthy of mention are *The Theory of the Planets (Theorica planetarum),* a lengthy astronomical work, written between 1261 and 1264 and dedicated to Pope Urban IV, which is extant in over sixty manuscripts and in early printed editions; *On the Quadrature of the Circle (De quadratura circuli); Computus major,* an astronomical treatise, written in 1268; *Tractatus de sphaera,* which cites both the *Computus* and *Theorica,* considers the precession of the equinoxes and the rival trepidation theory proposed by Thabit ibn Qurra; and *Tractatus de quadrante composite,* a short treatise concerned with the astronomical quadrant. Quite a number of other mathematical, astronomical, and astrological works are also ascribed to him.

Chalcidius (fl. early fourth century)

Almost nothing is known of Chalcidius' life, not even whether he was Christian, Jew, or Neoplatonist. But his translation of, and commentary on, approximately the first two thirds of Plato's *Timaeus* was of great intellectual significance for the Middle Ages, for it made available Plato's solitary physical treatise as a counterbalance to the predominant Aristotelian physics and cosmology.

Indeed, prior to the twelfth century it was the sole physical treatise available and was ardently studied at the Cathedral School of Chartres. The technical portions of Chalcidius' astronomical discourses are virtually translations from *A Manual of Mathematical Knowledge Useful for an Understanding of Plato* by Theon of Smyrna (fl. 125).

Claudius Ptolemy (fl. second century)

So little is known of Ptolemy's life that even the period of his activity can only be determined by examining his astronomical observations, which indicate that he observed at Alexandria between the years 127 and 145. The *Mathematical Syntaxis* (or *Mathematical Collection*), usually known as the *Almagest* from its Arabic title, is easily his greatest scientific work. Divided into thirteen books, it embodied, and added to, the achievements of Greek astronomy accumulated over a number of centuries. It also included a star catalogue for 1022 stars. Until Copernicus and even for some time after, it served as the most fundamental

astronomical treatise for Greek, Arabic, and Latin astronomers, for which reason alone Ptolemy must be considered among the greatest and most influential of scientists. But his greatness is confirmed by yet other high-level theoretical scientific treatises. For in addition to other astronomical works he wrote on mathematics, optics, music, and geography. In astrology his *Tetrabiblos* (or *Quadripartitum,* as it was known in its Latin translation) was a scientific attempt to determine the possible influences of celestial motions and configurations on the earth and on human life. Its subsequent impact, even to this day, has been enormous.

Coimbra Jesuit Commentators.

See Jesuit Commentators of Coimbra.

Constantine the African (d. 1087)

Born in Carthage, Constantine settled in Italy after much travel in the East. He resided in Monte Cassino and Salerno. Although he wrote some medical works, he is most famous for his translations of medical treatises (of both Greek and Arabic origin) from Arabic into Latin (indeed, he is the first of the major translators from Arabic to Latin). These works gave impetus to the study of medicine at Salerno.

Domingo Gundisalvo (Dominicus Gundissalinus) (fl. 1140)

As an important translator in Toledo, Gundisalvo collaborated with John of Seville, the latter perhaps translating from Arabic to Castilian and Domingo converting the Castilian to Latin. Many translations are ascribed to these two men. As an author Domingo wrote approximately five philosophical treatises, of which the most significant was *De divisione philosophiae (On the Division of Philosophy)*, treating of the classification of the sciences. While this work drew heavily upon al-Fārābi, his other works relied on Ibn Gabirol (known in Latin as Avicebron). In general, he incorporated into his own works the thoughts, ideas, and materials of the treatises he had translated.

Dominicus de Clavasio (fl. 1346)

Dominicus was probably born in Italy at Chivasso, near Turin. He studied at the University of Paris, where he was a teacher in arts from 1349 to 1350 and a member of the medical faculty in 1356–1357. Before 1368 he is said to have been a court astrologer to the French king (that is, either to John II or Charles V, since the latter succeeded to the throne in 1364). He wrote a treatise in applied, or practical, geometry *(Practica geometriae)*, a treatise on optics *(Questiones super perspectivam)*, a question on Sacrobosco's *Sphere*, and a commentary on Aristotle's *De caelo et mundo*.

Fibonacci. See Leonardo of Pisa.

Frederick II of Hohenstaufen (1197–1250)

Son of Henry VI, Holy Roman Emperor and King of Sicily, Apulia, and Calabria, and of Constance of Sicily, Frederick II, born in 1197 in Sicily, was orphaned in 1198 and left under the control of Pope Innocent III and a group of ecclesiastics. Through the maneuverings of Innocent III and the military defeat of the Welfs, Frederick, who was already King of Sicily, was confirmed as Holy Roman Emperor in 1215 and crowned at Rome in 1220. Never able to control the German princes, Frederick granted extensive powers to them in 1231, and, after putting down a rebellion led by his son Henry in 1235, he left Germany forever in 1237 to pursue his plans in Italy. Raised as ward of the Pope, Frederick was to become the greatest enemy of the papacy, which was solidly in control of central Italy. The initial difficulties arose from Frederick's promise to Pope Honorius III that he would lead a crusade to Jerusalem. A few years later, in 1227, when it seemed Frederick was deliberately stalling, Pope Gregory IX excommunicated him. During the year 1228–1229, when Frederick was actually on crusade and acquired the Kingdom of Jerusalem by negotiation, Gregory not only placed the Kingdom of Jerusalem under interdict but also attached Frederick's domains in southern Italy. Eventually absolved from excommunication, he was again excommunicated by the same Gregory in 1239, following Frederick's victory over the Lombard towns, which posed a serious threat to the Papal States and to Rome itself. A great propaganda war was launched by both sides. But the military struggle was also a desperate one, as Pope Innocent IV mustered all the powers of the Church and called for a crusade against Frederick. The latter's death from dysentery in 1250 brought the great struggle to an end. Although the Hohenstaufens fought on, the popes were eventually victorious when in 1268 the last of the King's line was beheaded. (For a brief description of Frederick's intellectual endowments, see the Introduction to Selection 85.)

Galileo Galilei (1564–1642)

Born in or near Pisa on February 15, 1564, Galileo received his early education in a monastery. In 1581 he entered the school of medicine at the University of Pisa, but turned to mathematics. Although he left Pisa without a degree, he became professor of mathematics at that university in 1589. It was about this time that his anti-Aristotelian treatise *De motu* was written. Galileo resigned from Pisa in 1592 and at the end of that year became professor of mathematics at the University of Padua, where he found intellectual stimulation and remained until 1610. During his Paduan period he formulated lectures on mechanics, made a thermoscope, and constructed a telescope with which he later observed the celestial bodies and discovered the four satellites of Jupiter, publishing his results in the *Siderius Nuncius*. In 1610 he was appointed Chief Mathematician and Philosopher to the Grand Duke of Tuscany and Chief Mathematician of the University of Pisa. Since the latter post did not require residence at Pisa, Galileo chose to live in Florence. In late 1615, Galileo, who had openly declared for the Copernican system, was warned not to hold or defend it. In 1616 the *De revolutionibus* was suspended until corrected and it was forbidden to reconcile its teachings with Scripture. In 1632, Galileo published in Florence his great *Dialogue Concerning the Two Chief World Systems*. Although he had church permission to publish the book, his enemies convinced the Pope that Galileo had defied the Church's ban against the teaching of Copernicanism. In 1633, after a trial by the Inquisition, Galileo was sentenced to life imprisonment, which took the form of house arrest. Although his books were banned and he was forbidden to write and publish others, he succeeded in writing his final and greatest work, a revolutionary treatise on physics entitled *Discourses on Two New Sciences,* which was published in Leyden in 1638. Totally blind during his last years, Galileo died in 1642 at the age of seventy-eight.

Gerard of Brussels (fl. thirteenth century)

Little more is known about Gerard than that his *De motu* was written sometime between 1187 and 1260 and that he may be the "Gernardus" who wrote a mathematical treatise titled *Algorithmus demonstratus.*

Gerard of Cremona (ca. 1114–1187)

This greatest of all translators of Arabic works into Latin was born in Cremona, Lombardy; he died in Toledo, Spain, in 1187. All the information about his life and translations is embodied in the biography and list of translations appended by his students to his translation of Galen's *Tegni*. This is reproduced in Selection 7.

Gilbert the Englishman (fl. 1250)

Almost nothing is known about Gilbert, but the fact that he was called "the Englishman" suggests that he lived outside England. He may have been chancellor of Montpellier in 1250. His major work was the *Compendium medicinae,* which drew upon Salernitan and Arabic medical sources. It includes discussions of fevers and of external and internal diseases of all kinds, and chapters on surgery.

Giles of Corbeil (d. between 1220 and 1224)

Giles was a Frenchman (active ca. 1200) who studied medicine in Salerno. After a period in Montpellier he went to Paris, becoming canon of Notre Dame. Most of his known medical works were written in verse, of which *On Urines (De urinis)* was perhaps the most popular textbook on uroscopy before the sixteenth century. Also popular was a work *On Pulses (De pulsibus).*

Guy de Chauliac (d. ca. 1368)

Of French peasant stock, Guy studied medicine in Toulouse and Montpellier; later he studied anatomy and surgery in Bologna. He was in Paris during the years 1315–1320 and also resided in

Lyon. However, much of his time was spent in Avignon, where he was physician to three popes, Clement VI (1342–1352), Innocent VI (1352–1362), and Urban V (1362–1370). His fame rests upon his *Chirurgia magna,* a treatise on surgery divided into seven tractates and completed in 1363 in Avignon.

He relied heavily on Galen, using new translations made in the fourteenth century, and Avicenna. The popularity of his work is attested by the translations of it into French, Italian, English, Dutch, Hebrew, and other languages.

Haly Rodohan ('Alī ibn Ridwān) (ca. 998–1061)

Born near Cairo, Haly Rodohan lived and died there. He was the author of many medical works and an important commentator on medical works of Galen and Hippocrates. In the Latin West his most popular treatise was a commentary on Galen's *Tegni* (or *Ars parva*), which was translated into Latin by Gerard of Cremona. As an astrological

author he was equally important, since his commentary on Ptolemy's *Quadripartitum* (or *Tetrabiblos*), the greatest of all astrological treatises, was translated into Latin in the mid-thirteenth century by Egidius de Tebaldis and exerted a wide influence.

Heinrich von Pfolspeundt (fl. 1460)

Heinrich was a German surgeon who was also a member of the Teutonic Order. In 1460 he wrote the *Buch der Bündth-Ertznei,* a treatise on rhinoplastic surgery, the techniques of which may have

been derived directly from Antonio Branca, a fifteenth-century Sicilian surgeon who left no writings but whose operations were described by others.

Henry of Harclay (ca. 1275–1317)

By 1296, Henry of Harclay was a master of arts, after which he became a Dominican in 1297 and chancellor of Oxford University in 1312. In the latter capacity he "presided at the condemnation of certain theological statements as heretical Feb. 1315" (A. B. Emden, *A Biographical Register of the University of Oxford to A.D. 1500* [Oxford: Clarendon Press, 1957–1959], II, 875, col. 1). He

died in Avignon in 1317 after arriving at the Roman Curia "as a proctor of the University for the settlement of its dispute with the Oxford Friars Preachers, 1317" *(ibid.).* Apart from recently discovered treatises on the infinite and the continuum, he left a *Commentary on the Sentences* and a series of *Questiones.*

Henry of Hesse (Heinrich von Langenstein) (1325?–1397)

Henry was born in Germany near Marburg (Hesse-Nassau) at Hainbuch, Langenstein. At the University of Paris he received a bachelor's degree on February 20, 1363, the licentiate in theology in 1375, and a doctorate in theology in 1376. Sometime before 1381 he was a vice-chancellor of the Sorbonne. Difficulties arising from the Great Schism made it expedient for Henry to leave Paris around 1383. He went to Germany and then to Vienna, where he became one of the founders of the University of Vienna, remaining in that city until his death in 1397.

Henry was a prolific author, writing on theology,

mysticism, religious reform, economics, the study of Hebrew, and science. Among the scientific works, mention should be made of his optical treatise *Questiones perspectivae,* and his anti-astrological treatises, *Questio de cometa,* which considered the comet of 1368, and *Tractatus contra astrologos coniunctionistas de eventibus futurorum,* written in 1373. Biological and physical ideas are contained in his *De reductione effectuum specialium in virtutes communes* and *De habitudine causarum et influxu naturae communis respectu inferiorum,* both composed before 1373.

Henry of Mondeville (d. ca. 1326)

Henry studied surgery at Montpellier with Lanfranc and in Italy under Theodoric Borgognoni,

who died in 1298 in Bologna. He served as a military surgeon with the armies of Philip the Fair

(1285–1314) and his son Louis X (1314–1316). He also taught anatomy and medicine in Montpellier in 1304 and then in Paris for a number of years commencing in 1306. His great work, *Chirurgia,* was intended to be in five parts. Begun in 1306, it was never completed, although Henry worked on it until around 1320. It was a vast work, notable for its advocacy of the new Italian methods of treating wounds. The *Chirurgia* never became very popular, because it was rivaled by Lanfranc's work and then superseded by the great surgical text of Guy de Chauliac.

Hugh of Saint-Victor (d. 1141)

Whether Hugh was born in Saxony or in the region of Ypres in Flanders is as yet unresolved. But from 1120 until his death he was the greatest master at the abbey school of Saint-Victor in Paris. Although Hugh's primary interest lay in theology and biblical exegesis, on which subjects he left a number of treatises, including a *Summa Sententiarum (A Summary of the Sentences)* and *De sacramentis christianae fidei (On the Sacraments of the Christian Faith),* he also composed a work on applied geometry *(Practica geometriae),* and in the *Didascalicon* included discussions on the classification of the sciences, the quadrivium, and other scientific topics. Philosophical issues and topics were touched upon in the *Didascalicon* and *Epitome Dindimi in philosophiam.*

Isidore of Seville (ca. 570–636)

Born into a prominent Spanish-Roman family, Isidore succeeded his brother as bishop of Seville around 600 and subsequently presided over church councils in Seville (619) and Toledo (633). Author of many theological treatises, he produced only two works of physical and scientific interest, *On the Nature of Things (De natura rerum)* and his later *Etymologies (Etymologiae),* an enormously influential encyclopedic work covering a wide range of topics. Isidore was a compiler utterly lacking in originality and frequently misunderstanding his unreliable sources.

Jacopo da Forlì (ca. 1350–1414)

Jacopo was a well-known physician and natural philosopher. In 1384–1385 he taught natural philosophy at the University of Bologna, and in the years 1400–1404 he taught at the University of Padua, to which, after withdrawing for a few years, he returned to teach medicine in 1407. He remained on the Padua faculty until his death in 1414. His medical works include commentaries on Galen's *Tegni,* the *Aphorisms* of Hippocrates, and Avicenna's *Canon of Medicine.* Like other Italian physicians, Jacopo was very interested in natural philosophy, as evidenced by his treatise *On the Intension and Remission of Forms (De intensione et remissione formarum),* in which he shows a good knowledge of the basic fourteenth-century physical discussions at Oxford and Paris.

Jesuit Commentators of Coimbra *(Conimbricenses)* (second half of sixteenth century)

A group of Jesuit authors in the second half of the sixteenth century who were associated with the University of Coimbra (hence the appellation *Conimbricenses)* and collectively produced a series of commentaries on the works of Aristotle. The leading member of the group was Peter de Fonseca (1548–1599).

Joannitius (Hunain ibn Isḥāq) (809–877)

Joannitius, or Abū Zaid Hunain ibn Isḥāq al-'Ibādī, was a Nestorian Christian physician who not only composed original works but was the greatest translator of Greek science into Syriac and Arabic. He seems to have made approximately 200 translations. Indeed, we learn from him that of the books of Galen he translated 95 into Syriac—five of them twice—and 39 into Arabic. In addition to the works of Galen he translated works by Hippocrates, Plato, Aristotle, Dioscorides, and Ptolemy. His methods were scholarly and his results excellent. Among his own compositions the most important from the standpoint of later medieval Latin science was the *Isagoge,* his medical introduction to Galen's *Tegni* (see Selection 90). After its translation into Latin in the late eleventh century

it was immensely popular for the next few centuries.

John Buridan (ca. 1300–ca. 1358)

Buridan was probably born in Béthune in the diocese of Arras in France. He is mentioned as rector of the University of Paris in a document dated February 9, 1328 (and mentioned as rector again in 1340). In another document of August 30, 1329, he is cited as a "celebrated philosopher" and recommended for a benefice (again recommended for a benefice in 1348). According to a papal register of November 2, 1330, he was awarded a benefice. His name is mentioned in a document in 1349, and he signed his name to a document on July 12, 1358. His date of death is unknown. Almost all of his extant works consist of commentaries or ques-

tions on the works of Aristotle, on almost all of which he left a commentary or set of problems or *questiones*. Of special importance are his *Questions on the De caelo* and three separate versions on the *Physics*. Buridan was a philosopher, not a theologian, which accounts for his singular concern for the works of Aristotle. His works were widely studied and discussed well into the sixteenth century throughout Europe. He undoubtedly ranks among the best thinkers who proposed novel and interesting solutions to a wide range of scientific problems that crystallized during the fourteenth century.

John Duns Scotus (ca. 1266–ca. 1308)

Only the most meager biographical data on Scotus are available. He was ordained to the priesthood in Northampton on March 17, 1291, and was at a convent in Oxford on July 26, 1300, at which time he was commenting on the *Sentences* of Peter Lombard. In 1302 he was sent to Paris, where he again lectured on the *Sentences*. In August of 1303, Scotus was probably compelled to leave France, since he had sided with Pope Boniface VIII against King Philip the Fair of France. Returning in 1304, he resumed his comments on the *Sentences* and probably received his doctorate in 1305. In February, 1308, he was in Cologne, probably teach-

ing. He is thought to have died there on November 8, 1308. His premature death left his major works incomplete. Disciples hastily edited and completed some of these and in the process produced great uncertainty as to the genuineness of many works and parts of works. Only in the last thirty years have scholars sought to establish the genuine works of Scotus. In addition to his *Sentence Commentaries*, which exist in different versions, mention should be made of his *Questiones quodlibetales*, *Collationes parisienses et oxonienses*, Questions on Aristotle's *Metaphysics*, and a number of logical works.

John Gaddesden (ca. 1280–1361)

John studied at Oxford, receiving the bachelor's degree by 1300, the master of arts degree by 1303, and a medical degree by 1309. At a later date he also took a theology degree and was, indeed, a priest. He acquired considerable fame as a medical practitioner in Oxford and then in London (Chau-

cer praised him in the Prologue to the *Canterbury Tales*). Sometime around 1314 he composed his *Rosa medicinae*, or *Rosa anglica* as it was more popularly known. Despite criticisms of this popular medical work, it included many fine observations and clinical descriptions.

John of Mirfeld (d. 1407)

John was an English medical writer whose year and place of birth are unknown, as is much else about him. Perhaps he was a priest associated with the hospital of St. Bartholomew in London. That he was a skilled university-trained physician at Oxford has been denied. Two books are ascribed

to him: *Breviarium Bartholomei* (written between 1380 and 1395), a medical treatise; and *Floriarium Bartholomei*, a theological treatise. The *Breviarium* is a nontheoretical scholastic encyclopedia (see Selection 101.3, n. 4).

John of Sacrobosco (d. ca. 1244–1256)

Very little is known of this author of three of the

most widely read elementary scientific treatises

written during the Middle Ages. Although he may have been born in England (Holywood or Halifax are often cited as possible birthplaces), it seems that his teaching career was spent at the University of Paris, where he is said to have composed his

Sphere (De sphera). Although a number of works are ascribed to him, only the *Sphere, Computus,* and *Algorismus* seem to be extant. All were e-normously popular well into the sixteenth century.

John of Saxony (fl. 1327–1355)

Although a German, as his name implies, John of Saxony flourished in Paris, where he worked with his master, John de Linières. Together they revised and helped introduce the Alfonsine Tables into Paris. Apparently it was John of Saxony who in 1327 added explanations and elaborations to his master's revision of the Alfonsine Tables. This version, of which a number of manuscript copies

exist, was published in Venice, 1483. The translation in this source book was made from it.

According to George Sarton (*Introduction to the History of Science,* Vol. III, Part I, p. 118), "E-phemerides, as well as astronomical and astrological commentaries, are ascribed to him; he was active as a professional astronomer from 1327 to 1355."

John Pecham (ca. 1230–1292)

Although he is known to be an Englishman, Pecham's year and place of birth are very uncertain. After studying in the faculties of arts at Paris and Oxford, he entered the Franciscan Order at Oxford, perhaps around 1250. Returning to Paris sometime between 1257 and 1259, Pecham began the study of theology and received his doctorate in that subject from the University of Paris in 1269. For the next two years he lectured in theology to Franciscans in Paris. During his stay at Paris, Pecham was not only active in philosophical and theological disputes against the Thomists, and a defender of the mendicants against the seculars, but he also composed many theological treatises. He returned to Oxford sometime during 1271–

1272 and again lectured to the Franciscans. In 1275 he became Provincial Minister of the English Franciscans, continuing his opposition to Thomism and Averroism. In 1279 he became Archbishop of Canterbury, a post he held until his death on December 8, 1292.

In the course of a busy administrative life Pecham found time to compose a number of scientific treatises, the most famous being his enormously popular optical treatise, *Perspectiva communis*. He also commented on Sacrobosco's *Sphere (Tractatus de sphera),* and wrote *Theorica plane-tarum, Tractatus de numeris, Tractatus de anima, Questiones de anima,* and perhaps a *Tractatus de animalibus*.

Jordan of Turre (fl. 1313–1335)

The little that is known of Jordan emerges from his connection with the University of Montpellier, where he was a master of medicine around 1313 and one of four consultants from the university who produced a regimen for an ailing bishop.

Among many medical treatises ascribed to Jordan are *De lepra (On Lepers,* from which Selection 101.2 has been translated), *De urina, De febribus, De experimentis,* and *Ad curam hydropisis.*

Jordanus of Nemore (fl. 1230–1260)

Biographical data are wholly lacking for Jordanus. He was formerly identified with Jordanus of Saxony, master general of the Dominican Order from 1222 to 1237, but this seems unlikely. It appears that he taught at the University of Toulouse, where the faculty of arts was not founded until 1229. This, and the fact that a number of his works were cited in the *Biblionomia,* a catalogue of books composed by Richard of Fournival sometime

between 1246 and 1260, makes it reasonable to conclude that his productive years fell between 1230 and 1260. Of the statical treatises ascribed to Jordanus, the *Elementa Jordani super demonstra-tionem ponderum (The Elements of Jordanus on the Demonstration of Weights)* is almost certainly genuine and the *De ratione ponderis (On the Theory of Weight)* is probably his composition. He was also the author of a number of mathematical treatises, including *Arithmetica, De numeris datis (On Given Numbers), De triangulis (On Triangles),* and probably *Planisphaerium,* or *De forma spere in plano (On the Form of a Sphere in a Plane).* A

number of other works, of uncertain attribution, have also been credited to him. He was an excellent mathematician and perhaps the leading mechanician of the Middle Ages.

Lanfranc (d. before 1306)

Lanfranc was an Italian surgeon, born in Milan, who studied under William of Saliceto. Banished from Milan around 1290, he went to France. After practicing in a few French provincial cities he went to Paris around 1295, where he became a successful surgeon and teacher. Between 1290 and 1295 he wrote his *Small Work on Surgery (Chirurgia parva)* and in 1296 completed his *Great Work on Surgery*

(Chirurgia magna), dedicated to Philip the Fair. The *Chirurgia magna,* from which Selection 114 has been translated, consisted of five treatises and contained many new ideas and techniques. It was translated into English, French, and Hebrew; the *Chirurgia parva* was translated into Spanish and German, the latter translation being by Otto Brunfels in the sixteenth century.

Leonardo of Pisa (Fibonacci) (b. ca. 1179; d. after 1240)

Leonardo was the greatest mathematician of the Latin Middle Ages. Our meager biographical information is derived largely from the prologue to his first work, *Liber abaci,* written in 1202 and revised in 1228. His youth was spent in the Pisan merchant compound in Bugia, Algeria, where his father served the Pisan merchants in a consular role. Here be became familiar with the Arabic positional number system developed by the Hindus. He learned more about this superior numerical system as he traveled about the Mediterranean to Egypt, Syria, Greece, and Sicily. Returning to Pisa, he eventually wrote five mathematical treatises, which reveal a good knowledge of Greek and Muslim mathematics, probably derived from Latin sources (there is no evidence that he ever learned

Arabic). In addition to the *Liber abaci,* which deals with number theory, square and cube roots, equations of the first degree, and a variety of other problems, Leonardo wrote *Practica Geometriae* in seven parts, containing theoretical and applied geometric propositions; *Flos super solutionibus quarumdam questionum ad numerum et ad geometriam pertinentium (The Flower of Solutions of Certain Questions Relevant to Number and Geometry),* which is concerned with determinate and indeterminate analysis of equations of the first degree; a letter containing two problems to Theodoric, astrologer at the court of Frederick II of Hohenstaufen; and finally, his last work, the *Liber quadratorum (Book of Square Numbers).*

Ludovicus (Luis) Coronel (fl. 1511)

Born in Segovia, Spain, Ludovicus studied in Paris and taught philosophy at the University of Paris. He wrote at least two works, one published at Paris in 1507 and called *Treatise on the Formation of Syllogisms (Tractatus de formatione syllo-*

gismorum) and the other written between 1506 and 1511 and published in Lyons (also, it seems, in Paris) in 1530, bearing the title *Physicae perscrutationes* but actually a commentary on the eight books of Aristotle's *Physics.*

Macrobius (fl. 400)

Virtually nothing is known of Macrobius' life. He was perhaps born in Africa, although it has not been determined whether he was Greek-speaking or not. He had a son, Eustachius, to whom he dedicated his two major works, *Saturnalia* and *Commentary on the Dream of Scipio.* The latter, a commentary on the closing portion of the text of Cicero's *On the Republic,* was an exposition of Neo-

platonism which was very influential in medieval scientific thought and was one of the main sources of medieval Platonism. Like most encyclopedists, however, Macrobius did not actually know at first hand the many works and authors he cited; rather, he relied on the numerous handbooks of the compilers of late antiquity.

Marsilius of Inghen (ca. 1340–1396)

Although his date of birth is unknown, Marsilius' birthplace was probably in the village of Ingen in the bishopric of Utrecht (some would place it near Nijmegen, however). The records of the English nation at the University of Paris list him as presenting an inaugural lecture as *magister* on September 27, 1362. A very popular teacher, he was also made rector in 1367 and 1371, and proctor of the English nation for the years 1362, and 1373 to 1375. In 1368 and 1376 he went to the papal court at Avignon to represent the University of Paris. Sometime before 1382 he left the University of Paris; he became first rector of the University of Heidelberg in 1386, a position he held six times. He died in Heidelberg on August 20, 1396, bequeathing his large book collection to the library of the University of Heidelberg.

In addition to a number of logical treatises and a commentary on the *Sentences* of Peter Lombard (dated ca. 1386) Marsilius wrote commentaries on Aristotle's *Physics* and *On Generation and Corruption*, probably while he was yet in Paris.

Martianus Capella (fl. 410–439)

Martianus Capella was probably a Latin-speaking lawyer who lived in Carthage (Africa). It is not known whether he was a Christian or a Neoplatonist. His fame rests on his allegorical encyclopedic handbook *On the Marriage of Philology and Mercury*. Modeled on Varro's *Nine Books of Disciplines*, it canonized the seven liberal arts, namely grammar, rhetoric, dialectic (the trivium), and arithmetic, geometry, astronomy, and music (the quadrivium). This extremely difficult, bombastic, and obscure book exerted a great influence on medieval authors. Like all such late encyclopedias, Capella's work was a collection of extant handbook materials, much of which he garbled, distorted, and misunderstood.

Master Nicholas (ca. 1200)

Although an anatomical treatise is ascribed to "Master Nicholas," nothing is known of him (see selection 95, n. 1).

Matthaeus Platearius (d. 1161)

Matthaeus was a Salernitan physician who wrote the earliest commentary on the *Antidotarium Nico-lai* and a treatise on simple medicines, *De simplici medicina*, or *Circa instans* after the opening words.

Moses Maimonides (Moses ben Maimon) (1135–1204)

A Spanish Jew born in Cordova, Maimonides and his family fled Spain when the Almohads took Cordova in 1148. After moves to Morocco and Palestine, they settled outside Cairo, where Maimonides became the recognized leader of Egyptian Jewry and the court physician to Saladin. Writing in Arabic and Hebrew, Maimonides composed treatises on the rabbinic tradition (here his greatest effort was the *Mishneh Torah,* written in Hebrew), on Aristotelian logic (*On the Terminology of Logic,* written when he was sixteen), and on philosophical problems generally, as in the *Guide of the Perplexed,* his best-known work, since it was translated from Arabic to Hebrew and from Hebrew to Latin. Thomas Aquinas and many other Latin scholastics were influenced by it.

Mundinus (Mondino de' Luzzi) (ca. 1265–1326)

Son of an apothecary, Mondino was born in Bologna, where he studied medicine at the university under Taddeo Alderotti. From sometime around 1290 until his death in 1326 he practiced in Bologna and taught at the university as professor of anatomy. Although his fame rests upon the *Anatomy* (see Selection 96, n. 1), he also wrote commentaries on classical medical texts by Galen, Hippocrates, and others, and composed a number of independent treatises on a variety of medical subjects.

Nicholas of Autrecourt (b. ca. 1300; d. after 1350)

Nicholas was born in Autrecourt, a village in the diocese of Verdun. Between 1320 and 1327 he was at the Sorbonne, probably as a student. During the years 1327 and 1340 he received the degrees of master of arts and bachelor of theology and laws and apparently lectured on the *Sentences* of Peter Lombard as well as on certain works of Aristotle. In 1338 he was made canon of the Cathedral of Metz, which was an ecclesiastical benefice enabling him to continue his studies. In December 1340, following a reaction against the teaching of certain doctrines by Ockham's followers, Nicholas was summoned to the papal curia at Avignon on charges of holding errors in theology and philosophy. The lengthy proceedings were concluded in 1346,

and Nicholas was compelled to make a public retraction of his errors on November 25, 1347. Not only was he deprived of his degrees and forbidden to receive a doctorate in theology but he was compelled to witness the public burning of his books. The last known mention of Nicholas was in 1350 and reveals that he was subsequently made a deacon at the Cathedral of Metz. The works which survived the burning are few and include nine letters to Bernard of Arezzo (only two are complete), a theological question, and a major work, virtually complete, known by the title *Exigit ordo executionis,* from which Selection 56.2 has been drawn.

Nicolaus Copernicus (1473–1543)

Copernicus was born in Thorn, on the banks of the Vistula, where his father was a merchant and town official. Following the death of his father, Copernicus, then ten years old, was raised by his maternal uncle, Lucas Watzelrode, who became bishop of Ermland in 1489. Lucas appointed young Copernicus canon of the Cathedral of Frauenburg, a post he held until his death in 1543 and from which he derived an income. In 1491, Copernicus entered the University of Cracow. A few years later in 1496, he departed for Italy, where he studied medicine, law, and the liberal arts at Bologna and Padua. In 1503, he received a doctorate in canon law from the University of Ferrara.

Copernicus was a man of many talents and interests. He knew Greek, was learned in medicine and canon law, and composed a treatise on problems of coinage. He even painted his own portrait. In astronomy, his fame rests on the epoch-making *De revolutionibus orbium caelestium,* in which he proclaimed the heliocentric theory in 1543. He formulated an earlier version of the heliocentric theory in the *Commentariolus,* which is of uncertain date and was unpublished during his lifetime. The only other extant astronomical work by Copernicus is the *Letter Against Werner,* written in 1524 and published after his death.

Nicole Oresme (ca. 1325–1382)

A native of Normandy, Oresme was probably born sometime between 1320 and 1330, perhaps in the village of Allemagne near Caen. His name appears on a list of masters of the Norman nation dated November 29, 1348, the year he entered the College of Navarre at the University of Paris. In 1356 he became grand master of the College of Navarre, presumably already having his doctorate in theology, since this was a prerequisite for the grand mastership. Many of his Latin treatises may already have been written by this time. A few years later Oresme came into association with the French royal family, serving as instructor to the dauphin, the future Charles V. His appointment as archdeacon of Bayeux in 1361 had to be relinquished because it conflicted with the grand mastership of

Navarre. On November 23, 1362, he was appointed canon of the Cathedral of Rouen (at this time he was apparently still an active teacher at the University of Paris); on February 10, 1363, he became canon at La Sainte Chapelle and on March 18, 1364, was elevated to the deanship of the Cathedral of Rouen, by which time he had probably resigned as grand master of Navarre. At the request of King Charles, Oresme made a series of translations from Latin into French of Aristotle's *Ethics* (1369–1370), *Politics, Economics* (both completed between 1372 and 1374), and *De caelo et mundo* (1377), the latter an especially important scientific treatise which helped formulate a new scientific terminology in the French language. In addition to monetary rewards for his labors Oresme received King Charles' support for the bishopric of Lisieux, which he held from August 3, 1377, until his death on July 11, 1382.

In addition to the major scientific treatises represented by selections in this source book *(Algorismus proportionum, De proportionibus proportionum, Questiones super geometriam Euclidis, De configurationibus qualitatum, De commensurabilitate vel incommensurabilitate motuum celi, Livre*

de divinacions, and *Le Livre du ciel et du monde)* and his other translations, Oresme wrote works in French and Latin on theology, money, natural magic, astrology, and cosmology, and Latin commentaries on some of the works of Aristotle. His broad intellectual interests and his extraordinary conceptual power mark him as one of the most penetrating philosophic and scientific thinkers of the Middle Ages.

Otto von Guericke (1602–1686)

Von Guericke was born in Prussian Saxony in the city of Magdeburg on November 20, 1602. He studied law in various German cities, but studied mathematics in Holland (Leiden). He was elected to a number of governmental offices (alderman and the mayor of Magdeburg in 1627 and 1646, respectively; in 1646 he also became a magistrate of Brandenburg). Much of his leisure was devoted to science and many of his ideas and experiments were incorporated into his famous work *Experimenta nova, ut vocant, Magdeburgica de vacuo spatio,* published in 1672. Here he not only expressed cos-

mological views on the size of the world, our planetary system, the existence of void within and without the world but also included his experiments with air pumps and vacua. For in 1650 he had invented the air pump, which enabled him in 1654 to demonstrate the great pressure of air on an evacuated pair of copper hemispheres (the "Magdeburg hemispheres"), which could not be pulled asunder by two teams of horses tugging in opposite directions. He was also interested in static electricity and produced charges by exciting a sulphur ball. On May 11, 1686, he died in Hamburg.

Petrus Bonus (fl. 1330)

Petrus Bonus (or Pietro Buono) was probably a physician of Ferrara, perhaps a menber of the Avogadrus family, although this is seriously questioned by some investigators. Next to nothing is known of him except that in 1323 he was in the city of Traú, in Dalmatia, and that in 1330 he wrote *The New Pearl of Great Price* (or simply, *The Precious New*

Pearl; Pretiosa Margarita Novella). In this prolix and rambling alchemical treatise the author, after boasting that the whole alchemical art can be taught in about an hour, admits at the end of the work that he himself does not yet know the secrets of alchemy (for further discussion, see the Introduction to Selection 75).

Petrus Peregrinus (Pierre de Maricourt) (fl. 1269)

Pierre de Maricourt of Picardy, or Petrus Peregrinus (Peter the Pilgrim), probably so-called because of service in a papal crusade in the Holy Land or Italy, was a Picard about whom little is known. He probably studied at the University of Paris, where he apparently received a master's degree (he is sometimes referred to as "Master" [*magister*]). Indeed, it may have been at the University of Paris that he met Roger Bacon, who speaks of him in glowing terms. The latter praised him as the only one capable of understanding optics and one who had actually undertaken the construction of a burning mirror. In his *Opus tertium* (ch. 13, p. 47 of the edition by J.S. Brewer, *Fr. Rogeri Bacon Opera quaedam hactenus inedita* [London, 1859], Vol. I), Bacon has this to say about Peregrinus:

"I know one only who can have praise for works of this science [i.e. optics]; for he does not trouble about discourses or quarrels over words, but follows the works of wisdom and keeps quietly to them. And so, though others strive blinkingly to see, as a bat in the twilight, the light of the sun, he himself contemplates it in its full splendour, on account of which he is a master of experiment *(magister experimentorum);* and therefore he knows by experiment natural history, and physic, and alchemy, and all things in the heavens and beneath them: indeed, he is shamed if any layman, or grandam, or soldier, or country bumpkin knows anything that he himself does not know. Wherefore he has inquired into all operations of metal-founding, and the working of gold and silver and other metals, and of all minerals; and he knows all things pertaining to the army and to arms and the chase: and he has examined all that relates to agriculture, the measurement of land, and earthworks, and he has even studied the experiments, devices, and incantations of witches and magicians, and likewise the illusions and tricks of jugglers; so that nothing is hidden from him which he ought to know, and he knows how to reprobate all things false and magical. And so it is impossible without him that philosophy should be exhausted, or be treated usefully and with certainty. But he is thus not worthy of price, nor does he estimate his own worth. For should he wish to stand well with kings and princes, he would find those who would honour him and would enrich him. Or, if he were to show in Paris by his works of wisdom all that he knows,

the whole world would follow him: yet because either way he would be hindered from the bulk of his experiments in which he most delights, so he neglects all honour and enrichment, the more since he might, whenever he wished it, attain to riches by his wisdom". (Translated by Silvanus P. Thompson, "Petrus Peregrinus de Maricourt and his Epistola de Magnete," *Proceedings of the British Academy*, II [1905–1906], 380.)

In the year 1269 Peregrinus was serving as an army engineer with the forces of Charles Anjou, who was then besieging the town of Lucera in southern Italy. It was during this military operation that he wrote his letter, which constitutes one of the finest scientific descriptions composed in the Middle Ages.

Pierre d'Ailly (1350–1420)

Born in Compiègne, France, d'Ailly entered the College of Navarre at the University of Paris in 1363, the same college at which Nicole Oresme had studied some fifteen to twenty years earlier. The latter's influence on d'Ailly was extensive, as evidenced partially by d'Ailly's not inconsiderable plagiarisms from Oresme's works, which he probably found readily available at Navarre. He became a master of arts at Navarre in 1368, began the study of theology in 1372, and commented on the *Sentences* of Peter Lombard in 1375. In 1381 he received his doctorate and became a professor of theology at the Sorbonne. He became in succession rector of the College of Navarre in 1384, chancellor of the University of Paris on October 7, 1389, chaplain to Charles VI in 1390, archdeacon

of Cambrai in 1391, bishop of Le Puy on April 2, 1395, and archbishop of Cambrai on June 5, 1397. For siding with the antipope John XXIII during the Great Schism, d'Ailly was made a cardinal by John in 1412. During the Council of Constance (1414–1418) he played the role of an opportunist, changing views when it seemed advantageous. He was a prodigious author (175 writings are credited to him), writing mostly theological treatises but also composing the *Ymago mundi* and philosophical treatises which reveal a generally nominalist position. Although hardly a deep thinker, he exerted a considerable influence on contemporary theologians and was also the teacher of Jean Gerson.

Richard of Wallingford (ca. 1292–1336)

Richard studied at Oxford University until approximately 1315 and became a Benedictine monk at the St. Albans abbey. After further study at Oxford (1318–1327) he became a bachelor of theology and lectured on the *Sentences* of Peter Lombard. On October 29, 1327, he was elected abbot of St. Albans, which was then declared invalid and validated again on February 1, 1328.

Leprosy caused a gradual loss of his physical powers until his death in 1336. In addition to his *Quadripartitum de sinibus demonstratis* he composed other astronomical works, as well as treatises describing astronomical instruments. A complicated astronomical clock was constructed by him for the church of St. Albans.

Robert Grosseteste (ca. 1168–1253)

Grosseteste is described as a *magister* (master) in a charter by Hugh, bishop of Lincoln, dated sometime between 1186 and 1189. He probably

received his early arts training at Oxford, but his theological training may have been received at Paris between 1209 and 1214 (by 1214 he was a doctor of theology). Sometime between 1214 and 1231 he became one of the first chancellors of Oxford University. In 1229 he was archdeacon of Leicester, and around 1232 he became first lecturer to the Franciscans at Oxford. Election as bishop of Lincoln came in 1235, in which office he remained until his death. Among his scientific works should be mentioned his commentaries on Aristotle's *Posterior Analytics* and *Physics* (the latter was written after 1230), and a commentary on the six days of creation, *Hexameron,* in which there is much cosmological discussion. A series of optical works was probably composed between 1230 and 1235 and includes *On Lines, Angles, and Figures, or On Refractions and Reflections of Rays (De lineis, angulis et figuris seu De fractionibus et reflexionibus radiorum); On the Rainbow (De iride); On the Nature of Places (De natura locorum);* and *On Color (De colore).* His "light metaphysics" was embodied in two earlier treatises, *On Light (De luce)* and *On Corporeal Motion and Light (De motu corporali et luce),* written perhaps between 1215 and 1220. His computus or calendrical treatises and a work *On the Sphere (De sphaera)* are also worthy of mention. During his lectureship to the Franciscans, Grosseteste studied Greek, and he later made a number of translations from Greek to Latin, including Aristotle's *De caelo.*

Roger Bacon (ca. 1219–1292)

Bacon was born perhaps near Ilchester in Dorset, England. He studied at the University of Paris, where, around 1237, he delivered some of the earliest lectures on Aristotle's natural books. Although studying in the faculty of theology, he never became a doctor of theology. He left teaching around 1247 and went to Oxford in order to learn about all sciences, so that he might apply them to theology. Between 1247 and 1267 he claims to have spent £2000 on secret books, experiments, and instruments. Around 1257 he entered the Franciscan order and soon after (around 1260) fell under suspicion of holding dangerous ideas, which led to his transfer to Paris. In 1263–1264 he met Gui de Foulquois, who became Pope Clement IV in 1265 and requested Bacon's format for applying science to theology. Thus inspired, Bacon wrote many of his better-known and important works between 1266 and 1268, namely the *Opus majus, Opus minus, De multiplicatione specierum (On the Multiplication of Species* or *On the Propagation of Forces),* and the *Opus tertium.* All were sent to the pope prior to the latter's death on November 29, 1268. Following Clement's death, whose opinion of the treatises is unknown, Bacon returned to Oxford, where he wrote additional theological and philosophical works, and Hebrew and Greek grammars (he was a strong advocate of the study of languages). Apparently because of attacks on both Dominicans and Franciscans, and perhaps also because of the Condemnation of 1277, Bacon was imprisoned for a time, possibly as long as from 1277 to 1292, being released shortly before his death in 1292. Always contentious and difficult, Bacon was a foe of authority and scornful of most of his contemporaries.

Roger of Salerno (fl. 1170)

Roger was a Salernitan physician who wrote a textbook on practical surgery *(Practica chirurgiae)* which was widely used and commented upon. Little more than this is known about him.

Rufinus (fl. 1287–1300)

In the introduction to his edition of *The Herbal of Rufinus* (Chicago: University of Chicago Press, 1946), Thorndike declares (p. xiv) that "Rufinus is called a holy doctor as well as master in the rubric or titulus of our text and later, in the course of the work, is spoken of as a monk and again as 'brother Rufinus, penitentiary of the lord archbishop of Genoa,' and as abbot of the monastery of Tyre, a position which he presumably held *in absentia.* In the Preface he tells us that he had pursued the seven liberal arts in the cities of Naples and Bologna and that, after studying astronomy and astrology, using the *Tables of Toledo,* he turned to the science concerning herbs, which he intended to investigate fully and perfectly. This investigation, as we shall see, was by no means to be limited to books and the classroom. Astrology, however, Rufinus taught, as well as studied, at Bologna. It is doubtful whether he may be identified with a 'brother Rufinus, of the order of Minorites, our chaplain and penitentiary,' to whom Urban IV addressed a letter on June 4, 1264, and to whom he again referred, on April 18, 1264, as called Columbo and as resigning from an archpresbytery."

Taddeo Alderotti (ca. 1223–ca. 1295)

A native of Florence, Alderotti studied medicine and philosophy in Bologna, where he later taught in 1260 and remained to do much of his writing. He produced many commentaries on texts by Hippocrates, Galen, and on the *Isagoge* of Hunain ibn Ishāq (Joannitius). The best of his clinical observations were included in his *Consilia medicinalia,* from which Selection 109.2 has been translated.

Themon, Son of the Jew (fl. 1349–1361)

Apparently associated with Münster, Westphalia, in his youth, Themon converted to Christianity and subsequently received his education at the Sorbonne, becoming a famous teacher at the University of Paris, where he incepted in arts in 1349. After teaching in Erfurt, Germany, for a few years, he returned to Paris in 1353, where he was elected three times as proctor of the English Nation at the University of Paris (1353, 1355, 1356) and was also its treasurer from 1357 to 1361, when he was replaced by Albert of Saxony, who was said to have been his teacher. Themon was author of a commentary on Aristotle's *Meteorologica,* a *Question on the Motion of the Moon,* a commentary and questions on the *Sphere* of Sacrobosco, and an analysis of a sophism. Apparently he also wrote some works on physics that are now lost.

Theodoric, Bishop of Cervia (1205–1298)

An Italian, Theodoric was "born at Lucca in 1205; died at Bologna on Christmas Eve, 1298. He assumed the Dominican habit; was penitentiary to Innocent IV (pope from 1243 to 1254); bishop of Bitonto (Bari delle Puglie) from 1262 to 1266, during which time he resided in Lucca; and bishop of Cervia (near Ravenna) from 1266 to 1298, with residence in Bologna" (Sarton, *Introduction to the History of Science,* Vol. II, Part II, pp. 654–655). Written in Latin around 1267, *The Surgery* of Theodoric was translated into many languages. He was also the author of other medical treatises, and of a work on horse medicine and one on falconry.

Theodoric of Freiberg (b. ca. 1250; d. shortly after 1310)

A German Dominican, Theodoric may have taught at a convent in Freiberg before 1275 and studied at the University of Paris around the years 1275–1277. In 1293, when he already had his doctorate in theology, Theodoric was made provincial of Germany for the Dominican Order. He was present at a General Chapter of the Dominicans held at Toulouse in 1304. The last known mention of him is in the Acts of the General Chapter in Piacenza in 1310. In addition to many treatises on logic, theology, and metaphysics he wrote a number of physical treatises, including *On the Rainbow (De iride,* written in 1304 at the request of Aymeric, master general of the Dominican Order), *On Time (De tempore), On the Elements (De elementis), On Light (De luce), On Celestial Bodies (De corporibus celestibus),* and a few others.

St. Thomas Aquinas (ca. 1225–1274)

An Italian born near Aquino between Naples and Rome, Thomas entered the Dominican order at an early age and studied under Albertus Magnus at Paris and Cologne. In 1256 he became a doctor of theology and assumed one of the two chairs of theology occupied by the Dominicans at Paris. He taught in various Italian cities between 1259 and 1269, after which he once again lectured in Paris, until 1272. For the next two years (1272–1274) he organized the Dominican house of theological studies in Naples. He died on March 7, 1274, on his way to the Council of Lyons. He was a prolific author, producing many commentaries on the works of Aristotle and numerous theological treatises, among which the best known are the *Summa contra Gentiles (Summary against the Gentiles)* and *Summa theologica (Summary of Theology)*.

Thomas Bradwardine (ca. 1290–1349)

Bradwardine himself tells us that he was born in Chichester. After earning his bachelor's (1321) and master's (1323) degrees he became proctor of Merton College, Oxford, in 1325. It was probably during his stay at Oxford University that he composed most of his philosophical, mathematical, and theological treatises. Around 1335, Richard of Bury, bishop of Durham and a great bibliophile, named Bradwardine his personal chaplain and called him to London. In 1338 he became chaplain-confessor to King Edward III and in 1347 was appointed archdeacon of Norwich. He died in England on August 26, 1349, shortly after his second election as archbishop of Canterbury on June 4, 1349. Although his fame was already widespread, his name was immortalized by Chaucer's flattering association of his name with those of Boethius and St. Augustine in the *Nun's Priest's Tale* of the *Canterbury Tales*. In addition to the *De Causa Dei* and *Treatise on Proportions,* from which selections have been included in this source book, Bradwardine wrote a *Speculative Geometry (Geometria speculativa), Speculative Arithmetic (Arithmetica speculativa),* and *Treatise On the Continuum (Tractatus de continuo)*. Other treatises of a logical and physical character are attributed to him. Early printed editions of many of Bradwardine's treatises testify to the continuing interest in his works well into the sixteenth and seventeenth centuries.

Trotula

For the little that can be said of Trotula see Selection 103, n. 1.

Ugo Benzi (Hugh of Siena) (1376–1439)

Born in Siena and studying at the universities of Florence (1392–1393), and Bologna (1393–1394), Ugo Benzi entered the University of Pavia in 1395 and "received his license at Pavia in 1396. He taught philosophy in the University of Pavia–Piacenza from 1396 to 1402, and at Bologna 1402–3. In 1403–4 his attempt to teach medicine at Bologna caused a scandal, and in the following academic year he went back to philosophy. He began his teaching of medicine, unchallenged, at Siena in 1405. Driven out of Siena in 1409 by his political enemies, he "visited" Perugia; and at Pisa he attached himself to Card. Cossa, who became Papal Legate at Bologna and in 1410 was elected Pope (John XXIII). Ugo taught at Bologna 1409–13, at Parma 1412–16, at Siena 1416–21, at Florence 1421–23, at Bologna 1423–25 (where he was physician to the Papal Legate, Gabriel Condulmer, later Pope Eugene IV), at Pavia 1425–29, at Padua 1429–31. From 1431 to 1439 he was court physician to Niccolo d'Este at Ferrara" (Dean P. Lockwood, *Ugo Benzi* [Chicago: University of Chicago Press, 1951], p. 159).

This philosopher-physician, friend of humanists though not himself a humanist, had a great reputation in his day in both philosophy and medicine. He wrote five commentaries on the fundamental medical texts studied at the universities, including commentaries on the *Aphorisms* of Hippocrates, the *Tegni* ascribed to Galen, and various parts of Avicenna's *Canon of Medicine*. He also wrote a series of independent questions and the *consilia,* or case histories, from which Selection 102 has been extracted. Among his philosophical works are included *An Introduction to Logic (De logicae artis ratione),* completed around 1399, a *Commentary on Aristotle's Parva Naturalia,* completed 1402–1405, and *Natural Questions,* written after 1405.

William Heytesbury (d. ca. 1373)

Heytesbury was a fellow of Merton College, Oxford, in 1330 and is mentioned as having been there again in 1338–1339. In 1341 he was named a foundation fellow of Queen's College, although he probably never took up the fellowship. By 1348 he had become a doctor of theology. Sometime

before 1363, and again in 1371, he was chancellor of Oxford University. He was primarily a logician and natural philosopher. A few of his works were used as textbooks in logic.

William of Alnwick (ca. 1270–1333)

William entered the Franciscan Order before 1290 and became a theologian. As a pupil of John Duns Scotus, he heard the latter's lectures at Oxford and Paris. Upon the death of his master, William edited Scotus' Paris and Oxford lectures on the *Sentences*. By 1314 he may have lectured on the *Sentences* at Paris. Around 1316 he became a doctor of theology at Oxford. He was a defender of the Franciscan doctrine of evangelical poverty and

ran afoul of Pope John XXII. As a consequence he was given asylum in Naples and became bishop of Giovinazzo by July 1330. He died in Avignon in March 1333. Aside from his Paris *Sentence Commentary,* he delivered *Questiones disputatae* and *Quodlibeta* at Oxford in which he was critical of a number of contemporaries, especially Henry of Harclay.

William of Moerbeke (ca. 1215–ca. 1286)

A Flemish Dominican, William was born at Moerbeke in East Flanders. He was a chaplain to a number of popes and eventually became archbishop of Corinth in 1278 until his death. The length of his residence in Greece is unknown, but

he was there on at least two separate occasions and learned the Greek language well enough to become the greatest translator from Greek into Latin (see also the Introduction to Selection 8).

William of Ockham (ca .1280–ca. 1349)

Little is known of William's early life, but it is most probable that he pursued his higher studies in theology at Oxford from about 1309 to 1315, lectured on the Bible from about 1315 to 1317, lectured on the *Sentences* from about 1317 to 1319, and was a *Baccalaureus Formatus* from about 1319 to 1323. It seems that Ockham was a *Magister Theologiae*—that is to say, he had fulfilled all the requirements necessary for that degree—but he never became a *Magister actu regens,* a Master occupying an official chair of theology; that is the reason why he was called *Inceptor,* and that is the meaning of this title so often misinterpreted in the sense of "originator of the nominalistic school." (*Philosophical Writings, A Selection: William of Ockham,* translated with an Introduction by Philotheus Boehner, O.F.M. [Edinburgh and London: Nelson, 1957], p. xii).

Charged with holding heretical opinions, Ockham was summoned to Avignon by the Pope. An examination of his views was carried on for three

years but was never completed, and none of his views were formally condemned. After siding with those Franciscans who favored evangelical poverty, Ockham opposed the Pope, and in 1328 he left Avignon secretly with Michael of Cesena, general of the Franciscan Order, fleeing to the Holy Roman Emperor, Ludwig of Bavaria. Ockham was promptly excommunicated. He lived in Munich for many years. After the death of Ludwig in 1347, Ockham's position became desperate and he sought a reconciliation with the Pope. He died around 1349, probably a victim of the black death.

Ockham was the author of many treatises in logic, physics, and theology, and the promulgator of numerous political tracts, many of which were directed against the Pope. Among the most important of his nonpolitical tracts are a tripartite work on logic *(Summa totius logicae),* an exposition of Aristotle's *Physics (Expositio super octo libros Physicorum),* and his Commentary on the Four Books of the *Sentences* of Peter Lombard.

Witelo (fl. 1250–1275)

A Pole born in Polish Silesia, Witelo (or Vitelo, Vitellio) was educated in Paris (ca. 1253), Padua

(ca. 1262–1268), and Viterbo (1269). He was the author of at least two theological treatises, but his major work was the *Perspectiva,* a treatise on optics in ten books dedicated to William of Moerbeke, whom he had met at Viterbo. Like Moerbeke, Witelo was a Neoplatonist and expressed his neoplatonic views in the preface to the *Perspectiva.* The theoretical portions of the work were based largely upon Alhazen's *Optics.* Witelo also included discussions on the psychological aspects of vision. The *Perspectiva* was influential into the seventeenth century, when Kepler himself read it and used it as a point of departure in his *Ad Vitellionem paralipomena quibus astronomiae pars optica traditur* (Frankfurt, 1604).

Index

Medieval figures are usually indexed under their first names unless they were better known by their last names, as, for example, Aquinas, Buridan, Bradwardine, Grosseteste, Ockham, Oresme, and so on. Both names should be tried before concluding that the name of a particular individual does not appear. Wherever appropriate, the Latinized forms of Arabic names are used (for example, Alhazen [not Ibn al-Haitham], Avicenna [not Ibn Sina], Averroes [not Ibn Rushd], and so on). Page citations followed by "n" signify the footnote on that page. The preface has not been indexed.

Index

Firmament, 442, 550, 632
Fishes: described in a bestiary, 651–54
Fissility, 624, 628
Fixation(s), 590, 597
Fleas, 653
Flesh, 682, 683
Flight (of birds), 675–78
Flint, 627
Flood, the, 466n
Florence, University of, 828
Flowers: on the nature and generation of, 697–98; two
    kinds distinguished, 697n; *mentioned*, 696
Fludd, Robert, 559n
Focilia, 672n
Fontaine, Jacques, 12n, 25n
Food(s): in regimen, 760–61; *mentioned*, 708
Force(s) : and motion in a vacuum, 340; natural,
    four in consimilar members of body, 728; inex-
    haustible,375n; maximum capacity of motive, 365n;
    measurement of, 360–67; minute, 359n; motive,
    infinite capacity of, 298n; vectorial analysis of,
    212
Forlì, Jacopo da: biography of, 818; on Galen's *Tegni,*
    720–22
Forma mixti, 607, 613
Forms: accidental, 604; intension and remission of,
    363n; substantial, 578, 579, 580, 613–14; sub-
    stantial and accidental, 581–82, 603, 607n; sub-
    stantial, and compounds, 604, 605, 609; sub-
    stantial, and elements, 608n; substantial, and
    generation and alteration, 609n; *mentioned,* 65
Forster, E.S., 674n, 689, 694n, 695n
Four Masters, 792, 793
Fournival (Furnival), Richard de, 646n, 820
Fox mange, 368
Franciscus de Mayronnis, 637n, 638n
Fractions, 134n
Fractures: of skull, 795–96
Frauenburg, longitudes of, 478
Frederick II of Hohenstaufen: biography of, 815; on the
    structure and habits of birds, 659–81; personality
    and intellect of, 657–59; *mentioned,* 114, 115n,
    128n, 821
Freestone, 627
Friction, 502
Friedlein, G., 17n, 23n, 137n
Frisk, Gösta, 779n, 780n
Fruit: and flowers, 698; and leaves, 697; in regimen,
    761; of trees, 694; *mentioned,* 694n, 695, 696
Fuga vacui, 356
Fulgeratus, 491
Fulton, Mr. (of British Museum), 757n, 758n
Fumigation(s): and menstruation, 764; *mentioned,* 767
Fumitory, 742
Functionality, 237
Functions: and faculties distinguished, 720; theory of,
    366n; trigonometric, 188n
Fungi, 684, 695
Furnace(s): on construction of, for alchemical opera-
    tions, 591; sublimation, 600
Fyfe, F. Marjorie: translation and annotation by, 659–
    81

Gabriel Condulmer, 828

Gaddesden, John: biography of, 819; on smallpox,
    770–72; *mentioned, 792*
Galen: advocates dessicative medicines, 805; and anato-
    my, 727; and apoplexy, 702n; and dissection, 724–
    25; and menstruation, 763; Avicenna's system of
    Galen's writings, 715n; great theriac of, 788–89;
    on centaury, 784; on conception, 765; on couching,
    738n; on crocodile dung, 650n; on faculties, 720;
    on pus, 802; on surgery, 792; on the acorn, 700;
    on the fetus, 765; on the pulse, 722n, 745; on thera-
    peutics, 730n; on treatment of wounds, 806; on
    ventricles, 735n, 736n; suppuration employed as
    last resort, 805; works of, translated from Arabic
    to Latin, 37–38; *mentioned,* 571n, 582, 681n, 703n,
    713, 722, 741, 742, 762n, 773, 781n, 793, 797, 801
    811, 817

WORKS:
Ars medica, 730n
Book of Sects, 793
Book on Tumors Contrary to Nature, 792
*Catageni (De compositione medicamentorum per
    genera),* 792, 794n
Commentary on the Aphorisms, 795, 805n
Commentary on the Regimen in Acute Diseases, 791,
    797
Definitiones medicae, 702n
De interioribus (De locis affectis), 733, 734, 735, 789
*De juvamentis membrorum (De usu partium corporis
    humani),* 731n, 736n, 737
De pulsuum differentiis, 722n
De simplici medicina, 790
Introduction, 794
Isagoge, 797
Megatechni, 805
Methodus medendi, 729–30
Miamir or *Compositions According to Plans (De
    compositione medicamentorum secundum locos),*
    794, 794n
Natural Faculties, 793
On Nutriments (De alimentis), 40
Pantegni, 781n
Tegni (Ars parva), 35, 38n, 805, 816
The Constitution of the Dogmatic Art, 793
Therapeutics, 740, 792, 793, 794, 795, 807
Galen, Pseudo-: *De anatomia vivorum,* 739n
Galenic physiology, 701n, 732n
Galenic system: as formulated by Avicenna, 715–20;
    as formulated by Joannitius (Hunain ibn Isḥāq),
    705–15
Galileo Galilei: abandons absolute heaviness and light-
    ness, 348–49; and experiments demonstrating
    nature's abhorrence of a vacuum, 328–29; and
    experiments on weight of air, 331n; and Joshua,
    508n; and mean speed theorem, 237; and the sepa-
    ration of plates, 359; attacks Aristotle's theories
    against motion in a void, 346–48; biography of,
    816; criticizes Gilbert, 373n; equality of velocities in
    a vacuum, 349–50; explanation of acceleration,
    283; interstitial vacua affirmed, 359–60; motion can
    arise when force less than resistance, 340n; on in-
    stantaneous motion, 239n, 346; on moment of rest
    between contrary motions, 289–92; on momentary
    existence of a vacuum, 329n; on motion in a void,

ment of rest between contrary motions, 285–89; mentioned, 52n, 275, 284, 287n, 288n, 291n, 329n, 339n, 340n, 358n, 530, 621n

Martianus Capella: biography of, 822; theory of two oceans, 31n; *mentioned,* 4, 29n, 30, 56n, 59n, 630n, 646

Martí-Ibañez, Félix, 743n, 767n

Martin, 496n

Marvin, F.S., 731n

Mary the Jewess, 598n

Māsawaih, Yūḥarrā ibn, 797

Mason-Hohl, Elizabeth: translation by, 761–67

Master Nicholas: biography of, 822; anatomy of, 727–29

Materia, 26

Mathematics, 90–198: and its parts, 65–67; and natural science, 200; mathematicians as astrologers, 75; as discipline, 66; as theoretical knowledge, 55; defined, 65; division of, 56; genus of, 66; in Aristotle's view, 53; in the quadrivium, 3–9; infinite series, 131–35; matter of, 66; notation, 150n; objective of, 58; on the importance of its study, 90–94; only its demonstrations are certain, 92–93; part of theoretical science, 54n; prior to other sciences, 93; probability, 130–31; proportion, defined, 132; proportional parts, 132–35; proportionality theory, 136–59; reality of points, lines, and planes in, 363n; required by every science, 91–94; seven arts of, 66; treats of concepts, 59. *See also* Algebra, Arithmetic, Geometry, Number theory, Numbers, Proportions, Ratios, Trigonometry.

Mathesis, 66

Matter: concept of community of, 356n; prime, 26; quantity of, 279; questions on, 200

Matthaeus Platearius. *See* Platearius, Matthaeus.

Matthews, Gareth B., 79n

Maudith, 188n

Maurus, 750

Maxima and minima: no single rule for assigning, 367n

Mayow, 727n

Mean(s): and extremes in nature, 604; between living and non-living, 684; center of the moon, 472; distinction between arithmetic, geometric, and musical, 7; geometric, 7, 9, 24; geometrical, compared to democratic state, 90; harmonic, 8n, 11n; musical, 7–8; proportionals, 101; speed theorem, 237–43, 251–53, 363n; state between health and disease, 711

Measure, units of, 72

Mechanics, 59. *See also* Dynamics, Kinematics, Statics.

Medical sects, 701, 793

Medicamentum, 69

Medicine(s), 700–808: a method of practice in, 760–61; among mechanical sciences, 54n; and astrology; and theology, 704n; compound, 787–91; definitions of, 700–701, 715–16; derivation of terms, 69, 701; dessicative, 805; divisions and parts of, 68–69, 714, 715; early medieval, 700–704; general instructions for the practitioner of, 742–45; operation of, 714; simple, 778–85; study of, 704; subject matter of, 716–18; theoretical and practical, 705, 716; theory of, 700–20; three methods of attacking disease in, 797; tools employed in treatment in, 775–91; translation of works on, 37; treatment of particular ailments in, 761–74

Medietas, 24n, 153n

Medium (media): and alteration of species, 393; and bending of lines of force, 386; and resistance to motion of species, 431–32; angles of refraction in different, 423–26; color and transparent, 391; density of, 366–67; function of, according to Averroes, 258n; illumination of, instantaneously and without resistance, 273, 364n; image formation in transparent, 427–30; location of images in spherical, 427–30; resistance of, 224–25; role of, in motion, 254–62, 264

Mehren, A.F., 491n

Melancholia, 702

Member(s): animal, 737–39; as symptoms of leprosy, 754–55; defined, 727; displacement of, 715; excess, diminution, and size of, 713; four kinds of, 706; generative, 732–34; of the body (official, animate, and natural), 731–32; spiritual, 734–37; threefold classification of, 728

Memory, 10

Menelaus (Mileus), 36

Menstruation: and use of wormwood, 782; Trotula on, 762–64; *mentioned,* 724

Mental concept (or content), 85, 86

Menut, Albert D.: translations by, 503–510, 548–54; *mentioned,* 504n, 537n, 547n, 548n, 550n, 553n

Mercadant (surgeon), 793

Mercury (planet): and Sun, 28, 458–59; deferent center of, 458; epicycle of, 458, 459; mean argument of, 466; mean motus of, 460; nature of, 632n; orbit of, 28n, 29–30; reflection of, perceptible, 464n; theory of motion of, 458–61; *mentioned,* 16, 27, 463, 518, 634

Mercury (metal): how sublimed, 598; one of four spirits, 591; preparation of, 598–600; properties and formation of, 571; red sulphide of, 592n; sign for, 587n; still, 598n; used for skin ailments, 769n, 770; what it is, 592; *mentioned,* 569, 586 (quicksilver), 602 (quicksilver)

Mergansers, 661n

Meridian: defined, 448, 541

Meroe, 13

Mersenne, 333n

Mertens, Joseph Charles, 368

Merton College, Oxford: mean speed theorem of, 237–43; *mentioned,* 132n, 235, 239n, 251n, 275, 828

Mesopontus, 13

Messehala, 36

Mesue, 789, 797n

Mesue junior (Johannes), 734

Metals: and planetary signs, 587n; and sulphur-mercury theory, 569n–570n; are imperfectly what gold is perfectly, 584–86; calcination of, 600–603; can be further developed, 583; cannot be transmuted, 572–73; cohesiveness of, 360; common, intermediate between quicksilver and gold, 586; differ only accidentally, 588; divination by means of, 492; formation of, 571–73, 588–89; may be transmuted into silver and gold, 591; reason against transmutation of, 575–80; *mentioned,* 209

Metaphysics, 53, 70, 92–93

Meteorological phenomena: questions on, 207–209

Meteorological sphere, 436n, 441n

Method, experimental, 440n